IN AUSTRALASIA OVER THE LAST 545 MILLION YEARS

PLANTS	Vegetation Ages	ANIMALS	Vertebrate Ages
ands	AGE OF FLOWERING PLANTS	Europeans arrive by 1606 Dingoes introduced 5,000–4,000 yrs BP	AGE OF MAMMALS
species (sclerophyll communities) e forests to low araucarian vine forest (Qld.)		Mass extinctions about 40,000 yrs BP Humans arrive 100,000–60,000 yrs BP	
op		First grazing mammals and Australian rodents	
comes progressively important			
forest			
ears		Very diverse fossil assemblages from Riversleigh, Qld.	
nd ears		Oldest Australian cetaceans (@ ~28 mya) Faunal immigration from S.E. Asia commences	
semblage (*Cinnamomum* flora) widely distributed		Most Gondwanan faunal migration routes disrupted by 50 mya	
appears		Oldest Australian marsupials, bats and world's oldest songbirds (@ 55 mya, from Qld.)	
appears (both Australia and New Guinea) of angiosperms throughout Gondwana		Mass extinctions @ 65 mya Dinosaur stampede trackway, Winton Qld.	
te Jurassic complexes	AGE OF GYMNOSPERMS & FERNS	Possibly largest dinosaurs known (tracks at Broome, W.A. @ 100 mya) First monotremes @ 110 mya from Lightning Ridge, N.S.W. Abundant ichthyosaurs @ ~115 mya, polar dinosaurs @ 120 mya	AGE OF REPTILES
		First undoubted birds (Eu.) by 150 mya First freshwater teleosts from Talbragar, N.S.W.	
	AGE OF SEED-FERNS	First mammals (N.A.) and first dinosaurs (S. America) @ ~ 210 mya Mammal-like reptiles diverse	
l transition n,deciduous forests in north, coniferous		Major extinction event @ 251 mya	AGE OF AMPHIBIANS
seed-ferns lubmosses, horsetails)		First reptiles @ 310 mya (N.A. and Eu.)	
-sized plants)		First amphibians (tracks from Victoria ~400 mya) Abundant and diverse fish faunas in Australasia	
ya)	AGE OF CLUBMOSSES	First land animals (arthropod tracks from W.A. @ 420 mya)	AGE OF FISHES
		First fish-like animals (from the N.T. @ 470 mya)	
		Early chordates @ 530 mya (N.A. and Asia). Australian marine invertebrate faunas diverse. First vertebrate (from S.A. @ 560 mya)	

d the boundaries of some intervals. For other geological sub-intervals (Ages or Australian Stages) see the more detailed table in each section.
ars ago; N.A. = North America; Qld. = Queensland; Eu. = Europe; N.S.W. = New South Wales; N.T. = Northern Territory; S.A. = South Australia; W.A. = Western Australia.

Sources for Notable Events Summary

ABC Science Online. 2003. Oldest vertebrate fossil found in Australia, scientists say. http://www.abc.net.au/science/news/stories/s973390.htm

Archer, M., Burnley, I., Dodson, J., Harding, R., Head, L. and Murphy, P. 1998. From plesiosaurs to people: 100 million years of Australian environmental history. *Australia: State of the Environment, Technical Paper Series (Portrait of Australia)*, pp.1–66. Department of the Environment, Canberra.

Archer, M. and Clayton, G. (eds). 1984. *Vertebrate Zoogeography and Evolution in Australasia*. Hesperian Press, Perth. 1203pp.

Frakes, L.A. 1999. Evolution of Australian environments. Pp.163–204. In Orchard, A.E. (ed). *Flora of Australia Volume 1, Introduction*. 2nd Edit. Australian Biological Resources Study, CSIRO Publishing, Melbourne. 670pp.

Hill, R.S., Truswell, E.M., McLoughlin, S. and Dettmann, M.E. 1999. Evo;ution of the Australian flora: fossil evidence. Pp.251–320. In Orchard, A.E. (ed). *Flora of Australia. Volume 1, Introduction*. 2nd Edit. Australian Biological Resources Study, CSIRO Publishing, Melbourne. 670pp.

Martin, H.A. 1998. Tertiary climatic evolution and the development of aridity in Australia. *Proceedings of the Linnean Society of New South Wales*, Vol. 119, pp.115–136.

White, M.E. 1998. *The Greening of Gondwana.*Third Edition. Kangaroo Press Pty. Ltd., Sydney. 256pp.

Young, G.C. and Laurie, J.R. (eds). 1996. *An Australian Phanerozoic Timescale*. Oxford University Press, Melbourne. 279pp.

Plate 1. One of the most familiar marine mammals, the bottlenose dolphin (*Tursiops aduncus*). This group is moving along the coast just out from the surf zone. Research in recent years has shown that these dolphins have very complex social relationships and group behaviour patterns; there are resident populations in large bays that have little interaction with groups in adjacent coastal waters. Photograph S. Allen.

Plate 2. An example of a poorly known endemic freshwater fish from inland Australia. This Barcoo grunter (*Scortum barcoo*) came from the Wilson River (Lake Eyre division). Photograph G. Schmida.

Plate 3. *Litoria bicolor* is a small, agile tree frog (Hylidae). Photograph M. Tyler.

Plate 4. This is one of about 35 species of freshwater turtle that occur in Australia and New Guinea. This long-necked form is from the genus *Chelodina*.
Photograph J. Cann.

Evolution and Biogeography of Australasian Vertebrates

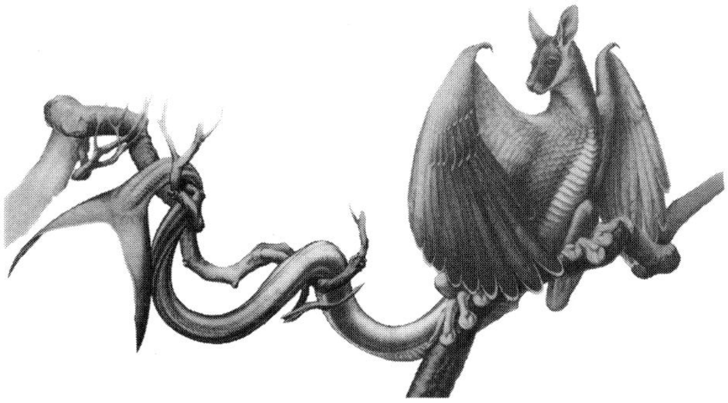

The financial contributions provided by the following organizations, towards preparation of this volume are gratefully acknowledged:

AUSTRALIA NEW GUINEA FISHES ASSOCIATION INC

THE LINNEAN SOCIETY OF NEW SOUTH WALES

WELLINGTON CAVES FOSSIL STUDIES CENTRE

Evolution and Biogeography of Australasian Vertebrates

Editors:

Merrick, J.R., Archer, M., Hickey, G.M. and Lee, M.S.Y.

auscipub
Australian Scientific Publishing

First published in 2006

Auscipub Pty Ltd
PO Box 68
Oatlands
NSW 2117
Australia
auscipub@btinternet.com
www.auscipub.com

National Library of Australia Cataloguing-in-Publication data

Evolution and biogeography of Australasian vertebrates
Bibliography.
Includes Index.
ISBN 0 9757790 0 1 (pbk)
0 9757790 1 X (hbk)

1. Vertebrates - Australasia - Evolution. 2. Vertebrates - Ecology.
3. Biogeography - Australasia. 4. Zoogeography - Australasia.
I. Merrick, J.R. (John Rodney), 1944–

596.09

Typeset by Chapter 8 Pty Ltd, Sydney, email: chapter8@btinternet.com
Cover Artwork by Professional Drafting and Photography, Sydney
(based on an original drawing by Peter Schouten)
Printed and bound by Southwood Press Pty Ltd, Sydney

Contents

About the Editors

John R. Merrick

Awarded an M.Sc. and a Ph.D. from the University of Sydney for studies on Australian freshwater fishes, John's research programmes have involved extensive fieldwork on streams, lakes and dams throughout Queensland, New South Wales, Victoria and the Northern Territory. Over 65 publications have included 6 books, many refereed papers and general articles on aquaculture and freshwater fisheries management, reproduction and taxonomy of a range of Australian native fishes as well as ecology and conservation management of endemic crayfishes.

In addition to teaching and postgraduate supervision in the Department of Biological Sciences and Graduate School of the Environment at Macquarie University, Dr. Merrick has undertaken a number of other activities related to publishing. These include: editing Scientific Journals and Conference Proceedings; editing and proofing of scientific or technical monographs; draft manuscript assessment; and compilation of specialist indexes.

Michael Archer

A graduate of Princeton University, Michael returned to Australia as a Fulbright Scholar and obtained his Ph.D. from the University of Western Australia. With colleagues he began work on the remarkable fossil deposits at Riversleigh (north-western Queensland) in 1976 and has retained a leading role in the continuing studies at Riversleigh.

As Professor of Biological Sciences at the University of New South Wales other research interests have included palaeontological investigations of early Tertiary mammals from Murgon (south-eastern Queensland). Mike's diverse and active research has resulted in many publications—he has received six Whitley awards for books.

During his appointment as Director of the Australian Museum in Sydney, Professor Archer continued to supervise postgraduate students and co-ordinate research activities—including a ground-breaking programme to reconstruct the genetic material of the Tasmanian tiger (the 'Thylacine Project'). Early in 2004 Mike returned to the University of New South Wales

Georgina Hickey

Georgina M. Hickey (née Clayton) is the Editor of *Nature Australia* magazine (published by the Australian Museum), where she has worked since 1984. She co-edited the original *Vertebrate and Zoogeography and Evolution in Australasia* (1984) after graduating from the University of New South Wales (Zoology and Mathematics) in 1980.

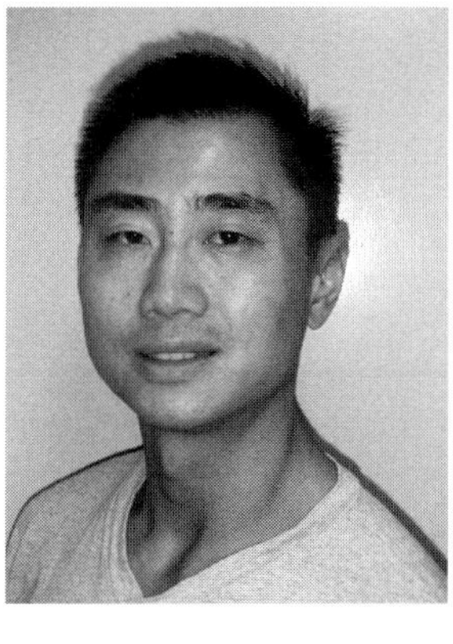

Michael Lee

Michael Lee's research interests include the evolution of fossil and living reptiles, molecular and anatomical evolution, methods for quantifying biodiversity, and the theory and practice of systematics. He studied at the University of Queensland and Cambridge University. Mike is currently Senior Research Scientist at the South Australian Museum and Associate Professor at the University of Adelaide

Perspectives

J.R. Merrick[1], M. Archer[2], G.M. Hickey[3] and M.S.Y. Lee[4]

[1] Graduate School of the Environment, Macquarie University, North Ryde. N.S.W. 2109 Australia
[2] Faculty of Science, University of New South Wales, Sydney. N.S.W. 2052 Australia;
[3] PO Box 4167, Manuka, A.C.T. 2603 Australia
[4] Department of Environmental Biology, Adelaide University, Adelaide. S.A. 5000 Australia
E-mail: jmerrick@gse.mq.edu.au

BACKGROUND

This volume is a direct development of the groundbreaking book *Vertebrate Zoogeography and Evolution in Australasia* (edited by Archer, M. and Clayton, G.) produced in 1984. That earlier work, affectionately dubbed 'the doorstop' because of its dimensions (A4, 1200 pp.), was enthusiastically received as it filled a very large gap—it comprehensively integrated, for the first time, information on extant Australasian vertebrates with the fossil record and geological and evolutionary factors. With an easily readable style and slightly irreverent approach, the doorstop sold well both locally and overseas. By 1988 stocks had sold out and work had already commenced on a revised edition.

The original book remained popular, becoming as 'rare as hens' teeth' and some copies were reportedly exchanged for high prices. With continuing demand, revised and new chapters were submitted for some years. It was soon clear that the huge amount of material collected could not all be included and the concept of a reduced volume was developed. While attempts have been made to retain continuity, where possible, so much has changed after 20 years that this has to be considered a new book on the same general topic—not just a revised edition.

For example, in systematics major developments include: (a) a rise in the use of numerical phylogenetics—combining a range of diverse data from newer techniques (electrophoresis, immunological reactions, chromosomal structures, DNA sequencing)—to determine lineages; and (b) attempts to introduce a new more consistent phylogenetic taxonomic system—to standardize nomenclatural rules and names as well as clarifying the relative status of taxa—especially above generic level. See pp.3–16 (Lee, 2006).

There have also been substantial changes in biogeographic (and zoogeographic) thought with the realization that observed patterns do not necessarily reflect gradual dispersal. The interaction of rapid and often extreme vicariance events (geological, climatic), combined with differential mutation rates within genomes and varying breeding strategies in small or isolated populations, has resulted in re-assessment of speciation processes and the resultant distributions. For a summary of the major influences on biogeography of regional vertebrates see the back endpapers; for details, readers are referred to recent books such as Humphries and Parenti (1999), Augee and Fox (2000), Cox and Moore (2000), MacDonald (2001) or Veevers (2001).

BOUNDARIES AND TERMS

To eliminate any confusion about exactly what region is covered and the distinction between Australasia and other regional terms such as Polynesia or Melanesia, readers are referred to Figure 1 where major zones are illustrated and briefly defined. A number of more detailed local 'bioregional' concepts, based on the occurrence of faunal assemblages or particular groups (e.g. Whiting *et al.*, 2000), will be introduced in subsequent chapters.

As Figure 1(a) shows a number of recognized geographical zones are contained within the larger Australasian zone. Technically Australasia covers a huge area of the central and western Pacific but, for practical purposes, the focus in this book is restricted to the three major land masses in the south-west (Australia, New Guinea and New Zealand) and adjacent coastal waters. The vertebrate faunas of the smaller islands are similar in many ways, but show some distinctive features. Unfortunately it has not been possible to discuss small island faunas in detail; although, in recent years, it has become increasingly apparent just how important islands are in terms of distinctive endemics, vulnerability to introduced taxa, and as conservation refuges.

CHANGES AND BASIC THEMES

Like its subject matter, this volume has undergone considerable evolution and this has led to altered emphases related to new data and new perspectives. For example, it is widely accepted that a human presence has been relatively recent in Australasia in comparison with other regions; colonization of some areas has only occurred in the last 4,000 years (see Figure 2). However, controversy rages about the extent of human impacts (Flannery, 1994; Kohen, 1995; Roberts, 1998; Wroe and Field, 2001; Wroe *et al.*, 2002). Recent archaeological studies have provided more evidence of just how significant particular hunting activities were in local regions

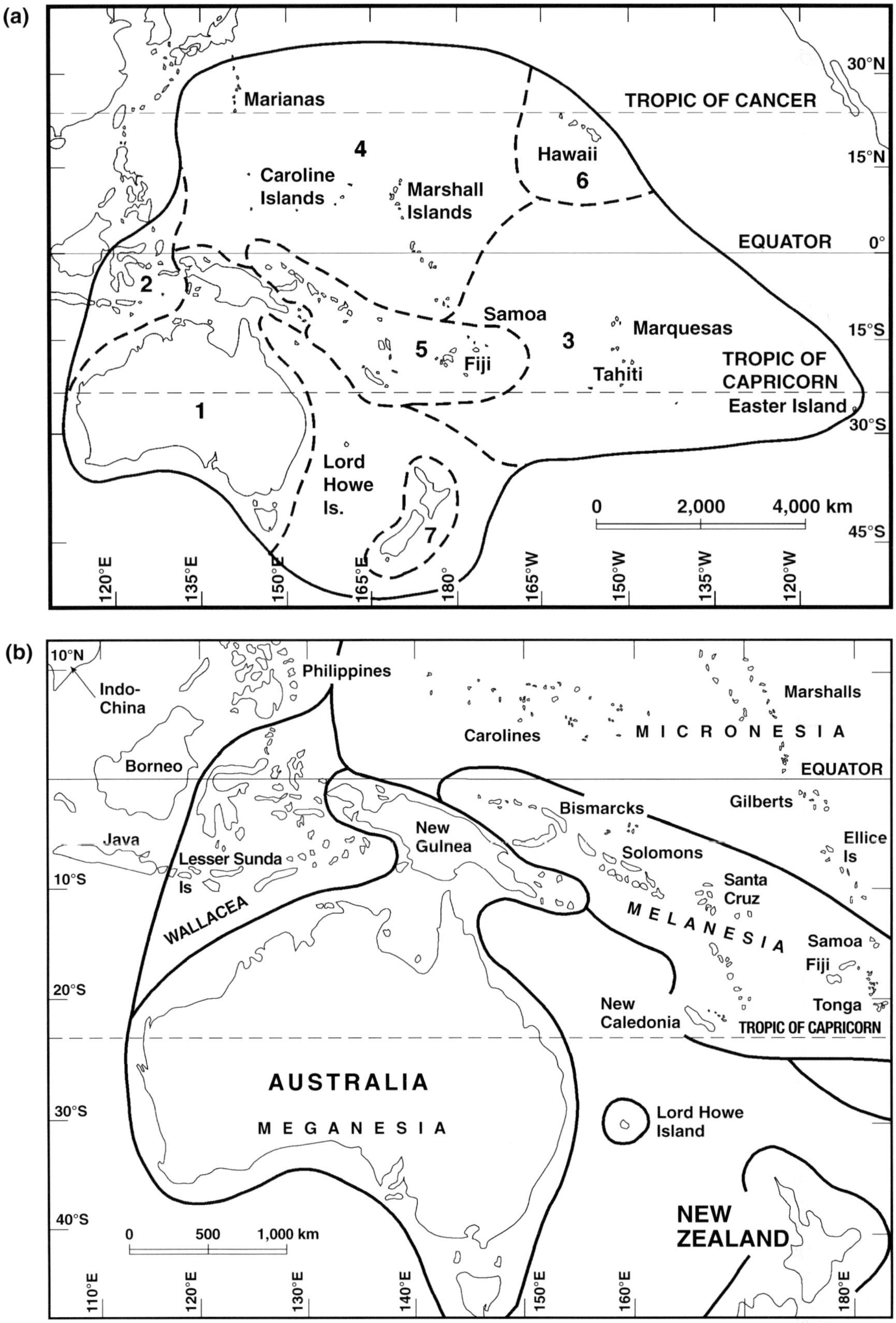

Figure 1. Major regional geographical zones with boundaries marked: **(a)** total area of Australasia outlined with sub-regions delineated; **(b)** details of Meganesia, New Zealand and adjacent zones (based on Illies, 1974; Whitmore, 1981). Code to zone numbering: 1 = Meganesia; 2 = Wallacea; 3 = Polynesia; 4 = Micronesia; 5 = Melanesia; 6 = Hawaii; 7 = New Zealand.

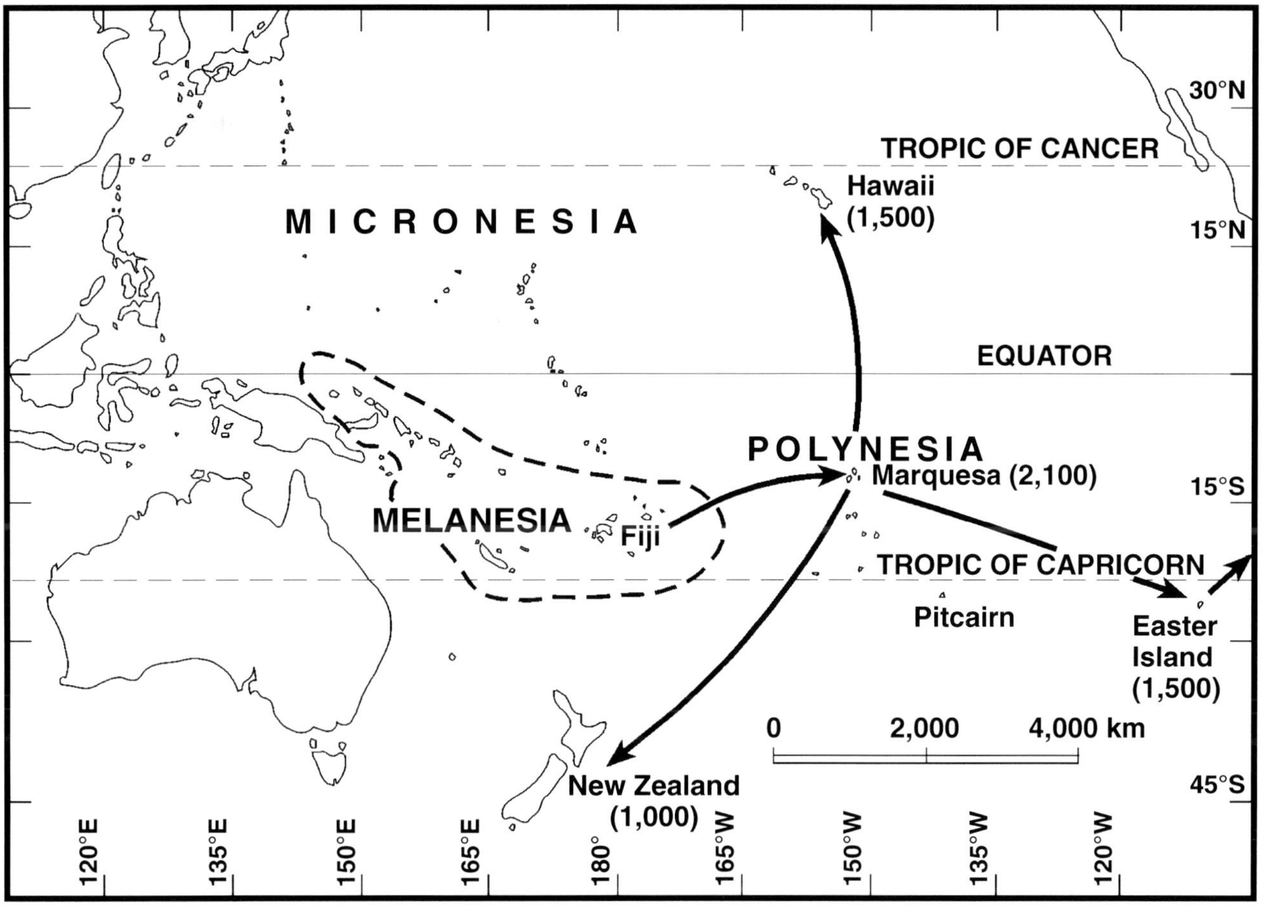

Figure 2. A summary of human colonization of the Pacific during the late Holocene (from 3,800–1,000 yrs BP). The Lapita expansion extended throughout Melanesia from 3,800–3,200 yrs BP; the arrows indicate subsequent phases of the Polynesian expansion with approximate arrival dates (yrs BP) in brackets. These two expansions are very recent in comparison with the initial arrival of *Homo sapiens* in New Guinea and Australia. Based on Roberts (1998, p.181).

(Flannery, 1994; Kohen and Merrick, 1998; Owen and Merrick, 1998), leading to conclusions of heavy utilization of selected resources and possible over-exploitation. Although in other situations human 'blitzkreig' has been challenged as a cause for megafaunal extinctions (Wroe *et al.*, 2002); climate change is at least as likely a significant cause of megafaunal extinction.

But it has only been in very recent years that the importance of two other factors (related to human activity) has become apparent. The first is the essential role of genetics in conservation—where populations are reduced and fragmented. Major findings from recent studies are that: small populations have a greatly increased risk of extinction (with significant inbreeding depression); endemics usually have reduced genetic variability; the use of isolated, refuge populations to re-stock other areas may not be appropriate (Frankham *et al.,* 2002). The second factor is the potential for problems with translocations and associated bio-invasions.

There has also been an enormous increase in general community awareness of environmental issues as well as pressure to urgently apply existing research data to alleviate current conservation problems and develop management strategies. The importance of maintaining biodiversity (Commonwealth of Australia, 1996), together with the physical environment which sustains it, is finally being politically acknowledged and new methods of storing and analysing biodiversity information are being developed. In addition, the implications of global warming and climate change are now being considered and debated (Brereton *et al.*, 1995; Hughes, 2000; Intergovernmental Panel on Climate Change, 2002; Chilcott *et al*., 2003).

The dominant theme in the following pages is still integration—integration of knowledge about the past evolution of the diversity of Australasian vertebrates with that of extant faunas, their distributions and future management. Several general comments about levels of integration are relevant.

First, some of the newer automated (computerized) biodiversity census techniques can be applied to many taxonomic groups and are able to synthesize huge amounts of disparate data in a short time. Also, molecular techniques mean that diversity in highly disparate groups with few morphological features in common can be measured using the common yardstick of sequence divergence. So it is essential to discuss these techniques as they afford opportunities never previously available—at a time when we are just beginning to grasp the enormity of the task of monitoring and managing existing biodiversity.

Second, the combined use of morphological and molecular techniques is revolutionizing systematics. In the previous volume, organismal relationships were nearly always based entirely on morphological information. Since then, the field of molecular systematics has grown remarkably. In the present volume, many chapters now discuss the implications of this new molecular data for the groups in question.

Third, the realization that extant biotas represent a tiny fraction of the total species that have existed has been pivotal in analysing biodiversity. It is clearly essential to know what has happened in the past in order to interpret current changes and predict possible trends (Archer *et al.*, 1998). For instance, the five great mass extinctions in the past are increasingly being scrutinized in order to help predict the consequences of the current, human-induced mass extinction. Also, despite the molecular revolution in phylogenetics, the vital contribution of fossil forms in understanding the evolutionary relationships of living taxa is continually being reaffirmed.

To achieve the prime integrative objective, we have:

(i) provided summaries of geological, physical and chemical conditions under which evolution has occurred locally;
(ii) included overviews of the major methods of classifying organisms and analysing their evolutionary relationships;
(iii) described the processes influencing the distributions of extinct and extant faunas and explained recently proposed systems of bioregionalization;
(iv) incorporated summaries of molecular and computerized techniques that will play an increasingly important role in analysis and decision-making related to management of biodiversity.

Clearly, it was not possible to include detailed chapters on every vertebrate group represented in Australasia; however, we have tried to compile a balanced coverage of most major groups. See Plates 1–11. Emphasis has been placed on Australasian coverage with comparisons included wherever feasible and relevant.

Australia has been recognized for some years as a megadiverse nation (Beattie, 1995, p.6, 81); furthermore, for a variety of geographical, historical, and other reasons, Australasia has a unique opportunity to retain a large part of its local biodiversity. Hopefully this reference volume will assist in that process. *Evolution and Biogeography of Australasian Vertebrates* provides a comprehensive synthesis of knowledge on Australasian vertebrates, explaining their evolution, relationships, zoogeography, prospects and management. Considerable effort has been taken to present often complex data or analyses in a way that is understandable by a broad readership.

The production of this book is also an example of what might happen more frequently in these times of 'economic rationalism', but success depends on:

(a) a medium- to long-term entrepreneurial approach; and
(b) direct interaction between the publishing industry and academic or research organizations.

ACKNOWLEDGEMENTS

For projects of this type, extending over many years, the list of people who have contributed in various ways soon becomes enormous. As the efforts of all current and previous authors, as well as their assistants, have been directly acknowledged in the relevant chapters, this list will be restricted to the smaller group who have played key roles in completing final production. A number of contributors have also acted as reviewers for related chapters; however, a number of additional researchers have kindly donated time to assist refereeing chapters in their area of expertise. These include: the late Don Adamson, Macquarie University, North Ryde; Per Ahlberg, British Museum (National History), London; Atholl Anderson, Australian National University, Canberra; Scott Baker, University of Auckland, New Zealand; John Bannister, Western Australian Museum, Perth; Andrew Beattie, Macquarie University, North Ryde; David Blair, James Cook University, Townsville; Richard Blewett, Australian Geological Survey Organisation, Canberra; W. Bock, Columbia University, New York; Marita Bradshaw, Australian Geological Survey Organisation, Canberra; Dave Briscoe, Macquarie University, North Ryde; Peter Brown, University of New England, Armidale; Keith Campbell, Australian National University, Canberra; Les Christidis, Museum Victoria, Melbourne; Bill Clemens, University of California, Berkeley; Merel Dalebout, University of Auckland, New Zealand; Chris Dickman, University of Sydney, Sydney; Steve Donnellan, South Australian Museum, Adelaide; Don Driscoll, University of Tasmania, Hobart; Michael Engelbretsen, Macquarie University, North Ryde; Tim Flannery, South Australian Museum, Adelaide; Nick Gales, Conservation and Land Management, Bentley; Fritz Geiser, University of New England, Armidale; Henk Godthelp, University of New South Wales, Sydney; Martin Gomon, Museum Victoria, Melbourne; Brett Gooden, Linden Park, South Australia; Allen Greer, Australian Museum, Sydney; Les Hall, University of Queensland, St. Lucia; Barry Hutchins, Western Australian Museum, Perth; Sandy Ingleby, Australian Museum, Sydney; Farish Jenkins, Harvard University, Massachusetts; Ben Kear, South Australian Museum, Adelaide; Scott Keogh, Australian National University, Canberra; Peter Kershaw, Monash University, Melbourne; Roger Kirkwood, Cowes, Victoria; Roger

Kitching, Griffith University, Nathan; Jim Kohen, Macquarie University, North Ryde; John Long, University of Western Australia, Nedlands; Wayne Longmore, Museum Victoria, Melbourne; Michael Mahony, University of Newcastle, Callaghan; Shane Maloney, University of Western Australia, Nedlands; Norm McKenzie, Conservation and Land Management, Wanneroo; Alan Newsome, C.S.I.R.O. Wildlife and Ecology, Canberra; Stewart Nicol, University of Tasmania, Hobart; Charles Oxnard, University of Western Australia, Nedlands; Dave Pollard, N.S.W. Fisheries, Sydney; Tom Rich, Museum Victoria, Melbourne; Greg Richards, Gungahlin, Australian Capital Territory; Glenn Shea, University of Sydney, Sydney; Roger Summons, Australian Geological Survey Organisation, Canberra; Sue Turner, Queensland Museum, Brisbane; John Veevers, Macquarie University, North Ryde; George (Buz) Wilson, Australian Museum, Sydney.

The person who has been instrumental in making this book happen is Julian Martell (Auscipub Pty. Ltd., Sydney). From the start Julian recognized the significance and potential of the work and organized basic support. Appreciation is expressed for his essential backing, supervision of all typesetting and advice at all stages of production. Several organizations are thanked for assistance with final manuscript preparation. The Linnean Society of New South Wales, Wellington Caves Fossil Studies Centre and Australia New Guinea Fishes Association all contributed on the basis of the relevance and interest of this book to their memberships; logos of these groups are displayed with publication details (reverse of title page).

We are grateful to Peter Schouten (Sydney, N.S.W.) for providing a colourful cover illustration, which embodies the notion of the continuity of vertebrates through time. John Cleasby (Sydney) is also thanked for assisting with drafting or standardizing of figures as well as scanning of photographic plates. All photographers are acknowledged in plate legends, but while most photographs were supplied by authors and/or their co-researchers, special thanks are also due to a group of other people who have provided a range of additional colour plates to more fully illustrate the diversity of Australasian vertebrates. These contributors include: Simon Allen (Macquarie University, North Ryde); Lars Bejder (Dalhousie University, Haliflax, Canada);John Cann (Sydney); B. Maryan (Queensland); Hamar Midgley (Bli Bli, Queensland); Gunther Schmida (Mudgeeraba, Queensland); G. Baker, R. Fidler, K. Ireland, G. and J. Little, P. Munchenberg, M. Seyfort, G. Taylor, G. Threlfo and B. Wells (Nature Focus, Australian Museum Sydney). Lastly appreciation is expressed to Heike Baumuller, Alison Cook, Helen Jolley, Michelle Novatin, Loren Schiavon and Kerstin Stuhr (Graduate School of the Environment, Macquarie University) for assistance with manuscript drafts and corrections. Sasiphan Bilmanoch helped with the final preparation of the index.

REFERENCES

Archer, M., Burnley, I., Dodson, J., Harding, R., Head, L. and Murphy, P. 1998. From plesiosaurs to people: 100 million years of Australian environmental history. *Australia: State of the Environment, Technical Paper Series (Portrait of Australia)*, pp.1–66. Department of the Environment, Canberra.

Archer, M. and Clayton, G. (eds). 1984. *Vertebrate Zoogeography and Evolution in Australasia.* Hesperian Press, Perth. 1203pp.

Augee, M.L. and Fox, M.D. 2000. *Biology of Australia and New Zealand.* Pearson Education, Sydney. 100pp.

Beattie, A.J. (ed). 1995. *Biodiversity, Australia's Living Wealth.* Reed Books, Sydney. 131pp.

Brereton, R., Bennett, S. and Mansergh, I. 1995. Enhanced greenhouse climate change and its potential effect on selected fauna of South-eastern Australia: a trend analysis. *Biological Conservation*, Vol. 72, pp.339–354.

Chilcott, C., Hilbert, D. and Howden, M. 2003. Chapter 5. Modelling biodiversity and climate change. Pp.63–66. In Howden, M., Hughes, L., Dunlop, M., Zethoven, I., Hilbert, D. and Chilcott, C. (eds). *Climate Change Impacts on Biodiversity in Australia. Outcomes of a Workshop Sponsored by the Biological Diversity Advisory Committee, 1–2 October 2002.* Commonwealth of Australia, Canberra. 98pp.

Commonwealth of Australia. 1996. *National Strategy for the Conservation of Australia's Biological Diversity.* Department of Environment, Sport and Territories, Canberra.

Cox, C.B. and Moore, P.D. 2000. *Biogeography: an Ecological and Evolutionary Approach.* 6th Edition. Blackwell Science, Malden, Massachusetts. 298pp.

Flannery, T. 1994. *The Future Eaters: an Ecological History of the Australasian Lands and People.* Reed Books, Sydney. 421pp.

Frankham, R., Ballou, J.D. and Briscoe, D.A. 2002. *Introduction to Conservation Genetics.* Cambridge University Press, Cambridge. 617pp.

Hughes, L. 2000. Biological consequences of global warming: is the signal already apparent? *Trends in Ecology and Evolution*, Vol. 15, pp.56–61.

Humphries, C.J. and Parenti, L.R. 1999. *Cladistic Biogeography: Interpreting Patterns of Plant and Animal Distributions.* 2nd Edition. Oxford University Press, Oxford. 187pp.

Illies, J. 1974. *Introduction to Zoogeography.* Macmillan, London. 120pp.

Intergovernmetal Panel on Climate Change. 2002. Climate change and biodervisity. *I.P.C.C. Technical Paper* V. http://www.ipcc.ch/pub/tpbiodiv.pdf

Kohen, J.L. 1995. *Aboriginal Environmental Impacts.* University of New South Wales Press, Kensington. 160pp.

Kohen, J.L. and Merrick, J.R. 1998. Limited usage of freshwater crayfishes (genus *Euastacus*) by Aborigines in eastern New South Wales: records and comments. *Proceedings of the Linnean Society of New South Wales*, Vol. 119, pp.101–105.

Lee. M.S.Y. 2006. Systematics and the Australasian vertebrate fauna. Pp.3–16. In Merrick, J.R., Archer, M., Hickey, G.M. and Lee, M.S.Y. (eds). *Evolution and Biogeography of Australasian Vertebrates.* Auscipub Pty. Ltd., Sydney. 942pp.

MacDonald, G. 2001. *Biogeography: Space, Time and Life.* 1st

Edition. J. Wiley and Sons, New York. 480pp.

Owen, J.F. and Merrick, J.R. 1998. Re-interpretation of remains of snapper (*Pagrus auratus*) from Holocene middens at Bass Point and Currarong, New South Wales. *Proceedings of the Linnean Society of New South Wales*, Vol. 120, pp.191–201.

Roberts, N. 1998. *The Holocene. An Environmental History*. Second Edition. Blackwell Publishers Ltd., Oxford. 316pp.

Veevers, J.J. 2001. *Atlas of Billion-year Earth History of Australia and Neighbours in Gondwanaland*. Gemoc Press, Sydney. 76pp.

Whiting, A.S., Lawler, S.H., Horwitz, P. and Crandall, K.A. 2000. Biogeographic regionalization of Australia: assigning conservation priorities based on endemic freshwater crayfish phylogenetics. *Animal Conservation*, Vol. 3, No. 2, pp.155–163.

Whitmore, T.C. (ed). 1981. *Wallace's Line and Plate Tectonics*. Clarendon Press, Oxford. 90pp.

Wroe, S. and Field, J. 2001. Mystery of megafaunal extinctions remains. *Australasian Science*, September, pp.21–25.

Wroe, S., Field, J. and Fullagar, R. 2002. Lost giants. *Nature Australia*, Vol.27, pp.54–61.

Background

GEOLOGICAL TIME SCALE FOR THE PHANEROZOIC

ERA	PERIOD or SUB-PERIOD	SERIES/EPOCH	STAGE	AGE m.y.a.	EXTINCTIONS (E) ICE AGE (I)
CAINOZOIC	QUATERNARY	PLEISTOCENE	Calabrian	1.8	E, I
CAINOZOIC	TERTIARY – NEOGENE	PLIOCENE	Piacenzian	3.6	I
CAINOZOIC	TERTIARY – NEOGENE	PLIOCENE	Zanclean	5.3	I
CAINOZOIC	TERTIARY – NEOGENE	MIOCENE Late	Messinian	7.12	
CAINOZOIC	TERTIARY – NEOGENE	MIOCENE Late	Tortonian	11.2	
CAINOZOIC	TERTIARY – NEOGENE	MIOCENE Mid	Serravalian	14.8	
CAINOZOIC	TERTIARY – NEOGENE	MIOCENE Mid	Langhian	16.4	
CAINOZOIC	TERTIARY – NEOGENE	MIOCENE Early	Burdigalian	20.52	
CAINOZOIC	TERTIARY – NEOGENE	MIOCENE Early	Aquitanian	23.8	
CAINOZOIC	TERTIARY – PALAEOGENE	OLIGOCENE Late	Chattian	28.5	E
CAINOZOIC	TERTIARY – PALAEOGENE	OLIGOCENE Early	Rupelian	33.7	
CAINOZOIC	TERTIARY – PALAEOGENE	EOCENE Late	Priabonian	37	
CAINOZOIC	TERTIARY – PALAEOGENE	EOCENE Mid	Bartonian	41.3	
CAINOZOIC	TERTIARY – PALAEOGENE	EOCENE Mid	Lutetian	49	
CAINOZOIC	TERTIARY – PALAEOGENE	EOCENE Early	Ypresian	54.8	
CAINOZOIC	TERTIARY – PALAEOGENE	PALEOCENE Late	Thanetian	57.9	
CAINOZOIC	TERTIARY – PALAEOGENE	PALEOCENE Late	Selandian	60.9	
CAINOZOIC	TERTIARY – PALAEOGENE	PALEOCENE Early	Danian	65	E
MESOZOIC	CRETACEOUS	LATE	Maastrichtian	73	
MESOZOIC	CRETACEOUS	LATE	Campanian	83	
MESOZOIC	CRETACEOUS	LATE	Santonian	87	
MESOZOIC	CRETACEOUS	LATE	Coniacian	89	
MESOZOIC	CRETACEOUS	LATE	Turonian	91	
MESOZOIC	CRETACEOUS	LATE	Cenomanian	97.5	
MESOZOIC	CRETACEOUS	EARLY	Albian	108	
MESOZOIC	CRETACEOUS	EARLY	Aptian	115	
MESOZOIC	CRETACEOUS	EARLY	Barremian	123	
MESOZOIC	CRETACEOUS	EARLY	Hauterivian	130	
MESOZOIC	CRETACEOUS	EARLY	Valanginian	135	
MESOZOIC	CRETACEOUS	EARLY	Berriasian	141	E
MESOZOIC	JURASSIC	LATE	Tithonian	146	
MESOZOIC	JURASSIC	LATE	Kimmeridgian	151	
MESOZOIC	JURASSIC	LATE	Oxfordian	159	
MESOZOIC	JURASSIC	MIDDLE	Callovian	165	
MESOZOIC	JURASSIC	MIDDLE	Bathonian	173	
MESOZOIC	JURASSIC	MIDDLE	Bajocian	180	
MESOZOIC	JURASSIC	MIDDLE	Aalenian	184	
MESOZOIC	JURASSIC	EARLY	Toarcian	190	E
MESOZOIC	JURASSIC	EARLY	Pliensbachian	195	
MESOZOIC	JURASSIC	EARLY	Sinemurian	202	
MESOZOIC	JURASSIC	EARLY	Hettangian	205	E
MESOZOIC	TRIASSIC	LATE	Rhaetian	207	
MESOZOIC	TRIASSIC	LATE	Norian	220	
MESOZOIC	TRIASSIC	LATE	Carnian	230	
MESOZOIC	TRIASSIC	MIDDLE	Ladinian	235	
MESOZOIC	TRIASSIC	MIDDLE	Anisian	241	
MESOZOIC	TRIASSIC	EARLY	Spathian	244	
MESOZOIC	TRIASSIC	EARLY	Nammalian	247	
MESOZOIC	TRIASSIC	EARLY	Griesbachian	251	

This detailed scale is based on Young and Laurie (1996) but Cambrian stages abbreviated ages are rounded to nearest 0.1 million years. More recent dating (e.g. United States Geological Survey) has resulted in some small adjustments to interval boundaries.

ERA	PERIOD or SUB-PERIOD	SERIES/EPOCH	STAGE	AGE m.y.a.	EXTINCTIONS (E) ICE AGE (I)
PALAEOZOIC	PERMIAN	LATE	Tatarian – Dorashamian	255	E
PALAEOZOIC	PERMIAN	LATE	Tatarian – Dzhulfian	259	I
PALAEOZOIC	PERMIAN	LATE	Tatarian – Midian	263	I
PALAEOZOIC	PERMIAN	LATE	Kazanian	267	I
PALAEOZOIC	PERMIAN	LATE	Ulfimian	270	I
PALAEOZOIC	PERMIAN	EARLY	Kungurian	274	I
PALAEOZOIC	PERMIAN	EARLY	Artinskian	285	I
PALAEOZOIC	PERMIAN	EARLY	Sakmarian	293	I
PALAEOZOIC	PERMIAN	EARLY	Asselian	298	I
PALAEOZOIC	CARBONIFEROUS	PENNSYLVANIAN – SILESIAN – STEPHANIAN	Gzhelian	302	I
PALAEOZOIC	CARBONIFEROUS	PENNSYLVANIAN – SILESIAN – STEPHANIAN	Kasimovian	305	I
PALAEOZOIC	CARBONIFEROUS	PENNSYLVANIAN – SILESIAN – WESTPHALIAN	Moscovian	311.5	
PALAEOZOIC	CARBONIFEROUS	PENNSYLVANIAN – SILESIAN – WESTPHALIAN	Bashkirian	314	
PALAEOZOIC	CARBONIFEROUS	SILESIAN – NAMURIAN	Serpukhovian	325	
PALAEOZOIC	CARBONIFEROUS	MISSISSIPPIAN – DINANTIAN – VISEAN	Brigantian	328.5	
PALAEOZOIC	CARBONIFEROUS	MISSISSIPPIAN – DINANTIAN – VISEAN	Asbian	330.5	
PALAEOZOIC	CARBONIFEROUS	MISSISSIPPIAN – DINANTIAN – VISEAN	Holkerian	335	
PALAEOZOIC	CARBONIFEROUS	MISSISSIPPIAN – DINANTIAN – VISEAN	Arundian	341.5	
PALAEOZOIC	CARBONIFEROUS	MISSISSIPPIAN – DINANTIAN – VISEAN	Chadian	344	
PALAEOZOIC	CARBONIFEROUS	MISSISSIPPIAN – DINANTIAN – TOURNAISIAN	Ivorian	348	
PALAEOZOIC	CARBONIFEROUS	MISSISSIPPIAN – DINANTIAN – TOURNAISIAN	Hastarian	354	
PALAEOZOIC	DEVONIAN	LATE	Famennian	364.5	E
PALAEOZOIC	DEVONIAN	LATE	Frasnian	369	
PALAEOZOIC	DEVONIAN	MIDDLE	Givetian	378	
PALAEOZOIC	DEVONIAN	MIDDLE	Eifelian	384	I
PALAEOZOIC	DEVONIAN	EARLY	Emsian	399.5	I
PALAEOZOIC	DEVONIAN	EARLY	Pragian	404.5	I
PALAEOZOIC	DEVONIAN	EARLY	Lochkovian	410	
PALAEOZOIC	SILURIAN	PRIDOLI		414	
PALAEOZOIC	SILURIAN	LUDLOW	Ludfordian	417	
PALAEOZOIC	SILURIAN	LUDLOW	Gorstian	420	
PALAEOZOIC	SILURIAN	WENLOCK	Homerian	422.5	
PALAEOZOIC	SILURIAN	WENLOCK	Sheinwoodian	425	
PALAEOZOIC	SILURIAN	LLANDOVERY	Telychian	428	I
PALAEOZOIC	SILURIAN	LLANDOVERY	Aeronian	431	I
PALAEOZOIC	SILURIAN	LLANDOVERY	Rhuddanian	434	E, I
PALAEOZOIC	ORDOVICIAN	ASHGILL	Bolindian	443	I
PALAEOZOIC	ORDOVICIAN	CARADOC	Eastonian	454.5	
PALAEOZOIC	ORDOVICIAN	CARADOC	Gisbornian	459	
PALAEOZOIC	ORDOVICIAN	LLANVIRN	Darriwilian	467	
PALAEOZOIC	ORDOVICIAN		Yapeenian	471	
PALAEOZOIC	ORDOVICIAN	ARENIG	Castlemainian	477	
PALAEOZOIC	ORDOVICIAN	ARENIG	Chewtonian	481	
PALAEOZOIC	ORDOVICIAN	ARENIG	Bendigonian	484.5	
PALAEOZOIC	ORDOVICIAN	TREMADOC	Lancefieldian	486	
PALAEOZOIC	ORDOVICIAN	TREMADOC	Warendan	490	
PALAEOZOIC	CAMBRIAN	LATE	Datsonian	491	
PALAEOZOIC	CAMBRIAN	LATE	Payntonian	492.5	
PALAEOZOIC	CAMBRIAN	LATE	Iverian	496	
PALAEOZOIC	CAMBRIAN	LATE	Idamean	497.5	E
PALAEOZOIC	CAMBRIAN	MIDDLE	Mindyallan	498.5	
PALAEOZOIC	CAMBRIAN	MIDDLE	Boomerangian	500	
PALAEOZOIC	CAMBRIAN	MIDDLE	Undillan	503	
PALAEOZOIC	CAMBRIAN	MIDDLE	Late Templetonian-Floran	506	
PALAEOZOIC	CAMBRIAN	MIDDLE	Ordian-Early Templetonian	509	
PALAEOZOIC	CAMBRIAN	EARLY	Atdabanian-Toyonian	528	
PALAEOZOIC	CAMBRIAN	EARLY	Tommotian	534	
PALAEOZOIC	CAMBRIAN	EARLY	Nemakit-Daldynian	545	

Chapter 1
Systematics and the Australasian Vertebrate Fauna

Michael S.Y. Lee
Department of Environmental Biology, Adelaide University, Adelaide S.A. 5000 Australia
and Department of Palaeontology, South Australia Museum, North Terrace, Adelaide S.A. 5000 Australia
E-mail: Lee.Mike@saugov.sa.gov.au

INTRODUCTION

Systematics is one of the unifying disciplines of biology, and accordingly pervades nearly all of this book. Indeed, the global importance of the Australasian vertebrate fauna lies not in its diversity, which is not higher than that of other continents, but in its taxonomic uniqueness. For instance, the fauna includes numerous relict forms or 'living fossils' representing lineages that were once highly diverse but which have gone largely or totally extinct everywhere else: the Queensland lungfish (the most primitive surviving lungfish), the echidnas and platypus (the only surviving egg-laying mammals) and the tuatara (a primitive surviving relative of lizards and snakes); see Holdaway and Worthy (2006), Merrick (2006) and Musser (2006). Similarly, the overall composition of the fauna is unique, with groups poorly represented elsewhere being very diverse, and vice versa: the mammal fauna is dominated by marsupials rather than placentals—see pp.551–574 (Wroe and Archer, 2006), and the snake fauna is dominated by highly venomous elapids rather than (generally harmless) colubrids (see Scanlon, 2006).

Knowledge of the relationships between these organisms is vital for answering important questions about their evolution, biogeography and conservation. For instance, it reveals the striking convergences between certain Australasian marsupials and their placental counterparts on other continents. Two of the most famous Australian animal icons fall into this category. The koala is an ecological analogue of a tree sloth (a slow-moving, low-energy arboreal herbivore), while kangaroos are ecological analogues of antelopes (fleet-footed grazers of open habitats). There are also marsupial versions of dogs, cats, moles, squirrels, badgers and rodents. Systematics has demonstrated how the Australian marsupial radiation has closely paralleled the placental radiations on other continents (Wroe and Archer, 2006).

Phylogenetic information allows one to pinpoint exactly where on the tree of life certain evolutionary changes occurred, and whether they coincided with other changes. In mammals, for instance, gradual changes in the feeding system appear to be accompanied by functionally-related changes in the locomotory systems, suggesting that mammals evolved from reptiles via 'correlated progression' (Kemp, 1982). Comparative methods using phylogenetic information are needed when investigating relationships among variables (life history, morphology, genetics) across different species, to account for the fact that closely related species will tend to be more similar in all variables due to shared ancestry rather than adaptation (Harvey and Pagel, 1991; Aplin, 2006; Groves, 2006). The two major processes of biogeography—vicariance and dispersal—also rely heavily on systematic information. The fusion and splitting of continents, and the appearance and disappearance of barriers within land masses (e.g. mountains, rivers, deserts) can also be revealed by concordant patterns of relationship among the affected taxa (Dawson and Dawson, 2006; Heinsohn and Hope, 2006). Indeed, early support for the theory of continental drift came from faunal similarities between Australia and South America, which suggested a previous land connection (White, 2006). Similarly, phylogenetic information can pinpoint examples of dispersal—for instance, how often rodents colonized Australia, and their source populations (Aplin, 2006). Finally, systematics also heavily influences conservation priorities, species with long independent histories, morphological uniqueness, or genetic divergence being of especially high conservation value—see Colgan and O'Meally (2006); the Australasian fauna in general scores heavily on all three criteria. The current global mass extinction has not spared Australasia, with the region having a particularly bad record for mammals: three frog, one bird, and 19 mammal species have disappeared from the Australian mainland since European colonization, and many others are endangered—pp.241–252 (Richards *et al.*, 2006).

What follows is an overview of systematics focusing on concepts and terms used commonly throughout the book and using mainly examples from the Australasian fauna. Systematics encompasses two major fields: phylogenetics and taxonomy. Phylogenetics is concerned with reconstructing evolutionary relationships between organisms: what is the shape of the tree of life? Its concerns are the external problem of how to define and recognize species, and the relationships within and between these species. Taxonomy is concerned with the classification of organisms: after we arrive at a particular tree, how do we name the branches? Its concerns include the naming of species, and of intraspecific and supraspecific taxa.

PHYLOGENETICS: INFERRING RELATIONSHIPS

SOURCES OF DATA

Historically, affinities between organisms were inferred on the basis of the features that were readily observed with the naked eye or light microscope (Mayr and Ashlock, 1991). These usually meant anatomical (morphological) traits, though behavioural and ecological features were also occasionally considered. Of these traits, skeletal anatomy was and remains one of the most utilized sources of information about vertebrate relationships. The reasons for this are both biological and pragmatic. The skeleton of vertebrates is highly complex, includes many easily recognizable ('homologous') landmarks, and exhibits large numbers of potentially illuminating similarities and differences between taxa. It is also relatively conservative (unlike, for instance, certain external features like fur colour), and thus likely to be informative about deeper divergences, such as those between higher taxa. Also, skeletons are easily prepared and stored, and as a result there are large collections in museums for systematists to study. In contrast, dissections showing 'soft' internal organs are difficult to prepare and store (especially for large animals, such as elephants and whales!), and there are consequently far fewer such specimens in most museums. Finally, skeletal features are usually the only features preserved in fossil taxa. Given that over 99% of species are extinct (Novacek and Wheeler, 1992) and known only from their fossilized skeletons (if at all), our understanding of the evolution of skeletal traits is potentially much better than behaviour, soft anatomy and molecules. It has been demonstrated that, without information from fossils, our interpretations of skeletal evolution can be different and probably incorrect (e.g. Gauthier *et al.*, 1988; Lee, 2001). This should make systematists wary of drawing sweeping conclusions from behaviour, soft anatomy or molecules, or any other traits that have only been sampled for limited taxa—if we only had a more complete knowledge of these traits, our interpretations of their evolution might be similarly revised.

Morphology remains an important source of phylogenetic information, and in some cases (most extinct species, i.e. the majority of life) the only potential source. However, progressively more genetic data has been accumulating rapidly, generated by a series of technological advances (see Hillis *et al.*, 1996; Page and Holmes, 1998). In the 1920s, new techniques allowed the determination of the shapes and bandings of chromosomes (the gross structures in the nucleus that contain the genes). Later, the presence of alternative alleles of protein-coding genes could be detected through electrophoresis. Also, the differences between the genotypes of organisms began to be investigated through immunological reactions and DNA hybridization. These were relatively crude techniques that revealed only the overall amount of divergence between two samples of DNA, and provided no information on precisely where those differences were located. More recently, however, cladistic analysis (see below) of allozyme data has improved this method somewhat. Finally, soon after the cracking of the genetic code (the famous double helix) by Watson and Crick (1953), the actual sequence of nucleotides along genes began to be used. Techniques were developed in the 1970s for cloning, amplifying and sequencing precise regions of an organism's genome, which allowed determination of the sequence of nucleotides at those regions. Thus, equivalent ('homologous') sequences from two organisms could be compared, allowing inference of both the overall amount of genetic divergence and also the precise location and nature of these differences.

It is now generally recognized that all heritable traits are potentially phylogenetically informative, and that different sources of data might be better at resolving particular problems. For instance, for very recent divergences between koala populations, allozymes, or rapidly evolving regions of 'junk' DNA (e.g. microsatellites), might be the most useful traits, while slowly-evolving structural genes and skeletal traits are unlikely to have changed at all. However, for more ancient divergences, such as those between the major lineages of marsupials, the latter traits will probably be more informative. Nevertheless, there has been a predictable but unjustified tendency for each new technique to arrive with much fanfare, its practitioners proclaiming that the new source of data was the most reliable yet and made all previous traits obsolete. Later, of course, it was invariably found that the new technique was not as foolproof as first thought, and that it complemented rather than superseded previous techniques. This is perhaps epitomized by the use of DNA sequences, which many once thought would completely resolve the tree of life by virtue of the immense number of nucleotides in each genome and their perceived immunity to large-scale, misleading convergence (e.g. Gould, 1985; Hedges and Maxson, 1996; Givnish and Sytsma, 1997). However, it is being increasingly realized that potentially confounding processes (such as functional convergence, base composition bias and rate heterogeneity) at the molecular level can be extensive, causing even very large molecular data sets such as complete mitochondrial genomes (16,000 bases) to converge on what appears to be the 'wrong' tree (Naylor and Brown, 1998; Page and Holmes, 1998; Curole and Kocher, 1999). Thus, no class of data is universally superior to all other classes. In combined analyses of both morphological and genetic data, the contributions of each type of data to tree resolution is often substantial (e.g. Baker *et al.*, 1998). Groupings of organisms supported by each and every data set are those in which we can have the most confidence.

However, while there is a consensus that all traits are relevant, there remains debate over exactly how to combine different classes of data in phylogenetic analyses. The 'simultaneous' or 'total evidence' approach involves constructing a single large matrix from all the data sets, and getting the optimal tree from this single combined analysis

Plate 5. One of the goannas (Genus *Varanus*). This group of reptiles has undergone extensive radiation in Australasia and includes species that are among the largest terrestrial predators in some areas. Photograph S. Allen.

Plate 6. This rare Oenpelli rock python (*Morelia oenpelliensis*) is from Kakadu in the Northern Territory. Photograph J. Cann.

Plate 7. The yellow-nosed albatross (*Diomedea chlororhynchos*) is among the smaller albatross species. It breeds on several southern ocean islands and regularly visits south-western, southern and south-eastern Australian coasts in winter. Photograph S. Allen.

Plate 8. The crested tern (*Sterna bergii*) is very common around the shores of Australia and Asia. Photograph S. Allen.

Plate 9. The Australian pelican (*Pelecanus conspicillatus*) is common on estuaries and inland waterways throughout Australia, but extends to New Guinea and other islands. Photograph S. Allen.

Plate 11. Kangaroos are often associated with inland or arid environments, but these macropods also live in coastal habitats. Photograph S. Allen.

Plate 10. The Australian brush-turkey (*Alectura lathami*) has an extensive range from Cape York south to the Hawkesbury River, N.S.W. It inhabits rainforest, dry scrubs and dense vegetation in coastal drainages—rarely occurs west of the Great Dividing Range. Photograph J. Cann.

(Kluge, 1989; Nixon and Carpenter, 1996). The 'taxonomic congruence' approach advocates analysing each data set separately, and seeing if there are groupings common to all the trees (Miyamoto and Fitch, 1995). The 'conditional combination' approach treads the middle ground (Bull *et al.*, 1993; Huelsenbeck *et al.*, 1996). It involves testing the separate data sets first for disagreement. If there is no statistically-significant conflict, the data sets are combined into a single analysis (the 'simultaneous' approach), while if there is severe conflict, the data sets should be analysed separately and the resultant trees compared for agreement (the 'taxonomic congruence' approach).

Finally, a fable presented by Avise (1994; see also Hillis, 1999) emphasizes very clearly how the apparent replacement of morphological data with new (and allegedly better) molecular techniques might largely be an accident of history. Imagine if humans were infinitesimally small and could only see chains of nucleotides, and not gross morphological features, with the naked eye. For decades systematists attempted to infer relationships between organisms through comparisons of nucleotide sequences alone, each organism being represented by a chain of the same four bases (A, T, C, G), with convergence and reversals being ubiquitous. Then a surveyor invented a new technique to map the shapes of morphological features in each organism. Each organism, it was now discovered, had its own unique set of morphological structures, and the amount of complexity and variation in these structures was apparently infinite, unlike the variation in the nucleotide sequences. It is likely that many systematists would hail morphology as the panacea to their problems, conclude that with such a large number of possible forms each morphological character must be almost immune from convergence, and proclaim that molecular techniques were now obsolete!

METHODS OF ANALYSIS

The differing opinions over which traits are best for phylogenetic analysis pale in comparison to the disputes over what analytic methods to use (Hull, 1988). What follows is a brief overview of the major methods; more detailed general reviews of the history of taxonomy can be found in Ridley (1986), Hull (1988), Mayr and Ashlock (1991) and Panchen (1992). Most of the systematic chapters in this book are concerned with groups of species ('higher taxa'), and so this discussion will focus on inferring the branching relationships between species, rather than on delimiting individual species and inferring the (often reticulating) relationships within them. The complexities of recognizing and naming individual species are discussed at length in three recent edited volumes (Claridge *et al.*, 1997; Howard and Berlocher, 1998; Wilson, 1999); see also Lee (2003).

Evolutionary Systematics and Phenetics

For a very long time inferred relationships between organisms were based largely on vague concepts of overall similarity, with emphasis on characters that systematists suspected of being 'more reliable'. This school of phylogenetics, which was prominent until the 1960s, came to be known as evolutionary systematics. However, the concept of a 'reliable' character was vague. When different experts focused on different characters and drew different phylogenies, there was no way to objectively settle the dispute. Nevertheless, much useful systematic work was done within this framework (or lack thereof), and many of the groupings inferred by these workers are still accepted today. Also, evolutionary taxonomists tended to focus (validly) on derived rather than primitive characters in inferring relationships, but this crucial distinction was never made explicit until the advent of cladistic analysis (see below).

Phenetics emerged partly as a reaction to the perceived subjectivity and authoritarianism of evolutionary taxonomy, and was facilitated by advances in computing. Rather than intuitive use of perceived similarity and 'reliable' characters to calculate phylogeny, pheneticists advocated a rigorous mathematical approach. Hence, as many characters as possible were scored for each taxon, and each was assigned the same unit weight. The overall similarity between each taxon was then calculated, and taxa clustered together based on their degree of similarity. It was initially proclaimed that this system of examining as many characters as possible and assigning them all the same weight led to more objective and reproducible results than evolutionary taxonomy. However, this apparent objectivity was undermined because there was no way to define a unit character. Furthermore, the same character data could yield many different trees, depending on what formula was used to calculate the overall similarity ('distance') between taxa, what algorithm was used to cluster these taxa, and even the order in which groups were joined to the growing tree (e.g. Farris *et al.*, 1996; de Queiroz and Good, 1997).

There was an additional problem with phenetics: it was concerned only with depicting similarity between organisms, and not with phylogeny. However, there was often the additional assumption that the patterns of overall similarity found in phenetic analyses also reflected phylogenetic relationships. This assumption will hold only if rates of evolution are constant—if so, taxa will accumulate differences at a constant rate and the amount of difference will indeed be directly related to time since divergence. However, if rates of evolution are highly variable, as is often the case, this assumption becomes dubious.

Cladistics

With the English translation of Hennig (1966), cladistic analysis began to supersede phenetics in the late 1960s. Like phenetics, it clusters organisms based on shared traits—however, while phenetics considers all traits indiscriminately, cladistics only groups taxa on the basis of shared *derived* traits (synapomorphies). Shared *primitive* traits (symplesiomorphies) and unique derived

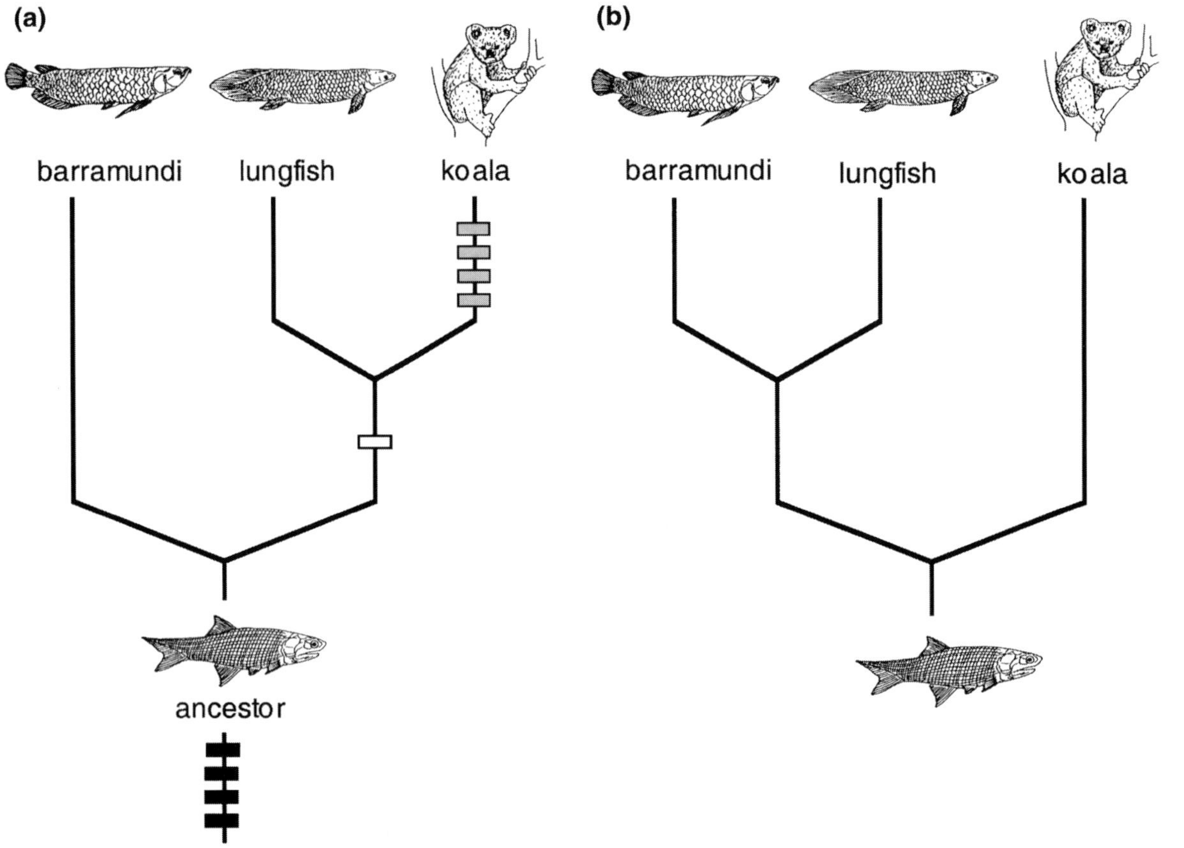

Figure 1. Illustration of why shared derived features, and not overall similarity, are accurate indicators of phylogenetic relationships. **(a)** This arrangement represents accepted relationships between the barramundi, lungfish and koala with character changes shown as bars on the branches. Shared primitive features between the barramundi and the lungfish are shown in black: these were retained from the common ancestor of all three taxa. Shared derived features uniting the lungfish and koala are shown in white, and unique derived features of the koala are shown in grey. If only shared derived features are considered, as in cladistic analysis, this topology is retrieved. **(b)** The arrangement obtained when all traits are considered indiscriminately, as in phenetic clustering.

traits (autapomorphies) are deliberately ignored. As the following example shows, when rates of evolution are highly variable, cladistics can often retrieve the 'correct' tree where phenetics fails.

We wish to work out evolutionary relationships between three taxa, the barramundi, the lungfish and the koala. The group consisting of all (three) taxa under investigation is termed the *ingroup*. Everyone agrees that the lungfish and the kangaroo are more closely related to one another than either is to the barramundi (Figure 1(a)). However, the barramundi and the lungfish are most similar—thus, a phenetic analysis would erroneously produce the arrangement in Figure 1(b). Why does similarity here not reflect affinity? The reason is that the ancestor of all three taxa was itself a fish, very similar to both the barramundi and lungfish (how we can infer this is discussed later). The barramundi and lungfish are similar because, although they diverged long ago, they did not evolve very much after this divergence. Thus, all their fishy similarities (e.g. gills, scales and fins) were already present in the common fish-like ancestor of the ingroup. They are shared primitive features (symplesiomorphies) of the barramundi and lungfish—see black bars in Figure 1(a). In contrast, although the koala has diverged from the lungfish quite recently, it has evolved very rapidly and lost most of its fishy characteristics within a short time. As a result, it differs greatly from the lungfish (and therefore also from the barramundi). However, the advanced traits in the koala (legs, ears and fur) are not found in either the barramundi, the lungfish, or the common ancestor of the ingroup. They are *unique* derived features (autapomorphies) of the koala—see grey bars in Figure 1(a). All these above traits would (erroneously) group the barramundi and lungfish, and separate off the koala. In contrast, only a few traits group (correctly) the lungfish and the koala—such as presence of lungs and fleshy appendages. These were not present in the fishy ancestor of the ingroup. These traits are therefore shared derived features (synapomorphies) of the lungfish and koala—see white bar in Figure 1(a).

Phenetics considers all traits indiscriminately, and thus groups the lungfish and the barramundi together, separating off the koala, because of the numerous shared primitive characters of the lungfish and barramundi, and unique derived characters of the koala. The shared derived features, which correctly group the lungfish and the koala, would also be considered but would be overwhelmed by the other characters. In contrast, cladistics explicitly excludes the latter traits, and would consider only the shared derived features. As a result, a cladistic analysis would correctly converge on the lungfish-koala grouping.

Of course, life (literally) isn't always so neat. Different shared derived characters might support different groups: for instance, we might later find a new shared derived trait that apparently unites the barramundi and koala. Disagreement between different derived traits is termed *homoplasy, character conflict,* or *incongruence.* In such cases, democracy prevails—the tree supported by the largest number of synapomorphies is provisionally accepted, and the (hopefully few) incongruent synapomorphies are consequently assumed to have evolved convergently (or to have reversed). This amounts to accepting the tree that implies the least homoplasy, termed the *shortest* or *most parsimonious* tree. The amount of homoplasy in an analysis is reflected in the consistency index (CI) and retention index (RI), which both decrease from a maximum of 1.0 as homoplasy increases. Groups that are supported by many characters and contradicted by few are the most convincing; such clades have greater branch support (Bremer, 1988; also termed 'Bremer support', 'support index' or 'decay index') and bootstrap frequencies (Felsenstein, 1985; Sanderson, 1995).

The above analysis assumed that we knew what the ancestor of our ingroup looked like, and used this knowledge to infer which traits were primitive (already present in this ancestor) and which were derived (not present in this ancestor). Determination of the direction of character evolution is called *polarity.* Of course, we will rarely, if ever, know actual ancestors, such as the actual ancestor of barramundis, lungfishes and koalas. However, we can make a good guess as to what it looked like using a method termed *outgroup comparison.* This involves examining taxa that are closely related to, but outside of, the clade under investigation (the ingroup). The character state present in all these outgroups was most probably present in the common ancestor of the ingroup. For instance, in attempting to infer whether fins or legs is primitive for our ingroup, we would look at the nearest outgroups. Sharks and extinct armoured fish called placoderms are closely related to, but outside of, our ingroup, and thus are appropriate outgroups. We would observe that they both had fins, not legs, and thus conclude that presence of fins was primitive for our ingroup, and presence of legs derived. (Note that we cannot use taxa such as kangaroos for 'outgroup comparison' in our analysis, since kangaroos do not lie outside our ingroup, but are actually very closely related to koalas). There are additional methods of inferring polarity of characters, such as through ontogeny (the character state that appears first in development is assumed to be primitive), or using stratigraphy (the character state that appears first in the fossil record is assumed to be primitive). However, both these methods are generally seen as less accurate than outgroup comparison, and cannot be employed to polarize molecular data.

The best recent overview of cladistic methology is Kitching *et al.* (1998); other useful summaries are Ridley (1986, 1993), Harvey and Pagel (1991) and Page and Holmes (1998). Also, the manuals for two of the most widely used computer programs for cladistic analyses, MacClade (Maddison and Maddison, 2000) and PAUP (Swofford and Begle, 1993; Swofford, 1999) also provide extensive background information. Cladistic methods for inferring trees are now used in almost all morphological studies, including those in this book. However, for molecular sequence data, methods known as maximum likelihood are increasingly being used.

Maximum Likelihood

Maximum likelihood and related Bayesian methods are the most recent methods of tree construction. Because they incorporate precise probability models of trait evolution, they are potentially the most powerful, but also entail the most assumptions. Currently, we do not have adequate models for morphological evolution, or for the certain types of genetic changes (e.g. deletions, insertions, and changes in gene order), so likelihood analyses can be performed only on certain types of molecular data (i.e. substitutional differences between sequences of nucleotides).

First we need a model of sequence evolution, i.e. the transformation probabilities between the four bases that are arranged linearly to form DNA sequences. These bases consist of two purines (A, G) and two pyrimidines (C, T). In the most simple model, all possible changes between these four bases are assigned the same probability. More complex models assign different probabilities to transitions and transversions. Transitions are purine-to-purine (A-G) or pyrimidine-to-pyrimidine (C-T) changes. Transversions are changes between purines and pyrimidines (A-C, A-T, G-C, G-T). For some molecules (e.g. protein-coding regions), functional constraints mean that transitions, where a base is replaced by another of the same type, are more probable than transversions. The most complex models assign separate probabilities to each of the six possible changes, and also consider other factors such as base composition bias (if there are already 'too' many Gs in a molecule, changes away from G might be more likely than changes to G). Finally, apart from different probabilities for various substitutions at each site, variations in the overall rate of evolution across different sites ('among-site rate heterogeneity') can be accommodated.

Once a model is adopted, we then take a tree and observe the distribution of character states for a particular site across all taxa. Under the model, the probability of getting the observed distribution can be determined. The same calculations are done for all other sites, and the likelihood of the observed data that tree topology is the product of the likelihoods for all sites (because these likelihoods are very small, they are usually expressed as logarithms). The likelihood of alternative possible tree topologies can be calculated in the same fashion, and the preferred tree topology is the one that confers the highest likelihood on the data. The strength of support for a particular clade can be evaluated using the Kishino-Hasegawa and other tests. If the optimal likelihood tree, which contains the clade of interest, is significantly ($p<0.05$) more likely than the best tree that lacks the clade, the clade can be assumed to be significantly supported.

Of course, rather than uncritically adopting a particular model, it is better to experiment with a range of models, and investigate which fits the data best. There is a problem, however, in that the more complex models (with more parameters free to vary) will inevitably fit the data more accurately than more simple models. However, it is possible to evaluate (by a likelihood ratio test) if the extra parameter improves the fit by just a small and insignificant amount, or a large and significant ($p<0.05$) amount. If the improvement is insignificant, the simpler model is all that is needed; if the improvement is significant, the more complex model is required.

Clear and balanced introductions to likelihood are Page and Holmes (1998) and Lewis (1998); other recent but more technical treatments are those of Yang *et al.* (1995), Huelsenbeck and Crandall (1997), and Nei and Kumar (2000). Useful introductions to Bayesian implementation of likelihood models include Huelsenbeck *et al.* (2001) and Hall (2001).

WHICH METHOD?

Intuitive methods (evolutionary taxonomy) are generally now seen as insufficiently rigorous, while phenetic methods are seldom used because they are extremely sensitive to rate heterogeneity, which appears to be common in both morphological and molecular data. Thus, the methods currently most commonly employed today are parsimony and likelihood.

Parsimony remains the only method readily applicable to morphological traits and some genetic traits. It is computationally fast and does not require many explicit assumptions (although it makes some implicit ones!). Thus, simultaneous analyses of hundreds of taxa are feasible (e.g. Soltis *et al.*, 1998; Kallersjö *et al.*, 1998). However, it has been demonstrated to fail under some circumstances: when *distantly* related taxa have each evolved very rapidly, parsimony tends to group them together erroneously, a phenomenon known as long-branch attraction (Felsenstein, 1978). Likelihood models have been argued to be more realistic, and less susceptible to long-branch attraction. However, it has been shown recently that when *closely* related taxa have each evolved very rapidly, likelihood tends to erroneously push them apart (Yang, 1998), a phenomenon dubbed long-branch repulsion (Siddall, 1998). In such cases, parsimony performs better and tends to correctly group these taxa together. Likelihood models also require knowledge of many parameters: different adopted values of these parameters can give different trees, and in practice one can never know which values are 'correct' (Siddall and Kluge, 1997; Page and Holmes, 1998; Lee, 1999a). They are also much more computationally intensive, and analyses of very large taxon sets are currently unfeasible. In defence, though, one can try a range of parameter values and adopt the one that seems to fit the data best (e.g. Yang *et al.*, 1995; Lewis, 1998), while computational limitations are undoubtedly only temporary.

More importantly, however, likelihood can only be readily applied at the moment to molecular sequences. Thus, simultaneous analyses of molecular sequence data with other types of molecular data (insertions and deletions, allozymes, restriction fragments) and with morphology can only be undertaken using parsimony. It is no coincidence that the strongest advocates for such simultaneous analyses are also advocates of parsimony! Because both parsimony and likelihood have their strengths and weaknesses, the debate over which one is universally better might not be resolvable. It might also be unnecessary: in practice, for most real data sets, both parsimony and likelihood give similar if not identical trees.

CLADOGRAMS AND EVOLUTIONARY TREES

Species usually diverge and seldom merge. Accordingly, the results of phylogenetic analyses are usually presented in the form of a branching diagram, where the lineages diverge and never merge. The most basic form of diagram is the cladogram, which presents information only on recency of common ancestry, and not on rates of evolution or on actual ancestor-descendant relationships. Thus, the cladogram in Figure 1(a) only shows that the lungfish and koala share a more recent ancestor than either does with the barramundi. It does not specify *how* lungfishes and koalas are related: lungfishes might be ancestral to koalas, koalas might be ancestral to lungfishes or (most likely) both might be descended from a different and as yet unknown ancestor. Stratigraphic information, and information on specializations (unique derived traits) of the koala and/or lungfish, must be used to evaluate these possibilities. Furthermore, taxa grouped together on a cladogram are not necessarily the most (or even very) similar, as exemplified by the koala-lungfish grouping: they are merely those that share the most recent common ancestor.

Phylogenetic trees are similar to cladograms except that presumed ancestor-descendant relationships are also depicted. An even more detailed method of presenting phylogenetic information is the phylogram, where each branch length is scaled to reflect the inferred amount of evolutionary change along it. Reticulated networks can also be used, to either depict fusion of lineages or character conflict. A good introduction to all these types of diagrams is Page and Holmes (1998). However, the simplest diagram, the cladogram, remains the most common and is the one used by most chapters in this book.

MONOPHYLY, PARAPHYLY AND POLYPHYLY

There are three major types of groupings that can be identified on cladograms: monophyletic, paraphyletic and polyphyletic. A monophyletic group, or *clade,* includes a node (hypothetical ancestor) on a cladogram and all taxa descended from that node (Figure 2(a)). Thus, all taxa in a monophyletic group share an exclusive common ancestor that no other taxa share. Monophyletic groups are readily recognized (diagnosed) by shared derived features (synapomorphies). A good example is the taxon Aves (birds), diagnosed by feathers, wings and a wishbone.

A paraphyletic group results when some highly aberrant taxa are excluded from an (otherwise) monophyletic group, and is diagnosed by shared primitive features (Figure 2(b)). A good example is reptiles, which is what is left over when birds and mammals are removed from an otherwise monophyletic Amniota. Reptiles are grouped together on features primitive for amniotes, i.e. scaly skin, ectothermy, sprawling gait.

A polyphyletic group results when distantly related taxa are grouped together, and is diagnosed by convergent features (Figure 2(c)). A good example is Haematothermia, which includes two distantly related groups (birds and mammals) that have separate origins and share only convergent features related to high metabolic rate (insulation, upright gait, endothermy). In practice the distinction between paraphyletic and polyphyletic taxa can sometimes be difficult, but both have in common the feature that they do *not* include an ancestor and all its descendants. Thus, members of paraphyletic or polyphyletic groups are not all each other's closest relatives. Some members are more closely related to members of *other* groups than to other members of their own group. Both types of assemblages are consequently are considered 'artificial' by most systematists.

Two monophyletic groups that are each other's closest relatives are termed sister-taxa. Sister taxa are important in biogeographic, stratigraphic, and evolutionary studies because they are equal in age and share an exclusive common ancestor.

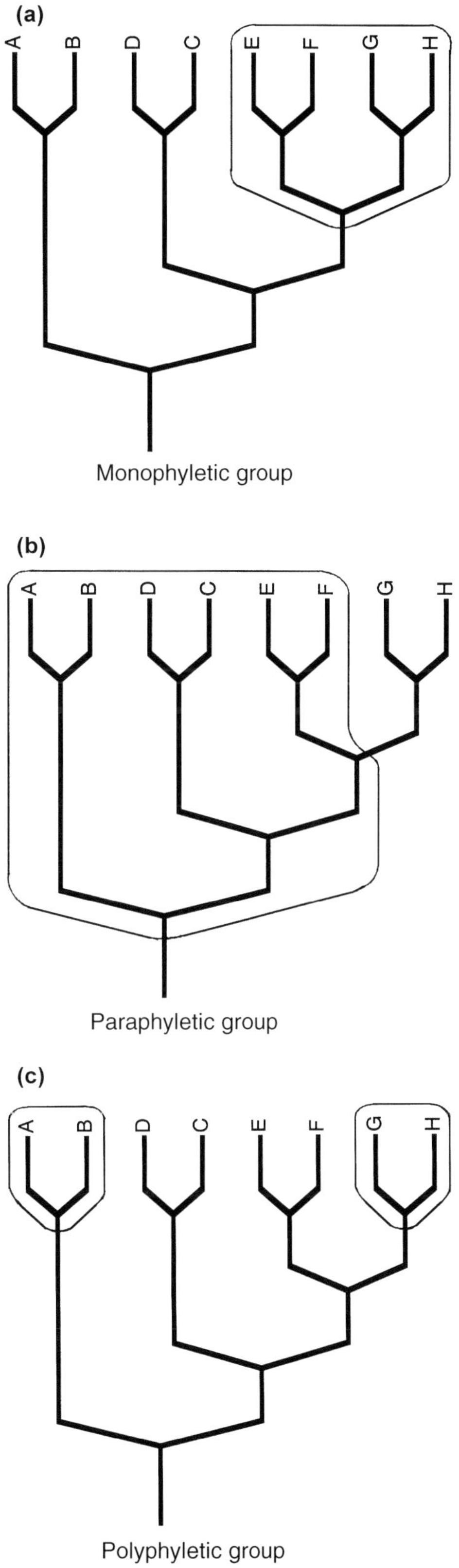

Figure 2. Illustrations of **(a)** monophyletic, **(b)** paraphyletic, and **(c)** polyphyletic groups. Note that only the monophyletic group includes a single common ancestor and all its descendants.

CLASSIFICATION: NAMING TAXA

Classification is the practice of attaching names to groups of organisms. Again, the focus here will be on naming groups of species ('higher taxa') rather than naming individual species, which is covered elsewhere

(ICZN Committee, 1999). In recent years, it has been realized that higher taxa should be clades, i.e. monophyletic groups of species diagnosed by shared derived traits. There has been an associated trend to eliminate paraphyletic and polyphyletic groups from formal classifications, although some paraphyletic groups continue to be used colloquially ('fishes', 'amphibians', 'reptiles'). Almost all the chapters in this book formally recognize only monophyletic higher taxa.

LINNEAN TAXONOMY

The Linnean system (Linnaeus, 1758) has been in use for centuries, and is a system for (1) attaching names to higher taxa (now usually clades) and (2) assigning each named higher taxon to a rank. However, despite being the 'official' approach to naming higher taxa (ICZN Committee, 1999), it is inadequate in many respects. Such shortcomings are not surprising, since the Linnean system was conceived a century before evolution (and thus phylogenies) became generally accepted, and was never intended to be a system for naming clades (de Queiroz, 1992, 1997).

The procedure of naming higher taxa in the Linnean system is largely arbitrary (de Queiroz, 1997; Lee, 1999b). Names are not explicitly tied to particular clades, but can be shifted up or down a tree at whim. A conventional definition of Hominidae, for instance, is the type species of the family and all species deemed sufficiently similar and/or closely related enough to be considered members of the same family. Once the taxonomist has (rather subjectively) decided on the boundaries of this clade, the diagnostic synapomorphies can be listed. However, another taxonomist might have more liberal (or stringent) criteria regarding the necessary degree of similarity or relationship to the type species. That person might therefore apply the same name to a more inclusive (or restricted) clade—if so, not only the included species, but also the diagnostic characters of Hominidae will change. There is no objective way to resolve these disputes—rather, the interpretation of Hominidae that is eventually adopted by the majority of scientists is the 'correct' one. If there is no consensus, instability and confusion will reign.

Once a higher taxon is named, it must be assigned to a Linnean rank. From lowest (least inclusive) to highest (most inclusive), these ranks are: species, genus, family, order, class, phylum, kingdom. Because seven primary ranks were often not enough to represent the complex fractal nesting of clades in large phylogenies (e.g. Figure 4;various prefixes (e.g. 'super' and 'infra') are often used to create additional, secondary ranks (e.g. superorder, infraclass). However, there are (surprisingly) few firm rules for determining the rank for a certain named clade, other than a general agreement that sister groups should be assigned the same rank and that higher-ranked taxa must be more inclusive than lower-ranked taxa. Apart from these loose guidelines, assigning ranks is largely based on tradition and intuition. A family, for instance, is often merely 'the clade a taxonomist deems distinctive or large enough to be called a family'. Thus, two sister clades might be dubbed separate families by one taxonomist, while another worker might lump both into a single family. There is no objective way to resolve such disputes, which create unnecessary instability in classification.

As a result of this subjectivity in ranking, higher taxa in different groups are often non-equivalent in terms of age of origin, species diversity or morphological and genetic distinctness. For instance, bird genera are, on average, less genetically distinct than reptile or mammal genera (e.g. Johns and Avise, 1998). Despite the fact that higher taxa are largely subjective constructs, it has been shown that counts of such higher taxa do mirror patterns in species and morphological diversity, suggesting that Linnean higher taxa are not totally arbitrary or devoid of information content (e.g. Balmford *et al.*, 1996; Lee, 1997). Clades that are more speciose or diverse do tend to be divided into more genera and families, although these divisions are far from objective and precise.

Proponents of Linnean ranks have also argued that ranks convey information about inclusiveness of higher taxa—for instance, if one knows that *Homo* is a genus and Hominidae is a family, then *Homo* must be nested within (i.e. a subgroup of) Hominidae, and thus be younger and less diverse. However, no such information is actually conveyed: without reference to a phylogeny that specifies that *Homo* is nested within Hominidae, there is no way that ranks alone can convey that *Homo* is a subgroup of Hominidae. The genus *Homo* might be a subgroup of a totally different family (e.g. Canidae), in which case it cannot be assumed to be younger or less diverse than Hominidae. The only information that ranks convey is that the higher-ranked taxon (e.g. the family Hominidae) cannot be a subgroup of a lower-ranked taxon, (e.g. the genus *Homo*), and that two names that have the same rank must refer to non-overlapping clades.

It has also been argued that ranks somehow make it easier for people to organize their knowledge of biological diversity. Large phylogenies (e.g. Figure 4) can contain dozens or even hundreds of nested clades, making them rather intimidating. It can be difficult to identify and remember 'important' clades, e.g. those that are particularly diverse, marked by major evolutionary innovations, or have special relevance to humans. However, the argument goes, if 'important' clades (e.g. Mammalia, Aves) are assigned to primary Linnean Ranks (e.g. class, family), and less important clades (e.g. Therapsida) are assigned to secondary ranks (e.g. superclass, suborder), then people's attention can be readily focused on certain key clades. This argument is difficult to justify rigorously. Most non-biologists already know the important clades of animals (crustaceans, insects, spiders, vertebrates, snakes, turtles, crocodiles, birds, mammals, marsupials, placentals) even if they have no knowledge of their Linnean ranks. Thus, names of 'important' groups will be used most frequently and will consequently be widely known and recognized—whether or not they are attached to primary ranks.

Conventional Linnean taxonomy is therefore unsatisfactory for two reasons. First, names are not explicitly tied to particular clades by some objective criterion, but can be shifted up or down a tree at whim. Second, even if there is agreement to attach a name to a particular clade, there might still be disagreement over what rank to assign that name. If ranks are to have standard endings (e.g. families ending in -idae), changes of rank will also necessitate changes in the names themselves. These problems can cause substantial instability even when biologists are in total agreement about the phylogeny!

PHYLOGENETIC TAXONOMY

A rigorous system for naming clades, called *phylogenetic taxonomy*, has recently been proposed that does not suffer from the above problems (de Queiroz and Gauthier, 1992, 1994). The new system is already increasingly being adopted, and a formal code has been released (Cantino *et al.*, 1999, 2000). Its proponents are optimistic that in future higher taxa (i.e. clades) will be named according to these rules, with traditional Linnean taxonomy used only for naming species, if at all. Phylogenetic taxonomy is a system of nomenclature explicitly designed for naming clades unambiguously. It attaches names to clades by marking the boundary of the named clade precisely using a taxon and a diagnostic synapomorphy, or two taxa. Reference taxa in this system are species and thus ultimately, their type specimens. The three most commonly used types of definition are as follows (Figure 3):

Apomorphy-based: the clade diagnosed by trait X homologous (=synapomorphic) with that in species A (i.e. the first species with trait X, and all its descendants).

Node-based: the least inclusive clade containing species A and B (i.e. the latest common ancestor of A and B, and all its descendants)

Stem-based: the most inclusive clade containing species A and not species C (i.e. the earliest ancestor of A but not C, and all its descendants).

The clades shown in **bold** in Figure 4 might be defined using these apomorphy-based definitions.

Theria: the clade diagnosed by live-bearing homologous with that in humans (Eutheria).

Mammalia: the clade diagnosed by hair homologous with that in humans (Eutheria).

Amniota: the clade diagnosed by an amniotic egg homologous with that in humans (Eutheria).

Tetrapoda: the clade diagnosed by dactyly (fingers) homologous with that in humans (Eutheria).

Sarcopterygii: the clade diagnosed by fleshy appendages homologous with that in humans (Eutheria).

Gnathostomata: the clade diagnosed by jaws homologous with that in humans (Eutheria).

Vertebrata: the clade diagnosed by vertebrae homologous with that in humans (Eutheria).

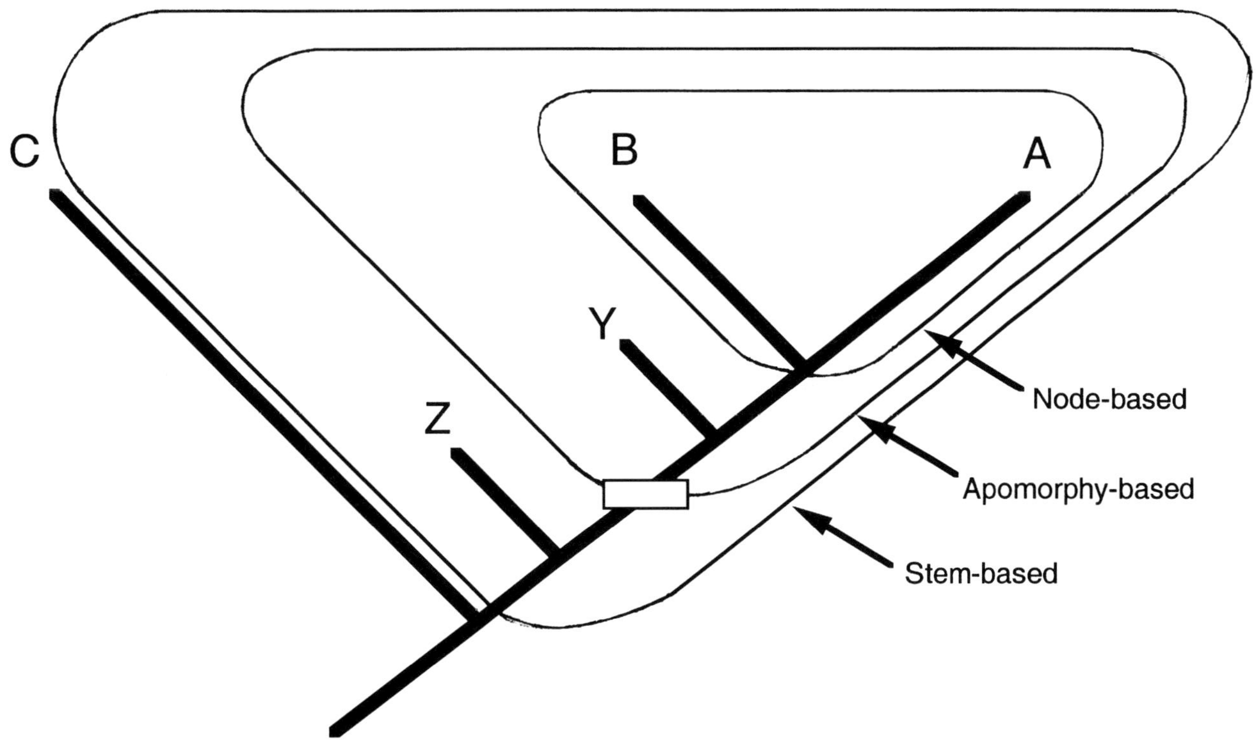

Figure 3. Illustrations of the three types of phylogenetic definitions of taxon names. See text for relevant definitions. Origin of the diagnostic apomorphy is depicted by the horizontal bar.

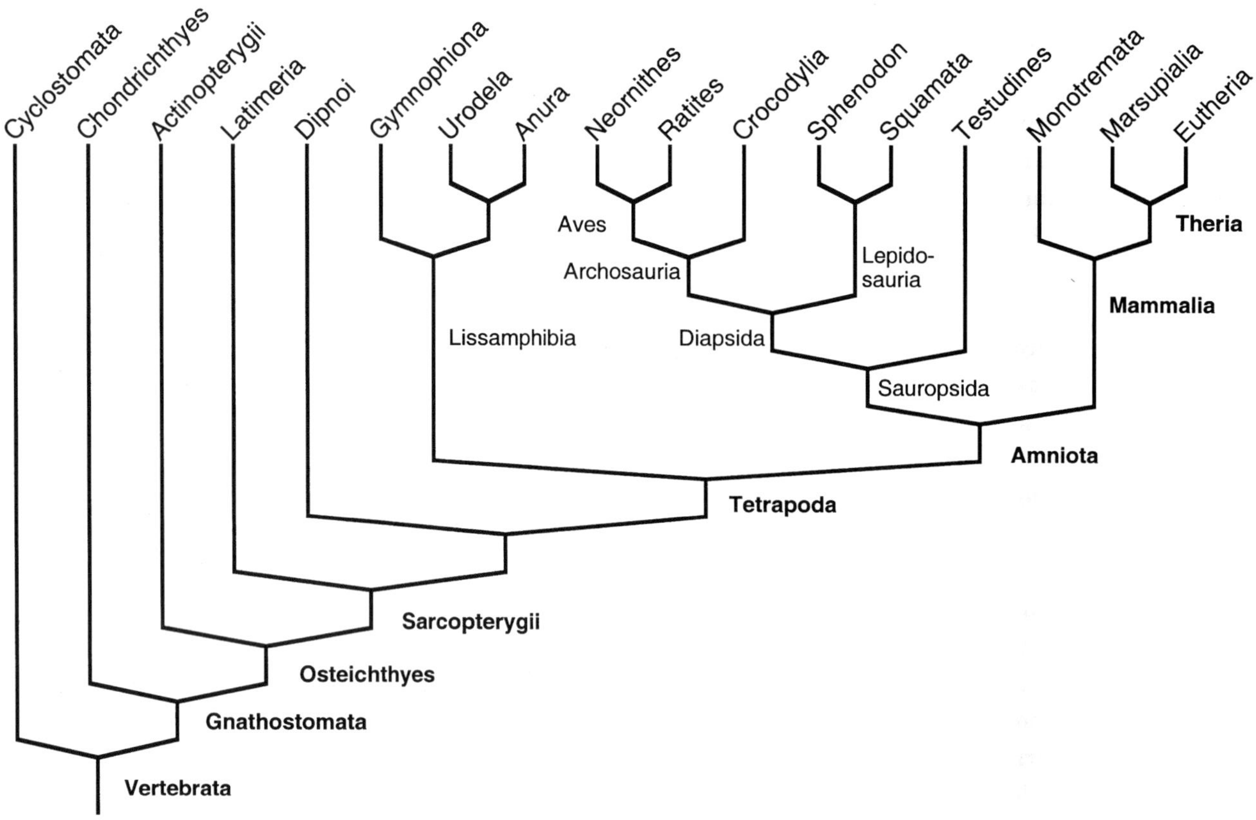

Figure 4. Traditionally accepted relationships between living vertebrates. For list of common names for these taxa, see Box 1. Note that new species are being discovered continuously, while (regrettably) some are going extinct, so numbers are always changing. Clade names in bold are those defined using examples of apomorphy-based and node-based definitions in the text.

These definitions use *Homo sapiens* as a 'reference' or 'specifying' taxon, and delineate a series of successively more-inclusive clades using successively more widespread synapomorphies (see Lee, 1999b). Alternatively, a series of clades might have been defined using node-based definitions:

Theria: the least inclusive clade containing humans (Eutheria) and koalas (Marsupialia).

Mammalia: the least inclusive clade containing humans (Eutheria) and echidnas (Monotremata).

Amniota: the least inclusive clade containing humans (Eutheria) and kookaburras (Neornithes).

Tetrapoda: the least inclusive clade containing humans (Eutheria) and cane toads (Anura).

Sarcopterygii: the least inclusive clade containing humans (Eutheria) and coelacanths (Latimeria)

Gnathostomata: the least inclusive clade containing humans (Eutheria) and great white sharks (Chondrichthyes).

Vertebrata: the least inclusive clade containing humans (Eutheria) and hagfishes (Cyclostomata).

These definitions again use *Homo sapiens* as a reference taxon, but now delineate successively more-inclusive clades by using successively more distantly related species as additional landmarks (see Lee, 1999b).

If the phylogeny is stable, such phylogenetic definitions will also be stable. However, if the areas of the phylogeny are uncertain, then taxa might shift into or out of the boundaries of the named clade (this problem, of course, also affects clades named according to Linnaean rules). There are, however, ways to 'buffer' phylogenetic definitions of taxon names in the face of unstable phylogenies, through judicious choice of reference taxa, the use of the right type of definition (node-, stem-, or apomorphy-based) for a given situation, and careful phrasing of the actual definition (e.g. Lee, 1998, 1999b; Sereno, 1999).

Because Linnean ranks cause instability, are arbitrary, and have been argued to convey little information, the named clades in phylogenetic taxonomy are deliberately unranked. Where necessary, information about inclusiveness is contained in a cladogram or an indented taxonomy (e.g. see pp.492–493, Figure 1). In such taxonomies, nesting of clades is conveyed by indenting. Clades defined under phylogenetic taxonomy, if desired, might be subsequently assigned Linnean ranks, but this extra step is not governed by the rules of phylogenetic taxonomy and is discouraged for the reasons above.

BOX 1. SCIENTIFIC AND COMMON NAMES FOR VARIOUS VERTEBRATE CLADES DEPICTED IN FIGURE 4.

Actinopterygii: ray-finned bony fishes.
Amniota: amniotes (reptiles, birds and mammals).
Anura: frogs and toads.
Archosauria: birds, crocodiles and their extinct relatives.
Cyclostomata: jawless fishes (lampreys and hagfishes).
Chondrichthyes: cartilaginous fishes (sharks, rays and chimaeras).
Crocodylia: crocodiles and alligators.
Diapsida: diapsid reptiles (lepidosaurs and archosaurs).
Dipnoi: lungfishes.
Eutheria: placental mammals.
Gymnophiona: caecilians (limbless worm-like amphibians).
Latimeria: coelacanth.
Lepidosauria: squamates and the tuatara.
Lissamphibia: modern amphibians.
Mammalia: monotremes, marsupials and placental mammals.
Marsupialia: marsupial mammals.
Monotremata: platypus, echidnas and their extinct relatives.
Neornithes: modern birds.
Osteichthyes: bony fishes (including tetrapods).
Ratites: primitive flightless birds (e.g. emus and ostriches, cassowaries, kiwis).
Sarcopterygii: lobe-finned bony fishes.
Sauropsida: living reptiles, birds and their extinct relatives.
Sphenodon: the tuatara.
Squamata: lizards, snakes and amphisbaenians.
Testudines: turtles and tortoises.
Tetrapoda: limbed vertebrates.
Urodela: salamanders and newts.

AUSTRALASIAN VERTEBRATES—A PHYLOGENETIC 'ROAD MAP'

The cladogram in Figure 4 outlines how each of the extant Australasian groups discussed in this volume fits into higher-level vertebrate phylogeny as a whole. The species diversity of each group is also shown. Thus, for instance, the cladogram shows the position of monotremes with respect to other vertebrates, and the clade is covered in Chapter 26 (Musser, 2006), which deals exclusively with the group; the status of monotremes is also briefly discussed in Chapter 25 (Johanson, 2006) which discusses mammals in general.

It should be emphasized that this cladogram portrays conventionally accepted relationships, ie. those based mainly on morphological evidence. Some of these relationships have recently been challenged by molecular studies. In particular, phylogenetic analyses of whole mitochondrial genomes have suggested that (1) cartilaginous fishes might not be the sister group to all other living vertebrates, but are nested within bony fishes (Rasmussen and Arnason, 1999), (2) turtles might not be the sister group to diapsids, but are related to archosaurs (Kumazawa and Nishida, 1999; see also Hedges and Poling, 1999), (3) ratite birds (struthioniforms) are not the sister group to all other birds, but are nested within Neornithes (Härlid and Arnason, 1999), (4) monotremes are not basal to other living mammals, but are the sister group to marsupials (Penny and Hasegawa, 1997).

Some workers have accepted these new molecular results and attempted to reinterpret morphological evolution in this new phylogenetic framework (e.g. Rieppel, 1999). However, sceptics have suggested that these heterodox groupings might instead demonstrate the fallibility of molecular, in particular mitochondrial, data (Naylor and Brown, 1998; Curole and Kocher, 1999; Lee, 1999c; Groth and Barrowclough, 1999;

Matthee *et al.*, 2001; Springer *et al.*, 2001). More data from all sources, and improved analytical methods, are needed to determine which of these interpretations is correct. Vertebrate systematics is therefore a highly dynamic field that continues to generate many intriguing results and questions; the chapters in this book attest to this.

ACKNOWLEDGEMENTS

This work was supported by the Australian Research Council.

REFERENCES

Aplin, K.P. 2006. Ten million years of rodent evolution in Australasia: phylogenetic evidence and a speculative historical biogeography. Pp.707–744. In Merrick, J.R., Archer, M., Hickey, G.M. and Lee, M.S.Y. (eds). *Evolution and Biogeography of Australasian Vertebrates*. Auscipub Pty Ltd., Sydney. 942pp.

Avise, J.C. 1994. *Molecular Markers, Natural History and Evolution.* Chapman and Hall, London. 511pp.

Baker, R.H., Yu, X. and deSalle, R. 1998. Assessing the relative contribution of molecular and morphological characters in simultaneous analysis trees. *Molecular Phylogenetics and Evolution,* Vol. 9, pp.427–436.

Balmford, A.M., Green, J.B. and Murray, M.G.1996. Using higher-taxon richness as a surrogate for species richness. 1: regional tests. *Proceedings of the Royal Society of London, Biological Sciences,* Series B, Vol. 263, pp.1267–1274.

Bremer, K. 1988. The limits of amino acid sequence data in angiosperm phylogenetic reconstruction. *Evolution*, Vol. 42, pp.795–803.

Bull, J.J., Huelsenbeck, J.P., Cunningham, C.W., Swofford, D.L. and Waddell, P.J. 1993. Partitioning and combining data in phylogenetic analysis. *Systematic Biology*, Vol. 42, pp.384–397.

Cantino, P.D., Bryant, H.N., de Queiroz, K., Donoghue, M.J., Eriksson, T., Hillis, D.M. and Lee, M.S.Y. 1999. Species names in phylogenetic nomenclature. *Systematic Biology*, Vol. 48, pp.790–807.

Cantino, P. and de Queiroz, K. (with Baum, D., Bryant, H.N., Donoghue, M.J., Eriksson, T., Gauthier, J.A., Hillis, D.M., Kron, K., Lee, M.S.Y., Minelli, A., Pleijel, F., Robeck, H., Rowe, T., Schander, C. and Thollesson, M.). 2000. *Phylocode: A Phylogenetic Code for Biological Nomenclature.* http://www.ohiou.edu/phylocode/index.html.

Claridge, M.F., Dawah, H.A. and Wilson, M.R. 1997. *Species: The Units of Biodiversity.* Chapman and Hall, New York. 439pp.

Colgan, D.J. and O'Meally, D. 2006. Molecular studies: implications for phylogeny, biogeography and conservation. Pp.851–871. In Merrick, J.R., Archer, M., Hickey, G.M. and Lee, M.S.Y. (eds). *Evolution and Biogeography of Australasian Vertebrates*. Auscipub Pty Ltd., Sydney. 942pp.

Curole, J.P. and Kocher, T.D. 1999. Mitogenomics: digging deeper with complete mitochondrial genomes. *Trends in Ecology and Evolution,* Vol. 14, pp.394–398.

Dawson, T.J. and Dawson, L. 2006. Evolution of arid Australia and consequences for vertebrates. Pp.51–70. In Merrick, J.R., Archer, M., Hickey, G.M. and Lee, M.S.Y. (eds). *Evolution and Biogeography of Australasian Vertebrates*. Auscipub Pty Ltd., Sydney. 942pp.

de Queiroz, K. 1992. Phylogenetic definitions and taxonomic philosophy. *Biology and Philosophy*, Vol. 7, pp.295–313.

de Queiroz, K. 1997. The Linnean hierarchy and the evolutionization of taxonomy, with emphasis on the problem of nomenclature. *Aliso*, Vol. 15, pp.125–144.

de Queiroz, K. and Gauthier, J. 1992. Phylogenetic taxonomy. *Annual Review of Ecology and Systematics*, Vol. 23, pp.449–480.

de Queiroz, K. and Gauthier, J. 1994. Toward a phylogenetic system of biological nomenclature. *Trends in Ecology and Evolution*, Vol. 9, pp.27–31.

de Queiroz, K. and Good, D.A. 1997. Phenetic clustering in biology: a critique. *Quarterly Review of Biology,* Vol. 72, pp.3–30.

Farris, J.S., Albert, V.A., Källersjö, M., Lipscomb, D. and Kluge, A.G. 1996. Parsimony jackkinfing outperforms neighbour-joining. *Cladistics,* Vol. 12, pp.99–124.

Felsenstein, J. 1978. Cases in which parsimony or compatability methods will be positively misleading. *Systematic Zoology*, Vol. 27, pp.401–410.

Felsenstein, J. 1985. Confidence limits on phylogenies: An approach utilizing the bootstrap. *Evolution*, Vol. 39, pp.783–791.

Gauthier, J.A., Kluge, A.G. and Rowe, T. 1988. Amniote phylogeny and the importance of fossils. *Cladistics*, Vol. 4, pp.105–209.

Givnish, T.J. and Sytsma, K.J. 1997. Homoplasy in molecular vs. morphological data: the likelihood of correct phylogenetic inference. Pp.55–101. In Givnish, T.J. and Sytsma, K.J. (eds). *Molecular Evolution and Adaptive Radiation.* Cambridge University Press, New York. 621pp.

Gould, S.J. 1985. A clock of evolution. *Natural History,* Vol. 4, pp.12–25.

Groth, J.G. and Barrowclough, G.F. 1999. Basal divergences in birds and the phylogenetic utility of the nuclear RAG-1 gene. *Molecular Phylogenetics and Evolution*, Vol. 12, pp.115–123

Groves, C.P. 2006. Humans among primates. Pp.779–831. In Merrick, J.R., Archer, M., Hickey, G.M. and Lee, M.S.Y. (eds). *Evolution and Biogeography of Australasian Vertebrates*. Auscipub Pty Ltd., Sydney. 942pp.

Hall, B.G. 2001. *Phylogenetic Trees Made Easy: A How-To Manual for Molecular Biologists*. Sinauer, Sunderland, Massachusetts.

Härlid, A. and Arnason, U. 1999. Analyses of mitochondrial DNA nest ratite birds within the Neognathidae: supporting a neotenous origin of ratite morphological features. *Proceedings of the Royal Society of London, Biological Sciences,* Series B, Vol. 266, pp.305–309.

Harvey, P.H. and Pagel, M.D. 1991. *The Comparative Method in Evolutionary Biology.* Oxford University Press, Oxford. 239pp.

Hedges, S.B. and Maxson, L.R. 1996. Molecules and morphology in amniote phylogeny. *Molecular Phylogenetics and Evolution*, Vol. 6, pp.312–314.

Hedges, S.B. and Poling, L.L. 1999. A molecular phylogeny of reptiles. *Science,* Vol. 283, pp.998–1001.

Heinsohn, T. and Hope, G. 2006. The Torresian connections: zoogeography of New Guinea. Pp.71–93. In Merrick, J.R., Archer, M., Hickey, G.M. and Lee, M.S.Y. (eds). *Evolution and Biogeography of Australasian Vertebrates.* Auscipub Pty Ltd., Sydney. 942pp.

Hennig, W. 1966. *Phylogenetic Systematics.* University of Illinois Press, Urbana. 263pp.

Hillis, D.M. 1999. SINEs of the perfect character. *Proceedings of the National Academy of Sciences of the United States of America,* Vol. 96, pp.9979–9981.

Hillis, D.M., Moritz, C. and Mable, B.K. 1996. *Molecular Systematics.* Sinauer Associates, Sunderland, Massachusetts. 588pp.

Holdaway, R.N. and Worthy, T.H. 2006. The evolution of New Zealand and its vertebrates. Pp.111–128. In Merrick, J.R., Archer, M., Hickey, G.M. and Lee, M.S.Y. (eds). *Evolution and Biogeography of Australasian Vertebrates.* Auscipub Pty Ltd., Sydney. 942pp.

Howard, D.J. and Berlocher, S.H. 1998. *Endless Forms: Species and Speciation.* Oxford University Press, Oxford. 470pp.

Huelsenbeck, J.P., Bull, J.J. and Cunningham, C.W. 1996. Combining data in phylogenetic analysis. *Trends in Ecology and Evolution,* Vol. 11, pp.152–158.

Huelsenbeck, J.P. and Crandall, K.A. 1997. Phylogeny estimation and hypothesis testing using maximum likelihood. *Annual Review of Ecology and Systematics,* Vol. 28, pp.437–466.

Huelsenbeck, J.P., Ronquist, F.R., Nielsen, R. and Bollback, J.P. 2001. Bayesian inference of phylogeny and its impact on evolutionary biology. *Science,* Vol. 294, pp.2310–2314.

Hull, D.L. 1988. *Science as a Process: An Evolutionary Account of the Social and Conceptual Development of Science.* University of Chicago Press, Chicago. 586pp.

ICZN Committee. 1999. *International Code of Zoological Nomenclature.* 4th Edition. International Trust for Zoological Nomenclature/The Natural History Museum, London.

Johanson, Z. 2006. Origins of mammals: morphology, molecules and a petrosal or two. Pp.491–522. In Merrick, J.R., Archer, M., Hickey, G.M. and Lee, M.S.Y. (eds). *Evolution and Biogeography of Australasian Vertebrates.* Auscipub Pty Ltd., Sydney. 942pp.

Johns, G.C. and Avise, J.C. 1998. A comparative summary of genetic distances in the vertebrates from the mitochondrial cytochrome *b* gene. *Molecular Biology and Evolution,* Vol. 15, pp.1481–1490.

Kallsersjö, M., Farris, J.S., Chase, M.W., Bremer, B., Fay, M.F., Humphries, C.J., Petersen, G., Seberg, O. and Bremer, K. 1998. Simultaneous parsimony jackknife analysis of 2538 *rbc*L DNA sequences reveals support for major clades of green plants, land plants, seed plants, and flowering plants. *Plant Systematics and Evolution,* Vol. 213, pp.259–287.

Kemp, T.S. 1982. *Mammal-like Reptiles and the Origin of Mammals.* Academic Press, London. 363pp.

Kitching, I.J., Forey, P.L., Humphries, C.J. and Williams, D.M. 1998. *Cladistics: The Theory and Practice of Parsimony Analysis.* Second Edition. Oxford University Press, Oxford.

Kluge, A.G. 1989. A concern for evidence and a phylogenetic hypothesis of relationships among *Epicrates* (Boidae, Serpentes). *Systematic Zoology,* Vol. 38, pp.7–25.

Kumazawa, Y. and Nishida, M. 1999. Complete mitochondrial DNA sequences of the green turtle and blue-tailed mole skink: statistical evidence for archosaurian affinity of turtles. *Molecular Biology and Evolution,* Vol. 16, pp.784–792.

Lee, M.S.Y. 1997. Documenting present and past biodiversity: conservation biology meets palaeontology. *Trends in Ecology and Evolution,* Vol. 12, pp.132–133.

Lee, M.S.Y. 1998. Phylogenetic uncertainty, molecular sequences, and the definition of taxon names. *Systematic Biology,* Vol. 47, pp.719–726.

Lee, M.S.Y. 1999a. Circularity, evolution, systematics ... and circularity. *Journal of Evolutionary Biology,* Vol. 12, pp.724–734.

Lee, M.S.Y. 1999b. Stability of higher taxa in traditional and phylogenetic taxonomy. *Zoologica Scripta,* Vol. 28, pp.361–366.

Lee, M.S.Y. 1999c. Molecular phylogenies become functional. *Trends in Ecology and Evolution,* Vol. 14, pp.177–178.

Lee, M.S.Y. 2001. Molecules, morphology and the monophyly of diapsid reptiles. *Contributions to Zoology,* Vol. 70, pp.1–22.

Lee, M.S.Y. 2003. Species concepts and species reality: salvaging a Linnean Rank. *Journal of Evolutionary Biology,* Vol.16, pp.179–188.

Lewis, P.O. 1998. Maximum-likelihood as an alternative to parsimony for inferring phylogeny using nucleotide sequence data. Pp.132–163. In Soltis, D.E., Soltis, P.S. and Doyle, J.J. (eds). *Molecular Systematics of Plants II: DNA Sequencing.* Kluwer Academic, Boston. 574pp.

Linnaeus, C. 1758. *Systema Naturae per Regna Tria Naturae.* Laurentii Salvii, Holmiae.

Maddison, W.P. and Maddison, D.R. 2000. *MacClade: Analysis of Phylogeny and Character Evolution.* Version 4. Sinauer Associates, Sunderland, Massachusetts.

Matthee, C.A., Burzlaff, J.D., Taylor, J.F. and Davis, S.K. 2001. Mining the mammalian genome for artiodactyl systematics. *Systematic Biology,* Vol. 50, pp.367–390.

Mayr, E. and Ashlock, P.D. 1991. *Principles of Systematic Zoology.* McGraw Hill, New York. 428pp.

Merrick, J.R. 2006. Australasian freshwater fish faunas: diversity, interrelationships, radiations and conservation. Pp.195–224. In Merrick, J.R., Archer, M., Hickey, G.M. and Lee, M.S.Y. (eds). *Evolution and Biogeography of Australasian Vertebrates.* Auscipub Pty Ltd., Sydney. 942pp.

Miyamoto, M.M. and Fitch, W.M. 1995. Testing species phylogenies and phylogenetic methods with congruence. *Systematic Biology,* Vol. 44, pp.64–76.

Musser, A.M. 2006. Furry egg-layers: monotreme relationships and radiations. Pp.523–550. In Merrick, J.R., Archer, M., Hickey, G.M. and Lee, M.S.Y. (eds). *Evolution and Biogeography of Australasian Vertebrates.* Auscipub Pty Ltd., Sydney. 942pp.

Naylor, G.J.P. and Brown, W.M. 1998. *Amphioxus* mitochondrial DNA, chordate phylogeny, and the limits of inference based on comparisons of sequences. *Systematic Biology,* Vol. 47, pp.61–77.

Nei, M. and Kumar, S. 2000. *Molecular Evolution and Phylogenetics.* Oxford University Press, Oxford.

Nixon, K.C. and Carpenter, J.M. 1996. On simultaneous analysis. *Cladistics*, Vol. 12, pp.221–241.

Novacek, M.J. and Wheeler, Q.D. 1992. Extinct taxa: accounting for 99.999...% of the earth's biota. Pp.1–16. In Novacek, M.J. and Wheeler, Q.D. (eds). *Extinction and Phylogeny.* Columbia University Press, New York.

Page, R.D.M. and Holmes, E.C. 1998. *Molecular Evolution: A Phylogenetic Approach.* Blackwell Science, Oxford. 346pp.

Panchen, A.L. 1992. *Classification, Evolution and the Nature of Biology.* Cambridge University Press, Cambridge. 403pp.

Penny, D. and Hasegawa, M. 1997. The platypus put in its place. *Nature*, Vol. 387, pp.549–550.

Rasmussen, A.S. and Arnason, U. 1999. Phylogenetic studies of complete mitochondrial DNA molecules place cartilaginous fishes within the tree of bony fishes. *Journal of Molecular Evolution*, Vol. 48, pp.118–123.

Richards, S.J., Alford, R.A. and Bradfield, K.S. 2006. The great frog decline in Australasia: causes, developments and conservation. Pp.241–252. In Merrick, J.R., Archer, M., Hickey, G.M. and Lee, M.S.Y. (eds). *Evolution and Biogeography of Australasian Vertebrates.* Auscipub Pty Ltd., Sydney. 942pp.

Ridley, M. 1986. *Evolution and Classification: The Reformation of Cladism.* Longman, London. 201pp.

Ridley, M. 1993. *Evolution.* Blackwell Scientific Publications, Cambridge, Massachusetts. 670pp.

Rieppel, O. 1999. Turtle origins. *Science*, Vol. 283, pp.945–946.

Sanderson, M.J. 1995. Objections to bootstrapping: a critique. *Systematic Biology,* Vol. 44, pp.299–320.

Scanlon, J.D. 2006. Origins and radiations of snakes in Australasia. Pp.309–330. In Merrick, J.R., Archer, M., Hickey, G.M. and Lee, M.S.Y. (eds). *Evolution and Biogeography of Australasian Vertebrates.* Auscipub Pty Ltd., Sydney. 942pp.

Sereno, P.C. 1999. Definitions in phylogenetic taxonomy: critique and rationale. *Systematic Biology*, Vol. 48, pp.329–352.

Siddall, M.E. 1998. Success of parsimony in the four-taxon case: long branch repulsion by likelihood in the Farris zone. *Cladistics*, Vol. 14, pp.209–220.

Siddall, M.E. and Kluge, A.G. 1997. Probabilism and phylogenetic inference. *Cladistics*, Vol. 13, pp.313–336.

Soltis, D.E., Soltis, P.S., Mort, M.E., Chase, M.W., Savolainen, V., Hoot, S.B. and Morton, C.M. 1998. Inferring complex phylogenies using parsimony: an empirical approach using three large DNA data sets for angiosperms. *Systematic Biology,* Vol. 47, pp.32–42.

Springer, M.S., DeBry, R.W., Douady, C., Amrine, H.M., Madsen, O., deJong, W.W. and Stanhope, M.J. 2001. Mitochondrial versus nuclear gene sequences in deep-level mammaliaan phylogeny reconstruction. *Molecular Biology and Evolution*, Vol. 18, pp.132–143.

Swofford, D.L. 1999. *PAUP*: Phylogenetic Analysis Using Parsimony (*and Other Methods).* Version 4. Sinauer Associates, Sunderland, Massachusetts.

Swofford, D.L. and Begle, D.P. 1993. *PAUP: Phylogenetic Analysis Using Parsimony.* Version 3.1.1. User's Manual. Smithsonian Institution, Washington D.C.

Watson, J.D. and Crick, F.H.C. 1953. A structure for Deoxyribose Nucleic Acid. *Nature*, Vol. 171, p.737.

White, M.E. 2006. Environments of the geological past. Pp.17–50. In Merrick, J.R., Archer, M., Hickey, G.M. and Lee, M.S.Y. (eds). *Evolution and Biogeography of Australasian Vertebrates.* Auscipub Pty Ltd., Sydney. 942pp.

Wilson, R.A. 1999. *Species: New Interdisciplinary Essays.* MIT Press, Cambridge, Massachusetts. 325pp.

Wroe, S. and Archer, M. 2006. Origins and early radiations of marsupials. Pp.551–574. In Merrick, J.R., Archer, M., Hickey, G.M. and Lee, M.S.Y. (eds). *Evolution and Biogeography of Australasian Vertebrates.* Auscipub Pty Ltd., Sydney. 942pp.

Yang, Z. 1998. On the best evolutionary rate for phylogenetic analysis. *Systematic Biology*, Vol. 47, pp.125–133.

Yang, Z., Goldman, N. and Friday, A. 1995. Maximum likelihood trees from DNA sequences: a peculiar statistical estimation problem. *Systematic Biology,* Vol. 44, pp.384–399.

Chapter 2
Environments of the Geological Past

Mary E. White
Falls Forest Retreat, 318 Isaacs Lane, Johns River, N.S.W. 2443 Australia

INTRODUCTION

Charles Darwin wrote in 1859 that, in the isolated and unique environment which is the Galapagos Islands, 'Both in space and time, we seem to be brought somewhat near to that great fact—that mystery of mysteries—the first appearance of new beings on this earth.' He recognized in those isolated ecosystems—such startlingly different vignettes in which simplification and clarity of detail are characteristic—the driving force of evolution. The initial inspiration led to the propounding of his theory, which so alarmed a world in which his concepts were considered blasphemous.

In rare instances, as in the Galapagos today, one can see the basic truth that it is the environment, through natural selection, that drives evolution. The mutual interdependence of life and environments is abundantly clear as we look around; the interactions between living things and their environment resulting in mutual modifications are obvious.

While the environments of the Phanerozoic Eon (the last 545 million years) are most relevant for this book, a brief summary of earlier conditions (from the first appearance of life in the fossil record) is provided as background. Ancestral photosynthetic life was the factor that prepared the world for the animal life forms that have characterized the last 500–600 million years. The contribution of green autotrophs, the nexus between them and the physical world today, is often taken for granted or ignored. The role of the plant kingdom as a moderator and controller of environmental factors, which are vital for the survival of animals, often goes unnoticed. The dependence of 99.9 percent of the animal kingdom on plant- and algal-based food chains emphasizes the importance of photosynthesis. The plant kingdom would have been able to evolve, albeit differently, without animals, but the reverse is not the case. A full account of the co-evolution of the Australian continent and its vegetation through 400 million years is given in *The Greening of Gondwana* (White, 1998).

The essential roles of micro-organisms (both aerobic and anaerobic), of all domains and established kingdoms (bacteria, archaea, eukarya) in helping to maintain conditions favourable for life, also remain little-noticed and poorly understood. These microscopic forms are involved in the chemical processes that: keep the proportions of gases in the atmosphere within the range of tolerance of living things; influence the levels of dissolved salts in freshwaters; regulate the removal of carbon from the system by the decay of organic matter; facilitate the recycling of trace elements, and a host of other functions. In summary, their activities maintain the out-of-equilibrium situation that is the characteristic of this living Earth (White, 2003). The concept of the biosphere actively maintaining and controlling environmental conditions suitable for its own maintenance is the basis of the Gaia hypothesis (Lovelock, 1987).

ENVIRONMENTS OF THE ARCHAEAN AND PROTEROZOIC EONS

Planet Earth is believed to have formed about 4,600 million years ago (m.y.a.). In its earliest days it was, indeed, elemental—earth, air, fire and water. For earth a primitive crust. For air a primitive atmosphere that escaped from the interior of the Earth when the primitive crust separated from the mantle. It contained essentially all of its nitrogen, much more carbon dioxide than now, hydrogen sulphide from volcanoes and traces of argon, but it lacked any significant amount of oxygen. The older concept of a methane-ammonia-hydrogen-water vapour mixture, such as might be expected from a solar dust cloud, is now in some doubt. Or, if such an atmosphere ever existed, it was blown away early in the life of the planet (Rush, 1957). For fire there was vulcanism, and for water, the early oceans. The composition of the early seas is uncertain, although over the last 2,000 million years there have been huge changes in ocean chemistry. There is recent evidence that hot springs on mid-ocean ridges contribute to salinity, but there are also balancing mechanisms as the salts precipitate out and become locked in bottom sediments.

It has been commonly thought that life started in the early seas; however, in recent years, there has been much controversy over particular sites, conditions and mechanisms that may have produced living matter. For example, the first organisms were able to synthesize organic molecules from ammonia, which was probably produced by hot springs related to submarine vulcanicity (Barns *et al.*, 1996; Stetter, 1996; Russell and Hall, 1997). But approxi-

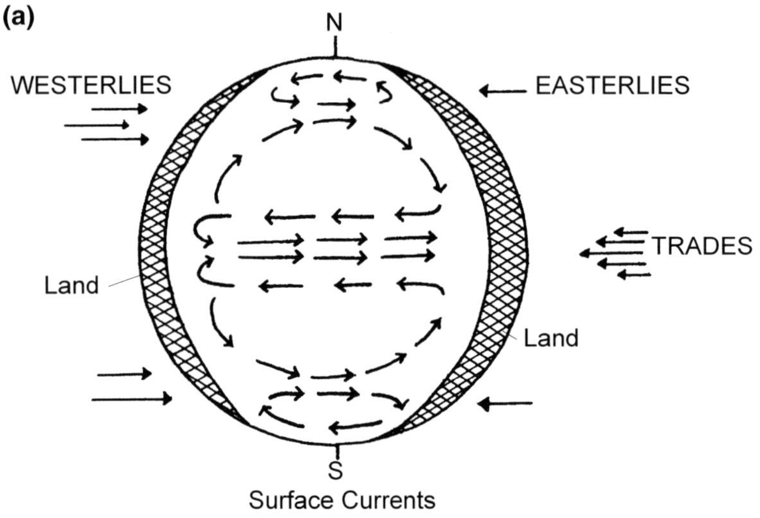

Figure 1. In a simple ocean surrounded by land **(a)**, surface currents are driven by the planetary winds. Two equatorial currents are blown westwards by the trades, and they turn back when they encounter the opposite shore, producing an equatorial countercurrent and two large revolving currents (gyres), one in each Hemisphere. Surface temperature distribution **(b)** follows the current pattern, with warmest waters at the end of the equatorial countercurrent (after van Andel, 1985).

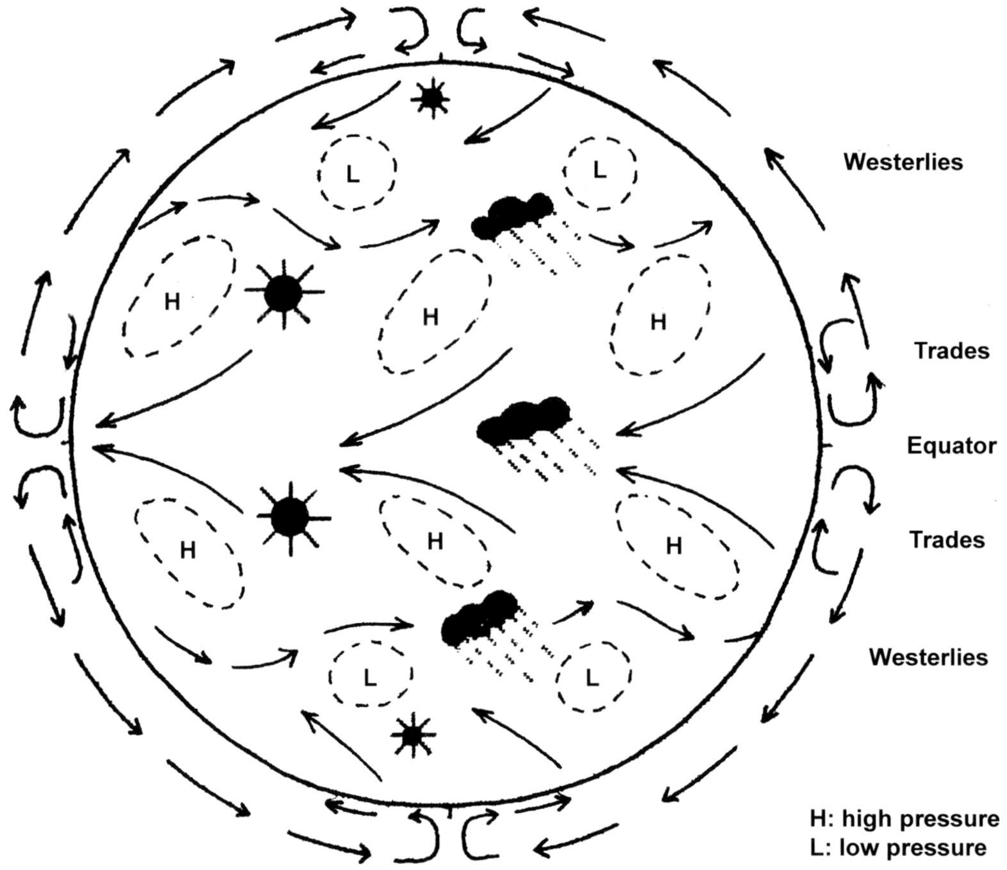

Figure 2. Planetary atmospheric circulation begins with warm air rising at the Equator and spreading towards the Poles. As it spreads, its moisture-holding capacity diminishes and tropical rains fall. In the subtropics, some of the now cooler air returns to the Equator. The Coriolis force turns the flow into easterly trade winds, and the sinking air is dry and results in a sunny climate. At mid-latitudes, a zone of westerly winds carries low-pressure systems and their rains eastward; its strongest expression is the jet stream at high altitude (after van Andel, 1985).

mately 3,500 million years ago cyanobacteria were abundant (Hill *et al.*, 1999). These ancestral prokaryotic organisms contained chlorophyll which enabled them to photosynthesize. By precipitating lime and trapping silt they formed stromatolite limestone reef structures in shallow water environments. Those at the North Pole locality in Western Australia are among the oldest fossils preserved anywhere on Earth (Hill *et al.*, 1999).

It was the activities of such organisms, over possibly 2,000 million years, that steadily generated oxygen as a product of the photosynthetic reactions. At first the oxygen was used up in oxidation of the many elements and minerals in the sea. For example, ironstone formations were laid down in the shallow epicontinental seas and in troughs along continental margins when the ferrous iron dissolved in sea-water was oxidized and precipitated out between 3,800 and 2,000 m.y.a. When the 'oxygen sink' in the seas had been satisfied, oxidized continental sediments such as redbeds or gypsum in marine evaporites suddenly became abundant, indicating that free oxygen had invaded the atmosphere. For oxygen to accumulate in air and sea it was necessary for carbon to be removed permanently from circulation. Thus burial and the transformation of organic matter into black shales and carbonates (as limestone or dolomites) were essential for the creation of an atmosphere suitable for oxygen-breathing organisms.

Oxygen concentrations rose in freshwaters, oceans and the atmosphere from about 1,800 m.y.a., and may be associated with the appearance of eukaryotes and their subsequent abundance. The origins of eukaryotic cells (with nuclei and other inclusions) involved symbiotic arrangements between cells of different types (Margulis and Sagan, 1987). Such advanced and efficient cells arose probably over 2,700 m.y.a., and it has been suggested that the vast explosion of primitive life forms led to depletion of the carbon dioxide in the atmosphere. This would have resulted in the opposite of the 'greenhouse' effect and could have contributed to the late Precambrian ice ages, which had several peaks (Hill *et al.*, 1999).

After the extinctions that the inordinately long and widespread ice ages had caused, the changed environments emerging as conditions improved and stabilized offered new opportunities, and prompted the evolutionary explosion that saw the establishment of most of the categories of invertebrate animals by Cambrian times (Glaessner, 1984)—545 million years ago (Young and Laurie, 1996; Walker, 2003). By the start of the Phanerozoic the concentration of oxygen was probably similar to its present value but the ozone layer, which shields the Earth from ultraviolet radiation, might have been less efficient. It is assumed that it was the harmful radiation that confined life to aquatic environments (mostly marine) for the first 200 million years of the Phanerozoic, until the continuing build-up of oxygen had provided a denser ozone layer.

From the beginning, conditions under which organisms existed have been determined by several categories of factors: by local factors like topography, latitude and microclimate; by global factors like primary currents and circulatory patterns and the effects on them of different arrangements of land masses; and by the interactions of the living things that occupy a habitat with each other and with their inanimate surroundings (Traube, 1985). This range of environmental influences is discussed throughout the remainder of this chapter with an emphasis on the Australasian region.

The global climate is an engine whose fuel is the heat energy of the sun, and whose parts are land, oceans and atmosphere. Basic circulatory patterns in sea and air (Figures 1–3) are a function of the differential heating from Equator to the poles, complicated by the Coriolis force caused by the rotation of the Earth. From the beginning, trade winds, westerlies, monsoons, jet streams, tropics and subtropics have been a feature of world climates. There have always been seasons with long periods of darkness in high latitudes in 'winter', even when temperature gradients were minimal. Currents and stratification of ocean waters by salinity and temperature differences have always been features of the oceans.

Oceans have long been considered to be biological deserts but this perception is changing. In terms of biodiversity, habitat areas and natural productivity it is clear that oceans dominate global biotic production and environmental diversity (Miller, 2000, pp.101, 198); however, it is clear that the shallow margins and sunlit surface waters (representing about 10% of ocean volume) are the most fertile areas. Shallow epicontinental seas, which were a feature of much of Palaeozoic time, have no modern counterpart. They were apparently even more important than terrestrial habitats at the time. The abundant life that evolved in them left deposits quite different from anything we find today.

It is only by having some understanding of the changing arrangements of land and sea through geological time that we can hope to interpret environments of the past. Evidence supplied by the fossil fauna and flora, by interpretation of sedimentation, structural geology and tectonics, added to logical deductions about climate and conditions implied by the global situation at any time, has to be synthesized to form a picture of past ages.

The very early history of the Earth, when large numbers of small land masses appear to have been moving about on the surface of the globe—colliding, joining, separating—does not concern us here. In that phase, land masses were probably small enough not to interfere with primary current patterns (McCulloch and Bennett, 1994); the uninterrupted flow of warm (or very warm) equatorial waters implies uniformly hot conditions worldwide, and high rainfall. High carbon dioxide levels, high radiation and volcanic hyperactivity contributed.

The plate tectonic model for the movement of land masses applies to the last 1,500 million years, by which time the volume of continental crust may have been about three-quarters of the present figure. Movement of plates and disposition of land and sea are well documented for the 545 million years of Phanerozoic time

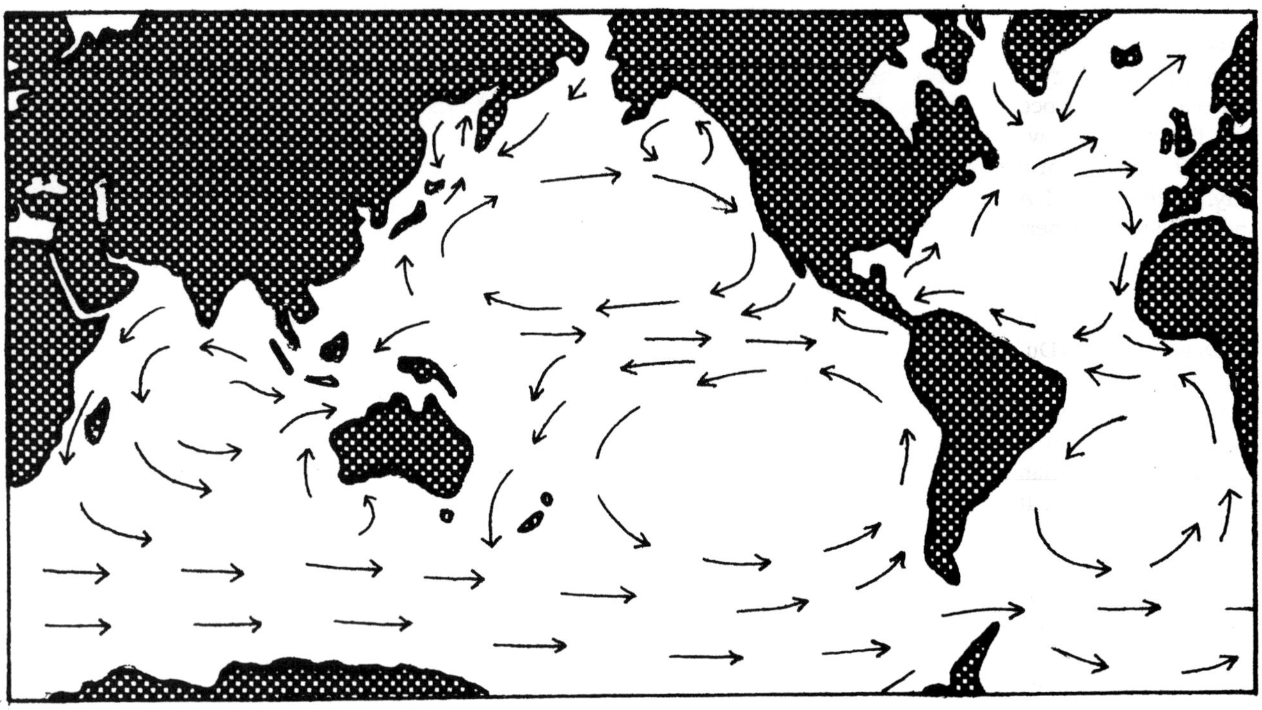

Figure 3. Circulation in topographically complex real oceans approximates that of the simple ideal ocean basin. Seaway connections between basins, such as the Drake Passage between Antarctica and South America and passages through the Indonesian Archipelago, are important for exchange and maintenance of patterns (after van Andel, 1985).

(Young and Laurie, 1996), helping us to assess the climate and conditions under which animals and plants lived in each geological period. See White (1999) for detailed discussion on these topics relating to Australia and New Zealand.

THE CHANGING FACE OF THE GLOBE

The idea that during past ages the continents might not have been in the positions they occupy today has been around for a long time. Early geographers had noted the neat fit of some coastlines, like those of Africa and South America. As knowledge of living flora increased in the 19th century, distribution patterns emerged that were hard to explain in the context of widely separated land masses. The British botanist Sir Joseph Hooker on voyages of scientific discovery saw the Antarctic beech flora of southern lands separated by wide oceans. He postulated that there might have been 'a great southern Continent' that included a vegetated Antarctica, and when he discussed his ideas with Charles Darwin they met with approval. Land-bridges were envisaged as the mechanisms that connected the vegetated Antarctica to the other lands to which the circum-Antarctic flora had migrated. Disjunct distribution patterns among primitive plants and animals were also being noticed increasingly as classification proceeded (Table 1).

Table 1. Examples of groups of living creatures showing disjunct distributions on southern continents that have been taken as evidence of plate tectonics (from Archer, 1984).

Group	South America	Australasia	Africa
Nothofagus (southern beeches)	+	+	–
Proteaceae	+	+	+
Marsupials	+	+	–
Galaxiid fish	+	+	+
Lungfish	+	+	+
Pipid frogs	+	–	+
Chelid turtles	+	+	–
Pelomedusid turtles	+	–	+
Amphisbaenids	+	–	+
Ratite birds	+	+	+
Acanthodriline worms	+	+	+
Banhamine worms	+	–	+
Onychophorans	+	+	+
Parastacid crustaceans	+	+	–
Mutelid molluscs	+	+	+
Bothriembryonid snails	+	+	–
Capniid insects	+	+	+
Macropathine insects	+	+	+

When the evidence of closely related fossils, in widely separated continents, was combined with the matching geology along the margins of lands on opposing sides of oceans, the theory of continental drift was formulated. It was first propounded by W.L. Green in 1857, but it went unnoticed by the scientific community. Then in 1912 Alfred Wegener first published his theories and is generally credited as being the father of continental drift (Wegener, 1929); however, it was Alex Du Toit who, in 1937, developed Wegener's ideas and in a study of geology of southern lands put drifting continents on the map (Du Toit, 1972).

While the continental drift theory had its adherents, there were also problems preventing its general acceptance. Initially no satisfactory explanation could be found for the mechanics involved in moving bits of the Earth's crust about. It was only when advances in technology enabled measurement of palaeomagnetism in rocks and detection of magnetic field reversals that theories could be tested. So the plate tectonic model for moving continents was developed, investigated and found to fit the facts.

Professor Harry Hess (1962) suggested that there might be movement of ocean floors in two directions away from mid-ocean ridges. Molten rock from the mantle came to the surface at positions marked by the ridges, spread laterally at right angles to their long axis and formed new crust. This sea-floor spreading was the force driving continents apart. To compensate for the appearance of new crust, he suggested that equivalent amounts of old crust were being subducted into deep-sea trenches. Vine and Matthew (1963) then proposed that if new ocean floor was being produced on the mid-ocean ridges it should, on cooling, have preserved a record of the Earth's magnetic field at the time of formation of the rock. This was because periodic reversals of the Earth's magnetic field were known to have occurred at frequent intervals during the Earth's history. When extensive magnetometer surveys of the Mid-Atlantic ridge were carried out, Vine and Matthew were proved right. The pattern that emerged was that each side of the ridge revealed a mirror image of a succession of magnetic reversals spreading away from the ridge. Radiometric dating showed the reversals to have occurred at approximately 450,000-year intervals. From this the spreading rate was calculated, and it was found that the Atlantic, for instance, was widening by three to five centimetres per year. The sea-floor of the South Atlantic has been shown to contain no rock older than 90 million years; implying that South America and Africa were joined until about that time. For a summary of the accepted mechanisms see Figure 4.

It has now been determined that the oldest sea-floor in any modern ocean basin is much younger than continental rocks may be. No sea-floor older than 200 million years has been dated, whereas continental rocks up to 3,800 million years old are known. While most spreading is on mid-oceanic ridges under water, rift zones also extend onto some continental masses. The Mid-Atlantic ridge extends across Iceland and is steadily splitting it apart; other rifts extend across eastern Africa and in the Baja California region of North America.

HOW PLATE TECTONICS WORKS

The Earth's crust consists of a number of plates (Figures 4, 5), which are rigid and practically undistorted slabs of lithosphere on which the continents ride. Continents are made of lighter material (SIAL)—mostly silica and alumina—with a density of 2.7. The rest of the plate is heavier (SIMA), which is mainly silica and magnesium (density 3.3). Convection currents are developed in the Earth's mantle of fluid or plastic rock, surfacing at mid-ocean ridges, fracturing and shifting crustal plates away from these ridges or rifts. These divergent or spreading zones where new crust is produced, moving plates apart at rates between five and ten centimetres per year, represent one type of boundary between plates. Others are subduction zones, where plates converge and the edge of one is forced down under the other and consumed; collision zones, which are former subduction zones, where the continents riding on the plates are colliding; and transform faults, where two plates are slipping past each other with no addition or subtraction of plate material.

Active zones of the Earth's crust, where earthquakes, volcanic activity and mountain building are taking place, lie on the boundaries of plates and are related to movements between them. The active volcanoes of today are primarily on the edges of plates. Those that occur within plates are associated with hot spots in the mantle, and chains of volcanoes (such as the Hawaiian Islands) result from the passage of a plate across a hot spot, with activity confined to the newest. Because the continents are made of lighter material than the plates on which they ride, it was thought that they were rarely sucked under at subduction zones. However, it is now known that about 1,000 kilometres of continental crust underlies the Himalayas, and about 500 kilometres lies below the Southern Uplands of Scotland where the Lakes District of northern England was partially subducted, causing the elevation of the Uplands. The Himalayas, Andes and New Guinea mountains developed at the leading edges of colliding plates as the lighter crust crumpled and sheared off when the heavier plate material nose-dived. In these cases, continental material was near the edge of plates and the boundaries were collision zones.

Generation of new sea-floor on rifts leads to formation of ocean basins, thus continents are separated as the process continues.

By the start of Phanerozoic time the southern continents, India and some other fragments were already a supercontinent—Gondwana (Figure 6). Land masses of the Northern Hemisphere were aggregating and would join up with that of North America, which was joining West

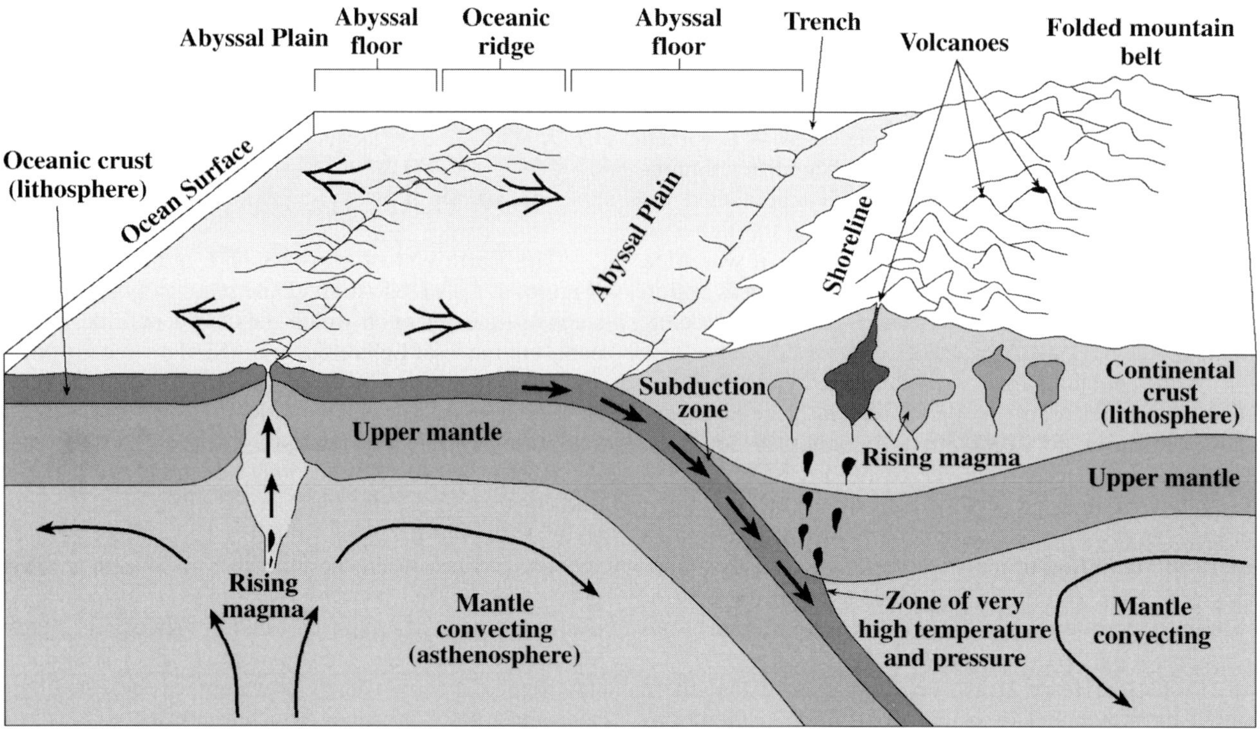

Figure 4. Diagrammatic summary of the mechanisms of plate tectonics: on the left showing general mantle convection with sea-floor spreading and continental drift; on the right a schematic model showing structures where an ocean plate under-rides a continental plate. Adapted from Sutton (1974) and Miller (2000).

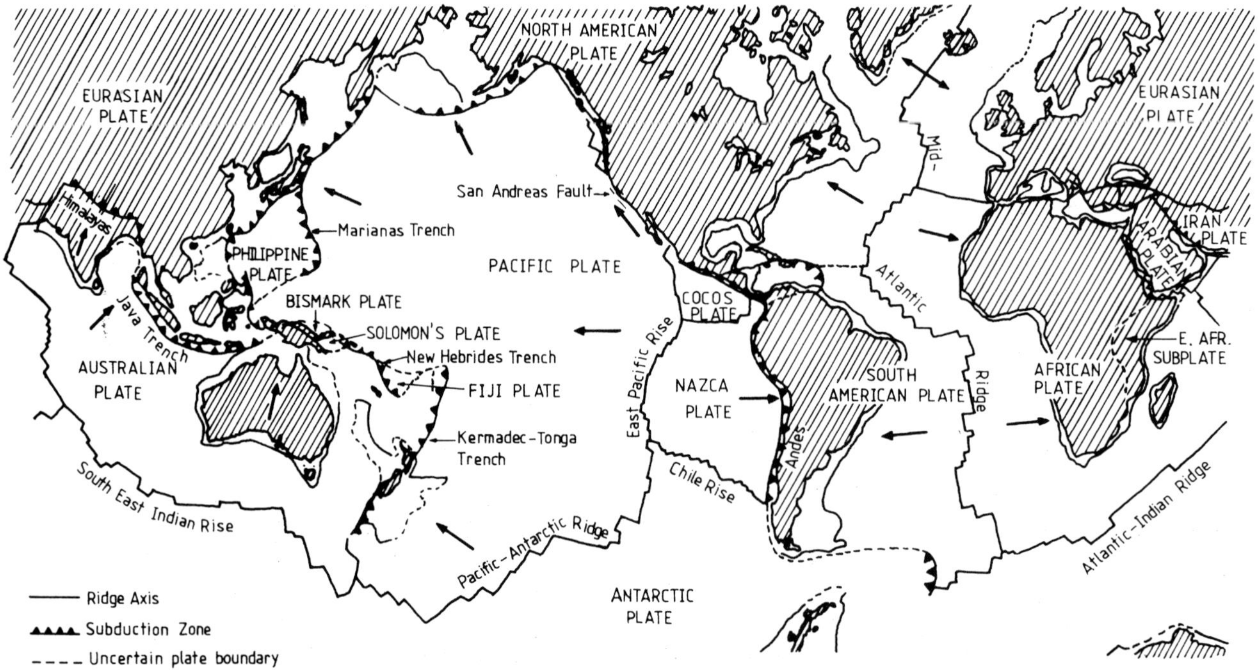

Figure 5. The major tectonic plates of the Earth's crust and significant associated features. Emergent portions of continents are cross-hatched; continental shelf margins are represented by the surrounding line. Arrows indicate direction of plate movements. New crustal material is being formed along the ridges (e.g. the South East Indian Rise). Older crustal material is being subsumed along subduction zones (e.g. the Java Trench). The Australian plate, which includes India, is sometimes called the Indo-Australian or Indian plate (from Archer, 1984).

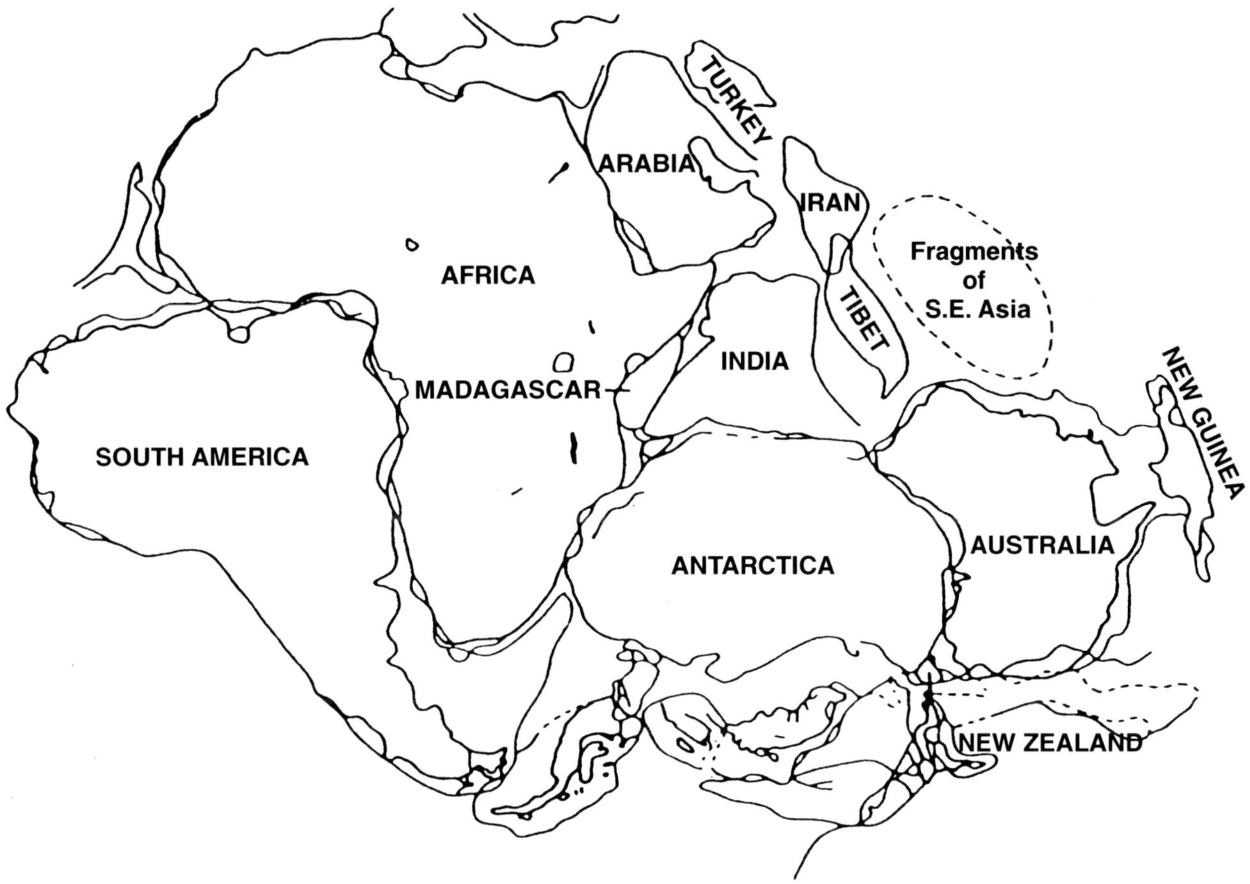

Figure 6. Gondwana, the great southern supercontinent, before its fragmentation began in the Jurassic (from White, 1998).

Gondwana. By 200 million years ago all the land masses were to form Pangaea, a single super-supercontinent. This coming together resulted in a C-shaped land mass stretching from pole to pole on one side of the globe, with the Tethys Sea, an embayment of the one ocean called Panthalassa, dividing the Northern Hemisphere lands (in their supercontinent Laurasia) and Gondwana.

Fragmentation of supercontinents with sea-floor spreading separating parts at the times shown in Figure 7, and development of ocean basins as the lands moved apart, has resulted in the globe as we find it today. All sea-floor prior to 200 million years ago has been subducted.

For recent summaries of regional events over the last billion years, as a context for the following detailed comments see Veevers (2000, 2001) and Metcalfe *et al.* 2001).

PHANEROZOIC ENVIRONMENTS

Environments in Australia, Tasmania and New Guinea are described in stratigraphical sequence from Cambrian time to the present; for further details on the weathering, erosional and drainage change processes that have shaped the Australian landscape readers are referred to Ollier (2001). Some key events relating to vegetation and faunas are listed in the summary table inside the front cover.

Environmental factors that affect the lives of animals are numerous. The most specialized creatures have very specific requirements and change of any sort puts their survival at risk. Others can tolerate a wide range of conditions and can adapt to gradual change. Most living things have difficulty in adapting to sudden and great alterations in conditions. When they do not have the flexibility and cannot survive, it is the opportunists exploiting the new conditions, with the genetic potential to change with the times, that take over. As it is with individuals, so it is with populations, and so it is also with whole dynasties of animal and plant types. Just as there is an in-built obsolescence in every living thing and it has a time to live, hopefully to thrive and breed, and to die, the same rule applies on a larger scale. Extinction is as much a part of life as is death. Efficiency under the prevailing conditions determines who dominates, competition keeps each individual at peak efficiency and innovation is essential if conditions change.

The enormous amounts of time involved when we look into the geological past can make the decline and extinction of a group of animals or plants appear sudden, when in fact it may have taken many millions of years.

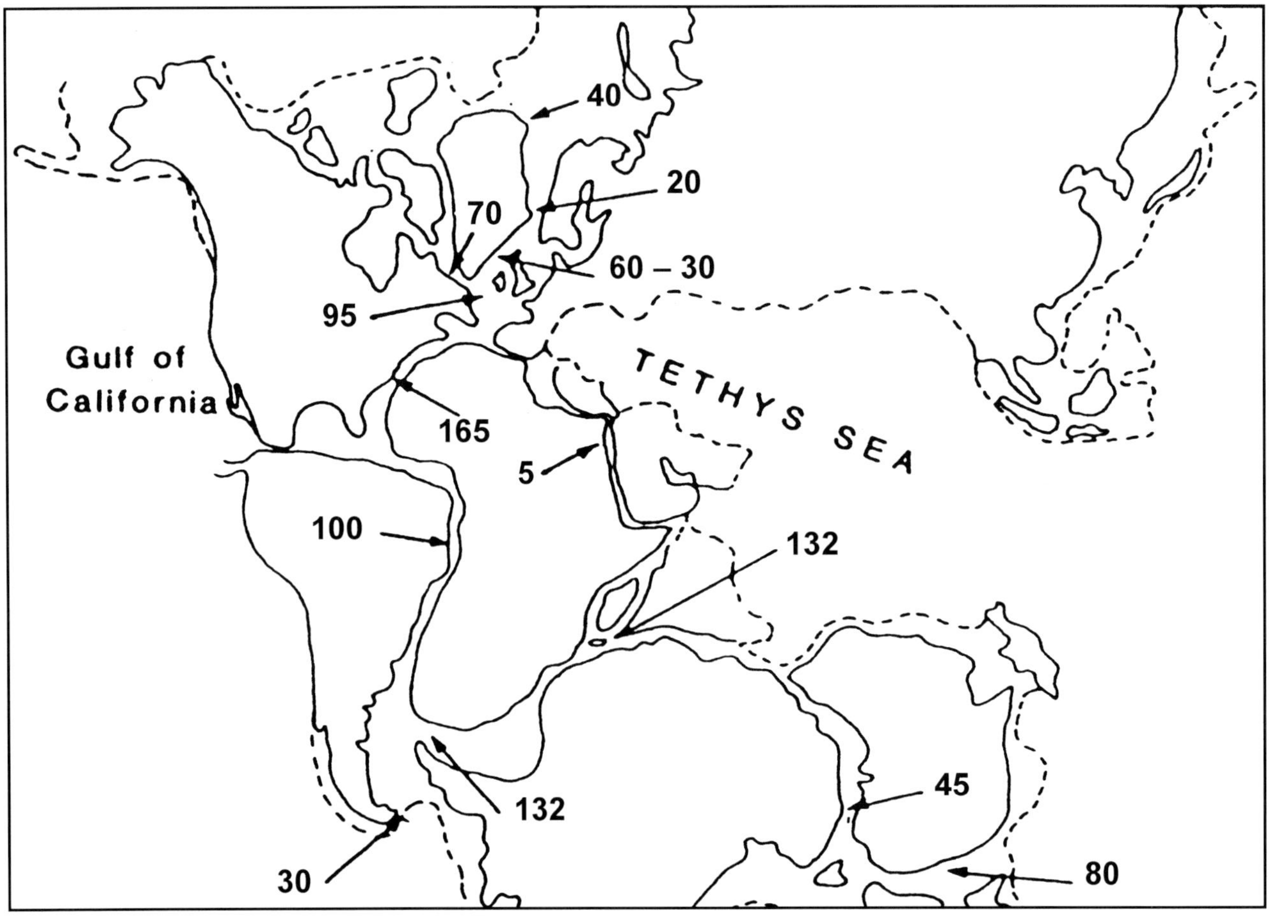

Figure 7. Pangaea 200 million years ago and the times of separation of its constituent parts (in m.y.a.).

Even one million years is a long time in terms of generations of living things. The fossil record is full of gaps, chance had a hand in selecting which organisms made it into the record, and erosion has removed much of the story originally recorded by the rocks (Figure 8). When the sparse collection of facts that can be elicited from all sources to form a picture of a specified time in the past is assembled, it can be no more than a generalization. As now, in the past a shifting mosaic of environments, vegetation types and animal populations was the pattern, and most of them have left no record.

Consider for a moment the last two million years, which is an insignificant fraction of the geological record, for which we have relatively complete information. We see the vast changes wrought by ice ages: on climate and landscapes worldwide; on fauna, with massive extinctions and the origin of *Homo sapiens*; and on vegetation, to a lesser extent. But if two million well-documented years are so important and significant, then the gaps in our knowledge of the more distant past are clearly important and our conclusions necessarily tentative.

The palaeogeography of succeeding geological periods is shown in a series of accompanying maps. They show the Australian continent's orientation as it was in the past, to emphasize the changes that have occurred through time. The north-south axis was at right angles to the present orientation for much of geological time. However, in description of relative movement, position of features and so on, present-day orientation is used. For example, if the sea is said 'to advance from the east into south-eastern Australia' that means present east and the region as now.

CAMBRIAN PERIOD

During Cambrian times (545–490 m.y.a.), Australia lay in the Northern Hemisphere. Part of the present-day coastline of Western Australia was on the Equator and the rest of the land lay between 0° and 30°N. Sea covered about a third of what is now land because globally sea levels were high throughout the early Palaeozoic. By the late Cambrian (Figure 9(a)), there was an embayment into central Australia. Thus shallow water environments were extensive, supporting an abundance of marine life. Climates worldwide were warm to hot, and seas were warm.

In Australian waters, archaeocyathids (early coral-like organisms now believed to be related to sponges) formed reefs, which imply warm waters off an arid hinterland. Trilobites and brachiopods dominated a diverse invertebrate fauna. Evolution of the shallow

Figure 8. The Australian terrestrial fossil record is largely contained in rocks laid down in the large sedimentary basins shown here (based on Long, 1991).

marine fauna proceeded apace with exploitation of all the niches—free-swimming, sedentary, burrowing—and by the late Cambrian large predators, such as cephalopods, had appeared.Early vertebrates were also present (ABC Science Online, 2003). The plant kingdom was represented by diverse algae (planktonic unicellular, multicellular, coralline and others) forming the basis of food chains for all the animals, but their fossil record is sparse.

Most life was confined to the seas, but sedimentation affects the fossil record and tectonic events influence the palaeogeography of later times, so some mention of conditions on land is relevant. Much of Australia was low-lying with hilly ground in the southern part of the Northern Territory and the Kimberley region of Western Australia. Under the arid conditions, shallow embayments were cut off from the open sea and phosphorites, gypsum and other evaporites of the early to mid-Cambrian are testimony to aridity, which affected much of the land. There were no land plants and, apart from the shorelines, no animals. The land would have looked barren.

Late in the Cambrian, seas over northern Australia from the Gulf of Carpentaria, and those in the Centre, retreated because there was uplift and folding in a zone across the continent north through Adelaide.

ORDOVICIAN PERIOD

During the Ordovician (490–434 m.y.a.), Australia still lay between 0° and 30°N, life was still confined to water—and the first vertebrates were to appear by the middle of the period (Figure 9(b)). Sea levels were as high as in the Cambrian, peaking at about 480 million years ago and then becoming somewhat lower during the Ordovician/Silurian ice age. They resumed their high and rising levels very early in Silurian times. The South Pole was on North Africa and there was glaciation there and in northern South America in the ice age, but Australia was little affected by cold. The equatorial current kept seas warm.

Following the retreat of inland seas in the late Cambrian, orogeny continued, accompanied by granite intrusion and metamorphosis along a zone from western Tasmania in a north-north-western direction through western Victoria to north-central South Australia (Delamerian Orogeny). Erosion of this high ground was rapid in the early Ordovician, implying high rainfall during this time.

The Larapintine Sea, a shallow, epicontinental sea joining an incursion in the Canning Basin to an embayment from the east, bisected the land in mid-Ordovician times. Once again there was a wealth of new shallow water environments and in them the rich and diverse shallow-water fauna abounded. Around 480 million years ago the first true fishes appear; *Arandaspis*, the first armoured, jawless fish, and three species of ostracoderms lived in the Larapintine Sea. The invertebrate fauna comprised members of many phyla; there were straight-cone cephalopods, graptolites, trilobites, gastropods, molluscs, echinoderms and crinoids. Fossil content of many of the strata suggests intertidal deposition. Corals indicate that sea temperatures were warm.

By late in the period (Figure 9(c)), seas had regressed and the Larapintine Sea disappeared from central Australia. Many of the deposits that had collected in it were uplifted and eroded. Deep water sedimentation continued along the east coast, and volcanic activity became widespread in what is now eastern Australia, building up a foundation of basic to intermediate volcanic rocks.

SILURIAN PERIOD

Although this period extended over 24 million years (from 434–410 m.y.a.) there was some gradation in climate during the earliest Silurian, from the north to the south of the continent. This gradient is thought to be related to the Ordovician/Silurian ice age, which was most intense when global sea-levels were lowest in the first few million years of the period. A counter-clockwise gyre had brought cooler water from polar regions off the North African coast towards northern Australia, modifying the warm to hot conditions that had applied elsewhere under the influence of equatorial currents. By mid-Silurian times, climates worldwide were warm to hot. Australian evidence supports warm waters on both sides of the continent.

Latest Ordovician and early Silurian time was marked by uplift and erosion of central Australia (Rodingan Movement) and the orogenic movements along the eastern oceanic margin (Benambran and Quidongan Orogenies). In the eastern area, offshore from a generally fairly narrow shelf, a complex pattern of troughs, shelves and elongate islands formed, their positions changing with time. Volcanic activity in the area resulted in the extrusion of volcanic rock in shallow waters, sometimes forming islands, and in some cases the magma cooled below the Earth's surface to form granite batholiths. Sedimentation continued in the deeper troughs, while shelf areas were sites for deposition of sandstones, shales and limestones. The limestones contain warm-water corals, brachiopods, trilobites and molluscan faunas. The *Baragwanathia* flora of Victoria, an assemblage of very early land plants, occurs in beds deposited inside the protection of elongate islands in a series of strata from late Silurian (Figure 9(d)) to early Devonian.

This *Baragwanathia* flora is the first evidence of successful adaptation of plants to life on dry land in this region and was a significant step in the evolutionary process. Colonization of land habitats by plants made it feasible for many more animals to leave the water, so initiating many new evolutionary processes.

While life was confined to the seas, the climate under which it operated was stable, fairly constant and predictable. On the bare and inhospitable land, there were variations of temperature (diurnal and seasonal) and exposure to heat and cold; radiation and drying became a fact of life. Having ancestors adapted to life in the water posed difficulties for early terrestrial invaders, exerting selection pressure to meet the challenges of the new environment. Adaptations of body structure and reproductive mechanisms had to be achieved for efficiency under the very different conditions. In the case of plants it meant the evolution of vascular plants, with specialization of the plant body with tissues for water absorption and transport, for support, photosynthesis and water conservation. The major evolutionary steps seen in vertebrates are from fish with gills and fins, shedding their eggs and sperm into the water for fertilization; to amphibians which have lungs and legs, and can breathe on land and may spend most of their life there, but have to return to the water to breed because their larval stages have gills; to reptiles, which perfected the amniote egg and cut the ties with water; and finally to mammals, which retain the egg within the body of the mother, protecting the embryo. These steps are mirrored in the plant kingdom where algae in water, shedding eggs and free-swimming sperm (haploid gametes), equate to fish; spore-producing land plants (ferns, lycopods, horsetails) have an alternation of separate haploid and diploid forms and are confined to limited habitats by their need for free water for swimming sperm, equating to amphibians; gymnospermous seed plants have eliminated the need for separate haploid and diploid plants as well as the requirement for free water, equating to reptiles; and angiospermous or flowering plants, which enclose their eggs in a vessel, equate to the mammals.

As plants established a foothold on the land, they not only provided food for land animals but they created micro-climates and micro-environments, modifying the harsh conditions. Early land plants, ill-adapted as they were to the dry land, created a new zone on the edges of the waters, in wetlands and swamps (see Figure 10). With the high sea levels of the times, suitable environments for them to conquer were abundant. Environmental pressures to evolve, to push the boundaries of their habitats ever inland, and the opportunity to do so, saw rapid

Figure 10. A reconstruction of an Australian Silurian landscape. Only the water margins were colonized by the first vascular land plants—rhyniophytes and zosterophylls. Higher dry areas were bare and eroded rapidly. Diarama by D. Dunphy, from White (1985, p. 11).

evolution. The increasing creation of new niches following their invasion enticed animals to fill them, and the race to inhabit the land was on.

Vegetation produced soil from the sands and clays, holding moisture, slowing erosion. From the beginning there was symbiosis between the land plants and bacteria or fungi that had colonized wet places long before. As communities were established, the interrelationships between all living things became more intertwined and more complex. Patterns were established, so that since those first terrestrial colonies, although the individuals changed through time, ecosystems maintained their characteristics.

DEVONIAN PERIOD

The Devonian lasted for about 56 million years (from 410–354 m.y.a.) and Australia was situated between 15°N and 15°S in early to mid-Devonian times (Figure 11(a)) and had slipped farther southwards to lie mainly between 30°S and 45°S, slightly differently orientated, by the late Devonian (Figure 11(b)).

World climates were warm to hot and there was no polar ice, so up to the middle of the period the Australian climate must have been hot, and in later times warm to hot. There is evidence of aridity, particularly in Western Australia, where the Canning Basin was a desert.

Volcanic activity on the eastern flank was rapidly building land onto the continent, and by the end of Devonian time the eastern continental margin approximated that of today. There were marine incursions into the Canning, Carnarvon and Bonaparte Basins in the west throughout the Devonian. In the early to mid-Devonian, the Canning Basin incursion spread across the continent, linking the western to the eastern seaboards by a wide ribbon of inland sea. The Tabberaberan Orogeny of the mid-Devonian saw shallow seas retreating eastward, and terrestrial deposits accumulating over large areas of the continent. These were laid down by exten-

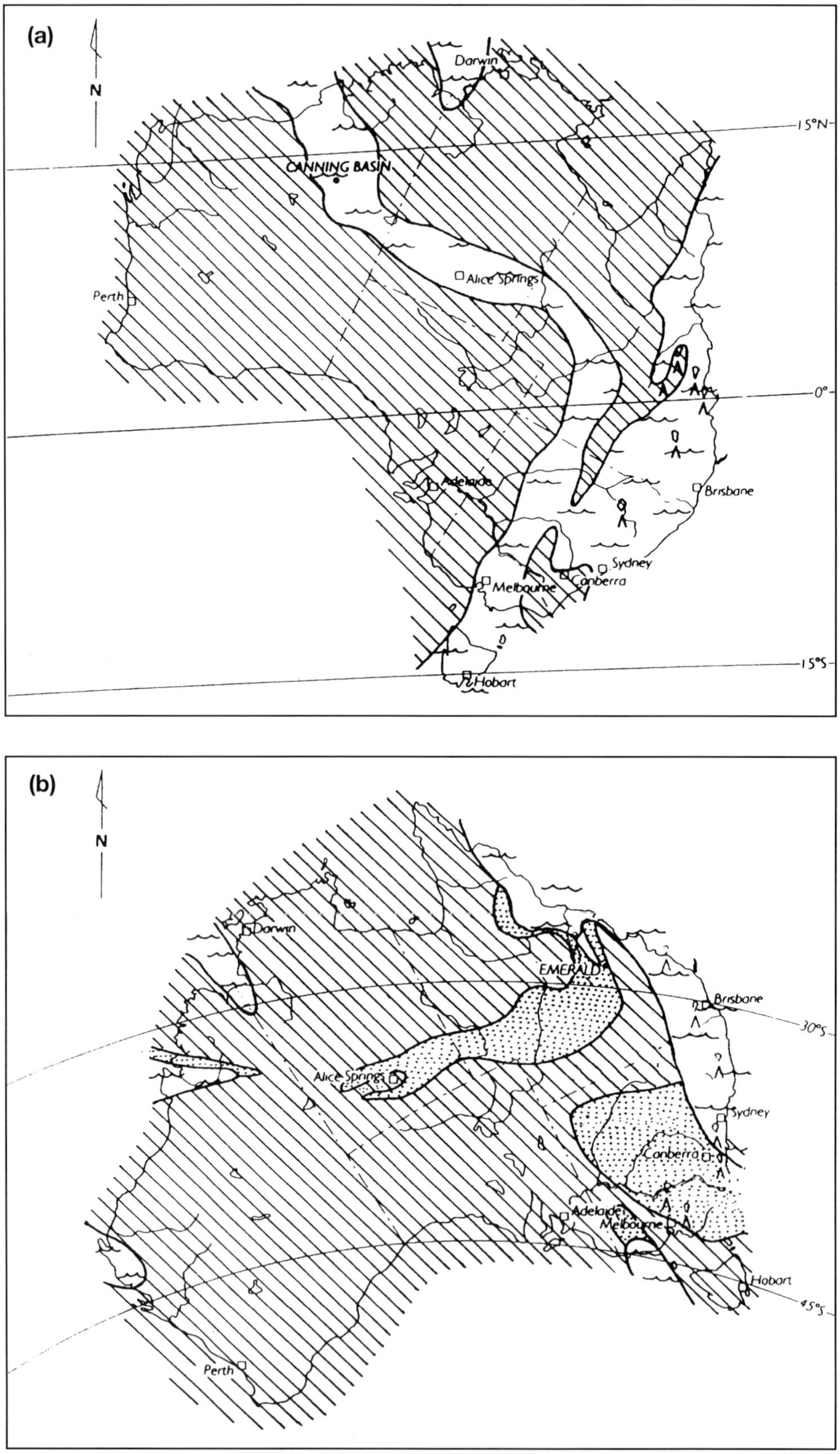

Figure 11. Palaeogeography of Australia during the Devonian period: **(a)** in the early to middle Devonian—an epicontinental sea again bisects the continent; **(b)** in the late Devonian (adapted from White, 1998). For key to symbols see Figure 9.

sive river systems, generally flowing eastward. Freshwater fish were abundant in rivers, probably in response to the food available, as vegetation thrived on river margins. Uplift in central Australia provided a source of thick sediments; these banked up against the emerging mountains and were also spread by the rivers. Good rainfall is implied for those episodes, although in a landscape where drier areas and hillsides were not vegetated, erosion would have been far more rapid than today.

Early Devonian plants were herbaceous and low-growing. The ability to produce secondary supporting and conducting tissues had not yet evolved, and plants were completely tied to watery environments. The *Baragwanathia* flora of Victoria, which originated in the late Silurian, persists into the early Devonian. It comprises zosterophylls, which are ancestral to lycopods (clubmosses); *Baragwanathia*, which is already a lycopod and similar to species of living *Lycopodium*; and the rhyniophytes, which are ancestral to ferns, horsetails and seed plants. The early Devonian horizons also contained more advanced flora such as trimerophytes; these are intermediate forms between rhyniophytes and higher plants.

By the mid-Devonian, land plants had acquired the ability to grow into large trees, and the age of giant clubmosses and horsetails dawned (Figure 12). Swamps supported forests; in Australia *Lepidosigillaria* appears in the mid-Devonian, *Leptophloeum* is the dominant tree of the late Devonian, and by the end of the period,

Figure 12. A Devonian swamp forest dominated by *Leptophloeum australe* (clubmoss trees with rhombic bark patterning). Plant life was still confined to swamps and areas with high water tables, and the dry slopes and hillsides were not vegetated. Diarama by D. Dunphy, from White (1985, p.13).

species of *Lepidodendron* appear. The swamps were no doubt home to thriving invertebrate populations, but so far no local fossils have been found.

The Devonian was the age of fishes of all sorts including the lungfish group—see pp.131–157 (Basden *et al.*, 2006) and pp.159–184 (Ritchie, 2006). Late in the period amphibians first make their appearance.

CARBONIFEROUS PERIOD

Early in the Carboniferous (354–298 m.y.a., duration 56 million years) Australia lay within almost the same latitudes as in the late Devonian, but it had rotated and was proceeding steadily towards the South Pole, its north-south axis at right angles to the present one. By late Carboniferous its southward movement had taken it into high latitudes with most of New South Wales and Queensland inside the 75°S circle (Figure 13).

In the early part of the period, world climates were warm, but by the middle Carboniferous, glaciation had commenced and, as polar ice caps developed, temperatures showed marked gradation from the Equator to the poles. Euramerica, in equatorial latitudes, supported a thriving swamp flora of giant clubmosses and horsetails, and the coal deposits of the Northern Hemisphere bear testimony to the luxuriance of the vegetation there. In the Southern Hemisphere, with the South Pole surrounded by Gondwana, effects of the continental ice sheets on the southern lands were profound.

Australia's giant clubmoss flora did not thrive under the deteriorating climatic conditions here, and was replaced by a pro-gymnosperm flora, the *Rhacopteris* flora, which seems to have grown under tundra conditions as the ice advanced (Figure 14). Mountain glaciers appeared on the volcanic highlands of eastern Australia in the late Carboniferous. At the peak of the ice age, at the very end of the period and beginning of Permian time, the continental ice sheet covered about half of Gondwana (Figure 15). The effect on flora and fauna of a continental ice sheet of such dimensions was dramatic. It cleared the land of all living things in its path, setting the stage for a great burst of evolution as new environments emerged when the climate improved and the ice retreated again.

The warm seas of the early Carboniferous in Australia supported a diverse fauna of corals, brachiopods, bryozoans, bivalves, gastropods, ammonoid cephalopods and rare trilobites. But as the journey south

Figure 13. Palaeogeography of Australia in the early Carboniferous (adapted from White, 1998). For key to symbols see Figure 9.

Figure 14. In the late Carboniferous, Australia lay close to the South Pole. An ice sheet spread across much of Gondwana during an ice age, and tundra with low-growing pro-gymnosperms (*Rhacopteris* spp.) existed near the ice. Diarama by D. Dunphy, from White (1985, p.15).

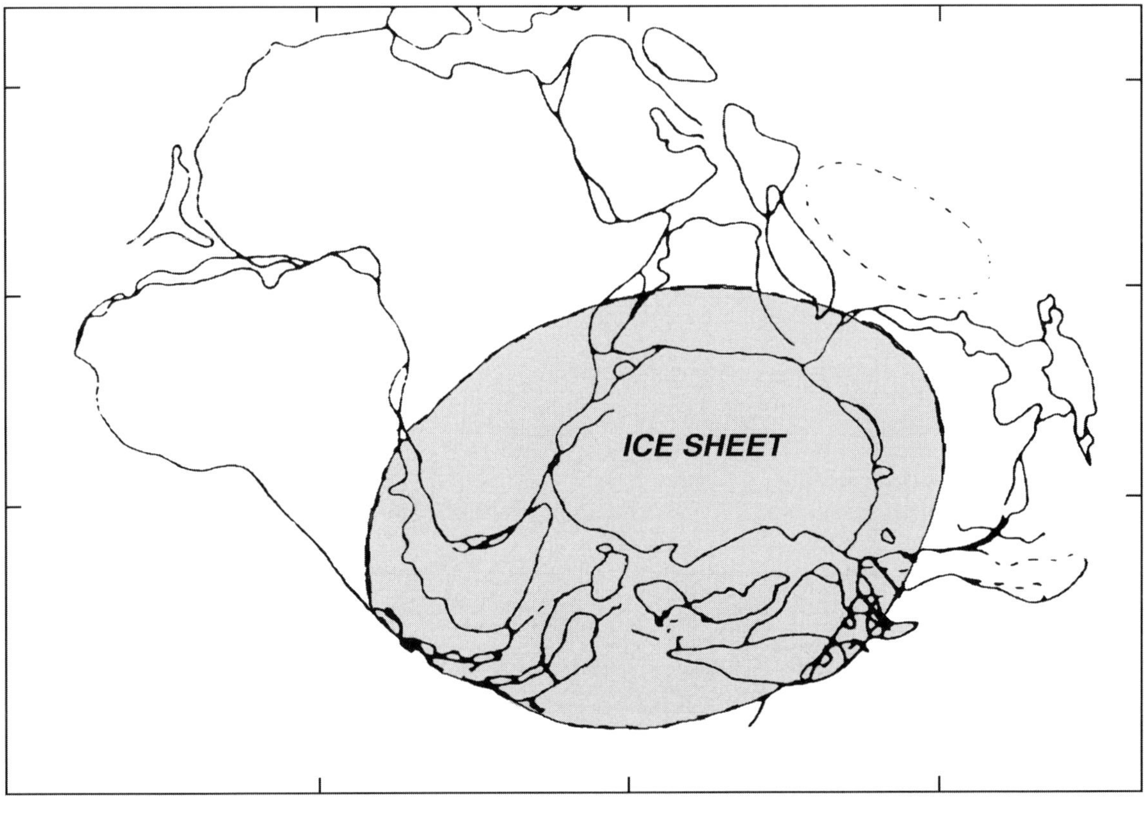

Figure 15. Extent of the ice sheet covering Gondwana in early Permian, 286 million years ago.

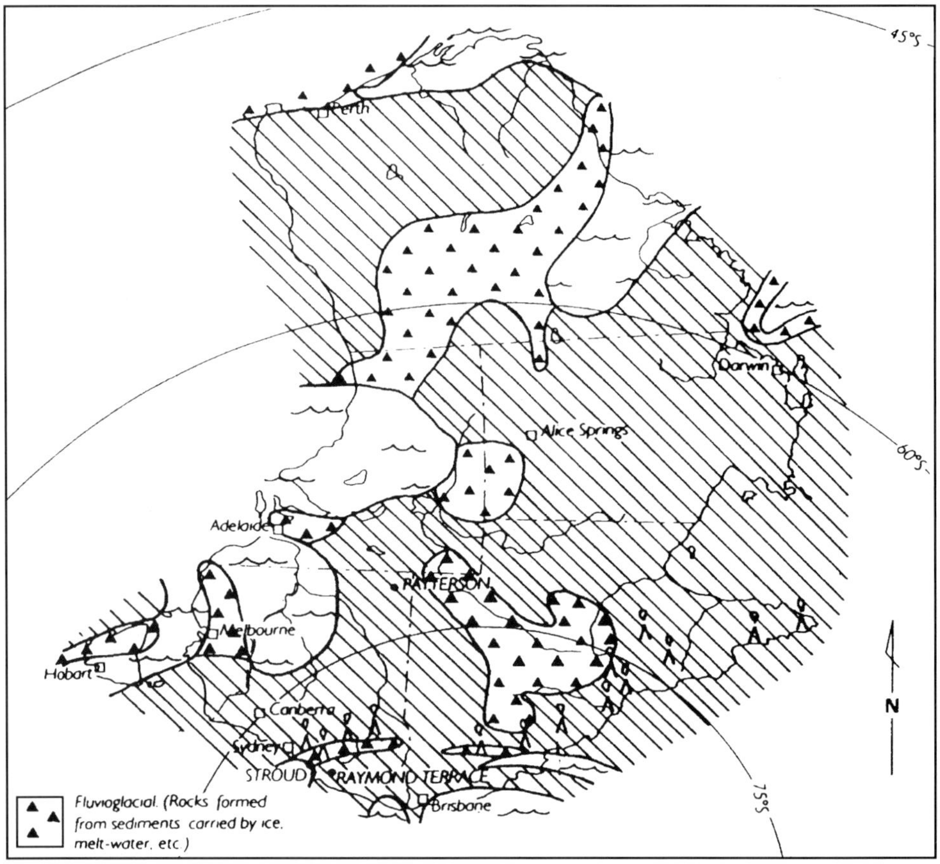

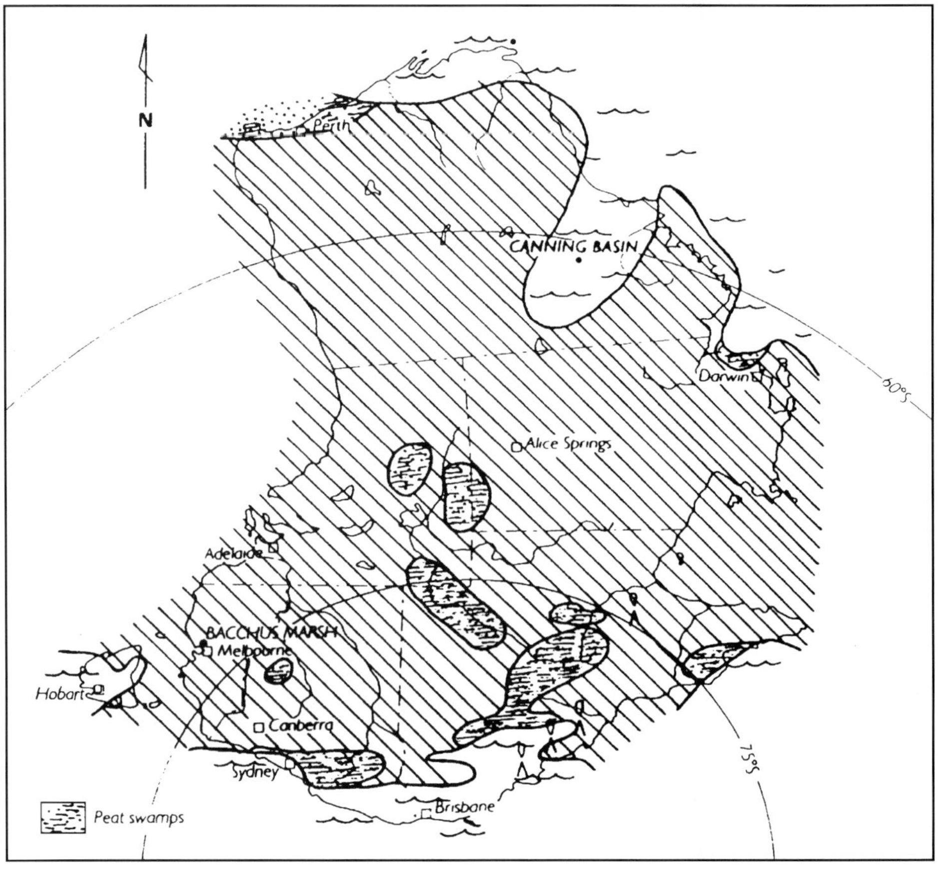

Figure 16. Palaeogeography of Australia from the late Carboniferous to early Permian periods: **(a)** the late Carboniferous to the earliest Permian times were characterized by a high latitude situation, extensive glacial sediments left by the ice sheets, vulcanism on the eastern margin, and marine penetration into some basins; **(b)** during the early Permian (adapted from White, 1998). For key to symbols see Figure 9.

proceeded and temperatures fell towards the ice age, aquatic faunas became increasingly impoverished and showed more endemism. On land, this was the age of amphibians, insects were abundant elsewhere but have not been found in the Australian record; the oldest known reptile also made its appearance (outside Australia).

During Carboniferous times, the geological evolution of the continent continued, with further volcanic activity and sedimentation adding to the eastern margin. Large-scale uplift in the Alice Springs Orogeny resulted in mountainous terrain in the area where the MacDonnell Ranges are today.

PERMIAN PERIOD

The Permian had a duration of 47 million years (298–251 m.y.a.). When the ice sheet retreated in the earliest Permian (Figure 16(a)) and sea levels rose worldwide with its melting, large areas of South Australia and Victoria were inundated and Tasmania was separated by an arm of the sea that penetrated inland on the Australia/Antarctica joint line. Other large encroachments of the sea occurred in the Canning Basin on the Western Australian coast. Glacial sediments were widely distributed after the ice age.

During the early Permian (Figure 16(b)), seas retreated from the south of the continent, but entered the Sydney and Bowen Basins. After the ice melted, climates improved, but remained cool to cold and alpine glaciers persisted on highlands in the east. Coal swamp conditions, perhaps similar to the peat bogs in Scotland now, existed in some areas in the early Permian. As conditions changed and became progressively warmer more typical coal swamps, with luxuriant growth related to warmer climates, were widespread. Pronounced growth rings in petrified tree trunks are strong evidence for distinct seasonality.

There was volcanic activity in the Sydney Basin and northwards on a line to Townsville. Rocks in the New England region were folded, intruded by granite and uplifted. Late in the period all the epicontinental seas had withdrawn. Australia still lay in high latitudes and, like all lands in polar regions, was subjected to several months of darkness each year.

The vegetation of Permian times, characterized by the

Figure 17. An Australian landscape in the Permian. Coal swamps, in which the *Glossopteris* flora flourished, were a feature of the times; the initially cool temperate conditions warmed throughout the period to become almost subtropical by the end. Early in the period winters were long, dark and cold, with residual glaciers on high ground. Early conifers, cycadophytes and ginkgophytes were growing on dry slopes away from swamps and water margins by mid-Permian. Temnospondyl amphibians were the dominant land vertebrates of the Permian coal swamps. Diarama by D. Dunphy, from White (1985, p.17).

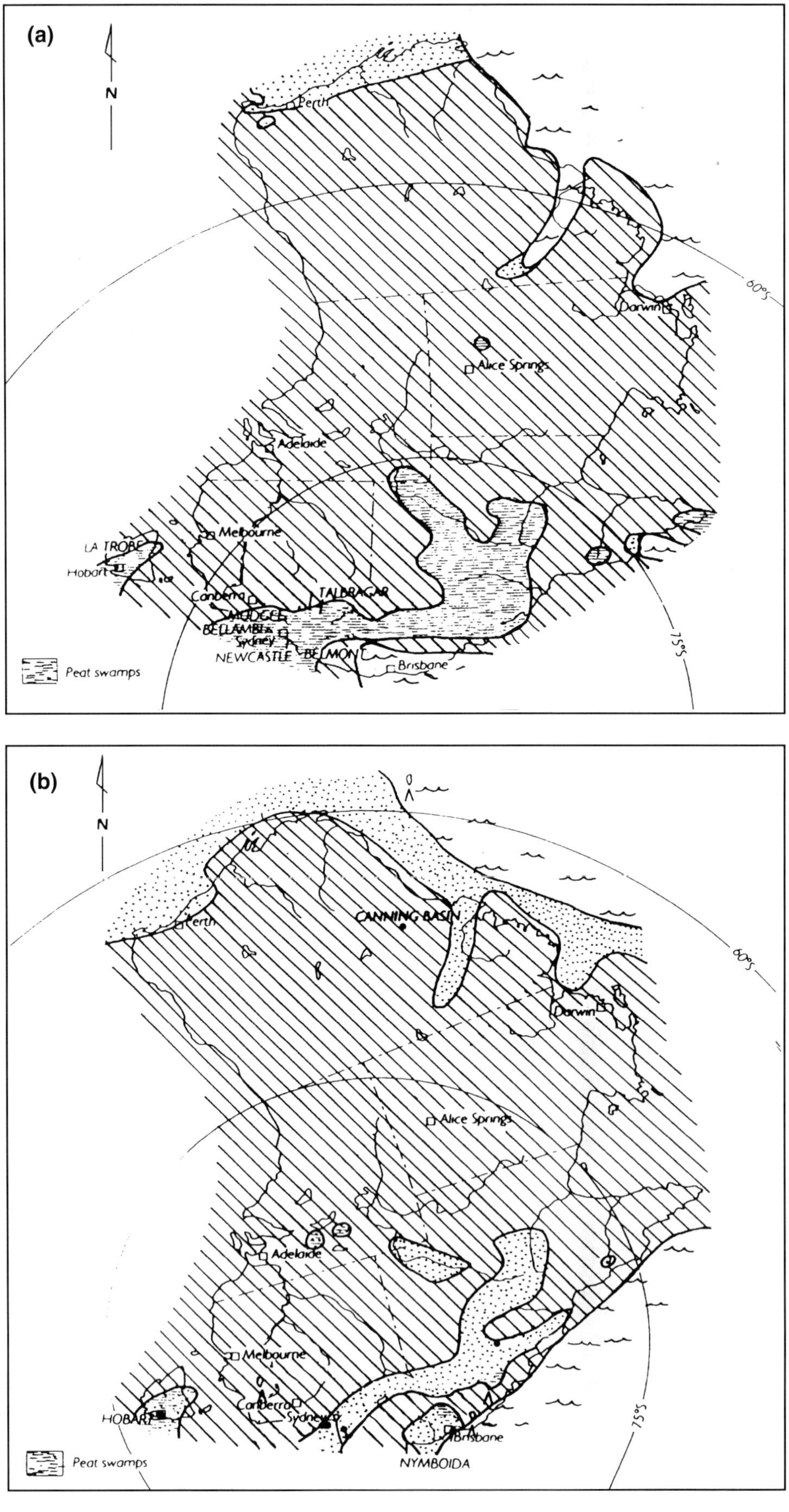

Figure 18. Palaeogeography of Australia in the late Permian and Triassic periods: **(a)** during the late Permian—major coal swamps were a feature of basins in the eastern part of the continent; **(b)** during the Triassic—the South Pole was in the vicinity of modern-day Bourke in western New South Wales, the world was warm without polar ice caps, and the *Dicroidium* flora characterized Gondwana (adapted from White, 1998). For key to symbols see Figure 9.

Glossopteris flora, thrived and evolved rapidly (Figure 17). The importance of glossopterids as possible ancestors of several later plant groups is dealt with fully in White (1998). The glossopterids were adapted to the wet soil of coal swamps. Their roots contained aerating tissues to overcome the anaerobic conditions in permanently wet places. By the end of the Permian, conifers, cycadophytes, ginkgophytes and some pteridosperms had broken their ties with water and were covering dry hillsides. An interval of warm to hot and wet climate on the Permian–Triassic boundary saw a floral change. Podocarp and Araucarian conifers (southern conifers that may have had a glossopterid ancestry) dominated instead of *Glossopteris*, and the first of the forked-frond seed ferns appeared.

The Permian marine fossil record is interesting. Increased shallow marine environments, due to high sea-levels after the ice age, and improving conditions resulted in rapid diversification of the impoverished fauna. By the middle of the period there were over 100 orders, 200 families and maybe 100,000 species globally. But at the end of the Permian (Figure 18(a)), with the loss of the shallow water environments, there was a massive 'collapse' and only about ten per cent of the once abundant marine species made the transition into the Triassic. This reduction was probably a reflection of loss of habitat rather than of global catastrophe. In contrast, the record for land plants at this time shows burgeoning communities colonizing new areas.

The Permian was the time when large amphibians were the dominant land animals. Insects were abundant, and there are some special localities where the insect fauna is particularly well preserved. One such site is the Insect Beds at Belmont, New South Wales, where volcanic ash settled on the surface of a lake on which insects and fragile bits of plants were floating, carrying them down and preserving them in such fine rock that minute structural detail is visible in the impressions that remain.

TRIASSIC PERIOD

After what appears to have been a warm and wet transition period on the Permian–Triassic boundary, the Triassic (251–205 m.y.a.) (Figure 18(b)) shows evidence

Figure 19. A landscape in the Sydney Basin in the Triassic where rivers spread sediments that formed the Narrabeen, Hawkesbury and Wianamatta sedimentary sequences. *Cylomeia* clubmosses grew like small palms in the mangrove niche; horsetails grew like rushes on water margins; *Dicroidium* forked-frond seed-ferns occupied the heath zone; and the dry slopes were colonized by conifers, cycadophytes and ginkgophytes. Diarama by D. Dunphy, from White (1985, p.19).

of being hot, often arid, with seasonality of rainfall. The early Triassic Narrabeen Group flora shows fewer aridity features than later Triassic floras; drought adaptation (xeromorphy) is pronounced in many plants of the middle and late Triassic. Conditions suitable for coal formation were confined to small areas in South Australia (Leigh Creek), Tasmania, north-eastern New South Wales and south-eastern Queensland. Red-bed formations confirm alternation of hot and dry periods with seasons of moderate precipitation in areas previously occupied by coal swamps. Mesozoic climates were largely influenced by the Pangaean Monsoon resulting from the type of global circulation suggested by Parrish (1993).

In the early Triassic in Western Australia, the sea encroached on the land spreading eastwards in a narrow embayment in the area between Geraldton and Carnarvon, before retreating in the mid-Triassic. On the North West Shelf, the marine transgression of the early Triassic was followed by the establishment of a river system and the formation of deltas near the Exmouth Plateau. Then a further transgression occurred at the end of the Triassic. Sandstone formed during the period constitutes some of the main reservoirs for the gas fields of the Shelf. It is interesting to note that in Australia many of the gas and oil deposits of post-Devonian age are terrestrial in origin, in contrast to those in other parts of the world, which have marine origins.

Freshwater sedimentation in the Bowen and Sydney Basins was a feature of the Triassic period. The massive Hawkesbury Sandstone and the Narrabeen Group sediments of the Sydney region are examples. Volcanoes were active on the continental shelf and on the land east of these basins.

The plants that characterize the Triassic in Gondwana are *Dicroidium* forked-frond seed ferns; their habit is still unknown, although much information has been gathered about their leaves and fruiting bodies. The abundance of *Dicroidium* in fossil deposits suggests that they grew locally in the areas in which the sediments were collecting. It seems likely that they formed a low-growing heath zone adjacent to damp areas, where horsetails and lycopods occupied the water margins (Figure 19). Conifers, ginkgophytes, cycadophytes, ferns and tree-ferns formed communities on the slopes and hillsides.

Temnospondyl amphibians left their footprints and bones in rocks containing *Dicroidium* flora. Insects were abundant. Triassic freshwater fish, including lungfish, have been found in quantity in Australia; rarer amphibians, crocodile-like and lizard-like primitive reptiles have also been found.

JURASSIC PERIOD

This period extended over 64 million years (205–141 m.y.a.) and in early Jurassic times major river systems developed in the east of the Australian continent, flowed generally eastward and entered the sea somewhere north of Brisbane (Figure 20(a)). They spread a sheet-like sand body across the area, which was subsequently buried and subjected to minor folding, and was to become one of the main aquifers of the Great Artesian Basin. In places it forms reservoirs for the accumulation of oil and gas. These active river systems indicate the high and consistent rainfall of the time, in accordance with the global climatic pattern, which was one of uniform warm to hot conditions and abundant rainfall. With the break-up of Pangaea the monsoonal cycle began to decline—especially in western areas—but coastal areas remained humid.

There were coal swamps in Western Australia north of Perth in the early Jurassic, and in north-eastern New South Wales and south-eastern Queensland in middle to late Jurassic times (Figure 20(b)), possibly along the lower reaches of an eastward-flowing river. Alluvial deposition continued in the eastern part of the continent throughout the period as the river system expanded.

Basalts were extruded near the western margin of the Sydney Basin late in the period. At that time a major geological event was getting under way with the start of fragmentation of Gondwana. Parts of Gondwana flanking the Western Australian coast started to move away, with the production of new sea-floor starting in the north-west. There had been faulting and rifting for some millions of years before the actual separation. Similar faulting and rifting on Australia's southern margin from mid-Jurassic was setting the stage for separation there, although this link with Antarctica was to be the last to be completely severed at the completion of the Gondwanan break-up. The extrusion of basalts on Kangaroo Island and west of Melbourne as well as over most of Tasmania was part of the process that heralded the break-up.

The Jurassic, with widespread warm and wet conditions, was the age of conifers, cycads, ginkgoes, ferns, seed-ferns and herbaceous horsetails and lycopods (Figure 21). A similar flora persisted into the early Cretaceous. This was the last age of 'prehistoric' vegetation, comprising all the ancient plant groups whose lineages went back into distant geological past. After this, the advent of modern floras, dominated by flowering plants, would relegate the old plants to relatively minor roles in vegetation patterns.

Among the conifers, kauri pine similar to that alive today, and monkey-puzzles like Norfolk and hoop pines were already here. There were insects, including cicadas, and abundant freshwater fishes. It was the age of reptiles, and dinosaurs and plesiosaurs appear in the fossil record.

There is evidence worldwide that the number of species of conifers, and of members of the other old plant groups that thrived under the benign conditions of the Jurassic, were declining as the period progressed. This may be evidence of senescence in the populations. Under very long periods of unchanging conditions, the genetic ability to change was being lost. Populations that reached this stage were headed for challenges to their dominance when conditions changed in the Cretaceous and new and

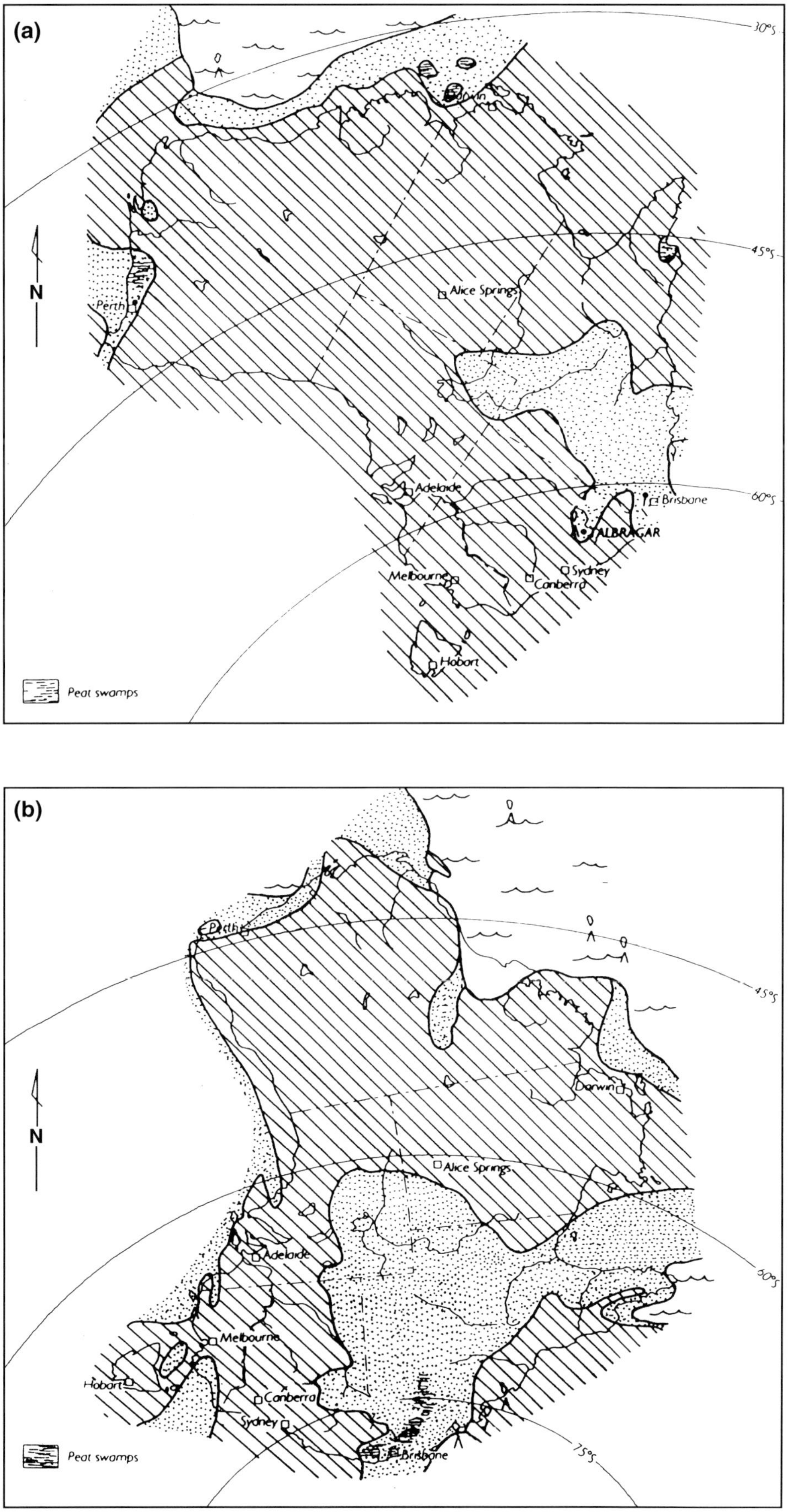

Figure 20. Palaeogeography of Australia during the Jurassic and early Cretaceous periods: **(a)** during the early Jurassic—freshwater sedimentation in the east began, creating the major aquifers of the Great Artesian Basin; **(b)** during mid-Jurassic to early Cretaceous times—greatly increased freshwater sedimentation occurred in a very large area of eastern Australia, completing the sequence of sandstones that are the major aquifers of the Great Artesian Basin (adapted from White, 1998). For key to symbols see Figure 9.

efficient forms (angiosperms) developed to meet those new challenges.

CRETACEOUS PERIOD

The Cretaceous period (141–65 m.y.a.) could well be called the 'Time of Change'. Dramatic events on the world scene included: supercontinents fragmenting and continents starting to move towards their present positions on the globe; the changeover from ancient-style vegetation to floras dominated by flowering plants; the end of the reign of the reptiles and start of the rise of the mammals. Australia remained locked to Antarctica, in polar latitudes.

There is evidence that very early in the period (Figure 22) there was considerable temperature fluctuation, from the warm to hot conditions of the Jurassic to winters in which there were possibly mountain glaciers on high ground and freezing over of lakes and shallow waters. Evidence for this is seen in the 'winter kill' of fish in the Koonwarra Fish Beds in Victoria (Waldman, 1984) and ice-rafted boulders in the Eromanga Sea. There is also evidence from marine plankton studies that seas were cool to cold off the south-eastern coast. After this cooler episode, temperatures rose steadily and for the latter part of the period may have been higher than any in the Mesozoic or since. Abyssal waters may have been 15°C where today they are 2°C, and ocean circulation would have been sluggish.

The ancient vegetation types (carried over from the Jurassic) were already in decline. In addition to climatic changes these floras were to be affected by extensive marine inundations of about half the exposed land. The rising sea levels accompanied the plate tectonic movements that were starting to break the supercontinents apart. In Australia flooding of land surfaces was most extensive in the early Cretaceous; sea levels reached a maximum height about 115 million years ago, then fell rapidly and by late Cretaceous most land was again exposed. In contrast to Africa, Europe and North America where the highest sea levels were in the late Cretaceous.

The fact that so much of Australia was inundated in early Cretaceous times indicates that it was already relatively flat. On the western side of the continent, sea

Figure 21. The Jurassic and early Cretaceous were uniformly warm and wet, and forests of conifers, cycads, ginkgophytes and ferns extended from pole to pole on Pangaea. In Australia, bones of the *Muttaburrasaurus* were preserved in sediments in the Eromanga Sea that covered much of the continent. Diarama by D. Dunphy, from White (1985, p. 23).

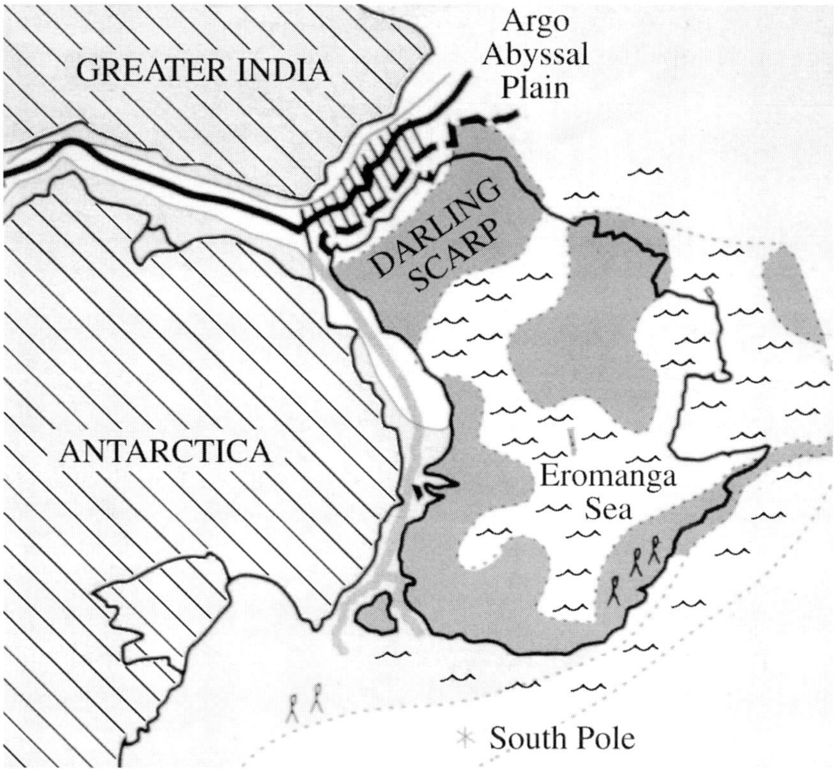

Figure 22. Palaeogeography of the region during early Cretaceous times (118 m.y.a.). India was moving away from the Western Australian coastline with the opening of the Indian Ocean and the sea was starting to enter the rift valley between Australia and Antarctica from the western end. Much of Australia was inundated by epicontinental seas. The Jurassic freshwater sedimentary strata in the Eromanga Basin were covered by marine sediments, completing the Great Artesian Basin's water-transporting and sealing sequence (from Veevers *et al.*, 1991).

invaded the rift between Australia and India 132 million years ago. Meanwhile, thinning of the crust between Australia and Antarctica as stretching occurred prior to rifting, and sagging of the thinned crust, led to the development of a wide rift valley between the two lands. The sea was able to invade this rift along Australia's southern margin before there was any sea-floor spreading there, and a seaway lay between the continents before any basaltic sea-floor appeared in the centre of the rift, 96 million years ago. A connection persisted via Tasmania and the South Tasman Rise until 45 m.y.a. (Veevers, 1987; Etheridge *et al.,* 1988).

Sea-floor spreading was freeing the southern part of Africa from Antarctica early in the period, and India was breaking loose from the Horn of Africa.

Ancestral flowering plants appeared on the scene as separation of parts of Gondwana was beginning. From the pollen record, which shows the oldest angiosperm pollen in West Gondwana, ancestral flowering plants could radiate into both Hemispheres (Figure 23). There was rapid evolution and establishment of many of the old families in all parts of Gondwana while radiation was still possible. A palm province became established in equatorial regions and members of this tropical flora were able to migrate into Australia across India while the connection remained. Separation of Africa from South America meant that some southern forms, notably *Nothofagus* which evolved outside Africa after that link was broken, are found in all southern lands to which they spread by the South America–Antarctica–Australia–New Zealand route, but not in Africa (or India). The route via Antarctica remained operative for much longer. By the time that Australia and New Zealand were isolated (Figure 24), each had a similar complement of modern-style ancestral flora, as had Antarctica at that time. The Eocene fossil flora of Australia is similar in its family content to the New Zealand flora of today.

Towards the late Cretaceous (Figure 25), Australia was freeing its eastern margin. The now mostly submerged part of Eastern Gondwana—the Lord Howe Rise—began drifting eastward, separating from the present coastline and opening up the Tasman Sea. The movement was accomplished between 80 and 60 million years ago. Formation of the Coral Sea Basin followed.

The Cretaceous was still the age of reptiles. Dinosaurs of all sizes and persuasions shared the shrinking land as epicontinental seas advanced; however, it is hard to assess the impact of grazers, like the sauropods, on the vegetation. Some of these giant reptiles (e.g. *Austrosaurus*, *Muttaburrasaurus*) could be tripedal; with short front legs and long back ones and a tail, some

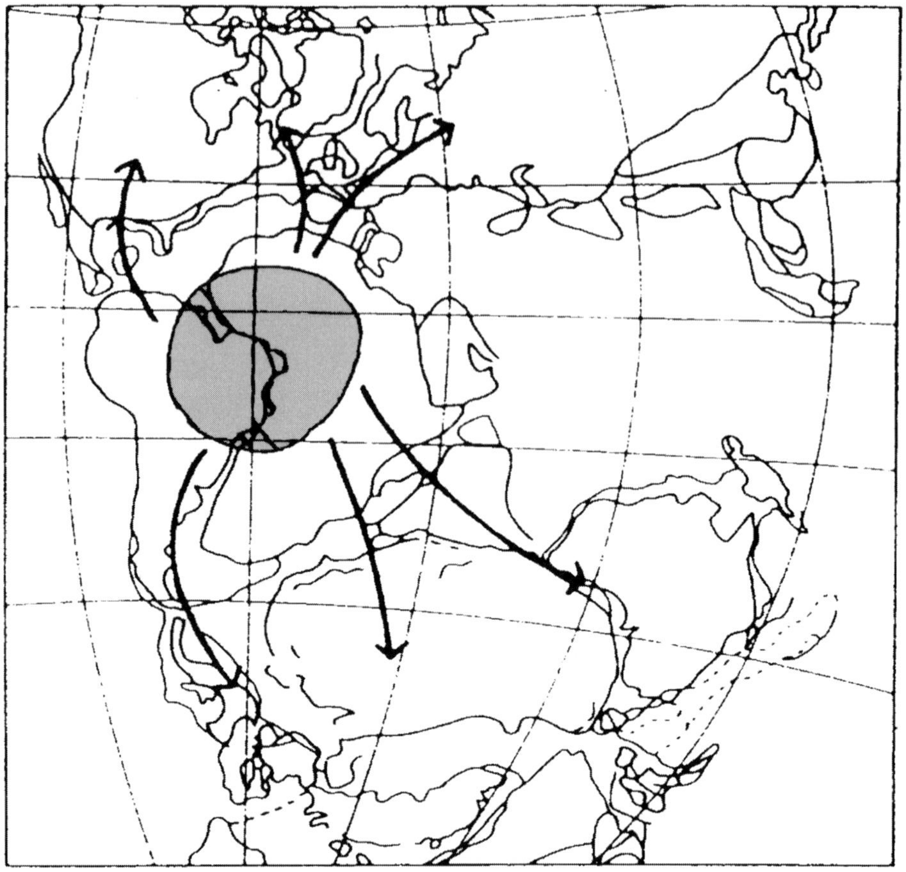

Figure 23. An early Cretaceous reassembly of the continents, showing West Gondwana as the centre where the angiosperms (flowering plants) originated, and from which the ancestral groups could radiate into both Hemispheres (from White, 1998).

could sit up on their haunches and graze the tops of tall conifers. They had gizzards full of well-worn stones to grind the hard and indigestible conifer, cycad or fern material and their stomachs (probably multiple) acted as digesting vats to break down the wood and cellulose. The large sauropods consumed huge amounts of plant material and, combined with a range of other herbivores of smaller size, must have exerted significant pressures—at least locally. As the vegetation at the time was slow to regenerate (in comparison with broad-leaf plants and grasses) there were probably stressful effects on preferred forage species at some times.

With the retreat of the epicontinental seas, conditions developed for another rapid phase of plant evolution. As the waters withdrew many new habitats, such as salt swamps and marshes, seabed sands, dunes and littoral flats, became available. In addition the rift

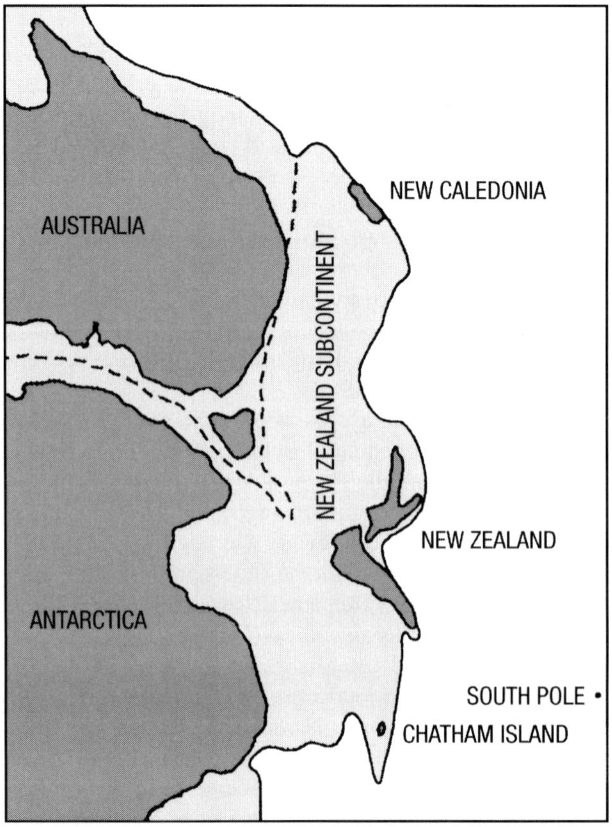

Figure 24. The relationship of Australia, Antarctica and New Zealand in the Jurassic period, 175 m.y.a.; the 'New Zealand Subcontinent' formed the outer (eastern) edge of Gondwana and Australia's present-day eastern margin was far from the ocean. The Subcontinent comprised the continental crust that forms the Lord Howe Rise, the Norfolk Rise, New Zealand and the undersea plateaux around it and New Caledonia.

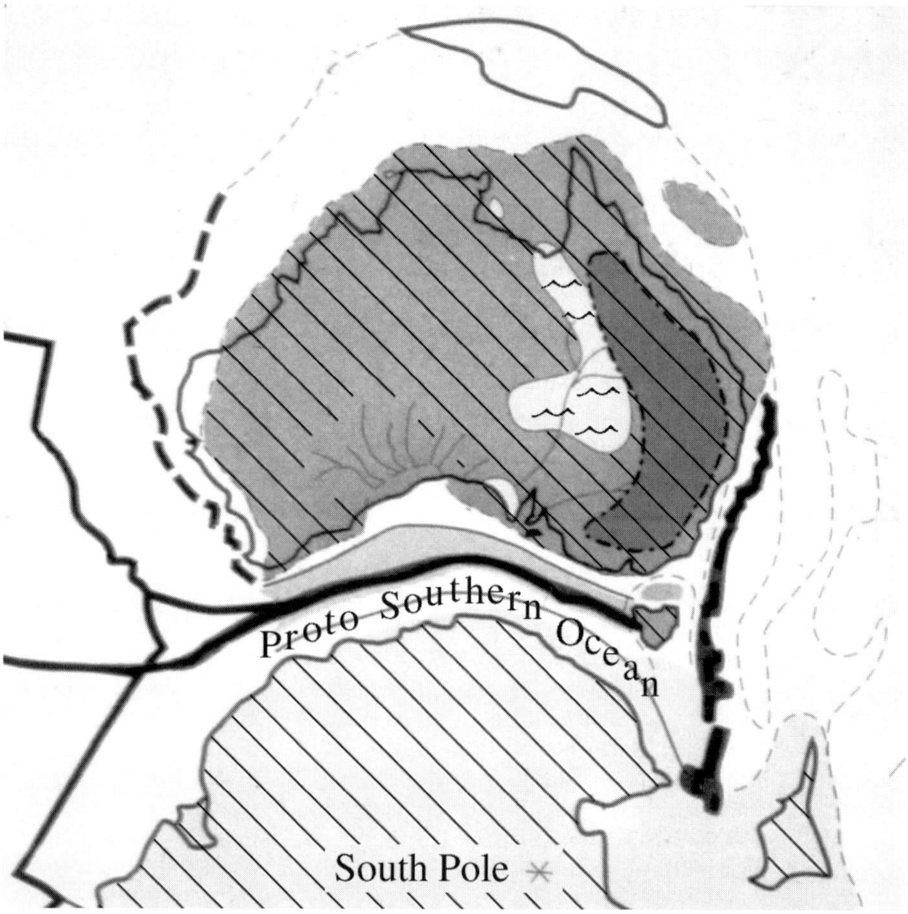

Figure 25. Palaeogeography of Australia during the late Cretaceous period (84 m.y.a.). With the opening of the Tasman Sea and the drifting away of the New Zealand subcontinent, the eastern margin of Australia was freed and a wide zone of adjacent land was elevated, while the centre of the continent started to sink. Australia was still connected to Antarctica through Tasmania and the South Tasman Rise but a wide seaway, the proto-Southern Ocean, lay between the continents (from Veevers *et al.*, 1991). For key to symbols see Figure 9.

valleys appearing between land masses, globally, provided many other new types of habitat. The existing vegetation could not utilize these newly available environments and so the newer, more efficient flowering plants colonized the exposed areas. They had the genetic capability and the changing landscapes provided the opportunity.

Whether some dinosaur digestive systems were so specialized that they were unable to utilize vegetative matter from flowering plants, is not known. Certainly, many of the new angiosperm families are noted for their alkaloid content, and significant changes in food resources may have contributed to the decline of some dinosaur groups (Archer and Molnar, 1984). But, at the end of the Cretaceous there was a dramatic extinction of plankton, which may have occurred in as little as 10,000 years. The dinosaurs also became extinct abruptly. The plant record in the Southern Hemisphere does not show an abrupt extinction event, although in the Northern Hemisphere an extinction peak occurs (Johnson *et al.*, 1989; Hill *et al.*, 1999).

There is general acceptance of a large meteorite impact in the Caribbean at the Cretaceous–Tertiary (K–T) boundary; but in addition to the immediate physical effects (such as ash clouds), a number of other factors have been suggested as contributing towards the world-wide faunal extinctions. The additional influences proposed include: global wildfires, release and dispersal of toxic sulphurous compounds, declining atmospheric oxygen concentrations leading to respiratory stress and suspended particulate matter in the atmosphere leading to 'nuclear winter'.

It is evident from the Australian plant fossil record that the vegetation type had changed and a modern style flora was established before the terminal Cretaceous event. The flowering plants were evolving rapidly and dispersing widely, and genera present before the event continued to diversify after the Cretaceous–Tertiary boundary. Traverse (1988) analyses the Northern Hemisphere information and concludes that the major faunal extinction events and times of floral change in the fossil record are out of step. Floral transformations have been due to gradual replacement and piecemeal, not mass, extinctions. The seed banks in the ground and the ability of plants to hybridize, to produce diploids and polyploids, means that plant populations have many variants that are ready to be selected as conditions change. While sudden climate change places animal species under increasing stress to the point where they can no longer thrive and they ultimately become extinct, the subtle changes in flora throughout the period of change result in new vegetation types without any sudden extinction events (see summary table inside front cover).

TERTIARY PERIOD

There are considerable data on the events over the 63.2 million years of the Tertiary (65–1.8 m.y.a.); in the first part of the period (Paleocene Epoch, 65–55 m.y.a.), Australia's climate was warm and wet. Sea-surface temperatures were relatively high, and erratic atmospheric circulation north of 60°S drove moisture-laden winds deep into central Australia, which was covered with rainforest. New Zealand was increasingly distanced from the Australian continent by continued widening of the Tasman Sea, and the Coral Sea was opening out to the north. The processes leading to separation of Australia from Antarctica continued. The proto-Southern Ocean filled most of the rift valley between the continents but connection to Antarctica was by a trailing edge through Tasmania and the South Tasman Rise (Figure 26).

The warm and wet conditions that favoured forests continued into the Eocene (55–34 m.y.a.) reaching a peak towards the end of that epoch. There were coal swamps in southern parts of the continent. In the mid-Eocene, about 45 million years ago, the Tasman Rise was breached and the sea between Australia and Antarctica started to widen. Warm ocean water transgressed from the Indian Ocean initially, forming an early Southern Ocean. There was a large marine transgression into the Eucla Basin (Nullarbor Plain) in the south, and into the Carnarvon Basin region of Western Australia. Global circulation patterns were changing with the development of a partial circumpolar current (Figure 27), but until the Drake Passage was opened in Oligocene times about 30 m.y.a., when a strong and widening circumpolar current was established, the effects were neither rapid nor severe.

The South Pole cooled steadily and by about 35 million years ago ice-age regimes can be deemed to have started, with glaciers appearing on Antarctica and worldwide cooling of surface and abyssal waters. In the mid-Eocene, India started boring its way into Asia, but while Gondwana was splitting, Laurasia was acting as one continent with land-bridges across the North Atlantic and North Pacific.

The eastern highlands of Australia continued to be uplifted and eroded throughout Tertiary times. An active drainage traversed most of the continent. The increased rainfall led to deep weathering. Active volcanoes along the Great Divide and outpourings of basalts in the south-east and also north along the ranges were features of the times, mostly related to the northward drift of the continent (Figure 28).

Truswell (1989) classifies as 'eucalypt forest' a vegetation type in the Lake Eyre and Lake Torrens area in which there is an abundance of gumnuts of *Eucalyptus* and other Myrtaceae, leaves with mesic characters (related to moderate moisture), and conifers like those of montane New Guinea. The age of this fossil flora, which is distributed quite widely in the present-day arid zone is

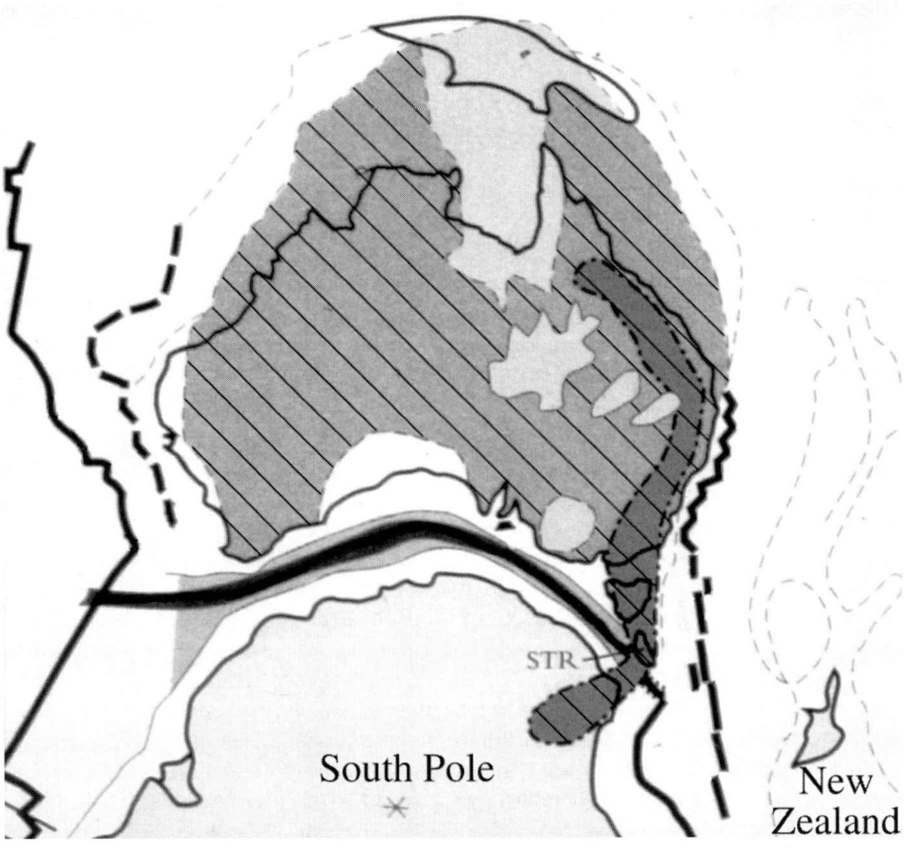

Figure 26. Palaeogeography of Australia during the early Tertiary period (60 m.y.a.). Australia was still connected to Antarctica, the Murray Basin was starting to sink (marine incursions occurred repeatedly throughout the Tertiary) and the Eucla Basin became an embayment of the Southern Ocean (from Veevers *et al.*, 1991). For key to symbols see Figure 9.

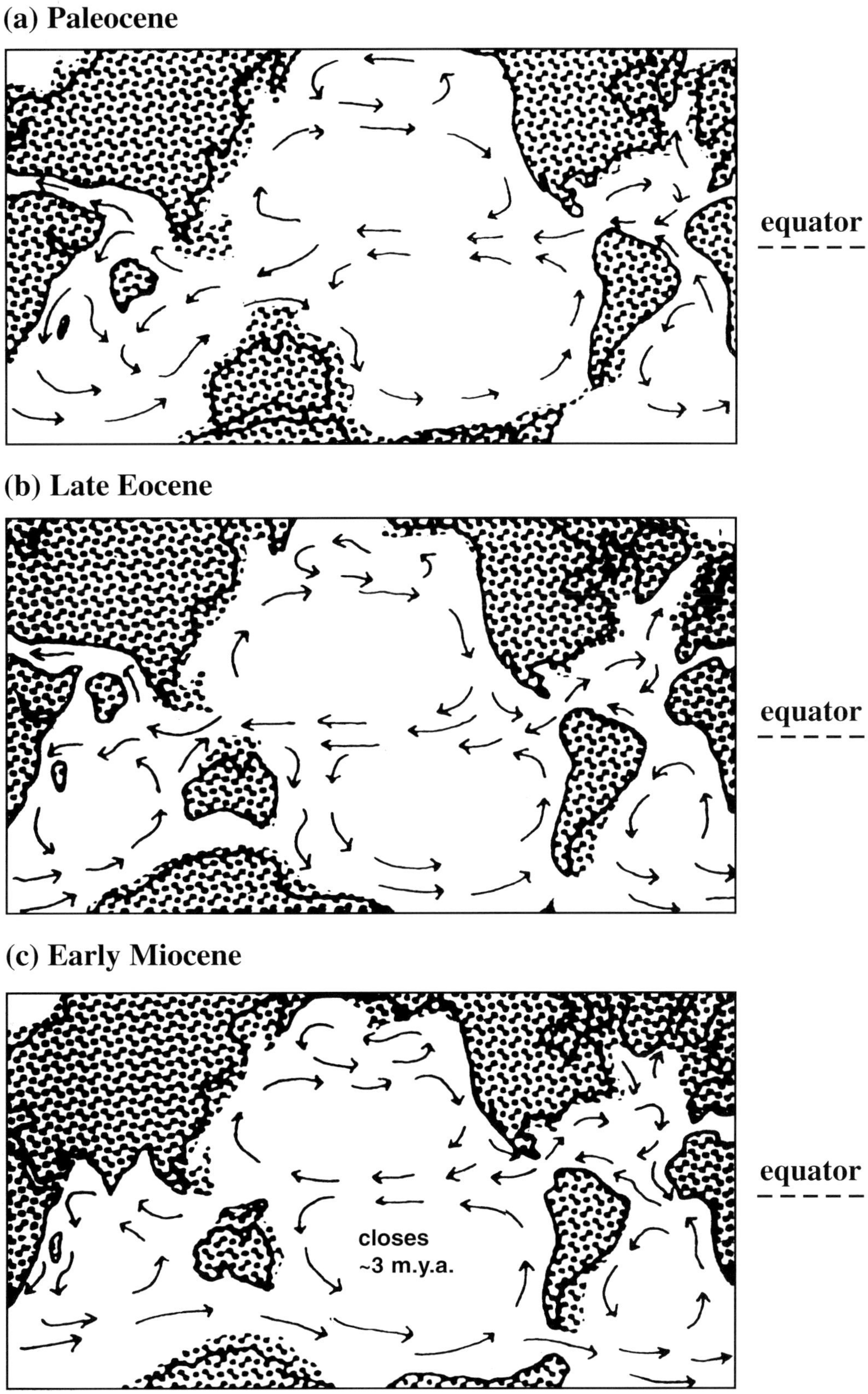

Figure 27. Patterns of oceanic currents during the Tertiary period, showing the major changes resulting in the development of a circumpolar current and establishment of modern circulation patterns: **(a)** Paleocene (60 m.y.a.)—Australia still attached to Antarctica; **(b)** mid Eocene (~40 m.y.a.) before Drake Passage opened; **(c)** early Miocene (20 m.y.a.) before closure of the seaway between South and North America.

White, M.E. 1998. *The Greening of Gondwana*. Third Edition. Kangaroo Press Pty. Ltd., Sydney. 256pp.

White, M.E. 1999. *Reading the Rocks. Animals and Plants in Prehistoric Australia and New Zealand.* Kangaroo Press Pty. Ltd., Sydney.256pp.

White, M.E. 2003. *Earth Alive! From Microbes to a Living Planet*. Rosenberg Publishing, Sydney. 192pp.

Young, G.C. and Laurie, J.R. (eds). 1996. *An Australian Phanerozoic Timescale*. Oxford University Press, Melbourne. 279pp.

Chapter 3
Evolution of Arid Australia and Consequences for Vertebrates

Terence J. Dawson and Lyndall Dawson
School of Biological, Earth and Environmental Sciences, University of New South Wales, Sydney. N.S.W. 2052 Australia
E-mail: t.dawson@unsw.edu.au

INTRODUCTION

The various groups of vertebrates that make up the Australian fauna have varying histories which reflect long past climatic and geographic events. Their ancestors have generally passed through numerous arid 'filters' (or climatic crises) and all families that have survived to Recent times contain many (in some, most) species that have adaptations to some degree of aridity. Groups of Gondwanan origin have experienced over 45 million years of evolution in relative isolation, during which time the continent drifted north from Antarctica and the climate and vegetation changed. Other elements of our fauna have immigrated from the north since Australia approached the Asian plate around 20 million years ago (m.y.a.).

The late Cretaceous and early Tertiary origins of most of Australia's endemic vertebrates are still poorly understood because of the sparse fossil record prior to the Late Oligocene epoch, about 25 m.y.a. (Vickers-Rich *et al.*, 1991; Archer *et al.*, 1999). However, in the last 20 years the fossil record of the past 25 million years has expanded dramatically, greatly improving our understanding of the previous diversity and evolutionary pathways of many taxa, especially the mammals (Archer *et al.*, 1999). We can now make 'educated guesses' about the most probable environmental and ecological forces (ancestry, climate, vegetation) that determined the nature of the vertebrates of this continent prior to European arrival. For further details of regional conditions and major processes influencing them, see pp.17–50 (White, 2006) as well as van Oosterzee (2006)—pp.95–110.

During the past 25 million years there have been two highly significant episodes of climate change, both of worldwide extent, the first at the end of the Miocene (~5–7 m.y.a.), the second in the late Pleistocene (< 0.5 m.y.a.). Both episodes involved cooling and drying of the climate. Australian vertebrates responded to this aridity in four major ways: some contracted their ranges to mesic climatic refugia (e.g. north Queensland and New Guinea); some more generalists were able to adjust their behaviour, diet, physiological responses or body size and so 'ride out the storm'; other groups responded by speciation (either allopatric speciation in refugia, or by *in situ* selection pressure on existing populations); while many species, often entire genera and sometimes even families, unable to adapt, became extinct.

Although all vertebrate groups have distinctive Australian representatives it is the mammals that have attracted special interest. This group also happens to have the most complete fossil record (see Table 1), and so has been most informative about change through time.

CLIMATE AND FAUNA BEFORE LATE MIOCENE

By the early Tertiary (approximately 45 m.y.a.) Australia had rifted free of Antarctica and its fauna began to diverge significantly from their Gondwanan ancestors. By the late Oligocene, when the detailed fossil record begins (Table 1), the marsupials had already diversified into distinctive Australian orders, and parts of the unique profile of the mammal fauna were firmly established. Non-mammalian vertebrates are not so well known, but the late Oligocene Ngapakaldi and Etadunna faunas from the Lake Eyre Basin of South Australia, and the System A faunas from Riversleigh indicate that the continent by then had endemic fishes (including up to ten species of lungfish), amphibians (hylid and myobatrachid frogs), reptiles (chelid and meiolaniid turtles, crocodiles, varanid and scincid lizards, snakes), birds (including passerines and dromornithids) and at least four families of bats (hipposiderids, megadermatids, mystacinids and mollosids) (Vickers-Rich *et al.*, 1991). However, no elements of the Oligocene to middle Miocene faunas suggest aridity—rather all suggest warm, moist, forested conditions. This is supported by the pollen record (Kershaw *et al.*, 1994; McPhail, 1997). The first hints of the climatic hardships to come occur in the later Miocene. Pollen data suggest a decrease in 'wet rainforest' taxa and an increase in 'open vegetation' taxa, including herbs and grasses, from late Miocene onwards (Kershaw *et al.*, 1994; Macphail, 1997).

Several marsupial families that characterized Oligo-Miocene faunas disappeared at or before the end of the Miocene. These were all leaf-eaters or browsers and included several possum-like families—the ektopodontids, miralinids, pilkipildrids; two 'wombat-like' families, the ilariids and wynardiids; and two of the earliest 'kangaroo-like' groups, the balbarid and bulungamyine macropodoids. Both of these 'kangaroo-like' groups had diversified into many species by the late Oligocene and

Table 1. Summary of the Australian fossil mammal record. Shading indicates periods with no fossil record. Adapted from Archer *et al.* (1999). Key to abbreviations: LF = Local Fauna; S.A. = South Australia, N.T. = Northern Territory; QLD. = Queensland; N.S.W. = New South Wales; VIC. = Victoria; W.A. = Western Australia; TAS. = Tasmania; PNG = Papua New Guinea.

Period/epoch with approx. age (m.y.a.)	FOSSIL RECORD						CLIMATE AND FAUNAS*
	S.A.	N.T.	QLD.	N.S.W.	VIC.	W.A/TAS./PNG	
PLEISTOCENE ~2	Many deposits, esp. Naracoorte & L. Eyre	Several deposits	Many deposits, esp. east Darling Downs	Many deposits, esp. Wellington Caves	Many deposits	Several deposits, esp. Mammoth Cave, W.A.	-late Pleistocene extinction
PLIOCENE 5	Kanunka LF 3.4 Palankarinna LF 3.9 Sunlands LF ~3		Chinchilla LF ~3.4 Bluff Downs LF ~3.8 Rackham's Roost	Fisherman's Cliff 2–3 Big Sink 3–5 Bow LF 3–5	Smeaton 2–3 Coimadai 3–6 Hamilton 4.46	Quanbun (W.A.) Awe (PNG) 3–4	-many modern mammal genera appear -first grazers evolve
MIOCENE 24	↑? Kutjamarpu LF ↓?	Ongeva LF ~6.5 Alcoota LF ~8.7 Bullock Crk LF ~12 ↑? Kangaroo Well LF	<u>Riversleigh Faunas</u> Encore LF Upper System C Lower System C System B ↓?	Canadian Lead ~17	Beaumaris LF 6–7	Wynyard (TAS.) ~21 Geilston Bay (TAS.)	-late Miocene extinctions -rainforests decline and woodlands increase -Aust. collides with Asia 15 m.y.a. -family-level extinctions
OLIGOCENE 38	Tarkaroola LF Ngama LF ~24.8 Ericmas LF Ngapakaldi LF Ditjimanka LF Oinpa LF		System A ↓?				-faunas show high species diversity, rainforest adaptations
EOCENE 55			Tingamarra LF ~55				-final separation from Antarctica (46–35 m.y.a.)
				Long gap in record			
65 CRETACEOUS					Lightning Rge ~100	Flat Rocks ~115	

*Details of faunas can be found in Rich (1991).

Plate 12. Numerous termite mounds in the spinifex grassland of the Tanami Desert, Northern Territory. Photograph T.J. Dawson.

Plate 13. Golden perch or yellowbelly (*Macquaria ambigua*), an adaptable widespread edible fish from inland Australia. Photograph T.J. Dawson.

Plate 15. Turtle frog (*Myobatrachus gouldii*), a burrowing termite-eating frog, often found associated with termite mounds. Photograph J.D. Roberts.

Plate 14. Cocooned frog (*Neobatrachus sutor*). The cocoon is derived from shed epidermal cells. Photograph P. Withers.

Plate 16. Thorny devil (*Moloch horridus*), a most unusual-looking desert lizard with unusual physiological features. It feeds only on ants. Photograph L. Dawson.

Plate 17. Emus (*Dromaius novaehollandia*). These large, widespread ratite birds have features that allow them to be successful in arid and mesic regions. Photograph T.J. Dawson.

Plate 19. A breeding group of red kangaroos (*Macropus rufus*), the common large native animal in much of inland Australia. The female mates a day or two after giving birth, the previous large young having been permanently excluded from the pouch days before the birth. The fertilized egg from the mating is held in embryonic diapause until required. Photograph T.J. Dawson.

Plate 18. Paucidont planigale (*Planigale gilesi*). One of the smallest of marsupials, this little arid zone dasyurid spends most time underground in regions with cracking soils. Photograph R. Williams.

Clockwise from top left

Plate 20. The southern cassowary (*Casuarius casuarius*) is widely distributed across New Guinea, Aru and the rainforests of north-east Queensland. A population on Seram in the Moluccas was probably introduced by human agency. Photograph T. Heinsohn.

Plate 21. A black-tailed giant rat (*Uromys anak*) in Kalangur Cave, West Papua. Weighing up to a kilogram, this montane species is one of New Guinea's old endemic rodents. Photograph G. Hope.

Plate 22. An adult female northern common cuscus (*Phalanger orientalis*) with pouch young. This species is found in northern New Guinea and surrounding islands. It was probably introduced to much of its oceanic island range in the Solomon Islands, Bismarck Archipelago, Moluccas and Timor by prehistoric human agency. It is one of the most common lowland cuscuses. Photograph T. Heinsohn.

Plate 23. Adult male white and red colour morph of the common spotted cuscus (*Spilocuscus maculatus maculatus*) subspecies from northern New Guinea and adjacent islands. This subspecies was probably introduced to the St. Matthias Group by prehistoric human agency and to New Ireland about 60–70 years ago. This is one of the common lowland cuscuses. Photograph T. Heinsohn.

Plate 24. Juvenile male grey and white colour morph of the common spotted cuscus (*Spilocuscus maculatus nudicaudatus*) subspecies, upper Cape York Peninsula. The ancestors of this essentially New Guinean possum probably entered Australia via the prehistoric Torres Strait land-bridge. Photograph T. Heinsohn.

Plate 25. An aerial view of the lowland swamps and forests of the Sepik River in northern Papua New Guinea. Photograph G. Hope.

Plate 26. Montane *Nothofagus* forest at 2,350 m, Otomona, West Papua. Photograph G. Hope.

PLATES 27–34

Clockwise from top right

Plate 27. Open subalpine vegetation including anthropogenic grassland in the vicinity of Lake Habbema in West Papua. Mount Trikora in the background is 4,730 m in elevation. Photograph G. Hope.

Plate 28. Glaciated peak of Puncak Jaya in the Snow Mountains of West Papua. Photograph G. Hope.

Plate 29. A giant long-beaked echidna (*Zaglossus bruijnii*) from Neon Basin, (3,050 m) in the Wharton Range, Papua New Guinea. Endemic to the Papuan subregion, this monotreme once had a wide range that extended across much of New Guinea and some continental satellite islands from sea level to over 4,000 metres in elevation. Due to its vulnerability to hunting by humans, it now has a much more restricted distribution and is regarded as an endangered species. Photograph G. Hope.

Plate 30. An adult female agile wallaby (*Macropus agilis*) with pouch young. This species exemplifies the Torresian connection by occurring in savanna-woodland habitats in northern Australia and southern New Guinea. It may have been prehistorically introduced to parts of its range in the South-east Papuan Islands. Photograph T. Heinsohn.

Plate 31. The female magnificent riflebird (*Ptiloris magnificus*) of New Guinea and upper Cape York Peninsula, Australia. This is one of the few essentially New Guinean birds of paradise with a range that extends to Australia. Photograph T. Heinsohn.

Plate 32. The green tree python (*Morelia viridis*) of New Guinea, Aru and upper Cape York Peninsula. This is an essentially New Guinean species that only just makes it to rainforest patches on upper Cape York Peninsula. Photograph T. Heinsohn.

Plate 33. The Javan rusa deer (*Cervus timorensis*) in the lowlands of Western Province, Papua New Guinea. This species was introduced to New Guinea by European colonists about a century ago. It is now widespread in West Papua and Papua New Guinea. Photograph R. Hide.

Plate 34. The cane toad (*Bufo marinus*) from tropical America was introduced to New Guinea, many adjacent islands and Australia in the 1930s in a misguided effort to control agricultural insect pests. It has since invaded vast areas of New Guinea and north-eastern Australia where it has become a major ecological pest as it tends to kill native species that ingest it. Photograph T. Heinsohn.

persisted through to middle Miocene, but left no representatives beyond the late Miocene climatic events.

Cerling *et al.* (1998) have noted a worldwide increase in C_4 plants (tropical and sub-tropical grasses and sedges) between 8 and 6 m.y.a., but suggested that these vegetation changes were caused by low atmospheric CO_2 levels rather than aridity. They have correlated this change in grasses with a dramatic change in mammalian herbivores in Europe, Africa and Asia, particularly an increase in those with hypsodont teeth suitable for grazing. However, in Australia herbivores of this period do not yet have specialized adaptations to grazing—in fact, there seems to be a discrepancy of about 2 million years between the evolution of grazing placentals and marsupials (Langer, 1994; Cerling *et al.*, 1998)! In Australia the earliest true grazer (exemplified by highly specialized unrooted molar teeth) is a vombatid genus, *Warendja*, from the late Miocene Encore Local Fauna of Riversleigh (Archer *et al.*, 1999). Grazing grade kangaroos do not appear until early Pliocene faunas less than 5 million years old, such as Bluff Downs in Queensland (Archer *et al.*, 1994). The reason for the late evolution of marsupial grazers in Australia is not understood.

CHANGE ACROSS THE MIOCENE–PLIOCENE BOUNDARY

During the latest Miocene and very early Pliocene, Australian marsupial faunas underwent profound change, with the essential character of Recent faunas (at least at the family level) becoming established by early Pliocene (5–6 m.y.a.). By this time Australia's close contact with the Indonesian archipelago of the Asian plate was established. At times of low sea level Australia and New Guinea were also connected by a land-bridge, enabling easy interchange of fauna. Conversely, New Guinea and the contiguous rainforest areas of north eastern Queensland became the refuge for several forest-dwelling Miocene taxa. Notable north Queensland and New Guinea 'refugees' include a ratite bird (the cassowary), a monotreme (*Zaglossus*), numerous 'possums', the peroryctid bandicoots, rat kangaroos (*Hypsiprymnodon*), tree kangaroos and small forest wallabies. New Guinea was also a refuge for several small diprotodontids and several browsing 'wallabies' of the genus *Protemnodon* (Archer *et al.*, 1991).

The marsupials are the best studied group and provide good insights into the magnitude of change at this time. Elements of many marsupial families did persist across the Miocene–Pliocene boundary within Australia, but in most families major changes occurred in generic diversity, habitat and dietary adaptation as well as geographic distribution. These 'survivors' fall into the three broad groups described below.

1. A few very conservative genera persisted from at least the middle Miocene into the modern fauna. These include taxa that occupy a specialized niche such as the marsupial moles (*Notoryctes*), and the carnivore (*Thylacinus*), as well as many of the forest-dwelling 'possums' including burramyids (*Burramys* and *Cercartetus*), pseudocheirids (*Pseudocheirops*), phalangerids (*Strigocuscus* and *Trichosurus*) and petaurids (*Dactylopsila*).
2. Several families persisted, but with reduced diversity and complete replacement at the generic level in the Plio-Pleistocene. These include the marsupial lions (Thylacoleonidae), the koalas (Phascolarctidae), the sloth-like palorchestids, and the 'bovine-like' zygomaturines and diprotodontines. The latter three groups all comprised relatively large browsers in the Miocene, but significantly larger species evolved as the Pliocene progressed. Of these, only the diprotodontines, which attained the largest body size, seem to have occurred in the most arid areas (Hope, 1982).
3. Several families increased in taxonomic diversity and abundance, with almost total replacement at the generic level during the Plio-Pleistocene. It is among this latter group that most of the arid-adapted marsupials are to be found. They include two bandicoot families, all omnivores, with many arid-adapted species (peramelids and thylacomyids); three sub-families of carnivorous marsupials (dasyurids); grazing wombats (vombatids); the rat kangaroos (potoroids); the short-faced browsing sthenurine 'kangaroos'; and the macropodines (the rest of the kangaroos and wallabies).

The best dated fossil faunas of the early Pliocene are dominated by mammals and tell us little about other vertebrate groups during this transition period. This is partly due to the nature of the deposits, which tells us something about the changing climate. Lacustrine deposits become rare, thus most Pliocene fossil faunas are from fluvial or cave deposits. In the lacustrine deposits of the Lake Eyre basin there is a notable decrease in lungfish diversity, from at least 10 species in the Oligo-Miocene to 3 or 4 in the Pliocene (Kemp, 1991). Unfortunately reptiles and birds are, as yet, poorly studied in deposits of late Miocene and Pliocene age.

THE PLIO-PLEISTOCENE—DEVELOPMENT OF THE MODERN PICTURE

While the late Miocene was the period when the Australian fauna made its most profound adjustment (in terms of family-level taxonomy) to increasing aridity, several more significant climatic 'events' were to follow, each having profound effects on vegetation and vertebrate habitats. Once again, it is the marsupial fossil faunas that have provided most insights into faunal change during this period.

Moderate Pliocene Climates

After the dramatic changes at the end of the Miocene, climate (for terrestrial vertebrates) settled down during the Pliocene (between 6 and 2 m.y.a.) to a seasonal regime of mainly summer rain and mild winter aridity (Macphail, 1997). Open savannas characterized the north of the continent, while south and central regions were characterized by woodland with many vestiges of cool temperate rainforest, particularly in the south-east (Kershaw *et al.*, 1994).

This period is relatively well represented by fossil deposits from eastern Australia. Mammal faunas still retain some genera that are reminiscent of Miocene taxa, but are dominated by new genera and species, particularly bandicoots, dasyurids, and small browsing-grade macropodines and sthenurines. While all early Pliocene faunas contain several 'wallabies' and some contain species of *Macropus*, these are reminiscent of the modern forest-dwelling species, which mainly browse. However, Queensland faunas (Bluff Downs and Chinchilla) have a diversity of highly specialized grazing kangaroos (various species of *Macropus*) by the early Pliocene (Archer *et al.*, 1994) and the 'grey kangaroo' lineage makes its first appearance around the mid-Pliocene in Victoria (Great Buninyong Estate Mine) (Rich, 1991). This distribution of grazing mammals reflects the pollen picture, indicating that while mesic climates and forest or woodland habitats prevailed over the more southern regions, grasslands had begun spreading from the north.

Change in the Early to Middle Pleistocene

From about 2.5–0.8 m.y.a. climates began to 'deteriorate' (i.e. become less stable) under the influence of a global 'rhythm' of low amplitude growth and recession of Arctic ice sheets (Williams *et al.*, 1998). In Australia Gondwanan rainforest plants, such as *Nothofagus* spp., retreated to isolated pockets along the eastern fringe of the continent (including Tasmania) while open woodland, herbs and grasslands expanded (Kershaw *et al.*, 1994). Confidently dated fossil faunas from this period are sparse (Tedford, 1994), but the record suggests significant changes in marsupial herbivore assemblages, with a decrease in diversity of diprotodontids and browsing forest wallabies and a sharp increase in grazing macropodines in all regions. Interestingly, the browsing sthenurines ('short-faced kangaroos') began to diversify during this period, particularly in southern regions, suggesting that a new adaptive niche may have opened up with the changes in vegetation.

For a summary of size and diversity of Quaternary herbivores see Table 2.

Late Pleistocene Climatic Roller-coaster

From about 0.8 m.y.a. the periodicity and amplitude of global climate swings changed from approximately 40,000 year (40 ka) cycles to approximately 100 ka cycles of glaciation (known as Milankovitch cycles—see Williams *et al.*, 1998). The start of 'true aridity' in Australia, when conditions for vertebrate fauna became extremely difficult, probably began around 370,000–350,000 years ago (Wasson, 1989; Hesse, 1994). Since then there have been four major glaciations, each probably as severe (or more

Table 2. Estimated body size of large living and extinct mammalian herbivores from the Quaternary of Australia. Species still extant are in bold, with weights (kg) in parentheses. Weights of extinct species are based on estimates of Murray (1991, p. 1156). Maximum observed weights of living species (males, unless otherwise indicated) from Strahan (1995) and Dawson (1995).

Wt Range (kg)	BROWSERS Diprotodontids	 Sthenurines	 Browsing Macropodines	GRAZERS Grazing Macropodines	 Vombatids
20–40		*Sthenurus gilli* (25)	*Wallabia bicolor* (20)	***Macropus agilis*** (27) ***M. rufogriseus*** (27) ***M. robustus*** (female 25) ***M. giganteus*** (female 35) ***M. rufus*** (female 40)	*Warendja wakefieldi* (30) ***Lasiorhinus latifrons*** (32) ***L. krefftii*** (32) ***Vombatus ursinus*** (39)
40–90		*Troposodon minor* (45) *Sthenurus atlas* (50) *S. andersoni* (50) *S. occidentalis* (50) *S. maddocki* (55) *S. oreas* (55) *S. brownei* (60) *S. orientalis* (60) *Procoptodon pusio* (75) *P. rapha* (90) *Sthenurus pales* (90) *S. tindalei* (90)	*Protemnodon anak* (50) *Protemnodon brehus* (75)	***M. robustus*** (male 52) [Several undescribed or poorly known fossil species of *Macropus* fall in this size range] *M. altus* (85) *M. titan* (85) *M. ferragus* (85) ***M. giganteus*** (male 90) ***M. rufus*** (male 90)	*'Phascolomys' medius* (45) *Ramsayia magna* (60)
Over 90	*Palorchestes parvus* (250) *Palorchestes azeal* (300) *Nototherium inerme* (400) *Zygomaturus trilobus* (500) *Diprotodon minor* (850) *Diprotodon optatum* (1150)	*Procoptodon goliah* (120)			*Phascolonus gigas* (150)

so) as the well documented last glacial maximum, which occurred between 20,000–17,000 years ago. In Australia 'glacial maxima' were exemplified by dune fields, dust and loess deposits. The evidence indicates extensive reduction in vegetation cover throughout the continent, particularly in the south-east (Wasson, 1989). Each of these episodes would have had a marked impact on the fauna, probably forcing most surviving species into peripheral refugia, promoting allopatric speciation. The best documented fossil faunas of this period, from Naracoorte Caves in South Australia, include abundant browsing herbivores (particularly sthenurines) but also a diverse macropodine assemblage (wallabies, kangaroos) including many modern species (Wells *et al.*, 1984).

Late Pleistocene Extinctions

As recently as 100,000 years ago the vertebrate fauna of Australia differed dramatically from that encountered by the first European settlers. An event known as 'the late Pleistocene megafaunal extinction' changed the nature of vertebrate faunas during that comparatively recent period. Our understanding of the timing and details of this phenomenon in Australia is clouded, mainly because we have not yet been able to accurately date significant influences (such as the arrival of humans). The fossil record, although abundant enough to identify the players, is still too scanty and poorly dated to write the script for the drama. However, the undeniable outcome was a drastic reduction in the diversity of larger vertebrates, mostly

Summary Box 1 MEGAFAUNAL EXTINCTIONS

The debate about late Pleistocene extinctions throughout the world has been dominated by the concept of 'the megafauna'. This has been so pervasive that it is generally understood that the Late Pleistocene faunal change simply involved removal of giant mammals from ecosystems throughout the world. To understand the dominance of this 'paradigm' and to put the Australian record in context, it is worth briefly exploring its origins. In North and South America and Europe many of the most spectacularly massive mammals survived throughout the Pleistocene only to disappear completely within a few thousand years around the early years of the Holocene (Anderson, 1984). The largest of these included most of the Edentates (the glyptodonts, ground sloths, megatheres), most proboscideans (the mastodons, gompotheres, mammoths and stegodonts), several large genera of deer, buffalo and antelope; and a suite of formidable carnivores including the European cave lion (*Panthera*), and the sabretooth tiger (*Smilodon*).

Because of their size, these mammals justifiably inspired their designation as 'megafauna', but the term has now been extended in common usage to include virtually all species (especially of mammals) that became extinct during the Pleistocene, even some not much larger than a rabbit (Anderson, 1984). A rough check on the late Pleistocene Rancholabrean deposits from California indicates that, while 24 giant genera became extinct a further 25 genera that were no larger than their modern equivalents, or were relatively small also succumbed at this time (Anderson, 1984).

The Australian Pleistocene vertebrate faunas also contained genera and species larger than those remaining today. However, they differed markedly from their North American and European equivalents, in being relatively much smaller. Murray (1991) has estimated that, although there was a similar distribution of 'weight categories' (i.e. smaller, medium, large, very large) among the larger mammals, the absolute weights of North American taxa were approximately two thirds greater than the equivalent Australian genera and species. Moreover, the Australian fauna was low in generic diversity and population density, but species rich, while the converse was true for North America. Murray (1991) suggests that these features are evidence for a strategy (in Australian faunas) where maximum control of the comparatively small amount of energy available is accomplished at the expense of maintaining high population densities.

As with the placental mammals in Northern Hemisphere continents, the marsupial herbivores included the largest bodied (and most diverse) elements in Australian Plio-Pleistocene faunas. The large herbivore niche also included dromornithid and ratite birds (both herbivorous in the Pleistocene). But despite the diversity of larger herbivores there were only two large marsupial carnivores (*Thylacinus* and *Thylacoleo*). Instead, the large carnivore niche was supplemented by a range of reptilian carnivores (crocodiles, *Megalania*, meiolaniid turtles and the snake *Wonambi*), completing the complement of 'megafauna' in the Australian Pleistocene.

While the Late Pleistocene extinctions in Australia did remove all the largest species (as in the rest of the world), a high proportion of small to mid-sized taxa also became extinct. In the trophic category most severely affected, the browsing herbivores, of a total of 23 extinct species only seven weighed more than an adult male red kangaroo (Table 2).

It is possible that over-use of the term 'megafauna' and over-emphasis on body size when considering late Pleistocene extinctions, may be masking more subtle ecophysiological and biogeographic constraints, that affected the ability of those taxa to adapt to climatic pressures (see Jablonski and Witfort, 1999). Whether or not the arrival of humans was one of the prime causes of extinction or merely 'the last straw', can only be determined when the date of human arrival is known and more precise and reliable dates for the last occurrence of extinct species become available (Roberts *et al.*, 2001).

mammals, and the extinction of some reptiles (including the giant lizard *Megalania*, the archaic snake *Wonambi* and the land turtle *Meiolania*) and the last dromornithid bird, *Genyornis*. See Summary Box 1 for further details of Pleistocene extinctions.

CONSEQUENCES OF ARIDITY AND VERTEBRATE RESPONSES

Aridity in Australia must be thought of not only in terms of average annual rainfall, or evaporation/precipitation ratios; unpredictability is also a key feature. Predictable seasonal rains usually occur in the extreme north (the monsoonal tropics) and the extreme south (the 'Mediterranean' climate of the south and south-western margins), but over much of the continent the overriding climatic paradigm is that of the Southern Oscillation El Niño/La Niña cycles. These 'cycles' are responsible for the fickle pattern of 'droughts and flooding rains'; this can mean a run of poor years when vegetation dies back and dust storms blow away the soil, only to be followed by a year of catastrophic floods—which fill the ephemeral lakes and streams, but often further erode the precious remnants of soil.

All modern Australian vertebrates, even the freshwater fishes, must either adapt to these precarious conditions or have access to refugia. Although the basic environmental conditions seem similar for all groups, their different habitat requirements and physiological endowments allow a wide range of adaptive solutions to overcome the problems of life in arid Australia.

BASIC CONSTRAINTS OF LIFE IN THE ARID ZONE

The problems for life in the Australian arid zone centre around the shortage of water, the difficulty of acquiring a predictable and adequate supply of food, and the need to cope with temperature extremes. Most importantly, water, through rain, determines plant growth and thus the primary production of the ecosystem. Because rainfall is sparse or erratic, primary production is low and the energy flow (food availability) in arid zone ecosystems is low. While absence of drinking water and temperature extremes provide immediate stresses to terrestrial arid zone vertebrates, it is the 'low energy' nature of the ecosystem that is the major constraint of all life in the arid zone.

Water

All vertebrates have an optimum concentration of body fluids (they are 65–75% water) and they must be able to achieve a continuing balance between inputs and outputs of water if they are to function effectively and survive. Some desert species, particularly some frogs, have evolved the ability to be significantly out of optimum water balance for extended periods as an adaptation for survival in extreme conditions, but there are limits to osmotic imbalance (Withers, 1992, pp.822–830).

There are three possible sources of water for vertebrates: free standing water, water available in food, and 'metabolic' water (i.e. water produced by the oxidation of food). Free standing drinking water is often not reliably available to arid and semi-arid zone vertebrates. Even though most animals will utilize free water when it is available, many are able to use alternative sources so successfully that they never drink.

Water obtained from food (either as free water or metabolic water) is the most important source for most arid zone vertebrates. Water in plants may vary from over 90% of weight for some succulent plants to less than 10% in dry plants and some seeds. Animal bodies consist of 65–75% water and many insectivores and carnivores rely on this water. Metabolic water may be only a small source, but it can provide sufficient extra water to enable some animals to stay in water balance without drinking.

When water is scarce, control of water lost by evaporation and excretion is crucial. Evaporative water loss occurs passively but also can be utilized as an active process for thermoregulation. Passive diffusion of water out of an animal occurs because skin is never completely impermeable to water and breathing exposes moist interior surfaces to the air. Active mechanisms that involve the cooling power of water evaporation for thermoregulation include sweating, panting or licking. These mechanisms enable some animals to dump heat in extreme conditions to avoid lethal high body temperatures. However, because a relatively large quantity of water is required, the flexibility provided by active evaporative heat loss is not available to all desert vertebrates. Excretory water loss is usually via faeces and urine. Strategies for reducing losses via these routes include concentrating urine, producing drier faeces and also reducing excretory residues by diet selection, thereby saving water. Some reptiles and birds have special salt excreting glands which also use a little water.

Food and Energy

Plants are at the base of the food chain and ultimately all vertebrates depend on plants for survival. Soil moisture is the principal determinant of plant growth and biomass levels in arid regions (Noy-Meir, 1973), but rainfall, and hence plant growth, is extremely variable in arid Australia. During a 12 year study of kangaroo, sheep and rabbit grazing interactions at Fowlers Gap Station in western New South Wales, Dawson and Ellis (1994) recorded changes in the cover of ground vegetation, through an El Niño/La Niña rainfall cycle, from a low of 9% to a high of 77%.

Such cycles of vegetation availability cannot be fully 'tracked' by vertebrate herbivore populations (McLeod, 1997) and much of the lush vegetation in a good season will remain uneaten. This eventually senesces to highly fibrous material (cellulose, hemicellulose and lignin), which cannot be digested by most vertebrates and is, therefore, not directly available as an energy source.

However, it can be digested by microbial action in some vertebrates and many invertebrates. Some highly specialized invertebrates have formed symbiotic associations with microbes to facilitate fibre consumption. This is particularly the case with termites (see Summary Box 2). These invertebrates, in turn, are utilized by numerous vertebrates as the energy base for life in the arid zone (Plate 12). Seeds are another major nutrient supply. As a source of food, seed is relatively more dependable from season to season; it can be available in ample amounts because many arid zone plants produce large quantities of seed in favourable times and this 'load' (seed bank) may be preserved in the soil for many years.

Summary Box 2 TERMITES AND FIBRE DIGESTION

Because of its structure fibre is not easily digested by common animal enzymes, this resistance and strength being part of plants' defences against herbivory. The ability of termites and other arthropods to digest cellulose and other plant fibre components makes them a fundamental food resource in the arid zone. In parts of Australia termites are so abundant (Plate 12) that their mass can equal that of all vertebrates (Watson *et al.*, 1973, 1978).

Termites, woodroaches and wood-boring beetles are highly specialized for fibre consumption, and the larvae of many other insects consume fibre. A few of these species have evolved special enzymes, such as cellulase, but many rely on symbiotic bacteria or protozoans to digest cellulose for them. For example, wood-eating termites have a rectal pouch that contains protozoans that efficiently convert food into usable products (Withers, 1992, p. 931). Most arthropods do not need to metabolize food to heat themselves and their metabolic rate is generally 5–10% of that of endotherms like birds and mammals. They are consequently able to allocate 10–40% of their utilizable energy intake to production of biomass (compared to ~1% for birds and insectivorous mammals)—consequently a large biomass of arthropods can be maintained in an arid ecosystem (Withers, 1992, pp.86–92).

Temperature

Arid zones are usually thought of as hot but winters may be cold, especially at night. The impact of temperature on an animal depends on whether it is an ectotherm or endotherm. For ectotherms, such as the reptiles, body temperature tracks environmental temperature and individuals may use behaviour to adjust body temperature; for example, it is common for lizards to shuttle from shade to sun to maintain a warm activity temperature (MacMillan *et al.*, 1989). Many vertebrates, both ectothermic and endothermic, seek refuge from temperature extremes in burrows. Soil temperatures are buffered all year and more closely reflect average seasonal temperatures (Figure 1) (Lovegrove and Knight-Eloff, 1988). The higher humidity underground can also be important in reducing water loss.

Many endotherms (birds and mammals) can cope with extremes of temperature, but must use water for cooling if temperatures are high. Effective environmental temperatures of about 70°C occur in the open on clear summer days in the inland. If these conditions cannot be avoided, a lot of water is required to maintain body temperature. Facing the cold requires energy (food) for heat production. Endotherms in deserts often avoid thermal extremes to save on the limited resources of water and energy.

ADAPTIVE STRATEGIES

Freshwater Fishes

Although not generally thought of as arid zone animals, particularly not as desert animals, fishes should not be overlooked in a review of arid zone vertebrates. The aquatic habitats of arid Australia are generally associated with riverine systems, though sometimes remotely; for example, Dalhousie Springs, deep in the deserts of northern South Australia (Richardson, 1994). The dispersed inland rivers, lakes and swamps are recharged by massive floods in the unpredictable wet years. These floods are essential for sustaining widespread fish populations, and many species have life cycles that make use of such events. Some respond to flooding with opportunistic upstream migration before spawning, which is made energetically feasible by the exceptionally low energy cost of swimming (Figure 2). This behaviour enhances survival of larvae and allows colonization of habitats that are only occasionally accessible. The golden perch (*Macquaria ambigua*)—shown in Plate 13—and the silver perch (*Bidyanus bidyanus*) migrate upstream; individual females of both species may release up to 500,000 pelagic eggs into a flood. This strategy allows precious extra time for the eggs to develop and hatch as they are moved downstream and dispersed by the floodwaters (Richardson, 1994).

Aestivation (the ability to lie dormant in the mud when streams and ponds dry up) is a survival adaptation adopted by some freshwater fishes, especially in Africa, but is relatively rare in the Australian fauna, reported to occur in only four species. The widespread spangled perch (spangled grunter) *Leiopotherapon unicolor*, is reported to aestivate. It appears to survive in bottom mud and emerge a few days after rain in areas previously dry and with no connections to permanent water (Merrick, 1996). Details of aestivation in the salamander fish (*Lepidogalaxias salamandroides*), from south-western Australia, were described by Pusey (1989); adults of the Tasmanian mudfish (*Neochanna cleaveri*) are known to aestivate over summer under logs or rocks and the related eastern little galaxias (*Galaxiella pusilla*) is reported to survive for several months in yabby burrows (McDowall and Fulton, 1996).

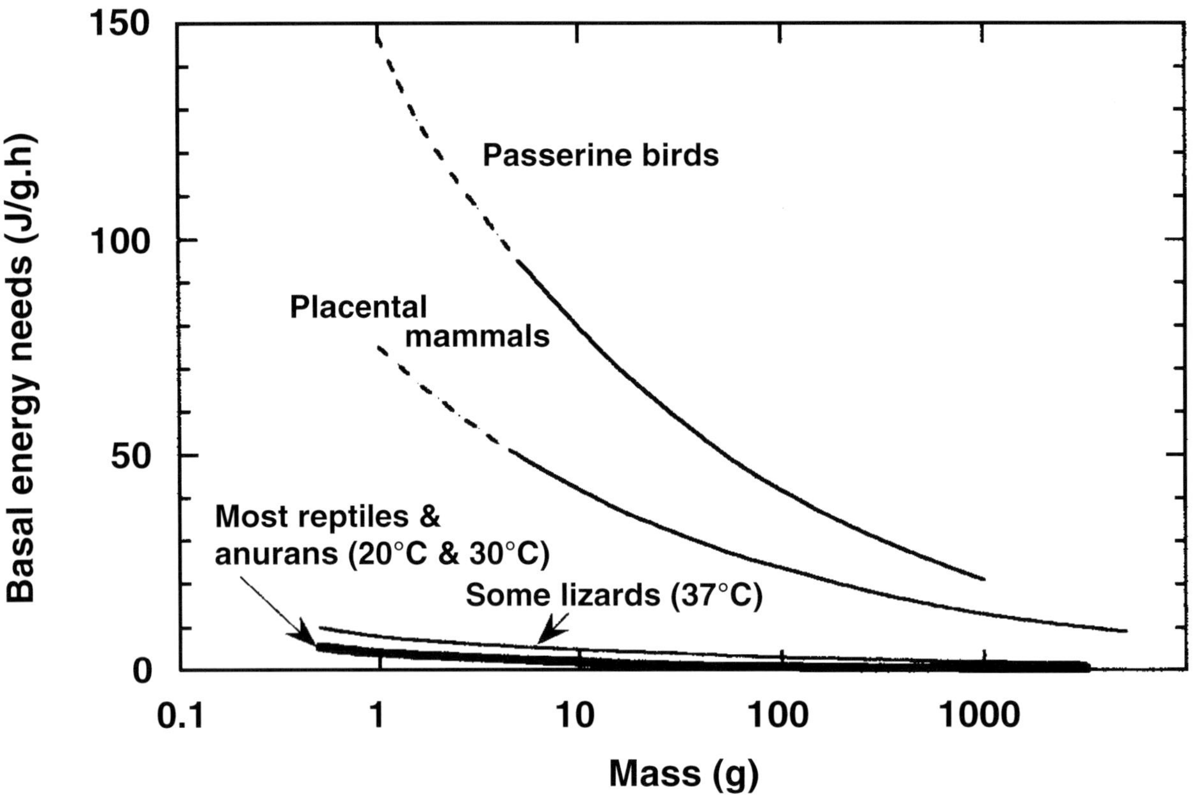

Figure 3. Resting metabolic rate as a function of size in terrestrial vertebrates. Values for frogs and toads fall within the general range for the reptiles; lizards with higher body temperatures have elevated metabolic rates. The dashed portions of the lines for birds and mammals indicate extensions from 5 g to 1 g, which are at the limit of the weight ranges for these groups (adapted from Pough, 1980).

1982). Two myobatrachid burrowing frogs, the sandhill frog (*Arenophryne rotunda*) and turtle frog (*Myobatrachus gouldii*)—see Plate 15, have adapted in a different way. Both live in seasonally dry, sandy country in the south-west of Western Australia and rarely experience free standing water (Withers and Roberts, 1993). Reproduction of these species does not require free water; the eggs develop directly into froglets. Mating takes place underground and the eggs in capsules are deposited (1–2 metres down) in damp sand (Roberts, 1981; 1984).

Reptiles

Australia has the most diverse reptilian fauna of any continent (Pianka, 1986). While the reasons for this are not fully understood, much of their success is undoubtedly related to the basic reptilian 'blueprint', which makes them almost 'pre-adapted' for life under arid conditions. As well as the basic attributes of ectothermy and low metabolic rate, reptiles have achieved control over water loss via the skin and also over water required for respiration and excretion. Most importantly, the ingenious cleidoic egg (see Withers, 1992, pp.655–657), a reptilian 'invention', freed them from needing a larval stage in water.

Relative to the amphibians, all amniotes (reptiles, birds and mammals) have achieved greater control over skin water loss (Table 3). Their skin is covered by a layer of dead, keratinized cells called the *stratum corneum*, which acts as a highly efficient diffusion barrier. Some arid adapted reptiles (e.g. the thorny devil, *Moloch horridus*) have skin that is up to 100 times more resistant to water loss than that of even 'waterproofed' frogs, and about five times as effective as the skin of birds and mammals (Table 3) (Withers, 1998).

To aid in water economy, reptiles use relatively insoluble uric acid as the principal excretory product for protein metabolism. Water can be resorbed from the urine while it is stored in the cloaca/rectum, and when this occurs most of the uric acid is precipitated in the excreta, thereby further conserving water (Withers, 1992, pp.881–882). This is important for insectivorous and carnivorous reptiles, in which most of the urinary products are nitrogenous metabolites. For many desert reptiles the salt load from herbivorous diets can also be a problem. To contend with this and still conserve water many desert lizards, for example, the shingleback (*Trachydosaurus rugosus*), have specialized salt glands.

Several factors may account for the high diversity of reptiles, especially lizards, in Australian arid zones. Underlying their ability to adapt is the fact that, where plant biomass production is unreliable, ectothermic lizards are favoured over birds and mammals because of their lower metabolic rate and lower energy needs.

Additionally, the relatively high vegetation cover, compared with deserts on other continents, provides adequate food and increases habitat complexity. Plant-eating invertebrates, especially termites, are abundant and form an important food source, especially for nocturnal lizards. The presence of sparse trees in Australian deserts provides habitat for a diversity of arboreal lizards. Studies from different habitats illustrate how lizards effectively utilize the microhabitat complexity of spinifex country (Morton and James, 1988), chenopod shrubland (Read, 1995) and mallee woodland (Pianka, 1994).

Ectothermy also allows reptiles access to small body size and specialized body shapes, which enable them to fill ecological niches that are not available to birds and mammals. Ectotherms can have small body size because they have fundamentally lower levels of metabolism. Their metabolic rates do not change directly with body mass; rather, they change at a varying rate in proportion to the 0.75 exponent of mass (i.e. $M^{0.75}$). This means that as body size decreases, the metabolic rates of ectotherms and endotherms diverge so that small birds and mammals need proportionally larger amounts of food (Figure 3) (Pough, 1980). Consequently, few adult birds and mammals weigh less than 10 g and the smallest species are rarely below 3 g. Reptiles are not faced with equivalent problems; more than 75% of reptiles are less than 10 g and about 20% are less than 1 g. Reptiles can also occupy some niches denied to mammals because they are not so restricted in their body shapes; for mammals, body forms that increase surface to mass ratios (such as elongate or very flattened shapes) cause problems with heat loss, which is not so much a concern of ectothermic reptiles.

Several arid zone reptiles have rather unexpected and ingenious adaptations to life in arid conditions. The most spectacular, the thorny devil (*Moloch horridus*) is a specialist ant-eater (Plate 16). Apart from its unusual appearance, the skin of the thorny devil can absorb water like a 'blotting paper' from rain, dew or even damp soil. However, the skin is impermeable and the water is taken into a network of capillary channels among the skin scales and eventually channelled into the mouth (Withers, 1993). The thorny devil can generally maintain energy and water balance solely on its diet of ants, but its water collecting abilities provide the reserve capabilities essential for a successful desert vertebrate.

Birds

Birds retain some of the physiological attributes of their reptilian ancestors, which enable them to cope well with aridity; they have even improved on the water conserving mechanisms of the terrestrial reptiles. However, they have also evolved endothermy, which is a high energy life strategy, but this is linked to flight. That birds can fly is the most significant feature facilitating their existence in the low energy environment of deserts. The energetics of flying are such that it is a very economical way to travel long distances (see Summary Box 3, Figure 2). So it is feasible for some birds to forage over vast distances in central Australia. Nevertheless, in summer when temperatures are extreme, free water may be needed within their daily flying range.

The physical and physiological limitations on flight constrain body size (weight). Basically, the relative costs of flight increase with body weight, such that flight by big birds (over 12 kg) is energetically too costly. To save energy in arid regions birds tend to be small or spend much time on the ground, even though they can fly (e.g. the Australian bustard, *Ardeotis australis*). Some larger birds (eagles and kites) overcome this energy problem by soaring and gliding, which uses much less energy than flapping flight.

The energetics of flight play a large part in determining the diets of arid zone birds. Because of their very high metabolic rate, birds need to process food quickly. This means they have limited ability to digest fibrous vegetation because this is a relatively slow process. Most birds that can utilize coarse plant fibre by using microbial fermentation are, by necessity, too large to fly (Foley and Cork, 1992). Some ratites (ostriches, rheas) are conventional hindgut fermenters, digesting plant fibre using microbial fermentation. But emus (*Dromaius novaehollandiae*) (Plate 17), are omnivores and normally eat relatively succulent food such as insects, berries, seeds and soft herbage. They have an unconventional small gut and are less reliant on microbial fermentation. However, when conditions deteriorate they are capable of altering their gut processes so they can digest some fibrous vegetation (Herd and Dawson, 1984). To acquire sufficient food to meet their considerable metabolic needs emus need to forage throughout the day and travel long distances. Although too large to fly, their lightweight bones and avian metabolic capacity mean they can run fast economically. They make excellent use of their feathers to protect them from cold and from the extreme solar heat load in summer, the feathers acting as a large insulating umbrella (Maloney and Dawson, 1995).

Seed eaters (granivores) are the most abundant birds in the Australian desert (Fisher *et al.*, 1972). They comprise 44% of all individuals, but only 17% of the total species observed. Because many granivorous species, such as parrots, cockatoos and pigeons are relatively large, the total biomass of seed eaters is much greater than that of insectivores, the majority of which are small. While seeds may be an abundant and reliable food supply, they are low in water. Since most birds are diurnal, they must face the desert's severe temperature extremes. Some seed-eating species like the budgerigar (*Melopsittacus undulatus*) and the zebra finch (*Taeniopygia guttata*) can exist without drinking when environmental temperatures are moderate, but in summer all seed eaters need to drink. Some of the most spectacular sights of the desert are of huge flocks of seed-eating birds—budgies, finches, corellas and galahs—coming to drink at waterholes on summer evenings. Fisher *et al.* (1972) recorded 47,000 budgerigars drinking at a waterhole on one day.

The other major food supply in the arid zone is invertebrates, largely insects and their larvae. In terms of total

Summary Box 3 ENERGETIC COSTS OF DIFFERENT TYPES OF LOCOMOTION

The various types of locomotion have very different costs and, additionally, these costs vary with body size (Figure 2). Swimming is the most economical form of locomotion, thus inland fish may make large upstream breeding migrations to disperse their eggs. Of non-aquatic forms of locomotion, flight is the most economical in terms of mass transported a given distance. The energy cost of flying per unit time is high, so birds must have a high metabolic capability to produce that energy. Travelling even short distances can require a high burst of energy use. So, if food is abundant and predators are few, birds may prefer to walk. However, flying becomes much more economical at higher optimum speeds. Flying has very different energetic costs to walking or running on two or four legs. With running and walking the cost of transport per unit of mass is higher for small walkers, as is cost relative to metabolic capabilities. Consequently, long-distance movements generally occur only in larger mammals. The hopping locomotion of kangaroos is unusual; at moderate speeds and above kangaroos travel long distances more economically than do runners.

individuals, Fisher *et al.* (1972) found that 61 insectivorous species constituted 38% of the desert avifauna. Carnivorous birds (including the raptors), which largely eat reptiles, make up another 6% of the arid zone birds. As insect and reptile tissues have a high water content, birds with these diets rarely or never drink. These birds largely excrete nitrogen as semi-solid uric acid/urates, using very little water. The bird kidney can also produce a moderately concentrated urine, but many water birds of the inland live on ephemeral lakes and rivers, which are often saline. Birds inhabiting these environments often have active salt glands, with some species being able to utilize salt water, excreting the salt and getting a net water gain (Maclean, 1976).

Birds are very tolerant of hot conditions and have some physiological 'tricks' to achieve this. Birds generally have higher body temperatures than mammals, with the spinifex pigeon (*Geophaps plumifera*) tolerating body temperatures near to 45°C. This means that gradients for dry heat loss (by radiation, conduction and convection) are high, and if environmental temperatures get above body temperature the gradient for heat inflow is smaller (Withers and Williams, 1990). The result of such a high body temperature is water saving. As distinct from mammals, the unique respiratory system of birds enables them to use respiratory cooling without dangerously affecting blood acid-base balance (Marder and Arad, 1989). To cushion the impact of water use, many arid zone species are particularly tolerant of dehydration; the zebra finch can tolerate a loss of water equivalent to 30% of its body mass (Maclean, 1996).

Because of the speed and efficiency of flight, nomadism is an option available to some arid zone birds; by moving away they can avoid the most extreme seasons and droughts (Davies, 1984). Few Australian arid zone birds are true migrants, but about a quarter are regularly nomadic (Keast, 1959). The number of species present in an arid region of Western Australia has been noted to vary from 10 in a drought year to over 90 in a wet year. Despite the availability of this option some desert species, such as the grass wrens (genus *Amytornis*) from central Australia, remain truly sedentary. These tiny birds have strong bills and utilize seeds as well as insects in their diet, and thus maximize the water and energy resources available to them (Schodde, 1982).

Mammals

While marsupials dominate, monotremes, rodents and bats are important components of the indigenous Australian mammal fauna and have successfully filled many niches in the arid zone. Mammals, like birds, live life in the 'energy fast lane' (Pough, 1980). In general, they have high body temperatures and high metabolic rates (Figure 3), which provide flexibility, but present problems maintaining food and water supplies. In adapting to arid conditions mammals have utilized many of the strategies used by the other vertebrates, but have also come up with a few of their own. There are mammals that burrow, use torpor or hibernate, subsist on water from food, modify their skin (fur covering), utilize a variety of forms of terrestrial locomotion, migrate and even fly. In small mammals metabolic rates may be lowered (Lovegrove, 2000) and lifespans may be lengthened in some desert species (Predavec, 1997). An increased lifespan will increase the chance that a good period for reproduction will be encountered by an individual within its lifetime.

Mammals have also exploited additional opportunities associated with increased body size. As explained, this attribute is not available to flying birds and while terrestrial birds such as the ratites have gone this route they generally have not been as prolific or successful as the mammals. Because of the effect of body size on potential digestive strategies and lifestyle options, the impact of aridity on the Australian mammal fauna is probably best considered relating to three size categories: small (below ~300 g); medium (300 g–8 kg); and large (above 8 kg). These size classes roughly correspond to the types of vegetation that herbivores can effectively use (Hume, 1999); carnivores, on the other hand, may be small or large depending on the size of their prey.

Small Mammals

Small terrestrial mammals cannot range far or move to distant water sources to drink; they must deal with the desert *in situ*, so small desert mammals avoid situations requiring thermoregulatory water loss. Even if water was available they have limited abilities to deal with heat, so they are usually nocturnal and use burrows.

Small mammals require readily digested food such as animal tissue or seed and soft vegetation. In Australia insectivorous/carnivorous dasyurid marsupials and the rodents (mainly granivorous) are the most successful groups of small terrestrial mammals. Arthropods and small lizards provide a fairly abundant food resource for the small dasyurids such as tiny planigales (genus *Planigale*) (Plate 18) and the many species of dunnarts (genus *Sminthopsis*). Larger dasyurid predators, such as the mulgara (*Dasycercus cristicaudata*) and the kowari (*Dasyuroides byrnei*) may even add small mammals and birds to their diet. These readily digestible prey have a high water content. Nitrogenous waste from protein digestion is excreted as soluble urea, not uric acid as in birds. The dasyurids overcome the problems of urea excretion by having kidneys that produce highly concentrated urine (Dawson *et al.*, 1989). To conserve energy these dasyurids also utilize daily torpor in the face of fluctuating food supplies. When utilizing such torpor individuals markedly drop body temperature and metabolism to low but regulated levels during the day (Song and Geiser, 1997). Additionally, several species store fat when prey is plentiful (e.g. the fat-tailed dunnart, *Sminthopsis crassicaudata*).

In some small mammals there is a tendency to a reduced (by about 25%) basal or resting metabolic rate in arid zone species (Predavec, 1997; Lovegrove, 2000). A lower basal metabolic rate generally suggests lower levels of daily metabolism in placental mammals. However, care is necessary when extrapolating to marsupials. They do have a low basal metabolism but they have high overall metabolic capabilities (Dawson and Dawson, 1982; Hinds *et al.*, 1993). Lovegrove emphasizes the unpredictability of the El Niño Southern Oscillation (ENSO) cycles as the evolutionary pressure behind the pattern seen in small desert mammals. But long cycles of very extreme aridity have also occurred with the glacial cycles throughout the Pleistocene, when these species were evolving; so it is suggested that, in Australia, they may have been pre-adapted to El Niño conditions.

The 'old endemic' murine rodents, which first entered Australia in the early Pliocene (Archer *et al.*, 1999) have been very successful in adapting to arid conditions. Again these are nocturnal and generally burrowing. They are seed-eaters but Murray *et al.* (1999) have shown that most Australian species also include substantial amounts of invertebrates and green plant material in their diet. These provide more protein and water than would be available from seeds alone. However, during droughts these rodents must survive on seeds and then water economy is essential. The hopping mice (genus *Notomys*) can produce the most concentrated urine of any mammal (MacMillan and Lee, 1969). The use of torpor to avoid extreme conditions had not been observed in Australian rodents until recently, although it is well known in rodents from the arid zones of other continents. Its occurrence in the sandy inland mouse *Pseudomys hermannsburgensis* (Predavec, 1997) is unlikely to be an isolated case among Australian species.

Microchiropteran bats are common in the night sky of arid Australia. Being insectivorous, they generally obtain sufficient water from their food, though they do fly to water. The key to their survival is the selection of favourable roost sites and the utilization of torpor as a major energy and water conservation mechanism. This is most important in winter when insect abundance is low and energy is needed for heat production (Hosken and Withers, 1997).

Medium-sized Mammals

Mammals in this size class have been diverse and successful, but have suffered a marked decline since European settlement. They include many desert adapted species, most of which have evolved since the late Miocene in response to the climatic changes associated with Plio-Pleistocene aridity. They include carnivores, omnivores and herbivores, with each feeding style utilizing somewhat different strategies to cope with arid conditions. They are usually nocturnal and some burrow, notably the rodents and the bilbies.

The only arid zone carnivores in this size class are the quolls (*Dasyurus geoffroii* and *D. hallucatus*). They obtain water from their prey and have good water conserving mechanisms and efficient thermoregulation (Arnold and Shield, 1970). Their decline in recent years may be related to a diminished supply of prey or to quolls themselves becoming prey to feral placental predators.

Two highly specialized but unrelated medium-sized mammals, the marsupial numbat (*Myrmecobius fasciatus*) and a monotreme, the short-beaked echidna (*Tachyglossus aculeatus*) eat an exclusive diet of termites and ants. The abundance of this food source seems to have ensured their success in the semi-arid zone. Both species have relatively low resting metabolic rates, as do larger mammalian anteaters elsewhere in the world (McNab, 1986). A low basal metabolism can considerably extend fat reserves. Also, water needs are consequently lower and the numbat and echidna can subsist on water obtained from their prey.

The omnivores have also been successful in adapting to arid conditions. At least six species of bandicoot inhabited the inland at the time of European settlement, but all are now extinct or endangered (Kemper, 1990). Information is available for two endangered species, the bilby (*Macrotis lagotis*) and the golden bandicoot (*Isoodon auratus*). Both species have low levels of resting metabolism and excellent water conserving abilities, but appear to employ quite different life strategies. The golden bandicoot does not burrow, but uses dense vegetation for surface cover and focuses on termites and other invertebrates for food; it has good thermoregulation (Bradshaw *et al.*, 1994). Bilbies also take invertebrates but more plant material is eaten, including seeds, bulbs and fungi, which often have a relatively high water content. The bilby is also strictly nocturnal and spends the days in deep burrows; it very much avoids the surface at thermally stressful times (Gibson and Hume, 2000). The only species of possum to inhabit the arid zone, the phalangerid brush-tailed possum, *Trichosurus vulpecula*, is also omnivorous. However, it tends to favour pockets of better habitat along watercourses. In the Alice Springs region it is found in more fertile valleys in the MacDonald Ranges.

The mid-sized herbivorous mammals of the dry country include a diverse range of taxa, including small 'rat kangaroos' (Potoroidae), several small macropodine wallabies (e.g. the nail-tailed wallabies (*Onychogalea*), rock wallabies (*Petrogale*), hare wallabies (*Lagorchestes*)), and some larger rodents. All these species have excellent water conserving abilities, and can survive without free drinking water. They shelter from environmental extremes by using a variety of avoidance tactics involving behavioural and lifestyle modifications. This group also contains many species that are now rare or endangered (some are extinct) and it may be that their dependence for survival on specific habitat niches has contributed to their vulnerability following European settlement (Dickman, 1994).

The diets of these mid-sized herbivores vary according to their size, the smaller species using the least fibrous vegetation. All use fermentation to some degree, but as size increases more reliance is placed on this and more fibrous vegetation is eaten. At the small end of the scale, the sticknest rat (*Leporillus conditor*) eats mostly succulent vegetation and has an enlarged stomach and expanded caecum to facilitate some fermentative digestion (Copley, 1999). The bettongs (genus *Bettongia*) have varied vegetable diets, with much emphasis on underground storage parts of plants such as tubers and bulbs. Bettongs also eat some seeds, green herbs and particularly underground fungal fruiting bodies. All these foods are of high water content and are reasonably easily digested. The hare wallabies (*Lagorchestes* spp.) at the larger end of the scale, survive (tenuously now) in the most arid deserts on a diet of seeds and nutritious forbs and grasses with high water content (when available), but unlike the smaller species, they can also digest high fibre grasses (Hume, 1999).

Large Mammals

In modern Australia the only large mammals (over 7–8 kg) are the dingo, koala, the wombats and macropodine kangaroos and wallabies. While koalas may enter semi-arid zones via some riverine habitats, only the dingo, hairy-nosed wombats (*Lasiorhinus* spp.) and four large kangaroo species in the genus *Macropus* (the red, the euro and two greys) inhabit really dry country today. The dingo is the only large carnivore of the arid zone, and the rest, the arid zone herbivores, all have one thing in common—they are specialized grazers. The rest of the larger herbivores of Australia, both grazers and browsers, all favour well-watered peripheral woodland or forest habitats.

The situation was remarkably different some time prior to the last glacial maximum when at least the semi-arid regions (we don't know about the deserts) were also inhabited by many large browsers (species of *Diprotodon, Procoptodon, Protemnodon, Sthenurus, Simosthenurus* etc.), plus the giant wombat *Phascolonus* and the carnivores *Thylacinus* and *Thylacoleo*. There were also more species of grazing kangaroo around, such as the very large *Macropus ferragus* and others, which were direct ancestors of the modern grey kangaroos and euro (*Macropus titan, M. altus*) and were larger than their modern descendants. The ancestry of the most arid-adapted kangaroo, the red kangaroo *Macropus rufus*, remains a mystery since it has not been confidently identified in fossil sites older than about 40,000 years.

The cause of the late Pleistocene extinctions (see Summary Box 1), which removed all the large browsers and completely changed the ecology of the entire continent, is hotly debated; with opinions ranging from climate change being the sole cause (see Horton, 2000), to human overkill as the sole cause (see Flannery, 1994; Roberts *et al.*, 2001). Many studies of late Pleistocene faunal change now focus on the complex responses of mammals to changing climates and other ecological parameters—including migratory scope, dietary preferences and mobility (Lister, 1997; Sher, 1997; Jablonski and Witfort, 1999). It is now clear that each continent differed in the timing of the final disappearance of the largest mammals and one cannot simply extrapolate from North American or European-based hypotheses to account for the Australian phenomenon (Martin and Klein, 1984). The search for an explanation for Australian extinctions must therefore be firmly based on the unique circumstances that pertained in this continent during the last 500,000 years. It is not possible here to offer an explanation for the late Pleistocene extinction of Australia's large mammals, rather, we will focus on the attributes of the 'megafauna' survivors. Their highly refined adaptations to life in the arid and semi-arid rangelands provide some insight into selective pressures that operated during the past 500,000 years of extreme environmental stress.

ADAPTIVE SIGNIFICANCE OF LARGE BODY SIZE

Large size provides various benefits for mammals. It gives relatively greater independence from the vagaries of the environment and allows access to a food source (fibrous vegetation) that only the invertebrates otherwise use significantly. These features occur because of the way the body's surface area and metabolic rate change as body size increases. Surfaces involved in water loss and heat exchange decrease markedly as body mass increases, making thermoregulation easier. Metabolic intensity (metabolism per unit mass) also decreases markedly as mammals get bigger. This means that food and water reserves can last much longer and food requirements relative to mass decrease, allowing the use of fibrous vegetation. The microbial fermentative digestion necessary to achieve this is a slow process, and bulky food must be retained in the stomach for long periods. The larger the mammal, the more fibrous the vegetation on which it can live (Hume, 1999). The largest of the Australian extinct marsupials, the palorchestids, diprotodontids and sthenurines, probably fed on the rough browse of semi-arid zone trees and shrubs (as African elephants and giraffes do today). The large kangaroos mainly eat grass, as do cattle and antelopes (Dawson, 1995).

Locomotion (walking and running) also becomes energetically cheaper with larger size and the speed at which animals can travel increases with size up to about 100 kg (Garland, 1983). The overall effect is that larger mammals have the flexibility to 'economically' range over large areas to obtain food or find water, which is obviously advantageous in arid regions. As Figure 2 shows the kangaroos have further improved on the economics of travel with the evolution of hopping (Dawson, 1995).

Given the benefits of large body size, it might be expected that immediately prior to European settlement the diversity and density of larger mammals would have been higher in arid and semi-arid regions of Australia. Today our rangelands support large numbers of introduced domestic mammals (sheep, goats, cattle, camels), as earlier Pleistocene environments supported the large marsupial browsers and grazers. Kangaroos and wombats alone somehow managed to pass through the final 'filter'. So what are the attributes that enabled them to survive?

For wombats, avoidance is the key; in the late Pleistocene, hairy-nosed wombats ranged across semi-arid eastern Australia, but are now endangered and confined to relict populations in Queensland and South Australia. They are specialized grazers, moderate in size (up to 35 kg) and their main strategy for survival is to avoid environmental extremes by being nocturnal and retreating to burrows during the day. Because this behaviour reduces their exposure to high temperatures, they can survive on the water they obtain from their food (Wells and Green, 1998). They are not particularly mobile and do not range far from their burrows. Study of the limb bones of extinct wombats also suggests that even the largest members of this family may have been burrowers (Murray, 1998).

The kangaroos have much more sophisticated adaptations; they include the largest native animals in arid Australia, with big males reaching almost 100 kg (Plate 19). They are highly mobile; their unique locomotion allows them to travel widely, quickly and economically to food and water. However, kangaroos are generally not nomadic and rely on intimate knowledge of their home range environment to enhance survival. The ability to digest grasses efficiently is the basis of kangaroos' ability to live in arid country (Dawson, 1995). They use fermentative digestion in a forestomach, analogous to the rumen of the bovids, such as African antelope and this can provide virtually all the required nutrients, even from poor quality vegetation. With this digestive system comes the ability to produce vitamins and to recycle waste nitrogenous compounds back into useable protein.

While kangaroos are largely nocturnal, in winter they seek the sun to save on energy for heat production. On summer days, to save water, they seek shade to avoid high solar heat loads. Some drinking water is needed, but water conservation is highly refined in arid zone kangaroos. The ability of red kangaroos to concentrate their urine surpasses all other large mammals (Blaney *et al.,* 2000). Kangaroos have a sophisticated pattern of thermoregulatory responses, which include an interplay of evaporative heat loss routes together with the cardiovascular mechanisms to reduce heat inflow in very hot conditions (Dawson *et al.*, 2000). The net effect is that the red kangaroo can survive summer temperatures with a minimum of water. Even in hot dry conditions most red kangaroos drink less than once a week and for much of the year rarely drink at all (Dawson *et al.,* 1975). Inland euros do well also, but rely on microhabitat selection to save thermoregulatory water use (Dawson and Denny, 1969). The reflective and insulative characteristics of the fur are important in water economy in summer. In winter kangaroos markedly increase the insulative thickness of their fur to reduce energy needs (Dawson, 1995).

The large semi-arid and arid zone kangaroos also have specialized reproductive strategies which are 'designed' to cope with the unpredictability of good seasons. This is illustrated by comparison of red and grey kangaroos (Dawson, 1995). The grey kangaroos favour the more predictable, mesic environments of the semi-arid zones, and are more seasonal in their breeding. Red kangaroos can cope with harsher conditions; they can breed continuously and their young grow more quickly. They wean a joey at one year, compared with 18 months for grey kangaroos. Unlike grey kangaroos, the red kangaroos can mate a day or two after a young is born. They then use a specialized development of embryonic diapause as part of the process of continuous production of young. In embryonic diapause, as seen in red kangaroos, a fertilized viable embryo is carried in the uterus for long periods (many months) with its development arrested at the stage of a blastocyst, which consists of some 70–100 cells. Development recommences when lactational inhibition is removed, either because of the loss of the pouch young, as can occur in drought, or by the reduced lactation toward the end of a normal pouch life. This process allows populations to recover quickly when droughts break (Figure 4).

It has been suggested that kangaroo numbers have increased in arid regions in recent times due to the provision of more stock watering sites. This observation has been made in semi-arid and arid pastoral country where sheep are grazed, particularly western New South Wales and south-central Queensland. Kangaroo numbers are high in these regions (Caughley, 1987), but the reason may be more complicated than just the provision of more water. In most of this region kangaroos, along with sheep, are also protected from dingos by the 'dog fence', which runs along large sections of the borders of New South

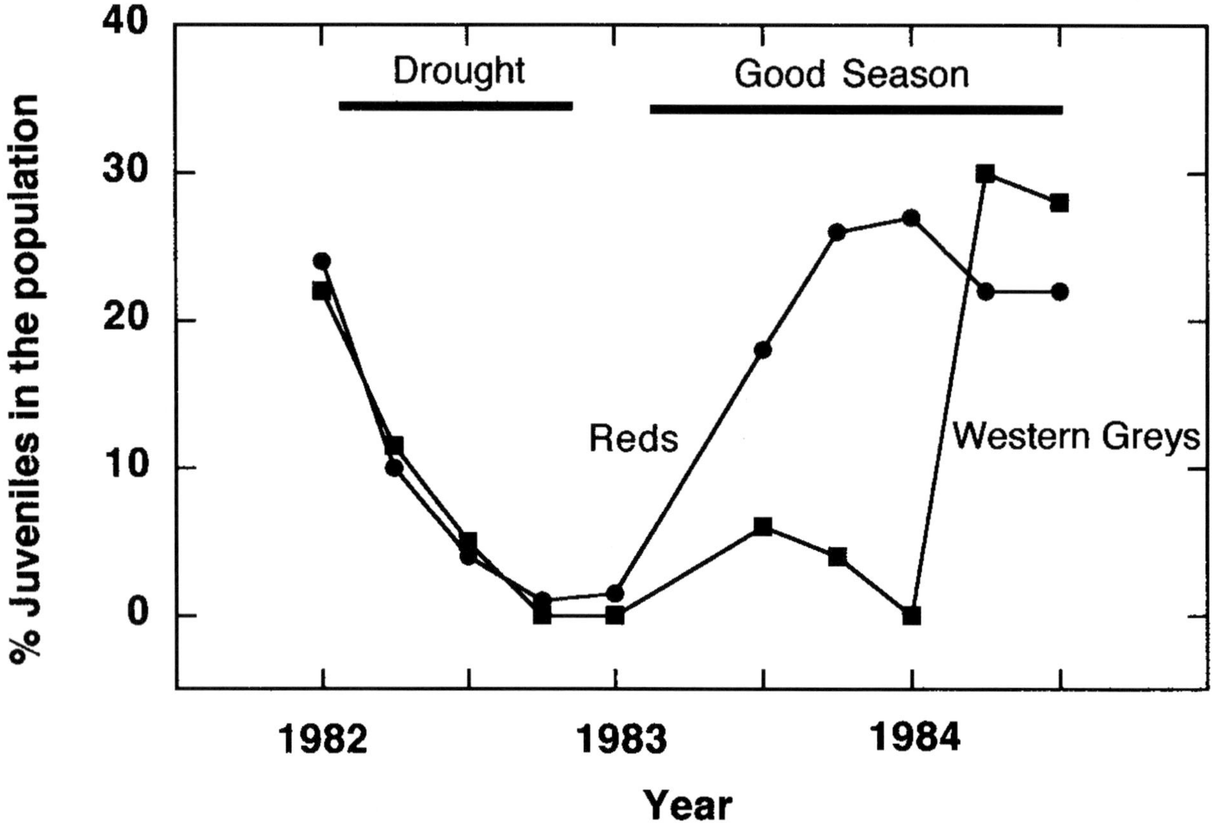

Figure 4. Percentage of juveniles in the populations of red kangaroos and western grey kangaroos through a drought and the following good season. The decrease in juveniles of both species was due to juvenile mortality and the cessation of breeding. The delayed increase in the western greys was due to the drought breaking after their breeding season (from Dawson, 1995).

Wales with South Australia and Queensland. Additionally, sheep have modified the grazing environment, perhaps to a greater extent than have cattle in other parts of Australia. It has been said that sheep produce 'marsupial lawns'. In recent travels outside the sheep rangelands we have been surprised at how sparse kangaroos generally are in arid Australia.

CONCLUSIONS

Since Australia drifted free from Antarctica (~ 45 m.y.a.) long isolation, low relief, poor soils and sluggish drainage patterns, combined with latitudinal change as it drifted north, have presented severe challenges for its vertebrate inhabitants. The modern endemic vertebrate faunas represent the survivors, largely those that responded to the biogeographic and climatic challenges over millennia by seeking refuge or adapting. The predominant challenge over the last 5 million years (over at least two thirds of the continent) has been increasing aridity and unpredictability of climate. As a consequence, a large proportion of modern Australian vertebrates are arid-adapted and many have evolved remarkable attributes to cope with the exigencies presented by an ENSO-influenced climate.

The early vertebrate inhabitants of the 'raft' that separated from Gondwana comprised a select subset of the vertebrates of the world. It appears that several major groups of modern vertebrates may have had their origins in Gondwana and left a legacy in Australia; examples include the myobatrachid frogs, the ratites, parrots and passerines and possibly even the bats. Gondwanan mammals present a unique and puzzling history. The monotremes left their only descendants on the raft. While early placental mammals may have inhabited Australia prior to its separation from Gondwana, they did not persist and the continent became the domain of marsupials. These radiated to fill most niches occupied by placentals elsewhere, using their distinctive physiological attributes to advantage in coping with extreme conditions, especially aridity. Additional immigrants arrived after Australia approached the Asian plate. These included frogs, reptiles, birds, bats and the rodents, all of which contributed to arid zone faunas.

Aridity brings particular problems for vertebrates, and their responses vary. There is the low metabolic strategy of ectotherms (fishes, amphibians, reptiles), which is contrasted with the high metabolic strategy of birds and mammals. Evolution has favoured both in Australian arid zones; for example, in the diversity of lizards on one hand, and birds on the other. Lizards have low food requirements and stored resources can be eked out over a long time, while birds have the flexibility of mobility.

Superimposed on these basic metabolic constraints, members of all groups have evolved remarkable behavioural, physiological and reproductive strategies to deal with particular climatic and environmental problems, particularly the extreme unpredictability of good seasons that characterizes Australian arid lands. Examples include the burrowing cocoon frogs, which can lie in 'suspended animation' for years waiting for rain; and the phenomenon of embryonic diapause in macropods, enabling them to repopulate rapidly after prolonged drought.

Throughout the history of the continent, extinctions have been part of the picture in Australia, as in the rest of the world; climates change, refuges disappear, adaptive niches or predation pressures may change beyond the response capabilities of the organisms living there. This process continues today, but Australia has a pattern of particular concern. In the 200 years since European settlement the ecosystems of semi-arid and arid regions of Australia have been drastically degraded by agriculture, grazing and the introduction of feral animals, particularly predators such as cats and foxes. It is in these zones that the highest levels of mammal extinctions have occurred. If we allow this degradation to continue, we will lose much more of our unique vertebrate fauna.

REFERENCES

Anderson, E. 1984. Who's who in the Pleistocene: A mammalian bestiary. Pp.40–89. In Martin, P. and Klein, R. (eds). *Quaternary Extinctions. A Prehistoric Revolution.* University of Arizona Press, Tucson. 892pp.

Archer, M., Arena, R., Bassarova, M.I., Black, K., Brammall, J., Cooke, B., Creaser, P., Crosby, K., Gillespie, A., Godthelp, H., Gott, M., Hand, S.J., Kear, B., Kirkmann, A., Macness, B., Muirhead, J., Musser, A., Myers, T., Pledge, N., Wang, Y. and Wroe, S. 1999. The evolutionary history and diversity of Australian mammals. *Australian Mammalogy*, Vol.21, No. 1, pp.1–45.

Archer, M., Hand, S.J. and Godthelp, H.J. 1991. *Riversleigh. The Story of Animals in Ancient Rainforests of Inland Australia.* Reed Books, Sydney. 264pp.

Archer, M., Hand, S.J. and Godthelp, H.J. 1994. Patterns in the history of Australia's mammals and inferences about palaeohabitats. Pp.80–103. In Hill, R.S. (ed). *History of the Australian Vegetation: Cretaceous to Recent.* Cambridge University Press, Cambridge. 433pp.

Arnold, J. and Shield, J. 1970. Oxygen consumption and body temperature of the Chuditch (*Dasyurus geoffroii*). *Journal of Zoology*, Vol. 160, pp.391–404.

Blaney, C.E., Dawson, T.J., McCarron, H.C.K. and Buffenstein, R. 2000. Water metabolism and renal function and structure in eastern grey kangaroos (*Macropus giganteus*): responses to water deprivation. *Australian Journal of Zoology*, Vol. 48, pp.335–345.

Blaylock, L.A., Ruibal, R. and Platt-Alola, K. 1976. Skin structure and wiping behaviour of phyllomedusine frogs. *Copeia*, 1976, pp.283–295

Bradshaw, S.D., Morris, K.D., Dickman, C.R., Withers, P.C. and Murphy, D. 1994. Field metabolism and turnover in the golden bandicoot (*Isoodon auratus*) and other small mammals from Barrow Island, Western Australia. *Australian Journal of Zoology*, Vol. 42, pp.29–41.

Caughley, G. 1987. Introduction to the sheep rangelands. Pp.1–13. In Caughley, G., Shepherd, N. and Short, J. (eds). *Kangaroos: Their Ecology and Management in Sheep Rangelands.* Cambridge University Press, Cambridge. 253pp.

Cerling, T.E., Ehlinger, J.R. and Harris, J.M. 1998. Carbon dioxide starvation, the development of C4 ecosystems, and mammalian evolution. *Philosophical Transactions of the Royal Society of London*, Vol. 353, pp.159–171.

Christian, K. and Parry, D. 1997. Reduced rates of water loss and chemical properties of skin secretions of the frogs *Litoria caerulea* and *Cyclorana australis*. *Australian Journal of Zoology*, Vol. 45, pp.13–20.

Copley, P. 1999. Natural history of Australia's stick-nest rats, genus *Leporillus* (Rodentia: Muridae). *Wildlife Research*, Vol. 26, pp.513–539.

Davies, S.J.J.F. 1984. Nomadism as a response to desert conditions in Australia. *Journal of Arid Environments*, Vol. 7, pp.183–196.

Dawson, T.J. 1995. *Kangaroos: the Biology of the Largest Marsupials.* University of New South Wales Press, Sydney. 162pp.

Dawson, T.J., Blaney, C.E., Munn, A.J, Krockenberger, A. and Maloney, S.K. 2000. Thermoregulation by kangaroos from mesic and arid habitats: influence of temperature on routes of heat loss in eastern grey kangaroos (*Macropus giganteus*) and red kangaroos (*Macropus rufus*). *Physiological and Biochemical Zoology*, Vol. 73, pp.374–381.

Dawson, T.J. and Dawson, W.R. 1982. Metabolic scope and conductance in response to cold of some dasyurid marsupials and Australian rodents. *Comparative Biochemistry and Physiology*, Vol. 71A, pp.59–64.

Dawson, T.J. and Denny, M.J.S. 1969. A bioclimatological comparison of the summer day microenvironments of two species of arid zone kangaroo. *Ecology*, Vol. 50, pp.329–332.

Dawson, T.J., Denny, M.J.S., Russell, E. and Ellis, B. 1975. Water usage and diet preferences of free ranging kangaroos, sheep and feral goats in the Australian arid zone during summer. *Journal of Zoology, London*, Vol. 177, pp.1–23.

Dawson, T.J. and Ellis, B.A. 1994. Did dietary competition play a role in the extinction of small macropodids in western New South Wales with the coming of European settlement? Pp.75–80. In Lunney, D., Hand, S., Reed, P. and Butcher, D. (eds). *Future of the Fauna of Western New South Wales.* Royal Zoological Society of New South Wales, Sydney. 246pp.

Dawson, T.J., Finch, E., Freedman, L., Hume, I.D., Renfree, M.B. and Temple-Smith, P.D. 1989. Morphology and physiology of the Metatheria. Pp.451–504. In Walton, D.W. and Richardson, B.J. (eds). *Fauna of Australia. Vol.*

1B. Mammalia. Australian Government Publishing service, Canberra. pp.401–1227.

Dickman, C.R. 1994. Native mammals of western New South Wales: past neglect, future rehabilitation? Pp.81–91. In Lunney, D., Hand, S., Reed, P. and Butcher, D. (eds). *Future of the Fauna of Western New South Wales.* Royal Zoological Society of New South Wales, Sydney. 246pp.

Fisher, C.D., Lindgren, E. and Dawson, W.R. 1972. Drinking patterns and behaviour of Australian desert birds in relation to their ecology and abundance. *Condor*, Vol. 74, pp.111–136

Flannery, T.F. 1994. *The Future Eaters: an Ecological History of the Australian Lands and People.* Reed Books, Sydney. 421pp.

Foley, W.J. and Cork, S.J. 1992. Use of fibrous diets by small herbivores: how far can the rules be 'bent'? *Tree*, Vol. 7, pp.159–162.

Garland, T. 1983. The relation between maximal running speed and body mass in terrestrial mammals. *Journal of Zoology, London,* Vol. 199, pp.157–170.

Gibson, L.A. and Hume, I.D. 2000. Seasonal field energetics and water influx rates of the greater bilby (*Macrotis lagotis*). *Australian Journal of Zoology*, Vol. 48, pp.225–239.

Guppy, M. and Withers, P. 1999. Metabolic depression in animals: physiological perspectives and biochemical generalizations. *Biological Reviews*, Vol. 74, pp.1–40.

Herd, R.M. and Dawson, T.J. 1984. Fibre digestion in the emu, *Dromaius novaehollandiae*, a large bird with a simple gut and high rates of passage. *Physiological Zoology*, Vol. 57, pp.70–84.

Hesse, P.P. 1994. The record of continental dust from Australia in Tasman Sea sediments. *Quaternary Science Reviews*, Vol. 13, pp.257–272.

Hinds, D.S., Baudinette, R.V., MacMillen, R.V. and Halpern, E.A. 1993. Maximum metabolism and the aerobic factorial scope of endotherms. *Journal of Experimental Biology,* Vol. 182, pp.41–56.

Hope, J. 1982. Late Cainozoic vertebrate faunas and the development of aridity in Australia. Pp.85–100. In Barker, W.R. and Greenslade, P.J.M. (eds). *Evolution of the Flora and Fauna of Arid Australia.* Peacock Publications, South Australia. 392pp.

Horton, D. 2000. *The Pure State of Nature. Sacred Cows, Destructive Myths and the Environment.* Allen and Unwin, Sydney. 192pp.

Hosken, D.J. and Withers, P.C. 1997. Temperature regulation and metabolism of an Australian bat, *Chainolobus gouldii* (Cheiroptera: Vespertilionidae) when euthermic and torpid. *Journal of Comparative Physiology,* Series B, Vol. 167, pp.71–80.

Hume, I.D. 1999. *Marsupial Nutrition.* Cambridge University Press, Cambridge. 434pp.

Jablonski, N.G. and Witfort, M.J. 1999. Environmental change during the Quaternary in East Asia and its consequences for mammals. Pp.307–316. In Baynes, A. and Long, J.A. (eds). *Papers in Vertebrate Palaeontology.* Records of the Western Australian Museum, Supplement No. 57.

Keast, A. 1959. Australian birds: their zoogeography and adaptations to an arid continent. Pp.89–114. In Keast, A., Crocker, R.I. and Christian, C.S. (eds). *Biogeography and Ecology in Australasia.* Monographiae Biologicae Vol. 8. W. Junk, The Hague.

Kemp, A. 1991. Australian Mesozoic and Cainozoic lungfish. Pp.465–489. In Vickers-Rich, P., Monaghan, J.M., Baird, R.F. and Rich, T.H. (eds). *Vertebrate Palaeontology of Australasia.* Pioneer Design Studio and Monash University Publications Committee, Melbourne. 1437pp.

Kemper, K. 1990. Status of bandicoots in South Australia. Pp.67–72. In Seebeck, J.H., Brown, P.R., Wallis, R.L. and Kemper, C.M. (eds). *Bandicoots and Bilbies.* Surrey Beatty and Sons, Sydney. 392pp.

Kershaw, A.P., Martin, H.A. and McEwan Mason, J.R.C. 1994. The Neogene: a period of transition. Pp.299–327. In Hill, R.S. (ed). *History of the Australian Vegetation: Cretaceous to Recent.* Cambridge University Press, Cambridge. 433pp.

Langer, P. 1994. Food and digestion of Cainozoic mammals in Europe. Pp.9–24. In Chivers, D.J. and Langer, P. (eds). *The Digestive System in Mammals: Food, Form, Function.* Cambridge University Press, Cambridge. 446pp.

Lister, A.M. 1997. The evolutionary response of vertebrates to Quaternary environmental change. Pp.287–302. In Huntley, B., Cramer, W., Morgan, A.V., Prentice, H.C. and Allen, J.R.M. (eds). *Past and Future Rapid Environmental Changes: the Spatial and Evolutionary Responses of Terrestrial Biota.* Springer-Verlag, Berlin. 523pp.

Long, J.A. 1995. *The Rise of Fishes: 500 Million Years of Evolution.* Johns Hopkins University Press, Baltimore. 223pp.

Lovegrove, B.G. 2000. The zoogeography of mammalian basal metabolic rate. *American Naturalist*, Vol. 156, pp.201–219.

Lovegrove, B.G. and Knight-Eloff, A. 1988. Soil and burrow temperatures, and the resource characteristics of the social mole-rat *Cryptomys damarensis* (Bathyergidae) in the Kalahari Desert. *Journal of Zoology, London*, Vol. 216, pp.403–416.

Maclean, G.L. 1976. A field study of the Australian dotterel. *Emu*, Vol. 76, pp.207–215.

Maclean, G.L. 1996. *The Ecophysiology of Desert Birds.* Springer-Verlag, Berlin. 181pp.

MacMillan, R.E., Augee, M.L. and Ellis, B.A. 1989. Thermal ecology and diet of some xerophilous lizards from western New South Wales. *Journal of Arid Environments*, Vol. 16, pp.193–201.

MacMillan, R.E. and Lee, A.K. 1969. Water metabolism of Australian hopping mice. *Comparative Biochemistry and Physiology*, Vol. 28, pp.439–514.

Macphail, M.K. 1997. Late Neogene climates in Australia: fossil pollen- and spore-based estimates in retrospect and prospect. *Australian Journal of Botany*, Vol. 45, pp.425–464.

Maloney, S.K. and Dawson, T.D. 1995. The heat load from solar radiation on a large, diurnally active bird, the emu (*Dromaius novaehollandiae*). *Journal of Thermal Biology*, Vol. 20, pp.381–387.

Marder, J. and Arad, Z. 1989. Panting and acid-base regulation in heat stressed birds. *Comparative Biochemistry and Physiology*, Vol. 94A, pp.395–400.

Martin, P.S. and Klein, R.G. (eds). 1984. *Quaternary Extinctions: A Prehistoric Revolution.* University of Arizona Press, Tucson. 892pp.

McDowall, R.M. and Fulton, W. 1996. Family Galaxiidae. Galaxiids. Pp.52–77. In McDowall, R. (ed). *Freshwater Fishes of South-eastern Australia.* Reed Books, Sydney. 247pp.

McLeod, S.R. 1997. Is the concept of carrying capacity useful in variable environments? *Oikos,* Vol. 79, pp.529–542.

McNab, B.K. 1986. Food habits, energetics and reproduction of marsupials. *Journal of Zoology, London,* Series A, Vol. 208, pp.595–614.

Merrick, J.R. 1996. Family Terapontidae. Freshwater grunters and perches. Pp.164–167. In McDowall, R. (ed). *Freshwater Fishes of South-eastern Australia.* Reed Books, Sydney. 247pp.

Morton, S.R. and James, C.D. 1988. The diversity and abundance of lizards in arid Australia: a new hypothesis. *American Naturalist*, Vol. 132, pp.237–256.

Murray, B.R., Dickman, C.R., Watts, C.H.S. and Morton, S.R. 1999. The dietary ecology of Australian desert rodents. *Wildlife Research*, Vol. 26, pp.421–437.

Murray, P. 1991. The Pleistocene megafauna of Australia. Pp.1071–1164. In Vickers-Rich, P., Monaghan, J.M., Baird, R.F. and Rich, T.H. (eds). *Vertebrate Palaeontology of Australasia.* Pioneer Design Studio and Monash University Publications Committee, Melbourne. 1437pp.

Murray, P.F. 1998. Palaeontology and palaeobiology of wombats. Pp.1–33. In Wells, R.T. and Pridmore, P.A. (eds). *Wombats.* Surrey Beatty and Sons, Sydney. 332pp.

Noy-Meir, I. 1973. Stability of grazing systems: an application of predator-prey graphs. *Journal of Ecology*, Vol. 63, pp.459–481.

Pianka, E.R. 1986. Ecology and natural history of desert lizards, an ecological perspective. *Copeia*, 1972, pp.127–144.

Pianka, E.R. 1994. Comparative ecology of *Varanus* in the Great Victoria Desert. *Australian Journal of Ecology*, Vol. 19, pp.395–408.

Pough, F.H. 1980. The advantages of ectothermy for tetrapods. *American Naturalist*, Vol. 115, pp.92–112.

Predavec, M. 1997. Variable energy demands in *Pseudomys hermannsburgensis*: possible ecological consequences. *Australian Journal of Zoology*, Vol. 45, pp.85–94.

Pusey, B.J. 1989. Aestivation in the teleost fish *Lepidogalaxias salamandroides* (Mees). *Comparative Biochemistry and Physiology*, Vol. 92A, pp.137–138.

Read, J.L. 1995. Subhabitat variability: a key to the high reptile diversity in chenopod shrublands. *Australian Journal of Ecology*, Vol. 20, pp.494–501.

Rich, T.H. 1991. Appendix 1. Australian Mesozoic and Tertiary Terrestrial Mammal Localities, Great Buniyong Estate Mine. P. 1044. In Vickers-Rich, P., Monaghan, J.M., Baird, R.F. and Rich, T.H. (eds). *Vertebrate Palaeontology of Australasia.* Pioneer Design Studio and Monash University Publications Committee, Melbourne. 1437pp.

Richardson, B.A. 1994. The human impacts on the ecology of freshwater fish in western New South Wales. Pp.169–176. In Lunney, D., Hand, S., Reed, P. and Butcher, D. (eds). *Future of the Fauna of Western New South Wales.* Royal Zoological Society of New South Wales, Sydney. 246pp.

Roberts, J.D. 1981. Terrestrial breeding in the Australian leptodactylid frog, *Myobatrachus gouldii* (Gray) (Leptodactylidae). *Australian Wildlife Research*, Vol. 8, pp.451–462.

Roberts, J.D. 1984. Terrestrial egg deposition and direct development in *Arenophryne rotunda* Tyler: a myobatrachid frog from coastal sand dunes at Shark Bay, Western Australia. *Australian Wildlife Research*, Vol. 11, pp.191–200.

Roberts, R.G., Flannery, T.F., Ayliffe, L.K., Yoshida, H., Olley, J.M., Prideaux, G.J., Laslett, G.F., Baynes, A., Smith, M.A., Jones, R. and Smith, B.L. 2001. New Ages for the last Australian megafauna: continent-wide extinction about 46,000 years ago. *Science*, Vol. 292, pp.1888–1892.

Schodde, R. 1982. Origin, adaptation and evolution of birds in arid Australia. Pp.191–224. In Barker, W.R. and Greenslade, P.J.M. (eds). *Evolution of the Flora and Fauna of Arid Australia.* Peacock Publications, South Australia. 392pp.

Sher, A.V. 1997. Late Quaternary extinction of large mammals in northern Eurasia: a new look at the Siberian contribution. Pp.319–342. In Huntley, B., Cramer, W., Morgan, A.W., Prentice, H.C. and Allen, J.R.M. (eds). *Past and Future Rapid Environmental Changes: the Spatial and Evolutionary Responses of Terrestrial Biota.* Springer-Verlag, Berlin. 523pp.

Song, X. and Geiser, F. 1997. Daily torpor and energy expenditure in *Sminthopsis macroura*: interactions between food and water availability and temperature. *Physiological Zoology*, Vol. 70, pp.331–337.

Strahan, R. (ed). 1995. *The Australian Museum Complete Book of Australian Mammals.* (Revised Edition). Reed Books, Sydney. 756pp.

Tedford, R.H. 1994. Succession of Pliocene through medial Pleistocene mammal faunas of southeastern Australia. *Records of the South Australian Museum*, Vol. 28, pp.263–284

Tyler, M.J. 1994. *Australian Frogs: A Natural History.* Reed Books, Sydney. 192pp.

Tyler, M.J. and Godthelp, H. 1993. A new species of *Lechoriodus* Boulenger (Anura, Leptodactylidae) from the early Eocene of Queensland. *Transactions of the Royal Society of South Australia*, Vol. 117, pp.187–189.

van Beurden, E. 1982. Desert adaptations of *Cyclorana platycephalus*: a holistic approach to desert-adaptation in frogs. Pp.235–240. In Barker, W.R. and Greenslade, P.J.M. (eds). *Evolution of the Flora and Fauna of Arid Australia.* Peacock Publications, South Australia. 392pp.

van Oosterzee, P. 2006. Drawing the Wallace Line. Pp.95–110. In Merrick, J.R., Archer, M., Hickey, G.M. and Lee, M.S.Y. (eds). *Evolution and Biogeography of Australasian Vertebrates.* Auscipub Pty. Ltd., Sydney. 942pp.

Vickers-Rich, P., Monaghan, J.M., Baird, R.F. and Rich, T.H. (eds). 1991. *Vertebrate Palaeontology of Australasia.* Pioneer Design Studio and Monash University Publications Committee, Melbourne. 1437pp.

Wasson, R.J. 1989. Desert dune building, dust raising and palaeoclimate in the southern hemisphere during the last 280,000 years. Pp.123–137. In Donnelly, T. and Wasson, R. (eds). *CLIMANZ 3. Proceedings of the Third Symposium on the Late Quaternary Climatic History of Australasia. Melbourne University 28–29 November 1987.* C.S.I.R.O., Division of Water Resources, Canberra.

Watson, J.A.L., Barrett, R.A. and Lendon, C. 1978. Termites. Pp.101–108. In Low, W.A. (ed). *The Physical and Biological Features of Kunoth Paddock in Central Australia.* Technical Paper 4, Division of Land and Resource Management. C.S.I.R.O., Canberra.

Watson, J.A.L., Lendon, C. and Low, B.S. 1973. Termites in mulga lands. *Tropical Grasslands*, Vol. 7, pp.121–126.

Wells, R.T. and Green, B. 1998. Aspects of water metabolism in the Southern Hairy-nosed Wombat *Lasiorhinus lat-*

ifrons. Pp.61–66. In Wells, R.T. and Pridmore, P.A. (eds). *Wombats*. Surrey Beatty and Sons, Sydney. 332pp.

Wells, R.T., Moriarty, K. and Williams, D.L.G. 1984. The fossil vertebrate deposits of Victoria Fossil Cave, Naracoorte: an introduction to the geology and fauna. *Australian Zoologist*, Vol. 21, pp.305–333.

White, M.E. 2006. Environments of the geological past. Pp.17–50. In Merrick, J.R., Archer, M., Hickey, G.M. and Lee, M.S.Y. (eds). *Evolution and Biogeography of Australasian Vertebrates*. Auscipub Pty. Ltd., Sydney. 942pp.

Williams, M., Dunkerley, D., De Decker, P., Kershaw P. and Chappell, J. (eds). 1998. *Quaternary Environments*. (2nd Edition). Arnold, London. 329pp.

Withers, P.C. 1992. *Comparative Animal Physiology*. Saunders College Publishing, Orlando, Florida. 949pp.

Withers, P.C. 1993. Cutaneous water acquisition by the thorny devil (*Molloch horridus*: Agamidae). *Journal of Herpetology*, Vol. 27, pp.265–270.

Withers, P.C. 1998. Evaporative water loss and the role of cocoon formation in Australian frogs. *Australian Journal of Zoology*, Vol. 46, pp.405–418.

Withers, P.C. and Richards, S.J. 1995. Cocoon formation by the treefrog *Litoria alboguttata* (Amphibia: Hylidae): a 'waterproof' taxonomic tool? *Journal of the Royal Society of Western Australia*, Vol. 78, pp.103–106.

Withers, P.C. and Roberts, J.D. 1993. Water balance response in the terrestrial Western Australian frog, *Myobatrachus gouldii* (Anura: Myobatrachidae). *Journal of the Royal Society of Western Australia*, Vol. 76, pp.49–52.

Withers, P.C. and Williams, J.B. 1990. Metabolic and respiratory physiology of an arid-adapted Australian bird, the spinifex pigeon. *Condor*, Vol. 92, pp.961–969.

Chapter 4
The Torresian Connections: Zoogeography of New Guinea

Tom Heinsohn and Geoffrey Hope
Department of Archaeology and Natural History, Research School of Pacific and Asian Studies,
Australian National University, Canberra, A.C.T. 0200 Australia
E-mail: tomh@coombs.anu.edu.au

INTRODUCTION

Leave the flat red northern coastline of Australia and cross a few hundred kilometres of water to the north or north-west and out of the shining tropical sea will rise a huge equatorial island, with higher peaks than occur in the whole Australian continent. In a biogeographical sense New Guinea differs from Australia's other tropical neighbours, Timor and even partial continental fragments such as New Caledonia, which are 'oceanic' islands, separated from Australia by deep oceans for millions of years. New Guinea is separated by the wide but very shallow Arafura Sea and Torres Strait.

This seaway has only existed during times of high sea level, so in fact New Guinea is an integral part of the Australian continent and forms its youngest geological province. Although it is often excluded from descriptions of Australia, this continental island or its precursor terranes and archipelagos has shared biota with Australia for at least ten million years—probably much longer. From 100,000 years to about 11,000 years ago, apart from a short swim across West Papua's very narrow Sagewin Strait, it would have been possible to trudge the 5,000 km from Waigeo Island, in the extreme north-west, to the hills of Maatsuyker, then a rocky peninsula of southernmost Tasmania.

ORIGINS

GROWING ISLANDS FROM THE SEA

New Guinea is mostly very young, the result of plate collision as Australia drifted northwards over the past 80 million years. Throughout the Tertiary it seems to have been a series of lowland islands barely emergent along the northern leading edge of a northward drifting Meganesia (also called Sahul or Greater Australia), the combined Tasmanian-Australian-New Guinean land mass of the Indo-Australian continental plate.

Prior to about 45 million years ago, Meganesia itself was joined to the Antarctic portion of the great southern super-continent of Gondwana. During the later Tertiary the Sahul plate drifted into closer proximity with the

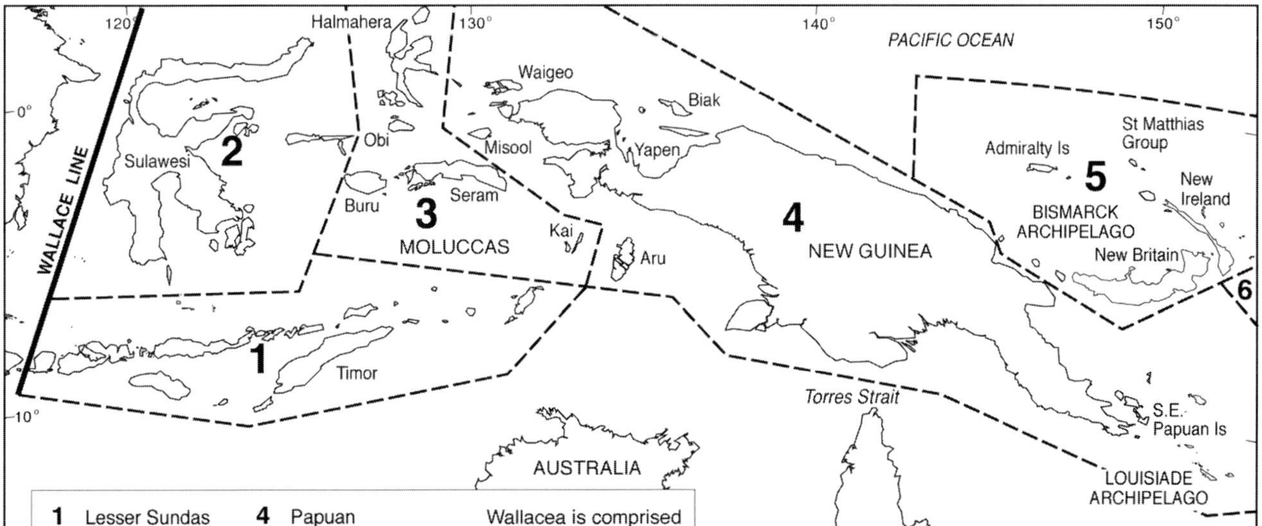

Figure 1. New Guinean region locations, with borders of zoogeographic subregions (east of Wallace's Line) marked.

Eurasian continent and island arcs formed that have gradually coalesced and welded into a composite land mass; for example, the Vogelkop Peninsula was probably an isolated island originally but became connected by a narrow neck of land to the rest of New Guinea. Towards the end of the Tertiary (3–4 m.y.a.) the collision with the Pacific Plate had initiated a continuing period of uplift, mountain building and accretion along the northern edge of Meganesia / Proto New Guinea. By the Quaternary New Guinea's high central cordillera and outlying northern ranges were in existence and tectonics had thrust up the Sunda Islands of Indonesia (Dow, 1977; Pigram and Davies, 1987; Flannery, 1995a; van Oosterzee, 1997; Aplin, 1998).

In the mid 19th century the renowned naturalist Alfred Russell Wallace visited the Aru Islands and New Guinea and was struck by the presence of birds of paradise, cassowaries and marsupials rather than the more familiar creatures of the Orient—farther west and north (Wallace, 1869). The vertebrate fauna appeared to have more in common with Australia, so he assigned it to the Australian (or Australasian) zoogeographic kingdom. Wallace concluded that biotas reflected the geological history of a region, but lacked the scientific knowledge needed to fully explain how the fauna of New Guinea had developed. For further information on the geology and faunal barriers between Asia, New Guinea and Australia see pp.95–110 (van Oosterzee, 2006).

More than a century later biologists consider that New Guinea falls into three biogeographic provinces delineated by their bioclimates, named Papua, Tumbuna and Torresia (Figure 2). Papua is the equatorial lowlands, Tumbuna the tropical montane, and Torresia the drier lowland habitats (Schodde and Calaby, 1972; Schodde, 1994). For mammals, Flannery (1995a) defines an additional province, Oceanic, which covers both mountain and lowlands of the Birds Head Peninsula and some areas to the east such as the Weyland Mountains as well as

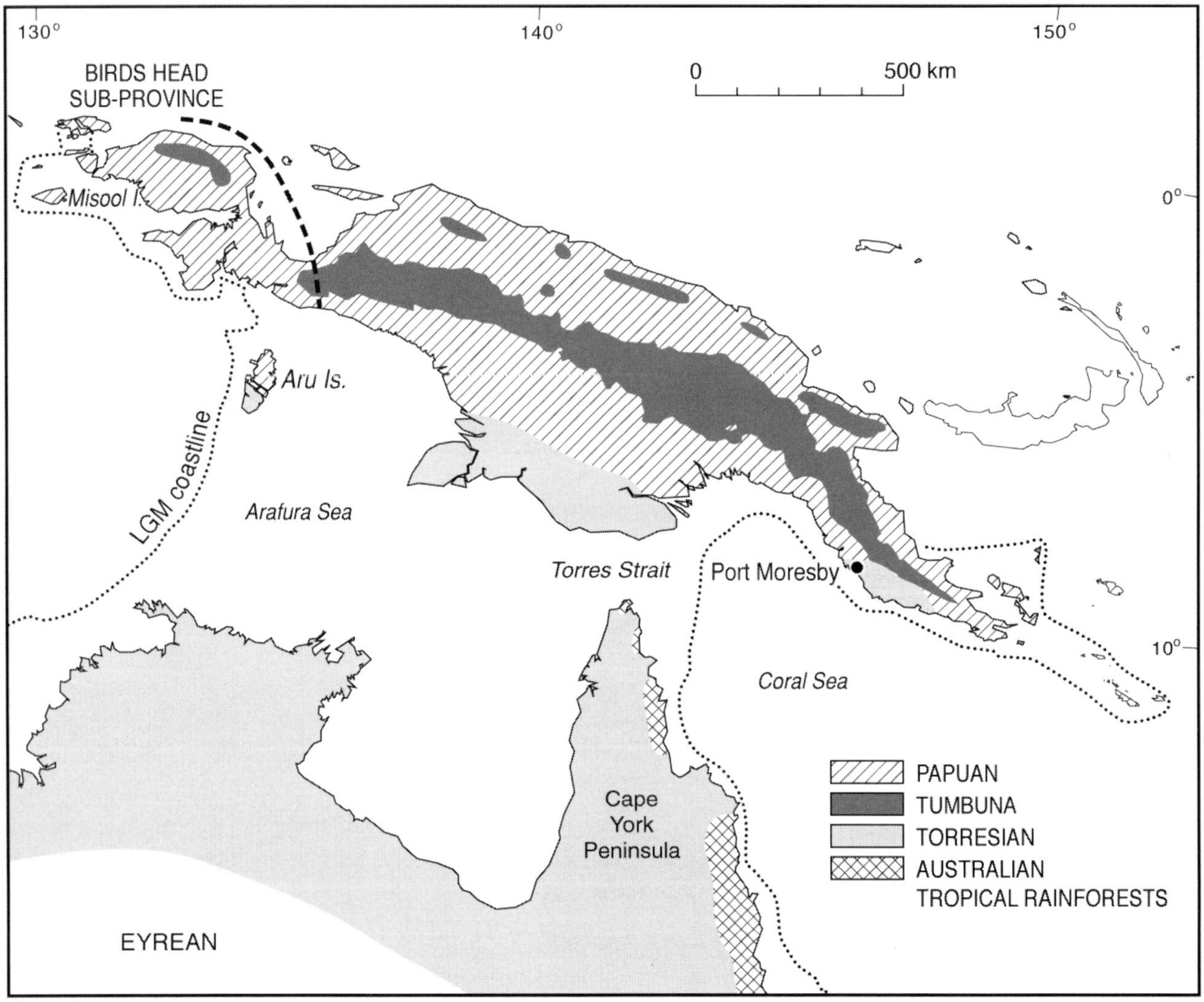

Figure 2. Currently recognized biogeographic provinces/vertebrate faunal divisions in New Guinea and northern Australia (adapted from Schodde, 1994).

north coast ranges such as the Cyclops and Bewani-Torricelli-Prince Alexander Mountains. He expands Tumbuna to include the lowlands and considers the seasonally dry southern grasslands to be a distinct province, Austral. However, the authors suggest that Austral is better seen as a sub-province of Torresia, as it shares so many characteristics (and a common history) with northern Australia it might well be linked at this level.

New Guinea itself is a part of the larger Papuan zoogeographic subregion (not to be confused with the aforementioned Papuan biogeographic province), which includes the main island plus surrounding continental satellites. This subregion is bounded on the west by the islands of Wallacea, to the north-east and east by the Northern Melanesian Islands, and to the south by Australia (see Figure 1).

COMPOSITION OF TERRESTRIAL VERTEBRATE GROUPS

At higher taxonomic levels, New Guinea has a terrestrial vertebrate fauna that is similar to that of Australia. The similarities decrease at the genus and species level, and for some groups there are also significant differences at the family level (see Tables 2–5). New Guinea also exhibits a terrestrial vertebrate diversity similar to that of Australia (summarized in Table 1); although a major difference is that while Australia is considerably more diverse in marsupials, lizards and snakes, New Guinea is more diverse in murid rodents and megachiropteran bats. Given its much smaller land area (825,000 km^2), but with vast tracts of equatorial humid forest, New Guinea is generally more species rich than Australia (7,682,000 km^2).

Frogs

The naturally occurring amphibians of Meganesia are limited to four families of frogs, the Myobatrachidae, Hylidae, Microhylidae and Ranidae. There are no non-anuran amphibians, such as salamanders or caecilians, as found in the adjacent Oriental region. Of these Australo-Papuan frog families, the myobatrachids and hylids appear to be Gondwanan, while the others are derived from Oriental stock that probably first began to arrive via the Malay Archipelago in the later Tertiary (Zweifel and Tyler, 1982; Allison, 1996). Other major frog families such as Rhacophoridae and Bufonidae, although pene-

Table 1. Comparison of mammal, bird and herpetofauna diversity in the Papuan subregion and Australia*. Recently extinct Holocene species, and some smaller introduced species such as murid rodents, birds and herpetofauna are included. Larger conspicuous introduced species such as ungulates, carnivores, lagomorphs and ostriches are excluded. Sources: Flannery (1995a, 1995b), Strahan (1995), Allison (1996), Pizzey and Knight (1997), Aplin (1998), Aplin *et al.* (1999), Cogger (2000), Coates and Peckover (2001).

Extant and Recently	New Guinea and Continental Satellites			Australia and Continental Satellites		
Occurring Groups	Families	Genera	Species	Families	Genera	Species
Amphibians						
Frogs	5	27	202	5	29	212
Reptiles						
Freshwater turtles	3	5	9	2	7	25
Snakes	6	25	70	5	33	155
Crocodiles	1	1	2	1	1	2
Lizards	6	34	189	5	78	614
Birds						
Ratite birds	1	1	3	1	2	4
Other birds	83	300	736	81	327	777
Mammals						
Monotremes	1	2	2	2	2	2
Marsupials	10	28	81	16	56	149
Murid rodents	1	28	80	1	15	65
Mega. bats	1	8	26	1	5	8
Micro. bats	5	21	54	6	20	57
Total	**123**	**480**	**1454**	**126**	**575**	**2070**

*Papuan subregion is defined as New Guinea and surrounding islands such as the West Papuan Islands, Aru Islands, islands of Geelvink Bay and South-east Papuan Islands, most of which are continental land-bridge islands, but some of which, such as Waigeo in the West Papuan Group and Biak-Supiori in Geelvink Bay, have remained cut off by sea channels. Adjacent non-continental islands belonging to Wallacea, the Bismarck Archipelago and Solomon Islands are excluded from the Papuan Subregion (Figure 1).

probably represent more recent arrivals mostly via the Malay Archipelago of Old World passerines, that despite possible ultimate Gondwanan origins, have radiated in Afro-Eurasia. In the Australo-Papuan region this smaller assemblage includes: true thrushes and Old World flycatchers (Muscicapidae); starlings and mynas (Sturnidae); swallows and martins (Hirundinidae); Old World warblers (Sylviidae); African warblers (Cisticolidae); white-eyes (Zosteropidae); Old World larks (Alaudidae); sunbirds (Nectariniidae); flowerpeckers (Dicaeidae); Old World wagtails and pipits (Motacillidae); and estrildine finches (Passeridae) (Sibley and Ahlquist, 1985; Sibley and Monroe, 1990; Pizzey and Knight, 1997; Coates and Peckover, 2001). Non-passerine birds with Oriental or Old World affinities make up another minor portion of the Australo-Papuan avifauna; while cosmopolitan non-passerines such as various water birds, sea birds, and raptors collectively make up a further significant component of the Meganesian bird fauna.

For further details of the extant bird fauna of the Papuan subregion see pp.413–458 (Schodde, 2006).

Mammals

In common with Australia, New Guinea lacks most of the eutherian mammal orders that characterize the adjacent Oriental region. The terrestrial mammal fauna consists only of a couple of surviving monotremes, and a considerable range of marsupials, murid rodents and bats. Among the latter, the microchiropteran bats are much more diverse than the megachiropterans. Of the Australo-Papuan mammals the monotremes and marsupials have ancient Gondwanan ancestry with either fossil or living relatives in South America. In contrast, the eutherians (murid rats and bats) appear to be largely of Indo-Malayan or Old World origins, having mostly arrived by over-water dispersal and island hopping in the latter part of the Tertiary (Flannery, 1996; Archer *et al.,* 1999). Curiously though, probable early Eocene remains of an archaeonycteridid microchiropteran bat (*Australonycteris clarkae*) and a possible condylarth mammal (*Tingamarra porterorum*), from the Murgon fossil site in Queensland indicate that ancient Gondwanan eutherian mammals did once inhabit the Meganesian continental area (Archer *et al.,* 1999), though prior to the consolidation of New Guinea. Endemism is high among the monotremes, marsupials and murid rodents, but decreases in the bats, with both microchiropterans and megachiropterans exhibiting a significant proportion of wide-ranging species, in addition to some endemic assemblages (Laurie and Hill, 1954; Smith and Hood, 1981; Corbet and Hill, 1992; Strahan, 1995; Flannery, 1996).

Among the monotremes or egg-laying mammals, New Guinea lacks the aquatic ornithorhynchid platypus (*Ornithorhynchus anatinus*), but in addition to the tachyglossid short-beaked echidna (*Tachyglossus aculeatus*) which it shares with Australia, it also possesses large long-beaked echidnas (genus *Zaglossus*) which are endemic to the Papuan zoogeographic subregion. Australia also appears to have once possessed a second large echidna genus (*Megalibgwilia* sp.) but this appears to have become extinct in the late Pleistocene (Musser, 1999). Extant Australian marsupial families that appear to be absent from New Guinea include the Myrmecobiidae (numbat), Phascolarctidae (koala), Vombatidae (wombats), Tarsipedidae (honey possum), Potoroidae (rat-kangaroos, potoroos and bettongs) and Notoryctidae (marsupial moles). The recently extinct Tasmanian thylacine (*Thylacinus cynocephalus*), also appears to have had congeners or conspecifics persisting in mainland Australia and New Guinea up into the latter half of the Holocene (Flannery, 1995a; Archer *et al.,* 1999).

On current evidence, although this could be partially due to a lack of palaeontological discovery, several extinct fossil marsupial families appear to be absent from New Guinea. Some salient examples include the ektopodontids, pilkipildrids, palorchestids and thylacoleonids. The pre-Holocene fossil marsupial megafauna known from New Guinea is limited to several genera of Pliocene and Pleistocene diprotodontids and macropodids, and these appear to represent narrow lineages of zygomaturines or *Protemnodon*-related macropodines, respectively. The recorded extinct Pliocene megafauna include diprotodontids such as *Kolopsis rotundus*, *'Kolopsis' watutense* and *Kolopsoides cultridens*; and the macropodines *Protemnodon buloloensis*, *P. otibandus* and *Watutia novaeguinea* (all apparently lowland browsers). This compares with a known Pleistocene diprotodontid fauna consisting of *Hulitherium thomasettii*, *'Kolopsis' watutense* and *Maokopia ronaldi*; and a macropodine megafauna of *Protemnodon hopei*, *P. nombe*, and *P. tumbuna* (all apparently mid-montane to sub-alpine browsers). Extinct species of *Dendrolagus*, *Thylogale* and *Thylacinus* are also recorded from the Pleistocene (Rich and Thompson, 1982; Hope *et al.,* 1993; Mountain, 1993; Menzies and Ballard, 1994; Flannery, 1995a).

The only Australian eutherian mammal family that is absent from New Guinea is the Megadermatidae, represented in Australia by the endemic ghost bat (*Macroderma gigas*) (Flannery, 1995a; Strahan, 1995). Given its northern Australian distribution, however, it is possible that the ghost bat may one day be discovered in a little surveyed southern New Guinea.

Among the murid rats the old endemic tribe Hydromyini has undergone a significant radiation in New Guinea with nine mostly endemic genera and 17 species, but is barely present in Australia, which only has two recorded species, the common water-rat (*Hydromys chrysogaster*) and false water-rat (*Xeromys myoides*), both of which are shared with New Guinea (Flannery, 1995a; Strahan, 1995; Hitchcock, 1998). A similar pattern is apparent for the old endemic tribe Uromyini, which has 13 genera and 38 species in New Guinea (Plate 21), but is only represented in Australia by three New Guinean genera and six species.

Conversely, the old endemic tribe Conilurini has undergone a significant radiation in Australia with eight mostly endemic genera and 40 extant species, but is only represented in New Guinea by two Australian species, the brush-tailed rabbit-rat (*Conilurus penicillatus*) and delicate mouse (*Pseudomys delicatulus*) that are restricted to southern areas (Waithman, 1979; Flannery, 1995a). The above pattern probably represents the arrival of different waves of ancestral murid stock in the late Tertiary. The ancestors of the Hydromyini and Uromyini found an opportunity to radiate into New Guinea's extensive rainforest habitats, whereas a drying Australian continent provided opportunities for the Conilurini to radiate into mesic and xeric habitats that were expanding from the later Tertiary into the Pleistocene.

In contrast to the murids discussed above the subfamily Murinae, represented by the genera *Stenomys* and *Rattus* in New Guinea but only the latter in Australia, probably represents a late arrival in both areas, but perhaps with a longer presence in New Guinea. Often referred to as 'new endemics', the 12 New Guinean and seven Australian species (some of which are shared) appear to be less specialized than many of their old endemic counterparts (Flannery, 1995a; Strahan, 1995).

For a detailed review of the evolution of rodents in Australasia see pp.707–744 (Aplin, 2006). A summary of the extant mammal fauna of the Papuan subregion is provided in Table 5.

BIOGEOGRAPHIC PROCESSES

MAJOR FACTORS INFLUENCING ZOOGEOGRAPHY

Today, although our knowledge is still far from perfect, we can suggest the major factors that have given rise to the peculiar zoogeography of New Guinea and the Papuan zoogeographic subregion to which it and its immediate continental satellites belong (see Figure 2). The fossil record of New Guinea is very poorly known so the processes have to be largely inferred from the relationships of the extant faunas. The six principal factors are listed below:

(1) New Guinea's Meganesian inheritance and the biotic legacy of the Proto New Guinea Archipelago;
(2) habitat diversification with the creation of altitudinal and latitudinal gradients, and geomorphic and climatic barriers within New Guinea;
(3) habitat resonance leading to considerable direct colonization from the Indo-Malayan region and Australia;
(4) connecting and isolating influences of fluctuating sea levels and habitats, particularly during the late Tertiary and Pleistocene;
(5) New Guinea's relative isolation, and adaptive radiation from a limited founding stock of vertebrates;
(6) ancient and recent human influences including species translocation, extinctions, disturbance and habitat modification.

Meganesian Inheritance

When Meganesia separated from Antarctica, it carried with it a strong Gondwanan biotic inheritance of southern plants and animals. Examples of Gondwanan-derived vertebrates include monotremes, marsupials, flightless casuariid birds, megapodes, chelid turtles, and myobatrachid and hylid frogs. During Meganesia's long drift northwards from Antarctic latitudes these Gondwanan organisms were joined by long-distance over-water colonizers from other land masses. Aided by increasing proximity to Asia, particularly in the last 20 million years, and island arc stepping stones, species with Asian affinities became the main over-water colonizers of Meganesia. Examples of these ancient Indo-Malayan colonizing influences include various megachiropteran and microchiropteran bats, murid rodents, a variety of bird families, several groups of snakes and lizards, and microhylid and ranid frogs. These ancient colonizers combined with pre-existing Gondwanan biota to create a mixed founding stock that underwent considerable evolution in isolation and adaptive radiation within a changing continent. This led to the formation of a distinctive and diverse endemic Meganesian vertebrate fauna, with Australia serving as a major source for species colonizing a nascent New Guinea.

In addition to a strong Meganesian influence, a further biotic inheritance that has probably left its mark on the zoogeography of New Guinea, is the legacy of Proto New Guinea and associated island arcs. Prior to its consolidation in the late Tertiary and early Pleistocene, New Guinea was a changing archipelago of islands in which Meganesian and newly colonizing Indo-Malayan and Oceanic elements underwent allopatric speciation. Flannery (1995a) considers that geological upheaval and drastically changing land areas probably led to extinctions that simplified the fauna, however, various Proto New Guinea islands retained endemic elements. Later when former Gondwanan terranes and oceanic islands of volcanic arc origin amalgamated into a consolidated New Guinea, they carried with them a biotic inheritance of local endemic assemblages that were to contribute to regional differences in New Guinea's zoogeography (Coates, 1990; Flannery, 1995a; Aplin, 1998).

Habitat Diversification—Gradients and Barriers

Environments

If New Guinea is not a 'real' island in a geological sense, it is largely an island of environmental difference from the rest of the continent. Reaching within 60 km of the equator, and over 4.8 km in altitude, several climatic zones occur that are found nowhere else on the Greater Australian continent. The geology differs from the rest of Australia as the island is rich in limestones, recent acid volcanics, deep sea sediments and ultramafic rocks. While active volcanoes (and their ash blankets) are restricted to the east of the island, rapid uplift and earthquakes occur everywhere in the ranges north of the flat

Table 5. Summary of the extant (and recently extinct) mammal fauna of the Papuan subregion (based on Honacki *et al.*, 1982; Nowak, 1991; Flannery, 1995a, b; Strahan, 1995; Aplin, 1998; Aplin *et al.,* 1999). Numbers of species per genus are in parentheses. Key to abbreviations: A = genus occurs in Australia; E = genus endemic to Papuan subregion; Int = introduced.

Common Name	Family	Genus
Monotremes		
	Tachyglossidae	
		Tachyglossus (1) A
		Zaglossus (1) E
Marsupials		
	Dasyuridae	
		'*Antechinus*' (5) ?E
		Dasyurus (2) A
		Murexia (2) E
		Myoictis (1) E
		Neophascogale (1) E
		Phacolosorex (2) E
		Planigale (1) A
		Sminthopsis (2) A
	Thylacinidae	
		Thylacinus [1 Extinct] A
	Peramelidae	
		Isoodon (1) A
	Peroryctidae	
		Echymipera (5) A
		Microperoryctes (3) E
		Peroryctes (2) E
	Burramyidae	
		Cercartetus (1) A
	Petauridae	
		Dactylopsila (4) [+1 ?Extinct] A
		Petaurus (3) A
	Pseudocheiridae	
		?Petauroides sp. [1 ?Extinct] A
		Pseudochirops (4) A
		Pseudochirulus (5) A
	Acrobatidae	
		Distoechurus (1) E
	Phalangeridae	
		Phalanger (8) A
		Spilocuscus (3) A
	Macropodidae	
		Dendrolagus (8) [+1 Extinct] A
		Dorcopsis (4) E
		Dorcopsulus (2) E
		Lagorchestes (1) A
		Macropus (1) A
		Protemnodon (5 Extinct) A
		Thylogale (4) [+2 Extinct] A
		Watutia (1 Extinct) E
	Diprotodontidae	
		Hulitherium [1 Extinct] E
		Kolopsis [1 Extinct] E
		'Kolopsis' [1 Extinct] E
		Kolopsoides [1 Extinct] E
		Maokopia [1 Extinct] E
Microchiropteran Bats		
	Emballonuridae	
		Emballonura (5)
		Saccolaimus (3) A
		Taphozous (1) A
	Hipposideridae	
		Aselliscus (1)
		Hipposideros (11) A
	Rhinolophidae	
		Rhinolophus (4) A
	Vespertilionidae	
		Chalinolobus (1) A
		Kerivoula (2) A
		Miniopterus (6) A
		Murina (1) A
		Myotis (1) A
		Nyctophilus (4) A
		Pharotis (1) E
		Philetor (1)
		Phoniscus (1)
		Pipistrellus (4) A
		Scotorepens (1) A
	Molossidae	
		Chaerephon (1) A
		Mormopterus (2) A
		Otomops (2)
		Tadarida (1)
Megachiropteran Bats		
	Pteropodidae	
		Aproteles (1) E
		Dobsonia (5) A
		Macroglossus (1) A
		Nyctimene (7) A
		Paranyctimene (1) E
		Pteropus (8) A
		Rousettus (1)
		Syconycteris (2) A
Rodents		
	Muridae	
Subfamily Hydromyinae		
	Tribe Conilurini	
		Conilurus (1) A
		Pseudomys (1) A
	Tribe Hydomyini	
		Crossomys (1) E
		Hydromys (4) A
		Leptomys (3) E
		Mayermys (1) E
		Microhydromys (2) E
		Neohydromys (1) E
		Parahydromys (1) E
		Paraleptomys (2) E
		Pseudohydromys (2) E
		Xeromys (1) A
	Tribe Uromyini	
		Abeomelomys (1) E
		Anisomys (1) E
		Chiruromys (3) E
		Coccymys (2) E
		Hyomys (2) E
		Lorentzimys (1) E
		Macruromys (2) E
		Mallomys (4) E
		Melomys (13) A
		Pogonomelomys (2) [+1 ?Extinct] E
		Pogonomys (4) A
		Uromys (4) A
		Xenuromys (1) E
	Subfamily Murinae	
		*Rattus (*14) [6 Int] A(9, 2 Int)
		Stenomys (5)
		Mus (1) [Int] A(Int)
Artiodactyls—Even-toed Ungulates		
	Suidae	
		Sus (1) [Int] A(Int)
	Cervidae	
		Cervus (1) [Int] A(Int)
		Axis (1) [Int] A(Int)
Carnivores		
	Canidae	
		Canis (2) [*C. dingo* and *C. familiaris*—Int] A(Int)
	Felidae	
		Felis (1) [Int] A(Int)

Oriomo Plateau. Constrictions to the seaway caused by the growth of the eastern Indonesian archipelago have pooled hot equatorial water in the western Pacific Ocean. Palaeoecology suggests that New Guinea has been the wettest place on Earth throughout the late Quaternary. In the mountains there are very high rainfalls throughout the year, which combined with the steep terrain causes erosion at rates that are two orders of magnitude higher than those affecting Australia (Gressitt, 1982; Hope, 1996b).

With more precipitation than evaporation over most of the island, true equatorial ever-wet environments with a cover of lowland and mountain rainforest has an extent of 700,000 km^2, much larger than the area of the modern Queensland rainforests. However seasonal tropical environments, similar to, but distinctly wetter than those of northern Australia, are found in the Torresian zone of southern New Guinea (south of 8°S latitude), where *Eucalyptus* and *Melaleuca* savanna woodlands are widespread. Other major biomes are mangrove and freshwater swamp forests, and an alpine complex of shrublands, grasslands and mires above 3,200 m (Figure 3).

Not surprisingly aquatic habitats are common as the huge river systems such as the Fly, Mamberamo and Sepik take the rainfall to the sea. There are several substantial lakes ranging from Lake Sentani and Lake Wanum near sea level through montane lakes such as Angi, Panai and Kutubu, the subalpine Lake Habbema and hundreds of glacial tarns. In addition the river systems have impounded floodplain lakes, the largest being Lake Murray on the Strickland River and the seasonally inundated Lakes Plain of the Mamberamo.

Topographic Controls

The 4,000 m high (isolated peaks to 4,884 m) central mountains that run west to east for most of the length of the island have had a strong influence on zoogeography by isolating the faunas of the northern lowlands from those of the south with massive mountain barriers. These also generated a peninsular effect by creating a long strip of montane habitat running west to east along the spine of the island and two long strips of lowland habitats running along the northern and southern lowlands respectively. The overall effect was to create either distinctive northern and southern faunas, west to east clines, or ring species that wrapped around the coastal lowlands.

Examples of lowland species that are largely restricted to the north of the central cordillera include the northern common cuscus (*Phalanger orientalis*), red-legged brush-turkey (*Talegalla jobiensis*) and Victoria crowned pigeon (*Goura victoria*). These are mirrored by congeneric counterparts that are largely restricted to the south of the main range, including the southern common cuscus (*Phalanger intercastellanus*), yellow-legged brush-turkey (*Talegalla fuscirostris*) and southern

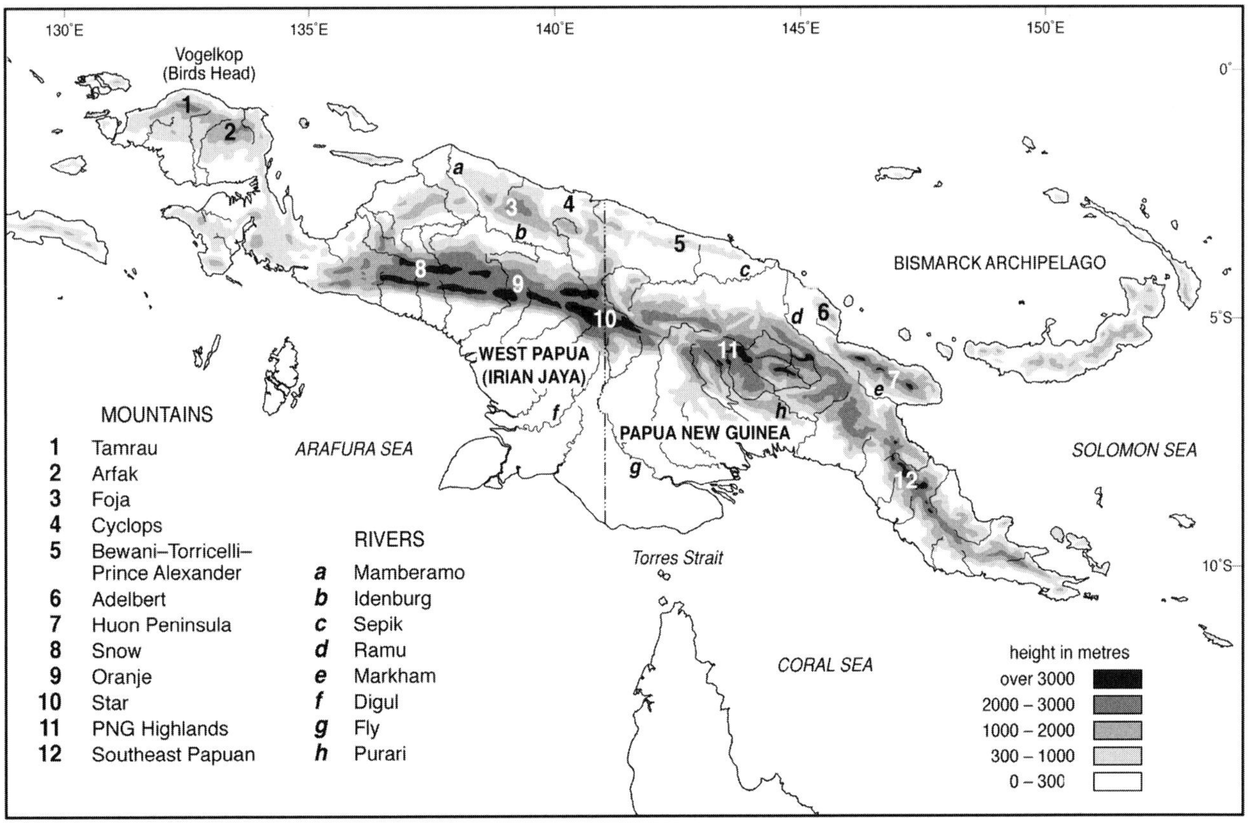

Figure 3. A physiographic map of New Guinea and associated islands showing the major mountain ranges and river systems referred to in the text.

crowned pigeon (*Goura scheepmakeri*) (Beehler *et al.*, 1986; Coates, 1990; Menzies, 1991; Flannery, 1995a).

The clinal phenomenon of ring populations that wrap all the way around the lowlands, is well illustrated by the different subspecies of the common spotted cuscus (*Spilocuscus maculatus*), a common possum of the lowlands up to about 1,200 m. Along much of the north coast one finds the subspecies *S. maculatus maculatus*, this gives way to *S. maculatus goldei* in the south-eastern lowlands, with *S. maculatus chrysorrhous* arising in the south-western lowlands (Flannery, 1994a; Singadan, 1996). A further subspecies *S. maculatus nudicaudatus* occurs on upper Cape York Peninsula in Australia. See Plates 22–24. The death adder (*Acanthophis praelongus*), is another element of lowland forest both north and south which also extends into the lower montane zone.

The long continuous west to east strip of montane habitat stretches 1,800 km from the Snow Mountains in the west to the Star Mountains, Central Highlands, Eastern Highlands and South-east Papuan mountains in the east. Here there are some clear examples of west to east clinal populations, but also some more complex patterns, particularly among birds, of disjunct, parapatric, and sympatric populations along a historically fluctuating but contiguous strip of montane habitat. Zoogeographer Jared Diamond (as cited by Beehler *et al.*, 1986) considers that the more complex pattern in birds is due to what he calls 'the drop out' phenomenon. This is where random extinctions along part of the once continuous cline have left some populations isolated, facilitating allopatric speciation. Some of these have later come back into contact to form parapatric and sympatric populations, while others have retained disjunct distributions. In some instances Pleistocene climate change may have played a role in these stochastic extinctions through ecological concatenations associated with habitat change. In other instances Coates (1990) and Flannery (1995a) and Aplin (1998) have speculated that some of the zoogeographic patchiness may be due to local suites of vertebrates inherited from the component land masses that eventually coalesced to form New Guinea.

The most marked peninsular effect can be seen in the suite of local endemic species (discussed later) that have formed on the relatively isolated Bird's Head Peninsula of far north-western New Guinea. This high mountainous area is only connected to the main body of New Guinea by a narrow and relatively low neck of land. This serves as both a constriction and ecological filter by causing general genetic attenuation or inhibiting the ready migration of montane species from New Guinea's main range. The Bird's Head mountain flora also has endemic elements and varying plant communities that distinguish it from the Central Cordillera. Conversely, the Bird's Neck region also serves as a faunal mixing zone where some of the northern and southern New Guinean lowland species come into contact. This has contributed to a composite aspect to the faunas of the Bird's Head region (Aplin, 1998). The same author also considers that both anthropogenic and natural extinctions may have simplified the mammal fauna of the Bird's Head, and this may exaggerate its apparent differences (Aplin, 1998; Aplin *et al.*, 1999).

For many vertebrate species, low intervening valleys and large rivers draining out of the mountains have acted as barriers, while isolated peaks and ranges have become virtual islands of montane or alpine habitats. This last effect is most marked in the isolated ranges of the north coast, which include the outlying Foja, Cyclops, Bewani-Torricelli-Prince Alexander, Adelbert, and Huon Ranges. These are cut off by the deep intervening river valleys of the Mamberamo, Idenburg, Grime, Sunggrum, Sepik, Ramu and Markham Rivers, respectively. A typical pattern with these northern ranges is to have a slightly depauperate subset of species from the main range, but with a suite of their own endemics.

The high mountains of the isolated Huon Peninsula for example, follow this pattern. They are lacking in some main range species but possess unique endemics such as the Huon tree-kangaroo (*Dendrolagus matschiei*), Huon astrapia (*Astrapia rothschildi*), emperor bird of paradise (*Paradisaea guilielmi*), spangled honey-eater (*Melipotes ater*), and Huon melidectes (*Melidectes foersteri*) (Diamond, 1985; Beehler *et al.*, 1986; Coates, 1990; Menzies, 1991; Flannery, 1995a). A further feature of the Huon Peninsula ranges, which exceed 4,000 m in altitude, is that in the absence of competition from some main range species, some vertebrates extend to higher altitudes than on the main range. For example, in the absence of competition from the silky cuscus (*Phalanger sericeus*), the mountain cuscus (*Phalanger carmelitae*) has been recorded at up to 3,660 m on the Huon Peninsula, compared with about 3,000 m on the Central Cordillera (George, 1987; Menzies, 1991; Flannery, 1995a).

Some further examples of endemics on outlying ranges include: the golden-fronted bowerbird (*Amblyornis flavifrons*) of the Foja Mountains; a frog (*Platymantis cheesmanae*) of the Cyclops Range; the tenkile (*Dendrolagus scottae*), northern glider (*Petaurus abidi*) and Musser's shrew-mouse (*Microhydromys musseri*) of the Bewani-Torricelli Mountains; and the fire-maned bowerbird (*Sericulus bakeri*) of the Adelbert Mountains (Zweifel and Tyler, 1982; Beehler *et al.*, 1986; Coates, 1990; Flannery, 1995a). Due to a combination of geological history, peninsular effect and its high isolated Arfak and Tamrau Mountains, the Vogelkop Peninsula is a considerable centre of regional endemism. Examples of some unique Vogelkop endemics include: the Vogelkop tree-kangaroo (*Dendrolagus ursinus*), the Arfak ringtail (*Pseudochirulus schlegeli*), Arfak astrapia (*Astrapia nigra*), grey-banded mannikin (*Lonchura vana*), and Vogelkop scrub-wren (*Sericornis rufescens*) (Beehler *et al.*, 1986; Flannery, 1995a; Aplin, 1998; Coates and Peckover, 2001).

Large rivers such as the Mamberamo, Idenburg and Sepik in the north, and the Digul, Fly, Strickland, Erave and Purari in the south, also seem to act as barriers that cause a degree of isolation for blocks of lowland habitat. This has probably contributed to the formation of distinct gene pools and varying faunal assemblages in different sectors of the New Guinean lowlands.

Altitudinal and Bioclimatic Zonations

The development of the largest habitat variety on the continent has encouraged niche diversification and speciation of vertebrates. New Guinea mountains provide an altitudinal zonation of habitats ranging from equatorial lowland rainforest and hill forest; a range of lower montane, upper montane and subalpine forests to the open alpine vegetation of grassland and mires, and glaciers (Plates 25–28). These environments are not symmetrical layers but have complex north-south gradients related to local climates. For example southern beech forests (*Nothofagus* spp.) are more common in the west and north of the island, but rare to absent on the southern slopes, due to narrow precipitation and temperature requirements.

Many faunal groups exhibit distinctive lowland tropical and montane sub-tropical to alpine temperate forms. This is well illustrated among the cuscuses, for example. On the Central Cordillera from sea level to about 1,200 or 1,500 m, the principal arboreal cuscuses are the common spotted cuscus (*Spilocuscus maculatus*) and northern common cuscus (*Phalanger orientalis*) or southern common cuscus (*Phalanger intercastellanus*). However, starting at 1,400–1,500 m these are replaced by the mountain cuscus (*Phalanger carmelitae*) and silky cuscus (*Phalanger sericeus*) which may extend up into the upper montane and sub-alpine zone, with recorded altitudinal limits of about 3,000 m and 3,900 m respectively (George, 1987; Menzies, 1991; Flannery, 1995a). A similar pattern can also be seen among reptiles such as pythons, with lowland species such as the amethystine python (*Morelia amethistina*) being replaced in the montane zone by Boelen's python (*Morelia boeleni*) (Parker, 1982; O'Shea, 1996). Among the frogs, ranids appear to be mostly found in the lowlands and lower montane zone, while the wide-ranging microhylids appear to be the only frogs that have colonized the alpine zone (Zweifel and Tyler, 1982).

Still other vertebrate species are restricted to very narrow altitudinal ranges within the hill forest to mid montane zone. The narrowness of these ranges indicates that competition as well as environment may be playing a role in altitudinal zonation of species. One example is Stein's cuscus (*Phalanger vestitus*) which on the main range typically occupies a narrow band of montane forest between 1,200 and 1,600 m, into which it may be squeezed by competition from other cuscuses (George, 1987; Flannery, 1995a). Although in the Tamrau and Arfak Mountains of the Bird's Head, where the montane cuscuses *P. carmetitae* and *P. sericeus* are absent, *P. vestitus* has been recorded at up to at least 2,200 m elevation (Aplin, 1998). The role of competition is emphasized in situations of almost complete competitive release, such as allopatric populations of New Guinea's northern common cuscus (*Phalanger orientalis*); on the island of New Britain or in the Solomons, where it is the sole possum, local people report that this essentially lowland phalangerid may be found all the way from sea level to more than 2,000 m at the tops of the highest mountains (Heinsohn, 2001b). A further case of probable competitive release is provided by Hope *et al.* (1993), who suggest that the arboreal, but partially terrestrial, coppery ringtail (*Pseudochirops cupreus*), the commonest possum of the subalpine zone on many of New Guinea's mountains, has expanded its range upwards as human predation has removed competition from other possums and wallabies in the grasslands.

Other species are subalpine to alpine specialists that are only found in the virtual habitat islands created by New Guinea's highest peaks. Vertebrates that are largely restricted to the high Snow Mountains of West Papua and some adjacent alpine areas, for example, include mammals such as the giant alpine woolly-rat (*Mallomys gunung*) and the glacier rat (*Stenomys richardsoni*); and birds such as the short-bearded melidectes (*Melidectes nouhuysi*); orange-cheeked honey-eater (*Lichenostomus chrysogenys*) and the Snow Mountain quail (*Anurophasis monorthonyx*) (Beehler *et al.,* 1986; Betz and Gruys, 1995; Flannery, 1995a; Coates and Peckover, 2001). All of these are typically found between 3,200 and 4,600 m in elevation. Equivalent but distant alpine habitats to the east may support other locally endemic alpine species such as the Giluwe rat (*Rattus giluwensis*) on Mt. Giluwe (Flannery, 1995a).

A few species seem to defy altitudinal zonation by occupying very wide altitudinal ranges. Examples include the long-beaked echidna (*Zaglossus bruijnii*)—sea level (SL) to 4,150 m; New Guinea quoll (*Dasyurus albopunctatus*) (SL to 3,500 m); ground cuscus (*Phalanger gymnotis*) (SL to 2,700 m); sugar glider (*Petaurus breviceps*) (SL to 3,000 m); uneven-toothed rat (*Anisomys imitator*) (SL to 3,500 m); common blossom-bat (*Syconycteris australis*) (SL to 3,000 m); and dwarf cassowary (*Casuarius bennetti*) (SL to 3,620 m) (Beehler *et al.,* 1986; Menzies, 1991; Flannery, 1995a; Coates and Peckover, 2001).

Habitat Resonance—Role as Refuge and Destination

The term 'habitat resonance' is defined here as similar habitats that make a new land mass receptive to colonizers from like habitats. A curious effect of mountain building in New Guinea, is that while much of the Australian portion of a northward-drifting Meganesia went through a drying trend that culminated in Pleistocene aridity, New Guinea's high mountains and blocking of the equatorial ocean circulation served to buffer the island against aridification. This enabled it to act as a refuge, particularly at altitude, for Gondwanan species with a preference for cool moist subtropical and temperate environments. Thus a relatively geologically young New Guinea was able to recreate

Gondwanan niches which preserve Gondwanan plant taxa such as *Dacrydium, Dacrycarpus* and *Nothofagus*. Pliocene faunas from southern Australia often overlap with modern mountain faunas in New Guinea, for example the 4.6 million-year-old site at Grange Burn, in western Victoria. Here mammals include cuscus-like phalangerines, tree-kangaroos and dorcopsis-like macropods, in a habitat of relatively open forest (Rich and Thompson, 1982; Flannery *et al.,* 1996). Similarly, the slightly younger Pliocene Otibanda fauna from New Guinea's Wau valley has some features in common with late Miocene faunas of Australia, including *Thylacinus*, *Protemnodon*, zygomaturine diprotodontids and an unidentified rodent (Rich and Thompson, 1982).

Some extinctions are known: several species of zygomaturine diprotodonts, some robust species of the macropod *Protemnodon* and a dwarf cassowary (*Casuarius lydekkeri*) died out in the Pleistocene, while thylacine and a pademelon (*Thylogale christenseni*) disappear in the Holocene (Rich and Thompson, 1982; Hope *et al.,* 1993; Menzies and Ballard, 1994; Flannery, 1995a). However, New Guinea never appears to have supported other marsupial or monotreme groups such as palorchestines, koalas, wombats, devils, *Thylacoleo* or platypus. On the other hand New Guinea did preserve a worm-eating monotreme, *Zaglossus bruijnii* (Plate 29); while Australia lost its second echidna genus (*Megalibgwilia* spp.). *Megalibgwilia* was once widespread throughout eastern areas until the group's gradual extinction in the Pleistocene (Musser, 1999), possibly due to drought and decimation of its prey.

New Guinea's high mountains probably also helped to preserve humid lowland habitats by catching moisture-laden winds and creating regions of high rainfall in lowland areas adjacent to mountains and escarpments. These probably acted as lowland rainforest refuges during the glacial maximum. Thus montane New Guinea served as a refuge for subtropical and temperate Meganesian biota; while lowland New Guinea served as a refuge for tropical Meganesian rainforest biota as well as a viable habitat for colonizing tropical Indo-Malayan species.

In contrast, New Guinea's Torresian lowland habitats collected differing suites of species. The extensive seasonally dry savanna-woodlands and patchy monsoon forests of the southern Trans-Fly region served as viable habitat for mesic-adapted Australian species entering via intermittent land-bridges or over-water dispersal; while sizeable areas of coastal mangroves collected Meganesian, Indo-Malayan and widespread mangrove species.

Pleistocene Fluctuations—Connections and Isolation

An historical influence on the altitudinal environments of New Guinea, is the legacy of Pleistocene temperature fluctuations, which brought about corresponding fluctuations in the altitudinal zonation of habitats. As temperatures fell during various glacial periods glaciers advanced and the tree-line retreated from 3,800 m down to about 2,200 m, only to advance back up to higher levels during warmer interglacial periods. Lower zonations responded less dramatically, the lower montane forest descending about 500 m, but with marked changes in composition. Depression of the tree-line during cool glacial periods enabled the spread of montane and alpine species normally isolated by habitat barriers. However, with the climatic amelioration at the end of the Pleistocene, many of these were left isolated in shrinking montane and alpine habitats on outlying ranges (Hope, 1996a).

The temperature fluctuations of the Pleistocene also resulted in fluctuating polar glaciation and associated sea level changes. Although sea levels fell by as much as 140 m, New Guinea and Meganesia have remained cut off from the land masses of the Asian Sunda Plate and the intervening Wallacean Islands by deep permanent sea channels, though a narrowing of these may have aided over-water dispersal (Hope, 1996b; van Oosterzee, 1997). Within Meganesia, however, vast areas of the Sahul continental shelf became exposed as a flat plain, adding an additional 1.4 million km^2 to the land area (New Guinea is 808,000 km^2 in area). The Sahul plain joined Australia, New Guinea, Aru and other continental islands into a contiguous land mass that facilitated an exchange of species. A saline lake formed west of Cape York and pollen data indicates that the plain was mostly open and arid over the last 250,000 years (Torgersen *et al.,* 1988). In New Guinea first Nix and Kalma (1972), and later Kaars (1995) as well as Kaars *et al.* (2000), suggested it is likely that the savanna-wet forest boundary extended northwards. This has been shown to have occurred on Aru Island on the Arafura shelf (O'Connor *et al.*, 2002). Faunal sequences from Lembudu Cave on Kobroor Island in the Aru Group reveal that savanna-woodland adapted agile wallabies (*Macropus agilis*) and some other species with a preference for mesic habitats once occurred on Aru (Plate 30). But, these appear to have been marginalized and eventually died out when much of Aru reverted to rainforest following climatic amelioration at the end of the Pleistocene (O'Connor *et al.*, 2002). Although savannas still occur on the Aru Islands, the reduced areas may have been too small to sustain populations of larger savanna-woodland adapted species, particularly if they were under hunting pressure from humans.

As sea levels rose during interglacial periods, populations became isolated on recreated islands and underwent allopatric speciation to form new species and subspecies. This cycle occurred a number times throughout the Pleistocene (and later Tertiary), with suites of species and subspecies owing their origins to different land-bridge events. The general rule being that closely related taxa either side of a temporary land-bridge are due to more recent connections, whereas more divergent sister taxa are due to events further back in time. Aplin *et al.* (1993), for example, use albumin immunological evidence and rates of molecular change to correlate approximate divergence dates with various late Tertiary and Pleistocene dispersal events for different groups of monotremes and marsupials. They estimate that one dispersal event, may have occurred in the early Miocene, some 20 million years ago. It is postulated that this may have been when the Australian ancestors of the peroryctid bandicoots, cuscuses (phalangerinae) and possibly the feather-tailed possums (*Distoechurus*) arrived in New Guinea or one of its precursor land masses. A second dispersal episode may have occurred 12–10 m.y.a. in the middle Miocene, with the Australian ancestors of the pseudocheirid ringtail possums, dactylopsiline striped possums, *Dorcopsis* wallabies, and dasyurids of the genera '*Antechinus*', *Phascolosorex* and *Neophascogale* entering New Guinea. A third dispersal event may have occurred 4.7–2.7 m.y.a., bringing the Australian ancestors of some additional pseudocheirids, tree-kangaroos (*Dendrolagus*) and the dasyurid genus *Myoictis* to New Guinea (Aplin *et al.,* 1993; Flannery, 1995a). Another major period of dispersal was during the dramatic sea level fluctuations of the Pleistocene with multiple episodes of faunal exchange across a fluctuating Torresian land-bridge.

A further dimension, to areas such as the Torres Strait and the broader Torresian plain, was that habitat on the ephemeral land-bridge varied from episode to episode. During some early episodes there may have been corridors of closed forest, although it seems clear that through most of the Quaternary there have been large disjunctions of savanna-woodland with only tenuous closed forest routes along watercourses (Nix and Kalma, 1972; Torgersen *et al.,* 1988; Flannery, 1995a). Thus the land-bridges have acted as ecological filters, that during one episode may allow the passage of some rainforest species but during another may mostly favour savanna-woodland adapted species. One more consideration regarding the exchange of fauna between New Guinea, Australia and other land-bridge satellites, was whether species could disperse to a stable and viable habitat, as opposed to an ephemeral ecological death trap (demonstrated by the extinct savanna-woodland adapted mammals of Aru mentioned above). This is because the climatic and sea level fluctuations of the Pleistocene also caused enormous shifts in the distribution of habitats.

Many exchanges, particularly among birds and bats, and even among some non-volant species are due to over-water dispersal rather than land-bridge events. While a general trend through geological time has been for an older Australia to serve as a major source of fauna for a nascent New Guinea, the latter has also served as a source for some New Guinea-radiated taxa that have re-invaded Australia. A general trend in these later species exchanges between Australia and New Guinea, is for mesic-adapted Australian species to colonize savanna-woodland habitats in southern New Guinea, and for rainforest-adapted New Guinean species to colonize patches of rainforest/monsoon forest habitat in northern or eastern Australia. There are few recent examples of Australian rainforest species colonizing New Guinean rainforests or of New Guinean savanna-woodland species colonizing Australia.

Some probable examples of later invasions of north-eastern Australia by New Guinean rainforest/monsoon forest species include: the long-nosed echymipera (*Echymipera rufescens*); common spotted cuscus (*Spilocuscus maculatus*); southern common cuscus (*Phalanger intercastellanus*); striped possum (*Dactylopsila trivirgata*); giant white-tailed rat (*Uromys caudimaculatus*); prehensile-tailed rat (*Pogonomys mollipilosus*); double-wattled cassowary (*Casuarius casuarius*); palm cockatoo (*Probosciger aterrimus*); eclectus parrot (*Eclectus roratus*); magnificent riflebird (*Ptiloris magnificus*) (Plate 31); trumpet manucode (*Manucodia keraudrenii*); green tree python (*Morelia viridis*) (Plate 32); and the wood frog (*Rana daemeli*).

Examples of Australian mesic species that have colonized savanna-woodland habitats in southern New Guinea include: the northern brown bandicoot (*Isoodon macrourus*); agile wallaby (*Macropus agilis*); spectacled hare-wallaby (*Lagorchestes conspicillatus*); brush-tailed rabbit-rat (*Conilurus penicillatus*); canefield rat (*Rattus sordidus*); delicate mouse (*Pseudomys delicatulus*); little red flying fox (*Pteropus scapulatus*); Australian bustard (*Ardeotis australis*); king brown snake (*Pseudechis australis*); sand monitor (*Varanus gouldii*); frilled lizard (*Chlamydosaurus kingii*); and desert tree frog (*Litoria rubella*). An example of an Australian rainforest species that has colonized a mosaic of monsoon forest, gallery forest and savanna-woodland in southern New Guinea, is the red-legged pademelon (*Thylogale stigmatica*); while the Papuan sheathtail-bat (*Saccolaimus mixtus*) is an example of a New Guinea open forest species that has colonized like habitats on the tip of Cape York Peninsula.

As well as exerting an influence on northern Australia, New Guinea has also had a strong influence on other continental satellites. Its major land-bridge satellite islands include: Waigeo, Batanta, Salawati, Misool, Japen, Aru Islands, D'Entrecasteaux Islands and the Louisiade Archipelago. Waigeo and Batanta are included in this list, because although their coalesced land mass remained separated from New Guinea during times of low sea levels, the narrowness of the channel enabled them to function as quasi-land-bridge islands. All these island groups possess a depauperate subset of the New Guinea mainland fauna, augmented by suites of their own endemic species. Together with New Guinea, these continental land-bridge islands form the Papuan zoogeographic subregion. Some examples of Papuan island endemics include: the black dorcopsis (*Dorcopsis atrata*) of Goodenough Island and Tate's triok (*Dactylopsila tatei*) on Fergusson Island in the D'Entrecasteaux Archipelago; and the Waigeo cuscus (*Spilocuscus papuensis*), red bird of paradise (*Paradisaea rubra*) and Wilson's bird of paradise (*Cicinnurus respublica*) of the Waigeo Archipelago (Beehler *et al.,* 1986; Menzies, 1991; Flannery, 1995b).

New Guinea has also exerted a significant, though attenuated, influence on surrounding non-continental satellites in the Moluccas, Biak Islands, Bismarck

Archipelago and Solomon Islands, all of which have remained cut off from the New Guinean mainland by deep, wide and permanent sea channels. In addition to widespread oceanic vertebrates and some local island arc endemics, these islands also typically have a depauperate subset of New Guinean species that have colonized through over-water dispersal. The herpetofauna of Moluccan islands such as Seram, for example, exhibit a strong New Guinean influence and include classic Papuan elements such as: the amethystine python (*Morelia amethistina*); northern death adder (*Acanthophis praelongus*); the Papuan skink (*Carlia novaeguineae*) and giant tree frog (*Litoria infrafrenata*). All these probably arrived through over-water dispersal (Monk *et al.*, 1997).

In other instances colonizing New Guinean elements have speciated into island endemic forms. Halmahera and some adjacent islands in the North Moluccas, for example, are home to two outlying birds of paradise, Wallace's standardwing (*Semioptera wallacii*) and the paradise crow (*Lycocorax pyrrhopterus*) both of which are probably derived from Papuan ancestors (Sibley and Monroe, 1990; Coates and Bishop, 1997); although these and some other North Moluccan endemics could be due to a more complex former geological connection with the Papuan subregion (Coates and Peckover, 2001; Heads, 2001). Biak-Supiori in the northern Geelvink Islands is home to the endemic Biak glider (*Petaurus biacensis*), a close relative of the common Australo-Papuan sugar glider (*Petaurus breviceps*), and the Biak giant-rat (*Uromys boeadii*) a relative of New Guinea's black-tailed giant rat (*Uromys anak*). Other examples include the New Britain water-rat (*Hydromys neobritanicus*) and Bismarck giant rat (*Uromys neobritannicus*) of New Britain in the Bismarck Archipelago, which are close relatives of the widespread Australo-Papuan common water rat (*Hydromys chrysogaster*) and *U. anak* respectively (Groves and Flannery, 1994; Flannery, 1995b). The Bismarck Archipelago is also home to a range of endemic bats, birds, and herpetofauna. While many of these are New Guinea derived, others are probably ancient Melanesian arc endemics. Generally though, New Guinea appears to be one of the principal sources for vertebrate taxa that have dispersed out into the islands of the south-west Pacific—particularly birds, bats, lizards, snakes and frogs (Koopman, 1979; Smith and Hood, 1981; Allison, 1996; Flannery, 1996; Keast, 1996; Coates and Peckover, 2001).

Isolation and Convergence in a Limited Founding Stock

Though New Guinea is recognized as one of the world's hot spots of biodiversity, its humid forests are not as species rich as comparable habitats in the Amazon or South-east Asia. This is probably due to a combination of four main factors. These include: New Guinea's relative isolation compared with the continental zoogeographic cross-roads of Amazonia and South-east Asia; its comparatively smaller land mass; its relatively young geological age; and a more limited founding stock. The latter factors of young geological age and limited founding stock, suggest that there may be a relative under-utilization of some niches.

Now that the remarkable diversity of rainforest marsupials in lowland Australia in the Miocene has been revealed by Mike Archer and colleagues at Riversleigh, north Queensland (Archer *et al.*, 1991), it seems that the Papuan refuge came too late, or was too difficult to reach, for more than a handful of Australian mammal species. For example possum diversity from single deposits at Riversleigh may be higher than that from a modern New Guinean mountain, i.e. a ratio of 16/13 (= 1.23) (Archer *et al.*, 1991; Flannery, 1994a). Had New Guinea directly inherited the Riversleigh faunas, we can speculate that the modern marsupial diversity would have been many times greater. This in turn would have severely limited the niche opportunities of the later immigrants, such as the rodents.

A parallel can be drawn with the Australian arid region, which is of comparable age to New Guinea because it also first becomes extensive in the Pliocene and expands greatly in the Pleistocene. The arid zone is much larger than New Guinea, but lacks great habitat diversity. Here rodents also form a large proportion of the mammal fauna. The New Guinean marsupial/rodent ratio prior to late Pleistocene extinctions is estimated to be 76/70 (=1.09) while the same figure for the Australian arid zone is 38/25 (=1.52). The arid zone index is intermediate between New Guinea and the total Australian index of 165/63 (=2.62), reflecting a longer history of interaction and few geographic barriers. However both new areas demonstrate the effect of new niche creation, which allowed both mammal groups to evolve new forms. The older Australian habitats were already occupied when rodents arrived less than 5 million years ago. The relatively young geological age of many New Guinea ecosystems probably explains, for example, why hydromyin rats were able to diversify into insectivorous shrew-like niches, with a radiation that at the small end of the size scale, vies with that of the insectivorous and carnivorous dasyurid marsupials.

Despite its limited stock, New Guinea does seem to exhibit some further remarkable and unexpected examples of convergence and adaptive radiation. Some convergent New Guinean species are so similar in body form, that they are difficult to distinguish from their morphological counterparts in other parts of the world; while in other instances the convergence is mostly ecological and occurs between species from completely different classes or orders of vertebrate. An example of the latter is the aquatic earless water-rat (*Crossomys monktonii*), which appears to have adopted a platypus-like niche in New Guinea's fast flowing montane rivers. It has greatly reduced ear pinnae, small eyes, a keel-like crest of hairs on the tail, webbed feet, immense whiskers and uses burrows with underwater entrances (Flannery, 1995a).

One of the most remarkable cases of convergence is that between the New Guinean green tree python (*Morelia viridis*) and the South American emerald tree boa (*Corallus canina*). Both are bright green arboreal constrictors that prey on birds, small mammals and other vertebrates, and it requires close examination to distinguish the two snakes (Stafford, 2000; Torr, 2000). There are also some cases of very close convergence among birds. Examples include the Papuan *Machaerirhynchus* boatbills, which in appearance and habits resemble the Neotropical *Todirorostrum* flycatchers; while New Guinea's tit berrypecker *Oreocharis* is strikingly similar to a Neotropical *Euphonia* tanager (Beehler *et al.,* 1986). Other convergences are more behavioural than morphological. The ground-feeding New Guinean megapodes occupy the Old World pheasant niche (Beehler *et al.,* 1986); Australo-Papuan death adders *Acanthophis*, in feeding behaviour and form, fill the Old World viper niche (Stafford, 2000); while some skinks of the genera *Tribolonotus* and *Sphenomorphus* live under rotting logs and other debris of the forest floor and have habits that are convergent with those of salamanders (Allison, 1982).

Among the mammals, the New Guinean trioks or striped possums resemble skunks with their striking black and white pelage colouration and in the production of a strong odour. Among these, the long-tailed triok (*Dactylopsila palpatator*) with its markedly elongated digit for fishing insect larvae out of wood tunnels is convergent with the feeding strategy of the Madagascan Aye Aye (*Daubentonia madagascariensis*). Furthermore, by ripping open wood burrows with their teeth to help in the extraction of larvae, trioks as mammals, may also be partly filling New Guinea's under-utilized avian woodpecker niche. Another example of mammalian convergence is the remarkable similarity of form between the New Guinean northern glider (*Petaurus abidi*) and various species of Old World and New World flying squirrels. Some of the New Guinean murid rats such as the one-toothed moss-mouse (*Mayermys ellermani*), are also remarkably similar to insectivorous shrews in habit and general appearance. Further examples of convergence among mammals include: tree-kangaroos, cuscuses and some ringtail possums, which are broadly convergent with folivorous monkeys, lemurs and sloths; while New Guinea's insect and worm-eating long-beaked echidna (*Zaglossus bruijnii*) is convergent with New Zealand's avian apterygid kiwis and the Neotropical mammalian myrmecophagid anteaters.

The under-utilization of niches is also highlighted by the remarkable radiation of the tree-kangaroos and bandicoots. New Guinea's eight species of tree-kangaroo appear to have evolved from a terrestrial macropodid ancestor (probably in Australia) that radiated into an under-utilized medium-size arboreal folivore niche (Flannery *et al.,* 1996). The apparent dearth of competition during this radiation is highlighted by the relative lack of refined morphological specialization and agility in tree-kangaroos compared with arboreal primates or possums. They lack specialized adaptations for arboreal life such as a developed hallux or opposable thumb-like toe on the hind feet as found in possums, or a prehensile tail as found in many arboreal monkeys and possums. The nine peroryctid bandicoots represent another major marsupial radiation in New Guinea. They range in size from the 100 gram mouse bandicoot (*Microperoryctes murina*) to the five kilogram giant bandicoot (*Peroryctes broadbenti*), although most species are 1–2 kg. They mostly feed on invertebrates and some small vertebrates, although some also consume significant amounts of plant matter. In body form and habits they show convergence with the macroscelid elephant-shrews of Africa. As bandicoots often dig for food and have some omnivorous tendencies they may also be partially filling some niche space elsewhere occupied by suids (Menzies, 1991; Nowak, 1991; Flannery, 1995a).

New Guinea has a very poor carnivorous mammal fauna, in comparison with South-east Asia, or even Australia. Other than a couple of small quolls (≤1 kg), it apparently lacks larger dasyurids. But as recently as the middle Holocene, New Guinea did have the dog-like thylacines or marsupial wolves. For further information on the Thylacinidae see pp.551–574 (Wroe and Archer, 2006) and pp.745–753 (Corbett, 2006). Ground cuscuses (*Phalanger gymnotis*) and a number of bandicoots may exhibit some opportunistic carnivorous tendencies; several arboreal rodents probably prey on nestlings; and some terrestrial and aquatic hydomyin murids probably include small vertebrates in their diets. The carnivore niche, however, is mainly filled by raptors and reptiles. Among these the Kapul eagle (*Harpyopsis novaeguineae*) will take prey up to the size of a tree-kangaroo; the amethystine python (*Morelia amethistina*) anything up to the size of a wallaby; while large varanids such as Salvadori's dragon (*Varanus salvadorii*) may take a variety of smaller vertebrates.

The South-east Asian-derived flora also seems out of balance, in lacking large mammalian dispersal agents such as monkeys. An *Aglaia* species produces a huge seed (> 1 kg in weight) and is presumably limited in its rate of colonization. The cassowaries are important medium and small fruit dispersers.

Humans

The initial human colonization of a naive New Guinea / Meganesia is thought to have occurred somewhere between 40,000 and 60,000 years ago (Groube *et al.,* 1986; Flannery, 1994c; Thorne *et al.,* 1999). A further major phase of human arrivals was that of the Austronesian-speaking agricultural and maritime peoples who colonized coastal New Guinea in the latter half of the Holocene, prior to contact with eastern Indonesian kingdoms. Then finally the brief colonial phase, by Dutch, British and Germans, occurred in the 19th and 20th centuries (Bellwood, 1995; Swadling, 1996).

Superimposed on the natural zoogeography of New

Guinea are some ancient and recent human influences. The five principal means by which humans influence zoogeography are: (1) human-induced habitat modification; (2) direct or indirect human-induced decline / extinction of animal species (e.g. through predation, competition or disturbance); (3) unintentional or deliberate animal translocation; (4) translocation of pathogens; (5) co-evolution and domestication (Heinsohn, 2001a).

In the New Guinea region, long-term human-induced habitat modification has probably had a significant impact on zoogeography. The main influence has been a reduction in the area of primary forest due to the combined impacts of swidden agriculture and anthropogenic fire. The greatest impacts have been in heavily populated and intensively cultivated parts of the highlands, or in fire-prone parts of the lowlands where there is a marked dry season. Swidden agriculture has created large areas of secondary forest, while the combined effects of agricultural clearing and repeated burning has produced large areas of anthropogenic grassland. The loss of primary forests may lead to contractions in ranges of species adapted to such vegetation and a relative increase in species adapted to secondary forests, disturbed environments and successional seres (Dwyer, 1984). Large areas of anthropogenic grassland may act as an ecological barrier to some closed forest species; or alternatively, it may facilitate the dispersal of some ecotone or savanna-adapted species. Such habitats then invite colonization by synanthropic vertebrates that follow in the wake of human disturbance. Some examples of synanthropic birds from South-east Asia that have recently colonized anthropogenic grasslands in New Guinea, include the pied chat (*Saxicola caprata*), Richard's pipit (*Anthus novaeseelandiae*) and long-tailed shrike (*Lanius schach*), with the last two targeting montane anthropogenic grasslands (Beehler *et al.,* 1986; Coates, 1990). Infestation by introduced weeds may also modify habitats and bring about changes in vertebrate faunas.

A further major impact on vertebrate zoogeography is hunting by humans. This has caused significant range contractions for a number of species and in some cases, local or total extinction. The first major impacts probably occurred with the arrival of humans in the late Pleistocene, although the exact degree to which humans may have contributed to the extinctions of New Guinea's Pleistocene megafauna (diprotodontids and macropodines) is the subject of much debate. As in Australia, resolution of the controversy is hampered by dating and stratigraphic uncertainties regarding the timing of events and difficulties in separating the impacts of climate change from those due to humans (Murray, 1984; Mountain, 1993; Flannery, 1994c, 1995a).

In addition to suspected Pleistocene extinctions, there are also some prehistoric Holocene extinctions that are probably due to humans. The disappearance of the New Guinean thylacine, for example, is most probably due to introduction of the semi-domesticated dog in the later Holocene; feral dogs probably both competed with and preyed on the thylacine. Deliberate hunting with trained dogs may have also contributed to the disappearance of the thylacine and some species of pademelon (*Thylogale* spp.), with species in the subalpine zone appearing to be especially vulnerable (Flannery, 1992a, b, 1995a; Hope *et al.,* 1993).

A particularly illuminating faunal sequence regarding Holocene extinctions comes from Kria and Toe Cave sites in the Ayamaru Lakes region, an undulating area a few hundred metres in altitude, in central Bird's Head Peninsula, West Papua (Aplin *et al.,* 1999). Discoveries at the site include: a diminutive new species of striped possum (*Dactylopsila kambuayai*); and an unusual form of pseudocheirid possum with no close relatives in New Guinea, but with similarities to Australia's greater glider (*Petauroides volans*). The site also includes a range of species persisting elsewhere in New Guinea, but apparently no longer extant on the Bird's Head Peninsula. This latter group includes two lowland phalangerids, the southern common cuscus (*Phalanger intercastellanus*) and black-spotted cuscus (*Spilocuscus rufoniger*) as well as two macropodids, Goodfellow's tree-kangaroo (*Dendrolagus goodfellowi*) and the small Dorcopsis (*Dorcopsulus vanheurni*)—both of which are normally regarded as montane species. A third group includes species extant in montane areas of the Bird's Head, but no longer present at lower altitude. Examples include Stein's cuscus (*Phalanger vestitus*), D'Albertis's ringtail (*Pseudochirops albertisii*), reclusive ringtail (*Pseudochirops coronatus*), Arfak ringtail (*Pseudochirulus schlegeli*), long-fingered triok (*Dactylopsila palpator*), ?De Vis's woolly-rat (*Mallomys* sp. cf. *M. aroaensis*), and waterside rat (*Parahydromys asper*).

Most of the regionally or totally extinct species appear to have disappeared by mid-Holocene times. Aplin *et al.* (1999) consider that these extinctions are due to a combination of hunting/disturbance by humans and environmental change brought about by the climatic amelioration of the terminal Pleistocene and early Holocene during which montane habitats retreated to higher elevations. Of the species mentioned above, the disappearance of lowland forms may be due mostly to humans, whereas the retreat or extinction of montane species may be largely due to environmental change, but exacerbated by humans (Aplin *et al.,* 1999). The history of the mammals of the Bird's Head region provides an illuminating example of how humans may have played a role in simplifying one of New Guinea's regional mammal faunas (Aplin, 1998). This apparent anthropogenic loss of biodiversity, may have parallels with some of the avifaunal extinctions described by Steadman (1995) as well as Steadman *et al.* (1999) for various Polynesian and Melanesian islands. Despite a range of regional and total extinctions, New Guinea still retains a very rich and diverse vertebrate fauna. It is also one of the last land masses to retain large areas of primary tropical forests.

In addition to prehistoric extinctions there are also a number of species that have undergone more recent human-induced decline and range contraction. For many, the process started in prehistoric times but has continued or even accelerated in recent historical times. Examples include: mammals such as Bulmer's fruit-bat (*Aproteles bulmerae*), the long-beaked echidna (*Zaglossus bruijnii*), the large black-spotted cuscus (*Spilocuscus rufoniger*), most species of tree-kangaroo and pademelon; and birds such as crowned pigeons (*Goura* spp.), palm cockatoo (*Probosciger aterrimus*), Blyth's hornbill (*Rhyticeros plicatus*), several birds of paradise (for example *Macgregoria pulchra*), and all three of New Guinea's cassowaries (Beehler *et al.,* 1986; Coates, 1990; Flannery, 1995a). Causes for very recent decline include hunting with firearms, the breakdown of traditional hunting taboos, excessive logging, habitat loss and general impacts of a growing human population.

A further significant human influence on the zoogeography of New Guinea is animal translocation, with a number of wild and domesticated vertebrates having been introduced since mid-Holocene, with most arriving via the Indonesian archipelago. New Guinea's main introduced terrestrial vertebrates include: pig (*Sus scrofa vittatus* x *Sus celebensis*), dog (*Canis lupus dingo/Canis familiaris*), domestic cat (*Felis catus*), Rusa deer (*Cervus timorensis*), Pacific rat (*Rattus exulans*), Himalayan rat (*Rattus nitidus*), ricefield rat (*Rattus argentiventer*), Asian house rat (*Rattus tanezumi*), black rat (*Rattus rattus*), brown rat (*Rattus norvegicus*), house mouse (*Mus musculus*), domestic chicken (*Gallus gallus*), common myna (*Acridotheres tristis*), common starling (*Sturnus vulgaris*), rock pigeon (*Columba livia*), sooty-headed bulbul (*Pycnonotus aurigaster*), house sparrow (*Passer domesticus*), tree sparrow (*P. montanus*), flowerpot snake (*Ramphotyphlops braminus*), Asian house gecko (*Hemidactylus frenatus*), and the American cane toad (*Bufo marinus*) (Groves, 1981; Flannery, 1995a; Coates and Peckover, 2001; Heinsohn, 2001a). Some of these, such as the domesticated triumvirate of dogs, pigs and chickens as well as the commensal Pacific rat (*Rattus exulans*), were introduced prehistorically in the middle to late Holocene; while many of the others are more recent arrivals. The Javan Rusa deer (*Cervus timorensis*), for example, was introduced into New Guinea about a century ago by European colonists (Plate 33). Further Asian species such as the Malay civet (*Viverra tangalunga*) and common palm civet (*Paradoxurus hermaphroditus*) have been introduced to various Moluccan Islands and Aru and will invade New Guinea, should they be afforded the chance (Heinsohn, 2001a). Such exotic predators pose a threat to New Guinea's unique insular fauna. The toxic American cane toad (*Bufo marinus*), which was introduced in the 1930s to control agricultural insect pests, is also a threat because it tends to kill predators that ingest it (Plate 34). It may be responsible for the recorded decline of the Papuan black snake (*Pseudechis papuanus*) in some areas (Parker, 1982).

In addition to receiving introduced species, New Guinea has also had some of its own vertebrates translocated to surrounding islands, and exerted a significant influence on regional zoogeography. As revealed by vertebrate sequences excavated by archaeologists of the Lapita Homeland Project, this is a process that appears to have begun in the late Pleistocene (~ 20,000 years ago) with the translocation of the northern common cuscus (*Phalanger orientalis*) to the Bismarck Archipelago (Flannery and White, 1991; Spriggs, 1997). Indeed, most of the non-continental distribution of this species from Timor, Sanana and the Moluccas to the Bismarck Archipelago and Solomons, is probably due to long-term human agency; the same thing applies to the common spotted cuscus, *Spilocuscus maculatus* (Heinsohn, 2001a, 2002a, b, c). Some of these may represent accidental introductions through escaped captives or the deliberate introduction of mammals to islands lacking in game (Heinsohn, 1998). Similar motives and a trade in live animals may be behind the apparent introduction of a number of other vertebrates (macropods, bandicoots, cassowaries) to various New Guinean satellites (Flannery and White, 1991; Flannery, 1995b; Spriggs, 1997; Heinsohn, 1998, 2001a). Other small New Guinean vertebrates may have been accidentally introduced to surrounding islands as stowaways in vessels and cargo (Heinsohn, 2001a); this is probably how the New Guinean skink (*Carlia fusca*) made it to the Solomon Islands (McCoy, 1980).

DISCUSSION

This overview of New Guinea establishes it as a major component of the Australian 'ark', carrying a unique faunal experiment through changing climates and increasing interactions with South-east Asia. Although present day New Guinea preserves some 'Gondwanan' habitats these have actually appeared in the recent past and so have not acted as complete refuges for mid-Tertiary or older Australian taxa. Torres Strait has remained a significant barrier through time, filtering the interchange with the rest of Australia; while far western New Guinea continues to serve as a limited zone of interchange with the Indo-Malayan faunas. Despite a partial Torres Strait barrier, some of the main divergences between the New Guinean and Australian vertebrate faunas may be due to the habitat disjunction between a mostly mesic Australia and a much wetter rainforest-dominated New Guinea. From the late Tertiary through to the Quaternary, while Australia's biota radiated into newly created mesic to xeric niches in a drying continent, New Guinea's founding stock radiated into a variety of lowland and montane rainforest niches. These radiations, however, are relatively young.

Whatever the future holds, there is a significant challenge to analyse the vertebrate zoogeography of New Guinea more completely, due to the current lack of systematic reviews and comprehensive faunal surveys. There are probably still many species among the herpetofauna that have not yet been discovered or described, and even some birds and mammals. Many published distribution maps may merely reflect where surveys have occurred; there are many gaps in distributional records, particularly in West Papua. In addition to the recent discovery of new species of large mammals such as the Dingiso tree-kangaroo (*Dendrolagus mbaiso*) (Flannery *et al.*, 1996), it is only in the last few years that the Australian spectacled hare-wallaby (*Lagorchestes conspicillatus*) and false water-rat (*Xeromys myoides*), have been discovered in southern New Guinea (Hitchcock, 1997, 1998). Furthermore, there have been very few archaeological faunal sequences reported from West Papua, despite an abundance of potential sites at all altitudes.

As further detailed systematic studies are carried out we may find that many vertebrates currently clumped into a single genus or species complex will be split into multiple genera, species and subspecies; this may further illuminate late Tertiary and Quaternary zoogeographic processes. Species now thought to be recent invaders across the Torres Strait or Arafura Sea, may turn out to be genetically distinct remnants of former widespread trans-Torresian populations with vicariance rather than dispersal explaining contemporary distributions. Added to this there are probably many ancient and some recent human influences such as anthropogenic range reductions and extinctions that have gone unrecognized and skew our view of New Guinea's 'natural' zoogeography.

Finally, as the New Guinean fossil record is so poorly investigated and a whole new genus of macropods (*Lagorchestes conspicillatus*) can suddenly turn up in southern areas, then future discoveries may result in substantial revisions of the current zoogeography. As demonstrated by Aplin *et al.* (1999), in relation to subfossil archaeological sequences, new investigations will potentially lead to the discovery of whole new faunal elements, such as the greater glider-like *Petauroides ayamaruensis* from middle Holocene deposits on the Bird's Head.

ACKNOWLEDGEMENTS

The authors are grateful for guidance over the years from many sources, particularly Ken Aplin, Mike Archer, Walter Boles, John Calaby, Brian Coates, Liz Dennis, Hobart van Deusen, Jared Diamond, Peter Dwyer, Yance de Fretes, Graeme George, Lin Gressitt, Tim Flannery, Colin Groves, Robin Hide, Jeannette Hope, Roy Mackay, Jim Menzies, Mary-Jane Mountain, Henry Nix, Sue O'Connor, Juliette Pasveer, Ron Petocz, Richard Schodde, Matthew Spriggs, Pam Swadling and Pat Woolley. Other biologists, palaeontologists and archaeologists too numerous to name, are also thanked. Work in New Guinea has been supported by ARC grants, Commonwealth scholarships, the Australian National University, and the Australian Museum. In Papua New Guinea support has been given by the Department of Environment and Conservation, the Institute of PNG Studies, provincial governments, and other institutions. Various Indonesian government departments and research institutions are thanked for their support of work in West Papua and adjacent areas.

Like other visitors, we owe our greatest debt to the farmers and hunters of New Guinea, whose knowledge of their local faunas so exceeds our own. We hope that this overview will encourage a new generation of home grown tertiary students and nature enthusiasts in Papua New Guinea and West Papua.

REFERENCES

Allison, A. 1982. Distribution and ecology of New Guinea lizards. Pp.803–813. In Gressitt, J.L. (ed). *Biogeography and Ecology of New Guinea*. W. Junk, The Hague. 983pp.

Allison, A. 1996. Biogeography of amphibians and reptiles of New Guinea and the Pacific region. Pp.407–436. In Keast, A. and Miller, S.E. (eds). *The Origin and Evolution of Pacific Island Biotas, New Guinea to Eastern Polynesia: Patterns and Processes*. SPB Academic Publishing, Amsterdam. 531pp.

Aplin, K.P. 1998. Vertebrate zoogeography of the Bird's Head of Irian Jaya, Indonesia. Pp.803–890. In Miedema, J., Ode, C. and Dam, R.A.C., (eds). *Perspectives on the Bird's Head of Irian Jaya, Indonesia*. Proceedings of the Conference, International Institute for Asian Studies, Leiden, 13–17 October 1997, Rodopi, Amsterdam.

Aplin, K.P. 2006. Ten million years of rodent evolution in Australasia: phylogenetic evidence and a speculative historical biogeography. Pp.707–744. In Merrick, J.R., Archer, M., Hickey, G.M. and Lee, M.S.Y. (eds). *Evolution and Biogeography of Australasian Vertebrates*. Auscipub Pty. Ltd., Sydney. 942pp.

Aplin, K.P., Baverstock, P.R. and Donnellan, S.C. 1993. Albumin immunological evidence for the time and mode of origin of the New Guinea terrestrial mammal fauna. *Science in New Guinea,* Vol. 19, No. 3, pp.131–145.

Aplin, K.P., Pasveer, J.M. and Boles, W.E. 1999. Late Quaternary vertebrates from the Bird's Head Peninsula, Irian Jaya, Indonesia, including descriptions of two previously unknown marsupial species. *Records of the Western Australian Museum, Supplement*, No. 57, pp.351–387.

Archer, M., Arena, R., Bassarova, M., Black, K., Brammall, J., Cooke, B., Creaser, P., Crosby, K., Gillespie, A., Godthelp, H., Gott, M., Hand, S.J., Kear, B., Krikmann, A., Mackness, B., Muirhead, J., Musser, A., Myers, T., Pledge, N., Wang, Y. and Wroe, S. 1999. The evolutionary history and diversity of Australian mammals. *Australian Mammalogy,* Vol. 21, No. 1, pp.1–45.

Archer, M., Hand, S. and Godthelp, H. 1991. *Australia's Lost World: Riversleigh, World Heritage Site*. Reed New Holland, Sydney. 264pp.

Beehler, B., Pratt T. and Zimmerman, D. 1986. *Birds of New Guinea*. Princeton University Press, Princeton. 293pp.

Bellwood, P. 1995. Austronesian prehistory in Southeast Asia: homeland, expansion and transformation. Pp.96–111. In Bellwood, P., Fox, J.J. and Tryon, D. (eds). *The Austronesians: Historical and Comparative Perspectives*. Anthropology Department, Research School of Pacific and

Asian Studies, Australian National University, Canberra. 359pp.

Betz, W.W. and Gruys, R.C. 1995. Range extension of the Snow Mountain quail (*Anurophasis monorthonyx*) to the Oranje Mountains, Irian Jaya. *Science in New Guinea,* Vol. 21, No. 2, pp.69–72.

Cann, J. 1998. *Australian Freshwater Turtles.* Beaumont Publishing, Singapore. 292pp.

Christidis, L. and Schodde, R. 1991. Relationships of Australo-Papuan songbirds—protein evidence. *Ibis,* Vol. 133, pp.277–285.

Christidis, L., Schodde, R. and Robinson, N.A. 1993. Affinities of the aberrant Australo-Papuan honeyeaters, *Toxorhamphus*, *Oedistoma*, *Timeliopsis* and *Epthianura*: protein evidence. *Australian Journal of Zoology,* Vol. 41, pp.423–432.

Coates, B.J. 1990. *The Birds of Papua New Guinea: Including the Bismarck Archipelago and Bougainville.* 2 Volumes. Dove Publications, Brisbane. 1040pp.

Coates, B.J. and Bishop, K.A. 1997. *A Guide to the Birds of Wallacea.* Dove Publications, Brisbane. 535pp.

Coates, B.J. and Peckover, W.S. 2001. *Birds of New Guinea and the Bismarck Archipelago: A Photographic Guide.* Dove Publications, Brisbane. 272pp.

Cogger, H.G. 2000. *Reptiles and Amphibians of Australia.* Sixth Edition. Reed, Sydney. 596pp.

Corbet, G.B. and Hill, J.E. 1992. *The Mammals of the Indomalayan Region: A Systematic Review.* Oxford University Press, Oxford. 488pp.

Corbett, L. 2006. The Australian dingo. Pp.745–753. In Merrick, J.R., Archer, M., Hickey, G.M. and Lee, M.S.Y. (eds). *Evolution and Biogeography of Australasian Vertebrates.* Auscipub Pty. Ltd., Sydney. 942pp.

Diamond, J.M. 1985. New distributional records and taxa from the outlying mountain ranges of New Guinea. *Emu,* Vol. 85, pp.65–91.

Dow, D.B. 1977. A geological synthesis of Papua New Guinea. *Bureau of Mineral Resources, Geology and Geophysics Bulletin*, No. 201.

Dwyer, P.D. 1984. From garden to forest: small rodents and plant succession in Papua New Guinea. *Australian Mammalogy,* Vol. 7, No. 1–2, pp.29–36.

Flannery, T.F. 1992a. Taxonomic revision of the *Thylogale brunii* complex (Macropodidae: Marsupialia) in Melanesia, with description of a new species. *Australian Mammalogy*, Vol. 15, pp.7–23.

Flannery, T.F. 1992b. New Pleistocene marsupials (Macropodidae, Diprotodontidae) from subalpine habitats in Irian Jaya, Indonesia. *Alcheringa,* Vol. 16, pp.321–331.

Flannery, T.F. 1994a. *Possums of the World: a Monograph of the Phalangeroidea.* Australian Museum and Geo Productions, Sydney. 240pp.

Flannery, T.F. 1994b. The fossil land mammal record of New Guinea: a review. *Science in New Guinea,* Vol. 20, No. 1, pp.39–48.

Flannery, T.F. 1994c. *The Future Eaters: an Ecological History of the Australian Lands and People.* Reed Books, Sydney. 421pp.

Flannery, T. 1995a. *Mammals of New Guinea.* Australian Museum and Reed Books, Sydney. 568pp.

Flannery, T. 1995b. *Mammals of the South-West Pacific and Moluccan Islands.* Australian Museum and Reed Books, Sydney. 464pp.

Flannery, T.F. 1996. Mammalian zoogeography of New Guinea and the Southwestern Pacific. Pp.399–406. In Keast, A. and Miller, S.E. (eds). *The Origin and Evolution of Pacific Island Biotas, New Guinea to Eastern Polynesia: Patterns and Processes.* SPB Academic Publishing, Amsterdam. 531pp.

Flannery, T.F., Martin, R. and Szalay, A. 1996. *Tree Kangaroos: a Curious Natural History.* Reed, Sydney. 202pp.

Flannery, T.F. and White, J.P. 1991. Animal translocation: zoogeography of New Ireland mammals. *National Geographic Research and Exploration,* Vol. 7, No. 1, pp.96–113.

George, G.G. 1987. Characterisation of the living species of cuscus (Marsupialia: Phalangeridae). Pp.507–526. In Archer, M. (ed). *Possums and Opossums: Studies in Evolution.* Surrey Beatty and Sons, Sydney. 788pp.

Georges, A. and Thomson, S. 2006. Evolution and zoogeography of Australian freshwater turtles. Pp.291–308. In Merrick, J.R., Archer, M., Hickey, G.M. and Lee, M.S.Y. (eds). *Evolution and Biogeography of Australasian Vertebrates.* Auscipub Pty. Ltd., Sydney. 942pp.

Gibbons, J.R.H. 1985. The biogeography and evolution of Pacific island reptiles and amphibians. Pp.125–142. In Grigg, G., Shine, R. and Ehmann, H. (eds). *Biology of Australasian Frogs and Reptiles.* Surrey Beatty and Sons, Sydney. 527pp.

Gressitt, J.L. (ed). 1982. *Biogeography and Ecology of New Guinea.* W. Junk, The Hague. 983pp.

Groube, L.J., Chappell, J., Muke, J. and Price, D. 1986. A 40,000 year-old occupation site at Huon Peninsula, Papua New Guinea. *Nature,* Vol. 324, pp.453–455.

Groves, C.P. 1981. *Ancestors for the pigs: taxonomy and phylogeny of the genus* Sus. Department of Prehistory, Research School of Pacific Studies, Australian National University, Canberra. 96pp.

Groves, C.P. and Flannery, T.F. 1994. A revision of the genus *Uromys* Peters 1867 (Muridae: Mammalia) with descriptions of two new species. *Records of the Australian Museum,* Vol. 46, pp.145–170.

Heads, M. 2001. Birds of Paradise, biogeography and ecology in New Guinea: a review. *Journal of Biogeography*, Vol. 28, pp.893–925.

Heinsohn, T. 1998. Captive ecology. *Nature Australia,* Vol. 26, No. 2, pp.36–43.

Heinsohn, T. 2000. Predation by the white-breasted sea eagle *Haliaeetus leucogaster* on phaleangerid possums in New Ireland, Papua New Guinea. *Emu,* Vol. 100, pp.245–246.

Heinsohn, T. 2001a. Human influences on vertebrate zoogeography: animal translocation and biological invasions across and to the east of Wallace's Line. Pp.153–170. In Metcalfe, I., Smith, J.M.B., Morwood, M. and Davidson, I. (eds). *Faunal and Floral Migrations and Evolution in SE Asia–Australasia.* A.A. Balkema, Lisse. 416pp.

Heinsohn, T. 2001b. Ecological plasticity of the common spotted cuscus (*Spilocuscus maculatus*) and northern common cuscus (*Phalanger orientalis*) in the Australasian Archipelago. *Paper prepared for the 47th Scientific Meeting of the Australian Mammal Society,* (*Possum and Glider Symposium*), *Brisbane, 2–6 July 2001.* 16pp.

Heinsohn, T. 2002a. Status of the common spotted cuscus

Spilocuscus maculatus and other wild mammals on Selayar Island, Indonesia, with notes on Quaternary faunal turnover. *Australian Mammalogy*, Vol. 24, pp.199–207.

Heinsohn, T. 2002b. Observations of probable camouflaging behaviour in a semi-commensal common spotted cuscus *Spilocuscus maculatus maculatus* (Marsupialia: Phalangeridae) in New Ireland, Papua New Guinea.. *Australian Mammalogy*, Vol. 24, pp.243–245.

Heinsohn, T. 2002c. Possum extinctions at the marsupial frontier: the status of the northern common cuscus *Phalanger orientalis* on Santa Ana Island, Makira Province, Solomon Islands. *Australian Mammalogy*, Vol. 24, pp.247–248.

Hitchcock, G. 1997. First record of a spectacled hare-wallaby, *Lagorchestes conspicillatus* (Marsupialia: Macropodidae), in New Guinea. *Science in New Guinea,* Vol. 23, No. 1, pp.47–51.

Hitchcock, G. 1998. First record of the false water-rat *Xeromys myoides* (Rodentia: Muridae), in New Guinea. *Science in New Guinea,* Vol. 23, No. 3, pp.141–144.

Honacki, J.H., Kinman, K.E. and Keoppl, J.W. 1982. *Mammal Species of the World: A Taxonomic and Geographic Reference.* Association of Systematics Collections and Allen Press, Kansas. 694pp.

Hope, G.S. 1996a. History of *Nothofagus* in New Guinea and New Caledonia. Pp.257–270. In Veblen, T.T., Hill, R.S. and Read, J. (eds). *The Ecology and Biogeography of Nothofagus Forests.* Yale University Press, New Haven. 403pp.

Hope, G.S. 1996b. Quaternary change and historical biogeography of Pacific Islands. Pp.165–190. In Keast, A. and Miller, S.E. (eds). *The Origin and Evolution of Pacific Island Biotas, New Guinea to Eastern Polynesia: Patterns and Processes.* SPB Academic Publishing, Amsterdam. 531pp.

Hope, G.S., Flannery, T.F. and Boeardi. 1993. A preliminary report of changing Quaternary mammal faunas in subalpine New Guinea. *Quaternary Research*, Vol. 40, pp.117–126.

Kaars, van der, S. 1995. Preliminary palynological results on the Pleistocene-Holocene transition, Seram Trench, offshore Irian Jaya, Indonesia. *Geologie en Mijnbouw,* Vol. 74, pp.285–286.

Kaars, van der, S., Wang, X., Kershaw, P., Guichard, F. and Setiabudi, D.A. 2000. A late Quaternary palaeoecological record from the Banda Sea, Indonesia: patterns of vegetation, climate and biomass burning in Indonesia and northern Australia. *Palaeogeography, Palaeoclimatology, Palaeoecology,* Vol. 155, pp.135–153.

Keast, A. 1996. Avian geography: New Guinea to the eastern Pacific. Pp.373–398. In Keast, A. and Miller, S.E. (eds). *The Origin and Evolution of Pacific Island Biotas, New Guinea to Eastern Polynesia: Patterns and Processes.* SPB Academic Publishing, Amsterdam. 531pp.

Koopman, K.F. 1979. Zoogeography of mammals from Islands off the northeastern coast of New Guinea. *American Museum Novitates,* No. 2690, pp.1–17.

Laurie, E.M.O. and Hill, J.E. 1954. *List of Land Mammals of New Guinea, Celebes and Adjacent Islands 1758–1952.* British Museum, London. 175pp.

McCoy, M. 1980. *Reptiles of the Solomon Islands.* Wau Ecology Institute, Wau. 80pp.

Menzies, J.I. 1991. *A Handbook of New Guinea Marsupials and Monotremes.* Kristen Press, Madang. 138pp.

Menzies, J.I. and Ballard, C. 1994. Some new records of Pleistocene megafauna from New Guinea. *Science in New Guinea,* Vol. 20, pp.113–139.

Menzies, J.I. and Dennis, E. 1979. *Handbook of New Guinea Rodents.* Wau Ecology Institute Handbook No. 6, Wau Ecology Institute, Wau. 68pp.

Monk, K.A., de Fretes, Y. and Reksodiharjo-Lilley, G. 1997. *The Ecology of Nusa Tenggara and Maluku.* Periplus Editions, Hong Kong. 966pp.

Mountain, M. 1993. Bones, hunting and predation in the Pleistocene of Northern Sahul. Pp.123–130. In Smith, M.A., Spriggs, M. and Frankhauser, B. (eds). *Sahul in Review: Pleistocene Archaeology in Australia, New Guinea and Island Melanesia.* Prehistory Department, Research School of Pacific Studies, Australian National University, Canberra. 312pp.

Murray, P. 1984. Extinctions downunder: a bestiary of extinct Australian Late Pleistocene monotremes and marsupials. Pp.600–628. In Martin, P.S. and Klein, R.G. (eds). *Quaternary Extinctions: A Prehistoric Revolution.* University of Arizona Press, Tucson. 892pp.

Musser, A.M. 1999. Diversity and relationships of living and extinct monotremes. *Australian Mammalogy,* Vol. 21, No. 1, pp.8–9.

Nix, H.A. and Kalma, J.D. 1972. Climate as a dominant control in the biogeography of northern Australia and New Guinea. Pp.61–91. In Walker, D. (ed). *Bridge and Barrier: the Natural and Cultural Heritage of the Torres Strait.* Department of Biogeography and Geomorphology, Research School of Pacific Studies, Australian National University, Canberra. 437pp.

Nowak, R.M. 1991. *Walker's Mammals of the World.* Fifth Edition. 2 volumes. Johns Hopkins University Press, Baltimore. 1629pp.

O'Connor, S., Aplin, K.P., Spriggs, M., Veth, P. and Ayliffe, L.K. 2002. From savannah to rainforest: changing environments and human occupation at Liang Lembudu, the Aru Islands, Maluku, Indonesia. Pp.279–306. In Kershaw, P., Trapper, N.J., David, B., Bishop, P.M. and Penny, D. (eds). *Bridging Wallace's Line: Environmental and Cultural History and Dynamics of the Southeast Asian-Australian Region.* Advances in Geology Series, Catena Verlag.

O'Shea, M. 1996. *A Guide to the Snakes of Papua New Guinea* Independent Publishing, Port Moresby. 239pp.

Parker, F. 1982. The snakes of Western Province. *Wildlife in Papua New Guinea,* Vol. 82, No. 1, pp.1–78.

Pigram, C.J. and Davies, H.L. 1987. Terranes and the accretion history of the New Guinea orogen. *Bureau of Mineral Resources, Journal of Australian Geology and Geophysics,* Vol. 10, pp.193–211.

Pizzey, G. and Knight, F. 1997. *Field Guide to the Birds of Australia.* Angus and Robertson, Sydney. 460pp.

Rich, P.V. and Thompson, E.M. 1982. *The Fossil Vertebrate Record of Australasia.* Monash University Press, Clayton, Victoria. 759pp.

Rich, P.V. and van Tets, G.F. 1982. Fossil birds of Australia and New Guinea: their biogeographic, phylogenetic, and biostratigraphic input. Pp.235–384. In Rich, P.V. and Thompson, E.M. (eds). *The Fossil Vertebrate Record of Australasia.* Monash University Press, Clayton, Victoria. 759pp.

Scanlon, J.D. 2006. Origins and radiations of snakes in Australasia. Pp.309–330. In Merrick, J.R., Archer, M., Hickey, G.M. and Lee, M.S.Y. (eds). *Evolution and Biogeography of Australasian Vertebrates.* Auscipub Pty. Ltd., Sydney. 942pp.

Schodde, R. 1994. Australian faunal divisions. Pp.30–31. In Bambrick, S. (ed). *The Cambridge Encyclopedia of Australia.* Cambridge University Press, Cambridge. 384pp.

Schodde, R. 2006. Australasia's bird fauna today--origins and evolutionary development. Pp.413–458. In Merrick, J.R., Archer, M., Hickey, G.M. and Lee, M.S.Y. (eds). *Evolution and Biogeography of Australasian Vertebrates.* Auscipub Pty. Ltd., Sydney. 942pp.

Schodde, R. and Calaby, J.H. 1972. The biogeography of the Australo-Papuan bird and mammal faunas in relation to Torres Strait. Pp.257–300. In Walker, D. (ed). *Bridge and Barrier: the Natural and Cultural Heritage of the Torres Strait.* Department of Biogeography and Geomorphology, Research School of Pacific Studies, Australian National University, Canberra. 437pp.

Sibley, C.G. and Ahlquist, J.E. 1985. The phylogeny and classification of the Australo-Papuan passerine birds. *Emu,* Vol. 85, No. 1, pp.1–14.

Sibley, C.G. and Ahlquist, J.E. 1990. *Phylogeny and Classification of Birds, a Study in Molecular Evolution.* Yale University Press, New Haven. 976pp.

Sibley, C.G. and Monroe, B.L.Jnr. 1990. *Distribution and Taxonomy of Birds of the World.* Yale University Press, New Haven. 1111pp.

Singadan, R.K. 1996. Notes on hybrid spotted cuscus, *Spilocuscus maculatus* X *Spilocuscus kraemeri* (Marsupialia: Phalangeridae). *Science in New Guinea,* Vol. 22, No. 2, pp.77–82.

Smith, J.D. and Hood, G.S. 1981. Preliminary notes on bats from the Bismarck Archipelago (Mammalia: Chiroptera). *Science in New Guinea,* Vol. 8, No. 2, pp.81–121.

Spriggs, M. 1997. *The Island Melanesians.* Blackwell, Oxford. 362pp.

Stafford, P. 2000. *Snakes.* The Natural History Museum, London. 112pp.

Steadman, D.W. 1995. Prehistoric extinctions of Pacific island birds: biodiversity meets zooarchaeology. *Science,* Vol. 267, pp.1123–1131.

Steadman, D.W., White, J.P. and Allen, J. 1999. Prehistoric birds from New Ireland, Papua New Guinea: extinctions on a large melanesian island. *Proceedings of the National Academy of Sciences of the United States of America,* Vol. 96, pp.2563–2568.

Strahan, R. (ed). 1995. *The Mammals of Australia.* Reed, Sydney. 756pp.

Swadling, P. 1996. *Plumes from Paradise.* Robert Brown and Associates, Brisbane. 352pp.

Thorne, A., Grun, R., Mortimer, G., Spooner, N.A., Simpson, J.J., McCulloch, M., Taylor, L. and Curnoe, D. 1999. Australia's oldest human remains: age of the Lake Mungo 3 skeleton. *Journal of Human Evolution,* Vol. 36, pp.591–612.

Torgersen, T., Luly, J., de Deckker, P., Jones, M.R., Searle, D.E., Chivas, A.R. and Ullman, W.J. 1988. Late Quaternary environments of the Carpentaria Basin, Australia. *Palaeogeography, Palaeoclimatology, Palaeoecology,* Vol. 67, pp.245–248.

Torr, G. 2000. *Pythons of Australia: a Natural History.* University of New South Wales Press, Sydney. 103pp.

Tyler, M.J. and Lee, M.S.Y. 2006. The origins of Australian frogs. Pp.237–240. In Merrick, J.R., Archer, M., Hickey, G.M. and Lee, M.S.Y. (eds). *Evolution and Biogeography of Australasian Vertebrates.* Auscipub Pty. Ltd., Sydney. 942pp.

van Oosterzee, P. 1997. *Where World's Collide: the Wallace Line.* Reed, Melbourne. 234pp.

van Oosterzee, P. 2006. Drawing the Wallace Line. Pp.95–110. In Merrick, J.R., Archer, M., Hickey, G.M. and Lee, M.S.Y. (eds). *Evolution and Biogeography of Australasian Vertebrates.* Auscipub Pty. Ltd., Sydney. 942pp.

Vincent, M. and Wilson, S. 1999. *Australian Goannas.* New Holland, Sydney. 152pp.

Waithman, J. 1979. A report on a collection of mammals from Southwest Papua, 1972–1973. *Australian Zoologist,* Vol. 20, pp.313–326.

Wallace, A.R. 1869. *The Malay Archipelago, the Land of the Orang-utan and the Bird of Paradise: a Narrative of Travel with Studies of Man and Nature.* Macmillan, London. 638pp.

Webb, G. and Manolis, C. 1989. *Australian Crocodiles: A Natural History.* Reed, Sydney. 160pp.

Welch, K.R.G. 1988. *Snakes of the Orient: a Checklist.* Krieger, Malabar. 183pp.

Welch, K.R.G., Cooke, P.S. and Wright, A.S. 1990. *Lizards of the Orient: a Checklist.* Krieger, Malabar. 162pp.

Whitaker, R and Whitaker, Z. 1982. Reptiles of Papua New Guinea. *Wildlife in Papua New Guinea,* Vol. 82, No. 2, pp.1–53.

Willis, P.M.A. 2006. Evolution and zoogeography of Australasian crocodilians. Pp.331–348. In Merrick, J.R., Archer, M., Hickey, G.M. and Lee, M.S.Y. (eds). *Evolution and Biogeography of Australasian Vertebrates.* Auscipub Pty. Ltd., Sydney. 942pp.

Wroe, S. and Archer, M. 2006. Origins and early radiations of marsupials. Pp.551–574. In Merrick, J.R., Archer, M., Hickey, G.M. and Lee, M.S.Y. (eds). *Evolution and Biogeography of Australasian Vertebrates.* Auscipub Pty. Ltd., Sydney. 942pp.

Zweifel, R.G. and Tyler, M.J. 1982. Amphibia of New Guinea. Pp.759–801. In Gressitt, J.L. (ed). *Biogeography and Ecology of New Guinea.* W. Junk, The Hague. 983pp.

Chapter 5
Drawing the Wallace Line

Penny van Oosterzee
EcOz, GPO Box 381, Darwin, N.T. 0801 Australia

'I think geologists are more converted to the concept of evolution than simple naturalists because they are more accustomed to reasoning'.
Charles Darwin to Alfred Russel Wallace, May 1860

INTRODUCTION

Biogeography is a science rooted in evolution, describing and explaining the distribution of individual species and of observed patterns of species diversity. Biogeography is dependent on an acceptance of evolution and, in turn, sheds light on the processes of evolution. The major objectives of this chapter are: first, to trace the discovery of the Wallace Line and outline regional geologic and climatic events that contributed to the formation of this faunal barrier; second, to discuss findings of select faunal studies in terms of biogeographic principles to explain current distribution patterns.

MYTH AS FACT

Unfortunately, Alfred Russell Wallace, the founder of biogeography, was born into an era where virtually all scientists accepted the traditional Christian concept of a Creator God. The Bible was regarded as a literal tome, a real account of the origins of the natural world. What was worse, Wallace was born (in 1823) into a family of no distinction, at a time when individuals were locked into their family's social status. Young Wallace went to school till the age of 14, when he was forced to leave because of his father's bankruptcy. Ironically for the future founder of biogeography and expert at Latinized designations, his two most painful subjects were Latin and Geography. Nevertheless he was an avid reader who devoured books on a multitude of topics and enjoyed way-out ideas such as socialism and evolution.

Mid-19th century naturalists generally scoffed at the concept of evolution. The idea of one species changing into another one was too crazy for words. With no evidence of how such a bizarre thing might happen and an historical belief in an ordered world, regulated by a 'Superior Wisdom', they resorted to creation—which, of course, required no evidence. When a species became extinct it was simply replaced by a new species by a very hands-on God. There was no essential variation between individuals of a species and hence species did not evolve. Natural history, itself, was the analysis of the thoughts of the Creator, as manifested in the animal and vegetable kingdom. An impious innovator with a free-ranging intellect, Wallace shunned these absurd creationist views; his one great advantage was that he knew what he was talking about.

THE SARAWAK LAW

Unlike most other natural philosophers of the time who spent their time theorizing in the comfort of their homes and social clubs, Wallace was out getting work experience. In the steamy sunlit jungles of the Amazon and Indonesia he found that instead of a static and neatly ordered world where animals formed distinct and clear species, always fixed in form, there were apparently chaotic outbursts of variation within species. He noticed endless modifications of structure, shape and colour, particularly in insects and innumerable adaptations to diverse environments. Clearly there was another principle regulating the infinite variety of life. Like his contemporary, Charles Darwin (1809–1882), Wallace was forever turning over the species question.

Unlike Darwin, who had been paralysed with concern for his reputation if his ideas on evolution were published, Wallace was not daunted by the possibility of ridicule by his peers; equally, he himself was ready to ridicule the often ludicrous metaphysical ideas of his time. As a result, in 1856, he reacted strongly to a paper by a colleague that suggested that life on Earth was obliterated and replaced by God from time to time through geological eras. This paper said that any similarities between eras was due to God's particular penchant for one design over another—beetles for instance. Annoyed with these absurd ideas when a simple, secular hypothesis could explain all the facts, Wallace sat down in Sarawak on the island of Borneo to write a paper titled 'On the Law Which has Regulated the Introduction of New Species' (Wallace, 1857); a law which came to be known as the Sarawak Law.

The essence of the Sarawak Law is that every species has come into existence in about the same place and time as its nearest relative. This breath-takingly simple observation was as close to an explanation of evolution as anyone had yet dared. Wallace pointed out that the distribution of animals and plants in space echoes their distribution in time. Higher taxa—classes and orders—are generally spread over the whole globe. Primates, for instance, are found everywhere except Australia and Antarctica. This is in effect an echo of their past distributions. Genera and species, more often than not, are specific to one region, often a very small area. In particular, closely allied species will be found geographically near to each other. For example, lemurs are found only in Madagascar and orangutans only in Borneo and Sumatra.

The Sarawak Law suggested the when and where of evolution: that evolving, from geographically nearby and closely related species, was occurring constantly. But the greatest mystery of the species question, the how, was still not answered.

One day in 1858, while Wallace was thrashing about in a malarial fever in Ternate (Maluku), he hit upon a mechanism for evolution. He was thinking about an essay written in 1798 by Thomas Malthus about the natural checks that kept populations (of humans) in control. Natural selection! As with humans, so it must be for the natural world. What had taken Darwin virtually all his adult life, Wallace had done in a sudden flash of insight. He thought it out during a feverish bout and when that subsided sat up quickly to write down a sketch of its various applications. He copied this on thin letter paper and titled it 'On The Tendency Of Varieties to Depart Indefinitely from The Original Type'. Wallace sent the manuscript to Charles Darwin, the only other person he knew who was interested in the species question, asking Darwin to publish it on his behalf. The letter jolted Darwin out of his 20-year inertia and made him forget the heretical implications of 'his' theory of evolution. For fear of being trumped by Wallace, the 'Theory of Evolution' by Charles Darwin and Alfred Russell Wallace was announced in the Linnean Society meeting in London of July 1858.

THE WALLACE LINE

The Wallace Line was drawn within the context of the Sarawak Law and the Theory of Evolution by natural selection, but Wallace realized that evolution alone was not a sufficient explanation for the diversity of life on Earth. While evolution transforms existing species there must also be a mechanism for creating new species. If a single species is divided in two, then evolution will proceed separately in each part and the two populations will eventually become so different as to be recognized as different species. To divide a species, there needs to be a barrier that will prevent the movement of individuals between the two populations. Populations on either side of that barrier will gradually change.

When Wallace travelled across the narrow, 25 kilometre strait of water from Bali to Lombok he expected to see more-or-less the same Asian fauna as he found in Bali. But when the giant swells of the Indian Ocean, squeezing through the deep strait, dumped him onto the steep beach of Ampanam Bay, Lombok, the first sound he heard was the loud screeching of the helmeted friarbird *Philemon buceriodes*, an Australian bird.

Where were the Oriental barbets, fruit-thrushes and woodpeckers, the same birds he had seen in Malaysia, Borneo and Bali, the latter still clearly visible across the strait? Instead the forests of Lombok echoed with the loud strangled screams of cockatoos, and honey-eaters flitted through the trees. For the first time he also met with the strange mound-making bird *Megapodius reinwardt*, an Australian megapode (literally 'big foot') that industriously uses its big feet to rake together 'cart-loads of rubbish' (dead leaves, sticks, stones, earth, rotten wood) till it forms a large mound perhaps two metres high and four metres wide. In the middle of this it lays its brick-red eggs, the rotting material of the mound producing heat, like an oven, to incubate the eggs.

Neither physical nor organic conditions could explain to Wallace the absence of prominent Asian birds and the presence of raucous Australian ones. In the midst of the Orient he was unprepared for echoes of Australia. This is because the difference between the fauna of Australia and Asia is greater than the difference between any other of the broad regions of the world. Australia naturally has no monkeys, cats, bears, or hyenas; or many of the other familiar types of mammals which exist in every other part of the world. Instead of these, it has marsupials (kangaroos and possums, wombats) and the duck-billed platypus. In birds it is almost as peculiar. There are no woodpeckers or pheasants, which exist elsewhere, but instead there are mound-making brush-turkeys, honey-eaters, cockatoos and the brush-tongued lories.

Robust barriers, such as wide oceans and high mountain ranges, usually separate major provinces—for example, the Himalayas separate the Palaeartic and the Oriental province—not a channel 25 kilometres wide. From an evolutionary point of view, a naturalist would not expect a string of close, physically similar islands to be populated with strikingly different animals on its western and eastern ends, let alone across a narrow channel.

Wallace found the differences between Sulawesi and Borneo to be even more striking. In Borneo, across the Makassar Straits, the forests abounded with monkeys of many kinds, wild-cats, civets, otters and squirrels; as well as oriental birds like woodpeckers, barbets, trogons, fruit-thrushes and leaf-thrushes. In Sulawesi he found many prehensile-tailed marsupial cuscuses as well as honeyeaters and parrots. But Sulawesi had even more surprises to spring. Its central position in the Indonesian archipelago, connected by small satellite islets and coral reefs to the Philippines to the north, Kalimantan to the west, the Moluccas to the east and the Lesser Sundas to the south, led him to expect the richness and variety of the whole archipelago. Instead the facts turned out to be just the reverse. Sulawesi 'was at once the poorest in the number of its species and the most isolated in the character of its productions of all the great islands of Indonesia'. With a land area nearly double that of Java, it had but half Java's numbers of mammals and land birds. Its 'productions', hemmed in on every side by islands, were either distantly related or were totally unrelated to those islands.

What Wallace saw at 'The Wallace Line' was not geographically close populations split in two by a narrow water barrier but two different faunal universes—see Figure 1. Later, other zoologists confirmed Wallace's observations—several examples are listed below.

Bernhard Rensch frequently travelled across the strait between Lombok and Bali in the 1920s. On first visiting Bali from Lombok he said 'What about the animal life? Is it really different from that of Lombok as has been claimed by so many other travellers? A strait which even the smallest bird could cross without difficulties… And the difference is indeed quite extraordinary! Much more conspicuous than I would have ever imagined. As soon as I entered the woods on a small native trail a whole chorus of strange bird songs greet me—in fact, among the real songsters there is not a single one with which I was familiar' (cited in Lincoln, 1975).

Ornithologist G.A. Lincoln (1975) studied birds in Java, Bali, Lombok and Sumbawa, on either side of the Wallace Line. He found that on Java and Bali the birds were nearly all Oriental and dominated by the oriental bulbul. On crossing the Wallace Line from Bali to Lombok the birds were predominantly Australian. The honeyeaters, in particular, attracted attention with their noisy calls and squabbling behaviour. Java and Bali had the most similar bird faunas. The least similar adjacent bird faunas were those between Bali and Lombok, astride the Wallace Line.

Zoologist R.C. Raven sailed back and forth across the Makassar Strait from Kalimantan to Sulawesi, from 1912 to 1923, collecting natural history material for American museums. Raven (1935) found that while a number of mammalian groups had indeed succeeded in passing the Wallace Line from the west (he was not concerned with movement from the east), a much greater number had been stopped by it. Although his data, based on 1935 systematics, are now out-of-date the general trends observed remain true. For instance, Raven found that: of 59 species of shrew only 13 crossed; only two of 21 deer species in the region had crossed; of 56 species

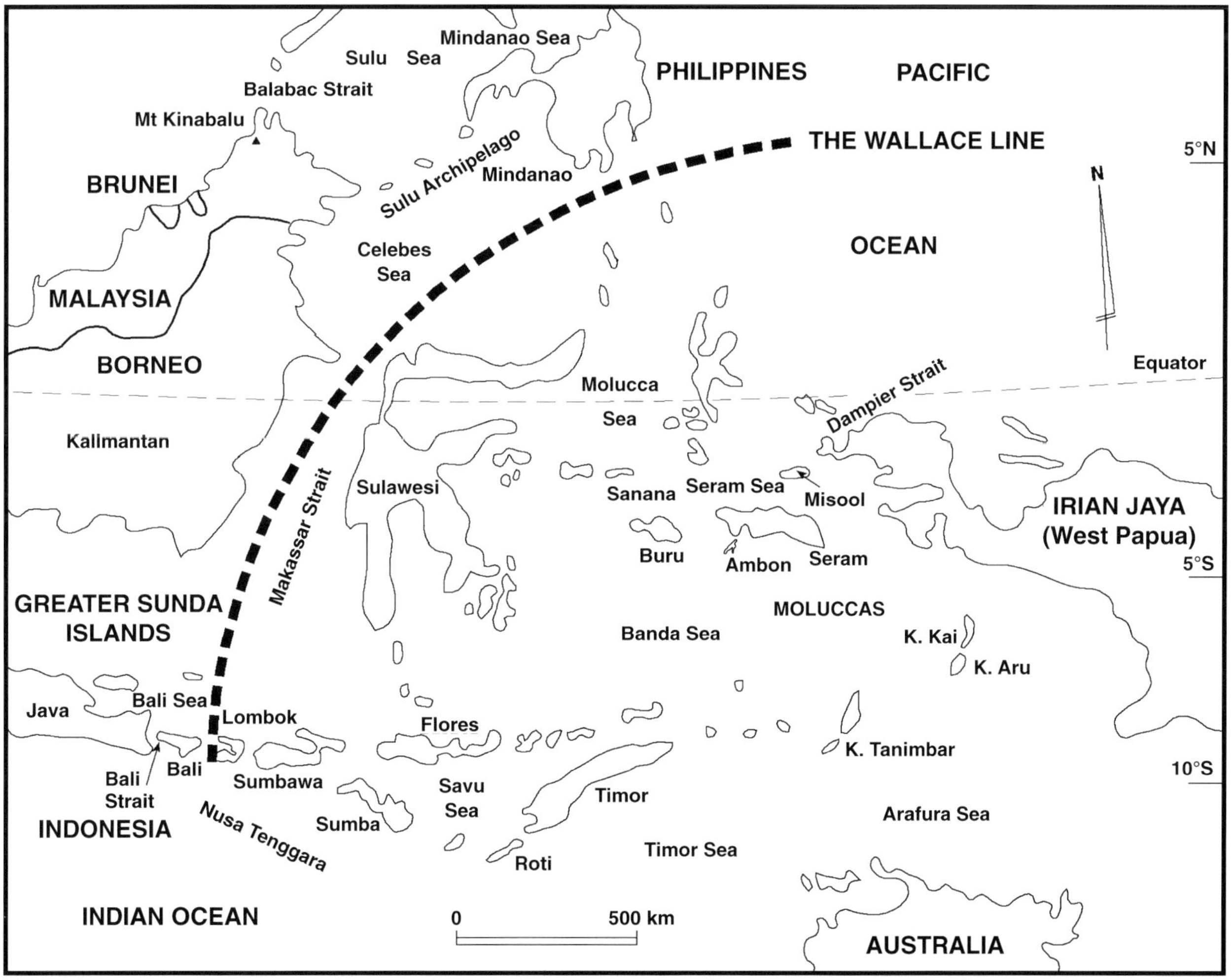

Figure 1. The position of the Wallace line showing the surrounding major land masses and oceanic islands as they exist today (adapted from van Oosterzee, 1997).

of civet cats identified, only seven species were found east of the Line; of 196 species of squirrel only 14 were found in Sulawesi, across the Line.

Raven (1935) also considered other faunal groups, including the following. The specialized pangolin, which ranges from India and China to Java and northward through Borneo to Palawan and the Philippines, does not naturally cross the Wallace Line. Tapirs and rhinos, flying lemurs and hedgehogs extend only as far east as Java or Borneo. None of the 16 species of insectivorous tree shrew (family Tupaiidae), which extend right to the Wallace Line, had crossed it and three species of the Asiatic mouse deer (chevrotain), extending through the Malayan region and Philippines, do not cross the Wallace Line. The muntjacs, or barking deer, are also limited to the Malayan region but do not extend east of Borneo. Only one canid, the dingo crosses the Wallace Line. The once widespread wild dog of Asia, *Cuon alpinus,* extends east to Sumatra and Java but no further. The Malayan sun bear (*Helarctos malayanus*) is found in Malaysia, Sumatra and Borneo but not east of the Wallace Line. Of 27 species of mustelids (weasels, badgers, skunks and otters) Raven identified, none had transgressed the Wallace Line.

Of the primates, *Tarsius*, macaques and humans occur to the east of the Wallace Line. The lemuroid slow loris extends east as far as Java, Borneo and the Philippines. The leaf-eating monkeys, which range over the Malay region including Borneo, do not transgress the Wallace Line. The highly specialized proboscis monkey, *Nasalis larvatus,* is characteristic of Borneo. Among the anthropoids, there are nine species of gibbons in the Malay Peninsula, Java and Sumatra, Borneo and Palawan; the orangutan is found in Borneo and Sumatra. All these primates are of east Asiatic derivation and none has transgressed the Wallace Line.

Recently Clode and O'Brien (1999) reassessed Wallace's Line using 2,517 avian specimens Wallace himself collected from Indonesia. Each species (excluding cosmopolitan genera) was classified into Asian or Australian groups on the basis of modern molecular research. Using cluster analysis their results confirm a sharp faunal break at the Wallace Line representing a change in the proportion of Asian to Australian species.

Recent work on butterflies (Kitching *et al.*, 1999) shows that the contrast in richness and faunal composition between Borneo and Sulawesi was strong across butterflies of the Papilionidae, Danainae and Satyrinae. Kitching *et al.*(1999) concluded that 'this segment of the Wallace Line is alive and well as far as the butterflies are concerned!'.

THE BIRTH

In January 1858, the year the joint theory of evolution was presented to the Linnean Society of London, a perplexed Wallace wrote a letter to his friend Henry Walter Bates, describing the existence of a near magical faunal boundary. 'In the Archipelago', he wrote, 'there are two distinct faunas rigidly circumscribed, which differ as much as those of South America and Africa, and more than those of Europe and North America. Yet there is nothing on the map or on the face of the islands to mark their limits. The boundary line often passes between islands closer than others in the same group.' (Marchant, 1916). This line became known as The Wallace Line.

Wallace had an explanation to account for the extraordinary juxtaposition of these two faunas which were poles apart. 'I believe the western part to be a separated portion of continental Asia, the eastern the fragmentary prolongation of a former Pacific continent' (Marchant, 1916). This sentence marks the birth of biogeography: Wallace had tried to explain where species come from and why they occur where they do realizing that ques-

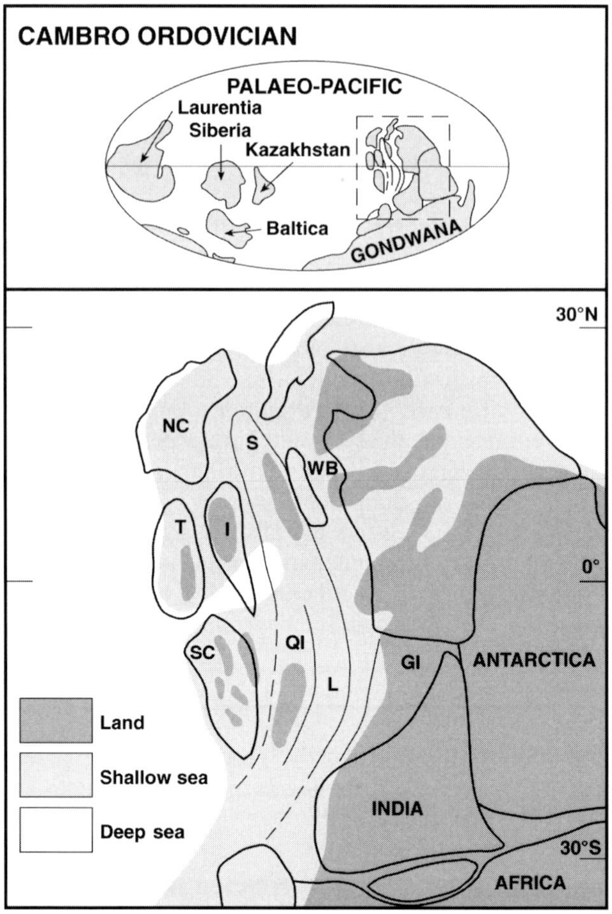

Figure 2. Reconstruction of eastern Gondwana from Cambro-Ordovician times, showing the postulated positions of the East and South-east Asian terranes, distribution of land and sea. Present day outlines are for reference only. Adapted from Metcalfe (1998). Key to abbreviations: GI = Greater India; I = Indochina; L = Lhasa; NC = North China; QI = Qiang tang; S = Sibumasu; SC = South China; T = Tarim; WB = West Burma.

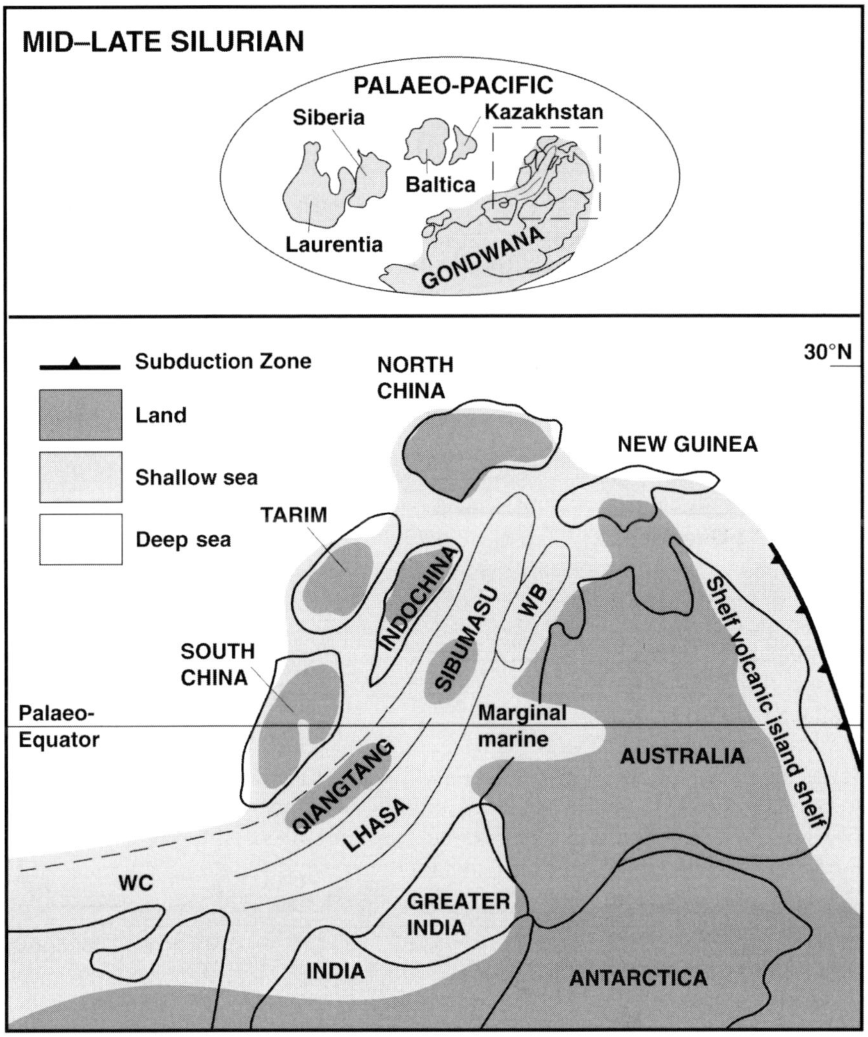

Figure 3. Reconstruction of eastern Gondwanaland for the mid–late Silurian, showing the postulated positions of the East and Southeast Asian terranes. Adapted from Metcalfe (1998). Key to abbreviations: WB = West Burma; WC = Western Cimmerian Continent.

tions of species' origin and species' distribution were inextricably linked. Wallace was not interested merely in placing boundaries around faunal regions. Such specific boundaries will inevitably be different for different taxonomic groups which have different evolutionary histories, different environmental requirements and hence different barriers to dispersal. Wallace was interested in the big picture. He was the first to realize that to really understand the distribution of species over the face of the Earth one had to appreciate not only the species' evolutionary history but also the geological history of the region where they occurred. His line underscored this important principle.

FANTASIA

To account for the formation of Indonesia, Wallace imagined the skyward heaving of Atlantic-deep ocean beds. Accompanied by a crescendo of earthquake shocks, volcanic action sent rivers of molten rock and sediment pouring down into the sea from the lands on either side, to build up the continents so that they spread toward each other. Islands rose, fantasia-like, in the intervening ocean channel, as continents strained toward each other, now connected, now disconnected. In this way, all the islands east of Java, Wallace imagined, once formed a part of a former Australian or Pacific continent, although some of them may never have been actually joined to it. This great continent had been shattered into bits over time with Sulawesi possibly marking its farthest western extension.

This fantastic vision was conjured within a world view that allowed movement in only one direction—up and down. Using this world view to explain the many faunal similarities between distant lands, whole continents had to rise from the ocean deeps, forming faunal highways. In this way the romantically named 'Lemuria' rose, streaming from the depths of the Indian ocean, like some ancient Atlantis, to levitate lemurs from Madagascar to India.

Wallace reasoned—incorrectly as it happens—that it would be impossible to tell the origins of the various fragments of land in this archipelagic jig-saw puzzle. The animals and plants, however, would reveal clues of their

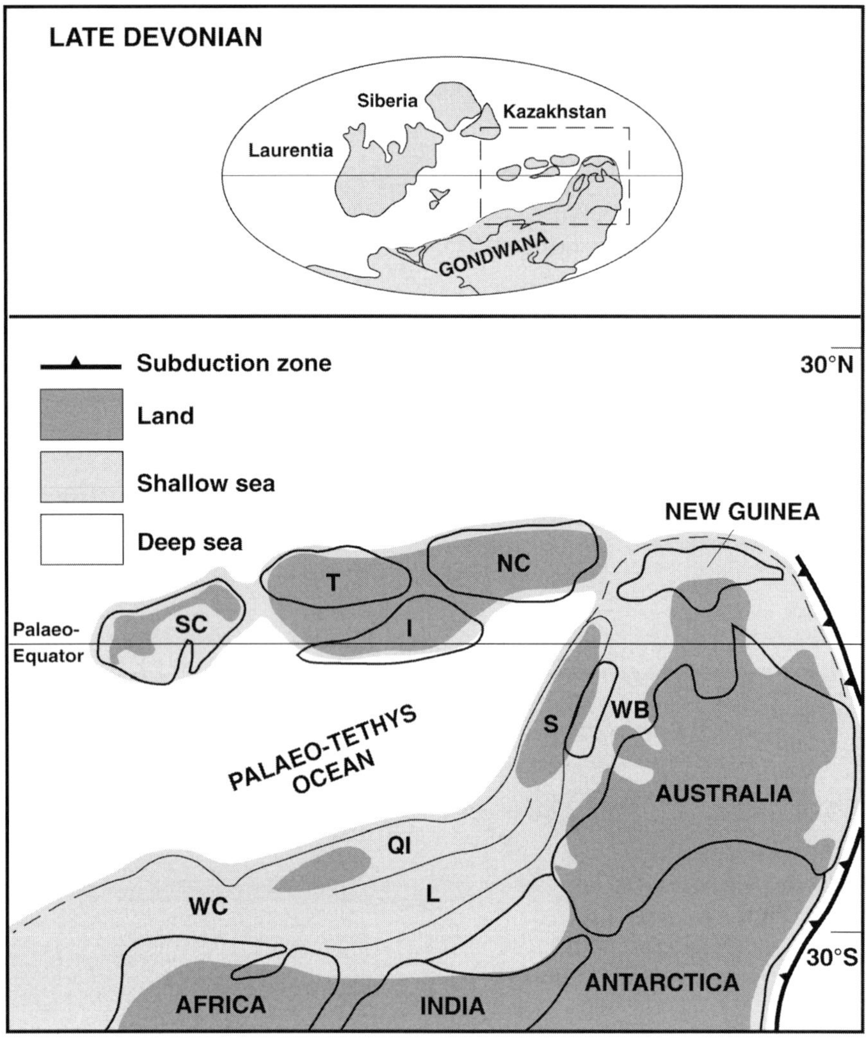

Figure 4. Reconstruction of eastern Gondwanaland for the late Devonian, showing the positions of the East and South-east Asian terranes and the opening of the Palaeo-Tethys Ocean. Present day outlines are included for reference and abbreviations are as for Figures 2 and 3. Adapted from Metcalfe (1998).

former history. 'Some portion of the upraised land might at different times have had a temporary connection with both continents, and would then contain a certain amount of mixture in its living inhabitants'—as with Sulawesi. 'Other islands, again, even though in such close proximity as Bali and Lombok,' Wallace said, 'might each exhibit an almost unmixed sample of the productions of the continents of which they had directly or indirectly once formed a part' (Wallace, 1859).

WORLDS COLLIDING

The extent to which species distributions are limited by underlying geology is a question that is as relevant today as it was in Wallace's time. As imaginative as Wallace was, he would surely have been agog to know that continents not only move up and down, but they also move laterally, rafting across the face of the globe like gargantuan armadas. The theory of plate tectonics says that the surface of the Earth is covered by a series of relatively thin but rigid, shell-like plates that slide over the Earth driven by gigantic convection cells of hot molten material within its core—see pp.17–50 (White, 2006).

Recent advances in plate tectonic theory indicate that faunal and floral links between South-east Asia and Australia extend back more than 500 million years. In the early Palaeozoic (Cambrian–Silurian), Indonesia, Malaysia, Indochina, South China, North China and Tibet existed off the India-Australian margin of the ancient supercontinent Gondwana (Metcalfe, 1998). See Figure 2. Over the period between 545 m.y.a. and 410 m.y.a. trilobites, archaeocyaths, brachiopods, gastropods, nautiloids, stromatoporoids and conodonts define distinctive Asia-Australian 'provinces'. Asian terranes rifted and separated from north-western Australian Gondwana as three continental slivers around 350, 270 and 200–140 million years ago in the Devonian, early Permian and late Triassic–late Jurassic respectively (Metcalfe, 1998). See Figures 2–5.

During the Cainozoic, the jostling of three major (Pacific, Australian and Eurasian) and numerous smaller plates determined the present configuration of land and sea in our region. This intense interaction has resulted in numerous examples of subduction and this has maintained volcanic arcs, which formed intermittent emergent island chains crossing the region. Rapid changes in topography and distribution of land and sea have provided multiple opportunities for dispersal and vicariance (see Figures 6–8).

What was happening at the region of the Wallace Line? Borneo formed part of Sundaland (South-east Asia) throughout the Cainozoic. Hall (2001) states that Sundaland was mainly land, or washed by very shallow seas and was probably linked with the rest of Asia for the whole period. The South China, Celebes and Sulu Seas opened from the Eocene, around 50 m.y.a., forming deep water barriers to the north and east. High mountains rising in Borneo from the early Miocene, around 25 m.y.a., expanded the area of land, and large deltas fanned out rapidly into the surrounding deep basins. The Makassar Strait section of The Wallace Line was, in fact, wider than at present, with deep water and wide marine shelves.

Hall's (2001) recent work in west Sulawesi indicates that emergence of land and uplift of mountains was quite recent—late Miocene or later—so that there was no direct way of crossing between Borneo and west Sulawesi. There may, however, have been routes into Sulawesi via Java by way of small islands, although west Sulawesi may itself have been little more than a set of small islands until the Pliocene, around 5 million years ago.

In eastern Indonesia, particularly Maluku, deformation through plate collision is particularly vigorous and active (Hall, 2001). Here high mountains on land are separated by deep ocean basins in a tectonically complex region which involves the collision of all three major plates. During the last few million years islands of north Maluku arrived from the east with island arcs of the Pacific plate. Other fragments in the Banda region glided into the region, along a series of splays of the Sorong fault system. New Guinea, at the collision zone between the Australian and Pacific plate, crumpled into mountains. Mountains have also risen in Seram and Timor within the last 3 million years.

Does the assembly of this breathtaking global jigsaw puzzle fit in with Wallace's observations? And does it provide any supporting evidence for the Wallace Line?

One of Wallace's greatest mysteries is resolved. Sulawesi and Borneo seem never to have been connected. In fact Sulawesi has always been an island. Wallace had suggested the possibility of a permanent separation but, at that time, such a thought was treated with scepticism. Ironically, biological confirmation of the fact comes, not from land animals, but from aquatic vertebrates.

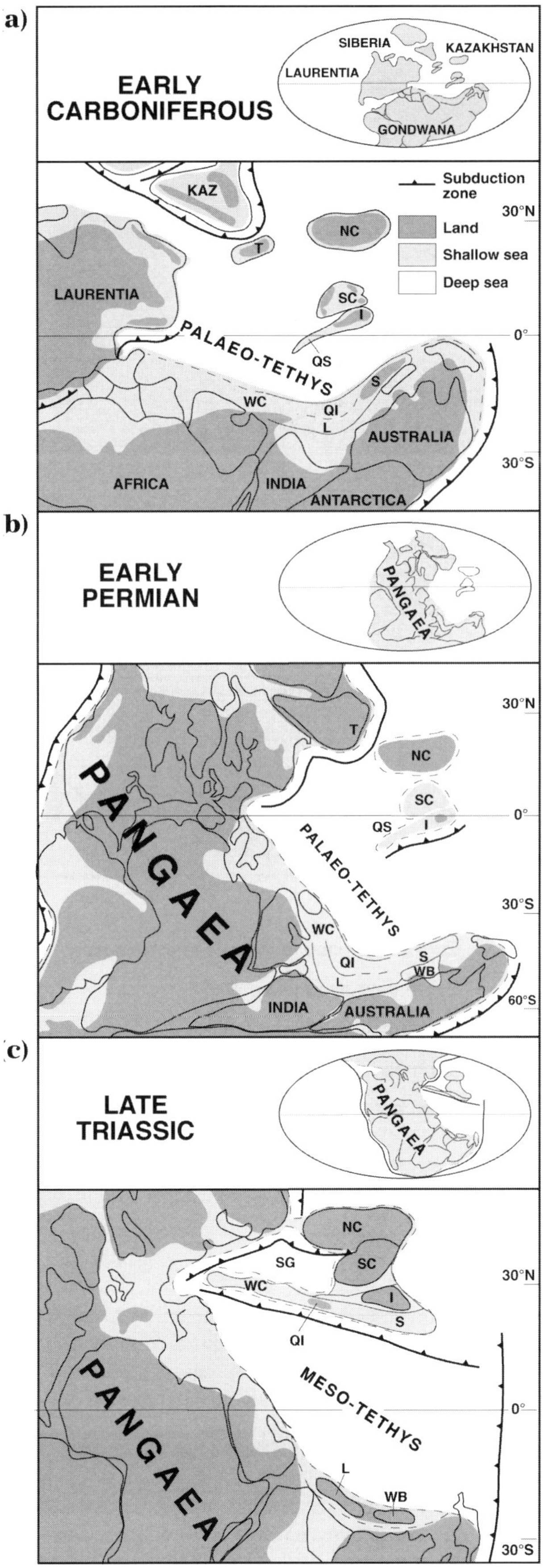

Figure 5. Reconstructions of the Tethyan region for: **(a)** early Carboniferous (340 m.y.a.); **(b)** early Permian (295 m.y.a.); **(c)** late Triassic (220 m.y.a.). Relative positions of East and South-east Asian terranes are shown with present day outlines included for reference and abbreviations as for Figures 2 and 3. Adapted from Metcalfe (1998).

Early Eocene (50 m.y.a.)

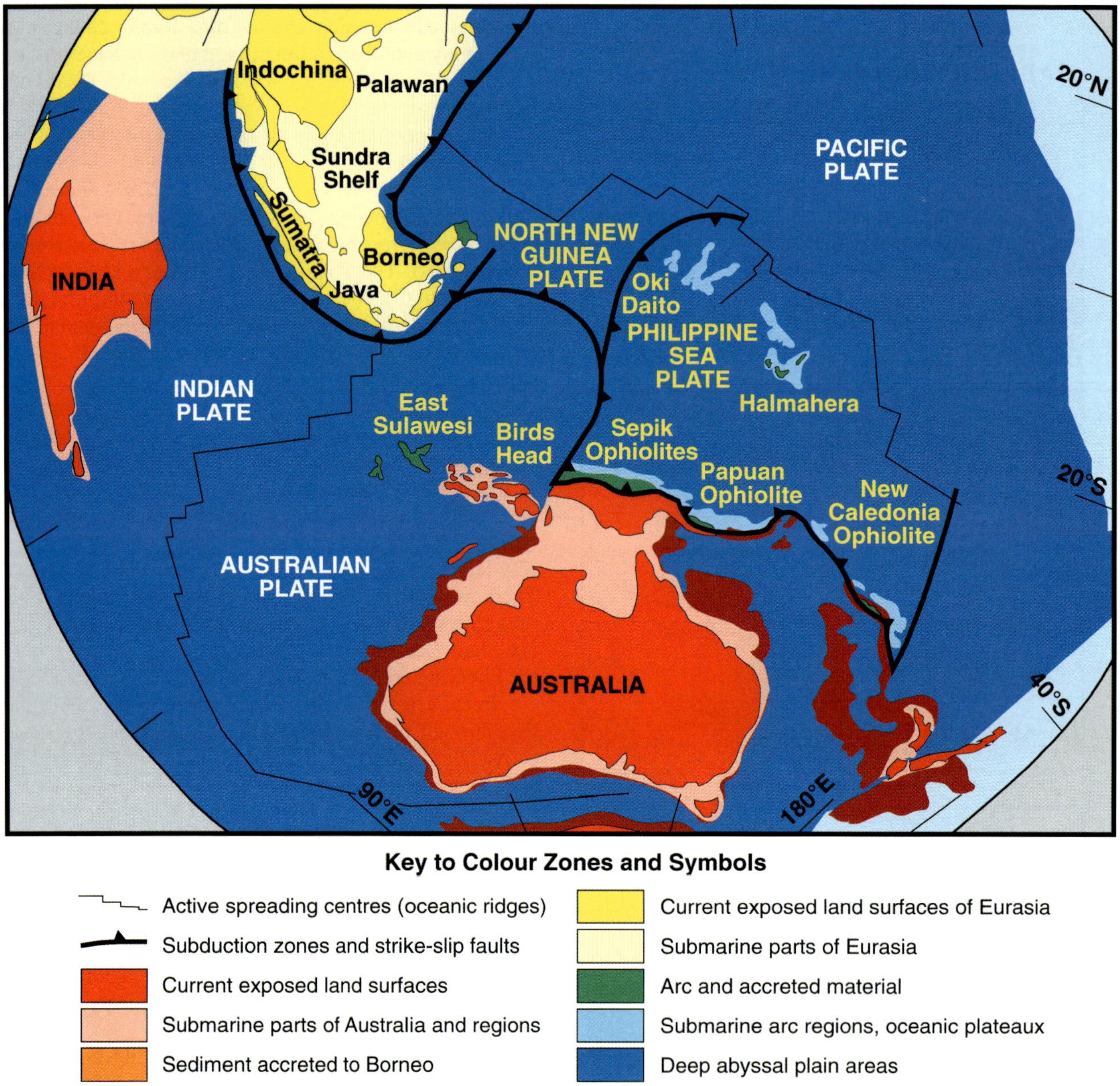

Figure 6. A reconstruction of South-east Asia and the western Pacific at the end of the early Eocene (~50 m.y.a.). Main features include the India-Asia collision and an arc-continent collision at the northern New Guinea margin, which eliminated old ocean crust. Massive volcanism occurred at the eastern edge of the Philippine Sea Plate, which rotated rapidly clockwise. Adapted from Hall (2000).

Freshwater fishes of the world are generally divided into two main categories depending on their ability to survive in salty water. Primary division families are those which evolved in freshwaters and have always been intolerant of salt water. Secondary division families are those which initially evolved in the sea, but now live in freshwater and are able to tolerate seawater for a short period. See pp.195–224 (Merrick, 2006). Natural occurrence on an island of primary division freshwater fishes indicates that some land connection must have existed between the island and the mainland, otherwise the fish could not have got to the island. The numbers of primary division species on Sumatra, Java and Borneo are 232, 104 and 340 respectively (Kottelat *et al.,* 1993), but Sulawesi has no primary division fishes.

Like a giant moat, the Makassar Strait, the most robust section of the Wallace Line, has always separated Borneo, indeed South-east Asia generally, from Sulawesi.

Of Maluku, in between Sulawesi and Australia/New Guinea, Wallace observed 'that the Moluccan fauna has

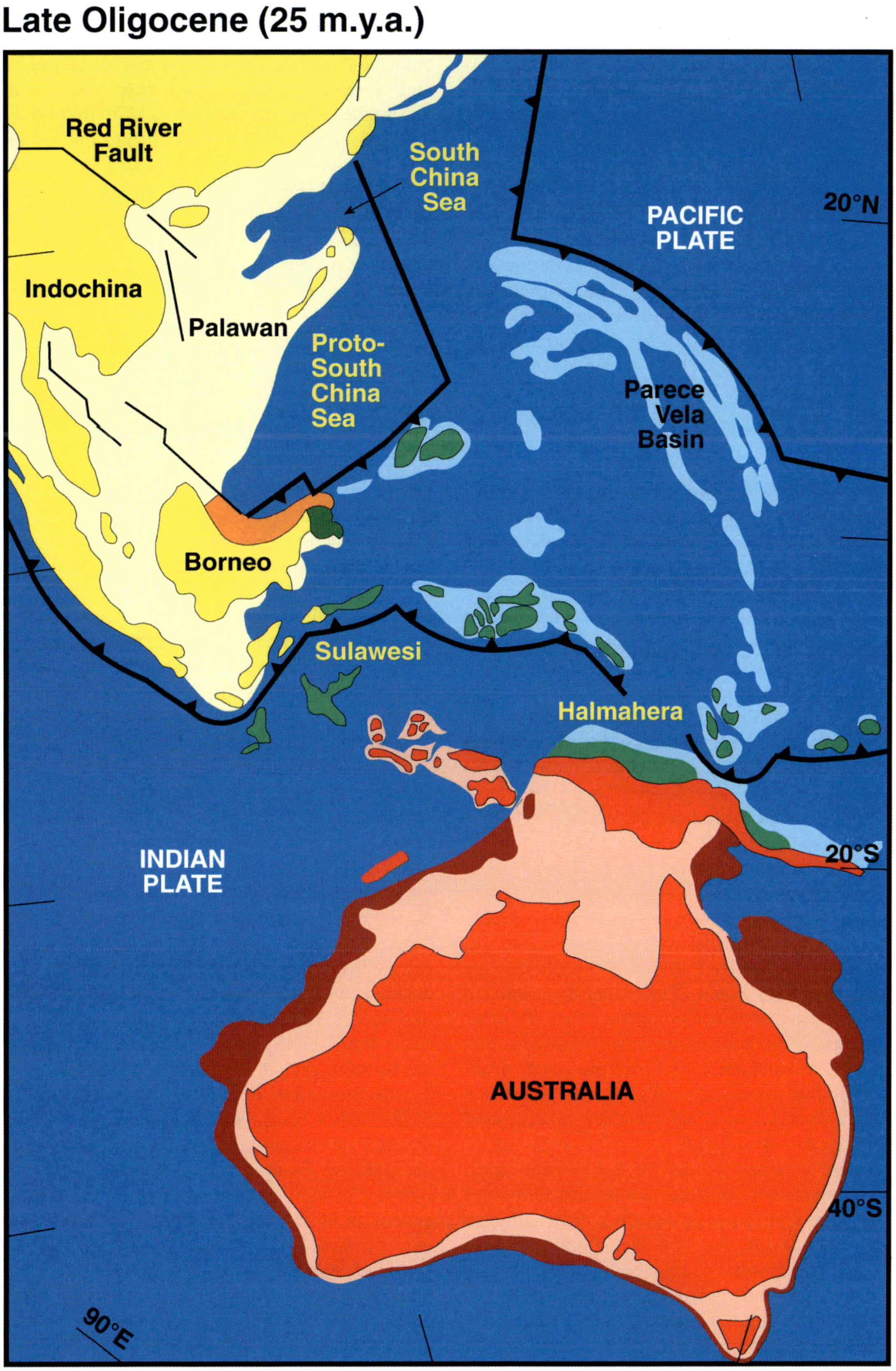

Figure 7. A reconstruction of South-east Asia and the western Pacific at the end of the Oligocene (~25 m.y.a.). The South China Sea opens and causes sediment to accrete on to Borneo. The northern New Guinea margin collides with another arc system. Adapted from Hall (2000). For key to colour zones and symbols see Figure 6.

been almost entirely derived from that of New Guinea'. Plate tectonics confirm that while the islands of Maluku have a rather chaotic geological history, their complicated relationships are all with New Guinea.

LOOKING-GLASS INSECTS

Wallace had said that to appreciate species' evolutionary history, one had to understand the geological history of the region where they occurred. Modern day biogeographers, aware of plate tectonic reconstructions, continue to search for common biogeographic patterns amongst taxa, to fine tune predictions of geological and climatic changes. They assume that animal and plant groups that have gone through the same geological and climatic events should largely reflect them in their distributions.

Groups of organisms that evolved in different periods will, in turn, show different patterns of area relationships. But the real problem for biogeographers is that different groups of organisms, forming the assemblage characteristic of an area, will almost certainly have responded differently to the same sequence of geographic, climatic and ecological changes. The first step in the search for general biogeographic patterns, then, is the search for an appropriate animal or plant group that will unambiguously reveal past geologic and climatic changes.

In a story of drifting armada-like continents and magical faunal boundaries, what could be more natural than 'looking-glass' insects to provide a more finely-tuned mirror to the scene; these insects are ancient, have a world-wide distribution, are poor dispersers and highly endemic—as such they mirror geological history.

Cicadas are hemipterans which grow up to 6.7 cm in length. Beloved of children who, in Australia, have given them names like the greengrocer, yellow Monday, double drummer and the floury baker, they are part of an old order of animals that evolved around 280 million years ago. At that time they could disperse easily around a world dominated by one supercontinent. As with other herbivorous insects we assume that their evolution was stimulated by the emergence of flowering plants from around 130 million years ago. With these plants the cicadas radiated extensively and there are now around 1,200 cicada species world-wide.

Cicadas are a perfect model for the biogeographic principle, championed by Wallace, that the distribution of animals and plants in space echoes their distribution in time. This principle predicts that higher taxa such as classes, orders and families are generally spread across the globe and, as a family, cicadas do indeed have this cosmopolitan distribution. Genera, and particularly species, should have more restricted distributions reflecting more recent regional, geologic and climatic configurations. Of the 38 Australian genera of cicadas, for instance, 28 are endemic; however, some generic distributions can still echo older patterns.

Some Oriental genera, for instance, appear to echo the geological entity known as the Sibumasu block (now embedded within Tibet, Myanmar or Burma, Malaysia and Sumatra) which rifted from the north-western Australian edge of Gondwana 270 million years ago (Duffels *et al.,* 1998). After rifting, the terranes had a long period of isolation during which the cicadas evolved and radiated. From Australia, at least one of the non-endemic genera is well represented in North America, a distribution that may echo a 250 million year old Pangaean pattern (Duffels and de Boer, 1990)

Evolution to species level is a more recent event still. In Australia this is reflected by the fact that all but four of the 250 species are endemic. These species show restricted geographical distributions within Australia, with closely allied species being found geographically near to each other.

Of course, not all animals and plants behave in such a model fashion for biogeographers. In the first place, many groups of animals and plants had not evolved in time to track wayward continents. Others have not diversified enough—there are not enough of them—to track distribution patterns since their evolution from a single ancestor. For instance it is not possible to determine the meandering paths of the continents by looking at distribution patterns of, say, elephants, because the family only evolved about 40 million years ago and there are currently only two extant species, hardly a statistically significant number. Other taxa, again, have widespread distributions, such as seabirds and some rats (and elephants). These animals just fly, raft, or swim over barriers that would stop any self-respecting cicada.

Cicadas are less flighty than butterflies (another group that has been studied), and many other insects. Being rather chunky they are also less likely to be transported elsewhere by birds or wind. That cicadas are homebodies is also indicated by their long subterranean nymphal phase lasting from two to six years, but even 17 years has been reported (Duffels and de Boer, 1990). While cicadas will invade new land, in general they are poor colonists. Those that live in an area are likely to be endemic to that area. Their distribution patterns, therefore, are likely to mirror the complex geological history of the region.

Duffels and de Boer (1990) have analysed the cicadas of the region east of the Wallace Line. Not surprisingly, Oriental cicadas show a distinct line of demarcation along the Makassar Strait section of Wallace's Line, where most Oriental genera reach their eastern limits. As expected, endemism of cicadas on Sulawesi is strong, even extending to generic level, which is a sure indication of a long period of isolation. From a cicada's point of view the geographic entity of Sulawesi includes those islands of the nearby Banggai and Sangihe Archipelagos, to the east and north respectively, another pattern that mirrors the plate tectonics of the region. See animations and maps on the web pages for SE Asia Research Group (2000).

Another example of cicadas' ability to mirror geology is the island of New Guinea. New Guinea is a geological composite of comparatively recent origin—see pp.71–93 (Heinsohn and Hope, 2006); it is not yet recognized by cicadas as a geographical entity. Instead they recognize the islands of Maluku and northern New Guinea as one endemic area. Across this area, which represents the pre-collisional Melanesian oceanic arc, cicadas are closely related at a high level. Cicadas from southern New Guinea are distinct and unrelated to the Melanesian forms. Indeed, it is only recently that cicadas have dispersed from northern New Guinea to the newly formed central New Guinea and vice versa. As yet no Australian cicadas have evolved to occupy habitats outside New Guinea and few Melanesian cicadas have evolved to occupy habitats in Australia.

Embedded within this high endemism area of Maluku to northern New Guinea are lower ranking areas of endemism. These sub-areas developed more recently and involve species or subspecies that have become restricted to the limited ranges. Perhaps the suspected close proximity of Seram, Buru and Sula several million years ago is being mirrored by the fact that three of the four southern Maluku endemics are found on these three islands. Some Maluku cicadas show a relationship with northern New Guinea including the Vogelkop Peninsula, while others show relationships with northern New Guinea but not the Vogelkop. This latter situation was possibly established before 3 million years ago when the islands of Maluku and northern New Guinea were adjacent—before the Vogelkop swung into its current position (see SE Asia Research Group, 2000).

The Vogelkop Peninsula itself comprises an endemic area of lower order with species from three more widespread genera evolving there. Cicadas of central New Guinea, literally the newest area on the map, are rapidly diversifying, so far mainly at the subspecies level, in the many habitats created by the geologically unstable central mountain ranges.

GLACIAL INDONESIA

Plate tectonics go at least part of the way toward unravelling the complex biogeography of Wallace's 'Malay Archipelago'. But what of the biogeographic barrier between Bali and Lombok, the southern section of the Wallace Line? Plate tectonic reconstructions do not suggest a tectonically-derived barrier between Java's offshore island, Bali, and Lombok, the westernmost island of Nusa Tenggara (the Lesser Sundas). In fact, Bali may not have existed as dry land until volcanic activity elevated it around 1–2 million years ago (Robert Hall, 1999, personal communication). Other islands of Nusa Tenggara, similarly, have a history of having been built by volcanic action upon ancient, but largely marine, rocks. What is it then that makes this section of the Wallace Line so potent?

Charles Darwin provided Wallace with a hint. Darwin wrote to Wallace pointing out that, elsewhere in the world, he had found a relationship between sea depth and known patterns of mammalian distribution; the deeper the sea the more distantly related the fauna, the shallower the sea the more closely related the fauna. Darwin had discovered that depth was a measure of time.

Shallow seas implied a recent connection. In England, for instance, with few exceptions, every mammal, bird, reptile, insect and plant is found on the adjacent continent. However in Sri Lanka, closer to India than Britain is to Europe—but with a deeper sea in between—many animals and plants are different and peculiar to the island. In the Galapagos near South America, and Madagascar near Africa, almost every living thing is endemic. Both are separated by deep sea trenches from the nearest continent.

This fitted in with Wallace's own observations that the (to this day) little known Aru Archipelago, in the Arafura Sea north of Australia, and the islands of Misool and Waigeo to the south and west of the Vogelkop Peninsula respectively 'agree with New Guinea in their species of mammalia and birds much more closely than they do with the Moluccas, and we find that they are all united to New Guinea by a shallow sea' (Wallace, 1859).

Borneo, Sumatra and Java clearly lay on the same shallow bank or continental shelf. Indeed so shallow was the sea that ships could anchor anywhere in the vast seas between the islands. That the faunas of these well known islands resembled each other was obvious. In fact Wallace thought the faunas were as similar as distant areas could be expected to be—even without an intervening sea. The sea must have been a recent invader, flooding the lowlands and deep valleys, and isolating the highlands and volcanic peaks that became Borneo, Java and Sumatra. The Wallace Line more or less marks the edge of the Asian continental shelf. Wallace had no explanation for the changing sea levels since even his fantastic imagination could not have concluded that he was in the temporary warm phase of an ice age.

Nothing could be further from the truth than to consider today's climate normal. The last time climatic conditions were similar to today's was around 125,000 years ago at the beginning of the last interglacial. The present ice age is characterized by regular glacials, punctuated by these interglacials (one of which we are currently in). Hominids (including *Homo sapiens*) are ice age animals having evolved during the last 2.5 million years, during which there have been scores of recognizable glacial to interglacial alternations, following a curious series of cycles, called Milankovich cycles. These cycles are triggered by small variations in the amount and distribution of solar radiation received by the Earth as a result of predictable changes in the Earth's orbit relative to the sun (Kershaw *et al.*, 2000). During these cycles the great continental ice sheets of the Northern Hemisphere and Antarctica wax and wane, in turn locking-up and releasing water so that far from being steady-state the sea levels rise and fall, from levels similar to those of today to levels as low as 130 metres below present levels.

At these glacial times Borneo, Sumatra, Java and its offshore island Bali as well as the Malay Peninsula have

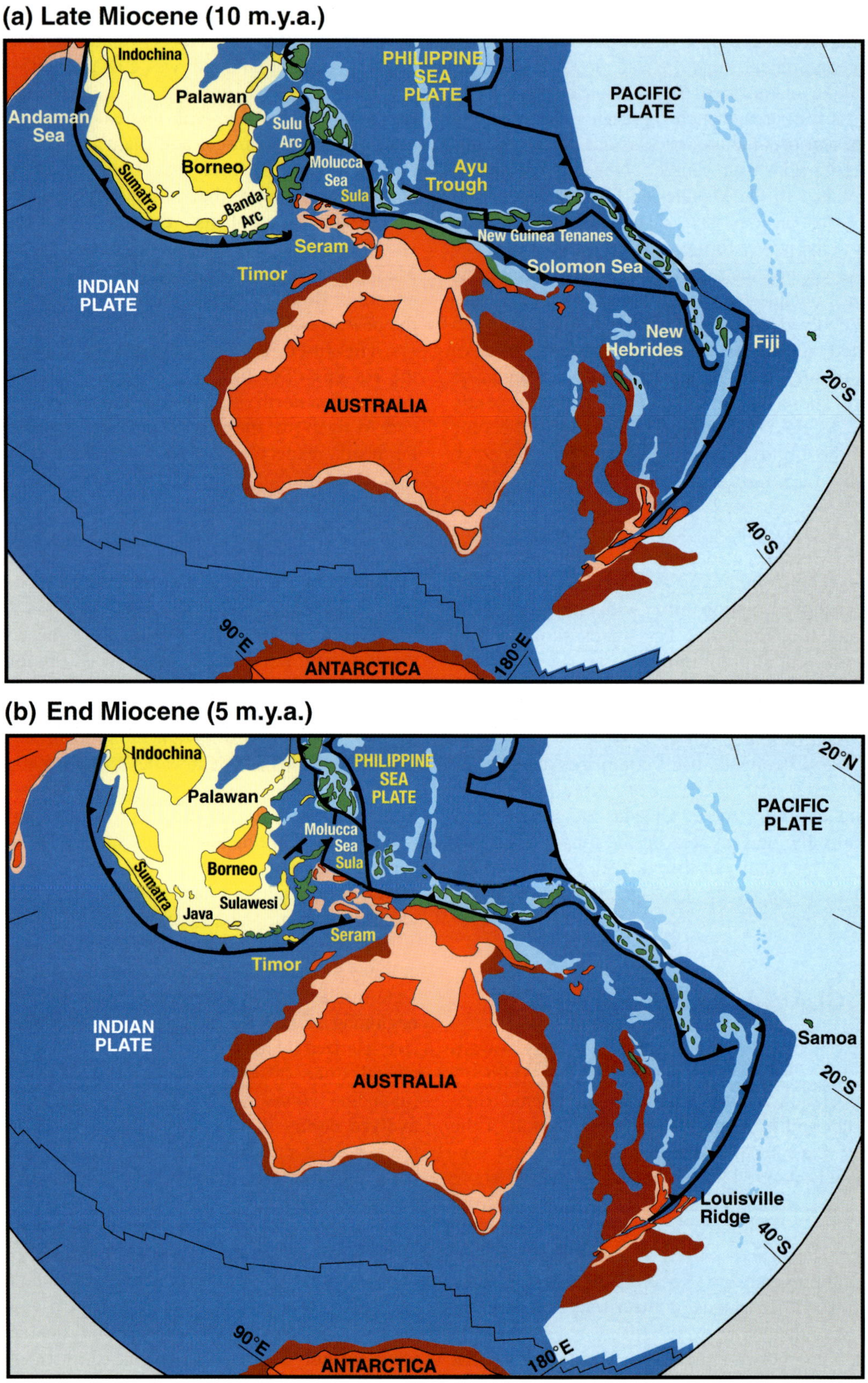

Figure 8. Reconstructions of South-east Asia and the western Pacific during the later Miocene, showing regional subduction zones and faults which have caused a complicated sequence of collisions resulting in the accretion of many of the islands now known as Indonesia. **(a)** Late Miocene (~10 m.y.a.); **(b)** end of Miocene or very early Pliocene (~5 my.a.). Adapted from Hall (2000). For key to colour zones and symbols see Figure 6.

been connected by dry land. The Lesser Sundas, or Nusa Tenggara, comprising Lombok, Sumbawa, Komodo, Flores and Alor emerge as a thin peninsula separated from Bali by a very narrow channel. New Guinea, Australia and Tasmania form a single land mass which includes the offshore islands of Misool, Waigeo and the Aru Archipelago. This vast land is separated from Timor by as little as 70 km. Timor itself is not much more than a stone's throw from Alor. A string of islands probably provides a tenuous link between Sulawesi and part of Nusa Tengarra. Borneo would probably be connected, through a thin strip of land (now Palawan), to the Philippines.

During glacials, rivers gouged out deep valleys which today appear as submarine channels on the sea-floor. A huge North Sunda river flowed northward to the present South China Sea. It collected its tributaries from eastern Sumatra, western Borneo and from Malaysia and other parts of the east coast of mainland Asia. Another river flowed eastwards throwing its fines into the sea north of Bali. The river embraced a catchment extending from south-east Sumatra, northern Java and southern Borneo. On greater Australia, two separate rivers flowed westwards. One from southern New Guinea entered the sea north of today's Aru Archipelago and the other drained an area that included today's Gulf of Carpentaria, Top End of the Northern Territory and Kimberley region.

Interglacials, characterized by sea levels similar to those of today, have been brief events lasting perhaps 10,000 years, with an oscillating build up towards full glacial conditions. If anything can be characterized as normal in the Pleistocene, the period in which *Homo sapiens* evolved, it is a climate tending toward glacial. The above comments should also be considered in the light of very recent studies of global climates 1–2 million years ago, which suggest that Quaternary glaciations are minor oscillations during a prolonged period of extreme aridity—see also pp.51–70 (Dawson and Dawson, 2006).

What sort of impact did these drastic changes have on the land? Ancient pollen provides a clue. Undisturbed by earth-movements or the activity of humans, an extinct crater on the Atherton Tablelands in northern Queensland, Australia has been steadily filling with sediment for over 125,000 years. The sediment, containing pollen from forests of the surrounding catchment, has been swept down and layered in a neat time-series. A drill-core from Lynch's Crater provides one of the few windows to check on modern interpretations of the region's past (Williams, 1991). Using the dustings of preserved pollen, a fingerprint of the landscape during the last interglacial to glacial to interglacial sequence appears.

Before 80,000 years ago, a rich tropical rainforest flourished on the volcanic soils of the Atherton Tablelands. Around 80,000 years ago this rich forest withered, under drying conditions, to become a cooler *Araucaria*-dominated rainforest. As the region entered the glacial peak of 18,000 years ago, rainforests were entirely obliterated. They were replaced by a sclerophyll forest dominated by *Eucalyptus* and *Casuarina*. It would have needed a rainfall plunging to 30 per cent of its former level to have caused this change. Only 10,000 years ago, as the climate ameliorated into that of the current interglacial, did the rich flowering forests return. This was only one of the many similar cycles that have occurred in the last 2.5 million years!

For these sorts of changes to have occurred, temperatures would have been cooler and rainfall half, in places only 20%, that of today. Falling sea levels and a greater land area help to explain these changes, as does the differing configuration of land and water. For instance, ocean currents that today enter the Malay Archipelago through the Torres Strait, the South China Sea, and south of Mindanao in the Philippines would have been severed. Their buffering effect, on a climate already sucked dry by polar icesheets, was removed. The climate would have become virtually continental.

That most of the Pleistocene was indeed characterized by these savanna environments is confirmed by the sequential palaeontological fossil record of Javan mammals (Sondaar, 1984). Though incomplete, the Javan mammalian biostratigraphy reveals a fascinating faunal succession through seven different ages. More in tune with scenes of Africa than tropical Asia, over the past 2 million years, stegodons, ancestral hippopotamuses, antelopes and deer were abundant as well as carnivorous large cats and ancient dogs. Later, tapirs, hyenas, rhinoceroses, modern elephants and large bovids arrived, and with the stegodons and pigs evolved into different forms. The nature and size of these animals indicate that, for most of the Pleistocene, Java was a wedge of land at the edge of a continent with herds of grazing and browsing mammals roaming large tracts of open forests. These were prey for tigers and a carnivorous assemblage characteristic of continents.

Of about seven different suites of fossil mammals from the region, only one that represents the fauna of an interglacial can be identified. Around 125,000 year ago, tropical Java was characterized by a high quantity of rainforest-loving orangutan. With the orangutans was a species of pig (still extant), which is a typical animal of the rainforest floor where it roots out much of its food from the moist soil. The modern elephant species *Elephas maximus* also makes its first appearance in the fossil record here.

At the height of Pleistocene glaciations land mammals, including humans, made their way from South-east Asia to what are now the islands of Sumatra, Java and Borneo. The Makassar Strait, the northern end of the Wallace Line, between Borneo and Sulawesi, remained a wide and deep barrier to the dispersal of most mammals from Asia. The orangutan, tiger, leopard, Malay bear, elephant, Malay tapir, giant pangolin, Javan rhinoceros, Sumatran rhinoceros and banteng have never been found in Sulawesi, either living or fossil. Even during the Pleistocene glaciations, the northern end of the Wallace Line maintained its integrity.

The Nusa Tenggara chain of islands extending from Bali to Timor, however, were accessible—at least to great travellers like elephants, stegodons, giant turtles, Asian crocodiles and giant pigs. These animals not only managed to hop the small water gap from Bali to Lombok, they also managed to transfer from Java and Flores to the south-west arm of Sulawesi, and from Alor to Timor during low sea levels. Certainly, modern species of elephants are very strong swimmers capable of swimming 45 km under extreme circumstances. They are also reported (Johnson, 1980) to be strongly inclined to swim shorter distances (several km) in search of food (Heaney, 1985).

Clearly the vicissitudes of the Pleistocene climate have had a major impact on the evolution of the flora and fauna of the Malay Archipelago. The periodic expansion and contraction of environments has driven dispersal and speciation and has largely created the distribution of plants and animals observed in Indonesia today. On present-day distributions the Wallace Line, in between Bali and Lombok, separates adjacent faunas that are quite different. Unlike the robust biogeographic barrier of the Makassar Strait, the difference is independent of origin. Java and Bali are part of the Asian continent and have a rich continental fauna whereas, even at the height of glacials, Lombok and the rest of Nusa Tenggara had an island (or oceanic) fauna limited by the size of the islands.

CHAPTERS OF ACCIDENTS

Perhaps surprisingly, the idea that there are indeed islands which have never been connected to a continental mainland was not acceptable in Wallace's time. 'This' Wallace said, explaining the impeccable logic of Victorian England, 'would imply that their animal population was a matter of chance; it has been termed the *flotsam and jetsam* theory, and it has been maintained that nature does not work by the 'chapter of accidents', (Wallace, 1859).

Wallace (1859) said that, on the contrary, 'we have the most positive evidence that such *has* been the mode of peopling the islands. Their productions *are* of that miscellaneous character which we should expect from such an origin and to suppose that they have been portions of Australia or of Java will introduce perfectly gratuitous difficulties, and render it quite impossible to explain those curious relations which the best-known group of animals (the birds) have been shown to exhibit'.

In fact, in Maluku, examples of haphazard colonization abound. The mountain tailorbird *Orthotomus cuculatus*, for instance, is present in Seram, Buru and Bacan, but not Halmahera, Sula or Obi, while the island Verditer flycatcher (*Eumyias panayensis*) is present in Seram and Obi but not in Buru, Bacan, Halmahera or Sula (Bowler and Taylor, 1993). The recent discovery of two new waterbirds for Seram, the comb-crested jacana (*Irediparra gallinacea*), toward the western limit of its spread from Australia and the white-breasted waterhen (*Amaurornis phoenicurus*) on the eastern limit of its spread from South-east Asia exemplifies the ongoing nature of colonization (Bowler and Taylor, 1993). Clearly the spread of species from west to east, or east to west, is a fluid process where small differences in the initial colonization of an island can result in chaotically different faunas.

An island is nothing like a continent. Compared to a continent's stable patchwork of interlocking habitats, an island provides small areas of different habitats, which may be unstable. The surrounding expanses of seawater act as a filter, so that only animals that can survive the crossing will make landfall in an environment that may not be suitable. Some large birds and bats can reach even distant islands. Smaller birds, bats and insects can be carried passively on high winds. Other animals such as rats make the journey on drifting rafts of vegetation (including human-made rafts and boats). Rarely, animals such as pigs, deer and elephants will swim to an island. Certain groups of animals are completely unable to disperse over water barriers; apart from notable exceptions, such as bats and rats, mammals are generally very poor dispersers. Not surprisingly, very old islands become evolutionary centres in their own right, developing their own species and—in time—even genera and families; examples are Madagascar and Sulawesi. For further discussion of the processes of colonization and development of island faunas see pp.459–476 (Ford, 2006).

While islands are depauperate in absolute numbers of species, they have extraordinarily high rates of endemism. Of the total of 385 butterfly species recorded in Maluku (Vane-Wright and Peggie, 1994), for instance, 82 are not known to occur elsewhere. Compared with the 2% endemism for butterflies of the mainland Malay Peninsula (for example), this 21% endemism is staggering. Indeed, Nusa Tenggara and Maluku together comprise a major centre of global bird diversity. Some 562 species have been recorded in the region and 144 of these are endemic. Maluku alone is home to 94 bird species found nowhere else on Earth—more endemic birds than for any other similar area (Bibby *et al.,* 1992). Mainland Malaysia, on the other hand, has a paltry 1% endemism.

The constellation of islands of Maluku, has maintained its isolation, even from a Pleistocene-expanded Australia, across water gaps rarely less than 70 km. Most of its birds are either endemic, derived from New Guinea, or derived from the Oriental stock of Sulawesi. But endemism in the birds of Nusa Tenggara and Maluku is dominated by a relatively small number of families: pigeons with 15 endemic species; parrots with 16 endemic species; honeyeaters with 15 endemic species; and white-eyes with 13 endemic species. In addition, Wallace found a preponderance of kingfishers and sunbirds. These are all families that have a remarkable capacity at colonizing oceanic islands, and then speciating (Bibby *et al.*, 1992).

In contrast, mammals show a remarkable incapacity at colonizing islands; no Australian marsupial has naturally survived the wide water crossing to Timor. Even the marsupial cuscus (*Phalanger orientalis*) that Wallace observed there, was likely to have been brought to Timor by people, since it is widespread and has the appealing habit of producing twins (Flannery, 1995). Lombok's impressive list of mammals also includes the long-tailed macaque, silvered leaf monkey, pangolin, Plantain squirrel, Sunda Island porcupine, leopard cat, little civet, common palm civet, barking deer and the Sunda shrew. All of these are known to be human commensals, kept as pets or to thrive in human-disturbed environments (Kitchener *et al.*, 1990) and are, therefore, likely to have been transported to Lombok by people. Without these animals, Lombok's mammal fauna would be reduced to populations of bats and rats.

The few mammals that have successfully dispersed to oceanic islands also show a remarkable capacity to speciate. In the oceanic islands of the Australasian region, for instance there are 80 species of fruitbat of which an extraordinary 73% are endemic (Flannery, 1995). In Nusa Tenggara, recent discoveries amount to no less than 12 new species of bats and rats (Kitchener *et al.*, 1990); of the known mammalian fauna of these islands, at least 10 bats and 7 murid rodents (rats and mice) are known to be endemic at the species level. Sulawesi and Flores, in particular, have large numbers of endemic genera of rodents. Tellingly, the speciation of rodents, at least to generic level, has been particularly rapid from about 8 million years ago, coinciding with the increasing number of island arcs appearing in the region (Kitchener *et al.,* 1991).

Patterns of speciation in Nusa Tenggara reflect the Pleistocene configuration of islands. During the last major glacial maximum four main islands existed: Lombok to Sumbawa; Komodo to Adonara; Timor, Semau and Roti; and the Tanimbar islands. Overlaying this is a more modern pattern of sub-speciation with different island species differing in appearance from each other. More isolated Maluku reflects the accidental or sweepstakes nature of colonization. Halmahera, for instance, has only one native rodent, while Seram has five endemic species, as well as an endemic bandicoot *Rhynchomeles prattorus*, the northern common cuscus *Phalanger orientalis*, and the common spotted cuscus *Spilocuscus maculatus*, though both these latter marsupials were probably introduced to Seram by people.

'The almost entire absence of Mammalia and of such widespread groups of birds as woodpeckers, thrushes, jays, tits, and pheasants' convinced Wallace that he was 'in a part of the world which has in reality but little in common with the great Asiatic continent, although an unbroken chain of islands seems to link them to it' (Wallace, 1859).

The region, equally, has little in common with the great Australian continent. For instance, eastern Indonesia does not have such widespread Australian groups of birds as fairy wrens, sitellas and treecreepers, whipbirds, wedgebills and quail-thrushes. In common with the Asian woodpeckers, thrushes, jays, tits and pheasants, they have a continental heritage. The strings and constellation of islands in the sea between Australia and Asia are neither Australian nor Asian but literally a biogeographic law unto themselves.

CONCLUSIONS

The sharp contrasts between vertebrate faunas on different Indonesian islands were reported by the naturalist Alfred Russel Wallace well over a century ago. These field observations contributed to the development of the concepts of natural selection and evolution. Although not correct in all aspects, Wallace's explanation of the faunal patterns was the first to combine long-term environmental changes (geological processes, climatic fluctuations) with evolutionary history. Various forms of evidence have since confirmed the existence of this unusual biotic barrier. Plate tectonics combined with more recent palaeoclimatological and palaeoecological data can now demonstrate how a number of processes have contributed to the current situation over many millions of years. These data can also be used to interpret apparently anomalous or discontinuous distribution patterns seen in the region today.

REFERENCES

Bibby, C.J., Collar, N.J., Crosby, M.J., Heath, M.F., Imboden, Ch., Johnson, T.H., Long, A.J., Stattersfield, A.J. and Thirgood, S.J. 1992. *Putting Biodiversity on the Map: Priority Areas for Global Conservation.* International Council for Bird Preservation, Cambridge, U.K.

Bowler, J. and Taylor, J. 1993. The avifauna of Seram. Cited in Edwards, I.D., MacDonald, A.A. and Proctor, J. 1995. *Natural History of Seram, Maluku, Indonesia.* Raleigh International, Intercept Ltd., Andover.

Clode, D. and O'Brien, R. 1999. The beginnings of biogeography: what made Wallace draw the line? *UNE Asia Centre Where Worlds Collide: Faunal and Floral Migrations and Evolution in SE Asia-Australasia. International Conference Abstracts.*

Dawson, T.J and Dawson, L. 2006. Evolution of arid Australia and consequences for vertebrates. Pp.51–70. In Merrick, J.R., Archer, M., Hickey, G.M. and Lee, M.S.Y. (eds). *Evolution and Biogeography of Australasian Vertebrates.* Auscipub Pty. Ltd., Sydney. 942pp.

Duffels, J.P. and de Boer, A.J. 1990. Areas of endemism and composite areas in east Malesia. In Baas, P., Kalkman, K. and Geesink, R. (eds). *The Plant Diversity of Malesia.* Kluwer Academic Publishers, Amsterdam. 420pp.

Duffels, J.P., de Boer, A.J. and Beuk P.L.Th. 1998. Distribution patterns of cicadas of Southeast Asia related to Plate Tectonics. In Hall, R. and Holloway, J.D. (eds). *Biogeography and Geological Evolution of SE Asia.* Backhuys Publishers, Leiden, Netherlands. 417pp.

Flannery, T.F. 1995. *Mammals of the South-West Pacific and Moluccan Islands.* Reed Books, Sydney. 464pp.

Ford, H.A. 2006. Island Biogeography: as illustrated by birds in the Australasian Region. Pp.459–476. In Merrick, J.R., Archer, M., Hickey, G.M. and Lee, M.S.Y. (eds). *Evolution and Biogeography of Australasian Vertebrates.* Auscipub Pty. Ltd., Sydney. 942pp.

Hall, R. 2000. Cenozoic plate reconstructions of SE Asia and the SW Pacific. Integrating onland geology with data from ocean basins. http://www.gl.rhbnc.ac.uk/seasia/welcome.html

Hall, R. 2001. Cenozoic reconstructions of SE Asia and SW Pacific: changing patterns of land and sea. Pp.35–56. In Metcalfe, I., Smith, J.M.B., Morwood, M. and Davidson, I. (eds). *Faunal and Floral Migrations and Evolution in SE Asia-Australasia.* A.A. Balkema, Lisse. 416pp.

Hall, R. and Holloway, J.D. (eds). 1998. *Biogeography and Geological Evolution of SE Asia.* Backhuys Publishers, Leiden, Netherlands. 417pp.

Heinsohn, T. and Hope, G. 2006. The Torresian connections: zoogeography of New Guinea. Pp.71–93. In Merrick, J.R., Archer, M., Hickey, G.M. and Lee, M.S.Y. (eds). *Evolution and Biogeography of Australasian Vertebrates.* Auscipub Pty. Ltd., Sydney. 942pp.

Johnson, D.L. 1980. Problems in the land vertebrate zoogeography of certain islands and the swimming powers of elephants. Cited in Heaney, L.R. 1985. Zoogeographic evidence for middle and late Pleistocene land bridges to the Philippine Islands. *Modern Quaternary Research in South East Asia,* Vol. 9, pp.127–143.

Kershaw, P., Quilty, P.G., David, B., van Huet, S. and McMinn, A. 2000. Palaeobiogeography of the Quaternary of Australasia. *Memoir of the Association of Australasian Palaeontologists*, Vol. 23, pp.471–516.

Kitchener, D.J., Aplin, K.P. and Boeadi. 1991. A new species of *Rattus* from Gunung Mutis, South West Timor Island, Indonesia. *Records of the Western Australian Museum*, Vol. 15, No. 2, pp.445–461.

Kitchener, D.J., Boeadi, Charlton, L. and Maharadatunkamsi. 1990. Wild mammals of Lombok. *Records of the Western Australian Museum, Supplement,* No. 33, pp.1–129.

Kitching, R.L., Eastwood, R. and Hurley, K. 1999. Butterflies and Wallace's Line: faunistic patterns and explanatory hypotheses within the south-east Asian butterflies. *UNE Asia Centre Where Worlds Collide: Faunal and Floral Migrations and Evolution in SE Asia-Australasia. International Conference Abstracts.*

Kottelat, M., Whitten, A.J., Kartikasari, S.N. and Wirjoatmodjo, S. 1993. *Freshwater Fishes of Western Indonesia and Sulawesi.* Periplus, Singapore.

Lincoln, G.A. 1975. Bird counts either side of Wallace's Line. *Journal of Zoology, London*, Vol. 177, pp.349–361.

Marchant, J. 1916. *Alfred Russell Wallace: Letters and Reminiscences.* Cassell, London.

Merrick, J.R. 2006. Australasian freshwater fish faunas: diversity, interrelationships, radiations and conservation. Pp.195–224. In Merrick, J.R., Archer, M., Hickey, G.M. and Lee, M.S.Y. (eds). *Evolution and Biogeography of Australasian Vertebrates.* Auscipub Pty. Ltd., Sydney. 942pp.

Metcalfe, I. 1998. Palaeozoic and Mesozoic geological evolution of the SE Asian region: multidisciplinary constraints and implications for biogeography. Pp.25–41. In Hall, R. and Holloway, J.D. (eds). *Biogeography and Geological Evolution of SE Asia.* Backhuys Publishers, Leiden, Netherlands. 417pp.

Raven, H.C. 1935. Wallace's Line and the distribution of Indo-Australian mammals. *Bulletin of the American Museum Natural History,* Vol.68, pp.179–293.

SE Asia Research Group. 2000. Watch SE Asia moving during the last 50 million years. http://www.gl.rhbnc.ac.uk/seasia/welcome.html 12 June.

Sondaar, P.Y. 1984. Fauna evolution and mammalian biostratigraphy of Java. *Courier Forschungs-Institut Senckenberg,* Vol. 69, pp.219–235.

van Oosterzee, P. 1997. *Where Worlds Collide: The Wallace Line.* Reed Books, Sydney. 234pp.

Vane-Wright, R.I. and Peggie, D. 1994. The butterflies of Northern and Central Maluku: diversity, endemism, biogeography, and conservation priorities. *Tropical Biodiversity,* Vol.2, No.1, pp.212–230.

Wallace, A.R. 1857. On the Law which has regulated the introduction of new species. *Annals and Magazine of Natural History*, Series 2, Vol.14, pp.184–196.

Wallace, A.R. 1859. *The Malay Archipelago: The Land of the Orang-utan and the Bird of Paradise.* Macmillan, London.

White, M.E. 2006. Environments of the geological past. Pp.17–50. In Merrick, J.R., Archer, M., Hickey, G.M. and Lee, M.S.Y. (eds). *Evolution and Biogeography of Australasian Vertebrates.* Auscipub Pty. Ltd., Sydney. 942pp.

Williams, M.A.J. 1991. Evolution of the landscape. In Haynes, C.D., Ridpath, M.G. and Williams, M.A.J. (eds). *Monsoonal Australia: Landscape, Ecology and Man in the Northern Lowlands.* A.A. Balkema, Rotterdam. 231pp.

Chapter 6
Evolution of New Zealand and its Vertebrates

Richard N. Holdaway[1] and Trevor H. Worthy[2]

[1] Palaecol Research, P.O. Box 16569, Christchurch, New Zealand
[2] Department of Earth and Envirinmental Sciences, University of Adelaide, S.A. 5005 Australia
E-mail: piopio@paradise.net.nz

ZOOGEOGRAPHIC HISTORY

New Zealand (in the political sense, excluding Macquarie Island, New Caledonia, Norfolk Island and Lord Howe Island, which are part of the same continental fragment), is an archipelago of hundreds of islands, stretching from Raoul in the Kermadec Group (29°15′S) to Campbell Island (52°30′S), encompassing a total of just 270,000 km^2 (Figure 1). The main islands (North, 114,740 km^2; South, 151,120 km^2; and Stewart, 966 km^2), span 13 degrees of latitude (34°S–47°S) and account for 98.8% of the land area. There are 254 small (>8 ha) islands within 50 km of the coast of the North and South islands, and over 400 of < 8 ha. The total land area is one thirtieth that of Australia, one third that of Madagascar, and comparable to the British Isles.

A BRIEF GEOLOGICAL SUMMARY

The geological history of New Zealand is complex and intimately related to plate tectonics; this process has already been discussed by White (2006, pp.17–50). At plate margins, one plate usually sinks (subducts) beneath another, or (as also occurs in New Zealand) the edges grind sideways past each other. Plate boundaries do at times form under continents, so may result in the buckling up or splitting apart of the continent. There is a zone of friction between the plates, which is revealed by numerous earthquakes generated as the stress at the junction is released suddenly. Friction, or other factors, melts some of the sinking rock, forming magma which on rising to the surface produces volcanoes. The process is slow (millimetres per year) but in the 80+ million years since New Zealand parted company with Australia and Antarctica, it amounts to over 1,000 km. New Zealand straddles the margins of the Pacific and Australian plates that are converging now at about 42 mm per year, about the same rate that fingernails grow; over a million years, that amounts to 42 km.

Long before New Zealand in any form took shape, all the world's continents were united in a single super-continent called Pangaea, and stayed conjoined until late in the Triassic Period some 251–205 million years ago (m.y.a.). Then, a split into northern Laurasia and southern Gondwana created the Tethys Ocean. About 190 m.y.a., the first separation of parts of Laurasia formed the Atlantic Ocean, when North America split from Europe and Africa. Beginning about 120 m.y.a., Gondwana began to rupture, with Africa separating from South America. By about 90 m.y.a., Antarctica was moving into its polar position, and India had started north on a solo journey. New Zealand as such still did not exist, but its parts were being formed elsewhere; major stages in its formation are illustrated in Figure 2.

Sediments eroded from the northern shore of Gondwana (from the region that split to become Australia and Antarctica) had been accumulating in a deep near-shore trough. About 100 m.y.a., a rupture (or plate boundary) formed along the margin of the Australian part of Gondwana, and with the onset of sea-floor spreading in the area, a tiny sliver of Gondwana and the huge pile of sediments accumulated in the trough moved east. Together these sediments formed the New Zealand mini-continent, which extends from New Caledonia in the north to the south of the present New Zealand archipelago. It includes the Lord Howe Rise, the Norfolk Ridge, the Chatham Rise, and the Campbell Plateau. The break from the parent continent was complete and New Zealand was on its own by 82 m.y.a. (Lawver *et al.,* 1992). Sea-floor spreading continued until ~60 m.y.a., by which time New Zealand was 1,200 km from Australia (at New Caledonia) and 1,500 km away at South Island. Australia remained attached to Antarctica all this time and started to break free only at about 55 m.y.a.; it is still heading north.

In the interval 38–21 m.y.a., the Tasman plate boundary became inactive and a new plate boundary formed near Antarctica. The new rupture spread northwards along what is now the Macquarie Ridge and so New Zealand found itself straddling the new plate boundary (Stevens, 1980, pp.458–459). The new edge of the Indian-Australian plate abutted the Pacific plate south of New Zealand, and formed the Macquarie Ridge. In the northern part of New Zealand, the Pacific plate sank under New Zealand off Hawke's Bay, and the deep Kermadec Trench formed along the subduction zone to the north. In between, the new New Zealand block was itself rent asunder, and the spectacular Alpine Fault was born about 20 million years ago. A twin for the San Andreas Fault, the Alpine Fault traverses the length of the South Island, and rocks displaced on either side indicate 450 km of lateral movement.

Before the development of the new plate boundary, New Zealand was carried along in a relatively stable tectonic setting. Without uplift the land mass of New Zealand was erod-

Antarctica, their remains have yet to be found. Given how recently—and almost accidentally—fossils of those large animals have come to light, there is a finite possibility that there are undiscovered remains of tiny mammals. The present absence of terrestrial non-volant mammals is hugely important in understanding the composition and evolution of the New Zealand fauna. Even for bats, only two families are represented, the almost cosmopolitan Vespertilionidae and the endemic Mystacinidae.

Of these bats, *Chalinolobus tuberculatus* (Vespertilionidae) is the New Zealand representative of a genus that is widely distributed in Australia, New Guinea, New Caledonia, and (formerly) Norfolk Island (Flannery, 1995). It is an aerial insectivore and although still widespread it is now thought to be the more endangered of the two surviving bat species in New Zealand (O'Donnell, 2000, personal communication). Fossils of *Chalinolobus* are rare and all are from caves and of Holocene age.

The unique assemblage of morphological features of *Mystacina* (Mystacinidae, 2 spp.) has puzzled systematists for more than a century, resulting in its placement at various times in seven of the 18 extant families of bats (King, 1990). Albumen immunology indicated to Pierson *et al.* (1986) that *Mystacina* was an early branch off the superfamily Phyllostomoidea and most closely related to the South American Noctilionidae, Mormoopidae and Phyllostomidae. This prompted the suggestion that *Mystacina* had dispersed to New Zealand more than 35 million years ago from South America, via Antarctica. In a classic case where fossils have overturned biochemical evidence, the recent discovery of early mystacinid fossils in Australia shows otherwise. Hand *et al.* (1998) described three new species of bat in a new genus *Icarops*, in the family Mystacinidae, from the early–middle Miocene of Riversleigh, and suggested that *Icarops* and *Mystacina* are sister taxa. It seems most likely that, because of the prevailing wind direction, *Mystacina* dispersed from Australia in the late Oligocene or early Miocene, when New Zealand was closer to Australia. Also, at this time, New Zealand was emerging (literally) from a prolonged period of widespread inundation by shallow seas.

Two species of *Mystacina* are known. Although *M. tuberculata* is considered to be endangered it still survives in suitable habitat in areas of both North and South islands, and on some offshore islets. *Mystacina robusta* was a much larger species that is now extinct, but is known from fossil and historical records (Worthy *et al.*, 1996); the last populations—on islands off Stewart Island—were exterminated in the 1960s by introduced rats. Neither of the New Zealand mystacinids has a pre-Pleistocene fossil record.

ENVIRONMENTAL INTERACTIONS INFLUENCING BIOTAS

As a continental fragment lying between the subtropics and the subantarctic, athwart the West Wind Drift, New Zealand has had a unique physical environment for much of its independent history. Surrounded by ocean, the relatively small, narrow land mass has a predominantly maritime climate, except high in the mountains and in central Otago and inland Canterbury (Mackenzie Basin) where the surrounding mountains allow continental extremes. A large latitudinal range means that the southern South Island enjoys a cool-temperate climate, whereas the northern North Island copes with reduced seasonality approaching a subtropical equability.

Geology and geophysics profoundly affect the climate pattern and with it the vegetation. The young mountains of the Southern Alps separate a narrow coastal plain in the west from broader plains and lower ranges to the east. Most of the moisture carried by the westerlies is deposited on the mountains, resulting in high erosion rates, soil leaching, and very humid conditions. The leaching compounds the almost universal problem of low natural soil fertility in New Zealand: even the young soils that might be thought to have high levels of nutrients are formed from rocks lacking in many trace elements. The rain shadow areas in the east rely on southerly rain and the occasional errant subtropical storm, but the great rivers that flow east from the Main Divide are fed by the westerlies as are the snowfields and glaciers in the mountains themselves.

Apart from the hard sandstones of the axial ranges in the south-east, the surficial rocks over much of the North Island are soft and erodible Tertiary marine sediments. The relatively low, but deeply dissected relief means that rainfall is not concentrated on either coast. The north and east are subject to intense rain bursts from cyclones, which exacerbate soil movement and leaching. The highest mountains in the North Island are volcanoes, and volcanoes of various lithologies are important parts of the landscape everywhere in the North Island except the south and east. Dominating the centre of the island is a plateau formed from the eruption products of several volcanic zones. Much of the volume of volcanic material has been erupted as extremely viscous rhyolite. Eruptions of rhyolite are so violent that the ejecta are spread to hundreds of kilometres from the vent, so, as with Lake Taupo, the volcano itself is not a mountain, but a large caldera. Devastating ignimbrite flows have repeatedly eliminated most of the vegetation within 100 km of the main vents.

The New Zealand biota was still recovering from the Oligocene marine transgressions (Cooper and Cooper, 1995) when tectonic uplift and astronomically-forced climate fluctuations came to dominate the New Zealand scene. The distribution of vegetation and fauna has ever since been governed by the interplay of the results of these global processes. In the South Island, the major factors have been the rise of the Southern Alps, the dramatic lateral displacement along the Alpine Fault, and the Quaternary glaciations. In the North Island, explosive volcanism has been more important than other tectonic processes and has resulted in relatively frequent (on a geological time scale) oscillations in the distribution of vegetation cover over much of the island. The factors were not unique to each island: glacial climates affected the North Island vegetation patterns over longer periods than each volcanic event (and the two were often superimposed), and some of the comparatively rare more violent explosions undoubtedly affected vegetation over much of the South Island. But, with the land area available, faunas and vegetation assemblages coped with most of the destructive influences, shifting with climate change and recovering rapidly from each volcanic episode. Only the human influence has been so pervasive, powerful and rapid as to leave many species and assemblages without a refuge or a chance of recovery.

ARCHAIC ELEMENTS

ORIGINALS

About 82 m.y.a. when New Zealand separated from Australia it carried with it a fauna that included dinosaurs and pterosaurs These survived until the end of the Cretaceous, (~ 65 m.y.a.), when presumably they met the same fate as their relatives elsewhere. After that, there is a large blank in the terrestrial fossil record. The composition of the rest of the fauna in the Cretaceous and much of the early Cainozoic is, however, hinted at by the Holocene-modern fauna. Over-water dispersal is most unlikely for some groups. As their respective evolutionary grades were in existence long before New Zealand became isolated, we can assume that sphenodontids and leiopelmatid frogs were part of the original fauna. Doubtless, it is the isolation from evolving forms elsewhere that allowed such taxa to survive on New Zealand. It seems likely that the archosaurs that gave rise to the mekosuchines were also present, although cross-channel swimming might have been easier with the narrower gap of the Paleocene and Eocene Tasman Sea.

A number of primitive birds would also have contributed to this original fauna, but few of these made it to the Holocene in any recognizable form. The examples most frequently quoted are the ratites (moas and kiwis) whose relatives have mainly a Gondwanan distribution. But the existence of volant ratites in the Northern Hemisphere in the Paleocene (Houde, 1986, 1988) raises other possibilities; mtDNA studies suggest kiwis are closer to cassowaries and emus than to moas and diverged from the former long after New Zealand became isolated (Cooper, 1997; Cooper *et al*., 1992). Without exploring the debate of ratite relationships, which is extensively covered by Sibley and Ahlquist (1990) and Worthy and Holdaway (2001), it seems reasonable to accept that moa ancestors were among the original fauna. Recent molecular evidence suggests many modern orders of birds evolved before the end of the Cretaceous (Cooper and Penny, 1997), and so several should have been in New Zealand from the outset. Strong contenders for members of this original fauna are the primitive passerines that gave rise to the family Acanthisittidae, of which the rifleman and rock-wrens survive. It is probable that even some parrots were present—the kakapo has ancient roots deep under the family tree that may pre-date the Tertiary (Cooper and Penny, 1997).

THE OLIGOCENE CRISIS

As we have noted, any changes in the New Zealand fauna after its isolation can only be conjecture. Certainly there must have been radiations of species in many groups to take advantage of the huge faunal gaps left when the dinosaurs went extinct, but so far, the fossil record is silent on the matter.

Much of New Zealand was covered by shallow seas in the Oligocene, as shown by the extensive deposits of limestone of that age. As little as 18% of present land area was above sea level (Cooper and Cooper, 1995); this alone would have reduced the terrestrial faunal diversity markedly. Paradoxically, as the inundation eliminated taxa that could not live in habitats that remained, it would have promoted speciation in those groups that could. Fragmentation into separate islands would have created isolated populations that could have diverged from each other and remained separate when the islands coalesced again. Then, as land emerged during the Miocene (20–12 m.y.a.), the diversity of habitats increased again. Some of the new species would have been lost, but the development of others would have been promoted as populations entered the new habitats. It could therefore be predicted that radiations in several modern groups may have started in the Oligocene or early Miocene. Molecular studies on several groups now support this hypothesis showing radiations commencing in the Miocene for *Leiopelma* frogs and *Leiolopisma* skinks (Daugherty *et al.,* 1982, 1990), as well as the radiations of moas, kiwis and acanthisittid wrens (Cooper, 1994; Cooper and Cooper, 1995).

LATE TERTIARY–EARLY QUATERNARY CHANGES

The absence of a rich Tertiary fossil record like that of Australia—in New Zealand represented by only a few bones of two ducks and one crocodilian—makes the late Pleistocene fauna of terrestrial vertebrates even more dramatic. The earlier Pleistocene record represented by the Marton site (at 1 million years) and a few fossil moa are all that are identifiable as Holocene taxa (see above) although it is clear that there has been substantial faunal turnover in the past million years. The discovery of further mid-Pleistocene faunas will allow the extent of this change to be documented.

LATE PLEISTOCENE–HOLOCENE CHANGES

In contrast to the extreme paucity of older material, extensive fossil deposits date from the last 50,000 years. The composition of the fauna is well known for parts of North and South islands in the glacial period (locally called the Otiran) including periods up to the coldest time at 18,000 (18 kyr) radiocarbon years BP, to the commencement of climatic amelioration at 14 kyr BP. The Holocene (10 kyr BP to the present) is even better known. One major difference between the New Zealand vertebrate fauna and that of other islands and continents populated by humans in the past 50,000 years is that all taxa known from the late Pleistocene (up to 10 kyr BP) survived to the late Holocene. The massive extinction event that characterized the end of the Pleistocene in North America and somewhat earlier in Australia happened much later in New Zealand, in the past 2,000 years, when 56 birds (Table 2), at least one or two lizards, a bat, three frogs and a fish became extinct.

While it is true that no species is known to have become extinct at the end of the New Zealand Pleistocene, there were local extirpations and changes in the distribution of faunas that matched climatic and vegetation changes (Worthy, 1999). Such faunal changes are most clearly exemplified by the moas. The 11 species were clearly separated ecologically into several groups (Worthy, 1990). Their large bones usefully provide the most easily found fossils and so, in areas where there is a clear history of vegetation change (shrubland/grasslands to closed forest) from the Pleistocene to the Holocene, the species present now provide strong evidence for the age of the deposits.

PRECIPITATING THE EXTINCTION

FLIGHTLESS TRADITION

The New Zealand fauna contained many flightless species and many others that fly poorly. While none may be better known than the New Zealand emblem, the kiwi, the trend to flightlessness transcended ordinal boundaries and some or all species in seven orders were flightless, and in those and others, individual species had reduced powers of flight. The ratites, with 11 moa (Dinornithiformes) and 3 kiwi (Apterygiformes) species were all flightless, but the condition is found in other groups. Examples are: ducks (Anseriformes) 9 spp., rails (Gruiformes) 13 spp., a species of parrot (Psittaciformes) and passerine birds (Passeriformes)—4 wrens in the Acanthisittidae. These 41 species represent 31% of the 132 land and freshwater birds breeding in New Zealand in the Holocene. Until the recent discovery of fossils of a flightless bunting in the Canary Islands (Rando *et al.*, 1999), the flightless passerines of New Zealand were the only such members of this order in the world.

Others, like the owlet-nightjar (*Aegotheles novaezealandiae*), all five members of the wattlebird family Callaeatidae, the piopios (*Turnagra* spp.) and two further members of the Acanthisittidae (*Xenicus* spp.) were poor fliers.

Of these flightless birds, the moas are notable as the only group of birds to have totally lost the wing skeleton. Only a single element of the pectoral girdle, the fused coracoid and scapula, remains. Most of the other flightless species have much reduced wing skeletons and the sternal carina (keel) is reduced in height. Ratites are, of course, named for this feature (the sternum is raft-like) and lack a carina altogether. The kakapo (*Strigops habroptilus*) and two of the wrens (*Traversia, Dendroscansor*) have the carina reduced to a line on the sternum.

Associated with flightlessness is a strong tendency to increased size. The only flightless parrot in the world, the New Zealand kakapo, is also the largest at 2–2.5 kg. At the extreme end of the scale, the moas are the stuff of legend with some forms reaching 250 kg and a physical reach of 3 m (although the normal height was a mere 2 m). The smallest moa was only a paltry 20 kg. Kiwi, which nest in holes in the ground, are smaller, and the species range in mass from 1 to 3 kg.

Like the parrots, the rails and their allies in New Zealand included some relative giants, especially the aberrant *Aptornis* spp., which were among the largest gruiforms at 15–20 kg. Of similar size there were among the waterfowl two species of primitive goose *Cnemiornis* spp. related to the Australian Cape Barren goose (*Cereopsis*) (Worthy *et al.,* 1997). The flightless rallids also included the world's largest swamphen *Porphyrio mantelli*, the largest coots *Fulica prisca* and *F. chathamensis*, and the largest derivative of the widespread genus *Gallirallus, Diaphorapteryx hawkinsi*.

However, the trend to increased size is not universal, especially in the waterfowl where the teals (*Anas* spp.) show a reverse trend, with flightless taxa being smaller than their flying congeners, both in New Zealand and on Macquarie Island, and on Amsterdam Island (*Anas marecula*), in the Indian Ocean (Olson and Jouventin, 1996). The flightless wrens remained diminutive (15–40 g).

Many of the flightless species are derivatives from volant Australian or mainland New Zealand ancestors. The various anatids demonstrate this well with New Zealand Australian pairs as follows: *Cnemiornis* spp.: *Cereopsis novaehollandiae*, *Biziura delautouri: B. lobata*, *Chenonetta finschi: Chenonetta jubata*. The story of the strongly volant Australian *Anas castanea*, the weakly volant mainland New Zealand *A. chlorotis*, and the flightless subantarctic *A. nesiotis* and *A. aucklandica* is more complex. Kennedy and Spencer (2000) have shown that the New Zealand species diverged from a common ancestor within New Zealand, independent of, and perhaps before, the split between the grey and chestnut teals within Australia, and that the New Zealand forms warrant species status as originally conferred, and later maintained by Oliver (1955) and Livezey (1990).

ABSENCE OF MAMMALIAN PREDATORS

Naivety

One attribute of species on islands that continental biologists often find difficult to believe is the often-extreme tameness or naivety of the vertebrates. Charles

Table 2. List of bird taxa (species, or taxa considered of specific status) that went extinct in the last 1,000 years in New Zealand. Author is author of taxon; if in CAPITAL LETTERS it is the reference advising of the undescribed status of the species. Status column indicates whether extinct globally (E), or locally extinct (LE). Key to distributions: K = Kermadec Islands; N = North Island; S = South Island; Ch = Chatham Islands; Sub = Subantarctic islands.

Species	Author *	Common name	Status	Distribution
Anomalopteryx didiformis	(Owen, 1844)	Little bush moa	E	N S
Emeus crassus	(Owen, 1846)	Eastern moa	E	S
Euryapteryx curtus	(Owen, 1846)	Coastal moa	E	N
Euryapteryx geranoides	(Owen, 1848)	Stout-legged moa	E	N S
Megalapteryx didinus	(Owen, 1843)	Upland moa	E	S
Pachyornis australis	Oliver, 1949	Crested moa	E	S
Pachyornis elephantopus	(Owen, 1856)	Heavy-footed moa	E	S
Pachyornis mappini	Archey, 1941	Mappin's moa	E	N
Dinornis struthoides	Owen, 1844	Slender bush moa	E	N S
Dinornis novaezealandiae	Owen, 1843	Large bush moa	E	N S
Dinornis giganteus	Owen, 1844	Giant moa	E	N S
Apteryx "East South Is"	WORTHY 1997	Eastern kiwi	E	S
Pterodroma alba	(Gmelin, 1789)	Phoenix petrel	LE	K?
Pterodroma sp. 1	MILLENER 1999	Unnamed Chatham petrel	E	Ch
Puffinus spelaeus	Holdaway & Worthy, 1994	Scarlett's shearwater	E	S
Oceanites maorianus	(Mathews, 1932)	New Zealand storm petrel	E	N S
Eudyptes "Chatham Island"	MILLENER, 1999	Chatham crested penguin	E	Ch
Ixobrychus novaezelandiae	(Potts, 1871)	New Zealand little bittern	E	N S Ch
Biziura delautouri	Forbes, 1892	New Zealand musk duck	E	N S
Cnemiornis calcitrans	Owen, 1865	South Island goose	E	S
Cnemiornis gracilis	Forbes, 1892	North Island goose	E	N
Chenonetta finschi	(Van Beneden, 1875)	Finsch's duck	E	N S
Malacorhynchos scarletti	Olson, 1977	Scarlett's duck	E	N S Ch
Mergus australis	Hombron & Jacquinot, 1841	New Zealand merganser	E	N S Ch Sub
Pachyanas chathamica	Oliver, 1955	Chatham Island duck	E	Ch
Tadorna "Chathams"	MILLENER 1999	Chatham Island shelduck	E	Ch
Circus eylesi	Scarlett, 1953	Eyles's harrier	E	N S
Harpagornis moorei	Haast, 1872	Haast's eagle	E	S
Coturnix novaezelandiae	Quoy & Gaimard, 1830	New Zealand quail	E	N S
Megapodius sp.	CHEESEMAN 1891	Megapode	?LE	K
Aptornis defossor	Owen, 1871	South Island adzebill	E	S
Aptornis otidiformis	(Owen, 1844)	North Island adzebill	E	N
Cabalus modestus	(Hutton, 1872)	Chatham Island rail	E	Ch
Capellirallus karamu	Falla, 1954	Snipe rail	E	N
Diaphorapteryx hawkinsi	(Forbes, 1892)	Hawkins' rail	E	Ch
Fulica chathamensis	Forbes, 1892	Chatham Island coot	E	Ch
Fulica prisca	Hamilton, 1893	New Zealand coot	E	N S
Gallinula hodgenorum	(Scarlett, 1955)	Hodgens' waterhen	E	N S
Porphyrio mantelli	(Owen, 1848)	North Island takahe	E	N
Coenocorypha chathamica	(Forbes, 1893)	Forbes' snipe	E	Ch
Coenocorypha iredalei	Rothschild, 1921	South Island snipe	E	S
Coenocorypha barrierensis	Oliver, 1955	North Island snipe	E	N
Nestor "Chatham Is"	MILLENER, 1999	Chatham Island kaka	E	Ch
Sceloglaux albifacies	(G.R. Gray, 1844)	Laughing owl	E	N S
Aegotheles novaezealandiae	(Scarlett, 1968)	New Zealand owlet-nightjar	E	N S
Dendroscansor decurvirostris	Millener & Worthy, 1991	Long-billed wren	E	S
Pachyplichas yaldwyni	Millener, 1988	Stout-legged wren	E	N S
Traversia lyalli	Rothschild, 1894	Lyall's wren	E	N S
Xenicus longipes	(Gmelin, 1789)	Bush wren	E	N S
Bowdleria rufescens	(Buller, 1869)	Chatham Island fernbird	E	Ch
Anthornis melanocephala	G.R. Gray, 1843	Chatham Island bellbird	E	Ch
Callaeas wilsoni	(Gmelin, 1788)	South Island kokako	E	S
Heteralocha acutirostris	(Gould, 1837)	Huia	E	N
Turnagra capensis	(Sparrman, 1787)	South Island piopio	E	S
Turnagra tanagra	(Schlegel, 1865)	North Island piopio	E	N
Corvus "mainland"		New Zealand raven	E	N S Sub?
Corvus moriorum	Forbes, 1892	Chatham Island raven	E	Ch

* Most of these original descriptions are not included in the reference list, but details can be quickly accessed in recent reference volumes listed.

Darwin remarked at length on the 'tameness' of birds in the Galapagos (Darwin, 1884, p.398), and, despite centuries of human contact, many species there are still very tolerant of humans (Grant, 1998). Significantly, however, they react to the presence of predators that have been on the islands for a much longer time. In New Zealand, species on the offshore and outlying islands are usually not afraid of humans: saddlebacks (*Philesturnus carunculatus* and *P. rufusater*) will feed from the hand. The same birds react to the presence of harriers (*Circus approximans*) and owls (*Ninox novaeseelandiae*).

Naivety of birds is not only manifested in contacts with humans. Where species evolved in the absence of any mammalian predators—as those on islands certainly did—then they do not react to introduced predators in the same way as they would to, say, an owl. Usually they sit and wait to be killed. A good example is the kakapo (*Strigops habroptilus*), one the largest surviving species, and now surviving only in managed populations. Kakapo ignore Pacific rats (*Rattus exulans*) even when they enter the nest cavity; these rats are known to take eggs and attack chicks. Rats, mustelids, people and other introduced predators were all totally unfamiliar to the endemic New Zealand species. It is very likely that moas and other birds would have stood and watched the first humans they saw, but taken cover at the approach of either Haast's eagle (*Harpagornis moorei*) or Eyles's harrier (*Circus eylesi*).

K-selection Attributes

Another aspect of the biology of many bird species on islands is that they are long-lived, their populations are maintained at or near the carrying capacity of the environment, and they have small clutches and low replacement rates. Species with these characteristics are termed K-selected, from the variable K, the carrying capacity in calculations of population changes. Even many of the small songbirds in New Zealand are K-selected; compared to songbirds elsewhere, even the smallest species survive for several years. The black robin (*Petroica traversi*) of the Chatham Islands cheated extinction because the famous female 'Old Blue' lived for over 12 years (Heather and Robertson, 1996) and provided chicks for the intensive management programme (Butler and Merton, 1992). Of the larger birds, a female northern royal albatross (*Diomedea sanfordi*) banded as an adult at Taiaroa Head near Dunedin lived for at least 62 years, finally disappearing in 1989 (Robertson and Nicholls, 2000).

Annual adult survivorship for large K-selected bird species is typically over 90%. Being K-selected makes a species vulnerable to predation on the adults: the reproductive rate is so low that adults are removed before they can replace themselves. For example, Croxall and Rothery (1991) found that juvenile survivorship must increase by 6% to offset a reduction of just 1% in the survival of adult wandering albatrosses. Indeed, at present the 'best' example of a group of K-selected species in danger because of predation on adults is that of the albatrosses breeding in the South Atlantic and southern Indian Oceans. Even extra mortality as low as 1% per year has resulted in catastrophic declines in several species (Weimerskirch *et al.*, 1987; Weimerskirch and Jouventin, 1987; Croxall *et al.*, 1990; Croxall and Rothery, 1991). There is no way that juvenile survivorship can increase to offset the losses. In terrestrial species, where predators also target juveniles and eggs, the declines are even faster. Lists of extinct and critically endangered birds are heavily biased towards K-selected species.

HUMANS—ARCH PREDATORS, ENVIRONMENTAL VANDALS

Polynesians

Pacific Island ecosystems, including those in New Zealand, were fully functional with all species living in balance and none likely to have been in threat of extinction until the ancestors of the Polynesians moved out into 'Remote Oceania' about 3,500 years ago (Green, 1994). In their travels, these people visited or colonized almost every island in the great triangle formed by Hawaii, Easter Island and New Zealand (Irwin, 1992). People changed the island ecosystems forever (Kirch, 1982, 1983, 1997; Steadman, 1997).

Most evidence suggests that settlement of New Zealand began in the late 13th century (Anderson, 1991; Higham *et al.*, 1999). Within about 100 years of contact, or by the close of the 14th century, almost all of the large birds on the main islands and the Chathams were either extinct or had been reduced to non-viable remnants (Holdaway, 1999b; Holdaway and Jacomb, 2000; McGlone *et al.*, 1994; McGlone *et al.*, 2001). It is possible that in some remote areas populations of such birds survived into the early 15th century but there is no evidence for later survival. Of the 39 species of bird that have become extinct in New Zealand (North and South islands) in the past 2,000 years, 34 have been found as food remains in archaeological sites (Worthy, 1999). The five other species were either recently described or are small species that are unlikely to have been used as food (Worthy, 1999). Some sites in the south-eastern South Island cover many hectares (Anderson, 1989) and provide some indication of the scale of the destruction of moas (Dinornithiformes). An analysis of the frequency of species in sites by size indicates that larger species were selected more often than small ones, but within a region species were probably taken in proportion to the frequency in which they were encountered (Worthy, 1999). Few sites provide evidence that any particular taxon was targeted: for example, the large faunas from Marfell's Beach indicate that swans, which may seem to be an obvious preferred prey, were not preferentially hunted, yet extinct coot were.

Other large taxa, such as the flightless geese (*Cnemiornis*) and adzebills (*Aptornis*) appear only in small numbers in the earliest archaeological sites and must have been early losses. The absence of remains of the large flightless species from sites with 'transitional' artefact assemblages suggests that most, if not all, of these species went extinct within about a century of settlement (Holdaway and Jacomb, 2000).

There are two forms of evidence suggesting that New Zealand's first colonizers may not have been the first visitors. The presence of Pacific rat bones dated to about 2,000 years BP from natural sites (Holdaway, 1996, 1999a; Holdaway and Beavan, 1999), and the fact that the mainland fauna was apparently missing some rat-susceptible species (Holdaway, 1999b), both suggest that people visited as early as 2,000 years ago. The rat dates have been disputed (Anderson, 1996; Smith and Anderson, 1998) on the basis that rat bones from archaeological sites gave dates that were older than other materials (shell, charcoal) in the same layers. Also, dates on different bones from non-articulated bones considered to be from a single skeleton gave quite disparate ages. However, Beavan Athfield *et al.* (1999) presented dates on Pacific rat bones from another archaeological site that were consistent with dates for other materials from the site. In addition, dates up to 600 years older than human settlement on rat gelatin from natural sites have been corroborated by stratigraphically consistent OSL dates and dates on collagen from other species (Holdaway and Beavan, 1999; Holdaway *et al.*, 2002).

If the old dates on rat bones from natural sites are accepted and corroborated by further evidence, it must be accepted that humans reached New Zealand about 2,000 years ago, but these first people did not stay and left the commensal Pacific rat (*Rattus exulans*) as a legacy. Carried throughout the Pacific by Polynesian voyagers (Roberts, 1991), the rat reached its southernmost limit near Stewart Island (Atkinson and Moller, 1990). Until recently, most biologists have regarded the Pacific rat as a harmless herbivore, and even as a natural part of the New Zealand environment, according to King (1990), being termed the 'native rat' by Hutton and Drummond (1905). However, the truth of the detrimental and often catastrophic effects of rat predation on plants, invertebrates and small vertebrates has now been recognized. The impact of Pacific rats on the birds is visible in the list of extinctions of small species before European arrival (Holdaway, 1999b).

An anomalous feature of the archaeological record, that may support the hypothesis of an early arrival of rats, is that archaeological deposits rich in the remains of petrels (Procellariiformes) are rare on the two main islands of New Zealand. Recent studies have shown that at least 10 species of petrel bred in large numbers on the South Island alone before human contact (Holdaway *et al.*, 2001). Anderson (1997) could point to only one site, at Tiwai Point, at the southern end of the South Island that included substantial remains of petrels. Elsewhere in the Pacific, such as on Norfolk Island (Holdaway and Anderson, 2001) and the Chatham Islands (Sutton, 1979), petrels down to the size of prions (100–200 g) were regular items of food for pre-European Polynesian inhabitants and the remains of these species are abundant in archaeological sites.

The absence of archaeological sites with abundant remains of petrels cannot be a result of depositional or excavation biases against small bones, because species as small as robins (*Petroica australis*) are regularly recorded, despite the routine use of excavation techniques that militated against the recovery of such material (Worthy, 1999). While 'muttonbirding' was an important historical activity in some areas, especially in southern New Zealand, Anderson (1997) was at a loss to explain the lack of petrel bones in middens and favoured a late development of the practice. He recorded *Puffinus* bones from 76 sites, and found that bones were infrequent or rare in most mainland sites with intensive exploitation suggested in only one North Island (Opito Bay) and one South Island site (Tiwai Point). Cultural practices that result in few petrel bones being in middens, as now practised in southern titi islands (Anderson, 1997), are unlikely to have been practised in the past; in sites on the Chathams, northern offshore islands (e.g. Red Mercury Island), Stewart Island (Old Neck) and the above mentioned mainland sites, bones are common (Anderson, 1997). The absence or rarity of petrel bones from mainland sites may best be explained by the suggestion that mainland colonies of petrels were small and minor during the period of human occupation. Both mainland sites with abundant shearwater bones are adjacent to islands where populations of petrels still survive today, and doubtless reflect exploitation of such colonies.

We therefore suggest that most of the smaller petrels were extinct, or at least economically so in terms of foraging, on the North and South Islands before permanent settlement began. Remaining petrel populations are still susceptible to rat predation and are in danger of extinction with the continued presence of the Pacific rat both in New Zealand (Booth *et al.*, 1996) and elsewhere (Brooke, 1995). The late Holocene abundance of petrels yet the absence of substantial evidence for human consumption of petrels on the main islands strongly supports the hypothesis of an early introduction of the rat.

While it is now clear that the Pacific rat irrevocably affected populations and species in the past and still continues to detrimentally do so today, there is little evidence that the Polynesian dog ever became feral before European settlement, and its use in moa hunting, if at all, was minor (Anderson, 1981).

Habitat Destruction

Pre-European Changes

Polynesian settlement brought with it widespread burning of the vegetation in the drier areas (McGlone, 1983, 1989). Most if not all of the destruction of forest and shrubland took place within a century of the arrival of Polynesian settlers (McGlone and Wilmshurst, 1999; Newnham *et al.*, 1998). One estimate places the area of predominantly woody vegetation in the South Island transformed into lowland and montane tussock grassland at about 50% of the land area (McGlone, 1989), most of the land east of the main ranges.

The removal of so much forest and shrubland created large areas of new habitats, particularly grassland and open swamp. Those new habitats were occupied by the few indigenous birds that could survive there, most importantly the pipit (*Anthus novaeseelandiae*) and quail (*Coturnix novaezelandiae*), which are both well represented in deposits up to 50,000 years old in New Zealand. The absence of a specialized grassland avifauna (Holdaway, 1990) and indeed even an extensive grassland fauna of invertebrates (Harrison and White, 1969) is strong evidence for the dominance of woody vegetation for much of the history of the land mass. The takahe, *Porphyrio hochstetteri*, has been argued to be a specialist grassland species (Mills *et al.*, 1984, 1988). However, we consider it best interpreted as one that lived in forest margin type habitats where sufficient monocotyledons were present, and at all altitudes (Beauchamp and Worthy, 1988; Trewick and Worthy, 2000). Thus, while in glacial times its range could expand into the increased grassland areas made available by reduced forests, it could not make use of the new anthropogenic grasslands in the late Holocene, because of human predation.

The partial ecological vacuum allowed new species to establish, from the continuous 'rain' of potential colonists that undoubtedly existed throughout the Tertiary and continued into the late Holocene, but few of which could find a place in the intact fauna and vegetation (Holdaway *et al.*, 2001). At present, at least 16 species of bird are thought to have colonized after the arrival of Polynesian settlers (Table 3), including species that have been considered to be long-term residents like the harrier (*Circus approximans*) and pukeko (*Porphyrio porphyrio*). The original fauna, where it is intact and not unduly ravaged by extinction, and living in mostly intact vegetation, as on Little Barrier Island, can resist the incursion of even such a vigorous colonist as the silvereye (*Zosterops lateralis*) (McCallum, 1982). In all, 230 bird species are known to have bred in New Zealand as defined here in the late Holocene, including extinct species and recent natural immigrants (Table 1).

Before the Polynesian deforestation, many if not most New Zealand swamps were swamp forests, with trees such as kahikatea (*Dacrycarpus dacrydioides*). Present swamp assemblages based on raupo (*Typha orientalis*) and New Zealand flax (*Phormium tenax*) were not extensive (McGlone, 1999, personal communication). Some immigrants such as the bittern (*Botaurus poiciloptilus*) seem to have taken advantage of the change, as did the pukeko when areas of grazing sward became available.

Post-European Changes

Europeans immediately added greatly to the list of exotic mammals, with 32 species now having populations somewhere in the archipelago (King, 1990). Successful introductions, both deliberate and accidental, date from the earliest European settlements in the late 18th and early 19th centuries. Explorers, whalers, and sealers followed deliberate policies of introducing mammals that they thought would assist both themselves and future occupants or shipwrecked sailors on the main and outlying islands (King, 1990). Populations of pigs (*Sus scrofa*) and goats (*Capra hircus*) were joined by Norway rats (*Rattus norvegicus*) and house mice (*Mus musculus*) that landed from the same ships. With the settlers came cats (*Felis catus*) and dogs (*Canis domesticus*), both of which rather rapidly went feral, and the European dog quickly replaced the Polynesian breed. At some time before 1860, the black rat (*Rattus rattus*) arrived, but when is a matter of debate. Atkinson (1973) suggested that they did not spread widely until the second half of the 19th century, but they probably arrived several decades before, at least in Northland, as Colenso (1843) explicitly describes their presence in the Bay of Islands in 1843.

Many species of mammals were introduced for sport, mostly deer and wallabies, but including the European rabbit (*Oryctolagus cuniculus*) and hare (*Lepus europaeus*). All the wallabies, ungulates, and the hare have caused damage to both the natural vegetation and to crops to the extent that control has been necessary. But two species—the rabbit and the red deer (*Cervus elaphus*)—have been subjects of extremely expensive control and eradication programmes because of the massive damage that they have caused in the grasslands and forests, respectively. Now attention is focused on the Australian brush-tail possum (*Trichosurus vulpecula*) whose numbers were estimated in the late 1990s as being about 70,000,000 animals. Possums are now recognized as being a threat not only to the forests and birds, but also to commercial farming because they carry tuberculosis and can transmit it to cattle and thence to humans.

Rats have been recognized relatively recently as major predators of New Zealand birds (Bell, 1978; Booth *et al.,* 1996; Holdaway, 1989, 1999b). Three species of mustelid (*Mustela erminea, M. furo, M. nivalis*), introduced in the 1880s in an attempt to control rabbits, were soon implicated in the destruction of birds. One species, the stoat (*M. erminea*) is now known to be the main cause of the dramatic reduction in the populations of kiwis (*Apteryx* spp.), takahe (*Porphyrio hochstetteri*), kaka (*Nestor meridionalis*), parakeet (*Cyanoramphus* spp.), and yellowhead (*Mohoua ochrocephala*) since the 1880s. *M. erminea,* with the black rat, has contributed to the extinction of laughing owls (*Sceloglaux albifacies*), piopios (*Turnagra* spp.) and the near extinction of the kokako (*Callaeas* spp.), and on the main islands extinction of saddlebacks (*Philesturnus* spp.) (O'Donnell, 1996; Heather and Robertson, 1996).

Populations of stoats are driven by population cycles in their main prey, the house mouse, which are in turn responses to the periodic mast seeding of the southern beech (*Nothofagus* spp.) trees. Before the end of the 19th century, plagues of Pacific rats were reported, which probably also were driven by the same cycle of food abundance, and which may have contributed to losses of smaller passerines on the main islands (Lovegrove, 1996a).

Removal of predators on small islands can have dramatic effects. For example, the recent removal of Norway

Table 3. List of bird species that have established breeding populations in the New Zealand region naturally in the last 500 years. Under 'notes' Pe means the population established in pre-European time (~ before 1800), but less than 500 years ago.

Species	Author *	Common name	Notes
Tachybaptus novaehollandiae	(Stephens, 1826)	Australian little grebe	
Thalassarche melanophris	(Temminck, 1828)	Black-browed albatross	
Pterodroma mollis	(Gould, 1844)	Soft-plumaged petrel	
Phalacrocorax sulcirostris	(Brandt, 1837)	Little black shag	Pe
Ardea novaehollandiae	Latham, 1790	White-faced heron	
Botaurus poiciloptilus	(Wagler, 1827)	Australasian bittern	Pe
Platalea regia	Gould, 1838	Royal spoonbill	
Circus approximans	Peale, 1848	Australasian harrier	Pe
Fulica atra	Gould, 1845	Eurasian coot	
Porphyrio melanotus	Temminck, 1820	Pukeko	Pe
Himantopus himantopus	(Linnaeus, 1758)	Australasian pied stilt	Pe
Charadrius melanops	Vieillot, 1818	Black-fronted dotterel	
Vanellus miles	(Boddaert, 1783)	Spur-winged plover	
Hirundo neoxena	(Gmelin, 1789)	Welcome swallow	
Zosterops lateralis	(Latham, 1801)	Silvereye	

* These original descriptions are not included in the reference list, but details can be quickly accessed in recent reference volumes listed.

and Pacific rats from Kapiti has led to a 160% increase in the population of parakeets within three years (Orange, 2000). Intensive trapping in small areas does lower predation rates and allow some recovery of numbers for yellowheads (Elliott, 1996; O'Donnell *et al.*, 1996), but not for yellow-crowned parakeets (Elliott *et al.*, 1996). In any event, such programmes are expensive, unlikely to be viable in the long term, and do not reach the core of the problem, which is the ubiquity and abundance of two of the three mustelids, and of the black rat, in New Zealand mainland forests. One of the mustelids, the ferret (*Mustela furo*) not only preys on ground birds, but also carries bovine tuberculosis (Ragg, 1998), so control programmes are more likely to be funded, developed and maintained.

ENDANGERED BIRDS—CURRENT SITUATION

Of the 216 species of bird breeding in New Zealand at the time of first human contact, 56 (25.9%) are now extinct (Holdaway *et al.*, 2001). Still others would be extinct were it not for remnant populations on islands that escaped inoculation with introduced mammalian predators. Indeed, the importance of the smaller offshore and outlying islands to conservation of the New Zealand fauna as a whole cannot be overstated. Even though few islands have an entirely natural fauna and flora (Adams and Disappointment Islands in the Auckland group, and The Snares are the largest), here and there, populations of species eliminated from the main islands of each group survive. See Plates 36 and 37.

In the Chathams, the shore plover (*Charadrius novaeseelandiae*), snipe (*Coenocorypha pusilla*), and other species, including many invertebrates, are extinct on main Chatham Island but survive on the islets surrounding the nearby Pitt Island. The Campbell Island duck (*Anas nesiotis*) survives on Dent Island off the north-west coast of Campbell Island itself, and an undescribed form of snipe (*Coenocorypha* sp.) was discovered in 1998 on Jacquemart Island off the south coast. Both species are absent from Campbell Island, and are presumed to have been exterminated by Norway rats that arrived with the first sealing and whaling ships to anchor in the island's sheltered bays.

Islands off North, South and Stewart Islands are now used for intensive programmes of translocations to provide alternative homes for species whose habitat is now untenable because of predators. See Plate 37. The modern translocations began as a means of saving the populations of saddleback (*Philesturnus carunculatus*), bush wren (*Xenicus longipes*), and snipe (*Coenocorypha iredalei*) on Big South Cape Island after *Rattus rattus* colonized from fishing boats in the late 1950s (Bell, 1978). The translocations for the wren and the snipe ultimately failed and these species are now extinct. The species that have benefited most from the translocations are the North and South Island saddlebacks (*P. rufusater* and *P. carunculatus*), both of which are now established on several new islands after 45 releases, not all of which have been successful (Lovegrove, 1996b).

Although most work until recently has concentrated on the birds, efforts are now being made to increase the number of populations of, for example, tuatara (*Sphenodon* spp.), lizards, and frogs, and captive breeding programmes are under way for the tuatara, frogs, some lizards, as well as for some of the larger invertebrates such as the giant wetas (Insecta: Orthoptera: *Deinacrida*).

Work has begun even more recently on the bats, but much has been learned in a short time. New populations of the surviving species of *Mystacina*, the lesser short-tailed bat (*M. tuberculata*) were found in 1998–1999, and its ecology and that of the long-tailed bat (*Chalinolobus tuberculatus*) are being studied by several groups in both main islands. Unfortunately, what was probably the last population of the greater short-tailed bat (*M. robusta*) was one of the species that could not be saved by removal from Big South Cape Island, its last known refuge.

Translocations began in the 1890s, but were not always successful. For example, the benefits of removing kakapo and kiwis to Resolution Island from the mainland

Table 4. Species of bird at risk in the New Zealand archipelago. Based on lists in (1) Molloy *et al.* (1994) and (2) Holdaway *et al.* (2001). Where nomenclature differs between sources, that in Appendix 1 of Holdaway *et al.* (2001) is used. Species for which one or more, but not all, subspecies are endangered, are not listed. Categories are as given in Molloy *et al.* (1994), with some changes indicated by dual derivation. Molloy *et al.* (1994) score taxa on taxonomic distinctiveness, status (number of populations, population size and largest population), threats (legal protection of habitat, habitat loss rate, predator or harvest impact, competition, other factors), vulnerability (habitat or diet specificity, reproductive or behavioural specializations, captive breeding), and values to the different ethnicities. Species with scores of over 47 are the highest priority threatened species (Category A), those scoring 39–47 are second priority (Category B). Category C species (those scoring 30–38) are not included below. As predation has been identified as the most important factor in the extinction of New Zealand birds, the lack of weighting of factors tends to over-accentuate factors of relatively low importance in this grading system, hence the different levels given to some taxa here.* = 'living dead', see text.

Category/Species	Common name	Source
Category A		
Anas chorotis	Brown teal	2
Anas nesiotis	Campbell Island duck	1
Apteryx australis	South Island brown kiwi	1, 2
Apteryx mantelli	North Island brown kiwi	1, 2
Apteryx "Okarito brown"	Okarito brown kiwi	1
Charadrius exilis	Auckland Islands banded dotterel	22
Charadrius obscurus	New Zealand dotterel	1,2
Cyanoramphus malherbi	Orange-fronted parakeet	1
Gallirallus australis	Weka	1, 2
Haematopus chathamensis	Chatham Island oystercatcher	1
Hemiphaga chathamensis	Chatham Island pigeon	1
Himantopus novaezelandiae	Black stilt	1
*Petroica traversi**	Black robin	1
*Porphyrio hochstetteri**	South Island takahe	1
Pterodroma axillaris	Chatham petrel	1
Pterodroma magentae	Taiko	1
*Strigops habroptilus**	Kakapo	1
Category B		
Charadrius frontalis	Wrybill	1
Apteryx haastii	Great spotted kiwi	1
*Apteryx owenii**	Little spotted kiwi	1
Charadrius novaeseelandiae	New Zealand shore plover	1, 2
Callaeas wilsoni	North Island kokako	1
Coenocorypha aucklandica	Auckland Island snipe	2
Coenocorypha huegeli	Snares Island snipe	1
Coenocorypha meinertzhagenae	Antipodes Island snipe	2
Coenocorypha pusilla	Chatham Island snipe	2
*Cyanoramphus forbesi**	Forbes' parakeet	1

Category/Species	Common name	Source
Category B (continued)		
Thalassarche sp.	Northern Buller's albatross	1
Thalassarche eremita	Chatham Island albatross	1
Thalassarche chrysostoma	Grey-headed albatross	1
Thalassarche impavida	Campbell Island albatross	1
Diomedea epomophora	Southern royal albatross	1
Diomedea sanfordi	Northern royal albatross	1
Eudyptes pachyrhynchus	New Zealand crested penguin	1
Eudyptes sclateri	Erect-crested penguin	1
Falco novaeseelandiae	New Zealand falcon	1
Hemiphaga novaeseelandiae	New Zealand pigeon	1
Hymenolaimus malacorhynchos	Blue duck	1
Leucocarbo carunculatus	New Zealand king shag	1
Leucocarbo onslowi	Chatham Island shag	1
Megadyptes antipodes	Yellow-eyed penguin	1
Mohoua ochrocephala	Yellowhead	1
Nestor meridionalis	Kaka	1
Nestor notabilis	Kea	1
Notiomystis cincta	Stitchbird	1
Pelagodroma albiclunis	Kermadec storm petrel	1
*Philesturnus carunculatus**	South Island saddleback	1
Philesturnus rufusater	North Island saddleback	2
Poliocephalus rufopectus	New Zealand dabchick	2
Procellaria parkinsoni	Black petrel	1
Procellaria westlandica	Westland petrel	1
Pterodroma cookii	Cook's petrel	2
Puffinus huttoni	Hutton's shearwater	1
Sterna albostriata	Black-fronted tern	1
Xenicus gilviventris	Rock wren	2

Plate 35. The primitive frog *Leiopelma hamiltoni* from Maud Island. Photograph T.H. Worthy.

Plate 36. Pycroft's petrel (*Pterodroma pycrofti*), Aorangi Island, Poor Knights group, northern New Zealand. Pycroft's petrel is a small species (160 g) that survives in relict populations on small islands off the north-eastern coast of the North Island. A very large colony existed on Norfolk Island before human contact, but was exterminated by a combination of exploitation by Polynesian settlers and predation by Pacific rats (*Rattus exulans*), which take eggs, chicks and adults. Photograph Paul M. Sagar.

Plate 37. Snares Island snipe (*Coenocorypha huegeli*) lives only on that small group to the south-west of Stewart Island. It is a representative of a genus that was once widespread in the south-west Pacific, from New Caledonia to Campbell Island, but which is now found only in the New Zealand subantarctic, on islands without the Pacific rat (*Rattus exulans*). It flies well in display, but rarely at other times, and is very tame. Populations of at least two species were present on the North, South, and Stewart Islands until the arrival of Pacific rats. The last population of the South Island snipe, on the Big South Cape Islands just south-west of Stewart Island was eliminated by ship rats (*Rattus rattus*) that arrived on fishing boats in the early 1960s. Photograph Paul M. Sagar

Plate 38. New Zealand sea lions (*Phocarctos hookeri*) at Otago. This group includes a pup and mature male. Two pinnipeds breed in New Zealand—the other species is the New Zealand fur seal (*Arctocephalus forsteri*). Photograph R. Harcourt.

Plate 39. A representative of the family Galaxiidae—the Southern Hemisphere equivalents of salmons and trouts. Galaxias are one of the most abundant and widespread species groups in south-eastern Australia, but in New Zealand they have undergone the most extensive radiation and comprise over 60% of the fauna. Photograph J. Merrick.

Plate 40. The short-beaked common dolphin (*Delphinus delphis*) usually travels in groups of 10–500 individuals, in offshore temperate and tropical waters. Common dolphins vary considerably world-wide and occur throughout Australasia. Adults are up to 2.2 m long and the diet consists mainly of small fishes and cephalopods. Photograph S. Allen.

Plate 41. The lungfish (*Neoceratodus forsteri*) is the only surviving species of lungfish in Australasia, although a number of fossil species are known . The natural range of *N. forsteri* when it was first described, was the Burnett River system. But it has since been successfully introduced to other adjacent drainages in south-eastern Queensland. Photograph J. Merrick.

Plate 42. One of two osteoglossid species in Australasia; the northern spotted barramundi (*Scleropages jardini*) has an extensive but patchy range in northern Australia and New Guinea. The congener, *S. leichhardti* has a more limited natural range further south, in the Fitzroy River system (North-east Coast drainage division), but has been translocated to adjacent drainages. Photograph J. Merrick.

Plate 43. A freshwater whaler (*Carcharhinus leucas*)—one of several marine carcharhinids that regularly inhabit freshwaters in Australia and New Guinea. Photograph S.H. Midgley.

Plate 44. Leichhardt's sawfish (*Pristis microdon*) is another elasmobranch inhabiting northern Australian and New Guinean freshwaters. *P. microdon* is known to breed in freshwaters. Photograph S.H. Midgley.

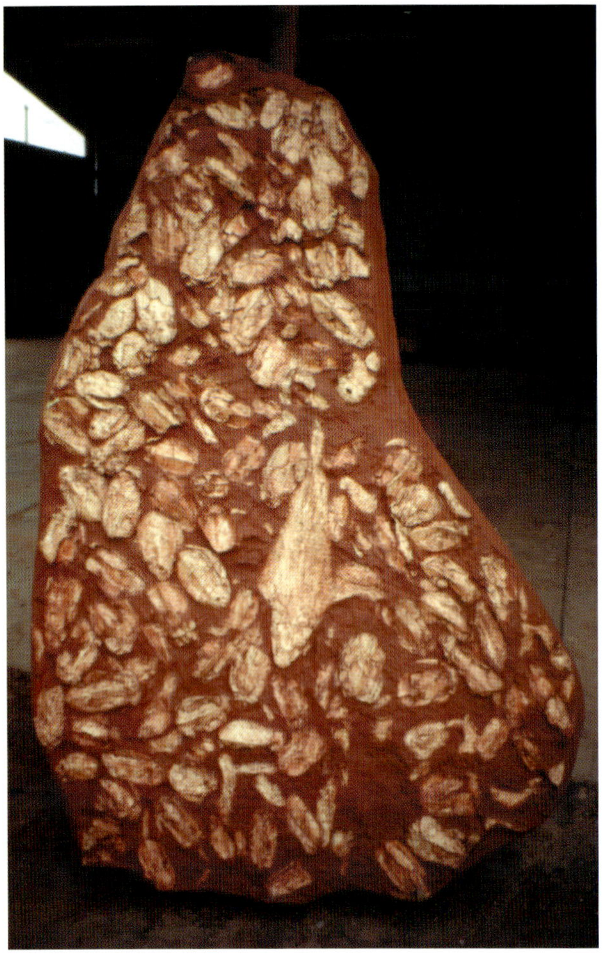

Plate 45. Replica of Devonian fish slab, recovered from near Canowindra in 1956, now on display in the Australian Museum, Sydney. Photograph A. Ritchie.

Plate 46. Council yard, Canowindra, July 1993. Alex Ritchie examines Devonian fish slabs recovered from the Gooloogong road site; these represent about half of the total number recovered in the 1993 dig. Photograph A. Ritchie.

Plate 47. The original (natural mould) and positive replica of the holotype of *Cabonnichthys burnsi* (AMF.96858), preserved on Slabs 11 (anterior end) and 12 (trunk and tail). Photograph A. Ritchie.

(Henry, 1903), were lost when stoats managed to swim to the island—before the end of the transfer programme. Other translocations of that period also failed because of the presence of predators on the new island (as for the kakapo moved to Kapiti Island) or were thwarted by the machinations of individuals, as when Sir Walter Buller diverted to England several live huia (*Heteralocha acutirostris*) collected expressly for movement to islands within New Zealand.

The importance of islands to the maintenance of endangered species in New Zealand is no better epitomized than by one island that was not protected at all at that time, although it was almost within sight of the MPs debating the subject of bird preservation. Stephens Island, at the northern end of Cook Strait, held, apart from the largest population of tuatara, viable populations of South Island kokako (*Callaeas cinerea*), South Island saddleback, South Island piopio (*Turnagra capensis*), and the only population of Lyall's wren (*Traversia lyalli*) and Hamilton's frog (*Leiopelma hamiltoni*) to survive into the European period. Of these, only the tuatara and the frog survived the replacement of forest with farmland when a lighthouse was built on the island in 1894.

The offshore and outlying islands have been important as places of origin of new taxa, and now they are pivotal in retaining the remnants of New Zealand's biodiversity, plants, animals, and other groups for which there is precious little information.

Tiritiri Matangi is an island in the Hauraki Gulf, near Auckland, which has come to epitomize island conservation in New Zealand. Totally cleared of natural forest and farmed until the 1970s, the island has become an important part of the conservation estate with the revegetation of the island and its restocking with a variety of rare and endangered species. Maintained in a mammalian predator-free state, the populations of birds such as takahe, saddleback, hihi, parakeets, kokako and little spotted kiwi are all easily accessible and visible to more than a third of the New Zealand population living in the Auckland region. The accessible nature of this island and its inhabitants is exceedingly important in maintaining the profile of conservation and hence making palatable the expense of conservation on other more remote places that are unlikely to be visited by most people.

One measure of the continuing plight of the New Zealand biota is that, in the best-known group of vertebrates—the birds—over 50 of the indigenous species are endangered (Table 4) and another 10–20 are potentially so (Molloy *et al.*, 1994). Included in the total of 54 species listed in Table 4 are five that are 'living dead' because there are no known natural populations; all existing birds have been translocated to safe islands. Two other species (black stilt and South Island takahe) are in that category because they are unlikely to have survived to the present without intensive management of the remaining populations, and in some instances, such as the species on South East Island in the Chathams, the sheer good luck that no predators arrived before the island became a reserve.

ACKNOWLEDGEMENTS

We thank Colin O'Donnell and Matt McGlone for sharing information cited here. We thank A. Anderson and another referee for comments on a draft text that improved the final product. This work has been supported in part by grants from the Foundation for Research Science and Technology to both authors—PLC801 (RNH) and TWO601 (THW).

REFERENCES

Anderson, A.J. 1981. Pre-European hunting dogs in the South Island, New Zealand. *New Zealand Journal of Archaeology,* Vol. 3, pp.15–20.

Anderson, A.J. 1989. *Prodigious Birds: Moas and Moa-hunting in Prehistoric New Zealand.* Cambridge University Press, Cambridge. 238pp.

Anderson, A.J. 1991. The chronology of colonization in New Zealand. *Antiquity,* Vol. 65, pp.767–795.

Anderson, A.J. 1996. Was *Rattus exulans* in New Zealand 2000 years ago? AMS radiocarbon ages from Shag River Mouth. *Archaeology in Oceania,* Vol. 31, pp.178–184.

Anderson, A.J. 1997. Historical and archaeological aspects of muttonbirding in New Zealand. *New Zealand Journal of Archaeology,* Vol. 17, pp.35–55.

Atkinson, I.A.E. 1973. Spread of the ship rat (*Rattus r. rattus* L.) in New Zealand. *Journal of the Royal Society of New Zealand,* Vol. 3, pp.457–472.

Atkinson, I.A.E. and Moller, H. 1990. Kiore, Polynesian rat. Pp.175–192. In King, C.M. (ed). *The Handbook of New Zealand Mammals.* Oxford University Press, Auckland. 600pp.

Beauchamp, A. and Worthy, T.H. 1988. The takahe's decline in distribution—a re-examination. *Journal of the Royal Society of New Zealand,* Vol. 18, pp.103–112.

Beavan Athfield, N.B., McFadgen, B.G. and Sparks, R. 1999. Reliability of bone gelatin AMS dating: *Rattus exulans* and marine radiocarbon dates from Pauatahanui midden sites in Wellington, New Zealand. *Radiocarbon*, Vol. 41, No. 2, pp.119–126.

Bell, B.D. 1978. The Big South Cape rat irruption. Pp.33–40. In Dingwall, P.R., Atkinson, I.A.E. and Hay, C. (eds). *The Ecology and Control of Rodents in New Zealand Nature Reserves.* Information Series No. 4, New Zealand Department of Lands and Survey.

Booth, A.M., Minot, E.O., Fordham, R.A. and Innes, J.G. 1996. Kiore (*Rattus exulans*) predation on the eggs of the little shearwater (*Puffinus assimilis haurakiensis*). *Notornis*, Vol. 43, pp.147–153.

Brooke, M. de L. 1995. The breeding biology of the gadfly petrels *Pterodroma* spp. of the Pitcairn Islands: characteristics, population sizes and controls. Pp.213–231. In Benton, T.G. and Spencer, T. (eds). *The Pitcairn Islands: Biogeography, Ecology and Prehistory*. Academic Press, London. 422pp.

Butler, D. and Merton, D. 1992. *The Black Robin: Saving the World's Most Endangered Bird.* Oxford University Press, Oxford. 294pp.

Cheeseman, T.F. 1891. On the birds of the Kermadec Islands.

Transactions and Proceedings of the New Zealand Institute, Vol. 23, pp.217–226.

Colenso, W. 1843. An account of some enormous fossil bones, of an unknown species of the class Aves, lately discovered in New Zealand. *Tasmanian Journal of Natural Science,* Vol. 2, No. VII, pp.81–107.

Cooper, A. 1994. Ancient DNA sequences reveal unsuspected phylogenetic relationships within New Zealand wrens (Acanthisittidae). *Experientia,* Vol. 50, pp.558–563.

Cooper, A. 1997. Studies of avian ancient DNA: from Jurassic Park to modern island extinctions. Pp.345–373. In Mindell, D.R. (ed). *Avian Molecular Evolution and Systematics.* Academic Press, San Diego. 382pp.

Cooper, A. and Cooper, R.A. 1995. The Oligocene bottleneck and New Zealand biota: genetic record of a past environmental crisis. *Proceedings of the Royal Society of London,* Series B, Vol. 261, pp.293–302.

Cooper, A., Mourer-Chauvire, C., Chambers, G.K., Haeseler, A., Wilson, A.C. and Pääbo, S. 1992. Independent origins of New Zealand moas and kiwis. *Proceedings of the National Academy of Sciences of the United States of America,* Vol. 89, pp.8741–8744.

Cooper, A. and Penny, D. 1997. Mass survival of birds across the Cretaceous-Tertiary boundary: molecular evidence. *Science,* Vol. 275, pp.1109–1113.

Croxall, J. P. and Rothery, P. 1991. Population regulation of seabirds: implications of their demography for conservation. Pp.272–296. In Perrins, C.M., LeBreton, J.D. and Hirons, G.M. (eds). *Bird Populations Studies: Relevance to Conservation and Management.* Oxford University Press, Oxford. 683pp.

Croxall, J.P., Rothery, P., Pickering, S.P.C. and Prince, P. A. 1990. Reproductive performance, recruitment and survival of wandering albatrosses *Diomedea exulans* at Bird Island, South Georgia. *Journal of Animal Ecology,* Vol. 59, pp.775–796.

Darwin, C. 1884. *A Naturalist's Voyage. Journal of Researches into the Natural History and Geology of the Countries Visited during the Voyage of H.M.S.* 'Beagle' *round the World under the Command of Capt. FitzRoy, R.N.* John Murray, London.

Daugherty, C.H., Maxon, L. R. and Bell, B.D. 1982. Phylogenetic relationships within the New Zealand frog genus *Leiopelma*—immunological evidence. *New Zealand Journal of Zoology,* Vol. 9, pp.239–242.

Daugherty, C.H., Patterson, G.B. and Hitchmough, R.A. 1994. Taxonomic and conservation review of the New Zealand herpetofauna. *New Zealand Journal of Zoology,* Vol.21, No. 4, pp.317–323.

Daugherty, C.H., Patterson, G.B., Thorn, C.J. and French, D.C. 1990. Differentiation of the members of the New Zealand *Leiolopisma nigriplantare* species complex (Lacertilia: Scincidae). *Herpetological Monographs,* Vol. 4, pp.61–76.

Elliott, G.P. 1996. Mohua and stoats: a population viability analysis. *New Zealand Journal of Zoology,* Vol. 23, pp.239–247.

Elliott, G.P., Dilks, P.J. and O'Donnell, C.F.J. 1996. The ecology of yellow-crowned parakeets (*Cyanoramphus auriceps*) in *Nothofagus* forest in Fiordland, New Zealand. *New Zealand Journal of Zoology,* Vol. 23, pp.249–265.

Flannery, T.F. 1995. *Mammals of the South-west Pacific and Moluccan Islands.* Reed Books, Sydney. 464pp.

Fordyce, R.E. 1991. A new look at the fossil vertebrate record of New Zealand. Pp.1191–1316. In Vickers-Rich, P., Monaghan, J.M., Baird, R.F. and Rich, T.H. (eds). *Vertebrate Palaeontology of Australasia.* Pioneer Design Studio and Monash University Publications Committee, Melbourne. 1437pp.

Fordyce, R.E. 2006. A southern perspective on cetacean evolution and zoogeography. Pp.755–777. In Merrick, J.R., Archer, M., Hickey, G.M. and Lee, M.S.Y. (eds). *Evolution and Biogeography of Australasian Vertebrates.* Auscipub Pty. Ltd., Sydney. 942pp.

Grant, P.R. 1998. Patterns on islands and microevolution. Pp.1–17. In Grant, P.R. (ed). *Evolution on Islands.* Oxford University Press, Oxford. 334pp.

Green, R.C. 1994. Changes over time; recent advances in dating human colonisation of the Pacific Basin area. Pp.19–51. In Sutton, D.G. (ed). *The Origins of the First New Zealanders.* Auckland University Press, Auckland. 269pp.

Hand, S.J., Murray, P., Megirian, D., Archer, M. and Godthelp, H. 1998. Mystacinid bats (Microchiroptera) from the Australian Tertiary. *Journal of Paleontology,* Vol. 72, No. 3, pp.538–545.

Harrison, R.A. and White, E.G. 1969. Grassland invertebrates. Pp.379–399. In Knox, G.A (ed). *The Natural History of Canterbury, 2nd Edition.* A.H. and A.W. Reed, Wellington. 620pp.

Heather, B.D. and Robertson, H.A.R. 1996. *The Field Guide to the Birds of New Zealand.* Viking, Auckland. 432pp.

Henry, R. 1903. *The Habits of the Flightless Birds of New Zealand; with Notes on other New Zealand Birds.* New Zealand Government Printer, Wellington. 88pp.

Higham, T., Anderson, A.J. and Jacomb, C. 1999. Dating the first New Zealanders: the chronology of Wairau Bar. *Antiquity,* Vol. 73, pp.420–427.

Holdaway, R.N. 1989. New Zealand's pre-human avifauna and its vulnerability. *New Zealand Journal of Ecology,* Vol. 12 (Supplement), pp.11–25.

Holdaway, R.N. 1990. Changes in the diversity of New Zealand forest birds. *New Zealand Journal of Zoology,* Vol. 17, No. 3, pp.309–321.

Holdaway, R.N. 1996. Arrival of rats in New Zealand. *Nature,* Vol. 384, pp.225–226.

Holdaway, R.N. 1999a. A spatio-temporal model for the invasion of the New Zealand archipelago by the Pacific rat (*Rattus exulans*). *Journal of the Royal Society of New Zealand,* Vol. 29, pp.91–105.

Holdaway, R.N. 1999b. Introduced predators and avifaunal extinction in New Zealand. Pp.189–238. In MacPhee, R.D.E. (ed). *Extinctions in Near Time: Causes, Contexts, and Consequences.* Kluwer Academic and Plenum Press, New York. 394pp.

Holdaway, R.N. and Anderson, A.J. 2001. The avifauna of the Emily Bay archaeological site, Norfolk Island, South-west Pacific: a preliminary account. *Records of the Australian Museum, Supplement,* No. 27, pp.87–102.

Holdaway, R.N. and Beavan, N.R. 1999. Reliable ^{14}C AMS dates on bird and Pacific rat *Rattus exulans* bone gelatin, from a $CaCO_3$-rich deposit. *Journal of the Royal Society of New Zealand,* Vol. 29, pp.185–211.

Holdaway, R.N. and Jacomb, C. 2000. Rapid extinction of the moas (Aves: Dinornithiformes): model, test, and implications. *Science,* Vol. 287, No. 5461, pp.2250–2254.

Holdaway, R.N., Roberts, R.G., Beavon-Athfield, N.R., Olley, J.M. and Worthy, T.H. 2002. Optical dating of quartz sediments and accelerator mass spectrometry ^{14}C dating of bone gelatin and moa eggshell: a comparison of age estimates for non-archaeological deposits in New Zealand. *Journal of the Royal Society of New Zealand,* Vol. 32, pp.463–505.

Holdaway, R.N., Worthy, T.H. and Tennyson, A.J.T. 2001. A working list of breeding bird species of the New Zealand region at first human contact. *New Zealand Journal of Zoology,* Vol. 28, No. 2, pp.119–187.

Houde, P. 1986. Ostrich ancestors found in the Northern Hemisphere suggest new hypothesis of ratite origins. *Nature,* Vol. 324, pp.563–565.

Houde, P. 1988. Palaeognathous birds from the early Tertiary of the Northern Hemisphere. *Publications of the Nuttall Ornithological Society,* Vol. 22, pp.1–148.

Hutton, F.W. and Drummond, J. 1905. *Animals of New Zealand: an Account of the Colony's Air-Breathing Vertebrates.* Whitcombe and Tombs, Christchurch. 375pp.

Irwin, G. 1992. *The Prehistoric Exploration and Colonisation of the Pacific.* Cambridge University Press, Cambridge. 240pp.

Kennedy, M. and Spencer, H.G. 2000. Phylogeny, biogeography and taxonomy of Australasian teals. *Auk,* Vol. 117, pp.154–163.

King, C.M. 1990. Introduction. Pp.3–21. In King, C.M. (ed). *The Handbook of New Zealand Mammals.* Oxford University Press, Auckland. 600pp.

Kirch, P.V. 1982. Transported landscapes. *Natural History,* Vol. 91, pp.32–35.

Kirch, P.V. 1983. Man's role in modifying tropical and subtropical Polynesian ecosystems. *Archaeology in Oceania,* Vol. 18, pp.26–31.

Kirch, P.V. 1997. The environmental history of oceanic islands. Pp.1–21. In Kirch, P.V. and Hunt, T.L. (eds). *Historical Ecology in the Pacific Islands: Prehistoric Environmental and Landscape Change.* Yale University Press, New Haven. 331pp.

Lawver, L.A., Gahagan, L.M. and Coffin, M.F. 1992. The development of paleoseaways around Antarctica. *American Geophysical Union, Antarctic Research Series,* Vol. 56, pp.7–30.

Lewis, K.B., Carter, L. and Davey, F.J. 1994. The opening of Cook Strait: interglacial tidal scour and aligning basins at a subduction to transform plate edge. *Marine Geology,* Vol. 116, pp.293–312.

Livezey, B.C. 1990. Evolutionary morphology of flightlessness in the Auckland Islands teal. *Condor,* Vol. 92, pp.639–673.

Long, J.A. 1998. *Dinosaurs of Australia and New Zealand and Other Animals of the Mesozoic.* Harvard University Press, Cambridge, Massachusetts. 188pp.

Lovegrove, T.G. 1996a. A comparison of the effects of predation by Norway (*Rattus exulans*) and Polynesian rats (*R. exulans*) on the saddleback (*Philesturnus carunculatus*). *Notornis,* Vol. 43, pp.91–112.

Lovegrove, T.G. 1996b. Island releases of saddlebacks *Philesturnus carunculatus* in New Zealand. *Biological Conservation,* Vol. 77, pp.151–157.

McCallum, J. 1982. The penetration of exotic passerines into modified forests on Little Barrier Island, northern New Zealand. *Tane,* Vol. 28, pp.37–51.

McDowall, R.M. 1990. *New Zealand Freshwater Fishes: a Guide and Natural History.* Heinemann Educational, Auckland. 230pp.

McDowall, R.M. 2000. "Rivers destitute of fish"—another look at New Zealand's freshwater fish fauna. *Species 2000: New Zealand. Millenial Symposium, A Review and Inventory of New Zealand's Biodiversity.* Museum of New Zealand Te Papa Tongarewa, Wellington, 1–5 February 2000. (abstract only).

McDowall, R.M. and Chadderton, W.L. 1999. *Galaxias gollumoides* (Teleostei: Galaxiidae), a new fish species from Stewart Island, with notes on other non-migratory freshwater fishes present on the island. *Journal of the Royal Society of New Zealand,* Vol. 29, No. 1, pp.77–88.

McDowall, R.M. and Pole, M. 1997. A large galaxiid fossil (Teleostei) from the Miocene of Central Otago, New Zealand. *Journal of the Royal Society of New Zealand,* Vol. 27, No. 2, pp.193–198.

McGlone, M.S. 1983. Polynesian deforestation of New Zealand: a preliminary synthesis. *Archaeology in Oceania,* Vol. 18, pp.11–25.

McGlone, M.S. 1989. The Polynesian settlement of New Zealand in relation to environmental and biotic changes. *New Zealand Journal of Ecology,* Vol. 12 (Supplement), pp.115–129.

McGlone, M.S., Anderson, A.J. and Holdaway, R.N. 1994. An ecological approach to the Polynesian settlement of New Zealand. Pp.136–163. In Sutton, D.G. (ed). *The Origins of the First New Zealanders.* Auckland University Press, Auckland. 269pp.

McGlone, M.S., Holdaway, R.N. and Wilmshurst, J. 2001. Ecological dimensions to Polynesian settlement of New Zealand: cataclysm and accommodation. In Sutton, D.G. (ed). *Origins of the First New Zealanders, 2nd Edition.* Auckland University Press, Auckland.

McGlone, M.S. and Wilmshurst, J.M. 1999. Dating initial Maori environmental impact in New Zealand. *Quaternary International,* Vol. 59, pp.5–16.

Merrick, J.R. 2006. Australasian freshwater fish faunas: diversity, interrelationships, radiations and conservation. Pp.195–224. In Merrick, J.R., Archer, M., Hickey, G.M. and Lee, M.S.Y. (eds). *Evolution and Biogeography of Australasian Vertebrates.* Auscipub Pty. Ltd., Sydney. 942pp.

Millener, P.R. 1999. The history of the Chatham Islands' bird fauna of the last 7000 years—a chronicle of change and extinction. Proceedings of the 4th International meeting of the Society of Avian Paleontology and Evolution. Washington, USA, June 1996. *Smithsonian Contributions to Paleobiology,* pp.85–109.

Mills, J.A., Lavers, R.B. and Lee, W.G. 1984. The takahe—a relict of the Pleistocene grassland avifauna of New Zealand. *New Zealand Journal of Ecology,* Vol. 7, pp.57–70.

Mills, J.A., Lavers, R.B. and Lee, W.G. 1988. The post-Pleistocene decline of takahe (*Notornis mantelli*): a reply. *Journal of the Royal Society of New Zealand,* Vol. 18, pp.112–118.

Molloy, J., Davis, A. and Tisdall, C. 1994. *Setting Priorities for the Conservation of New Zealand's Threatened Plants and Animals. 2nd Edition.* New Zealand Department of Conservation, Wellington.

Molnar, R.E. and Pole, M. 1997. A Miocene crocodilian from New Zealand. *Alcheringa,* Vol. 21, pp.65–70.

Molnar, R.E. and Wiffen, J. 1994. Polar dinosaur faunas from

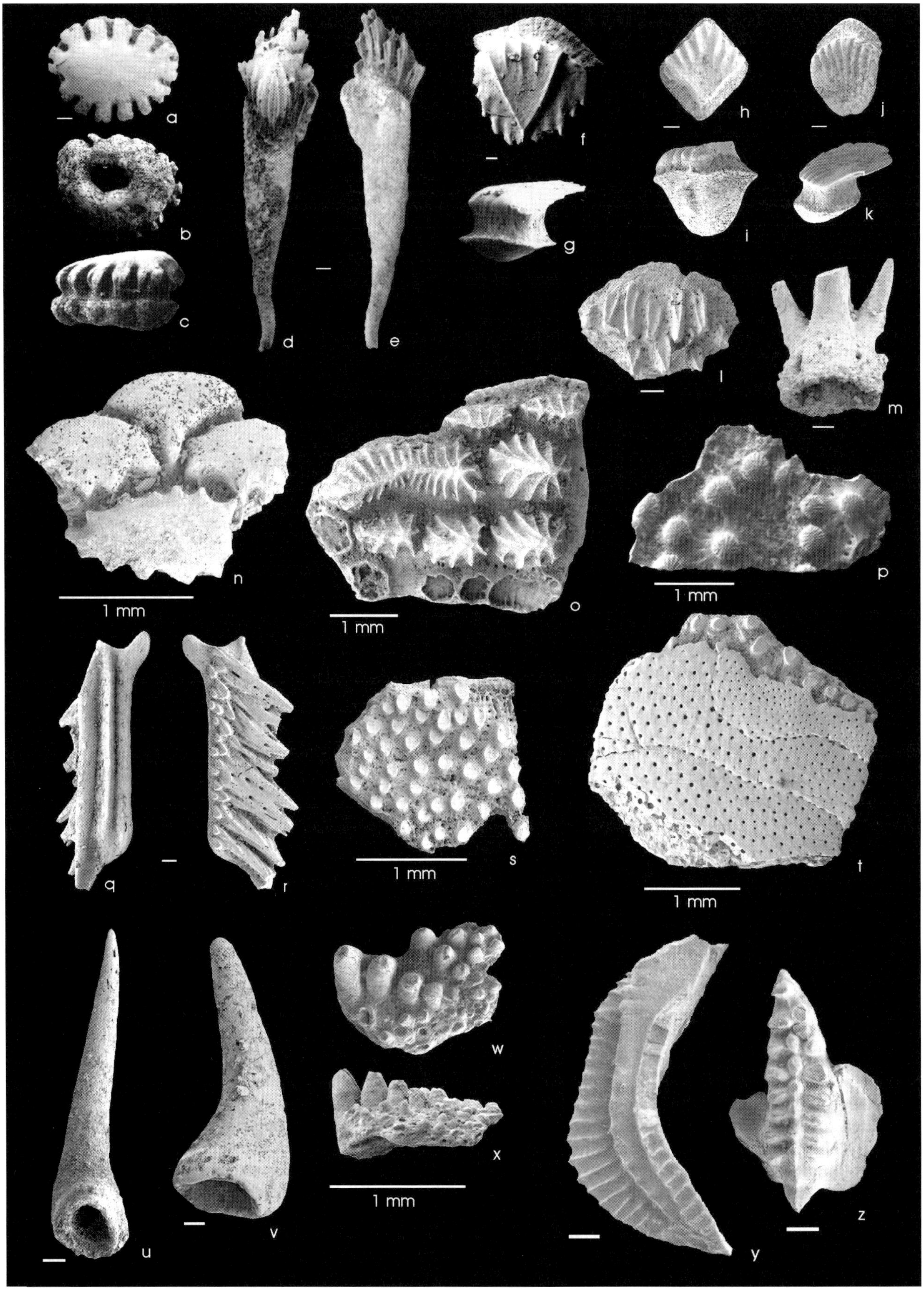
a
b
c
d
e
f
g
h
i
j
k
l
m
n
1 mm
o
1 mm
p
1 mm
q
r
s
1 mm
t
1 mm
u
v
w
x
1 mm
y
z

of early vertebrates, support a vertebrate affinity (Aldridge, 1987; Aldridge *et al.*, 1993; Sansom *et al.*, 1992, 1994). Recognition of what appears to be wear on the surface of the conodont elements has led to the suggestion that conodont animals were macrophagous feeders, with the conodont elements functioning as teeth designed for grasping, shearing and crushing (Purnell, 1993, 1994, 1995). However, other workers disagree with this interpretation, considering the histology of the conodont elements to be more consistent with protochordate affinities instead (Kemp and Nicoll, 1995, 1996).

Regardless of their phylogenetic position, conodonts are an ideal basis for a stratigraphic scheme because of their wide geographic distribution and often short-ranging species. Sedimentary rocks can be precisely dated according to the conodonts they contain, to within a few hundred thousand years in some cases. By recovering vertebrate microremains and conodonts simultaneously from long, closely-sampled stratigraphic sections, the vertebrates can be precisely dated by the accompanying conodonts and can, in turn, be used to establish a biostratigraphic scheme based solely on the vertebrate microfossils. This is useful because, unlike conodonts which are marine, some fish are euryhaline (can tolerate different salinities) at different stages of their life cycle. Consequently, fish fossils have potential as a biostratigraphic tool in marginal environments, such as brackish or freshwater, where conodonts are not preserved.

MICROVERTEBRATE STUDIES

Limestone samples collected for conodonts at Taemas (Burrinjuck), Buchan and the area near Tyers in eastern

Figure 4 (*opposite page*). Devonian vertebrate microfossils and conodonts from eastern Australia. Scale bar = 100 µm, unless indicated otherwise.

(a)–(e) Thelodont scales, *Turinia* sp.

(a)–(c) Head scale AMF101203 from YWJ117.5, Cavan Formation, *dehiscens* Conodont Zone. a. crown view, b. basal view, c. lateral view.

(d)–(e) Body scale AMF101177 from SLO51m, Taravale Formation, *dehiscens* Conodont Zone. d. crown view, e. basal view.

(f)–(k) Acanthodian scales.

(f)–(g) *Nostlepoides platymarginata* scale AMF101183 from SLO8.5m, Taravale Formation, *dehiscens* Conodont Zone. f. crown view, g. lateral view.

(h)–(i) cf. *Gomphonchus bogongensis* scale from locality 63/66 (NHM London 1963 expedition) at Taemas, *Spirifer yassensis* Limestone. h. crown view, i. lateral view.

(j)–(k) *Cheiracanthoides* sp. scale from YWJ117.5, Cavan Formation, *dehiscens* Conodont Zone. j. crown view, k. lateral view.

(l)–(m) Shark microremains.

(l) Crown view of *Ohiolepis* sp. scale from locality 63/76 (NHM London 1963 expedition) at Taemas, Warroo Limestone.

(m) Tooth AMF106412 from TQ7.3m, Coopers Creek Limestone, Tyers Quarry, eastern Victoria.

(n)–(p) Placoderm microremains.

(n) *Brindabellaspis* sp. platelet AMF106387 from BOO10.6, Coopers Creek Limestone, Boola Quarry, eastern Victoria, *kindlei* Conodont Zone.

(o) *Buchanosteus* sp. fragment AMF101197 from Gelantipy Road 15.2, Taravale Formation, near Buchan, *inversus* Conodont Zone.

(p) Acanthothoracid fragment locality 63/76 (NHM London 1963 expedition) at Taemas, Warroo Limestone.

(q)–(x) Osteichthyan microremains.

(q)–(r) Palaeoniscid scale—*Ligulalepis* sp. from locality 55/34 (NHM London 1955 expedition) at Taemas, sample P33686, *Receptaculites* Limestone. q. outer surface, r. inner surface.

(s) Porolepiform fragment from locality 55/33 (NHM London 1955 expedition) at Taemas, sample P33675, boundary of *Receptaculites* and Bloomfield limestones.

(t) Dipnoan or porolepiform fragment with cosmine on surface, from locality 63/79 (NHM London 1963 expedition) at Taemas, *Receptaculites* Limestone.

(u)–(v) Crossopterygian teeth. u. from locality 55/34 (NHM London 1955 expedition) at Taemas, sample P33686, *Receptaculites* Limestone. v. from YWJ196.2 at Taemas, Majurgong Formation.

(w)–(x) Dipnoan toothplate QMF31845 from SD131/174.6 Storm Dam section, Papilio Formation, middle *varcus* Conodont Zone at Broken River, north Queensland. w. upper view, x. side view.

(y)–(z) Conodonts.

Middle Devonian conodonts from Broken River Group, north Queensland (precise locality unknown).

Key to institutions where specimens are deposited: AMF = Australian Museum, Sydney; NHM = Natural History Museum, London; QMF = Queensland Museum, Brisbane.

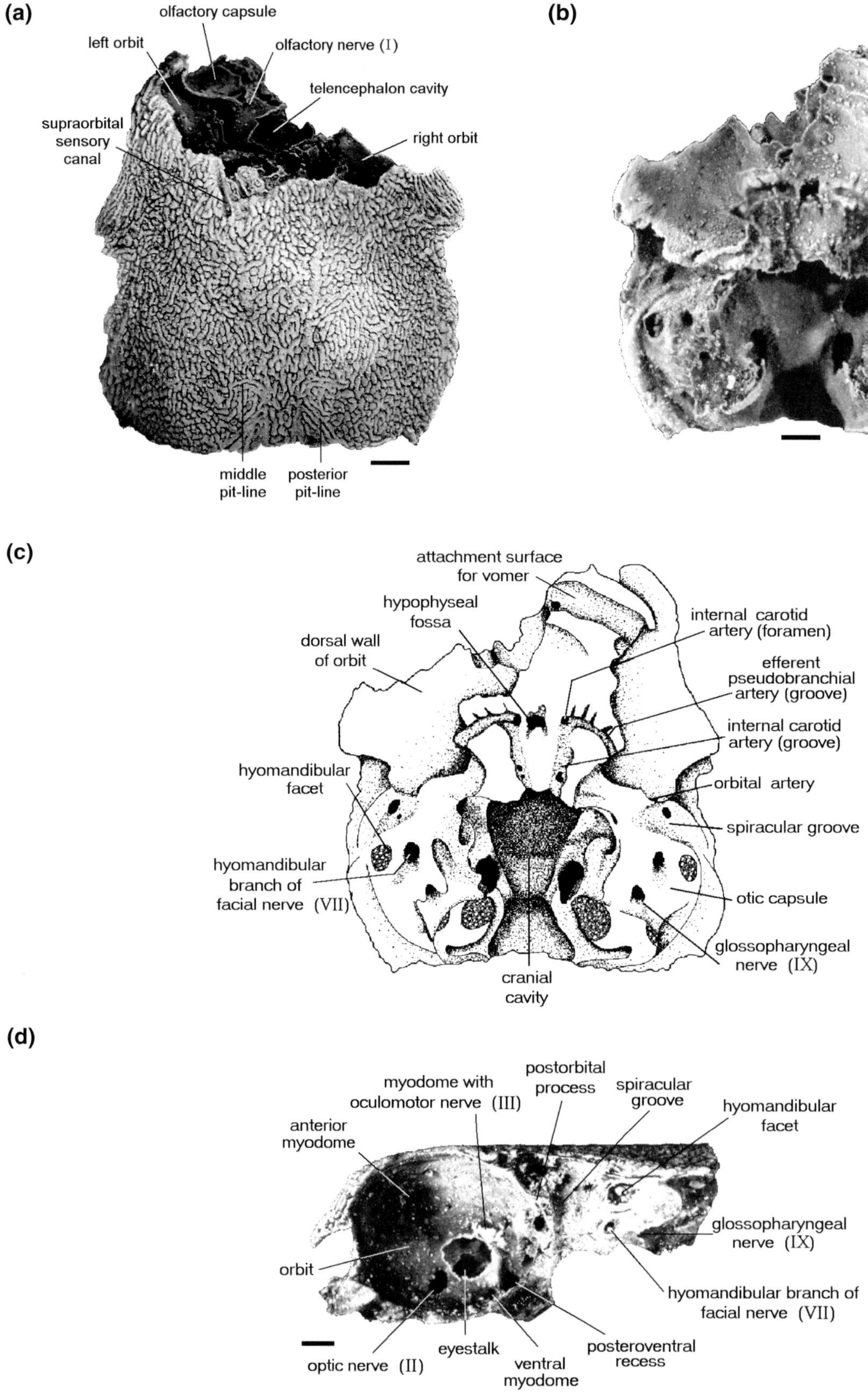

Figure 5. Primitive actinopterygian braincase (*Ligulalepis* sp.) from Wee Jasper: **(a)** dorsal view showing skull roof, with part of the underlying braincase structure visible through broken anterior margin; **(b)** ventral view; **(c)** ventral interpretation; **(d)** lateral view. Scale bar = 1 mm.

Victoria (Mawson, 1987; Mawson *et* al., 1992; Mawson and Talent, 1994) have provided both accurate dating of the limestones and an abundance of vertebrate microremains for study (Basden, 1999a; Basden *et al.*, 2000a). See Figure 4. As well as being potential biostratigraphic indicators, the vertebrate microfossils can be useful in other ways. Burrow (1994) studied the arrangement and degree of interlocking of isolated scales of *Ligulalepis toombsi* from central N.S.W., and has drawn conclusions about the mobility, swimming style and probable ecological range of that particular palaeoniscid. As discussed later in this chapter, Trinajstic (1999b, c) has undertaken similar studies of scales and articulated specimens of palaeoniscids *Mimia toombsi* and *Moythomasia durgaringa* from Western Australia, and identified the changing scale morphology from juvenile through subadult to adult stages of growth, which again gives clues about the palaeoecology. Comparison of abundance and composition of microvertebrate, conodont and invertebrate faunas from shallow marine limestones of similar ages allows inferences about the depositional environment, for example, water depth or flow (Basden, 1999b) and predator-prey relationships.

The Devonian Burrinjuck fauna is particularly diverse and abundant, and has been fairly widely studied; see references in Long (1995). A variety of invertebrate groups have been reported—sponges (Pickett, 1969), corals (Hill, 1940; Pedder *et al.*, 1970), trilobites and brachiopods (Chatterton, 1971, 1973), ostracodes (Reynolds, 1978), gastropods (Tassell, 1982), and bivalves (Johnston, 1993)—in addition to both macro- and micro-vertebrate remains. Two extensive collecting expeditions in 1955 and 1963, mounted by the Natural History Museum (London), resulted in the transport of several hundred fish specimens to London for preparation with acetic acid. White and Toombs (1972) and White (1978) subsequently described some of the larger arthrodire (placoderm) bones, and Schultze (1968) erected a new form genus, *Ligulalepis*, for palaeoniscid scales recovered from the residues. Little else has been done, but some of the hundreds of residue samples in the Museum collections are currently being examined and two features have become clear. First is the diversity of the Burrinjuck microfossil fauna, especially the acanthodians, and second is the similarity between early Devonian faunas from Burrinjuck and Victoria as well as central western New South Wales.

The Burrinjuck fauna comprises all major groups of jawed fishes plus the jawless thelodonts, but is dominated by placoderms. Work on this group in recent decades by Young (1978, 1979, 1980, 1981, 1985, 1986a) has shown what a bizarre and varied bunch they were. A number of new genera have been erected for specimens recovered from Burrinjuck: arthrodires *Taemasosteus* and *Williamsaspis* (White, 1952), *Burrinjucosteus, Goodradigbeeon* and *Toombsosteus* (White, 1978), *Arenispiscis* and *Errolosteus* (Young, 1981); acanthothoracids *Weejasperaspis* (White, 1978) and *Brindabellaspis* (Young, 1980); petalichthyids *Notopetalichthys* (Woodward, 1941), *Wijdeaspis* (Young, 1978) and *Shearsbyaspis* (Young, 1985); and a second lungfish, *Speonesydrion* (Campbell and Barwick, 1983). Recent work by Lindley, (2000, 2001, 2002) has added several new species and the genera *Cavanacanthus* and *Cambaracanthus* to the list of described acanthodians.

OSTEICHTHYAN BRAINCASE (FROM WEE JASPER) AND ITS LINKS

One of the most recent additions to the list of Burrinjuck fossils, and probably one of the most controversial, is a tiny, one-centimetre long osteichthyan braincase collected during an Australian Museum field trip to Wee Jasper. The specimen consists of an ornamented, anteriorly incomplete skull roof (Figure 5(a)), attached to the surface ossification of the cartilaginous braincase. The details of braincase structure are preserved in fine detail and include openings for cranial nerves, vessels and sensory canals as well as the central cavity for the brain (Figures 5 (b)–(d)). The controversy surrounding this specimen stems not only from its unusual anatomy, but also from its strange combination of features (Basden *et al.*, 2000b; Basden and Young, 2001).

The specimen displays a mélange of characters. While the ornamentation of the dermal bone (Figure 5(a)), the skull roof pattern, and the general proportion and structure of the endocranium are typically actinopterygian, other features more closely resemble endocrania of other fish groups. For example, the curved pit lines, the widely-separated orbital walls, and the short, broad cavity for the telencephalon (front section of the brain) (Figure 5(a)) are typical of sarcopterygians (e.g. Yu, 1998; Zhu *et al.*, 1999); whereas other known primitive osteichthyans tend to have short, stubby pit lines, orbits placed closer together, and a longer telencephalon. The position and orientation of the postorbital

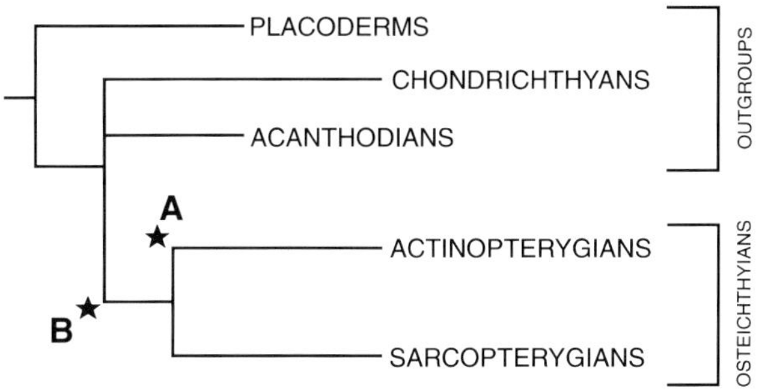

Figure 6. Diagrammatic interpretation of inferred interrelationships of major gnathostome groups, based on morphological characters, showing two possible phylogenetic positions for the Wee Jasper braincase (Specimen AMF101607). A = primitive actinopterygian; B = stem-group osteichthyan.

Figure 8. Microvertebrate remains from the Gneudna Formation. Thelodonti: **(a)** *Australolepis seddoni*; **(b)** *Turinia hutkensis*. Placodermi: **(c)** *Holonema westolli*. Chondrichthyes: **(d)** shark tooth; **(e)** *Ctenacanthus* sp. indet.; **(f)** *Protocrodus* sp. indet. Acanthodii: **(g)** *Cheiracanthus* sp.; **(h)** Acanthodidae gen. et sp. indet.; **(i)–(j)** *Moythomasia durgaringa* type B scales; **(k)** *Mimia toombsi* type B scale; **(l)** *Mimia toombsi* type D scale. Sarcopterygii scales: **(m)** *Gogonaseus*; **(n)** osteolepiform scale; **(o)** dipnoan toothplate (juvenile).

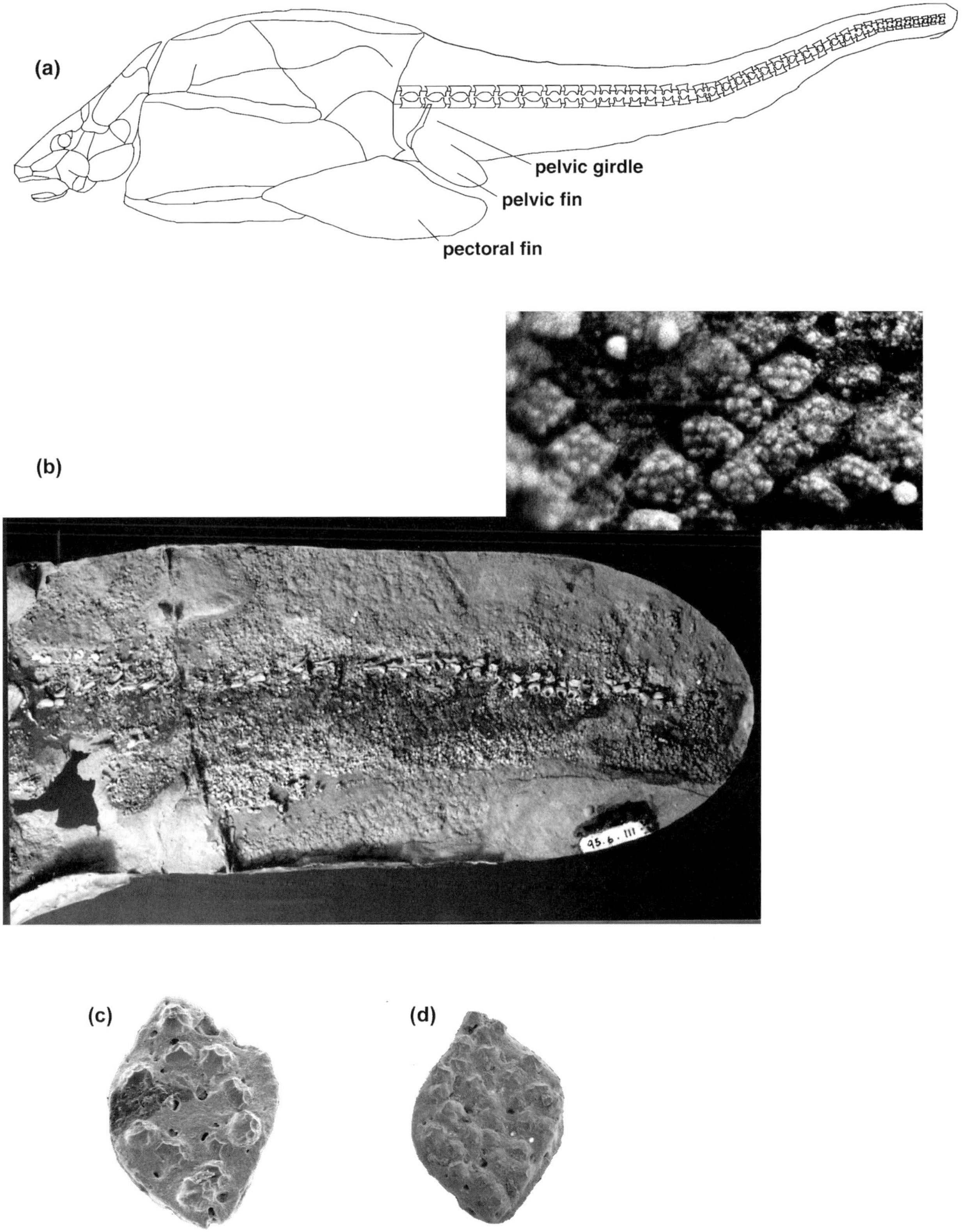

Figure 9. The placoderm *Holonema westolli* from the Gogo Formation with scales from the Gneudna Formation: **(a)** reconstruction of *H. westolli*; **(b)** tail region of *H. westolli*, showing vertebral elements and the arrangement of scales (insert); **(c)** *H. westolli* scale from the Gogo Formation; **(d)** isolated scale (identified as *H. westolli*) from the Gneudna Formation.

Palaeozoic sequences in Australia (Turner, 1993; Young, 1993a, b; De Pomeroy, 1996; Burrow, 1996; Trinajstic, 1999a, b, c, 2001a, b). The data obtained from these works is helping to correlate marine and non-marine sequences throughout Australia, and to integrate microvertebrate and other biostratigraphies (Turner, 1999). Analysis of major variation in biodiversity and changes in biogeographic differentiation along the Gondwanan continental margin during the mid Palaeozoic would not have been possible without the inclusion of data obtained from microvertebrate fossils.

IDENTIFYING MICROVERTEBRATE REMAINS

Microvertebrates are common and easily obtained from small amounts of sedimentary rock. But despite their abundance, the accurate identification of species is a major obstacle in working with microvertebrate fossils, because of the considerable variation in scale shape and in the pattern of crown ornamentation observed in many fishes. Phylogenetic relationships, biocorrelation and biogeography all rely on the accurate identification of genera and/or species, and this has resulted in microvertebrates being often overlooked as a potential research tool by palaeontologists. The comparison of isolated teeth and scales with more complete, often articulated specimens has provided a framework for the accurate identification of microvertebrate fossils. It was by direct comparison to articulated specimens of the placoderm *Holonema westolli* (Figure 9(a)) from the Gogo Formation (Miles, 1971) that isolated scales recovered from the Gneudna Formation could be identified (Trinajstic, 1999a)—see Figures 9(b)–(d).

There are many other taxa where morphological variation has been studied on articulated specimens. In thelodonts several different classifications of scale varieties within a species have been used. Märss (1986) studied *Phlebolepis elegans* and recognized five distinct morphological areas, thereby refining the traditional head, trunk and body scale classification. These studies have made it possible to determine species based on isolated scales (Figures 10(a)–(e)), and to utilize these scales in solving biostratigraphic problems (Turner, 1997, 1999; Bloom, 1999; Trinajstic, 2001a).

V.T. Young (1995) carried out one of the most comprehensive studies of scale variation in acanthodians. In that study, differences in scale morphology were shown to be present in nine taxa of early and middle Devonian articulated acanthodian fishes from the British Isles. Comparison of isolated scales from the Gneudna Formation to articulated specimens from Europe showed that the nine scale types from the Gneudna Formation represented just two acanthodian genera (Figure 8(g), (h)). Studies of this type assist in recognizing the true species diversity of the taxa present by preventing scales from different body regions being categorized as different species.

It has long been recognized that the morphology of shark scales is influenced by both ontogenetic and regional variation (Applegate, 1967; Reif, 1974, 1978, 1985; Karatajute-Talimaa, 1992). Williams (1985) showed a continuum of simple scales, ranging from single lepidomoria to complex scales, consisting of multiple odontodes on the body of *Phoebodus heslerorum*. A large

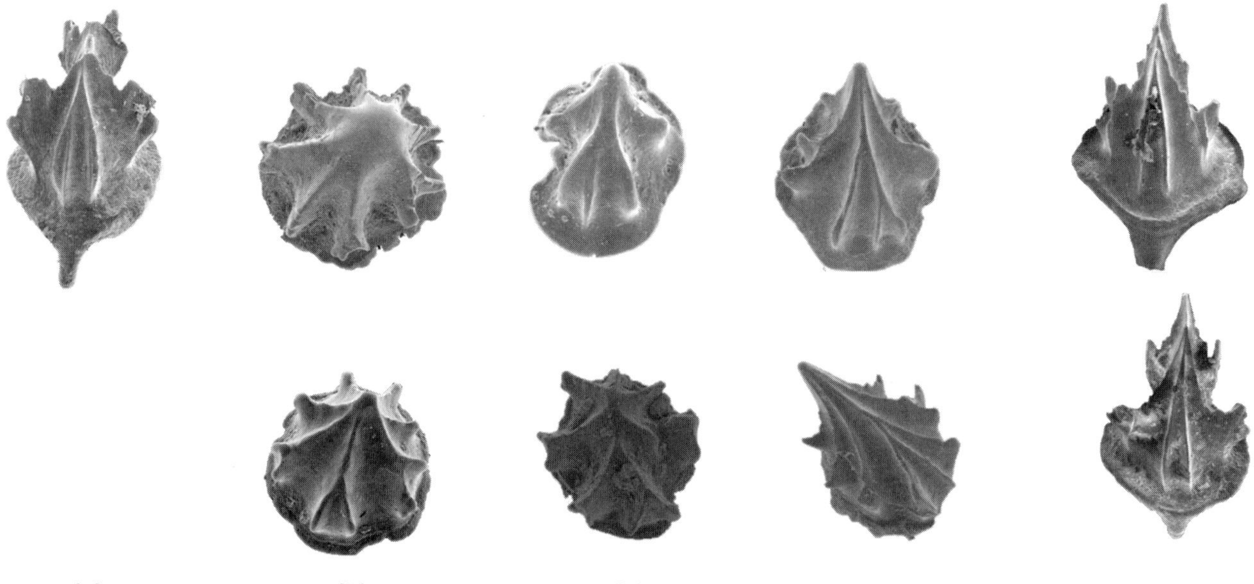

(a) (b) (c) (d) (e)

Figure 10. Isolated scales identified as coming from different body regions of *Australolepis seddoni*: **(a)** pineal; **(b)** oral; **(c)** cephalo-pectoral; **(d)** post-pectoral; **(e)** precaudal.

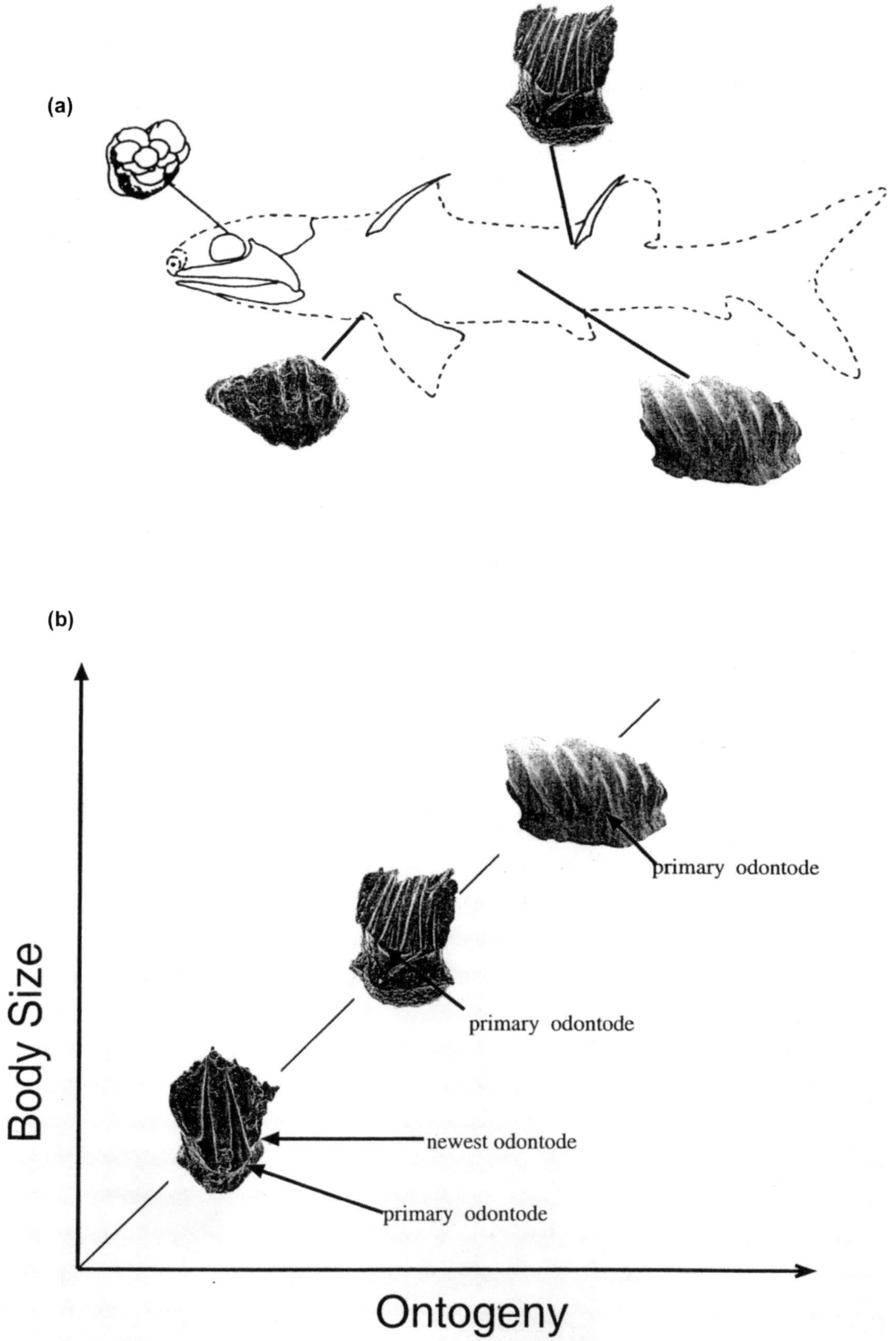

Figure 11. Ctenacanthid scales: **(a)** outline of a ctenacanthid shark indicating the position that isolated scales originated from; **(b)** ctenacanthid scales at different stages of growth. Odontodes were added from alternating sides as the scale increased in size.

diversity of scale types on single individuals from the shark family Ctenacanthoidea has also been recognized (Reif, 1978; Williams, 1985; Turner, 1993). During skin growth new scales are added to the shark's body, resulting in scales of various ontogenetic stages being present on a single individual (Reif, 1978, 1980; Karatajute-Talimaa, 1992). To make the assignment of isolated scales more reliable several methods of identification and classification have been proposed, based on the recognition of different morphogenic types of squamation in elasmobranchs (Stensiö, 1961; Ørvig, 1967; Reif, 1978, 1980; Karatajute-Talimaa, 1992). By using these methods, seven scale types, identified as *Ctenacanthus* scales from the Gneudna Formation, were considered to represent scales from only one species, but originating from different body regions (Figure 11(a)) and representing different stages of growth (Figure 11(b)).

The usefulness of palaeoniscoid scales in biocorrelation has so far been limited despite their occurrence in both marine and freshwater facies. This has primarily been the result of difficulties in identifying taxa from isolated scales. To overcome this difficulty, Esin (1990) described a pattern of squamation based on nine distinct morphological areas, providing a uniform reference for the identification of isolated scales. Esin's method was successfully used by Burrow (1994) to attribute *Ligulalepis toombsi* (Schultze, 1968) scales, of various forms, to particular body areas, and by Trinajstic (1999b, c) to refer isolated scales from the Gneudna and Virgin Hills Formations to *Moythomasia durgaringa* (Figure 12) and *Mimia toombsi.*

The recognition of variation in scale morphology, both ontogenetic and regional variation on articulated macrovertebrates, has made the identification of isolated scales to generic and, in some cases, species level possible. This has increased the known range of some species and also enabled the greater use of microvertebrate taxa in cladistic analyses, biocorrelation and biogeographic studies. This has also helped to limit the overestimation of species present, a real problem when dealing with partial fossil remains.

BIOCORRELATION BASED ON MICROVERTEBRATES

Biocorrelations between local successions are usually based on similar species occurring in different localities. The most useful fossils for correlation are those that

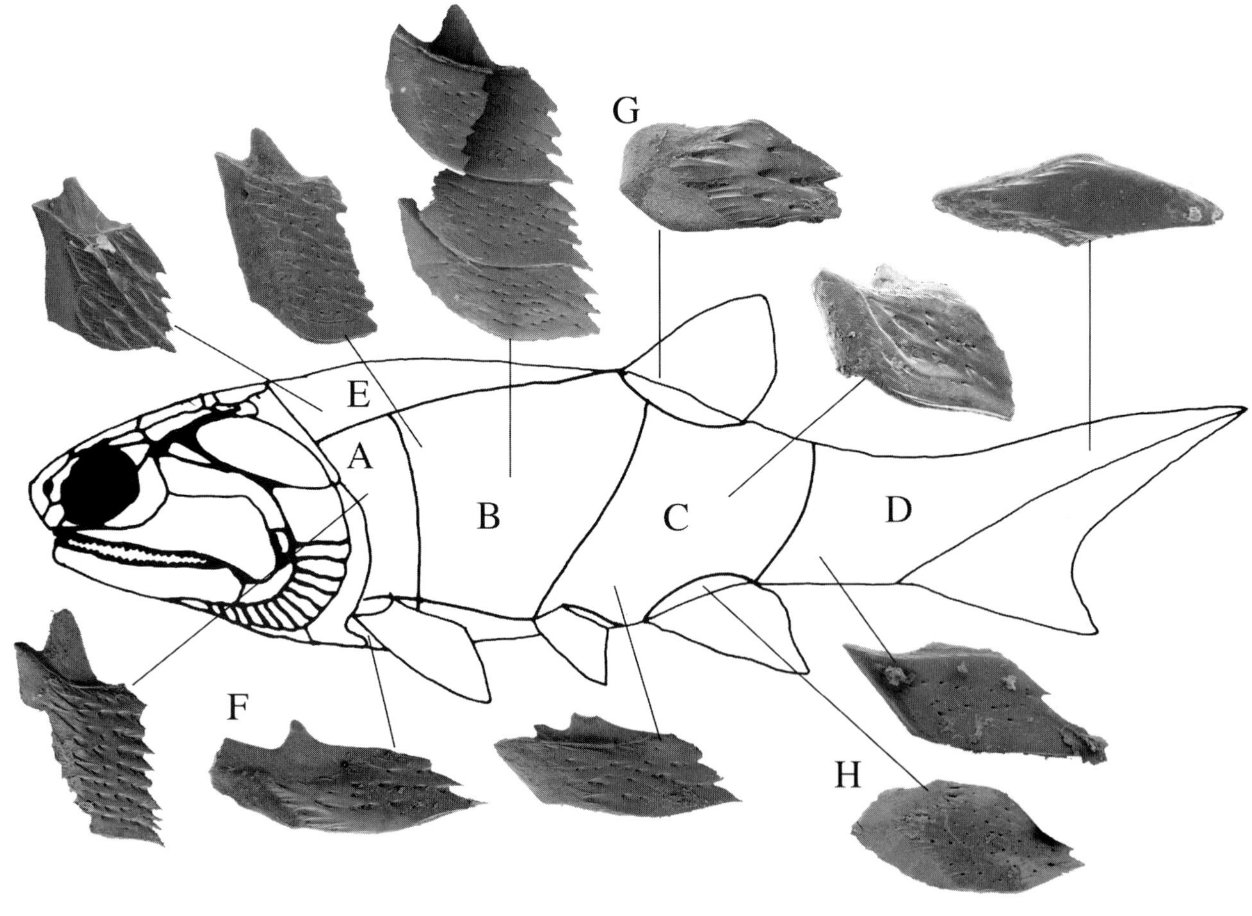

Figure 12. *Moythomasia durgaringa* indicating the variation in scales found on a single individual. Zones A–H indicate the primary scale type present, although it is clear there is some variation within zones.

are abundant, widespread, rapidly evolving and are distributed in a variety of facies. In continental rocks, microvertebrate remains such as teeth, scales and fin-spines are often the only age indicators preserved. Standard index fossils such as ammonoids, goniatites and conodonts occur only in marine facies, whereas many fossil fish species, such as sharks, palaeoniscoids and acanthodians, occurred in transitional environments. In some instances particular species, such as phoebodont sharks, may have inhabited deep marine, near shore and/or non-marine environments. Phoebodonts seem to have preferably inhabited open deep water marine facies, in association with the conodont *Palmatolepis*; but they are also found in nearshore environments associated with the conodont *Icriodus* and very shallow, restricted facies where there are no conodonts present (Ginter, 1991). In this situation phoebodont teeth take on an independent biostratigraphic role.

Where fishes have been used for biostratigraphy, such as in the Lower Old Red Sandstone of England (White, 1938, 1950, 1961; Stubblefield, 1985), the biozones have not always been precisely correlated to marine invertebrate-dominated sections. However, some units bearing microvertebrates are extremely well dated through correlating faunas to standard conodont zonations (Turner, 1993, 1997; Ginter and Ivanov, 2000). For example, in the East Baltic there is a nearly complete Silurian–Devonian succession in which the vertebrate fauna is correlated with the standard graptolite and conodont biozonations (Valiukevicius, 1985; Blieck *et al.*, 1988; Rzhonsnitskaya, 1988; Märss, 1989).

Many fossil groups that are widely used for age control of Devonian successions overseas have not been studied in detail for Australian sequences, and refinement of the biostratigraphic zonation requires more detailed systematic and biostratigraphic palaeontology (Young, 1995). Vertebrate correlation between major sedimentary basins in different states of Australia has largely been based on macrovertebrate assemblages of genera, or similar faunas (Ritchie, 1973; Young, 1993a, b, 1995). Recently Young (1995) and Turner (1993, 1995, 1997) have devised microvertebrate zones based on index fossils.

The work of correlating the Palaeozoic basins of Western Australia, especially the Frasnian and Fammenian sections using both micro- and macro-remains of fish, to the standard conodont and ammonite zones is currently being undertaken. In addition to the description of microvertebrates from the Gneudna Formation, Trinajstic (2000a,b) has now reported *Australolepis seddoni* in residue of limestone samples taken from the Gogo Formation, the Virgin Hills Formation and the middle Frasnian division of the Horse Springs section. The presence of conodonts with thelodonts in the Canning Basin has enabled more precise dating of the fauna from the Gneudna Formation. Turner (1997) has used thelodont scales to determine zonations for the Devonian of East Gondwana, and these zones have been successfully correlated to the standard conodont zones already known. The use of the East Gondwanan thelodont zonations (Turner, 1997) was crucial in dating the Gneudna Formation, where only rare and long ranging conodonts have been found (Trinajstic, 2001a).

Ginter (1991, 1997, 1999), Ginter and Ivanov (1996, 2000) and Turner (1993) have been working on using phoebodont teeth as zone fossils. Phoebodonts were cosmopolitan sharks and show great potential for global correlation. Phoebodont zones have been determined for the middle to late Devonian and early Carboniferous. The recent discovery of phoebodont teeth from the Virgin Hills Formation, the Gogo Formation and the Gneudna Formation of Western Australia (Trinajstic, 2000b) has permitted their inclusion in these schemes. Recent work suggests that with continuing research, these fossils have the potential to provide more refined biostratigraphic data.

Currently work is being undertaken in Australia and overseas to determine zonations for other vertebrate taxa. The holonematid arthrodires are one of the most widespread and common placoderm families, and are known from many localities in the middle Devonian and Frasnian. The association of *Holonema* with placoderms such as *Groenlandaspis* and *Bothriolepis*, which also have a wide geographic distribution, may provide a tool for long-distance correlation between marine and non-marine stratigraphic sequences (Leliévre and Goujet, 1986). Recent discoveries of placoderm scales with placoderm body plates (Burrow, 1996; Burrow and Turner, 1998, 1999; Trinajstic, 1999a) present the possibility that in the future these scales will be useful biostratigraphic indicators for Devonian sequences (Trinajstic, 1999a).

Vertebrate remains are proving useful in biocorrelation. They are particularly valuable because they occur across a number of different facies types. In addition, microvertebrates can make an important contribution to the dating of near shore and restricted marine facies where other index fossils are often absent. To date thelodont, chondrichthyan and acanthodian scales have proved the most useful but, as research continues on other taxa, it appears that they too will be useful age indicators.

BIOGEOGRAPHY OF FRASNIAN VERTEBRATES

Comparisons between faunas can be used to recognize biogeographic relationships. However, there are many uncertainties associated with faunal comparisons. These range from phylogenetic status to geographic range and are compounded when only fossil groups are considered. Further, fossils are rare and this raises difficulties in obtaining data of similar age and in suitable amounts to give an adequate sample size from more than one location. In palaeontology it is never expected that the complete record of biota for a given time will be obtained (Janvier and Blieck, 1993).

Table 2. Comparative similarity indexes (based on 13 reference taxa present) between the Gneudna Formation fauna (100% similar to itself) and 11 other Frasnian faunas from around the world.

	Frasnian Fish Faunas											
Reference Genera	**Gneudna**	**Gogo**	**Poland**	**Iran**	**Germany**	**France**	**USA**	**Bolivia**	**Antarctica**	**South Africa**	**Eastern Australia**	**South China**
Australolepis	x			x								
Groenlandaspis	x								x	x	x	
Holonema	x	x	x	x		x	x					
Bothriolepis	x	x	x				x	x	x	x	x	x
Ctenacanthus	x		x	x	x	x	x	x				
Protocrodus	x		x	x	x	x	x		x			
Cheiracanthus	x		x	x	x	x			x	x		x
Moythomasia	x	x	x	x	x	x		x				
Mimia	x	x	x									
Onychodus	x	x		x	x	x	x	x				
Chirodipterus	x	x		x	x							
Holodipterus	x	x	x									
Adololphus	x	x										
Total Similarity (%)	**100**	**62**	**62**	**62**	**46**	**46**	**38**	**31**	**31**	**23**	**15**	**15**

With regard to Gondwanan vertebrate faunas, the scarcity of data is a major obstacle to the discovery of significant palaeobiological patterns. In describing the late Devonian, the single land mass of Gondwana is often divided into West Gondwana, consisting primarily of what is now South America, Africa and the Arabic Peninsula, and East Gondwana, consisting primarily of what is now Australia, Antarctica and India (Figure 13). In addition, a number of small terranes such as Armorica (now the Iberian Peninsula and southern France), parts of Bohemia, western Turkey, southern Caucasus, northern and central Iran and South-east Asia extended along what was the northern margin of Gondwana (Scotese and McKerrow, 1990). However, there are competing hypotheses as to how close to the coast of Gondwana these smaller terranes were located and how expansive the Givetian/Frasnian ocean basins were (Young, 1993a, b).

In addition to the problem of small data sets are the reputedly cosmopolitan global vertebrate faunas, in particular marine fish faunas, during the late Devonian (Young, 1990). Typically, endemic freshwater fishes have been used in biogeographic studies because of their limited dispersal capabilities (Darlington, 1957; Young, 1990, 1993a, b). However, the environment of the Gneudna Formation has been interpreted as nearshore marine, and, to date, only one endemic species has been identified. A biogeographic analysis based on wide-ranging marine taxa is possible but requires good phylogenetic analyses at the species level (Leliévre *et al.,* 1993). Unfortunately these conditions can rarely be met, with Long's (1983) and Young's (1988) analyses of antiarchs being the only global biogeographic studies of Devonian vertebrates to date. When considering microvertebrates, phylogenetic analysis is even harder, as few specimens can be identified to the species level. This is particularly true of the vertebrate fauna of the Gneudna Formation, with few taxa identified to species. When using higher taxonomic groups, the problems pertaining to phylogenetic status and geographic range (vertebrate mobility varies between species) are compounded (Schultze and Cloutier, 1996).

Despite these difficulties, the inclusion of microvertebrates in biogeographic analyses permits a more accurate interpretation of faunal links between continents. The microvertebrate fauna from the Gneudna Formation supports tectonic data indicating a connection, via the Northwest Shelf, between the Carnarvon Basin and the Canning Basin (Stuckmeyer and Totterdell, 1992). There appear to be closer links between the marine faunas of Western Australia and those of Western Gondwana and Armorica than with the faunas of eastern Australia (Table 2). Faunas from the Holy Cross Mountains, Poland and central Iran (Janvier, 1974) show equal similarities in numbers of shared taxa to the Gneudna Formation. The similarity of the Kielce Polish fauna to the Gneudna fauna supports Young (1986b, 1997), who suggested the close affinities of Gondwana, Armorica and the Middle East. Ongoing research into Western Australian microvertebrates provides, for the first time, crucial evidence as to biogeographic relationships of Western Australia within the Gondwanan land mass during the late Devonian (Trinajstic, 2001a).

The faunal assemblages found in East Gondwana are less similar to those of the Gneudna Formation than the

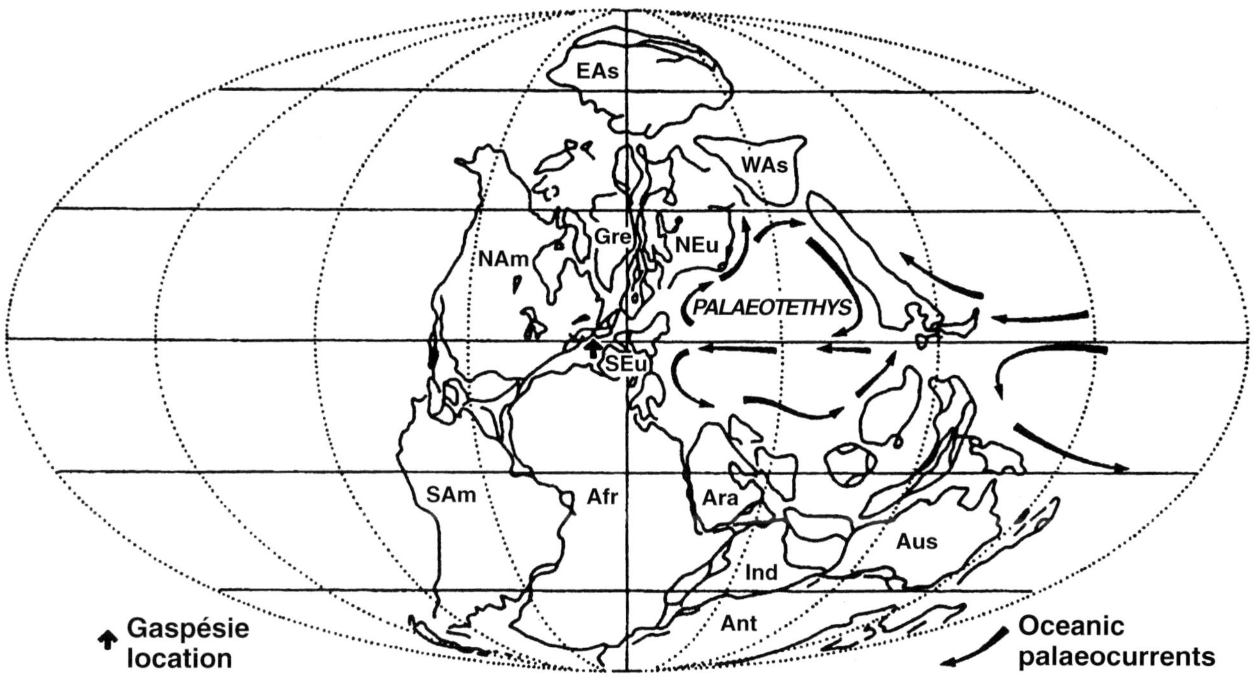

Figure 13. Palaeogeographic map of the Frasnian indicating oceanic palaeocurrents. Note that the currents that affected eastern Australia were different from those around Western Australia (based on Scotese and McKerrow, 1990). Key to abbreviations: Afr = Africa; Ant = Antarctica; Ara = Armorica; Aus = Australia; EAs = Eastern Asia; Gre = Greenland; Ind = India; NAm = North America; NEu = Northern Europe; SAm = South America; SEu = Southern Europe; WAs = Western Asia.

faunas of Western Gondwana and Armorica (Table 2). There are two possible causes for this lack of similarity. The first is that the faunas of eastern Australia come mainly from non-marine facies and that the discontinuity of non-marine conditions possibly prevented faunal exchange between eastern and western regions of Australia. The second possibility is that eastern Australia was influenced by different ocean currents, which favoured migration to other regions than Western Australia (Figure 13). Klapper (1980) found a similar pattern in the biogeographic relationship of conodonts, and concluded migration of cosmopolitan species (both off-shore and near shore conodont species) was affected mainly by oceanic currents. A palaeogeographic map indicating the main palaeocurrents supports this view, with the coast of Western Australia influenced by different currents than the shores of what is now eastern Australia (Figure 13).

MESOZOIC TO PRESENT

All the major fish groups had developed in the Palaeozoic and several, such as the heterostracans and placoderms and acanthodians, had flourished—undergoing extensive radiations; but a number of both jawless and jawed assemblages had become extinct by the end of the Permian (see Figure 1).

TRIASSIC TO CRETACEOUS

From the groups that persisted in the Mesozoic and Cainozoic two major new lineages diverged. The rays (Batoidea) first appeared in the early Jurassic—derived from the Neoselachii (true sharks). Cytogenetic studies by Schwartz and Maddock (2002) suggest that most families of extant elasmobranchs had diverged by late Jurassic (~155 m.y.a.); dasyatid rays are known from late Cretaceous.

But the most important development had occurred earlier in the early to mid Triassic, when teleosts first appeared. Teleosteans, derived from an advanced group of actinopterygians (Neoptergyii), have four key features:

(a) uroneural bones in the tail skeleton, which stiffen the dorsal tail lobe, increasing the swimming power;
(b) jaw modifications, including free premaxillae;
(c) ventral gill arch bones (basibranchials) with unpaired toothplates on them;
(d) specializations of head musculature.

The teleosts diversified rapidly in the Jurassic (see Long, 1995, pp.148–151) and were first recorded in Australia (at Talbragar) during this period (Long, 1991). As Figure 3 shows, Cretaceous fish remains are known principally from sites in Queensland and Victoria—and most of these Cretaceous fishes (elasmobranchs, holocephalans, actinopterygians and dipnoans) were widely distributed species; for example, mid Cretaceous sharks from W.A. have affinities with European forms (Siverson, 1996, 1997). But by late Cretaceous, Australian fish faunas were beginning to show some distinctness—in the form of groups such as the palaeoniscoids and ceratodontid dipnoans (Henderson *et al.*, 2000).

By the Cretaceous teleosts were well established in both marine and freshwater environments; for example, herring-like teleosts occurred in a freshwater lake in Victoria in early Cretaceous times, 135 m.y.a. (Waldman, 1984). Teleosts have become the most successful ray-finned fishes and have diversified to dominate extant fish faunas.

Unlike lungfishes, where the Australian fossil record is very complete throughout their history (Campbell and Barwick, 1987; Long, 1987, 1995, pp.162–163; Kemp, 1993, 1994, 1997a, b, 1998), the records for Mesozoic teleosts are fragmentary; similarly the chondrichthyan record, during Mesozoic and Tertiary, is very incomplete.

CAINOZOIC

The trend towards endemism referred to above, which began in late Cretaceous, continued in mid to high latitude marine faunas throughout the Cainozoic (Henderson *et al.,* 2000). Little work has been done on Australasian chondrichthyan faunas of the Cretaceous (Siverson, 1996, 1997); but more has been published on the Tertiary fauna—on the basis of extensive tooth collections. The first fossil Carcharhinidae are known from mid Eocene (Schwartz and Maddock, 2002) and extant genera of both sharks and rays are known from Oligocene and Miocene (Kemp, 1991).

Likewise, the records for Cainozoic teleosts are fragmentary; but from studies completed several general conclusions can be drawn. First, by the beginning of the Cainozoic many of the families of extant fishes (including percomorphs) had evolved (Long, 1995, p.138; Li *et al.*, 1997). Second, some fish families appear to have changed very little in 40 million years (Long, 1995, pp.156–157). Third, the limited evidence suggests that many extant genera had developed by the Oligocene, with a number of both marine and freshwater species virtually unchanged since the Miocene (Merrick and Schmida, 1984; Long, 1991).

Further details of Cainozoic fossils of extant groups inhabiting freshwaters are included below; but surveys of literature (Long and Turner, 1984; Kemp, 1991; Long, 1991) show that, although there are a number of isolated records of Tertiary or Quaternary fossils, there are few major studies of Cainozoic fish faunal assemblages in Australia. This situation is the result of factors such as limited resources (human or financial), access to potential sites, research priorities or collecting bias, and it still persists (Molnar and Kurz, 1997). Even in the few well publicized, supported and documented excavations of recent years, such as Riversleigh, the fishes identified (Archer *et al.*, 1991) are minor components of a diverse vertebrate fauna, which includes some conspicuous and spectacular tetrapods.

Freshwater Faunas

Extant Australian faunas clearly have diverse origins and fossils of 15 genera in 14 families represented have been described from Tertiary and Quaternary deposits, although several lineages extend much further back into the Mesozoic (Merrick and Schmida, 1984; Rowland, 1989; Allen *et al.*, 2002, pp.3–8). These 14 families include elasmobranchs as well as examples of the major subdivisions of bony fishes such as osteoglossids, salmoniforms, ostariophysans and percoids (Plates 41–44).

Fossil identification to generic level is based on conspicuous osteological features, but many of the smaller characteristics used to differentiate species are not adequately preserved—so often it is not possible to determine if the fossilized individual is a member of a living species or an extinct congener; however, occasionally specimens are preserved in exceptionally good condition and can be definitely identified to species level. An example of this is a series of Murray cod remains, from diatomaceous earth in northern New South Wales, dated at 7–13 million years. It is clear that even some of the more derived teleosts have changed little for many millions of years.

A few isolated studies continue to be released on Pleistocene teleost remains in Australia (Molnar and Kurz, 1997), but the other regional faunas are very poorly known. At least two galaxiids have been identified from Miocene deposits in New Zealand (McDowall and Pole, 1997); however, no freshwater fish records have been found for New Guinea (Allen, 1996).

Throughout Australasia there is little information on vertebrates from the earlier part of the Quaternary; but, for a variety of reasons (including duration, extreme climate oscillations and associated sea level changes), little significant evolutionary change is expected in faunas over most of the period (Kershaw *et al.,* 2000).

CONCLUSIONS

This chapter outlines the long evolutionary history of fishes and introduces the enormous diversity of the total fish assemblage (within a systematic framework). By contrast, in the following chapters, discussion focuses on relationships of particular groups of fossil fishes—often below order level—or on diversity of extant faunas.

The Devonian period is known as 'the age of fishes' because it was the time of the first major radiation of the jawed vertebrates. All major fish groups are known from the early Devonian, and in the early Devonian limestones of south-eastern Australia agnathans (thelodonts), placoderms, chondrichthyans, acanthodians and osteichthyans are all represented, although not in equal abundance or diversity. The fossil record based on microvertebrate material tends to mirror the macrovertebrate record, showing various cosmopolitan forms with a significant component of East Gondwanan endemics. The use of microremains, such as scales, teeth and bone fragments, together with conodont elements is proving increasingly valuable for precise inter-regional stratigraphic alignment, and will assist in establishing a microvertebrate-based biozonation for the Devonian of Australia. In addition, the unusual combination of features of the recently described osteichthyan braincase from Wee Jasper provide new insight into the origin and radiation of the bony fishes. It is the first specimen from the major osteichthyan clade to show previously unknown primitive conditions, especially an opening for a cartilaginous eyestalk, suggesting possible links to both ancestral ray-finned fishes and the lineage ancestral to lobe-finned fishes (and other vertebrates).

The range and distribution of microvertebrate remains in Western Australia have been clarified over the last ten years. These studies have shown that, using the Gneudna and Gogo Formations, many of the fishes from the Gogo Formation that were previously thought to be endemic, had a much greater geographic range along the coast of Western Australia. Microvertebrate data from the Frasnian of Western Australia have also made comparisons between faunas from eastern Australia, the Middle East and Europe possible, because previously unreported taxa such as thelodonts (with the exception of *Australolepis seddoni*), phoebodont sharks and acanthodians have now been found. These discoveries are important in determining biogeographic differentiation along the continental margins during the mid Palaeozoic.

Nearly all the major fish taxa (thelodonts, placoderms, chondrichthyans and acanthodians) have now been used in vertebrate zonations on either a regional or global scale (Blieck and Turner, 2000). The discovery of conodonts, thelodonts and phoebodonts at Horse Springs and Virgin Hills Formation has confirmed the age range of *Australolepis seddoni* in East Gondwana, permitting a zonation for the later Frasnian in the northern Canning Basin.

The global vertebrate fauna is still dominated by fishes (>23,000 species are currently recognized), but the history of many living groups is not as well documented as the extinct Palaeozoic groups mentioned. Except for lungfishes, the Australasian fossil fish record since the Devonian is not well known. The main events demonstrated in Mesozoic faunas are the appearance and diversification of teleosts; furthermore it is clear, from the few studies of Cainozoic fish faunas, that a number of extant taxa have remained little changed for tens of millions of years.

ACKNOWLEDGEMENTS

Dr. J.A. Long (Western Australian Museum, Perth) is thanked for unpublished observations on the Gogo Fauna and appreciation is expressed to Dr. Z. Johanson (Australian Museum, Sydney), as well as Dr. P. Ahlberg (British Museum (Natural History), London), for critical comment on the manuscript.

REFERENCES

ABC Science Online, 2003. Oldest vertebrate fossil found in Australia, scientists say. http://www.abc.net.au/science/news/stories/s973390.htm

Aldridge, R.J. (ed.) 1987. *Palaeobiology of Conodonts*. Ellis Horwood, Chichester. 180pp.

Aldridge, R.J., Briggs, D.E.G., Smith, M.P., Clarkson, E.N.K. and Clark, N.D.L. 1993. The anatomy of conodonts. *Philosophical Transactions of the Royal Society of London,* Series B, Vol. 340, pp.405–421.

Allen, G.R. 1996. Freshwater fishes of Irian Jaya. Pp.15–21. In Kitchener, D.J. and Suyanto, A. (eds). *Proceedings of the First International Conference on Eastern Indonesia–Australian Vertebrate Fauna, Manado, Indonesia, November 22–26, 1994.* Western Australian Museum, Perth. 174pp.

Allen, G.R., Midgley, S.H. and Allen, M. 2002. *Field Guide to the Freshwater Fishes of Australia.* Western Australian Museum, Perth. 394pp.

Applegate, S.P. 1967. A survey of shark hard parts. In Gilbert P.W., Mathewson, R.F. and Rall, D.P. (eds). *Sharks, Skates and Rays.* Johns Hopkins University Press, Baltimore. 624pp.

Archer, M., Hand, S.J. and Godthelp, H. 1991. *Riversleigh. The Story of Animals in Ancient Rainforests of Inland Australia.* Reed Books, Sydney. 264pp.

Basden, A.M. 1999a. Emsian (early Devonian) microvertebrates from the Buchan and Taemas areas of southeastern Australia. *Records of the Western Australian Museum, Supplement,* No. 57, pp.15–21.

Basden, A.M. 1999b. Early Devonian microvertebrates from the Tyers-Boola area of central Victoria, Australia. *Bollettino della Società Paleontologica Italiana*, Vol. 37(2–3), pp.527–541.

Basden, A.M., Burrow, C., Hocking, M., Parkes, R. and Young, G.C. 2000a. Siluro-Devonian microvertebrates from southeastern Australia. *Courier Forschungsinstitut Senckenberg,* Vol. 223, pp.201–222.

Basden, A.M., Young, G.C., Coates, M.I. and Ritchie, A. 2000b. The most primitive osteichthyan braincase? *Nature*, Vol. 403, pp.185–188.

Basden, A.M. and Young, G.C. 2001. A primitive actinopterygian neurocranium from the early Devonian of southeastern Australia. *Journal of Vertebrate Paleontology*, Vol. 21, No. 4, pp.754–766.

Blieck, A., Mark-Kurik, E. and Marss, T. 1988. Biostratigraphical correlations between Siluro-Devonian invertebrate-dominated and vertebrate-dominated sequences: the East Baltic example. Pp.579–587. In McMillan, N.J., Embry, A.F. and Glass, D.J. (eds). *Devonian of the World: Proceedings of the Second International Symposium on the Devonian System.* Canadian Society of Petroleum Geologists, Calgary.

Blieck, A. and Turner, S. 2000. Palaeozoic vertebrate biochronology and global marine/non-marine correlation. Final report of IGCP 328 (1991–1996). *Courier Forschungsinstitut Senckenberg,* Vol. 223, pp.1–68.

Bloom, H. 1999. *Loganellia* (Thelodonti, Agnatha) from the lower Silurian of northern Greenland. *Acta Geologica Polonica,* Vol. 49, No.2, pp.97–104.

Burrow, C. 1994. Form and function in scales of *Ligulalepis toombsi* Schultze, a palaeoniscoid from the early Devonian of Australia. *Records of the South Australian Museum*, Vol. 27, No. 2, pp.175–185.

Burrow, C. 1996. Placoderm scales from the lower Devonian of New South Wales, Australia, *Modern Geology,* Vol. 20, pp.351–370.

Burrow, C. and Turner, S. 1998. Devonian placoderm scales from Australia. *Journal of Vertebrate Paleontology,* Vol. 18, pp.677–695.

Burrow, C. and Turner, S. 1999. A review of placoderm scales, and their significance in placoderm phylogeny. *Journal of Vertebrate Paleontology,* Vol. 19, pp, 204–219.

Campbell, K.S.W. and Barwick, R.E. 1983. Early evolution of dipnoan dentitions and a new genus *Speonesydrion. Memoir of the Association of Australasian Palaeontologists*, Vol. 1, pp.17–49.

Campbell, K.S.W. and Barwick, R.E. 1987. Palaeozoic lungfishes—a review. *Journal of Morphology, Supplement,* Vol. 1, pp.93–131.

Chang, M.M. 1982. *The braincase of* Youngolepis, *a lower Devonian crossopterygian from Yunnan, south-western China.* Ph.D. Thesis (unpublished), University of Stockholm, Sweden.

Chatterton, B.D.E. 1971. Taxonomy and ontogeny of Siluro-Devonian trilobites from near Yass, New South Wales. *Palaeontographica* A, Vol. 137, pp.1–108.

Chatterton, B.D.E. 1973. Brachiopods of the Murrumbidgee Group, Taemas, New South Wales. *Bureau of Mineral Resources, Geology and Geophysics, Bulletin,* No. 137, pp.1–146.

Coates, M.I. and Sequeira, S.E.K. 1998. The braincase of a primitive shark. *Transactions of the Royal Society of Edinburgh: Earth Sciences,* Vol. 89, pp.63–85.

Darlington, P.J. 1957. *Zoogeography: The Geographic Distribution of Animals.* John Wiley and Sons, New York. 210pp.

De Pomeroy, A. 1996. Biostratigraphy of Devonian microvertebrates from Broken River, North Queensland. *Records of the Western Australian Museum.* Vol. 17, No. 2, pp.417–438.

Dring, R. 1980. *Palaeobiology of the Gneudna Formation, late Devonian (Frasnian), Carnarvon Basin, Western Australia.* Ph.D. Thesis (unpublished), University of Western Australia, Perth.

Esin, D. 1990. The scale cover of *Amblypterina costata* (Eichwald) and palaeoniscoid taxonomy based on isolated scales. *Paleontological Journal,* Vol. 2, pp.90–98.

Gardiner, B.G. 1984. The relationships of the palaeoniscid fishes, a review based on new specimens of *Mimia* and *Moythomasia* from the upper Devonian of Western Australia. *Bulletin of the British Museum of Natural History (Geology)*, Vol. 37, pp.173–428.

Ginter, M. 1991. Late Famennian shark teeth from the Holy Cross Mountains, Poland. *Ichthyolith Issues,* No. 6, pp.6–24.

Ginter, M. 1997. *Sharks from the Devonian/Carboniferous boundary beds of Thuringia.* The Amadeus Grabau Symposium, Rochester.

Ginter, M. 1999. Famennian–Tournaisian chondrichthyan microremains from the Eastern Thuringian Slate Mountains. *Abhandlungen und Berichte Fur Naturkunde,* Vol. 21, pp.25–47.

Ginter, M. and Ivanov, A. 1996. Relationships of *Phoebodus. Modern Geology,* Vol. 20, pp.263–274.

Ginter, M. and Ivanov, A. 2000. Stratigraphic distribution of chondrichthyans in the Devonian on the East European Platform margin. *Courier Forschungsinstitut Senckenberg,* Vol. 223, pp.325–339.

Goujet, D. 1984. Placoderm interrelationships: a new interpretation, with a short review of placoderm classification. *Proceedings of the Linnean Society of New South Wales*, Vol. 107, pp.211–243.

Gross, W. 1937. Das Kopfskelett von *Cladodus wildungensis* Jaekel. *Senckenbergiana*, Vol. 19, pp.80–107.

Gross, W. 1969. *Lophosteus superbus* Pander, ein Teleostome aus dem Silur. *Lethaia*, Vol. 2, pp.12–47.

Gross, W. 1971. Unterdevonische Thelodontier- und Acanthodier-Schuppen aus Westaustralien. *Palontologisches Zeitschrift,* Vol. 45, pp.97–106.

Henderson, R.A., Crampton, J.S., Dettmann, M.E., Douglas, J.G., Haig, D., Shafik, S., Stilwell, J.D. and Thulborn, R.A. 2000. Biogeographical observations on the Cretaceous biota of Australasia. *Memoir of the Association of Australasian Palaeontologists*, Vol. 23, pp.355–404.

Hill, D. 1940. The lower Middle Devonian rugose corals of the Murrumbidgee and Goodradigbee Rivers, NSW. *Journal of the Royal Society of New South Wales*, Vol. 74, pp.247–276.

Janvier, P. 1974. Preliminary report on late Devonian fishes from central Iran. *Geological Survey of Iran Report,* No. 31, pp.5–48.

Janvier, P. and Blieck, A. 1993. The Silurian–Devonian agnathan biostratigraphy of the Old Red Continent. Pp.67–86. In Long, J.A. (ed.). *Palaeozoic Vertebrate Biostratigraphy and Biogeography*. Bellhaven Press, London. 369pp.

Johnston, P.A. 1993. Lower Devonian Pelecypoda from south-eastern Australia. *Memoir of the Association of Australasian Palaeontologists*, Vol. 14, p. 134.

Karatajute-Talimaa, V. 1992. The early stages of the dermal skeleton formation of chondrichthyans. Pp.223–243. In Mark-Kurik, E. (ed.). *Fossil Fishes as Living Animals*. Academy of Sciences of Estonia. Tallinn.

Kemp, A. 1993. *Ceratodus diutinus,* a new ceratodont from Cretaceous and late Oligocene-medial Miocene deposits in Australia. *Journal of Paleontology,* Vol. 67, No. 5, pp.883–888.

Kemp, A. 1994. Australian Triassic lungfish skulls. *Journal of Paleontology*, Vol.68, No.3, pp.647–654.

Kemp, A. 1997a. A revision of the Australian Mesozoic and Cenozoic lungfish of the family Neoceratodontidae (Osteichthyes: Dipnoi), with a description of four new species. *Journal of Paleontology*, Vol. 71, No.4, pp.713–733.

Kemp, A. 1997b. Four species of *Metaceratodus* (Osteichthyes: Dipnoi, Family Ceratodontidae) from Australian Mesozoic and Cenozoic deposits. *Journal of Vertebrate Paleontology*, Vol. 17, No.1, pp.26–33.

Kemp, A. 1998. Skull structure in post-Paleozoic lungfish. *Journal of Vertebrate Paleontology*, Vol. 18, No.1, pp.43–63.

Kemp, A. and Nicoll, R.S. 1995. Protochordate affinities of conodonts. *Courier Forschungsinstitut Senckenberg,* Vol. 182, pp.235–245.

Kemp, A. and Nicoll, R.S. 1996. A histochemical analysis of biological residues in conodont elements. *Modern Geology*, Vol. 20, No. 3, pp.287–302.

Kemp, N.R. 1991. Chondrichthyans in the Cretaceous and Tertiary of Australia. Pp.497–568. In Vickers-Rich, P., Monaghan, J.M., Baird, R.F. and Rich, T.H. (eds). *Vertebrate Palaeontology of Australasia*. Pioneer Design Studio and Monash University Publications Committee, Melbourne. 1437pp.

Kershaw, P., Quilty, P.G., David, B., Van Huet, S. and McMinn, A. 2000. Palaeogeography of the Quaternary of Australasia. *Memoir of the Association of Australasian Palaeontologists*, Vol. 23, pp.471–516.

Klapper, G. 1980. The Montagne Noire Frasnian (Upper Devonian) conodont succession. *Canadian Society of Petroleum Geologists*, Vol. 14, No. 3, pp.449–468.

Lelièvre, H. and Goujet, D. 1986. Biostratigraphic significance of some uppermost Devonian placoderms. *Annales de la Société Géologique de Belgique*, Vol. 109, No. 1, pp.55–59.

Leliévre, H., Janvier, P. and Blieck, A. 1993. Silurian-Devonian vertebrate biostratigraphy of Western Gondwana and related terranes (South America, Africa, Armorica—Bohemia, Middle East.). Pp.139–173. In Long, J. A. (ed). *Palaeozoic Vertebrate Biostratigraphy and Biogeography.* Bellhaven Press, London. 369pp.

Li, G.-Q., Grande, L. and Wilson, M.V.H. 1997. The species of *Phareodus* (Teleostei: Osteoglossidae) from the Eocene of North America and their phylogenetic relationships. *Journal of Vertebrate Paleontology*, Vol. 17, No.3, pp.487–505.

Lindley, I.D. 2000. Acanthodian fish remains from the lower Devonian Cavan Bluff Limestone (Murrumbidgee Group), Taemas district, New South Wales. *Alcheringa*, Vol. 24, pp.11–35.

Lindley, I.D. 2001. Lower Devonian ischnacanthid fish (Gnathostomata: Acanthodii) from the Taemas Limestone, Lake Burrinjuck, New South Wales. *Alcheringa*, Vol. 25, pp.269–291.

Lindley, I.D. 2002. Acanthodian, onychodontid and osteolepidid fish from the middle–upper Taemas Limestone, (early Devonian), Lake Burrinjuck, New South Wales. *Alcheringa*, Vol. 26, pp.103–126.

Long, J. 1991. The long history of Australian fossil fishes. Pp.337–428. In Vickers-Rich, P., Monaghan, J.M., Baird, R.F. and Rich, T.H. (eds). *Vertebrate Palaeontology of Australasia*. Pioneer Design Studio and Monash University Publications Committee, Melbourne. 1437pp.

Long, J. and Turner, S. 1984. A checklist and bibliography of Australian fossil fish. Pp.235–254. In Archer, M. and Clayton, G. (eds). *Vertebrate Zoogeography and Evolution in Australasia.* Hesperian Press, Perth. 1203pp.

Long, J.A. 1983. New bothriolepid fishes from the late Devonian of Victoria, Australia. *Palaeontology,* Vol. 26, pp.295–320.

Long, J.A. 1987. A description of lungfish *Eoctenodus microsoma* (Hills 1929) with a reassessment of genus *Dipterus* Sedgwick & Murchison 1828 in Australia. *Records of the Western Australian Museum,* Vol.13, Pt. 2, pp.297–314.

Long, J.A. 1995. *The Rise of Fishes. 500 Million Years of Evolution.* University of New South Wales Press, Sydney. 223pp.

Long, J.A. and Trinajstic, K. 2000. Devonian microvertebrate faunas of Western Australia. *Courier Forschungsinstitut Senckenberg,* Vol. 223, pp.471–485.

Märss, T. 1986. [Silurian vertebrates of Estonia and West Latvia], *Fossilia Baltica,* pp.1–104. *Akademia Nauk Estonia SSR, Instituut Geologii, "Valgus", Tallinn.* (in Russian).

Märss, T. 1989. Vertebrates. Pp.284–289. In Holland, C. H. and Bassett, M. G. (eds). *A Global Standard for the Silurian System.* Geological Series, Vol. 9, National Museum of Wales. Cardiff.

Mawson, R. 1987. Early Devonian conodont faunas from Buchan and Bindi, Victoria, Australia. *Palaeontology*, Vol. 30, pp.251–297.

Mawson, R. and Talent, J.A. 1994. Age of an early Devonian carbonate fan and isolated limestone clasts and megaclasts, east-central Victoria. *Proceedings of the Royal Society of Victoria*, Vol. 106, pp.31–70.

Mawson, R., Talent, J.A., Brock, G.A. and Engelbretsen, M.J. 1992. Conodont data in relation to sequences about the Pragian-Emsian boundary (early Devonian) in south-eastern Australia. *Proceedings of the Royal Society of Victoria*, Vol. 104, pp.23–56.

McCoy, F. 1876. Prodromus of the palaeontology of Victoria. *Prodromus of the Palaeontology of Victoria*, Vol. 3, pp.1–40.

McDowall, R.M. and Pole, M. 1997. A large galaxiid fossil (Teleostei) from the Miocene of Central Otago, New Zealand. *Journal of the Royal Society of New Zealand*, Vol. 27, No. 2, pp.193–198.

Merrick, J.R. and Schmida, G.E. 1984. *Australian Freshwater Fishes. Biology and Management.* J.R. Merrick, Sydney. 409pp.

Miles, R. S. 1971. The Holonematidae (Placodern Fishes): a review based on new specimens of *Holonema* from the upper Devonian of Western Australia. *Philosophical Transactions of the Royal Society of London,* Series B, Vol. 263, pp.101–234.

Miles, R.S. 1973. Relationships of acanthodians. Pp.63–103. In Greenwood, P.H., Miles, R.S. and Patterson, C. (eds). *Interrelationships of Fishes.* Academic Press, London. 536pp.

Molnar, R.E. and Kurz, C. 1997. The distribution of Pleistocene vertebrates on the eastern Darling Downs, based on the Queensland Museum collections. *Proceedings of the Linnean Society of New South Wales*, Vol. 117, pp.107–134.

Ørvig, T. 1967. Phylogeny of tooth tissues:evolution of some calcified tissues in early vertebrates. Pp.45–110. In Miles, A.E. (ed). *Structural and Chemical Organisation of Teeth.* Vol. 1. Academic Press, London.

Pedder, A.E.H., Jackson, J.H. and Philip, G.M. 1970. Lower Devonian biostratigraphy of the Wee Jasper region, New South Wales. *Journal of Paleontology*, Vol. 44, pp.206–251.

Pickett, J.W. 1969. Middle and upper Palaeozoic sponges from New South Wales. *Memoirs of the Geological Survey of New South Wales, Palaeontology*, Vol. 16, pp.1–24.

Purnell, M.A. 1993. Feeding mechanisms in conodonts and the function of the earliest vertebrate hard tissues. *Geology,* Vol. 21, pp.375–377.

Purnell, M.A. 1994. Skeletal ontogeny and feeding mechanisms in conodonts. *Lethaia*, Vol. 27, pp.129–138.

Purnell, M.A. 1995. Microwear on conodont elements and macrophagy in the first vertebrates. *Nature*, Vol. 374, pp.798–800.

Reif, W.-E. 1974. Morphogenese und Musterbildung im Hautzähnchen-Skelett von *Heterodontus. Lethaia,* Vol. 7, pp.25–42.

Reif, W.-E. 1978. Types of morphogenesis of the dermal skeleton in fossil sharks. *Palontologisches Zeitschrift,* Vol. 52, pp.110–128.

Reif, W.-E. 1980. A model of morphogenetic processes in the dermal skeleton of elasmobranchs. *Neues Jahrbuch für Geologie und Paläontologie, Abhandlungen,* Vol. 159, pp.339–359.

Reif, W.-E. 1985. Squamation and ecology of sharks. *Courier Forschungsinstitut Senckenberg*, Vol, 78, pp.1–255.

Reynolds, L.R. 1978. The taxonomy and palaeoecology of ostracodes from the Devonian *Receptaculites* Limestone, Taemas, New South Wales, Australia. *Palaeontographica* A, Vol. 162, pp.144–203.

Ritchie, A. 1973. *Wuttagoonaspis* gen. nov., an unusual arthrodire from the Devonian of western New South Wales, Australia. *Palaeontographica* A, Vol. 143, pp.58–72.

Ritchie, A. 2006. The great Devonian fish kill at Canowindra. Pp.159–184. In Merrick, J.R., Archer, M., Hickey, G. and Lee, M.S.Y. (eds). *Evolution and Biogeography of Australasian Vertebrates.* Auscipub Pty. Ltd., Sydney. 942pp.

Rowland, S.J. 1989. Aspects of the history and fishery of the Murray cod, *Maccullochella peeli* (Mitchell) (Percichthyidae). *Proceedings of the Linnean Society of New South Wales,* Vol. 111, pp.201–213.

Rzhonsnitskaya, M.A. 1988. Biostratigrafiya Devona okrain Kuznetskogo Bassyina, Tom.1 Stratigrafiya. *Trudy vses Nauchno-Issledovatel'skiy, Institut Geologii*, pp.1–286.

Sansom, I.J., Smith, M.P., Armstrong, H.A. and Smith, M.M. 1992. Presence of the earliest vertebrate hard tissues in conodonts. *Science*, Vol. 256, pp.1308–1311.

Sansom, I.J., Smith, M.P. and Smith, M.M. 1994. Dentine in conodonts. *Nature*, Vol. 368, p. 591.

Schultze, H.-P. 1968. Palaeoniscoidea-Schuppen aus dem Unterdevon Australiens und Kanadas und aus dem Mitteldevon Spitsbergen. *Bulletin of the British Museum of Natural History (Geology)*, Vol. 16, pp.343–368.

Schultze, H.-P. and Cloutier, R. 1996. Comparision of the Escuminac Formation ichthyofauna with other late Givetian/early Frasnian ichthyofauna. Pp.348–368. In Schultze, H.-P. and Cloutier, R. (eds). *Devonian Fishes and Plants of Miguasha, Quebec, Canada.* Dr. F. Pfeil Verlag, München.

Schwartz, F.J. and Maddock, M.B. 2002. Cytogenetics of the elasmobranchs: genome evolution and phylogenetic implications. *Marine and Freshwater Research*, Vol. 53, No. 2, pp.491–502.

Scotese, C.R. and McKerrow, W.S. 1990. Revised World maps an introduction. Pp.1–21. In Scotese, C.R. and McKerrow, W.S (eds). *Palaeozoic Palaeogeography and Biogeography*. Memoir Vol.12. Geological Society, London.

Seddon, G. 1969. Conodont and fish remains from the Gneudna Formation, Carnarvon Basin, Western Australia. *Proceedings of the Royal Society of Western Australia,* Vol. 52, pp.21–30.

Siverson, M. 1996. Lamniform sharks of the mid Cretaceous Alinga formation and Beedagong Claystone, Western Australia. *Palaeontology*, Vol. 39, Pt.4, pp.813–849.

Siverson, M. 1997. Sharks from the mid-Cretaceous Gearle siltstone, southern Carnarvon Basin, Western Australia. *Journal of Vertebrate Paleontology*, Vol. 17, No. 3, pp.453–465.

Stensiö, E. 1961. Permian vertebrates. Pp.231–247. In Raasch, G. O. (ed). *Geology of the Arctic*. Toronto.

Stiassny, M.L.J., Parenti, L.R. and Johnson, G.D. (eds). 1996. *Interrelationships of Fishes.* Academic Press, London. 496pp.

Stubblefield, J. 1985. Errol Ivor White 1901–1985, elected FRS 1956. *Biographical Memoir of Fellows of the Royal Society,* Vol. 31, pp.635–651.

Stuckmeyer, H.I.M. and Totterdell, J.M. 1992. *Australia, Evolution of a Continent.* Australian Government Publishing Service, Canberra. 96pp.

Tassell, C.B. 1982. Gastropods from the early Devonian '*Receptaculites*' Limestone, Taemas, New South Wales. *Records of the Queen Victoria Museum*, Vol. 77, pp.1–30.

Teichert, C. 1940. *Helicoprion* in the Permian of Western Australia. *Journal of Paleontology*, Vol. 14, pp.140–149.

Thomas, G. 1957. Lower Carboniferous deposits in the Fitzroy Basin, Western Australia. *Australian Journal of Science,* Vol. 19, pp.160–161.

Thomas, G. 1959. The lower Carboniferous Laurel Formation of the Fitzroy Basin. *Bureau of Mineral Resources Report,* No.38, pp.31–36.

Thomson, K.S. and Campbell, K.S.W. 1971. The structure and relationships of the primitive Devonian lungfish *Dipnorhynchus süssmilchi* Etheridge. *Bulletin of the Peabody Museum of Natural History*, Vol. 38, pp.1–109.

Trinajstic, K. 2001a. A description of additional variation seen in the scale morphology of the Frasnian thelodont *Australolepis seddoni* Turner and Dring 1981. *Records of the Western Australian Museum,* Vol. 20, No. 2, pp.200–237.

Trinajstic, K. 2001b. Acanthodian microremains from the Frasnian Gneudna Formation, Western Australia. *Records of the Western Australian Museum,* Vol. 20, No. 2, pp.187–198.

Trinajstic, K.M. 1999a. New anatomical information on *Holonema* (Placodermi) based on material from the Gogo Formation and the Givetian-Frasnian Gneudna Formation, Western Australia. *Geodiversitas,* Vol. 21, No.1, pp.69–84.

Trinajstic, K. 1999b. Scale morphology of the late Devonian palaeoniscoid *Moythomasia durgaringa* Gardiner and Bartram, 1977. *Alcheringa,* Vol. 23, pp.9–19.

Trinajstic, K.M. 1999c. Scales of palaeoniscoid fishes (Osteichthyes: Actinopterygii) from the late Devonian of Western Australia. *Records of the Western Australian Museum, Supplement,* No. 57, pp.93–106.

Trinajstic, K.M. 2000a. Microvertebrates from the Frasnian Gneudna Formation, Australia. *Early and Lower Vertebrates Abstracts,* No.4, p. 20.

Trinajstic, K.M. 2000b. Conodonts, thelodonts and phoebodonts—together at last. *Palaeontology Down Under Abstracts, Geological Society of Australia,* No. 61, p.121.

Turner, S. 1991. Palaeozoic vertebrate microfossils in Australasia. Pp.429–464. In Vickers-Rich, P., Monaghan, J.M., Baird, R.F. and Rich, T.H. (eds). *Vertebrate Palaeontology of Australasia.* Pioneer Design Studio and Monash University Publications Committee, Melbourne. 1437pp.

Turner, S. 1993. Early Carboniferous microvertebrates from the Narrien Range, central Queensland. *Memoir of the Association of Australasian Palaeontologists,* Vol. 15, pp.289–304.

Turner, S. 1995. Defining Devonian microvertebrate faunal assemblages or zones for Eastern Gondwana. *Ichthyolith Issues Special Publication*, No. 1, pp.43–52.

Turner, S. 1997. Sequence of Devonian thelodont scale assemblages in East Gondwana. *Geological Society of America Special Paper,* No. 321, pp.295–315.

Turner, S. 1999. Early Silurian to early Devonian thelodont assemblages and their possible ecological significance. Pp.42–78. In Boucot, A.J. and Lawson, J. (eds). *Palaeocommunities: case studies in International Geological Correlation 53, Programme Project Ecostratigraphy, Final Report.* Cambridge University Press, Cambridge.

Turner, S. and Dring, R.S. 1981. Late Devonian thelodonts (Agnatha) from the Gneudna Formation, Carnarvon Basin, Western Australia. *Alcheringa*, Vol. 5, pp.39–48.

Valiukevicius, J.J. 1985. *Acanthodians from the Narva Regional Stage of the Main Devonian Field.* Mosklas, Vilnius. 143pp. [In Russian with English summary].

Waldman, M. 1984. The fossil lake fauna of Koonwarra, Victoria. Pp.231–233. In Archer, M. and Clayton, G. (eds). *Vertebrate Zoogeography and Evolution in Australasia.* Hesperian Press, Perth. 1203pp.

Warren, A. 2006. Origins and early radiations of the amphibians. Pp.227–236. In Merrick, J.R., Archer, M., Hickey, G. and Lee, M.S.Y. (eds). *Evolution and Biogeography of Australasian Vertebrates.* Auscipub Pty. Ltd., Sydney. 942pp.

White, E.I. 1938. The ostracoderm *Pteraspis* Kner and the relationship of the agnathan vertebrates. *Philosophical Transactions of the Royal Society of London,* Vol. 527B, pp.381–457.

White, E.I. 1950. The vertebrate faunas of the Lower Old Red Sandstones of the Welsh Borders. *Bulletin of the British Museum (Natural History), Geology*, Vol. 1, pp.51–67.

White, E.I. 1952. Australian arthrodires. *Bulletin of the British Museum of Natural History (Geology)*, Vol. 1, pp.249–304.

White, E.I. 1961. The Old Red Sandstone of Brown Clee Hill and the adjacent area. II. Palaeontology. *Bulletin of the British Museum (Natural History), Geology,* Vol. 5, pp.243–310.

White, E.I. 1978. The larger arthrodiran fishes from the area of the Burrinjuck Dam, NSW. *Transactions of the Zoological Society of London*, Vol. 34, pp.149–262.

White, E.I. and Toombs, H.A. 1972. The buchanosteid arthrodires of Australia. *Bulletin of the British Museum of Natural History (Geology)*, Vol. 22, pp.379–419.

Williams, M.E. 1985. The "Cladodont level" sharks of the Pennsylvanian Black Shales of central North America. *Palaeontographica* A, Vol. 190, pp.83–158.

Woodward, A.S. 1941. The headshield of a new macropetalichthyid fish (*Notopetalichthys hillsi*, gen. et sp. nov.) from the middle Devonian of Australia. *Annals and Magazine of Natural History,* Series 2, Vol. 8, pp.91–96.

Young, G.C. 1978. A new early Devonian petalichthyid fish from the Taemas/Wee Jasper region of New South Wales. *Alcheringa*, Vol. 2, pp.103–116.

Young, G.C. 1979. New information on the structure and relationships of *Buchanosteus* (Placodermi: Euarthrodira) from the early Devonian of New South Wales. *Zoological Journal of the Linnean Society*, Vol. 66, pp.307–352.

Young, G.C. 1980. A new early Devonian placoderm from New South Wales, Australia, with a discussion of placoderm phylogeny. *Palaeontographica* A, Vol. 167, pp.10–76.

Young, G.C. 1981. New early Devonian brachythoracids (placoderm fishes) from the Taemas-Wee Jasper region of New South Wales. *Alcheringa*, Vol. 5, pp.245–271.

Young, G.C. 1985. Further petalichthyid remains (placoderm fishes, early Devonian) from the Taemas-Wee Jasper region of New South Wales. *Bureau of Mineral Resources, Journal of Australian Geology and Geophysics*, Vol. 9, pp.121–131.

Young, G.C. 1986a. The relationships of placoderm fishes. *Zoological Journal of the Linnean Society*, Vol. 88, pp.1–57.

Young, G.C. 1986b. Cladistic methods in Palaeozoic continental reconstruction. *Journal of Geology,* No. 94, pp.523–537.

Young, G.C. 1988. Antiarchs (placoderm fishes) from the Devonian Aztec Siltstone, southern Victoria Land, Antarctica, *Palaeontographica* A, Vol. 202, pp.1–125.

Young, G.C. 1990. Devonian vertebrate distribution patterns, and cladistic analysis of palaeogeographic hypotheses. Pp.243–255. In McKerrow, W.S. and Scotese, C.R. (eds). *Palaeozoic Palaeogeography and Biogeography.* Memoir, Vol. 12. Geological Society, London.

Young, G.C. 1993a. Middle Palaeozoic macrovertebrate biostratigraphy of eastern Gondwana. Pp.208–251. In Long, J.A. (ed). *Palaeozoic Vertebrate Biostratigraphy and Biogeography.* Bellhaven Press, London. 369pp.

Young, G.C. 1993b. Vertebrate faunal provinces in the middle Palaeozoic. Pp.293–323. In Long, J.A. (ed). *Palaeozoic Vertebrate Biostratigraphy and Biogeography.* Bellhaven Press, London. 369pp.

Young, G.C. 1995. Timescales 4: Devonian, Australian Phanerozoic Timescales, biostratigraphic charts and explanatory notes, Second Series. *Australian Geological Survey Organisation Record,* 1995/33, pp.1–47.

Young, G.C. 1997. Ordovician microvertebrate remains from the Amadeus Basin, central Australia, *Journal of Vertebrate Paleontology,* Vol. 17, pp.1–25.

Young, V.T. 1995. Micro-remains from early and middle Devonian acanthodian fishes from the U.K. and their biostratigraphic possibilities. *Ichthyolith Issues, Special Publication,* No.1, pp.65–68.

Yu, X. 1998. A new porolepiform-like fish, *Psarolepis romeri*, gen. Et sp. Nov. (Sarcopterygii, Osteichthyes) from the lower Devonian of Yunnan, China. *Journal of Vertebrate Paleontology*, Vol. 18, pp.261–274.

Zhu, M., Yu, X. and Janvier, P. 1999. A primitive fossil fish sheds light on the origin of bony fishes. *Nature*, Vol. 397, pp.607–610.

Chapter 8
The Great Devonian Fish Kill at Canowindra

Alex Ritchie

Australian Museum, 6 College Street, Sydney, N.S.W. 2010 Australia
E-mail: Alex.Ritchie@austmus.gov.au

'The driver of the bulldozer wiped his brow and took a second look at a huge piece of quartzite which had slaked off from the road level. He had been levelling a dangerous curve on the Gooloogong Road near Canowindra, 200 miles west of Sydney. As he looked down from his cabin he could see the unmistakable impression of a fish-like creature, about 15 inches long, centred in the rock.

When he looked closer he could see other smaller impressions in the rock, some of them clearly showing scales and what looked like armoured joints. "Might be of interest," he thought as he shovelled the slab carefully to the side of the road where it remained for 12 months before its importance was recognised' (Hastings, 1956).

BACKGROUND

These opening paragraphs of a newspaper report announced one of the world's great fossil discoveries to the Australian public. The sequel, in which the author was heavily involved, came 37 years later in 1993. But the Canowindra story really began more than 360 million years ago, in late Devonian times, when a severe drought trapped, killed and, most importantly, entombed thousands of fish—preserving a dramatic record of an ancient disaster.

This unique fossil site was discovered in central west New South Wales, near the small town of Canowindra. About 10 km west of Canowindra the unsealed road to Gooloogong skirts the south end of a low sandstone range, constrained on its south side by an overflow channel of the Belubula River. Until 1955 a hard sandstone ridge had formed a dangerous hump on a bend and had been the site of several accidents. The local council (then Boree Shire) hired a contractor (Jack Burge) to carry out road improvements. His bulldozer driver, Charlie Stevens, intrigued by strange marks on one massive sandstone slab, saved it by moving it clear of the road against the fence.

We don't know how long the slab lay there, certainly less than the 12 months cited by Peter Hastings. But then, as in many major fossil finds, serendipity played a major role. Early in 1956 the Australian Museum in Sydney received a letter informing them of a rock near Canowindra 'with what looks like a fossil reptile on it'. This letter was sent by William (Bill) Simpson, a local beekeeper and carpenter from Holmwood, near Cowra, in New South Wales. While checking his hives in an adjacent field, Bill Simpson noticed the marks on the rock and was intrigued.

Simpson died in 1965 and unfortunately his version of the discovery was not recorded; furthermore, it was assumed that Simpson had been alone when he found the fossil slab. Then, in 1999, the author tracked down a living eye-witness to the 1956 discovery. When Simpson visited his hives that day he was accompanied by his foster son, Robert (Bob) Scott. Bob Scott remembers that it was very hot and ants were raiding the hives and disturbing the bees. After poisoning the ants, Simpson and Scott left the field quickly, leapt the fence and sat on a large rock slab waiting to be picked up. As he rolled himself a smoke, Simpson looked down, recognized the strange marks as fossils and pushed Bob Scott off the rock. When he got home he wrote to inform the Australian Museum of his discovery.

Harold Fletcher, the Museum's palaeontologist at the time and Ted Rayner (from the N.S.W. Geological Survey) had planned to visit Condobolin, west of Canowindra in March 1956 and decided to check Simpson's report on the way. When they saw the slab and realized its scientific importance, they cancelled the rest of their trip and arranged for the slab to be removed to Sydney (Fletcher, 1956).

The original Canowindra slab (2m x 1m) bears the impressions of at least 115 late Devonian fishes, most of them complete (Figure 1, Plate 45). Four types of fish are now known to be present on this slab but the fauna is dominated by two armoured antiarchs, *Bothriolepis* (23 specimens) and *Remigolepis* (89 specimens). Near the middle of the slab lay a single beautifully-preserved specimen of a choanate, lobe-finned, sarcopterygian fish; this was later described by Thomson (1973) and named *Canowindra grossi*. *C. grossi* was the first complete sarcopterygian discovered in Devonian rocks of the Southern Hemisphere. The fourth type of fish was only recognized in 1980, when the author identified two small specimens of an armoured fish, an arthrodire called *Groenlandaspis*, lying amongst the larger specimens. It was this discovery that stimulated attempts to relocate the Canowindra site.

By the time the original slab was recovered in 1956, the roadworks had been completed and the position of the

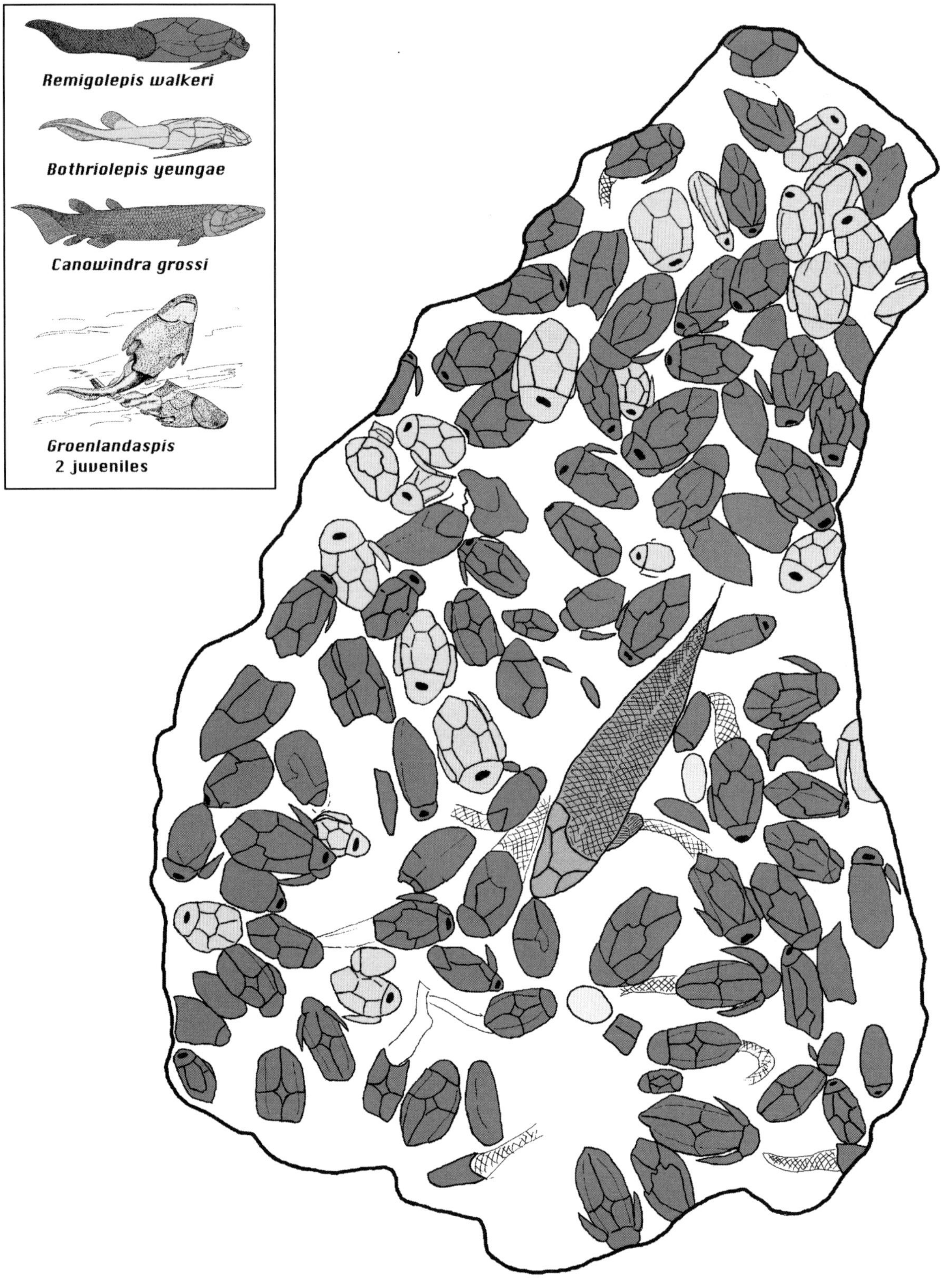

Figure 1. Schematic diagram of the Devonian fish slab collected from near Canowindra in 1956. The distribution of 115 specimens representing four taxa are mapped. The only known example of the sarcopterygian, *Canowindra grossi* Thomson (1973), lies near the centre of the slab, surrounded by antiarchs, *Remigolepis* (89 specimens) and *Bothriolepis* (23 specimens). Most show only the armoured head and trunk shield but some *Remigolepis* also display the scale-covered body and tail. The later discovery of two small specimens of the arthodire, *Groenlandaspis,* on this slab led to the 1993 excavation of the Canowindra site and the creation of an Age of Fishes Museum at Canowindra.

fossil layer could not be identified. After succeeding Fletcher as the Australian Museum's Palaeontologist in 1968, I visited Canowindra at least six times between 1973 and 1990, trying to relocate the fish layer but without success. Heavy earth-moving equipment was clearly needed, but funds to hire such assistance were not available.

REDISCOVERY

In September 1992 I was invited to talk to Canowindra Rotary Club on the 'Great Canowindra Fish Kill' and seized the opportunity to stimulate local interest. Some local councillors heard the talk and were intrigued. Shortly afterwards the council (now Cabonne Shire) offered the use of earth-moving equipment, free of charge or conditions, to try to relocate and excavate the 1956 fossil fish site.

In January 1993, with a 22 tonne excavator and its skilled operator (Fred Fewings), the site was reopened with a preliminary trial dig. Within three hours we had located the fossil layer at the side of the road and traced it for 20 metres up the bank. Every slab turned over was covered with articulated fish fossils, indicating that the site still contained thousands of superbly preserved specimens. The prospect of spectacular new discoveries was dramatically confirmed when parts of three very large sarcopterygian fishes (1.5–2 m long) were recovered.

We returned to excavate the site in July 1993, using the same Cabonne Council excavator and driver and with remarkable personal and logistic support from the Canowindra community—business people, farmers, lorry drivers, teachers, high-school students and local residents. Results exceeded expectations! Some 60 tonnes of fossil slabs were recovered; the largest weighed more than 2 tonnes. All the slabs were stored at Canowindra (Plate 46) where they have been systematically cleaned, prepared and cast for study by monthly groups of volunteers, under the supervision of the author and Zerina Johanson (Australian Museum).

To date more than 3,700 fish specimens have been identified, representing at least eight species, most of them new to science. This fossil layer apparently represents the remains of a late Devonian fish assemblage; all specimens (which represent only part of the original deposit) come from a single layer within the Mandagery Sandstone. Details of these fishes are provided in the following sections. The Canowindra fauna is dominated (> 97%) by two types of antiarch (*Bothriolepis* and *Remigolepis*) but includes another armoured fish, the arthrodire *Groenlandaspis* and at least five different types of sarcopterygian.

The recovery of these remarkable fossils is only the last link in a chain of events that began 360 million years ago when a lake, river or billabong dried up during a severe drought. The fish, tightly concentrated in a small area, were rapidly buried after they died and decomposed but before they disintegrated. Hundreds of millions of years later, this evidence of a mass mortality was brought back to the surface by earth movements and natural erosion. This occurred just where the fossil layer was intersected by a country road; it was spotted by a sharp-eyed bulldozer driver and saved for posterity.

MATERIAL

In 1993 some 195 rock slabs were recovered from the Canowindra locality but, because of the speed of the recovery, attempts to record how the rock slabs originally fitted together were largely unsuccessful. Only the upper surface of the fossil fish layer was removed—the undersurface remains *in situ*, preserving the ventral impressions of thousands of fish specimens and invaluable information on how they were distributed in the deposit. At the end of the dig (12–22 July), the undersurface layer was cleaned, protected with plastic sheets as well as layers of hay and temporarily reburied to prevent vandalism. Until the site can be safely reopened for scientific study, information on the distribution and preservation of this fauna can only be obtained from the recovered slabs.

The largest slabs (up to 3 m by 1.5 m) are covered with tightly packed and mainly adult antiarchs. Although *Bothriolepis* and *Remigolepis* are almost equally abundant they are not evenly distributed throughout the deposit. For example, Slab 4 has 164 *Bothriolepis* and 91 *Remigolepis*, while Slab 6, has far fewer *Bothriolepis* than *Remigolepis* (Figure 2). Although the various taxa are not randomly distributed in the final deposit, reasons for this variation are unknown at present. There are several slabs that appear to represent the outer margins of the fish deposit—perhaps the edges of the final pool. Instead of densely packed adults these contain mainly juvenile individuals; they are less crowded and terminate along a well defined line (see Figure 3).

How the Canowindra fish fauna perished has been investigated by considering events and processes occurring immediately before and after death. But other general ecological data can be gained from these remains. For example, population profiles of the antiarchs can indicate relative abundances of juveniles and adults in the community, or if the mortality was sudden rather than over an extended period. Differences in population profiles can also indicate timing of reproduction, growth rates and levels of predation.

PRESERVATION AND PREPARATION

Because most of the original bone had weathered away, the Canowindra specimens were mostly preserved as three dimensional natural moulds of the dermal surfaces, infilled with sandstone casts and separated by a greasy clay, probably representing the weath-

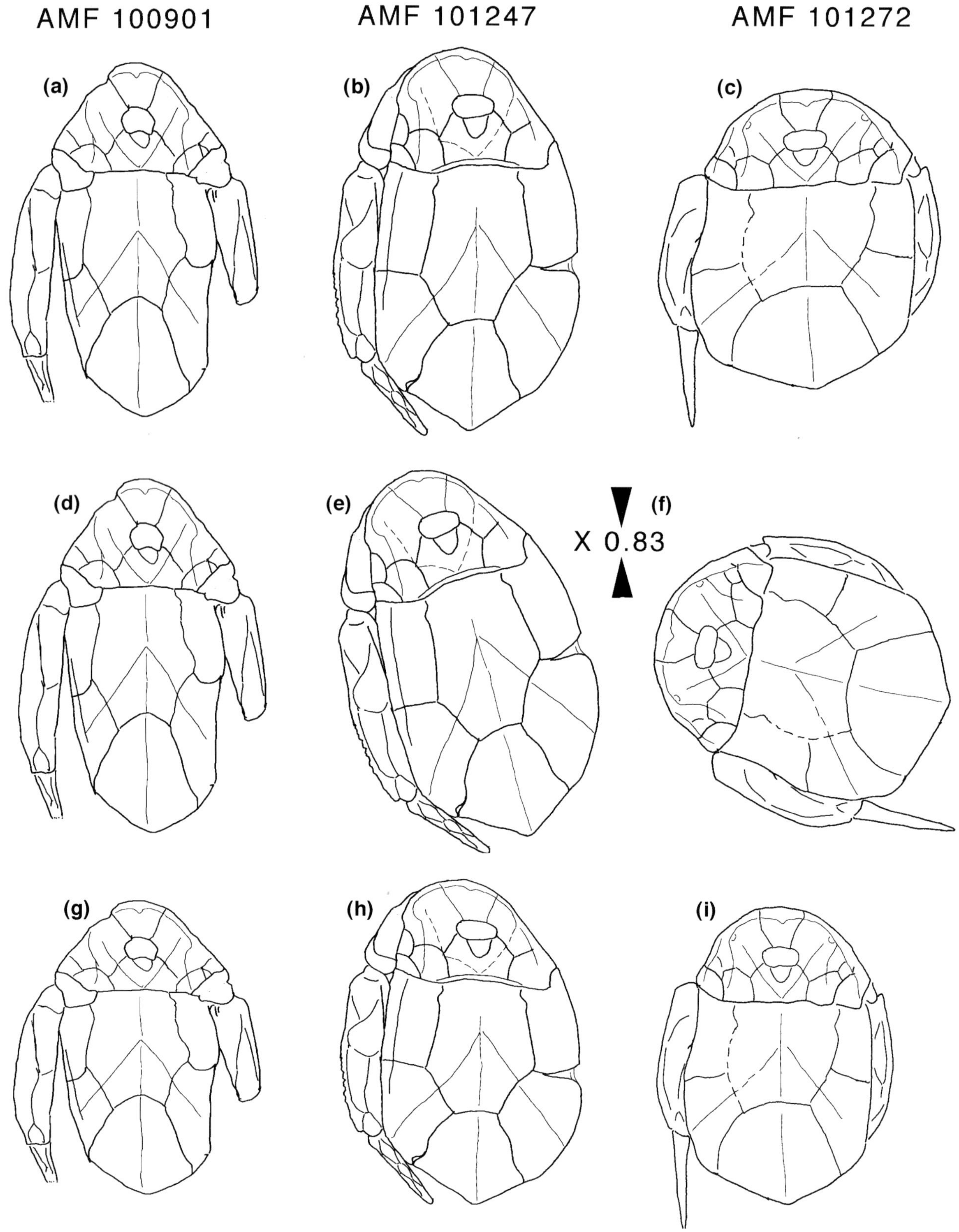

Figure 4. Three *Bothriolepis yeungae* specimens showing variation in shape and proportions caused by tectonic deformation. **(a)–(c)** AMF.100901 (Johanson, 1998); AMF.101247, (Johanson, 1998); AMF.101272 (Johanson, 1998). **(d)–(f)** Same specimens, as originally oriented on the slab, with degree and direction of correction indicated. **(g)–(i)** Same specimens, reoriented for comparison with (a)–(c).

Table 1. Summary of the systematics of fossil fishes found at Canowindra, N.S.W.

Taxa (to generic level) and Original Authors	
Subclass	Placodermi (McCoy, 1848)
	Order Antiarchi (Cope, 1885)
	Suborder Asterolepidoidei (Miles, 1968)
	Family Remigolepididae (Stensiö, 1931)
	Genus *Remigolepis* (Stensiö, 1931)
	Remigolepis walkeri (Johanson, 1997b)
	Suborder Bothriolepidoidei (Miles, 1968)
	Family Bothriolepidae (Cope, 1886)
	Genus *Bothriolepis* (Eichwald, 1840)
	Bothriolepis yeungae (Johanson, 1998)
	Order Arthrodira (Woodward, 1891)
	Suborder Phlyctaenoidei (Miles, 1973)
	Family Groenlandaspididae (Obruchev, 1964)
	Genus *Groenlandaspis* (Heintz, 1932)
Class	Sarcopterygii (Romer, 1955)
	Rhipidistia (Cope, 1889)
	Tetrapodomorpha (Ahlberg, 1991)
	Family Canowindridae (Young, Long and Ritchie, 1992)
	Canowindra grossi (Thomson, 1973)
	Family Tristichopteridae (Cope, 1889)
	Cabonnichthys loomesi (Ahlberg and Johanson, 1997)
	Mandageria fairfaxi (Johanson and Ahlberg, 1997)
	Family Rhizodontida (Andrews and Westoll, 1970)
	Gooloogongia loomesi (Johanson and Ahlberg, 1998)
	Order Dipnomorpha (Ahlberg, 1991)
	Dipnoi (Muller, 1844)
	Genus *Soederberghia* (Lehman, 1955)
	Soederberghia simpsoni (Ahlberg, Johanson and Daeschler, 2001).

Antiarchi

Antiarchs were small to medium-sized placoderms, most of which lived in deltaic or marginal marine environments. They first appear in late Silurian rocks of China and had spread world-wide by the middle Devonian and reached their peak of diversity during the late Devonian when they were probably the most abundant placoderm fishes.

As in most placoderms, the head and anterior trunk region of antiarchs was encased in, and protected by, overlapping bony plates. In antiarchs, the head shield is relatively short and the eyes, nostrils and pineal plate were all housed in a small oval opening on the skull roof. The main feature distinguishing antiarchs from other placoderms is the presence of an elongate trunk shield with two median dorsal plates (one in arthrodires), one median ventral plate (two in most arthrodires), and by the peculiar construction of their pectoral fins (or appendages). The pectoral fin is entirely enclosed in a tubular armour of small polygonal plates. In most antiarchs, this articulates with the trunk armour in a ball-and-socket arrangement and could either be extended laterally or tucked back against the body armour.

The most successful and widespread antiarchs belong to two groups, the asterolepidoids and bothriolepidoids, which flourished world-wide in middle and late Devonian times. At Canowindra these are represented by *Bothriolepis* and *Remigolepis* respectively (Figures 5, 6).

Antiarch Population Characteristics

Over several years, all the antiarch specimens on several fully prepared slabs were measured for analysis. In the absence of other features used for age determination, size was used as an indication for age of individuals. Length measurements of the combined head and trunk-shield were taken from a total of 343 individuals, 233 specimens of *Bothriolepis yeungae* and 110 specimens of *Remigolepis walkeri*. Although both *Bothriolepis* and *Remigolepis* are present in approximately equal numbers (47.8%:49.7%) throughout the deposits, the *Remigolepis* sample was smaller on these particular slabs.

The size-frequency distributions for both antiarchs were skewed; *Remigolepis walkeri* towards larger individuals and *Bothriolepis yeungae* towards smaller individuals (Johanson, 1998). They also reveal that the Canowindra fauna includes relatively few small individuals, particularly of *Remigolepis*. Juveniles under 70 mm are rare for both taxa but in *Bothriolepis yeungae* specimens 50–100 mm long make up some 27% of the sample while in *Remigolepis walkeri* specimens in the 50–100 mm size range are very rare (only 0.06% of the sample).

Smaller juveniles are missing from both population profiles, which may reflect biological or physical factors. Predation by sarcopterygians in the Canowindra fauna could certainly have been important in removing smaller individuals from the fauna in preference to larger antiarchs. Physical processes resulting from active transport (breakage, dissolution and abrasion) can also remove smaller individuals from a fauna.

In a mass or catastrophic mortality of fish (where a total population is sampled in a short period), the profile should include distinct year-classes represented by multiple peaks in size-frequency distribution. These peaks should be more distinct for younger individuals, since growth slows in older fish and peaks become less distinct. The profiles of *Bothriolepis* and *Remigolepis* show possible age classes for medium to large individuals but not for smaller juveniles (Johanson, 1998).

Bothriolepis

The antiarch *Bothriolepis* (meaning 'pitted-scale') was the most successful armoured fish, or placoderm, that ever lived (Figure 5). It first appears in middle Devonian rocks and flourished until the end of the late Devonian period. Its fossilized remains have now been found, often in abundance, on every continent. More than 100 species of *Bothriolepis* have been described but this number may be artificially high. Many species of *Bothriolepis* have been described on the basis of isolated (often distorted) plates and small sample sizes.

The pectoral appendages of *Bothriolepis* were relatively long, usually extending beyond the posterior mar-

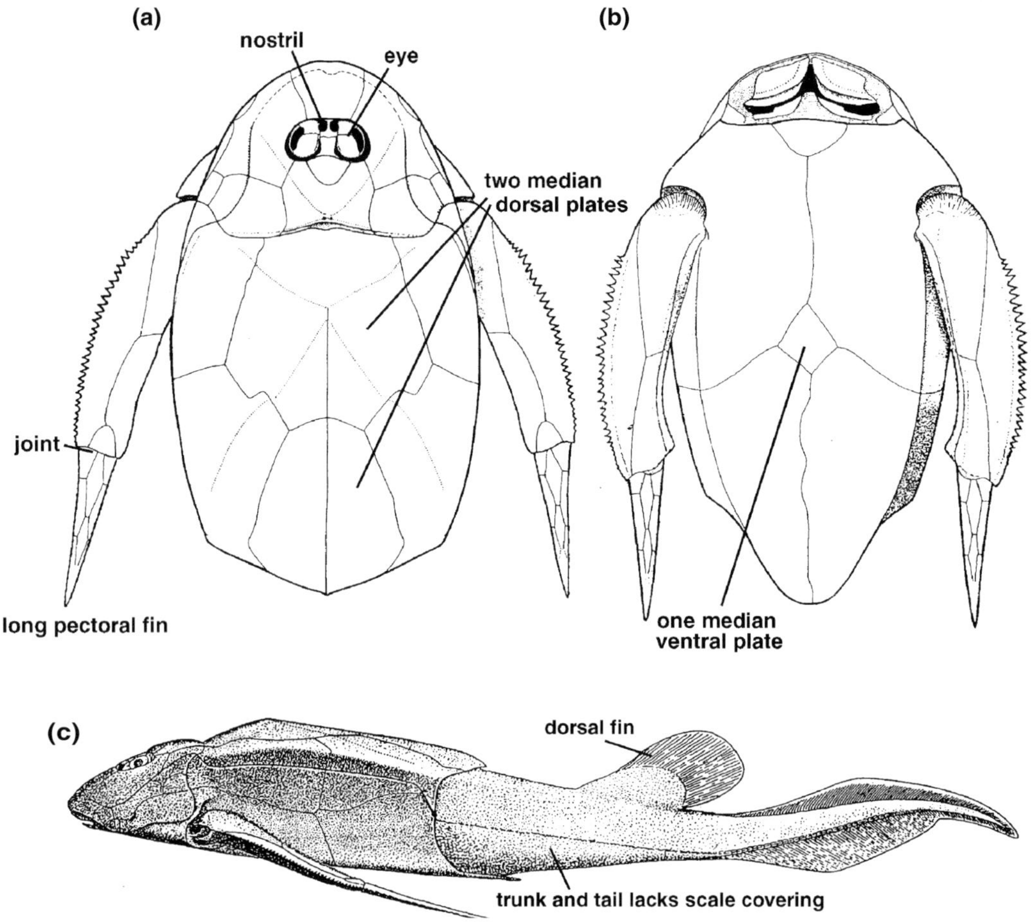

Figure 5. Generalized reconstructions of *Bothriolepis* sp.: **(a)** dorsal view; **(b)** ventral view; **(c)** lateral view. Modified from Stensiö (1963) and Miles (1968).

gin of the trunk shield when folded back against the body. The pectoral fin is composed of two segments, a longer wider proximal portion, and a shorter narrower distal segment. The trunk and tail of *Bothriolepis* is seldom preserved because, in most species, it lacked a scale covering. Only a few specimens at Canowindra show faint impressions of the trunk and tail but these indicate that it was developed essentially as in *Bothriolepis canadensis* from Miguasha, Canada (Figure 5(c)).

Bothriolepis yeungae

Over 1,700 specimens of *Bothriolepis* have been identified so far from Canowindra. Combined head+trunk shield lengths range from 50 mm to 220 mm, indicating that adults reached around 500 mm in length when fully grown. Using a variety of measurements a Principal Components Analysis was performed, to test the hypothesis that only one species of *Bothriolepis* was present; the range of variation in head and trunk plates of *Bothriolepis* from Canowindra supported the presence of a single species (Johanson, 1998). The species was named in honour of Monica Yeung, of Canberra; her ecotourism organization (Gondwana Dreaming) has provided the Canowindra project with over 1,000 volunteer workers since 1993 (Cribb, 1996; Yeung *et al.,* 1998).

Remigolepis

The Late Devonian antiarch, *Remigolepis,* has been recorded from East Greenland (Stensiö, 1931), China (Pan *et al.*, 1980; 1987), the Tula region of Russia (Luksevics, 1991), Kirgizia (Panteleyev, 1992), and possibly Scotland (Andrews, 1978). The presence of this genus in Australia has been known for some time, mostly from disarticulated head and trunkshield plates (Young, 1993a), but the first Australian species were not described until recently. The first, *Remigolepis hunterensis* (Johanson, 1997a) from Hunter Siltstone near Grenfell, N.S.W. is known predominantly from isolated plates. In contrast the second, *Remigolepis walkeri* (Johanson, 1997b) from Canowindra, is represented by at least 1,780 specimens—most of them articulated and including a wide size range.

Remigolepis walkeri

Remigolepis walkeri (Johanson, 1997b) is a medium-sized species (Figure 6). In 70 measured specimens the combined head + trunk shield lengths ranged from 78–185

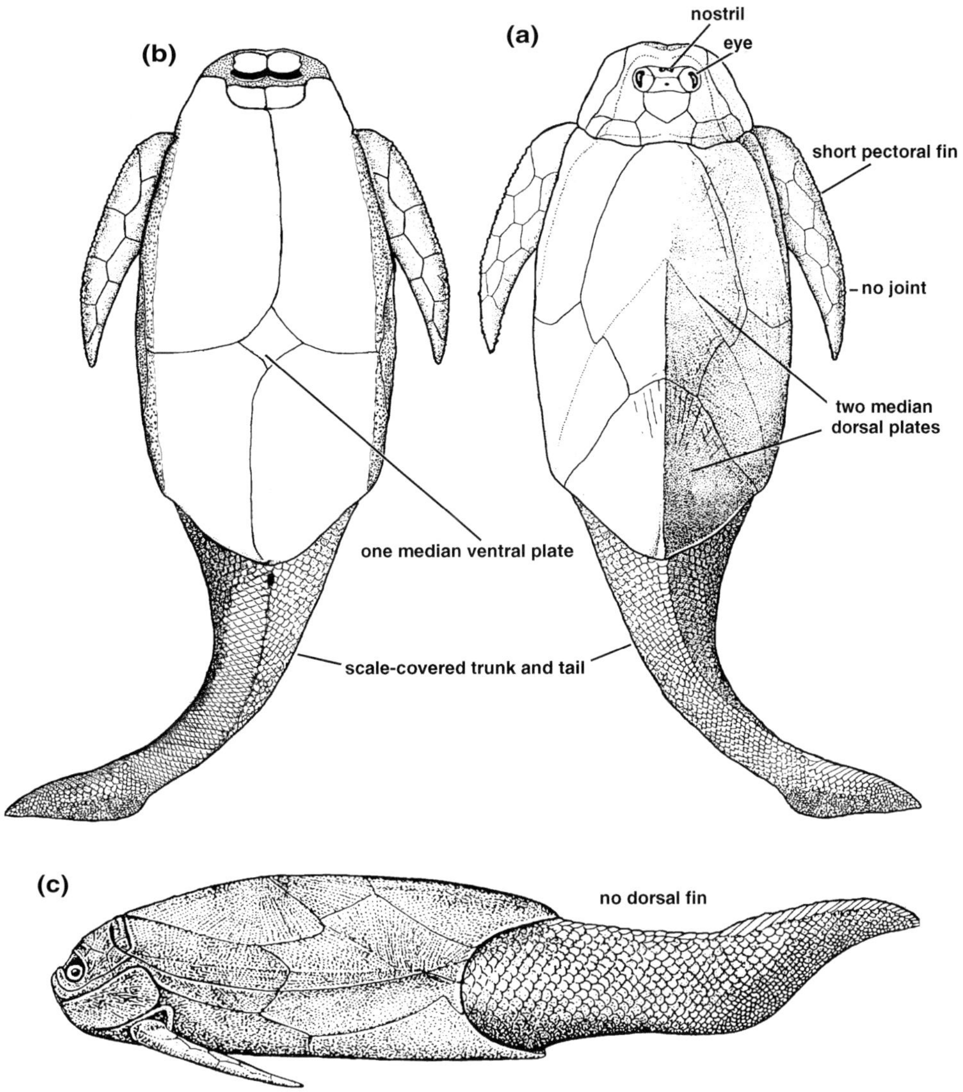

Figure 6. Reconstruction of whole *Remigolepis walkeri*: **(a)** dorsal view; **(b)** ventral view; and **(c)** lateral view.

mm suggesting that fully grown adults probably reached a length of around 350 mm. Most of the Canowindra specimens of *Remigolepis* are preserved in life position, with little disarticulation of skeletal elements. This suggests a rapid but gentle burial of the fauna with no predation or subsequent reworking. Incoming sediment not only covered the craniothoracic armour but also partly infilled it prior to burial, producing detailed internal moulds. Although most of the antiarch shields were crushed to various degrees after burial, in *Remigolepis*, where the trunk armour was subtriangular in cross section with a prominent dorsal ridge, dorso-ventral compaction was often less extreme than in the lower profile *Bothriolepis* shields.

The articulated specimens of *Remigolepis walkeri* from Canowindra preserve many features poorly known or preserved in overseas finds—examples are the internal surfaces of the head and trunk shield, pectoral fins, the scale-covered tail and portions of the mouth region including the internal surface of the suborbitals, possible attached palatoquadrate, infragnathals and Meckel's cartilage (Johanson, 1995, 1997b). The internal surface of the suborbital plate shows an unusual feature, a large oval structure at the anteromesial edge of the plate which probably represents an attachment for the autopalatine portion of the palatoquadrate (Johanson, 1997b). This feature is consistently present on all specimens in the fauna where suborbitals are preserved. In the absence of evidence to the contrary, it suggests a single species of *Remigolepis* is present at Canowindra. Among asterolepidoids, this morphology appears most similar to *Pterichthyodes* from the middle Devonian of Scotland.

The pectoral fins of *R. walkeri* were unjointed, flattened and short, terminating midway along the trunk shield. Although they are most often preserved lying close to the body it is thought that they could probably extend laterally at nearly right angles. Canowindra specimens provide the best evidence to date on the trunk and caudal fin in *Remigolepis,* which was shorter than the head and trunk shield (about 75%) and covered in small rhomboid scales (Johanson, 1997b). There is no trace of dorsal, ventral or anal fins. The caudal fin was gently heterocercal and its dorsal margin was fringed with a single row of elongate, sloping scales. The scale-covering of the caudal fin web decreases in size towards the ventral margin. The pectoral and caudal fin morphology of *Remigolepis walkeri* indicate that it was probably a slow-swimming, bottom-dwelling fish whereas *Bothriolepis yeungae* may have been better able to generate sufficient lift from the craniothoracic armour and pectoral fins to enter the water column on a regular basis.

R. walkeri is similar to *Remigolepis* sp. from near Eden (N.S.W.) and *Asterolepis ornata,* but differs from *Remigolepis* sp. from China (Pan *et al.*, 1987) which appears to lack the dorsal row of elongate scales present in *R. walkeri.*

Arthrodira

Unlike antiarchs, the pectoral fins of arthrodires are not encased in bone and the arthrodiran trunk shield has only one median dorsal plate and two median ventral plates.

Groenlandaspis

The rarest of the armoured fishes at Canowindra, *Groenlandaspis*, has an interesting history of discovery. The name of this armoured fish (meaning 'Greenland shield') tells us where it was first found. Scandinavian expeditions to East Greenland in the early 1930s collected fragmentary remains of *Groenlandaspis* from late Devonian rocks; enough to name it but not to reconstruct it (Heintz, 1932). In 1970–71 the author discovered the first *Groenlandaspis* remains from the Southern Hemisphere, in the late Devonian Aztec Siltstone of Southern Victoria Land (Antarctica) and the whole animal could be reconstructed for the first time (Figures 7, 8).

Then in 1972, the first two *Groenlandaspis* species from Australia, were discovered from separate sites near Forbes and Grenfell in central west N.S.W. These were followed by a chain of discoveries, some in the field, others in museum collections. *Groenlandaspis* species are now known also from Europe, the Middle East, South Africa and North America (Ritchie, 1974, 1975; Janvier and Ritchie, 1977; Long, 1995a; Long *et al.*, 1997; Anderson *et al.,* 1999).

It was the unexpected recognition, in 1980, of two small *Groenlandaspis* specimens on the original 1956 Canowindra slab in the Australian Museum (Plate 45) that renewed the search to relocate the Canowindra site. The 60 specimens of *Groenlandaspis* now recovered from Canowindra include the finest specimens known of this ancient armoured fish from anywhere in the world. Many of the Canowindra individuals are remarkably complete, with the head and trunk shields preserved relatively uncrushed, in three dimensions (Figure 9). An almost complete growth series has been discovered, ranging from juveniles (craniothoracic shields 50 mm long) to adults

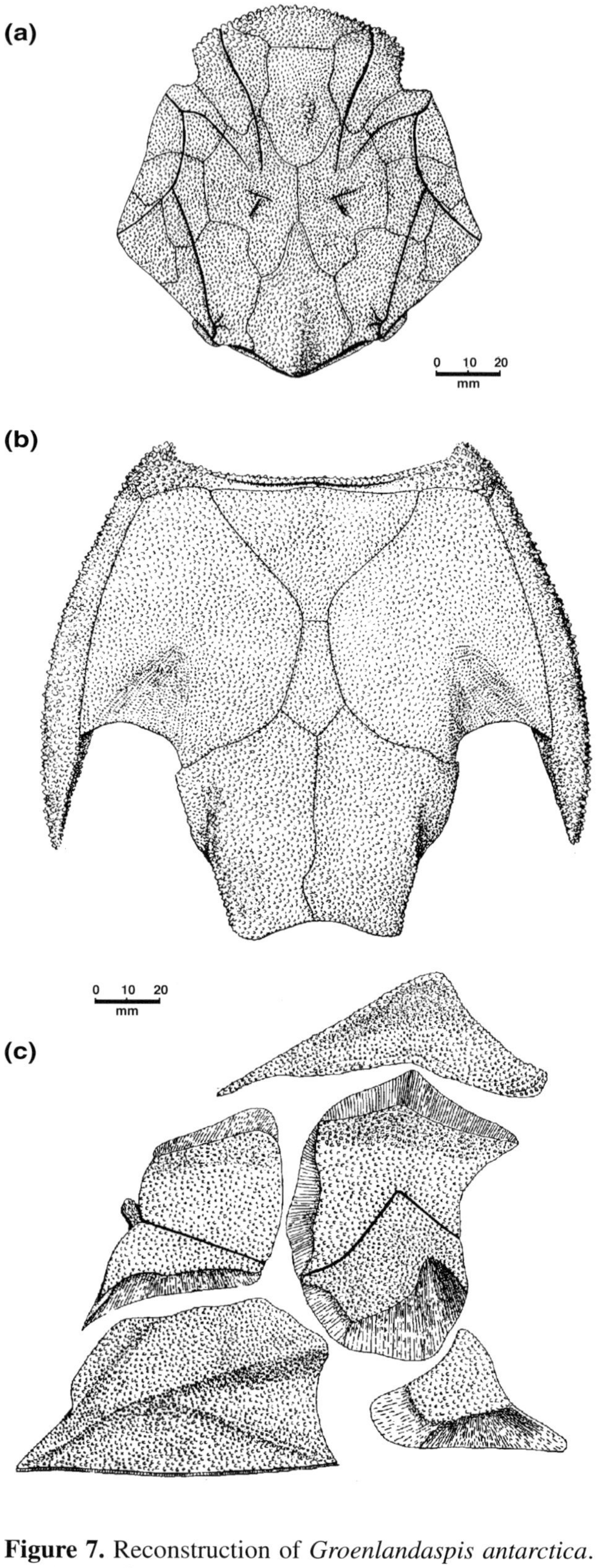

Figure 7. Reconstruction of *Groenlandaspis antarctica*. Apart from minor differences in proportions the Canowindra species of *Groenlandaspis* is similar to *G. antarctica*: **(a)** head shield in dorsal view; **(b)** trunk shield in ventral view; **(c)** lateral and median dorsal trunk plates, showing overlap relationships.

Figure 8. *Groenlandaspis*. Reconstruction, with tail and fins based on material from Mt. Howitt, Victoria.

with shield lengths to 210 mm. This suggests that adults grew to at least 400–450 mm in length. The head shield was sub-pentagonal, with the eyes situated anterolaterally. The trunk shield, with which the head articulated by a double ball-and-socket joint, was triangular in cross section. The ventral shield was broad and flat and the lateral walls rose steeply to meet in the midline under a single long, and very narrow median dorsal plate with a pointed, backwardly directed crest. All of the dermal plates are covered with a tubercular ornament (Figures 7, 8).

Unlike the antiarchs, the pectoral fins of arthrodires were not encased in bony paddles; they were flexible flaps which emerged through openings at the posterolateral corners of the trunk armour, protected laterally by a long, fixed bony spinal plate. The trunk, fins and tail of *Groenlandaspis* are known from a few flattened, but articulated, specimens from the late Devonian of Mt. Howitt in Victoria, where they are covered with small scales. Faint traces of the trunk and tail are occasionally visible in smaller *Groenlandaspis* individuals at Canowindra, preserved near what appear to be the edges of the fish deposit (Figure 3). The species of *Groenlandaspis* from Canowindra closely resembles *G. antarctica* (Ritchie, 1975) in general proportions and ornament.

SARCOPTERYGII

The rarest, but the most interesting, fishes found at Canowindra are the sarcopterygians, all referable to the taxon Rhipidistia. Both living and fossil sarcopterygians (literally 'fleshy-fin'), are characterized by a muscular basal lobe in their paired fins, supported internally by a monobasal internal bony skeleton that articulates with the shoulder girdle.

Recent discoveries from the Devonian of Europe and North America, as well as Australia, have thrown new light on the interrelationships of sarcopterygian fishes and the earliest tetrapods, but their classification remains in a state of flux. Ahlberg's proposal (1991) to recognize two major groups of rhipidistians, the Tetrapodomorpha and Dipnomorpha, has received wide, but not universal, acceptance.

Tetrapodomorpha

The Tetrapodomorpha, which includes the tetrapods (amphibians and their descendants), are represented at Canowindra by four genera from three families. The sole representative of the Dipnomorpha is a long-snouted dipnoan, *Soederberghia*. Other members of the Dipnomorpha (Porolepiformes) have not so far been discovered at Canowindra. This is a little surprising because the easily recognizable cosmopolitan porolepiform, *Holoptychius,* is known from other Late Devonian sites in N.S.W. near Grenfell, Forbes and Eden.

Devonian tetrapodomorphs are also known from several other localities in Australia including Mount Howitt, Victoria (Long, 1985b, 1987); Gogo Formation, Western Australia (Long, 1985c) and Hatchery Creek, New South Wales (Young and Gorter, 1981).

The sarcopterygians also include the Actinistia (or coelacanths) and Onychodontiformes, with the former

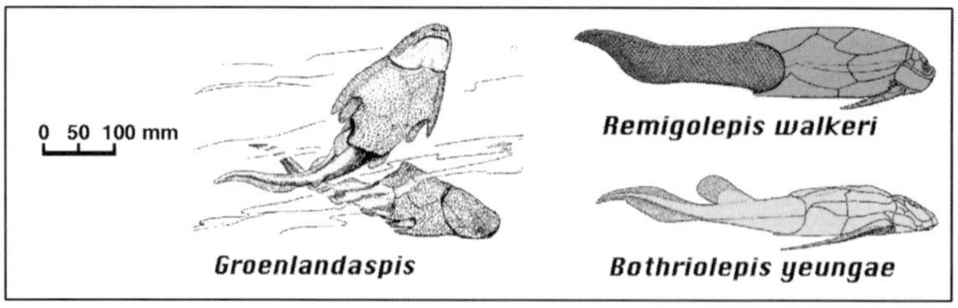

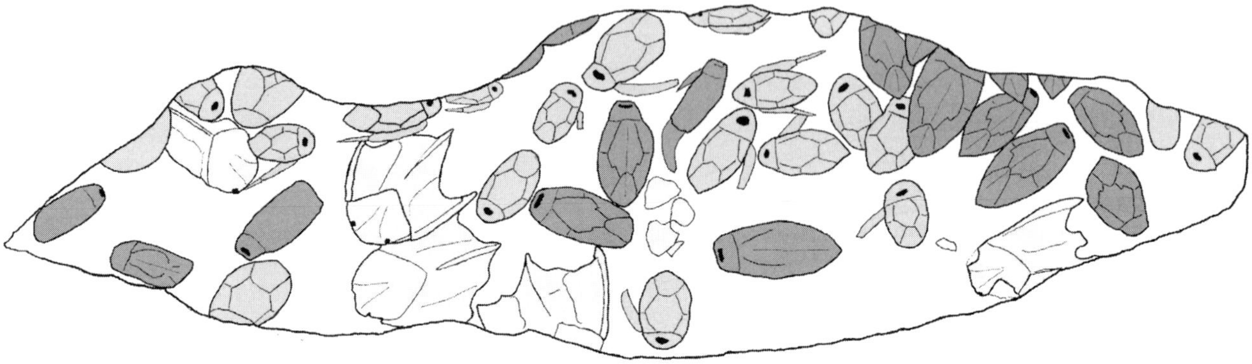

Figure 9. Slab 9 (1.67 x 0.47 m), one of the most important finds from the 1993 dig, contains 5 adult *Groenlandaspis* specimens. These relatively uncrushed specimens from Canowindra are some of the finest examples of *Groenlandaspis* found anywhere in the world. The slab also contains 17 *Remigolepis* and 21 *Bothriolepis.*

having living representatives as well as several groups known only from fossils. Both the Actinistia and Onychodontiformes are poorly known from Australia. Actinistians have been described from Victoria (Long, 1999) and onychodonts are currently being described from the Gogo Formation of Western Australia. Neither group is known from any late Devonian sites in New South Wales, although Long (1999) has identified a possible fossil coelacanth from this state.

Canowindra grossi

One of the most unexpected aspects of our new discoveries is the rarity of the fish that first put Canowindra on the palaeontological map. Despite the discovery of so many new, and beautifully preserved, sarcopterygian specimens from the site in 1993, *Canowindra grossi* itself is still known only from the unique holotype on the 1956 slab (AMF.47153), illustrated in Plate 45.

Keith Thomson (1973) named *Canowindra grossi* after the town and in honour of Walter Gross, a highly respected German fossil fish worker. Thomson, unfortunately, did not have access to the original specimen of *Canowindra* and had to work from an old latex cast prepared many years earlier. The holotype is preserved as a natural mould of the dorsal surface, with the head region preserved in considerable relief. With only a cast of the dorsal surface available, the exact shape and size of the dorsal, pelvic, anal and caudal fins remains conjectural but the original total length is estimated at 550–600 mm. *Canowindra grossi* displayed a strange mixture of diagnostic characters leaving Thomson unable to determine whether it belonged to an osteolepiform or porolepiform sarcopterygian. The scales of *C. grossi* display a distinctive tuberculate ornament quite different from any of the other sarcopterygian specimens found at Canowindra in 1993.

In the early 1980s, the original specimen and new discoveries from the late Devonian of Victoria were restudied. From these, and a new cast of *C. grossi*, Long (1985a) reinterpreted the head region and concluded that *Canowindra* was an osteolepiform sarcopterygian, not a porolepiform (Figure 10). Long (1987) described a new genus, *Beelarongia*, from the Devonian of Victoria and showed that it was related to *Canowindra.*

Young *et al.* (1992) described another new taxon, *Koharalepis,* from the Aztec Siltstone of southern Victoria Land, Antarctica which shares many of the same characters—very small orbits, large postorbitals and jugals, and a broad postparietal shield with an extratemporal. On this basis Young *et al.* (1992) assigned *Canowindra*, *Beelarongia* and *Koharalepis* to an endemic middle–late Devonian Gondwanan family, the Canowindridae.

Tristichopteridae

The two most abundant sarcopterygians in the Canowindra fauna, *Mandageria fairfaxi* (Johanson and Ahlberg, 1997) and *Cabonnichthys burnsi* (Ahlberg and Johanson, 1997), both belong to the family Tristichopteridae (= Eusthenopteridae of Cloutier and Ahlberg, 1996). They are the only unambiguous trichichopterids described from Australia, although *Eusthenodon*, the sister taxon to *Mandageria* (Ahlberg

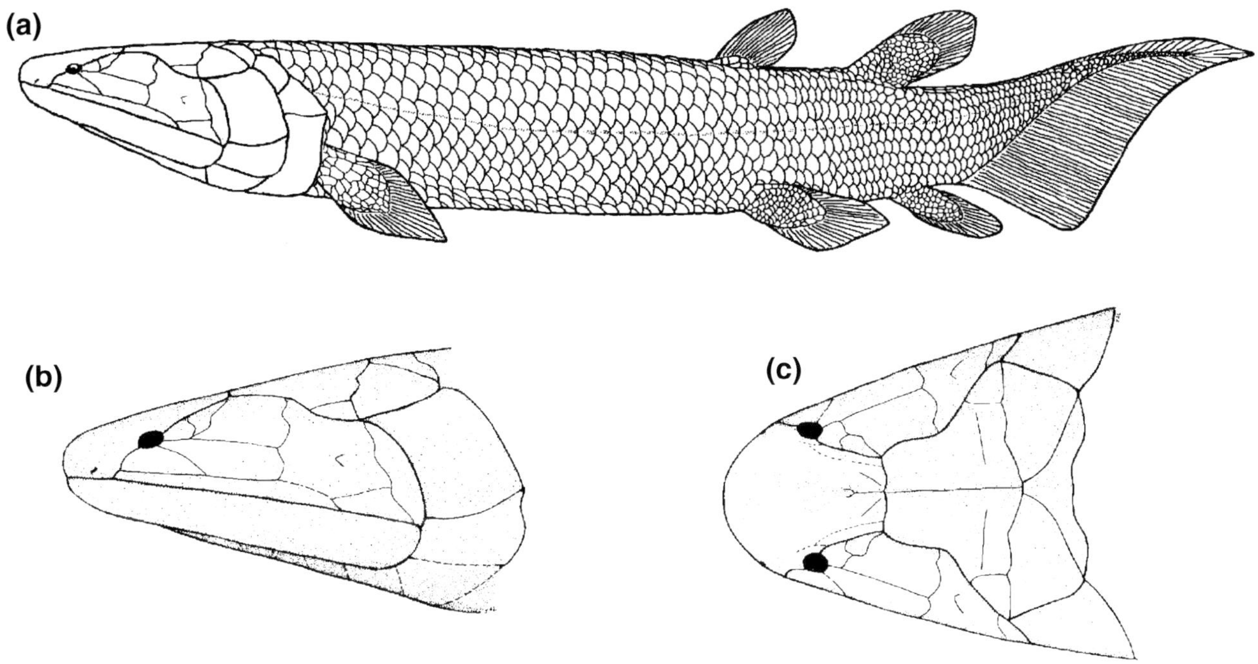

Figure 10. Reconstructions of *Canowindra grossi*: **(a)** whole fish in lateral view. Body, fins and tail after Thomson (1973) with head region after Long (1985a); **(b)** head in lateral view; **(c)** head in dorsal view. After Long (1985a).

and Johanson, 1997), may be present at Grenfell (Johanson and Ritchie, 2000) and near Eden, in south-eastern New South Wales.

The Tristichopteridae, a widespread family, is characterized by the possession of postspiracular bones. In addition to the two new taxa from Canowindra the family also includes: *Tristichopterus* Egerton (1861), Givetian, Scotland; *Eusthenopteron* Whiteaves (1881), Frasnian, Canada, Latvia and Russia; *Platycephalichthys* Vorobyeva (1959) and *Jarvikina* Vorobyeva (1977), Frasnian, Baltic; *Eusthenodon* Jarvik (1952), Famennian of Greenland, Russia, Pennsylvania (and possibly Australia); also *Spodichthys* Jarvik (1985), late Devonian, Greenland; and *Hyneria* Thomson (1968), Famennian, Pennsylvania. The more primitive tristichopterids (*Tristichopterus, Eusthenopteron, Spodichthys, Jarvikina, Platycephalichthys*) all come from the Laurussian continent.

Mandageria fairfaxi

Mandageria fairfaxi (Figure 11) is by far the largest fish in the fauna, possibly exceeding 1.6 m in length. It is named after the Mandagery Sandstone and in honour of James Fairfax, who has sponsored research on Canowindra fishes since 1993.

Mandageria fairfaxi is known from remains of 10–12 individuals, including several complete or partial heads and bodies, preserved as natural moulds. Because of its large size, parts of the same individual were often recovered on two or more slabs, the original associations of which will remain uncertain until the original site is reopened. The original undersurface of the Canowindra fish layer (reburied in 1993), still contains detailed ventral impressions of at least eight complete or partial specimens of *Mandageria,* lying *in situ.* When the site is re-excavated these ventral impressions can be used to identify and reassemble their dorsal counterpart surfaces preserved on slabs recovered in 1993.

Mandageria had a narrow, triangular head with an acutely pointed snout. The junction between the parietal and postparietal shields (and also, presumably, the underlying intracranial joint) is set relatively far back on the skull roof. Unusually, the head region of many of the *Mandageria* specimens found at Canowindra is preserved almost in three dimensions. It would appear that after death the soft tissues rotted quickly but the endocranial and branchial skeletons remained attached, and still associated, probably held together by desiccated ligaments and other soft tissues.

Behind the head it was a different story. As the flesh rotted, the scale covering of the trunk and tail collapsed onto the ventral squamation which in turn usually lies in a shallow trough obviously caused by the weight of the heavy fish settling in the sediment. In several *Mandageria* specimens the vertebral column and very short ribs, which were apparently poorly ossified, became trapped between the dorsal and ventral squamation which draped over and around them and preserved their shape, albeit rather faintly. This was aided by the unusually small size of the scales in *Mandageria* which are of normal tristichopterid type. *Mandageria* had approximately 120 scale rows between the shoulder girdle and caudal fin, compared to 50 scale rows in *Eusthenopteron.*

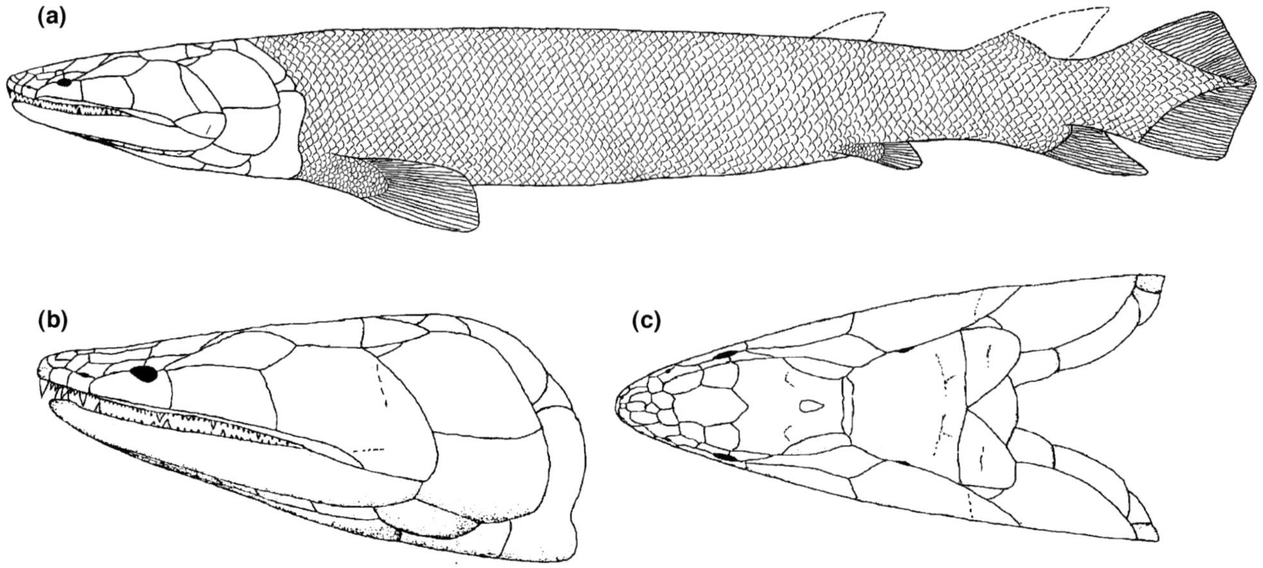

Figure 11. Reconstructions of *Mandageria fairfaxi:* **(a)** whole fish in lateral view (after Johanson and Ahlberg, 1997), with squamation and fin details added by Ritchie; **(b)** head region in lateral view; **(c)** head region in dorsal view. After Johanson and Ahlberg (1997).

The complete collapse of the skin and scales once attached to the head and gill region probably created openings through which fine-grained sediment was then able to infiltrate, filling up most of the cranial and buccal cavities and surrounding the basibranchial skeleton. In normal circumstances such three dimensional specimens, encased in solid quartzite, would be virtually impossible to prepare out of the rock but at Canowindra nature, and luck, has come to our assistance.

During the thousands of years that the fossil fish layer has lain near the surface at Canowindra, percolating groundwater, natural acids and subsurface weathering have combined to reduce the once-solid bone to a sticky orange clay. It was a relatively simple matter to clean out this clay to reveal beautifully detailed natural moulds in the sandstone endocasts. These were then cast in latex and moulded using silicone rubbers to produce positive resin replicas for scientific study, display and exchange purposes.

Some idea of the anatomical details preserved can be obtained from one superb specimen (AMF.98593b), that displays all of the branchial arches in the floor of the mouth, in position and uncrushed. In every respect it compares closely with the branchial skeleton of another tristichopterid, *Eusthenopteron*, from the Devonian of Miguasha, Quebec, Canada which was reconstructed by Jarvik (1954, 1980) from specimens that had been serial sectioned, a laborious process that took many years and thousands of man-hours.

Interestingly, the branchial arches preserved in AMF 98593b have helped to resolve a problematical interpretation of the morphology of the second basibranchial and the articulating arches in *Eusthenopteron.* Jarvik (1980) reconstructed an elongate second basibranchial with posterior arches articulating separately to this basibranchial. However this region was incomplete, and AMF 98593b clearly showed that this interpretation was incorrect, probably for the Tristichopteridae as a whole. In fact, the second basibranchial was short, and the last branchial arch articulated to the arch in front, rather than to the basibranchial. This is a common morphology throughout the Rhipidistia, occurring in taxa such as the osteolepid *Medoevia* (Lebedev, 1995), as well as porolepiforms and lungfishes (Rosen *et al.,* 1981; Johanson and Ahlberg, 1997).

Mandageria fairfaxi has recently been discovered to possess an unusual feature, a functional neck joint, a feature characteristic of post-Devonian tetrapods but hitherto unknown in sarcopterygian fishes (Johanson *et al.*, 2003). *Mandageria* indicates that the osteolepiform fishes may have produced two quite separate and mechanically distinct solutions to the problem of building a mobile neck; this probably took place in relation to a 'snapping' form of food capture rather than as an adaptation to terrestrialization.

Cabonnichthys burnsi

Cabonnichthys burnsi (Ahlberg and Johanson, 1997) was a medium-sized tristichopterid (Figure 12) and the second to be described from the Canowindra fauna. The genus is named for Cabonne Shire Council and the species honours Bruce Burns, Sydney, whose invitation to me to address Rotary International in Canowindra in late 1992 led to the offer of council logistic support.

Cabonnichthys is known from at least 11 individuals, many of which have the head region preserved in considerable relief. The most complete specimen, the holotype (AMF 96858) preserved on Slabs 11 and 12 (Figure 13, Plate 47), is 740 mm long but other less complete specimens indicate it grew slightly larger. The head and snout

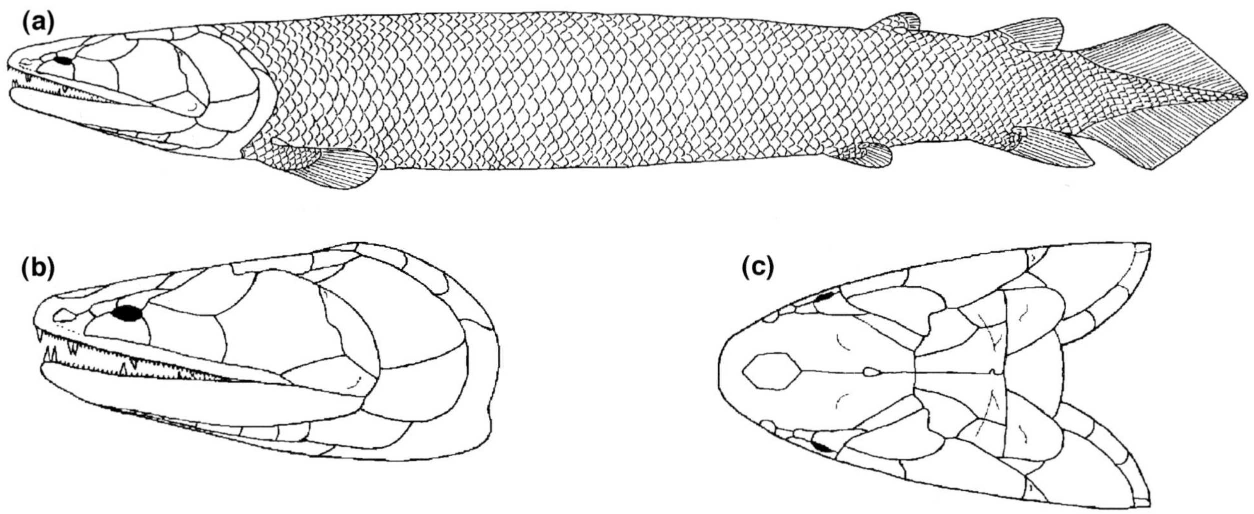

Figure 12. Reconstructions of *Cabonnichthys burnsi*: (**a**) whole fish in lateral view (after Ahlberg and Johanson, 1997), with squamation and fin details added by Ritchie; (**b**) head in lateral view; (**c**) head in dorsal view. After Ahlberg and Johanson (1997).

are broader than in *Mandageria* and the postparietal shield is proportionately wider, but the intracranial joint is in approximately the same position as in *Mandageria* and *Eusthenodon* although farther back than in *Eusthenopteron*.

Cabonnichthys burnsi resembles both *Mandageria* and the Laurussian *Eusthenodon* in possessing: a posterior supraorbital that extends ventrally behind the orbit to exclude the postorbital from the orbital margin; a posteriorly positioned pineal surrounded by kite or teardrop shaped bones; and an enlarged fang-like tooth at the anterior end of the premaxillary tooth row. *Cabonnichthys* differs from *Mandageria* and *Eusthenodon* in retaining contact between the intertemporal and posterior supraorbital, and in possessing 'horn-like' processes on the postparietal shield that notch into the posterior margin of the postorbital.

In more primitive tristichopterids the median fins were more anteriorly placed and the caudal fin had a high aspect-ratio with a narrow peduncle. The tail was strongly asymmetrical in *Tristichopterus*, less so in *Eusthenopteron*. In *Mandageria* and *Cabonnichthys* the tail is almost completely symmetrical with a much lower aspect ratio than in *Tristichopterus* and *Eusthenopteron* and the median fins are much smaller and positioned closer to the tail. This suggests a change from a rather active free-swimming mode of life towards an ambush-predator role with good acceleration but little sustained swimming ability.

Rhizodontida

Until recently, this group was known primarily from teeth, scales and robust postcranial material (Andrews, 1985). Nevertheless, they have featured prominently in discussions on the origins of tetrapod limbs. *Gooloogongia* (discussed below) has provided important new information as to the relationships, both within the Rhizodontida and between the rhizodonts and other tetrapodomorphs.

Gooloogongia loomesi

Gooloogongia loomesi (Johanson and Ahlberg, 1998) from Canowindra, represents the most primitive, and by far the most complete, rhizodont sarcopterygian discovered so far. It is very rare in the fauna, represented by three incomplete specimens, only two of which have the head region preserved, and several fragments. It was named for Gooloogong, the nearest town to the west of the fossil site, and for Bruce Loomes of Canowindra who helped organize the 1993 dig and has been heavily involved in the project since then.

Gooloogongia loomesi was about 800 mm long (Figure 14). The head and body were broader than deep, the eyes faced dorsally, and the lower jaw projected forwards of the snout and was furnished with needle-sharp, backward-directed fangs. The skull roof of *Gooloogongia* (Johanson and Ahlberg, 1998) is similar to that of *Barameda* from the early Carboniferous of Victoria (Long, 1989), the only other rhizodont where it is preserved in articulation.

Some published phylogenetic analyses have been based on poorly substantiated characters, such as the alleged presence of two external nostrils in *Barameda*. *Gooloogongia,* which has several bones in the nasal series, appears less derived than *Barameda.* However it is clear that *Gooloogongia* has only one nostril (Johanson and Ahlberg, 1998), which has the same relationship to surrounding bones as in other tetrapodomorphs. Subsequent re-examination of the *Barameda* material by Long and Ahlberg (1999) indicated that it too conforms to this pattern. In tetrapodomorphs including elpistostegids and tetrapods, the single external nostril is associat-

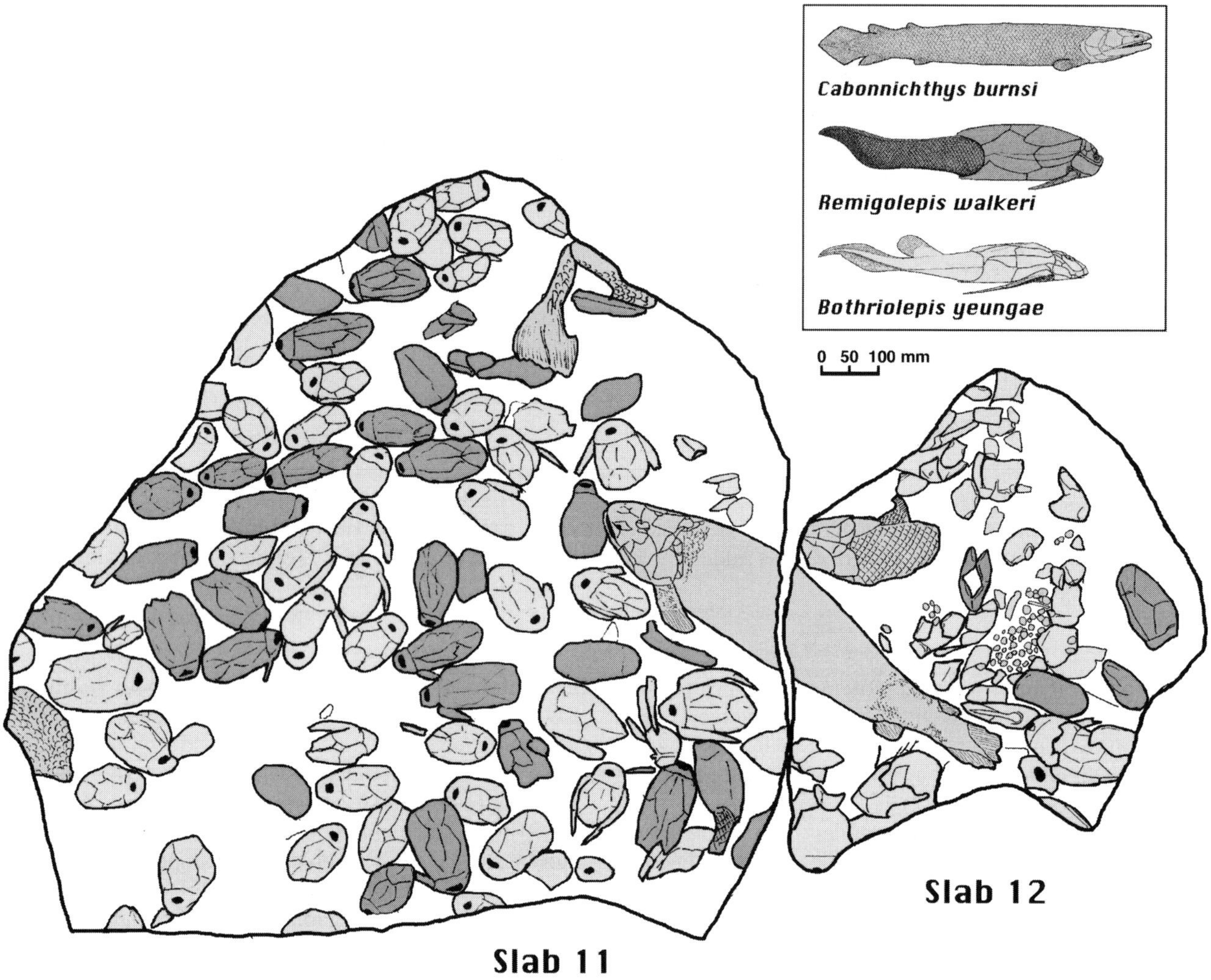

Figure 13. Slabs 11 and 12, which fit together, jointly contain the holotype (AMF.96858) and finest example of *Cabonnichthys burnsi* Ahlberg and Johanson (1997) from Canowindra. The head and anterior trunk region are on Slab 11 (1.17 x 1.25 m) and the posterior trunk and tail on Slab 12 (0.83 x 0.67 m). The head of a second *Cabonnichthys burnsi* lies nearby and parts of three other individuals are present elsewhere on the same slabs. Slab 11 also carries 52 *Bothriolepis* and 32 *Remigolepis* while Slab 12 has 5 *Bothriolepis* (+ scattered plates) and 4 *Remigolepis;* most of these are adults.

ed with a palatal nostril or choana but, unfortunately, the palate is not preserved in *Gooloogongia*.

The cheek, previously largely unknown in rhizodonts is tetrapodomorph-like in *Gooloogongia*. However, it has a broad preopercular, an apparently primitive feature shared with porolepiforms and onychodonts, but absent in the Tetrapodomorpha where this bone is narrow. The new evidence from *Gooloogongia* indicates that rhizodont sarcopterygians are resolved near the base of the tetrapodomorph clade (Johanson and Ahlberg, 2001) and are more remote from the Tetrapoda than are 'osteolepiform' and elpistostegid lobe-fin fishes.

The phylogenetic position of rhizodonts is important because the pectoral fin of the late Devonian rhizodont, *Sauripteris,* has featured prominently in discussions about the origin of tetrapod limbs. The preserved elements of the pectoral fin endoskeleton in *Gooloogongia* (Johanson and Ahlberg, 1998) resemble those of *Sauripteris* (Ahlberg, 1989; Daeschler and Shubin, 1998) and other rhizodonts, indicating that this basic pattern characterizes the Rhizodontida. However, the pectoral fin lepidotrichia of *Gooloogongia* have basal segments equal in length to the jointed distal portions (Johanson and Ahlberg, 1998). In other rhizodonts the basal segments are much longer than the jointed parts, whereas the generalized osteichthyan condition (e.g. as in actinopterygians and osteolepiforms) is to have very short basal segments. The posterior fins of *Gooloogongia* show this generalized condition, whereas other rhizodonts have long basal lepidotrichial segments in all fins (Andrews, 1985). Thus *Gooloogongia* only shows the derived condition of the rhizodont lepi-

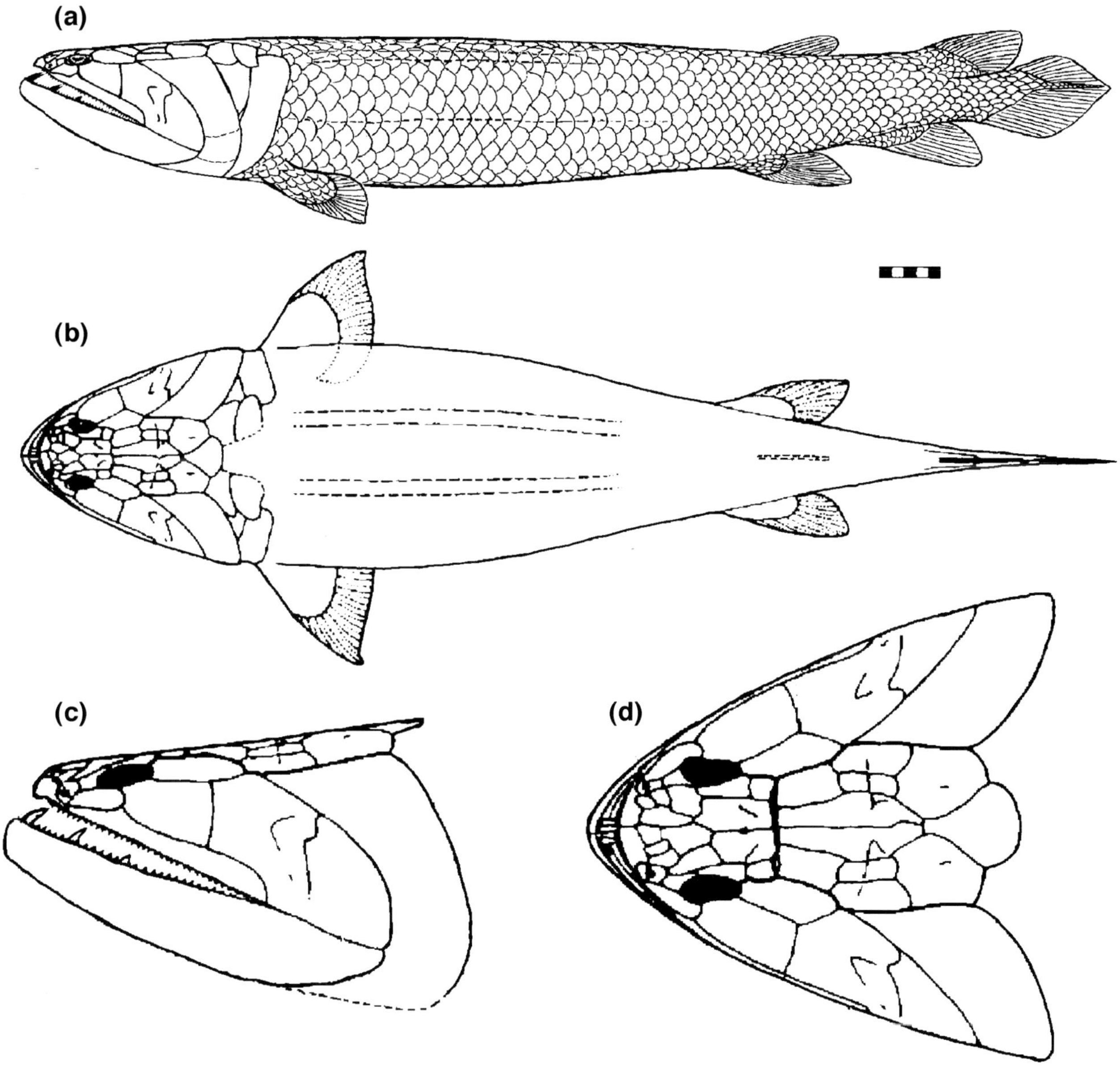

Figure 14. Reconstructions of *Gooloogongia loomesi*: **(a)** whole fish in lateral view; **(b)** whole fish in dorsal view (after Johanson and Ahlberg, 1998), with squamation and fin details added by Ritchie; **(c)** head in lateral view; **(d)** head in dorsal view (after Johanson and Ahlberg, 1998).

dotrichia in the pectoral fin, and even here it approaches the more typical tetrapodomorph condition.

Given the basal phylogenetic position of the Rhizodontida within the Tetrapodomorpha (Ahlberg and Johanson, 1998) similarities between rhizodont fins and tetrapod limbs are thus probably convergent, and the pectoral fin of *Sauripteris* should not be used as a model for the precursor of tetrapod limbs. Indeed, the rhizodont skull morphology illustrated by *Gooloogongia* and *Barameda* differs from the elpistostegid and tetrapod pattern in lacking frontal bones and having a short snout, widely separated eyes, a well-developed intracranial joint between parietals and postparietals, and extratemporals. These characters of rhizodonts are shared with porolepiforms, which are basal stem-group lungfishes; they are therefore probably shared primitive characters of the lungfish and the Tetrapodomorpha. They also occur in osteolepid (but not tristichopterid) tetrapodomorphs, which may indicate that osteolepids are paraphyletic relative to elpistostegids and tetrapodomorphs (Ahlberg and Johanson, 1998).

With its broad head, dorsally facing eyes and small caudal fin (Figure 14) the body form of *Gooloogongia* indicates that it may have been a benthic ambush predator. It differs considerably from the only other known rhizodont body, that of the Carboniferous genus *Strepsodus* (Andrews, 1985), which has a symmetrical tail and reduced median fins. Superficially *Strepsodus* resembles an elpistostegid or basal tetrapod, whereas *Gooloogongia* is closer to the generalized sarcopterygian condition exemplified by *Osteolepis*. *Strepsodus* also has multiple lateral lines while *Gooloogongia*, which has rather large rounded scales, displays a simpler pattern with a double dorso-lateral line.

Dipnomorpha

Dipnoans, or lungfishes, form a significant element in many middle and late Devonian fish faunas (Achanarras, Caithness, Scotland; Miguasha, Quebec, Canada; Mount Howitt, Victoria, Australia; Gogo, Western Australia) where they occur in association with antiarchs, arthrodires and osteolepiform sarcopterygians.

Strangely, of more than 3,600 fish specimens recovered from Canowindra, only two dipnoan specimens have been identified to date (although this may well be an artefact of preservation). However, the scientific implications of these two specimens far outweigh their rarity in the fauna. Ahlberg *et al.* (2001) identified both specimens as a new, small species of a late Devonian long-snouted dipnoan genus, *Soederberghia.*

The type species of this genus, *Soederberghia groenlandica* Lehman (1955) from the late Devonian (Famennian, Remigolepis Series) of East Greenland, is known from many specimens, mainly partial skulls and isolated cranial plates. It was a large species in which the skull alone reached at least 280 mm in length (Lehman, 1959). The second discovery of *Soederberghia* came, surprisingly, from the other side of the world, from south-eastern Australia. Campbell and Bell (1982) identified an isolated long-snouted dipnoan skull roof from the late Devonian Cloghnan Shale of the Jemalong Range, west of Forbes N.S.W. as *Soederberghia* sp. Although smaller than Greenland specimens it was still a substantial animal, with a skull roof 150 mm long. It was associated with antiarchs (*Bothriolepis*, *Remigolepis*), arthrodires (*Groenlandaspis* and a phyllolepid) and a mandible of *Metaxygnathus* (Campbell and Bell, 1977), the only known late Devonian tetrapod body fossil from the Southern Hemisphere.

Soederberghia simpsoni

Ahlberg, Johanson and Daeschler (2001) described new finds of *Soederberghia* from Pennsylvania (U.S.A.) and Canowindra and, at the same time, reviewed the earlier Greenland and Australian finds. They concluded that *Soederberghia* sp. from the Jemalong Range (N.S.W.) is specifically indistinguishable from *S. groenlandica* from Greenland. *Soederberghia* from the Catskill Formation of Pennsylvania is known (like the Jemalong specimen) only from an isolated and incomplete skull roof (~80 mm long) of a small individual. It is therefore about the same size as the larger of the two Canowindra specimens but, despite minor differences, it falls within the range of variation of *S. groenlandica* from East Greenland and so has been referred to that species together with the Jemalong skull.

The new Canowindran species has been named *S. simpsoni* in honour of Bill Simpson, the local beekeeper, who reported the original find to the Australian Museum in 1956. The smaller of the two dipnoan specimens from the Mandagery Sandstone (AMF 102819) represents an articulated but incomplete individual, around 210 mm long, displaying parts of the right side of the skull, most of the postcranial squamation and parts of the median fins; the paired fins are not preserved. *Soederberghia simpsoni* is thus the only species of this genus in which the postcranial skeleton is known, even if incompletely (Figure 15(b)). The second specimen of *S. simpsoni*, the holotype (AMF 102820, Figure 15(a)), is a natural mould of a skull roof and left cheek. The skull is about 70 mm long, indicating an original length of about 320 mm. It differs from *S. groenlandica* in the relative size of some cranial bones at the rear of the skull, in the presence of one postorbital (instead of two) and the fact that the jugal forms part of the orbit.

Relationships and Palaeobiogeography

Soederberghia groenlandica from Greenland and *Rhynchodipterus elginensis* Säve-Söderbergh (1937) from Scotland have long been regarded as sister taxa and placed in the same family, the Rhynchodipteridae. Until recently *Griphognathus,* another long-snouted dipnoan from the Gogo Formation of Western Australia, was thought to be related to *Soederberghia* and *Rhynchodipterus,* but Ahlberg *et al.* (2001) now exclude *Griphognathus* from the Rhynchodipteridae based on differences in the dermal bone pattern of the skull and the absence of a stalked parasphenoid and cranial ribs (thought to be air-breathing

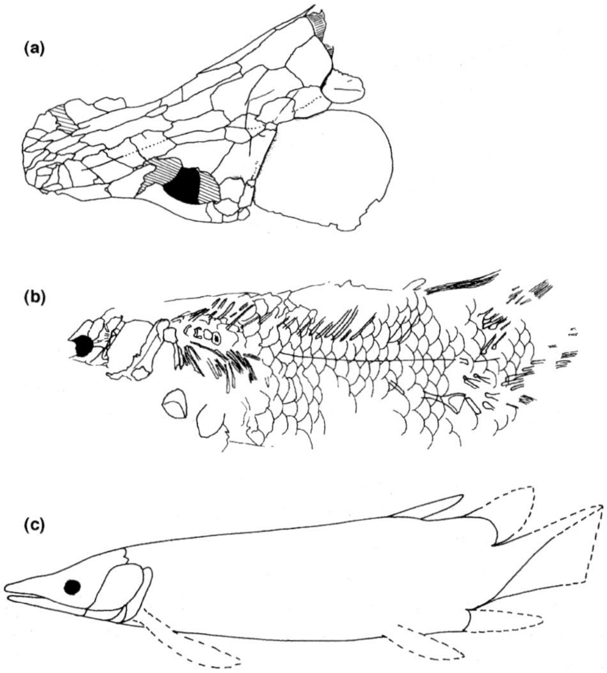

Figure 15. *Soederberghia simpsoni* n. sp. (Ahlberg, Johanson and Daeschler, 2001): **(a)** holotype (AMF.102820), skull roof and left cheek (70 mm long); **(b)** articulated but incomplete specimen (AMF.102819) (210 mm long); **(c)** provisional reconstruction of body profile and fins.

adaptations). Ahlberg *et al.* (2001) suggest that rhynchodipterids were a group of derived air-breathing late Frasnian–Famennian dipnoans, apparently restricted to Old Red Sandstone environments and possibly related to another dipnoan group, the fleurantiids.

The distribution of *Soederberghia,* found in both the Northern and Southern Hemispheres in association with tetrapods, tristichopterid sarcopterygians, arthrodires and antiarchs, must be taken into account for late Devonian biostratigraphy and palaeogeographic reconstructions. The occurrence of species of *Soederberghia* in the Famennian of Greenland, Pennsylvania and New South Wales, together with tetrapods, indicates close faunal connections between these areas in the late Devonian (Young, 1993a, b). Decreasing endemicity and increasing cosmopolitanism in vertebrate faunas towards the end of the Devonian may be indicated. These points have considerable bearing on the palaeomagnetic reconstructions considered below.

COMPARING CANOWINDRA AND OTHER DEVONIAN FAUNAS

For comparative comments on early Devonian fish faunas in Australia see pp.131–157 (Basden *et al.*, 2006).

BIOGEOGRAPHIC CATEGORIES

The Canowindra vertebrate assemblage can be divided into the three categories listed below for biogeographical purposes (Ahlberg *et al.,* 2001):

(1) Endemic Gondwanan Groups—only represented by *Canowindra grossi*, which belongs to the Canowindridae, an Australian–Antarctic family of sarcopterygians.

(2) Endemic Genera Belonging to Cosmopolitan Groups —represented by *Mandageria* and *Cabonnichthys,* which belong to the Tristichopteridae, a cosmopolitan family that originated in Laurussia; and by *Gooloogongia* (Family Rhizodontida), another cosmopolitan clade, but which appears to be of Gondwanan origin.

(3) Cosmopolitan Genera—represented by antiarchs *Bothriolepis* and *Remigolepis*, the arthrodire *Groenlandaspis* and the lungfish *Soederberghia;* are all known from Laurussia as well as Gondwana.

Phylogenetic analysis of widely distributed taxa such as *Bothriolepis*, *Remigolepis* and *Groenlandaspis* has provided valuable information on Devonian palaeobiogeography and, with improved resolution of species-level relationships, this is likely to increase. *Bothriolepis* provides a good example; with recent discoveries from South America (Janvier and Villarroela, 1998; Young and Moody, 2002), *Bothriolepis* species are now known from every modern continent. Cladograms of *Bothriolepis* species relationships have been prepared (Long, 1983; Young, 1988; Johanson and Young, 1999), but it is expected that these can be further refined, with implications for biogeography.

It is interesting to compare the Canowindra fauna with two other late Devonian faunas from south-east Australia.

CLOGHNAN SHALE FORMATION

In the Jemalong Range, west of Forbes, N.S.W., *Bothriolepis, Remigolepis, Groenlandaspis* and a phyllolepid occur with *Soederberghia* represented by the Greenland species *S. groenlandica,* not *S. simpsoni*. The Jemalong assemblage also contains *Holoptychius*, a porolepiform sarcopterygian which is widespread in the late Devonian of Laurussia, but also occurs in eastern Australia at Grenfell (Johanson and Ritchie, 2000) and near Eden N.S.W. as well as South America (Janvier and Villarroela, 1998). The Cloghnan Shale has also produced *Metaxygnathus*, a tetrapod similar to *Ventastega* from Latvia and to an unnamed tetrapod jaw from the Catskill Formation (Ahlberg *et al.,* 1994; Ahlberg and Clack, 1998).

WORANGE POINT FORMATION

The assemblage at Worange Point, near Eden in south-eastern N.S.W., contains *Bothriolepis, Remigolepis, Groenlandaspis* and *Holoptychius* (Young, 1993a), as well as a large undescribed tristichopterid assignable to the cosmopolitan genus *Eusthenodon* (Ahlberg and Johanson, 2000, personal communication), but the same sites have not yet yielded *Soederberghia.* The Worange Point Formation is thought to be the lateral equivalent of the Genoa River Formation (Young, 1993a), from which Devonian tetrapod trackways were described by Warren and Wakefield (1972).

Although new discoveries could well change this picture, the Canowindra assemblage appears somewhat less cosmopolitan than the others; this may support Young's contention (1999) that the Canowindra fauna is somewhat older, possibly late Frasnian. Fish faunas of the Cloghnan Shale (N.S.W.), Catskill Formation (eastern U.S.A.), the Remigolepis Series (East Greenland) and other late Devonian Old Red Sandstone localities, are dominated by taxa that are restricted to the late Devonian.

PALAEOMAGNETIC EVIDENCE

This strongly suggests that late Devonian distribution patterns of these taxa reflect continental positions and dispersal routes during the late Devonian, not the continent positions of an earlier time. But the current dispute on relative positions of Laurussia and Gondwana during the late Devonian makes it difficult to assess current palaeogeographic models.

deposit and have not yet been collected. The relatively late recognition (in 1997) of two specimens of *Soederberghia simpsoni* on the slabs illustrates how relatively delicate specimens can easily be masked by the crowded skeletal remains of robust antiarchs.

Estimates of total numbers of each species identified on slabs recovered from Canowindra are summarized in Table 2. These population figures are regularly revised and updated as preparation work on the slabs is completed and latex casts are prepared for detailed study. From experience, estimates based on examination of the rock slabs alone usually underestimate the total numbers, especially of the antiarchs where they are so tightly packed together.

ROLES OF INDIVIDUAL FISH SPECIES

The non-dipnoan sarcopterygians, with their strongly developed jaws and dentitions, were almost certainly the predators in the fauna, preying predominantly on the placoderms. As mentioned previously, caudal fin morphology suggests the sarcopterygians were slow or burst swimmers, rather than steady swimmers that were capable of moving constantly at moderate to fast speeds (Webb, 1982; Johanson and Ahlberg, 1997; Ahlberg and Johanson, 1997). These sarcopterygians probably waited in ambush, accelerating and attacking prey as it passed. It may seem unusual that all predators in the Canowindra fauna were relatively weak or short duration swimmers, but this appears characteristic of many Devonian sarcopterygians (Belles-Isles, 1992).

In contrast, both *Bothriolepis* and *Remigolepis* possess ventral mouths with small, weak lower jaws (Stensiö, 1931, 1948; Johanson, 1997b), suggesting that they were bottom feeders, taking small particles from the sediment surface, or perhaps ingesting and processing the sediment itself for food (Denison, 1941; Weems *et al.,* 1981). However, Johanson (1997b) noted that *Bothriolepis* and *Remigolepis* differ in jaw structure, as well as in pectoral and caudal fin morphology, suggesting differences in swimming ability and feeding. From its relatively longer tail and pectoral fins *Bothriolepis* appears to have been the better and more manoeuvrable swimmer, able to feed within the water column. *Remigolepis,* with its relatively shorter pectoral fins (and reduced capacity for generating lift), was probably the poorer swimmer and more restricted to bottom feeding.

One aspect of the Canowindra fauna that requires further investigation is the comparative rarity of very small specimens or juveniles of the dominant antiarchs. One reason could be sampling bias, with smaller specimens overlooked during collection. However, the Canowindra fauna was removed *en masse* on large slabs and cleaned later under supervision. Smaller individuals are much rarer, and more difficult to recognize, on rock slabs crowded with adult specimens, but this bias is not believed to be large. Most of the juvenile antiarchs discovered at Canowindra occur on slabs that appear to represent the margins of the assemblage, perhaps representing the shallow water at the edge of the final pool as it dried up. In these, the individuals are less crowded and predominantly represented by juveniles.

Another possible influence on the abundance of juveniles is timing of reproduction. Reproductive patterns and seasons vary widely among fishes and are influenced by environmental factors, food availability and presence of predators. The relative scarcity of very small *Remigolepis* specimens might suggest differences in timing of reproduction relative to *Bothriolepis*. Different spawning times in the two dominant species would certainly lessen the pressure on local habitat resources.

Although the mass mortality could have occurred during a non-reproductive period for both *Bothriolepis* and *Remigolepis*, the composition of the fauna is such that another explanation is possible. For example, many fish species tend to aggregate (or school) before and during spawning. This was clearly the case in the fauna studied by Wilson (1996), where one species of paracanthopterygian fish dominated the fauna (1,700 specimens) to the near exclusion of the other two present (69 osmerids and 2 osteoglossids). Wilson (1996) described the diversity in this fauna as being low, and considered that this might be due to an aggregation of spawning species in a pool of shallow water during a high-water event. Wilson (1996) was further able to suggest that the fishes in the Joffre Bridge fauna aggregated to spawn because he was able to recognize breeding tubercles (on the males). No such structures have been observed at Canowindra, indeed, it is at present impossible to distinguish the sexes in the Canowindra fauna.

CONCLUSIONS

All the evidence suggests that the Canowindra fish remains represent a single late Devonian palaeocommunity, which was trapped in a pool that dried up over a restricted period of time. A most unusual set of circumstances just after the mortality (mummification, rapid but gentle burial by fine sand, limited disintegration and an absence of reworking) has resulted in an extremely high level of preservation—with fine detail of both external and internal features still discernible. Sub-surface weathering has destroyed most of the original bone, leaving behind detailed three-dimensional natural moulds of the fossils in sandstone.

Although less diverse (in numbers of species) than some other Australian Devonian deposits, the Canowindra fauna is highly significant for several reasons. First, the *Groenlandaspis* specimens recovered have increased the knowledge of this arthrodire as they are among the finest specimens of this taxon known. Second, the distribution patterns of this arthrodire and the dominant antiarchs provide indications of how continents were grouped before the break-up of Gondwana. Third, prior to 1993 the Southern Hemisphere had produced very few well-preserved rhipidistians, but the discoveries reported here are making a major contribution to understanding sarcopterygian relationships and biogeography. For example, the evidence now suggests that a derived clade within the Tristichopteridae (including *Cabonnichthys* and *Mandageria*) extended over a wide range, including Gondwana. The discovery of *Gooloogongia* has greatly improved knowledge of rhizodont anatomy and confirms that rhizodonts are not as closely related to tetrapods as taxa such as tristichopterids; its occurrence also suggests that Rhizodontida may have originated in Eastern Gondwana in the Devonian. Fourth, the Canowindra fauna includes many taxa that are either cosmopolitan in distribution or related to Laurussian forms.

The ecological comments possible at this stage are limited but the following three conclusions can be drawn: differences in body structure and swimming abilities, of the two dominant antiarchs, may be related to differing life cycle strategies that permit partitioning of local resources; the frequency of juveniles of the dominant antiarchs is low; the large sarcopterygians almost certainly predated on juvenile placoderms.

Re-opening of the site will enable a number of studies to proceed. These include: investigating the total extent of the layer; documenting the sedimentology above and below the fossil layer; expanding searches for other groups (plants, invertebrates, tetrapods); surveying distribution patterns of different taxa; examining ventral impressions of exposed individuals and re-assessing orientation or alignment on existing slabs.

This fossil site and associated Age of Fishes Museum (AOFM) provide a major rural tourist attraction with considerable benefits for the local community, tour operators and professional scientists. In the author's opinion, if management follows original recommendations (with Australian Museum scientific supervision), when fully excavated and developed the locality would have the potential for nomination as a World Heritage Fossil Site.

ACKNOWLEDGEMENTS

I am indebted to my colleagues, Zerina Johanson and Per Ahlberg, for permission to extract much of the material used in this article from their papers on the Canowindra fauna. Cabonne Shire Council's logistic support, and Canowindra's community support, made possible the 1993 rediscovery and excavation of the Canowindra site. The on-going success of the Canowindra project owes much to the efforts of monthly teams of 'paying' volunteers organized by Monica Yeung (Gondwana Dreaming, Canberra) and Bruce Loomes, Canowindra. Since 1993 research work on the Canowindra fauna (by the author and Zerina Johanson) has been generously funded by James Fairfax, of Bowral, N.S.W.

REFERENCES

Ahlberg, P.E. 1989. Paired fin skeletons and relationships of the fossil group Porolepiformes (Osteichthyes: Sarcopterygii). *Zoological Journal of the Linnean Society,* Vol. 96, pp.119–166

Ahlberg, P.E. 1991. A re-examination of sarcopterygian interrelationships, with special reference to the Porolepiformes. *Zoological Journal of the Linnean Society,* Vol. 103, pp.241–287.

Ahlberg, P.E. and Clack, J.A. 1998. Lower jaws, lower tetrapods—a review based on the Devonian genus *Acanthostega. Transactions of the Royal Society of Edinburgh: Earth Sciences,* Vol. 89, pp.11–46.

Ahlberg, P.E. and Johanson, Z. 1997. The second Tristichopterid (Sarcopterygii, Osteolepiformes) from the upper Devonian of Canowindra, New South Wales, Australia. *Journal of Vertebrate Paleontology,* Vol. 17, pp.653–673.

Ahlberg, P.E. and Johanson, Z. 1998. Osteolepiforms and the ancestry of tetrapods. *Nature,* Vol. 395, pp.792–794.

Ahlberg, P.E., Johanson, Z. and Daeschler, E.B. 2001. The late Devonian lungfish *Soederberghia* (Sarcopterygii, Dipnoi) from Australia and North America, and its biogeographical implications. *Journal of Vertebrate Paleontology*, Vol. 21, No. 1, pp.1–12.

Ahlberg, P.E., Luksevics, E. and Lebedev, O. 1994. The first tetrapod finds from the Devonian (upper Famennian) of Latvia. *Philosophical Transactions of the Royal Society of London*, Series B, Vol. 343, pp.303–328.

Anderson, M.E., Long. J.A., Evans, F.J., Almond, J.E., Theron, J.N. and Bender, P. 1999. Biogeographic affinities of middle and late Devonian fishes of South Africa. *Records of the Western Australian Museum, Supplement,* No. 57, pp.157–168.

Andrews, S.M. 1978. A possible occurrence of *Remigolepis* in the topmost Old Red Sandstone of Berwickshire. *Scottish Journal of Geology,* Vol. 14, pp.311–315.

Andrews, S. M. 1985. Rhizodont crossopterygian fish from the Dinantian of Foulden, Berwickshire, with a re-evaluation of this group. *Transactions of the Royal Society of Edinburgh: Earth Sciences,* Vol. 76, pp.67–95.

Andrews, S.M. and Westoll, T.S. 1970. The postcranial skeleton of rhipidistian fishes excluding *Eusthenopteron. Transactions of the Royal Society of Edinburgh,* Vol. 68, pp.391–489.

Australian Museum 2002. Field sites—the Canowindra story. http://www.amonline.net.au/palaeontology/field_sites/canowindra.htm

Basden, A.M., Trinajstic, K.M. and Merrick, J.R. 2006. Eons of fishy fossils. Pp.131–157. In Merrick, J.R., Archer, M., Hickey, G.M. and Lee, M.S.Y. (eds). *Evolution and Biogeography of Australasian Vertebrates.* Auscipub Pty. Ltd., Sydney. 942pp.

Belles-Isles, M. 1992. The modes of swimming of sarcopterygians. Pp.117–130. In Mark-Kurik, E. (ed). *Fossil Fishes as Living Animals.* Academy of Sciences of Estonia, Tallinn.

Campbell, K.S.W. and Bell, M.W. 1977. A primitive amphibian from the late Devonian of New South Wales. *Alcheringa,* Vol. 1, pp.369–381.

Campbell, K.S.W. and Bell, M.W. 1982. *Soederberghia*

(Dipnoi) from the late Devonian of New South Wales. *Alcheringa*, Vol. 6, pp.143–149.

Cloutier, R. and Ahlberg, PE. 1996. Morphology, characters and the interrelationships of basal sarcopterygians. Pp.445–479. In Stiassny, M.L.J., Parenti, L. and Johnson, G.D. (eds). *Interrelationships of Fishes.* Academic Press, New York. 496pp.

Conolly, J.R. 1965. The stratigraphy of the Hervey Group in central New South Wales. *Journal and Proceedings of the Royal Society of New South Wales,* Vol. 98, pp.37–83.

Cope, E.D. 1885. The position of *Pterichthys* in the system. *American Naturalist,* Vol. 19, pp.289–291.

Cope, E.D. 1886. An interesting connecting genus of Chordata. *American Naturalist,* Vol. 20, pp.1027–1031.

Cope, E.D. 1889. Synopsis of the families of Vertebrata. *American Naturalist,* Vol. 23, No. 2, pp.1–29.

Cribb, J. 1996. A prize catch. *Australian Geographic,* No. 43 (July-Sept.), pp.100–115.

Daeschler, E.B. and Shubin, N.H. 1998. Fish with fingers? *Nature,* Vol. 391, p.133.

Denison, R.H. 1941. The soft anatomy of *Bothriolepis*. *Journal of Paleontology,* Vol. 15, pp.553–561.

Denison, R.H. 1978. Placodermi. Pp.1–128. In Schultze, H-P. (ed). *Handbook of Paleoichthyology, 2.* Gustav Fischer Verlag, Stuttgart.

Egerton, P. de M.G. 1861. *Tristichopterus alatus*; pp.51–55 in Figures and Descriptions Illustrative of British Organic Remains. *Memoirs of the Geological Survey of the United Kingdom, Decade X.*

Eichwald, E.I. von, 1840. Die Thier-und Pflanzenreste des alten rothen Sandsteins und Bergkalks im Novgorodschen Gouvernement. *Bulletin de l'Academie des Sciences de Saint-Petersbourg,* Vol. 7, pp.78–91.

Elder, R.L. and Smith, G.R. 1988. Fish taphonomy and environmental inference in paleolimnology. *Palaeogeography, Palaeoclimatology, Palaeoecology,* Vol. 62, pp.577–592.

Fletcher, H.O. 1956. Ancient armoured fishes discovered at Canowindra, N.S.W. *Australian Museum Magazine,* Vol. 12, pp.37–40

Hastings, P. 1956. Beside a N.S.W. bush road… he found a fish 350 million years old. *Sunday Telegraph*, 1 April, p.13.

Heintz, A. 1932. Beitrag zur Kenntnis der Devonischen Fischfauna Ost-Gronlands. *Skrifter Skrifter om Svalbard og Ishavet,* Vol. 42, pp.1–27.

Janvier, P. 1996. Early vertebrates. *Oxford Monographs on Geology and Geophysics,* 33, pp.1–393, *Clarendon Press, Oxford.*

Janvier, P. and Ritchie, A. 1977. Le genre *Groenlandaspis* Heintz (Pisces, Placodermi, Arthrodira) dans le Devonien d'Asie. *Colloques Researches Academie des Sciences de Paris*, Series D, Vol. 284, pp.1385–1388.

Janvier, P. and Villarroela, C. 1998. Los peces devónicos del Macizo de Floresta (Boyacá, Colombia). Consideraciones taxonómicas, bioestratigráficas, biogeográficas y ambientales. *Geología Colombiana,* Vol. 23, pp.3–18.

Jarvik, E. 1952. On the fish-like tail in the ichthyostegid stegocephalians with descriptions of a new stegocephalian and a new crossopterygian from the upper Devonian of East Greenland. *Meddelelser om Gronland,* Vol. 114, pp.1–90.

Jarvik, E. 1954. On the visceral skeleton in *Eusthenopteron*, with a discussion of the parasphenoid and palatoquadrate in fishes. *Kungliga Svenska Vetenskapsakademiens Handlingar,* Vol. 4, pp.1–104

Jarvik, E. 1980. *Basic Structure and Evolution of Vertebrates.* Volume 1. Academic Press, London.

Jarvik, E. 1985. Devonian osteolepiform fishes from East Greenland. *Meddelelser om Gronland, Geoscience,* Vol. 13, pp.1–52.

Johanson, Z. 1995. New information on jaw elements of *Remigolepis* (Placodermi, Antiarchi) from Canowindra, NSW, Australia (upper Devonian). *Geobios, Memoire Special,* Vol. 19, pp.103–108.

Johanson, Z. 1997a. New antiarchs (Placodermi) from the Hunter Siltstone (Famennian) near Grenfell, New South Wales. *Alcheringa,* Vol. 21, pp.191–217.

Johanson, Z. 1997b. New *Remigolepis* (Placodermi; Antiarchi), from Canowindra, NSW, Australia. *Geological Magazine,* Vol. 133, pp.813–846.

Johanson, Z. 1998. The upper Devonian fish *Bothriolepis* (Placodermi: Antiarchi) from near Canowindra, New South Wales, Australia. *Records of the Australian Museum,* Vol. 50, pp.315–348.

Johanson, Z. and Ahlberg, P.E. 1997. A new tristichopterid (Osteolepiformes, Sarcopterygii) from the Mandagery Sandstone (late Devonian, Famennian) near Canowindra, NSW, Australia. *Transactions of the Royal Society of Edinburgh: Earth Sciences,* Vol. 88, pp.39–68.

Johanson, Z. and Ahlberg, P.E. 1998. A complete primitive rhizodont from Australia. *Nature,* Vol. 394, pp.569–573.

Johanson, Z. and Ahlberg, P.E. 2001. Devonian rhizodontids and tristichopterids (Sarcopterygii; Tetrapodomorpha) from East Gondwana. *Transactions of the Royal Society of Edinburgh: Earth Sciences,* Vol. 92, Issue 1, pp.43–74.

Johanson, Z. Ahlberg, P.E. and Ritchie, A. 2003. The braincase and palate of the tetrapodomorph sarcopterygian *Mandageria fairfaxi*: morphological variability near the fish-tetrapod transition. *Palaeontology*, Vol. 46, No. 2, pp.271–293.

Johanson, Z. and Ritchie, A. 2000. Rhipidistians (Sarcopterygii) from the Hunter Siltstone (late Famennian) near Grenfell, NSW, Australia. *Geowissenschaftliche Reihe*, Vol. 3, pp.111–136.

Johanson, Z. and Young, G.C. 1999. New *Bothriolepis* (Antiarchi; Placodermi) from the Braidwood region, New South Wales, Australia (middle-late Devonian). *Records of the Western Australian Museum, Supplement,* No. 57, pp.55–75.

Kent, D.V. and Van der Voo, R. 1990. Palaeozoic palaeogeography from palaeomagnetism of the Atlantic-bordering continents. Pp.49–56. In Scotese, C.R. and McKerrow, W.S.(eds). *Palaeozoic Palaeogeography and Biogeography.* Memoir, Vol. 12. Geological Society, London.

Lebedev, O.A. 1995. Morphology of a new osteolepidid fish from Russia. Pp.287–342. In Arsenault, M., Lelievre, H. and Janvier, P. (eds). *Etudes sur les Vertebres Inferieurs.* Vol. 17(C). Bulletin du Museum National d'Histoire Naturelle, Paris.

Lehman, J-P. 1955. Les Dipneustes du Dévonien supérieur du Groenland. *Comptes Rendues des Séances Academie des Sciences,* Vol. 240, pp.995–997.

Lehman, J-P. 1959. Les Dipneustes du Devonien superieur du Groenland. *Meddelelser om Gronland,* Vol. 160, pp.1–58.

Li, Z.X., Powell, C. McA. and Trench, A. 1993. Palaeozoic global reconstructions. Pp.25–53. In Long, J.A. (ed). *Palaeozoic Vertebrate Biostratigraphy and Biogeography*. Bellhaven Press, London. 369pp.

Long, J.A. 1982. The history of fishes on the Australian continent. Pp.53–86. In Rich, P.V. and Thompson, E.M. (eds). *The Fossil Vertebrate Record of Australasia*. Monash University Press, Clayton, Victoria. 759pp.

Long, J.A. 1983. New bothriolepid fish from the late Devonian of Victoria, Australia. *Palaeontology,* Vol. 26, pp.295–320.

Long, J.A. 1985a. New information on the head and shoulder girdle of *Canowindra grossi* Thomson, from the late Devonian Mandagery Sandstone, New South Wales. *Records of the Australian Museum,* Vol. 37, pp.91–99.

Long, J.A. 1985b. The structure and relationships of a new osteolepiform fish from the late Devonian of Victoria, Australia. *Alcheringa,* Vol. 9, pp.1–22.

Long, J.A. 1985c. A new osteolepid fish from the upper Devonian Gogo Formation, Western Australia. *Records of the Western Australian Museum,* Vol. 12, pp.361–377.

Long, J.A. 1987. An unusual osteolepiform fish from the late Devonian of Victoria, Australia. *Palaeontology,* Vol. 30, pp.839–852.

Long, J.A. 1989. A new rhizodontiform fish from the early Carboniferous of Victoria, Australia, with remarks on the phylogenetic position of the group. *Journal of Vertebrate Paleontology,* Vol. 9, pp.1–17.

Long, J.A. 1995a. New groenlandaspidid arthrodire (Pisces, Placodermi) from the middle Devonian Aztec Siltstone, southern Victoria Land, Antarctica. *Records of the Western Australian Museum,* Vol. 17, pp.35–41.

Long, J.A. 1995b. *The Rise of Fishes: 500 Million Years of Evolution*. University of New South Wales Press, Sydney. 223pp.

Long, J.A. 1999. A new genus of fossil coelacanth (Osteichthyes: Coelacanthiformes) from the middle Devonian of southeastern Australia. *Records of the Western Australian Museum, Supplement,* No. 57, pp.37–54.

Long, J.A. and Ahlberg, P.E. 1999. New observations on the snouts of rhizodont fishes (Palaeozoic Sarcopterygii). *Records of the Western Australian Museum, Supplement,* No. 57, pp.169–173.

Long, J.A., Anderson, M.E., Gess, R. and Hiller, N. 1997. New placoderm fishes from the late Devonian of South Africa. *Journal of Vertebrate Paleontology,* Vol. 17, No. 2, pp.253–268.

Luksevics, E.V. 1991. New *Remigolepis* (Pisces, Antiarchi), from the Famennian deposits of the Central Devonian field (Russia, Tula region). *Daba un Muzejs,* Vol. 3, pp.51–56 (in Russian).

McCoy, F. 1848. On some new fossil fish of the Carboniferous period. *Annals and Magazine of Natural History,* Vol. 2, pp.1–10.

Miles, R.S. 1968. The Old Red Sandstone antiarchs of Scotland. Family Bothriolepididae[sic]. *Palaeontographical Society Monographs,* Vol. 122, pp.1–130.

Miles, R.S. 1973. An actinolepid arthrodire from the lower Devonian Peel Sound Formation, Prince of Wales Island. *Palaeontographica, Stuttgart, (A),* Vol. 143, pp.109–118.

Muller, J. 1844. Uber den Bau und die Grenzen der Ganoiden, and uber das naturlich System der Fische. *Akademie der Wissenschaften Berlin,* 1844, pp.67–85

Obruchev, D.V. 1964. Agnathan and fish. In Orlov, Y.U. (ed). *Fundamentals of Paleontology.* Vol. 11. Moscow. 522pp.

Pan, J., Huo, F., Cao, J., Gu, Q., Liu, S., Wang, J., Gao, L. and Liu, C. 1987. *Continental Devonian System of Ningxia and its Biotas.* Geological Publishing House, Beijing.

Pan, J., Wang, S.T., Liu, S.-Y., Gu, Q. and Hang, J. 1980. Discovery of Devonian *Bothriolepis* and *Remigolepis* in Ninxia. *Acta Geologica Sinica*, Vol. 54, pp.176–185.

Panteleyev, N. 1992. New remigolepids and high armoured antiarchs of Kirgizia. Pp.185–192. In Mark-Kurik, E. (ed). *Fossil Fishes as Living Animals*. Academy of Sciences, Estonia, Tallinn.

Ritchie, A. 1974. From Greenland's icy mountains … a detective story in stone. *Australian Natural History,* Vol. 18, pp.28–35.

Ritchie, A. 1975. *Groenlandaspis* in Antarctica, Australia and Europe. *Nature,* Vol. 254, pp.569–573.

Romer, A.S. 1955. Fish origins—fresh or salt water? Pp.262–280. In *Papers in Marine Biology and Oceanography*. Pergamon Press, London.

Rosen, D.E., Forey, PL., Gardiner, B.G. and Patterson. C. 1981. Lungfishes, tetrapods, paleontology, and plesiomorphy. *Bulletin of the American Museum of Natural History,* Vol. 167, pp.159–276.

Säve-Söderbergh, G. 1937. On *Rhynchodipterus elginensis* n.g., n.sp., representing a new group of dipnoan-like Choanata from the upper Devonian of East Greenland and Scotland. *Arkiv Zoologica,* Vol. 29B, pp.1–8.

Schäfer, W. 1972. *Ecology and Palaeoecology of Marine Environments*. University of Chicago Press, Chicago. 568pp.

Scotese, C.R. and McKerrow, W.S. 1990. Revised world maps and introduction. Pp.1–21. In Scotese, C.R. and McKerrow, W.S. (eds). *Palaeozoic Palaeogeography and Biogeography.* Memoir, Vol.12. Geological Society, London.

Stensiö, E.A. 1931. Upper Devonian vertebrates from East Greenland, collected by the Danish Greenland expedition in 1929 and 1930. *Meddelelser om Gronland,* Vol. 86, pp.1–212.

Stensiö, E.A. 1948. On the Placodermi of the upper Devonian of East Greenland. 2. Antiarchi: subfamily Bothriolepinae. With an attempt at a revision of the previously described species of that family. *Meddelelser om Gronland,* Vol. 139, pp.1–622.

Stensiö, E.A. 1963. Anatomical studies on the arthrodiran head. Part 1. *Kunglica Svenska Vetenskapsakademiens Handlingar,* Vol. 9, pp.1–419.

Thomson, K.S. 1968. A new Devonian fish (Crossopterygii: Rhipidistia) considered in relation to the origin of the amphibia. *Postilla,* Vol. 124, pp.1–13.

Thomson, K.S. 1973. Observations on a new rhipidistian fish from the upper Devonian of Australia. *Palaeontographica* A, Vol. 143, pp.209–220.

Trewin, N.H. 1986. Palaeoecology and sedimentology of the Achanarras fish bed of the Middle Old Red Sandstone, Scotland. *Transactions of the Royal Society of Edinburgh: Earth Sciences,* Vol. 77, pp.21–46.

Van der Voo, R. 1988. Paleozoic paleogeography of North America, Gondwana and displaced terranes: comparisons of paleomagnetism with paleoclimatology and biogeographical patterns. *Geological Society of America Bulletin,* Vol. 100, pp.311–324.

Vorobyeva, E.I. 1959. A new crossopterygian fish genus *Platycephalichthys*—from the upper Devonian of the Lovat River. *Paleontologicheskii Zhurnal,* Vol. 3, pp.95–106. (In Russian).

Vorobyeva, E.I. 1977. Morphology and nature of evolution of crossopterygian fish. *Trudy Paleontologicheskogo Instituta,* Vol. 163, pp.1–239. (In Russian).

Warren, J.W. and Wakefield, N.A. 1972. Trackways of tetrapod vertebrates from the late Devonian of Victoria, Australia. *Nature,* Vol. 238, pp.469–470.

Webb, P.W. 1982. Locomotor patterns in the evolution of Actinopterygian fishes. *American Zoologist,* Vol. 22, pp.329–342.

Weems, R.E., Beem, K.A. and Miller, T.A. 1981. A new species of *Bothriolepis* (Placodermi: Bothriolepidae) from the upper Devonian of Virginia (USA). *Proceedings of the Biological Society of Washington,* Vol. 94, pp.984–1004.

Whiteaves, J.F. 1881. On some remarkable fossil fishes from the Devonian rocks of Scaumenac Bay, Province of Quebec, with descriptions of a new genus and three new species. *Canadian Naturalist,* Vol. 10, pp.27–35.

Wilson, M.V.H. 1996. Taphonomy of a mass-death layer of fishes in the Paleocene Paskapoo Formation at Joffre Bridge, Alberta, Canada. *Canadian Journal of Earth Sciences,* Vol. 33, pp.1487–1498.

Woodward, A.S. 1891. *Catalogue of the Fossil fishes in the British Museum (Natural History), Pt. II. Elasmobranchii.* British Museum (Natural History), London. 567pp.

Yeung, M., Stephenson, T. and Ritchie, A. 1998. The Canowindra Age of Fishes Museum project—Matrix Ecotourism at work. Pp.35–39. In Eastburn, D. and Milligan, A. (eds). *Reading the Land Workshop Proceedings, Canberra, 21–22 May 1997.* Murray Darling Basin Commission, Canberra.

Young, G.C. 1988. Antiarchs (placoderm fishes) from the Devonian Aztec siltstone, southern Victoria Land, Antarctica. *Palaeontographica Abteilung* A, Vol. 202, pp.1–125.

Young, G.C. 1993a. Middle Palaeozoic macrovertebrate biostratigraphy of eastern Gondwana. Pp.208–251. In Long, J.A. (ed). *Palaeozoic Vertebrate Biostratigraphy and Biogeography.* Bellhaven Press, London. 369pp.

Young, G.C. 1993b. Vertebrate faunal provinces in the middle Palaeozoic. Pp.292–323. In Long, J.A. (ed). *Palaeozoic Vertebrate Biostratigraphy and Biogeography.* Bellhaven Press, London. 369pp.

Young, G.C. 1995. Timescales 4: Devonian, Australian Phanerozoic Timescales, biostratigraphic charts and explanatory notes. Second Series. *Australian Geographic Survey Organisation Record,* 1995/33, pp.1–47.

Young, G.C. 1999. Preliminary report on the biostratigraphy of new placoderm discoveries in the Hervey Group (upper Devonian) of central New South Wales. *Records of the Western Australian Museum, Supplement,* No. 57, pp.139–150.

Young, G.C. and Gorter, J.D. 1981. A new fish fauna of middle Devonian age from the Taemas/Wee Jasper region of New South Wales. *Bureau of Mineral Resources, Geology and Geophysics Bulletin,* No. 209, pp.83–147.

Young, G.C., Long, J.A. and Ritchie, A. 1992. Crossopterygian fishes from the Devonian of Antarctica: systematics, relationships and biogeographic significance. *Records of the Australian Museum, Supplement,* No. 14, pp.1–77.

Young, G.C. and Moody, J.M. 2002. A middle–late Devonian fish fauna from the Sierra de Perijà, Western Venezuela, South America. *Mitteilungen aus dem Museum für Naturkunde in Berlin. Geowissenschaftliche Reihe,* Vol. 5, pp.155–206.

Chapter 9
Australian Marine Fishes: Zoogeography, Endemics and Conservation

John R. Paxton[1], Gerald R. Allen[2] and Douglass F. Hoese[1]
[1] Australian Museum, 6 College Street, Sydney, N.S.W. 2010 Australia
[2] Western Australian Museum, Francis Street, Perth, W.A. 6000 Australia
E-mail: John.Paxton@austmus.gov.au

INTRODUCTION

In 1790, 11 new species of fishes were described and figured from Botany Bay in White's *Journal of a Voyage to New South Wales*. From that inauspicious beginning, the known Australian fish fauna has grown, through 1,281 species in Macleay (1881a, b, c, d, 1884), 2,023 species in McCulloch (1929–1930), 2,447 species in Whitley (1964), 3,600 estimated species in Paxton *et al.* (1989) to an estimated 4,500+ species today (Hoese *et al.*, in press). Approximately 200 of those species are predominantly freshwater species—see Allen *et al.* (2002) Merrick (2006)—while another 200+ species are recognized as distinct, but have not been described as new or in some cases conclusively identified. The aims of the present chapter are: to summarize the diversity of the named marine fish fauna, presently some 4,135 species, by areas and habitats; to compare the fishes of one habitat, coral reefs, in Australia with those of other selected countries in the Australasian region; to identify areas and groups of endemism; to consider conservation problems of marine fishes; and to identify areas for future research.

Australia has an immense marine environment covering approximately 9 million km^2, with more than 36,000 km of coastline in its Exclusive Economic Zone that includes external territorial islands, but not the Australian Antarctic Territory (Kailola *et al.*, 1993, p.1). Numerous habitats are included, from shallow estuaries to deep sea waters in excess of 4,000 m. Coastal waters include sandy beaches, rocky shores and coral reefs, while the end of the outer shelf and beginning of the open ocean and deep sea are defined by the 200 m depth contour. The Australian continental shelf is approximately 2.2 million km^2 in area and less than 10 km wide off northern N.S.W., southern Qld. and north-west W.A. to more than 300 km wide off northern Australia (Kailola *et al.*, 1993, pp.22–23). Equally significant in terms of fish diversity is the latitudinal extent of the marine environment, from 10° S to 44° S around the continent.

It is not the intent of this article to undertake a detailed analysis of fish distributions on a regional basis in terms of zoogeography; this must await the species compilation with distributions for the second fish volume of the *Zoological Catalogue of Australia* currently nearing completion (Hoese *et al.*, in press). Nonetheless, a brief review of previous works on the subject is of value.

HISTORICAL REVIEW

Whitley's (1932) scheme of marine zoogeographic provinces of Australia, a refinement of Hedley (1926), also used information from fish distributions. He recognized six provinces, three tropical northern (Damperian west from the tip of Cape York to about Geraldton, W.A.; Solanderian east from the tip of Cape York for the Great Barrier Reef proper; and Banksian for the Queensland coast in the same area south to Fraser Island), one subtropical south-eastern (Peronian from Fraser Island to the middle of Bass Strait), and two southern temperate (Flindersian from Geraldton to the middle of Bass Strait and Maugean for Tasmania).

Ekman (1953), in his classic *Zoogeography of the Sea*, mainly followed Clark (1946) based on echinoderm distributions, which in turn was mostly based on an earlier scheme of Hedley (1916). He recognized two tropical provinces (Damperian and Solanderian), one eastern subtropical province (Peronian) and a single warm temperate (South Australian) province. The boundaries were similar to those of Whitley (1932) in the west, with the southern limit of the Damperian near 29° S at the Houtman-Abrolhos Islands and Geraldton on the mainland. A separate Banksian province was not recognized, and the southern limit of the eastern Solanderian was near Sydney on the mainland, and included Lord Howe Island. The southern limit of the subtropical Peronian was also further south to include all of the east coast of Tasmania, while Tasmania's west coast was included in the South Australian province; a separate Maugean province was not recognized.

Mead (1970) presented 'a history' of south Pacific fishes, with an emphasis on historical biogeography. Provinces and boundaries were not the focus, attempting rather to discern common elements or trends in the origins, dispersals and colonization patterns of the fishes. This erudite essay was limited to analyses of the major shore-fish superorders. While Australian distributions were not detailed, the conclusions reached will be of value to future studies on the zoogeography and evolution of Australian fishes.

Briggs' (1974) *Marine Zoogeography* was based primarily on fish distributions, as the author is an ichthyologist. He also recognized two tropical provinces in Australia with a break at the tip of Cape York. He placed the southern boundary of the western Damperian province a little further north at Shark Bay, and considered the eastern tropical area of Queensland to be part of a large Indo-Polynesian province, the southern boundary of which was at the north end of Fraser Island. Separate provinces were recognized if at least 10% of the species were endemic to the area. He recognized a separate tropical Lord Howe–Norfolk province that included Elizabeth and Middleton Reefs, based on a study of Waite (1916) with 22% endemic fishes, plus supporting evidence from echinoderms (Clark, 1946). Briggs did not include subtropical regions as provinces, but recognized two warm temperate provinces: a South-western Australian province from Shark Bay, W.A. to Robe, S.A. and a South-eastern Australian province from Fraser Island, Qld. to Bermagui, N.S.W. Following Bennett and Pope's (1953, 1960) studies based on intertidal invertebrates, plus fish distributions, Briggs recognized a cold temperate area that included the states of Victoria and Tasmania termed the Tasmanian region, which appears to be a higher category than a province.

Wilson and Allen (1987) reviewed the major components and distribution of the Australian marine fauna. Their valuable survey included elements of historical review, as well as summaries of geological history and the effects of ocean currents. Historical zoogeography was emphasized. Distributional patterns for Australian fishes, molluscs, echinoderms and corals were presented, with a series of maps illustrating different zoogeographic schemes. The most general pattern showed only a northern tropical region and a southern temperate region, with eastern and western overlap zones. The concluding map (see Figure 1) illustrated a complex series of 13 overlapping areas that represent the common distribution patterns of shallow water benthic animals. These authors found little evidence, in terms of endemicity, for the recognition of a distinct cool temperate, or Maugean province in Tasmania. They concluded with a brief, but cogent, discussion of province names and boundaries.

Paxton *et al.* (1989) did not analyse the Australian zoogeographic patterns of the 2,000+ fish species in their first volume of the *Zoological Catalogue of Australia*, because it only covered half the fauna. However, they did categorize the 1,780 named marine species as inshore and

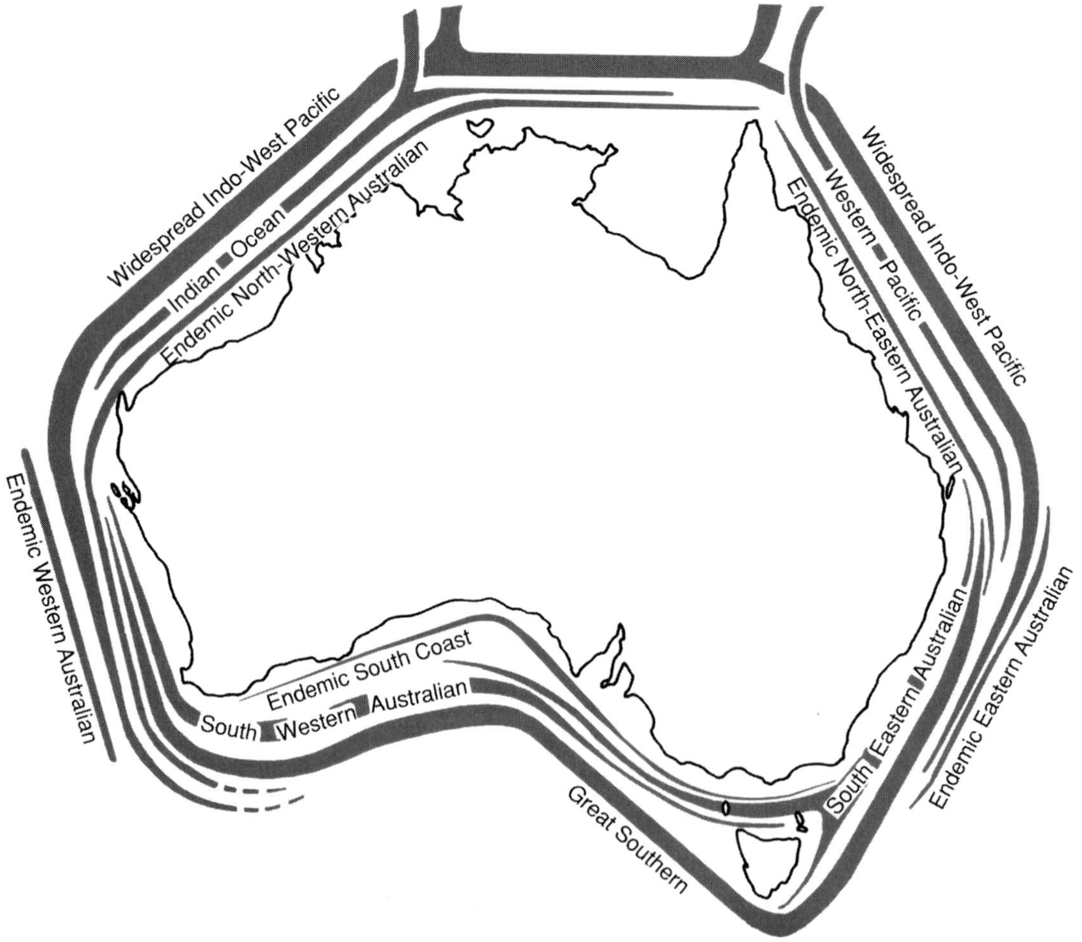

Figure 1. Common distribution patterns of shallow water benthic animals on the Australian coast, showing approximate boundaries of the Indo-West Pacific Region (northern Australian sector) and southern Australian Region and the wide Eastern and Western Overlap Zones between them (from Wilson and Allen, 1987, Figure 3.13). Courtesy Australian Biological Resources Study, Environment Australia.

offshore (beyond the 200 m depth contour) and listed the numbers of each as tropical, temperate, and tropical/temperate, as well as by state. In addition each species was placed in one of 28 mutually exclusive world zoogeographic patterns, such as endemic, introduced, Indo-west Pacific, etc. Inshore fishes made up 69% of the fauna, while tropical species constituted 58%.

As a basis for the establishment of representative marine protected areas (MPAs) for the conservation of biodiversity, Environment Australia is developing an Interim Marine and Coastal Regionalisation for Australia (IMCRA). A recent version (IMCRA, 1998) included nine separate demersal provinces of shelf environment separated by broad transition zones, based on distributions of benthic fishes, sponges, corals, sea grasses and physical characteristics such as bathymetry, sediments, currents, water chemistry and temperature. The provinces include three tropical provinces (North Western and Northern west of Cape York and North Eastern on the east coast), a single subtropical Central Western province, three warm temperate provinces (South Western, Gulfs, and Central Eastern), and two cold temperate provinces (Bassian and Tasmanian).

While not dealing specifically with Australian marine fishes, the following references should be consulted at the beginning of any study on the zoogeography of Australian tropical fishes. Springer (1982) summarized an immense amount of data in a masterful study of Pacific Plate biogeography. Randall (1998) presented a personal review of the zoogeography of Indo-Pacific shore fishes.

SOURCES AND SITES

The numbers of named species and families used in this chapter come from Paxton *et al.* (1989), the first half of the Australian fish fauna covered in the *Zoological Catalogue of Australia,* and the second half recently submitted (Hoese *et al.*, in press). The 200+ species of Australian freshwater fishes that have been excluded include diadromous species that spend part of their life in marine waters (e.g. Anguillidae, Galaxiidae).

Also, the numbers in this chapter only include the species recorded from the continental Australian Exclusive Economic Zone. Fishes from waters around most of the Australian territorial islands have been the subjects of separate publications, as follows: Lord Howe Island (Allen *et al.,* 1976; Francis, 1993); Norfolk Island (Francis, 1993); Elizabeth and Middleton Reefs (Gill and Reader, 1992); Christmas Island (Allen and Steene, 1988); Cocos-Keeling Island (Allen and Smith-Vaniz, 1994); and Maquarie Island (Williams, 1988).

The numbers of coral reef fish species (Table 1) have been determined by counting the individual species in the references noted for each country.

Table 1. World's top 10 countries for coral reef fishes.

Locality	No. spp.	Reference
Indonesia	2,057	Allen and Adrim, 2003
Philippines	1,600	Allen, 2001
Australia	1,573	Hoese *et al.*, in press
Papua New Guinea	1,494	Kailola, 1987–1991; Allen, 1998
Japan	1,315	Masuda *et al.*, 1984
Palau	1,278	Myers, 1999
Taiwan	1,172	Shen *et al.*, 1993
New Caledonia	1,097	Rivaton *et al.*, 1989
Fiji	919	Myers, 1999
Maldives	894	Randall and Anderson, 1993

SURVEYS AND ANALYSES

FAUNAL SUMMARY

With some 4,300+ marine fish species, Australia has one of the largest marine fish faunas in the world, larger than that of Japan (Nakabo, 2002) and perhaps of the same order as those of Indonesia and the Philippines, if the sampling and study efforts in those countries were equivalent. With some 14,500 species in the world found only in the oceans (Nelson, 1994), Australia has about 30% of the total marine fish fauna. Comparisons between countries are not directly equivalent because sampling efforts are uneven, coastal areas vary and latitudinal spreads in each country differ. Australia and Indonesia also have two ocean basins bordering their coasts, while most countries have only one.

The most obvious major divisions of Australia's marine fish fauna are tropical in the north and temperate in the south, as well as inshore and offshore. While most species can be easily categorized (tropical 72%, temperate 29%, inshore 78%, offshore 22%, Hoese *et al.*, in press), there are some with such broad distributions that they fit in both categories. As a result 8% of the species are classified as tropical/temperate, while the total of freshwater + inshore + offshore species is 6% more than the true total due to multiple classifications of some species (Hoese *et al.*, in press).

Offshore fishes include epipelagic species that live in surface waters (< 200 m depth) and deepsea species living below 200 m in benthic and meso- and bathypelagic habitats, as well as those that live above and below the boundary depth. More than 900 species, or more than 20% of Australia's named marine fishes, are offshore species. As a rule offshore species have broader distributions than inshore species, as there are fewer barriers to distribution. For example, 15% of the offshore species of the Australian fish fauna are classified as tropical/temperate, more than twice that of either the freshwater or inshore species. As a result, fewer offshore species, only 10% are endemic to Australia (Hoese *et al.*, in press). This is especially true for pelagic offshore families in Australia, as the Myctophidae with some 100 species has no endemic species. With the exception of south-eastern Australia, most Australian waters below 200 m are seriously undersampled. For instance, all rocky habitats below 200 m, most tropical waters below 200 m, and all waters below 1000 m, except a few localities off Tasmania need further sampling. As a result, detailed analyses of our deepsea fish fauna are premature. Except for a brief comment on conservation of seamount species, the offshore fauna will not be considered further.

Numerous inshore habitats occur in Australia: sandy beaches, rocky reefs, seagrass beds, mud flats, mangroves, the continental shelves, coral reefs. Boundaries between these habitats are not always clear cut, and many species live in two or more habitats at some stage of their lives. The most distinctive of these habitats are coral reefs, which are particularly important in Australia. These fishes are considered in more detail.

CORAL REEF FISHES

For the purpose of this discussion, coral reef fishes are defined as species that are wholly or mainly confined to coral reefs and intermingled habitats (such as sand and rubble patches or weed beds) in depths of less than 60 m. This depth approximates the lower limit of safe scuba diving surveys, and also that of most reef-building corals (Veron, 2000).

Northern Australia is an integral part of the Indo-Australian Archipelago, sometimes referred to as the 'Coral Triangle', especially by conservationists (Werner and Allen, 1998). The region, which includes Australia, Indonesia, Malaysia, Papua New Guinea, and the Philippines, forms the world's richest province for marine biodiversity due to its overwhelming wealth of coral reef habitats (Allen, 1997).

Australia, Indonesia and the Philippines boast the highest number of coral reef fishes in the world, with between 1,500–2,100 species. Accurate figures are lacking for the Philippines. Herre (1953) included about 1,300 coral reef species in his comprehensive checklist of Philippine fishes, but allowing for considerable collecting activity in recent decades the total is probably closer to 1,600 (Allen, 2001). Although extensive surveys have been conducted in both the Philippines and Indonesia, there is a critical need for comprehensive faunal reviews that integrate historical records and more recent collecting efforts.

The world's most speciose coral reef fish faunas are listed in Table 1, based on countries where up to date figures are available. Australia is currently ranked number 3, surpassed only by Indonesia and the Philippines.

Owing to the wide dispersal capabilities of coral reef fishes in general (due to their pelagic larval stage) the faunal composition on Australian coral reefs is similar to that of other regions in the Indo-Pacific. The most abundant families in terms of number of species are gobies (Gobiidae), wrasses (Labridae), damselfishes (Pomacentridae), gropers (Serranidae), blennies (Blenniidae), and cardinalfishes (Apogonidae). The 10 most speciose families, which collectively account for 54% of the fauna, are shown in Figure 2.

The Australian coral reef ichthyofauna is composed mainly of widely distributed elements that are recruited as postlarvae, after a variable pelagic stage. The total species present at a particular locality is ultimately dependent on the availability of food and shelter and the diversity of substrata. Coral and rocky reefs exposed to periodic strong currents are by far the richest habitat in terms of fish diversity. These areas provide an abundance of shelter for fishes of all sizes and the currents are vital for supporting numerous planktivores, most of which provide food for larger predators. Silty bays (often relatively rich in corals), mangroves, seagrass beds, and pure sand-rubble areas are consistently the poorest areas for fish diversity. The highest numbers of fish species are usually recorded at sites incorporating the following features: predominantly coral or rock reef substratum; relatively clear water; periodic strong currents; and presence of additional habitats (sand-rubble, seagrass, mangroves) in close proximity. In other words these sites incorporate a high level of habitat heterogeneity (Allen, 1998).

Endemic Coral Reef Species

Australia belongs to the overall Indo-west Pacific faunal community. Its reef fishes are similar to those inhabiting other areas within this vast region, stretching eastward

Table 2. Coral reef fishes endemic to the combined Australia–New Guinea region.

Family, Genus, Species	Family, Genus, Species
Orectolobidae	**Apogonidae**
Eucrossorhinus dasypogon	*Apogon brevicaudata*
Orectolobus ornatus	*Apogon ruppellii*
Hemiscylliidae	**Nemipteridae**
Hemiscyllium ocellatum	*Pentapodus porosus*
Carcharhinidae	**Haemulidae**
Carcharhinus cautus	*Plectorhinchus unicolor*
Antennariidae	**Chaetodontidae**
Tathicarpus butleri	*Chaetodon aureofasciatus*
Atherinidae	**Labridae**
Hypoatherina tropicalis	*Choerodon monostigma*
Syngnathidae	*Choerodon vitta*
Haliichthys taeniophorus	*Cirrhilabrus condei*
Scorpaenidae	**Blenniidae**
Cheroscorpaena tridactyla	*Ecsenius aequalis*
Serranidae	**Tripterygiidae**
Pseudanthias engelhardi	*Enneapterygius larsonae*
Pseudochromidae	**Gobiidae**
Congrogadus amplimaculatus	*Gobiodon unicolor*
Congrogadus malayanus	**Microdesmidae**
Plesiopidae	*Aioliops novaeguineaensis*
Assessor macneilli	**Ephippidae**
Opistognathidae	*Zabidus novemaculeatus*
Opistognathus papuensis	

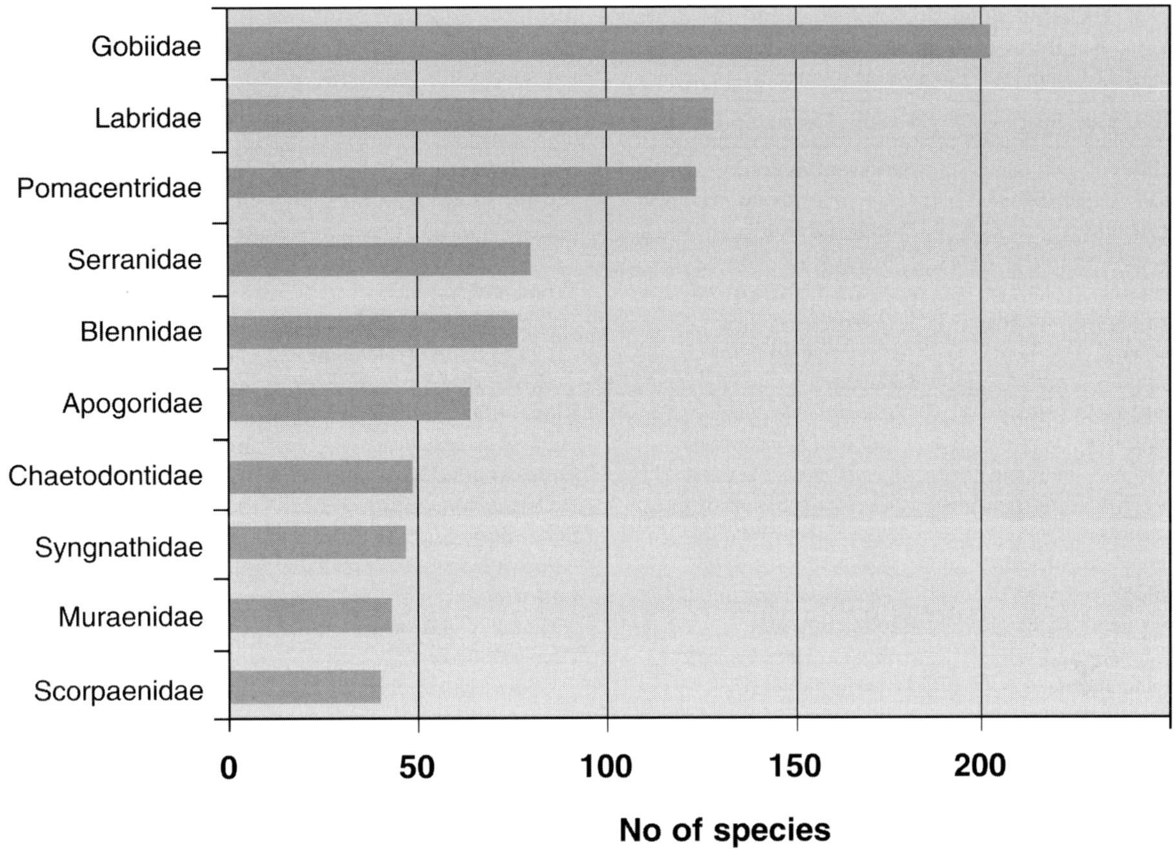

Figure 2. Australia's 10 most speciose coral reef fish families.

from East Africa and the Red Sea to the islands of Micronesia and Polynesia. Although most families and many genera are consistently present across the region, the species composition varies greatly according to locality.

While a significant portion of Australia's coral reef fishes range widely across the Indo-west Pacific, from East Africa to Oceania, most species occur in neighbouring regions such as New Guinea and Indonesia. The coral reef fish fauna is most closely linked with that of its nearest neighbour, New Guinea (Papua New Guinea and the Indonesian province of Irian Jaya). They share numerous wide-ranging species, as well as a select group of 26 species that are endemic to the combined Australia–New Guinea region (Table 2).

While the high proportion of endemism in Australian temperate fishes is well known (Wilson and Allen, 1987), there is also a significant tropical endemic element consisting of 101 species or about 6.4% of the overall coral fish fauna. The largest contingent of endemics occurs in north-western seas or the region between the Abrolhos Islands, Western Australia and the Gulf of Carpentaria, Queensland, with 49 species (Table 3.) The less extensive region covered by Australia's eastern coast supports 29 endemics (Table 4). In addition, 23 species are more or less widely distributed across both regions (Table 5; Figure 3).

The families with the most number of endemic Australian coral fish species are Labridae (12–13), Blenniidae (9), Pomacentridae (9), Pseudochromidae (10), Gobiidae (7), Opistognathidae (6), and Syngnathidae (6).

Tropical endemic Australian fishes are not restricted to coral reefs. Endemic species in families such as Gobiidae and Syngnathidae are found in other tropical habitats as well. When the endemic totals are finalized, those of other tropical habitats will probably exceed, on a percentage basis, the coral reef endemics.

ENDEMIC TEMPERATE FISHES

While the number of endemic coral reef species is somewhat surprising, the high number of temperate Australian endemic fishes has been well documented. Wilson and Allen (1987, p.51) estimate 96% of the temperate inshore species are either endemic (85%) or shared with New Zealand (11%). Hoese *et al.* (in press) include 78% endemic and 19% shared with New Zealand for the temperate inshore fishes, as the offshore shelf species above 200 m are included in the inshore total. Some 24% of Australian marine fish species are now considered endemic.

Of the eight endemic Australian fish families (Paxton *et al.*, 1989), three are freshwater and all the rest are temperate marine families. These five families include the monotypic Dinolestidae, Enoplosidae and Gnathanacanthidae, as well as the Brachionichthyidae with four named species and the

Table 3. Endemic coral reef fishes of north-western Australia (from the Abrolhos Islands to the western coast of Cape York Peninsula).

Family, Genus, Species	Family, Genus, Species
Narcinidae	**Pomacanthidae**
Narcine westraliensis	*Chaetodontoplus personifer*
Gobiesocidae	**Pomacentridae**
Lepadichthys sandaracatus	*Amphiprion rubrocinctus*
Muraenidae	*Dischistodus darwiniensis*
Gymnothorax mccoskeri	*Pomacentrus limosus*
Plotosidae	*Pomacentrus milleri*
Paraplotosus butleri	**Labridae**
Paraplotosus muelleri	*Anampses lennardi*
Syngnathidae	*Choerodon cauteroma*
Acentronura larsonae	*Choerodon rubescens*
Choeroichthys latispinosus	*Cirrhilabrus morrisoni*
Scorpaenidae	*Cirrhilabrus randalli*
Dampierosa daruma	*Conniella apterygia*
Serranidae	*Thalassoma septemfasciatum*
Epinephelus bilobatus	**Pholidichthyidae**
Pseudanthias sheni	*Pholidichthys anguis*
Pseudochromidae	**Tripterygiidae**
Assiculoides desmonotus	*Enneapterygius namarrgon*
Assiculus punctatus	**Blenniidae**
Labracinus lineatus	*Ecsenius alleni*
Pseudochromis howsoni	*Cirripectes alleni*
Opistognathidae	*Cirripectes hutchinsi*
Opistognathus darwiniensis	*Meiacanthus naevius*
Opistognathus inornatus	**Callionymidae**
Opistognathus reticulatus	*Pterosynchiropus occidentalis*
Stalix flavida	**Gobiidae**
Apogonidae	*Eviota bimaculata*
Archamia melasma	*Eviota inutilis*
Pterapogon mirifica	*Feia notata*
Nemipteridae	*Stonogobius larsonae*
Pentapodus vitta	*Sueviota atrinasa*
Scaevius milii	**Siganidae**
Chaetodontidae	*Siganus trispilos*
Chaetodon assarius	
Chelmon marginalis	

Pataecidae with three species. All families are represented in the southern states of Tasmania, Victoria, South Australia and Western Australia. Other temperate families with a number of endemic species include the Odacidae, with eight of 10 Australian species endemic and the other two in New Zealand. Other families endemic to Australia and New Zealand together include the Arripidae, Chironemidae, and Leptoscopidae (Paxton *et al.*, 1989).

The widespread families Syngnathidae, Monacanthidae and Antennariidae have undergone significant radiation in temperate Australia, with numerous endemic genera and species (Wilson and Allen, 1987). Other families with significant numbers of Australian species include the Urolophidae, Trachichthyidae, Platycephalidae, Sillaginidae, Pempheridae, Pentacerotidae and Uranoscopidae.

Table 4. Endemic coral reef fishes of eastern Australia (from Cape York south to the Capricorn Group, Great Barrier Reef).

Family, Genus, Species	Family, Genus, Species
Syngnathidae	**Pomacentridae**
Campichthys tryoni	*Chromis nitida*
Corythoicthys paxtoni	*Parma oligolepis*
Doryrhamphus negrosensis malus	*Pomacentrus wardi*
Pseudochromidae	*Pomacentrus australis*
Ogilbyina novaehollandiae	*Stegastes apicalis*
Ogilbyina queenslandiae	**Labridae**
Pseudochromis flammicauda	*Choerodon frenatus*
Plesiopidae	*Choerodon venustus*
Plesiops genaricus	*Coris aurilineata*
Opistognathidae	**Pinguipedidae**
Opistognathus eximus	*Parapercis stricticeps*
Serranidae	**Blenniidae**
Epinephelus undulostriatus	*Ecsenius australianus*
Lethrinidae	*Ecsenius mandibularis*
Gymnocranius audleyi	*Ecsenius tigris*
Chaetodontidae	*Meiacanthus lineatus*
Chaetodon rainfordi	*Petroscirtes fallax*
Pomacanthidae	**Tripterygiidae**
Chaetodontoplus meredithi	*Ceratobregma acanthops*
	Gobiidae
	Trimma necopinna
	Microdesmidae
	Aioliops tetrophthalmus

CONSERVATION ISSUES

Marine fishes contribute significantly to the gross national product of Australia. Descriptions of the most important commercial species are provided in Kailola *et al.* (1993), while Caton (2002) described the current status of most fisheries. The vast majority of Australia's fisheries are considered overfished, fully fished or of uncertain status.

Clearly overfishing will eventually result in a threat to biodiversity. Until the last decade, conservation of fish biodiversity dealt with the serious problems facing freshwater fishes where significant numbers of extinctions have occurred (Almada-Villela, 1998). More recently attention has focused on threats to marine fishes. Pogonoski *et al.* (2002) completed a first survey of threatened marine fishes in Australia. In this overview, 114 species were analysed in detail for their conservation status. Species were chosen for analysis based on their presence in state or overseas lists, evidence of decline, or nomination by Australian fish or fisheries experts.

Of the 114 species analysed, 17 species were placed in the legally threatened categories of Critically Endangered (3), Endangered (6) and Vulnerable (8). Other International Union for the Conservation of Nature (IUCN) categories included Lower Risk conservation dependent (16), Lower Risk near threatened (13), Data Deficient (53) and Lower Risk least concern (15). An additional 19 species of sharks and rays, included in the IUCN Species Survival Commission's forthcoming publication on the conservation status of chondrichthyan fishes (Fowler *et al.*, in press), were very briefly analysed and determined to be Lower Risk least concern in Australian waters. The 17 species at highest risk in Australia include 7 sharks, 6 rays (including 3 sawfishes), 3 handfishes of the family Brachionichthyidae, and one rockcod of the family Serranidae. The 3 Critically Endangered species include a carcharhinid shark, a pristid sawfish and a handfish.

Sharks and rays world-wide make up a disproportionate number of threatened species because of their reproductive mode of internal fertilization and resultant low number of young (Camhi *et al.*, 1998). In Australia and other areas they are target species of both commercial and recreational fishing, caught as bycatch in other fisheries, and killed in protective beach-meshing programmes. The increasing demands for fins, flesh, jaws and cartilage have put a number of these species at risk.

The endemic handfish family Brachionichthyidae includes a few small, benthic species with limited distributions and reproductive potential. An introduced starfish has been implicated in egg predation and habitat modification for one species, although development and water pollution also may be involved (Bruce *et al.*, 1999).

Threatening processes for our marine species include overfishing by both commercial and recreational fishers (including non-target bycatch), habitat degradation (water pollution, sedimentation, trawling, dredging, development impacts) and exotic species introductions. Species with reduced reproductive potential and/or restricted distributions are most likely to be effected by these threatening processes. Previously commercial species were considered unlikey to be at risk of extinction due to overfishing, on the premise that the economics of fishing would result in fishery closure before the target species became threatened. However, the extremely high prices paid for select species, like the southern bluefin tuna, could result in continued fishing for small numbers of individuals. Increasingly commercial fishing is seen as a conservation threat to vulnerable species. Previous assumptions that commercial fishes would recover rapidly from prolonged declines have also been questioned (Hutchings, 2000, 2001).

Endemic species should be those of greatest concern, for if they are threatened in Australia there is no other population as a backup. The numerous species that currently are listed as Data Deficient require study, as the lack of information may hide species under threat. Future research on Australian fish conservation should survey all of our endemic species in a risk assessment analysis.

SEAMOUNTS

Marine environments below 200 m generally are not subjected to the level of human impacts that inshore habitats sustain, and the fishes below 200 m generally have broader distributions than their inshore counterparts. However, the faunas of seamounts do not appear to follow these generalizations. Richer de Forges *et al.* (2000) reported potential endemism levels of 29–34% based on the number of new species found in 850+ species of macro-invertebrates and fishes from 24 seamounts on the Norfolk Ridge, Lord Howe Rise and south of Tasmania. The number of fish species was not mentioned in this brief article. However, Koslow and Gowlett-Holmes (1998) listed 36 species of fishes and 242 species of invertebrates from the 14 seamounts south of Tasmania. Of those, three species were mesopelagic fishes caught over the seamount. Of the remaining 33 species, 14 were new records from Australia and 12 species, or 36% of the benthic

Table 5. Endemic coral reef fishes that are widely distributed in northern Australia.

Family, Genus, Species	Family, Genus, Species
Orectolobidae	**Apogonidae**
Orectolobus wardi	*Apogon pallidofasciatus*
Hemiscylliidae	*Gymnapogon anona*
Hemiscyllium trispeculare	**Haemulidae**
Carcharhinidae	*Plectorhinchus multivittatus*
Carcharhinus fitzroyensis	**Pempheridae**
Carcharhinus tilstoni	*Pempheris ypsilychnus*
Muraenidae	**Chaetodontidae**
Gymnothorax longinquus	*Chelmon muelleri*
Syngnathidae	**Labridae**
Choeroichthys suillus	*Choerodon cephalotes*
Serranidae	*Choerodon cyanodus*
Rainfordia opercularis	*Choerodon sugillatum*
Pseudochromidae	**Tripterygiidae**
Blennodesmus scapularis	*Enneapterygius clea*
Pseudochromis quinquedentatus	*Enneapterygius gracilis*
Pseudochromis wilsoni	**Callionymidae**
Opistognathidae	*Calliurichthys grossi*
Opistognathus latitabundus	**Gobiidae**
	Valenciennea alleni

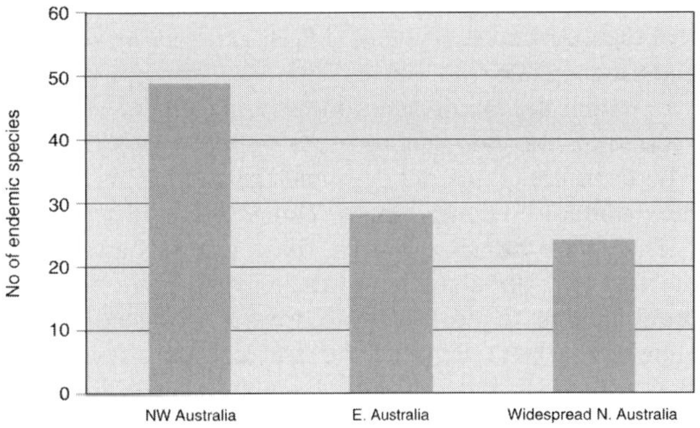

Figure 3. Regional distribution of Australia's endemic coral reef fishes.

seamount fish species, were either confirmed (2) or possible (10) undescribed species and potential endemics.

Seamounts around the world are commercially fished (Rogers, 1994) and the damage to the benthic epifaunal community by bottom trawling is significant (Koslow and Gowlett-Holmes, 1998). Protection of selected seamounts from bottom trawling, as has been done for the 14 Tasmanian seamounts, is justified by the levels of endemism found and the long-term damage done to the epifauna by bottom trawling.

MARINE PROTECTED AREAS

When numerous species are data deficient in terms of abundance and distribution, such as many invertebrates and small, cryptic fishes (e.g. gobiids, blennids, muraenids), taxon-based conservation measures are likely to be inadequate. Establishment of marine protected areas (MPAs) such as marine parks and reserves is one of the best alternative solutions (Randall, 1982). Placements of these protected areas are often centred on 'hot spots', areas with the greatest number of endemic species and/or overall diversity (Roberts *et al*., 2002). Although it is a relatively new idea for marine conservation, the 'hot spot' concept is rapidly gaining popularity as an important priority-setting tool for terrestrial ecosystems (Myers *et al*., 2000; Flemons and Cassis, 2006).

Within the marine environment, placement of MPAs centred on hot spots will be of less value in those regions where currents transport eggs and larvae to areas outside their normal breeding ranges. Unless the adults migrate back to their breeding ranges, these vagrants are lost to the breeding population, and their inclusion in an MPA does nothing to protect those species from extinction. Hutchins (1991) detailed the tropical fish species in Western Australia transported south by the Leeuwin Current and provided examples of tropical species carried into southern N.S.W. by the East Australian Current. Based on a survey of Sydney Harbour captures in the 1970s (Paxton, 2001, unpublished data), most east coast juvenile tropical fishes arrive in the summer and disappear by the middle of winter. Predation by local predators when metabolism slows due to lowered temperatures is probably the most important factor in their disappearance. These vagrant individuals clearly are lost from the gene pool of the species, and their subtropical/temperate distribution records should not be utilized when attempting to determine the placement of marine protected areas. It is possible to discount vagrant distributions based on size of individuals and temporal variation of occurrences; however, information on the reproductive distributions of Australian marine fishes, as distinct from non-breeding distributions, will be a better basis for the establishment of future protected areas for conservation. Greater emphasis on the description of breeding distributions is needed in future studies of fish ranges.

DISCUSSION AND CONCLUSIONS

Detailed analyses of both historical biogeography and faunal boundaries of Australian marine fishes have yet to be done. It is only in the last decade that increased collecting of shore fishes, begun in 1965, has provided the databases in the various Australian fish collections to allow meaningful results. An example of spurious results due to incomplete data is evident in Briggs (1974). Briggs recognized a separate Lord Howe–Norfolk province, citing the study of Waite (1916) with 22% endemic species. Allen *et al.* (1976) more than doubled the known species from Lord Howe Island and listed 3.8% of the shore fishes as endemic (including those species also known from Norfolk Island), while Randall (1976) found 12% endemic species when Lord Howe, Norfolk, and Elizabeth and Middleton Reefs were combined. Francis (1993) decreased the level of endemism to less than 5%, even with the Kermadec Islands added, due to taxonomic revisions and additional collections from those islands and extralimitally.

While some gaps in the details of individual distributions remain, particularly in the far north, the Australian shore fish fauna is better known today than ever before. However, that does not mean that an analysis of faunal boundaries will be without problems. Whittaker (1967) emphasized that individual species have independent limiting factors and the distribution limits of numerous species are a continuum. Boundaries are discernible only where environmental parameters rapidly change so the distributional limits of a number of species coincide. The community/continuum argument is still debated by botanists (e.g. Austin and Smith, 1989), but seems rarely considered in zoogeographic studies. As Wilson and Allen (1987, p.66) stated 'The problem with these classifications is that the location of the boundaries between the provinces varies according to the group of organisms studied and that there is so much overlap that exceptions outnumber the species which obey the rules.' Francis (1993) described additional problems with locating faunal boundaries on the basis of endemics, including clinal variation in faunal species numbers and annual variations in the distributions of vagrants. While faunal boundaries may have a role in conservation, as a guide to the distribution of MPAs given the lack of other information, understanding present distributions in terms of ecology and evolution should prove more meaningful.

Most Australian fish communities below 50 m have been so undersampled that analyses would not be very productive. For instance, Williams *et al.* (1996) reported on the first survey of W.A. slope fishes to 1,500 m, based on 150 bottom trawls. Of 388 species recorded, some 100 species or more than 25% were new records from Australia or undescribed species. Clearly more sampling is required before detailed analytical studies will be worth the effort.

ACKNOWLEDGEMENTS

Rudie Kuiter caught most of the Sydney Harbour tropical fishes discussed under Conservation Issues above. Jeffrey Leis and Winston Ponder aided with the references. The Australian Biological Resources Study (ABRS), Canberra kindly supplied a copy of Figure 1. Martin Gomon improved the manuscript with a thorough review. We are grateful to all.

REFERENCES

Allen, G.R. 1997. *Marine Fishes of Tropical Australia and South-east Asia.* Western Australian Museum, Perth. 292 pp

Allen, G.R. 1998. Reef and shore fishes of Milne Bay Province, Papua New Guinea. Pp.39–49. In Werner, T.B and Allen, G.R (eds). *A Rapid Biodiversity Assessment of the Coral Reefs of Milne Bay Province, Papua New Guinea.* Rapid Assessment Program Working Papers, No. 11. Conservation International, Washington, DC. 109pp.

Allen, G.R. 2001. Reef and shore fishes of the Calamianes Islands, Palawan Province, Philippines. Pp.31–44. In Werner, T.B and Allen, G.R. (eds). *A Rapid Biodiversity Assessment of the Coral Reefs of the Calamianes Islands, Palawan Province, Philippines.* Rapid Assessment Program Bulletin of Biological Assessment, Conservation International, Washington, DC. 127pp.

Allen, G. R. and Adrim, M. 2003. Coral reef fishes of Indonesia. *Zoological Studies (Taiwan),*Vol. 42, pp.1–72.

Allen, G.R., Hoese, D.F., Paxton, J.R., Randall, J.E., Russell, B.C., Starck, W. A., Talbot, F.H. and Whitley, G.P. 1976. Annotated checklist of the fishes of Lord Howe Island. *Records of the Australian Museum,* Vol. 30, pp.365–454.

Allen, G.R., Midgely, S.H. and Allen, M. 2002. *Field Guide to Freshwater Fishes of Australia.* Western Australian Museum, Perth. 394pp.

Allen, G.R. and Smith-Vaniz, W.F. 1994. Fishes of Cocos (Keeling) Islands. *Atoll Research Bulletin,* No. 412, pp.1–21.

Allen, G.R. and Steene, R.C. 1988. *Fishes of Christmas Island, Indian Ocean.* Christmas Island Natural History Association, Christmas Island. 197pp.

Almada-Villela, P. C. 1998. Endangered species. Pp.48–53. In Paxton, J. R. and Eschmeyer, W. N. (eds). *Encyclopedia of Fishes.* (Edit. 2). Academic Press, San Diego. 240pp.

Austin, M.P. and Smith, T.M. 1989. A new model for the continuum concept. *Vegetatio,* Vol. 83, pp.35–47.

Bennett, I. and Pope, E.C. 1953. Intertidal zonation of the exposed rocky shores of Victoria, together with a rearrangement of the biogeographical provinces of temperate Australian shores. *Australian Journal of Marine and Freshwater Research,* Vol. 4, pp.105–159.

Bennett, I. and Pope, E.C. 1960. Intertidal zonation of the exposed rocky shores of Tasmania and its relationship with the rest of Australia. *Australian Journal of Marine and Freshwater Research,* Vol. 11, pp.182–221.

Briggs, J.C. 1974. *Marine Zoogeography.* McGraw-Hill, New York. 475pp.

Bruce, B.D., Green, M.A. and Last, P.R. 1999. Aspects of the biology of the endangered spotted handfish (*Brachionichthys hirsutus*: Lophiiformes) off southern Australia. Pp.369–380. In Seret, B. and Sire, J.-T. (eds). *Proceedings of the 5th Indo Pacific Fish Conference, Noumea 1997.* Societe Francaise d'Ichthyologie and Institut de Recherche pour Developpement, Paris. 866pp.

Camhi, M., Fowler, S., Musick, J., Brautigam, A. and Fordham, S. 1998. Sharks and their relatives. Ecology and conservation. *Occasional Paper of the IUCN Species Survival Commission,* No. 20, pp.1–39.

Caton, A. (ed.). 2002. *Fishery Status Reports, 2000–2001.* Bureau of Rural Sciences, Canberra. 252pp.

Clark, H.L. 1946. The echinoderm fauna of Australia. *Publications of the Carnegie Institute,* No. 566, pp.1–567.

Ekman, S. 1953. *Zoogeography of the Sea.* Sidgwick and Jackson, London. 417pp.

Flemons, P. and Cassis, G. 2006. The role of GIS in Systematic Conservation Planning. Pp.873–887. In Merrick, J.R., Archer, M., Hickey, G.M. and Lee, M.S.Y. (eds). *Evolution and Biogeography of Australasian Vertebrates.* Auscipub Pty. Ltd., Sydney. 942pp.

Fowler, S.L., Camhi, M., Burgesss, G.H., Cailliet, G., Fordham, S.V., Cavanagh, R.D., Simpfendorfer, C.A. and Musick, J.A. *Sharks, Rays and Chimaeras: the Status of the Chondrichthyan Fishes.* IUCN [International Union for the Conservation of Nature] SSC [Species Survival Commission] Shark Specialist Group. IUCN, Gland, Switzerland. (in press).

Francis, M.P. 1993. Checklist of the coastal fishes of Lord Howe, Norfolk, and Kermadec Islands, southwest Pacific Ocean. *Pacific Science,* Vol. 47, pp.136–170.

Gill, A.C. and Reader, S.E. 1992. Fishes. Pp.90–93, 193–228. In Hutchings, P.A. (ed). *Reef Biology, a Survey of Elizabeth and Middleton Reefs, South Pacific. Kowari* 3. Australian National Parks and Wildlife Service, Canberra.

Hedley, C. 1916. Mollusca. *Australasian Antarctic Expedition 1911/14, Science Report C,* Vol. 4, pp.1–80.

Hedley, C. 1926. Zoogeography. P.743. In *Australian Encyclopedia.* Vol. 2. Angus and Robertson, Sydney.

Herre, A.W. 1953. Check list of Philippine fishes. *United States Fish and Wildlife Service Research Report,* No. 20, pp.1–977.

Hoese, D.F., Bray, D.J., Allen, G.R., Allen, C., Cross, N. and Paxton, J.R. *Zoological Catalogue of Australia. Vol. 7, Pisces Part 2.* Australian Biological Resources Study, Canberra. (in press).

Hutchings, J.A. 2000. Collapse and recovery of marine fishes. *Nature,* Vol. 406, No. 6798, pp.882–885.

Hutchings, J.A. 2001. Conservation biology of marine fishes: perceptions and caveats regarding assignment of extinction risk. *Canadian Journal of Fisheries and Aquatic Sciences,* Vol. 58, pp.108–121.

Hutchins, J.B. 1991. Dispersal of tropical fishes to temperate seas in the southern hemisphere. *Journal of the Royal Society of Western Australia,* Vol. 74, pp.79–84.

IMCRA Technical Group. 1998. *Interim Marine and Coastal Regionalisation for Australia: An Ecosystem-based Classification for Marine and Coastal Environments.* Version 3.3. Environment Australia, Commonwealth Department of the Environment, Canberra. 114pp.

Kailola, P.J. 1987–1991. The fishes of Papua New Guinea. A revised and annotated checklist. Vol. 1. Myxinidae to Synbranchidae (1987), Vol 2. Scorpaenidae to Callionymidae (1987), Vol. 3. Gobiidae to Molidae (1991). *Papua New Guinea Department of Fisheries, Marine Resources Bulletin,* Vol. 41, pp.1–572.

Kailola, P.J., Williams, M.J., Stewart, P.C., Reichelt, R.E., McNee, A. and Grieve, C. 1993. *Australian Fisheries Resources.* Bureau of Resource Sciences and Fisheries Research and Development Corporation, Canberra. 422pp.

Koslow, J.A. and Gowlett-Holmes, K. 1998. *The Seamount Fauna off Southern Tasmania: Benthic Communities, their Conservation and Impacts of Trawling.* Fisheries Research and Development Corporation and Environment Australia Final Report, FRDC Project 95/058. 104pp.

Macleay, W. 1881a. Descriptive catalogue of the fishes of Australia. Part I. *Proceedings of the Linnean Society of New South Wales,* Series 1, Vol. 5, Pt. 3, pp.302–444.

Macleay, W. 1881b. Descriptive catalogue of the fishes of Australia. Part II. *Proceedings of the Linnean Society of New South Wales,* Series 1, Vol. 5, Pt. 4, pp.510–529.

Macleay, W. 1881c. Descriptive catalogue of the fishes of Australia. Part III. *Proceedings of the Linnean Society of New South Wales,* Series 1, Vol. 6, Pt. 1. pp.1–138.

Macleay, W. 1881d. Descriptive catalogue of the fishes of Australia. Part IV. *Proceedings of the Linnean Society of New South Wales,* Series 1, Vol. 6, Pt. 2, pp.202–387.

Macleay, W. 1884. Supplement to the descriptive catalogue of the fishes of Australia. *Proceedings of the Linnean Society of New South Wales,* Series 1, Vol. 9, Pt. 1, pp.2–64.

Masuda, H., Amaoka, K., Araga, C., Uyeno, T. and Yoshino, T. 1984. *The Fishes of the Japanese Archipelago.* 3 Vols. Tokai University Press, Tokyo. 456pp.

McCulloch, A.R. 1929–1930. A checklist of the fishes recorded from Australia. *Memoirs of the Australian Museum,* No. 5, Pts. 1–4, pp.1–534.

Mead, G.W. 1970. A history of South Pacific fishes. Pp.236–251. In Wooster, W. S. (ed). *Scientific Exploration of the South Pacific.* National Academy of Sciences, Washington DC. 257pp.

Merrick, J.R. 2006. Australasian freshwater fish faunas: diversity, interrelationships, radiations and conservation. Pp.195–224. In Merrick, J.R, Archer, M., Hickey, G.M. and Lee. M.S.Y. (eds). *Evolution and Biogeography of Australasian Vertebrates.* Auscipub Pty. Ltd., Sydney. 942pp.

Myers, N., Mittermeier, R.A., Mittermeier, C.G., da Fonseca, G.A.B. and Kent, J. 2000. Biodiversity hotspots for conservation priorities. *Nature,* Vol. 403, No. 6772, pp.853–858.

Myers, R. F. 1999. *Micronesian Reef Fishes.* Coral Graphics, Guam. 298pp.

Nakabo, T. (ed.) 2002. *Fishes of Japan with Pictorial Keys to the Species, English Edition 2.* 2 vols., Tokai University Press, Tokyo. 1719pp.

Nelson, J.S. 1994. *Fishes of the World.* (Edit. 3). John Wiley and Sons, New York. 600pp.

Paxton, J.R., Hoese, D.F., Allen, G.R. and Hanley, J.E. 1989. *Zoological Catalogue of Australia. Vol. 7. Pisces. Petromyzontidae to Carangidae.* Australian Government Publishing Service, Canberra. 665pp.

Pogonoski, J.J., Pollard, D.A. and Paxton, J.R. 2002. *Conservation Overview and Action Plan for Australian Threatened and Potentially Threatened Marine and Estuarine Fishes.* Environment Australia, Canberra. 375pp.

Randall, J.E. 1976. The endemic shore fishes of the Hawaiian Islands, Lord Howe Island and Easter Island. *Travaux et Document ORSTOM,* No. 47, pp.49–73.

Randall, J.E. 1982. Tropical marine sanctuaries and their significance in reef fisheries research. *National Oceanic and Atmospheric Administration Technical Memorandum NMFS-SEFC,* No. 80, pp.167–177.

Randall, J.E. 1998. Zoogeography of shore fishes of the Indo-Pacific region. *Zoological Studies (Taiwan),* Vol. 37, pp.227–268.

Randall, J.E. and Anderson, C.R. 1993. Annotated checklist of the epipelagic and shorefishes of the Maldive Islands. *J. L. B. Smith Institute of Ichthyology Bulletin,* No. 59, pp.1–47.

Richer de Forges, B., Koslow, J.A. and Poore, G.C.B. 2000. Diversity and endemism of the benthic seamount fauna. *Nature,* Vol. 405, No.6789, pp.944–947.

Rivaton, J., Fourmanoir, P., Bourret, P. and Kulbicki, M. 1989. Catalogue des poissons de Nouvelle-Calédonie. *Catalogues Sciences de la Mer Biologie Marine.* ORSTOM, Centre de Noumea. 170pp.

Roberts, C.M., McClean, C.J., Veron, J.E.N., Hawkins, J.P., Allen, G.R., McAllister, D.E., Mittermeier, C.G., Schueler, F.W., Spalding, M., Wells, W., Vynne, W. and Werner, T.B. 2002. Marine biodiversity hotspots and conservation priorities for tropical reefs. *Science,* Vol. 295, No. 8885, pp.1280–1284.

Rogers, A.D. 1994. The biology of seamounts. *Advances in Marine Biology,* Vol. 30, pp.305–350.

Shen, S-C., Shao, K.T. Lee, S.C., Mok, H.K., Chen, C.T. and Chen, C.H. 1993. *Fishes of Taiwan.* National Taiwan University Press, Taipei. 960pp.

Springer, V.G. 1982. Pacific Plate biogeography, with special reference to shorefishes. *Smithsonian Contributions to Zoology,* No. 367, pp.1–182.

Veron, J.E.N. 2000. *Corals of the World.* 3 Vols. Australian Institute of Marine Science, Townsville. 1332pp.

Waite, E.R. 1916. A list of the fishes of Norfolk Island, and indication of their range to Lord Howe Island, Kermadec Island, Australia, and New Zealand. *Transactions of the Royal Society of South Australia,* Vol. 40, pp.452–458.

Werner, T.B. and Allen, G.R. (eds). 1998. *A Rapid Biodiversity Assessment of the Coral Reefs of Milne Bay Province, Papua New Guinea.* Rapid Assessment Program Working Papers, No. 11. Conservation International, Washington, DC. 109pp.

White, J. 1790. *Journal of a Voyage to New South Wales.* Debrett, London. 299pp.

Whitley, G.P. 1932. Marine zoogeographical regions of Australia. *Australian Naturalist,* Vol. 8, pp.166–167.

Whitley, G.P. 1964. A survey of Australian ichthyology. *Proceedings of the Linnean Society of New South Wales,* Vol. 89, pp.11–127.

Whittaker, R.H. 1967. Gradient analysis of vegetation. *Biological Reviews,* Vol. 42, pp.207–264.

Williams, A., Last, P.R., Gomon, M.F. and Paxton, J.R. 1996. Species composition and checklist of the demersal ichthyofauna of the continental slope off Western Australia (20–35°S). *Records of the Western Australian Museum,* Vol. 18, pp.135–155.

Williams, R. 1988. The nearshore fishes of Macquarie Island. *Papers and Proceedings of the Royal Society of Tasmania,* Vol. 122, Pt. 1, pp.233–245.

Wilson, B.R. and Allen, G.R. 1987. Major components and distribution of marine fauna. Pp.43–68. In Dyne, G.R. and Walton, D.W. (eds). *Fauna of Australia, Vol. 1A. General Articles.* Australian Government Publishing Service, Canberra. 339pp.

Chapter 10
Australasian Freshwater Fish Faunas: Diversity, Interrelationships, Radiations and Conservation

John R. Merrick
Graduate School of the Environment, Macquarie University, North Ryde, N.S.W. 2109 Australia
E-mail: jmerrick@gse.mq.edu.au

INTRODUCTION

As Paxton *et al.* (2006) have already explained, the marine fish faunas of Australasia are among the most diverse in the world (pp.185–194), but the main freshwater ichthyofaunas (Australia, New Guinea and New Zealand) are all depauperate relative to freshwater fish faunas elsewhere (Filewood, 1984; Flannery, 1984; Paxton, 1994: Allen *et al.,* 2002).

DEPAUPERATE BUT DIVERSE

Australia had been estimated to have 180–190 native species (Merrick and Schmida, 1984, p.13; Allen, 1989; McDowall, 1996b, p.9) and, more recently, Allen *et al.* (2002) have reported over 200 species. Many of the 214 species reported from New Guinea (Allen, 1991, p.10) also occur in northern Australia, but only 27–37 non-marine species have been recorded from New Zealand (Flannery, 1984; McDowall, 1990; National Institute of Water and Atmospheric Research, 2001). By contrast, 1400 species have been estimated for African freshwaters (Allen *et al.,* 2002, p.2), over 2000 species are estimated for the Amazon Basin (Allen, 1991, p.9), the Mekong River system in South-east Asia is thought to support between 500–1200 species (Dudgeon, 2000) and Borneo is reported to have 340 freshwater species (van Oosterzee, 2006).

Although low in numbers of species, the taxa present in Australasia are diverse ranging from jawless lampreys, to sharks and the sawfish, lungfish, gobies, minnows and soles. Only a few are considered primary freshwater fishes (i.e. having completed their evolution in freshwater); all the others are more recent marine invaders (Bond, 1979, pp.249–250, 261–262; Allen *et al.*, 2002, p.7). The timing of freshwater fish invasions is unknown and, as explained previously, a number of marine transgressions are known to have occurred since mid Cretaceous times (Holdaway and Worthy, 2006; White, 2006). Indeed, Crowley (1990a, b) hypothesized invasion patterns related to these marine inundations and subsequent regressions. For example, the divergent stocks within the atherinid genus *Craterocephalus* were thought to have resulted from two separate invasions (mid Cretaceous and Oligocene–early Miocene) of ancestral marine stocks (Crowley, 1990a); it was suggested that other groups, such as percichthyids, probably invaded in early Tertiary, whereas ambassids and ariids were thought to have arrived in more recent Pliocene–Pleistocene times (Crowley, 1990b).

However, regardless of arrival times, it is clear that the unique characteristics of Australasian species are at least partly related to the relative isolation of the region for over 40 million years, as well as the geologic and climatic conditions prevailing during that time (Frakes, 1999; Heinsohn and Hope, 2006; Holdaway and Worthy, 2006; White, 2006). For a summary see the front endpaper table.

ORIGINS AND ZOOGEOGRAPHIC PATTERNS

The evolution of pre-Tertiary fish groups in Australian freshwaters has been referred to previously (Long, 1995, pp.150–151; Basden *et al.*, 2006; Ritchie, 2006). Although details are still unknown, it is clear that the extant Australasian faunas have diverse origins, being the products of the overlap of Indo-Pacific, South-east Asian and southern temperate faunas (Frankenberg, 1974; McDowall and Whitaker, 1975; McDowall, 1981; Gressitt, 1982).

Cainozoic biogeography in Australasia has been influenced by five major categories of physico-chemical processes (McGowran *et al.*, 2000). These are:

(1) Tasman and Coral Sea sea-floor spreading (terminated early Eocene) with accelerated Australia-Antarctica divergence (commenced mid Eocene) and topographic relief variation in late Neogene (mountain chains in New Guinea and New Zealand, but subdued relief in Australia);
(2) Global cooling, steepened oceanic gradients and increased aridification of Australia, which all began in late Palaeogene;
(3) Isolation of land masses from eastern Gondwana (commenced Paleocene, clear gap south of Australia since Oligocene);
(4) Interactions of warm currents with cool waters south of the Subtropical Convergence;
(5) Strong long-term influence of El Niño-Southern Oscillation (ENSO) on Australia.

But Quaternary events (local uplift or volcanism, high amplitude climatic oscillations) have been particularly important for freshwater faunas, with most ecosystems only developing in the last 14,000 years (the current interglacial interval).

It has been suggested (Kershaw *et al.*, 2000) that evolutionary change, at a specific or higher level, is unlikely to have been significant in the Quaternary because of the short timescale; however, a number of other processes and pressures have contributed to the development of extant aquatic faunas (see back endpaper).

Clearly many factors affect the distribution of individual fish species, but for over 30 years the Australasian freshwater fish faunas have been related to major geographic barriers (in Australia these are usually the boundaries of drainage divisions). It is considered that geographic barriers, such as those separating drainages, are of prime importance in restricting most aquatic faunal exchange and thus likely to reflect biotic boundaries. For details of the drainage division system, descriptions of habitat types, and other factors known to affect distribution of Australian freshwater fishes see Merrick and Schmida (1984, pp.15–22), Hurwood and Hughes (1998) and Unmack (2001a,b).

The current biotic concepts of biomes or biogeographic regions (for terrestrial vegetation and ecosystem types) don't correspond to drainage divisions—although they may include part of a division or largely correspond to a major drainage basin (or river system) (Beattie, 1995, p.69; Department of Environment, Sport and Territories, 1996). It should also be noted that, as freshwater habitats are relatively isolated, a high level of endemism is common, so regional biodiversity can be high even if local diversity is low (Beattie, 1995, pp.73–74). On the Australian mainland the absence of large, permanent lakes has no doubt contributed to lower diversity among freshwater faunas generally, nonetheless some high diversity 'hotspots', such as south-eastern Australia (and Tasmania), north-western and south-western Australia are recognized (World Conservation Monitoring Centre, 2000b). Several river systems in New Guinea have also been identified as important for fish diversity (Allen *et* al., 1992; W.C.M.C., 2000b). An extensive analysis of zoogeography for Australian freshwater fishes was completed by McDowall (1981) and Unmack (2001a), in the most recent comprehensive survey, proposed a number of biogeographic provinces and subprovinces for Australian freshwater ichthyofaunas. Additional zoogeographic subprovinces, related to fish faunas in New Guinea, were briefly discussed by Allen (1991, pp.13–15; 1996a).

Further information on the evolution and radiations of select groups is included in later sections; however, before considering the details of diversity within Australasian faunas it is appropriate to briefly discuss the differences between local and Asian faunas.

A number of families, widespread in Asia, are not naturally present in Australasia; these include the Cyprinidae, Homalopteridae, Bagridae, Ophiocephalidae and Anabantidae (Filewood, 1984). There are also some families with overall ranges extending into both Australasian and Asian regions, but which have developed characteristic freshwater forms entirely restricted to Australia and/or New Guinea. Examples of this second group are the Clupeidae, Ariidae, Atherinidae, Terapontidae, Kuhliidae and Eleotrididae. Finally, there are a few groups that are apparently endemic to Australasia; these include the Ceratodontidae, Retropinnidae, Lepidogalaxiidae, Gadopsidae, Melanotaeniidae and Pseudomugilidae (Thoney *et al.,* 2003, p.57).

Using the findings of recent published surveys, specific taxonomic studies and broader systematic revisions, the diversity, interrelationships and zoogeography of extant Australasian freshwater ichthyofaunas are briefly reviewed in subsequent sections. The suggested phylogenetic grouping of families, family names, generic and specific names are generally based on Paxton *et al.* (1989); however, recent studies have recommended a number of modifications. These changes include a revision of the relationships of galaxioid families, placement of gadopsid and kuhliid species with percichthyids and the status of the gudgeon group. Some of these changes are reflected in Table 1, but others continue to be the subject of controversy or on-going research, and so the original separate family designations are retained.

The objectives of this chapter are: to briefly summarize the main factors considered to have influenced the evolution of freshwater fishes in Australasia, highlighting the differences between local and Asian faunas; to discuss interrelationships within extant native faunas and demonstrate diversity in a family key with comparative generic lists; to describe recent speciation and radiations that have been documented in select groups of native fishes; and to briefly outline the impacts of introduced fishes, their significance in management strategies and implications for conservation.

DIVERSITY AND INTERRELATIONSHIPS

As Figure 1 illustrates, there are basically four disparate groups comprising Australasian freshwater fish faunas. These are the jawless lampreys, the elasmobranchs (sharks, rays, sawfish), the lungfish (Plates 41–44) and teleosts. The early evolution and possible interrelationships of these major groupings are not discussed further here, and Figure 1 includes them as four distinct assemblages with separate lineages; indeed, even relationships between many of the Recent teleosts are unclear. As the non-teleostean groups only involve a few taxa, which have remained virtually unchanged for long periods of time, most discussion focuses on the teleost families—some of which have clearly undergone recent speciation.

FAMILY EVOLUTION

The non-teleostean families are all ancient groups—the lampreys (Petromyzontiformes) represent relicts of a group dating from the mid Carboniferous, about 330 m.y.a. (Potter *et al.,* 1986; Thoney *et al.,* 2003). The Petromyzontiformes are monophyletic and

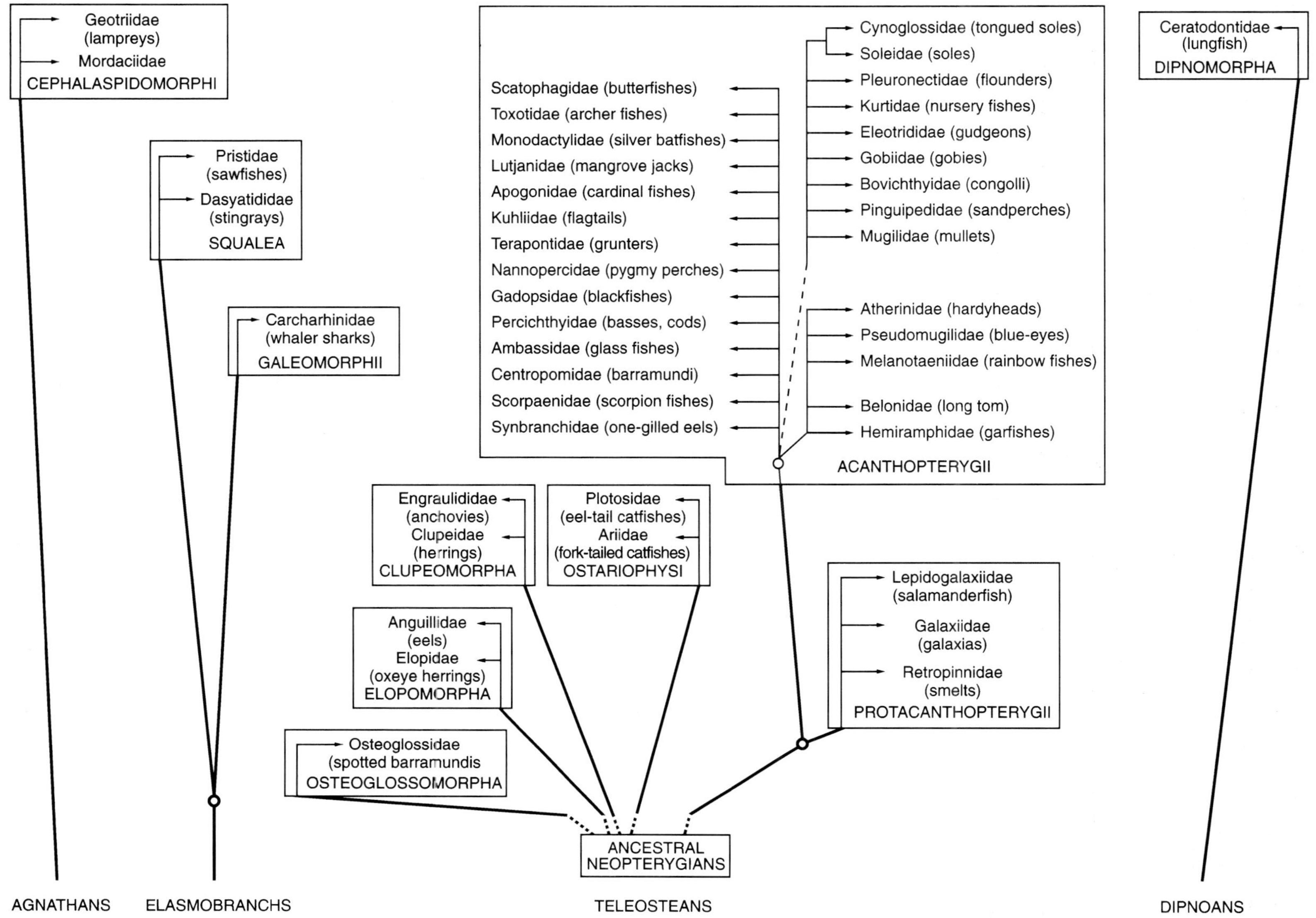

Figure 1. This diagram summarizes the suggested systematic relationships of native Australasian freshwater fishes. The lampreys, lungfish and elasmobranchs have separate histories with distinct lineages from the 38 teleostean families discussed. It is acknowledged that many interrelationships, at various levels, are under review; here the generally accepted divisional, supraordinal and family names are retained. Uncertain relationships are indicated by a dotted line. Based on Greenwood *et al.* (1966), Greenwood *et al.* (1973), Carvalho (1996), Cloutier and Ahlberg (1996), Forey *et al.* (1996), Li and Wilson (1996), Johnson and Patterson (1996), Lecointre and Nelson (1996), Parenti and Song (1996) and Waters *et al.* (2002).

most species occur in the Northern Hemisphere (family Petromyzontidae), but four species of Southern Hemisphere lampreys are placed in two families, Mordaciidae and Geotriidae (Potter, 1996a). Recent studies of the chemical structure of lamprey insulins suggest that these two southern families have been separated from each other for a long period—possibly since pre-Tertiary (Conlon *et al.*, 2001).

Major modern groups of elasmobranchs had evolved by the late Jurassic (~155 m.y.a) (Schwartz and Maddock, 2002); although rays had evolved from sharks earlier (Basden *et al.*, 2006). Dasyatid ray fossils are reported from late Cretaceous elsewhere (Schwartz and Maddock, 2002), but the oldest Australian specimen assigned to the family Dasyatididae is reported as early Miocene in age (Long and Turner, 1984; Kemp, 1991). Sawfish fossils are known from mid Eocene times (~45 m.y.a.) and an Australian species closely related to the Pristidae has been dated as being early to middle Miocene in age (Long and Turner, 1984). Although fossils representing a number of extant shark genera are known from Australia the earliest *Carcharhinus* dates from the Oligocene (in Victoria); teeth referrable to an extant carcharhinid (and dated from early Pliocene) have been recorded at a number of Victorian and Tasmanian sites (Kemp, 1991).

Due to their unique structure, and interest in their exact relationship to other vertebrate groups, considerable study has been completed on all extant dipnoans, as well as fossil remains. Australian lungfishes are relatively well documented; and, although only one species (*Neoceratodus forsteri*) survives, the almost complete fossil record indicates how diverse the group has been (Long, 1995, p.162; Molnar and Kurz, 1997; Ritchie, 2006). The family to which *N. forsteri* belongs has been revised and enlarged by Kemp (1997), who reports *N. forsteri* toothplates from early Cretaceous deposits.

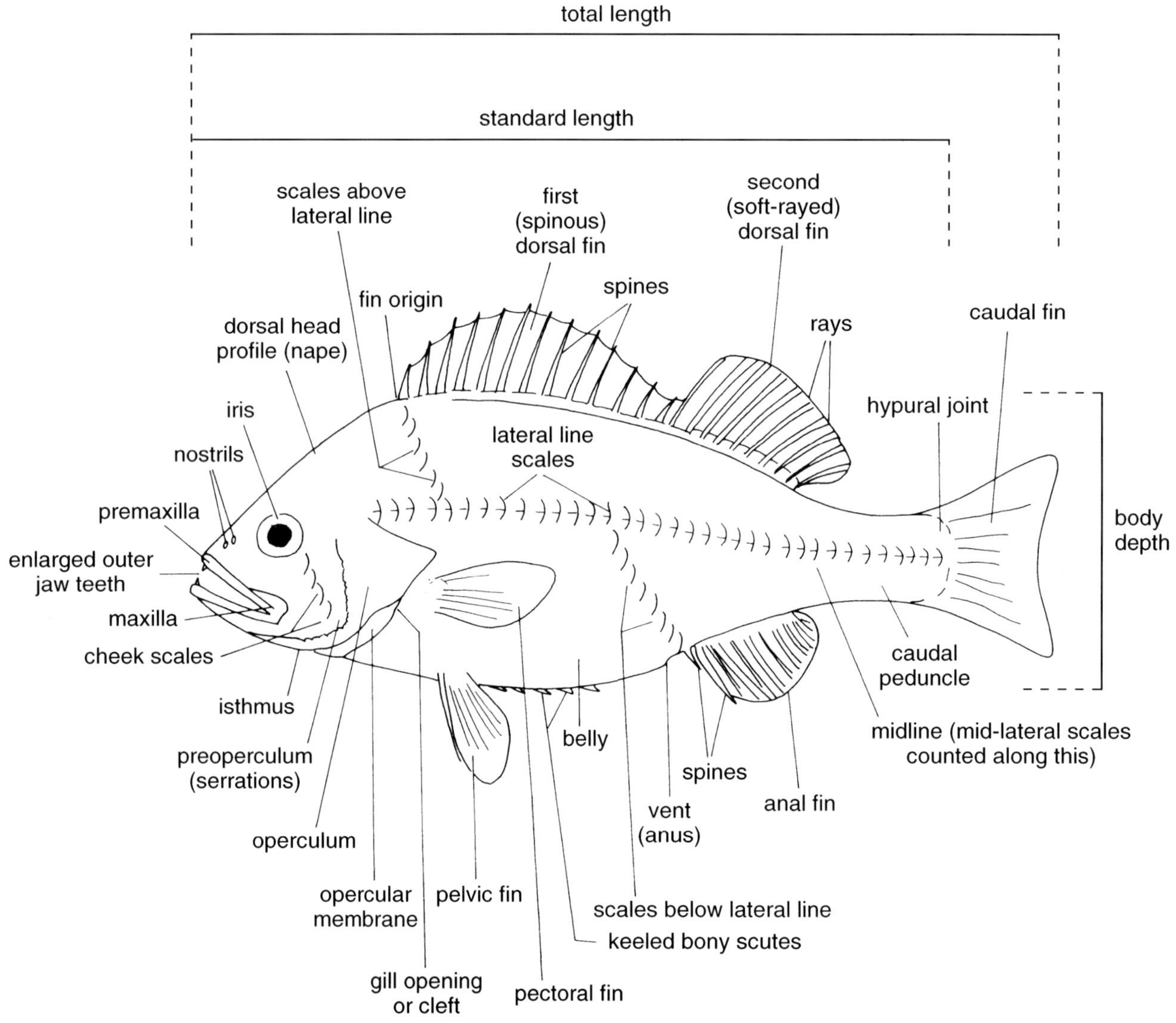

Figure 2. Diagram of a hypothetical bony fish showing many of the external features used in identification. Adapted from McDowall (1996b).

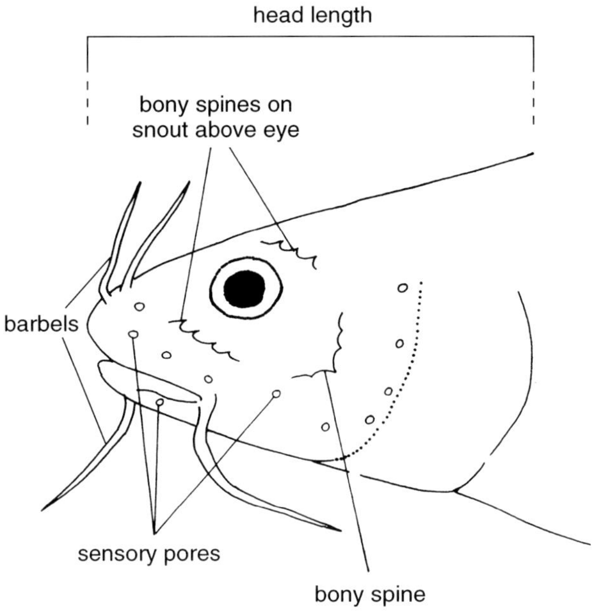

Figure 3. Selected features of the head of a bony fish used in identification. Adapted from McDowall (1996b).

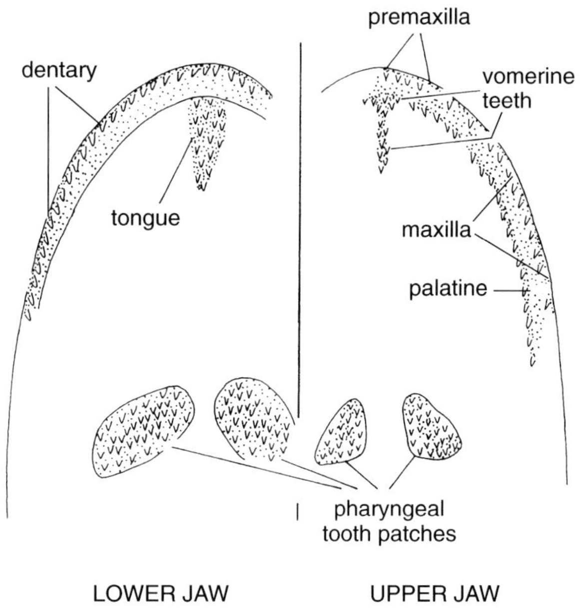

Figure 4. Features of the dentition of a bony fish used in identification. Adapted from McDowall (1996b).

Within the teleostean assemblage a number of broad groupings of closely related families can be recognized; these associations are illustrated in Figure 1 and discussed briefly below. The Osteoglossomorpha have been considered among the most primitive living teleosts (Lauder and Liem, 1983; Nelson, 1984) and derivation of the family Osteoglossidae has been extended back to early Cretaceous or late Jurassic (~120–150 m.y.a.); but a number of extant genera, including *Scleropages,* are known from mid Oligocene (Forey *et al.*, 1996; Li and Wilson, 1996). Likewise, the elopomorph family Elopidae (Megalopidae) dates from the early Cretaceous (108–97.5 m.y.a.), whereas fossil Anguillidae are not known until early Eocene (54.8–49 m.y.a.) (Forcy *et al.*, 1996).

The inter- and intrafamilial relationships of the five osmeroid families (Retropinnidae, Prototroctidae, Galaxiidae, Aplochitonidae and Lepidogalaxiidae) have been extensively studied (McDowall, 1984, 1997, 2002; Johnson and Patterson, 1996; McDowall and Wallis, 1996; Waters and White, 1997; Williams, 1997; Waters *et al.*, 2000a, b); all these groups represent a Southern Hemisphere radiation comparable to the osmerid and salmonid radiations in temperate freshwaters of the Northern Hemisphere.Recent molecular studies by Waters *et al.* (2002) have indicated that prototroctids should be included in the Retropinnidae and that the aplochitonid genus should be assigned to the Galaxiidae. *Lepidogalaxias salamandroides* is distinct and its affinities have been the subject of considerable discussion; however,the data have confirmed that this relict species is related to galaxioids (Allen and Berra, 1989; Williams, 1997; Waters *et al.*, 2002). It has been suggested that the lepidogalaxiids first evolved up to 180 m.y.a. (Merrick and Schmida, 1984, pp.78–79).

Among the ostariophysans, both the families Ariidae and Plotosidae (dating from Eocene–Oligocene times) have undergone their largest radiations in Australia and New Guinea (Merrick and Schmida, 1984, pp.115, 123; Allen *et al.*, 2002); however, much of the previous uncertainty about the taxonomy of these groups has now been clarified. Similarly, there has been much confusion about the Melanotaeniidae and related forms, but recent revisionary studies and the erection of the family Pseudomugilidae have clarified the situation (Saeed *et al.*, 1989). Unlike the catfishes these latter families are restricted to Australia and New Guinea.

The largest single grouping—in terms of both families and species—comprises the fishes in the order Perciformes; these percoids include small families (e.g. Toxotidae) as well as large families (e.g. Terapontidae) that have undergone major radiations in Australasia and may well have polyphyletic origins.

It is emphasized that the key and tables below are limited to taxa reported from the three major faunas. With the exceptions of major island groups like New Caledonia or Hawaii, where studies by Watson (and colleagues) and McDowall have been important, the freshwater ichthyofaunas of most islands are very poorly documented (Watson and Poellabauer, 1998; Watson *et al.*, 2000, 2001, 2002; McDowall, 2003). Almost certainly future surveys will detect additional genera from isolated or remote localities. Very recent studies on Micronesian island fish faunas indicate that species richness is greater on large high islands (with more habitat complexity) than on small low islands; but other factors, such as geographical location, dispersal capacity or vicariance events, also affect diversity (Donaldson and Myers, 2002).

FAMILY KEY

The key below is of a standard dichotomous type and is adapted from a section in McDowall (1996b) with most diagrams from Merrick and Schmida (1984). The construction of this kind of identification tool necessitates the use of a number of technical structural terms; however, it is not possible to include glossaries here, so readers are referred to McDowall (1996b) for detailed definitions or explanatory diagrams. Most of the features used in the keys are illustrated in Figures 2–4.

Although recent studies have incorporated microscopic, chemical or molecular criteria to delineate taxa, the emphasis here is on the traditional, macroscopic features in order to maximize general usage by non-specialists. Likewise, new computer-generated keys (such as Lucid) can accommodate multi-state characters with multiple alternatives (Centre for Pest Information Technology and Transfer, 2001); however, for simplicity and familiarity to users, the dichotomous format is retained.

1a. No jaws, varying numbers of teeth on oral disc, no paired fins 2
b. Jaws and paired fins present (sometimes no pelvics) 5

2a. Eyes present 3
b. Eyes absent 4

3a. Two large tricuspid teeth on oral disc above mouth, other teeth on disc few in number in radial rows; eyes dorsolateral .... Mordaciidae (adults)
b. Single tooth on oral disc above mouth, other teeth on disc numerous in irregular spirals; eyes distinctly lateral .... Geotriidae (adults)

3a

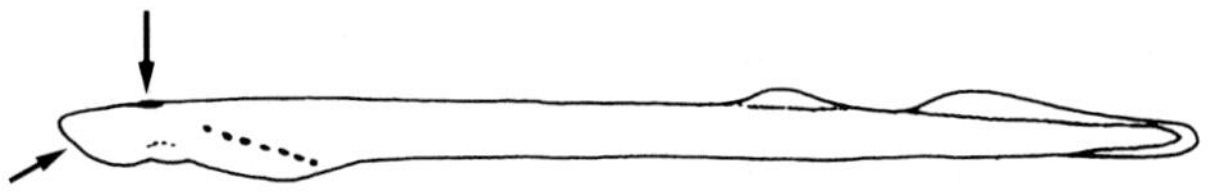

3b

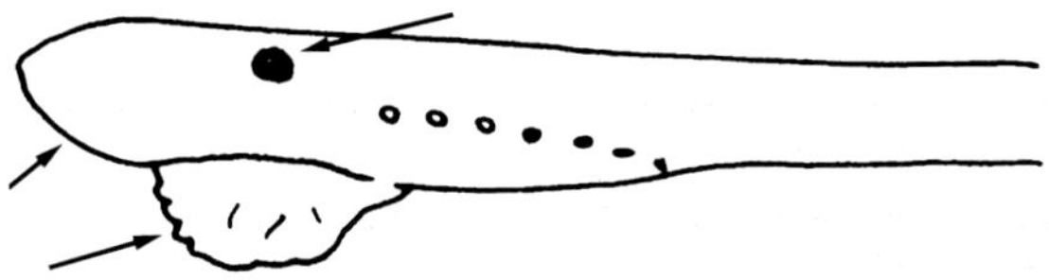

4a. Snout to vent (anterior margin) distance 83% or more of total length .... Mordaciidae (juvenile ammocoete)
b. Snout to vent (anterior margin) distance 80% or less of total length .... Geotriidae (juvenile ammocoete)

4a

5a. Five pairs of external gill openings; no operculum (gill cover) 6
b. One pair of external gill openings; an operculum (gill cover) present 8

6a. Body strongly dorso-ventrally flattened 7
b. Body not dorso-ventrally flattened .... Carcharhinidae

6b

7a. Snout produced as an elongate, flattened, strongly toothed saw; a typical shark-like heterocercal tail without spines .... Pristidae
b. Snout not elongate and without teeth; tail elongate and whip-like with serrated spines .... Dasyatididae

7a

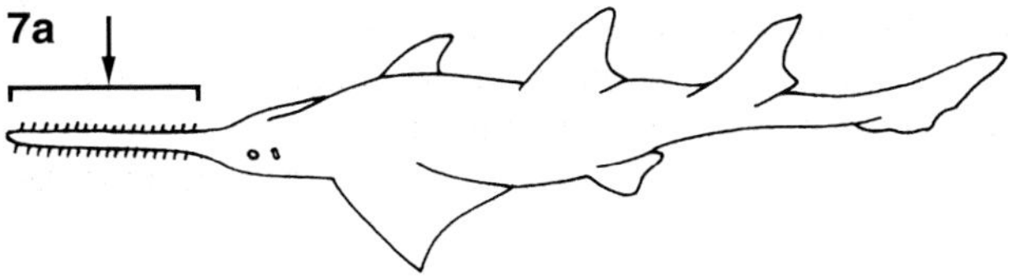

7b

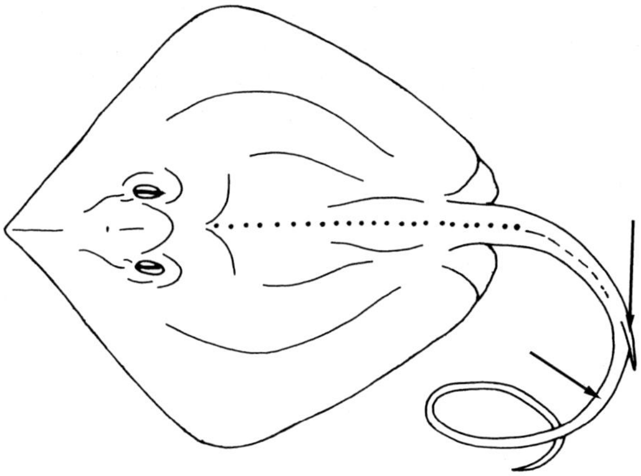

8a. Dorsal, caudal and anal fins continuous and inseparable; fish elongate and slender, tubular 9
b. Dorsal, caudal and anal fins quite distinct; fish stocky; laterally compressed or dorso-ventrally flattened 11

9a. Gill openings narrow slits, paired or single; no pelvic fins 10
b. Gill openings more extensive, paired and extending below head; large pelvic fins prop or leg-like .... Ceratodontidae

9b

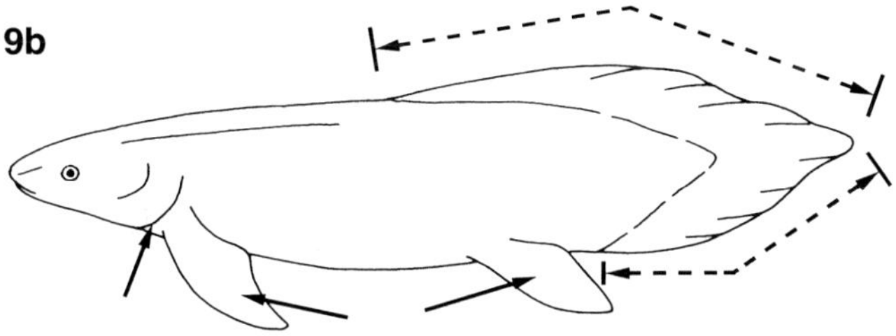

10a. Paired vertical slits in front of pectoral fin bases; eyes medium to large .... Anguillidae
b. Single slit across throat, no pectorals; eyes small when present and covered with skin .... Synbranchidae

10a

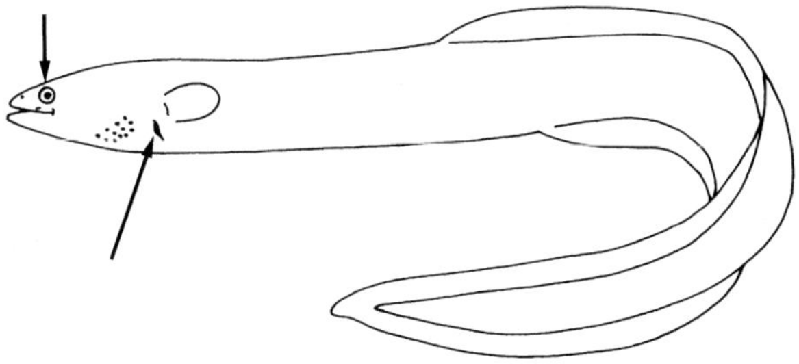

10b

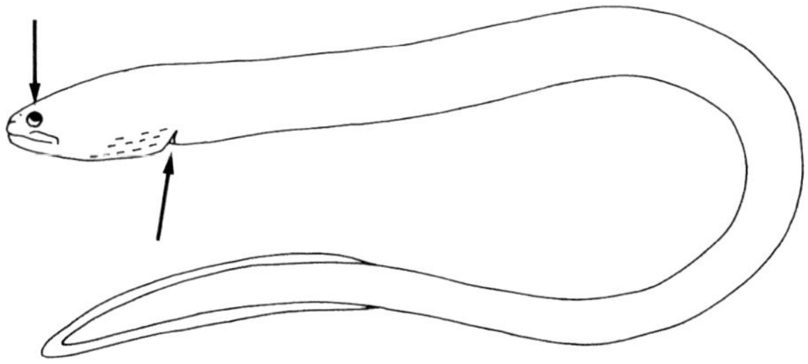

11a. Adipose fin present (may be very small) 12
b. No adipose fin 14

12a. Stout spines present in leading edges of dorsal and pectoral fins; barbels present around mouth .... Ariidae
b. Fins entirely soft-rayed; no barbels 13

12a

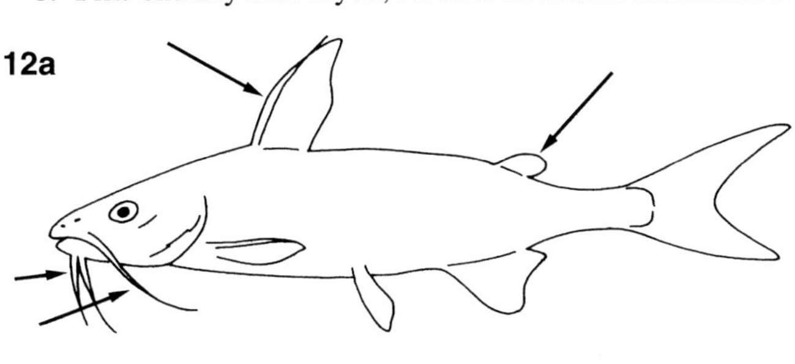

13a. Dorsal fin base wholly in front of anal fin origin and just behind pelvic fin bases; horny shelf surrounds lower jaw which is shorter than upper ....Retropinnidae (including Prototroctidae)
b. Dorsal fin partly or largely above anal fin, well behind pelvic bases; no horny shelf around lower jaw, jaws about equal in length ....Retropinnidae

13a

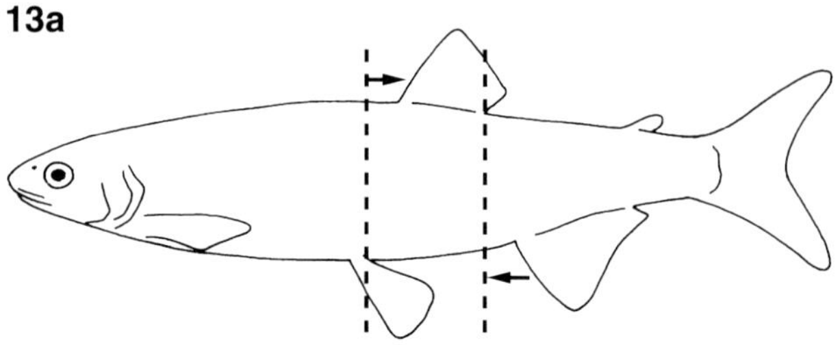

13b

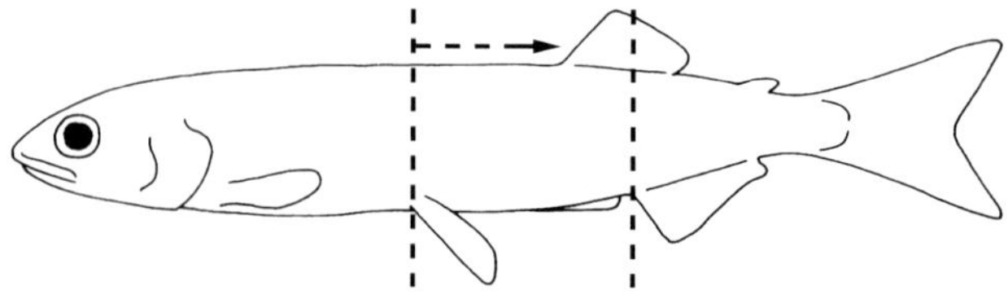

14a. Ventral surface of abdomen serrate owing to the presence of scaly scutes 15
b. Ventral surface of abdomen not serrate 16

15a. Large inferior mouth with long upper jaw bone (maxilla) extending well back behind eye............Engraulididae
b. Small to moderate mouth with maxilla extending to beneath eye..Clupeidae

15a

15b

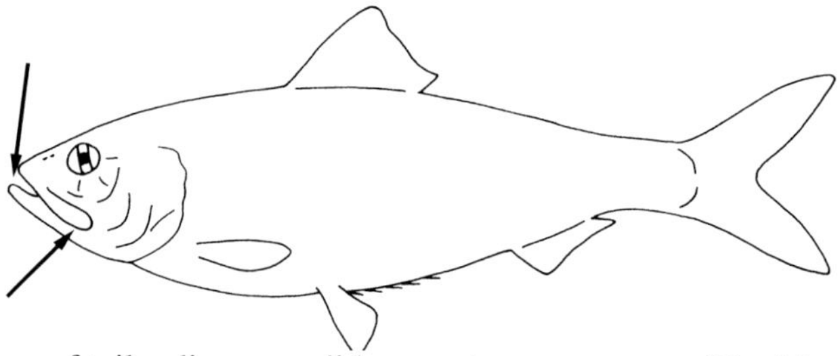

16a. Last ray of dorsal fin an elongate filament reaching base of tail; adipose eyelid present........................Elopidae
b. Last ray of dorsal fin not elongate; adipose eyelid rarely present 17

16a

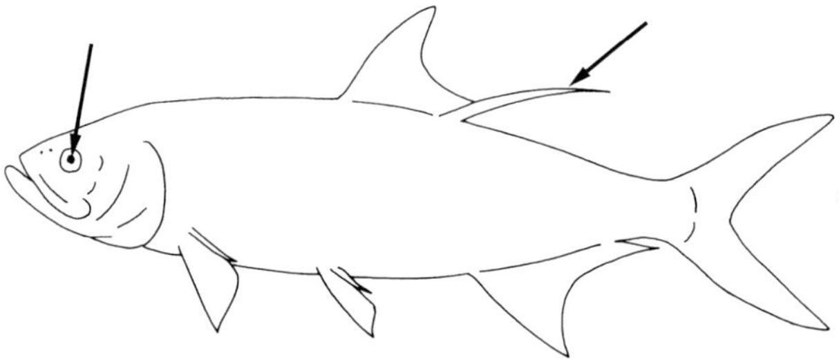

17a

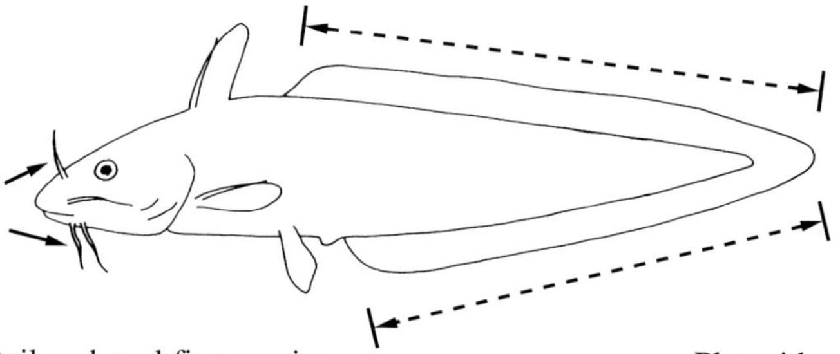

17a. Several long barbels around mouth; second dorsal, tail and anal fins continuous ....Plotosidae
b. No barbels around mouth (occasionally one short one beneath chin); second dorsal fin (if more than one), caudal and anal fins quite distinct 18

18a. Scales thick and bony, head scaleless ....Osteoglossidae
b. Scales not thick and bony 19

18a

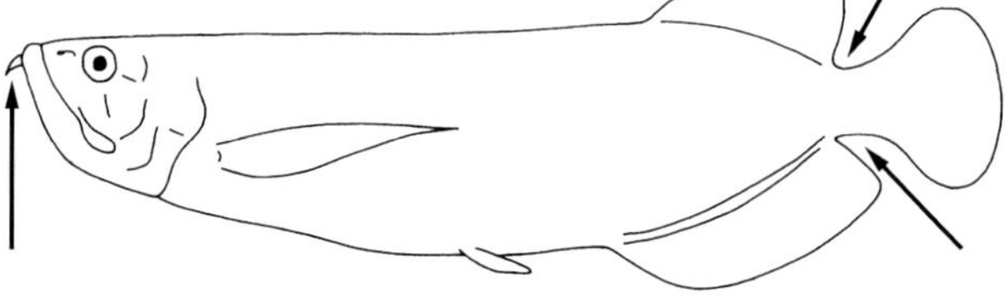

19a. Head with a very slender and elongate beak 20
b. Head without a slender elongate beak 21

20a. Beak formed from an elongation of both upper and lower jaws; jaws about equal in length .... Belonidae
b. Beak formed from an elongation of only lower jaw which is many times longer than upper ...... Hemiramphidae

20a

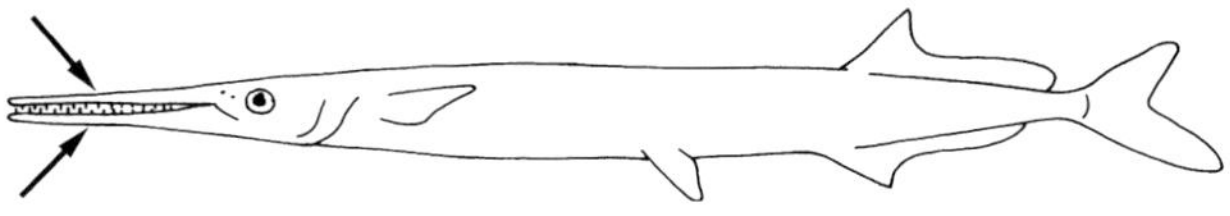

20b

21a. Pelvic fins reduced to a slender, indistinct spine and a single, divided ray .... Gadopsidae
b. Pelvic fins, if present, with two or more rays 22

21a

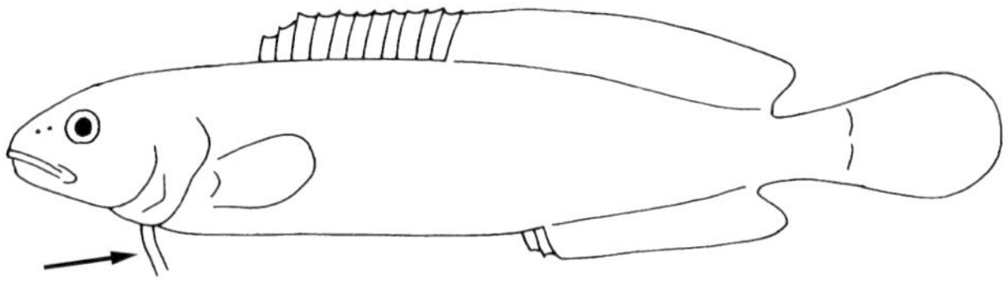

22a. Fins lack spines, only soft rays 23
b. Fins with spines present, although these often slender 27

23a. Body very compressed, fish lies on one side, two eyes on opposite side of head 24
b. Body more or less tubular to compressed; fish swims upright with an eye on each side of head 26

24a. Fish lies on right side, two eyes on left side .... Cynoglossidae
b. Fish lies on left side, two eyes on right side 25

24a

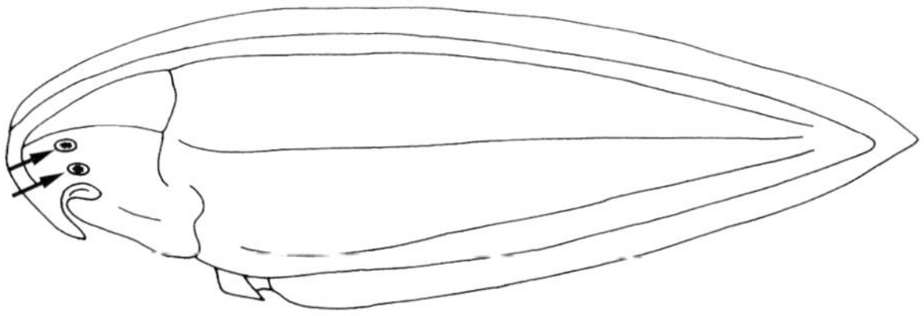

25a. Caudal fin continuous with dorsal and anal, pectorals reduced or absent .... Soleidae
b. Caudal fin not continuous with dorsal and anal, pectorals not reduced .... Pleuronectidae

25a

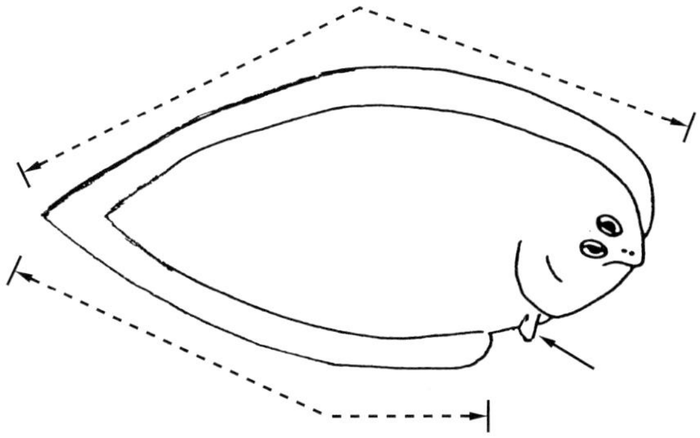

25b

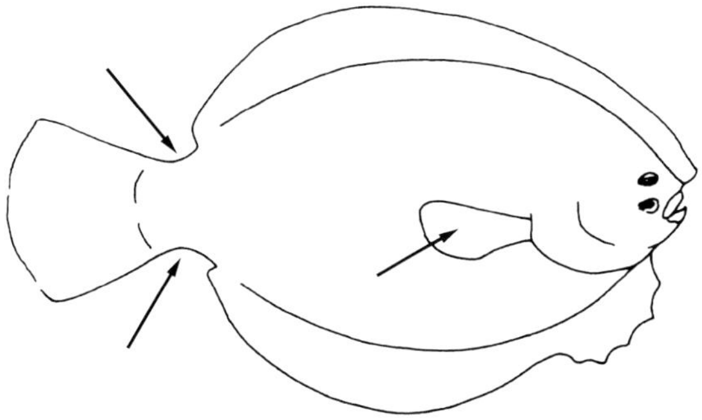

26a. Thin cycloid scales present, no lateral line .... Lepidogalaxiidae
b. Scales absent; lateral line presen .... Galaxiidae (including Aplochitonidae)

26a

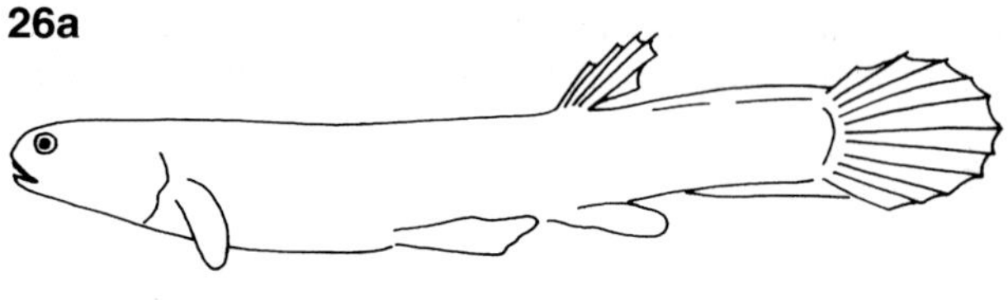

26b

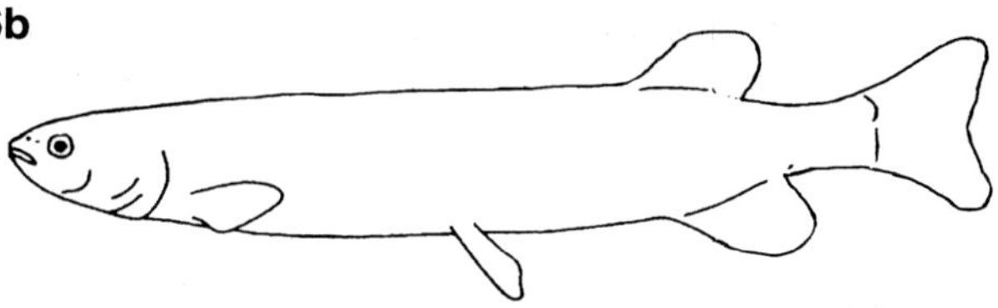

27a. A small to moderate-sized genital papilla between vent and anal fin origin; one slender spine at front of anal fin; head with many rows of small papillae 28
b. Usually no genital papilla; usually more than one spine at front of anal fin; usually few or no papillae on head 29

28a. Pelvic fins joined to form a small cup .... Gobiidae
b. Pelvic fins separate, not forming a cup .... Eleotrididae

28a

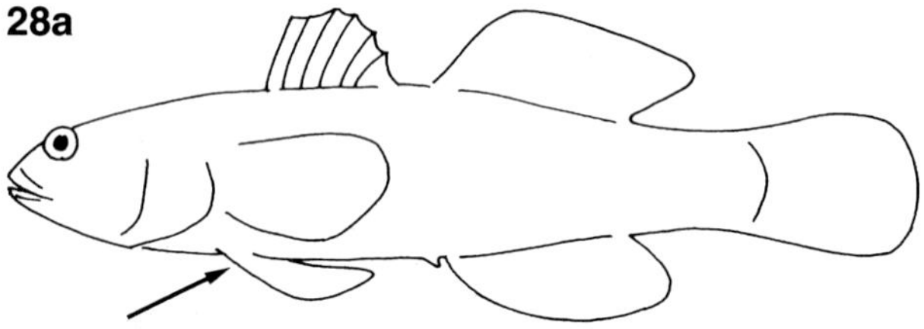

28b

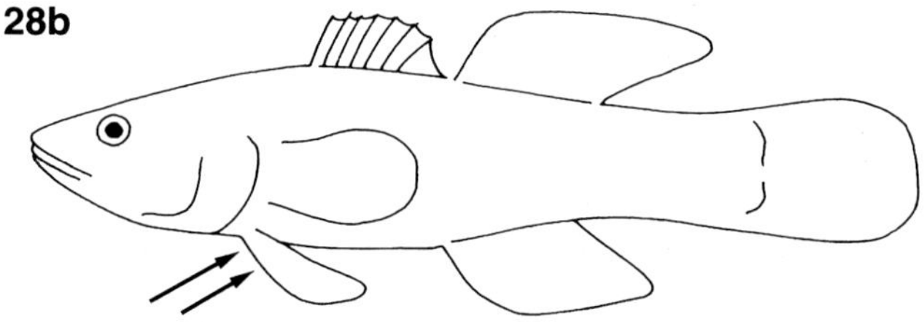

29a. Head armed with numerous sharp, bony spines on snout, above eyes, on cheeks and gill covers ...Scorpaenidae
b. Head not armed with numerous sharp, bony spines 30

29a

30a. Spinous and soft-rayed dorsal fins quite distinct, usually separated by a distinct gap 31
b. Spinous and soft-rayed dorsal fins continuous, although there may be a deep but incomplete notch 36

31a. Single nostril on each side of head; pelvics originating well forward of pectoral fin bases, dorsal usually behind pectoral base .... Bovichthyidae
b. Two nostrils on each side of head; pelvic and dorsal fin origins level with or behind pectoral fin bases 32

32a. Lateral line present; scales usually ctenoid and very thick; mouth large, reaching below eyes ....Apogonidae
b. Lateral line absent; scales cycloid and often very thin; mouth small, not reaching below eyes 33

31a

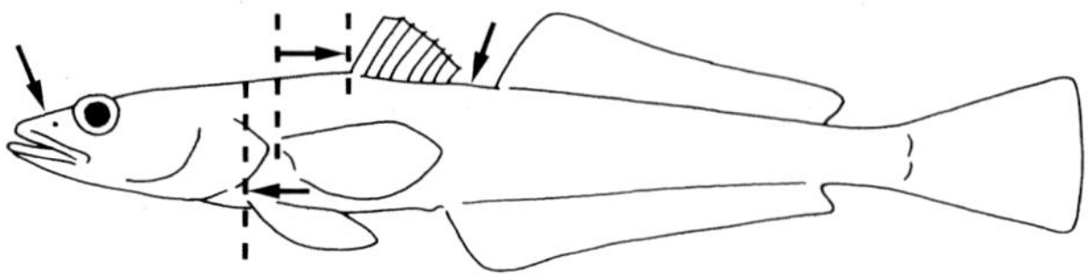

32a

Plate 48. Rainbow fishes (Family Melanotaeniidae) are colourful small species that have been extensively studied and appear to have originated in the New Guinea region. Photograph J. Merrick.

Plate 49. A fork-tailed catfish (Family Ariidae) from northern Australia. This species, *Arius graeffei*, has a wide range and is a mouth-brooder; males incubate the eggs. Photograph G. Schmida.

Plate 50. The eel-tail catfishes (Family Plotosidae) have also radiated in Australasian freshwaters. The *Tandanus tandanus* shown here is one of the larger and more widespread species. Photograph G. Schmida.

Plate 51. The Macquarie perch (*Macquaria australasica*) is a representative of the family Percichthyidae; other members of this family (including the Murray cod and golden perch) have radiated extensively in Australian freshwaters. The range and abundance of *M. australasica* has greatly declined in recent years: factors contributing to this include stream modifications and introduced trouts. Photograph J. Merrick.

Plate 52. The spangled grunter (*Leiopotherapon unicolor*) is one of the most common and abundant species in Australian freshwaters. Some regional variation has been noted in this widespread species and *L. unicolor* is very hardy. A member of the family Terapontidae which is another group known to have radiated extensively in Australasia. Photograph J. Merrick.

Plate 53. The sleepy cod (*Oxyeleotris lineolatus*) which is one of the larger gudgeon species. Eleotridids have also radiated extensively in Australasian freshwaters. Photograph. J. Merrick.

33a. Soft dorsal and/or anal fins long-based, dorsal with up to 16 rays, anal with 7–22 rays 34
b. Soft dorsal and anal fins short-based, dorsal with less than nine rays, anal with less than 18 rays 35

34a. Body moderately compressed and elongate, first dorsal vestigial or with 3–8 flexible spines, eyes blue or green ....Pseudomugilidae
b. Body compressed, often deep, first dorsal not vestigial, eyes not blue or green ....Melanotaeniidae

34a

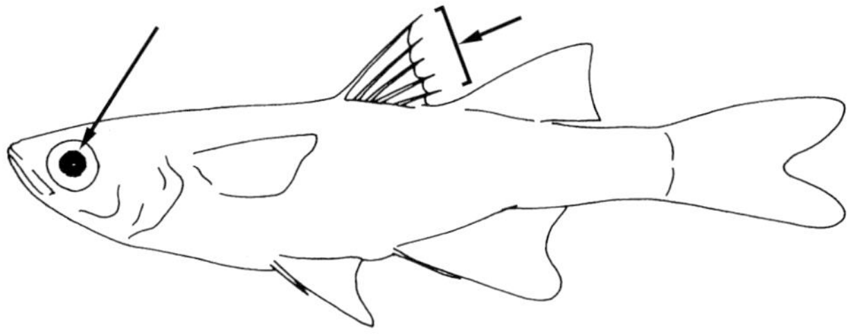

34b

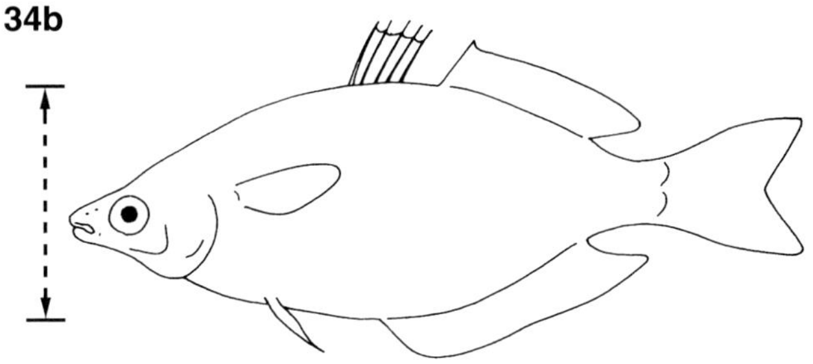

35a. Spines in vertical fins stout and rigid; more than one anal spine; preorbital bone serrate in adults; gill rakers very long and slender, very numerous, more than 50 ....Mugilidae
b. Spines in vertical fins slender and flexible; one anal spine; preorbital bone smooth; gill rakers usually short and stubby but sometimes long and slender, less than 50 ....Atherinidae

35a

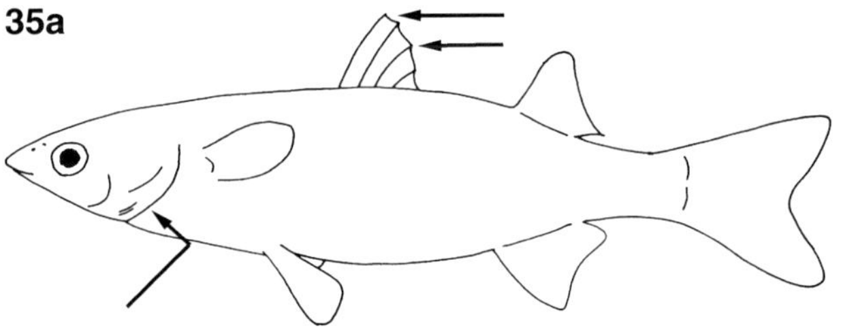

35b

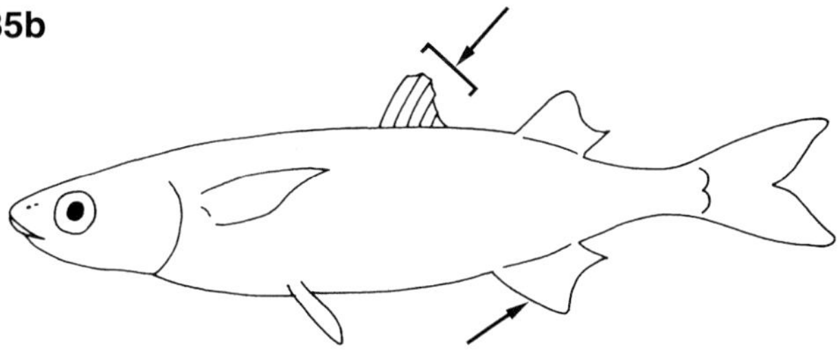

36a. Total rays and spines in anal fin more than 35...37
b. Total rays and spines in anal fin less than 30 38

37a. Very deep, strongly compressed body, small head and mouth; four short spines at front of dorsal grading abruptly into much longer spines and rays; pelvic fins reduced in adults, with one spine and 2–3 rays ....Monodactylidae
b. Deep compressed body, medium to large head with large protractile mouth; reduced dorsal spines grading abruptly into much longer rays; pelvic fins not reduced, with one spine and five rays; bony crest or hook on forehead of males ....Kurtidae

37a

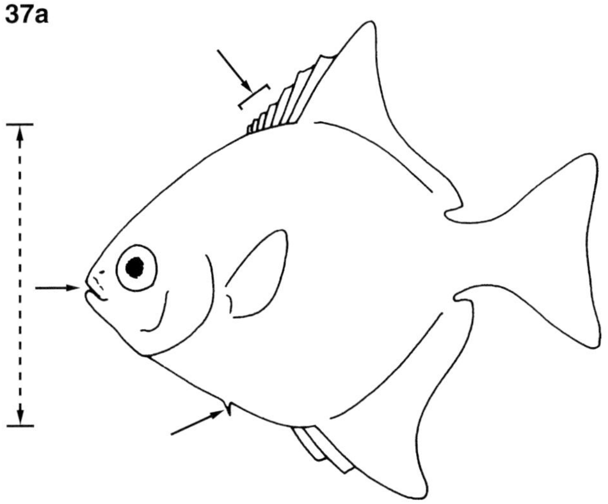

37b

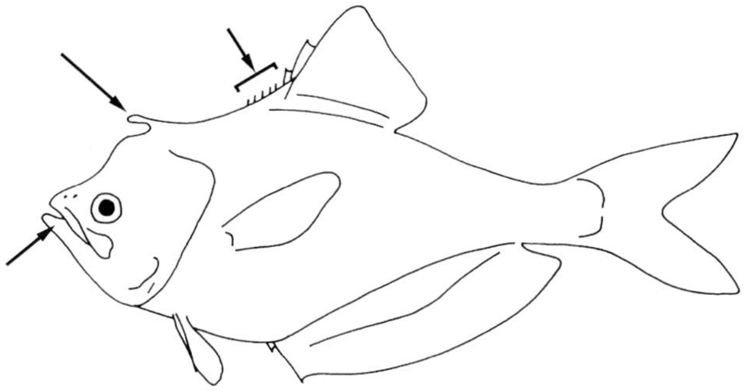

38a. Anal fin with no spines, 18 rays .... Pinguipedidae
b. Anal fin with spines and rays 39

39a. Gill membranes joined to each other across isthmus, free from isthmus; scales very small, 80 or more along lateral line .... Scatophagidae
b. Gill membranes free from each other and isthmus; scales larger, less than 80 along lateral line 40

38a

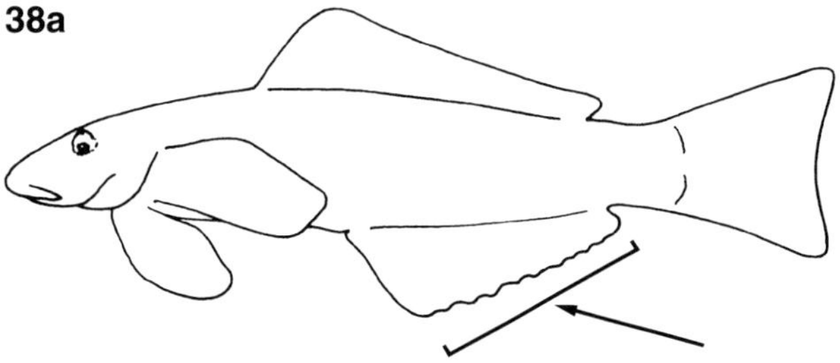

39a

40a. Dorsal and anal fin bases sheathed with large scales .... Ambassidae
b. Dorsal and anal fin bases not sheathed with large scales (small scales may spread on to fin bases) 41

41a. Lateral line absent or represented by an incomplete series of pored scales; about 25–45 mid lateral scales .... Kuhliidae
b. Lateral line present and continuous; scales may be smaller and more than 50 along side 42

40a

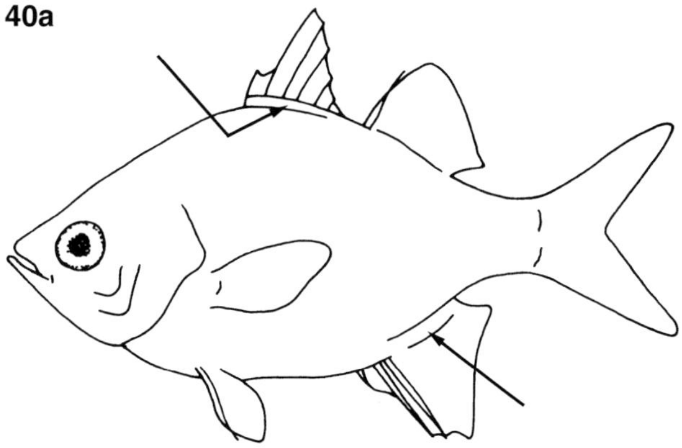

41a

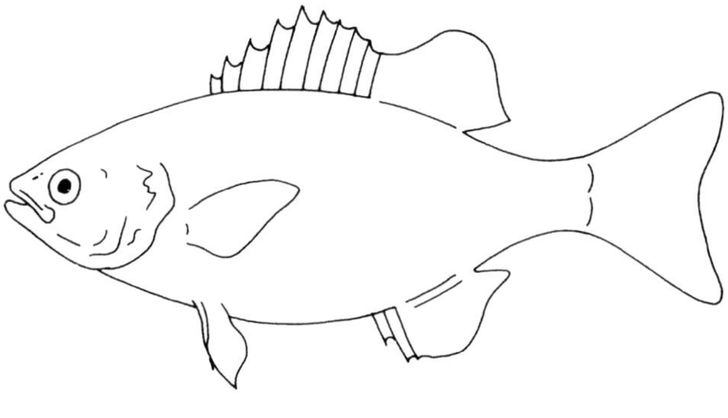

42a. Deep compressed body, snout to dorsal profile straight or slightly convex; dorsal posteriorly positioned above anal .... Toxotidae
b. Compressed body, snout to dorsal profile convex or concave; dorsal not posteriorly positioned 43

42a

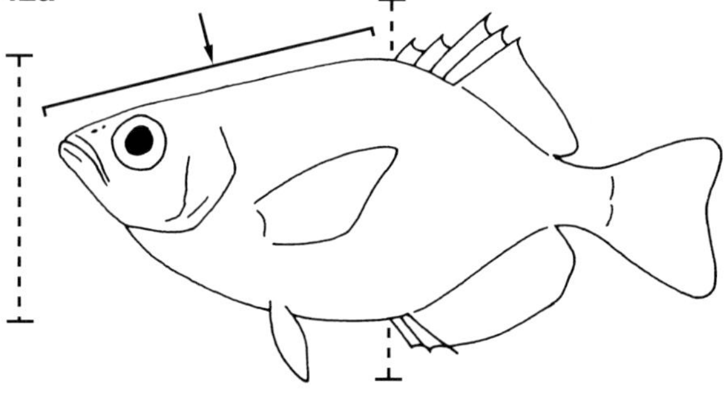

43a

43a. Teeth in jaws usually in a few rows, outer row enlarged, conical, sharp and stout; axillary process present at base of pelvic fins .... Lutjanidae
b. Teeth in jaws small, numerous, several rows wide; no axillary process at pelvic fin base 44

44a. Dorsal fin deeply divided 45
b. Dorsal fin not deeply divided 46

45a. Lateral line scales continuing on caudal .... Centropomidae
b. Lateral line scales not continuing on caudal .... Nannopercidae

45a

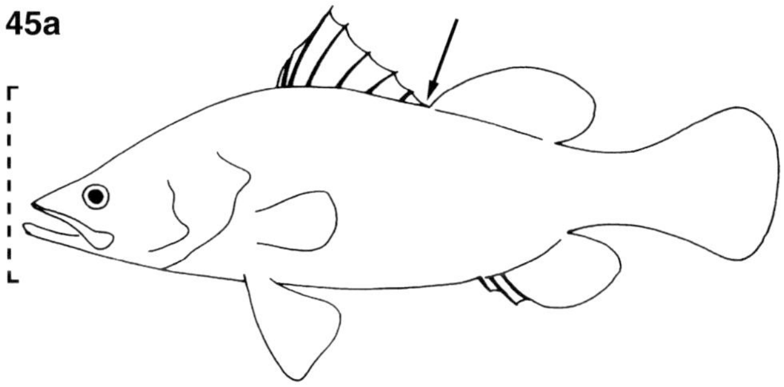

45b

46a. Pelvic fin origin below or in front of pectoral fin bases; mouth large, usually reaching to hind margin of eye; tail usually rounded .... Percichthyidae
b. Pelvic fin origin distinctly behind pectoral fin bases; mouth smaller, usually not reaching middle of eye; tail forked or emarginate .... Terapontidae

46a

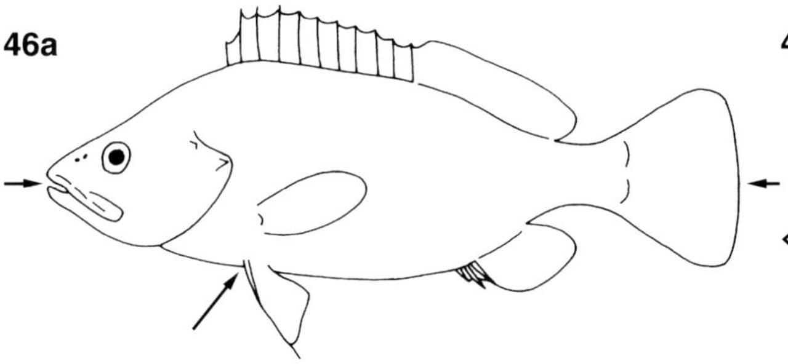

46b

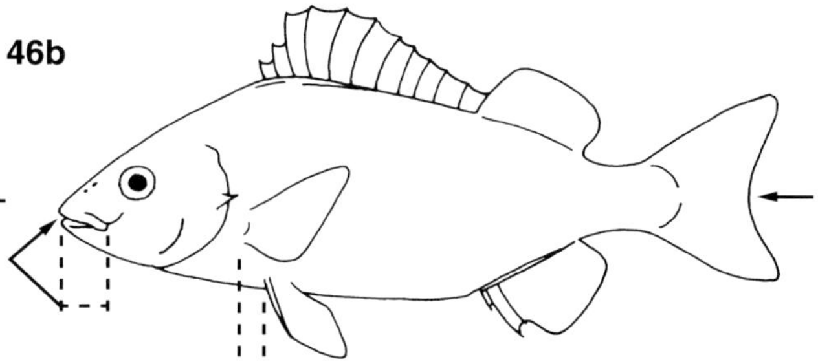

GENERIC AND SPECIFIC LONGEVITY

Like the families mentioned on p.198, a number of genera and species have remained unchanged for long periods of time. Fishes identical to the modern *Neoceratodus* are present in early Cretaceous deposits (Kemp, 1997). This indicates that *N. forsteri* has remained unchanged for over 100 million years and represents one of the oldest vertebrate genera (Merrick and Schmida, 1984, p.46). Including the exceptional *Neoceratodus,* fossil remains of 13 genera represented in extant faunas have been dated variously from 20–100 million years of age; these genera are from 12 families. Another two genera, from two additional families, have been dated at 5–15 million years of age (Merrick and Schmida, 1984; Rowland, 1989; McDowall and Pole, 1997).

While some groups, such as the melanotaeniids, have undergone considerable recent speciation the longevity of species has been the subject of much discussion. The one consistent trend is that as more data come to hand the estimated ages for groups increase.

COMPARISONS AND CONTRASTS

Direct comparisons of species numbers from previous ichthyofaunal surveys—as indicators of relative diversity—have to be treated with considerable caution for several reasons. First, criteria used by researchers for inclusion in the different surveys have varied; second, knowledge of both northern Australian and New Guinean faunas has been acknowledged as incomplete; and third, taxonomy of some groups remains confused with recent molecular studies revealing genetic structuring within taxa and natural hybridization (Musyl and Keenan, 1992; Douglas *et al.*, 1995; Jerry and Woodland, 1997; Hughes *et al.*, 1999; Jerry *et al.*, 1999; Bertozzi *et al.*, 2000; McGlashan and Hughes, 2002). In addition, recent studies have indicated how adaptive life cycle strategies, such as amphidromy, can influence numbers of taxa present. Among very speciose eleotridid or gobiid lineages, amphidromy is common but considerable variation is known to occur in time spent in freshwaters and local distributions may be ephemeral or intermittent (McDowall, 2003).

Table 1. A summary of the diversity of the freshwater fish faunas of Australasia; compiled from the following sources, listed alphabetically (Allen, 1985, 1987, 1991, 1993, 1996a, b, 1997, 1998, 2001; Allen and Boeseman, 1982; Allen and Burgess, 1990; Allen and Coates, 1990a, b; Allen and Feinberg, 1998; Allen and Hoese, 1986, 1991; Allen and Ivantsoff, 1985; Allen and Jebb, 1993; Allen and Jenkins, 1999; Allen and Pusey, 1999; Allen *et al.*, 1993, 1998, 2002; Allen and Renyaan, 1996a, b, 1998, 1999; Aplin, 1998; Crowley and Ivantsoff, 1990a, b, 1992; Crowley *et al.*, 1991, 1995; Filewood, 1984; Hoese and Allen, 1987, 1990; Ivantsoff *et al.*, 1987a, b, 1997; Jenkins *et al.*, 2000; Kailola, 1983, 1989,1990, 2000; Last and Stevens, 1994; McDowall, 1996b, 1997; McDowall and Whitaker, 1975; Merrick and Schmida, 1984; NIWA, 2001; Paulin *et al.*, 1989; Paxton *et al.*, 1989; Potter, 1996a, b; Pusey and Kennard, 2001; Rowland, 1993; Saeed *et al.*, 1989;Waters *et al.*, 2002; Watson, 1994, 1995; Watson and Allen, 1999; Watson *et al.*, 1998).

Family	Genus	Number of Species*		
		Aust	NG	NZ
Mordaciidae	*Mordacia*	2	–	–
Geotriidae	*Geotria*	1	–	1
Dasyatididae	*Himantura*	1	1	–
Pristidae	*Pristis*	1	1	–
Carcharhinidae	*Carcharhinus*	1	1	–
	Glyphis	2	1	–
Ceratodontidae	*Neoceratodus*	1	–	–
Osteoglossidae	*Scleropages*	2	1	–
Elopidae	*Megalops*	1	1	–
Anguillidae	*Anguilla*	4	5	3
Clupeidae	*Clupeoides*	–	2	–
	Nematalosa	1	3	–
	Potamalosa	1	–	–
Engraulididae	*Thryssa*	1	2	–
Retropinnidae	*Retropinna*	2	–	1
	Stokellia	–	–	1
	Prototroctes	1	–	–
Galaxiidae	*Galaxias*	14	–	17
	Galaxiella	3	–	–
	Neochanna	1	–	4
	Paragalaxias	4	–	–
	Lovettia	1	–	–
Lepidogalaxiidae	*Lepidogalaxias*	1	–	–
Ariidae	*Arius*	5	15	–
	Cinetodus	1	1	–
	Cochlefelis	–	2	–
	Doiichthys	–	1	–
	Nedystoma	–	1	–
	Tetranesodon	–	1	–
Plotosidae	*Anodontiglanis*	1	–	–
	Neosiluroides	1	–	–
	Neosilurus	6	7	–
	Oloplotosus	–	3	–
	Plotosus	–	1	–
	Porochilus	3	2	–
	Tandanus	3	–	–
Hemiramphidae	*Arrhamphus*	1	1	–
	Zenarchopterus	2	6	–
Belonidae	*Strongylura*	1	1	–
Melanotaeniidae	*Cairnsichthys*	1	–	–
	Chilatherina	–	9	–
	Glossolepis	–	7	–
	Iriatherina	1	1	–
	Melanotaenia	12	36	–
	Pelangia	–	1	–
	Rhadinocentrus	1	–	–
Pseudomugilidae	*Pseudomugil*	5	11	–
	Kiunga	–	1	–
Atherinidae	*Craterocephalus*	13	6	–
	Leptatherina	1	–	–
Synbranchidae	*Ophisternon*	3	2	–
	Monopterus	1	–	–
Scorpaenidae	*Notesthes*	1	–	–
Centropomidae	*Lates*	1	1	–
Ambassidae	*Ambassis*	6	5	–
	Denariusa	1	1	–
	Parambassis	1	3	–
	Tetracentrum	–	3	–
Percichthyidae	*Bostockia*	1	–	–
	Guyu	1	–	–
	Maccullochella	3	–	–
	Macquaria	5	–	–
Gadopsidae	*Gadopsis*	2	–	–
Nannopercidae	*Edelia*	1	–	–
	Nannatherina	1	–	–
	Nannoperca	4	–	–
Terapontidae	*Amniataba*	1	1	–
	Bidyanus	2	–	–
	Hannia	1	–	–
	Hephaestus	5	10	–
	Leiopotherapon	3	–	–
	Pingalla	3	1	–
	Scortum	4	–	–
	Syncomistes	4	–	–
	Variichthys	1	2	–
Kuhliidae	*Kuhlia*	1	1	–
Apogonidae	*Glossamia*	1	7	–
Lutjanidae	*Lutjanus*	1	2	–
Monodactylidae	*Monodactylus*	1	1	–

Family	Genus	Number of Species* Aust	NG	NZ
Toxotidae	*Toxotes*	3	3	–
Scatophagidae	*Scatophagus*	1	1	–
	Selenotoca	1	–	–
Mugilidae	*Cestraeus*	–	1	–
	Crenimugil	–	1	–
	Liza	1	1	–
	Mugil	1	1	1
	Myxus	1	–	–
	Valamugil	–	1	–
Pinguipedidae	*Cheimarrichthys*	–	–	1
Bovichthyidae	*Pseudaphritis*	1	–	–
Gobiidae	*Afurcagobius*	1	–	–
	Awaous	1	3	–
	Chlamydogobius	5	–	–
	Glossogobius	7	23	–
	Gobius	–	1	–
	Lentipes	–	4	–
	Mugilogobius	–	1	–
	Pseudogobiopsis	–	1	–
	Pseudogobius	1	–	–
	Redigobius	1	1	–
	Schismatogobius	1	–	–
	Sicyopterus	1	5	–
	Sicyopus	–	5	–
	Stenogobius	–	6	–
	Stiphodon	1	7	–
	Tasmanogobius	1	–	–
Eleotrididae	*Bostrichthys*	1	3	–
	Bunaka	1	1	–
	Butis	–	3	–
	Eleotris	–	2	–
	Giurus	1	–	–
	Gobiomorphus	2	–	7
	Hypseleotris	10	2	–
	Kimberleyeleotris	2	–	–
	Milyeringa	1	–	–
	Mogurnda	6	22	–
	Ophiocara	1	1	–
	Oxyeleotris	5	11	–
	Philypnodon	2	–	–
	Prionobutis	1	1	–
	Tateurndina	–	1	–
Kurtidae	*Kurtus*	1	1	–
Pleuronectidae	*Rhombosolea*	–	–	1
Soleidae	*Aseraggodes*	1	1	–
	Brachirus	2	1	–
Cynoglossidae	*Cynoglossus*	1	1	–

* Good regional specific identification keys are available in Fulton (1990), Larson and Martin (1990), Herbert and Peeters (1995) and McDowall (1996b).

In contrast to some previous studies, this review has used common criteria for inclusion of species in all faunas. Only fishes that are limited to freshwater or that must enter freshwater to complete the life cycle or that regularly spend prolonged periods in freshwaters are considered as freshwater species; however, there are many additional marine species that visit freshwater habitats for varying amounts of time (Parker, 1980; Merrick and Schmida, 1984, p.13; Allen, 1991; Allen *et al.*, 2002). Taxonomic work is continuing on a number of groups. However, Table 1 summarizes the present knowledge of the Australasian faunas in terms of currently recognized genera and species. General comments about each of the faunas and brief discussions of particular problems are included in the subsections below.

Australian Fauna

This is the fauna with the widest diversity (42 families, 100 genera and 231 species). As mentioned previously, very few primary species are represented—those groups identified so far include the lungfish, osteoglossids and the lepidogalaxiid.

Although comprising 42 families, 17 of these are only represented by a single species; a further 13 families contain 2–4 species in up to three genera. Another five families include between five and ten species in 2–4 genera. Four families contain between 11 and 20 species (Plotosidae with five genera and 14 species, Atherinidae with two genera and 14 species, Melanotaeniidae with four genera and 15 species, Gobiidae with ten genera and 20 species) and only three families contain more than 20 species. Galaxiidae has five genera comprising 23 species, Terapontidae with nine genera is represented by 24 species and Eleotrididae is the largest (12 genera and 33 species).

Only three families (Ceratodontidae, Lepidogalaxiidae, Gadopsidae) are endemic to Australia. However, generic and specific endemism levels are more significant. Over 40 genera (40% of the fauna) and 155 species (67%) are unique to the Australian mainland.

The fauna is generalized in that most species have wide physiological tolerances, varied diets and standard reproductive strategies (Merrick, 1990). The extant assemblage is the result of interactions of southern, north-eastern and Asian faunas although in some parts of the continent there are local peculiarities. For example, the Tasmanian ichthyofauna consists largely of galaxiids.

New Guinean Fauna

This is the richest fauna (32 families, 79 genera and 290 species)—related to geographic area—but still relatively impoverished compared with Asian faunas. The most recent field guide (Allen, 1991) records 316 native species from New Guinean freshwaters, but acknowledges that many of these are short-term estuarine or marine visitors. *Scleropages jardinii* is the only primary freshwater species present (Allen and Boeseman, 1982).

Twelve families are only represented by a single

species; a further nine families contain 2–5 species in one to five genera. Three families contain between six and ten species in one or two genera; four families contain between 11 and 20 species (Pseudomugilidae with two genera and 11 species, Plotosidae with four genera and 13 species, Ambassidae with four genera and 13 species, Terapontidae with four genera and 14 species). Only four contain more than 20 species. Ariidae has six genera with 21 species, Eleotrididae has nine genera with 37 species, Melanotaeniidae with five genera is represented by 54 species and Gobiidae has eleven genera and 57 species.

None of the families is endemic to New Guinea; however 24 genera (about 30%) and 223 species (over 76%) are restricted to New Guinea. Further taxonomic work is proceeding on a number of families including the Melanotaeniidae, Atherinidae, Gobiidae and Eleotrididae.

Although this is the most poorly known fauna it has been noted that there are relatively large numbers of large species present. The current assemblage is the result of interactions between Asian and Indo-Pacific faunas, combined with a number of geologically recent processes contributing to the current drainage and topographic barriers. This is discussed below and more fully—in relation to other vertebrate groups—on pp.71–93 (Heinsohn and Hope, 2006).

New Zealand Fauna

This is a small fauna by any standards (eight families, 10 genera and 37 species); it is considered, in part, to be the result of the relatively long period during which New Zealand has been isolated from other land masses by oceanic barriers (N.Z.N.F.F.S., 2003a). No primary species are present and, as over 50% of the fauna retains a diadromous life cycle, it is thought that many species arrived via marine dispersal (N.Z.N.F.F.S., 2003a). Four families are represented by a single species; a further two families contain two or three species from one or two genera. The Eleotrididae includes seven species from one genus and the Galaxiidae is the most speciose (with two genera and 21 species).

None of the families is endemic, but three genera (30%) and 31 species (over 80%) are unique to New Zealand; this high degree of endemism at species level, but not in higher taxonomic groupings, suggests that the fauna is not very old. Further taxonomic work is required to fully determine the status of several galaxiid and eleotridid populations which have been recognized as possibly additional species (McDowall and Whitaker, 1975; N.Z.N.F.F.S., 2003b). Dietary studies of this relatively well documented fauna indicate that most species are generalized, opportunistic feeders. The current assemblage is the result of interactions between southern temperate faunas and northern Indo-Pacific faunas (McDowall and Whitaker, 1975; NIWA, 2001).

RADIATIONS

Many mechanisms have been hypothesized to explain distributions of particular species or groups, with much discussion of concepts such as vicariance, dispersal and generalized tracks (Croizat *et al.,* 1974; Rosen, 1976; McDowall, 1981; Nelson and Rosen, 1981; Kirsch, 1984); however, regardless of individual phenomena or local events, it is clear that a number of groups of fishes have undergone major radiations in Australasian freshwaters but not elsewhere.

Unmack (2001a) focussed on factors affecting the distribution and differentiation of aquatic faunas in Australia. Based on levels of endemism, his analyses indicated that: there are some faunal differences within drainage divisions; many species have been present for long periods; climate is the most important influence in most areas; and obvious zones of faunal similarity coincide with low drainage divides or Tertiary volcanism.

Although speciation has been minimal in many areas, it is emphasized that regional geological change and extreme climatic oscillations—especially in the Quaternary—have been of prime importance in determining ranges and movements of extant taxa in limited freshwater habitats (McGowran *et al.*, 2000). For example, arid zones are generally depauperate, for both fishes and other vertebrate groups (Unmack, 2001b); opportunities for radiation are limited and biotas are characterized by diverse genera (many monospecific). The types of ichthyofaunas present are considered to be the result of: (a) a long history of culling and extinction; and (b) few and ephemeral niches in limited pockets of relictual habitat. While the above situation applies to large areas of Australia, by contrast, tectonic movement has been the most important influence on species radiation in New Guinea; extensive volcanism (or other processes) has caused local disruption, separation or isolation of stocks. Furthermore, it is now clear that: evolutionary response in some vertebrates can be very rapid (Frankham *et al.*, 2002); divergence in some isolated populations can be significant in a short time with significant in-breeding depression in small populations (Frankham *et al.*, 2002). Any human impacts (prior to European settlement) would have been localized and very recent (in the Holocene) (Kohen, 1997).

GALAXIIDS

Perhaps the best documented group, both in terms of individual ranges and specific interrelationships, is the family Galaxiidae. Although this family comprises about 50 species extending over a wide range in the Southern Hemisphere, some 22 species occur in southern Australia; Tasmania has the most diverse fauna with 16 species, including 11 endemics (McDowall and Fulton, 1996, p.52). Frankenberg (1974) suggested that the galaxiids originated in south-eastern Australia and that the radiation in Tasmania has been continuing since early Tertiary times. He hypothesized that the radiation and speciation now observed in Tasmania could occur for two reasons. First, conditions for galaxiids in Tasmanian freshwaters were more favourable—no hot, erratic mainland climate. Second, there was reduced competition from other freshwater percoids, due to the relative isolation of Tasmania from the mainland. Unmack (2001a) related high levels of endemism in southern Tasmania to a peninsular effect (being at the extreme southern end of Australia) with the coldest extremes during climatic fluctuations and very effective ocean barriers due to a narrow continental shelf.

The specific relationships established in Tasmania can be related to life cycles. Populations of the diadromous coastal species, such as *Galaxias brevipinnis* and *G. truttaceus,* have progressively become established in lakes or swamps on the Central Plateau and have diverged morphologically from parent species. These derivatives, adapted to quiet waters of a permanent type, are often restricted to one locality or a very limited range. More recent studies on the widespread and variable mainland species *Galaxias olidus* have also indicated some divergence of populations in small isolated catchments in Victoria and eastern highlands further north (Rich, 1987; Raadik, 2001a). A review of the systematics of the *G. olidus* species complex is underway (Raadik, 2001a; Raadik and Kuiter, 2002).

The galaxiids have also radiated in New Zealand where, in numbers of species, they comprise about 60% of the fauna. A number of recent studies have indicated that the current fauna has resulted from:

(a) periodic oceanic dispersal of long-lived marine stages of diadromous species such as *Galaxias maculatus, G. brevipinnis* and *Neochanna cleaveri* from Australia; followed by
(b) landlocking of local populations or loss of marine juvenile stages (to become non-diadromous) and genetic differentiation among populations of widespread species, such as *Galaxias vulgaris* (Allibone *et al.*, 1996);
(c) or regional isolation of particular river systems (altered drainage patterns) where non-diadromous stocks have radiated locally giving rise to species with very restricted distributions. For example, recent revisionary taxonomic studies have resulted in five distinct species being described—of these, four are only found within one river system (McDowall and Wallis, 1996; McDowall, 1997).

In addition to demonstrating genetic structuring within and between galaxiid populations, recent molecular studies have also provided estimates of divergence times for particular stocks. Estimated divergence time for trans-Tasman stocks of *Neochanna* is ~ 8–9 m.y.a. and for *G. brevipinnis* is up to 13 m.y.a.; whereas separation times for Australian species are 0.8–4.3 m.y.a. for *G. truttaceus* and *G. auratus,* and 3–10 m.y.a. for *Paragalaxias mesotes* and *P. julianus* (Waters and White, 1997; Waters *et al.*, 2000b).

Of course, any interpretation of the biogeography of the New Zealand ichthyofauna must consider the violent, regional tectonic history (McDowall, 1996). Although early Tertiary dispersals cannot be discounted, it has to be emphasized that there was little emergent land during Oligocene suggesting that virtually the entire New Zealand biota has resulted from post-Oligocene dispersal. Indeed, McDowall (1997) concludes that the biogeography of both Australian and New Zealand galaxiid species is consistent with post-Oligocene geology—especially events since the start of the Pleistocene.

MELANOTAENIIDS

Melanotaeniidae and the closely related Pseudomugilidae are the only two freshwater fish families solely confined to the northern Australia–New Guinea region; melanotaeniids have been subject to considerable study in recent years (Plate 48). No rainbowfish fossils are known, but there is little doubt that the group originated in Australian freshwaters because there are no closely related forms occurring elsewhere (Allen and Cross, 1982, p.20; Tappin, 2001). Melanotaeniids are considered to be a specialized offshoot of the atherinid evolutionary line and it has been suggested that they evolved from a pseudomugilid ancestor, originally adapted to brackish conditions; over 60 species are described—of the 15 that occur in Australia, three species also have a New Guinean range (Allen and Cross, 1982, p.20; McGuigan *et al.*, 2000; Allen *et al.*, 2002, p.142).

In Australia the majority of extant melanotaeniids occur from northern Western Australia eastwards throughout the Timor Sea, Gulf of Carpentaria and North-east Coast Drainage Divisions, where there is higher rainfall and more suitable habitat. Movement north into New Guinea could have been facilitated by (a) Lake Carpentaria, until about 12,000 years ago—bounded on the east by the Torres Strait land-bridge; and (b) the exposed Western Arafura Shelf (with drainages flowing west to the Timor Sea) which allowed interchange between West New Guinea, Arnhem Land and the Kimberley via streams quite separate from Lake Carpentaria (Tappin, 2001).

From the most recent comprehensive study of melanotaeniid relationships and biogeography, based on both molecular and morphological data, by McGuigan *et al.* (2000) the following conclusions can be drawn:

(a) six monophyletic clades can be identified, which are strongly consistent with morphology;
(b) the distributions of these clades correspond well with Northern and Southern Drainage Divisions (separated by the central ranges) in New Guinea; but do not correlate with divisions on the Australian mainland where, in three clades, sub-ranges are disjunct;
(c) three biogeographic provinces should be recognized for melanotaeniids (North New Guinea, Western New Guinea, Southern New Guinea and Australia);

(d) molecular divergence values suggest that all three provinces were separated ~5 m.y.a., which corresponds with central highland uplift in New Guinea;
(e) current populations of Australian species may be remnants of formerly widespread ancestral taxa, where large-scale extinctions of intervening populations have occurred—perhaps as a result of aridification or Pleistocene climatic oscillations.

Two endemic genera inhabit northern New Guinea and the three species shared with northern Australia are all south of the ranges. The only endemic genera in Australia occur in the far north-east where topography and conditions are similar to those in much of New Guinea. Lower divergences between southern New Guinean and Australian clades and between local Western New Guinea stocks suggest recent speciation and range expansion (McGuigan *et al.*, 2000). The presence of several closely allied rainbows on either side of Torres Strait led Allen (1991, p.12) to suggest speciation within the last 8,000 years. Over 80% of the melanotaeniids now described are found in New Guinea (Tappin, 2001).

Several major factors can be identified as contributing to the rapid, recent speciation of the Melanotaeniidae in New Guinea. These include: evolving a freshwater life cycle; being generalist omnivores with little ecological or morphological specialization; the lowered competition in freshwaters due to the general depauperate nature of fluviatile ichthyofaunas in the region; and the formation of many isolated drainages (resulting from volcanic events or sea level fluctuations). It cannot be assumed that this phase of rapid radiation and speciation has been completed.

ARIIDS

The family Ariidae is another group that has undergone an extensive radiation into Australasian freshwaters, especially in New Guinea. Members of this large group (estimated at 80 species) inhabit freshwater, estuarine and marine environments in tropical and sub-tropical regions throughout the world (Kailola, 1990, 2000; Allen *et al.*, 2002, p.69). The local ariid fauna has not been studied to the same degree as the galaxiids or melanotaeniids, but 23 species of fork-tailed catfishes have been reported from freshwaters. Five *Arius* species are endemic to northern New Guinea, but there are only two endemics (*Arius midgleyi, A. paucus*) in Australia. The other four species occurring in Australian rivers also extend to New Guinea (Kailola, 2000). See Plate 49.

The same geologic events influencing melanotaeniid speciation in New Guinea would also affect ariids; some additional general points are relevant in considering current ranges. First, it is postulated that the ancestral southern fauna originally extended into northern New Guinea; the northern fauna then evolved from species that became isolated by formation of the central mountains (5–6 m.y.a.). Second, the central mountainous region is still mostly devoid of fishes and this is attributed to recent violent geologic events as well as climatic conditions; volcanism and glaciation were concurrent up to 200,000 years ago. Widespread glaciation, with lowered temperatures, persisted until 15,000 years ago. Third, the mountain barrier is almost complete, but a possible valley corridor exchange route, near Mount Hagen, has been identified; this may have been disrupted only 200,000 years ago (Allen, 1991, pp.12–13).

Roberts (1978) points out that the majority of the Fly River ariids occur in the lower to middle reaches and probably represent several ancestral stocks. Many factors contributing to radiation and speciation in this group remain unknown, but it is clear that broad physiological tolerances, very generalized diets and reproduction involving oral incubation would not disadvantage these catfishes in freshwaters. Roberts (1978) lists dietary specializations of Fly River species (to feed on snails, shrimp and aquatic insect larvae) as adaptations to riverine conditions. There is no evidence that the radiation/speciation process in the Ariidae has been as rapid as that in Melanotaeniidae and events such as the dramatic Miocene marine cooling may have accelerated the ariid penetration inland (Archer, 1984).

OTHER GROUPS AND RECENT INTRODUCTIONS

Other families that have undergone extensive radiation and speciation in Australasian freshwaters are the Terapontidae and Plotosidae (Australia and New Guinea), Percichthyidae (Australia) and Eleotrididae (all regions including New Zealand). See Plates 50–53.

Elements of the New Zealand fauna may have originated in Gondwana but transoceanic dispersal is considered very important in the derivation of the existing ichthyofauna there. Comparison shows that New Zealand species have strong affinities with those of south-eastern Australia and Tasmania. It is suggested that the East-Australian Current—running down the east coast past Tasmania, turning east across the Tasman Sea to the western New Zealand coast—has been the main mechanism by which euryhaline Australian freshwater fishes have been carried to New Zealand where speciation then occurred (McDowall and Whitaker, 1975; Waters and White, 1997; McDowall, 2002).

As noted above, most species now inhabiting Australasian freshwaters are marine invaders and there is no reason to think that marine invasions are not continuing. Furthermore, although the above discussion has focussed on radiations of native species, another kind of rapid and short-term dispersal is now occurring as a result of introduced species throughout Australasia. Some researchers have considered that because the local faunas are small and relatively unspecialized they are especially vulnerable to specialized competitive introductions such as *Tilapia* species (McKay, 1977; Kawei, 1989;

McDowall, 1989; Courtenay, 1990; Allen, 1991, p.209, 211; Miller *et al.*, 1999; Heijden, 2002a).

The impacts of introductions or translocations on indigenous species have been widely documented elsewhere (Pollard, 1990; Arthington and McKenzie, 1997; Burrows, 2002; NSW Fisheries, 2003) and detailed discussion is neither appropriate nor possible here. But, as the management of non-indigenous stocks is now such an important element of conservation strategies for indigenous freshwater fishes in Australasia, a comparative summary of introduced species currently established in the wild (Table 2) and some general comments on the situation are warranted.

Historical and Current Exotics

The history of introduced fish species in Australasia can be broadly divided into three phases. Initial attempts (from the 1860s onwards) to establish a small number of salmons, trouts and other European sport fishes (Clements, 1988). A second phase, commencing in the 1950's, to import and establish a small number of sport or food fishes such as common carp, tilapia, quinnat salmon, Atlantic salmon, snakehead gudgeon (Heijden, 2002a, b). A final continuing phase, which commenced in the early 1970s, based on regular importation of large numbers of smaller, ornamental species for the aquarium trade.

The immediate impacts of the initial trout and European species introductions were not studied and are now impossible to estimate, although research on impoundment fisheries suggested that large trout populations exerted direct predation pressures on small, indigenous species such as galaxiids (Tilzey, 1976; McIntosh *et al.*, 1992, 1994; Lintermans, 2000; McDowall, 2003b); declines in other larger native fishes (such as Macquarie perch) were also noted (Jackson, 1981; NSW Fisheries, 2003). Results of the second phase of introductions varied widely. Quinnat salmon have not established despite repeated stockings, but the most recently introduced (new genetic) stock of European carp has caused major problems in south-eastern Australia (Shearer and Mulley, 1978; Merrick and Schmida, 1984, pp.333–335; NSW Fisheries, 2003). More recently, large feral populations of tilapia have developed in eastern Queensland (Bluhdorn et al., 1990; Faragher and Harris, 1994; Raadik, 2003b); but two African cichlids (*Oreochromis mossambicus*, *Tilapia rendalli*) have become an important food source for people in the Sepik and Ramu River systems of northern New Guinea (Allen, 1991, p.209; Miller *et al.*, 1999; Heijden, 2002a, b).

The importation of ornamental species for the aquarium trade has, unfortunately, led to accidental releases and is thought to be associated with the occurrence of disease problems. Some of these released species (e.g. Oriental weatherloach) have now established large feral populations in natural waterways (Lintermans and Burchmore, 1996; Pratt, 1999, personal communication; Allen *et al.*, 2002, p.363; Koster *et al.*, 2002).

If ornamental species are included, it has to be acknowledged that it has been legally possible to import many hundreds of exotic fishes over the years; indeed a recent national inventory indicated that 1181 exotic species have been recorded in Australia in the last 40 years (Raadik, 2003a). The following discussion is confined to those species that have established wild populations. Without going into specific details, or recent quarantine (regulatory) initiatives, the current situation is summarized below.

In Australia some 31 non-indigenous species (representing 7 families) have been documented in the wild in various parts of the continent (see Table 2); another nine cichlids and two cyprinids have also been recorded, but are not known to be established (Raadik, 2004). Several salmonids are the bases of valuable recreational fisheries in restricted cooler areas, but other small species (e.g. *Gambusia* spp.) have spread over wide areas (Allen *et al.*, 2002). In New Guinea some 15 non-indigenous species (from 9 families) have been recorded (Table 2), including two cichlids which are now important in small lake fisheries in the highlands and northern drainages (Allen, 1991; Miller *et al.*, 1999; Heijden, 2002a, b). Werry (1998) reported that 30 species had been introduced between 1930 and 1998. The list includes seven Australian native species which have not survived. The status of the remaining species is uncertain, but there is no evidence that they have established in the wild. The New Zealand situation is similar to Australia—with 21 species from 5 families reported (NIWA, 2001; Raadik, 2003a); several salmonids are important in recreational fisheries but, as Table 2 shows, other exotics have been introduced for varying reasons.

The interactions between native and non-indigenous species are poorly known and this aspect of research is now being acknowledged and afforded a higher priority, especially as introduced species are significant components of the faunas in many areas (Cashner *et al.,* 1999; Pratt, 1999, personal communication; NSW Fisheries, 2003). Just as it is clear that many introductions are now permanent features of local fish faunas, so their relative ecological importance in particular areas will also change with time. Many freshwater fisheries, based on exotic species, are centred on large lakes or impoundments and their associated feeder streams, most of which have been degraded to some degree. In formulating future management of these populations, with ongoing impacts or remedial habitat strategies, it is important to consider the resilient capacity now documented in aquatic ecosystems (Cottingham and Carpenter, 1994; Carpenter and Cottingham, 1997).

MANAGEMENT STRATEGIES

Components of all three indigenous faunas discussed have suffered declines to varying degrees as a result of environmental modifications. The various physical processes and chemical changes to Australian aquatic environments have been well documented (Merrick and

Table 2. Non-indigenous fishes documented as established in the wild in Australasia (based on Merrick and Rimmer, 1984; Merrick and Schmida, 1984; Clements, 1988; Allen, 1991, 1996a; Miller *et al.*, 1999; Barlow, 2000; NIWA, 2001; Allen *et al.*, 2002, pp.357–371; Heijden, 2002a, b; Raadik, 2001b, 2003a, b, 2004).

Family, Genus, Species	Range			Comments
	AU	NG†	NZ	
Salmonidae (*)				
Oncorhynchus mykiss	x	x	x	Recreational angling, aquaculture; native to western North America and eastern Siberia
Oncorhynchus nerka	–	–	x	Recreational angling; native to western North America
Oncorhynchus tshawytscha	x	–	x	Recreational angling; limited range in Victoria—replenished annually from hatchery stock; native to western North America and north-east Asia.
Salmo salar	x	–	x	Recreational angling, populations NOT self-sustaining.
Salmo trutta	x	x	x	Recreational angling, aquaculture; native to Europe
Salvelinus fontinalis	x	–	x	Recreational angling; native to eastern North America.
Salvelinus namaycush	–	–	x	Recreational angling; limited to single population; native to North America.
Ictaluridae	–	–	x	Native to North America
Cyprinidae				
Carassius auratus	x	x	x	Ornamental, aquaculture; native to eastern Asia.
Ctenopharygnodon idella	–	–	x	Water weed control; NOT self-sustaining.
Cyprinus carpio	x	x	x	Recreational angling, food; native to Asia.
Hypophthalmichthys molitrix	–	–	x	Phytoplankton control; NOT self sustaining.
Leuciscus idus	–	–	x	Recreational angling; native to northern Europe.
Puntius conchonuis	x	–	–	Ornamental; native to northern India.
Scardinius erythrophthalmus	–	–	x	Native to Europe.
Rutilus rutilus	x	–	–	Recreational angling; native to Europe.
Tanichthys albonubes	x	–	–	Ornamental; reported at single site; native to China.
Tinca tinca	x	–	x	Recreational angling; native to Europe.
Cobitidae				
Misgurnus anguillicaudatus	x	–	–	Ornamental; native to Asia.
Clariidae				
Clarias batrachus	–	x	–	Native to S.E. Asia.
Poeciliidae				
Gambusia dominicensis	x	–	–	Mosquito control; native to West Indies.
Gambusia holbrooki	x	x	x	Mosquito control, native to Central America

Schmida, 1984; Cullen and Lake, 1995). In Papua New Guinea widespread degradation of waterways has been reported due to logging and mining (Mowbray, 1996, personal communication); similarly, in western New Guinea, researchers have also expressed concern about conservation of ichthyofaunas threatened by these activities (Allen, 1996). The reductions in abundance and ranges are often accompanied by fragmentation of distribution, which poses a new set of management problems related to genetics. In addition to localized habitat modification, in the last decade, it has become clear that human-induced climate change will have profound effects on freshwater habitats and ichthyofaunas throughout Australasia. Known impacts include: earlier reproduction, due to rising temperatures; reduction in river flows; extensive salt intrusion into low-lying swamps with a small increase in sea level (Chilcott *et al.*, 2003; Howden, 2003).

Although the genetics of Australasian native fishes is generally poorly known, three conclusions can be drawn from current information: (a) considerable genetic variation has been demonstrated between regional wild populations of some widespread species; (b) more sensitive DNA sequencing will undoubtedly show more intra-populational variation; (c) significant in-breeding depression may occur rapidly in some isolated small populations. These factors have to be considered when decisions relating to captive breeding, stock enhancement or re-stocking are required (NSW Fisheries, 2003).

Many of the factors which influence conservation of freshwater fishes in Australia (and elsewhere) have been summarized in Merrick (1997) as well as Morris *et al.* (2001) and Pusey and Arthington (2003). With increased research as well as public pressure and awareness, the native faunas are being assessed on a regular basis for conservation status. Categories have varied locally (Crook, 2000) but the IUCN series (see Table 3) is now internationally accepted.

At present in Australia 10 freshwater species are considered Critically Endangered, another 10 Endangered, 19

Table 2 *continued.*

Family, Genus, Species	Range			Comments
	AU	NG†	NZ	
Poeciliidae *continued*				
Phalloceros caudimaculatus	x	–	x	Ornamental; native to eastern South America.
Poecilia latipinna	x	–	x	Ornamental; native to southern North America
Poecilia reticulata	x	x	x	Ornamental; native to northern South America.
Xiphophorus helleri	x	x	x	Ornamental; native to eastern Central America.
Xiphophorus maculatus	x	–	–	Ornamental; native to eastern Central America.
Percidae				
Perca fluviatilis	x	–	x	Recreational angling; native to temperate Europe.
Cichlidae				
Aequidens pulcher	x	–	–	Ornamental; native to Central America
Cichlasoma nigrofasciatum	x	–	–	Ornamental; native to Central America.
Amphilophus (Cichlasoma) citrinellum	x	–	–	Ornamental; native to Central America
Amphilophus (Cichlasomz) labiatus	x	–	–	Ornamental; native to Central America
Astronotus ocellatus	x	–	–	Ornamental; native to South America
Haplochromis (Astatotilapia) burtoni	x	–	–	Ornamental; native to northern Africa
Hemichromis guttatus	x	–	–	Ornamental; native to Africa
Labeotropheus sp.	x	–	–	Ornamental; native to East Africa
Oreochromis mossambicus	x	x	–	Recreational, food species; native to East Africa.
Tilapia mariae	x	–	–	Ornamental; native to west Africa.
Tilapia rendalli	–	x	–	Food species; native to southern Africa.
Anabantidae				
Anabas testudineus	–	x	–	Native to south-east Asia.
Belontiidae				
Trichogaster pectoralis	–	x	–	Food fish; native to south-east Asia.
Trichogaster trichopterus	x	x	–	Ornamental with limited range around Port Moresby and in northern Queensland; native to south-east Asia.
Osphronemidae				
Osphronemus goramy	–	x	–	Food fish; widely distributed; native to western Indonesia.
Channidae				
Channa striata	–	x	–	Food fish; native to south-east Asia.

(*) Some other European trout stocks are known to have been stocked in Australia but have not been reported in recent years, although they may persist in small numbers in a few restricted upper catchments.

† Unlike the situation in Australia and New Zealand where a number of the larger species were initially introduced as sport, in New Guinea the prime objective of a number of introductions was to enhance subsistence fisheries.

Vulnerable and 27 at Lower Risk; with 10 considered Data Deficient (Crook and Pogonoski, 2003). Figures are not as precise for New Guinea but at least 12 species are considered threatened (W.C.M.C., 2000a, b). In New Zealand the southern grayling (*Prototroctes oxyrhynchus*) has been declared Extinct. *Galaxias cobitinis* is considered Critically Endangered, with another two galaxiids listed as Endangered; a further five galaxiids and an eleotridid are assessed as Lower Risk (N.Z.N.F.F.S., 2003b). As more research becomes available on native species so there will be changes in the numbers in the respective conservation categories over time. For example, at one stage a few years ago the Australian grayling was considered endangered, but then detailed surveys revealed a number of additional populations and the status has been downgraded to Lower Risk (Farragher and Harris, 1994; Crook and Pogonoski, 2003). In another initiative, advances in captive breeding techniques have enabled stocks of the endangered trout cod (*Maccullochella macquariensis*) to be re-introduced to managed environments, where natural recruitment has now been reported (Douglas and Brown, 2000); in the longer term this may contribute to a change in conservation category. Likewise the large number of species currently placed in the indeterminate category will fall as more data become available; a number of New Zealand galaxiids (both recognized species as well as potentially new species) are still poorly documented, but this will change as new taxa are investigated (N.Z.N.F.F.S., 2003b).

The management situation is further complicated by translocations of native species between drainages, often to areas outside their known natural ranges (Leggett and Merrick, 1997; Anon., 1999; Low, 1999; Rowland, 2001; NSW Fisheries, 2003); but, in recent years, many recommendations have been made about the rehabilitation and

Table 3. Summary of conservation status categories and prime criteria for Australian fishes. Ratings are based on the IUCN system (Crook, 2000).

Classification*	Criteria
EXTINCT	Taxa where there is no reasonable doubt that the last individual has died.
EXTINCT IN THE WILD	Taxa known only to survive in cultivation, in captivity or as a naturalized population (or populations) well outside the past range.
CRITICALLY ENDANGERED	Taxa facing an extremely high risk of extinction in the wild in the immediate future.
ENDANGERED	Taxa facing a very high risk of extinction in the wild in the near future.
VULNERABLE	Taxa facing a high risk of extinction in the wild in the medium-term future.
LOWER RISK	Taxa that when evaluated do not satisfy the criteria for any of the above categories; there are three sub-categories:
1) CONSERVATION DEPENDENT	Taxa which are the focus of a continuing taxon-specific or habitat-specific conservation programme targeted towards the taxon in question, the cessation of which would result in the taxon qualifying for one of the threatened categories above within five years.
3) NEAR THREATENED	Taxa which do not qualify for Conservation Dependent, but which are close to qualifying for Vulnerable.
3) LEAST CONCERN	Taxa which do not qualify for Conservation Dependent or Near Threatened.
DATA DEFICIENT	Taxa where there is inadequate information to make a direct or indirect assessment of their risk of extinction based on distribution and /or population status.
NOT EVALUATED	Taxa that have not yet been assessed against the criteria.

* Similar series of categories (with some variation in names or definitions) have been, or still are, used in some regions/organizations.

maintenance of water quality, aquatic habitats and biota (Merrick, 1995, 1997; Carpenter and Cottingham, 1997; NSW Fisheries, 2003). While some uncertainty and controversy continues, relating to specific strategies in particular areas, the following general findings are agreed:

(a) waterways and the immediate riparian zones are, for many reasons (e.g. drought refuges, concentrated biodiversity, migration routes, repositories of all catchment run-off), key focus areas in any integrated catchment management;
(b) in limited areas where active restoration followed by protective management and monitoring has been implemented, significant improvements in water quality and regeneration or re-colonization have occurred within a short time (a few years);
(c) representative groups of aquatic faunas (e.g. omnivorous or carnivorous fishes), are integrative in terms of habitat processes and, can be used as indicators of system health;
(d) effective conservation has to be system-based (single species or species-by-species management is not feasible), with predictive programmes utilizing a wide range of biotic and abiotic criteria.

CONCLUSIONS

The main Australasian freshwater fish faunas are depauperate but diverse—both in morphology and origin. Four major lineages are present but the overwhelming majority of species are teleosts that invaded from marine environments during the Tertiary. Very few primary species are represented, some common Asian families are absent and other widespread families have undergone major radiations in Australasian freshwaters. Using a common set of criteria over 480 species from 44 families have been identified; many species are shared between northern Australia and New Guinea but only *Anguilla reinhardtii* and *Mugil cephalus* are common to New Zealand and both other faunas.

The family key demonstrates the diversity within Australasian freshwater ichthyofaunas, with the comparative list of taxa highlighting the many monospecific or small genera present. While further research may result in re-assessment of the significance of particular distinguishing features and inter-generic relationships, the validity of most of these genera is not seriously questioned at present. But recent studies, in both Australia and New Guinea, demonstrate that further significant change may be necessary in several families. For example, the revision of the *Chlamydogobius eremius* complex (Family Gobiidae), the discovery of new eleotridid species in northern New Guinea and the additional species of melanotaeniids as surveys of more remote areas are completed .

The main events affecting the speciation and distribution of Australasian freshwater fishes are: the detachment of New Zealand and gradual splitting of Australia from Gondwana with subsequent complete separation of Tasmania from Antarctica; the opening of the Tasman Sea and subsequent evolution of island groups such as Fiji; the relative position of the Indian subcontinent to Australia until 50 million years ago; the world-wide marine regression in Late Miocene times—many Pacific shallow shelf areas became inland lakes; the recent development of New Guinea with considerable volcanicity and dramatic uplift of central and northern areas; the prolonged, repeated emergence and drowning of extensive land-bridges formed by the Sahul Shelf between Australia and New Guinea; oceanic current patterns and prevailing wind regimes; and Pleistocene oscillations with glaciations, lowered temperatures and regional aridity.

The interrelationships of early teleosts are still uncertain, but it is clear that the origins of many groups in the extant faunas extend back to the early Tertiary period or beyond; some of the more derived species appear to have changed little over millions of years. Molecular studies may help to clarify the sequence of derivation, age and relationships of some groups. Although familial endemism is low, generic and specific endemism is significant. Only the Anguillidae, Mugilidae and Eleotrididae are common to all Australasian regions; in each case a single genus with an extensive range is involved. Of the 126 genera represented, 53 are common to New Guinea and Australia—over 60 species are shared. The Pinguipedidae and Pleuronectidae from New Zealand are not represented in either of the other faunas. All the other families are shared with Australia. Seven genera are common to Australia and New Zealand and six species have trans-Tasman ranges.

The importance of geographic barriers to aquatic faunas is explained and a number of major Australasian radiations are briefly discussed. It is suggested that galaxiids originated in south-eastern Australia with the subsequent Tasmanian radiation and speciation related to a favourable climate and reduced competition; the derived species, adapted to the quiet waters of the Central Plateau, generally have restricted ranges. Although vicariance may have played some role the New Zealand fauna is considered to be largely the result of oceanic dispersal of pelagic juvenile marine stages of diadromous species from south-eastern Australia—with subsequent colonization and local radiations as postulated for Tasmania. The true extent of galaxiid speciation (in both Australia and New Zealand) has only become clearer in recent years, with the aid of genetic techniques.

The development of a freshwater life cycle, broad physiological tolerances, relatively low competition in Australasian freshwaters and the recent formation of many isolated drainages are all contributing factors to the sometimes rapid speciations that have occurred in groups such as the Melanotaeniidae and Ariidae, especially in New Guinea. Unlike the situation in New Guinea, the geomorphological and climatological information suggests that once a species entered Australian mainland freshwaters, rapid dispersal over wide areas was possible. The slow erection of geographical barriers during the Miocene and Pliocene would have caused a gradual reduction in exchange routes; faunal disruption could be expected to be minimal. Population exchange prior to the Pleistocene and the more recent fluctuating conditions would have minimized speciation.

In fact the composition of the Australian ichthyofauna is consistent with the combination of processes listed above. A number of monospecific or small genera are present, indicating age where ecological niches are occupied (for survival) but conditions are not conducive to radiation. The one area where considerable sub-speciation seems to have occurred is the extremely arid centre of Australia, where exchange would have been more limited for the longest time (e.g. *Chlamydogobius* species group).

In contrast to the Plotosidae, which is considered to have northern origins, it has been suggested that the Percichthyidae and Terapontidae could have originated in the south. As mentioned previously transoceanic dispersal by the East-Australian Current is considered important in the derivation and subsequent speciation of New Zealand freshwater fishes.

In the current world climate of environmental destruction and modification it is essential that faunal diversity be documented urgently to assist in future management. This is particularly important locally because Australasia is probably one of the few regions where considerable biotic diversity should be able to be maintained in the forseeable future. Relatively large numbers of exotic fish species have been introduced and a few, such as carp and the mosquito fish, have become very widespread. Because of their long-term ecological effects and economic or recreational fishery importance, the management policies for introduced species must be afforded higher priority and integrated with conservation strategies for indigenous ichthyofaunas.

Indigenous fish faunas have declined due to widespread environmental degradation and at least 35 species are currently considered endangered or threatened; however, system-based active catchment management is the key to the future.

ACKNOWLEDGEMENTS

Appreciation is expressed to Gerald Allen (Perth), David Mowbray (Port Moresby) and Bryan Pratt (Canberra) who have provided unpublished information; John Cleasby (Sydney) prepared the figures. Thanks also to Dave Pollard (Sydney) and Tarmo Raadik (Melbourne) for critical comments on the manuscript.

REFERENCES

Allen, G.R. 1985. Description of two new species of freshwater catfishes (Plotosidae) from Papua New Guinea. *Records of the Western Australian Museum,* Vol. 12, Pt. 3, pp.247–256.

Allen, G.R. 1987. *Melanotaenia iris,* a new freshwater rainbow fish (Melanotaeniidae) from Papua New Guinea with notes on the fish fauna in head waters. *Japanese Journal of Ichthyology,* Vol. 34, pp.15–20.

Allen, G.R. 1989. *Freshwater Fishes of Australia.* T.F.H. Publications, Neptune City, New Jersey. 240pp.

Allen, G.R. 1991. *Field Guide to the Freshwater Fishes of New*

Guinea. Christensen Research Institute, Madang, Papua New Guinea. 268pp.

Allen, G.R. 1993. *Variichthys*, a replacement name for the tetrapontid fish genus *Varia* and first record of *V. lacustris* from Australia. *Records of the Western Australian Museum*, Vol. 16, Pt. 3, pp.459–460.

Allen, G.R. 1996a. Freshwater fishes of Irian Jaya. Pp.15–21. In Kitchener, D.J. and Suyanto, A. (eds). *Proceedings of the First International Conference on Eastern Indonesian—Australian Vertebrate Fauna, Manado, November 22–26, 1994.* Western Australian Museum, Perth. 174pp.

Allen, G.R. 1996b. Two new species of rainbowfishes (*Melanotaenia*: Melanotaeniidae), from the Kikori River system, Papua New Guinea. *Revue Francaise d'Aquariologie*, Vol. 23, Nos. 1–2, pp.9–16.

Allen, G.R. 1997. *Lentipes watsoni*, a new species of freshwater goby (Gobiidae) from Papua New Guinea. *Ichthyological Exploration of Freshwaters,* Vol. 8, No. 1, pp.33–40.

Allen, G.R. 1998. A new genus and species of rainbowfish (Melanotaeniidae) from fresh waters of Irian Jaya, Indonesia. *Revue Francaise d'Aquariologie*, Vol. 25, Nos. 1–2, pp.11–16.

Allen, G.R. 2001. *Lentipes multiradiatus*, a new species of freshwater goby (Gobiidae) from Irian Jaya, Indonesia. *Aqua, Journal of Ichthyology and Aquatic Biology,* Vol. 4, No. 3, pp.121–124 .

Allen, G.R. and Berra, T.M. 1989. Life history aspects of the West Australian salamanderfish. *Records of the Western Australian Museum,* Vol. 14, Pt. 3, pp.253–267.

Allen, G.R. and Boeseman, M. 1982. A collection of freshwater fishes from western New Guinea with descriptions of two new species (Gobiidae and Eleotridae). *Records of the Western Australian Museum,* Vol. 10, Pt. 2, pp.67–103.

Allen, G.R. and Burgess, W.E. 1990. A review of the glassfishes (Chandidae) of Australia and New Guinea. *Records of the Western Australian Museum, Supplement*, No. 34, pp.139–206.

Allen, G.R. and Coates, D. 1990a. An ichthyological survey of the Sepik River, Papua New Guinea. *Records of the Western Australian Museum, Supplement*, No. 34, pp.31–116.

Allen, G.R. and Coates, D. 1990b. A new species of freshwater eleotridid fish from northern Papua New Guinea. *Records of the Western Australian Museum, Supplement*, No. 34, pp.131–137.

Allen, G.R. and Cross, N.J. 1982. *Rainbowfishes of Australia and Papua New Guinea*. Angus and Robertson, Sydney. 141pp.

Allen, G.R. and Feinberg, M.N. 1998. Descriptions of a new genus and four new species of freshwater catfishes (Plotosidae) from Australia. *Aqua, Journal of Ichthyology and Aquatic Biology*, Vol. 3, No. 1, pp.9–18.

Allen, G.R. and Hoese, D.F. 1986. The eleotrid fishes of Lake Kutubu, Papua New Guinea with descriptions of four new species. *Records of the Western Australian Museum,* Vol. 13, Pt. 1, pp.79–100.

Allen, G.R. and Hoese, D.F. 1991. A review of the genus *Mogurnda* (Pisces: Eleotrididae) from New Guinea with descriptions of three new species. *Ichthyological Exploration of Freshwaters,* Vol. 2, No. 1, pp.31–46.

Allen, G.R. and Ivantsoff, W. 1985. Deux nouvelles especes de blue-eyes (*Pseudomugil:* Melanotaeniidae) de Nouvelle-Guinee. *Revue Francaise d'Aquariologie,* Vol. 12, pp.85–88.

Allen, G.R., Ivantsoff, W., Shepherd, M.A. and Renyaan, S.J. 1998. *Pseudomugil pellucidus* (Pisces: Pseudomugilidae), a newly discovered blue-eye from Timika-Tembagapura region, Irian Jaya. *Aqua, Journal of Ichthyology and Aquatic Biology,* Vol. 3, No. 4, pp.1–8.

Allen, G.R. and Jebb, M. 1993. A collection of fishes from the upper Purari River system, Papua New Guinea, with descriptions of two new species (Terapontidae and Eleotrididae). *Ichthyological Exploration of Freshwaters,* Vol. 4, No. 3, pp.233–240.

Allen, G.R. and Jenkins, A.P. 1999. A review of the Australian freshwater gudgeons, genus *Mogurnda* (Eleotridae) with descriptions of three new species. *Aqua, Journal of Ichthyology and Aquatic Biology*, Vol. 3, No. 4, pp.141–156.

Allen, G.R., Larson, H.K. and Midgley, S.H. 1993. A new species of *Scortum* Whitley (Pisces: Terapontidae) from the Northern Territory, Australia. *The Beagle (Occasional Papers of the Northern Territory Museum of Arts and Sciences)*, Vol. 10, No. 1, pp.71–74.

Allen, G.R., Midgley, S.H. and Allen, M. 2002. *Field Guide to the Freshwater Fishes of Australia.* Western Australian Museum, Perth. 394pp.

Allen, G.R., Parenti, L.R. and Coates, D. 1992. Fishes of the Ramu River, Papua New Guinea. *Ichthyological Exploration of Freshwaters,* Vol. 3, No. 4, pp.289–304.

Allen, G.R. and Pusey, B.J. 1999. *Hephaestus tulliensis* De Vis, a valid species of grunter (Pisces: Terapontidae) from fresh waters of north-eastern Queensland, Australia. *Aqua, Journal of Ichthyology and Aquatic Biology,* Vol. 3, No. 4, pp.157–162.

Allen, G.R. and Renyaan, S.J. 1996a. *Chilatherina pricei,* a new species of rainbowfish (Melanotaeniidae) from Irian Jaya. *Revue Francaise d'Aquariologie,* Vol. 23, Nos. 1–2, pp.5–8.

Allen, G.R. and Renyaan, S.J. 1996b. Eleotrid fishes of the Triton Lakes, Irian Jaya, with descriptions of four new species. *Revue Francaise d'Aquariologie,* Vol. 23, Nos. 1–2, pp.47–56.

Allen, G.R. and Renyaan, S.J. 1998. Three new species of rainbowfishes (Melanotaeniidae) from Irian Jaya, Indonesia. *Aqua, Journal of Ichthyology and Aquatic Biology*, Vol. 3, No. 2, pp.69–80.

Allen, G.R. and Renyaan, S.J. 1999. *Pseudomugil ivantsoffi* (Pseudomugilidae), a new species of blue-eye from the Timika region, Irian Jaya. *Aqua, Journal of Ichthyology and Aquatic Biology,* Vol. 3, No. 4, pp.163–168.

Allibone, R.M., Crowl, T.A., Holmes, J.M., King, T.M., McDowall, R.M., Townsend, C.R. and Wallis, G.P. 1996. Isozyme analysis of *Galaxias* species (Teleostei: Galaxiidae) from the Taieri River, South Island, New Zealand: a species complex revealed. *Biological Journal of the Linnean Society*, Vol. 57, pp.107–126.

Anon. 1999. *National Policy for the Translocation of Live Aquatic Organisms. Issues, Principles and Guidelines for Implementation.* Ministerial Council on Forestry, Fisheries and Aquaculture, Canberra. 31pp.

Aplin, K.P. 1998. Vertebrate zoogeography of the Bird's Head of Irian Jaya, Indonesia. Pp.803–890. In Miedema, J., Odé,

C. and Dam, R.A.C. (eds.). *Perspectives on the Bird's Head of Irian Jaya, Indonesia.* Natural History Museum, Leiden.

Archer, M. 1984. Earth-shattering concepts for historical zoogeography. Pp.45–59. In Archer, M. and Clayton, G. (eds). *Vertebrate Zoogeography and Evolution in Australasia.* Hesperian Press, Perth. 1203pp.

Arthington, A. and McKenzie, F. 1997. Review of impacts of displaced/introduced fauna associated with inland waters. http://www.ea.gov.au/soe/techpapers/series1/14fauna.html

Barlow, G.W. 2000. *Cichlid Fishes: Nature's Grand Experiment in Evolution.* Perseus Publishing, Cambridge, Massachusetts. 335pp.

Basden, A., Trinajstic, K. and Merrick, J.R. 2006. Eons of fishy fossils. Pp.131–157. In Merrick, J.R., Archer, M., Hickey, G.M. and Lee, M.S.Y. (eds). *Evolution and Biogeography of Australasian Vertebrates.* Auscipub Pty. Ltd., Sydney. 942pp.

Beattie, A.J. (ed). 1995. *Biodiversity, Australia's Living Wealth.* Reed Books, Sydney. 131pp.

Bertozzi, T., Adams, M. and Walker, K.F. 2000. Species boundaries in carp gudgeons (Eleotrididae: *Hyseleotris*) from the River Murray, South Australia: evidence for multiple species and extensive hybridization. *Marine and Freshwater Research,* Vol. 51, No. 8, pp.805–815.

Bluhdorn, D.R., Arthington, A.H. and Mather, P.B. 1990. The introduced cichlid, *Oreochromis mossambicus* in Australia: a review of distribution, population genetics, ecology, management, management issues and research priorities. Pp.83–92. In Pollard, D.A. (ed). Translocated fishes and their ecological effects. *Australian Society for Fish Biology Workshop: Bureau of Rural Resources Proceedings 8,* Canberra. 181pp.

Bond, C.E. 1979. *Biology of Fishes.* W.B. Saunders, Philadelphia. 514pp.

Burrows, D. 2002. Managing the spread of translocated native fish into aquatic protected areas in Queensland. P.52. In *World Congress on Aquatic Protected Areas, What Works Best and How do we Know? 14-17 August 2002, Cairns, Queensland, Australia.* Australian Society for Fish Biology (abstract only).

Carpenter, S.R. and Cottingham, K.L. 1997. Resilience and restoration of lakes. http://www.consecol.org/vol1/iss1/art2

Carvalho, M.R. de. 1996. Higher-level elasmobranch phylogeny, basal squaleans and paraphyly. Pp.35–62. In Stiassny, M.L.J., Parenti, L.J. and Johnson, G.D. (eds). *Interrelationships of Fishes.* Academic Press, San Diego. 496pp.

Cashner, R.C., Hawkes, G.P., Gartside, D.F. and Marsh-Matthews, E. 1999. Fishes of the Nymboida, Mann and Orara Rivers of the Clarence River drainage, New South Wales, Australia. *Proceedings of the Linnean Society of New South Wales,* Vol. 121, pp.89–100.

Centre for Pest Information Technology and Transfer, University of Queensland. 2001. *Lucid Professional Version 2.1. Tools for Identification and Diagnosis.* CSIRO Publishing.

Chilcott, C., Hilbert, D. and Howden, M. 2003. Chapter 5. Modelling biodiversity and climate change. Pp.63–66. In Howden, M., Hughes, L., Dunlop, M., Zethoven, I., Hilbert, D. and Chilcott, C. (eds). *Climate Change Impacts on Biodiversity in Australia. Outcomes of a Workshop Sponsored by the Biological Diversity Advisory Committee, 1–2 October 2002.* Commonwealth of Australia, Canberra. 98pp.

Clements, J. 1988. *Salmon at the Antipodes. A History and Review of Trout, Salmon and Char and Introduced Coarse Fish in Australasia.* J. Clements, Victoria. 391pp.

Cloutier, R. and Ahlberg, P.E. 1996. Morphology, characters and interrelationships of basal sarcopterygians. Pp.445–479. In Stiassny, M.L.J., Parenti, L.J. and Johnson, G.D. (eds). *Interrelationships of Fishes.* Academic Press, San Diego. 496pp.

Conlon, J.M., Wang, Y. and Potter, I.C. 2001. The structure of *Mordacia mordax* insulin supports the monophyly of the Petromyzontiformes and an ancient divergence of Mordaciidae and Geotriidae. *Comparative Biochemistry and Physiology,* Part B, Vol. 129, Issue 1, pp.65–71.

Cottingham, K.L. and Carpenter, S.R. 1994. Predictive indices of ecosystem resilience in models of north temperate lakes. *Ecology,* Vol. 75, pp.2127–2138.

Courtenay, W.R. Jnr. 1990. Fish introductions and translocations and their impacts in Australia. Pp.171–179. In Pollard, D.A. (ed). Translocated fishes and their ecological effects. *Australian Society for Fish Biology Workshop: Bureau of Rural Resources Proceedings 8,* Canberra. 181pp.

Croizat, L., Nelson, G. J. and Rosen, D.E. 1974. Centres of origin and related concepts. *Systematic Zoology,* Vol. 23, pp.265–287.

Crook, D. 2000. Threatened fishes committee report. *Australian Society for Fish Biology Newsletter,* Vol. 30, No. 2, pp.41–50.

Crook, D. and Pogonoski, J. 2003. Threatened fishes committee report December 2002. *Australian Society for Fish Biology Newsletter,* Vol. 32, No. 2, pp.45–52.

Crowley, L.E.L.M. 1990a. Biogeography of the endemic freshwater fish *Craterocephalus* (Family Atherinidae). *Memoirs of the Queensland Museum,* Vol. 28, Pt. 1, pp.89–98.

Crowley, L.E.L.M. 1990b. *Biogeography and the systematics of the endemic freshwater fish genus* Craterocephalus *(Family Atherinidae).* Ph.D. Thesis (unpublished), School of Biological Sciences, Macquarie University, North Ryde.

Crowley, L.E.L.M. and Ivantsoff, W. 1990a. A second hardyhead, *Craterocephalus gloveri* (Pisces: Atherinidae), from Dalhousie Springs, central Australia. *Ichthyological Exploration of Freshwaters,* Vol. 1, No. 2, pp.113–122.

Crowley, L.E.L.M. and Ivantsoff, W. 1990b. A review of species previously identified as *Craterocephalus eyresii* (Pisces: Atherinidae). *Proceedings of the Linnean Society of New South Wales,* Vol. 112, pp.87–103.

Crowley, L.E.L.M. and Ivantsoff, W. 1992. Redefinition of the freshwater fish genus *Craterocephalus* (Teleostei: Atherinidae) of Australia and New Guinea with an analysis of three species. *Ichthyological Exploration of Freshwaters,* Vol. 3, No. 3, pp.273–287.

Crowley, L.E.L.M., Ivantsoff, W. and Allen, G.R. 1991. Freshwater fishes of the genus *Craterocephalus* (Atherinidae) from the southern drainages of Papua New Guinea and Irian Jaya with reference to *C. stercusmuscarum* from Australia. *Records of the Western Australian Museum,* Vol. 15, Pt. 1, pp.33–52.

Crowley, L.E.L.M., Ivantsoff, W. and Allen, G.R. 1995. Description of a new species of hardyhead, *Craterocephalus fistularis* (Pisces: Atherinidae) from Irian Jaya. *Records of the Western Australian Museum*, Vol. 17, Pt. 3, pp.325–329.

Cullen, P. and Lake, P.S. 1995. Water resources and biodiversity: past, present and future problems and solutions. Pp.115–125. In Bradstock, R.A., Auld, T.D., Keith, D.A., Kingsford, R.T., Lunney, D. and Sivertsen, D.P. (eds). *Conserving Biodiversity: Threats and Solutions.* Surrey Beatty and Sons, Sydney. 428pp.

Department of Environment, Sport and Territories. 1996. *Australia: State of the Environment 1996.* CSIRO Publishing, Melbourne.

Donaldson, T.J. and Myers, R.F. 2002. Insular freshwater fish faunas of Micronesia: patterns of species richness and similarity. *Environmental Biology of Fishes,* Vol. 65, pp.139–149.

Douglas, J.P. and Brown, P. 2000. Notes on successful spawning and recruitment of a stocked population of the endangered Australian freshwater fish, trout cod, *Maccullochella macquariensis* (Cuvier) (Percichthyidae). *Proceedings of the Linnean Society of New South Wales,* Vol. 122, pp.143–147.

Douglas, J.W., Gooley, G.J., Ingram, B.A., Murray, N.D. and Brown, L.D. 1995. Natural hybridization between Murray cod, *Maccullochella peelii peelii* (Mitchell) and trout cod, *Maccullochella macquariensis* (Cuvier) (Percichthyidae), in the Murray River, Australia. *Marine and Freshwater Research,* Vol. 46, No. 4, pp.729–734.

Dudgeon, D. 2000. Conservation of freshwater biodiversity in Oriental Asia: constraints, conflicts, and challenges to science and sustainability. *Limnology,* Vol. 1, pp.237–243.

Faragher, R.A. and Harris, J.H. 1994. The historical and current status of freshwater fish in New South Wales. *Australian Zoologist*, Vol. 29, No. 3–4, pp.166–176.

Filewood, W. 1984. The Torres Strait connection: zoogeography of New Guinea. Pp.1121–1131. In Archer, M. and Clayton, G. (eds). *Vertebrate Zoogeography and Evolution in Australasia.* Hesperian Press, Perth. 1203pp.

Flannery, T.F. 1984. New Zealand: a curious zoogeographic history. Pp.1089–1094. In Archer, M. and Clayton, G. (eds). *Vertebrate Zoogeography and Evolution in Australasia.* Hesperian Press, Perth. 1203pp.

Forey, P.L., Littlewood, D.T.J., Ritchie, P. and Meyer, A. 1996. Interrelationships of elopomorph fishes. Pp.175–191. In Stiassny, M.L.J., Parenti, L.J. and Johnson, G.D. (eds). *Interrelationships of Fishes.* Academic Press, San Diego. 496pp.

Frakes, L.A. 1999. Evolution of Australian environments. Pp.163–203. In Orchard, A.E. (ed). *Flora of Australia, Introduction.* Volume 1. 2nd Edn. Australian Biological Resources Study, CSIRO Publishing, Melbourne. 670pp.

Frankenberg, R. 1974. Native freshwater fish. Pp.113–140. In Williams, W.D. (ed). *Biogeography and Ecology in Tasmania.* W. Junk, The Hague. 498pp.

Frankham, R., Ballou, J.D. and Briscoe, D.A. 2002. *Introduction to Conservation Genetics.* Cambridge University Press, Cambridge. 617pp.

Fulton, W. 1990. *Tasmanian Freshwater Fishes*. University of Tasmania, Hobart. 80pp.

Greenwood, P.H. 1975. *A History of Fishes*. E. Benn, London. 467pp.

Greenwood, P.H., Miles, R.S. and Patterson, C. (eds). 1973. *Interrelationships of Fishes.* Academic Press, London. 536pp.

Greenwood, P.H., Rosen, D.E., Weitzmann, S.H. and Myers, G.S. 1966. Phyletic studies of teleostean fishes with a provisional classification of living forms. *Bulletin of the American Museum of Natural History,* Vol. 131, Pt. 4, pp.341–455.

Gressitt, J.L. 1982. Zoogeographical summary. Pp.897–918. In Gressitt, J.L. (ed). *Biogeography and Ecology of New Guinea.* W. Junk, The Hague. 983pp.

Heijden, P.G.M. van der. 2002a. The artisanal fishery in the Sepik-Ramu catchment area, Papua New Guinea. *Science in New Guinea*, Vol. 27, Nos. 1–3, pp.101–119.

Heijden, P.G.M. van der. 2002b. Fisheries in Yonki Reservoir, Papua New Guinea. *Science in New Guinea*, Vol. 27, Nos. 1–3, pp.120–130.

Heinsohn, T. and Hope, G. 2006. The Torresian connections: zoogeography of New Guinea. Pp.71–93. In Merrick, J.R., Archer, M., Hickey, G. and Lee, M.S.Y. (eds). *Evolution and Biogeography of Australasian Vertebrates.* Auscipub Pty. Ltd., Sydney. 942pp.

Herbert, B. and Peeters, J. 1995. *Freshwater Fishes of Far North Queensland.* Information Series Q195018, Department of Primary Industries, Queensland. 74pp.

Hoese, D.F. and Allen, G.R. 1987. Australian fishes. Part 10. A new genus and two new species of freshwater eleotridid fishes (Gobioidei) from the Kimberley region of Western Australia. *Memoirs of the Museum of Victoria,* Vol. 48, pp.35–42.

Hoese, D.F. and Allen, G.R. 1990. Descriptions of two new freshwater *Glossogobius* (Pisces: Gobiidae) from northern Papua New Guinea. *Records of the Western Australian Museum, Supplement,* No. 34, pp.117–129.

Holdaway, R. and Worthy, T. 2006. Evolution of New Zealand and its vertebrates. Pp.111–128. In Merrick, J.R., Archer, M., Hickey, G. and Lee, M.S.Y. (eds). *Evolution and Biogeography of Australasian Vertebrates.* Auscipub Pty. Ltd., Sydney. 942pp.

Howden, M. 2003. Chapter 2. Climate trends and climate change scenarios. Pp.8–13. In Howden, M., Hughes, L., Dunlop, M., Zethoven, I., Hilbert, D. and Chilcott, C. (eds). *Climate Change Impacts on Biodiversity in Australia. Outcomes of a Workshop Sponsored by the Biological Diversity Advisory Committee, 1–2 October 2002.* Commonwealth of Australia, Canberra. 98pp.

Hughes, J.M., Ponniah, M.H., Hurwood, D.A., Chenoweth, S. and Arthington, A.H. 1999. Strong genetic structuring in a habitat specialist, the Oxleyan pygmy perch *Nannoperca oxleyana. Heredity,* Vol. 83, pp.5–14.

Hurwood, D.A. and Hughes, J.M. 1998. Phylogeography of the freshwater fish, *Mogurnda adspersa*, in streams of northeastern Queensland, Australia: evidence for altered drainage patterns. *Molecular Ecology*, Vol. 7, pp. 1507–1517

Ivantsoff, W., Aarn, Shepherd, M.A. and Allen, G.R. 1997. *Pseudomugil reticulatus* (Pisces: Pseudomugilidae) a review of the species originally described from a single specimen, from Vogelkop Peninsula, Irian Jaya with further evaluation of the systematics of Atherinoidea. *Aqua, Journal of Ichthyology and Aquatic Biology,* Vol. 2, No. 4,

pp.53–64.

Ivantsoff, W., Crowley, L.E.L.M. and Allen, G.R. 1987a. Descriptions of three new species and one subspecies of freshwater hardyhead (Pisces: Atherinidae: *Craterocephalus*) from Australia. *Records of the Western Australian Museum,* Vol. 13, Pt. 2, pp.171–188.

Ivantsoff, W., Crowley, L.E.L.M. and Allen, G.R. 1987b. Description of a new species of freshwater hardyhead *Craterocephalus kailolae* (Pisces: Atherinidae) from Safia, northeastern Papua New Guinea. *Proceeding of the Linnean Society of New South Wales,* Vol. 109, pp.331–337.

Jackson, P.D. 1981. Trout introduced into south-eastern Australia: their interaction with native fishes. *Victorian Naturalist*, Vol. 98, No. 1, pp.18–24.

Jenkins, A.P., Buston, P.M. and Allen, G.R. 2000. Two new species of freshwater gudgeons (Eleotrididae: *Mogurnda*) from Lake Kutubu, Papua New Guinea. *Ichthyological Exploration of Freshwaters,* Vol.11, No.1, pp.47–54.

Jerry, D.R., Raadik, T.A., Cairns, S.C. and Baverstock, P.R. 1999. Evidence for natural interspecific hybridization between the Australian bass (*Macquaria novemaculeata*) and estuary perch (*M. colonorum*). *Marine and Freshwater Research,* Vol. 50, No. 7, pp.661–666.

Jerry, D.R. and Woodland, D.J. 1997. Electrophoretic evidence for the presence of the undescribed 'Bellinger' catfish (*Tandanus* sp.) (Teleostei: Plotosidae) in four New South Wales mid-northern coastal rivers. *Marine and Freshwater Research,* Vol. 48, No. 3, pp.235–240.

Johnson, G.D. and Patterson, C. 1996. Relationships of lower euteleostean fishes. Pp.251–332. In Stiassny, M.L.J., Parenti, L.J. and Johnson, G.D. (eds). *Interrelationships of Fishes.* Academic Press, San Diego. 496pp.

Kailola, P.J. 1983. *Arius graeffei* and *Arius armiger,* valid names for two common species of Australo-Papuan fork-tailed catfishes (Pisces: Ariidae). *Transactions of the Royal Society of South Australia,* Vol. 107, Pt. 3, pp.187–196.

Kailola, P.J. 1989. Fork-tailed catfishes. Family Ariidae. Pp.47–55. In Allen, G.R. *Freshwater Fishes of Australia.* T.F.H. Publications, Neptune City, New Jersey. 240 pp.

Kailola, P.J. 1990. A review of the freshwater fork-tailed catfishes (Pisces: Ariidae) in northern New Guinea, with descriptions of two new species. *Records of the Western Australian Museum, Supplement,* No. 34, pp.1–30.

Kailola, P.J. 2000. Six new species of fork-tailed catfishes (Pisces, Teleostei, Ariidae) from Australia and New Guinea. *The Beagle (Records of the Museums and Art Galleries of the Northern Territory),* Vol. 16, pp.127–144.

Kawei, M.H. 1989. Review on introduced fresh water fish in Papua New Guinea. *Australian Society for Fish Biology Preconference Workshop: Introduced and Translocated Fishes and their Ecological Effects* (abstract only).

Kemp, A. 1991a. Australian Mesozoic and Cainozoic lungfish. Pp.465–496. In Vickers-Rich, P., Monaghan, J.M., Baird, R.F. and Rich, T.H. (eds). *Vertebrate Palaeontology of Australasia.* Pioneer Design Studio and Monash University Publications Committee, Melbourne. 1437pp.

Kemp, A. 1997. A revision of Australian Mesozoic and Cenozoic lungfish of the family Neoceratodontidae (Osteichthyes: Dipnoi), with a description of four new species. *Journal of Paleontology*, Vol. 71, No. 4, pp.713–733.

Kemp, N.R. 1991b. Chondrichthyans in the Cretaceous and Tertiary of Australia. Pp.497–568. In Vickers-Rich, P., Monaghan, J.M., Baird, R.F. and Rich, T.H. (eds). *Vertebrate Palaeontology of Australasia.* Pioneer Design Studio and Monash University Publications Committee, Melbourne. 1437pp.

Kershaw, P., Quilty, P.G., David, B., Van Huet, S. and McMinn, A. 2000. Palaeogeography of the Quaternary of Australasia. *Memoir of the Association of Australasian Palaeontologists*, Vol. 23, pp.471–516.

Kirsch, J. 1984. Vicariance biogeography. Pp.109–112. In Archer, M. and Clayton, G. (eds). *Vertebrate Zoogeography and Evolution in Australasia.* Hesperian Press, Perth. 1203pp.

Kohen, J.L. 1997. *Aboriginal Environmental Impacts.* University of New South Wales Press, Sydney. 160 pp.

Koster, W., Raadik, T.A. and Clunie, P. 2002. Scoping study of the potential spread and impact of the exotic fish Oriental weatherloach in the Murray-Darling Basin, Australia: a resource document. *Agriculture, Fisheries and Forestry Australia Report*, pp. 1–78. *Arthur Rylah Institute, Melbourne.*

Larson, H.K. and Hoese, D.F. 1996. Family Gobiidae, subfamilies Eleotridinae and Butinae. Gudgeons. Pp.200–228. In McDowall, R.M. (ed). *Freshwater Fishes of South-eastern Australia.* Second Edition. Reed Books, Sydney. 247pp.

Larson, H.K. and Martin, K.C. 1990. *Freshwater Fishes of the Northern Territory.* Handbook Series No. 1, Northern Territory Museum of Arts and Sciences, Darwin. 102pp.

Last, P.R. and Stevens, J.D. 1994. *Sharks and Rays of Australia.* C.S.I.R.O. Australia. 513pp.

Lauder, G.V. Jnr. and Liem, K.F. 1983. The evolution and interrelationships of the actinopterygian fishes. *Bulletin of the Museum of Comparative Zoology, Harvard University,* Vol. 150, pp.95–197.

Lecointre, G. and Nelson, G. 1996. Clupeomorpha, sister-group of Ostariophysi. Pp.193–207. In Stiassny, M.L.J., Parenti, L.J. and Johnson, G.D. (eds). *Interrelationships of Fishes.* Academic Press, San Diego. 496pp.

Leggett, R. and Merrick, J.R. 1997. Australia's Lake Eacham rainbowfish—lessons and outlook. *Aquarium Sciences and Conservation*, Vol. 1, No. 1, pp.37–43.

Li, G.-Q. and Wilson, M.V.H. 1996. Phylogeny of Osteoglossomorpha. Pp.163–174. In Stiassny, M.L.J., Parenti, L.J. and Johnson, G.D. (eds). *Interrelationships of Fishes.* Academic Press, San Diego. 496pp.

Lintermans, M. 2000. Recolonization by the mountain galaxias *Galaxias olidus* of a montane stream after the eradication of rainbow trout *Oncorhynchus mykiss. Marine and Freshwater Research,* Vol. 51, No. 8, pp.799–804.

Lintermans, M. and Burchmore, J. 1996. Family Cobitidae. Loaches. Pp.114–115. In McDowall, R.M. (ed). *Freshwater Fishes of South-eastern Australia.* Second Edition. Reed Books, Sydney. 247pp.

Long, J.A. 1995. *The Rise of Fishes. 500 Million Years of Evolution.* University of New South Wales Press, Sydney. 223pp.

Long, J. and Turner, S. 1984. A checklist and bibliography of Australian fossil fish. Pp.235–254. In Archer, M. and Clayton G. (eds). *Vertebrate Zoogeography and Evolution in Australasia.* Hesperian Press, Perth. 1203pp.

Low, T. 1999. *Feral Future. The Untold Story of Australia's Exotic Invaders.* Viking Penguin Books, Victoria. 380pp.

McDowall, R.M. 1981. The relationships of Australian freshwater fishes. Pp.1253–1273. In Keast, A. (ed). *Ecological Biogeography of Australia.* W. Junk, The Hague. 1434pp.

McDowall, R.M. 1984. Southern Hemisphere freshwater salmoniforms: development and relationships. Pp.150–153. In Moser, H.G., Richards, W.J., Cohen, D.M., Fahay, M.P., Kendall, A.W. Jr and Richardson, S.L. (eds). Ontogeny and systematics of fishes. *American Society of Ichthyologists and Herpetologists, Special Publication,* Vol. 1, pp.1–760.

McDowall, R.M. 1989. Filling in the gaps – the introduction of exotic fishes into New Zealand. *Australian Society for Fish Biology Preconference Workshop: Introduced and Translocated Fishes and their Ecological Effects* (abstract only).

McDowall, R.M. 1990. *New Zealand Freshwater Fishes. A Natural History and Guide.* Heinemann Reed, Auckland. 553pp.

McDowall, R.M. 1996a. Volcanism and freshwater fish biogeography in the northeastern North Island of New Zealand. *Journal of Biogeography,* Vol. 23, No. 2, pp.139–148.

McDowall, R.M. (ed). 1996b. *Freshwater Fishes of South-eastern Australia.* Second Edition. Reed Books, Sydney. 247pp.

McDowall, R.M. 1997. Affinities, generic classification and biogeography of the Australian and New Zealand mudfishes (Salmoniformes: Galaxiidae). *Records of the Australian Museum,* Vol. 49, No. 2, pp.121–137.

McDowall, R.M. 2002. Accumulating evidence for a dispersal biogeography of southern cool temperate freshwater fishes. *Journal of Biogeography,* Vol. 29, No. 2, pp.207–219.

McDowall, R.M. 2003a. Hawaiian biogeography and the islands' freshwater fish fauna. *Journal of Biogeography,* Vol. 30, No. 5, pp.703–710.

McDowall, R.M. 2003b. Impacts of introduced salmonids on native galaxiids in New Zealand upland streams: a new look at an old problem. *Transactions of the American Fisheries Society,* Vol.132, pp.229–238.

McDowall, R.M. and Fulton, W. 1996. Family Galaxiidae. Galaxiids. Pp.52–77. In McDowall, R.M. (ed). *Freshwater Fishes of South-eastern Australia.* Second Edition. Reed Books, Sydney. 247pp.

McDowall, R.M. and Pole, M. 1997. A large galaxiid fossil (Teleostei) from the Miocene of Central Otago, New Zealand. *Journal of the Royal Society of New Zealand,* Vol. 27, No. 2, pp.193–198.

McDowall, R.M. and Wallis, G.P. 1996. Description and redescription of *Galaxias* species (Teleostei: Galaxiidae) from Otago and Southland. *Journal of the Royal Society of New Zealand,* Vol. 26, No. 3, pp.401–427.

McDowall, R.M. and Whitaker, A.H. 1975. The freshwater fishes. Pp.277–299. In Kuschel, G. (ed). *Biogeography and Ecology in New Zealand.* W. Junk, The Hague. 689 pp.

McGlashan, D.J. and Hughes, J.M. 2002. Extensive genetic divergence among populations of the Australian freshwater fish, *Pseudomugil signifer* (Pseudomugilidae), at different hierarchical scales. *Marine and Freshwater Research,* Vol. 53, No. 5, pp.897–907.

McGowran, B., Archer, M., Bock, P., Darragh, T.A., Godthelp, H., Hageman, S., Hand, S.J., Hill, R., Li, Q., Maxwell, P.A., McNamara, K.J., Macphail, M., Mildenhall, D., Partridge, A.D., Richardson, J., Shafik, S., Truswell, E.M. and Warne, M. 2000. Australasian palaeobiogeography: the Palaeogene and Neogene record. *Memoir of the Association of Australasian Palaeontologists,* Vol. 23, pp.405–470.

McGuigan, K. Zhu, D., Allen, G.R. and Moritz, C. 2000. Phylogenetic relationships and historical biogeography of melanotaeniid fishes in Australia and New Guinea. *Marine and Freshwater Research,* Vol. 51, No. 7, pp.713–723.

McIntosh, A.R. Crowl, T.A. and Townsend, C.R. 1994. Size-related impacts of introduced brown trout on the distribution of native common river galaxias. *New Zealand Journal of Marine and Freshwater Research,* Vol. 28, pp.135–144.

McIntosh, A.R., Townsend, C.R. and Crowl, T.A. 1992. Competition for space between introduced brown trout (*Salmo trutta* L.) and a native galaxiid (*Galaxias vulgaris* Stokell) in a New Zealand stream. *Journal of Fish Biology,* Vol. 41, pp.63–81.

McKay, R.J. 1977. The Australian aquarium industry and the possibility of the introduction of exotic fish species and diseases. *Fisheries Division of the Department of Primary Industry, Canberra, Fisheries Paper,* No. 25.

Merrick, J.R. 1990. Freshwater fishes. Pp.7–15. In Hand, S.J. (ed). *Care and Handling of Australian Native Animals.* Surrey Beatty and Sons and Royal Zoological Society of New South Wales, Sydney. 210pp.

Merrick, J.R. 1995. Diversity, distribution and conservation of freshwater crayfishes in the eastern highlands of New South Wales. *Proceedings of the Linnean Society of New South Wales,* Vol. 115, pp.247–258.

Merrick, J.R. 1997. The conservation of freshwater fishes in Australia: politics, bureaucracy and potential roles for aquarists. *Aquarium Sciences and Conservation*, Vol. 1, No. 1, pp.29–35.

Merrick, J.R. and Rimmer, M.A. 1984. Reservoir fishes and water quality monitoring systems. Baseline studies 1981–1983. *Metropolitan Water Sewerage and Drainage Board Report,* pp.1–242. *Sydney.*

Merrick, J.R. and Schmida, G.E. 1984. *Australian Freshwater Fishes. Biology and Management.* J.R. Merrick, Sydney. 409pp.

Miller, S., Hyslop, E., Kula, G. and Burrows, I. 1999. Status of biodiversity in Papua New Guinea. http://ww3.datec.com.pg/CRC/CHAPTER6statusofbiodiversity.html

Molnar, R.E. and Kurz, C. 1997. The distribution of Pleistocene vertebrates on the eastern Darling Downs, based on the Queensland Museum collections. *Proceedings of the Linnean Society of New South Wales,* Vol. 117, pp.107–134.

Morris, S.A., Pollard, D.A., Gehrke, P.C. and Pogonoski, J.J. 2001. Threatened and potentially threatened freshwater fishes of coastal New South Wales and the Murray-Darling Basin, *NSW Fisheries Final Report Series*, No. 33, pp.1–166.

Musyl, M.K. and Keenan, C.P. 1992. Population genetics and zoogeography of Australian freshwater golden perch, *Macquaria ambigua* (Richardson 1845) (Teleostei: Percichthyidae), and electrophoretic identification of a new species from the Lake Eyre Basin. *Australian Journal of Marine and Freshwater Research,* Vol. 43, No. 6,

pp.1585–1601.

National Institute of Water and Atmospheric Research (NIWA). 2001. *Atlas of New Zealand Freshwater Fishes.* http://www.niwa.cri.nz/rc/freshwater/fishatlas/

Nelson, G. and Rosen, D.E. 1981. *Vicariance Biogeography: a Critique.* Columbia University Press, New York. 576pp.

Nelson, J.S. 1984. *Fishes of the World.* 2nd Edition. Wiley, New York. 523pp.

New Zealand Native Freshwater Fish Society (N.Z.N.F.F.S.). 2003a. Facts about the native freshwater fishes of New Zealand. http://www.nzfreshwater.org/index_wildlife.html

New Zealand Native Freshwater Fish Society (N.Z.N.F.F.S.). 2003b. Threatened freshwater fish list. http://www.nzfreshwater.org/index_resources.html

NSW Fisheries. 2003. *Environmental Impact Statement on Freshwater Fish Stocking in NSW.* 3 Vols. NSW Fisheries, Cronulla, New South Wales.

Parenti, L.R. and Song, J. 1996. Phylogenetic significance of the pectoral-pelvic fin association in acanthomorph fishes: a re-assessment using comparative neuroanatomy. Pp.427–444. In Stiassny, M.L.J., Parenti, L.J. and Johnson, G.D. (eds). *Interrelationships of Fishes.* Academic Press, San Diego. 496pp.

Parker, P. 1980. Marine fishes that may enter fresh water. Pp.193–194. In McDowall R.M. (ed). *Freshwater Fishes of South-eastern Australia.* A.H. and A.W. Reed, Sydney. 208pp.

Paulin, C., Stewart, A., Roberts, C. and McMillan, P. 1989. New Zealand fish. A complete guide. *National Museum of New Zealand, Miscellaneous Series,* Vol. 19, pp.1–279.

Paxton, J.R. 1994. Habitats and adaptations. Pp.32–41. In Paxton, J.R. and Eschmeyer, W.N. (eds). *Encyclopedia of Fishes.* University of New South Wales Press, Sydney. 240pp.

Paxton, J.R., Allen, G.R. and Hoese, D.F. 2006. Australian marine fishes: zoogeography, endemics and conservation. Pp.185–194. In Merrick, J.R., Archer, M., Hickey, G.M. and Lee, M.S.Y. (eds). *Evolution and Biogeography of Australasian Vertebrates.* Auscipub Pty. Ltd., Sydney. 942pp.

Paxton, J.R., Hoese, D.F., Allen, G.R. and Hanley, J.E. 1989. *Zoological Catalogue of Australia. Vol. 7. Pisces. Petromyzontidae to Carangidae.* Australian Government Publishing Service, Canberra. 665pp.

Pollard, D.A. (ed) 1990. *Introduced and Translocated Fishes and their Ecological Effects.* Australian Society for Fish Biology Workshop, Magnetic Island, 24-25 August 1989. Bureau of Rural Resources Proceedings No.8. Australian Government Publishing Service, Canberra. 181pp.

Potter, I.C. 1996a. Family Mordaciidae. Shortheaded lampreys. Pp.32–35. In McDowall, R.M. (ed). *Freshwater Fishes of South-eastern Australia.* Second Edition. Reed Books, Sydney. 247pp.

Potter, I.C. 1996b. Family Geotriidae. Pouched lamprey. Pp.36–38. In McDowall, R.M. (ed). *Freshwater Fishes of South-eastern Australia.* Second Edition. Reed Books, Sydney. 247pp.

Potter, I.C., Hilliard, R.W. and Neira, F.J. 1986. The biology of Australian lampreys. Pp.207–230. In De Deckker, P. and Williams, W.D. (eds). *Limnology in Australia.* C.S.I.R.O., Melbourne and Junk, Dordrecht. 671pp.

Pusey, B.J. and Arthington, A.H. 2003. Importance of the riparian zone to the conservation and management of freshwater fish: a review. *Marine and Freshwater Research,* Vol. 54, No. 1, pp.1–16.

Pusey, B. and Kennard, M. 2001. *Guyu wujalwujalensis,* a new genus and species (Pisces: Percichthyidae) from north-eastern Queensland, Australia. *Ichthyological Exploration of Freshwaters,* Vol. 12, No. 1, pp.17–28.

Raadik, T. 2001a. When is a mountain galaxias not a mountain galaxias ? *Fishes of Sahul,* Vol. 15, pp.785–789.

Raadik, T. 2001b. Alien zone. Report of the Exotic Fishes Sub-Committee—to May 2001. *Australian Society for Fish Biology Newsletter,* Vol. 31, No.1, pp.26–32.

Raadik, T. 2003a. Alien zone (II). Exotic Fishes Committee Report – to June 2002. *Australian Society for Fish Biology Newsletter,* Vol. 32, No. 2, pp.60–67.

Raadik, T. 2003b. Alien zone (III). Exotic Fishes Committee Report—to June 2003. *Australian Society for Fish Biology Newsletter,* Vol. 33, No. 1, pp.45–53.

Raadik, T. 2004. Alien zone (No.5). Report of the Exotic Fishes Committee—to June 2004. *Australian Society for Fish Biology Newsletter,* Vol. 34, No. 1, pp.51–57.

Raadik, T.A. and Kuiter, R.H. 2002. Koscuiszko galaxias. *Fishes of Sahul,* Vol. 16, No. 2, pp.829–835.

Rich, C. 1987. Biogeographic patterns in *Galaxias olidus. Australian Society of Fish Biology Newsletter,* Vol. 17, p.30 (abstract only).

Ritchie, A. 2006. The great Devonian fish kill at Canowindra. Pp.159–184. In Merrick, J.R., Archer, M., Hickey, G.M. and Lee, M.S.Y. (eds). *Evolution and Biogeography of Australasian Vertebrates.* Auscipub Pty. Ltd., Sydney. 942pp.

Roberts, T.R. 1978. An ichthyological survey of the Fly River in Papua New Guinea with descriptions of new species. *Smithsonian Contributions to Zoology,* No. 281, pp.1–70.

Rosen, D.E. 1976. A vicariance model of Caribbean biogeography. *Systematic Zoology,* Vol. 24, pp.431–464.

Rowland, S.J. 1989. Aspects of the history and fishery of the Murray cod, *Maccullochella peeli* (Mitchell) (Percichthyidae). *Proceedings of the Linnean Society of New South Wales,* Vol. 111, pp.201–213.

Rowland, S.J. 1993. *Maccullochella ikei,* an endangered species of freshwater cod (Pisces: Percichthyidae) from the Clarence River system, NSW and M. *peelii mariensis,* a new subspecies from the Mary River system, Qld. *Records of the Australian Museum.* Vol. 45, pp.121–145.

Rowland, S.J. 2001. Record of the banded grunter *Amniataba percoides* (Teraponidae) from the Clarence River, New South Wales. *Australian Zoologist,* Vol. 31, pp.603–607.

Saeed, B., Ivantsoff, W. and Allen, G.R. 1989. Taxonomic revision of the family Pseudomugilidae (Order Atheriniformes). *Australian Journal of Marine and Freshwater Research,* Vol. 40, No. 6, pp.719–787.

Schwartz, F.J. and Maddock, M.B. 2002. Cytogenetics of the elasmobranchs: genome evolution and phylogenetic implications. *Marine and Freshwater Research,* Vol. 53, No. 2, pp.491–502.

Shearer, K.D. and Mulley, J.C. 1978. The introduction and distribution of the carp, *Cyprinus carpio* Linnaeus, in Australia. *Australian Journal of Marine and Freshwater Research,* Vol. 29, No. 5, pp.551–563.

Tappin, A.R. 2001. Rainbowfish habitats. http://members.optushome.com.au/chelmon.Intro.htm

Thoney, D.A., Loiselle, P.V. and Schlager, N. (eds). 2003.

Grzimek's Animal Life Encyclopedia. Volume 4. Fishes I. Second Edition. Gale Group Inc., Farmington Hills, Michigan. 517pp.

Tilzey, R.D.J. 1976. Observations on interactions between indigenous Galaxiidae and introduced Salmonidae in the Lake Eucumbene catchment, New South Wales. *Australian Journal of Marine and Freshwater Research*, Vol. 27, No. 4, pp.551–564.

Unmack, P.J. 2001a. Biogeography of Australian freshwater fishes. *Journal of Biogeography*, Vol. 28, No. 9, pp.1053–1089.

Unmack, P.J. 2001b. Fish persistence and fluvial geomorphology in central Australia. *Journal of Arid Environments*, Vol. 49, pp.653–669.

van Oosterzee, P. 2006. Drawing the Wallace Line. Pp.95–110. In Merrick, J.R., Archer, M., Hickey, G.M. and Lee, M.S.Y. (eds). *Evolution and Biogeography of Australasian Vertebrates.* Auscipub Pty. Ltd., Sydney. 942pp.

Waters, J.M., Dijkstra, L.H. and Wallis, G.P. 2000a. Biogeography of a southern hemisphere freshwater fish: how important is marine dispersal ? *Molecular Ecology,* Vol. 9, pp.1815–1821.

Waters, J.M., Lopez, J.A. and Wallis, G.P. 2000b. Molecular phylogenetics and biogeography of galaxiid fishes (Osteichthyes: Galaxiidae) : dispersal, vicariance, and the position of *Lepidogalaxias salamandroides. Systematic Biology,* Vol. 49, No. 4, pp.777–795.

Waters, J.M., Saruwatari, T., Kobayashi, T., Oohara, I., McDowall, R.M. and Wallis, G.P. 2002. Phylogenetic placement of retropinnid fishes: data set incongruence can be reduced by using asymmetric character state transformation costs. *Systematic Biology*, Vol. 51, No. 3, pp.432–449.

Waters, J.M. and White, R.W.G. 1997. Molecular phylogeny and biogeography of the Tasmanian and New Zealand mudfishes (Salmoniformes: Galaxiidae). *Australian Journal of Zoology,* Vol. 45, No.1, pp.39–48.

Watson, R.E. 1994. *Awaous (Awaous) acritosus* a new species of freshwater goby from southern New Guinea and northeastern Australia (Teleostei: Gobiidae). *Ichthyological Exploration of Freshwaters,* Vol. 5, No. 4, pp.371–376.

Watson, R.E. 1995. A new species of *Sicyopus* from Papua New Guinea with a redescription of *Sicyopus multisquamatus* (Teleostei: Gobiidae: Sicydiinae). *Ichthyological Exploration of Freshwaters,* Vol. 6, No. 3, pp.276–278.

Watson, R.E. and Allen G.R. 1999. New species of freshwater gobies from Irian Jaya, Indonesia (Teleostei: Gobiidae). *Aqua, Journal of Ichthyology and Aquatic Biology,* Vol. 3, No. 3, pp.113–118.

Watson, R.E., Allen, G.R. and Kottelat, M. 1998. A review of *Stiphodon* from Halmahera and Irian Jaya, Indonesia, with descriptions of two new species (Teleostei: Gobiidae). *Ichthyological Exploration of Freshwaters,* Vol. 9, No. 3, pp.293–304.

Watson, R.E., Keith, P. and Marquet, G. 2001. *Sicyopus (Smilosicyopus) chloe*, a new species of freshwater goby from New Caledonia (Sicydiinae). *Cybium,* Vol. 25, No. 1, pp.41–52.

Watson, R.E., Keith, P. and Marquet, G. 2002. *Lentipes kaaea,* a new species of freshwater goby (Teleostei: Gobioidei: Sicydiinae) from New Caledonia. *Bulletin Francais de la Peche et de la Pisciculture,* No. 364, pp.173–185.

Watson, R.E., Marquet, G. and Poellabauer, C. 2000. New Caledonia fish species of the genus *Sicyopterus* (Teleostei: Gobioidei: Sicydiinae). *Aqua, Journal of Ichthyology and Aquatic Biology,* Vol. 4, No. 1, pp.5–34.

Watson, R.E. and Poellabauer, C. 1998. A new genus and species of freshwater goby from New Caledonia with a complete lateral line (Pisces: Teleostei: Gobioidei). *Senckenbergiana Biologica,* Vol. 77, No. 2, pp.147–153.

Werry, L.P. 1998. A review of freshwater fish introductions in Papua New Guinea. *Science in New Guinea*, Vol. 24, No. 1, pp.33–38.

White, M.E. 2006. Environments of the geological past. Pp.17–50. In Merrick, J.R., Archer, M., Hickey, G.M. and Lee, M.S.Y. (eds). *Evolution and Biogeography of Australasian Vertebrates.* Auscipub Pty. Ltd., Sydney. 942pp.

Williams, R.R.G. 1997. Bones and muscles of the suspensorium in the galaxioids and *Lepidogalaxias salamandroides* (Teleostei: Osmeriformes) and their phylogenetic significance. *Records of the Australian Museum*, Vol. 49, No. 2, pp.139–166.

World Conservation Monitoring Centre (W.C.M.C.). 2000a. Freshwater biodiversity: a preliminary global assessment. 3. Status, trends and threats. http://www.unep-wcmc.org/information_services/publications/freshwater/3.htm.

World Conservation Monitoring Centre (W.C.M.C.). 2000b. Freshwater biodiversity: a preliminary global assessment. 4. Important areas for freshwater diversity. http://www.unep-wcmc.org/information_services/publications/freshwater/4.htm.

Primitive Tetrapods: Amphibians

GEOLOGICAL TIME SCALE FOR THE PHANEROZOIC

ERA	PERIOD or SUB-PERIOD	SERIES/EPOCH	STAGE	AGE m.y.a.	EXTINCTIONS (E) ICE AGE (I)
CAINOZOIC	QUATERNARY	PLEISTOCENE	Calabrian	1.8	E, I
	TERTIARY – NEOGENE	PLIOCENE	Piacenzian	3.6	I
			Zanclean	5.3	I
		MIOCENE	Messinian	7.12	
		Late	Tortonian	11.2	
			Serravalian	14.8	
		Mid	Langhian	16.4	
			Burdigalian	20.52	
		Early	Aquitanian	23.8	
	TERTIARY – PALAEOGENE	OLIGOCENE Late	Chattian	28.5	E
		Early	Rupelian	33.7	
		EOCENE Late	Priabonian	37	
			Bartonian	41.3	
		Mid	Lutetian	49	
		Early	Ypresian	54.8	
		PALEOCENE	Thanetian	57.9	
		Late	Selandian	60.9	
		Early	Danian	65	E
MESOZOIC	CRETACEOUS	LATE	Maastrichtian	73	
			Campanian	83	
			Santonian	87	
			Coniacian	89	
			Turonian	91	
			Cenomanian	97.5	
		EARLY	Albian	108	
			Aptian	115	
			Barremian	123	
			Hauterivian	130	
			Valanginian	135	
			Berriasian	141	E
	JURASSIC	LATE	Tithonian	146	
			Kimmeridgian	151	
			Oxfordian	159	
		MIDDLE	Callovian	165	
			Bathonian	173	
			Bajocian	180	
			Aalenian	184	
		EARLY	Toarcian	190	E
			Pliensbachian	195	
			Sinemurian	202	
			Hettangian	205	E
	TRIASSIC	LATE	Rhaetian	207	
			Norian	220	
			Carnian	230	
		MIDDLE	Ladinian	235	
			Anisian	241	
		EARLY	Spathian	244	
			Nammalian	247	
			Griesbachian	251	

ERA	PERIOD or SUB-PERIOD	SERIES/EPOCH	STAGE	AGE m.y.a.	EXTINCTIONS (E) ICE AGE (I)
PALAEOZOIC	PERMIAN	LATE	Tatarian – Dorashamian	255	E
			Tatarian – Dzhulfian	259	I
			Tatarian – Midian	263	I
			Kazanian	267	I
			Ulfimian	270	I
		EARLY	Kungurian	274	I
			Artinskian	285	I
			Sakmarian	293	I
			Asselian	298	I
	CARBONIFEROUS	PENNSYLVANNIAN – SILESIAN – STEPHANIAN	Gzhelian	302	I
			Kasimovian	305	I
		WESTPHALIAN	Moscovian	311.5	
			Bashkirian	314	
		NAMURIAN	Serpukhovian	325	
		MISSISSIPPIAN – DINANTIAN – VISEAN	Brigantian	328.5	
			Asbian	330.5	
			Holkerian	335	
			Arundian	341.5	
		TOURNAISIAN	Chadian	344	
			Ivorian	348	
			Hastarian	354	
	DEVONIAN	LATE	Famennian	364.5	E
			Frasnian	369	
		MIDDLE	Givetian	378	I
			Eifelian	384	I
		EARLY	Emsian	399.5	I
			Pragian	404.5	I
			Lochkovian	410	
	SILURIAN	PRIDOLI		414	
		LUDLOW	Ludfordian	417	
			Gorstian	420	
		WENLOCK	Homerian	422.5	
			Sheinwoodian	425	
		LLANDOVERY	Telychian	428	I
			Aeronian	431	I
			Rhuddanian	434	E, I
	ORDOVICIAN	ASHGILL	Bolindian	443	I
		CARADOC	Eastonian	454.5	
			Gisbornian	459	
		LLANVIRN	Darriwilian	467	
			Yapeenian	471	
		ARENIG	Castlemainian	477	
			Chewtonian	481	
			Bendigonian	484.5	
		TREMADOC	Lancefieldian	486	
			Warendan	490	
	CAMBRIAN	LATE	Datsonian	491	
			Payntonian	492.5	
			Iverian	496	
			Idamean	497.5	E
		MIDDLE	Mindyallan	498.5	
			Boomerangian	500	
			Undillan	503	
			Late Templetonian-Floran	506	
			Ordian-Early Templetonian	509	
		EARLY	Atdabanian-Toyonian	528	
			Tommotian	534	
			Nemakit-Daldynian	545	

This detailed scale is based on Young and Laurie (1996) but Cambrian stages abbreviated ages are rounded to nearest 0.1 million years. More recent dating (e.g. United States Geological Survey) has resulted in some small adjustments to interval boundaries.

Chapter 11
Origins and Early Radiations of the Amphibians

Anne Warren
Department of Zoology, La Trobe University, Melbourne, Vic. 3086 Australia
E-mail: a.warren@latrobe.edu.au

INTRODUCTION

The word amphibian literally means 'both lives', that is, adapted for life both in the water and on land and so can be applied to many vertebrates. The group of vertebrates known as Amphibia are best described as anamniote tetrapods, or four-footed vertebrates which do not have the advanced characters found in the Amniota (reptiles, birds and mammals). Phylogenetic research has shown that the anamniote tetrapods are not a monophyletic, or natural, group. The most widely supported hypothesis is that the Lissamphibia, the living frogs, salamanders and caecilians, may be closely related to the largest group of ancient anamniotic tetrapods, the Temnospondyli, in which case the Temnospondyli plus the Lissamphibia can be termed Amphibia (Figure 1(a)). On the other hand, Laurin and Reisz (1997) have hypothesized that the Lissamphibia are more closely related to primitive members of the amniote lineage than to the temnospondyls with the result that the name Amphibia may have to be restricted to the Lissamphibia.

The subject of this chapter is the Australian anamniote tetrapods that are not lissamphibians. They ranged in time from the Devonian to the Cretaceous and included the temnospondyls, the group to which most anamniotic tetrapods belonged. The older and more widely known name for these early tetrapods is 'labyrinthodonts', a term that refers to the labyrinthine folding of the dentine in their teeth. The 'labyrinthodonts' included a number of early tetrapods whose relationships are as yet poorly understood, as well as the temnospondyls and the early members of the amniote lineage, the anthracosaurs.

TETRAPOD ORIGINS

There is now little doubt that the tetrapods originated from the bony 'osteolepiform' fishes, which are first known in the fossil record of the middle Devonian (Figure 1(a)). The process of that transition is now well-documented so that Carroll (1996) has used it as an example of large scale evolutionary change, or macroevolution. The fish closest morphologically to tetrapods is the Russian *Panderichthys*, which has a tetrapod 'look', with a flattened skull, more posteriorly placed orbits, separate frontal bones and no dorsal or anal fins (Carroll, 1995, 1996).

The time of divergence of *Panderichthys* plus tetrapods from the osteolepiforms must have been no later than early Middle Devonian; the time of divergence between *Panderichthys* and tetrapods is also unknown, although presumably no later than middle Devonian (Thomson, 1993). Curiously, no indication of digit-like bones is present in the paired fins of any osteolepiform. These bones are present in the Rhizodontida (the sister group to 'osteolepiforms' plus tetrapods) which exhibit an additional tetrapod character, the presence of a single external nostril (Johanson and Ahlberg, 1998). Osteological characters of the tetrapods therefore appear to have developed in mosaic fashion, with digit equivalents and a single nostril appearing in rhizodonts, skull roof characters in panderichthyids (which lack digits), and braincase and further limb transformation in the earliest tetrapod taxa.

THE EARLIEST TETRAPODS

The earliest material now identifiable as tetrapod comes from the late Frasnian (early Late Devonian) and consists of two European taxa (described originally as fishes) and the early Famennian *Metaxygnathus* from Australia. *Elginerpeton*, from Scotland, was recognized as tetrapod especially by its humerus and tibia which carries facets for the ankle bones, a feature found only in tetrapods (Ahlberg, 1995). *Elginerpeton* and the closely related *Obruchevichthys* from Latvia and Russia (Vorobyeva, 1977) are thought to represent an early radiation not closely related to any known Devonian or Carboniferous tetrapods (Carroll, 1995). Later in the Devonian (Famennian), well-preserved, articulated material of *Ichthyostega* (Jarvik, 1995) and *Acanthostega* is plentiful in Greenland (Coates and Clack, 1990; Coates, 1996; Clack, 2000a), while fragmentary tetrapods have been found in Russia (Lebedev and Coates, 1995) and eastern North America (Daeschler, 2000). The precise phylogenetic relationships between these Devonian tetrapods and those from the Carboniferous are not yet established.

The best known Devonian tetrapods have been shown to be primarily (if not fully) aquatic, with remains of the opercular series of bones still present behind the skull, fish-like tails, polydactylous paddles and, in *Acanthostega* at least, internal gills (Coates and Clack, 1990). Although the pectoral limb of *Ichthyostega* is robust it is not seen as

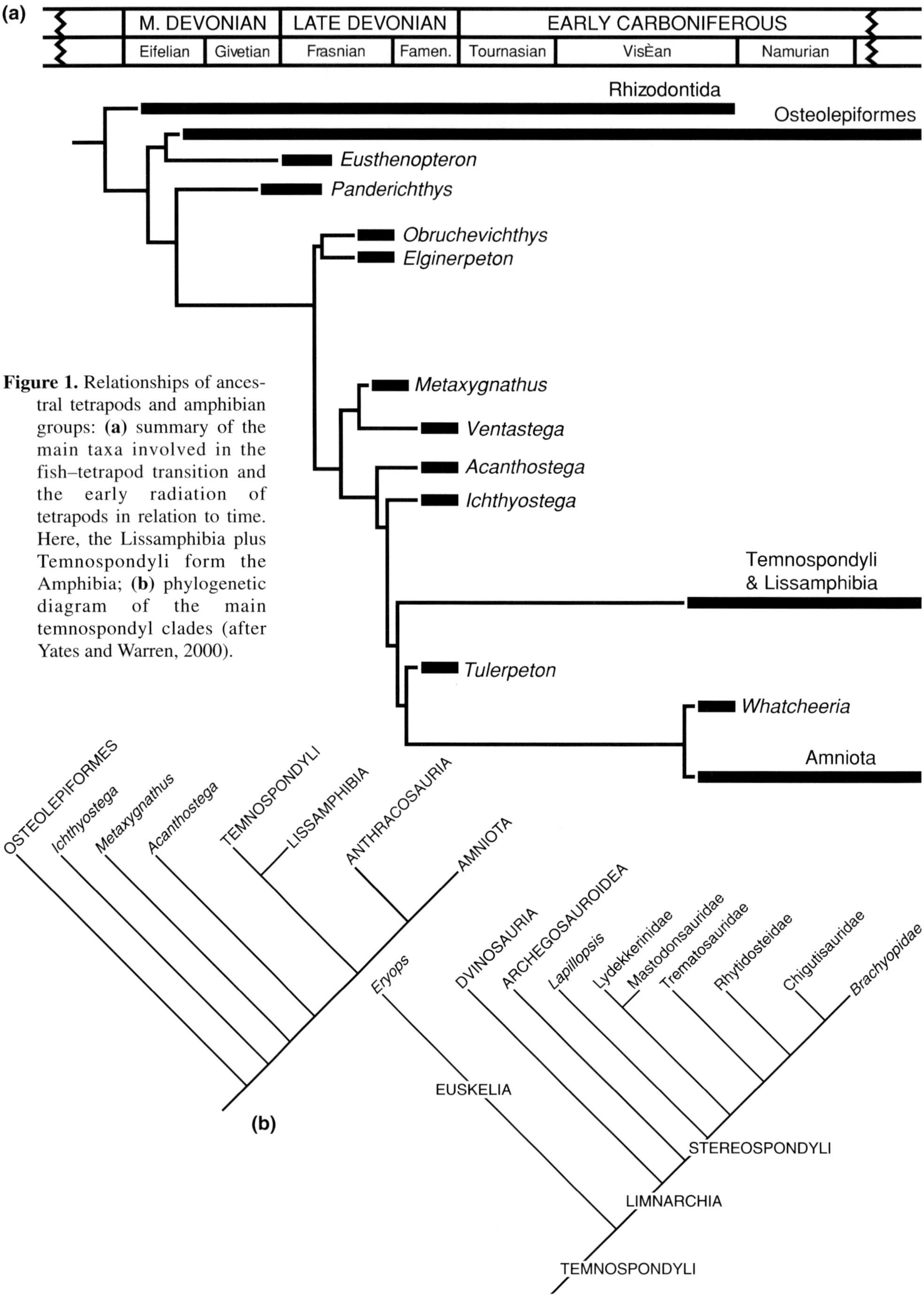

Figure 1. Relationships of ancestral tetrapods and amphibian groups: **(a)** summary of the main taxa involved in the fish–tetrapod transition and the early radiation of tetrapods in relation to time. Here, the Lissamphibia plus Temnospondyli form the Amphibia; **(b)** phylogenetic diagram of the main temnospondyl clades (after Yates and Warren, 2000).

a propulsive organ. Both *Ichthyostega* and *Acanthostega* (whose pectoral limb is tiny) were visualized as adopting a locomotory pattern involving hind limb and tail propulsion (Coates and Clack, 1990) which was unlikely to have been efficient on land. Nevertheless the Devonian tetrapods had already developed the fundamental skeletal framework needed for land-based locomotion.

The remaining skeletal transformations associated with adopting terrestrial habitats presumably came during the first 20 million years of the Carboniferous; by the late Early Carboniferous fully terrestrial tetrapod faunas are known (Wood *et al.*, 1985). Over that time period a functional stapes developed, preopercular bones were lost, digit number was reduced to five plus five in the amniotes and four plus five in the temnospondyls and lissamphibians, the forelimb enlarged in relation to the hindlimb, and fin rays were lost from the tail fin.

Changes in soft morphology and in physiological processes associated with the move on to land will have occurred in this time period also. Thomson (1993) has reviewed many of these. Some changes were adaptations of features already present in the Devonian tetrapods, and perhaps the panderichthyids. Lungs are primitive for bony fishes, and a circulatory system able to service both gills and lungs was presumably primitive for the lobe-finned fishes. The problem of a sudden change in osmoregulatory needs in a terrestrial habitat is lessened if the early tetrapods are seen as living an anadromous life in a coastal wetland. As pointed out by Thomson, all lineages of Devonian lobe-finned forms had marine members, and unless the continents had contiguous freshwater drainage systems, the immediate ancestors of the early tetrapods must have been able to tolerate marine conditions. Living in a salt-rich environment is much like living on the land as both environments can be seen as physiological deserts (Thomson, 1993).

With many of the structural changes needed for a terrestrial life established in the earliest tetrapods, it is surprising that the transition to land was not made earlier. The Devonian terrestrial ecosystem offered abundant food in the form of invertebrates and fishes, shelter and safety from aquatic predators. Perhaps, as suggested by Thomson (1993), the transition from an aquatic to an amphibious way of life was in response to the evolution of a rich, semi-aquatic ecosystem resulting from the emergence of vegetation on to the land.

THE AUSTRALIAN RECORD

Australian fossil anamniote tetrapods are of particular interest in four time periods covering approximately 250 million years. We know that they were present here from almost the beginning of the tetrapod radiation in the late Devonian and that these earliest tetrapods survived here into the early Carboniferous during the critical transition to a terrestrial habitat. A burst of radiation occurred during the late Permian or early Triassic, resulting in the chief clade of anamniote tetrapods from the Mesozoic, the Stereospondyli. The last remnants of the stereospondyls survived only in southern Australia until towards the end of the early Cretaceous.

DEVONIAN

The first evidence that Devonian tetrapods were present in Gondwana comes from trackways preserved in Famennian sediments in the Genoa River Beds near the N.S.W.–Victorian border (Warren and Wakefield, 1972). A redescription by Clack (1996) affirmed the tetrapod nature of two of the three trackways with characteristic overstepping of the fore foot by the hind foot on alternate sides of the body. At least five digits impressed the sediment but overstepping could have concealed further impressions so that we do not know whether the maker of these tracks was polydactylous like the Devonian tetrapods from the Northern Hemisphere.

Another trackway was described from the Red Man Bluff Subgroup of the Grampians Group, Victoria (Warren *et al.*, 1986) but was considered by Clack (1996) to be unlikely to have been generated by a tetrapod. Clack's conclusion is supported by a revision of the geology of the Grampians which maintains that the youngest possible age of the Grampians Group is early Silurian (Cayley and Taylor, 1997), some 70 million years before the first tetrapod body fossils are found.

In Australia a single body fossil is known from the Devonian. *Metaxygnathus* (Campbell and Bell, 1977), a right lower jaw ramus from the Cloghnan Shale (early Famennian) near Forbes, New South Wales, is considered certainly tetrapod, appearing between *Ichthyostega* and *Acanthostega* on the tetrapod stem in a phylogenetic analysis of lower jaws (Ahlberg and Clack, 1998). The Genoa River footprints and *Metaxygnathus* remain the only evidence for the presence of Devonian tetrapods outside Euramerica.

CARBONIFEROUS

Despite their presence in the Devonian, no tetrapods were known from Australia between the late Devonian and the latest Permian. That tetrapods survived, at least in northern Australia, was shown by the discovery of a new fauna from the Ducabrook Formation near Emerald in central Queensland (Thulborn *et al.*, 1996). This is the only early Carboniferous tetrapod fauna from Gondwana and contains taxa not found previously outside Euramerica. Because this fauna is from the early Carboniferous it should help to clarify relationships between the Devonian and later Carboniferous tetrapods and to detail some of the morphological changes that took place in the tetrapod body during the transition from fully aquatic to amphibious and terrestrial tetrapods.

Material collected to date includes well preserved (but disarticulated) parts of the postcranial skeleton of a

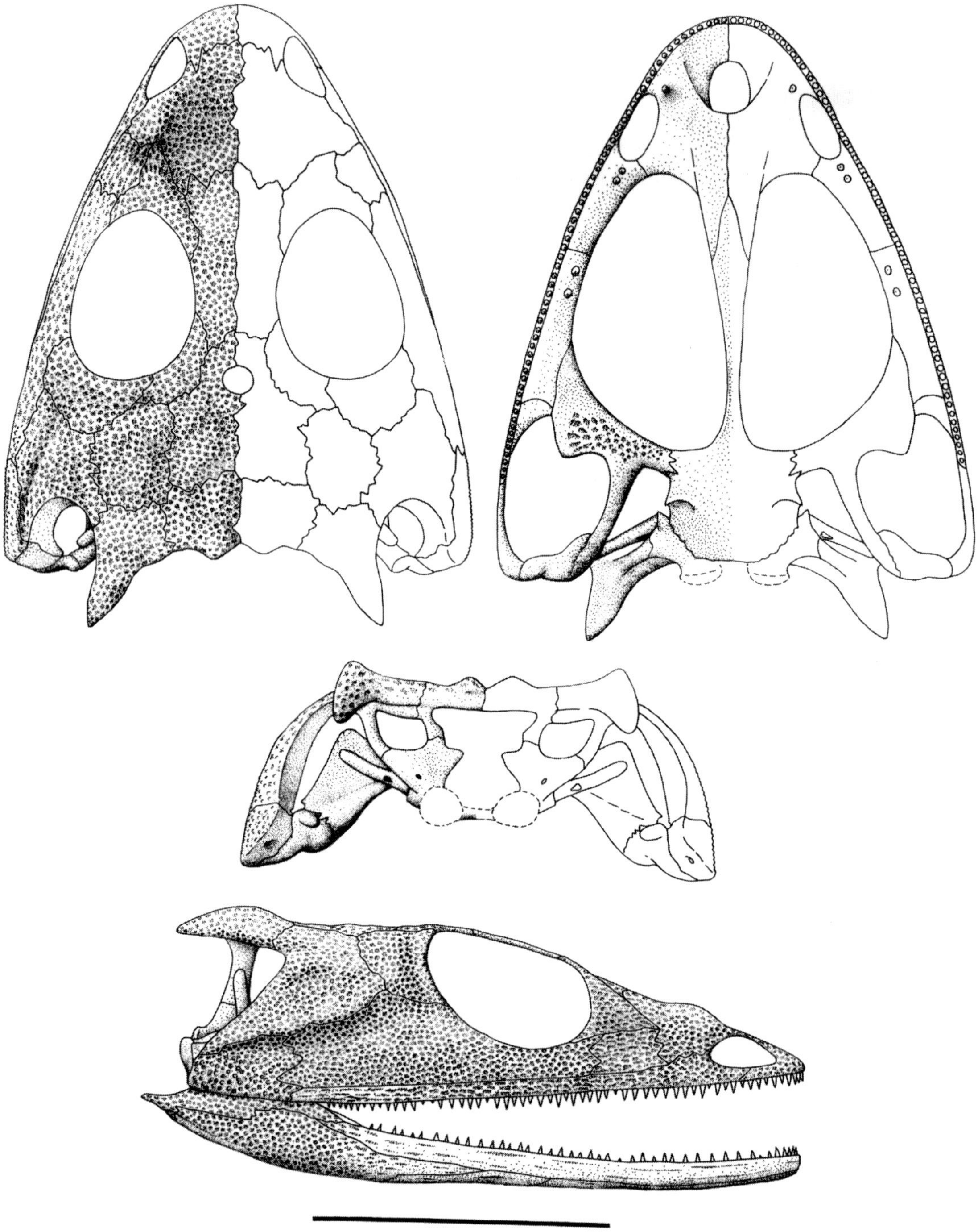

Figure 3. Reconstruction of the skull and lower jaw of *Lapillopsis*, the tiny terrestrial insectivorous stereospondyl from the Arcadia Formation, Queensland. Note the absence of sensory canals, which are usually associated with an aquatic way of life. Scale bar = 10 mm. From Yates (1999).

been found at the base of the Triassic in the Sydney Basin. *Bulgosuchus* (Mastodonsauridae), the only known vertebrate from the early Triassic fluvial Bulgo Sandstone of the Narrabeen Group, is of interest because the specimen is the largest described mastodonsaurid mandible, whereas early Triassic mastodonsaurids are usually of small size (Damiani, 1999).

From the Terrigal Formation, which was deposited by low to mid sinuosity streams spanning the late Early Triassic and early Middle Triassic, come two small stereospondyl skeletons, the juvenile brachyopid *Platycepsion* (Warren and Marsicano, 1998) and a small species of the mastodonsaurid, *Parotosuchus* (Damiani and Warren, 1997); Damiani (1999) has now described a fragmentary

Figure 4. Restoration of the mastodonsaurid *Paracyclotosaurus* hyperventilating beside a lake in the middle Triassic of the Sydney Basin. The animal was about two metres in length. Drawn by Adam Yates.

mandibular ramus from a second larger mastodonsaurid taxon. An articulated brachyopoid was recovered in 1997 from the Terrigal Formation. This specimen is of interest because it is unusually well preserved with most parts of the skeleton present and because it appears to have carbonaceous impressions of an outer body covering. This specimen and the small articulated stereospondyls were preserved in their natural environment in overbank or lacustrine sediments.

Subcyclotosaurus, a single mastodonsaurid skull impression (Watson, 1958), and some shoulder girdle material are known from the diverse fluvial environment of the middle Triassic Hawkesbury Sandstone which overlies the Terrigal Formation.

The later Middle Triassic sediments with stereospondyls in the Sydney Basin were deposited by a fresh water lake which formed the Ashfield Shale of the Wianamatta Group. For many years the Wianamatta Group was dated as late Triassic based solely on the presence of a supposedly late Triassic stereospondyl, *Paracyclotosaurus* (Watson, 1958), but evidence from pollen and plant macrofossils now agrees with a middle Triassic position (Balme and Foster, 1996). The large fish-eater, *Paracyclotosaurus* (Figure 4) is the most completely preserved of the Mastodonsauridae and one of the best known stereospondyls (Watson, 1958). A small brachyopid skull impression, *Notobrachyops*, is also known from the Ashfield Shale, while a fragmentary large brachyopoid was associated with *Paracyclotosaurus*.

Leigh Creek Coal Measures

South Australia had no known stereospondyl fossils until a probable member of the Chigutisauridae was found in the late Triassic Leigh Creek Coal Measures in 2000 (Pledge, 2000, personal communication).

POST-TRIASSIC STEREOSPONDYLS

Until 1983 no stereospondyls had been recovered from sediments later than the Triassic, the assumption being that they had become extinct at the close of that period. *Siderops* (Figure 5) was the first known post-Triassic stereospondyl (Warren and Hutchinson, 1983). This specimen was preserved in a concretion from the Westgrove Ironstone Member of the early Jurassic Evergreen Formation, near Wandoan, south central Queensland. *Siderops* is the most complete chigutisaurid stereospondyl with close relatives in Argentina, India and South Africa. Subsequently, more fragmentary remains of an early Cretaceous chigutisaurid,

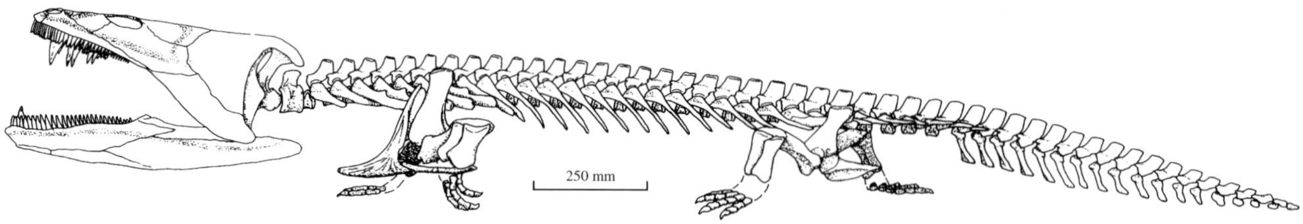

Figure 5. Skeletal reconstruction of *Siderops*, a Jurassic member of the Chigutisauridae from the Evergreen Sandstone, Queensland. Scale bar = 250mm. Drawn by Mark Hutchinson.

Koolasuchus, were found associated with dinosaurs in the fluvial Strzelecki Group exposed on the southern coast of Victoria (Warren *et al.,* 1997). *Koolasuchus* remains the last known stereospondyl, although several specimens, all referred to the Brachyopidae, have now been found in post-Triassic formations of Asia. The recent discovery of the late Triassic Leigh Creek chigutisaurid makes the Australian record for that taxon the longest and most complete for a stereospondyl, stretching from the stem chigutisaurid *Keratobrachyops* of the early Triassic through the Leigh Creek specimen to the Jurassic *Siderops* and the Cretaceous *Koolasuchus*.

PALAEOBIOGEOGRAPHY

Tetrapods have been present in the Australian region almost from the time that their remains first appeared in the fossil record, in the early part of the late Devonian. This appearance is presumed to be synchronous in Euramerica and Australia but it is not yet known where they originated, or whether that event happened more than once. A Gondwanan origin for the tetrapods has been proposed, based on a biogeographic and cladistic analysis of bony fishes (Long, 1990; Milner, 1993). Long cited the presence of plesiomorphic osteolepiform fishes as well as the earlier appearance of tetrapods in Australia in support of his theory, but more recent establishment of the closest relatives to the tetrapods as Euramerican panderichthyid fish, and discoveries of Frasnian tetrapods in Scotland and Latvia (Ahlberg, 1995) have now weakened Long's argument. As yet, no pre-Frasnian tetrapods have been found either in Gondwana or Euramerica so we have no more evidence as to their geographic origin. Nevertheless their widespread distribution in the Frasnian–Famennian suggests a pre-Frasnian origin, as does the presumed time of their divergence from the panderichthyids.

The aquatic adaptations of the Famennian tetrapods, especially the abundant *Ichthyostega* and *Acanthostega* from Greenland, suggests that invasion of the land by early tetrapods must have taken place somewhat later, during the early Carboniferous. Unfortunately, the first 20 million years of the Carboniferous represents a hiatus in the tetrapod record world-wide, with two Tournaisian records only: fragmentary material has been collected from Horton Bluff, Nova Scotia (Carroll, 1997) and a single articulated specimen has been found in Scotland (Clack, 2000b). From the mid Viséan increasingly diverse tetrapod faunas are known in the eastern United States of America, Scotland and Ireland. Many of the tetrapod taxa from these faunas can be related phylogenetically to later members of the early anamniote tetrapods or the amniote clade but as yet the relationships between these later Carboniferous tetrapods and the Devonian forms is unclear.

It has been suggested that the Carboniferous stem tetrapods were confined to Euramerica, forming the Mississippian Tetrapod Province (Milner, 1993). The new Ducabrook fauna from the Queensland early Carboniferous has shown that tetrapods were present also at almost the opposite side of the early Carboniferous globe (Thulborn *et al.*, 1996). We do not know as yet whether the Mississippian Tetrapod Province fauna was cosmopolitan, extending across the whole of the tropical early Carboniferous land masses, or whether the Australian Devonian tetrapods were the origin of a separate East Gondwanan Tetrapod Province, represented as yet only in the Ducabrook Formation. A separate radiation of tetrapods in the Australian region could also help to explain some sudden appearances of taxa in Europe (Milner, 1996). If these forms were a part of an Australian early Carboniferous radiation and drifted northwards on land masses detached from the Gondwanan mainland, they would appear in the European fauna only after collision of their land mass with the Euramerican block (Milner, 1996).

Following the Ducabrook fauna, the Australian region saw a period of 60 million years during which no tetrapod fossils have been found. It is probable that the reasons for the supposed absence of early Carboniferous tetrapods from Gondwana, that of increasing glaciation, did hold for southern Australia in the early Carboniferous and for the whole of East Gondwana until the late Permian. Evidence for this view is that fish faunas of the type associated with tetrapods in Euramerica and in the Drummond Basin of Queensland (Turner, 1993) were present in southern Australia in the early Carboniferous, for example in the Devil's Plain Formation at Mansfield (Warren *et al.*, 2000), but no tetrapods have been found associated with them.

Whether tetrapods remained in Australia during the late Carboniferous and Permian or re-invaded in the later stages of the Permian can only be resolved by the discovery of tetrapod remains from those periods.

Fragments of tetrapods are next recorded from

Australia in the late Permian to be followed by an explosion of diversity within the Stereospondyli in the early Triassic. It has been suggested (Milner, 1994) that the Mesozoic radiation of stereospondyls originated in eastern Gondwana. Further, Yates and Warren (2000) speculated that there may have been an almost complete elimination of anamniote tetrapods from Laurasia during the major extinction event at the end of the Permian. As a result, the earliest Stereospondyli as well as relict populations of earlier temnospondyl radiations, which are now known to have invaded southern Africa (Warren, 1999), survived in a safe haven in Gondwana to re-invade the north in the earliest Triassic. Although Australia's late Permian amphibian record is meagre, the presence of the basal stereospondyl, *Lapillopsis*, in the Arcadia Formation accompanied by a great diversity of other stereospondyls (including primitive members of several higher stereospondyl taxa), suggests that the stereospondyl radiation may have occurred in or close to the area of Gondwana that is now Australia.

Later in the Triassic, stereospondyls appear to have declined both in diversity and abundance in Australia, with only the Mastodonsauridae and Brachyopidae present in the middle Triassic of the Sydney Basin. The Chigutisauridae were most likely present also as they occurred in post-Triassic sediments. The new find from the late Triassic of South Australia should bridge the time gap in Australia between the early Triassic and early Jurassic chigutisaurids.

The survival of large members of the Chigutisauridae into the early Jurassic in Queensland and the early Cretaceous in Victoria is not yet fully explained. These stereospondyls were large (about 2.5 metres long in *Siderops*), suction-gulp feeders and, at least in the Victorian Cretaceous, existed alongside dinosaurs such as *Allosaurus*. The Strzelecki Group is noted for the presence of a number of endemic and relict vertebrate taxa and is thought to have constituted a polar safe haven at the end of the East Gondwana Peninsula (Warren *et al.,* 1997).

CONCLUSIONS

Australia has two of the earliest records of tetrapods in the early Famennian, *Metaxygnathus* and the Genoa River footprints; but suggestions that the tetrapods may have originated here have been weakened by the discovery of panderichthyid fishes in Canada and Europe as well as Frasnian tetrapods in Scotland, Latvia and Russia. The recent discovery of a mid-Viséan tetrapod fauna in Queensland is highly significant as it may help to bridge the morphological and phylogenetic gap between archaic Devonian tetrapods and the more terrestrial forms from the Carboniferous of Euramerica. Following the mid-Viséan, no anamniote or amniote tetrapods are known from the Australian fossil record until the window into the early radiation of the Stereospondyli afforded by the early Triassic Arcadia Formation. Despite their abundance in the earliest Triassic, stereospondyl diversity dwindled in Australia later in the Mesozoic; however, conditions in this portion of Gondwana favoured the survival of one higher stereospondyl taxon (Chigutisauridae), into the later part of the early Cretaceous.

ACKNOWLEDGEMENTS

I am grateful to the Australian Research Council for a series of grants supporting my work in the Mesozoic and Carboniferous of Australia. N. Pledge, South Australian Museum, provided unpublished observations, Lisa Masini drafted Figure 1 and Adam Yates drew Figures 2–4.

REFERENCES

Ahlberg, P.E. 1995. *Elginerpeton pancheni* and the earliest tetrapod clade. *Nature*, Vol. 373, pp.420–425.

Ahlberg, P.E. and Clack, J.A. 1998. Lower jaws, lower tetrapods-a review based on the Devonian genus *Acanthostega. Transactions of the Royal Society of Edinburgh: Earth Sciences*, Vol. 89, pp.11–46.

Balme, B. and Foster, C.B. 1996. Triassic (Chart 7). In Young, G.C. and Laurie, J.R. (eds). *An Australian Phanerozoic Timescale.* Oxford University Press, Melbourne. 279pp.

Campbell, K.S.W. and Bell, M.W. 1977. A primitive amphibian from the late Devonian of New South Wales. *Alcheringa*, Vol. 1, pp.369–381.

Cantrill, D.J. and Webb, J.A. 1998. Permineralised pleuromeid lycopsid remains from the early Triassic Arcadia Formation, Queensland, Australia. *Review of Palaeobotany and Palynology*, Vol. 102, pp.189–211.

Carroll, R.L. 1995. Between fish and amphibian. *Nature*, Vol. 373, pp.389–390.

Carroll, R.L. 1996. Revealing the pattern of macroevolution. *Nature*, Vol. 381, pp.19–20.

Carroll, R.L. 1997. The earliest Carboniferous tetrapods from North America. *Journal of Vertebrate Paleontology,* Vol. 17, Supplement to No. 3, pp.36A.

Cayley, R.A. and Taylor, D.H. 1997. Grampians special map area geological report. *Geological Survey of Victoria Report*, Vol. 107, pp.1–141.

Clack, J.A. 1996. Devonian tetrapod trackways and trackmakers; a review of the fossils and footprints. *Palaeogeography, Palaeoclimatology, Palaeoecology*, Vol. 130, pp.227–250.

Clack, J.A 2002a. The dermal skull roof of *Acanthostega*, an early tetrapod from the late Devonian. *Transactions of the Royal Society of Edinburgh: Earth Sciences,* Vol. 93, pp.17–33.

Clack, J.A 2002b. An early tetrapod from 'Romer's Gap'. *Nature*, Vol. 418, pp.72–76.

Coates, M.I. 1996. The Devonian tetrapod *Acanthostega gunnari* Jarvik: postcranial anatomy, basal tetrapod interrelationships and patterns of skeletal evolution. *Transactions of the Royal Society of Edinburgh: Earth Sciences*, Vol. 87, pp.363–421.

Coates, M.I. and Clack, J.A. 1990. Polydactyly in the earliest known tetrapod limbs. *Nature*, Vol. 347, pp.66–69.

Cosgriff, J.W. 1965. A new genus of Temnospondyli from the Triassic of Western Australia. *Journal of the Royal Society* *of Western Australia*, Vol. 48, pp.65–95.

Cosgriff, J.W. 1974. Lower Triassic Temnospondyli of Tasmania. *The Geological Society of America, Special Paper* 149, pp.1–131.

Daeschler, E.B. 2000. Early tetrapod jaws from the late Devonian of Pennsylvania, USA. *Journal of Paleontology*, Vol. 74, pp.301–308.

Damiani, R.J. 1999. Giant temnospondyl amphibians from the early to middle Triassic Narrabeen Group of the Sydney Basin, New South Wales, Australia. *Alcheringa*, Vol. 23, pp.87–109.

Damiani, R.J. 2001. A systematic revision and phylogenetic analysis of Triassic mastodonsauroids (Temnospondyli: Stereospondyli). *Zoological Journal of the Linnean Society*, Vol. 133, pp.379–482.

Damiani, R.J. and Warren, A.A. 1996. A new look at members of the superfamily Brachyopoidea (Amphibia, Temnospondyli) from the early Triassic of Queensland and a preliminary analysis of brachyopoid relationships. *Alcheringa*, Vol. 20, pp.277–300.

Damiani, R.J. and Warren, A.A. 1997. Re-interpretation of *Parotosuchus wadei* Cosgriff from the Triassic Narrabeen series at Gosford, New South Wales. *Alcheringa*, Vol. 21, pp.281–289.

Jarvik, E. 1995. The Devonian tetrapod *Ichthyostega*. *Fossils and Strata*, Vol. 40, pp.1–213.

Johanson, Z. and Ahlberg, P.E. 1998. A complete primitive rhizodont from Australia. *Nature*, Vol. 394, pp.569–573.

Laurin, M. and Reisz, R.R. 1997. A new perspective on tetrapod phylogeny. Chapter 2. In Sumida, S.S. and Martin, K. L.M. (eds). *Amniote Origins: Completing the Transition to Land*. Academic Press, San Diego. 510pp.

Lebedev, O.A. and Coates, M.I. 1995. The postcranial skeleton of the Devonian tetrapod *Tulerpeton curtum* Lebedev. *Zoological Journal of the Linnean Society*, Vol. 114, pp.307–348.

Long, J.A. 1990. Heterochrony and the origin of tetrapods. *Lethaia*, Vol. 23, pp.157–166.

Marsicano, C.A. and Warren, A.A. 1998. The first Paleozoic rhytidosteid record: *Trucheosaurus major* Watson 1956 from the late Permian of Australia, and a reassessment of the Rhytidosteidae (Amphibia, Temnospondyli). *Bulletin of the Natural History Museum of London. Geology*, Vol. 54, pp.147–154.

Milner, A.R. 1993. Biogeography of Palaeozoic tetrapods. Chapter 3. In Long, J. (ed). *Palaeozoic Biostratigraphy and Biogeography*. Bellhaven Press, London. 369pp.

Milner, A.R. 1994. Late Triassic and Jurassic amphibians: fossil record and phylogeny. Pp.5–22. In Fraser, N.C. and Sues, H.-D. (eds). *In the Shadow of the Dinosaurs: Early Mesozoic Tetrapods*. Cambridge University Press, Cambridge. 435pp.

Milner, A.R. 1996. Early amphibian globetrotters? *Nature*, Vol. 381, pp.741–742.

Northwood, C. 1997. *Palaeontological interpretation of the early Triassic Arcadia Formation, Queensland*. Ph.D. Thesis (unpublished), La Trobe University, Melbourne.

Thomson, K.S. 1993. The origin of tetrapods. *American Journal of Science*, Vol. 293-A, pp.33–62.

Thulborn, R.A., Warren, A.A., Turner, S. and Hamley, T. 1996. Carboniferous tetrapods from Australia. *Nature*, Vol. 381, pp.777–780.

Turner, S. 1993. Early Carboniferous microvertebrates from the Narrien Range, central Queensland. *Memoir of the Association of Australasian Palaeontologists*, Vol. 15, pp.289–304.

Vorobyeva, E.I. 1977. Morphology and nature of evolution of crossopterygian fishes. *Trudy Paleontological Institute Akademie Nauk SSSR*, Vol. 163, pp.1–239.

Warren, A.A. 1985. Two long snouted temnospondyls (Amphibia, Labyrinthodontia) from the Triassic of Queensland. *Alcheringa*, Vol. 9, pp.293–295.

Warren, A.A. 1997. A tetrapod fauna from the Permian of the Sydney Basin. *Records of the Australian Museum*, Vol. 49, pp.25–33.

Warren, A.A. 1999. Karroo tupilakosaurid: a relict from Gondwana. *Transactions of the Royal Society of Edinburgh: Earth Sciences*, Vol. 89, pp.145–160.

Warren, A.A., Currie, B.P., Burrow, C. and Turner, S. 2000. A redescription of *Gyracanthides murrayi* Woodward 1906 (Acanthodii, Gyracanthidae) from the lower Carboniferous of the Mansfield Basin, Victoria, Australia. *Journal of Vertebrate Paleontology*, Vol. 20, pp.225–242.

Warren, A.A. and Hutchinson, M.N. 1983. The last labyrinthodont? A new Brachyopoid (Amphibia, Temnospondyli) from the early Jurassic Evergreen formation of Queensland, Australia. *Philosophical Transactions of the Royal Society of London,* B, Vol. 303, pp.1–62.

Warren, A.A. and Hutchinson, M.N. 1988. A new capitosaurid (amphibian) from the early Triassic of Queensland, and the ontogeny of the capitosaurid skull. *Palaeontology*, Vol. 31, pp.857–876.

Warren, A.A., Jupp, R. and Bolton, B. 1986. Earliest tetrapod trackway. *Alcheringa*, Vol. 10, pp.183–186.

Warren, A. and Marsicano. C. 1998. Revision of the Brachyopidae (Temnospondyli) from the Triassic of the Sydney, Carnarvon and Tasmania Basins, Australia. *Alcheringa*, Vol. 22, pp.329–342.

Warren, A.A., Rich, T.H. and Vickers-Rich, P.V. 1997. The last last labyrinthodonts? *Palaeontographica* A, Vol. 247, pp.1–24.

Warren, A.A. and Schroeder, N. 1995. Changes in the capitosaur skull with maturity: an extension of the growth series of *Parotosuchus aliciae*. *Alcheringa*, Vol. 19, pp.41–46.

Warren, J.W. and Wakefield, N.A. 1972. Trackways of tetrapod vertebrates from the upper Devonian of Victoria, Australia. *Nature*, Vol. 238, pp.469–470.

Watson, D.M.S. 1958. A new labyrinthodont (*Paracyclotosaurus*) from the upper Trias of New South Wales. *Bulletin of the British Museum (Natural History Geology)*, Vol. 3, pp.235–263.

Wood, S.P., Panchen, A.L. and Smithson, T.R. 1985. A terrestrial fauna of the Scottish lower Carboniferous. *Nature*, Vol. 314, pp.355–356.

Yates, A.M. 1999. The Lapillopsidae a new family of small temnospondyls from the early Triassic of Australia. *Journal of Vertebrate Paleontology*, Vol. 19, pp.302–320

Yates, A.M. 2000. A new tiny rhytidosteid (Temnospondyli: Stereospondyli) from the early Triassic of Australia and the possibility of hidden temnospondyl diversity. *Journal of Vertebrate Paleontology*, Vol. 20, pp.484–489.

Yates, A.M. and Warren, A.A. 2000. The phylogeny of the 'higher temnospondyls' (Vertebrata, Choanata). *Zoological Journal of the Linnean Society*, Vol. 128, pp.77–121.

Plate 54. *Litoria bicolor* is a small, agile tree frog (Hylidae). Photograph M. Tyler.

Plate 55. *Austrochaperina fryi* has direct development entirely within the egg capsule (Microhylidae). Photograph M. Tyler.

Plate 56. *Limnodynastes interioris* is a large ground-dwelling frog (Limnodynastinae). Photograph M. Tyler.

Plate 57. *Pseudophryne semimarmorata* a small, so-called toadlet (Myobatrachinae). Photograph M. Tyler.

Plate 58. The green-eyed tree frog, *Litoria genimaculata,* from the Atherton Tableland (north-eastern Queensland). Photograph S. Richards.

Plate 59. Populations of the torrent frog, *Litoria eucnemis,* from northern Queensland have suffered mass mortality. Photograph S. Richards.

Plate 60. Populations of this rainforest frog, *Nyctimystes dayi*, have declined at upland sites across its range in northern Queensland. Photograph S. Richards.

Plate 61. This upland rainforest frog, *Taudactylus acutirostris,* from the Windsor Tableland is another species that has been subjected to mass mortality. Photograph S. Richards.

Chapter 12
The Origins of Australian Frogs

Michael J. Tyler[1] and Michael S.Y. Lee[1, 2]

[1] Department of Environmental Biology, University of Adelaide, S.A. 5005 Australia
[2] Department of Palaeontology, South Australian Museum, North Terrace, Adelaide, S.A. 5000 Australia
E-mail: michael.tyler@adelaide.edu.au

INTRODUCTION

Australian frogs consist of at least five distinct lineages (Plates 54–57): the Hylidae, Ranidae, Microhylidae, Limnodynastinae, and Myobatrachinae. The hylids (3 genera, 76 species) are commonly known as tree frogs, and include such familiar forms as the green tree frogs (*Litoria caerulea*). All members of two of the genera (*Litoria* and *Nyctimystes*) possess toe pads and an extra toe cartilage or bone, enabling many to be adept climbers. They are diverse and distributed across most of Australia including the extremely arid centre. The ranids are the typical frogs abundant throughout most of the world. In Australia, however, they are poorly represented (a single genus and species) and restricted to the tropical north. The sole Australian ranid (*Rana daemeli*) can be recognized by a skin fold extending from the eye to the hindlimb. The microhylids (2 genera, 18 species) are small frogs which lay eggs on land that develop directly into froglets; they are also relatively poorly represented and restricted to the tropical north (Burton, 1993).

The limnodynastines (10 genera, 46 species) are distributed across the entire continent with the exception of the Nullarbor Plain and are represented by water-holding frogs (*Notaden* and *Neobatrachus*) in the arid centre; several genera share the habit of foam nesting (Tyler and Davies, 1979) and they can be diagnosed by a derived arrangement of the muscles of the lower jaw (the intermandibularis muscle underlies the submentalis muscle: Figure 1). This group used to be named the Cycloraninae (Parker, 1940); however, following the transfer of the oldest included genus (*Cyclorana*) to the Hylidae (Tyler, 1972, 1978), the name reverted to being based on the next oldest genus (*Limnodynastes*). The myobatrachines (12 genera, 80 species) are a large assemblage of small forms that include such species as the bizarre turtle frog (*Myobatrachus*), a mole-like form that occasionally inhabits termite nests, and the pouched frog (*Assa*) where males rear larvae inside lateral pockets. Myobatrachines are diverse and distributed across most of Australia. However, the evidence for their monophyly remains weak (Farris *et al.,* 1982), and the problem deserves further investigation.

The Limnodynastinae and Myobatrachinae were previously grouped together into the Leptodactylidae and subsequently into the Myobatrachidae. This was a grouping based entirely on geography rather than intrinsic character evidence (e.g. morphology, molecules or behaviour), and consisted of Australian frogs that vaguely resembled the South American leptodactylids (Lynch, 1971). The two included lineages, the Limnodynastinae and Myobatrachinae, share no derived characters permitting the interpretation that they are one another's closest relatives. In contrast, the myobatrachines possess a few characters linking them to the Sooglossidae, an obscure group restricted to the Seychelles Archipelago (Cannatella and Hillis, 1993). The 'Myobatrachidae' is therefore probably an artificial (polyphyletic) assemblage and its two discrete components (Myobatrachinae and Limnodynastinae) are treated separately. Some authors attribute family status to these groups (Zug *et al.*, 2001).

ORIGINS AND BIOGEOGRAPHY

Until about 1965 the Australian frog fauna was perceived to have originated elsewhere and to have entered the continent from the north (Darlington, 1957). The Australian continental plate was assumed to be static, and the only plausible route of entry was therefore island-hopping across Indonesia and New Guinea. With the acceptance of continental drift, other mechanisms of faunal origin became apparent. Tyler (1979) suggested that the Australian herpetofauna (including frogs) comprised three distinct elements generated by three processes. These are listed below.

(1) Ancient lineages that were distributed across the remains of the supercontinent Gondwana (Australia, Antarctica and South America) and were subsequently isolated in Australia and South America, during the breakup of Gondwana and the freezing of Antarctica.
(2) Recent lineages that represent invasions from the north that became possible when Australia collided with the Asian plate.
(3) Autochthonian (endemic) diversification of each lineage which came to be isolated in Australia via either mechanism described above.

Here, the focus will be on the two methods by which frog lineages could have colonized Australia. The mechanisms of subsequent endemic diversification, and relationships within each endemic lineage, are discussed in Roberts and Watson (1993), Littlejohn *et al.* (1993), Tyler and Davies (1993), Schauble *et al.* (2000) and Read *et al.* (2001).

Frogs usually do not tolerate salt water very well (the introduced cane toad, *Bufo marinus*, is a rare and notorious exception). Apart from the prospect of highly unlike-

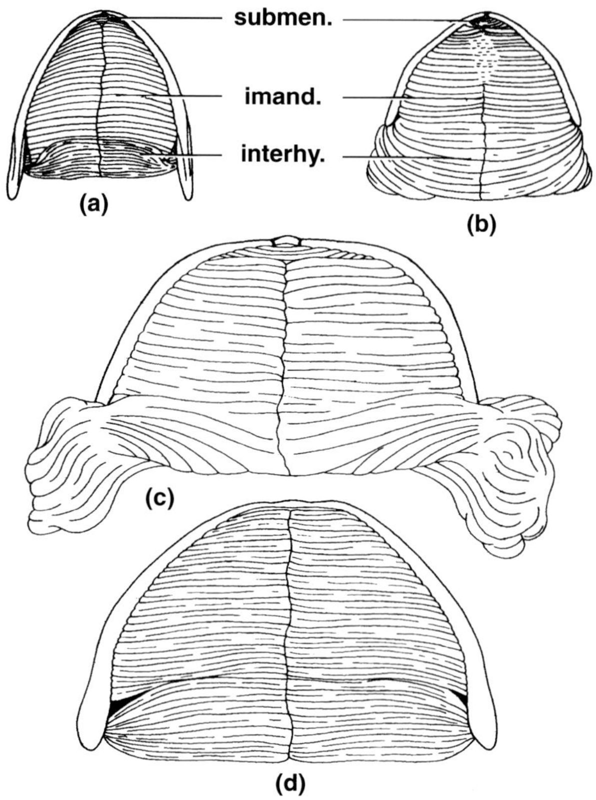

Figure 1. Superficial mandibular musculature in four species (from Tyler, 1972). **(a)** *Adelotus brevis*; **(b)** *Myobatrachus gouldii*; **(c)** *Notaden nichollsi*; **(d)** *Neobatrachus centralis*. Note that all except *Myobatrachus* are limnodynastines, exhibiting various degrees of the intermandibularis underlying the submentalis. Key to abbreviations: submen. = musculus submentalis; imand. = musculus intermandibularis; interhy. = musculus interhyoideus.

ly long-distance (transoceanic) dispersal, the Gondwanan or Asian routes are the only two terrestrial entry modes which frogs and other terrestrial animals could have used; for further discussion of migratory routes, relating to reptiles, see pp.309–330 (Scanlon, 2006). Anurans could have entered from South America via Antarctica, when all three continents were connected as the remains of Gondwana. This route was severed (~45 m.y.a.), when Australia separated from Antarctica and drifted north (Metcalfe, 2001). For the next 20 million years Australia was isolated but, commencing ~25 m.y.a., it began to collide with the Asian plate, raising the New Guinea massif and creating a new connection for migration from Asia across the Indonesian archipelago (Hall, 2001). For further details of the formation of New Guinea, associated island arcs and land-bridge connections see pp.71–93 (Heinsohn and Hope, 2006) and pp.95–110 (van Oosterzee, 2006).

It can be predicted that lineages with South American origins would have been in Australia, and isolated from their nearest relatives (ancestors), for at least 45 million years. This long period of isolation and endemic evolution should result in the following characteristics: wide distributions across the continent; morphological diversity; and distinctness due to extensive evolution. The hylids, limnodynastines and myobatrachines seem to meet most of these criteria. They are each distributed across most of Australia (with some forms occupying even the arid centre). They are also highly diverse, each consisting of several very distinct genera. While hylids currently only include three genera, one of these (*Litoria*) consists of over 60 diverse species in Australia and a further 50 confined to New Guinea; when phylogenetic relationships become better known it will be split into several discrete genera. Two new myobatrachine genera have also been recently described, increasing that lineage's known diversity: *Bryobatrachus* Rounsevell *et al.* (1994) and *Spicospina* Roberts *et al.* (1997). The hylids, limnodynastines and myobatrachines are very distinct from their nearest relatives—in fact, so distinct that these relatives have yet to be identified with confidence.

Bagnara and Ferris (1975) found the red skin pigment rhodomelanochrome in South American frogs of the endemic subfamily Phyllomedusinae and in some species of Australian *Litoria*. This led them to suggest that those *Litoria* species were more closely related to Phyllomedusinae than to other Australian *Litoria* species. Tyler and Davies (1978) investigated this novel proposal and concluded that the Australian *Litoria* species possessing rhodomelanochrome were nevertheless more closely related to other *Litoria* than to phyllomedusine frogs: the skin pigment had evolved in each group independently. These observations suggest a long period of endemic evolution.

There are, however, lingering difficulties with this interpretation. First, the affinities between Australian and South American frogs remain unconfirmed, and is currently restricted to vague similarities between the Australian limnodynastines and myobatrachines on the one hand and the South American leptodactylids on the other. The evidence for affinities between the myobatrachines and the sooglossids from the Seychelles (Cannatella and Hillis, 1993), if correct, would be inconsistent with this simple scenario. However Tyler (1985) demonstrated, in a small suite of characters, closer affinity of sooglossids to ranoids and suggested an African origin for the ancestors of this family. The close proximity of the Seychelles to Africa renders this interpretation attractive. Also, the South American frog fauna seems to show greater similarity to that of Africa, reflecting the long period of direct physical contact of these continents (Laurent, 1979). For instance, both continents possess a diverse extant radiation and fossil record of pipids (totally aquatic frogs that include the laboratory clawed toad, *Xenopus laevis*), indicating that there was an ancient diversification of pipids that occurred before the separation of South America and Africa (Baez and Pugener, 1998). If so, the total absence of pipids from Australia is puzzling; extinction is not a satisfactory explanation since no pipid fossils have been found here either, though abundant remains of other frog lineages have been unearthed.

The earliest fossil frogs in Australia have also recently been discovered, from the Eocene-Paleocene boundary at Murgon in south-eastern Queensland (Tyler and Godthelp, 1993). The antiquity and morphological conservativeness of these specimens also challenge the widespread belief that hylids, limnodynastines and myobatrachines each diversified subsequent to Australia's isolation (White, 1984). Modern hylid (*Litoria*) and limnodynastine (*Limnodynastes, Lechriodus*) genera have been found at Murgon, along with the extinct *Australobatrachus*. This fact suggests that hylids and limnodynastines had already radiated by this time (55 m.y.a.). However, Australia was still connected to the rest of Gondwana (Crook, 1981; Metcalfe, 2001), so that the endemic diversification of each of these three groups should not have commenced. These fossils suggest very strongly that at least two endemic radiations (hylids and limnodynastines) commenced before Australia had broken away from Western Antarctica. Immunological distances suggest an even earlier divergence time for limnodynastines (Roberts and Maxson, 1986), although this conclusion needs to be confirmed with nucleotide sequences. Even if confirmed, such early divergence times would still not destroy the accepted model: it is possible that Antarctica formed a partial barrier (due to cool climate and a distance or 'peninsula' effect) to faunal interchange for some time before the final separation of Australia. This might have allowed some endemic Australian radiations to commence earlier.

In contrast to Gondwanan forms, lineages with Asian origins could only have arrived within the last 30 million years, and probably much later. Their recent arrival would have several results. They are unlikely to be widely distributed, but should be restricted to, or at least most diverse, in northern habitats that are adjacent and environmentally similar to their most likely entry point (tropical Cape York). They should not be very diverse, nor very distinct from their ancestral stocks, and thus should have clear affinities with Asian forms.

In relation to the Ranidae, Menzies (1987) observed, 'There is no fossil or even subfossil evidence to support any theories on the evolution of Papuan *Rana*'. Equally, there are no morphological or biochemical data. Allison (1996) observed about the New Guinean frog fauna in general 'Most of the initial New Guinea colonists would have come from the Oriental region and these probably included all the major families now represented on that island except for hylid frogs... which were widespread in Australia...'. Unfortunately for both the Ranidae and Microhylidae there remains no supporting evidence for this perfectly reasonable assumption. Knowledge of the families in Australia is improving and, rather than being confined to Cape York (as previously thought), both groups are now known to be more widely distributed, extending up to 1,000 km west to the Northern Territory (Zweifel, 1985; Ingram, 1993).

Several other chapters contain material relating to Australasian frogs. The diversity of extant faunas is demonstrated in keys—pp.253–262 (White, 2006). Comparative comment on frog evolution or biogeography in New Guinea and New Zealand is included in Heinsohn and Hope (2006) and Holdaway and Worthy (2006) respectively. Readers are also referred to pp.95–110 (van Oosterzee, 2006) for a summary of the geologic and climatic processes that have contributed to the isolation of Australia from Asia and influenced faunal exchange from New Guinea.

CONCLUSIONS

While the dual origin model proposed by Tyler (1979) has not yet been seriously questioned, more work needs to be done to confirm its predictions. In particular, the presumed relationships between the Australian and South American frog lineages need to be established. The two possible entry routes not only predict different relationships between Australian and overseas forms (South America versus Asia), but also very different divergence times from their overseas sister-groups (>45 versus <25 million years). Detailed and well-corroborated phylogenies (based on morphological and molecular traits), in conjunction with distributional and temporal (fossil and molecular clock) information, will therefore allow these biogeographic hypotheses to be formulated and tested. Also, details of the model need to be revised; for instance, it is likely that some of the endemic Australian radiations pre-dated rather than post-dated the total isolation of Australia, perhaps due to earlier physical or ecological barriers forming in Antarctica.

REFERENCES

Allison, A. 1996. Zoogeography of amphibians and reptiles of New Guinea and the Pacific region. Pp.407–436. In Keast, A. and Miller, S.E. (eds). *The Origin and Evolution of Pacific Island Biotas, New Guinea to Eastern Polynesia: Patterns and Processes.* SPB Academic Publishing, Amsterdam. 531pp.

Baez, A.M. and Pugener, L.A. 1998. A new Paleogene pipid frog from northwestern Patagonia. *Journal of Vertebrate Paleontology*, Vol. 18, No. 3, pp.511–524.

Bagnara, J.F. and Ferris, W. 1975. The presence of phyllomedusine melanosomes and pigments in Australian hylids. *Copeia*, 1975, No. 3, pp.592–595.

Burton, T.C. 1993. Family Microhylidae. Pp.64–66. In Glasby, C.J., Ross, G.J.B. and Beesley, P.L. (eds). *Fauna of Australia. Vol 2A. Amphibia and Reptilia.* Australian Government Publishing Service, Canberra. 439pp.

Cannatella, D.C. and Hillis, D.M. 1993. Amphibian relationships: phylogenetic analysis of morphology and molecules. *Herpetological Monographs,* Vol. 7, pp.1–7.

Crook, K.A.W. 1981. The break-up of the Australian-Antarctic segment of Gondwanaland. Pp.1–14. In Keast, A. (ed). *Ecological Biogeography of Australia.* W. Junk, The Hague.

Darlington, P.J. 1957. *Zoogeography: the Geographical Distribution of Animals.* John Wiley and Sons, New York. 675pp.

Farris, J.S., Kluge, A.G. and Mickevich, M.F. 1982. Phylogenetic analysis, the monothetic group method, and myobatrachid frogs. *Systematic Zoology,* Vol. 31, pp.317–327.

Hall, R. 2001. Cenozoic reconstructions of SE Asia and the SW Pacific: changing patterns of land and sea. Pp.35–56. In Metcalfe, I., Smith, J.M.B., Morwood, M. and Davidson, I. (eds). *Faunal and Floral Migrations and Evolution in SE Asia–Australasia*. Balkema Publishers, Exton, Pennsylvania. 416pp.

Heinsohn, T. and Hope, G. 2006. The Torresian connections: zoogeography of New Guinea. Pp.71–93. In Merrick, J.R., Archer, M., Hickey, G.M. and Lee, M.S.Y (eds). *Evolution and Biogeography of Australasian Vertebrates*. Auscipub Pty. Ltd., Sydney. 942pp.

Holdaway, R.N and Worthy, T.H. 2006. The evolution of New Zealand and its vertebrates. Pp.111–128. In Merrick, J.R., Archer, M., Hickey, G.M. and Lee, M.S.Y (eds). *Evolution and Biogeography of Australasian Vertebrates*. Auscipub Pty. Ltd., Sydney. 942pp.

Ingram, G.J. 1993. Family Ranidae. In Glasby, C.J., Ross, G.J.B. and Beesley, P.L. (eds). *Fauna of Australia. Vol. 2A. Amphibia and Reptilia*. Australian Government Publishing Service, Canberra. 439pp.

Laurent, R.F. 1979. Herpetofaunal relationships between Africa and South America. Pp.55–71. In Duellman, W.E. (ed). *The South American Herpetofauna: its Origin, Evolution, and Dispersal*. Monograph of the Museum of Natural History, University of Kansas, No. 7, pp.1–485.

Littlejohn, M.J., Roberts, J.D., Watson, G.F. and Davies, M. 1993. Family Myobatrachidae. Pp.41–57. In Glasby, C.J., Ross, G.J.B. and Beesley, P.L. (eds). *Fauna of Australia. Vol 2A. Amphibia and Reptilia.* Australian Government Publishing Service, Canberra. 439pp.

Lynch, J.D. 1971. Evolutionary relationships, osteology and zoogeography of leptodactyloid frogs. *University of Kansas Museum of Natural History, Miscellaneous Publication,* No. 53, pp.1–238.

Menzies, J.I. 1987. A taxonomic revision of Papuan *Rana* (Amphibia: Ranidae). *Australian Journal of Zoology*, Vol. 35, pp.373–418.

Metcalfe, I. 2001. Palaeozoic and Mesozoic tectonic evolution and biogeography of SE Asia–Australasia. Pp.15–34. In Metcalfe, I., Smith, J.M.B., Morwood, M. and Davidson, I. (eds). *Faunal and Floral Migrations and Evolution in SE Asia–Australasia.* Balkema Publishers, Exton, Pennsylvania. 416pp.

Parker, H.W. 1940. Australasian frogs of the family Leptodactylidae. *Novitates Zoologicae*, Vol. 42, No. 1, pp.1–106.

Read, K., Keogh, J.S., Scott, I.A.W., Roberts, J.D. and Doughty, P.D. 2001. Molecular phylogeny of the Australian frog genera *Crinia, Geocrinia*, and allied taxa (Anura: Myobatrachidae). *Molecular Phylogenetics and Evolution*, Vol. 21, pp.294–308.

Roberts, J.D., Horwitz, P., Wardell-Johnson, G., Maxson, L.R. and Mahony, M.J. 1997. Taxonomy, relationships and conservation of a new genus and species of myobatrachid frog from the high rainfall region of southwestern Australia. *Copeia,* 1977, pp.373–381.

Roberts, J.D. and Maxson, L.R. 1986. Phylogenetic relationships in the genus *Limnodynastes* (Anura: Myobatrachidae): a molecular perspective. *Australian Journal of Zoology,* Vol. 34, pp.561–574.

Roberts, J.D. and Watson, G.F. 1993. Biogeography and phylogeny of the Anura. Pp.35–40. In Glasby, C.J., Ross, G.J.B. and Beesley, P.L. (eds). *Fauna of Australia. Vol 2A. Amphibia and Reptilia*. Australian Government Publishing Service, Canberra. 439pp.

Rounsevell, D.E., Ziegler, D., Brown, P.B., Davies, M. and Littlejohn, M.J. 1994. A new genus and species of frog (Anura: Leptodactylidae) from southern Tasmania. *Transactions of the Royal Society of South Australia*, Vol. 118, No. 3, pp.171–185.

Scanlon, J.D. 2006. Origins and radiations of snakes in Australasia. Pp.309–330. In Merrick, J.R., Archer, M., Hickey, G.M. and Lee, M.S.Y. (eds). *Evolution and Biogeography of Australasian Vertebrates.* Auscipub Pty. Ltd., Sydney. 942pp.

Schauble, C., Moritz, C. and Slade R.W. 2000. A molecular phylogeny for the frog genus *Limnodynastes* (Anura: Myobatrachidae). *Molecular Phylogenetics and Evolution,* Vol. 16, pp.379–391.

Tyler, M.J. 1972. Superficial mandibular musculature, vocal sacs and the phylogeny of Australo-Papuan leptodactylid frogs. *Records of the South Australian Museum*, Vol. 16, No. 9, pp.1–20.

Tyler, M.J. 1978. *Amphibians of South Australia.* South Australian Government Printer, Adelaide. 84pp.

Tyler, M.J. 1979. Herpetofaunal relationships of South America with Australia. Pp.73–106. In Duellman, W.E. (ed). *The South American Herpetofauna: its Origin, Evolution, and Dispersal.* Monograph of the Museum of Natural History, University of Kansas, No. 7, pp.1–485.

Tyler, M.J. 1985. Phylogenetic significance of the superficial mandibular musculature and vocal sac structure of sooglossid frogs. *Herpetologica*, Vol. 41, No. 2, pp.173–176.

Tyler, M.J. and Davies, M. 1978. Phylogenetic relationships of Australian hyline and Neotropical phyllomedusine frogs of the family Hylidae. *Herpetologica*, Vol. 34, No. 2, pp.219–224.

Tyler, M.J. and Davies, M. 1979. Foam nest construction by Australian leptodactylid frogs (Amphibia, Anura, Leptodactylidae). *Journal of Herpetology*, Vol. 13, pp.509–510.

Tyler, M.J. and Davies, M. 1993. Family Hylidae. Pp.58–63. In Glasby, C.J., Ross, G.J.B. and Beesley, P.L. (eds). *Fauna of Australia. Vol 2A. Amphibia and Reptilia.* Australian Government Publishing Service, Canberra. 439pp.

Tyler, M.J. and Godthelp, H. 1993. A new species of *Lechriodus* Boulenger (Anura: Leptodactylidae) from the early Eocene of Queensland. *Transactions of the Royal Society of South Australia*, Vol. 117, pp.187–189.

van Oosterzee, P. 2006. Drawing the Wallace Line. Pp.95–110. In Merrick, J.R., Archer, M., Hickey, G.M. and Lee, M.S.Y (eds). *Evolution and Biogeography of Australasian Vertebrates*. Auscipub Pty. Ltd., Sydney. 942pp.

White, A. 1984. Origins and early radiations of the amphibians. Pp.225–261. In Archer, M. and Clayton, G. (eds). *Vertebrate Zoogeography and Evolution in Australasia.* Hesperian Press, Perth. 1203pp.

White, A.W. 2006. Keys to families and genera of frogs in Australia, New Zealand, New Guinea and neighbouring islands. Pp.253–262. In Merrick, J.R., Archer, M., Hickey, G.M. and Lee, M.S.Y (eds). *Evolution and Biogeography of Australasian Vertebrates.* Auscipub Pty. Ltd., Sydney. 942pp.

Zug, G.R., Vitt, L.J. and Caldwell, J.P. 2001. *Herpetology. An Introductory Biology of Amphibians and Reptiles.* Second Edition. Academic Press, San Diego. 630pp.

Zweifel, R.G. 1985. Australian frogs of the family Microhylidae. *Bulletin of the American Museum of Natural History,* Vol. 182, Pt. 3, pp.265–388.

Chapter 13
The Great Frog Decline in Australasia: Causes, Developments and Conservation

Stephen J. Richards[1], Ross A. Alford and Kay S. Bradfield
School of Tropical Biology, James Cook University, Townsville, Qld. 4811 Australia
[1] Current address: Vertebrates Department, South Australian Museum, North Terrace, Adelaide, S.A. 5000 Australia
E-mail: richards.steve@saugov.sa.gov.au

INTRODUCTION

Although the first anecdotal reports of widespread declines in amphibian populations were made public over ten years ago (Barinaga, 1990), individual frog declines have been reported for over a century (Bury, 1999). In the past decade an intensive global effort has been made to determine the extent and causes of these declines. Australian researchers have been at the forefront of this effort, responding to evidence that a large number of Australian frogs have suffered range contractions since at least 1979. Localized disappearances of frogs in Australia were first noticed in south-eastern Queensland, where populations of the southern gastric brooding frog *Rheobatrachus silus* and the sympatric day frog *Taudactylus diurnus* crashed during the late 1970s (Czechura and Ingram, 1990). Subsequently, the Eungella gastric brooding frog *Rheobatrachus vitellinus* vanished in the mid-1980s, and several other species suffered population declines in the Wet Tropics of northern Queensland during the early 1990s (Richards *et al.*, 1993). A large number of population declines and range contractions have now been documented across Australia, concentrated largely in Queensland and New South Wales.

The situation in Australia was first summarized by Tyler (1991), who reported that 23 species had suffered population declines. Since that time extensive data have been accumulated on the status of Australian frogs, as shown in *The Action Plan for Australian Frogs* (Tyler, 1997). As a number of comprehensive literature reviews on Australian and global amphibian declines have also recently been published (Ferraro and Burgin, 1993; Mahony, 1996; Waldman and Tocher, 1998; Alford and Richards, 1999), we will not present another review of the general problem. Our approach will be to describe progress in understanding of the situation in Australia and how emphases of research programmes, on causes of amphibian declines, have changed. We will attempt to place the Australian situation in a global context, comment on the most important current research foci, and suggest areas for fruitful research in the future.

WHAT WE KNOW

When declines were first reported in Australia intensive efforts were made to quantify the status of frog species considered to be at risk, through surveys and the establishment of monitoring programmes (Richards *et al.*, 1993; Trenerry *et al.*, 1994; Gillespie and Hollis, 1996; Ehmann, 1997; Osborne *et al.*, 1999). These data allowed rigorous assessment and refinement of the species considered to be of conservation concern, as demonstrated by a comparison between the species of concern listed by Tyler (1991) and Tyler (1997).

In the interim between these publications, the population declines of 12 species listed (Tyler, 1991) were shown to represent local, small-scale population fluctuations and were eliminated from Tyler's (1997) updated list. However, the availability of more data led to the inclusion of an additional 16 species in the 1997 list, bringing the total number of frog species that have been identified as suffering from population declines to 27 or 13% of the Australian frog fauna. While we have added another three species (*Litoria genimaculata, L. pearsoniana, Adelotus brevis*) to our table of declining species and postulated causes (Table 1), we have excluded four species from the Tyler (1997) list. *Litoria olongburensis* was excluded because, like the other 'acid' frogs (species that inhabit low nutrient soils near coastal lowlands and sand islands), its population declines are localized and due to habitat loss as well as fragmentation (Hines *et al.*, 1999); the other three species (*Geocrinia vitellina, Spicospina flammocaerulea* and *Taudactylus pleione*) are at risk solely because they have extremely restricted geographic distributions but are not known to have declined.

A number of other Australian species may also be at risk because of their restricted distributions. Although localized population declines or extinctions have occurred in some of these species, our discussion is restricted to species that have exhibited dramatic declines over a considerable portion of their range. We are not suggesting, however, that species with restricted distributions should not be of concern.

Since 1991 substantially more information has become available on likely threats to frog populations and two general categories of factors have been identified—anthropogenic and unknown. First, declines in populations of nine species appear to be linked directly to anthropogenic causes such as habitat loss or degradation of habitat quality, pollution and introduced predators (Ferraro and Burgin, 1993; Thumm and Mahony, 1999). Although significant from the perspective of amphibian conservation, the effects of these factors on amphibian populations and how to mitigate them are relatively well understood. The issue for conservation of these taxa is political will rather than scientific investigation and innovation.

The second category involves a group of 18 species. Although a number of theories have been proposed, no causal mechanism has yet been identified. Most of these species share a number of ecological characteristics including: a closer association with streams than non-declining species (Richards *et al.*, 1993; Mahony, 1996; Williams and Hero, 1998; McDonald and Alford, 1999); occurrence in protected habitats at relatively high altitudes; having aquatic larvae. Declines of upland stream-dwelling frogs have been reported from a number of sites scattered throughout eastern Australia (Czechura and Ingram, 1990; McDonald, 1990; Ingram and McDonald, 1993; Richards *et al.*, 1993; Mahony, 1996).

Species that have declined in, or disappeared from, relatively pristine upland stream habitats are phylogenetically diverse, occurring in two families (Hylidae, Myobatrachidae) and five genera; see Table 1 and Plates 58–61. Furthermore, declining species invariably occur in sympatry with other species with stable populations (Richards *et al.*, 1993). The species with stable populations tend to spend more time away from streams, breed in lentic water bodies, or have a reproductive strategy that excludes an aquatic larval stage (e.g. Family Microhylidae). Research has been initiated in northern Queensland to further compare the ecology and behaviour of species that have declined with those of sympatric species that have not, in an attempt to identify differences in behaviour or habitat use that could expose declining species to environmental influences that may cause or accelerate population declines.

While populations of many stream-associated frogs in eastern Queensland declined to disappearance, populations of at least a few other species appeared to undergo moderate or severe population declines followed by varying degrees of recovery. This effect was striking in *Litoria genimaculata*, in which closely monitored populations declined near Paluma and at Big Tableland (McDonald and Alford, 1999) but recovered over several succeeding years to pre-decline numbers. No populations of *Taudactylus eungellensis* were found between 1986 and 1992, but since then, they have been found at several sites (McDonald and Retallick, 2001, personal communication); this probably reflects some degree of recovery in numbers combined with the discovery of previously unknown populations. Populations of *Taudactylus rheophilus*, an upland rainforest frog from northern Queensland, declined markedly during the early 1990s but small, apparently remnant, populations of this species were subsequently discovered in 1996 (Marshall, 1998). More recently (McDonald and Alford, 1999) it appears that several other species, including *Litoria rheocola, L. nannotis* and *Nyctimystes dayi*, may be recolonizing a few sites from which they disappeared in the early 1990s. Thus far, however, only a few individuals have been sighted at a few locales where these species were formerly common, and in several cases the small numbers of apparent recolonizers have subsequently disappeared. This suggests that there may be some hope for natural recovery of populations, but by no means assures that it is occurring.

Mysterious declines in rainforest habitats are not restricted to Australia. A number of declines showing remarkable parallels with Australian species have occurred in the mountains of Central America. The best known of these involved the golden toad, *Bufo periglenes*, which disappeared from montane cloud forests in Costa Rica (Pounds and Crump, 1994). Pounds *et al.* (1999) argued that the disappearance of 20 anurans from Costa Rican montane forests is linked to changing patterns of dry-season mist frequency due to global warming. Similar patterns of decline were reported for amphibian assemblages in montane Panama and Costa Rica (Lips, 1998, 1999). Although no obvious cause has been identified, there have been suggestions that infection by a virulent chytrid fungus played some role in these declines (Berger *et al.*, 1998; Lips, 1999).

The status of frog populations in South-east Asia remains poorly known. In Borneo there is no evidence for population declines of frogs along lowland rainforest streams (Inger and Voris, 1993; Voris and Inger, 1995). Allison (1993) noted the absence of several species of stream-dwelling frogs (*Litoria* and *Nyctimystes*) from an upland rainforest in Papua New Guinea, and postulated that increased stream sedimentation from mining was responsible for their disappearance. Ecologically similar *Litoria* and *Nyctimystes* species have declined in northern Queensland but insufficient data are available to determine whether other populations of stream-dwelling New Guinean frogs have declined. The diverse assemblages of *Litoria* and *Nyctimystes* in montane New Guinea are clearly at risk of decline in the future. Bickford (1998) reported that during a severe drought in Papua New Guinea in 1997 rainforest frogs changed their behavioural patterns, presumably in response to stress resulting from the lack of moisture in this normally saturated environment. It is not known if any species of New Guinean frogs disappeared during this severe drought, but populations of several species of microhylid frogs in the Star Mountains appeared to occur at much lower densities after the drought than in preceding years (Richards, 2001, unpublished data).

Although extensive surveys continue to be an integral part of many research programmes, in the past five years there has been a major shift in emphasis away from anecdotal and observational data towards rigorous experimental approaches. Current research on amphibian declines, both in Australia and elsewhere, is focused on three major areas: (1) experimental ecology including the impact of UV-B radiation on embryonic survival and the impact of exotic predators on native species; (2) the role of diseases, particularly emerging infectious diseases; and (3) long-term population monitoring and autecological comparisons.

Table 1. Summary of Australian amphibian species (listed alphabetically) that have experienced significant declines, and postulated causes of these declines. A question mark (?) denotes that the postulated cause is entirely speculative; while a tick (✓) indicates that there is evidence implicating the postulated cause in the decline of the species, but does not indicate that the postulated cause is definitely responsible, either wholly or in part, for the decline. It should be noted that, within species, different populations may have declined due to different causes, and multiple factors may have contributed to the decline of some populations, and that there is currently disagreement among amphibian researchers as to the relative significance of potential causal factors for many species. A number of species have ? in the Disease column, because, although chytrids have been identified in specimens of these species, the specimens were either captive or collected after population declines occurred and/or are from areas where declines have not been recorded.

SPECIES	POSTULATED CAUSE/S OF DECLINE								
	Disease	Habitat Disturbance	Pollution	Ultraviolet Radiation	Introduced Predators	Unknown	Other	Recovering/ Recovered	Key Sources
Adelotus brevis	?					✓			1,2
Litoria aurea		✓		?	✓		drought		3,4,5,6,7
Litoria castanea		✓	?	?		✓	drought		5,8
Litoria genimaculata	?					✓		✓	2,9
Litoria lorica	?					✓			2,9
Litoria nannotis	✓					✓		possibly	9,10,11
Litoria nyakalensis	?					✓			2,9,10
Litoria pearsoniana	?					✓			1,2
Litoria piperata		?	?		?	✓			12,13
Litoria raniformis	?	✓	?	?	?		drought		2,5,14,15
Litoria rheocola	✓					✓		possibly	9,10,11
Litoria spenceri	?	✓			✓				2,11,13,16,17
Litoria subglandulosa		?	?		?	✓			18
Litoria verreauxii alpina				✓			drought		19,20
Nyctimystes dayi	?					✓		possibly	2,9,10
Geocrinia alba		✓							21
Mixophyes balbus		?	?		?	✓			13,18,22
Mixophyes fleayi	?	?				✓			11,23,24
Mixophyes iteratus		?	?			✓			24,25,26
Philoria frosti		?	?	?	?	✓	drought		15,20
Pseudophryne corroboree		✓	?			✓	drought		15,20,27
Rheobatrachus silus	?					✓	climate change?		2,28,29
Rheobatrachus vittelinus	?					✓			2,9,30
Taudactylus acutirostris	✓					✓			9,10,11
Taudactylus diurnus	?					✓	climate change?		2,28
Taudactylus eungellensis	?					✓		✓	9,11,30
Taudactylus rheophilus	?					✓		✓	2,9,10

Key to Sources: 1) Ingram and McDonald, 1993; 2) Berger *et al.*, 1999; 3) Lewis and Goldingay, 1999; 4) Morgan and Buttemer, 1996; 5) Osborne *et al.*, 1996; 6) van de Mortel *et al.*, 1998; 7) White and Pyke, 1996; 8) White and Ehmann, 1997; 9) McDonald and Alford, 1999; 10) Richards *et al.*, 1993; 11) Berger et al., 1998; 12) Mahony *et al.*, 1997a; 13) Gillespie and Hines, 1999; 14) Ehmann and White, 1997; 15) Tyler, 1997; 16) Gillespie and Hollis, 1996; 17) Gillespie, 1997; 18) Anstis, 1997; 19) Broomhall *et al.*, 2000; 20) Osborne *et al.*, 1999; 21) Roberts *et al.*, 1999; 22) Mahony et al., 1997b; 23) Mahony *et al.*, 1997c; 24) Hines *et al.*, 1999; 25) Mahony *et al.*, 1997d; 26) Lemckert, 1999; 27) Osborne, 1989; 28) Ingram, 1990; 29) Tyler and Davies, 1985; 30) McDonald, 1990.

CHANGING EMPHASES—CURRENT RESEARCH

EXPERIMENTAL ECOLOGY

UV-B Radiation

Embryos of some amphibian species have significantly lower survival and higher rates of deformities when exposed to ambient UV-B radiation (Blaustein *et al.*, 1998; Starnes *et al.*, 2000). Other studies have demonstrated more indirect impacts of UV-B including reduced hatchling size (Pahkala *et al.*, 2000), damage to eyes (Fite *et al.*, 1998) and shifts in activity patterns (Blaustein *et al.*, 2000).

The results of several studies comparing the susceptibility to UV-B of declining species with sympatric species that have 'stable' populations hint that UV-B may play a role in the declines of at least some amphibian populations. In Australia, embryonic mortality of montane *Litoria verreauxi alpina* and *Crinia signifera* populations was higher in experimental treatments exposed to ambient UV-B than in treatments where UV-B was filtered. Furthermore, mortality of *L. verreauxi alpina*, a declining species, was significantly higher than that of *C. signifera*, a non-declining species, in high altitude treatments where UV-B effects are likely to be greatest and where *L. verreauxi alpina* populations no longer persist (Broomhall *et al.*, 2000).

Other studies, both in Australia and elsewhere, have produced more equivocal results. Van de Mortel and Buttemer (1996) found no evidence that UV-B is affecting survivorship of pre-hatching embryos of *Litoria aurea*, a species that has suffered a large range reduction in eastern Australia. The ability of adult and larval *L. aurea* to detect and avoid peaks in UV-B radiation appears to be no different from that of sympatric species with stable populations (Van de Mortel and Buttemer, 1998). However this species does have lower photolyase activity than two sympatric non-declining frogs, and there is a trend for lower hatching success than the sympatric non-declining frogs under UV-B exposure (van de Mortel *et al.*,1998).

Although there is a large and rapidly increasing literature documenting the impacts of UV-B on the survival rates of aquatic stages of amphibians, there is no clearly emerging trend to support the contention that increased UV-B is affecting amphibian populations at a broad scale. Many species, such as *Rana temporaria*, do not exhibit enhanced survivorship in the absence of UV-B (Merilä *et al.,* 2000) and do not suffer increased mortality when exposed to UV-B levels substantially higher than ambient (Cummins *et al.,* 1999). Indeed no studies have demonstrated convincingly that increased UV-B is directly responsible for any population decline; both the protocols and interpretation of UV-B experiments have been criticized (Licht and Grant, 1997; Alford and Richards, 1999; Cummins *et al.,* 1999).

It is difficult to compare the results of experiments undertaken in Australia with those reported from elsewhere in the world due to widely varying experimental protocols. Future approaches to the experimental evaluation of UV-B, as a factor in amphibian declines, must include appropriate and consistent UV-B dosimetry to allow comparisons among experiments (Cummins *et al.,* 1999). In addition, the ecological relevance of mortality from exposure to UV-B, at levels predicted from increasing ozone depletion, must be rigorously assessed (Alford and Richards, 1999).

Disease

Infectious diseases have been associated with mass mortality of a number of amphibian populations. Pathogens causing die-offs of amphibians include bacteria (Nyman, 1986; Bradford, 1991; Carey, 1993), fungi (Taylor *et al.,* 1999) and iridoviruses (Cunningham *et al.,* 1996). Although a virus has been proposed as an agent of population declines in north Queensland rainforests (Laurance *et al.,* 1996) there is little evidence to suggest that viruses are responsible for any population declines in Australia (Alford and Richards, 1997; Hero and Gillespie, 1997).

Many dead and dying stream-dwelling frogs found in upland rainforests in Central America and Australia were infected with a virulent chytrid fungus, *Batrachochytrium dendrobatidis* (Berger *et al.*, 1998). Chytrid fungi were not previously known to infect vertebrates, and the precise mechanism by which *B. dendrobatidis* kills amphibians remains to be determined (Pessier *et al.,* 1999). *B. dendrobatidis* resides in the skin where it is closely associated with keratin. It is possible that the fungus releases enzymes or other toxic compounds into the skin, or that it disrupts water regulation and electrolyte balance (Berger *et al.*, 1999; Pessier *et al.*, 1999). The fungus does not appear to cause mortality in tadpoles, in which keratin is restricted to the mouthparts.

Chytrid fungi appear to be widely present in amphibian populations throughout the world, but their virulence appears to differ markedly among populations and species (Speare, 2000). The observation that mortality associated with fungal infestation has occurred in relatively pristine habitats and involves a previously unknown disease raises several interesting questions about its origin.

It is possible that *B. dendrobatidis* is causing mortality in amphibian populations because the immune systems of amphibians are being increasingly compromised by anthropogenic stressors or changing climatic conditions (Carey *et al.*, 1999). These stressors may shift existing, stable host-parasite relationships (Pounds and Crump, 1994; Pounds *et al.*, 1999). Alternatively, it is possible that this disease has evolved recently or has recently entered many amphibian populations, and is capable of killing amphibians without any associated anthropogenic or environmental stresses on their immune systems. In pristine rainforest habitats it is possible that this 'new' pathogen has been introduced by humans or other agents and is infecting novel hosts, or that it evolved *in situ* in these habitats (Daszak *et al.*, 1999).

Carey (2000) has noted that there are similarities between mass mortalities involving amphibians, and new diseases recently reported for both coral reefs and humans. In all of these groups outbreaks of disease involving new pathogens appear to have been increasing since the 1970s. Whether their emergence is enhanced by human activities is unclear (Carey, 2000), but it is evident that the increasing incidence of emerging infectious diseases is a threat not only to amphibians but to the Earth's biodiversity as a whole (Daszak *et al.*, 2000).

The past five years have seen a substantial increase in knowledge about the diseases of amphibians in general, and about the potential role of disease in population declines. Future research in Australia appears likely to initially concentrate on determining whether disease outbreaks are caused by the appearance of a novel pathogen or by outbreaks of an established pathogen. Once this has been determined, the focus will shift to either understanding the mechanisms and causes of the spread of the disease, if a novel pathogen is implicated, or understanding what factors have contributed to outbreaks if the pathogen is not novel.

A major international research programme, funded by the U.S. National Science Foundation and including over 40 researchers from several countries, commenced in late 1999 with the objective of examining the interactions of amphibians and pathogens. One group of Australian participants, at the Australian Animal Health Laboratories, is focusing on developing new techniques for diagnosing chytridiomycosis in animals, and surveying for its presence in habitats. In collaboration with the Amphibian Research Centre (Melbourne) they are carrying out experiments examining the disease process in a variety of species. The Animal Health Group is also co-operating with James Cook University in surveys and monitoring of disease in wild populations. A second group of staff (including the authors) and students at James Cook is examining the dynamics of the chytrid in natural populations—conducting experiments on the development and transmission of chytrid infections. Other workers are continuing to focus on the autecology of species that have declined and those that have not, enabling more detailed comparisons of their biology that may help to shed light on what factors make some species vulnerable to declines.

Species Interactions

In Australia the impacts of introduced amphibians (the cane toad, *Bufo marinus*) and fishes on amphibian populations have been examined in some detail. *Bufo marinus* was introduced into Australia from Hawaii in 1935. It has subsequently spread throughout all suitable habitat in Queensland, and has penetrated into northern New South Wales and is rapidly extending its range in the Northern Territory (Freeland and Martin, 1985). Negative effects of cane toads on a variety of native fauna have been documented (Covacevich and Archer, 1975; Crossland and Alford, 1998). Perhaps surprisingly, adult cane toads do not appear to be significant predators of native frogs (Freeland *et al.*, 1986; Freeland and Kerin, 1988; Alford, 2001, unpublished observation). Assemblages of native frogs did not change significantly when censused before and immediately after the arrival of *Bufo marinus* (Freeland and Kerin, 1988). However, larval cane toads occur at high densities in many habitats (Hearnden, 1991; Alford *et al.* 1995) and exert substantial intra- and inter-specific competitive effects (Hearnden, 1991; Alford, 1994; Alford *et al.*, 1995; Crossland, 1997). The eggs and hatchlings of cane toads are toxic and are consumed by the larvae of many native frogs (Crossland and Alford, 1998). The competitive and toxic effects of *Bufo marinus* on native frogs may take several years to appear after toads invade, because most negative effects are exerted on the larval stages of frogs. Although it appears unlikely that cane toads have led directly to any of the mysterious frog declines that have occurred in Australia since 1979, they may be negatively affecting populations of native species within their expanding range.

Introduced predatory fishes have been implicated in the declines or disappearances of amphibian species in a number of locations (Petranka, 1983; Bradford *et al.*, 1993; Fellers and Drost, 1993; Brönmark and Edenhamn, 1994; Fisher and Schaffer, 1996; Gamradt and Kats, 1996; Funk and Dunlap, 1999). Larvae without an evolutionary history of exposure to a particular species of predator may not possess appropriate mechanisms to avoid predation (Gamradt and Kats, 1996; Kiesecker and Blaustein, 1997).

At least 24 exotic fish species, including mosquito fishes and three species of trout, have established populations in Australia (Gillespie and Hero, 1999). However, only the effects of trout and mosquito fish on native anurans have received any attention. Introduced trout have been implicated in declines of the endangered treefrog *Litoria spenceri* (Gillespie, 1997) and occur in streams where populations of *Litoria barringtonensis, L. booroolongensis* (Anstis, 1997), *L. piperata* (Gillespie and Hines, 1999), *L. subglandulosa* and *Mixophyes balbus* (Anstis, 1997) have declined or disappeared. The mosquito fish *Gambusia affinis holbrooki* is an effective predator of Australian anuran larvae (Morgan and Buttemer, 1996; Webb and Joss, 1997; Komak and Crossland, 2000). There is particular concern about its impact on populations of the endangered green and golden bell frog, *Litoria aurea* (Mahony, 1993; Morgan and Buttemer, 1996; White and Pyke, 1996). Experimental investigations have demonstrated that *Gambusia* is capable of significantly reducing larval populations of *L. aurea* (Morgan and Buttemer, 1996). A negative correlation between the presence of predatory fish (*Gambusia* being the most prevalent) and breeding by the green and golden bell frog has also been reported (Pyke and White, 1996). Fortunately *L. aurea* is significantly more likely to breed in ephemeral ponds than in more permanent ones containing fishes (Pyke and White, 1996) and, in some habitats, persists successfully in the presence of high densities of *Gambusia* (van de Mortel and Goldingay, 1998).

Clearly the evidence linking fishes to frog population declines remains equivocal. However, it seems reasonable to conclude that introduced predatory fishes pose a threat to at least some Australian frogs, particularly because strong evidence exists that introduced fishes have had a major impact on frog populations elsewhere (Fisher and Shaffer, 1996).

POLLUTANTS

Environmental pollutants such as insecticides, herbicides and heavy metals can have serious negative impacts on amphibians, including sterility, retarded growth, abnormalities and mortality (see Ferraro and Burgin, 1993). Pesticides may have played a role in the declines of *Rana muscosa* and *Bufo canorus* in the Sierra Nevada Mountains in California (Bradford *et al.*, 1994). There is also a strong association between declines of *Rana aurora draytoni* in the Sierra Nevada and areas upwind of agriculture, suggesting that pesticide drift may be involved (Davidson *et al.*, 2001). Relatively small quantities of pollutants deposited in remote areas may be concentrated by physical processes and adversely affect frog populations (Pounds and Crump, 1994). It has also been speculated that environmental pollutants may adversely affect amphibians by suppressing their immune systems, thus increasing their susceptibility to pathogens (Mann and Bidwell, 1999).

In Australia, environmental toxicology is a relatively new field of research and the effects of environmental pollutants on Australian amphibians have received little attention (Mann and Bidwell, 1999). Mann and Bidwell (1999) located just 18 studies (including unpublished theses) concerning the effects of pollutants on native amphibians; none of these involved species that have exhibited significant declines. Although pollution has been suggested as a potential cause of decline for several native species (Table 1), no studies have been undertaken to determine its exact role.

ARE AMPHIBIANS REALLY DECLINING?

The problem of separating natural population fluctuations from population declines was demonstrated by Pechmann *et al.* (1991). Pechmann and Wilbur (1994) argued that insufficient data were available to determine whether observed amphibian declines simply represent stochastic events and are just one component of a general biodiversity crisis. They argued that null models of the behaviour of amphibian populations are required to test the significance of observed declines. Pounds *et al.* (1997) used such an approach to argue that simultaneous declines of amphibian populations in Costa Rica were unlikely to be coincidental. Alford and Richards (1999) used a model of amphibian population dynamics to generate null hypotheses about population behaviour. They examined whether 85 time series of amphibian population dynamics in the literature conformed to predictions about population behaviour, and found that more amphibian populations declined between 1951–1997 than the model predicted. Subsequently, Houlahan *et al.* (2000) examined trends in amphibian populations using data from 936 populations studied between 1940 and 1997. Despite significant spatial and temporal variation, they found that on a global scale amphibian populations have been declining since the late 1950s. Alford *et al.* (2001) re-analysed these data and concluded that there is little evidence for a general decline in amphibian populations globally prior to the 1990s.

LONG-TERM MONITORING

Prior to 1991 the long-term population dynamics of Australian frogs were poorly known. Few studies exceeded three years (Table 2) so the extent to which populations fluctuate under natural conditions remained undocumented. Since 1991 a number of new monitoring programmes have been initiated (Table 2). These have provided valuable insights into how individual populations behave, as well as data which support the contention that many frog populations in Australia are undergoing declines rather than exhibiting normal population fluctuations. Despite these studies, and a general acceptance that amphibian populations are now declining at a rate substantially higher than normal 'background' extinction, it remains to be determined whether taxa other than amphibians are suffering similarly increased levels of population declines.

Developmental Stability Analysis as a Tool

Although long-term monitoring has greatly increased our knowledge of how frog populations change over time, monitoring cannot predict or prevent declines. Declines typically happen rapidly (Richards *et al.*, 1993; McDonald and Alford, 1999), such that by the time they can be confirmed from monitoring data the population has already undergone a severe reduction or has disappeared. One technique that holds promise for improving the ability to detect problems before populations crash is developmental stability analysis or DSA (Alford *et al.*, 1999). In DSA, levels of developmental instability (DI) are assessed by careful measurements of the degree of asymmetry of normally bilaterally symmetrical body parts of individual animals (Alford *et al.*, 1999). Studies on a variety of taxa have shown that declining populations (Tsubaki, 1998) or populations in stressful environments (Valentine *et al.*, 1973; Møller, 1996; Sarre, 1996; Siikamäki and Lammi, 1998) exhibit increased levels of DI. These increases can occur before numbers of individuals change, providing advance warning of population declines (Alford *et al.*, 1999). Although recently incorporated into several monitoring programmes, it is still too early to determine whether DSA is fulfilling its promise as a monitoring technique (McDonald, 2000, personal communication).

Table 2. Long-term monitoring programmes for Australian frogs.

Location	Species Monitored	Monitoring Period	Declines Recorded?	References
Birthday Creek, Mt Spec, northern Qld.	*Litoria genimaculata, Litoria nannotis, Nyctimystes dayi, Mixophyes schevilli*	1987–1998	yes	Richards *et al.*, 1993 McDonald and Alford, 1999
Big Tableland, ~30 km S of Cooktown, northern Qld.	*Litoria genimaculata, L. nannotis, L. rheocola, Nyctimystes dayi, Taudactylus acutirostris*	1992–present	yes	Richards *et al.*, 1993 McDonald and Alford, 1999
Tully, northern Qld.	*Litoria genimaculata, L. nannotis, L. rheocola* and *Nyctimystes dayi*	1992–present	no	McDonald, 2000, personal communication
McIlwraith Range, Cape York Peninsula, northern Qld.	*Litoria eucnemis, L. longirostris, Rana daemeli*	1993–present	no	McDonald and Alford, 1999
Eungella National Park, central Qld.	*Rheobatrachus vitellinus, Taudactylus eungellensis, T. liemi*	1984–1990	yes	McDonald, 1990
Eungella National Park, central Qld.	*Litoria chloris, L. lesueuri, L. revelata, Mixophyes fasciolatus, Taudactylus eungellensis, T. liemi*	1994–1998	no	Retallick, 2001, personal communication
Australian Alps, N.S.W. and A.C.T.	*Pseudophryne corroboree, Pseudophryne pengilleyi*	1986–present	yes	Osborne *et al.*, 1999
Australian Alps, Vic.	*Philoria frosti*	1983–1984, and 1993–1997	yes	Osborne *et al.*, 1999
South-eastern N.S.W. and eastern Vic.	*Litoria spenceri*	1994–present	yes	Gillespie and Hines, 1999
Various National Parks, State Forests and Reserves, north-eastern N.S.W.	various, including *Litoria pearsoniana, Mixophyes fleayi, M. iteratus*	1996–present	no	Goldingay *et al.*, 1999

CONTINGENCY STRATEGIES

Animal husbandry and assisted reproductive technologies are techniques that can be used as 'insurance' against future losses, and to retain a species' evolutionary potential (Mahony *et al.*, 1999). Given that the cause or causes of decline have not been identified for numerous frog species and that, even if they are identified, there may be little that can be done to alleviate their adverse effects (particularly in the short term), these techniques may be crucial to the continued survival of many species (Mahony *et al.*, 1999).

Husbandry techniques can be used for the conservation of rare and endangered species in a number of ways. These include enhancing recruitment by increasing the survival rate of a particular stage of the life cycle beyond the rate of survival in the field, providing adults and/or young for experimental research into causes of population declines, and producing animals for release into the wild to re-establish populations that have disappeared, or to enhance existing ones (Marantelli, 1999). In 1997, a population augmentation (recruitment enhancement) project was initiated for *Pseudophryne corroboree*. This project aims to determine whether (1) it is possible to use captive rearing techniques together with active field management to prevent tadpole mortality in the field, thus increasing recruitment through to metamorphosis, and (2) increased recruitment through to metamorphosis translates to increased numbers of breeding adults in remnant populations (Hunter *et al.*, 1999). In the initial year of the project, Hunter *et al.* (1999) significantly increased recruitment through to metamorphosis via captive rearing, which is a promising result. It remains to be seen, however, whether increased survival to metamorphosis translates to increased numbers of breeding adults.

Assisted reproductive technologies (ART) can be used to introduce additional genetic diversity into populations, both wild and captive-bred, that are experiencing inbreeding depression (due to small population size). They can also be used to maintain captive-bred populations and to re-establish extinct populations (Clulow *et al.*, 1999). Technologies being developed include hormonal inductions of sperm release and ovulation, IVF, cryopreservation of sperm, androgenesis (creating an organism from sperm only) and cloning (Clulow *et al.*, 1999). Although these procedures have the potential to contribute significantly to management of endangered species or populations, their development for use in amphibian conservation programmes is in its infancy and techniques for the cryopreservation of eggs, stem cells and embryos are not presently available (Clulow *et al.*, 1999). Successful establishment of sperm, egg and embryo cryobanks would provide the potential for species that have gone extinct in the wild to be 'retrieved' from these banks (Clulow *et al.*, 1999). But even if these technologies prove to be viable, there is considerable debate about the long-term success of frog (and reptile) populations repatriated into the wild (Burke, 1991; Dodd and Seigel, 1991). The ultimate success of these technologies will still depend on an understanding of the causal mechanisms involved in population declines or extinctions and consideration of the suitability of target species for reintroduction or long-term captive-breeding programmes.

FUTURE DIRECTIONS

Although it is tragic that more than 10% of the Australian frog fauna has declined or disappeared in the past 25 years, it is encouraging that these declines and disappearances have been a huge stimulus to increasing our knowledge of our frogs and their biology. We also know much more about what information is lacking and how to go about acquiring it. Future research is likely to be more focused on gaining the information needed to understand the status of Australian frogs and to maintain the integrity of their populations.

One major area in which a great deal of work is needed is in refining our knowledge of species distributions, abundances, and habitat requirements. At present, it is not unusual for fauna surveys to lead to major range extensions for species, or to their discovery in habitats they were not known to use. We cannot preserve the diversity of the Australian anuran fauna without knowing how it is constituted.

Another area that should be fruitful for future research is clearly the complex interactions between species, their diseases and parasites, and the external environment. It is critical to understand, as soon as possible, the nature of the outbreaks of chytrid fungus infection that have been associated with large-scale declines. If these result from the invasion of a novel pathogen, means to slow its spread and reduce its effects on frog populations are urgently required. If the outbreaks have occurred because an endemic pathogen has been potentiated by anthropogenic environmental effects, we need to understand the natural host-parasite biology of chytrids and frogs and how it has been disrupted, so that such disruptions can be prevented from having drastic effects on frog populations in the future.

In conjunction with work specifically on diseases, it will be necessary to learn a great deal more about the population biology of frogs. At present, we have very little idea of what rates of growth, survival and reproduction are within local frog populations, and know almost nothing of how, and in which species, local populations are connected to form regional metapopulations. This makes it impossible to predict the true effects of habitat fragmentation and changes in rates of survival and reproduction on species. Work on captive maintenance and more exotic technological solutions to the artificial preservation of species diversity should continue, because in some cases that may be the only approach capable of assuring species survival in the short term.

Finally, it would be fruitful to continue to explore more efficient techniques for assessing the status and health of frog populations. Early work on developmental stability analysis shows promise, but additional information, for example data on the mean and variance of developmental instability in non-declining reference populations of many species, are needed before this technique can be fully incorporated into surveying and monitoring programmes. Techniques for more accurate rapid assessments of frog distribution, abundance and habitat use (Heyer *et al.*, 1994; Alford and Richards, 1999) would also contribute greatly to the utility of research as a tool for frog conservation.

SUMMARY AND CONCLUSIONS

We have learned a great deal about the status and biology of Australia's frogs since the alarm about declining amphibian populations was first raised in the late 1980s. We now know that at least 10% of the fauna has declined in the past 25 years, but still do not know the causes of more than half of those declines. We do know that they occurred largely in protected habitats, making it clear that we cannot simply set aside reserves and assume that this will maintain the diversity of our native fauna. We have identified those groups of frogs that are most likely to be at risk in the future, including diverse assemblages of upland stream-dwelling frogs in the rainforests of New Guinea. Research foci have shifted from documenting declines and their patterns of occurrence, to carefully designed rigorous experimental programmes to reveal processes responsible for declines. These studies have shown several potential causal agents including a chytrid fungus that has been implicated in a number of declines around the world. Despite these advances more research is urgently required in order to fully understand these declines, so that similar reductions in frogs and possibly in other fauna can be prevented in future.

ACKNOWLEDGEMENTS

We would like to thank K.R. McDonald and R. Retallick for access to their unpublished data and for acting as valued collaborators in our declining frog research. Our work has been supported by grants from the Australian Research Council, James Cook University, the

Rainforest Cooperative Research Centre, the United States National Science Foundation, and the Wet Tropics Management Authority.

REFERENCES

Alford, R.A. 1994. Interference and exploitation competition in larval *Bufo marinus*. Pp.297–306. In Mishra, P.C., Behera, N., Senapati, B.K. and Guru, B.C. (eds). *Advances in Ecology and Environmental Sciences*. Ashish Publishing House, New Delhi.

Alford, R., Bradfield, K. and Richards, S. 1999. Measuring and analysing developmental instability as a tool for monitoring frog populations. Pp.34–43. In Campbell, A. (ed). *Declines and Disappearances of Australian Frogs*. Environment Australia, Canberra. 234pp.

Alford, R.A., Cohen, M.P., Crossland, M.R., Hearnden, M.N. and Schwarzkopf, L. 1995. Population biology of *Bufo marinus* in northern Australia. Pp.173–181. In Finlayson, C.M. (ed). *Proceedings of the Wetland Research in the Wet–Dry Tropics of Australia Workshop*. Supervising Scientist Report 101, Canberra.

Alford, R.A., Dixon, P.M. and Pechmann, J.H.K. 2001. Global amphibian population declines. *Nature*, Vol. 412, pp.499–500.

Alford, R.A. and Richards, S.J. 1997. Lack of evidence for epidemic disease as an agent in the catastrophic decline of Australian rain forest frogs. *Conservation Biology*, Vol. 11, pp.1026–1029.

Alford, R.A. and Richards, S.J. 1999. Global amphibian declines: a problem in applied ecology. *Annual Review of Ecology and Systematics*, Vol. 30, pp.133–165.

Allison, A. 1993. Biodiversity and conservation of the fishes, amphibians, and reptiles of Papua New Guinea. Pp.157–225. In Beehler, B.M. (ed). *Papua New Guinea Conservation Needs Assessment*. Volume 2. The Biodiversity Support Program, Washington, D. C.

Anstis, M. 1997. *Litoria subglandulosa*, glandular tree frog. Pp.213–221. In Ehmann, H. (ed). *Threatened Frogs of New South Wales: Habitats, Status and Conservation*. Frog and Tadpole Study Group of New South Wales, Sydney. 237pp.

Barinaga, M. 1990. Where have all the froggies gone? *Science*, Vol. 247, pp.1033–1034.

Berger, L., Speare, R., Daszak, P., Green, D.E., Cunningham, A.A., Goggin, C.L., Slocombe, R., Ragan, M.A., Hyatt, A.D., McDonald, K.R., Hines, H.B., Lips, K.R., Marantelli, G. and Parkes, H. 1998. Chytridiomycosis causes amphibian mortality associated with population declines in the rain forests of Australia and Central America. *Proceedings of the National Academy of Sciences of the United States of America*, Vol. 95, pp.9031–9036.

Berger, L., Speare, R. and Hyatt, A. 1999. Chytrid fungi and amphibian declines: overview, implications and future directions. Pp.23–33. In Campbell, A. (ed). *Declines and Disappearances of Australian Frogs*. Environment Australia, Canberra. 234pp.

Bickford, D. P. 1998. Effects of the El Niño drought on the frogs of Papua New Guinea. *Froglog*, Vol. 27, p.1.

Blaustein, A.R., Chivers, D.P., Kats, L.B. and Kiesecker, J.M. 2000. Effects of ultraviolet radiation on locomotion and orientation in roughskin newts (*Taricha granulosa*). *Ethology*, Vol. 106, pp.227–234.

Blaustein, A.R., Kiesecker, J.M., Chivers, D.P., Hokit, D.G., Marco, A., Belden, L.K. and Hatch, A. 1998. Effects of ultraviolet radiation on amphibians: field experiments. *American Zoologist*, Vol. 38, pp.799–812.

Bradford, D.F. 1991. Mass mortality and extinction in a high-elevation population of *Rana muscosa*. *Journal of Herpetology*, Vol. 25, pp.174–177.

Bradford, D.F., Gordon, M.S., Johnson, D.F., Andrews, R.D. and Jennings, W.B. 1994. Acidic deposition as an unlikely cause for amphibian population declines in the Sierra Nevada, California. *Biological Conservation*, Vol. 69, pp.155–161.

Bradford, D.F., Tabatabai, F. and Graber, D.M. 1993. Isolation of remaining populations of the native frog, *Rana muscosa*, by introduced fishes in Sequoia and King's Canyon National Parks, California. *Conservation Biology*, Vol. 7, pp.882–888.

Brönmark, C. and Edenhamn, P. 1994. Does the presence of fish affect the distribution of tree frogs (*Hyla arborea*)? *Conservation Biology*, Vol. 8, pp.841–845.

Broomhall, S.D., Osborne, W.S. and Cunningham, R.B. 2000. Comparative effects of ambient ultraviolet-B radiation on two sympatric species of Australian frogs. *Conservation Biology*, Vol. 14, pp.420–427.

Burke, R.L. 1991. Relocations, repatriations, and translocations of amphibians and reptiles: taking a broader view. *Herpetologica*, Vol. 47, pp.350–357.

Bury, R.B. 1999. A historical perspective and critique of the declining amphibian crisis. *Wildlife Society Bulletin*, Vol. 27, pp.1064–1068.

Carey, C. 1993. Hypothesis concerning the causes of the disappearance of boreal toads from the mountains of Colorado. *Conservation Biology*, Vol. 7, pp.355–362.

Carey, C. 2000. Infectious disease and worldwide declines of amphibian populations, with comments on emerging diseases in coral reef organisms and in humans. *Environmental Health Perspectives*, Vol. 108, Supplement 1, pp.143–150.

Carey, C., Cohen, N. and Rollins-Smith, L. 1999. Amphibian declines: an immunological perspective. *Developmental and Comparative Immunology*, Vol. 23, pp.459–472.

Clulow, J., Mahony, M., Browne, R., Pomering, M. and Clark, A. 1999. Applications of assisted reproductive technologies (ART) to endangered anuran amphibians. Pp.219–225. In Campbell, A. (ed). *Declines and Disappearances of Australian Frogs*. Environment Australia, Canberra. 234pp.

Covacevich, J. and Archer, M. 1975. The distribution of the cane toad, *Bufo marinus*, in Australia and its effects on indigenous vertebrates. *Memoirs of the Queensland Museum*, Vol. 17, pp.305–310.

Crossland, M.R. 1997. *Impact of the eggs, hatchlings, and tadpoles of the introduced toad* Bufo marinus *(Anura: Bufonidae) on native aquatic fauna in northern Queensland, Australia.* Ph.D. Thesis (unpublished), James Cook University, Townsville.

Crossland, M.R. and Alford, R.A. 1998. Evaluation of the toxicity of eggs, hatchlings and tadpoles of the introduced toad *Bufo marinus* (Anura: Bufonidae) to native Australian aquatic predators. *Australian Journal of Ecology*, Vol. 23, pp.129–137.

Cummins, C.P., Greenslade, P.D. and McLeod, A.R. 1999. A test of the effect of supplemental UV-B radiation on the common frog, *Rana temporaria* L., during embryonic development. *Global Change Biology,* Vol. 5, pp.471–479.

Cunningham, A.A., Langton, T.E.S., Bennett, P.M., Lewin, J.F., Drury, S.E.N., Gough, R.E. and MacGregor, S.K. 1996. Pathological and microbiological findings from incidents of unusual mortality of the common frog (*Rana temporaria*). *Philosophical Transactions of the Royal Society of London,* Series B, Vol. 351, pp.1539–1557.

Czechura, G.V. and Ingram, G.J. 1990. *Taudactylus diurnus* and the case of the disappearing frogs. *Memoirs of the Queensland Museum,* Vol. 29, pp.361–365.

Daszak, P., Berger, L., Cunningham, A.A., Hyatt, A.D., Green, D.E. and Speare, R. 1999. Emerging infectious diseases and amphibian population declines. *Emerging Infectious Diseases,* Vol. 5, pp.735–748.

Daszak, P., Cunningham, A.A. and Hyatt, A.D. 2000. Emerging infectious diseases of wildlife—threats to biodiversity and human health. *Science,* Vol. 287, pp.443–449.

Davidson, C., Shaffer, H.B. and Jennings, M.R. 2001. Declines of the California red-legged frog: climate, UV-B, habitat, and pesticides hypotheses. *Ecological Applications,* Vol. 11, pp.464–479.

Dodd, C.K. and Seigel, R.A. 1991. Relocation, repatriation, and translocation of amphibians and reptiles: are they conservation strategies that work? *Herpetologica,* Vol. 47, pp.336–350.

Ehmann, H. (ed). 1997. *Threatened Frogs of New South Wales: Habitats, Status and Conservation.* Frog and Tadpole Study Group of New South Wales, Sydney. 237pp.

Ehmann, H. and White, A.W. 1997. *Litoria raniformis*, southern bell frog. Pp.195–202. In Ehmann H. (ed). *Threatened Frogs of New South Wales: Habitats, Status and Conservation.* Frog and Tadpole Study Group of New South Wales, Sydney. 237pp.

Fellers, G.M. and Drost, C.A. 1993. Disappearance of the cascades frog *Rana cascadae* at the southern end of its range, California, USA. *Biological Conservation,* Vol. 65, pp.177–181.

Ferraro, T.J. and Burgin, S. 1993. Review of environmental factors influencing the decline of Australian frogs. Pp.205–218. In Lunney, D. and Ayers, D. (eds). *Herpetology in Australia: A Diverse Discipline.* Transactions of the Royal Zoological Society of New South Wales, Mosman. 414pp.

Fisher, R.N. and Shaffer, H.B. 1996. The decline of amphibians in California's Great Central Valley. *Conservation Biology,* Vol. 10, pp.1387–1397.

Fite, K.V., Blaustein, A., Bengston, L. and Hewitt, H.E. 1998. Evidence of retinal light damage in *Rana cascadae*: a declining amphibian species. *Copeia,* Vol. 1998, pp.906–914.

Freeland, W.J., Delvinquier, B.L.J. and Bonnin, B. 1986. Food and parasitism of the cane toad, *Bufo marinus*, in relation to time since colonization. *Australian Wildlife Research,* Vol. 13, pp.489–499.

Freeland, W.J. and Kerin, S.H. 1988. Within-habitat relationships between invading *Bufo marinus* and Australian species of frog during the tropical dry season. *Australian Wildlife Research,* Vol. 15, pp.293–305.

Freeland, W.J. and Martin, K.C. 1985. The rate of range expansion by *Bufo marinus* in northern Australia, 1980–1984. *Australian Wildlife Research,* Vol. 12, pp.555–559.

Funk, W.C. and Dunlap, W.W. 1999. Colonization of high-elevation lakes by long-toed salamanders (*Ambystoma macrodactylum*) after the extinction of introduced trout populations. *Canadian Journal of Zoology,* Vol. 77, pp.1759–1767.

Gamradt, S.C. and Kats, L.B. 1996. Effect of introduced crayfish and mosquitofish on California Newts. *Conservation Biology,* Vol. 10, pp.1155–1162.

Gillespie, G. and Hero, J-M. 1999. Potential impacts of introduced fish and fish translocations on Australian amphibians. Pp.131–144. In Campbell, A. (ed). *Declines and Disappearances of Australian Frogs.* Environment Australia, Canberra. 234pp.

Gillespie, G. and Hines, H. 1999. Status of temperate riverine frogs in south-eastern Australia. Pp.109–130. In Campbell, A. (ed). *Declines and Disappearances of Australian Frogs.* Environment Australia, Canberra. 234pp.

Gillespie, G.R. 1997. *The biology of the spotted tree frog* (Litoria spenceri) *and examination of factors responsible for population declines.* Report (unpublished) to the Biodiversity Group, Environment Australia, Canberra. Arthur Rylah Institute, Department of Natural Resources and Environment, Heidelberg, Victoria.

Gillespie, G.R. and Hollis, G.J. 1996. Distribution and habitat of the spotted tree frog, *Litoria spenceri* Dubois (Anura: Hylidae), and an assessment of potential causes of population declines. *Wildlife Research,* Vol. 23, pp.49–75.

Goldingay, R., Newell, D. and Graham, M. 1999. The status of rainforest stream frogs in north-eastern New South Wales: decline of recovery? Pp.64–71. In Campbell, A. (ed). *Declines and Disappearances of Australian Frogs.* Environment Australia, Canberra. 234pp.

Hearnden, M.N. 1991. *Reproductive and larval ecology of* Bufo marinus *(Anura: Bufonidae).* Ph.D. Thesis (unpublished), James Cook University, Australia.

Hero, J-M. and Gillespie, G.R. 1997. Epidemic disease and amphibian declines in Australia. *Conservation Biology,* Vol. 11, pp.1023–1025.

Heyer, W.R., Donnelly, M.A., McDiarmid, R.W., Hayek, L.C. and Foster, M.S. 1994. *Measuring and Monitoring Biological Diversity: Standard Methods for Amphibians.* Smithsonian Institution Press, Washington D.C. 364pp.

Hines, H., Mahony, M. and McDonald, K. 1999. An assessment of frog declines in wet subtropical Australia. Pp.44–63. In Campbell, A. (ed). *Declines and Disappearances of Australian Frogs.* Environment Australia, Canberra. 234pp.

Houlahan, J.E., Findlay, C.S., Schmidt, B.R., Meyer, A.H. and Kuzmin, S.L. 2000. Quantitative evidence for global amphibian population declines. *Nature,* Vol. 404, pp.752–755.

Hunter, D., Osborne, W., Marantelli, G. and Green, K. 1999. Implementation of a population augmentation project for remnant populations of the Southern Corroboree Frog (*Pseudophryne corroboree*). Pp.158–167. In Campbell, A. (ed). *Declines and Disappearances of Australian Frogs.* Environment Australia, Canberra. 234pp.

Inger, R.F. and Voris, H.K. 1993. A comparison of amphibian communities through time and from place to place in Bornean forests. *Journal of Tropical Ecology,* Vol. 9,

pp.409–433.

Ingram, G. 1990. The mystery of the disappearing frog. *Wildlife Australia,* Vol. 27, No. 3, pp.6–7.

Ingram, G.J. and McDonald, K.R. 1993. An update on the decline of Queensland's frogs. Pp.297–303. In Lunney, D. and Ayers, D. (eds). *Herpetology in Australia: A Diverse Discipline.* Transactions of the Royal Zoological Society of New South Wales, Mosman. 414pp.

Kiesecker, J.M. and Blaustein, A.R. 1997. Population differences in responses of red-legged frogs (*Rana aurora*) to introduced bullfrogs. *Ecology,* Vol. 78, pp.1752–1760.

Komak, S. and Crossland, M.R. 2000. An assessment of the introduced mosquitofish (*Gambusia affinis holbrooki*) as a predator of eggs, hatchlings and tadpoles of native and non-native anurans. *Wildlife Research,* Vol. 27, pp.185–189.

Laurance, W.F., McDonald, K.R. and Speare, R. 1996. Epidemic disease and the catastrophic decline of Australian rain forest frogs. *Conservation Biology,* Vol. 10, pp.406–413.

Lemckert, F. 1999. Impacts of selective logging on frogs in a forested area of northern New South Wales. *Biological Conservation,* Vol. 89, pp.321–328.

Lewis, B. and Goldingay, R. 1999. A preliminary assessment of the status of the green and golden bell frog in north-eastern NSW. Pp.94–98. In Campbell, A. (ed). *Declines and Disappearances of Australian Frogs.* Environment Australia, Canberra. 234pp.

Licht, L.E. and Grant, K.P. 1997. The effects of ultraviolet radiation on the biology of amphibians. *American Zoologist,* Vol. 37, pp.137–145.

Lips, K.R. 1998. Decline of a tropical montane amphibian fauna. *Conservation Biology,* Vol. 12, pp.106–117.

Lips, K.R. 1999. Mass mortality and population declines of anurans at an upland site in western Panama. *Conservation Biology,* Vol. 13, pp.117–125.

Mahony, M.J. 1993. The status of frogs in the Watagan Mountains area of the Central Coast of New South Wales. Pp.257–264. In Lunney, D. and Ayers, D. (eds). *Herpetology in Australia: A Diverse Discipline.* Transactions of the Royal Zoological Society of New South Wales, Mosman. 414pp.

Mahony, M. 1996. The decline of the green and golden bell frog *Litoria aurea* viewed in the context of declines and disappearances of other Australian frogs. *Australian Zoologist,* Vol. 30, pp.237–247.

Mahony, M., Clulow, J., Browne, R., and Pomering, M. 1999. Declines and disappearances of frogs: risk assessment and contingency strategies. Pp.202–211. In Campbell, A. (ed). *Declines and Disappearances of Australian Frogs.* Environment Australia, Canberra. 234pp.

Mahony, M., Knowles, R. and Pattinson, L. 1997a. *Litoria piperata,* peppered tree frog. Pp.189–194. In Ehmann, H. (ed). *Threatened Frogs of New South Wales: Habitats, Status and Conservation.* Frog and Tadpole Study Group of New South Wales, Sydney. 237pp.

Mahony, M., Knowles, R. and Pattinson, L. 1997b. *Mixophyes balbus,* stuttering barred frog. Pp.65–71. In Ehmann, H. (ed). *Threatened Frogs of New South Wales: Habitats, Status and Conservation.* Frog and Tadpole Study Group of New South Wales, Sydney. 237pp.

Mahony, M., Knowles, R. and Pattinson, L. 1997c. *Mixophyes fleayi,* silverblue-eyed barred frog. Pp.72–77. In Ehmann, H. (ed). *Threatened Frogs of New South Wales: Habitats, Status and Conservation.* Frog and Tadpole Study Group of New South Wales, Sydney. 237pp.

Mahony, M., Knowles, R. and Pattinson, L. 1997d. *Mixophyes iteratus,* gold-eyed barred frog. Pp.78–84. In Ehmann, H. (ed). *Threatened Frogs of New South Wales: Habitats, Status and Conservation.* Frog and Tadpole Study Group of New South Wales, Sydney. 237pp.

Mann, R. and Bidwell, J. 1999. Toxicological issues for amphibians in Australia. Pp.185–201. In Campbell, A. (ed). *Declines and Disappearances of Australian Frogs.* Environment Australia, Canberra. 234pp.

Marantelli, G. 1999. Husbandry: science or art?—Are captive technologies ready to contribute to recovery processes for Australian frogs? Pp.168–176. In Campbell, A. (ed). *Declines and Disappearances of Australian Frogs.* Environment Australia, Canberra. 234pp.

Marshall, C.J. 1998. The reappearance of *Taudactylus* (Anura: Myobatrachidae) in north Queensland streams. *Pacific Conservation Biology,* Vol. 4, pp.39–41.

McDonald, K. and Alford, R. 1999. A review of declining frogs in northern Queensland. Pp.14–22. In Campbell, A. (ed). *Declines and Disappearances of Australian Frogs.* Environment Australia, Canberra. 234pp.

McDonald, K.R. 1990. *Rheobatrachus* Liem and *Taudactylus* Straughan and Lee (Anura: Leptodactylidae) in Eungella National Park, Queensland: distribution and decline. *Transactions of the Royal Society of South Australia,* Vol. 114, pp.187–194.

Merilä, J., Laurila, A. and Pahkala, M. 2000. Effects of ambient UV-B radiation on early development of the common frog (*Rana temporaria*) embryos in the subarctic. *Annales Zoologici Fennici,* Vol. 37, pp.51–58.

Møller, A.P. 1996. Development of fluctuating asymmetry in tail feathers of the barn swallow *Hirundo rustica. Journal of Evolutionary Biology,* Vol. 9, pp.677–694.

Morgan, L.A. and Buttemer, W.A. 1996. Predation by the non-native fish *Gambusia holbrooki* on small *Litoria aurea* and *L. dentata* tadpoles. *Australian Zoologist,* Vol. 30, pp.143–149.

Nyman, S. 1986. Mass mortality in larval *Rana sylvatica* attributable to the bacterium, *Aeromonas hydrophila. Journal of Herpetology,* Vol.20, pp.196–201.

Osborne, W.S. 1989. Distribution, relative abundance and conservation status of Corroboree frogs, *Pseudophryne corroboree* Moore (Anura: Myobatrachidae). *Australian Wildlife Research,* Vol. 16, pp.537–547.

Osborne, W.S., Hunter, D. and Hollis, G. 1999. Population declines and range contraction in Australian alpine frogs. Pp.145–157. In Campbell, A. (ed). *Declines and Disappearances of Australian Frogs.* Environment Australia, Canberra. 234pp.

Osborne, W.S., Littlejohn, M.J. and Thomson, S.A. 1996. Former distribution and apparent disappearance of the *Litoria aurea* complex from the Southern Tablelands of New South Wales and the Australian Capital Territory. *Australian Zoologist,* Vol. 30, pp.190–198.

Pahkala, M., Laurila, A. and Merilä, J. 2000. Ambient ultraviolet-B radiation reduces hatchling size in the common frog, *Rana temporaria. Ecography,* Vol. 23, pp.531–538.

Pechmann, J.H.K., Scott, D.E., Semlitsch, R.E., Caldwell,

J.P., Vitt, L.J. and Gibbons, J.W. 1991. Declining amphibian populations: the problem of separating human impacts from natural fluctuations. *Science,* Vol. 253, pp.892–895.

Pechmann, J.H.K. and Wilbur, H.M. 1994. Putting declining amphibian populations in perspective: natural fluctuations and human impacts. *Herpetologica,* Vol. 50, pp.65–84.

Pessier, A.P., Nichols, D.K., Longcore, J.E. and Fuller, M.S. 1999. Cutaneous chytridiomycosis in poison dart frogs (*Dendrobates* spp.) and White's tree frogs (*Litoria caerulea*). *Journal of Veterinary Diagnostic Investigation,* Vol. 11, pp.194–199.

Petranka, J.W. 1983. Fish predation: A factor affecting the spatial distribution of a stream-breeding salamander. *Copeia,* 1983, pp.624–628.

Pounds, J.A. and Crump, M.L. 1994. Amphibian declines and climate disturbance: the case of the golden toad and the harlequin frog. *Conservation Biology,* Vol. 8, pp.72–85.

Pounds, J.A., Fogden, M.P.L. and Campbell, J.H. 1999. Biological response to climate change on a tropical mountain. *Science,* Vol. 398, pp.611–615.

Pounds, J.A., Fogden, M.P.L., Savage, J.M. and Gorman, G.C. 1997. Tests of null models for amphibian declines on a tropical mountain. *Conservation Biology,* Vol. 11, pp.1307–1322.

Pyke, G.H. and White, A.W. 1996. Habitat requirements for the green and golden bell frog *Litoria aurea* (Anura: Hylidae). *Australian Zoologist,* Vol. 30, pp.224–232.

Richards, S.J., McDonald, K.R. and Alford, R.A. 1993. Declines in populations of Australia's endemic tropical rainforest frogs. *Pacific Conservation Biology,* Vol. 1, pp.66–77.

Roberts, D., Conroy, S. and Williams, K. 1999. Conservation status of frogs in Western Australia. Pp.177–184. In Campbell, A. (ed). *Declines and Disappearances of Australian Frogs.* Environment Australia, Canberra. 234pp.

Sarre, S. 1996. Habitat fragmentation promotes fluctuating asymmetry but not morphological divergence in two geckos. *Researches on Population Ecology,* Vol. 38, pp.57–64.

Siikamäki, P. and Lammi, A. 1998. Fluctuating asymmetry in central and marginal populations of *Lychnis viscaria* in relation to genetic and environmental factors. *Evolution,* Vol. 52, pp.1285–1292.

Speare, R. 2000. Amphibian diseases home page. http://www.jcu.edu.au/school/phtm/PHTM/frogs/ampdis.htm

Starnes, S.M., Kennedy, C.A. and Petranka, J.W. 2000. Sensitivity of embryos of southern Appalachian amphibians to ambient solar UV-B radiation. *Conservation Biology,* Vol. 14, pp.277–282.

Taylor, S.K., Williams, E.S., Thorne, T., Mills, K.W., Withers, D.I. And Pier, A.C. 1999. Causes of mortality of the Wyoming toad. *Journal of Wildlife Diseases,* Vol. 35, pp.49–57.

Thumm, K. and Mahony, M. 1999. Loss and degradation of red-crowned toadlet habitat in the Sydney region. Pp.99–108. In Campbell, A. (ed). *Declines and Disappearances of Australian Frogs.* Environment Australia, Canberra. 234pp.

Trenerry, M.P., Laurance, W.F. and McDonald, K.R. 1994. Further evidence for the precipitous decline of endemic rainforest frogs in tropical Australia. *Pacific Conservation Biology,* Vol. 1, pp.150–153.

Tsubaki, Y. 1998. Fluctuating asymmetry of the Oriental fruit fly (*Dacus dorsalis*) during the process of its extinction from the Okinawa Islands. *Conservation Biology,* Vol. 12, pp.926–929.

Tyler, M.J. 1991. Declining amphibian populations—a global phenomenon? An Australian perspective. *Alytes,* Vol. 9, pp.43–50.

Tyler, M.J. 1997. *The Action Plan for Australian Frogs.* Wildlife Australia, Canberra. 77pp.

Tyler, M.J. and Davies, M. 1985. The gastric brooding frog *Rheobatrachus silus.* Pp.469–470. In Grigg, G., Shine, R. and Ehmann, H. (eds). *The Biology of Australasian Frogs and Reptiles.* Royal Zoological Society of New South Wales, Surrey Beatty and Sons, Sydney. 527pp.

Valentine, D.W., Soulé, M.E. and Samollow, P. 1973. Asymmetry analysis in fishes: A possible statistical indicator of environmental stress. *Fishery Bulletin,* Vol. 71, pp.357–370.

van de Mortel, T.F. and Buttemer, W.A. 1996. Are *Litoria aurea* eggs more sensitive to ultraviolet-B radiation than eggs of sympatric *L. peronii* or *L. dentata*? *Australian Zoologist,* Vol. 30, pp.150–157.

van de Mortel, T.F. and Buttemer, W.A. 1998. Avoidance of ultraviolet-B radiation in frogs and tadpoles of the species *Litoria aurea, L. dentata* and *L. peronii. Proceedings of the Linnean Society of New South Wales,* Vol. 119, pp.173–179.

van de Mortel, T., Buttemer, W., Hoffman, P., Hays, J. and Blaustein, A. 1998. A comparison of photolyase activity in three Australian tree frogs. *Oecologia,* Vol. 115, pp.366–369.

van de Mortel, T. and Goldingay, R. 1998. Population assessment of the endangered green and golden bell frog *Litoria aurea* at Port Kembla, New South Wales. *Australian Zoologist,* Vol. 30, pp.398–404.

Voris, H.K. and Inger, R.F. 1995. Frog abundance along streams in Bornean forests. *Conservation Biology,* Vol. 9, pp.679–683.

Waldman, B. and Tocher, M. 1998. Behavioral ecology, genetic diversity, and declining amphibian populations. Pp.394–443. In Caro, T. (ed). *Behavioural Ecology and Conservation Biology.* Oxford University Press, New York. 582pp.

Webb, C. and Joss, J. 1997. Does predation by the fish *Gambusia holbrooki* (Atheriniformes: Poeciliidae) contribute to declining frog populations? *Australian Zoologist,* Vol. 30, pp.316–324.

White, A.W. and Ehmann, H. 1997. *Litoria castanea,* New England bell frog. Pp.165–170. In Ehmann, H. (ed). *Threatened Frogs of New South Wales: Habitats, Status and Conservation.* Frog and Tadpole Study Group of New South Wales, Sydney. 237pp.

White, A.W. and Pyke, G.H. 1996. Distribution and conservation status of the green and golden bell frog *Litoria aurea* in New South Wales. *Australian Zoologist,* Vol. 30, pp.177–189.

Williams, S.E. and Hero, J-M. 1998. Rainforest frogs of the Australian Wet Tropics: guild classification and the ecological similarity of declining species. *Proceedings of the Royal Society of London,* Series B, Vol. 265, pp.597–602.

Chapter 14

Keys to Families and Genera of Frogs in Australia, New Zealand, New Guinea and Neighbouring Islands

A.W. White

Frog Ecology and Behaviour Group, Australian Museum, 6 College Street, Sydney, N.S.W. 2010 Australia
E-mail: awhite@integral.net.au

INTRODUCTION

The geographic regions covered by this key include Australia, New Zealand, New Guinea, and neighbouring islands. Six families of frogs occur in this region: (1) family Hylidae, from Australia, New Guinea, Timor, Moluccas, New Britain, Solomon Islands and recently introduced into New Zealand; (2) family Myobatrachidae, from Australia and New Guinea; (3) family Microhylidae, from northern Australia, New Guinea, Timor, Moluccas and the islands around eastern New Guinea; (4) family Ranidae, from northern Australia, New Guinea, Moluccas, New Britain, Solomon Islands and Fiji; (5) family Bufonidae, introduced into northern and eastern Australia, New Guinea, Solomon Islands and Fiji; and (6) family Leiopelmatidae, occurring only in New Zealand. For further discussion of New Guinean frog faunas see Heinsohn and Hope (2006); New Zealand frogs have been considered by Holdaway and Worthy (2006).

Early in the evolution of frogs (see Tyler and Lee, 2006) a body form developed that was well suited to both swimming and hopping, and since then frogs have remained basically conservative in body structure. Such conservatism often makes it difficult to find external features that distinguish frogs at higher taxonomic levels; this has necessitated the use of internal or skeletal characters, in addition to external characters at some point in this key. Figure 1 below shows the external and skeletal structure of a typical frog.

To use these keys, begin at the first couplet of alternative statements. Choose the statement that best describes the frog specimen you are trying to identify. This will give you either the number of the next couplet to which you should proceed, or the family or genus to

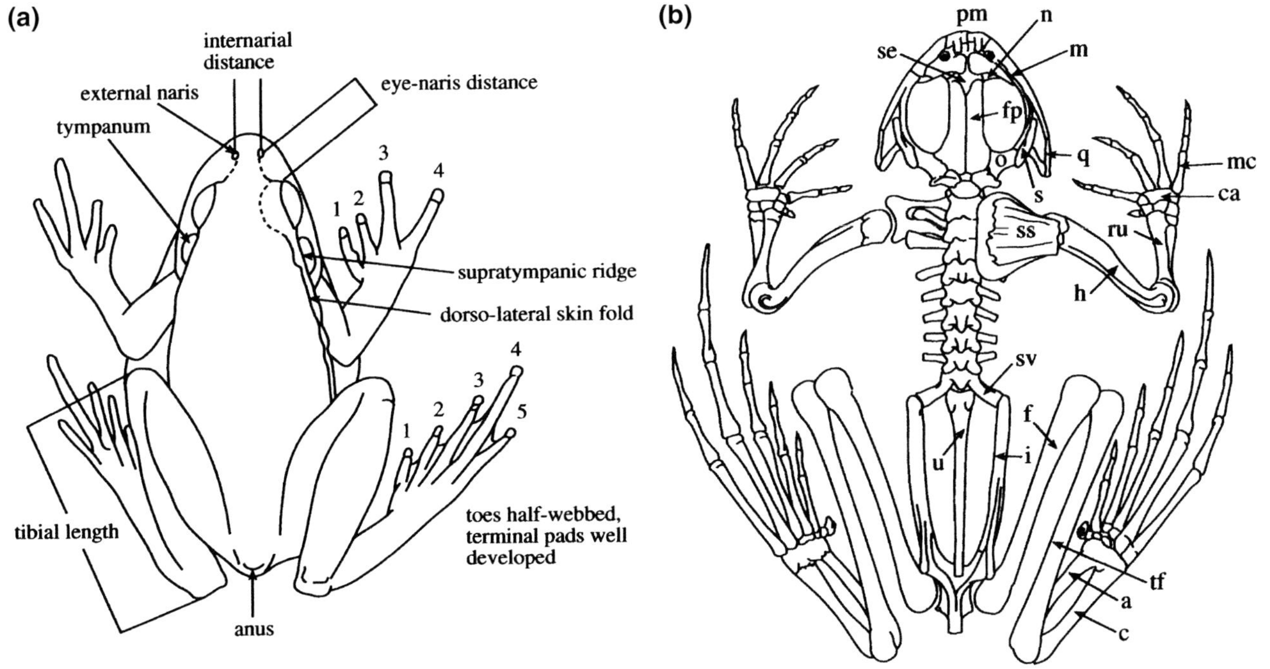

Figure 1. Dorsal views of a hylid frog: (**a**) external structure; (**b**) skeleton. Key to labels: 1, first digit; 2, second digit etc. (fingers and toes are numbered, beginning with the digit that is closest to the body); n, external naris; t, tympanic membrane; pm, premaxilla; m, maxilla; n, nasal; se, sphenethmoid; fp, frontoparietal; o, otoccipital; s, squamosal; qj, quadratojugal; ss, suprascapular; h, humerus; ru, radio-ulna; c, carpal bone; mc, metacarpal; sv, sacral vertebra; u, urostyle; l, ilium; f, femur; tf, tibio-fibula; a, astragalus; c, calcaneum; mt, metatarsal.

which the specimen belongs. Repeat the procedure until the latter is reached. No attempt is made here to identify frogs beyond the generic level and, in the case of microhylid frogs, as far as subfamily level. The taxonomic backgrounds of these frog families are varied and the generic and specific status of frogs in the families Ranidae and Microhylidae, in particular, are not well understood.

Abbreviations used in the key are as follows: sp. = species (singular); spp. = species (plural); HW = head width; S-V = snout to vent length; ED = eye diameter; ED:S-V = ratio of eye diameter to snout-vent length; < = less than.

KEY TO FAMILIES

1a. Epicoracoid cartilages freely overlap, not fused along midline 3
b. Epicoracoid cartilages united along the midline 2

1a

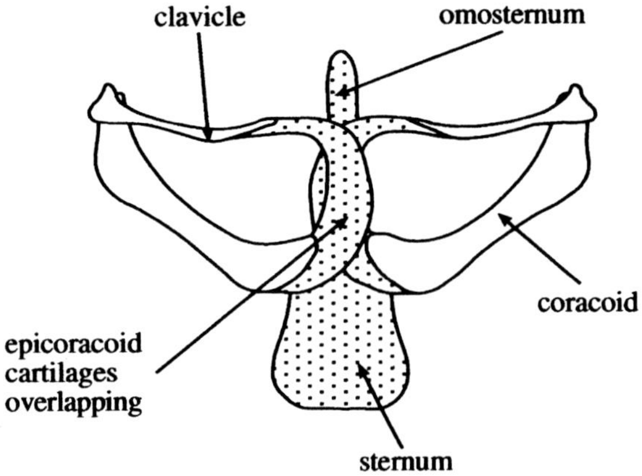

1b

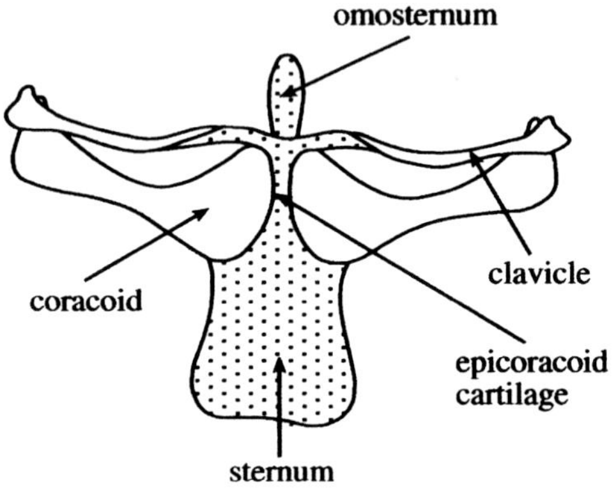

2a. Vomerine teeth present near the choanae; maxillary teeth always present .... Ranidae
b. Vomerine teeth absent, or if present, maxillary teeth absent (in *Xenobatrachus* spp.) .... Microhylidae

2b

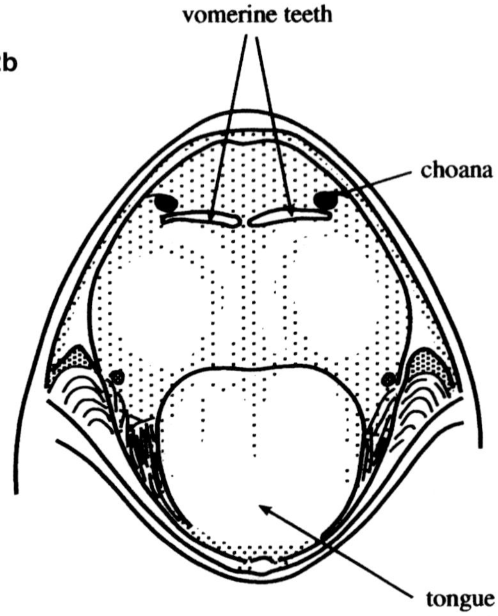

3a

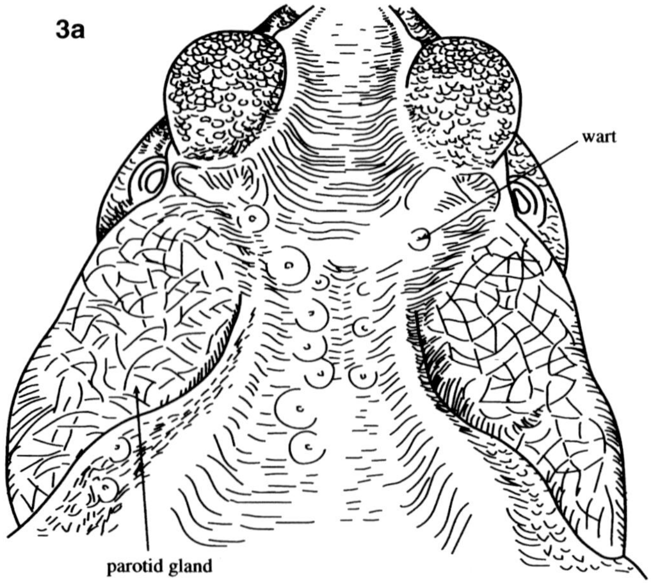

3a. Large poison-secreting glands located behind eyes; thick skin ridge above the eye .... Bufonidae
b. Other than the above 4

4a. Sacral vertebrae articulate with the urostyle by a single condyle; found only in New Zealand ...... Leiopelmatidae
b. Other than the above 5

4a

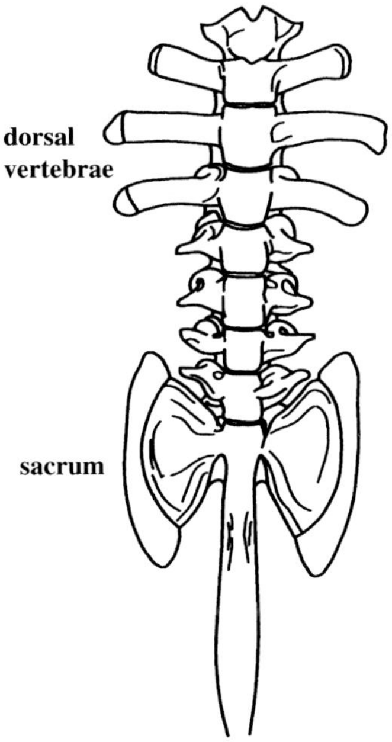

5a

5b

5a. Fingers simple in profile .... Myobatrachidae
b. Fingers with distinct notch between the fingertip and disc .... Hylidae

KEYS TO GENERA (BY FAMILY)

RANIDAE

This family is represented in the region considered here by two subfamilies (Raninae and Platymantinae) and by two to four genera depending on systematic opinions. The genera *Rana* and *Platymantis* are particularly well represented throughout New Guinea and its surrounding islands with 40 species currently described (Menzies, 1987). The validity of the genera *Cornufer* and *Hylarana* in this geographic region is disputed (see Zweifel, 1965, 1969; Menzies, 1976; Burton, 1986).

1a. Toes extensively webbed; paired vocal sacs in males .... *Rana*
b. Toes poorly webbed; single vocal sac in males .... *Platymantis*

1a

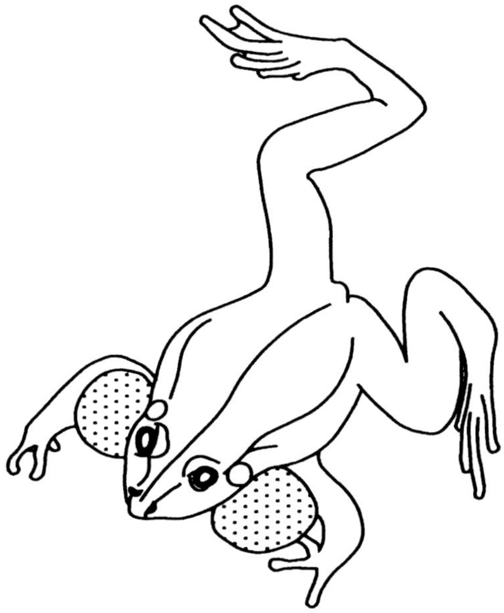

1b

MICROHYLIDAE

Microhylid frogs are represented in this region by two extensively sympatric subfamilies (Asterophryninae and Genyophryninae). Asterophrynines are distributed from the Moluccas, across New Guinea to the Louisiade Archipelago. Genyophrynines are also widely distributed across New Guinea; they occur in northern Australia and the Pacific islands (including New Britain). Asterophrynines are divided into eight genera (Burton, 1986). The genyophrynines are placed in eleven genera (Zweifel and Parker, 1989; Burton and Zweifel, 1995; Zweifel, 2000). The number of species in each genus is given in brackets after the name.

The subfamily Genyophryninae has been under considerable review for the past 15 years. Two Australian genera (*Sphenophryne* and *Cophixalus*) have long been recognized (Cogger, 2000) but recently all Australian species previously ascribed to *Sphenophryne* were reassigned to *Austrochaperina* (Zweifel, 2000). The genus *Cophixalus* has also been redefined (Zweifel and Parker, 1989; Burton and Zweifel, 1995; Menzies, 1999; Kraus and Allison, 2001) and many new species have been described (Allison and Kraus, 2000).

1a. Vertebrae procoelous (centrum concave at the front end and convex at the rear); tongue relatively free (only half or less of the area of the tongue attached to the floor of mouth)....................subfamily Genyophryninae 9
b. Vertebrae displasiocoelous (all vertebrae procoelous except the last, which is amphicoelous); tongue broadly fixed to mouth (about three quarters of the area of the tongue adhering to the floor of the mouth) ....subfamily Asterophryninae 2

1a

2a. Anterior ends of maxillae not touching and separated by premaxillae 3
b. Anterior ends of the maxillae touching or very close 4

3a. Fingers long with small discs and prominent sub-articular tubercles....................*Hylophorbus* (1 sp.)
b. Fingers short, lacking discs and sub-articular tubercles ....*Barygenys* (8 spp.)

2a

2b

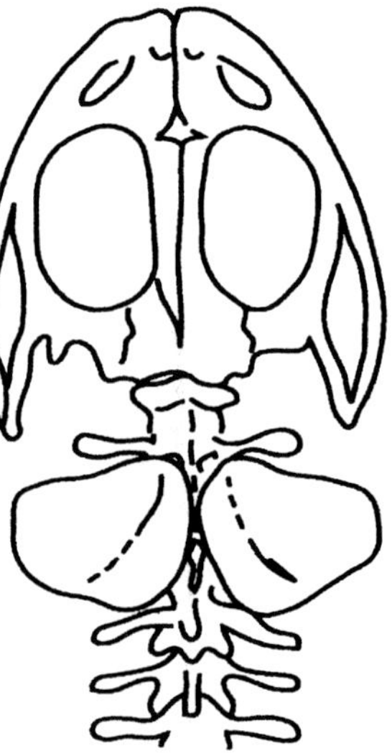

3a

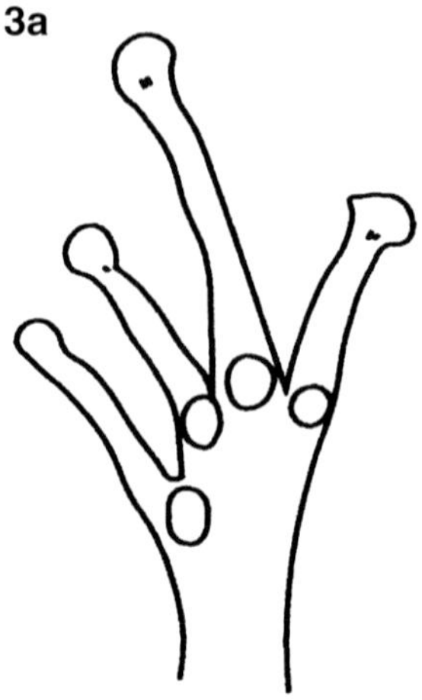

3b

4a. One or two prominent spikes on each vomer .... *Xenobatrachus* (13 spp.)
b. No projections of the vomer visible from the inside of the mouth 5

5a. Eyelids have spiky, cutaneous projections; head very broad (HW:S-V = 0.5) .... *Asterophrys* (2 spp.)
b. Eyelids lack spiky projections; head relatively narrow (HW:S-V = 0.46) 6

4a

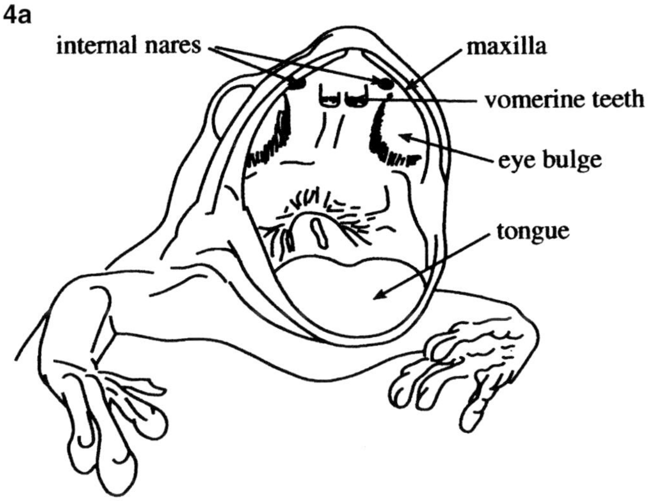

5a

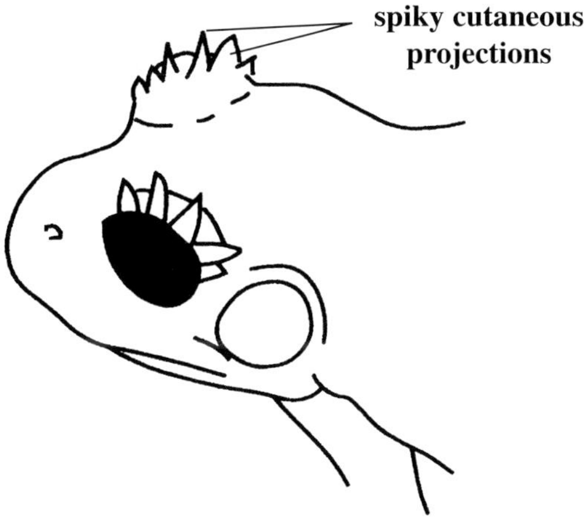

6a. Area between squamosal and maxilla free of bone 7
b. Area behind the eye (between squamosal and maxilla) filled by bone .... *Pherohapsis* (1 sp.)

7a. Eyes relatively small (ED: S-V length = 0.09 or less) .... *Xenorhina* (7 spp.)
b. Eyes relatively large (ED: S-V = 0.1 or greater) 8

8a. Ilial shaft lacks a dorsal crest .... *Callulops* (7 spp.)
b. Ilial shaft bears dorsal crest .... *Mantophryne* (3 spp.)

9a. Head broad, head width half the snouth-vent length .... *Genyophryne* (1 sp.)
b. Head width less than half the snout-vent length...10

6a

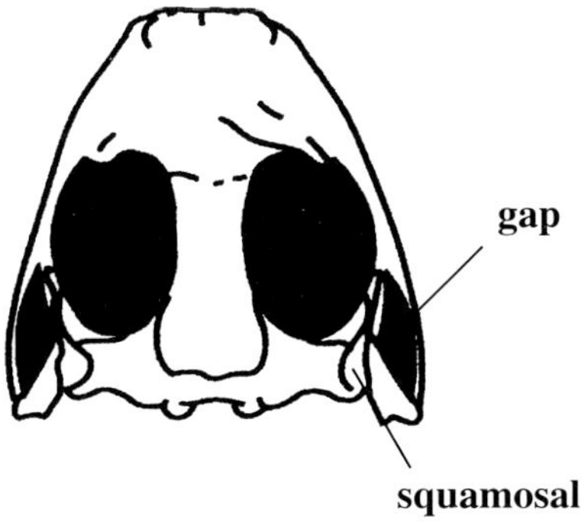

6b

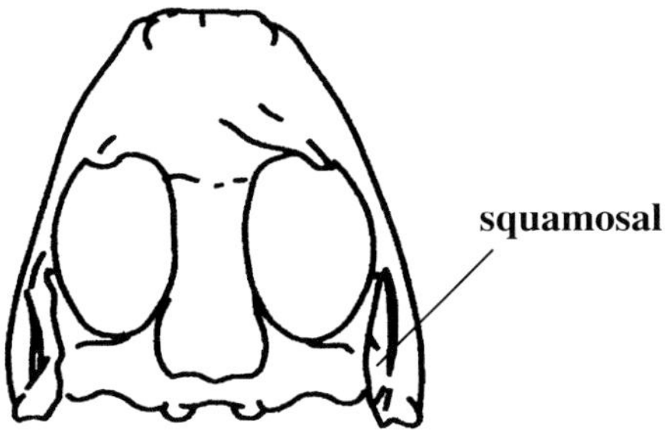

9a

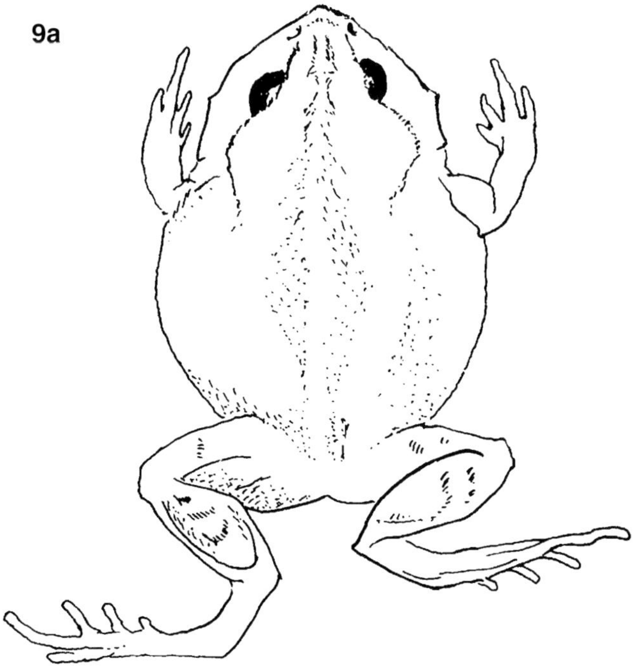

10a. 5th toe longer than the 3rd toe 11
b. 3 rd toe longer than 5th toe 12

11a. No webbing between toes .... *Albericus* (10 spp.)
b. Webbing present between toes .... *Oreophryne* (15 spp.)

12a. 8 pre-sacral vertebra .... *Aphantophryne* (2 spp.)
b. 7 pre-sacral vertebra 12

13a. Nasals elongated to form prominent snout .... *Choerophryne* (3 spp.)
b. Nasals not excessively elongated, snout rounded 13

14a. Pectoral girdle complete (clavicles and procoracoid cartilages present) 15
b. Clavicles and procoracoid cartilages absent .... *Cophixalus* (18 spp.)

13a

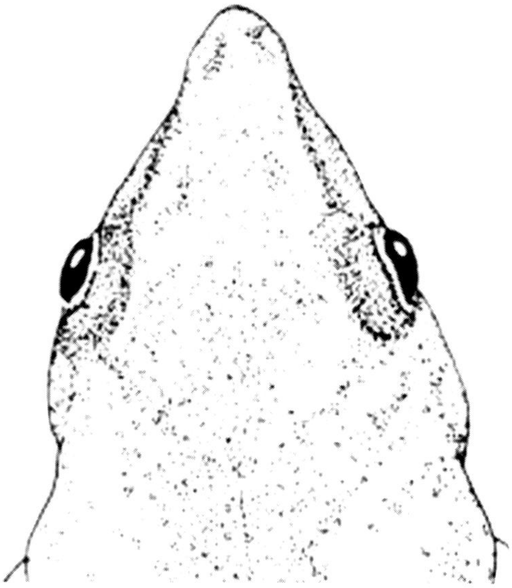

16a

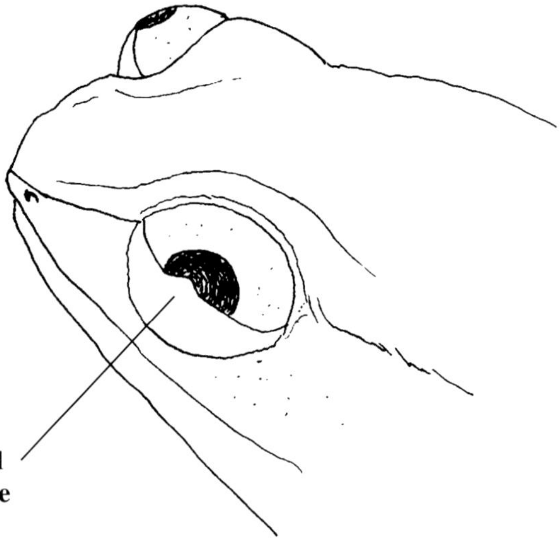

15a. Tympanum unpigmented, nostrils lateral .... *Copiula* (3 spp.)
b. Tympanum pigmented, nostrils anterolateral 16

16a. Single, pointed tubercle on each eyelid, tips of digits (except first finger) with obvious discs .... *Sphenophryne* (1 sp.)
b. No tubercle on eyelids, fingers discs small or absent 17

17a. Tips of fingers pointed or rounded, not flattened .... *Oxydactyla* (5 spp.)
b. Tips of fingers with small discs (only as wide as the finger stem) 18

18a. Legs long, half the snout-vent length .... *Liophryne* (6 spp.)
b. Legs clearly less than half the snout-vent length .... *Austrochaperina* (23 spp.)

BUFONIDAE

Toads are not endemic to the Australasian region. However, one species, the cane toad (*Bufo marinus*) is now widespread throughout the area. Originally found only in South and Central America, this species was introduced into Queensland in 1935 in an attempt to control beetle pests infesting sugar cane plantations. The cane toad has dispersed rapidly from its original release sites and is now commonly found in northern Australia, southern New Guinea, the Solomon Islands and Fiji. Accounts of the introduction and spread of this species can be found in Tyler (1975, 1989) and Menzies (1976). Recently, the presence of a second toad species, *B. melanostictus,* was reported from the Arfak Mountains in the far west of New Guinea (Menzies and Tapilatu, 2000).

1a. The tips of the dorsal warts are black and pointed .... *Bufo melanostictus*
b. The dorsal warts are brown, not pointed and the tips are not black .... *Bufo marinus*

LEIOPELMATIDAE

This family is confined to the North Island of New Zealand and surrounding smaller islands (Gill and Whitaker, 1996). There are only four species, all of which belong to the genus *Leiopelma.* Only one species (*L. hochstetteri*) can be regarded as common, being distributed across the northern coast and Coromandel area of the North Island. It is more aquatic than the other three (*L. hamiltoni*, *L. pakeka* and *L. archeyi*) and can be readily recognized by the extensive webbing between its toes. The other three species are fully terrestrial and lack toe webbing. *Leiopelma hamiltoni* is confined to Stephens Island in Cook Strait while *L. pakeka* (previously included in *L. hamiltoni*) is confined to Maud Island in Marlborough Sound. The distribution of *L. archeyi* overlaps that *of L. hochstetteri* and is confined to the Coromandel area. No key provided.

MYOBATRACHIDAE

These frogs have only recently been accorded familial status (Cogger, 2000), previously being regarded as a subfamily of the Leptodactylidae. Myobatrachids are widespread throughout Australia and New Guinea. There are 23 genera in Australia, five of which (*Crinia, Lechriodus, Limnodynastes, Mixophyes* and *Uperoleia*) are also represented in New Guinea. *Cyclorana* had previously been included in this family but has been removed and placed in the Hylidae.

Three subfamilies are recognized: the Rheobatrachinae (with one genus); the Limnodynastinae (with 11 genera); the Myobatrachinae (with 11 genera). Cogger (2000) gives generic and specific descriptions of members of this family. For the purposes of this key, *Megistolotus* is recognized as a valid genus, although it has been argued that it should be included in *Limnodynastes* (Schauble *et al.*, 2001).

1a. Tongue adhering entirely to the floor of the mouth 2
b. Tongue free of the floor of the mouth at least at the posterior edge of the tongue 3

1a

1b

2a. Inside of the jaw feels smooth, not serrated (maxillary teeth absent), toes poorly webbed .... *Notaden* (4 spp.)
b. Inside of jaw feels serrated and rough (maxillary teeth present); toes fully webbed ...... *Rheobatrachus* (2 spp.)

3a. Vomerine teeth obvious on the roof of the mouth 4
b. Vomerine teeth absent or not obvious 12

4a. Toes fully webbed, webbing extending between the outer metatarsals .... *Mixophyes* (6 spp.)
b. Toes not fully webbed, webbing not reaching the outer metatarsals 5

3a

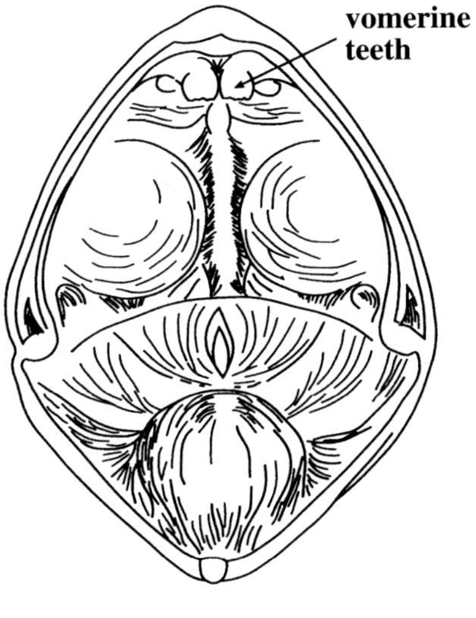

4a

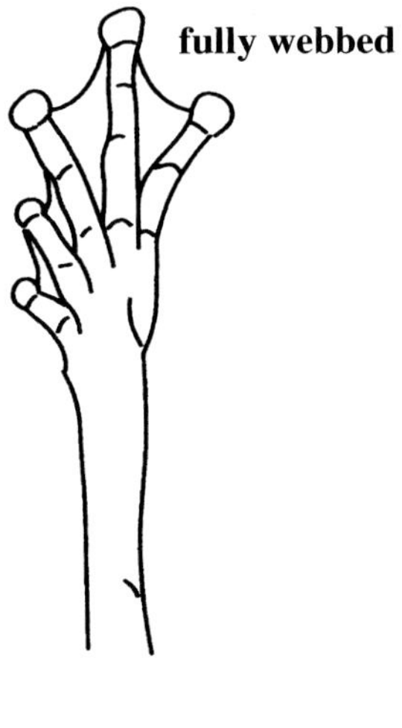

5a

5a. Inside of the groin bears bright red/orange patches; males develop paired, fang-like pseudo-teeth in the lower jaw .... *Adelotus* (1 sp.)
 b. No red/orange patches on the inside of the groin 6

6a. Fingers have small terminal discs .... *Lechriodus* (4 spp.)
 b. Fingers simple, lacking terminal discs 7

7a. Vomerine teeth located between the choanae 8
 b. Vomerine teeth behind the level of the choanae 9

7a **7b**

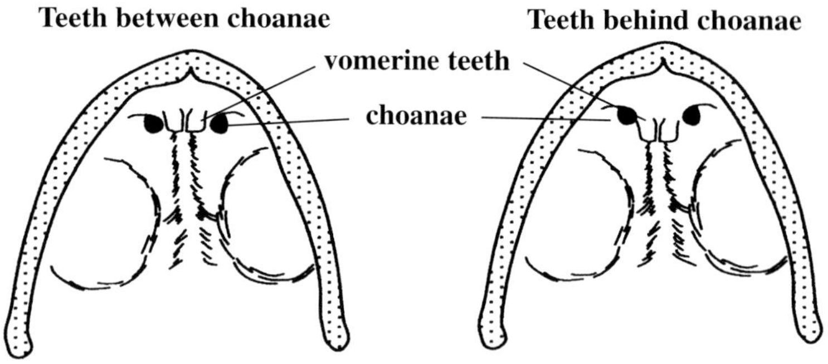

9a **9b**

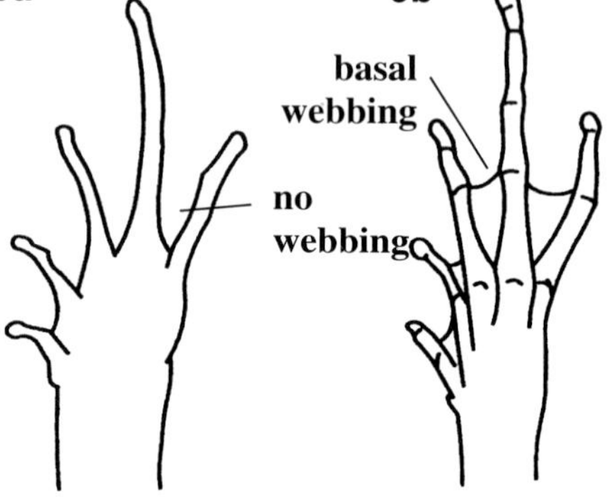

8a. Metatarsal tubercles brown or black .... *Neobatrachus* (10 spp.)
 b. Metatarsal tubercles unpigmented .... *Heleioporus* (6 spp.)

9a. Toes lack any trace of webbing 10
 b. Toes possess some webbing: may be only basally webbed 11

10a. Large parotoid glands present .... *Philoria* (1 sp.)
 b. Parotoid glands not obvious .... *Kyrranus* (6 spp.)

11a. A distinct flap of skin in the corner of the eye .... *Megistolotus* (1 sp.)
 b. No skin flaps present in the anterior corner of the eye .... *Limnodynastes* (12 spp.)

12a. Inner and outer metatarsal tubercles similar in size .... *Uperoleia* (24 spp.)
 b. Inner metatarsal tubercles much bigger than outer metatarsal tubercles (which may be absent altogether) 13

11a

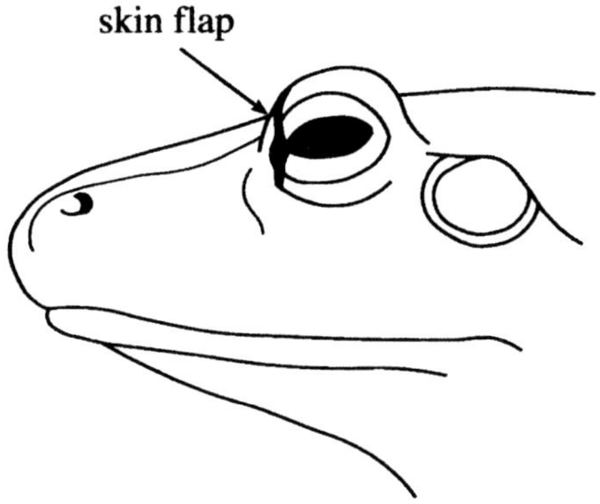

12a

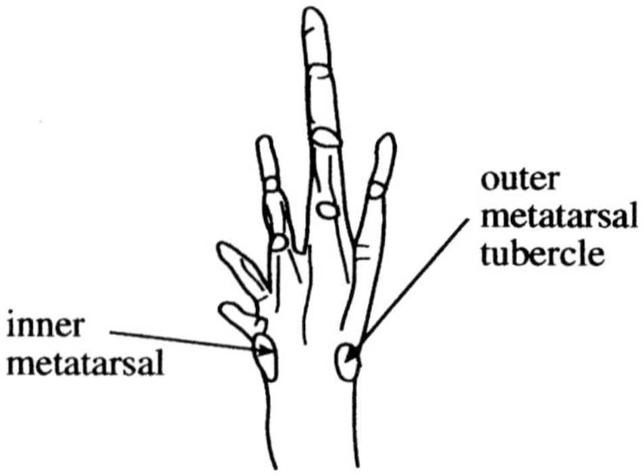

Ventral view of foot

13a. Maxillary teeth absent (jaw feels smooth not serrated) 14
 b. Maxillary teeth present (jaw surface feels serrated) 16

14a. Belly is granular; small (S-V length < 25 mm) dark frog from Western Australia .... *Metacrinia* (1 sp.)
 b. Belly smooth, no lumps or warts present 15

15a. An unusual, turtle-shaped frog with short stumpy limbs ....*Myobatrachus* (1 sp.)
b. Small, dark toadlet with a black and white mottled pattern on the belly ....*Pseudophryne* (11 spp.)

15a

16a

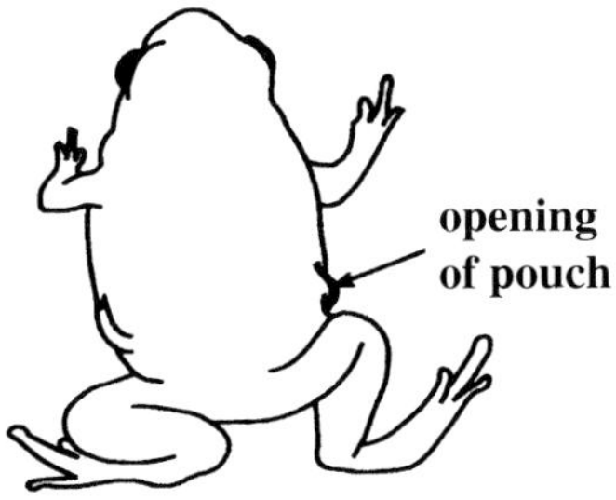

20a

21a

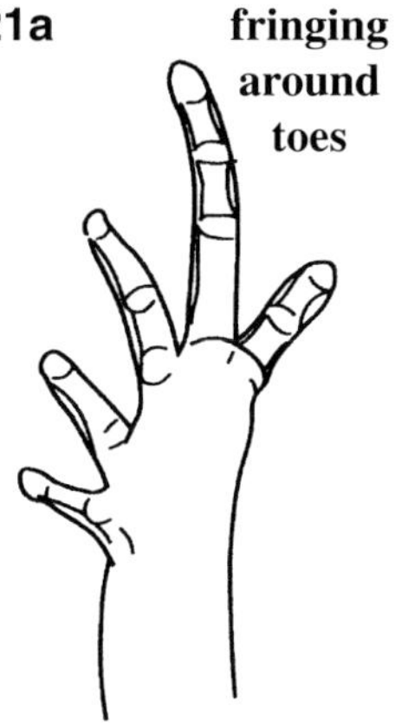

16a. Males develop brood pouches on their flanks; first finger small (vestigial) ....*Assa* (1 sp.)
b. Males lack brood pouches; first finger normal...17

17a. Tips of fingers with small, but discernible T-shaped discs..*Taudactylus* (6 spp.)
b. Fingers lack discs...18

18a. Belly smooth (not granular); vomerine teeth present ....*Geocrinia* (7 spp.)
b. Belly granular, vomerine teeth may be absent 19

19a. Prominent glands on neck or on dorsum 20
b. No enlarged glands on side of neck or on dorsum 21

20a. Large glands evident on neck and flanks; frog rotund, from Geraldton–Shark Bay
area of Western Australia ....*Arenophryne* (1 sp.)
b. Small, raised glands over dorsal area. Dorsum purple, throat yellow and belly mostly blue ......*Spicospina* (1 sp.)

21a. Toes fringed...*Paracrinia* (1 sp.)
b. Toes simple. Lack fringe 22

22a. No webbing between fingers or toes ....*Crinia* (14 spp.)
b. Traces of webbing between fingers and toes ....*Bryobatrachus* (1 sp.)

HYLIDAE

Some of the members of this family are often called `tree frogs'. The digits are usually equipped with prominent finger and toe pads (except for *Cyclorana*). The Australian representatives of this family fall into three distinct genera. Savage (1973) proposed that the Australian hylid frogs should be placed in their own family (Pelodryadidae) but this proposal has not found support. The Australian hylids are recognized, however, as a unique subfamily, the Pelodryadinae.

Hylid diversity is highest in Australia and New Guinea but representatives occur in the Moluccas, New Britain and the Solomon Islands.

1a

2b

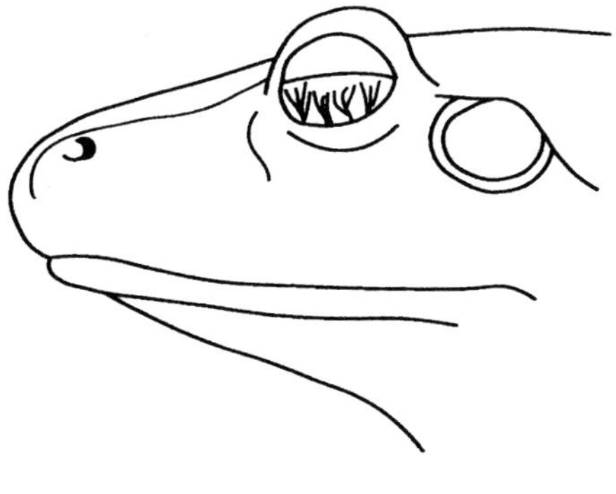

1a. Tips of the fingers have a distinct notch (and usually a prominent terminal disc) 2
b. Notch not apparent, discs absent, first finger is opposable to the rest ....*Cyclorana* (13 spp.)

2a. Pupil horizontal, lower eyelid lacks pigmentation ....*Litoria* (103 spp.)
b. Pupil vertical, lower eyelid with vein-like pattern ....*Nyctimystes* (19 spp.)

CONCLUSIONS

The evolution, distributions and conservation of Australasian frog faunas have been discussed in a number of other chapters (Heinsohn and Hope, 2006; Holdaway and Worthy, 2006; Richards *et al.*, 2006; Tyler and Lee, 2006), but this chapter summarizes the diversity of extant frogs (also see Plates 35, 54–61) throughout the region.

Including two introduced toads the six family keys include 49 genera and over 430 recognized species. The Microhylidae, Myobatrachidae and Hylidae are by far the most speciose families (including 386 species); but of the 45 genera represented most are small—34 contain between one and ten species. Another ten genera have between 11–24 species and the largest genus is *Litoria* with 103.

It is hoped that these combined keys will supplement and complement the detailed field keys (down to specific level) currently available for regional faunas (Gill and Whitaker, 1996; Cogger, 2000).

REFERENCES

Allison, A. and Kraus, F. 2000. A new species of frog of the genus *Xenorhina* (Anura:Microhylidae) from the north coast ranges of Papua New Guinea. *Herpetologica,* Vol. 56, pp.285–294.

Burton, T.C. 1986. A re-assessment of the Papuan sub-family Asterophryne (Anura:Microhylidae). *Records of the South Australian Museum.* Vol. 19, pp.405–450.

Burton, T.C. and Zweifel, R.G. 1995. A new genus of genyophrynine microhylid frogs from New Guinea. *American Museum Novitates,* No. 3129, pp.1–7.

Cogger, H.G. 2000. *Reptiles and Amphibians of Australia.* 6th Edit. Reed Books, Sydney. 596pp.

Gill, B.J. and Whitaker, A.H. 1996. *New Zealand Frogs and Reptiles.* David Bateman Ltd, Auckland.

Heinsohn, T. and Hope, G. 2006. The Torresian connections: zoogeography of New Guinea. Pp.71–93. In Merrick, J.R., Archer, M., Hickey, G.M. and Lee, M.S.Y. (eds). *Evolution and Biogeography of Australasian Vertebrates.* Auscipub Pty. Ltd., Sydney. 942pp.

Holdaway, R.N. and Worthy, T.H. 2006. The evolution of New Zealand and its vertebrates. Pp.111–128. In Merrick, J.R., Archer, M., Hickey, G.M. and Lee, M.S.Y. (eds). *Evolution and Biogeography of Australasian Vertebrates.* Auscipub Pty. Ltd., Sydney. 942pp.

Kraus, F. and Allison, A. 2001. A review of the endemic New Guinea microhylid frog genus *Choerophryne. Herpetologica,* Vol. 57, No. 2, pp.214–232.

Menzies, J.I. 1976. *Handbook of Common New Guinea Frogs.* Wau Ecology Institute Handbook No. 1, Wau, Papua New Guinea. 75pp.

Menzies, J.I. 1987. A taxonomic revision of Papuan *Rana* (Amphibia:Ranidae). *Australian Journal of Zoology,* Vol. 35, pp.373–418.

Menzies, J.I. 1999. A study of *Albericus* (Anura:Microhylidae) of New Guinea. *Australian Journal of Zoology,* Vol. 47, pp.327–360.

Menzies, J.I. and Tapilatu, R.F. 2000. The introduction of a second species of toad (Amphibia: Bufonidae) into New Guinea. *Science in New Guinea,* Vol. 25, pp.70–73.

Richards, S.J., Alford, R.A. and Bradfield, K.S. 2006. The great frog decline in Australasia: causes, developments and conservation. Pp.241–252. In Merrick, J.R., Archer, M., Hickey, G.M. and Lee, M.S.Y. (eds). *Evolution and Biogeography of Australasian Vertebrates.* Auscipub Pty. Ltd., Sydney. 942pp.

Savage, J.M. 1973. The geographic distributions of frogs: patterns and predictions. Pp.352–445. In Vial, J.V. (ed). *Evolutionary Biology of the Anurans; Contemporary Research on Major Problems.* University of Missouri Press, Columbia. 470pp.

Schauble,C.S., Moritz, C. and Slade, R.W. 2000. A molecular phylogeny for the frog genus *Limnodynastes* (Anura: Myobatrachidae). *Molecular Phylogenetics and Evolution.* Vol. 16, pp.379–391.

Tyler, M.J. 1975. *Frogs.* William Collins and Sons, Melbourne. 256pp.

Tyler, M.J. 1989. *Australian Frogs.* Viking O'Neil, Melbourne. 220pp.

Tyler, M.J. and Lee, M.S.Y. 2006. The origins of Australian frogs. Pp.237–240. In Merrick, J.R., Archer, M., Hickey, G.M. and Lee, M.S.Y. (eds). *Evolution and Biogeography of Australasian Vertebrates.* Auscipub Pty. Ltd., Sydney. 942pp.

Zweifel, R.G. 1965. Revisionary notes on Australian microhylid frogs of the genus *Sphenophryne*. *American Museum Novitates*, No. 2214, pp.1–9.

Zweifel, R.G. 1969. Frogs of the genus *Platymantis* (Ranidae) in New Guinea, with the description of a new species. *American Museum Novitates*, No. 2374, pp.1–19.

Zweifel, R.G. 2000. Partition of the Australopapuan microhylid frog genus *Sphenophryne* with descriptions of new species. *Bulletin of the American Museum of Natural History*, Vol. 253, pp.1–130.

Zweifel, R.G. and Parker, F. 1989. New species of microhylid frog from the Owen Stanley Mountains of Papua New Guinea and the resurrection of the genus *Aphantophryne. American Museum Novitates,* No. 2954, pp.1–19.

Primitive Amniotes: Reptiles

GEOLOGICAL TIME SCALE FOR THE PHANEROZOIC

ERA	PERIOD or SUB-PERIOD	SERIES/EPOCH	STAGE	AGE m.y.a.	EXTINCTIONS (E) ICE AGE (I)
CAINOZOIC	QUATERNARY	PLEISTOCENE	Calabrian	1.8	E, I
CAINOZOIC	TERTIARY – NEOGENE	PLIOCENE	Piacenzian	3.6	I
CAINOZOIC	TERTIARY – NEOGENE	PLIOCENE	Zanclean	5.3	I
CAINOZOIC	TERTIARY – NEOGENE	MIOCENE Late	Messinian	7.12	
CAINOZOIC	TERTIARY – NEOGENE	MIOCENE Late	Tortonian	11.2	
CAINOZOIC	TERTIARY – NEOGENE	MIOCENE Mid	Serravalian	14.8	
CAINOZOIC	TERTIARY – NEOGENE	MIOCENE Mid	Langhian	16.4	
CAINOZOIC	TERTIARY – NEOGENE	MIOCENE Early	Burdigalian	20.52	
CAINOZOIC	TERTIARY – NEOGENE	MIOCENE Early	Aquitanian	23.8	
CAINOZOIC	TERTIARY – PALAEOGENE	OLIGOCENE Late	Chattian	28.5	E
CAINOZOIC	TERTIARY – PALAEOGENE	OLIGOCENE Early	Rupelian	33.7	
CAINOZOIC	TERTIARY – PALAEOGENE	EOCENE Late	Priabonian	37	
CAINOZOIC	TERTIARY – PALAEOGENE	EOCENE Mid	Bartonian	41.3	
CAINOZOIC	TERTIARY – PALAEOGENE	EOCENE Mid	Lutetian	49	
CAINOZOIC	TERTIARY – PALAEOGENE	EOCENE Early	Ypresian	54.8	
CAINOZOIC	TERTIARY – PALAEOGENE	PALEOCENE Late	Thanetian	57.9	
CAINOZOIC	TERTIARY – PALAEOGENE	PALEOCENE Late	Selandian	60.9	
CAINOZOIC	TERTIARY – PALAEOGENE	PALEOCENE Early	Danian	65	E
MESOZOIC	CRETACEOUS	LATE	Maastrichtian	73	
MESOZOIC	CRETACEOUS	LATE	Campanian	83	
MESOZOIC	CRETACEOUS	LATE	Santonian	87	
MESOZOIC	CRETACEOUS	LATE	Coniacian	89	
MESOZOIC	CRETACEOUS	LATE	Turonian	91	
MESOZOIC	CRETACEOUS	LATE	Cenomanian	97.5	
MESOZOIC	CRETACEOUS	EARLY	Albian	108	
MESOZOIC	CRETACEOUS	EARLY	Aptian	115	
MESOZOIC	CRETACEOUS	EARLY	Barremian	123	
MESOZOIC	CRETACEOUS	EARLY	Hauterivian	130	
MESOZOIC	CRETACEOUS	EARLY	Valanginian	135	
MESOZOIC	CRETACEOUS	EARLY	Berriasian	141	E
MESOZOIC	JURASSIC	LATE	Tithonian	146	
MESOZOIC	JURASSIC	LATE	Kimmeridgian	151	
MESOZOIC	JURASSIC	LATE	Oxfordian	159	
MESOZOIC	JURASSIC	MIDDLE	Callovian	165	
MESOZOIC	JURASSIC	MIDDLE	Bathonian	173	
MESOZOIC	JURASSIC	MIDDLE	Bajocian	180	
MESOZOIC	JURASSIC	MIDDLE	Aalenian	184	
MESOZOIC	JURASSIC	EARLY	Toarcian	190	E
MESOZOIC	JURASSIC	EARLY	Pliensbachian	195	
MESOZOIC	JURASSIC	EARLY	Sinemurian	202	
MESOZOIC	JURASSIC	EARLY	Hettangian	205	E
MESOZOIC	TRIASSIC	LATE	Rhaetian	207	
MESOZOIC	TRIASSIC	LATE	Norian	220	
MESOZOIC	TRIASSIC	LATE	Carnian	230	
MESOZOIC	TRIASSIC	MIDDLE	Ladinian	235	
MESOZOIC	TRIASSIC	MIDDLE	Anisian	241	
MESOZOIC	TRIASSIC	EARLY	Spathian	244	
MESOZOIC	TRIASSIC	EARLY	Nammalian	247	
MESOZOIC	TRIASSIC	EARLY	Griesbachian	251	

This detailed scale is based on Young and Laurie (1996) but Cambrian stages abbreviated ages are rounded to nearest 0.1 million years. More recent dating (e.g. United States Geological Survey) has resulted in some small adjustments to interval boundaries.

ERA	PERIOD or SUB-PERIOD	SERIES/EPOCH	STAGE	AGE m.y.a.	EXTINCTIONS (E) ICE AGE (I)
PALAEOZOIC	PERMIAN	LATE	Tatarian – Dorashamian	255	E
PALAEOZOIC	PERMIAN	LATE	Tatarian – Dzhulfian	259	I
PALAEOZOIC	PERMIAN	LATE	Tatarian – Midian	263	I
PALAEOZOIC	PERMIAN	LATE	Kazanian	267	I
PALAEOZOIC	PERMIAN	LATE	Ulfimian	270	I
PALAEOZOIC	PERMIAN	EARLY	Kungurian	274	I
PALAEOZOIC	PERMIAN	EARLY	Artinskian	285	I
PALAEOZOIC	PERMIAN	EARLY	Sakmarian	293	I
PALAEOZOIC	PERMIAN	EARLY	Asselian	298	I
PALAEOZOIC	CARBONIFEROUS	PENSYLVANNIAN – SILESIAN – STEPHANIAN	Gzhelian	302	I
PALAEOZOIC	CARBONIFEROUS	PENSYLVANNIAN – SILESIAN – STEPHANIAN	Kasimovian	305	I
PALAEOZOIC	CARBONIFEROUS	PENSYLVANNIAN – SILESIAN – WESTPHALIAN	Moscovian	311.5	
PALAEOZOIC	CARBONIFEROUS	PENSYLVANNIAN – SILESIAN – WESTPHALIAN	Bashkirian	314	
PALAEOZOIC	CARBONIFEROUS	SILESIAN – NAMURIAN	Serpukhovian	325	
PALAEOZOIC	CARBONIFEROUS	MISSISSIPPIAN – DINANTIAN – VISEAN	Brigantian	328.5	
PALAEOZOIC	CARBONIFEROUS	MISSISSIPPIAN – DINANTIAN – VISEAN	Asbian	330.5	
PALAEOZOIC	CARBONIFEROUS	MISSISSIPPIAN – DINANTIAN – VISEAN	Holkerian	335	
PALAEOZOIC	CARBONIFEROUS	MISSISSIPPIAN – DINANTIAN – VISEAN	Arundian	341.5	
PALAEOZOIC	CARBONIFEROUS	MISSISSIPPIAN – DINANTIAN – TOURNAISIAN	Chadian	344	
PALAEOZOIC	CARBONIFEROUS	MISSISSIPPIAN – DINANTIAN – TOURNAISIAN	Ivorian	348	
PALAEOZOIC	CARBONIFEROUS	MISSISSIPPIAN – DINANTIAN – TOURNAISIAN	Hastarian	354	
PALAEOZOIC	DEVONIAN	LATE	Famennian	364.5	E
PALAEOZOIC	DEVONIAN	LATE	Frasnian	369	
PALAEOZOIC	DEVONIAN	MIDDLE	Givetian	378	
PALAEOZOIC	DEVONIAN	MIDDLE	Eifelian	384	I
PALAEOZOIC	DEVONIAN	EARLY	Emsian	399.5	I
PALAEOZOIC	DEVONIAN	EARLY	Pragian	404.5	I
PALAEOZOIC	DEVONIAN	EARLY	Lochkovian	410	
PALAEOZOIC	SILURIAN	PRIDOLI		414	
PALAEOZOIC	SILURIAN	LUDLOW	Ludfordian	417	
PALAEOZOIC	SILURIAN	LUDLOW	Gorstian	420	
PALAEOZOIC	SILURIAN	WENLOCK	Homerian	422.5	
PALAEOZOIC	SILURIAN	WENLOCK	Sheinwoodian	425	
PALAEOZOIC	SILURIAN	LLANDOVERY	Telychian	428	I
PALAEOZOIC	SILURIAN	LLANDOVERY	Aeronian	431	I
PALAEOZOIC	SILURIAN	LLANDOVERY	Rhuddanian	434	E, I
PALAEOZOIC	ORDOVICIAN	ASHGILL	Bolindian	443	I
PALAEOZOIC	ORDOVICIAN	CARADOC	Eastonian	454.5	
PALAEOZOIC	ORDOVICIAN	CARADOC	Gisbornian	459	
PALAEOZOIC	ORDOVICIAN	LLANVIRN	Darriwilian	467	
PALAEOZOIC	ORDOVICIAN		Yapeenian	471	
PALAEOZOIC	ORDOVICIAN	ARENIG	Castlemainian	477	
PALAEOZOIC	ORDOVICIAN	ARENIG	Chewtonian	481	
PALAEOZOIC	ORDOVICIAN	ARENIG	Bendigonian	484.5	
PALAEOZOIC	ORDOVICIAN	TREMADOC	Lancefieldian	486	
PALAEOZOIC	ORDOVICIAN	TREMADOC	Warendan	490	
PALAEOZOIC	CAMBRIAN	LATE	Datsonian	491	
PALAEOZOIC	CAMBRIAN	LATE	Payntonian	492.5	
PALAEOZOIC	CAMBRIAN	LATE	Iverian	496	
PALAEOZOIC	CAMBRIAN	LATE	Idamean	497.5	E
PALAEOZOIC	CAMBRIAN	MIDDLE	Mindyallan	498.5	
PALAEOZOIC	CAMBRIAN	MIDDLE	Boomerangian	500	
PALAEOZOIC	CAMBRIAN	MIDDLE	Undillan	503	
PALAEOZOIC	CAMBRIAN	MIDDLE	Late Templetonian-Floran	506	
PALAEOZOIC	CAMBRIAN	MIDDLE	Ordian-Early Templetonian	509	
PALAEOZOIC	CAMBRIAN	EARLY	Atdabanian-Toyonian	528	
PALAEOZOIC	CAMBRIAN	EARLY	Tommotian	534	
PALAEOZOIC	CAMBRIAN	EARLY	Nemakit-Daldynian	545	

Chapter 15
Dinosaurs and Other Mesozoic Reptiles of Australasia

John D. Scanlon
Department of Zoology, University of Queensland, Brisbane Qld. 4072 Australia
Department of Palaeontology, South Australian Museum, North Terrace, Adelaide S.A. 5000 Australia
E-mail: Scanlon.John@saugov.sa.gov.au

INTRODUCTION

The number of dinosaur-bearing deposits and total amount of dinosaur material known from Australasia are still very small in comparison to other continents, except Antarctica. There have been several thorough reviews in recent years (Molnar, 1982a, 1991; Long, 1998; Long and Buffetaut, 2001); there is also a series of papers from the Symposium on Gondwanan Dinosaurs published in 1996 (*Memoirs of the Queensland Museum*, Vol. 39). This chapter attempts to briefly summarize the biological features of major dinosaurian lineages, and the history of each lineage in Australasia, incorporating evidence from both body fossils and trackways. Because of the paucity of well-preserved body fossils of terrestrial vertebrates, including dinosaurs, the record of their tracks provides a vital complementary source of information (Gillette and Lockley, 1989; Thulborn, 1990a; Lockley, 1991). The main focus will be an interpretation of the evolutionary history and biogeography of Australasian dinosaurs and a key to published sources, rather than a catalogue of material or detailed taxonomic assessment.

As well as dinosaurs, this chapter treats their aquatic counterparts, ichthyosaurs and sauropterygians, which are equally characteristic of Mesozoic faunas. Groups such as turtles, pterosaurs, mosasaurs, and early branches of the amniote radiation (Figure 1) that did not survive the Triassic, have a more limited presence in the Australasian Mesozoic record and will be considered only in passing.

TRACES OF THE TRIASSIC TERRESTRIAL FAUNA

The end of the Permian and beginning of the Mesozoic was marked by a marine regression from margins of the single supercontinent Pangaea (margins including the eastern and north-western parts of what is now Australia), and the most severe mass extinction of the Phanerozoic, estimated to have wiped out over 90% of marine and nearly as many terrestrial animal species in well under a million years (Benton, 2003). During the Triassic, climates became progressively warmer and drier, and the adaptive radiation of the surviving lineages, including reptiles, produced a steep increase in diversity which was punctuated by further waves of extinction, minor only by comparison to the end-Permian event (Tucker and Benton, 1982). Due to the contiguous land masses, and despite large areas of desert in lower latitudes, there were few significant barriers to animal migration. As a result, the Triassic (the shortest period of the Mesozoic) is characterized by low endemism and frequent global changes in the composition of faunas, with a succession of distinct vertebrate faunas providing correlation across widely separated regions.

Because of Australia's then near-polar location, Triassic fossil-bearing deposits formed under a cool temperate climate represent mainly freshwater lakes, streams, and swamps dominated by fishes and temnospondyl amphibians; no equivalent faunas are known from New Zealand. There is also a sparse record of some terrestrial amniote groups from the early Triassic: therapsids, procolophonids, archosauromorphs, and lepidosauromorphs, representing a small sample of the fauna typical of Scythian–Anisian terrestrial sediments elsewhere in the world—for example, Karoo beds in South Africa, or Lower Fremouw Formation in the Transantarctic Mountains (Vickers-Rich and Rich, 1993). This limited representation of a widespread fauna may be partly due to the (relatively) extreme climate, but perhaps mainly to the small number of exposures of terrestrial fossil-bearing deposits of this age; almost all of the non-marine reptilian material comes from a single locality, 'The Crater', in the early Triassic Arcadia Formation near Rewan, Qld. (see Table 1). The skeletal material of these terrestrial animals is often very limited, enough to show that a given group was represented but giving little information about the morphology or palaeobiology of most of the Australian species. Complementing the fossil bones to some extent, fossilized tracks of terrestrial reptiles are also known in the early Triassic Narrabeen Group and middle Triassic Hawkesbury Sandstone near Sydney, N.S.W., and the Blackstone Formation and other late Triassic beds in Queensland.

For a small number of species, the Rewan reptiles are remarkably diverse at a high taxonomic level—no two species belong to the same order (as traditionally classified—but of course, reptilian 'orders' shared a much closer common ancestry then than they do now).

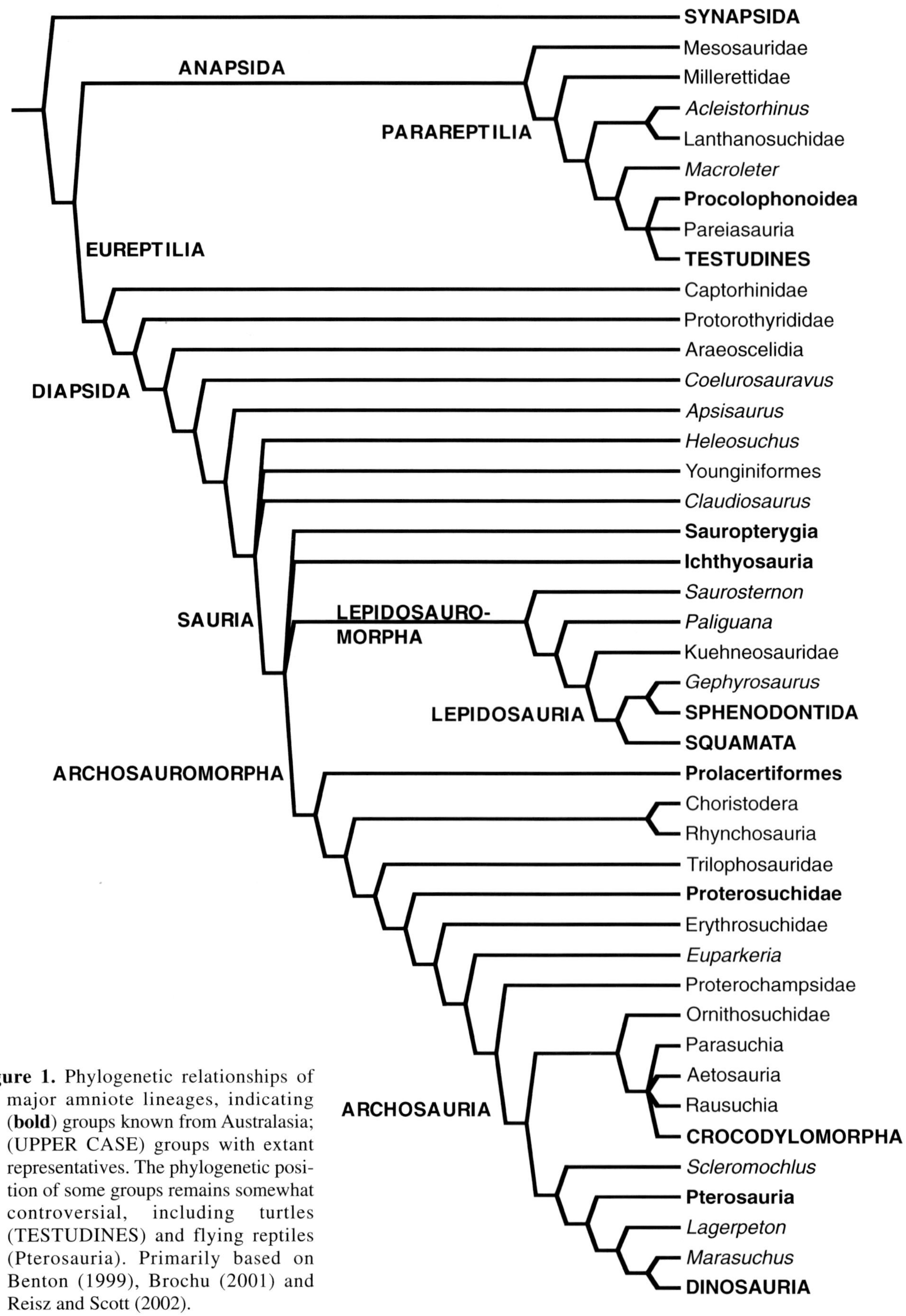

Figure 1. Phylogenetic relationships of major amniote lineages, indicating (**bold**) groups known from Australasia; (UPPER CASE) groups with extant representatives. The phylogenetic position of some groups remains somewhat controversial, including turtles (TESTUDINES) and flying reptiles (Pterosauria). Primarily based on Benton (1999), Brochu (2001) and Reisz and Scott (2002).

The only therapsid yet detected is a sheep-sized, herbivorous dicynodont. This unnamed taxon (for a long time represented by a single skull bone, though more has been found recently) is similar to the important indicator taxa *Lystrosaurus* or *Kannemeyeria*. Other therapsid lineages such as cynodonts, more closely related to mammals and including herbivores as well as the dominant land carnivores of this time, have not been detected but might also have been present. Early Triassic tracks from south of Sydney have been described as *Dicynodontipus*, presumed to belong to a *Lystrosaurus*-type animal close to a metre long (Retallack, 1996). Procolophonids were chunky, lizard-like herbivores (about the size and shape of a blue-tongue or shingle-back skink), members of the 'parareptile' group also containing pareiasaurs, mesosaurs and probably turtles. The Rewan procolophonid material (Bartholomai, 1979) has yet to be described in full, but includes dozens of skulls, partial skeletons and a complete skeleton of a species of *Procolophon*, thus also showing close links with Triassic faunas of other continents (Thulborn, 1986b; Hamley, 1997).

Three partial skulls from the same site (with some associated postcranial material) belong to almost equally divergent lineages of diapsid reptiles: *Kudnu*, a primitive lepidosaur; *Kadimakara*, a prolacertid archosauromorph; and *Kalisuchus* (Figure 2(a)), a proterosuchid archosaur, the oldest archosaur known in Australia (Bartholomai, 1979; Thulborn, 1986a). *Kudnu* and *Kadimakara* would have been lizard-like in their body form and habits, and probably insectivorous, despite the prolacertiforms such as *Kadimakara* being more closely related to archosaurs (Gow, 1975; Benton, 1985). The early Mesozoic record of lepidosaurs is poor worldwide, mainly because of the very small body size typical of lizard biology. Apart from *Kudnu* no other taxa have been identified in Australasia until the appearance of mosasaurs (giant marine varanoid lizards) in the late Cretaceous.

Proterosuchians included several distinct lineages by the early Triassic, and may represent a paraphyletic stem-group of the other more derived archosaur lineages. The Australian forms are proterosuchids, a group best known from South African and Chinese material of *Chasmatosaurus*, thought to have been semi-aquatic and ecologically similar to small crocodiles (Cruikshank, 1972). *Kalisuchus* and *Tasmaniosaurus* (a related form from the similar-aged Knocklofty Sandstone in Tasmania) reached about 2 m in maximum length, barely larger than some of the amphibians in the same deposits, their likely prey (Camp and Banks, 1978; Thulborn, 1986a). Other families of primitive archosauromorphs were more terrestrial and included the dominant land predators of the late Triassic, but have yet to be detected in Australia. Footprints in the Blackstone Formation (Qld.) include some referred to the ichnogenus *Plectropterna*, representing a slender quadruped over a metre in length, possibly a rauisuchian.

Table 1. Stratigraphic distribution of Australasian Mesozoic reptiles (modified from Molnar, 1991; Fordyce, 1991; Long, 1998). For each geological formation producing reptile fossils or traces, the table gives the name of the collecting locality (or nearby town or feature) and references, then lists the reptile fauna (including birds).

Period/Region	Formation and Fauna
TRIASSIC	
Western Australia	
	Blina Shale, early Triassic (Canning Basin) (McKenzie 1961, Cosgriff and Garbutt 1972)
	?ichthyosaur indet.
Tasmania	
	Knocklofty Sandstone, early Triassic, Scythian-Anisian (Crisp and Gunn's Quarry, Hobart) (Camp and Banks 1978, Thulborn 1986a)
	proterosuchid thecodont *Tasmaniosaurus triassicus*
New South Wales	
	Narrabeen Group, early Triassic (Bellambi Colliery, Wollongong) (Retallack 1996)
	therapsid(?) *Dicynodontipus bellambiensis* (tracks)
	Hawkesbury Sandstone, middle Triassic (Berowra Creek) (Fletcher 1948)
	quadrupedal reptile (tracks)
Queensland	
	Arcadia Formation (Rewan Group), early Triassic, Scythian-Anisian (The Crater, Duckworth Creek, Rewan, Rolleston) (Bartholomai and Howie 1970, Jensen 1975, Bartholomai 1979, Thulborn 1979, 1983, 1985b, 1986a, b, 1990b, Molnar 1991, Hamley 1997)
	dicynodont therapsid aff. *Kannemeyeria*
	procolophonid indet.
	prolacertid archosauromorph *Kadimakara australiensis*
	'paliguanid' (?)lepidosaur *Kudnu mackinlayi*
	lepidosaur indet.
	proterosuchian thecodont *Kalisuchus rewanensis*
	Blackstone Formation, late Triassic (Rhondda Colliery, Dinmore; Bergin Hill Quarry, Goodna) (Bartholomai 1966a, Molnar 1982b)
	thecodont or lepidosaur (tracks) *Plectropterna* sp.
	large tetrapod tracks (Striped Bacon Seam)
	large theropod (tracks) cf. ***Eubrontes***
	unnamed beds, late Triassic, Rhaetian (Callide Basin) (Molnar 1991)
	small theropod(?) tracks
New Zealand	
	formation not stated, early Triassic (Anisian) (Cruikshank *et al.* 1999)
	pachypleurosaur sauropterygian
	Hokonui Series, late Triassic, Otamitan (Otamita Stream, Nugget Point, South Island) (Campbell 1965, Fordyce 1991)
	ichthyosaurs
	shastasaurid(?)
	ichthyosaur indet.

Torlesse Supergroup, upper Triassic, Oretian (Mt Potts, Rangitata Valley, South Island) (Hector 1874, Campbell 1965, Fleming *et al.* 1971, Campbell and Fordyce 1972, Fordyce 1991)
  ichthyosaurs indet. (including '*Ichthyosaurus australis*' Hector)
formation not stated, Triassic(?) (Houghton Bay, Wellington, North Island) (Stevens 1974, Fordyce 1991)
  ichthyosaur indet.

JURASSIC
Western Australia

Colalura Sandstone, middle Jurassic, Bajocian (Bringo Cutting, Geraldton) (Long 1992, Long and Molnar 1998, Long and Cruikshank 1998)
  elasmosaurid(?) plesiosaur
  sauropod indet. (caudal vertebra)
  theropod (family indet.) ***Ozraptor subotai*** (tibia)

Queensland

Evergreen Formation, early Jurassic (Kolane, Springsure) (Thulborn and Warren 1980)
  plesiosaur indet. (elasmosaurid or cryptocleidid)
Precipice Sandstone, early Jurassic (Carnarvon Gorge) (Molnar 1991, Thulborn 1994)
  ornithopod tracks, ***Anomoepus*** cf. *A. gracillimus*
Razorback beds, early Jurassic (Mt Morgan) (Bartholomai 1966a, b, Molnar 1991, Cruikshank *et al.* 1999)
  pachypleurosaur(?) sauropterygian ("Mount Morgan 'plesiosaur'")
  large, primitive theropod (tracks, manus with 5 digits)
Hutton Sandstone, middle Jurassic, ?Bajocian (Injune Creek Beds, Taloona, Roma) (Longman 1926, 1927, Thulborn 1985a, Molnar 1991, Hunt *et al.* 1994, Rich 1996)
  cetiosaurid(?) sauropod ***Rhoetosaurus brownei*** (partial skeleton)
Walloon Group, middle Jurassic, ?Bajocian (Westvale, Balgowan, Lanefield, Oakleigh, and New Hope Collieries; Rosewood, Oakey, Jeebropilly) (Hill *et al.* 1966, Haubold 1971, Molnar 1991, Thulborn 1994, 1999)
  theropods
    large theropod tracks ***Changpeipus bartholomaii***
    large theropod tracks cf. *Eubrontes*
    small theropod tracks
  ornithopods
    small bipedal ornithopod tracks cf. *Anomoepus*
    large bipedal ornithopod tracks (iguanodontian)
    small quadrupedal ornithopod tracks cf. *Wintonopus*
  quadrupedal ornithischian (thyreophoran?) tracks

CRETACEOUS
Western Australia

Birdrong Sandstone, early Cretaceous, Hauterivian-Barremian (Kalbarri, Cardabia Station, Murchison House Station) (Long and Cruikshank 1996, Cruikshank and Long 1997, Long 1998, Choo 1999)
  pliosauroid *Leptocleidus clemai*
  ichthyosaur *Platypterygius* sp.
  theropod indet. (caudal vertebra)
Broome Sandstone, early Cretaceous, Barremian-Aptian (Broome, Gantheaume Point, Prices Point) (Glauert 1952, Colbert and Merrilees 1967, Long 1990, 1993, Dayton 1991, Thulborn *et al.* 1994)
  theropods
    Megalosauropus broomensis (tracks)
    theropods indet. (tracks)
  sauropods
    large sauropod ***Brontopodus*** sp. (tracks)
    other sauropods (tracks)
  ornithopods
    small ornithopod ***Wintonopus*** sp. (tracks)
    large ornithopods (tracks)
  thyreophoran (stegosaur?) (tracks)
Molecap Greensand, late Cretaceous, ?Cenomanian-Santonian (Gingin, Dandaragan) (Teichert and Matheson 1944, Lundelius and Warne 1960, Long 1995, Long and Cruikshank 1998, Choo 1999)
  ichthyosaur
  elasmosaurid plesiosaurs
  mosasaurid lizard cf. *Platecarpus*
  theropod indet., cf. *Allosaurus* (pedal phalanx)
Alinga Formation, late Cretaceous, Cenomanian (Thyridine Point, Murchison House Station, Kalbarri) (Choo 1999)
  ichthyosaur
Miria Formation, late Cretacous, Maastrichtian (Giralia Range, Exmouth) (Bennett and Long 1991, Long 1992, 1998)
  mosasaurid lizard indet.
  azhdarchid(?) pterosaur
  tetanuran theropod (humerus)

Northern Territory

Bathurst Island Formation, early Cretaceous, Albian (Mullamann Beds; Fannie Bay, Nightcliff, Casuarina Beach, Darwin) (Teichert and Matheson 1944, Murray 1985, 1987, Long 1998)
  ichthyosaur *Platypterygius* sp.
  elasmosaurid plesiosaur

South Australia

Bulldog Shale, Maree Sub-group (formerly Maree Formation), early Cretaceous, Aptian-Albian (Andamooka, Coober Pedy, Neales River) (Persson 1960, Pledge 1980, Molnar and Pledge 1980, Vickers-Rich and Rich 1985, Wade 1990, Molnar 1991, Long 1998, Schroeder 1998)
  plesiosaurs
    pliosaurid *Leptocleidus* sp. ('Eric')
    elasmosaurid *Woolungasaurus* sp.
    other plesiosaur(s), e.g. 'Andamooka sea monster' (elasmosaurid?)
  ichthyosaur
  theropods
    ?coelurid ***Kakuru kujani*** (tibia; phalanx, fibula fragment)
    theropod indet. (pedal phalanx, possibly also *Kakuru*)
  ornithopod indet.

Victoria

Merino Group, early Cretaceous (Casterton) (Warren 1969, Gaffney 1981, 1991)
- cryptodire turtle *Chelycarapookus arcuatus*

Strzelecki Group, early Cretaceous, ?Valanginian-Aptian (except Koonwarra locality, Albian) (Wonthaggi Formation, Strzelecki Ranges; Cape Paterson, Eagles Nest, Flat Rocks, Inverloch, Koonwarra) (Woodward 1906, Talent *et al.* 1966, Waldman 1970, Molnar 1980c, Molnar *et al.* 1981, Flannery and Rich 1981, Rich 1987, 1996, Vickers-Rich *et al.* 1988, Rich and Vickers-Rich 1989, 1994, 1999, Vickers-Rich 1991, 1996, Currie *et al.* 1996, Chinsamy *et al.* 1998)
- cryptodire turtles indet.
- theropods
  - allosaurid ***Allosaurus*** sp. (astragalus)
  - theropod indet. ('Cape Paterson claw')
  - ornithomimid indet. (vertebrae)
  - ?bird indet. (Koonwarra feathers)
- hypsilophodontid ornithopods:
  - ***Leaellynasaura amicagraphica*** (skull, partial skeleton, etc.)
  - ***Atlascopcosaurus loadsi*** (jaws, teeth)
  - ?***Fulgurotherium*** sp. (femora)
  - ***Qantassaurus intrepidus*** (jaws, teeth)
  - other undescribed hypsilophodontids
- ceratopsian(?) cf. *Leptoceratops* (ulna)

Otway Group, early Cretaceous, Albian (Middle Eumeralla Formation, Otway Ranges; Knowledge Creek, Dinosaur Cove, Cape Otway, Apollo Bay, Lorne) (Flannery and Rich 1981, Rich 1987, 1996, Vickers-Rich *et al.* 1988, Rich and Vickers-Rich 1989, 1994, 1999, Gross *et al.* 1993, Vickers-Rich 1996, Currie *et al.* 1996, Chinsamy *et al.* 1997, 1998)
- turtles indet.
- plesiosaur indet.
- pterosaur indet.
- theropods
  - ornithomimid ***Timimus hermani*** (femur, pubis, vertebrae)
  - oviraptorosaurian(?) indet. (surangular, vertebra)
  - dromaeosaurid indet. (teeth)
  - small theropod tracks
- hypsilophodontid ornithopods:
  - ***Leaellynasaura amicagraphica***
  - ***Atlascopcosaurus loadsi***
  - ?***Fulgurotherium*** sp.
  - other undescribed hypsilophodontids*

New South Wales

Coreena Formation, early Cretaceous, Aptian-Albian (White Cliffs) (Etheridge 1897, 1904, Persson 1960, Molnar 1982a, 1991)
- ichthyosaur cf. *Platypterygius longmani*
- plesiosaurs
  - cimoliasaurid *Cimoliasaurus maccoyi*
  - pliosaurid *Trinacromeron leucoscopelus* (=?*Dolichorhynchops* sp.)
- theropod indet.

Griman Creek Formation, early Cretaceous, Albian (Wallangulla Sandstone Member, Lightning Ridge) (Etheridge 1917, von Huene 1932, Molnar 1980a, c, 1982a, 1991, Bartholomai and Molnar 1981, Gaffney 1981, 1991, Molnar and Galton 1986, Ritchie 1988, Molnar 1996a, Smith 1997, 1999, Long 1998)
- primitive eucryptodire turtles indet.
- pliosaurid plesiosaur cf. *Leptocleidus*
- crocodilians
  - eusuchian '*Crocodylus*' *selaslophensis*
  - other crocodilians indet.
- ornithopods
  - hypsilophodontids
    - ***Fulgurotherium australe*** (femora, scapula)
    - other hypsilophodont material (attributed to *Fulgurotherium* sp., *Atlascopcosaurus loadsi*)
  - large ornithopod ***Muttaburrasaurus*** sp. (isolated bones, teeth)
  - large ornithopod (tracks)
- sauropod indet.
- theropods
  - coelurid(?) ***Rapator ornitholestoides*** (metacarpal)
  - family indet. ***Walgettosuchus woodwardi*** (caudal vertebra)
- bird indet.

Queensland

Normanton Formation (Molnar 1991)
- large reptile indet.

Bungil Formation, early Cretaceous, Aptian (Mack Gully, Roma) (Molnar 1980b, 1985b, Molnar and Frey 1987)
- plesiosaur
- ankylosaur ***Minmi paravertebra*** (vertebrae, ribs, pes, ventral armour)

Wallumbilla Formation, early Cretaceous, Aptian-Albian (Richmond, 'Army Downs', Grampian Valley, 'Rainscourt') (Romer and Lewis 1960, Persson 1960, 1982, Molnar 1987a, 1991)
- ichthyosaur
- plesiosaurs
  - pliosaurid *Kronosaurus* sp.
  - elasmosaurid *Woolungasaurus glendowerensis*

Toolebuc Formation, early Cretaceous, Aptian-Albian (Hughenden, Boulia, Hamilton River, Nelia, Julia Creek, 'Telemon', 'Dunluce', 'Sylvania', 'Dunraven') (Owen 1882, Longman 1915, 1924, 1930, Molnar and Thulborn 1980, Gaffney 1981, 1991, Molnar 1985a, 1986, 1987b, 1991, 1994, 1996b, Wade 1984, 1990, Vickers-Rich and Rich 1993, Kurochkin and Molnar 1997)
- desmatochelyid turtles
  - *Notochelone costata*
  - *Cratochelone berneyi*
- ichthyosaur *Platypterygius longmani*
- pliosaurid plesiosaur *Kronosaurus queenslandicus*
- crocodilian indet.
- ankylosaur ***Minmi*** sp.
- pteranodontid pterosaurs

?*Ornithocheirus* sp.
?*Pteranodon* sp.
enantiornithine birds
Nanantius eos (tibiotarsus; cervical vertebra?)
Nanantius sp. (tibiotarsus)

Allaru Mudstone, early Cretaceous, Albian (Maxwelton, Richmond, 'Marathon', 'Dunluce', Hughenden, Flinders River, Barcaldine) (Longman 1933, Molnar and Thulborn 1980, Gaffney 1981, Molnar 1982a, 1991, 1994, 1996a, b, Wade 1984, 1990, Thulborn 1987, Hunt *et al.* 1994)
turtle ?*Cratochelone berneyi*
ichthyosaur *Platypterygius longmani*
pliosaurid plesiosaur (complete skeleton)
large ornithopod ***Muttaburrasaurus*** sp. (good skull)
titanosaurid? sauropod ***Austrosaurus mckillopi*** (vertebrae, rib fragments)
ankylosaur ***Minmi*** sp. (mostly complete skeleton)
anhanguerid pterosaur cf. *Anhanguera*

Mackunda Formation, early Cretaceous, Albian (Muttaburra) (Bartholomai and Molnar 1981, Molnar 1991, 1996a)
plesiosaur
pliosaur cf. *Trinacromerum*
large ornithopod ***Muttaburrasaurus langdoni*** (partial skeleton)

unknown horizon, early Cretaceous, Albian (Hughenden) (Molnar 1982a, 1991)
large sauropod cf. *Brachiosaurus* (cervical vertebra; possibly *Austrosaurus*)

Griman Creek Formation, early Cretaceous, Albian (Molnar 1991)
plesiosaur
?dinosaur

Winton Formation, late Cretaceous, Cenomanian (Winton, Lark Quarry, Isisford) (Thulborn and Wade 1979, 1984, Coombs and Molnar 1981, Molnar 1982a, 1991, Molnar and Willis 1996)
turtle indet.
crocodilian cf. *Bernissartia*
theropods
large theropod ***Tyrannosauropus*** sp. (tracks)
small theropod ***Skartopus australis*** (tracks)
sauropod ?***Austrosaurus*** sp. (limb bones, shoulder girdle, vertebrae)
ornithopods
small ornithopod ***Wintonopus latomorum*** (tracks)
large ornithopods (tracks, braincase fragments)

New Zealand

Makirikiri Formation, early Cretaceous, Motuan (Tinui, North Island) (Fleming *et al.* 1971)
ichthyosaurs indet.

unknown horizon, late(?) Cretaceous (Dargaville, North Auckland, North Island) (Grant-Mackie in Fordyce 1991)
ichthyosaur indet.

Maungataniwha Sandstone, late Cretaceous, Piripauan-Haumurian (=Campanian-Maastrichtian) (Mangahouanga Stream, Hawke's Bay, North Island) (Molnar 1980c, 1981, Wiffen 1981a, b, 1986, Molnar and Scarlett 1984, Wiffen and Moisley 1986, Wright 1989, Wiffen and Molnar 1988, 1989, Molnar and Wiffen 1994, Wiffen 1996, Hiller *et al.* 1999, Long 1998)
cryptodire (desmatochelyid?) turtle
plesiosaurs
elasmosaurids
Tuarangisaurus keyesi
Mauisaurus haasti ?
pliosauroids indet.
mosasaurid lizard *Moanasaurus* (?=*Mosasaurus*) *mangahouangae*
theropods
2 or 3 spp. indet. (vertebrae, phalanges)
sauropod indet. (rib)
ornithopod cf. *Dryosaurus* (ilium)
ankylosaur (nodosaurid?) indet. (vertebrae, rib)
pterosaur indet. (ulna)

Laidmore Formation, late Cretaceous, Haumurian (=Maastrichtian) (Waipara River, Cheviot, Haumuri Bluff, South Island) (Owen 1861, Clark 1861, Hood 1870, Haast 1870, Hector 1870, 1874, McKay 1877, Hutton 1894, Persson 1963, Welles and Gregg 1971, Warren and Spedden 1977)
plesiosaurs
pliosauroids indet.
elasmosaurid *Mauisaurus haasti*
elasmosaurid indet.
polycotylid indet.
mosasaurid lizards
Tylosaurus haumuriensis
Taniwhasaurus oweni
Prognathodon waipuraensis
Mosasaurus mokoroa
mosasaur indet.
bird (femur)

Conway Formation late Cretaceous, Haumurian (North Canterbury, South Island) (Hiller *et al.* 1999)
elasmosaurid plesiosaur *Mauisaurus haasti*

Katiki Formation (Shag Point, North Otago, South Island) (Fordyce 1983, 1991, Cruikshank and Fordyce 1996, 1998, Cruikshank *et al.* 1999)
plesiosaurs
elasmosaurid indet.
cryptoclidid cf. *Cryptoclidus*
mosasaurid lizard indet.

'*Lingula* bed', late Cretaceous, Maastrichtian (Otarama, Waimakariri River, South Island) (Fordyce 1991)
elasmosaurid plesiosaurs indet.
mosasaurid lizards indet.

---

*Additional material of hypsilophodontids, other ornithopods, and theropods, recently collected from Lightning Ridge, some privately owned and still undescribed, is held at the Australian Museum (Ritchie 1988, Long 1998, Smith 1999).

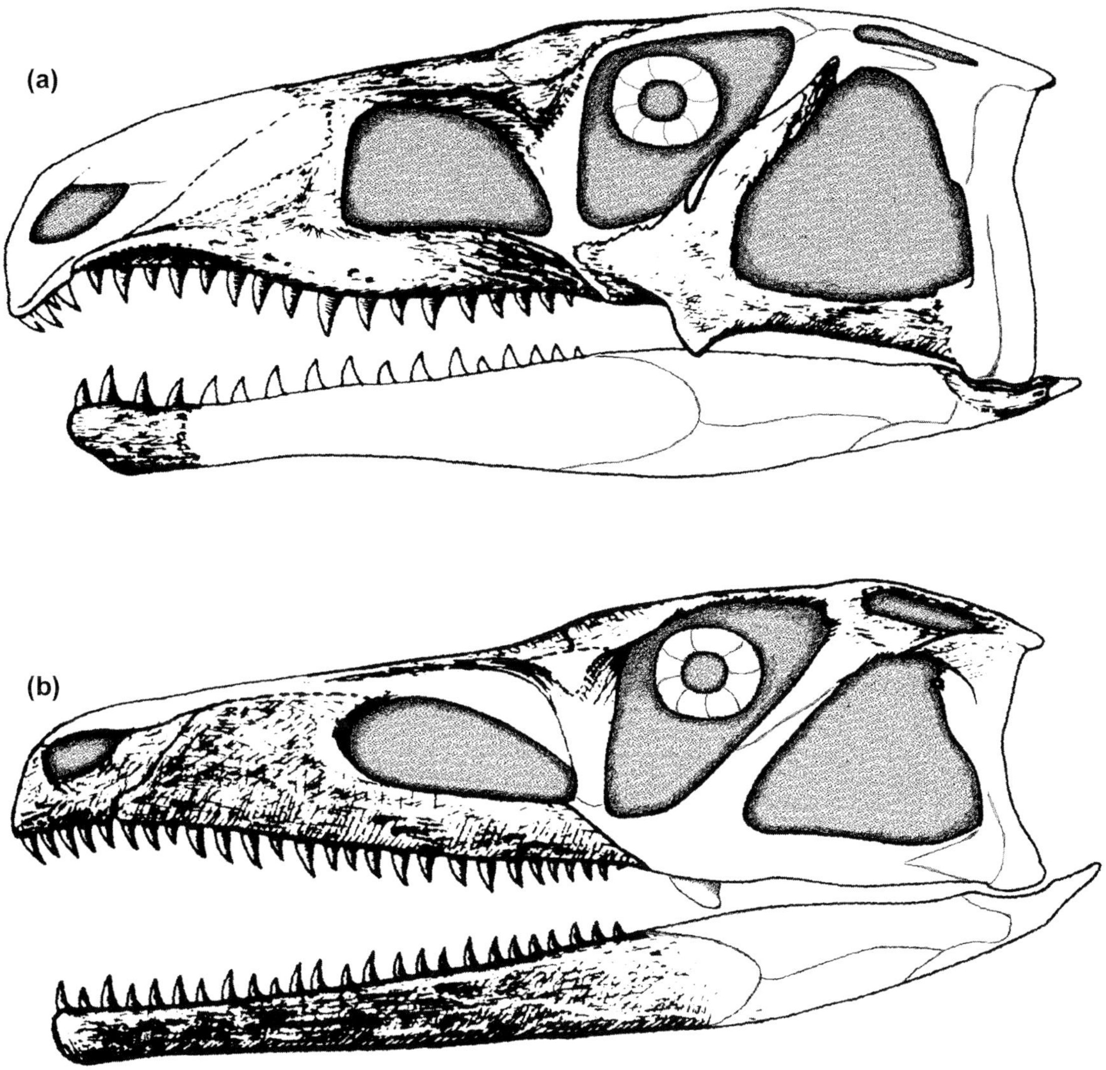

Figure 2. Reconstructed skulls of the two known Australian proterosuchians. **(a)** *Kalisuchus rewanensis*; **(b)** *Tasmaniosaurus triassicus.* Modified from Thulborn (1986a) and Cruikshank (1972).

DINOSAURIA

Another one of the terrestrial archosaurian lineages that appeared during the Triassic was the dinosaurs. These were characterized by: bipedal posture (with elongate hindlimb, strengthened sacrum, shortened forelimb, and numerous associated changes in details of the pelvis, limb bones and joints); features of the posterior skull roof (loss of the postfrontal bone, near-closure of the post-temporal opening) and palate (Sereno *et al.*, 1993). In other respects they were typical archosaurs, with a total length of about 1 m (already much larger than most reptiles), a relatively long neck, the teeth compressed, serrated and recurved but heterodont (varying in shape as well as size along the tooth row), a large antorbital fossa in the side of the snout, and the pubic bones projecting forward and downward from the hip joint (primitive 'saurischian' condition, as in squamates and crocodiles). The earliest known dinosaur fossils, from early in the late Triassic of Argentina (mid-Carnian, about 230 m.y.a.) show that the divergence between (what were to become) three major lineages had already occurred (Rogers *et al.*, 1993). *Eoraptor* and *Herrerasaurus* were basal theropods, becoming more specialized as bipedal predators, while the jaws and dentition of *Pisanosaurus* show characteristics of the herbivorous ornithischians. The first sauropodomorphs, also herbivorous (Galton, 1985) but retaining the saurischian hip structure, had also originated by this time, and are represented by *Saturnalia* in similar-aged deposits in Brazil (Langer *et al.*, 1999) and *Isanosaurus* in Thailand (Buffetaut *et al.*, 2000). Over the next 15 million years, dinosaurs increased in diversity but remained subordinate in size and abundance to other amniote lineages. Only 'by accident' (Sereno, 1999), with the extinction of most of the remaining therapsids, rhynchosaurs and other archosauromorphs, did dinosaurs

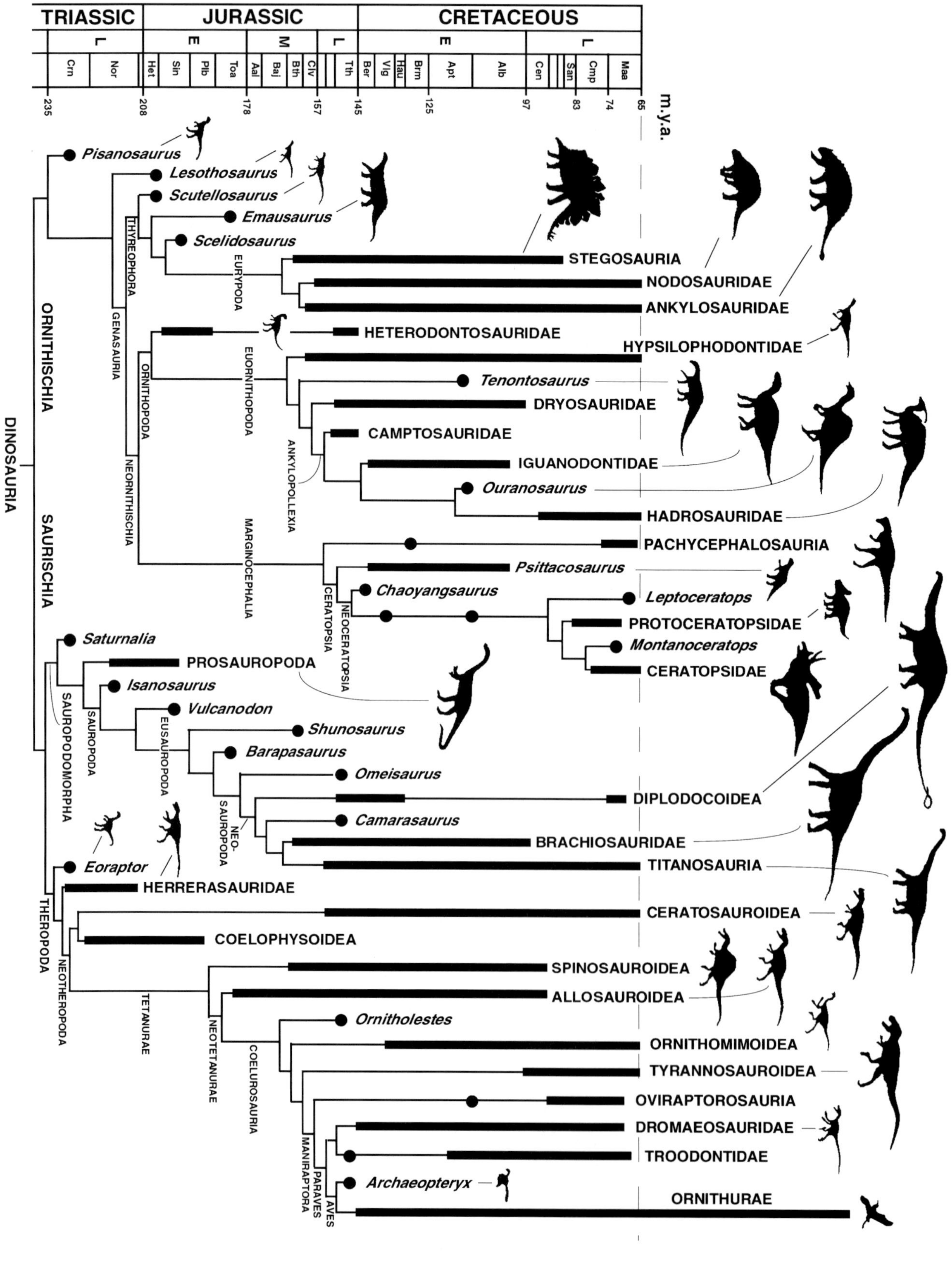

Figure 3. Phylogenetic relationships and stratigraphic ranges of dinosaurian lineages. Modified from Sereno (1999).

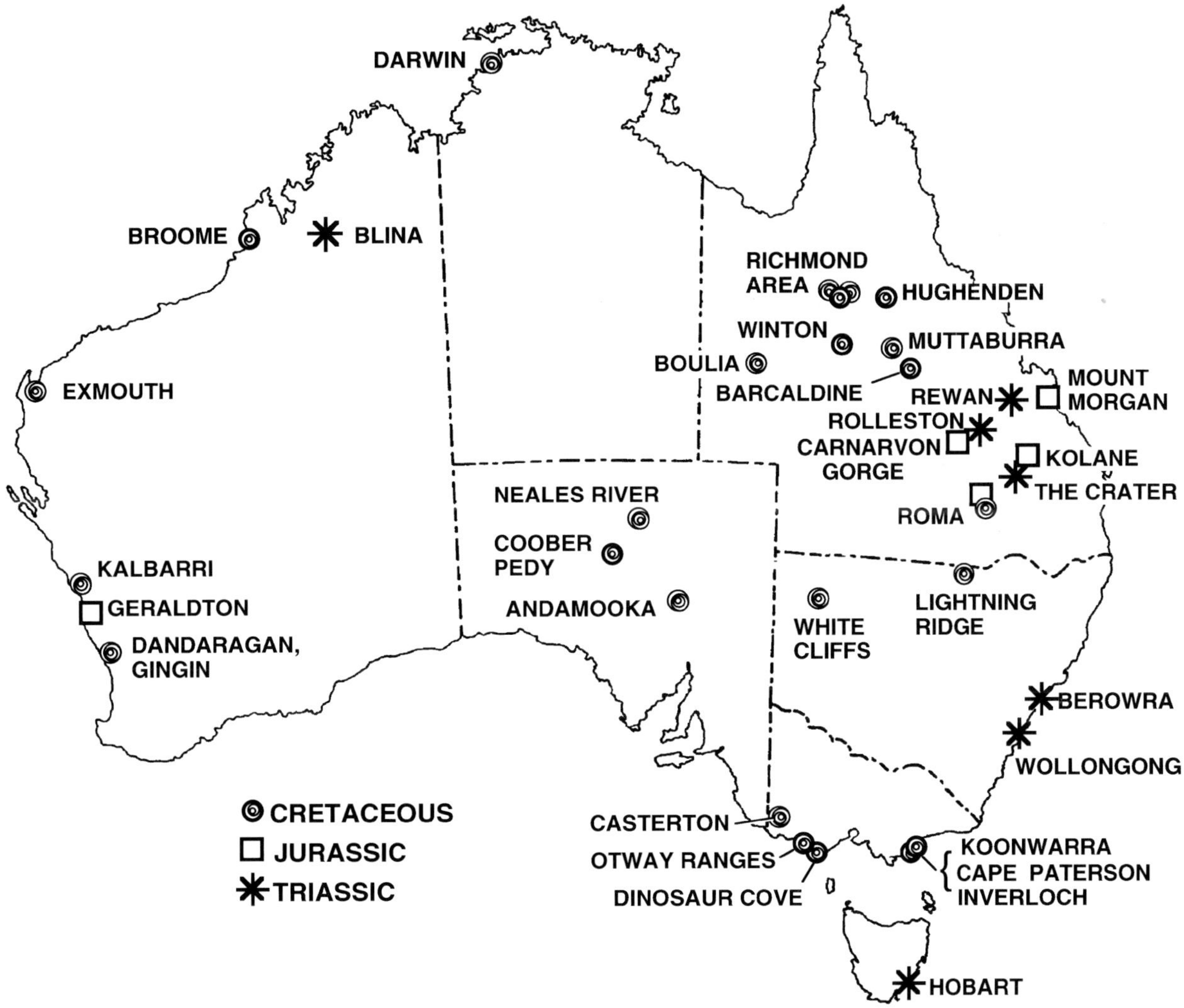

Figure 4. Map showing Mesozoic reptile localities in Australia with sites of Triassic, Jurassic and Cretaceous age distinguished. Based on Molnar (1991) and Long (1998).

come to dominate terrestrial faunas towards the end of the Triassic (mid-Norian). For summaries of dinosaur relationships and Mesozoic reptile localities see Figures 3 and 4.

THEROPODS: BIPEDAL PREDATORS

Specializations that appeared in early theropods (*Herrerasaurus* and/or *Eoraptor*) include: hollowing of the long bones and vertebral centra to lighten the skeleton; reduction of the outer two digits of the hand, and enlargement of the inner three digits for grasping prey; presence of a sliding joint between the front and back halves of the lower jaw (similar to the intramandibular joint of varanoid lizards and snakes), probably an adaptation for seizing and swallowing large prey; and stiffening of the posterior part of the tail by overlapping vertebral processes (Sereno and Novas, 1992; Sereno *et al.*, 1993).

The old classification of theropods as large 'carnosaurs' and small 'coelurosaurs' is now obsolete. Above the primitive grade of herrerasaurids, the neotheropods comprise two lineages that persisted to the end of the Cretaceous: ceratosaurians (coelophysoids and ceratosauroids), and tetanurans (including allosauroids and coelurosaurians). The coelurosaurians dominated both small and large carnivore niches in the Cretaceous, including various groups best known from Mongolia and North America such as dromaeosaurs, ornithomimids, oviraptorids and tyrannosaurids, and also gave rise to birds before the end of the Jurassic (Sereno, 1999). Unfortunately, few of the Australasian theropods can yet be assigned their precise places in the phylogeny, but the presence of several coelurosaurian lineages (see below) shows that no major groups were excluded by biogeographic barriers.

The oldest records of theropods (or any dinosaurs) in Australia are trackways from the Blackstone Formation (late Triassic of southern Queensland), and unnamed Rhaetian beds in the Callide Basin (Figure 4). Some of these are a large form comparable to the ichnogenus *Eubrontes* (common in the early Jurassic of North America), estimated to have been made by a theropod about 6 m long. The early Jurassic Razorback Beds around Mount Morgan (Qld.), have also produced tracks attributed to large theropods. Remarkably, these are said to preserve impressions of five-toed 'hands' as well as three-toed hind feet (Bartholomai, 1966a; Molnar, 1991); this is unusual for these generally bipedal animals, and also because the outer digits of the hand are vestigial in even the most primitive known theropods (Sereno *et al.*, 1993). See Figure 5.

From the middle Jurassic, *Ozraptor subotai* has been described based on the distal end of a tibia from the Colalura Sandstone near Geraldton, W.A. (Long and Molnar, 1998). This is a fairly informative element, preserving information about the shape of the fibula, astragalus and calcaneum, which vary significantly in theropods (Welles and Long, 1974), but does not indicate a close relationship between *Ozraptor* and any other known theropod. Thus, it suggests this lineage had already been evolving independently in Gondwana for some time. *Ozraptor* would have been about 2 m in total length.

No other skeletal remains are yet known from the Jurassic, but trackways of several kinds of theropods appear in the middle Jurassic Walloon Group in southern Queensland, impressed into coal seams (Bartholomai, 1966a; Hill *et al.*, 1966; Haubold, 1971; Molnar, 1991). These include: a large form like the North American track-maker *Eubrontes* (from a dinosaur up to about 12 m long, one of the largest Jurassic predators); another distinctive large theropod track attributed to *Changpeipus bartholomaii*—similar to a Chinese footprint fossil of the same age (Haubold, 1971); and other footprints about 12 cm long, from a theropod smaller than *Ozraptor*. The large forms might be from animals related to the Antarctic allosauroid *Cryolophosaurus* of about the same age (Hammer and Hickerson, 1994, 1998). See Figure 5.

Australasian Cretaceous deposits have produced skeletal material of quite a few theropods, but in almost all cases they are represented by single elements. The earliest is a small caudal vertebra from the marine Birdrong Sandstone near Kalbarri, W.A. (Hauterivian–Barremian), but the affinities of the animal, or even whether it is definitely a theropod, are uncertain. Slightly younger are the trackway sites around Broome (Barremian–Aptian), where the diverse (ichno)fauna included several theropod trackmakers (Thulborn *et al.*, 1994); the only track yet formally described is *Megalosauropus broomensis* (Colbert and Merrilees, 1967), from an animal that probably reached about 9 m in length (Thulborn *et al.*, 1994; Long, 1998).

Most of the Cretaceous theropods come from the more extensive Aptian–Albian deposits. The first of these specimens to be described was a claw from Cape Paterson (Strzelecki Group, Vic; Woodward, 1906), and attempts in the late 1970s to relocate the site of its collection led to the discovery of extensive coastal exposures of fossils in the Strzelecki and Otway Ranges east and west of Melbourne (Flannery and Rich, 1981; Rich and Vickers-Rich, 1989). The claw itself is about 25 mm long, and was thought by Woodward to come from a dinosaur similar to *Megalosaurus*, but does not provide much information. The first more diagnostic specimen, from less than 200 m away, is an astragalus—an ankle bone, fitting a characteristically shaped recess on the tibia (Welles and Long, 1974) nearly identical to that of *Allosaurus*; it is a little smaller but more robust than the well-known *A. fragilis* from the Jurassic of North America (Molnar *et al.*, 1981). Other identifiable material—still only fragments or isolated elements—has come mostly from Dinosaur Cove in the slightly younger Otway Group, and represents a diverse fauna of smaller theropods. The remains include: an ornithomimid *Timimus hermani* (a complete femur; also a pubis and vertebrae possibly of the same taxon); the distinctive teeth of a small dromaeosaurid (a relative of *Velociraptor*); a peculiar jaw bone (probable surangular) and a vertebra, which appear to belong to a primitive oviraptorosaur (though also similar to dromaeosaurids; Currie *et al.*, 1996).

Ornithomimids, cursorial ostrich-like dinosaurs, are now thought to have been mainly herbivorous (Kobayashi *et al.*, 1999); *Timimus* was also about ostrich-sized (about 2.5 m long including tail), though other ornithomimids reached about 4 m in length. While the *Allosaurus* is the last of its genus and considered a relict, the ornithomimosaur and slightly smaller oviraptorosaur are among the earliest known members of their groups, suggesting a possible Gondwanan origin. However, all four Victorian theropod lineages presumably diverged during the Jurassic (they form successive outgroups to birds, so must be older than *Archaeopteryx*) and could have dispersed widely before Laurasia and Gondwana separated, wherever they originated (Sereno, 1999).

Also in the Strzelecki Group, but slightly younger, fossil feathers at the lacustrine Koonwarra site indicate the presence of birds (Talent *et al.*, 1966; Waldman, 1970; Rogers, 1987; Vickers-Rich, 1991). They appear to differ morphologically from feathers of any modern birds, and it is now known that feather-like structures, or even true feathers, were present more widely in tetanuran theropods and can no longer be regarded as strictly diagnostic of birds (usually defined as the clade including *Archaeopteryx* and modern birds, and all descendants of their last common ancestor; Sereno, 1999). However, the small size of the Koonwarra specimens is consistent with birds such as the enantiornithine *Nanantius* from Queensland, which was about the size of a European blackbird (Molnar, 1986) and smaller than any known non-avian theropod.

Of similar age to the Rift Valley deposits in Victoria are the marine sediments of S.A., N.S.W. and Qld.

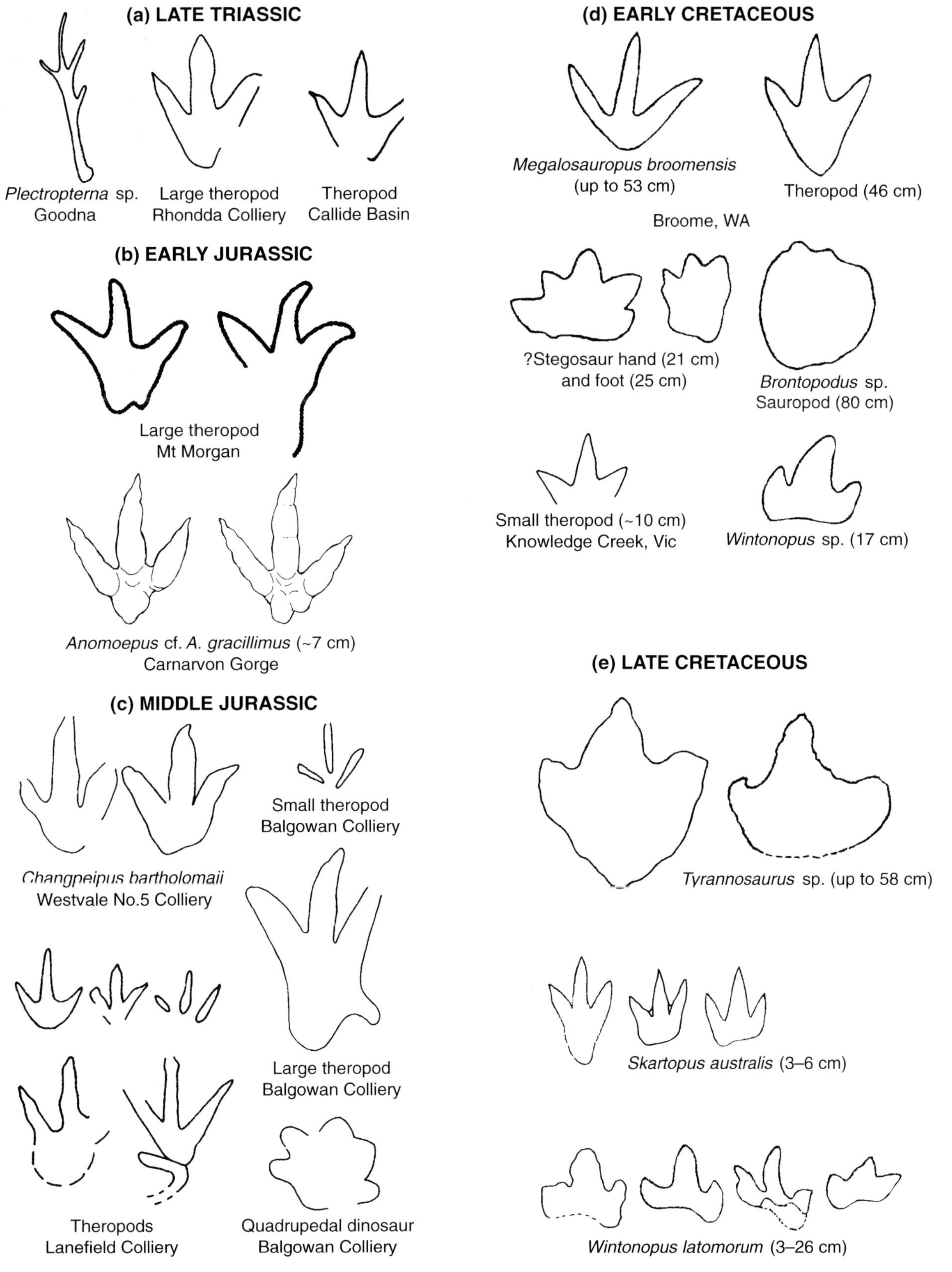

Figure 5. Dinosaur footprints from the Triassic, Jurassic, and Cretaceous of Australia: **(a)** late Triassic (south-eastern Queensland); **(b)** early Jurassic (south-eastern Queensland); **(c)** middle Jurassic (south-eastern Queensland); **(d)** early Cretaceous (Western Australia and Victoria); **(e)** late Cretaceous (north-western Queensland). Based on Thulborn and Wade (1984), Molnar (1991), Thulborn (1994), Thulborn *et al.* (1994) and Long (1998).

Theropod remains occur in some of these deposits near the palaeoshorelines, and *Rapator ornitholestoides*, *Walgettosuchus woodwardi* (both from Lightning Ridge, N.S.W.; von Huene, 1932), and *Kakuru kujani* (Andamooka, S.A.; Molnar and Pledge, 1980) have been formally described. *Rapator* resembles the small North American 'coelurid' (basal coelurosaurian) *Ornitholestes* in the shape of its first metacarpal and may have been related to it, but would have been about 9 m in length; other fragments from Lightning Ridge, such as vertebral centra, may also belong to this form. *Walgettosuchus*, based on a poorly preserved caudal vertebra, could represent an allosaurid, coelurid, or ornithomimid, so it is doubtful whether it is actually a distinct or recognizable taxon. *Kakuru*, known from a distal tibia and possibly a few other opalized fragments (such as a metatarsal from Coober Pedy), was a slender beast about 2–3 m in length, similar to a Mongolian theropod (*Avimimus*) in proportions but apparently not closely related to it, or any of the theropods whose ankle bones were studied by Welles and Long (1974).

The extensive marine deposits of Queensland have produced surprisingly little theropod material, only a few tiny specimens obtained by acid-processing of Toolebuc limestone from near the Hamilton River, on the western edge of the basin. These include tibiotarsi of the oldest enantiornithine birds, *Nanantius eos* and *Nanantius* sp., and a cervical vertebra possibly from *N. eos* (Molnar, 1986, 1991; Kurochkin and Molnar, 1997). In the younger (Cenomanian) Winton Formation, the famous Lark Quarry trackway site records the activities of two kinds of theropods. One large and heavily built individual (about 9–10 m long) whose tracks are classified as *Tyrannosauropus* sp. (possibly an abelisaur, giant allosaur or some other large tetanuran predator, not necessarily a tyrannosaurid) apparently caused a stampede by a mixed-age flock of small coelurosaurs (ichnospecies *Skartopus australis*), no more than 22 cm tall at the hip, running the other way in company with somewhat larger ornithopods (Thulborn and Wade, 1979, 1984).

A theropod roughly 4 m long, and comparable to *Allosaurus*, is represented by a toe bone from the Molecap Greensand near Gingin, W.A. (Cenomanian or Santonian; Long, 1995). The youngest theropod specimen from the mainland is a partial humerus from the Miria Formation (late Maastrichtian) in the Giralia Range, W.A., which would have been about 35 cm long when complete. In describing it, Long (1992) noted similarities to various theropod lineages including ornithomimosaurs, oviraptorids and dromaeosaurs, but then excluded them from consideration on biogeographic grounds; so it is ironic that these same three groups have since been reported from much older deposits in Victoria. Relative size of the humerus varies greatly among theropod lineages, so the specimen could represent either a medium-sized dromaeosaur (with the long arms characteristic of this group) or a larger member of another tetanuran lineage with smaller arms.

At about the same time (Campanian–Maastrichtian) theropods are represented by isolated vertebrae and phalanges in the Mangahouanga Stream deposits in the North Island of New Zealand. These indicate at least 2 or 3 species ranging in size up to that of *Allosaurus* (Wiffen, 1996), but their relationships are uncertain due to the lack, so far, of more diagnostic elements. See Figure 6.

SAUROPODOMORPHS: LONG-NECKED GIANTS

The oldest known sauropodomorph, *Saturnalia* from the Carnian (late Triassic) of Brazil, was a gracile herbivore about 1.5 m long (Langer *et al.*, 1999) but a few million years later (Norian) the semi-bipedal prosauropods, such as *Plateosaurus,* were the first dinosaurs to attain large body size (about 8 m). *Agrosaurus macgillivrayi* was the first dinosaur described as coming from Australia; it is represented by a tibia, a claw and some other fragments, and has been regarded as Australia's only known Triassic dinosaur. The Triassic age and prosauropod nature of the material can be accepted, but not the presumed locality; based on the chemical composition of the matrix, the specimens were almost certainly collected in England, and are probably referable to *Thecodontosaurus antiquus* (Vickers-Rich *et al.*, 1999). Thus, there is no real evidence that prosauropods existed here, although this may be an artefact of the few bone-bearing deposits from the late Triassic and early Jurassic known in Australia (Figure 4).

Sauropods, obligate quadrupeds with elephantine limbs, evidently form the sister group to Prosauropoda rather than their descendants, since prosauropods have a number of synapomorphies including the twisted shape of the thumb, with an enlarged claw that points inward (Upchurch, 1995; Sereno, 1999). The latter feature also occurs in some theropods and one ornithischian, and has sometimes been considered a dinosaurian synapomorphy reversed several times, rather than a multiple convergence (Bakker and Galton, 1974). The absence of recorded sauropods until well into the early Jurassic thus meant they were a 'ghost lineage' until the recent description of the first Triassic sauropod, *Isanosaurus*, from the Rhaetian of Thailand (Buffetaut *et al.,* 2000), and the even older (Norian) *Antetonitrus* from South Africa (Yates and Kitching, 2003). Sauropods remain rare until the later part of the middle Jurassic (Hunt *et al.*, 1994), so that the single partial skeleton of *Rhoetosaurus brownei* from the mid-Jurassic Hutton Sandstone near Roma, Qld. (Longman, 1927; Thulborn, 1985a; Molnar, 1991; Long, 1998) still ranks as one of the earliest known. Although the skull and forelimbs were not found, it is still one of the most complete dinosaur fossils discovered in Australia, and represents a primitive grade of eusauropods (referred to Cetiosauridae by Hunt *et al.*, 1994), however it should be remembered that this group is poorly defined and likely paraphyletic (Upchurch,

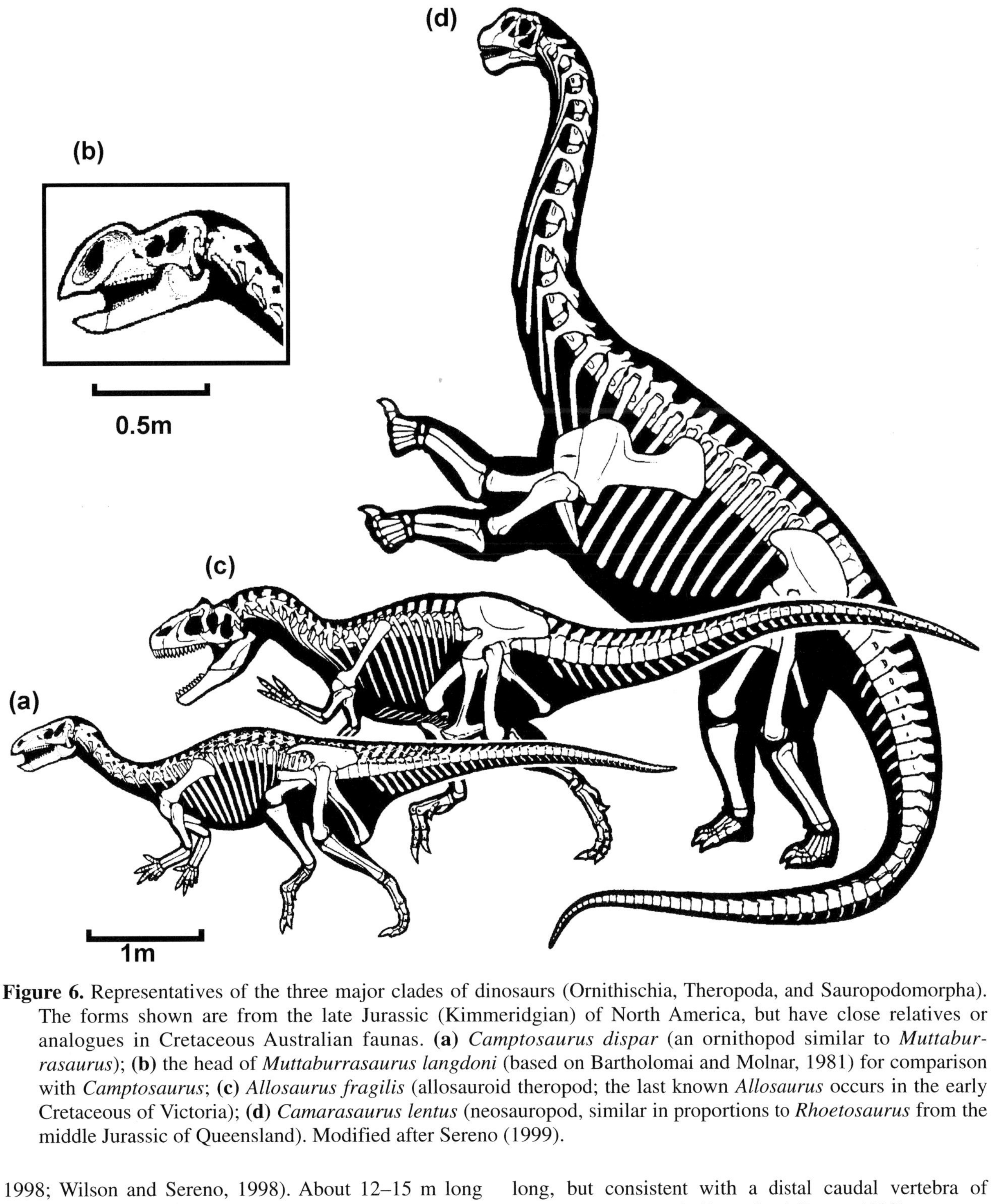

Figure 6. Representatives of the three major clades of dinosaurs (Ornithischia, Theropoda, and Sauropodomorpha). The forms shown are from the late Jurassic (Kimmeridgian) of North America, but have close relatives or analogues in Cretaceous Australian faunas. **(a)** *Camptosaurus dispar* (an ornithopod similar to *Muttaburrasaurus*); **(b)** the head of *Muttaburrasaurus langdoni* (based on Bartholomai and Molnar, 1981) for comparison with *Camptosaurus*; **(c)** *Allosaurus fragilis* (allosauroid theropod; the last known *Allosaurus* occurs in the early Cretaceous of Victoria); **(d)** *Camarasaurus lentus* (neosauropod, similar in proportions to *Rhoetosaurus* from the middle Jurassic of Queensland). Modified after Sereno (1999).

1998; Wilson and Sereno, 1998). About 12–15 m long and weighing something like 20 tonnes, *R. brownei* was massively built and probably resembled *Camarasaurus* in its proportions, lacking the extreme elongation of neck and/or tail seen in many advanced forms like *Diplodocus*, *Brachiosaurus* or *Mamenchisaurus*.

Of about the same age (Bajocian, middle Jurassic) is a vertebral centrum from Bringo Cutting near Geraldton, W.A. (Colalura Sandstone); this specimen is only 63 mm long, but consistent with a distal caudal vertebra of *Rhoetosaurus* or a similar animal (Long, 1992).

The next evidence of sauropods does not appear until the early Cretaceous, where they are represented in the great Barremian–Aptian trackway sites near Broome, W.A. (Thulborn *et al.*, 1994). At least two species of sauropods appear to be recorded in these heavily trafficked sediments, with some of the individual footprints (up to 1.5 m diameter) indicating truly enormous animals walking near a coastal lagoon. The shape and arrangement of hind and fore foot impressions allows the largest form to be included in the ichnogenus *Brontopodus*, but the number of track making species, and much else about their behaviour and interactions, remains to be revealed by further study.

More tantalizing skeletal remains of sauropods come from the Albian (and possibly Cenomanian) of N.S.W. and Qld. Fragments of sauropod bone are known from the opal fields of Lightning Ridge (Molnar, 1991)—part of the Griman Creek Formation. Some associated dorsal vertebrae and ribs from the Allaru Mudstone (near Maxwelton, Qld.), were described as *Austrosaurus mckillopi* by Longman (1933). These vertebrae seem remarkably primitive for a Cretaceous sauropod, the broad centra composed of spongy bone within a thin cortex, as in *Rhoetosaurus* and other mid-Jurassic forms, and with only rudimentary pleurocoels (the weight-saving hollows between buttresses that become increasingly complex in advanced sauropod lineages). Somewhat younger remains from near Winton (Winton Formation, Qld.), have also been referred to *Austrosaurus* sp. and include massive but elongate limb bones, girdle elements, and simple caudal vertebrae corresponding to the dorsals in the type material (Coombs and Molnar, 1981). These remains are too incomplete to reconstruct either the body proportions or relationships of *Austrosaurus* with confidence, but it could have been about 15 m long and may be related either to the earlier cetiosaurs, or to the titanosaurids which were the dominant Cretaceous sauropods of other Gondwanan lands.

One other specimen, suggesting that a more advanced sauropod coexisted with this seemingly archaic form, is a partial cervical vertebra from near Hughenden; this is probably from the Toolebuc Formation or Allaru Mudstone and hence most likely Albian (Molnar, 1991). It possessed well-developed pleurocoels, defined by a complex arrangement of crests on either side of the centrum and neural arch, and in these details closely resembles an African species of *Brachiosaurus*. If truly a brachiosaur, it would have been about 20 m long. However, the possibility cannot yet be ruled out that it represents the neck of *Austrosaurus* (despite the great difference in 'style' of vertebrae), because no cervicals are associated with the material of that form.

ORNITHISCHIA: 'BIRD-HIPPED', BEAKED HERBIVORES

Appearing at the same time as the earliest known theropods, the most basal known ornithischian *Pisanosaurus* is poorly known but already shows specializations for herbivory such as a high coronoid process and tooth-to-tooth wear facets (Bonaparte, 1976; Sereno *et al.*, 1993). By the beginning of the Jurassic, ornithischians such as *Lesothosaurus* (=*Fabrosaurus* of Thulborn, 1992) had evolved a cropping beak formed by the horn-covered premaxilla (toothless in most advanced forms) and a unique new bone (the predentary) at the front of the lower jaw, as well as fleshy cheeks enclosing a buccal cavity lateral to the tooth row. With the development of cheeks, and shift of jaw adductor muscles for chewing and grinding, ornithischians tend to reduce or lose the antorbital fenestrae, where major adductors attach in theropods and other archosaurs. They retained a bipedal stance, small body size, and five-clawed hands, but the body cavity was elongated backward between the hindlimbs by rotation of the pubic bones, making room for a longer herbivorous gut (rotation of the pubis occurred independently, and presumably for other reasons, in theropods on the bird lineage). By early Jurassic times, three main ornithischian lineages had diverged: Thyreophora, Ornithopoda and Marginocephalia (Sereno, 1999).

Thyreophora: Armoured Quadrupeds

Beginning with early Jurassic forms, such as *Scutellosaurus* and *Scelidosaurus*, this lineage made the transition back to a quadrupedal way of life. They retained relatively short forelimbs from their bipedal ancestors, and except for a few basal genera their necks were also quite short, so they were restricted to feeding mainly or exclusively on low-growing plants. These habits would tend to increase their vulnerability to predators, but were countered by an early investment in bony dermal armour, such as parasagittal rows of plates and spines. It has also been suggested that stegosaurs could adopt a bipedal stance for feeding and/or defence (Bakker, 1986). Several later lineages also evolved large body size (to 10 m) and lethal defensive weapons, such as massive knobs or long horn-covered spikes on the end of the heavily muscled tail (Greene, 1988). Stegosaurs, and two main groups of ankylosaurs (nodosaurids and ankylosaurids) appeared by mid-Jurassic; stegosaurs were especially common in the late Jurassic but had disappeared from most areas during the early Cretaceous, while the ankylosaurian groups flourished throughout the Cretaceous.

The first hint of thyreophorans in this region is a single broad footprint in the middle Jurassic Walloon Group at Balgowan Colliery, south-east Qld. (Hill *et al.*, 1966; Molnar, 1991), which seems to show the outlines of five short, hoof-like toes. All the saurischians and ornithopods have fewer functional digits, and most groups are ruled out in any case by their longer digits and claws, so it seems likely that the track was made by a thyreophoran manus—a neoceratopsian could make a similar track, but they almost certainly did not evolve till much later (Lockley, 1991; Sereno, 1999). This could represent a relict of the early Jurassic scelidosaurs, an early stegosaur, or a primitive ankylosaurian.

Other traces, and body fossils, of thyreophorans are found in Australia's early Cretaceous. One of the numerous dinosaur species that left their marks in the Broome Sandstone was a thyreophoran, most likely a stegosaur, with broad five-toed hands and narrower hind feet with three functional toes (Long, 1990, 1998). However, this is the only evidence currently available for stegosaurs in

Australia, and like the rest of the Broome trackways it is still incompletely described (Thulborn *et al.*, 1994; Long, 1998). Few Cretaceous stegosaurs are known anywhere, but *Dravidosaurus* survived into the late Cretaceous (Coniacian) in India (Yadagiri and Ayyasami, 1979) and, given the sparse fossil record here, it is possible that they also had a richer or longer history in our part of Gondwana.

The most completely known of all Australasian dinosaurs is an ankylosaur. *Minmi paravertebra* was discovered in the Aptian Bungil Formation near Roma, Qld. (Molnar, 1980b, 1985b; Molnar and Frey, 1987), the type material consisting of partly articulated portions of the postcranial skeleton indicating a relatively small, lightly armoured ankylosaur of uncertain affinities. A number of similar partial specimens or fragments (possibly all *Minmi*) were later recovered from the Toolebuc Formation (Aptian–Albian) and Allaru Mudstone (Albian) and, in 1990, the most complete skeleton of any Australasian dinosaur was collected from the Allaru on 'Marathon' Station near Richmond; this is embedded in a limestone nodule which provides excellent preservation and can be dissolved with acid (Molnar, 1991, 1996b). Except for one that may be a distinct species, these specimens are all similar in size (2.5–3.5 m) and hence probably mature despite the lack of closure of many cranial sutures (fused cranial bones being usual in ankylosaurids and nodosaurids). They come from marine deposits near the eastern palaeoshorelines of the Eromanga Sea, so *Minmi* apparently inhabited the eastern Queensland land mass (an island in the Aptian, but joined to south-eastern Australia during the Albian); readers are referred to pp.17–50 for a summary of changes in Mesozoic climates and sea levels in Australia (White, 2006).

Possibly the relatively small size of *Minmi* and its light armour reflect a dwarfing and release from predatory pressure, as part of an island fauna (Molnar, 1996b); but a much better fossil record would be needed to document such a process. The taphonomy of the near-complete specimen is interesting: the articulated specimen was found on its back and details, such as the way the dermal ossicles have sunken in between the ribs, indicate that it was 'mummified' before being washed out to sea and sinking to the bottom (Molnar, 1996b). This suggests that it died, either on a floodplain or a beach, without being preyed upon or significantly scavenged, and that an intense dry spell was followed by flooding (by rain, storm, or tide). Oddly, quite a few ankylosaurian skeletons around the world have been found preserved in the same way, upside down in marine sediments. Molnar (1996b) mentions another Queensland skeleton discovered more recently (probably also *Minmi*) but not yet described.

The domed, rather than flat, snout of *Minmi* is characteristic of ankylosaurids, and it appears to be a primitive member of that lineage rather than a nodosaurid or more basal ankylosaurian (Molnar, 1996b; Sereno, 1999). Rather than a continuous armour of heavy bone plates on the back as in many later ankylosaurids and nodosaurids, *Minmi* had dorsal rows of scutes a few centimetres across, surrounded by a 'pavement' of tiny ossicles embedded in the skin, and also forming flexible 'chainmail' on the belly and limbs. There were also enlarged, spine-bearing scutes in rows across the neck and shoulders, as well as from the hips along both sides of the tail; but *Minmi* apparently lacked the kind of weapons that other ankylosaurians used for offensive defence—such as horns on the skull, long shoulder spikes, or a tail-club. The most distinctive feature was the presence of obliquely overlapping bony rods (paravertebrae) embedded in the dorsal muscles beside the vertebrae. These may have compensated for reduced dermal armour in stiffening the trunk, providing support for the great weight of guts needed to digest vegetation only coarsely chopped by the turtle-like beak and small narrow teeth. Traces of gut contents, fossilized as a 'colo-lite', within the 'Marathon' specimen (Molnar and Clifford, 2000) indicate a diet of soft vegetation including foliage, small stems and fruiting bodies.

Ankylosaur fragments are also reported from the Aptian–Albian Rift Valley deposits (either Strzelecki or Otway Group) in Victoria (Rich, 1996; Vickers-Rich, 1996), but have yet to be described. If *Minmi* was really a dwarfed insular form, the Victorian material might be expected to represent a larger, more heavily armoured taxon.

Across the Tasman and much later in the Cretaceous (Campanian or Maastrichtian), but also in a polar environment like the Victorian Albian, more ankylosaur remains have been found at Mangahouanga Stream, N.Z. (Wiffen, 1996). Some articulated tail vertebrae and a rib are attributed to a nodosaurid, which accords with the recent record of a nodosaurid from Antarctica (Gasparini *et al.*, 1996). If this is correct then both ankylosaurian families, as well as stegosaurs, existed in Australasia.

Ornithopods: Bipedal Browsers

Ornithopods are first known in the early Jurassic, represented by heterodontosaurids in southern Africa. These small, kangaroo-like herbivores exaggerated the heterodont dentition of ancestral dinosaurs to produce small upper 'incisors' that bit against the predentary beak, large interlocking 'canines', and a row of shearing/grinding 'molars'. They also had strong-clawed hands, and it is not hard to imagine them using their claws and canines in intraspecific struggles for dominance and mating rights (Santa Luca, 1980). Other ornithopod lineages (which dispensed with the raking claws and canines and, mostly, the premaxillary teeth) are not identified from skeletal material until the late Middle Jurassic (Sereno, 1999), but may have begun radiating earlier in some part of the world with a poor fossil record—such as Australia.

The first evidence of ornithopods here consists of tracks in the early Jurassic Precipice Sandstone of Carnarvon Gorge, Qld., which are referred to the widespread ichnogenus *Anomoepus* (Thulborn, 1994). Although they used an alternating gait rather than hopping, the small ornithopods that made *Anomoepus* tracks were also like kangaroos in the way they sometimes shifted between quadrupedal and bipedal locomotion (Lockley, 1991). Trackways from the Walloon Coal Measures of south-east Qld. (middle Jurassic, ?Bajocian) indicate an increasingly diverse ornithopod radiation, including not only small, bipedal *Anomoepus*-like tracks but larger bipedal ones (iguanodontians), and a form similar to the later *Wintonopus*, in quadrupedal gait (Thulborn, 1999).

The combination of *Wintonopus* and large ornithopod (iguanodontian) tracks also appears in the Broome Sandstone (Long, 1990; Thulborn *et al.*, 1994); it persists into the Cenomanian where *Wintonopus latomorum* was described from the famous 'stampede' trackway at Lark Quarry, near Winton, Qld. (Thulborn and Wade, 1979, 1984). It is only in the Aptian–Albian deposits of eastern Australia that well-preserved skeletal remains of both small and large ornithopods are found.

Hypsilophodontidae has long been recognized as a family comprising numerous small (mostly less than 4 m long), bipedal, primitive ornithopods; these are more derived than *Heterodontosaurus* but lack the distinctive dental and jaw modifications, flattened snouts, crests, or larger body sizes characteristic of iguanodontids and hadrosaurs. Complete skeletons of some forms are known (Galton, 1974), and several synapomorphies of the group have been recognized (Sereno, 1986). However, given the taxonomic diversity within hypsilophodonts and their basal position relative to more advanced lineages, it is tempting to regard this as a paraphyletic group, and there have been proposals to split it up (Scheetz, 1998; Winkler *et al.*, 1998). Australian forms, known from incomplete remains such as their distinctive teeth or femora and of uncertain relationship to *Hypsilophodon*, should be classified as 'Euornithopoda incertae sedis' (Rich and Vickers-Rich, 1999), but 'hypsilophodont' can serve as an informal term. See Figure 7 for skeletal features of this group.

The first hypsilophodont to be described from Australia was *Fulgurotherium australe*, based on a femur from Lightning Ridge originally thought to be from a theropod (Griman Creek Formation, N.S.W.; von Huene, 1932; Molnar and Galton, 1986), but hypsilophodonts were surprisingly diverse here, particularly in the Aptian–Albian polar environments of Victoria. Three genera and species of small to medium-sized hypsilophodonts have been described based on jaw and tooth characters; these are *Leaellynasaura amicagraphica*, *Atlascopcosaurus loadsi*, and *Qantassaurus intrepidus*. The first two were described from Dinosaur Cove near Cape Otway (Rich and Vickers-Rich, 1989), the last only recently, from the Flat Rocks site near Inverloch, Strzelecki Ranges. This locality is now prominent as the most—in fact, only—concentrated lode of Mesozoic mammals known in Australia (Rich and Vickers-Rich, 1999). Because most of the material is broken and disarticulated, it is difficult to assign postcranial elements to particular taxa, but femora have been assigned to *Leaellynasaura* based on small size. Femora from these localities are currently allocated to *L. amicagraphica* and *Fulgurotherium australe*. The Victorian femora of *F. australe* show some diversity in size and shape suggesting that, when material assigned to this form species is better known, it could well be divided into three or more genera. Rich and Vickers-Rich (1994, 1999) estimate a total of six hypsilophodont taxa in the Otway and Strzelecki deposits, and at least two of them also occur at Lightning Ridge.

Among various dinosaur faunas reported from (formerly) polar regions with near-freezing temperatures and months of winter darkness, the Victorian hypsilophodonts are the only ones appearing to show special adaptations for the extreme environment (Rich, 1996). The type of *Leaellynasaura amicagraphica* is a partial skull showing relatively large orbits and a natural endocast with (compared to several other hypsilophodonts where the internal surface of the braincase is known) especially prominent optic lobes (Rich and Vickers-Rich, 1989; Vickers-Rich and Rich, 1993). This hypertrophy of the visual system is suggested to have facilitated activity during the several months of darkness each winter. Evidence that the Victorian hypsilophodonts were indeed active and growing year-round is provided by their bone histology; an ornithomimid theropod from the same deposit, in contrast, shows cyclic bone growth implying periods of reduced feeding and growth (Chinsamy *et al.*, 1998).

An articulated partial skeleton from Dinosaur Cove, lacking a skull but possibly belonging to *Atlascopcosaurus*, also provides information on the biology of these animals. Its left tibia shows gross pathology resulting from a long-term infection, chronic osteomyelitis (Gross *et al.*, 1993). Though it may eventually have died from the infection, the animal is inferred to have persisted in a somewhat incapacitated state for many months; this suggests that encounters with predators were relatively infrequent, and their population density was low. This observation tends to support Vermeij's (1987) view that polar regions are relatively 'safe places' from predation.

Apart from *Fulgurotherium*, finds at Lightning Ridge include isolated bones and teeth of a large ornithopod, as well as tracks in the ceiling of a mine gallery, which might represent a single taxon (Molnar, 1991; Smith, 1999). The teeth—high-crowned, ridged, and with wear facets indicating a shearing action as in hypsilophodonts, rather than the fancier close-packed teeth and grinding surfaces of iguanodontids and hadrosaurs—belong to a primitive iguanodontian (*Muttaburrasaurus*) represented by a number of specimens from central Queensland (Figure 6). Only one species, *M. langdoni*, has been named so far, based on a partial skeleton from the Albian Mackunda Formation at Muttaburra (Bartholomai and Molnar, 1981); but differences in the material from Lightning Ridge, and especially a skull from the Allaru Mudstone, suggest the presence of two or three species, perhaps an ancestor-descendant sequence (Molnar, 1996a). Sexual dimorphism is another possible explanation for some differences in skull shape, especially in the degree of development of the swollen nasal 'bulla' char-

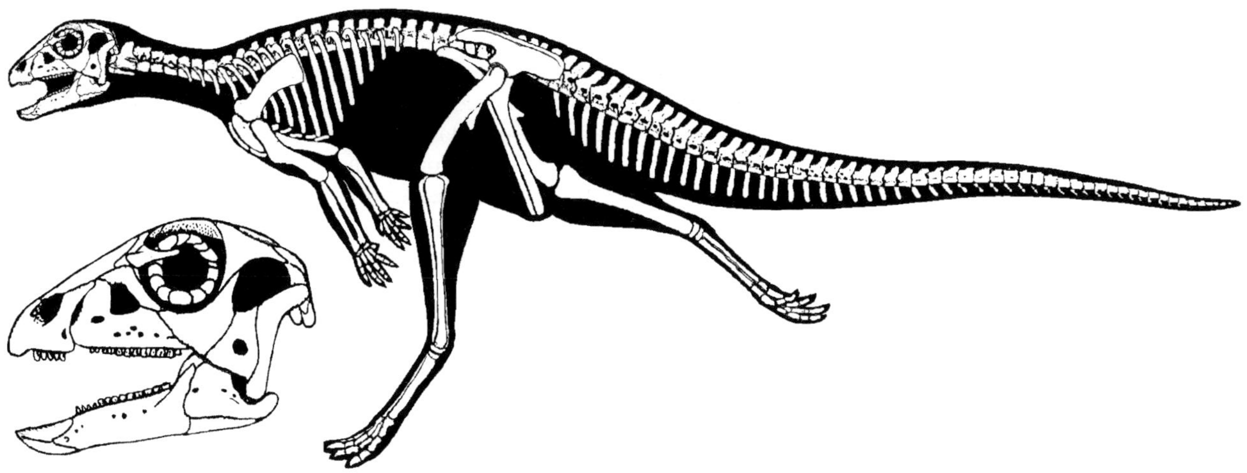

Figure 7. Skull, skeleton and body outline of a hypsilophodont dinosaur—based on *Hypsilophodon foxii.* Adapted from Galton (1974) and Long (1998).

acteristic of the genus, and possibly even for differences in the teeth. Reaching about 7 m in length, similar to the well-known *Iguanodon* in proportions, and probably at least partly quadrupedal, *Muttaburrasaurus* was originally considered either an iguanodontid or camptosaurid (Bartholomai and Molnar, 1981; Rich, 1987). Recently, more detailed character analysis of ornithopods suggests it diverged from the ancestry of these iguanodontians at an earlier stage. Sereno (1999) places it at the same level as the North American *Tenontosaurus* (these two forming an unresolved trichotomy with the dryosaur-iguanodont-hadrosaur lineage). But Molnar (1996a) suggests, based on the partial isolation of the Australian fauna and similarities in tooth morphology, that *Muttaburrasaurus* could represent a descendant of the much smaller 'hypsilophodont' taxon *Atlascopcosaurus*. This would imply that ornithopods evolved large size repeatedly, not only on different continents but (given the Jurassic trackway evidence) perhaps twice or more in Australia.

Small ornithopods are also found in late Cretaceous deposits in New Zealand, at Mangahouanga Stream (Wiffen and Molnar, 1989). A partial ilium from this site is similar to that of dryosaurids, a group which was widespread in the mid- to late Jurassic (Galton, 1981). Although structurally more like the large iguanodontians (e.g. lacking premaxillary teeth), these resembled hypsilophodonts in their small size and bone histology, suggesting a similar pattern of growth and general biology (Chinsamy, 1995; Chinsamy *et al.*, 1998). It has been suggested that hypsilophodonts were predominantly an upland group at lower latitudes, and hence tend to be underrepresented in fossil faunas, whereas in high latitudes they ranged down to sea level and were more likely to be fossilized (Rich, 1996).

Marginocephalia: Spiky-headed Herbivores

Marginocephalia is a weakly diagnosed group comprising two ornithischian lineages that are unknown before the Cretaceous: pachycephalosaurs and ceratopsians. They share a spiny shelf of bone across the rear of the skull, and evolutionary trends in both lineages imply they also shared the use of the head as a weapon in defensive and aggressive interactions. Pachycephalosaurs and the basal ceratopsian *Psittacosaurus* remained bipedal, resembling ornithopods such as hypsilophodonts except for their domed, horned and/or beaked heads. Both groups are best known from East Asia and North America; reports of both from Gondwanan areas are subject to doubt. Sues (1980) reported remains of a pachycephalosaur skull from Madagascar, but it has since been reinterpreted as a theropod (Sampson *et al.*, 1997). Two possible neoceratopsians have been reported from Cretaceous deposits in Gondwana, one in Argentina and one in the Strzelecki Group near Inverloch, Victoria (Tapia, 1918; Rich and Vickers-Rich, 1994). The massive-beaked ceratopsians developed a unique rostral bone in front of the premaxilla—opposing the predentary in the lower jaw (Coombs, 1982); neoceratopsians (more derived than *Psittacosaurus*) were quadrupedal, with very large beak-like snout and jaws balanced by a long bony frill over the neck and shoulders. The only material attributed to this group from Victoria is an ulna, which closely matches that of a relatively small (1.5 m total length) and primitive neoceratopsian, *Leptoceratops*, that lived in cool, upland areas of western North America towards the end of the Cretaceous (Lehman, 1987). Because of the possibility of convergence, additional material will be required to confirm the identification of Australasia's only marginocephalian. However, given the long temporal gap between the first known marginocephalians and the first appearance of their sister taxon Ornithopoda (i.e. almost the entire duration of the Jurassic), it is plausible that this group had its first radiation in Gondwana and only later dispersed to the northern continents (Rich and Vickers-Rich, 1994).

EURYAPSID MARINE REPTILES

Early in the Triassic, several groups of aquatic reptiles appear that share a pattern of temporal openings described as euryapsid (derived from early diapsids by loss of the lower temporal, or quadratojugal, bar). Their phylogenetic relationships are poorly resolved, but ichthyosaurs are apparently only distantly related to sauropterygians, so that the euryapsid condition is convergent (Callaway, 1997; Rieppel, 1997; Motani *et al.*, 1998). Each of these groups initially had a crocodile- or salamander-like body form, with short neck, clawed limbs, and long, rudder-like tail. Each of them independently developed webbed, then paddle-like limbs as they became progressively more adapted to aquatic life, but in other respects they diverged greatly. Placodonts were the least predatory, with dentition specialized for crushing shellfish, and some of them were strikingly convergent with sea turtles; they did not survive beyond the Triassic, and are not known in our region. The two more important groups are ichthyosaurs and plesiosaurs.

ICHTHYOSAURIA: ICHTHYOSAURS ('FISH-LIZARDS')

The ichthyosaurs are characterized by a long, tapering snout and practically no neck, a design enabling them to disable prey by slashing the snout rapidly through the water using the inertia and power of the whole body (as in swordfish, and long-snouted crocodiles and dolphins). Ichthyosaurs all have four fin-like limbs (the anterior pair usually longest), but some of the earliest (mid-Triassic Shastasauridae) still had an oar-like tail, and their flexible body form resembled that of the late Cretaceous mosasaurs. However, later shastasaurids and all later ichthyosaurs had developed a stiff, barrel-shaped trunk and vertical tail-fin, with the vertebral column kinked downwards into the lower fluke; they usually swam by rapid oscillation of the tail-fin on its narrow peduncle, although some of them may also have used the forelimbs as paddles for slow swimming. Ichthyosaurs thus converged biomechanically with tuna and pelagic sharks, and were among the fastest-moving predators on the planet. In many respects they are analogous to dolphins, except that cetaceans swim with up-and-down rather than lateral oscillation of the tail, and lack hindlimbs. Also unlike cetaceans, ichthyosaurs did not develop sonar, but were visual hunters; many had enormous eyes, suggesting an ability to hunt either at night, or at considerable depth in the ocean. Callaway (1997) reviews the history of discovery and classification, and the phylogeny of the group has recently been analysed by Motani (1999).

Australia and New Zealand have a long but depauperate record of ichthyosaurs. There are a few poorly preserved remains from the late Triassic of New Zealand, indeterminate apart from a probable shastasaurid from Otamita Stream. In addition, Cosgriff and Garbutt (1972) tentatively identified what would have been the oldest known ichthyosaur remains from the Blina Shale, (early Triassic of W.A.), but this has not been confirmed yet. During the Jurassic, the period of greatest abundance and diversity for ichthyosaurs in other parts of the world, most of continental Australia was dry land and few marine deposits were formed, while New Zealand has both marine and terrestrial deposits of this age but, for reasons which are not clear, they have yet to produce fossil vertebrates (Fordyce, 1991). Diversity was much lower during the Cretaceous and only a single genus, *Platypterygius* (Ophthalmosauria, *sensu* Motani, 1999) is currently recognized from Australasia. The cool, clear, shallow, Eromanga Sea of the Aptian and Albian Stages, with abundant fish and cephalopods (mainly belemnites), has been described as an 'ichthyosaur feeding-lot', and contains a majority of the Cretaceous ichthyosaur fossils known from anywhere in the world (Wade, 1985, 1990). Numerous partial to complete skulls and some fairly complete skeletons from the Allaru Mudstone and Toolebuc Formation of central Qld. (mainly Albian) are assigned to the endemic species *P. longmani*—see Figure 8 (Wade, 1990), but older (Hauterivian–Barremian, or possibly Aptian Stage) specimens from W.A. appear to be

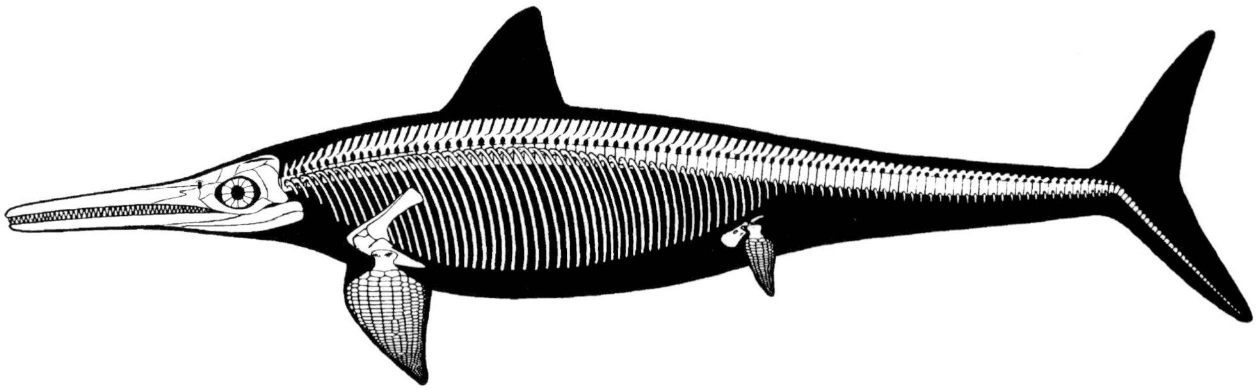

Figure 8. Skeletal reconstruction of the Australian Cretaceous ichthyosaur *Platypterygius longmani*. Modified from a drawing in Molnar (1991).

distinct, and more like some European species in the shape of the humerus (Choo, 1999). Ichthyosaurs died out during the Cenomanian, and some of the latest records are in the Alinga Formation near Kalbarri, W.A.

SAUROPTERYGIA: PLESIOSAURS AND THEIR KIN

Primitive sauropterygians of the Triassic were formerly grouped together as 'nothosaurs', but these are now recognized as a heterogeneous (paraphyletic) assemblage of primitive relatives of plesiosaurs (Taylor, 1989; Rieppel, 1997). The most basal, pachypleurosaurids, were mostly small, inhabiting lagoons and shallow seas, and swam mainly using the tail. Simosaurids and nothosaurids had shorter tails and more paddle-like limbs, using both for marine locomotion. *Pistosaurus* had attained the open-marine habits, hyperphalangy (increased numbers of bones in the digits), reduced tail, and hydrodynamic design of later plesiosaurs, and is regarded as the sister group to Plesiosauria. The four paddle-like limbs of plesiosaurs, articulated to massive pectoral and pelvic bones (which form characteristic fenestrated plates across the ventral side of the body), were shaped like aerofoils and used for underwater 'flight', generating thrust on both upward and downward strokes (as in penguins, seals and sea-turtles, though these modern analogues use only a single pair of limbs for propulsion). In contrast to the 'neckless' ichthyosaurs, early sauropterygians developed a somewhat elongated neck, allowing them to grab prey in motion without turning their whole body.

Most of the fossils of the primitive groups (non-plesiosaurians) are from the Triassic of Europe and China, but very recently, probable pachypleurosaurs have been reported in Australasia (Cruikshank *et al.*, 1999). These specimens, not yet fully described, are from the early Triassic of New Zealand (Anisian Stage; locality not named), and early Jurassic of Queensland (Razorback Beds at Mount Morgan). The Jurassic example, earlier considered to be a pliosauroid (Bartholomai, 1966b; Molnar, 1991) may represent a freshwater relict population of this formerly widespread group, and is much larger than most pachypleurosaurs (Molnar, 1991).

Plesiosauria includes Pliosauroidea and Plesiosauroidea. Pliosaurs developed relatively large heads and shorter necks than their ancestors, while plesiosauroids tend to have small heads and some have greatly elongated necks (see Figure 9). The Jurassic record includes a possible elasmosaurid plesiosauroid (Colalura Sandstone, W.A.) and another long-necked plesiosauroid, either elasmosaurid or cryptocleidid (Evergreen Formation, near Springsure, Qld.); the latter, like the Mount Morgan pachypleurosaur, is from a freshwater deposit. As noted above (in reference to ichthyosaurs), there are few marine records for the Jurassic in our region.

The Cretaceous marine record is much more extensive—especially for the Aptian–Albian Stages in eastern Australia, and Campanian–Maastrichtian (or Haumurian) in New Zealand. Cretaceous plesiosaurs were diverse and abundant, and include some of the most spectacular fossils of both Australia and New Zealand, with a number of relatively complete skeletons, some of them composed of precious opal, and one of the largest predators ever to have lived. The pliosauroids include: species of *Leptocleidus* (*L. clemai* from the Birdrong Sandstone, W.A.; a new species from the Bulldog Shale at Coober Pedy, S.A., represented by the famous opalized skeleton 'Eric'; possibly another *Leptocleidus* sp. from the Griman Creek Formation at Lightning Ridge, N.S.W.); *'Trinacromeron' leucoscopelus* from the Coreena Formation at White Cliffs, N.S.W.; possibly a species of *Dolichorhynchops* and a possible *Trinacromeron* from the Mackunda Formation (near Muttaburra, Qld.); and the giant *Kronosaurus queenslandicus* (originally described from the Toolebuc Formation, but now known from numerous partial skeletons spanning a considerable temporal range in the Eromanga Basin).

Cretaceous plesiosaurs include numerous elasmosaurids; examples include: indeterminate forms from the Molecap Greensand near Dandaragan, W.A.; Bathurst Island Formation around Darwin, N.T.; *Woolungasaurus glendowerensis* from the Wallumbilla Formation near Richmond, Qld.; *Woolungasaurus* sp. from the Bulldog Shale at Andamooka, S.A.; *Tuarangisaurus keyesi* and *Mauisaurus haasti* from the Maungataniwha Sandstone, Laidmore and Conway Formations, N.Z.; indeterminate elasmosaur material from the Katiki Formation and '*Lingula* bed', N.Z. and representatives of several other families. Unidentified plesiosaurian material (pliosaurs and plesiosaurs, including partial or complete skeletons still to be described) occurs in the same marine or estuarine deposits, and others in the Otway Group (Vic.), which (as with the Mount Morgan specimen) may represent freshwater species.

GENERAL SUMMARY

Dodson (1990) used statistical models to estimate the total diversity of dinosaurs at 900–1,200 genera, suggesting that the fossil record of dinosaurs (at the generic level) was then about 25% complete. The world figures have moved forward (new dinosaur taxa are described every few weeks, many of them valid), but the geological history of Australasia is such that only a very small fraction of the actual Mesozoic diversity here will ever be detected. In order to form an overall view of dinosaurian and marine reptile evolution in this region, we must leap over inordinately large gaps in the record, in both space and time.

At the beginning of the Mesozoic when there was only a single supercontinent (Pangaea) terrestrial animals were able to disperse relatively freely to all parts of it. The part that was to become Australasia already occupied a relatively remote position, close to the South Pole and

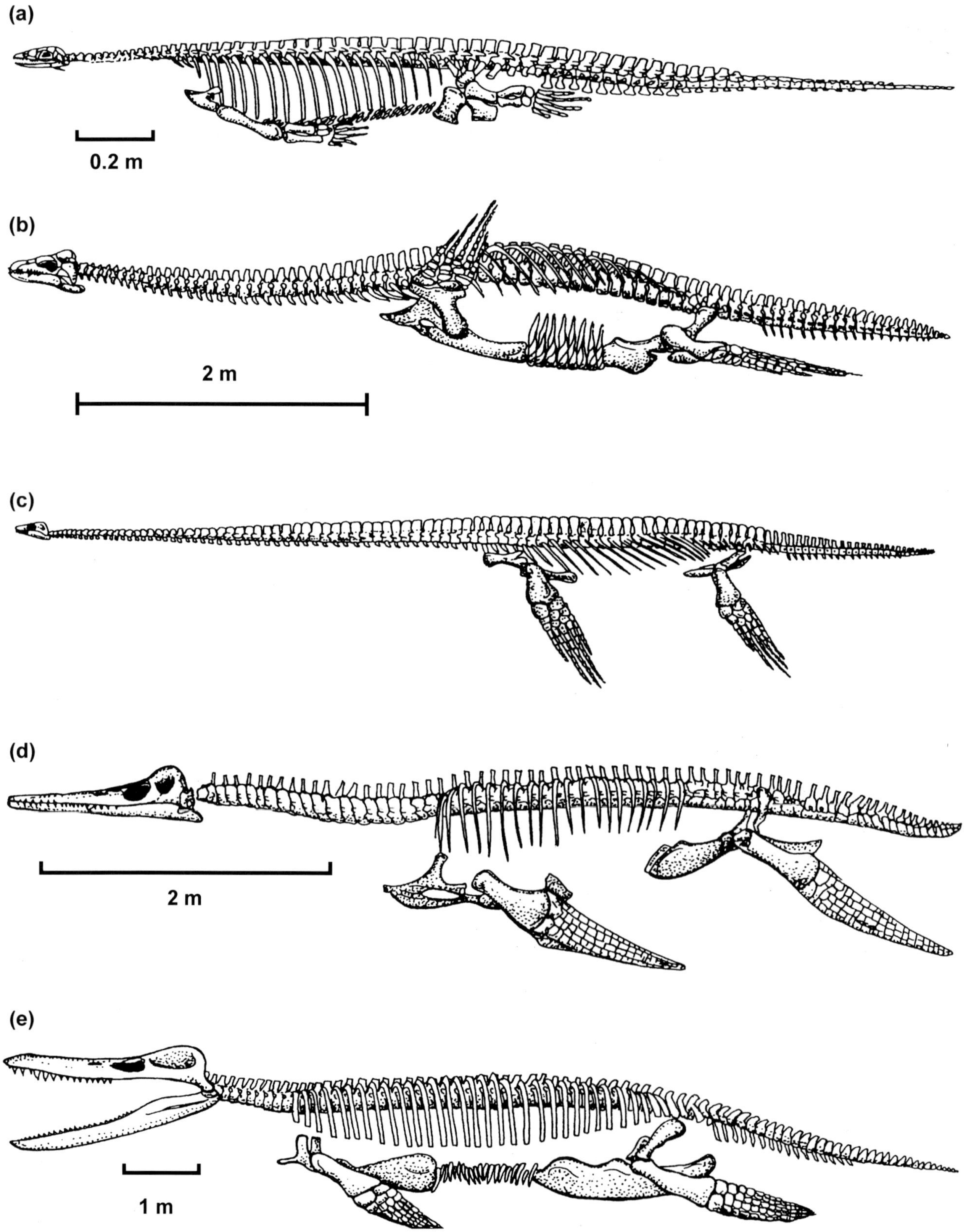

Figure 9. Representative skeletons of sauropterygian lineages: **(a)** a pachypleurosaurid (*Pachypleurosaurus*); **(b)** a short-necked plesiosaur (*Cryptoclidus*, Cryptoclididae); **(c)** a long-necked plesiosaur (*Hydrotherosaurus*, Elasmosauridae); **(d)** a long-necked pliosaurid (*Dolichorhynchops*); **(e)** the giant short-necked pliosaurid *Kronosaurus*. Based on Cruikshank *et al.* (1999).

partly surrounded by ocean. Distance and climate thus may have acted as barriers to dispersal for many taxa, and it could be expected that the fauna would already differ from that of more centrally located regions. However, what is known of the early Triassic amniotes suggests that endemism, if it existed at all, was at a fairly low level—though this is not equally the case for the temnospondyls (Thulborn, 1986a, b).

In the remainder of the Triassic and whole Jurassic the body-fossil record is so sparse that comparisons with other continents are difficult to make at the 'fauna' level; but the trackway evidence suggests there was a diverse assemblage of theropods and ornithopods here about as early as could be expected. *Rhoetosaurus* may be a fairly typical early sauropod, but the affinities of *Ozraptor* remain uncertain and it is not clear whether it represents an endemic theropod lineage or part of a more cosmopolitan group. The reported pachypleurosaur from Mount Morgan (Cruikshank *et al.*, 1999), a freshwater form and one of the last of its kind, parallels later freshwater relicts here at the edge of the world such as the temnospondyls in the early Cretaceous of Victoria, or ceratodont lungfish in the Tertiary and Recent of Queensland and Central Australia.

The fragmentation of Gondwana was under way by early Cretaceous, but the fauna of this time—the best known part of the Mesozoic in Australia—still suggests only the relative isolation of distance and climate. Some groups seem to be relicts (*Austrosaurus*, *Allosaurus*, possibly the stegosaur in the Broome trackway assemblage), others are only slightly different from relatives elsewhere (*Minmi*, the hypsilophodonts, *Muttaburrasaurus*, most of the marine reptiles). Still others seem to be among the first members of groups that flourished during the late Cretaceous in 'Asiamerica' (*Timimus*, the possible oviraptorid, the possible ceratopsian).

The affinities of most of the theropods, the sauropods, and even the hypsilophodonts are too poorly known to assess the amount of faunal exchange with other continental areas during the Jurassic and Cretaceous, but the proportion of endemic taxa does support a considerable level of isolation (Molnar, 1991). Molnar also suggested that the closest similarity in general was with the South American fauna; while the grounds for this comparison are not really clear (in the Mesozoic fauna), it may nevertheless be borne out by ongoing discoveries, particularly in the Otway and Strzelecki sites in Victoria and the similar-aged (Aptian) Santana Formation in Brazil.

ACKNOWLEDGEMENTS

I thank J.R. Merrick for encouraging me to write this chapter, and B. Kear, M.S.Y. Lee, J. Long, R.E. Molnar, N. Pledge and P.M.A. Willis for discussion of drafts or of Mesozoic reptiles in general. I thank the Australian Research Council for financial support.

REFERENCES

Bakker, R.T. 1986. *The Dinosaur Heresies*. William Morrow and Co., New York. 481pp.

Bakker, R.T. and Galton, P.M. 1974. Dinosaur monophyly and a new class of vertebrates. *Nature,* Vol. 248, pp.168–172.

Bartholomai, A. 1966a. Fossil footprints in Queensland. *Australian Natural History,* Vol. 15, pp.147–150.

Bartholomai, A. 1966b. The discovery of plesiosaurian remains in freshwater sediments in Queensland. *Australian Journal of Science,* Vol. 28, p. 437.

Bartholomai, A. 1979. New lizard-like reptiles from the early Triassic of Queensland. *Alcheringa,* Vol. 3, pp.225–234.

Bartholomai, A. and Howie, A. 1970. Vertebrate fauna from the lower Trias of Australia. *Nature,* Vol. 225, p. 1063.

Bartholomai, A. and Molnar, R.E. 1981. *Muttaburrasaurus langdoni*, new genus new species of iguanodontid (Ornithischia: Ornithopoda) dinosaur from the lower Cretaceous of Queensland, Australia. *Memoirs of the Queensland Museum,* Vol. 20, pp.319–350.

Bennett, S.C. and Long, J.A. 1991. A large pterodactyloid pterosaur from the late Cretaceous (late Maastrichtian) of Western Australia. *Records of the Western Australian Museum,* Vol. 15, pp.435–444.

Benton, M.J. 1985. Classification and phylogeny of the diapsid reptiles. *Zoological Journal of the Linnean Society, London,* Vol. 84, pp.97–164.

Benton, M.J. 1999. *Scleromochlus taylori* and the origin of dinosaurs and pterosaurs. *Philosophical Transactions of the Royal Society of London,* Series B, Vol. 354, pp.1423–1446.

Benton, M.J. 2003. *When Life Nearly Died: the Greatest Mass Extinction of All Time*. Thames and Hudson, London.

Bonaparte, J.F. 1976. *Pisanosaurus mertii* Casimaquela and the origin of the Ornithischia. *Journal of Paleontology,* Vol. 50, pp.808–820.

Brochu, C.A. 2001. Progress and future directions in archosaur phylogenetics. *Journal of Paleontology,* Vol. 75, pp.1185–1201.

Buffetaut, E., Suteethorn, V., Cuny, G., Tong, H., Le Loeuff, J., Khansubha, S. and Jongautchariyakul, S. 2000. The earliest known sauropod dinosaur. *Nature,* Vol. 407, pp.72–74.

Callaway, J.M. 1997. Ichthyosauria. Introduction. Pp.3–16. In Callaway, J.M. and Nicholls, E.L. (eds). *Ancient Marine Reptiles*. Academic Press, San Diego. 501pp.

Camp, C.L. and Banks, M.R. 1978. A proterosuchian reptile from the early Triassic of Tasmania. *Alcheringa,* Vol. 2, pp.143–158.

Campbell, J.D. 1965. New Zealand Triassic saurians. *New Zealand Journal of Geology and Geophysics,* Vol. 8, pp.505–509.

Campbell, J.D. and Fordyce, E.R. 1972. Stratigraphy of the Mount Potts Group at Rocky Gully, Rangitata Valley, Canterbury. *New Zealand Journal of Geology and Geophysics,* Vol. 15, pp.157–167.

Chinsamy, A. 1995. Ontogenetic changes in the bone histology of the late Jurassic ornithopod *Dryosaurus lettowvorbecki*. *Journal of Vertebrate Paleontology,* Vol. 15, pp.96–104.

Chinsamy, A., February, E.C., Harley, E., Rich, T.H. and Vickers-Rich, P. 1997. Wood within a polar dinosaur bone. *South African Journal of Science,* Vol. 93, p. 48.

Chinsamy, A., Rich, T.H. and Vickers-Rich, P. 1998. Polar dinosaur bone histology. *Journal of Vertebrate Paleontology,* Vol. 18, pp.385–390.

Choo, B. 1999. Cretaceous ichthyosaurs from Western Australia. *Records of the Western Australian Museum, Supplement,* No. 57, pp.207–218.

Clark, W.B. 1861. *Recent Geological Discoveries in Australasia.* 2nd Edit. Joseph Cook, Sydney.

Colbert, E.H. and Merrilees, D. 1967. Cretaceous dinosaur footprints from Western Australia. *Journal of the Royal Society of Western Australia,* Vol. 50, pp.21–25.

Coombs, W.P. Jnr. 1982. Juvenile specimens of the ornithischian dinosaur, *Psittacosaurus mongoliensis. Palaeontology,* Vol. 25, pp.89–108.

Coombs, W.P. Jnr. and Molnar, R.E. 1981. Sauropods (Reptilia, Saurischia) from the Cretaceous of Queensland. *Memoirs of the Queensland Museum,* Vol. 20, pp.351–373.

Cosgriff, J.W. and Garbutt, N.K. 1972. *Erythrobatrachus noonkanbaliensis*, a trematosaurid species from the Blina Shale. *Journal of the Royal Society of Western Australia,* Vol. 55, pp.5–18.

Cruikshank, A.R.I. 1972. The proterosuchian thecodonts. Pp.89–119. In Joysey, K.A. and Kemp, T.S. (eds). *Studies in Vertebrate Evolution: Essays Presented to F.R. Parrington.* Oliver and Boyd, Edinburgh. 284pp.

Cruikshank, A.R.I. and Fordyce, R.E. 1996. A late Cretaceous plesiosaur (Plesiosauria, Cryptoclididae?) from Shag Point, North Otago, New Zealand. *Geological Society of New Zealand, Miscellaneous Publication,* No. 91a, p. 62.

Cruikshank, A.R.I. and Fordyce, R.E. 1998. High latitude late Cretaceous plesiosaurs in Gondwana. *Journal of African Earth Sciences,* Vol. 27, pp.50–51.

Cruikshank, A.R.I., Fordyce, R.E. and Long, J.A. 1999. Recent developments in Australasian sauropterygian palaeontology (Reptilia: Sauropterygia). *Records of the Western Australian Museum, Supplement*, No. 57, pp.201–205.

Cruikshank, A.R. and Long, J.A. 1997. A new species of pliosaurid reptile from the early Cretaceous Birdrong Sandstone of Western Australia. *Records of the Western Australian Museum,* Vol. 18, pp.263–276.

Currie, P.J., Vickers-Rich, P. and Rich, T.H. 1996. Possible oviraptorosaur (Theropoda, Dinosauria) specimens from the early Cretaceous Otway Group of Dinosaur Cove, Australia. *Alcheringa,* Vol. 20, pp.73–79.

Dayton, L. 1991. Missing dinosaurs turn up in Australia. *New Scientist,* Vol. 131, p.14.

Dodson, P. 1990. Counting dinosaurs: How many kinds were there? *Proceedings of the National Academy of Sciences of the United States of America,* Vol. 87, pp.7608–7612.

Etheridge, R. Jnr. 1897. An Australian sauropterygian (*Cimoliasaurus*), converted into precious opal. *Records of the Australian Museum,* Vol. 3, pp.19–29.

Etheridge, R. Jnr. 1904. A second sauropterygian converted into opal, from the upper Cretaceous of White Cliffs, New South Wales. *Records of the Australian Museum,* Vol. 5, pp.306–316.

Etheridge, R. Jnr. 1917. Reptilian notes: *Megalania prisca*, Owen, and *Notiosaurus dentatus*, Owen; lacertilian dermal armour, opalized remains from Lightning Ridge. *Proceedings of the Royal Society of Victoria,* Vol. 29, pp.127–133.

Flannery, T.F. and Rich, T.H. 1981. Dinosaur digging in Victoria. *Australian Natural History,* Vol. 20, pp.195–198.

Fletcher, H.O. 1948. Footprints in the sands of time. *Australian Museum Magazine,* Vol. 9, pp.247–251.

Fleming, C.A., Gregg, D.R. and Welles, S.P. 1971. New Zealand ichthyosaurs—a summary, including new records from the Cretaceous. *New Zealand Journal of Geology and Geophysics,* Vol. 14, pp.734–741.

Fordyce, R.E. 1983. The North Otago plesiosaur. *Geological Society of New Zealand Newsletter,* 61, pp.47–48.

Fordyce, R.E. 1991. A new look at the fossil vertebrate record of New Zealand. Pp.1191–1316. In Vickers-Rich, P., Monaghan, J.M., Baird, R.F. and Rich, T.H. (eds). *Vertebrate Palaeontology of Australasia.* Pioneer Design Studio and Monash University Publications Committee, Melbourne, Victoria. 1437pp.

Gaffney, E.S. 1981. A review of the fossil turtles of Australia. *American Museum Novitates,* No. 2720, pp.1–38.

Gaffney, E.S. 1991. The fossil turtles of Australia. Pp.704–720. In Vickers-Rich, P., Monaghan, J.M., Baird, R.F. and Rich, T.H. (eds). *Vertebrate Palaeontology of Australasia.* Pioneer Design Studio and Monash University Publications Committee, Melbourne, Victoria. 1437pp.

Galton, P.M. 1974. The ornithischian dinosaur *Hypsilophodon* from the Wealden of the Isle of Wight. *Bulletin of the British Museum of Natural History (Geology),* Vol. 25, pp.1–125.

Galton, P.M. 1981. *Dryosaurus*, a hypsilophodontid dinosaur from the upper Jurassic of North America and Africa: postcranial skeleton. *Palaeontologische Zeitschrift,* Vol. 55, pp.271–312.

Galton, P.M. 1985. Diet of prosauropod dinosaurs from the late Triassic and early Jurassic. *Lethaia,* Vol. 18, pp.105–124.

Gasparini, Z., Pereda-Suberbiola, X. and Molnar, R.E. 1996. New data on the ankylosaurian dinosaur from the late Cretaceous of the Antarctic Peninsula. *Memoirs of the Queensland Museum,* Vol. 39, pp.583–594.

Gillette, D. and Lockley, M.G. (eds). 1989. *Dinosaur Tracks and Traces.* Cambridge University Press, Cambridge. 454pp.

Glauert, L. 1952. Dinosaur footprints near Broome. *Western Australian Naturalist,* Vol. 3, pp.82–83.

Gow, C.E. 1975. The morphology and relationships of *Youngina capensis* Broom and *Prolacerta broomi* Parrington. *Palaeontologia Africana,* Vol. 18, pp.89–131.

Greene, H.W. 1988. Antipredator mechanisms in reptiles. Pp.1–152. In Gans, C. and Huey, R.B. (eds). *Biology of the Reptilia.* Vol. 16. Alan R. Liss Inc., New York.

Gross, J.D., Rich, T.H. and Vickers-Rich, P. 1993. Dinosaur bone infection. *National Geographic Research and Exploration,* Vol. 9, pp.286–293.

Haast, J. 1870. Notes on a collection of saurian remains from the Waipara River, Canterbury, in the possession of J.H. Cockburn Hood, Esq. *Transactions and Proceedings of the New Zealand Institute,* Vol. 2, pp.186–189.

Hamley, T. 1997. Biogeography of the procolophonids. *Geological Society of Australia Abstracts,* 48, p. 34.

Hammer, W.R. and Hickerson, W.J. 1994. A crested theropod dinosaur from Antarctica. *Science,* Vol. 264, pp.828–830.

Hammer, W.R. and Hickerson, W.J. 1998. Antarctic

dinosaurs: a glimpse into the early Jurassic of Gondwana. *Journal of Vertebrate Paleontology,* Vol. 18, No.3 (supplement), p. 48A.

Haubold, H. 1971. Ichnia Amphibiorum et Reptiliorum fossilum. In Kuhn, O. (ed). *Handbuch der Paläoherpetologie* 18. G. Fischer, Stuttgart.

Hector, J. 1870. *Catalogue of the Colonial Museum, Wellington, New Zealand.* Government Printer, Wellington.

Hector, J. 1874. On the fossil Reptilia of New Zealand. *Transactions and Proceedings of the New Zealand Institute,* Vol. 6, pp.333–358.

Hill, D., Playford, G. and Woods, J.T. 1966. *Jurassic Fossils of Queensland.* Queensland Palaeontological Society, Brisbane. 32pp.

Hiller, N., Mannering, A. and Jones, C.M. 1999. *Mauisaurus haasti*—a new look at an old species of plesiosaur. *Records of the Western Australian Museum, Supplement,* No. 57, p. 402 (abstract only).

Hood, T.H.C. 1870. Geological observations on the Waipara River, New Zealand. *Quarterly Journal of the Geological Society of London,* Vol. 26, pp.409–413.

Hunt, A.P., Lockley, M.G., Lucas, S.G. and Meyer, C.A. 1994. The global sauropod fossil record. *Gaia,* Vol. 10, pp.261–279.

Hutton, F.W. 1894. On a new plesiosaur from the Waipara River. *Transactions of the New Zealand Institute,* Vol. 26, pp.354–358.

Jensen, A.R. 1975. Permo-Triassic stratigraphy and sedimentation in the Bowen Basin, Queensland. *Bureau of Mineral Resources, Geology and Geophysics Bulletin,* No. 154, pp.1–187.

Kobayashi, Y., Lu, J.-C., Dong, Z.-M., Barsbold, R., Azuma, Y. and Tomida, Y. 1999. Herbivorous diet in an ornithomimid dinosaur. *Nature,* Vol. 402, pp.480–481.

Kurochkin, E.N. and Molnar, R.E. 1997. New material of enantiornithine birds from the early Cretaceous of Australia. *Alcheringa,* Vol. 21, pp.291–297.

Langer, M.C., Abdala, F., Richter, M. and Benton, M.J. 1999. A sauropodomorph dinosaur from the upper Triassic (Carnian) of southern Brazil. *Comptes Rendus de L'Academie des Sciences, Sciences de la Terre et des Planetes,* Series 2, Vol. 329, pp.511–517.

Lehman, T.M. 1987. Late Maastrichtian paleoenvironments and dinosaur biogeography in the western interior of North America. *Palaeogeography Palaeoclimatology Palaeoecology,* Vol. 60, pp.189–218.

Lockley, M.G. 1991. *Tracking Dinosaurs: A New Look at an Ancient World.* Cambridge University Press, Cambridge. 238pp.

Long, J.A. 1990. *Dinosaurs of Australia and Other Animals of the Mesozoic Era.* Reed Books, Sydney. 86pp.

Long, J.A. 1992. First dinosaur bones from Western Australia. *The Beagle (Records of the Northern Territory Museum of Arts and Sciences),* Vol. 9, pp.21–28.

Long, J.A. 1993. *Dinosaurs of Australia and Other Animals of the Triassic, Jurassic and Cretaceous Periods.* Reed Books, Sydney.

Long, J.A. 1995. A theropod dinosaur bone from the late Cretaceous Molecap Greensand, Western Australia. *Records of the Western Australian Museum,* Vol. 17, pp.143–146.

Long, J.A. 1998. *Dinosaurs of Australia and New Zealand, and Other Animals of the Mesozoic Era.* University of New South Wales Press, Sydney. 188pp.

Long, J.A. and Buffetaut, E. 2001. A biogeographic comparison of the dinosaurs and associated vertebrate faunas from the Mesozoic of Australia and Southeast Asia. Pp.97–103. In Metcalfe, I., Smith, J.M.B., Morwood, M. and Davidson, I. (eds) *Faunal and Floral Migrations and Evolution in SE Asia-Australasia.* A.A. Balkema, Lisse. 416pp.

Long, J.A. and Cruickshank, A.R.I. 1996. First record of an early Cretaceous theropod dinosaur bone from Western Australia. *Records of the Western Australian Museum,* Vol. 18, pp.219–221.

Long, J.A. and Cruickshank, A.R.I. 1998. Further records of plesiosaurian reptiles from the Jurassic and Cretaceous age from Western Australia. *Records of the Western Australian Museum,* Vol. 19, pp.47–55.

Long, J.A. and Molnar, R.E. 1998. A new Jurassic theropod dinosaur from Western Australia. *Records of the Western Australian Museum,* Vol. 19, 121–129.

Longman, H.A. 1915. On a giant turtle from the Queensland lower Cretaceous. *Memoirs of the Queensland Museum,* Vol. 3, pp.24–29.

Longman, H.A. 1924. A new gigantic marine reptile from the Queensland Cretaceous, *Kronosaurus queenslandicus* new genus and species. *Memoirs of the Queensland Museum,* Vol. 8, pp.26–28.

Longman, H.A. 1926. A giant dinosaur from Durham Downs, Queensland. *Memoirs of the Queensland Museum,* Vol. 8, pp.183–194.

Longman, H.A. 1927. The giant dinosaur *Rhoetosaurus brownei. Memoirs of the Queensland Museum,* Vol. 9, pp.1–18.

Longman, H.A. 1930. *Kronosaurus queenslandicus*, a gigantic Cretaceous pliosaur. *Memoirs of the Queensland Museum,* Vol. 10, pp.1–7.

Longman, H.A. 1933. A new dinosaur from the Queensland Cretaceous. *Memoirs of the Queensland Museum,* Vol. 13, pp.133–144.

Lundelius, E. Jnr. and Warne, S.S. 1960. Mosasaur remains from the upper Cretaceous of Western Australia. *Journal of Paleontology,* Vol. 34, pp.1215–1217.

McKay, A. 1877. On the reptilian beds of New Zealand. *Transactions of the New Zealand Institute,* Vol. 9, pp.581–590.

McKenzie, K.H. 1961. Vertebrate localities in the Triassic Blina Shale of the Canning Basin, Western Australia. *Journal and Proceedings of the Royal Society of Western Australia,* Vol. 44, pp.69–76.

Molnar, R.E. 1980a. Procoelous crocodile from the lower Cretaceous of Lightning Ridge, New South Wales. *Memoirs of the Queensland Museum,* Vol. 20, pp.65–75.

Molnar, R.E. 1980b. An ankylosaur (Ornithischia: Reptilia) from the lower Cretaceous of southern Queensland. *Memoirs of the Queensland Museum,* Vol. 20, pp.77–87.

Molnar, R.E. 1980c. Reflections on the Mesozoic of Australia. *Mesozoic Vertebrate Life,* Vol. 1, pp.47–60.

Molnar, R.E. 1981. A dinosaur from New Zealand. Pp.91–95. In *Proceedings, 5th International Gondwana Symposium, Wellington, New Zealand, 11–16 February 1980.*

Molnar, R.E. 1982a. Australian Mesozoic reptiles. Pp.170–225. In Vickers-Rich, P.V. and Thompson, E.M. (eds). *The Fossil*

Vertebrate Record of Australasia. Monash University Offset Printing Unit, Melbourne. 759pp.

Molnar, R.E. 1982b. A catalogue of fossil amphibians and reptiles in Queensland. *Memoirs of the Queensland Museum*, Vol. 20, pp.613–633.

Molnar, R.E. 1985a. *Notochelone costata*, the southern turtle; ?*Ornithocheirus*, the western flying dragon. Pp.126–131. In Vickers-Rich, P. and van Tets, G.F. (eds). *Kadimakara: Extinct Vertebrates of Australia*. Pioneer Design Studio, Melbourne. 284pp.

Molnar, R.E. 1985b. *Minmi paravertebra* the minmi, an armoured dinosaur. Pp.172–176. In Vickers-Rich, P. and van Tets, G.F. (eds). *Kadimakara: Extinct Vertebrates of Australia*. Pioneer Design Studio,Melbourne. 284pp.

Molnar, R.E. 1986. An enantiornithine bird from the lower Cretaceous of Queensland, Australia. *Nature*, Vol. 322, pp.736–738.

Molnar, R.E. 1987a. A pterosaur pelvis from western Queensland. *Alcheringa*, Vol. 11, pp.87–94.

Molnar, R.E. 1987b. A sea-serpentine sauropterygian, *Woolungasaurus glendowerensis*. Pp.33–35. In Hand, S. and Archer, M. (eds). *The Antipodean Ark: Creatures from Prehistoric Australia*. Angus and Robertson, Sydney. 90pp.

Molnar, R.E. 1991. Fossil reptiles in Australia. Pp.605–702. In Vickers-Rich, P., Monaghan, J.M., Baird, R.F. and Rich, T.H. (eds). *Vertebrate Palaeontology of Australasia*. Pioneer Design Studio, and Monash University Publications Committee, Melbourne. 1437pp.

Molnar, R.E. 1994. *Minmi*, all tanked up and ready to grow. *Dinonews* (*Western Australian Museum*), Vol. 7, pp.7–9.

Molnar, R.E. 1996a. Observations on the Australian ornithopod dinosaur, *Muttaburrasaurus*. *Memoirs of the Queensland Museum*, Vol. 39, pp.639–652.

Molnar, R.E. 1996b. Preliminary report on a new ankylosaur from the early Cretaceous of Queensland, Australia. *Memoirs of the Queensland Museum*, Vol. 39, pp.653–668.

Molnar, R.E. and Clifford, H.T. 2000. Gut contents of a small ankylosaur. *Journal of Vertebrate Paleontology*, Vol. 20, pp.194–196.

Molnar, R.E., Flannery, T.F. and Rich, T.H. 1981. An allosaurid theropod dinosaur from the early Cretaceous of Victoria, Australia. *Alcheringa*, Vol. 5, pp.141–146.

Molnar, R.E. and Frey, E. 1987. The paravertebral elements of the Australian ankylosaur *Minmi* (Reptilia: Ornithischia, Cretaceous). *Neues Jahrbuch für Geologie und Paläontologie, Abhandlungen*, Vol. 175, pp.19–37.

Molnar, R.E. and Galton, P.M. 1986. Hypsilophodontid dinosaurs from Lightning Ridge, New South Wales, Australia. *Geobios*, Vol. 19, pp.231–244.

Molnar, R.E. and Pledge, N.S. 1980. A new theropod dinosaur from South Australia. *Alcheringa*, Vol. 4, pp.281–287.

Molnar, R.E. and Scarlett, R.J. 1984. Terrestrial bird or dinosaur phalanx from the New Zealand Cretaceous. *New Zealand Journal of Zoology*, Vol. 11, pp.271–275.

Molnar, R.E. and Thulborn, R.A. 1980. First pterosaur from Australia. *Nature*, Vol. 288, pp.361–363.

Molnar, R.E. and Wiffen, J. 1994. A late Cretaceous polar dinosaur fauna from New Zealand. *Cretaceous Research*, Vol. 15, pp.689–706.

Molnar, R.E. and Willis, P.M.A. 1996. A neosuchian crocodile from the Queensland Cretaceous. *Journal of Vertebrate Paleontology*, Vol. 16, No. 3 (supplement), p. 27A (abstract only).

Motani, R. 1999. Phylogeny of the Ichthyopterygia. *Journal of Vertebrate Paleontology*, Vol. 19, pp.473–496.

Motani, R., Minoura, N. and Ando, T. 1998. Ichthyosaurian relationships illuminated by new primitive skeletons from Japan. *Nature*, Vol. 393, pp.255–257.

Murray, P.F. 1985. Ichthyosaurs from Cretaceous Mullamann Beds near Darwin, Northern Territory. *The Beagle (Records of the Northern Territory Museum of Arts and Sciences)*, Vol. 2, pp.39–55.

Murray, P.F. 1987. Plesiosaurs from Albian-aged Bathurst Island Formation siltstones near Darwin, Northern Territory. *The Beagle (Records of the Northern Territory Museum of Arts and Sciences)*, Vol. 4, pp.95–102.

Owen, R. 1861. On the remains of a plesiosaurian reptile (*Plesiosaurus australis*) from the Oolitic Formation in the Middle Island of New Zealand. *Geologist*, Vol. 4, pp.444–445.

Owen, R. 1882. On an extinct chelonian reptile (*Notochelys costata*, Owen) from Australia. *Quarterly Journal of the Geological Society of London*, Vol. 38, pp.178–183.

Persson, P.O. 1960. Lower Cretaceous plesiosaurians (Rept.) from Australia. *Lunds Universitets Arsskrift*, Series 2. Vol. 56, pp.1–21.

Persson, P.O. 1963. A revision of the classification of the Plesiosauria with a synopsis of the stratigraphical and geographical distribution of the group. *Lunds Universitets Arsskrift*, Series 2. Vol. 59, pp.1–60.

Persson, P.O. 1982. Elasmosaurid skull from the lower Cretaceous of Queensland (Reptilia: Sauropterygia). *Memoirs of the Queensland Museum*, Vol. 20, pp.647–655.

Pledge, N. 1980. *Vertebrate Fossils of South Australia*. South Australian Museum, Adelaide. 11pp.

Reisz, R.R. and Scott, D. 2002. *Owenetta kitchingorum*, sp. nov., a small parareptile (Procolophonia: Owenettidae) from the lower Triassic of South Africa. *Journal of Vertebrate Paleontology*, Vol. 22, pp.244–256.

Retallack, G.J. 1996. Early Triassic therapsid footprints from the Sydney Basin, Australia. *Alcheringa*, Vol. 20, pp.301–314.

Rich, T.H. 1987. Cretaceous kangaroo mimics: Hypsilophodontidae and Camptosauridae species. Pp.40–42. In Hand, S. and Archer, M. (eds). *The Antipodean Ark: Creatures from Prehistoric Australia*. Angus and Robertson, Sydney. 90pp.

Rich, T.H. 1996. Significance of polar dinosaurs in Gondwana. *Memoirs of the Queensland Museum*, Vol. 39, pp.711–717.

Rich, T.H. and Vickers-Rich, P. 1989. Polar dinosaurs and biotas of the early Cretaceous of southeastern Australia. *National Geographic Research*, Vol. 5, pp.15–53.

Rich, T.H. and Vickers-Rich, P. 1994. Neoceratopsians and ornithomimosaurs: dinosaurs of Gondwana origin? *National Geographic Research & Exploration*, Vol. 10, pp.129–131.

Rich, T.H. and Vickers-Rich, P. 1999. The hypsilophodontidae from southeastern Australia. Pp.167–180. In Tomida, Y., Rich, T.H. and Vickers-Rich, P. (eds). *Proceedings of the Second Symposium on Gondwanan Dinosaurs*. National Science Museum of Japan, Tokyo.

Rieppel, O. 1997. Sauropterygia. Introduction. Pp.107–119. In Callaway, J.M. and Nicholls, E.L. (eds). *Ancient Marine Reptiles*. Academic Press, San Diego. 501pp.

Ritchie, A. 1988. Who should pay for Australia's past? *Australian Natural History,* Vol.22, pp.368–370.

Rogers, D. 1987. *Description and interpretation of a lower Cretaceous fossil feather from Koonwarra, South Gippsland, Victoria.* Third Year Thesis (unpublished), Monash University, Melbourne. 70pp.

Rogers, R.R., Swisheri, C.C., Sereno, P.C., Monetta, A.M., Forster, C.A. and Martinez, R.N. 1993. The Ischigualasto tetrapod assemblage (late Triassic, Argentina) and argon-40/argon-39 dating of dinosaur origins. *Science,* Vol. 260, pp.794–797.

Romer, A.S. and Lewis, A.D. 1960. A mounted skeleton of the giant plesiosaur, *Kronosaurus*. *Breviora,* Vol. 112, pp.1–15.

Sampson, S.D., Witmer, L.M., Forster, C.A. and O'Connor, P.M. 1997. Discovery of a complete theropod skull from the late Cretaceous of Madagascar yields new information on *Majungasaurus*, *Majungatholus*, and Abelisauridae. *Journal of Vertebrate Paleontology,* Vol. 17, No. 3 (supplement), p. 73A (abstract only).

Santa Luca, A.P. 1980. The postcranial skeleton of *Heterodontosaurus tucki* (Reptilia, Ornithischia) from the Stormberg of South Africa. *Annals of the South African Museum,* Vol. 79, pp.159–211.

Scheetz, R.D. 1998. Phylogeny of basal ornithopod dinosaurs and the dissolution of the Hypsilophodontidae. *Journal of Vertebrate Paleontology,* Vol.18, No.3 (supplement) p. 75A. (abstract only)

Schroeder, N. 1998. *A review of order Pliosauria, and the description of a new, opalised pliosauroid, [name deleted], from the early Cretaceous of Coober Pedy, South Australia.* M.Sc. thesis (unpublished), Monash University, Melbourne.

Sereno, P.C. 1986. Phylogeny of the bird-hipped dinosaurs (Order Ornithischia). *National Geographic Research,* Vol. 2, pp.234–256.

Sereno, P.C. 1999. The evolution of dinosaurs. *Science,*Vol. 284, pp.2137–2147.

Sereno, P.C., Forster, C.A. Rogers, R.R. and Monetta, A.M. 1993. Primitive dinosaur skeleton from Argentina and the early evolution of Dinosauria. *Nature,* Vol. 361, pp.64–66.

Sereno, P.C. and Novas, F.E. 1992. The complete skull and skeleton of an early dinosaur. *Science,* Vol. 258, pp.1137–1140.

Smith, E. 1997. Lightning Ridge fossils: hunting on a new frontier. *Riversleigh Notes,* Vol. 36, pp.6–8.

Smith, E. 1999. *Black Opal Fossils of Lightning Ridge: Treasures from the Rainbow Billabong.* Kangaroo Press, Sydney. 112pp.

Stevens, G.R. 1974. *Rugged Landscape*. A.H. and A.W. Reed, Wellington. 286pp.

Sues, H.D. 1980. A pachycephalosaurid dinosaur (*Majungatholus atopus*, new record), from the upper Cretaceous of Madagascar and its paleobiogeographical implications. *Journal of Paleontology,* Vol. 54, pp.954–962.

Talent, J.A., Duncan, P.M. and Handley, P.L. 1966. Early Cretaceous feathers from Victoria. *Emu,* Vol. 66, pp.81–86.

Tapia, A. 1918. Una mandibula de dinosaurio procedente de Patagonia. *Physis,* Vol. 4, pp.369–370.

Taylor, M.A. 1989. Sea-saurians for sceptics. *Nature,* Vol. 338, pp.625–626.

Teichert, C. and Matheson, R.S. 1944. Upper Cretaceous ichthyosaurian and plesiosaurian remains from Western Australia. *Australian Journal of Science,* Vol. 6, pp.167–170.

Thulborn, R.A. 1979. A proterosuchian thecodont from the Rewan Formation of Queensland. *Memoirs of the Queensland Museum,* Vol. 19, pp.331–355.

Thulborn, R.A. 1983. A mammal-like reptile from Australia. *Nature,* Vol. 303, pp.330–331.

Thulborn, R.A. 1985a. *Rhoetosaurus brownei* the giant Queensland dinosaur. Pp.166–171. In Vickers-Rich, P.V. and van Tets, G.F. (eds). *Kadimakara: Extinct Vertebrates of Australia*. Pioneer Design Studio, Melbourne. 284pp.

Thulborn, R.A. 1985b. Dicynodontia. The Queensland dicynodont. Pp.177–182. In Vickers-Rich, P. and van Tets, G.F. (eds). *Kadimakara: Extinct Vertebrates of Australia*. Pioneer Design Studio, Melbourne. 284pp.

Thulborn, R.A. 1986a. The Australian Triassic reptile *Tasmaniosaurus triassicus* (Thecodontia: Proterosuchia). *Journal of Vertebrate Paleontology,* Vol. 6, pp.123–142.

Thulborn, R.A. 1986b. Early Triassic tetrapod faunas of southeastern Gondwana. *Alcheringa,* Vol. 10, pp.297–313.

Thulborn, R.A. 1987. An enigmatic brontosaur, *Austrosaurus mckillopi*. Pp.44–46. In Hand, S. and Archer, M. (eds). *The Antipodean Ark: Creatures from Prehistoric Australia*. Angus and Robertson, Sydney. 90pp.

Thulborn, R.A. 1990a. *Dinosaur Tracks*. Chapman and Hall, London. 410pp.

Thulborn, R.A. 1990b. Mammal-like reptiles of Australia. *Memoirs of the Queensland Museum,* Vol. 28, p. 169.

Thulborn, R.A. 1992. Taxonomic characters of *Fabrosaurus australis*, an ornithischian dinosaur from the lower Jurassic of southern Africa. *Geobios,* Vol. 25, pp.283–292.

Thulborn, R.A. 1994. Ornithopod dinosaur tracks from the lower Jurassic of Queensland. *Alcheringa,* Vol. 18, pp.247–258.

Thulborn, R.A. 1999. New dinosaur tracks from the Walloon Coal Measures (middle Jurassic) of Queensland. *Records of the Western Australian Museum, Supplement*, No. 57, p. 419 (abstract only).

Thulborn, R.A., Hamley, T. and Foulkes, P. 1994. Preliminary report on sauropod dinosaur tracks in the Broome Sandstone (lower Cretaceous) of Western Australia. *Gaia,* Vol. 10, pp.85–96.

Thulborn, R.A. and Wade, M. 1979. Dinosaur stampede in the Cretaceous of Queensland, Australia. *Lethaia,* Vol. 12, pp.275–279.

Thulborn, R.A. and Wade, M. 1984. Dinosaur trackways in the Winton Formation (mid-Cretaceous) of Queensland (Australia). *Memoirs of the Queensland Museum,* Vol. 21, pp.413–518.

Thulborn, R.A. and Warren, A.A. 1980. Early Jurassic plesiosaurs from Australia. *Nature,* Vol. 285, pp.224–225.

Tucker, M.E. and Benton, M.J. 1982. Triassic environments, climates and reptile evolution. *Palaeogeography Palaeoclimatology Palaeoecology,* Vol. 40, pp.361–379.

Upchurch, P. 1995. The evolutionary history of sauropod dinosaurs. *Philosophical Transactions of the Royal Society of London,* Series B, Vol. 349, pp.365–390.

Upchurch, P. 1998. The phylogenetic relationships of sauropod dinosaurs. *Zoological Journal of the Linnean Society,* Vol. 124, pp.43–103.

Vermeij, G.J. 1987. *Evolution and Escalation: an Ecological History of Life*. Princeton University Press, Princeton, New Jersey. 527pp.

Vickers-Rich, P. 1991. The Mesozoic and Tertiary History of birds on the Australian plate. Pp.721–808. In Vickers-Rich, P., Monaghan, J.M., Baird, R.F. and Rich, T.H. (eds). *Vertebrate Palaeontology of Australasia*. Pioneer Design Studio and Monash University Publications Committee, Melbourne, Victoria. 1437pp.

Vickers-Rich, P. 1996. Early Cretaceous polar tetrapods from the Great Southern Rift Valley, Southeastern Australia. *Memoirs of the Queensland Museum,* Vol. 39, pp.719–723.

Vickers-Rich, P. and Rich, T.H. 1985. Plesiosauridae. The Andamooka sea monster. Pp.143–146. In Rich, P.V. and van Tets, G.F. (eds). *Kadimakara: Extinct Vertebrates of Australia*. Pioneer Design Studio, Melbourne. 284pp.

Vickers-Rich, P. and Rich, T.H. 1993. *Wildlife of Gondwana*. Reed Books, Sydney. 276pp.

Vickers-Rich, P., Rich, T.H., McNamara, G.C. and Milner, A. 1999. Is *Agrosaurus macgillivrayi* Australia's oldest dinosaur? *Records of the Western Australian Museum, Supplement*, No. 57, pp.191–200.

Vickers-Rich, P., Rich T.H., Wagstaff, B.E., Mason, J.M., Douthitt, C.B., Gregory, R.T. and Felton, E.A. 1988. Evidence for low temperature and biologic diversity in Cretaceous high latitudes of Australia. *Science,* Vol. 242, pp.1403–1406.

von Huene, F. 1932. Die fossile Reptil-Ordnung Saurischia, ihre Entwicklung und Geschichte. *Monographs in Geologie und Paläontologie,* Vol. 1, pp.1–361.

Wade, M. 1984. *Platypterygius australis*, an Australian Cretaceous ichthyosaur. *Lethaia,* Vol. 17, pp.99–113.

Wade, M. 1985. *Platypterygius australis:* a Cretaceous marine reptile. Pp.137–142. In Rich, P.V. and van Tets, G.F. (eds). *Kadimakara: Extinct Vertebrates of Australia.* Pioneer Design Studio, Melbourne. 284pp.

Wade, M. 1990. A review of the Australian Cretaceous longipinnate ichthyosaur *Platypterygius* (Ichthyosauria, Ichthyopterygia). *Memoirs of the Queensland Museum,* Vol. 28, pp.115–152.

Waldman, M. 1970. A third specimen of a lower Cretaceous feather from Victoria, Australia. *Condor,* Vol. 72, p. 377.

Warren, G. and Speden, I.G. 1977. The Piripauan and Haumurian stratotypes (Mata Series, upper Cretaceous) and correlative sequences in the Haumuri Bluff district, South Marlborough. *New Zealand Geological Survey Bulletin,* Vol. 92, pp.1–60.

Warren, J.W. 1969. A fossil chelonian of probably lower Cretaceous age from Victoria. *Memoirs of the National Museum of Victoria,* Vol. 29, pp.23–28.

Welles, S.P. and Gregg, D.R. 1971. Late Cretaceous marine reptiles of New Zealand. *Records of the Canterbury Museum,* Vol. 9, pp.1–111.

Welles, S.P. and Long, R.A. 1974. The tarsus of theropod dinosaurs. *Annals of the South African Museum,* Vol. 64, pp.191–218.

White, M.E. 2006. Environments of the geological past. Pp.17–50. In Merrick, J.R., Archer, M., Hickey, G.M., and Lee, M.S.Y. (eds). *Evolution and Biogeography of Australasian Vertebrates.* Auscipub Pty. Ltd., Sydney. 942pp.

Wiffen, J. 1981a. *Moanasaurus*, a new genus of marine reptile (Family Mosasauridae) from the upper Cretaceous of North Island, New Zealand. *New Zealand Journal of Geology and Geophysics,* Vol. 23, pp.507–528.

Wiffen, J. 1981b. The first late Cretaceous turtles from New Zealand. *New Zealand Journal of Geology and Geophysics,* Vol. 24, pp.292–300.

Wiffen, J. 1986. Flying reptiles from the late Cretaceous. Geological Society of New Zealand Annual Conference 1986, Programme and Abstracts. *Geological Society of New Zealand Miscellaneous Publications* 35A, p. 113 (abstract only).

Wiffen, J. 1996. Dinosaurian palaeobiology: a New Zealand perspective. *Memoirs of the Queensland Museum,* Vol. 39, pp.725–731.

Wiffen, J. and Moisley, W.L. 1986. Late Cretaceous reptiles (families Elasmosauridae, Pliosauridae) from the Mangahouanga Stream, North Island, New Zealand. *New Zealand Journal of Geology and Geophysics,* Vol. 29, pp.205–252.

Wiffen, J. and Molnar, R.E. 1988. First pterosaur from New Zealand. *Alcheringa,* Vol. 12, pp.53–59.

Wiffen, J. and Molnar, R.E. 1989. An upper Cretaceous ornithopod from New Zealand. *Geobios,* Vol. 22, pp.531–536.

Wilson, J.A. and Sereno, P.C. 1998. Early evolution and higher-level phylogeny of sauropod dinosaurs. *Society of Vertebrate Paleontology Memoir* 5, pp.i–vi, 1–68 (supplement to *Journal of Vertebrate Paleontology,* Vol. 18, No. 2).

Winkler, D.A., Murry, P.A and Jacobs, L.L. 1998. The new ornithopod dinosaur from Proctor Lake, Texas, and the deconstruction of the family Hypsilophodontidae. *Journal of Vertebrate Paleontology,* Vol. 18, No. 3 (supplement), p.87A.

Woodward, A.S. 1906. A tooth of *Ceratodus* and a dinosaurian claw from the lower Jurassic of Victoria, Australia. *Records of the Geological Society of Victoria,* Vol. 2, pp.135–137.

Wright, K.R. 1989. On the taxonomic status of *Moanasaurus mangahouangae* Wiffen (Squamata: Mosasauridae). *Journal of Palaeontology,* Vol. 63, pp.126–127.

Yadagiri, P. and Ayyasami, K. 1979. A new stegosaurian dinosaur from upper Cretaceous sediments of south India. *Journal of the Geological Society of India,* Vol. 20, pp.521–530.

Yates, A.M. and Kitching, J.W. 2003. The earliest known sauropod dinosaur and the first steps towards sauropod locomotion. *Proceedings of the Royal Society of London,* Series B. DOI:10.1098/rspb.2003.2417.

Chapter 16
Evolution and Zoogeography of Australian Freshwater Turtles

Arthur Georges and Scott Thomson
Applied Ecology Research Group and Cooperative Research Centre for Freshwater Ecology, University of Canberra, A.C.T. 2601 Australia
E-mail: georges@aerg.canberra.edu.au

TURTLE DIVERSITY AND ORIGINS

The turtle shell is one of the most enduring morphological innovations in vertebrate history. It is the single unifying theme for the great diversity of turtle and tortoise species that have come to occupy our oceans, forests, swamps, lakes, rivers and deserts. An estimated 220 species of freshwater turtle alone occupy the waters of all vegetated continents, from the tropics to just outside the Arctic Circle. Centres of biodiversity occur in eastern North America and in southern Asia.

Turtles first appear in the fossil record 210 million years ago (late Triassic), as a very early side-branch of reptilian evolution. No transitional forms exist. Turtles appear abruptly in the fossil record as *Proganochelys quenstedti* (Jaeckel, 1914, 1918) (Figure 1). This earliest known turtle has distinctive and well-developed chelonian features of the carapace and plastron (Figure 2), and the shoulder girdle is located within the rib cage (Gaffney, 1990).

The search for turtle ancestry has relied heavily on osteological data, usually taken from the skull, to compare *bona fide* turtles with groups of reptiles at or near the base of the amniote stock. Gaffney and Meylan (1988) link turtles with the Captorhinidae (anapsids with a generalized tetrapod body plan), but this would require the turtle lineage to have diverged in the early Permian or even the late Carboniferous, some 100 million years before *Proganochelys* appears in the fossil record. Reisz and Laurin (1991) argued that the procolophonids, a group of small tetrapod reptiles (or 'parareptiles') from the Permian of South Africa, are the ancestors of turtles. They found one postcranial and nine cranial characters unique to the two groups. Their hypothesis implies an origin for turtles in the late Permian, closer to the appearance of *Proganochelys* in the late Triassic. Unfortunately, neither of these hypotheses address the transition from a generalized tetrapod body plan to that of a turtle.

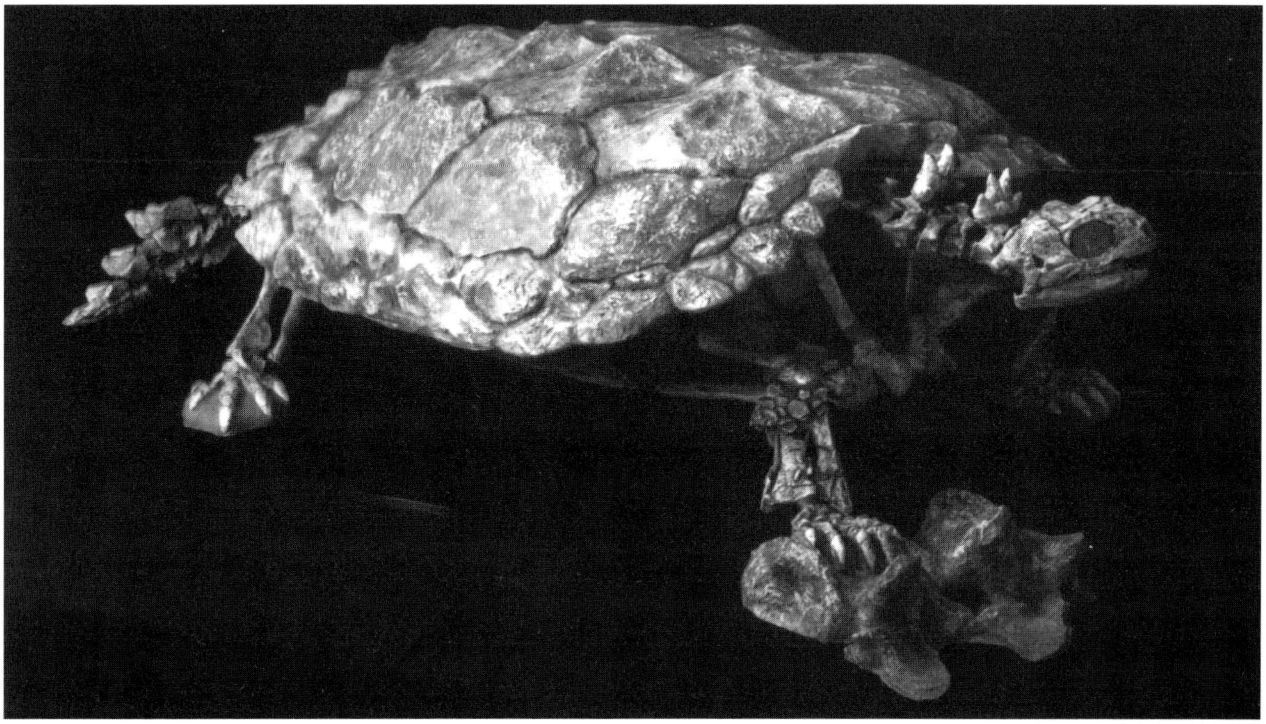

Figure 1. A cast of a reconstructed skeleton of the first known turtle, *Proganochelys quenstedti,* from the late Triassic Norian of Germany (with permission from the American Museum of Natural History).

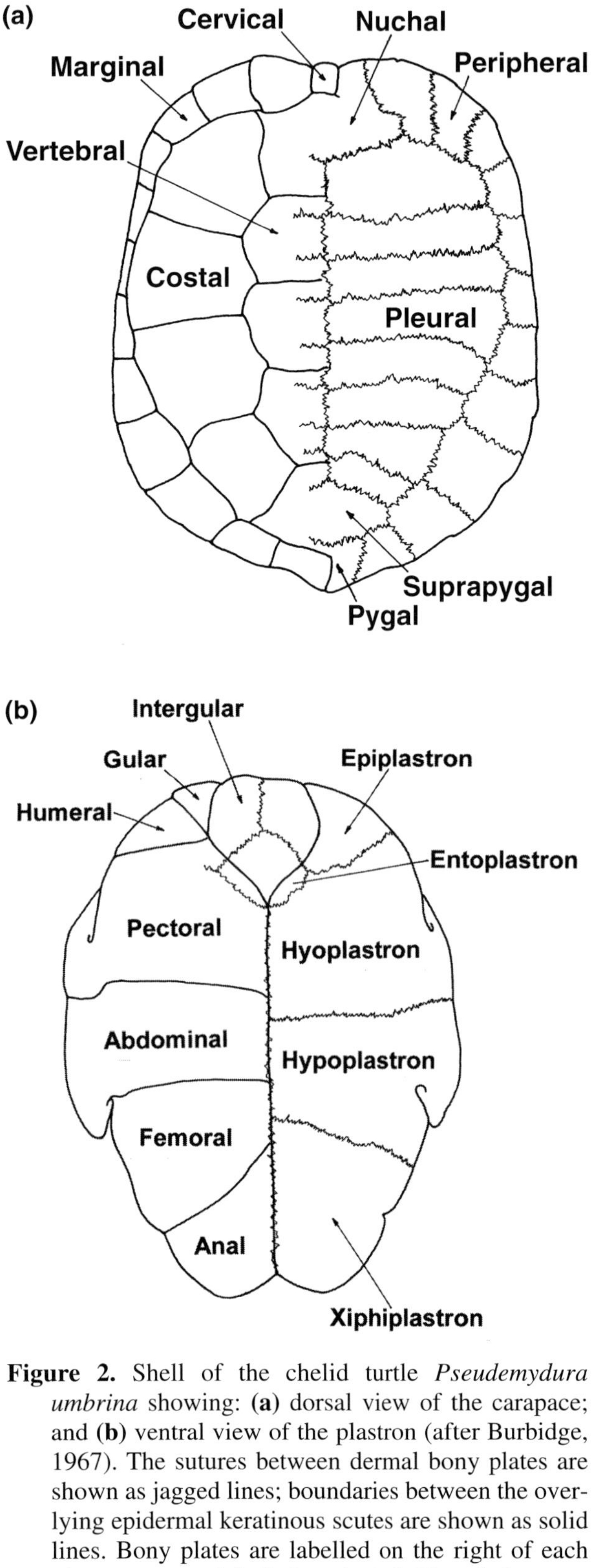

Figure 2. Shell of the chelid turtle *Pseudemydura umbrina* showing: **(a)** dorsal view of the carapace; and **(b)** ventral view of the plastron (after Burbidge, 1967). The sutures between dermal bony plates are shown as jagged lines; boundaries between the overlying epidermal keratinous scutes are shown as solid lines. Bony plates are labelled on the right of each view; scutes are labelled on the left. Neural plates are present only as subsurface elements in most Australian chelid turtles (Thomson and Georges, 1996) and so are not shown. Nomenclature follows that of Zangerl (1969).

Lee (1993) suggests that turtles have evolved from pareiasaurs (strictly, the Pareiasauridae). Pareiasaurs are large and clumsy herbivorous reptiles prominent in the late Permian (Araujo, 1984). Having established this relationship with an impressive array of shared derived characters, Lee speculates on the origins of the turtle shell. The dorsal surface, flanks and sometimes the belly of pareiasaurs were covered in longitudinal rows of bony plates. Lee argues that the extensive dermal ossifications in these animals and the dermal elements of the armour of turtles are homologous—'the precursors of the chelonian carapace and plastron can now be seen in the osteoderms of pareiasaurs' (Lee, 1993, p.1719). Fusion of these dermal osteoderms with the unmodified neural arches of the vertebrae and the ribs resulted in what we know as the turtle carapace. The migration of the pectoral girdle within the shell was possible because the long ribs of pareiasaurs extend laterally from the vertebral column with only slight curvature.

RELATIONSHIPS AND DISTRIBUTION

Proganochelys had unspecialized cervical vertebrae (Gaffney, 1990) and presumably was unable to retract its head beneath the shell. Head retraction was later achieved in two ways. Turtles of the Suborder Cryptodira flex their necks in the vertical plane when the head is retracted beneath the protection of the shell (Plate 62). The Cryptodira comprises nine extant families, and most of the living species of turtles (Pritchard, 1979). Most occur in temperate regions of the Northern Hemisphere—in Europe, Asia and North America, and in the tropics. Notable radiations include the soft-shelled turtles of Asia, Africa and North America (Trionychidae), the terrestrial tortoises of Asia, Europe, Africa and the Americas (Testudinidae), the pond and river turtles of all vegetated continents except Australia (Emydidae), and the marine turtles (Cheloniidae) (Ernst and Barbour, 1989). Marine turtles of the families Cheloniidae and Dermochelyidae, and the freshwater pig-nosed turtle, *Carettochelys insculpta* (Carettochelydidae) are the only living Australian cryptodires.

Turtles of the Suborder Pleurodira flex their neck in the horizontal plane when it is withdrawn into the shell (Plate 63). These are the so-called side-necked turtles. The suborder comprises two families. The Pelomedusidae includes five genera: *Erymnochelys* of Madagascar, *Pelomedusa* and *Pelusios* of Africa, and *Peltocephalus* and *Podocnemis* of South America. Pelomedusids were once much more widespread, with fossils known from deposits in Europe, North and South America, Africa, India and Asia. Their fossil record dates back to the Cretaceous (Pritchard, 1979). *Stupendemys geographicus*, from the Miocene of Venezuela, is the largest turtle known to have existed—with a carapace length of 2.3 m (Wood, 1976).

The second family, Chelidae, is represented by 13 genera—*Acanthochelys, Batrachemys, Chelus,*

Hydromedusa, Mesoclemmys, Phrynops and *Platemys* of South America and *Chelodina, Emydura, Elseya, Elusor, Pseudemydura* and *Rheodytes* of the Australasian region. Their fossil record extends back to the late Cretaceous (de Broin, 1987) of South America and the Miocene of Australia (Gaffney *et al.*, 1989), but no fossil chelids are known from outside the present range of the family (Williams, 1953a, b; Gaffney, 1991). Their Gondwanan origins would appear beyond dispute. All but one living species of Australian freshwater turtle are chelids.

AUSTRALASIAN FAUNAS

FOSSIL RECORD

The fossil turtle fauna provides an essential backdrop to the interpretation of the living turtle fauna of Australia. Four of the six living genera of Australian freshwater turtles are monotypic (*Pseudemydura*, *Rheodytes*, *Elusor* and *Carettochelys*), represented by species that are distinct morphologically (Ramsay, 1886; Siebenrock, 1901; Legler and Cann, 1980; Cann and Legler, 1994) and biochemically (Georges and Adams, 1996). Each of these genera arises from lineages of considerable antiquity. Many distinctive species, including those in the genera above, are restricted to one or a few drainages. Many of these are threatened or endangered by virtue of their restricted natural distributions, coupled with the alterations to habitat brought about by human activities. Although today Australia is the driest vegetated continent on Earth, this was not always so. It is clear from the fossil record, that the extant freshwater fauna of Australia should be considered a relict of a much more diverse fauna of wetter times past (Table 1). There is no doubt that palaeontology provides a fertile field in which many more of our freshwater turtle ancestors await discovery.

The fossil record reveals that the land mass we now call Australia was once home to a number of families of turtle, both pleurodire and cryptodire (Table 1). Pleurodires were the dominant non-marine turtles in the Recent Australian fauna, and were the dominant turtles throughout the Tertiary (Gaffney, 1981), but only cryptodires have been found in the Australian Mesozoic (Gaffney, 1979, 1981; Gaffney *et al.*, 1998).

The oldest turtles from Australia with definitive stratographic data are marine, the chelonioids *Cratochelone* and *Notochelone* from the early Cretaceous Toolebuc Limestones of Julia Creek (Gaffney, 1981, 1991). The oldest freshwater turtle from Australia is *Otwayemys cunicularius*, a cryptodire from the early Cretaceous Eumeralla Formation of Cape Otway in Victoria. It has been assigned to the family Sinemydidae, found elsewhere in the Cretaceous of Mongolia (Brinkman and Peng, 1993).

Table 1. Fossil species of Chelonia from Australia and New Guinea. The list is restricted to fossils that have been assigned to specific species.

Current Name	Family	Original Reference	Horizon and Site
Carettochelys insculpta	Carettochelydidae	Glaessner (1942)	Late Miocene, Papua New Guinea.
Birlimarr gaffneyi	Chelidae	Megirian and Murray (1999)	Miocene Camfield Beds, N.T.
Chelodina insculpta	Chelidae	de Vis (1897)	Plio–Pleistocene, Darling Downs, Qld.
Chelodina longicollis	Chelidae	Thomson (2001, unpublished data)	Pleistocene, Naracoorte Caves, S.A.
Chelodina novaeguineae	Chelidae	Thomson and Mackness (1999)	Early Pliocene, Bluff Downs, Qld.
Elseya lavarackorum	Chelidae	White and Archer (1994)	Pleistocene, Riversleigh Station, Qld.
Elseya nadibajagu	Chelidae	Thomson and Mackness (1999)	Early Pliocene, Bluff Downs, Qld.
Elseya uberima	Chelidae	de Vis (1897)	Plio-Pleistocene, Darling Downs, Qld.
Emydura macquarii	Chelidae	Thomson and Mackness (1999)	Early Pliocene, Bluff Downs, Qld.
Rheodytes devisi	Chelidae	Thomson (2000b)	Plio-Pleistocene, Darling Downs, Qld.
Meiolania brevicollis	Meiolaniidae	Megirian (1992)	Middle Miocene, Bullock Creek, N.T.
Meiolania mackayi	Meiolaniidae	Anderson (1925)	Late Pleistocene, Walpole Island
Meiolania platyceps	Meiolaniidae	Owen (1886)	Late Pleistocene, Lord Howe Island, N.S.W.
Nynjemys oweni	Meiolaniidae	Woodward (1901)	Late Pleistocene, King's Creek, Qld.
Warkalania carinaminor	Meiolaniidae	Gaffney *et al.* (1992)	Miocene, Riversleigh Station, Qld.
Cratochelone berneyi	Chelonioidea	Longman (1915)	Early Cretaceous, Hughenden, Qld.
Notochelone costata	Chelonioidea	Owen (1882)	Early Cretaceous, Julia Creek, Qld.
Chelycharapookus arcuartus	Sinemydidae	Warren (1969)	Early Cretaceous, Merino Group, Vic.
Otwayemys cunicularius	Sinemydidae	Gaffney *et al.* (1998)	Early Cretaceous, Cape Otway, Vic.

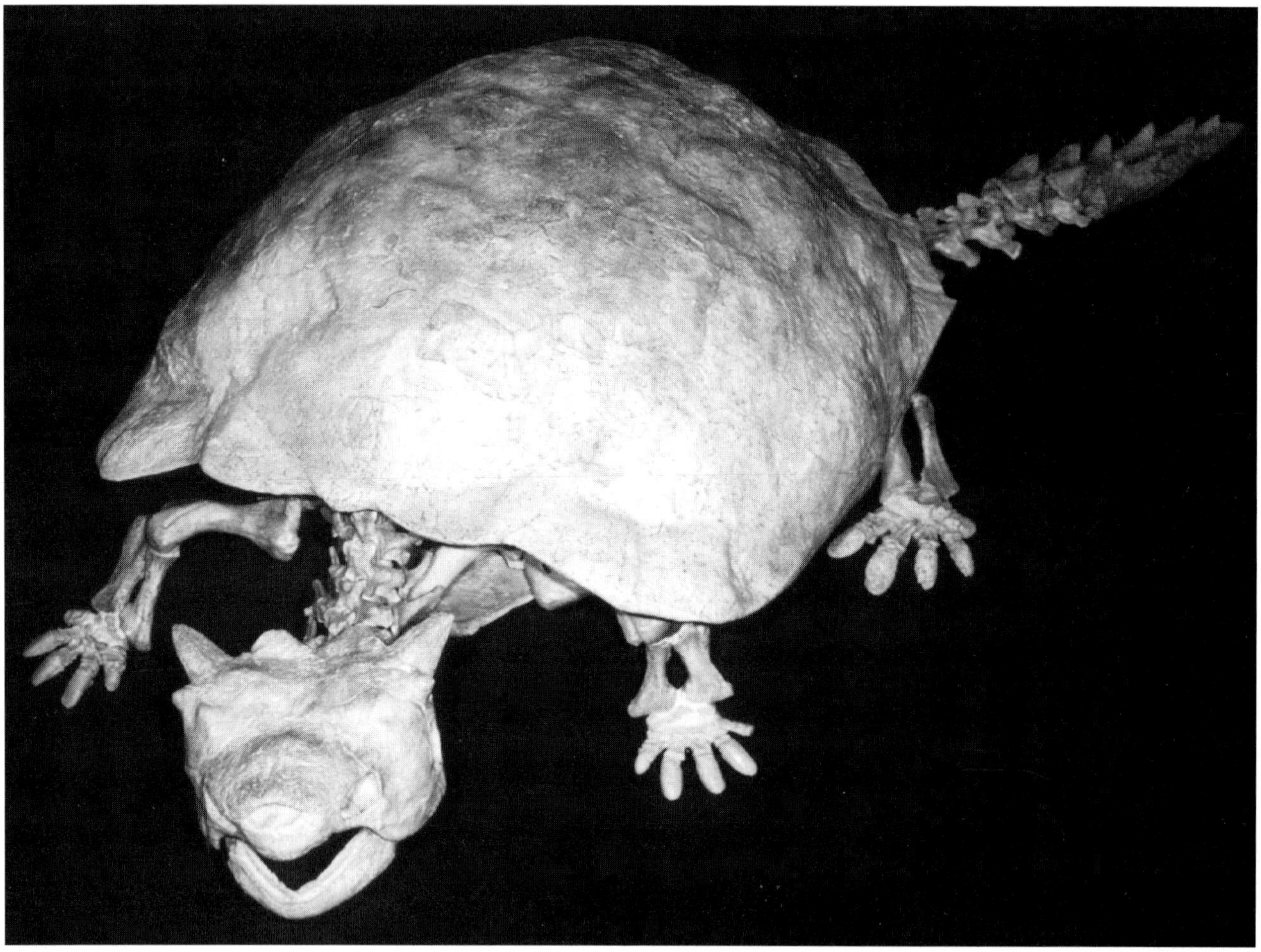

Figure 3. Reconstructed skeleton of *Meiolania platyceps*, from Pleistocene calcarenites of Lord Howe Island (with permission from the American Museum of Natural History).

Another Mesozoic cryptodire, *Chelycarapookus*, is known as an internal mold (Warren, 1969), and as fossil fragments from Lightning Ridge, New South Wales (Gaffney, 1981; Molnar, 1991). *Chelycarapookus* is of uncertain affinity, but it is similar in many features to the sinemydids (Gaffney *et al.*, 1998), to which it may one day be assigned.

The meiolaniids are the real novelties of our fossil record. They occur from the Miocene to the Pleistocene on mainland Australia, in the Pleistocene of Lord Howe Island and New Caledonia, and in the Cretaceous and Eocene of South America (Gaffney, 1981). The best known fossil specimens are of *Meiolania platyceps* from Lord Howe Island (Figure 3). They were large terrestrial turtles with cranial horns and frills, cervical ribs, a bony club at the end of their tails and numerous epidermal ossicles (Gaffney, 1983, 1985). Their bizarre appearance led to much confusion when early palaeontologists endeavoured to identify the fossil remains. Their ecology is unknown, but it is thought likely that they had habits similar to the testudinids (giant tortoises) of the Galapagos and Aldabra Islands (Gaffney, 1985). Phylogenetically, they form an unresolved trichotomy with the sinemydids and the modern cryptodires (Gaffney *et al.*, 1998).

Fossil trionychid material from Australian deposits is fragmentary, but has been diagnosed as Trionychidae on the basis of surface texture pattern and a natural distal edge to the costals, rather than a suture, suggesting that peripheral bones were absent (Gaffney and Bartholomai, 1979). Trionychids were widespread through Queensland (White, 1994), distinct from the extant *Pelochelys* of New Guinea (White, 1990), and range from the Eocene of Murgon (White, 2000, personal communication) to the Pliocene of the Darling Downs and Chinchilla (de Vis, 1897, p.127; Mackness and Thomson, 2000, unpublished data). However, they are absent from the Oligo-Miocene deposits of Riversleigh (Gaffney *et al.*, 1989).

The chelid fossil fauna of Australia is poorly known. The fossil record for chelids dates to the Cretaceous, but the earliest Australian chelid is from the Eocene (White, 2000, personal communication). They dominate the fauna from the Miocene to the present. Good comparative morphological data on living forms are lacking, so *Emydura* and *Chelodina* have been used as broad receptacles for a range of fossil taxa in the Chelidae. Recent advances in our knowledge of the postcranial osteology of chelids (Thomson and Georges, 1996; Thomson *et al.*,

1997), and particularly the identification of characters that distinguish *Elseya* and *Emydura*, have permitted a re-examination of a number of fossil forms. They include *Elseya uberima* (de Vis, 1897) from Chinchilla in Queensland, *Rheodytes devisi* Thomson (2000b) from the Darling Downs, and *Elseya lavarackorum* (White and Archer, 1994) from Riversleigh in the Gulf country (Figure 4). The latter species was first described as a fossil (White and Archer, 1994) but is now known also from living populations in the rivers that flow into the Gulf of Carpentaria (Thomson *et al.*, 1997). Other recent work has uncovered a number of additional forms, such as *Birlimarr gaffneyi* (Megirian and Murray, 1999), from the Miocene Camfield Beds of the Northern Territory and *Elseya nadibajagu* Thomson and Mackness (1999) from Bluff Downs near Charters Towers in Queensland.

Riversleigh Station is a rich source of early to middle Miocene turtle fossils (Gaffney *et al.*, 1989). A specimen of *Chelodina* is the first Miocene record and oldest fossil of this living Australian genus. The discovery there of a fossil *Pseudemydura* is particularly notable as it dramatically extends the range of a genus now extremely rare and restricted to a single species, *Pseudemydura umbrina*, from the south-west corner of Western Australia. These discoveries, together with that of a third genus of freshwater turtle (*Emydura/Elseya*) and perhaps also a new genus at Riversleigh (White, 1993), extend the temporal and geographic range of living Australian turtle taxa and give a better glimpse of the Australian fauna during the Tertiary (Gaffney *et al.,* 1989).

DIVERSITY OF LIVING FORMS

The classification of Australasian chelid turtles is poorly founded, and in drastic need of review (Cogger, 2000). This has been a major impediment to conservation and has limited our understanding of how freshwater turtle biodiversity is distributed across the Australasian landscape. The situation has changed recently, with detailed surveys and analysis of species boundaries using allozyme electrophoresis (Georges and Adams, 1996; Georges *et al.*, 2002) and the subsequent formal description of a number of new species (Cann, 1997a, b, c, d; Thomson *et al.*, 1997; Thomson *et al.*, 2000). Further descriptions are imminent.

Emydura

Species of *Emydura* are widespread and abundant in eastern and northern Australia, where they occupy rivers and the larger, permanent waterholes and billabongs of their floodplains. They are omnivorous short-necked species and are general in their habitat requirements, pro-

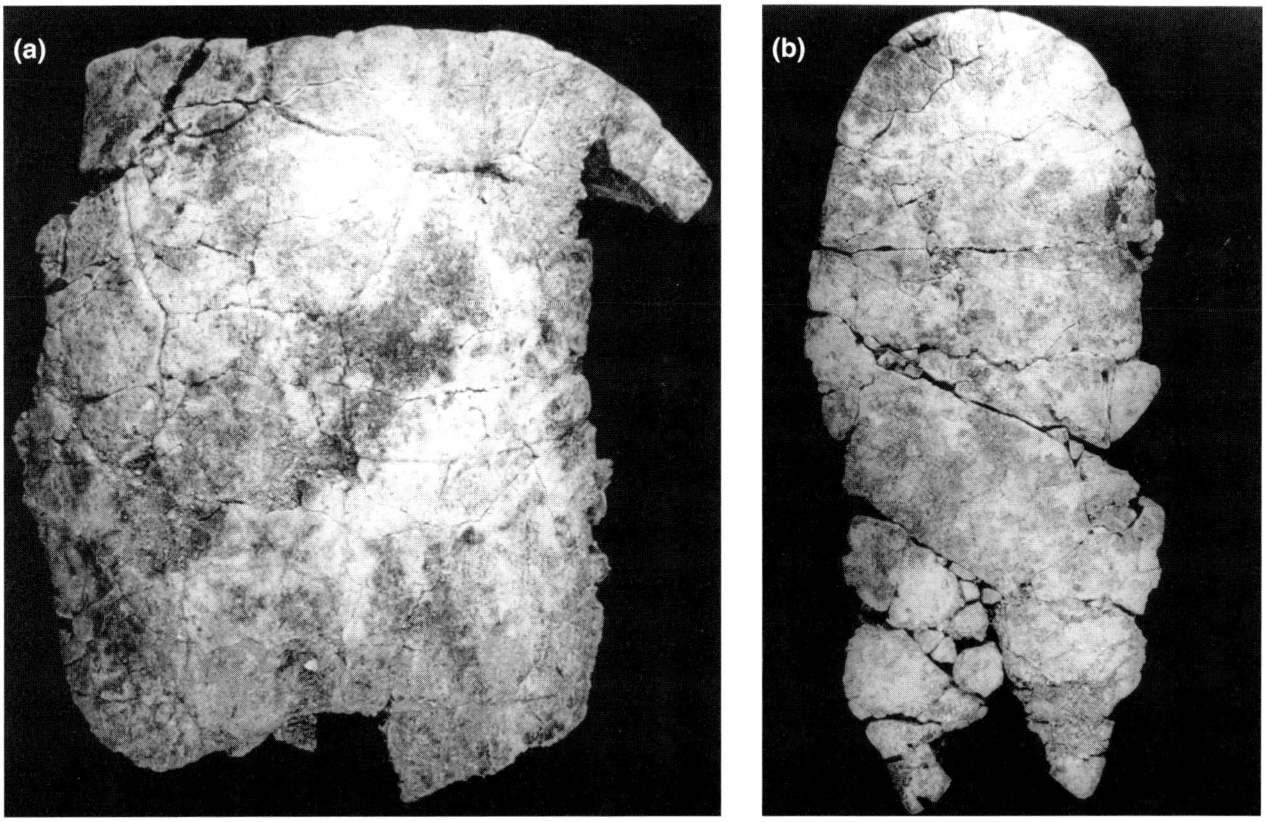

Figure 4. Shell features of a fossil of *Elseya lavarackorum* from the Pleistocene deposits at Riversleigh Station (Queensland). This species has since been found alive and well in the same region (Thomson *et al.*, 1997). **(a)** Dorsal view of the carapace; **(b)** ventral view of the plastron.

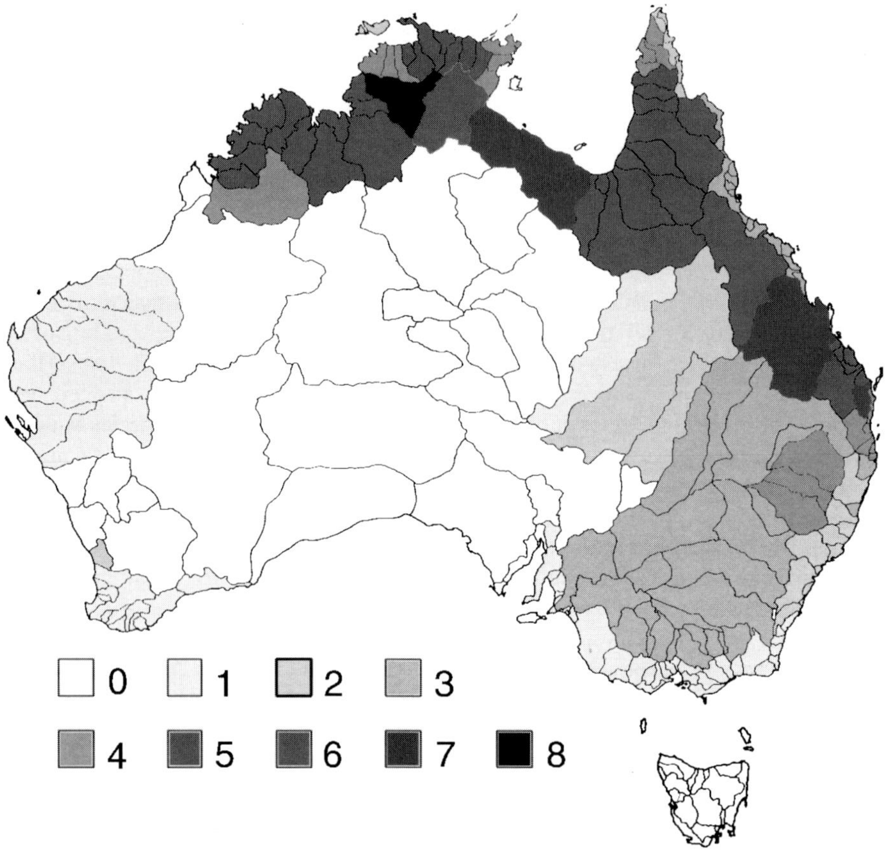

Figure 5. A map of drainage basins showing the distribution of turtle species richness across Australia. There are 'hot spots' in east coastal Australia, centred on the Fitzroy-Dawson drainage (6–7 species) and the 'Top End' centred on the Daly drainage (8 species). Lack of knowledge of the taxonomy prevents a similar map for New Guinea being presented.

vided the water is permanent. Electrophoretic surveys found no clear distinctions between *Emydura signata*, *E. krefftii* and *E. macquarii*. They are found nowhere in sympatry despite abutting distributions, which complicates assessment of species status for very closely related forms. In the absence of clear morphological characters to separate them, they should be regarded as a single biological species, *Emydura macquarii* (Georges and Adams, 1992, 1996). A case can be made for sub-specific distinctions using morphology, colouration and geographic range. Populations of the Murray-Darling drainage and coastal rivers from the Brisbane River south to the Hawkesbury-Nepean are recognized as *Emydura macquarii macquarii*, whereas populations from coastal Queensland north from the Mary River to the vicinity of Princess Charlotte Bay are *Emydura macquarii krefftii.* The population from Fraser Island, and from Cooper Creek in central Australia can each be regarded as subspecies, *Emydura macquarii nigra* McCord *et al.* (2003) and *E. macquarii emmotti* Cann *et al.* (2003) respectively. A number of other subspecies have been suggested for this widespread southern *Emydura* (Cann, 1998), and several workers still prefer to regard *Emydura krefftii* as a distinct species, though without clear diagnosis.

In the north, there are also a number of closely related species. The northern red-faced turtle, *Emydura victoriae*, ranges from the Fitzroy River of Western Australia east to the Daly River of the Northern Territory. The diamond-head or Worrell's turtle is found in the rivers from the Daly River in the west to the rivers flowing into the Gulf of Carpentaria along the west coast of Cape York. This form is electrophoretically indistinguishable from populations of *Emydura subglobosa* in New Guinea, but lacks the red suffusion of the New Guinean populations. It is best referred to as *Emydura subglobosa worrelli. Emydura subglobosa subglobosa* is found in the Jardine River at the tip of Cape York and is widespread in the southern flowing rivers of New Guinea. The northern yellow-faced turtle, *Emydura tanybaraga*, is widespread across northern Australia. All three species of *Emydura* are found in the Daly River.

These species of *Emydura* are all very closely related, differing at most by a few fixed allozyme differences, and are presumably a recent radiation. This is not the case for the genus *Elseya*.

Elseya

What was until recently regarded as a single widespread species, the northern snapping turtle (*Elseya den-*

Table 2. Described species of living freshwater turtle from Australia and New Guinea. Several other species that await description include *Chelodina* sp. (aff. *expansa*) from the Cadell River of the Northern Territory and undescribed species identified by Georges and Adams (1996).

Current Name	Original Name	Original Reference	Geographic Range
Carettochelys insculpta	*Carettochelys insculptus*	Ramsay (1886)	N.T. and southern New Guinea
Pelochelys bibroni	*Trionyx (Gymnopus) bibroni*	Owen (1853)	New Guinea, north of the central highlands
Pelochelys cantori	*Trionyx cantori*	Gray (1864)	New Guinea, south of the central highlands
Chelodina burrungandjii	*Chelodina burrungandjii*	Thomson *et al.* (2000)	Arnhem Land plateau and escarpment, N.T.
Chelodina canni	*Chelodina canni*	McCord and Thompson (2002)	North of Rockhampton, Qld, westwards to the Roper River, N.T.
Chelodina oblonga	*Chelodina colliei*	Gray (1856)	South-west W.A.
Chelodina expansa	*Chelodina expansa*	Gray (1857)	Murray-Darling and south coastal Qld.
*Chelodina kuchlingi**	*Chelodina kuchlingi*	Cann (1997c)	Kimberley, W.A.
Chelodina longicollis	*Testudo longicollis*	Shaw (1794)	Eastern Australia north to the Burdekin, Cooper Creek in central Australia
Chelodina mccordi	*Chelodina mccordi*	Rhodin (1994b)	Roti Island, Indonesia; East Timor
Chelodina novaeguineae	*Chelodina novae-guineae*	Boulenger (1888)	Southern New Guinea
Chelodina parkeri	*Chelodina parkeri*	Rhodin and Mittermeier (1976)	Fly/Strickland Rivers of New Guinea
Chelodina pritchardi	*Chelodina pritchardi*	Rhodin (1994a)	Kemp/Welch Rivers of New Guinea
Chelodina reimanni	*Chelodina reimanni*	Philippen and Grossman (1990)	Merauke region of west New Guinea
Chelodina rugosa	*Chelodina rugosa*	Ogilby (1890)	Northern Australia; Kimberley to Cape York
Chelodina steindachneri	*Chelodina steindachneri*	Siebenrock (1914)	Middle west of W.A.
Elseya belli	*Phrynops belli*	Gray (1841)	Namoi/Gwydir and Border Rivers (N.S.W.)
Elseya branderhorstii	*Emydura branderhorstii*	Ouwens (1914)	South-west New Guinea to Fly River
Elseya dentata	*Chelymys dentata*	Gray (1863)	North-west N.T. to Kimberley of W.A.
Elseya georgesi	*Elseya georgesi*	Cann (1997a)	Bellinger River, N.S.W.
Elseya irwini	*Elseya irwini*	Cann (1997b)	Burdekin River, Qld.
Elseya latisternum	*Elseya latisternum*	Gray (1867)	Richmond River N.S.W., coastal area to South Alligator River, N.T.
Elseya lavarackorum	*Emydura lavarackorum*	White and Archer (1994)	Nicholson River, Qld., to Roper River, N.T.
Elseya novaeguineae	*Platemys Novae Guineae*	Meyer (1874)	Vogelkopf Region of west New Guinea
Elseya purvisi	*Elseya purvisi*	Wells and Wellington (1985)	Manning River, N.S.W.
Elseya schultzei	*Emydura schultzei*	Vogt (1911)	Sepik River region of New Guinea
Elusor macrurus	*Elusor macrurus*	Cann and Legler (1994)	Mary River, Qld.
Emydura macquarii	*Chelys (Hydraspis) macquarii*	Gray (1830)	Murray-Darling system, eastern Qld. to Cooktown, coastal N.S.W. south to Sydney
Emydura subglobosa	*Euchelymys subglobosa*	Krefft (1876)	New Guinea and onshore islands, Cape York to Daly River, N.T.
Emydura tanybaraga	*Emydura tanybaraga*	Cann (1997d)	Cape York, Qld. to Kimberley region W.A.
Emydura victoriae	*Hydraspis victoriae*	Gray (1841)	Daly River, N.T. to Kimberley region W.A.
Pseudemydura umbrina	*Pseudemydura umbrina*	Seibenrock (1901)	Perth region of W.A.
Rheodytes leukops	*Rheodytes leukops*	Legler and Cann (1980)	Fitzroy River of Qld.

**Chelodina kuchlingi* was described from a single specimen of uncertain origin. It was held for a substantial period in captivity and its distinction from *C. rugosa* may be questioned.

tata) is now known to be a series of highly divergent allopatric species (Georges and Adams, 1996). They are river turtles and largely herbivorous. *Elseya dentata* is restricted to the northern rivers west of, but not including, the Mary River of the Northern Territory. A distinct undescribed species occurs in the Mary River, the Alligator Rivers region and the rivers flowing north from the Arnhem Land plateau (*Elseya* sp. [Sth Alligator] of Georges and Adams, 1996). The Gulf snapping turtle, *Elseya lavarackorum*, is poorly known, but its distribution is thought to extend from the Roper River of the Northern Territory in the west to at least as far east as the Gregory-Nicholson systems in Queensland. There is an undescribed species in the Johnstone Rivers region near Cairns, *Elseya irwini* is restricted to the Burdekin and there is a third undescribed species in the Mary, Burnett and Fitzroy Rivers of south-eastern Queensland (Georges and Adams, 1996). A related form, the New Guinea snapping turtle *Elseya novaeguineae*, is widespread in New Guinea; however, *E. branderhorsti* is found only in southern rivers. It is likely that many more species will be discovered in New Guinea and adjacent islands (Rhodin and Genorupa, 2000; Sademi and Iskandar, 2000).

A second lineage of carnivorous and omnivorous species is currently grouped within *Elseya*. They tend to be most abundant in the upper reaches of rivers and their tributaries. The common sawshell, *Elseya latisternum*, is the most widespread. It occurs in coastal rivers from the Richmond River of northern New South Wales, to the Jardine River on the tip of Cape York, as well as rivers discharging into the Gulf of Carpentaria, and the headwaters of Arnhem Land plateau rivers. By contrast, other species in this group are very restricted. The Bellinger River turtle, *Elseya georgesi*, and Manning River turtle, *Elseya purvisi*, are each restricted to the respective coastal river systems (New South Wales) for which they are named. They are of particular interest because they are a cryptic species pair, morphologically difficult to distinguish on external examination, but deeply divergent genetically (Georges and Adams, 1996). Once thought to be a single species, the genetic examination prompted a closer look at their morphology and revealed that *Elseya purvisi* has a well-developed series of exposed neural bones, consistently lacking in *Elseya georgesi* (Thomson and Georges, 1996). Neural bones in most chelid turtles are reduced subsurface bony elements of the carapace lying immediately above the vertebral column (Thomson and Georges, 1996). A fourth species in this group, *Elseya belli*, is found in the granitic headwaters of the Namoi, Gwydir and Severn tributaries of the Darling River in inland New South Wales.

Other Short-necked Genera

The remaining short-necked chelid genera are monotypic. *Rheodytes leukops* is restricted to the Fitzroy-Dawson drainage, *Elusor macrurus* is found only in the Mary River of coastal Queensland, and *Pseudemydura umbrina* is restricted to coastal swamps near Perth.

Chelodina

Australia is well known for its snake-necked chelid turtles (genus *Chelodina*) because their necks can be spectacularly long in relation to their bodies, an innova-

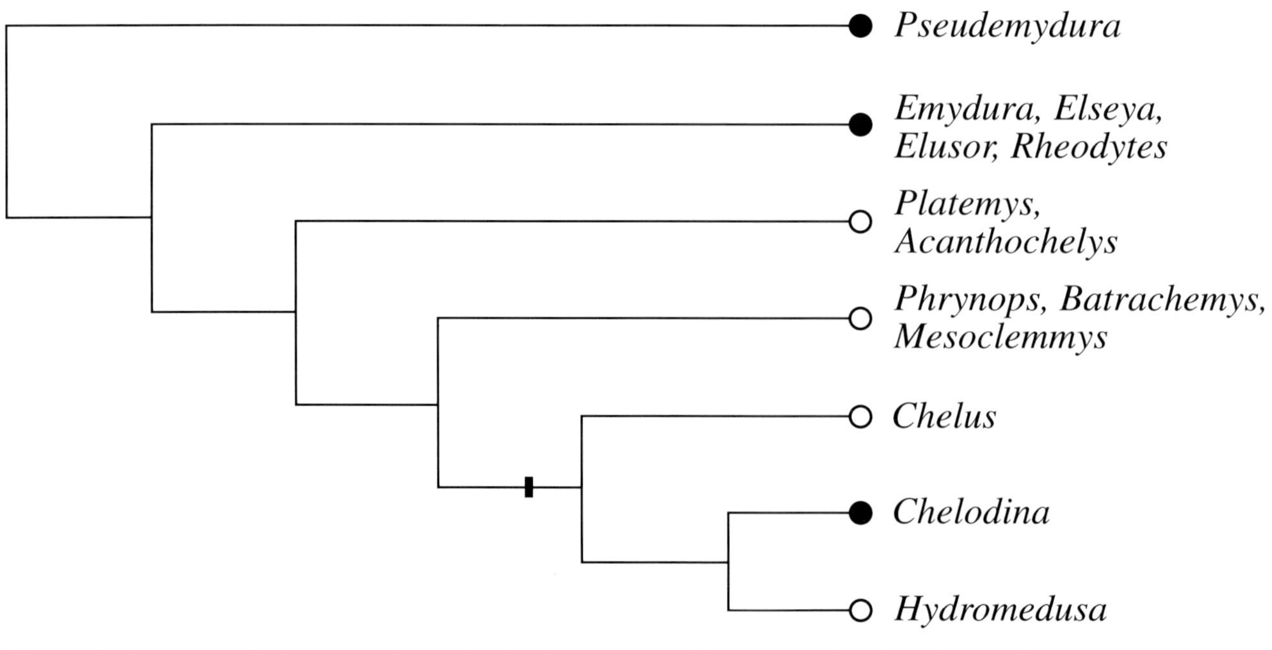

Figure 6. The most widely accepted hypothesis of the relationships among chelid turtles (Gaffney, 1977). Note that the three long-necked genera (*Chelus*, *Hydromedusa*, *Chelodina*) form a monophyletic group. This phylogeny, based on morphological characters, is at odds with the best molecular hypothesis (Georges *et al.*, 1998). Key to symbols: ▮ = long-necked; ● = Australian genera; ❍ = South American genera.

tion rare among the cryptodiran turtles that dominate the turtle fauna of the Northern Hemisphere. The evolution of long necks confers benefits of attenuated strike and gape feeding that enables these animals to secure fast-moving prey not available to their short-necked cousins. When feeding, the head and neck are thrust out against the inertia of the body. At the same time, the floor of the mouth is lowered causing an in-rushing of water to suck in the prey (Parmenter, 1976).

Species of *Chelodina* fall into three sub-generic groups (Burbidge *et al.*, 1974). *Chelodina longicollis* and *C. steindachneri* of Australia, *C. pritchardi, C. novaeguineae* and *C. reimanni* of New Guinea, *C. mccordi* of Roti and *C. canni* of Australia and New Guinea belong to Group A. They are carnivorous foragers with relatively narrow heads, shorter thinner necks and broader plastrons (Goode, 1967; Rhodin, 1994a, b). *Chelodina expansa, C. burrungandjii, C. rugosa, C. kuchlingi* and an undescribed species (*Chelodina* sp. aff. *expansa* [Cadell]) of Australia, and *C. siebenrocki* and *C. parkeri* of New Guinea belong to Group B. They are ambush predators with relatively broad heads, longer thicker necks and narrower plastrons (Goode, 1967; Rhodin and Mittermeier, 1976; Thomson *et al.*, 2000). *C. rugosa* and *C. siebenrocki* have been variously regarded as distinct species (Wermuth and Mertens, 1961; Burbridge *et al.*, 1974; Cogger, 1983) or the same (Siebenrock, 1909, 1915; Goode, 1967). Morphological evidence in support of their separation is scant. Burbidge *et al.* (1974) cite only 'consistent, if minor, differences' between the two as their basis for recognizing them as separate species. Rhodin and Mittermeier (1976) failed to reliably distinguish *C. rugosa* and *C. siebenrocki* following a morphological analysis, and chose to refer to them collectively as the *C. rugosa* complex. In the absence of any substantive diagnostic morphological differences or any substantive allozyme differences (Georges *et al.*, 2002), we synonymize *C. siebenrocki* with *C. rugosa* (the latter name has precedence). *Chelodina kuchlingi* was described from a single specimen of uncertain origin. It was held for a substantial period in captivity and, following a preliminary morphological analysis, we also have doubts about its separate identity from *C. rugosa*.

Chelodina oblonga of south-western Australia (= *C. colliei*, of Thomson, 2000a) is superficially similar to species of the Group B, and has often been placed in that group (Goode, 1967; Legler, 1981). However we follow Burbidge *et al.* (1974) and place it in a third subgeneric group, Group C. It is distinguished from other described *Chelodina* by a consistent set of well developed neural bones (Burbidge *et al.*, 1974; Thomson and Georges, 1996).

Carettochelys

Finally, the cryptodire, *Carettochelys insculpta*, is found in the Victoria, Fitzmaurice, Daly, South Alligator, East Alligator and Goomadeer Rivers of the Northern Territory.

In all there are 24 species of freshwater turtle currently described in Australia and 12 in New Guinea

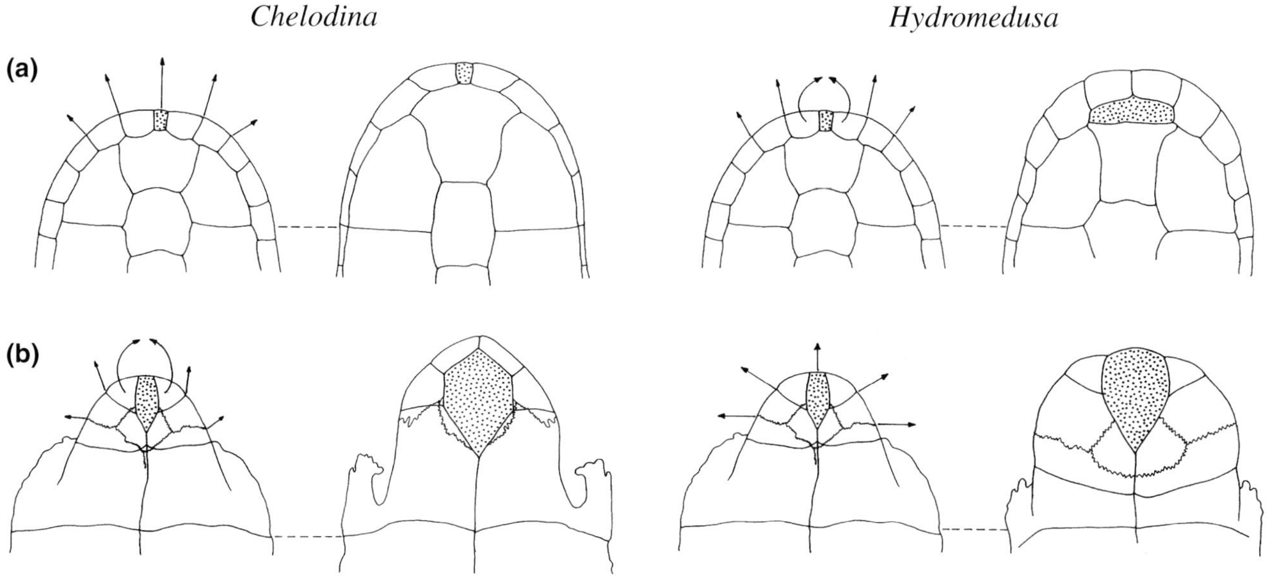

Figure 7. Shell and scute expansion in the evolution of *Chelodina* and *Hydromedusa* to accommodate the head and long neck within the shell. **(a)** Carapace. In *Chelodina*, expansion is achieved by elongation of the anterior vertebral and costal scutes, and the cervical scute (stippled) remains largely unaltered. In *Hydromedusa*, the cervical scute is expanded laterally and recedes behind the marginals. **(b)** Plastron. In *Chelodina*, the pectorals have elongated, and the gulars and humerals have migrated around the intergular scute (stippled). In *Hydromedusa*, the entoplastron and intergular scute are enlarged. Pritchard (1984) believes this argues against a close relationship between the two long-necked genera. Dashed line: common reference line. For scute names, refer to Figure 2. (After Pritchard, 1984).

(Table 2). As a result of recent molecular and morphological work we are, for the first time, able to map the distribution of freshwater turtle biodiversity across the Australian landscape (Figure 5). This map shows a hot spot of biodiversity in the Northern Territory. The Daly River alone has eight species and the broader region supports four endemics (*Carettochelys insculpta*, *Chelodina burrungandjii*, *Chelodina* sp. [Cadell], *Elseya* sp. [Sth Alligator] of Georges and Adams, 1996). There is a second hot spot in southern coastal Queensland, in the Mary, Burnett and Fitzroy drainages (Figure 5). The Fitzroy drainage basin alone supports six species and the broader region supports three endemics (*Rheodytes leukops*, *Elusor macrurus*, *Elseya* sp. [Burnett] of Georges and Adams, 1996). A seventh species, *Chelodina canni*, may also occur in the Fitzroy, as it is present in the small adjacent coastal drainage, the Styx. A third hot spot occurs in the south-western corner of Western Australia, based not on species richness, but on its two species, both endemic (*Chelodina oblonga*, *Pseudemydura umbrina*).

PHYLOGENY AND ZOOGEOGRAPHIC HYPOTHESES

Living and fossil chelid turtles are restricted to South America and Australasia, which raises the question of how the two faunas relate to each other.

SUGGESTED RELATIONSHIPS

The very early literature splits the Chelidae into two groups that do not recognize continental boundaries—those with a neck shorter than the shell (*Phrynops*, *Platemys*, *Elseya* and *Emydura*) and those with longer necks (*Chelus*, *Hydromedusa* and *Chelodina*) (Boulenger, 1888). This arrangement is reflected in the most widely accepted phylogeny for the chelids (Gaffney, 1977) based on a numerical cladistic analysis of a wide range of cranial and post-cranial characters (Figure 6). It is also supported by the more recent work on fossil material from Argentina by de Broin and Fruente (1993), and the morphological analysis

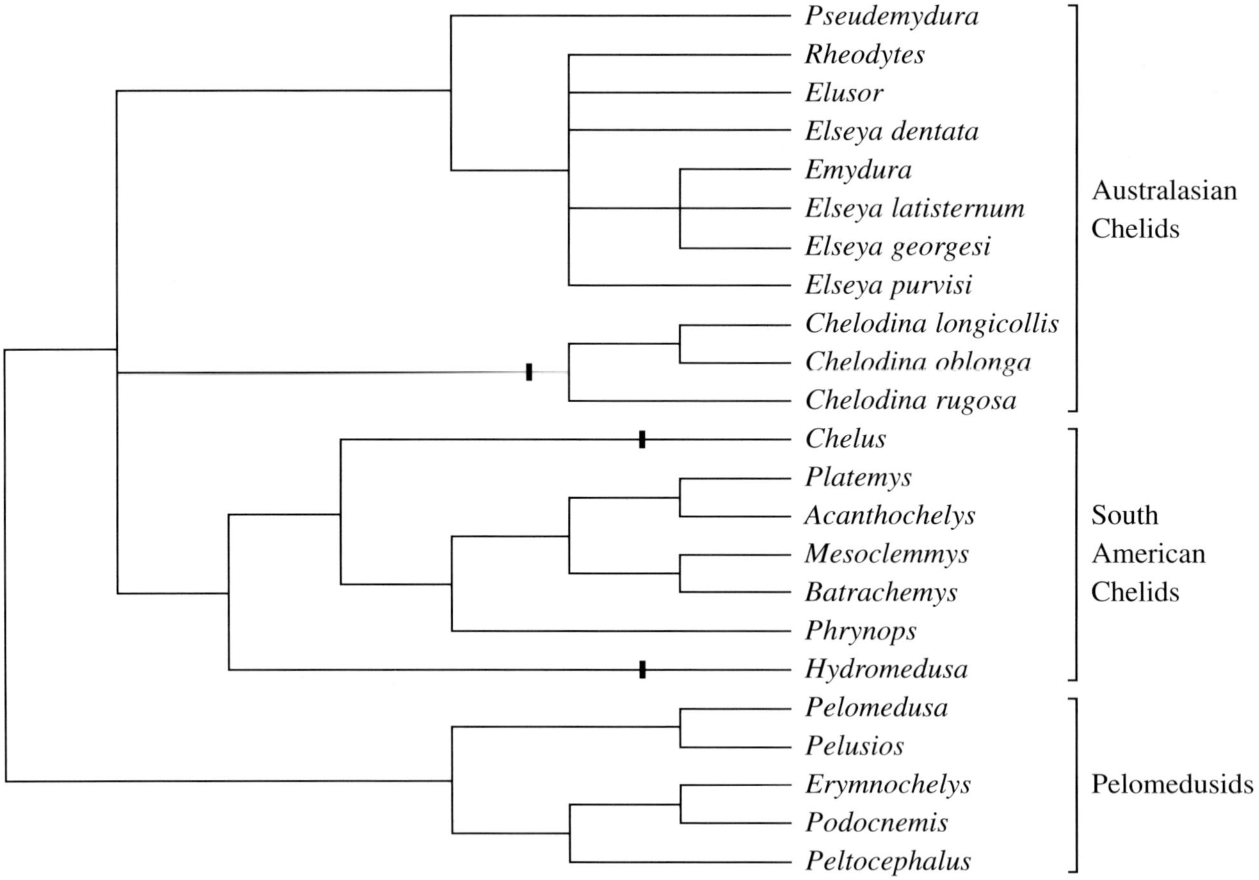

Figure 8. A consensus phylogeny for the Pleurodira, based on analysis of sequence data from three mitochondrial genes and one nuclear gene (Georges *et al.*, 1998). Bootstrap support for all resolved nodes exceeded 70%. The pelomedusids were used as an outgroup for the chelids. Note that the South American chelids are monophyletic, and that none of the long-necked genera (*Chelus*, *Hydromedusa*, *Chelodina*) are particularly closely related. Most of the molecular analyses had the Australian and South American chelids as reciprocally monophyletic, and the Australasian clade consisting of short-necked and long-necked lineages. Key: ▮ = long-necked.

Plate 62. The mode of withdrawal of the head and neck in an Australian hidden-necked turtle (Cryptodira: Carettochelyidae), the pig-nosed turtle *Carettochelys insculpta*. Photograph J. Cann.

Plate 63. The mode of withdrawal of the head and neck in an Australian side-necked turtle (Pleurodira: Chelidae), the Gulf snapper *Elseya lavarackorum*. Photograph A. Georges.

Plate 64. This broad-shelled turtle (*Chelodina expansa*) has a broad range in the Murray–Darling and southern North-east Coast drainage divisions. This secretive species can be found in both rivers and still impoundments, such as farm dams. Photograph J. Cann.

Plate 65. Another freshwater turtle from the Namoi River, New South Wales. This is one of the *Elseya latisternum* group from mid-eastern Australia. Photograph J. Cann.

of Shaffer *et al.* (1997) which consistently grouped *Chelodina* with *Chelus*. These analyses support a common derived origin of the long neck and associated habits of the South American *Hydromedusa*, *Chelus* and the Australian *Chelodina*, that pre-dated the separation of Australia and Antarctica from South America.

Although the above scheme summarized the prevalent view, it was not universally accepted. Pritchard (1984) took a more functional approach to his interpretation of characters than did Gaffney (1977). He argues that the development of a long neck requires expansion of the anterior plastral lobe and the anterior carapace to accommodate and protect the neck when withdrawn. Pritchard shows that turtles in the three genera have achieved this in fundamentally different ways (Figure 7), suggesting independent origins of the long neck, with supporting evidence from a range of other characters. Pritchard does not regard any of the long-necked genera as being particularly closely related, but regards *Hydromedusa* as more closely related to *Chelus* than to *Chelodina.*

The tensions that developed as a result of these conflicting views erupted in an interesting, if unproductive, debate between those who adhered to a numerical cladistic approach to phylogenetic reconstruction and Pritchard, who would admit functional interpretations of characters as evidence of close relationship or convergence (Pritchard, 1994; De Queiroz, 1995; Frost, 1995; Kluge, 1995; Meylan, 1995).

Pritchard's conclusion of independent origins of the long necks of *Chelodina* and the South American forms receives strong support from recent molecular studies (Seddon *et al.*, 1997; Shaffer *et al.*, 1997; Georges, 1997). The molecular phylogeny, based on DNA sequences from four mitochondrial genes (*12S rRNA*, *16S rRNA*, *CO1* and *cytochrome b*) and a nuclear gene (c-*mos*), has all South American chelids as a monophyletic assemblage to the exclusion of *Chelodina* and the other Australian chelids (Figure 8). Bootstrap values were in the range 76–82% (Georges *et al.*, 1998). Our reasons for preferring the molecular phylogeny over that generated by applying numerical cladistics to morphological data are presented by Georges *et al.* (1998). It would seem from the molecular analyses that the living Australian chelid turtles represent radiations that no longer have representatives in South America, though a strict analysis has the South American chelids, *Chelodina* and the remaining Australasian chelids in an unresolved trichotomy (Georges *et al.*, 1998). This caveat aside, much of the current freshwater turtle diversity in Australia most likely arose through radiations that occurred after the separation of Australia and Antarctica from South America, perhaps from a very limited surviving stock.

Carettochelydids are not of Gondwanan origin, as the fossil record for the family shows that it was a diverse group covering much of Laurasia by the Eocene (Meylan, 1987). *Carettochelys* is thought to have entered the Australasian region via South-east Asia.

Detailed molecular phylogenies are available for each of the short-necked and long-necked chelid turtles of Australasia (Georges and Adams, 1992; Georges *et al.*, 1998; Georges *et al.*, 2002), an essential prerequisite for generating zoogeographic hypotheses. The short-necked phylogeny is interesting in many respects (Figure 9). First, the genus *Elseya* is paraphyletic, as the closest common ancestor to the species of *Elseya* has *Emydura* among its descendants. This is an unacceptable arrangement for a genus, and is soon to be resolved by erecting a new genus for *Elseya latisternum*, *E. georgesi*, *E. purvisi* and *E. belli* (Thomson *et al.*, 2000, unpublished data).

A second interesting feature of the short-neck phylogeny is that the cryptic species pair, *Elseya georgesi* and *E. purvisi*, are not sister taxa (Georges and Adams, 1992; Georges *et al.*, 1998). *Elseya latisternum* and *E. belli* have arisen from a common ancestor with *E. georgesi* after the divergence of *E. georgesi* from *E. purvisi*. The rivers that they occupy are similar—clear, continuously flowing, over a stony bed—and in an area of long-term geological stability. There are two evolutionary scenarios to explain this aspect of the phylogeny. Either *Elseya georgesi* and *Elseya purvisi* have evolved independently but in parallel to maintain their striking similarity since they diverged, or they have jointly retained the primitive features of their common ancestor. The latter interpretation is much more likely, and if true, we have a rare window in time to view the ancestral form of *Elseya latisternum* and *E. belli*.

The *Chelodina* phylogeny is not contentious in most respects (Figure 10). One trend evident in the phylogeny is toward progressively greater robustness in the jaws of turtles most closely related to *Chelodina longicollis* (Group A species), culminating in the robust jaws of the molluscivorous *Chelodina reimanni*. Long robust necks and other modifications associated with an attenuated strike and suction mode of feeding, in response to a piscivorous diet (Pritchard, 1984), have evolved in species related to *Chelodina expansa* (Group B species). Traditionally, *Chelodina oblonga* was placed in this latter group, their similarity of form taken to indicate close phylogenetic relationship (Goode, 1967; Legler, 1981); however, recent molecular phylogenies (Georges and Adams, 1992; Georges *et al.*, 1998; Georges *et al.*, 2002) indicate that the adaptations associated with piscivory in this group have arisen twice independently. *Chelodina oblonga* has its closest relatives among the *Chelodina longicollis* group (Figure 10). Indeed, it would now appear that the exceptionally long necks and attenuated strike and suction mode of feeding have arisen four times in the Chelidae, in response to adopting a piscivorous diet (Pritchard, 1984), and perhaps because side-necked turtles are pre-disposed to do so. It has arisen in *Chelus* (sister to *Phyrnops*), *Hydromedusa* (sister to the remaining South American turtles), the *Chelodina expansa* group (sister to the remaining *Chelodina*), and *Chelodina oblonga* (sister to the *Chelodina longicollis* group) (Figures 8 and 10).

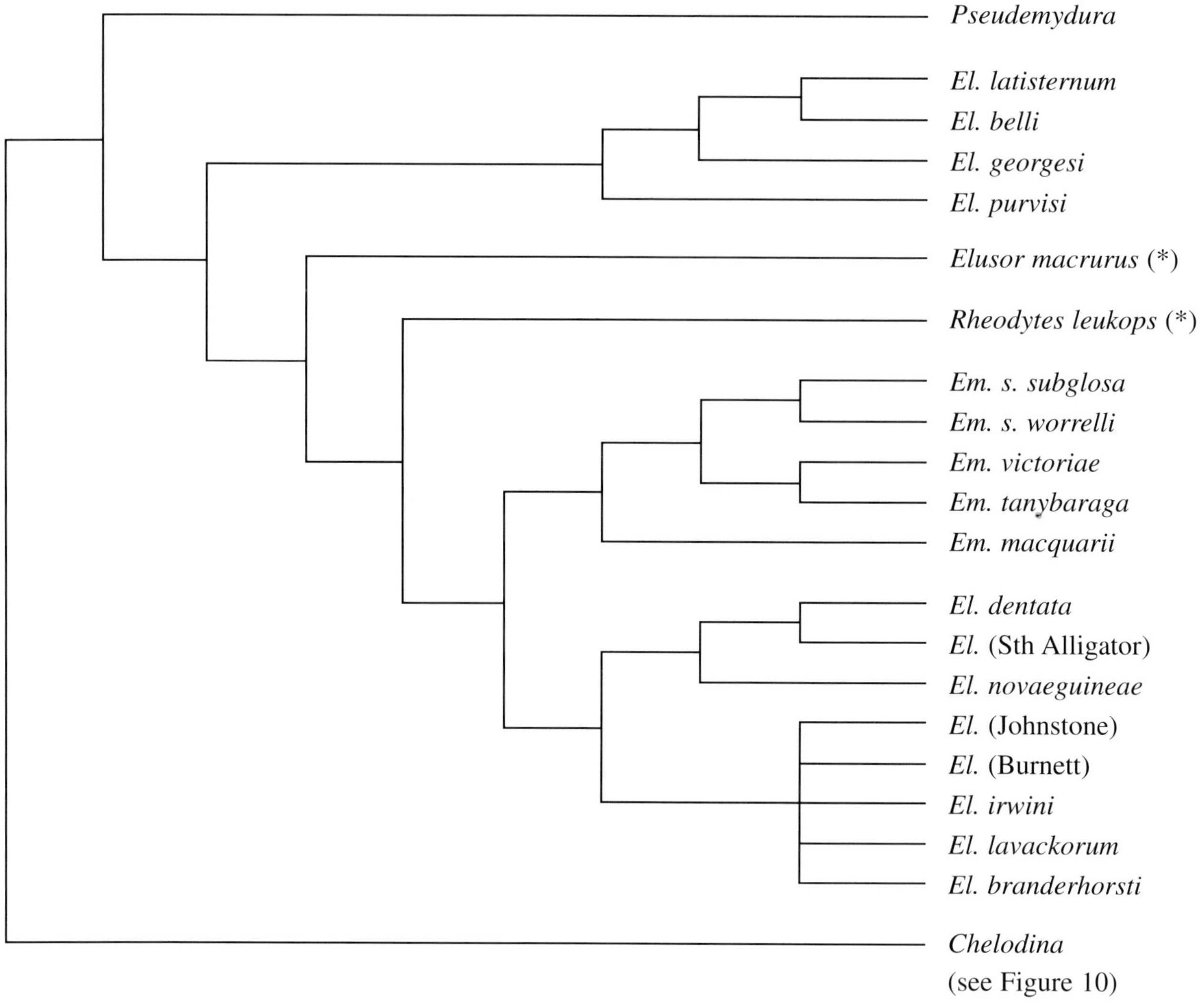

Figure 9. A phylogeny for the extant Australian chelids (outgroup = *Chelodina*), combining molecular (Georges and Adams, 1992; Georges *et al.*, 1998) and morphological results (Megirian and Murray, 1999). *Pseudemydura* was placed as sister to the remaining short-necks, to the exclusion of *Chelodina*, by the DNA sequence data (Georges *et al.*, 1998). The relationships of *Rheodytes leukops* and *Elusor macrurus* (*) are based entirely on the morphological data of Megirian and Murray (1999). Note that the genus *Elseya* is paraphyletic. Key to name abbreviations: *El.* = *Elseya*; *Em.* = *Emydura.*

DISTRIBUTION PATTERNS AND INFLUENCES

With the species boundaries well established and well supported phylogenies available for these species, it is possible to speculate on zoogeography within Australia. A number of climatic and geomorphological factors are thought to have influenced freshwater turtle distribution in Australia and Burbidge *et al.* (1974) first presented detailed arguments on the subject. They interpreted high endemism in the turtle fauna of south-west Western Australia to be a consequence of the sea that divided Australia from the Gulf of Carpentaria to the south coast in the Cretaceous and to the progressively increasing aridity and fragmentation of suitable habitat in the Tertiary. Discovery of a species of *Pseudemydura* from the Miocene of north-eastern Australia weakens the link to a sea barrier, but there is no doubt that increasing aridity in the Tertiary fragmented the landscape and limited dispersal opportunities. Turtles are absent from much of Australia's interior, and the western species *Chelodina steindachneri*, *Chelodina oblonga*, and *Pseudemydura* are isolated by aridity that extends to the coast.

Sea level change too has had an important influence. There have been 17 distinct glacial-interglacial cycles of at least 100,000 years duration during the last 1.8 million years of the Quaternary (Williams, 1984). These cycles have affected global eustatic sea levels (discounting local and regional tectonic and isostatic effects). Glacial episodes have resulted in low global sea levels, while

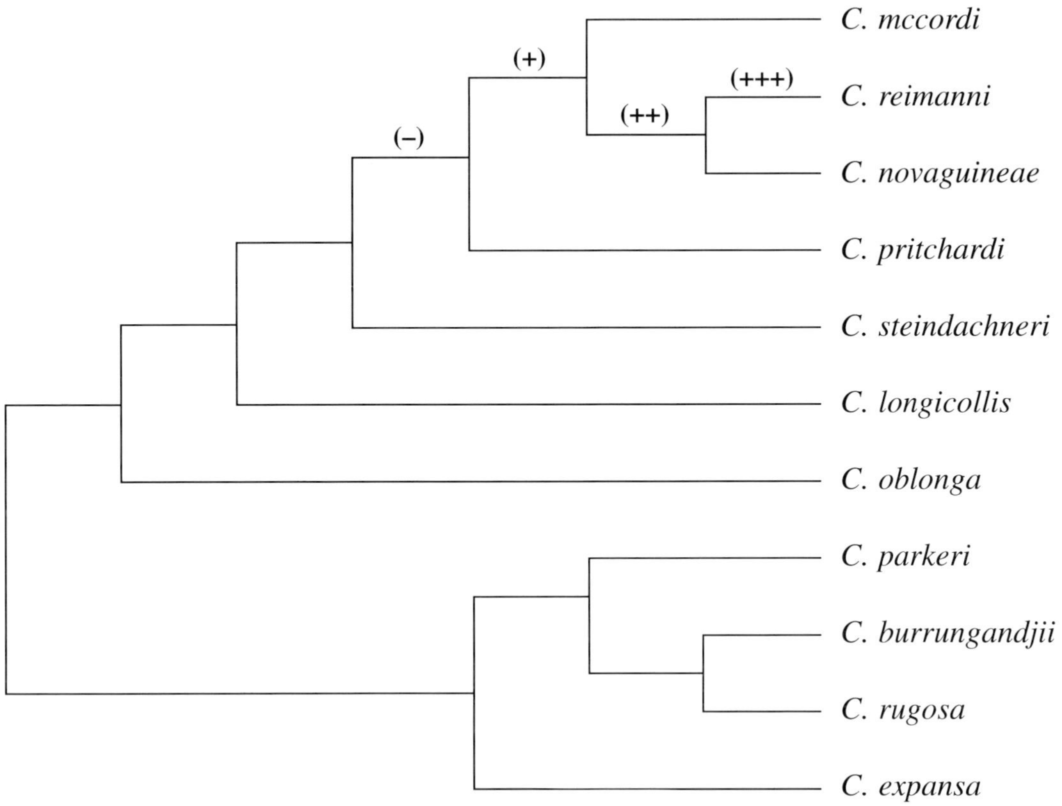

Figure 10. A phylogeny for the extant long-necked Australasian chelids (genus *Chelodina*) based on allozyme electrophoresis (Georges and Adams, 1992). No instances of paraphyly arise. Note that *Chelodina oblonga*, superficially similar to turtles in the *Chelodina expansa* group, has its strongest affinities with the *Chelodina longicollis* group. This arrangement is supported by the DNA sequence data (Georges *et al.*, 1998). Note that *C. reimanni* and *C. novaeguineae* (including *C. canni*) could not be separated electrophoretically, but we retain them as separate on the basis of morphological evidence (Philippen and Grossman, 1990; Rhodin, 1994a). Key to symbols: + = the progressive development of robusticity in both the skull and triturating surfaces; – = not robust.

glacial melting (interglacials) has resulted in high eustatic levels. Interglacial episodes with high global shorelines, such as those being experienced at present, have only accounted for about 10% of each cycle (Williams, 1984). Global sea levels are currently higher than at anytime during the last 120,000 years, separating Australia and New Guinea; however, Torres Strait has acted consistently as a land-bridge since the last interglacial about 118,000 years ago up until about 6,000 years ago, when marine transgression closed it (Chapell, 1983, 1994). This land-bridge is almost certainly the avenue by which *Emydura subglobosa subglobosa* entered Australia; the role of the exposed Sahul continental shelf and its implications for other New Guinean vertebrates has been discussed by Heinsohn and Hope (2006)—see pp.71–93.

About 12,000 years ago, sea levels were low enough that the Arafura Sill (53 m below current sea level) was exposed, and 20,000 years ago sea levels were 120 m below present levels (Chapell, 1994). The Arafura Sill is a broad feature nearly 100 km wide, with a total relief of less that 2 m (Torgerson *et al.*, 1983) that extends from the north-eastern extent of Arnhem Land to New Guinea east of Pulau Dolak and west of the mouth of the Fly. Between 12,000 and 55,000 years ago, the Gulf of Carpentaria was a large inland lake (Chapell, 1994) (Figure 11). The mighty Fly River flowed into the lake before being diverted to the east (Blake and Ollier, 1971; Torgersen *et al.*, 1988; McCulloch *et al.*, 1989). Pollen studies have shown that the vegetation surrounding the lake was very much as it is today in the open savanna country. The lake would have been fresh or brackish for much of its existence. If it connected to the sea, and no channel is evident from bathometry, then it would have done so to the west.

Evidence from deep core drilling reveals a pattern of establishment and marine inundation of Lake Carpentaria that appears to have been repeated. It was a freshwater lake in the Jurassic then inundated by a marine transgression (in limestone deposits), and there was a further freshwater episode in the Miocene, followed by another marine transgression (McConneckie, 1996, personal communication).

The current distribution of a number of northern turtle species can be explained by the opportunities the lake and the exposed Arafura Sill provided. *Chelodina rugosa* of Australia and *Chelodina siebenrocki* of New Guinea

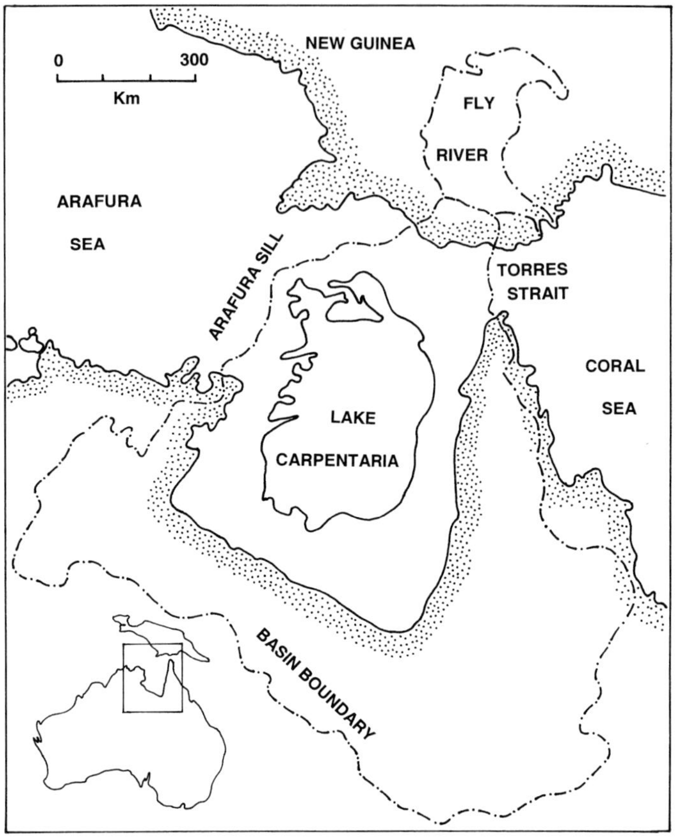

Figure 11. A map showing the relationship between Australia and New Guinea in the late Quaternary (55,000–12,000 years ago). The Gulf of Carpentaria was then a large inland freshwater or brackish lake. The Arafura Sill is 53 m below current sea levels, and Torres Strait is 12 m below current sea levels. The lake boundary corresponds to the –53 m contour, which corresponds to the maximum extent of Lake Carpentaria (Torgersen *et al.*, 1983; Torgersen *et al.*, 1985; Torgersen *et al.*, 1988). The Fly River flowed into the lake prior to its diversion into the Coral Sea in the late Quaternary. At the glacial maximum, most of what is shown as the Arafura Sea would have been terrestrial (after McCulloch *et al.*, 1989).

are indistinguishable electrophoretically (Georges *et al.*, 2002) and poorly differentiated morphologically (Rhodin and Mittermeier, 1976), and we regard them as a single taxon once forming a contiguous series of populations that circled Lake Carpentaria. Lack of variation shown by *Emydura subglobosa subglobosa* from southern Papua New Guinea and *Emydura subglobosa worrelli* from the Australian rivers discharging into the Gulf of Carpentaria (Georges and Adams, 1996) can be similarly interpreted. So too can the distribution of *Chelodina novaeguineae* and the closely related *C. canni*, and preliminary morphological analyses indicate that *Elseya lavarackorum* from the Gulf country and *Elseya branderhorsti* from southern New Guinea are very closely related. It seems that a number of contiguous turtle distributions have been disrupted in the last few tens of thousands of years by the inundation of Lake Carpentaria.

The Arafura Sill that defined the western boundary of Lake Carpentaria would also have provided a land-bridge to New Guinea presumably with drainages flowing west to the Timor Sea. This would have allowed potential interchange of forms between West Papua, Arnhem Land and the Kimberley via coastal rivers and associated habitat quite different from that provided by Lake Carpentaria. It would also have isolated the turtle fauna of these western and west-central rivers from those flowing into the eastern seaboard of Australia and south eastern New Guinea. This may explain the high level of endemism for the region from the Kimberley and western Arnhem Land—*Emydura victoriae*, *Carettochelys insculpta*, *Elseya dentata*, *Elseya* sp. [Sth Alligator], *Chelodina* sp. [Cadell] and *Chelodina burrungandjii*. *Carettochelys* presumably entered Australia from New Guinea via the opportunities that arose from the Arafura Sill land-bridge. These zoogeographic hypotheses, and others to do with Lake Carpentaria, could be tested by molecular comparisons between the Australian and New Guinean fauna.

In south-eastern Australia, the primary driving force behind current turtle distributions appears to be climatic. Australia's Pleistocene and Holocene climatic history is reasonably well documented (Frakes *et al.*, 1987). The last glacial build-up started about 30,000 years ago with the glacial maximum achieved about 18,000 years ago (Augustinus and Macphail, 1997). Ice sheets and glaciers were nowhere near as extensive as in the Northern Hemisphere. In Tasmania, where natural populations of freshwater turtles no longer exist, the ice cap was 700 m thick and covered an area of at least 6,000 km^2 (Kiernan 1990), whereas on mainland Australia, the glacial area was small, perhaps no more than 25 km^2. The area affected by frost and winter snowfalls was far more extensive however. Snowlines dropped as much as 1,000 m below current levels, and air temperatures were at least 6.5°C lower at the glacial maximum. Maximal aridity occurred about 12,000 years ago over south coastal and central Australia—81% of sampled lakes were low or dry (Harrison, 1993). Between 18,000 and 12,000 years ago in the Murray–Darling catchment the situation was similarly arid (Fried, 1993).

During these dry cold times, many of the more southerly distributed plant and animal species, including turtles, would have shifted their ranges north, suffering local extinction in the south, only to recolonize these areas after the retreat of glaciers around 10,000 years ago (Frakes *et al.*, 1987). *Emydura macquarii* currently ranges and breeds only as far into the colder regions of Australia as Burrinjuck Dam, near the border of the Australian Capital Territory. Its distribution in the Murray-Darling at the time of the glacial maximum would have contracted substantially, perhaps as far north as Brisbane, though to a lesser degree on the coast. *Chelodina expansa* would have been more severely affected, as it breeds in the autumn and winter, and its range extends less into colder regions than *Emydura macquarii*. Breeding populations of *Chelodina longicollis* occur near Cooma, which regularly receives winter snowfalls, so it would have been least affected. Accidental introductions of *Chelodina longicollis* in the north-west of Tasmania have shown that the climate is suitable for their establishment much further south than its current natural distribution.

Zoogeographic hypotheses relating to Pleistocene contractions of freshwater turtle populations, followed by rapid reinvasion during the current interglacial, are currently being tested using fine-scale molecular techniques of mtDNA and nuclear microsatellites (Shaffer *et al.*, 1999).

CONCLUSIONS

In summary, the zoogeography of the Australasian freshwater turtles can be understood on a number of temporal scales. Cryptodires dominated the Mesozoic turtle fauna of what we now call Australia, shifting to a chelid dominated fauna in the Tertiary. The Gondwanan origins of the Chelidae are not in doubt, as chelids are not known outside their present range, even as fossils. The Australian fauna was separated from that of South America in the Cretaceous, and the two chelid faunas now are probably reciprocally monophyletic (Georges *et al.*, 1998).

Progressively increasing aridity during the Tertiary reduced opportunities for freshwater turtles (now absent from much of central Australia) and fragmented the coastal regions, isolating in particular the fauna of the extreme south west (*Chelodina oblonga*, *Pseudemydura umbrina*). Recent rises and falls in sea level resulted in intermittent interconnection of Australia and New Guinea, providing opportunities for interchange between the two via a Torres Strait land-bridge, Lake Carpentaria and the rivers flowing west from the land-bridge formed by the exposure of the Arafura Sill. The terrestrial components of the land-bridges between New Guinea and Australia would have formed a barrier between the north-western fauna and the east coast fauna.

In the south, Pleistocene climate changes have resulted in the contraction and reinvasion of south-eastern Australia, particularly in the Murray-Darling drainage, and have resulted in isolated pockets of endemism in coastal New South Wales. These, coupled with Recent sea level rises, have prevented the reinvasion of Tasmania, although accidental introductions of *Chelodina longicollis* in the north-west of the island have shown that the climate is suitable for their establishment there. The Asian influence on our freshwater turtle fauna is represented by the cryptodires *Pelochelys bibroni* and *P. cantori* which have reached New Guinea but not yet Australia, and *Carettochelys insculpta*, which is found in both regions.

ACKNOWLEDGEMENTS

The authors wish to express their appreciation to Gene Gaffney for his interest and efforts in providing figures; the assistance of the American Museum of Natural History (New York) in providing illustrations is acknowledged. Arthur White (Sydney) is also thanked for permission to use unpublished information, and Jan Martin kindly reviewed the literature on Lake Carpentaria.

REFERENCES

Anderson, C. 1925. Notes on the extinct chelonian *Meiolania*, with a record of a new occurrence. *Records of the Australian Museum*, Vol. 14, pp.223–242.

Araujo, D.C.F. 1984. Sistematica e taxonomia dos pareiasaurios: historico prespectivas atuais. *Pesquisas, Porto Alegre,* Vol. 16, pp.227–249.

Augustinus, P.C. and Macphail, M.K. 1997. Early Pleistocene stratigraphy and timing of the Bulgobac Glaciation, western Tasmania, Australia. *Palaeogeography, Palaeoclimatology, Palaeoecology*, Vol. 128, pp.153–167.

Blake, D.H. and Ollier, C.D. 1971. Alluvial plains of the Fly River, Papua. *Zeitschrift für Geomorphologie, Supplement* 12, pp.1–17.

Boulenger, G.A. 1888. On the chelydoid chelonians of New Guinea. *Annali del Museo Civico di Storia Naturale di Genova*, Vol. 6, No. 2, pp.449–452.

Brinkman, D.B. and Peng, J.H. 1993. Ordosemys leios, n.gen., n.sp., a new turtle from the early Cretaceous of the Ordos Basin, Inner Mongolia. *Canadian Journal of Earth Sciences,* Vol. 30, pp.2128–2138.

Broin, F. de 1987. The late Cretaceous fauna of Los Alamitos, Patagonia, Argentina. Part IV. Chelonia. Revista del Museo Argentino de Ciencias Naturales, Bernardino Rivadavia. *Paleontologia,* Vol. 3, No. 3, pp.131–139.

Broin, F. de and Fuente, M. 1993. Les tortues fossiles d'Argentine: synthese. *Annales de Paleontologie,* Vol. 79, pp.169–232.

Burbidge, A.A. 1967. *The biology of south-western Australian tortoises*. PhD Thesis (unpublished). University of Western Australia, Perth, W.A.

Burbidge, A.A., Kirsch, J.A.W. and Main, A.R. 1974. Relationships within the Chelidae (Testudines: Pleurodira) of Australia and New Guinea. *Copeia*, 1974, pp.392–409.

Cann, J. 1997a. Georges short-necked turtle. *Monitor (Victorian Herpetological Society, Melbourne),* Vol. 9, pp.18–23, 31, 32.

Cann, J. 1997b. Irwin's turtle. *Monitor (Victorian Herpetological Society, Melbourne),* Vol. 9, No. 1, pp.36–40, 31–32.

Cann, J. 1997c. Kuchling's long-neck turtle. *Monitor (Victorian Herpetological Society, Melbourne),* Vol. 9, No. 1, pp.41–44, 31–32.

Cann, J. 1997d. The northern yellow-faced turtle. *Monitor (Victorian Herpetological Society, Melbourne),* Vol. 9, No. 1, pp.24–29, 31–32, 34–35.

Cann, J. 1998. *Australian Freshwater Turtles.* Beaumont Publishing, Singapore. 292pp.

Cann, J. and Legler, J. 1994. The Mary River tortoise: a new genus and species of short-necked chelid from Queensland, Australia (Testudines: Pleurodira). *Chelonian Conservation and Biology,* Vol. 1, pp.81–96.

Cann, J., McCord, W.P. and Joseph-Ouni, N. 2003. Emmett's short-necked turtle, *Emydura macquarii emotti* spp. nov. *Reptilia,* Vol. 27, pp.60–61.

Chapell, J. 1983. A revised sea-level record for the last 300,000 years from Papua New Guinea. *Search,* Vol. 14, pp.99–100.

Chapell, J. 1994. Upper quaternary sea levels, coral terraces, oxygen isotopes and deep sea temperatures. *Journal of Geography,* Vol. 103, pp.828–840.

Cogger, H.G. 1983. *Reptiles and Amphibians of Australia.* A.H. and A.W. Reed, Sydney. 640pp.

Cogger, H.G. 2000. *Reptiles and Amphibians of Australia.* Reed New Holland, Sydney. 808pp.

De Queiroz K. 1995. Phylogenetic approaches to classification and nomenclature, and the history of taxonomy (an alternative interpretation). *Herpetological Review,* Vol. 26, pp.79–81.

Ernst, C.H. and Barbour, R.W. 1989. *Turtles of the World.* Smithsonian Institution Press, London.

Frakes, L.A., McGowran, B. and Bowler, J.M. 1987. Evolution of Australian environments. Pp.1–16. In Dyne, G.R. and Walton, D.W. (eds). *Fauna of Australia. Volume 1A. General Articles.* Australian Government Publishing Service, Canberra. 338pp.

Fried, A.W. 1993. Late Pleistocene river morphological change, south-eastern Australia: the conundrum of sinuous channels during the Last Glacial Maximum. *Palaeogeography, Palaeoclimatology, Palaeoecology,* Vol. 101, pp.305–316.

Frost, D. 1995. Response to Peter Pritchard. *Herpetological Review,* Vol. 26, pp.81–83.

Gaffney, E.S. 1977. The side-necked turtle family Chelidae: a theory of relationships using shared derived characters. *American Museum Novitates,* No. 2620, pp.1–28.

Gaffney, E.S. 1979. Fossil chelid turtles of Australia. *American Museum Novitates,* No. 2681, pp.1–23.

Gaffney, E.S. 1981. A review of the fossil turtles of Australia. *American Museum Novitates,* No. 2720, pp.1–38.

Gaffney, E.S. 1983. The cranial morphology of the extinct horned turtle, *Meiolania platyceps,* from the Pleistocene of Lord Howe Island, Australia. *Bulletin of the American Museum of Natural History,* Vol. 175, pp.361–480.

Gaffney, E.S. 1985. *Meiolania platyceps* Owen 1886. The Lord Howe Island horned turtle. Pp.132–136. In Rich, P.V. and van Tets, G.F. (eds). *Kadimakara. Extinct Vertebrates of Australia.* Pioneer Design Studio, Melbourne. 284pp.

Gaffney, E.S. 1990. The comparative osteology of the Triassic turtle *Proganochelys. Bulletin of the American Museum of Natural History,* Vol. 194, pp.1–263.

Gaffney, E.S. 1991. The fossil turtles of Australia. Pp.704–720. In Vickers-Rich, P., Monaghan, J.M., Baird, R.F. and Rich, T.H. (eds). *Vertebrate Palaeontology of Australasia.* Pioneer Design Studio and Monash University Publications Committee, Melbourne. 1437pp.

Gaffney, E.S., Archer, M. and White, A. 1989. Chelid turtles from the Miocene freshwater limestones of Riversleigh Station, Northwestern Queensland, Australia. *American Museum Novitates,* No. 2959, pp.1–10.

Gaffney, E.S., Archer, M. and White, A. 1992. *Warkalania,* a new meiolaniid turtle from the Tertiary Riversleigh deposits of Queensland, Australia. *The Beagle (Records of the Museums and Art Galleries of the Northern Territory),* Vol. 9, No. 1, pp.35–48.

Gaffney, E.S. and Bartholomai, A. 1979. Fossil trionychids of Australia. *Journal of Paleontology,* Vol. 53, pp.1354–1360.

Gaffney, E.S., Kool, L., Rich, T., Vickers-Rich, P., Pledge, N., Archer, M. and White, A. 1998. *Otwayemys,* a new cryptodiran turtle from the early Cretaceous of Australia. *American Museum Novitates,* No. 3233, pp.1–28.

Gaffney, E.S. and Meylan, P.A. 1988. A phylogeny of turtles. Pp.157–219. In Brenton, M.J. (ed). *The Phylogeny and Classification of the Tetrapods.* (Systematics Association Special Volume 35A). Clarendon Press, Oxford.

Georges, A. and Adams, M. 1992. A phylogeny for Australian chelid turtles based on allozyme electrophoresis. *Australian Journal of Zoology,* Vol. 40, pp.453–476.

Georges, A. and Adams, M. 1996. Electrophoretic delineation of species boundaries within the short-necked chelid turtles of Australia. *Zoological Journal of the Linnean Society,* Vol. 118, pp.241–260.

Georges, A., Adams, M. and McCord, W. 2002. Delineation of species boundaries and a phylogeny for the snake-necked freshwater turtles of Australasia (Testudines: Chelidae: Chelodina). *Zoological Journal of the Linnean Society,* Vol. 134, pp.401–421.

Georges, A., Birrell, J., Saint, K., McCord, W.P. and Donnellan, S. 1998. A phylogeny for side-necked turtles (Chelonia: Pleurodira) based on mitochondrial and nuclear gene sequences. *Biological Journal of the Linnean Society,* Vol. 67, pp.213–246.

Glaessner, M.F. 1942. The occurrence of the New Guinea turtle (*Carettochelys*) in the Miocene of Papua. *Records of the Australian Museum,* Vol. 21, pp.106–109.

Goode, J. 1967. *Freshwater Tortoises of Australia and New Guinea (in the Family Chelidae).* Lansdowne Press, Melbourne. 154pp.

Gray, J.E. 1830. A synopsis of the species of the class Reptilia. Pp.1–110. In Griffith, E. (ed). *The Animal Kingdom. Arranged in Conformity with its Organization by the Baron Cuvier.* Whittaker, Treacher, London.

Gray, J.E. 1841. A catalogue of the species of reptiles and amphibia hitherto described as inhabiting Australia, with a description of some new species from Western Australia. Pp.422–449. In Grey, G. (ed). *Journals of Two Expeditions of Discovery in North-west and Western Australia, during the years 1837, 38 and 39, under the Authority of Her Majesty's Government.* T. and W. Boone, London.

Gray, J.E. 1856. On some new freshwater tortoises from North America, Ceylon and Australia, in the collection of the British Museum. *Proceedings of the Zoological Society of London,* 1856, pp.197–202.

Gray, J.E. 1857. Description of a new species of *Chelodina* from Australia. *Proceedings of the Zoological Society of London,* 1856, pp.369–371.

Gray, J.E. 1863. On the species of *Chelymys* from Australia, with the description of a new species. *Annals and Magazine of Natural History*, Vol. 12, pp.98–99, 246.

Gray, J.E. 1864. Revision of the species of Trionychidae found in Asia and Africa, with the descriptions of some new species. *Proceedings of the Zoological Society of London,* 1864, No. 6, pp.76–98.

Gray, J.E. 1867. Description of a new Australian tortoise (*Elseya latisternum*). *Annals and Magazine of Natural History*, Vol. 20, pp.43–45.

Harrison, S.P. 1993. Late quaternary lake-level changes and climates in Australia. *Quaternary Science Reviews*, Vol. 12, pp.211–231.

Heinsohn, T. and Hope, G. 2006. The Torresian connections: zoogeography of New Guinea. Pp.71–93. In Merrick, J.R., Archer, M., Hickey, G.M. and Lee, M.S.Y (eds). *Evolution and Biogeography of Australasian Vertebrates*. Auscipub Pty. Ltd., Sydney. 942pp.

Jaekel, O. 1914. Ueber die Wirbeltierfunde in der oberen Trias von Halberstadt. *Palaeontologica Zhurnal,* Vol. 1, pp.155–215.

Jaekel, O. 1918. Die Wirbeltierfunde aus dem Keuper von Halberstadt. Series II. Testudinata. Teil 1. *Stegochelys dux*, n.g., n.sp. *Palaeontologica Zhurnal*, Vol. 2, pp.88–214.

Kiernan, K. 1990. The extent of late Cainozoic glaciation in the central highlands of Tasmania, Australia. *Arctic and Alpine Research*, Vol. 22, pp.341–358.

Kluge, A.G. 1995. Parsimony. *Herpetological Review*, Vol. 26, pp.77–78.

Krefft, G. 1876. Notes on Australian animals in New Guinea with description of a new species of fresh water tortoise belonging to the genus *Euchelymys* (Gray). *Annali di Museo Civico di Storia Naturale di Giacomo Doria*, Vol. 1, pp.390–394.

Lee, M.S.Y. 1993. The origin of the turtle body plan: bridging a famous morphological gap. *Science*, Vol. 261, pp.1716–1720.

Legler, J.M. 1981. The taxonomy, distribution, and ecology of Australian freshwater turtles (Testudines: Pleurodira: Chelidae). *National Geographic Society Research Reports*, Vol. 13, pp.391–404.

Legler, J.M. and Cann, J. 1980. A new genus and species of chelid turtle from Queensland, Australia. *Contributions to Science (Natural History Museum of Los Angeles County)*, Vol. 324, pp.1–18.

Longman, H.A. 1915. On a giant turtle from the Queensland lower Cretaceous. *Memoirs of the Queensland Museum*, Vol. 3, pp.24–29.

McCord, W.P., Cann, J. and Joseph-Ouni, N. 2003. Fraser Island short-neck turtle, *Emydura macquarii nigra* spp. nov. *Reptilia*, Vol. 27, pp.62–63.

McCord, W.P. and Thompson, S.A. 2002. A new species of *Chelodina* (Testudinata: Pleurodira: Chelidae) from northern Australia. *Journal of Herpetology*. Vol. 36, pp.255–267.

McCulloch, M.T., DeDeckker, P. and Chivas, A.R. 1989. Strontium isotope variations in single ostracod valves from the Gulf of Carpentaria, Australia. A paleoenvironmental indicator. *Geochimica et Cosmochemica Acta*, Vol. 53, No. 7, pp.1703–1710.

Megirian, D. 1992. *Meiolania brevicollis* sp. nov. (Testudines: Meiolaniidae): a new horned turtle from the Australian Miocene. *Alcheringa*, Vol. 16, pp.93–106.

Megirian, D. and Murray, P. 1999. Chelid turtles (Pleurodira, Chelidae) from the Miocene Camfield Beds, Northern Territory of Australia, with a description of a new genus and species. *The Beagle (Records of the Museums and Art Galleries of the Northern Territory*), Vol. 15, pp.75–130.

Meyer, A.B. 1874. *Platemys novae-guineae* sp. nov. Eine Mittheilung von Hrn. Adolf Bernhard Meyer uber die von ihm auf Neu-Guinea und den Inseln Jobi, Mysore und Mafoor im Jahre 1873 gesammelten Amphibien. *Monatsberichte der Koniglich Preussischen Akademie der Wissenschaften zu Berlin*, Vol. 39, pp.128–140.

Meylan, P.A. 1987. The phylogenetic relationships of soft-shelled turtles (Family Trionychidae). *Bulletin of the American Museum of Natural History*, Vol. 186, pp.1–101.

Meylan, P. 1995. Cladistics: a response to Pritchard. *Herpetological Review*, Vol. 26, pp.83–85.

Molnar, R.E. 1991. Fossil reptiles in Australia. Pp.605–702. In Vickers-Rich, P., Monaghan, J.M., Baird, R.F. and Rich, T.H. (eds). *Vertebrate Palaeontology of Australasia.* Pioneer Design Studio and Monash University Publications Committee, Melbourne. 1437pp.

Ogilby, J.D. 1890. Description of a new Australian tortoise. *Records of the Australian Museum*, Vol. 1, pp.56–59.

Ouwens, P.A. 1914. List of Dutch East Indian Chelonians in the Buitenzong Zoological Museum. *Contributions a la Faune des Indes Neelandaises*, Vol. 1, pp.29–32.

Owen, R. 1853. *Descriptive Catalogue of the Osteological Series contained in the Museum of the Royal College of Surgeons of England. Volume 1. Pisces, Reptilia, Aves, Marsupalia.* Taylor and Frances, London.

Owen, R. 1882. On an extinct chelonian reptile (*Notochelone costata*, Owen) from Australia. *Journal of the Geological Society of London*, Vol. 38, pp.178–183.

Owen, R. 1886. Description of fossil remains of two species of a Megalanian genus (*Meiolania*) from "Lord Howe Island". *Transactions of the Royal Society of London*, Vol. 177, pp.471–480.

Parmenter, C.J. 1976. *The natural history of the Australian freshwater turtle* Chelodina longicollis *Shaw (Testudinata, Chelidae).* PhD Thesis (unpublished), University of New England, Armidale, N.S.W.

Philippen, H.D. and Grossman, P. 1990. Eine neue Schlangenhalsschildkrote von Neuguinea: *Chelodina reimanni* sp. n. (Reptilia, Testudines, Pleurodira, Chelidae). *Zoologische Abhandlungen, Staatliches Museum für Tierkunde, Dresden*, Vol. 46, No. 5, pp.95–102.

Pritchard, P.C.H. 1979. Taxonomy, evolution, and zoogeography. Pp.1–42. In Harless, M. and Morlock, H. (eds). *Turtles: Perspectives and Research.* John Wiley, New York. 695pp.

Pritchard, P.C.H. 1984. Piscivory in turtles, and evolution of the long-necked Chelidae. *Symposia of the Zoological Society of London*, Vol. 52, pp.87–110.

Pritchard, P.C.H. 1994. Cladism: the great delusion. *Herpetological Review*, Vol. 25, pp.103–110.

Ramsay, E.P. 1886. On a new genus and species of fresh water turtle from the Fly River, New Guinea. *Proceedings of the Linnean Society of New South Wales*, Vol. 1, pp.158–162.

Reisz, R.R. and Laurin, M. 1991. *Owenetta* and the origin of turtles. *Nature*, Vol. 349, pp.324–326.

Rhodin, A.G.J. 1994a. Chelid turtles of the Australasian Archipelago: I. A new species of *Chelodina* from south-eastern Papua New Guinea. *Breviora (Museum of Comparative Zoology)*, No. 497, pp.1–36.

Rhodin, A.G.J. 1994b. Chelid turtles of the Australasian Archipelago: II. A new species of *Chelodina* from Roti Island, Indonesia. *Breviora (Museum of Comparative Zoology)*, No. 498, pp.1–31.

Rhodin, A.G.J. and Genorupa, V.R. 2000. Conservation status of freshwater turtles in Papua New Guinea. *Chelonian Research Monographs*, Vol. 2, pp.129–136.

Rhodin, A.G.J. and Mittermeier, R.A. 1976. *Chelodina parkeri*, a new species of chelid turtle from New Guinea, with a discussion of *Chelodina siebenrocki* Werner, 1901. *Bulletin of the Museum of Comparative Zoology*, Vol. 147, No. 11, pp.465–488.

Sademi and Iskandar, D.T. 2000. Freshwater turtle and tortoise conservation and utilization in Indonesia. *Chelonian Research Monographs*, Vol. 2, pp.106–111.

Seddon, J.M., Georges, A., Baverstock, P.R. and McCord, W. 1997. Phylogenetic relationships of chelid turtles (Pleurodira: Chelidae) based on mitochondrial 12S rRNA gene sequence variation. *Molecular Phylogenetics and Evolution*, Vol. 7, pp.55–61.

Shaffer, H.B., Meylan, P. and McKnight, M.L. 1997. Tests of turtle phylogeny: molecular, morphological, and paleontological approaches. *Systematic Biology*, Vol. 46, pp.235–268.

Shaffer, H.B., Storz, B.L., Georges, A. and Starkey, D. 1999. Pleistocene glaciation, population substructure, and genetic differentiation in long-generation time vertebrates: a case study in eastern Australian turtles. *Evolution*, Vol. 99.

Shaw, G. 1794. *Zoology of New Holland.* J. Davis, London.

Siebenrock, F. 1901. Beschreibung einer neuen Schildkrotengattung aus der Familie Chelydidae aus Australien: Pseudemydura. *Anzeiger der Akademischen Wissenschaften Wien,* Vol. 38, pp.248–251.

Siebenrock, F. 1909. Synopsis der rezenten Schilderkroten. *Zoologische Jahrbucher, Supplement* No. 10, pp.427–618.

Siebenrock, F. 1914. Eine neue Chelodina Art aus Westaustralien. *Anzeiger der Akademischen Wissenschaften Wien*, Vol. 17, pp.386–387.

Siebenrock, F. 1915. Die Schildkrotensammlung *Chelodina* Fitz. *Sitzungsberichte der Kaiserlichen Akademie der Wissenschaften, Mathmatisch-Naturwissenschaftliche Classe, Wien,* Vol. 124, pp.13–35.

Thomson, S.A. 2000a. On the identification of the holotype of *Chelodina oblonga* (Testudinata: Chelidae) with a discussion of the taxonomic implications. *Chelonian Conservation and Biology,* Vol. 3, No. 4, pp.745–749.

Thomson, S.A. 2000b. A revision of the fossil chelid turtles (Pleurodira) described by C.W. DeVis (1897). *Memoirs of the Queensland Museum*, Vol. 43, pp.593–598.

Thomson, S.A. and Georges, A. 1996. Neural bones in chelid turtles. *Chelonian Conservation and Biology*, Vol. 2, pp.82–86.

Thomson, S.A., Kennett, R. and Georges, A. 2000. A new species of long necked turtle (Chelidae:Chelodina) from the Arnhem Land plateau, Northern Territory, Australia. *Chelonian Conservation and Biology*, Vol. 3, No. 4, pp.675–685.

Thomson, S.A. and Mackness, B. 1999. Fossil turtles from the early Pliocene Bluff Downs Local Fauna, with a description of a new species of *Elseya. Transactions of the Royal Society of South Australia*, Vol. 123, pp.101–105.

Thomson, S.A., White, A. and Georges, A. 1997. Re-evaluation of *Emydura lavarackorum*: identification of a living fossil. *Memoirs of the Queensland Museum*, Vol. 42, pp.327–336.

Torgersen, T., Hutchinson, M.F., Searle, D.E. and Nix, H.A. 1983. General bathymetry of the Gulf of Carpentaria and the Quaternary physiography of Lake Carpentaria. *Palaeogeography, Palaeoclimatology, Palaeoecology*, Vol. 41, pp.207–225.

Torgersen, T., Jones, M.R., Stephens, D.E. and Ullman, W.J. 1985. Late Quaternary hydrological changes in the Gulf of Carpentaria. *Nature*, Vol. 313, pp.785–787.

Torgersen, T., Luly, J., De Decker, P., Jones, M.R., Searle, D.E., Chivas, A.R. and Ullman, W.J. 1988. Late Quaternary environments of the Gulf of Carpentaria, Australia. *Palaeogeography, Palaeoclimatology, Palaeoecology*, Vol. 67, pp.245–261.

Vis C.W. de 1897. The extinct fresh-water turtles of Queensland. *Annals of the Queensland Museum,* Vol. 3, pp.1–7.

Vogt, T. 1911. *Emydura schultzei*, sp. nov. Reptilien und Amphibien aus Neu Guinea. *Sitzgunsberichte der Gesellschaft der naturforschenden Freunde, Berlin*, Vol. 9, pp.410–412.

Warren, J.W. 1969. A fossil chelonian of probable lower Cretaceous age from Victoria. *Memoirs of the National Museum, Victoria*, Vol. 29, pp.23–28.

Wells, R. and Wellington, R. 1985. A classification of the Amphibia and Reptilia of Australia. *Australian Journal of Herpetology, Supplementary Series*, No. 1, pp.1–61.

Wermuth, H. and Mertens, R. 1961. *Schildkroten, Krokodile, Bruckenechsen.* Gustav Fischer Velag, Jena.

White, A. 1990. 'Murgon's Salad Bowl'. The only thing missing is the dressing. *Riversleigh Notes,* Vol. 10, pp.6–7.

White, A. 1994. The fossil history of soft-shelled turtles. *Riversleigh Notes*, Vol. 14, pp.2–4.

White, A.W. 1993. Riversleigh researchers. Part 3—Arthur White reveals much about aspects of his "Turtlemania". *Riversleigh Notes*, Vol. 17, pp.4–6.

White, A.W. and Archer, M. 1994. *Emydura lavarackorum*, a new Pleistocene turtle (Pleurodira: Chelidae) from fluvatile deposits at Riversleigh, northwestern Queensland. *Records of the South Australian Museum*, Vol. 27, pp.159–167.

Williams, E.E. 1953a. Fossils and the distribution of Chelyid turtles (1). *Breviora (Museum of Comparative Zoology),* No. 13, pp.1–8.

Williams, E.E. 1953b. Fossils and the distribution of Chelyid turtles (2). *Breviora (Museum of Comparative Zoology),* No. 32, pp.1–6.

Williams, M.A.J. 1984. Paleoclimates and paleoenvironments. Pp.42–47. In Veevers, J.J. (ed). *Phanerozoic Earth history of Australia*. Clarendon Press, Oxford. 418pp.

Wood, R.C. 1976. *Stupendemys geographicus*, the world's largest turtle. *Breviora (Museum of Comparative Zoology),* No. 436, pp.1–31.

Woodward, A.S. 1901. On some extinct reptiles from Patagonia, of the genera *Meiolania, Dinilysia*, and *Genyodectes. Proceedings of the Zoological Society of London* 1901, pp.169–184.

Zangerl, R. 1969. The turtle shell. Pp.311–339. In Gans, C., Bellairs, A.d'A. and Parsons, T.S. (eds). *Biology of the Reptilia.* Academic Press, New York.

Chapter 17
Origins and Radiations of Snakes in Australasia

John D. Scanlon

Department of Palaeontology, South Australian Museum, North Terrace, Adelaide, S.A. 5000 Australia
E-mail: Scanlon.John@saugov.sa.gov.au

INTRODUCTION

Like so much of the rest of our fauna, Australasia's complement of snakes is very distinctive compared to those of the other continents; distinctive not only in the groups that are present and have radiated widely, but in the absence or poor representation of other groups that are dominant elsewhere (Cadle, 1987; Hutchinson and Donnellan, 1993; Greer, 1997).

In the modern Australasian fauna, the Typhlopidae (blindsnakes), Pythoninae (pythons) and especially Elapidae (elapids, or fixed-front-fanged venomous snakes) are represented by diverse radiations. The elapids include two groups of sea-snakes (*Laticauda*, sea kraits, and Hydrophiini, true sea-snakes), which both probably originated in this region but are now widely distributed in tropical seas. Other lineages present are *Cylindrophis* (Cylindrophiidae, small, primitive, burrowing snakes); *Candoia* (Boinae, true boas); *Acrochordus* (Acrochordidae, aquatic filesnakes); and several groups within 'Colubridae' in its broad sense, including non-venomous and rear-fanged colubroid snakes (Colubrinae, Natricinae and Homalopsinae). Many other widespread colubroid groups are absent, including vipers and pit-vipers (Viperidae).

The few extensive radiations of snakes seen in Australia today are mainly products of the Neogene, the last 25 million years or so, which has been called 'The Age of Snakes' (Stanley, 1979), although the precise time when ancestors of these groups arrived is uncertain. As in Cogger and Heatwole (1981), it is convenient—in principle—to recognize three broad groupings according to biogeographic origins of Australasian lineages: an old, Mesozoic or Gondwanan fauna, the intermediate, post-Gondwanan or Tertiary fauna and the modern, or Quaternary fauna. Until the Paleocene, about 60 m.y.a., Australia still had land connections with South America via Antarctica, and thus was effectively still part of Gondwana (Audley-Charles, 1987; Woodburne and Case, 1996). There followed a period of isolation as the Australian plate (preceded by various micro-plates or terranes) drifted north from the Antarctic; the isolation began to end (~30 m.y.a.) as the tectonic collision with Southeast Asia produced the Sunda and Melanesian island arcs, thus forming a filter-bridge allowing dispersal of terrestrial animals from the north (Hall, 2001). The hiatus defines a dichotomy between an old, southern biota, and the intermediate and modern groups of northern origin. Some of the lineages of snakes found here (living or extinct) can be assigned confidently to one of these groups, but uncertainty remains in several cases due to large gaps in Australasia's fossil record of land vertebrates.

Even for those groups where an Asian (i.e. Laurasian) origin can be most confidently asserted—elapids, colubrids and acrochordids—direct evidence for the time of first entry to the Australasian region is hard to obtain. Cogger and Heatwole (1981) suggested that the 'intermediate' lineages evolved in relative isolation here for some 30–35 million years with 'little or no modification by later migrations of the same families until well into the Quaternary'; whereas Molnar (1991) suggested that Australia's complete isolation from reptilian migration lasted from Eocene to Pliocene. There is some evidence from the patchy fossil record, phylogenetic analysis of extant forms and molecular clocks, but we must also consider not only continental drift, but global climatic effects on sea level. Hutchinson and Donnellan (1993) note that the geological history of the Asian-Australasian 'gap' remains poorly understood, and 'geology is not likely to provide rigid constraints on biogeographic hypotheses'; this still appears to be true (Hall, 2001). Nevertheless, we must suppose that the progressive approximation of the continents makes dispersal from the north more likely at later dates than earlier, but superimposed on this trend are several intervals of reduced sea level, exposing greater land areas and narrowing water gaps. For example, Pleistocene glaciations and interglacials have clearly been involved in dispersal and isolation of human and some animal populations (Cogger and Heatwole, 1981; Archer *et al.*, 1995). In addition, sea levels were low through most of the late Oligocene, due to global 'icehouse' conditions beginning before 30 m.y.a. (Frakes *et al.*, 1987), corresponding well with the time scale suggested by Cogger and Heatwole (1981) for the mid-Tertiary dispersals.

The last few years have been exciting times in snake systematics, in Australia and the rest of the world, with the publication of numerous important reviews and new fossil discoveries. This chapter attempts to place the Australian snake fauna in context by reviewing recent literature on the phylogenetic relationships, fossil history and biogeography of ophidian families that occur here, and to indicate the current state of knowledge (and of controversy) of many of the questions involved.

Palaeontological studies of snakes are usually based mainly on isolated vertebrae, for much the same reasons as palaeomammalogy has been largely concerned with teeth—they are abundant, durable, and can provide information about function as well as taxonomic or phylogenetic relationships at various levels, sometimes down to species. Until recent decades, comparisons were often based on features and proportions of supposedly 'mid-trunk' vertebrae; this neglected actual variation within the column, and resulted in numerous synonyms for some fossil species (Simpson, 1933; Auffenberg, 1963; Hoffstetter and Gasc, 1969; Holman, 1979; Rage, 1984). Articulated fossil skeletons of snakes would be much more informative if they could be completely described, but are not only rare, but much more difficult to prepare and study than isolated bones. So in practice, our best information usually comes from very incomplete, broken and disarticulated skeletons (Hoffstetter, 1939; Rage, 1984; LaDuke, 1991; Holman, 2000).

ORIGINS AND PHYLOGENY

In discussing the origin of snakes, it is necessary to differentiate between two kinds of questions and explanations. First, when considering phylogenetic patterns, it is necessary to ask the following: what are the nearest extant and/or fossil relatives of snakes, and exactly how are they related? Second, when analysing adaptive processes it is necessary to ask: what was the ecological context for the particular derived features diagnosing snakes, in what order, and under what selective pressures were they acquired? These two domains have long been recognized (Hoffsetter, 1955; Rieppel, 1988) but not always adequately distinguished, although modern cladistic methods make it easier to do so. To soundly infer what adaptive processes actually took place in the early history of snakes, it is necessary first to analyse the phylogenetic relationships among snakes and between them and other squamates (lizards)—see pp.349–366 (Hutchinson and Smith, 2006). The actual features of ancestors or intermediates can, at least potentially, be derived from those of known taxa by parsimony-based character optimization, once a preferred phylogenetic hypothesis is adopted. However, neither phylogenetic analysis nor inference of ancestral habits can proceed effectively while ignoring the fossil record (Gauthier *et al.*, 1988; Donoghue *et al.*, 1989; Bryant and Russell, 1992).

Because snakes are only one of many groups of elongate, more or less limbless squamates, most attempts at 'process' explanations of snake origins have emphasized things that most other limbless lizards have in common with snakes: elongate body, reduction and loss of limbs, reduced ears and eyes, and modified skull structure. These can be explained by burrowing, use of crevices, small size of the body and, particularly, of the head (Janensch, 1906; Bellairs and Underwood, 1951; Rieppel, 1988), so analogy suggests that snakes may have evolved from a small fossorial lizard. However, there was clearly not just one fossorial stage in the early history of snakes, but rather various lineages of snakes have entered or emerged from the burrowing 'zone' at various times. All extant families include more or less specialized burrowers, but it is clear that most, if not all, of these fossorial groups are descended from more generalized terrestrial snakes. Even if all extant snakes have a common burrowing ancestry, it does not follow that a unique subterranean phase necessarily corresponds to the origin of the snake body form or particular skeletal modifications.

Several 'pattern' theories of snake relationships have also been held in the past century. One is that snakes evolved very long ago from a common ancestor with lizards, i.e. they are the living sister group of lizards (Boulenger, 1891; Hoffstetter, 1939, 1955, 1968; Underwood, 1970; Rage, 1982), but this idea has been almost entirely abandoned, in part because the fossil record of lizards is considerably longer than that of snakes. The case for lizards being monophyletic was always weak, being based on 'Dollo's law' (irreversibility of evolution) applied to very few characters. Most of the morphological (and recently, molecular) evidence has indicated that snakes are quite deeply nested within the 'lizard' radiation, and in particular there is strong support for relationship with the varanoid lizards, which include the living goannas, *Varanus* spp. (Camp, 1923; Nopcsa, 1923; Bellairs and Underwood, 1951; McDowell and Bogert, 1954; McDowell, 1972a; Rieppel, 1988; Estes *et al.*, 1988; Schwenk, 1988, 1993; Forstner *et al.*, 1995; Reeder, 1995; Lee, 1997b, 1998).

On the other hand, some workers have suggested phylogenetic hypotheses in which snakes are related to extant lineages of more or less limbless, elongate fossorial squamates such as anguids, pygopodids, and especially amphisbaenians and dibamids (Senn and Northcutt, 1973; Rage, 1982; Greer, 1985; Cundall *et al.*, 1993; Hallermann, 1998; Evans and Barbadillo, 1998). The 'subterranean' theory of snake origin has usually been based on analogy with many groups of sheltering or fossorial lizards rather than homology (i.e. evidence of common descent) with any of them, and the most thorough reviews have concluded that the extant burrowing lizard groups are poor candidates to be relatives of snakes (Janensch, 1906; Bellairs and Underwood, 1951; Underwood, 1967; Rage, 1987; Rieppel, 1988).

Nevertheless, these elongate and limbless squamates do tend to cluster together in phylogenetic analyses of morphology—at least in analyses which do not include fossils (Estes *et al.*, 1988; Cundall *et al.*, 1993; Hallermann, 1998; Lee, 1998). The most basal lineages of extant snakes are small and fossorial (Scolecophidia and Anilioidea), so that a fossorial origin has remained a popular idea even though there are no primitive burrowing snakes in the fossil record. A burrowing varanoid, *Lanthanotus borneensis*, has been interpreted as a close relative or 'structural ancestor' of snakes (Bellairs and

Underwood, 1951; McDowell and Bogert, 1954; McDowell, 1972a), but is now considered to be an offshoot of terrestrial varanids rather than a proto-snake (Rieppel, 1980; Lee, 1997b). The late Cretaceous *Dinilysia patagonica*, until recently the best-known fossil snake, resembles anilioids in many respects (Estes *et al.*, 1970; Frazzetta, 1970; Hecht, 1982; Rage and Albino, 1989) but its phylogenetic position has been much debated without a consensus emerging (e.g. Estes *et al.*, 1970; Rage, 1977; Hecht, 1982; McDowell, 1987; Kluge, 1991), and in this sense the early fossil record of snakes has been uninformative.

The oldest fossils identified as snakes are two tiny vertebrae from a lake deposit in the early Cretaceous of Spain (Barremian, 124.5–131.8 m.y.a.; Rage and Richter, 1994). However, this identification is now doubtful because 'dolichosaurs' (elongate, aquatic varanoid lizards) such as *Coniasaurus* have vertebrae that are just as snake-like, while retaining four limbs (Caldwell and Cooper, 1999; Rage and Werner, 1999). In fact, recent fossil discoveries and phylogenetic analyses imply that snakes are most closely related to Cretaceous marine varanoid lizards including dolichosaurs and the giant mosasaurs (Figure 1). This in turn, suggests that they originated as relatively large aquatic predators—about a metre long—rather than small fossorial insectivores (Scanlon, 1996; Lee, 1997b, 1998; Lee and Caldwell, 1998, 2000; Caldwell, 1999; Scanlon *et al.*, 1999; Scanlon and Lee, 2000; Lee and Scanlon, 2002a, b). This corresponds approximately to views of snake origins and relationships proposed earlier by Cope (1869), Nopcsa (1908, 1923), and McDowell (McDowell and Bogert, 1954; McDowell, 1972a, 1987), though a relationship between snakes and varanoids has often been supported on morphological grounds without considering the marine forms (Camp, 1923; Bellairs and Underwood, 1951; Estes *et al.*, 1988; Rieppel, 1988; Schwenk, 1988, 1993).

If marine varanoids had formed a single sister group to snakes, and early marine snakes also formed a single basal clade, it would be equally parsimonious to assume the 'origin' of snakes took place on land or in the water; in either case there would be two habitat shifts (Scanlon *et al.*, 1999). It now appears that marine varanoids, and possibly also early marine snakes, form several successive clades along the stem lineage of extant snakes, so that there is stronger support for a marine origin (Lee and Caldwell, 2000; Lee and Scanlon, 2002a, b). Note that the aquatic hypothesis, as a 'scenario' or theory of evolu-

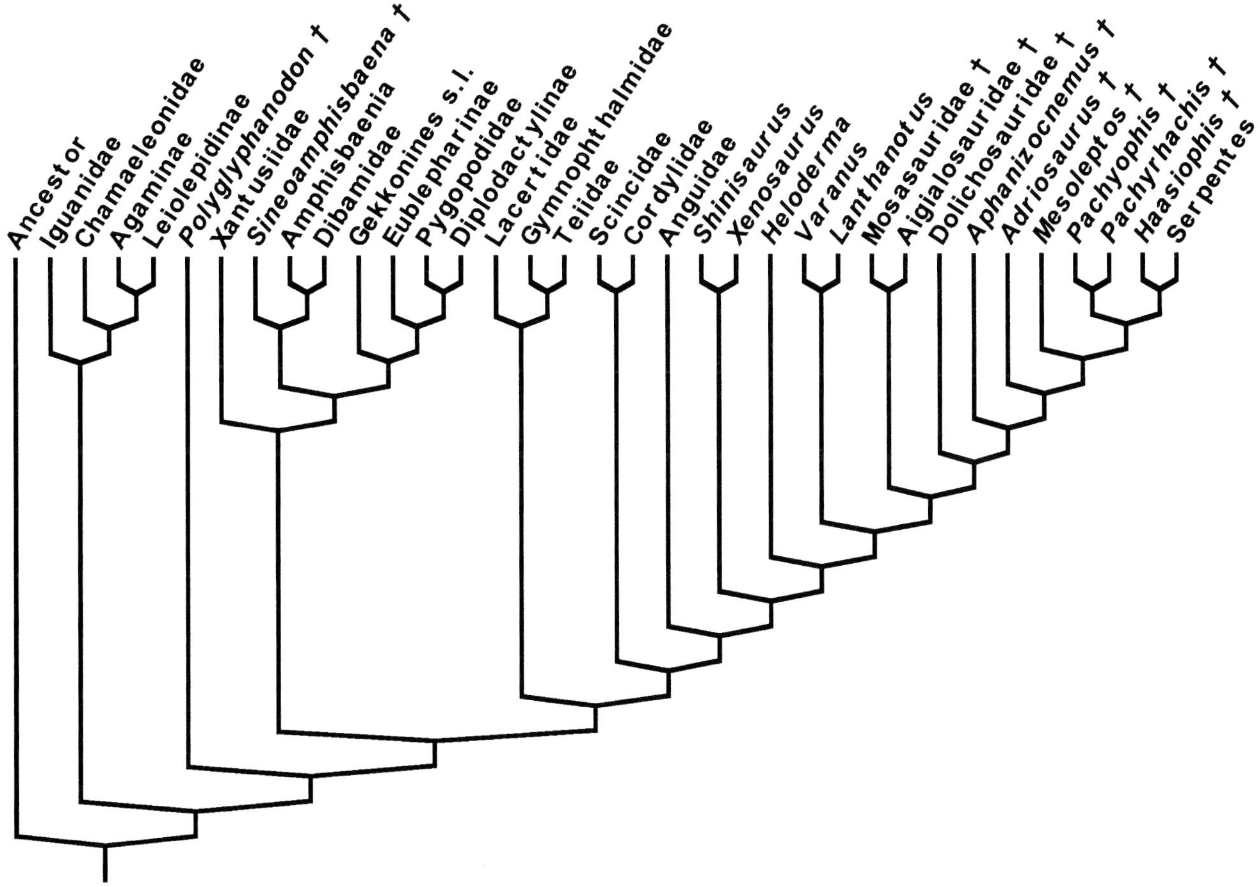

Figure 1. Phylogenetic hypothesis for squamate lineages based on morphological data, after Lee and Caldwell (2000) and Lee and Scanlon (2002a). If extinct groups (marked †) are excluded from the analysis, the same set of characters tends to support a (Serpentes, (Dibamidae, Amphisbaenia)) clade.

lineage to be compared with living taxa and included in a phylogenetic analysis (Figure 2), supporting the view of madtsoiids—along with pachyophiids and *Dinilysia*—as archaic snakes rather than alethinophidians (McDowell, 1987; Scanlon and Lee, 2000). Another species, *W. barriei*, is also known in Oligo-Miocene deposits at Riversleigh, but was less than half as long as the Plio-Pleistocene form at under 3 m (Scanlon and Lee, 2000); another medium-sized madtsoiid remains undescribed (Scanlon, 1996, 1997b). Two small species of *Nanowana* (under a metre) apparently fed on small, tough-scaled lizards, both showing different adaptations of the dentition associated with a diet of skinks in living snakes (Scanlon, 1997a). The medium-sized and large madtsoiids probably took a wide range of vertebrate prey, especially small and medium-sized marsupials such as bandicoots, possums and small macropods, but may also have tangled with reptilian prey such as larger lizards which were present at Riversleigh, including species of *Tiliqua* (Shea and Hutchinson, 1991), *Physignathus* (Covacevich *et al.*, 1990) and *Varanus* (Scanlon, 2002, unpublished data), or even small crocodilians (Willis, 1997).

SCOLECOPHIDIA

TYPHLOPIDAE

Scolecophidians (literally 'worm-snakes', known as blindsnakes and threadsnakes) are small, burrowing, mainly insectivorous snakes with highly reduced eyes, gape and dentition (McDowell, 1967a, 1974, 1987; Ehmann and Bamford, 1993; Greer, 1997). The only scolecophidian family present in the Australian region is Typhlopidae, which also occurs in South and Central America, Africa, Madagascar, southern Europe and southern Asia. Leptotyphlopidae is nearly as widespread, while the third (and least derived) scolecophidian family, Anomalepididae, is restricted to South and Central America. Scolecophidians (presumably the typhlopid *Ramphotyphlops*; see below) were present in Australia by the late Oligocene, and fossil vertebrae and ribs have been found at Riversleigh (Scanlon, 1996; Archer *et al.*, 1991, 1997).

It has often been presumed that scolecophidians exhibit the primitive snake condition in many of their fea-

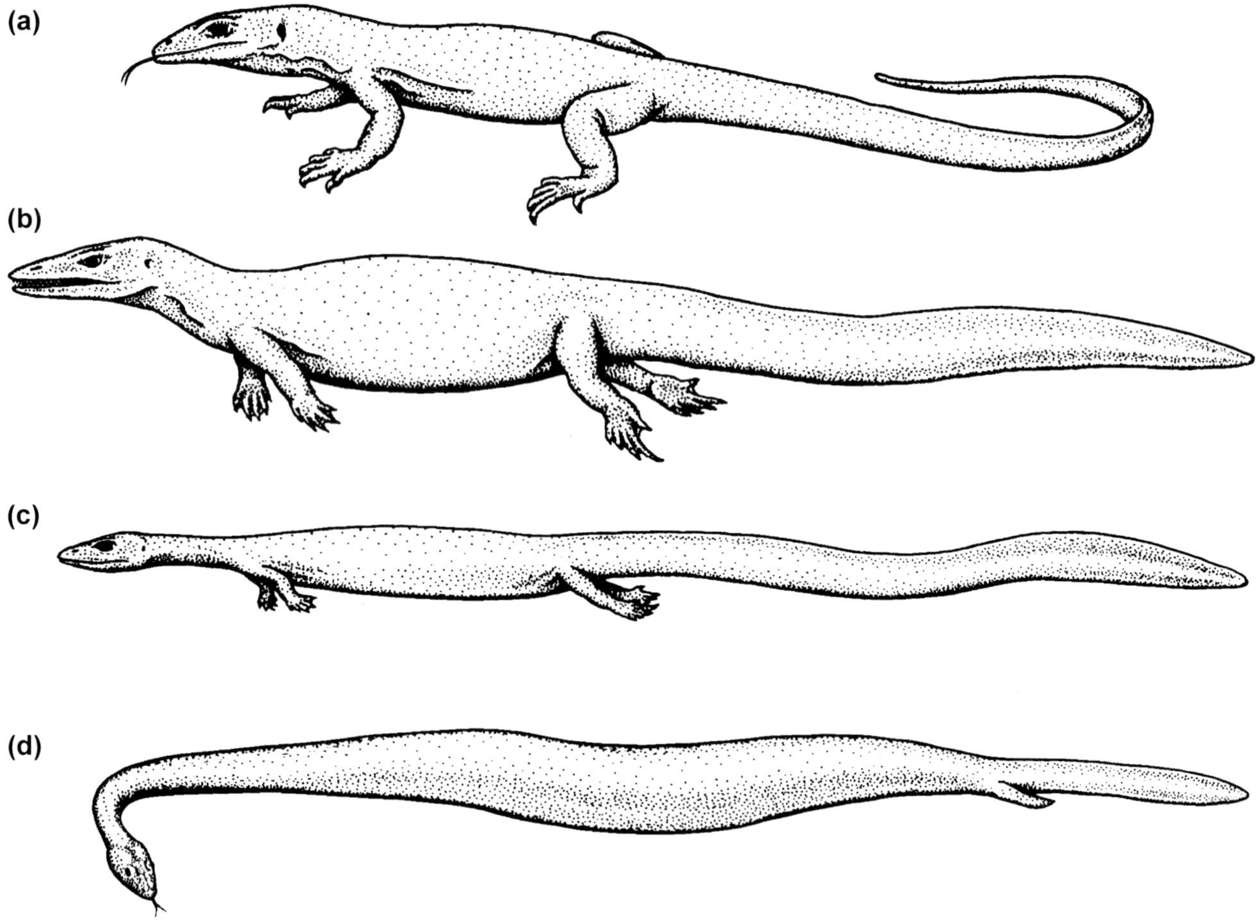

Figure 3. Body form in a terrestrial varanid and Cenomanian marine varanoids: **(a)** extant *Varanus* sp.; **(b)** primitive mosasauroid (*Opetiosaurus bucchichi*, based on skeletal reconstruction by Caldwell *et al.*, 1995); **(c)** 'dolichosaur' (*Adriosaurus suessi*, modified after Lee and Caldwell, 2000); **(d)** early snake with hindlimbs (*Pachyrhachis problematicus*), based on Haas (1979, 1980). Not to same scale: (a), (b), (d) total length approx. 1.2 m, (c) 0.3 m.

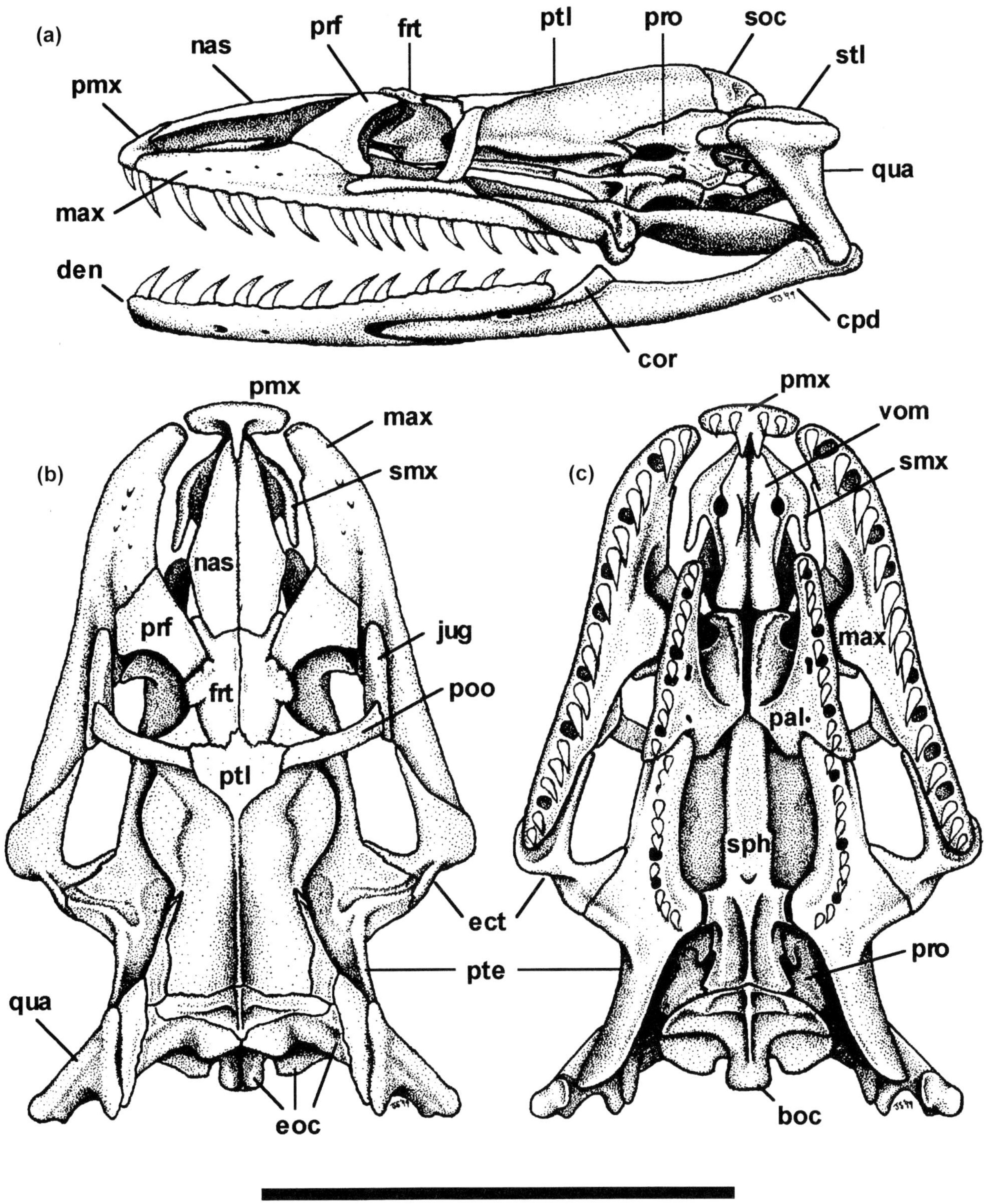

Figure 4. Reconstruction of the skull of *Wonambi naracoortensis*, a large madtsoiid snake known from Pliocene and Pleistocene cave deposits in southern Australia: **(a)** lateral view; **(b)** dorsal view; **(c)** palatal view. Key to abbreviations: **boc** = basioccipital; **cor** = coronoid; **cpd** = compound mandibular element (articular-prearticular-surangular); **den** = dentary; **ect** = ectopterygoid; **eoc** = exoccipital (including opisthotic); **frt** = frontal; **jug** = jugal; **max** = maxilla; **nas** = nasal; **pal** = palatine; **pmx** = premaxilla; **poo** = postorbital; **prf** = prefrontal; **pro** = prootic; **pte** = pterygoid; **ptl** = parietal; **qua** = quadrate; **smx** = septomaxilla; **soc** = supraoccipital; **stl** = supratemporal; **vom** = vomer. Scale = 100 mm. Based on the largest known specimen (Scanlon and Lee, 2000).

tures, including small gape and lizard-like, 'microphagous' feeding habits—feeding frequently on small prey, mostly immature stages of ants or termites (Underwood, 1967; Greene, 1983, 1997; Cundall and Greene, 2000; Webb *et al.*, 2000). However, if snakes are closely related to varanoids—which often feed on large items, including vertebrates—it is most parsimonious to assume instead that a capacity for 'macrophagous' habits is actually primitive for snakes. Thus, Lee *et al.* (1999b) interpret snake feeding mechanisms as essentially continuous with those in mosasaurs, the large marine varanoids of the Cretaceous; this was previously suggested by McDowell and Bogert (1954) and McDowell (1972a). The fact that some scolecophidians eat infrequent meals of many small items (e.g. some *Leptotyphlops*) or single large items (*Acutototyphlops*) shows that scolecophidian diets are not necessarily as 'lizard-like' as sometimes thought (Webb *et al.*, 2000). The reinforced braincase of snakes is usually considered evidence for early burrowing adaptation, but if macrophagy is primitive an enclosed braincase could have initially been favoured to prevent brain injury while subduing or swallowing large prey (Hoffstetter, 1968).

The world fossil record of scolecophidians begins in the Eocene of Europe (Rage, 1987). Only vertebrae are known, and unfortunately there are no vertebral characters known to distinguish the three families—or lower taxa—among modern forms (List, 1966; Rage, 1987), so the fossils are currently of little phylogenetic significance. If the basal dichotomy between Scolecophidia and Alethinophidia is real, both groups should be equally old, so that the lack of fossil blindsnakes before the Eocene might seem puzzling. However, it is possible that some early 'alethinophidian' fossils are actually basal scolecophidians (e.g. the supposed 'aniliid' *Coniophis*, mid-Cretaceous to Eocene; Gardner and Cifelli, 1999).

Blindsnakes are sometimes stated to have shown slow rates of evolution, an idea based on the overall morphological homogeneity of the group, but this does not necessarily imply a slow rate of speciation (in any case, recent studies of scolecophidians find they are not so homogeneous after all). The large, widespread genus *Typhlops* is represented on the northern fringe of the Australian region by several species, placed by McDowell (1974) in three more widely distributed species groups, so presumably resulting from at least three separate invasions from the Indonesian region. The presence of four species of *Typhlops* in the region (Wallach, 1997) appears to be analogous to that of Colubridae (see below) in that it represents multiple invasions by a relatively large cosmopolitan group. These invasions were presumably quite recent, since such limited radiation has occurred.

Most of the New Guinean typhlopids, and all Australian species, belong to a group (usually considered a single genus *Ramphotyphlops*) diagnosed by exceptionally long, coiled 'awns' on the hemipenes, and retrocloacal sacs, both characters observable only in males (Robb, 1966; McDowell, 1974; Aplin and Donnellan, 1993; Wallach, 1995, 1996). *Ramphotyphlops* is concentrated in Australia and New Guinea, but a number of species extend to Borneo, the Philippines and New Caledonia. Many species have restricted geographic ranges, and new Australian species are still being described, often from only one or a few specimens. But *R. braminus* (most similar to a northern Papuan species, *R. erycinus*) is one of the most widespread and abundant snakes in the world, being common on the coasts and islands of the Indian and Pacific Oceans from South Africa to China and as far as Mexico. This distribution is a recent result of association with humans (as a 'flowerpot snake') and its ability, as the only known all-female, parthenogenetic snake, to found colonies with a single individual (McDowell, 1974); it is still expanding its range in Australia, having recently become established in Townsville, Qld. (Calvert, 1999, personal communication) and towns in the Kimberley and Pilbara regions of W.A. (Coate, 1997; Aplin and Smith, 2001). A distinctive group of species in the Solomons, Bismarck Archipelago and eastern New Guinea has recently been distinguished as a separate genus *Acutotyphlops* (Wallach, 1995); these differ from other scolecophidians in feeding on relatively large, elongate prey (earthworms) rather than small insects (Webb and Shine, 1993; Webb *et al.*, 2000).

McDowell's (1974) arrangement of New Guinea *Ramphotyphlops* into species groups was neither based on cladistic methods nor aimed at a monophyletic classification; he explicitly sought groups based on overall homologous similarity, not distinguishing monophyly from paraphyly (see McDowell, 1987). Greer (1997) has made a start on phylogenetic (cladistic) analysis of some Australian *Ramphotyphlops* using 10 morphological characters. Unfortunately both analyses used geographically limited sets of species as well as different methodological approaches, and there are also conflicts where they overlap. However, both agree that the most primitive or basal lineages within the genus are found in warm, mesic northern and eastern areas, and the species found in southern and arid areas represent a small number of invasions of new habitat.

Tyler (1979) considered the distribution of *Ramphotyphlops* to indicate an Oriental origin for the genus; but a local origin from Gondwanan *Typhlops* does not seem to be ruled out by any evidence. The presence of typhlopids in the Oligo-Miocene deposits of Riversleigh is not conclusive evidence of Gondwanan origin, since elapids certainly arrived from Asia and pythons may also have done, and all three families are found together at Riversleigh. This question could be solved either by much older typhlopid fossils in Australia (indicating a Gondwanan origin, as for madtsoiids), or if some lineage of *Typhlops* were found to represent the living sister-group of *Ramphotyphlops*, and to have a restricted distribution in South America (Gondwana) or southern Asia (Laurasia). So far, either seems possible.

ALETHINOPHIDIA

Alethinophidians (literally 'true snakes') include all the largest and most familiar groups such as boas and pythons, tree-snakes, and all the venomous families such as elapids and vipers. This group is thought to be monophyletic and the sister group to Scolecophidia, but is far more diverse. Most alethinophidians belong to the clade Macrostomata, while the remaining taxa are grouped together as Anilioidea (Figure 2).

ANILIOIDEA, ANILIIDAE

In recent decades, anilioids (pipesnakes, shield-tail snakes) have often been considered the most primitive living snakes; they are small, more or less specialized burrowers (but also partly aquatic in *Anilius* and some *Cylindrophis*), with some adaptations for swallowing large prey but lacking others found in more advanced alethinophidians (McDowell, 1972a; Cundall and Rossman, 1993; Cundall, 1995). Anilioids have often been considered a paraphyletic assemblage, the stem-group of either Scolecophidia (Dowling, 1959; Scanlon, 1996) or Macrostomata (Rage, 1998). Recent molecular studies confirm *Anilius* is a basal alethinophidian (i.e. on the macrostomatan stem) but the relationships of cylidrophiids and uropeltids remain labile (Slowinski and Lawson, 2002; Wilcox *et al.*, 2002; Vidal and Hedges, 2002). The anilioids *Anomochilus* and Uropeltidae, in particular, show many similarities with scolecophidians (e.g. Cundall *et al.*, 1993), which can be interpreted as either primitive for snakes, derived in parallel, or indicating that scolecophidians evolved from anilioids; Scanlon and Lee (2000) found more evidence supporting parallelism than the other possibilities (see Figure 2).

A number of fossils have been assigned to Anilioidea, the oldest being *Coniophis*—mid-Cretaceous to Eocene of North and South America, Europe and Africa (Gardner and Cifelli, 1999; Rage and Werner, 1999). Tertiary anilioids are known from western Europe (*Eoanilius* and *Michauxophis*, late Eocene to middle Pliocene) and South America (*Colombophis* and *Hoffstetterella*, Paleocene to middle Miocene; Rage, 1984, 1998; Szyndlar and Schleich, 1993). All are known only from isolated vertebrae, and it has been suggested that *Coniophis* could be related to *Dinilysia* or Scolecophidia rather than living anilioids (McDowell, 1987). Apart from a single species, *Anilius scytale* of South America, living anilioids are almost entirely restricted to southern Asia. Species of *Cylindrophis* occur throughout the Sunda Islands of Indonesia, and one species (*C. aruensis*) at the extreme edge of this range is endemic to the Aru Islands, close to the mainland of New Guinea (McDowell, 1975). This distribution is presumably the result of relatively recent dispersal across Wallacea, so that anilioids have probably never been members of the continental Australasian fauna.

MACROSTOMATA

The 'large-mouthed' snakes form the sister group to Anilioidea, but many characters such as their large gape and (mostly) surface-active lifestyles are likely to be retained from early snakes rather than derived (Scanlon and Lee, 2000). Apart from booids and caenophidians (see below), there are several currently or formerly widespread lineages of macrostomatans not found in this region (Figure 2): Xenopeltidae (two genera, South-east Asia and Central America); Tropidophiidae and Ungaliophiidae (each with two extant genera from Central and South America and the Antilles, and Eocene fossils from North America and Europe); and Bolyeriidae (two genera, Round Island and subfossils from neighbouring Mauritius, Indian Ocean). Some of these have been considered Gondwanan relicts (Underwood, 1976; Rage, 1987; Barrie, 1990), but given their actual distributions (including fossils) this view seems dubious. They are relicts certainly, but possibly as much Laurasian as Gondwanan, and there is no reason to believe they ever reached our end of Gondwana. Recent molecular studies suggest that tropidophiids are basal snakes only distantly related to other macrostomatans, implying homoplasy in the evolution of large gape (Slowinski and Lawson, 2002; Wilcox *et al.*, 2002; Vidal and Hedges, 2002).

BOOIDEA, BOIDAE

Pythons and boas have also sometimes been regarded as primitive snakes (Boulenger, 1893; Dowling, 1959), and their relatively large size, constricting habits, and common association with water may indeed be retained ancestral features. Of the three broadly distributed booid lineages, most of the pythonines and one genus of boines occur in our region (Ehmann, 1993a; Greer, 1997). The third lineage, erycine 'sand boas', may once have reached South America (Albino, 1993) but is now restricted to mostly semi-arid areas of North America, northern Africa, and western and central Eurasia (Kluge, 1993a).

The three species of small Pacific boas (*Candoia*) range from Sulawesi in the west, through New Guinea and the Solomons to Samoa. This kind of distribution is not unusual for groups of skinks, geckoes or frogs (Austin, 2000), but *Candoia* presents a true biogeographic puzzle since the only other extant boas are found in South America and Madagascar. In the absence of a fossil record or any near relatives in this part of the world, it is not clear whether this genus is an old relict in Australasia (i.e. Gondwana), or has dispersed from another part of the world. With the view that boines are basically a South American radiation, several authors have suggested that *Candoia* arrived relatively recently across the Pacific (via the Equatorial Current, i.e. 'Kon-Tiki' dispersal); the same mechanism has been used to explain the origin of the Fijian iguanas (*Brachylophus* spp.) (Underwood, 1976; McDowell, 1979; Gibbons, 1985). However, phylogenetic studies do not support this idea: an analysis of morphology implied that *Candoia* is the most basal boine lineage, the sister group to the clade comprising extant taxa in Madagascar as well as South America (Kluge, 1991); whereas a recent molecular study groups *Candoia* with the Madagascan genus *Sanzinia* (Austin, 2000; see also Vences *et al.*, 2001). With a molecular clock assumption, Austin's results suggest a split from New World boas before 40 m.y.a., from *Sanzinia* only slightly later, and that divergences within *Candoia* itself are nearly as old. Thus, these small boas probably *are* old relicts, but not necessarily Gondwanan. While there is no evidence of Tertiary boines in Australasia, they were present in Europe from Eocene to early Miocene times (*Palaeopython* and *Bavarioboa*; Rage, 1987; Szyndlar and Schleich, 1993). The fossil record in Asia is poorly known, so it is perhaps more likely that *Candoia* is a relict of an otherwise extinct Laurasian lineage, rather than a Gondwanan one.

Pythonines seem to have a stronger claim as old residents of our region: most of the world's species and all but one genus occur in Australia and New Guinea, and a recent phylogenetic analysis indicates that the most basal lineages are the Australian endemics *Aspidites* and *Antaresia* (Kluge, 1993b). See Figure 5. Earlier analyses had suggested that the African and Asian genus *Python* was the sister group, or even stem group, of the Australasian python radiation (McDowell, 1975; Underwood and Stimson, 1990), but Kluge's results imply instead that *Python* is derived from within the Australian group. The fossil record does not exactly contradict this view, but it also does not support it (Scanlon, 2001); molecular phylogenies currently under construction will provide an independent test.

Fossil pythons are known from Australia, most material coming from the Oligo-Miocene deposits at Riversleigh. Several partly articulated and relatively complete skeletons of pythons have yet to be fully prepared (Scanlon, 1996, 2001), but nearly all the material studied so far can be referred to *Morelia riversleighensis*, described as *Montypythonoides reiversleighensis* by Smith and Plane (1985). This may be conspecific with a form found in the somewhat younger deposits of Bullock Creek, N.T. (*Morelia antiqua*, described in the same paper). The supposedly basal lineages *Aspidites* and *Antaresia* are not yet known as fossils, and pythons referable to *Liasis* (in the broad sense; the paraphyletic 'mid-

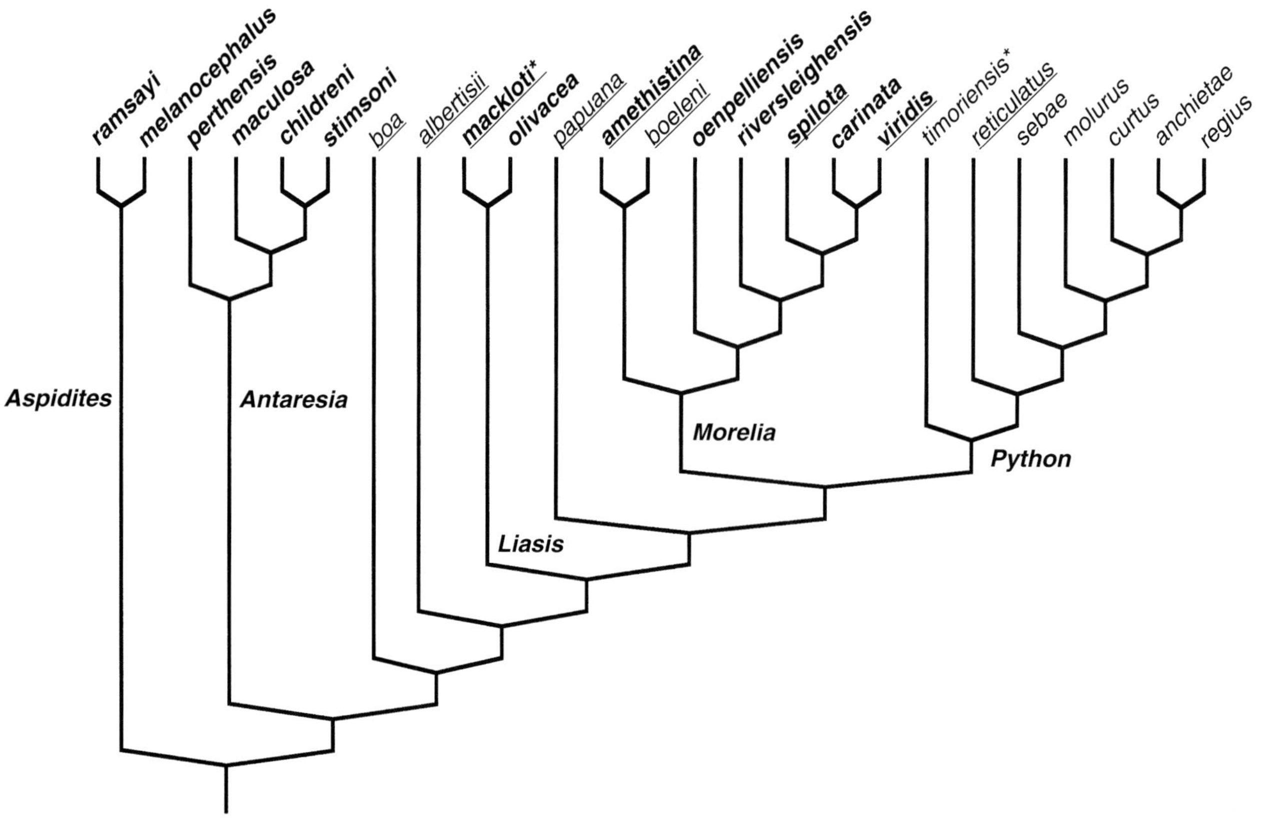

Figure 5. Phylogenetic hypothesis for Pythoninae, from Kluge (1993b), with position of *Morelia riversleighensis* based on Scanlon (2001). Taxa shown in **bold** occur in Australia; underlined, Melanesia (New Guinea, Bismarck Archipelago, Solomons; *P. reticulatus* only on islands to west of New Guinea); *, Timor and Lesser Sunda Islands; plain italic, found only outside the Australasian region. Generic names omitted for *Bothrochilus boa, Leiopython albertisii, Apodora papuana*.

dle' of most of Kluge's cladograms) are not known until the Pliocene (Scanlon, 2001; Scanlon and Mackness, 2002). On the other hand, the Eocene of Europe contains python-like snakes (Szyndlar and Böhme, 1993; Augé *et al.*, 1997; Duffaud and Rage, 1997), and if these are found to be true pythonines, an Australian origin for the group would be difficult to maintain.

Understanding of the phylogeny and biogeography of booids would receive a large boost from detailed description of some articulated fossil skeletons from the Eocene of Europe—such as those from Messel and Geiseltal in Germany (Barnes, 1926; Kuhn, 1939; Keller and Schaal, 1988; Szyndlar and Böhme, 1993); as well as Oligocene material from Wyoming in North America (Breithaupt and Duvall, 1986). So far, however, these very well-preserved skeletons have not been adequately described (e.g. Kluge, 1993a; Szyndlar and Böhme, 1993; Baszio and Schaal, 1997).

CAENOPHIDIA

Caenophidians (literally 'recent snakes') comprise the large clade of 'advanced' or 'typical' snakes, Colubroidea, together with a peculiar aquatic side-branch Acrochordidae. Though often regarded as highly specialized colubroids in older classifications, the acrochordids are now generally recognized as the sister group to colubroids (Rieppel, 1988; Cundall *et al.*, 1993; Scanlon and Lee, 2000). Caenophidians evolved within a cluster of 'boa-like' macrostomatan lineages (Figure 2), but are so diverse and widespread that it is difficult to suggest where or how they acquired their many distinctive characters. Most authors have regarded Caenophidia as basically a Cainozoic radiation, but vertebrae of fossil families referred to Acrochordoidea and Colubroidea (Nigerophiidae and Russellophiidae respectively) are now known from the Cenomanian of Sudan as well as terminal Cretaceous and Tertiary deposits (Rage and Werner, 1999). But, in the absence of cranial material, it cannot be confirmed that these early fossils are actually caenophidian.

ACROCHORDIDAE

File snakes are one of four living lineages of highly aquatic snakes, though none of these four is a holdover from the hypothetical aquatic radiation associated with the origin of snakes; thus, they may demonstrate 'palinosis', or return to an ancestral habitat or mode of life (Morgan, 1994). Similar reversals to and from a burrowing existence have also occurred, and surface-active terrestrial snakes must also be palinotic if either hypothesis of the environment of snake origin is correct.

McDowell (1975, 1979, 1987) and Rage (1984, 1987) recognize a superfamily Acrochordoidea to include several extinct aquatic groups as well as *Acrochordus*, but the composition and suggested affinities of their proposed groups differ greatly between the two authors. As none of the extinct taxa included is well known, or found in this region, they will not be discussed further here.

There are three extant species of file snake in the genus *Acrochordus* (McDowell, 1979; Shine and Houston, 1993; Greer, 1997). *A. arafurae* is restricted to northern Australia and southern New Guinea, *A. granulatus* is widely distributed along coasts from India to the Solomons. *A. javanicus* does not occur in the Australian region, but like *arafurae* is a freshwater species. McDowell (1979) suggests that marine adaptations are primitive for this family and shared with Palaeophiidae, so that freshwater habits would be derived, but it is doubtful whether this would be a synapomorphy of the two larger species. *Acrochordus* was present from middle Miocene to Pliocene times in the Siwalik deposits (India and Pakistan); the earliest of these is an extinct species, *A. dehmi*, but late Miocene and Pliocene vertebrae are similar to *A. javanicus* (Rage, 1987). Despite previous tentative identifications (Smith and Plane, 1985), no acrochordid fossils have yet been found in Australian deposits of any age. Archer *et al.* (1991) suggest that *Acrochordus* might already have been in the Australian region in the Miocene, but did not reach Riversleigh because the depositional basin was not yet open to the sea; evidence from other sites and other taxonomic groups (such as anadromous fish) is required to test this idea. If *A. dehmi* is an outgroup for living species, large size and a relatively long and high neural spine are primitive; *A. javanicus* retains these, but the neural spine is low and short in both other species (Hoffstetter and Gayrard, 1965; McDowell, 1979).

COLUBROIDEA

Colubridae: Homalopsinae, Colubrinae and Natricinae

Colubrids are often regarded as typical snakes, reflecting the fact that they are the most numerous and species-diverse element in modern snake faunas on all continents—except Australia. While often of small size and low trophic level, and including many invertebrate feeders (rare in most other families), this does not explain the diversity of the group as many of them are also relatively large and feed only on vertebrates. The family 'Colubridae', in the words of Hoffstetter (1968), 'constitutes a highly ramified assemblage containing more than 300 aglyphous and opisthoglyphous genera, the classification of which [is] a real nightmare for the herpetologist'. Within such a large group, many of the morphological or behavioural specializations of anilioid, booid, viperid or elapid groups are duplicated or mimicked; there are moderately large constrictors, aquatic fish-eaters, highly venomous tree-snakes, brightly-coloured 'shovel-snouted' burrowers, insectivorous and 'goo-eating' forms, and many more (Greene, 1997). There are a number of recognizable subfamilial groups, but also a large remaining

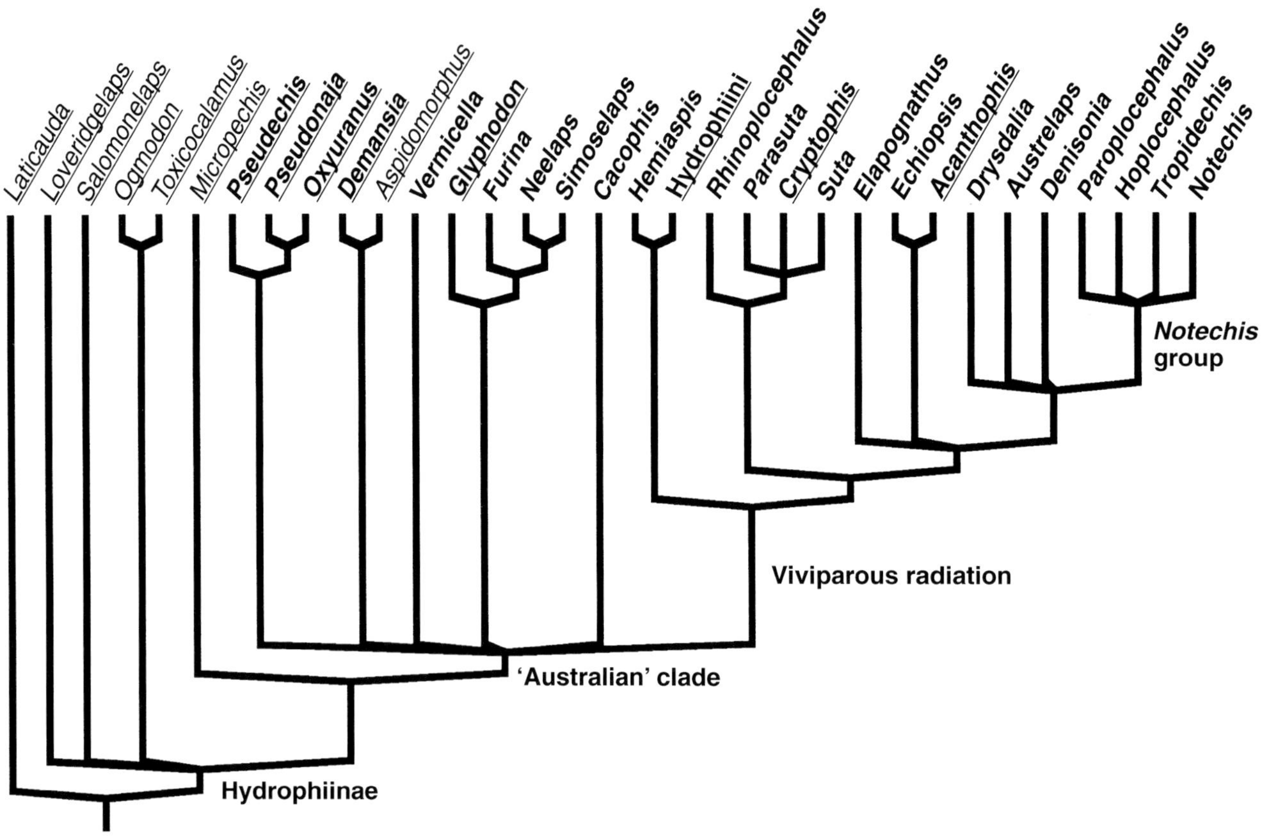

Figure 7. Phylogenetic hypothesis for Australasian Elapidae, based on various sources including Keogh (1998), Keogh *et al.* (1998, 2000), and unpublished analyses by the author; see also Scanlon and Lee (2003). *Laticauda* (sea kraits) and Hydrophiini (true sea snakes; numerous genera, not shown) are littoral or marine, and widely distributed in Indo-Pacific tropics; all others are terrestrial. Taxa shown in **bold** are wholly or mainly Australian; underlined, Melanesian (i.e. New Guinea, except *Loveridgelaps* and *Salomonelaps*, endemic to the Solomons, and *Ogmodon*, endemic to Fiji. *Acanthophis* also extends to a few islands west of New Guinea). *Parapistocalamus* (endemic to Bougainville), has uncertain relationships near the base of this tree, but is omitted for clarity.

estimates of divergence dates for Neotropical colubrid lineages, to argue that the basal radiations leading to Elapidae and Viperidae must have occurred over 30 m.y.a. and possibly 40–60 m.y.a.

The distribution and apparent relationships of modern forms are thought to be most consistent with an origin of Elapidae in eastern Asia (Hoffstetter, 1939; Cadle, 1987), and there is some support for monophyly of the Asian, African and American taxa as the sister taxon to Australasian and marine groups (Slowinski and Keogh, 2000). For example, the New World elapids are monophyletic but nested within the Asian coral snake group (Slowinski *et al.*, 2001). However, Africa has also been a centre of radiation and dispersal, since Asian cobras of the *Naja naja* species complex (Wüster and Thorpe, 1992) appear to be a monophyletic offshoot of the African cobra radiation (Szyndlar and Rage, 1990; Szyndlar and Zerova, 1990; Slowinski and Keogh, 2000).

Storr (1964) regarded the Australian elapids as an old and declining group. Cogger and Heatwole (1981) suggested that (along with other intermediate lineages) they had radiated here in isolation for 30–35 million years. Schwaner *et al.* (1985) doubt the implication that this radiation is now being overtaken by further reptilian invasions from the north; they contrast this with Stanley's (1979) view of the Neogene as the 'Age of Snakes', and suggestion that elapids, like most other snake lineages, are a dynamic, speciating group. Using transferrin immunological distances and a 'molecular clock' assumption, Schwaner *et al.* (1985) calculated a date of 12 million years for the oldest divergences within the Australian group (for example, between *Laticauda* and other genera). This was more recent than previous estimates, and Schwaner *et al.* suggested that the actual date could have been somewhat earlier, but probably not before 15 m.y.a. See Figure 7.

Until recently, the fossil history of elapids in Australia has been extremely limited and of little practical relevance to questions of the age and pattern of the radiation. Pleistocene elapid fossils have been described from Victoria Cave at Naracoorte, South Australia (referred to the extant genera *Pseudonaja*, *Pseudechis* and *Notechis*; Smith, 1975) and reported from the fluviatile Wyandotte Formation, north Queensland (McNamara, 1990). Further elapid remains, not yet described, are known from Pleistocene deposits in Queensland (Darling Downs, Floraville and Riversleigh), New South Wales (Wellington Caves) and Victoria (Bacchus Marsh); most of this material represents large species similar to *Pseudechis* or *Pseudonaja* (Scanlon, 1996 and unpublished data).

Early to middle Pliocene elapid remains are known from several sites including Bluff Downs (Allingham Formation, north Queensland; Archer and Wade, 1976; Scanlon, 1996), Chinchilla (Scanlon, unpublished data) and Corra-Lynn Cave, Yorke Peninsula, South Australia (Pledge, 1992) but have not yet been described.

Recent discoveries at Riversleigh, north-west Queensland, extend the record of elapids in Australia back to the early Miocene (Scanlon, 1995a, 1996; Scanlon *et al.,* 2003). A fragmentary vertebra is also known from the middle Miocene Bullock Creek Local Fauna, Northern Territory (Scanlon, 1992). This suggests that the molecular clock may have been running slow; Schwaner *et al.* (1985) estimated the divergence between Australasian and Asian lineages as taking place at 20 m.y.a., but by that time elapids were already here. As suggested above, the earliest time when dispersal of terrestrial animals from Asia became likely was a period of low global sea level during the late Oligocene, so elapids presumably arrived not much earlier than 30 m.y.a. (the same is likely for other squamate lineages such as pythons, typhlopids, varanids, and agamids). Although the skeletal morphology of living forms is not yet well enough known to interpret the phylogenetic significance of this fossil record, further study of available material should provide tests of both molecular clock dates. The 12 or 15 million year dates (Schwaner *et al.*, 1985) for the basal divergences in the extant Australasian radiation could be accurate if early (basal) branches in the Australian radiation all went extinct.

External morphology, soft anatomy, chromosomes, ecological characters and particularly molecular evidence have provided partial resolution of phylogeny in extant Australasian elapids (McDowell, 1967b, 1970, 1985, 1987; Wallach, 1985; Mengden, 1985a, b; Mengden *et al.*, 1986; Shine, 1985a; Greer, 1997; Keogh, 1998, 1999; Keogh *et al.*, 1998, 2000; Slowinski and Keogh, 2000; Scanlon and Lee, 2003); see Figure 7. Several lines of evidence suggest that the most basal lineages include *Laticauda* (sea kraits), generalized terrestrial forms from the Solomon Islands (*Loveridgelaps* and *Salomonelaps*) and several burrowing forms from the Solomons, New Guinea and Fiji (*Parapistocalamus, Toxicocalamus*, and *Ogmodon*). The other terrestrial forms are mainly Australian (except for the New Guinean endemics *Micropechis* and *Aspidomorphus*). Relationships among various lineages within the Australian group remain poorly resolved, with much disagreement among analyses (compare Wallach, 1985; Mengden, 1985a; Greer, 1997; Keogh *et al.*, 1998). But most recent studies agree that there is a large lineage of viviparous forms (including the death adders *Acanthophis*, tiger snakes *Notechis* and numerous others) which in turn gave rise to the 'true' (viviparous) sea-snakes (Shine, 1985a; Keogh *et al.*, 2000), and this is supported by combined analysis of morphology and DNA sequences(Scanlon and Lee, 2003). Thus Australian elapids, by returning to the marine environment which may have been the source of the very first snakes, have given rise to yet another radiation of more than 50 species, and spread across half the world.

SUMMARY

The grouping of Australian reptile lineages into three waves of arrival—Gondwanan, mid-Tertiary and modern (Cogger and Heatwole, 1981)—appears to remain valid, though there are still large gaps in the fossil record and phylogenetic uncertainties bearing on the origin of some groups. Despite a widely held view that the early radiations of snakes were basically Gondwanan, only one family of snakes is known to have been present in Australia as part of an old fauna shared with South America and Africa: the extinct Madtsoiidae. Of the many lineages known as Cretaceous and Palaeogene fossils in the other Gondwanan continents, there is no evidence that any others ever reached Australasia. This negative evidence suggests that the early radiations of snakes, on or adjacent to the Gondwanan land mass, were remote from the Australian part of Gondwana and mostly separated by effective dispersal barriers. A climatic 'filter' barrier involving the position of Antarctica (already over the South Pole in the early Cretaceous) has been mentioned above.

At the other extreme, three groups of *Typhlops*, a species of *Cylindrophis*, *Acrochordus*, and approximately nine distinct colubrid lineages are present on the northern periphery of the region, and are likely to have expanded their range from South-east Asia into New Guinea in the last few million years. No precise date for the beginning of this wave of snakes is available, but it may correspond to the relatively well-dated arrival of murid rodents in the early Pliocene (Rich *et al.*, 1991); see also pp.707–744 (Aplin, 2006).

Three lineages occupy the interesting intermediate position: pythons, elapids, and probably *Ramphotyphlops*, which were present in northern Australia by the late Oligocene or early Miocene, and are still here in force. Cogger and Heatwole (1981) interpreted these as Asian lineages with a 30–35 million-year history in Australasia, thus effectively predicting the Oligo-Miocene fossils that have now been found. Smith and Plane (1985) and Molnar (1991) did not envisage such early dispersal from Asia and considered the presence of pythons in Australia's middle Miocene to imply Gondwanan origin. However, the hypothesis of a Gondwanan origin for Australian elapids has been refuted by DNA-based phylogenetic analyses (Keogh *et al.*, 1998); Asian origins for pythons and *Ramphotyphlops* are therefore also possible and this interpretation will be tested by further phylogenetic studies of living species. Precise knowledge of when these dispersals occurred may be very difficult to obtain, due to the paucity of terrestrial vertebrate deposits in Australasia before the late Oligocene.

The ground boas *Candoia* remain a puzzle, with neither a fossil record nor a nearby sister group to indicate how they came to be here. The time scale and relationships implied by Austin's (2000) molecular study seem to contradict both the Kon-Tiki and Gondwanan-relict hypotheses, which leaves the possibility that this is another 'intermediate', Eurasian lineage whose relatives have yet to be identified as fossils.

ACKNOWLEDGEMENTS

I thank M. Archer who supervised my Ph.D. research at the University of New South Wales, and originally suggested that I write this chapter (or something like it); R. Shine, A.E. Greer, H.G. Cogger, J. Covacevich, R.E. Molnar, M.N. Hutchinson, J.C. Rage, K.P. Aplin, J. Coventry, N. Pledge and J. Long for encouragement, discussion, and access to collections; M.S.Y. Lee and J.S. Keogh for discussions and fruitful collaborations; H. Godthelp and B.S. Mackness for bringing several snake-bearing fossil localities to my attention; G. Calvert for unpublished observations; and the Australian Research Council for financial support in recent years.

REFERENCES

Albino, A.M. 1986. Nuevos Boidae Madtsoiinae en el Cretacico tardio de Patagonia (Formacion Los Alamitos, Rio Negro, Argentina). Pp.15–21. In Bonaparte, J.F (ed). *Simposio Evolucion de los Vertebrados Mesozoicos.* Actas IV Congreso Argentino de Paleontología y Bioestratigrafía. Mendoza, Argentina.

Albino, A.M. 1993. Snakes from the Paleocene and Eocene of Patagonia (Argentina): paleoecology and coevolution with mammals. *Historical Biology,* Vol. 7, pp.51–69.

Alfaro, M.E. and Arnold, S.J. 2001. Molecular systematics and evolution of *Regina* and the thamnophiine snakes. *Molecular Phylogenetics and Evolution,* Vol. 21, pp.408–423.

Aplin, K. P. 2006. Ten million years of rodent evolution in Australasia: phylogenetic evidence and a speculative historical biogeography. Pp.707–744. In Merrick, J.R., Archer, M., Hickey, G.M. and Lee. M.S.Y. (eds). *Evolution and Biogeography of Australasian Vertebrates.* Auscipub Pty. Ltd., Sydney. 942pp.

Aplin, K.P. and Donnellan, S.C. 1993. A new species of blindsnake, genus *Ramphotyphlops* (Typhlopidae, Squamata), from northwestern Western Australia, with a redescription of *R. hamatus*, Storr 1981. *Records of the Western Australian Museum,* Vol. 16, pp.243–256.

Aplin, K.P. and Smith, L.A. 2001. Checklist of the frogs and reptiles of Western Australia. *Records of the Western Australian Museum, Supplement,* No. 63, pp.51–74.

Archer, M., Hand, S.J. and. Godthelp, H. 1991. *Riversleigh, the Story of Animals in Ancient Rainforests of Inland Australia.* Reed Books, Sydney. 256pp.

Archer, M., Hand, S.J. and Godthelp, H. 1995. Tertiary environmental and biotic change in Australia. Pp.77–90. In Vrba, E.S., Denton, G.H., Partridge, T.C. and Burckle, L.H. (eds). *Paleoclimate and Evolution, with Emphasis on Human Origins.* Yale University Press, New Haven. 547pp.

Archer, M., Hand, S.J., Godthelp, H. and Creaser, P. 1997. Correlation of the Cainozoic sediments of the Riversleigh World Heritage fossil property, Queensland, Australia. Pp.131–152. In Aguilar, J.-P., Legendre, S. and Michaux, J. (eds). *Actes du Congrès BiochroM'97.* École Pratique des Hautes Études Institut de Montpellier, France.

Archer, M. and Wade, M. 1976. The results of the Ray E. Lemley Expedition Part 1. The Allingham Formation and a new Pliocene vertebrate fauna from northern Queensland. *Memoirs of the Queensland Museum,* Vol. 17, pp.379–397.

Audley-Charles, M.G. 1987. Dispersal of Gondwanaland: relevance to evolution of the angiosperms. Pp.5–25. In Whitmore, T.C. (ed). *Biogeographical Evolution of the Malay Archipelago.* Clarendon Press, Oxford. 147pp.

Audley-Charles, M.G. 1991. Tectonics of the New Guinea area. *Annual Reviews in Earth and Planetary Sciences,* Vol. 19, pp.17–41.

Auffenberg, W. 1959. *Anomalophis bolcensis* (Massalongo), a new genus of fossil snake from the Italian Eocene. *Breviora, Museum of Comparative Zoology,* Vol. 114, pp.1–16.

Auffenberg, W. 1963. The fossil snakes of Florida. *Tulane Studies in Zoology,* Vol. 10, pp.131–216.

Augé, M., Duffaud, S., Lapparent de Broin, F. de, Rage, J.C. and Vasse, D. 1997. Les amphibiens et les reptiles de Prémontré (Cuisien, Bassin de Paris); une herpétofaune de référence pour l'Éocène inférieur. *Géologie de la France,* 1997, pp.23–33.

Austin, C.C. 2000. Molecular phylogeny and historical biogeography of Pacific Island Boas (*Candoia*). *Copeia,* 2000, pp.341–352.

Averianov, A.O. 1997. Paleogene sea snakes from the eastern part of Tethys. *Russian Journal of Herpetology,* Vol. 4, pp.128–142.

Barnes, B. 1926. Eine eozäne Wirbeltier-Fauna aus der Braunkohle des Geiseltales. *Jahrbuch des Halleschen Verbandes für die Erforschung der mitteldeutschen Bodenschätze, Neue Folge,* Vol. 6, pp.5–24.

Barrie, D.J. 1990. Skull elements and associated remains of the Pleistocene boid snake *Wonambi naracoortensis. Memoirs of the Queensland Museum,* Vol. 28, pp.139–151.

Baszio, S. and Schaal, S. 1997. The Eocene snake fauna from the Messel Pit fossil site. P. 15. In Rocek, Z. and Hart, S. (eds). *Herpetology '97: Abstracts of the Third World Congress of Herpetology.* Third World Congress of Herpetology, Prague, Czech Republic.

Bellairs, A.d'A. and Underwood, G. 1951. The origin of snakes. *Biological Reviews,* Vol. 26, pp.193–237.

Boles, W.E. 1999. Early Eocene shorebirds (Aves: Charadriiformes) from the Tingamarra Local Fauna, Murgon, Queensland, Australia. *Records of the Western Australian Museum, Supplement,* No. 57, pp.229–238.

Bolkay, S.J. 1925. *Mesophis nopcsai* n. g., n. sp.; ein neues, schlangenähnliches Reptil aus der unteren Kreide

(Neocom) von Bilek-Selista (Ost-Hercegovina). *Glasnika Zemaljskog Museja u Bosni i Hercegovini,* Vol. 37, pp.125–136.

Boulenger, G.A. 1891. Notes on the osteology of *Heloderma horridum* and *H. suspectum*, with remarks on the systematic position of the Helodermatidae and on the vertebrae of the Lacertilia. *Proceedings of the Zoological Society of London,* Vol. 59, pp.109–118.

Boulenger, G.A. 1893. *Catalogue of the Snakes in the British Museum (Natural History).* Vol. 1. Taylor and Francis Ltd., London.

Breithaupt, B.H. and Duvall, D. 1986. The oldest record of serpent aggregation. *Lethaia,* Vol. 19, pp.181–185.

Bryant, H.N. and Russell, A.P. 1992. The role of phylogenetic analysis in the inference of unpreserved attributes of extinct taxa. *Philosophical Transactions of the Royal Society of London.* Series B *(Biological Sciences),* Vol. 337, pp.405–418.

Cadle, J.E. 1987. Geographic distribution: problems in phylogeny and zoogeography. Pp.77-105. In Seigel, R.A., Collins, J.T.C. and Novak, S.S. (eds). *Snakes: Ecology and Evolutionary Biology*. MacMillan, New York.

Cadle, J.E. 1988. Phylogenetic relationships among advanced snakes: a molecular perspective. *University of California Publications in Zoology,* Vol. 119, pp.1–77.

Cadle, J.E. 1994. The colubrid radiation in Africa (Serpentes: Colubridae): phylogenetic relationships and evolutionary patterns based on immunological data. *Zoological Journal of the Linnean Society,* Vol. 110, pp.103–140.

Caldwell, M.W. 1999. Squamate phylogeny and the relationships of snakes and mosasauroids. *Zoological Journal of the Linnean Society,* Vol. 125, pp.115–147.

Caldwell, M.W., Carroll, R.L. and Kaiser, H. 1995. The pectoral girdle and forelimb of *Carsosaurus marchesetti* (Aigialosauridae), with a preliminary phylogenetic analysis of mosasauroids and varanoids. *Journal of Vertebrate Paleontology,* Vol. 15, pp.516–531.

Caldwell, M.W. and Cooper, J.A. 1999. Redescription, palaeobiogeography and palaeoecology of *Coniasaurus crassidens* Owen, 1850 (Squamata) from the Lower Chalk (Cretaceous; Cenomanian) of SE England. *Zoological Journal of the Linnean Society,* Vol. 127, pp.423–452.

Caldwell, M.W. and Lee, M.S.Y. 1997. A snake with legs from the marine Cretaceous of the Middle East. *Nature,* Vol. 386, pp.705–709.

Calligaris, R. 1988. I rettili fossili degli "Strati calcarei ittiolotici di Comen" e dell'Isola di Lesina. *Atti del Museo Civico di Storia Naturale Trieste,* Vol. 41, pp.85–125.

Camp, C.L. 1923. Classification of the lizards. *Bulletin of the American Museum of Natural History,* Vol. 48, pp.289–482.

Carroll, R.L. and deBraga, M. 1992. Aigialosaurs: mid-Cretaceous varanoid lizards. *Journal of Vertebrate Paleontology,* Vol. 12, pp.66–86.

Coate, K.H. 1997. First records of the blind snake *Ramphotyphlops braminus* from Western Australia. *Western Australian Naturalist,* Vol. 21, p. 181.

Coates, M. and Ruta, M. 2000. Nice snake, shame about the legs. *Trends in Ecology and Evolution,* Vol. 15, pp.503–507

Cogger, H.G. and Heatwole, H. 1981. The Australian reptiles: origins, biogeography, distribution patterns and island evolution. Pp.1333–1373. In Keast, A. (ed). *Ecological Biogeography of Australia.* W. Junk, The Hague.

Cope, E.D. 1869. On the reptilian orders Pythonomorpha and Streptosauria. *Proceedings of the Boston Society of Natural History,* Vol. 12, pp.250–266.

Covacevich, J., Couper, P., Molnar, R.E., Witten, G. and Young, W. 1990. Miocene dragons from Riversleigh: new data on the history of the family Agamidae (Reptilia: Squamata) in Australia. *Memoirs of the Queensland Museum,* Vol. 29, pp.339–360.

Cundall, D. 1995. Feeding behaviour in *Cylindrophis* and its bearing on the evolution of alethinophidian snakes. *Journal of Zoology, London,* Vol. 237, pp.353–376.

Cundall, M. and Greene, H.W. 2000. Feeding in snakes. Pp.293–333. In Schwenk, K. (ed). *Feeding: Form Function and Evolution in Tetrapod Vertebrates.* Academic Press, San Diego.

Cundall, D. and Rossman, D.S. 1993. Cephalic anatomy of the rare Indonesian snake *Anomochilus weberi. Zoological Journal of the Linnean Society,* Vol. 109, pp.235–273.

Cundall, D., Wallach, V. and Rossman, D.S. 1993. The systematic relationships of the snake genus *Anomochilus. Zoological Journal of the Linnean Society,* Vol. 109, pp.275–299.

Dal Sasso, C. and Pinna, G. 1997. *Aphanizocnemus libanensis* n. gen. n. sp., a new dolichosaur (Reptilia, Varanoidea) from the upper Cretaceous of Lebanon. *Paleontologia Lombarda della Società Italiana di Scienze Naturali e del Museo Civico di Storia Naturale di Milano, Nuova serie,* Vol. 7, pp.1–31.

de Buffrénil, V. and Rage, J.C. 1993. La "pachyostose" vertébrale de *Simoliophis* (Reptilia, Squamata): données comparatives et considérations fonctionelles. *Annales de Paléontologie (Vert.-Invert.),* Vol. 79, pp.315–335.

Donoghue, M.J., Doyle, J.A., Gauthier, J., Kluge, A.G. and Rowe, T. 1989. The importance of fossils in phylogeny reconstruction. *Annual Review of Ecology and Systematics,* Vol. 20, pp.431–460.

Dowling, H.G. 1959. Classification of the Serpentes: a critical review. *Copeia,* 1959, pp.38–52.

Duffaud, S. and Rage, J.C. 1997. Les remplissages karstiques polyphasés (Éocène, Oligocène, Pliocène) de Saint-Maximin (Phosphorites du Gard) et leur apport à la connaissance des faunes européennes, notamment pour l'Éocène moyen (MP13). Pp.729–735. In Aguilar, J.-P., Legendre, S., Michaux, J. (eds). *Actes du Congrès BiochroM'97.* 'École Pratique des Hautes Études Institut de Montpellier, France.

Ehmann, H.F.W. 1993a. Family Boidae. Pp.284–289. In Glasby, C.J., Ross, G.J.B. and Beesley, P.L. (eds). *Fauna of Australia. Vol. 2A. Amphibia and Reptilia.* Australian Government Publishing Service, Canberra. 439pp.

Ehmann, H.F.W. 1993b. Family Colubridae. Pp.290–294. In Glasby, C.J., Ross, G.J.B. and Beesley, P.L. (eds). *Fauna of Australia. Vol. 2A. Amphibia and Reptilia.* Australian Government Publishing Service, Canberra. 439pp.

Ehmann, H.F.W. and Bamford, M.J. 1993. Family Typhlopidae. Pp.280–283. In Glasby, C.J., Ross, G.J.B. and Beesley, P.L. (eds). *Fauna of Australia. Vol. 2A. Amphibia and Reptilia.* Australian Government Publishing Service, Canberra. 439pp.

Estes, R., De Queiroz, K. and Gauthier, J.A. 1988. Phylogenetic relationships within Squamata. Pp.119–281. In Estes R. and Pregill, G. (eds). *Phylogenetic Relationships of the Lizard Families: Essays Commemorating Charles L. Camp*. Stanford University Press, Stanford, California.

Estes, R., Frazzetta, T.H. and Williams, E.E. 1970. Studies on the fossil snake *Dinilysia patagonica* Woodward: Part 1. cranial morphology. *Bulletin of the Museum of Comparative Zoology, Harvard,* Vol. 140, pp.25–74.

Evans, S.E. and Barbadillo, J. 1998. An unusual lizard (Reptilia: Squamata) from the early Cretaceous of Las Hoyas, Spain. *Zoological Journal of the Linnean Society,* Vol. 124, pp.235–265.

Forstner, M.R., Davis, S.K. and Arévalo, E. 1995. Support for the hypothesis of anguimorph ancestry for the suborder Serpentes from phylogenetic analysis of mitochondrial DNA sequences. *Molecular Phylogenetics and Evolution,* Vol. 4, pp.93–102.

Frakes, L.A., McGowran, B. and Bowler, J.M. 1987. Evolution of Australian environments. Pp.1–16. In Dyne, G.R. and Walton, D.W. (eds). *Fauna of Australia, Vol. 1A, General Articles*. Australian Government Publishing Service, Canberra. 339pp.

Frazzetta, T.H. 1970. Studies on the fossil snake *Dinilysia patagonica* Woodward. II. Jaw machinery in the earliest snakes. *Forma et Functio*, Vol. 3, pp.205–221.

Gardner, J.D. and Cifelli, R.L. 1999. A primitive snake from the Cretaceous of Utah. *Special Papers in Palaeontology,* Vol. 60, pp.87–100.

Gauthier, J., Kluge, A.G. and Rowe, T. 1988. Amniote phylogeny and the importance of fossils. *Cladistics,* Vol.4, pp.105–209.

Gibbons, J.R.H. 1985. The biogeography and evolution of Pacific island reptiles and amphibians. Pp.125–142. In Grigg, G., Shine, R. and Ehmann, H. (eds). *Biology of Australasian Frogs and Reptiles*. Surrey Beatty and Sons, Sydney. 527pp.

Godthelp, H., Archer, M., Cifelli, R., Hand, S.J. and Gilkeson, C.F. 1992. Earliest known Australian Tertiary mammal fauna. *Nature,* Vol. 359, pp.514–516.

Greene, H.W. 1983. Dietary correlates of the origin and radiation of snakes. *American Zoologist,* Vol. 23, pp.431–441.

Greene, H.W. 1997. *Snakes: the Evolution of Mystery in Nature*. University of California Press, Berkeley. 351pp.

Greene, H.W. and Cundall, D. 2000. Snake origins: response. *Science,* Vol. 288, pp.1344–1345.

Greer, A.E. 1985. The relationships of the lizard genera *Anelytropsis* and *Diabamus*. *Journal of Herpetology,* Vol. 19, pp.116–156.

Greer, A.E. 1997. *The Biology and Evolution of Australian Snakes*. Surrey Beatty and Sons, Sydney. 358pp.

Grigg, G., Shine, R. and Ehmann, H. (eds). 1985. *Biology of Australasian Frogs and Reptiles*. Surrey Beatty and Sons, Sydney. 527pp.

Gyi, K.K. 1970. A revision of colubrid snakes of the subfamily Homalopsinae. *University of Kansas Publications of the Museum of Natural History,* Vol. 20, pp.47–223.

Haas, G. 1979. On a new snakelike reptile from the lower Cenomanian of Ein Jabrud, near Jerusalem. *Bulletin du Muséum National d'Histoire Naturelle, Paris,* Series 4, Section C, Vol. 1, pp.51–64.

Haas, G. 1980a. *Pachyrhachis problematicus* Haas, snakelike reptile from the lower Cenomanian: ventral view of the skull. *Bulletin du Muséum National d'Histoire Naturelle, Paris,* Series 4, Vol. 2, pp.87–104.

Haas, G. 1980b. Remarks on a new ophiomorph reptile from the lower Cenomanian of Ein Jabrud, Israel. Pp.177–192. In Jacobs, L.L. (ed). *Aspects of Vertebrate History,* Museum of Northern Arizona Press, Flagstaff, Arizona.

Hall, R. 2001. Cenozoic reconstructions of SE Asia and the SW Pacific: changing patterns of land and sea. Pp.35–56. In Metcalfe, I., Smith, J.M.B., Morwood, M. and Davidson, I. (eds). *Faunal and Floral Migrations and Evolution in SE Asia-Australalasia.* A.A. Balkema, Lisse. 416pp..

Hallermann, J. 1998. The ethmoidal region of *Dibamus taylori* (Squamata: Dibamidae), with a phylogenetic hypothesis on dibamid relationships within Squamata. *Zoological Journal of the Linnean Society,* Vol. 122, pp.385–426.

Heatwole, H. and Cogger, H.G. 1993. Family Hydrophiidae. Pp.310–318. In Glasby, C.J., Ross, G.J.B. and Beesley, P.L. (eds). *Fauna of Australia. Vol. 2A. Amphibia and Reptilia.* Australian Government Publishing Service, Canberra. 439pp.

Heatwole, H. and Guinea, M.L. 1993. Family Laticaudidae. Pp.319–321. In Glasby, C.J., Ross, G.J.B. and Beesley, P.L. (eds). *Fauna of Australia. Vol. 2A. Amphibia and Reptilia.* Australian Government Publishing Service, Canberra. 439pp.

Hecht, M.K. 1982. The vertebral morphology of the Cretaceous snake, *Dinilysia patagonica* Woodward. *Neues Jahrbuch für Geologie und Paläontologie Monatshefte,* Vol. 1982, pp.523–532.

Hoffstetter, R. 1939. Contribution a l'étude des Elapidae actuels et fossiles et de l'ostéologie des ophidiens. *Archives du Muséum d'Histoire Naturelle de Lyon,* Vol. 15, pp.1–78.

Hoffstetter, R. 1955. Squamates de type moderne. Pp.606–662. In Piveteau, J. (ed). *Traité de Paléontologie*. Masson, Paris.

Hoffstetter, R. 1959. Un serpent terrestre dans le Cretacé inférieur du Sahara. *Bulletin du Societé Géologique de France,* Vol. 7, pp.897–902.

Hoffstetter, R. 1968. A contribution to the classification of snakes, by Garth Underwood. *Copeia,* 1968, pp.201–213.

Hoffstetter, R. and Gasc, J.P. 1969. Vertebrae and ribs of modern reptiles. Pp.201–310. In Gans, C., Bellairs, A.d'A. and Parsons, T.S. (eds). *Biology of the Reptilia.* Vol. 1. Academic Press, London. 373pp.

Hoffstetter, R. and Gayrard, Y. 1965. Observations sur l'osteologie et la classification des Acrochordidae (Serpentes). *Bulletin du Muséum National d'Histoire Naturelle, Paris,* Series 2, Vol. 36, pp.677–696.

Holman, J.A. 1979. A review of North American Tertiary snakes. *Publications of the Museum—Michigan State University, Paleontological Series,* Vol. 1, pp.200–260.

Holman, J.A. 2000. *Fossil Snakes of North America: Origin, Evolution, Distribution, Paleoecology*. Indiana University Press, Bloomington, Indiana.

Hutchinson, M.N. and Donnellan, S.C. 1993. Biogeography and phylogeny of the Squamata. Pp.210–220. In Glasby, C.J., Ross, G.J.B. and Beesley, P.L. (eds). *Fauna of Australia. Vol. 2A. Amphibia and Reptilia.* Australian Government Publishing Service, Canberra. 439pp.

Hutchinson, M.N. and Smith, S.A. 2006. Australian lizards—relationships and origins. Pp.349–366. In Merrick, J.R., Archer, M., Hickey, G.M. and Lee. M.S.Y. (eds). *Evolution and Biogeography of Australasian Vertebrates.* Auscipub Pty. Ltd., Sydney. 942pp.

Janensch, W. 1906. Über *Archaeophis proavus* Mass., eine Schlange aus dem Eocän des Monte Bolca. *Beiträge zur Paläontologie und Geologie Österreich-Ungarns und des Orients,* Vol. 19, pp.1–33.

Keller, T. and Schaal, S. 1988. Schuppenechsen—Reptilien auf Erfolgskurs. Pp.121–133. In Schaal, S. and Ziegler, W. (eds). *Messel—Ein Schaufenster in die Geschichte der Erde und des Lebens.* Senckenbergischen Naturforschenden Gesellschaft, Frankfurt am Main.

Keogh, J.S. 1998. Molecular phylogeny of elapid snakes and a consideration of their biogeographic history. *Biological Journal of the Linnean Society,* Vol. 63, pp.177–203.

Keogh, J.S. 1999. Evolutionary implications of hemipenial morphology in the terrestrial Australian elapid snakes. *Zoological Journal of the Linnean Society,* Vol. 125, pp.232–278.

Keogh, J.S., Scott, I.A.W. and Scanlon, J.D. 2000. Molecular phylogeny of viviparous Australian elapid snakes: affinities of *'Echiopsis' atriceps* (Storr, 1980) and *'Drysdalia' coronata* (Schlegel, 1837), with description of a new genus. *Journal of Zoology, London,* Vol. 252, pp.317–326.

Keogh, J.S., Shine, R. and Donellan, S. 1998. Phylogenetic relationships of terrestrial Australo-papuan elapid snakes (subfamily Hydrophiinae) based on Cytochrome b and 16S rRNA sequences. *Molecular Phylogenetics and Evolution,* Vol. 10, pp.67–81.

Kluge, A.G. 1991. Boine snake phylogeny and research cycles. *Miscellaneous Publications of the Museum of Zoology, University of Michigan,* Vol. 178, pp.1–58.

Kluge, A.G. 1993a. *Calabaria* and the phylogeny of erycine snakes. *Zoological Journal of the Linnean Society,* Vol. 107, pp.293–351.

Kluge, A.G. 1993b. *Aspidites* and the phylogeny of pythonine snakes. *Records of the Australian Museum, Supplement,* No. 19, pp.1–77.

Kraus, F. and Brown, W.M. 1998. Phylogenetic relationships of colubroid snakes based on mitochondrial DNA sequences. *Zoological Journal of the Linnean Society*, Vol. 122, pp.455–487.

Kuhn, O. 1939. Die Schlangen (Boidae) aus dem Mitteleozän des Geiseltales. *Nova Acta Leopoldina*, Neue Folge, Vol. 7, pp.119–133.

LaDuke, T.C. 1991. The fossil snakes of Pit 91, Rancho La Brea, California. *Natural History Museum of Los Angeles County Contributions in Science,* Vol. 424, pp.1–28.

Lee, M.S.Y. 1997a. Phylogenetic relationships among Australian elapid snakes: the soft anatomical data reconsidered. *The Herpetological Journal,* Vol. 7, pp.93–102.

Lee, M.S.Y. 1997b. The phylogeny of varanoid lizards and the affinities of snakes. *Philosophical Transactions of the Royal Society of London,* Series B, Vol. 352, pp.53–91.

Lee, M.S.Y. 1998. Convergent evolution and character correlation in burrowing reptiles: towards a resolution of squamate phylogeny. *Biological Journal of the Linnean Society*, Vol. 65, pp.369–453.

Lee, M.S.Y., Bell, G.L. and Caldwell, M.W. 1999. The origin of snake feeding. *Nature,* Vol. 400, pp.655–659.

Lee, M.S.Y. and Caldwell, M.W. 1998. Anatomy and relationships of *Pachyrhachis problematicus*, a primitive snake with hindlimbs. *Philosophical Transactions of the Royal Society of London,* Series B, Vol. 353, pp.1521–1552.

Lee, M.S.Y. and Caldwell, M.W. 2000. *Adriosaurus* and the affinities of mosasaurs, dolichosaurs, and snakes. *Journal of Paleontology,* Vol. 74, pp.915–937.

Lee, M.S.Y., Caldwell, M.W. and Scanlon, J.D. 1999. A second primitive marine snake: *Pachyophis woodwardi* from the Cretaceous of Bosnia-Herzegovina. *Journal of Zoology, London,* Vol. 248, pp.509–520.

Lee, M.S.Y. and Scanlon, J.D. 2002a. The Cretaceous marine squamate *Mesoleptos* and the origin of snakes. *Bulletin of the British Museum (Natural History) Zoology*, Vol. 68, pp.131–142.

Lee, M.S.Y. and Scanlon, J.D. 2002b. Snake phylogeny based on osteology, soft anatomy and behaviour. *Biological Reviews,* Vol. 77, pp.333–402.

List, J.C. 1966. Comparative osteology of the snake families Typhlopidae and Leptotyphlopidae. *Illinois Biological Monographs,* No. 36, pp.1–112. *University of Illinois Press, Urbana.*

Long, J.A. 1998. *Dinosaurs of Australia and New Zealand, and Other Animals of the Mesozoic Era.* University of New South Wales Press, Sydney. 188pp.

Mackness, B.S. and Scanlon, J.D. 1999. The first Pliocene record of the madtsoiid snake genus *Yurlunggur* Scanlon, 1992 from Queensland. *Memoirs of the Queensland Museum,* Vol. 43, Pt. 2, pp.783–785.

Malnate, E.V. and Underwood, G. 1988. Australasian natricine snakes of the genus *Tropidonophis*. *Proceedings of the Academy of Natural Sciences of Philadelphia,* Vol. 140, pp.59–201.

McDowell, S.B. 1967a. Osteology of the Typhlopidae and Leptotyphlopidae: a critical review. *Copeia,* 1967, pp.686–692.

McDowell, S.B. 1967b. *Aspidomorphus*, a genus of New Guinea snakes of the family Elapidae, with notes on related genera. *Journal of Zoology, London,* Vol. 151, pp.497–543.

McDowell, S.B. 1970. On the status and relationships of the Solomon Island elapid snakes. *Journal of Zoology, London,* Vol. 161, pp.145–190.

McDowell, S.B. 1972a. The evolution of the tongue of snakes, and its bearing on snake origins. Pp.191–273. In Dobzhansky, T., Hecht, M.K. and Steere, W.(eds). *Evolutionary Biology.* Vol. 6. Appleton-Century-Crofts, New York. 445pp.

McDowell, S.B. 1972b. The species of *Stegonotus* (Serpentes, Colubridae) in Papua New Guinea. *Zoologische Mededelingen,* Vol. 47, pp.6–26.

McDowell, S.B. 1974. A catalogue of the snakes of New Guinea and the Solomons, with special reference to those in the Bernice P. Bishop Museum. Part I. Scolecophidia. *Journal of Herpetology,* Vol. 8, pp.1–57.

McDowell, S.B. 1975. A catalogue of the snakes of New Guinea and the Solomons, with special reference to those in the Bernice P. Bishop Museum. Part II. Anilioidea and Pythoninae. *Journal of Herpetology,* Vol. 9, pp.1–80.

McDowell, S.B. 1979. A catalogue of the snakes of New Guinea and the Solomons, with special reference to those in the Bernice P. Bishop Museum. Part III. Boinae and Acrochordoidea (Reptilia, Serpentes). *Journal of Herpetology,* Vol. 13, pp.1–92.

McDowell, S.B. 1984. Results of the Archbold Expeditions. No. 112. The snakes of the Huon Peninsula. *American Museum Novitates,* No. 2775, pp.1–28.

McDowell, S.B. 1985. The terrestrial Australian elapids: general summary. Pp.261–264. In Grigg, G., Shine, R. and

Ehmann, H. (eds). *The Biology of Australasian Frogs and Reptiles*. Surrey Beatty and Sons, Sydney. 527pp.

McDowell, S.B. 1987. Systematics. Pp.1–50. In Seigel, R.A., Collins, J.T.C. and Novak, S.S. (eds). *Snakes: Ecology and Evolutionary Biology*. MacMillan, New York.

McDowell, S.B. and Bogert, C.M. 1954. The systematic position of *Lanthanotus* and the affinities of the anguimorphan lizards. *Bulletin of the American Museum of Natural History,* Vol. 105, pp.1–154.

McNamara, G.C. 1990. The Wyandotte Local Fauna: a new, dated, Pleistocene vertebrate fauna from northern Queensland. *Memoirs of the Queensland Museum,* Vol. 28, pp.285–297.

Mengden, G.A. 1985a. Australian elapid phylogeny: a summary of the chromosomal and electrophoretic data. Pp.185–192. In Grigg, G., Shine, R. and Ehmann, H. (eds). *Biology of Australasian Frogs and Reptiles*. Surrey Beatty and Sons, Sydney. 527pp.

Mengden, G.A. 1985b. A chromosomal and electrophoretic analysis of the genus *Pseudonaja*. Pp.193–208. In Grigg, G., Shine, R and Ehmann, H. (eds). *Biology of Australasian Frogs and Reptiles*. Surrey Beatty and Sons, Sydney. 527pp.

Mengden, G.A., Shine, R. and Moritz, C. 1986. Phylogenetic relationships among venomous Australasian snakes of the genus *Pseudechis* (Elapidae). *Herpetologica,* Vol. 42, pp.215–229.

Molnar, R.E. 1991. Fossil reptiles in Australia. Pp.605–702. In Vickers-Rich, P., Monaghan, J.M., Baird, R.F. and Rich, T.H. (eds). *Vertebrate Palaeontology of Australasia.* Pioneer Design Studio and Monash University Publications Committee, Melbourne. 1437pp.

Morgan, E. 1994. *The Descent of the Child: Human Evolution from a New Perspective*. Souvenir Press, London.

Nessov, L.A., Zhegallo, V.I. and Averianov, A.O. 1998. A new locality of late Cretaceous snakes, mammals and other vertebrates in Africa (western Libya). *Annales de Paleontologie,* Vol. 84, pp.265–274.

Nopcsa, F. 1908. Zur Kenntnis der fossilen Eidechsen. *Beiträge zur Paläontologie und Geologie Österreich-Ungarns und des Orients,* Vol. 21, pp.33–62.

Nopcsa, F. 1923. *Eidolosaurus* und *Pachyophis*. Zwei neue Neocom-Reptilien. *Palaeontographica,* Vol. 65, pp.99–154.

Nopcsa, F. 1925. Ergebnisse der Forschungsreisen Prof. E. Stromers in den Wüsten Ägyptens. II. Wirbeltier-Reste der Baharîje-Stufe (unterstes Cenoman). 5. Die *Symoliophis*-Reste. *Abhandlungen der Bayerischen Akademie der Wissenschaften, Mathematisch-naturwissenschaftliche Abteilung,* Vol. 30, pp.1–27.

Pascual, R., Archer, M., Ortiz Jaureguizar, E., Prado, J.L., Godthelp, H. and Hand, S.J. 1992. The first non-Australian monotreme: an early Paleocene South American platypus (Monotremata, Ornithorhynchidae). Pp.1–14. In Augee, M.L. (ed). *Platypus and Echidnas*. Royal Zoological Society of New South Wales, Sydney. 296pp.

Pledge, N.S. 1992. The Curramulka Local Fauna: a new late Tertiary fossil assemblage from Yorke Peninsula, South Australia. *The Beagle (Records of the Northern Territory Museum of Arts and Sciences),* Vol. 9, pp.115–142.

Polcyn, M.J., Tchernov, E. and Jacobs, L.L. 1999. The Cretaceous biogeography of the eastern Mediterranean with a description of a new basal mosasauroid from 'Ein Yabrud, Israel. Pp.259–290. In Tomida, Y., Rich, T.H. and Vickers-Rich, P. (eds). *Proceedings of the Second Gondwanan Dinosaur Symposium*. National Science Museum Monograph, No. 15, Tokyo.

Rage, J.C. 1977. La position phylétique de *Dinilysia patagonica*, serpent du Cretacé supérieur. *Comptes Rendus de l'Académie des Sciences, Paris,* Vol. 284, pp.1765–1768.

Rage, J.C. 1982. La phylogénie des Lépidosauriens (Reptilia): une approche cladistique. *Comptes Rendus de l'Académie des Sciences, Paris,* Vol. 294, pp.399–402.

Rage, J.C. 1984. *Handbuch der Paläoherpetologie. 11. Serpentes*. G. Fischer, Stuttgart.

Rage, J.C. 1987. Fossil History. Pp.57–76. In Seigel, R.A., Collins, J.T.C. and Novak, S.S. (eds). *Snakes: Ecology and Evolutionary Biology*. MacMillan, New York.

Rage, J.C. 1998. Fossil snakes from the Paleocene of São José de Itaboraí, Brazil. Part I. Madtsoiidae, Aniliidae. *Palaeovertebrata,* Vol. 27, No. 3–4, pp.109–144.

Rage, J.C. and Albino, A.M. 1989. *Dinilysia patagonica* (Reptilia, Serpentes): matériel vertébral additionel du Crétacé supérieur d'Argentine. Étude complémentaire des vertèbres, variations intraspécifiques et intracolumnaires. *Neues Jahrbuch für Geologie und Paläontologie Monatshefte,* 1989, pp.433–447.

Rage, J.C., Buffetaut, E., Buffetaut-Tong, H., Chaimanee, Y., Ducrocq, S., Jaeger, J.-J. and Suteethorn, V. 1992. A colubrid snake in the late Eocene of Thailand: the oldest known Colubridae (Reptilia, Serpentes). *Comptes Rendus de l'Académie des Sciences, Paris,* Series 2, Vol. 314, pp.1085–1089.

Rage, J.C. and Escuillié, F. 2000. Un nouveau serpent bipède du Cénomanien (Cretacé). Implications phyletiques. *Comptes Rendus de l'Academie des Sciences, Paris, Sciences de la Terre et des Planètes,* Vol. 330, pp.513–520.

Rage, J.C. and Escuillié, F. 2002. *Eupodophis*, new name for the genus *Podophis* Rage and Escuillié, 2000, an extinct bipedal snake, preoccupied by *Podophis* Wiegmann, 1834 (Lacertilia, Scincidae). *Amphibia-Reptilia*, Vol. 23, pp.232–233.

Rage, J.C. and Richter, A. 1994. A snake from the lower Cretaceous (Barremian) of Spain: the oldest known snake. *Neues Jahrbuch für Geologie und Paläontologie Monatshefte,* 1994, pp.561–565.

Rage, J.C. and Werner, C. 1999. Mid-Cretaceous (Cenomanian) snakes from Wadi Abu Hashim, Sudan: the earliest snake assemblage. *Palaeontologia Africana,* Vol. 35, pp.85–110.

Reeder, T.W. 1995. Phylogenetic placement of snakes within the Squamata: evidence from molecules and morphology. P. 166. *American Society of Ichthyologists and Herpetologists, Program and Abstracts*. University of Alberta, Edmonton.

Rich, T.H., Archer, M., Hand, S.J., Godthelp, H., Muirhead, J., Pledge, N.S., Flannery, T.F., Woodburne, M.O., Case, J.A., Tedford, R.H., Turnbull, W.D., Lundelius, E.L. Jnr., Rich, L.S.V., Whitelaw, M.J., Kemp, A. and Vickers-Rich, P. 1991. Australian Mesozoic and Tertiary terrestrial mammal localities. Pp.1005–1058. In Vickers-Rich, P., Monaghan, J.M., Baird, R.F. and Rich, T.H. (eds). *Vertebrate Palaeontology of Australasia*. Pioneer Design Studio and Monash University Publications Committee, Melbourne. 1437pp.

Rieppel, O. 1980. The phylogeny of anguinomorph lizards.

Denkschriften der Schweizerischen Naturforschenden Gesellschaft, Vol. 94, pp.1–86.

Rieppel, O. 1988. A review of the origin of snakes. *Evolutionary Biology,* Vol. 22, pp.37–130.

Rieppel, O. and Zaher, H. 2000a. The intramandibular joint in squamates, and the phylogenetic relationships of the fossil snake *Pachyrhachis problematicus* Haas. *Fieldiana (Geology),* New Series, No. 43, pp.1–69.

Rieppel, O. and Zaher, H. 2000b. The braincases of mosasaurs and *Varanus,* and the relationship of snakes. *Zoological Journal of the Linnean Society,* Vol, 129, pp 489–514.

Rieppel, O. and Zaher, H. 2001. Re-building the bridge between mosasaurs and snakes. *Neues Jahrbuch für Geologie und Paläontologie Abhandlungen,* Vol. 221, pp.111–132.

Robb, J. 1966. The generic status of the Australian typhlopids (Reptilia: Squamata). *Annals and Magazine of Natural History,* Vol. 13, pp.675–679.

Romer, A.S. 1956. *Osteology of the Reptiles.* University of Chicago Press, Chicago.

Scanlon, J.D. 1992. A new large madtsoiid snake from the Miocene of the Northern Territory. *The Beagle (Records of the Northern Territory Museum of Arts and Sciences),* Vol. 9, pp.49–60.

Scanlon, J.D. 1993. Madtsoiid snakes from the Eocene Tingamarra Fauna of eastern Queensland. *Kaupia: Darmstädter Beiträge zur Naturgeschichte,* Vol. 3, pp.3–8.

Scanlon, J.D. 1995a. An elapid snake (Hydropheinae) from the middle Miocene Encore Site, Riversleigh. P. 20. *Abstracts of 5th Conference on Australasian Vertebrate Evolution, Palaeontology and Systematics, Canberra.*

Scanlon, J.D. 1995b. First records from Wellington Caves, New South Wales, of the extinct madtsoiid snake *Wonambi naracoortensis* Smith, 1976. *Proceedings of the Linnean Society of New South Wales,* Vol. 115, pp.233–238.

Scanlon, J.D. 1996. *Studies in the palaeontology and systematics of Australian snakes.* Ph.D. thesis (unpublished), University of New South Wales, Sydney.

Scanlon, J.D. 1997a. *Nanowana* gen. nov., small madtsoiid snakes from the Miocene of Riversleigh: sympatric species with divergently specialised dentition. *Memoirs of the Queensland Museum,* Vol. 41, pp.393–412.

Scanlon, J.D. 1997b. Mysterious origins of Australian snakes. *Riversleigh Notes,* Vol. 33, pp.2–6.

Scanlon, J.D. 2001. *Montypythonoides*: the Miocene snake *Morelia riversleighensis* (Smith and Plane, 1985) and the geographical origin of pythons. *Memoir of the Association of Australasian Palaeontologists,* Vol. 25, pp.1–35.

Scanlon, J.D. and Lee, M.S.Y. 2000. The Pleistocene serpent *Wonambi* and the early evolution of snakes. *Nature,* Vol. 403, pp.416–420.

Scanlon, J.D. and Lee, M.S.Y. 2004. Phylogeny of Australasian venomous snakes (Colubroidea, Elapidae, Hydrophiinae) based on phenotypic and molecular evidence. *Zoologica Scripta,* Vol. 33, No. 4, pp.335–367.

Scanlon, J.D., Lee, M.S.Y. and Archer, M. 2003. Mid-Tertiary elapid snakes (Squamata, Colubroidea) from Riversleigh, northern Australia: early steps in a continent-wide adaptive radiation, *Geobios,* Vol. 36, No. 5, pp.573–601.

Scanlon, J.D., Lee, M.S.Y., Caldwell, M.W. and Shine, R. 1999. The palaeoecology of the primitive snake *Pachyrhachis. Historical Biology,* Vol. 13, pp.127–152.

Scanlon, J.D. and Mackness, B.S. 2002. A new giant python from the Pliocene Bluff Downs Local Fauna of northeastern Queensland. *Alcheringa,* Vol. 25, pp.425–437.

Schwaner, T.D., Baverstock, P.R., Dessauer, H.C. and Mengden, G.A. 1985. Immunological evidence for the phylogenetic relationships of Australian elapid snakes. Pp.177–184. In Grigg, G., Shine, R. and Ehmann, H. (eds). *Biology of Australasian Frogs and Reptiles.* Surrey Beatty and Sons, Sydney. 527pp.

Schwenk, K. 1988. Comparative morphology of the lepidosaur tongue and its relevance to squamate phylogeny. Pp.569–598. In Estes, R. and Pregill, G. (eds). *Phylogenetic Relationships of the Lizard Families: Essays Commemorating Charles L. Camp.* Stanford University Press, Stanford, California.

Schwenk, K. 1993. The evolution of chemoreception in squamate reptiles: a phylogenetic approach. *Brain Behaviour and Evolution,* Vol. 41, pp.124–137.

Senn, D.G. and Northcutt, R.G. 1973. The forebrain and midbrain of some squamates and their bearing on the origin of snakes. *Journal of Morphology,* Vol. 140, pp.135–152.

Shea, G.M. and Hutchinson, M.N. 1991. A new species of *Tiliqua* (Lacertilia, Scincidae) from the Oligo-Miocene deposits of Riversleigh, northwestern Queensland. *Memoirs of the Queensland Museum,* Vol. 32, pp.303–310.

Shea, G., Shine, R. and Covacevich, J.C. 1993. Family Elapidae. Pp.295–309. In Glasby, C.J., Ross, G.J.B. and Beesley, P.L. (eds). *Fauna of Australia. Vol. 2A. Amphibia and Reptilia.* Australian Government Publishing Service, Canberra. 439pp.

Shine, R. 1985a. Ecological evidence on the phylogeny of Australian elapid snakes. Pp.255–260. In Grigg, G. Shine, R. and Ehmann, H. (eds) *Biology of Australasian Frogs and Reptiles.* Surrey Beatty and Sons, Sydney. 527pp.

Shine, R. 1985b. The evolution of viviparity in reptiles: an ecological analysis. Pp.605–694. In Gans, C. and Billett, F. (eds). *Biology of the Reptilia.* Vol. 15. John Wiley and Sons, New York.

Shine, R. 1991a. Strangers in a strange land: ecology of the Australian colubrid snakes. *Copeia,* 1991, pp.120–131.

Shine, R. 1991b. *Australian Snakes: a Natural History.* Reed Books, Sydney. 223pp.

Shine, R. and Houston, D. 1993. Family Acrochordidae. Pp.322–324. In Glasby, C.J. Ross, G.J.B. and Beesley, P.L. (eds). *Fauna of Australia. Vol. 2A. Amphibia and Reptilia.* Australian Government Publishing Service, Canberra. 439pp.

Shine, R. and Keogh, J.S. 1996. Food habits and reproductive biology of the endemic Melanesian elapids: are tropical snakes really different? *Journal of Herpetology,* Vol. 30, pp.238–247.

Simpson, G.G. 1933. A new fossil snake from the *Notostylops* beds of Patagonia. *Bulletin of the American Museum of Natural History,* Vol. 67, pp.1–22.

Slowinski, J.B., Boundy, J. and Lawson, R. 2001. The phylogenetic relationships of Asian coral snakes (Elapidae: *Calliophis* and *Maticora*) based on morphology and molecular characters. *Herpetologica,* Vol. 57, pp.233–245.

Slowinski, J.B. and Keogh, J.S. 2000. Phylogenetic relationships of elapid snakes based on Cytochrome *b* mtDNA sequences. *Molecular Phylogenetics and Evolution,* Vol. 15, pp.157–164.

Slowinski, J.B. and Lawson, R. 2002. Snake phylogeny: evidence from nuclear and mitochondrial genes. *Molecular*

Phylogenetics and Evolution, Vol. 24, pp.194–202.

Smith, M.J. 1975. The vertebrae of four Australian elapid snakes (Squamata: Elapidae). *Transactions of the Royal Society of South Australia,* Vol. 96, pp.125–137.

Smith, M.J. 1976. Small fossil vertebrates from Victoria Cave, Naracoorte, South Australia. IV. Reptiles. *Transactions of the Royal Society of South Australia,* Vol. 100, pp.39–51.

Smith, M.J. and Plane, M. 1985. Pythonine snakes (Boidae) from the Miocene of Australia. *Bureau of Mineral Resources, Journal of Australian Geology and Geophysics,* Vol. 9, pp.191–195.

Stanley, S.M. 1979. *Macroevolution: Pattern and Process*. W. H. Freeman, San Francisco.

Storr, G.M. 1964. Some aspects of the geography of Australian reptiles. *Senckenbergiana Biologica,* Vol. 45, pp.577–589.

Szyndlar, Z. and Böhme, W. 1993. Die fossilen Schlangen Deutschlands: Geschichte der Faunen und ihrer Erforschung. *Mertensiella,* Vol. 3, pp.381–431.

Szyndlar, Z. and Rage, J.C. 1990. West Palearctic cobras of the genus *Naja* (Serpentes: Elapidae): interrelationships among extinct and extant species. *Amphibia-Reptilia,* Vol. 11, pp.385–400.

Szyndlar, Z. and Schleich, H.H. 1993. Description of Miocene snakes from Petersbuch 2 with comments on the lower and middle Miocene ophidian faunas of southern Germany. *Stuttgarter Beiträge zur Naturkunde, Serie B (Geologie und Paläontologie),* Vol. 192, pp.1–47.

Szyndlar, Z. and Zerova, G.A. 1990. Neogene cobras of the genus *Naja* (Serpentes: Elapidae) of east Europe. *Annalen des Naturhistorischen Museums in Wien, Serie A, Mineralogie und Petrographie, Geologie und Paläontologie, Anthropologie und Prähistorie,* Vol. 91, pp.53–61.

Tatarinov, L.P. 1988. Structure of the skull in the sea snake *'Archaeophis' turkmenicus* from the lower Eocene of Turkmeniya. *Paleontologischeskii Zhurnal*, 1988, pp.75–82.

Tchernov, E., Rieppel, O., Zaher, H., Polcyn, M.J. and Jacobs, L.L. 2000. A fossil snake with limbs. *Science*, Vol. 287, pp.2010–2012.

Tyler, M.J. 1979. Herpetofaunal relationships of South America with Australia. Pp.73–106. In Duellman, W.E. (ed). *The South American Herpetofauna: its Origin, Evolution and Dispersal.* Monograph of the Museum of Natural History, University of Kansas, No. 7, pp.1–485.

Underwood, G. 1967. *A Contribution to the Classification of Snakes*. British Museum (Natural History), London.

Underwood, G. 1970. The eye. Pp.1–97. In Gans, C. and Parsons, T.S. (eds). *Biology of the Reptilia.* Vol. 2. Academic Press, London. 374pp.

Underwood, G. 1976. A systematic study of boid snakes. Pp.151–175. In Bellairs, A.d'A. and Cox, C.B. (eds). *Morphology and Biology of Reptiles.* Academic Press, London.

Underwood, G. and Stimson, A.F. 1990. A classification of pythons (Serpentes, Pythoninae). *Journal of Zoology, London,* Vol. 221, pp.565–603.

Vences, M., Glaw, F., Kosuch, J., Böhme, W. and Veith, M. 2001. Phylogeny of South American and Malagasy boine snakes: molecular evidence for the validity of *Sanzinia* and *Acrantophis* and biogeographic implications. *Copeia,* 2001, pp.1151–1154.

Vidal, N. and Hedges, S.B. 2002. Higher-level relationships of snakes inferred from four nuclear and mitochondrial genes. *Comptes Rendues, Biologies,* Vol. 325, pp.977–985.

Wallach, V. 1985. A cladistic analysis of the terrestrial Australian Elapidae. Pp.223–253. In Grigg, G., Shine, R. and Ehmann, H. (eds). *Biology of Australasian Frogs and Reptiles*. Surrey Beatty and Sons, Sydney. 527pp.

Wallach, V. 1995. A new genus for the *Ramphotyphlops subocularis* species group (Serpentes: Typhlopidae), with description of a new species. *Asiatic Herpetological Research,* Vol. 6, pp.132–150.

Wallach, V. 1996. The systematic status of the *Ramphotyphlops flaviventer* (Peters) complex (Serpentes: Typhlopidae). *Amphibia-Reptilia,* Vol. 17, pp.341–359.

Wallach, V. 1997. Two new blind snakes of the *Typhlops ater* species group from Papua New Guinea (Serpentes: Typhlopidae). *Russian Journal of Herpetology,* Vol. 3, pp.107–118.

Webb, J.K. and Shine, R. 1993. Prey-size selection, gape limitation and predator vulnerability in Australian blindsnakes. *Animal Behaviour,* Vol. 45, pp.1117–1126.

Webb, J.K., Shine, R., Branch, W.R. and Harlow, P.S. 2000. Life-history strategies in basal snakes: reproduction and dietary habits of the African thread snake *Leptotyphlops scutifrons* (Serpentes: Leptotyphlopidae). *Journal of Zoology, London,* Vol. 250, pp.321–327.

Werner, C. and Rage, J.C. 1994. Mid-Cretaceous snakes from Sudan. A preliminary report on an unexpectedly diverse snake fauna. *Comptes Rendus de l'Académie des Sciences, Paris*, Series 2, Vol. 319, pp.247–252.

Wiffen, J. 1990. New mosasaurs (Reptilia, family Mosasauridae) from the upper Cretaceous of North Island. *New Zealand Journal of Geology and Geophysics,* Vol. 33, pp.67–85.

White, A.W. 2001. A new Eocene soft-shelled turtle (Trionychidae) from Murgon, south-eastern Queensland. *Memoir of the Association of Australasian Palaeontologists,* Vol. 25, pp.37–43.

Wilcox, T.P., Zwickl, D.J., Heath, T.A. and Willis, D.M. 2002. Phylogenetic relationships of the dwarf boas and a comparison of Bayesian and bootstrap measures of phylogenetic support. *Molecular Phylogentics and Evolution,* Vol. 25, pp.361–371.

Willis, P.M.A. 1997. New crocodilians from the late Oligocene White Hunter Site, Riversleigh, northwestern Queensland. *Memoirs of the Queensland Museum,* Vol. 41, pp.423–438.

Woodburne, M.O. and Case, J.A. 1996. Dispersal, vicariance, and the late Cretaceous to early Tertiary land mammal biogeography from South America to Australia. *Journal of Mammalian Evolution,* Vol. 3, pp.121–161.

Wüster, W. and Thorpe, R.S. 1992. Asiatic cobras: population systematics of the *Naja naja* species complex (Serpentes: Elapidae) in India and Central Asia. *Herpetologica,* Vol. 48, pp.69–85.

Zaher, H. 1998. The phylogenetic position of *Pachyrhachis* within snakes (Squamata, Lepidosauria). *Journal of Vertebrate Paleontology,* Vol, 18, pp.1–3.

Zaher, H. and Rieppel, O. 1999. Tooth implantation and replacement in squamates, with special reference to monasaur lizards and snakes. *American Museum Novitates,* No. 3271, pp.1–19.

Zaher, H. and Rieppel, O. 2000. A brief history of snakes. *Herpetological Review,* Vol, 31, pp.73–76.

Chapter 18
Evolution and Zoogeography of Australasian Crocodilians

Paul M.A. Willis
Quinkana Pty Ltd, 1/662 Elizabeth Street, Waterloo, N.S.W. 2017 Australia
E-mail: pwillis@ozemail.com.au

INTRODUCTION

Contrary to what might be expected crocodilians have not always been lethargic creatures lying around in swamps looking like a cross between a dinosaur and a lizard. There are some weird and wonderful animals in the crocodilian family tree and crocodiles are tenacious survivors that have lasted 210 million years.

The fossil record of crocodilians is pretty good, which is not surprising. Many crocodilians lived in or near environments where fossilization is likely to occur. Within Australia, crocodiles have one of the best fossil records of any vertebrate group but it has only been during the last couple of decades that this material has received the study needed to identify taxa and reconstruct phylogenies.

WHAT IS A CROCODILIAN?

Crocodiles are quadrupedal archosaurs closely related to the dinosaurs (Romer, 1956; Benton, 1985; Benton and Clark, 1988). The term crocodilian is applied to members of the order Crocodilia and includes all members of the suborder Eusuchia (which includes all living crocodilians) as well as a huge diversity of extinct forms. Details of the morphology of various extant crocodilian skulls are included in Iordansky (1973) while Langston (1973) takes an historical perspective; Rowe *et al.* (1999) provide a more detailed description of the skull of *Alligator mississippiensis*.

Crocodilian features include: a depressed skull; external nares in a terminal position; some development of a bony palate; choanae posterior to the external nares; parietals solidly united along the midline; postorbital skull roof broad and flat, forming a 'skull deck'; antorbital fenestrae reduced or absent; quadrate strongly inclined, bordered anteriorly by a long, slender quadratojugal; palate akinetic with pterygoids and quadrates fused to the braincase; prefrontals extend ventrally to reach the palatines; pterygoids with wide, deep wings; interpterygoid vacuity absent; many cranial bones and the articular more or less pneumatic; eustachian tubes more or less enclosed; posttemporal fenestrae reduced; squamosal, quadrate and paraoccipital processes combine to form an otic meatus; mandible deepened posteriorly with a well-developed retroarticular process; interclavicles present but clavicles reduced or absent and; pubis excluded from the acetabulum. Another feature observed in crocodilian skulls and mandibles as well as the scutes is a sculptured texture of pits and grooves on most external surfaces.

A BRIEF HISTORY OF THE GROUP AND AUSTRALIAN FORMS

Benton and Clark (1988) place crocodiles (their Crocodylomorphia) within the Pseudosuchia and it appears that the crocodiles have a thecodont archosaurian ancestor somewhere in the mid Triassic. It's possible that, like the dinosaurs, the ancestral crocodilian was at least partly bipedal (Walker, 1970; Neill, 1971).

The earliest recognizable crocodiles are members of the Protosuchidae. This family includes dog-like animals such as *Protosuchus richardsoni* from the late Triassic of Arizona (Colbert and Mook, 1951), but it is not until the early Jurassic that crocodilians recognizably similar to extant species evolved. The teleosaurids from Europe had the characteristically crocodilian long snout, long body with relatively short limbs and long tail for swimming. Their relatives, the metriorhynchids, were fully marine adapted with limbs modified into paddles and a tail bent ventrally to support a vertical fluke, also seen in ichthyosaurs.

Since the Jurassic there has always been some form of crocodile that represents the generalist crocodilian form: a semi-aquatic hunter with a broad, flat snout. Another crocodilian form that has repeatedly and convergently evolved is the longirostrine form: long, thin snouted crocodilians that specialize in piscivory. Both forms can be found in crocodilians ranging from a few metres (m) in length up to 10 metres or more. The largest crocodilian ever is *Deinosuchus*, a 15 metre long terror from the late Cretaceous of Montana and Texas. This generalist crocodilian could have taken on its contemporary, *Tyrannosaurus rex,* and won.

There has been an array of unusual crocodilians throughout their history. Ziphodont crocodilians have teeth that are medio-laterally compressed, usually with serrate carinae on both the anterior and posterior edges set into a deep snout. This pattern has been independently derived half a dozen times within the crocodilians. The teeth of some forms, such as *Sebecus* spp., are so similar to those of theropod dinosaurs that isolated finds of teeth from South America were taken as evidence that some dinosaurs had survived the Cretaceous extinction (Colbert, 1946). Ziphodonty is recognized as an adaptation to terrestrial carnivory and, apart from the distinctive teeth, this form is characterized by having deep snouts with laterally directed orbits (Langston, 1973, 1975).

Another form recurrent among the crocodilians is the small dog-like animal with box-like head. Some are virtually toothless but in others the teeth are well developed and may have multiple cusps (Clark *et al.*, 1989; Wu *et al.*, 1995; Golmani, 1997). Again, this seems to be an adaptation to terrestrial life. However, all box-heads are small creatures (< 1.5 m in length) suggesting some niche separation from the larger ziphodonts.

Extant crocodilians represent only a shadow of the phylogenetic and morphological diversity of their ancestors and extinct kin. They are divided into three groups: the Crocodyloidea (true crocodiles), the Alligatoriodea (alligators and caymans) and the Gavialoidea (gavials). All are semi-aquatic and are either generalists or longirostrine piscivores.

Today there are only two species of crocodiles in Australia, both in the same genus (*Crocodylus porosus* and *C. johnstoni*). However, the fossil record of crocodilians in Australia extends back to the early Cretaceous and includes a number of endemic forms representing a variety of ecomorphic forms discussed below. It has only been through recent analyses that the monophyly of most extinct Australian crocodilians has been recognized and the subfamily Mekosuchinae has been erected to include them.

RESEARCH AND EARLY FINDS IN AUSTRALIA

Crocodilian fossils have been recovered from Australia for more than 130 years, but it has only been since the late 1970s that this group has been studied in detail. This developing interest in fossil crocodilians is the result of several factors. First, more crocodile fossils recovered during new palaeontological digs at sites such as Riversleigh, Alcoota, Bullock Creek and Murgon. Second, the abundance of new material permitted comparative reassessment of previously known material. Third, as there was enough work for a Ph.D. or two, the momentum for study of Australian fossil crocodilians was underway.

The first record of a fossil crocodilian from Australia is given by Clarke (1869) who refers to bones of crocodiles in Crinum Creek near Peak Downs in Queensland. These were sent to Thomas Huxley in England but he seems to have made no comment on the material. Daintree (1872) mentions crocodile teeth from Maryvale Creek in north-eastern Queensland that he named *Crocodilus australis*, although Jack and Etheridge (1892) refer all crocodile material from this site to *Crocodilus porosus*. It is relevant to note at this point that the archaic spelling of *Crocodilus* has been retained where employed by the original authors. It is also emphasized that the specific name '*australis*' has also been applied to South American forms now recognized as species of *Proalligator* (Steel, 1973).

Typical of most British colonies of last century, important fossil specimens from Australia were sent to the major collections in England, usually the collections of the British Museum of Natural History. Thus, when Lydekker (1888) published the *Catalogue of the Fossil Reptilia and Amphibia in the British Museum (Natural History)*, it contained the first published records of many Australian specimens and localities. Lydekker records *Crocodilus porosus* from Clifton Plains, Gowrie and the Condamine River in Queensland. These specimens had been donated by several collectors, including Daintree (1872); some of the material has now been reassessed and may be assigned to *Pallimnarchus* (Willis and Molnar, 1997a).

Jack and Etheridge (1892) record a number of occurrences of crocodilian fossils in Queensland. They note the occurrences mentioned by Clark (1869), Daintree (1872) and Lydekker (1888) and list both *Crocodilus porosus* and *Pallimnarchus pollens* from the Darling Downs. In a later section of the same publication they list *C. porosus* from Maryvale Creek, and from Gowrie and Clifton Plains on the Darling Downs. *P. pollens* from Chinchilla, Weimbilla Creek, Gowrie Creek, Condamine River and Eight-mile Plains (near Brisbane) are also included.

CRETACEOUS CROCODILIANS

There are now three published records of crocodiles from Cretaceous sediments of Australia (Figure 1). These occurrences are fragmentary and uninformative but probably contemporaneous and possibly conspecific. Three other, previously unpublished Cretaceous crocodilian specimens are also known from Australia.

Etheridge (1917) referred a small opalized fragment of reptilian mandible from Albian–Aptian deposits at Lightning Ridge, New South Wales to *Crocodylus (Bottosaurus) selaslophensis*. Molnar (1980) referred a procoelous vertebra and several other pieces from the same locality to this taxon and concluded that it represented an eusuchian.

A quadratojugal with characteristic crocodilian sculpture and two scutes have been recovered from early Cretaceous sites at Dinosaur Cove in the Otway Ranges of Victoria. This undescribed material is fragmentary and not very informative but indicates a small (< 2 m long) crocodile within the Dinosaur Cove fauna (Rich and Rich, 1989).

Two dentary fragments representing the one dentary and two posterior maxillary fragments from Lightning Ridge were described by Molnar and Willis (2000). These pieces appear to belong to the same taxon but lack adequate

Plate 66. The rough scale python (*Morelia carinata*) is reported to be rare. Photograph J. Cann.

Plate 67. Tiger snakes (genus *Notechis*) have a wide but disjunct range in south-western Western Australia, south-eastern Australia, Tasmania and offshore islands. This specimen is from a mainland population, but considerable variation in colour and size has been noted between island populations. These elapids are highly venomous and viviparous (produce live young). Photograph J. Cann.

Plate 68. A saltwater crocodile (*Crocodylus porosus*) in a small floodplain pool. Since commercial hunting ceased about 30 years ago, populations of this large predatory species have increased and the total range has expanded. Management problems have now increased, especially around regional centres such as Darwin and Cairns, so the relevant departments are considering re-introducing a limited annual cull. Photograph J. Cann.

Plate 69. The head of an adult saltwater crocodile (*Crocodylus porosus*) showing the broad jaw and tooth pattern. Photograph J. Cann.

Plate 70. A freshwater crocodile (*Crocodylus johnstoni*) basking. This species is much smaller than *C. porosus* and has a more limited total range. Note the narrow jaw and needle-like teeth that can grip slippery struggling prey, such as fishes, which are the main part of the diet. Photograph J. Cann.

Plate 71. This lace monitor (*Varanus varius*) is common at the edges of forests and woodlands. The species is considered to be arboreal and is often seen basking (or sunning) on the trunks of trees. Photograph J. Cann.

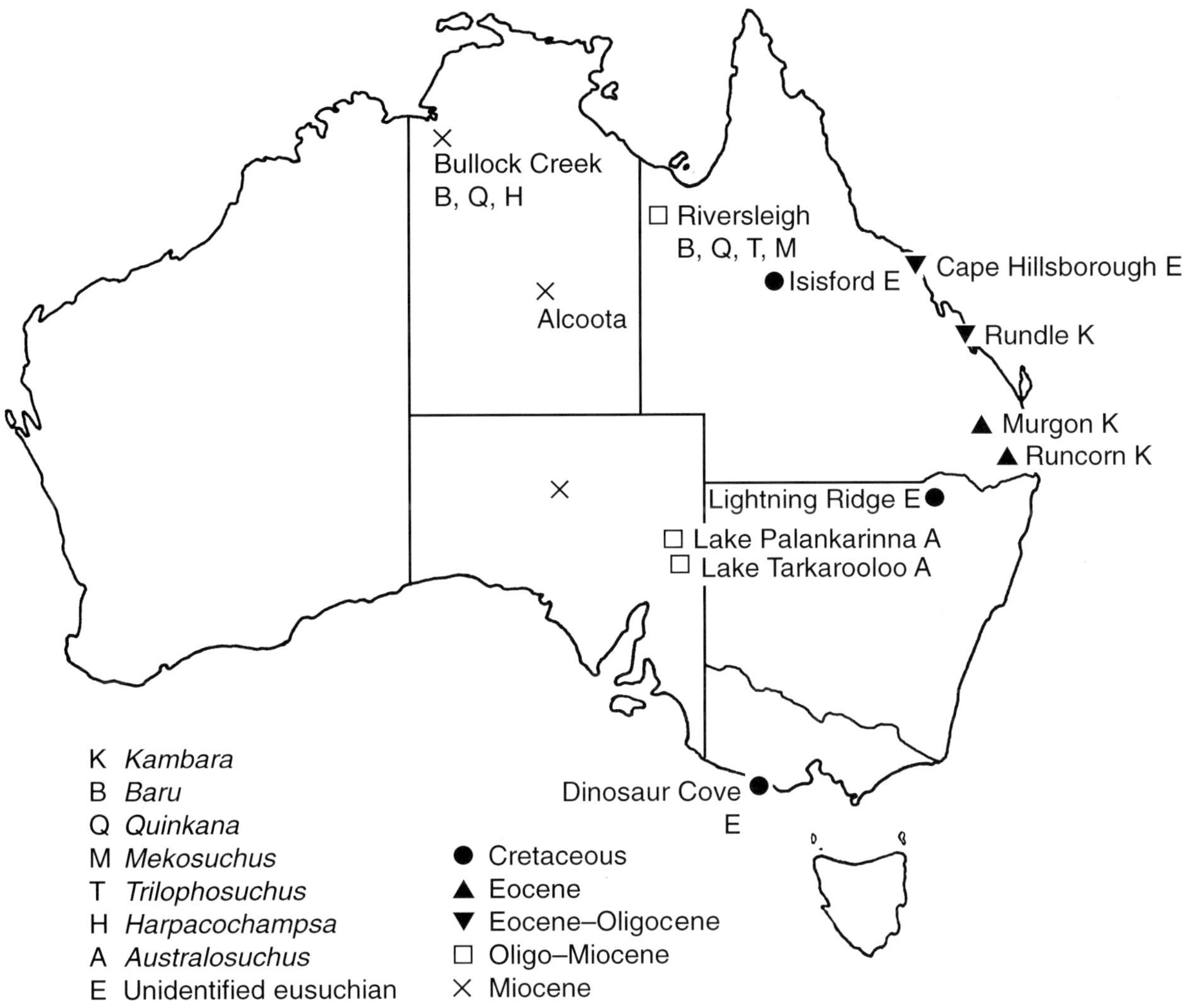

Figure 1. Location of Mesozoic, early and mid Tertiary crocodile-bearing deposits in Australia. Sources: Hills, 1943; Molnar, 1980, 1991; Rich and Rich, 1989; Willis *et al.*, 1990, 1993; Megirian *et al.*, 1991, 1993; Willis and Molnar 1991a, b; Murray and Megirian, 1992; McNamara, 1993; Murray *et al.*, 1993; Willis, 1993, 1995; Megirian, 1994.

features required for formal description but this material represents a different crocodilian taxon from the dentary referred to *Crocodylus* (*Bottosaurus*) *selasophensis*. The undescribed crocodilian is small, has a dorsoventrally compressed snout, homodont dentition and the anterior teeth appear to have been procumbent. A posterior skull fragment in the collections of the Australian Museum and a maxillary fragment from yet another crocodilian in the collections of the Queensland Museum are currently being studied by the author.

Two almost complete crocodilian skeletons have been recovered from Cenomanian deposits near Isisford in Central Queensland. These two specimens appear to represent the same taxon of eusuchian crocodile but both are encased in a firmly cemented matrix that is difficult to remove. Preparation is proceeding on these specimens while Steve Salisbury, Dino Frey and Ralph Molnar are conducting the formal study and description of this material.

TERTIARY AND QUATERNARY CROCODILES

A number of crocodilian specimens have been recovered from Paleocene to Oligocene deposits and all come from Queensland (Figure 1). Most of this material is fragmentary and uninformative but material from the Tingamarra deposits (near Murgon) and the Rundle Oil shales near Gladstone is more complete. Most of this material can be referred to the genus *Kambara*.

Until the 1980s, crocodilian material from Miocene deposits was rare and poorly understood (e.g. Woodburne, 1967), but subsequently there has been a rapid increase in the number of specimens from Riversleigh in north-western Queensland, Bullock Creek and Alcoota in the Northern Territory and Miocene deposits in the Lake Eyre Basin of South Australia (Figure 1). The Riversleigh and Bullock Creek localities have produced particularly diverse and interesting crocodilian faunas with some sites containing up to four crocodilian species (Murray and Megirian, 1992; Willis, 1992, 1993, 1995, 1997, 2001).

The first records of crocodilians from Australia were from Plio-Pleistocene deposits of Queensland (Clarke, 1869; Daintree, 1872; Lydekker, 1888). Subsequent excavations have revealed Plio-Pleistocene crocodilians from northern New South Wales, northern Western Australia and central South Australia (Figure 2). Quite large crocodilians (5 m or more in length) inhabited the inland drainage basins of Australia as well as the coastal basins until comparatively recently (late Pleistocene in most cases).

As mentioned previously phylogenetic analyses indicate that most extinct crocodilians from Australia belong to an endemic subfamily, the Mekosuchinae (Willis, 1986, 1992, 1993, 1995, 1997, 2001; Willis *et al.*, 1990, 1993; Megirian *et al.*, 1991; Willis and Molnar, 1991a, b, 1997a; Megirian, 1994; Salisbury and Willis, 1996; Willis and Mackness, 1996). Members of this group first appear in the Eocene and the last representatives may only have become extinct in the Holocene. Characters that identify mekosuchines are given in Table 1. *Crocodylus* species either arrived in Australia during, or subsequent to, the Pliocene or evolved here during the Pleistocene (Willis, 1986, 1995; Willis and Archer, 1990; Willis and Molnar, 1997b).

KAMBARA

The early Tertiary genus *Kambara* was first described by Willis *et al.* (1993), who recognized a single species, *K. murgonensis* (Figure 3), from the Eocene Tingamarra deposits in south-eastern Queensland. Subsequently, a second species, *Kambara implexidens*, has been recognized from this site (Salisbury and Willis, 1996). Material from the Eocene-Oligocene oil shale deposits of Rundle, eastern Queensland, briefly described by Hills (1943) and Molnar (1991), may represent a third, as yet undescribed species of *Kambara*. The anterior den-

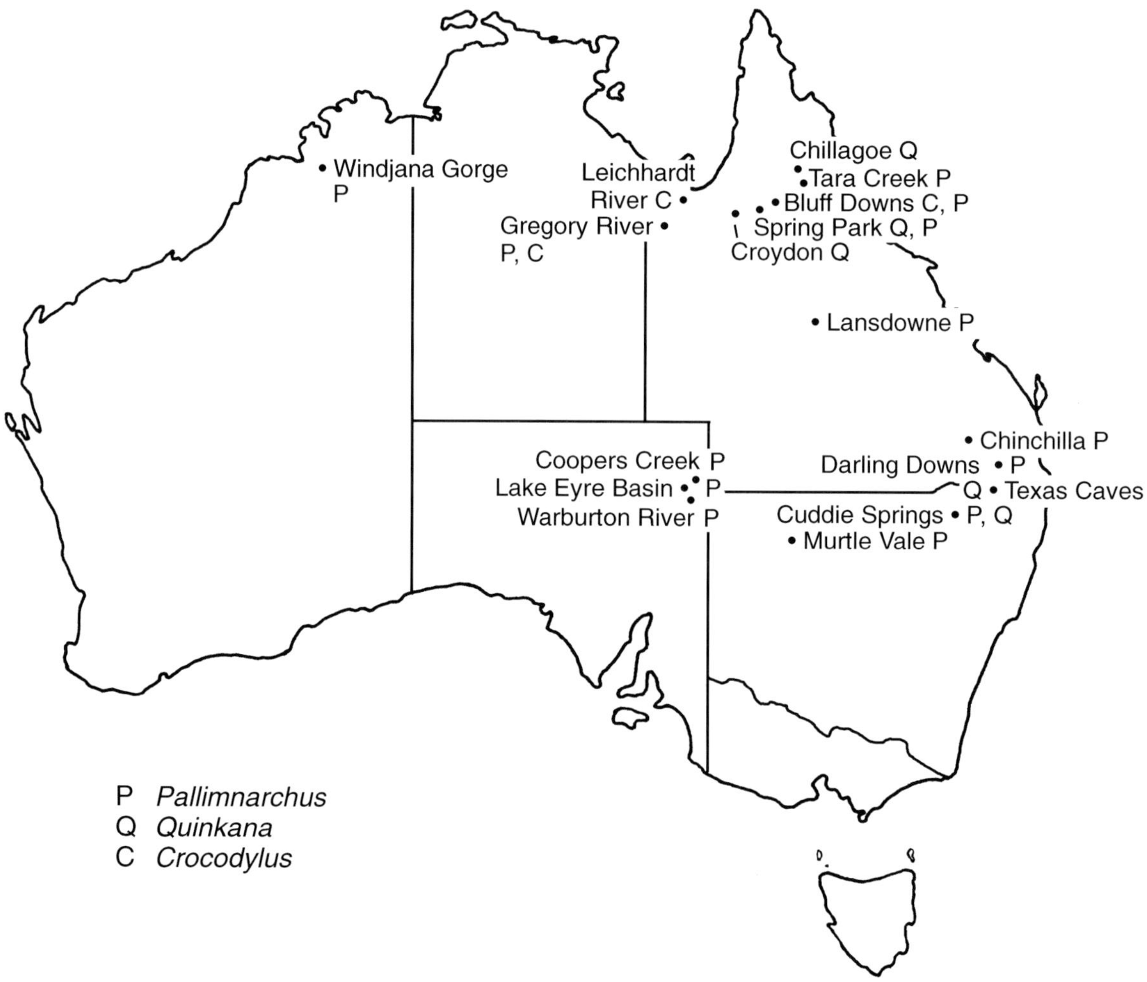

Figure 2. Location of Pliocene and Pleistocene crocodile-bearing deposits in Australia. Sources: de Vis, 1886, 1907; Longman, 1924, 1925a, b; Anderson and Fletcher, 1934; Anderson, 1937; Woodburne, 1967; Hecht and Archer, 1977; Molnar, 1977, 1978a, b, 1979, 1981, 1982b; Gorter and Nicoll, 1978; Thompson, 1980; McNamara, 1990; Tedford and Wells, 1990; Willis and Archer, 1990; Dodson *et al.*, 1993; Megirian *et al.*, 1993; Willis and Mackness, 1996.

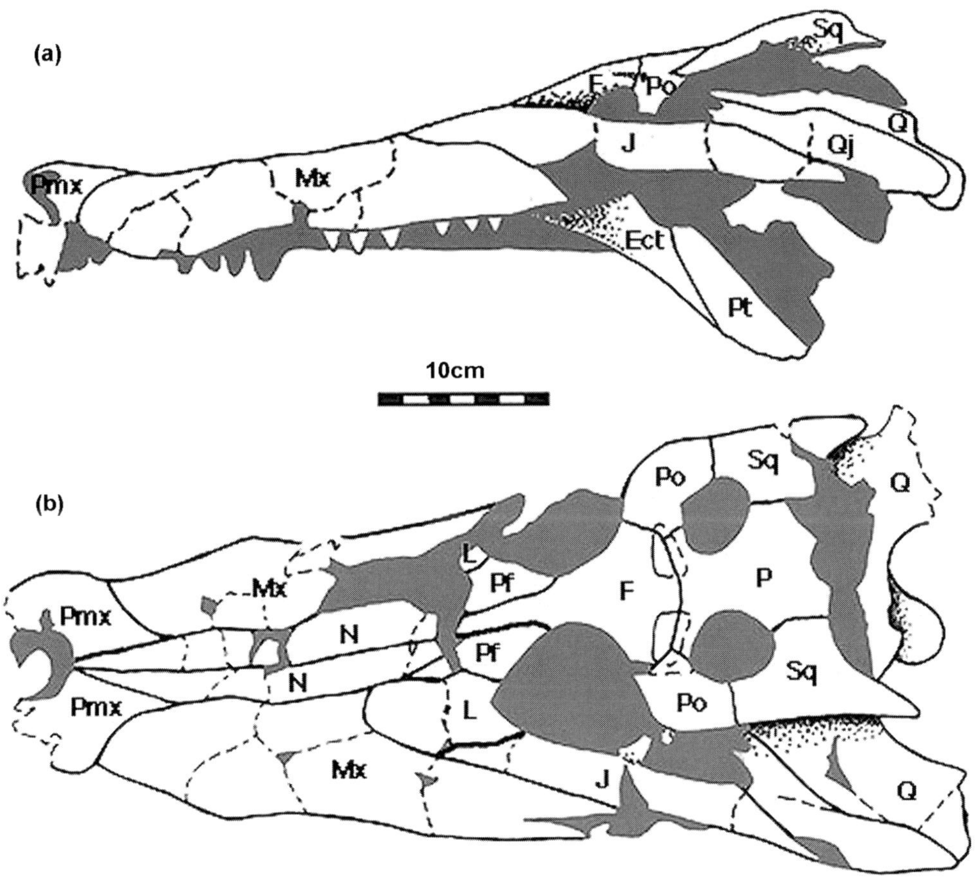

Figure 3. *Kambara murgonensis* skull, line interpretation of the type specimen (QM F21115): **(a)** lateral view; **(b)** dorsal view. Key to abbreviations: Ect = ectopterygoid; F = frontal; J = jugal; L = lachrymal; Mx = maxilla; N = nasal; P = parietal; Pf = prefrontal; Pmx = premaxilla; Po = postorbital; Pt = pterygoid; Q = quadrate; Qj = quadratojugal; Sq = squamosal.

tary from Murgon described by Molnar (1982b) has been referred to *Kambara* (Willis *et al.*, 1993).

Kambara is a moderately broad-snouted, generalist crocodile notable for the disparity in size between the smallest and largest teeth. This tooth disparity is also seen in other mekosuchines but *Kambara* is the earliest member of the group to display it. Distinctive sculptural patterns, such as large sculptured pits between the orbits and the supratemporal fenestrae, and a deep excavation anterior to the internal nares, also characterize this genus. The two species differ in a number of features, the most significant being the nature of the tooth occlusion. *Kambara murgonensis* had an overbite dentition while *K. implexidens* had interlocking teeth.

Another possible species of *Kambara* has been recorded by Willis and Molnar (1991a) from Eocene sediments near Brisbane. This material consists of an anterior dentary fragment and an isolated vertebral centrum. This is a more longirostrine form than the other *Kambara* species but it displays a similar tooth disparity.

An intriguing recent discovery of Eocene-Oligocene crocodile scutes has come from Cape Hillsborough, northern Queensland (McNamara, 1993). These small, squarish scutes probably belong to an advanced neosuchian (*sensu* Benton and Clark, 1988) but otherwise this material is uninformative.

Riek (1952) records a piece of skin and some bone fragments possibly from a crocodile found in the early Tertiary Redbank Plains Series near Brisbane. The bone fragments cannot be found and it is not clear from Riek's paper that these fragments were actually collected. Known skin impressions from this site represent lepidosaurs and lungfish but not crocodilians (Molnar, 2000, personal communication).

BARU

Relatively complete cranial and mandibular material from Riversleigh and Bullock Creek formed the basis of the new genus and species *Baru darrowi* (Willis *et al.*, 1990). This is a moderately large (around 4 m long) crocodile with a particularly massive head, enormous caniniform teeth that are moderately compressed and a

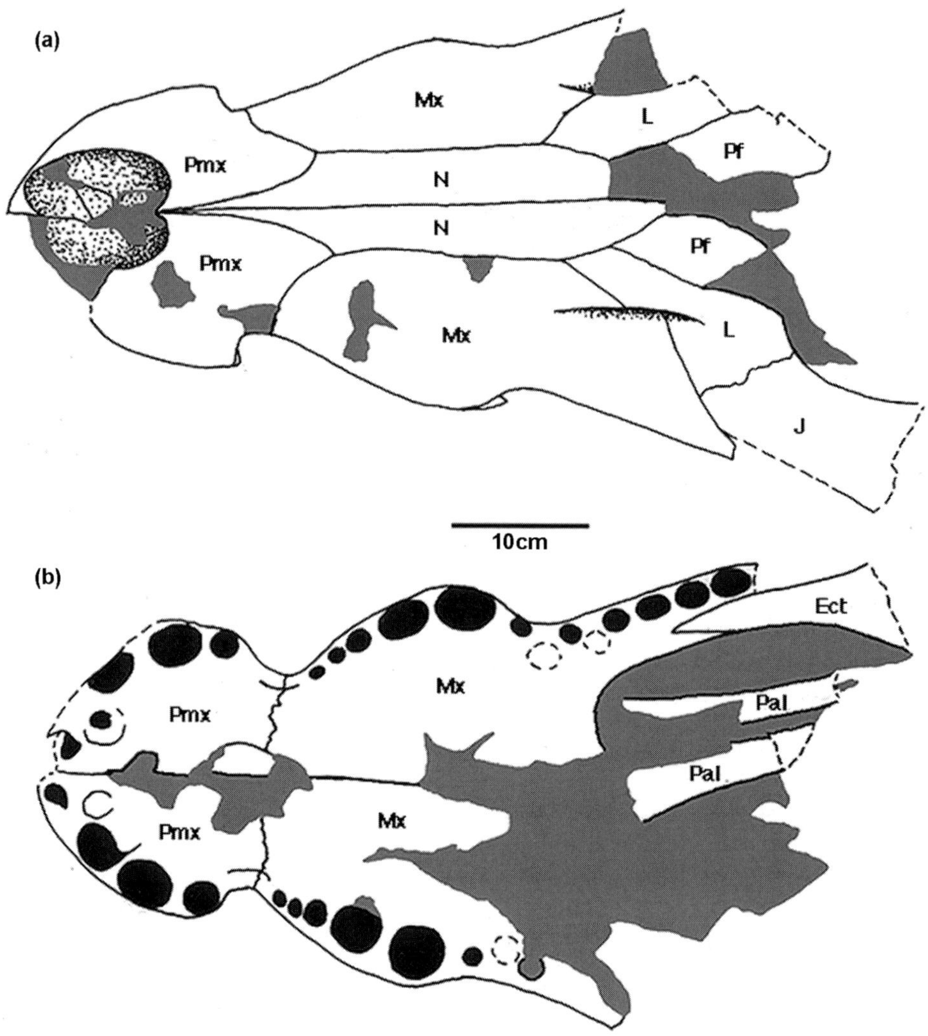

Figure 4. *Baru wickeni* skull, line interpretation of the type specimen (QM F16822): **(a)** dorsal view, **(b)** ventral view. Abbreviations as for Figure 3, except: Pal = palatine.

broad posterior region of the skull allowing large areas for massive jaw musculature. Subsequent work has shown that the Riversleigh material represents a second species, *B. wickeni* (Figure 4), separate from *B. darrowi* (Figure 5) of Bullock Creek (Willis, 1992, 1997). A smaller, third species, *B. huberi* (Figure 6), also exists at White Hunter Site, Riversleigh (Willis, 1997). Species of *Baru* are now recognized in many older sites at Riversleigh and the larger species are restricted to some of the very earliest sites (Willis, 1997). Similarly, another smaller species of *Baru* may be present at Bullock Creek (Willis, 1995).

Crocodilian material from mid to late Miocene sediments at Alcoota was originally described by Woodburne (1967) as *Crocodylus* sp.; however, Murray and Megirian (1992) examined new material from the site and assigned it to a species of *Baru*, possibly *B. darrowi*. It is likely that the material Woodburne (1967) identified as *Crocodylus* actually belongs to *Baru*.

QUINKANA

Quinkana is a genus of ziphodont crocodilians that is demonstrably mekosuchine and endemic to Australia. Thus their acquisition of ziphodont characters is independent from and convergent on other ziphodont crocodilians from around the world. *Quinkana* is also thought to have been fully terrestrial based on comparisons with pristichampsines and sebecosuchians and some aspects of the taphonomy of the Pleistocene species *Q. fortirostrum* (Molnar, 1977, 1978a, b, 1981). The early presence in Australia of ziphodont crocodilians and other large reptilian carnivores, combined with the absence of large mammalian carnivores, indicates that the occupation of the large carnivore guild noted for Australian Pleistocene faunas (Murray, 1984) probably has a much more extensive history.

The first reference to ziphodont crocodilian material from Meganesia is by Plane (1967) who recorded ziphodont teeth in the Awe Fauna from New Guinea.

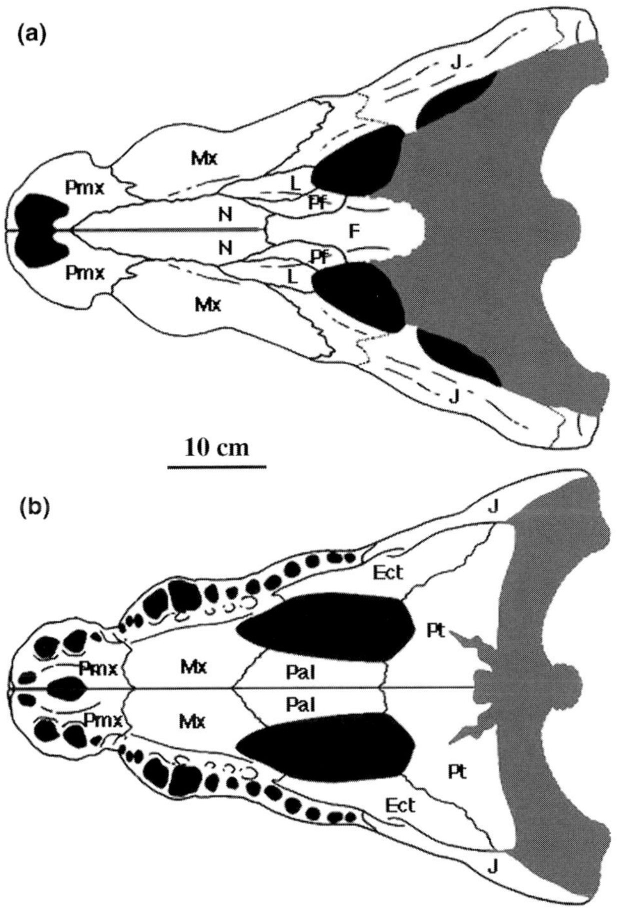

Figure 5. *Baru darrowi*, line interpretation for a skull reconstruction based on various specimens: **(a)** dorsal view; **(b)** ventral view. Abbreviations as for Figures 3 and 4.

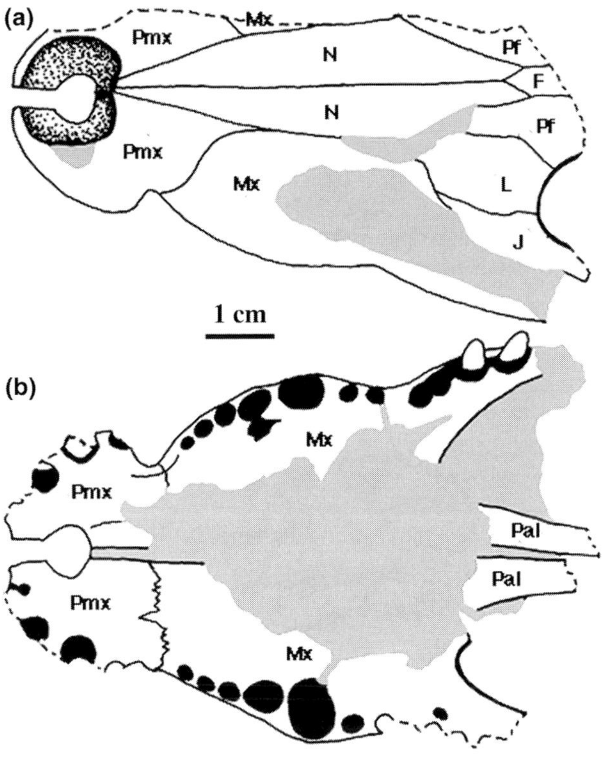

Figure 6. *Baru huberi* skull, line interpretation based on the type specimen (QM F31060): **(a)** dorsal view; **(b)** ventral view. Abbreviations as for Figures 3 and 4.

Hecht and Archer (1977) described two maxillary fragments with ziphodont teeth, one from the Lake Eyre Basin, South Australia, the other from Texas Caves, Queensland, and referred them to an unknown sebecosuchian. Meanwhile Molnar (1977, 1978a, b), working with more complete ziphodont material from Chillagoe Caves, north Queensland, recognized that this ziphodont form was eusuchian with similarities to the pristichampsines. Molnar (1981) described and defined this form as the new genus and species *Quinkana fortirostrum*, and referred to this genus the material described by Hecht and Archer (1977) from Texas Caves. Unfortunately the specimen from the Lake Eyre Basin has been lost and its referral to, or exclusion from, *Quinkana* cannot be corroborated. However, until evidence to the contrary can be produced, previous claims for sebecosuchians in Australia (Hecht and Archer, 1977) should be discounted.

Several other species of *Quinkana* have since been described. The Pliocene species *Quinkana babarra* from the Spring Park locality, north Queensland (Willis and Mackness, 1996), differs from *Q. fortirostrum* in having a shorter snout with mild festooning (completely suppressed in other species of *Quinkana*) and is known from a partial maxilla and isolated teeth. *Quinkana meboldi* from White Hunter Site, Riversleigh, is quite small in comparison to previously known species (Willis, 1997) as are elements referable to *Quinkana* from a number of other Riversleigh sites (Willis, 1995). Megirian (1994) described the *Quinkana timara* from Bullock Creek, Northern Territory as having a significantly narrower snout than that of the Pleistocene *Q. fortirostrum*; however, a specific determination of the *Quinkana* in the Ongeva Local Fauna of Alcoota (Megirian *et al.*, 1993; Murray *et al.*, 1993) cannot be made based on material currently available.

Isolated ziphodont crocodilian teeth from Pleistocene localities in Meganesia should tentatively be referred to *Quinkana* sp. cf. *Q. fortirostrum* (Willis and Molnar, 1997b). This includes the ziphodont teeth referred to by Plane (1967) from the Awe Fauna of Papua New Guinea, as well as teeth from Croydon, north Queensland and various sites on the Queensland Darling Downs reported by Molnar (1981) and Cuddie Springs in New South Wales as reported by Dodson *et al.* (1993).

MEKOSUCHUS

This distinctive genus of small (less than 1m long) crocodilians is unusual in that the maxilla participates in the orbit thus excluding the lacrimal from contact with

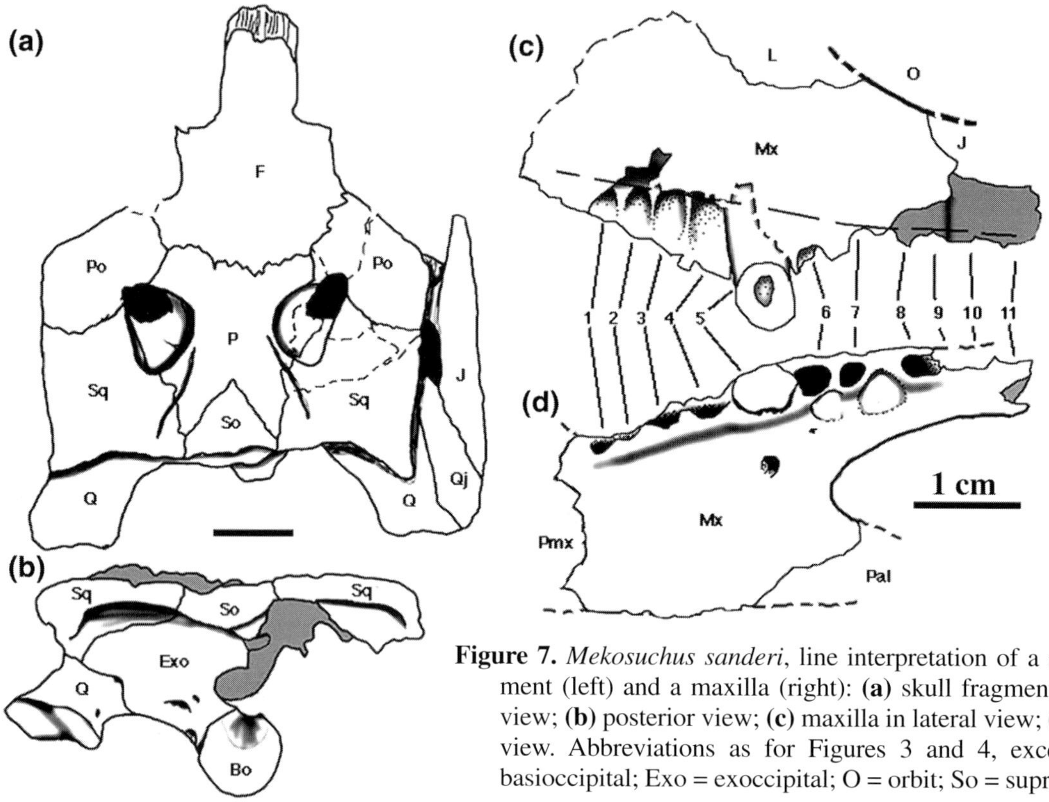

Figure 7. *Mekosuchus sanderi*, line interpretation of a skull fragment (left) and a maxilla (right): **(a)** skull fragment in dorsal view; **(b)** posterior view; **(c)** maxilla in lateral view; **(d)** ventral view. Abbreviations as for Figures 3 and 4, except: Bo = basioccipital; Exo = exoccipital; O = orbit; So = supraoccipital.

the jugal. *Mekosuchus* species have very short, deep heads and may have been terrestrial or even arboreal. This conclusion is based on postcranial material from the Pleistocene species and comparison with Mesozoic crocodilians such as atoposaurids, libycosuchids and notosuchids which have a similar cranial morphology.

Balouet and Buffetaut (1987) described a small, deep-snouted crocodile from cave deposits in New Caledonia as the new genus and species *Mekosuchus inexpectatus*. These deposits were initially dated as Recent (Balouet, 1991), but it is now thought that the deposits are more likely to be Pleistocene (Flannery, 2000, personal communication). Being an unusual, short-faced, deep-headed form, initial phylogenetic analyses erroneously suggested that *Mekosuchus* formed a sister-group relationship to all other eusuchian crocodilians (Balouet and Buffetaut, 1987). More recent analyses have interpreted some of the supposed plesiomorphies of *Mekosuchus* as characters which reversed as part of a functional complex, similar to the morphology of *Osteolaemus* and *Paleosuchus* (Willis, 1993, 1995).

The appearance of two species of *Mekosuchus* in Miocene deposits of Riversleigh (Willis, 1992, 1995, 1997, 2001) was surprising considering that the only previous record for this genus was in New Caledonia. *Mekosuchus whitehunterensis* from White Hunter Site and *M. sanderi* (Figure 7) from Ringtail Site, both at Riversleigh, are smaller than *M. inexpectatus* and differ from each other in cranial sculpture among other features.

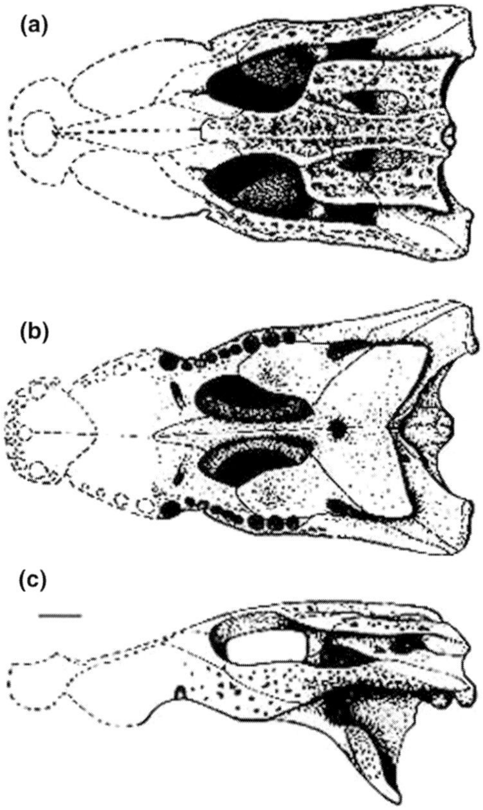

Figure 8. *Trilophosuchus rackhami*, line interpretations of a skull fragment based on the type specimen (QM F16856): **(a)** dorsal view; **(b)** ventral view; **(c)** lateral view.

TRILOPHOSUCHUS

The unusual species *Trilophosuchus rackhami* (Figure 8) was described by Willis (1993) from Ringtail Site at Riversleigh. Similar to *Mekosuchus* species in being small and having a short deep head, *Trilophosuchus* has the striking feature of three longitudinal crests along the skull roof. Low bumps indicate a poorly developed additional pair of crests lateral to the other three. The supratemporal fenestrae are elongate and laterally compressed, which is either a juvenile or paedomorphic feature.

Busbey (1977) analysed the neck musculature of the North American and European ziphodont *Pristichampsus* and showed that it was significantly different from that of other crocodilians; these differences may reflect adaptations to a terrestial habitat. Willis (1993) found several similarities between the morphology of the region of the skull of *Trilophosuchus* where neck muscles attached and the same region of *Pristichampsus*. These similarities were offered as tentative evidence for a terrestrial habit for *Trilophosuchus*.

HARPACOCHAMPSA

The poorly known *Harpacochampsa camfieldensis* (Figure 9) from Bullock Creek (Northern Territory) is based on a posterior cranial fragment, an anterior snout fragment and a small piece of the left dentary, all probably from the same individual, as well as some postcranial material. Megirian *et al.* (1991) recognized that certain longirostrine characters, such as the extremely large and cavernous supratemporal fenestrae and the globose anterior of the snout, were reminiscent of gavialids. But their phylogenetic analysis indicated mekosuchine or crocodyline affinities. More recent phylogenetic analyses (Willis, 1995; Salisbury and Willis, 1996) indicate that *Harpacochampsa* is likely to be a mekosuchine or crocodyloid where many key characters have been masked by the longirostrine condition. *Harpacochampsa* is a moderately large crocodilian (probably reaching lengths of 4–5 m) and its longirostrine condition indicates a piscivorous diet.

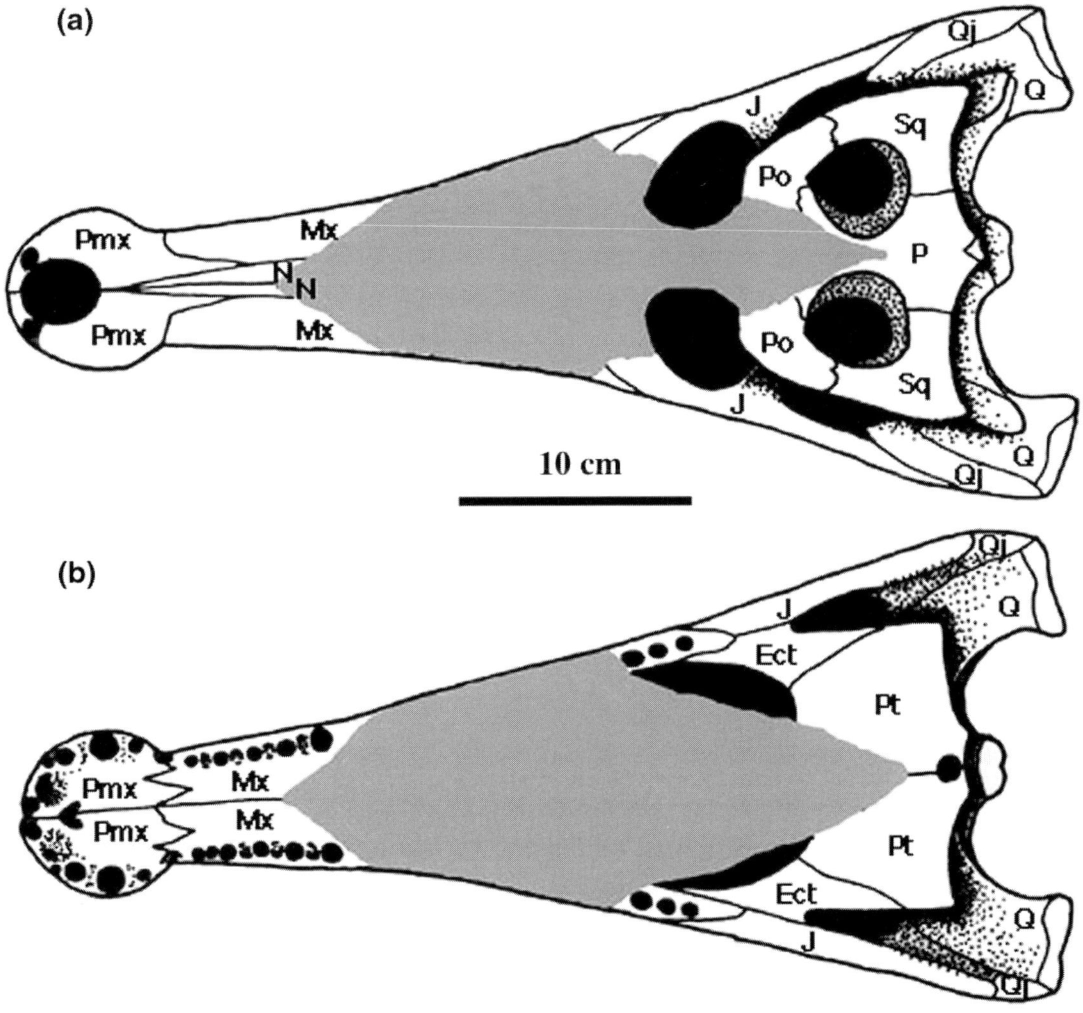

Figure 9. *Harpacochampsa camfieldensis*, line interpretation of a skull reconstruction based on type specimens (NTM P87106-5, NTM P87106-1): **(a)** dorsal view; **(b)** ventral view. Abbreviations as for Figure 3.

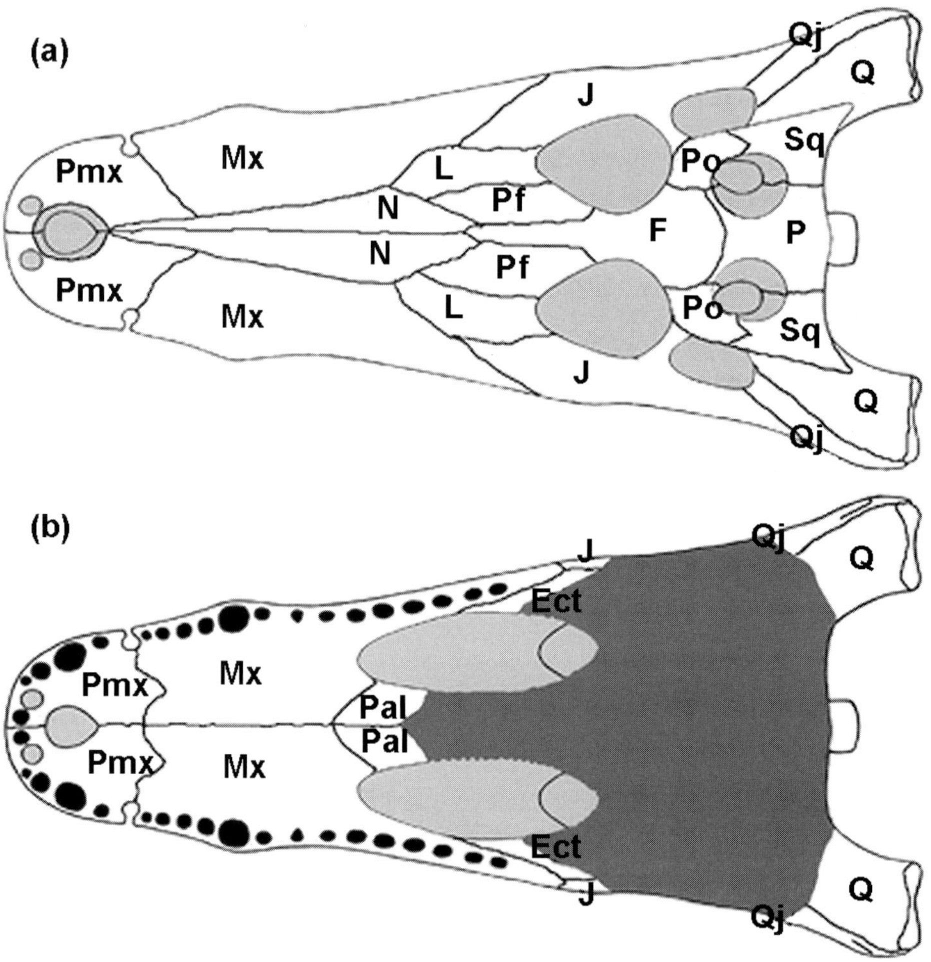

Figure 10. *Australosuchus clarkae*, line interpretation of the skull based on a reconstruction from various specimens: (**a**) dorsal view; (**b**) ventral view. Abbreviations as for Figures 3 and 4.

AUSTRALOSUCHUS

The medium sized (~3 m in length) generalized crocodilian *Australosuchus clarkae* (Figure 10) is known from abundant material from a number of sites in the Lake Eyre Basin, South Australia (Willis and Molnar, 1991b). Being a moderately broad-snouted mekosuchine, *Australosuchus* has the unusual condition of the fourth dentary tooth notch in the snout being almost totally enclosed. Complete skulls and mandibles are known of *A. clarkae* as well as abundant postcranial elements, dermal armour and an articulated partial skeleton. Reports of *A. clarkae* from younger sediments (Willis and Molnar, 1991b; Megirian *et al.*, 1993) are based on specimens that have probably reworked out of older deposits.

PALLIMNARCHUS

Pallimnarchus is a large crocodilian with a particularly broad and flat snout. The first detailed discussion of crocodilian fossils from Australia was by de Vis (1886), when he named *Pallimnarchus pollens* and noted that it was a heavily built, large crocodilian. Later de Vis (1907) described more fossil material from the Gulf of Carpentaria and assigned it to *P. pollens*. De Vis never formally defined *Pallimnarchus* or *P. pollens* but the names gained a *de facto* validity through use in popular and scientific literature.

Molnar (1982b) reviewed the Cainozoic crocodilians of Queensland with particular attention to *Pallimnarchus*. Although the concepts *Pallimnarchus* and *P. pollens* had received widespread common usage, it was not until Molnar's revision that these concepts were properly defined and lectotype material nominated. Molnar's generic and specific definition was based on features of the anterior dentary, but referred material included other mandibular elements, premaxillae and quadrates. Molnar also described an interorbital fragment and assigned it to *P. pollens*. Molnar (1982b) showed that *C. nathani* (Longman, 1924) was an invalid taxon and variously assigned the *C. nathani* material to either *C. porosus* or *P. pollens*.

A second species of *Pallimnarchus, P. gracilis*, (Figure 11) was described by Willis and Molnar (1997a).

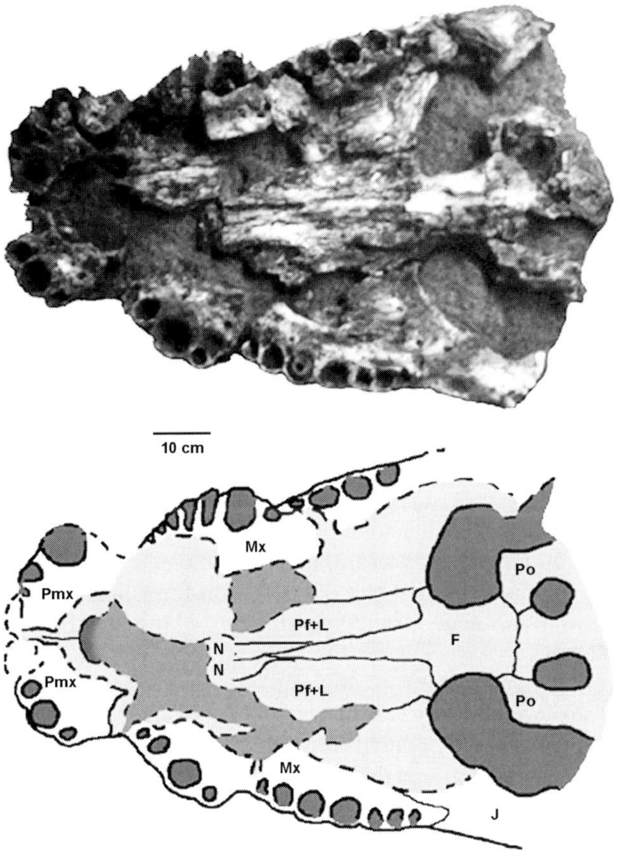

Figure 11. *Pallimnarchus gracilis*, photo and line interpretation of the Mirani Museum skull (MM89-1072), in ventral view with most of the palate missing. Abbreviations as for Figure 3.

Pallimnarchus gracilis is very similar to *P. pollens* in outline and size but differs in having a much lower snout. Several specimens previously designated as *P. pollens* were reassigned to *P. gracilis* by Willis and Molnar (1997a). Longman (1925a) had described a large crocodilian snout from Lansdowne Station, central Queensland, referring it to *P. pollens*. Molnar (1982b) re-examined this specimen and identified it as *Crocodylus porosus*. Subsequently Willis and Molnar (1997a) reassigned the Lansdowne snout to *Pallimnarchus*, but to the new species *P. gracilis*. Longman (1925b) also described a large, heavily built crocodilian mandible from Macalister, south-eastern Queensland, referring it to *P. pollens*.

Mackness and Sutton (2000) report on a crocodilian metatarsal, from the Pliocene Bluff Downs locality, that displays trauma they suggest might be due to intraspecific combat. The bone has a series of puncture marks consistent in spacing and form with having been inflicted by another crocodile. Based on size and likely habits of the three crocodilian taxa known from this site, it is suggested that this element is most likely to belong to *Pallimnarchus* (Mackness and Sutton, 2000).

Based on isolated teeth identified by Willis and Molnar (1997b), *Pallimnarchus* is probably present at Cuddie Springs in northern New South Wales (Anderson and Fletcher, 1934, Dodson *et al.*, 1993) as well as the Wyandotte Local Fauna of northern Queensland (McNamara, 1990). The report of *Crocodylus* from Windjana Gorge, Western Australia (Gorter and Nicoll, 1978), is based on a premaxilla and scutes. Willis and Molnar (1997a) suggest that it should be referred to *Pallimnarchus*.

Thompson (1980) reported two Pleistocene river bank deposits along the Darling River in western New South Wales, one of which (Myrtle Vale) contained a crocodilian tooth. Thompson describes the conical tooth as being 15mm across the base. This could represent a large *C. porosus* but would more likely represent *Pallimnarchus* sp. because of its size. The present location of the fossils Thompson collected is unknown and the Myrtle Vale site has been submerged under the waters of a weir. Thus there appears to be little chance of correctly determining the identity of this crocodilian or of collecting more material. However, similar deposits may occur elsewhere along the Darling River (Hope, 2000, personal communication). Myrtle Vale and Cuddie Springs are the most southern records of Pleistocene crocodilians in Australia.

CROCODYLUS

The genus *Crocodylus* has a confused history due to many fossil taxa being included within the genus that probably do not belong there (Brochu, 2000). This situation has not been helped by the genus being poorly diagnosed and many fossils being fragmentary. The genus has yet to be satisfactorily revised. Undoubted *Crocodylus* species first appeared in the late Miocene (Brochu, 2000). The genus has subsequently dispersed across the equatorial regions of the world probably through saltwater tolerant species crossing the major oceanic boundaries (Taplin, 1984, 1989). The extant saltwater crocodile, *Crocodylus porosus*, appears to be a recent immigrant to Australia probably arriving from South-east Asia in the Pliocene (Willis, 1986, 1995). It is poorly represented in the fossil record within Australia.

Molnar (1979) identified the snout of a small crocodilian from Bluff Downs, north Queensland, reported by Archer and Wade (1976), as *Crocodylus porosus*. This is the first time fossil crocodilian material from Australia was correctly identified as belonging to this species. The dating of Bluff Downs is based on Potassium-Argon (K/Ar) dating of an overlying basalt, originally thought to be 4–4.5 million years old. But this age has been revised to a significantly younger date within the Pliocene (Mackness, 2000, personal communication). This makes this specimen the second oldest fossil record of any extant species of crocodilian (the oldest being for a specimen of *Gavialis gangeticus*).

Being an endemic species, the apparent absence of *C. johnstoni* from fossil deposits of Australia seemed enigmatic until Willis and Archer (1990) described a dentary from Pleistocene deposits at Riversleigh, north-western Queensland. In a note added in proof to Willis and Archer (1990), a second dentary was identified from Pleistocene deposits of Floraville Downs in north-western Queensland. Both specimens appeared to be somewhat large for this species but were otherwise unremarkable.

Some identifications of material as being *Crocodylus* sp. should, on review, be assigned to various mekosuchine taxa. Swinton (1924) identified crocodilian remains collected from the Lake Eyre Basin, South Australia, by Gregory and referred them variously to *Crocodilus* or *C. porosus*. This material may be referred to species of *Pallimnarchus* (Willis, 2000, unpublished observations) which would be consistent with the findings of Megirian *et al.* (1993). These workers recorded a *Crocodylus*-like specimen from the Kutjitara Formation of the Lake Eyre Basin which has possibly been reworked from other deposits. A jugal possibly referrable to *Crocodylus* is also reported from the Katipiri Formation of the Lake Eyre Basin (Megirian *et al.*, 1993). However, Megirian (2000, personal communication) cautions that designations from these sites are based on fragmentary material with few diagnostic features.

Gorter and Nicoll (1978) referred a poorly preserved premaxillary fragment and some scutes from Windjana Gorge, Western Australia, to *Crocodylus* sp. but this material should now be referred to *Pallimnarchus* (see above). Woodburne (1967) described crocodilian fossils from Alcoota (Northern Territory) and referred them to *Crocodylus* sp. but this material should probably be referred to *Baru*. Plane and Gatehouse (1967) noted the presence of 'nondescript' crocodile teeth in the Awe Fauna from New Guinea. These can probably be referred to a species of *Crocodylus* but may represent a species of *Pallimnarchus*. Resolution of this matter requires a re-examination of the material.

GAVIALIS PAPUENSIS

De Vis (1905) described fragmentary mandibular material from Woodlark Island in the Solomon Sea as the new species *Gavialis papuensis*. Molnar (1982a) reviewed this material and assigned it to a *Euthecodon*-like crocodile. Subsequently Aoki (1988) and Molnar (1993) suggested that it could belong to a malformed individual of *Gavialis bengawanicus*. A popular account of this taxon is given by Willis (1987).

ISOLATED REMAINS

Isolated crocodilian teeth are often found in fossil deposits. Crocodilian teeth are regularly replaced throughout the life of the individual and so old teeth that have dropped out tend to be abundant. In many sites, only isolated crocodile teeth are known and identification of the genus and species is very difficult. In an attempt to address this identification problem for teeth found in Plio-Pleistocene sites of Australia, Willis and Molnar (1997b) have produced a diagnostic key identifying the teeth of the major known crocodilian taxa and the varanid *Megalania* (which are similar to those of some ziphodont crocodilians).

The only crocodilian fossil known from New Zealand is an angular bone from the Miocene of central Otago (Molnar and Pole, 1997); although the taxon has not been determined it is thought to be mekosuchine.

Further crocodilian material has recently been recovered from this site and will be described in the near future.

A new mekosuchine taxon, *Volia*, has been described from presumed Pleistocene deposits of Fiji by Molnar *et al.* (2002). This crocodilian is represented by several specimens from two cave deposits on the island of Viti Levu. It is a generalized crocodilian reaching lengths of 2 metres.

A new species of *Mekosuchus* has been described from Vanuatu by Mead *et al.* (2002). *Mekosuchus kalpokasi* is only known from a maxilla recovered from a cave on Efate and is dated at 3,000 years old. It is heavily built for a member of this genus but represents a small animal.

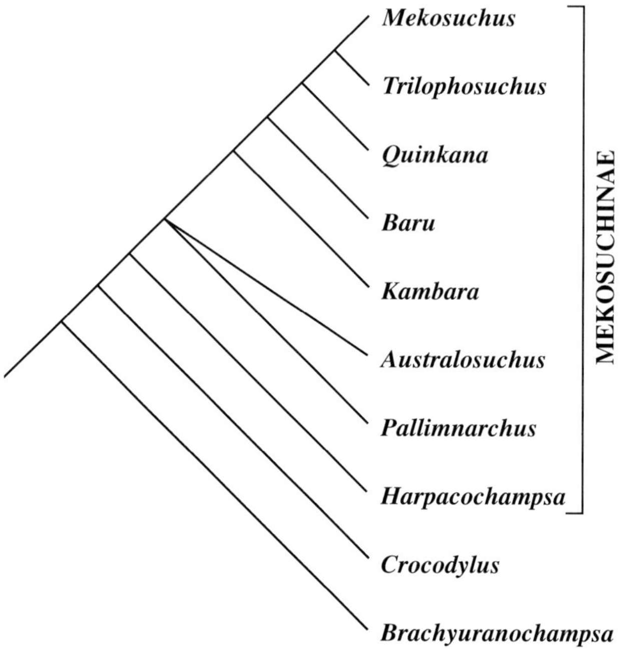

Figure 12. Cladogram showing the relationships of the mekosuchines based on an analysis of 15 characters. Data were analysed using Paup 3.0; ordered parsimony was used and a heuristic search using general options. *Brachyuranochampsa* was nominated as the outgroup. Characters unweighted. After Willis (1995).

Table 1. Character distribution for analysis of the Mekosuchinae.

Generic Group	1	2	3	4	5	6	7	8	9	10	11	12	13	14	15	Number of unknowns
Mekosuchus	1	2	1	1	1	1	2	1	1	?	1	0	1	1	0	1
Trilophosuchus	0	1	?	1	1	?	2	1	1	?	1	0	2	1	0	3
Quinkana	1	2	1	0	?	1	2	1	1	1	1	1	?	0	?	3
Baru	1	2	1	1	?	1	1	1	1	1	0	1	0	0	1	1
Kambara	1	1	1	1	1	1	0	0	0	0	0	0	0	0	0	1
Australosuchus	1	1	1	1	1	0	0	0	0	0	0	0	0	0	0	0
Pallimnarchus	1	1	1	1	1	0	0	0	0	0	0	0	0	0	1	0
Harpacochampsa	?	1	1	1	0	0	0	?	0	0	0	0	0	0	0	2
Crocodylus	0	0	0	0	0	0	0	0	0	0	0	0	0	0	0	0
Brachyuranochampsa	1	0	0	0	0	0	0	0	0	0	0	0	0	0	0	0
Unknowns	1	0	1	0	2	1	0	1	0	2	0	0	1	0	1	

(Columns 1–15: Character and State*)

*Characters and states:

1) Presence or absence of an anterior process of the palatines.
   Present (0)
   Absent (1)
2) Presence or absence of an alveolar process.
   Absent (0)
   Low alveolar process (1)
   Well developed alveolar process (2)
3) Size of the fifth premaxillary alveolus relative to the first maxillary alveolus.
   Fifth premaxillary tooth smaller than the first maxillary tooth (0)
   Fifth premaxillary tooth larger than the first maxillary tooth (1)
4) Presence or absence of a great disparity in tooth size.
   Small tooth disparity (0)
   Large tooth disparity (1)
5) Presence or absence of a supraoccipital wedge.
   Absent (0)
   Present (1)
6) Shape of the external nares.
   Oval or round nares (0)
   Trapezoidal nares (1)
7) Depth of the snout.
   Shallow snout (0)
   Moderately deep snout (1)
   Very deep snout (2)
8) Location of the palatal fenestrae.
   Posterior to 8th alveoli (0)
   Reach or anterior to the 8th alveoli (1)
9) Location of the orbits.
   Posteriorly located orbits (0)
   Anteriorly located orbits (1)
10) Form of the snout notch.
   Symmetrical notch (0)
   Inclined notch (1)
11) Direction of the orbits.
   Dorsally facing orbits (0)
   Laterally facing orbits (1)
12) Degree of alveolar compression.
   Rounded section (0)
   Compressed section (1)
13) Degree of lateral compression of the supratemporal fenestrae.
   Uncompressed supratemporal fenestrae (0)
   Moderately compressed (1)
   Strongly compressed (2)
14) Degree of participation of the pterygoids in the narial passage.
   Restricted pterygoid participation (0)
   Extensive participation (1)
15) Size of the supratemporal fenestrae.
   Moderate sized supratemporal fenestrae (0)
   Small fenestrae (1)

PHYLOGENETIC SYSTEMATICS

Early work on crocodilian taxa from Australia recognized various taxa as either crocodyline or alligatorine (de Vis, 1886), although crocodyline affinities were favoured (Longman, 1925a, b). *Quinkana* was initially thought to be sebecosuchian (Hecht and Archer, 1977) or eusuchian, possibly pristichampsine (Molnar, 1981).

It was not until more recent work that similarities were recorded for Australia's endemic crocodilian taxa (Willis, 1986; Willis *et al.*, 1990; Willis and Molnar, 1991b). Initially, this group was termed the Australian Tertiary radiation of crocodiles (Willis *et al.*, 1990). Cladistic analyses (Figure 12) supported the monophyly of these taxa (Willis, 1993, 1995; Salisbury and Willis, 1996). The subfamily Mekosuchinae was erected by Willis *et al.* (1993) to encompass the pre-existing concept of an Australian Tertiary radiation of crocodilians, redefining the taxonomic concept of Mekosuchidae originally erected by Balouet and Buffetaut (1987).

Mekosuchine characters include: a great disparity of alveolar size, between the smallest and largest alveoli; the development of a wedge of the supraoccipital on the dorsal surface of the skull; and reduction or loss of the anterior process of the palatines. The Mekosuchinae includes *Baru*, *Quinkana*, *Mekosuchus*, *Trilophosuchus*, *Pallimnarchus*, *Kambara* and *Australosuchus*. The very derived, longirostrine genus *Harpacochampsa* (Megirian *et al.*, 1991) is a problematic taxon previously of uncertain phylogenetic affinities. However, Willis (1995) presents an argument for its inclusion in the Mekosuchinae.

The Mekosuchinae forms a sister group to the Crocodylinae, which includes species of *Crocodylus* and *Osteolaemus*. Character distributions are shown in Table 1 and the affinities of the Mekosuchinae are shown in Figure 12.

EVOLUTIONARY HISTORY

The phylogenetics of the Mekosuchinae appear to fit neatly into the zoogeographic history of Australia. The earliest representatives of the group (*Kambara* spp.) were present in Australia prior to the rifting from Antarctica and while Antarctica was still attached to South America. Such a land-bridge may have offered a potential dispersal route for mekosuchines into or out of Australia, although only through very high latitudes. Similarly, alligatorines and other crocodilians present in South America at or prior to this time (Simpson, 1937a, b; Langston, 1965; Gasparini, 1971, 1981, 1984; Gasparini and Buffetaut, 1980) were also presented with dispersal opportunities into Australia. However, although some mammal taxa appear to have migrated via this route (Archer *et al.*, 1993, 1994), neither the mekosuchines nor South American crocodilian taxa appear to have dispersed through Antarctica.

Whatever the origin of the mekosuchines, they were present in Australia before its isolation from Antarctica and provided the rootstock of subsequent crocodilian faunas throughout Tertiary until the Pliocene. Possibly as a response to nutrient-poor Australian environments, favouring large ectothermic predators over endothermic predators (Flannery, 1994), mekosuchines evolved diverse faunas throughout the Tertiary including some forms that occupied terrestrial habitats. Reduction of mekosuchine diversity and their eventual extinction may be related to the decline of the associated megafauna as well as increased competition with varanids, madtsoiids, marsupial carnivores and newly arrived crocodylines.

Two species of *Crocodylus* are currently found in Australia and they are not mekosuchines. The sudden appearance of *Crocodylus porosus* in the early Pliocene record of Australia (Molnar, 1979) is compatible with a dispersal event, probably from South-east Asia, sometime in the early Pliocene or possibly the latest Miocene (Willis, 1986; Willis *et al.*, 1990). The salt-tolerance of this species allows it to disperse across oceanic barriers (Taplin, 1984). The subsequent appearance of *C. johnstoni* in Australia suggests speciation from a *C. porosus* ancestor (Willis and Archer, 1990), which is broadly consistent with phylogenetic analyses of the ingroup relationships of *Crocodylus* (Brooks, 1979; Brooks and O'Grady, 1989; Densmore and Owen, 1989; Brochu, 2000).

An interesting palaeobiogeographic problem is presented by the distribution of contemporaneous crocodilian taxa in Australia during the Oligo-Miocene. Although some of the sites in the Lake Eyre Basin in South Australia and Riversleigh in Queensland are demonstrably contemporaneous (Woodburne *et al.*, 1993) and there are large collections from both localities representing many crocodilian specimens, no common crocodilian taxa are shared between the two areas. While this disparity in distributions, for semi-aquatic taxa (*Baru*, *Australosuchus*), could be due to separate and isolated drainage basins the same explanation cannot extend to taxa that are terrestrial or, at least, competent in terrestrial environments (*Quinkana*, *Mekosuchus*, *Trilophosuchus*). Either the barriers to distribution for small to medium-sized terrestrial crocodilian taxa were formidable, or their abilities to cross terrestrial environments were not as developed as has been supposed. Perhaps the 'terrestrial' adaptations of *Quinkana*, *Mekosuchus* and *Trilophosuchus* represent the terrestrial limit of the operating potential of mostly aquatic species. For example, the ziphodont characters may well have allowed *Quinkana* to take prey and process food on land but the species were still tied to permanent water bodies for protection, reproduction or thermoregulation.

The biogeographic distribution of species of *Mekosuchus* (Miocene species in north Queensland, Pleistocene species in New Caledonia) also presents an intriguing problem. Postcranial material known for *M. inexpectatus* indicates a fully terrestrial habit. Indeed, the form of the humerus is so similar to that of some *Varanus* species, that it is likely to have had a similar habit, possibly being arboreal (Willis, 2001). Such a terrestrial species is unlikely to have swum between Australia and New Caledonia. It is also unlikely that the distribution of *Mekosuchus* species is the result of a vicariant event, the last land-bridge between the two land masses occurred 80 million years ago in the late Cretaceous (Balouet, 1991). Being small (no more than 1 m snout-to-vent length) and possibly arboreal, *Mekosuchus* may have rafted from Australia to New Caledonia.

ECOMORPHIC DIVERSITY

Mekosuchines display a broad range of ecomorphs from small taxa such as *Mekosuchus* and *Trilophosuchus* (< 1 m long) to comparative giants such as *Pallimnarchus* (> 5 m long). Even more striking are the differences in head shape from small deep 'box-heads' (*Mekosuchus* spp. and *Trilophosuchus rackhami*), to long but broad-snouted ziphodonts (*Quinkana* spp.), to massive-headed, 'hatchet-mouthed' taxa (*Baru* spp.), to moderately broad snouted, generalized forms (*Australosuchus clarkae* and *Kambara* spp.) to large, very broad snouts (*Pallimnarchus* spp.) and the longirostrine condition shown in *Harpacochampsa camfieldensis*.

Meyer (1984) demonstrated that more than one species of crocodilian can live sympatrically if they have different head shapes. The difference in head shape reflects differing diets and thus resources can be divided. This rule holds true for most Australian localities where more than one species of crocodile has been found. For example in Bullock Creek there is a large species of *Baru* and possibly a second, smaller species, a ziphodont species of *Quinkana* and a longirostrine species of *Harpacochampsa*. But Ringtail Site at Riversleigh is more difficult to explain; both *Mekosuchus sanderi* and *Trilophosuchus rackhami* occur here and are similar in size as well as cranial morphology. The slight differences in skull morphology between these two species may reflect significantly different diets, but local habitat utilization (slow versus fast flowing water) or disparate behavioural patterns (diurnal versus nocturnal) may have permitted sympatry. However, the occurrence of the remains together does not necessarily mean that the species were sympatric in life. Recognition of two species of *Kambara,* identical in size and with similar skull morphology, in deposits at Murgon raises the same problem.

CONCLUSIONS

Over the last 20 years newly excavated material has renewed interest and stimulated research on fossil crocodiles of Meganesia. These new studies have revealed the presence of diverse fossil crocodilian faunas in many parts of Australia since Cretaceous times (up to 115 m.y.a.). Including *Crocodylus* (represented by two extant species), specimens of some ten genera are now recognized in Australasian deposits.

It is now accepted that the Australian endemic taxa (subfamily Mekosuchinae) are monophyletic and radiated during the Tertiary. Cladistic analyses indicate that *Australosuchus*, *Baru*, *Kambara*, *Mekosuchus*, *Pallimnarchus*, *Quinkana* and *Trilophosuchus* should be included in the Mekosuchinae, but the affinities of the more derived genus, *Harpacochampsa,* are less certain; the sister group, Crocodylinae, includes species of *Crocodylus* and *Osteolaemus*.

The evolution and distribution of mekosuchines can be related to tectonic history, land-bridge dispersal, isolation, regional habitat utilization, competition and other ecological changes. In many fossil crocodilian faunas ecomorphic diversity (in size and head shape) can be related to varying ecological roles and differential resource usage, although a few sites have two apparently similar species that would, potentially, compete directly.

On-going research into the fossil crocodilians of Australia is being conducted by the author and Steve Salisbury (University of Queensland), Ralph Molnar (formerly of the Queensland Museum), Dirk Megirian (Northern Territory Museum of Arts and Sciences) and Neville Pledge (South Australian Museum).

REFERENCES

Anderson, C. 1937. Palaeontological notes No. IV. *Records of the Australian Museum*, Vol. 20, pp.73–77.

Anderson, C. and Fletcher, H.O. 1934. The Cuddie Springs Bone Bed. *Australian Museum Magazine*, Vol. 5, pp.152–158.

Aoki, R. 1988. Protruded toothed fossil crocodilian from the Solomon Islands—*Gavialis papuanus* (sic). *Kiiyoh to Seibutu (Aquabiology),* Vol. 10, p.60.

Archer, M., Hand S.J. and Godthelp H. 1993. Early Eocene marsupial from Australia. *Kaupia Darmstädter Beiträge zur Naturgeschichte*, Vol. 3, pp.193–200.

Archer, M., Hand S.J. and Godthelp H. 1994. Tertiary environmental and biotic change in Australia. In Vrba, E. (ed). *Environmental Change and Evolution*. Yale University Press, New Haven.

Archer, M. and Wade, M. 1976. The results of the Ray E. Lemley Expedition Part 1. The Allingham Formation and a new Pliocene vertebrate fauna from northern Queensland. *Memoirs of the Queensland Museum*, Vol. 17, pp.379–397.

Balouet, J.C. 1991. The fossil vertebrate record of New Caledonia. Pp.1383–1409. In Vickers-Rich, P., Monaghan, J.M., Baird, R.F. and Rich, T.H. (eds). *Vertebrate Palaeontology of Australasia*. Pioneer Design Studio and Monash University Publications Committee, Melbourne. 1437pp.

Balouet, J.C. and Buffetaut, E. 1987. *Mekosuchus inexpectatus n. g. n. sp.* Crocodilian nouveau de l'Holocene de Nouvelle Caledonie. *Comptes Rendus de l'Académie des Sciences, Paris*, Vol. 304, pp.853–857.

Benton, M.J. 1985. Classification and phylogeny of the diapsid reptiles. *Zoological Journal of the Linnean Society of London*, Vol. 79, pp.163–186.

Benton, M.J., and Clark, J.M. 1988. Archosaur phylogeny and the relationships of the crocodilia. Pp.295–338. In Benton, M.J. (ed). *The Phylogeny and Classification of the Tetrapods. Vol 1: Amphibians, Reptiles, Birds.* Clarendon Press, Oxford.

Brochu, C.A. 2000. Phylogenetic relationships and divergence timing of *Crocodylus* based on morphology and the fossil record. *Copeia*, 2000, pp.675–683.

Brooks, D.R. 1979. Testing hypotheses of evolutionary relationships among parasites: the digeneans of crocodilians. *American Zoologist*, Vol. 19, pp.1225–1238

Brooks, D.R. and O'Grady, R.T. 1989. Crocodilians and their helminth parasites: macroevolutionary considerations. *American Zoologist,* Vol. 29, pp.873–883

Busbey, A.B. 1977. *Functional morphology of the head of* Pristichampsus vorax *(Crocodilia, Eusuchia) from the Eocene of North America*. M.Sc. Thesis (unpublished), University of Texas.

Clark, J.M., Jacobs, L.L. and Downs, W.R. 1989. Mammal-like dentition in a Mesozoic crocodylian. *Science,* Vol. 244, pp.1064–1066.

Clarke, W.B. 1869. *Dinornis* an Australian genus. *Geological Magazine*, Vol. 6, pp.383–384.

Colbert, E.H. 1946. Sebecus, representative of a peculiar sub-order of fossil Crocodilia from Patagonia. *Bulletin of the American Museum of Natural History*, Vol. 87, pp.217–270.

Colbert, E.H. and Mook, C.C. 1951. The ancestral crocodilian *Protosuchus*. *Bulletin of the American Museum of Natural History*, Vol. 97, pp.146–182.

Daintree, R. 1872. Notes on the geology of the Colony of Queensland. *Quarterly Journal of the Geological Society of London*, Vol. 28, pp.271–317.

Densmore, L.D. and Owen, R.D. 1989. Molecular systematics of the Order Crocodilia. *American Zoologist*, Vol. 29, pp.831–841.

Dodson, J., Fullagar, R., Furby, J., Jones, R. and Prosser, I. 1993. Human and megafauna in a late Pleistocene environment from Cuddie Springs, northwestern New South Wales. *Archaeology in Oceania*, Vol. 28, pp.94–99.

Etheridge, R. 1917. Reptilian Notes; *Megalania prisca*, Owen, and *Notiosaurus dentatus*, Owen, lacertian dermal armour; opalised remains from Lightning Ridge. *Proceedings of the Royal Society of Victoria*, Vol. 29, pp.127–133.

Flannery, T.F. 1994. *The Future Eaters: an Ecological History of Australasian Lands and People*. Reed Books, Sydney. 423pp.

Gasparini, Z. 1971. Los Notosuchia del Cretacicode America del sur como un nuevo infraden de los Mesosuchia (Crocodilia). *Ameghiniana*, Vol. 8, pp.83–103.

Gasparini, Z. 1981. Los Crocodylia fosiles de la Argentina. *Ameghiniana,* Vol. 28, pp.177–205.

Gasparini, Z. 1984. New Tertiary Sebecosuchia (Crocodylia: Mesosuchia) from Argentina. *Journal of Vertebrate Paleontology*, Vol. 4, pp.85–95.

Gasparini, Z. and Buffetaut, E. 1980. *Dolichochampsa minima*, n. g. n. sp., a representative of a new family of eusuchian crocodiles from the late Cretaceous of northern Argentina. *Neues Jahrbuch für Geologie und Paläeontologie: Monatshefte*, 1980, pp.257–271.

Golmani, E.M. 1997. A crocodyliform from the early Cretaceous Dinosaur Beds, Northern Malawi. *Journal of Vertebrate Paleontology*, Vol. 17, pp.280–294.

Gorter, J.D. and Nicoll, R.S. 1978. Reptilian fossils from Windjana Gorge, Western Australia. *Journal of the Royal Society of Western Australia*, Vol. 60, pp.97–104.

Hecht, M.K. and Archer, M. 1977. Presence of xiphodont crocodilians in the Tertiary and Pleistocene of Australia. *Alcheringa,* Vol. 1, pp.383–385.

Hills, E.S. 1943. Tertiary fresh-water fishes and crocodilian remains from Gladstone and Duaringa, Queensland. *Memoirs of the Queensland Museum*, Vol. 12, pp.96–100.

Iordansky, N.N. 1973. The skull of the crocodile. Pp.201–202. In Gans, C. and Parsons, T.S. (eds). *Biology of the Reptilia. 4. Morphology D*. Academic Press, London.

Jack, R.L. and Etheridge, R. 1892. *The Geology and Palaeontology of Queensland and New Guinea*. Government Printer, Brisbane.

Langston, W. 1965. Fossil crocodilians from Colombia and the Cenozoic history of the crocodile in South America. *University of California Publications in Geological Science*, Vol. 52, pp.1–157.

Langston, W. 1973. The crocodilian skull in historical perspective. In Gans, C. and Parson, T.S. (eds). *Biology of the Reptilia. 4. Morphology D*. Academic Press, London.

Langston, W. 1975. Xiphodont Crocodilia. *Fieldiana (Geology),* Vol. 33, pp.291–314.

Longman, H.A. 1924. Some Queensland fossil vertebrates. *Memoirs of the Queensland Museum*, Vol. 8, pp.16–24.

Longman, H.A. 1925a. A crocodilian fossil from Lansdowne Station. *Memoirs of the Queensland Museum*, Vol. 8, pp.103–108.

Longman, H.A. 1925b. A large jaw of *Pallimnarchus pollens*. *Memoirs of the Queensland Museum*, Vol. 9, pp.158–159.

Lydekker, R. 1888. *Catalogue of the Fossil Reptilia and Amphibia in the British Museum (Natural History). Part 1.* Trustees of the British Museum (Natural History), London.

Mackness, B. and Sutton, R. 2000. Possible evidence for intraspecific aggression in a Pliocene crocodile from north Queensland. *Alcheringa,* Vol. 24, pp.55–62.

McNamara, G.C. 1990. The Wyandotte Local Fauna: a new, dated, Pleistocene vertebrate fauna from northern Queensland. *Memoirs of the Queensland Museum*, Vol. 28, pp.285–297.

McNamara, G.C. 1993. Cape Hillsborough: a dated Eocene-Oligocene vertebrate fossil site from Northeastern Queensland. *CAVEPS '93; Conference on Australasian Vertebrate Evolution, Palaeontology and Systematics. Adelaide, 19–21 April 1993.* (abstract only).

Mead, J.I., Steadman, D.W., Bedford, S.H., Bell, C.J. and Spriggs, M. 2002. New extinct mekosuchine crocodile from Vanuatu, South Pacific. *Copeia*, 2002, pp.632–641.

Megirian, D. 1994. A new species of *Quinkana* Molnar (Eusuchia: Crocodylidae) from the Miocene Camfield Beds of Northern Australia. *The Beagle (Records of the Northern Territory Museum of Arts and Sciences)*, Vol. 11, pp.145–166

Megirian, D., Murray, P.F. and Wells, R.T. 1993. The late Miocene Ongeva Local Fauna from the Waite Formation of central Australia. *CAVEPS '93; Conference on Australasian Vertebrate Evolution, Palaeontology and Systematics. Adelaide, 19–21 April 1993.* (abstract only).

Megirian, D., Murray, P. F. and Willis, P. 1991. A new crocodile of the gavial ecomorph morphology from the Miocene of northern Australia. *The Beagle (Records of the Northern Territory Museum of Arts and Sciences),* Vol. 8, pp.135–158.

Meyer, E.R. 1984. Crocodilians as living fossils. In Etheridge, N. and Stanley, S.M. (eds). *Living Fossils*. Springer Verlag, New York. 291pp.

Molnar, R.E. 1977. Crocodile with laterally compressed snout; first find in Australia. *Science,* Vol. 197, pp.62–64.

Molnar, R.E. 1978a. Age of the Chillagoe crocodile. *Search,* Vol. 9, pp.156–158.

Molnar, R.E. 1978b. The crocodile from Tea Tree Cave and ziphodont crocodiles in Australia. *Journal of the Sydney Speleological Society*, Vol. 22, pp.3–10.

Molnar, R.E. 1979. *Crocodylus porosus* from the Pliocene Allingham Formation of north Queensland. *Memoirs of the Queensland Museum*, Vol. 19, pp.357–365.

Molnar, R.E. 1980. Procoelous crocodile from lower Cretaceous of Lightning Ridge, N.S.W. *Memoirs of the Queensland Museum*, Vol. 20, pp.65–75.

Molnar, R.E. 1981. Pleistocene ziphodont crocodilians of Queensland. *Records of the Australian Museum*, Vol. 33, pp.803–834.

Molnar, R.E. 1982a. A longirostrine crocodilian from Murua (Woodlark), Solomon Sea. *Memoirs of the Queensland Museum*, Vol. 20, pp.675–685.

Molnar, R.E. 1982b. *Pallimnarchus* and other Cenozoic crocodiles of Queensland. *Memoirs of the Queensland Museum*, Vol. 20, pp.657–673.

Molnar, R.E. 1991. Fossil reptiles in Australia. Pp.605–701. In Vickers-Rich, P., Monaghan, J.M., Baird, R.F. and Rich, T.H. (eds). *Vertebrate Palaeontology of Australasia.* Pioneer Design Studio and Monash University Publications Committee, Melbourne. 1437pp.

Molnar, R.E. 1993. Biogeography and phylogeny of the Crocodylia. In Glasby, C.J., Ross, G.J.B. and Beesley, P.L. (eds). *Fauna of Australia. Vol. 2A: Amphibia and Reptilia.* Australian Government Publishing Service, Canberra. 439pp.

Molnar, R.E. and Pole, M. 1997. A Miocene crocodilian from New Zealand. *Alcheringa*, Vol. 21, pp.65–70.

Molnar, R.E. and Willis, P.M.A. 1996. A neosuchian crocodile from the Queensland Cretaceous. *Fifty-sixth Annual Meeting, Society of Vertebrate Paleontology, American Museum of Natural History, New York, New York: 54A.* (abstract).

Molnar, R.E. and Willis, P.M.A. 2000. New crocodyliform material from the early Cretaceous Griman Creek Formation, at Lightning Ridge, New South Wales. In Grigg, G.C., Seebacher, F. and Franklin, C.E. (eds). *Crocodilian Biology and Evolution.* Surrey Beatty and Sons, Sydney. 446pp.

Molnar, R.E., Worth, T. and Willis, P.M.A. 2002. An extinct Pleistocene endemic mekosuchine crocodylian from Fiji. *Journal of Vertebrate Paleontology*, Vol. 22, pp.612–628.

Murray, P. 1984. Extinctions downunder: a bestiary of extinct Australian late Pleistocene monotremes and marsupials. In Martin, P.S. and Klein, R.G. (eds). *Quaternary Extinctions: a Prehistoric Revolution.* University of Arizona Press, Tucson. 892pp.

Murray, P. and Megirian, D. 1992. Continuity and contrast on middle and late Miocene vertebrate communities from the Northern Territory. *The Beagle (Records of the Northern Territory Museum of Arts and Sciences)*, Vol. 9, pp.195–218.

Murray, P., Megirian, D. and Wells, R. 1993. *Kolopsis yperus* sp. nov. (Zygomaturinae, Marsupialia) from the Ongeva Local Fauna: new evidence for the age of the Alcoota fossil beds of Central Australia. *The Beagle (Records of the Northern Territory Museum of Arts and Sciences)*, Vol. 10, pp.155–172.

Neill, W.T. 1971. *The Last of the Ruling Reptiles; Alligators, Crocodiles and their Kin.* Columbia University Press, New York. 486pp.

Plane, M.D. 1967. Stratigraphy and vertebrate fauna of the Otibanda Formation, New Guinea. *Bureau of Mineral Resources, Geology and Geophysics Bulletin*, No. 86, pp.1–64.

Plane, M.D. and Gatehouse, C.G. 1967. A new vertebrate fauna from the Tertiary of northern Australia. *Australian Journal of Science*, Vol. 30, pp.272–273.

Rich, T.H.V. and Rich, P.V. 1989. Polar dinosaurs and biotas of the early Cretaceous of southeastern Australia. *National Geographic Research,* Vol. 5, pp.15–53.

Riek, E.F. 1952. The fossil insects of the Tertiary Redbank Plains Series. Part I: an outline of the fossil assemblage with descriptions of the fossil insects of the orders Mecoptera and Neuroptera. *University of Queensland Papers, Department of Geology*, Vol. 4, pp.3–14.

Romer, A.S. 1956. *Osteology of the Reptiles.* University of Chicago Press, Chicago. 722pp.

Rowe, T., Brochu, C.A., Colbert, M., Merck, J.W. Jnr, Kishi, K., Saglamer, E. and Warren, S. 1999. Introduction to Alligator: Digital atlas of the skull. *Society of Vertebrate Paleontology, Memoir*, No. 6.

Salisbury, S.W. and Willis, P.M.A. 1996. A new crocodylian from the early Eocene of southeastern Queensland and a preliminary investigation into the phylogenetic relationships of crocodyloids. *Alcheringa,* Vol. 20, pp.179–226.

Simpson, G.G. 1937a. An ancient eusuchian crocodile from Patagonia. *American Museum Novitates*, No. 965, pp.1–20.

Simpson, G.G. 1937b. New reptiles from the Eocene of South America. *American Museum Novitates*, No. 927, pp.1–3.

Steel, R. 1973. Crocodilia. *Handbuch der Paleontologie*, Vol. 16, pp.1–116.

Swinton, W.E. 1924. *Description of the vertebrate remains collected by Professor J.W. Gregory, D.Sc., PRS, in the Lake Eyre district of South Australia.* Hunterian Museum, University of Glasgow.

Taplin, L. 1984. Evolution and zoogeography of crocodilians; a new look at an ancient order. In Archer, M. and Clayton, G. (eds). *Vertebrate Zoogeography and Evolution in Australasia.* Hesperian Press, Perth. 1203pp.

Taplin, L. 1989. Historical zoogeography of the eusuchian crocodilians: a physiological perspective. *American Zoologist*, Vol. 29, pp.885–901.

Tchernov, E. 1986. Evolution of the Crocodiles in East and North Africa. *Éditions du Centre National de la Recherche Scientifique, Paris.*

Tedford, R.H. and Wells, R.T. 1990. Pleistocene deposits and fossil vertebrates from the "Dead Heart of Australia". *Memoirs of the Queensland Museum*, Vol. 28, pp.263–284.

Thompson, P. 1980. *Report on investigation of Myrtle Vale and Redbank; two fossil bone sites; Darling River Weir No 20 North-west N.S.W.* N.S.W. National Parks and Wildlife Service.

Vis, C.W. de. 1886. On the remains of an extinct saurian. *Proceedings of the Royal Society of Queensland,* Vol. 2, pp.181–191.

Vis, C.W. de. 1905. Fossil vertebrates from New Guinea. *Annals of the Queensland Museum*, Vol. 6, pp.26–31.

Vis, C.W. de. 1907. Fossils of the Gulf Watershed. *Annals of the Queensland Museum*, Vol. 7, pp.1–7.

Walker, A.D. 1970. A revision of the Jurassic reptile *Hallopus victor* (Marsh), with remarks on the classification of crocodiles. *Philosophical Transactions of the Royal Society of London,* Series B, (*Biological Sciences*), Vol. 257, pp.323–372.

Willis, P.M.A. 1986. *The Riversleigh crocodiles.* B.Sc. (Hons.) Thesis (unpublished), University of New South Wales.

Willis, P.M.A. 1987. A fish eating crocodile from the South Pacific. In Hand, S. and Archer, M. (eds). *The Antipodean Ark.* Angus and Robertson, Sydney. 90pp.

Willis, P.M.A. 1992. Four new crocodilians from the early Miocene sites at Riversleigh Station, northwestern Queensland. *The Beagle (Records of the Northern Territory Museum of Arts and Sciences),* Vol. 9, p.629.

Willis, P.M.A. 1993. *Trilophosuchus rackhami*, gen. et sp. nov., a new crocodilian from the early Miocene Limestones of Riversleigh, northwestern Queensland. *Journal of Vertebrate Paleontology,* Vol. 13, pp.90–98.

Willis, P.M.A. 1995. *The phylogenetic systematics of Australian crocodilians.* Ph.D. Thesis (unpublished), University of New South Wales.

Willis, P.M.A. 1997. New crocodilians from the Oligo-Miocene White Hunter Site, Riversleigh Station, northwestern Queensland. *Memoirs of the Queensland Museum*, Vol. 41, Pt. 2, pp.423–438.

Willis, P.M.A. 2001. New crocodilian material from Ringtail Site, Riversleigh, northwestern Queensland. In Grigg, G.C. Seebacher, F. and Franklin, C.E. (eds). *Crocodilian Biology and Evolution.* Surrey Beatty and Sons, Sydney. 446pp.

Willis. P.M.A. and Archer, M. 1990. A Pleistocene longirostrine crocodilian from Riversleigh: first fossil occurrence of *Crocodylus johnstoni* Krefft. *Memoirs of the Queensland Museum*, Vol. 28, pp.159–163.

Willis, P.M.A. and Mackness, B. 1996. *Quinkana babarra*, a new species of ziphodont mekosuchine crocodile from the early Pliocene Bluff Downs Local Fauna, northern Australia with a revision of the genus. *Proceedings of the Linnean Society of New South Wales*, Vol. 116, pp.143–151.

Willis, P.M.A. and Molnar, R.E. 1991a. A longirostrine crocodile from the early Tertiary of southeastern Queensland. *Alcheringa*, Vol. 15, pp.229–233.

Willis, P.M.A. and Molnar, R.E. 1991b. A new middle Tertiary crocodile from Lake Palankarinna, South Australia. *Records of the South Australian Museum*, Vol. 25, pp.39–55.

Willis, P.M.A. and Molnar, R.E. 1997a. A review of the Plio-Pleistocene crocodilian genus *Pallimnarchus. Proceedings of the Linnean Society of New South Wales*, Vol. 117, No. 1, pp.223–242.

Willis, P.M.A. and Molnar, R.E. 1997b. Identification of large reptilian teeth from Plio-Pleistocene deposits of Australia. *Journal and Proceedings of the Royal Society of New South Wales*, Vol. 130, No. 3–4, pp.79–92.

Willis, P.M.A., Molnar, R.E. and Scanlon, J.D. 1993. An early Eocene crocodilian from Murgon, southeastern Queensland. *Kaupia Darmstädter Beiträge zur Naturgeschichte*, Vol. 3, pp.25–32.

Willis, P.M.A., Murray, P.F. and Megirian, D. 1990. *Baru darrowi* gen. et. sp. nov., a large, broad-snouted crocodyline (Eusuchia: Crocodylidae) from mid-Tertiary freshwater limestones in Northern Australia. *Memoirs of the Queensland Museum*, Vol. 29, pp.521–540.

Woodburne, M.O. 1967. The Alcoota Fauna, central Australia; an intergrated palaeontological and geological study. *Bureau of Mineral Resources, Geology and Geophysics Bulletin,* No. 87, pp.1–187.

Woodburne, M.O., Macfadden, B.J., Case, J.A., Springer, M.S., Pledge, N.S., Power, J.D., Woodburne, J.M. and Springer, K.B. 1993. Land mammal biostratigraphy of the Etadunna Formation (late Oligocene) of South Australia. *Journal of Vertebrate Paleontology*, Vol. 13, No. 4 pp.483–515.

Wu, X., Sues, H-D. and Sun, A. 1995. A plant-eating crocodyliform reptile from the Cretaceous of China. *Nature*, Vol. 376, pp.678–680.

Chapter 19
Australian Lizards—Relationships and Origins

Mark N. Hutchinson[1] and Sarah A. Smith[2]
[1] South Australian Museum, North Terrace, Adelaide, S.A. 5000 Australia
[2] Department of Biology, Villanova University, 800 Lancaster Ave., Villanova, Pennsylvania 19085 U.S.A.
E-mail: Hutchinson.Mark@saugov.sa.gov.au

BACKGROUND

Lizards and snakes are a recent success story in the evolution of reptiles, with nearly 7,000 living species. New species are still being discovered world-wide at the rate of dozens per year, so the actual total is not going to be clear for some time yet. From the evolutionary point of view lizards and snakes belong to a single lineage, the Squamata. Snakes are essentially just one of several groups of lizards which have evolved a limbless body shape coupled with the adoption of undulatory locomotion, and are thus part of the adaptive radiation of lizards. However, snakes are by far the most diverse legless squamates and so common usage tends to treat lizards and snakes as distinct, as does this book. Chapter 17 by John Scanlon (2006) discussed the snakes in detail; this chapter looks at the other squamates, the lizards.

The term 'lizards' really means squamates that are not snakes; they are not exclusively defined and so constitute a paraphyletic group. They are merely those squamates that lack the specializations that distinguish snakes.

In thinking about the evolution of life, it is more productive to concentrate on monophyletic lineages, which are those that include all the descendants from a common ancestor. Within the lizards there are four distinct monophyletic lineages (excluding snakes), some more closely related to snakes than they are to other lizard lineages. These lineages are the Iguania, the Scincomorpha, the Gekkota and the Anguimorpha (Estes *et al.*, 1988); the last group might include snakes. As it happens, Australia has one family from each of these major lineages; the Agamidae (Iguania), Scincidae (Scincomorpha), Gekkonidae (Gekkota) and Varanidae (Anguimorpha). 'Lizards' can be a useful vernacular term (a shorthand for non-snake squamates), but evolutionary and biogeographic scenarios will need to deal with each component lineage as an independent entity.

Lizards are in some ways a conservative group in their anatomy and biology. They essentially retain the body plan of the earliest known reptiles, such as the captorhinomorphs of the Carboniferous (Carroll, 1988). True lizards, that is, undoubted members of the Squamata, do not appear in the fossil record until the Jurassic although the date of their origins is still uncertain. This is due to the fact that the early diapsid reptiles, which gave rise to the Squamata, are themselves very lizard-like in general appearance and have independently evolved some lizard-like features (Fraser, 1986, 1988; Rieppel, 1994). Which brings us to the question: what makes a squamate?

LIZARD CHARACTERISTICS

Living members of the Squamata share several unique features that indicate their descent from a common ancestor. Some of the important features are soft anatomical structures, such as the paired copulatory organs termed hemipenes, which are unique to male squamates. These cannot be assessed in fossils, so we must rely on skeletal features. Some diagnostic derived features are discussed below.

SKULL ANATOMY

Squamates have kinetic skulls, a term which means that various bony units within the skull can move with respect to one another. Squamate kinesis evolved via a unique anatomical change, the loss of the quadratojugal bone. In most diapsid reptiles, including birds and crocodiles, the quadratojugal is part of the lower temporal arch, a bony strut that formes a brace for the quadrate bone. The absence of this bone in squamates means that the quadrate is free to swing forwards and back, and because it is the articulation point for the lower jaw, this motion protracts or retracts the mandible. In most squamates, similar movements also occur in the bones of the snout, including the upper jaw, which can be raised or lowered with respect to the braincase and rear of the skull. In lizards the motion is made possible by a distinctive hinge between the frontal and parietal bones; snakes have the hinge in a new position, between the nasals and frontals. This attribute enables the animal to greatly improve seizing and handling prey, to better apply force to the food item with its jaws, and allows the mouth to open more widely and engulf larger food items than would otherwise be possible (Frazetta, 1983; Iordansky, 1990).

In other respects, the squamate skull is specialized with respect to that of early reptiles in lacking tabular and postparietal bones; the palatines also separate the pterygoids from the vomers on the midline of the palate. In the

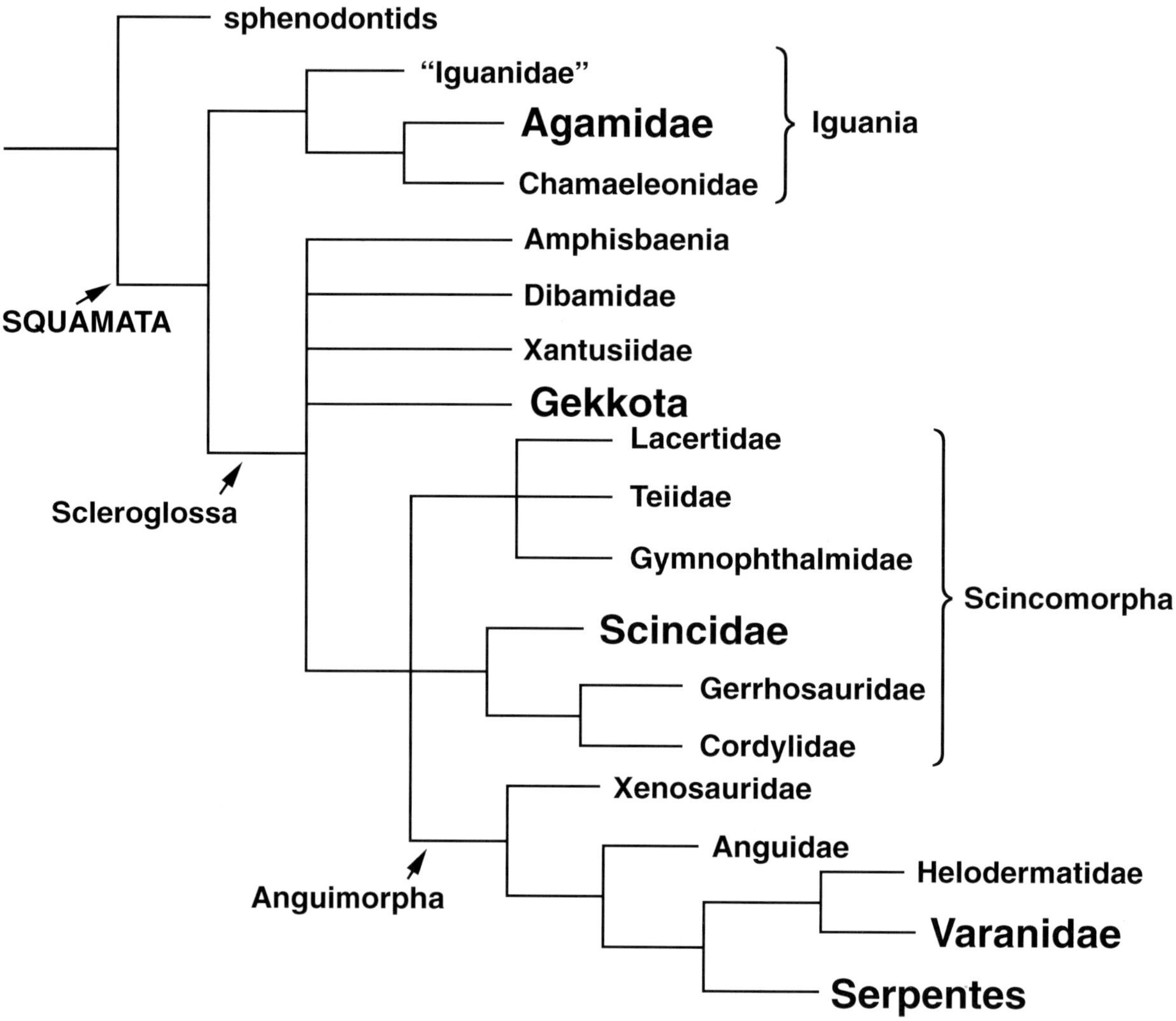

Figure 1. Major living lineages of the Squamata, and their phylogenetic relationships. Relationships shown are a subjective summary taken from Wu *et al.* (1996), Lee (1998), Honda *et al.* (2000b) and Harris *et al.* (2001). Iguanidae is treated as of uncertain monophyly and Agamidae as the monophyletic sister lineage of Chamaeleonidae. Relationships of Gekkota, Amphisbaenia, Dibamidae and Xantusiidae are shown as scleroglossan but otherwise unresolved. Following Lee (1998) traditional Scincomorpha are shown as paraphyletic with respect to Anguimorpha; Serpentes is treated as a varanoid offshoot. Clades occurring in Australia shown in larger typeface.

lower jaw, the articular and prearticular are fused (the single bone resulting is usually called the articular) and there is a well-developed, complex retroarticular process projecting caudal to the articular facet (Estes, 1983).

TEETH

Most squamates have teeth that are replaced throughout life; the teeth are attached by temporary bone to the inner margins of the jaw bones (pleurodonty), usually supported lingually by a longitudinal ridge, the subdental shelf (Edmund, 1969). An exception to this is seen in some lineages (Agamidae, Chamaeleonidae, some Teiidae) in which teeth attach to the far edges of the jaw bones (acrodonty) and are not replaced, although additional teeth are added to the jaw as long as it is growing (Robinson, 1976).

VERTEBRAE AND RIBS

Most squamate vertebrae are procoelous, with an anterior cotyle (socket) and a posterior condyle (ball). But some of the earliest lizards show a plesiomorphic (primitive) vertebral structure, a persistent notochord, with the vertebral centrum pierced lengthwise by a canal. There is little development of the bony condyle, making the vertebral centra hollow at both ends (amphicoelous). Most living gekkotans also have amphicoelous vertebrae, regarded by most recent workers as a consequence of paedomorphic reversal rather than being truly plesiomorphic. Lizards have determinate growth; growth ceases once separate long bone epiphyses fuse to their shafts. Squamate ribs have a single head, ending in a cotyle, which forms a ball-and-socket joint with the pleurapophysis of the vertebral centrum.

MAJOR LINEAGES

Much of the morphological evidence that had been accumulated on the relationships of lizards was compiled and analysed by Estes *et al.* (1988). This tabulation of data has since been re-analysed by several workers, and has been augmented by them (Wu *et al.*, 1996; Evans and Barbadillo, 1997; Macey *et al.*, 1997; Lee, 1998, 2000; Reynoso, 1998). It is notable how the consensus on the composition of major lineages is not matched by strong agreement on the phylogenetic relationships between lineages. The data set is acknowledged as highly homoplastic, with considerable evidence of parallel and reversal of evolution. Lee's (1998) study demonstrated that correlated suites of character states (associated with limbless locomotion) heavily influenced the placement of clades on a phylogenetic tree. The major lineages and their included groups are shown in Figure 1.

IGUANIA

Includes the 'Iguanidae', sometimes subdivided into five related lineages (Frost and Etheridge, 1989), the Agamidae and the highly specialized Chamaeleonidae. Chamaeleons and agamids share acrodont dentition and the study by Frost and Etheridge (1989) found that some analyses nest chamaeleons within agamids, casting doubt on the monophyly of the Agamidae. Many subsequent authors have treated the Agamidae as of uncertain monophyly, but this uncertainty has only begun to be addressed using molecular methods (Honda *et al.*, 2000b; Macey *et al.*, 2000); so far these tend to favour a monophyletic Agamidae as the sister lineage to the Chamaeleonidae.

In Australia, the Iguania is represented by the Agamidae—see Figure 2 for skeletal features. The skull is short and deep, with broad upper temporal fenestrae bordered by a laterally compressed postfrontal-squamosal arch, and medially by a similarly shaped supratemporal process of the parietal. The jugal is also broad and flattened and in broad contact with the squamosal. The palate has broad, plate-like palatines with only lightly impressed choanal grooves. The parietal foramen is large and lies on the fronto-parietal suture. The agamid braincase has a distinctive pair of openings, the epiotic foramina, on the dorsal surface. The mandible has a very well-developed, dorso-caudally directed retroarticular process, and a distinctive medial prearticular process. The characteristic acrodont teeth are triangular in profile and enlarge towards the posterior end of the tooth row. The premaxillary teeth and the anteriormost two to three teeth on the maxilla and dentary are pleurodont, cylindrical and sharply pointed. The best osteological reference has not been published (Moody, 1980) but useful information is presented by Frost and Etheridge (1989) and Covacevich *et al.* (1990).

SCLEROGLOSSA

All remaining squamates, including snakes, are now regarded as forming a sister lineage to the Iguania, usually termed the Scleroglossa (redefined by Lee, 1998). Within Scleroglossa are several very distinctive lineages whose relationships remain unknown or controversial. The snakes (Serpentes) are one such problem group, as are the Amphisbaenia, limbless burrowers that do not occur in Australia. Smaller groups that are equally in doubt are the nocturnal Xantusiidae of North America and the limbless, burrowing Dibamidae of South-east Asia, New Guinea and Central America (Underwood, 1957; Rieppel, 1984a, b; Lee, 2000; Vicario *et al.*, 2003). These groups present problems due to convergence involving limbless burrowing adaptations or evolutionary reversals linked to paedomorphosis. These aside, the remaining squamates are generally placed in three scleroglossan lineages.

GEKKOTA

This lineage includes the Gekkonidae and Pygopodidae as its core living groups. Relationships of these two to one another are strongly supported by anatomy (Estes, 1983; Kluge, 1987), but each lineage has numerous specializations that make relationships to other squamates difficult to establish. Gekkonids show considerable paedomorphosis and apparently primitive traits exhibited by them may therefore be evolutionary reversals (Moffat, 1973). Pygopods are problematic because of their highly developed snake-like anatomy, with many characters that would have potential in establishing relationships lost or highly modified. Traditionally regarded as the sister group of geckos, morphology (Kluge, 1987) and molecules (Donnellan *et al.*, 1999) now show that pygopods are nested within geckos as the sister group of the Australian Diplodactylinae. Most analyses place the Gekkota at the base of the Scleroglossa. In fact, the modification of so much of gekkotan anatomy makes their relationships to other squamates very uncertain.

Gekkonid skulls (see Figure 3) are lightly ossified, with no upper temporal arch and upper temporal fenestra; the chevron-shaped postfrontal and very reduced squamosal are well separated from one another. The frontal has a characteristic tubular canal on its ventral surface formed by fused descending processes. The parietals are usually paired in Australian species, and there is never a parietal foramen. The jugal has no ascending process so that there is no bony posterior orbital margin. The braincase is peculiar in having the opening for the trigeminal nerve surrounded by bone, rather than emerging from a

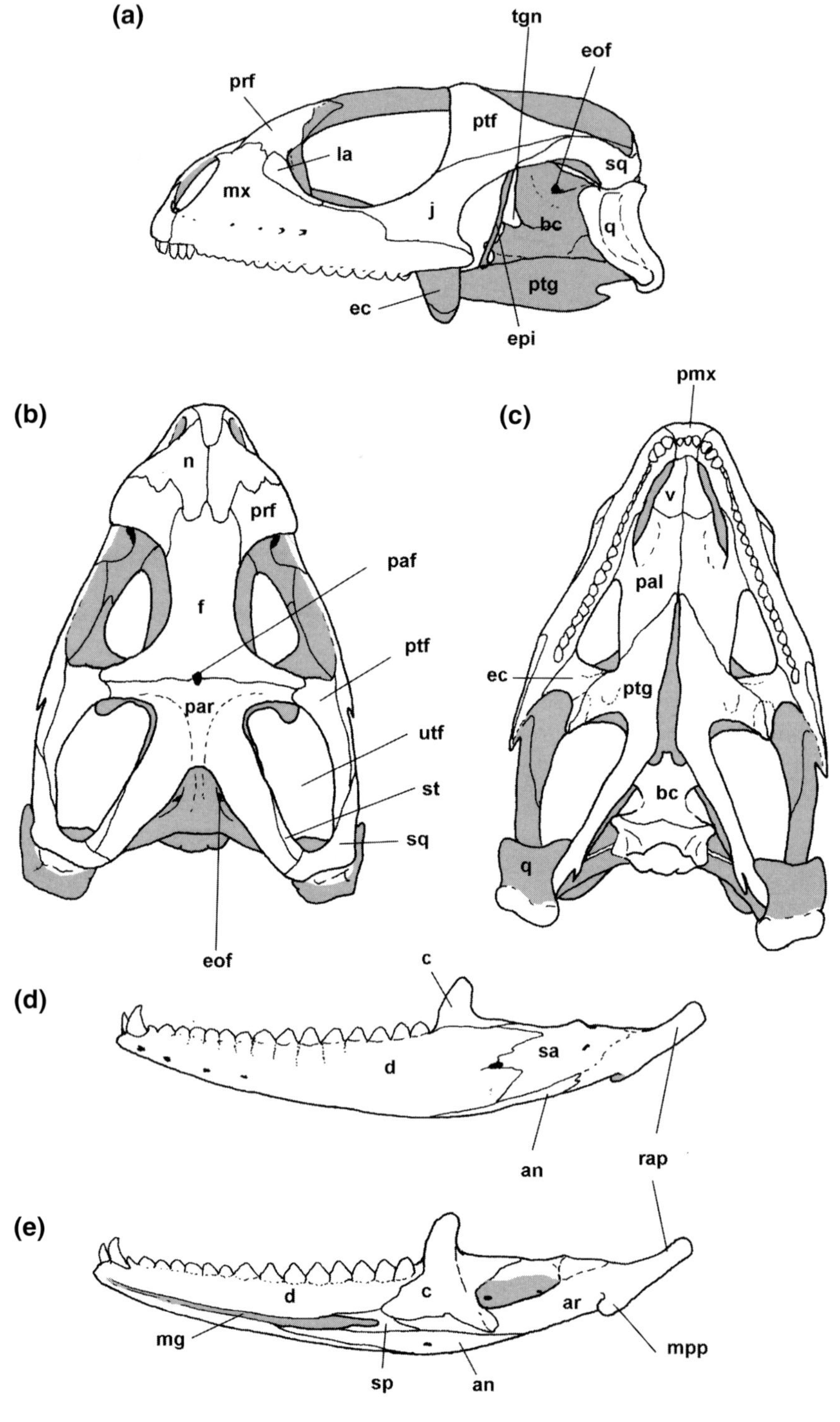

Figure 2. Skull and mandible of the Agamidae: **(a)** *Physignathus lesueurii* skull, left lateral view; **(b)** *P. lesueurii* skull, dorsal view (after Witten, 1993); **(c)** *P. cocincus* skull, ventral view (based on Moody, 1980). The large lacrimal bone is absent in most Australian agamids; **(d)** *Ctenophorus pictus* mandible, labial view; **(e)** *C. pictus* mandible, lingual view. Drawings not to scale. Key to abbreviations: an = angular; ar = articular; bc = braincase (usually a single fused unit in mature individuals; comprises lateral prootics and epiotics, ventral basisphenoid and basioccipital and dorsal supraoccipital); c = coronoid; d = dentary; ec = ectopterygoid; eof = epiotic foramen; epi = epipterygoid; f = frontal; j = jugal; la = lacrimal; mgr = Meckelian groove; mx = maxilla; n = nasal; paf = parietal foramen; pal = palatine; par = parietal; pmx = premaxilla; po = postorbital; ptf = postfrontal; ptg = pterygoid; q = quadrate; rap = retroarticular process; sa = surangular; sm = septomaxilla; sp = splenial; sq = squamosal; st = supratemporal; tgf = trigeminal foramen; tgn = trigeminal notch; utf = upper temporal fenestra; v = vomer.

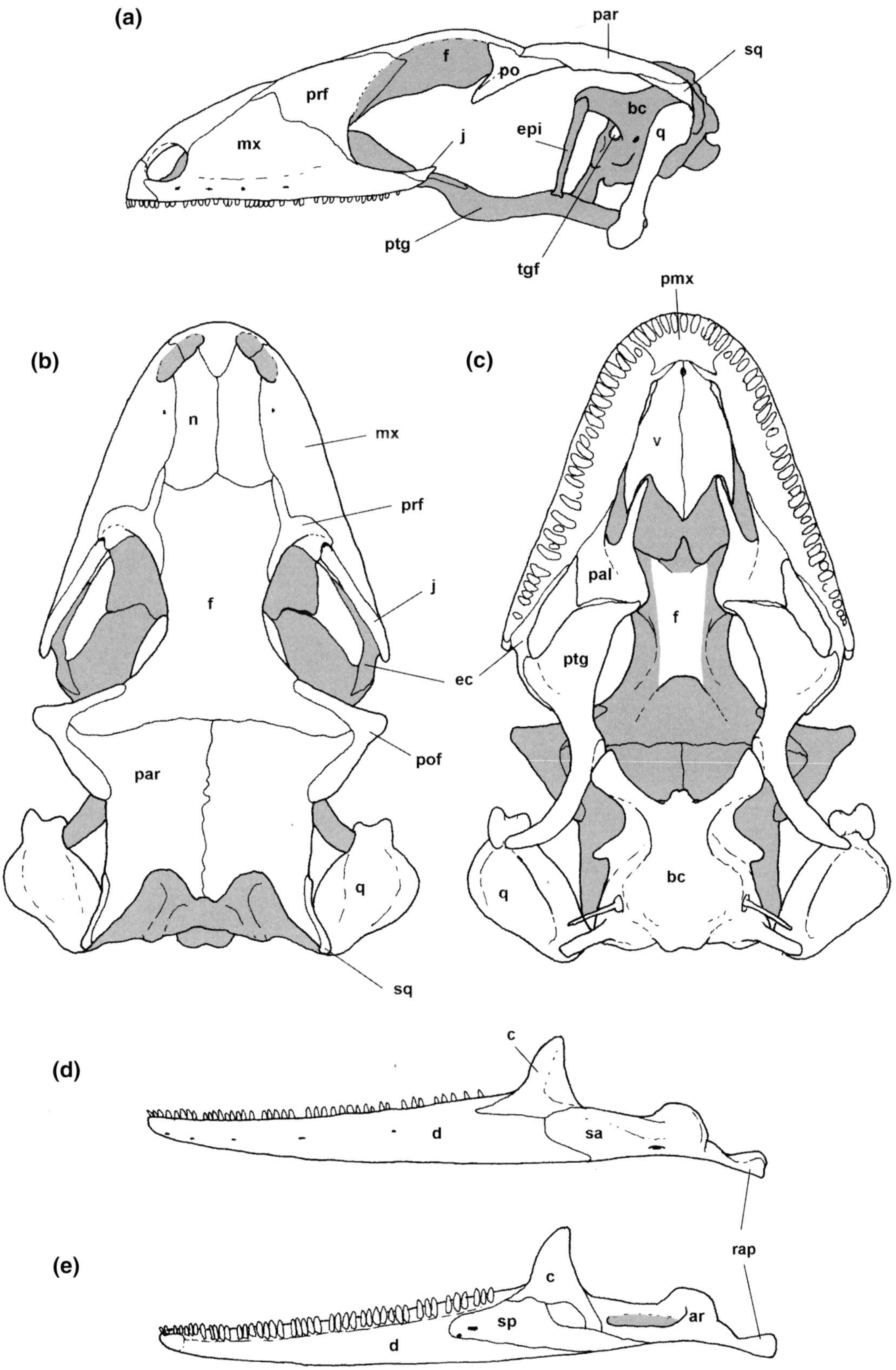

Figure 3. Skull and mandible of the Gekkota: **(a)** *Carphodactylus laevis* skull, left lateral view (after Greer, 1989); **(b)** *Gekko vittatus* skull, dorsal view; **(c)** *G. vittatus* skull, ventral view; **(d)** *Carphodactylus laevis* mandible, labial view; **(e)** *C. laevis* mandible, lingual view. Drawings not to scale, for label abbreviations see legend for Figure 2.

notch, bordered anteriorly by cartilage, as in other lizards. The palate is very lightly ossified with the right and left halves widely separated. The mandible has a low-set, narrow-based retroarticular process, the Meckelian groove is completely concealed by dentary overgrowth; in Australian species, the angular is absent, apparently fused with the splenial. Gekkonid teeth are characteristically very small and numerous. Osteological references include Stephenson (1960) and Rieppel (1984b).

Pygopods are similar to those of typical gekkonids, but more elongate and more robust in their degree of ossification. The braincase has a trigeminal notch as in other lizards, rather than the foramen characteristic of 'geckos'. Pygopod teeth are very variable, much more so than those of geckos (Hutchinson, 1997).

SCINCOMORPHA

The major living lizard lineage recognized by most workers has been the Scincomorpha, comprising several large, diverse and widely distributed families, the Scincidae, Cordylidae, Gerrhosauridae, Lacertidae, Teiidae, Gymnophthalmidae and possibly the Xantusiidae. Within the Scincomorpha, the Cordylidae and Gerrhosauridae are regarded as closely related, as Cordyliformes (Lang, 1991), in turn being close to the Scincidae—the Scincoidea of Estes (1983). Teiids and gymnophthalmids are also treated as closely related, and have only recently been regarded as distinct families (Presch, 1988). Lacertids are usually placed as the sister group of teiids plus gymnophthalmids (Estes *et al.*, 1988). Lee's (2000) reanalysis and initial evidence from *c-mos* sequences (Saint *et al.,* 1998; Harris *et al.*, 2001) have cast doubt on the monophyly of the Scincomorpha. The problem areas are particularly in the relationships of lacertids, teiids and possibly xantusiids to one another and to the scincoids (Vicario *et al.*, 2003).

Australia's scincomorphs are the Scincidae—see Figure 4 for skeletal features. Scincid skulls have broad flat roofs, often embossed by the fused osteoderms of the dorsal head scales. The parietal is broad, often partly or completely roofing over the upper temporal fenestra. The underside of the parietal has a pair of descending processes that articulate with the epipterygoids. The frontal (single in all Australian species) is often almost as broad anteriorly as posteriorly. In a few species the lacrimal and postorbital bones are present, if small, but in most Australian species both bones are absent. The palatines are distinctive, each with a broad ventral plate that separates the internal airway from the mouth. On the mandible the Meckelian groove may be exposed, but in many it is concealed by overgrowth of the dentary (Greer, 1970, 1979). The dentary has a posterodorsal coronoid process that overlaps the coronoid labially. Many skinks are fossorial, and have more elongate skulls, with shorter tooth rows and reduced, sometimes incomplete orbital and upper temporal arches (Greer and Cogger, 1985).

ANGUIMORPHA

The anguimorphs have long been recognized as a lineage due to their distinctive system of tooth replacement, in which new teeth lie beside functional teeth, rather than growing up into the bases of the functional teeth as in scincomorphs, gekkotans and iguanians (McDowell and Bogert, 1954). Anguimorphs conventionally are divided into the relatively plesiomorphic anguoids (Anguidae and Xenosauridae) and the highly modified, carnivorous Platynota (living Varanidae and Helodermatidae plus the extinct mosasaurs) (Rieppel, 1980). Serpentes have been treated as derived from anguimorphs in general (McDowell and Bogert, 1954) or as members of the Platynota with a relationship to mosasaurs (Lee and Caldwell, 1998).

In Australia the anguimorphs are represented by the Varanidae (goannas), always bearing in mind the likelihood that the Serpentes belong here as well. Varanid skulls are elongate and lightly built—see Figure 5 (Rieppel, 1980). The snout has extremely elongate external nostrils that extend posteriorly to separate the maxillae from the small, fused nasals. The widely open snout exposes the septomaxilla, concealed within the nasal cavity in other lizard lineages. The premaxilla is elongate, forming a medial rod connecting the nasal and tip of the snout. The frontals are paired, with well-developed descending processes that contact medially. The upper temporal arch is slender but complete and the supratemporal is relatively large. The jugal is very slender and fails to contact the postfrontal. The mandible has an internal hinge running ventrally from the coronoid, the splenial plus dentary articulating with the surangular plus articular plus angular. The teeth are characteristically, pointed and recurved in profile, laterally flattened and with a sharp cutting edge.

EARLY FOSSIL HISTORY

The squamate fossil record has been poor, as squamates are largely a terrestrial group, and mostly small, so that fossilization is unlikely (reviewed by Estes, 1983). Palaeoherpetology is also limited in that most fossils are from the Northern Hemisphere; the fossil record of the southern continents since the Cretaceous is relatively poorly known. But, despite these limitations, several recent reviews have greatly expanded the number of described fossil lizards—pushing the origins of the major lineages back to the start of the Jurassic (Evans, 1993, 1998; Evans and Milner, 1994); considerable diversity has been documented in some regions, especially the Gobi Desert deposits of the late Cretaceous (approximately Campanian) of Mongolia (Borsuk-Bialynicka, 1984; Borsuk-Bialynicka and Moody, 1984; Gao and Hou, 1995; Gao and Norell, 1998, 2000).

The oldest fossils that appear to belong to true lizards are Jurassic in age (Evans, 1993). What is striking about

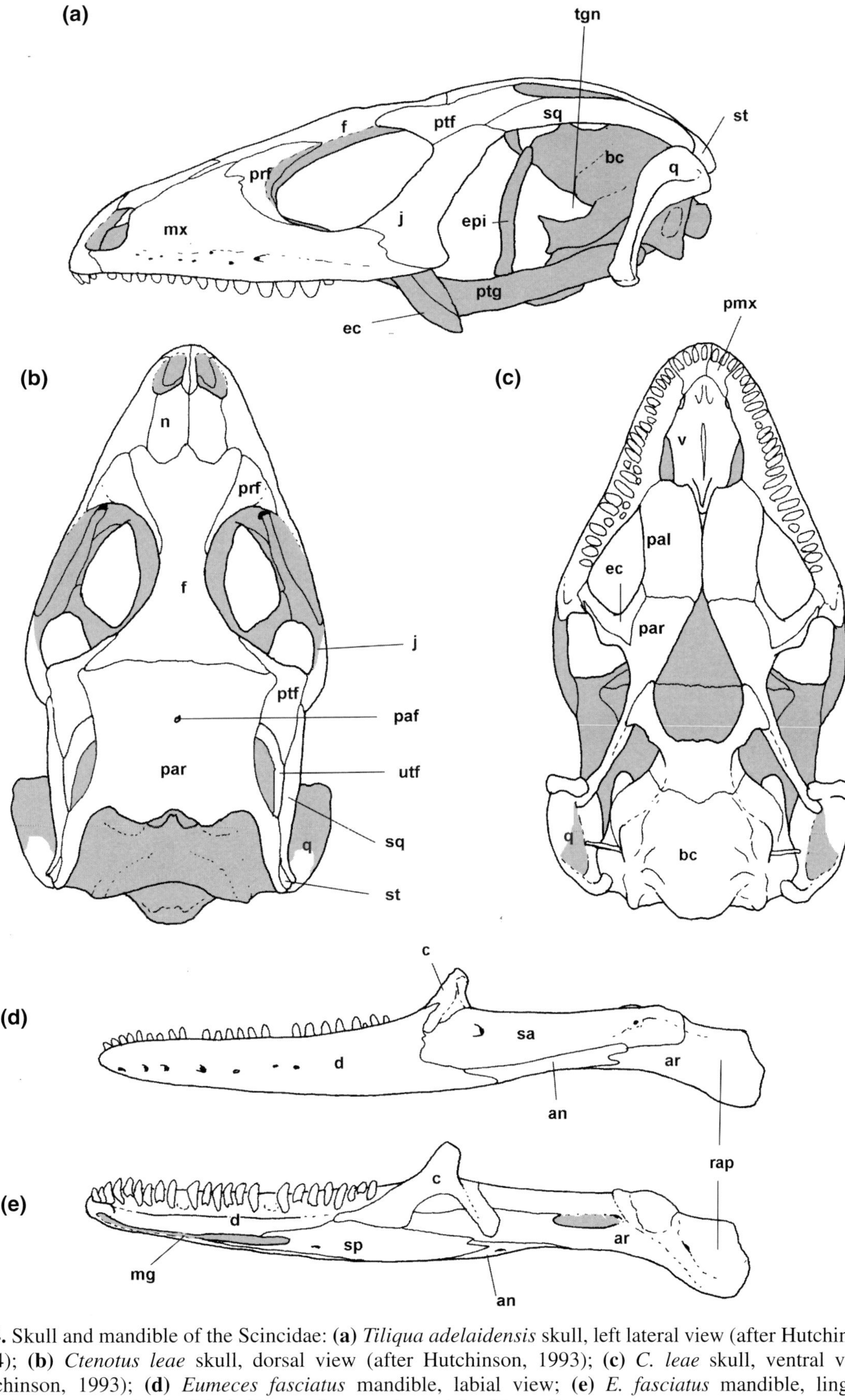

Figure 4. Skull and mandible of the Scincidae: **(a)** *Tiliqua adelaidensis* skull, left lateral view (after Hutchinson *et al.*, 1994); **(b)** *Ctenotus leae* skull, dorsal view (after Hutchinson, 1993); **(c)** *C. leae* skull, ventral view (after Hutchinson, 1993); **(d)** *Eumeces fasciatus* mandible, labial view; **(e)** *E. fasciatus* mandible, lingual view. Drawings not to scale, for label abbreviations see legend for Figure 2.

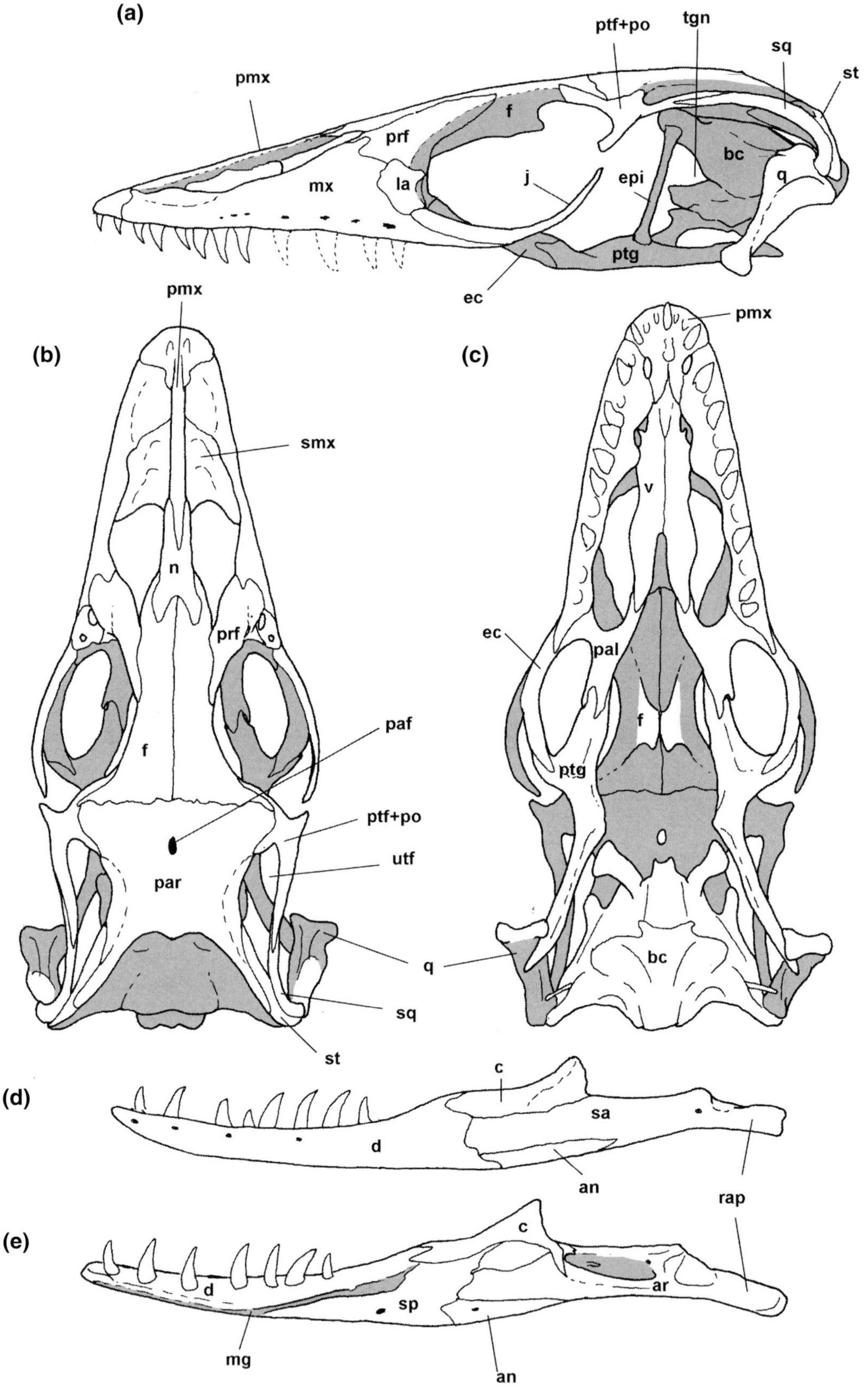

Figure 5. Skull and mandible of the Varanidae: **(a)** *Varanus salvator* skull, left lateral view; **(b)** *V. salvator* skull, dorsal view; **(c)** *V. salvator* skull, ventral view; **(d)** *V. salvator*, mandible, labial view; **(e)** *V. salvator* mandible, lingual view. From Rieppel (1980). Drawings not to scale, for label abbreviations see legend for Figure 2.

these finds is that they imply a markedly earlier origin for lizards as a group, as these first fossils include taxa already assignable to living lineages. Fossils from Europe and North America include fragmentary remains of several lizards regarded as probable scleroglossans. Evans (1998) recognizes early anguimorphs (*Dorsetisaurus*, *Parviraptor*) and early scincomorphs (*Paramacellodus*, *Saurillodon*), as well as much more doubtfully, early gekkotans in the middle Jurassic Forest Marble Formation of England. Some of the same taxa (*Parviraptor*, *Paramacellodus*) are also present in the similarly aged Morrison Formation of North America. *Parviraptor* has been suggested as a possible early platynotan (Evans, 1994b) while *Paramacellodus* has been interpreted as related to living cordyliforms (Estes, 1983), or to scincoids (cordyliforms plus skinks). These interpretations imply that differentiation within scincomorphs and anguimorphs had already begun, pushing their origins back towards the early Jurassic and late Triassic. However, it may be that these fossil forms simply show a few of the synapomorphies of living groups but they pre-date this later differentiation.

The domination of the earliest lizard fossil record by scleroglossans is in contrast to the complete absence of any iguanians, universally regarded as a more archaic squamate group. The implication is that the iguanians and scleroglossans were already separate lineages by the middle Jurassic. An alternative not yet seriously considered is that the scleroglossan body plan is actually plesiomorphic for squamates, with iguanians representing a later-derived group characterized by numerous reversals. Recent applications of molecular techniques to lizard relationships suggest that this idea might be worth further consideration as studies such as sequence data from *c-mos* (Saint *et al.*, 1998; Harris *et al.*, 2001) place iguanians within the scleroglossan radiation, not basal to it. Evans *et al.* (2002) may have filled this early gap in iguanian history with the discovery of a probable agamid from the middle Jurassic of India.

Some later fossils appear to be late survivors of the ancestral true lizards. *Huehuecuetzpalli* (Reynoso, 1998) from the early Cretaceous of Mexico, surely one of the most frightening names in all of palaeontology, is known from two almost complete skeletons (an adult and an immature individual). It was a slender, long-nosed predator, almost certainly a squamate, but lacking any obvious relationship with a living lineage. The details of its skull and vertebrae combine a few unique features with otherwise plesiomorphic squamate character states.

IGUANIA

Iguanians are first known relatively late in the fossil record. The only Mesozoic iguanians are from the late Cretaceous of Mongolia, where both pleurodont 'iguanids' and acrodont 'agamids' have been found (Borsuk-Bialynicka and Moody, 1984; Gao and Hou, 1997; Gao and Norell, 2000). The most complete Mesozoic agamid is *Priscagama*, from the late Cretaceous of Mongolia (Borsuk-Bialynicka and Moody, 1984; Gao and Norell, 2000). Priscagamines (the group includes at least two other genera) are an early lineage of acrodont iguanians that appear to be distinct from living lineages. They retain several plesiomorphic features (large splenial, poorly developed acrodonty in some taxa, absence of an epiotic foramen) but show some derived features—for example, labial overlap of the coronoid by the dentary.

GEKKOTA

For much of the 20th century, some of the oldest and most complete lizard fossils known were regarded as gekkotans. The late Jurassic *Ardeosaurus, Bavarisaurus, Eichstaettisaurus* and *Yabeinosaurus* were all described from almost complete skeletons and, after various attributions, were generally interpreted as gekkotans (Mateer, 1982; Estes, 1983). Review of this material has resulted in altered allocations, as they were initially made on the basis of plesiomorphic character states (Evans, 1994a). These fossils are now regarded as probable scleroglossans, but possess too few gekkotan synapomorphies to be unequivocally placed with living gekkotans (Evans, 1994a). Even so, one of these, *Eichstaettisaurus*, is very gecko-like in the proportions and layout of the skull bones (Mateer, 1982). At present, many of these gecko-like attributes have not been coded for phylogenetic analysis, and so this genus like the others still branches basally due to its many retained plesiomorphic character states.

The oldest fossil that appears to be unequivocally gekkotan is *Hoburogekko* (Alifanov, 1989) from the early Cretaceous (Aptian–Albian) of Mongolia. The fossil consists of bones of the snout that are very gecko-like, including a T-shaped frontal with its descending processes forming a completely closed canal for the stem of the forebrain, a broad facial process on the maxilla, small circular external naris and vomers notched posteriorly. It retains an open Meckelian groove in the dentary, a plesiomorphic feature not seen in any living gekkotan. Gao and Norell (2000) included a newly described late Cretaceous Mongolian lizard, *Myrmecodaptria*, as a basal gekkotan; but the list of supposed gekkotan synapomorphies is by no means unique to the Gekkota, being found elsewhere especially in reduce-limbed fossorial lizards. The proportions of the skull strongly suggest that *Myrmecodaptria* was a burrower.

SCINCOMORPHA

Most Jurassic to early Cretaceous lizard finds are regarded as scincomorphan. Scincoids are represented by a number of Mesozoic genera from across Eurasia, including *Becklesius*, *Paramacellodus*, *Saurillus* and *Pseudosaurillus* from western Europe and North Africa (Hoffstetter, 1967;

Estes, 1983), *Sharovisaurus* from Kazakhstan (Hecht and Hecht, 1984) and *Mimobecklesisaurus* from China—erroneously described as a scincid (Li, 1985). Although similar to living cordyliforms in many respects, they are placed in their own family, the Paramacellodidae (Estes, 1983). Paramacellodids were like living gerrhosaurids in having heavy osteoderms underlying the body scales (Richter, 1994).

Uncertainty on the relationships of living forms makes attribution of other Mesozoic scincomorphs problematic. Lacertoids (i.e. lacertid- and teiid-like forms) have been identified as early as the late Jurassic, with the recent suggestion by Evans and Barbadillo (1997), that *Meyasaurus* was teiid-like. Several genera regarded as teiid-like (*Polyglyphanodon, Macrocephalosaurus, Darchansaurus*) are known from well preserved skulls and skeletons from the Cretaceous of Mongolia and North America (Sulimski, 1975). They retain some plesiomorphic features compared with living teiids and this has prompted their placement within the teiids by Estes (1983). According to Estes (1983), they are not strongly differentiated from living teiids, and some of their specializations, including transversely oriented, subacrodont cheek teeth, are shared with some living teiids (e.g. the genus *Teiius*). By contrast, Wu *et al.* (1996) found that polyglyphanodontines were distinct from teiids and lacertids, although still a sister lineage, and were closer to amphisbaenians.

Other scincomorphan families are more doubtfully known prior to the Tertiary. Paradoxically the largest living family, the Scincidae, has the worst fossil record. Several specimens from the Mesozoic have been identified as truly scincid, rather than merely scincomorphan, but none of the evidence is compelling. Gao and Fox (1996) reviewed the meagre history of supposed scincids from the Mesozoic, and noted that several taxa from North America initially identified as skinks probably belonged to other families (*Contogenys*, a possible xantusiid, and *Sauriscus,* possibly a cordylid). Gao and Fox accepted as true skinks the durophagous lizard described by Rowe *et al.* (1992), Richter's (1994) unnamed early Cretaceous skink, and their own three new taxa (*Penemabuya*, *Orthrioscincus* and *Aocnodromeus*), all from the late Cretaceous of western Canada. Gao and Fox greatly understated the variation that occurs within scincids and the similarities that are shared by scincids and cordyliforms. The conclusion that their evidence points to is thus the more limited one—i.e. that scincoids, possibly gerrhosaurids, were present in North America and western Europe in the Cretaceous, but there are no definite records of true scincids.

ANGUIMORPHA

An early anguimorph from the same area as *Parviraptor*, *Dorsetisaurus* (Jurassic to early Cretaceous), represents a distinct lineage, possibly closer to the Anguioidea (Borsuk-Bialynicka, 1984) than the Platynota. The anguioids are common in the late Cretaceous of North America (Gao and Fox, 1976), with both xenosaurs and slightly more derived anguids recorded; in the late Cretaceous of Mongolia anguioids are represented by an extinct lineage related to xenosaurs, the carusiids (Gao and Norell, 1998).

Mesozoic platynotans were more diverse than the living varanids and helodermatids (the venomous North American gila monster and bearded lizard). Their large size and marine habits of the mosasaurs and aigialosaurs have helped to create a good fossil record, which has been reviewed several times (Borsuk-Bialynicka, 1984; Carroll and de Braga, 1992; Norell and Gao, 1997; Caldwell, 1998). Fossils show that the modern Varanidae and Helodermatidae were distinct lineages by the end of the Cretaceous, and that they were widespread across the Northern Hemisphere, occurring in North America, western Europe, North Africa and eastern Asia. True varanids are not known from the Mesozoic, but similar taxa such as the genus *Saniwa* (Gilmore, 1928), show that the varanid body plan (relatively large, elongate predators) was well established in the Cretaceous.

MODERN RADIATIONS

The major lineages of lizards appear to stem from a late Jurassic period of initial diversification, while some modern families appeared early in the Cretaceous. Until recently there was little information of any sort that could shed any light on the origins of Australia's modern lizard fauna. The only speculations possible were of a superficial sort, where Australian taxa were seen to be broadly similar in appearance to Asian taxa, leading to the conclusion that Australia had essentially received its lizard fauna from Asia by island hopping. The difficulty of this crossing and Australia's essential isolation were considered as sufficient to explain the distinctive features of the Australian genera *vis-à-vis* their presumed Asian ancestors.

Fossil data are limited to an Eocene limb bone, probably scincid, from the Rundle Formation, near Rockhampton, Queensland (Hocknull, 2000), and the remains from the Oligo-Miocene of the Etadunna Formation, South Australia (Estes, 1984) and Riversleigh, Queensland (Covacevich *et al.,* 1990; Hutchinson, 1992). Skinks are recorded from the earliest Late Oligocene beds, while gekkonids and agamids are known from the early Miocene. Where identifiable (e.g. Shea and Hutchinson, 1992; Hutchinson, 1992, 1997), these Tertiary fossils belong to extant Australian lineages.

Much of the recent information on the relationships and origins of Australian lizards has come not from morphological analysis but through the application of molecular methods, especially DNA sequencing. DNA sequencing has provided a new source of data to establish phylogenetic frameworks for Australian taxa and their possible extralimital relatives. In some cases we now have suggestions for where the sister lineages of the

Australian radiations are found. The primarily divergent nature of molecular sequence evolution can also make some estimation possible on when the Australian taxa diverged from these sister lineages. Some work has been done for all four lizard lineages.

IGUANIA

Agamids and chamaeleons have long been regarded as closely associated based on their distinctive acrodont dentition, as well as general cranial and other anatomical similarities. A major review of the Iguania (Frost and Etheridge, 1989) found some support for chamaeleons being nested within the traditional Agamidae, but also found an equivalent amount of support for the traditional view of chamaeleons and agamids as sister taxa. In the former scenario, the most basal split was not between chamaeleons and agamids, but between the archaic agamid genera *Leiolepis* and *Uromastyx* and the rest of the acrodont lizards. Within Australia (Witten, 1993), agamid genera are still defined for the most part on typological grounds—species that look different are placed in different genera. The only major phylogenetic split recognized within the major Australian radiation is based on the difference in preanal and femoral pore structure, whether these pores are situated within a scale or between scales. The latter arrangement is probably apomorphic and suggests that species in the genera *Ctenophorus*, *Pogona*, *Moloch* and *Tympanocryptis* (including *Rankinia*, sometimes regarded as distinct) form a distinct lineage within the Australian fauna (Greer, 1989). The water dragon (*Physignathus*) and the rainforest dragons (*Hypsilurus*) are regarded as Australian, but of earlier divergence from the main Australian radiation.

Macey *et al.* (2000) presented a phylogenetic analysis of all acrodont lizards (the families Agamidae and Chamaeleonidae) based on sequence variation of a large fragment (1685–1778 BP) of the mitochondrial genome. This study is notable for its authors' strong inferences on origins. Macey *et al.* identified several major monophyletic radiations within these lizards, one being the Australo-Papuan radiation that they termed the Amphibolurinae. Their data were equivocal regarding the position of chamaeleons, but confirmed a long-standing biogeographic puzzle, the close relationship of the Asian water dragon, *Physignathus cocincinus*, to the Australian radiation—although not specifically to the Australian water dragon alone. While they acknowledged that dispersal may account for this relationship, they preferred a vicariant explanation. They proposed that accretion of Gondwanan fragments with Laurasia, notably India, brought successive groups of acrodontans to the Asian region, while other acrodontans (such as Madagascan chamaeleons, Australian agamids) differentiated in isolated Gondwanan fragments. Also see Schulte *et al.* (2003).

If the suggested mechanism of Macey *et al.* (2000) is correct, then the times of origins of these lizard families are pushed back; much of the accretion of Gondwanan plate fragments (terranes) was complete by the early Jurassic (Metcalfe, 1998). Unfortunately for this thesis, their mitochondrial sequence data are not ideal for resolving either the relationships between the major lineages, nor the times of origin, even in relative terms. This is because the mitochondrial gene fragment being used is relatively rapidly evolving so that mutations begin to saturate after 10 million years (Moritz *et al.*, 1987), with the relationship between divergence in sequence and time levelling off as a consequence. This means that the data are unlikely to resolve Mesozoic or early Tertiary divergences (>50 m.y.a), yet these authors relate some divergences to vicariant events that are Jurassic or earlier. The phylogenetic trees obtained by Macey *et al.* (2000) are statistically robust with respect to the existence of major lineages, such as the Australian amphibolurines; but are poorly supported with respect to deeper branching order among lineages.

Honda *et al.* (2000b) also used mitochondrial sequence data to probe acrodont lizard relationships,

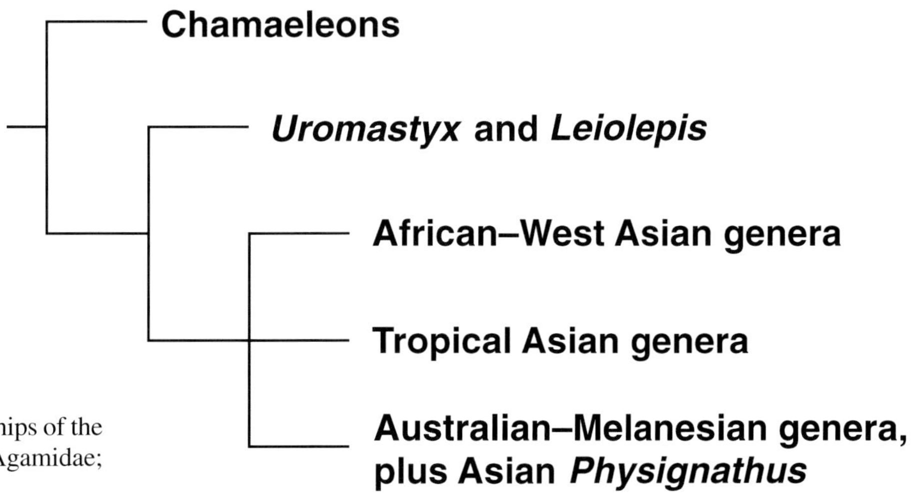

Figure 6. Phylogenetic relationships of the major lineages within the Agamidae; from Honda *et al.* (2000b).

using the more slowly evolving 12S and 16S ribosomal RNA genes. They concluded that chamaeleons are the sister lineage of all agamids, thus supporting the continued recognition of the two traditional families. They also found the same major lineages within the agamids as Macey *et al.* (2000), including a distinctive Australian lineage which included the Asian *P. cocincinus* (Figure 6). Honda *et al.* have better support for some of the basal branches within the agamids, including a strong sister-group relationship between South-east Asian agamids and those from Africa plus Western Asia. This cluster then has uncertain relationships to two other major assemblages, the Australian taxa plus *P. cocincinus* and the leiolepidines (*Leiolepis* and *Uromastyx*). Honda *et al.* (2000b) supported a northern origin for agamids, implying that the Australian radiation is the result of dispersal southwards across the Indonesian (or rather, proto-Indonesian) region.

GEKKOTA

Relationships among the Australian gekkonids (King and Horner, 1993), including the pygopod legless lizards (Shea, 1993), have been examined by Kluge in a series of papers. Kluge (1967) established a subfamilial scheme for geckos that is still in use, and revised the Australian endemic subfamily Diplodactylinae. Diplodactylines were regarded as falling into two tribes, the Diplodactylini (entirely Australian) and Carphodactylini, more ecologically and morphologically diverse with radiations in New Caledonia and New Zealand as well as Australia. Kluge (1974, 1976) then revised the pygopodid lizards and produced a synthesis and review of relationships among gekkonids (Kluge, 1987). In this last study, Kluge concluded that pygopods were nested within the Gekkonidae, close to or even within the Australian Diplodactylinae. Bauer (1990) revised Kluge's Carphodactylini, although without testing its monophyly with respect to the Diplodactylini.

The relationships among gekkonine geckos have proven to be more difficult to resolve. Kluge (1987) provided a general phylogenetic framework for this large subfamily, but this left the relationships of several lineages to one another unresolved. Within Australia little phylogenetic work has been carried out among the Australian genera or between these genera and their possible non-Australian relatives. Bauer *et al.* (1997) reviewed the relationships of the leaf-toed geckos ('*Phyllodactylus*') and confirmed the generic distinctiveness of members from the Australian region, recognizing them as a separate genus *Christinus*. The relationships of *Christinus* appear to be with the geographically remote African and Mediterranean species formerly placed in *Phyllodactylus*. Data are needed regarding the nearest relative of *Christinus* as well as the timing of the split between this genus and our other gekkonines

Molecular data on diplodactyline gekkonid relationships were presented by Donnellan *et al.* (1999) (Figure 7). This relied mainly on sequence data from the nuclear regulatory gene, *c-mos*, which has emerged as a useful phylogenetic tool for taxa diverging during the past 100 million years (Graybeal, 1994; Saint *et al.*, 1998). Donnellan *et al.* found good support for the origin of pygopods from Australian diplodactyline geckos; they also suggested that the Australian padless genera (such as *Carphodactylus*, *Phyllurus* and *Nephrurus*) were at the base of the diplodactyline radiation. The long-established subdivision of the diplodactylines into two tribes, Carphodactylini and Diplodactylini, was not strongly supported. Instead, the best trees suggested that the 'Carphodactylini' was a paraphyletic assemblage, and that the 'Diplodactylini' was therefore just one of several distinctive lineages within the Diplodactylinae.

King's (1987a,b) models of karyotypic evolution within the Diplodactylinae and Pygopodidae were not supported by the trees generated from sequence data. The presence of distinctive diplodactyline radiations on New Caledonia and in New Zealand implies the presence of diplodactylines in Australia during the late Cretaceous (at least), and pushes the divergence of the diplodactyline and gekkonine geckos even further back in time.

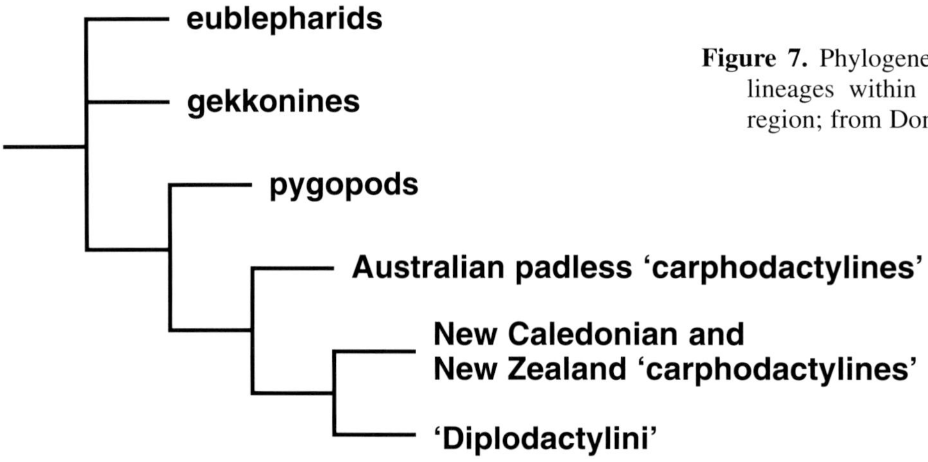

Figure 7. Phylogenetic relationships of the major lineages within Gekkota from the Australian region; from Donnellan *et al.* (1999).

SCINCOMORPHA

The skinks are the largest group of Australian squamates (Hutchinson, 1993), the approximate total of 400 species outnumbering all other families put together; see the generic keys and species list in the next chapter (Couper *et al*., 2006). The currently accepted subfamilial classification of the Scincidae is that proposed by Greer (1970). Under this classification the Scincidae is made up of four subfamilies, distinguished from each other on the basis of a combination of cranial osteology and scalation characteristics. The Lygosominae, the most diverse and geographically widespread skink subfamily, contains over 80% of the living species. The Scincinae includes 29 genera distributed throughout east and south central Asia. The Acontinae and Feylininae occur in Africa and include three and two genera respectively. Very little work following that of Greer has addressed the relationships between the scincid subfamilies. Molecular data are starting to be accumulated to relationships among scincomorphs and among squamates, with recent data from the nuclear gene *c-mos* presented by Saint *et al.* (1998) and Harris *et al.* (2001).

The lack of an independent, robust phylogenetic framework makes it difficult to frame hypotheses about scincid origins. Biogeographical considerations were an integral part of Greer's (1970) division of the Scincidae; Greer considered the Scincinae to be the living ancestors to the other subfamilies and hypothesized that the African distribution of the Feylininae and Acontinae is the result of separate origins of each subfamily from African scincines. He went on to suggest that the contemporary distribution of the Lygosominae and Scincinae in the Pacific is the result of lygosomines originating from east Asian Scincinae, dispersing into the scincine-free Australian region, diversifying there and radiating back to replace the Scincinae throughout south and east Asia. What is needed to confirm Greer's hypothesis is an explicit phylogeny showing that (1) the nearest relatives of lygosomines are Asian scincines, implying an Asian origin for lygosomines, and (2) the most basal lygosomines are Australian, implying that lygosomines radiated from Australia northwards.

The Lygosominae is the only skink subfamily to occur in Australia and has been the focus of significant taxonomic and systematic study (Couper *et al.,* 2006). The Australian Lygosominae have been informally subdivided into three lineages, termed the *Eugongylus*, *Egernia* and *Sphenomorphus* groups (Greer, 1979). However the monophyly of some of these groups and even their placement within the Lygosominae are still open questions. The remaining lygosomines (all of which are non-Australian) have been termed for convenience the *Mabuya* group (Hutchinson, 1981). Although this name has never been thought to define a monophyletic group, genera of the Australian *Egernia* group have been included in it by some authors (Greer, 1989)

The relationships of the characteristically Australian *Egernia* Group (*Egernia*, *Tiliqua*, *Cyclodomorphus* and *Corucia*) are central to the current uncertainty regarding skink evolution. Greer (1979) and Hutchinson (1981) suggested a closer relationship of the *Egernia* group to the *Eugongylus* group, but Rawlinson (1974) had queried whether *Egernia* and its relatives were even lygosomines. Recently, Honda *et al.* (2000a) have presented sequence data (12S and 16Sr RNA) bearing on broad relationships among the Lygosominae (see Figure 8). They emphasized the '*Mabuya*' group in their sampling, and confirmed that the *Egernia* group is a distinct lineage in its own right and that the '*Mabuya*' group comprises at least two monophyletic lineages. This study confirms the existence of three lygosomine lineages in Australia, and suggests the existence of two other lineages not found here; these can be termed the *Mabuya* Group (now restricted to *Mabuya* and its immediate relatives) and a *Lygosoma* Group.

Honda *et al.* (2000a) have only weak support for the branching order of their tree, and no strong consideration of time-frames or alternative origins of any of the lineages were presented. The mitochondrial data support Greer (1979) in so far as the *Sphenomorphus* Group is the sister lineage of the *Eugongylus* and *Egernia*

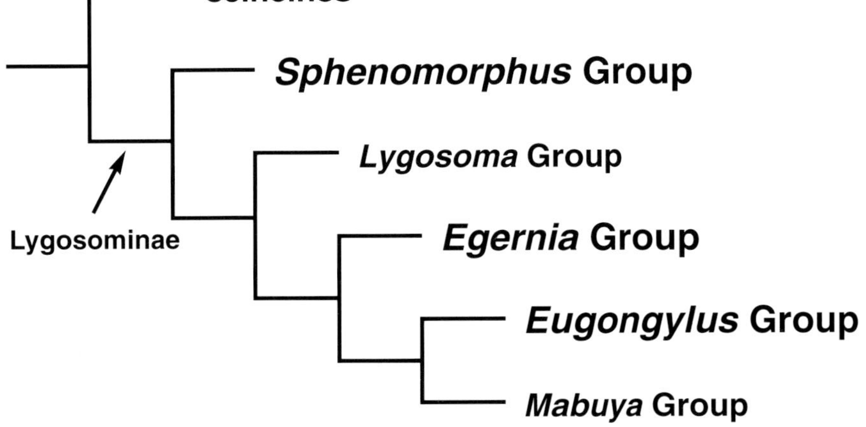

Figure 8. Phylogenetic relationships of the major lineages of lygosomine skinks; from Honda *et al.* (2000a). Lineages represented in Australia are in larger typeface.

Publications, No. 81, pp.1–65.

Fuller, S., Baverstock, P. and King, D. 1998. Biogeographic origins of goannas (Varanidae): a molecular perspective. *Molecular Phylogenetics and Evolution,* Vol 9, pp.294–307.

Gao, K. and Fox, R.C. 1996. Taxonomy and evolution of late Cretaceous lizards (Reptilia: Squamata) from western Canada. *Bulletin of Carnegie Museum of Natural History,* No. 33, pp.1–107.

Gao, K. and Hou, L. 1995. Iguanians from the upper Cretaceous Djadochta Formation, Gobi Desert, China. *Journal of Vertebrate Paleontology,* Vol. 15, pp.57–78.

Gao, K. and Norell, M.A. 1998. Taxonomic revision of *Carusia* (Reptilia: Squamata) from the late Cretaceous of the Gobi Desert and phylogenetic relationships of anguimorphan lizards. *American Museum Novitates,* No. 3230, pp.1–51.

Gao, K. and Norell, M.A. 2000. Taxonomic composition and systematics of late Cretaceous lizard assemblages from Ukhaa Tolgod and adjacent localities, Mongolian Gobi Desert. *Bulletin of the American Museum of Natural History,* Vol. 249, pp.1–118.

Gilmore, C.W. 1928. The fossil lizards of North America. *Memoirs of the National Academy of Sciences,* Vol. 22, pp.1–210.

Graybeal, A. 1994. Evaluating the phylogenetic utility of genes: a search for genes informative about deep divergences among vertebrates. *Systematic Biology,* Vol. 43, pp.174–193.

Greer, A.E. 1970. A subfamilial classification of scincid lizards. *Bulletin of the Museum of Comparative Zoology, Harvard,* Vol. 139, pp.151–184.

Greer, A.E. 1979. A phylogenetic subdivision of Australian skinks. *Records of the Australian Museum,* Vol. 32, pp.339–371.

Greer, A.E. 1989. *The Biology and Evolution of Australian Lizards.* Surrey Beatty and Sons, Sydney. 264pp.

Greer, A.E. and Cogger, H.G. 1985. Systematics of the reduce-limbed and limbless skinks currently assigned to the genus *Anomalopus* (Lacertilia: Scincidae). *Records of the Australian Museum,* Vol. 37, pp.11–54.

Harris, D.J., Marshall, J.C. and Crandall, K.A. 2001. Squamate relationships based on C-*mos* nuclear DNA sequences: increased taxon sampling improves bootstrap support. *Amphibia-Reptilia,* Vol. 22, pp.235–242.

Hecht, M.K. and Hecht, B.M. 1984. A new lizard from Jurassic deposits of Middle Asia. *Paleontological Journal,* Vol. 18, pp.133–136.

Hocknull, S.A. 2000. Remains of an Eocene skink from Queensland. *Alcheringa,* Vol. 24, pp.63–64.

Hoffstetter, R. 1967. Coup d'oeil sur les sauriens (=lacertiliens) des couches de Purbeck (Jurassique Supérieur d'Angleterre). Pp.349–371. In *Problèmes Actuels de Palontologie (Evolution des Vertébrés).* Colloques Internationaux du Centre National de la Recherche Scientifique, Paris.

Honda, M., Ota, H., Kobayashi, M., Nabhitabhata, J., Yong, H-S. and Hikida, T. 2000a. Phylogenetic relationships, character evolution and biogeography of the subfamily Lygosominae (Reptilia: Scincidae) inferred from mitochondrial DNA sequences. *Molecular Phylogenetics and Evolution,* Vol. 15, pp.452–461.

Honda, M., Ota, H., Kobayashi, M., Nabhitabhata, J., Yong, H-S, Sengoku, S. and Hikida, T. 2000b. Phylogenetic relationships of the family Agamidae (Reptilia: Iguania) inferred from mitochondrial DNA sequences. *Zoological Science,* Vol. 17, pp.527–537.

Hutchinson, M.N. 1981. The systematic relationships of the genera *Egernia* and *Tiliqua* (Lacertilia: Scincidae). A review and immunological reassessment. Pp.176–193. In Banks, C.B. and Martin, A.A. (eds). *Proceedings of the Melbourne Herpetological Symposium.* Zoological Board of Victoria, Melbourne.

Hutchinson, M.N. 1992. Origins of the Australian scincid lizards: a preliminary report on the skinks of Riversleigh. *The Beagle (Records of the Northern Territory Museum of Arts and Sciences),* Vol. 9, pp.61–70.

Hutchinson, M.N. 1993. Scincidae. Pp.261–279. In Glasby, C. J., Ross, G. J. B and Beesley, P. L. (eds). *Fauna of Australia. Vol. 2A: Amphibia and Reptilia.* Australian Government Publishing Service, Canberra. 439pp.

Hutchinson, M.N. 1997. The first fossil pygopod (Squamata: Gekkota), and a review of mandibular variation in living species. *Memoirs of the Queensland Museum,* Vol. 41, pp.355–366.

Hutchinson, M.N., Milne, T. and Croft, T. 1994. Redescription and ecological notes on the pygmy bluetongue, *Tiliqua adelaidensis* (Squamata: Scincidae). *Transactions of the Royal Society of South Australia,* Vol. 118, pp.216–227.

Iordansky, N.N. 1990. Evolution of cranial kinesis in lower tetrapods. *Netherlands Journal of Zoology,* Vol. 40, pp.32–54.

King, D. and Green, B. 1993. Varanidae. Pp.253–260. In Glasby, C.J., Ross, G.J.B. and Beesley, P.L. (eds). *Fauna of Australia. Vol. 2A: Amphibia and Reptilia.* Australian Government Publishing Service, Canberra. 439pp.

King, M. 1987a. Chromosomal evolution in the Diplodactylinae (Gekkonidae: Reptilia). I. Evolutionary relationships and patterns of change. *Australian Journal of Zoology,* Vol. 35, pp.507–531.

King, M. 1987b. Monophyleticism and polyphyleticism in the Gekkonidae: a chromosomal perspective. *Australian Journal of Zoology,* Vol. 35, pp.641–654.

King, M. and Horner, P. 1993. Gekkonidae. Pp.221–233. In Glasby, C.J., Ross, G.J.B. and Beesley, P.L. (eds). *Fauna of Australia. Vol. 2A: Amphibia and Reptilia.* Australian Government Publishing Service, Canberra. 439pp.

Kluge, A.G. 1967. Higher taxonomic categories of gekkonid lizards and their evolution. *Bulletin of the American Museum of Natural History,* Vol. 135, pp.1–60.

Kluge, A.G. 1974. A taxonomic revision of the lizard family Pygopodidae. *Miscellaneous Publications of the Museum of Zoology, University of Michigan,* No. 147, pp.1–221.

Kluge, A.G. 1976. Phylogenetic relationships in the lizard family Pygopodidae: an evaluation of theory, methods and data. *Miscellaneous Publications of the Museum of Zoology, University of Michigan,* No. 152, pp.1–72.

Kluge, A.G. 1987. Cladistic relationships in the Gekkonoidea (Squamata, Sauria). *Miscellaneous Publications of the Museum of Zoology, University of Michigan,* No.173, pp.1–54.

Lang, M. 1991. Generic relationships within Cordyliformes (Reptilia: Squamata). *Bulletin de l'Institut Royal des*

Sciences Naturelles de Belgique, Biologie, Vol. 61, pp.121–188.

Lee, M.S.Y. 1998. Convergent evolution and character correlation in burrowing reptiles: towards a resolution of squamate relationships. *Biological Journal of the Linnean Society,* Vol. 65, pp.369–453.

Lee, M.S.Y. 2000. Soft anatomy, diffuse homoplasy, and the affinities of lizards and snakes. *Zoologica Scripta,* Vol. 29, pp.101–130.

Lee, M.S.Y. and Caldwell, M.W. 1998. Anatomy and relationships of *Pachyrhachis problematicus*, a primitive snake with hindlimbs. *Philosophical Transactions of the Royal Society of London,* Series B, Vol. 353, pp.1521–1552.

Li, J. 1985. A new lizard from late Jurassic of Subei, Gansu. *Vertebrata Palasiatica,* Vol. 23, pp.13–18.

Macey, J.R., Larson, A., Ananjeva, N.B., Fang, Z. and Papenfuss, T.J. 1997. Two novel gene orders and the role of light stranded replication in rearrangement of the vertebrate mitochondrial genome. *Molecular Biology and Evolution,* Vol. 14, pp.91–104

Macey, J.R., Schulte, J.A., Larson, A., Ananjeva, N., Wang, Y., Pethiyagoda, R., Rastegar-Pouyani, N. and Papenfuss, T.J. 2000. Evaluating trans-Tethys migration: an example using acrodont lizard phylogenetics. *Systematic Biology,* Vol. 49, pp.233–256.

Mateer, N. 1982. Osteology of the Jurassic lizard *Ardeosaurus brevipes* (Meyer). *Palaeontology,* Vol. 25, pp.461–469.

McDowell, S.B. and Bogert, C.M. 1954. The systematic position of *Lanthanotus* and the affinities of the anguimorphan lizards. *Bulletin of the American Museum of Natural History,* Vol. 105, pp.1–142.

Mertens, R. 1942. Die Familie der Warane (Varanidae). Dritter Teil: Taxonomie. *Abhandlungen der Senckenberg Naturforschungs Gessellschaft,* No. 466, pp.235–391.

Metcalfe, I. 1998. Palaeozoic and Mesozoic geological evolution of the SE Asian region: multidisciplinary constraints and implications for biogeography. Pp.25–41. In Hall, R. and Holloway, J.D. (eds). *Biogeography and Geological Evolution of SE Asia.* Backhuys Publishers, Leiden. 417pp.

Moffat, L.A. 1973. The concept of primitiveness and its bearing on the phylogenetic classification of the Gekkota. *Proceedings of the Linnean Society of New South Wales,* Vol. 97, pp.275–301.

Moody, S.M. 1980. *Phylogenetic and historical biogeographical relationships of the genera in the family Agamidae.* Ph. D. thesis (unpublished), University of Michigan. 373pp.

Moritz, C., Dowling, T.E. and Brown, W.M. 1987. Evolution of animal mitochondrial DNA: relevance for population biology and systematics. *Annual Review of Ecology and Systematics,* Vol. 18, pp.269–292.

Norell, M.A. and Gao, K. 1997. Braincase and phylogenetic relationships of *Estesia mongoliensis* from the late Cretaceous of the Gobi Desert and the recognition of a new clade of lizards. *American Museum Novitates,* No. 3211, pp.1–25.

Presch, W. 1988. Phylogenetic relationships of the Scincomorpha. Pp.471–492. In Estes, R. and Pregill, G. (eds). *Phylogenetic Relationships of the Lizard Families. Essays Commemorating Charles L. Camp.* Stanford University Press, Stanford.

Rawlinson, P.A. 1974. Biogeography and ecology of the reptiles of Tasmania and the Bass Strait area. Pp.291–338. In Williams, W.D. (ed). *Biogeography and Ecology in Tasmania.* W. Junk Publishers, The Hague. 498pp.

Reynoso, V-H. 1998. *Huehuecuetzpalli mixtecus* gen. et sp. nov: a basal squamate (Reptilia) from the early Cretaceous of Tepexi de Rodriguez, Central Mexico. *Philosophical Transactions of the Royal Society of London,* Series B, Vol. 353, pp.477–500.

Richter, A. 1994. Lacertilia aus der Unteren Kreide von Uña und Galve (Spanien) und Anoual (Marokko). *Berliner Geowissenschaftliche Abhandlungen,* Vol. 14, pp.1–147.

Rieppel, O. 1980. The phylogeny of anguinomorph lizards. *Denkschriften der Schweizerischen Naturforschenden Gesellschaft,* Vol. 94, pp.1–86.

Rieppel, O. 1984a. The cranial morphology of the fossorial lizard genus *Dibamus* with a consideration of its phylogenetic relationships. *Journal of Zoology, London,* Vol. 204, pp.289–327.

Rieppel, O. 1984b. The structure of the skull and jaw adductor musculature in the Gekkota, with comments on the phylogenetic relationships of the Xantusiidae (Reptilia: Lacertilia). *Zoological Journal of the Linnean Society*, Vol. 82, pp.291–318.

Rieppel, O. 1994. The Lepidosauromorpha: an overview with special emphasis on the Squamata. Pp.23–37. In Fraser, N.C. and Sues, H.-D. (eds). *In the Shadow of the Dinosaurs: Early Mesozoic Tetrapods.* Cambridge University Press, Cambridge. 435pp.

Robinson, P.L. 1976. How *Sphenodon* and *Uromastyx* grow their teeth and use them. Pp.43–64. In Bellairs, A.d'A and Cox, C.B. (eds). *Morphology and Biology of Reptiles.* Linnean Society Symposium Series No. 3. Academic Press, London.

Rowe, T., Cifelli, R.L., Lehman, T.M. and Weil, A. 1992. The Campanian Terlingua Local Fauna, with a summary of other vertebrates from the Aguja Formation, Trans-Pecos Texas. *Journal of Vertebrate Paleontology,* Vol. 12, pp.472–493.

Saint, K.M., Austin, C.C., Donnellan, S.C. and Hutchinson, M.N. 1998. *C-mos*, a nuclear marker useful for squamate phylogenetic analysis. *Molecular Phylogenetics and Evolution,* Vol. 10, pp.259–263.

Scanlon, J.D. 2006. Origins and radiations of snakes in Australasia. Pp.309–330. In Merrick J.R., Archer, M., Hickey, G.M. and Lee, M.S.Y. (eds). *Evolution and Biogeography of Australasian Vertebrates.* Auscipub Pty. Ltd., Sydney. 942pp.

Schulte, J.A., Melville, J. and Larson, A. 2003. Molecular phylogenetic evidence for ancient divergence of lizard taxa on either side of Wallace's line. *Proceedings: Biological Sciences*, Vol. 270, pp.597–603.

Shea, G.M. 1993. Pygopodidae. Pp.234–239. In Glasby, C.J., Ross, G.J.B. and Beesley, P.L. (eds). *Fauna of Australia. Vol. 2A: Amphibia and Reptilia.* Australian Government Publishing Service, Canberra. 439pp.

Shea, G.M. and Hutchinson, M.N. 1992. A new species of *Tiliqua* (Lacertilia: Scincidae) from the Miocene of Riversleigh, Queensland. *Memoirs of the Queensland Museum,* Vol. 32, pp.303–310.

Stephenson, N.G. 1960. The comparative osteology of Australian geckoes and its bearing on their morphological status. *Zoological Journal of the Linnean Society,* Vol. 44, pp.278–299.

Sulimski, A. 1975. Macrocephalosauridae and Polyglyphanodontidae (Sauria) from the late Cretaceous of Mongolia. *Palaeontologia Polonica,* Vol. 33, pp.25–102.

Underwood, G. 1957. On the lizards of the family Pygopodidae, a contribution to the morphology and phylogeny of the Squamata. *Journal of Morphology,* Vol. 100, pp.207–268.

Vicario, S., Caccone, A. and Gauthier, J. 2003. Xantusiid "night" lizards: a puzzling phylogenetic problem revisited using likelihood-based Bayesian methods on mtDNA sequences. *Molecular Phylogenetics and Evolution*, Vol. 26, pp.243–261.

Witten, G.J. 1993. Agamidae. Pp.240–252. In Glasby, C.J., Ross, G.J.B. and Beesley, P.L. (eds). *Fauna of Australia. Vol. 2A: Amphibia and Reptilia.* Australian Government Publishing Service, Canberra. 439pp.

Wu, X., Brinkman, D.B. and Russell, A.P. 1996. *Sineoamphisbaenia hexatabularis*, an amphisbaenian (Diapsida: Squamata) from the upper Cretaceous redbeds at Bayan Mandahu (Inner Mongolia, People's Republic of China), and comments on the phylogenetic relationships of the Amphisbaenia. *Canadian Journal of Earth Sciences,* Vol. 33, pp.541–577.

Chapter 20
The Genera of Skinks (Family Scincidae) of Australia and its Island Territories: Diversity, Distribution and Identification

P.J. Couper, J.A. Covacevich, A.P. Amey and A.M. Baker
Queensland Museum, Box 3300, South Bank, Qld. 4101 Australia

BACKGROUND AND RECENT DEVELOPMENTS

The history of discovery and description of Australia's skinks has been documented by Hutchinson (1993) and major reviews of this family in Australia have been published recently (Greer, 1989; Hutchinson, 1993). Comprehensive guides and a catalogue to identification of the Australian species are also available (Cogger, 1975, 1979, 1983, 1986, 1992, 1996, 2000; Cogger *et al.*, 1983; Wilson and Knowles, 1988; Ehmann, 1992; Wilson and Swan, 2003). The 37 genera recognized from Australia and its island territories in this work are, largely, endemic to Australia. *Tiliqua, Lygisaurus, Carlia, Emoia, Ctenotus, Egernia, Glaphyromorphus, Eugongylus* and *Cryptoblepharus* occur in Papua New Guinea and, in a few cases, in Indonesia and other islands of the Pacific; *Oligosoma,* of Lord Howe Is. and Philip and Nepean Islands (near Norfolk Is.) is essentially a New Zealand genus. Greer's (1989) synthesis of the Australian skinks (all members of the subfamily Lygosominae) recognizes three groups (Table 1).

The Australian skinks are fairly well known. This notwithstanding, the words of G.A. Boulenger, one of herpetology's early 'giants', remain true. Boulenger (1887) compiled the first review (in English), to include the Australian skinks. This work remains a valuable guide to their identification. He wrote … 'I have met with great difficulty in arranging the genera of this family. The majority of characters hitherto employed for the distinction of genera, such as the degree of development of the limbs, the presence or absence of a transparent disk in the lower eyelid, the presence or absence of keels on scales … are in many cases not even of specific value …'. Today, despite the availability of comprehensive guides and two major generic reviews for the family world-wide (Smith, 1937; Mittleman, 1952), some generic boundaries within the Australian skinks remain ill-defined and blurred by exceptions to the rule. For example, *Glaphyromorphus*, despite the definitions of Wells and Wellington (1984) and Cogger (2000) and wide acceptance, remains best defined by what is not assigned to it, rather than by a clear-cut or unique combination of characters. Of *Glaphyromorphus,* Greer (1989) notes … 'There is an assemblage of 13 species which lack traits distinctive enough to ally them either with each other or with other genera in the *Sphenomorphus* group…'

Other examples further emphasize some of the problems associated with both devising identification keys and using them. Amongst the 5/5 (i.e. those having 5 fingers and 5 toes) genera, separating species formerly assigned to '*Leiolopisma*', and choosing new generic names for these, has presented considerable difficulty. Detailed revisionary work on species complements for many genera separated from *Leiolopisma* has yet to be completed. Until this is done, both clear-cut generic boundaries and correct names remain elusive, because decisions cannot be based simply on either chronology or precedence. The frequently consulted, standard texts differ in nomenclature, concepts and generic allocations: Wilson and Knowles (1988) maintain *Leiolopisma;* Greer (1989) largely follows the morphology-based review of Wells and Wellington (1985) and incorporates some concepts of Fuhn (1967); Cogger (2000) combines the names of Hutchinson *et al.* (1990) with those of Fuhn (1967); Wilson and Swan (2003) largely follow Cogger (2000), but substitute some of the generic allocations of Hutchison *et al.* (1990) for those of Wells and Wellington (1985) and also incorporate the findings of Stuart-Fox *et al.* (2002).

But this case is not straightforward because at least three names proposed by Wells and Wellington (1985)—*Carinascincus, Litotescincus, Tasmascincus*—are available for *Niveoscincus* (Hutchinson *et al.*, 1990; Shea, 2000, personal communication). In using some of the names proposed by Wells and Wellington (1985), with *Pseudemoia* Fuhn 1967, to 'cover' species previously assigned to *Leiolopisma*, we do so to adhere to the Principle of Priority (ICZN, 1999; Wells and Wellington, 1999). We treat *Acritoscincus, Carinascincus, Harrisoniascincus* and *Techmarscincus* as senior synonyms of *Bassiana, Niveoscincus, Cautula* and *Bartleia* respectively of Hutchinson *et al.* (1990). These genera, despite definitions by Fuhn (1967), Wells and Wellington (1985) and Hutchinson *et al.* (1990), remain hard to separate.

This problem results largely from the fact that, while biochemical and skeletal distinctions between them are often strong and well-defined, distinguishing external characters are subtle, even elusive. To illustrate this, the distinction between the monotypic genera *Techmarscincus* Wells and Wellington, 1985 and *Harrisoniascincus* Wells and Wellington, 1985 serves well. *Techmarscincus* (as *Bartleia*) is said to be distinct in having supraciliaries as wide as long. For *Harrisoniascincus* (as *Cautula*), this character is not defined, but is illustrated as longer than wide (Hutchinson *et al.,* 1990, p.541). The distinction appears to be clear, but

Table 1. Comparative summary of the nomenclature of Australian genera of skinks, since the first popular, general guide to their identification was published. Authors of the genera are listed below the generic history. Genera asterisked (*) are listed in the Appendix but not included in generic keys of Cogger (2000).

Worrell, 1963	**Couper, Covacevich, Amey and Baker, 2000**	**Cogger, 2000**
		***Egernia* Group**
Egernia	*Egernia*	*Egernia*
Trachydosaurus	*Tiliqua*	*Tiliqua*
Tiliqua		*Trachydosaurus*
	Cyclodomorphus	*Hemisphaeriodon*
		Cyclodomorphus
Lygosoma (Lygosoma)	*Lygosoma*	*Lygosoma*
		***Eugongylus* Group**
	Cryptoblepharus	*Cryptoblepharus*
Ablepharus	*Menetia*	*Menetia*
	Morethia	*Morethia*
	Proablepharus	*Proablepharus*
Emoia	*Emoia*	*Emoia*
	Acritoscincus	*Bassiana**
	Carlia	*Carlia*
	Carinascincus	*Niveoscincus**
	Eroticoscincus	*Eroticoscincus*
	Harrisoniascincus	*Cautula**
Lygosoma (Leiolopisma)	*Lampropholis*	*Lampropholis*
	Lygisaurus	*Lygisaurus*
	Nannoscincus	*Nannoscincus*
	Oligosoma	*Cyclodina*
	Pseudemoia	*Pseudemoia*
	Saproscincus	*Saproscincus*
	Techmarscincus	*Bartleia**
Riopa	*Eugongylus*	*Eugongylus*
		***Sphenomorphus* Group**
	Hemiergis	*Hemiergis*
	Saiphos	*Saiphos*
	Anomalopus	*Anomalopus*
	Ophioscincus	*Ophioscincus*
	Coeranoscincus	*Coeranoscincus*
	Glaphyromorphus	*Glaphyromorphus*
Rhodona	*Lerista*	*Lerista*
	Calyptotis	*Calyptotis*
	Ctenotus	*Ctenotus*
	Eremiascincus	*Eremiascincus*
Lygosoma (Sphenomorphus)	*Eulamprus*	*Eulamprus*
	Notoscincus	*Notoscincus*
	Coggeria	*Coggeria**
	Nangura	*Nangura**
Tropidophorus	*Gnypetoscincus*	*Gnypetoscincus*

is not. It is confounded by some specimens of *Harrisoniascincus zia* that have the '*Techmarscincus* state' (i.e. supraciliaries as wide as long, rather than longer than wide). However, when *Techmarscincus jigurru* and *Harrisoniascincus zia* are examined side by side, other, more superficial characters, like colour pattern, tail length and body form separate them easily. In our key, colour pattern has been used to separate them.

A major feature of *Ctenotus* is the presence of conspicuous, pointed, anterior ear lobules. This, again, would seem to be an easily defined and easily used character. However, one subspecies of *Ctenotus*, *C. essingtoni brevipes*, lacks ear lobules. *Saproscincus* is another genus with one 'outsider'; all *Saproscincus* species, except *S. tetradactylus,* have five fingers. The 4/5 digital formula of *S. tetradactylus* accounts for the presence of this genus in both keys below.

Carlia spp. and *Lygisaurus* spp. all have the 4/5 digital formula, but present another problem. The latter was described early (de Vis, 1884), but was treated as a synonym of the former for many years. Ingram and Covacevich (1988) resurrected *Lygisaurus* for what seemed to be a distinct 'cluster' of closely related small, brown, shiny, litter dwellers. A key character in the separation of these genera was the number of supradigital lamellae. Cogger (1993, 2000) has perpetuated the use of number of supradigital lamellae to distinguish these genera. Specimens of *Lygisaurus* spp. usually have fewer than 10 supradigital lamellae on the fourth toe, while those of *Carlia* spp. usually have 10 or more. We have found that this character is not invariably reliable. Occasional (2 of 55 individuals examined in Queensland Museum collections) specimens of *Lygisaurus* spp. have 10 supradigital lamellae.

Further, recent biochemical studies of *Carlia* spp. and *Lygisaurus* spp., as defined by Ingram and Covacevich (1988, 1989), have indicated another blurring of the boundaries between these genera (Stuart-Fox *et al.*, 2002). These facts notwithstanding, *Carlia* spp. and *Lygisaurus* spp. do seem to form easily recognized, morphological groups. Size, ornamentation of dorsal scales and (to a lesser degree) colour, are good distinguishing characters. *Carlia* spp. tend to be larger, with a maximum snout-to-vent length (max SVL) of 69 mm; whereas *Lygisaurus* spp. have a max SVL of 39 mm. *Carlia* spp. generally have keeled or carinate dorsal scales versus striate or smooth dorsal scales in *Lygisaurus* spp.; and *Carlia* spp. tend to be more colourful than *Lygisaurus* spp., which are usually essentially brown. *Menetia*, another of the 4/5 digital formulae genera, also appears twice in the key to genera with a digital formula of less than 5/5, to overcome another variation. Six out of nine *Menetia* species have very distinctive oblique supraoculars, a character long used to diagnose this genus. In the key, *M. koshlandae, M. sadlieri* and *M. timlowi* (the species lacking this character) are separated from *Lygisaurus* spp. by numbers of paravertebral scales.

The first popular guide to Australia's reptiles was that of Worrell (1963). In that work, nine genera comprised the family Scincidae in Australia. Now, as a result of resurrections of old genera from synonymy, descriptions of new genera to more properly reflect relationships or to accommodate totally new taxa, and name changes made for a variety of taxonomic or nomenclatural reasons, there are many more. Although 37 Australian genera are recognized in this work, 39 are recognized by Cogger (2000). Table 1 summarizes the nomenclatural histories of Australian skink genera between 1963 and June, 2003, and provides a list of authors of all generic names.

There is a strong correlation between the genera recognized in this work and those recognized by Cogger (2000). In using *Cyclodomorphus* and *Tiliqua* only, we follow Shea (1990) not Cogger (2000), who recognized *Hemisphaeriodon* and *Trachydosaurus* also. Cogger's (2000) most recent guide is very comprehensive and includes generic changes made subsequent to the 1992 and 1996 editions, but the generic key (pp.380–382) was not updated from Cogger (1992, 1996). Thus *Techmarscincus, Acritoscincus* and *Harrisoniascincus* Wells and Wellington, 1985, *Coggeria* Couper *et al.,* 1996, *Lygosoma* Hardwicke and Gray, 1827, *Oligosoma* O'Shaughnessy, 1874, *Nangura* Covacevich *et al.*, 1993 and *Carinascincus* Wells and Wellington, 1985 are not included. To supplement the work of Cogger (2000) and summarize other recent research we have:

(a) presented two new, comprehensive keys (one for skinks with a digital formula of 5/5, the second for those with reduced digit numbers);
(b) illustrated the continental and island ranges of the 37 Australian genera;
(c) listed the species of skinks currently known from Australia and its island territories (Table 2); and
(d) commented briefly on diversity and conservation of Australian skinks.

Since this chapter went to press, Stuart-Fox *et al.* (2002) have completed a major biochemical review of species assigned to *Carlia* and *Lygisaurus*. *Lygisaurus* has been re-synonymized with *Carlia* and one species, present-

Table 1 (opposite). Original sources: *Ablepharus* Fitzinger, 1824; *Acritoscincus* Wells and Wellington, 1985; *Anomalopus* Duméril and Duméril, 1851; *Bartleia* Hutchinson, Donnellan, Baverstock, Krieg, Simms and Burgin, 1990; *Bassiana* Hutchinson, Donnellan, Baverstock, Krieg, Simms and Burgin, 1990; *Calyptotis* De Vis, 1886; *Carinascincus* Wells and Wellington, 1985; *Carlia* Gray, 1845; *Cautula* Hutchinson, Donnellan, Baverstock, Krieg, Simms and Burgin, 1990; *Coeranoscincus* Wells and Wellington, 1984; *Coggeria* Couper, Covacevich, Marsterson and Shea, 1996; *Cryptoblepharus* Wiegmann, 1834; *Ctenotus* Storr, 1964; *Cyclodomorphus* Fitzinger, 1843; *Egernia* Gray, 1838; *Emoia* Gray, 1845; *Eremiascincus* Greer, 1979; *Eroticoscincus* Wells and Wellington, 1984; *Eugongylus* Fitzinger, 1843; *Eulamprus* Fitzinger, 1843; *Glaphyromorphus* Wells and Wellington, 1984; *Gnypetoscincus* Wells and Wellington, 1984; *Harrisoniascincus* Wells and Wellington, 1985; *Hemiergis* Wagler, 1830; *Hemisphaeriodon* Peters, 1867; *Lampropholis* Fitzinger, 1843; *Lerista* Bell, 1833; *Leiolopisma* Duméril and Bibron, 1837; *Lygisaurus* De Vis, 1884; *Lygosoma* Hardwicke and Gray, 1827; *Menetia* Gray, 1845; *Morethia* Gray, 1845; *Nangura* Covacevich, Couper and James, 1993; *Nannoscincus* Günther, 1872; *Niveoscincus* Hutchinson, Donnellan, Baverstock, Krieg, Simms and Burgin, 1990; *Notoscincus* Fuhn, 1969; *Oligosoma* Girard, 1857; *Ophioscincus* Peters, 1874; *Proablepharus* Fuhn, 1969; *Pseudemoia* Fuhn, 1967; *Rhodona* Gray, 1839; *Riopa* Gray, 1839; *Saiphos* Gray, 1831; *Saproscincus* Wells and Wellington, 1984; *Sphenomorphus* Fitzinger, 1843; *Techmarscincus* Wells and Wellington, 1985; *Tiliqua* Gray, 1825; *Trachydosaurus* Gray, 1825; *Tropidophorus* Duméril and Bibron, 1839.

ly assigned to *Menetia* (*M. timlowi*), should probably now either be transferred to *Carlia* or, with two congeners (*koshlandae* and *sadlieri*), be assigned to a new genus.

However, these changes have little impact on our keys. Stuart-Fox *et al.* (2002) reinforce the fact that most '*Lygisaurus*' species form a single clade within *Carlia.* As they are easily recognized, although not by synapomorphies, it is logical to retain '*Lygisaurus*' as a unit in the key. Given that Stuart-Fox *et al.* (2002) make no firm recommendation about the generic status of *timlowi*, *koshlandae* and *sadlieri*, their maintenance in *Menetia* also seems reasonable for the present.

GENERIC KEYS

The first key (to skinks with a digital formula of 5/5) includes, among others, all the large, conspicuous, frequently encountered skinks (e.g. *Tiliqua, Egernia* and *Ctenotus*). In the second key are all genera containing burrowing species, along with a host of small, rarely encountered and even more rarely captured species (e.g. *Anomalopus, Coggeria, Lerista* and *Eroticoscincus*). The keys have been devised for use with live or preserved whole specimens, and thus external characters only are used. However, some of these are not easily seen without the aid of a hand lens or microscope. Each key is made up of a series of couplets. A statement in each will lead the user to either a generic identification or to another couplet.

No key can ever be perfect or absolutely reliable for two prime reasons. First, new discoveries automatically render keys partly obsolete. Second, an occasional exception to a rule or a 'fuzzy' generic boundary will always confound the best efforts of devisers and users of keys. Thus, as a precaution, particularly when an identification is needed for more than passing interest, these keys should be used always in association with, or checked against, more detailed descriptions. Further, identifications can be checked against reference or research collections held in state museums, where accuracy is important.

KEY TO GENERA WITH A DIGITAL FORMULA OF 5/5

Letters in parentheses at the end of statements refer to adjacent figures illustrating features referred to in the couplet.

1. Parietal scales (sometimes fragmented) widely separated by interparietal **(a)** **2**
   Parietals forming a suture behind the interparietal, which may be fused to frontoparietals **(b)** **7**

2. Toe 4 longer than toe 3 **(c)** **3**
   Toe 3 and 4 subequal, or toe 3 slightly longer than toe 4 .... **5**

(a)

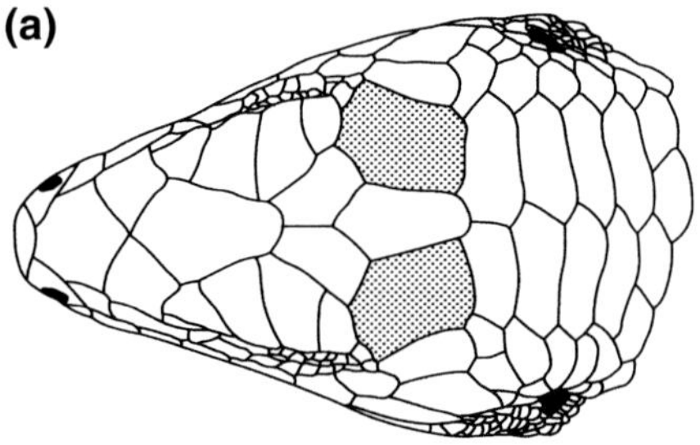

(b)

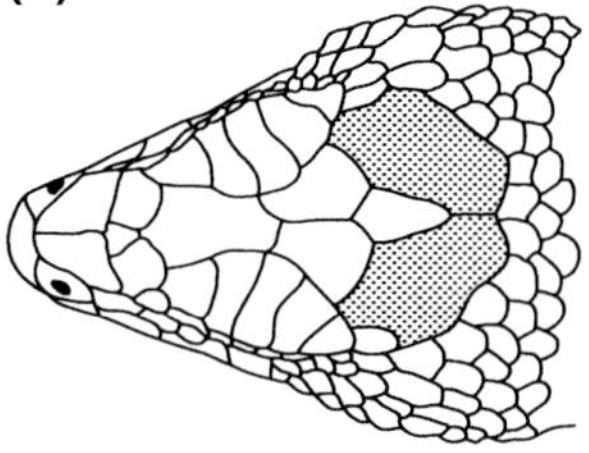

(c)

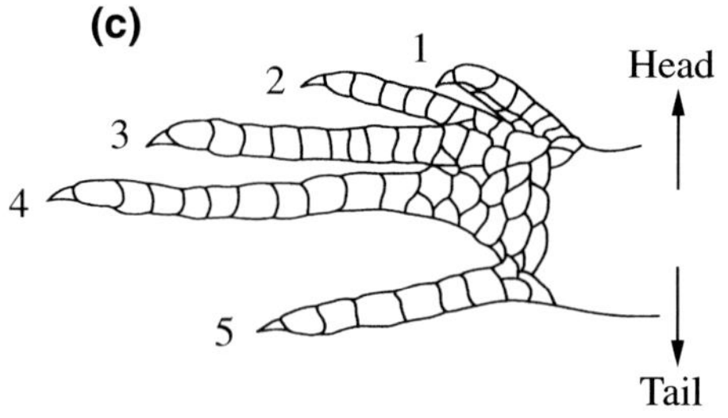

3. Anterior ear lobules present **(d)** .... *Egernia*
   Anterior ear lobules absent **4**

4. Dorsal scales strongly keeled .... *Nangura*
   Dorsal scales smooth .... *Eulamprus*

5. Post narial groove absent .... *Cyclodomorphus*
   Post narial groove present **(e)** **6**

6. Prefrontals in medial contact **(f)** .... *Tiliqua*
   Prefrontals separated .... *Cyclodomorphus*

(d)

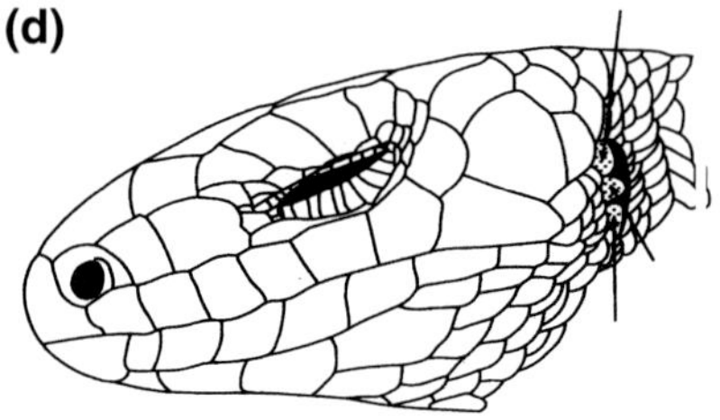

(e)

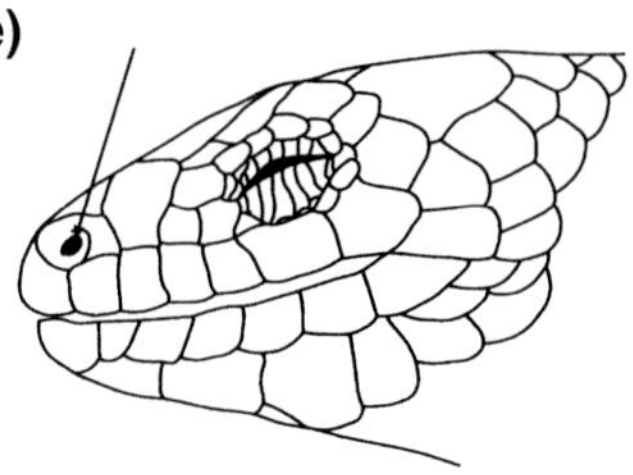

(f)

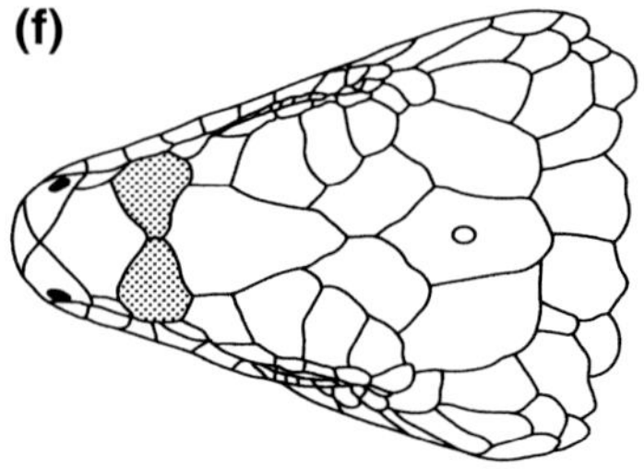

7. Dorsal and lateral scales strongly keeled .... *Gnypetoscincus*
Dorsal and lateral scales smooth, rugose or weakly keeled **8**

8. Medial pair of preanal scales overlapped by adjacent preanal scales **(g)** **9**
Medial pair of preanal scales overlap (sometimes completely) adjacent preanal scales **(h)** **26**

9. Lower eyelid immovable, the disc forming a transparent spectacle covering the eye **(i)** **10**
Lower eyelid movable **(j)** **12**

(g)

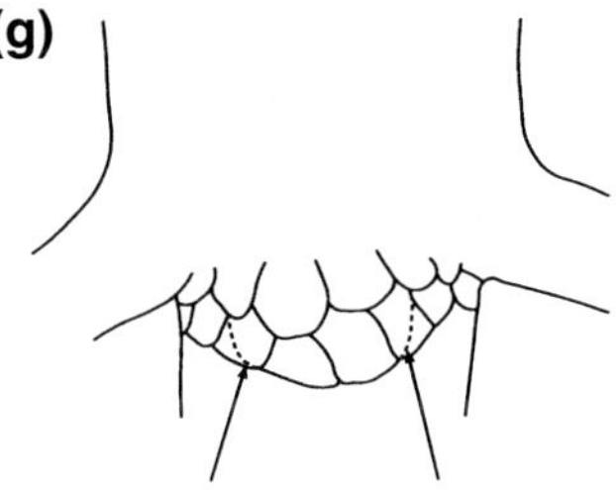

(h)

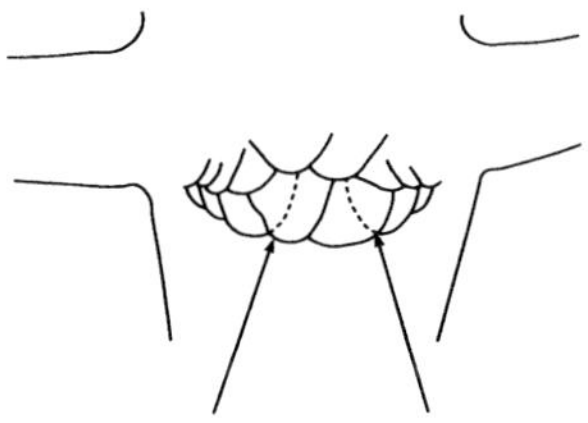

(i)

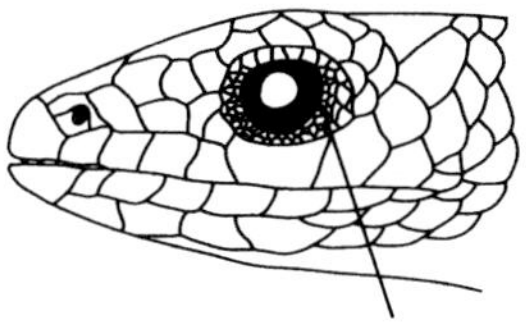

10. Frontoparietals fused to interparietal to form single shield **(k)** **11**
Frontoparietals not fused to interparietal, which is distinct **(l)** .... *Proablepharus*

11. Enlarged scales (3–4) present along upper edge of eye, immediately below supraciliaries **(m)** ... *Cryptoblepharus*
Uniformly small scales along upper edge of eye, immediately below supraciliaries .... *Morethia*

(j)

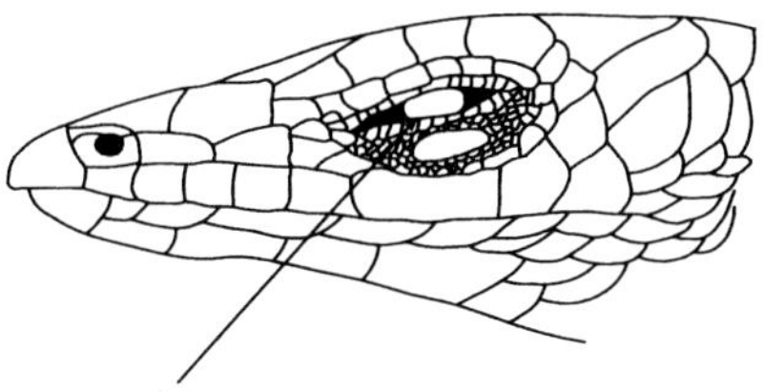

(k)

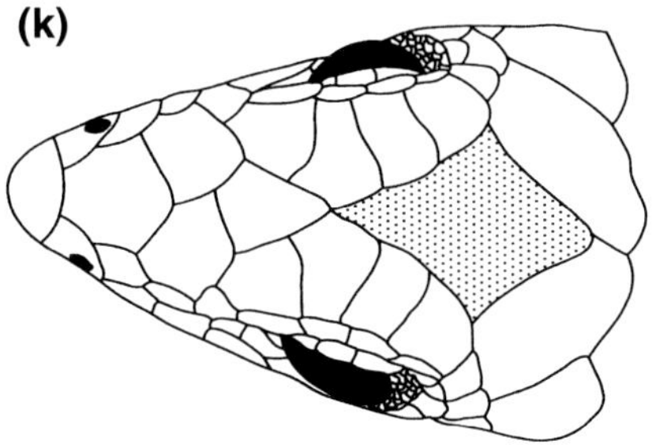

(l)

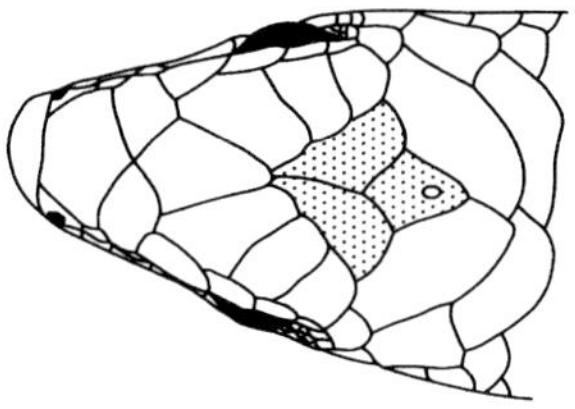

12. Frontoparietals fused to interparietal to form single shield (interparietal free or fused) **(k)** **13**
Frontoparietals paired and separate from interparietal **(l)** **17**

13. Supranasals present **(n)** **14**
Supranasals absent **15**

14. Subdigital lamellae under 4th toe < 30 **(o)** .... *Carinascincus*
Subdigital lamellae under 4th toe ≥ 32 .... *Emoia*

(m)

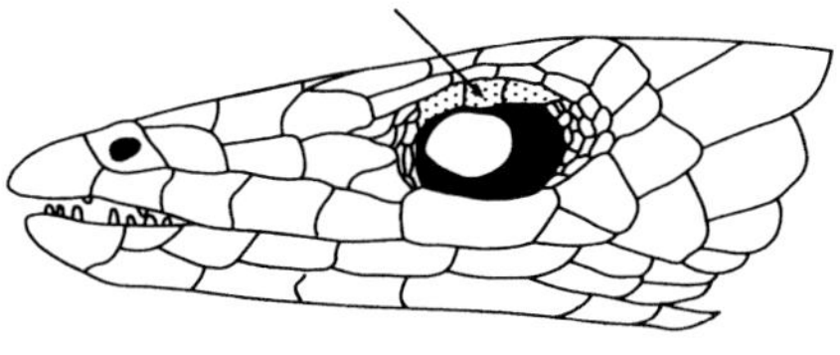

(n)

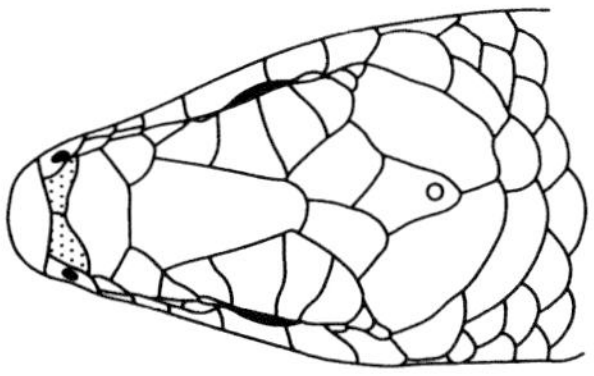

(o)

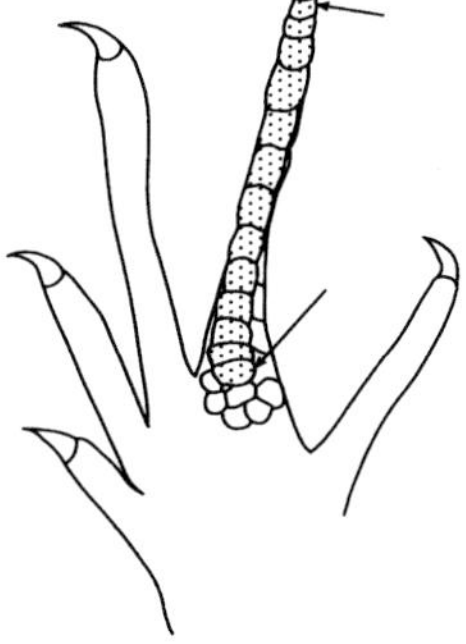

15. Suture between rostral and frontonasal about as wide as frontal scale **(p)** .... *Lampropholis*
Suture between rostral and frontonasal narrower than frontal scale **(q)** **16**

16. Supraciliaries (usually) 5 **(r)**; a broad, dark lateral stripe with sharply defined edges; throat flushed with pink/red in life .... *Acritoscincus*
Supraciliaries (usually) > 5; dark lateral stripe without sharply defined edges; throat not flushed with pink/red in life .... *Carinascincus*

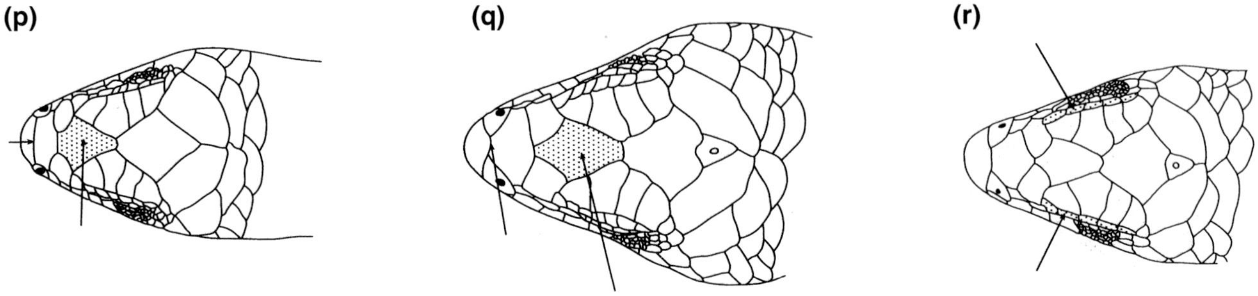

17. Supranasals present **(n)** **18**
Supranasals absent **20**

18. Anterior ear lobules present **(d)** .... *Eugongylus*
Anterior ear lobules absent **19**

19. Subdigital lamellae under 4th toe < 16 **(o)** .... *Lygosoma*
Subdigital lamellae under 4th toe > 16 .... *Pseudemoia*

20. Midbody scale rows > 35 .... *Oligosoma*
Midbody scale rows < 35 .... **21**

21. Prefrontals absent .... *Nannoscincus*
Prefrontals present **(f)** **22**

22. Posterior base of thigh with white spot **(s)** .... *Saproscincus*
Posterior base of thigh without white spot **23**

23. Mid-lateral line broken into a series of prominent, pale blotches .... *Techmarscincus*
Mid-lateral line continuous, or absent **24**

24. Subdigital lamellae smooth **(t)** .... *Pseudemoia*
Subdigital lamellae rough **(u)** **25**

25. Subdigital lamellae under 4th toe < 18 **(o)** .... *Harrisoniascincus*
Subdigital lamellae under 4th toe > 18 .... *Carinascincus*

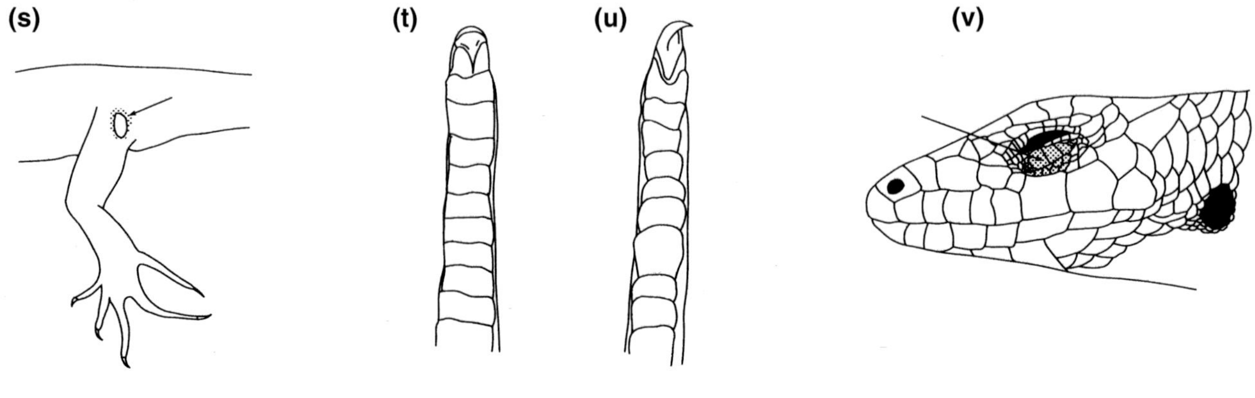

Plate 72. Another goanna (*Varanus glebopalma*) which is restricted to Australia. Photograph J. Cann.

Plate 73. The head of *Varanus prasinus*; this specimen is from New Guinea but the species also occurs in Australia. It has been proposed to assign this species to a subgenus *Euprepiosaurus*, on the basis of molecular data. Photograph J.Cann.

Plate 74. Rosenberg's monitor (*Varanus rosenbergii*) has been studied in detail. This specimen is a juvenile, but adults can grow to 1.0 m and weigh 2.0 kg. *V. rosenbergii* occurs along much of the southern coast of Australia and is very abundant on Kangaroo Island. This medium-sized goanna has a varied diet that includes: mammals, birds and reptiles (and their eggs) as well as invertebrates. Photograph J. Cann.

Plate 75. The Jacky lizard (*Amphibolurus muricatus*) from south-eastern Australia. The range extends from southern Queensland along the coast and ranges to western Victoria and into South Australia. Inhabits coastal heath, rocky ridges and dry sclerophyll forests; is an opportunistic carnivore—takes insects and other small invertebrates. Photograph S. Allen.

Plate 76. This skink (*Ctenotus colletti*), from the Port Hedland area of Western Australia, also occurs in arid western central parts of Australia. It inhabits red sand dunes and interdune areas dominated by spinifex. Several subspecies are recognized. Photograph B. Maryan.

Plate 77. Another skink (*Ctenotus vertebralis*) from northern Australia. This specimen is from Nourlangie Rock in the Northern Territory. The species occurs mainly in woodland habitats, where stony hills and escarpments meet the plain areas, in western Arnhem Land. Photograph B. Maryan.

Plate 78. A representative of a different genus of skinks (Family Scincidae). This specimen (*Lerista muelleri*) is from the Cloncurry area, Queensland, but the species has an extensive continental range from drier interiors of eastern states to central coastal regions of Western Australia. *L. muelleri* utilizes a variety of habitats from dry sclerophyll forests and woodlands to river floodplains and arid scrublands. Photograph B. Maryan.

26. Lower eyelid immovable, the disc forming a transparent spectacle covering the eye **(i)** .... *Notoscincus*
    Lower eyelid movable **(j)** **27**

27. Lower eyelid with transparent disc **(j)** **28**
    Lower eyelid scaly **(v)** **29**

28. Ear opening present .... *Lerista*
    Ear opening absent .... *Hemiergis*

29. Ear opening with anterior lobules **(d)** .... *Ctenotus*
    Ear opening without anterior lobules; or absent **30**

30. Scales on rump and base of tail keeled .... *Eremiascincus*
    Scales on rump and base of tail not keeled **31**

31. Toe 4 slightly longer than toe 3 .... *Calyptotis*
    Toe 4 markedly longer than toe 3 **(c)** **32**

32. Flanks with a sharp-edged, pale mid-lateral stripe .... *Ctenotus*
    Flanks without pale mid-lateral stripe **33**

33. Adpressed limbs overlapping **(w)** .... *Eulamprus*
    Adpressed limbs not overlapping .... *Glaphyromorphus*

(w)

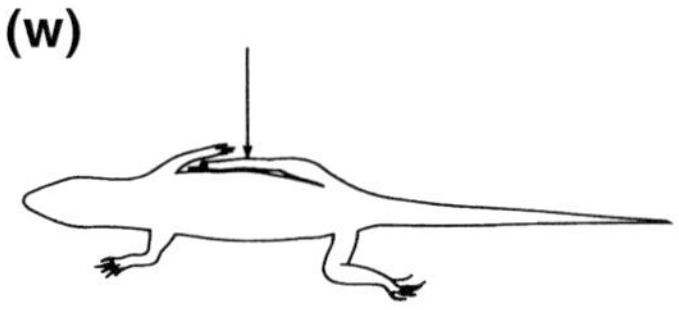

(x)

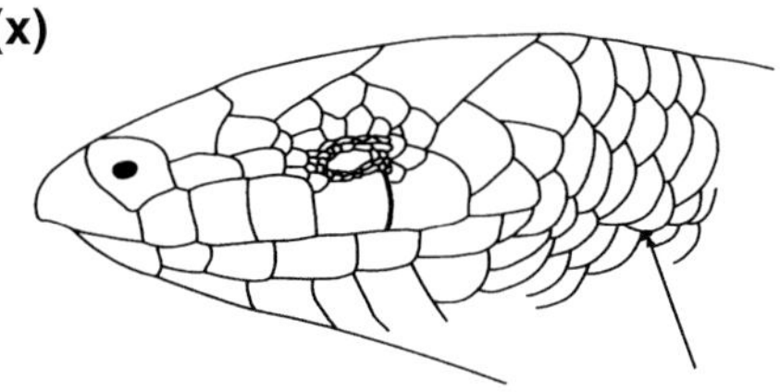

(y)

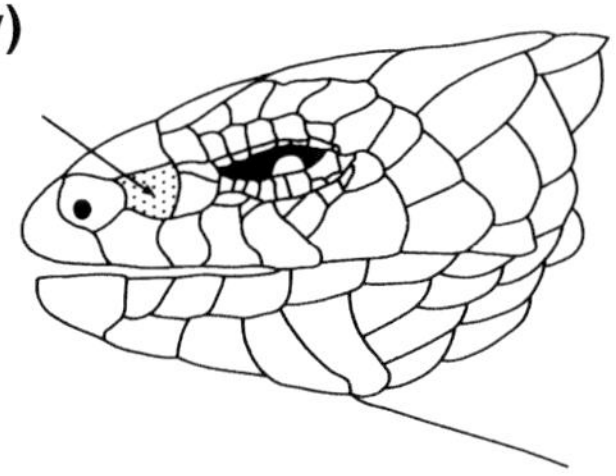

KEY TO GENERA WITH A DIGITAL FORMULA OF LESS THAN 5/5

Letters in parentheses at the end of statements refer to adjacent figures illustrating features referred to in the couplet.

1. Medial pair of preanal scales overlap adjacent preanal scales **(h)** **2**
   Medial pair of preanal scales overlapped by adjacent preanal scales **(g)** **12**

2. Limbs present **3**
   Limbs absent **9**

3. Ear opening present **(x)** .... *Lerista*
   Ear opening absent **4**

4. Lower eyelid with transparent disc **(j)** .... *Hemiergis*
   Lower eyelid scaly **(v)** **5**

5. Digital formula 3/3 .... **6**
   Not as above **8**

6. Loreal scale 1 **(y)**, belly yellow to orange .... *Saiphos*
   Loreal scales 2 **(z)**, belly variegated with grey **7**

7. First two supraoculars contact frontal **(aa)**; snout bluntly conical in profile .... *Coeranoscincus*
   Only first supraocular contacts frontal **(bb)**; snout sharp in profile .... *Coggeria*

(z)

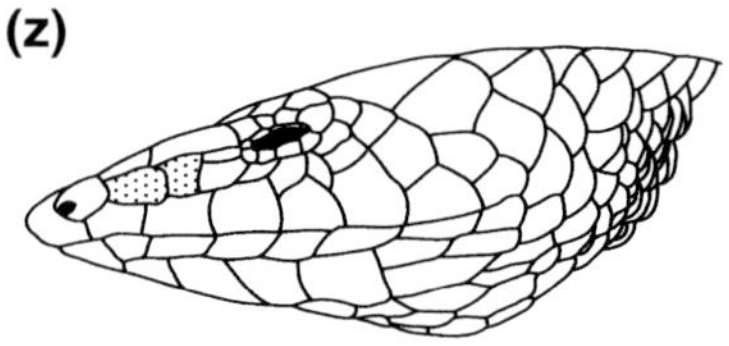

(aa)

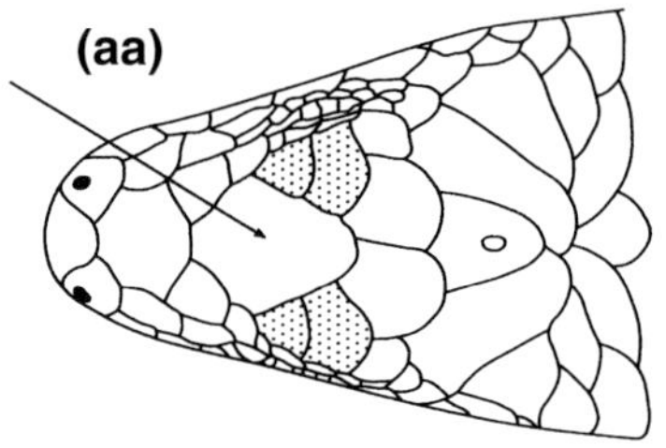

(bb)

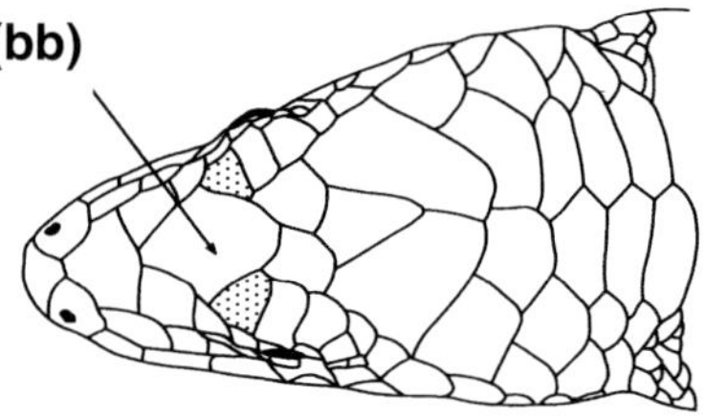

8. Forelimb without digits, or with one digit .... *Ophioscincus*
   Forelimb with more than one digit .... *Anomalopus*

9. Ear opening present, minute .... *Lerista*
   Ear opening absent **10**

10. Paravertebral scales (between parietal scales and preanal scales) >150 **(cc)** .... *Coeranoscincus*
    Paravertebral scales <140 .... **11**

11. Lateral stripe, broad and continuous .... *Ophioscincus*
    Laterally unstriped or, at most, flecked .... *Anomalopus*

12. Supraoculars obliquely oriented **(dd)** .... *Menetia*
    Supraoculars transversely oriented **(ee)** **13**

(cc)

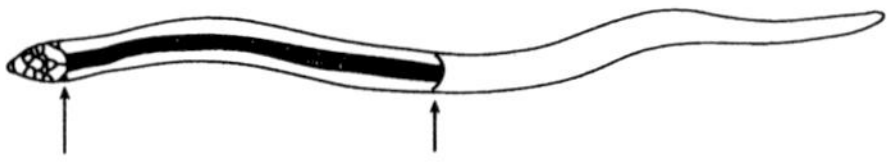

(dd)

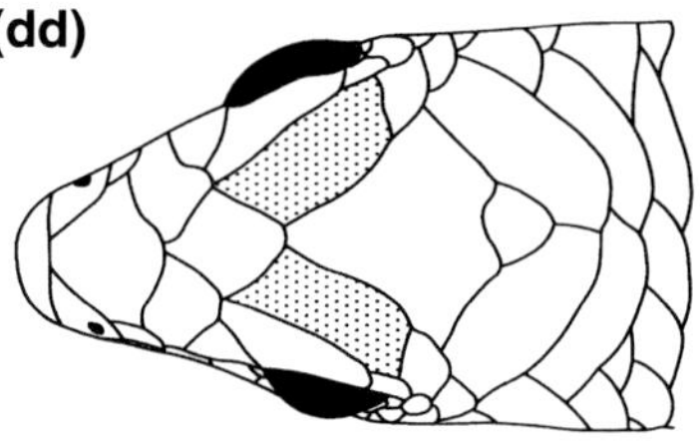

(ee)

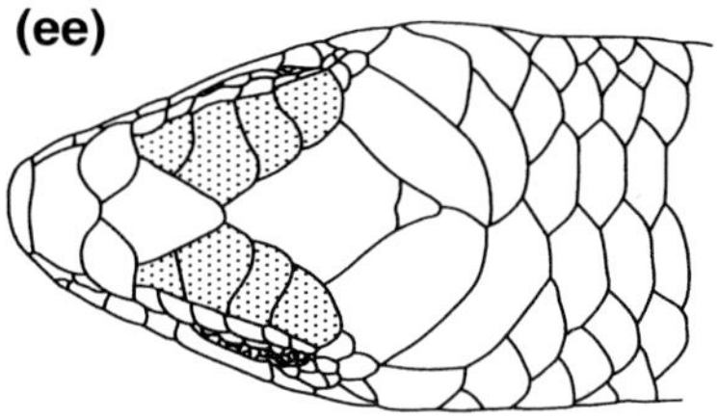

13. Prefrontals absent........ *Eroticoscincus*
    Prefrontals present **(f)**........ **14**

14. Frontoparietals paired **(l)** .... *Saproscincus*
    Frontoparietals fused to interparietal forming single shield **(k)** **15**

15. Supradigital scales on 4th toe ≥ 10 **(ff)** .... *Carlia*
    Supradigital scales on 4th toe <10........ **16**

16. Paravertebral scales (between parietals and posterior edge of thigh) 39–50 **(gg)** .... *Lygisaurus*
    Paravertebral scales 52–66 .... *Menetia*

(ff)

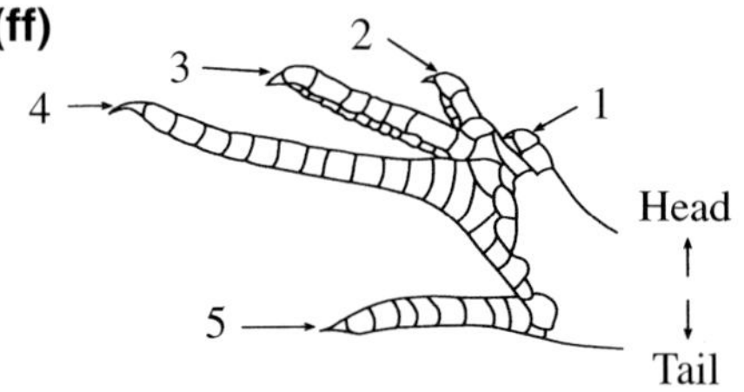

(gg)

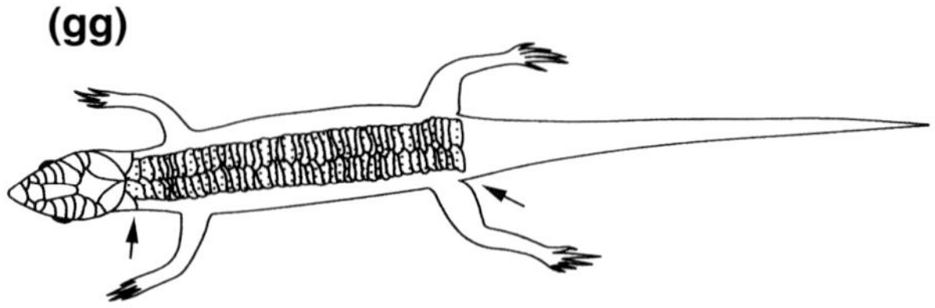

GENERIC DISTRIBUTIONS

The total geographic range of each of the 37 skink genera from Australia and its island territories is shown in Figures 1–5. Maps are arranged alphabetically by genus and are modified from Cogger (2000), Covacevich and Couper (1991), Greer (1983), Hutchinson *et al.* (1990), Shea and Sadlier (1999), Storr *et al.* (1999) and Wells and Wellington (1984, 1985).

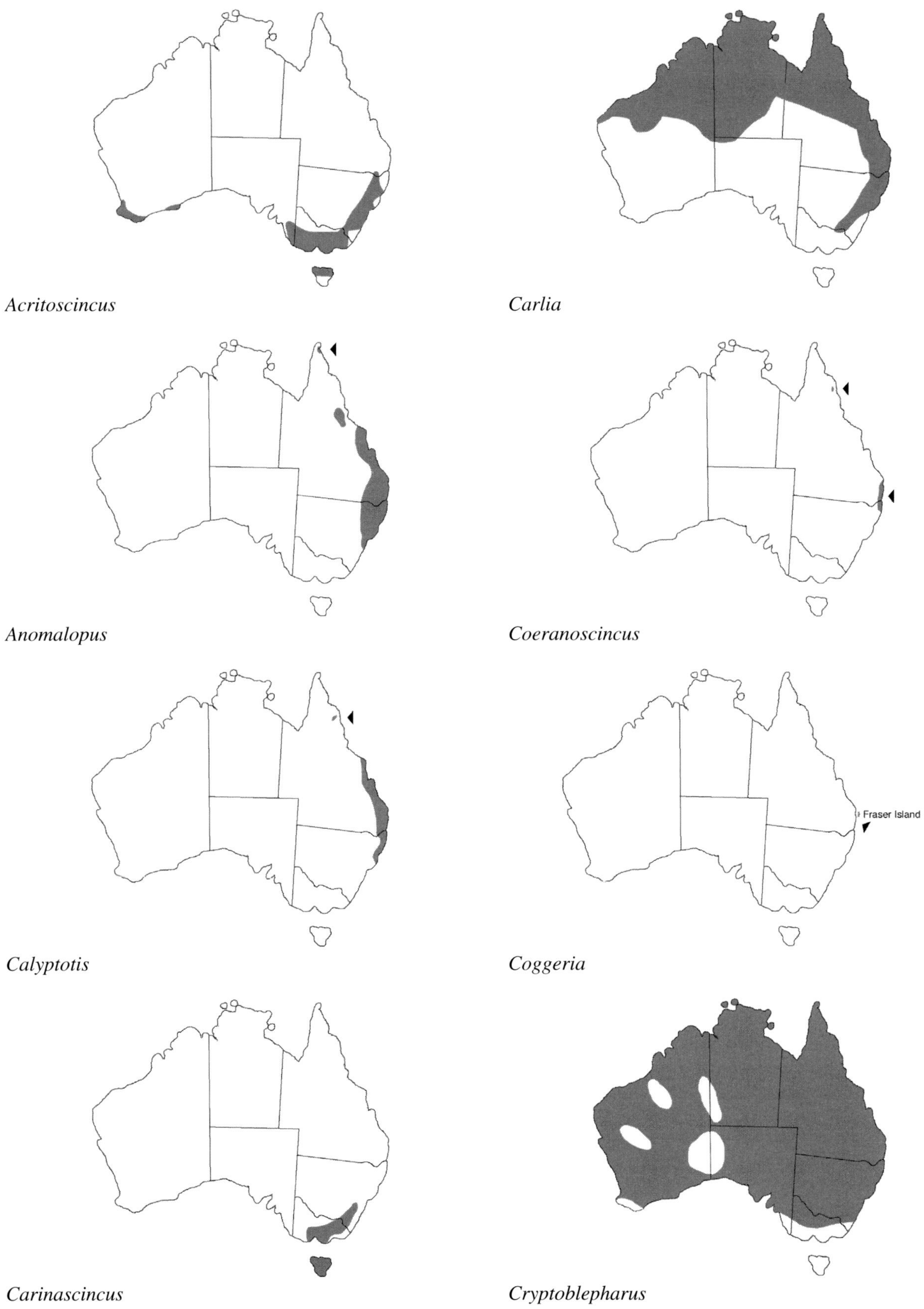

Figure 1. Total generic ranges for eight genera of Australian skinks (*Acritoscincus to Cryptoblepharus*). Based on data to September 2004.

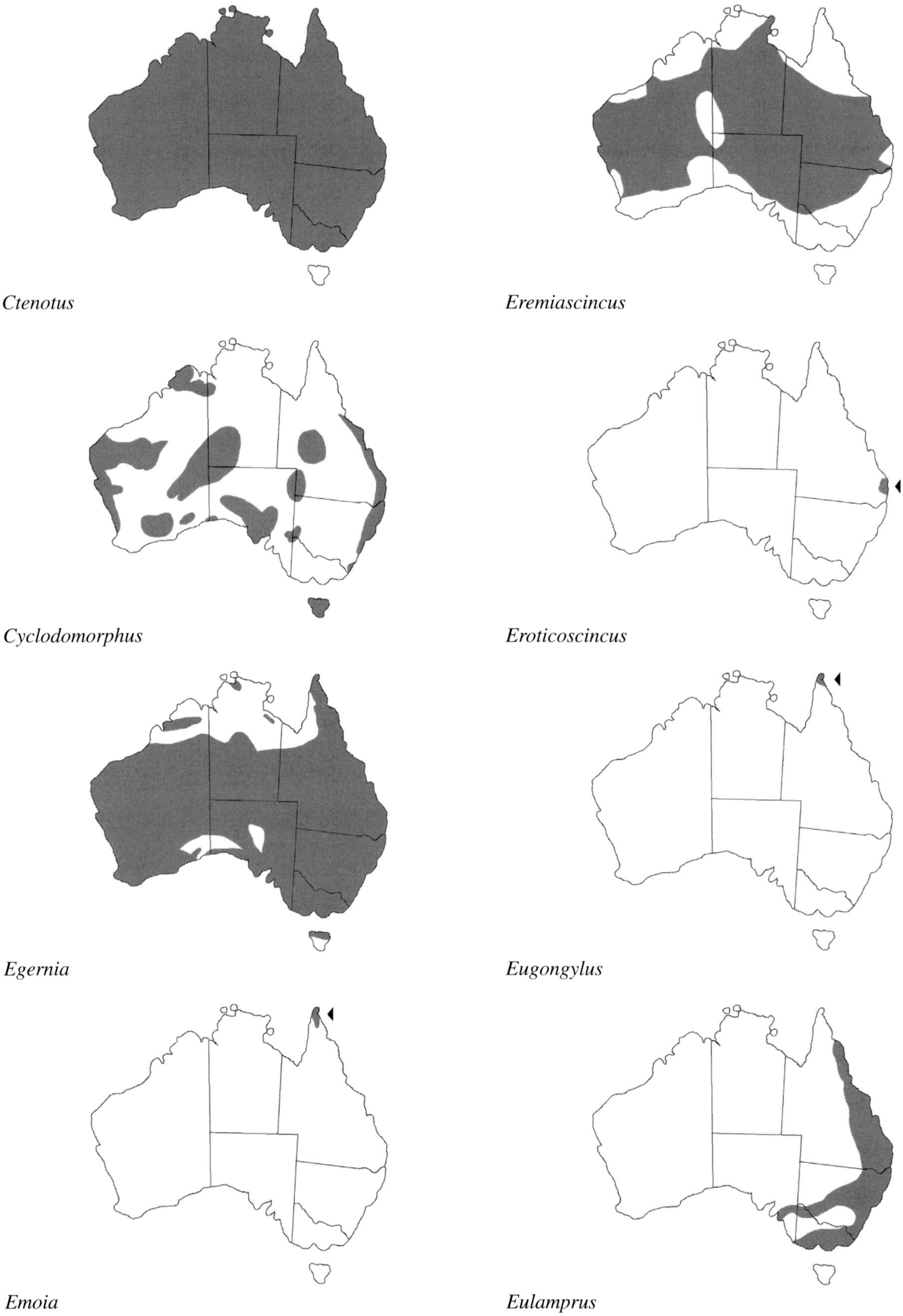

Figure 2. Total generic ranges for eight genera of Australian skinks (*Ctenotus to Eulamprus*). Based on data to September 2004.

Glaphyromorphus

Lampropholis

Gnypetoscincus

Lerista

Harrisoniascincus

Lygosoma (Christmas Island)

Lygisaurus (North and East Coast)

Hemiergis

Menetia

Figure 3. Total generic ranges for nine genera of Australian skinks (*Glaphyromorphus to Menetia*). Based on data to September 2004.

Morethia

Oligosoma

Nangura

Ophioscincus

Nannoscincus

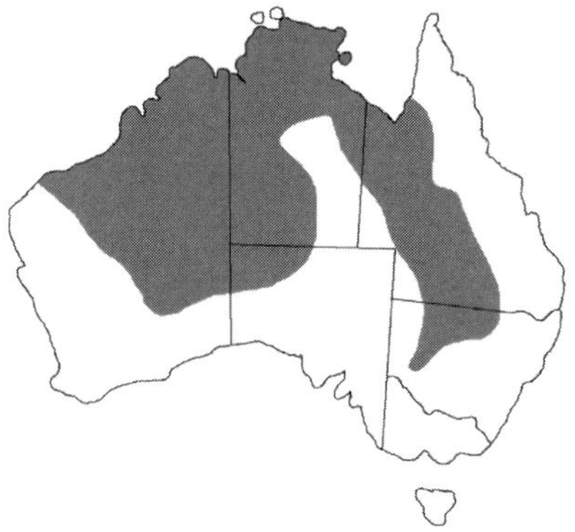

Proablepharus

Notoscincus

Pseudemoia

Figure 4. Total generic ranges for eight genera of Australian skinks (*Morethia to Pseudemoia*). Based on data to September 2004.

Saiphos

Techmarscincus

Saproscincus

Tiliqua

Figure 5. Total generic ranges for four genera of Australian skinks (*Saiphos* to *Tiliqua*). Based on data to September 2004.

CURRENTLY KNOWN SPECIES OF AUSTRALIAN SKINKS

Table 2 is a list of all species known to September, 2004. Genera (and species within them) are arranged alphabetically. Authors and dates of description are listed, but most are not included in the reference list; details of species descriptions and synonymy can be quickly accessed in recent reference volumes listed.

DISCUSSION

DIVERSITY

Three hundred and eighty-eight (388) species of skinks are currently recognized from Australia and its island territories. *Ctenotus,* with species occurring throughout continental Australia, is the most speciose genus (98 species) and the most widely distributed. *Ctenotus* spp. are most numerous in drier habitats, as are those of *Lerista*, another widely distributed and speciose (81 species) genus. Amongst narrowly occurring continental genera are three monotypic Queensland genera (*Coggeria, Nangura* and *Techmarscincus*), from single localities or a handful of sites close together. Some genera that are monotypic or contain only two or three species range more widely. These (*Coeranoscincus, Eroticoscincus, Gnypetoscincus, Harrisoniascincus* and *Saiphos*) are usually confined to rainforests. *Emoia,* in continental Australia, is narrowly restricted to Cape York Peninsula. Yet this genus is widespread and speciose on … 'islands of south-east Asia east through the Indo-Australian Archipelago, the Philippines, New Guinea, the Bismarcks and north into Micronesia, south into Melanesia, and east into Polynesia' (Ingram, 1979). One species of *Emoia* occurs on Australia's Indian Ocean territory, Christmas Island.

CONSERVATION AND THE FUTURE

Cogger *et al.* (1993) identified 16 species or subspecies of skinks as threatened, i.e. endangered or vulnerable. Species of nine genera (*Acritoscincus, Anomalopus, Coeranoscincus, Ctenotus, Egernia, Eulamprus, Lerista, Oligosoma* and *Tiliqua*) are threatened and many more are of conservation concern as 'rare or insufficiently known'. Since completion of that work, *Lerista allanae* Longman, 1937 has been shown to be probably extinct (Covacevich *et al.,* 1996). This species, formerly very narrowly restricted to black soil plains in the vicinity of Capella (MEQ) is the first reptile species presently believed to have been 'lost' since European settlement. Its putative extinction has been attributed to habitat destruction in the face of increased grazing and intensive, broad scale grain farming. Bridgewater (1993) observed that, '...As a group, reptiles seem to have survived the perturbations of the last 200 years (of European settlement of Australia) better than the mammals and birds...'. It is true that only one Australian species of reptile is believed to be extinct (Covacevich *et al.*, 1996) while many species of mammals and one species of bird are extinct (Garnett, 1992; Lee, 1995; Maxwell *et al.,* 1996). However, such an apparently rosy picture for the reptiles does not tell their full story.

Reptiles are less well known, less frequently studied and less 'popular' with the public and politicians than mammals and birds. They do not fare well in competition for slices of the research money cake. Czech *et al.* (1998) have shown recently that, in the United States (often a model for Australia in many environmental matters)...' a myriad of social and political values and influences converge to favour birds, mammals, fish and plants in the allocation of endangered species funds'. Bridgewater (1993) noted that ...Reptiles were neglected by scientists, planners and decision-makers.

Despite the ever-present need for more information and research money to spend in the quest for solutions to conservation problems, a lot has been done with the publication of the *Action Plan for Australian Reptiles* (Cogger *et al.,* 1993). One reviewer noted, it '... takes the problem (of reptile species conservation) out of the airy-fairy world of conservation philosophy and into the cold, hard world of facts and figures. It provides exact locations, precise data on destructive impacts and, most importantly, costs in terms of manpower and money...' (Covacevich, 1995). At that time, expenditure of about $5 million could have been used to start the implementation of recovery plans to conserve all 50+ threatened species of Australian reptiles. Costs of land acquisition to protect habitat for these species were not included in the cost sheet at the time. Only 16 species of skinks are known to be threatened so the costs, of ensuring their maximum survival prospects, are not great when measured beside the cost of just one military aircraft, one naval vessel or the cost of each gold medal Australia won in the 2004 Olympic Games!

Plans for how to dramatically improve skink (and other reptile) conservation in Australia have been available since 1993. Costs of doing so have been fairly well known as well. Yet little, if anything, has changed since that time. It seems reasonable to suggest that the absence of real advances results from a lack of political pressure, which translates via political will/action, into properly-funded implementation of habitat protection and management of the species. Such political 'pull' does not exist for reptiles. This problem re-emphasizes the findings of Czech *et al.* (1998) who wrote that 'ecological importance and rarity are the most important factors in prioritizing species for conservation, ... Plants, birds, mammals and fish have distinctly more positive social construction than reptiles, amphibians, invertebrates and micro-organisms'.

Given these facts, in our view, prospects for Australia's presently threatened skink (and other reptile) species seem unlikely to improve, unless community and political perceptions of their value change. Further, with ever-increasing environmental pressures, it seems reasonable to expect that more species will slip from common to rare to threatened. One ray of hope for promoting necessary changes may rest with Bridgewater's (1993) 'eager cohort' of herpetologists, professional and amateur.

ACKNOWLEDGEMENTS

Dr. D. Stuart-Fox contributed unpublished observations and Drs. Glenn Shea (Department of Veterinary Anatomy, University of Sydney) and John Merrick (Graduate School of the Environment, Macquarie University, Sydney) generously helped prepare this chapter. Ms Lauren Keim and Mr Steve Wilson checked the keys and assisted us to improve them.

REFERENCES

Boulenger, G.A. 1887. *Catalogue of the Lizards in the British Museum (Natural History).* Vol III. Trustees of the British Museum, London. 575pp.

Bridgewater, P. 1993. Foreword. In Cogger, H.G., Cameron, E.E., Sadlier, R.A. and Eggler, P. *The Action Plan for Australian Reptiles.* Australian Nature Conservation Agency, Canberra. 254pp.

Cogger, H.G. 1975. *Reptiles and Amphibians of Australia.* A.H. and A.W. Reed, Sydney. 584pp.

Cogger, H.G. 1979. *Reptiles and Amphibians of Australia.* A.H. and A.W. Reed, Sydney.

Cogger, H.G. 1983. *Reptiles and Amphibians of Australia.* A.H. and A.W. Reed, Sydney.

Cogger, H.G. 1986. *Reptiles and Amphibians of Australia.* A.H. and A.W. Reed, Sydney.

Cogger, H.G. 1992. *Reptiles and Amphibians of Australia.* Reed Books Australia, Sydney. 775pp.

Cogger, H.G. 1993. General description and definition of the Order Squamata. Pp.157–171. In Glasby, C.J., Ross, G.J.B. and Beesley, P.L. (eds). *Fauna of Australia. Vol 2A. Amphibia and Reptilia.* Australian Government Publishing Service, Canberra. 439pp.

Cogger, H.G. 1996. *Reptiles and Amphibians of Australia.* Reed Books Australia, Port Melbourne. 796pp.

Cogger, H.G. 2000. *Reptiles and Amphibians of Australia.* Reed New Holland, Sydney. 808pp.

Cogger, H.G., Cameron, E.E. and Cogger, H.M. 1983. *Zoological Catalogue of Australia 1. Amphibia and Reptilia.* Australian Government Publishing Service, Canberra. 313pp.

Cogger, H.G., Cameron, E.E., Sadlier, R.A. and Eggler, P. 1993. *The Action Plan for Australian Reptiles.* Australian Nature Conservation Agency, Canberra. 254pp.

Covacevich, J. 1995. Review, 'The Action Plan for Australian Reptiles'. *Australian Natural History*, Vol. 24, No. 12, p.72.

Covacevich, J.A. and Couper, P.J. 1991. The reptile records.

Pp.45–140. In Ingram, G.J. and Raven, R.R. (eds). *An Atlas of Queensland's Frogs, Reptiles, Birds and Mammals.* Board of Trustees, Queensland Museum, Brisbane. 391pp.

Covacevich, J.A. Couper, P.J. and McDonald, K.R. 1996. *Lerista allanae* (Scincidae: Lygosominae): 60 years from exhibition to extinction? *Memoirs of the Queensland Museum,* Vol. 39, Pt. 2, pp.247–256.

Czech, B., Krausman, P.R. and Borkhataria, R. 1998. Social construction, political power, and the allocation of benefits to endangered species. *Conservation Biology,* Vol. 12, No. 5, pp.1103–1112.

De Vis, C.W. 1884. On new Queensland lizards. *Proceedings of the Royal Society of Queensland,* Vol. 1, No. 2, pp.77–78.

Ehmann, H. 1992. *Encyclopedia of Australian Animals. The National Photographic Index of Australian Wildlife. Reptiles.* Collins Angus and Robertson Publishers, Sydney. 495pp.

Fuhn, I.E. 1967. *Pseudemoia,* eine neue monotypische Gattung aus Südaustralien *Ablepharus/Emoia spenceri* (Lucas and Frost, 1894). *Zoologischer Anzeiger,* Vol. 179, No. 3/4, pp.243–247.

Garnett, S. 1992. *The Action Plan for Australian Birds.* Australian National Parks and Wildlife Service, Canberra. 262pp.

Greer, A.E. 1983. The Australian scincid lizard genus *Calyptotis* De Vis: resurrection of the name, description of four new species and discussion of relationships. *Records of the Australian Museum,* Vol. 35, No. 1, pp.29–59.

Greer, A.E. 1989. *The Biology and Evolution of Australian Lizards.* Surrey Beatty and Sons, Sydney. 264pp.

Hutchinson, M.N. 1993. Family Scincidae. Pp.261–279. In Glasby, C.J., Ross, G.J.B. and Beesley, P.L. (eds). *Fauna of Australia. Vol 2A. Amphibia and Reptilia.* Australian Government Publishing Service, Canberra. 439pp.

Hutchinson, M.N., Donellan, S.C., Baverstock, P., Krieg, H., Simms, S. and Burgin, S. 1990. Immunological relationships and generic revision of the Australian lizards assigned to the genus *Leiolopisma* (Scincidae: Lygosominae). *Australian Journal of Zoology,* Vol. 38, No. 5, pp.535–554.

ICZN. 1999. *International Code of Zoological Nomenclature.* Fourth Edition. International Trust for Zoological Nomenclature, c/- The Natural History Museum, London. 306pp.

Ingram, G.J. 1979. The occurrence of lizards of the genus *Emoia* (Lacertilia, Scincidae) in Australia. *Memoirs of the Queensland Museum,* Vol. 19, Pt. 3, pp.431–437.

Ingram, G. and Covacevich, J. 1988. Revision of the genus *Lygisaurus* de Vis (Scincidae: Reptilia) in Australia. *Memoirs of the Queensland Museum,* Vol. 25, Pt. 2, pp.335–354.

Ingram, G. and Covacevich, J. 1989. Revision of the genus *Carlia* (Reptilia: Scincidae) in Australia with comments on *Carlia bicarinata* of New Guinea. *Memoirs of the Queensland Museum,* Vol. 27, Pt. 2, pp.443–490.

Lee, A.K. 1995. *The Action Plan for Australian Rodents.* Australian Nature Conservation Agency, Canberra. 62pp.

Maxwell, S., Burbidge, A.A. and Morris, K. (eds) 1996. *The 1996 Action Plan for Australian Marsupials and Monotremes.* Wildlife Australia, Commonwealth Department of the Environment, Sport and Territories, the Australian Nature Conservation Agency and the Australian Heritage Commission, Canberra. 234pp.

Mittleman, M.B. 1952. A generic synopsis of the lizards of the subfamily Lygosominae. *Smithsonian Miscellaneous Collection,* Vol. 117, No. 17, pp.1–35.

Shea, G.M. 1990. The genera *Tiliqua* and *Cyclodomorphus* (Lacertilia: Scincidae): generic diagnoses and systematic relationships. *Memoirs of the Queensland Museum,* Vol. 29, Pt. 2, pp.495–519.

Shea, G.M. and Sadlier, R.A. 1999. A catalogue of the non-fossil amphibian and reptile type specimens in the collection of the Australian Museum: types currently, previously, and purportedly present. *Technical Reports of the Australian Museum,* No. 15, pp.1–91.

Smith, M.A. 1937. A review of the genus *Lygosoma* (Scincidae: Reptilia) and its allies. *Records of the Indian Museum,* Vol. 39, No. 3, pp.213–234.

Storr, G.M., Smith, L.A. and Johnstone, R.E. 1999. *Lizards of Western Australia. 1 Skinks.* Western Australian Museum, Perth. 291pp.

Stuart-Fox, D.M., Hugall, A.F. and Moritz, C. 2002. A molecular phylogeny of rainbow skinks (Scincidae: *Carlia*): taxonomic and biogeographic implications. *Australian Journal of Zoology*, Vol. 50, pp.39–51.

Wells, R. and Wellington, C.R. 1999. A response to Ken Aplin's article on herpetology in Australia. *Monitor, Journal of the Victorian Herpetological Society,* Vol. 10, No. 2/3, pp.110–112.

Wells, R.W. and Wellington, C.R. 1984. A synopsis of the Class Reptilia in Australia. *Australian Journal of Herpetology,* Vol. 1, No. 3–4, pp.73–129.

Wells, R.W. and Wellington, C.R. 1985. A classification of the Amphibia and Reptilia of Australia. *Australian Journal of Herpetology Supplementary Series,* No. 1, pp.1–61.

Wilson, S. and Swan, G. 2003. *A Complete Guide to Reptiles. of Australia.* Reed New Holland, Sydney. 480pp.

Wilson, S.K. and Knowles, D.G. 1988. *Australia's Reptiles. A Photographic Reference to the Terrestrial Reptiles of Australia.* William Collins, Sydney. 447pp.

Worrell, E. 1963. *Reptiles of Australia. Crocodiles, Turtles, Tortoises, Lizards, Snakes.* Angus and Robertson, Sydney. 207pp.

Birds

GEOLOGICAL TIME SCALE FOR THE PHANEROZOIC

ERA	PERIOD or SUB-PERIOD	SERIES/EPOCH	STAGE	AGE m.y.a.	EXTINCTIONS (E) ICE AGE (I)
CAINOZOIC	QUATERNARY	PLEISTOCENE	Calabrian	1.8	E, I
	TERTIARY – NEOGENE	PLIOCENE	Piacenzian	3.6	I
			Zanclean	5.3	I
		MIOCENE Late	Messinian	7.12	
			Tortonian	11.2	
		MIOCENE Mid	Serravalian	14.8	
			Langhian	16.4	
		MIOCENE Early	Burdigalian	20.52	
			Aquitanian	23.8	
	TERTIARY – PALAEOGENE	OLIGOCENE Late	Chattian	28.5	E
		OLIGOCENE Early	Rupelian	33.7	
		EOCENE Late	Priabonian	37	
		EOCENE Mid	Bartonian	41.3	
			Lutetian	49	
		EOCENE Early	Ypresian	54.8	
		PALEOCENE Late	Thanetian	57.9	
			Selandian	60.9	
		PALEOCENE Early	Danian	65	E
MESOZOIC	CRETACEOUS	LATE	Maastrichtian	73	
			Campanian	83	
			Santonian	87	
			Coniacian	89	
			Turonian	91	
			Cenomanian	97.5	
		EARLY	Albian	108	
			Aptian	115	
			Barremian	123	
			Hauterivian	130	
			Valanginian	135	
			Berriasian	141	E
	JURASSIC	LATE	Tithonian	146	
			Kimmeridgian	151	
			Oxfordian	159	
		MIDDLE	Callovian	165	
			Bathonian	173	
			Bajocian	180	
			Aalenian	184	
		EARLY	Toarcian	190	E
			Pliensbachian	195	
			Sinemurian	202	
			Hettangian	205	E
	TRIASSIC	LATE	Rhaetian	207	
			Norian	220	
			Carnian	230	
		MIDDLE	Ladinian	235	
			Anisian	241	
		EARLY	Spathian	244	
			Nammalian	247	
			Griesbachian	251	

This detailed scale is based on Young and Laurie (1996) but Cambrian stages abbreviated ages are rounded to nearest 0.1 million years. More recent dating (e.g. United States Geological Survey) has resulted in some small adjustments to interval boundaries.

ERA	PERIOD or SUB-PERIOD	SERIES/EPOCH	STAGE	AGE m.y.a.	EXTINCTIONS (E) ICE AGE (I)
PALAEOZOIC	PERMIAN	LATE	Tatarian – Dorashamian	255	E
			Tatarian – Dzhulfian	259	I
			Tatarian – Midian	263	I
			Kazanian	267	I
			Ulfimian	270	I
		EARLY	Kungurian	274	I
			Artinskian	285	I
			Sakmarian	293	I
			Asselian	298	I
	CARBONIFEROUS	PENNSYLVANIAN – SILESIAN – STEPHANIAN	Gzhelian	302	I
			Kasimovian	305	I
		WESTPHALIAN	Moscovian	311.5	
			Bashkirian	314	
		NAMURIAN	Serpukhovian	325	
		MISSISSIPPIAN – DINANTIAN – VISEAN	Brigantian	328.5	
			Asbian	330.5	
			Holkerian	335	
			Arundian	341.5	
		TOURNAISIAN	Chadian	344	
			Ivorian	348	
			Hastarian	354	
	DEVONIAN	LATE	Famennian	364.5	E
			Frasnian	369	
		MIDDLE	Givetian	378	
			Eifelian	384	I
		EARLY	Emsian	399.5	I
			Pragian	404.5	I
			Lochkovian	410	
	SILURIAN	PRIDOLI		414	
		LUDLOW	Ludfordian	417	
			Gorstian	420	
		WENLOCK	Homerian	422.5	
			Sheinwoodian	425	
		LLANDOVERY	Telychian	428	I
			Aeronian	431	I
			Rhuddanian	434	E, I
	ORDOVICIAN	ASHGILL	Bolindian	443	I
		CARADOC	Eastonian	454.5	
			Gisbornian	459	
		LLANVIRN	Darriwilian	467	
			Yapeenian	471	
		ARENIG	Castlemainian	477	
			Chewtonian	481	
			Bendigonian	484.5	
		TREMADOC	Lancefieldian	486	
			Warendan	490	
	CAMBRIAN	LATE	Datsonian	491	
			Payntonian	492.5	
			Iverian	496	
			Idamean	497.5	E
		MIDDLE	Mindyallan	498.5	
			Boomerangian	500	
			Undillan	503	
			Late Templetonian-Floran	506	
			Ordian-Early Templetonian	509	
		EARLY	Atdabanian-Toyonian	528	
			Tommotian	534	
			Nemakit-Daldynian	545	

Chapter 21
The Avian Fossil Record of Australia: An Overview

Walter E. Boles

Division of Vertebrate Zoology, Australian Museum, 6 College Street, Sydney N.S.W. 2010 Australia; and School of Biological, Earth and Environmental Sciences, University of New South Wales N.S.W. 2052 Australia

E-mail: walterb@austmus.gov.au

REGIONAL RECORDS

This overview of the Australian avian fossil record benefits from following the in-depth reviews by Vickers-Rich (1991)—of Mesozoic and Tertiary data—and Baird (1991a), who discussed the Quaternary. Some aspects raised by those authors are updated here; however, the level of detail in their reviews is beyond this summary. Emphasis is placed on studies that have appeared since those reviews. For more thorough discussions and references not cited in the text, consult these accounts; for a global overview, see Olson (1985) and Feduccia (1999).

Australasia has a patchy record, both in temporal and faunistic representation and in palaeontological research. Although the record is richer and longer for Australia than it is for New Zealand, New Caledonia and New Guinea, it is poor compared with that of Europe or North America.

The early fossil record of birds in New Zealand is scanty. Although the record extends from at least the Paleocene, other than for penguins it is sparse and uninformative. Material from the late Pleistocene forms the vast majority of the known record and much can be referred to living taxa; see Fordyce (1991), Millener (1991) and especially Worthy and Holdaway (2002) for recent reviews. The only New Guinean sites, of Plio-Pleistocene age, have provided a handful of avian fossils, most of them cassowaries *Casuarius* (Rich *et al.*, 1988; Vickers-Rich, 1991; Aplin *et al.*, 1999). Published deposits from New Caledonia are late Quaternary, mostly late Holocene, and indicate a greater diversity of non-passerine taxa than that on the island today (Balouet and Olson, 1989); information on the passerines has not yet been published. Avian fossils from other, small islands in the Australasian region are restricted to the Quaternary (Meredith, 1991).

Australia has by far the longest avian fossil record in Australasia, but it is not without its problems. The earliest record of birds comes from the early Cretaceous (110–100 m.y.a.) of western Queensland, north central New South Wales and southern Victoria (Figure 1(a)). There is a large, almost empty gap between the Cretaceous fossils and the considerably richer record that starts in the late Oligocene. Within this span, the appearance of birds is almost non-existent other than for a few specimens from the early Eocene, recovered from south-east Queensland and the south-eastern coast of the mainland (Figure 1(a)).

Of the only two major intervals with rich, diverse avian records, the first, starting in the late Oligocene and continuing through the middle Miocene, is represented predominantly in the present Great Artesian Basin (Lake Eyre and Lake Tarkarooloo Basins) of central Australia and in northern Australia (Riversleigh, Bullock Creek) (Figure 1(b)). The record through the late Miocene is much sparser, with only a single Northern Territory site (Alcoota) of this age yielding avian fossils. Pliocene-aged localities in northern, central and eastern Australia (Figure 1(c)) are not as rich as those of the late Oligocene–middle Miocene. Important sites include Riversleigh, Bluff Downs, Lake Kanunka, Lake Palankarinna and Darling Downs. The second major fossil-bearing interval consists mainly of late Quaternary cave and aeolian-fluvial deposits, principally across southern Australia (Figure 1(d)).

Australia's Mesozoic and Tertiary fossil record suffers from several deficiencies. As already mentioned, the most profound is its incomplete nature, with gaps at critical times. The different taphonomic forces of the Quaternary have yielded more complete remains than the usually incomplete and disassociated Tertiary material. Differences in the types of preservation and accumulating factors between pre-Quaternary and Quaternary deposits often render them incomparable. Another drawback of the pre-Quaternary record is the paucity of radiometrically dated sites; most dates have been assigned through tenuous sequences correlating remains between a series of sites (Rich and Baird, 1986). Much of the continent remains to be explored for suitable fossiliferous sites, with many potential areas in Western Australia yet to be exploited. The record exhibits a strong bias towards certain taxonomic groups, which is particularly evident in the Tertiary.

HISTORY OF BIRDS IN AUSTRALIA

The following section builds upon the reviews of Vickers-Rich (1991) and Baird (1991a), incorporating published investigations in the past decade. Since Rich and van Tets (1982) presented a graphic representation of the taxonomic and temporal distributions of birds in this record, there has been a steady increase in discovery, study and description of specimens, both new and old; compare the figures from Rich and van Tets (1982, fig. 27) and Boles (1991, fig. 2) with Figure 2 in this chapter.

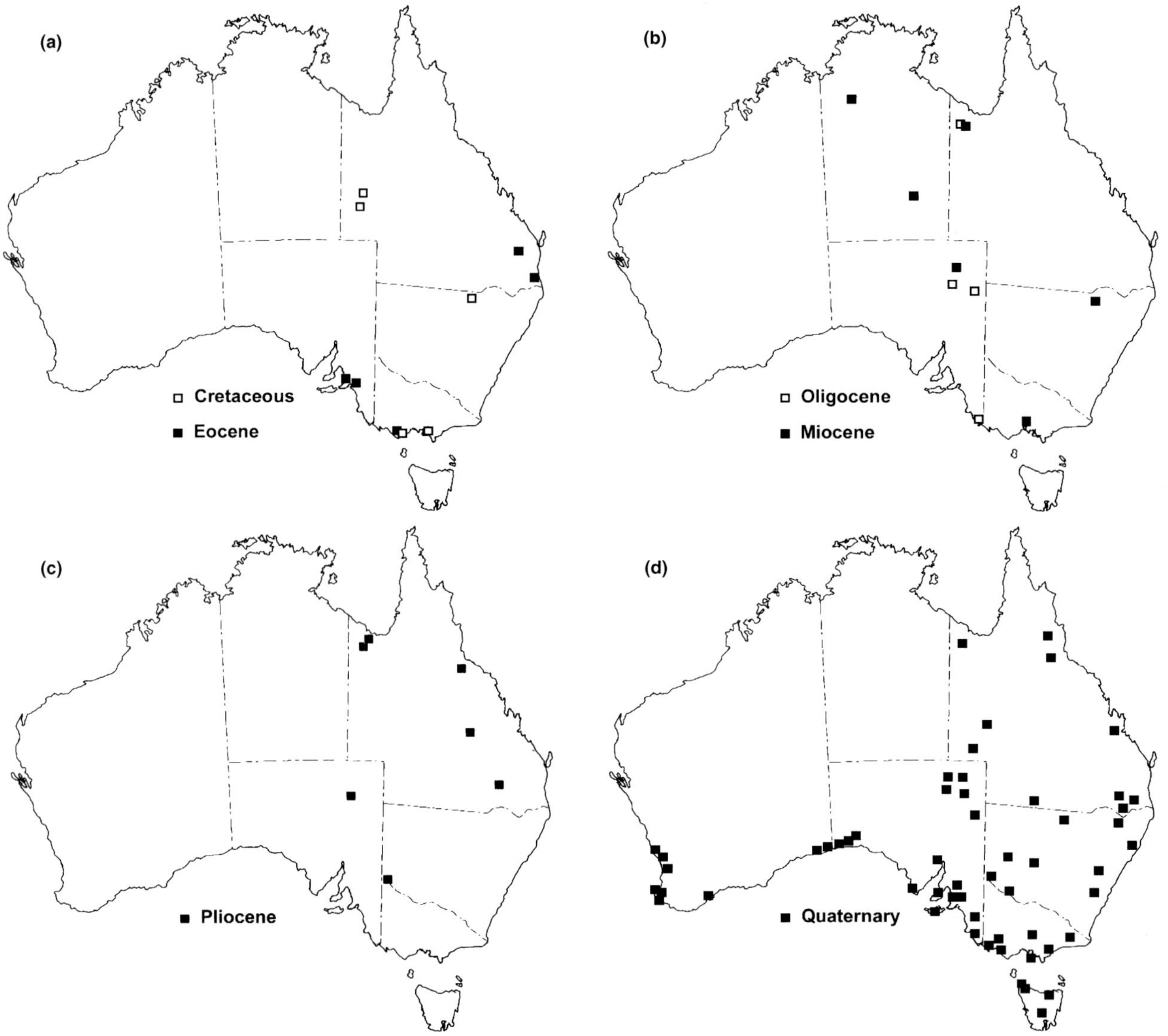

Figure 1. Major Australian localities that have produced fossil birds: **(a)** Cretaceous and Eocene sites; **(b)** Oligocene and Miocene sites; **(c)** Pliocene sites; **(d)** Quaternary sites. Modified from Rich and Baird (1986) and Vickers-Rich (1991). Quaternary localities are taken mainly from Baird *et al.* (1991), but are not exhaustive.

Early attempts to explain the distribution of animals on different continents used the premise that the positions of the land masses were, and had been, static through time. Using such an assumption, only one means was available to explain the mutual occurrence of a taxon at different localities: dispersal from one area to another, including the crossing of intervening water bodies. This approach characterized the initial attempts of biogeographers to account for the characteristic Australasian birds (and other biota) and the presence of Wallace's Line—see pp.95–110 (van Oosterzee, 2006). The most influential of these efforts was that of Mayr (1944), which assumed that birds moved into the Australian region from the north by way of the Indo-Malayan Archipelago.

The emergence of ideas of continental drift and plate tectonics and their eventual acceptance dramatically altered the available explanatory mechanisms for animal distributions. Rather than a static system, the arrangement of the continents was now seen as dynamic, with the configurations and positions of continental land masses varying through time. At different times, the land masses were joined into one supercontinent, Pangaea, or two—Laurasia, comprising the northern continents, and Gondwana, the southern ones. As the plates moved, the arrangements of the supercontinents changed, and component land masses broke off, drifting to new positions, sometimes making contact with other land masses. Through time, the locations and orientations of the continents changed, as did their proximity and connections, and the amount of land that was above water; for further details about regional history and tectonic events see pp.17–50 (White, 2006).

Plate 79. Australian magpies (*Gymnorhina tubicen*) are among the most familiar birds, because they have adapted to urban areas. The species is widespread and common throughout Australia and extends to New Guinea; it has also been introduced to New Zealand. Photograph J. Cann.

Plate 80. A black swan (*Cygnus atratus*) landing on an inland waterway. At times hundreds congregate on low-lying coastal wetlands. Photograph S. Allen.

Plate 81. The Cape petrel (*Daption capense*) has a huge range, extending from the Antarctic ice to Lat. 25°S; some regional variation in colour, from islands near New Zealand, has been noted. *D. capense* breeds in summer on Antarctic shores and nearly all sub-Antarctic Islands, then moves north in winter following cold-water currents. It is a common visitor from June to December to southern Australian and all New Zealand waters. Photograph S. Allen.

Plate 82. A great egret (*Ardea alba*) in flight. Widespread in shallow freshwater habitats, estuaries and coastal lagoons. Rarely present in large numbers, but breeds in treetop colonies overhanging water. Photograph S. Allen.

Plate 83. The tawny frogmouth (*Podargus strigoides*) is most active at dusk or in the evening. It inhabits forests, woodlands and urban areas throughout Australia. Can be common locally. Photograph S. Allen.

Plate 84. The Australian white ibis or sacred ibis (*Threskiornis molucca*) is one of three ibis species common and widespread in eastern Australia. *T. molucca* also occurs in New Guinea. Most types of wetland are utilized and although locally abundant it is nomadic. Photograph S. Allen.

It is not known when birds first appeared on the Australian continent. The earliest records are from the early Cretaceous, but without doubt the initial occurrence substantially preceded this. One group that would have been a major component of a Mesozoic avifauna was the subclass Enantiornithes. This group of birds was first reported by Walker (1981), based on strange bones from South America. They were morphologically diverse and widespread—specimens have now been found from every continent except Africa and Antarctica—and appear to have been the dominant lineage of birds until the end of the Cretaceous, when they became extinct (Chiappe, 1995; Martin, 1995; Kurochkin, 1996; Chiappe and Walker, 2002). Chiappe (1991) considered that the enantiornithines originated in Gondwana and moved from there into Laurasia, whereas Zhou (1995) proposed the opposite route. Bocheński (1999) believed that the stage of development of the flight apparatus precluded the crossing of extensive stretches of open water, thus limiting these birds to dispersal by land. Given these constraints, the Enantiornithes would have had to differentiate by the middle Jurassic in order to achieve their known distribution across continents, including Australia.

Australia was once part of Gondwana, which also included Antarctica, South America, Africa, India and New Zealand. Starting about 100 m.y.a., these components began to gradually separate from the main land mass, until the last remaining connection was that between Australia and Antarctica. This began to break about 55 m.y.a. and was completed about 45–38 m.y.a.. From this point, Australia was isolated from other continents for 20 million years or more. In this period of isolation, there was little interchange of non-marine vertebrate animals between different continents. The animals then on the Australian continent evolved in isolation for millions of years, developing their unique characteristics. In the late Miocene, Australia approached and eventually collided with the Asian plate. New Guinea, which had previously been no more than some islands along the edge of the Asian plate, rose from the ocean along the line of contact between the plates. For further details on the formation of New Guinea see Aplin (2006), Heinsohn and Hope (2006) and van Oosterzee (2006). The approach of Australia to Asia allowed new animals to invade from the north, using the Indo-Malayan islands and New Guinea as stepping stones.

It is reasonable to assume that some lineages of neornithine birds with pan-Gondwanan distributions before the separation of the southern continents had representatives on two or more of the component land masses after separation. An ongoing source of speculation and debate is the identification of those components of the Australian avifauna that had vicariant (Gondwanan) origins and those that had arrived from dispersal from the north (Indo-Malayan). Identification of Gondwanan forms has been attempted in part by noting which taxa were endemic to Australasia and had few, if any, close relatives in the Old World. Additionally, the presence of related groups on other southern continents strengthened the support.

Potential Gondwanan elements of the Australian avifauna have been nominated by a range of workers (Cracraft, 1972, 1976; Mayr, 1972; Serventy, 1972; Rich, 1975a,b, 1976; Schodde, 1982; Olson, 1988; Vickers-Rich, 1991) with varying degrees of conservatism and uncertainty. The groups that are most often cited as of likely southern origin have been the ratites (Casuariidae), megapodes (Megapodiidae) and several groups of parrots (Platycercinae, Loriinae, Cacatuidae). Others forwarded on more than one occasion include: groups of distinctive waterfowl, particularly *Anseranas*; flamingos (Phoenicopteridae); several groups of charadriiforms; frogmouths (Podargidae); owlet-nightjars (Aegothelidae) and suboscine passerines (previously thought to include lyrebirds, Menuridae, and scrub-birds, Atrichornithidae).

At the same time, Olson (1988) cautioned that, based on the fossil record, all the major groups of living birds that are now considered characteristic of the Southern Hemisphere were actually widespread in the Tertiary of the northern continents. Mourer-Chauviré (1999) proposed a 'basal continental avifauna', shared by South America, North America and Europe (see references in Mourer-Chauviré, 1999). Because some of these avian groups are characterized by reduced or lost volancy, this distributional pattern of fossil forms required a pathway between South America and Europe via North America, using probable sequential connections between continents. Differential extinctions of groups from these continents produced the dissimilar distributional patterns that characterize modern avifaunas. Olson (1988) considered that the present-day avifaunas of the classic biogeographic realms are not autochthonous, but rather aggregations of relictual taxa that were unsuccessful elsewhere.

For example, Mourer-Chauviré (1982) announced the presence of megapodes, frogmouths and owlet-nightjars in the Eo-Oligocene 'Phosphorites du Quercy', France. Mourer-Chauviré (1988) subsequently created the genus *Quercypodargus* for the frogmouths, placing this in the same family, Podargidae, as the Australo-Papuan *Podargus* (see also Peters, 1991). The owlet-nightjar material has not yet been further described. The putative megapodes, however, were later (Mourer-Chauviré, 1992) placed in the genus *Quercymegapodius* in their own family, the Quercymegapodiidae, considered to be the sister-group of all Recent Galliformes. This family is now known from both Europe and South America, represented by the same genus, *Ameripodius* (Alvarenga, 1995; Mourer-Chauviré, 1999). Olson (1999) placed the genus *Anatalavis* from the late Cretaceous–early Palaeogene of the Northern Hemisphere into the Anseranatidae, a family until then restricted to the Australian endemic magpie goose *Anseranas semipalmatus*. This allocation was rejected by Dyke (2001), however, who considered *Anatalavis* closer to the true ducks (Anatidae). As these examples illustrate, caution must be exercised with the interpretation of northern occurrence of southern groups pending closer systematic examination.

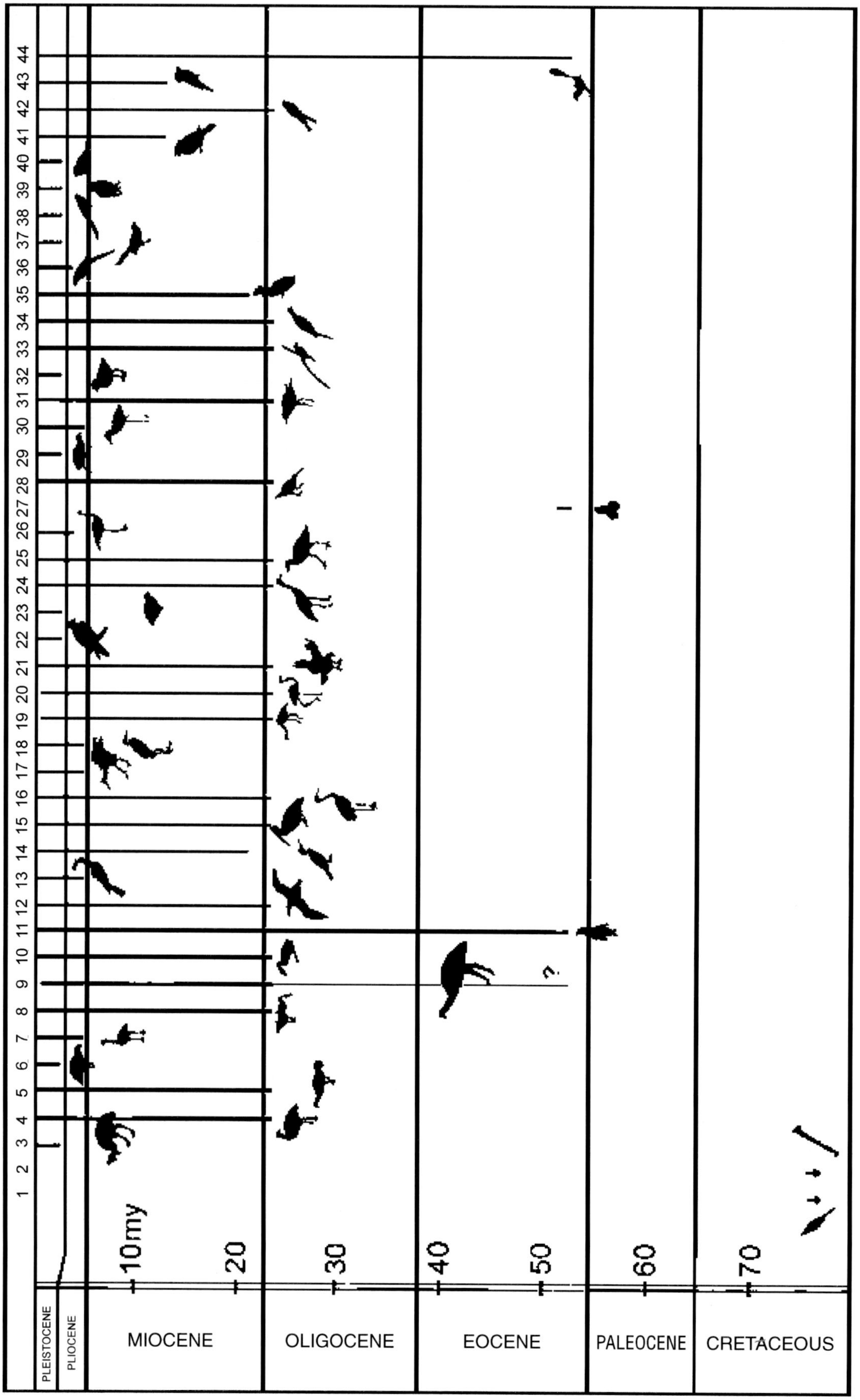
PLEISTOCENE
PLIOCENE
MIOCENE
OLIGOCENE
EOCENE
PALEOCENE
CRETACEOUS
10my
20
30
40
50
60
70
1 2 3 4 5 6 7 8 9 10 11 12 13 14 15 16 17 18 19 20 21 22 23 24 25 26 27 28 29 30 31 32 33 34 35 36 37 38 39 40 41 42 43 44

Among the families that Mourer-Chauviré (1999) nominated as components of her continental basal avifauna are two known from the Australian record. The Phoenicopteridae and Palaelodidae have been found from late Oligocene into the Pleistocene. The early age at which these are first recognized in Australia suggests that both families were present on the continent before its last connections with other Gondwanan land masses were severed.

Growing fossil information demonstrates that modern distributions of avian families do not reflect those of the past. It is thus risky to infer Gondwanan histories on the basis of current occurrence, as has been frequently done. Many of these issues cannot be resolved until the fossil histories of critical groups are better documented. Table 1 gives the oldest records of avian families from Australia. Because references to the respective records are given in the table, they are not always cited in the text.

CRETACEOUS

The first indication of Cretaceous birds in Australia comprised several small feather impressions from Koonwarra, Victoria (Talent *et al.*, 1966; Waldman, 1970). The enantiornithine *Nanantius eos* was described from western Queensland (Molnar, 1986); additional specimens have been subsequently found (Kurochkin and Molnar, 1997; Kear *et al.*, 2003). Non-enantiornithine birds have been recovered from north central New South Wales (Molnar, 1999; Smith, 1999), and indeterminate avian remains come from southern Victoria (Rich *et al.*, 1999). While verifying the presence of enantiornithine birds in Australia, the Cretaceous record offers nothing about the origins of the neornithine radiation in Australia.

The critical period from the end of the Cretaceous to the start of the Eocene, when Australia was still connected to Antarctica, is unrepresented in the record, so it is unknown what lineages were already established by that time.

EOCENE

The meagre early Eocene record confirms the existence of few groups of birds. Remains of penguins, mostly indeterminate forms, have been found along the coasts of south-eastern South Australia and south-western Victoria. Among these was the giant penguin *Anthropornis nordenskjoeldi*, which had been originally discovered at Seymour Island; it may have reached 1.35 m and 90 kg (Wiman, 1905; Jenkins, 1974, 1985).

The foot impression of a large bird from south-eastern Queensland has been referred to the Dromornithidae, a group of large, flightless birds endemic to Australia (Vickers-Rich and Molnar, 1996). If the fossil is correctly assigned, then it is the earliest record of this family, but

Figure 2 *(opposite)*. Geologic ranges of major groups of Australasian birds—based on Boles (2000), originally modified from Rich and van Tets (1982) for Australia to indicate new finds and accommodate revisions in ages of sites. Families without fossil records or ones restricted to the Holocene are omitted. Light thin line and question mark indicate possible range of Dromornithidae based on non-osteological evidence. Key to diagram numbers:

1. feathers from Koonwarra, Vic.;
2. *Nanantius eos* and other enanthiornithine material, and indeterminate specimens from Lightning Ridge, N.S.W. and Flat Rock, Vic.;
3. Casuariinae (cassowaries);
4. Dromaiinae (emus);
5. Megapodiidae (megapodes);
6. Phasianidae (quail);
7. Anseranatidae (magpie geese);
8. Anatidae (waterfowl);
9. Dromornithidae (dromornithid birds);
10. Podicipedidae (grebes);
11. Sphenisciformes (penguins);
12. Diomedeidae (albatrosses);
13. Anhingidae (darters);
14. Phalacrocoracidae (cormorants);
15. Pelecanidae (pelicans);
16. Ciconiidae (storks);
17. Threskiornithidae (spoonbills, ibises);
18. Ardeidae (herons);
19. Palaelodidae (*Palaelodus*);
20. Phoenicopteridae (flamingoes);
21. Accipitridae (hawks, eagles);
22. Falconidae (falcons);
23. Turnicidae (button-quail);
24. Gruidae (cranes);
25. Rallidae (rails);
26. Otididae (bustards);
27. Graculavidae ('transitional Charadriiformes');
28. Charadriiformes (indeterminate waders);
29. Pedionomidae (plains wanderers);
30. Scolopacidae (sandpipers);
31. Burhinidae (stone-curlews);
32. Charadriidae (plovers/lapwings);
33. Laridae (gulls);
34. Columbidae (pigeons);
35. Cacatuidae (cockatoos);
36. Psittacidae (parrots/lorikeets);
36. Cuculidae (cuckoos);
37. Centropodidae (coucals);
38. Tytonidae (barn owls);
39. Strigidae (typical owls);
40. Podargidae (frogmouths);
41. Eurostopodidae (eared nightjars);
42. Aegothelidae (owlet-nightjars);
43. Apodidae (swifts);
44. Halcyonidae (land kingfishers);
45. Passeriformes (songbirds).

Table 1. Earliest records of higher taxonomic groups of Australian birds. Taxa designated as 'unnamed' are awaiting detailed study; those designated 'indeterminate' cannot be identified to a greater taxonomic resolution. The taxonomy of modern taxa follows Christidis and Boles (1994), except for the use of Eupetidae instead of Cinclosomatidae (Bock, 1994). The allocation of the Dromornithidae to the Anseriformes follows Murray and Megirian (1998) and the recognition of the form-family Graculavidae is after Olson and Parris (1987). Names preceded by '†' designate extinct taxa.

Higher taxon	Genus/species	Age	Formation	Locality	Fauna/LF	References
Aves	indeterminate (feather impressions)	E Cretaceous		Vic: Koonwarra		Talent *et al.* (1966), Waldman (1970)
	indeterminate	E Cretaceous	Griman Creek	NSW: Lightning Ridge		Molnar (1999), Smith (1999)
	indeterminate	E Cretaceous	Wonthaggi	Vic: Flat Rocks		Rich *et al.* (1999)
†Enantiornithes	indeterminate	E Cretaceous	Toolebuc	Qld: Canary Station		Kurochkin and Molnar (1997), Kear *et al.* (2003)
†family indeterminate	†*Nanantius eos*	E Cretaceous	Toolebuc	Qld: Warra Station		Molnar (1986), Kurochkin and Molnar (1997)
Neornithes Struthioniformes						
Casuariidae	†*Emuarius gidju*	L Oligocene		Qld: Riversleigh	LSO, Hiatus South, White Hunter	Patterson and Rich (1987), Boles (1992, 1997c, 2001a)
	†*Emuarius guljaruba*	L Oligocene	Etadunna	SA: Lake Palankarinna	Ngama	Boles (2001a)
Galliformes						
Megapodiidae	†*Ngawupodius minya*	L Oligocene	Namba	SA: Lake Pinpa	Ericmas	Boles and Ivison (1999)
Phasianidae	*Coturnix* sp.	Pleistocene	unnamed cave deposits	Vic: Pyramids Cave	unnamed	Baird (1993)
Anseriformes						
Anseranatidae	*Anseranas* cf *A. semipalmatus*	E/M Pliocene	Chinchilla Sands	Qld: Chinchilla	Chinchilla	Rich *et al.* (1991)
Anatidae	unnamed	L Oligocene	Etadunna	SA: Lake Palankarinna	Ngapakaldi, Ngama	Vickers-Rich (1991), Rich *et al.* (1991)
				SA: Lake Kanunka	Ngama	
			Namba	SA: Lake Pinpa SA: Lake Yanda	Pinpa Yanda	
				Qld: Riversleigh	White Hunter	Boles (1997d)
†Dromornithidae	?indeterminate (foot impression)	E Eocene	Redbank Plains	Qld: Redbank Plains		Vickers-Rich and Molnar (1996)
	†*Barawertornis tedfordi*	L Oligocene		Qld: Riversleigh	various	Rich (1979), Boles (2000)
	unnamed genus and species	L Oligocene		Qld: Riversleigh	various	Boles (2000)

Table 1. (continued)

Higher taxon	Genus/species	Age	Formation	Locality	Fauna/LF	References
Podicipediformes						
Podicipedidae	unnamed	L Oligocene	Namba	SA: Lake Pinpa	Pinpa	Vickers-Rich (1991), Rich *et al.* (1991)
Sphenisciformes						
Spheniscidae	†*Anthropornis nordenskjoeldi*	L Eocene	Blanche Point	SA: Blanche Point		Wiman (1905), Jenkins (1974, 1985)
	indeterminae	L Eocene	Blanche Point	SA: Blanche Point Vic: Otway Ranges: Brown Creek		Finlayson (1938), Glaessner (1955), Simpson (1957), Jenkins (1985), Vickers-Rich (1991)
Procellariiformes						
Procellariidae	numerous species	Holocene	unnamed cave and dune deposits	SA, Tas, Vic, offshore islands, Lord Howe Is, Macquarie Is, Norfolk Is, Nepean Is		Baird (1991a and references therein), Meredith (1991)
Diomedeidae	†*Diomedea thyridata*	L Miocene	Black Rock	Vic: Beaumaris	Beaumaris	Wilkinson (1969)
Hydrobatidae	*Pelagodroma marina*	Holocene	unnamed dune deposits	Lord Howe Is, Norfolk Is, Nepean Is		Meredith (1991)
	Fregetta grallaria			Lord Howe Is		
Pelecaniformes						
Phaethontidae	*Phaethon rubricauda*	Holocene	unnamed dune deposits	Norfolk Is		Meredith (1991)
Sulidae	*Sula dactylatra*	Holocene	unnamed dune deposits	Lord Howe Is, Norfolk Is, Nepean Is		Meredith (1991)
Anhingidae	†*Anhinga malagurala*	E Pliocene	Allingham	Qld: Bluff Downs	Bluff Downs	Mackness (1995)
Phalacrocoracidae	unnamed	E Miocene		Qld: Riversleigh		Boles (1997d)
Pelecanidae	†*Pelecanus tirarensis*	L Oligocene	Etadunna Namba	SA: Lake Palankarinna SA: Lake Pinpa	Ngapakaldi Pinpa	Miller (1966), Rich and van Tets (1981)
Fregatidae	no record					
Ciconiiformes						
Ciconiidae	†*Ciconia* new species	L Oligocene		Qld: Riversleigh	White Hunter	Boles (unpublished data)
Threskiornithidae	*Threskiornis* cf. *T. molucca*	E Pliocene	Allingham	Qld: Bluff Downs	Bluff Downs	Boles and Mackness (1994)
Ardeidae	cf. *Ardea picata*	E Pliocene	Allingham	Qld: Bluff Downs	Bluff Downs	Boles and Mackness (1994)

Table 1. (continued)

Higher taxon	Genus/species	Age	Formation	Locality	Fauna/LF	References
Phoenicopteriformes						
Phoenicopteridae	†*Phoeniconotius eyrensis*	L Oligocene	Etadunna	SA: Lake Palankarinna	Ngapakaldi	Miller (1963a)
	†*Phoenicopterus novaehollandiae*	L Oligocene	Etadunna	SA: Lake Pitikanta	Ngapakaldi	Miller (1963a)
†Palaelodidae	†*Palaelodus pledgei*	L Oligocene	Etadunna	SA: Lake Palankarinna	Minkina, Ditjimanka	Baird and Rich (1998)
	†*Palaelodus wilsoni*	L Oligocene	Etadunna	SA: Lake Palankarinna	Minkina, Ditjimanka, Ngama	Baird and Rich (1998)
			Namba	SA: Lake Yanda SA: Lake Pinpa	Ericmas Pinpa	
Falconiformes						
Accipitridae	†*Pengana robertbolesi*	L Oligocene		Qld: Riversleigh	Sticky Beak	Boles (1993c)
	unnamed	L Oligocene	Etadunna	SA: Lake Palankarinna	Ngapakaldi, Ngama	Rich *et al.* (1991), Gaff (2002)
Falconidae	*Falco berigora*	Pleistocene	unnamed cave deposits	SA: Green Waterhole Cave		Rich *et al.* (1982)
Gruiformes						
Gruidae	indeterminate	L Oligocene	Etadunna	SA: Lake Palankarinna, Lake Ngapakaldi	Ngapakaldi	Vickers-Rich (1991), Rich *et al.* (1991)
Rallidae	†*Gallinula* new species	L Oligocene		Qld: Riversleigh		Boles (unpublished data)
	unnamed	L Oligocene	Etadunna	SA: Lake Palankarinna	Ngapakaldi, Ngama	Vickers-Rich (1991), Rich *et al.* (1991)
			Namba	SA: Lake Pinpa	Pinpa	
Otididae	cf. *Ardeotis australis*	L Pliocene	Tirari	SA: Lake Kanunka	Kanunka	Vickers-Rich (1991), Rich *et al.* (1991)
Turniciformes						
Turnicidae	*Turnix varia*	Pleistocene	unnamed cave deposits	Vic: Pyramids Cave		Baird (1993)
	indeterminate/unnamed	L Oligocene	Etadunna	SA: Lake Palankarinna	Ngama	Vickers-Rich (1991), Rich *et al.* (1991)
Charadriiformes						
†Graculavidae	unnamed	E Eocene		Qld: Murgon	Tingamarra	Boles (1999a)
Pedionomidae	*Pedionomus* cf. *P. torquatus*	Plio-Pleistocene	unnamed cave deposits	SA: Victoria Fossil Cave		Rich and McEvey (1980)
Scolopacidae	cf. *Numenius* sp.	E Pliocene	Allingham	Qld: Bluff Downs	Bluff Downs	Boles and Mackness (1994)
Rostratulidae	no record					

Table 1. (continued)

Higher taxon	Genus/species	Age	Formation	Locality	Fauna/LF	References
Jacanidae	no record					
Burhinidae	unnamed	L Oligocene	Etadunna	SA: Lake Palankarinna	Ngapakaldi, Ngama	Vickers-Rich (1991), Rich *et al.* (1991)
			Namba	SA: Lake Pinpa	Pinpa	
Haematopodidae	no record					
Recurvirostridae	no record					
Charadriidae	*Charadrius australis*	Pleistocene	unnamed cave deposits	Vic: Victoria Fossil Cave		van Tets and Smith (1974)
Glareolidae	no record					
Laridae	unnamed	L Oligocene	Namba	SA: Lake Pinpa	Pinpa	Rich and van Tets (1982)
Columbiformes						
Columbidae	unnamed	L Oligocene	Etadunna Namba	SA: Lake Palankarinna SA: Lake Pinpa	Ngapakaldi Ericmas	Vickers-Rich (1991), Rich *et al.* (1991)
Psittaciformes						
Cacatuidae	*Cacatua* indeterminate	E Miocene		Qld: Riversleigh	RSO	Boles (1993a)
Psittacidae	*Melopsittacus undulatus*	E Pliocene		Qld: Riversleigh: Rackham's Roost	Rackham's Roost	Boles (1998)
Cuculiformes						
Cuculidae	*Chrysococcyx* sp.	Pleistocene	unnamed cave deposits	Vic: Pyramids Cave		Baird (1993)
Centropodidae	†*Centropus colossus*	Quaternary	unnamed cave deposits	SA: Green Waterhole Cave		Baird (1985)
Strigiformes						
Tytonidae	*Tyto novaehollandiae*	Pleistocene	unnamed cave deposits	Vic: Pyramids Cave		Baird (1993)
Strigidae	*Ninox connivens*	Holocene	unnamed cave deposits	Vic: McEachern's Cave		McNamara (1991)
Caprimulgiformes						
Podargidae	*Podargus strigoides*	Holocene	unnamed cave deposits	Vic: McEachern's Cave		McNamara (1991)
Caprimulgidae	no record					
Eurostopodidae	*Eurostopodus mystacalis*	Pleistocene	unnamed cave deposits	Vic: Pyramids Cave		Baird (1993)
Aegothelidae	†*Quipollornis konberi*	M Miocene		NSW: Bugaldi		Rich and McEvey (1977)

Table 1. (continued)

Higher taxon	Genus/species	Age	Formation	Locality	Fauna/LF	References
Apodiformes						
Apodidae	†*Collocalia buday*	L Oligocene		Qld: Riversleigh	Gillespie's Gully	Boles (2001b)
Coraciiformes						
Alcedinidae	*Alcedo azurea*	Holocene	unnamed cave deposits	Vic: Curran's Creek Cave		Baird (1991a)
Halcyonidae	unnamed	M Miocene		Qld: Riversleigh	Last Minute	Boles (1997b)
Meropidae	no record					
Coraciidae	no record					
Passeriformes						
	indeterminate	E Eocene		Qld: Murgon	Tingamarra	Boles (1995a, 1997a)
Pittidae	no record					
Menuridae	†*Menura tyawanoides*	E Miocene		Qld: Riversleigh	Upper	Boles (1995b)
Atrichornithidae	*Atrichornis rufescens*	Pleistocene	unnamed cave deposits	Vic: Pyramids Cave		Baird (1993)
Climacteridae	*Climacteris* sp.	Pleistocene	unnamed cave deposits	Vic: Pyramids Cave		Baird (1993)
Maluridae	*Amytornis textilis*	Holocene	unnamed cave deposits	WA: Madura Cave		Baird (1991a)
Pardalotidae	*Dasyornis broadbenti*	Pleistocene	unnamed cave deposits	SA: Green Waterhole Cave		Baird (1985)
Meliphagidae	unnamed	M Miocene		Qld: Riversleigh	Last Minute, Ringtail, AL'90?	Boles (unpublished data)
Petroicidae	*Petroica* sp.	Holocene	unnamed cave deposits	Vic: Clogg's Cave		Baird (1991a)
Orthonychidae	†*Orthonyx kaldowinyeri*	M Miocene		Qld: Riversleigh	Last Minute	Boles (1993b)
Pomatostomidae	*Pomatostomus superciliosus*	Holocene	unnamed cave deposits	WA: Madura Cave		Baird (1991a)
Eupetidae	*Cinclosoma punctatum* *Psophodes olivaceus*	Pleistocene	unnamed cave deposits	Vic: Pyramids Cave		Baird (1993)
Neosittidae	*Daphoenositta chrysoptera*	Holocene	unnamed cave deposits	Vic: Mabel Cave		Baird (1991a)
Pachycephalidae	*Oreoica gutturalis*	Pleistocene	unnamed cave deposits	WA: Koonalda Cave		Baird (1991a)
Dicruridae	*Grallina cyanoleuca*	Pleistocene	unnamed cave deposits	SA: Victoria Fossil Cave		Baird (1991a)
Campephagidae	*Lalage* cf. *L. leucopyga*	Holocene	unnamed dune deposits	Norfolk Is, Nepean Is		Meredith (1991)
Oriolidae	†*Longmornis robustirostrata*	E Miocene		Qld: Riversleigh	Neville's Garden	Boles (1999b)
Artamidae	*Strepera versicolor*	Pleistocene	unnamed cave deposits	WA: Devil's Lair		Baird (1991a)

Table 1. (continued)

Higher taxon	Genus/species	Age	Formation	Locality	Fauna/LF	References
Paradisaeidae	no record					
Corvidae	*Corvus* sp.	Pleistocene	unnamed cave deposits	SA: Green Waterhole Cave		Baird (1985)
Corcoracidae	*Corcorax melanorhamphos*	Quaternary	unnamed cave deposits	NSW: Wellington Caves		Boles (unpublished data)
Ptilonorhynchidae	*Ptilonorhynchus violaceus*	Holocene	unnamed cave deposits	Vic: Amphitheatre Cave		Baird (1992)
Alaudidae	no record					
Motacillidae	*Anthus novaeseelandiae*	Pleistocene	unnamed cave deposits	WA: Devil's Lair		Baird (1991a)
Passeridae	indeterminate	Pleistocene	unnamed cave deposits	WA: Devil's Lair		Baird (1991a)
Nectariniidae	no record					
Dicaeidae	no record					
Hirundinidae	indeterminate	Pleistocene	unnamed cave deposits	SA: Green Waterhole Cave		Baird (1985)
Sylviidae	*Cincloramphus cruralis*	Pleistocene	unnamed cave deposits	WA: Koonalda Cave		Baird (1991a)
Zosteropidae	*Zosterops* sp.	Holocene	unnamed cave deposits	Vic: Clogg's Cave		Baird (1991a)
Muscicapidae	*Turdus poliocephalus*	Holocene	unnamed cave deposits	Norfolk Is		Meredith (1991)
Sturnidae	*Aplonis* cf. *A. fusca*	Holocene	unnamed cave deposits	Norfolk Is		Meredith (1991)

gives no other clues about these birds; their diversity is documented later in the record.

The Graculavidae is a potentially unnatural assemblage of unrelated forms with similar postcranial skeletons, which resemble that of the living stone-curlews (Burhinidae) (Olson and Parris, 1987). It is now known that this morphology also characterized early waterfowl (Anseriformes) (Ericson, 1997; Livezey, 1997). The occurrence in the Tingamarra Local Fauna of Murgon, south-eastern Queensland, of birds assigned to the Graculavidae (Boles, 1999a) may say more about the distribution of this morphology than it does about phylogeographic relationships between land masses.

The important finding at this time was the discovery of the Passeriformes also in the Tingamarra Local Fauna (Boles, 1995a, 1997a). The origin of this order, numerically the most successful among birds, has been the subject of much speculation. The traditional, pre-continental drift theory that passerines arose in the Northern Hemisphere and colonized Australasia by a series of invasions (e.g. Mayr, 1944) has been challenged by the conclusions of molecular studies (e.g. Sibley and Ahlquist, 1985) and more recent interpretations of the fossil record (Feduccia and Olson, 1982; Olson, 1988). A southern origin for the songbirds is now considered likely by a number of authors (e.g. Feduccia and Olson, 1982; Olson, 1988; Maclean, 1990; Christidis, 1991; Cracraft, 2001; Barker *et al.*, 2002; Ericson *et al.*, 2002). The early Eocene occurrence of passerines in Australia, 25 million years older than the oldest Northern Hemisphere record (Mourer-Chauviré *et al.*, 1989), adds support for this idea.

Between these early Eocene records and the late Oligocene is a particularly significant gap. It was during this time period that the final connection between Australia and Antarctica was severed, leaving Australia isolated from other land masses.

OLIGOCENE–MIOCENE

The growing understanding of birds during the late Oligocene and early Miocene is a result of new fossil deposits and revised dating of previously exploited ones. This was an important period for interchange of taxa between the north and south in other parts of the world (Olson, 1988) and, in some places (e.g. New Zealand; Cooper and Cooper, 1995; Holdaway and Worthy, 2006) of extreme ecological crises. Australia, however, would have been still isolated. Families now known to have been present during that time seem likely to have been on the continent before the final connection with Antarctica was severed. The alternative is that they flew here, but the only land masses that would have been realistic sources for such birds were other Gondwanan ones. It is possible that some lineages may be of sufficient antiquity that they occurred in both Gondwana and Laurasia, when these were in closer proximity or even in contact (as Pangaea).

The presence of several taxa of the palaeognathous Casuariidae (*Emuarius*, *Dromaius*) at the same time is strongly indicative that this lineage had been established well before it began to diversify in Australia (Patterson and Rich, 1987; Boles, 1992, 2001a). Results from DNA sequencing imply that the closest relatives to the Casuariidae among living birds are the kiwis (Apterygidae) of New Zealand (Cooper *et al.*, 1992; van Tuinen *et al.*, 1998; Haddrath and Baker, 2001). Both families exhibit structural modifications consistent with having been flightless for extended periods. Assuming that they evolved from a flightless common ancestor, then the alternative scenarios of how kiwis got to New Zealand depend largely on the date of the cassowary/emu–kiwi divergence. If this occurred before New Zealand split from Australia, then vicariance is a satisfactory explanation. Alternatively, if the divergence post-dated the separation event, then kiwis must have reached New Zealand by a land-bridge across the Norfolk Arc or other colonization means. The possibility that kiwis evolved in an isolated New Zealand from a flying, over water colonist that also gave rise to the casuariids in Australia, seems unlikely.

Similarly, the late Oligocene–early Miocene record of the other major group of flightless birds—the Dromornithidae (*Barawertornis* and an undescribed genus)—points to a long history of these birds in Australia (Rich, 1979; Boles, 2000). The phylogenetic analysis of Murray and Megirian (1998) concluded that the dromornithids were the sister-group of the South American screamers (Anhimidae). A distal fragment of a large tarsometatarsus from the middle Miocene–early Oligocene La Meseta Formation, Seymour Island, Antarctica (Tambussi *et al.*, 1994), is of appropriate size for the Dromornithidae, but requires detailed comparisons before anything useful can be said about its affinities.

Families with more global distributions are represented at this time, mainly from the central Australian Etadunna and Namba Formations, and to a lesser extent, sites at Riversleigh. The central Australian deposits are dominated by water birds, and thus the record at this time includes the Anatidae (waterfowl), Podicipedidae (grebes), Phalacrocoracidae (cormorants), Pelecanidae (pelicans), Phoenicopteridae (flamingos), Palaelodidae, Gruidae (cranes), Rallidae (rails) and Burhinidae (stone-curlews). Non-aquatic forms are limited to the Accipitridae (birds of prey) and Columbidae (pigeons). Only a few of these specimens have been studied beyond the level of family allocation and been formally named. The occurrences of grebes, pelicans, palaelodids, stone-curlews and pigeons are earlier than any from the Northern Hemisphere, albeit sometimes only marginally so. The oldest Australian records of rails are roughly comparable with those elsewhere in the world. The Northern Hemisphere histories of waterfowl, cormorants, flamingos, cranes and birds of prey exceed in age those in Australia.

The pelicans and palaelodids have been reviewed in depth (Miller, 1966; Rich and van Tets, 1981; Baird and Rich, 1998); several species are represented, all belonging to widespread genera. The flamingos exhibited a remarkable Australian radiation, documented from this time period through into the Quaternary (Miller, 1963a; Rich *et al.*, 1987). Six species in four genera are known, a diversity rivalling or exceeding that anywhere else in the world. The only rail thus far studied in detail is a species of *Gallinula* from Riversleigh (Boles, unpublished data). This was a flightless form that evolved on the mainland in the presence of mammalian carnivores. Many Quaternary rails have lost the power of flight, but almost all are confined to small islands free of mammalian predators.

Many of these families were established in both hemispheres by the start of the Neogene. This suggests an original Pangaean distribution for the lineages or early dispersal from one supercontinent to the other after origination. Many of the families include birds capable of extensive, long distance flight, so the latter alternative would have remained feasible for some time after the initial split between Laurasia and Gondwana. At this point, a choice between these possibilities cannot be made, and both may be valid for different families. The Phoenicopteridae and Palaelodidae formed part of the basal continental avifauna of Mourer-Chauviré (1999).

Other avifaunal elements appearing in the record during the early Miocene are more restricted in their fossil and modern distributions. The oldest records of the Cacatuidae (cockatoos) (Boles, 1993a) and Apodidae: Collocaliinae (swiftlets) (Boles, 2001b) come from the Riversleigh deposits. Also found here are numerous remains of passerines. Most are unidentified below order, but there are named species in the Menuridae (lyrebirds) (Boles, 1995b), Orthonychidae (logrunners) (Boles, 1993b) and Oriolidae (orioles and kin) (Boles, 1999b). The first two distinctive families are considered autochthonous to Australia. They are represented by the modern genera *Menura* and *Orthonyx*, respectively, known to have evolved by at least the middle Miocene. The Oriolidae is also part of the corvoid radiation that is centred on the Australo-Papuan region.

It is uncertain when it first became feasible for birds from outside Australia to colonize the continent. Olson (1988) proposed that it was when the continent reached its current position during the Miocene. Because of birds' ability to fly, their initial colonization may have taken place substantially before that of non-volant vertebrates. Avian families appearing in the record by the middle Miocene may have had several origins.

If the record of owlet-nightjars (Aegothelidae) from the Eo-Oligocene of France (Mourer-Chauviré, 1988) is correct, and in the absence of a comparable record from Australia, then it suggests that this family may have entered from the north by the middle Miocene. The French material requires more documentation before its identification can be confirmed and no conclusions should be made on this basis.

Nominal taxa in the Ciconiidae: Ciconiinae (ciconiine storks) have been reported from the early Miocene of Europe and more widespread localities from younger deposits elsewhere in the world. Because of the relative scrappiness of the material representing several named Tertiary species, and the conservative trend to refer most poorly known small storks to the genus *Ciconia*, it is difficult to interpret Riversleigh fossils of this genus in a global perspective. The current diversity of recognized species in this subfamily may reflect taxonomic practice as much as, or more than, actual generic relationships.

Fry (1980) speculated that the Halcyonidae (land kingfishers) diversified in the rainforests of Malesia after its present geographical configuration had been attained. A fossil from the middle Miocene of Riversleigh retains a feature found in living primitive genera (Boles, 1997b). While Fry's scenario may be correct as far as the location of the halcyonid radiation is concerned, there are some problems with the timing of this event. During the early to middle Miocene Australia had not yet approached the position, relative to Asia, that it occupies today and sections of the Indo-Malaysian island arc were still undergoing formation.

Olson (1988) suggested that it was possible at this time for megapodes and frogmouths to arrive in Australia from the north. Megapodes have been shown to be established in Australia by that time (Boles and Ivison, 1999), but there is no evidence of frogmouths to challenge the suggestion of a northern origin for this group.

The Australian and Asian plates had collided by the late Miocene, forcing the New Guinean land mass above the ocean. There are few sites of this age that have yielded bird remains. The only new family appearing in the Australian record at this time is the Diomedeidae (albatrosses) (Wilkinson, 1969). Because of the extreme vagility of these birds and their dependence on oceanic wind systems, their presence reveals little about intercontinental faunal connections.

By this time, the Dromornithidae had diversified into several genera (Rich, 1979). Alcoota, Northern Territory, has yielded two species of *Ilbandornis*, comparatively gracile ostrich-like forms, and the gigantic *Dromornis stirtoni*, which may have been the heaviest bird to have ever lived (3 m, 450–500 kg). The ecological role of dromornithid birds is under debate. Some had massive skulls, leading to questions about the type of foods that they ate (Murray and Megirian, 1998; Wroe, 1999).

The low taxonomic diversity of bird fossils from the late Miocene is unfortunate. Potential colonization routes between Australia and Asia were becoming firmly established, and northern groups of birds would have moved along them into Australia. The actual and relative timing of these events for different families cannot be established in the absence of a suitable record.

Most of the passerine families within the Corvida, as originally recognized by Sibley and Ahlquist (1990), are centred in Australo-Papua and are endemic or near endemic to this region. They are considered on molecular grounds (Sibley, 1976; Sibley and Ahlquist, 1985, 1990; Sibley *et al.*, 1988; Baverstock *et al.*, 1991; Christidis, 1991; Christidis and Schodde, 1991) to have autochtho-

Table 3. Named avian genera known from the fossil record that no longer occur in Australia. The major references to each taxon in Australia are given.

Status	Range (Genus)	Family	Sources
Extinct			
	Endemic to Australia		
	Nanantius	Family indeterminate	Molnar (1986), Kurochkin and Molnar (1997)
	Emuarius	Casuariidae	Boles (1992, 1997a, 2001a)
	Ngawupodius	Megapodiidae	Boles and Ivison (1999)
	Progura	Megapodiidae	van Tets (1974)
	Barawertornis	Dromornithidae	Rich (1979); Boles (2000)
	Bullockornis	Dromornithidae	Rich (1979)
	Ilbandornis	Dromornithidae	Rich (1979)
	Dromornis	Dromornithidae	Owen (1872), Rich (1979)
	Genyornis	Dromornithidae	Stirling (1913), Stirling and Zietz (1896, 1900, 1905), Rich (1979)
	Anthropodytes	Spheniscidae	Simpson (1959)
	Pseudoaptenodytes	Spheniscidae	Simpson (1970)
	Tasidyptes	Spheniscidae	van Tets and O'Connor (1983)
	Ocyplanus	Phoenicopteridae	De Vis (1905), Rich *et al.* (1987)
	Phoeniconotius	Phoenicopteridae	Miller (1963a), Rich *et al.* (1987)
	Xenorhynchopsis	Phoenicopteridae	De Vis (1905), Rich *et al.* (1987)
	Pengana	Accipitridae	Boles (1993c)
	Quipollornis	Aegothelidae	Rich and McEvey (1977)
	Longmornis	Oriolidae	Boles (1999b)
	Not endemic to Australia		
	Anthropornis	Spheniscidae	Wiman (1905), Jenkins (1985)
	Palaelodus	Palaelodidae	Baird and Rich (1998)
Extant, no longer in Australia			
	Ciconia	Ciconiidae	Boles (unpublished data)
	Phoenicopterus	Phoenicopteridae	Miller (1963a), Rich *et al.* (1987)

the scrappy nature of the fossils and a conservative approach to describing them. The number of named fossil forms can be expected to increase as more of the existing specimens from Murgon, central Australia and Riversleigh are studied. Even granted the preceding contributing factors, the pattern of apparent conservatism at higher taxonomic levels contrasts with the record from the rest of the world. The compilation by Unwin (1993) of avian families in the fossil record was used as the basis for comparisons, with some modification. There have been additions to this record since it was published, and there is considerable lack of agreement on the relationships and higher taxonomic allocation of many taxa. Nonetheless, it serves for presenting a broad overview of neornithine family and ordinal diversity during the Tertiary.

Although palaeofamilies are known from throughout the Tertiary, the overwhelming preponderance is from the Paleocene and Eocene, with a number extending into the Oligocene (Martin, 1983). Their occurrence in deposits of this age alone is adequate to explain the absence of comparable palaeofamilies in the Australian record: there are no Australian fossil deposits of Paleocene age that have yielded birds.

The hiatus between the early Eocene and late Oligocene sites encompasses the time during which many modern families are thought to have arisen, replacing the more archaic families. The antiquity of many modern lineages is starting to be demonstrated, however, with putative records from the Cretaceous (Olson, 1992; Kurochkin, 1995; Chiappe, 1996; Chatterjee, 1997; Stidham, 1998; Hope, 2002); but some of these must be treated with caution. The proposed turnover in more primitive families seems supported by the fossil record. Undoubtedly the lack of such archaic groups in the Australian avian record reflects in part the influence of the age of known fossil deposits. The lack of a residue of archaic families into the late Oligocene suggests that substantial differences existed between the taxonomic diversity in Australia and elsewhere throughout early Tertiary.

Palaeofamilies known from the Miocene are considerably fewer in number throughout the world. Some of these comprised large, flightless, terrestrial forms, and these may be taken as essentially local evolutionary events, comparable to the Dromornithidae and Casuariidae in Australia. The Australian Miocene record, although much richer than in the Palaeogene, albeit still patchy, does not yet show any novel families of volant birds, all also being known from elsewhere in the world. Because Australia was still isolated from other land masses for most of this time, it would have received little

Table 4. Named avian species known from the fossil record that no longer occur in Australia. The major references to each taxon in Australia are given.

Status	Range, Species	Family	Source(s)
Extinct			
	Endemic to Australia		
	Extinct genera		
	Emuarius gidju	Casuariidae	Patterson and Rich (1987); Boles (1992, 1997a)
	Emuarius guljaruba	Casuariidae	Boles (2001a)
	Ngawupodius minya	Megapodiidae	Boles and Ivison (1999)
	Progura gallinacea	Megapodiidae	De Vis (1888a,b, 1889, 1892), van Tets (1974)
	Progura naracoortensis (may be conspecific with *P. gallinacea*)	Megapodiidae	van Tets (1974), Boles (unpublished data)
	Barawertornis tedfordi	Dromornithidae	Rich (1979), Boles (2000)
	Bullockornis planei	Dromornithidae	Rich (1979), Murray and Megirian (1998)
	Ilbandornis lawsoni	Dromornithidae	Rich (1979)
	Ilbandornis woodburnei	Dromornithidae	Rich (1979)
	Dromornis stirtoni	Dromornithidae	Rich (1979), Jennings (1990)
	Dromornis australis	Dromornithidae	Owen (1872), Rich (1979)
	Genyornis newtoni	Dromornithidae	Stirling (1913), Stirling and Zietz (1896, 1900, 1905), Rich (1979), Miller *et al.* (1999)
	Anthropodytes gilli	Spheniscidae	Simpson (1959)
	Pseudoaptenodytes macraei	Spheniscidae	Simpson (1970)
	Pseudoaptenodytes minor	Spheniscidae	Simpson (1970)
	Tasidyptes hunteri	Spheniscidae	van Tets and O'Connor (1983)
	Phalacrocorax serventyorum	Phalacrocoracidae	van Tets (1994)
	Ocyplanus proeses	Phoenicopteridae	De Vis (1905), Miller (1963a), Rich *et al.* (1987)
	Phoeniconotius eyrensis	Phoenicopteridae	Miller (1963a), Rich *et al.* (1987)
	Xenorhynchopsis minor	Phoenicopteridae	De Vis (1905), Rich *et al.* (1987)
	Xenorhynchopsis tibialis	Phoenicopteridae	De Vis (1905), Rich *et al.* (1987)
	Palaelodus pledgei	Palaelodidae	Baird and Rich (1998)
	Palaelodus wilsoni	Palaelodidae	Baird and Rich (1998)
	Pengana robertbolesi	Accipitridae	Boles (1993c)
	Quipollornis koniberi	Aegothelidae	Rich and McEvey (1977)
	Longmornis robustirostrata	Oriolidae	Boles (1999b)
	Extant genera, currently Australian		
	Casuarius lydekkeri	Casuariidae	Lydekker (1891), Rothschild (1911), Miller (1962), Rich *et al.* (1988)
	Dromaius ocypus	Casuariidae	Miller (1963b), Patterson and Rich (1987)
	Diomedea thyridata	Diomedeidae	Wilkinson (1969)
	Anhinga malagurala	Anhingidae	Mackness (1995)
	Pelecanus tirarensis	Pelecanidae	Miller (1966), Rich and van Tets (1981)
	Pelecanus cadimurka	Pelecanidae	Rich and van Tets (1981)
	Pelecanus proavus (possibly conspecific with extant *P. conspicillatus*)	Pelecanidae	De Vis (1892), Rich and van Tets (1981)
	Gallinula new species	Rallidae	Boles (unpublished data)
	Centropus colossus	Centropodidae	Baird (1985)
	Collocalia buday	Apodidae	Boles (2001b)
	Menura tyawanoides	Menuridae	Boles (1995b)
	Pycnoptilus fordi	Pardalotidae	Baird (1993)
	Orthonyx kaldowinyeri	Orthonychidae	Boles (1993b)
	Orthonyx hypsilophus	Orthonychidae	Baird (1985)
	Orthonyx wakefieldi	Orthonychidae	Baird (1993)
	Extant genera, formerly Australian		
	Ciconia nana	Ciconiidae	De Vis (1888b, 1905), Boles (unpublished data)
	Ciconia new species	Ciconiidae	Boles (unpublished data)
	Phoenicopterus novaehollandiae	Phoenicopteridae	Miller (1963a), Rich *et al.* (1987)
	Not endemic to Australia		
	Anthropornis nordenskjoeldi	Spheniscidae	Wiman (1905), Jenkins (1974, 1985)
Extant, no longer in Australia			
	Phoenicopterus ruber	Phoenicopteridae	Miller (1963a), Rich *et al.* (1987)

Table 5. Taxa proposed by C.W. De Vis that are currently under study or require further examination. Those in the Accipitridae were reviewed by Gaff (2002), but this work is not yet published.

Genera and Species	Family	Publication
Necrastur	Accipitridae	De Vis (1892)
Taphaetus	Accipitridae	De Vis (1890, 1891)
Palaeolestes	Accipitridae? possibly non-avian	De Vis (1911)
Phalacrocorax gregorii	Phalacrocoracidae	De Vis (1905)
Phalacrocorax vetustus	Phalacrocoracidae	De Vis (1905)
Baza gracilis	Accipitridae	De Vis (1905)
Necrastur alacer	Accipitridae	De Vis (1892)
Taphaetus brachialis	Accipitridae	De Vis (1890)
Taphaetus lacertosus	Accipitridae	De Vis (1905)
Palaeolestes gorei	Accipitridae? possibly non-avian	De Vis (1911)

trans-oceanic colonization by new families. Any evolutionary experimentation would have had to occur from within the existing Australian Gondwanan avifauna. There is no fossil evidence of this.

The marsupial fauna would have been constrained in the same way, although with even less chance of colonization from outside. A comparison of the mid to late Tertiary record of Australian birds and mammals exhibits some noticeable differences between these groups. Physical constraints imposed by flight restrict the limits of morphological diversification of birds. Allowing for these limitations, several groups of Australian birds, for example, the columbiforms, psittaciforms and passeriforms, exhibit as much diversity as seen anywhere in the world. It can be argued that this is comparable to that seen in Australia's modern marsupial radiation.

The fossil records of Australian birds and mammals present a different picture from that evident today (mammalian record based on Archer *et al.*, 1999 and Long *et al.*, 2002). The Tertiary record of mammals includes two now extinct orders (Yalkaparadontoidea; unnamed, includes Yingabalaridae; Archer *et al.*, 1988). The marsupials exhibit striking radiations and the diverse emergence of a number of lineages no longer alive. In extant orders, there are several palaeofamilies (e.g. Wynyardiidae, Iliariidae, Balbaridae), some with quite divergent morphologies, and more than 85 named palaeogenera. Possums (Phalangeriformes) alone have some three extinct families and 11 extinct genera named to date, in addition to those still extant. Some of these novel groups did not persist much beyond the early Miocene, but most were not evolutionary experiments restricted to the early Tertiary. In contrast, there is an absence of marked radiations among birds beyond those groups still represented today. To some degree, these contrasts may reflect differences between mammalogists and ornithologists in philosophies and practices of taxonomy and classification; however, this seems inadequate to explain the degree of variation between the fossil records of the two classes. It appears that during the Australian Tertiary there were fundamental differences between the pattern of avian diversification and that exhibited by marsupials on this continent; although the reasons for such discrepancies are unclear.

PATTERNS WITHIN LINEAGES

Ancestor-descendant

Few possible ancestor-descendant pairs can be identified from these fossil and living species. There are frequently long time periods between the occurrence of members of a lineage and the radiation of most of these groups is poorly appreciated. The best documented fossil history is that of the Dromornithidae; however, much of the diversity of these birds is yet to be discovered. No direct ancestry of known taxa has been demonstrated, although preliminary analysis that an unnamed taxon from the late Oligocene–early Miocene of Riversleigh is the sister-taxon to *Bullockornis* from the middle Miocene of Bullock Creek raises the possibility of direct ancestry (Boles, 2000); additional specimens are needed to test this idea rigorously.

The lack of crucial specimens hampers interpretation of relationships within the Dromaiinae, the lineage comprising emus *Dromaius* and its extinct sister-group *Emuarius*. The latter genus is recognized in part because of its possession of a primitive *Casuarius*-like femur combined with an advanced *Dromaius*-like tarsometatarsus (documented in *Emuarius gidju*) (Boles, 1992). Whether the recently described *E. guljaruba*, tentatively placed in this genus, is actually closer to *Dromaius* cannot be assessed because the femur is unknown. Dromaiine specimens exist from the period between *E. guljaruba* (late Oligocene) and *Dromaius ocypus* (Pliocene), but were considered indeterminate by Patterson and Rich (1987) because they were inadequate for identification below generic level or still required detailed study. Evaluation of a possible ancestor-descendant relationship between late Oligocene forms and *D. ocypus* and/or *D. novaehollandiae*, both known from Pliocene-aged sites, awaits more evidence.

The lack of specimens from critical intervening times is also pronounced in the megapodes. *Ngawupodius* (late Oligocene) has a tarsometatarsus with proportions most similar to those of *Progura* and *Leipoa* among other megapode taxa; a direct relationship between these genera should be considered (Boles and Ivison, 1999). The flightless Riversleigh *Gallinula* cannot be the direct ancestor of the living volant *G. ventralis*, but the possibility that it gave rise to the similarly flightless *G. mortierii* cannot be ruled out (Boles, unpublished data). Avian candidates as megafauna are few. The giant megapode *Progura gallinacea*, coucal *Centropus colossus* and large accipitrids (Gaff, 2002) are among the few palaeospecies. Boles has speculated that *G. mortierii* may be a direct megafaunal descendant of the Riversleigh *Gallinula*.

Other possible ancestor-descendant series are in the songbird genera *Orthonyx* and *Menura*. The diversification within *Orthonyx* was greater than it is at present, with three palaeospecies and two neospecies (Baird, 1985, 1993; Boles, 1993b). *Orthonyx kaldowinyeri* (middle Miocene) may have been in the direct lineage of some or all of these Quaternary-aged taxa. Likewise, *Menura tyawanoides* may have been ancestral to the two living species (Boles, 1995b). The likelihood of a direct ancestry to modern forms is increasingly probable between Pliocene–Recent congeners, such as *Anhinga malagurala* and *A. melanogaster*, respectively.

Morphological Diversification

There is little evidence of morphologies that diverged markedly within a lineage beyond that expected among related taxa. Perhaps the only significant example is the development of the hyperflexible tarsal joint in the bird of prey *Pengana* (Boles, 1993c). Even this cannot be considered a radical innovation, however, as it has appeared convergently in two unrelated modern genera and the extinct Horusornithidae (Mourer-Chauviré, 1991).

Morphological trends can only be documented in a few lineages. One of the best illustrated is the adaptation of the hindlimb to an increasingly cursorial lifestyle. The best represented skeletal element in the dromaiines is the tarsometatarsus. From the late Oligocene species of *Emuarius*, through Pliocene *Dromaius ocypus* to the modern *D. novaehollandiae*, the trends of overall lengthening of the element and reduction in digit II can be followed. A similar although less pronounced pattern is discernible in the Dromornithidae. Both families exhibit other locomotory-related changes with time.

In the Aegothelidae a different transformation of the hindlimb is evident. The Miocene aged *Quipollornis koniberi* had proportionally longer hindlimbs than the living *Aegotheles*, suggesting that early owlet-nightjars had a greater reliance on the legs for locomotion than do modern ones (Rich and McEvey, 1977).

There are substantial differences in size between early and late members of some lineages. Several Eocene penguins were substantially larger than any living today (Jenkins, 1985). This contrast appears real because the number of known fossil and Recent taxa in the Spheniscidae is large enough for meaningful analysis. In other groups, the sample size is too small to ensure that it is a true reflection of a trend and not an artefact of an insufficient record. *Longmornis* is larger than any living member of the Oriolidae. *Ngawupodius minya* and *Orthonyx kaldowinyeri* are the smallest taxa in their respective families.

EXTINCTION

Reasons for the loss of taxa before the arrival of Europeans may be attributed to (1) climatic shifts and concomitant changes in habitats, (2) direct and indirect effects of human activities, including predation, (3) effects of non-human predators, (4) competition from other species or (5) a combination of two or more of these.

Australia has undergone a number of icehouse-greenhouse cycles. The two that probably had the greatest influence on the avifauna were the onset of drying in the late Middle Miocene, when the environment began to dry and rainforests started their retreat from central parts of the continent, and the height of aridification in the Pleistocene at the peak of the 'ice age', with the final loss of the permanent lake systems in central Australia and the spread of grasslands. No doubt, many rainforest lineages were extinguished by the retraction of rainforests, although among birds these have yet to be identified. The loss of flamingos and palaelodids from Australia has been attributed to the loss of suitable shallow lakes.

As more remains are found and identified, and earlier misidentifications corrected (e.g. van Tets and Rich, 1990), the extent of Pleistocene extinctions can be more accurately assessed. Older bird groups (e.g. Dromornithidae) that radiated in the Palaeogene were already showing a post-Tertiary reduction in diversity and may have also been terminally affected (Rich and van Tets, 1982). Large-bodied taxa, such as the dromornithid *Genyornis*, were prominent among the Pleistocene extinctions. Perhaps the impact on megafaunal forms is over emphasized because of the preferential preservation and relative ease of finding large bones and a predilection of palaeo-ornithologists for studying them. The growing record of small birds may show that the emphasis on gigantism is somewhat artefactual.

After reaching their peak, megafaunal forms exhibited either of two responses. Many became extinct. Those that survived underwent dwarfing—decreasing in size from this megafaunal peak to achieve the dimensions now seen in the extant form. Among marsupials, this phenomenon is well demonstrated in the grey kangaroo; the very large *Macropus* (*giganteus*) *titan* of the Pleistocene gave rise to the smaller *Macropus giganteus* of today. Within birds, a similar occurrence is suspected only for the megapodes. In this group, current studies by the author suggest that the giant megapode *Progura gallinacea* is the megafaunal form of the living *Leipoa ocellata*.

Whether dromornithid birds, and indeed most of the megafauna, went extinct because of the Pleistocene climatic changes or the activities of humans is currently a subject of considerable debate. Anthropogenic effects may not necessarily have been direct, as in hunting for food; indirect influences, such as habitat alteration through burning, may have had a greater and more widespread impact. Whatever relative impacts humans and climate had on dromornithids may have been exacerbated by the sudden rise of macropodine kangaroos. Rich

(1985) speculated that dromornithids had limited success moving into the expanding grasslands, unlike the rapidly radiating grazing macropodines.

Non-human predators as a cause of extinction have only been invoked for the loss of *Gallinula mortierii* from the mainland. Baird (1991b) proposed that the arrival of the dingo within the last 4,000 years might have caused this flightless bird to become restricted to Tasmania, where it still lives in the absence of this predator.

Only a few cases of possible competitive exclusion have been proposed. It has been suggested that the large penguins of the Eocene and Oligocene were eventually displaced by the seals and small cetaceans. There may have also been a relationship with climatic events resulting from the separation of Australia and Antarctica and subsequent strengthening of the Circum-Antarctic Current (Jenkins, 1985).

It might be expected that invading taxa from the north provided competition with ecologically similar counterparts from the late Miocene onwards. The fossil record of this period is insufficient to track the appearance of new forms and disappearance of previously occurring ones in a manner that allows correlation of these events. The genus of small storks *Ciconia* extended from the late Oligocene–early Miocene until the Pleistocene (Boles, unpublished data). There would have been considerable temporal overlap between *Ciconia nana* and the invading and much larger *Ephippiorhynchus asiaticus*; however, size and probable behavioural differences argue against competition between these storks. Closer to *Ciconia* in ecology are ibises *Threskiornis* and *Plegadis*, also likely invaders co-occurring with *C. nana* for some time. The loss of *C. nana* may have been due to climatic and environmental changes, but any such pressures could have been intensified through competition with ibises.

CONCLUSIONS

What can be said about the fossil record of birds in Australia? Much is an indication of what it lacks. Despite these deficiencies, the fossil record makes a useful contribution toward the understanding of Australasia's avifauna. It provides basic documentation needed for building a broad picture of avian evolution on these southern land masses by documenting changing distributions and early faunal connections with other continents; questioning or corroborating previous ideas of Australian endemicity; testing theories on origins, colonization routes and divergence times of various bird groups; and aiding the interpretation of changes in climate and environment. The record is still scanty, but will assume growing importance. Greater accessibility to known and future fossil localities, increased funding for field work and preparation, and the growing number of Australian-based palaeo-ornithologists cannot but help build on the existing knowledge of the avian fossil record.

ACKNOWLEDGEMENTS

I offer my gratitude to many colleagues for discussions over the years that helped me focus my thoughts, to the referees for constructive comments that helped clarify some of the arguments presented here, and to Lynne Ho, who assisted with the artwork in its preliminary stages.

REFERENCES

Alvarenga, H.M.F. 1995. Um primitivo membro da ordem Galliformes (Aves) do Terciario médio da Bacia de Taubaté, Estado de São Paulo, Brasil. *Anais da Academia Brasiliera de Ciências*, Vol. 67, pp.33–44.

Aplin, K.P. 2006. Ten million years of rodent evolution in Australasia: phylogenetic evidence and a speculative historical biogeography. Pp.707–744. In Merrick, J.R., Archer, M., Hickey, G.M. and Lee, M.S.Y. (eds). *Evolution and Biogeography of Australasian Vertebrates.* Auscipub. Pty. Ltd., Sydney. 942pp.

Aplin, K.P., Pasveer, J. and Boles, W.E. 1999. Late Quaternary vertebrates from the Bird's Head Peninsula, Irian Jaya, Indonesia, including descriptions of two previously unknown marsupial species. *Records of the Western Australian Museum, Supplement,* No. 57, pp.351–387.

Archer, M., Arena, R., Bassarova, M., Black, K., Brammall, J., Cooke, B., Creaser, P., Crosby, K., Gillespie, A., Godthelp, H., Gott, M., Hand, S.J., Kear, B., Krikmann, A., Mackness, B., Muirhead, J., Musser, A., Myers, T., Pledge, N., Wang, A. and Wroe, S. 1999. The evolutionary history and diversity of Australian mammals. *Australian Mammalogy,* Vol. 21, pp.1–45.

Archer, M., Hand, S. and Godthelp, H. 1988. A new order of Tertiary zalambdodont marsupials. *Science*, Vol. 239, pp.1528–1531.

Baird, R.F. 1985. Avian fossils from Quaternary deposits in 'Green Waterhole Cave', south-eastern South Australia. *Records of the Australian Museum,* Vol. 37, pp.353–370.

Baird, R.F. 1991a. Avian fossils from the Quaternary of Australia. Pp.809–870. In Vickers-Rich, P., Monaghan, J.M., Baird, R.F. and Rich, T.H. (eds). *Vertebrate Palaeontology of Australasia.* Pioneer Design Studio and Monash University Publications Committee, Melbourne. 1437pp.

Baird, R.F. 1991b. The dingo as a possible factor in the disappearance of *Gallinula mortierii* from the Australian mainland. *Emu*, Vol. 91, pp.121–122.

Baird, R.F. 1992. Fossil avian assemblage of pitfall origin from Holocene sediments in Amphitheatre Cave (G-2), south-western Victoria, Australia. *Records of the Australian Museum,* Vol. 44, pp.21–44.

Baird, R.F. 1993. Pleistocene avian fossils from Pyramid Cave (M-89), eastern Victoria, Australia. *Alcheringa,* Vol. 17, pp.383–404.

Baird, R.F. and Rich, P.V. 1998. *Palaelodus* (Aves: Palaelodidae) from the middle to late Cenozoic of Australia. *Alcheringa*, Vol. 22, pp.135–151.

Baird, R.F., Rich, P.V. and van Tets G.F. 1991. Appendix II. Localities yielding avian assemblages of Quaternary age in Australia. Pp.850–870. In Vickers-Rich, P., Monaghan,

J.M., Baird, R.F. and Rich, T.H. (eds). *Vertebrate Palaeontology of Australasia*. Pioneer Design Studio and Monash University Publications Committee, Melbourne. 1437pp.

Balouet, J.C. and Olson, S.L. 1989. Fossil birds from late Quaternary deposits in New Caledonia. *Smithsonian Contributions to Zoology,* No. 469, pp.1–38.

Barker, F.K., Barrowclough, G.F. and Groth, J.G. 2002. A phylogenetic hypothesis for passerine birds: taxonomic and biogeographic implications of an analysis of nuclear DNA sequence data. *Proceedings of the Royal Society of London,* Series B, Vol. 269, pp.295–308.

Baverstock, P.R., Schodde, R., Christidis, L., Kreig, M. and Sheedy, C. 1991. Microcomplement fixation: preliminary results from the Australasian avifauna. Pp.611–618. In Bell, B.D., Cossee, R.O., Flux, J.E.C., Heather, B.D., Hitchmough, R.A., Robertson, C.J.R. and Williams, M.J. (eds). *Acta XX Congressus Internationalis Ornithologici, Canterbury, 1990.* New Zealand Ornithological Trust Board, Wellington. 2568pp.

Bocheński, Z. 1999. Enantiornithes: earlier birds than *Archaeopteryx*? *Smithsonian Contributions to Paleobiology*, No. 88, pp.285–288.

Bock, W.J. 1994. History and nomenclature of avian family-group names. *Bulletin of the American Museum of Natural History*, Vol. 222, pp.1–281.

Boles, W.E. 1991. The origin and radiation of Australasian birds: perspectives from the fossil record. Pp.383–391. In Bell, B.D. Cossee, R.O., Flux, J.E.C., Heather, B.D., Hitchmough, R.A., Robertson, C.J.R. and Williams, M.J. (eds). *Acta XX Congressus Internationalis Ornithologici, Canterbury, 1990.* New Zealand Ornithological Trust Board, Wellington. 2568pp.

Boles, W.E. 1992. Revision of *Dromaius gidju* Patterson and Rich 1987 from Riversleigh, northwestern Queensland, Australia, with a reassessment of its generic position. *Natural History Museum of Los Angeles County, Science Series* 36, pp.195–208.

Boles, W.E. 1993a. A new cockatoo (Psittaciformes: Cacatuidae) from the Tertiary of Riversleigh, northwestern Queensland, and an evaluation of rostral characters in the systematics of parrots. *Ibis*, Vol. 135, pp.8–18.

Boles, W.E. 1993b. A logrunner *Orthonyx* (Passeriformes: Orthonychidae) from the Miocene of Riversleigh, north-western Queensland. *Emu*, Vol. 93, pp.44–49.

Boles, W.E. 1993c. *Pengana robertbolesi*, a peculiar bird of prey from the Tertiary of Riversleigh, northwestern Queensland, Australia. *Alcheringa*, Vol. 17, pp.19–25.

Boles, W.E. 1995a. The world's oldest songbird. *Nature*, Vol. 374, pp.21–22.

Boles, W.E. 1995b. A preliminary analysis of the Passeriformes from Riversleigh, northwestern Queensland, Australia, with the description of a new species of lyrebird. *Courier Forschungsinstitut Senckenberg*, Vol. 181, pp.163–170.

Boles, W.E. 1997a. Fossil songbirds (Passeriformes) from the early Eocene of Australia. *Emu*, Vol. 97, pp.43–50.

Boles, W.E. 1997b. A kingfisher (Halcyonidae) from the Miocene of Riversleigh, northwestern Queensland, with comments on the evolution of kingfishers in Australo-Papua. *Memoirs of the Queensland Museum*, Vol. 41, pp.229–234.

Boles, W.E. 1997c. Hindlimb proportions and locomotion of *Emuarius gidju* (Patterson and Rich, 1987) (Aves: Casuariidae). *Memoirs of the Queensland Museum,* Vol. 41, Pt. 2, pp.235–240.

Boles, W.E. 1997d. Riversleigh birds as palaeoenvironmental indicators. *Memoirs of the Queensland Museum,* Vol. 41, Pt. 2, pp.241–246.

Boles, W.E. 1998. A budgerigar *Melopsittacus undulatus* from the Pliocene of Riversleigh, north-western Queensland. *Emu*, Vol. 98, pp.32–35.

Boles, W.E. 1999a. Early Eocene shorebirds (Aves: Charadriiformes) from the Tingamarra Local Fauna, Queensland, Australia. *Records of the Western Australian Museum, Supplement,* No. 57, pp.229–238.

Boles, W.E. 1999b. A songbird (Aves: Passeriformes: Oriolidae) from the Miocene of Riversleigh, northwestern Queensland, Australia. *Alcheringa,* Vol. 23, pp.51–56.

Boles, W.E. 2000. *Investigations on the Tertiary avifauna of Australia, with emphasis on the birds of Riversleigh.* Ph.D. Thesis (unpublished), School of Biological Sciences, University of New South Wales. 643pp.

Boles, W.E. 2001a. A new emu (Dromaiinae) from the late Oligocene Etadunna Formation. *Emu,* Vol. 101, pp.317–321.

Boles, W.E. 2001b. A swiftlet (Apodidae: Collocaliini) from the Miocene of Riversleigh, northwestern Queensland. *Memoir of the Association of Australasian Palaeontologists*, Vol. 25, pp.45–52.

Boles, W.E. and Ivison, T.J. 1999. A new genus of dwarf megapode (Galliformes: Megapodiidae) from the late Oligocene of central Australia. *Smithsonian Contributions to Paleobiology,* Vol. 88, pp.199–206.

Boles, W.E. and Mackness, B. 1994. Birds from the Bluff Downs Local Fauna, Allingham Formation, Queensland. *Records of the South Australian Museum*, Vol. 27, pp.139–149.

Chatterjee, S. 1997. *The Rise of Birds: 225 Million Years of Evolution*. Johns Hopkins University Press, Baltimore. 312pp.

Chiappe, L.M. 1991. Cretaceous birds of Latin America. *Cretaceous Research,* Vol. 12, pp.55–63.

Chiappe, L.M. 1995. The first 85 million years of avian evolution. *Nature,* Vol. 378, pp.349–355.

Chiappe, L.M. 1996. Early avian evolution in the southern hemisphere: the fossil record of birds in the Mesozoic of Gondwana. *Memoirs of the Queensland Museum,* Vol. 39, pp.533–555.

Chiappe, L.M. and Walker, C.A. 2002. Skeletal morphology and systematics of the Cretaceous Euenantiornithes (Ornithothoraces: Enantiornithes). Pp.240–267. In Chiappe, L.M. and Witmer, L.M. (eds). *Mesozoic Birds: Above the Heads of Dinosaurs*. University of California Press, Berkeley. 520pp.

Christidis, L. 1991. Biochemical evidence for the origins and evolutionary radiations in Australasian avifauna: the song-birds. Pp.392–397. In Bell, B.D., Cossee, R.O., Flux, J.E.C., Heather, B.D., Hitchmough, R.A., Robertson, C.J.R. and Williams, M.J. (eds). *Acta XX Congressus Internationalis Ornithologici, Canterbury, 1990.* New Zealand Ornithological Congress Trust Board, Wellington. 2568pp.

Christidis, L. and Boles, W.E. 1994. *Taxonomy and Species of Birds of Australia and its Territories*. Royal Australasian Ornithologists Union, Melbourne. 112pp.

Christidis, L. and Schodde, R. 1991. Relationships of the

Australo-Papuan songbirds—protein evidence. *Ibis*, Vol. 133, pp.277–285.

Cooper, A. and Cooper, R.A. 1995. The Oligocene bottleneck and New Zealand biota: genetic record of a past environmental crisis. *Proceedings of the Royal Society of London*, Series B, Vol. 261, pp.293–302.

Cooper, A., Mourer-Chauviré, M., Chambers, G.K., von Haeseler, A., Wilson, A.C. and Pääbo, S. 1992. Independent origins of New Zealand moas and kiwis. *Proceedings of the National Academy of Sciences of the United States of America*, Vol. 89, pp.8741–8744.

Cracraft, J.A. 1972. Continental drift and Australian avian biogeography. *Emu,* Vol. 72, pp.171–174.

Cracraft, J. 1976. Avian evolution on southern continents: influences of palaeogeography and palaeoclimatology. Pp.40–52. In Frith, H.J. and Calaby, J.H. (eds). *Proceedings of the 16th International Ornithological Congress, Canberra, 1974.* Australian Academy of Sciences, Canberra. 765pp.

Cracraft, J. 2001. Avian evolution, Gondwana biogeography and the Cretaceous-Tertiary mass extinction event. *Proceedings of the Royal Society of London,* Series B, Vol. 268, pp.459–469.

De Vis, C.W. 1888a. Australian ancestry of the crowned pigeon of New Guinea. *Proceedings of the Royal Society of Queensland*, Vol. 5, pp.127–131.

De Vis, C.W. 1888b. A glimpse of the post-Tertiary avifauna of Queensland. *Proceedings of the Linnean Society of New South Wales,* Series 2, Vol. 3, pp.1277–1290.

De Vis, C.W. 1889. Additions to the list of fossil birds. *Proceedings of the Royal Society of Queensland*, Vol. 6, pp.55–58.

De Vis, C.W. 1890. On a bone of a giant eagle. *Proceedings of the Royal Society of Queensland*, Vol. 6, pp.161–162.

De Vis, C.W. 1891. Note on an extinct eagle. *Proceedings of the Linnean Society of New South Wales*, Vol. 6, pp.123–125.

De Vis, C.W. 1892. Residue of the extinct birds of Queensland as yet detected. *Proceedings of the Linnean Society of New South Wales*, Series 2, Vol. 6, pp.437–445.

De Vis, C.W. 1905. A contribution to the knowledge of the extinct avifauna of Australia. *Annals of the Queensland Museum*, Vol. 6, pp.3–25.

De Vis, C.W. 1911. *Palaeolestes gorei* n. sp. An extinct bird. *Memoirs of the Queensland Museum*, Vol. 10, pp.15–17.

Dyke, G. J. 2001. The fossil waterfowl (Aves: Anseriformes) from the Eocene of England. *American Museum Novitates*, No. 3354, pp.1–15.

Ericson, P.G.P. 1997. Systematic relationships of the palaeogene family Presbyornithidae (Aves: Anseriformes). *Zoological Journal of the Linnean Society*, Vol. 121, pp.429–483.

Ericson, P.G.P., Christidis, L., Cooper, A., Irestadt, M., Jackson, J., Johansson, U.S. and Norman, J.A. 2002. A Gondwanan origin of passerine birds supported by DNA sequences of the endemic New Zealand wrens. *Proceedings of the Royal Society of London,* Series B, Vol. 269, pp.235–241.

Feduccia, A. 1999. *The Origin and Evolution of Birds*. 2nd Edition. Yale University Press, New Haven. 466pp.

Feduccia, A. and Olson, S.L. 1982. Morphological similarities between the Menurae and the Rhinocryptidae, relict passerine birds of the South Hemisphere. *Smithsonian Contributions to Zoology,* No. 366, pp.1–22.

Finlayson, H.H. 1938. On the occurrence of a fossil penguin in Miocene beds in South Australia. *Transactions of the Royal Society of South Australia*, Vol. 62, pp.14–17.

Fordyce, R.E. 1991. A new look at the fossil vertebrate record of New Zealand. Pp.1191–1316. In Vickers-Rich, P.V., Monaghan, J.M., Baird, R.F. and Rich, T.H. (eds). *Vertebrate Palaeontology of Australasia*. Pioneer Design Studios and Monash University Publications Committee, Melbourne. 1437pp.

Fordyce, R.E. and Jones, C.M. 1990. Penguin history and new fossil material from New Zealand. Pp.419–446. In Davis, L.S. and Darby, J.T. (eds). *Penguin Biology*. Academic Press, San Diego. 467pp.

Fry, C.H. 1980. The evolutionary biology of kingfishers (Alcedinidae). *Living Bird*, Vol. 18, pp.113–160.

Gaff, P. 2002. *The fossil history of the family Accipitridae in Australia*. M.Sc. Thesis (unpublished), Monash University, Clayton. 206pp.

Glaessner, M.F. 1955. Pelagic fossils (*Aturia*, penguins, whales) from the Tertiary of South Australia. *Records of the South Australian Museum*, Vol. 11, pp.353–372.

Haddrath, O. and Baker, A.J. 2001. Complete mitochondrial DNA genome sequences of extinct birds: ratite phylogenetics and the vicariance biogeography hypothesis. *Proceedings of the Royal Society of London,* Series B, Vol. 268, pp.939–945.

Heinsohn, T. and Hope, G. 2006. The Torresian connections: zoogeography of New Guinea. Pp.71–93. In Merrick, J.R., Archer, M., Hickey, G.M. and Lee, M.S.Y. (eds). *Evolution and Biogeography of Australasian Vertebrates.* Auscipub Pty. Ltd., Sydney. 942pp.

Holdaway, R.N. and Worthy, T.H. 2006. Evolution of New Zealand and its vertebrates. Pp.111–128. In Merrick, J.R., Archer, M., Hickey, G.M. and Lee, M.S.Y. (eds). *Evolution and Biogeography of Australasian Vertebrates.* Auscipub Pty. Ltd., Sydney. 942pp.

Hope, S. 2002. The Mesozoic radiation of Neornithes. Pp.339–388. In Chiappe, L.M. and Witmer, L.M. (eds). *Mesozoic Birds: Above the Heads of Dinosaurs*. University of California Press, Berkeley. 520pp.

Jenkins, R.J.F. 1974. A new giant penguin from the Eocene of Australia. *Palaeontology,* Vol. 17, pp.291–310

Jenkins, R.J.F. 1985. *Anthropornis nordenskjoeldi* Wiman, 1905. Nordenskjoeld's Giant Penguin. Pp.183–187. In Rich, P.V. and van Tets, G.F. (eds). *Kadimakara, Extinct Vertebrates of Australia*. Pioneer Design Studio, Lilydale, Victoria. 284pp.

Jennings, S.F. 1990. *The musculoskeletal anantomy* [sic], *locomotion and posture of the dromornithid* Dromornis stirtoni *from the Late Miocene Alcoota Local Fauna.* Hons. Thesis (unpublished), School of Biological Sciences, Flinders University, South Australia. 302pp.

Kear, B.P., Boles, W.E. and Smith, E.T. 2003. Unusual gut contents in an early Cretaceous ichthyosaur. *Proceedings of the Royal Society of London*, Series B, (submitted).

Kurochkin, E.N. 1995. Synopsis of Mesozoic birds and early evolution of class Aves. *Archaeopteryx*, Vol. 3, pp.47–66.

Kurochkin, E.N. 1996. A new enantiornithid of the Mongolian late Cretaceous, and a general appraisal of the infraclass Enantiornithes (Aves). Special Issue. Russian Academy of Sciences, Moscow.

Kurochkin, E.N. and Molnar, R.E. 1997. New material of

enantiornithine birds from the early Cretaceous of Australia. *Alcheringa,* Vol. 21, pp.291–297.

Livezey, B.C. 1997. A phylogenetic analysis of basal Anseriformes, the fossil *Presbyornis*, and the interordinal relationships of waterfowl. *Zoological Journal of the Linnean Society,* Vol. 121, pp.361–428.

Long, J., Archer, M., Flannery, T. and Hand, S. 2002. *Prehistoric Mammals of Australia and New Guinea.* University of New South Wales Press, Sydney. 244pp.

Lydekker, R. 1891. *Catalogue of the Fossil Birds in the British Museum (Natural History).* British Museum (Natural History), London. 368pp.

Mackness, B.S. 1995. *Anhinga malagurala*, a new pygmy darter from the early Pliocene Bluff Downs Local Fauna, north-eastern Queensland. *Emu,* Vol. 95, pp.265–271.

Maclean, G.L. 1990. Evolution of the passerines in the Southern Hemisphere. *Transvaal Museum Bulletin,* Vol. 22, Supplement, pp.1–11.

Martin, L.D. 1983. The origin and early evolution of birds. Pp.291–338. In Brush, A.H. and Clark, G.A., Jnr. (eds). *Perspectives in Ornithology. Essays Presented for the Centenary of the American Ornithologists Union.* Cambridge University Press, Cambridge. 560pp.

Martin, L.D. 1995. The Enantiornithes: terrestrial birds of the Cretaceous. *Courier Forschungsinstitut Senckenberg*, Vol. 181, pp.23–36.

Mayr, E. 1944. Timor and the colonization of Australia by birds. *Emu,* Vol. 44, pp.113–130.

Mayr, E. 1972. Continental drift and the history of the Australian bird fauna. *Emu*, Vol. 72, pp.26–28.

McNamara, G.F. 1991. *Fossil avifauna from the late Quaternary deposits of McEachern's Cave, Lower Glenelg National Park, Victoria.* 3rd Year Report (unpublished), Department of Earth Sciences, Monash University.

Meredith, C. 1991. Vertebrate fossil faunas from islands in Australasia and the southwest Pacific. Pp.1345–1382. In Vickers-Rich, P.V., Monaghan, J.M., Baird, R.F. and Rich, T.H. (eds). *Vertebrate Palaeontology of Australasia.* Pioneer Design Studio and Monash University Publications Committee, Melbourne. 1437pp

Millener, P.R. 1991. The Quaternary avifauna of New Zealand. Pp.1317–1344, In Vickers-Rich, P.V., Monaghan, J.M., Baird, R.F. and Rich, T.H. (eds). *Vertebrate Palaeontology of Australasia.* Pioneer Design Studio and Monash University Publications Committee, Melbourne. 1437pp.

Miller, A.H. 1962. The history and significance of the fossil *Casuarius lydekkeri. Records of the Australian Museum,* Vol. 10, pp.235–238.

Miller, A.H. 1963a. The fossil flamingoes of Australia. *Condor,* Vol. 65, pp.289–299.

Miller, A.H. 1963b. Fossil ratite birds of the late Tertiary of South Australia. *Records of the South Australian Museum,* Vol. 14, pp.413–420.

Miller, A.H. 1966. The fossil pelicans of Australia. *Memoirs of the Queensland Museum*, Vol. 14, pp.181–190.

Miller, G.H., Magee, J.W., Johnson, B.J., Fogel, M.L., Spooner, N.A., McCulloch, M.T. and Ayliffe, L.K. 1999. Pleistocene extinction of *Genyornis newtoni*: human impact on Australian megafauna. *Science,* Vol. 283, pp.205–208.

Molnar, R. 1986. An enantiornithine bird from the lower Cretaceous of Queensland, Australia. *Nature*, Vol. 322, pp.736–738.

Molnar, R. 1999. Avian tibiotarsi from the early Cretaceous of Lightning Ridge, New South Wales. *National Science Museum Monographs,* Vol. 15, pp.197–209.

Mourer-Chauviré, C. 1982. Les oiseaux fossiles des Phosphorites du Quercy (Eocène supérieur à Oligocène supérieur): Implications paléobiogéographiques. *Géobios (Mémoire Spécial*), Vol. 6, pp.413–426.

Mourer-Chauviré, C. 1988. Les Caprimulgiformes et les Coraciiformes de l'Éocene et du l'Oligocene des phosphorites du Quercy et description de deux genres nouveaux de Podargidae et Nyctibiidae. Pp.2047–2055. In Ouellet, H. (ed). *Acta XIX Congressus Internationalis Ornithologici, Ottawa, 1986.* National Museum of Natural Sciences, Ottawa. 2815pp.

Mourer-Chauviré, C. 1991. Les Horusornithidae fam. nov., Accipitriformes (Aves) a articulation intertarsienne hyperflexible de l'Éocène du Quercy. *Géobios*, Vol. 13, pp.183–192.

Mourer-Chauviré, C. 1992. The Galliformes (Aves) from the Phorphorites due Quercy (France): systematics and biostratigraphy. *Natural History Museum of Los Angeles Country, Science Series,* 36, pp.67–95.

Mourer-Chauviré, C. 1999. Les relations entre les avifaunes du Tertiaire inférieur d'Europe et d'América du Sud. *Bulletin de la Société Géologique de France*, Vol. 170, pp.85–90.

Mourer-Chauviré, C., Hugueney, M. and Jonet, P. 1989. Découverte de Passeriformes dans l'Oligocène supérior de France. *Comptes Rendus de l'Académie des Sciences de Paris,* Series 2, Vol. 309, pp.843–849.

Murray, P.F. and Megirian, D. 1998. The skull of dromornithid birds: anatomical evidence for their relationship to Anseriformes. *Records of the South Australian Museum*, Vol. 31, pp.51–97.

Olson, S.L. 1975. The extinct rails of C. W. De Vis, being mainly an extinct form of *Tribonyx mortierii* from Queensland. *Emu*, Vol. 75, pp.49–54.

Olson, S.L. 1977. The identity of the fossil ducks described from Australia by C.W. de Vis. *Emu*, Vol. 77, pp.127–132.

Olson, S.L. 1985. The fossil record of birds. Pp.79–238. In Farner, D.S., King, J.R. and Parkes, K.C. (eds). *Avian Biology. Volume 8.* Academic Press, New York. 256pp.

Olson, S.L. 1988. Aspects of global avifaunal dynamics during the Cenozoic. Pp.2023–2029. In Ouellet, H. (ed). *Acta XIX Congressus Internationalis Ornithologici, Ottawa, 1986.* National Museum of Natural Sciences, Ottawa. 2815pp.

Olson, S.L. 1992. *Neogaeornis wetzeli* Lambrecht, a Cretaceous loon from Chile (Aves: Gaviidae). *Journal of Vertebrate Paleontology,* Vol. 12, pp.122–124.

Olson, S.L. 1999. The anseriform relationships of *Anatalavis* Olson and Parris (Anseranatidae), with a new species from the early Eocene London Clay. *Smithsonian Contributions to Paleobiology*, No. 88, pp.231–243.

Olson, S.L. and Parris, D.C. 1987. The Cretaceous birds of New Jersey. *Smithsonian Contributions to Paleobiology*, No. 63, pp.1–22.

Owen, R. 1872. [Untitled]. *Proceedings of the Zoological Society of London*, 1872, pp.682–683.

Patterson, C. and Rich, P.V. 1987. The fossil history of the emus, *Dromaius* (Aves: Dromaiinae). *Records of the South Australian Museum,* Vol. 21, pp.85–117.

Peters, D.S. 1991. Zoogeographical relationships of the Eocene avifauna from Messel (Germany). Pp.572–577. In Bell, B.D., Cossee, R.O., Flux, J.E.C., Heather, B.D., Hitchmough, R.A., Robertson, C.J.R. and Williams, M.J. (eds). *Acta XX Congressus Internationalis Ornithologici, Canterbury, 1990*. New Zealand Ornithological Congress Trust Board, Wellington. 2568pp.

Rich, P.V. 1975a. Antarctic dispersal routes, wandering continents, and the origin of Australia's non-passeriform avifauna. *Memoirs of the National Museum of Victoria,* Vol. 36, pp.63–125.

Rich, P.V. 1975b. Changing continental arrangements and the origin of Australia's non-passerine continental avifauna. *Emu*, Vol. 5, pp.97–112.

Rich, P.V. 1976. The history of birds on the island continent Australia. Pp.53–65. In Frith, H.J. and Calaby, J.H. (eds). *Proceedings of the 16th International Ornithological Congress, Canberra, 1974*. Australian Academy of Sciences, Canberra. 765pp.

Rich, P.V. 1979. The Dromornithidae, an extinct family of large ground birds endemic to Australia. *Bureau of Mineral Resources, Geology and Geophysics Bulletin,* No. 184, pp.1–196.

Rich, P.V. 1985. *Genyornis newtoni* Stirling and Zietz, 1896. A mihirung. Pp.188–194. In Rich, P.V. and van Tets, G.F. (eds). *Kadimakara, Extinct Vertebrates of Australia*. Pioneer Design Studio, Lilydale, Victoria. 284pp.

Rich, P.V. and Baird, R.F. 1986. History of the Australian avifauna. *Current Ornithology,* Vol. 4, pp.97–139.

Rich, P.V. and McEvey, A.R. 1977. A new owlet-nightjar from the early to mid-Miocene of eastern New South Wales. *Memoirs of the National Museum of Victoria*, Vol. 38, pp.247–253.

Rich, P.V. and McEvey, A.R. 1980. A fossil plains wanderer (Aves: Pedionomidae) from Fire-hole deposits, Morwell, southeastern Victoria, Australia. *Natural History Museum of Los Angeles County, Contributions in Science,* No. 330, pp.251–255.

Rich, P.V., Plane, M. and Schroeder, N. 1988. A pygmy cassowary (*Casuarius lydekkeri*) from late Pleistocene bog deposits at Pureni, Papua New Guinea. *Bureau of Mineral Resources, Journal of Australian Geology and Geophysics*, Vol. 10, pp.377–389.

Rich, P.V. and van Tets, G.F. 1981. The fossil pelicans of Australasia. *Records of the South Australian Museum*, Vol. 18, pp.235–264.

Rich, P. and van Tets, J. 1982. Fossil birds of Australia and New Guinea: their biogeographic, phylogenetic and biostratigraphic input. Pp.235–384. In Rich, P.V. and Thompson, E.M. (eds). *The Fossil Vertebrate Record of Australasia*. Monash University Offset Printing Unit, Melbourne. 759pp.

Rich, P.V., van Tets, G.F. and McEvey, A.R. 1982. Pleistocene records of *Falco berigora* from Australia and the identity of *Asturaetus furcillatus* De Vis (Aves: Falconidae). *Memoirs of the Queensland Museum*, Vol. 20, pp.687–693.

Rich, P.V., van Tets, G.F., Rich, T.H.V. and McEvey, A.R. 1987. The Pliocene and Quaternary flamingoes of Australia. *Memoirs of the Queensland Museum*, Vol. 25, pp.207–225.

Rich, T.H., Archer, M., Hand, S.J., Godthelp, H., Muirhead, J., Pledge, N.S., Flannery, T.F., Woodburne, M.O., Case, J.A., Tedford, R.H., Turnbull, W.D., Lundelius, E.L., Jnr., Rich, L.S.V., Whitelaw, M.J., Kemp, A. and Rich, P.V. 1991. Appendix I. Australian Mesozoic and Tertiary terrestrial mammal localities. Pp.1005–1058. In Vickers-Rich, P., Monaghan, J.M., Baird, R.F. and Rich, T.H. (eds). *Vertebrate Palaeontology of Australasia*. Pioneer Design Studio and Monash University Publications Committee, Melbourne. 1437pp.

Rich, T.H., Vickers-Rich, P., Constantine, A., Flannery, T.F., Kool, L. and van Klaveren, N. 1999. Early Cretaceous mammals from Flat Rocks, Victoria, Australia. *Records of the Queen Victoria Museum*, Vol, 106, pp.1–35.

Rothschild, W. 1911. On the former and present distribution of the so-called Ratitae or ostrich-like birds. Pp.144–169. In Schalow, H. (ed). *Verhandlungen des V. Internationalen Ornithologen-Kongresses in Berlin. 30.Mai bis 4.Juni 1910.* Deutsche Ornithologische Gesellschaft, Berlin. 1186pp.

Schodde, R. 1982. Origin, adaptation and radiation of birds in arid Australia. Pp.191–224. In Barker, W.R. and Greenslade, P.J.M. (eds). *Evolution of the Flora and Fauna of Arid Australia.* Peacock Publications, South Australia. 392pp.

Schodde, R. 2006. Australasia's bird fauna today—origins and evolutionary development. Pp.413–458. In Merrick, J.R., Archer, M., Hickey, G.M. and Lee, M.S.Y. (eds). *Evolution and Biogeography of Australasian Vertebrates.* Auscipub Pty. Ltd., Sydney. 942pp.

Serventy, D.L. 1972. Causal ornithogeography of Australia. Pp.574–584. In Voous, K.H. (ed). *Proceedings of the XV International Ornithological Congress, The Hague, 1970.* E.J. Brill, Leiden. 745pp.

Sibley, C.G. 1976. Protein evidence of the relationships of some Australian passerine birds. Pp.557–570. In Frith, H.J. and Calaby, J.H. (eds). *Proceedings of the 16th International Ornithological Congress, Canberra, 1974.* Australian Academy of Sciences, Canberra. 765pp.

Sibley, C.G. and Ahlquist, J.E. 1985. The phylogeny and classification of the Australo-Papuan songbirds. *Emu,* Vol. 85, pp.1–14.

Sibley, C.G. and Ahlquist, J.E. 1990. *Phylogeny and Classification of Birds: A Study in Molecular Evolution.* Yale University Press, New Haven, Connecticut. 976pp.

Sibley, C.G., Ahlquist, J.E. and Monroe, B.L., Jnr. 1988. A classification of the living birds of the world based on DNA-DNA hybridization studies. *Auk*, Vol. 105, pp.409–423.

Simpson, G.G. 1957. Australian fossil penguins, with remarks on penguin evolution and distribution. *Records of the South Australian Museum*, Vol. 13, pp.51–70.

Simpson, G.G. 1959. A new fossil penguin from Australia. *Proceedings of the Royal Society of Victoria,* Vol. 71, pp.113–119.

Simpson, G.G. 1970. Miocene penguins from Victoria, Australia and Chubut, Argentina. *Memoirs of the National Museum of Victoria*, Vol. 31, pp.17–24.

Smith, E. 1999. *Black Opal Fossils of Lightning Ridge. Treasures form the Rainbow Billabong*. Kangaroo Press, Sydney. 112pp.

Stidham, T.A. 1998. A lower jaw from a Cretaceous parrot. *Nature*, Vol. 396, pp.29–30.

Stirling, E.C. 1913. Fossil remains of Lake Callabonna. Part IV. 1. Description of some further remains of *Genyornis* *newtoni*, Stirling and Zietz. *Memoirs of the Royal Society of South Australia,* Vol. 1, pp.111–126.

Stirling, E.C. and Zietz, A.H.C. 1896. Preliminary notes on *Genyornis newtoni*: a new genus and species of fossil struthious bird found at Lake Callabonna, South Australia. *Transactions of the Royal Society of South Australia*, Vol. 20, pp.171–190.

Stirling, E.C. and Zietz, A.H.C. 1900. Fossil remains of Lake Callabonna. Part II. 1. *Genyornis newtoni*. A new genus and species of fossil struthious bird. *Memoirs of the Royal Society of South Australia,* Vol. 1, pp.41–80.

Stirling, E.C. and Zietz, A.H.C. 1905. Fossil remains of Lake Callabonna. Part III. Description of the vertebrae of *Genyornis newtoni*. *Memoirs of the Royal Society of South Australia*, Vol. 1, pp.81–110.

Talent, J.A., Duncan, P.M. and Handley, P.L. 1966. Early Cretaceous feathers from Victoria. *Emu*, Vol. 66, pp.81–86.

Tambussi, C.P, Noriega, J.I., Gazdzicki, A., Tatur, A., Reguero, M.A. and Vizcaino, S.F. 1994. Ratite bird from the Paleogene La Meseta Formation, Seymour Island, Antarctica. *Polish Polar Research*, Vol. 15, pp.15–20.

Unwin, D.M. 1993. Aves. Pp.717–736. In Benton, M.J. (ed). *The Fossil Record 2*. Chapman and Hall, London. 845pp.

van Oosterzee, P. 2006. Drawing the Wallace Line. Pp.95–110. In Merrick, J.R., Archer, M., Hickey, G.M. and Lee, M.S.Y. (eds). *Evolution and Biogeography of Australasian Vertebrates*. Auscipub Pty. Ltd., Sydney. 942pp.

van Tets, G.F. 1974. A revision of the fossil Megapodiidae (Aves), including a description of a new species of *Progura* De Vis. *Transactions of the Royal Society of South Australia,* Vol. 98, pp.213–224.

van Tets, G.F. 1994. An extinct new species of cormorant (Phalacrocoracidae, Aves) from a Western Australian peat swamp. *Records of the South Australian Museum,* Vol. 27, pp.135–138.

van Tets, G.F. and O'Connor, S. 1983. The Hunter Island Penguin, an extinct new genus and species from a Tasmania midden. *Records of the Queen Victoria Museum*, Vol. 81, pp.1–13

van Tets, G.F. and Rich, P.V. 1980. A review of the de Vis fossil pigeons of Australia. *Memoirs of the Queensland Museum*, Vol. 20, pp.89–93.

van Tets, G.F. and Rich, P.V. 1990. An evaluation of De Vis' fossil birds. *Memoirs of the Queensland Museum*, Vol. 28, pp.165–168.

van Tets, G.F. and Smith, M.J. 1974. Small fossil vertebrates from Victoria Cave, Naracoorte, South Australia. III. Birds (Aves). *Transactions of the Royal Society of South Australia*, Vol. 98, pp.225–227.

van Tuinen, M., Sibley, C.G. and Hedges, S.B. 1998. Phylogeny and biogeography of ratite birds inferred from DNA sequences of the mitochondrial ribosomal genes. *Molecular and Biological Evolut*ion, Vol. 15, pp.370–376.

Vickers-Rich, P. 1991. The Mesozoic and Tertiary history of birds on the Australian plate. Pp.721–808. In Vickers-Rich, P., Monaghan, J.M., Baird, R.F. and Rich, T.H. (eds). *Vertebrate Palaeontology of Australasia*. Pioneer Design Studio and Monash University Publications Committee, Melbourne. 1437pp.

Vickers-Rich, P. and Molnar, R.E. 1996. The foot of a bird from the Eocene Redbank Plains Formation of Queensland, Australia. *Alcheringa,* Vol. 20, pp.21–29.

Waldman, M. 1970. A third specimen of a lower Cretaceous feather from Victoria, Australia. *Condor*, Vol. 72, p.377.

Walker, C.A. 1981. New subclass of birds from the Cretaceous of South America. *Nature*, Vol. 292, pp.51–53.

White, M.E. 2006. Environments of the geological past. pp.17–50. In Merrick, J.R., Archer, M., Hickey, G.M. and Lee, M.S.Y. (eds). *Evolution and Biogeography of Australasian Vertebrates*. Auscipub Pty. Ltd., Sydney. 942pp.

Wilkinson, H.E. 1969. Description of an upper Miocene albatross from Beaumaris, Victoria, Australia, and a review of fossil Diomedeidae. *Memoirs of the National Museum of Victoria,* Vol. 29, pp.41–51

Wiman, C. 1905. Vorläufige Mitteilung über die alteriären Vertebraten der Seymourinsel. *Bulletin of the Geological Institution of the University of Upsala,* Vol. 6, pp.247–256.

Worthy, T.N. and Holdaway, R.N. 2002. *The Lost World of the Moa. Prehistoric Life in New Zealand.* Indiana University Press, Bloomington. 718pp.

Wroe, S. 1999. The bird from hell? *Nature Australia,* Vol. 26, No.7, pp.56–63

Zhou, Z. 1995. The discovery of early Cretaceous birds in China. *Courier Forschungsinstitut Senckenberg,* Vol. 181, pp.9–22.

Chapter 22
Australasia's Bird Fauna Today—Origins and Evolutionary Development

Richard Schodde
Australian Biological Resources Study, G.P.O. Box 787, Canberra. A.C.T. 2601 Australia
E-mail: RichardSchodde@aol.com

DIVERSITY OF THE FAUNA

Where did it come from? How did it diversify? These two questions sum up the issues—still controversial—surrounding the sources and subsequent radiation of Australasia's modern bird fauna. The answers to these questions are important in global zoogeography because the bird life of Australia and its sister island New Guinea is as spectacular as that of any continent and the most distinct and 'different' in the world. A multitude of ancient forms are either endemic or centred there: emus and cassowaries, mound-building megapodes, phabine and fruit pigeons, cockatoos, lories and platycercine parrots, masked owls, frogmouth-like birds, land kingfishers, lyrebirds and scrub-birds, tree-creepers, and, indeed, the central corvidan assemblage of the world's songbirds. The composition of the bird fauna for the entire region, including New Zealand, the archipelagos fringing east New Guinea and surrounding seas, is summarized in Table 1. World figures (Sibley and Monroe, 1990) are included in parentheses for perspective. Australia itself is home to some 690 species, including migrants; and there are another 970 in the rest of Australasia, mainly in Papuasia. The combined Australasian bird fauna comprises some 1,300 species of birds, allowing for about 360 shared among regions.

ECOLOGICAL AND FAUNAL CHARACTERISTICS

Except in New Zealand, the regional bird fauna is ecologically diverse and harmonic. New Zealand is rich in marine (95 species), freshwater (33 species) and shore birds (~32 regular species), but impoverished in land birds: only about 40 were present at European settlement. Around Australia and Papuasia, the marine faunas are just as rich and, except for most shearwaters (*Puffinus*) and terns (Sternini), are partitioned as well at the tropical convergence (25–30°S) into:

(i) temperate sea assemblages to the south: penguins (Spheniscidae), albatrosses (*Diomedea*), prions (*Pachyptila*), fulmarine petrels (*Fulmarus, Macronectes, Aphodroma*), procellariine petrels (*Procellaria*), diving petrels (*Pelecanoides*), the storm petrels *Oceanites*, *Garrodia* and *Pelagodroma*, gannets (*Morus*), and gulls and skuas (Larinae, Stercorariinae); and
(ii) tropical sea assemblages to the north: many gadfly petrels (*Pterodroma*), the storm petrels *Fregetta* and *Oceanodroma*, tropicbirds (Phaethontidae), boobies (*Sula*), frigatebirds (Fregatidae), and noddies (Megalopterini).

Many members of the temperate sea assemblage reach Australian seas only in winter, on a northward migratory shift from south polar waters. On the west coast, a number of tropical terns (*Sterna dougalii, S. fuscata, S. anaethetus, Anous stolidus* and *A. tenuirostris*) breed further south (Houtman Abrolhos) than in the east, due to the warm inshore Leeuwin current. Curiously, several temperate species (e.g. *Larus pacificus, Sterna nereis*) nest north of their eastern range there as well.

In freshwater and shore birds, Australasia possesses a diversity of ducks and shore birds, as well as numerous cormorants, herons, ibises and spoonbills, rails and several storks and cranes. It is an assemblage familiar on any continent; and its members fill the same niches as their vicars world-wide: dabbling and grazing, cursorial combing of beach and lake shores, under-water fishing, fishing by stalking, probing and sifting, and skulking to peck and glean. The scolopacid wader group (~35 species) is uniquely migratory, spending the austral summer in Australasia away from its frozen breeding grounds in the holarctic; some plovers (*Charadrius, Pluvialis*) are also similarly migratory. In New Guinea, where wetlands are more densely vegetated and expose less muddy open shore, rails are richest: 18 species, including a radiation in rainforest rails, *Rallina* (5 species). Shore birds, however, are richer in Australia: 55 species, including several endemic genera, compared to just over 40 in New Guinea. It reflects the different ecological opportunities afforded historically by dry open and rainforested environments, a schism still clearer in the regional land bird fauna.

It is the immense land bird fauna of Australia and New Guinea, comprising over 70% of the total regional avifauna, that gives Australasia's bird life its special character. Missing from it are the woodpeckers, jays and nine-primaried songbirds (true finches, buntings and allies) so prevalent in the Northern Hemisphere. Specialized carrion-feeders such as vultures (*Gyps* and allies, Cathartidae) are absent too and nocturnal birds-of-prey are few, particularly strigid owls. Nor are there any of the humming birds or suboscinine passerines that have radiated so diversely in the Americas. In their stead is a suite of 36 land bird families that are either endemic to or centred in Australasia (Table 1), all replacing Eurasian and New World families in equivalent ecological niches.

Nectar-feeding here is not the province of sunbirds or humming birds, but of honeyeaters (Meliphagidae) and lorikeets (Loriinae). Fruit-pigeons (Treroninae), parrots (Psittacidae), bowerbirds (Ptilonorhynchidae) and birds of paradise (Paradisaeidae) are the dominant arboreal fruit- and seed-eaters, not hornbills, toucans, trogons, barbets and bulbuls. On the ground, seed-eating parrots, cockatoos and phabine pigeons, plus a handful of grass finches, Estrildidae, replace fringillid finches, pheasants and allies. Among insectivores, the undershrubbery and foliage gleaning niche is filled by acanthizid warblers (Acanthizidae) and malurid wrens (Maluridae), not sylviid warblers and Old World babblers or South American ant-thrushes and ovenbirds; the perch-pouncers and sallying flycatchers are Australasian robins (Petroicidae), whistlers (Pachycephalidae), and monarchs and fantails (Monarchidae), not holarctic thrushes, or muscicapid and tyrant flycatchers, puff-birds and bee-eaters; and treecreepers and sittellas (Climacteridae, Neosittidae) are the scansorial tree-feeders, not certhiid creepers and nuthatches. Even the predatory shrikes and jays of the Northern Hemisphere are replaced by butcherbirds and currawongs (Artamidae). These last groups are all songbirds—passerines with complex intrinsic muscling of the syrinx—and they dominate the bird fauna, comprising over two thirds of land bird species in the region.

These faunistic contrasts highlight the distinctiveness of the Australasian bird fauna and, reflecting enormous endemic radiation, pose the still more fascinating question: how did it happen?

Geological periods and epochs are explained elsewhere and time scales, taken from Ludbrook (1980) and Harland *et al.* (1990), are similar to the standard intervals used throughout this volume.

Summary Box 1.
TERMS AND DEFINITIONS

In this chapter a number of terms are used with the meanings listed below:

Afro-Eurasia: Africa and Eurasia.
Allotaxa: phylogenetically sister taxa that replace one another in different geographical or altitudinal areas.
Australia: the Australian continent and its islands, including Tasmania.
Australasia: Australia, Papuasia and New Zealand
Autochthon: a native in terms of origin.
Causal zoogeography: synthesized zoogeographic patterns explained by historical (palaeogeographical and evolutionary) processes.
Craton: a principal continental plate of the Earth's crust.
Eurasia: the Europe-Asia continent, including India and the islands of the Sunda shelf (*e.g.* Sumatra, Java, Borneo).
Functional zoogeography: synthesized zoogeographic patterns determined as they exist now.
Papuasia: New Guinea and satellite islands, Bismarck Archipelago and Solomon Islands.
Scleromorphic, sclerophyll: refers to the hard-leaved vegetation (forest, woodland, heath) of Australia, dominated by *Eucalyptus*, *Acacia* and diverse Proteaceae.
Species: taxa delimited by the biological species concept.
Subspecies: regional forms of species with differentiated genetic bases.

ORIGINS

In the ensuing text, Australian avian taxonomy and nomenclature follows Stanger *et al.* (1998), modified for songbirds by Schodde and Mason (1999). Sources for Papuasia and New Zealand are Beehler and Finch (1985), Sibley and Monroe (1990) and the Checklist Committee, Ornithological Society of New Zealand (1990).

TRADITIONAL THEORIES AND FLAWS

Researchers from Mayr (1944a, b; 1965) to Keast (1981,1984), considered that Australia received its ancestral stocks of birds from Eurasia, in a series of immigrant waves across Indonesian island stepping stones over the last 30–40 million years. Those that came first radiated furthest, and those that came last were thought to have changed little; among these are Australia's handful of species of such prominent Afro-Eurasian families as bee-eaters, pipits, larks, starlings, thrushes, flowerpeckers and sunbirds. Older stocks coming via the Lesser Sundas were thought to have produced the sclerophyll-inhabiting bird faunas that occupy most of Australia today. Rainforest assemblages on the east coast, however, were explained by more recent arrival from New Guinea.

This immigrant wave hypothesis rested on three premises now known to be flawed. One was the steady state of the Earth's crust, its continents and archipelagos in supposedly fixed positions (Audley-Charles, 1966). Another was the close resemblance of the main stocks of Australasian songbirds to those in Eurasia in outward form, notably in syringeal ('voice box') structure and a ten-primaried wing. The third circumstance was the fit that an older, sclerophyll-inhabiting bird fauna and a recently-arrived rainforest avifauna made with traditional geographic perceptions of their habitat, Australia's vegetation (Hooker, 1860; Burbidge, 1960).

Continental Drift

The first chinks in the immigrant wave construct were exposed by continental drift. When its first Eurasian bird stocks were supposedly arriving, Australia was nowhere near its present position; instead it was thousands of kilometres south, part of the supercontinent Gondwana (Cracraft, 1973; McKenna, 1973; Rich, 1975; Stevens, 1991). This knowledge affected the Eurasian immigration hypothesis in two ways. First, it drew attention to a number of primitive groups of birds centred today on the southern continents, such as the flightless ratites, parrots and cockatoos, megapodes and South American curassows, magpie-geese and South American screamers, whistling ducks, button-quail, plains wanderers and seed-snipe, pigeons, cuculine cuckoos, and frogmouth-like birds. These now came to be thought of as Gondwanan, forming an initial—and meagre—avifaunal base in Australia (Schodde, 1982a, Appendix 5; Rich and van Tets, 1984). Second, the position of Australia so far south then led to speculation that the Eurasian stocks had arrived later and radiated faster, overcoming much of the original Gondwanan bird fauna as they did (Mayr, 1976). Proto-Australia was still perceived as an avian vacuum, into which birds had poured along one pathway or another (Rich, 1975); the possibility that whatever stocks Australia got from Gondwana it inherited *in situ* was not recognized.

Fossils and Palaeogeography

The fossil record provided its own anomalies. It reveals an end-of-Cretaceous origin for modern neornithine birds globally, followed by their radiation through the early Tertiary, in parallel with mammals, to supplant the enantiornithine and early ornithurine birds of the late Mesozoic (Feduccia, 1995, 1996). It evidently occurred rapidly 65–30 m.y.a., when the root lineages of all modern bird families were established. This questions a Gondwanan origin for any southern group of weak dispersers, such as the ratites, which today are shared among Africa, India and New Zealand, lands that had split from one another before then (Feduccia 1995, 1996). Yet most birds fly and, although some such as the Australasian malurid wrens, treecreepers and bowerbirds never appear to have spread from the Australian continental plate, others traverse hemispheres annually on migration. Even such weak, round-winged fliers as rails (Rallidae) and fantails (Monarchidae) cross significant water barriers each year.

Immigration from outside is the simplest explanation for the sudden appearance of passerines bursting into the Northern Hemisphere fossil record from the late Oligocene (25–30 m.y.a.) (Olson, 1985). In Australia, where the record is far more patchy, there are much older passerine deposits, dating back to the early Eocene (about 50 m.y.a.); at the very time that Australia began to break away from the remnants of Gondwana (Boles, 1995, 1997). Whether these fossils are of songbirds or sub-oscine passerines is unclear, but by the late Oligocene or very early Miocene (~30 m.y.a.), deposits at Riversleigh, (north-west Queensland), reveal a diverse songbird fauna of lyrebirds, logrunners (Orthonychidae), honeyeaters and other modern Australian families (Boles, 1991). That passerine or perching birds are crucial in resolving origins and radiations in modern birds may be appreciated from their prevalence today: they comprise over half the global bird fauna. There are two groups: (i) the sub-oscines, with a massive radiation in South and Central America into at least six families and over 1000 species, plus outliers of just over 50 species in New Zealand (xenicid wrens) and the palaeotropics (pittas, broadbills and Madagascan asities); (ii) the songbirds which, with over 4,500 species including Australia's lyrebirds and scrub-birds (Bock, 1972; Feduccia, 1975; Raikow, 1985; Sibley and Ahlquist, 1990), are cosmopolitan and dominate the bird faunas on all continents except South America.

Global sequences in the passerine fossil record and the vicariant disposition of Australian songbird and South American suboscine radiations suggest a Gondwanan origin for the passerines (Olson, 1988; McLean, 1990; Schodde, 1991). Except for the ratites and dendrocygnine ducks, other southern non-passerine groups in Australia are also shared prevailingly with South America, which was connected archipelagically to proto-Australia (East Gondwana) via West Antarctica until *after* the beginning of neornithine radiation. This connection is reflected as well in the habitat—the vegetation—in which the nascent Gondwanan land bird faunas were evolving. At that time, Australia, like southern South America and much of Antarctica, was substantially covered not by a sclerophyll eucalypt flora but subtropical and temperate rainforests dominated by Antarctic beech (*Nothofagus*), myrtles (Myrtaceae), laurels and monimiads (Laurales), cunoniads, escalloniads and podocarp gymnosperms (Kemp, 1978; Martin, 1982, 1984; Hill *et al.*, 1999).

In Australo-Papuasia, such forests survive today along the east coast of Australia and in montane New Guinea where, as stressed by Schodde and Calaby (1972) and Schodde and Faith (1991), they hold a multitude of root elements of the Australasian bird fauna split vicariantly between the two lands: the megapodes *Alectura* and *Aepypodius*, king parrots (*Alisterus*), logrunners (*Orthonyx*), scrubwrens (*Sericornis*), caligavine honeyeaters (*Lichenostomus* subgenus *Caligavis*), treecreepers (*Cormobates*), Australasian robins (*Tregellasia*, *Poecilodryas*) and tower-building bowerbirds (*Prionodura*, *Amblyornis*). Other non-shared members of these forests seem similarly old: the tiger parrots (*Psittacella*), ifrits and melampittas (*Ifrita*, *Melampitta*), berrypeckers (Melanocharitidae), satin-birds (Cnemophilidae), pygmy drongo (*Chaetorhynchus*), core elements of the birds of paradise (Paradisaeidae) in New Guinea, and the fruit pigeon (*Lopholaimus*) and lyrebirds and scrub-birds (*Menura*, *Atrichornis*) in Australia. If this fauna and its habitat are co-eval for

Nothofagus alone (Vuilleumier and Kikkawa, 1991), their association fits a Gondwanan, not northern immigrant origin.

Molecular and Morphological Studies

The similarities between Eurasian and Australian songbirds in external form, long taken as evidence of Eurasian immigration, have been shown to be otherwise by molecular data and structural morphology. The first breakthrough, from DNA-DNA hybridization (Sibley and Ahlquist, 1985, 1990) was followed by protein electrophoresis (Christidis and Schodde, 1991), micro-complement fixation (Baverstock *et al.*, 1991) and mitochondrial DNA sequencing (Cracraft *et al.*, 1998). Then the osteological survey of Schodde and Mason (1999), revealed that the core Australasian songbird families were not related to Eurasian look-alikes but to one another. Malurid wrens and acanthizid warblers were allied to one another and to honeyeaters, not Eurasian troglodytine wrens or sylviid warblers; fantails and monarch-flycatchers were allied to cracticid crow-shrikes and birds of paradise; and the Australasian robins, babblers and bowerbirds stood apart from one another *within* the Australasian assemblage. Australasia's 'flycatchers', 'wrens', 'robins', 'thrushes', 'babblers' and even 'magpies' merely resembled Eurasian birds of the same name through convergent adaptation to the same niches and modes of foraging. It was a radiation of marsupial-like proportions. DNA-DNA hybridization and the structure of the humeral fossa in the wing indicate, in fact, that modern songbirds comprise two primary groups (Figure 1): the Corvida of 28 families and about 1,100 species centred in Australasia, and the Passerida (32 families and approximately 3,600 species) well diversified throughout the Northern Hemisphere and Africa (Sibley and Monroe, 1990, as modified by Schodde and Mason, 1999). Within Australia itself, moreover, corvidan families comprise 85% of songbird species, and in the rest of Australasia 77%. Australia-New Guinea is the locus of global corvidan radiation.

But what is its source? Numerical dominance in songbirds would suggest Eurasia, as tradition has held. Does genic distance between families, which equates to relative age of divergence and potentially sorts ancestral groups from derived, confirm it? Sibley and Ahlquist's (1990) data are non-committal, recording three root lineages in the massive Northern Hemisphere passeridan radiation, and

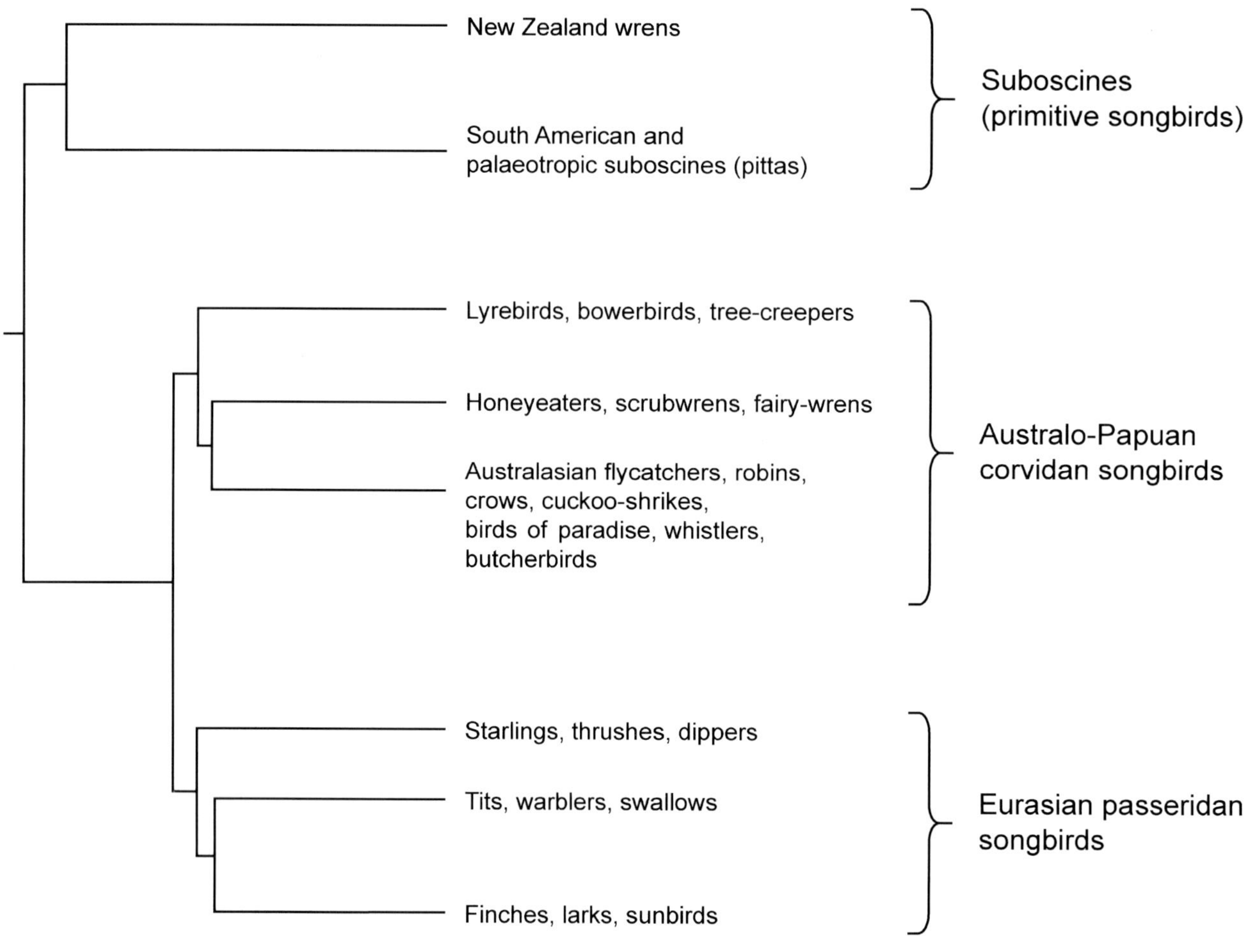

Figure 1. A simplified phylogeny for passerine or perching birds showing the split between Australasian corvidan songbirds and Eurasian passeridan songbirds, according to the DNA-DNA hybridization data of Sibley and Ahlquist (1990).

another three of at least equal depth in the much less speciose corvidan radiation in Australasia: (i) lyrebirds, scrub-birds, bowerbirds, treecreepers; (ii) malurid wrens, acanthizid warblers, honeyeaters and (iii) the remaining corvidan families of monarch flycatchers, robins, whistlers, cracticids, cuckoo-shrikes and birds of paradise. However, current evidence from DNA sequencing (Barker *et al.*, 2001; Ericson *et al.*, 2001) not only corroborates much of this, but also finds (i) the New Zealand xenicid wrens (Acanthisittidae) ancestral to all perching birds or passerines (which include songbirds), and (ii) the entire passeridan cluster of families nested in, and so derived from, the Corvida. Ancestral morphological conditions, too, prevail in corvidans: (i) variability of the syrinx in lyrebirds, scrub-birds and treecreepers (Ames, 1971, 1987), (ii) a single pneumatized humeral fossa that is shared with suboscines, and (iii) a greater than usual number of secondaries in lyrebirds and bowerbirds (Schodde and Mason, 1999). Far from supporting a Eurasian source, these data indicate (i) an East Gondwanan origin for the songbirds, with initial (corvidan) radiation on the Australian plate, and (ii) derivation of the Northern Hemisphere passeridan fauna from the Corvida, with initial dispersal in the reverse direction, from Australia to the Old World, followed by an explosion of speciation (Figure 2).

SYNTHESIS OF EVIDENCE

Reconciling tectonic events with the fossil record and current phylogeographic information recasts an origin for Australasia's bird fauna that is almost the antithesis of Eurasian immigration theory. It now seems *more likely* that the basic stocks of Australasia's modern land bird fauna—its ratites, megapodes, cockatoos and parrots, phabine and fruit pigeons, cuculine cuckoos, possibly its button-quail, tytonid owls, frogmouth-like birds and eurostopodine nightjars, and certainly its major corvidan songbird lineages—collectively summing to over 80% of present species—were inherited *in situ* from Gondwana. All belong to families or subfamilies that are prevailingly endemic, indicating massive indigenous radiation. Other lineages of waterfowl (*Anseranas*, *Stictonetta*, *Dendrocygna*, *Biziura*), raptors (*Hamirostra*, *Lophoictinia* and possibly *Erythrotriochis*), waders (*Pedionomus*, *Cladorhynchus*, vanelline plovers) and the now-extinct flamingos (Phoenicopteridae) are probably Gondwanan as well (Olson, 1988). Most probably arose in East Gondwana, many vicariantly with West Gondwanan stocks that gave rise in parallel to the arine parrots, cracids, screamers, columbinine pigeons (Johnson and Clayton, 2000), coccyzine cuckoos, seed-snipe and suboscinine passerines in South America today. For the land birds, *Nothofagus*-associated subtropical and temperate rainforest was the prevailing ancestral habitat.

Eurasian immigration, like Australasian emigration, did occur, but perhaps no earlier than early Miocene times (20–25 m.y.a.). Then, judged by the explosive advent of passerines in the Eurasian fossil record, the northward drift of Australia and its Indonesian chards (Audley-Charles, 1987; Stevens, 1991; Hall, 1996) had brought the two land masses close enough to exchange dispersive elements. Even so, few Eurasian immigrants appear to have gained a foothold in Australasia since, contributing less than 20% of total modern species. The likely reasons are twofold: (i) the faunal nexus that the Indonesian archipelagos still pose between Wallace's and Weber's Lines (Mayr, 1944c), inhibiting free interchange and sifting out groups reluctant to cross water barriers; and (ii) an already established land bird fauna in Australia-New Guinea that had filled most available niches. Not surprisingly, most of the immigrants, judged from modern and fossil distributions, appear to be water birds (anatine ducks, herons, ibis, rails and summering scolopacid waders and plovers), as well as raptors (kites, falcons and perhaps strigid owls) and coraciiforms (land kingfishers, bee-eaters and rollers). The few passerines are mostly of passeridan families—one lark, two pipits, two or three thrushes, a chat, several sunbirds, flowerpeckers, cisticolas, swallows, glossy starlings and mynas, and three moderate radiations in estrildid finches, white-eyes and grassbirds. Several corvidans—the Schach Shrike in New Guinea and *Corvus* itself, with eight or nine regional species—have also returned to Australasia secondarily.

Lack of endemic diversification among the immigrants may reflect relatively recent arrival or dearth of opportunity to radiate in an environment where most niches were already filled. Both circumstances probably applied, the latter explaining the anomalous radiation of grassbirds (Megalurinae) and estrildid finches in grass-field habitats that were themselves newly adventive in desiccating Australia towards the end of the Tertiary (Hill *et al.*, 1999). The overall picture that emerges is of a bird fauna built by layers of different age and different diversification. At the base is the primary Gondwanan fauna, filling all treed and shrub-field niches on land. Superimposed upon it are successive but minor waves of Eurasian immigrants that are today most diverse in grasslands and wetlands, habitats to which Gondwanan stocks are poorly adapted or from which they have been culled by extinction. It is a picture, nevertheless, that is only as clear as its underpinning data, and these in turn gain their strength not from any single set but from corroboration between sets—molecular and morphological phylogeny, palaeontology, palaeoecology and plate tectonics. However strong that corroboration may appear, it is still circumstantial and so susceptible to the same shifts that have befallen the ideas of the mid 20th century.

EVOLUTIONARY DEVELOPMENT

How did the Australasian bird fauna adapt, change and develop from its origins and ancestral environments in the mid Tertiary (20–30 m.y.a.) to its present expression?

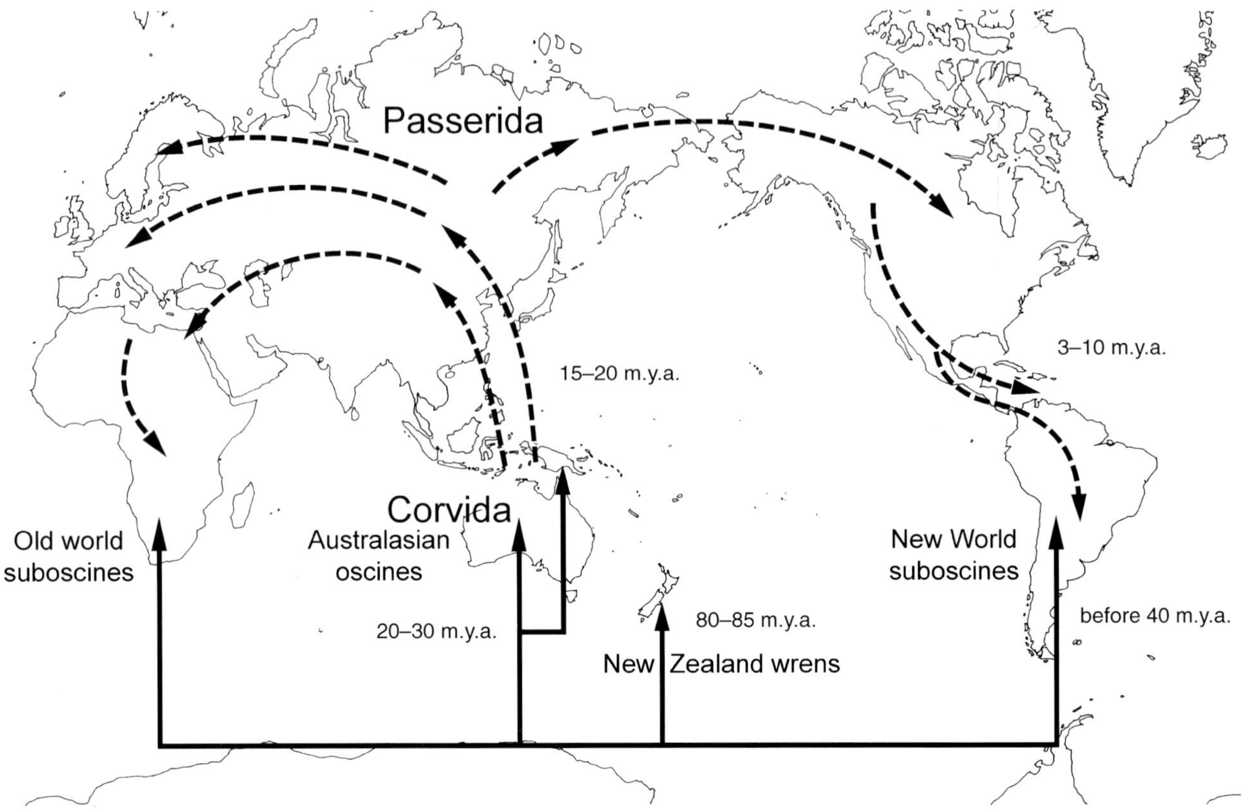

Figure 2. Hypothetical dispersal of passerine or perching birds out of Gondwanan continents. Based on phylogenetic data in Barker *et al.* (2001) and Ericson *et al.* (2001). Straight arrows indicate vicariant lineages probably inherited *in situ* from Gondwana; curved arrows indicate likely dispersal events.

TRADITIONAL APPROACHES—REGION BY REGION

The foundations of Australian bird geography were laid by Baldwin Spencer (1896) when he defined southern wet-winter Bassian, northern wet-summer Torresian, and arid, central Eyrean regions for the Australian fauna. Because the regions reflected a historic zoning of environment and habitat, Serventy (1948, 1953, 1972), Gentilli (1949), Keast (1961) and others began to interpret them as source areas for constructing patterns of speciation. The ecological regions had become regional faunulas; and out of them grew the rudiments of a *causal* geography for Australian birds because the great bulk of the Australian land and freshwater bird fauna outside rainforest was found to approximate to a Bassian, Torresian or Eyrean 'type' distribution. Yet the Spencer model also had its limitations. In its conventional form, it could only provide a geographic frame for interpreting modern speciation. Underlying its simplicity was the perception that it only had to explain radiation in one bird fauna, the old, sclerophyll-inhabiting bird fauna of supposed Eurasian origin (Mayr, 1944a, b; Keast, 1981, 1984). Even then it made no connection with the serial advent of that fauna; nor did it acknowledge the rainforest elements on the east coast—these it submerged in the Torresian avifauna; and it took no account of faunal connections with New Guinea.

An alternative model was computed by Kikkawa and Pearse (1969), based on a numerical analysis of taxon densities. Its regions had a quantitative rather than qualitative basis, and, taken together, were an excellent example of what they were intended to produce: a *functional* bird geography for Australia. In it, however, the factor of historic time was missing, making it inappropriate for interpreting evolutionary development in the bird fauna (Vuilleumier, 1975). Elsewhere in New Zealand, the much simpler bird fauna there was perceived as a layered one, built largely by immigrants from Australia (Fleming, 1962; Bull and Whitaker, 1975; Schodde and Faith, 1991). For Papuasia, zoogeographic syntheses by Schodde and Hitchcock (1972) and Pratt (1982) summarized distribution patterns in relation to terrain, habitat and physical barriers, and so processes driving faunal diversification and impoverishment. These, like the Australian studies, limited themselves to their own region and focused on contemporary speciation. It led Keast (1981, 1984) to view the great Australian and New Guinean bird faunas as 'twin' faunas, one sclerophyll vegetation-adapted, the other rainforest-adapted.

HOLISTIC APPROACHES—THE AUSTRALIA-NEW GUINEA CONNECTION

To draw the evolutionary threads of the whole Australasian bird fauna together, a more integrative approach has been needed; one that connects zoogeographic patterns and addresses deeper phyletic linkages rather than just recent speciation. Schodde and Calaby's (1972) analysis of avifaunal connections between Australia and New Guinea was a first step. It concluded that: (i) the sclerophyll woodland bird fauna in southern New Guinea was an outlier of the Torresian bird fauna of northern Australia, (ii) the rainforest bird fauna of Cape York Peninsula alone was derived from the tropical lowland rainforests of New Guinea, and (iii), most significantly, the bird faunas of the subtropical rainforests further south along the east Australian coast were vicariant with those isolated in montane New Guinea.

Further clarification followed from Kikkawa *et al.* (1981) and Pratt (1991), together with Schodde and Faith's (1991) cladistic analysis of the entire Australasian region at both generic and species level (Figure 3). These holistic analyses reversed earlier 20th century perceptions about directions of evolutionary radiation in Australasia's bird fauna. First, as stressed by both Schodde (1991) and Pratt (1991), Australia and New Guinea hold not twin faunas (Keast 1981,1984) but a single fauna of different elements adapted ecologically over time to different habitat: rainforest mainly in New Guinea and sclerophyll woodland mainly in Australia. Second, the subtropical Australian rainforest bird fauna and its vicar in montane New Guinea is ancestral Gondwanan, rich in root Australian land bird lineages and occupying rainforests of laurels, mytles, Winteraceae, *Nothofagus*, cunoniads, escalloniads and podocarps that were widespread in Australia in the mid Tertiary (Kemp, 1978; Truswell and Harris, 1982; Martin, 1984; Hill *et al.*, 1999). As Archer and Fox (1984) paraphrased elegantly, 'a walk up a New Guinea mountain is like a walk back into time'. Third, by corollary, the bird faunas of the tropical jungles of lowland Papuasia and sclerophyll Australian woodlands are derived, and not of old immigrant stocks from Eurasia except for an anomaly or two (e.g. Blyth's hornbill (*Rhyticeros*) in Papuasia and the Australian bustard (*Ardeotis*) in Australia).

The three zoogeographic units of Spencer (Bassian, Torresian and Eyrean) had become inadequate as a model for reflecting pattern and process in the development of the Australasian bird fauna as a whole. Accordingly, Schodde and Calaby (1972) and Schodde and Faith (1991) added a further two: the Irian for the avifaunula of the lowland tropical rainforests of Papuasia, and the Tumbunan, from Neo-Melanesian pidgin 'tumbuna' meaning ancestor, for the vicariant rainforest avifaunula of montane New Guinea and subtropical east Australia. All were perceived, not as static regions, but as dynamic

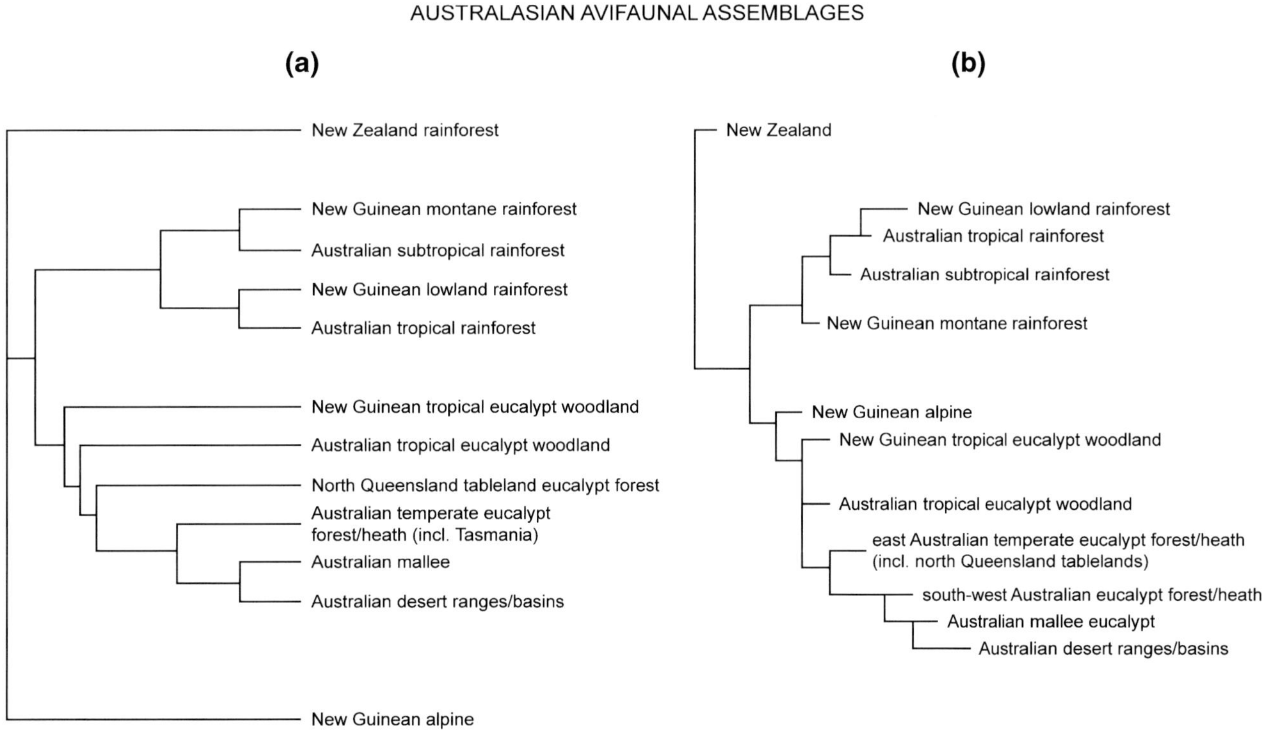

Figure 3. Strict consensus area cladograms of Australasian faunas: **(a)** cladogram of assemblages for genera, based on five PAUP trees; **(b)** cladogram of assemblages for species, based on six HENNIG 86 trees. Modified from Schodde and Faith (1991).

faunas, co-adapted to habitat and responsive to environmental change. It reset the conceptual framework for interpreting the development of Australasia's modern bird fauna.

EARLIER EVENTS (MID-TERTIARY TO PLEISTOCENE)

More than any other group of vertebrates, land birds are tied to vegetation niches, dependent on them for shelter, food, and nest sites. So palaeogeographic events that shaped their diversification in Australasia are tied here to those in the flora.

Later Tertiary Environments in Australia and Water Birds

The roots of Australasia's contemporary bird fauna trace back to the mid-Tertiary (20–30 m.y.a.). By that time, New Zealand was well separated from Australia on the Campbell Plateau, and New Guinea had not begun to coalesce; it comprised two sections, a southern bloc that formed the northern edge of the exposed Australian Continental Plate and a northern complex of Australian continental fragments and a fringing island arc. Australia itself, some 1,000 km south of its present position, had nevertheless assumed much of its present form. For further details of topography and conditions over the continent then see pp.17–50 (White, 2006), but key points relating to bird faunas include: a stable continent of low relief except for an embryonic Great Dividing Range on its eastern rim; extensive lacustrine environments throughout the eastern central region (Great Artesian Basin) and a cool, moist climate with subtropical rainforests covering much of the landscape.

It was a time when rainforest and freshwater bird faunas flourished (Rich and van Tets, 1984; Boles, 1991). The fossil record indicates that a rich water and shore bird fauna had diversified on the lake systems of the Great Artesian Basin: grebes, the short-limbed *Pelecanus tirarensis*, cormorants, ducks, cranes, rails, several species of *Burhinus*, flamingo-like palaelodid waders and two or three species of flamingos (Miller, 1963a,b, 1966a,b; Rich, 1975; Rich and van Tets, 1984). A high number of old autochthons in the Eyrean avifaunula (monotypic genera) may trace their origins to this fauna. Examples are: freckled duck (*Stictonetta*), pink-eared duck (*Malacorhynchus*), plains wanderer (*Pedionomus*), banded stilt (*Cladorhynchus*), black-fronted dotterel (*Elseyornis*), red-kneed dotterel (*Erythrogonys*), and Australian praticole (*Stiltia*). No other Australasian avifaunula is as rich in old freshwater or shore autochthons today (Schodde, 1982a).

The primary environmental force influencing avian diversification in Australasia from then on has been the desertification of Australia. It began locally in central Australia in the Oligocene (~30 m.y.a.), and gradually intensified, spreading out towards the rim of the continent to culminate in the cool-dry glacial episodes of the Plio-Pleistocene (Bowler, 1976; Kemp, 1978; Truswell and Harris, 1982; Frakes, 1999: Dawson and Dawson, 2006). Rainforest gave way first to sclerophyll woodlands and heaths dominated today by *Eucalyptus*, then to savannas and grasslands, and finally to deserts (Martin, 1984; Truswell, 1993). At first conditions favoured large cursorial birds and raptors, such as giant dromornithids (6 genera, 8 species), giant megapodes (*Progura*) and giant eagles (*Harpagornis*), which reached their zenith then with the rest of the megafauna (van Tets, 1974; Rich, 1979; Rich and van Tets, 1984). Then the arrival of man and fluctuations in climate coincided with megafaunal extinctions 50,000–60,000 years ago—see pp.51–70 (Dawson and Dawson, 2006). Much of the freshwater bird fauna of the inland basins suffered the same fate. According to the fossil record, a small pelican (*Pelecanus cadimurka*), a pygmy cormorant, a broad-headed darter (now synonymized with living *Anhinga novaehollandiae*), a very large heron, a second species of stork, an eagle (*Taphaetus lacertosus*), and the last of the flamingos died out (Rich and van Tets, 1984). Even in climatically more stable parts of south-east and south-west Australia, Hope *et al.* (1977), Merrilees (1979) and Baird (1985) document local extinction in many birds. Judged as well by the present diversity of relictual elements in rainforested montane New Guinea, Australia's bird fauna today may be little more than half of its size in later Tertiary times.

Compounding loss by extinction was loss by exodus. As Australia desiccated, the subtropical Gondwanan rainforests and their bird faunas were compressed progressively coastwards. To the east, the developing Great Dividing Range (Jennings and Mabbutt, 1977; Ollier, 1978) provided refuges along its coastward scarps (Figure 4). To the north-east, the biota also found refuge in the enormous central cordillera of New Guinea that began to rise coherently right along the northern rim of the Australian continental plate from the later Miocene into the Quaternary, within the last 10 million years (Visser and Hermes, 1962; Dow, 1977; Pigram and Davies, 1987; Audley-Charles, 1991; Struckmeyer *et al.*, 1993).

The Development of New Guinea and its Avifaunulas

In contrast to the tectonic stability of Australia, the palaeogeographic changes to New Guinea in later Tertiary times were catastrophic. At the beginning of the Miocene, New Guinea lay in disconnected fragments off north Australia. Details of the formation of New Guinea have been discussed previously—see pp.71–93 (Heinsohn and Hope, 2006) and pp.95–110 (van Oosterzee, 2006)—but the important conclusions, in relation to birds, are: that the island of New Guinea, as we know it, only began to form 4–5 m.y.a.; and that the land link with Australia was only finally broken (intermittently) by eustatic rises in sea level in late Pleistocene.

Climate not only became drier but also warmer as Australia drifted north. So the rising of the cooler New

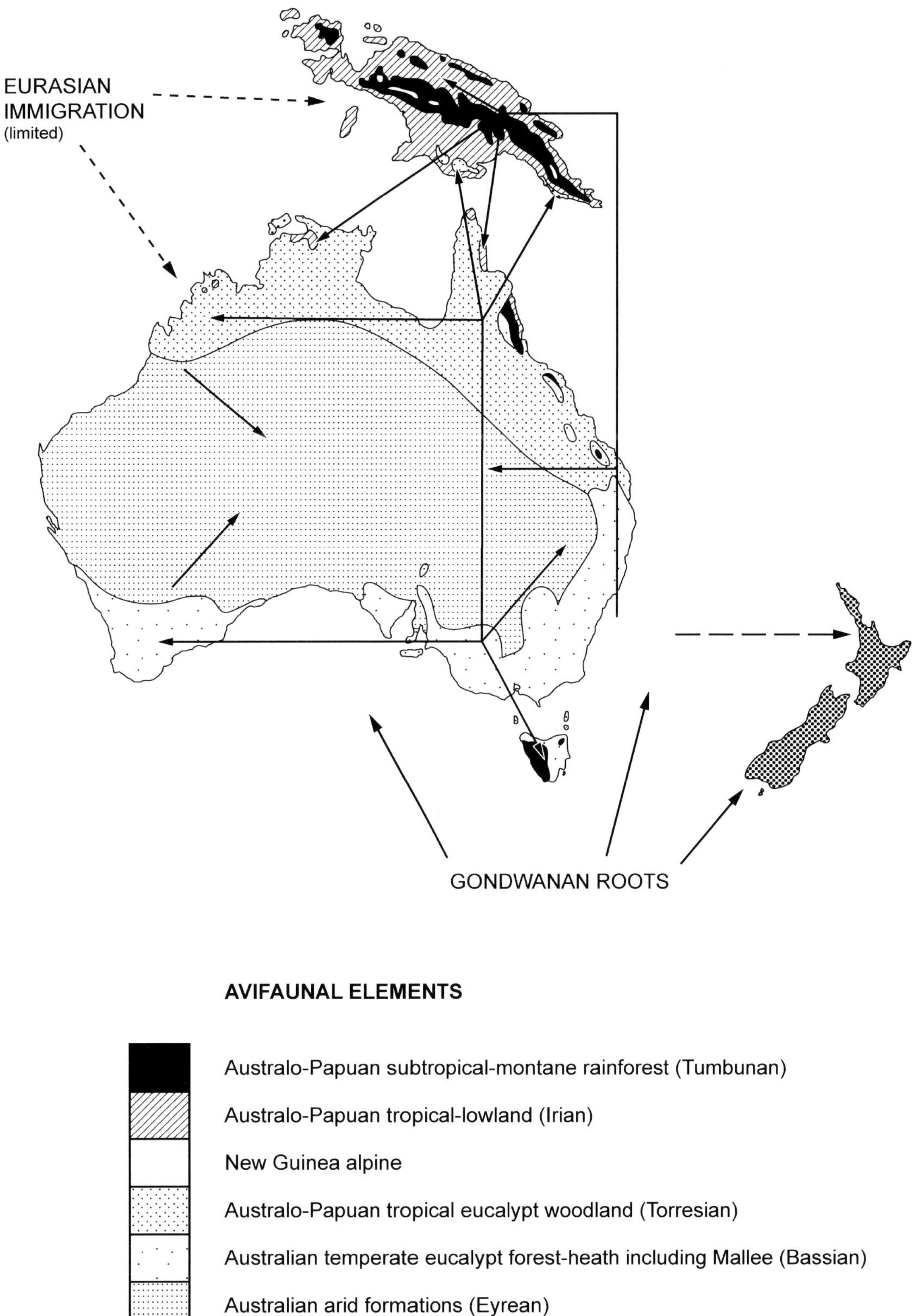

Figure 4. Suggested developmental radiation in the Australasian bird fauna through the later Tertiary to Quaternary, indicated by arrows on a base of established avifaunal elements. Modified from Schodde (1991). Solid arrows indicate likely *in situ* Gondwanan sources and regional vicariant connections; dashed arrows indicate immigrant oversea dispersal.

fringes (Schodde and Calaby, 1972). From such forms arose the red-winged parrot group (*Aprosmictus*) in the Torresian north and *Polytelis* itself in the Bassian south, both co-evolving with sclerophyll woodlands in inland Australia through Miocene times. *Aprosmictus* has probably changed little to this day, except for budding off a second species in Timor. *Polytelis*, however, whose populations were fragmented by Plio-Pleistocene climatic oscillations around the periphery of the arid zone in southern Australia, speciated and subspeciated. Its modern Eyrean representative, the Princess parrot (*P. alexandrae*), probably arose either (i) from an early phase of isolation in south-west Australia, to later shift adaptively inland during pluvial episodes, or (ii) from refuging and speciating vicariantly in central ranges during dry (glacial) periods.

In the other example, the whipbirds (*Psophodes*), the pathway is similar but, being intra-generic, possibly more recent (Figure 6 (b)). From stocks represented today by *Psophodes olivaceus* and *Androphobus* in the Tumbunan rainforests of east Australia and montane New Guinea, a southern Bassian heath group (*Psophodes* subgenus *Phodopses*) first budded off vicariantly in the south-west of Australia (Schodde, 1982a; Schodde and Mason, 1991). From this group the eremic-adapted wedgebills (*Psophodes occidentalis*, *P. cristatus*) then evolved by one or other of the mechanisms suggested above for *Polytelis alexandrae* before they themselves speciated vicariantly, split by the Pleistocene Eyrean Barrier.

These sequences and other distribution patterns traced by Serventy (1953, 1972), Keast (1961), Harrison (1973, 1975), Ford (1974) and Schodde (1982a) show that the adaptive trail runs consistently from mesic to eremic, the Bassian and Torresian biotas serving as 'transition zones'. There appear to have been two principal routes for adaptive colonization: one via Bassian south-west Australia, illustrated as above, the other via Torresian north-west Australia (e.g. *Amytornis*) (Figure 4). Climatic gradients in both regions, unaffected by the Great Dividing Range, were broad and more conducive to adaptive shifts to eremic environments than any in the east (Schodde, 1982a).

The guilds fostered by the two routes were different. In the Bassian south-west, mallee and sandplain heaths run structurally, if not floristically, into the low *Acacia* woodlands and hummock (spinifex) grasslands of arid Australia. Diverse shrubbery and ground-foraging groups as well as arboreal foliage feeders entered the arid zone by this track, providing the major contribution to the Eyrean fauna of the family Acanthizidae (7 species in 3 genera) except *Gerygone*, quail-thrushes *Cinclosoma*, lichenostomatine and phylidonyrine honeyeaters, platycercine parrots, emu-wrens (*Stipiturus*) and others. Since the end of the Tertiary, Torresian northwest Australia has been covered by a simpler eucalypt woodland over a tussock grass understorey. Guilds coming from that source appear to have been predominantly tree-feeders: a number of honeyeaters (*Conopophila*, *Certhionyx*), bowerbirds (*Chlamydera*), *Gerygone fusca*, and perhaps the eremic woodswallows (*Artamus*), plus several ground-foraging seed-eaters—the pigeons *Geophaps*, *Ocyphaps* and *Geopelia*, finch *Taeniopygia castanotis* and part-insectivorous grasswrens (*Amytornis*).

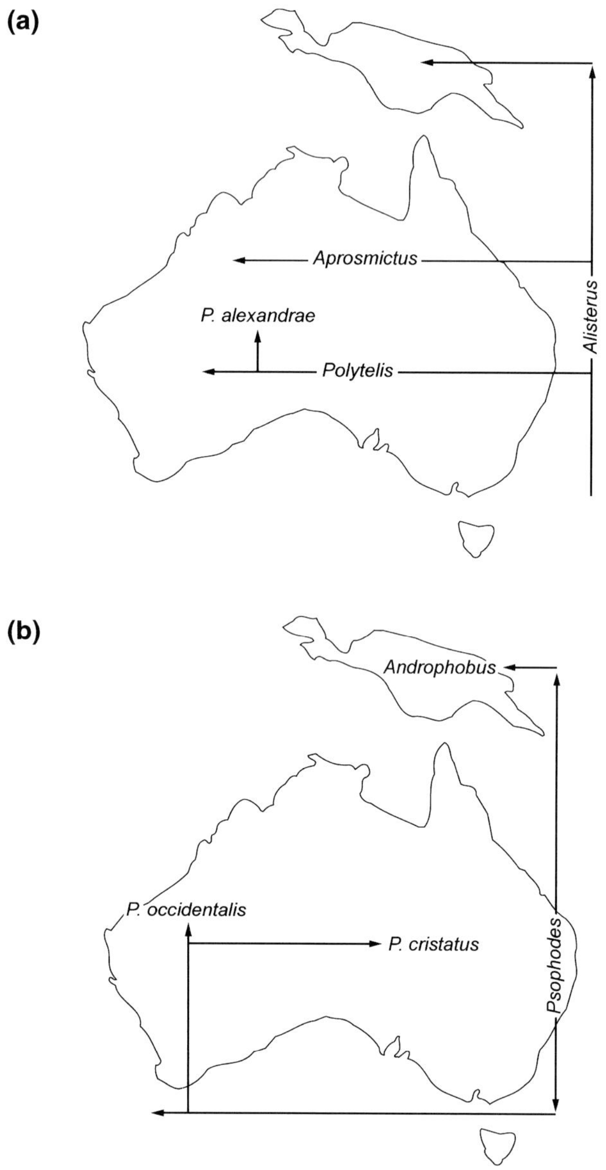

Figure 6. Hypothetical area phylogenies (evolutionary tracks) in: **(a)** the polyteline parrots (Polytelinae) and; **(b)** the Australo-Papuan whipbirds (*Psophodes*, *Androphobus*) in Australasia. Modified from Schodde (1982a).

QUATERNARY RADIATIONS

The term 'speciation' employed here to describe the Quaternary radiations is used in its broad sense, covering the successive processes of both speciation and subspeciation. Speciation for modern Australasian birds may extend back into the later Pliocene, but probably rarely beyond about 4 m.y.a. and often much less, judged by DNA distances in the logrunners, *Orthonyx* (Joseph *et al.*, 2001) and speciation in Australian grasswrens amid dune deserts of Pleistocene age.

Centres of Endemism and Faunal Barriers

By the end of the Tertiary, New Guinea had taken on its present form with a massive central mountain spine, Australia was substantially desiccated, and climatic sifting had established the five core regional avifaunulas of today. Plio-Pleistocene glacial maxima progressively denuded much of New Zealand and brought the snow line to within 300 m of present sea level in Tasmania; and sea level dropped episodically by up to 200 m to maintain broad land connections between Australia, Tasmania and New Guinea and effectively raise the tops of Australian ranges by equivalent altitude. Life zones and their avifaunulas shifted in concert with the intensifying cycling of climate, from warm-wet to cold-dry and back. In New Guinea they moved altitudinally up and down mountain ranges by 500–1,500 m (van der Kaars, 1991; Hope, 1996); and in cool dry episodes, savannas encroached to separate blocs of rainforest in the lowlands (Nix and Kalma, 1972). In Australia, the zones expanded and contracted latitudinally, making and breaking connections among more mesic avifaunulas (Tumbunan, Bassian and Torresian) around the rim of Australia, driving avifaunal accretion and extinction in the Eyrean centre. Pockets of humid-adapted biota were isolated in refugial ranges inland and around the coast during cold-dry episodes as well (Crocker and Wood, 1947; Gentilli, 1949; Serventy, 1953, 1972; Keast, 1961; Kikkawa and Pearse, 1969; Schodde, 1982a). The scene had been laid for contemporary speciation.

The cyclical development of concentric faunal zones and montane refugia in both Australia and New Guinea, however different the scale, generated: (i) successive rings of replacement species (allospecies) and subspecies around the lowland rims of both land masses; and (ii) chains of such taxa along their cordilleras. Simple examples of such rings are the white-cheeked rosellas (*Platycercus adscitus* superspecies), treecreepers (*Climacteris*), shrike-tits (*Falcunculus*) and white-naped honeyeaters (*Melithreptus* subgenus *Melithreptus*) in Australia, and crowned pigeons (*Goura*), streaked lories (*Chalcopsitta*), frilled monarchs (*Arses*) and greater/lesser birds of paradise (*Paradisaea apoda* superspecies) in New Guinea. Those of chains are the caligavine honeyeaters (*Lichenostomus* subgenus *Caligavis*), and megasubspecies of striated pardalotes (*Pardalotus striatus*) in east Australia and Tasmania, and the ptiloprorine and core melidectine honeyeaters (*Ptiloprora, Melidectes*) and flag and astrapia birds of paradise (*Parotia, Astrapia*) in New Guinea.

The core ranges of such taxa today were their refugial foci and source in past glacial cycles; and their present limits are the sites of past barriers, environmental as well as physical. For a great many members in all Australasian avifaunulas, distributional cores and limits are coincident, identifying centres of regional endemism and major faunal barriers respectively. Obvious centres of endemism in Australia are Tasmania, the Arnhem Land Plateau, the Atherton uplands in north-east Queensland, and the humid corner of the far south-west; others in New Guinea include the mountains of the Vogelkop and Huon Peninsula, and the Mamberamo-Sepik basin. The best known barriers in Australia are those that lack obvious physiographic or environmental disjunction today, namely: (i) Carpentarian, a shaft of savanna at the south-east head of the Gulf of Carpentaria that separates Arnhem Land-Kimberley and Cape York Peninsula sectors of the Torresian avifaunula (Macdonald, 1969); (ii) Torresian which, between the Endeavour and Burdekin Rivers in north-east Queensland, separates the rainforest and sclerophyll avifaunas of Cape York Peninsula from central east Queensland (Schodde and Mason, 1980; Schodde, 1982b, p. 107; Ford, 1986); and (iii) Eyrean, a broad gulch running north from the head of Spencer Gulf to Lake Eyre that divides east and west Bassian and Eyrean avifaunulas. Schodde and Mason (1999, pp.787–788) identified 22 such barriers in Australia. These, together with others in New Guinea and regional centres of endemism, will be elaborated further below.

Both refugial centres and faunal barriers operated serially through the Pleistocene glacial cycles; this is indicated by layered differentiation within the centres, at both species and subspecies levels. The last phase of subspeciation on mainland Australia and lowland New Guinea probably took place in refugia created in the last glacial cycle (12,000–20,000 years ago). For forms isolated since by warming climate on New Guinean mountain tops and rising sea level on off-shore islands everywhere, differentiation may be no older than 6,000–11,000 years.

New Zealand

The land bird fauna of the New Zealand region is depauperate today, culled by tectonic events and the Pleistocene glaciations. It is a layered fauna, with a uniquely Gondwanan base of kiwis, the recently extinct moas, nestorine and owl parrots, and xenicid wrens. Building on it have not been endemic radiations, but successive waves of immigrants, mostly from Australia (Fleming, 1962). Earlier Tertiary immigrants probably included the ancestors of the cyanoramphine parrots (*Cyanoramphus*), laughing owl (*Sceloglaux*), mohouines (*Finschia, Mohoua*), the New Zealand thrush (*Turnagra*), crow-honeyeaters (*Prosthemadera, Anthornis*) and callaeatids (*Callaeas, Huia*)—unless these too were inherited *in situ* from Gondwana. Then came modern Australian elements (*Gerygone, Petroica, Rhipidura*), and lastly species such as the black-faced cuckoo-shrike (*Coracina novaehollandiae*) and silvereye (*Zosterops lateralis*) which have established themselves since European settlement. For further discussion of the evolution of New Zealand birds and recent impacts of human colonization see pp.111–128 (Holdaway and Worthy, 2006).

Table 2. Taxa characteristic of regional avifaunulas present in Australasia today. The lists for regional avifaunulas are **not** complete, and commonly omit taxa ranging into more than one avifaunula, particularly in Australia. As a rule, monospecific genera are indicated by their names without qualification. Heavy lines between rows of families separate,

Family	TUMBUNAN		IRIAN		
	Australia	New Guinea	Australia	Papuasia	Mangrove obligates
Casuariidae (cassowaries, emus)	–	*Casuarius bennetti*	*Casuarius casuarius*	*Casuarius casuarius, C. unappendiculatus*	–
Megapodiidae (mound-builders)	*Alectura*	*Aepypodius arfakianus*	*Megapodius reinwardt*	*Talegalla* spp., *Megapodius* spp.	–
Anseranatidae (magpie-geese)	–	–	–	–	–
Anatidae (ducks)	–	*Anas waigiuensis*	*Tadorna radjah, Nettapus* spp.	*Tadorna radjah, Nettapus* spp.	–
Podicipedidae (grebes)	–	–	–	–	–
Ardeidae (herons)	–	–	*Ixobrychus flavicollis, Ardea picata, A. sumatrana*	*Ixobrychus flavicollis, Zonerodius, Ardea picata, A. sumatrana*	*Butorides striatus* (Australia)
Threskiornithidae (ibis)	–	–	–	–	–
Accipitridae (hawks)	–	*Accipiter melanochlamys, Erythrotriorchis buergersi*	*Aviceda subcristata,*	*Aviceda subcristata Henicopernis* spp., *Megatriorchis, Accipiter* spp., *Aquila gurneyi*	–
Falconidae (falcons)	–	–	–	–	–
Gruidae (cranes)	–	–	–	–	–
Rallidae (rails)	–	*Rallina* subg. *Rallicula*	*Rallina tricolor, Amaurornis moluccanus, Poliolimnas*	*Rallina tricolor, Megacrex, Amaurornis moluccanus, Nesoclopeus, Poliolimnas*	*Eulabeornis*
Turnicidae (button-quail)	*Turnix melanogaster*	–	–	–	–
Pedionomidae (plains wanderers)	–	–	–	–	–
Jacanidae (lily-trotters)	–	–	*Irediparra gallinacea*	*Irediparra gallinacea*	–
Recurvirostridae (stilts)	–	–	–	–	–
Charadriidae (plovers)	–	–	–	–	–
Glareolidae (pratincoles)	–	–	–	–	–

in descending order: non-passerines from passerines or perching birds, suboscines from songbirds and corvidan (presumed Gondwanan) songbirds from passeridan (presumed Palaearctic) songbirds; bracketed taxa in corvidan songbirds are presumed secondary invaders from Palaearctic sources.

Family	TORRESIAN		BASSIAN		EYREAN
	Australia	New Guinea	Australian sclerophyll/ swamplands	Mallee	
Casuariidae (cassowaries, emus)	–	–	–	–	–
Megapodiidae (mound-builders)	–	–	–	*Leipoa ocellata*	–
Anseranatidae (magpie-geese)	*Anseranas*	*Anseranas*	–	–	–
Anatidae (ducks)	*Dendrocygna eytoni*	–	*Cereopsis, Biziura, Tadorna tadornoides, Anas castanea, Chenonetta, Oxyura* sp., *Cygnus atratus*	–	*Stictonetta, Malacorhynchus*
Podicipedidae (grebes)	–	–	*Poliocephalus*	–	–
Ardeidae (herons)	–	–	*Botaurus poiciloptilus*	–	*Ardea pacifica*
Threskiornithidae (ibis)	–	–	*Platalea flavipes*	–	–
Accipitridae (hawks)	*Hamirostra melanosternon, Erythrotriorchis radiatus*	–	–	–	*Elanus scriptus, Circus assimilis*
Falconidae (falcons)	–	–	–	–	*Falco subniger, F. hypoleucos*
Gruidae (cranes)	*Grus* spp.	*Grus rubicunda*	–	–	–
Rallidae (rails)	–	–	*Gallinula mortierii*	–	*Gallinula ventralis*
Turnicidae (button-quail)	*Turnix maculosa, T. olivii T. castanota*	*Turnix maculosa*	*Turnix varia*	*Turnix varia*	*Turnix velox*
Pedionomidae (plains wanderers)	–	–	–	–	*Pedionomus*
Jacanidae (lily-trotters)	–	–	–	–	–
Recurvirostridae (stilts)	–	–	–	–	*Cladorhynchus leucocephalus*
Charadriidae (plovers)	–	–	–	–	*Charadrius australis, Vanellus tricolor, Elseyornis, Erythrogonys*
Glareolidae (pratincoles)	–	–	–	–	*Stiltia isabella*

Table 2. (continued)

Family	TUMBUNAN		IRIAN		
	Australia	New Guinea	Australia	Papuasia	Mangrove obligates
Columbidae (pigeons)	*Lopholaimus*	*Otidiphaps, Gymnophaps, Gallicolumba beccarii, Ptilinopus rivoli, Ducula chalconota*	*Chalcophaps, Ptilinopus, Ducula, Macropygia amboinensis*	*Goura* spp., *Microgoura, Reinwardtoena* spp., *Chalcophaps* spp., *Trugon, Henicophaps, Ducula* spp., *Ptilinopus* spp., *Gallicolumba* spp.	–
Cacatuidae (cockatoos)	–	–	*Probosciger*	*Probosciger*	–
Psittacidae (parrots)	*Alisterus scapularis*	*Neopsittacus, Psittrichas, Alisterus* spp., *Psittacella*	*Geoffroyus geoffroyi, Eclectus*	*Chalcopsitta* spp., *Eos* spp., *Pseudeos, Lorius* spp., *Psittaculirostris* spp., *Micropsitta* spp., *Geoffroyus* spp., *Eclectus*	–
Cuculidae (cuckoos)	–	*Chalcites ruficollis, Cacomantis flabelliformis*	*Chalcites minutillus, Eudynamys, Scythrops*	*Rhamphomantis, Chalcites minutillus, C. meyerii, Eudynamys* sp., *Microdynamis, Caliechthrus*	–
Centropodidae (coucals)	–	–	–	*Centropus* spp.	–
Strigidae (hawk-owls)	–	–	*Ninox rufa*	*Ninox rufa, Uroglaux*	–
Tytonidae (masked-owls)	*Tyto tenebricosa, T. multipunctata*	*Tyto tenebricosa*	–	–	–
Podargidae (frogmouths)	–	–	*Podargus papuensis*	*Podargus* spp.	–
Aegothelidae (owlet-nightjars)	–	*Aegotheles insignis, A. albertisii*	–	*Aegotheles tatei, A. bennetti*	–
Caprimulgidae (nightjars)	–	*Eurostopodus archboldi*	–	*Eurostopodus papuensis*	–
Apodidae, Hemiprocnidae (swifts, tree-swifts)	–	*Aerodramus papuensis*	*Aerodramus terraereginae*	*Aerodramus vanikorensis, A.spodiopygius, Chaetura novaeguineae, Hemiprocne mystacea*	–
Alcedinidae (kingfishers)	–	*Clytoceyx, Syma megarhyncha*	*Syma torotoro, Tanysiptera sylvia, Alcedo pusilla*	*Dacelo gaudichaud, Todiramphus* spp., *Melidora, Tanysiptera* spp., *Syma torotoro, Ceyx lepidus Alcedo pusilla*	*Todiramphus chloris* (Australia)
Bucerotidae (hornbills)	–	–	–	*Aceros plicatus*	–
Pittidae (pittas)	–	–	*Pitta* spp.	*Pitta* spp.	–

Table 2. (continued)

Family	TORRESIAN		BASSIAN		EYREAN
	Australia	New Guinea	Australian sclerophyll/ swamplands	Mallee	
Columbidae (pigeons)	*Geopelia humeralis, G. placida, Geophaps scripta, G. smithii, Petrophassa*	*Geopelia humeralis, G. placida*	*Phaps elegans, Leucosarcia*	*Ocyphaps lophotes*	*Phaps histrionica, Geopelia cuneata, Geophaps plumifera, Ocyphaps lophotes*
Cacatuidae (cockatoos)	*Calyptorhynchus banksii, Cacatua galerita, C. sanguinea*	*Cacatua sanguinea*	*Calyptorhynchus* subg. *Zanda, Calyptorhynchus lathami, Callocephalon, Cacatua tenuirostris*	*Cacatua leadbeateri, Eolophus, Nymphicus*	*Cacatua leadbeateri, Eolophus, Nymphicus*
Psittacidae (parrots)	*Psitteuteles versicolor, Trichoglossus rubritorquis, T. chlorolepidotus, Aprosmictus* spp., *Platycercus venustus,* subg. *Psephotellus*	–	*Glossopsitta* spp., *Polytelis swainsonii, Pezoporus wallicus, Lathamus, Neophema, Purpureicephalus, Platycercus elegans, Psephotus haematonotus*	*Glossopsitta porphyrocephala, Polytelis anthopeplus, Barnardius, Psephotus varius*	*Polytelis alexandrae, Pezoporus occidentalis, Melopsittacus, Northiella, Barnardius, Psephotus varius, Neopsephotus*
Cuculidae (cuckoos)	–	–	*Cacomantis flabelliformis* (Australia)	–	*Chalcites osculans*
Centropodidae (coucals)	*Centropus phasianinus*	*Centropus phasianinus*	–	–	–
Strigidae (hawk-owls)	–	–	*Ninox strenua*	–	–
Tytonidae (masked-owls)	–	–	–	–	–
Podargidae (frogmouths)	–	–	–	–	–
Aegothelidae (owlet-nightjars)	–	*Aegotheles cristatus*	–	–	–
Caprimulgidae (nightjars)	–	–	–	–	–
Apodidae, Hemiprocnidae (swifts, tree-swifts)	–	–	–	–	–
Alcedinidae (kingfishers)	*Dacelo leachii, Todiramphus macleayii*	*Dacelo leachii, Todiramphus macleayii*	*Dacelo novaeguineae*	*Todiramphus pyrrhopygia*	*Todiramphus pyrrhopygia*
Bucerotidae (hornbills)	–	–	–	–	–
Pittidae (pittas)	–	–	–	–	–

Table 2. (continued)

Family	TUMBUNAN		IRIAN		
	Australia	New Guinea	Australia	Papuasia	Mangrove obligates
Ptilonorhynchidae (bowerbirds)	*Ailuroedus crassirostris, Scenopoeetes, Prionodura, Sericulus, Ptilonorhynchus (pars)*	*Ailuroedus crassirostris, Archboldia, Amblyornis, Sericulus*	–	*Ailuroedus buccoides*	–
Turdidae (thrushes)	*Zoothera heinei*	*Zoothera heinei*	–	–	–
Sturnidae (starlings)	–	–	*Aplonis metallica*	*Aplonis* spp., *Mino*	–
Hirundinidae (swallows)	–	–	–	*Hirundo tahitica*	–
Zosteropidae (white-eyes)	–	*Zosterops* spp.	–	*Zosterops* spp.	*Zosterops lutea*
Sylviidae (warblers)	–	*Phylloscopus trivirgatus*	–	*Megalurulus* spp.	–
Dicaeidae (flowerpeckers)	–	–	–	*Dicaeum* spp.	–
Nectariniidae (sunbirds)	–	–	*Nectarinia jugularis*	*Nectarinia jugularis, N. aspasia*	–
Estrildidae (grass finches)	–	*Oreostruthus*	*Erythrura trichroa*	*Erythrura* spp.	–

Tumbunan Avifaunula

The Tumbunan avifaunula is confined today to subtropical and temperate rainforests of Gondwanan origin in montane mainland New Guinea (above 1,000–1,500 m a.s.l.) and along the east Australian coast, from Tasmania to the foot of Cape York Peninsula (Big Tableland)—Figures 7, 8. Corvidan songbirds are by far the dominant group today, comprising 170 species (77%) of the total fauna of 225 species (Table 2), most of them New Guinean where 130 are endemic. Prevalent families are honeyeaters (30 species), birds of paradise, including *Macgregoria* and cnemophiline satin-birds of uncertain affinity (26 species), acanthizid scrubwrens and allies (21), Australasian robins (16 species), whistlers and allies (16 species), bowerbirds (12 species) and the endemic New Guinean berrypeckers (8 species). The same families are also rich in endemic Tumbunan genera—4, 9, 2, 4, 3, 6 and 4 respectively—reflecting a long, continuous history of diversification. Non-passerine representation is patchy, and includes a scattering of families dominated by pigeons (9 species) and parrots (11 species). Adventive elements (e.g. *Accipiter melanochlamys*, *Scolopax saturata*, *Syma megarhyncha* and *Zosterops* spp. in New Guinea and *Columba leucomela* and *Zoothera* spp. in Australia) are few and comprise less than 15 % of species. None has diversified to match the old endemic corvidan or parrot groups, except perhaps the four or five forest rails (*Rallina*) in montane New Guinea.

Tumbunan Avifaunula in Australia

Quaternary climatic cycles fragmented Australia's Tumbunan avifaunula into four main blocs: (i) the Atherton between the Big Tableland and Paluma Range-Mt. Elliot, north-east Queensland; (ii) the Eungella on the Clarke-Connors Ranges inland from Mackay, Queensland; (iii) the central-south-east coast between the Dawes, Jimna and Bunya Ranges, south-east Queensland, and the Illawarra-Cambewarra district, New South Wales, with outlying pockets south and west through Gippsland to the Otway Range, Victoria; and (iv) the Tasmanian in montane and west Tasmania (Figure 7). Connections between the blocs probably pre-dated the last glacial cycle, at ages no younger than 25,000 years and perhaps rather older.

The richest (in both taxa and endemics) is the Atherton bloc; of the 32 species 11 are endemics, includ-

Plate 85. Female ospreys (*Pandion haliaetus*) may have a wing-span up to 1.7m. This large hawk inhabits coastal areas (including islands), bays and estuaries throughout Australia, although is now rare in Victoria and Tasmania. *P. haliaetus* will also move up large rivers and extend far inland. Photograph J. Cann.

Plate 86. Another bird of prey—this is the Pacific baza (*Aviceda subcristata*). Photograph K. Ireland.

Plate 87. Regent bowerbirds (*Sericulus chrysocephalus*) at the bower constructed by the male (foreground); the female of this species (background) is much less colourful. The total range extends throughout rainforest and eucalypt forests of eastern coastal Australia—from north-eastern Queensland to the Hawkesbury River, N.S.W. Photograph G. Threlfo.

Plate 88. A red-backed kingfisher (*Todiramphus pyrrhopygia*). Photograph B. Wells.

Plate 89. A white-eared honeyeater (*Lichenostomus leucotis*) at the nest. Photograph G. and J. Little.

Plate 90. The fairy martin (*Hirundo ariel*). Photograph M. Seyfort.

Table 2. (continued)

Family	TORRESIAN		BASSIAN		EYREAN
	Australia	New Guinea	Australian sclerophyll/ swamplands	Mallee	
Ptilonorhynchidae (bowerbirds)	*Chlamydera nuchalis, C. cerviniventris*	*Chlamydera nuchalis, C. cerviniventris, C. lauterbachi*	–	–	*Chlamydera maculata, C. guttata*
Turdidae (thrushes)	–	–	*Zoothera lunulata*	–	–
Sturnidae (starlings)	–	–	–	–	–
Hirundinidae (swallows)	–	–	*Hirundo neoxena, Petrochelidon nigricans*	*Petrochelidon nigricans*	*Cheramoeca, Petrochelidon ariel*
Zosteropidae (white-eyes)	–	–	*Zosterops lateralis*	–	–
Sylviidae (warblers)	*Megalurus timoriensis, Cisticola exilis*	*Megalurus timoriensis, Cisticola exilis*	*Megalurus gramineus*	–	*Cincloramphus* spp., *Eremiornis*
Dicaeidae (flowerpeckers)	–	–	–	–	–
Nectariniidae (sunbirds)	–	–	–	–	–
Estrildidae (grass finches)	*Neochmia* spp., *Poephila, Taeniopygia bichenovii, Erythrura gouldiae, Lonchura* spp., *Heteromunia*	*Neochmia phaeton, Lonchura* spp.	*Stagonopleura* spp., *Neochmia temporalis*	–	*Emblema, Taeniopygia guttata*

ing a sooty owl (*Tyto multipunctata*), the fernwren (*Oreoscopus*), honeyeaters *Xanthotis macleayana* and *Lichenostomus frenatus*, a logrunner (*Orthonyx spaldingii*), shrike-thrush (*Colluricincla boweri*) and riflebird (*Ptiloris victoriae*), and two bowerbirds (*Scenopooetes*, *Prionodura*)—see Table 2. Another 12 subspecies are endemic as well: ultrataxic endemism is thus 70%, higher than in any other centre of endemism in Australia. Many species are zoned altitudinally, as in New Guinea, and most occur above 500 m. One barrier, the Black Mountain Barrier, has been found within the bloc; it runs from the Daintree lowlands to Julatten-Mt. Molloy (Joseph and Moritz, 1994), separating forms of *Orthonyx spaldingii* and *Poecilodryas albispecularis*, as well as tree kangaroos, *Dendrolagus* spp. With only 17 species and one endemic (*Lichenostomus hindwoodi*), the Eungella bloc is depauperate, reflecting limited niches in limited pockets of relictual habitat (Table 2). Two subspecies are also endemic: *Cormobates leucophaeus intermedius* and *Gerygone mouki amalia*. Despite the proximity of the Atherton bloc, zoogeographic links are strongest with the central-south-east bloc: *Sericulus chrysocephalus*, *Tyto tenebricosa* and most subspecies are contaxic with populations in the southern bloc, e.g. *Alisterus s. scapularis*, *Sericornis m. magnirostris* and *Psophodes o. olivaceus* (Schodde and Mason, 1997, 1999).

The central-south-east coast bloc, with 29 species, becomes increasingly impoverished north to south as habitat tails off (Table 2). Five species—black-breasted button-quail, double-eyed fig-parrot, marbled frogmouth, Albert's lyrebird and white-eared monarch—do not extend south of the Clarence River; another six drop out by around the Hunter River valley (rufous scrub-bird, pale-yellow robin, barred cuckoo-shrike, paradise riflebird, regent bowerbird and russet-tailed thrush). A further two—a logrunner and green catbird—stop at the Illawarra-Cambewarra Ranges. Five species are endemic: *Turnix melanogaster*, *Menura alberti*, *Atrichornis rufescens*, *Orthonyx temminckii* and *Ptiloris paradiseus*. Although several zoogeographic barriers cut through the bloc, infra-specific differentiation is limited, evident in only *Atrichornis rufescens*, *Cormobates leucophaeus*, *Sericornis citreogularis*, *S. magnirostris* and *Pachycephala olivacea* (Schodde and Mason, 1999). Their budding may be no older than the last glacial cycle (12,000–22,000 years ago). The Tasmanian bloc is poorest, with only four species: two endemic acanthizids

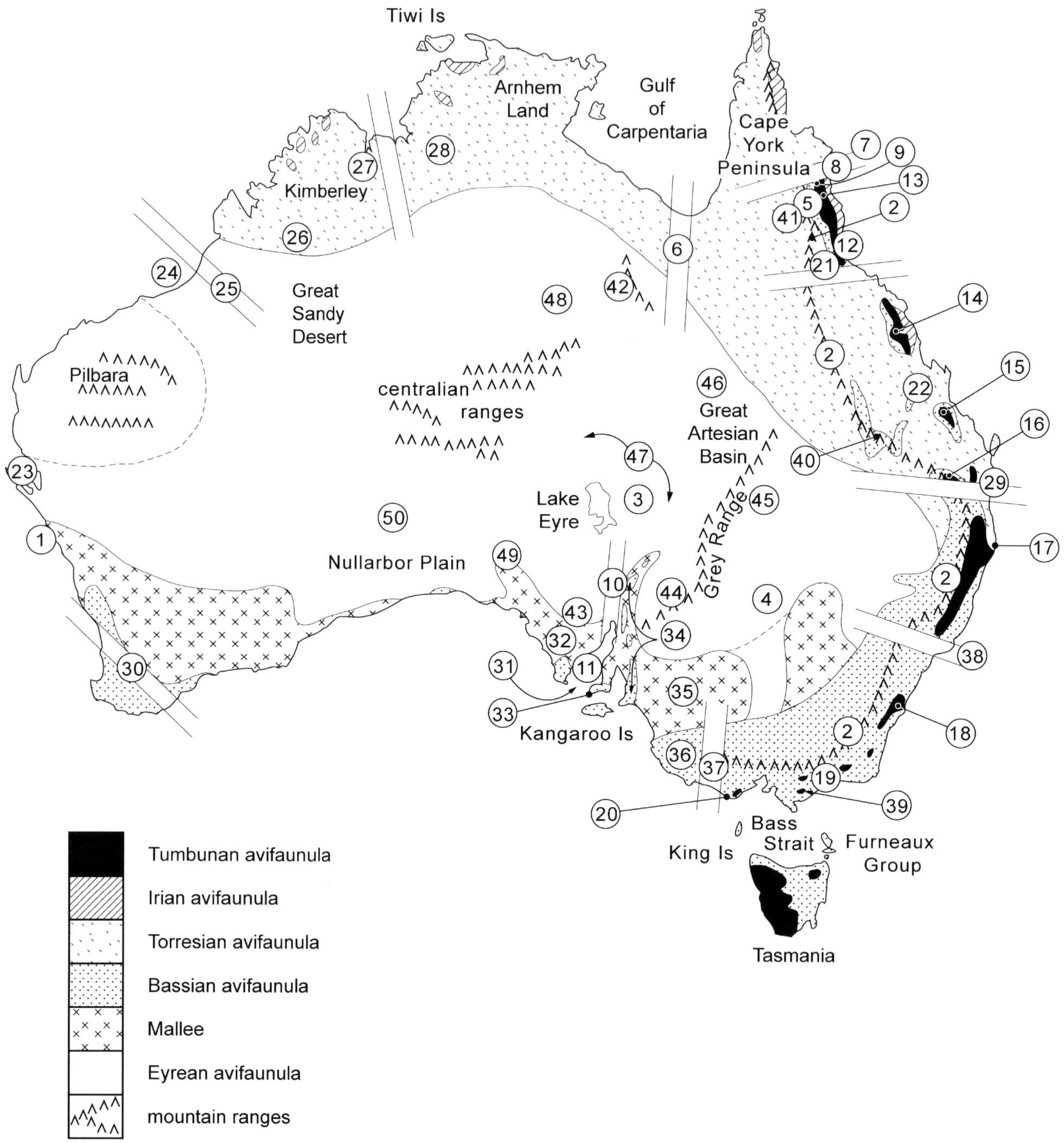

Figure 7. Distribution of avifaunulas in Australia, their refugia, barriers and localities cited in the text. Key to numbers: 1 = Houtman Abrolhos; 2 = Great Dividing Range; 3 = Lake Eyre Basin; 4 = Murray-Darling Basin; 5 = Atherton uplands; 6 = Carpentaria Barrier; 7 = Torresian Barrier; 8 = Endeavour River (Cooktown); 9 = Big Tableland; 10 = Eyrean Barrier; 11 = Spencer Gulf; 12 = Paluma Range (Ingham); 13 = Black Mountain Barrier; 14 = Eungella Plateau; 15 = Dawes Range-Kroombit Tops; 16 = Bunya Mountains; 17 = Clarence River; 18 = Illawarra-Cambewarra district; 19 = Gippsland; 20 = Otway Ranges; 21 = Burdekin Gap (River); 22 = Dawson-Mackenzie drainage; 23 = Shark Bay; 24 = Eighty Mile Beach; 25 = Canning Barrier; 26 = Fitzroy Basin; 27 = Ord River; 28 = Victoria River; 29 = Mary-Burnett-Brisbane drainage (Granite Belt, Border Ranges); 30 = Darling Barrier; 31 = South Australian gulfs; 32 = Eyre Peninsula; 33 = Yorke Peninsula; 34 = Mt Lofty-Flinders Ranges; 35 = Murravian Gulf; 36 = South East of South Australia; 37 = Bassian Volcanic Barrier; 38 = Hunter River (Barrier); 39 = Strzelecki Ranges; 40 = Carnarvon-Chesterton Ranges; 41 = Lolworth-Annan uplands; 42 = Selwyn Ranges; 43 = Gawler Ranges; 44 = Olary Spur; 45 = Bulloo Basin; 46 = Eromanga Basin; 47 = Simpson-Strzelecki Deserts; 48 = Sandover Desert; 49 = Barton sandhills; 50 = Great Victoria Desert.

(*Acanthornis* and *Acanthiza ewingii*, which also occurs in Bassian eucalypt forests), the pink robin (*Petroica rodinogaster*) and olive whistler (*Pachycephala olivacea*), also present in Bassian forest.

With little over 50 species, Australia's Tumbunan avifaunula is impoverished today (Table 2). Little species-budding and high proportions of monospecific genera (*Alectura*, *Lopholaimus*, *Oreoscopus*, *Scenopooetes*, *Prionodura* and *Ptilonorhynchus*) as well as ancient-lineage species (*Alisterus scapularis*, *Menura alberti*, *Atrichornis rufescens*, *Cormobates leucophaeus*, *Orthonyx* spp.) reflect a long history of culling and contraction. Atherton and central-south-east coast blocs are the primary residual centres. Vicariance at specific (*Tyto multipunctata-T. tenebricosa, Acanthiza katherina-A. pusilla, Lichenostomus frenatus-L. hindwoodi, Orthonyx spaldingii-O. temminkii* and *Ptiloris victoriae-P. paradiseus*) and subspecific levels (*Alisterus scapularis, Cormobates leucophaeus, Sericornis citreogularis, Gerygone mouki, Meliphaga lewinii, Tregellasia capito, Psophodes olivaceus, Ailuroedus crassirostris* and *Ptilonorhynchus violaceus*) shows the centres linked closer to one another than to montane New Guinea. This situation reflects a series of make-and-break connections along the Great Dividing Range deep into the Pliocene. Where species pairs overlap today, the southern member is the invader, displaced into sclerophyll habitats in the process, e.g. *Ninox b. boobook*, *Lichenostomus chrysops*(?), *Sericornis frontalis* and *Eopsaltria australis*. Such competitive interaction may well be the basis of adaptation to drier environments in other Australian birds.

Tumbunan Avifaunula in Papuasia

The modern bastion of the Tumbunan avifaunula is montane New Guinea: the enormous central cordillera and smaller fringing ranges of the Vogelkop, north coast and Huon Peninsula (Figure 8). Except for a handful of honeyeaters (*Vosea*, *Meliarchus*, *Stresemannia*), montane whistlers and fantails (e.g. *Pachycephala implicata*, *Rhipidura drownei*) and the pygmy parrot (*Micropsitta bruijnii*) on outlying peaks in the Bismarck Archipelago and Solomons, all of ~180 Tumbunan species in Papuasia are confined to the mainland New Guinean ranges (Table 2). Several families of corvidan songbirds are endemic: the berrypeckers (Melanocharitidae, Paramythiidae) and quite possibly the satin birds of paradise (Cnemophilinae), ifrits (*Ifrita*) and melampittas (*Melampitta*). All genera and most species occur on the central cordillera, a number of the genera are endemic: *Oreornis*, *Androphobus*, *Eulacestoma*, *Cnemophilus*, *Loboparadisea*, *Macgregoria* and *Pteridophora*, the bowerbird *Archboldia*, and moss forest finch *Oreostruthus*.

Three features stand out. One is a drop-off in species diversity with altitude (Kikkawa and Williams, 1971), steepening above 2,500 m where it reflects reduction in niches and habitat. Another is allospecific replacement at different altitudes, primarily on the central cordillera, which stratifies the bird fauna altitudinally. Replacement duos and trios are found in lorikeets (*Neopsittacus*), scrubwrens (*Sericornis*), honeyeaters (*Ptiloprora* and *Melidectes*), ground-robins (*Amalocichla*), the golden whistler species complex (*Pachycephala*), birds of paradise (*Epimachus*) and bowerbirds (*Amblyornis*), across both families and foraging guilds (Schodde and Hitchcock, 1972; Diamond, 1972a; Christidis *et al.*, 1988). Sequences extend into the Irian lowlands (e.g. jewel-babblers, *Ptilorrhoa*), indicative of adaptive routes between faunulas.

The third feature is the faunistic make-up of the poorer Tumbunan outliers in the Vogelkop, north coast and Huon Peninsula ranges; they consist almost exclusively of allotaxa—species and subspecies—present in the central cordillera. Obvious representation from the West Melanesian Island Arc, whence the north coast and Huon Peninsula ranges arose, is missing. Instead, these ranges appear to have been colonized by past dispersal from adjacent sectors of the central cordillera (Diamond, 1985): the 8 species and 68 subspecies endemic to the mountains of the Vogelkop (70% of the bird fauna) and 5 species and 28 subspecies in montane Huon Peninsula are consistently most closely related to allotaxa on respective sectors of the central cordillera opposite (data from Mayr, 1941; Rand and Gilliard, 1967). Moreover, all non-endemic Huon Peninsula taxa (59) except for *Casuarius bennetti hecki*, *Tregellasia leucops melanogenys* and *Parotia wahnesi* are also shared with adjacent sectors of the central cordillera, many exclusively (Mayr, 1941; Rand and Gilliard, 1967). Colonization of the north coast range probably occurred along the low bridges of hills separating the Waponga and Tariku drainages in the west, Mamberamo-Sepik drainages in the centre, and, in the case of the Huon Peninsula, even by direct flight across the narrow Ramu-Markham gulch. Judged by differentiation at specific and subspecific levels, it happened in waves, probably in cooler glacial cycles when life zones were depressed and no earlier than the very late Pliocene (2–3 m.y.a.) because neither the coast ranges nor their connecting bridges were in place before then. Sequential speciation there today is reflected in the forest rails (*Rallina*, *Rallicula*) and the geography of differentiation in the lorikeets *Charmosyna papou* and *Neopsittacus musschenbroekii* and bowerbirds *Amblyornis macgregoriae* and *A. flavifrons*.

Links between the mountains of the Vogelkop and central cordillera, via the low but hilly terrain at the head of Geelvink Bay, are similar in pattern but complicated by: (i) sequences of reverse colonization of the western end of the cordillera by Vogelkop forms of e.g. *Otidiphaps nobilis, Charmosyna josefinae, Alisterus amboinensis, Peneothello cryptoleucus, Ptiloprora erythropleura, Sericornis arfakianus, Monarcha axillaris* and *Rhamphocharis crassirostris*; and (ii) the presence on the Vogelkop of old Tumbunan elements otherwise isolated today at each end of the central cordillera: e.g. *Cormobates*, *Melidectes ochromelas*, *Pachycephalopsis* and *Orthonyx* (the 'drop-out' phenomenon of Diamond, 1972a, 1985). If such fragmentation dates back to the early–mid Pliocene, as genetic distances indicate in

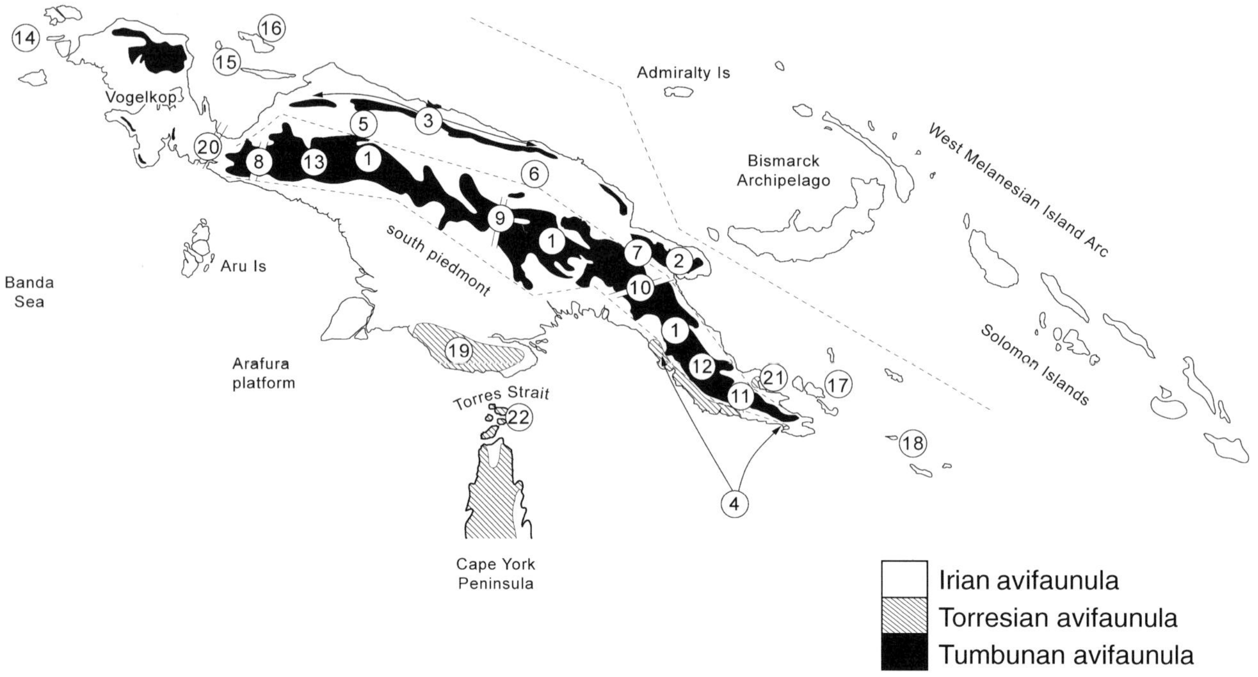

Figure 8. Distribution of avifaunulas in Papuasia, their refugia, barriers and localities cited in the text. Key to numbers: 1 = central cordillera; 2 = Huon Peninsula; 3 = north coast ranges; 4 = Papuan Peninsula (tail); 5 = Mamberamo Basin; 6 = Sepik-Ramu Basin; 7 = Ramu-Markham Gulch; 8 = Wissel Lakes Depression; 9 = Strickland-Sepik Divide; 10 = Kukukuku Barrier; 11 = Domara-Liba Divide; 12 = Owen Stanley Range; 13 = Snow Mountains; 14 = West Papuan Islands; 15 = Geelvink Bay Islands; 16 = Biak-Supiori; 17 = D'Entrecasteaux Archipelago; 18 = Louisiade Archipelago; 19 = Oriomo Plateau; 20 = Zoogeographer's Gap (see Stresemann, 1923); 21 = Collingwood Bay.

Orthonyx (Joseph *et al.*, 2001), then vicariant divergence on the old Vogelkop and Papuan Peninsula continental blocs becomes implicated as well. The central cordillera was formed by progressive east to west uplift from the Papuan Peninsula, suggesting that its Tumbunan stocks spread from east to west as uplift occurred. This pattern is reflected in the geographic affinities of those stocks at the western end of the cordillera today; at specific and generic level, 38 of 56 taxa endemic to the central cordillera are common to west and east sectors, compared to only 11 out of 19 common to the west sector and Vogelkop.

Speciation in the Central New Guinean Cordillera

Two models have been proposed to explain diversification in the cordilleran bird fauna and the altitudinal layering of allotaxa. The first, the well-known 'drop-out' model of Diamond (1972a, 1985), draws on the MacArthur and Wilson (1967) concept of cyclical turnover in island taxa. Yet in the continental-scale bird fauna of the cordillera, there is no tangible evidence for the process on which this model rests; so its presumption seems unrealistic.

The second model is the palaeoclimatic model of Schodde and Hitchcock (1972) and is based on palaeogeographic change and the climatic fluctuations that pushed life zones up and down the cordillera over the last several million years. In the present post-glacial period, life zones are at their highest and most expanded: the change from hill and fagaceous oak forest on lower slopes to mixed myrtle-laurel and *Nothofagus* forest begins at 1,200–1,600 m a.s.l. and is complete by 2,000 m, a mixed podocarp-myrtle-laurel forest replaces it at about 3,000 m and this in turn gives way to natural alpine meadow at around 4,000 m, the tree-line. During the last glacial (~18,000 years ago) much of the cordillera was covered by ice sheets, the tree-line was depressed to 2,500 m altitude, and montane forest zones compressed to between 500 and 2,000 m (van der Kaars, 1991, 1995; Hope, 1996). The Schodde-Hitchcock model sees these fluctuations effecting the union and separation of intra-zone populations in alternating sequence: during periods of altitudinal depression, montane populations would merge on lower, less-dissected terrain, and during warmer times of upward expansion would separate on higher, less-connected massifs, there to commence differentiating in isolation (Figure 5). After a cycle of such separation, re-depressed and re-meeting populations would either intergrade secondarily as in *Melidectes* (Mayr and Gilliard, 1952), or remain discrete, depending on the extent of speciation.

Evidence of intergradation today is probably reflected in stepped latitudinal (e.g. *Crateroscelis robusta*, *Lophorina superba*) and altitudinal (e.g. *Sericornis nouhuysi*, *Rhipidura albolimbata*) character gradients. Evidence of speciation comes from latitudinal species chains (*Melidectes* and the birds of paradise *Astrapia* and

Parotia) and the altitudinal species pairs quoted previously. How did altitudinal separation develop? Because the life zones on massifs of different size are at different levels at any one time (higher on larger massifs, lower on smaller or outlying ranges), populations from high central and lower peripheral ranges would have impinged on one another at different mean altitudes whenever and wherever they re-met. Such pre-adaptation, together with competitive interaction, is then thought to enable impinging species to 'slide' over or under one another altitudinally. This idea is supported by the fact that lower altitude members of such species pairs occur consistently on lower or outlying mountain massifs.

Coincident steps in latitudinal character gradients and species chains identify three major and five minor faunal barriers on the central cordillera, all at low points where summit ridges dip below 2,000 m a.s.l. (Schodde and Hitchcock, 1972)—see Figure 8. The three major barriers are:

(i) Wissel Lakes depression which separates the Weyland-Charles Louis horst from the rest of the cordillera at its western extremity. Of the 104 species localized or differentiated within the central cordillera, 40 either reach their eastern or western limits there or are represented by different forms on either side.

(ii) Strickland-Sepik divide along the Strickland River catchment, which separates the eastern central highlands from the entire western half of the cordillera. About 70 species either reach their limits in the region or are represented by different forms on either side.

(iii) Kukukuku barrier (Tauri-Watut divide) between the Wharton-Owen Stanley Ranges and central eastern highlands. Another 55 species either reach eastern or western limits there or are represented by different forms on either side.

The five minor barriers are: (i) Domara-Liba divide, east Owen Stanley Range (Schodde and Hitchcock, 1972); (ii) breaks around the west end of the Wahgi valley which mark the eastern and western limits of a number of birds of paradise and bowerbirds (*Astrapia*, *Parotia*, *Pteridophora*, *Cnemophilus*, *Archboldia*); (iii) valleys separating the northern Hunstein-Schraderberg outliers in north-west Papua New Guinea; (iv) the central valley dividing the 'Snow Mountains' (Baliem Valley); and (v) watersheds between the north and south ranges of the Snow Mountains composite. Most species affected are lower altitude forms, such as the robins *Tregellasia leucops* and *Pachycephalopsis poliosoma*, and jewel-babbler *Ptilorrhoa castanonota*.

Irian Avifaunula

The lowland mesophyll rainforests of Papuasia and northern Australia are the home of this avifaunula. In Australia, these forests are limited to pockets in the Kimberley, Arnhem Land and east Cape York Peninsula (north-east Queensland); in Papuasia they cover the lowlands up to Eurasian oak (Fagaceae) and hill forests on ranges at 1,000–1,500m a.s.l (Figures 7, 8). This avifaunula is much the richest and most diverse in Australasia, with about 410 species (Table 2). Outnumbering songbirds, non-passerines are themselves dominated by 50+ species of pigeons, with major radiations in *Ptilinopus* (17 species), *Ducula* (14 species) and macropygiines (2–3 genera, 6 species). Parrots are the next most diverse group, with 36 species (over half of them lorikeets), followed by moderate radiations in kingfishers (9 genera, 21 species), raptors (17 diurnal species, including 10 of *Accipiter*, and 11 nocturnal), rails (9 species), megapodes (8 species), and swifts and tree-swifts (8 species). Significant radiations in songbirds are limited to the three most dispersive corvidan families in Australasia: honeyeaters (47 species, dominated by *Myzomela* (14), *Meliphaga* (11) and *Philemon* (7)), monarchs and fantails (38 species) and cuckoo-shrikes and trillers (13 species).

Irian Avifaunula in Australia

Australia holds the tail-ends of the Irian avifaunula, despite its 95 species there. Few are endemic: a swiftlet (*Aerodramus terraereginae*), fairy-wren (*Malurus amabilis*), pitta (*Pitta iris*), two honeyeaters (*Meliphaga*), two robins (*Poecilodryas*) and three monarchs (*Arses* spp., *Monarcha leucotis*), as well as two mangrove specialists (*Gerygone tenebrosa* and *Pachycephala lanioides*). The Cape York Peninsula bloc north of the Endeavour River (Figure 7) is richest with 71 species (Table 2), and includes the only Australian populations of the New Guinean palm cockatoo (*Probosciger*), eclectus and red-cheeked parrots (*Eclectus*, *Geoffroyus*), yellow-billed kingfisher (*Syma torotoro*), red-bellied pitta (*Pitta erythrogaster*), and trumpet manucode and magnificent riflebird (*Phonygammus*, *Ptiloris magnificus*).

Diversity drops off southwards along the coast in steps, clearest at breaks in habitat at the lower Burdekin drainage (e.g. *Rallina tricolor*, *Podargus papuensis*, *Malurus amabilis*, *Meliphaga gracilis*, *Arses*, *Oriolus flavocinctus*) and the southern limit of complex tropical rainforest around Repulse Bay (e.g. *Ninox rufa*, *Aerodramus terraereginae*, *Poecilodryas superciliosa*, *Philemon buceroides*, *Cracticus quoyi* and *Aplonis metallica*). Only a handful reach the rainforests of eastern New South Wales, mostly dispersive insectivores and frugivores adapted to forest fringes and disturbance; examples are *Ptilinopus* spp., *Macropygia amboinensis*, *Chalcophaps indica*, *Monarcha* spp., *Lalage leucomela* and *Sphecotheres* (Table 2). Barely half the Australian complement (49 species) occurs in the fragmentary pockets of rainforest in north-west Arnhem Land and north-west Kimberley, all their members found also on Cape York Peninsula except the chestnut rail (*Eulabeornis*), banded fruit-dove (*Ptilinopus cinctus*), rainbow pitta (*Pitta iris*) and white-lined honeyeater (*Meliphaga albilineata*)—see Table 2. Attenuation westwards ends in monsoonal scrubs in the Kimberley, which lack seven species found in Arnhem Land.

Balancing losses through habitat attenuation, habitat fragmentation in dry Pleistocene climatic cycles fostered moderate refuging and speciation. This is clearest on the north coast, where 16 species are exclusive to either Cape York Peninsula or Arnhem Land-Kimberley, and 17 more are represented by differentiated pairs of subspecies (e.g. in *Megapodius reinwardt, Ptilinopus regina, Chalcites minutillus, Ninox rufa, Gerygone magnirostris, Philemon buceroides, Cracticus quoyi, Lalage leucomela* and *Sphecotheres vieilloti*) (Table 2). Even within the Arnhem Land-Kimberley bloc, there has been divergence, in *Pitta iris, Gerygone chloronotus, Meliphaga albilineata, Philemon buceroides* and *Lalage leucomela*. Such differentiation dates back at least to the last glacial cycle when continuity of rainforest along the northern rim of Australia last broke up. Although the Arafura platform was exposed then, assisting faunal interchange, drier climate would have driven the rainforest into fragmentary refugia on higher outcrops, such as the west Arnhem Land and north-west Kimberley Plateaux, as exemplified in subspeciation in *Philemon buceroides* (Schodde *et al*., 1979). Speciation in the pittas and robins (*Poecilodryas*), as well as the relictual occurrence there of *Ptilinopus cinctus, Gerygone chloronotus* and *Meliphaga albilineata*, reflects older cycles of connection and isolation with New Guinea and the Banda island arc to the north and west.

On the east coast, differentiation is focused at two faunal barriers (Figure 7). One is the Torresian Barrier on south-east Cape York Peninsula, where two species of the monarch *Arses* and subspecies within a further 15 species are split (e.g. *Megapodius reinwardt, Ptilinopus magnificus, Ninox rufa, Podargus papuensis, Alcedo azurea, Pitta versicolor, Meliphaga* spp., *Colluricincla megarhyncha, Machaerirhynchus flaviventer, Cracticus quoyi* and *Oriolus flavocinctus*). Another 11 species differentiate subspecifically at different points across the second barrier system, the dry Burdekin Gap–Dawson-Mackenzie drainage in central east Queensland (Schodde and Mason, 1999, pp.787–788). The few Irian taxa that occur further south are often quite distinct forms, for example *Ptilinopus magnificus, Podargus ocellatus, Colluricincla megarhyncha* and *Sphecotheres*. Fragmentation of habitat by recent dry climatic cycles is again the likely driver of subspeciation, centering differentiation in three main refugial foci: central east Cape York Peninsula, coastal humid north-east Queensland between Cooktown and Ingham, and the central east Australian coast (Figure 7).

Irian Avifaunula in Papuasia

Tectonic movement, rather than climatic change, has driven species radiation in Papuasia. It formed divisive mountain ranges in New Guinea and built, moved and separated islands—and their birds—in the west Papuan, Geelvink Bay, south-east Papuan, Bismarck, Admiralty and Solomons archipelagos. Dispersal and the taxon cycle may explain species turn-over and presence-absence among the islands (Greenslade, 1968; Diamond, 1970a, b, 1972b), but it does not account for overall diversity, which follows differentiation among isolated island populations.

The Papuasian archipelagos are together almost as species-rich as New Guinea itself (data from Rand and Gilliard, 1967); some 115 species are endemic and nearly another 200 shared with New Guinea (Table 2). Diversification on islands follows two main patterns. One, of a rich, 'continental' type New Guinean fauna but with relatively few endemics, is found on the inner satellite islands of New Guinea; examples are: west Papuan groups of Batanta, Waigeo and Misool (165 species, 5 endemics); Geelvink Bay group (135 species, 10 endemics); Aru Islands (130 species, none endemic); and south-east Papuan groups, including the D'Entrecasteaux and Louisiade Archipelagos (100 species, 7 endemics—see Figure 8, Table 2). Endemic subspecies in these islands number 55, 65, 40 and 70 respectively. Factors used to explain this pattern include proximity to mainland New Guinea and, particularly for the Aru Islands and Yapen in Geelvink Bay, episodic connection by land through much of the Pleistocene, facilitating faunal exchange. Significantly, the island with greatest specific endemism, Biak-Supiori in Geelvink Bay with 9, has probably never been connected to New Guinea, its stocks arriving over water to differentiate in greater isolation. Species-budding there was rapid, gauged by the rising of the island complex above sea at the Pliocene-Quaternary boundary, about 2–3 m.y.a. (Aplin, 1998).

The other pattern of diversification is shared by the archipelagos of the Melanesian Island Arc further out from New Guinea: the Bismarck Archipelago (including Mussau group), Admiralty Islands and Solomons (Figure 8). In contrast to New Guinea's satellite islands, all are relatively poor in species despite their greater area but are much richer in endemics; for example, with about 120 species, there are 38 endemics plus 110 endemic subspecies in the Bismarck Archipelago and 110, 49 and 156 respectively in the Solomons (Table 2). The tiny Admiralty Islands have only 43 Irian species, yet 6 are endemic plus another 17 at subspecific level (53%). Australasian families that respect water barriers are missing as well (Table 2). In contrast, the four most speciose families in the Bismarck Archipelago and Solomons—Columbidae, Psittacidae, Monarchidae and Alcedinidae—are also the most dispersive.

Evidently, the modern bird faunas of these archipelagos were built largely by oversea dispersal among themselves and from the late Tertiary east-west continental blocs of New Guinea (Greenslade, 1968; Diamond, 1970a,b; Schodde, 1977). Yet high endemism reflects a brake on dispersal; and niche pre-occupation may be a significant contributor, preventing new colonists from replacing established populations that are differentiating. It does not account, however, for a number of apparently old endemics: *Accipiter imitator, Nesoclopeus, Microgoura, Cacatua ducorpsii, Nesasio, Actenoides bougainvillei, Pitta superba*, and the honeyeaters *Stresemannia, Guadalcanaria, Meliarchus* and *Vosea*. All but two are endemic to the Solomons. In contrast to the Bismarck Archipelago, which has only emerged within the last 1–2 million years (Exon and Marlow, 1990), elements of the Solomons have remained above sea level since Miocene times (Schodde, 1977). They are thus the likely source for their older endemics, and of differentiation in the local sea-eagle *Haliaeetus sanfordi* and frogmouth, *Podargus inexpectatus*.

The geography of speciation in lowland New Guinea itself—its centres and the operation of barriers—is still poorly understood. One reason is a seriously oversplit infra-specific taxonomy, many elements of which are intergradient populations between real forms. An example of this is the Pinon Imperial pigeon in which the subspecies '*Ducula pinon rubiensis*' occurs at either end of New Guinea, between *D. p. pinon* south of the central cordillera and *D. p. jobiensis* to the north (Mayr, 1941; Rand and Gilliard, 1967). Even with these complex patterns three main foci of radiation have been identified: the Vogelkop with 50 endemic forms; the lowlands south of the central cordillera between the neck of the Vogelkop and south-east tail with about 105; and the Sepik-Mamberamo basin between Geelvink Bay and Huon Peninsula with approximately 80 (Pratt, 1982, 1991)—see Figure 8, Table 2. Within these centres there are further sub-foci in the south and far east: the central south piedmont between the neck of the Vogelkop and Gulf of Papua with 50 endemics, the Oriomo Plateau (see below), and the south-east Papuan tail (51 endemics); a further 14 subspecies are endemic just to the north coast of the tail (e.g. *Ptilorrhoa caerulescens geislerorum* and *Paradisaea augustaevictoriae intermedia*). The endemics in regional centres are usually allotaxa that replace one another in rings around the central cordillera. Thus in the tawny-breasted honeyeater (*Xanthotis flaviventer*), there are five valid regional forms in mainland New Guinea: greenish *flaviventer* on the Vogelkop, brown and patterned *saturatior* on the southern piedmont between the neck of the Vogelkop and Gulf of Papua, rufous and patterned *visi* around the south-east tail, pallid *madaraszi* around the Huon Peninsula, and plain grey *meyeri* in the Sepik-Mamberamo Basin.

Differentiation (at the infra- or allo- specific level) was probably driven by the interaction of the rising central cordillera, which separated lowland populations north and south (Schodde and Hitchcock, 1972), with Pleistocene climatic cycles which made and broke connections at east and west ends of the cordillera and parts of the southern piedmont (Pratt, 1982, 1991). The climatically-induced barriers lie at the neck of land between the Vogelkop and the rest of New Guinea, the north rim of the Oriomo Plateau, the west end of the Papuan tail and the rain-shadowed Ramu-Markham gulch, the driest zones in New Guinea today. Projections by Nix and Kalma (1972) indicate that these zones were still drier—reduced substantially to savannas—during the last Pleistocene glacial cycles, isolating the main centres of lowland rainforest biota from one another then. These events evidently date speciation there, because nearly all differentiation today is at infra- or allo-specific level, indicating that the Irian avifaunula ranged rather homogeneously throughout lowland New Guinea beforehand. Older endemics are few, most of them occurring in the Sepik-Mamberamo Basin (e.g. *Philemon brassi*, *Monarcha rubiensis, Epimachus bruijnii* and *Corvus fuscicapillus*), which is richer too in endemic allospecies (*Megapodius affinis, Talegalla jobiensis, Goura victoriae, Chalcopsitta duivenbodei, Psittaculirostris edwardsii, Arses insularis, Lalage atrovirens, Poecilodryas brachyura* and *Paradisaea minor*)—see Table 2. These taxa may well have differentiated in isolation on the lower slopes of the north coast ranges before the uplifting of the Successor Basin joined them to New Guinea at the beginning of the Quaternary.

Arafura Platform—Development of a Mangrove Avifauna

The Arafura Platform is the bridge between Australia and New Guinea and the historic meeting ground for Irian and Torresian avifaunulas. It is a terrane of low relief. For much of the Plio-Pleistocene it was exposed (Voris, 2000), but today it is largely submerged, with only the Aru Islands, Cape York Peninsula, the Oriomo Plateau and intervening Torres Strait Islands remaining exposed (Figures 7 and 8). Eucalypt and paperbark (*Melaleuca*) woodland occupied by the Torresian avifaunula fringe it on Cape York Peninsula, Irian elements surviving there in restricted pockets of rainforest (see above). The Torres Strait Islands are a conduit for New Guinea–Australia migration today, but not important in regional avifaunal interaction historically (Kikkawa *et al.*, 1981; Draffan *et al.*, 1983). Only on the Aru Islands and Oriomo Plateau are Irian elements prominent, predominating on the first and interdigitating with Torresian on the second.

There are two facies to the Irian avifaunula on the Platform: dryland/swamp rainforest and mangrove. The dryland/swamp rainforest assemblage is much more speciose, with over 160 taxa, but is proportionally poorer in endemics (Table 2). Only four species are Arafura-centred (*Ptilinopus wallacii, Dacelo tyro, Tanysiptera hydrocharis* and *Arses lorealis*), the first three shared exclusively by the Aru Islands and Oriomo Plateau. Endemic subspecies and subspecies-groups are more diverse (e.g. *Gallicolumba rufigula, Probosciger aterrimus, Trichoglossus haematodus, Cyclopsitta diophthalma, Cacomantis castaneiventris, Xanthotis polygramma, Glycichaera fallax, Tregellasia leucops, Coracina caeruleogrisea, Cracticus quoyi* and *Ailuroedus crassirostris*). Most are again shared by the Aru Islands and Oriomo Plateau, and some with Cape York Peninsula as well. Affinities otherwise lie with Irian south New Guinea, indicative of recent land connection; but layered differentiation at both species and subspecies levels also reflects episodic isolation by sea back beyond the last post-glacial rise in sea level 10,000–12,000 years ago (Phipps, 1970; Jennings, 1972; Voris, 2000).

Though small, the Australasian mangrove bird fauna is still large by world standards and rich in endemics: 12

species are confined to mangroves (obligate taxa), another 16 are semi-obligate, limited to them for part of their range (e.g. *Todiramphus chloris*, *Microeca flavigaster tormenti* in the Kimberley), and another 80 or 90 use them regularly (Schodde *et al.*, 1982; Ford, 1982, 1983; Johnstone, 1990). The heartland of this fauna today is the coast of south New Guinea and north Australia, on opposite rims of the Arafura Platform (Figures 7, 8). The Australian component, which disappears by Shark Bay in the west and the Hunter River in the east, is the richer. Of its obligate taxa, four are endemic (*Pachycephala lanioides, Gerygone tenebrosa, Lichenostomus fasciogularis* and *Zosterops lutea*) and the remainder are shared with New Guinea and the fringing Banda Basin (e.g. the rail *Eulabeornis* and songbirds *Peneonanthe pulverulenta*, *Rhipidura phasiana*, *R. dryas*, *Myiagra ruficollis* and *Gerygone levigaster*). The diversity of these taxa, the dearth of old generic endemics, and the patchy representation of obligate and facultative species together indicate that this Arafura mangrove fauna has been built locally and recently, in piecemeal fashion from mainly Irian sources.

Clues to its development come from two patterns of distribution. One is the presence of mangrove birds not so much in the impressive mangrove forests of *Bruguiera* and stilt-rooted *Rhizophora* in deltaic south New Guinea and Cape York Peninsula as in depauperate, strand-like stands mixed with *Avicennia* and *Ceriops* both there and along the estuaries of the Gulf of Carpentaria, Arnhem Land, Kimberley and the Pilbara. The other is the marked disjunction in diversity at Cape York Peninsula, with a decline down the east coast: four obligates range commonly along the north Australian coast to the west peninsula or just around its east tip (e.g. *Pachycephala lanioides, Rhipidura dryas, Myzomela erythrocephala*); another five extend in pockets down the east Australian coast as well, as though they have just reached it (e.g. *Pachycephala melanura, Myiagra ruficollis* and *Zosterops lutea*); only one obligate, *Lichenostomus fasciogularis*, is endemic to the east coast.

Underlying the break at Cape York Peninsula is the projection of the Great Dividing Range across Torres Strait as the east watershed of the Arafura Platform. When the Platform was exposed at times of low sea level in the Pleistocene, its eastern rim became an abrupt barrier to the Coral Sea. Across its broad western drainages, however, rivers from both northern Australia and southern New Guinea meandered into the centre and then west out to the Banda Sea (Voris, 2000). Climate was dry tropical and seasonal, as in northern Australia today (Nix and Kalma, 1972), favouring the development of mangroves of the same composition and structure as now found there. This was the crucible in which Australasia's modern mangrove bird fauna evidently radiated (Schodde *et al.*, 1982). Cyclical changes in climate and sea level made and broke continuity of habitat to promote speciation (e.g. *Gerygone tenebrosa-G. magnirostris*, *Pachycephala lanioides-P. monarcha* and Banda Sea forms of *Myiagra ruficollis* and *Rhipidura dryas*). The last marine flooding (10,000–12,000 years ago) probably initiated subspecies-budding in *Geopelia humeralis, Todiramphus chloris, Gerygone tenebrosa, Pachycephala lanioides, P. melanura* and *Zosterops lutea* around the Eighty Mile Beach in north-west Australia (Figure 7). Obligates in pockets on the east coast probably reached those pockets then, when Torres Strait was last breached (Jennings, 1971). Endemics there, such as *Butorides striatus macrorhynchus*, *Alcedo pusilla halli*, *Gerygone levigaster cantator* and *Lichenostomus fasciogularis*, probably reflect earlier invasions.

Torresian Avifaunula

The Torresian avifaunula occupies the monsoonal eucalypt and paperbark (*Melaleuca*) woodlands that today range across all northern Australia, with outliers in the Pilbara and south and south-east New Guinea (Figures 7, 8). It is a rather uniform vegetation, of sparse to closely-spaced trees over a dense 0.75–1.5 m high sward of annual grasses, except on outcropping rocky plateaux which are covered by *Triodia* and *Plectrachne* hummock grassland. Its avifaunula is correspondingly simple, with only about 90 species centred in it (Table 2). Honeyeaters (16 species) dominate, followed by estrildine finches (11), butcherbirds and woodswallows (6) and phabine pigeons (7), their maximum radiation. Together with the estrildine finches, four platycercine parrots, *Aprosmictus*, several cockatoos and three button-quail (*Turnix*), they comprise the most diverse guild of granivores in Australasia. Except for birds of paradise and eupetids, most songbird families are present, though acanthizids and monarchs, elsewhere so diverse, are represented by only one or two species each. Endemism, nevertheless, is high, much of it at generic level indicative of antiquity: the magpie goose (*Anseranas*), two raptors (*Hamirostra*, *Erythrotriorchis*), the ground-pigeons *Petrophassa* and *Geophaps*, the *Psephotellus* group of grass parrots, the honeyeaters *Entomyzon*, *Conopophila*, *Ramsayornis* and *Trichodere*, mudnester *Struthidea*, bowerbird *Chlamydera*, and grass-finches *Poephila* and *Heteromunia*.

Torresian Avifaunula in Australia

Including the Pilbaran outlier, the Australian Torresian avifaunula ranges today across the north of the continent south to the eucalypt-*Acacia* transition, inland deserts and upper Murray-Darling Basin in the east (Figure 7). Adaptation to specific habitat and the operation of historic faunal barriers have driven moderate radiations. All habitat specialists except the honeyeater *Trichodere cockerelli* (heath) and fairy-wren *Malurus coronatus* (stream-side galleries) are limited to spinifex *Triodia-Plectrachne* savannas on patchy sandstone outcrops between the Kimberley and Gulf of Carpentaria (Table 2). The patchiness has driven full speciation in rock pigeons (*Petrophassa*—2 species) and grasswrens (*Amytornis*—3 species), subspeciation in one rock pigeon and a fairy-wren (*Malurus lamberti*), and neither in a shrike-thrush (*Colluricincla woodwardi*). Their diverse affinities and disparate differentiation indicate uncoordinated histories of speciation, and so independent adaptation to sandstone spinifex habitat in each group.

Historic faunal barriers have created four major centres of differentiation and endemism. These are the Pilbara outlier, north-west Australia (the Kimberley and Arnhem Land), Cape York Peninsula, and eastern Queensland; each is discussed below.

(i) The Pilbara, separated from north-west Australia by the Great Sandy Desert (Canning Barrier of Schodde and Mason, 1999), has a depauperate Torresian complement mixed with Eyrean elements. Examples are the doves *Geopelia humeralis* and *G. placida,* corella *Cacatua sanguinea*, coucal *Centropus phasianinus*, kookaburra *Dacelo leachii*, treecreeper *Climacteris melanura*, honeyeaters *Lichmera indistincta* and *Conopophila whitei*, babbler *Pomatostomus temporalis*, woodswallow *Artamus minor*, and finch *Neochmia ruficauda* (Table 2). Differentiation has been limited subspecifically to the doves, corella, kookaburra and treecreeper, and may well date back to the forming of the Great Sandy Desert 500,000–600,000 years ago (Bowler, 1982). Pilbaran forms of *Geopelia humeralis*, *G. placida*, *Geophaps plumifera*, *Cacatua sanguinea*, *Podargus strigoides*, *Dacelo leachii* and *Malurus lamberti*—or their intergrades—also occur relictually in the Fitzroy Basin, north-west of the desert today (Schodde and Mason, 1997, 1999). It suggests that the Canning Barrier has been breached by episodic gene flow along the coast since and that Pilbaran subspeciation may actually be younger.

(ii) The north-west Australian bloc of the Kimberley and Arnhem Land extends east to the south Gulf of Carpentaria. It holds a full specific complement of the Torresian avifaunula today, and most of its 17 endemics are either spinifex habitat specialists (see above) or allotaxa of species groups found in other blocs (e.g. *Turnix castanota, Geophaps smithii, Trichoglossus rubritorquis, Platycercus venustus, Psephotus dissimilis, Cracticus argenteus,* and *Poephila longicauda*)—see Table 2. These and subspecies of another 42 species, such as *Calyptorhynchus banksii, Todiramphus macleayii, Manorina flavigula* and *Entomyzon cyanotis*, are separated from, or intergrade with, eastern allotaxa around the head of the Gulf of Carpentaria mostly at the Carpentaria Barrier. Today the barrier coincides with the confluence of a number of major streams that, at times of low sea level during Plio-Pleistocene glacial cycles, opened a broad channel through the centre of the Gulf (Galloway and Löffler, 1972; Voris, 2000). This low, mangrove-lined estuarine system evidently operated as a significant serial barrier to trans-Torresian connections, and did so serially judged by differentiation at specific and subspecific levels. In those drier times, most Torresian elements isolated in the north-west probably differentiated in Kimberley and Arnhem Land refugial uplands, then another 200–300 m a.s.l; the last episode of subspeciation probably coincided with the last glacial cycle, 12,000–22,000 years ago. Two minor barriers evidently operated within the north-west bloc then as well (Figure 7). One was the Ord-Victoria River systems between the Kimberley and Arnhem Land uplands, which operated like the Carpentaria Barrier and differentiated species of *Petrophassa* and subspecies of *Geophaps smithii, Platycercus venustus* and others on either side (Ford, 1978). The other minor barrier is the strait between Arnhem Land and the Tiwi Islands; the last flooding of this, about 10,000–12,000 years ago probably initiated subspeciation in *Tyto novaehollandiae*, *Dacelo leachii*, *Manorina flavigula* and *Pachycephala rufiventris*, among others.

(iii) The Cape York Peninsula bloc is separated from the rest of Torresian Australia by the Carpentarian Barrier to the south-west and combined Torresian Barrier and Burdekin Gap in the south (Figure 7). Despite this, it is as much a meeting ground for north-west Australian and eastern Queensland elements as a centre of endemism and radiation, emphasizing its unique position as the axle of the Australasia's modern avifaunal wheel. Allowing for New Guinean outliers, only five species are endemic (*Turnix olivii, Psephotus chrysopterygius, Trichodere cockerelli, Cracticus mentalis* and *Chlamydera cerviniventris*) and 22 subspecies—rather fewer than in north-west Australia (Table 2). Yet the Peninsula shares with north-west Australia 10 species not found in the eastern Queensland bloc plus another 22 subspecies. It also shares with east Queensland 11 species missing from north-west Australia (e.g. *Trichoglossus chlorolepidotus, Platycercus adscitus, Dacelo novaeguineae, Lichenostomus flavus, Philemon corniculatus* and *Poephila cincta*), plus another 10 subspecies (e.g. in *Calyptorhynchus banksii, Todiramphus macleayii, Lichmera indistincta, Pomatostomus temporalis* and *Taeniopygia bichenovii*).

Between the Torresian Barrier (Endeavour River) and Burdekin Gap (Burdekin River), Cape York Peninsula elements drop out or intergrade at different points. Some 11 species (e.g. *Philemon argenticeps, Cracticus mentalis*) and 14 subspecies around the Endeavour River, as well as 5 species and 9 subspecies around the Burdekin (e.g. *Lichenostomus unicolor, Rhipidura rufiventris*; subspecies of *Geophaps scripta, Centropus phasianinus, Lichenostomus flavus*). Subspecies of another 18 species intergrade broadly between the two (e.g. in *Climacteris picumnus, Malurus melanocephalus* and *Daphoenositta chrysoptera*). In all, 57 species comprising some 65% of the Torresian avifaunula are stopped or differentiated at and between these paired barriers (Ford, 1986). The region between them today comprises a complex belt of high montane spurs and tablelands (400–1,500 m) that extends inland across much of the foot of Cape York Peninsula. These uplands separate the Peninsula from Gulf of Carpentaria lowlands and eastern Queensland basins, and are clothed with rainforest and cool (Bassian) sclerophyll forest on eastern slopes and summit ridges. During past glacial cycles, the mountains were effectively several hundred metres higher and even colder, presenting a still more formidable barrier to Torresian birds and promoting differentiation on either side (Schodde, 1982b, p. 107).

(iv) The east Queensland bloc extends today from the lower Burdekin Basin and Burdekin-Flinders Rivers divide throughout eastern Queensland south to north-eastern New South Wales and the central Murray-Darling Basin (Figure 7). Little over half of Torresian species (~48) occur in it, and only three endemics: *Psephotus pulcherrimus*, *Myzomela sanguinolenta* and *Aidemosyne modesta* (Table 2). Its 36 endemic subspecies, moreover, are nearly all allotaxa of Cape York Peninsula forms differentiated at the Torresian-Burdekin barriers (e.g. in *Geophaps scripta*, *Platycercus adscitus*, *Malurus melanocephalus*, *Melithreptus albogularis*, *Coracina papuensis* and *Poephila cincta*).

Their derivation probably dates back to the last glacial cycles which cut off ancestral elements in coastal eastern Queensland basins such as the Burdekin, Dawson-Mackenzie and Mary-Burnett (Short *et al.*, 1983b). The present ranges of *Pardalotus striatus melanocephalus, Daphoenositta chrysoptera leucocephala, Microeca flavigaster laetissima* and *Artamus cinereus dealbatus* (Schodde and Mason, 1999) may reflect refugial distribution then.

Torresian Avifaunula in New Guinea

With only 47 species, the Torresian avifaunula tails off in New Guinea in two main outliers, one in the *Melaleuca-Sinoga-Eucalyptus* woodlands of the Oriomo Plateau and surrounding plains, the other in the eucalypt woodlands of south-eastern New Guinea between the Lakekamu River and Cape Rodney, with pockets in the lower sectors of the Owen Stanley Range through to Collingwood Bay on the north Papuan Peninsula (Figure 8). These enclaves are relicts of a more widespread Torresian biota in drier glacial cycles. No species are endemic, except for several mannikins (*Lonchura*) and the swamp-inhabiting *Megalurus albolimbatus*, which are recent adventives. Links with Torresian Australia lie via Cape York Peninsula and the east Arafura corridor (Torres Strait), as indicated by exclusively shared species (*Cracticus mentalis, Chlamydera cerviniventris*) and subspecies (*Gerygone olivacea cinerascens, Microeca flavigaster flavissima, Colluricincla harmonica superciliosa, Coracina papuensis oriomo* and *Neochmia phaeton evangelinae*).

Differences in species composition as well as subspecific differentiation suggest that the two outliers have been isolated from one another historically and that their respective connections with Cape York Peninsula may have been independent of one another. Eight of the 25 resident species in the south-east outlier (e.g. *Todiramphus macleayii*, *Gerygone olivacea*, *Microeca fascinans*, *Lichenostomus flavescens*, *Cracticus mentalis, Lonchura castaneothorax*) are missing from the Oriomo, though all except one finch (*Lonchura caniceps*) are shared with Cape York Peninsula. Conversely, 19 of the 39 residents in the Oriomo outlier are missing from the south-east (e.g. *Anseranas semipalmata, Aquila audax, Ardeotis australis, Cacatua sanguinea, Aprosmictus erythropterus, Ninox boobook* and *Philemon citreogularis*), though all except three species of *Lonchura* are again shared with Cape York Peninsula (Table 2). Of subspecies exclusive to the south-east New Guinea enclave, six are endemic and two shared with Cape York Peninsula (*Gerygone olivacea cinerascens, Microeca flavigaster flavissima*); conversely, of those exclusive to the Oriomo, 12 are endemic and six shared with Cape York Peninsula (e.g. *Tyto novaehollandiae galei, Pomatostomus temporalis temporalis, Neochmia phaeton evangelinae*). Such differentiation may well be attributable to patchwork isolation of south-east and Oriomo outliers from one another and Cape York Peninsula by the flooding of Torres Strait and inner Gulf of Papua at the close of the last glacial cycle 10,000–12,000 years ago.

Bassian Avifaunula

Endemic to Australia, the Bassian avifaunula is the most diverse of those adapted to scleromorphic vegetation. It occupies the sclerophyllous forests and woodlands that today rim the temperate south from south-west Australia to Tasmania and the south-east and penetrates north along the cooler summit ridges of the Great Dividing Range to tablelands in north-east Queensland (Figure 7). The 145 core species (Table 2) are rich in nectarivores, ground- and foliage- feeding insectivores, and granivores.

The corvidan families Meliphagidae (22 species) and Acanthizidae (19 species) dominate, with radiations in the honeyeater genera *Lichenostomus*, *Anthochaera*, *Melithreptus* and *Phylidonyris*, in *Acanthiza* (7 species) and the sericornithine scrubwrens (8 species in 4 genera). Platycercine parrots (13 species in 6 genera) and cockatoos (7 species in 3 genera) are also prominent; evidently, the Bassian biota has been the cradle of their radiation. *Leipoa* (Megapodiidae), *Turnix varia* (Turnicidae) and *Phaps* and *Leucosarcia* (Columbidae) are also characteristic.

Most Australasian songbird families are well represented except the bowerbirds, birds of paradise and cuckoo-shrikes, for which frugivory seems crucial. In contrast, passeridan representation is trivial, the most significant being four grass finches, three in the Bassian-centred genus *Stagonopleura*. As well as developing the mallee biota, the climatic cycles of the Plio–Pleistocene separated the Bassian avifaunula around the south and east coast into a series of isolated centres: south-west Australia, the South Australian gulfs, Tasmania and Bass Strait islands, south-east Australia, and the uplands of the Great Dividing Range in Queensland (Figure 7).

South-west Australia

This centre comprises the south-west corner of the continent inland to the mallee line (Figure 7) and holds 65 core Bassian species. Specific endemism is higher than in any other centre: 15 species and 22 subpecies plus 1 and 11 respectively if forms also present in the western mallee are included (Table 2). At least three cycles of isolation from Bassian east Australia have driven differentiation if the endemic white-breasted robin, *Eopsaltria georgiana* is sister to the Australian yellow robins, *Eopsaltria australia* and *E. griseogularis*. The first cycle budded off ancestral *E. georgiana*, the next ancestral *E. griseogularis* and the third the two subspecies of *griseogularis* (nominotypical *griseogularis* and *E. g. rosinae*), the last probably in the round of glacial and post-glacial climatic shifts within the last 25,000 years. Such events have been linked to the Darling Barrier, which runs from the west coast north of Perth south-east to Albany and marks the meeting zone of three pairs of allospecies (*Calyptorhynchus baudinii-C. latirostris*, *Psophodes nigrogularis-P. leucogaster*, and *Malurus elegans-M. pulcherrimus*) and another 11 pairs of allosubspecies (Schodde and Mason, 1991, 1999). The barrier (see Figure 7) probably manifests the interface between a more static wet *Eucalyptus marginata-E. diversicolor* forest refuge in the far south-west and a more mobile mallee-heath belt on sand plains across southern Australia from the early Quaternary. In drier glacial times, both forest and mallee were confined to refugial pockets, the mallee withdrawing south along the south coast as far east as the South Australian gulfs. Then, when climate ameliorated, the pockets expanded, the mallee more than the forest because of edaphic factors and its capacity to respond to lesser increases in rainfall. This may account for the older endemics in the far south-west (e.g. *Purpureicephalus spurius, Calyptorhynchus baudinii* species group, *Atrichornis clamosus* and *Eopsaltria georgiana*), and the differentiation between forest and mallee-heath forms there. It does not, however, explain the differentiation in taxa confined to forests within the south-west (e.g. subspecies of *Platycercus icterotis*), nor east Bassian allotaxa on Eyre Peninsula (e.g. *Calyptorhynchus funereus*, *Malurus cyaneus*), or apparently undifferentiated subspecies of Bassian species in both south-west and south-east Australian forests (e.g. in *Chalcites lucidus* and *Pardalotus punctatus*).

South Australian Gulfs

This centre comprises Eyre and Yorke Peninsulas, the Mt. Lofty-south Flinders Range system and Kangaroo Island (Figure 7); it is the meeting ground for east and west Bassian Australia. The Gulfs are relatively species-rich (with 78 species) yet poor in endemics (Table 2). Geographic affinities of Kangaroo Island birds are mixed, most lying with south-east Australia (e.g. *Calyptorhynchus* spp., *Platycercus elegans, Malurus cyaneus, Acanthiza pusilla, Anthochaera chrysoptera, Acanthorhynchus tenuirostris, Corvus* spp. and *Neochmia temporalis*) and some with the south-west (e.g. *Neophema* spp., *Calamanthus cautus, Sericornis frontalis, Lichenostomus cratitius, Psophodes leucogaster* and *Pachycephala pectoralis*); others (in *Petroica boodang* and *Zosterops lateralis*) are intergradient.

Eyre Peninsula is stronger in western affinities, in accord with its geographic position. A few eastern forms reach western limits there (e.g. *Calyptorhynchus funereus, Glossopsitta concinna, Malurus cyaneus, Corcorax melanorhamphos* and *Stagonopleura guttata*) and, conversely, several south-western taxa their eastern limits (e.g. *Climacteris rufa, Malurus pulcherrimus, Eopsaltria griseogularis*). Most Bassian forms on Eyre Peninsula, nevetheless, are of western-centred taxa that extend still further east in mallee, dry sclerophyll or coastal heath habitat (e.g. *Glossopitta porphyrocephala, Neophema elegans, Pardalotus punctatus xanthopygus, Calamanthus cautus, Lichenostomus cratitius, Drymodes brunneopygia,* and *Pachycephala inornata*).

In contrast, the Mt. Lofty-south Flinders Ranges east of the gulfs are an outlier of Bassian south-east Australia. Not only are they the western limits for 20 east Bassian species (e.g. *Platycercus elegans, Psephotus haematonotus, Calamanthus pyrrhopygius, Acanthiza reguloides, Manorina melanocephala* and *Falcunculus frontatus*), but their subspecies of pan-Bassian species are of prevailingly eastern affinity (e.g. in *Sericornis frontalis, Phylidonyris novaehollandiae, Melithreptus lunatus, Microeca fascinans* and *Colluricincla harmonica*).

Marking the division between eastern and western biotas is the Eyrean Barrier at the head of Spencer Gulf. Its historic role in separating Bassian birds was enhanced by the isolation of Kangaroo Island by sea 10,000–11,000 years ago (Thom and Chappell, 1975; Hope *et al.*, 1977), the development of minor barriers to west and east (Nullarbor Plain, Murravian Gulf), and the refuge offered by the Mt. Lofty-Flinders Ranges. All these factors have combined to differentiate 40 endemic subspecies: 16 on Kangaroo Island, 2 on Eyre-Yorke Peninsulas (in *Stipiturus malachurus, Strepera versicolor*), 12 in the Mt. Lofty-Flinders Ranges, and 8–9 shared (Condon, 1975; Schodde and Mason, 1999). Despite the presence of Mt. Lofty-Flinders Ranges refuges since the late Tertiary (Callen and Tedford, 1976), differentiation has not proceeded beyond subspecific level. This and the mix of east and west forms suggests that the gulf environments have been too marginal through Quaternary climatic cycles for more than temporary refuge; the current round of subspeciation may be no older than the last glacial cycle 12,000–22,000 years ago.

Tasmania and Bass Strait Islands

The Furneaux group and King Island hold 66 Bassian species, including 12 endemics and 45 endemic subspecies (see Table 2 and Figure 7). Cyclical isolation by sea from the Australian mainland during Plio-Pleistocene interglacials has been the primary driver of differentiation, judging by divergence at both specific and subspecific levels. But barriers of habitat probably also came into play, whenever Tasmania was linked to Australia by land during glacial periods of lowered sea level. For example, during the most recent land-bridges (12,000–50,000 years ago), and probably well before, dune systems from central south Australia extended south-east across Bass Strait as far as northern Tasmania, covering the lowlands of the strait with heaths and dry sclerophyll woodlands (Bowler, 1982). This may explain why most of the older endemic species are inhabitants of tall, wet forest and most subspecies adapted to heathy shrubberies and drier woodlands. Ancestral stocks of the last were able to filter across the land-bridge with relative ease, most of them budding off only since the final flooding of Bass Strait (~14,000 years ago). Examples include *Phaps elegans, Platycercus eximius, Stipiturus malachurus, Manorina melanocephala, Sericornis frontalis, Cinclosoma punctatum, Cracticus torquatus, Gymnorhina tibicen* and *Zosterops lateralis*.

Poorer in species yet richer in older endemics, the wetter eucalypt forests have had a longer history of isolation and probably extinction. During extreme glacial cycles, they were confined to a thin skein around Tasmania itself when the tree-line dropped to 200 m a.s.l. (Davies, 1967). Many species-groups abundant today in mainland forests on the opposite side of Bass Strait are surprisingly absent, e.g. wonga pigeon, Australian king parrot, powerful owl, superb lyrebird, treecreepers and striated thornbill.

The Bass Strait islands are connected by ridges to Tasmania, not the mainland, and remained bridged to Tasmania until 10,000–12,000 years ago (King Island) and 8,000–9,000 years ago (Furneaux group). This would account for exclusively Tasmanian allospecies there, e.g. *Platycercus caledonicus, Acanthiza ewingii, Lichenostomus flavicollis, Melithreptus validirostris, M. affinis, Anthochaera paradoxa, Melanodryas cucullata* and *Strepera fuliginosa* (Table 2). Subspecific budding probably did not begin until the last post-glacial flooding, which cut off King Island earlier and accounts for its greater endemism (8 subspecies compared to 3 in the Furneaux group).

South-east Australia

From the south-east of South Australia east and north along both flanks of the Great Dividing Range to the 'granite belt' of south-east Queensland, is the core of the modern Bassian avifaunula (Figure 7). It holds 95 Bassian species and, although only 10 are endemic, they include a number of old relictual forms such as the wonga pigeon (*Leucosarcia*), gang-gang cockatoo (*Callocephalon*), powerful owl (*Ninox strenua*), superb lyrebird (*Menura novaehollandiae*), red-browed treecreeper (*Climacteris erythrops*) and two dasyornithine acanthizids (*Dasyornis brachypterus, Pycnoptilus*)—see Table 2. At least 55 subspecies have budded off as well. Three internal barriers—the Great Dividing Range, Bassian Volcanic Barrier in western Victoria, and dry Hunter River valley—have created four minor refuges: the west slopes of the Great Dividing Range, a south-west bloc (south-east of South Australia-far west Victoria), a central bloc (central Victoria -Hunter River valley), and a north-east bloc (north-east New South Wales-south-east Queensland)—see Figure 7. The central bloc is richest, holding all species; the old relictual forms *Callocephalon, Pycnoptilus* and rock warbler (*Origma*) are confined to it. Other species reach their northern limits (e.g. *Strepera versicolor, Stagonopleura bella*) and south-western limits (e.g. *Pyrrholaemus sagittatus, Lichenostomus melanops*) at its respective boundaries. At least six species diverge in separate subspecies on east and west flanks of the Dividing Range (e.g. *Pardalotus striatus, Acanthiza nana, Lichenostomus melanops* and *Cracticus torquatus*). Still other species hardly cross the summit ridges, e.g. *Polytelis swainsonii, Neophema pulchella* and *Corvus mellori* on the west slopes and *Glossopsitta pusilla, Manorina melanophrys* and *Strepera versicolor* on the east. Where the mallee biota abuts on western outliers of the range, a number of allotaxa approach or meet one another—examples are *Polytelis anthopeplus-P. swainsonii, Malurus splendens-M. cyaneus, Acanthiza apicalis-A. pusilla,* and subspecies in *Pardalotus punctatus* and *Microeca fascinans*. This interface, termed by Ford (1987a,b) the Mallee Barrier, is not so much a 'barrier' as the frontline of the mallee following its expansion eastwards during wetter interglacial periods.

The Bassian Volcanic Barrier (Figure 7), in an area where local vulcanism and lava flows apparently split land bird populations over much of the late Plio-Pleistocene (Baird, 1985; Schodde *et al.*, 1993), generated subspeciation in 22 species (e.g. *Calyptorhynchus lathami, Stipiturus malachurus, Dasyornis broadbenti, Acanthiza reguloides, Pachycephala olivacea, Strepera versicolor, Zosterops lateralis*)—see Table 2. Some south-west subspecies (e.g. in *Podargus strigoides, Pardalotus striatus, Pachycephala pectoralis* and *Daphoenositta chrysoptera*) may nevertheless have reached the bloc with the expansion of the mallee 8,000–10,000 years ago. Outlying southern ranges on the Victorian Bassian plain—the Otway and Strzelecki Ranges—have also served as minor refuges in recent times, judging by endemic differentiation in *Dasyornis broadbenti, Sericornis frontalis* and *Melithreptus brevirostris* (Schodde and Mason, 1999). The Hunter Barrier (Figure 7) seems to have effected differentiation in at least another 11 Bassian species (e.g. *Platycercus eximius, Menura novaehollandiae, Eopsaltria australis, Pachycephala pectoralis, Zosterops lateralis, Cormobates leucophaeus* and *Coracina papuensis*) (Schodde and Mason, 1999).

Queensland Uplands

These are high, cool eucalypt-forested tablelands and ridges that outcrop along the spine of the Great Dividing Range in four main areas, south to north: the Dawes Range-Kroombit Tops coastal outlier, the Carnarvon-Chesterton Ranges (including Blackdown Tableland), the coastal Clarke-Connors Range (Eungella Plateau) inland from Mackay, and the Lolworth Range-Annan River uplands between Charters Towers, Hughenden and Cooktown (Figure 7). Only 48 core Bassian species, barely more than half those in the South-east bloc, reach these uplands, in as few as 20 endemic subspecies; there are no endemic species (Table 2). The break from the south-east centre lies along the Brisbane-Burnett-upper Dawson Rivers drainages and south through the northern 'granite belt' of south-east Queensland. Some 13 south-east Bassian species stop there (e.g. *Neophema pulchella, Menura novaehollandiae, Climacteris erythrops, Stipiturus malachurus, Calamanthus pyrrhopygius* and *Petroica boodang*); and another 11 are differentiated into subspecies on either side (e.g. in *Calyptorhynchus lathami, Sericornis frontalis* and *Acanthiza pusilla*), in addition to a number of pan-Australian or Torresian species. Subspecies that are separated in south-east Australia by the Great Dividing Range also meet there and intergrade across the divide (e.g. in *Malurus lamberti, Acanthiza nana, Lichenostomus leucotis* and *Cracticus torquatus*). Although Short *et al.* (1983a) discounted a Granite Belt (= Border Ranges) barrier, the extent of differentiation, species drop-off and change-over there seems to be too great to be chance coincidence, as appreciated by Ford (1980); but how the barrier operated historically remains unclear.

Species drop-out northwards is proportional to distance from south-east Australian sources and inversely proportional to endemic differentiation. Some 45 or so Bassian species are present in the southern Dawes-Kroombit bloc, but only two sub-endemic subspecies that are shared also with the Carnarvon-Chesterton Ranges bloc and south-east Queensland 'granite belt': *Acanthiza reguloides nesa* and *A. lineata alberti* (Table 2). Species numbers drop to 30 in the northern Lolworth-Annan bloc, but 8 are represented by endemic subspecies (e.g. in *Pardalotus punctatus, Acanthiza nana, Acanthorhynchus tenuirostris* and *Strepera graculina*); another 4 subspecies are shared exclusively with the Eungella bloc immediately south. At these latitudes, 15–20°S, core Bassian elements are limited to cool tablelands above 700 m a.s.l., underscoring an important control on avian distribution commonly overlooked in Australia: temperature. Because life-zones were altitudinally depressed in glacial cycles, isolation that led to subspecific differentiation on the Eungella and Lolworth-Annan blocs may well have come from climatic warming since the close of the last cycle (12,000 years ago) when the Bassian biota was pushed above the low divides that separate the blocs today.

Mallee

This is a unique biome of low close woodland of multi-stemmed eucalypts and associated sand-plain heath that occurs across semi-arid southern Australia from the westralian wheatbelt to the south-west outliers of the Great Dividing Range (Figure 7). Floristic and avifaunal composition indicate that it developed out of south-west Australia (Hill, 1990; Schodde, 1990). It interfaces the Eyrean avifaunula on its northern fringes and holds many species that today are more widespread in Eyrean regions. Of its 130 species only 52 are core Bassian taxa (Table 2). Another 28 are widespread through arid Australia have Bassian affinities at specific or subspecific level and appear to be derived from the Bassian sources.

In all, 65% of the mallee bird fauna is Bassian in origin (Schodde, 1990), the rest being of pan-Australian forms (e.g. *Aquila audax, Grallina cyanoleuca, Gymnorhina tibicen*), arid zone specialists of unclear source (e.g. *Eolophus roseicapilla, Oreoica gutturalis, Cheramoeca leucosternum, Artamus personatus/superciliosus*), or Eyrean elements of Torresian origin (e.g. *Ocyphaps, Taeniopygia guttata*). Thus the mallee bird fauna is treated here as a Bassian outlier.

Despite its diverse sources and ecotonal position between Bassian and Eyrean avifaunulas, the mallee bird fauna is distinctive. Fourteen species are endemic or almost so (e.g. *Leipoa ocellata, Polytelis anthopeplus, Malurus pulcherrimus, Calamanthus cautus, Lichenostomus cratitius, Drymodes brunneopygia, Pachycephala inornata* and *Cinclosoma castanotus*), and this indicates some age. Yet only 19 subspecies have budded off (e.g. in *Polytelis anthopeplus, Pardalotus punctatus, Calamanthus cautus, Manorina flavigula, Melithreptus brevirostris, Microeca fascinans* and *Cinclosoma castanotus*)—see Schodde (1990). Given the transcontinental range of the mallee and its spanning of the Eyrean Barrier, intrinsic subspeciation is significantly less than in most other Bassian centres. Differentiation of 10 involves the Eyrean Barrier itself (e.g. *Calamanthus cautus, Melithreptus brevirostris* and *Cinclosoma castanotus*); and the eastwards spread of another, *Eopsaltria griseogularis rosinae*, is stopped there. Otherwise most mallee forms range from Western Australia to the Murray Basin without variation (e.g. *Glossopsitta porphyrocephala, Pardalotus punctatus xanthopyge, Lichenostomus cratitius, Microeca fascinans assimilis, Drymodes brunneopygia* and *Pachycephala inornata*). Lack of intrinsic differentiation indicates that the mallee fauna, adapted to a drier climatic regime, has been less fragmented by Pleistocene glacial cycles, and for shorter periods, than other Bassian enclaves in south-west and south-east Australia.

Speciation in the *Timixos* group of *Pachycephala* (whistlers) indicates the development of the mallee bird fauna from south-west Australia in Plio-Pleistocene times. An early dry climatic episode first split an ancestral trans-Bassian population of *Timixos* into proto-*Pachycephala olivacea* in south-east Australia and proto-*P. inornata-P. rufogularis* in the south-west, there adapted incipiently to drier sclerophyll vegetation, including mallee. In a subsequent pluvial interglacial, the ancestral *inornata-rufogularis* group spread east with its mallee biome to the Murray Basin, approaching ancestral *olivacea* still confined to the abruptly wetter heights of the Great Dividing Range. The next dry glacial cycle then cut the *inornata-rufogularis* mallee lineage itself at the Nullarbor Plain or Eyrean Barrier (or both), vicariantly budding off ancestral *inornata* in the west and *rufogularis* in the east. During the most recent interglacial, the mallee once more expanded out of its refuges across southern inland Australia, and with it *inornata* spread east, reaching western outliers of the Great Dividing Range. There it now overlapped *rufogularis* which, characteristic of eastern Bassian taxa, remained largely contained in pockets of mallee in the Murray–Darling Basin.

This sequential pattern of differentiation in isolation in south-west refugia, followed by eastward spread to the Murray-Darling Basin, is repeated again and again in mallee-adapted birds (e.g. *Hylacola* group of *Calamanthus*, *Lichenostomus* group of *Lichenostomus*, *Eidopsarus* group of *Melithreptus* and *Morganornis* group of *Pomatostomus*). On a wider continental scale, such steps trace the path of eremic adaptation in Australian land birds in general.

Eyrean Avifaunula

This avifaunula inhabits the immense mulga (*Acacia*) woodlands, spinifex (*Triodia*) dune deserts and dry stony ranges that today cover all inland Australia out to the mid westralian coast (Figures 4, 7). In proportion to its range, it is depauperate: its environment offers diverse niches for survival but few opportunities for radiation. The 82 core species (Table 2) include a surprisingly rich waterfowl and shore bird fauna of 3 and 7 species respectively and a prevalence of ground-feeding seed- and insect-eaters. Even so, it resembles the Bassian in its core elements: honeyeaters (13 species), acanthizid warblers (8 species) and parrots (8 species, 7 of them platycercine) are the dominant families, but malurid wrens with 9 species, 7 of them grasswrens (*Amytornis*), are also prominent. Except for monarchid flycatchers, elements of most Australasian families are represented, including three phabine pigeons, three diverse lineages of cockatoos (*Nymphicus, Eolophus, Cacatua leadbeateri*), two campephagids (the eclipse-plumaged *Lalage tricolor* and ground-feeding *Coracina maxima*), and two bowerbirds (*Chlamydera*). Under environmental constrictions imposed by prolonged drought and erratic rains, adaptations to nomadism and opportunistic breeding are common. A boom-and-bust mode of survival, missing from other avifaunulas, is common to many Eyrean species (e.g. *Elanus scriptus*, *Gallinula ventralis*, *Phaps histrionica*, *Melopsittacus undulatus* and *Certhionyx variegatus*).

Many Eyrean taxa, their centres of distribution to the south of the Tropic of Capricorn, appear to be of Bassian origin (e.g. all platycercine parrots, most acanthizid warblers, the honeyeaters *Acanthagenys, Phylidonyris albifrons* and most of the genus *Lichenostomus*). Others seem to be originally Torresian (e.g. the pigeons *Geopelia cuneata* and *Ocyphaps lophotes*, black cockatoo *Calyptorhynchus banksii*, kingfisher *Todiramphus pyrrhopygius*, fairy-wren *Malurus leucopterus*, two bowerbirds (*Chlamydera*) and possibly the grasswrens (*Amytornis*)). Diverse pan-Australian species (e.g. emu, wedge-tailed eagle, weebill, grey shrike-thrush, magpie-lark, willie wagtail and Australian magpie) are also present.

Other species are autochthons of no clear source (e.g. the endemic water and shore bird groups (*Stictonetta, Malacorhynchus, Cladorhynchus, Elseyornis, Erythrogonys*), two eremic falcons (*Falco*), galah (*Eolophus*), cockatiel (*Nymphicus*), budgerigar (*Melopsittacus*), desert chats (*Epthianura, Ashbyia*), crested bellbird (*Oreoica*), ground cuckoo-shrike (*Coracina* subgenus *Pteropodoccys*), the migratory woodswallows (*Artamus* subgenus *Campbellornis*) and white-backed swallow (*Cheramoeca*). Many of these taxa are 'old' monospecific genera, emphasizing the minimal species-level radiation in the Eyrean avifaunula. Apart from seven grasswrens (*Amytornis*) and three whitefaces (*Aphelocephala*), there has been little intrinsic differentiation through the later Plio–Pleistocene; some 70% of genera are represented by single species and only 35 out of 60 subspecies have arisen *in situ* from Eyrean stocks (Schodde, 1982a). Widespread nomadism is a contributing factor; lack of internal physical and climatic barriers during Plio-Pleistocene glacial cycles is another. What radiation there is, nevertheless, reflects the overriding influence of those cycles, leading to generic differentiation in the epthianurine chats and radiation in 10 species-groups, budding off over 20 species and 35 subspecies. The three physical factors that have facilitated differentiation in this avifaunula are discussed below.

Refugia

These areas consisted mostly of isolated mountain ranges that offered refuge during glacial cycles of extreme aridity (Harrison, 1973, 1975; Schodde, 1982a)—see Figure 7. Species and subspecies that probably arose in them are: *Conopophila whitei*, *Chlamydera guttata* and western subspecies of *Geophaps plumifera, Eolophus roseicapilla, Amytornis striatus,* and *Rhipidura albiscapa* in the Pilbara uplands; *Polytelis alexandrae, Amytornis purnelli* and subspecies of *Calyptorhynchus banksii*, *Malurus splendens* and *Gerygone fusca* in the centralian ranges; *Amytornis merrotsyi* and subspecies of *Barnardius zonarius* and *Lichenostomus plumulus* in the Flinders-Gawler Ranges; and *Amytornis ballarae*, *Pomatostomus halli*, *Chlamydera maculata* and subspecies of *Barnardius zonarius, Malurus splendens, Acanthiza apicalis* and *Cinclosoma castaneothorax* in the Selwyn-Grey Ranges (Table 2). Other species that probably budded off in one or other of these range systems were *Ocyphaps lophotes, Eolophus roseicapilla, Stipiturus ruficeps, Pyrrholaemus brunneus, Acanthagenys rufogularis, Lichenostomus plumulus* and *Emblema picta*.

Following periods of climatic amelioration, populations that had diverged in isolation in the ranges tended to spread, many of them approaching the coast and overlapping or intergrading with allotaxa. This may explain the present patchy sympatry between red-capped and scarlet robins (*Petroica goodenovii, P. multicolor*), variegated and blue-breasted fairy-wrens (*Malurus lamberti, M. pulcherrimus*), and grey-fronted and yellow-plumed honeyeaters (*Lichenostomus plumulus, L. ornatus*), as well as the various zones of subspecific intergradation in *Barnardius zonarius*, *Malurus splendens*, *Pardalotus rubricatus* and *Manorina flavigula*. A refuge of a different nature is the Lake Eyre Basin itself. Among land birds, the grey grasswren (*Amytornis barbatus*), chestnut-breasted whiteface (*Aphelocephala pectoralis*), gibberbird (*Ashbyia lovensis*) and chestnut-crowned babbler (*Pomatostomus ruficeps*) are endemic to it, leading Serventy (1972) to regard it as a centre of speciation. But these species, like most Eyrean autochthons, have the attributes of 'adversity-selection', suggesting that the Lake Eyre Basin is more an area of faunal accumulation (and extinction) than a cradle of evolution.

Zoogeographic Barriers

There are at least two barriers of this type. One is the Grey Range Barrier (Ford, 1987; Schodde and Mason, 1999), a chain of low ranges that runs from the Olary Spur in South Australia to central Queensland and separates the Eromanga (Lake Eyre) and Murray-Darling Basins. Its effect has been minor, involving recent subspeciation in *Psephotus haematonotus*, *Malurus splendens*, *Amytornis barbatus* and *Acanthiza apicalis,* among core Eyrean taxa. The other, the Eyrean Barrier (see above), has played a more significant role, affecting the more sedentary tree- and shrub-living species most. One generic pair, five pairs of species and at least eight pairs of subspecies in core Eyrean taxa are separated by it (Table 2). The most recent differentiates have been east-west replacement subspecies in *Cacatua leadbeateri, Barnardius zonarius, Climacteris affinis, Amytornis textilis* and *Lichenostomus plumulus*, which probably budded off during the most recent glacial cycles. Differentiation of species, notably in the *Aphelocephala nigricincta*, *Pomatostomus superciliosus*, *Psophodes cristatus* and *Artamus personatus* species-groups, is not only older but also more complex, and often involves secondary overlap, as in the case of *Pomatostomus superciliosus* over *P. ruficeps* and *Artamus personatus* over *A. superciliosus*. Wherever this occurs, it is the western form that overlaps the eastern, paralleling historic movements in the mallee biota. Generic differentiation in the epthianurine chats may have followed a similar course (Serventy, 1972), *Epthianura* arising on the western side of the barrier and *Ashbyia* on the east. *Epthianura* has since radiated complexly, whereas *Ashbyia* has not diversified at all after adapting to terrestrial life on the gibber plains.

Terrain

This refers to the interpolation of desert sand fields among hard pan and gibber plains (from early Pleistocene) (King, 1960; Bowler, 1982) which created new and fragmented distributions for ground-dwelling birds (see Figure 7). Differentiation in response has happened in at least four groups of species and four pairs of subspecies (Table 2). Unlike the Eyrean Barrier, however, barriers of terrain developed in different regions at the same time, causing contemporaneous isolation in multiple populations of the one ancestral stock. This is evident in the unparallelled radiation in grasswrens, *Amytornis*, in which seven species and five subspecies have budded off. At least three major lineages are present. One, represented by the grey grasswren (*Amytornis barbatus*) is a relict, limited to flood plain tussock grassland and lignum (*Muehlenbeckia*) in the lower Eromanga and Bulloo Basins where isolates have subspeciated (Schodde and Christidis, 1987). Another, the striated grasswren group, comprises at least two species; *A. merrotsyi* in stony hummock (*Triodia*) grass fields in the Flinders and Gawler Ranges, and *A. striatus* which, though widespread in hummock grass fields throughout the Australian arid zone, has differentiated into three subspecies on different substrates of sand plain and stony hillside. Another species **possibly** related to this group is the Eyrean grasswren (*A. goyderi*) which is allopatric on the dune systems of the Simpson-Strzelecki Deserts. The third lineage is again adapted to different substrates. Two species, *A. purnelli* and *A. ballarae*, are vicariant on stony *Triodia*-covered hills in the centralian and Selwyn Range systems respectively, separated by the Sandover Desert; and *A. textilis*, which comprises two very distinct subspecies on either side of the Eyrean Barrier, occupies hard pan shrubberies patchily across the southern inland. Populations of *A. textilis* west of the Eyrean Barrier are also differentiated, but more slightly so, in contrast to the marked differences between forms of the cinnamon quail-thrush (*Cinclosoma cinnamomeum*) separated by the same terrain: the Barton sandhills-Great Victoria Desert.

Either evolutionary rates have varied greatly among Eyrean birds or barriers of terrain have operated in more complex ways over a longer span of time than is presently understood. The latter seems the more likely. Despite evidence suggesting that nearly all dune building in eremic Australia took place towards the end of the Pleistocene, within the last 500,000–600,000 years (Twidale, 1972; Bowler, 1982; Wasson, 1982), the extent of speciation in ground-living Eyrean birds implies that the inland Australian sand deserts, or their beginnings, are older.

Migrants, Adventives and Opportunists

Completing the development of Australasia's contemporary bird fauna are elements best described as migrants, adventives and opportunists. Among the migrants, the major contribution comes from intercontinental, trans-hemispheric migration, and this, aside from sea birds, is limited—the great North and South American and Afro-Eurasian migrations in songbirds, waterfowl and raptors are missing. Apart from several raptors in New Guinea, the Oriental cuckoo (*Cuculus saturatus*), several swifts and swallows (*Hirundapus, Apus, Hirundo, Cecropis*) and Old World warblers (e.g. *Locustella fasciolata*), the principal intercontinental migrants are shore birds. Like other migrants here they are Eurasia-centred and use Australasia as a non-breeding feeding refuge over summer. Migratory shore birds that reach Australasia regularly include 6 plovers (Charadriidae), some 33 curlews, snipe, godwits, sandpipers and phalaropes (Scolopacidae) and 1 pratincole (Glareolidae). For a number of species Australia-New Guinea are the main, if not sole wintering grounds.

The adventives are 'new Australians', members of indisputably Eurasian-centred families that have dispersed to Australasia within the last 15 million years, and often within the last five. They comprise just over 270 species, some 75 of which are freshwater and shore birds and another 50 are nocturnal and diurnal raptors. Significantly, most adventive land birds occupy grasslands, savannas, reed beds and chenopodiaceous shrub steppes, habitats as Eurasian as their occupants (Barlow, 1981; Crisp *et al.*, 1999). In Australasia, they are occupied by quail (Phasianidae), raptors (*Elanus*, *Circus*, *Haliastur*, *Tyto capensis*), sylviid warblers (Sylviidae), larks (Alaudidae), pipits (Motacillidae), swallows (Hirundinidae), grass finches (Estrildidae) and, in New Guinea, thrushes and chats (Turdidae) and a shrike (Laniidae).

The most distinctive Eurasian land bird community in Australasia occurs in natural alpine grasslands crowning the New Guinean cordilleras above 3,000 m (Figure 3). Endemic to it are two grass finches (*Lonchura*), a pipit (*Anthus gutturalis*) and a huge alpine quail (*Anurophasis*), as well as subspecies of the brown quail (*Coturnix ypsilophora*), buff-banded rail (*Gallirallus philippensis*), pectoral rail (*Rallus pectoralis*), spotless crake (*Porzana tabuensis*), island blackbird (*Turdus poliocephalus*) and a grassbird, *Megalurus macrurus*. No equivalent assemblage has developed in alpine Australia or New Zealand, probably because of the severity of southern winters.

The 'opportunists' are Australasian taxa that, for distributional and adaptational reasons, are no longer unambiguous elements of one Australasian avifaunula or another. Papuasian examples are the hill forest species *Gallicolumba jobiensis, Geoffroyus simplex, Chalcites meyerii, Lichenostomus obscurus, Ptilorrhoa castanonota* and *Coracina caeruleogrisea* that span the zone between Tumbunan and Irian avifaunulas. Others such as *Cacomantis flabelliformis* and the *Daphoenositta chrysoptera* superspecies are present in one avifaunula in New Guinea and another in Australia. Examples within Australia are widespread taxa (e.g. *Lophoictinia isura, Phaps chalcoptera, Podargus strigoides, Todiramphus sanctus, Smicrornis brevirostris, Rhipidura leucophrys, Grallina cyanoleuca, Gymnorhina tibicen* and *Coracina novaehollandiae*) that range through all sclerophyllous avifaunulas.

These opportunists have commonly modified their ancestral niche envelopes, enabling them to exploit new foraging zones free from ancestral competitors. Thus altitudinal shifts in New Guinean *Gallicolumba jobiensis, Geoffroyus simplex* and *Ptilorrhoa castanonota* have probably been driven, at least partly, by competition from altitudinally parapatric congeners (Diamond, 1972a). In many other species, adaptive change in morphology and behaviour has occurred. Thus the kite *Lophoictinia* has developed a harrier-like mode of hunting over forest canopy; the very stout-billed *Coracina caeruleogrisea* presumably takes larger fruits and harder-shelled insects than its congeners; the frogmouth *Podargus strigoides* has become grey-plumaged and relatively long-winged and short-tailed for camouflage and sallying in open space; and the acanthizid weebill, *Smicrornis brevirostris,* has developed a stubby pardalote-like bill fitted for foliage-gleaning. In Australia itself, a common thread runs through these shifts in foraging zone: adaptation to life in drier, more open environments.

CONCLUSIONS

Building on foundations laid by Schodde and Calaby (1972), perceptions of the origin and evolutionary development of Australasia's bird fauna have become the antithesis of mid 20th century views. The old concepts, summarized by Keast (1984), saw dispersal—from primarily Asian sources—as a panacea for explaining all avifaunal distribution patterns, despite its fuzzy correlations. Instead, it now seems that the core of the bird fauna has arisen out of Gondwana, to develop in vicarious isolation from the Eurasian. But, although root Australasian bird families—the parrots and cockatoos, corvidan songbirds, pigeons, megapodes, turnicid quail, magpie-geese and cuculine cuckoos, and even old divergent groups such as the eurostopodine nightjars, and the raptor genera *Hamirostra*, *Lophoictinia* and *Eurythrotriorchis*—may have Gondwanan origins, there is no absolute proof, and there may never be. The Australasian bird fauna is not a vehicle for testing dispersal and vicariance as competing paradigms because **both** have contributed to its development.

What does seem clear is that Australasia's modern bird fauna—particularly its land bird fauna—radiated in isolation out of root stocks hardly found today on other continents. Furthermore, Australia itself was the source of this radiation, that began in subtropical rainforests of Gondwanan origin and early to middle Tertiary age which survive today in montane New Guinea, New Zealand and refugia along the east Australian coast. Asian immigrants did not begin to reach Australasia until the later Tertiary, gaining a limited foothold in new habitats created by changing climate.

Adaptation and climatic sifting of the autochthonous bird fauna, driven by increasing warmth and intensifying aridity as Australasia entered the tropics in the later Tertiary, interacted to develop five main regional faunulas:

(i) the Tumbunan, preserving many ancestral elements of the mid Tertiary bird fauna, in the subtropical Gondwanan rainforests of montane New Guinea, New Zealand and east coastal Australia;

(ii) the Irian in tropical rainforests of Malaysian origin throughout lowland New Guinea and Papuasian archipelagos, with limited outliers along the north coast of Australia;

Plate 91. An Australian owlet-nightjar (*Aegotheles cristatus*). Photograph G. Baker.

Plate 92. The Australian (Mallee) ringneck parrot (*Barnardius zonarius*). Photograph G. Taylor.

Plate 93. The plumed pigeon (*Geophaps lophotes*). Photograph R. Fidler.

Plate 94. A willie wagtail adult (*Rhipidura leucophrys*) at the nest with young. Occurs in open forest, cultivated farmland as well as parks and gardens throughout Australia; also extends to Indonesia and the Solomon Islands. Photograph P. Munchenberg.

(iii) the Torresian in the tropical eucalypt woodlands around the rim of northern Australia, with outliers in southern New Guinea;
(iv) the Bassian in the temperate eucalypt forests and woodlands around the rim of southern Australia, including Tasmania; and
(v) the Eyrean in arid central Australia, the end point of radiation and adaptation in the Australasian bird fauna as a whole.

Throughout, the broad direction of radiation for land birds seems to have been from cool rainforests into more open and drier habitats, particularly the scleromorphic forests and woodlands of Australia itself.

This interpretation of the development of Australasia's bird fauna is only as good as the systematics upon which it is based. Advances in taxonomic information explain much of the altered perceptions of sources and directions of regional radiation; a critical factor has been the re-framing of the phylogeny of Australasia's diverse songbird fauna by molecular research, indicating a radiation in corvidan families.

Change in the immediate future will come from human impact on the Australasian environment. This process, from its beginnings with Aboriginal man within the last 50,000 years, has accelerated dramatically in the last century or two with the introduction of European technology. Its result, from land clearing of various kinds, has been to fragment and diminish forest and woodland faunas, while increasing habitat for those few species such as the crested pigeon, galah, Australian magpie and Eurasian adventives that are adapted to grasslands and open ground. In historical context, it is probably little different in effect from a Pleistocene glacial cycle, though more selective for such regions as arable basins; and it has the same potential to generate differentiation and cause extinction. How much we act to influence the process will depend as much on our philosophical attitude towards the natural environment as our knowledge about how to manage it. This is an issue of public education that can only be helped by a better understanding of the diversity of Australasia's bird life, its historic habitats and its global distinctiveness. We need to know **what** we are losing.

ACKNOWLEDGEMENTS

I am indebted to Brigitte Kuchlmayr for preparing the figures.

REFERENCES

Ames, P.L. 1971. The morphology of the syrinx in passerine birds. *Bulletin of the Peabody Museum of Natural History*, Vol. 37, pp.1–194.

Ames, P.L. 1987. The unusual syrinx morphology of Australian treecreepers *Climacteris*. *Emu*, Vol. 87, pp.192–195.

Aplin, K.P. 1998. Vertebrate zoogeography of the Bird's Head of Irian Jaya, Indonesia. Pp.803–890. In Miedama, J., Odé, C. and Dam, R.A.C. (eds). *Perspectives on the Bird's Head of Irian Jaya, Indonesia*. Natural History Museum, Leiden.

Archer, M. and Fox, B. 1984. Background to vertebrate zoogeography in Australia. Pp.1–15. In Archer, M. and Clayton, G. (eds). *Vertebrate Zoogeography and Evolution in Australasia*. Hesperian Press, Perth. 1203pp.

Audley-Charles, M.G. 1966. Mesozoic palaeogeography of Australasia. *Palaeogeography, Palaeoclimatology, Palaeoecology*, Vol. 2, pp.1–25.

Audley-Charles, M.G. 1987. Dispersal of Gondwanaland: relevance to the dispersal of the angiosperms. Pp.32-49. In Whitmore, T.C.(ed). *Biogeographical Evolution of the Malay Archipelago*. Clarendon Press, Oxford. 147pp.

Audley-Charles, M.G. 1991. Tectonics of the New Guinea area. *Annual Review of Earth and Planetary Sciences*, Vol.19, pp.17–41.

Baird, R.F. 1985. Avian fossils from Quaternary deposits in 'Green Waterhole Cave', south-eastern South Australia. *Records of the Australian Museum*, Vol. 37, pp.353–370.

Barker, F.K., Barrowclough, G. and Groth, J.G. 2001. A phylogenetic hypothesis for passerine birds: taxonomic and biogeographic implications of an analysis of nuclear DNA sequence data. *Proceedings of the Royal Society of London*, Series B, Vol. 269, pp.295–308.

Barlow, B.A. 1981. The Australian flora: its origin and evolution. Pp.25–75. In George, A.S. (ed.). *Flora of Australia*, 1. Australian Biological Resources Study, Canberra.

Baverstock, P.R., Schodde, R., Christidis, L., Krieg, M. and Sheedy, C. 1991. Microcomplement fixation: preliminary results from the Australasian avifauna. Pp.611–618. In Bell, B.D., Cossee, R.O., Flux, J.E.C., Heather, B.D., Hitchmough, R.A., Robertson, C.J.R. and Williams, M.J. (eds). *Acta XX Congressus Internationalis Ornithologici, Canterbury, 1990.* New Zealand Ornithological Trust Board, Wellington. 2568pp.

Beehler, B.M. and Finch, B.W. 1985. *Species-Checklist of the Birds of New Guinea*. Royal Australasian Ornithologists Union, Melbourne.

Bock, W.J. 1972. Review of "P.L. Ames. The morphology of the syrinx in passerine birds". *Auk*, Vol. 89, pp.899–903.

Boles, W.E. 1991. The origin and radiation of Australasian birds: perspectives from the fossil record. Pp.383–391. In Bell, B.D., Cossee, R.O., Flux, J.E.C., Heather, B.D., Hitchmough, R.A., Robertson, C.J.R. and Williams, M.J. (eds). *Acta XX Congressus Internationalis Ornithologici, Canterbury, 1990.* New Zealand Ornithological Trust Board, Wellington. 2568pp.

Boles, W.E. 1995. The world's oldest songbird. *Nature,* Vol. 374, pp.21–22.

Boles, W.E. 1997. Fossil songbirds (Passeriformes) from the early Eocene of Australia. *Emu*, Vol. 97, pp.43–50.

Bowler, J.M. 1976. Aridity in Australia: age, origins and expression in aeolian landforms and sediments. *Earth-Science Reviews,* Vol. 12, pp.279–310.

Bowler, J.M. 1982. Aridity in the late Tertiary and Quaternary of Australia. Pp.35–45. In Barker, W.R. and Greenslade, P.J.M. (eds). *Evolution of the Flora and Fauna of Arid Australia*. Peacock Publications, South Australia. 392pp.

Bull, P.C. and Whitaker, A.H. 1975. The amphibians, reptiles, birds and mammals. Pp.231-276. In Kuschel, G. (ed). *Biogeography and Ecology in New Zealand.* Monographiae Biologicae, Vol. 27. W. Junk, The Hague.

Burbidge, N.T. 1960. The phytogeography of the Australian region. *Australian Journal of Botany,* Vol. 8, pp.75–209.

Callen, R.A. and Tedford, R.H. 1976. New late Cainozoic rock units and depositional environments, Lake Frome area, South Australia. *Transactions of the Royal Society of South Australia,* Vol. 100, pp.125–167.

Checklist Committee, Ornithological Society of New Zealand. 1990. *Checklist of the Birds of New Zealand and the Ross Dependency, Antarctica.* Edit. 3. Ornithological Society of New Zealand, Auckland.

Christidis, L. and Schodde, R. 1991. Relationships of Australo-Papuan songbirds—protein evidence. *Ibis,* Vol. 133, pp.277–285.

Christidis, L., Schodde, R. and Baverstock, P.R. 1988. Genetic and morphological differentiation and phylogeny in the Australo-Papuan scrubwrens (*Sericornis,* Acanthizidae). *Auk,* Vol. 105, pp.616–629.

Condon, H.T. 1975. *Checklist of the Birds of Australia* Part1 Non-Passerines. Royal Australasian Ornithologists Union, Melbourne.

Cracraft, J. 1973. Continental drift, paleoclimatology, and the evolution and biogeography of birds. *Journal of Zoology, London,* Vol. 169, pp.455–545.

Cracraft, J., Espinoza, A., Nunn, G., Helm-Bychowski, K. and Feinstein, J. 1998. Phylogenetic relationships and radiation of the corvine oscines. *XXII International Ornithological Congress, Durban.* (Poster).

Crisp, M.D, West, J.G. and Lindner, H.P. 1999. Biogeography of the terrestrial flora. Pp.321–368. In Orchard, A.E. (ed). *Flora of Australia 1, Introduction.* 2nd Edit. Australian Biological Resources Study, CSIRO Publishing, Melbourne. 670pp.

Crocker, R.L. and Wood, J.G. 1947. Some historical influences on the development of the South Australian vegetation communities and their bearing on concepts and classification in ecology. *Transactions of the Royal Society of South Australia,* Vol. 71, pp.91–136.

Davies, J.L. 1967. Tasmanian landforms and Quaternary climates. Pp.1–25. In Jennings, J.N. and Mabbutt, J.A. (eds). *Landform Studies from Australia and New Guinea.* Australian National University Press, Canberra.

Dawson, T.J. and Dawson, L. 2006. Evolution of arid Australia and consequences for vertebrates. Pp.51–70. In Merrick, J.R., Archer, M., Hickey, G.M. and Lee, M.S.Y. (eds). *Evolution and Biogeography of Australasian Vertebrates.* Auscipub Pty. Ltd., Sydney. 942pp.

Diamond, J.M. 1970a. Ecological consequences of island colonization by southwest Pacific birds, I. Types of niche shifts. *Proceedings of the National Academy of Sciences of the United States of America,* Vol. 67, pp.529–536.

Diamond, J.M. 1970b. Ecological consequences of island colonization by southwest Pacific birds, II. The effect of species diversity on total population density. *Proceedings of the National Academy of Sciences of the United States of America,* Vol. 67, pp.1715–1721.

Diamond, J.M. 1972a. *Avifauna of the Eastern Highlands of New Guinea.* Nuttall Ornithological Club, Cambridge, Massachusetts.

Diamond, J.M. 1972b. Biogeographic kinetics: estimation of relaxation times for avifaunas of southwest Pacific islands. *Proceedings of the National Academy of Sciences of the United States of America,* Vol. 69, pp.3199–3203.

Diamond, J.M. 1985. New distributional records and taxa from the outlying mountain ranges of New Guinea. *Emu,* Vol. 85, pp.65–91.

Dow, D.B. 1977. A geological synthesis of Papua New Guinea. *Bureau of Mineral Resources, Geology and Geophysics Bulletin,* No. 201.

Draffan, R.D.W., Garnett, S.T. and Malone, G.J. 1983. Birds of the Torres Strait: an annotated list and biogeographical analysis. *Emu,* Vol. 83, pp.207–234.

Ericson, P.G.P., Christidis, L., Cooper, A., Irestedt, M., Jackson, J., Johansson, U.F. and Norman, J.A. 2001. A Gondwanan origin of passerine birds supported by DNA sequences of the endemic New Zealand wrens. *Proceedings of the Royal Society of London,* Series B, Vol. 269, pp.235–241.

Exon, N.F. and Marlow, M.S. 1990. The New Ireland Basin: a frontier basin in Papua New Guinea. Pp.513–532. In Carman, G.J. and Carman, Z. (eds). *Petroleum Exploration in Papua New Guinea.* PNG Chamber of Mines and Petroleum, Port Moresby, Papua New Guinea. 597pp.

Feduccia, A. 1975. Morphology of the bony stapes in the Menuridae and Acanthisittidae: evidence for oscine affinities. *Wilson Bulletin,* No. 87, pp.418–420.

Feduccia, A. 1995. Explosive evolution in Tertiary birds and mammals. *Science,* Vol. 267, pp.637–638.

Feduccia, A. 1996. *The Origin and Evolution of Birds.* Yale University Press, New Haven, Connecticut. 420pp.

Fleming, C.A. 1962. History of the New Zealand bird fauna. *Notornis,* Vol. 9, pp.270–274.

Ford, J. 1974. Speciation in Australian birds adapted to arid habitats. *Emu,* Vol. 74, pp.161–168.

Ford, J. 1978. Geographical isolation and morphological and habitat differentiation between birds of the Kimberley and Northern Territory. *Emu,* Vol. 78, pp.25–35.

Ford, J. 1980. Hybridization between contiguous subspecies of the varied sittella. *Emu,* Vol. 80, pp.1–12.

Ford, J. 1982. Origin, evolution and speciation of birds specialized to mangroves in Australia. *Emu,* Vol. 82, pp.12–23.

Ford, J. 1983. Taxonomic notes on some mangrove-inhabiting birds in Australasia. *Records of the Western Australian Museum,* Vol. 10, pp.381–415.

Ford, J. 1986. Avian hybridisation and allopatry in the region of the Einasleigh uplands and Burdekin-Lynd divide, north-eastern Queensland. *Emu,* Vol. 86, pp.12–22.

Ford, J. 1987. Minor isolates and minor geographical barriers in avian speciation in continental Australia. *Emu,* Vol. 87, pp.90–102.

Frakes, L.A. 1999. Evolution of Australian environments. Pp.163–204. In Orchard, A.E. (ed). *Flora of Australia 1, Introduction.* 2nd Edit. Australian Biological Resources Study, CSIRO Publishing, Melbourne. 670pp.

Galloway, R.W. and Löffler, E. 1972. Aspects of geomorphology and soils in the Torres Strait region. Pp.11–28. In Walker, D. (ed). *Bridge and Barrier: the Natural and Cultural History of Torres Strait.* Department of Biogeography and Geomorphology, Research School of Pacific Studies, Australian National University, Canberra. 437pp.

Gentilli, J. 1949. Foundations of Australian bird geography. *Emu,* Vol. 49, pp.85–129.

Greenslade, P.J.M. 1968. Island patterns in the Solomon Islands bird fauna. *Evolution,* Vol. 22, pp.751–761.

Hall, R. 1996. Reconstructing Cenozoic SE Asia. Pp.153-184. In Hall, R. and Blundell, D. (eds). *Tectonic Evolution in Southeast Asia*. Special Publication 106. The Geological Society, London.

Harland, W.B., Armstrong, R.L., Cox, A.V., Craig, L.E., Smith, A.G. and Smith, D.G.1990. *A Geologic Time Scale 1989*. Cambridge University Press, Cambridge. 263pp.

Harrison, C.J.O. 1973. The origin of apparent speciation in eastern central Australia. *Emu,* Vol. 73, pp.129–133.

Harrison, C.J.O. 1975. Speciation in arid habitats: some further comments. *Emu,* Vol. 75, p.157.

Heinsohn, T. and Hope, G. 2006. The Torresian connections: zoogeography of New Guinea. Pp.71–93. In Merrick, J.R., Archer, M., Hickey, G.M. and Lee, M.S.Y. (eds). *Evolution and Biogeography of Australasian Vertebrates*. Auscipub Pty. Ltd., Sydney. 942pp.

Hill, K.D. 1990. Biogeography of the mallee eucalypts. Pp.16–20. In Noble, J.C., Joss, P. and Jones, G.K. (eds). *The Mallee Lands: A Conservation Perspective*. CSIRO Publishing, Melbourne.

Hill, R.S., Truswell, E.M., McLoughlin, S. and Dettmann, M.E. 1999. Evolution of the Australian flora: fossil evidence. Pp.251–320. In Orchard, A.E. (ed). *Flora of Australia 1, Introduction. 2nd Edit*. Australian Biological Resources Study, CSIRO Publishing, Melbourne. 670pp.

Holdaway, R.N. and Worthy, T.H. 2006. Evolution of New Zealand and its vertebrates. Pp.111–128. In Merrick, J.R., Archer, M., Hickey, G.M. and Lee, M.S.Y. (eds). *Evolution and Biogeography of Australasian Vertebrates*. Auscipub Pty. Ltd., Sydney. 942pp.

Hooker, J.D. 1860. Introductory essay. Pp.I-cxxviii. In *Botany of the Antarctic Voyage of H.M. Discovery Ships* Erebus *and* Terror *in the Years 1839–1843, III. Flora Tasmaniae*. Lovell Reeve, London.

Hope, G.S. 1996. Quaternary change and the historical biogeography of Pacific Islands. Pp.165–190. In Keast, A. and Miller, S.E. (eds). *The Origin and Evolution of Pacific Islands Biotas. New Guinea to Eastern Polynesia: Patterns and Processes*. SPB Academic Publishing, Amsterdam. 531pp.

Hope, J.H., Lampert, R.J., Edmondson, E., Smith, M.J. and van Tets, G.F. 1977. Late Pleistocene faunal remains from Seton rock shelter, Kangaroo Island, South Australia. *Journal of Biogeography*, Vol. 4, pp.363–385.

Jennings, J.N. 1971. Sea level changes and land links. Pp.1–13. In Mulvaney, D.J. and Golson, J. (eds). *Aboriginal Man and Environment in Australia*. Australian National University Press, Canberra. 389pp.

Jennings, J.N. 1972. Some attributes of Torres Strait. Pp.29–38. In Walker, D. (ed). *Bridge and Barrier: the Natural and Cultural History of Torres Strait*. Department of Biogeography and Geomorphology, Research School of Pacific Studies, Australian National University, Canberra. 437pp.

Jennings, J.N. and Mabbutt, J.A. 1977. Physiographic outlines and regions. Pp.38–52. In Jeans, D.N. (ed). *Australia: a Geography*. Sydney University Press, Sydney.

Johnson, K.P. and Clayton, D.H. 2000. Nuclear and mitochondrial genes contain similar phylogenetic signal for pigeons and doves (Aves: Columbiformes). *Molecular Phylogenetics and Evolution*, Vol. 14, pp.141–151.

Johnstone, R.E. 1990. Mangroves and mangrove birds of Western Australia. *Records of the Western Australian Museum, Supplement,* No. 32, pp.1–120.

Joseph, L. and Moritz, C. 1994. Mitochondrial DNA phylogeography of birds in eastern Australian rainforests: first fragments. *Australian Journal of Zoology,* Vol. 42, pp.385–403.

Joseph, L., Slikas, B., Alpers, D. and Schodde, R. 2001. Molecular systematics and phylogeography of New Guinea logrunners (Orthonychidae). *Emu,* Vol. 101, pp.273–280.

Keast, A. 1961. Bird speciation on the Australian continent. *Bulletin of the Museum of Comparative Zoology, (Harvard)*, Vol. 123, pp.307–495.

Keast, A. 1981. The evolutionary biogeography of Australian birds. Pp.1585–1635. In Keast, A. (ed). *Ecological Biogeography of Australia*. Monographiae Biologicae, Vol. 41. W. Junk, The Hague.

Keast, A. 1984. Contemporary ornithogeography: the Australian avifauna, its relationships and evolution. Pp.457–468. In Archer, M. and Clayton, G. (eds). *Vertebrate Zoogeography and Evolution in Australasia*. Hesperian Press, Perth. 1203pp.

Kemp, E.M. 1978. Tertiary climatic evolution and vegetation history in the southeast Indian Ocean region. *Palaeogeography, Palaeoclimatology, Palaeoecology*, Vol. 24, pp.169–208.

Kikkawa, J., Ingram, G.J. and Dwyer, P.D. 1979. The vertebrate fauna of Australian heathlands—an evolutionary perspective. Pp.231–279. In Doodall, D.W. (ed). *Ecosystems of the World* 9, *Heathlands and Related Shrublands*. Elsevier, Amsterdam.

Kikkawa, J., Monteith, G.B. and Ingram, J.G. 1981. Cape York Peninsula: major region of faunal interchange. Pp.1695–1742. In Keast, A. (ed). *Ecological Biogeography of Australia*. Monographiae Biologicae, Vol. 41. W. Junk, The Hague.

Kikkawa, J. and Pearse, K. 1969. Geographical distribution of land birds in Australia—a numerical analysis. *Australian Journal of Zoology,* Vol. 17, pp.821–840.

Kikkawa, J. and Williams, W.T. 1971. Altitudinal distribution of land birds in New Guinea. *Search,* Vol. 2, pp.64–65.

King, D. 1960. The sandridge deserts of South Australia and related aeolian landforms of the Quaternary arid cycles. *Transactions of the Royal Society of South Australia*, Vol. 83, pp.93–108.

Klicka, J. and Zink, R.M. 1977. The importance of Recent ice ages in speciation: a failed paradigm. *Science,* Vol. 277, pp.1666–1669.

Lange, R.T. 1982. Australian Tertiary vegetation. Pp.44–89. In Smith, J.M.B. (ed). *A History of Australasian Vegetation*. McGraw-Hill, Sydney. 168pp.

Ludbrook, N.H. 1980. *A Guide to the Geology and Mineral Resources of South Australia*. Government Printer, Adelaide. 230pp.

MacArthur, R.H. and Wilson, E.O. 1967. *The Theory of Island Biogeography*. Princeton University Press, Princeton, New Jersey. 203pp.

Macdonald, J.D. 1969. Notes on the taxonomy of *Neositta*. *Emu,* Vol. 69, pp.169–174.

Martin, A.R.H. 1982. Proteaceae and the early differentiation of the central Australian flora. Pp.77–84. In Barker, W.R. and Greenslade, P.J.M. (eds). *Evolution of the Flora and Fauna of Arid Australia*. Peacock Publications, South including a description of a new species of *Progura* De Vis. *Transactions of the Royal Society of South Australia,* Vol. 98, pp.214–224.

Visser, W.A. and Hermes, J.J. 1962. Geological results of the exploration for oil in Netherlands New Guinea. In *Verhandelingen van het Koninklijk Nederlands geologisch-mijnbouwkundig genootschap*. Geologie ser. XX.

Voris, H.K. 2000. Maps of Pleistocene sea levels in Southeast Asia: shorelines, river systems and time durations. *Journal of Biogeography,* Vol. 27, pp.1153–1167.

Vuilleumier, F. 1975. Zoogeography. Pp.421–496. In Farner, D.S., King, J.R. and Parkes, K.C. (eds). *Avian Biology*. Vol. 5. Academic Press, New York.

Vuilleumier, F. and Kikkawa, J. 1991. Reconstructing the history of *Nothofagus* forest avifaunas. Pp.578–586. In Bell, B.D., Cossee, R.O., Flux, J.E.C., Heather, B.D., Hitchmough, R.A., Robertson, C.J.R. and Williams, M.J. (eds). *Acta XX Congressus Internationalis Ornithologici*, *Canterbury, 1990*. New Zealand Ornithological Trust Board, Wellington. 2568pp.

Wasson, R.J. 1982. Landform development in Australia. Pp.23–34. In Barker, W.R. and Greenslade, P.J.M. (eds). *Evolution of the Flora and Fauna of Arid Australia*. Peacock Publications, South Australia. 392pp.

White, M.E. 2006. Environments of the geological past. Pp.17–50. In Merrick, J.R., Archer, M., Hickey, G.M. and Lee, M.S.Y. (eds). *Evolution and Biogeography of Australasian Vertebrates*. Auscipub Pty. Ltd., Sydney. 942pp.

Chapter 23
Island Biogeography: as Illustrated by Birds in the Australasian Region

Hugh A. Ford
Zoology, University of New England, Armidale, N.S.W. 2351 Australia
E-mail: hford@metz.une.edu.au

BACKGROUND

Island biogeography deals with the distribution of animals and plants on islands. It is concerned with the origin of organisms on the islands, the numbers of species present as well as their evolution and ecology. Islands have fascinated biologists for at least 200 years, not only because they are remote and the animals and plants on them are often strange, but also because they represent simple ecosystems in which ecological and evolutionary problems can be tackled without the overwhelming complexity of continental ecosystems. Alfred Russell Wallace's extensive travels and writings on the Malay Archipelago provided an early comprehensive study of island fauna and flora (Wallace, 1869). The massive archipelago between South-east Asia and New Guinea now bears the name of Wallacea. Good accounts of island biogeography are given in MacArthur (1972), Gorman (1979) and Spellerberg *et al.* (1999).

Organisms on islands show classic examples of evolutionary change. Charles Darwin visited the Galapagos Islands in the early 19th century. At first, the common, drab finch-like birds that he saw did not make much of an impression and he didn't even keep specimens from different islands separate. Later though he realized that, whereas many of these small birds superficially looked different, they had similar calls and their basic plumage patterns were the same. In fact, they differed chiefly in size and in the size and shape of their bills. Darwin (1958, p. 328) realized that they must be closely related but had changed and diverged since reaching the islands. In his own words 'seeing this gradation and diversity of structure in one small, intimately related group of birds, one might really fancy that from an original paucity of birds in this archipelago, one species had been taken and modified for different ends'. In more modern language, one species had undergone an adaptive radiation into several species that occupied a variety of ecological niches.

Subsequent research has shown that Darwin was right. David Lack visited the Galapagos Islands before World War 2 and carried out the first detailed study of Darwin's finches. His classic book (*Darwin's Finches*) has been reprinted with extensive notes to update the information (Lack, 1983). More recently Peter Grant and his students have extended our knowledge of how Darwin's finches have evolved (Grant, 1986). Weiner (1994) gives a highly readable account of the finches and much more about evolution; see also the video by Parer and Parer-Cook (1999).

All of Darwin's finches are biochemically similar and have recently evolved from a mainland species that is still common on the neighbouring coast. However, they are ecologically very diverse (Figure 1). Several species, the ground finches *Geospiza*, have changed little since colonization. They eat seeds found on the ground and from plants. The tree finches *Camarhynchus* have less heavy beaks and take many insects as well as seeds. The warbler finch *Certhidea* has a fine beak and is entirely insectivorous, gleaning from leaves. The remarkable woodpecker finch (*Cactospiza pallida*) fills the niche of the absent woodpeckers. It has a relatively long beak like a woodpecker's, but lacks the long tongue that woodpeckers use to locate prey in holes in branches. Instead, the woodpecker finch uses a tool to locate such prey. Often it uses a cactus spine to probe into rotten wood to find burrowing insects, on other occasions it uses a small twig to prise beetles and larvae out of crevices in the bark. The most unusual source of food is exploited by the aptly named sharp-beaked or vampire finch, *Geospiza difficilis*. It is a blood sucker. These finches land on sea birds and peck vigorously at the base of their wing and tail feathers until they bleed. This behaviour probably arose from the finches first visiting sea birds and other animals to pick off ticks and other ectoparasites. Now they go straight for the primary source.

Isolation can give rise to bizarre animals, such as the giant tortoises of the Galapagos and Aldabra Islands, the Komodo dragon of Indonesia, or the giant extinct elephant birds of Madagascar. Although the island giants are best known (Worthy *et al.,* 1999), other animals become miniaturized on islands; for example, the islands of the Mediterranean not only had giant dormice and hedgehogs, but also pig-sized hippos and elephants one metre high (Davis, 1985). Plants too can grow to abnormal sizes, such as the daisies on Galapagos and New Zealand, which have evolved into trees. Another common adaptation is the loss of flight in birds and insects; for discussion of the role of loss of flight in the decline of New Zealand birds see pp.111–128 (Holdaway and Worthy, 2006).

Sadly, although these animals have evolved to cope with the unusual conditions of their island homes, they are poorly adapted to the direct or indirect effects of human settlement. Many of them have become extinct through

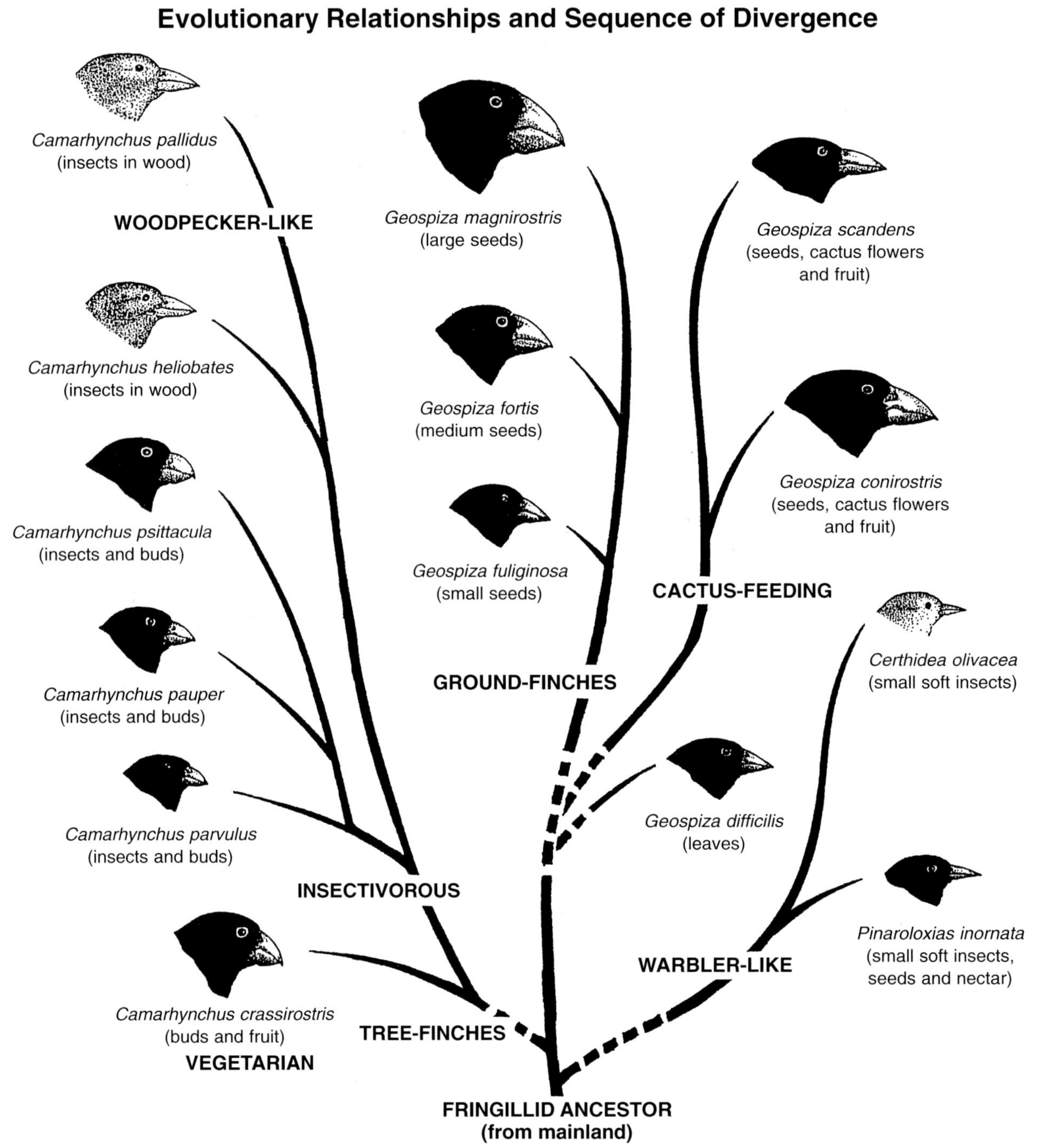

Figure 1. Suggested evolutionary relationships and sequence of divergence of Darwin's finches from a single colonizing stock; prime dietary items are indicated (in brackets) under species names. Based on Lack (1983).

hunting, loss of habitat or the impacts of introduced predators; many more species are threatened with extinction.

Extinction is a natural phenomenon, just as much as immigration and speciation are. Species disappear often from islands and are replaced by others. Occasionally, such disappearances mean that a species has gone globally extinct. However, human activities over the past few centuries have greatly increased the rate of extinction on islands. It is estimated that some 200 species or subspecies of birds have gone extinct since 1680, and 90% of these inhabited islands, including such famous species as the dodo (*Raphus cucullatus*) of Mauritius.

Collar *et al.* (1994) identified 1,111 species of birds (11% of the world's species) as being currently threatened with extinction. A further 10% are near-threatened or information is inadequate to decide whether they are threatened. A disproportionate number of these species occurs on islands. Indonesia has more threatened bird species (104) than any other country, narrowly beating Brazil (103 species). The Philippines are third with 86 species and although Australia is far behind, many of our threatened species are also found on small islands.

A number of reasons make island species especially prone to extinction. First, they often have small distributions and total population sizes. Habitat loss and degradation mean that population sizes soon become very small with random extinctions quite likely. Over half of the threatened bird species suffer from reduced or altered habitat. Second, many island species are highly specialized, and may have existed without predators, parasites or competitors for a long time. They may have lost the ability to fly and to respond to terrestrial predators. Introduced predators and herbivores can be devastating to such island specialists. The tameness of many New Zealand birds was probably a significant factor in their demise due to terrestrial predators (Holdaway and Worthy, 2006).

Studies on the fauna and flora of islands, and especially the relationship between the numbers of species on an island and its size, shape and location, have been applied to the conservation of nature reserves. This has led to altered insights into management, though there are many differences between reserves and real islands. More recently the field of landscape ecology has emerged from the disciplines of island biogeography and community ecology, and is providing useful guidelines for managing the small remnants of native vegetation in many regions.

In this chapter I shall give an outline of island biogeography, using examples from birds in the Australo-Pacific region. I shall also discuss how island biogeography theory has been applied to conservation in reserves in Australia.

ORIGINS OF ISLANDS

Islands form in a number of ways. Some emerge from beneath the sea due to volcanic activity, typically when a tectonic plate passes over a hot spot. These are called oceanic islands and have never been in contact with continental areas. Most of the Pacific Islands, such as the Galapagos and Hawaii, are volcanic in origin and some continue to have active volcanoes. Oceanic islands are lifeless at first, but gradually accumulate land plants and animals by dispersal across the sea.

Continental islands form when sea levels rise and isolate parts of the mainland, as happened in the case of Tasmania and the Bass Strait islands. These islands start life with a complex and diverse fauna and flora. Over time species go extinct, others colonize and new subspecies or even species evolve. Some endemics on continental islands are the surviving populations of species, originally more widespread, that have disappeared everywhere except the island; the Tasmanian devil (*Sarcophilus harrisii*) is a good example of this pattern.

A few of the larger and older islands do not fall conveniently into either category. These were formed from the fragmentation of ancient continents. The main islands of New Zealand broke away from the part of Gondwana that became Australia some 82 million years ago. Although they are continental in origin and a few remnants of the ancient Gondwanan plants and animals survive, most of the fauna and flora have reached New Zealand across the sea (see Holdaway and Worthy, 2006). Madagascar is another ancient land mass, separated from Africa over 150 m.y.a., retaining a diversity of primitive mammals, such as the lemurs. New Caledonia is a remnant of the Australian plate, with a distinctly Gondwanan flora and a few animals of ancient origin. One example is the kagu (*Rhynochetos jubatus*), a strange, flightless bird, which probes the forest floor for worms and snails (Létocart and Salas, 1997). It is the only member of its family and what its nearest relatives are we do not know. Like so many island endemics it is now endangered due to habitat loss and the activities of introduced predators.

New Guinea and many of the islands of Wallacea emerged as the Australian plate collided with the Southeast Asian region. Hall (1999) provides an excellent computer reconstruction of the appearance, movement and collision of islands in this region. Some islands, such as Sulawesi, result from the coalition of several islands of different origin, which has allowed the mixing and sorting of different biotas. See pp.71–93 (Heinsohn and Hope, 2006) and pp.95–110 (van Oosterzee, 2006).

ORGANISMS REACHING ISLANDS

Oceanic islands receive all of their land animals and plants by chance dispersal. Continental islands also receive potential colonists to add to the species already present. The most frequent method of dispersal in most groups is being carried by air currents. Many plants have very tiny seeds (such as orchids) or spores (ferns), that can be swept for long distances. Others have seeds attached to parachutes, which assist wind dispersal. The composites, such as thistles and dandelions, use this method of dispersal and have been very successful in reaching islands. Many insects fly, and others, as well as spiders, can be carried aloft in updrafts to be deposited far away.

Most birds can fly and islands are the recipients of vagrants of numerous species, though only a few successfully colonize. Surprisingly, the rails, which fly little during the day, have colonized almost all islands. Many of these have later evolved flightlessness. It is possible that many flying birds are carried unwillingly between islands. Stronger fliers, as David Lack once said, may use their wings to stay where they are! The loss of flight possibly reduces the risk of being blown off an island and also probably reduces the demand for energy and protein to be allocated to feathers, bone and muscle. The Lord Howe woodhen (*Gallirallus sylvestris*) is a large flightless rail, restricted to Lord Howe Island. Its numbers declined due to predation, but it is now recovering after protection and captive breeding. A related species may still survive in New Caledonia.

Arrival on a new island is of course only the first step. Successful colonization depends on survival and reproduction. The island may well have a very different habitat from the place of origin and important food or other resources may be absent. There may already be ecologically similar species present, which have become adapted to local conditions and able to prevent colonization by new species. Generalists, which can eat a wide variety of foods, typically establish more successfully on islands than do specialists, whose requirements cannot easily be met. A pair or group of individuals needs to arrive for breeding to be successful; the fact that birds often travel in flocks assists colonization. Small founder populations will contain only a limited range of the genetic variability in the parent population, and intense inbreeding may reduce the gene pool even further. So there is a high chance of extinction soon after colonization; probably only a tiny proportion of organisms that arrive on isolated islands becomes established. A few well-studied islands in Europe have received a steady rain of migrants and vagrants, though almost none colonizes. For the lucky few though, islands provide great opportunities, as some foods may be abundant and competitors or predators may be few.

More recently, habitat changes caused by humans have provided the opportunity for colonization by new species that may previously have arrived but been unable to overcome competition from species already there. Silvereyes (*Zosterops lateralis*) colonized New Zealand after the forests were opened up and native birds had succumbed to introduced predators. Several birds from South-east Asia have colonized the highland grasslands of New Guinea that result from forest clearance—see pp.71–93 (Heinsohn and Hope, 2006).

ISLAND BIOGEOGRAPHY THEORY

As more animals and plants colonize a new island, a biological community gradually forms. Island communities typically have fewer species than the mainland, because many specialists cannot survive. As they have to be less abundant than their prey there may only be a few predators; furthermore, there may only be a few prey species, reducing the opportunity for the predator to switch between targets when one species becomes rare.

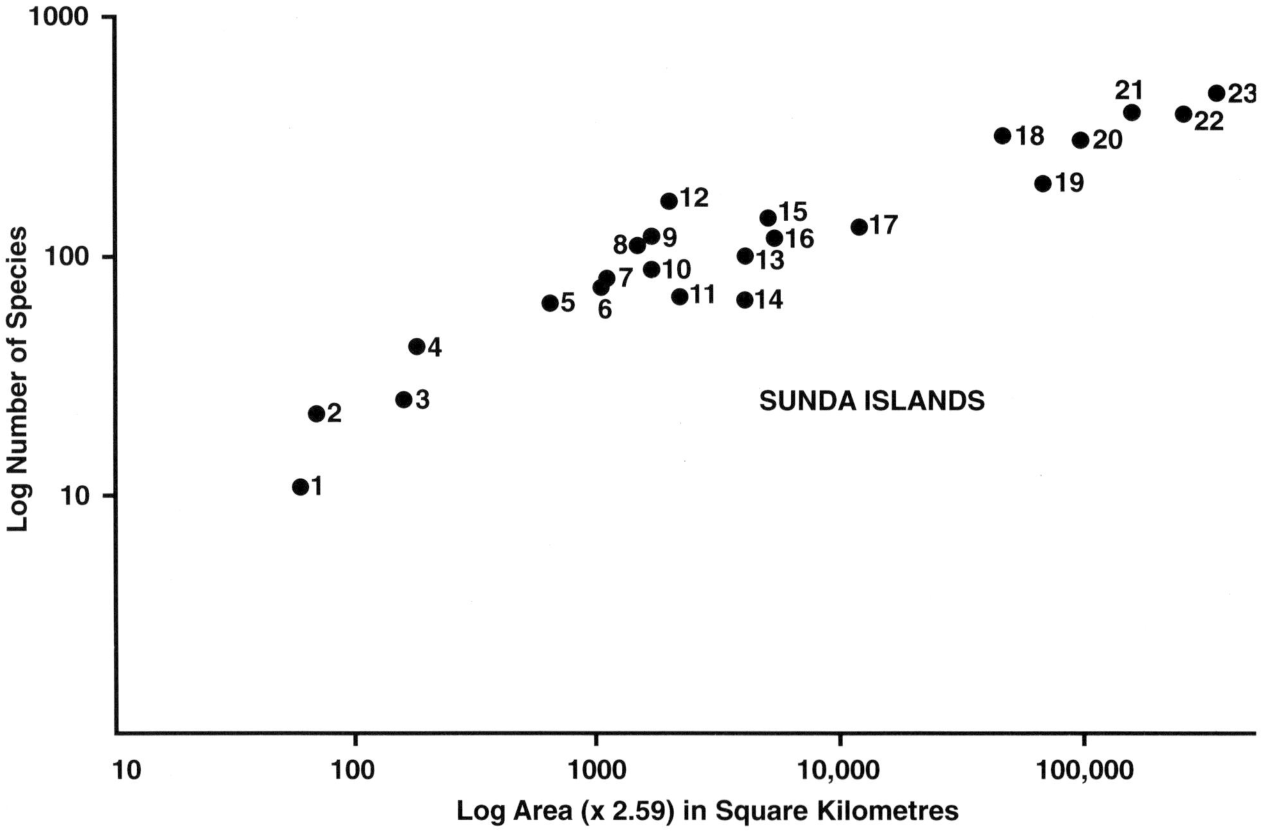

Figure 2. The relationship between the number of species of land and freshwater birds and individual island area in Indonesia, the Philippines and New Guinea. Key to island identification numbers: 1 = Christmas, 2 = Bawean, 3 = Engano, 4 = Savu, 5 = Simalur, 6 = Alor, 7 = Wetar, 8 = Nias, 9 = Lombok, 10 = Billiton, 11 = Mentawei, 12 = Bali, 13 = Sumba, 14 = Bangka, 15 = Flores, 16 = Sumbawa, 17 = Timor, 18 = Java, 19 = Sulawesi, 20 = Philippines, 21 = Sumatra, 22 = Borneo, 23 = New Guinea. From MacArthur and Wilson (1967).

SIZE, DEGREE OF ISOLATION, SPECIES EQUILIBRIUM

One of the most obvious relationships found on islands is the species-area relationship. In general, larger islands have more species of invertebrates, vertebrates and plants than smaller islands. In fact, the relationship between the logarithm of area and the logarithm of number of species is linear (Figure 2). As a simple rule of thumb, as the area increases tenfold the number of species doubles. This is partly a matter of straightforward sampling. A larger area has more individuals and even within mainland communities there is a direct relationship between the number of individuals and number of species.

There are two other main reasons why larger islands have more species. First, each species requires a certain minimum area to satisfy its requirements for survival. For instance, some small songbirds have a territory of only about one hectare whereas eagles have territories of several square kilometres; so increasing the area increases the number of species whose basic requirements are satisfied. Second, as the area increases there will be an increase in the number of habitats and this too will increase the chance that any particular species will find the island suitable. But despite the above observations, it is emphasized that there is continuing controversy about the relative importance of area and its exact relationship to taxa present.

In 1967 MacArthur and Wilson published a book entitled *The Theory of Island Biogeography*, which was to revolutionize the way in which we thought about species on islands. They proposed that the number of species of a particular taxon on an island was in equilibrium. This was because islands of a particular size and location had characteristic immigration and extinction rates. As the number of species increases, the number of potentially new immigrants will decline as most animals and plants arriving will already have colonized. At the same time, the chance of extinction will increase because, with more species, there will be fewer individuals of each species and there are more competitive and predatory interactions between the species.

Isolation, the distance to the nearest source of new species, is important. Islands close to the mainland, or aided by favourable winds and currents, receive a rain of animals and plants. Islands far offshore may be out of range for many forms that only survive for short periods. As one travels from New Guinea to Tahiti in the Pacific Ocean, the number of species in almost all groups declines rapidly. However, it is now recognized that the fauna of most oceanic islands had been drastically reduced before Europeans arrived (Steadman, 1995). So it is clear that the paucity of species on isolated islands and archipelagos has been overstated.

Larger islands will have lower extinction rates than small islands, because, on average, population sizes will be higher. Furthermore, larger islands may also have higher immigration rates than small islands; they represent larger targets and are more likely to have suitable and sufficient habitat. MacArthur and Wilson (1967) constructed a figure plotting immigration and extinction rates against number of species (Figure 3). Where the number of new species from immigration equals the number being lost through extinction, this leads to an equilibrium number of species. Larger and near islands should have more species as the immigration and extinction lines cross at a higher equilibrium number of species. Conversely, small and distant islands have their immigration/extinction curves crossing at a lower number of species.

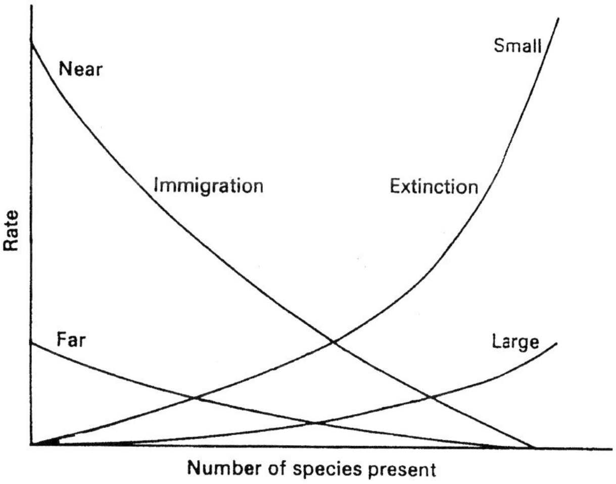

Figure 3. Relationship between immigration and extinction rates and number of species on large and small islands that are near to, or distant from, continental areas. Adapted from MacArthur and Wilson (1967).

It is important to realize that the equilibrial number of species on an island is not a static one but a dynamic one. Even though the number of species may stay more or less the same for a period of time, the actual species present may change. This is usually called species turnover. The evidence that island communities are in equilibrium is in fact not very strong. Remote oceanic islands may change very slowly, but those nearer to continents frequently show increases or decreases in the number of species. For instance, islands in the Åland Archipelago of Finland have been colonized by 8–18 new species of birds since the 1920s, whereas only three have become extinct. Even more remote islands off New Zealand and southern Australia have shown marked changes in the number of bird species that occupy them (Abbott and Grant, 1976). Unfortunately, much of this apparent lack of equilibrium has been associated with human activities; which have caused the extinction of endemic fauna, the introduction of exotic species and habitat destruction or alteration. We know little about the dynamics of species composition on islands that are not affected by humans.

One way to test whether the community is in equilibrium is to alter the number of species and see if it returns to the same level. An extreme version of this is to demolish all the species! This can only be done by divine the centre of the Pacific. These islands have no native terrestrial mammals, and currently have a depauperate bird fauna, at least partly due to their extreme isolation. However, a variety of hawks, pigeons, parrots, flycatchers, honeyeaters and other passerines have managed to reach these islands. Europeans in the late 18th century brought 'civilization', along with rats, mice, cattle, sheep, goats, horses, dogs, cats as well as venereal diseases, leprosy, tuberculosis, alcohol and opium! The former decimated the remaining native birds, whereas the latter were devastating to the human population. Six species of birds collected in early expeditions have not been seen since. Twenty of the remaining species are very rare and endangered.

In recent decades it has become clear that the impact of Europeans on the avifauna of the Pacific islands pales into insignificance compared with that of earlier settlers (Steadman, 1989, 1995, 1997; James, 1995; MacPhee, 1999; White, 2002). Humans settled Polynesia from about 3,000 to 1,000 years ago (see Roberts, 1998), brought pigs, rats and chickens with them, and removed and altered the forests. Subfossil remains indicate quite a rich pre-human avifauna, especially of rails, pigeons, doves, parrots and passerines. Many of these disappeared within a few centuries of human settlement. For instance, of 27 species of land bird present on Tonga before humans arrived, only five remain. The most extreme case of extinction was on Easter Island, which now has no indigenous land birds, but had at least six species before humans arrived about 1,500 years ago and destroyed the forests. There is little evidence of past extinctions on the Galapagos Islands, which were not settled until less than 200 years ago. Similar patterns of high extinction to those in Polynesia apparently occurred in Micronesia and Melanesia. The scale of extinction was immense, considering the number of islands and the degree of endemism. Steadman (1995) suggested that perhaps 8,000 species or populations of land birds have been lost, with possibly 2,000 species of rails. Not only is this a devastating decline in species diversity, but also the longer-term ecological effects may be profound. Many extinct species were important pollinators and seed dispersers, so that the whole forest community on most islands is now threatened.

HAWAII

The amazing adaptive radiation of the Hawaiian honeycreepers has already been described (Figure 8). However, a range of other land birds, including flycatchers, thrushes and honeyeaters reached this most isolated of the world's archipelagos (James and Olson, 1991; Olson and James, 1991). Of 84 endemic species of birds, 44 went extinct after colonization by humans (Scott *et al.*, 1988; MacPhee, 1999). They are known only from subfossils. The 29 of the remainder currently threatened, include sea birds, a goose, ducks, a hawk, a crow, and 14 of the honeycreepers. The last individual kaui oo (*Moho braccatus*), a honeyeater, probably died a few years ago. The major causes are habitat loss and modification, but also avian malaria, which is carried by the night mosquito. This disease was introduced by exotic birds (Collar *et al.,* 1994) and has virtually wiped out all native birds in habitats below 600 m above sea level.

APPLYING ISLAND BIOGEOGRAPHY THEORY TO CONSERVATION

MacArthur and Wilson (1967) stated that the patterns and processes of island biogeography could apply also to isolated patches of habitat. These could occur naturally, for instance in the case of alpine habitats on mountaintops. They could also result from human activity, in particular the clearing of natural vegetation. In many parts of the world, our nature reserves and national parks are effectively islands of habitat in a sea of farmland or suburbs. They behave more like islands the more that they become isolated from other natural areas and the more that their area is eroded. We would predict that immigration rates should fall and extinction rates should rise. Consequently, the number of species should gradually decline to a new equilibrium.

RESERVE SIZE, SHAPE AND DISTRIBUTION

Diamond (1975b) as well as Wilson and Willis (1975) not only extended the theory of island biogeography to conservation, but also predicted which sorts of reserves should be most effective. Obviously, larger reserves are better. A close relationship between area and the number of species is shown by bird species in small patches of eucalypt forests near Walcha and rainforests near Dorrigo, in New South Wales (Howe *et al.,* 1981; Howe, 1984)—see Figure 9. The same is true of birds, mammals and lizards from reserves in the wheatbelt of Western Australia (Kitchener, 1982).

In fact, very large reserves may be required to be sure of maintaining viable populations of species with large home ranges, such as top predators (Belovsky, 1987). An area of 10,000 km^2 has been suggested for tigers (*Panthera tigris*), grizzly bears (*Ursus horribilis*) and red goshawks (*Erythrotriorchis radiatus*). Large areas are also needed to include the habitats of specialists. Not only do larger reserves have more species, but they also tend to retain their species for longer. A study of mammal species in the United States found that proportionally more species had been lost from smaller reserves (Newmark, 1987). In fact, only the largest reserve (of over 20,000 km^2) had retained its entire mammal fauna. The loss of bird species has been noted in many isolated patches of native vegetation in Australia. For instance, Traill *et al.* (1996) described the loss of the crested bellbird (*Oreoica gutturalis*) from a 4,200 ha remnant in Victoria. In highly fragmented landscapes, such local losses can become

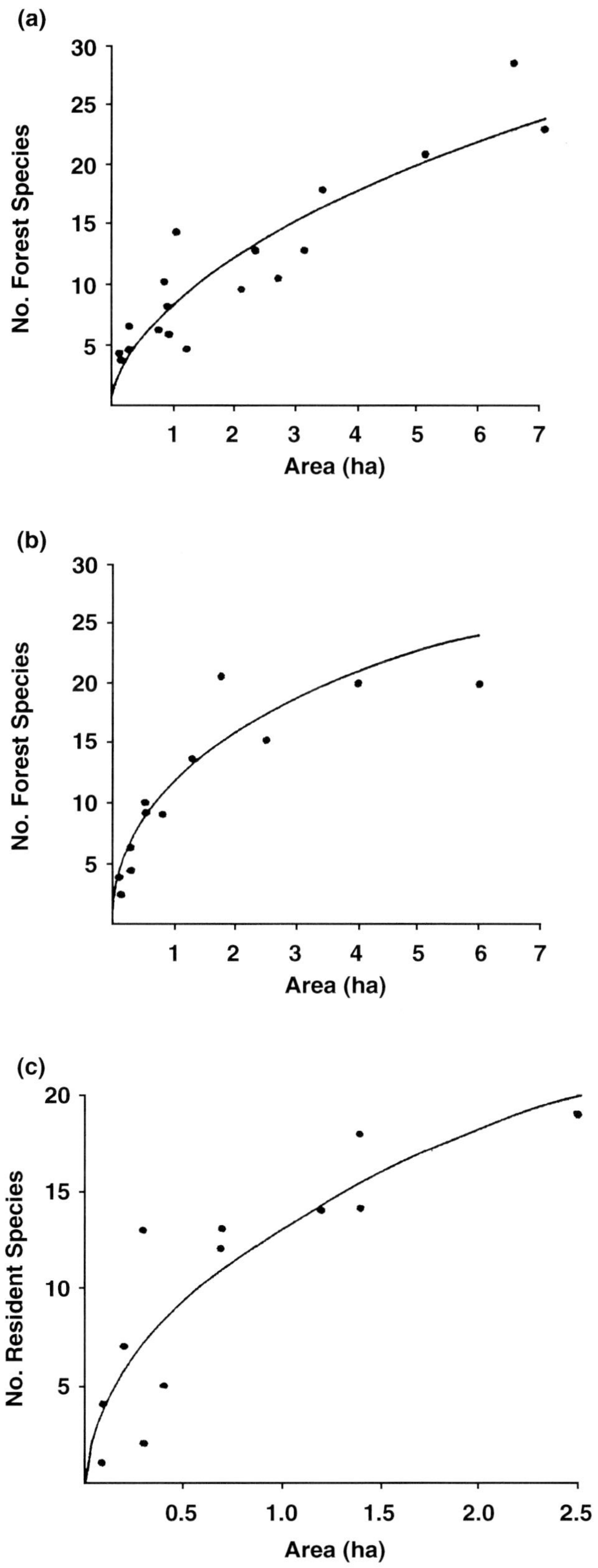

Figure 9. Numbers of bird species in: **(a)** 'islands' of eucalypt forest; **(b)** control plots in continuous eucalypt forest; and **(c)** rainforest 'islands', in north-eastern New South Wales. From Ford (1989).

regional losses as Saunders (1989) described for the Western Australian wheatbelt. The proposal then that the larger the reserve the better is now accepted by all.

Wilson and Willis (1975) made several further suggestions (see Figure 10):

(1) one large reserve is better than several small reserves of the same total area (Figure 10(a));
(2) reserves should be circular to maximize the core area and minimize the edge, which often contains different species and provides a source for alien species and other forms of degradation (Figure 10(b));
(3) where there are several reserves these should be as close together as possible (Figure 10(c));
(4) reserves should be linked with corridors (Figure 10(d)).

The last three suggestions have been generally accepted, but the first was more controversial. This led to the SLOSS (Single Large or Several Small) debate. Simberloff and Abele (1984) argued that there was no theoretical reason why one large reserve should hold more species, in fact the available data indicated that the opposite was often true. For instance, Kitchener *et al.* (1980) found a large number of lizard species in a collection of small patches of scrub in the wheatbelt of Western Australia. This is because most regions are heterogeneous and species of lizards frequently have specific habitat requirements and consequently restricted distributions. The numerous small reserves had sampled a wide variety of the available habitat. For a single reserve to include the full range of habitats and lizard species, it would have had to occupy an area 600 times that of the present reserves. But lizards are small and have rather modest energy requirements, as they are ectothermic. Far larger areas would probably be required to sustain viable populations of birds.

INFLUENCES ON SPECIES DECLINE

Data on the numbers of species in reserves may be misleading, because it may take a long time for a new equilibrium on an island to be reached. Diamond (1975a) showed that a group of islands off New Guinea, which had only become isolated by recent rises in sea level, have more species of birds than would be predicted by their area alone. He argued that the diversity on these islands is in the process of declining to a new equilibrium. The North American reserves mentioned earlier that had lost mammal species had mostly been dedicated 70 to 90 years ago, but extinctions are still taking place.

Ford and Howe (1980) looked at the number of species of birds in reserves in the Mount Lofty Ranges in South Australia. These ranges contain an isolated patch of eucalypt forest, cut off from more continuous forest in south-eastern Australia. After Europeans arrived much of the forest was cleared. Only about 10% remains and only about 2% of the original forest is now conserved, mostly in parks of 1,000 ha or less. The avifauna is well known both for the Mount Lofty Ranges as a whole and for individual parks. The number of birds that originally bred in be expensive relative to their effectiveness, and money may be better spent in setting up new reserves, improving the quality of intervening habitat generally or even by physically moving animals between reserves (Simberloff *et al.*, 1992). In the Australian landscape corridors typically consist of a few rows of trees along roadsides. Although these attract birds, they also increase their risk of death through collision with vehicles (Andrews, 1990). Furthermore, narrow strips of vegetation may be inhabited by the aggressive noisy miner (*Manorina melanocephala*), which may inhibit movement by dispersing birds (Catterall *et al.*, 1991). Major *et al.* (1999a) found fewer red-capped robins (*Petroica goodenovii*) in linear strips than in large remnants, and more of the latter were young birds, which suggests that strips were less preferred habitat. In addition, there were more nest predators in linear strips, and higher nest predation rates (Major *et al.*, 1999b).

CONCLUSIONS

Islands in the Australasian region consist of land-bridge islands, such as Tasmania, which were recently attached to the mainland, and oceanic islands, as in Polynesia, that have never been connected to continents. A few islands, such as New Zealand and New Caledonia, are fragments from ancient continents.

Birds have been excellent colonizers of even the most isolated islands, where they frequently evolved into new species. Evolution to the level of endemic species is often rapid and has occurred in Tasmania despite its recent connection with the mainland. Speciation has been most prolific on the Hawaiian Islands, with the adaptive radiation of the honeycreepers.

The number of species on an island increases through immigration (and in the long term through evolution) and decreases due to extinction. MacArthur and Wilson's theory of island biogeography suggests that a dynamic equilibrium is eventually reached between these processes. Interspecific competition and predation play a role in determining the composition and structure of the island community.

Bird species on islands are very prone to extinction after the arrival of humans, with their attendant feral mammals. Catastrophic extinctions of birds occurred throughout the Pacific as each island group was colonized. European settlement has extended this pattern of losses with the majority of recently extinct and currently endangered bird species occurring on islands.

Extending MacArthur and Wilson's theory of island biogeography to nature reserves has led to some useful insights, but also to controversy. Some have argued that it has provided little of value; however, it has served to: (a) widen the focus to all of the species rather than just a few threatened ones; (b) highlight the fact that dedication of nature reserves and management by benign neglect is inadequate. Active protection/intervention in parks (or other reserves) is clearly required if all species are to be retained; (c) emphasize the need for large parks; (d) indicate the value of corridors.

We have now moved beyond a close adherence to the theory of island biogeography and recognized the need to:

(1) consider the whole landscape; not just focus on the remnants;
(2) increase the representativeness of reserves as well as increasing the area of existing parks;
(3) address off-park conservation, which may be more beneficial to a wide range of species than extending parks;
(4) evaluate the cost of providing corridors against other measures (e. g. incentives to farmers to protect existing patches of vegetation);
(5) integrate autecological (individual species) studies, with studies on ecological processes at the level of the ecological community.

REFERENCES

Abbott, I. 1973. Birds of Bass Strait. Evolution and ecology of the avifaunas of some Bass Strait islands, and comparisons with those of Tasmania and Victoria. *Proceedings of the Royal Society of Victoria,* Vol. 85, pp.197–223.

Abbott, I. 1974. The avifauna of Kangaroo Island and causes of its impoverishment. *Emu,* Vol. 74, pp.124–134.

Abbott, I. and Grant, P.R. 1976. Nonequilibrial bird faunas on islands. *American Naturalist,* Vol. 110, pp.507–528.

Andrews, A. 1990. Fragmentation of habitat by roads and utility corridors: a review. *Australian Zoologist*, Vol. 26, pp.130–141.

Baird, R.F. 1991. Holocene avian assemblage from Skull Cave (AU-8), south-western Western Australia. *Records of the Western Australian Museum*, Vol. 15, pp.267–286.

Baird, R.F. 1993. Pleistocene avian fossils from Pyramid Cave (M-89), eastern Victoria, Australia. *Alcheringa*, Vol. 17, pp.383–404.

Beier, P. and Noss, R.F. 1998. Do habitat corridors provide connectivity? *Conservation Biology*, Vol. 12, pp.1241–1252.

Belovsky, G.E. 1987. Extinction models and mammalian persistence. Pp.35–58. In Soulé, M.E. (ed). *Viable Populations for Conservation.* Cambridge University Press, Cambridge. 189pp.

Butler, D. and Merton, D. 1992. *The Black Robin: Saving the World's Most Endangered Bird.* Oxford University Press, Auckland. 294pp.

Catterall, C.P., Green, R.J. and Jones, D.N. 1991. Habitat use by birds across a forest suburb interface in Brisbane: implications for corridors. In Saunders, D.A. and Hobbs, R.J. (eds). *Nature Conservation 2: The Role of Corridors.* Surrey Beatty and Sons, Sydney. 442pp.

Collar, N.J., Crosby, M.J. and Stattersfield, A.J. 1994. *Birds to Watch 2. The World List of Threatened Birds.* BirdLife International, Cambridge. 407pp.

Connor, E.F. and Simberloff, D. 1986. Competition, scientific method and null models in ecology. *American Scientist,* Vol. 74, pp.155–162.

Darwin, C. 1958. *The Voyage of the Beagle*. Bantam Books, New York.

Davis, S. 1985. Tiny elephants and giant mice. *New Scientist,* 3 January, pp.25–27.

Diamond, J.M. 1975a. Assembly of species communities. Pp.342–444. In Cody, M.L. and Diamond, J.M. (eds). *Ecology and Evolution of Communities.* Harvard University Press, Cambridge, Massachusetts. 545pp.

Diamond, J.M. 1975b. The island dilemma: lessons of modern biogeographic studies for the design of nature reserves. *Biological Conservation*, Vol. 7, pp.129–146.

Diamond, J.M. and Gilpin, M.E. 1982. Examination of the 'null' model of Connor and Simberloff for species co-occurrences on islands. *Oecologia,* Vol. 52, pp.64–74.

Ford, H.A. 1989. *The Ecology of Birds: an Australian Perspective*. Surrey Beatty and Sons, Sydney. 288pp.

Ford, H.A., Barrett, G.W. and Recher, H.F. 1995. Birds in a degraded landscape—safety nets for capturing regional biodiversity. Pp.43–50. In Saunders, D.A., Craig, J.L. and Mattiske, E.M. (eds). *Nature Conservation: the Role of Networks.* Surrey Beatty and Sons, Sydney. 684pp.

Ford, H.A., Barrett, G.W., Saunders, D.A. and Recher, H.F. 2001. Why have birds in the woodlands of southern Australia declined? *Biological Conservation,* Vol. 97, pp.71–88.

Ford, H.A. and Howe, R.W. 1980. The future of birds in the Mt Lofty Ranges. *South Australian Ornithologist,* Vol. 28, pp.85–89.

Ford, H.A. and Paton, D.C. 1975. Impoverishment of the avifauna of Kangaroo Island. *Emu,* Vol. 75, pp.155–156.

Garnett, S.T. and Crowley, G.M. 2000. *The Action Plan for Australian Birds 2000.* Environment Australia, Canberra. 673pp.

Gilpin, M.E. and Diamond, J.M. 1982. Factors contributing to non-randomness in species co-occurrences on islands. *Oecologia,* Vol. 52, pp.75–84.

Gorman, M.L. 1979. *Island Ecology.* Chapman and Hall, London. 79pp.

Grant, P.R. 1986. *Ecology and Evolution of Darwin's Finches*. Princeton University Press, Princeton, New Jersey. 458pp.

Hall, R. 1999. http://www.gl.rhbnc.ac.uk/seasia/welcome.html.

Heinsohn, T. and Hope, G. 2006. The Torresian connections: zoogeography of New Guinea. Pp.71–93. In Merrick, J.R., Archer, M., Hickey, G.M. and Lee, M.S.Y. (eds). *Evolution and Biogeography of Australasian Vertebrates.* Auscipub Pty. Ltd., Sydney, 942pp.

Holdaway, R.N. 1999. Introduced predators and avifaunal extinction in New Zealand. Pp.189–238. In MacPhee, R.D.E. (ed). *Extinctions in Near Time: Causes, Contexts, and Consequences.* Kluwer Academic/Plenum, New York. 394pp.

Holdaway, R.N and Worthy, T.H. 2006. Evolution of New Zealand and its vertebrates. Pp.111–128. In Merrick, J.R., Archer, M., Hickey, G.M. and Lee, M.S.Y. (eds). *Evolution and Biogeography of Australasian Vertebrates*. Auscipub Pty. Ltd., Sydney. 942pp.

Howe, R.W. 1984. Local dynamics of bird assemblages in small forest habitat islands in Australia and North America. *Ecology*, Vol. 65, pp.1585–1601.

Howe, R.W., Howe, T.D. and Ford, H.A. 1981. Bird distributions on small rainforest remnants in New South Wales. *Australian Wildlife Res*earch, Vol. 8, pp.637–653.

James, H.F. 1995. Prehistoric extinctions and ecological changes on oceanic islands. *Ecological Studies,* Vol. 115, pp.87–102.

James, H.F. and Olson, S.L. 1991. Descriptions of thirty-two new species of birds from the Hawaiian Islands. Part II. Passerines. *Ornithological Monographs,* No. 46, pp.1–88.

Jones, D.N., Dekker, R.W.R.J. and Roselaar, C.S. 1995. *The Megapodes: Megapodiidae.* Oxford University Press, Oxford. 262pp.

Kakapo. 2000. http//:www.kakaporecovery.org.nz/index1.html.

Keast, J.A. 1968. Competitive interactions and the evolution of ecological niches as illustrated by the honeyeater genus *Melithreptus* (Meliphagidae). *Evolution,* Vol. 22, pp.762–784.

Kitchener, D.J. 1982. Predictors of vertebrate species richness in nature reserves in the Western Australian wheat belt. *Australian Wildlife Res*earch, Vol. 9, pp.1–7.

Kitchener, D.J., Chapman, A., Dell, J., Muir, B.G. and Palmer, M. 1980. Lizard assemblage and reserve size and structure in the Western Australian wheatbelt—some implications for conservation. *Biological Conservation,* Vol. 17, pp.25–62.

Lack, D. 1983. *Darwin's Finches.* Cambridge University Press, Cambridge. 208pp.

Lambeck, R.J. 1997. Focal species: a multi-species umbrella for nature conservation. *Conservation Biology,* Vol. 11, pp.849–856.

Létocart, Y. and Salas, M. 1997. Spatial organisation and breeding of Kagu *Rhynochetos jubatus* in Rivière Bleue Park, New Caledonia. *Emu*, Vol. 97, pp.97–107.

MacArthur, R.H. 1972. *Geographical Ecology. Patterns in the Distribution of Species.* Harper and Row, New York. 269pp.

MacArthur, R.H. and Wilson, E.O. 1967. *The Theory of Island Biogeography.* Princeton University Press, Princeton, New Jersey. 203pp.

MacPhee, R.D.E. (ed). 1999. *Extinction in Near Time: Causes, Contexts, and Consequences.* Kluwer Academic/Plenum, New York. 394pp.

Major, R.E., Christie, F.J., Gowing, G. and Ivison, T.J. 1999a. Age structure and density of red-capped robin populations vary with habitat size and shape. *Journal of Applied Ecology,* Vol. 36, pp.901–908.

Major, R.E., Christie, F.J., Gowing, G. and Ivison, T.J. 1999b. Elevated rates of predation on artificial nests in linear strips of habitat. *Journal of Field Ornithology,* Vol. 70, pp.351–364.

Mayr, E. 1964. *Systematics and the Origin of Species*. Dover Publications, New York. 334pp.

McIntyre, S. and Barrett, G.W. 1992. Habitat variegation, an alternative to fragmentation. *Conservation Biology*, Vol. 6, pp.146–147.

Morris, R. and Smith, H. 1988. *Wild South: Saving New Zealand's Endangered Birds*. TVNZ, Auckland.

Moulton, M.P. and Pimm, S.L. 1987. Morphological assortment in introduced Hawaiian passerines. *Evolutionary Ecology,* Vol. 1, pp.113–124.

Newmark, W.D. 1987. A land-bridge island perspective on mammalian extinctions in western North American parks. *Nature,* Vol. 325, pp.430–432.

Olsen, P. 2000. A time to choose: setting conservation priorities. *Wingspan,* Vol. 10, No. 4, pp.18–24.

Olson, S.L. 1975. The fossil rails of C.W. De Vis, being mainly an extinct form of *Tribonyx mortierii* from Queensland. *Emu,* Vol. 75, pp.49–54.

Olson, S.L. and James, H.F. 1991. Descriptions of thirty-two new species of birds from the Hawaiian Islands: Part I. Non-Passeriformes. *Ornithological Monographs,* No. 45, pp.1–88.

Parer, D. and Parer-Cook, E. 1999. Islands of the Vampire Birds. Video, Australian Broadcasting Commission, Sydney.

Possingham, H. 2000. The Extinction Debt: the future of birds in the Mount Lofty Ranges. *Environment SA,* Vol. 8, p.10.

Raikow, R.J. 1976. The origin and evolution of the Hawaiian honeycreepers (Drepanididae). *Living Bird,* Vol. 15, pp.95–117.

Roberts, N. 1998. *The Holocene. An Environmental History.* Second Edition. Blackwell Publishers Ltd., Oxford. 316pp.

Rohde, K. 1992. Latitudinal gradients in species diversity: the search for the primary cause. *Oikos,* Vol. 65, pp.514–527.

Rosenzweig, M. 1995. *Species Diversity in Space and Time.* Cambridge University Press, Cambridge. 436pp.

Saunders, D.A. 1989. Changes in the avifauna of a region, district and remnant as a result of fragmentation of native vegetation: the wheatbelt of Western Australia. A case study. *Biological Conservation,* Vol. 50, pp.99–135.

Saunders, D.A. and Hobbs, R.J. (eds). 1991. *Nature Conservation 2: the Role of Corridors.* Surrey Beatty and Sons, Sydney. 442pp.

Scott, J.M., Kepler, C.B., van Riper, C. and Fefer, S.I. 1988. Conservation of Hawaii's vanishing avifauna. *BioScience,* Vol. 38, pp.238–253.

Simberloff, D. and Abele, L.G. 1984. Conservation and obfuscation: subdivision of reserves. *Oikos,* Vol. 42, pp.339–401.

Simberloff, D.S., Farr, J.A., Cox, J. and Mehlman, D.W. 1992. Movement corridors: conservation bargains or poor investments? *Conservation Biology*, Vol. 6, pp.493–504.

Spellerberg, I.P., Sawyer, J.W.D. and Whitten, T. 1999. *An Introduction to Applied Biogeography.* Cambridge University Press, New York. 257pp.

Steadman, D.W. 1989. Extinction of birds in eastern Polynesia: a review of the record, and comparisons with other Pacific Island groups. *Journal of Archaeological Science,* Vol. 16, pp.177–205.

Steadman, D.W. 1995. Prehistoric extinctions of Pacific Island birds: biodiversity meets zooarchaeology. *Science,* Vol. 267, pp.1123–1131.

Steadman, D.W. 1997. Extinction of Polynesian birds: reciprocal impacts of birds and people. Pp.51–79. In Kirch, P. V. and Hunt, T.L. (eds). *Historical Ecology in the Pacific Islands: Prehistoric Environmental and Landscape Change.* Yale University Press, New Haven. 331pp.

Thornton, I.W.B., Zann, R.A., Rawlinson, P.A., Tidemann, C.R., Adikerana, A.S. and Widjoya, A.H.T. 1988. Colonization of the Krakatau islands by vertebrates and possible delayed extinction. *Proceedings of the National Academy of Sciences of the United States of America,* Vol. 85, pp.515–518.

Traill, B.J., Collins, E., Peake, P. and Jessup, S. 1996. Current and past status of birds of Chiltern—a box-ironbark forest in north-eastern Victoria. *Australian Bird Watcher,* Vol. 16, pp.309–326.

van Oosterzee, P. 2006. Drawing the Wallace Line. Pp.95–110. In Merrick, J.R., Archer, M., Hickey, G.M. and Lee, M.S.Y. (eds). *Evolution and Biogeography of Australasian Vertebrates.* Auscipub Pty. Ltd., Sydney. 942pp.

Wallace, A.R. 1869. *The Malay Archipelago.* Dover, New York.

Walpole, S. 1999. Assessment of the economic and ecological impacts of remnant vegetation on pasture productivity. *Pacific Conservation Biology,* Vol. 5, pp.28–35.

Weiner, J. 1994. *The Beak of the Finch: a Story of Evolution in Our Time.* Knopf, New York. 332pp.

White, P. 2002. The human creation of Pacific faunal patterns. *Linnean Society of New South Wales Newsletter,* No. 103, pp.4–6.

Whittaker, R.J. 1995. Disturbed island ecology. *TREE,* Vol. 10, pp.421–425.

Wiens, J.A. 1989. *The Ecology of Bird Communities. Volume 1. Foundations and Patterns.* Cambridge University Press, Cambridge. 539pp.

Wilson, E.O. 1961. The nature of the taxon cycle in the Melanesian ant fauna. *American Naturalist,* Vol. 95, pp.169–193.

Wilson, E.O. and Willis, E.O. 1975. Applied biogeography. Pp.522–534. In Cody, M.L. and Diamond, J.M. (eds). *Ecology and Evolution of Communities.* Belknap Press, Cambridge, Massachussetts. 545pp.

Worthy, T., Anderson, A.J. and Molnar, R.E. 1999. Megafaunal expression in a land without mammals—the first fossil faunas from terrestrial deposits in Fiji (Vertebrata: Amphibia, Reptilia, Aves). *Senckenbergiana Biologica,* Vol. 79, pp.237–242.

Chapter 24
Key to the Families of Australasian Birds

Walter E. Boles

Division of Vertebrate Zoology, Australian Museum, 6 College Street, Sydney. N.S.W. 2010 Australia and School of Biological, Earth and Environmental Sciences, University of New South Wales N.S.W. 2052 Australia
E-mail: walterb@austmus.gov.au

INTRODUCTION

This key has been designed to allow identification of Australasian birds to family level for non-passerines and ordinal level for passerines. It is intended to work with fully grown adult birds only, which have not lost key feathers through moult or have abnormalities. External characters employed are those that can be examined in the hand on a live bird or museum study skin. While these should permit separation of Australasian taxa, they are not guaranteed for extralimital members of higher taxa.

Australasia is here used to encompass Australia (including its external territories), New Zealand and New Guinea. The taxonomic coverage is all families of non-passerine birds (including introduced ones) and the order Passeriformes. The ordinal and familial classifications of Australian birds follow Christidis and Boles (1994), except for changes indicated by recent systematic studies: the recognition of two orders of large ratites, rather than one, i.e. the Struthioniformes (ostriches) and Casuariiformes (cassowaries, emus) are kept separate; and the recognition of family Eurostopodidae (eared nightjars) separate from the Caprimulgidae (regular nightjars) (Sibley and Ahlquist, 1990; Mariaux and Braun, 1996). The key also includes several taxa that do not occur in Australia, but that are found elsewhere in Australasia. These comprise one order (Apterygiformes) and three families (Apterygidae—kiwis; Hemiprocnidae—whiskered swifts; Bucerotidae—hornbills).

KEY TO FAMILIES

Important terms are defined and characters described and illustrated throughout the key. These are indicated in the couplets by italics and a letter reference to the appropriate illustration.

1. Wings modified as *flippers* [A] .... Spheniscidae: Sphenisciformes (penguins)
   Wings not modified as flippers 2

   [A] Flipper: forelimb modified for swimming, characterized by being broad and flat, and having the joints rigid, permitting little movement; feathering scale-like in appearance

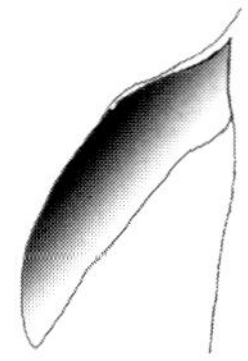

2. *Hallux* [B] present 3
   Hallux absent 55

   [B] Hallux: the hind toe or digit 1 or toe I; may be large or reduced, elevated or level with other toes, functional or vestigial

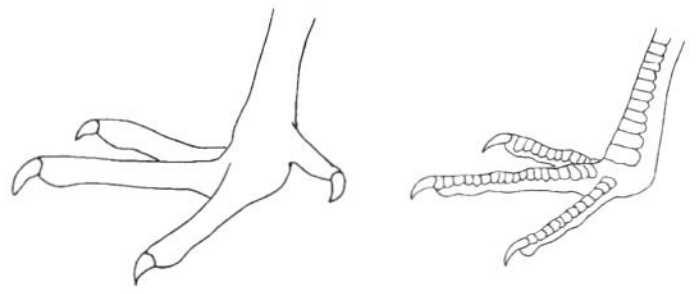

3. *Wings vestigial* [C]; *aftershaft* [D] almost as long as contour feather; upper *mandible with 'segments'* [E] .... Apterygidae: Apterygiformes (kiwis)
   Wings developed; aftershaft short relative to contour feather; upper mandible lacking 'segments' 4

   [C] Wings vestigial: forelimbs greatly reduced in size, or absent, and no longer capable of flight

[HH] Claw straight: weakly arched on top and bottom surfaces

Claw down-curved: claw noticeably arched on both top and bottom surfaces

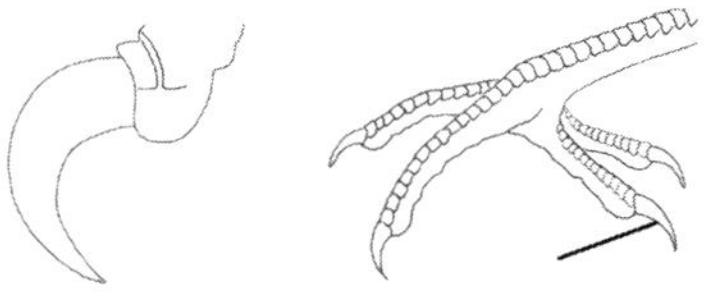

22. Foot fully *palmate* [II] 23
Foot not fully palmate (at most with brief basal webs only) 26

[II] Palmate: having the toes joined by webbing; fully palmate with webs extending to end of toes (cf. semipalmate [J]); usually front three toes joined, as in families below, but sometimes all four (totipalmate [F])

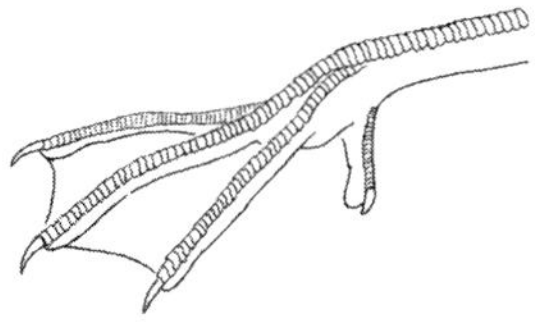

23. Distal half of *bill bent* abruptly downwards [JJ] .... Phoenicopteridae: Phoenicopteriformes (flamingos)
Bill straight or up-curved 24

[JJ] Down-bent mandibles: bill bent just anterior to nostrils, with most of bill directed ventrally (downwards)

24. Bill *recurved* [KK] and *dorsoventrally thin* [KK] .... Recurvirostridae (part): Charadriiformes (avocets)
Bill straight and upper mandible broad or *bilaterally compressed* [SSS] 25

[KK] Recurved: Distal portion of bill curved dorsally (upwards)
Dorsoventrally thin: compressed top to bottom

25. Bill deeper than broad; lower mandible not flattened .... Laridae: Charadriiformes (gulls, terns, skuas)
Bill not deeper than broad (breadth equal to or greater than depth);
lower mandible *dorsoventrally flattened* [LL] .... Anatidae: Anseriformes (ducks, geese, swans)

[LL] Dorsoventrally flattened: much broader from side to side than deep

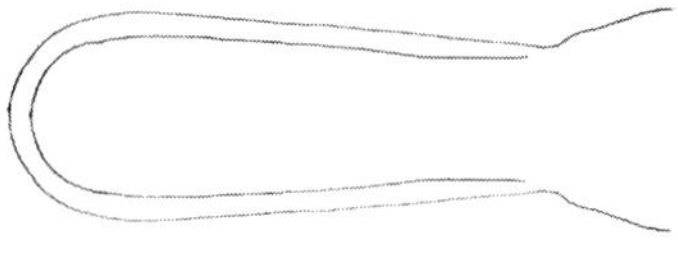

26. *Hallux functional* [MM] 27
Hallux vestigial [NN] 47

[MM] Hallux functional: length >60% of that of middle toe, including claw; able to grasp

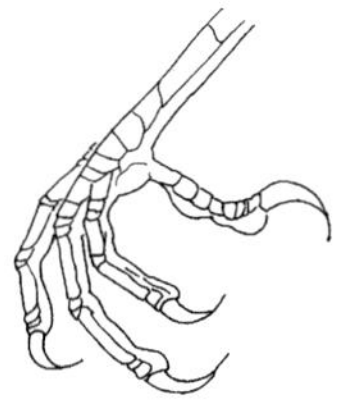

[NN] Hallux vestigial: length <50% of that of middle toe, including claw; unable to grasp

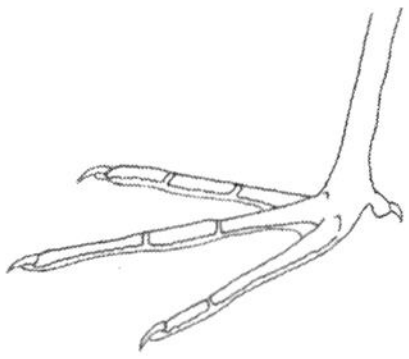

27. Bill with fleshy *cere* [DD] .... Columbidae: Columbiformes (pigeons, doves)
Bill without fleshy cere 28

28. Bill very broad and short; *commissural point* [OO] extended below eye 29
Bill rather long and often pointed; commissural point in front of eye 34

[OO] Commissural point: point at which the upper and lower mandibles meet

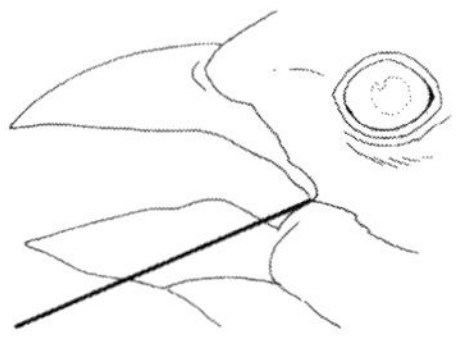

29. Claws not strongly curved 30
Claws strongly curved 33

30. Middle toe nail *pectinate* [X] 31
Middle toe nail not pectinate 32

31. Long, stiff, prominent *rictal bristles* [PP] present .... Caprimulgidae: Caprimulgiformes (nightjars)
Rictal bristles absent .... Eurostopodidae: Caprimulgiformes (eared nightjars)

[PP] Rictal bristles: stiffened, hair-like feathers bordering the gape; see hair-like bristles [W]

32. Nostrils at base of bill; *uropygial gland* [QQ] absent .... Podargidae: Caprimulgiformes (frogmouths)
Nostrils near tip of bill; uropygial gland present .... Aegothelidae: Caprimulgiformes (owlet-nightjars)

[QQ] Uropygial gland: oil gland located on the dorsal surface at the posterior end of the body, just above the insertion of the rectrices; overlain by feathers

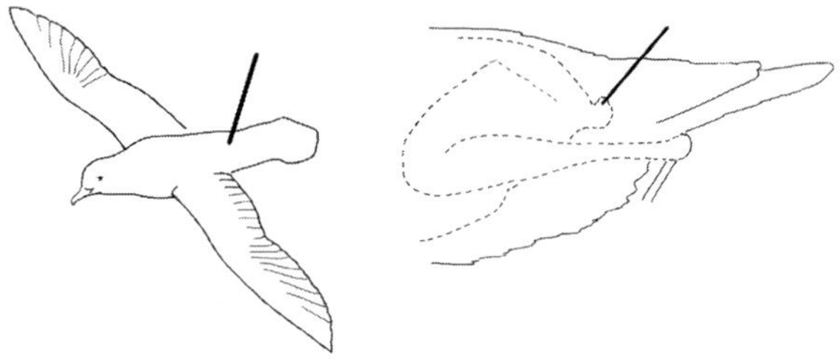

33. Face with elongated *moustachial* and *superciliary feathers* [RR] .... Hemiprocnidae: Apodiformes (whiskered swifts)
Face plain, without pluming .... Apodidae: Apodiformes (swifts)

[RR] Moustachial feathers: feathers extending from the base of the lower mandible along the cheek like a moustache

Superciliary feathers: feathers extending over the top of the eye like an eyebrow

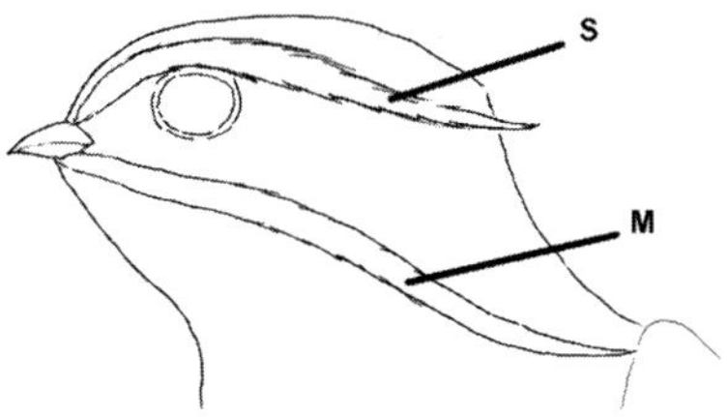

34. Toes *semipalmate* [J] .... Anseranatidae: Anseriformes (magpie geese)
Toes without webbing 35

35. Toes and claws greatly elongated; claw of hallux longer than hallux .... Jacanidae: Charadriiformes (jacanas)
Toes and claw of hallux not elongated 36

36. Bill and tarsus both elongate 37
Either bill or tarsus elongate, but not both 40

37. Nostrils *perforate* [SS] ....Gruidae: Gruiformes (cranes)
Nostrils *imperforate* [SS]..38

[SS] Perforate: septum between nostrils absent, making it possible to see through nostrils from one side of bill to the other

Imperforate: septum absent, not possible to see through bill

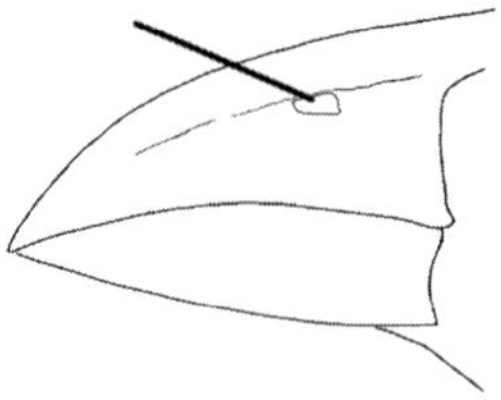

38. Bill *decurved* [TT] or *dorsoventrally flattened* [LL] and *laterally expanded* at tip [UU] ....
....Threskiornithidae: Ciconiiformes (ibises, spoonbills)
Bill straight 39

[TT] Decurved: end of bill curves ventrally (downwards)

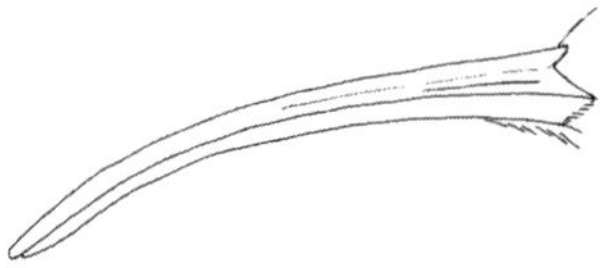

[UU] Laterally expanded: end of bill is extended laterally into flattened, spoon-like shape

39. Middle toe *pectinate* [X]; *powder down patches* [VV] on chest present ....Ardeidae: Ciconiiformes (herons)
Middle toe not pectinate; powder down patches on chest absent ....Ciconiidae: Ciconiiformes (storks)

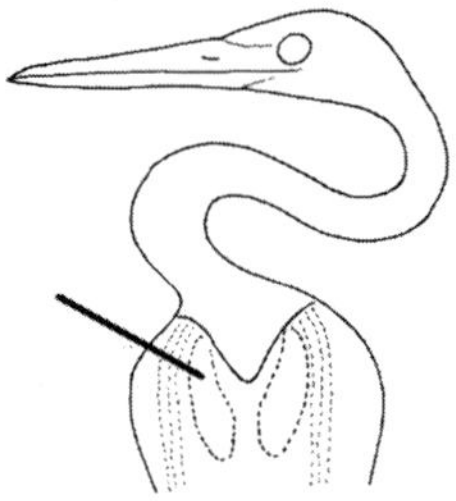

[VV] Powder down patches: dense areas of specialized feathers lying beneath contour feathers on each side of breast. Often yellowish in colour, these break down into fine powder used for grooming the plumage.

40. *Toes* [Z] II and III free from one another (toes III and IV usually so) 41
Toes III and IV *syndactyl* [WW], usually past first joint, and toes II and III syndactyl to lesser extent, sometimes to first joint...43

[WW] Syndactyl: having basal parts of the toes fused

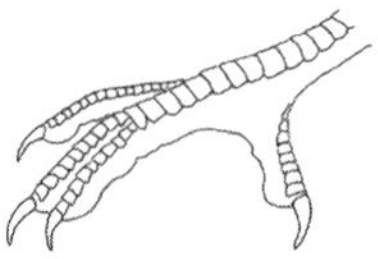

41. *Claw lengths* [XX] of *toes* [Z] I and IV not both > 16 mm; claws often strongly *curved* [HH] and *acute* [YY] ...
....Passeriformes (songbirds) (part)
Claw lengths of toes I and IV elongate (both > 16 mm); claws relatively *straight* [HH], not strongly curved, somewhat *obtuse* [ZZ]..42

[XX] Claw length: the straight line distance from the base of the claw on the dorsal side to the tip

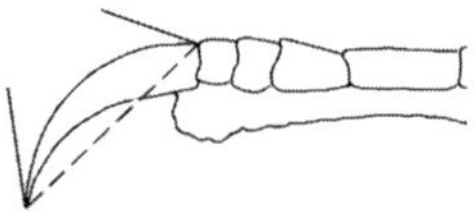

[YY] Claws acute: tip pointed

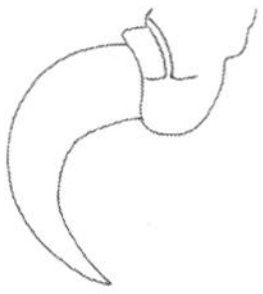

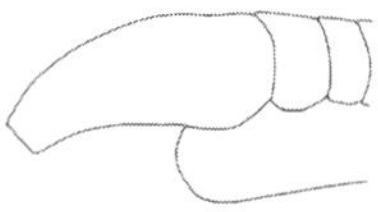

[ZZ] Claws obtuse: tip blunt or rounded off

42. *Tail length* [AAA] > 400 mm ....Megapodiidae: Galliformes (megapodes)
Tail length < 300 mm ....Passeriformes (songbirds) (part)

[AAA] Tail length: distance from insertion of central rectrices to tip of longest rectrix

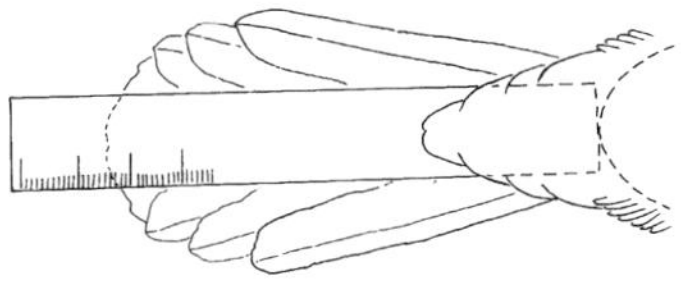

43. Bill *broad at gape* [BBB], about as wide as long; upper mandible hooked [G] ......Coraciidae: Coraciiformes (rollers)
Bill narrow, at least twice as long as wide; upper mandible not hooked 44

[BBB] Bill broad at gape: width across gape is large relative to length of bill, forming broad base to triangle

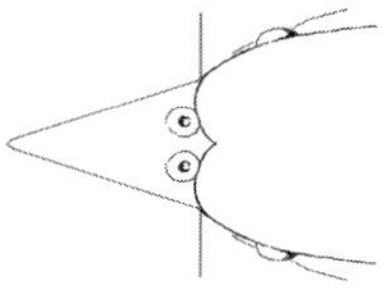

44. Tips of both mandibles *decurved* [TT] 45
Tips of both mandibles not decurved (lower straight) 46

45. *Mandibles massively enlarged* [CCC]; eyes with eye lashes ....Bucerotidae: Coraciiformes (hornbills)
Mandibles not greatly enlarged; eyes without eyelashes ....Meropidae: Coraciiformes (bee-eaters)

[CCC] Mandibles massively enlarged: massive in size relative to head; both mandibles considerably elongate and deep

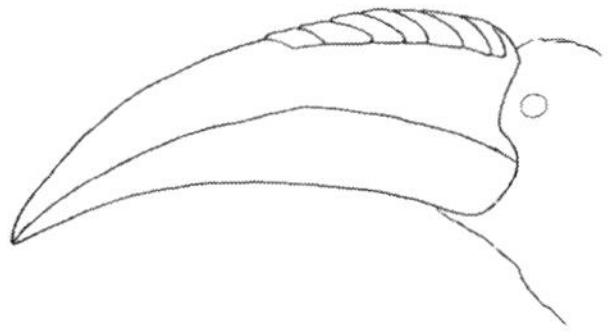

46. *Bill depth* [DDD] greater than *bill width* [EEE]; tail very short, < 75% of *exposed culmen length* [L] ....
....Alcedinidae: Coraciiformes (water kingfishers)
Bill width greater than bill depth; tail of moderate length, > 75% of exposed culmen length.............................
....Halcyonidae: Coraciiformes (land kingfishers)

[DDD] Bill depth: distance from dorsal side of upper mandible to ventral side of lower mandible measured at front of nostrils

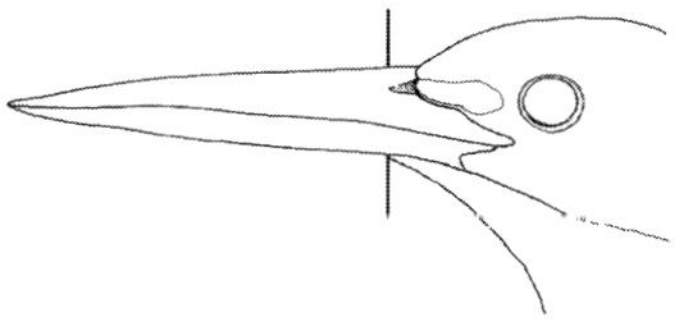

[EEE] Bill width: distance measured across bill at posterior margin of nostrils

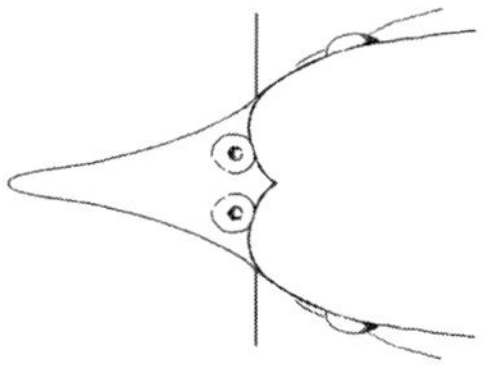

47. *Wing pointed* [FFF] 48
Wing rounded [GGG] 49

[FFF] Wing pointed: longest primaries (major flight feathers) are outermost (last feather may be greatly reduced in size, thereby making the next adjacent feather the functional outermost)

[GGG] Wing rounded: longest primaries are not outermost; they are usually 3rd–5th, counting inwards from last primary

48. Folded wings very long, extending well beyond ends of feet; lateral profile of upper mandible swollen toward tip relative to *constricted mid-bill* [HHH] ....Glareolidae: Charadriiformes (pratincoles)
Folded wings not extending beyond ends of feet; upper mandible variably long and slender, lacking a markedly swollen tip ....Scolopacidae: Charadriiformes (sandpipers)

[HHH] Constricted mid-bill: distal section of bill sloping anteroventrally from base, before reaching expanded or swollen proximal tip (dertrum), creating dorsoventrally pinched section near midpoint of bill

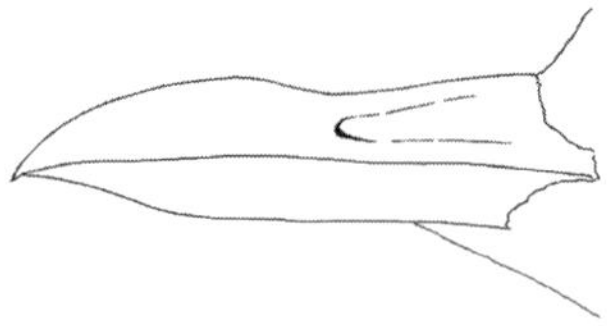

49. *Rhamphotheca complex* [III]; *nostrils sheathed* [JJJ] ....Chionididae: Charadriiformes (sheathbills)
Rhamphotheca simple; nostrils not sheathed 50

[III] Rhamphotheca: horny covering of bill

Complex: comprised of several well demarcated sections

[JJJ] Nostrils sheathed: sheath overlying base of upper mandible, partially covering nostrils

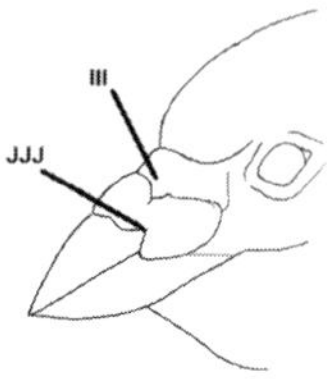

50. Tip of upper *mandible swollen relative to constricted mid-bill* [HHH] 51
Tip of upper mandible narrowed, acute or attenuated 52

51. *Exposed culmen* [L] more than twice as long as *bill width* at rear of nostrils [EEE]
....Vanellinae: Charadriidae: Charadriiformes (lapwings)
Exposed culmen less than twice as long as bill width at rear of nostrils ....
....Pedionomidae: Charadriiformes (plains-wanderer)

52. Nostrils *perforate* [SS] 53
Nostrils *imperforate* [SS]..54

53. *Tip of upper mandible droops* [KKK] over end of lower mandible; buff strap marks across *scapular region* [LLL]
....Rostratulidae: Charadriiformes (painted snipe)
Tip of upper and lower mandibles meeting; buff strap marks missing from scapular region ....
....Rallidae: Gruiformes (rails)

[KKK] Tip of upper mandible droops: end of otherwise straight bill slightly decurved

[LLL] Scapular region: shoulder area of dorsal and dorsolateral surfaces anterior to bend of wing

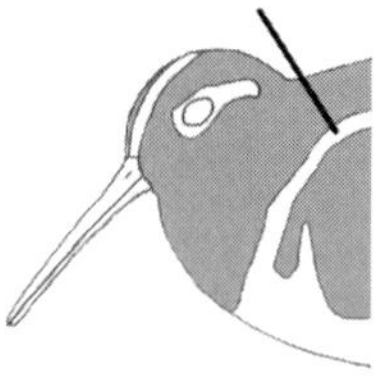

54. Hallux *incumbent* [MMM]..Odontophoridae: Galliformes (New World quail)
Hallux *elevated* [NNN] ....Phasianidae: Galliformes (Old World quail and allies)

[MMM] Incumbent: at same level as other toes on the foot

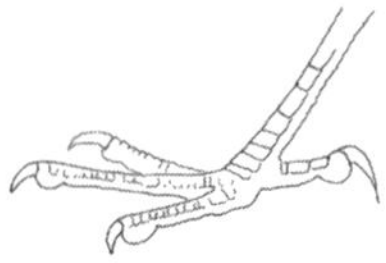

[NNN] Elevated: situated above (more proximally) than other toes on the foot

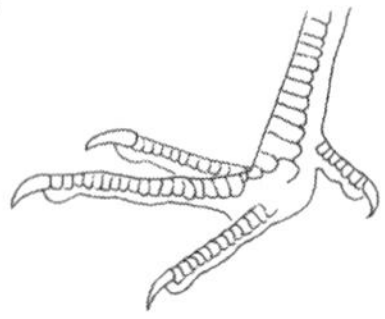

55. *Wings vestigial* [C]; *aftershaft* [D] almost as long as contour feather; upper *mandible with 'segments'* [E] 56
Wings developed; aftershaft short relative to contour feather; upper mandible lacking 'segments' 57

56. *Toes 2* [OOO] ....Struthionidae: Struthioniformes (ostriches)
Toes 3 ....Casuariidae: Casuariiformes (cassowaries, emus)

[OOO] Toes 2: only forward toes III and IV are present

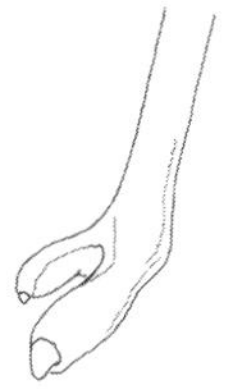

57. Tip of upper mandible swollen relative to *constricted mid-bill* [HHH] ....Charadriidae: Charadriiformes (plovers)
Tip of upper mandible narrowed, acute or elongate 58

58. Bill, *tibia* [PPP] and tarsus long, thin ....Recurvirostridae (part): Charadriiformes (stilts)
Bill not thin dorsoventrally; tibia and tarsus not markedly elongate 59

[PPP] Tibia: segment of the hindlimb proximal to the tarsus

59. Tarsus *reticulate* [QQQ] 60
Tarsus *scutellate* [RRR] 61

[QQQ] Reticulate: covered with fine, irregularly shaped scales, which are not arranged in rows

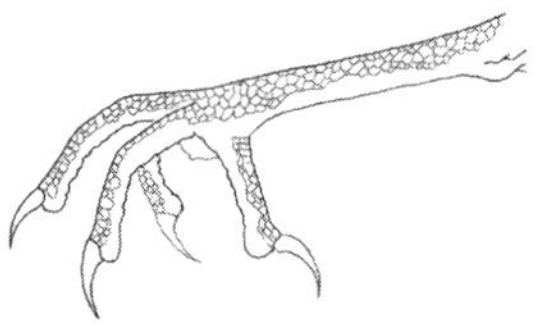

[RRR] Scutellate: anterior (front) side covered with row of large, flat scales

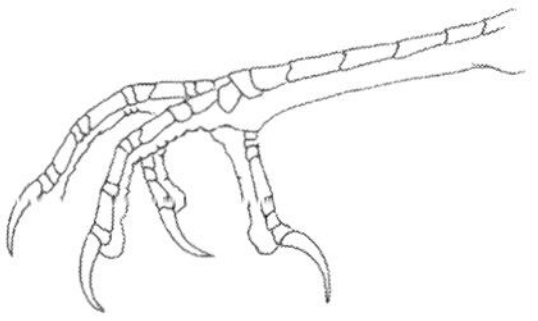

60. Bill *bilaterally compressed* [SSS]; *exposed culmen* [L] longer than *tarsus* [K] ....Haematopodidae: Charadriiformes (oystercatchers)
Bill not so compressed; exposed culmen shorter than tarsus ....Burhinidae: Charadriiformes (stone-curlews)

[SSS] Bilaterally compressed: bill thin from side to side; width considerably less than depth

61. *Uropygial gland* [QQ] present, feathered ....Turnicidae: Turniciformes (button-quail)
Uropygial gland absent ....Otididae: Gruiformes (bustards)

CONCLUSION

This key demonstrates the diversity of the extant Australasian bird faunas. It is a practical aid that: supplements the regional and national field guides, providing identification to a specific level; and complements the earlier chapters in this section, which discuss bird fossils (Boles, 2006), overall avifaunal composition (Schodde, 2006) and biogeography of bird faunas (Ford, 2006). For further details on evolution, distributions or biology of birds in particular areas, see pp.71–93 (Heinsohn and Hope, 2006) and pp.111–128 (Holdaway and Worthy, 2006).

ACKNOWLEDGEMENTS

I thank the reviewers for their useful comments and suggestions and L. Ho for the artwork.

REFERENCES

Boles, W.E. 2006. The avian fossil record of Australia: an overview. Pp.387–411. In Merrick, J.R., Archer, M., Hickey, G.M. and Lee, M.S.Y. (eds). *Evolution and Biogeography of Australasian Vertebrates.* Auscipub Pty. Ltd., Sydney. 942pp.

Christidis, L. and Boles, W.E. 1994. *Taxonomy and Species of Birds of Australia and its Territories.* Royal Australasian Ornithologists Union, Melbourne.

Ford, H.A. 2006. Island biogeography: as illustrated by birds in the Australasian region. Pp.459–476. In Merrick, J.R., Archer, M., Hickey, G.M. and Lee, M.S.Y. (eds). *Evolution and Biogeography of Australasian Vertebrates.* Auscipub Pty. Ltd., Sydney. 942pp.

Heinsohn, T. and Hope, G. 2006. The Torresian connections: zoogeography of New Guinea. Pp.71–93. In Merrick, J.R., Archer, M., Hickey, G.M. and Lee, M.S.Y. (eds). *Evolution and Biogeography of Australasian Vertebrates.* Auscipub Pty. Ltd., Sydney. 942pp.

Holdaway, R.N. and Worthy, T.H. 2006. Evolution of New Zealand and its vertebrates. Pp.111–128. In Merrick, J.R., Archer, M., Hickey, G.M. and Lee, M.S.Y. (eds). *Evolution and Biogeography of Australasian Vertebrates.* Auscipub Pty. Ltd., Sydney. 942pp.

Mariaux, J. and Braun, M.J. 1996. A molecular phylogenetic survey of the nightjars and allies (Caprimulgiformes) with special emphasis on the potoos (Nyctibiidae). *Molecular Phylogenetics and Evolution,* Vol. 6, pp.228–244.

Schodde, R. 2006. Australasia's bird fauna today—origins and evolutionary development. Pp.413–458. In Merrick, J.R., Archer, M., Hickey, G.M. and Lee, M.S.Y. (eds). *Evolution and Biogeography of Australasian Vertebrates.* Auscipub Pty. Ltd., Sydney. 942pp.

Sibley, C.G. and Ahlquist, J.E., Jnr. 1990. *Phylogeny and Classification of Birds. A Study in Molecular Evolution.* Yale University Press, New Haven, Connecticut. 976pp.

Mammals

GEOLOGICAL TIME SCALE FOR THE PHANEROZOIC

ERA	PERIOD or SUB-PERIOD	SERIES/EPOCH		STAGE	AGE m.y.a.	EXTINCTIONS (E) ICE AGE (I)
CAINOZOIC	QUATERNARY	PLEISTOCENE		Calabrian	1.8	E, I
	TERTIARY – NEOGENE	PLIOCENE		Piacenzian	3.6	I
				Zanclean	5.3	I
		MIOCENE		Messinian	7.12	
			Late	Tortonian	11.2	
				Serravalian	14.8	
			Mid	Langhian	16.4	
				Burdigalian	20.52	
			Early	Aquitanian	23.8	
	TERTIARY – PALAEOGENE	OLIGOCENE	Late	Chattian	28.5	E
			Early	Rupelian	33.7	
		EOCENE	Late	Priabonian	37	
				Bartonian	41.3	
			Mid	Lutetian	49	
			Early	Ypresian	54.8	
		PALEOCENE		Thanetian	57.9	
			Late	Selandian	60.9	
			Early	Danian	65	E
MESOZOIC	CRETACEOUS	LATE		Maastrichtian	73	
				Campanian	83	
				Santonian	87	
				Coniacian	89	
				Turonian	91	
				Cenomanian	97.5	
		EARLY		Albian	108	
				Aptian	115	
				Barremian	123	
				Hauterivian	130	
				Valanginian	135	
				Berriasian	141	E
	JURASSIC	LATE		Tithonian	146	
				Kimmeridgian	151	
				Oxfordian	159	
		MIDDLE		Callovian	165	
				Bathonian	173	
				Bajocian	180	
				Aalenian	184	
		EARLY		Toarcian	190	E
				Pliensbachian	195	
				Sinemurian	202	
				Hettangian	205	E
	TRIASSIC	LATE		Rhaetian	207	
				Norian	220	
				Carnian	230	
		MIDDLE		Ladinian	235	
				Anisian	241	
		EARLY		Spathian	244	
				Nammalian	247	
				Griesbachian	251	

This detailed scale is based on Young and Laurie (1996) but Cambrian stages abbreviated ages are rounded to nearest 0.1 million years. More recent dating (e.g. United States Geological Survey) has resulted in some small adjustments to interval boundaries.

ERA	PERIOD or SUB-PERIOD	SERIES/EPOCH		STAGE	AGE m.y.a.	EXTINCTIONS (E) ICE AGE (I)
PALAEOZOIC	PERMIAN	LATE	Tatarian	Dorashamian	255	E, I
			Tatarian	Dzhulfian	259	I
			Tatarian	Midian	263	I
				Kazanian	267	I
				Ulfimian	270	I
		EARLY		Kungurian	274	I
				Artinskian	285	I
				Sakmarian	293	I
				Asselian	298	I
	CARBONIFEROUS	PENSYLVANNIAN – SILESIAN	STEPHANIAN	Gzhelian	302	I
				Kasimovian	305	I
			WESTPHALIAN	Moscovian	311.5	
				Bashkirian	314	
			NAMURIAN	Serpukhovian	325	
		MISSISSIPPIAN – DINANTIAN	VISEAN	Brigantian	328.5	
				Asbian	330.5	
				Holkerian	335	
				Arundian	341.5	
				Chadian	344	
			TOURNAISIAN	Ivorian	348	
				Hastarian	354	
	DEVONIAN	LATE		Famennian	364.5	E
				Frasnian	369	
		MIDDLE		Givetian	378	I
				Eifelian	384	I
		EARLY		Emsian	399.5	I
				Pragian	404.5	I
				Lochkovian	410	I
	SILURIAN	PRIDOLI			414	
		LUDLOW		Ludfordian	417	
				Gorstian	420	
		WENLOCK		Homerian	422.5	
				Sheinwoodian	425	
		LLANDOVERY		Telychian	428	I
				Aeronian	431	I
				Rhuddanian	434	E, I
	ORDOVICIAN	ASHGILL		Bolindian	443	I
		CARADOC		Eastonian	454.5	
				Gisbornian	459	
		LLANVIRN		Darriwilian	467	
				Yapeenian	471	
		ARENIG		Castlemainian	477	
				Chewtonian	481	
				Bendigonian	484.5	
		TREMADOC		Lancefieldian	486	
				Warendan	490	
	CAMBRIAN	LATE		Datsonian	491	
				Payntonian	492.5	
				Iverian	496	
				Idamean	497.5	E
		MIDDLE		Mindyallan	498.5	
				Boomerangian	500	
				Undillan	503	
				Late Templetonian-Floran	506	
				Ordian-Early Templetonian	509	
		EARLY		Atdabanian-Toyonian	528	
				Tommotian	534	
				Nemakit-Daldynian	545	

Chapter 25
Origins of Mammals: Morphology, Molecules and a Petrosal or Two

Zerina Johanson
Palaeontology, Australian Museum, 6 College Street, Sydney, N.S.W. 2010 Australia

INTRODUCTION

We are mammals, and so the origin of the Mammalia is of interest to us. What was the first mammal? Where did the group evolve? What features did the first mammals possess? To answer these questions, we need an accurate phylogeny of the Mammalia. Aspects of mammalian origins examined in this review will focus primarily on innovations in higher-level mammalian phylogeny that have occurred since 1984, when this volume's predecessor *Vertebrate Zoogeography and Evolution in Australasia* was published (Archer and Clayton, 1984).

First, the definition and diagnosis of the taxon Mammalia are discussed. Rowe (1988, 1993), using the concepts of phylogenetic taxonomy as outlined by de Quieroz and Gauthier (1990) and Lee (2006), highlighted difficulties with the traditional character-based definition of Mammalia. Rowe defined Mammalia based on common ancestry rather than on possession of particular characters, and sought to stabilize Mammalia by restricting this name to extant taxa, the common ancestor of these taxa, and all descendants (both fossil and extant). One consequence of Rowe's work was a change in traditional taxonomic assignments to Mammalia. Taxa no longer assigned to the Mammalia included *Morganucodon* (late Triassic–early Jurassic of Europe, early Jurassic of Asia and North America (Kühne, 1949)), *Sinoconodon* (early Jurassic of Asia (Patterson and Olson, 1961)) and *Adelobasileus* (Late Triassic of North America (Lucas and Hunt, 1990)). Following Rowe's revision, these taxa were assigned to the new groups Mammaliaformes and Mammaliamorpha.

Second, a short list of taxa relevant to discussions of higher-level mammalian phylogeny is given. Taxa selected include those used in recent phylogenetic analyses of mammalian relationships (see Figure 1), as well as those important in more traditional classifications, for example, describing dental evolution within the Mammalia (*Kuehneotherium* (Kermack *et al.*, 1968) and *Aegialodon* (Kermack *et al.*, 1965)). The classification of the list loosely follows the recent mammalian classification provided by McKenna and Bell (1997).

Third, the characters or character systems currently being used to diagnose Mammalia are discussed. Traditionally, characters such as morphology of the dentition and the sidewall of the braincase have been utilized, but aspects of the base of the skull (basicranium) are providing an abundance of new characters for phylogenetic analyses. For example, *Vincelestes neuquenianus* (early Cretaceous La Amarga Formation, Argentina (Bonaparte, 1986b)), is closely related to tribosphenic therian mammals (marsupials and placentals) and is supplying important cranial and postcranial characters for 'prototribosphenic' mammals as a whole. As well, beautifully preserved fossils recently discovered in China (Hu *et al.*, 1997; Ji *et al.*, 1999) are providing postcranial characters for taxa previously known largely from teeth.

Some more traditional characters, such as dental characters, are proving to be homoplasious in their phylogenetic distribution (not diagnosing monophyletic groups). For example, groups like the Triconodonta (postcanine teeth with three anteroposteriorly aligned cusps) are no longer considered monophyletic (Rougier *et al.*, 1996; Ji *et al.*, 1999). As will be seen, however, it may be difficult to find any character system completely free of homoplasy.

Fourth, the increasing use of molecular analyses to estimate the timing of the origin of mammalian groups will be briefly discussed. Controversially, this work suggests that **modern** orders of mammals (work thus far has focused on placental mammals) originated in the Mesozoic. However, the fossil evidence suggests that the Mesozoic holds the key to the origins of the 'Mammalia' as a whole, not just modern mammalian orders. This is true even if Rowe's (1988) definition of the Mammalia as the last common ancestor of monotremes, placentals and marsupials and all descendants is followed, because fossil monotremes such as *Steropodon galmani* (Archer *et al.*, 1985) and *Kollikodon ritchiei* (Flannery *et al.*, 1995) occur in the early Cretaceous of Australia. Conversely, the majority of modern placental orders make their first appearance in the fossil record much later, in the early Tertiary.

The final aspect discussed in this review involves the 'where' of mammalian origins. If we use Rowe's Mammaliamorpha (defined as the last common ancestor of Mammalia and the cynodont group Tritylodontidae, and all descendants; this more or less reflects the more traditional concept of Mammalia), can we pinpoint the origins of this group to a particular place? Does the oldest fossil record of a mammaliamorph necessarily indicate the origins of the group? Or, should we interpret the

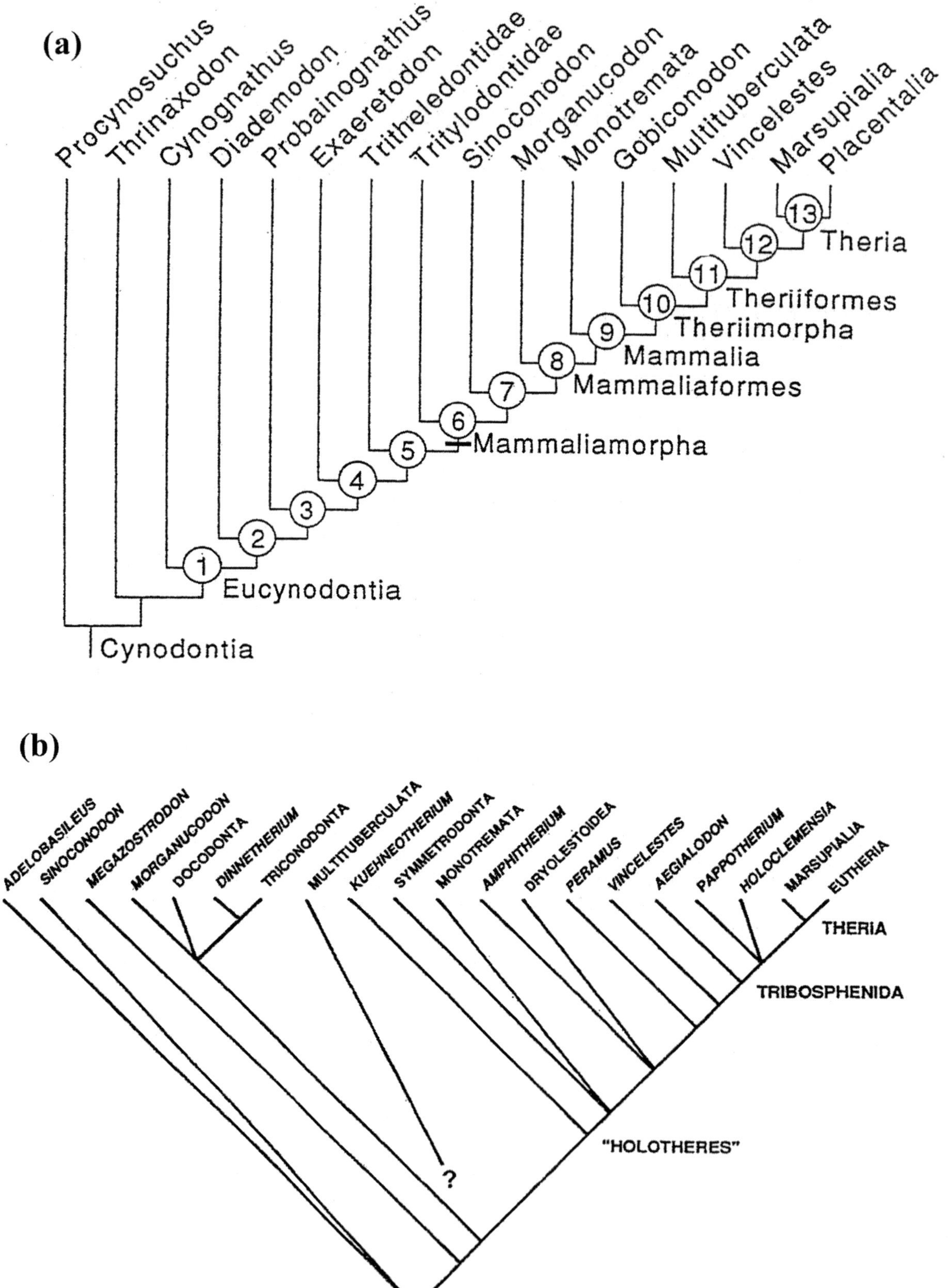

Figure 1. Cladograms summarizing the results from several recent studies of mammaliamorph relationships: **(a)** Rowe (1993); **(b)** Hopson (1994); **(c)** Wible *et al.* (1995); **(d)** Rougier *et al.* (1996).

(c)

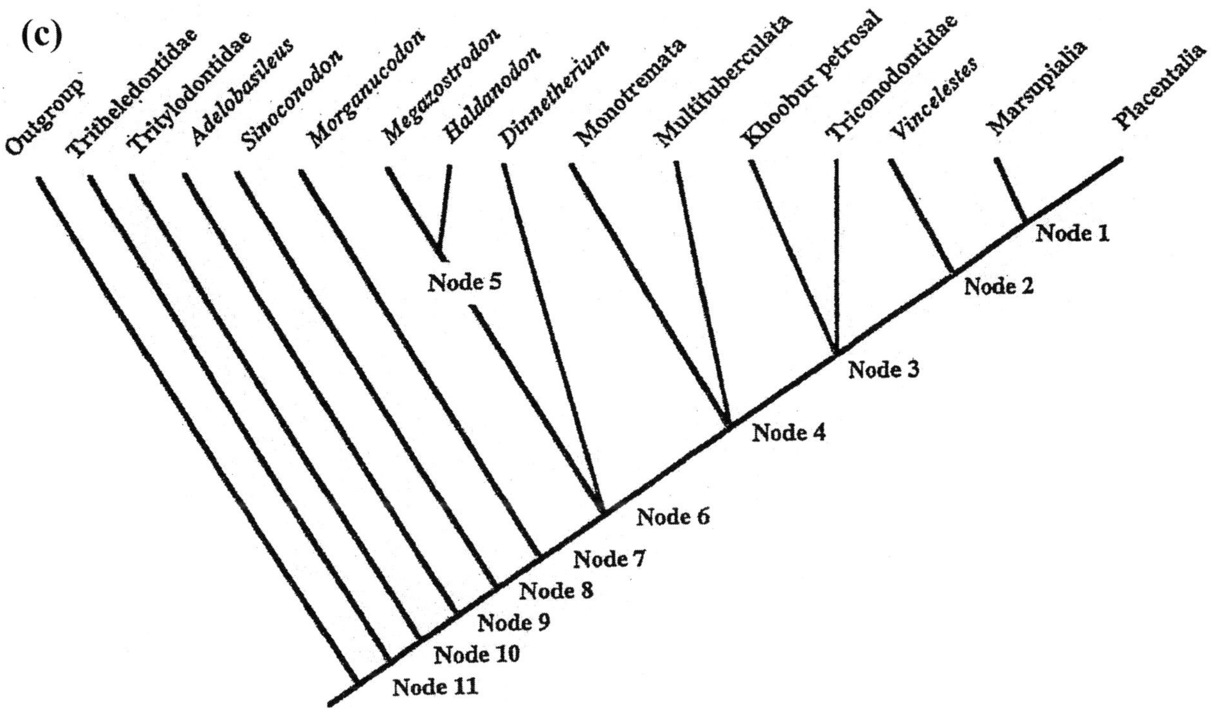
Outgroup
Tritheledontidae
Tritylodontidae
Adelobasileus
Sinoconodon
Morganucodon
Megazostrodon
Haldanodon
Dinnetherium
Monotremata
Multituberculata
Khoobur petrosal
Triconodontidae
Vincelestes
Marsupialia
Placentalia
Node 1
Node 2
Node 3
Node 4
Node 5
Node 6
Node 7
Node 8
Node 9
Node 10
Node 11

(d)

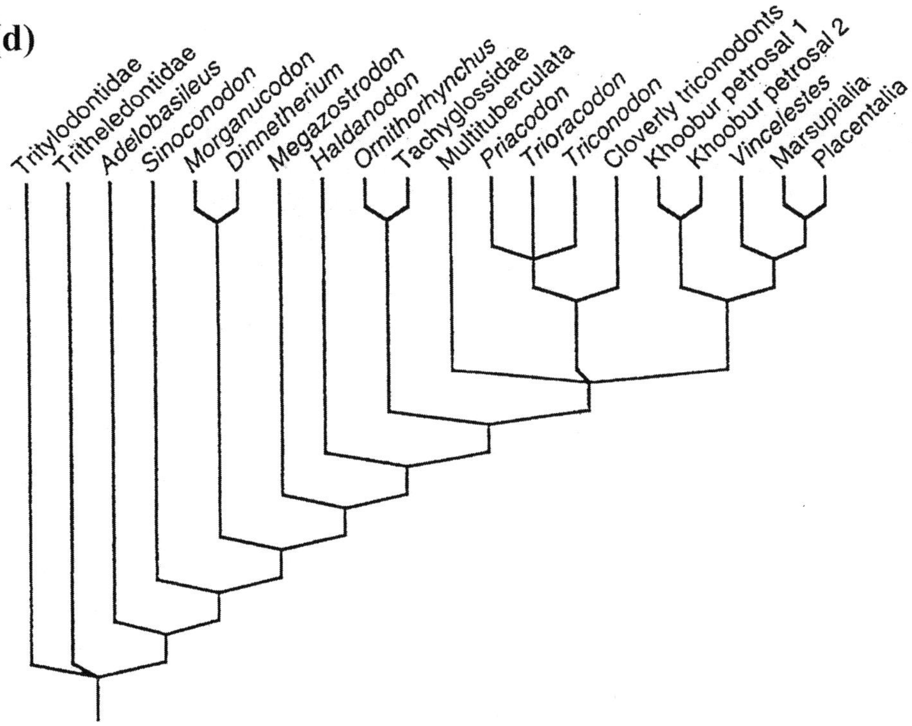
Tritylodontidae
Tritheledontidae
Adelobasileus
Sinoconodon
Morganucodon
Dinnetherium
Megazostrodon
Haldanodon
Ornithorhynchus
Tachyglossidae
Multituberculata
Priacodon
Trioracodon
Triconodon
Cloverly triconodonts
Khoobur petrosal 1
Khoobur petrosal 2
Vincelestes
Marsupialia
Placentalia

origins of the Mammaliamorpha from cladograms? Should we also consider distributions of the sister group taxa like the tritylodont and/or tritheledont cynodonts? For example, recent discoveries of tritylodont cynodonts in North America (Clark and Hopson, 1985; Lewis, 1986; Sues, 1986) provide some evidence for the origin of the mammaliamorphs in this region. Interestingly, North America is also home to one of the phylogenetically basal taxa within the mammaliamorph clade, *Adelobasileus* (Lucas and Hunt, 1990). Australia's possible, but contentious contribution to these issues is also discussed.

ORIGINS OF MAMMALIA, MAMMALIAFORMES OR MAMMALIAMORPHA?

Before we can begin to investigate the origins of the mammals, we need to determine just what animals are included within the clade or taxon Mammalia. Most people would recognize a mammal when they see one. Things become more difficult, however, when we try to extract mammalian origins from the fossil record. Most obvious is the problem of non-preservation, in the fossil record, of easily recognizable mammal characters such as hair, mammary glands or endothermy.

Also, the fragmentary nature of fossil material (often only teeth, one of the most durable parts of the body) means that many taxa lack vital morphological information on the skull and postcranial skeleton. However, these materials have become available across a broader range of taxa in recent years.

Another problem concerns the identification of mammalian synapomorphies, or characters uniquely shared by the Mammalia. Prior to the revisions of Rowe (1988), mammals were generally thought to have originated in the late Triassic. Therefore, skeletal and dental characters were the focus of the search for mammalian synapomorphies.

These 'hard' characters included a jaw articulation between the dentary (lower jaw) and squamosal (a skull bone) and a diphyodont dentition, in which teeth anterior to the molars were only replaced once (and molars not replaced at all). Differentiation of the postcanine teeth into premolars and molars was also thought to be characteristic of the Mammalia (Kemp, 1983; Hopson and Barghusen, 1986; Rowe, 1988). The uniqueness of these and a variety of other characters were evaluated by comparison to the Cynodontia (Figure 3). Cynodonts are part of a larger group of 'mammal-like reptiles' assigned to the Synapsida, and among synapsids, were identified as the most closely related non-mammalian group to the mammals (Figure 2). Two cynodont groups have variously been identified as the closest outgroups (sister groups), the Tritylodontidae and Trithelodontidae (contrast Figure 1(a), (c) and Figures 1(d), 2(b)).

It soon became apparent, however, that there were problems in using morphological characters to define Mammalia. Certain cynodonts possessed what were thought to be mammalian characters (Crompton and Jenkins, 1968), while early mammals often retained non-mammalian characteristics (Hopson and Barghusen, 1986). For example, a variety of early taxa (*Morganucodon*, *Sinoconodon*, the docodont *Haldanodon* (Kühne and Krusat, 1972) and even the early holotherian *Kuehneotherium*) possess a trough on the inner posterior surface of the lower jaw (Crompton and Jenkins, 1979; Lillegraven and Krusat, 1991; Wible and Hopson, 1993; Hopson, 1994). The dentary dominates the mammalian lower jaw, but in these taxa, the trough ('postdentary trough') houses remnants of bones more commonly found in non-mammalian jaws (Crompton, 1963; Crompton and Parker, 1978; Crompton and Hylander, 1986). This is important because some of these bones (e.g., the articular and angular) are homologous to bones found in the middle ear of more derived mammals (malleus and tympanic bones).

Mammalian features occurring among the Cynodontia include a reduction in the rate of tooth replacement (a condition approaching diphyodonty), double rooted molars, differentiation of the dentition and a variety of postcranial characters (Crompton and Jenkins, 1979; Kemp, 1983; Hopson and Barghusen, 1986). Hopson and Barghusen (1986) recognized 39 characters shared between *Morganucodon* and cynodonts.

These problems are not new, nor are they unique to the Mammalia. However, these difficulties led Rowe (1988, 1993) to redefine the Mammalia based on **common ancestry** rather than possession of particular characters. Specifically, he defined Mammalia as the monophyletic group including the most recent common ancestor of the living mammal groups Monotremata and Theria (marsupials and placentals), and all descendants of this common ancestor (Figure 1(a)). Defining taxa based on ancestry incorporates the methodology of phylogenetic taxonomy (Lee, 2006). For examples of recent studies on mammalian phylogeny see Murphy *et al.* (2001) and Luo *et al.* (2002)

Of course, characters are used to produce the phylogenies on which this classification is based, but Rowe made the distinction between **defining** the Mammalia (based on common ancestry) and **diagnosing** the Mammalia (based on characters). The definition of Mammalia was linked to **extant** mammalian taxa (monotremes, placentals and marsupials), in order to provide a stability to the name 'Mammalia' that had previously been lacking. Extant taxa are morphologically complete (compared to fossils) and provide a larger number of characters for analysis. New analyses are less likely to overturn the relationships of these living groups. Fossil taxa are accommodated, if they share common ancestry with the living taxa.

The goals of phylogenetic taxonomy (promotion of taxonomic stability) are laudable, and problems using characters to define taxa have long been recognized. However, these methods have required that fossil taxa previously assigned to the Mammalia, but not nested within the clade that unites monotremes and therians

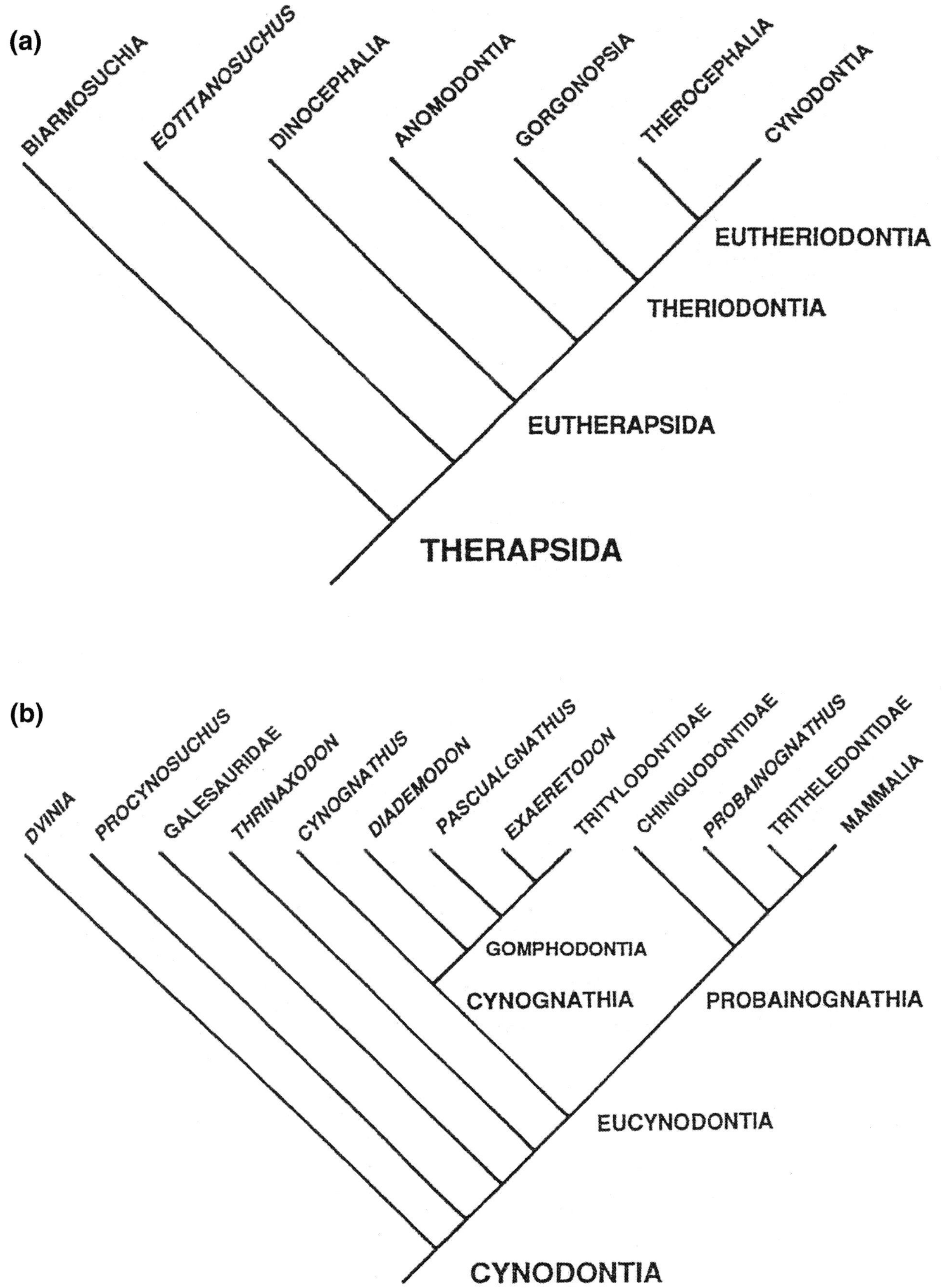

Figure 2. Cladograms of **(a)** synapsid and **(b)** cynodont relationships (from Hopson, 1994).

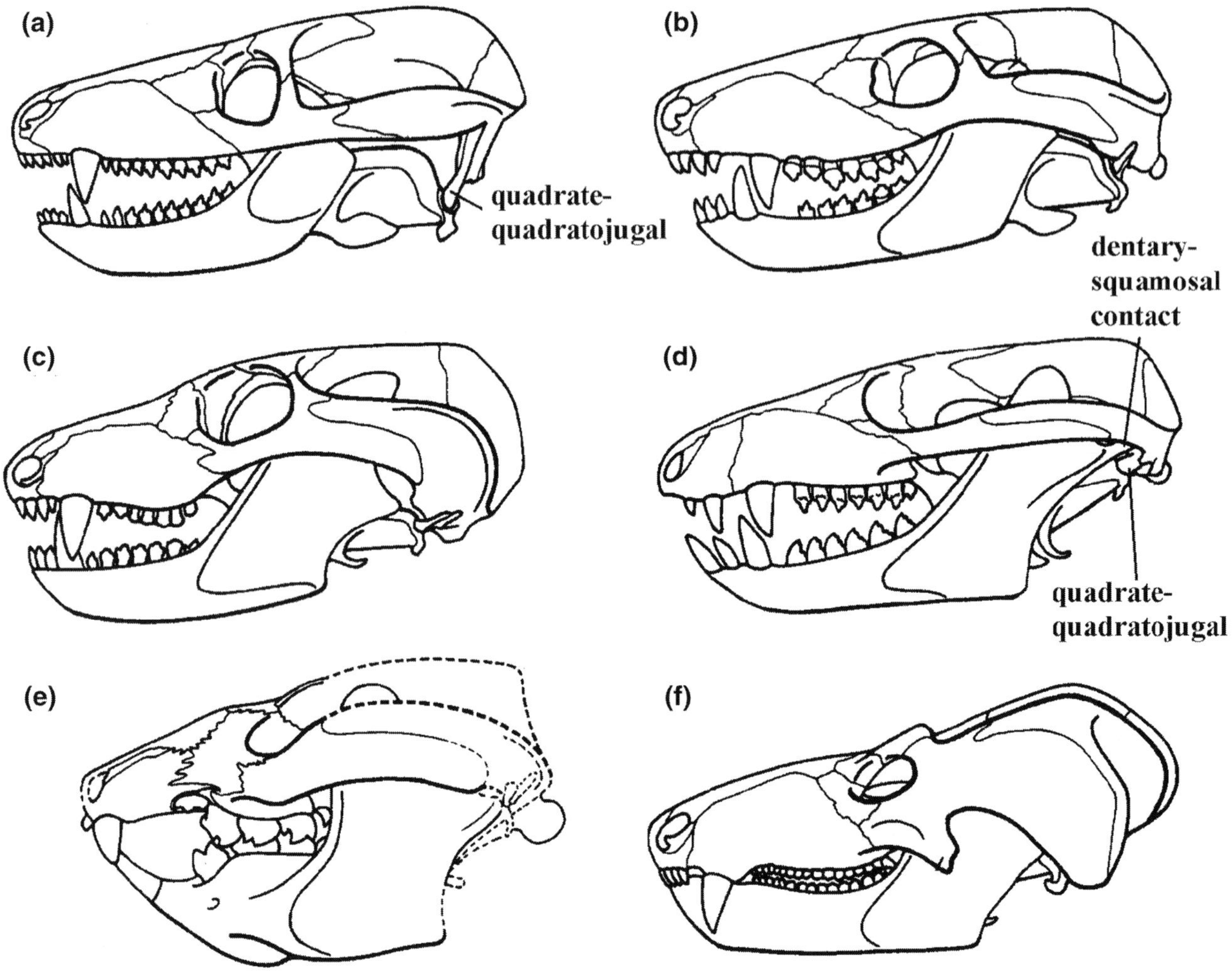

Figure 3. Lateral views of cynodont skulls (after Hopson, 1994): **(a)** *Procynosuchus*; **(b)** *Thrinaxodon*; **(c)** *Probainognathus*; **(d)** *Diademodon*; **(e)** *Bocatherium* (a tritylodontid); **(f)** *Pachygenelus* (a trithelodontid).

(e.g., the tricododontid *Morganucodon*) be assigned to new taxa. Thus, Mammaliaformes was defined by Rowe (1988, 1993) as 'the last common ancestor of *Morganucodon* and Mammalia and all its descendants'.

Some researchers (Lucas and Luo, 1993) have questioned whether having to provide new names for taxa traditionally assigned to the Mammalia (like Mammaliaformes) actually helps to promote taxonomic stability. Others have noted (Wible and Hopson, 1995) that, depending on the phylogeny used, a variety of fossil taxa would be included in the Mammalia as defined by Rowe. More specifically, Wible *et al.* (1995) have pointed out that the scope of the Mammalia is dependent on the phylogenetic position of the Monotremata. In turn, this depends on molar homologies of fossil monotremes like *Steropodon* and *Kollikodon* with holotherian and tribosphenic mammal molars. Some researchers (Pascual *et al.*, 1992b; Archer *et al.*, 1993; Wible *et al.*, 1995) deny these homologies exist, so that a variety of taxa are more closely related to the Holotheria than are monotremes. This would significantly expand 'Mammalia'. Other researchers (Hopson, 1994) homologize fossil monotreme molars with holotherian molars, suggesting monotremes are closely related to the Holotheria, and restricting 'Mammalia'.

Proponents of phylogenetic taxonomy would suggest these concerns are instead one of the strengths of the method. The taxa included may change (as new taxa and specimens are discovered), but the definition of Mammalia will remain constant.

These ideas are still controversial, but appear to have been accepted and used in several recent publications (Wible *et al.*, 1995; Rougier *et al.*, 1996; McKenna and Bell, 1997). The definition of Mammalia proposed by Rowe (1988, 1993) will be followed here as well, but many interesting evolutionary changes occur prior to the origin of Rowe's Mammalia (Lucas and Luo, 1993; Luo, 1994; Wible and Hopson, 1993; Wible *et al.*, 1995; Rougier *et al.*, 1996), within the more traditional concept of 'Mammalia' and its outgroups.

Rowe's original designation of Mammaliaformes as 'the last common ancestor of *Morganucodon* and

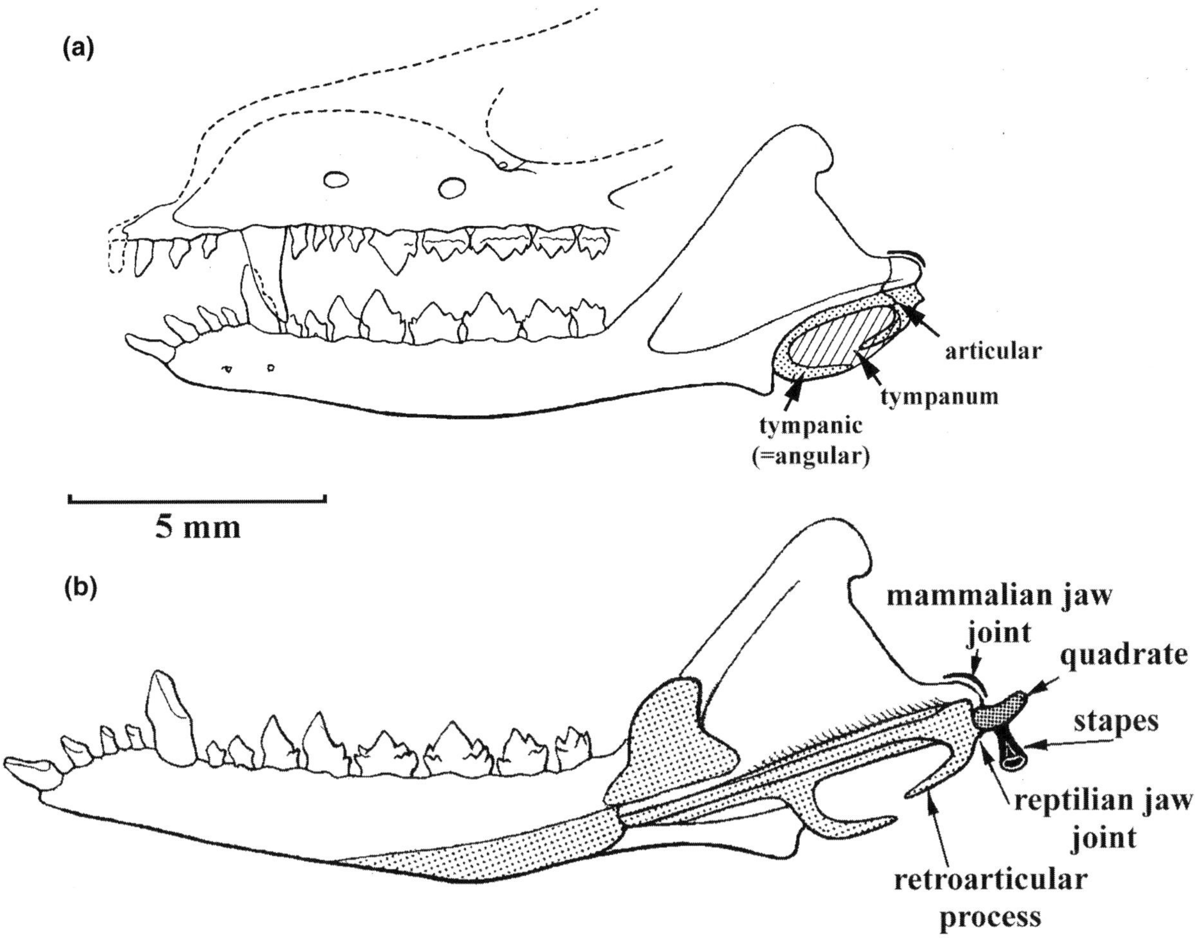

Figure 4. Reconstruction of skull and jaws of a morganucodontid: **(a)** lateral view; **(b)** medial view. Adapted from Crompton and Jenkins (1979).

Mammalia and all its descendants' circumscribed taxa assigned to the traditional Mammalia, as it was then understood. Since then, new taxa and phylogenetic analyses indicate that two 'mammalian' taxa (*Adelobasileus* and *Sinoconodon*) are phylogenetically basal to *Morganucodon* and the Mammaliaformes (Lucas and Hunt, 1990; Rowe, 1993; Wible *et al.*, 1995). The next taxon defined by Rowe (1988, 1993) was Mammaliamorpha, for the last common ancestor of the cynodont Tritylodontidae and Mammalia, and all descendants (Figure 1(a)). *Adelobasileus* was not included in Rowe's cladogram, but his Mammaliamorpha includes *Sinoconodon*.

The chapter sections below will begin with cynodont outgroups and move crownwards (see Lee (2006) for definitions of crown and stem groups) to the Mammalia (Rowe, 1988; Lee, 2006).

MAMMALIAMORPHA: WHO'S WHO

The discussion below lists and briefly discusses taxa incorporated in more recent phylogenetic analyses (Rowe, 1988, 1993; Lucas and Luo, 1993; Hopson, 1994; Luo, 1994; Wible *et al.*, 1995; Rougier *et al.*, 1996; Hu *et al.*, 1997; Ji *et al.*, 1999). These analyses broadly agree, but differ on several points, for example, the position of the multituberculates and monotremes. Recently described taxa, including new, nearly complete forms discovered in China (Hu *et al.*, 1997; Ji *et al.*, 1999) are also discussed. Higher-level classification follows Rowe (1988, 1993) and McKenna and Bell (1997). However, strict Linnean categories (Family, Order, Class, etc.) are omitted. A more complete listing of taxa assigned to the higher-level groupings below can be obtained from McKenna and Bell (1997).

CYNODONTIA

The Cynodontia is a phylogenetically derived group within the Synapsida, or the 'mammal-like' reptiles, and is considered to comprise those animals most closely related to the traditional concept of 'Mammalia' (Hopson and Barghusen, 1986); see Figure 2. Various physiological, morphological and ecological changes during the transition from these 'mammal-like reptiles' to the 'Mammalia' have already been examined in some detail (Archer, 1984;

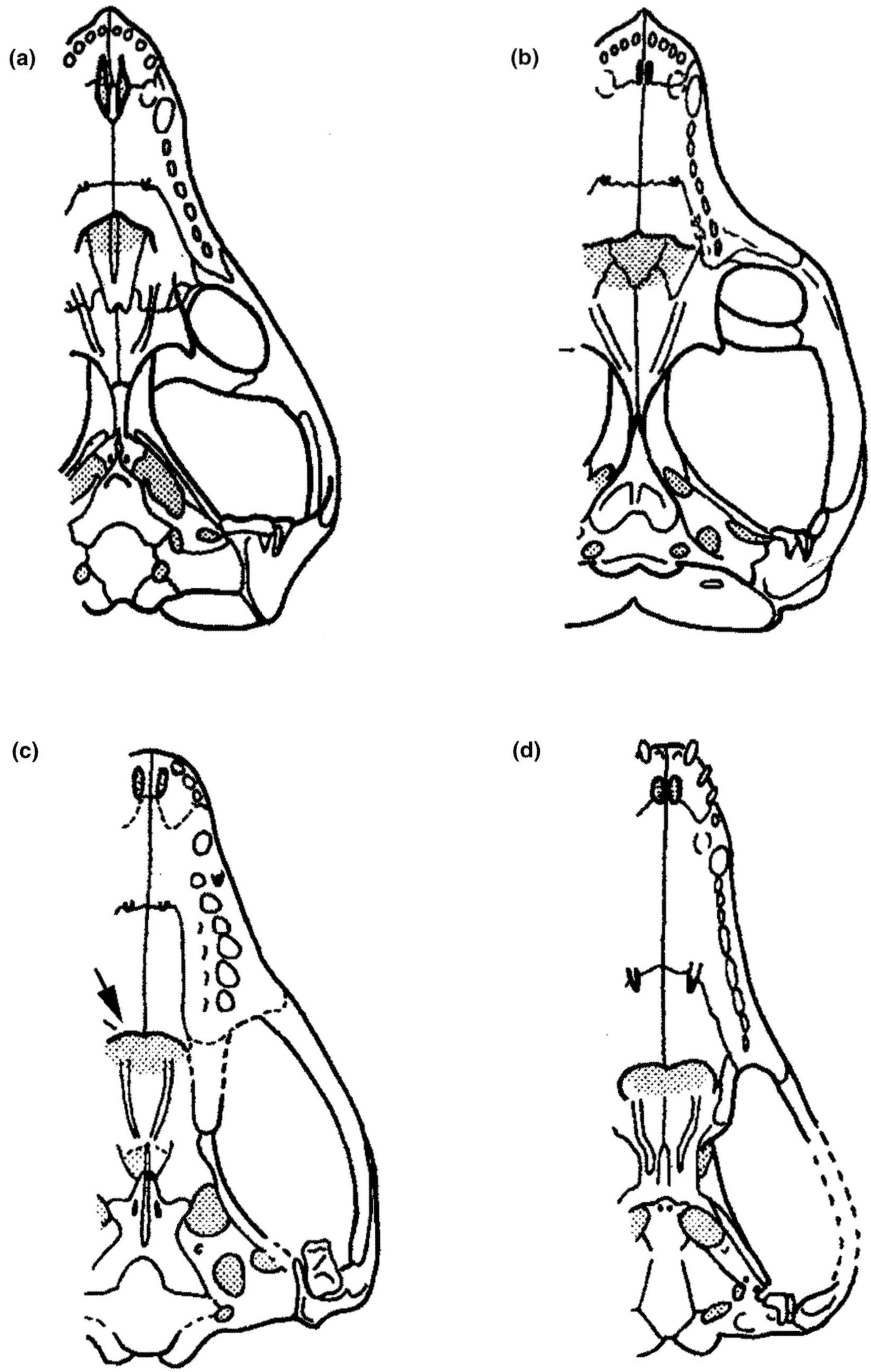

Figure 5. Palatal views of cynodonts and the 'mammal' *Morganucodon*: **(a)** thrinaxodontid cynodont; **(b)** probainognathid cynodont; **(c)** trithelodontid cynodont—note extension of palate to end of tooth row marked by arrow; **(d)** mammal. Based on Hopson and Barghusen (1986).

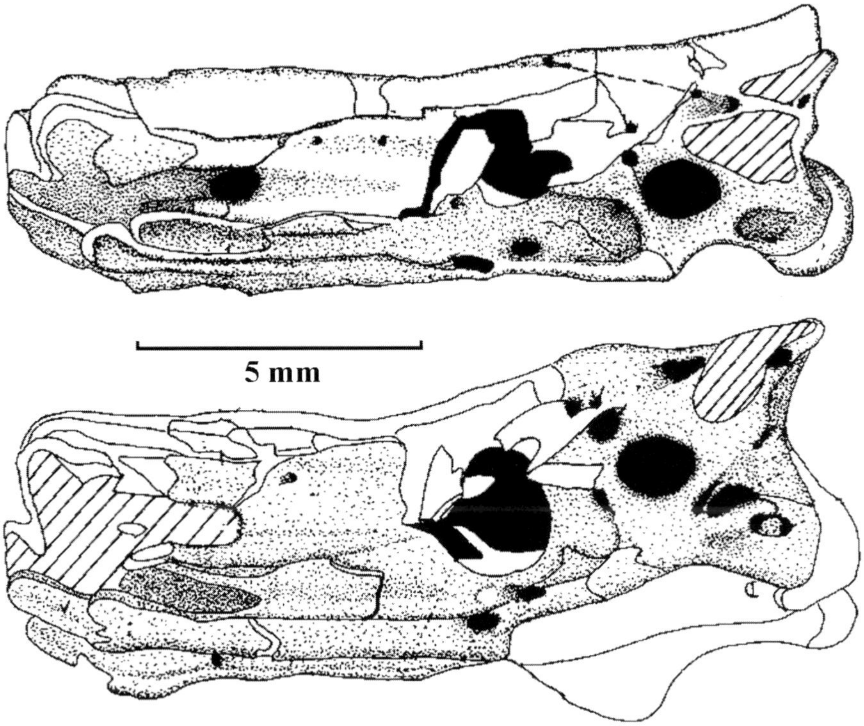

Figure 6. Skull of *Adelobasileus cromptoni* (from Lucas and Luo, 1993).

Hotton *et al.,* 1986; Hopson, 1994). Important morphological changes involve the lower jaw and the dentition, the middle ear, the basicranial region of the skull, and postcranial skeleton (Sues, 1986). The non-mammaliform Synapsida range in age from the Permian to Jurassic; they are exceptionally well known, particularly from the Karoo Beds of South Africa, but also North America, Russia, China and other areas (Parrish *et al.*, 1986).

Within the Cynodontia, the Tritylodontidae and Tritheledontidae are generally considered most closely related to the traditional Mammalia (Kemp, 1983; Sues, 1985; Rowe, 1988, 1993; Shubin *et al.*, 1991; Crompton and Luo, 1993; Lucas and Luo, 1993; Hopson, 1994; Luo, 1994; Luo and Crompton, 1994; Wible *et al.*, 1995). Kemp (1983) identified several similarities shared between the Tritylodontidae (late Triassic–late Jurassic) and the 'Mammalia' (e.g., postcanine teeth with multiple roots), although a dentary-squamosal joint does not seem to have been developed (Crompton and Sun, 1985). Rowe (1988) recognized 62 characters diagnosing the Mammaliamorpha, and Wible (1991) recognized 20, the majority of these in the postcranial skeleton. Wible *et al.* (1995) also resolved the tritylodonts as the most closely related cynodonts, in an analysis based on the petrosal bone of the basicranial region.

Sues (1985) questioned many of the similarities identified by Kemp (1983), and Hopson (1994) noted that a sister group relationship between the tritylodonts and mammals implied convergence in cranial and dental characters shared by tritylodonts and the cynodont group Gomphodontia (Figure 2(b)). Hopson (1994) also pointed out that Rowe's (1988, 1993) analyses focused primarily on postcranial rather than cranial and dental characters.

The tritheledontids may or may not possess a dentary-squamosal joint homologous to that in 'Mammalia' (Crompton and Sun, 1985; Shubin *et al.*, 1991; Hopson, 1994), but Kemp (1983) noted that trithelodontid teeth possess prismatic enamel (also Grine and Vrba, 1980; Hopson and Barghusen, 1986; Shubin *et al.*, 1991). This type of enamel is said to be characteristic of 'Mammalia', but not present in the tritylodontids. Hopson and Barghusen (1986) also identified five characters shared by tritheledontids and 'mammals' not shared by tritylodontids, including a secondary palate that extends to the end of the tooth row (Figure 5(c)). Shubin *et al.* (1991) listed seven trithelodontid-'mammal' synapomorphies, and discussed a shared pattern of unilateral occlusion where the lower jaw moves dorsomedially relative to the upper, producing wear on the lingual surface of the upper teeth and buccal wear on the lowers. Shubin *et al.* (1991) said this type of occlusion is absent from the Tritylodontidae. Other researchers have also supported a sister group relationship between the Trithelodontidae and a traditional Mammalia (Hopson and Barghusen, 1986; Crompton and Luo, 1993; Hopson, 1994; Luo, 1994; Luo and Crompton, 1994). However, Rowe (1993) places the Trithelodontidae as the sister taxon to his Mammaliamorpha (including the Tritylodontidae), as do other recent researchers (Wible, 1991; Wible and Hopson, 1993; Wible *et al.*, 1995).

It appears that the problems resolving the most closely related cynodont have to do with the poorly known postcranial skeleton in tritheledonts, and the prevalence

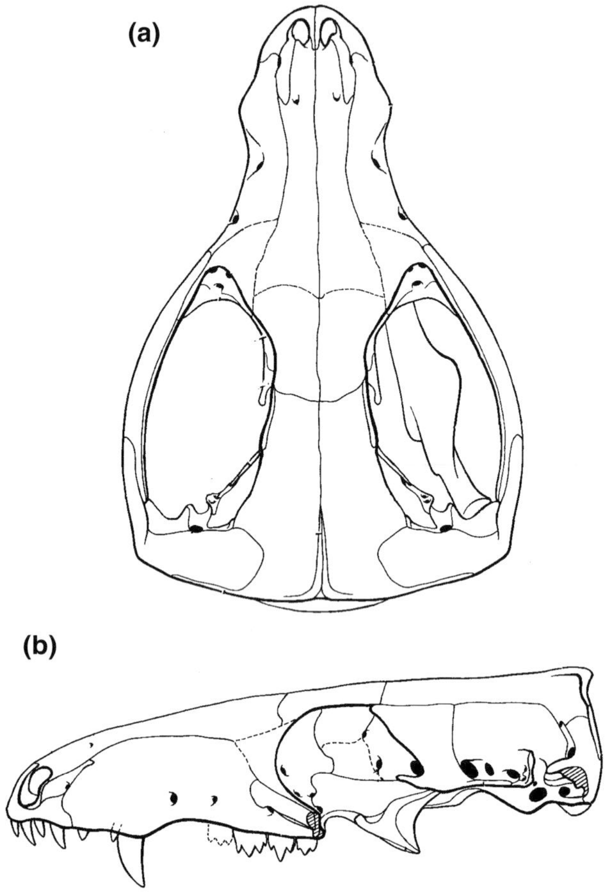

Figure 7. *Sinocodon* sp. skull: **(a)** dorsal view; **(b)** lateral view. Based on Crompton and Luo (1993).

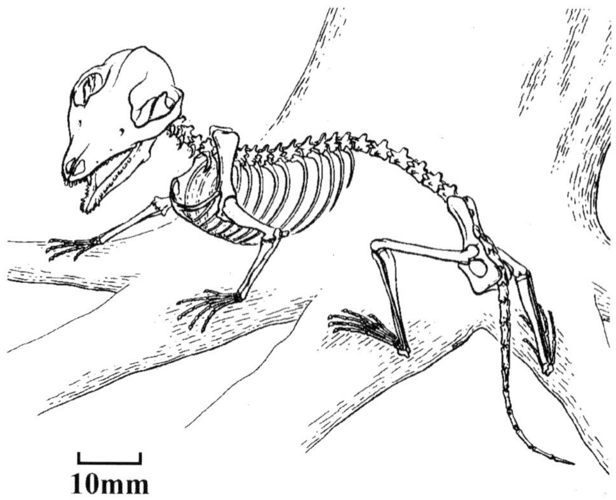

Figure 8. *Megazostrodon rudnerae* (from Jenkins and Crompton, 1979).

of these characters in favouring the Tritylodontidae (Wible, 1991). Cranial and dental characters appear to resolve the Tritylodontidae well away from the Mammaliamorpha, and favour the Trithelodontidae (Hopson, 1994).

MAMMALIAMORPHA

ADELOBASILEUS

Adelobasileus cromptoni (Lucas and Hunt, 1990), from the Tecovas Formation of Texas (late Carnian [late Triassic, 225 m.y.a.] in age), was originally described as the oldest known mammal (in the traditional sense), approximately 10 million years older than *Sinoconodon* (Patterson and Olson, 1961; Crompton and Sun, 1985) and other late Triassic–early Jurassic ('Rhaeto-Liassic') mammals (Lucas and Hunt, 1990; Lucas and Luo, 1993). These mammaliamorphs are illustrated in Figures 6 and 7. See also Rauhut *et al.* (2002).

Adelobasileus appears near the base of the mammaliamorph clade in phylogenetic analyses of Wible *et al.* (1995) and Rougier *et al.* (1996; Figure 1(c), (d)) and in a dichotomy with *Sinoconodon* in others (Lucas and Luo, 1993; Hopson, 1994; Luo, 1994). *Adelobasileus* is known from a single skull. A mandible and dentition were not originally described, but the dentition is similar to that of *Sinoconodon* (Lucas, 2000, personal communication). Lucas and Hunt (1990) assigned *Adelobasileus* to 'Order and Family incertae sedis'. McKenna and Bell (1997) did not assign *Adelobasileus* to any family within the Mammaliaformes, and noted that its status is questionable. However, evidence from the basicranium appears to establish the phylogenetic position of *Adelobasileus* between the cynodont outgroups and taxa like *Sinoconodon* and *Morganucodon* (see below).

The biogeographic implications of *Adelobasileus* being the oldest but also one of the phylogenetically most basal mammaliamorphs are discussed further below.

SINOCONODON

Sinoconodon (Patterson and Olson, 1961; Crompton and Sun, 1985) was described from the Rhaeto-Liassic Lufeng Series of Yunnan Province, China (Figures 7 and 10). The postdentary groove or trough on the internal surface of the lower jaw is present, but reduced in *Sinoconodon* (lower jaws unknown in *Adelobasileus*). The articular bone still forms a joint with the quadrate, medial to the dentary-squamosal joint (Hopson, 1994). A distinction between premolars and molars is not readily recognized, either on the basis of morphology or replacement pattern. *Sinoconodon* also lacks the dental occlusion characteristic of early Mammaliaformes such as *Morganucodon* or *Dinnetherium* (Jenkins *et al.*, 1983), and *Kuehneotherium* (Crompton and Sun, 1985; Wible, 1991; Hopson, 1994). However, *Sinoconodon* is derived relative to *Adelobasileus* in features of the basicranium described below.

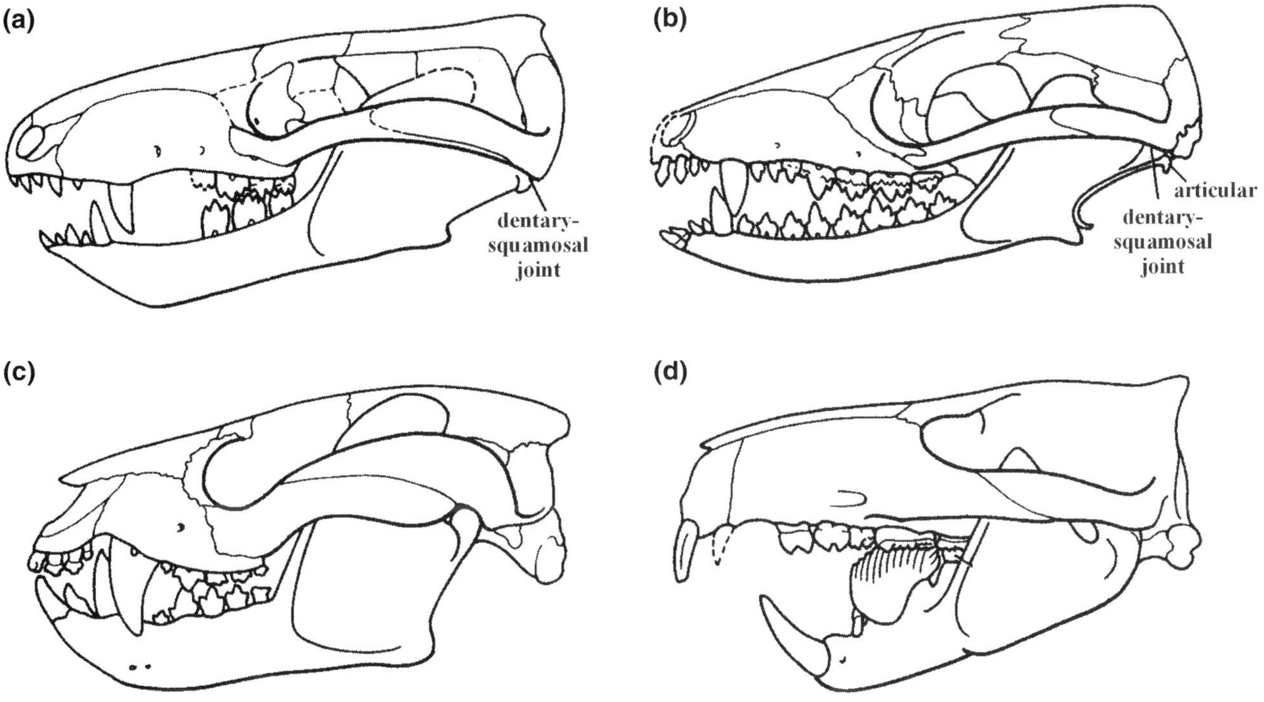

Figure 9. Skulls in lateral view (from Hopson, 1994): **(a)** *Sinoconodon*; **(b)** *Morganucodon*; **(c)** *Vincelestes*; **(d)** multituberculate *Ptilodus*.

MORGANUCODONTA/DOCODONTA

The morganucodontids (Kermack *et al.*, 1973), including the well known *Morganucodon* (Figure 9(b)), have been described from the late Triassic to the middle Jurassic of Asia, Europe, North America and South Africa (McKenna and Bell, 1997). *Morganucodon* is known from the late Triassic of Britain, and as such represents one of the oldest known mammals in the traditional sense (Jenkins and Crompton, 1979).

Morganucodon and *Megazostrodon* (Crompton and Jenkins, 1968), shown in Figure 8, are known from both cranial and postcranial material (Kermack *et al.*, 1973, 1981; Jenkins and Parrington, 1976; Jenkins and Crompton, 1979; Gow, 1986). The morphology of the *Morganucodon* braincase (Figure 10(b)) played an important role in separating mammals into the Prototheria and Theria. This is discussed in more detail below. *Morganucodon* and other morganucodontids are characterized by a triconodont dentition (Figures 11, 12), but the Triconodonta is not currently resolved as a monophyletic group (Rougier *et al.*, 1996; Ji *et al.*, 1999).

Recent phylogenetic analyses resolve members of the Docodonta (Kretzoi, 1946), such as *Haldanodon*, in unresolved polytomies with the morganucodontids and other triconodont taxa such as *Dinnetherium* (Wible, 1991; Hopson, 1994; Wible *et al.*, 1995). Docodonts (Figure 13) are of middle–late Jurassic age, from North America and Europe (Kron, 1979; Lillegraven and Krusat, 1991). Wible *et al.* (1995; Figure 1(c)) separated *Morganucodon* from a *Dinnetherium*/*Haldanodon* group, but include *Megazostrodon* as a sister taxon to *Haldanodon* (Lillegraven and Krusat, 1991). McKenna and Bell (1997) also classify *Megazostrodon* in the Order Docodonta. According to Wible *et al.* (1995), the relationship of these two genera is only weakly supported (a single opening for the maxillary and mandibular nerves in the side wall of the braincase and a ventral crest on the promontorium of the petrosal bone), but supports the views of other researchers such as Butler (1988). Conversely, Lillegraven and Krusat (1991) place the docodont *Haldanodon* at what would be the base of the Mammaliamorpha clade, which Wible *et al.* (1995) note requires four extra steps to their shortest tree to obtain.

Teeth of the Docodonta are an unusual rectangular shape (Figures 12, 13), and a postdentary trough is present on the internal surface of the mandible for the articular and angular bones (Lillegraven and Krusat, 1991; Wible and Hopson, 1993).

MAMMALIA

MONOTREMATA

Traditionally, Mammalia was divided between the Prototheria (monotremes, triconodonts, multituberculates, docodonts) and the Theria (marsupials and placentals). This separation was based, in part, on the morphology of the side wall of the braincase (Hopson, 1970;

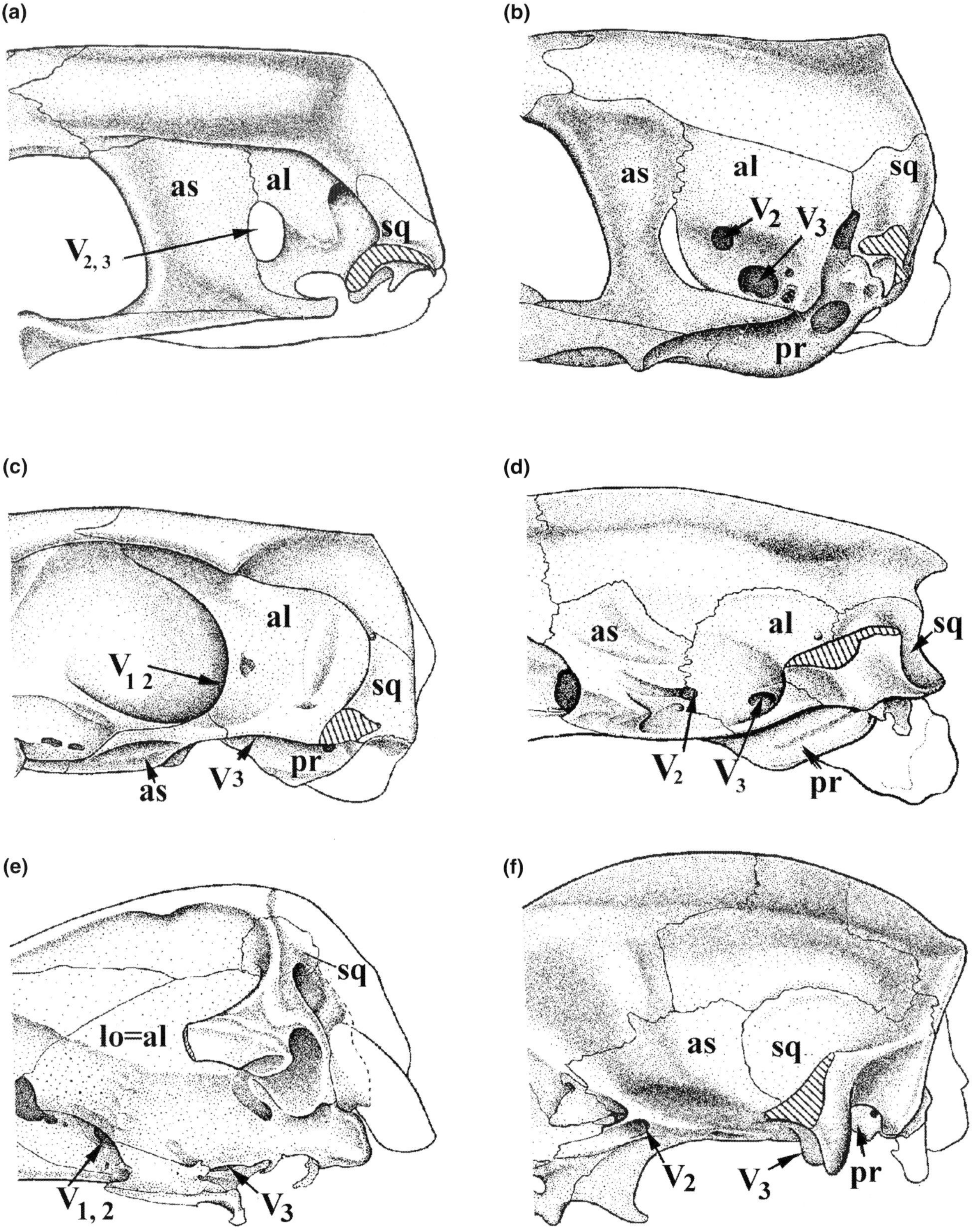

Figure 10. Lateral views of the side wall of the braincase (adapted from Wible and Hopson, 1993): **(a)** *Pachygenelus* (trithelodontid); **(b)** *Morganucodon*; **(c)** *Chusanbataar* (multituberculate); **(d)** *Vincelestes*; **(e)** *Ornithorhynchus*; **(f)** *Didelphis* (marsupial). Key to labels: **as** = alisphenoid; **al** = anterior lamina of petrosal; **lo** = lamina obturans; **os** = orbitosphenoid; **pr** = promontorium of the petrosal; **sq** = squamosal; **V1, V2, V3** = foramen for maxillary and mandibular branches of the trigeminal nerve.

Kemp, 1983; Hopson and Rougier, 1993) and on dental morphology.

However, most recent analyses indicate that the Prototheria is not a monophyletic group, with certain triconodonts and multituberculates being more closely related to the Theria than are monotremes and docodonts (Rowe, 1993; Rougier *et al.*, 1996). The phylogenetic position of the Monotremata within the Mammalia (*sensu* Rowe) is also problematic as researchers try to resolve homologies of the cusps and crests on fossil monotreme molars relative to the Holotheria (possessing upper and lower molars with a cusp pattern of opposing triangles, for example, in *Kuehneotherium* (Figure 11)). Some researchers (Archer *et al.*, 1993; Flannery *et al.*, 1995; Wible *et al.*, 1995) suggest these homologies are not recognizable, while others assign the Monotremata to the Holotheria on this basis (Hopson, 1994).

Until recently, monotremes (families Ornithorhynchidae, Tachyglossidae, Steropodontidae and Kollikodontidae) were thought to be restricted to the Australian land mass, with discoveries of the Cretaceous monotremes *Steropodon galmani* and *Kollikodon ritchei* from opal deposits at Lightning Ridge, N.S.W. (Archer *et al.*, 1985; Flannery *et al.*, 1995). *Teinolophus trusleri* represents a third Australian Cretaceous monotreme recently described from the Flat Rocks fossil locality, Victoria (Rich *et al.*, 1999, 2001). These taxa represent a major range extension of the Monotremata in geological time, although other fossil forms were known from the middle and later parts of the Tertiary (reviewed in Pascual *et al.*, 1992b), including *Obdurodon insignis* (Woodburne and Tedford, 1975) and *O. dicksoni* (Archer *et al.*, 1992). Although *Obdurodon insignis* was known only from isolated teeth and a piece of postcranial skeleton, *O. dicksoni*, from the Riversleigh deposits of northwestern Queensland (Archer *et al.*, 1991, 1992), included associated skull and dental material. This firmly established *Obdurodon* as a monotreme, assigned to the

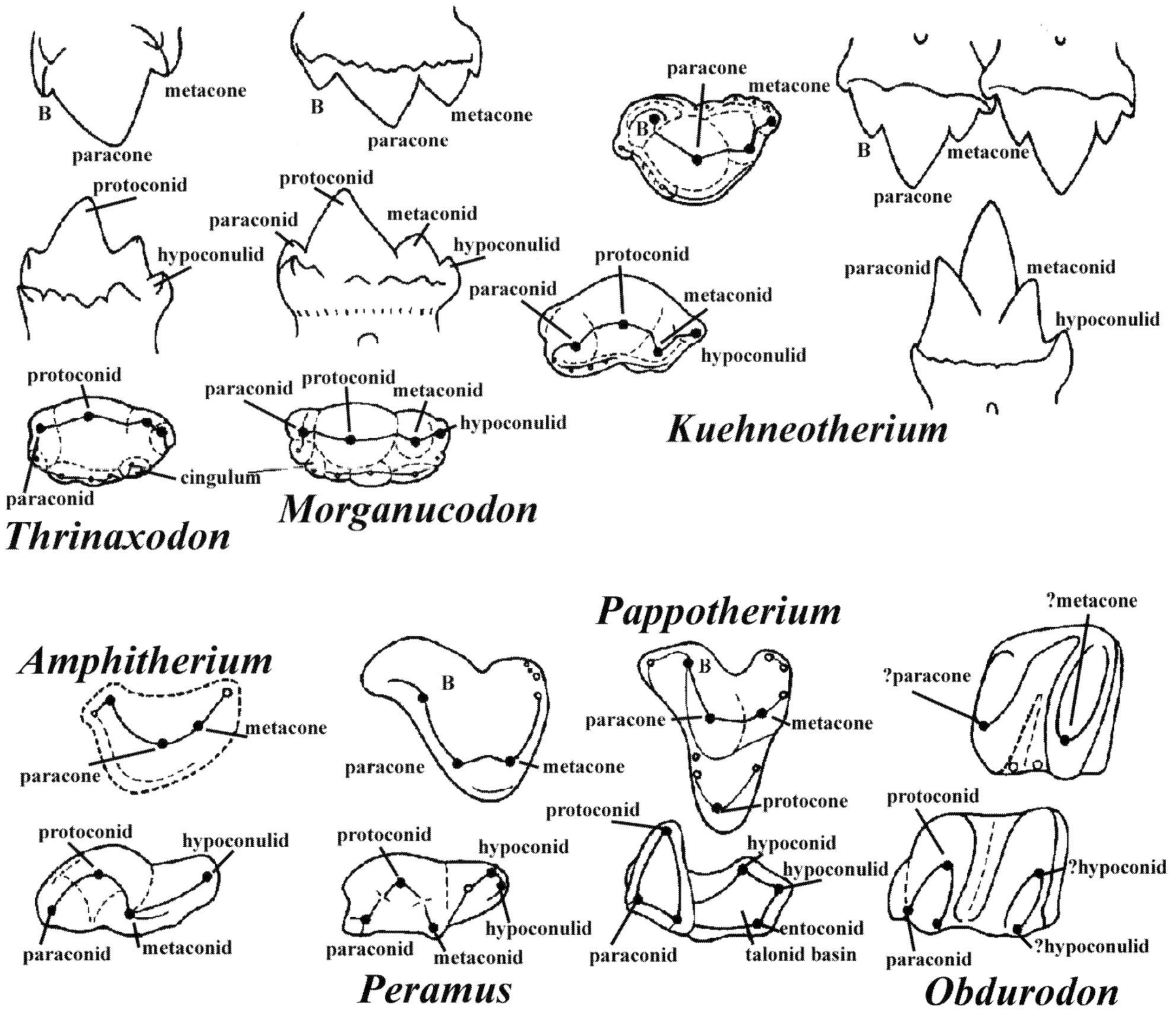

Figure 11. Upper and lower molar dentitions of seven taxa. Based on Hopson (1994).

ters. For example, the scapulae show infra- and supraspinous fossae of approximately equal size, more characteristic of extant therians (Ji *et al.*, 1999) and also *Gobiconodon* (Jenkins and Schaff, 1988). Nevertheless, in the phylogenetic analysis of Ji *et al.* (1999) *Jeholodens* is resolved as crownward of the morganucodontids, but not closely related to the therians. This results in convergence of these derived postcranial characters.

HOLOTHERIA

Hopson (1994) used the informal term 'holothere' to characterize mammals possessing a reversed triangle dentition, where cusps on the upper and lower molars formed opposing triangles (Figure 11). Holotheria was more formally defined by Wible *et al.* (1995) and Flynn *et al.* (1999) as the common ancestor of *Kuehneotherium* and therians plus all descendants. Multituberculates and triconodonts are not assigned to the Holotheria because they lack the reversed-triangle pattern of molar cusps and crests characteristic of this group. On this basis, Wible *et al.* (1995) also question monotreme and dryolestid affinities with the Holotheria.

Kuehneotheria

Kuehneotherium

Kuehneotherium (Kermack *et al.*, 1968) is a poorly known taxon from the late Triassic–early Jurassic of Europe, but its importance lies in its age (similar to *Morganucodon*, and previously one of the oldest mammals in the traditional sense) and in its possession of a derived reversed triangle dental morphology, relative to the triconodonts (Figure 11). From this triconodont and reversed triangle morphology, the tribosphenic condition could be derived by the addition of the lingual protocone and a basined talonid (Crompton, 1971; Cassiliano and Clemens, 1979; Hopson, 1994).

As noted by Cassiliano and Clemens (1979), *Kuehneotherium* was traditionally assigned to the Symmetrodonta, based on similarity between *Kuehneotherium* teeth and those of other symmetrodonts. However, they also note that these similarities were probably plesiomorphic, and that *Kuehneotherium* lacks characters seen in more derived symmetrodonts. McKenna (1975) assigned *Kuehneotherium* and other kuehneotherids to their own 'Superlegion Kuehneotheria', separate from the Symmetrodontida (also Prothero, 1981; McKenna and Bell, 1997). This separation is maintained here, but without the Linnean category.

Kuehneotherium is also problematic as its position within the Holotheria results in homoplasious character distributions for the Mammalia (*sensu* Rowe) as a whole. *Kuehneotherium* combines the derived reversed triangle dentition with a lower jaw preserving a postdentary trough (Figure 16). As described above, this suggests that the full complement of mammalian middle ear bones may not have been present. The trough is absent in monotremes and multituberculates, so that the position of these taxa in most cladograms relative to *Kuehneotherium* and the

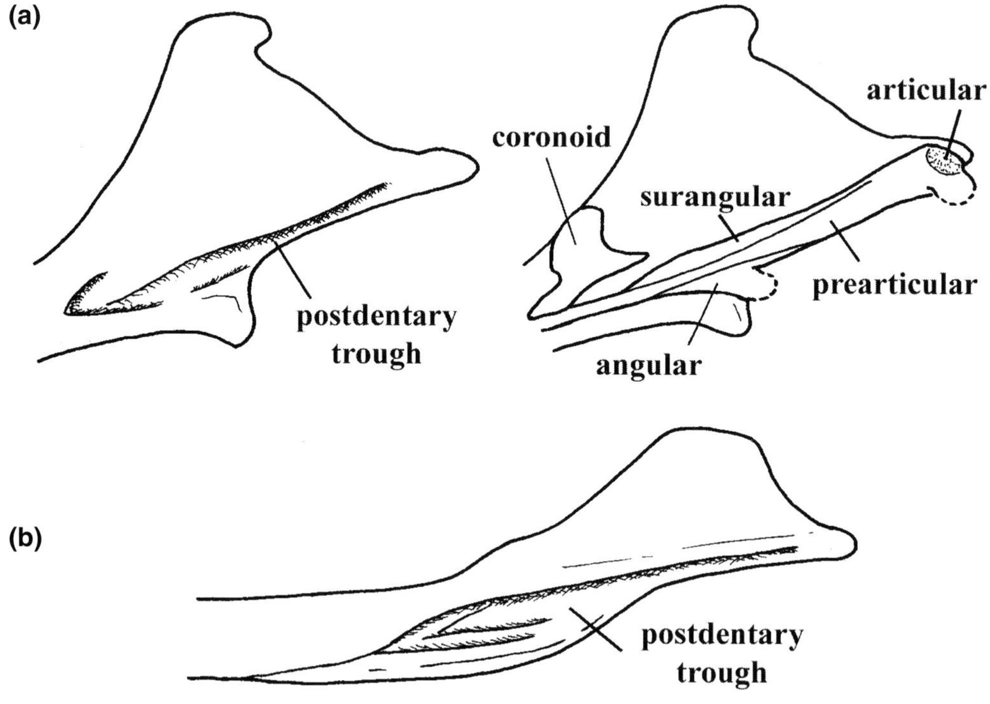

Figure 16. Lower jaws of: **(a)** a morganucodontid; and **(b)** the holothere *Kuehneotherium praecursoris*, showing the postdentary trough (from Kemp, 1983).

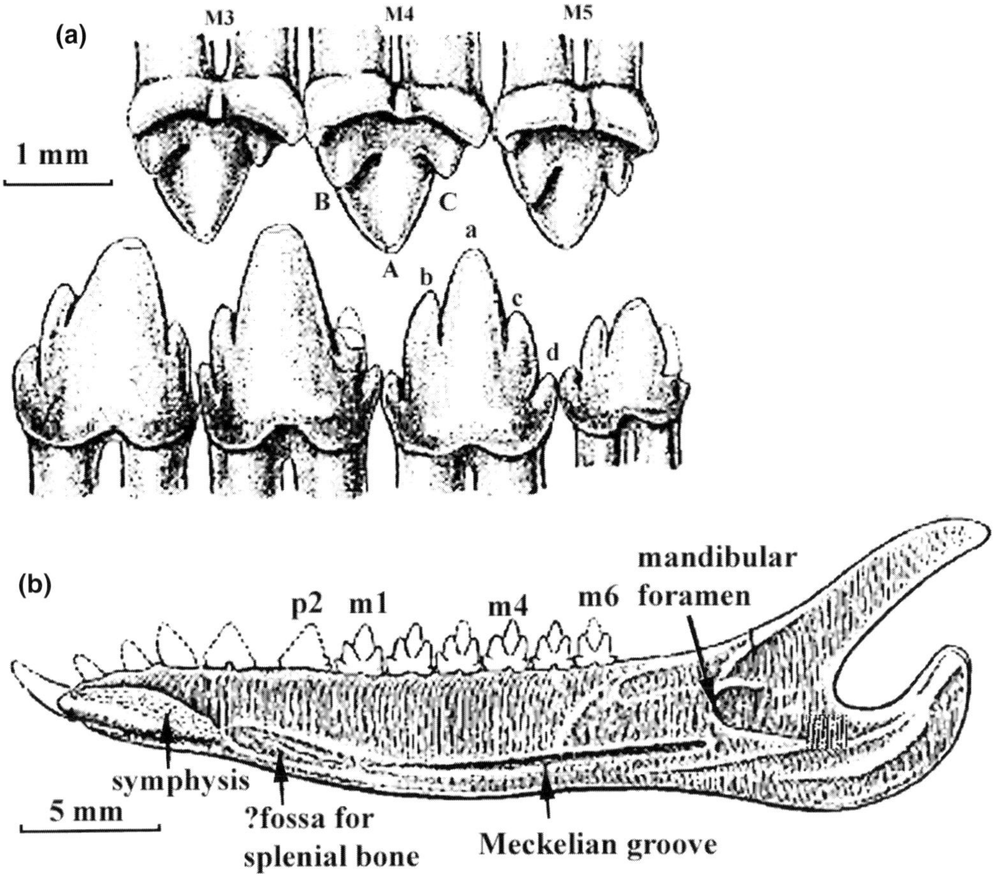

Figure 17. Teeth and a medial view of the lower jaw of *Zhangheotherium*, showing the Meckelian groove separated from the mandibular foramen. Adapted from Hu *et al.* (1997).

Holotheria implies that the middle ear bones evolved twice within the Mammalia.

Symmetrodonta

The Symmetrodonta (Simpson, 1925) includes a variety of taxa, known primarily from dental material (e.g., Cassiliano and Clemens, 1979; Fox, 1968, 1985; Cifelli and Madsen, 1999). The validity of the group Symmetrodonta is discussed in some detail by Cifelli and Madsen (1999). An exception to the poor preservation of the symmetrodonts is *Zhangheotherium* from the late Jurassic or early Cretaceous of China (Hu *et al.*, 1997). Interestingly, *Zhangheotherium* is known from a nearly complete animal, but characters of the dentition, lower jaw and postcranial skeleton still create problems of homoplasy within the Mammalia. Phylogenetically, *Zhangheotherium* is resolved as more closely related to the Theria than are monotremes and multituberculates (Hu *et al.*, 1997). Again, the problems with homoplasy have to do with the resolution of the monotremes and multituberculates at the base of the Mammalia (*sensu* Rowe) clade.

Hu *et al.* (1997) observed that *Zhangheotherium* shows a cylindrical and finger-shaped promontorium on the basicranial region of the skull, similar to *Sinoconodon*, *Morganucodon*, triconodontids, and multituberculates. They noted that this morphology is linked to an uncoiled cochlea of the inner ear, and differs from more derived therians possessing an oval and bulbous promontorium and a cochlea that is coiled at least 270 degrees. The latter is characteristic of monotremes, suggesting that the cochlear coiling is homoplasious. Along with *Morganucodon* (Kermack *et al.*, 1973) and *Kuehneotherium* (Patterson, 1956; Mills, 1964), lower jaws of *Zhangheotherium* retain a Meckelian groove (Hu *et al.*, 1997), absent in monotremes and multituberculates. For further details about the development of the middle ear see Wang *et al.* (2001).

The Meckelian groove in *Zhangheotherium* is separated from the mandibular foramen at the back of the jaw (Figure 17), resembling the condition in advanced symmetrodonts, eupantotheres and tribosphenic mammals (Hu *et al.*, 1997; Kielan-Jaworowska *et al.*, 1998). However, in *Kuehneotherium* and other early taxa such as *Morganucodon* and *Sinoconodon*, the groove contacts the mandibular foramen. This character also has some bearing on the newly discovered, possibly placental mammal from the early Cretaceous of Australia. This is discussed further below.

Homoplasy indicated by the postcranial skeleton of *Zhangheotherium* is also discussed in a following section.

Yinotheria

The higher-level group Yinotheria (Chow and Rich, 1982) was established for the late Jurassic *Shuotherium dongi* from China, a taxon with a superficially tribosphenic dentition, represented by lower molars with a basined talonid. However, the preservation of these lower molars in the mandible indicated that the talonid basin was on the anterior part of the tooth, rather than the trigonid. This reversed the positions of the trigonid and talonid in the typical tribosphenic condition. A phylogenetic analysis by Wang *et al.* (1998) established the Yinotheria as the sister taxon to the Cladotheria, the latter including both prototribosphenic and tribosphenic groups. Thus, the 'pseudotribosphenic' dentition of *Shuotherium* was derived independently from the true tribosphenic condition. Other taxa possessing this unusual condition and assigned to the Yinotheria include a second species of *Shuotherium* from the same unit as *S. dongi* and a second taxon from the Jurassic of England (Sigogneau-Russell, 1998). Interestingly, *Shuotherium* has been resolved as sister taxon to the new clade Australosphenida, a Gondwanan group of mammals in which a true tribosphenic condition has been derived separately from Laurasian forms (the new taxon Boreosphenida). This analysis (Luo *et al.,* 2001) is discussed in further detail at the end of this chapter.

Cladotheria

In the classification of McKenna and Bell (1997), Cladotheria is classified at the same taxonomic level ('Legion') as the Symmetrodonta. The Legion Cladotheria, in fact, includes all remaining taxa discussed in this section. Within the Cladotheria, non-tribosphenic taxa were previously referred to the taxon Eupantotheria, including the Peramuridae (*Peramus*) and the Dryolestidae. The eupantotheres possess an angular process on the lower jaw that symmetrodonts lack (Kraus, 1979).

Dryolestoidea

The Dryolestoidea (Butler, 1939) are assigned to the Holotheria based on the possession of the reversed triangle dentition (Figure 18(a)). Dryolestids are well known from the late Jurassic of North America and Europe (Martin, 1999), but less well during the Cretaceous (Kraus, 1979; Bonaparte, 1990; Martin, 1999). However, in the Cretaceous of Gondwana (Campanian, Los Alamitos Formation of Argentina), the dryolestoids are represented by the Family Dryolestidae and two new families, the Mesungulatidae and the Regitheriidae (Bonaparte, 1990).

In the Northern Hemisphere, the Campanian is characterized by a diversity of tribosphenic mammals (Bonaparte, 1990; Flynn and Wyss, 1998). Conversely, tribosphenids have not been identified in the Los Alamitos fauna. Bonaparte (1990) attributes the absence of dryolestoids in the Northern Hemisphere to extinction

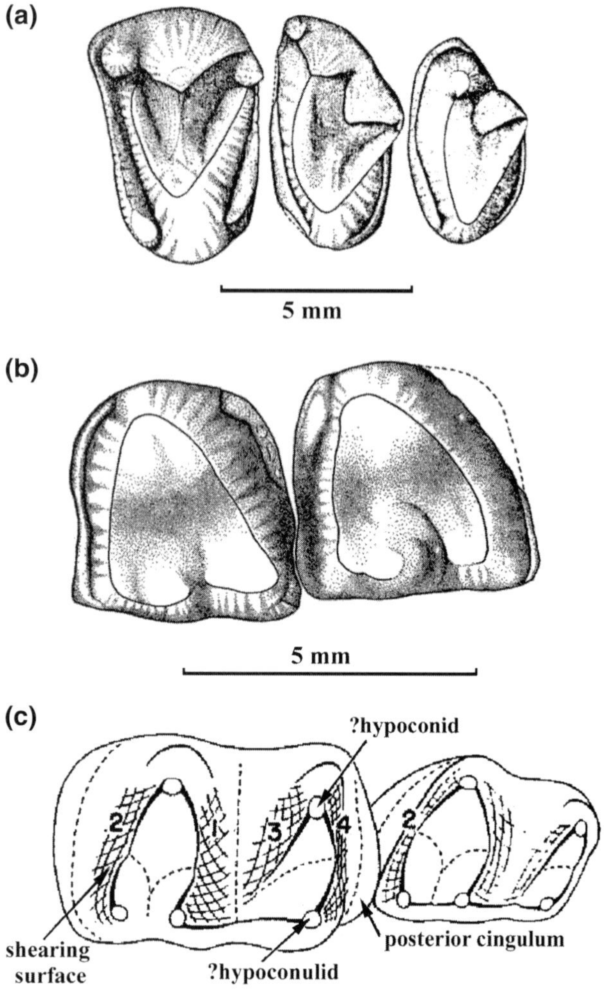

Figure 18. Molars:**(a)** upper (top) and **(b)** lower (bottom) molars of the South American dryolestid *Mesungulatum houssayi* (from Bonaparte, 1986b); **(c)** lower molar of *Steropodon galmani* (from Kielan-Jaworowska *et al.*, 1987).

caused by tribosphenic taxa, and suggests the presence of dryolestoids in the Cretaceous of Gondwana is due to the absence of tribosphenic competitors.

Additionally, Bonaparte (1990) and other researchers (Pascual *et al.*, 1992a, b; Archer *et al.*, 1993) suggested the trigonid basin on the lower molars of *Obdurodon* and *Steropodon* (Archer *et al.*, 1985) resembled the dryolestid molar as a whole. Bonaparte postulated a derivation of *Steropodon* from an ancestral dryolestid condition, such that the large talonid basin of *Steropodon* (Bonaparte, 1990) represented a further development of the dryolestoid condition. Hopson (1994), however, suggested that the dryolestids have secondarily lost the talonid basin.

Prototribosphenida

Prototribosphenida is the group consisting of the last common ancestor of *Vincelestes* and the Theria and all descendants. As noted by McKenna and Bell (1997), the name Prototribosphenida was originally erected in an unpublished Ph.D thesis (Rougier, 1993), and they use the term 'Zatheria' instead. However, Prototribosphenida is used by Wible *et al.* (1995) and Rougier *et al.* (1996), and is also retained here. This group is also similar to the taxon Eupantotheria of previous researchers (Kraus, 1979).

Vincelestidae

Vincelestes neuquenianus (Bonaparte, 1986b) comes from the early Cretaceous La Amarga Formation, Argentina (Bonaparte and Rougier, 1987), and is a crucial fossil taxon in terms of therian phylogeny (Figure 1). *Vincelestes* possesses the reversed triangle dentition with a small protocone on the upper molars and a small talonid 'heel' (Figure 19). A true 'basined' talonid was lacking, such that the occluding protocone and talonid characteristic of the tribosphenic dentition was absent. More importantly, skull, mandible and postcranial material of *Vincelestes* was also recovered, providing a wealth of characters for phylogenetic analyses (the postcranial material has not yet been fully described). Taxa at this grade of evolution ('pre-tribosphenic') were most commonly known only from teeth and jaws (Kraus, 1979). Information from *Vincelestes* provides and will provide a unique insight into the evolution of the cranial and postcranial skeletons just prior to the origin of the Tribosphenida and the Theria.

Based on this character information, *Vincelestes* is generally resolved as the sister taxon to the Tribosphenida (Bonaparte and Rougier, 1987; Wible, 1990; Rowe, 1993; Wible and Hopson, 1993; Hopson, 1994; Hu *et al.*, 1997). Importantly, the cochlea of the inner ear is coiled more than 270 degrees, supporting the more crownward position of *Vincelestes*.

Tribosphenida

Flynn *et al.* (1999) proposed defining the Tribosphenida (McKenna, 1975) as 'the first mammal possessing a dentition in which the protocone occludes with the talonid basin, plus all of its descendants'. This is an 'apomorphy-based definition' (de Queiroz and Gauthier, 1990), in which a derived character is used to define a group. Flynn *et al.* (1999) could have used a definition based on common ancestry, for example, 'the last common ancestor of *Aegialodon* and the Theria, and all descendants.' However, the cladogram presented by Flynn *et al.* (1999) indicates that *Aegialodon* forms an unresolved polytomy with three other fossil tribosphenid taxa at the base of the Tribosphenida (*Kermackia* (Slaughter, 1971); *Slaughteria* (Butler, 1978); *Potamotelses* (Fox, 1972)). Flynn *et al.* (1999) also described the oldest known tribosphenic mammal, *Ambondro mahabo*, from the middle Jurassic of Madagascar, but phylogenetically, it too formed a polytomy at the base of the Tribosphenida (Flynn *et al.*, 1999). Most recently, Luo *et al.* (2001) resolved *Ambondro* in a clade of Gondwanan taxa (including the monotremes), resulting in the convergent acquisition of a tribosphenic dentition in this clade and in a Laurasian mammal clade. This contradicts the 'Tribosphenida' as defined by Flynn *et al.* (1999).

Aegialodontida

Aegialodon dawsoni (Kermack *et al.*, 1965) was based on a single lower molar from the early Cretaceous of England. Even so, this molar was believed to mark an important evolutionary step in the development of the mammalian dentition. The lower molar of *Aegialodon* possessed a well-developed talonid with a wear pattern consistent with the presence of an equally well-developed, occluding protocone on the upper molar (Crompton, 1971). Thus, *Aegialodon* possessed a tribosphenic dentition, something taxa like *Vincelestes* (with no true talonid basin) lacked.

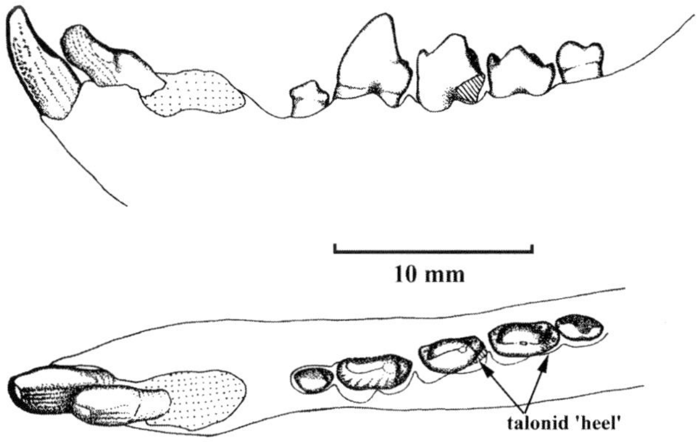

Figure 19. Lower jaw and teeth of *Vincelestes neuquenianus* (from Bonaparte, 1986a).

THERIA

The Theria represent the group Marsupialia + Placentalia. These groups are featured in more detail in other chapters in this book, and will not be discussed further here.

TRENDS IN MORPHOLOGY

A recent trend in higher-level phylogenetic analyses of Mammalia/Mammaliamorpha has been the shift away from the use of dental characters and the morphology of the lateral braincase wall to other cranial (particularly basicranial) and postcranial characters. Although cranial and postcranial material has been known from *Morganucodon*, *Megazostrodon* and other 'triconodont' taxa for many years (Kermack, 1963; Jenkins and Parrington, 1976; Crompton and Jenkins, 1979; Kermack *et al.*, 1981), new specimens of 'triconodont' (*Jeholodens*), symmetrodont (*Zhangheotherium*) and prototribosphenic taxa (*Vincelestes*) are providing a wealth of new data, particularly for taxa phylogenetically closely related to the Theria.

SIDE WALL OF THE BRAINCASE

As noted above, one of the primary divisions of the Mammalia traditionally was between the Prototheria or 'non-therians' (monotremes, triconodonts, multituberculates, docodonts) and the Theria (marsupials and placentals). This separation was based, in part, on the morphology of the side wall of the braincase (Hopson, 1964, 1970; Hopson and Crompton, 1969; Crompton and Jenkins, 1979; Kermack and Kielan-Jaworowska, 1971; Hopson and Rougier, 1993).

The braincase wall of the Prototheria included a lateral extension of the petrosal called the anterior lamina, and in the case of *Morganucodon*, an alisphenoid was also present (labelled 'as' in Figure 10). In the Theria, such as *Didelphis* (Figure 10(f)), this wall was formed solely by the alisphenoid bone. The anterior lamina of the petrosal ('al') was absent in the Theria, possibly resulting from the expansion of the squamosal ('sq') onto the braincase (Kemp, 1983).

However, work by Presley and Steel (1976) and Presley (1981) indicated that both the anterior lamina of the petrosal and the alisphenoid formed in the same membranous wall of the braincase, called the sphenoobturator membrane. Thus, the two bones were homologous, and differences only occurred when the bone forming in the membrane either fused to the petrosal, or became part of the alisphenoid. The division of Mammalia into Prototheria and Theria could not then be maintained on this basis.

Hopson and Rougier (1993) have recently challenged these interpretations, based on information provided by *Vincelestes* (Bonaparte, 1986b; Bonaparte and Rougier, 1987). *Vincelestes* possesses both an alisphenoid and anterior lamina of the petrosal (Figure 10(d)), separated by a distinct, digitating suture. This suggests the presence of two separate, and separately forming bones within the sphenoobturator membrane that subsequently sutured to one another. The maxillary and mandibular branches of the trigeminal nerve (V2, V3) pierce the anterior lamina, and the maxillary nerve opens on the suture between the two bones (Hopson and Rougier, 1993).

The presence of two separate bones is also seen in *Morganucodon* and *Sinoconodon* (Hopson and Rougier, 1993), suggesting that this condition is plesiomorphic for the Mammaliamorpha. Hopson and Rougier (1993) noted that the presence of an anterior lamina of the petrosal pierced by the V2 and V3 nerves is the plesiomorphic condition and is not a synapomorphy of the Prototheria. From this condition:

(1) the anterior lamina is lost and the alisphenoid expands backwards in marsupials and placentals and any taxon that shares a common ancestor with these (like the deltatheroidians with the marsupials);
(2) the alisphenoid is lost and the anterior lamina of the petrosal extends anteriorly to replace it in monotremes, multituberculates and possibly some triconodonts (but not *Morganucodon* and *Sinoconodon*).

Hopson and Rougier (1993) describe the first condition as a synapomorphy for the marsupials and placentals but noted that whether the second condition is a synapomorphy for Monotremata + Multituberculata or is convergent is extremely uncertain. Recent phylogenetic analyses (Rowe, 1993; Rougier *et al.*, 1996) suggest the multituberculates are more closely related to the Theria than are monotremes, based largely on the postcranial skeleton. Indeed, Wible and Hopson (1993) noted that the triconodont and multituberculate postcranial skeleton is more similar to Theria than is the monotreme postcranial skeleton.

However, researchers such as Meng and Wyss (1995) presented evidence from the morphology of the middle bones of the inner ear suggesting monotremes and multituberculates may form a monophyletic group; Miao (1993) lists references back to Cope (1888) that have also proposed this group. Wible and Hopson (1993) list two additional synapomorphies for the Monotremata + Multituberculata, including the replacement of the jugal in the zygomatic arch by an elongate process of the maxilla (Hopson *et al.*, 1989). The results of Hopson and Rougier (1993) may provide more evidence for this relationship, although Miao (1993), analysing cranial morphology, was unable to propose any unequivocal synapomorphy. In any event, the grouping of triconodonts, multituberculates, docodonts, and monotremes as the Prototheria cannot be maintained on the morphology of the braincase wall.

DENTITIONS

Teeth have also traditionally played an important role dividing the Mammalia into groups, and largely paralleled divisions based on braincase morphology. The Theria or Holotheria (*sensu* Hopson, 1994; Wible *et al.*, 1995) were distinguished from the 'triconodont' condition by the possession of a 'reversed triangle' pattern of cusps and crests on opposing upper and lower molars (Wible *et al.*, 1995). Researchers were able to trace an evolutionary sequence of dental development, from the straight-line 'triconodont' dentition through to the reversed-triangle symmetrodont dentition.

The next step involved taxa adding a cingulum to the upper molar (in the position of the protocone) and small additional cusps to their lowers; examples are *Peramus* and *Amphitherium*, variously called the pantotheres or eupantotheres, and referred to the Cladotheria above. This development extended through to *Aegialodon* and other tribosphenic mammals with a well-developed protocone and talonid basin (Kermack *et al.*, 1965; Crompton and Jenkins, 1968; Crompton and Sita-Lumsden, 1970; Crompton, 1971; Hopson, 1994).

In the absence of either upper and lower molars, the presence of a tribosphenic condition could be evaluated by the recognition of appropriate wear surfaces. Crompton (1971) established a developmental series of dental wear surfaces between occluding upper and lower molars from the triconodont through to the tribosphenic condition. In the latter, the protocone produced recognizable anterior and posterior wear surfaces in the opposing talonid basin. This work has had some implication for Australian early Cretaceous mammalian fossils, for example, the monotreme *Steropodon* (Figure 18(c)). Lower molars originally thought to be tribosphenic appeared to lack the requisite wear surfaces within the talonid (Kielan-Jaworowska *et al.*, 1987).

Interpretation of the early Cretaceous mammal *Ausktribosphenos* (Rich *et al.*, 1997, 1998, 1999) as the first reported Mesozoic placental from Australia was based, in part, on wear surfaces within the talonid basin. These wear surfaces are not the distinct anterior and posterior surfaces produced by the margins of a protocone within the confines of the talonid basin, but this may be due to extensive wear (Rich *et al.*, 1997, 1999).

Although the dentition of the Holotheria was well-defined, that of the Prototheria was a bit of a grab bag of all other dental types—basically, taxa lacking the reversed triangle dentition of the Holotheria. Cusps on the teeth of triconodonts like *Morganucodon* (Figure 11) form a straight line, but evidence from the basicranium indicates that taxa possessing this dentition are spread throughout the mammaliamorph cladogram, rather than grouping as a monophyletic taxon (Wible *et al.*, 1995; Rougier *et al.*, 1996). Other prototheres appeared to have unusual squarish molars with many cusps, defying a neat interpretation relating to the reversed triangle condition. These taxa included the multituberculates (Clemens and Kielan-Jaworowska, 1979) and the docodonts (Kron, 1979; Lillegraven and Krusat, 1991), illustrated in Figures 12–14.

A group such as the Prototheria cannot be characterized as monophyletic by the lack of a derived character (reversed triangle dentition), and perhaps more importantly, there are difficulties in homologizing the rows of multituberculate cusps and the blade systems of monotremes with the cusps and crests of other mammals. This is unfortunate when some of the most promising mammal specimens, for example, the Cretaceous monotremes, are known only from teeth. Although the Holotheria can be diagnosed based on dental characters, other taxa may not be. Indeed, the analysis of Luo *et al.* (2001) indicates that the taxon Tribosphenida, based on the presence of an upper molar protocone and lower molar occluding basin, may not be a monophyletic group—with these characters occurring convergently in Gondwanan and Laurasian groups.

PETROSALS

Recent analyses of the Mammaliamorpha/Mammalia have incorporated characters from the basicranial region of the skull. The robust petrosal bone in this region represents the bony otic capsule, surrounding the inner ear ('pr' in Figure 20). Isolated petrosals of taxa like the triconodonts *Trioracodon*, *Triconodon* and *Morganucodon* have been known for some years (Kermack, 1963), but it seems that investigators have been more confident recently in assigning these petrosals to a wider variety of higher-level Mesozoic taxa (e.g., Luo, 1994; Meng and Fox, 1995; Wible *et al.*, 1995; Rougier *et al.*, 1996), and using them in phylogenetic analyses. Additionally, the skulls of a number of phylogenetically basal mammaliamorphs are available for study (e.g., *Adelobasileus*, *Sinocodon* and *Morganucodon*).

Lucas and Luo (1993), Luo (1994) and Luo *et al.* (1995) have developed an evolutionary sequence of petrosal development from non-mammalian cynodonts such as *Probainognathus* and *Thrinaxodon* to taxa such as *Adelobasileus* and *Morganucodon* (Figure 20). In most non-mammalian cynodonts, the otic capsule is composed of separate bones including the prootic, basisphenoid, basioccipital and opisthotic. In the cynodont *Thrinaxodon* (Luo, 1994; Luo *et al.*, 1995) shown in Figure 20(a), a projection ('bsw') of the basisphenoid bone ('bs') reaches backwards to contact the oval window ('ow'), and the prootic ('po') and opisthotic bones ('ops') are not fused. However, in the two putative sister taxa to the Mammaliamorpha, these bones are either fused (Tritylodontidae) or fused in adults with sutures between the two bones in juveniles (Trithelodontidae)—as demonstrated by Luo *et al.* (1995). In *Adelobasileus*, the prootic and opisthotic bones are also fused and the basisphenoid projection ('wing') is reduced, while in *Morganucodon*, the projection is lost completely.

Other phylogenetically important information from the petrosal involves the cochlea (the auditory sensory organ of the inner ear) and the development of the bones housing this structure (Figure 20). As with the petrosal as a whole, Luo *et al.* (1995) were able to develop an evolutionary sequence of the development of the cochlea from non-mammals to the Mammaliamorpha. Luo *et al.* (1995) recognized two mammaliamorph synapomorphies: the elongation of the cochlea, and the development of the promontorium to house the cochlea (also shown by Crompton and Luo, 1993). The presence of a floor to the cavum epiptericum housing the ganglion of the trigeminal nerve was suggested as an additional synapomorphy (Lucas and Luo, 1993).

Among the basal mammaliamorph taxa, the skull of *Adelobasileus* shows an 'incipient promontorium' at the posteroventral edge of the skull (Figure 20(b)), and a partially floored cavum epiptericum (Lucas and Luo, 1993; Luo *et al.*, 1995; Wible *et al.*, 1995). A large basisphenoid wing contributes to the promontorium of *Adelobasileus*, as in non-mammalian cynodonts. The size of the promontorium of *Adelobasileus* is also less inflated than that of *Morganucodon* (Figure 20(c)) (Luo *et al.*, 1995).

Lucas and Luo (1993) considered the state of development of the promontorium in *Adelobasileus* to be intermediate between cynodonts and other mammaliamorph taxa. Additionally, the floor of the cavum epiptericum complete fossil materials. Whether this can be extended to recent results provided by molecular data is debatable, and will be discussed briefly below.

MOLECULES

Recent molecular analyses of interest in this chapter concern studies involving the molecular clock. Under clock models, rates of evolutionary change in certain molecules and amounts of genetic differentiation between taxa can be combined to estimate the divergence times of taxa from one another. Rates ('substitution rates', or amount of molecular change per unit time) are calibrated by using estimated divergence times from the fossil record. These rates of change may not be similar between different higher-level taxa, or between different molecules, but attempts are made to account for these differences in the analyses. Problems may also arise in the calibration date used, as different calibration dates can affect substitution rates (Bromham *et al.*, 1999); also see Jow *et al.* (2002).

MOLECULAR CLOCK—DIVERGENCE TIMES

Work by Springer *et al.* (1997) and Kumar and Hedges (1998) focused on divergence of extant placental groups from one another, while Hedges *et al.* (1996) and Waddell *et al.* (1999) considered divergence times between birds and placentals. Phylogenetic analyses using morphological characters have not been entirely successful in resolving placental relationships (reviewed by Novacek, 1992), so it seems reasonable to also use information provided by molecules. However, molecular evidence suggests that various placental orders diverged some time in the Cretaceous. For example, Kumar and Hedges (1998) have five placental groups diverging more than 100 million years ago (early Cretaceous) and most modern orders of placentals diverging before the end of the Cretaceous. Springer *et al.* (1997) suggest a divergence of their 'African clade' (including hyraxes, manatees, elephants, aardvarks) from other placental mammals at about 91 million years. Hedges *et al.* (1996) have Rodentia-Artiodactyla, Primates-Rodentia and Rodentia-Artiodactyla divergences near or after 100 million years, and certainly well prior to the Cretaceous-Tertiary boundary (~65 m.y.a); see also Springer *et al.* (2003). Waddell *et al.* (1999) also provide divergence dates between many placental groups earlier than this boundary, up to 169 m.y.a (middle Jurassic) for the split between the armadillo (Order Edentata) and the remainder of the placentals tested. This rivals the age of the oldest tribosphenic mammal at approximately 167 million years in Madagascar (Flynn *et al.*, 1999). *Ambondro* was resolved by Flynn *et al.* (1999) as part of a polytomy at the base of the Tribosphenida, but possessed certain characters which prevented an assignment to the Placentalia.

CONTRADICTORY FOSSIL EVIDENCE

These molecular divergence results are contrary to the fossil evidence, because the Cretaceous placentals present around this time include taxa such as *Asioryctes* and *Kennalestes* from Asia (Kielan-Jaworowska, 1975), *Gypsonictops* (Simpson, 1927) from North America, and *Lainodon* (Gheerbrant and Astibia, 1994) from Europe; see Rich *et al.* (1997) for a summary. McKenna and Bell (1997) have classified the majority of these as 'Superorder Lepticida', separate from Insectivora and perhaps near the base of the placental clade (Novacek, 1992). In fact, the majority of modern or extant placental orders have their first fossil records in the Paleocene and Eocene, and no higher-level placental orders are represented by Cretaceous fossils (Foote *et al.*, 1999).

As noted in the first section of this chapter, the results of these molecular clock analyses are problematic, whether the more traditional concept of Mammalia is used (including taxa like *Adelobasileus*, *Sinoconodon* or *Morganucodon*) or that of Rowe (1988). If modern placental orders diverged in the Cretaceous, then the rest of the traditional Mammalia must have evolved very quickly after their first appearance (*Adelobasileus*, late Triassic). Or, the origins of the traditional Mammalia could be much earlier than indicated by *Adelobasileus*, with the vagaries of preservation in the fossil record accounting for what would have to be a substantial amount of missing diversity.

Rowe's concept of Mammalia includes the common ancestor of monotremes, placentals and marsupials and all descendants, but Flannery *et al.* (1995), commenting on the diversity shown by fossil monotremes in the early Cretaceous, place monotreme origins in the Jurassic. This still rivals the divergence times for armadillos and rodents from the rest of placentals suggested by Waddell *et al.* (1999).

EXPLAINING DISCREPANCIES

Basically, fossils and molecular clock models conflict, with evidence from the latter being too old for the evidence provided by the former. Either the fossil record is too incomplete to preserve the history of placentals and by implication other mammals, or the molecular dates overestimate times of divergence.

Foote *et al.* (1999) examine the first problem, an incomplete fossil record, by estimating how low the rate of fossilization had to be for all species in a group to escape detection, over a specified interval of geologic time. With regards to the modern placental orders, they considered the observed diversity of these groups in the Tertiary and their hypothesized divergence time in the Cretaceous, and noted this implied a period of missing placental history of about 64 million years. They estimated the fossilization or preservation rate for the Cretaceous that would correspond to this missing history and then compared this rate to preservation rates of other late Cretaceous mammals. The estimated fossilization rate for placental taxa in the Cretaceous would have had to have been very low, about an order of magnitude lower than for other Cretaceous mammals. Because of these discrepancies, Foote *et al.* (1999) concluded that the Cretaceous placental diversity suggested by the molecular evidence is not simply missing due to an incomplete fossil record (also Flynn *et al.*, 1999). Estimates of rates of molecular evolution in different placental orders may need to be further revised.

Although the molecular estimates for divergence times currently appear to be overestimates, recently described fossils, like the tribosphenic mammals from the Jurassic of Madagascar (Flynn *et al.*, 1999) are themselves pushing back previous estimates of taxon origin. In the case of the Tribosphenida, the oldest forms previously known were early Cretaceous in age (McKenna and Bell, 1997).

One intriguing suggestion is that the origins of modern mammalian taxa, such as the various placental orders, occur in areas with a very poor fossil record (Bromham *et al.*, 1999; Foote *et al.*, 1999). For example, the 'African clade' of Springer *et al.* (1997) suggests an origin for several groups, including elephants and some insectivores, in the Cretaceous. As Bromham *et al.* (1999) noted, late Cretaceous terrestrial deposits are poorly known in Africa, so these origins may simply not be preserved. In fact, Rich *et al.* (1997, 1999) note up to three Cretaceous mammal localities in Africa, perhaps the best known of which preserves an early Cretaceous form (*Tribotherium africanum* (Sigogneau-Russell, 1991)) lacking placental and marsupial characteristics. This matches, to some extent, the better known early Cretaceous record in North America, where several taxa also lack these characters, and have been referred to as 'Theria of meta-eutherian grade' (Slaughter, 1971; Kielan-Jaworowska *et al.*, 1979). Even though these African localities are early Cretaceous in age, rather than late, the absence of recognizable placental mammals makes an African origin of placental groups in the Cretaceous seem unlikely.

Other more recent studies that may help to resolve the relative ages of different groups include Archibald *et al.* (2001), Lou *et al.* (2001) and Ji *et al.* (2002).

MAMMALIA, MAMMALIAFORMES, MAMMALIAMORPHA—WHERE DID THEY ORIGINATE?

Of all the taxa discussed in the sections above, *Adelobasileus* is the oldest (Lucas and Hunt, 1990). Described from the latest Triassic of Texas, *Adelobasileus* was said to be 10 million years older than taxa from the Triassic of Europe. Lucas and Hunt (1990) suggested that this shifts the search for mammaliamorph (as used in this chapter) origins from the Old World to the New World, but this is not strictly correct.

In fact, the question of mammaliamorph origins should be answered through an examination of the phylogenetic relationships of the group. The results of these analyses indicate which taxa are phylogenetically closest to the base of the mammaliamorph clade and so closest to the origins of the clade. The sister taxa to the Mammaliamorpha, such as the Trithelodontidae and Tritylodontidae, can also provide evidence.

If *Adelobasileus* had been resolved in phylogenetic analyses as a member of the Tribosphenida, then it would say little about the origin of the clade Mammaliamorpha, regardless of its age. Quite conveniently, however, *Adelobasileus* has been resolved as the most basal mammaliamorph taxon in some of the phylogenetic analyses discussed in the sections above. Other analyses (Hopson, 1994; Luo, 1994) show *Adelobasileus* and *Sinoconodon* at an unresolved polytomy at the base of the mammaliamorph radiation.

Thus, the origins of the Mammaliamorpha could reside near` *Adelobasileus* in Texas, or perhaps equally, with *Sinoconodon* in China. Both taxa are late Triassic in age, although *Adelobasileus* is older (Lucas and Hunt, 1990; Lucas and Luo, 1993). In this instance, the fossil record corresponds closely with evidence provided by phylogenetic relationships.

Further evidence on biogeographic origins can be provided by the distribution of outgroup taxa. The presence of an outgroup taxon and a basal ingroup taxon in an area indicates their common ancestor (marking the origin of the ingroup clade) may have also occurred in the region. Nearly all tritylodont taxa are Jurassic in age, and have a worldwide distribution including southern Africa, China, Great Britain, Arizona and Mexico (Clark and Hopson, 1985). The trithelodonts are slightly more restricted, being known predominantly from the Jurassic of South Africa, but also from eastern Canada (Shubin *et al.*, 1991).

Adelobasileus occurs in Texas, while the outgroup tritylodont taxa *Oligokyphus* and *Kayentatherium* (Sues, 1986) are known from the Kayenta Formation in Arizona (Jenkins *et al.*, 1983); *Nearctylodon* (Lewis, 1986) is now considered a synonym of *Kayentatherium*. The Kayenta Formation is younger than the Tecovas (early Jurassic; Sues, 1986), but still establishes the presence of tritylodont outgroups with basal mammaliamorphs like *Adelobasileus* in the southern United States.

Sinoconodon occurs, in China, with tritylodonts such as *Bienotheroides*, *Bienotherium* and *Lufengia* (Sun, 1984). It may be therefore difficult to pinpoint the specific area in which the traditional Mammalia or Mammaliamorpha *sensu* Rowe (1988, 1993) originated, but it appears that this origin occurred in the Northern Hemisphere, if the tritylodonts are used as the outgroup. However, the morganucodontid *Megazostrodon* (Jenkins and Crompton, 1979) occurs in South Africa, and in the same formation (the Stormberg Group) as the trithelodontids mentioned above (Shubin *et al.*, 1991). This might suggest a Gondwanan origin for the Mammaliamorpha, however, *Megazostrodon* is derived within this group, compared to *Adelobasileus* and *Sinoconodon* (Rougier *et al.*, 1996).

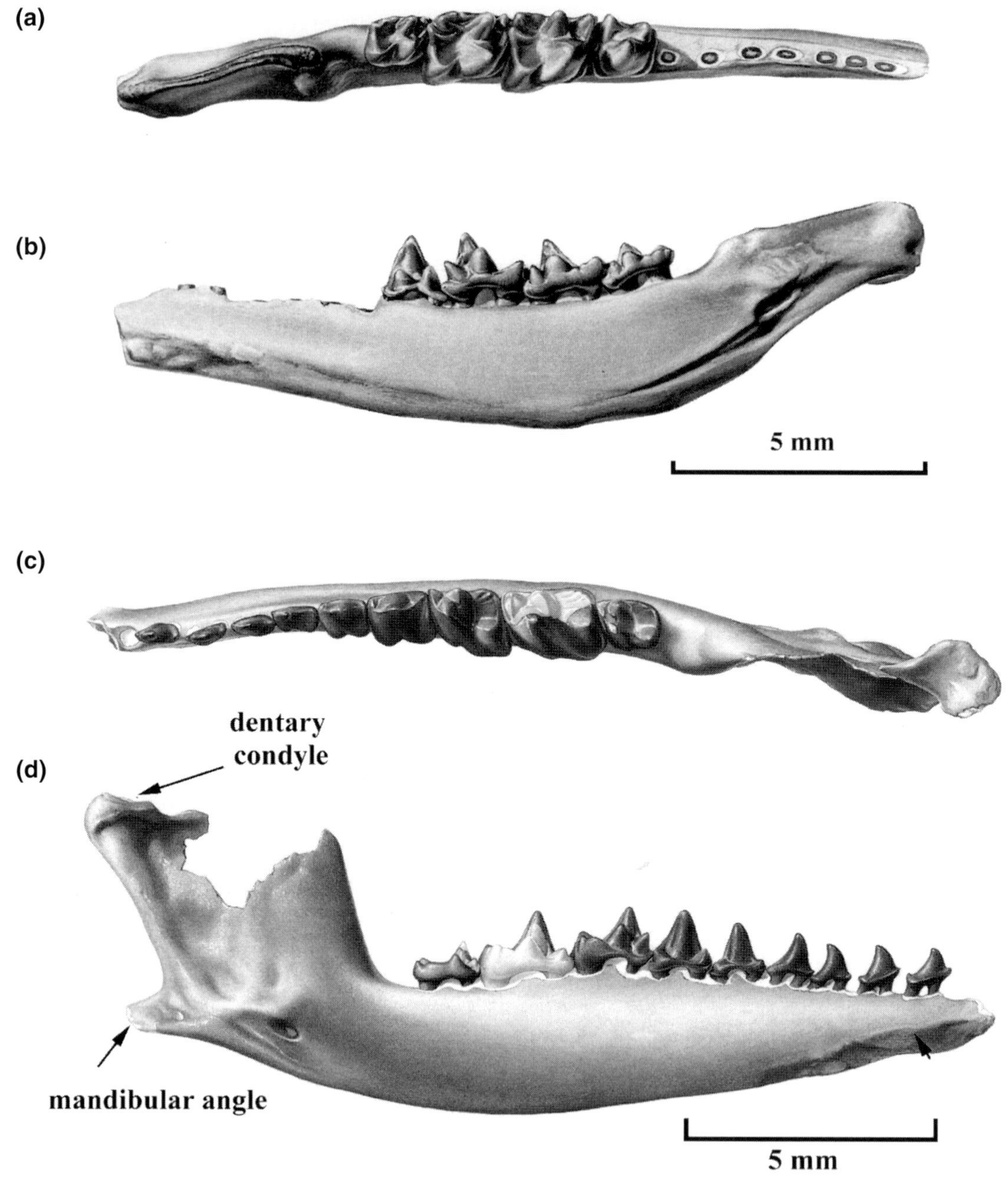

Figure 21. Mandibles of early Cretaceous Australian ?placental mammals. *Ausktribosphenos nyktos*: **(a)** occlusal view; **(b)** lingual view (from Rich *et al.*, 1999). *Bishops whitemorei*: **(c)** occlusal view; **(d)** lingual view (from Rich *et al.*, 2001).

EARLY MAMMALS IN AUSTRALIA

Australia, thus far, appears to contribute little to the question of mammaliamorph origins, but it has much to say about the origins of Mammalia *sensu* Rowe. Cynodonts have not been described, although a small number of specimens have been found in the Arcadia Formation of south central Queensland, and identified as a close relative of the dicynodont *Kannemeyeria* (Thulborn, 1983, 1990). *Kannemeyeria* is known from the Triassic of South Africa. The dicynodonts are a separate clade from the derived Cynodontia (Hopson and Barghusen, 1986). Animals comparable to *Adelobasileus* and *Sinoconodon* are unknown from Australia, so the very basal nodes of the Mammaliamorpha (or Mammalia in the traditional sense) and its outgroups are not represented in Australia.

Compared to the Northern Hemisphere, Australian Mesozoic mammals are rare. Until recent discoveries at the Flat Rocks and Dinosaur Cove localities, in Victoria (Rich *et al.*, 1997, 1999, 2001), these mammals were limited to early Cretaceous monotremes from the Lightning

Ridge region of New South Wales (Archer *et al.*, 1985; Flannery *et al.*, 1995). The distribution of monotremes outside Australia and New Guinea is restricted to the Paleocene *Monotrematum sudamericanum* (Patagonia; Pascual *et al.*, 1992a, b). This distribution indicates some degree of biogeographic continuity between Australia and South America. A variety of other taxa, including lungfishes, are also shared between these areas (Pascual *et al.*, 1992a). However, the non-therian and non-tribosphenic mammals dominating late Cretaceous deposits of the Patagonian region of South America (Pascual *et al.*, 1992a, b), are so far unknown from the Australian Cretaceous (Flynn and Wyss, 1998; Rich *et al.*, 1999; Archibald, 1999).

Two taxa have recently been described from the Flat Rocks locality, Victoria, and assigned to the Placentalia although their relationships are controversial. The first taxon, *Ausktribosphenos nyktos* (Rich *et al.*, 1997; Figure 21(a, b)), was described as a tribosphenic mammal based on the presence of a tribosphenic wear pattern in the talonid basin. This implies that the upper molars would possess an equally well-developed protocone. Rich *et al.* (1997) differentiated *A. nyktos* from monotremes in the possession of this wear pattern and considered *A. nyktos* to be a placental based on the presence of three molars in the lower jaw and a submolariform last premolar. Marsupials and the sister group to the marsupials, the Deltatheroida (Rougier *et al.*, 1998) have four molars and the most posterior premolar is simple, not submolariform in appearance.

Controversy surrounded the identification of *Ausktribosphenos* for several reasons. One was the absence of an angular process on the holotype lower jaw, which is more characteristic of non-tribosphenic taxa such as *Kuehneotherium* (Kraus, 1979; Kielan-Jaworowska *et al.*, 1998; Archibald, 1999). Also, the postdentary trough on the internal surface of the holotype lower jaw of *Ausktribosphenos* contacted the mandibular foramen, a character seen in *Kuehneotherium* and other early taxa such as *Morganucodon* and *Sinoconodon* (Rich *et al.*, 1998, 1999). Kielan-Jaworowska *et al.* (1998) suggested a derivation of *Ausktribosphenos* from primitive Gondwanan symmetrodonts.

Rich *et al.* (1998) countered these suggestions with new specimens from the Flat Rocks site and direct comparisons to the basal placental mammal *Prokennalestes* from the early Cretaceous of Asia (Kielan-Jaworowska and Dashveg, 1989). Specifically, new lower jaws showed that *Ausktribosphenos* did in fact possess an angular process on the lower jaw, but that this was not inflected (the marsupial condition). Further, this process was located under the dentary jaw joint. This mirrored the condition in *Prokennalestes* and more derived fossil placentals like *Asioryctes*, also from the late Cretaceous of Asia (Rich *et al.*, 1998), but differed from non-mammalian cynodonts and *Morganucodon* (Jenkins *et al.*, 1983; Rich *et al.*, 1998). In the latter, the angular process is anterior to the condyle (Rich *et al.*, 1998; Figure 16(a)).

A second reason for the controversy involved the presence of wear facets in the talonid basin of *Ausktribosphenos*. Rich *et al.* (1998) reiterated their identification of these tribosphenic wear surfaces, noting that absence of distinct and separate shearing surfaces within the talonid was due to an advanced state of wear. Recently, new mandibles from the Flat Rocks locality were referred to a second placental taxon, *Bishops whitmorei*, possessing three lower molars showing wear within the talonid basin, indicating the presence of an occluding protocone on the upper molars. Interestingly, the hypoconulid and entoconulid cusps on the lower molar talonid are separated rather than 'twinned' (Rich *et al.*, 2001). The latter is more characteristic of the Marsupialia.

If *Ausktribosphenos* and *Bishops* do prove to be placental mammals, then the biogeographic implications will be important. Rich *et al.* (1999) have explored these issues in some detail, focusing on the absence of tribosphenic mammals in the Cretaceous of South America. Rich *et al.* (1998, 1999) suggested that placental mammals may have originated in Australia and rafted northward with south-east Asian terranes in the Mesozoic. Using information provided by Metcalfe (1996), Rich *et al.* (1999) noted that the West Burma microplate is part of Eastern Gondwana in the late Jurassic, but part of the Eurasian land mass at the end of the Cretaceous. This would account for the presence of early Cretaceous placentals, such as *Prokennalestes*, in Asia.

More intriguingly, the phylogenetic implications of the identification of *Ausktribosphenos* (*Bishops* was not included in the analysis) as a mammal possessing a tribosphenic dentition have recently been addressed by Luo *et al.* (2001). Using a cladistic analysis of 55 dental and mandibular characters and 21 taxa (including the Southern Hemisphere forms *Ambrondo*, *Ausktribosphenos* and *Steropodon*), they identified two major mammalian clades, Australosphenida and Boreosphenida. Biogeographically, Australosphenida represented a Gondwanan mammal clade (including all monotremes), and Boreosphenida, a Laurasian clade. Most significantly, the analysis indicated that the tribosphenic dentition arose separately within the Australophenida (absent in *Steropodon*, but present in *Ambondro* and *Ausktribosphenos*) and in the Boreosphenida. As well, *Ausktribosphenos* is separated from *Prokennalestes* and *Asioryctes*, and from the Placentalia as a whole. One character separating Australosphenida from Boreosphenida was the posterior placement of the mandibular angle to the level of the dentary condyle, or jaw joint (Luo *et al.*, 2001). As noted above, certain mandibles of *Ausktribosphenos* possess a posterior angle (Rich *et al.* 1998, 1999), as does *Bishops* (Rich *et al.*, 2001). In *Bishops* (Figure 21(d)), the angle appears to be positioned beneath the dentary condyle, although perhaps the angle is not as posteriorly directed compared to undoubted placentals, such as *Prokennalestes* (see comparison in Kielan-Jaworowska *et*

Simmons, N. B. 1993. Phylogenetic systematics and the early history of mammals. Pp.146–164. In Szalay, F.S., Novacek, M.J. and McKenna, M.C. (eds). *Mammal Phylogeny, Volume I.* Springer-Verlag, New York. 249pp.

Simpson, G.G. 1925. Mesozoic Mammalia III. Preliminary comparison of Jurassic mammals except multituberculates. *American Journal of Science*, Series 5, Vol. 10, pp.559–569.

Simpson, G.G. 1927. Mammalian fauna of the Hell Creek Formation of Montana. *American Museum Novitates*, No. 267, pp.1–7.

Simpson, G.G. 1928. *A Catalogue of the Mesozoic Mammalia in the Geological Department of the British Museum.* British Museum (Natural History), London. 215pp.

Slaughter, B.H. 1971. Mid-Cretaceous (Albian) therians of the Butler Farm local fauna, Texas. Pp.131–144. In Kermack, D.M. and Kermack, K.A. (eds). *Early Mammals. Zoological Journal of the Linnean Society*, Vol. 50.

Springer, M.S., Cleven, G.C., Madsen, O., de Jong, W.W., Waddell, V.G., Amrine, H.M. and Stanhope, M.J. 1997. Endemic African mammals shake the phylogenetic tree. *Nature*, Vol. 388, pp.61–64.

Springer, M.S., Murphy, W.J., Eizirik, E. and O'Brien, S.J. 2003. Placental mammal diversification and the Cretaceous-Tertiary boundary. *Proceedings of the National Academy of Sciences of the United States of America*, Vol. 100, pp.1056–1061.

Sues, H.D. 1985. The relationships of the Tritylodontidae (Synapsida). *Zoological Journal of the Linnean Society*, Vol. 85, pp.205–217.

Sues, H.D. 1986. The skull and dentition of two tritylodontid synapsids from the lower Jurassic of western North America. *Bulletin of the Museum of Comparative Zoology (Harvard)*, Vol. 151, pp.217–268.

Sun, A.L 1984. The skull of the tritylodontid *Bienotheroides* (Tritylodontia, Reptilia) from Sichuan Province. *Scientia Sinica,* Series B, 1984 No.3, pp.257–268.

Thulborn, R.A. 1983. A mammal-like reptile from Australia. *Nature*, Vol. 303, pp.330–331.

Thulborn, R.A. 1990. Mammal-like reptiles of Australia. *Memoirs of the Queensland Museum*, Vol. 28, Pt. 1, p.169.

Trofimov, B.A. 1978. The first triconodonts (Mammalia, Triconodonta) from Mongolia. *Reports, Academy of Sciences of the U.S.S.R.*, Vol. 243, pp.213–216.

Waddell, P.J., Cao, Y., Hasegawa, M. and Mindell, D.P. 1999. Assessing the Cretaceous superordinal divergence times within birds and placental mammals by using whole mitochondrial protein sequences and an extended statistical framework. *Systematic Biology*, Vol. 48, pp.119–137.

Wang, Y., Clemens, W.A., Hu, Y. and Li, C. 1998. A probable pseudo-tribosphenic upper molar from the late Jurassic of China and the early radiation of the Holotheria. *Journal of Vertebrate Paleontology*, Vol. 18, pp.777–787.

Wang, Y., Hu, Y., Meng, J. and Li, C. 2001. An ossified Meckel's cartilage in two Cretaceous mammals and origin of the mammalian middle ear. *Science*, Vol. 294, pp.357–361.

Wible, J.R. 1990. Petrosals of late Cretaceous marsupials from North America and a cladistic analysis of petrosal features in therian mammals. *Journal of Vertebrate Paleontology*, Vol. 10, pp.183–205.

Wible, J.R. 1991. Origin of Mammalia: the craniodental evidence reexamined. *Journal of Vertebrate Paleontology*, Vol. 11, pp.1–28.

Wible, J.R. and Hopson, J. 1993. Basicranial evidence for early mammal phylogeny. Pp.45–62. In Szalay, F.S., Novacek, M.J. and McKenna, M.C. (eds). *Mammal Phylogeny, Volume I.* Springer-Verlag, New York. 249pp.

Wible, J.R. and Hopson, J. 1995. Homologies of the prootic canal in mammals and non-mammalian cynodonts. *Journal of Vertebrate Paleontology*, Vol. 15, pp.331–356.

Wible, J.R., Rougier, G.W., Novacek, M.J., McKenna, M.C. and Dashzeveg, D. 1995. A mammalian petrosal from the early Cretaceous of Mongolia: implications for the evolution of the ear region and mammaliamorph interrelationships. *American Museum Novitates*, No. 3149, pp.1–19.

Woodburne, M.O. and Tedford, R.H. 1975. The first Tertiary monotreme from Australia. *American Museum Novitates*, No. 2588, pp.1–11.

Chapter 26
Furry Egg-layers: Monotreme Relationships and Radiations

Anne M. Musser
Palaeontology, Australian Museum, 6 College Street, Sydney, N.S.W. 2010 Australia
E-mail: annem@austmus.gov.au

INTRODUCTION

Living monotremes—the egg-laying platypus and echidnas—have intrigued and confounded zoologists and palaeontologists since they were first brought to the attention of Western science in the late 1700s. No other living mammals are oviparous, and none share other monotreme features, which include a bony bill or beak, the ability to detect electrical fields of prey, and an ancient type of shoulder girdle that would be at home in long-extinct reptiles (cynodonts) that were the ancestors of living mammals. Monotremes today survive only in Australia, New Guinea and some offshore islands, but they have an ancient history that almost certainly predates the Cretaceous and, geographically, spans at least the eastern part of the great supercontinent of Gondwana.

'Monotremata' comes from the Greek 'mono', meaning 'one', and 'tremos', or 'hole', referring to the single orifice (or cloaca) in monotremes for passage of both reproductive products (sperm and eggs) and excretory products (urine and faeces). Although there are other mammals that can be considered cloacate, none has the comparatively primitive 'reptilian' reproductive system that monotremes do. Living monotremes share numerous primitive (or plesiomorphic) characters; however, monotremes can't be said to represent primitive, or Mesozoic, mammals in any general way. Both the amphibious platypus and terrestrial, insectivorous echidnas are so specialized for their particular niches that they bear little resemblance to an archetypal Mesozoic mammal.

At present three monotreme families are recognized: an extinct early Cretaceous family (Steropodontidae) that includes *Steropodon galmani* from Lightning Ridge and (provisionally) *Teinolophos trusleri* from Flat Rocks, Victoria; the platypuses (Ornithorhynchidae); and the echidnas (Tachyglossidae). A second early Cretaceous mammal, *Kollikodon ritchiei* (Kollikodontidae; Flannery *et al.* 1995), is described as a highly derived bunodont monotreme; however, further analysis casts doubt on this interpretation and suggests that, for the present, *K. ritchiei* should be considered a basal mammal or mammaliaform of uncertain affinities.

In a recent classification of all mammals above the species level, extant or extinct, monotremes were elevated from an order (Monotremata) to a subclass (Prototheria). Platypus-like monotremes are grouped together at the ordinal level (order Platypoda; families Steropodontidae and Ornithorhynchidae) as are the echidnas (order Tachyglossa; family Tachyglossidae) (McKenna and Bell 1997).

This recent revision has not yet been incorporated by many researchers in the field and, unfortunately, dispensed with the well-known term Monotremata—using the term Prototheria. While an elevation in rank is justified, my objection to Prototheria rests on the fact that Monotremata is recognized as a monophyletic group and is a widely-used, descriptive term; Prototheria is not descriptive and has traditionally included a wide range of unrelated, non-therian mammals. The term Monotremata is used in this account.

There are many questions raised by the enigmatic monotremes. How ancient are they? How far did they roam? Without a dental record, can we determine where the edentate echidnas came from? To which of the known mammal groups, living or extinct, are monotremes most closely related? In fact, there is debate as to whether monotremes are really mammals at all, and about their phylogenetic position; placement in the mammalian family tree differs greatly depending on the researcher's particular focus. This chapter discusses the morphology and palaeontology of monotremes, reviews the historical debates on phylogenetic affinities and considers developments of the past 15 years.

Many aspects of monotreme morphology, relationships and biogeography have been studied previously. For example, Griffiths (1978) discussed morphology and physiology in detail, while Grant (1995, 2004), Augee and Gooden (1997) and Grant *et al.* (2004) provide excellent summaries of anatomy, physiology and lifestyles of extant species.

LIVING MONOTREMES

THE PLATYPUS

The semi-aquatic platypus, *Ornithorhynchus anatinus*, is the only surviving member of the Ornithorhynchidae. Ornithorhynchids are considered to be more primitive than tachyglossids because they have a much longer palaeontological record and appear to have more ancestral character states for a number of features—see Musser and Archer (1998).

The geographical range of *Or. anatinus* extends along the east coast of Australia from Cooktown (south of Cape York) to Tasmania. The platypus primarily feeds on a range of aquatic invertebrates, sifted from bottom sediments, but will on occasion take larger prey such as frogs. Platypuses are generally crepuscular, usually feeding at dawn, dusk or through much of the night. They shelter and raise their young in burrows dug into banks along rivers and streams.

The platypus has often been described as an odd collection of disparate animal parts. No one that has held a living platypus or watched one swim in a stream or billabong could fail to be impressed by the natural beauty and elegant design of this small, amphibious mammal.

The body is streamlined, almost torpedo-shaped, with short but powerful limbs, a flat, tapered tail and, of course, the inimitable bill. Propulsion is accomplished by alternate strokes of the forelimbs while the hindlimbs and tail act as rudders. Each hand, or manus, is equipped with extensive, fan-shaped webbing that aids in swimming but which is retracted under the forepaw when the platypus is moving overland or burrowing. Short, dense, waterproof fur, ranging from light to very dark brown, covers the body. Adult males have hollow ankle spurs connected to venom glands over the thighs although female platypuses lose their spurs on reaching adulthood. The sharp spurs are primarily used by males in combat during the breeding season.

The bill of the platypus is an extraordinary sense organ. The underlying structure is formed by the bones of the snout and shape is given to the bill by an extensive sheet of cartilage. The bill is covered by soft, sensitive skin. Platypuses, which feed and mate underwater, habitually keep normal senses 'shut down.' The eye and ear are housed together in a horizontal groove that can be tightly shut, keeping water from entering but effectively rendering the platypus blind and deaf. Flaps of skin seal the dorsally located nostrils when swimming, shutting down the normal mammalian sense of smell (although the platypus does have Jacobson's Organ in the roof of the mouth, a more reptilian odour-detecting system). The bill is imbued with thousands of tiny specialized pores (or rod organs) that are either electroreceptors or touch (mechano-) receptors and together create a unique dual sensory system (Pettigrew *et al.*, 1998). Using these cutaneous receptors, the platypus can react to both touch and to the small electric stimuli given off by active prey like a crayfish. This highly sophisticated system—shared with echidnas but most developed in the platypus (Scheich *et al.*, 1986)—has been perfected over millions of years and has surely been instrumental in the success of this ancient group. It is a completely different sensory system from those of living therian mammals, which have developed vibrissae (or whiskers) on the face but not electroreceptive capabilities.

Like the edentate echidnas, the platypus has developed a unique solution to the loss of true teeth in the adult. Juveniles have vestigial molar teeth that are retained until the platypus begins to feed on its own soon after leaving the burrow. Adults masticate their food with flat dental pads of keratin that form over the molar toothbeds. The dental pads mimic the teeth they've replaced: the upper dental pads are subdivided into two sections, corresponding to the two upper molars in Tertiary ornithorhynchids (see Figure 9), while the lower dental pads additionally have a small posterior segment that corresponds to the small, button-like lower third molar in Tertiary platypuses. The platypus has an unusual tongue structure; two keratinous spikes, at the anterior end of a round pad at the base of the tongue, assist mastication. Two anterior, longitudinal keratinous ridges substitute for an anterior dentition, acting as grasping structures to hold active prey.

Platypuses are still fairly common over their range (see Figure 13) although there have been local declines and extinctions since European settlement. Much of their habitat has been altered or destroyed through clearing and grazing, however, and they are currently classed as 'potentially vulnerable'.

ECHIDNAS

Echidnas (family Tachyglossidae), sometimes called spiny anteaters or porcupines, have taken a very different evolutionary path. They have lost all vestiges of a dentition (lacking even a dental lamina). Their rounded, bird-like crania end in tubular bony beaks and their long tongues are coated with sticky saliva to trap their invertebrate prey. Echidnas are covered by keratinous spines as well as by fur. They have dome-like bodies and their skeletons are much more robust than that of the platypus. The forefeet have wide, spade-like claws used in digging. The hind feet are rotated outward, giving the echidnas a peculiar stance. Long claws on the hind feet are used for grooming between the spines rather than for digging. Echidnas have spurs on the ankles, but the venom apparatus is non-functional and distribution of the spurs between males and females is less consistent (female echidnas occasionally have spurs while they are sometimes absent in males). Like the platypus, echidnas have both electroreceptors and touch receptors in the skin of their beaks.

The phenomenon of morphological convergence can be seen in the way tachyglossids have independently evolved many of the specializations seen in placental anteating mammals. Particularly striking are the elongation of the rostrum into a bony, toothless beak and the correlated reduction of the lower jaw seen in both tachyglossids and in South American anteaters (xenarthrans). Both groups have long, extensible tongues coated with saliva and large front claws to dig into ant and termite mounds. In an interesting parallel, both monotremes and some xenarthrans have retained the septomaxilla on the snout, a small, splint-like bone otherwise seen only in reptiles and some early mammals.

Echidnas use their tongues for both the capture and mastication of prey. The short-beaked echidna, *Tachyglossus aculeatus*, feeds primarily on ants and termites. The distal portion of the sticky tongue is highly mobile, darting in and out of termite mounds or ant nests at lightning speed. *Zaglossus* species, long-beaked echidnas from New Guinea, feed primarily on oligochaete worms and insect larvae although they also feed occasionally on ants and termites. *Zaglossus* species have spines in a deep groove on the distal end of the tongue; these sharp hooks are used to snare worms. The long palates in both types of echidna have transverse keratinous spines that work against opposing spines on the posterior part of the tongue to grind the invertebrate prey into a paste, which is then swallowed.

The two living genera are distinguished primarily by the length and shape of the beak and by the distribution of spines covering the body. *T. aculeatus* is found in both Australia and New Guinea. It has, as its name implies, a relatively short beak that may be either straight or slightly upturned but does not have a marked downward curvature. Its spines are long, sharp and usually black-tipped, and cover the back and tail. Long-beaked echidnas today are found only in New Guinea (Flannery and Groves, 1998) although similar echidnas are known as fossils from the Australian mainland. *Zaglossus* species have proportionately longer beaks with downward curvatures of varying degrees, and have smaller, less numerous spines than *T. aculeatus*. In some *Zaglossus* the spines do not cover the back and some *Z. bruijnii* have spines on the belly as well as on the back. *T. aculeatus* is a small echidna of about 40 cm in length. *Zaglossus* species are much more variable: the single known specimen of *Z. attenboroughi* is about the size of a typical *T. aculeatus*; *Z. bruijnii* is of medium size; and the variable *Z. bartoni* reaches nearly one metre in length in the eastern part of its range (Flannery and Groves, 1998).

T. aculeatus is quite possibly the most widely distributed Australasian mammal. It is found in suitable habitat throughout Australia (including Tasmania) and also occurs on offshore islands such as Kangaroo Island and King Island. In New Guinea, unfortunately, neither its range nor its habits are well documented. The status of the several named subspecies is also questionable; Augee and Gooden (1997) believe that two forms—the mainland form (which includes New Guinean *Tachyglossus*) and the distinctive Tasmanian form—probably deserve subspecific status.

Figure 1. The eastern New Guinea long-beaked echidna *Zaglossus bartoni* illustrates the basic monotreme body plan: a bony beak or bill covered by sensitive skin, small eyes, compact body with short, splayed limbs and hind feet that are rotated outwards. Echidnas have long, tubular beaks, vertical aural slits rather than external ear pinnae and spines covering the body. Long-beaked echidnas have long, downcurved beaks and a sparse covering of spines (from Augee and Gooden, 1997).

Taxonomy of New Guinea *Zaglossus* has recently been revised by Flannery and Groves (1998), who base their study on re-examination of disputed features (e.g. the number of claws on the feet and certain cranial features) and on statistical analyses of these features plotted against the geographical ranges of the known specimens. Prior to this study it was generally accepted that *Z. bruijnii* was the sole living, although variable, species of *Zaglossus*. In a nutshell, the species from western New Guinea (including Salawati) that normally has three clawed digits on each forefoot is considered to be *Zaglossus bruijnii,* the type species for the genus. *Z. bruijnii* occurs from mountain peaks to sea level (on the low-lying island of Salawati) although *Zaglossus* is usually considered a mountain-dweller. A second species of *Zaglossus*, normally with five clawed digits on each forefoot, has been put into the resurrected species *Z. bartoni*, with four subspecies (distinguished on size and proportional differences along the east-west morphocline) proposed by Flannery and Groves (1998). *Z. bartoni* occupies the mountain summits of the central cordillera and the Huon Peninsula. The rarest *Zaglossus*, the newly discovered *Z. attenboroughi* from the high, remote peaks of the Cyclops Mountains, differs from *Z. bruijnii* and *Z. bartoni* in having: a relatively short, straight beak; small size; and fur of a different colour and texture. It has five claws on the forefoot. Superficially *Z. attenboroughi*, with its small size and short, straight beak, resembles a cross between *Zaglossus* and *Tachyglossus*.

T. aculeatus appears to be an echidna success story; the ants and termites it usually feeds on occur all over the Australian continent, and its numbers currently appear to be stable. Its status in New Guinea is not known, although it may be hunted for food. Species of *Zaglossus*, however, are in greater danger and *Z. attenboroughi* may already be extinct. *Z. bruijnii* and *Z. bartoni* are often taken by hunters and some subspecies of *Z. bartoni* in particular should be considered endangered. Much of New Guinea, however, remains as wilderness, raising hopes for survival of these unique mammals.

MONOTREME MORPHOLOGY

Monotremes have a mix of primitive features in combination with very specialized features called, appropriately, a mosaic; this confounding but fascinating situation in the monotremes has made the phylogenetic placement of these ‘living fossils’ an ongoing and hotly debated exercise. The concept of mosaic evolution is intriguing; a group of mammals may have branched off the tree at a very early stage but may have evolved highly sophisticated adaptations. In addition, they may have convergently acquired traits that define more advanced mammal groups. Certain recent fossil discoveries display such mosaic complexes; for example the eutriconodont *Jeholodens jenkinsi* (described by Ji *et al.*, 1999) has an advanced shoulder girdle but primitive pelvis and hindlimb.

The phenomenon of mosaic evolution underscores the need to look at monotreme morphology as a unit, not limiting examinations to any single system such as dental structure or genetic data. Investigations on specific areas such as the ear region (Meng and Wyss, 1995), basicranium (Wible and Hopson, 1993) and braincase (Kielan-Jaworowska, 1971) in addition to molecular studies (e.g. Westerman and Edwards, 1992) have been of enormous help in clarifying the relative placement of these features in comparison to other mammal groups. However, these studies have also resulted in a blizzard of cladograms with placement of monotremes in every conceivable position along the line—see pp.491–522 (Johanson, 2006). The situation is complicated by the fact that many of the earliest mammals are known primarily from teeth and jaws—critical features such as the development of the skull or shoulder girdle are unknown for many key groups.

SKULL STRUCTURE

Ornithorhynchids and tachyglossids share many cranial features, although they differ in details such as bill or beak shape. The majority of these characteristics are primitive features although certain specializations are shared between the two. To explain the many plesiomorphic features in monotremes, a palaeontologist must see these as either retained (never lost) within the lineage, or as ‘reversions’—features secondarily reverted back to after having been dispensed with by the group.

A partial list of plesiomorphic cranial features in living monotremes includes: a post-temporal canal linking the temporal fossa to the area above the ear (also present in stem reptiles, mammal-like reptiles and in some extinct mammals); an open ear region without an auditory bulla, but with primitive configurations of the incus, malleus and stapes and a semi-coiled cochlea (about 270°; in contrast to the cochlea of living therians, which is fully coiled at 360°); a septomaxilla in the snout (a large bone on the rostrum that is found in reptiles, therapsids, some early mammals and some living xenarthrans); and ectopterygoid bones on the ventral surface of the skull (possibly remnants of the reptilian pterygoids). Specializations include: the early fusion (and obliteration) of skull sutures; elongation of the snout into a bony bill or beak; extensive development of the trigeminal nerve to service the sensitive, electroreceptive beak (reflected in numerous foramina through the skull and rostrum); and loss of functional teeth in adult living monotremes.

Features of the jawbone (dentary) in living monotremes are difficult to assess because of their extreme specializations. The angular and coronoid processes are either reduced or absent. An internal mylohyoid process just above the medial mandibular foramen, however, appears to be a unique development in ornithorhynchids. The dental canal carrying nerves to the snout is large, although it is more developed in ornithorhynchids than in tachyglossids.

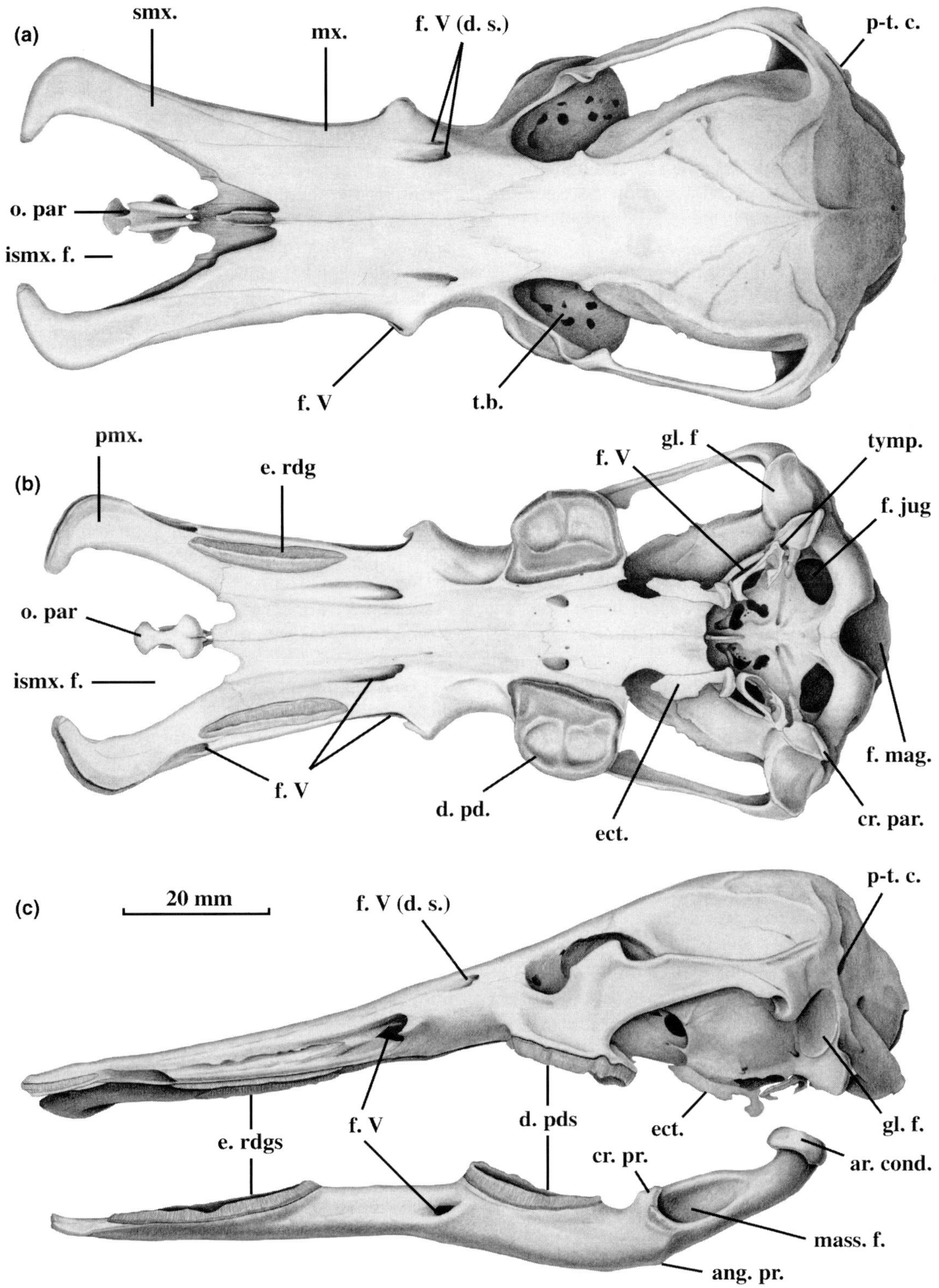

Figure 2. Skull and dentary of the living platypus *Ornithorhynchus anatinus* (from Musser and Archer, 1998): **(a)** dorsal view; **(b)** ventral view; **(c)** lateral view. Key to abbreviations: ang. pr. = (vestigial) angular process; ar. cond. = articular condyle; cr. par. = crista parotica; cr. pr. = coronoid process; d. pd(s). = dental pad(s); ect. = ectopterygoid; e. rdg(s). = epithelial ridge(s); f. jug. = foramen jugulare; f. mag. = foramen magnum; f. V = foramina for V (trigeminal nerve); f. V (d. s.) = foramina for V innervating dorsal shield; gl. f. = glenoid fossa; ismx. f. = interseptomaxillary fenestra; mass. f. = masseteric fossa; mx. = maxilla; o. par. = os paradoxum; pmx. = premaxilla; p-t.c. = post-temporal canal; smx. = septomaxilla; t.b. = tooth bed; tymp. = tympanic bone.

The skull of the platypus (Figure 2) does not look like that of any other mammal, past or present. The wide, flat bill is formed by rostral projections of the septomaxilla, maxilla and the premaxilla. A small, dumbbell-shaped bone (os paradoxum) is sandwiched between the prongs of the bill. Its purpose isn't clearly understood and it isn't present in any other mammal, but the bilateral openings of Jacobson's Organ lie in the constriction between the expanded sections of this odd bone. The dorsal nostrils open anterolaterally to the os paradoxum. Large foramina on the bill carry branches of the trigeminal nerve that service the electroreceptors and mechanoreceptors. The braincase is large and inflated although proportionately it is smaller than in echidnas. The zygomatic arches are thin and close to the skull, reflecting the reduced jaw musculature in this 'edentulous' mammal. Flat, perforated 'toothbeds' formed by expansion of the maxillae form the bases of the dental pads.

The ventral aspect of the platypus skull illustrates a number of specialized features. The palate is very long and perforated by large foramina for the trigeminal nerve, as is the basicranium. Keratinous dental pads substitute for a functional molar row. Two longitudinal ridges of hardened epithelium, midway between the dental pads and the tip of the bill, take the place of an anterior dentition and are used to hold and masticate prey. Some skull features are plesiomorphic: the tympanic region is completely open and no bulla is formed; in life, a cartilaginous ear funnel terminates over the horse-shoe-shaped tympanic bone supporting the tympanum as it may have in therapsid reptiles (Crompton, 2001, personal communication). Thin, plate-like ectopterygoids, sutured to the palatines, abut the tympanic bones.

The ephemeral molars of *Or. anatinus* were first described by Simpson (1929) and in greater detail by Green (1937). Green determined the dental formula to be I0/1-?5 C1/1 P1-2/1-2 M1-3/1-3 and that the last premolar, first upper molar and all three lower molars erupt. The small molars, which do not fill the tooth bed, vary in structure between individuals. The unique triangular blade systems, cuspule development and shallow roots seen clearly in fossil ornithorhynchids can just be discerned in the degenerate *Or. anatinus*, which is distinguished from Tertiary ornithorhynchids by its lower crowns, greater development of cuspules, more obliquely-oriented lophs, more dentate cingula and loss of the anterior cingulum on M1/ (Woodburne and Tedford, 1975).

In lateral view (Figure 2 (c)) the flattening of the skull and the downward deflection of the bill are clearly seen. The dentary has both anterior epithelial ridges and posterior dental pads that 'occlude' with their counterparts on the skull. The articular condyle is fairly well-developed, but the coronoid process has been lost and the angular process may or may not be present in *Or. anatinus*. The masseteric fossa is developed into a deep pocket. The small eye is placed high on the skull, an adaptation common in aquatic animals that float at the water's surface. The reptilian post-temporal canal passes between the skull and the posteriormost part of the zygomatic arch.

The short-beaked echidna (*Tachyglossus aculeatus*) has a comparatively short, straight beak without a marked curvature, while long-beaked echidnas have proportionately longer and more recurved beaks. The dorsal nostrils open through the median space (interseptomaxillary fenestra) between the rostral prongs. The foramina for branches of the trigeminal nerve to the skin of the bill are more numerous than in the platypus but much smaller because of the reduced role that electro- and mechanoreceptors play in the sensory apparatus of echidnas. The surprisingly large braincase is usually teardrop-shaped in *T. aculeatus* (Figure 3).

The palate in echidnas is even longer than in the platypus; the palate of *Z. bruijnii* is said to be the longest palate known in mammals. Robust ectopterygoids fused to the posterior palatines overhang the ear regions to create a protective shelf, although this can't be considered a substitute for an auditory bulla. Tympanic bones connected to the ear ossicles (missing from Figure 3) complete the middle ear. The footplate for the stapes is circular in monotremes, a plesiomorphic feature. There is no postglenoid process since the weak lower jaws have no need for buttressing.

The small, subterminal mouth opens ventrally and the mandible is bound to the upper jaw by ligaments in life, creating a 'tunnel' for the extensible tongue. The jaws cannot open any wider than the circumference of the tongue. The echidna mandible, perhaps comparatively the weakest lower jaw in Mammalia, has been reduced to two thin, tapered rods joined loosely at the symphysis. The angular process and articular condyle are poorly developed and the coronoid process is vestigial. The masseteric fossa has been lost entirely; the lower jaws do not operate in an up-and-down orthal movement. Instead, they twist about their long axes to open the mouth in a unique jaw-opening mechanism known only in echidnas (Murray, 1978a).

POSTCRANIAL SKELETON

The postcranial skeletons of the platypus and echidnas are similar in many ways. Most of the shared features are plesiomorphic, although living monotremes share specializations for digging (fossorial adaptations). The platypus is a more lightly-built animal with some aquatic specializations (especially in the flattened, streamlined body form) while the echidnas are robustly built diggers (Figure 4).

Monotremes have a sprawling stance where the limbs project horizontally from the body. The humerus projects from the body at least 45° and the hand (manus) normally contacts the ground lateral to the shoulder girdle (Sereno and McKenna, 1995). This posture is probably a fossorial adaptation developed from a primitive

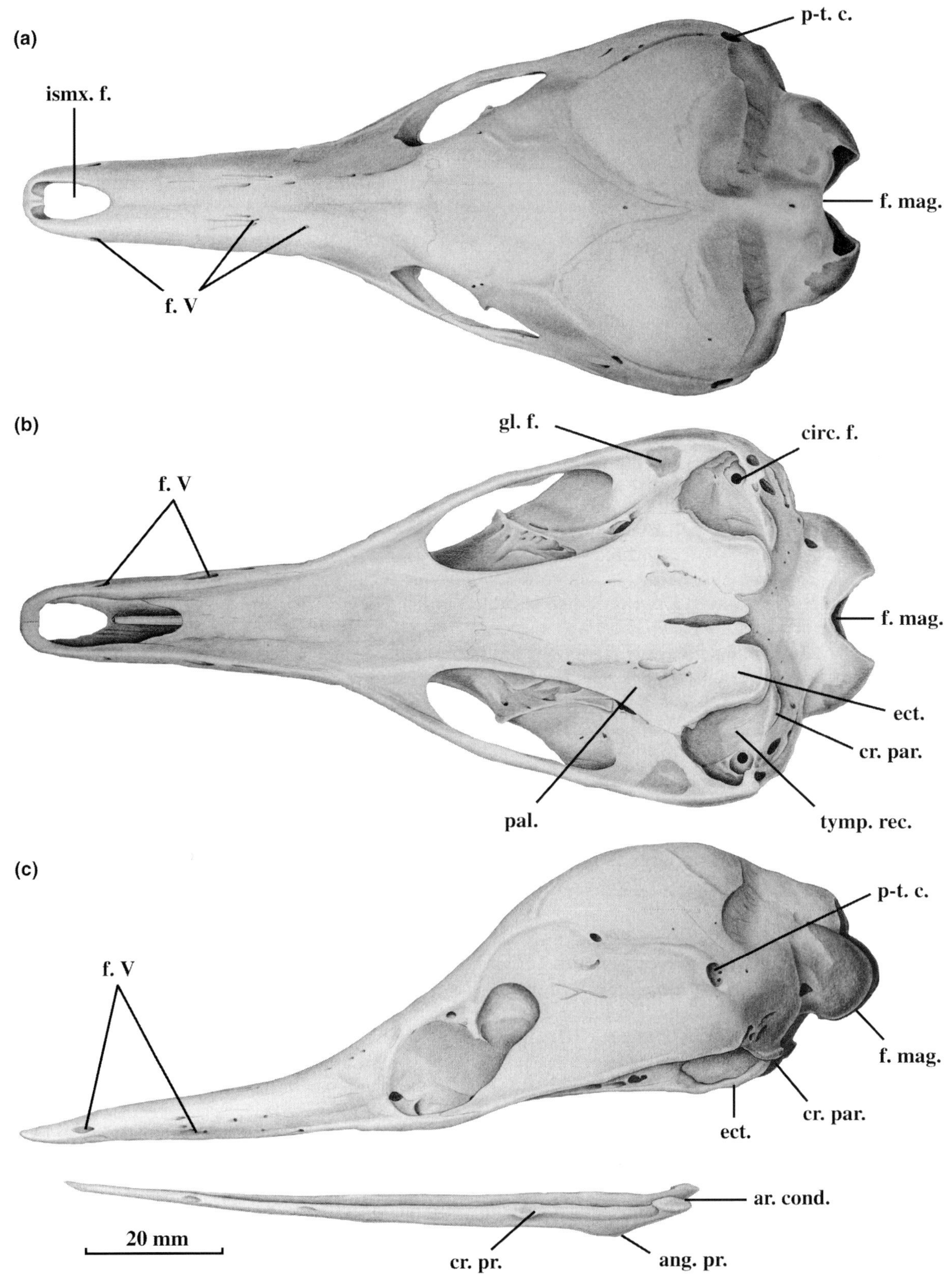

Figure 3. Skull and dentary of the short-beaked echidna *Tachyglossus aculeatus:* **(a)** dorsal view; **(b)** ventral view; **(c)** lateral view. Key to abbreviations: ang. pr. = (vestigial) angular process; ar. cond. = articular condyle; circ. f. = circular footplate for stapes; cr. par. = crista parotica; cr. pr. = (vestigial) coronoid process; ect. = ectopterygoid; f. mag. = foramen magnum; f. V = foramina for V (trigeminal nerve); gl. f. = position of glenoid fossa; ismx. f. = interseptomaxillary fenestra; pal. = palatine; p-t.c. = post-temporal canal; smx. = septomaxilla; tymp. rec. = tympanic recess.

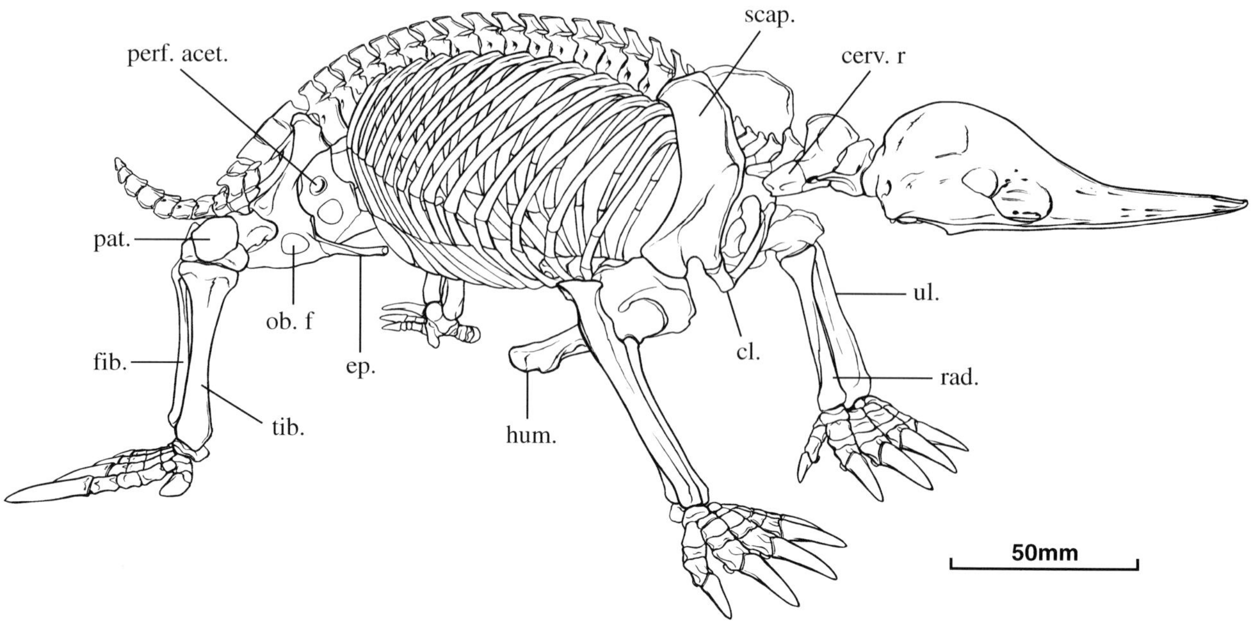

Figure 4. The skeleton of the short-beaked echidna *Tachyglossus aculeatus* (from Augee and Gooden, 1997). The robustly-built *T. aculeatus* has a bird-like skull, compact, domed body, large hands with spatulate digging claws, short tail, and an opening in the acetabulum. Like the platypus, echidnas have cervical ribs (on the vertebrae of the neck), a splayed posture, a therapsid-like shoulder girdle, sternal (costal) plates, epipubic bones, and outwardly-rotated hind feet. Key to abbreviations: cerv. r. = cervical rib; cl. = clavicle; ep. = epipubic bone; fib. = fibula; hum. = humerus; ob. f. = obturator foramen; pat. = patella; perf. acet. = perforated acetabulum; rad. = radius; scap. = scapula; tib. = tibula; ul. = ulna.

mammalian body plan rather than a reptilian stance (Jenkins, 1973). Limbs held at such an angle do result in a more reptilian mode of locomotion where the body undulates from side to side. In contrast, many Mesozoic mammals and all modern mammals (unless specialized) hold their limbs closer to the midline of the body (a parasagittal posture) which permits a faster, more 'mammalian' gait.

The monotreme vertebral column is plesiomorphic in several respects: the cervical vertebrae bear ribs (as in reptiles); the vertebrae are variable in number; and more of the lumbar vertebrae bear ribs than in therian mammals. The ribs do not have a head or tubercle but attach directly to the bodies of the vertebrae. Monotremes have large, flat, overlapping sternal ribs (ossified as in birds) that are separated by cartilage from the ribs attaching to the vertebrae. The sternal ribs, therefore, do not attach to the sternum by cartilage, as in other quadrupeds, but by bone. Echidnas have a xiphisternum while the platypus does not.

The shoulder girdle in monotremes (Figure 5) is similar to that of therapsid forebears and basal mammals although some elements (i.e. the bones of the sternum) are similar to those of more advanced mammals (Klima, 1973). A series of extra bones—a median, T-shaped interclavicle, procoracoids (also called precoracoids or epicoracoids, derived from the anterior coracoid of therapsids) and metacoracoids (derived from the posterior coracoid of therapsids)—are present (Klima, 1973). These bones buttress the clavicle and scapula, the only elements of the therapsid shoulder girdle still retained by other living mammals (at least as adults; the chondral part of the interclavicle can be seen during embryological development in marsupials and placentals) (Klima, 1973). The scapula is plesiomorphic in that the flat anterior plate of bone (the supraspinous fossa) for origin of m. supraspinatus and a central spinous process, as in therian mammals, has not been developed. The addition of the supraspinous fossa and its attached musculature in therians keeps the shoulder stable when running, an asymmetrical gait not possible with the rigid shoulder structure of monotremes (Bramble and Jenkins, 1989). The lateral orientation of the glenoid and the buttressing of the coracoid, interclavicle and clavicle are considered to be fossorial specializations.

Interclavicle, procoracoids and metacoracoids appear to have been lost or greatly reduced in nearly every mammal group for which the shoulder girdle is known. Recently discovered skeletal material for certain key early mammal groups have shown surprisingly advanced shoulder girdle morphologies (Sereno and McKenna, 1995; Hu *et al.*, 1997; Ji *et al.,* 1999). Although details differ, all of these mammals have jettisoned most of the extra bones and rigid structure retained in monotremes; loss of these bones allows for more mobile shoulder girdle articulation. The monotreme shoulder girdle is probably their most perplexing anatomical feature. However, it has been a useful retention in the fossorial monotremes: the extra bones provide extensive areas for attachment of pectoral musculature that assists in both digging and swimming, and the rigidity of the girdle helps keep the animal stable during these activities.

The forelimbs in both living monotreme families are well-developed, particularly in tachyglossids. The short, broad humerus is twisted across the shaft and is expanded at both ends for the attachment of robust forelimb musculature, as in therapsids, many archaic mammals and placental moles. The proximal end of the ulna is expanded and the ulna articulates with the humerus via a condyle; this structure, modified from the spiral ulnar condyle of cynodonts, is most similar to the elbow of some Jurassic mammals (Jenkins, 1973). Therian mammals have developed a groove between the ulnar and radial condyles, forming a trochlea (an indication of parasagittal posture). A humerus of early Cretaceous age—the first mammal from the Victorian Cretaceous—appears to be that of a monotreme; interestingly, the as yet undescribed humerus appears to have a pit (olecranon fossa) at the base of the humerus for the ulna, which would indicate that this monotreme, at least, may have had a more upright stance (Rich and Rich, 2000).

The pelvis in monotremes, although short and broad, is basically like that of primitive therian mammals and some advanced therapsids but with some specializations. Monotremes have epipubic bones, along with advanced therapsids, many early mammals, marsupials and archaic placentals (Novacek *et al.,* 1997); oddly, echidnas have a perforation in the acetabulum through which the head of the femur projects. Many features of the hindlimb are plesiomorphic while other features are specializations (the knee joint structure, elongation of the fibular process). The fibula is twisted posteriorly, which accounts in part for the posterolateral orientation of the hindfeet. The tarsus (ankle) is a very specialized version of the plesiomorphic mammalian (non-therian) tarsus.

The ankle (crural) spurs have traditionally been considered a uniquely monotreme feature. However, recent discoveries of well-preserved Mesozoic mammals with similar crural spurs (Jenkins and Schaff, 1988; Hu *et al.*, 1997) suggest that these structures are plesiomorphic.

SOFT ANATOMY AND PHYSIOLOGY

The most 'reptilian' anatomical system in monotremes is the reproductive system, designed, as in reptiles, for the very 'un-mammalian' purpose of laying eggs. The female reproductive tract is basically lizard-like: monotremes have paired oviducts, paired uteri and an unpaired urogenital sinus but no median uterus in which young develop. The eggs, which are coated by shell glands, pass through the urogenital sinus and out through the cloaca. The monotreme egg is meroblastic; this large yolked egg is similar to bird and reptile eggs but not to other mammals, which have eggs with small yolks.

Unlike most other mammals the males have internal testes and the penis is held internally, only everted when the occasion demands. The penis in both platypus and echidnas is subdivided at the head (bifid in the platypus and divided into four 'rosettes' in the echidnas). The penis is only used for copulation as urine passes out through the cloaca in both sexes.

A hatchling monotreme must extricate itself from its egg, and does so in a very reptilian way. A sharp egg tooth, shed after birth, is used to pick through the parchment-like shell with the aid of a pointed bump on the snout made of hardened epithelium (the os caruncle). Monotremes have functional mammary glands that can reach enormous size when the female is lactating. However, monotremes have no teats, an interesting and arguably primitive feature. The young suck the milk directly from special mammary hairs from an areolar patch on the mother's belly.

Does such a seemingly archaic system fundamentally separate monotremes from other living mammals? It was undoubtedly the mode of birth for the earliest mammals (Jenkins, 1990). However, although the marsupial neonate is at a similar stage of development as the monotreme hatchling when born, the reproductive apparatus—without development of a placenta, birth canal or teats—does appear to differ at a very basic level from the reproductive systems of living therian mammals.

Body temperature (TB) in monotremes is low for mammals, about 32°C (compared to 35–38° for marsupials and placentals respectively). This low TB may be similar to that of Jurassic mammals and is probably a primitive condition retained by monotremes (Jenkins,

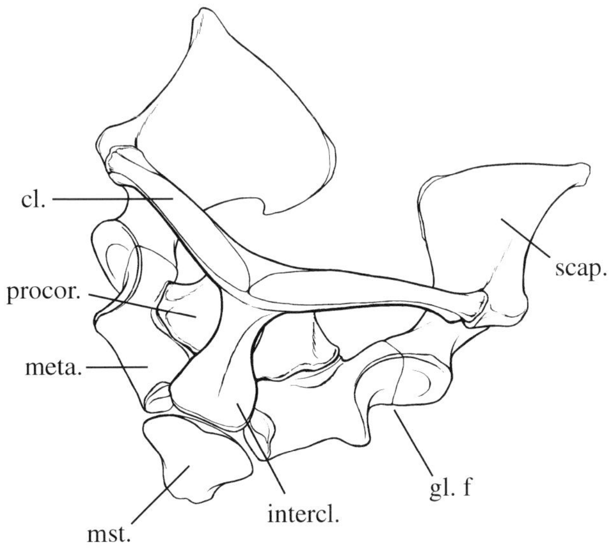

Figure 5. Shoulder girdle of the short-beaked echidna, *Tachyglossus aculeatus* (adapted from Augee and Gooden, 1997). Key to abbreviations: cl. = clavicle; gl. f. = glenoid fossa; intercl. = interclavicle; meta. = metacoracoid; mst. = manubrium sterni; procor. = procoracoid; scap. = scapula.

1990). It had been believed that, like reptiles, monotremes could not regulate their TBs efficiently but TBs do remain constant under normal circumstances. Echidnas may hibernate in colder regions, dropping their TBs to a little above ambient temperature (just as in therians) although platypuses are not known to hibernate. Neither platypus nor echidnas tolerate high temperatures well. Basal metabolism differs between the platypus and echidnas; the metabolic rate of the platypus approaches that of therian mammals while the echidnas, subsisting on low-energy insect diets, have the low metabolic rates seen in other insectivores (Jenkins, 1990).

Like other living mammals (and possibly advanced therapsids), monotremes have a four-chambered heart, muscular diaphragm and non-nucleated red blood cells although, as always, they have their own peculiarities. The kidneys are basically mammalian in structure, as are the lungs.

The monotreme brain is therian-like and is proportionately quite large (the echidna brain is proportionately similar to that of a rat). The platypus has a lissencephalic brain like an opossum (*Didelphis*) without creases, or convolutions, on its surface. In contrast, the surface of the echidna cerebrum is highly convoluted. Interestingly, the pattern of convolutions on the echidna brain differs from the pattern of convolutions on the brains of therian mammals (Griffiths, 1978). The platypus neocortex (site of higher mental activities) is quite large, reflecting the role it plays in use of the bill. However, the neocortex of echidnas is hugely expanded and is the largest of any mammal, including humans. Although echidnas do not lack in intelligence (at about the level of a rat in standard tests), the exact purpose of this impressive neural apparatus is one of life's little mysteries.

The musculature of monotremes, built onto their mosaic body plan, reflects the primitive/specialized skeletal structure. The detrahens mandibulae is a single-bellied slip of the jaw-closing muscle group that has migrated to the underside of the jaw to become the jaw-opener in monotremes. It is innervated by the trigeminal nerve. The presence of the detrahens in monotremes has been used as an argument for an ancient origin and that the detrahens may be a holdover from the time, far back in mammalian history, when the jaw joint was evolving from the articular-quadrate joint to the dentary-squamosal joint.

The facial musculature in monotremes is also distinctly different from that of living therians. The well-developed superficial facial musculature extends over the head and back and, remarkably, inserts onto the forelimbs in monotremes; this specialized musculature contrasts with the superficial musculature of therian mammals, which share the general plan seen in both primitive marsupials and placentals (Huber, 1930).

The shoulder girdle musculature is almost certainly close in form to the pectoral musculature of the cynodonts and the earliest mammals that had a similar shoulder girdle arrangement. Certain developments are fossorial specializations to create powerful musculature for digging and some muscles appear to be reptilian holdovers. The unique arrangement of the musculature in monotremes makes it difficult to homologize the muscles of the pectoral girdle with those in living mammals and suggests a separate line of evolution (Shrivastava, 1962).

FOSSIL MONOTREMES

CRETACEOUS FORMS

The first Mesozoic mammal from Australia—in fact from Gondwana—was 'discovered' by palaeontologist Alex Ritchie in a tray of opalized early Cretaceous fossils from Lightning Ridge, N.S.W. Perhaps not surprisingly, it was a monotreme. The specimen is a jaw fragment with three molars *in situ* and an alveolus for the last premolar (Figure 6) (Archer *et al.*, 1985). It has been christened *Steropodon galmani* because of its opalized nature ('sterope' is Greek for 'flash of lightning') and to honour the Galman brothers who procured the jaw.

The dentition of *S. galmani* is a more plesiomorphic version of the distinctive blade/cusp structure found in ornithorhynchids (Figure 9). The rectangular molars of *S. galmani* differ from ornithorhynchids in that diakidrepanids (two cusps linked by a crest: Every, 1974), rather than triakididrepanids (three cusps linked by two crests to form a triangle: Every, 1974), occur on the posterior halves of M/1 and M/2; M/3 is well-developed (and not vestigial); the ridges between cusps are weakly developed (incipient) rather than well-developed shearing crests as in ornithorhynchids and buccal cingula are present. The numbers of roots are difficult to determine, but roots of the wide molars (particularly on the posterior halves) are either very large or have subdivided into multiple roots. The presence of three well-developed lower molars in *S. galmani* suggests that there were three well-developed, occluding upper molars in *S. galmani*.

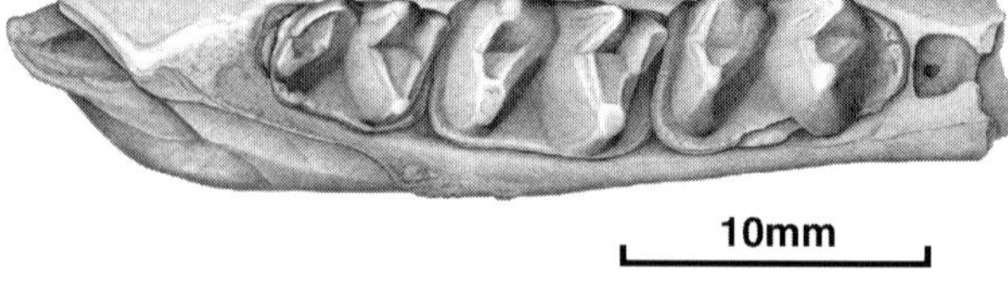

Figure 6. Dentary and dentition of *Steropodon galmani*: occlusal view of right dentary fragment with M/1-M/3 in place and an alveolus for the last premolar (anterior to right). The first lower molar (M/1) has diakidrepanid-like structures anteriorly and posteriorly; M/2 and the smaller M/3 have triakididrepanid-like structures anteriorly and diakidrepanid-like structures posteriorly. M/1 and M/2 have well-developed anterior, posterior and buccal cingula; M/3 has a prominent anterior and buccal cingulum but the posterior cingulum is not well-developed. M1 is damaged along the anterobuccal margin.

The fragmentary dentary of *S. galmani* is missing almost all normally diagnostic features: coronoid process, articular condyle, angular (or pseudangular) process, mandibular foramen, and all of the dentary anterior to the posterior alveolus for the ultimate premolar. However, *S. galmani* retains archaic features including a faint meckelian groove and what appear to be comparatively large facets for one or more postdentary bones (at least a splenial bone) anterior to the mandibular foramen (as in mammal-like reptiles and archaic mammals). Although detailed comparisons are not yet complete, these facets suggest that *S. galmani* had a jaw form more archaic than that of therian mammals or in fact holotherians (mammals included in the lineage leading to therians).

Luo *et al.* (2002) illustrate a mandibular foramen and possible postdentary trough posterior to it in their interpretation of the *S. galmani* jaw. In cynodonts and some basal mammals this trough, on the inside of the lower jaw behind the mandibular foramen, houses several postdentary bones. However, since a mandibular foramen in *S. galmani* is not present (and neither is a postdentary trough posterior to this), the depressed area on the jaw may instead be space for postdentary bones as described above. As well, the meckelian groove in *S. galmani* is weakly developed and not a deep sulcus, as suggested by Luo *et al.* (2002). Any suggestion of postdentary bones or remains of a meckelian groove or sulcus are missing in the more advanced ornithorhynchids. Additionally, several other features distinguish *S. galmani* from ornthorhynchids. The dentary is proportionately deeper and more compressed in *S. galmani* than it is in ornithorhynchids, which have a shallower dentary and widened areas for the large, rectangular teeth. The dental canal is large in *S. galmani*, but not as capacious as it is in ornithorhynchids because of the deeper molar roots in *S. galmani*, which extend into the dental canal. This morphological evidence supports molecular evidence (Flannery *et al.*, 1995) for a family-level distinction between *S. galmani* and ornithorhynchids.

When *S. galmani* was described, Archer *et al.* (1985) interpreted the teeth as tribosphenic (and therefore therian) rather than non-therian, upsetting the prevailing view of monotreme affinities. They cited loss of postdentary bones, tribosphenid-like cusp patterns and wide talonid basins that would indicate the presence of an occluding protocone (a defining tribosphenic feature) to link monotremes to tribosphenic therian mammals (Archer *et al.*, 1985). Kielan-Jaworowska *et al.* (1987) re-examined the evidence and proposed that *S. galmani*, because it had no wear on the talonid to positively indicate it had an occluding protocone and probably had no protocone or entoconid, had not reached the tribosphenic level of organization. However, they did consider *S. galmani* therian, possibly derived from a form similar to the pretribosphenic therian *Peramus*, which has a talonid with a hypoconid and hypoconulid but no protocone (and so no basined talonid).

S. galmani was initially placed in Ornithorhynchidae because of its platypus-like molars and the large mandibular canal suggesting a bill-like structure (Archer *et al.*, 1985). But *S. galmani* is now in its own family, Steropodontidae because, according to some molecular evidence, the split between the platypus and echidna lineages probably happened close to the Cretaceous–Tertiary boundary (Westerman and Edwards, 1992), which would exclude an early Cretaceous monotreme from membership in the family (Flannery *et al.*, 1995). Results from other molecular studies (Messer *et al.*, 1998) give an earlier date for divergence; it is suggested that morphological differences—primarily in jaw structure and possibly function—may also be great enough for separation of *S. galmani* from ornithorhynchids at the family level. These are discussed in detail below.

In addition to the archaic features of the jaw described above, the deeper and more compressed jaw in *S. galmani* differs considerably in form from the shallower and wider dentary in later ornithorhynchids. The shallow, boat-shaped jaw in ornithorhynchids is correlated with the flattening of the skull and reduction in length of the molar roots, shown in the *Obdurodon dicksoni* skull, which is a specialization within the Ornithorhynchidae (and synapomorphy for the family) to the probable exclusion of steropodontids. The dentition in *S. galmani* certainly shares many features with

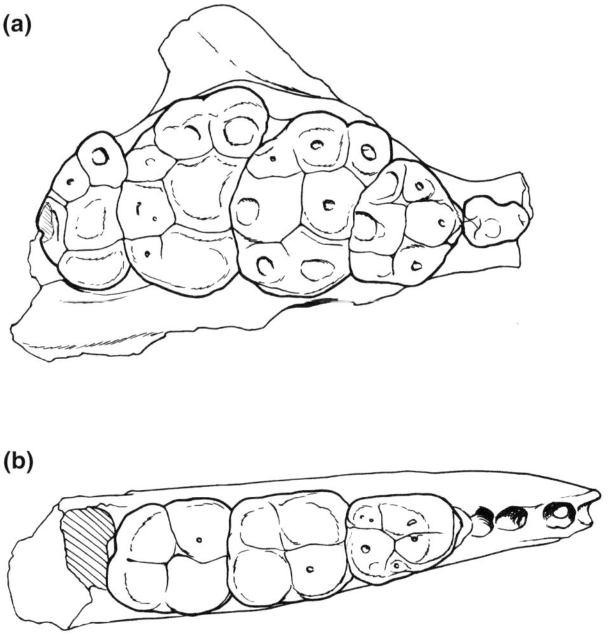

Figure 7. Dentition of *Kollikodon ritchiei.* **(a)** Occlusal view of right maxillary fragment, with last premolar and M/1–M/4 in place. The wide crescentic upper molars are pitted on the mamelonated cusps. **(b)** Occlusal view of right dentary fragment with M/1–M/3 in place (the anterior alveolus for M/4 has been preserved but the molar is missing). The rectangular lower molars are simpler in form than upper molars, but also exhibit pitting.

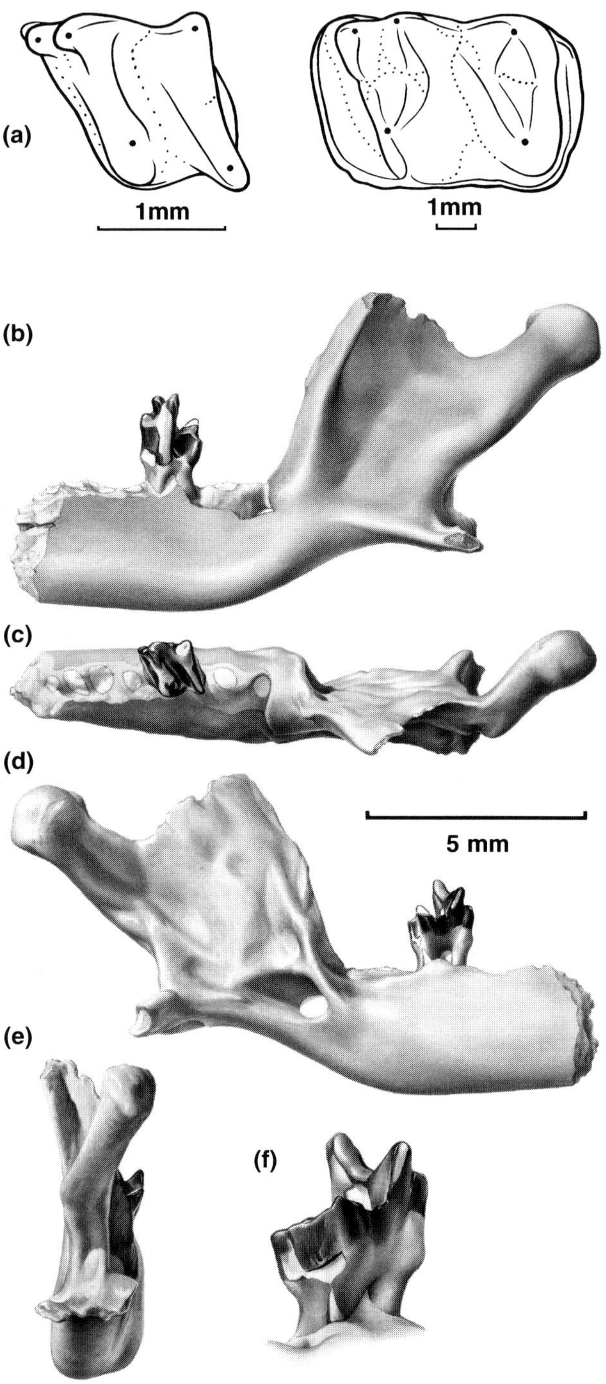

Figure 8. Dentary and molar structure in *Teinolophos trusleri* (from Rich *et al.,* 2001): **(a)** comparison between the single preserved molar of *T. trusleri,* left and the M/2 of *S. galmani,* right, illustrating similarities in molar form between the two molar teeth; **(b)** lateral view of the left mandibular fragment with condyle, base of the ascending ramus, angular process, and ultimate molar; **(c)** occlusal view; **(d)** medial view showing the angular process and large medial mandibular foramen; **(e)** posterior view; and **(f)** oblique-labial view of the ultimate molar.

ornithorhynchids, but the much stronger development of transverse shearing crests in ornithorhynchids, the loss of buccal cingula, and the possible inclusion of hardened epithelial tissue in the dental battery suggests that platypuses were using their dentition in a fundamentally different way from earlier steropodontids.

S. galmani was extraordinarily large for a Mesozoic mammal, measuring about 35 cm from head to tail (similar in size to female *Or. anatinus* in the more northerly parts of its range). This large size can be attributed to the cool-to-cold climate of Lightning Ridge that, at the time of deposition, was close to the inland Eromanga Sea just north of the Antarctic Circle. Lightning Ridge deposits are Aptian–Albian in age (~ 110 m.y.a.).

One of Australia's most surprising fossil finds was that of its second Mesozoic mammal, *Kollikodon ritchiei* (Flannery *et al.,* 1995). *K. ritchiei,* also found at Lightning Ridge, was described as a very specialized, bunodont monotreme on the basis of the holotype lower jaw (Flannery *et al.,* 1995). However, recovery of a partial maxilla and reanalysis of the lower jaw strongly suggest that *K. ritchiei* is not monotreme but instead represents a hitherto-unknown order of mammals.

Teinolophos trusleri, a tiny early Cretaceous mammal from Flat Rocks in Victoria, was found as a lower jaw fragment with a single molar (Figure 8) covered by hard matrix (Rich *et al.*, 1999). *T. trusleri* was originally described as a eupantothere based on what could be seen of the molar structure, but when the single, delicate tooth was more thoroughly cleaned the uniquely monotreme triakididrepanid structure could be seen (Rich *et al.*, 2001). With a head length of about 25 mm, *T. trusleri* might have been about the size of a small dasyurid marsupial which makes it by far the smallest monotreme known. At about 115 million years of age, it is also the oldest known monotreme.

T. trusleri has provisionally been assigned to Steropodontidae because of similarities in molar structure. The single preserved molar of *T. trusleri* has a triakididrepanid (trigonid) anteriorly and a diakidrepanid (talonid) posteriorly, as in *S. galmani*, but both trigonid and talonid are very compressed and angled obliquely. Anterior and posterior cingula are present in *T. trusleri*, but are more poorly developed than in *S. galmani*.

The most surprising aspect of the *T. trusleri* material is that the jaw structure is unlike that of any known monotreme. The articular condyle is well above the horizontal ramus of the dentary and, remarkably, the well-developed angular process has both medial and lateral flanges. This development is in stark contrast to ornithorhynchid and tachyglossid jaws in which the small, variably developed angular process, when present, projects only medially. As discussed above, the posterior sections of the jaw in both *S. galmani* and *K. ritchiei* are missing. The jaw structure of *T. trusleri* may be quite specialized and would have required muscle attachments unlike those in ornithorhynchids and, by inference, in an animal directly ancestral to them.

Possible dietary preferences for *T. trusleri* haven't been proposed, but Rich *et al.* compare the jaw structure to that of some triconodonts and spalacotheriids as well as to Geomyidae (the placental pocket gophers). The raised condyle suggests a more powerful bite than in the primarily insectivorous ornithorhynchids or tachyglossids. Further study may show what sort of dietary specializations *T. trusleri* had.

ORNITHORHYNCHIDS

Ornithorhynchids are the 'true' platypuses. Three genera are currently included in the family: *Monotrematum*, *Obdurodon* and *Ornithorhynchus*. Ornithorhynchid fossil material dates from the earliest Tertiary and consists primarily of isolated molar teeth referred to *Obdurodon*. However, a beautifully preserved skull was found in 1985 which gives us a clear look at an archaic, toothed platypus (Archer *et al.*, 1992; 1993; Musser and Archer, 1998); lower jaw fragments and part of an ilium have also been found.

An incomplete dental formula for *Obdurodon* species can be inferred from the skull and dentaries. The *Obdurodon dicksoni* skull is missing all molar teeth but has alveoli for M1/ and M/2 in addition to retaining P1/ and P2/ in the skull, which gives us the number of teeth in the adult skull. There are no canines or incisors, or alveoli for them. Unfortunately, none of the dentary fragments (referred to *Ob. dicksoni* and *Ob. insignis*) are complete anteriorly and the number of lower premolars and presence of incisors or canines in the jaw can't be determined. This is of interest in part because of Green's (1937) determination that *Or. anatinus* has toothbuds for five lower incisors although there are no anlages (buds) for upper incisors. There is no indication in this adult skull of Green's third upper molar and no molar fitting this position is known. Both *Ob. dicksoni* and *Ob. insignis* have a small, single root for a very reduced third lower molar. The dental formula for *Obdurodon* species, as far as can be determined, is therefore I0/? C0/?0 P1–2/?1–2 M1–2/1–03.

Premolars have only been recovered for *Ob. dicksoni* and are briefly described by Musser and Archer (1998) although not figured. These premolars are unicusped and well-developed, with basal cingula that nearly encircle the base of the teeth. The second upper premolar has three roots, which is unusual in mammals where two premolar roots are the norm. Like multiple molar roots, multiple roots in the premolars may be a specialization in *Obdurodon* species (Musser and Archer, 1998). There are almost certainly only two lower premolars, matching the uppers, although confirmation must wait until additional material is procured.

Molar teeth have been recovered for all fossil ornithorhynchid taxa although no species has a complete set of upper and lower molars (Figure 9). Ornithorhynchid molars, like those of *S. galmani*, have triakididrepanon (-id) and diakidrepanon (-id) blade systems and embrasures between the blade systems. They differ from the (lower) molars of *S. galmani* in elaboration of the blades into transverse shearing crests; addition of triakididrepanids to M/1 and M/2 (replacing the posterior diakidrepanids); loss of buccal cingula; reduction of M/3; and development of shallow, multiple roots. The multiple roots are interesting; because they are not set deeply into the dentary or skull, flattening of the skull and dentary in ornithorhynchids was possible (not an option for an animal with a deeper skull and jaws). The teeth sit in flat toothbeds, possibly buttressed by hardened gum tissue, and may have been shed during life (as the shallow-rooted vestigial molars are in *Or. anatinus*). Upper molars are less elongate (rectangular) than lower molars, and both lingual and buccal cuspules are more prominent. The two small, button-like M/3s recovered for *Ob. dicksoni* differ in form (to be expected in a vestigial tooth) and functionally would have contributed very little to the molar row.

The oldest known ornithorhynchid (and only early Tertiary monotreme) is not Australian. *Monotrematum sudamericanum* (Figure 9) was first described from a single molar recovered from early Paleocene deposits in Patagonia, Argentina (Pascual *et al.*, 1992a, b). The importance of this find cannot be exaggerated; it confirms the Gondwanan nature of monotremes as well as the antiquity of Ornithorhynchidae. Based on three teeth recovered to date (two worn upper second molars and the less-worn posterior half of the first lower molar) the robust *M. sudamericanum* would have been the largest member of the family.

Molar morphology is typically ornithorhynchid and although *M. sudamericanum* has been given separate generic status, there are few differences of any consequence between *M. sudamericanum* and species of *Obdurodon*. This is a striking example of dental conservatism considering that these taxa are separated by about 35 million years. The molars of *M. sudamericanum* are larger and more robust than those of *Obdurodon* species, but other features that were originally thought to separate the two genera are either variable in *Obdurodon* species or present in both taxa. *M. sudamericanum* was recovered from the Banco Negro site just west of Golfo de San Jorge (central Patagonia). The fossil site today is a bleak and windswept rocky plain but when *M. sudamericanum* was alive it was a lush lowland rainforest near the widening Atlantic Ocean (Pascual *et al.*, 1992a). The early Paleocene site, dated at 63.2–61.8 m.y.a., supported a very diverse vertebrate fauna (Pascual *et al.*, 1992a; Archer, 1995).

The oldest Australian ornithorhynchid, *Obdurodon insignis*, was originally described from two molar teeth discovered separately near Lake Eyre (Etadunna Formation) and Lake Frome (Namba Formation) in central Australia (Woodburne and Tedford, 1975). These late Oligocene sites (26–24 m.y.a.) are part of one of the oldest and best known Tertiary vertebrate fossil assemblages

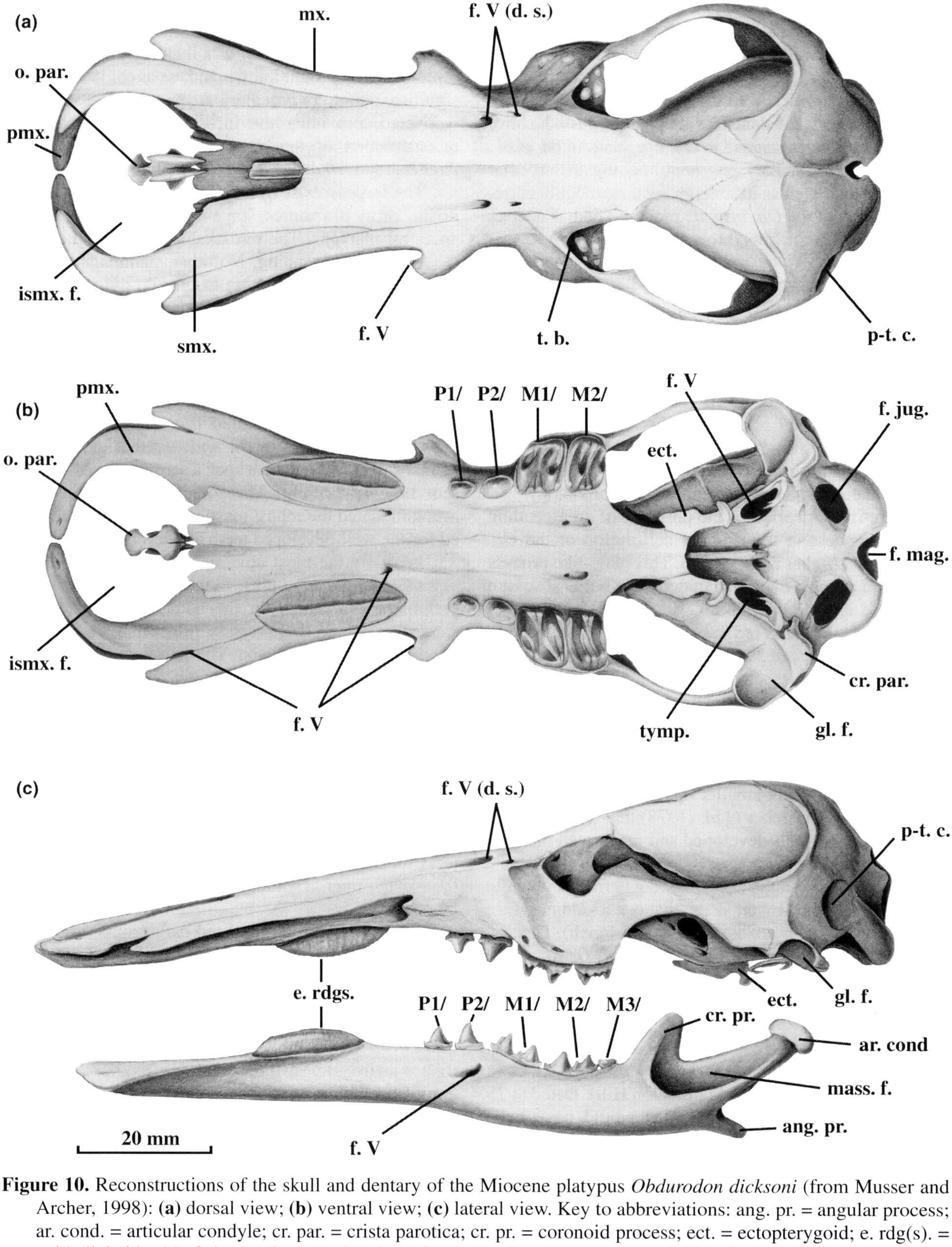

Figure 10. Reconstructions of the skull and dentary of the Miocene platypus *Obdurodon dicksoni* (from Musser and Archer, 1998): **(a)** dorsal view; **(b)** ventral view; **(c)** lateral view. Key to abbreviations: ang. pr. = angular process; ar. cond. = articular condyle; cr. par. = crista parotica; cr. pr. = coronoid process; ect. = ectopterygoid; e. rdg(s). = epithelial ridge(s); f. jug. = foramen jugulare; f. mag. = foramen magnum; f. V = foramina for V (trigeminal nerve); f. V (d. s.) = foramina for V innervating dorsal shield; gl. f. = glenoid fossa; mass. f. = masseteric fossa; mx. = maxilla; o. par. = os paradoxum; pmx. = premaxilla; p-t.c. = post-temporal canal; smx. = septomaxilla; t.b. = tooth bed; tymp. = tympanic bone.

just above the mandibular foramen, and large mandibular foramen and canal. The dentary is quite flat, corresponding to the flattened nature of the skull. It is proportionately deeper than in *Or. anatinus* but not as deep as in *S. galmani*. The presence of both an ascending coronoid process and an internal coronoid process in *Ob. dicksoni* is proof that the internal process is not the vestigial ascending coronoid process, as had been presumed (Musser and Archer, 1998).

It had been thought that bill size in ornithorhynchids gradually increased through time and that the wide, rectangular bill of the living *Or. anatinus* developed from the smaller bill of a platypus like *Ob. insignis*. The size and shape of the bill in *Ob. dicksoni* led Musser and Archer (1998) to suggest that the Riversleigh platypus might actually have been even more specialized than *Or. anatinus*, at least in development of the bill, and that perhaps *Ob. dicksoni* was well off the main evolutionary line leading to *Or. anatinus*. *Ob. insignis*, with a smaller bill and lighter build, may have been closer to the ancestor of today's lightly-built *Or. anatinus*.

The only ornithorhynchid material reported from the Pliocene is that of a limb bone from the Bow Local Fauna of N.S.W., referred to the genus *Ornithorhynchus* (Rich *et al.*, 1991). Numerous platypus cranial and postcranial elements from the Pleistocene and Holocene have been recovered, the majority of this material from cave deposits in Tasmania (Marshall, 1992). Pleistocene platypus material appears to be identical to *Or. anatinus* and all Pleistocene ornithorhynchid material recovered is assigned to this species (Archer *et al.*, 1978; Marshall, 1992).

The oldest Pleistocene record, from Bunya Siding, N.S.W., dates from about 100,000 years ago (Davis, 1996). All Pleistocene sites fall within the range of the living *Or. anatinus* except for a record from Lake Tandou west of the Darling River (Hope, 2001, personal communication).An edentulous jaw and tibia from King's Creek, Qld (Archer *et al.*, 1978), an undescribed shoulder girdle element from the Darling Downs, Qld. (Wilkinson, 2001, personal communication) and a juvenile skull from the Snowy River, N.S.W. (Marshall, 1992) represent natural deposition in streams, rivers or caves. Two archaeological sites have produced fossil platypus material: the late Holocene Brisbane River Platypus Rock Shelter site and the late Pleistocene cave deposits from Tasmania, dating from 14,000–29,000 years BP (Marshall, 1992). The large number of skeletal elements from the Tasmanian deposits suggests that the platypus may have been used as a source of dietary fat during glacial periods when Tasmanian Aborigines were using the caves (Marshall, 1992).

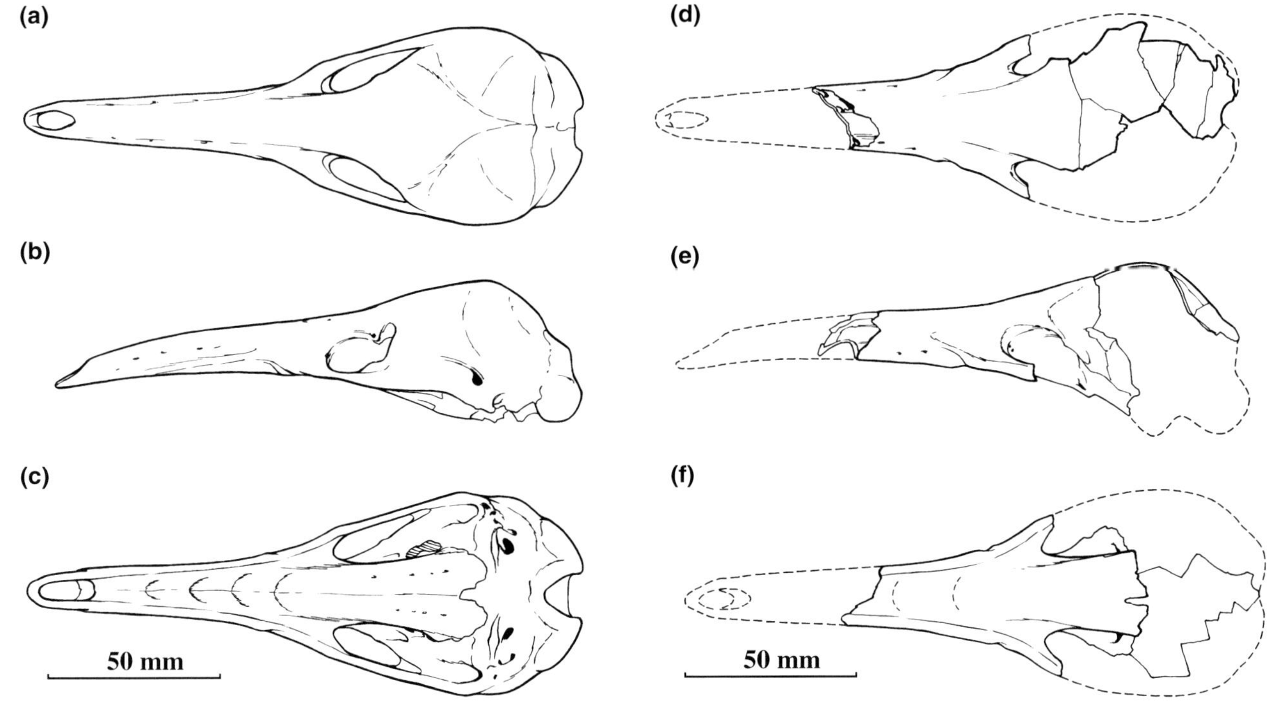

Figure 11. Skulls of fossil Australian long-beaked echidnas. The Pleistocene long-beaked echidna *Megalibgwilia ramsayi,* known from complete skulls as well as postcranial material, had a long, fairly straight beak and flattened, *Zaglossus*-like cranium (redrawn from Griffiths *et al.,* 1991): **(a)** dorsal view; **(b)** lateral view; and **(c)** ventral view. '*Zaglossus' robusta,* possibly a much older tachyglossid of Miocene age, is only known from a single specimen but appears to have been similar to *M. ramsayi* in most respects (redrawn from holotype AMF51451, Australian Museum): **(d)** dorsal view; **(e)** lateral view; and **(f)** ventral view.

TACHYGLOSSIDS

Animals without true teeth are at a disadvantage in the fossilization process because teeth, with an outer coat of durable enamel, are often preserved when other elements are not. The edentate echidnas are no exception to this general rule. The echidna fossil record is almost unknown prior to the relatively recent Pleistocene. The oldest known echidna, possibly as old as middle Miocene (*'Zaglossus' robusta*), is edentate and can tell us nothing about dental evolution in tachyglossids.

Traditionally, classification of fossil echidnas was relatively simple: small, short-beaked echidnas were assigned to the genus *Tachyglossus* and a more diverse group of larger, long-beaked echidnas (known from both New Guinea and Australia) were put into the genus *'Zaglossus'*. However, Griffiths *et al.* (1991) suggest that the story may be more complex and that the diversity of long-beaked echidnas should be recognized at the generic level, reflecting differences in feeding strategies and, therefore, in niche partitioning.

'Zaglossus' robusta (Figure 11) comes from a now-defunct gold mine near Gulgong in New South Wales. The site, originally thought to be Pliocene, may instead be middle Miocene in age—at least 13 or 14 million years old (Woodburne *et al.*, 1985). The argument over the age of the site—a difference of over 10 million years—depends in part on whether the associated fauna found in the mine shaft is archaic or is more recent in age. Woodburne *et al.* (1985) describe the assemblage from the mine as taxonomically similar to Miocene faunas, while the marsupial fauna and the type of preservation suggest to Murray (1978b) that the material may be Pliocene or Pleistocene. Preservation is similar to that of the nearby Wellington Caves Bone Cave deposits, which have a 'rubbery' date of about 250,000 years (Augee, 2001, personal communication).

A partial skull, complete humerus, atlas vertebra and clavicular fragment were recovered from the mine shaft (Dun, 1895). The skull is typically tachyglossid and is similar to that of the better known Pleistocene long-beaked echidna *Megalibgwilia ramsayi*. Unfortunately, the anterior section of the beak is missing which prevented Griffiths *et al.* from formally re-assigning *'Z.' robusta* to the genus *Megalibgwilia* (Griffiths *et al.*, 1991). The humerus was originally described by Dun (1895) as that of a giant platypus and given the name *Ornithorhynchus maximus*. It is undoubtedly tachyglossid, however, and is most likely from the same animal as the skull (Mahoney and Ride, 1975).

There are no Pliocene cranial remains but there are a few postcranial elements. Two well-preserved, complementary tibiae were found near Cooma, N.S.W. and a partial humerus and scapular fragment were recovered from Chinchilla in south-east Queensland. The humerus (*'Echidna' owenii* Krefft, 1868) was the first record of a large, long-beaked echidna from either Australia or New Guinea; this discovery pre-dated the description of the living New Guinea long-beaked echidna by eight years (*'Tachyglossus' bruijnii* Peters and Doria, 1876). The form of the humerus closely resembles that of *Megalibgwilia ramsayi*, described below. Although Krefft described the Chinchilla Sands deposit as Pleistocene, most material from this deposit is now believed to be Pliocene in age.

Pleistocene Echidnas

The vast bulk of fossil tachyglossid material is Pleistocene and most of this has been excavated from various cave and swamp deposits in eastern and southern Australia. Henschke's Cave, at Naracoorte in South Australia, is particularly rich in tachyglossid material, with three contemporaneous species of echidna recovered to date (Barrie, 2001, personal communication).

There are at least three distinct genera of Pleistocene echidnas: the long-beaked *Megalibgwilia* or *Zaglossus* species and the huge *'Zaglossus' hacketti* as well as the living short-beaked echidna *Tachyglossus aculeatus*. These three types are separated by size, by differences in the beak and palate (indicating differences in dietary preferences) and by proportional differences in the limbs. By the Pleistocene, then, echidnas had successfully radiated out into various niches, much like the anteaters of South America, reflecting the patchwork of wet and dry environments (and their respective invertebrate faunas) that continue to characterize Australia today.

Taxonomy of mainland long-beaked echidnas has long been confusing. There are many synonyms, with specific status given to limb bones and other fragmentary remains. The first comprehensive study of fossil long-beaked echidnas was by Peter Murray (1978b). Most of the material, from the Naracoorte cave deposits and from Tasmania, appeared to represent a single, variable species that he assigned to *Zaglossus ramsayi* (the name *Z. ramsayi* having priority over other names given to some of the fossil material). *Zaglossus ramsayi* (Figures 11, 12) was described as a large, robustly built echidna with a fairly straight, strong beak and long, well-muscled limbs. These features indicated a powerful digger with a beak that could act as a lever in prying up logs and rocks in search of insect larvae. Murray thought *Z. ramsayi* might have been less specialized than either *Z. bruijnii* or *T. aculeatus*; skull morphology suggests *Z. ramsayi* probably lacked the acanthid groove on the tongue of *Z. bruijnii* and probably had not developed the complex tongue musculature of *T. aculeatus*. In a twist on the debate over relationships, Murray thought that *Z. ramsayi* was actually closer in form to *T. aculeatus* than to *Z. bruijnii*, although *Z. bruijnii* had probably evolved from *Z. ramsayi* or an echidna very much like it.

Restudy of *'Zaglossus'* cranial material prompted Griffiths *et al.* (1991) to propose a separate genus (*Megalibgwilia*) for *Zaglossus ramsayi*. Its relatively short, straight, untapered beak and the shape and conformation of its palate suggested a difference in diet, a distinction that in many mammal groups is considered fundamental enough to separate taxa at the generic level. The diet of *M. ramsayi* was probably not primarily oligochaete worms, as is usually the case in New Guinea *Zaglossus*, but perhaps large insect larvae such as those of beetles or moths; these grubs would have been plentiful in the cool, moist grasslands of Pleistocene south-eastern Australia. To Griffiths *et al.* the cranial morphology of *M. ramsayi* appeared to be intermediate between that of *Tachyglossus* and *Zaglossus* and they proposed that *Megalibgwilia* may have been the ancestral form for these later genera.

Fossil echidnas resembling the New Guinea *Zaglossus* forms are known from the late Pleistocene of mainland Australia (Murray, 1978b) and from Pleistocene deposits in New Guinea (Flannery, 1995). Both *'Z.' robusta* and *M. ramsayi* are described as being larger and more robust than these *Z. bruijnii*-like forms (Murray, 1984) although since Murray's diagnosis, larger New Guinean *Zaglossus* species (up to 1.0 metre in length) have been recorded (Flannery, 1995). Murray sees the *Z. bruijnii/Z. bartoni* species, with their long, recurved beaks and less robust build, as the specialized, wet-forest end-forms of the *M. ramsayi* lineage. The slender, recurved beak is perfectly designed for probing leaf litter of rainforests for worms but is ill-equipped to pry up the rocks or logs that litter the drier woodlands that were the habitat for *M. ramsayi*.

The Pleistocene fluctuations in sea levels and extensive land-bridges between Australia and New Guinea—with associated opportunities for faunal exchange—have been documented in detail in several other chapters (Aplin, 2006; Heinsohn and Hope, 2006; van Oosterzee, 2006). For *Zaglossus* species, the mountainous terrain of New Guinea preserved ideal habitat. During climate fluctuations the vertical zonation that characterizes mountain slopes would have ensured access to suitable habitat regardless of climate shifts (Murray, 1978b).

The mid-Miocene *'Z.' robusta* is similar to *M. ramsayi*. Both Murray (1978b) and Griffiths *et al.* (1991) believe that it should either be included in *M. ramsayi* or be considered as an ancestral species. However, as mentioned, the rostrum of the skull, which would help in positive identification, is missing. A femur of a long-beaked echidna, described as a separate species (*Z. harrissoni*), is indistinguishable from the *M. ramsayi* material and should be assigned to that species (Murray, 1978b).

Some researchers feel that there isn't a generic-level distinction between the New Guinea species of *Zaglossus* and the forms assigned to *Megalibgwilia*. I've examined the *'Z.' robusta* material and tend to agree, especially in light of the morphological diversity within New Guinea *Zaglossus*. If the taxonomy proposed by Griffiths *et al.* is upheld, the genus *Megalibgwilia* would include most of the Miocene to Pleistocene/Holocene mainland long-beaked echidnas and the genus *Zaglossus* would be confined to Pleistocene to present forms close to the living New Guinean species. If the forms now assigned to

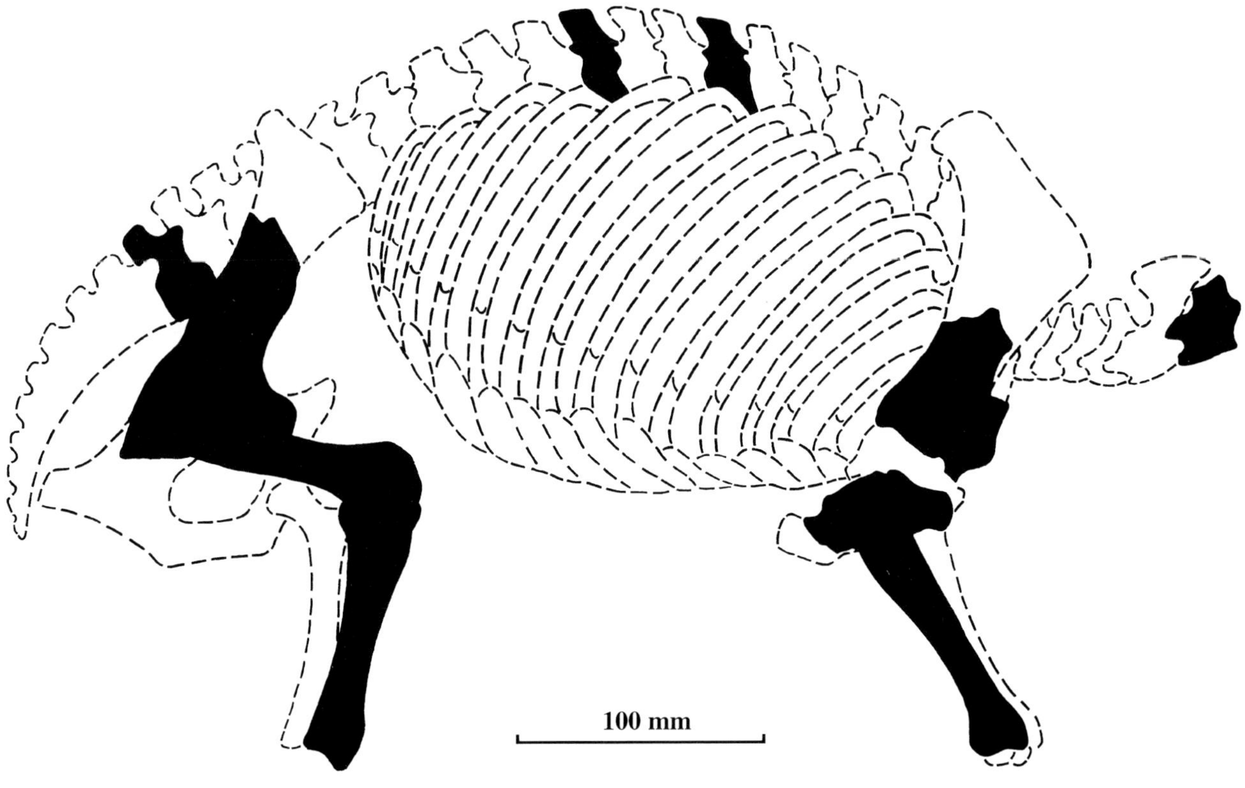

Figure 12. Restoration of the skeleton of the giant Western Australian echidna *'Zaglossus' hacketti* (from Murray, 1984). Shaded areas indicate the skeletal elements recovered to date. *'Z' hacketti* had relatively long limbs, but features of the hip suggest it may have been able to rear up on its hindlegs.

Megalibgwilia are re-assigned back to the genus *Zaglossus*, most long-beaked types would be considered species of *Zaglossus*. The fossil history of the genus *Zaglossus*, would then date from the ?Miocene to the present and distribution would include both mainland Australia and New Guinea. A wealth of recently excavated tachyglossid material from Naracoorte, S. A., may help to settle the debate.

The most distinctive Pleistocene echidna, *'Zaglossus' hacketti* (Figure 12) is an enormous echidna recovered from two caves in Western Australia. Pelvic fragments, hindlimb bones, a partial shoulder girdle, front limb elements and vertebrae were recovered but, unfortunately, there was no cranium that would help to answer questions about its relationships to other echidnas. A complete cranium would also allow a taxonomist to assign *'Z.' hacketti* to a new genus, a status it almost certainly deserves. The skeletal elements recovered indicate this echidna was about 1.0 m long and had a mix of both *Tachyglossus* and *Zaglossus/Megalibgwilia* features. A fragment of a large humerus recovered from Western Australian cave deposits may belong to *'Z.' hacketti* (Murray, 1984). The entepicondylar area is more like that of the older *'Z.' robusta* than either *M. ramsayi* or *Z. bruijnii*, although Murray describes this part of the humerus as variable in echidnas (Murray, 1978b).

Limb proportions of *'Z.' hacketti*, surprisingly, are closer to those of *T. aculeatus* than to any of the long-beaked echidnas. The lengths of the tibia and femur are nearly equal in *'Z.' hacketti* and in *T. aculeatus*, while in both *M. ramsayi* and in the New Guinea *Zaglossus* species the tibia is much longer than the femur. Perhaps the proportionately longer limbs aided in movement over logs and litter in forest environments (Murray, 1984). Some of the Naracoorte tachyglossid fossils resemble the *'Z.' hacketti* material although these elements are not especially large (Barrie, 2001, personal communication). It is probable that this giant among echidnas was either a rare form or that it was part of an earlier radiation that had largely disappeared by the Pleistocene.

Long-beaked echidnas from the Australian mainland have a much longer pedigree than the short-beaked echidna, *T. aculeatus*, which only dates back to the Pleistocene. The *'Z.' robusta* record is evidence that long-beaked echidnas may have been present at least in New South Wales from the middle Miocene to the Pleistocene. Records from Pleistocene deposits in western, southern and eastern Australia indicate that long-beaked echidnas had a wide distribution during this period. These Pleistocene environments were primarily woodlands or mosaic grasslands (Murray, 1978b). Speciation into *Tachyglossus* and *Zaglossus* types may have resulted as the Miocene and Pliocene forests retreated. Drier environments, with their new suites of invertebrate prey (particularly ants and termites), would have favoured *T. aculeatus* while *Zaglossus* species would have become increasingly restricted to remnant forest habitats.

Long-beaked echidnas appear to have survived into the Holocene on mainland Australia. Extinction times can be guessed at from the spotty fossil record as well as from some interesting but circumstantial evidence. The latest record for *M. ramsayi* comes from Holocene deposits in Tasmania. This island state may have been a refuge for *M. ramsayi* as well as for a number of other Australian mammals like the thylacine. The single record for *'Z.' hacketti* from the late Pleistocene indicates the species became extinct at some point after this (Pledge, 1980). Intriguingly, Holocene rock art (possibly as old as Pleistocene) from western Arnhem Land appears to depict a long-beaked echidna (Murray, 1984). Northern Australia and New Guinea, Murray believes, may have, like Tasmania, been refugia for long-beaked echidnas as the interior dried out.

What caused long-beaked echidnas to disappear from mainland Australia? Short-beaked echidnas are commonly eaten by aborigines in Australia and long-beaked echidnas are a favourite food item in New Guinea, now that taboos against eating them have broken down. Murray (1984) suggests that hunting pressure may have been responsible for the demise of at least *'Z.' hacketti* and overhunting may have contributed to the extinction of *M. ramsayi* as well. Loss of the extensive inland forests and of the moist grasslands may have been a factor in the reduction of scarab beetle numbers (with their fat-rich larvae).

Tachyglossus aculeatus first makes its appearance in the Pleistocene. It had a wide distribution during the Pleistocene, as it does today, with records from Wombeyan Caves, N.S.W., Lake Menindee, N.S.W., Naracoorte, S.A. and Mammoth Cave, W. A. However, it is not as common in fossil deposits as long-beaked echidnas are. Some *T. aculeatus* from Naracoorte were around 10% larger than their living counterparts, overlapping in size with some of the smaller long-beaked echidnas at the site (Pledge, 1980).

The relationship between echidnas and other monotremes remains a mystery. The lack of a dental 'trail' means that there are no echidna teeth to compare to the teeth of any of the other two monotreme families. The platypus and echidnas are similar to each other at the molecular level (Westerman and Edwards, 1992; Gemmell *et al.*, 1994). Pascual *et al.* (1992a) proposed that echidnas may have diverged from the platypus line, an idea also proposed by Gregory (1947). Pascual *et al.* base their hypothesis on genetic data, suggesting the platypus and echidna lines diverged from each other at some point near the Cretaceous–Tertiary boundary (Westerman and Edwards, 1992), well after the platypus line was established. Oddly, this would make echidnas derived platypuses, which are themselves highly specialized. If the genetic results are underestimates (Messer *et al.*, 1998) and if the echidna-platypus split dates from the early to middle Cretaceous, it is quite probable that echidnas descended from an as-yet-unknown monotreme. In this scenario, ancestral echidnas might have been terrestrial and more generalized than platypuses.

BIOGEOGRAPHY

The discovery of an ornithorhynchid monotreme in South America confirmed that monotreme mammals were Gondwanan and not uniquely Australian. The Gondwanan pieces—South America, India, Africa, Antarctica and Australia—were joined throughout the Mesozoic, separating sequentially from west to east before the final breach between Antarctica and the South Tasman Rise ~ 30 m.y.a. (Lawver *et al.*, 1992). With this history, it is reasonable to assume that monotremes may have been widespread over Gondwana when travel between the land masses was possible.

To construct a biogeographical scenario, we have to make some assumptions about evolutionary paths and dispersal opportunities. Because the earliest monotremes hail from Australia it is probable (but by no means certain) that they originated either on this continent or perhaps in Antarctica. *T. trusleri* and *S. galmani* are both specialized (in particular, with heavily innervated faces that may indicate mechano- or electroreceptors), suggesting that their lineages had been evolving for a long time. Many researchers, myself included, accept the possibility of a Jurassic origin for monotremes, but the Jurassic mammalian fossil record for Gondwanan continents is minute. There are only a handful of Jurassic mammal sites for the Gondwanan continents (none from Australia) and the Cretaceous record is not much better.

By 120 m.y.a., West Gondwana (Africa, Madagascar and India) had separated from the East Gondwanan land masses (South America, Antarctica, Australia and New Zealand) clustered around the South Pole. During the Aptian and Albian, Victoria lay within the Antarctic Circle and Lightning Ridge, on the shores of the inland Eromanga Sea, was situated just to the north. Temperatures were low although there is debate about whether the climate was actually polar (with development of permafrost) or merely cool-to-cold, perhaps like the cooler parts of Tasmania today; however, it is clear that Victoria and Lightning Ridge would have experienced the long winter nights and summer days of higher latitudes.

Ornithorhynchid monotremes may have spread across Antarctica to southern South America during the latter part of the Cretaceous. By the late Cretaceous, temperatures had warmed considerably and even the most southerly regions were traversable. From the late Cretaceous to Eocene, south-eastern Australia, New Zealand (up to 80 m.y.a.), Antarctica and southern South America formed a continuous and fairly homogeneous cool temperate forest environment dominated by southern beech (*Nothofagus*) (Case, 1989). This corridor would have been the dispersal route for, among other taxa, late Cretaceous ornithorhynchids travelling from Australia to South America as well as for ancestral South American marsupials entering Australia via Antarctica. Pascual *et al.* (1992a) believe that ornithorhynchids, along with gondwanatheres, may have been in South America since early Cretaceous and that southern South America may have been isolated from the early Cretaceous to the early Tertiary. Some geological evidence, however, indicates that South America remained connected to Antarctica through an archipelagic connection throughout this period (Lawver *et al.*, 1992).

M. sudamericanum, recovered from deposits that date from just 2-3 million years after the end of the Cretaceous, may represent either a single dispersal of monotremes to South America or a relict of a greater South American monotreme radiation. Pascual *et al.* (1992a, b) surmise that *M. sudamericanum* and the gondwanatheres represent the 'last stand' for archaic, non-therian mammals in South America. Contemporaneous faunas further to the north have neither monotremes nor gondwanatheres but do record diverse marsupial and placental mammals, immigrants and competitors from North America (Pascual *et al.*, 1992b). Monotremes, perhaps surprisingly, haven't been recorded from the late Cretaceous Los Alamitos site to the south of Banco Negro (gondwanatheres and many other non-tribosphenid mammals have been found; however, no therian mammals have yet been recovered (Bonaparte, 1990). Ornithorhynchids probably went extinct in South America as marsupial and placental mammals became established and faunal compositions changed. Australia, entering a long period of isolation, provided a less competitive environment for continued survival of ornithorhynchids.

Marsupial and placental mammals have been recorded from Seymour Island, an Eocene site off the Antarctic Peninsula (dated to ~40 m.y.a.). These mammals are endemic Antarctic forms of South American taxa, suggesting that Antarctica by this time was isolated from South America. Ornithorhynchids have not yet been recorded from Antarctica and, conversely, none of the distinctive Antarctic mammals have been found in Australia. Current estimates for the onset of glaciation over the Antarctic continent and for the complete separation of Antarctica from Australia vary. Antarctic floral records suggest that most terrestrial plant life (and, by inference, the animal life dependent on it) had become extinct by the late Oligocene or early Miocene (25–20 m.y.a.) (Askin, 1992). This agrees with an estimation for establishment of a strong circumpolar current around Antarctica (30–20 m.y.a.), effectively isolating the continent from ameliorating climatic influences (Lawver *et al.*, 1992).

The oldest Australian ornithorhynchid fossils are late Oligocene in age (about 25 m.y.o.). As Australia drifted north the continent experienced many climatic fluctuations; however, the onset of arid conditions is a comparatively recent event (White, 2006). This drying had a devastating effect on inland river systems, most of which have now disappeared along with the vast inland forests.

These events have reduced the range of ornithorhynchids to a considerable extent, and platypuses are 'out on a limb' in evolutionary terms. During the early Tertiary they occupied three continents; during the

Although molecular techniques are new and useful tools to investigate phylogenetic questions, in the debate over monotreme origins, some recent molecular findings are in conflict with morphological data and the palaeontological record. Some recent studies have strongly suggested that monotremes and marsupials share a close relationship; these investigations using mtDNA and DNA hybridization (Janke *et al.*, 1996, 1997; Kirsch and Mayer, 1998) are in agreement that an early divergence of monotremes from the stem leading to living therian mammals is the least likely outcome (Penny and Hasegawa, 1997). Alternative techniques have supported great genetic distance between monotremes and other living mammals (Westerman and Edwards, 1992; Retief *et al.*, 1993; Killian *et al.*, 2000; Belov *et al.*, 2002). Of particular interest is work using nuclear, rather than mitochondrial, DNA, which clearly supports the morphological conclusion that monotromes form a group far removed from therian (marsupial and placental) mammals (Killian *et al.*, 2001).

Molecular genetics can also be used to estimate the point chronologically where one group may have diverged from another. This is especially helpful where there are gaps in the fossil record. However, divergence times estimated in this way vary greatly depending on the techniques used. Estimates of a split between monotremes and the line leading to living therians range from early Jurassic (Messer *et al.*, 1998) to the early Cretaceous (Janke *et al.*, 1996, 1997; Kirsch and Mayer, 1998), a difference of about 60 million years.

THE GONDWANAN RADIATION

It may be that researchers, particularly those in the Northern Hemisphere, have been 'looking in the wrong place' for answers to the question of monotreme origins. Fossil mammals of Jurassic and Cretaceous age are now appearing on southern continents and evidence is mounting for an endemic southern mammalian radiation during the Mesozoic. This was first suggested by Bonaparte (1990) who excavated the late Cretaceous Los Alamitos site in Argentina and was struck by the unusual nature of the mammalian fauna, most of them unique, non-tribosphenic mammals with older Northern Hemisphere relatives. He proposed that the Northern Hemisphere (Laurasia) had been dominated by tribosphenic mammals during the later Mesozoic but that the Southern Hemisphere (Gondwana) had been predominantly dominated by these derived non-tribosphenic types.

Luo *et al.* (2001, 2002) take a different approach, prompted by the discovery of Mesozoic mammals from Madagascar (*Ambondro mahabo* Flynn *et al.*, 1999); Patagonia (*Asfaltomylos patagonicus* Rauhut *et al.*, 2002), and Australia (*Ausktribosphenos nyktos* Rich *et al.*, 1997, 1999; *Bishops whitmorei* Rich *et al.*, 1999). The middle Jurassic *Am. mahabo* is described as the oldest tribosphenic mammal to date (older by 25 million years than any of the northern tribosphenic mammals) and is considered by Flynn *et al.* (1999) to be the first Jurassic Gondwanan holothere known. *Au. nyktos*, from the early Cretaceous Victorian locality that produced *T. trusleri*, is described as an archaic placental mammal by Rich *et al.* (1997) although its affinities are highly controversial (Kielan-Jaworowska *et al.,* 1998; Musser and Archer, 1998). Luo *et al.* group *Au. nyktos*, *Am. mahabo* and monotremes together as part of an endemic radiation of Gondwanan mammals with tribosphenic dentitions (or dentitions derived from this pattern in the case of monotremes) separate from the northern tribosphenic radiation.

Characters of the teeth and jaws (the only material recovered for *Am. mahabo, As. patagonicus*, *Au. nyktos* and *B. whitmorei*) were analysed together with craniodental and postcranial characters for monotremes and all other known mammalian taxa. Results suggested a division of 'tribosphenic' mammals into two morphologically and geographically distinct groups—a northern group and a southern group. Luo *et al.* have proposed a new classification: a southern subclass Australosphenida (including *Am. mahabo*, *Au. nyktos* and monotremes) and a northern subclass Boreosphenida (including the Northern Hemisphere tribosphenic therians). This concept counters a northern origin for tribosphenic mammals and suggests that mammals with advanced, tribosphenic teeth appeared independently in the Southern Hemisphere before they appeared and radiated in the Northern Hemisphere.

EVOLUTIONARY TRENDS

What are we to make of the basic monotreme body plan: tribosphenic teeth, primitive shoulder girdle and modern pelvic girdle? The total monotreme design is a complex of features, some of which have probably developed convergently with those of therian mammals.

There is no consensus among researchers about which anatomical systems are so unique and complex that they can only have evolved once. The middle ear ossicular system, a mobile shoulder girdle and tribosphenic dentition have all been called too complex to have arisen more than once in mammalian evolution. However, Ji *et al.* (1999) have concluded that homoplasies are common in the postcranial skeleton (especially in the pectoral girdle and forelimbs but less so in the pelvic girdle and hindlimb), and that many seemingly shared, specialized features may be convergent, evolving at least twice in Mesozoic mammals. Several studies are in general agreement: that dental characters may be more highly homoplasic than those of the postcranium, which may explain the 'reinvention' of a tribosphenid-like tooth form.

A modern shoulder appears to have had an ancient origin; monotremes in this respect are at a very basal level, along with cynodonts and the earliest mammals, as discussed above. On morphological grounds, evidence points to an ancient origin—from the early to middle Jurassic—for monotremes. Their unique combination of features is most likely the result of early branching, a specific suite of inherited features, and a long period in which to refine specializations.

SUMMARY

Is it possible to come to a consensus about relationships and origins? The short answer may be 'no' at present, but for some very interesting reasons. First, the conflicting results between morphological and molecular studies should not be dismissed, but should be seen as a catalyst for further investigations. Second, the relationships of Mesozoic mammals need further resolution; taxonomy of early to middle Jurassic mammals, 'triconodonts' and multituberculates, are currently in flux. New fossil finds will help to determine the suites of features in these early groups, which will in turn help to clarify the position of monotremes. Third, we need to keep an open mind about biological evolutionary processes, that may have combined to create the unique combinations of features found in a number of mammalian groups. If reversions rather than retentions can explain the primitive features of monotremes, there must be an evolutionary mechanism to account for this reinvention and perhaps to allow a close relationship between monotremes and more recent mammals, particularly marsupials. Studies into such mechanisms—the switching on of hox genes, for example—are ongoing and may help to provide answers.

We may find that the most satisfying answer to the question of monotreme origins is the view promoted for most of the last 200 years, since their discovery by western science. Living monotremes may be the last surviving representatives of the earliest mammals that have miraculously persisted for perhaps 150 million years. The long-term survival of these ancient mammals can only be assured by addressing the issues of habitat preservation, hunting pressures and protection of populations from disease or predation.

ACKNOWLEDGEMENTS

I would like to thank Mike Archer, Mike Augee, John Barrie, Tim Flannery and Peter Murray for their insightful comments and suggestions. A.W. Crompton, Jeanette Hope and Joanne Wilkinson generously provided unpublished information and personal observations. Special thanks are due to Colin Macgregor for providing valuable advice and the opinion of a non-specialist.

REFERENCES

Aplin, K.P. 2006. Ten million years of rodent evolution in Australasia: phylogenetic evidence and a speculative historical biogeography. Pp.707–744. In Merrick, J.R., Archer, M., Hickey, G.M. and Lee, M.S.Y. (eds). *Evolution and Biogeography of Australasian Vertebrates.* Auscipub Pty. Ltd., Sydney. 942pp.

Archer, M. 1995. Prehistoric platypus fits the bill. *Australian Geographic*, Vol. 38, pp.87–103.

Archer, M., Flannery, T.F., Ritchie, A. and Molnar, R.E. 1985. First Mesozoic mammal from Australia—an early Cretaceous monotreme. *Nature,* Vol. 318, pp.363–366.

Archer, M., Hand, S., Godthelp, H. and Creaser, P. 1997. Correlation of the Cainozoic sediments of the Riversleigh World Heritage Fossil property, Queensland, Australia. Pp.131–152. In Aguilar, J.-P., Legendre, S. and Michaux, J. (eds). *Actes du Congrès BiochroM'97.* École Pratique des Hautes Etudes, Institut de Montpellier, Montpellier, France.

Archer, M., Jenkins, F.A. Jnr., Hand, S.J., Murray, P. and Godthelp, H. 1992. Description of the skull and non-vestigial dentition of a Miocene platypus (*Obdurodon dicksoni* n. sp.) from Riversleigh, Australia, and the problem of monotreme origins. Pp.15–27. In Augee, M.L. (ed). *Platypus and Echidnas.* Royal Zoological Society of New South Wales, Sydney. 296pp.

Archer, M., Murray, P., Hand, S.J. and Godthelp, H. 1993. Reconsideration of monotreme relationships based on the skull and dentition of the Miocene *Obdurodon dicksoni* (Ornithorhynchidae) from Riversleigh, Queensland, Australia. Pp.75–94. In Szalay, F.S., Novacek, M.J. and McKenna, M.C. (eds). *Mammal Phylogeny, Volume 1.* Springer-Verlag, New York. 249pp.

Archer, M., Plane, M.D. and Pledge, N.S. 1978. Additional evidence for interpreting the Miocene *Obdurodon insignis* Woodburne and Tedford, 1975, to be a fossil platypus (Ornithorhynchidae: Monotremata) and a reconsideration of the status of *Ornithorhynchus agilis* De Vis, 1885. *Australian Zoologist*, Vol. 20, pp.9–27.

Askin, R.A. 1992. Late Cretaceous-early Tertiary Antarctic outcrop evidence for past vegetation and climates. *Antarctic Research Series*, Vol. 56, pp.61–73.

Augee, M. and Gooden, B. 1997. *Echidnas of Australia and New Guinea.* New South Wales University Press, Kensington. 76pp.

Belov, K., Hellman, L. and Cooper, D.W., 2002. Characterisation of echidna IgM provides insights into the time of divergence of extant mammals. *Developmental & Comparative Immunology*, Vol. 26, pp.831–839.

Bonaparte, J.F. 1990. New late Cretaceous mammals from the Los Alamitos Formation, northern Patagonia. *National Geographic Research*, Vol. 6, pp.63–93.

Bramble, D.M. and Jenkins, F.A. Jnr. 1989. Structural and functional integration across the reptile-mammal boundary: the locomotor system. Pp.133–146. In Wake, D.B. and Roth, G. (eds). *Complex Organismal Functions: Integration and Evolution in Vertebrates.* John Wiley and Sons, New York. 451pp.

Case, J.A. 1989. Antarctica: the effect of high latitude heterochroneity on the origin of the Australian marsupials. Pp.217–226. In Crane, J.A. (ed). *Origins and Evolution of the Antarctic Biota.* Geological Society Special Publication, No. 47. Bath.

Crompton, A.W. and Jenkins, F.A. Jnr. 1973. Mammals from reptiles: a review of mammalian origins. *Annual Review of*

Earth and Planetary Science, Vol. 1, pp.131–155.

Davis, A.C. 1996. *Quaternary mammal faunas and their stratigraphy in the northern Monaro region, southeastern Australia*. Ph.D. thesis (unpublished), Australian National University, Canberra.

Dun, W.S. 1895. Notes on the occurrence of monotreme remains in the Pliocene of New South Wales. *Records of the Geological Survey of New South Wales*, Vol. 4, pp.118–126.

Every, R.G. 1974. Thegosis in prosimians. Pp.579–619. In Martin, R.D., Doyle, G.A. and Walker, A.C. (eds). *Prosimian Biology*. Duckworth, London.

Flannery, T.F. 1995. *Mammals of New Guinea*. Australian Museum and Reed Books, Sydney. 568pp.

Flannery, T.F., Archer, M., Rich, T.H. and Jones, R. 1995. A new family of monotremes from the Cretaceous of Australia. *Nature*, Vol. 377, pp.418–420.

Flannery, T.F. and Groves, C.P. 1998. A revision of the genus *Zaglossus* (Monotremata, Tachyglossidae), with description of new species and subspecies. *Mammalia*, Vol. 62, pp.367–396.

Flynn, J.J., Parrish, J.M., Rakotosamimanana, B., Simpson, W.F. and Wyss, A.R. 1999. A middle Jurassic mammal from Madagascar. *Nature*, Vol. 401, pp.57–60.

Gemmell, N.J., Janke, A., Western, P.S., Watson, J.M., Paabo, S. and Graves, J.A.M. 1994. Cloning and characterisation of the platypus mitochondrial genome. *Journal of Molecular Evolution,* Vol. 39, pp.200–205.

Grant, T. 1995. *The Platypus: a Unique Mammal.* University of New South Wales Press Ltd., Sydney. 92pp.

Grant, T.R. 2004. Captures, capture mortality, age and sex ratios of platypuses, *Ornithorhynchus anatinus*, during studies over 30 years in the upper Shoalhaven River in New South Wales. *Proceedings of the Linnean Society of New South Wales*, Vol. 125, pp.217–226.

Grant, T.R., Griffiths, M. and Temple-Smith, P.D. 2004. Breeding in a free-ranging population of platypuses, *Ornithorhynchus anatinus* in the upper Shoalhaven River, New South Wales—a 27 year study. *Proceedings of the Linnean Society of New South Wales*, Vol. 125, pp.227–234.

Green, H.L.H.H. 1937. The development and morphology of the teeth of *Ornithorhynchus*. *Philosophical Transactions of the Royal Society of London,* Series B, Vol. 228, pp.367–420.

Gregory, W.K. 1947. The monotremes and the palimpsest theory. *Bulletin of the American Museum of Natural History*, Vol. 88, pp.5–52.

Griffiths, M. 1978. *The Biology of the Monotremes*. Academic Press, New York. 367pp.

Griffiths, M., Wells, R.T. and Barrie, D.J. 1991. Observations on the skulls of fossil and extant echidnas (Monotremata: Tachyglossidae). *Australian Mammalogy*, Vol. 14, pp.87–101.

Heinsohn, T. and Hope, G. 2006. The Torresian connections: zoogeography of New Guinea. Pp.71–93. In Merrick, J.R., Archer, M., Hickey, G.M. and Lee, M.S.Y. (eds). *Evolution and Biogeography of Australasian Vertebrates.* Auscipub Pty. Ltd., Sydney. 942pp.

Hopson, J.A. and Crompton, A.W. 1969. Origin of mammals. Pp.15–72. In Dobzhansky, T., Hecht, M.K. and Steere, W.C. (eds). *Evolutionary Biology*. Vol. 3. Appleton-Century-Crofts, New York.

Hopson, J.A. and Rougier, G.W. 1993. Braincase structure in the oldest known skull of a therian mammal: implications for mammalian systematics and cranial evolution. *American Journal of Science*, Vol. 293, pp.268–299.

Hu, Y., Wang, Y., Luo, Z. and Li, C. 1997. A new symmetrodont mammal from China and its implications for mammalian evolution. *Nature*, Vol. 390, pp.137–142.

Huber, E. 1930. Evolution of facial musculature and cutaneous field of trigeminus. *Quarterly Review of Biology*, Vol. 5, pp.133–188.

Janke, A., Gemmell, N.J., Feldmaier-Fuchs, G., von Haesler, A. and Pääbo, S. 1996. The mitochondrial genome of a monotreme—the platypus (*Ornithorhynchus anatinus*). *Journal of Molecular Evolution*, Vol. 42, pp.153–159.

Janke, A., Xu, X. and Arnason, U. 1997. The complete mitochondrial genome of the wallaroo (*Macropus robustus*) and the phylogenetic relationship among Monotremata, Marsupialia and Eutheria. *Proceedings of the National Academy of Sciences of the United States of America,* Vol. 94, pp.1276–1281.

Jenkins, F.A. Jnr. 1973. Functional anatomy of the humero-ulnar joint. *American Journal of Anatomy*, Vol. 137, pp.281–298.

Jenkins, F.A. Jnr. 1990. Monotremes and the biology of Mesozoic mammals. *Netherlands Journal of Zoology*, Vol. 40, pp.5–31.

Jenkins, F.A. Jnr. and Parrington, F.R. 1976. The postcranial skeleton of the Triassic mammals *Eozostrodon*, *Megazostrodon* and *Erythrotherium. Philosophical Transactions of the Royal Society of London*, Series B, Vol. 273, pp.387–431.

Jenkins, F.A. Jnr. and Schaff, C.R. 1988. The early Cretaceous mammal *Gobiconodon* (Mammalia, Triconodonta) from the Cloverly Formation in Montana. *Journal of Vertebrate Paleontology*, Vol. 6, pp.1–24.

Ji, Q., Luo, Z. and Ji, S. 1999. A Chinese triconodont and mosaic evolution of the mammalian skeleton. *Nature*, Vol. 398, pp.326–330.

Johanson, Z. 2006. Origins of mammals: morphology, molecules and a petrosal or two. Pp.491–522. In Merrick, J.R., Archer, M., Hickey, G.M. and Lee, M.S.Y, (eds). *Evolution and Biogeography of Australasian Vertebrates.* Auscipub Pty. Ltd., Sydney. 942pp.

Kemp, T.S. 1983. The interrelationships of mammals. *Zoological Journal of the Linnean Society*, Vol. 77, pp.353–384.

Kielan-Jaworowska, Z. 1971. Skull structure and affinities of the Multituberculata. *Palaeontologica Polonica*, Vol. 25, Results of the Polish-Mongolian Palaeontological Expeditions, Part III: pp.5–41.

Kielan-Jaworowska, Z., Cifelli, R. and Luo, Z. 1998. Alleged Cretaceous placental from down under. *Lethaia,* Vol. 31, pp.267–268.

Kielan-Jaworowska, Z., Crompton, A.W. and Jenkins, F.A. Jnr. 1987. The origin of egg-laying mammals. *Nature*, Vol. 326, pp.871–873.

Killian, J.K., Buckley, T.R., Stewart, N., Mundy, B.L. and Jirtle, R.L. 2001. Marsupials and eutherians reunited: genetic evidence for the Theria hypothesis of mammalian evolution. *Mammalian Genome*, Vol. 12, pp.513–517.

Killian, J.K., Burd, J.C., Jirtle, J.V., Mundy, B.L. 2000. M6P/IGF2R imprinting evolution in mammals. *Molecular*

Cell, Vol. 5, pp.707–716.

Kirsch, J.A.W. and Mayer, G.C. 1998. The platypus is not a rodent. *Philosophical Transactions of the Royal Society of London,* Series B, Vol. 353, pp.1221–1237.

Klima, M. 1973. Die Frühentwicklung des Schultergürtels und des Brustbeins bei den Monotremen (Mammalia: Prototheria). *Advances in Anatomy Embryology and Cell Biology*, Vol. 47, pp.1–80.

Krebs, B. 1991. Das skelett von *Henkelotherium guimarotae* gen. et sp. nov. (Eupantotheria, Mammalia) aus dem Oberen Jura von Portugal. *Berliner Geowissenschaftliche Abhandlungen*, Vol. 133, pp.1–121.

Krefft, G. 1868. On a new and gigantic species of echidna. *Annals of the Magazine of Natural History*, Vol. I, Series 4, pp.113–114.

Lawver, L.A., Gahagan, L.M. and Coffin, M.F. 1992. The development of paleoseaways around Antarctica. *Antarctic Research Series*, Vol. 56, pp.7–30.

Luo, Z., Cifelli, R.L. and Kielan-Jaworowska, Z. 2001. Dual origin of tribosphenic mammals. *Nature*, Vol. 409, pp.53–57.

MacIntyre, G.T. 1967. Foramen pseudovale and quasi-mammals. *Evolution*, Vol. 21, pp.834–841.

Mahoney, J.A. and Ride, W.D.L. 1975. Index to the genera and species of fossil Mammalia described from Australia and New Guinea between 1838 and 1969. *Western Australian Museum Special Publication*, Vol. 6, pp.1–247.

Marshall, B. 1992. Late Pleistocene human exploitation of the platypus in southern Tasmania. Pp.268–276. In Augee, M.L. (ed). *Platypus and Echidnas*. Royal Zoological Society of New South Wales, Sydney. 296pp.

Martin, H.A., 1998. Tertiary climate evolution and the development of aridity in Australia. *Proceedings of the Linnean Society of New South Wales*, Vol. 119. pp.115–136.

McKenna, M.C. and Bell, S.K., 1997. *Classification of Mammals Above the Species Level.* Columbia University Press, New York. 631pp.

Meng, J. and Wyss, A.R. 1995. Monotreme affinities and low-frequency hearing suggested by multituberculate ear. *Nature*, Vol. 377, pp.141–144.

Merrick, J.R. 2006. Australasian freshwater fish faunas: diversity, interrelationships, radiations and conservation. Pp.195–224. In Merrick, J.R., Archer, M., Hickey, G.M. and Lee, M.S.Y. (eds). *Evolution and Biogeography of Australasian Vertebrates.* Auscipub Pty. Ltd., Sydney. 942pp.

Messer, M., Weiss, A.S., Shaw, D.C. and Westerman, M. 1998. Evolution of the monotremes: phylogenetic relationship to marsupials and eutherians, and estimation of divergence dates based on alpha-lactalbumin amino acid sequences. *Journal of Mammalian Evolution*, Vol. 5, pp.95–105.

Murray, P. 1978a. A Pleistocene spiny anteater from Tasmania (Monotremata: Tachyglossidae, *Zaglossus*). *Papers and Proceedings of the Royal Society of Tasmania*, Vol. 112, pp.39–67.

Murray, P.F. 1978b. Late Cenozoic monotreme anteaters. Pp.29–55. In Augee, M. (ed). *Monotreme Biology*. Special Symposium Issue of Australian Zoologist, Vol. 20, Part 1.

Murray, P.F. 1984. Furry egg-layers: the monotreme radiation. Pp.571–583. In Archer, M. and Clayton, G. (eds). *Vertebrate Zoogeography and Evolution in Australasia.* Hesperian Press, Perth. 1203pp.

Musser, A.M. 1998. Evolution, biogeography and palaeoecology of the Ornithorhynchidae. *Australian Mammalogy*, Vol. 20, pp.147–162.

Musser, A.M. and Archer, M. 1998. New information about the skull and dentary of the Miocene platypus *Obdurodon dicksoni*, and a discussion of ornithorhynchid relationships. *Philosophical Transactions of the Royal Society of London,* Series B, Vol. 353, pp.1063–1079.

Novacek, M.J., Rougier, G.W., Wible, J.R., McKenna, M.C., Dashzeveg, D. and Horovitz, I. 1997. Epipubic bones in eutherian mammals from the late Cretaceous of Mongolia. *Nature*, Vol. 389, pp.483–486.

Pascual, R., Archer, M., Ortiz-Jaureguizar, E., Prado, J.L., Godthelp, H. and Hand, S.J. 1992a. First discovery of monotremes in South America. *Nature*, Vol. 356, pp.704–706.

Pascual, R., Archer, M., Ortiz-Jaureguizar, E., Prado, J.L., Godthelp, H. and Hand, S.J. 1992b. The first non-Australian monotreme: an early Paleocene South American platypus (Monotremata, Ornithorhynchidae). Pp.1–14. In Augee, M.L. (ed). *Platypus and Echidnas*. Royal Zoological Society of New South Wales, Sydney. 296pp.

Penny, D. and Hasegawa, M. 1997. The platypus put in its place. *Nature*, Vol. 387, pp.549–550.

Peters, W. and Doria, G. 1876. Descrizione di una nuova specie di Tachyglossus proveniente dalla Nuova Guinea settentrionale. *Annali del Museo Civico di Storia Naturale di Genova,* Vol. 9, pp.183–185.

Pettigrew, J.D., Manger, P.R. and Fine, S.L.B. 1998. The sensory world of the platypus. *Philosophical Transactions of the Royal Society of London,* Series B, Vol. 353, pp.1199–1210

Pledge, N. 1980. Giant echidnas in South Australia. *South Australian Naturalist*, Vol. 55, pp.27–30.

Retief, J.D., Winkfein, R.J. and Dixon, G.H. 1993. Evolution of the monotremes: the sequences of the protamine P1 genes of platypus and echidna. *European Journal of Biochemistry*, Vol. 218, pp.457–461.

Rich, T.H., Archer, M., Hand, S., Godthelp, H., Muirhead, J., Pledge, N., Flannery, T.F., Woodburne, M.O., Case, J.A., Tedford, R.H., Turnbull, W.D., Lundelius, E.L.Jnr., Rich, L.S.V., Whitelaw, M.J., Kemp, A. and Rich, P.V. 1991. Monotremes, placentals, and marsupials: their record in Australia and its biases. Pp.892–1069. In Vickers-Rich, P., Monaghan, J.M., Baird, R.F. and Rich, T.H. (eds). *Vertebrate Palaeontology of Australasia.* Pioneer Design Studio and Monash University Publications Committee, Melbourne. 1437pp.

Rich, T.H. and Rich, P.V. 2000. *Dinosaurs of Darkness.* Indiana University Press, Bloomington. 223pp.

Rich, T.H., Vickers-Rich, P., Constantine, A., Flannery, T.F., Kool, L. and Van Klaveren, N. 1997. A tribosphenic mammal from the Mesozoic of Australia. *Science*, Vol. 278, pp.1438–1442.

Rich. T.H., Vickers-Rich, P., Constantine, A., Flannery, T.F., Kool, L. and van Klaveren, N. 1999. Early Cretaceous mammals from Flat Rocks, Victoria, Australia. *Records of the Queen Victoria Museum*, No. 106. pp.1–35.

Rich. T.H., Vickers-Rich, P., Trusler, P., Flannery, T.F., Cifelli, R., Constantine, A., Kool, L. and van Klaveren, N. 2001. Monotreme nature of the Australian early Cretaceous mammal *Teinolophos*. *Acta Palaeontologica Polonica*, Vol. 46, pp.113–118.

Rougier, G.W. 1993. Vincelestes neuquenianus *Bonaparte*

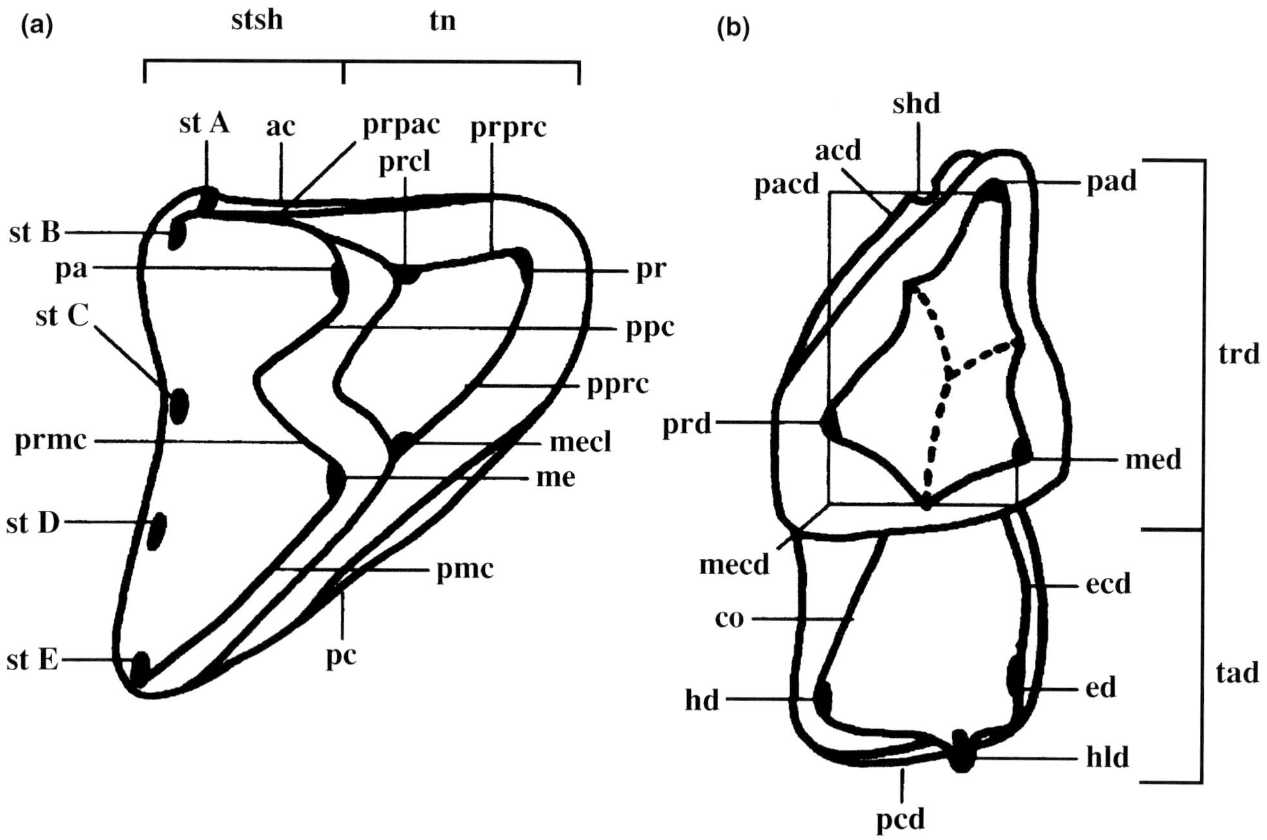

Figure 1. Dental terminology used in this chapter. Key to abbreviations: ac = anterior cingulum; acd = anterior cingulid; co = cristid obliqua; ecd = entocristid; ed = entoconid; hd = hypoconid; hld = hypoconulid; me = metacone; mecd = metacristid; mecl = metaconule; med = metaconid; pa = paracone; pacd = paracristid; pad = paraconid; pc = posterior cingulum; pcd = posterior cingulid; pmc = postmetacrista; ppc = postparacrista; pprc = postprotocrista; pr = protocone; prcl = protoconule; prd = protoconid; prmc = premetacrista; prpac = preparacrista; shd = sulcus for hypoconulid of preceding molar; st A = stylar cusp A; st B = stylar cusp B; st C = stylar cusp C; st D = stylar cusp D; st E = stylar cusp E; stsh = stylar shelf; ta = talonid; trd = trigonid; tn = talon. After Wroe (1999a).

using 156 cranio-dental features and 42 species, many important taxa are represented by very incomplete material. For some, more than 90% of the input data matrix is unknown. Novacek *et al.* (1988) consider that unless at least 30% of features are represented for a taxon, realistic elucidation of phylogeny cannot be expected. Consequently, these results have to be interpreted cautiously. Still, for the time being it's the best we've got.

On the issue of definitions, there are also various interpretations of dental and cranial terminology. Here dental nomenclature follows Flower (1867) and Luckett (1993) regarding the molar–premolar boundary, where the adult (unreduced) postcanine cheektooth formula of marsupials is P1–3 and M1–4 and dP3 is the only postcanine tooth to be replaced (Figure 3). Dental terminology follows Wroe (1999a) and is given in Figure 1. Basicranial terminology follows Archer (1976a), excepting that for the V^3 foramina (see Gaudin *et al.*, 1996; Wroe, 1997). Systematic terminology largely follows Aplin and Archer (1987), but with that used for borhyaenoids following Marshall *et al.* (1990) as amended by Muizon (1999) and that for Dasyuromorphia as amended by Krajewski *et al.* (1994; 1997) and Wroe (1997).

That out of the way, we ask again, what is a marsupial? Going back to the 11 potential synapomorphies forwarded by Archer (1984) it is now clear that several must now be cast aside. On the other hand, some features, treated as plesiomorphic by Archer (1984), may be marsupial synapomorphies, e.g., presence of a medially inflected angle of the lower jaw. Furthermore, some characters not treated by Archer (1984) may represent marsupial synapomorphies, e.g., absence of vascular sulci on the promontorium of the periotic and presence of a horizontally oriented prootic canal connected to the postglenoid venous system (Wible, 1990; Rougier *et al.*, 1998).

Starting from the top, most investigators would now concur that the 'ancestral' marsupial did not possess stylar cusp C. Clemens (1979) observed that stylar cusp C may not have been present in the ancestral marsupial. Fox (1987) further supported this idea, as did Marshall *et al.* (1990), who pointed out that stylar cusp C was absent

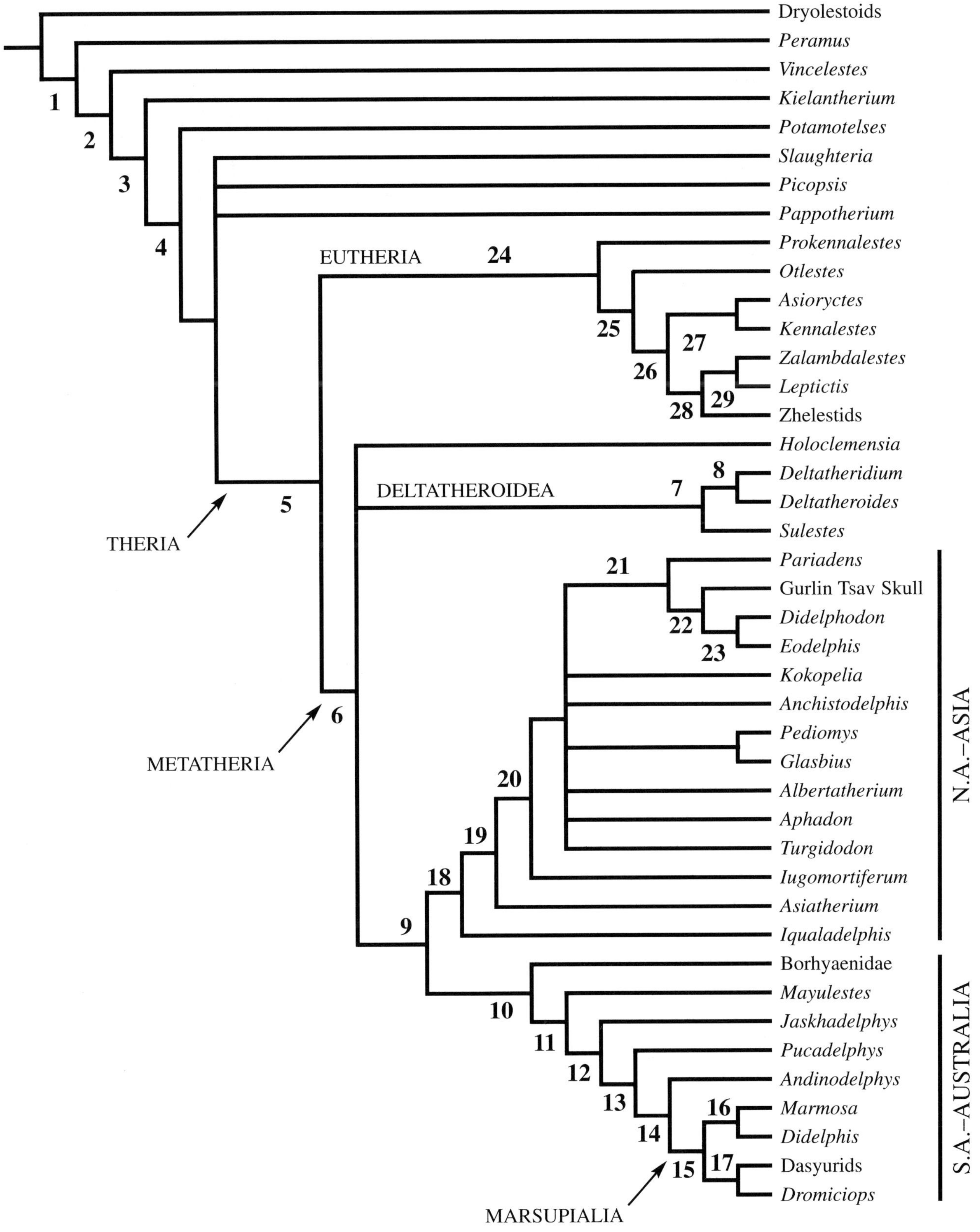

Figure 2. Results of the cladistic analysis of marsupial relationships performed by Rougier *et al.* (1998). Note that these authors apply a far more restricted definition of the term 'marsupial' than is used here, i.e., our Marsupialia equals Rougier *et al.*'s Metatheria. Key to abbreviations: N.A. = North America; S.A. = South America.

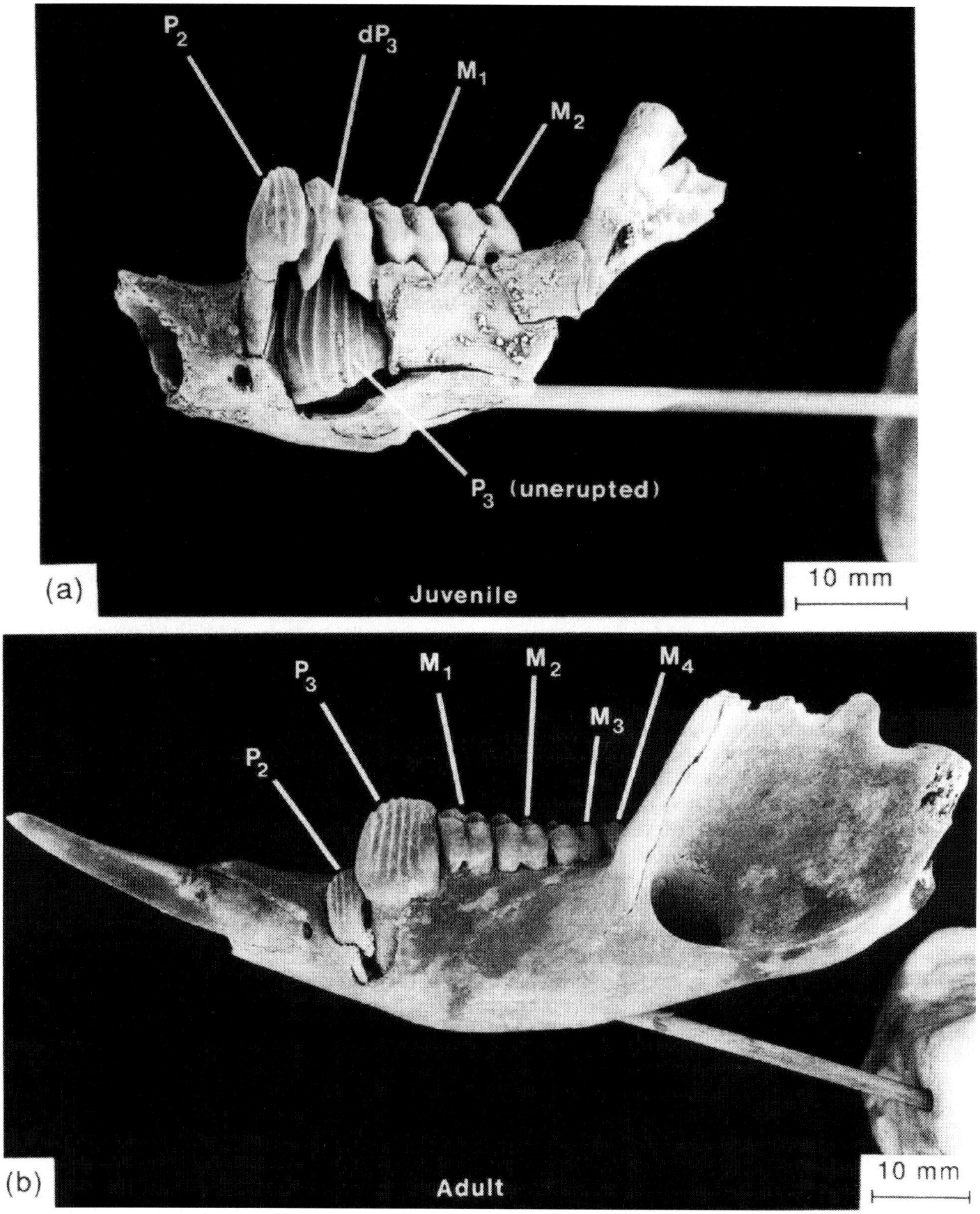

Figure 3. Now generally regarded as perhaps the single strongest defining synapomorphy of marsupials is their unique pattern of dental replacement, whereby only a single postcanine tooth is replaced. Although very derived, the giant extinct rat-kangaroo, *Ekaltadeta ima*, shows the same tooth replacement pattern evident in all marsupials, with only dP3 replaced by P3. Another curious phenomenon regarding *Ekaltadeta* is that the one organ, P2, serves two completely different purposes in the course of the individual's lifetime. In the juvenile **(a)** P2 forms part of a shearing blade, but in the adult **(b)** P2 is used as a buttress for the massive P3. After Wroe and Archer (1995).

in material previously assigned to the late Cretaceous North American genus *Alphadon*. This taxon they referred to as '*Alphadon*' *creber* [-C]. Cifelli (1990a) observed that stylar cusp C was variably present in '*Alphadon*' *creber* and named it *Protalphadon creber*. Johanson (1996) later divided *P. creber* into two species under the new generic name, *Varalphadon*. Marshall *et al.* (1990) argued that stylar cusp C may have arisen independently as many as five times within Marsupialia. The distribution of this feature, among taxa historically considered as marsupials, is further summarized by Godthelp *et al.* (1999) and these authors also agree that the value of this feature at such high taxonomic levels is questionable. It seems that lack of stylar cusp C is a poor basis for the exclusion of taxa from the marsupial fold.

This brings us to a point where we should more rigorously examine a couple of taxa not traditionally placed within Marsupialia. We'll start with deltatheroidans (Figure 4), which, incidentally, don't have a stylar cusp C. Whether or not this clade shares a special relationship with marsupials remains a debated issue (Rougier *et al.*, 1998; Cifelli, 1993, 1999). Deltatheroidans are known

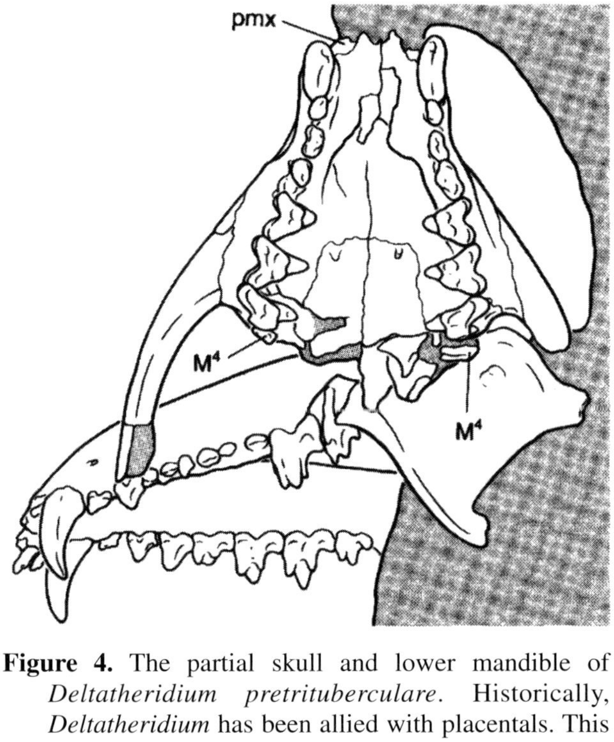

Figure 4. The partial skull and lower mandible of *Deltatheridium pretrituberculare*. Historically, *Deltatheridium* has been allied with placentals. This has been due, in part, to the fact that based on the fragmentary remains available, it was thought that they had an upper molar count of three, a feature typical of placentals. This recent find and others from the late Cretaceous of Mongolia show that not only did *Deltatheridium* have the marsupial dental formula of three premolars and four molars, but that it also showed the marsupial dental replacement pattern, i.e., only dP3 is evicted by a tooth from the same tooth family. After Rougier *et al.* (1998).

from the late Cretaceous of Asia and maybe North America (Cifelli, 1990b) and were originally thought to be basal placentals. However, support for a marsupial affinity has grown in some quarters (Kielan-Jaworowska and Nessov, 1990; Marshall and Kielan-Jaworowska, 1992; Rougier *et al.*, 1998). Several important potential marsupial synapomorphies are present in recently described material referred to *Deltatheridium*. These include the replacement of only one deciduous postcanine tooth, dP3 by P3 (Rougier *et al.*, 1998).

Another taxon that has recently figured prominently in discourse over marsupial origins is *Kokopellia juddi* (Figure 5). Hailing from the early Cretaceous of Utah, this species was originally described by Cifelli (1993) on the basis of a mandible containing alveoli for C1 and P1, as well as well-preserved P2–3 and M1–4. Further material, including a M3, was assigned to the species by Cifelli and Muizon (1997). This tooth shows no trace of stylar cusp C either, but *K. juddi* does have the marsupial postcanine dental formula of three premolars and four molars. Similarly, *Aenigmadelphys archeri*, from the very late Campanian (about 72 m.y.a.) of southern Utah, also lacks stylar Cusp C (Cifelli and Johanson, 1994).

A taxon that deserves special mention is *Asiatherium reshetovi*, from the late Cretaceous Udan Sayr Local Fauna of Mongolia (Szalay and Trofimov, 1996). In addition to a typical marsupial dental formula, *Asiatherium* also exhibits a twinned hypoconulid and entoconid, but again lacks stylar cusp C. Szalay and Trofimov (1996) were highly sceptical of the phylogenetic value of presence or absence of stylar cusps in general. With the demise of stylar cusp C as a cladistically useful tool at higher taxonomic levels, the case for the inclusion of an important prior contender for world's oldest marsupial title, *Holoclemensia texana* (middle Cretaceous Butler farm Locality of Texas), is diminished. But more on this later. To sum up, at present, stylar cusp C has been sidelined as a marsupial synapomorphy.

Regarding the second feature alluded to by Archer (1984), presence of a metacone subequal to or taller than the paracone, this character state is open to interpretation. Just how much smaller than the paracone does the metacone have to be before it's no longer subequal? Nonetheless, there is no doubt that a paracone taller than the metacone is plesiomorphic and many authors, including ourselves (Rougier *et al.*, 1998; Szalay and Trofimov, 1996; Godthelp *et al.*, 1999) have accepted that relative enlargement of the metacone is a useful feature, although it remains rather subjective.

The third feature, twinning of the hypoconulid and entoconid on the lower molars, is also effectively sidelined as a marsupial synapomorphy if deltatheroidans and *Kokopellia juddi* are included within the clade. In the cladistic analysis performed by Rougier *et al.* (1998), this feature is treated as a synapomorphy uniting all marsupials except *Holoclemensia texana* and deltatheroidans. However, this requires that lack of twinning of these cusps in *K. juddi* constitutes an apomorphic reversal to a

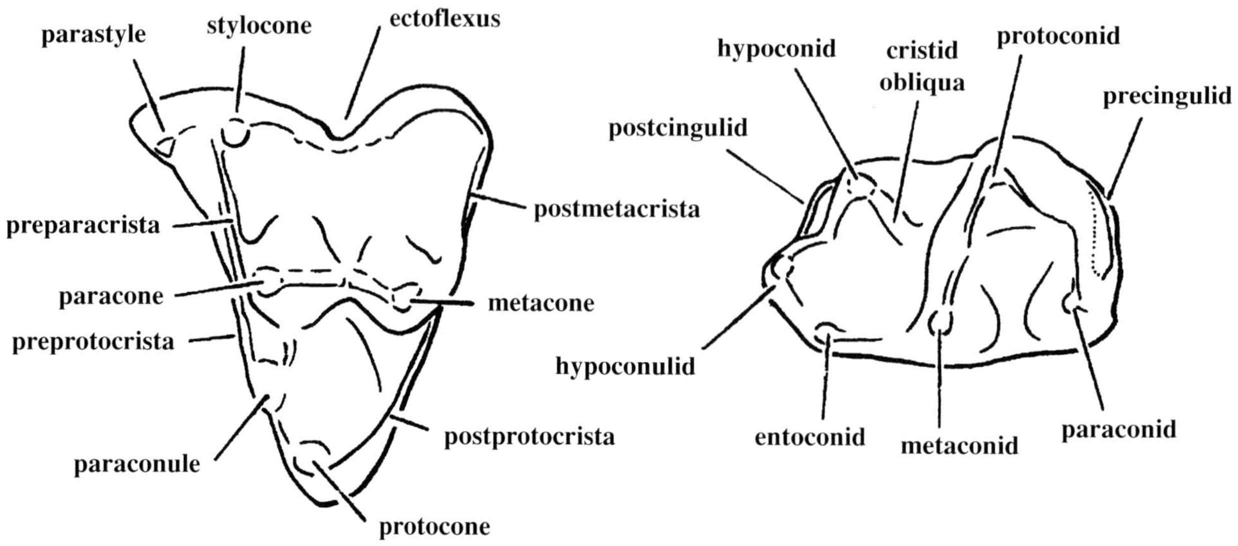

Figure 5. Left upper and lower molars of *Kokopellia juddi*. Note the absence of stylar cusp C and the 'untwinned' hypoconulid and entoconid. After Cifelli and Muizon (1997).

plesiomorphic state. Szalay and Trofimov (1996) consider approximation of the hypoconulid and entoconid as part of a character complex with enlargement of the metacone. If so, then the two should be treated as a single character in phylogenetic analysis.

The dubious nature of character 4 is already highlighted by Archer (1984) and we won't expand on it here, but regarding the next two interrelated features, 5 and 6, things have changed. Berkovitz (1967) and Archer (1974, 1978) presented evidence that no postcanine teeth were replaced in marsupials, based on studies of tooth replacement in *Didelphis* and dasyurids respectively. These analyses indicated that the tooth formerly regarded as dP3 (the dP4 of some) developed in a distinct position along the dental lamina and was not from the same tooth family as P3. Moreover, based on the timing of embryological initiation, Archer (1978) deduced that it formed part of the molar series. Hence, he concluded that dP3 was not a deciduous tooth, but a M1. Consequently, it was posited that primitively, the postcanine tooth formula of marsupials was P1–3 and M1–5. Archer's (1978) construal was widely accepted by subsequent authors, but still more recent evidence (Luckett, 1993; Luckett and Woolley, 1996) contradicts this interpretation, supporting the earlier proposition of Flower (1867) that the shed tooth is a dP3 and not an M1. Most have accepted this latter finding.

Where does this leave us with respect to the elucidation of marsupial synapomorphies? Well, it is now generally acknowledged that although marsupials are plesiomorphic in retaining four molars, they are derived among therians in that they have only three premolars (Martin, 1996) and in having only dP3 replaced among post canine teeth (Luckett and Woolley, 1996). The validity of this tooth replacement pattern as a marsupial synapomorphy has been further supported by studies of tooth replacement in undoubted early marsupial fossil taxa, such as *Alphadon* and *Pucadelphys* (Cifelli *et al.*, 1996; Cifelli and Muizon, 1998*a*). The presence of both these features constitutes a strong basis for the inclusion of deltatheroidans within Marsupialia (Rougier *et al.*, 1998).

This brings us to the next of our potential synapomorphies, feature 7: flooring of the middle ear by the alisphenoid. In 1994 Christian de Muizon described the complete skull of the borhyaenoid, *Mayulestes ferox*, from the early Paleocene Tiupampan Local Fauna of Bolivia and more recently near-complete postcranial remains have been recovered (Figure 6). *Mayulestes* lacks even the slightest hint of a contribution of the alisphenoid to the tympanic bulla (Figure 7). Similarly, this feature is lacking in two didelphoid species from the same locality, *Pucadelphys andinus* and *Andinodelphys cochabambensis* (Marshall and Muizon, 1995; Muizon *et al.*, 1997). These three taxa are the oldest marsupials for which well-preserved, complete skulls are known. Collectively, they constitute persuasive evidence backing the hypothesis that the ancestral marsupial lacked an alisphenoid tympanic wing. It seems highly unlikely that this feature could have been secondarily lost in these otherwise very generalized marsupials. Muizon (1994) argues that an alisphenoid tympanic wing has independently appeared several times in the course of marsupial evolution.

The validity of features 8 and 10 have already been treated as equivocal by Archer (1984) and feature 11 was considered to require further substantiation. So, at the end of the day what are we left with? Well, there does not seem to be any doubt that the presence of a pseudovaginal canal (feature 9) is a bona fide marsupial synapomorphy, although this of course is of zero value to palaeontologists. Incidentally, whether or not a prolonged period of nipple fixation was also present in early placentals, as

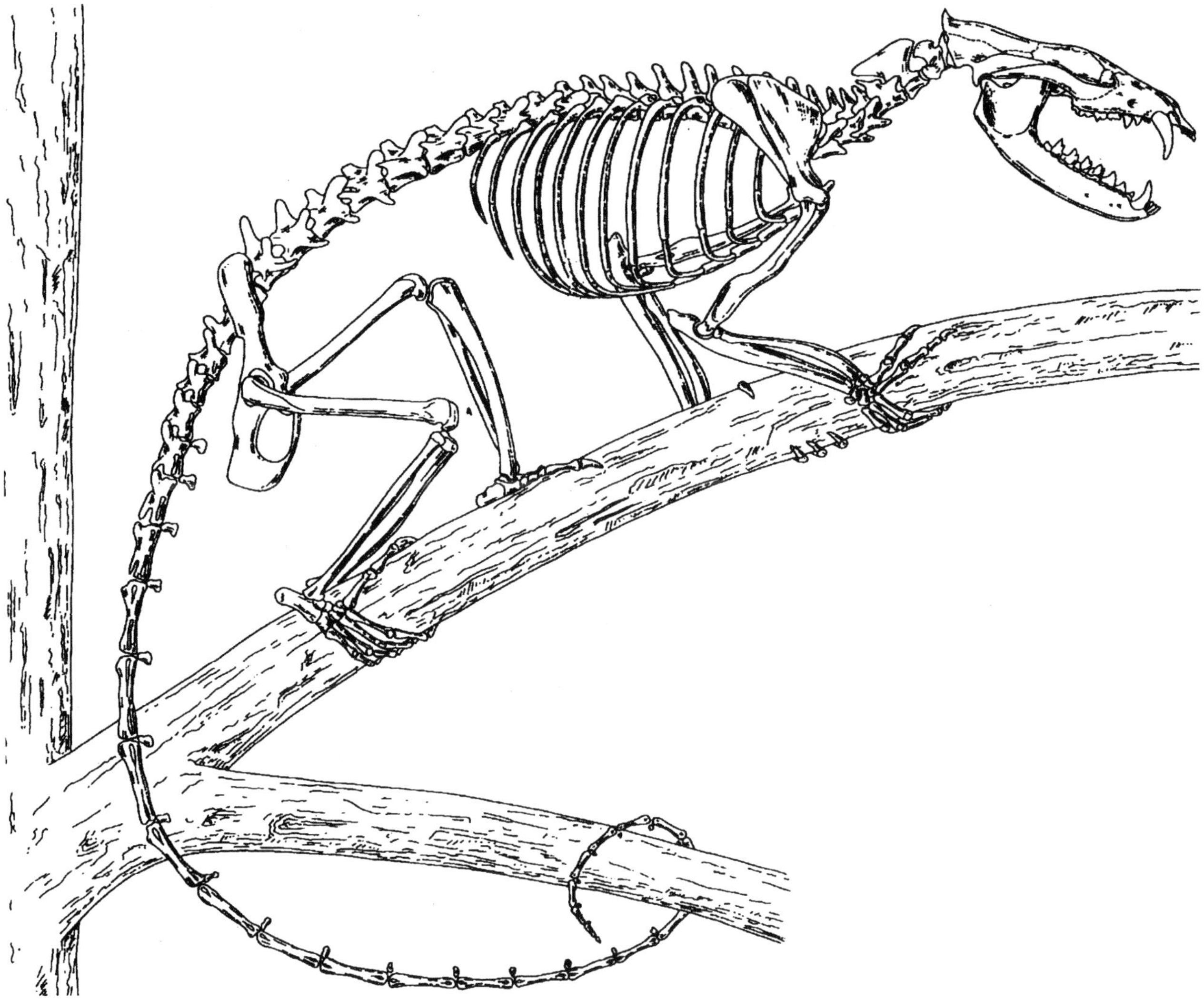

Figure 6. Reconstruction of the skeleton of *Mayulestes ferox*, from the Paleocene Tiupampan Local Fauna of Bolivia. As with other borhyaenoids that are known from near-complete postcranial material, as well as the Tasmanian 'tiger', *Mayulestes* lacks ossified epipubic or 'marsupial bones'. After Muizon (1998).

well as in living and extinct marsupials, has recently been called into question. Novacek *et al.* (1997) cite the presence of epipubics in two late Cretaceous placentals, *Ukhaatherium* and *Zalambdalestes*, as possible evidence for the prolonged external attachment of altricial young, on the grounds that epipubics in marsupials have been correlated with their reproductive mode. This is hard to prove or disprove. Certainly the case isn't watertight. Ossified epipubic bones are absent in some marsupials, i.e., *Thylacinus cynocephalus* and at least some borhyaenoids (Archer, 1984).

Back to the issue at hand, of the 11 features posited as potential marsupial synapomorphies by Archer (1984) only three seem to be regarded as useful from a palaeontological perspective: reduction of premolar number to three, non-replacement of dP1 and dP2 and enlargement of the metacone relative to the paracone. *Hang on!*, we hear you exclaim, there must have been some major advances since 1984 that allow us to pin down at least a few more shared derived features for the clade. Well, indeed many additional marsupial synapomorphies have been flagged in recent years. For example, in subjective treatments, Marshall *et al.* (1990) forward 23 synapomorphies for Marsupialia, while Marshall and Kielan-Jaworowska (1992) posit 12 [the apparent discrepancy here may be explained by the fact that Marshall and Kielan-Jaworowska (1992) treat deltatheroidans as marsupials, whereas Marshall *et al.* (1990) do not].

The cladistic analysis of Rougier *et al.* (1998) produced 18 potential shared derived features. Highlighted as most significant among these were non-replacement of dP1 and dP2, loss of vascular sulci for the stapedial arterial system and presence of a small, horizontally oriented prootic canal connected to the postglenoid venous system.

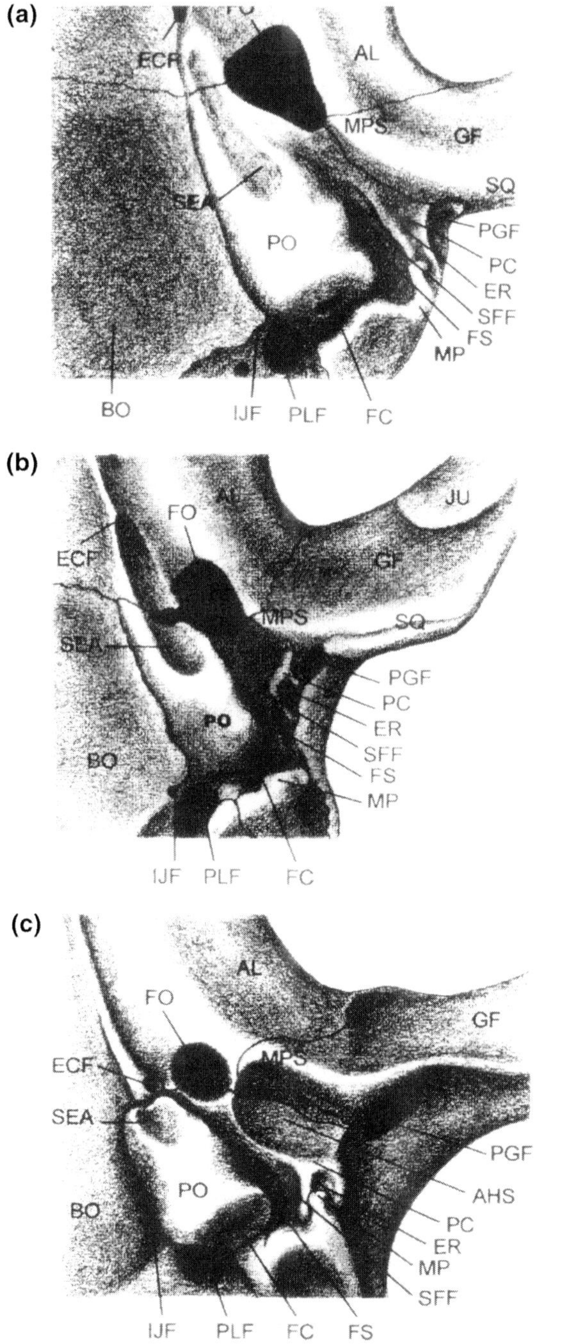

Figure 7. Comparative reconstructions of the ear regions of: **(a)** *Pucadelphys*; **(b)** *Andinodelphys*; **(c)** *Mayulestes*. Key to abbreviations: AHS = alisphenoid hypotympanic sinus; AL = alisphenoid; BO = basioccipital; ECF = entocarotid foramen; ER = epitympanic recess; FC = fenestra cochlearis; FO = foramen ovale; FS = fascial sulcus; GF = glenoid fossa; IJF = inferior jugular foramen; MP = mastoid process; MPS = medial process of the squamosal; PC = petrosal crest; PGF = postglenoid foramen; PLF = posterior lacerate foramen; PO = promontorium; SEA = sulcus for the entocarotid artery; SFF = secondary facial foramen; SQ = squamosal. After Muizon *et al.* (1997).

Two other features were forwarded as likely marsupial synapomorphies by Rougier *et al.* (1998), although they were not supported by parsimony-based analysis: presence of a medially inflected angle of the lower jaw and a posteriorly directed process of the premaxilla that reaches the alveolus of the canine. Of these latter two, the first was considered to be a plesiomorphy by Archer (1984). Certainly the claim that a medially inflected angle of the lower jaw is absent in Mesozoic outgroups, monotremes and placentals (Rougier *et al.*, 1998) may not be, strictly speaking, correct. Sanchez-Villagra and Smith (1997) argue that this process is not present in any placental, fossil or extant and postulate that it represents a marsupial synapomorphy. This position is at odds with that of Kielan-Jaworowska and Nessov (1990), who note the presence of a medially inflected angle in some Cretaceous placentals. Moreover, it is variably present in at least one monotreme, the living platypus (Musser, 2000, personal communication). Whether this feature is present in Mesozoic monotremes is unknown, as the posterior portion of the jaw is not preserved in any known specimen.

At any rate, having sifted through the more recent literature, it appears on the basis of consensus, that there are only two common denominators accepted by all in the 'marsupial synapomorphy stakes': non-replacement of dP1 and dP2 and reduction of premolar number to three. So, again dependent on interpretation and having accounted for additional information, the field has narrowed further still with regard to cladistically valid marsupial features. On the other hand, the number of taxa included within Marsupialia has expanded to include deltatheroidans and *Asiatherium*.

THE FIRST MARSUPIAL: WHO, WHEN AND WHERE?

Just which taxon wears the mantle of oldest marsupial has become a sticky question indeed. Moreover, we should preface any further discussion with the rider that the location of the oldest known specimen and Marsupialia's actual point of origin are not necessarily one and the same, although palaeontologists commonly make this assumption on the basis of parsimony. Before delving into the palaeontological record we should also make a note of parameters thus far flagged on the basis of molecular studies. Numbers denoting the potential marsupial-placental split range from 107 to 128 million years ago (Richardson, 1988; Hope *et al.*, 1990; Westerman *et al.*, 1990; Kirsch *et al.*, 1997).

Marsupials are now known from every continent, but historically North America has most commonly been advanced as the clade's point of origin. However, Kirsch (1984) has formerly argued for an Australian origin, while Szalay (1994) has posited that Asia may represent the marsupial motherland. For many years the North American middle Cretaceous *Holoclemensia texana*, described by Slaughter (1968), was thought to be the oldest marsupial title holder (Archer, 1984). This species, known only from a partial M3, a M4 and a lower left molar, possessed three features then thought to be marsupial synapomorphies: a large stylar cusp C; a hypoconulid that sits closer to the entoconid than the hypoconid; and a metacone only slightly smaller than the paracone. But, as we have seen, the first of these is no longer considered synapomorphic for the clade, while the second would appear to represent a possible synapomorphy uniting *H. texana* with marsupials more derived than the basal marsupial taxon Deltatheroidea, or *Kokopellia juddi*. Meanwhile, although at least potentially useful the third feature remains open to subjective interpretation.

Regardless of where *H. texana* fits within the greater scheme of marsupial phylogeny, other Cretaceous taxa now rival or exceed it in age. Marshall and Kielan-Jaworowska (1992) note that undescribed Mongolian material of late Early Cretaceous age has been referred to Deltatheroidea by Reshetov and Trofimov (1984). At this point it's worth restating that not all investigators accept marsupial affinity for deltatheroidans. Cifelli (1999) notes that principle components analysis of tooth shape places *Deltatheridium* among primitive therians. Interestingly, the same analysis places *H. texana* within the placental fold. On this basis *K. juddi*, of late Early Cretaceous age (around 110–100 m.y.a.), would be the most ancient of known marsupials, giving credence to the classical view that North America gave rise to the first marsupial.

This evidence aside, there are other factors that complicate the interpretation of deltatheroidan phylogeny in particular. Although deltatheroidans do sport a marsupial tooth formula and replacement pattern, they are in many respects quite specialized animals. For example, it is generally accepted that the 'prototypical' marsupial had five upper and four lower incisors, but in *Deltatheridium* this count is reduced to 4/3. Moreover, in this taxon the ultimate lower and upper molars are greatly reduced in overall size. Thus, although *Deltatheridium* may be a marsupial, it certainly wasn't *directly* ancestral to New World taxa. On the other hand, new evidence presented by Cifelli and Muizon (1998b) casts doubt on whether the taxon that for decades was cited as the archetypal marsupial, i.e., *Alphadon*, had a primitive dental formula. These authors observe that in a new species, *Alphadon eatoni* (Figure 8), it seems likely that only three lower incisors were present. Additionally, the second incisor was very large and lanceolate in shape. The anterior-most portion of the jaw is not preserved in any other specimens of *Alphadon*, allowing the possibility that these features were typical of the genus. Their presence in this otherwise dentally plesiomorphic taxon seems incongruous and thickens the plot further.

To sum up on the issues of identity and location of the oldest marsupial, we can only point out that potential can-

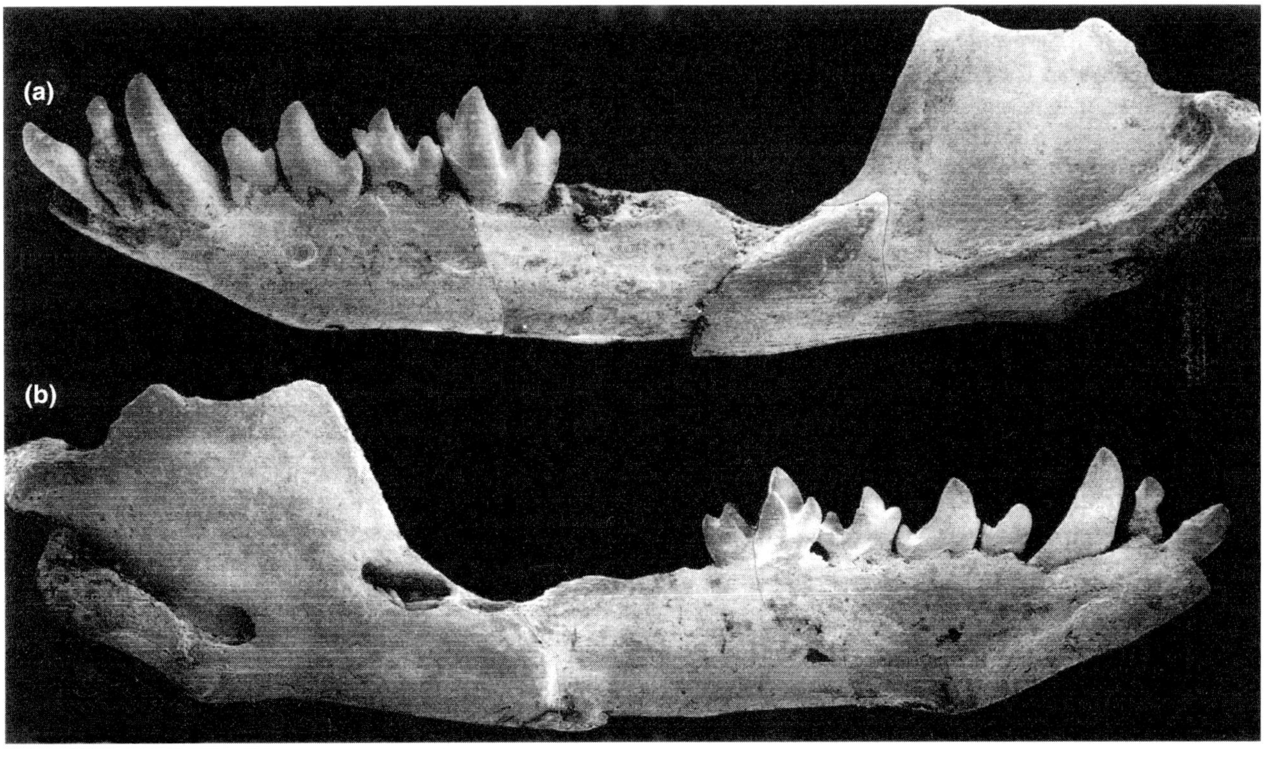

Figure 8. *Alphadon eatoni* from the late Cretaceous of Utah in buccal **(a)** and lingual **(b)** views. Although, like other peradectids, this species is plesiomorphic among marsupials for most dental features, it apparently had reduced its lower incisor count from four to three. Moreover, I2 was a specialized lanceolate tooth. Such derived morphology was unexpected in such an otherwise 'primitive' marsupial. Scale = 2 mm. After Cifelli and Muizon (1998).

America. However, precisely when they jumped ship is less clear. A lower limit is apparently set by the Los Alamitos Local Fauna of Patagonia, at around 73 to 83 million years in age (Campanian). This fauna has yielded a variety of non-tribosphenic mammals including triconodonts, multituberculates, dryolestids and symmetrodonts, but no marsupials (Bonaparte, 1990). Mammal bearing South American local faunas of earliest Tertiary age are known from Argentina, Bolivia, Peru and Brazil. The most diverse of these is the Tiupampan assemblage from the early Paleocene of Bolivia, with at least 11 marsupial species representing four orders alongside a wide variety of placental taxa. Because the marsupials of this fauna are diverse and more divergent from North American relatives of comparable age, than are its placentals, Muizon (1991) and Muizon *et al.* (1997) suggest that the arrival of marsupials and placentals in South America was diachronous. Thus, marsupials may have entered the continent sometime well before the K/T event. But, precisely when this might have been remains another open question.

Exactly when vicariance effectively halted dispersal between North and South America is also unknown. At any rate, regardless of when and how marsupials got to South America one thing is clear… they really liked it! Next to Australia no other continent has ever boasted such a rich marsupial presence. The curious and glaring contrast between Australia and South America is that in the former, marsupials dominated all terrestrial mammalian niches while in South America marsupials seem only to have conquered the insectivore, omnivore and carnivore roles, leaving the herbivorous niches for placentals. That is until North and South America again became chummy in the middle Pliocene. The culpability of placentals in the extinction of South American predaceous marsupials during the late Tertiary is another open question. Climatic change, together with competition from giant, terrestrial phorusrhacoid birds may also have played significant roles in the demise of the continent's large marsupial carnivores (Marshall, 1978).

Anyway… let's dwell on the positives… in the interval of isolation preceding this unfortunate late Tertiary tragedy, South American marsupials really took off, with around seven superfamilies represented, providing a record almost as spectacular as Australia's nine or so. Even a brief summary of all the South American families that have existed is way beyond the scope of this chapter. What we'll try and do is very briefly consider South American taxa that have been linked with Antarctic and Australian radiations. The radiation of Australian marsupials is discussed in more detail in the next chapter (Archer and Hand, 2006).

DIDELPHIMORPHIA

This taxon contains at least three families, Didelphidae (Figure 11), Pucadelphydae and Sparassocynidae and has been forwarded as structurally, if not actually ancestral to Australian dasyuromorphians, which have themselves been seen as archetypal Australian marsupials (Wroe *et al.*, 2000, 2003). At least some authors also place the carnivorous borhyaenoids (sparassodontans) within Didelphimorphia (Muizon *et al.*, 1997). As previously mentioned, historically Didelphidae has been something of a paraphyletic basket-case, used to dispose of anything too plesiomorphic to fit anywhere else.

Fortunately, attempts have been made to address this issue and the Didelphidae has been redefined several times in recent decades. Notably, all recent authors seem to concur on the removal of two taxa formerly treated as didelphids, i.e., Peradectidae and Microbiotheriidae. Still, it remains difficult to pin convincing synapomorphies on the family. Marshall *et al.* (1990) cite three tarsal features as didelphid synapomorphies. However, the same three 'synapomorphies' are used by these authors to denote the monophyletic clade comprising Didelphidae and Sparassocynidae. This doesn't hold water in a cladistic sense. Rougier *et al.* (1998) found two synapomorphies for Didelphidae (Stylar cusp E distal to D, second lower premolar larger than third). But this analysis only included two didelphoid taxa, *Didelphis* and *Marmosa*. Certainly neither of these features is common to all didelphoids, living or fossil. Marshall *et al.* (1990) list a total of 34 genera of didelphid falling within five subfamilies. Of these, 22 genera and three subfamilies are extinct. Although often cited as 'living fossils', at least some didelphids show quite derived morphology (Goin, 1993). Thus, among basicranial features for example, although there is not much in the way of alisphenoid or petrosal tympanic wings in most species, in some, such as *Caluromys*, these processes are well-developed. Similarly, a derived marsupial feature, the secondary foramina ovale, is also present in a number of didelphids (Wroe, 1997).

Because Didelphimorphia is essentially generalized, or at least inclusive of some very morphologically conservative taxa, unravelling the clade's origins has proven perennially problematical. Fortunately, a shaft of phylogenetic light has beamed down on the question in the form of two exquisitely well-preserved new fossil species from the Tiupampan Local Fauna: *Pucadelphys andinus* and *Andinodelphys cochabambensis*. Both taxa were originally described as didelphids by Marshall and Muizon (1988). More recently, Marshall *et al.* (1990) and Woodburne and Case (1996) advocated placement of *A. cochabambensis* as the sister taxon to the Australian radiation. But, as more complete material has been discovered (Marshall and Muizon, 1995; Muizon *et al.*, 1997; Muizon, 1998) it has become clear that these two should be placed in the same new family (Pucadelphydae) and that no special relationship exists between *A. cochabambensis* and Australian marsupials. These two taxa are plesiomorphic among marsupials for almost all cranial features. Not only do they lack alisphenoid tympanic wings,

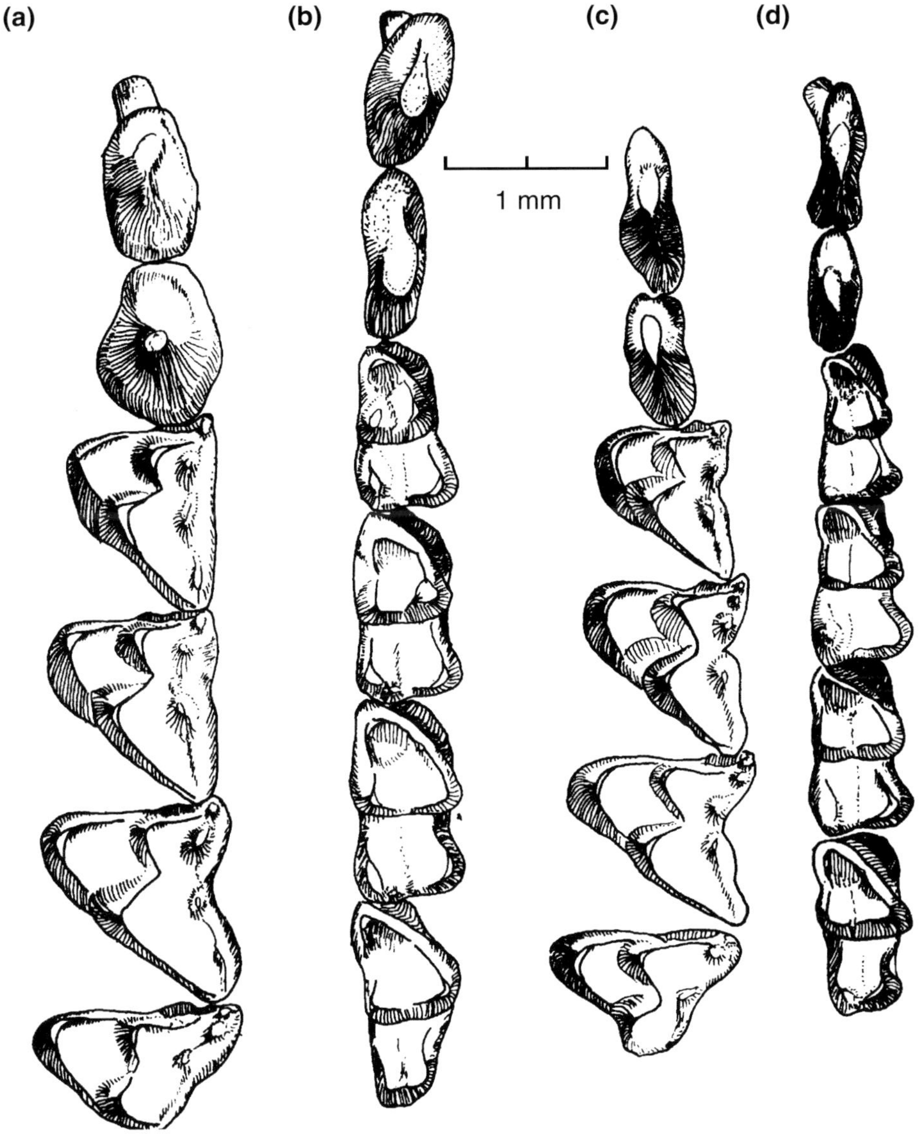

Figure 11. Occlusal views of two didelphid taxa: *Philander opossum*, **(a)** upper P2–M4 and **(b)** lower P2–M4; and *Metachirus nudicaudatus*, **(c)** upper P2–M4 and **(d)** lower P2–M4. Among living marsupials didelphids are generally held to have the least specialized dentition. After Reig *et al.* (1987).

but they have no hypotympanic sinuses, a feature found in all other marsupials for which the middle ear region is known.

SPARASSODONTA

Known from a relatively complete fossil record spanning most of the Tertiary, the five sparassodontan families (also known as borhyaenoids) represent the single greatest radiation of marsupial carnivores ever to have stalked the Earth [although advances in Australian palaeontology are closing the gap (Wroe, 1999b)]. They showed a wide range of morphologies, ranging from huge bear-sized beasts and marsupial sabre-tooths (Figure 12), to small, probably arboreal forms. The palaeontology and evolution of sparassodontans has been extensively covered by Marshall (1976, 1977, 1978, 1979, 1981). Marshall *et al.* (1990) included stagodontids and hondadelphids as members of this clade, a position not widely held by others. The five families we recognize are Mayulestidae, Hathylacinidae, Borhyaenidae, Proborhyaenidae and Thylacosmilidae.

The smallest, oldest and least specialized of sparassodontan taxa was *Mayulestes ferox* from the Tiupampa Local Fauna of Bolivia. Together with another relatively plesiomorphic taxon, *Allqokirus australis*, it has been placed in a family, Mayulestidae, treated as the plesiomorphic sister taxon to all other members of the superfamily by Muizon (1999). Until recently, one of the most solid synapomorphies uniting the group was the reduction of the mastoid part of the petrosal, which does not appear on the occipital shield at the rear of the skull in any, excepting *M. ferox*. With the discovery of *M. ferox* defining features of the Sparassodonta have had to be rewritten. In a cladistic analysis of relationships, Muizon (1999) forwarded eight synapomorphies for the superfamily.

The precise relationship of sparassodontans to other marsupials remains the topic of debate. Both pediomyids and deltatheroideans have previously been flagged as possible sister taxa by Marshall (1977) and Marshall and Kielan-Jaworowska (1992), respectively, although both interpretations have been rejected by Muizon *et al.* (1997) and Muizon (1999). Until quite recently, several authors had also argued the case for a special relationship between sparassodontans and thylacinids (Archer, 1976b). Similarities, on the basis of dentitions at least, are striking. Indeed, Sinclair (1906) actually referred borhyaenid material from the Santa Cruz Locality of Patagonia to the family Thylacinidae. However, scholarly push and shove over this issue was effectively hit on the head in the early 1980s by the results of molecular (Lowenstein *et al.*, 1981) and morphological (Szalay, 1982) investigations. These convincingly demonstrated a dasyuromorphian origin for thylacinids. Still more recent studies have put this finding beyond reasonable doubt (Krajewski *et al.*, 1997; 2000; Muirhead and Wroe, 1998). Most now appear to agree that sparassodontans are part of, or closely allied to, the didelphimorphian radiation (Archer, 1984; Muizon *et al.*, 1997).

POLYDOLOPIMORPHIA

Polydolopimorphians are a largely South American group that has also been the subject of some controversy. Four families are included within the superfamily by Marshall *et al.* (1990): Polydolopidae, Bonapartheriidae, Prepidolopidae and Protodidelphidae. Marshall *et al.* (1990) determined Polydolopimorphia to be the sister taxon to Didelphoidea based on a suite of 11 synapomor-

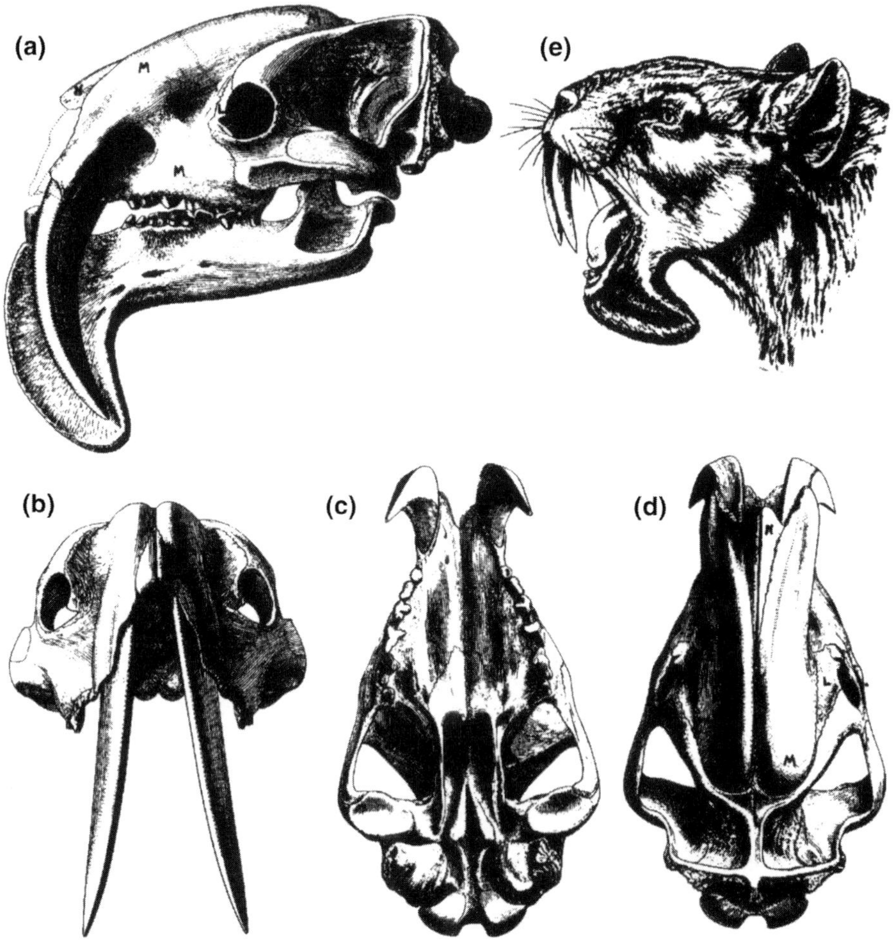

Figure 12. The skull of *Thylacosmilus atrox*, a late Tertiary South American marsupial convergent on sabre-toothed cats, indeed in some ways it was even more specialized, e.g., it has completely lost its incisor teeth. Australia's answer to the big felid niche, *Thylacoleo carnifex*, was also super-specialized, but in different ways. At up to around 164 kg, *T. carnifex* was also around 50 % larger (Wroe *et al.*, 1999, Wroe *et al.*, 2003). After Riggs (1934).

phies. Certainly this is not at odds with the consensus view advocating that the group is descended from a 'didelphoid-like' ancestor (Archer, 1984). However, it's worth bearing in mind that some have argued for a special relationship between polydolopids and Australian diprotodonts (Pascual and Herrera, 1973). On the face of it, the two clades do share some possible synapomorphies, including diprotodonty, enlargement of P3, reduction of M4 and presence of a quadritubercular, omnivorous- herbivorous molar morphology. A development that might be seen as tacitly supporting this notion has been the discovery of two species of polydolopid from Antarctica (Case *et al.*, 1988; Woodburne and Zinsmeister, 1984).

PAUCITUBERCULATA

Two superfamilies comprise Paucituberculata, Caenolestoidea and Argyrologoidea. Hayman *et al.* (1971) and Kirsch (1977) demonstrated that living caenolestids are distinct from all other extant marsupials. Refreshingly, there appears to be little in the way of discord over the affinity of fossil taxa placed within the superfamily. All caenolestoids have a procumbent lower incisor, a series of small teeth between the canine and the P_3 and a steep molar gradient.

Although most seem reasonably comfortable with the intrasuprafamilial phylogeny of caenolestoids, the same can not be said of intersuperfamilial relationships. Historically, it has been argued that they are closely related to Australian diprotodonts or bandicoots, but in more recent years an affinity with didelphoids has been cautiously accepted by many (Archer, 1984). This position is supported by the presence of epididymal sperm-pairing in the two clades, although pairing is head to head in caenolestoids versus side by side in didelphoids (Temple-Smith, 1987). At least some molecular evidence (Gemmell and Westerman, 1994) also supports caenolestid-didelphid affinity, however whole serum-serology (Kirsch, 1977), placed living caenolestids outside of all other living marsupials, Australian and American alike. In short, the debate over caenolestoid relationships ain't over yet.

MICROBIOTHERIIDOIDEA

Small, mousy and inconspicuous, a couple of decades ago no one could have foreseen that the South American Monito del Monte, *Dromiciops australis* (Figure 13) to its friends, was about to be catapulted into the centre of an intercontinental cladistic bun fight of mammoth proportions. Controversy is inversely proportional to body size when it comes to *D. australis*, traditionally once considered as part of the South American didelphoid radiation. In recent decades, on the basis of both tarsal morphology and molecular evidence, this species and its various putative fossil relatives have been placed as the sister taxon to dasyuromorphians (Szalay 1994), diprotodontians (Kirsch *et al.*, 1997), diprotodontians, peramelemorphians and dasyuromorphians (Colgan, 1999) or the entire Australian marsupial radiation plus *Andinodelphys cochabambensis* (Marshall *et al.*, 1990; Woodburne and Case, 1996).

While most investigators now seem to have accepted some kind of special relationship between microbiotheriids and the Australian marsupial radiation, or parts there-

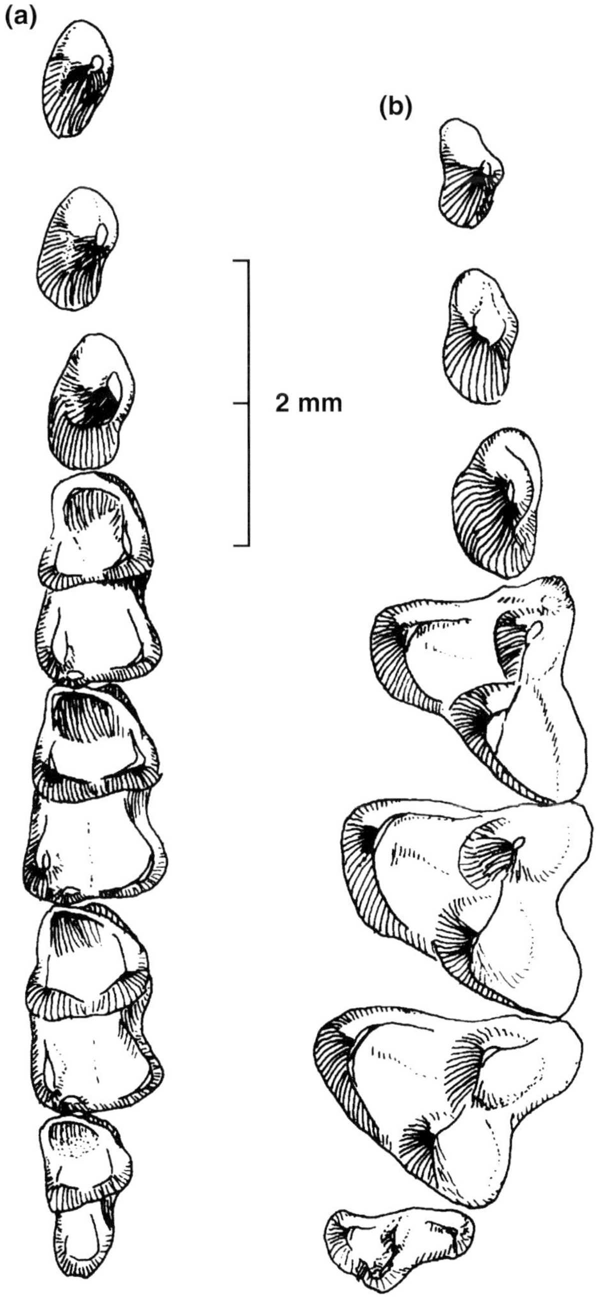

Figure 13. *Dromiciops australis*, sole surviving member of the Microbiotheriidae: **(a)** lower postcanine dentition, P1–M4 and **(b)** upper postcanine dentition P1–M4. After Reig *et al.* (1987).

of, at least one author has taken a radically different view. Hershkovitz (1992, 1995, 1999) treats *Dromiciops* as the sister taxon to Ameridelphia-Australidelphia. Central to Hershkovitz's interpretation is the absence of a derived feature for marsupials in this taxon that is present in ameridelphians and australidelphians, a staggered I3.

Just what constitutes a micobiotheroid is in itself a point of at best marginal consensus. Although all authors seem to concur on the close relationship of *D. australis* to the Miocene South American taxon *Microbiotherium tehuelchum*, agreement appears to end here. Monophyly has been proposed for these two and various other fossil taxa, including *Mirandatherium* and the Australian taxon *Ankotarinja tirarensis* (Marshall, 1987), as well as *Eomicrobiotherium*, *Khasia*, *Monodelphopsis*, *Pachybiotherium*, and Pedyomyidae (Marshall *et al.*, 1990). Also at issue is the putative microbiotherian affinity of fossil material from Seymour Island, Antarctica (Goin and Carlini, 1995; Marenssi *et al.*, 1994; Goin *et al.*, 1999). But again, more on this later. A thorough review of the ins and outs of microbiotherian phylogeny is beyond the scope of this chapter. But, before leaving this gritty topic, we observe that because Microbiotheriidae, *sensu* either Marshall (1987) or Marshall *et al.* (1990), is not well-defined on the basis of dental data, we concur along with Szalay (1994) and Wroe *et al.* (2000), that, excepting *Microbiotherium*, the referral of any fossil taxa to the clade should be ingested with a strong dose of salt.

ANTARCTICA

From at least late Cretaceous times until the opening of the Drake Passage about 36 m.y.a., South America and the Antarctic Peninsula were contiguous (Woodburne and Case, 1996). Dispersal of marsupials to Antarctica from South America had certainly occurred by the middle Eocene (Woodburne and Case, 1996). The polydolopid, *Antarctodolops dailyi*, described by Woodburne and Zinsmeister (1984), was the first of eight marsupials described to date from the middle Eocene La Meseta Formation of Seymour Island. Subsequently, Case *et al.* (1998) described another polydolopid, *Euydolops seymouriensis*. The classification of these two species is not questioned. However, in our view the affinity of a third taxon described from this formation is debatable.

Marenssi *et al.* (1994) informally, and Goin and Carlini (1995) formally, refer an edentulous dentary from Seymour Island to Microbiotheriidae. This specimen was originally referred to Didelphimorphia by Vizcaino *et al.* (1988). Goin and Carlini (1995, p. 205) cite two synapomorphies of the dentary, originally raised by Marshall (1982), in support of a microbiotheriid affinity for MLP 88–I- 1–1, i.e., '...large mental foramen below P2 and depth of ramus relatively constant from back to front'. As pointed out by Wroe *et al.* (2000), a well-developed mental foramen below P2 is present in various non-microbiotherian taxa and the degree of confidence that can be attributed to the second feature is difficult to assess without examining the specimen at first hand. Moreover, Marshall *et al.* (1990, p. 438) use neither of these characters to define Australidelphia, Microbiotheria or Microbiotheriidae in their cladogram.

Very recently, Goin *et al.* (1999) described five new marsupial taxa from the same deposits: three didelphimorphians, *Derorhynchus minutus* and *Pauladelphys juanjoi* and *Xenostylus peninsularis* (Derorhynchidae); the polydolopimorphian *Perrodelphys coquinense* (Prepidolopidae); an unnamed taxon of uncertain ordinal affinities; and the alleged microbiotheriid, *Marambiotherium gracilis*. Goin *et al.* (1999) refer MLP 88–I 1–1 to this species. The holotype and only other specimen of *M. gracilis* comprises a fragmentary lower jaw containing M4 and the posterior alveolus of M3.

Goin *et al.* (1999, p.354) cite three features of *M. gracilis* that link it with microbiotheriids: '...the M4 is reduced relative to M3, the hypoconid is reduced relative to the entoconid, and its crest forms a sharp angle'. Consequently, it is unclear as to whether the allocation of *M. gracilis* to Microbiotheriidae is based on synapomorphy or overall similarity. Either way, as argued by Szalay (1994) and Wroe *et al.* (1999), we maintain that the most convincing synapomorphies uniting microbiotheriids are those of the middle ear, an area of the anatomy only known in *Dromiciops* and *Microbiotherium*. Consequently, we reiterate the position of Wroe *et al.* (2003), i.e, that the use of putative microbiotherian taxa, not known from cranial data, in the construction of phylogenetic and biogeographic scenarios should be strongly qualified.

AUSTRALIA

In a review of the literature, Woodburne and Case (1996) conclude that overland dispersal from Antarctica to Australia was probably possible up to 64 million years ago, but progressively less likely since. These authors further posit that by 52 m.y.a., at the latest, the two continents were separated by a shallow seaway. Similarly, Goin *et al.* (1999) suggest that Australia and Antarctica were separated by the earliest Paleocene. Archer *et al.* (1993) argue that marsupials had dispersed into Australia by the latest Paleocene via a route that probably involved filter barriers. Archer *et al.* (1998), citing Veevers (1991), maintain that the last links via Tasmania and the South Tasman Rise were severed sometime before 38 million years ago. Differences of opinion between these authors do not centre around when the first marsupials could have entered Australia via Antarctica, but when the last may have made the trip.

Prior to 1992 the total of our 'hard' knowledge on the early radiation of marsupials into Australia could have been very easily summed up as a big fat zero. There simply were no marsupial-bearing deposits known that exceeded late Oligocene in age (~26 m.y.a.). Fortunately, things have taken a turn for the better. Australia now has its very own early Tertiary mammal-bearing assemblage, the 55 million year old Tingamarra Local Fauna of Murgon, south-eastern Queensland (Godthelp *et al.*, 1992). To date, among terrestrial mammals, two marsupials and a ?condylarth placental have been formally described from this fauna [another notable find has been one of the world's oldests bats (Hand *et al.*, 1994)].

The first of these early Eocene Murgon marsupials to make a formal debut was *Thylacotinga bartholomaii* (Archer *et al.*, 1993). *Thylacotinga* (Figure 14) is a curious beast and its relationships were initially a complete enigma. Archer *et al.* (1993) treated it as order and family *incertae sedis* and observed that in terms of overall similarity, *Thylacotinga* was closest to some early Tertiary South American bunodont taxa such as *Protodidelphis vanzolinii*, *Reigia punae* (Pascual, 1983), *Robertbutleria mastodontoidea* (Marshall, 1987), *Roberthoffstetteria nationalgeographica* (Marshall *et al.*, 1983) and glasbiids. Potential synapomorphies between *Thylacotinga* and these taxa include: bunodont crowns, reduction of the stylar shelf to two large stylar cusps (B and D) positioned buccal to the paracone and metacone respectively, distinctive parastyle and a high protocone. Archer *et al.* (1993) concluded that if *Thylacotinga* does share a special relationship with any of these early Tertiary South American taxa, then relatively recent contiguity with western Gondwanan faunas was implied.

The mystery of *Thylacotinga*'s place among marsupials has recently been at least partially illuminated following the description of a new species, *Chulpasia mattaueri*, from Paleocene deposits of Peru by Crochet and Sigé, 1993. Combined investigation of *Thylacotinga* and *Chulpasia* by Sige *et al.* (1996) led these authors to the conclusion that the two were sister taxa. A further undescribed Murgon species may share even more synapomorphies with *Chulpasia mattaueri* than *Thylacotinga* (Sigé *et al.*, 1996). This may play out as an argument in support of the notion that South America, Antarctica and Australia were more or less intimately connected until well after 62–64 million years ago, contradicting the conclusions of Woodburne and Case (1996) and Goin *et al.* (1999).

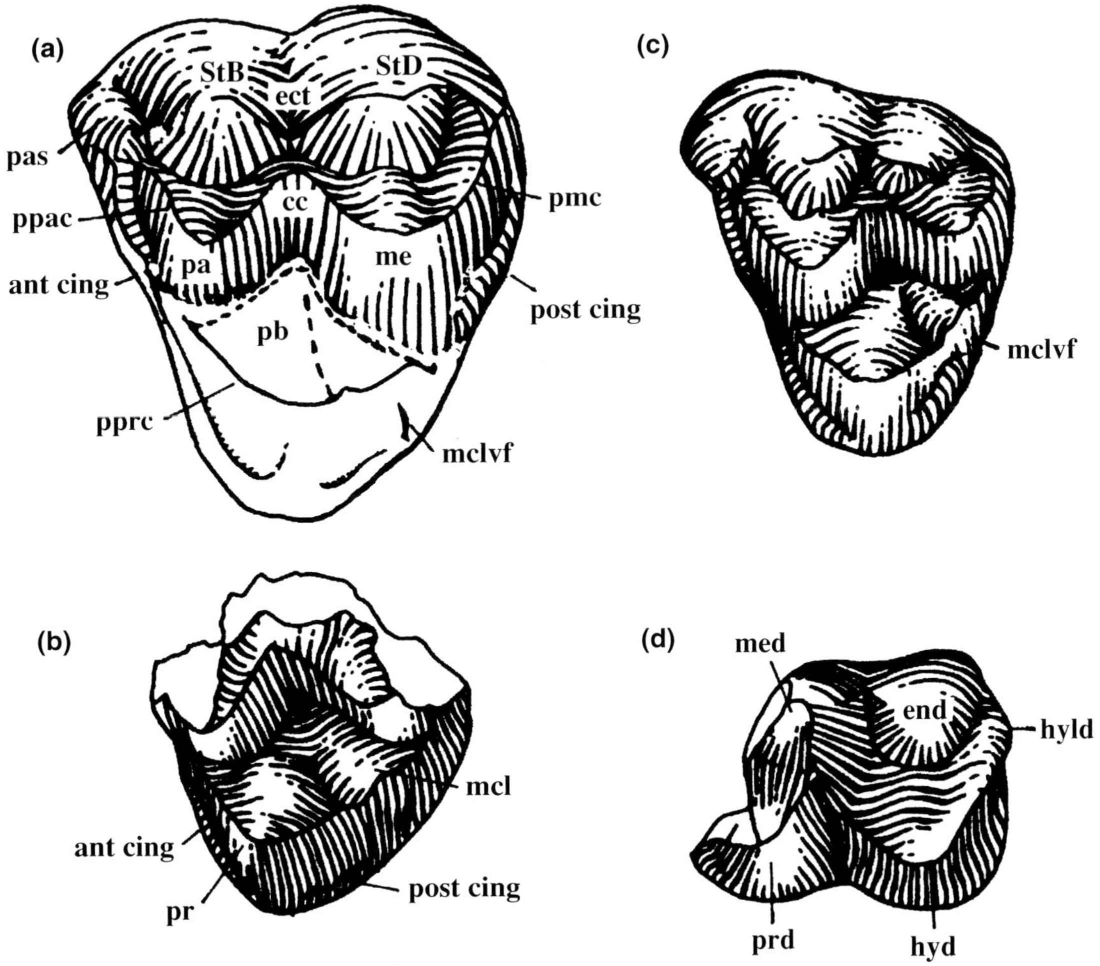

Figure 14. *Thylacotinga bartholomaii*, the first marsupial to be described from the early Tertiary Tingamarra Local Fauna of south-eastern Queensland. After Archer *et al.* (1993).

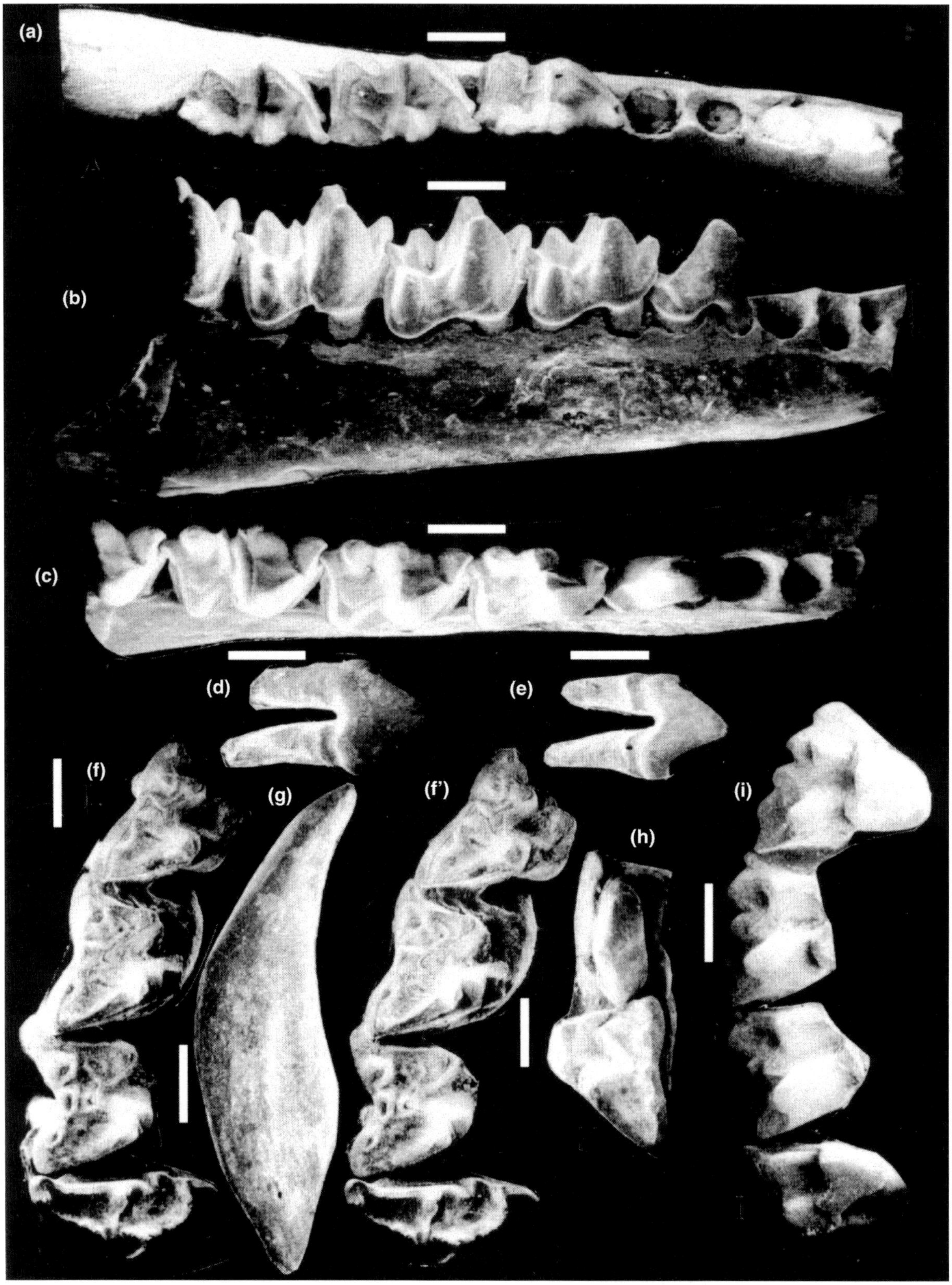

Figure 15. *Djarthia murgonensis*, from the 55 m.y.o. Tingamarra Local Fauna, is Australia's least specialized marsupial. In fact it is so generalized that it can not be placed within either the Australidelphian or Ameridelphian clades with any confidence. (a, d–i) QM F31458. **(a)** M1–3, occlusal view. **(b, c)** QM F 31460, P3, M1–3 and M4 trigonid in buccal **(b)** and occlusal **(c)** views. **(d)** right P3, buccal view. **(e)** P1 or P2, buccal view. **(f f')** right M1–4, stereo occlusal view. **(g)** partial right C1, buccal view. **(h)** left maxillary fragment with P3 and M1, occlusal view. **(i)** right M1–4, lingual-occlusal view. Scale = 1 mm. After Godthelp *et al.* (1999).

The second formally described marsupial taxon from Murgon is the small and very generalized *Djarthia murgonensis* (Godthelp *et al.*, 1999). In fact, *Djarthia* is undoubtedly the least derived Australian marsupial known. Among 18 features considered in the character analysis of Godthelp *et al.* (1999), *Djarthia murgonensis* (Figure 15) was plesiomorphic among marsupials for at least eight. A ninth character present in *Djarthia*, twinning of the hypoconid and entoconid, may represent a derived feature within Marsupialia, as we have seen. But this feature only unites the species within a taxon that includes all marsupials to the exclusion of deltatheriodans and *Kokopellia juddi*. Among remaining features examined by these authors, seven features present in *D. murgonensis* were probably derived among marsupials. However, at least six were present in at least some representatives of both Ameridelphia and Australidelphia.

Only one derived feature was shared with any Australian taxa to the exclusion of all other marsupials, presence of a 'central cusp' at the apex of the centrocrista. This apomorphy is also evident in the two late Oligocene Central Australian marsupicarnivore taxa, *Keeuna woodburnei* and *Ankotarinja tirarensis*. Although both these taxa have previously been classified as dasyurids Archer (1976c), Godthelp *et al.* (1999) and Wroe *et al.* (2003) argue that they should be treated as Marsupialia *incertae sedis* because no derived marsupial features present in either unambiguously unite them with any Australian or American orders, or representatives thereof.

To sum up, on the basis of known (i.e., dental) evidence, *Djarthia* could be considered as an unspecialized ameridelphian or australidelphian. Moreover, because the only non-dental character state currently given any weight as a potential synapomorphy for Australidelphia, presence of a continuous lower ankle joint (Szalay, 1994), is unknown for this species, it cannot be placed even in Australidelphia with any confidence. As pointed out by Godthelp *et al.* (1999, p. 307) 'There can be little doubt that most investigators would treat this taxon as a didelphoid were it found in South America'. These problems in taxonomically pigeon-holing *D. murgonensis* draw attention to the difficulties that confront us, as students of Australian marsupial origins. In our view, with one possible exception (reduction of lower incisor number for an Australian clade), no synapomorphies of the cranium or dentition clearly define either the australidelphian or Australian marsupial radiations to the exclusion of ameridelphians. Even this feature is not exclusive to Australian marsupials.

It has been argued that the marsupial faunas of South America and Australia are fundamentally disjunct, with only one exception, Microbiotheria (Woodburne and Case, 1996). Both Woodburne and Case (1996) and Marshall *et al.* (1990) suggest that the prototypical australidelphian possessed recognizably microbiotherian rather than didelphimorphian dental features. While this interpretation is of course a possibility, it is not accepted as writ by Godthelp *et al.* (1999); Wroe *et al.* (2000) or Wroe *et al.* (2003).

The bottom line is that while the many subjectively determined synapomorphies for Australidelphia, the Australian marsupial radiation, or Dasyuromorphia forwarded by Marshall *et al.* (1990) are undoubtedly derived within Marsupialia, many also apply at basal nodes within Ameridelphia. Among dental features these include: V-shaped (dilambdodont) centrocrista; paracone smaller and lower than metacone; cristid obliqua meets posterior wall of trigonid buccal to protocristid notch; and hypoconulid lower than entoconid. Furthermore, a cranial feature advanced as synapomorphic for the Australian marsupial radiation by Marshall *et al.* (1990), i.e., presence of a squamosal epitympanic sinus, also seems shaky. This feature is absent in the following Australian taxa: *Yalkaparidon coheni* (Archer *et al.*, 1988), basal extant peramelemorphians such as *Echymiperra kalubu* and *Peroryctes longicauda*, and at least one dasyuromorphian, *Badjcinus turnbulli* (Muirhead and Wroe, 1998; Wroe *et al.*, 1998). Thus, if this feature is an Australian marsupial synapomorphy we have to accept that its absence in the above, largely unspecialized taxa, constitute reversals to a plesiomorphic state... unlikely in our view.

One last Murgon taxon deserves a mention, even though it's probably not a marsupial. *Tingamarra porterorum* (see Figure 16), the Murgon ?condylarth, represents a truly unexpected find. Previously, it had generally been assumed that the absence of placentals in Australian faunas could be attributed simply to their failure to reach eastern Gondwana prior to the estrangement of Australia and Antarctica. Clearly, this view is turned on its head if *Tingamarra* is confirmed as a placental, i.e., the lack of placentals would have to be seen as a product of unsuccessful competition with marvellous marsupials... so much for innate placental superiority. But before we get too carried away here it must be conceded that not all authors are convinced that *T. porterorum* is a placental (Woodburne and Case, 1996) and until more complete material is recovered the identity of this species will remain contentious.

The Tingamarra Local Fauna continues to produce marsupials of great interest. Among these are a possible microbiotheriid and peramelemorphian (Archer *et al.*, 1998) and further descriptions and phylogenetic analyses are in the pipeline. However, until this material is formally described we will resist the temptation to further speculate on their affinities. Curiously, to date, clear relationships have yet to be demonstrated between Murgon taxa and the rest of the great Australian marsupial radiation.

ACKNOWLEDGEMENTS

Funding has been provided to S. Wroe by grants from the following institutions: University of Sydney (U2000 Postdoctoral Research Fellowship), Centre for Research into the Evolution of Australia's Terrestrial Ecosystems

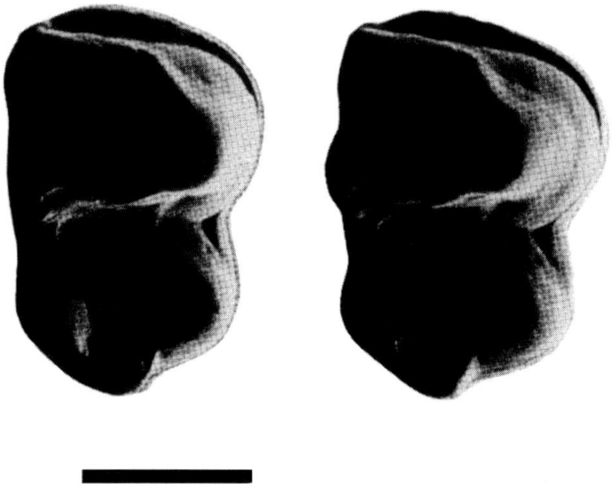

Figure 16. Stereo view of lower right molar of *Tingamarra porterorum* (occlusal view), from the 55 million-year-old Tingamarra Local Fauna was a real shock discovery. This animal appears to be a placental ?condylarth. If so, then it dismisses the notion that Australia's marvellous marsupial radiation only managed to make it by default. It seems our marsupials were just too good, completely out-competing placentals before they really got off the ground. Scale = 1 mm. After Godthelp *et al.* (1992).

(Australian Museum), French Ministry of Foreign Affairs, Linnean Society of New South Wales, Australian Geographic Society, Institute of Wildlife Research, and the University of New South Wales. Support has also been given by the Australian Research Council (to M. Archer); the National Estate Grants Scheme (Queensland) (grants to M. Archer and A. Bartholomai); the Department of Environment, Sports and Territories; the Queensland National Parks and Wildlife Service; the Commonwealth World Heritage Unit (Canberra); the University of New South Wales; ICI Australia Pty. Ltd.; the Australian Geographic Society; the Queensland Museum; the Australian Museum; Century Zinc Pty. Ltd.; Mt. Isa Mines Pty. Ltd.; Surrey Beatty and Sons Pty. Ltd.; the Riversleigh Society Inc.; the Royal Zoological Society of New South Wales and many private supporters.

REFERENCES

Aplin, K. and Archer, M. 1987. Recent advances in marsupial systematics, with a new syncretic classification. Pp.xv–lxxii. In Archer, M. (ed). *Possums and Opossums: Studies in Evolution.* Vol. 1. Surrey Beatty and Sons, Sydney. pp.1–400.

Archer, M. 1974. The development of the cheek-teeth in *Antechinus flavipes* (Marsupialia, Dasyuridae). *Journal of the Royal Society of Western Australia*, Vol. 57, pp.54–63.

Archer, M. 1976a. The basicranial region of marsupicarnivores (Marsupialia), interrelationships of carnivorous marsupials, and affinities of the insectivorous marsupial peramelids. *Zoological Journal of the Linnean Society*, Vol. 59, pp.217–322.

Archer, M. 1976b. The dasyurid dentition and its relationship to that of didelphids, thylacinids, borhyaenids (Marsupicarnivora) and peramelids (Peramelina: Marsupialia). *Australian Journal of Zoology, Supplementary Series,* No. 39, pp.1–34.

Archer, M. 1976c. Miocene marsupicarnivores (Marsupialia) from Central South Australia, *Ankotarinja tirarensis* gen. et. sp. nov., *Keeuna woodburnei* gen. et. sp. nov., and their significance in terms of early marsupial radiations. *Transactions of the Royal Society of South Australia*, Vol. 100, pp.53–73.

Archer, M. 1978. The nature of the molar premolar boundary in marsupials and a reinterpretation of the homology of marsupial cheek teeth. *Memoirs of the Queensland Museum*, Vol. 18, pp.157–164.

Archer, M. 1984. Origins and early radiations of marsupials. Pp.585–625. In Archer, M. and Clayton, G. (eds). *Vertebrate Zoogeography and Evolution in Australasia.* Hesperian Press, Perth. 1203pp.

Archer, M., Arena, R., Bassarova, M., Black, K., Brammall, J., Cooke, B., Creaser, P., Crosby, K., Gillespie, A., Godthelp, H., Gott, M., Hand, S.J., Kear, B., Krikman, A., Mackness, B., Muirhead, J., Musser, A., Myers, T., Pledge, N., Wang, Y. and Wroe, S. 1999. The evolutionary history and diversity of Australian mammals. *Australian Mammalogy*, Vol. 21, pp.1–45.

Archer, M., Burnley, I., Dodson, J., Harding, R., Head, L. and Murphy, P. 1998. From plesiosaurs to people: 100 million years of Australian environmental history. *Australia: State of the Environment, Technical Paper Series (Portrait of Australia)*, pp.1–66. *Department of the Environment, Canberra.*

Archer, M., Godthelp, H. and Hand, S.J. 1993. Early Eocene marsupial from Australia. *Kaupia, Darmstädter Beiträge zur Naturgeschichte*, Vol. 3, pp.193–200.

Archer, M. and Hand, S.J. 2006. The Australian marsupial radiation. Pp.575–646. In Merrick, J.R., Archer, M., Hickey, G.M. and Lee, M.S.Y. (eds). *Evolution and Biogeography of Australasian Vertebrates.* Auscipub Pty. Ltd., Sydney. 942pp.

Archer, M., Hand, S. and Godthelp, H. 1988. A new order of zalambdodont marsupials. *Science*, Vol. 239, pp.1528–1531.

Berkovitz, B.K.B. 1967. The dentition of a 25-day pouch-young specimen of *Didelphis virginia* (Didelphidae: Marsupialia). *Archives of Oral Biology*, Vol. 12, pp.1211–1212.

Bonaparte. J. 1990. New late Cretaceous mammals from the Los Alamitos Formation, northern Patagonia. *National Geographic Research*, Vol. 6, pp.63–93.

Bown, T.M. and Kraus, M.J. 1979. Origin of the tribosphenic molar and metatherian and eutherian dental formulae. Pp.172–181. In Lillegraven, J.A., Kielan-Jaworowska, Z. and Clemens, W.A. (eds). *Mesozoic Mammals: the First Two Thirds of Mammalian History.* University of California Press, Berkeley. 311pp.

Butler, P.M. 1990. Early trends in the evolution of tribosphenic molars. *Biological Reviews,* Vol. 65, pp.529–552.

Butler, P.M. and Kielan-Jaworowska, Z. 1973. Is *Deltaderidium* a marsupial? *Nature*, Vol. 245, pp.105–106.

Case, J.A., Woodburne, M.O., and Chaney, D.S. 1988. A new genus of polydolopid marsupial from Seymour Island, Antarctic Peninsula; Paleogene differentiation of land mammal fauna; land mammal dispersal in the southern hemisphere. Pp.505–521. In Feldmann, R.M. and Woodburne, M.O. (eds). *Geology and Paleontology of Seymour Island, Antarctic Peninsula*. Geological Society of America Memoirs, Vol. 169.

Cifelli, R.L. 1990a. Cretaceous mammals of southern Utah. I. Marsupials from the Kaiparopwits Formation. *Journal of Vertebrate Paleontology*, Vol. 10, pp.295–319.

Cifelli, R.L. 1990b. Cretaceous mammals of southern Utah. III. Therian mammals from the Turonian (early Late Cretaceous). *Journal of Vertebrate Paleontology*, Vol. 10, pp.332–345.

Cifelli, R.L. 1993. Theria of metatherian-eutherian grade and the origin of marsupials. Pp.205–215. In Szalay, S.F., Novacek, M.J. and McKenna, M.C. (eds). *Mammal Phylogeny; Mesozoic Differentiation, Multituberculates, Monotremes, Early Therians and, Marsupials*. Springer-Verlag, New York.

Cifelli, R.L. 1999. Tribosphenic mammal from the North American early Cretaceous. *Nature*, Vol. 401, pp.363–366.

Cifelli, R.L. and Johanson, Z. 1994. New marsupial from the upper Cretaceous of Utah. *Journal of Paleontology*, Vol. 14, pp.292–295.

Cifelli, R.L. and Muizon, C. de. 1997. Dentition of *Kokopellia juddi*, a primitive marsupial or near-marsupial from the medial Cretaceous of Utah. *Journal of Mammalian Evolution*, Vol. 4, pp.241–258.

Cifelli, R.L. and Muizon, C. de. 1998a. Tooth eruption and replacement in early marsupials. *Earth and Planetary Sciences*, Vol. 326, pp.215–220.

Cifelli, R.L. and Muizon, C. de. 1998b. Marsupial mammal from the upper Cretaceous North Horn Formation, Central Utah. *Journal of Paleontology*, Vol. 72, pp.532–538.

Cifelli, R.L., Rowe, T.B., Luckett, W.P., Banta, J., Reyes, R. and Howes, R.I. 1996. Fossil evidence for the origin of the pattern of marsupial tooth replacement. *Nature*, Vol. 379, pp.715–718.

Clemens, W.A. 1966. Fossil mammals of the type Lance Formation, Wyoming. Part II, Marsupialia. *University of California Publications in Geological Science*, Vol. 62, pp.1–122.

Clemens, W.A. 1979. Marsupialia. Pp.192–220. In Lillegraven, J.A., Kielan-Jaworowska, Z. and Clemens, W.A. (eds). *Mesozoic Mammals: The First Two Thirds of Mammalian History*. University of California Press, Berkeley. 311pp.

Clemens, W.A., Lillegraven, J.A., Lindsay, E.H. and Simpson, G.G. 1979. Where, when, and what—a survey of known Mesozoic mammal distribution. Pp.7–58. In Lillegraven, J.A., Kielan-Jaworowska, Z. and Clemens, W.A. (eds). *Mesozoic Mammals: the First Two Thirds of Mammalian History*. University of California Press, Berkeley. 311pp.

Colgan, D.J. 1999. Phylogenetic studies of marsupials based on phosphoglycerate kinase DNA sequences. *Molecular Phylogenetics and Evolution*, Vol. 11, pp.13–26.

Crochet, J.-Y. 1979. Diversite systematique des Didelphidae (Marsupialia) Europeens Tertiaires. *Geobios*, Vol. 12, pp.365–378.

Crochet, J.-Y. 1980. *Les Marsupiaux du Tertiare d'Europe*. Editions Foundation Singer-Polinac, Paris. 279pp.

Crochet, J.-Y. and Sigé, B. 1993. Les mammiferes de Chulpas (Formation Umayo, Transition Cretace-Tertiaire, Perou): donees préliminaires. *Documents du Laboratoire de Geologie de Lyon*, Vol. 125, pp.97–107.

Eaton, J.G. and Cifelli, R.L. 1987. Preliminary report on late Cretaceous mammals of the Kaiparowits Plateau, southern Utah. *Contributions to Geology, University of Wyoming*, Vol. 26, pp.45–55.

Flower, W.H. 1867. On the development and succession of teeth in the Marsupialia. *Philosophical Transactions of the Royal Society of London*, Vol. 157, pp.631–641.

Fox, R.C. 1971. Marsupial mammals from the early Campanian Milk River Formation, Alberta, Canada. Pp.145–164. In Kermack, D.M. and Kermack, K.A. (eds). *Early Mammals*. Academic Press, London. 203pp.

Fox, R.C. 1979. Ancestory of the "dog-like" marsupials. *Journal of Paleontology*, Vol. 53, pp.733–735.

Fox, R.C. 1987. Palaeontology and the early evolution of marsupials. Pp.161–169. In Archer, M. (ed). *Possums and Opossums: Studies in Evolution*. Vol. 1. Surrey Beatty and Sons, Sydney. pp.1–400.

Gaudin, T.J., Wible, J.R., Hopson, J.A. and Turnbull, W.D. 1996. Reexamination of the morphological evidence for the cohort Epitheria (Mammalia, Eutheria). *Journal of Mammalian Evolution*, Vol. 3, pp.31–79.

Gauthier, J., Kluge, A.G. and Rowe, T. 1988. Amniote phylogeny and the importance of fossils. *Cladistics*, Vol. 4, pp.105–209.

Gemmell, N.J. and Westerman, M. 1994. Phylogenetic relationships within the class Mammalia: a study using mitochondrial 12S RNA sequences. *Journal of Mammalian Evolution*, Vol. 2, pp.3–23.

Godthelp, H., Archer, M., Cifelli, R., Hand, S.J. and Gilkeson, C.F. 1992. Earliest known Australian Tertiary mammal fauna. *Nature*, Vol. 356, pp.514–516.

Godthelp, H., Wroe, S. and Archer, M. 1999. A new marsupial from the early Eocene Tingamarra Local Fauna of Murgon, southeastern Queensland: a prototypical Australian marsupial? *Journal of Mammalian Evolution*, Vol. 6, pp.289–313.

Goin, F.J. 1993. Living South American opposums are not living fossils. P. 112. In Augee, M. (ed). *Abstracts of the Sixth International Theriological Conference*. University of New South Wales, Sydney.

Goin, F.J. and Carlini, A.A. 1995. An early Tertiary microbiotheriid marsupial from Antarctica. *Journal of Vertebrate Paleontology*, Vol. 15, pp.205–207.

Goin, F.J., Case, J.A., Woodburne, M.O., Vizcaino, S.F. and Reguero, M.A. 1999. New discoveries of "Opposum-like" marsupials from Antarctica (Seymour Island, medial Eocene). *Journal of Mammalian Evolution*, Vol. 6, pp.335–365.

Gregory, W.K. 1910. The orders of mammal. *Bulletin of the American Museum of Natural History*, Vol. 27, pp.1–524.

Haines, R.W. 1958. Arboreal or terrestrial ancestry of placental mammals. *Quarterly Review of Biology*, Vol. 33, pp.1–23.

Hand, S., Novacek, M., Godthelp, H. and Archer, M. 1994. First Eocene bat from Australia. *Journal of Vertebrate Paleontology*, Vol. 14, pp.375–381.

Hayman, D.L., Kirsch, J.A.W., Martin, P.G. and Waller, P.F. 1971. Chromosomal and serological studies of the Caenolestidae and their implications for marsupial evolution. *Nature*, Vol. 231, pp.194–195.

Hershkovitz, P. 1992. Ankle bones: the Chilean oppossum

Wroe, S. 1997. A re-examination of proposed morphology-based synapomorphies for the families of Dasyuromorphia (Marsupialia): Part I, Dasyuridae. *Journal of Mammalian Evolution*, Vol. 4, pp.19–52.

Wroe, S. 1999a. The geologically oldest dasyurid (Marsupialia), from the middle Miocene of Riversleigh, northwestern Queensland. *Palaeontology*, Vol. 42, pp.501–527.

Wroe, S. 1999b. Killer kangaroos and other murderous marsupials. *Scientific American*, Vol. 285, pp.68–74.

Wroe, S. 2003. Australian marsupial carnivores: an overview of recent advances in palaeontology and phylogeny. Pp.102–123. In Jones, M., Dickman, C. and Archer, M. (eds). *Predators with Pouches: the Biology of Carnivorous Marsupials*. CSIRO, Melbourne. 486pp.

Wroe, S. and Archer, M. 1995. Extraordinary diphyodonty-related change in dental function for a tooth of the extinct marsupial *Ekaltadeta ima* (Propleopinae, Hypsiprymnodontidae). *Archives of Oral Biology*, Vol.40, Issue 7, pp.597–603.

Wroe, S., Brammall, J. and Cooke, B.N. 1998. The skull of *Ekaltadeta ima* (Marsupialia: Hypsiprymnodontidae?): An analysis of some cranial features among marsupials and a re-investigation of propleopine phylogeny, with notes on the inference of carnivory in mammals. *Journal of Paleontology*, Vol. 72, pp.738–751.

Wroe, S., Ebach, M., Ahyong, S., Muizon, C. de. and Muirhead, J. 2000. Cladistic analysis of dasyuromorphian (Marsupialia) phylogeny using cranial and dental data. *Journal of Mammalogy*, Vol. 81, pp.1008–1024.

Wroe, S., Myers, T., Seebacher, F., Kear, B., Gillespie, A., Crowther, M. and Salisbury, S. 2003. An alternative method for predicting body-mass: the case of the marsupial lion. *Paleobiology*, Vol. 29, pp.404–412.

Wroe, S., Myers, T.J., Wells, R.T. and Gillespie, A. 1999. Estimating the weight of the Pleistocene marsupial lion (*Thylacoleo carnifex*: Thylacoleonidae): implications for the ecomorphology of a marsupial super-predator and hypotheses of impoverishment of Australian marsupial carnivore faunas. *Australian Journal of Zoology*, Vol. 47, pp.489–498.

Chapter 28
The Australian Marsupial Radiation

Michael Archer and Suzanne J. Hand[1]

[1] School of Biological, Earth and Environmental Sciences, University of New South Wales, Sydney. N.S.W. 2052 Australia
E-mail: m.archer@unsw.edu.au

INTRODUCTION

This chapter represents a radically abridged but updated version of Archer's 1984 chapter 'The Australian Marsupial Radiation' in *Vertebrate Zoogeography and Evolution in Australasia* (Archer and Clayton, 1984). Looking back it's delightful to note how many of the unknowns at that time have become 'knowns' and how the fossil records, for many groups previously a complete blank, have now begun to fill out. Even whole new orders, superfamilies and families have been discovered (e.g., Order Yalkaparidontia, Superfamily Yaraloidea and Family Ilariidae). That said, there are still four groups of Australian mammals, the echidnas, numbats, honey possums and fruit bats, that have so far stubbornly refused to reveal archaic representatives, leaving their early evolutionary history a provocative mystery. Tomorrow's challenge! In this chapter, we will discuss the fossil record and a bit about diversity in all Australian marsupial groups through time.

In terms of the time periods that have produced new information about the older record of Australian marsupials, the biggest breakthrough since 1984 has been the plethora of fascinating creatures found in the early Eocene, 55-million-year-old Tingamarra faunal assemblage from south-eastern Queensland (Godthelp *et al.*, 1992). This is the only early Tertiary mammal fauna from the whole continent. Hence the record of Australian mammals now starts in the early Cretaceous (only monotremes and other strange beasts—no marsupials) at about 115 m.y.a., then enters a dark age until the early Eocene Tingamarra deposit, and then slips back again into a second dark age that doesn't brighten until the late Oligocene about 25 m.y.a. From that time on, the record becomes much better known.

In terms of new information for clarifying intergroup relationships, one of the largest contributions since 1984 has been biomolecular systematics (e.g., Richardson *et al.*, 1986; Westerman *et al.*, 1990; Springer *et al.*, 1994; Messer *et al.*, 1998; Krajewski *et al.*, 1994, 2000; Krajewski and Westerman, 2003). While not always in agreement with each other, or with all of the conclusions reached on the basis of morphological studies of living and extinct groups, from these studies there is a general picture emerging that has broad agreement from many areas of independent research.

And there have been significant advances in understanding about the links between the evolution of Australia's marsupials and the relatively newly-understood icehouse/greenhouse cycles of climate change that have shaped Australian environments for the last 55 million years (McGowran and Li, 1994).

In the discussions that follow, we use the terms: 'faunal assemblage' to mean a group of species that were found together in one spot; 'Local Fauna' (abbreviated LF) to mean all known species interpreted to have lived together at the same time in one local area (it can be documented by fossils from only one quarry; hence sometimes it is used interchangeably with 'faunal assemblage'); and 'Fauna' to mean all the species that lived at the same time across a broad region. In most cases we stick to LFs because we are rarely certain how geographically widespread extinct species were when they were alive.

For more about the radiation of Australian marsupials, their evolutionary history, history of discovery and geological context see Archer (1984), Archer and Hand (1984), Rich *et al.* (1984, 1991), Archer *et al.* (1999), McGowran *et al.* (2000) and most recently Long *et al.* (2002).

AUSTRALIAN FOSSIL MAMMALS

Australia's mammals, living and extinct, are among the most distinctive in the world. All other continents, apart from Antarctica, have interchanged terrestrial mammal communities at times of low sea levels, which is why South America, North America, Eurasia and Africa, but not Australia, have shared camels, lions, edentates and elephants. Of continents permanently inhabitable by terrestrial mammals, only Australia has been significantly distant from the rest of the world during most of the Cainozoic, an independent laboratory where evolutionary processes have operated to produce a second, very different creation. The history of Australia's mammals is the other side of evolution's mirror. When adequately documented, the independent Australian record should significantly increase understanding about the forces that produce evolutionary change.

In the past some researchers have suggested, on the basis of the high diversity of marsupials in Australia, that this group might have originated here. However, subsequent discovery of the most 'primitive'-known marsupials in early Cretaceous deposits in North America and Asia, suggests that this group probably originated in the Northern Hemisphere (see Wroe and Archer, 2006). All marsupials from both South America and Australia are far more 'derived'. The oldest undoubted marsupials (and placentals for that matter) from South America are latest Cretaceous to earliest Paleocene in age.

The Australian fossil record for mammals could be fairly described as patchy. While the clade Mammaliformes is at least 230 million years old (see Johanson, 2006), the Australian record spans only half that period—the last 115 million years. Hence there is a *first* frustrating 'dark age' of ignorance about Australian mammals that spans that interval from 230 to about 115 m.y.a. The record of Australia's mammals begins with rare early Cretaceous monotremes from Lightning Ridge, New South Wales (Archer *et al.*, 1985; Flannery *et al.,* 1995) and rare monotremes (Rich *et al.,* 2001) and the controversial (Archer *et al.,*1999; Long *et al.*, 2002; Johanson, 2006) ausktribosphenids from southern Victoria (Rich *et al.*, 1997, 1998, 1999, 2001; Musser, 2006). Other Mesozoic mammal-producing sites (e.g., Hazel Creek in northern Queensland; Godthelp *et al.*, unpublished data) are only now being assessed for their potential.

The *second* infuriating dark age then spans the late Cretaceous to late Paleocene interval, from about 100 to 56 m.y.a. When the curtain lifts, we find Australia's oldest marsupial and placental mammals, discovered in 1983, in freshwater clays on a property called Tingamarra on the outskirts of the town of Murgon, south-eastern

(a)

Figure 1. Localities of mammal-bearing sediments in Australia: **(a)** Tertiary sites; **(b)** Pleistocene sites. After Archer and Hand (1984).

Queensland (Godthelp *et al.*, 1992; see Figure 1). Illites in clays containing the Tingamarra fauna have been dated as >55.6 million years old (early Eocene or latest Paleocene depending on time scale used), on the basis of potassium/argon ratios (Godthelp *et al.*, 1992). This age has been confirmed by the presence in the fauna of many other animals that are about this age in other parts of the world such as graculavid birds (Boles, 1999) and archaeonycterid bats (Hand *et al.*, 1994; Sigé *et al.*, 2004). The Tingamarra mammals also include *Tingamarra*, a possible condylarth-like placental (Godthelp *et al.*, 1992) and many marsupials including several with affinities to Casamayoran (early Eocene) as well as Tiupampian (Paleocene) South American groups (Archer *et al.*, 1993; Godthelp *et al.*, 1999; Sigé *et al.*, 2004). One of these marsupials appears to be a microbiotheriid, a group represented today only by *Dromiciops australis*, a tiny hunter of the southern beech forests of Chile. Microbiotheriids are widely regarded as the likely evolutionary source for most, if not all, of Australia's living marsupial groups. Thus it seems probable that sometime between the late Cretaceous and early Eocene (70–55 m.y.a.), archaic marsupials with South American ties first colonized Australia.

Between the early Eocene and late Oligocene (the *third* frustrating dark age of information about Australian mammals), Australia finally completely separated from the rest of Gondwana (Antarctica plus South America)—sometime between 46 and 35 m.y.a. (Veevers, 1991; McGowran *et al.*, 2000) and began drifting northwards.

Figure 1 *(continued).*

South America pulled away from Antarctica around 35 m.y.a. (McGowran *et al.*, 2000) resulting in establishment of the cold circum-Antarctic current, which was to have a major impact on climates in the Southern Hemisphere—see pp.17–50 (White, 2006). Fossil marsupials spanning this middle Eocene to middle Oligocene interval are known from Antarctica and South America (Wroe and Archer, 2006) but they shed little light on what was happening during this interval in Australia—other than to confirm that marsupials, including microbiotheriids, were present in Antarctica.

Fortunately, when the curtain again lifts at the end of the third dark age, Australian mid Tertiary mammals turn up in abundance (Figure 1). This follows discoveries of many rich late Oligocene and Miocene deposits from central and northern Australia (Stirton, 1955; Archer *et al.*, 1989, 1994, 1999; Murray and Megirian, 1992; Long *et al.,* 2002; Megirian e*t al.,* 2003). By the late Oligocene (26–24 m.y.a.), marsupials had differentiated into five orders (two of which are now extinct) and many different families including most that still survive. Climates were cool to mild, indicative of the 'icehouse' conditions that were affecting much of the continent at that time. From 23 to 16 m.y.a., in the early Miocene, rainfall appears to have markedly increased supporting inland rainforests in at least northern, eastern and possibly central Australia. High marsupial species diversity seems to have been a hallmark of this period, with marsupials representing about 85% of mammal species-level diversity in faunas, for example, from Riversleigh, north-western Queensland.

About 15 m.y.a. the world entered another icehouse climatic phase with a dramatic decrease in rainfall. The late Miocene (~11–5 m.y.a.) was a critical period in the climatic and biotic history of Australia (McGowran *et al.,* 2000; White, 2006). It was during this time that the structure and nature of most of the marsupial groups, that dominate the continent now, began to take shape and to replace some of the older groups that dominated the once forested continent. But unfortunately, perhaps because of the increasingly drier conditions, there are very few late Miocene mammal sites from the continent. The best-known include Alcoota in the Northern Territory and Riversleigh in Queensland (Murray and Megirian, 1992; Myers *et al.*, 2001).

In the early Pliocene, about 5 m.y.a., there was a brief respite of warmer, wetter conditions before the continued onslaught of icehouse conditions culminated in the 'ice ages' of the Pleistocene. The Pliocene is better known palaeontologically than the late Miocene, with mammal assemblages recorded from all states except Tasmania and the Northern Territory. Nevertheless, absolute ages are well-understood for only two (Hamilton LF in Victoria and Bluff Downs LF, Queensland; Turnbull and Lundelius, 1970; Mackness *et al.*, 2000a). Rackham's Roost at Riversleigh contains a diverse assemblage of Pliocene rodents, bats and fewer marsupials (Archer *et al.*, 1994).

In the Pleistocene, the world entered a period of rapid and extreme climatic changes as the polar ice caps grew and retreated. In the Northern Hemisphere, huge areas were covered in glaciers, and world sea levels dropped by 200 metres. In Australia, there were no great ice sheets but the waxing and waning of the ice caps resulted in rapid cycles of cold, dry conditions alternating with warmer, wetter periods, resulting in a constantly changing pattern of forests, grasslands and deserts. During periods of low sea level, temporary land-bridges were exposed between mainland Australia and Tasmania in the south, and Australia and New Guinea in the north, facilitating the dispersal of mammals, including humans, between land masses. Quaternary deposits (Figure 1) document aspects of the origin and controversial demise of Australia's megafauna (Murray, 1991a; Dawson and Dawson, 2006), and the interactions of the biota with Australia's first people (Furby *et al.*, 1993; Flannery, 1994a; Miller *et al.,* 1999; Roberts *et al.*, 2001).

Periodic reviews (Rich *et al.*, 1982; Woodburne *et al.,* 1985, 1994; Archer *et al.*, 1994, 1995, 1997, 1999) of understanding about the biostratigraphic framework for Australia's Mesozoic and Cainozoic mammals have been based primarily on the evidence of stratigraphic superposition, stage of evolution arguments, palynology, invertebrate index fossils and, less commonly, radiometric dates and magnetostratigraphy. Woodburne *et al.* (1993) and Archer *et al.* (1999) have stressed the need for greater understanding about the absolute and relative ages of the assemblages that contain these mammals. Riversleigh has more than 200 species-rich assemblages (Creaser, 1997) whose apparent relationships to each other and to those from other areas of Australia (Myers and Archer, 1997), as well as intercontinentally (Hand *et al.*, 1997; Hand and Kirsch, 1998), need to be tested because at this point they are based primarily on superposition and biocorrelation.

DIVERSITY AND EVOLUTION OF AUSTRALIAN MARSUPIALS

Now that we have given a brief background of the Australian marsupial fossil record, we will briefly consider the major groups involved, both living and extinct. It is important to realize that the relationships of some, particularly in fossil, groups are often controversial—in most cases because there is not enough information known. For example, one of the most enigmatic groups is the Yingabalanaridae, known only from Riversleigh and only from two isolated teeth! Fortunately, most other groups are much better represented so more can be said about them with confidence.

DASYUROMORPHIANS—THINGS THAT BITE IN THE NIGHT

Dasyuromorphians for our present purposes include such diverse groups as Australia's quolls, antechinuses, dunnarts, planigales, phascogales, the numbat, thylacines and some other extinct marsupicarnivores whose relation-

Figure 2. Two young, hand-reared Tasmanian devils. The Tasmanian devil (*Sarcophilus laniarius*), Australia's largest living dasyurid, is now restricted to Tasmania, but was once also common on the mainland. Photograph M. Archer.

ships to other dasyuromorphians is not at all clear. For relevant papers on all topics about dasyuromorphians, readers should consult *Carnivorous Marsupials* (Archer, 1982) and *Pouched Predators* (Jones *et al*., 2003).

DASYURIDS

The Dasyuridae is a large family of mostly nocturnal, insectivorous and carnivorous marsupials. There are at least 64 currently recognized species in Australia and New Guinea (Flannery, 1995a, b; Strahan, 1995). They range in weight from about 4 g to about 8 kg (Figures 2–4, Plate 95), and include the quolls, devils, antechinuses, phascogales, planigales, dunnarts, ningauis, dibblers, mulgaras, kowaris, murexias and many others. Today they are the dominant, terrestrial (non-flying) insectivores and carnivores of Australia and are found all over Meganesia and adjacent islands, in habitats ranging from lush rainforest to the driest deserts.

All dasyurids have the adult tooth formula: I1–4/1–3 C1/1 P1–2 or 3/1–2 or 3 M1–4/1–4 (Figure 5). Aplin and Archer (1987), Wroe (1997a, 1999) and Wroe *et al.* (2000) have reviewed the characteristic features of dasyurids, including the presence of a distinct tubal foramen for the eustachian tube, presence of a foramen pseudovale (or secondary foramen ovale), no posterolateral palatal foramen, commonly reduced P3, a paraoccipital sinus, and a petrosal tympanic wing and hypotympanic sinus.

There is one extinct subfamily of dasyurids, the Barinyainae, and three currently recognized extant subfamilies: the Dasyurinae, Sminthopsinae and Phascogalinae (Figure 6). The latter are defined on the basis of both molecular and morphological data.

The oldest undoubted dasyurid is *Barinya wangala* Wroe, 1999, a species that is relatively common in early to middle Miocene deposits at Riversleigh. It is placed in its own subfamily the Barinyainae (Wroe, 1999). *Barinya* species are distinguished from other dasyurids by their more primitive dentition and skull morphology (Wroe, 1999).

Most other named extinct dasyurids are late Miocene or younger in age and belong to extant dasyurid subfamilies (but see the discussion on Dasyuromorphia *incertae sedis* below). Of these, the oldest is *Ganbulanyi djadjinguli* Wroe, 1998, an early Late Miocene species from Encore Site, at Riversleigh, which is referred to the subfamily Dasyurinae. Dasyurines include the living quolls, dibblers, pseudantechinuses, the kowari, mulgara, kaluta and devils. Wroe (1998) described *Ganbulanyi* as the world's smallest mammalian bone-cracker. It is characterized by a dome-like, powerfully-built posterior premolar.

Also referred to the subfamily Dasyurinae are the following early Pliocene taxa: *Archerium chinchillaensis* Wroe and Mackness, 2000, from the Chinchilla LF, south-eastern Queensland; *Dasyurus dunmalli* Bartholomai, 1971, also from Chinchilla as well as the Bluff Downs LF in north-eastern Queensland and Bow LF in central New South Wales (Wroe and Mackness, 1998); '*Dasycercus*' *worboysi* Dawson, Muirhead and Wroe, 1999, from the Big Sink LF, Wellington Caves, central New South Wales; and *Glaucodon ballaratensis* Stirton, 1957a, a carnivore from Ballarat, western Victoria.

Archerium is a relatively unspecialized dasyurine but shares a few features with species of *Dasyuroides* or *Dasycercus* such as reduced entoconids on the lower molars (Wroe and Mackness, 2000), a feature often found in dasyurids living in dry areas.

Figure 3. The arboreal brush-tailed phascogale (*Phascogale tapoatafa*), a rat-sized carnivorous species that occurs in coastal woodlands and forests around most of Australia (drawing by R. Goldingay).

Figure 4. The tinier end of the dasyurid range, a smaller than mouse-sized insectivorous long-tailed planigale (*Planigale ingrami*) from the Kimberley region of Western Australia; the matchbox gives an indication of scale. Most species of this genus have flattened heads and during the heat of day survive by staying deep within the cracks that develop in the clayey soils that characterize their habitats. Photograph M. Archer.

Quolls or 'native cats', species of *Dasyurus*, are the largest marsupial predators left on the Australian mainland, reaching the size of a large domestic cat (see Strahan, 1995). The one extinct species, *Dasyurus dunmalli*, is not particularly informative about the evolutionary history of the group, other than demonstrating an intermediate stage in loss of its last premolar.

'Dasycercus' worboysi Dawson *et al.*, 1999, from the early Pliocene of Wellington Caves, was larger than the living mulgara, *D. cristicauda,* which is a desert-dweller in central Australia. Its assignment to *Dasycercus* is in doubt because only a partial dentition is known.

Dasyuroides achilpatna Archer, 1982b, of either late Pliocene or early Pleistocene age, from the Lake Victoria area of south-western New South Wales, resembles its living relative *D. byrnei* which is also a desert-dweller.

Glaucodon ballaratensis Stirton, 1957a, came from Pliocene deposits near Ballarat, western Victoria. The species was described on the basis of a lower jaw containing two teeth. It is, in several features, intermediate between quolls (*Dasyurus* spp.) and devils (*Sarcophilus* spp.).

The Tasmanian devil *Sarcophilus laniarius* (Owen, 1838; formerly regarded by most workers as *S. harrisii*) is the largest living dasyurid (Figure 2). The fossil record for this genus includes the early Pleistocene *Sarcophilus moornaensis* Crabb, 1982, from the Lake Victoria area of south-western New South Wales. Late Pleistocene representatives of *S. laniarius* may have been 16% larger than modern Tasmanian devils (Werdelin, 1987) and it is not clear whether the living animal is the same or a distinct, descendant species of the fossil form (Dawson, 1982).

The dasyurid subfamily Phascogalinae includes phascogales (Figure 3) and antechinuses, as well as *Murexia*, the latter being only known from New Guinea. *Antechinus puteus* Van Dyck, 1982, from late Pleistocene cave deposits near Texas, south-eastern Queensland, is the oldest official phascogaline species, but an unnamed *Antechinus*-like beast from the early Miocene of Riversleigh may represent an ancestral group that gave rise to both *Antechinus* and *Murexia* (Van Dyck, 2003, personal communication).

The subfamily Sminthopsinae includes dunnarts (*Sminthopsis*), ningauis (*Ningaui*), planigales (*Planigale*) and the species of *Antechinomys* (Archer, 1981). *Planigale* sp. (Archer, 1982b) from the Bluff Downs LF in north-eastern Queensland, is the oldest member of this group of tiny dasyurids. *Sminthopsis floravillensis* Archer, 1982a, from the Floraville LF of northern Queensland, may be either late Pliocene or Pleistocene in age and again is the oldest species of the genus. See Figures 3, 4 and 5.

Now that the Riversleigh fossil dasyurid and dasyurid-like taxa, spanning the period 24–5 m.y.a., have been studied (Wroe, 1997a, b, 1998, 1999, 2003), it is clear that something profound happened to this group in the late Miocene, between about 10 and 5 million years ago (Figure 6). Prior to that time there were very few 'recognizable' lineages but by the early Pliocene many of the modern genera had made an appearance. The drying out of the continent in the late Miocene probably triggered not only a wave of extinctions of the older types, but stimulated a new evolutionary radiation of the modern lineages. The close relationships of living dasyurids to each other may reflect this relatively recent radiation after the group emerged from a late Miocene 'bottleneck'. Probably related to this, many dasyurid groups show independent evolution of arid-adapted features such as evacuated palates, long tails, fat tails, granulated soles on the feet, digit loss and large alisphenoid tympanic wings. This convergence (or parallelism, if the similar end product results from modification of the same, i.e. homologous, starting structure) probably reflects a widespread response to the late development of aridity in Australia.

As a parting consideration, let's think about what should have been our first consideration: the basis for the

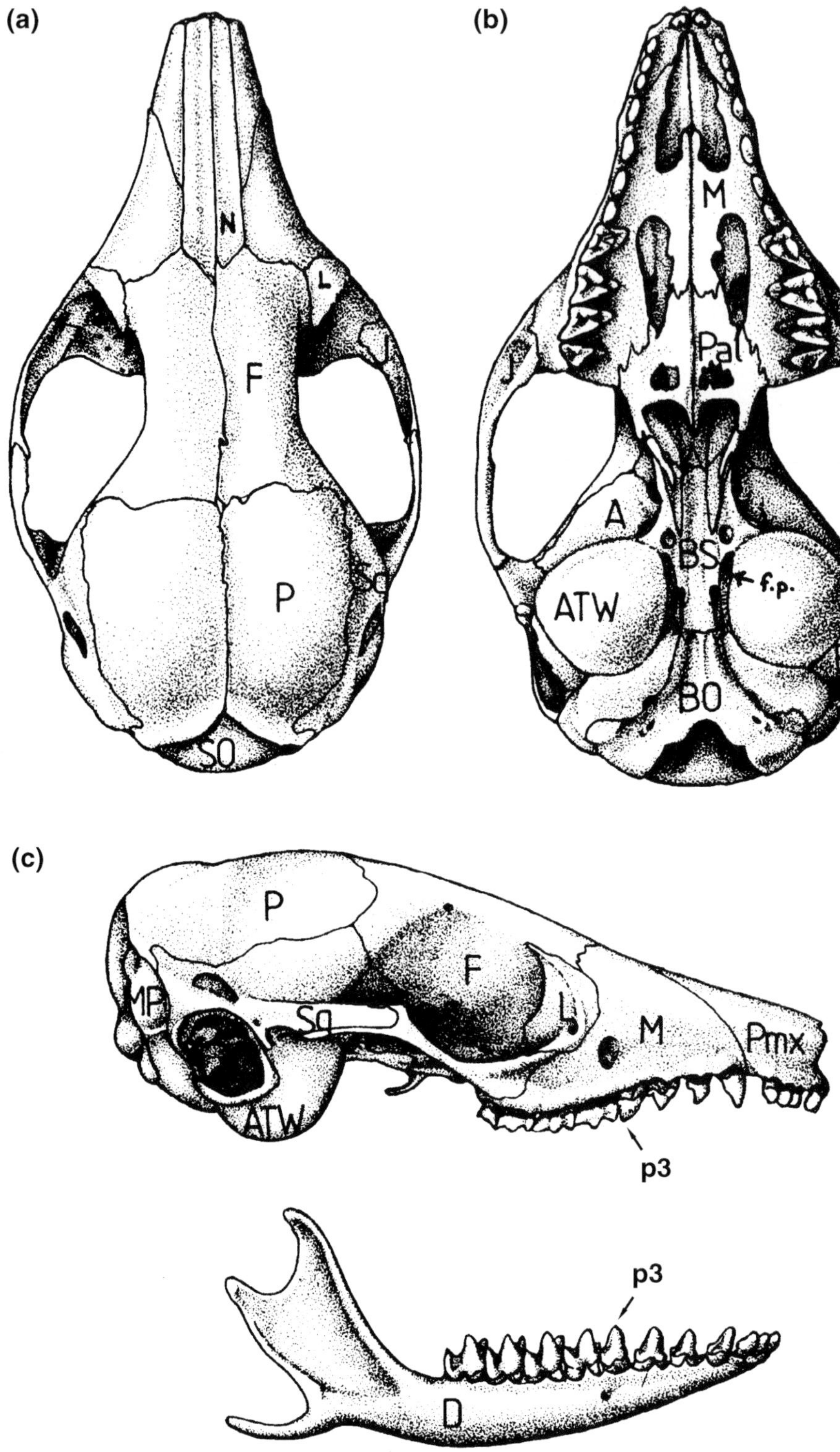

Figure 5. The skull and right dentary of the large mouse-sized kultarr (*Antechinomys laniger*): **(a)** dorsal view; **(b)** ventral view; **(c)** lateral view. The three sets of large vacuities in the palate are (from front to back) the incisive or premaxillary vacuities, the maxillary vacuities and the palatine vacuities. The alisphenoid, periotic and mastoid tympanic wings form a bony floor for the middle ear and its associated sinuses (modified from Archer, 1984, from a drawing by J.M. Thomas). Key to abbreviations: Pmx = premaxilla; MP = mastoid process; M = maxilla; L = lachrymal; N = nasal; J = jugal; F = frontal; P = parietal; Sq = squamosal; A = alisphenoid; ATW = alisphenoid tympanic wing; E = ectotympanic; SO = supraoccipital; Pal = palatine; BS = basisphenoid; BO = basioccipital; f.p. = foramen pseudovale; p3 = third premolar; D = dentary.

conclusion of dasyurid monophyly. As a group, dasyurids exhibit a number of seemingly synapomorphic features (compared, e.g., with didelphoids) including: a large stylar cusp D, small (or absent) stylar cusp C4/3 incisors, loss of the foramen ovale and elaboration of the foramen pseudovale for passage out of the braincase of the trigeminal nerve, a well-developed petrosal tympanic wing, well-developed squamosal epitympanic sinus, petrosal hypotympanic sinus and paraoccipital sinus, tendency towards suppression of P3 and loss of the posterolateral palatal foramen. Many of these features and others are discussed in Archer (1976b, c; 1982b), Kirsch and Archer (1982), Aplin and Archer (1987) and Wroe (1999). Sticking to this list of features, most of the early Eocene, late Oligocene and early to middle Miocene dasyurid-like beasts in the fossil record, such as *Ankotarinja* and *Keeuna*, cannot be confidently allocated to the Dasyuridae, which is why they are considered separately below.

NUMBATS

Numbats are among Australia's most distinctive marsupials. There is, in fact, just one species (*Myrmecobius fasciatus*) in the family Myrmecobiidae, although different forms have been named (e.g., the rusty numbat *M. f. rufus* from central Australia).

Even less is known about the fossil record of numbats. In fact, there are no pre-Pleistocene records for the group and even the Pleistocene records (e.g., from deposits in Nullarbor caves) represent the living species. See Figure 7 and Plate 96.

Understanding the relationships of the numbat has had a long history. In part, the uncertainty about its relationships has been the result of its diet. In specializing as a termite- (and, to a lesser extent, ant-) consumer, its molar teeth have lost most of their basic pattern and become somewhat vestigial (Figure 8). And because

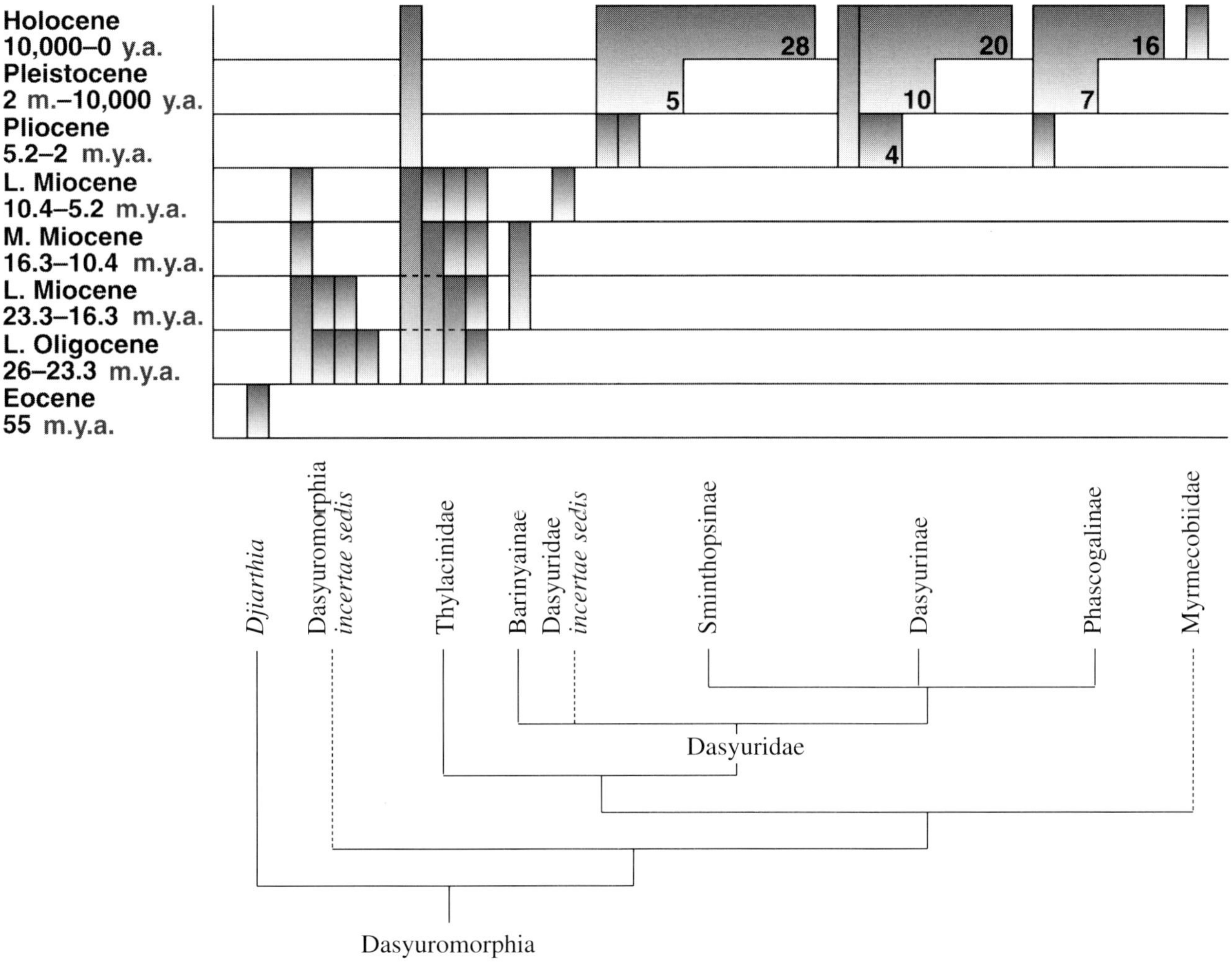

Figure 6. Australian marsupicarnivore diversity through time and a cladogram of probable relationships within the group. Uncertain relationships are indicated by dashed lines. Each block represents a distinct species, species diversity is indicated by a number, and blocks that extend without breaks through more than one time period indicate apparently long-surviving species. The Eocene *Djiarthia murgonensis* is presented in this context, but its ordinal relationships are not necessarily within the Dasyuromorphia. After Wroe and Muirhead (1999).

mammalogists have long relied on molar tooth morphology as a guide to interpreting an animal's phylogenetic relationships, the numbat caused confusion. Many 19th century zoologists thought it was a direct descendant of one or another group of Mesozoic mammals, partly because the molar teeth were simple and almost tricon-odont-like and partly because it seemed to have eight postcanine cheek teeth, one more than was known to occur in any other living terrestrial mammal.

Gradually, the idea that it was part of the dasyuroid radiation became accepted (Kirsch, 1968, 1977; Aplin and Archer, 1987). Although it now appears to be appropriately placed beside dasyurids and thylacinids as part of a monophyletic dasyuroid radiation, its distinctive features were argued by Archer and Kirsch (1977) to rationalize it being placed within its own family, the Myrmecobiidae. However, Wroe *et al.* (2000) and Wroe and Musser (2002) have expressed doubts about its relationships within Dasyuromorphia. Apomorphic features that make the numbat unique among dasyuroids include the following: it appears to either retain dP3 (despite eruption of P3) so that an adult can have three functional premolars and five functional molars all at the same time (Figure 8)—an otherwise unique condition among marsupials; or it has an M5 which would be an equally distinctive feature. Other distinctive features include: splint-like dentaries; the molar teeth display a series of conical cusps on a low base; the palate is extended posteriorly; the alisphenoid tympanic wing forms virtually the whole floor of the middle ear; there is an incomplete bony bar behind the orbit formed by processes from the frontal as well as jugal; the frontal bone develops an orbital process out over the orbit; the lachrymal bone is very large and extends a long way out onto the face; the animal is diurnal; and it has a very long protrusible tongue for invading the galleries of termite mounds. Like dasyurids, numbats have lost the foramen ovale but unlike dasyurids they have retained the posterolateral palatal foramen. Other distinctive aspects of its morphology are noted by Archer (1976b, c) and Jones (1923).

The adult dental formula of the numbat is rather variable in regard to the number of molariform teeth but characteristically is I1–4/1–3 C1/1 ?retained dP3/3 P1–3/1–3 M1–4 or 5/1–4 or 5.

As Archer (1984) pointed out, it is curious that there is just one species of numbat. Why, in a continent festooned by untold billions of diverse termites aren't there

Figure 7. The origins of the insectivorous, rabbit-sized numbat (*Myrmecobius fasciatus*) are obscure. It is the only known species in its genus and family and it has no pre-Pleistocene record (from Gould, 1863).

different sorts of numbats? But then again, perhaps there was a diverse radiation and we don't know about it yet. But if so, why is there only one species now? Today, the numbat's stronghold is a small area of Wandoo woodland in south-western Western Australia. Here it sleeps by night in hollow logs and hunts termites by day in rotten wood or termite mounds. However, when Europeans first arrived on this continent, numbats extended from their present range, north into central Western Australia, east through central Australia and even into the eastern states. Their decline has been catastrophic and foxes are undoubtedly one of the major factors in reducing their numbers. Current captive breeding and release programs carried out in Western Australia appear to be having success, particularly when baiting for foxes is conducted at the same time. With outright extermination of the thylacine and the toolache wallaby on our conscience, as well as indirect responsibility for the extinction of many other species, everyone wants to be sure that everything possible is being done to ensure that the beautiful numbat, the last of its kind, survives.

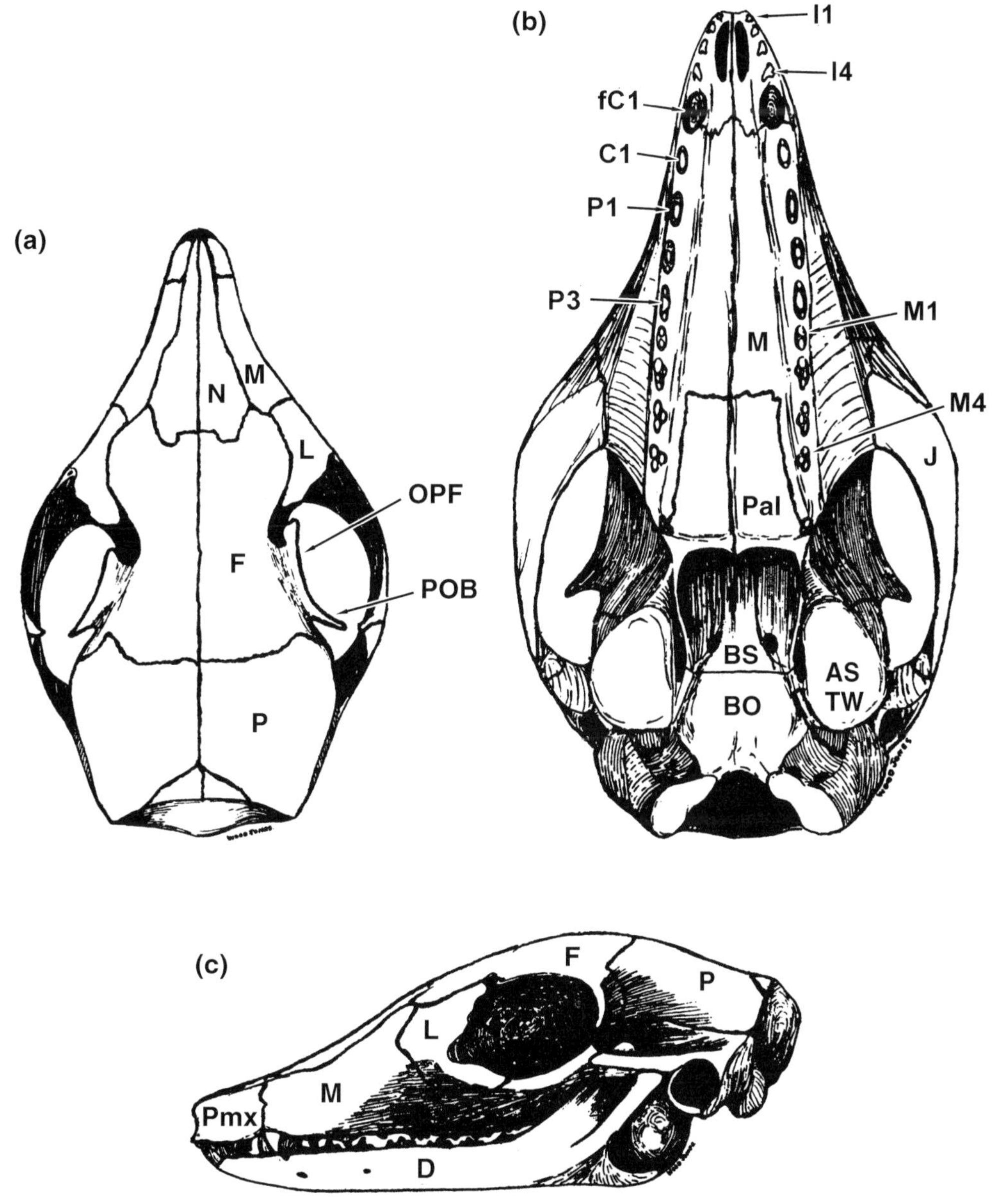

Figure 8. The skull and left dentary of the numbat (modified from Jones, 1923): **(a)** dorsal view; **(b)** ventral view; **(c)** lateral view. Key to abbreviations: I1–4 = upper incisors; fC1 = depression in the premaxilla for the tip of the C1; C1 =upper canine; P1–3=upper premolars; M1–4 = upper molars; M = maxilla; N = nasal; Pal.=palatine; J = jugal; AS TW = alisphenoid tympanic wing; BS = basisphenoid; BO = basioccipital; Pmx = premaxilla; L = lachrymal; F = frontal; P = parietal; D = dentary; OPF = orbital process of the frontal; POB = postorbital bar of frontal.

THYLACINES

The now-extinct wolf-sized thylacine or Tasmanian tiger (*Thylacinus cynocephalus*) was the last surviving member of the family Thylacinidae. Until its extermination by Europeans last century (Figure 9), the thylacine was the largest carnivorous marsupial in the world. Members of this family are characterized by a scrotal pouch in males, cartilaginous (vestigial) epipubic bones, missing hallux on the hind foot, the deciduous third premolar lost before birth and a wide and tubular ectotympanic bone in the auditory canal. All have the adult tooth formula: I1–4/1–3 C1/1 P1–3/1–3 M1–4/1–4 (Figure 10).

Thylacines have been an integral part of the Australian landscape for the last 24 million years. Twenty million years ago at least four different kinds of thylacines inhabited the rainforests of at least northern Australia. Some were the size of rats while others were the size of foxes. By 8 m.y.a. only three kinds of thylacines existed, including the largest of them all—the powerful thylacine (*Thylacinus potens*). By 5 m.y.a., there was just one kind left, the Tasmanian thylacine (*T. cynocephalus*), although it was distributed throughout Australia and New Guinea.

The last remnant of this ancient line, the Tasmanian thylacine, survived on the Australian mainland until about 4,000 years ago (Plate 97), when the dingo (*Canis lupus dingo*) was introduced to Australia by peoples unknown. Why did it go extinct on the mainland? Perhaps dingoes were more adept in competition for food and shelter. The two sorts of carnivorous mammals were, after all, markedly convergent in size and body form, a fact that suggests they might have competed with each other. Further, given the short, wide heads of the oldest known dingoes, it is clear that they were part of a domesticated population. Perhaps they had an added advantage because they were being looked after by people? After their extinction on the mainland, thylacines survived in Tasmania—until the 1930s, when Europeans finally hunted and/or hounded them to extinction. Johnson and Wroe (2003) argue that intensification of land use by Aborigines, which coincided with the dingo's introduction, may have been another significant factor in the extinction of the thylacine on the mainland. The last known thylacine died in the Beaumaris Zoo, Hobart, in 1936.

One of the most interesting aspects of the thylacine is the history of its interpreted relationships. It was first described as a species of *Didelphis* (an American genus of didelphids) but was soon regarded to be part of the Australian marsupial radiation. Bensley (1903), however, studying its teeth and feet, felt that it was not comfortably placed among Australian marsupials. Sinclair (1906) describing the middle Miocene borhyaenids from Patagonia, declared them to be most closely related to thylacinids. This conclusion was elaborated by Woods (1924) who noted many striking apparent synapomorphies between the two groups including their reduced epipubic bones and many striking dental features. Simpson (1945), however, concluded that thylacinids were merely specialized dasyurids and in this basic conclusion was followed by Marshall (1981). In a study of basicranial morphology (Archer, 1976c), it was concluded that borhyaenids and thylacinids shared no distinctive features. However, in a comparative study of tooth morphology (Archer, 1976b), it was concluded that thylacinids and borhyaenids could have shared a common ancestor that had the specialized dental morphology but a plesiomorphic basicranium (Figure 11). However, in two 1982 studies, one by Sarich *et al.* (1982) on albumin and one by Szalay (1982) on tarsal morphology, Bensley's original view was confirmed and is now broadly accepted. Most likely thylacinids are the plesiomorphic sister-group of the dasyurids and together these two are the sister-group of the numbats (Archer *et al.*, 1999); but Wroe *et al.* (2000) and Wroe and Musser (2001) have expressed doubts about the relationships of the three families (indicated in Figure 6).

The fossil record of thylacinids is now much better than it was when Archer sat down to write a similar review in 1984. Since then seven extinct genera and 11 species have been described from late Oligocene to late Miocene deposits (Wroe, 1996, 2001; Muirhead, 1997; Muirhead and Wroe, 1998; Murray, 1998; Murray and Megirian, 2000).

The oldest thylacine known is *Badjcinus turnbulli* Muirhead and Wroe, 1998, from late Oligocene deposits at Riversleigh, north-western Queensland. The genus was described on the basis of a partial skull and almost complete upper and lower jaws. Other extinct forms include: the primitive, fox-sized *Muribacinus gadiyuli* Wroe, 1996a, from the middle Miocene of Riversleigh; the very primitive *Nimbacinus dicksoni* Muirhead and Archer, 1990, from the late Oligocene to middle Miocene of Riversleigh (known from a complete skeleton and skull (Plate 98)); and the middle Miocene of Bullock Creek; the specialized *Wabulacinus ridei* Muirhead, 1997, from the early Miocene of Riversleigh, which appears closely related to *Thylacinus*; the small dog-sized *Ngamalacinus timmulvaneyi* Muirhead, 1997, from the early Miocene of Riversleigh; *Maximucinus muirheadae* Wroe, 2001, from middle Miocene sediments at Riversleigh, one of the largest thylacinids known for its age with estimates of body weight being 18 kg (comparable to small Tasmanian thylacines); *Mutpuracinus archibaldi* Murray and Megirian, 2000, from the middle Miocene Bullock Creek LF of the Northern Territory, the smallest thylacinid yet discovered; *Tyarrpecinus rothi* Murray and Megirian, 2000, from the late Miocene Alcoota LF of the Northern Territory, interpreted to be more closely related to species of *Thylacinus*; *Thylacinus macknessi* Muirhead, 1992, from early Miocene sites at Riversleigh the oldest species of this modern genus; and *T. potens* Woodburne, 1967a, from the Alcoota LF, the largest thylacine known, slightly larger but more robust than the Tasmanian thylacine.

Thylacines are symbolic of the gap between our scientific curiosity and our greed. They had survived in a seemingly healthy way in Tasmania long after their

Figure 9. The Tasmanian tiger or thylacine (*Thylacinus cynocephalus*), the last of its long and distinctive lineage, was exterminated by Europeans in the 1930s. Photograph Tasmanian Museum and Art Gallery.

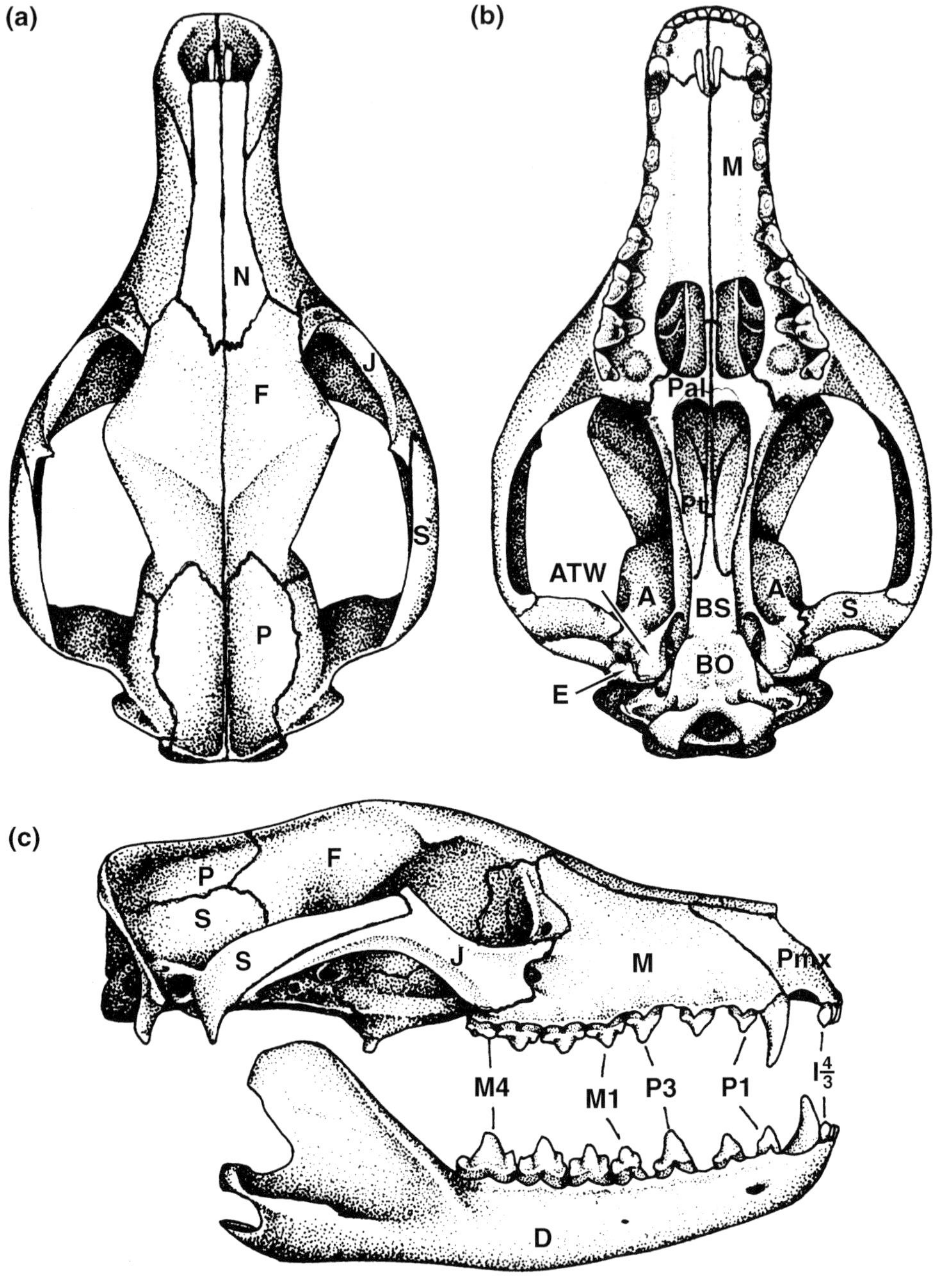

Figure 10. The skull and right dentary of the Tasmanian thylacine (*Thylacinus cynocephalus*) (from drawings by J.L. Rainbird in Green, 1983): **(a)** dorsal view; **(b)** ventral view; **(c)** lateral view. Key to abbreviations: I4/3 = incisors; P1–4 = premolars; M1–4 = molars; Pmx = premaxilla; M = maxilla; J = jugal; F = frontal; S = squamosal; N = nasal; P = parietal; Pal.= palatine; Pt = pterygoid; A = alisphenoid; ATW = alisphenoid tympanic wing; BS = basisphenoid; BO = basioccipital; E = ectotympanic; D = dentary.

demise on the Australian mainland. Not until we came, was there any problem. Aborigines and thylacines coexisted for well over 60,000 years. Europeans couldn't tolerate thylacines (or for that matter the Tasmanian Aborigines) for more than 200 years. It seems to be a trademark of European humans that, wherever we go, if we can, we exterminate the dominant carnivores in every ecosystem. There are also always apologists. In the case of the thylacine, some have tried to claim that disease was responsible for their extinction, not the continuous predation by Europeans, nor the bounties paid year after year (even after the population crash of the thylacines around 1910) by the Tasmanian Government of the day. Combined with the snares all over the bush by graziers, the total assault was all but over when the last thylacine fell over in the Beaumaris Zoo in Hobart. If any survived after that time, not a single shred of solid evidence supports the claim. In fact this happened for all the wrong

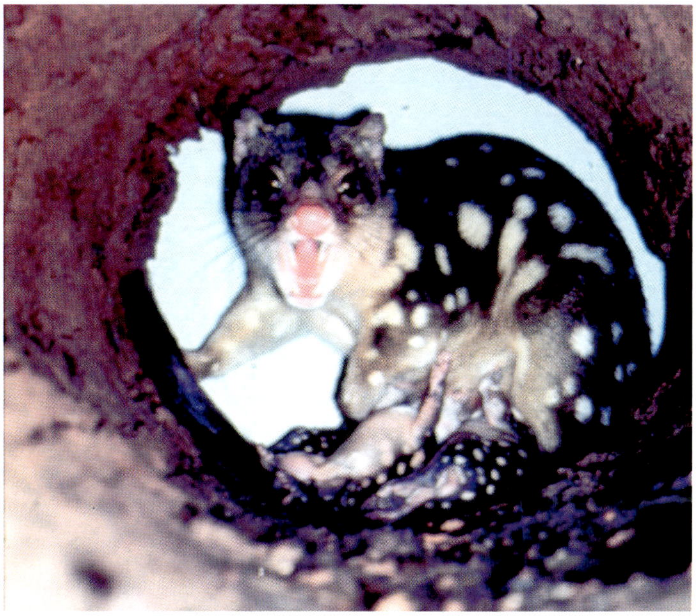

Plate 95. A characteristic dasyurid, the cat-sized tiger quoll (*Dasyurus maculatus*) with young. This marsupial carnivore inhabits forests in eastern Australia, from Queensland to Tasmania. It is one of the largest still living on the Australian mainland. Photograph G. Settle.

Plate 96. Juvenile numbats (*Myrmecobius fasciatus*) from Western Australia. Numbats are diurnal marsupials that hunt termites by day in rotten wood or termite mounds and sleep by night in hollow logs. Photograph D. Whitford.

Plate 97. Mummified remains of the thylacine *Thylacinus cynocephalus* from Thylacine Hole, a cave on the Western Australian Nullarbor Plain. Photograph M. Archer.

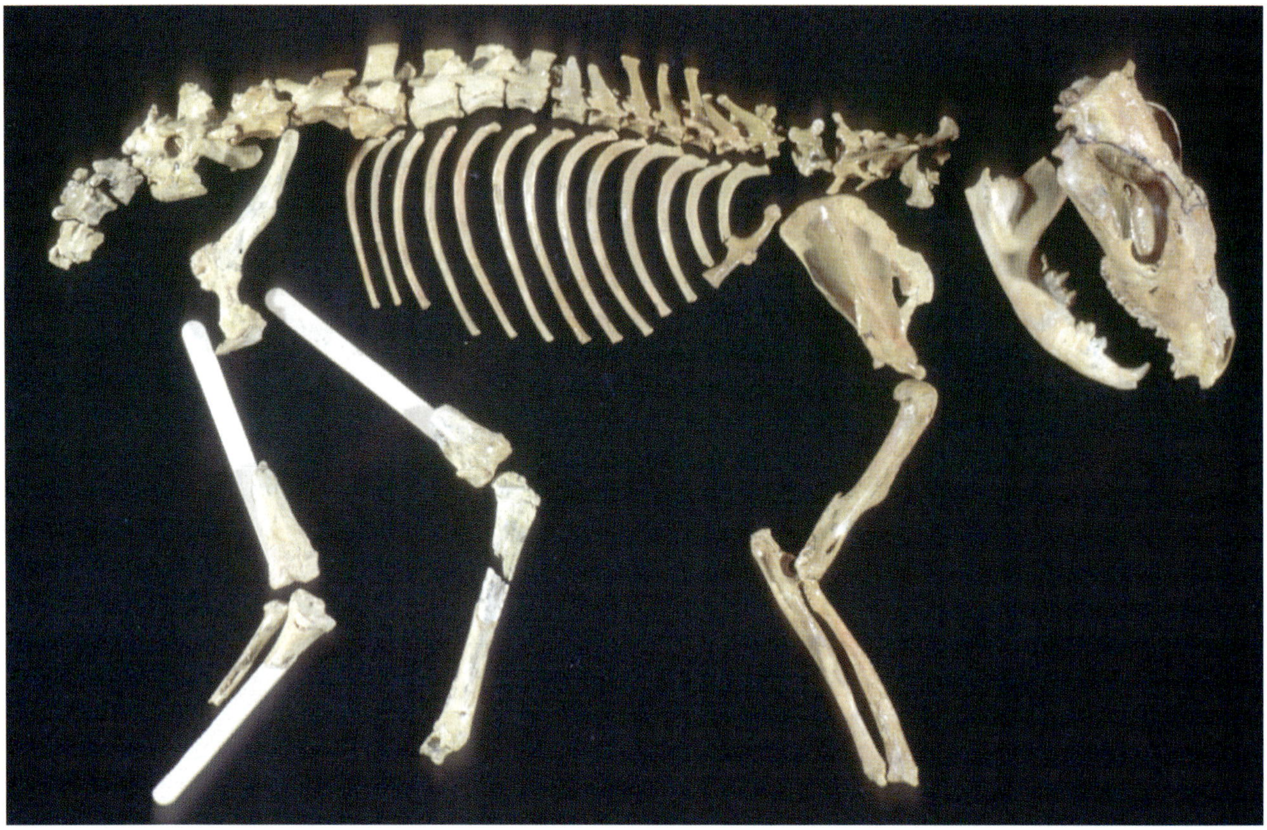

Plate 98. The skull and skeleton of *Nimbacinus dicksoni* from the Miocene of Riversleigh. This species was one of six thylacine species, ranging from the size of a rat to that of a dog, recovered from Riversleigh's Oligo-Miocene deposits. Photograph K. Black.

Plate 99. The skull and lower jaw of a species of *Yarala* emerges from a block of Riversleigh limestone during acid-processing. The extinct yaraloids appear to be the most plesiomorphic of all bandicoots. Photograph K. Atkinson.

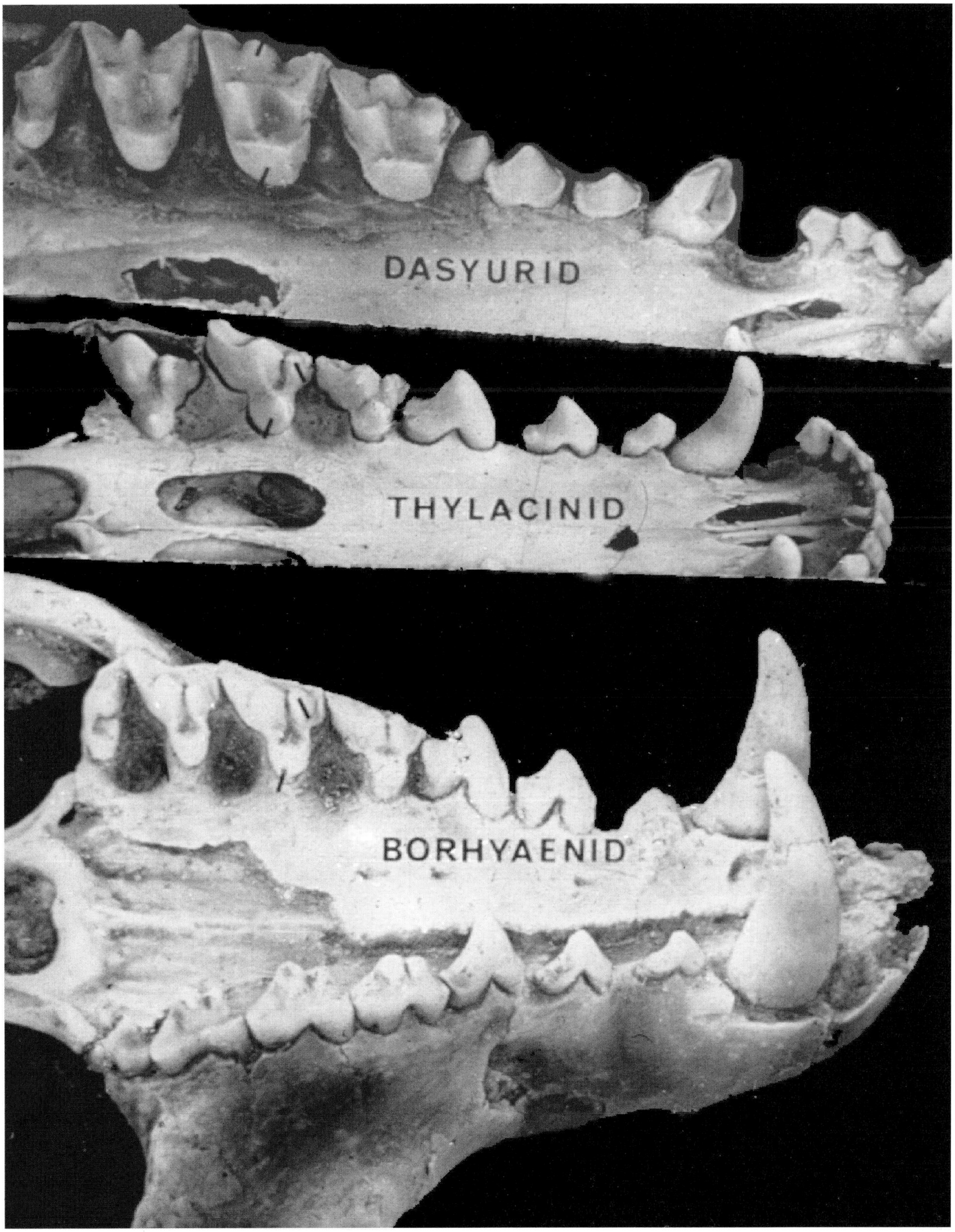

Figure 11. A comparison of the upper right dentitions of a young adult dasyurid (Australia), a juvenile thylacinid (with M4 still erupting), and an adult borhyaenid (South America; lack of incisors due to poor preservation). The markedly greater similarity between the thylacinid and borhyaenid is clear, but evidently the result of convergence. Photograph M. Archer.

reasons; wild dogs were killing Tasmanian sheep but thylacines were the scapegoat—see Paddle (2000).

But there is one, albeit remote hope: the Australian Museum's Thylacine Project. Scientific breakthroughs in genetic engineering and cloning over the last 10 years have raised the possibility of bringing extinct animals back to life—literally 'raising the dead'. Museum geneticists have recovered large quantities of thylacine DNA from teeth, bones and other tissues of museum specimens and work is underway to characterize, repair and reassemble the genome with the ultimate aim of bringing back a viable, breeding population of Tasmanian thylacines to be reinstalled in suitable Tasmanian natural habitat. It's a dream fraught with challenges, and naturally this is a very controversial project (see the Australian Museum's website for a discussion of some of the issues and challenges), but if per chance it works, we may get a second chance to restore part of the balance of nature which should never have been tipped over in the first place.

OTHER DASYUROMORPHIAN PUZZLES

A number of Oligo-Miocene marsupial taxa have been placed by Wroe (1996a, 1997a) in Dasyuromorphia *incertae sedis* (that is, in an uncertain position) because they lack enough key information to enable them to be placed with confidence in any particular modern family. Wroe (1996a) argued that their relationships were so unclear that they should be regarded as *incertae sedis* although conservatively left in Dasyuromorphia. Without complete skulls, or at least basicrania, their familial relationships remain enigmatic.

Taxa of this kind include: the mouse-size *Ankotarinja tirarensis* Archer, 1976a, from the late Oligocene Ditjimanka LF of South Australia which has been argued variously to be either a dasyurid, peradectid (archaic marsupials known from the Northern Hemisphere) or microbiotheriid (mostly South American marsupials); *Apoktesis cuspis* Campbell, 1976a, from the late Oligocene,

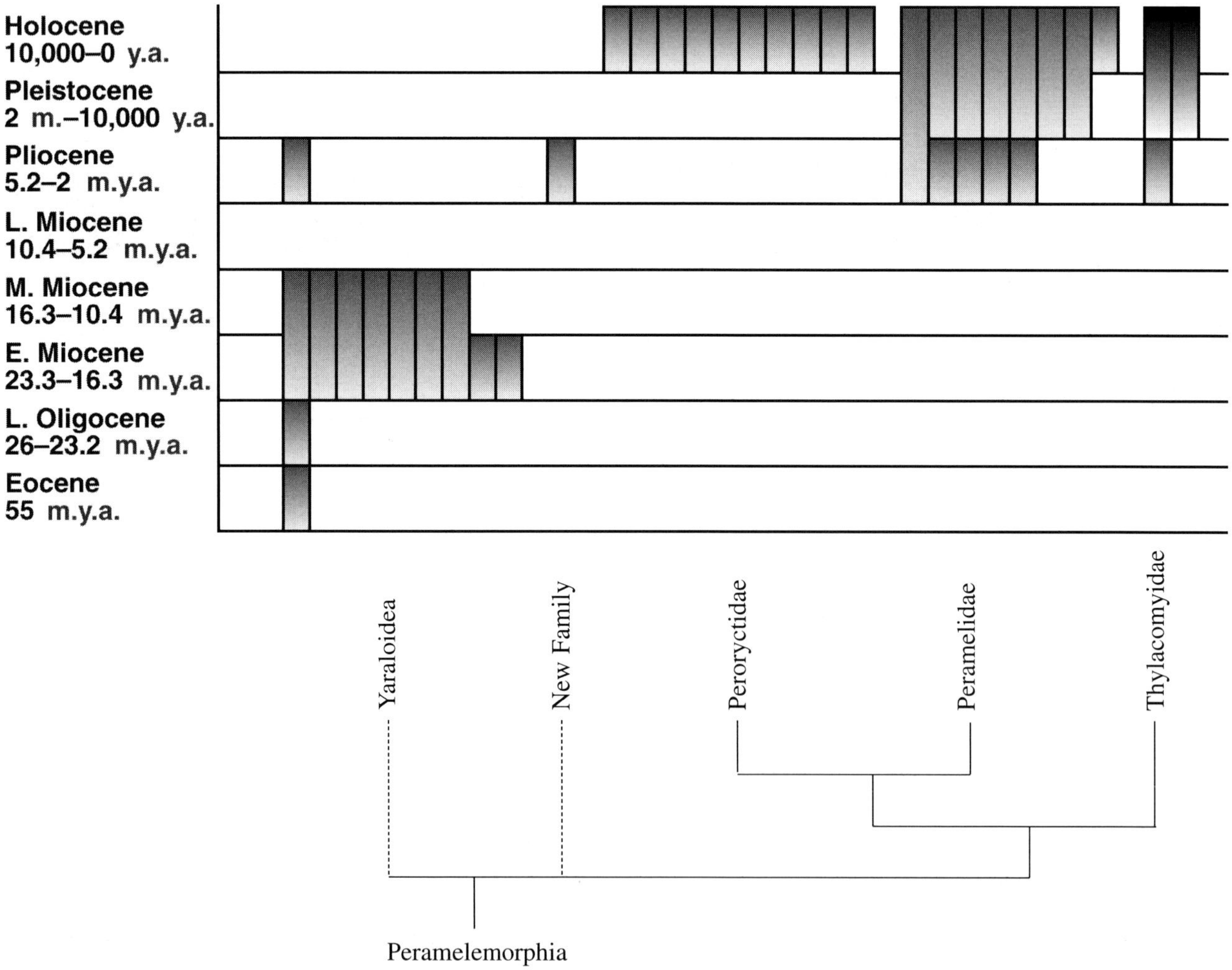

Figure 12. Bandicoot diversity through time and a cladogram of probable relationships within the group. Uncertain relationships are indicated by dashed lines. Each block represents a distinct species, and blocks that extend without breaks through more than one time period indicate apparently long-surviving species. The Eocene (Murgon) taxon is placed here among the Yaraloidea although its affinities are uncertain. After Muirhead (1999).

Ngapakaldi LF, South Australia, which was described as a dasyurid most closely allied to *Dasyurus* but is not clearly part of any known group; *Dasylurinja kokuminola* Archer, 1982b, from the middle Miocene Yanda LF of South Australia; and *Keeuna woodburnei* Archer, 1976a, from the late Oligocene, Ditjimanka LF of South Australia. In the case of *Keeuna,* Archer (1982b) noted that the referred maxilla has alveoli for just four incisors. Didelphimorphians normally have five upper incisors but all dasyuromorphians have four.

Wakamatha tasselli Archer and Rich, 1979, may be either late Oligocene or Pleistocene (provenance uncertain), as are its relationships. It was regarded to be part of the Billeroo Creek LF, South Australia, but was found in sediment washed down from nearby Oligocene and Pleistocene formations. As a result, its exact age is uncertain. Further, although its teeth resemble in some respects those of the dasyurid genus *Sminthopsis*, Wroe (1996a) has questioned whether this creature could not alternatively represent a very primitive bandicoot. The jury's still out on this one.

Mayigriphus orbus Wroe, 1997b, from the early Late Miocene Encore LF of Riversleigh, is another enigmatic beastie. It is the smallest dasyuromorphian known from the Oligocene or Miocene of Australia and differs from all dasyurids in many features including reduced metaconids and metacristids on M1–4, features of thylacinids. But it also differs from thylacinids in many other features including a greatly reduced paraconid on M1 but only moderately reduced metaconid, so its relationships remain a puzzle (Wroe, 1997b).

As a group, dasyuromorphians have been regarded (Bensley, 1903) to be plesiomorphic within the Australian marsupial radiation. This may be true. Certainly they are the only Australian group to retain, as a combination of features, more than one lower incisor on each side, relatively unspecialized molar tooth morphology, an unfused second and third hind toe (i.e. they are not syndactylous), and so on. In this combination of retained plesiomorphic features they are, among Australian groups, most similar to some of the South American marsupials. It is important to note that the family considered as most plesiomorphic within the order, the Dasyuridae, is in fact derived among marsupials for many cranial features. On the other hand, thylacinids, although derived in their dentition, retain many plesiomorphic cranial features (Wroe and Musser, 2001).

Figure 13. The greater bilby *Macrotis lagotis* is the only surviving thylacomyid. It is a desert-adapted, omnivorous bandicoot that lives in burrows during the heat of the day and forages for food in the cool of the night (from Gould, 1863).

In addition, some bandicoots have five upper incisors (as do many South American marsupials but none of the other Australian marsupials) and a more plesiomorphic middle ear and basicranial construction than dasyurids. To further shake the ancestral pedestal of dasyurids, Szalay (1982) has shown striking synapomorphic features of the tarsal bones shared by South American microbiotheriids and some Australian possums. He visualizes the dasyurid hind foot to be a secondarily derived structure, one that descended from an ancestral condition similar to that found in microbiotheriids. So, we're delighted to say that there are still more questions than

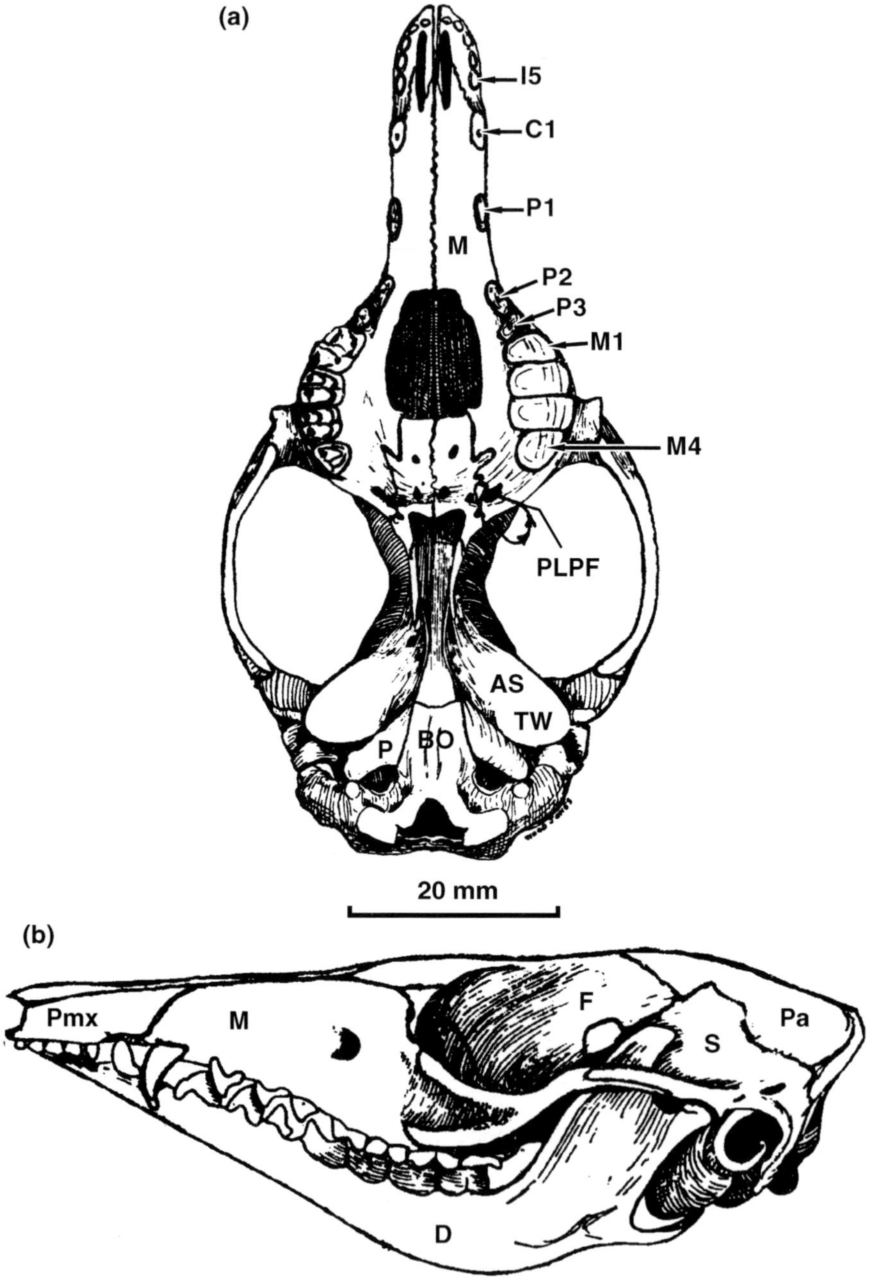

Figure 14. The skull and dentary of thylacomyids: **(a)** ventral view of the skull of *Macrotis lagotis*; **(b)** left lateral view of the skull and left dentary of *M. leucura*. The right molar row of **(a)** is shown as it would appear when little worn. The left molar row is shown as these teeth appear when heavily worn—with flat and wide, almost featureless, grinding surfaces. The enlarged and pear-shaped alisphenoid tympanic wings (ASTW), marked no-gradual change in width of the rostrum anterior to the molar row, very large upper canine and molar cusp arrangement are distinctive features of species of this genus (modified from Jones, 1924). Key to abbreviations: M = maxilla; PLPF = posterolateral palatal foramen; P = periotic tympanic wing; BO = basioccipital; Pmx = premaxilla; F = frontal; Pa = parietal; S = squamosal; D = dentary.

answers and lots for future generations of palaeontologists and phylogeneticists to sort out!

PERAMELEMORPHIANS—BANDICOOTS AND BILBIES

Bandicoots and bilbies have always fascinated zoologists and phylogeneticists, in part, because they refuse to be easily pigeon-holed in terms of their relationships to other marsupials and to each other. Readers should refer to *Bandicoots and Bilbies* (Seebeck *et al.*, 1990) for research into and overviews about their biology, history and phylogenetic relationships, although our family level classification follows Long *et al.* (2002).

As currently recognized there are three families of living bandicoots and bilbies, all placed in the superfamily Perameloidea (Archer and Kirsch, 1977); one family of extinct bandicoots, the Yaralidae, in the superfamily Yaraloidea (Muirhead, 2000); one mystery group from the Pliocene Bluff Downs LF in Queensland; and a single very archaic perameloid-like marsupial from the early Eocene Tingamarra LF of Queensland (Figure 12).

In the Thylacomyidae there are two known genera of bilbies (living species of *Macrotis* and a Pliocene species of *Ischnodon*); see Figures 13 and 14. They have a dental formula of I1–4/1–3 C1/1 P1–3/1–3 M1–4/1–4. There are three genera of ordinary bandicoots placed in the family Peramelidae (*Perameles, Isoodon* and *Chaeropus*, the last being extinct). They have an adult dental formula of I1–5/1–3 C1/1 P1–3/1–3 M1–4/1–4. There are in addition several unnamed genera of extinct peramelids. Species of *Peroryctes*, *Microperoryctes* and *Rhynchomeles* (the last probably now extinct) are all placed in the family Peroryctidae, or forest bandicoots, and are all known from New Guinea and surrounding islands—no living species of these genera occur in Australia. The peroryctid genus *Echymipera*, however, has Australian as well as New Guinean representatives (Figure 15). Peroryctids have an adult dental formula of I1–4/1–3 C1/1 P1–3/1–3 M1–4/1–4. The peramelid genus *Isoodon*, which is abundant in Australia, is also represented in New Guinea (Figure 16). The plesiomorphic yaralids, which encompass many Oligocene and Miocene genera including *Yarala* Muirhead and Filan, 1995, have a dental formula of I1–5/1–3 C1/1 P1–3/1–3 M1–4/1–4.

The phylogenetic interrelationships of bandicoots as a whole among marsupials in general, and of the individual families of bandicoots to each other, are controversial. Perameloids have, as a synapomorphic feature, syndactyly and are presumed to be monophyletic. However, this monophyly has yet to be demonstrated. Thylacomyids are monophyletic and have, among other synapomorphic features, reduced paraconids on the lower molars, lingually expanded metacones, no metaconules and tubular, markedly enlarged alisphenoid hypotympanic sinuses.

Considering the origins of bandicoots, Archer (1976b, c) suggested that dental and basicranial structure seemed to ally them more closely with certain South American than Australian groups. But Szalay (1982) has given reasons, based on pedal morphology, for believing that perameloids are more like other Australian marsupials than any American marsupial group (except the microbiotheriids); and Kirsch (1977) found that perameloids and dasyuroids were each other's closest sister-group based on studies of comparative serology. Then again, based on DNA-DNA hybridization data, bandicoots appear to be quite distinct from other Australian marsupials and perhaps closer to some South American marsupial groups, or even outside all other living marsupials (Springer *et al.*, 1994; Retief *et al.*, 1995). It's all up in the air. The data from basicranial and dental morphology may be reconcilable with the pedal and serological data if the more archaic features of perameloids are retentions of plesiomorphic features lost in other groups of Australian marsupials. Perameloids may have retained features present in the, as yet, unknown ancestral stock for the Australian marsupial radiation.

The other side of the bandicoot phylogenetic coin focuses on their possession of the apomorphic state of syndactyly—a condition in which the second and third toes of the hind foot are reduced and fused at their bases. At least superficially, this feature resembles that found in diprotodontian marsupials and is commonly used as the basis for recognition of Syndactyla (Szalay, 1982). However, the idea that the bandicoot and diprotodontian syndactyly is monophyletic is not without its detractors. Kirsch (1977) considers bandicoot syndactyly to be convergent because of the close dasyuroid-perameloid serological relationship.

Figure 17 further illustrates the apparently intermediate condition of bandicoots because it also considers the distribution of the apomorphic state of diprotodonty. Most 'primitive' marsupial groups are, besides being didactylous (i.e., the toes of the hind foot are unfused),

Figure 15. The cat-sized rufous spiny bandicoot (*Echymipera rufescens*) from Cape York, north-eastern Australia. All bandicoots have elongated snouts and are insectivorous to omnivorous. Photograph University of New South Wales.

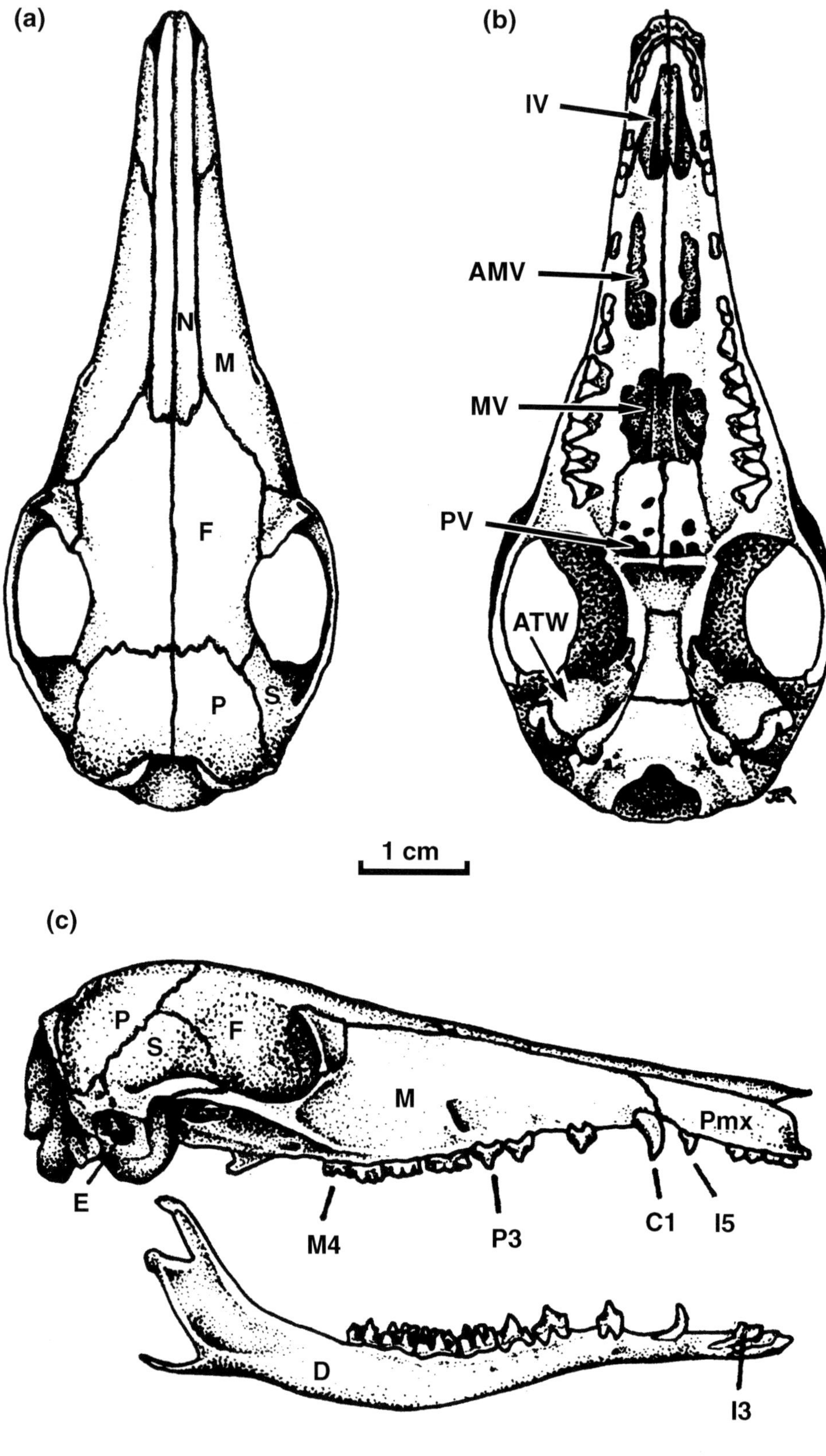

Figure 16. The skull and right dentary of the eastern barred bandicoot (*Perameles gun*) (from drawings by J.L. Rainbird in Green, 1983): **(a)** dorsal view; **(b)** ventral view; **(c)** lateral view. Key to abbreviations: I5 = fifth incisor; C1 = canine; P3 = third upper premolar; M4 = fourth upper molar; Pmx = premaxilla; M = maxilla; F = frontal; S = squamosal; P = parietal; E = ectotympanic;; N = nasal; IV = incisive vacuity; AMV = anterior maxilliary vacuity; MV = maxilliary vacuity; PV = palatine vacuity; ATW = alisphenoid tympanic wing.

polyprotodont (i.e., they have more than one lower incisor on each side and the median one is not markedly enlarged or procumbent). Bandicoots are polyprotodont (and hence in this regard plesiomorphic). Diprotodontians are both syndactylous and diprotodont. This could place bandicoots in a structurally intermediate position between other polyprotodonts and the syndactylous diprotodontians.

If this structurally intermediate condition is indicative of phylogenetic history, bandicoots may well have been the base for the Australian diprotodont radiation. But if the syndactyly is convergent, all bets are off—bandicoots could have nothing to do with the diprotodontian radiation.

In terms of their adaptive radiation, peramelemorphians are spread from one habitat extreme to the other. Some occur in rainforests (e.g., some species of *Perameles* and *Microperoryctes*) and others occur in the most arid parts of Australia (e.g., other species of *Perameles* and *Chaeropus*). Some are recently extinct (e.g. *Chaeropus ecaudatus*), the result of European activities such as land-clearing and the introduction of cats and foxes, while others can be commonly found in backyards (e.g., other species of *Perameles* and *Isoodon*). In the latter case, the future of at least three species appears secure.

EXTINCT BANDICOOTS AND BILBIES

The early Pliocene *Perameles allinghamensis* Archer, 1976d, has been described from the Bluff Downs LF of Queensland and *P. bowensis* Muirhead *et al.*, 1997, from the Bow LF of New South Wales. Although only recently extinct, the pig-footed bandicoot (*Chaeropus ecaudatus*) is unlike other peramelids in having a specialized forefoot with only two functional toes and a hindfoot with only one supporting toe. The result is a pig-like appearance of the forefeet and a kangaroo-like appearance of the hindfeet. It is also the only bandicoot thought to have been a grazer. Muirhead and Godthelp (1996) report a species of *Chaeropus* (probably *C. ecaudatus*) from a Quaternary fossil deposit near Chillagoe in north-eastern Queensland.

No extinct species of the family Peroryctidae have yet been described from Australia or New Guinea.

The extinct bilby *Ischnodon australis* Stirton, 1955, was described on the basis of a single right dentary from the possibly late Pliocene Palankarinna LF of South Australia. Archer and Kirsch (1977) noted the features of *I. australis* that ally it with species of *Macrotis* rather than any other bandicoot genus. This conclusion supports Stirton's original assessment. The greater reduction of the paraconids in species of *Macrotis* even make it possible that *Ischnodon australis* could be close to the ancestral stock for *Macrotis*.

Thylacomyids have not been doing terribly well of late—at least since the arrival of Europeans and their cats, sheep and rabbits. At the time of settlement by Europeans, the greater bilby was common in the drier areas of all states except Tasmania. In the late Pleistocene it was even in the Wellington area of New South Wales. Today it is confined to a few very small areas in Western Australia and the Northern Territory. The lesser bilby (*M. leucura*) appears to have gone extinct around the turn of last century. Like the greater bilby, it was a desert-adapted, omnivorous marsupial, living in burrows during the heat of the day and foraging for food in the cool of the night.

The plesiomorphic yaralids are all extinct but most have yet to be described, a work in progress by Jeanette Muirhead. They are the most numerous of all marsupials in Riversleigh's Oligo-Miocene sediments (Plate 99), and closely related forms are known from late Oligocene sites in the Tirari Desert, South Australia, and from Kangaroo Well in the Northern Territory. Yaralids are believed to be close to the base of the bandicoot family tree (Figure 12). They range in size from that of a mouse to some of the largest living bandicoots. Riversleigh's early Miocene *Yarala burchfieldi* Muirhead and Filan, 1995 was smaller than any living bandicoot—about the size of an antechinus.

The bizarre perameloid-like taxon from the Bluff Downs LF is so strange that its seeming perameloid affinities may be misleading. Even its premolars are bicuspid. The very plesiomorphic perameloid from the early Eocene Tingamarra LF is so far represented by a single tooth. More excavation will hopefully reveal a lot more about this potentially very significant and oldest perameloid, doing bandicootish things while Australia was still connected to Antarctica and via that land to South America. Perhaps similar perameloids will also turn up in some of the early Tertiary deposits of Argentina or Antarctica?

MOLES AND OTHER ENIGMAS

We have grouped here some of the most enigmatic of Australian marsupials—the marsupial moles, yalkiparidontians ('thingodontans'), yingabalanarids ('bizarrodontans') and chulpasinines because there are no undoubted relationships to other groups of marsupials.

MARSUPIAL MOLES

There are those (even the odd modern zoologist) who claim that a mole is a mole is a mole. Not so! There are indeed many mole-like animals in the world such as the marsupial moles (*Notoryctes typhlops* and *N. caurinus*), the placental Cape golden mole (a chrysochlorid), other zalambdodont insectivores and even South America's enigmatic *Necrolestes patagonicus*, *whatever* that thing is. Although they are all indeed mole-like, there seems abundant evidence that this is the result of convergence. Any attempt to formalize a taxonomic group for all moles would result in a polyphyletic group, and that's a no-no.

Australia's marsupial moles are delightful but puzzling little creatures (Figure 18). They are blind, almost

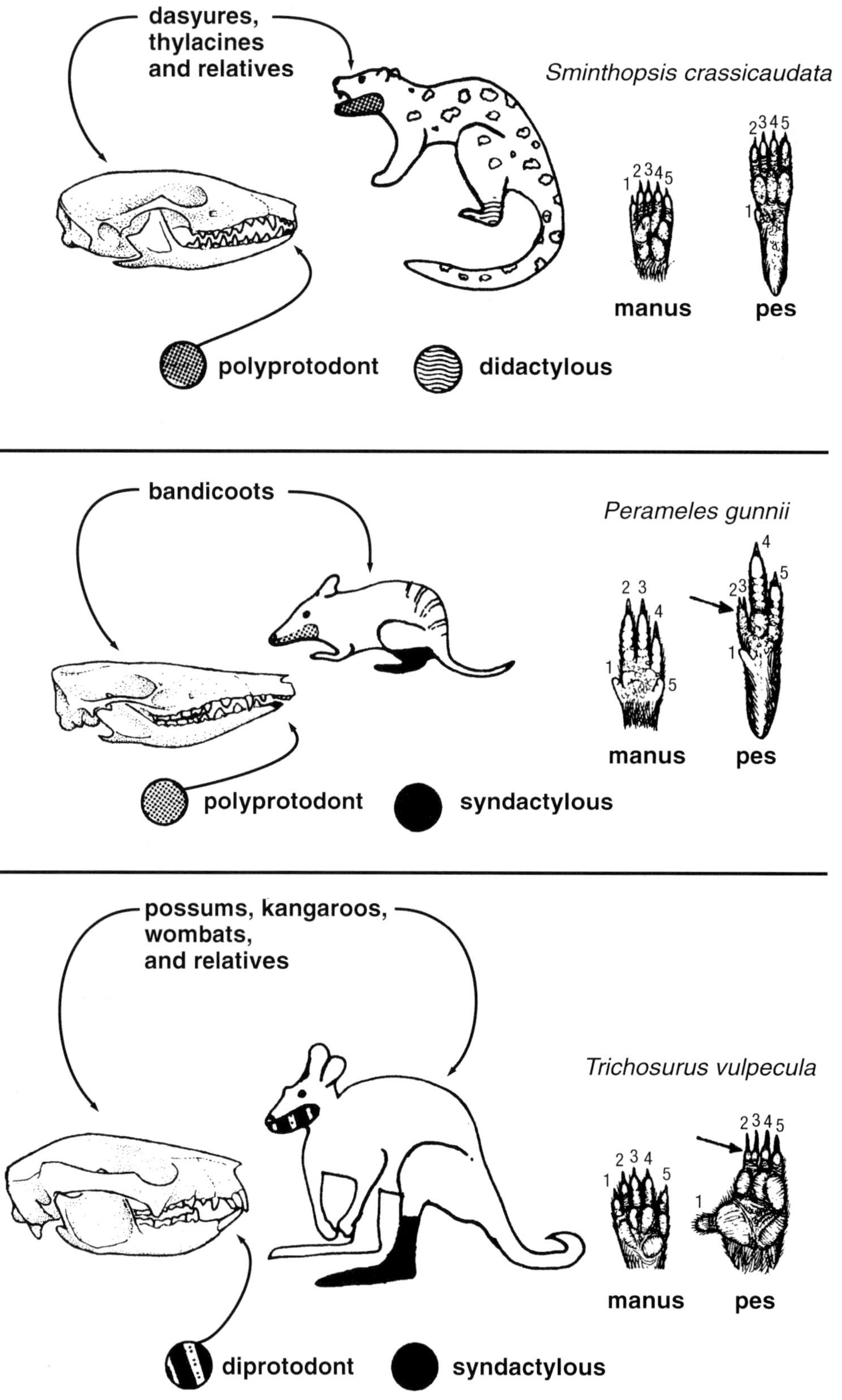

Figure 17. Bandicoots occupy a position structurally between other polyprotodonts (marsupials with more than one lower incisor on each side) and diprotodontians (Australian marsupials with only one lower incisor on each side). The bandicoots are, like the former, polyprotodont (i.e., they are plesiomorphic in this feature), but like the latter, syndactylous (an apomorphic condition in which the second and third toes of the hind foot are reduced in size and joined by soft tissues (from Archer, 1984).

earless (there is no pinna), they have highly specialized forefeet for digging, as well as a well-developed shield for the nose and a short stumpy tail. Two species are now recognized on the basis of molecular data and a slight difference in size and distribution. Fossils found in the early Miocene deposits at Riversleigh (Gott, 1988; Archer *et al.*, 1994, 1997; Warburton *et al.*, 2003) represent a far more primitive genus within the family Notoryctidae (Archer *et al.*, 1999).

It is not at all clear which family of Australian marsupials is most closely related to the notoryctids. Their middle ear construction (Archer, 1976c) and aspects of their soft anatomy, suggest they may have something to do with the diprotodontian radiation but their dentitions (Figure 19) show no trace of the diprotodont condition that seems to be the single most distinctive synapomorphy of the Diprotodontia. It is not even clear whether or not the hind foot specializations are superimposed on a didactylous or a syndactylous condition. Molecular biology has given mixed signals which all add up to the reason why Aplin and Archer (1987) placed them in their own order, the Notoryctemorphia, a conclusion commonly followed by others (e.g., Long *et al.*, 2002).

Regarding the lifestyle of the living marsupial moles, moles of course dig. And to dig you must have a 'digupable' substrate. Today, they are found in the sandy deserts of central Australia where they hike along just under the surface digging burrows that collapse behind them. They may eat insect larvae or worms. No one's sure. To catch a mole you evidently should wait for rain. Indigenous people watch for a mole's distinctive tracks on the surface after rain, follow them to the point where they disappear into the sand and then dig in the direction of the trackway, watching for traces of the collapsed burrow.

Until the fossil discoveries at Riversleigh, we assumed that their specialized features reflected a long evolutionary history of adapting to deserts. Now that we know ancestral forms were living in the ancient forests of Riversleigh, it seems more likely that they evolved their burrowing adaptations in response to chasing edible

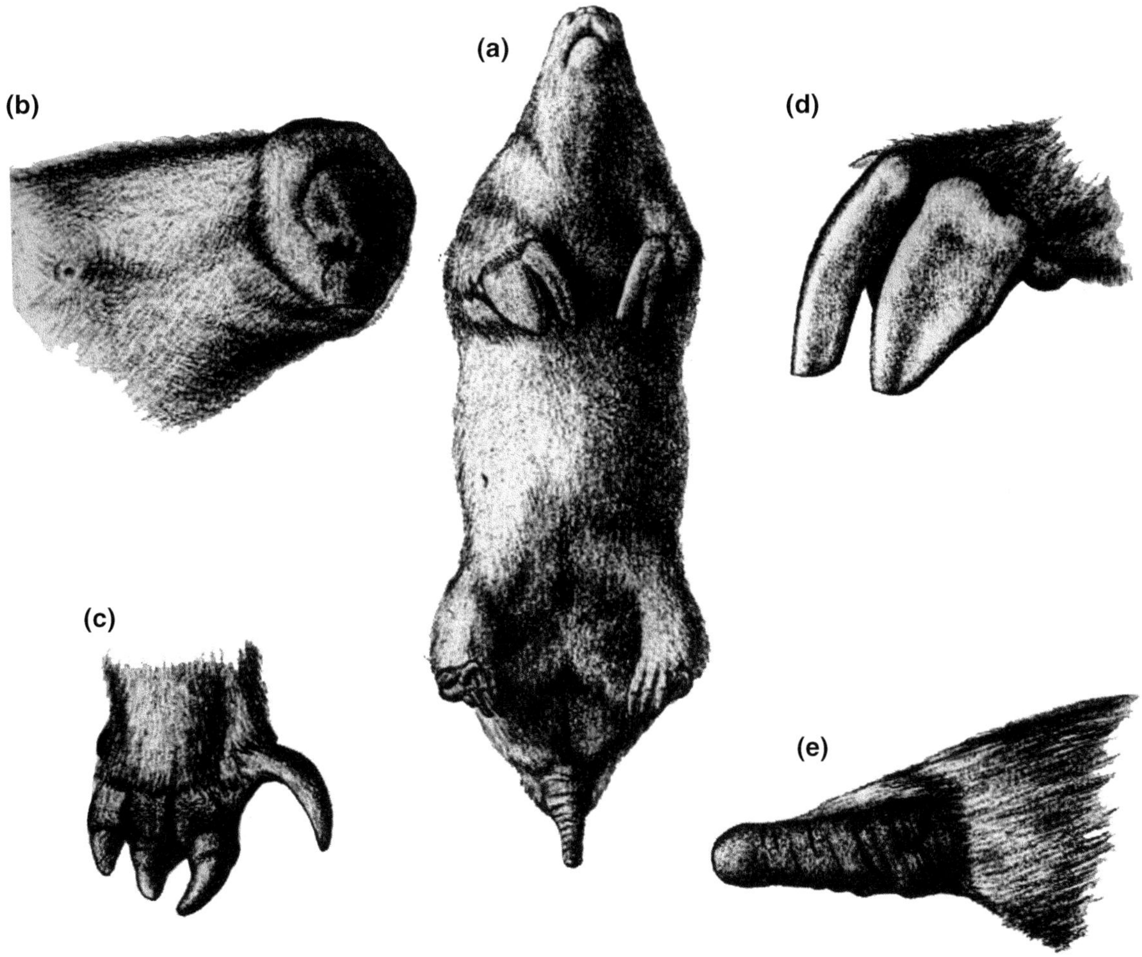

Figure 18. External features of a marsupial mole (*Notoryctes typhlops*): **(a)** the ventral surface of a female; **(b)** the left side of the head and thorax. Note feet, lack of eyes, the well developed nose shield, the ear opening (which has been exposed by parting the hair); **(c)** the highly specialized hind foot; **(d)** the forelimb; and **(e)** the short, stumpy tail. Adapted from Jones (1924).

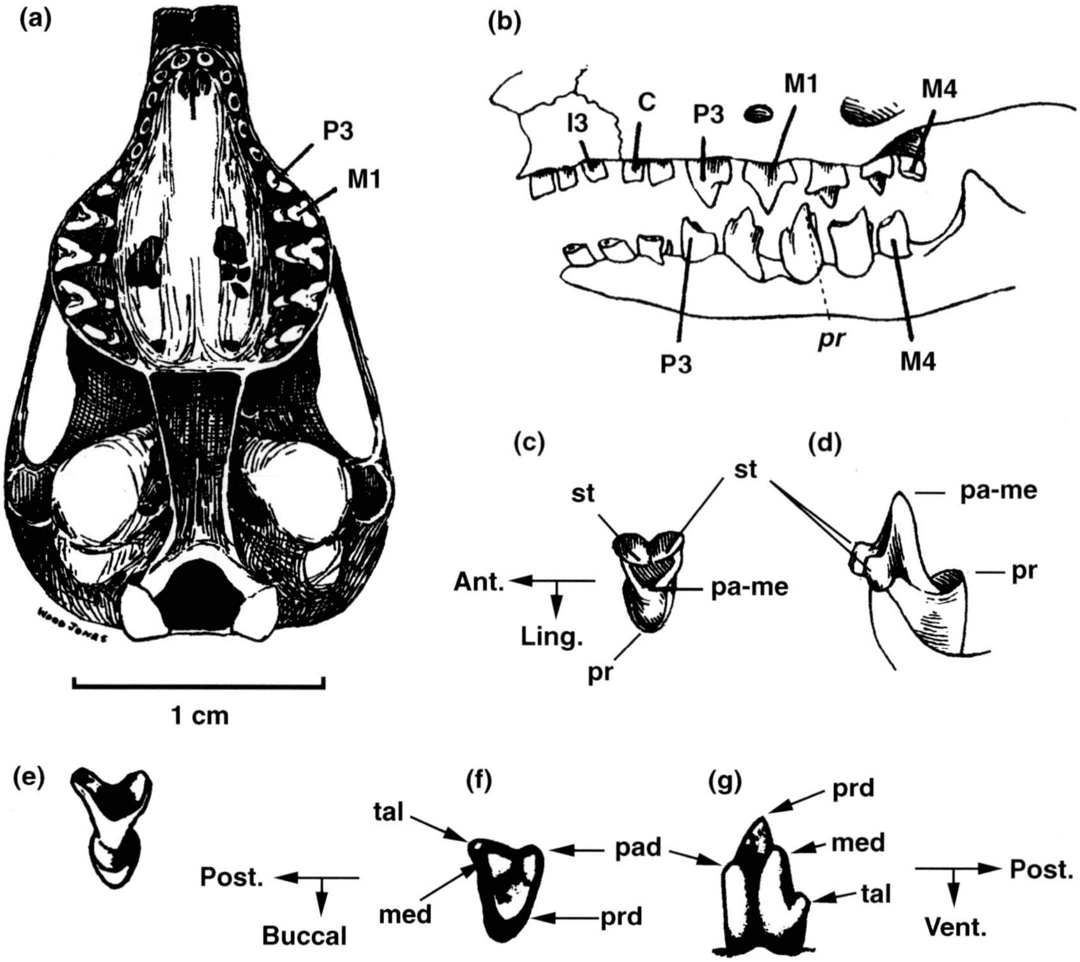

Figure 19. The skull and dentition of a marsupial mole. Although the exact number of teeth seems to vary between individuals, there may be up to I1–4/1–3 C1/1 P2–3/1–3 M1–4/1–4. From Osborne (1907) and Jones (1924): **(a)** ventral view; **(b)** lateral view of left upper and lower dentition (dentary); **(c)–(g)** views of individual teeth. Key to abbreviations: I3 = third incisor; C = canine; P3 = third premolar; M1 = first molar; M4 = fourth molar; st = stylar cusp; pa-me = parametacone; pr = protocone; ta = talonid; pad = paraconid; med = metaconid; prd = protoconid; Ant. = Anterior; Ling. = lingual; Post. = posterior; Vent. = ventral.

things in the soft soils or even moss floors of forests and just kept burrowing as the vegetation overhead slowly changed over the last 20 million years.

THINGODONTANS

Every now and then, an extinct mammal is found that reminds us of just how much we still have to learn about the significant features of Australia's extraordinary past. The order Yalkaparidontia Archer, Hand and Godthelp, 1988 was erected for an extraordinarily odd creature (Plate 100) found in the early to middle Miocene sediments of Riversleigh. It is so odd that it was almost like finding—*for the first time*—a whale, a bat or a human, all representatives of three other orders of mammals.

The interordinal affinities of these 'Thingodontans', as we refer to them unofficially, are a tantalizing mystery. Their cheek teeth are simple, almost moon-shaped crescents, a condition called zalambdodonty, while their lower incisor is huge and ever-growing (Figure 20). Zalambdodont teeth occur in a few other groups around the world, presumably evolved independently, such as the placental tenrecids (tenrecs), chrysochlorids (e.g., Cape golden mole), solenodontids (solenodons) and even, to a lesser extent, Australia's marsupial moles. The adult dental formula of yalkaparidontians is: I1–3/1 C1/0 P1–3/1 or 2–3 M1–3/1–3.

The combination of features in their dentition indicates that they could not shear food horizontally—only vertically. Combined with the huge incisor, we have speculated that they may have eaten eggs, caterpillars or worms (Figure 21). *Whatever* they were doing millions of years ago in Riversleigh's ecosystems, nothing today is doing the same thing. They must have occupied a niche that is either no longer available or no longer filled by any living mammal.

The family Yalkaparidontidae contains one genus, *Yalkaparidon*, which has two species: *Yalkaparidon coheni*

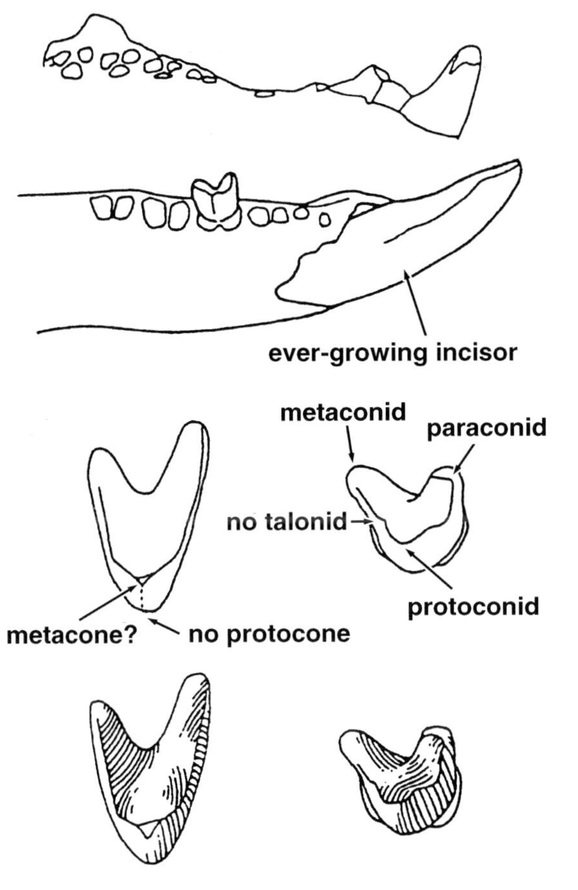

Figure 20. Thingodonta's teeth are quite unlike those of any other living mammal. Their simple, crescent shape provides few clues about the origins and relationships of this order of mammals to all others. Drawing J. Muirhead.

Archer, Hand and Godthelp, 1988, named after a long-time supporter of Riversleigh and Australian environments, Barry Cohen, and *Yalkaparidon jonesi* Archer, Hand and Godthelp, 1988, named after another long-time Riversleigh supporter and political supporter of science, Barry Jones. The name *Yalkaparidon* is actually based on the Waanyi Aboriginal word *yalkapari* which means 'boomerang', an allusion to the distinctive shape of the molar teeth.

OTHER EXTRAORDINARY MARSUPIALS

Thylacotinga bartholomaii Archer, Godthelp and Hand, 1993, from the early Eocene Tingamarra LF of south-eastern Queensland is known only from rather strange isolated teeth, so its relationships are difficult to work out. However, it shares some strikingly similar features with a similar-aged marsupial from Peru known as *Chulpasia*, a relationship that could be possible given the fact that South America, Antarctica and Australia were still connected to each other at that time. The broad, swollen molars suggest that it was probably omnivorous.

Djarthia murgonensis Godthelp, Wroe and Archer, 1999, also from the early Eocene Tingamarra LF, is probably the most primitive marsupial ever found in Australia. It is so primitive, in fact, that it cannot be referred to any of the living orders of Australian marsupials. For that matter, this tiny, insectivorous marsupial can't unambiguously be referred to any other marsupial group anywhere, although its simple dental structure resembles a number of groups known from the early Cainozoic record of South America. It is not impossible that it could be part of the stock that gave rise to some of

Figure 21. A reconstruction of the early Miocene thingodontan *Yalkaparidon coheni*. What thingodontans ate and how they lived remain mysteries (illustration by D. Dunphy).

the Australian groups, but much more information is needed before this possibility could be tested.

Yingabalanara richardsoni Archer, Every, Godthelp, Hand and Scally, 1990, from the early Miocene of Riversleigh, was appropriately nick-named 'Bizarrodonta'. It was placed in its own family, the Yingabalanaridae. Perhaps more than any other fossil mammal found in Australia, this one defies all attempts to pigeon-hole it into any known group. It is not even clear whether it is a marsupial or not (Figure 22). It's known from two isolated teeth (Archer *et al.*, 1999), each of which came from a different deposit at Riversleigh. When we first published this creature, we had to leave as possibilities that it *could* be a marsupial, perhaps similar to 'thingodontans', *or* a bat, *or* a primate, or something utterly different from *anything* anyone has ever seen before. So now we anxiously await more information about this intriguing beast!

DIPROTODONTIANS—BEASTS BEARING TWO SPECIAL FRONT TEETH

What follows is a look at an enormous pile of marsupials, all placed in the order Diprotodontia, that do not appear, at first glance, to have much in common. But on second glance, they do share what appear to be at least three distinctive synapomorphic features: diprotodonty, the fasciculus aberrans (a structure in the brain) and a superfi-

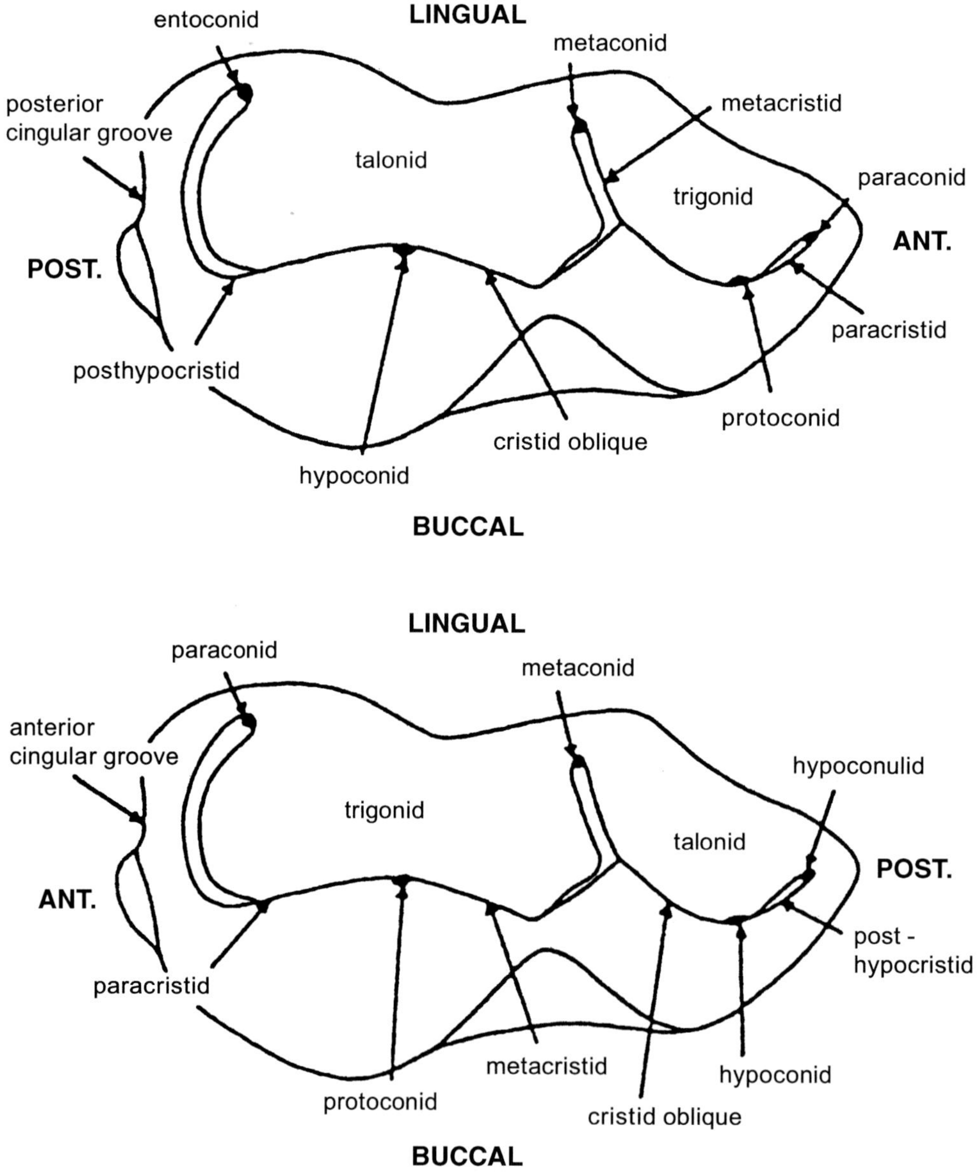

Figure 22. Two interpretations of the cusps of the lower molar of Riversleigh's early Miocene Bizarrodonta *Yingabalanara richardsoni.* Note the reversal of anterior and posterior in the two cases proposed. Drawing J. Muirhead.

cial thymus gland. In addition, if syndactyly in diprotodontians was not inherited from bandicoots (and there is controversy about this), then it would be a fourth conspicuous synapomorphy that unites all members of this order.

The condition of diprotodonty is such that the two central lower incisors are extremely large and procumbent (more-or-less horizontal), while most if not all of the other lower incisors are lost. There is also often a diastema (gap) between the enlarged incisors and the teeth behind it. Upper incisor numbers are reduced to three, two or one on each side. Diprotodonty probably evolved as a carnivorous adaptation, the procumbent teeth being useful as swords to stab prey. Even sweet things like sugar gliders have been seen to stab mice to death with these teeth. Kirsch (1979) demonstrated that living South American caenolestid marsupials also kill prey using their procumbent lower incisors. Alternatively, diprotodonty may have evolved in conjunction with the diastema as an aid in avoiding predation. Ardran *et al.* (1958) have suggested that the reason for diastemata in herbivores may be that they enable long stems of grass or branches to project through this space

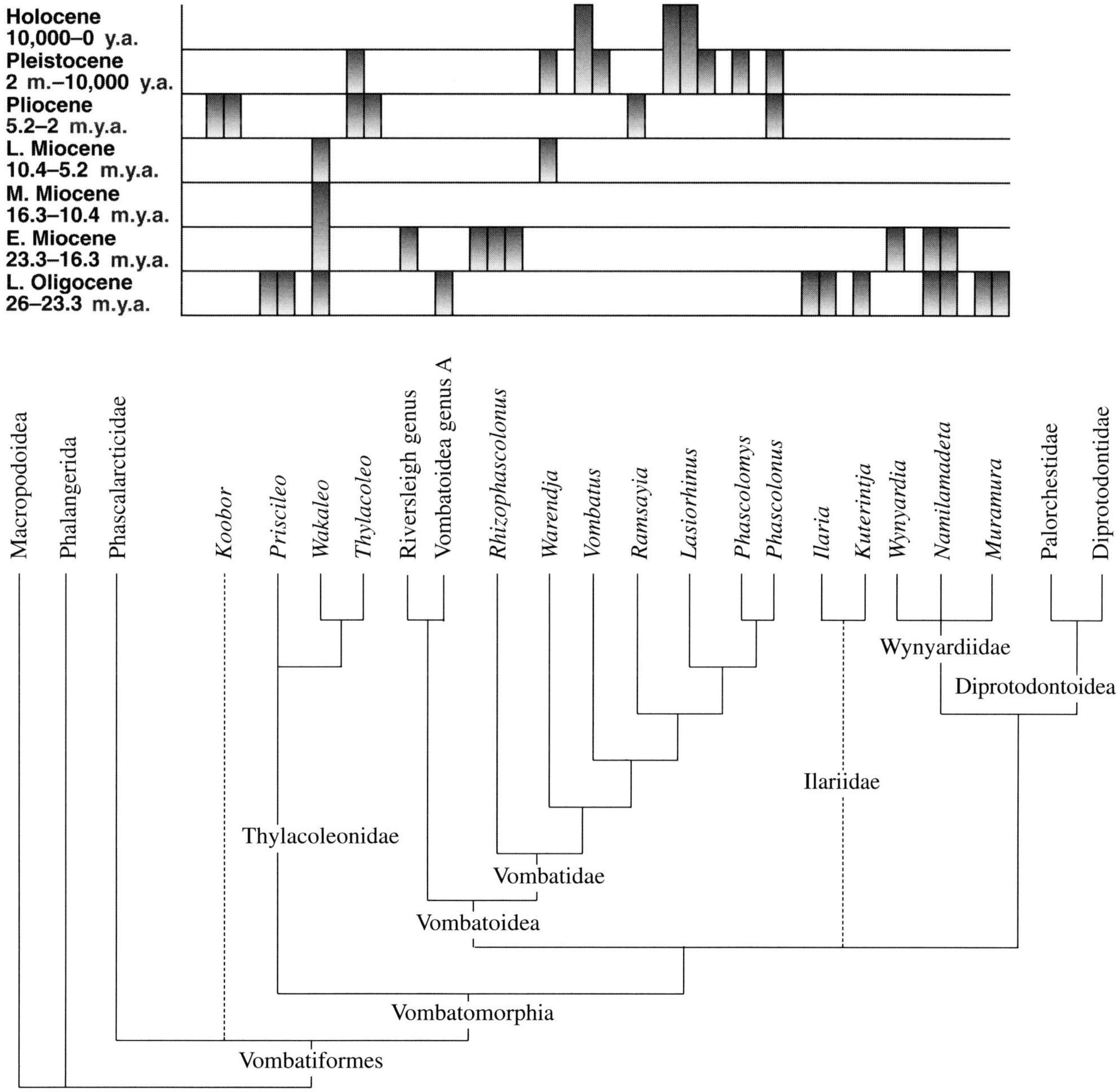

Figure 23. Diversity through time and a cladogram of probable interfamilial and intergeneric relationships of vombatiform marsupials. Uncertain relationships are indicated by dashed lines. Each block represents a distinct species, and blocks that extend without breaks through more than one time period indicate apparently long-surviving species. After Myers *et al.* (1999).

while being fed by the tongue into the cheek tooth battery on the opposite side. This ability would reduce the time the herbivore spent with its head down, unaware of approaching marsupial lions. In some bandicoots, the lower incisors, although three in number on each side, are virtually procumbent with the medial one enlarged. This could be evidence for an evolutionary relationship between bandicoots and diprotodontians.

The functional significance of the other two apparently synapomorphic features of diprotodontians, the fasciculus aberrans and the superfical thymus, are less easily interpreted. The fasciculus aberrans is a bundle of nerve fibres that connects the two hemispheres of the brain. There are many such transverse bundles in mammalian brains but this particular one appears to be confined to diprotodontians (Abbie, 1937). The significance of the superficial thymus is similarly obscure. But it does appear (Yadav, 1973) to be a synapomorphic feature.

So, equipped with two pokey lower incisors, a new brain hook-up and a fancy gland, enter the magnificent diprotodontians.

Figure 24. The modern koala (*Phascolarctos cinereus*) is a folivore, living on the leaves of some gum trees (*Eucalyptus* spp.), as well as other non-eucalypts. The forcipate hand, in which the two first digits oppose the other three, is evident here. Other characteristic features include a deep head, relatively short face, backward opening pouch and lack of a well developed tail. The last feature is curious in an arboreal animal and suggests that ancestral phascolarctids may not have been arboreal (drawing by R. Goldingay).

VOMBATIFORMES—WOMBAT-LIKE DIPROTODONTIANS

In this suborder are placed two living groups, the koalas and wombats, and many bizarre extinct families (Figure 23). The living vombatiforms share hook-shaped spermatozoa, many other features of the skeleton and musculature and masses of biomolecular similarities. In fact, if you squint, a koala walking on the ground can look more than a bit like a wombat.

Koalas

Until rather recently, the modern koala (Figures 24, 25) was interpreted as '...a phylogenetically senile and over specialised Phalanger, and most nearly related originally to the Ring-tailed Opossums' (Jones, 1924). Very little of this interpretation now seems defensible.

In the first place, although the only surviving species may be vulnerable, it is not at all clear that it is phylogenetically senile. For the whole of our well known fossil record back to the late Oligocene (Figure 26), there appear to have been only one or at most two phascolarctids at any given time—and over the last six million years their numbers have increased dramatically. Further, as discussed in more detail below, their closest living relatives appear to be wombats, not possums. In fact, of diprotodontian groups with living representatives, phascolarctids appear to be among the most plesiomorphic. This conclusion is reached for two reasons: the structure of the molars of koalas, while distinctive and called selenodont (Figure 27), is in fact readily derived from an ancestral insectivore pattern. Further, this pattern or some simpler version of it becomes more common in all diprotodontian lineages the further back in time we go. This suggests that it could be basic to the whole of the diprotodontian radiation. Unfortunately for this hypothesis, the modern koala is so very specialized for its arboreal leaf-eating life that it makes the mind verily boggle to think of such a weird beast at the base of the diprotodontian family tree. But that is of course the point—it isn't. It is a modern very specialized twig of the basal phascolarctid branch (Figure 26).

There are five genera of extinct phascolarctids (including one Riversleigh taxon yet to be named) and one living genus and species, *Phascolarctos cinereus*. Modern koalas are highly specialized tree-dwelling marsupials that prefer the leaves of some of the most toxic plants in Australia—gum trees. Perhaps related to this diet of toxins, they have a brain that apparently doesn't even fill up the endocranial space available to it! All koalas for which a full dentition is known (and many are only known from a few teeth) have a dental formula of: I1–3/1 C1 or 0/0 P3/3 M1–4/1–4.

The most archaic genus is an undescribed taxon from Riversleigh (Black, 1999), while the most derived genus appears to be *Litokoala*. In between are a plethora of taxa that span the last 24 million years, most being known from the early to middle Miocene.

Litokoala kutjamarpensis Stirton, Tedford and Woodburne, 1967, from the early Miocene Kutjamarpu LF of South Australia was first described on the basis of a single tooth. However, other species (e.g., Springer, 1987 and Black and Archer, 1997a) are now adding significantly to understanding about this group of small koalas.

Madakoala devisi Woodburne, Tedford, Archer and Pledge, 1987a, from the late Oligocene, Pinpa, Ericmas and Tarkarooloo LFs of South Australia, appears to be the primitive sister taxon to all other named koalas (but is more derived than the unnamed genus from Riversleigh; Black, 1999). It was about the same size as the living koala but would have had a conspicuously shorter face with an uninflated narial region. Other late Oligocene species from South Australia are known including *Madakoala wellsi* and *Madakoala* sp. (Woodburne *et al.*, 1987a).

Nimiokoala greystanesi Black and Archer, 1997a, from the early Miocene of Riversleigh represented a very distinctive group of koalas, with high-crowned teeth, and hyper-developed features (Figures 25, 27) that characterize normal koalas. *Nimiokoala* sp. Black and Archer (1997a) from the late Oligocene, South Prospect B LF of South Australia indicates that this group was widespread across the continent.

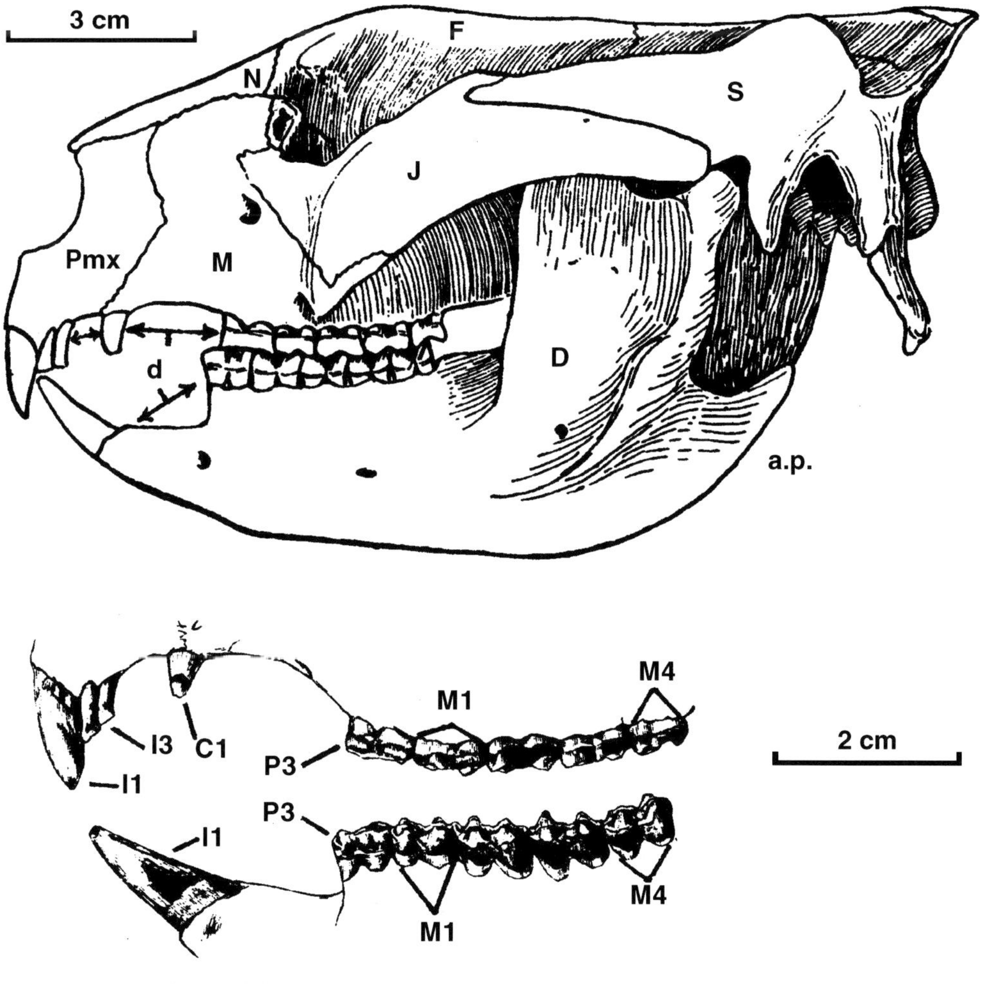

Figure 25. The skull and dentition of the koala (*Phascolarctos cinereus*) in left profile. The dental formula is I1–3/l Cl/0 P3/3 M1–4/1–4. The Ml is vestigial and resorbed without becoming functional. Diastemata (d) separate the I3 from C1, C1 from P3 and Il from P3. The very large angular process of the dentary reflects the large area for attachment of the muscles that move the dentary sideways. This action is more essential for a herbivore than a carnivore because of the greater need to grind up coarse plant material. The skull also exhibits a distinctive feature not found in polyprotodont marsupials: the squamosal bone forms a significant part of the bony chamber surrounding the middle ear. This feature also characterizes wombats, diprotodontoids and thylacoleonids and is part of the justification for recognition of the diprotodontan superfamily Vombatoidea. Koalas are, in contrast, unlike wombats in that they have reduced the degree of inflection of the angular process. Key to abbreviations: Pmx = premaxilla; N = nasal; M = maxilla; J = jugal; F = frontal; S = squamosal; D = dentary; d = diastema; a.p. = angular process of the dentary.

Perikoala palankarinnica Stirton, 1957b, from the late Oligocene Ditjimanka LF of South Australia, was the first Tertiary koala to be found in central Australia. Species of *Perikoala* appear to be the sister-group of a combined *Nimiokoala, Litokoala, Phascolarctos* clade (Black and Archer, 1997a; Black, 1999).

There are several extinct species of the living genus *Phascolarctos* including *Phascolarctos yorkensis* (Pledge, 1992a) from the late Miocene or early Pliocene Curramulka LF of South Australia. Pledge (1992a) originally placed this species in its own genus as *Cundokoala*, but Black (1999) thinks it should be placed in *Phascolarctos*. This koala is far and away the largest known, being probably twice the weight of the living koala. Gum trees must have been stronger in those days! *Phascolarctos maris* Pledge, 1987b, from the early Pliocene Sunlands LF of South Australia may be the same (Black, 1999) as *P. stirtoni* Bartholomai, 1968, from the Pleistocene, Cement Mills LF of Queensland. In fact, it may be questionable whether either is distinct from the living koala, *P. cinereus.*

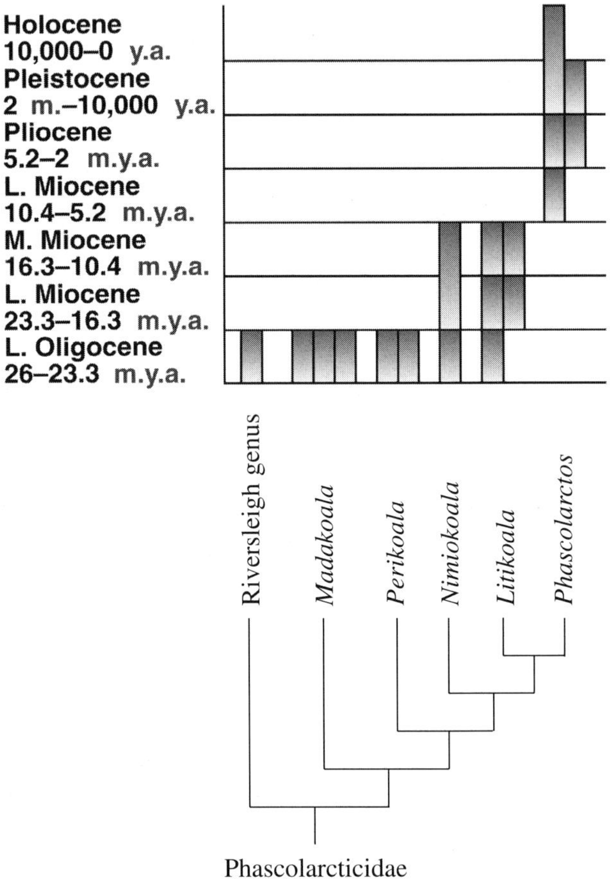

Figure 26. Koala diversity through time and a cladogram of probable relationships within the group. Uncertain relationships are indicated by dashed lines. Each block represents a distinct species, and blocks that extend without breaks through more than one time period indicate apparently long-surviving species (after Black, 1999).

The early Pliocene species of *Koobor* Archer, 1976e, from the Chinchilla and Bluff Downs LFs of Queensland, although originally regarded to be koalas (a placement tentatively ratified by Black and Archer 1997a), remain the subject of controversy. Pledge (1987a) has suggested that they might be ilariids, a conclusion supported by Marshall *et al.* (1990). Myers and Archer (1997) did not agree and Black (1999) has concluded that, until more complete material is known, affinities of this group will remain unclear.

The modern koala has been the subject of much research, a lot of it summarized and expanded on in Smith and Hume (1994) and Martin and Handasyde (2000).

Wombats

Wombats in the loose sense (i.e., vombatids and vombatid-like beasts) include nine genera of late Oligocene to Recent species. There are three living species: the common wombat (*Vombatus ursinus;* Figure 28), the southern hairy-nosed wombat (*Lasiorhinus latifrons*) and the northern hairy-nosed wombat (*L. krefftii*). And then there are an army of extinct taxa, some based on a lot of information, some based on little.

As noted above, most diprotodont marsupial groups seem to show at least distant relationships to phascolarc-

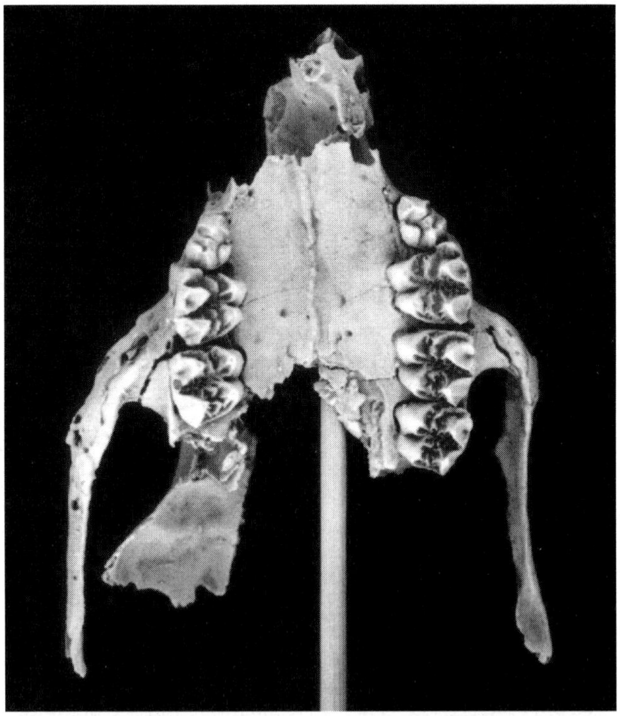

Figure 27. The snout of the early Miocene *Nimiokoala greystanesi* found at Neville's Garden Site, Riversleigh displays the distinctive selenodont pattern found in the molars of all koalas. The upper and lower molars have four linked crescent V-shaped or moon-shaped blades developed on each molar (called selenodonty), well suited for slicing up the leaves of trees. Photograph K. Black.

tids. Of living groups, this statement is most applicable to wombats. In terms of molecular data, sperm morphology, skeletal morphology and so on, wombats are closest to (share the most apparent synapomorphies with) koalas. In this sense, koalas could be seen as vombatoids that have secondarily become arboreal. But perhaps not. Could ancestral phascolarctids have been also arboreal? Probably not. If ancestral phascolarctids were arboreal, why would they have lost their tails and have backwards (downwards) opening pouches (one sneeze and you've lost junior)? As almost all other groups of arboreal animals clearly demonstrate, tails are decidedly useful things in trees. It seems to make more sense to suggest that ancestral phascolarctids were terrestrial, sharing the reduced-tail character state and the backward opening pouch, with at least wombats, and that wombats were never arboreal.

Modern wombats have teeth that grow throughout life. They have no roots, just crowns (Figures 29, 30). One Miocene fossil form (*Rhizophascolonus crowcrofti*), until recently known from a single tooth (Stirton *et al.*, 1967; much more is now known from Riversleigh), was obviously a rooted wombat as are many other genera and species that have now been found in the Oligocene to Miocene deposits of Riversleigh and central Australia. Rootless wombats evidently evolved this condition of constantly growing teeth as an adaptation to abrasive vegetation. *Rhizophascolonus crowcrofti* could be seen as representative of an early stage in wombat evolution, one in which the teeth still retained roots but the crown heights were much taller than in other Miocene diprotodontians.

Some late Tertiary and Pleistocene wombats were giants (e.g. *Phascolonus gigas*), over three times the size of the largest modern species (Figure 31). Did they, like all modern wombats, burrow? If so, those would have been mighty tunnels. Indeed, Woolnough and Steele (2001) have argued that, due to the physical and physiological limitations of burrows, giant wommies probably didn't tunnel.

Warendja wakefieldi from Pleistocene deposits in McEachern's Cave in western Victoria, despite its relatively young age and although plesiomorphic in many features (Hope and Wilkinson, 1982), has rootless teeth. At first it seemed peculiar that this plesiomorphic form should appear

Figure 28. The common wombat (*Vombatus ursinus*) is one of Australia's three surviving wombat species. All three have probably been around since at least the late Pleistocene. Photograph M. Archer.

weighed perhaps 200 kg (Murray, 1991a). Its wide, strap-like, curved and ever-growing two upper first incisors are among the most distinctive teeth of any of Australia's marsupials. *Ramsayia magna* (Owen, 1872) from the Pleistocene of many areas, and related species from the Pliocene of Queensland, were also huge (although smaller than *P. gigas*), probably twice as heavy as the living common wombat (*Vombatus ursinus*). *Rhizophascolonus crowcrofti* Stirton, Tedford and Woodburne, 1967, from the early Miocene Kutjamarpu LF of South Australia, and now also possibly Riversleigh, was the first-known pre-Pliocene wombat. Materials from Riversleigh appear to include most of the upper and lower molars which confirm its intermediate condition in terms of its molars evolving from a rooted to rootless state.

There are several fossil species of the modern genera *Lasiorhinus* and *Vombatus*. All are reviewed in Dawson (1983) and Long *et al.* (2002). The oldest may be represented by an early Pliocene tooth from the Hamilton LF, Victoria (Rich, 1982). Most of the later Pliocene and Pleistocene species of these two genera differ in relatively minor dental or craniomandibular ways from the living species. Triggs (1996) and Woodford (2001) provide excellent accounts of the natural history of living wombats.

One of the most incredible aspects of modern wombats is the way they shape their teeth through deliberate thegosis ('tooth grinding') before they have eaten any solid food. The unworn molar crowns of juvenile wombats show some complexity and look a little bit like degenerate selenodont teeth. No animal with selenodont teeth in Australia can eat grass—they are all in the trees munching on softer leaves (e.g., koalas and ringtail possums). With grasslands developing only very late in Australia's history, perhaps only within the last 3,000,000 years, wombats may not have had time to evolve the transversely lophodont teeth, like those of grazing kangaroos, that *are* suitable for eating grasses. But by thegosing their molars while still in the pouch, before they have touched solid food, they have ground away their 'ancestral' crown and carved transverse ridges in the molar surfaces that do enable them to deal with grass. This may be the only example of an animal anywhere in the world using a behavioural instinct to deliberately reshape the morphology of its teeth to suit an otherwise challenging diet.

Gigantic and Not So Gigantic Cousins

Although, of living groups, koalas appear to be the sister-group of the wombats, the wombats have even more immediate sister-groups that are totally extinct including the ilariids, wynyardiids and diprotodontoids.

DIPROTODONTOIDS—BIG, BIGGER, BIGGEST

One of the earliest, awe-inspiring discoveries in Australian palaeontology were the bones of *Diprotodon optatum* (Figures 32, 33). At nearly 3 tonnes in weight and much larger than a rhinoceros, this was the biggest marsupial that ever lived. But it is only one of many kinds that have been found since the early 1800s spanning the last 24 million years. The generic-level diversity of these giants is shown in Figure 34. Some (among the diprotodontids) may have had cheek pouches and resembled gigantic chipmunks (Murray, 1992). Others (among the palorchestids) were tapir-like with overslung schnozzles and giraffe-like ribbon tongues.

Two families of diprotodontoids are recognized: the Palorchestidae and the Diprotodontidae. Both have had chequered taxonomic histories and there must, even now, remain some doubt about the monophyly within and between these groups. For the moment, however, we accept that the whole lot is probably monophyletic but are much less confident about the monophyly of each family.

Both groups appear to have terminated in the late Pleistocene without leaving any phylogenetic successors. Much less can be said with confidence about the origins of the diprotodontoids as a whole. Although they may be close relatives of wombats, they may be more closely related to the extinct wynyardiids (see below).

Before we consider each family in more detail, there is another matter that may be relevant here. This is a curious ecological problem that may explain the origin of the famous Australian bush salute. Diprotodontoids must have laid rather colossal poops. And, poops of course are the stuff of dreams for flies (the maggots are laid in the stuff by adult flies) and dung beetles (including coprinids). In Africa, where poops come in all sizes and shapes, there are incredibly diverse sorts of dung beetles. Many can dismantle a good-sized poop before the flies know it hit the ground. In fact, rumour has it that some of the more enterprising wait right at the herbivore's poop escape hatch and ride the stuff down. You can't get there much quicker than that! Anyway, the result of this competition is that Africa apparently doesn't have much of a fly problem. It is most reasonable to presume that Australia had its own coprinids that dined on the finest output that the diprotodontoids could produce. But what happened? The giant poop-works shut down when the diprotodontoids went extinct and those specialized dung beetles would have been retired from the game. But not the flies. Being more adaptable, any poop would do, even the less commodious roo poo. Now, enter Europeans with cattle, camels and horses. Giant poops were up for grabs once again but now only the deliriously happy little bush flies were doing the grabbing. So perhaps we should spare a thought for the poor defunct giant diprotodontoids while we whisk away those delightful little flies from our faces. Things might have been different.

The diprotodontoid radiation as a whole reflects an evolutionary reaction to the progressively changing Australian environment. The late Oligocene to middle Miocene forms were sheep-sized. By Pliocene times, many were the size of small cows and by Pleistocene times, they ranged from bull-size to bigger than rhinoceroses. As the Australian continent began to dry out, larger herbivores were probably better able to cope with poorer

Figure 32. At 2.8 tonnes, *Diprotodon optatum* is the largest marsupial the world has seen. Reasons for its extinction in the late Pleistocene remain controversial (illustration by P. Murray).

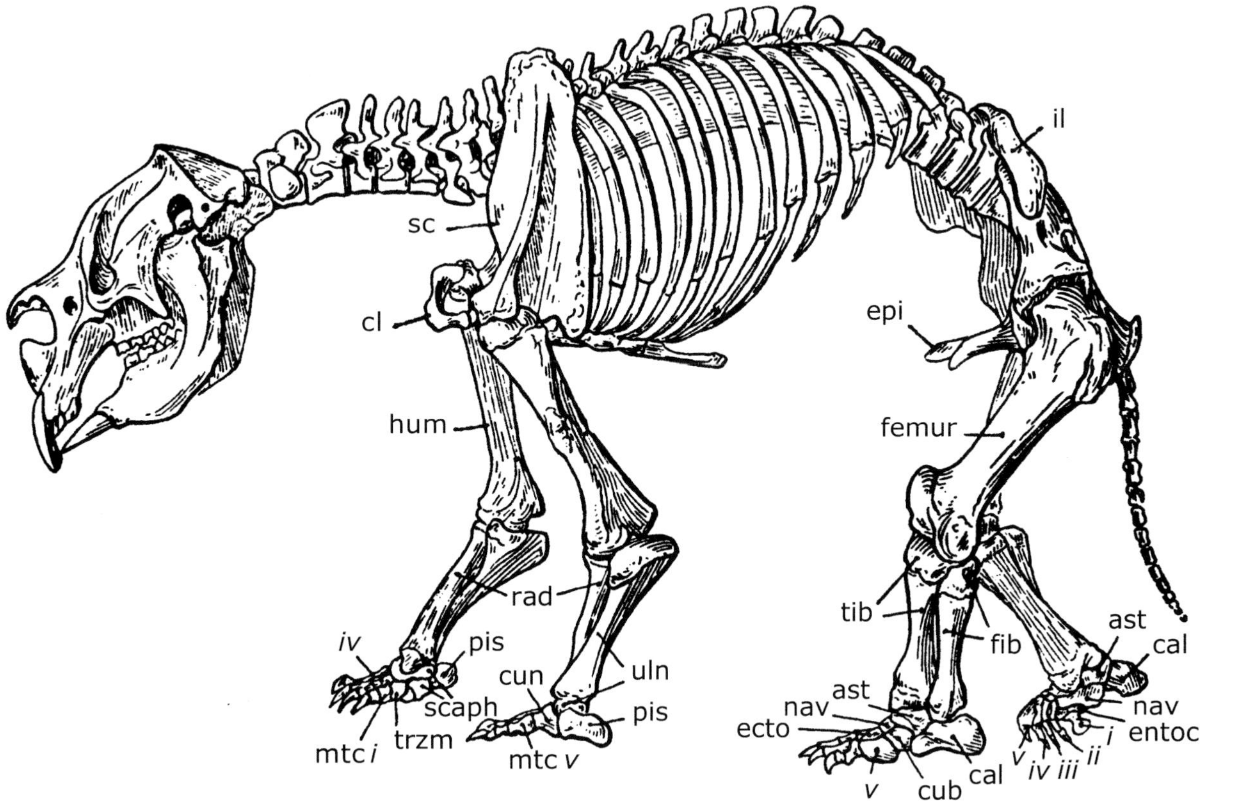

Figure 33. The skeleton of the Pleistocene diprotodontine *Diprotodon optatum*. The pillar-like legs, broad, flattened feet and steeply arched back reflect the great mass of this animal (from Gregory, 1951). Key to abbreviations: sc = scapula; cl = clavicle; hum = humerus; rad = radius; uln = ulna; pis = pisiform; scaph = scaphoid; trzm = trapezium; mtcI = metacarpal 1; cun = cuneiform; il = ilium bone of the pelvis; epi = epipubic; tib = tibia; fib = fibula; ast = astragalus; nav = navicular; ecto = ectocuneiform; cub = cuboid; cal = calcaneum; ento = entocuneiform; I–V, toes one to five.

quality food—they could eat more of it. Therefore, as Australia dried there may have been increasing selective pressure for an increase in body size. But there were hidden clauses in this evolutionary contract. In tiny letters at the bottom it said that once you became very large your absolute (per individual) water requirement increases. That may have been no problem until the late Pleistocene when the bottom dropped out of the Australian water bucket. A second problem resulting from large size was the increase in time necessary to successfully reproduce. The larger the animal, the more investment it must place in raising its young and the less capable it is of successfully reproducing during the shorter and unpredictable 'good seasons' of the Pleistocene. A third problem is that, all else being equal, larger body size means smaller population size. These trick clauses may ultimately have been the diprotodontoids' undoing. Perhaps, unlike some lineages of kangaroos, they could not adapt fast enough to the Pleistocene's changing demands (i.e. by reducing their size), and they found themselves out of the game. Or perhaps the arrival of fire-wielding humans in Australia had a significant impact (Wroe *et al.*, 2004).

Palorchestids

When the first palorchestid was described in 1873 it was thought to be some type of giant kangaroo. Richard Owen named *Palorchestes* on the basis of the anterior portion of a skull found in Victoria. The molar teeth of species of *Palorchestes* are rather like those of kangaroos in having longitudinal links between and in front of the transverse lophs. In fact, this basis for the conclusion that species of *Palorchestes* were kangaroos was not chal-

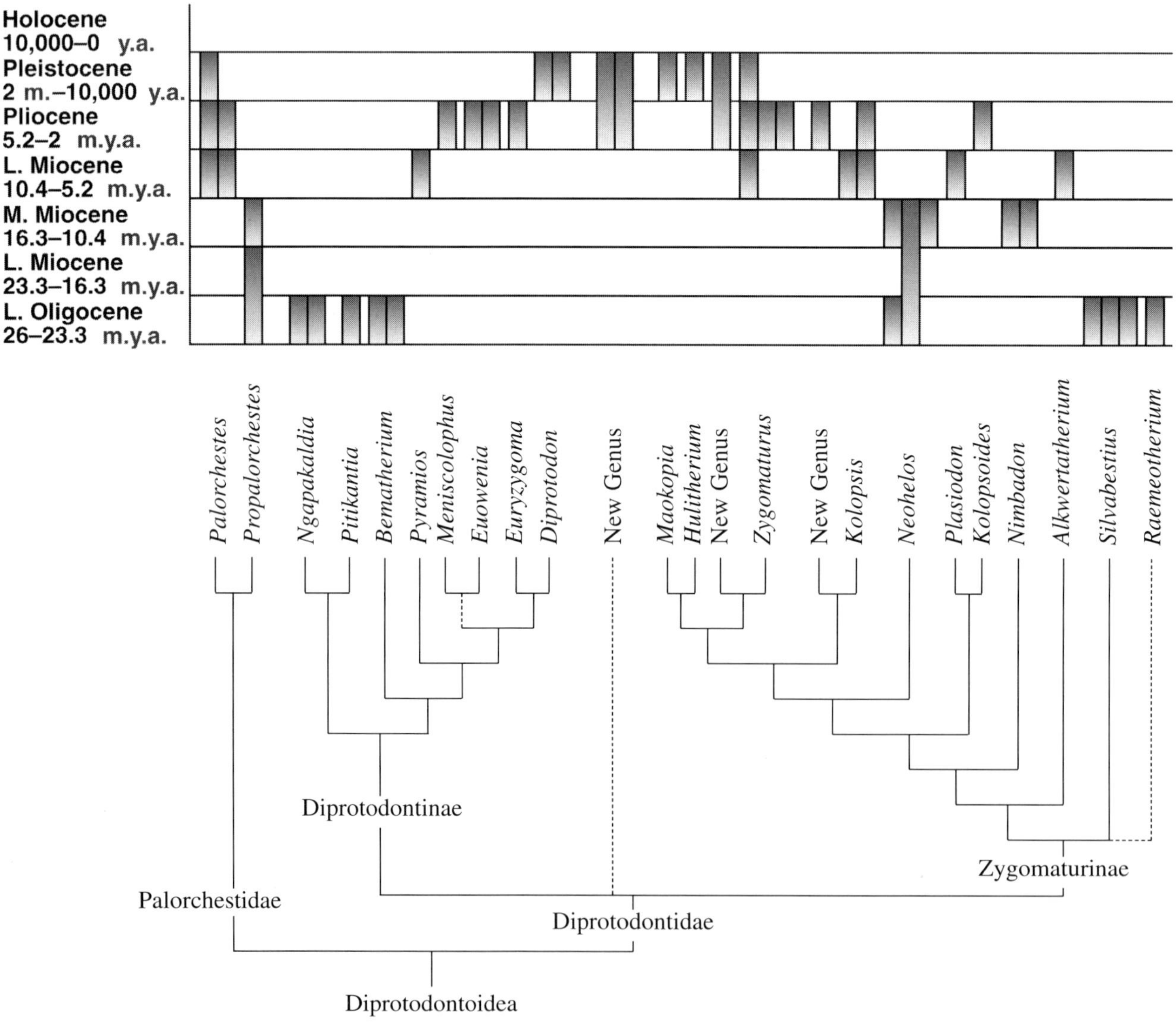

Figure 34. Diprotodontoid diversity through time and a cladogram of probable interfamilial and intergeneric relationships. Uncertain relationships are indicated by dashed lines. Each block represents a distinct species, and blocks that extend without breaks through more than one time period indicate apparently long-surviving species (after Black and Mackness, 1999).

lenged until 1958 when the genus was revised (Woods, 1958). Woods noted that the dentary lacked the large mandibular foramen and masseteric canal that were features of all other kangaroos. Consequently, he suggested that it might be more appropriate to place these odd marsupials within the Diprotodontidae, the only other family of lophodont marsupials known from Australia.

Archer and Bartholomai (1978) placed the species of *Palorchestes* and *Ngapakaldia* (see below) in the Palorchestidae as a family distinct from the Diprotodontidae and placed both in the superfamily Diprotodontoidea.

Anyway, whatever their phylogenetic relationships, discovery of an almost complete skull of the late Miocene *Palorchestes painei* from Alcoota began to reveal just how odd this group was. This skull showed three unusual features: retracted nasal bones, markedly enlarged infraorbital foramina and a very narrow, elongate rostrum. The first feature suggested that the animal may have had a tapir-like trunk. This view was put forward by Bartholomai (1978). The infraorbital foramen, which carries a major portion of the nerves and blood vessels that serve the rostrum, suggested that the animal might have had a rather larger, perhaps even prehensile trunk (Figure 35). The narrow rostrum and lower jaw suggest that this animal had a long, narrow, protrusible giraffe-like tongue that might have been used to reach up into trees to pull leaves down. Other even more bizarre features are less easy to explain, such as the enormous, laterally compressed koala-like claws on the fore and hind feet and the very powerful forelimbs. What did herbivores with equipment like this do for a living? Could they have been partly arboreal, hanging upside down like sloths? At this point, we are waiting for an ambitious anatomically-minded student to carry out a more detailed functional interpretation of the postcranial skeleton.

All palorchestids have the following adult tooth formula: I1–3/l P3/3 Ml–4/1–4. The deciduous Ml is also unusually large.

There are only two genera and a number of species of palorchestids: *Palorchestes azael* Owen, 1873, from Pleistocene deposits over much of the continent; *Palorchestes painei* Woodburne, 1967, from the late Miocene Alcoota LF of the Northern Territory; *Palorchestes parvus* De Vis, 1895, from the early Pliocene Chinchilla LF and elsewhere in Australia; *Palorchestes selestiae* Mackness, 1995, from the early Pliocene Bluff Downs LF of Queensland; *Palorchestes annulus* Black, 1997b, from the late Miocene of Riversleigh, Queensland; *Propalorchestes novaculacephalus* Murray, 1986, from the middle Miocene Bullock Creek LF, Northern Territory; and *Propalorchestes ponticulus* Murray, 1990a, from the late Oligocene to middle Miocene of Riversleigh, Queensland.

Diprotodontids

We are not sure what a diprotodontid is: that is to say, we do not know which, if any, synapomorphic features tie this group together. They are clearly more similar to each other when compared with the palorchestids, but of course simple similarity is not a good guide to phylogenetic relationship. That said, let's assume monophyly and consider the component groups of this family. There are two subfamilies: the Zygomaturinae and the Diprotodontinae. Primitively, all diprotodontids have the tooth formula quoted above.

The zygomaturines appear to be monophyletic but again the definitive studies in this superfamily (currently being investigated by Karen Black) have yet to be completed. Certainly most share a distinctive complex P3 that has between three and five cusps, a distinct parastyle and commonly a hypocone. A paracone and metacone are indistinguishable in the older members and hence the buccal cusp in this region is called the parametacone. In later zygomaturines, a paracone and metacone are distinguishable and the hypocone is much larger.

The middle Miocene, small cow-sized species of *Neohelos* were common throughout much of central and northern Australia. In some sites where they occur, they are the most common of the larger mammals. It is probable that they lived in herds (or 'mobs' to borrow a term applied to kangaroos). They were almost certainly browsers. Judging by the variation shown in large samples such as those from the middle Miocene Bullock Creek LF of the Northern Territory, males were probably larger than females.

The late Miocene is also the time of first appearance of zygomaturines referable to the genus *Zygomaturus*. These were the giants of the subfamily, one species of which (*Z. trilobus*) survived into the late Pleistocene. It was common in many of the more coastal regions of the continent. It had a skull with peculiar prominences on the nasal and frontal bones. These suggest the possibility that there were protuberances on the face.

In the late Pleistocene deposits of Mammoth Cave in south-western Western Australia, *Z. trilobus* was found associated with bones interpreted to have been butchered or at least interfered with by Aboriginal people (Archer *et al.,* 1980). Possibly relevant to this suggested culinary relationship with humans is an artifact given to Kim Akerman by Kimberley Aborigines (Akerman, 1973). It is a well handled premolar of *Z. trilobus* mounted in spinifex gum. Did the tooth come from a Kimberley fossil deposit as yet unknown to Europeans? Or was it traded, as an obviously interesting object, by Aboriginals from its original source in a fossil deposit somewhere else in Australia? Or, perhaps more implausibly, has it survived as an ancient relict of a banquet held 30,000 years ago, passed down from one generation to the next?

Other notable zygomaturines include: *Alkwertatherium webborum* Murray, 1990b, from the late Miocene Alcoota LF, Northern Territory; *Hulitherium tomasettii* Flannery and Plane, 1986, from late Pleistocene Pureni LF of New Guinea; *Kolopsis torus* Woodburne, 1967b, from the late Miocene Alcoota LF, Northern Territory; *Kolopsoides cultridens* Plane, 1967, from the early Pliocene Awe LF of New Guinea; *Maokopia ronaldi* Flannery, 1992, from late Pleistocene cave deposits in New Guinea; *Neohelos tirarensis* Stirton, 1967a, from the early Miocene Kutjamarpu LF of South Australia; *Nimbadon lavarackorum* Hand *et al.*, 1993, from the middle Miocene of Riversleigh, Queensland (Plate 101); *Plaisiodon centralis* Woodburne, 1967b, from the late Miocene Alcoota LF, Northern territory; *Raemeotherium yatkolai* Rich, Archer and Tedford, 1978, from the late Oligocene Ericmas LF of South Australia; *Silvabestius johnnilandi* Black and Archer, 1997b, from the late Oligocene of Riversleigh, Queensland; and *Zygomaturus trilobus* Macleay, 1858, from the Pleistocene of all states of Australia.

The other major group of diprotodontids is the Diprotodontinae. Here there is little doubt about monophyly. They range in type from the structurally simple, late Oligocene *Ngapakaldia* (Figure 36) to the highly specialized Pleistocene giant, *Diprotodon optatum* (Figures 32, 33).

Of these, *Euryzygoma dunense* is one of the best known. A well preserved skull of *E. dunense* reveals, in section, another extraordinary feature of most (if not all) diprotodontids: a cranium full of bone bubbles covering an amazingly tiny brain. When the skull of this bullock-sized animal was being prepared, it was found that a tree root had fractured and separated the rear half from the front half. The inside of the cranium, which should have been the braincase, was found to be almost totally taken up by thin-walled bony sinuses. The brain was restricted to a small recess in the base of the cranium barely as large as a human fist. The skull, however, had to be large enough to provide areas of attachment for the massive chewing muscles.

The most famous nototheriine is *Diprotodon optatum* (Figures 32 and 33). It was almost 3 m long and 2 m high at the shoulder and weighed between two to three tonnes (Wroe *et al.*, 2003). It roamed much of central Australia

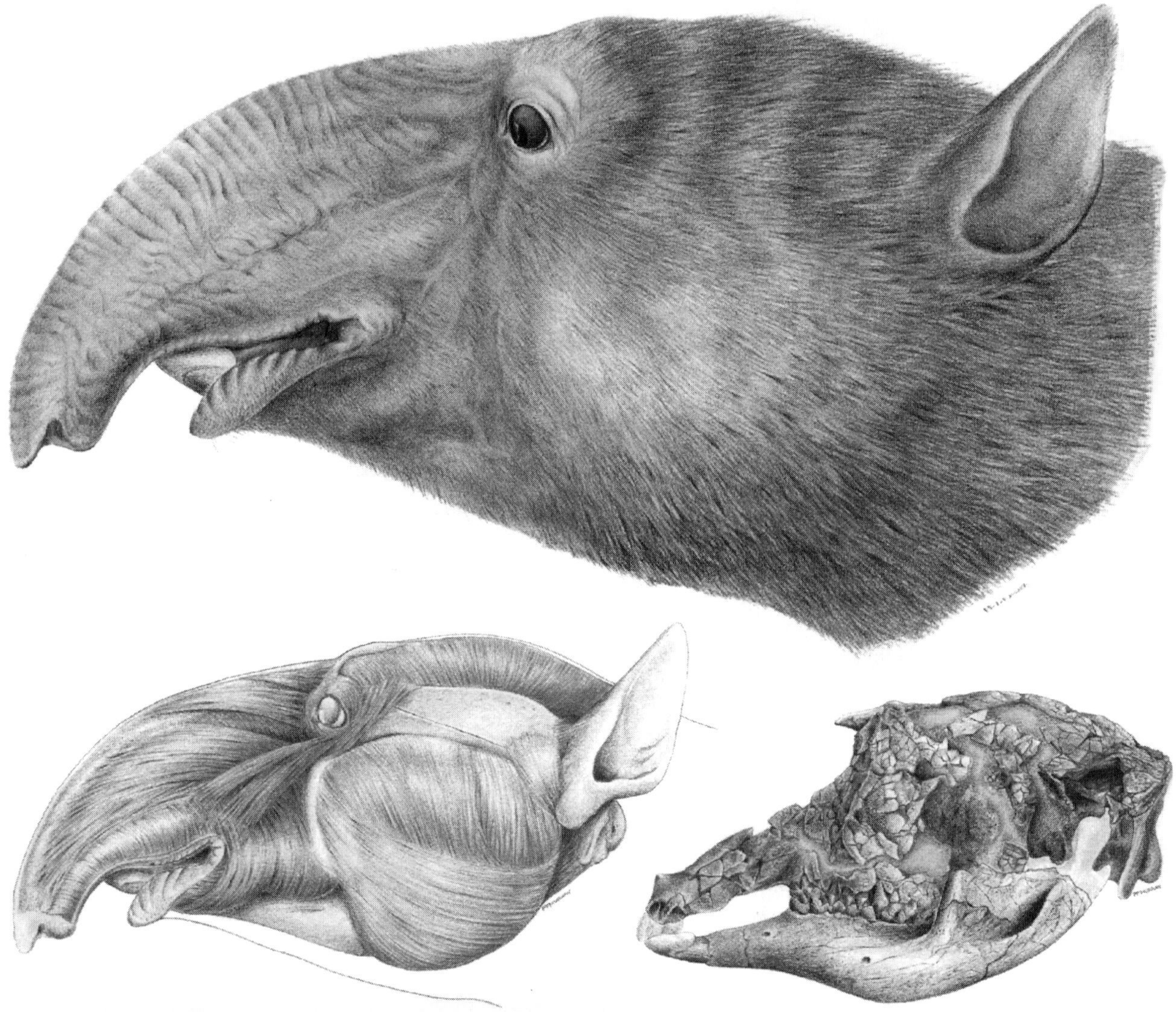

Figure 35. Palorchestids, like the Miocene *Palorchestes painei*, appear to have sported tapir-like trunks which they may have used to manipulate their plant food (illustrations by P. Murray).

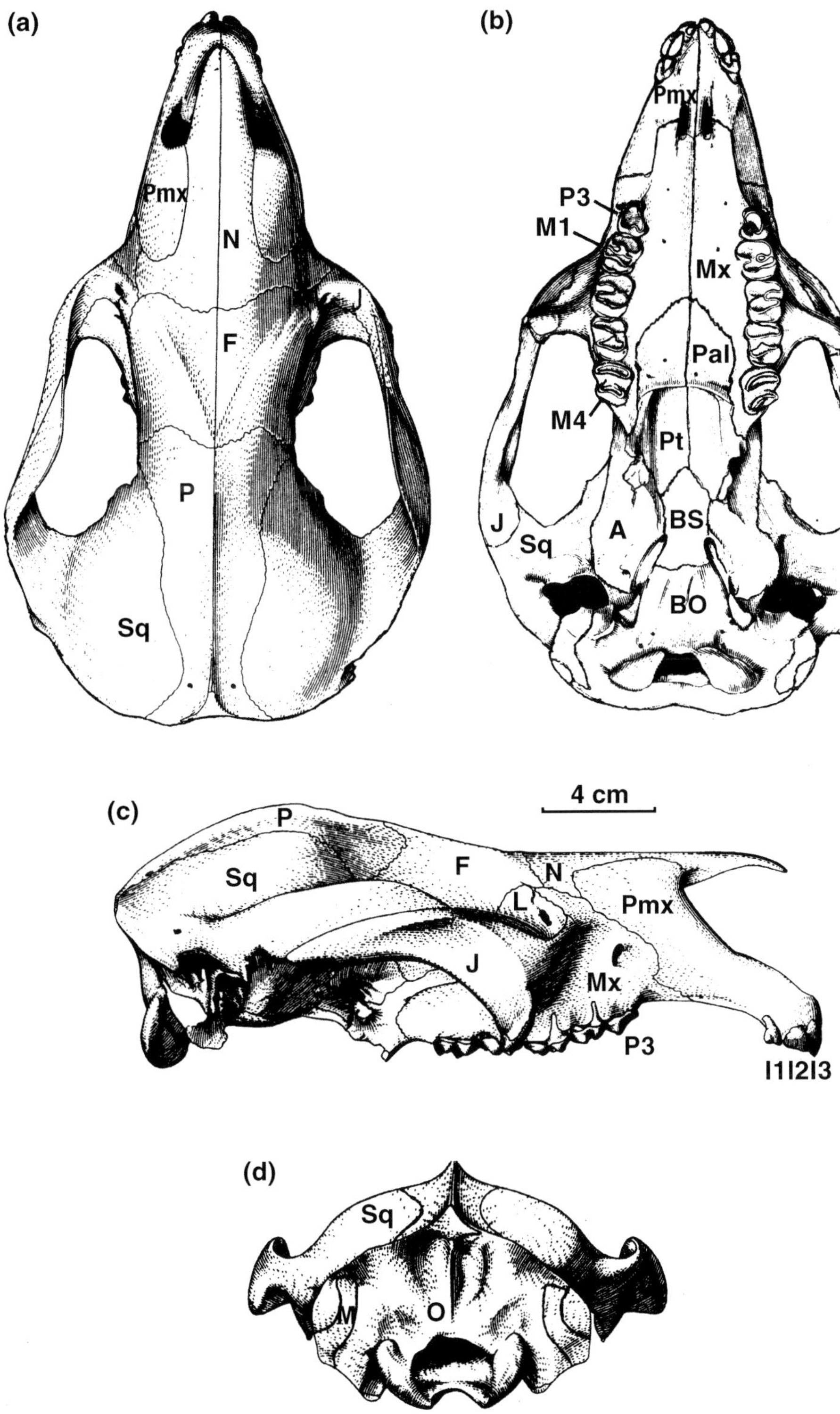

Figure 36. The skull of the of the sheep-sized, browsing diprotodontine *Ngapakaldi tedfordi* from late Oligocene deposits of the Tirari Desert, South Australia (modified from Stirton, 1967a). **(a)** dorsal view; **(b)** ventral view; **(c)** right lateral view; **(d)** posterior view. It lacks almost all of the apomorphic specializations shown in the Miocene to Pleistocene species of *Palorchestes* and is, in this sense, much more plesiomorphic. Conspicuous features include the large epitympanic recesses (the black spaces immediately behind the squamosals) and the tympanic wing of the squamosal (but not alisphenoid) bone. Key to abbreviations: Pmx = premaxilla; N = nasal; F =frontal; P = parietal; Pal = palatine; Pt = pterygoid; Mx = maxilla; BS = basisphenoid; A = alisphenoid; BO = basioccipital; J = jugal; L = lachrymal; Sq = squamosal; M = mastoid component of the periotic bone; 0 = occipital.

during Pleistocene times, seemingly avoiding only some of the continentally peripheral areas occupied by *Zygomaturus trilobus*. It has been suggested (Rich, 1983) that *Diprotodon optatum* was adapted more to the open woodlands and grasslands, while species of *Zygomaturus* were adapted to forested areas. This would explain the occurrence of species of *Zygomaturus* in central Australia in the Pliocene and the appearance and spread of *Diprotodon optatum* only in the Pleistocene, a spread coincident with the establishment of arid conditions throughout central Australia.

With the death of the last species of *Diprotodon*, sometime in the late Pleistocene, a magnificent era in marsupial evolution passed into history. Not only were the real giants gone, but a whole family disappeared. And although Europeans were not there to observe this twilight and passing, Aborigines were. In fact, it is becoming apparent that Aborigines ate these animals when the opportunity afforded itself. But *were* early Aborigines significant factors in the extinction? No one yet knows.

Large accumulations of *Diprotodon optatum* skeletons in the late Pleistocene clays of Lake Callabonna in South Australia may represent one of the 'last stands' of this species. The seeming need on their part to cross the muddy and evidently treacherous surface of a drying lake may be an indication of a non-human critical factor in their extinction—reduction in the availability of fresh water. They may have needed to reach the shrinking body of water in the lake. This need for water, combined with the probable overgrazing of surrounding vegetation, the difficulty of effectively replacing their numbers when faced with unpredictable and short 'good' seasons, and possibly the presence of humans, all finally put the lid on the diprotodontoid radiation.

Representative diprotodontines include the following: *Bematherium angulum* Tedford, 1967, from the late Oligocene of Riversleigh, Queensland; *Diprotodon optatum* Owen, 1838, from the Pleistocene of most areas of Australia; *Euowenia grata* De Vis, 1887, from the early Pliocene Chinchilla LF, Queensland; *Euryzygoma dunense* De Vis, 1888, from the early Pliocene Chinchilla LF, Queensland; *Meniscolophus mawsoni* Stirton, 1955, from the early Pliocene Palankarinna LF, South Australia; *Ngapakaldia tedfordi* Stirton, 1967b, from the late Oligocene Ngapakaldi LF, South Australia; *Nototherium watutense* Anderson, 1937, from the Pliocene Awe LF, New Guinea; *Pitikantia dailyi* Stirton, 1967b, from the late Oligocene Ngapakaldi LF, South Australia; and *Pyramios alcootense* Woodburne, 1967b, from the late Miocene Alcoota LF, Northern Territory.

In passing, as noted above, diprotodontoids left no descendants; just fossils and a fertile field for human contemplation about the rise and fall of mighty beasts.

Wynyardiids

Since its discovery at the foot of a cliff at Wynyard, Tasmania, the first known Tertiary mammal from Australia, *Wynyardia bassiana*, has had more rhetoric written about it than almost any other fossil marsupial. Why? Because it slid down that cliff on its palate and neatly ground off all of its information-filled teeth. Without teeth, most mammalian palaeontologists feel hamstrung. So battles about interpretation have been fought based on the remaining bits of its skeleton. It was first described by Baldwin Spencer in 1901. He concluded that it represented a group structurally intermediate between polyprotodonts (e.g. dasyurids) and diprotodontians (e.g. phalangerids). Jones (1930) came down on the side of diprotodontian affinity (and specifically phalangerids). Ride (1964) also came down on the side of diprotodontian affinity but concluded that its middle ear structure was more plesiomorphic than that of any other known diprotodontian. Haight and Murray (1981) concluded, on the basis of an endocast of the cranial cavity, that it was clearly a phalangeroid and one most closely resembling the living phalangerid *Trichosurus vulpecula*. Aplin (1987) on the basis of basicranial information, placed it firmly in the Vombatiformes, but suggested wynyardiids may be the primitive sister-group to the other members of the group. Its relationships remain controversial. Such is the uncertain fate of a beast with no teeth.

Now wynyardiids have turned up in the late Oligocene deposits of central Australia and in the late Oligocene to early Miocene deposits of Riversleigh—all with teeth and skulls (Rich and Archer, 1979; Pledge, 1987c; Myers and Archer, 1997). And these teeth have caused a bit of a stir among the masses of devotees of marsupial tooth evolution. The adult tooth formula of these middle Miocene wynyardiids is: I1–3/1 C1/? P3/3 M1–4/1–4.

Namilamadeta snideri, a late Oligocene wynyardiid from the Tarkarooloo LF of South Australia (Rich and Archer, 1979), was the first of the non-Tasmanian materials to be described. To those of us afflicted with the compulsion to categorize unknowns, this beast presented a headful of intrigue. Notable features included the enlarged I1, persistent canine, elongate and sectorial P3, the cusp pattern of P3 and the peculiar molar morphology.

The P3 looks a bit like the sectorial premolars of tree-kangaroos (species of *Dendrolagus*). There is perhaps also a resemblance to some Tertiary phascolarctids. But the resemblance to the P3 of certain Miocene diprotodontids impresses us the most. There are what appear to be analogues of the parastyle, parametacone and posterolingual protocone of zygomaturine diprotodontids. It would not be difficult to derive a *Neohelos*-like molar from the same tooth in archaic wynyardiids. Despite their distinct cusps, with wear, wynyardiid teeth appear and probably functioned as transversely lophodont teeth. The unworn teeth indicate, however, that simple lophodonty is not what is going on here. The paracones and metacones were positioned midway along the transverse lophs. The buccal ends of the lophs were in fact stylar cusps C and D. It has become the custom to regard the ends of the lophs in lophodont mammals to represent the positions of the principle cusps (protocone, paracone, metacone and hypocone in uppers; and protoconid, metaconid (or paraconid), entoconid and hypoconid in lowers). But if a selenodont pattern was in fact ancestral to the lophodont pattern in diprotodontians, the principal buccal cusps on the upper molars would have been stylar cusps. By alteration of some of the minor crests linking these cusps it would be easy to produce a lophodont molar of the sort exhibited by diprotodontids. The wynyardiid molar pattern is therefore tantalizingly intermediate between the selenodont and the lophodont pattern, both functionally and structurally. This observation, combined with the basically similar incisor and premolar morphology, is at least good reason for considering that Wynyardiidae may be the plesiomorphic sister-group of Diprotodontoidea, a conclusion accepted by Myers *et al.*(1999). Anyway, it is possibly appropriate that the name *Namilamadeta* comes from Aboriginal words meaning "changeling tooth" (Rich and Archer, 1979).

All wynyardiids, as such, snuffed it sometime before the middle Miocene, the last ones being known from the early Miocene local faunas of Riversleigh (Figure 23).

Ilariids

The first ilariid to turn up was a lower jaw found at Lake Palankarinna in the Lake Eyre Basin by Mike Plane. This lake was one of the first explored by Reuben Stirton in the 1950s, in a southern part of the Simpson Desert known as the Tirari Desert. At first glance, it looked like a monstrous koala jaw, with huge, selenodont teeth, but with lots of added extra cusps and crests that made it clear to Mike that this creature belonged to yet another distinct and previously unrecognized group of diprotodontian marsupials. Hot on the heels of that discovery, Dick Tedford and his colleagues working in the dry lake systems of the Frome Basin on the other side of the Flinders Ranges found crushed skeletons with teeth identical to those found by Mike in the Lake Eyre Basin. Mike Woodburne teemed up with Dick to eventually publish the new group as the family Ilariidae. They recognized that it was somehow related to vombatomorphian marsupials, but weren't clear about which group in particular might have been most similar.

Shortly thereafter, Neville Pledge found a single tooth of a much smaller ilariid in the late Oligocene Mammalon Hill LF. Pledge further argued that this thing might have some relationship to the species of *Koobor* (see koalas above), hence that the latter might be an ilariid rather than a koala. Mike Archer and his colleagues then found whole dentitions and jaws that appear to be the same ilariid described by Neville, in the late Oligocene deposits of Riversleigh, and used this overlap (Myers and Archer, 1997) to help correlate the fossil deposits between central and northern Australia.

Now that we have whole dentitions, it is possible to determine the adult dental formula to be: I1–3/1 C1/0 P3/3 M1–4/1–4 (Figure 37).

Known ilariids include: *Ilaria illumidens* Tedford and Woodburne, 1987, from the late Oligocene Pinpa LF, South Australia; *Ilaria lawsoni* Tedford and Woodburne, 1987, from the late Oligocene Ditjimanka LF, South Australia (the one found by Mike Plane); *Kuterintja ngama* Pledge, 1987a, from the late Oligocene Ngama LF, South Australia and the White Hunter LF, Riversleigh, Queensland (Myers and Archer, 1997).

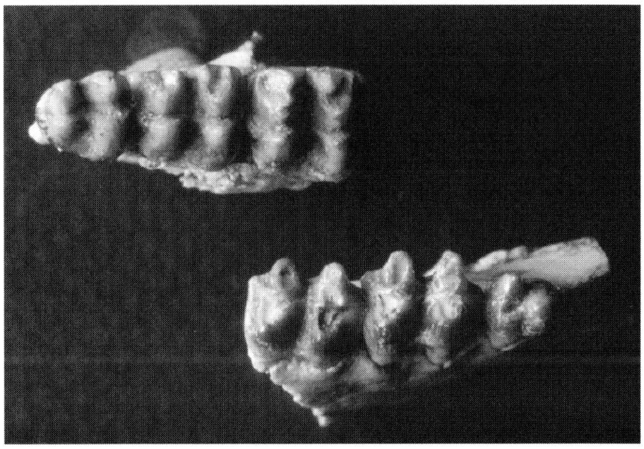

Figure 37. The cheekteeth (upper above; lower below) of the ilariid *Kuterintja ngama*, these specimens are from the late Oligocene of Riversleigh. The same species is known from late Oligocene deposits at Lake Palankarinna, South Australia. Photograph T. Myers.

Thylacoleonids—Marsupial Lions

These include the beasts that, if survivors of the late Pleistocene, might have made us all sleep much less easily in the Australian bush. *Thylacoleo carnifex*, an animal as large as sabre-toothed lions and perhaps smarter, has to be one of Australia's most fascinating extinct creatures.

There are three genera of thylacoleonids collectively spanning the late Oligocene to late Pleistocene placed in two subfamilies: the Thylacoleoninae (*Thylacoleo*) and Wakaleoninae (*Prisciléo* and *Wakaleo*). See Figure 38. The oldest known come from central Australia and Riversleigh; the youngest survived long enough to have overlapped with and terrorized Aborigines.

The phylogenetic relationships of the family are somewhat obscure, although this may soon be sorted by Anna Gillespie who's currently researching complete skulls of the tiniest and most primitive found at Riversleigh. Pending further clarity, it was once commonly presumed that they were bizarrely specialized descendants of some group of possums (e.g. burramyids or phalangerids). Recent independent studies, however, are reaching different conclusions. Suffice it to say here that they appear to be more closely related to the vombatiforms than to any of the possums or kangaroos (e.g. Aplin and Archer, 1983; Archer, 1987; Archer *et al.*, 1999).

As for descendants, fossils and growing intrigue appear to be all that's left. So let's consider these extraordinarily toothsome beasts in a bit more detail. The oldest thylacoleonids are represented by a maxillary fragment from the late Oligocene Ngapakaldi LF called *Prisciléo pitikantensis* and a skull and other materials

Figure 38. Reconstruction of the leopard-sized, middle Miocene marsupial lion *Wakaleo vanderleueri* from Bullock Creek in the Northern Territory (illustration by P. Murray).

from Riversleigh named *Priscileo roskellyae*. The next oldest thylacoleonids are a variety of species of *Wakaleo* from central and northern Australia, including a skull from the Bullock Creek LF, Northern Territory, spanning the late Oligocene to late Miocene (Figure 39). The species of *Thylacoleo* don't appear before the early Pliocene and extend to the late Pleistocene.

In all of them there are certain distinctive features that earmark the whole group. The most striking feature is the evolution of the third premolar into an increasingly elongate, shearing blade, while at the same time the molar row begins to degenerate from the rear forwards (Figure 39). This trend sees the oldest having three premolars (the last of which is becoming rather long) and four molars (the last of which is reduced in size), while the late Pleistocene forms have a third premolar that is simply enormous and just one vestigial molar left behind the premolar. These are the dentitions of renderers of flesh.

It has been noted (Van Valen, 1969) that carnivorous specializations in the dentition of *Thylacoleo carnifex* represent an extreme, unmatched in any other lineage of carnivorous mammals. This extreme is the ratio of length of the carnassial tooth (the specialized meat-cutting tooth; in this case P3) to the length of the whole tooth row. In other words, it is a coefficient of carnassiality, a measure of the extent to which a single tooth in the dentition carries out the function of meat-cutting. As thylacoleonids elaborated the length and function of P3, they reduced the rest of the cheek tooth row on either side of P3. The posterior molars were lost and the premolars anterior to P3 were reduced.

The huge P3 could have functioned more efficiently than bolt-cutters because, as a result of the opposing curvatures of the blades in the upper and lower P3, the object that was being cut (such as a bone) was forced into and held in a suitable position for the slicing stroke (Wells *et al.*, 1982; Wroe *et al.*, 2003). From an edible kangaroo's point of view, this was indeed a set of teeth to be respected.

Thylacoleo carnifex was quite variable in size. Specimens from Balladonia (Western Australia), for example, are very small, while some from eastern Australia are very large. Further, there may have been a marked difference in the size of males and females. This conclusion is based on the size range of specimens from single sites such as the Wellington Caves. The skeletons in a cave on the Western Australian Nullarbor recently discovered by John Long and his team from the Western Australian Museum should soon shed much light on the size range of marsupial lions.

What can we say about the thylacoleonid way of life? It has been accused of being everything from a nutcracker, to a melon-eater to an eater of bones. Recent studies have come down heavily on the side of meat-eating, with the odd bit of bone-chewing on the side (e.g. Wells *et al.*, 1982). There have also been a number of studies of chewed bones that have concluded that *T. carnifex* was the probable culprit (De Vis, 1884; Horton and Wright, 1981; Runnegar, 1983), although traces left on the bones may be incidental products of meat removal rather than osteophagy (Wroe *et al.*, 2003).

How could it have performed this grizzly business? Most likely it used its large-clawed thumb (see Wells and Nichol, 1977) and hand to hold its intended prey and then its lower incisors as lances to pierce and tear the throat. These caniniform teeth would have been most effective knives. Anyone who has ill-advisedly thrust his hand amongst a pile of sweet cuddly squirrel gliders (*Petaurus norfolcensis*) can attest to the effectiveness of diprotodontian lower incisors as highly effective stabbing weapons.

Another debate has centred around the size of the late Pleistocene *Thylacoleo carnifex*. Tim Flannery and many others have assumed they were about leopard-size and hence not awe-inspiringly fearsome in size. Steve Wroe and his colleagues, however, have been exhaustively re-analysing all known materials and exploring body ratios in living marsupials (Wroe *et al.*, 2003). Their conclusions are that marsupial lions ranged from 90 to 160 kg in weight, hence comparable in size to African lionesses, larger than some sabre-toothed lions, and had relatively large brains. 'Respect' has suddenly been returned to this fellest carnivore of the southlands.

Another point of curiosity surrounds their movements. They were at first generally assumed to be terrestrial, before their hind feet were known. Then Wells and Nichol (1977) described a composite hind foot, noting that it was very like that of the common brushtail possum (*Trichosurus vulpecula*), a decidedly (but not obligate) arboreal mammal. The idea that a carnivorous marsupial lion might have used trees in much the way a leopard does (cache prey away from other predators) was not, therefore, a great leap of faith (Wroe *et al.*, 2003). And although it still seems plausible that they used trees when available, it is also the case that thylacoleonids occur in a number of more inland faunas that do not provide any other indication of an arboreal component (e.g. Kanunka, Malkuni and Spring Creek LFs). Perhaps they were satisfied to hunt and live on the ground when trees were absent.

Considering species diversity, the following forms have been named: *Priscileo pitikantensis* Rauscher, 1987, from the late Oligocene Ngapakaldi LF of South Australia; *Priscileo roskellyae* Gillespie, 1997, from the early Miocene of Riversleigh, Queensland; *Thylacoleo carnifex* Owen, 1858, from many Pleistocene localities throughout Australia; *Thylacoleo crassidentatus* Bartholomai, 1962, from the early Pliocene Chinchilla and other LFs in Queensland and New South Wales; *Thylacoleo hilli* Pledge, 1977, from the late Miocene to Pliocene Curramulka LF and the Pliocene Bow LF of New South Wales; *Wakaleo vanderleueri* Clemens and Plane, 1974, from the middle Miocene Bullock Creek LF of the Northern Territory; *Wakaleo oldfieldi* Clemens and Plane, 1974, from the early Miocene Kutjamarpu LF, South Australia. Many additional unnamed species are now known from Riversleigh.

PHALANGERIDANS—POSSUMS AND KANGAROOS

Although some would argue, with some justification, that kangaroos and possums should be treated as separate suborders, we follow Aplin and Archer (1987) in combining the two into the suborder Phalangerida. The oft-cited similarities, both anatomical and molecular, of the two groups makes this arrangement seem not unreasonable. This is despite the historical presumption by some (e.g. Bensley, 1903) that possums and koalas were closely related, an idea that gradually gave way to the realization that the dental similarities between these two groups of leaf munchers were the result of either symplesiomorphy or convergence and that koalas were actually far more closely related to wombats (see above).

There are many groups of possums, but unlike some of the vombatiforms, in most cases there are more living groups than extinct ones with only three families restricted to the fossil record.

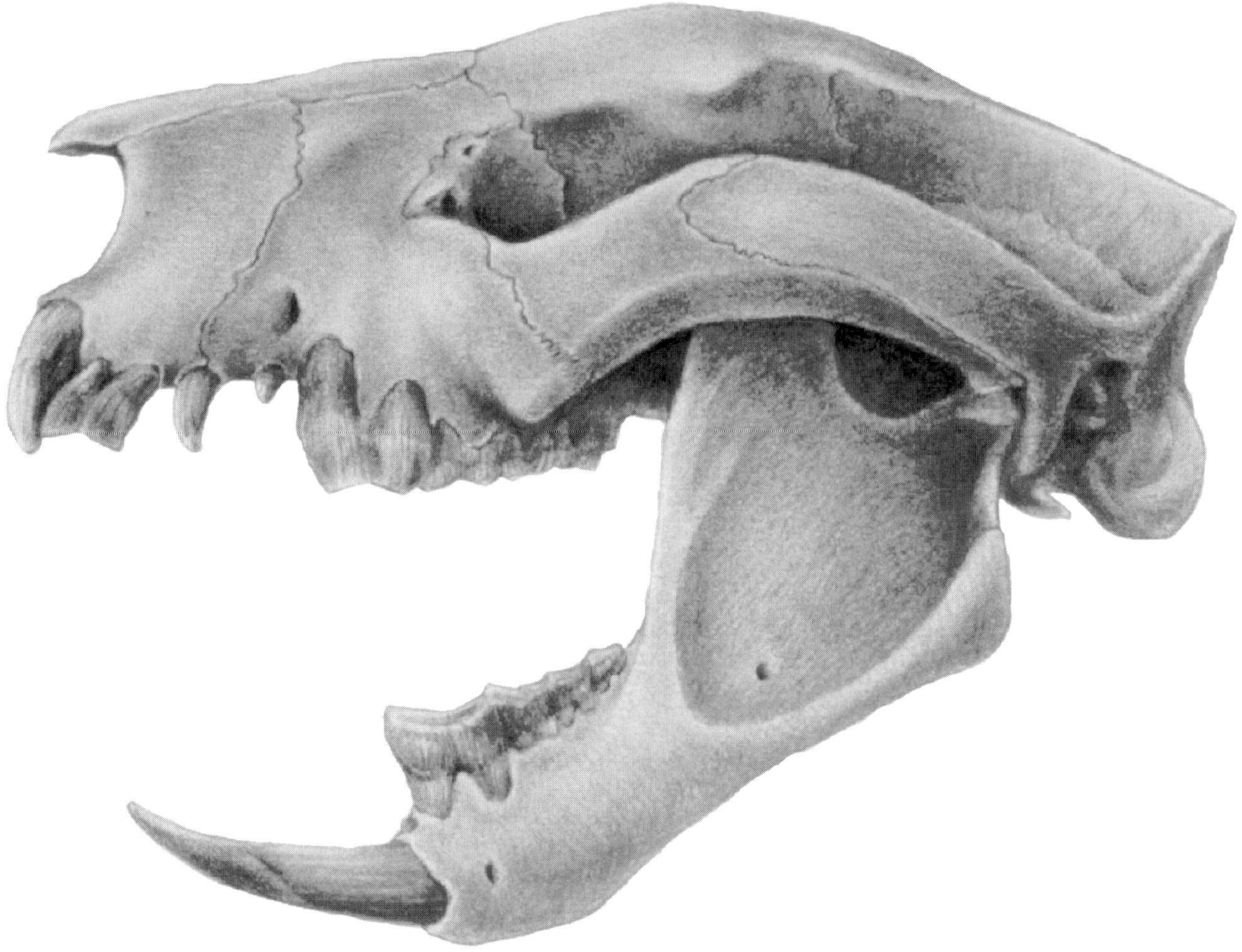

Figure 39. The reconstructed skull and dentary of *Wakaleo vanderleueri* showing the characteristic thylacoleonid dental specializations for flesh-eating (illustration by P. Murray).

BURRAMYOIDEA

Although there is only one family in this group, Burramyidae, Aplin and Archer (1987) considered its distinction from other phalangeriforms merited superfamily status. This sort of judgement can be subjective but it is again not an unreasonable view—they are indeed a very special group of possums!

Mountain Pygmy-possums

Opinion varies about whether there are four, three or two genera of these tiny pygmy-possums as they are collectively known. At the lower extreme, which we accept here, there is the monotypic *Burramys parvus*, or mountain pygmy-possum, and four species of *Cercartetus* or 'ordinary' pygmy-possums (Figures 40, 41). At the other extreme, there is *B. parvus*, *Cercartetus concinnus*, *C. nanus*, *Dromiciola lepida* and *Eudromicia caudata*. Again, decisions about whether to separate a species off as a separate genus can be subjective and is frequently subject to change. So here we will take the conservative option and leave all the ordinary pygmy-possums in the genus *Cercartetus*.

Figure 40. The eastern pygmy-possum *Cercartetus nanus* from the forests of coastal eastern Australia. The fossil record for this genus probably extends from at least the Miocene, but extinct species have yet to be described. Photograph D. Whitford.

Primitively, burramyids have a tooth formula of I1–3/1 or 2 C1/C? or 0 P1–3/1–3 M1–4/1–4. The tooth questionably regarded as ?I2 or ?C is of decidedly uncertain homology. Odds are, however, that it is homologous with the similarly situated tooth in the members of the *Acrobates-Distoechurus* group (acrobatids) and in a number of other diprotodont groups (e.g. some petaurids, pseudocheirids, phalangerids and potoroids). Our guess is that it is an I2.

Since its original description in 1896, no one has doubted the generic-level distinction of the mountain pygmy-possum, *Burramys parvus* (Figure 42). This lack of doubt, however, has been based on the large number of autapomorphic features exhibited by this possum such as its extraordinary circular-saw-like P3, a shape called plagiaulacoid.

Burramys parvus has had a rather curious systematic career. After discovery as a Pleistocene fossil near the Wombeyan Caves in New South Wales, the animal suffered a long period of misunderstanding. Broom (1896a) at first thought it was a link between possums and kangaroos. Although he later (1896b) changed his mind regarding it to be a specialized possum, other workers maintained the view that it was at least near the base of the kangaroo family tree. The misunderstanding was also fed by the failure to appreciate the real antiquity of the marsupial radiation. Tate (1947), for example, considered that all except a few dasyurid lineages had developed within the Pleistocene. So it was probably not such an extraordinary thought at the time that *Burramys parvus* could be a fossil form near the base of the whole kangaroo family tree.

The idea that *B. parvus* might have something to do with kangaroos persisted in various forms until David Ride began his doctoral research at Oxford in 1952. He restudied the Wombeyan fossil material and, on the basis of broader comparisons, concluded (Ride, 1956), as it turned out correctly, that it was a possum phylogenetically close to pygmy-possums.

Then, in 1966, a live and distinctly odd possum turned up in a woodpile in a ski hut on Mount Hotham, Victoria. While it was being studied by Bob Warneke (Fisheries and Wildlife Department of Victoria), its mouth was pried open by an intensely curious Norman Wakefield. He saw and recognized a large plagiaulacoid P3 that he knew characterized *Burramys parvus*. Much of this story of the discovery of *Burramys parvus* and aspects of its biology are recorded in Epstein (1981) and Mansergh and Broome (1994).

The fossil record of *Burramys*-like forms contributes little to help clarify their phylogenetic position within the family. There are, however, now many more species known from the fossil record (Figure 43). These include: *Burramys brutyi* Brammall and Archer, 1997, from the late Oligocene to middle Miocene of Riversleigh; *Burramys wakefieldi* Pledge, 1987d, from the late

Oligocene Ngama LF of South Australia; and *B. triradiatus* Turnbull, Rich and Lundelius, 1987a, from the early Pliocene Hamilton LF of Victoria. An unnamed burramyid was reported from the early Miocene Geilston Bay LF of Tasmania based on an isolated lower incisor (Tedford *et al.*, 1975) but others have questioned its identity as a pygmy-possum.

Extant specimens of *Burramy parvus* have only been found in two main areas, one being 20 x 10 sq. km in the Mount Hotham area and the other about 30 x 10 sq. km in the Kosciusko National Park in New South Wales. Its preferred habitat appears to be low shrubs amongst boulders, above the winter snow line. For at least three months of the year they are active beneath a thick layer

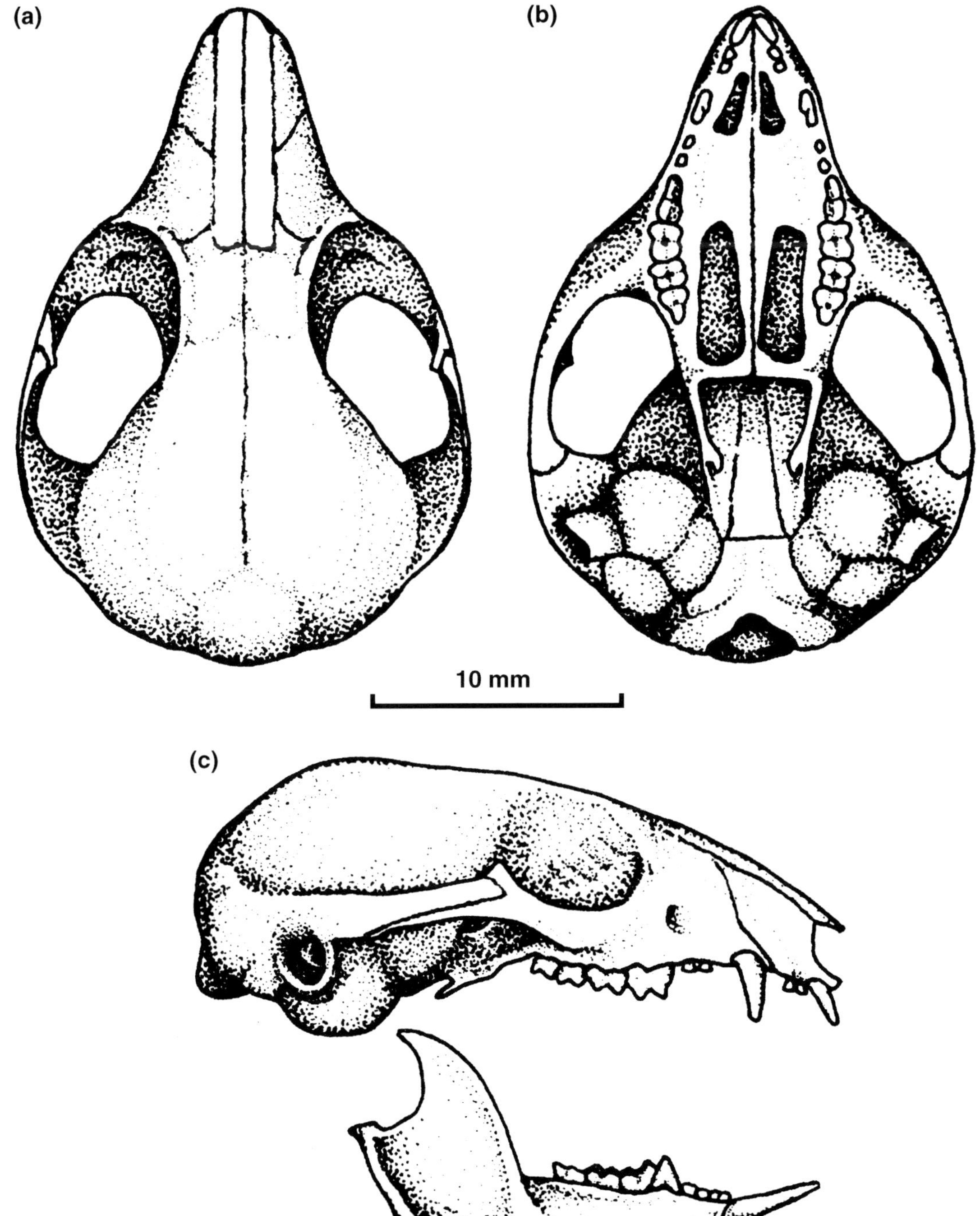

Figure 41. The skull and right dentary the eastern pygmy-possum (*Cercartetus nanus*), from Tasmania: **(a)** dorsal view; **(b)** ventral view; **(c)** lateral view. Note that M4/4 are not present and P3 is distinctly bifid while P3 is not (from drawings by J.L. Rainbird in Green, 1983).

of snow. The life style of *Burramys parvus* is now the subject of considerable research. For those interested, consult Smith and Hume (1984), Mansergh and Broome (1994) and Strahan (1995).

Because there is now concern about what will happen to its habitat with climate change, Archer, Hand and Godthelp (1992) suggested we should pay attention to the fossil record. Species of this genus have always lived as parts of diverse mammal communities in lowland rainforest communities such as those of Riversleigh and Hamilton, and perhaps central Australia. Why shouldn't we see, by controlled experiment, if it can survive, for example, in the lowland rainforests of the Daintree or the lowland temperate rainforests in the Barrington Tops area which is now at about the latitude Riversleigh would have been at 23 million years ago when its forests were filled with *Burramys*? Before we let its habitat disappear off the top of the mountains where it now occurs, should we at least think about these options?

The interrelationships of the species of *Cercartetus*, based mainly on dental morphology, have been reviewed by Wakefield (1978) and Turnbull and Schram (1973). They note that all of the species are quite distinct from one another. This has been used by others to support the view that each should represent a monotypic genus. But if all of the species of *Cercartetus* are in fact monophyletic (i.e., no other non-*Cercartetus* taxon lies within *Cercartetus*), there is really no compelling reason for recognizing each of them as a separate genus.

Before abandoning burramyid phylogenetic considerations, we should consider the group as a whole, if only to note yet another unresolved problem. It again concerns

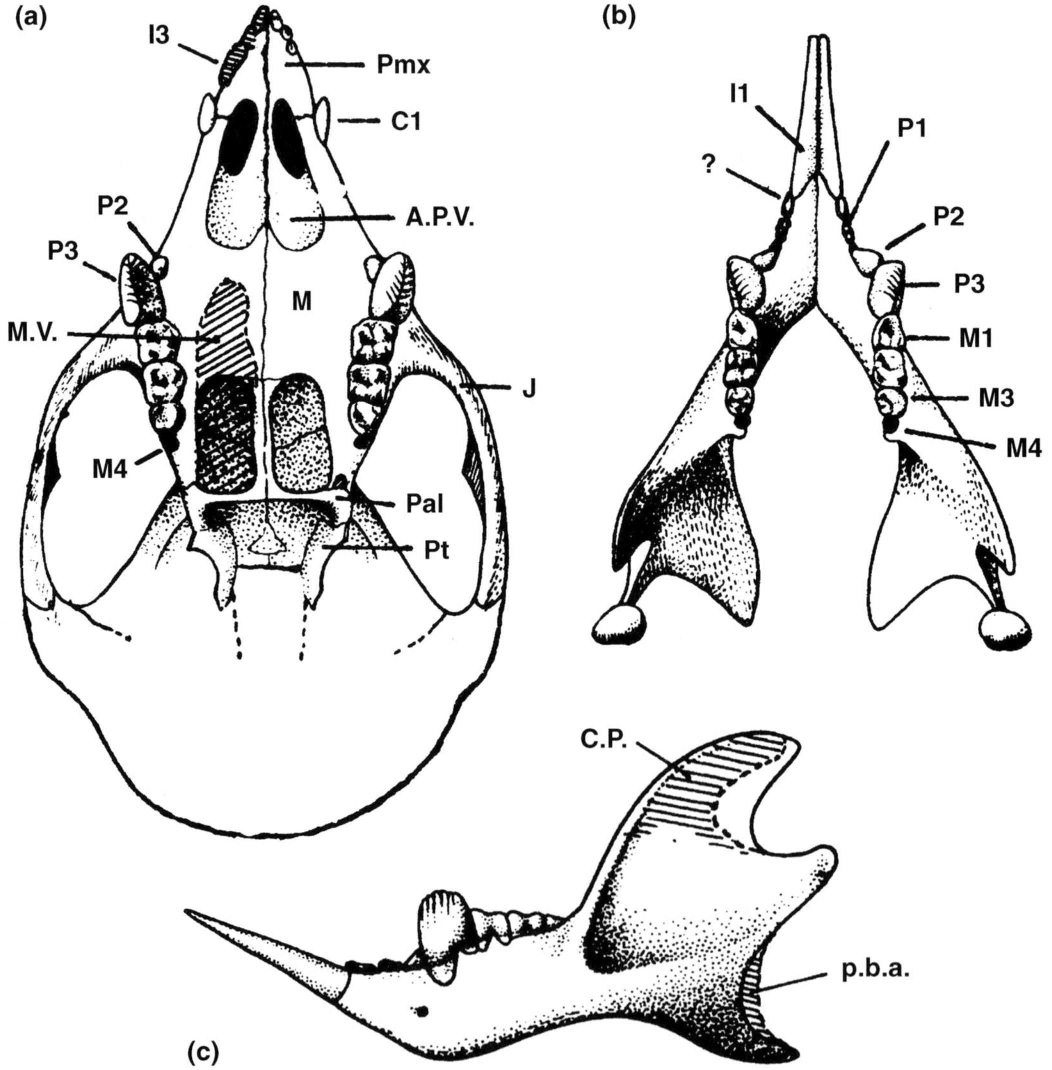

Figure 42. The skull and dentition of the modern mountain pygmy-possum, *Burramys parvus* (after Ride, 1956): **(a)** ventral view of the skull; **(b)** dorsal view of dentaries; **(c)** lateral view of the left dentary. Conspicuous features are the greatly enlarged premaxillary vacuities, steeply reducing molar gradient and sectorial P3. Key to abbreviations: Pmx = premaxilla; A.P.V. = anterior palatal vacuity; M = maxilla; M.V. = maxillary vacuity; J = jugal; Pal = palatine; Pt = pterygoid; C.P. = coronoid process; p.b.a. = posterior border of the angular process.

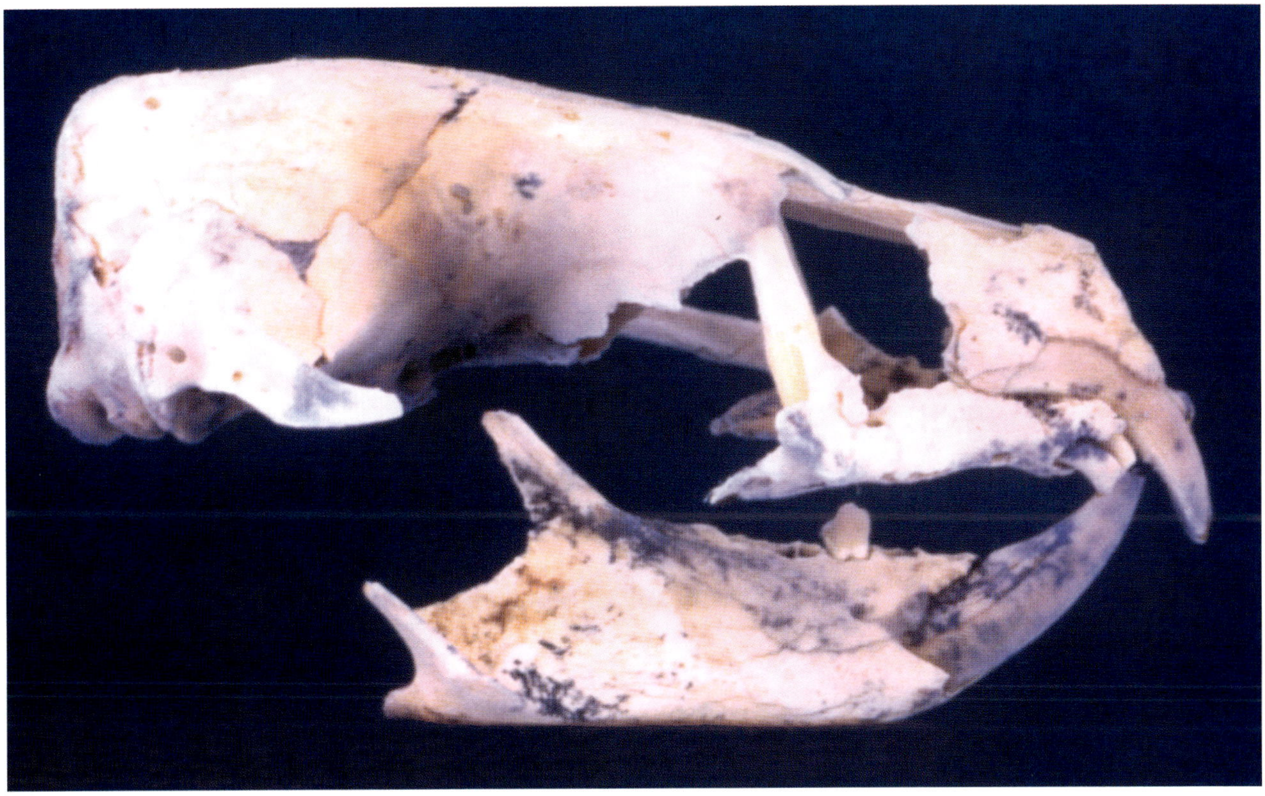

Plate 100. *Yalkaparidon coheni* is known from a complete skull and lower jaw, but nearly everything else about it remain mysteries, including its diet, life-style and ancestors. Photograph R. Arnett.

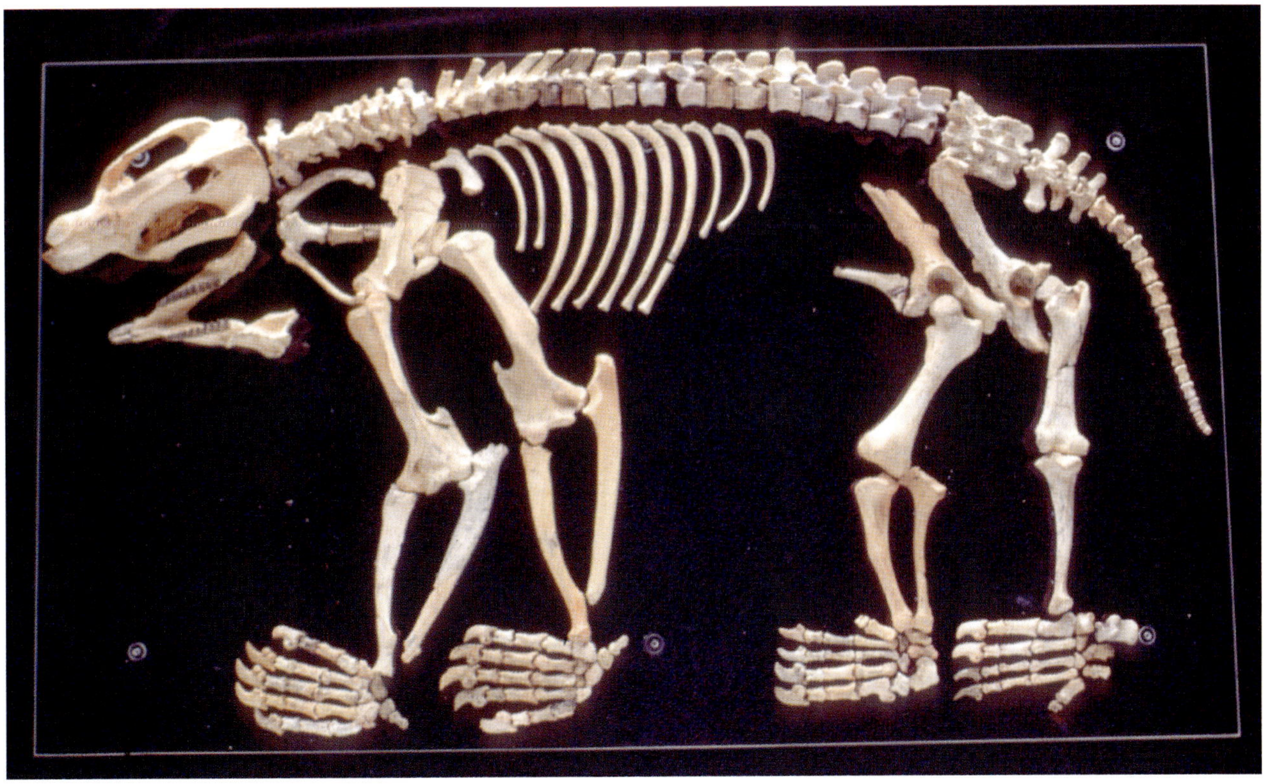

Plate 101. The skeleton of *Nimbadon lavarackorum* from the middle Miocene of Riversleigh, one of the best preserved zygomaturines known. Photograph C. Bento, Australian Museum.

Plate 102. Australia's feathertail glider *Acrobates pygmaeus* is a tiny omnivorous possum that eats nectar, pollen and insects. A row of long hairs extending from each side of its tail helps this possum glide. Photograph D. Andrew.

Plate 103. Living species of *Petaurus*, like this one, are charming, gliding possums that inhabit mainly sclerophyll forests and woodlands. Photograph A. Easton, Queensland Museum.

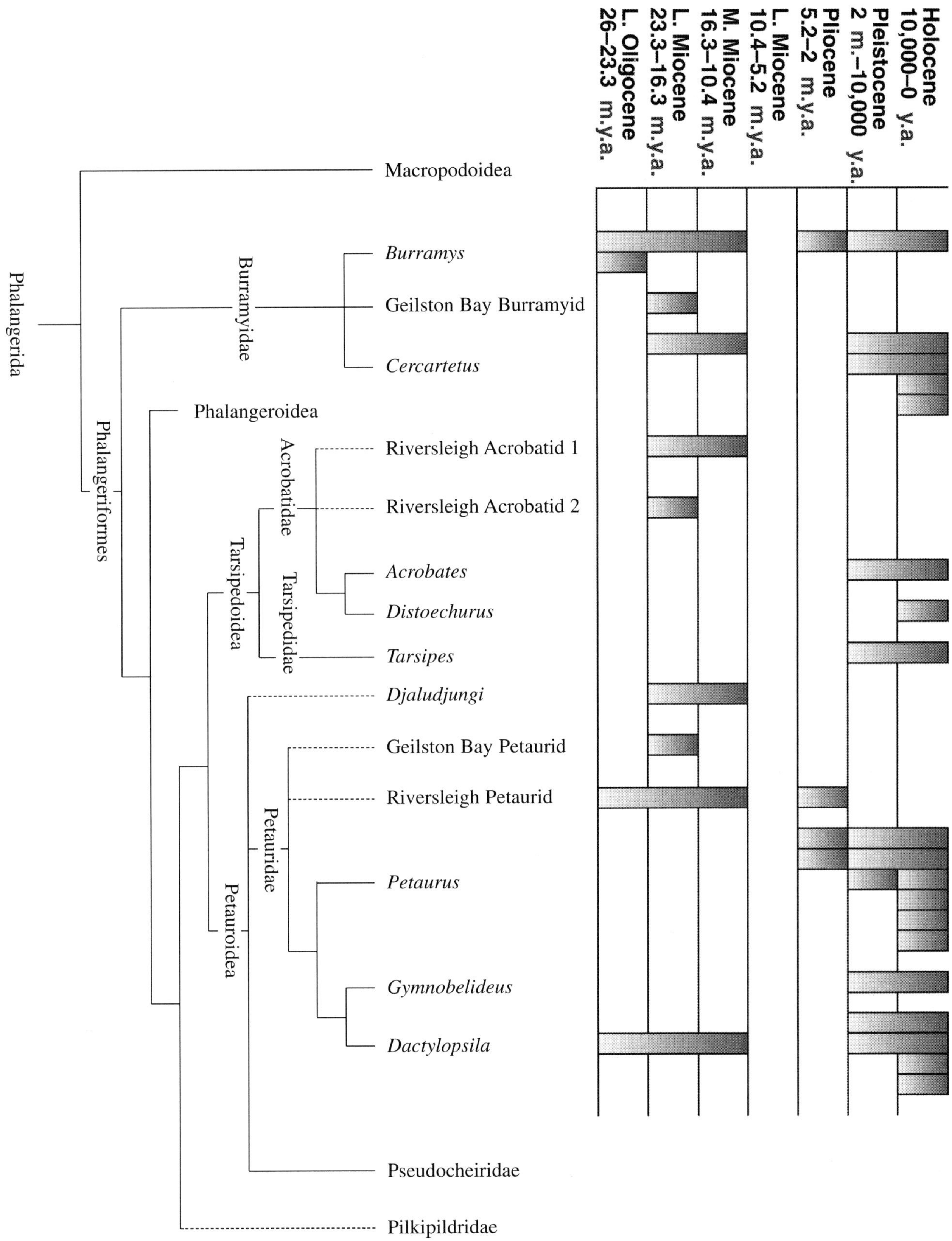

Figure 43. Diversity through time and a cladogram of probable interfamilial and intergeneric relationships of acrobatids, tarsipedids and burramyids. Uncertain relationships are indicated by dashed lines. Each block represents a distinct species, and blocks that extend without breaks through more than one time period indicate apparently long-surviving species (after Brammall and Archer, 1999).

the position of pygmy-possums within the marsupial radiation as a whole. Bensley (1903) regarded them to be structurally ancestral to phalangerids and petaurids (i.e., the petaurids in the sense used here, not pseudocheirids which he considered to be relatives of phascolarctids). Other workers have placed them in an even more plesiomorphic position relative to other diprotodont groups. In contrast, Archer (1976e) suggested that they might instead be a derived group descended from selenodont ancestors. At present, these and other competing ancestor-descendant hypotheses are difficult to test.

The fossil record of *Cercartetus*-like forms is poor. Pleistocene deposits, particularly those from caves, abound with burramyids (Wakefield, 1972) but all of these are modern species. A new *Cercartetus*-like species is represented in the early to middle Miocene sediments of Riversleigh, but this has yet to be described (Brammall and Archer, 1999).

About the life styles of the species of *Cercartetus* all are insectivorous, nectarivorous and ferocious predators of pollen grains. *Cercartetus caudatus* lives in rainforest in north-eastern Queensland and New Guinea. *Cercartetus nanus* lives in sclerophyll forest and rainforest throughout much of coastal eastern Australia. *Cercartetus lepidus* has been found in a wide range of habitats in south-eastern Australia from mallee scrub in north-western Victoria to wet sclerophyll forest (but not rainforest) in Tasmania. *Cercartetus concinnus* occurs in mallee heath and dry sclerophyll forest in south-western Western Australia, southern South Australia and north-western Victoria. Much about the biology of these species is summarized in Smith and Hume (1984) and Flannery (1994b).

TARSIPEDOIDS—HONEY POSSUMS AND FEATHER-TAIL POSSUMS

In this group are placed some of the most delightful of the possums, the diminutive, long-snouted tarsipedids and the exquisitely beautiful acrobatids.

Honey Possums

There is only one known species, *Tarsipes rostratus* (or 'Noolbenger' as it was called by Aborigines of south-western Western Australia); see Figure 44. It has many claims to fame besides being the only member of its family. It is one of only two mammals (the quokka, *Setonix brachyurus*, being the other) endemic to the south-western corner of the continent. The spermatozoan is also the largest known of any species of mammal (Woolley and Scarlett, 1984) and the testicles make up nearly 30% of its body weight! One irritating fact is that it has no known pre-Pleistocene record. This, combined with its vestigial dentition, has frustrated determination of its phylogenetic position within the diprotodont marsupials.

Until recently it was generally assumed to be strictly arboreal. In fact, one of the most common sources for specimens was the swimming pool at Yanchep National Park, supposedly because the honey possums kept falling out of the overhanging banksia trees. More recently it has been found that the best way to capture them in the heaths of the south-western coast is to set pit-fall traps in the ground. This surprising discovery made by a field survey party from the Western Australian Museum, has enabled population studies to be carried out—something that would have been very difficult if the animals were confined to trees.

The flower-fixation and singularly appropriate shape of these blossom-probing creatures raises another interesting matter. It is gradually becoming apparent that some plants count on more than birds, bats and insects for pollination. In the heaths of south-western Western Australia there are even some species of *Banksia* (e.g. *B. petiolaris*) that develop flower stalks near the ground that give off a musky scent most noticeable at night. It has been suggested (Turner, 1982) that a number of groups of Australian plants co-evolved with marsupials. That is, the morphology, time of flowering and chemistry of the plants probably evolved to suit changes in the needs of the evolving marsupials upon which they came to depend for efficient pollination.

Distinctive features (mostly autapomorphic) of honey possums include the longitudinal stripes, the feathery tongue, the vestigial dentition, the broad nasal-lachrymal contact, the large ventral squamosal flange on the skull, the posteriorly incomplete palate, the tiny pterygoids, the lack of an inturned angular process on the dentary and almost complete suppression of the coronoid process of the dentary.

The adult dentition is somewhat variable (Figure 45) but appears to be I1–2/1 C1/0 P–M (undifferentiated) (1–4)/(1–3). The lower incisor is procumbent and lanceolate (lance-like), presumably for inserting into insect bodies or the bases of flowers. The other teeth are minute, peg-like gadgets possibly used to remove pollen and nectar from the tongue as it slides past.

The relationships of this group have long been the subject of controversy, partly because their vestigial teeth gave no clear indication of who their closest cousins were. Even molecular biology and other soft-tissue studies gave conflicting indications (Archer, 1984; Aplin, 1987; Smith and Hume, 1984). Aplin (1987) reached the conclusion we follow here that they are most closely related to the feather-tail gliders, and with these to the petauroids (other gliders, ringtails and their relatives). See Figure 43.

In the meantime, our mysterious little friend goes about the business of sticking his nose into other organisms' nectaries, without seeming to care at all about his enigmatic pedigree. He just is, and for him that seems to be quite enough.

Feather-tails and Pen-tails

The feather-tail possums (e.g. *Acrobates pygmaeus;* Plate 102) and pen-tail possums (e.g. *Distoechurus pen-*

Figure 44. Three dark-eyed honey possums (*Tarsipes rostratus*) depicted by Gould (1863). The long protrusible tongue is shown by the individual on the left.

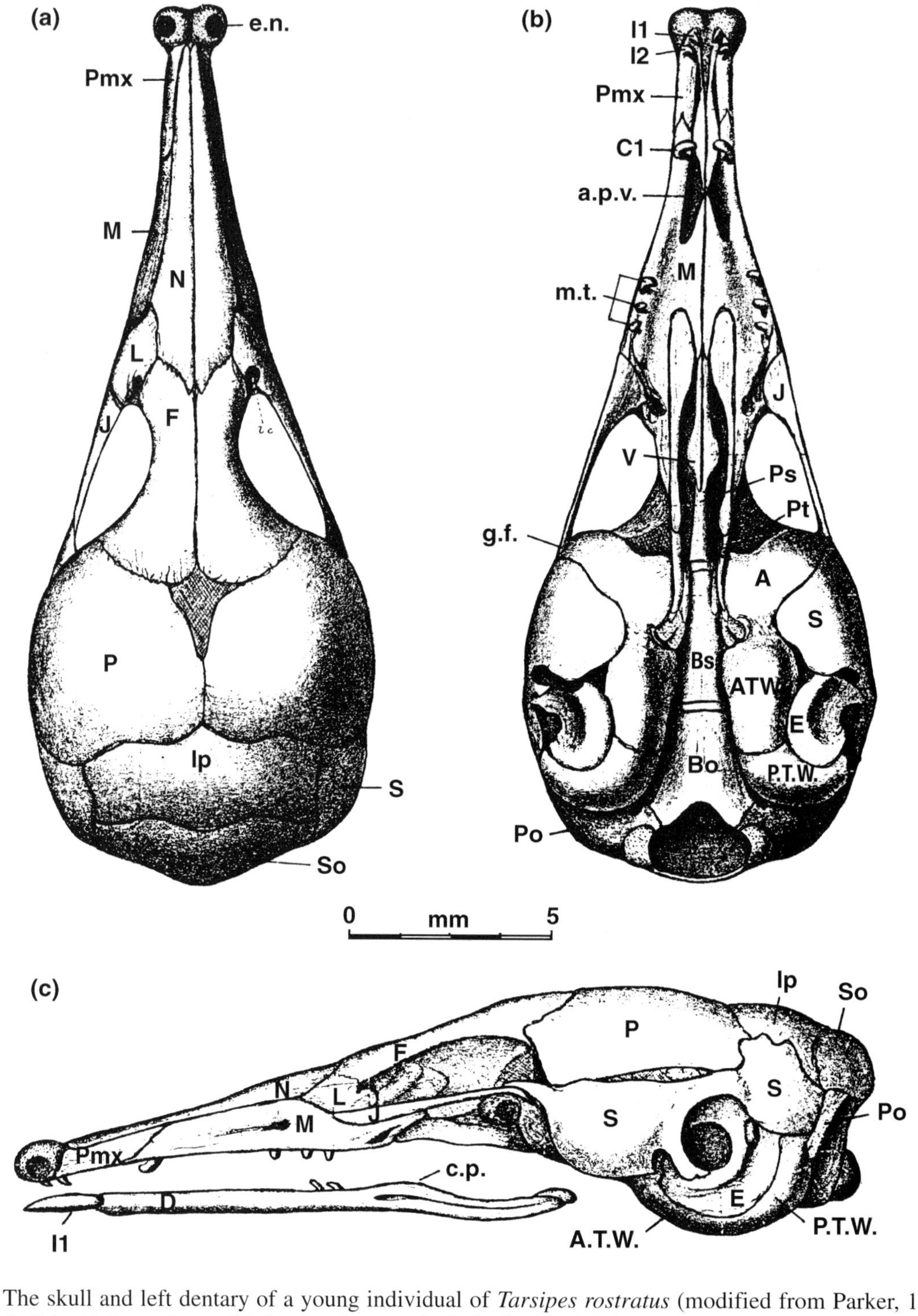

Figure 45. The skull and left dentary of a young individual of *Tarsipes rostratus* (modified from Parker, 1890). The whole skull is 12.5 mm long. Distinctive features include the vestigial dentition, the virtual lack of a coronoid process (c.p.) on the dentary, the elongate snout, nasal-lachrymal contact, relatively small frontal, palatine and vomer bones, very large interparietal and squamosal bones, large ventral flange of the squamosal, splint-like jugal (and zygomatic arch) and incomplete posterior margin of the palate. Key to abbreviations: e.n. = external naris; Pmx = premaxilla; N = nasal; M = maxilla; L = lachrymal; F = frontal; J = jugal; P = parietal; Ip = interparietal; So = supraoccipital; S = squamosal; Po = paraoccipital; E = ectotympanic; A.T.W. = alisphenoid tympanic wing; P.T.W. = tympanic wing of the periotic (?mastoid portion); D = dentary; (c.p.) = position where coronoid process develops in other mammals; a.p.v. = anterior palatal vacuity; m.t. = posterior maxillary teeth (not clearly differentiated into premolars and molars); V = vomer; Ps = presphenoid; Pt = pterygoid; g.f. = glenoid fossa for articulation of the dentary; A = alisphenoid; Bs = basisphenoid; Bo = basioccipital.

natus), once thought to be some kind of pygmy-possums, are now understood to be quite separate from this group (Figure 43). In the course of finishing a review of the burramyids, Mike Archer (1984) concluded that the species of these two genera stood well apart from the other burramyids. Ken Aplin then had a detailed look at the basicranium (Aplin, 1987) and came down on the side of peeling these two genera off at a family level (a view locked in by Aplin and Archer, 1987). Molecular biology has since generally confirmed this view.

So much for the systematic relationships within the possum group. What about possible ancestors or descendants? We don't know. But there are some tantalizing things to think upon here. Szalay (1982) demonstrated that the tarsal bones of what he then regarded to be burramyids shared apomorphic states with those of the South American marsupial *Dromiciops australis*. Among the Australian forms, he noted in particular the marked similarity involving *Distoechurus pennatus*. However, it is also clear that acrobatids are diprotodont, syndactylous, have a superficial thymus, and so on, all features that indicate they are part of the diprotodontian radiation. Hence the possible relationship to *Dromiciops australis* might be broader and to the Diprotodontia as a whole. For sure maybe.

Did the *Distoechurus-Acrobates* radical give rise to any other group of possums? We very much doubt it. The middle ear structure of *A. pygmaeus* and *D. pennatus* is so highly autapomorphic (Archer, 1984) that to derive from it any of the other more plesiomorphic possum basicranial morphologies would require a major redesign of the whole region. The dental formula of the two known living species is: Il–3/1 or 1–2 Cl/1 or 0 Pl–3/1–2 or 3 Ml–4/1–4. The Ml is inferred but has not yet been seen.

In Australia, *A. pygmaeus* is an inhabitant of coastal forests. This species is also known from one New Guinean specimen (as *Acrobates oulchellus*) but this may have been an escaped Australian pet (Honacki *et al.*, 1982) and a synonym of *A. pygmaeus*. Feathertail gliders live on nectar, insects and perhaps fruits. *Distoechurus pennatus* is a larger, non-gliding possum that lives mainly in small trees and undergrowth in rainforests and moss forests (Flannery, 1995a). It probably feeds on nectar, pollen, fruit and insects.

Although it is a highly distinctive group, the known fossil record of the group is poor. We have a dentary from the Leaf Locality (middle Miocene Kutjamarpu LF) and at least two different kinds of acrobatids from the Oligo-Miocene deposits of Riversleigh, which have yet to be named.

PHALANGEROIDEA—FROM RAINFORESTS TO CONCRETE JUNGLES

In this group are placed four families: Phalangeridae, Pilkipildridae, Miralinidae and Ektopodontidae, the last three have only extinct species. The position of pilkipildrids in this group remains highly controversial—they may belong with the petauroids or an entirely different group. There is too little known at this point to place them with certainty anywhere—so for convenience we'll drop them here, following Long *et al.* (2002).

Cuscuses, Brushtails and their Relatives

The most familiar marsupial, even to city-bound Australians, must be the common brushtail possum (*Trichosurus vulpecula*), a ubiquitous member of the Phalangeridae. It is also one of the most widespread marsupials occurring from wet coastal forests (and some northern rainforests) around Australia to the arid woodlands of Alice Springs (Kerle, 2001). Its relatives are a bit fussier. The cuscuses (*Phalanger*, *Spilocuscus* etc.) are rainforest beasts of Australia, New Guinea and surrounding islands (Plates 22, 23, 24), and the scaly-tailed possum (*Wyulda squamicaudata*) which lives only in the woodlands of northern Western Australia.

The members of this family are almost certainly monophyletic (the only possible fly in the ointment here being the phylogenetic position of the ektopodontids, which could make the group paraphyletic; see below). All phalangerids have a dental formula of Il–3/1 or 1–2 C1/1 or 0 Pl–3/1,3 M1–4/1–4. The dP3 is large and molariform. The P2 is missing in some species. The possible I2 or C1, whatever it actually is, is almost certainly homologous with the similarly positioned tooth in members of many other diprotodont families. The adult molars are semilophodont with transverse crests extending lingually from the paracones and metacones and buccally from the metaconids (except on M1) and entoconids.

The phylogenetic position of the Phalangeridae within the diprotodont radiation remains controversial. Previously phalangerids have commonly been regarded as representative of 'possums' as a whole, perhaps because they are not clearly specialized in many ways of their own. The MC'F studies of Baverstock (1984) are relevant here. He found phalangerids to be the sister-group of the Burramyidae (in the restricted sense noted above). This conclusion has not been reached by most morphologists, although Bensley (1903) suggested that phalangerids were descendants of burramyids.

There is also the question of possible phalangerid descendants. Elsewhere (Archer, 1976e, 1981, 1984; Archer *et al.*, 1999) it has been suggested that ektopodontids were possible descendants of ancestral phalangerids. Rich (1982), however, suggested they may instead be descendants of phascolarctids. Until whole skulls are found, with intact basicrania, the question of relationships will remain unanswered.

As noted above, cuscuses are the most diverse of the phalangerid groups. There are two species known from the rainforests of north-eastern Queensland, *Spilocuscus maculatus* and *Phalanger orientalis*. Both of these species and at least nine others (Flannery, 1995a, b) are also known from New Guinea and adjacent islands. They are one of the most common types of arboreal mammals

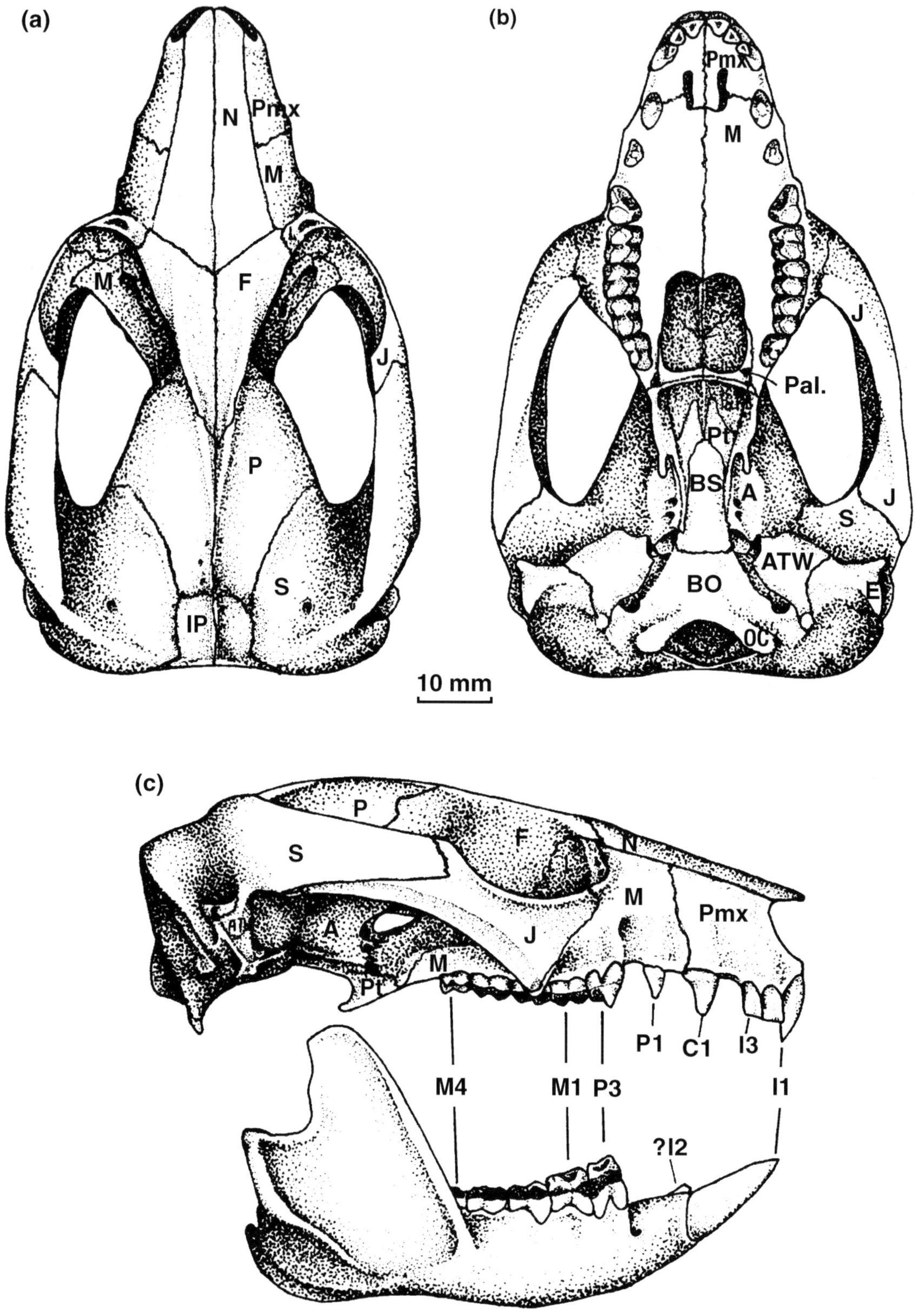

Figure 46. The skull and right dentary of the common brushtail possum, *Trichosurus vulpecula*. This individual comes from Tasmania where the largest individuals occur (subspecies *Trichosurus vulpecula fuliginosa*): **(a)** dorsal view; **(b)** ventral view; **(c)** lateral view. Key to abbreviations: Pmx = premaxilla; N = nasal; M = maxilla; J = jugal; F = frontal; L = lachrymal; Pt = pterygoid; A = alisphenoid; P = parietal; S = squamosal; E = ectotympanic; ATW = alisphenoid tympanic wing; IP = interparietal; Pal = palatine; BS = basisphenoid; BO = basioccipital; OC = occipital condyle. Drawings by J.L. Rainbird from Green (1983).

known from the New Guinean rainforests. A little is known about the biology of Australian populations (Strahan, 1995), but even less about the New Guinean ones (Flannery, 1995a).

There is also a curious zoogeographic question here. It concerns Australia's two species. How did they get to Australia? In the past, when species distribution patterns were almost invariably seen as the result of simple dispersals from centres of origin, this distribution pattern would have been interpreted as a classic example of dispersal. Obviously, two species originally restricted to New Guinea (the living Australian cuscuses) dispersed at some recent time, south to Queensland. Maybe yes.

But maybe no. We can think of at least two equally good reasons for why this distribution may not be a simple case of dispersal—rather one of vicariance. First, consider that there is, at present, no suitable rainforest habitat linking the populations of New Guinea and Australian cuscuses. Even when the sea levels were lowered at intervals of glaciation during the Pleistocene, it would have been at a time of lowered rainfall and colder temperatures, hardly the conditions to develop a continuous tropical rainforest corridor between Papua and Queensland. So, simple dispersal from New Guinea to Queensland along such a corridor during the Pleistocene seems improbable. This is more the case for *P. orientalis* than *S. maculatus* because, although *S. maculatus* will occasionally wander into mangrove forests, *P. orientalis* rarely leaves rainforest.

Second, the fossil record indicates that species of *Phalanger* were present in eastern Australia at least as long ago as the early Miocene (Riversleigh). In fact, there is an enormously greater probability that Australia, not New Guinea, was the 'centre of origin' for cuscuses. Is it not, therefore, equally probable that populations now in north-eastern Queensland and New Guinea represent remnants of once continuous and more extensive New Guinean-Australian populations? If so, their present disjunct distribution would probably have resulted from the loss of the rainforest that once linked these now isolated populations; such loss probably having occurred sometime at the end of the Pliocene or earliest Pleistocene with the development of Australia's arid conditions. This would be a classic example of a vicariance rather than a dispersal event.

The other two genera of phalangerids, *Trichosurus* (Figure 46) and *Wyulda*, contain only Australian species. We suspect that these represent offshoots of the cuscus radiation, parts of it that became adapted to the drier conditions that developed in Australia in the late Tertiary.

There are three species of brushtail possums currently recognized in the genus *Trichosurus*: the widespread *T. vulpecula*, the northern *T. arnhemensis* (which may not be specifically distinct from *vulpecula*) and the eastern Australian *T. caninus*. A lot of research has been done on *T. vulpecula*, particularly of late because of its introduction to New Zealand and subsequent declaration there as a pest.

Wyulda squamicaudata is much less well known (Strahan, 1995). Its distinctively scaly tail, the basis for the vernacular name scaly-tailed possum, is strongly prehensile. This omnivorous possum is so far known only from very rugged and rocky areas in the Kimberley division of Western Australia. *Wyulda asherjoli*, Crosby, Nagy and Archer, 2001, from the early Miocene of Riversleigh, is the oldest and only extinct member of this genus.

Strigocuscus reidi Flannery and Archer, 1987, from middle Miocene LFs of Riversleigh, was originally considered to be a relative of ground cuscuses such as *Strigocuscus gymnotis*. But the huge P3 and other *gymnotis*-like features it shows may have convergently evolved. *Strigocuscus notialis*, Flannery, Turnbull, Rich and Lundelius, 1987, from the early Pliocene Hamilton LF in Victoria, is similar to the Riversleigh taxon—*whatever* that is.

Trichosurus dicksoni Flannery and Archer, 1987, from the middle Miocene of Riversleigh, is more plesiomorphic than living species but the whole group is under review by Kirsten Crosby who now recognizes many more species and even genera from Riversleigh than were recognized by Flannery and Archer (1987)—science lurches ahead! *Trichosurus hamiltonensis* Flannery, Turnbull, Rich and Lundelius, 1987, an early Pliocene taxon from the Hamilton LF, Victoria, may be similarly in need of revision.

Other phalangerids are known from the Pliocene Bow LF, of New South Wales (Archer and Hand, 1984) and from the early Miocene Geilston Bay LF, Tasmania (Tedford *et al.*, 1975). Tedford and Kemp (1998), however, concluded that the Geilston Bay material is actually referable to the Petauroidea. Then again (!), Kirsten Crosby and Jenny Brammall have expressed doubts about this interpretation.

Miralinids

When Mike Woodburne and Mike Archer were beavering away in central Australia at a place called Tedford Locality (one of the locations for the Ditjimanka LF), they kept finding teeth that were more than a bit bizarre—not as strange as ektopodontids (see below) but certainly teeth that represented another group of possums no one had seen before. Once we began to work out how all of the isolated teeth went together to form a whole dentition, the most striking aspects of the dentition were the huge, serrated premolar and high blade-like ridges on the first molars. Also striking were the razor sharp cutting edges of the molars which had been honed in life to be far sharper than they would have been when the teeth had first erupted. Our best guess at the time was that it was a herbivore of some kind, but perhaps one not averse to the odd bit of beast. Certainly it was cutting up food that required very sharp teeth. The adult dental formula for miralinids appears to be: I?/1–2 C1/0 P1,3/1,3 M1–4/1–4.

While we couldn't be certain, miralinids appeared to retain features that reminded us all of cuscus teeth, but

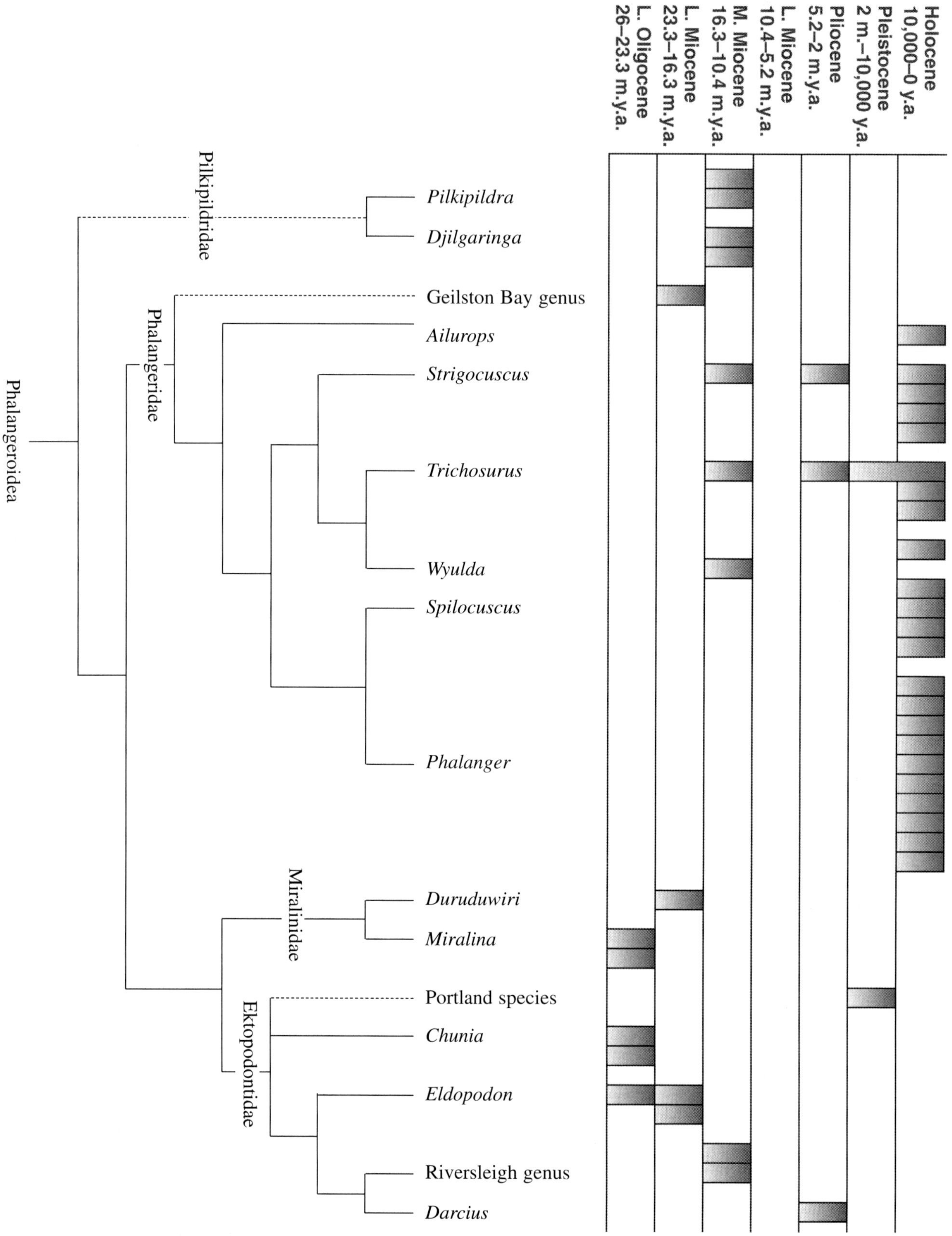

Figure 47. Diversity through time and a cladogram of probable interfamilial and intergeneric relationships of phalangerid, ektopodontid, miralinid and pilkipildrid possums. Uncertain relationships are indicated by dashed lines. Each block represents a distinct species, and blocks that extend without breaks through more than one time period indicate apparently long-surviving species. After Crosby *et al.* (1999).

had other features, like many complex ridges within the molars, that appeared to anticipate the ektopodontids—hence we nicknamed the first one found 'Protoektopodon' in our field notes. The family contains two subfamilies (Miralininae and Durudawirinae) and two described genera (*Miralina* and *Durudawiri*), all of them extinct, from the late Oligocene to Miocene—another group that filled some kind of a niche that presumably has either disappeared or remains vacant.

Kirsten Crosby and Mike Archer placed the very distinctive form that was found at Riversleigh in a new subfamily, the Durudawirinae, at the same time allocating the central Australian forms to the subfamily Miralininae (Figure 47).

The only durudawirine that has been described is *Durudawiri inusitatus* Crosby and Archer, 2000, from the early Miocene Wayne's Wok LF of Riversleigh. It is smaller and, although younger, seemingly more plesiomorphic than the late Oligocene species of *Miralina*.

There are possibly three miralinines: *Miralina doylei* Woodburne, Pledge and Archer, 1987, from the late Oligocene Ditjimanka LF of South Australia, which was about the size of a common brushtail possum; *Miralina minor* Woodburne, Pledge and Archer, 1987, also from the late Oligocene, Ditjimanka LF; and another possible distinct species (Woodburne, Pledge and Archer, 1987) from the late Oligocene Yanda, Tarkarooloo and Wadikali LFs of South Australia.

Ektopodontids

When the first teeth of these beasts were found by Reuben Stirton and his crew in 1961 at the early Miocene Leaf Locality in the Tirari Desert of South Australia, they were thought to be evidence for Australian multituberculates, a group of archaic, somewhat rodent-like mammals common in the Mesozoic and early Cainozoic deposits of the Northern Hemisphere. There was excitement in the field camp! But just as the champagne corks were about to pop, so went the multituberculate idea. One of the more critical palaeontologists noted that wear marks on one of the teeth indicated that the rows of cusps must have been oriented transversely in the jaw, not longitudinally as they are in multituberculates. So from the depths of despair, but still undaunted, they asked themselves: 'Well then, what is this thing?'

What indeed. Let's consider an ecological as well as a phylogenetic answer to this question. What did ektopodontids do? One guess is that they were giant rodent-like possums (Archer, 1981a). Rodents have similar teeth with lots of small cusps in flat rows (albeit in longitudinal rather than transverse rows) and are similarly diprotodont with two procumbent lower incisors. There are no other marsupials with rodent-like teeth. Perhaps most suggestive of rodent-like habits is the fact that ektopodontids declined soon after rodents arrived. The last undoubted ektopodontid occurs in the early Pliocene Hamilton local fauna of Victoria and the first placental rodent appears in the early Pliocene Bluff Downs local fauna of north-eastern Queensland—see pp.707–744 (Aplin, 2006). There is a possible ektopodontid known from the early Pleistocene of Hamilton in Victoria, but that has yet to be confirmed as a last survivor of this strange group.

Pledge (1982), who found a nearly complete lower jaw (Figure 48), confirmed their nature as possums of some kind and noted that a skull fragment had large and forward pointing orbits—suggesting a nocturnal animal with well developed stereoscopic vision. Stereoscopic vision is commonly found among arboreal animals. He further suggested that ektopodontid teeth would have been

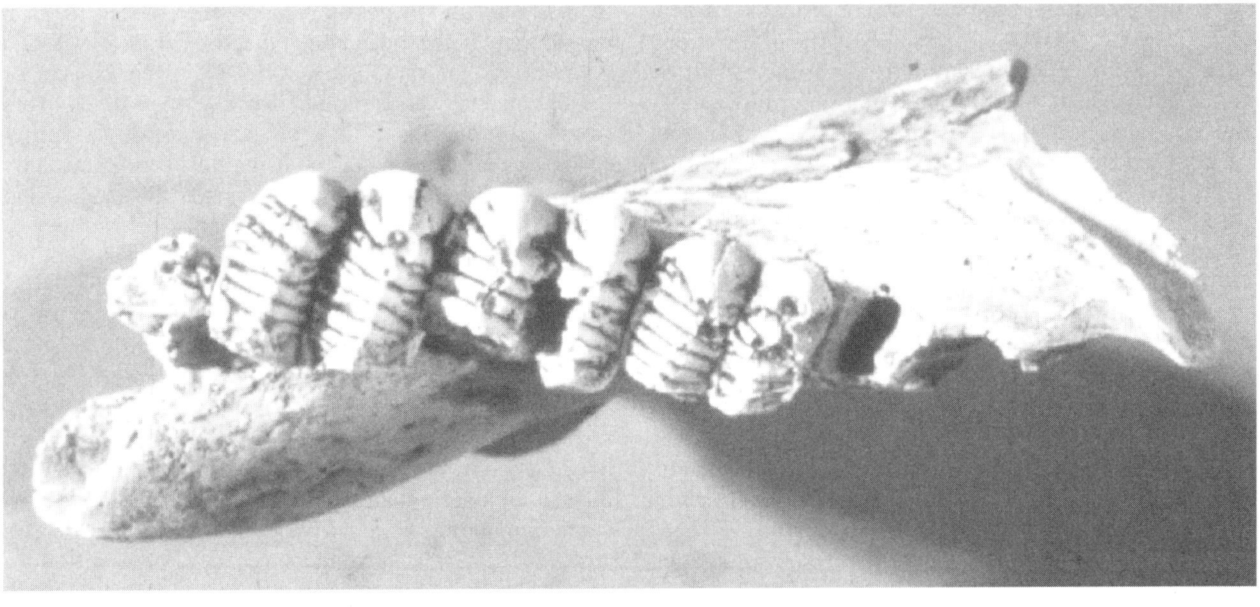

Figure 48. The lower jaw of the ektopodontid *Chunia illuminata* from Tedford Locality in South Australia. Photograph N. Pledge.

suitable for munching a number of different foods including leaves, flowers, fruits, seeds and invertebrates. The lower jaw and skull fragment enabled much of the adult dentition to be determined: I?/1–2 C1/0 P3/3 M1–4/1–4.

Rich (1982) suggested that the ektopodontid P3 and I1 are koala-like rather than phalangerid-like and accordingly proposed that the relationships of ektopodontids lie with phascolarctids. We remain open-minded about this principally because the evidence for both hypotheses is thin. But, because the complex molar morphology of miralinids appears to be so neatly intermediate between that of phalangerids and ektopodontids, we consider it a bit more likely that ektopodontids are highly autapomorphic descendants of ancestral phalangerids (Figure 43).

Species named so far include: *Chunia illuminata* Woodburne and Clemens, 1986a, from the late Oligocene, Ditjimanka LF of central Australia; *Chunia omega* Woodburne and Clemens, 1986, from the late Oligocene Tarkarooloo LF of central Australia; *Darcius duggani* Rich, 1986, from the early Pliocene Hamilton LF of Victoria; *Ektopodon serratus* Stirton, Tedford and Woodburne, 1967, from the early Miocene Kutjamarpu LF of central Australia and early Miocene LFs of Riversleigh; *Ektopodon stirtoni* Pledge, 1986, from the late Oligocene Ngama LF of central Australia; and *E. litolophus* Pledge *et al.,* 1999, from the early Miocene Kutjamarpu LF of central Australia. There is also an undescribed genus of ektopodontids, related to *Darcius*, known from the early Miocene of Riversleigh.

Pilkipildrids

When the first teeth of pilkipildrids turned up in late Oligocene deposits in central Australia, Mike Woodburne and Mike Archer thought they were looking at gigantic pygmy-possums (gigantic that is for pygmy-possums—they were still very tiny!). And that's the way they entered the literature, before they were officially named, as giant pygmy-possums. It wasn't until a whole jaw turned up at Riversleigh that the really distinctive nature of this group began to be apparent. In the meantime, other isolated teeth had turned up in other areas of central Australia and in 1987 they were all published as members of the new family Pilkipildridae.

Unfortunately too little is known about them to clarify their relationships to other families of possums (Figure 47). We could have placed them with the petauroids with almost as much justification as the phalangeroids—if a skull eventually turns up, the question can be settled.

Pilkipildrids, so far as known, are characterized by having a steeply inclined I1, huge, powerful P3s, prominent, isolated metastylids, low-crowned molars, and prominent, swollen buccal cusps on the upper molars with reduced paracones and metacones. They were probably omnivores, perhaps specializing in fruits or seeds (Archer *et al.*, 1987). The dental formula, so far as is known, is: I?/1–? C?/0 P?,3/1–3 M?/1–4.

Taxa identified, some known from lower jaw fragments, others from isolated teeth, include: *Pilkipildra handae* Archer, Tedford and Rich, 1987, from the late Oligocene Billeroo Creek LF of South Australia; *Pilkipildra taylorae* Archer, Tedford and Rich, 1987, from the late Oligocene Ditjimanka LF of South Australia; *Djilgaringa gillespiae* Archer, Tedford and Rich 1987; from the middle Miocene Last Minute LF of Riversleigh; and *Djilgaringa thompsonae* Archer, Tedford and Rich, 1987, from the late Oligocene Yanda LF of South Australia.

PETAUROIDS—RINGTAILS, GLIDERS AND THEIR RELATIVES

In this group are the pseudocheirids or ringtail possums, and the petaurids which include the *Petaurus* gliders as well as Leadbeater's possum and the striped possums.

Ringtail Possums

The possums in this group are the ringtails (a paraphyletic vernacular group name including living species referable to the genera *Pseudocheirus*, *Pseudochirulus, Pseudochirops*, *Petroseudes* and *Hemibelideus*) and the greater glider (*Petauroides volans*). All pseudocheirids have (as far as can be determined) selenodont molars, buccally situated paracones and metacones, offset postprotocristae and prehypocristae, paracristids on the M2–4 that extend to the lingual sides of the teeth and the following dental formula: Il–3/1 or l–2 Cl/1 or 0 Pl–3/ l–3 Ml–4/1–4.

There has been general agreement that at least the living forms (Figure 49) placed in this group are monophyletic. But there has been much less agreement about where pseudocheirids should be placed within the marsupial radiation. Most of the early systematists regarded them simply to be 'possums' and hence placed them in the broadly conceived Phalangistidae. Bensley (1903) was among the first to suggest that they were cousins of the koala (the koala supposedly having evolved from ringtails) and hence in his phalangerid subfamily Phascolarctinae (all other 'phalangerids' being placed by him in the Phalangerinae). He reached this conclusion mainly because, like koalas, the ringtails had well developed selenodont molars and forcipate hands. Even Tate (1945a, b) continued to regard ringtails and greater gliders as phascolarctines.

Kirsch (1968, 1977) suggested quite a different relationship. Partly on the basis of his studies in serology, he decided that these possums were in fact most closely related to species of *Petaurus*, *Gymnobelideus* and possibly *Dactylopsila*. For these two possum groups he proposed the family name Petauridae. Koalas were, he concluded, much more closely related to wombats. But Archer (1984) treated ringtails as a separate family from the petauroids and this has been followed by most subsequent workers, including Long *et al.* (2002). That said, it is also clear from most molecular studies that petaurids and pseudocheirids are most closely related to each other, even if their dentitions differ.

The dental formula of pseudocheirids is, however, primitive within the Diprotodontia, and certainly more plesiomorphic than that of the only other conspicuously selenodont group, the phascolarctids, which (in common with wombats and diprotodontoids) have lost P1–2 and all lower teeth between I1 and P3 (Figure 50). It is possible that petaurids (Petauridae) are descendants of pseudocheirids but in a cladistic sense they may be each other's immediate sister-group (Brammall and Archer, 1999).

Alright then, we don't know for certain where pseudocheirids came from or even what came from them. But we can be more positive about relationships within the family (Figure 51). Points that are now absolutely clear are: (1) the rock ringtail possum, *Petroseudes dahli* is most closely related to the species of *Pseudochirops*; (2) the lemuroid ringtail possum, *Hemibelideus lemuroides*, is most closely related (among living forms) to the greater glider, *Petauroides volans*. All tissue systems so far examined seem to support these findings.

Less certain but highly probable are the following conclusions: (1) the Pliocene *Pseudokoala erlita* represents a third distinct part of the pseudocheirid radiation; (2) the Pliocene *Petauroides stirtoni* (described as *Pseudocheirus stirtoni*) is part of the *Hemibelideus-Petauroides* radiation; and (3) the species of *Pseudochirops* are monophyletic. The intrafamilial relationships of the species of *Pseudochirulus* and *Pseudocheirus* are less clear but, as suggested by Bassarova and Archer (1999), they may be each other's closest sister-group, and together the sister-group of the *Petauroides/Hemibelideus* group.

What do we know about the warm, furry pseudocheirids that adorn the modern forests of Australia and New Guinea? In fact, only the common ringtail pos-

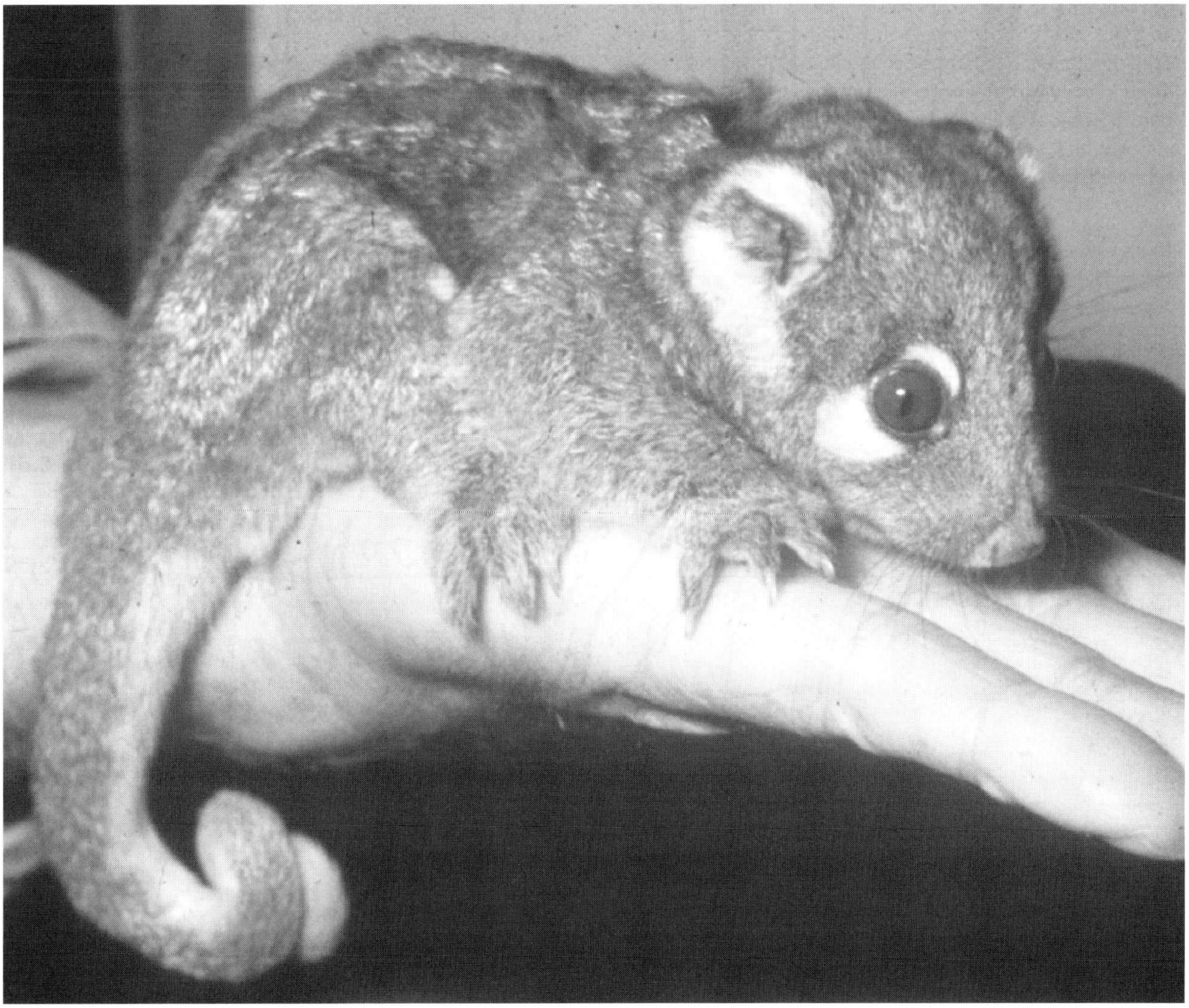

Figure 49. The green ringtail possum (*Pseudochirops archeri*) is found in the rainforests of north-eastern Queensland and is a close relative of several extinct woolly ringtail possums recovered from Oligo-Miocene sediments at Riversleigh. Photograph M. Archer.

able to the *Hemibelideus-Petauroides* lineage appear at least as early as the early Pliocene (in the Hamilton LF; *Petauroides stirtoni*). Fifth, there were giant forms in the Pleistocene (e.g., taxa in the Gore and Portland LFs), as there were in other marsupial families.

Among named extinct species are the following: *Marlu kutjamarpensis* Woodburne, Tedford and Archer, 1987, from the early Miocene Kutjamarpu LF of South Australia and possibly Miocene LFs from Riversleigh, Queensland; *Marlu praecursor* Woodburne, Tedford and Archer, 1987, from the late Oligocene Wadikali LF of South Australia; *Paljara tirarense* Woodburne, Tedford and Archer, 1987, from the early Miocene Kutjamarpu LF of South Australia (other species occur in the Riversleigh LFs); *Petauroides stirtoni* (Turnbull and Lundelius, 1970), from the early Pliocene Hamilton LF; *Petauroides ayamaruensis* Aplin in Aplin, Pasveer and Boles, 1999, from the early Holocene Kria Cave LF of Irian Jaya; *Petauroides marshalli* (Turnbull and Lundelius, 1970) from the early Pliocene Hamilton LF of Victoria; *Pildra antiquus* Woodburne, Tedford and Archer, 1987, from the late Oligocene Pinpa LF from South Australia; *Pildra secundus* Woodburne, Tedford and Archer, 1987, from the late Oligocene Ditjimanka and Ericmas LFs of South Australia; *Pildra tertius* Woodburne, Tedford and Archer, 1987, from the early Miocene Kutjamarpu LF from South Australia; *P. magnus* Pledge, 1987e, from the late Oligocene Ngama LF of South Australia; *Pseudokoala erlita* Turnbull and Lundelius, 1970, from the early Pliocene Hamilton LF of Victoria; *Pseudokoala cathysantamaria* Archer, Black and Nettle, 1997, from the early Pleistocene Portland LF; *Pseudokoala curramulkensis* (Pledge, 1992a) from the late Miocene to Pliocene Curramulka LF of South Australia.

Archer (1992), Bassarova and Archer (1999) and Bassarova *et al.* (2001) reported more than 12 pseudocheirids from Oligo-Miocene sediments at Riversleigh. These include species of *Pildra, Paljara, Marlu, Pseudochirops* and at least two unnamed genera. Archer and Bartholomai (1978) noted a species of *Pseudochirops* from the late Miocene Alcoota LF (Northern Territory) and Mackness and Archer (2002) reported another species of *Pseudochirops* from the early Pliocene Bluff Downs LF of Queensland. *Petropseudes* has a record extending back to the early Pliocene of Riversleigh (Archer *et al.*, 1994a). Clearly the fossil record of pseudocheirids is very rich and getting richer.

Gliders and Relatives

This is the other half of the older (1968) concept of Pseudocheiridae, regarded (e.g., by Kirsch, 1968, 1977)

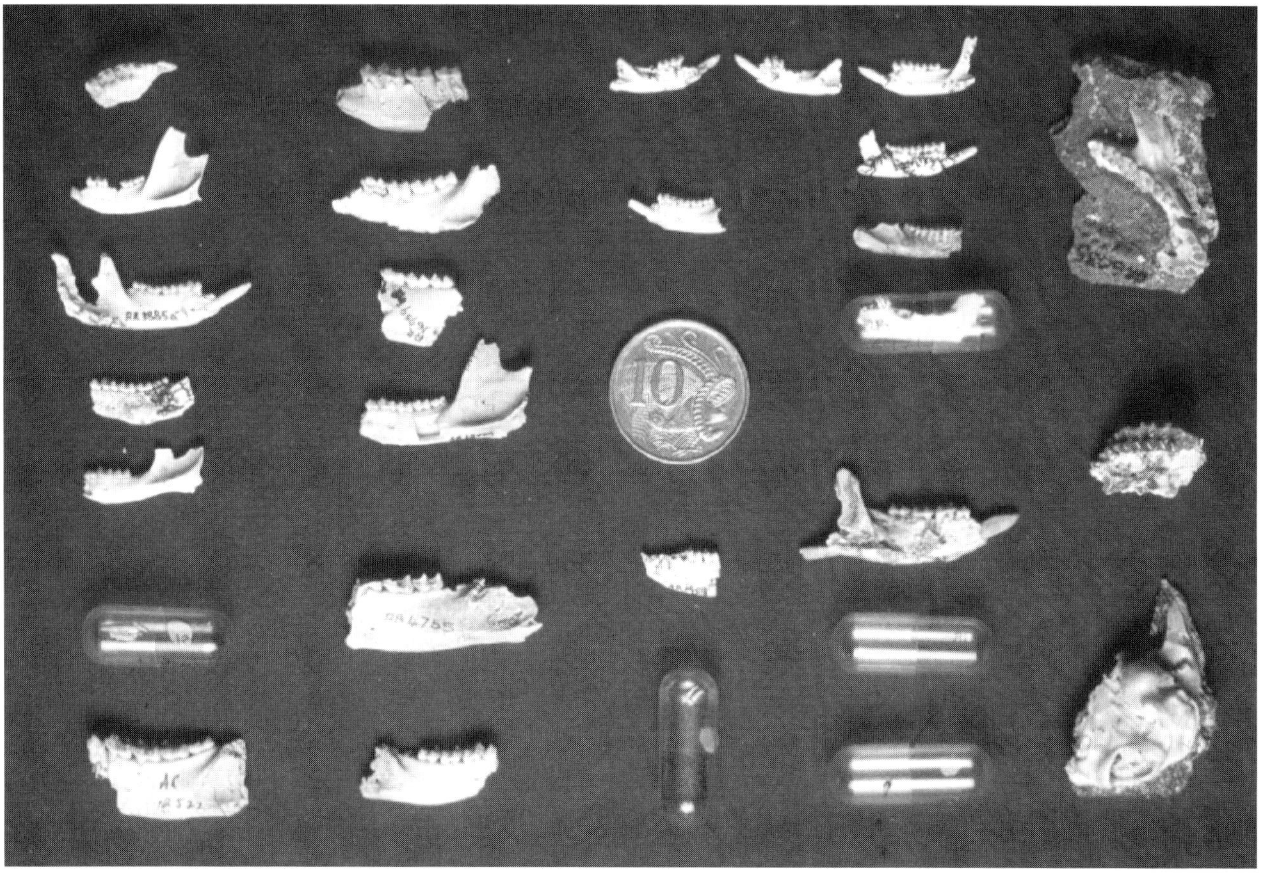

Figure 52. Many different kinds of ringtail possums are known from Australian Oligo-Miocene deposits. From Riversleigh alone, 18 species have been recorded. Photograph M. Archer.

to be the Petaurinae within the Petauridae. The reasons for peeling it off at a family level are explained above. As it stands here, Petauridae contains Leadbeater's possum (*Gymnobelideus leadbeateri*), the four species of *Petaurus*, and four species of *Dactylopsila*. See Figure 53 and Plate 103.

The monophyly of this group is in little doubt, but what little there is focuses on the species of *Dactylopsila*. Kirsch (1977b) proposed Dactylopsilinae as a new subfamily for the striped possums which he regarded to be related to the petaurines. However, he regarded them to be also closely related to pseudocheirids (in the sense used here). The overall features of the dentition (not considering the many autapomorphic features) also appear to ally the striped possums with petaurids rather than pseudocheirids. Petaurids all also share the following dental formula: Il–3/1 or l–2 Cl/1 or 0 Pl–3/ 1–3 Ml–4/1–4.

Within the Petauridae, we recognize two subfamilies: the Petaurinae (species of *Gymnobelideus* and *Petaurus*) and the Dactylopsinae (species of *Dactylopsila*). Let's have a brief look at each.

Within Petaurinae, Leadbeater's possum (*Gymnobelideus leadbeateri*), a non-gliding, arboreal omnivore, has had a curious modern history. It was discovered in July, 1867, in the Bass River valley of South Gippsland in Victoria and named by Professor F. McCoy in the same year (Wilkinson, 1961). From that time until 1909, only four specimens were known. In 1909 a fifth and supposedly last was presented to the National Museum of Victoria. By 1921, the species was presumed to be extinct, the victim of the destruction of the scrub and forests along the Bass River. Then, in 1961, H.E. Wilkinson, R.F. Lees and A. Sonnenberg, while surveying an area of bush in the Cumberland Valley of Victoria (~115 km from Melbourne), saw a very live and acrobatic Leadbeater's possum appear and vanish in a blackwood tree. Puzzlement gave way to jubilation at the discovery. Quite oblivious to the opinion of humans, Leadbeater's possum had risen from the ranks of the extinct (Wilkinson, 1961). Now, although it is clear that it occupies only a very small area of north-eastern Victoria, its future looks a bit brighter than it did in 1921. The field and laboratory studies of Andrew Smith and others should help to ensure its survival if the Forestry Department can ensure, among its other needs, the preservation of mature *Eucalyptus regans* trees which develop the appropriate sorts of spouts and hollows needed by this possum as nest sites.

The other petaurines are all species of *Petaurus*. All are accomplished gliders and devastatingly beautiful. There are now five species recognized: the large yellow-bellied glider (*P. australis*) from eastern Australia (Figure 54); the diminutive sugar glider (*P. breviceps*) from most of coastal Australia (except western and southern Western Australia) and New Guinea; the squirrel glider (*P. norfolcensis*) from eastern Australia (Figure 55); the recently described *P. abidi* from New Guinea; and the mahogany glider (*P. gracilis*) from mid-eastern Queensland. All species of *Petaurus* are highly social beasts. A pet orphan yellow-bellied glider raised in Brisbane used to intrigue Mike Archer by her habit of grabbing the hair on the back of his head in her hands and vigorously rubbing the crown of her head against his head. Although this ritual was repeated every night, its significance was unclear until Russell (1984) noted that this species uses crown and post-anal gland scents to knit 'family-level' groups together. Mutual head to tail-base rubbing is the way in which the dominant male ensures its scent is spread to its females and any other subordinates. That beguiling ball of fluff was simply telling Mike in its own instinctual way that he was 'family'. Schultze-Westrum's (1965) study of the social behaviour of the Papuan sugar gliders is another masterful demonstration of the complex social behaviour of these animals. Other more current studies of field behaviour and ecology in species of *Petaurus* are noted in Smith and Hume (1984).

Figure 53. Leadbeater's possum (*Gymnobelideus leadbeateri*) is a relatively plesiomorphic petaurid. This nimble, arboreal omnivore of the forests of north-eastern Victoria was thought to be extinct until rediscovered in 1961 in Victoria's Cumberland Valley (drawing by R. Goldingay).

The dactylopsilines are much less well known. They are distinguished from other petaurids by their lack of a gliding membrane, darkly coloured striped fur and elongated fourth digit on the hand. At present four living species are recognized (Honacki *et al.*, 1982;

Flannery, 1994b,1995a): *Dactylopsila trivirgata* (Australia and New Guinea); *D. palpator* (New Guinea); *D. megalura* (New Guinea); and *D. tatei* (New Guinea). What little is known about the biology of these species is available in Rand (1937), Moeller (1976), Winter *et al.* (1983), Smith (1982a) and Flannery (1994b, 1995a). Moeller's (1976) account (in German) is particularly interesting because it documents some previously unsuspected uses for the elongated fourth finger of species of this genus. As has long been known, they use it to withdraw, in fish-hook fashion, wood-boring insect larvae from their tunnels; behaviour that is remarkably convergent on that of the similarly long-fingered aye aye (*Daubentonia madagascariensis*). But Moeller (1976) has also shown that they may grip branches with their elongate fourth fingers, in forceps-like fashion, for support while the other fingers are manipulating food in front of the mouth.

The fossil record of petaurids is relatively poor although some new species have been described (see below). Most have been recorded but not fully described from the middle Miocene Tarkarooloo, early Pliocene Hamilton, and many Riversleigh LFs. The reasons for this relative scarcity are not clear but we can hazard one suggestion. At least as far as species of the genus *Petaurus* are concerned, their apparent lack of appearance prior to the Pliocene may reflect the fact that open forests (the habitat of gliders) were not as widespread as they are today until the very end of the Tertiary. In closed forests there would be little point in evolving the capacity to glide. Perhaps it was not until the mid to late Miocene, when the central Australian closed forests were

Figure 54. The yellow-bellied glider (*Petaurus australis*) in a characteristic pose—upside down in a tree. These magnificent gliders feed mainly on nectar, sap, pollen and some insects found under bark. They are highly social and like all members of this genus communicate with a complex array of sounds and scents (drawing by R. Goldingay).

beginning to open out that there was selection pressure on arboreal possums to develop the ability to glide.

Fossil taxa include: *Dactylopsila kambuayai* Aplin in Aplin, Pasveer and Boles, 1999, from the early Holocene, Cave sediments of Kria Cave, Irian Jaya; Brammall (1998) and Brammall and Archer (1999) reported that at least one species of *Dactylopsila* is present in the late Oligocene to middle Miocene limestones of Riversleigh; Turnbull *et al.* (1987) reported *Petaurus* sp. cf. *P. norfolcensis* and *P.* sp. cf. *P. australis* from the early Pliocene Hamilton LF of Victoria; *Djaludjangi yadjana* Brammall, 1998, comes from the early to ?early–Late Miocene Dirk Towers LF and many other early and middle Miocene sites at Riversleigh (it lacks some characteristic features of petaurids, however, and may represent a new family of petauroids; Brammall, 1998).

Figure 55. The squirrel glider (*Petaurus norfolcensis*) is a voracious consumer of small animals as well as the products of plants—nectar, sap, pollen and some insects found under bark (drawing by R. Goldingay).

From whence came the petaurids and to what if anything were they ancestral? The answers to these questions are tied to considerations of the sister-group of the Petauridae. As noted above, there are two current views: first that Petauridae and Pseudocheiridae are each other's sister-group, the view we share here; second, that they may be even more closely related to tarsipedids and species of the *Acrobates-Distoechurus* radical (Brammall and Archer, 1999).

KANGAROOS—NON-ROOTINE ROOBOUNDING ROOMINANTS

Because Bernie Cooke has presented a whole chapter (Cooke, 2006) on this interesting group of marsupials, we will not attempt to cover the same ground here. This is a fascinating group and Bernie has done a superb job of summarizing the salient and most interesting aspects of this great radiation (Plates 104, 105). It's worth contemplating that at the same time funny little primates were coming out of the forests into the grasslands of Africa to stand up on two legs, freeing their arms to manipulate things, kangaroos were doing exactly the same thing. The only difference is that kangaroos didn't then set out to subjugate the world!

ACKNOWLEDGEMENTS

We acknowledge with intense gratitude the many colleagues and students who have worked so very hard to turn Riversleigh's limestones and the clays of central Australia into soluble mud and taxonomic treasures! In particular, Henk Godthelp, Anna Gillespie, Kirsten Crosby, Jeanette Muirhead, Phil Creaser, Karen Black, Steve Wroe, Mina Bassarova, Troy Myers, Rick Arena, Arthur and Karen White, Georgina Hickey, Mike Woodburne, Dick Tedford, Rod Wells, Peter Murray, Dirk Megirian, Tom and Pat Rich, Neville Pledge, Steve Van Dyck, Scott Hocknull, Alex Baynes, Tim Flannery, John Long, Lloyd Campbell, the various managers of Riversleigh, Etadunna and Frome Downs Stations, the University of New South Wales, Australian Research Council, Queensland Museum, Queensland National Parks and Wildlife Service, World Heritage Unit of Environment Australia, Frank Nissan (surveyor), Rod and Michelle Low Mow, an army of PhD and Honours students, many volunteers from the public, the Riversleigh Society, the Riversleigh Interpretive Centre in Mount Isa, the Waanyi aboriginal people of the Riversleigh region, Pasminco, Earthwatch, the Australian Museum Society and so many other individuals and organizations who have devoted so much of their lives to developing the understanding that unpins what we are able to summarize above.

REFERENCES

Abbie, A.A. 1937. Some observations on the major subdivisions of the Marsupialia, with especial reference to the position of the Peramelidae and Caenolestidae. *Journal of Anatomy*, Vol. 71, pp.426–436.

Akerman, K. 1973. Two Aboriginal charms incorporating fossil giant marsupial teeth. *Western Australian Naturalist*, Vol. 12, pp.139–141.

Anderson, C. 1937. Palaeontological notes. No. IV. Fossil marsupials from New Guinea. *Records of the Australian Museum,* Vol. 20, pp.73–78.

Aplin, K.P. 1987. Basicranial anatomy of the early Miocene diprotodontian *Wynyardia bassiana* (Marsupialia: Wynyardiidae) and its implications for wynyardiid phylogeny and classification. Pp.369–391. In Archer, M (ed). *Possums and Opossums: Studies in Evolution*. Vol. 1. Surrey Beatty and Sons, Sydney. pp.1–400.

Aplin, K.P. 2006. Ten million years of rodent evolution in Australasia: phylogenetic evidence and a speculative historical biogeography. Pp.707–744. In Merrick, J.R., Archer, M., Hickey, G.M. and Lee, M.S.Y. (eds). *Evolution and Biogeography of Australasian Vertebrates.* Auscipub Pty. Ltd., Sydney. 942pp.

Aplin, K.P. and Archer, M. 1987. Recent advances in marsupial systematics with a new syncretic classification. Pp xv–lxxii. In Archer, M (ed). *Possums and Opossums: Studies in Evolution*. Vol. 1. Surrey Beatty and Sons, Sydney. pp.1–400.

Aplin, K.P., Pasveer, J.M. and Boles, W.E. 1999. Late Quaternary vertebrates from the Bird's Head Peninsula, Irian Jaya, Indonesia, including descriptions of two previously unknown marsupial species. *Records of the Western Australian Museum, Supplement,* No. 57, pp.351–387.

Archer, M. 1976a. Miocene marsupicarnivores (Marsupialia) from central South Australia, *Ankotarinja tirarensis* gen. et sp. nov., *Keeuna woodburnei* gen. et sp. nov., and their significance in terms of early marsupial radiations. *Transactions of the Royal Society of South Australia*, Vol. 100, pp.53–73.

Archer, M. 1976b. The dasyurid dentition and its relationships to that of didelphids, thylacinids, borhyaenids (Marsupicarnivora) and peramelids (Peramelina: Marsupialia). *Australian Journal of Zoology*, Supplementary Series, Vol. 39, pp.1–34.

Archer, M. 1976c. The basicranial region of marsupicarnivores (Marsupialia), inter-relationships of the carnivorous marsupials, and affinities of the insectivorous marsupial peramelids. *Zoological Journal of the Linnean Society*, Vol. 59, pp.217–322.

Archer, M. 1976d. Bluff Downs local fauna. *Memoirs of the Queensland Museum*, Vol. 17, pp.383–398.

Archer, M. 1976e. Phascolarctid origins and the potential of the selenodont molar in the evolution of diprotodont marsupials. *Memoirs of the Queensland Museum*, Vol. 17, pp.367–371.

Archer, M. 1977. *Koobor notabilis* (de Vis), an unusual koala from the Pliocene Chinchilla Sand. *Memoirs of the Queensland Museum*, Vol. 18, pp.31–35.

Archer, M. 1981. A review of the origins and radiations of Australian mammals. Pp.1437–1488. In Keast, A. (ed). *Ecological Biogeography in Australia.* Vol. 3.W. Junk, The Hague. pp.1435–2142.

Archer, M. (ed). 1982a. *Carnivorous Marsupials.* 2 Vols. Surrey Beatty and Sons, Sydney. 804pp.

Archer, M. 1982b. Review of the dasyurid (Marsupialia) fossil record, integration of data bearing on phylogenetic interpretation, and suprageneric classification. Pp.397–443. In Archer, M. (ed). *Carnivorous Marsupials.* Vol. 2. Surrey Beatty and Sons, Sydney. pp.397–804

Archer, M. 1984. The Australian marsupial radiation. Pp.633–808. In Archer, M and Clayton, G. (eds). *Vertebrate Zoogeography and Evolution in Australasia.* Hesperian Press, Perth. 1203pp.

Archer, M. (ed). 1987. *Possums and Opossums: Studies in Evolution.* 2 Vols. Surrey Beatty and Sons, Sydney. 788pp.

Archer, M. 1992. Ringtail possums (Pseudocheiridae, Marsupialia) from the Tertiary deposits of Riversleigh. *The Beagle (Records of the Northern Territory Museum of Arts and Sciences)*, Vol. 9, No. 1, pp.257–258.

Archer, M., Arena, D.A., Bassarova, M., Black, K., Brammall, J.R., Cooke, B.N., Creaser, P.H., Crosby, K., Gillespie, A.K., Godthelp, H.J., Gott, M., Hand, S.J., Kear, B.P., Krikmann, A., Mackness, B.S., Muirhead, J., Musser, A.M., Myers, T.J., Pledge, N.S., Wang, Y. and Wroe, S. 1999. The evolutionary history and diversity of Australian mammals. *Australian Mammalogy*, Vol. 21, pp.1–45.

Archer, M. and Bartholomai, A. 1978. Tertiary mammals of Australia: a synoptic review. *Alcheringa*, Vol. 2, pp.1–19.

Archer, M., Black, K. and Nettle, K. 1997. Giant ringtail possums (Marsupialia, Pseudocheiridae) and giant koalas (Phascolarctidae) from the late Cainozoic of Australia. *Proceedings of the Linnean Society of New South Wales*, Vol. 117, pp.3–15.

Archer, M. and Clayton, G. (eds). 1984. *Vertebrate Zoogeography and Evolution in Australasia*. Hesperian Press, Perth. 1203pp.

Archer, M., Crawford, I.M. and Merrilees, D. 1980. Incisions, breakages and charring, some probably manmade, in fossil bones from Mammoth Cave, Western Australia. *Alcheringa*, Vol. 4, pp.115–131.

Archer, M., Every, R.G., Godthelp, H.J., Hand, S.J. and Scally, K.B. 1990. Yingabalanaridae, a new family of enigmatic mammals from Tertiary deposits of Riversleigh, northwestern Queensland. *Memoirs of the Queensland Museum*, Vol. 28, pp.193–202.

Archer, M., Flannery, T.F., Ritchie, A. and Molnar, R.E. 1985. First Mesozoic mammal from Australia—an early Cretaceous monotreme. *Nature*, Vol. 318, pp.363–366.

Archer, M. and Godthelp, H. 1992. Wombat-like marsupials from the Oligo-Miocene faunal assemblages of Riversleigh. *The Beagle (Records of the Northern Territory Museum of Arts and Sciences)*, Vol. 9, No. 1, pp.256–257.

Archer, M., Godthelp, H.J., Gott, M., Wang, Y. and Musser, A.M. 1999. The evolutionary history of notoryctids, yingabalanarids, yalkaparidontids and other enigmatic groups of Australian mammals. *Australian Mammalogy*, Vol. 21, pp.13–15, 34–45.

Archer, M., Godthelp, H.J. and Hand, S.J. 1993. Early Eocene marsupial from Australia. Pp.193–200. In Schrenk, F and Ernst, K. (eds). *Kaupia: Darmstädter Beiträge zur Naturgeschichte. Monument Grube Messel—Perspectives and Relationships. Part 2.* Hessisches Landesmuseum Darmstadt, Darmstadt, Germany.

Archer, M., Godthelp, H.J., Hand, S.J. and Megirian, D. 1989. Fossil mammals of Riversleigh, northwestern Queensland: preliminary overview of biostratigraphy, correlation and environmental change. *Australian Zoologist*, Vol. 25, pp.29–65.

Archer, M. and Hand, S.J. 1984. Background to the search for Australia's oldest mammals. Pp.517–565. In Archer, M. and Clayton, G. (eds). *Vertebrate Zoogeography and Evolution in Australasia.* Hesperian Press, Perth. 1203pp.

Archer, M., Hand, S.J. and Godthelp, H.J. 1988. A new order of Tertiary zalambdodont marsupials. *Science*, Vol. 239, pp.1528–1531.

Archer, M., Hand, S.J. and Godthelp, H.J. 1992. Back to the future: the contribution of palaeontology to the conservation of Australian forest faunas. Pp.67–80. In Lunney, D. (ed). *Conservation of Australia's Forest Fauna.* Royal Zoological Society of New South Wales, Sydney. 415pp.

Archer, M., Hand, S.J. and Godthelp, H.J. 1994. *Riversleigh: the Story of Animals in Ancient Rainforests of Inland Australia.* (Reprint) Reed Books, Sydney. 264pp.

Archer, M., Hand, S.J. and Godthelp, H.J. 1995. Tertiary environmental and biotic change in Australia. Pp.77–90. In Vrba, E., Denton, G.H., Partridge, T.C. and Burckle, L.H. (eds). *Palaeoclimate and Evolution, with Emphasis on Human Origins.* Yale University Press, New Haven.

Archer, M., Hand, S.J., Godthelp, H. and Creaser, P. 1997. Correlation of the Cainozoic sediments of the Riversleigh World Heritage fossil property, Queensland, Australia. Pp.131–152. In Aguilar, J.-P., Legendre, S. and Michaux, J. (eds). *Actes du Congrès BiochroM'97.* École Pratique des Hautes Études, Institut de Montpellier, Montpellier, France.

Archer, M. and Kirsch, J.A.W. 1977. The case for the Thylacomyidae and Myrmecobiidae, Gill, 1872, or why are marsupial families so extended? *Proceedings of the Linnean Society of New South Wales*, Vol. 102, pp.18–25.

Archer, M. and Rich, T.H. 1979. *Wakamatha tasselli* gen. et sp. nov., a fossil dasyurid (Marsupialia) from South Australia convergent on modern *Sminthopsis*. *Memoirs of the Queensland Museum*, Vol. 19, pp.309–317.

Archer, M. and Rich, T.H. 1982. Results of the Ray E. Lemley expeditions. *Wakaleo alcootaensis* n. sp. (Thylacoleonidae, Marsupialia), a new marsupial lion from the Miocene of the Northern Territory with a consideration of early radiation in the family. Pp.495–502. In Archer, M. (ed). *Carnivorous Marsupials.* Vol. 2. Surrey Beatty and Sons, Sydney. pp.397–804.

Archer, M., Tedford, R.H. and Rich, T.H. 1987. The Pilkipildridae, a new family and four new species of ?petauroid possums (Marsupialia: Phalangerida) from the Australian Miocene. Pp.607–627. In Archer, M. (ed). *Possums and Opossums: Studies in Evolution.* Vol. 2. Surrey Beatty and Sons, Sydney. pp.401–788

Ardran, G.M., Kemp, F.H. and Ride, W.D.L. 1958. A radiographic analysis of mastication and swallowing in the domestic rabbit *Oryctolagus cuniculus* (L). *Proceedings of the Zoological Society of London*, Vol. 130, pp.257–274.

Bartholomai, A. 1962. A new species of *Thylacoleo* and notes on some caudal vertebrae of *Palorchestes azael*. *Memoirs of the Queensland Museum*, Vol. 14, pp.33–40.

Bartholomai, A. 1968. A new fossil koala from Queensland and a reassessment of the taxonomic position of the problematic species, *Koalemus ingens* DeVis. *Memoirs of the Queensland Museum*, Vol. 15, pp.65–71.

Bartholomai, A. 1971. *Dasyurus dunmalli*, a new species of fossil marsupial (Dasyuridae) in the upper Cainozoic deposits of Queensland. *Memoirs of the Queensland Museum*, Vol. 16, pp.19–26.

Bartholomai, A. 1978. The rostrum in *Palorchestes* Owen (Marsupialia: Diprotodontidae); results of the Ray E. Lemley expedition, Part 3. *Memoirs of the Queensland Museum*, Vol. 18, pp.145–149.

Bassarova, M. and Archer, M. 1999. Living and extinct pseudocheirids (Marsupialia, Pseudocheiridae): phylogenetic relationships and changes in diversity through time. *Australian Mammalogy*, Vol. 21, pp.25–27, 34–45.

Bassarova, M., Archer, M. and Hand, S.J. 2001. New species of ringtail possums (Pseudocheiridae, Marsupialia) referable to the Oligo-Miocene genus *Paljara* from deposits of the Riversleigh World Heritage property, northwestern Queensland. *Memoir of the Association of Australasian Palaeontologists,* Vol. 25, pp.61–75.

Baverstock, P.R. 1984. The molecular relationships of Australasian possums and gliders. Pp.1–8. In Smith, A.P. and Hume, I.D. (eds). *Possums and Gliders*, Surrey Beatty and Sons, Sydney. 598pp.

Baverstock, P.R., Archer, M., Adams, M. and Richardson, B.J. 1982. Genetic relationships among 32 species of Australian dasyurid marsupials. Pp.641–650. In Archer, M. (ed). *Carnivorous Marsupials.* Vol. 2. Surrey Beatty and Sons, Sydney. pp.397–804.

Baverstock, P.R., Birrell, J. and Krieg, M. 1987. Albumin immunologic relationships among Australian possums: a progress report. Pp.229–234. In Archer, M. (ed). *Possums and Opossums: Studies in Evolution.* Vol. 1. Surrey Beatty and Sons, Sydney. pp.1–400.

Bensley, B.A. 1903. On the evolution of the Australian Marsupialia; with remarks on the relationships of the marsupials in general. *Transactions of the Linnean Society of London, Zoology*, Vol. 9, pp.83–217.

Black, K. 1997. A new species of Palorchestidae (Marsupialia) from the late middle to early late Miocene Encore Local Fauna, Riversleigh, northwestern Queensland. *Memoirs of the Queensland Museum*, Vol. 41, pp.181–185.

Black, K. 1999. Diversity and relationships of living and extinct koalas (Phascolarctidae, Marsupialia). *Australian Mammalogy*, Vol. 21, pp.16–17, 34–45.

Black, K. and Archer, M. 1997a. *Nimiokoala* gen. nov. (Marsupialia, Phascolarctidae) from Riversleigh, northwestern Queensland, with a revision of *Litokoala*. *Memoirs of the Queensland Museum*, Vol. 41, pp.209–228.

Black, K. and Archer, M. 1997b. *Silvabestius* gen. nov., a primitive zygomaturine (Marsupialia, Diprotodontidae) from Riversleigh, northwestern Queensland. *Memoirs of the Queensland Museum*, Vol. 41, pp.193–208.

Black, K. and Mackness, B.S. 1999. Diversity and relationships of diprotodontoid marsupials. *Australian Mammalogy*, Vol. 21, pp.20–21, 34–45.

Boles, W.E. 1999. Early Eocene shorebirds (Aves: Charadriiformes) from the Tingamarra Local Fauna, Murgon, Queensland, Australia. *Records of the Western Australian Museum, Supplement*, No. 57, pp.229–238.

Brammall, J.R. 1998. A new petauroid possum from the Oligo-Miocene of Riversleigh, northwestern Queensland. *Alcheringa*, Vol. 23, pp.31–50.

Brammall, J.R. and Archer, M. 1997. A new Oligocene–Miocene species of *Burramys* (Marsupialia, Burramyidae) from Riversleigh, northwestern Queensland. *Memoirs of the Queensland Museum*, Vol. 41, pp.247–268.

Brammall, J.R. and Archer, M. 1999. Living and extinct petaurids, acrobatids, tarsipedids and burramyids (Marsupialia): relationships and diversity through time. *Australian Mammalogy*, Vol. 21, pp.24–25, 34–45.

Broom, R. 1896a. On a small fossil marsupial with large grooved premolars. *Proceedings of the Linnean Society of New South Wales,* (Series 2), Vol. 10, pp.563–567.

Broom, R. 1896b. On a small fossil marsupial allied to *Petaurus. Proceedings of the Linnean Society of New South Wales* (Series 2), Vol. 10, pp.568–570.

Broom, R. 1898. On the affinities and habits of *Thylacoleo. Proceedings of the Linnean Society of New South Wales*, Vol. 23, pp.57–74.

Campbell, C. 1976a. Tertiary Dasyuridae and Peramelidae from the Tirari Desert, South Australia. *Dissertation Abstracts International B. Science and Engineering*, Vol. 37, No. 9, p. 4375.

Campbell, C. 1976b. *Tertiary Dasyuridae and Peramelidae (Marsupialia) from the Tirari Desert, South Australia.* Ph.D. dissertation, University of California, Berkeley. 220pp.

Clemens, W.A. and Plane, M.D. 1974. Mid–Tertiary Thylacoleonidae (Marsupialia, Mammalia). *Journal of Paleontology*, Vol. 48, pp.652–660.

Cooke, B. 2006. Kangaroos. Pp.647–672. In Merrick, J.R., Archer, M., Hickey, G.M. and Lee, M.S.Y. (eds). *Evolution and Biogeography of Australasian Vertebrates.* Auscipub Pty. Ltd., Sydney. 942pp.

Crabb, P.L. 1982. Pleistocene dasyurids (Marsupialia) from south-western New South Wales. Pp.511–516. In Archer, M. (ed). *Carnivorous Marsupials.* Vol. 2. Surrey Beatty and Sons, Sydney. pp.397–804.

Creaser, P. 1997. Oligocene–Miocene sediments of Riversleigh: the potential significance of topography. *Memoirs of the Queensland Museum*, Vol. 41, pp.303–314.

Crosby, K. and Archer, M. 2000. *Durudawiri inusitatus*, a new phalangeroid marsupial from the Miocene of Riversleigh, northwestern Queensland. *Journal of Paleontology*, Vol. 74, pp.327–335.

Crosby, K., Godthelp, H.J., Archer, M. and Pledge, N.S. 1999. Diversity and evolution of phalangerid, ektopodontid, miralinid and pilkipildrid marsupials. *Australian Mammalogy*, Vol. 21, pp.22–23, 34–45.

Crosby, K., Nagy, M. and Archer M. 2001. *Wyulda asherjoeli*, a new phalangerid (Diprotodontia: Marsupialia) from the early Miocene of Riversleigh, northwestern Queensland. *Memoir of the Association of Australasian Palaeontologists*, Vol. 25, pp.77–82.

Dawson, L. 1982. Taxonomic status of fossil devils (*Sarcophilus*, Dasyuridae, Marsupialia) from late Quaternary eastern Australian localities. Pp.517–525. In Archer, M. (ed). *Carnivorous Marsupials.* Vol. 2. Surrey Beatty and Sons, Sydney. pp.397–804.

Dawson, L. 1983. The taxonomic status of small fossil wombats (Vombatidae: Marsupialia) from the Quaternary deposits, and of related modern wombats. *Proceedings of the Linnean Society of New South Wales*, Vol. 107, pp.99–121.

Dawson, T.J. and Dawson, L. 2006. Evolution of arid Australia and consequences for vertebrates. Pp.51–70. In Merrick, J.R., Archer, M., Hickey, G.M. and Lee, M.S.Y. (eds). *Evolution and Biogeography of Australasian Vertebrates.* Auscipub Pty. Ltd., Sydney. 942pp.

Dawson, L., Muirhead, J. and Wroe, S. 1999. The Big Sink Local Fauna: a lower Pliocene mammalian fauna from the Wellington Caves complex, Wellington, New South Wales. *Records of the Western Australian Museum, Supplement*, No. 57, pp.265–290.

De Vis, C.W. 1884. On tooth-marked bones of extinct marsupials. *Proceedings of the Linnean Society of New South Wales*, Vol. 8, pp.187–190.

De Vis, C.W. 1887. In: Anonymous. *The Brisbane Courier*, No. 9224, p. 6.

De Vis, C.W. 1888. In: Anonymous. On a supposed new species of *Nototherium. Abstract of the Proceedings on the Linnean Society of New South Wales* for December 28th 1887, p. v.

De Vis, C.W. 1895. A review of the fossil jaws of the Macropodidae in the Queensland museum. *Proceedings of the Linnean Society of New South Wales*, Vol. 2, No. 10, pp.75–133.

Epstein, J. 1981. *The Friends of Burramys.* Oxford University Press, Melbourne.

Flannery, T.F. 1992. New Pleistocene marsupials (Macropodidae, Diprotodontidae) from subalpine habitats in Irian Jaya, Indonesia. *Alcheringa*, Vol. 16, pp.321–331.

Flannery, T.F. 1994a. *The Future Eaters. An Ecological History of the Australasian Lands and People.* Reed Books, Sydney. 423pp.

Flannery, T.F. 1994b. *Possums of the World.* Geo Productions, Sydney. 240pp.

Flannery, T.F. 1995a. *Mammals of New Guinea.* Revised Edition. Reed Books, Sydney. 568pp.

Flannery, T.F. 1995b. *Mammals of the Southwest Pacific and Moluccan Islands.* Reed Books, Sydney. 464pp.

Flannery, T.F. and Archer, M. 1987. *Strigocuscus reidi* and *Trichosurus dicksoni*, two new fossil phalangerids (Marsupialia: Phalangeridae) from the Miocene of northwestern Queensland. Pp.527–536. In Archer, M. (ed). *Possums and Opossums: Studies in Evolution.* Vol. 2. Surrey Beatty and Sons, Sydney. pp.401–788.

Flannery, T.F., Archer, M., Rich, T.H. and Jones, R. 1995. A new family of monotremes from the Cretaceous of Australia. *Nature*, Vol. 377, pp.418–420.

Flannery, T.F. and Plane, M.D. 1986. A new late Pleistocene diprotodontid (Marsupialia) from Pureni, Southern Highlands Province, Papua New Guinea. *Bureau of Mineral Resources, Journal of Australian Geology and Geophysics*, Vol. 10, pp.65–76.

Flannery, T.F., Turnbull, W.D., Rich, T.H. and Lundelius, E.L. Jnr. 1987. The phalangerids (Marsupialia: Phalangeridae) of the early Pliocene Hamilton Local Fauna, south-western Victoria. Pp.537–546. In Archer, M. (ed). *Possums and Opossums: Studies in Evolution.* Vol. 2. Surrey Beatty and Sons, Sydney. pp.401–788.

Furby, J.H., Fullagar, R.K.L., Dodson, J.R. and Prosser, I. 1993. The Cuddie Springs bone bed revisited, 1991. Sahul in Review: Pleistocene archaeology in Australia, New Guinea and island Melanesia. *Occasional Papers in Prehistory*, No. 24, pp.204–210.

Gillespie, A.K. 1997. *Priscileo roskellyae* sp. nov. (Thylacoleonidae, Marsupialia) from the Oligocene–Miocene of Riversleigh, northwestern Queensland. *Memoirs of the Queensland Museum*, Vol. 41, pp.321–327.

Godthelp, H., Archer, M., Cifelli, R.L., Hand, S.J. and Gilkeson, C.F. 1992. Earliest known Australian Tertiary mammal fauna. *Nature*, Vol. 356, pp.514–516.

Godthelp, H., Wroe, S. and Archer, M. 1999. A new marsupial from the early Eocene Tingamarra Local Fauna of Murgon, southeastern Queensland: a prototypical Australian marsupial? *Journal of Mammalian Evolution*, Vol. 6, pp.514–516.

Gott, M. 1988. *A Tertiary marsupial mole (Marsupialia: Notoryctidae) from Riversleigh, north-eastern Australia and its bearing on notoryctemorphian phylogenetics.* B. Sc. (Hons.) Thesis (unpublished), School of Biological Science, University of New South Wales, Sydney.

Gould, J. 1863. *The Mammals of Australia*. Reprinted in 1983 by Macmillan, Melbourne.

Green, R.H. 1983. *Skulls of the Mammals in Tasmania*. Queen Victoria Museum and Art Gallery, Launceston.

Gregory, W.K. 1951. *Evolution Emerging: a Survey of Changing Patterns From Primeval Life to Man*. Macmillan, New York.

Haight, J.R. and Murray, P.F. 1981. The cranial endocast of the early Miocene marsupial, *Wynyardia bassiana*: an assessment of taxonomic relationships based upon comparisons with Recent forms. *Brain, Behaviour, Evolution*, Vol. 19, pp.17–36.

Hand, S.J., Archer, M. and Godthelp, H.J. 1997. First record of *Hydromops* (Microchiroptera: Molossidae) from Australia: its biocorrelative significance. Pp.153–162. In Aguilar, J.–P., Legendre, S. and Michaux, J. (eds). *Actes du Congrès BiochroM'97*. École Pratique des Hautes Études, Institut de Montpellier, Montpellier, France.

Hand, S.J., Archer, M., Godthelp, H.J., Rich, T.H. and Pledge, N.S. 1993. *Nimbadon*, a new genus and three new species of Tertiary zygomaturines (Marsupialia: Diprotodontidae) from northern Australia, with a reassessment of *Neohelos*. *Memoirs of the Queensland Museum*, Vol. 33, pp.193–210.

Hand, S.J. and Kirsch, J.A.W. 1998. A southern origin for the Hipposideridae (Microchiroptera)? Evidence from the Australian fossil record. Pp.72–90. In Kunz, T. and Racey, P. (eds). *Bat Biology and Conservation*. Smithsonian Institution, Washington, D.C. 365pp.

Hand, S.J., Novacek, M.J., Godthelp, H.J. and Archer, M. 1994. First Eocene bat from Australia. *Journal of Vertebrate Paleontology*, Vol. 14, pp.375–381.

Harding, H.R. 1987. Interrelationships of the families of the Diprotodonta—a view based on spermatozoan ultrastructure. Pp.195–216. In Archer, M. (ed). *Possums and Opossums: Studies in Evolution*. Vol. 1. Surrey Beatty and Sons, Sydney. pp.1–400.

Honacki, J.H., Kinman, K.E. and Koeppl, J.W. (eds). 1982. *Mammal Species of the World*. Allen Press Inc. and Association of Systematic Collections, Lawrence, Kansas. 694pp.

Hope, J.H. and Wilkinson, H.E. 1982. *Warendja wakefieldi*, a new genus of wombat (Marsupialia, Vombatidae) from Pleistocene sediments in McEacherns Cave, western Victoria. *Memoirs of the National Museum Victoria*, Vol. 43, pp.109–120.

Horton, D.R. and Wright, R.V.S. 1981. Cuts on Lancefield bones: carnivorous *Thylacoleo*, not humans, the cause. *Archaeology in Oceania*, Vol. 16, pp.73–80.

Johanson, Z. 2006. Origins of mammals: morphology, molecules and a petrosal or two. Pp.491–522. In Merrick, J.R., Archer, M., Hickey, G.M. and Lee, M.S.Y. (eds). *Evolution and Biogeography of Australasian Vertebrates*. Auscipub Pty. Ltd., Sydney. 942pp.

Johnson, C.N. and Wroe, S. 2003. Causes of extinctions of vertebrates during the Holocene of mainland Australia: arrival of the dingo or human impact? *Holocene*, Vol. 13, Issue 6, pp.941–948.

Jones, F. W. 1923–25. *The Mammals of South Australia*. Parts 1–3. Government Printer Adelaide, (Reprinted 1968).

Jones, F. W. 1930. A re-examination of the skeletal characters of *Wynyardia bassiana*, an extinct Tasmanian marsupial. *Papers and Proceedings of the Royal Society of Tasmania* 1930, pp.96–115.

Jones, M., Dickman, C. and Archer, M. (eds). 2003. *Predators with Pouches: the Biology of Carnivorous Marsupials*. CSIRO Publishing, Melbourne. 486pp.

Kerle, A. 2001. *Possums: the Brushtails, Ringtails and Greater Glider*. UNSW Press, Sydney. 128pp.

Kirsch, J.A.W. 1968. Prodromus of the comparative serology of Marsupialia. *Nature*, Vol. 217, pp.418–420.

Kirsch, J.A.W. 1977. The classification of marsupials. Pp.1–50. In Hunsaker, D. (ed). *The Biology of Marsupials*, Academic Press, New York,

Kirsch, J.A.W. 1979. Notes on the trapping and behaviour of the Caenolestidae (Marsupialia). *Journal of Mammalogy*, Vol. 60, pp.390–437.

Kirsch, J.A.W. and Archer, M. 1982. Polythetic cladistics, or, when parsimony's not enough: the relationships of carnivorous marsupials. Pp.595–619. In Archer, M. (ed). *Carnivorous Marsupials*. Vol. 2. Surrey Beatty and Sons, Sydney. pp.397–804.

Krajewski, C., Painter, J., Buckley, L. and Westerman, M. 1994. Phylogenetic structure of the marsupial family Dasyuridae. *Journal of Mammalian Evolution*, Vol. 2, pp.25–35.

Krajewski, C. and Westerman, M. 2003. Molecular systematics of Dasyuromorphia. Pp.3–20. In Jones, M., Dickman, C. and Archer, M. (eds). *Predators with Pouches: the Biology of Carnivorous Marsupials*. CSIRO Publishing, Melbourne. 486pp.

Krajewski, C., Wroe, S. and Westerman, M. 2000. Molecular evidence for the timing and pattern of cladogenesis in dasyurid marsupials. *Zoological Journal of the Linnean Society*, Vol. 130, pp.375–404.

Long, J., Archer, M., Flannery, T.F. and Hand, S.J. 2002. *Prehistoric Mammals of Australia and New Guinea*. UNSW Press, Sydney. 244pp.

Mackness, B. 1995. *Palorchestes selestiae*, a new species of palorchestid marsupial from the early Pliocene Bluff Downs Local Fauna, northeastern Queensland. *Memoirs of the Queensland Museum*, Vol. 38, pp.603–609.

Mackness B.S., and Archer M. 2002. *Pseudochirops winteri* n. sp, a new petaurid possum from the Pliocene Bluff Downs Local Fauna of northern Queensland. *Alcheringa*, Vol. 25, pp.439–444.

Mackness B.S., Whitehead, P. and McNamara, G. 2000a. A new potassium-argon date in relation to the Pliocene Bluff

Downs Local Fauna, northern Australia. *Australian Journal of Earth Sciences*, Vol. 47, pp.807–811.

Mackness, B. S., Wroe, S., Muirhead, J., Wilkinson, C. and Wilkinson, D. 2000b. First fossil bandicoot from the Pliocene Chinchilla Local Fauna. *Australian Mammalogy*, Vol. 22, pp.133–136.

Macleay, W.S. 1858. Owen in Anonymous. *Proceedings of the Geological Society of London*, No. 12, pp.49–50 (abstract only).

Mansergh, I. and Broome, L. 1994. *The Mountain Pygmy-possum of the Australian Alps*. UNSW Press, Sydney.114pp.

Marshall, L.G. 1981. The families and genera of Marsupialia. *Fieldiana (Geology)*, New Series, No. 8, pp.1–65.

Marshall, L.G., Case, J.A. and Woodburne, M.O. 1990. Phylogenetic relationships of the families of marsupials. Pp.433–502. In Genoways, H.H. (ed). *Current Mammalogy.* Vol. 2. Plenum Press, New York.

Martin, R. and Handasyde, K. 2000. *The Koala: Natural History, Conservation and Management*. UNSW Press, Sydney.

McGowran, B. and Li, Q. 1994. The Miocene oscillation in southern Australia. *Records of the South Australian Museum*, Vol. 27, pp.197–212.

McGowran, B., Archer, M., Bock, P., Darragh, T. A., Godthelp, H., Hageman, S., Hand, S. J., Hill, R., Li, Q., Maxwell, P. A., McNamara, K. J., Macphail, M., Mildenhall, D., Partridge, A. D., Richardson, J., Shafik, S., Truswell, E. M. and Warne, M. 2000. Australasian palaeobiogeography: the Palaeogene and Neogene record. *Memoir of the Association of Australasian Palaeontologists*, Vol. 23, pp.405–470.

Megirian, D., Murray, P., Scwartz, L. and Von Der Borch, C. 2003. The Kangaroo Well Local Fauna from the Uta Limestone of the Lake Eyre Basin. *CAVEPS 2003; Conference on Australasian Vertebrate Evolution, Palaeontology and Systematics, Brisbane Abstracts*, pp.10–11.

Messer, M., Weis, A.S., Shaw, D.C. and Westerman, M. 1998. Evolution of the monotremes: phylogenetic relationship to marsupials and eutherians, and estimation of divergence dates based on lactalbumin amino-acid sequences. *Journal of Mammalian Evolution*, Vol. 5, pp.95–105.

Miller, C.H., Magee, J.W., Johnson, B.J., Fogel, M.L., Spooner, M.A., McCullock, M.T. and Ayliffe, L.K. 1999. Pleistocene extinction of *Genyornis newtoni*: human impact on Australian megafauna. *Science*, Vol. 283, pp.205–208.

Moeller, H. 1976. *Dactylopsila trivirgata* (Phalangeridae) Klettern unde Fressen. *Encyclopedia Cinematographica Institute fur den Wissenschlaftlichen Film: Gottingen Film* E2267: 3–12.

Muirhead, J. 1992. A specialised thylacinid, *Thylacinus macknessi*, (Marsupialia: Thylacinidae) from Miocene deposits of Riversleigh, northwestern Queensland. *Australian Mammalogy*, Vol. 15, pp.67–75.

Muirhead, J. 1997. Two new early Miocene thylacines from Riversleigh, northwestern Queensland. *Memoirs of the Queensland Museum*, Vol. 41, pp.367–377.

Muirhead, J. 1999. Bandicoot diversity and evolution (Peramelemorphia, Marsupialia): the fossil evidence. *Australian Mammalogy*, Vol. 21, pp.11–13, 34–45.

Muirhead, J. 2000. Yaraloidea (Marsupialia, Peramelemorphia), a new superfamily of marsupial and a description and analysis of the cranium of the Miocene *Yarala burchfieldi*. *Journal of Paleontology*, Vol. 74, pp.512–523.

Muirhead, J. and Archer, M. 1990. *Nimbacinus dicksoni*, a plesiomorphic thylacine (Marsupialia: Thylacinidae) from Tertiary deposits of Queensland and the Northern Territory. *Memoirs of the Queensland Museum*, Vol. 28, pp.203–221.

Muirhead, J., Dawson, L. and Archer, M. 1997. *Perameles bowensis*, a new species of *Perameles* (Peramelemorphia, Marsupialia) from Pliocene faunas of Bow and Wellington Caves, New South Wales. *Proceedings of the Linnean Society of New South Wales*, Vol. 117, pp.163–173.

Muirhead, J. and Filan, S.L. 1995. *Yarala burchfieldi*, a plesiomorphic bandicoot (Marsupialia, Peramelemorphia) from Oligo-Miocene deposits of Riversleigh, north-western Queensland. *Journal of Paleontology*, Vol. 69, pp.127–134.

Muirhead, J. and Godthelp, H.J. 1996. Fossil bandicoots of Chillagoe (north-eastern Queensland) and the first known specimens of the pig–footed bandicoot *Chaeropus* Ogilby, 1838 from Queensland. *Australian Mammalogy*, Vol. 19, pp.73–76.

Muirhead, J. and Wroe, S. 1998. A new genus and species, *Badjcinus turnbulli* (Thylacinidae: Marsupialia) from the late Oligocene of Riversleigh, northern Australia, and an investigation of thylacinid phylogeny. *Journal of Vertebrate Paleontology*, Vol. 18, pp.612–626.

Murray, P.F. 1986. *Propalorchestes novaculacephalus* gen. et sp. nov., a new palorchestid (Diprotodontoidea: Marsupialia) from the middle Miocene Camfield Beds, Northern Territory, Australia. *The Beagle (Occasional Papers of the Northern Territory Museum of Arts and Sciences)*, Vol. 3, No. 1, pp.195–211.

Murray, P.F. 1990a. Primitive marsupial tapirs (*Propalorchestes novaculacephalus* Murray and *P. ponticulus* sp. nov.) from the mid–Miocene of north Australia (Marsupialia: Palorchestidae). *The Beagle (Records of the Northern Territory Museum of Arts and Sciences)*, Vol. 7, No. 1, pp.39–51.

Murray, P.F. 1990b. *Alkwertatherium webbi*, a new zygomaturine genus and species from the late Miocene Alcoota Local Fauna, Northern Territory (Marsupialia: Diprotodontidae). *The Beagle (Records of the Northern Territory Museum of Arts and Sciences)*, Vol. 7, No. 1, pp.53–80.

Murray, P.F. 1991. The Pleistocene megafauna of Australia. Pp.1071–1163. In Rich, P.V., Monaghan, J.M., Baird R.F. and Rich, T.H. (eds). *Vertebrate Palaeontology of Australasia.* Pioneer Design Studio and Monash University Publications Committee, Melbourne. 1437pp.

Murray, P.F. 1992. Thinheads, thickheads and airheads—functional craniology of some diprotodontian marsupials. *The Beagle (Records of the Northern Territory Museum of Arts and Sciences)*, Vol. 9, No. 1, pp.71–87.

Murray, P.F. 1998. *Thylacinus megiriani*, a new species of thylacine (Marsupialia: Thylacinidae) from the Ongeva Local Fauna of central Australia. *Records of the South Australian Museum*, Vol. 30, pp.43–61.

Murray, P.F. and Megirian, D. 1992. Continuity and contrast in middle and late Miocene vertebrate communities from the Northern Territory. *The Beagle (Records of the Northern Territory Museum of Arts and Sciences)*, Vol. 9, No. 1, pp.195–217.

Murray, P.F. and Megirian, D. 2000. Two new genera and three new species of Thylacinidae (Marsupialia) from the Miocene of the Northern Territory, Australia. *The Beagle (Records of the Northern Territory Museum of Arts and Sciences)*, Vol. 16, pp.145–162.

Murray, P.F., Wells, R.T. and Plane, M.D. 1987. The cranium of the Miocene thylacoleonid, *Wakaleo vanderleuri*: click go the shears—a fresh bite at thylacoleonid systematics. Pp.433–466. In Archer, M. (ed). *Possums and Opossums: Studies in Evolution*. Vol. 2. Surrey Beatty and Sons, Sydney. pp.401–788.

Myers, T.J. and Archer, M. 1997. *Kuterintja ngama* (Marsupialia, Ilariidae): a revised systematic analysis based on material from the late Oligocene of Riversleigh, northwestern Queensland. *Memoir of the Queensland Museum*, Vol 41, pp.379–392.

Myers, T.J., Archer, M., Krikmann, A. and Pledge, N.S. 1999. Diversity and evolutionary relationships of ilariids, wynyardiids, vombatids and related groups of marsupials. *Australian Mammalogy*, Vol 21, pp.18–19, 34–45.

Myers, T. J., Crosby, K., Archer, M. and Tyler, M. 2001. The Encore Local Fauna, a late Miocene assemblage from Riversleigh, northwestern Queensland. *Memoir of the Association of Australasian Palaeontologists*, Vol. 25, pp.147–154.

Osborne, H.F. 1907. *Evolution of Mammalian Molar Teeth*. Macmillan Co., London.

Owen, R. 1838. Fossil remains from Wellington Valley, Australia. Marsupialia. Pp.359–369. Appendix to Mitchell, T.L. *Three Expeditions into the Interior of Eastern Australia, with Descriptions of the Recently Explored Region of Australia Felix, and of the Present Colony of New South Wales.* 2 Vols, T. and W. Boone, London.

Owen, R. 1858. Odontology. Teeth of mammals. *Encyclopaedia Britannica, (8th edition)*, Vol. 16, p. 447.

Owen, R. 1872. The fossil mammals of Australia. *Nature*, Vol. 5, pp.503–504.

Owen, R. 1873. On the fossil mammals of Australia. Family Macropodidae. *Proceedings of the Royal Society*, Vol 21, pp.386–387.

Paddle, R. 2000. *The Last Tasmanian Tiger*. Cambridge University Press, Cambridge. 273pp.

Parker, W.K. 1890. On the skull of *Tarsipes rostratus*. *Studies from the Museum of Zoology in University College, Dundee*, pp.79–83.

Plane, M.D. 1967. Two new diprotodontids from the Pliocene Otibanda Formation, New Guinea. *Bureau of Mineral Resources, Geology and Geophysics Bulletin*, No. 85, pp.105–128.

Pledge, N.S. 1977. A new species of *Thylacoleo* (Marsupialia: Thylacoleonidae) with notes on the occurrences and distribution of Thylacoleonidae in South Australia. *Records of the South Australian Museum*, Vol. 17, pp.277–283.

Pledge, N.S. 1982. Enigmatic *Ektopodon*: a case history of palaeontological interpretation. Pp.480–488. In Rich, P.V. and Thompson, E.M. (eds.). *The Fossil Vertebrate Record of Australasia.* Monash University Offset Printing Unit, Clayton, Victoria. 759pp.

Pledge, N.S. 1986. A new species of *Ektopodon* (Marsupialia; Phalangeroidea) from the Miocene of South Australia. Pp.43–67. In Woodburne, M.O. and Clemens, W.A. (eds). *Revision of the Ektopodontidae (Mammalia: Marsupialia; Phalangeroidea) of the Australian Neogene*. University of California Publications in Geological Sciences, Vol. 131, California. 114pp.

Pledge, N.S. 1987a. *Kuterintja ngama*, a new genus and species of primitive vombatoid marsupial from the medial Miocene Ngama Local Fauna of South Australia. Pp.419–422. In Archer, M. (ed). *Possums and Opossums: Studies in Evolution*. Vol. 2. Surrey Beatty and Sons, Sydney. pp.401–788.

Pledge, N.S. 1987b. *Phascolarctos maris*, a new species of koala (Marsupialia: Phascolarctidae) from the early Pliocene of South Australia. Pp.327–330. In Archer, M. (ed). *Possums and Opossums: Studies in Evolution*. Vol. 1. Surrey Beatty and Sons, Sydney. pp.1–400.

Pledge, N.S. 1987c. *Muramura williamsi*, a new genus and species of ?wynyardiid (Marsupialia: Vombatoidea) from the middle Miocene Etadunna Formation of South Australia. Pp.393–400. In Archer, M. (ed). *Possums and Opossums: Studies in Evolution*. Vol. 1. Surrey Beatty and Sons, Sydney. pp.1–400.

Pledge, N.S. 1987d. A new species of *Burramys* Broom (Marsupialia: Burramyidae) from the middle Miocene of South Australia. Pp.725–728. In Archer, M. (ed). *Possums and Opossums: Studies in Evolution*. Vol. 2. Surrey Beatty and Sons, Sydney. pp.401–788.

Pledge, N.S. 1987e. Pseudocheirids (Marsupialia: Pseudocheiridae) from the middle Miocene Ngama Local Fauna of northern South Australia. Pp.681–688. In Archer, M. (ed). *Possums and Opossums: Studies in Evolution*. Vol. 2. Surrey Beatty and Sons, Sydney. pp.401–788.

Pledge, N.S. 1992. The Curramulka local fauna: a new late Tertiary fossil assemblage from Yorke Peninsula, South Australia. *The Beagle (Records of the Northern Territory Museum of Arts and Sciences)*, Vol. 9, No. 1, pp.115–142.

Pledge, N.S., Archer, M., Hand, S.J. and Godthelp, H. 1999. Additions to knowledge about ektopodontids (Marsupialia: Ektopodontidae): including a new species *Ektopodon litolophus*. *Records of the Western Australian Museum, Supplement*, No. 57, pp.255–264.

Rand, A.L. 1937. Results of the Archbold Expeditions. No. 17. Some original observations of the habits of *Dactylopsila trivirgata* Gray. *American Museum Novitates*, No. 957.

Rauscher, B. 1987. *Priscileo pitikantensis*, a new genus and species of thylacoleonid marsupial (Marsupialia: Thylacoleonidae) from the Miocene Etadunna Formation, South Australia. Pp.423–432. In Archer, M. (ed). *Possums and Opossums: Studies in Evolution*. Vol. 2. Surrey Beatty and Sons, Sydney. pp.401–788.

Retief, J.D., Krajewski, C., Westerman, M., Winkfien, R.J. and Dixon, G.H. 1995. Molecular phylogeny and evolution of marsupial protoamine P1 genes. *Proceedings of the Royal Society London*, Vol. 259, pp.7–14.

Rich, T.H. 1982. Monotremes, placentals, and marsupials: their record in Australia and its biases. Pp.386–477. In Rich, P.V. and Thompson, E.M. (eds). *The Fossil Vertebrate Record of Australasia,* Monash University Offset Printing Unit, Clayton, Victoria. 759pp.

Rich, T.H. 1983. The largest marsupial. *Diprotodon optatum*. Pp.62–63. In Quirk, S. and Archer, M. (eds). *Prehistoric*

Sciences, Vol. 131, California. 114pp.

Woodburne, M.O. and Clemens, W.A. (eds). 1986b. *Revision of the Ektopodontidae (Mammalia; Marsupialia; Phalangeroidea) of the Australian Neogene.* University of California Publications in Geological Sciences, Vol. 131, California. 114pp.

Woodburne, M.O., MacFadden, B.J., Case, J.A., Springer, M.S., Pledge, N.S., Power, J.D., Woodburne, J.M. and Springer, K.B. 1993. Land mammal biostratigraphy and magnetostratigraphy of the Etadunna Formation (late Oligocene) of South Australia. *Journal of Vertebrate Paleontology*, Vol. 13, No. 4, pp.483–515.

Woodburne, M.O., Pledge, N.S. and Archer, M. 1987b. The Miralinidae, a new family and two new species of phalangeroid marsupials from Miocene strata of South Australia. Pp.581–602. In Archer, M. (ed). *Possums and Opossums: Studies in Evolution.* Vol. 2. Surrey Beatty and Sons, Sydney. pp 401–788.

Woodburne, M.O., Tedford, R.H. and Archer, M. 1987c. New Miocene ringtail possums (Marsupialia: Pseudocheiridae) from South Australia. Pp.639–679. In Archer, M. (ed). *Possums and Opossums: Studies in Evolution.* Vol. 2. Surrey Beatty and Sons, Sydney. pp.401–788.

Woodburne, M.O., Tedford, R.H., Archer, M. and Pledge N.S. 1987a. *Madakoala*, a new genus and two species of Miocene koalas (Marsupialia: Phascolarctidae) from South Australia, and a new species of *Perikoala*. Pp.293–317. In Archer, M. (ed). *Possums and Opossums: Studies in Evolution.* Vol. 1. Surrey Beatty and Sons, Sydney. pp.1–400.

Woodburne, M.O., Tedford, R.H., Archer, M., Turnbull, W.D., Plane, M.D. and Lundelius, E.L. 1985. Biochronology of the continental mammal record of Australia and New Guinea. *Special Publication, South Australian Department of Mines and Energy*, Vol. 5, pp.347–363.

Woodford, J. 2001. *The Secret Life of Wombats.* Text Publishing, Melbourne. 226pp.

Woods, H.E. 1924. The position of the "sparassodonts": with notes on the relationships and history of the Marsupialia. *Bulletin of the American Museum of Natural History*, Vol. 51, pp.77–101.

Woods, J.T. 1958. The extinct marsupial genus *Palorchestes* Owen. *Memoirs of the Queensland Museum*, Vol. 13, pp.177–193.

Woolley, P.A. and Scarlett, G.L. 1984. Observations on the reproductive anatomy of male *Tarsipes rostratus* (Marsupialia: Tarsipedidae). Pp.445–450. In Smith, A.P. and Hume, I.D. (eds). *Possums and Gliders.* Surrey Beatty and Sons, Sydney. 598pp.

Woolnough, A.P. and Steele, V.R. 2001. The palaeoecology of the Vombatidae: did giant wombats burrow? *Mammal Review*, Vol.31, pp.33–45.

Wroe, S. 1996. *Muribacinus gadiyuli*, (Thylacinidae: Marsupialia), a very plesiomorphic thylacinid from the Miocene of Riversleigh, northwestern Queensland, and the problem of paraphyly for the Dasyuridae (Marsupialia). *Journal of Paleontology*, Vol. 70, pp.1032–1044.

Wroe, S. 1997a. A re-examination of proposed morphology-based synapomorphies for the families of Dasyuromorphia (Marsupialia). Part I. Dasyuridae. *Journal of Mammalian Evolution*, Vol. 4, pp.19–52.

Wroe, S. 1997b. *Mayigriphus orbus* gen. et sp. nov., a Miocene dasyuromorphian from Riversleigh, northwestern Queensland. *Memoirs of the Queensland Museum*, Vol. 41, pp.439–448.

Wroe, S. 1998. A new 'bone-cracking' dasyurid (Marsupialia), from the Miocene of Riversleigh, northwestern Queensland. *Alcheringa*, Vol. 22, pp.277–284.

Wroe, S. 1999. The geologically oldest dasyurid, from the Miocene of Riversleigh, north-west Queensland. *Palaeontology*, Vol. 42, pp.501–527.

Wroe, S. 2001. *Maximucinus muirheadae*, gen. et sp. nov. (Thylacinidae: Marsupialia), from the Miocene of Riversleigh, north-western Queensland, with estimates of body weights for fossil thylacinids. *Australian Journal of Zoology*, Vol. 49, pp.603–614.

Wroe, S. 2003. Australian marsupial carnivores: an overview of recent advances in palaeontology and phylogeny. Pp.102–123. In Jones, M., Dickman, C. and Archer, M. (eds). *Predators with Pouches: the Biology of Carnivorous Marsupials.* CSIRO Publishing, Melbourne. 486pp.

Wroe, S. and Archer, M. 2006. Origins and early radiations of marsupials. Pp.551–574. In Merrick, J.R., Archer, M., Hickey, G.M. and Lee, M.S.Y. (eds). *Evolution and Biogeography of Australasian Vertebrates.* Auscipub Pty. Ltd., Sydney. 942pp.

Wroe, S., Crowther, M., Dortch, J. and Chong, J. 2003. The size of the largest marsupial and why it matters. *Proceedings of the Royal Society of London,* Series B *(Supplement)*, Vol. 270.

Wroe, S., Ebach, M., Ahyong, S., Muizon, C. de and Muirhead, J. 2000. Cladistic analysis of dasyuromorphian (Marsupialia) phylogeny using cranial and dental features. *Journal of Mammalogy*, Vol. 81, pp.1008–1024.

Wroe, S., Field, J., Fullagar, R. and Jermiin, L.S. 2004. Megafaunal extinction in the late Quaternary and the global overkill hypothesis. *Alcheringa*, Vol. 28, No. 1, pp.291–331.

Wroe, S. and Mackness, B.S. 1998. Revision of the Pliocene dasyurid, *Dasyurus dunmalli* (Dasyuridae, Marsupialia). *Memoirs of the Queensland Museum*, Vol. 42, pp.605–612.

Wroe, S. and Mackness, B. 2000a. A new genus and species of dasyurid from the Pliocene Chinchilla Local Fauna of south-eastern Queensland. *Alcheringa*, Vol. 24, pp.319–326.

Wroe, S. and Mackness, B. 2000b. Additional material of *Dasyurus dunmalli* from the Pliocene Local Fauna of south-eastern Queensland and its phylogenetic implications. *Memoirs of the Queensland Museum*, Vol. 42, pp.605–612.

Wroe, S. and Muirhead, J. 1999. Evolution of Australia's marsupicarnivores: Dasyuridae, Thylacinidae, Myrmecobiidae, Dasyuromorphia *incertae sedis* and Marsupialia *incertae sedis*. *Australian Mammalogy*, Vol. 21, pp.10–11, 34–45.

Wroe, S. and Musser, A. 2001. The skull of *Nimbacinus dicksoni* (Thylacinidae: Marsupialia). *Australian Journal of Zoology*, Vol. 49, pp.487–514.

Wroe, S., Myers, T.J., Seebacher, F., Kear, B., Gillespie, A., Crowther, M. and Salisbury, S. 2003. An alternative method for predicting body mass: the case of the Pleistocene marsupial lion. *Paleobiology*, Vol. 29, pp.404–412.

Yadav, M. 1973. The presence of the cervical and thoracic thymus lobes in marsupials. *Australian Journal of Zoology*, Vol. 21, pp.285–301.

Chapter 29
Kangaroos

Bernard Cooke

School of Natural Resource Sciences, Queensland University of Technology, G.P.O. Box 2434, Brisbane, Qld. 4001 Australia and Vertebrate Palaeontology Laboratory, School of Biological, Earth and Environmental Sciences, University of New South Wales, Sydney. N.S.W. 2052 Australia

INTRODUCTION

Following the final liberation of Australia from Gondwana, the continent was used by nature for tens of millions of years as an isolated proving ground for evolutionary experiments that relied heavily on the use of marsupials as test subjects. In the opinion of us macropodiphiles, the most successful products of these experiments were kangaroos. While the aboriginal people of Australia had no doubt frequently dined with kangaroos during the tens of thousands of years of their acquaintance, the existence of these quintessentially Australian marsupials remained unknown to the wider world. That began to change as 17th and 18th century European explorers of the Australian region started to publish descriptions of their encounters with kangaroos. Frith and Calaby (1969) provide an excellent overview of meetings between kangaroos and these and later explorers. Kangaroos did not necessarily benefit from these encounters. In 1606, for instance, Don Diego d Prado y Tovar, while sailing along the south-eastern coast of New Guinea in company with the explorer Torres, was presented with what was probably a pademelon (*Thylogale* sp.). Pausing briefly to make notes regarding its scaly tail and pendulous testicles, he ate it.

Early voyagers to the spice islands of the East Indies, either out of curiosity or geographical embarrassment, sometimes landed (occasionally inadvertently and rather forcefully) in Western Australia and they too provided accounts of local kangaroos. Francisco Pelsaert, in his 1647 account of his ill fated voyage in the 'Batavia,' wrecked in Houtman's Abrolhos off Western Australia in 1629, describes 'a species of cats', now known to be Tammar wallabies (*Macropus eugenii*). These too served as food for the rescuers and survivors of the shipwreck and the mutiny and massacre which followed it.

Samuel Volkerson landed on a Western Australian island in 1658 and saw a 'wild cat resembling a civet-cat'—in fact, the quokka, *Setonix brachyurus*. Willem de Vlamingh landed on the same island in 1696, but thought the quokkas were rats and so named the island 'Rottnest'. William Dampier, buccaneer captain and later explorer, first landed in Western Australia in 1688. Undeterred, he returned in 1699, this time as the respectable commander of an English naval vessel. His report of a kind of raccoon (actually the banded hare wallaby, *Lagostrophus fasciatus*), seen on this voyage, included a note that it was very tasty.

James Cook's account of his 1770 voyage along the Australian east coast included descriptions of a kangaroo from the Endeavour River. Cook thought the animal resembled a jerboa, as did the naturalist Joseph Banks, who also took notes on the beast. Whatever it was (and there is still doubt about the species—see Frith and Calaby, 1969) Cook thought the animal provided 'excellent food', but Banks was far less complimentary.

Cook's was a scientific expedition as well as a voyage of exploration and included trained naturalists who made written observations, prepared illustrations and collected specimens, not all of which were eaten. It marks the change from curiosity or mere culinary interest to genuine scientific interest in kangaroos. Just as early European explorers had difficulty in relating kangaroos to any animals they were familiar with, kangaroos also posed difficulties for systematists of the time, being first grouped with a variety of different rodents and later with opossums. Information regarding 18th and 19th century attempts to elucidate the affinities of kangaroos can be found in Ride (1978).

The Darwin-inspired revolution in scientific thinking about the processes of evolution stimulated attempts to not just classify organisms, but to determine their evolutionary origins and relationships. There have been numerous attempts to do this for kangaroos using data drawn from many different kinds of morphological, anatomical, cytological, biochemical and molecular studies. Unfortunately kangaroos are phylogenetically frustrating animals, reluctant to give away too much evolutionary information. They leave false clues about relationships through an indulgence in evolutionary convergence, which can make it difficult to distinguish truly homologous structures within the group. Tim Flannery emphasized this tendency when he subtitled his 1989 treatise on kangaroo phylogeny 'a study in convergence'. They also leave lamentable gaps in crucial parts of their fossil record and fossils have an important role to play in phylogenetic analyses.

Before reviewing attempts to unravel kangaroo phylogeny, we should have a brief look at the characteristics of the group and of the major kinds of kangaroos, take some time to examine the current status of kangaroo systematics and examine the fossil legacy that kangaroos have left us.

KANGAROO CHARACTERISTICS AND SYSTEMATICS

The toes and teeth of kangaroos place them firmly within the order Diprotodontia—they have syndactylous second and third toes and retain only a single pair of functional lower incisors. Some Oligo-Miocene kangaroos had a second pair of lower incisors or analogues of these teeth, but these were tiny, closely resembling the deformed caniniform second incisors of phalangerid possums. Aplin and Archer (1987) include all kangaroos within the superfamily Macropodoidea, within the suborder Phalangerida which also includes all families of possums. This reflects the general consensus that kangaroos share a common ancestor with possums.

In the past macropodoids have been phylogenetically linked to wynyardiids (Archer and Bartholomai, 1978; Bartholomai, 1978; Rich and Archer, 1979). The work of Aplin (1987) on basicranial anatomy of wynyardiids has demonstrated a closer relationship between wynyardiids and vombatiform diprotodontians than with phalangeridans, effectively declaring wynyardiids innocent of involvement in kangaroo origins. Within Phalangerida, phalangerids (brushtails, cuscuses) are probably chief suspects as a potential sister-group to kangaroos, but burramyids (mountain pigmy possums) have not yet been completely exonerated (Bensley, 1903; Wroe *et al.*, 1998). Definitive conclusions regarding the ancestral relations of kangaroos are frustrated by the lack of an earlier Oligocene fossil record.

Aplin and Archer (1987) divide Macropodoidea into two families: Macropodidae, including all (herbivorous) kangaroos and wallabies, and Potoroidae, encompassing all of the more or less omnivorous rat kangaroos. Why

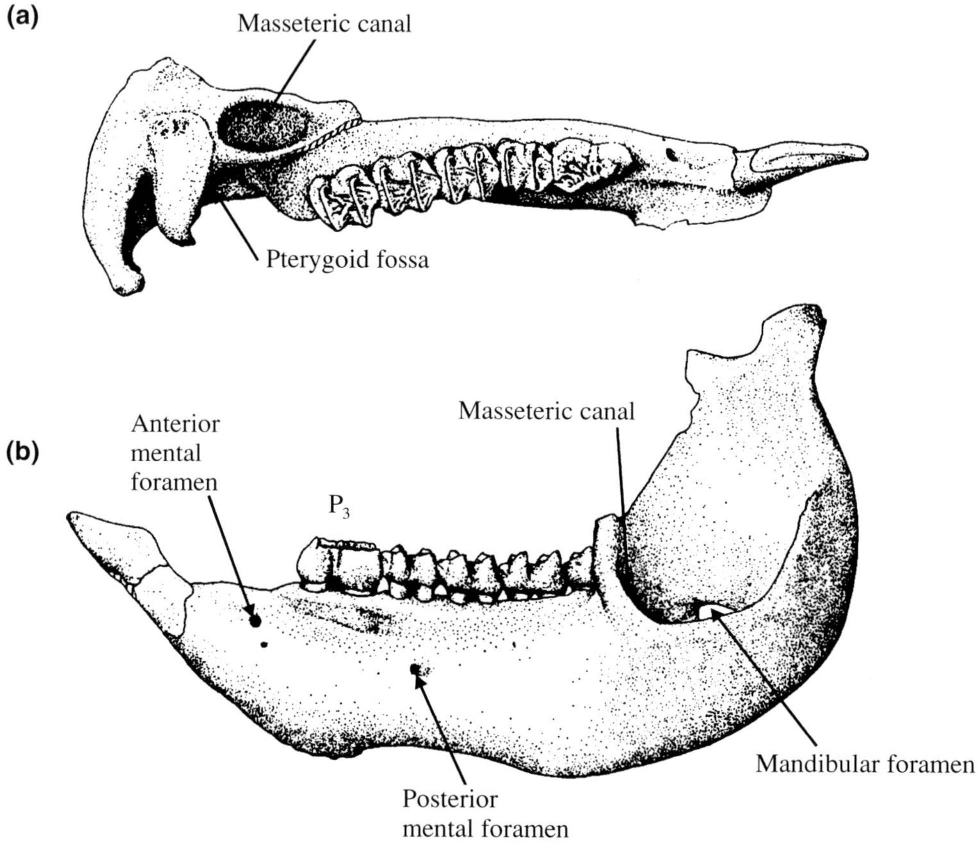

Figure 1. The mandible of a sthenurine kangaroo showing some of the distinctive features of kangaroos. **(a)** occlusal view; and **(b)** buccal view. P3 is the only premolar tooth: its deciduous precursor, (dP3) and P2 were both evicted as it erupted. P1 was lost early in kangaroo phylogeny. The masseteric canal allows penetration of the deep layer of the masseter muscle within the body of the mandible. The large mandibular foramen which opens between the masseteric canal and the pterygoid fossa is a comparatively recent development. In plesiomorphic Oligo-Miocene kangaroos, the mandibular foramen is the posterior opening into the pterygoid fossa of the inferior dental canal (IDC). A short, posterior length of the canal opens into the masseteric canal via a masseteric foramen. Over time, the posterior portion of the IDC decreased in length, allowing overlap, then merging of the masseteric and mandibular foramina. The posterior mental foramen, common in sthenurines but absent in modern kangaroos, appears to be a plesiomorphic character since it is present also in the majority of Oligo-Miocene species. After Tedford (1966).

the common name of potoroids should infer some kind of relationship with a particularly scruffy group of placentals is hard to understand. No potoroid had even so much as seen a murid until the Pliocene. A full classification of macropodoids is presented in Flannery (1989). Like previous macropodoid classifications, this too is subject to modification as research continues.

Macropodoids are distinguished from all other diprotodontians, indeed from all other mammals, by several synapomorphies including: the use of a bipedal hopping gait; loss of P1 and the eviction of both P2 and the deciduous third premolar, dP3, by the eruption of P3. A third synapomorphy is unique among mammals; this is a peculiarity of attachment of one of the main chewing muscles, the masseter, to the mandible. In the vast majority of mammals the masseter runs from the zygomatic arch to insert on the buccal surface of the posterior portion of the mandible, whereas in kangaroos the deep layer of the masseter muscle penetrates into the mandible (Figures 1 and 2).

Strangely, this peculiar kangaroo adaptation was not commented on in scientific literature until described formally by Abbie (1939), who also noticed differences between potoroids and macropodids in the manner of this penetration. In potoroids part of the masseter runs forward inside the mandible within a masseteric canal below the molar tooth row (Figure 2). The masseteric canal is also confluent with the inferior dental canal, carrying nerves and blood vessels along the length of the jaw. In macropodids the degree of penetration is much less—into a masseteric fossa which is anteriorly partitioned from the inferior dental canal by a bony lamina. Because any such insertion of the masseter within the jaw is extremely odd to say the least, the shallow penetration seen in macropodids has been considered to be plesiomorphic for macropodoids and the deep penetration in potoroids to be apomorphic. Tuck this away in the back of your brain because we'll come back to it later.

Pearson (1950) noted another difference between potoroids and macropodids relating to skull structure. In potoroids the frontal and squamosal bones make contact on the lateral wall of the braincase. In macropodids these bones are separated by the dorsal extension of the alisphenoid, which makes contact with the parietal (Figure 3). Because alisphenoid-parietal contact is usual among other diprotodontians, this condition has been considered plesiomorphic for macropodoids and the potoroid squamosal-frontal contact to be a derived condition. More about this later too.

In potoroids the basicranium lies in about the same horizontal plane as the palate. In derived macropodids the

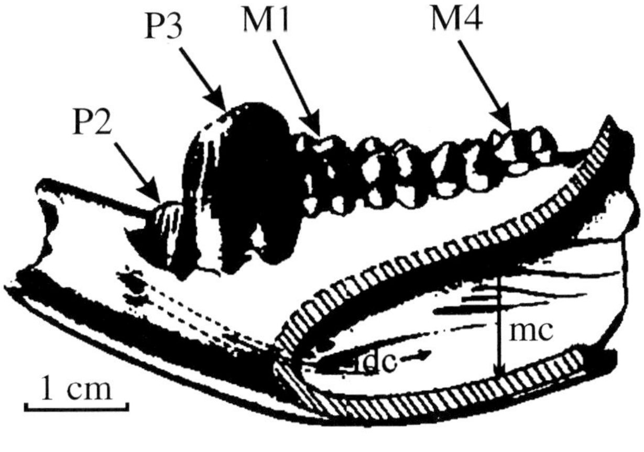

Figure 2. Left dentary of the propleopine potoroid, *Ekaltadeta*. The jaw has been sectioned to show the extensive masseteric canal (mc), which is confluent over much of its length with the inferior dental canal (idc). Retention of P2 after the eruption of P3 is apparently a propleopine synapomorphy. After Archer (1984).

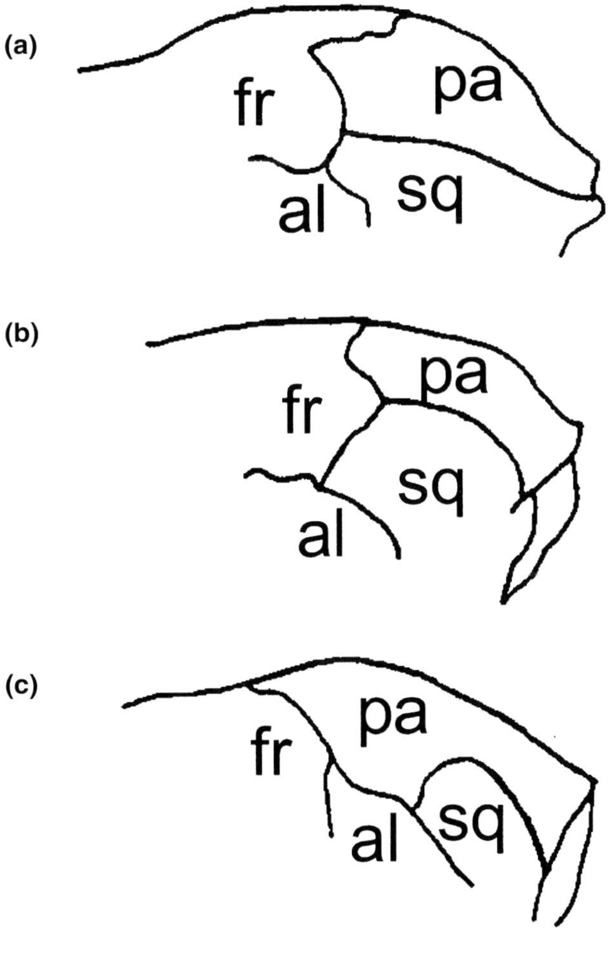

Figure 3. Bone contacts in the cranial wall of different kinds of kangaroos: **(a)** *Hypsiprymnodon moschatus*: short length of contact between the frontal (fr) and squamosal (sq) bones; **(b)** normal potoroid condition: extensive contact between frontal and squamosal; **(c)** normal macropodid condition: contact between the alisphenoid (al) and parietal (pa) bones. After Burk *et al.* (1998).

skull exhibits marked flexion—the basicranium is inclined antero-dorsally and the palate antero-ventrally. The zygomatic arch of potoroids makes a smooth transition to the cheek region and there is little development of a masseteric process. In more derived macropodids the anterior end of the zygomatic arch stops rather abruptly with a marked anteorbital sulcus between it and the cheek region. The masseteric process of the arch is also produced ventrally and may extend below the level of the molar alveoli (Figure 4). These differences are likely related to differences in diet and associated differences in masseteric muscle bulk required for food processing. Potoroid mandibles are usually ventrally convex, those of macropodids are straighter or even slightly undulating. The differences relate to the position and size of the digastric eminence to which attaches the digastric muscle involved in opening the jaw. The ascending ramus of the dentary is more recumbent in potoroids than in macropodids.

There are also differences in dental morphology between potoroids and macropodids. Potoroid lower incisors rise from the jaw at a steeper angle when compared to the mostly procumbent lower incisors of macropodids. They retain upper canine teeth (as do some plesiomorphic macropodids) and have low-crowned, bunodont or bunolophodont molars with four distinct cusps, partially linked by transverse ridges in bunolophodont forms. Macropodids have bilophodont molar teeth—two transverse enamel crests or lophs on each molar. Posterior molar teeth in macropodids tend be larger than anterior ones. The reverse is usually the case among potoroids.

As far as differences in squishy bits between marcropodids and potoroids are concerned, Pearson (1950) commented upon differences between the female reproductive organs of potoroids and macropodids, including an unusual anterior vaginal expansion or vaginal caecum in potoroids. Rose (1978, 1989) supported the validity of this as a taxonomic characteristic. There are also differences in stomach structure discussed further below.

POTOROID SUBFAMILIES

Hypsiprymnodontinae

Hypsiprymnodontine kangaroos have never been a diverse group and are represented by a single extant species. This is the diurnal, omnivorous/frugivorous (Johnson and Strahan, 1982; Dennis, 1995) musky rat kangaroo, *Hypsiprymnodon moschatus* of north Queensland rainforests. Like other extant potoroids, *H. moschatus* construct nests and can carry nesting material in their curled, partly prehensile, scaly tails. These shy and retiring beasts must surely be the arch-conservatives among kangaroos. They have much simpler stomachs than any other extant kangaroos. They have steadfastly

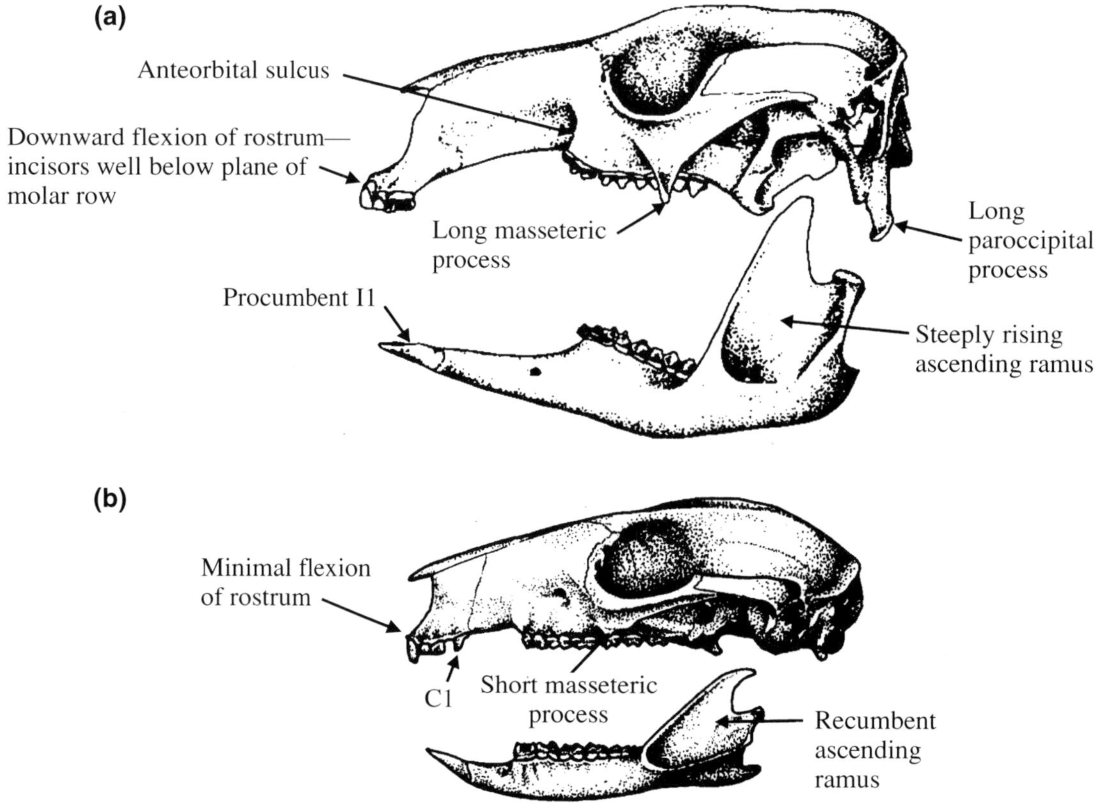

Figure 4. Some features of the skull and mandible. **(a)** *Macropus giganteus* (adult); **(b)** *Bettongia gaimardi* (juvenile).

resisted adopting the new-fangled hopping method of getting about, and stick to using all four feet, even if it might result in the embarrassment of being mistaken for a bandicoot, grounded possum, or other lesser marsupial. They have also stubbornly retained their halluxes (big toes), appendages that seem to be the kangaroo equivalent of the safari suit and long since relegated to the back of the evolutionary wardrobe by other, trendier, modern kangaroos. The living *Hysiprymnodon* is the most plesiomorphic of all living kangaroos. By remaining in its ancestral rainforest home, it has managed to avoid the need for too much evolutionary change. As a result it is similar to its Miocene antecedent, *H. bartholomaii* described from Gag Site at Riversleigh by Flannery and Archer (1987a). Pliocene *Hysiprymnodon* fossils are also known from as far south as Hamilton in Victoria and from the Bow Fauna of the N.S.W. Hunter Valley.

Propleopinae

Propleopine history stretches back as far as Oligo-Miocene times. They are represented in the Oligo-Miocene fossil record at Riversleigh by two species of *Ekaltadeta*: *E. ima* and *E. jamiemulvaneyi*. All Pleistocene species are members of the genus *Propleopus*, and Ride (1993) has added a monotypic Pliocene genus, *Jackmahoneyia*. Propleopines share some dental characteristics with hypsiprymnodontines, notably in having impressive, plagiaulacoid premolars and bunodont molars. They have the unusual characteristic of retaining P2 after the eruption of P3 (Wroe and Archer, 1995). Since P2 is sometimes briefly similarly retained in *Hypsiprymnodon*, Wroe (1996) regarded this as a plesiomorphic character, a view supported by observation of similar retention of P2 among some species of fossil balbarine kangaroos from Riversleigh (Cooke, unpublished data). Partly because of their large size, comparative rarity in the fossil record and their impressive premolars, there has been speculation that some propleopine species may have been at least partly carnivorous (Archer and Flannery, 1985; Ride *et al.*, 1997); but Wroe *et al.* (1998) express doubts about carnivory among propleopines.

Ride (1993) placed propleopines and hysiprymnodontines in a new, third family of kangaroos, Hypsiprymnodontidae. Wroe *et al.* (1998) indicate cranial features which suggest that propleopines may have a closer relationship with the extinct, Oligo-Miocene balbarine kangaroos than with hypsiprymnodontines.

Potoroinae

Potoroines are the most diverse of the potoroids. There are four extant genera: *Potorous*, *Bettongia*, *Caloprymnus* and *Aepyprymnus*. Flannery (1989) groups these into two tribes, Potoroini (including *Potorous*) and Bettongini (*Bettongia*, *Caloprymnus* and *Aepyprymnus*). Unfortunately Potoroines have not always fared well since the arrival of Europeans in Australia. The sole species of *Caloprymnus*, the arid-dwelling desert rat kangaroo (*C. campestris*), has not been seen since 1935 (Smith, 1983) and is presumed extinct, as is the broad-face potoroo, *Potorous platyops*. Most others have suffered range reductions and some have become extinct on mainland Australia

All extant potoroines exhibit architectural skills. Most construct 'nests' or shelters in which they rest during the day. *Bettongia lesueuri*, the burrowing bettong, goes a step further and excavates deep burrows which can develop into elaborate warrens and shelter many individuals. All species carry grass and plant material for these purposes using their partly prehensile tails.

The potoroine stomach is more complex than that of *Hysiprymnodon*. Like macropodids, potoroines have a stomach with three distinct regions—a sacculated forestomach and a tubiform midstomach, regions associated with temporary storage and fermentation, and an acid-secreting hindstomach. Because of their nocturnality, potoroines spend less time feeding than do most macropodids. They therefore have a relatively large sacculated forestomach for storage of food, which can be digested while they rest during the day. See Freudenberger *et al.* (1989) for further information on the potoroine digestive system. Potoroids have a fondness for subterranean fungi which are the principal food source for *Potorous* and some of the species of *Bettongia* (Seebeck *et al.*, 1989).

Potoroines are known from as far back as Oligo-Miocene times. Flannery *et al.* (1983) described *Gumardee pascuali* from Site D at Riversleigh, now considered to be late Oligocene in age. Shortly thereafter two new genera and species were recorded from South Australia with Woodburne (1984) describing *Wakiewakie lawsoni* from the Kutjumarpu Local Fauna of the Wipajiri Formation, and Case (1984) describing *Purtia mosaicus* from the Ngapakaldi Local Fauna of the Etadunna Formation. Woodburne *et al.* (1993) indicate a late Oligocene age for these taxa; they also report a further new Oligocene genus and species from the Etadunna Formation, which is said to be the geologically oldest potoroine yet known.

Flannery and Archer (1987b) described *Bettongia moyesi* from Two trees Site at Riversleigh, the first known pre-Pliocene occurrence of this genus. This species, like *Hysiprymnodon bartholomaii*, described by the same authors, exhibited alisphenoid-parietal contact on the lateral wall of the cranium, prompting them to the conclusion that frontal squamosal contact could no longer be considered a synapomorphy for Potoroidae. They indicated that this derived condition had evolved independently four times within Potoroidae. We'll revisit this a bit later.

The Curramulka Local Fauna from Cora Lynne Cave on Yorke Peninsula, South Australia, described by Pledge (1992) and regarded by him as early Pliocene or late Miocene in age, includes a species of *Potorous*. Tedford (1994) supports a late Miocene age for this fauna. If correct, this would represent the first known pre-Pleistocene occurrence for this genus. A further new potoroine genus and species, *Milliyowi bungantitj* was described by Flannery *et al.* (1992) from the early Pliocene Hamilton Local Fauna of Victoria.

Because of its deeply-penetrating masseteric canal, confluent with the inferior dental canal, Woodburne (1967) originally included the late Miocene *Dorcopsoides fossilis* from Alcoota in Potoroinae. Because of the similarity of this species with those of the New Guinean forest wallabies, *Dorcopsis* and *Dorcopsulus*, he included these too in Potoroinae. Bartholomai (1978) argued for synonymy of *Dorcopsoides* with *Wallabia*. Although this has not been adopted, *Dorcopsoides* is now regarded as a macropodine.

Bulungamayinae

This subfamily was created by Flannery *et al.* (1983) to accommodate *Wabularoo naughtoni* described by Archer (1979) from D Site at Riversleigh, and a new Riversleigh species, *Bulungamaya delicata*. These were weird potoroids. While having long, finely-ridged although crescentic premolars and other, normal potoroid characteristics, their teeth were lophodont, a condition hitherto regarded as a macropodid prerogative (Figure 5). Indeed, both Woodburne (1984) and Case (1984) argued that, because of their lophodonty, these species were macropodids. Their inclusion within Potoroidae was maintained by Flannery *et al.* (1984) on the basis of possession of a number of what were regarded as potoroid synapomorphies: ventral margin of the mandible convex below the middle of the molar row; a buccally expanded, deeply penetrating masseteric canal; elongate, finely-ridged premolars.

Cooke (1997b) described two additional genera and species of bulungamayines, *Nowidgee matrix*, a distinctly bunolophodont form, and *Ganguroo bilamina*, an

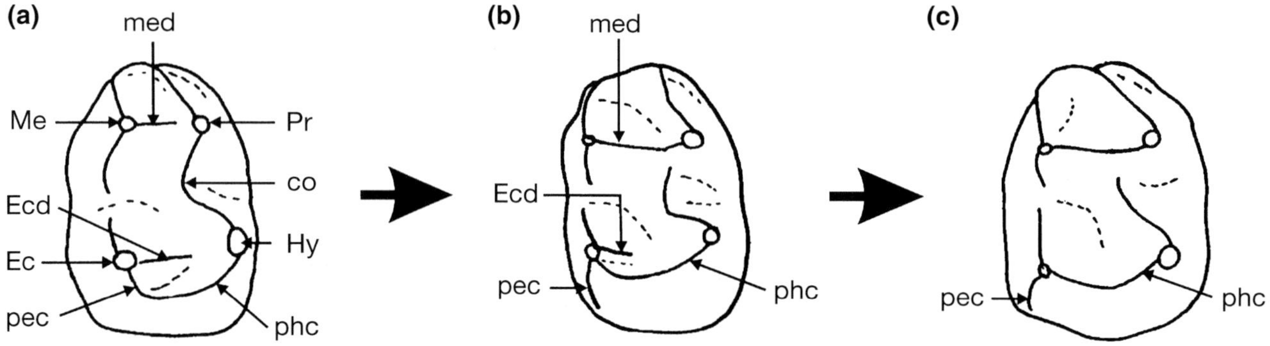

Figure 5. Lophid evolution in lower molars of bulungamayines. **(a)** RM1 of the bunolophodont *Nowidgee matrix*. The longitudinal crest formed by paracristid and cristid obliqua is prominent. Transverse partial links between lingual and buccal cuspids are formed by cristids originating from the metaconid and entoconid respectively. The post hypocristid and postentocristid unite low on the posterior face of the crown. **(b)** RM1 of *Bulungamaya delicata*. The longitudinal crest is de-emphasized. Protoconid and metaconid are linked by the buccally-directed crest originating from the metaconid (with the probable contribution of a short, lingually-directed crest from the protoconid). The posthypocristid is higher on the posterior face of the crown and links the apices of hypoconid and entoconid. The buccally-directed crest from the entoconid is reduced and the postentocristid is retained. **(c)** RM1 of the completely lophodont *Ganguroo bilaminar*. All trace of the buccally-directed crest from the entoconid is lost. The hypolophid is formed by the elevated and transversely-oriented posthypocristid. Key to abbreviations: co = cristid obliqua; Ec = entoconid; Ecd = entocristid; Hy = hypoconid; Me = metaconid; med = metacristid; pec = postentocristid; phc = posthypocristid; Pr = protoconid.

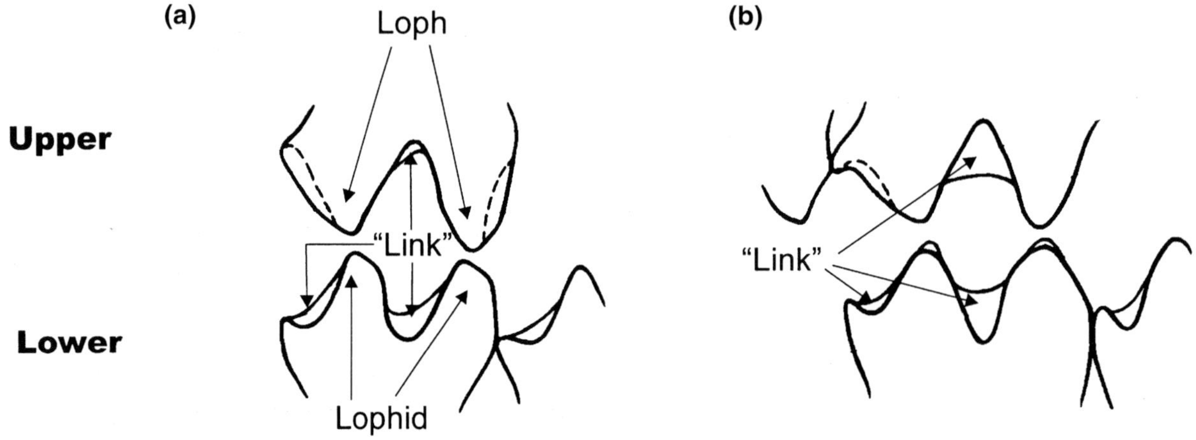

Figure 6. Differences in the development of longitudinal 'links' in **(a)** a browsing macropodine, and **(b)** a grazing macropodine. After Dawson (1995).

Plate 104. A common kangaroo along Australia's east coast in southern Queensland and northern New South Wales is the whiptail or 'pretty-faced' wallaby (*Macropus parryi*). Photograph M. Archer.

Plate 105. Fossil remains of the Miocene kangaroo *Ekaltadeta ima*, whose skull and teeth indicate its diet ranged from fruit to flesh. Photograph R. Arnett.

Plate 106. Many bats roost in the exposed foliage of trees. Flying-foxes may form camps numbering hundreds or thousands of bats in tropical and subtropical areas of the Old World. Australia's little red flying-fox *Pteropus scapulatus* is one of 13 living pteropodids found in Australia. Despite the fact that megabats have a long fossil history elsewhere, their record in Australia is relatively short, extending back only to the late Pleistocene. Photograph M. Archer.

Plate 107. *Mystacina tuberculata* is one of only two mammals native to New Zealand (both are bats) and the only living member of the Mystacinidae. It is a terrestrial, arboreal and burrowing bat that has a unique way of folding and protecting its wings while on the ground. A number of fossil mystacinids are known from Australia, the first pre-Pleistocene records for the family. Photograph R. Morris.

Plate 108. Northern Australia's orange leaf-nosed bat *Rhinonicteris aurantius* is a cave-dwelling hipposiderid whose ancestry can be traced back to the early Miocene in limestone deposits at Riversleigh. Photograph L. Hall.

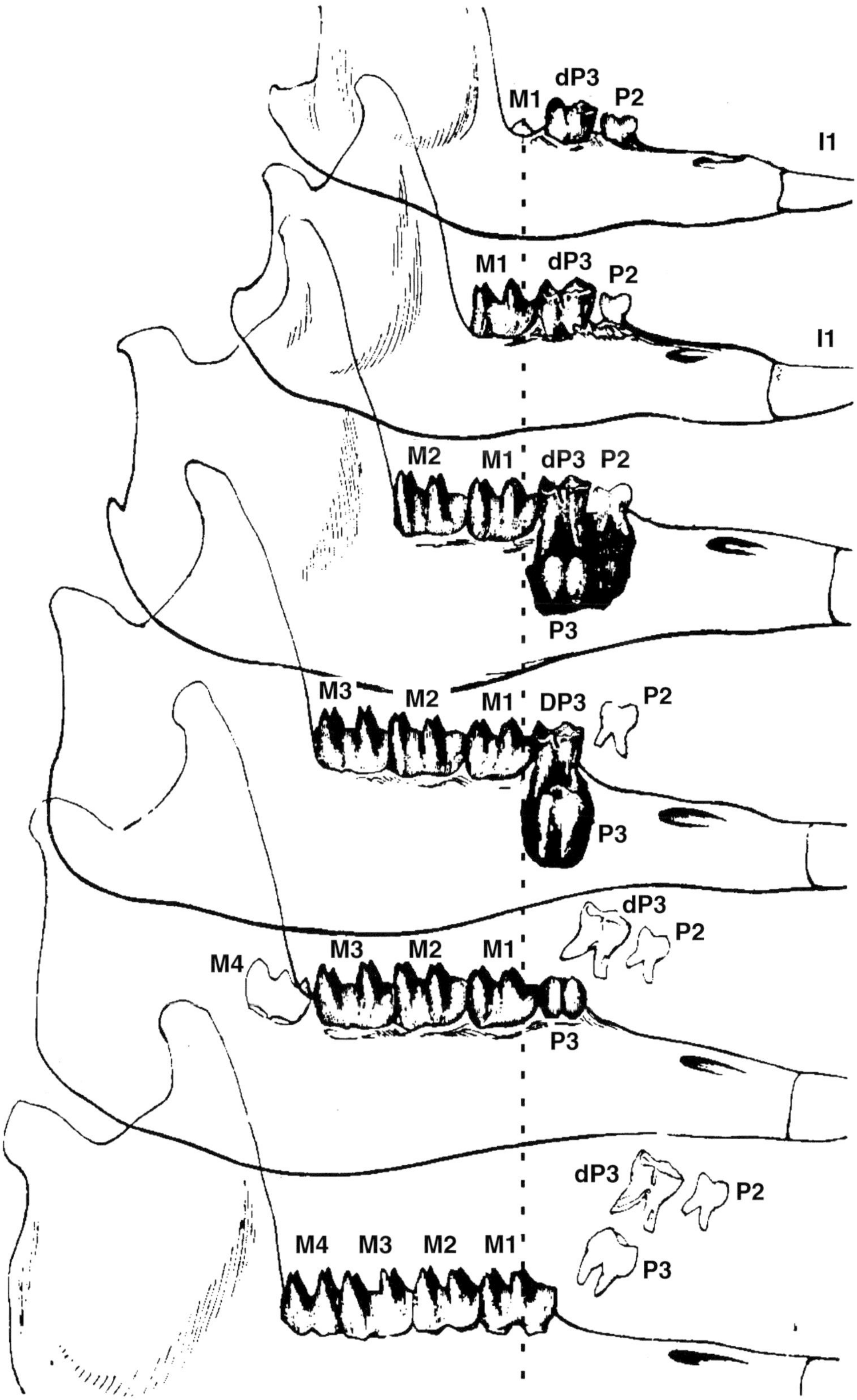

Figure 7. Sequence of eruption of cheek teeth and molar progression in a modern grazing kangaroo. The eruption of the third permanent premolar (P3), results in the eviction of P2 and the molariform deciduous third premolar (dP3). Molar teeth erupt in sequence from anterior to posterior and the cheek tooth row moves anteriorly, resulting in the progressive loss of the worn anterior teeth. After Owen (1845).

indubitably completely lophodont form. By then much more material of *Bulungamaya delicata* had become available from Riversleigh. Together the three species show a nice evolutionary sequence of lower molar evolution from a bunolophodont to a completely lophodont form. Bulungamayines achieved this in a manner different from that described by Flannery and Rich (1986) as occurring in Oligo-Miocene balbarine kangaroos, considered by Flannery (1989) to be ancestral to later macropodids (Figures 6 and 7). Cooke (1999) described *Wanburoo hilarus*, yet another lophodont bulungamayine taxon, from early to middle Miocene at Riversleigh and suggested that bulungamayines should be given consideration as potential ancestors of lophodont macropodids.

Several Riversleigh bulungamayines have also considerately provided skull material for study. These include specimens referable to *Bulungamaya delicata* and *Wabularoo naughtoni*. More recently beautifully preserved skulls of an adult and a juvenile *Ganguroo* have also been recovered. Perhaps these result from a tragic Miocene accident that took the lives of a mother and child? Hang on while I find a tissue...sorry! Just as in *Hypsiprymnodon bartholomaii* and *Bettongia moyesi*, all of these skulls exhibit alisphenoid-parietal contact, providing even more food for thought regarding the polarity of this character and the systematic placement of bulungamayines (Figures 8, 9, 10). Let's leave bulungamayines in Potoroidae for the moment and return to consider these topics later.

MACROPODID SUBFAMILIES

Macropodinae

This subfamily includes the largest of our native marsupials, the grazing kangaroos. These forms are comparatively recent evolutionary novelties, arising as grasslands began to spread dramatically from about early Pliocene times. In grazing macropodines the teeth are high-crowned or hypsodont to allow for wear. They also have strongly developed longitudinal crests or 'links' on their molars. Browsing macropodids, including macropodines such as the swamp wallaby, *Wallabia bicolor* have lower crowned teeth with less developed links (Figure 11). See Dawson (1995). Browsing kangaroo molars emphasize crushing of plant material to release cell contents, while those of grazers are adapted for cutting grass into small pieces to speed up the process of microbial fermentation in the gut. Longitudinal links also occur on the molars of the large sized, large-clawed palorchestids whose faces were probably also equipped with short trunks. Because of their molar links palorchestids were long considered to be kangaroos, until Woods (1958) pointed out that their dentaries did not have the distinctive kangaroo character of a masseteric canal.

Grazing kangaroos have evolved an unusual method of coping with the heavy wear to which their molar teeth are subjected while processing grass, which is not only fibrous but contains the mineral silica as well. The latter is probably a defence that grasses have evolved precisely to wear down the teeth of grazers and discourage their grass-eating habit. Kangaroos, have evolved a defence against the defence—molar progression. As each of their four molars erupts in sequence from front to back, the anterior, more heavily worn molars and the premolars are pushed forwards, eventually being evicted from the front of the tooth row. The Nabarlek or little rock wallaby, *Peradorcas concinna*, one of the smallest living macropodines, has taken molar progression a step further by producing extra molars beyond the normal four to which the rest of the kangaroos are limited. Molar rows in grazing macropodines are also arched so that the maximum biting force is concentrated on one or two pairs of molars at any one time. Sanson (1989) reviews macropodoid dental adaptations to diet.

Modern macropodids, particularly the largest species, are famous for their graceful bipedal bounding or

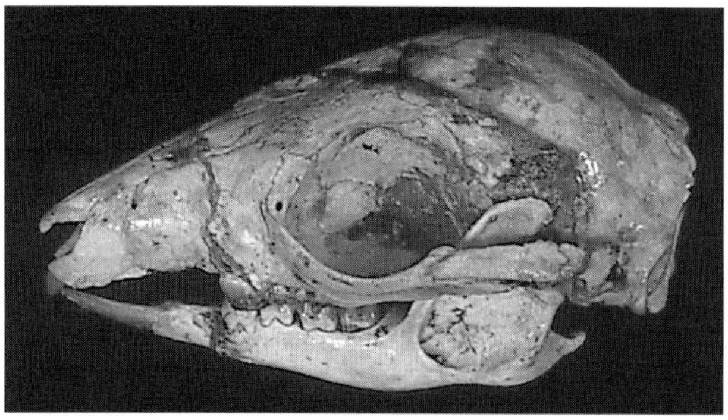

Figure 8. Skull and mandible of a juvenile of *Ganguroo* sp., a Miocene bulungamayine kangaroo from Riversleigh.

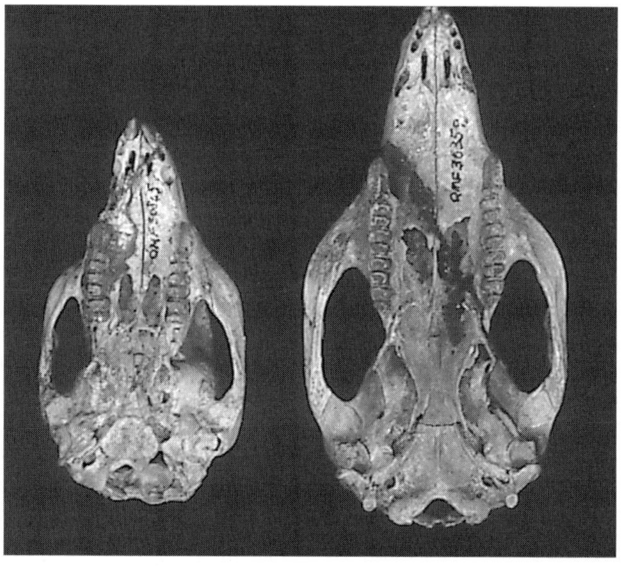

Figure 9. Palatal views of the skulls of juvenile (left) and adult (right) skulls of Miocene bulungamayine kangaroos. Both specimens, members of a species of *Ganguroo*, were recovered from AL90 Site at Riversleigh and may be from a mother and her young.

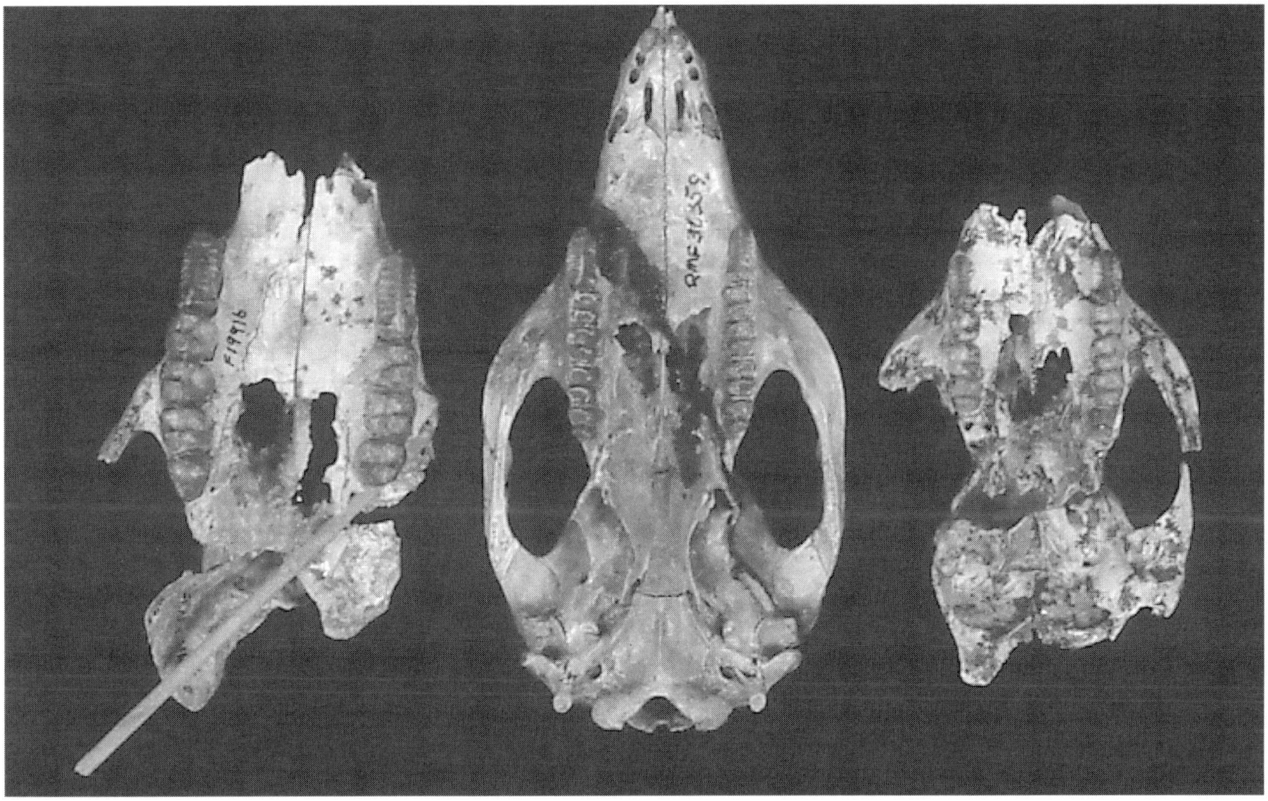

Figure 10. Palatal views of skulls of bulungamayine kangaroos from Riversleigh. From left to right: ?*Wabularoo naughtoni*, *Ganguroo* sp. and ?*Bulungamaya delicata*.

saltatorial gait, used at other than very low speeds. No other mammals of comparable size utilize such a means of locomotion. At moderate speeds, hopping turns out to be more efficient than running on all fours, conserving energy and using less oxygen (Dawson and Taylor, 1973; Dawson, 1977, 1995).

Macropodines are an extremely diverse group with 50 extant species distributed among 14 genera, and a continually growing number of extinct genera and species. Their sizes range from gigantic forms such as the extinct *Macropus titan* through smaller, extant species such as *M. rufus* or *M. giganteus* and the shorter but built-like-a-brick-shipyard *M. robustus*, down to diminutive species such as the warabi (*Petrogale burbidgei*), the smallest of all extant species weighing in at only around a kilo. Macropodines are also highly adaptable, with modern species living in environments as diverse as the arid inland plains, eucalypt forests, rainforests, tall grasslands, dense thickets and rocky escarpments. One group, the tree kangaroos, belonging to the genus *Dendrolagus*, has resumed an arboreal way of life.

Nor are macropodines confined to Australia. New Guinea is the centre of diversity of tree kangaroos with eight species, while only two species, *Dendrolagus lumholtzi* and *D. bennettianus*, survive in the north Queensland rainforest canopies. Those interested in these amazing kangaroos will be delighted by perusal of Flannery *et al.* (1996). Also in New Guinea are the most plesiomorphic of the living macropodines, the forest wallabies belonging to the genera *Dorcopsis* and *Dorcopsulus*. Both tree kangaroos and *Dorcopsis* (*D. wintercookorum*) inhabited early Pliocene rainforests as far south in Australia as Hamilton in Victoria (Flannery *et al.*, 1992). A number of influences contributed to the survival of these kinds of kangaroos including: the tropical latitudes; the insular status of regional environments; and upthrust of the New Guinean highlands during mid-Miocene, Pliocene and Pleistocene—a result of the collision of the north-going Australian plate with the Asian plate. These factors perpetuated the habitats to which these kangaroos were already adapted but which were fast disappearing in Australia. In addition New Guinea has endemic species of pademelons (*Thylogale*) as well as populations of one Australian pademelon (*T. stigmatica*) and the agile wallaby, *Macropus agilis* (Flannery, 1990).

Macropodines are well documented in the Pliocene and Pleistocene but their earlier history is not known with certainty beyond the late Miocene *Dorcopsoides fossilis* mentioned earlier. Pledge (1992) also indicates a possible late Miocene presence of macropodines in the Curramulka Local Fauna of South Australia—two species of *Baringa*, a genus described by Flannery and Hahn (1984) from early Pleistocene deposits of Victoria, and a species of *Protemnodon*.

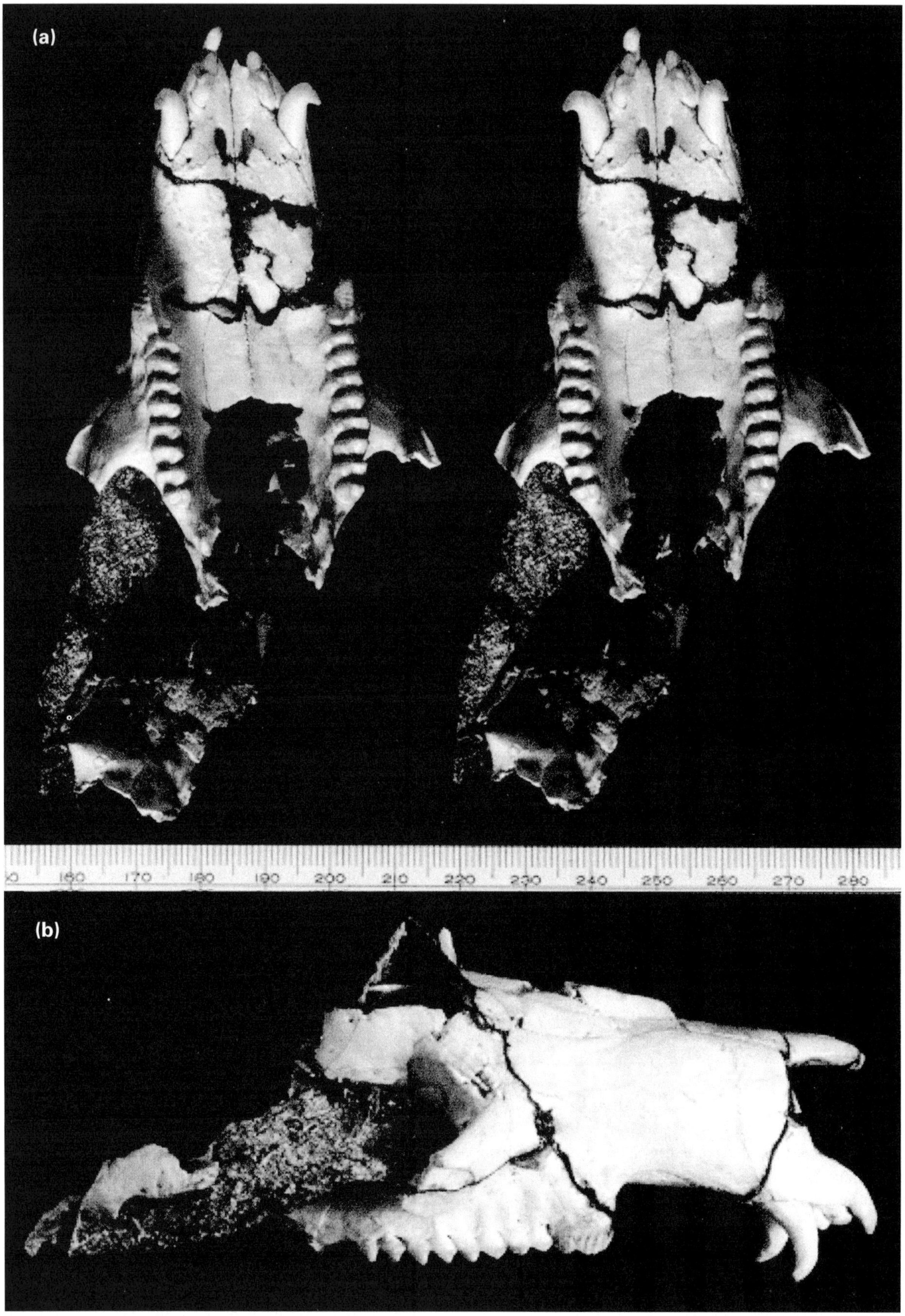

Figure 11. The paratype of *Balbaroo fangaroo*. **(a)** stereoimage of palatal view; **(b)** right lateral view. Scale in mm.

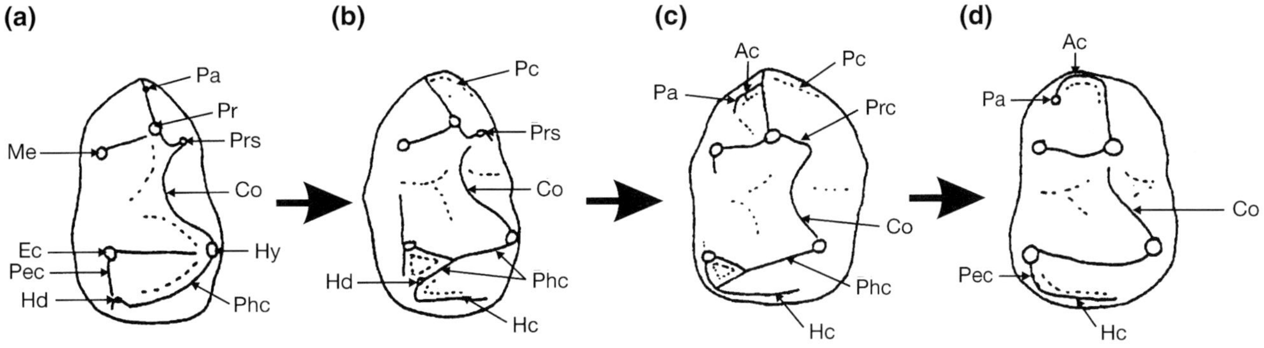

Figure 12. Lophid evolution in lower molars of balbarines. **(a)** RM1 of a plesiomorphic balbarine, *Nambaroo couperi*. The paracristid runs directly anteriorly from the apex of the centrally positioned protoconid. The protolophid is formed by a buccally directed crest from the metaconid. The cristid obliqua terminates anteriorly at a protostylid. Hypoconid and entoconid are linked by a buccal crest from the entoconid. The posthypocristid descends the posterior face of the crown to a hypoconulid which is also joined by the posthypocristid. **(b)** RM1 of a more derived *Nambaroo* (c.f. *N. saltavus*). The hypoconulid is reduced in size and a posterior shelf, the hypocingulid (= posterior cingulid), is formed low on the posterior face of the crown. The proximal portion of the posthypocristid is linked to the entoconid crest to form a neomorphic hypolophid. The distal portion of the posthyocristid and the postentocristid border a triangular depression on the posterior, lingual surface of the hypolophid. Anteriorly a precingulid is formed buccal to the paracristid. **(c)** RM1 of a more derived balbarine, *Wururoo dayamayi*. The protostylid is lost but the protoconid crest linking protoconid and protostylid is retained and is joined by the cristid obliqua. The paraconid is lingually displaced and contacted by a branch of the paracristid which forms a lingual anterior cingulid. **(d)** RM1 of a derived balbarine, *Balbaroo fangaroo*. The anterior cingulid is more developed. All trace of the protostylid is lost and the cristid obliqua terminates anteriorly on the posterior face of the protoconid. The distal part of the posthypocristid is lost and the postentocristid is continuous with the hypocingulid. Key to abbreviations: Ac = anterior cingulid; Hc = hypocingulid (posterior cingulid); Hd = hypoconulid; Pa = paraconid; Pc = precingulid; Prs = protostylid; others in Figure 5.

The first description of a fossil macropodine was of the giant species, *Macropus titan,* by Owen (1838) on the basis of material collected by Sir Thomas Mitchell from the Pleistocene deposits of Wellington Cave in New South Wales. Owen described further species of Pleistocene kangaroos and the sthenurine genera *Procoptodon*, and *Sthenurus*, as well as the macropodine, *Protemnodon* in a continuing series of papers (1873, 1874a, b, 1876, 1877). Between 1883 and 1895, Charles de Vis described many species of macropodine and sthenurine kangaroos from Pleistocene deposits of the Darling Downs of Queensland; however, many de Vis species have since been synonymized.

Many new genera of fossil macropodines have been described since those times. Stirton (1955) described *Prionotemnus palankarinnicus,* from the then newly discovered Pliocene Palankarinna Fauna of the Tirari Formation of South Australia. Flannery and Archer (1984) added the genus, *Kurrabi*, and two new species, *K. mahoneyi* and *K. merriwaensis*, from the early Pliocene Bow Fauna of New South Wales. Bartholomai (1973b) described the new macropodine genus and species, *Fissurodon pearsoni*, from Pleistocene deposits of the eastern Darling Downs. Later Flannery (1980) proposed the inclusion of *F. pearsoni* within *Macropus*. Flannery and Archer (1982) indicate an extensive range for this species, recording its occurrence in the early Pleistocene Katipiri Sands of Lake Kanunga, South Australia, and at Glen Garland on Cape York Peninsula. Flannery and Szalay (1982) described a new genus and species of giant tree kangaroo, *Bohra paulae*, from the Wellington Caves. Flannery and Hahn (1984) added the new genus and species, *Baringa nelsonensis*, from the early Pleistocene Lower Nelson Bay Formation of Victoria. Most recently McNamara (1994) added a further new genus and species, *Congruus congruus*, from a Pleistocene cave fill near Naracoorte, South Australia.

Additional Pliocene macropodine species are now known from New Guinea. Plane (1967) described *Protemnodon otibandus* and *P. buloloensis* from the Pliocene Awe Local Fauna of the Otibanda Formation in New Guinea. Flannery *et al.* (1989) described a new plesiomorphic macropodine genus and species, *Watutia novaeguineae*, from the same local fauna.

There have been a number of revisions made of macropodine genera, and for some, revisions of the revisions. Stirton (1963) reviewed the genus *Protemnodon*, revising the generic diagnosis and questioning the validity of the six species described by Owen. He suggested that two species may have been present in faunas of the Darling Downs, but regarded even that number as questionable. Bartholomai (1973a) reviewed the occurrence of the genus *Protemnodon* in later Cainozoic deposits of Queensland. He recognized five Queensland species of the genus: *P.*

anak, *P. brehus* and *P. roechus* restricted to Pleistocene sediments, and two new species, *P. chinchillaensis* and *P. devisi*, from the Pliocene Chinchilla Sands, noting the similarity of the last species to *P. otibandus* Plane (1967).

Bartholomai (1975) recognized three subgenera of *Macropus* within later Cainozoic deposits of Queensland: *M. (Macropus)* including *M. titan* and a new species, *M. rama*, both known only from Pleistocene deposits; *M. (Osphranter)* including *M. atlas*, and *M. ferragus* from Pleistocene deposits, and *M. pan* and a new species, *M. woodsi*, both occurring in the Pliocene Chinchilla Sands; *M. (Prionotemnus)* including *M. agilis siva* de Vis (1895), *M. gouldii*, *M. thor* and a new species, *M. piltonensis*, all restricted to Pleistocene sediments, and *M. dryas* and *M. palankarinnicus*, both from the Chinchilla Sands. In their revision of the genus *Macropus,* Dawson and Flannery (1985), disagreed with Bartholomai's inclusion of *M. (Prionotemnus) palankarinnicus* within the genus on the grounds that it represents a much more plesiomorphic member of Macropodinae. They also proposed a new subgeneric name, *Notamacropus*, for those species other than the subgeneric type included by Bartholomai within *Macropus (Prionotemnus).*

Bartholomai (1976) reviewed the occurrence of the genus *Wallabia* in later Cainozoic deposits of Queensland. He included *W. indra* de Vis (1895) within this genus and synonymized '*Halmaturus*' *vishnu* de Vis (1895) with *W. indra*, noting the occurrence of this species in both Pleistocene and Pliocene deposits on the Darling Downs.

Sthenurinae

Sthenurinae contains the largest and some of the most bizarre kangaroos that ever lived. They come in two forms: those with long heads (dolichocephalic forms) and those with short heads and flatter faces (brachycephalic forms). The brachycephalic species of *Procoptodon* had long arms, forward-facing eyes and an upright stance, parallels with certain hominids that did not go unnoticed by Flannery (1983a, 1994, p.123). *Procoptodon goliah* was the largest kangaroo ever, weighing in at around 200 kg. One wonders what might have happened if its brawn had been matched with a brain capacity similar to that of hominids. But for the Pleistocene extinction we could have fielded, in both rugby codes, invincible teams made up of real wallabies and kangaroos.

Sthenurines, particularly the larger species, have a number of distinctive features that set them apart from other kinds of kangaroos. One of these is the trend towards monodactyly of the hind foot in which only the fourth toe is functional. There are also modifications of the ankle joint that tend to limit flexion of the foot to a fore and aft motion, probably associated with a shift of the body weight over the fourth digit. Interestingly, Kear (1998) reports similar adaptations in the bulungamayine genus, *Wanburoo*. The dentaries of sthenurines are also usually strongly fused or ankylosed. This is in contrast to all other kangaroos in which left and right dentaries are capable of limited independent movement. Flannery (1983b, 1989) considered that the latter character constituted a synapomophy for sthenurines. He noted also that a 'boss' or 'chin' is frequently present on the dentary. Other synapomorphies listed by Flannery included: a well-developed postlink (neometaconule crista) on upper molars; an overlap between the postmetacrista and posthypocrista (the latter is now termed the postmetaconule crista); a well-developed premetacristid and a sinuous shaped occlusal edge on I1.

On the basis of these synapomorphies Flannery (1983b) added the genera *Troposodon* and *Lagostrophus* to the three sthenurine genera already recognized at that time: *Sthenurus*, *Simosthenurus* and *Procoptodon*. Bartholomai (1967) had proposed *Troposodon* as a new genus for *Sthenurus minor* Owen (1877) and its synonym, *Halmaturus vinceus* de Vis (1895). Additional species were added to the genus by Campbell (1973) and Flannery and Archer (1983). *Lagostrophus* includes only the extant (barely!—its range is now restricted to two islands in Shark Bay, W.A.) species, *L. fasciatus,* the diminutive banded hare wallaby. Since then Murray (1991) has indicated sthenurine affinities for the late Miocene Alcoota macropodid, *Hadronomas puckridgi*, described by Woodburne (1967). *Hadronomas* could thus be the oldest known sthenurine and *Lagostrophus* the sole surviving and smallest member of the lineage. Size apparently isn't everything.

Balbarinae

The subfamily Balbarinae was erected by Flannery *et al.* (1983) to accommodate two newly discovered fossil species of kangaroos: *Balbaroo camfieldensis* from the middle Miocene Bullock Creek Local Fauna of the Northern Territory, and *B. gregoriensis* from Oligo-Miocene deposits at Riversleigh. These were species with lophodont molars (Figure 12) but with a first molar whose trigonid region was markedly laterally compressed. Flannery and Rich (1986) added a new genus, *Nambaroo* (Figures 14 and 15), and three new species from late Oligocene deposits at Lake Tarkarooloo in South Australia. Because of the presence of a protostylid on M1 and other aspects of molar morphology in these species, they were considered to be the most plesiomorphic macropodids known. The species of *Nambaroo* were smaller than those of *Balbaroo* and had much more gracile premolars. Flannery (1989) added considerable phylogenetic significance to the subfamily by proposing balbarines as the common ancestors of macropodines and sthenurines.

Additional balbarine genera and species have since been added. Cooke (1992) described a new genus, *Ganawamaya*, and three new species from Riversleigh. These had similar premolars to those of *Nambaroo*, but were somewhat more derived in molar morphology. Cooke (1997a) added a new species of *Nambaroo* (*N. couperi*) and a new genus and species, *Wururoo dayamayi*, both from Riversleigh. *N. couperi* retains an addi-

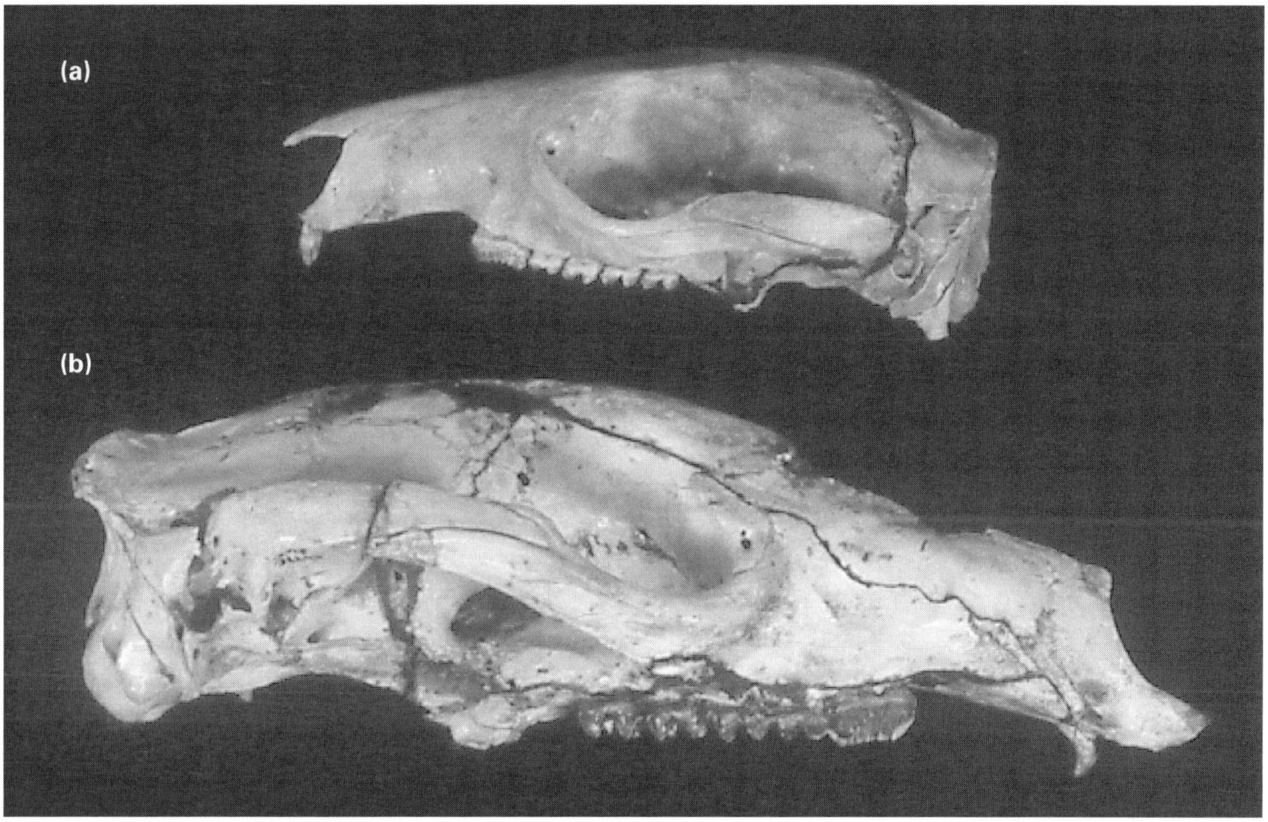

Figure 13. Lateral views of the skulls of lophodont, Miocene kangaroos from Riversleigh: **(a)** a bulungamayine, *Ganguroo* sp.; **(b)** a balbarine, *Balbaroo* sp.

tional plesiomorphic cusp, the hypoconulid on its M1. *W. dayamayi* retains some trace of a protostylid on M1, but its premolar morphology suggests a closer affinity with the species of *Balbaroo* than with *Nambaroo* or *Ganawamaya*.

Cooke (2000) described cranial remains of a new and startling species, *Balbaroo fangaroo*. The specific epithet, 'fangaroo', whose origin can be laid at the door of one Michael Archer, is a reference to the very large upper canine teeth present in this species (Figure 16). The function of hypertrophied canines in an Oligo-Miocene, rainforest-dwelling, herbivorous kangaroo is cause for speculation. Possession of upper canines is clearly plesiomorphic for macropodoids: they occur in all phalangeridans, all potoroids, all balbarines in which the condition can be ascertained and are retained in some of the more plesiomorphic macropodines such as *Dorcopsis* and *Dendrolagus*. But why are they so big in 'fangaroo'?

The answer may lie in the observation that enlarged canines also occur in extinct and extant ungulates. Among the latter are species of *Tragulus*, *Moschus*, *Hydropotes*, *Cervulus* and *Elaphodus*, inhabitants of dense vegetation, and in which canine size is sexually dimorphic. Enlarged canines may be similarly sexually dimorphic in *B. fangaroo* and may have been useful for close-range visual signalling of sexual identity and probably also in agonistic displays within a visually restrictive habitat. Physical conflicts between many modern kangaroos involve an upright stance, the use of forelegs and claws in grappling and scratching (frequently with the head averted to avoid eye damage) and hind legs for kicking. Fangaroo's canines may well have been its chief weapons during conflicts if it could not employ its forelimbs for combat (see below).

Balbarines were an extremely diverse group during the latter part of the Oligocene and up until about the middle Miocene. Woodburne *et al.* (1993) indicate the presence of at least six species from late Oligocene deposits in the Tirari Desert. Cooke (1997c) recognizes a further 11 species of balbarines from Riversleigh. Significantly, all but one such species are older than mid Miocene. The dramatic decline in balbarine numbers may have been the result of competition with the other lophodont kangaroos of the time, the bulungamayines.

Among the Riversleigh balbarines are seven additional species of *Nambaroo*, one of which is known from very well preserved cranial and postcranial remains—amazingly complete for a probable late Oligocene fossil kangaroo. Studies of the postcranial remains by Kear (1998) suggest that this animal did not hop but used a quadrupedal gait as does *Hypsiprymnodon*—the same may well be true for *Balbaroo fangaroo*. Like *Hypsiprymnodon*, this *Nambaroo* species also retained its first toe and may have had similar limited climbing abil-

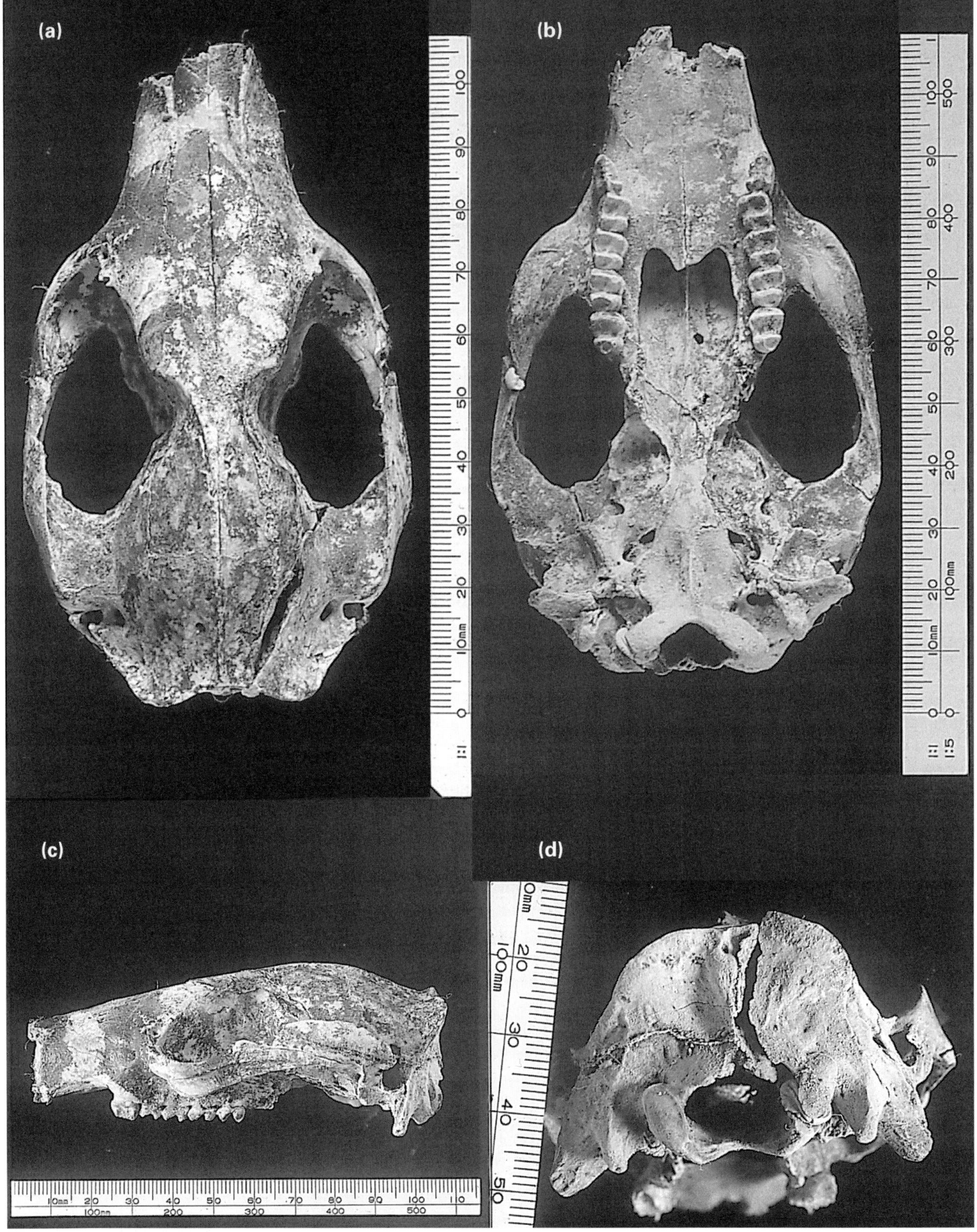

Figure 14. The skull of an undescribed species of *Nambaroo* from Quantum Leap site (Riversleigh). This is estimated to be late Oligocene in age and the specimen may be the oldest kangaroo skull known: **(a)** dorsal view; **(b)** palatal view; **(c)** lateral view; **(d)** posterior view showing the hypertrophied mastoid processes—a probable balbarine synapomorphy.

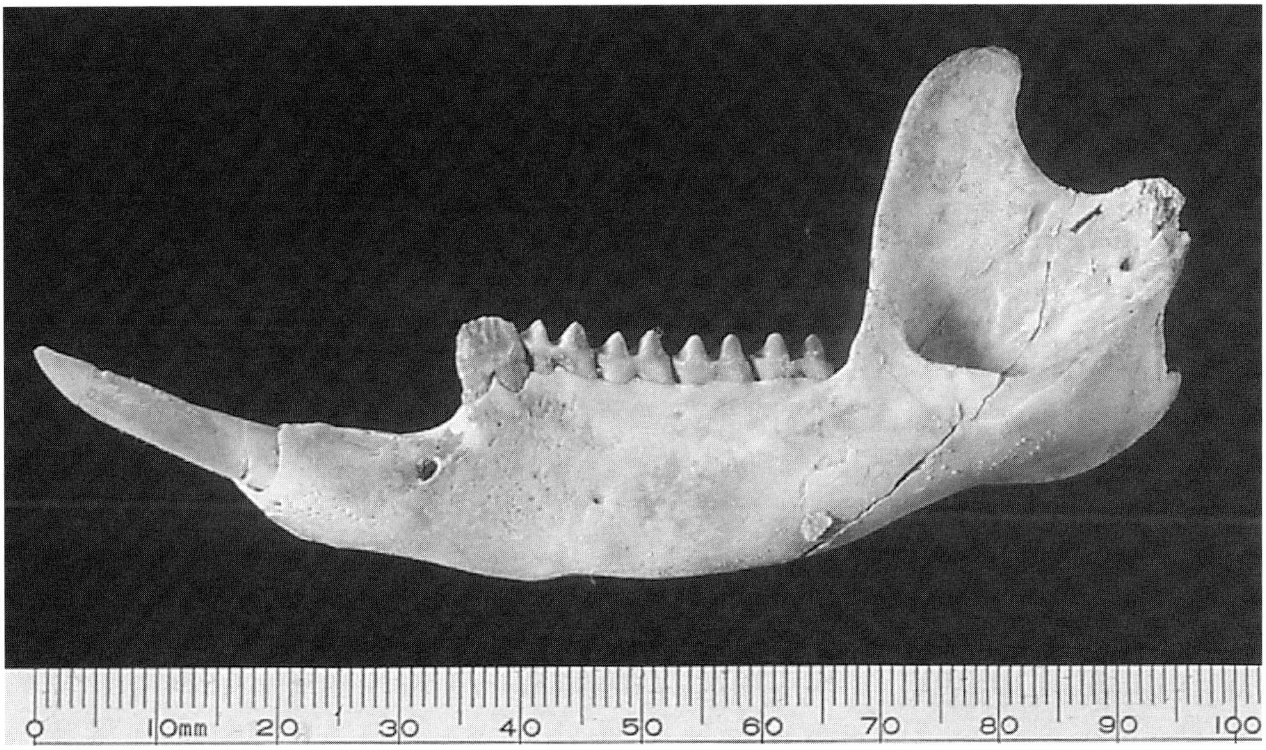

Figure 15. Lateral view of the dentary associated with the Quantum Leap *Nambaroo*. The dentary has a deeply penetrating masseteric canal confluent with the inferior dental canal, and a posterior mental foramen. Both are apparently plesiomorphic for macropodoids.

ity. The cranial remains of this animal, like those of *B. fangaroo* and of other undescribed balbarines, suggest that these were somewhat fat-faced animals with inflated frontal sinuses, marked post-orbital constriction of the skull, hypertrophied mastoid processes and well-developed nuchal crests, indicating a powerful neck musculature. Surprisingly, all have extensive frontal-squamosal contact on the wall of the braincase, a character studiously avoided by all other macropodids. This would represent the fifth instance of independent evolution of this character among macropodoids, perhaps beginning to stretch the bounds of evolutionary coincidence a little too far. Cooke (1997a) has argued, on the grounds of parsimony, that frontal-squamosal contact is plesiomorphic for Macropodoidea.

Balbarine dentaries also have extensive masseteric canals which, in the plesiomorphic species, are confluent with the inferior dental canal. The same condition occurs in hysiprymnodontines, potoroines, propleopines, bulungamayines and plesiomorphic macropodines such as *Dorcopsoides*. Its widespread occurrence suggests that this too is plesiomorphic for Macropodoidea.

Apart from being lophodont, molar morphologically speaking, more plesiomorphic dental characters are retained in species of balbarines than in any other macropodoids. Flannery and Rich (1986) outlined an evolutionary sequence leading to hypolophid formation in balbarines. Cooke (1997a) suggested modifications to this and then (1997b) showed that the hypolophid of bulungamayines, evolution's second attempt at building a plant-eating kangaroo, was achieved via a different evolutionary pathway.

Cooke (1997a) proposed that balbarines are in fact not macropodids at all, but basal macropodoids that diverged early from the main stem of macropodoid evolution, independently and probably rapidly evolving lophodonty to exploit a herbivore niche. Wroe *et al.* (1998) also support separation of Balbarinae from Macropodidae, but indicate potential synapomorphies with Propleopinae. If monophyly of balbarines as suggested by Cooke (1997d) and Kear (1998) can be demonstrated conclusively, it would result in the elevation of Balbarinae to familial status. That still leaves the problem of the origins of macropodids. On the basis of cranial, dental and postcranial morphology, Cooke (1997a, b, d, 1999) and Cooke and Kear (1999) argue that bulungamayines, not balbarines, are directly ancestral to macropodids. As you can see, all has not yet been made clear in kangaroo phylogeny.

KANGAROO PHYLOGENY

Winge (1893) made one of the earlier post-Darwinian attempts to derive a phylogeny for kangaroos (Figure 18(a)). He placed macropodoids and all modern possums within the family Phalangistidae (possums),

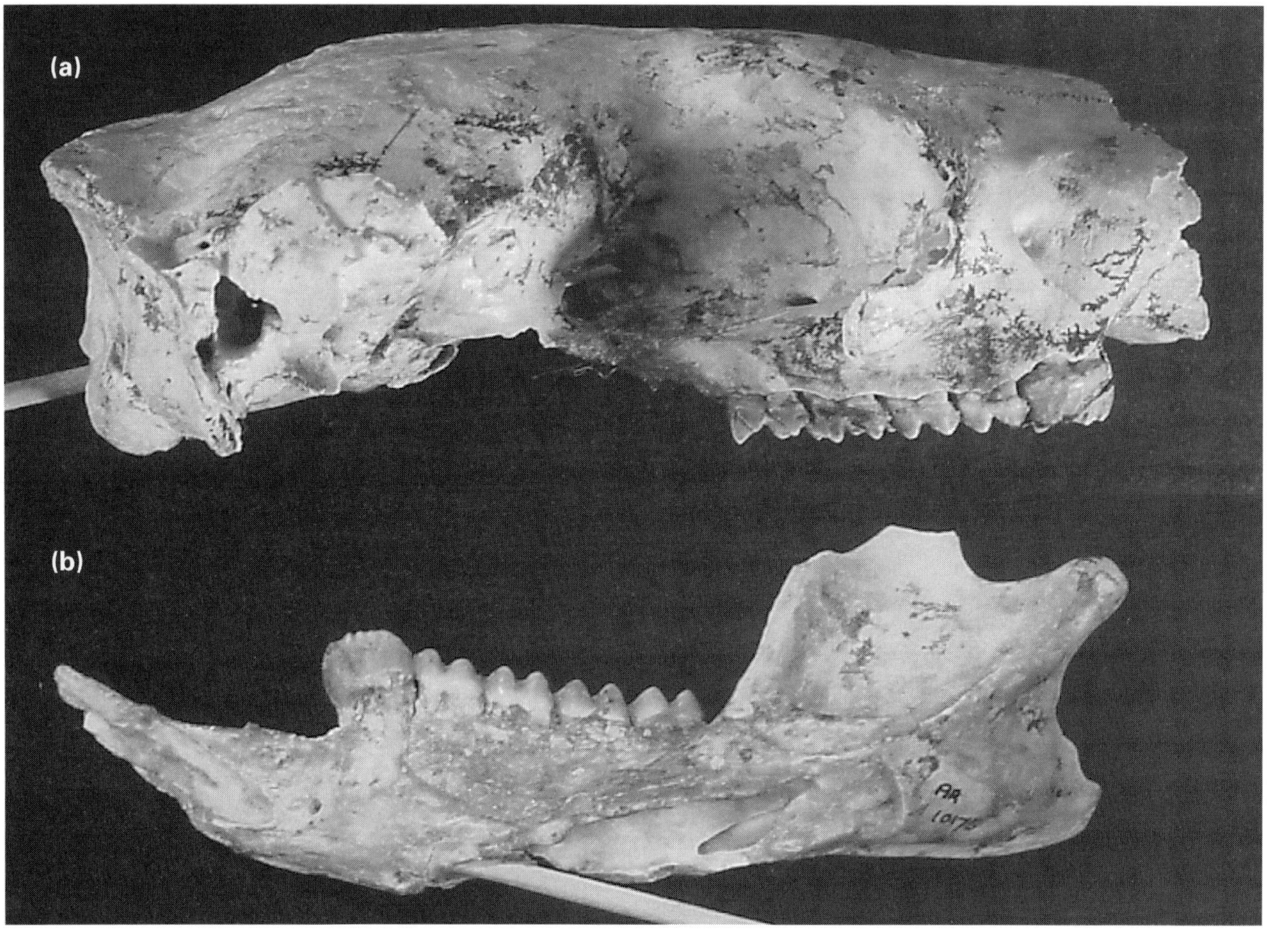

Figure 16. The holotype of *Balbaroo fangaroo*. Like all known balbarine skulls in which the character is visible, the skull exhibits contact between the frontal and squamosal bones, suggested herein to be the plesiomorphic condition for macropodoids. **(a)** lateral view of the skull; **(b)** dentary.

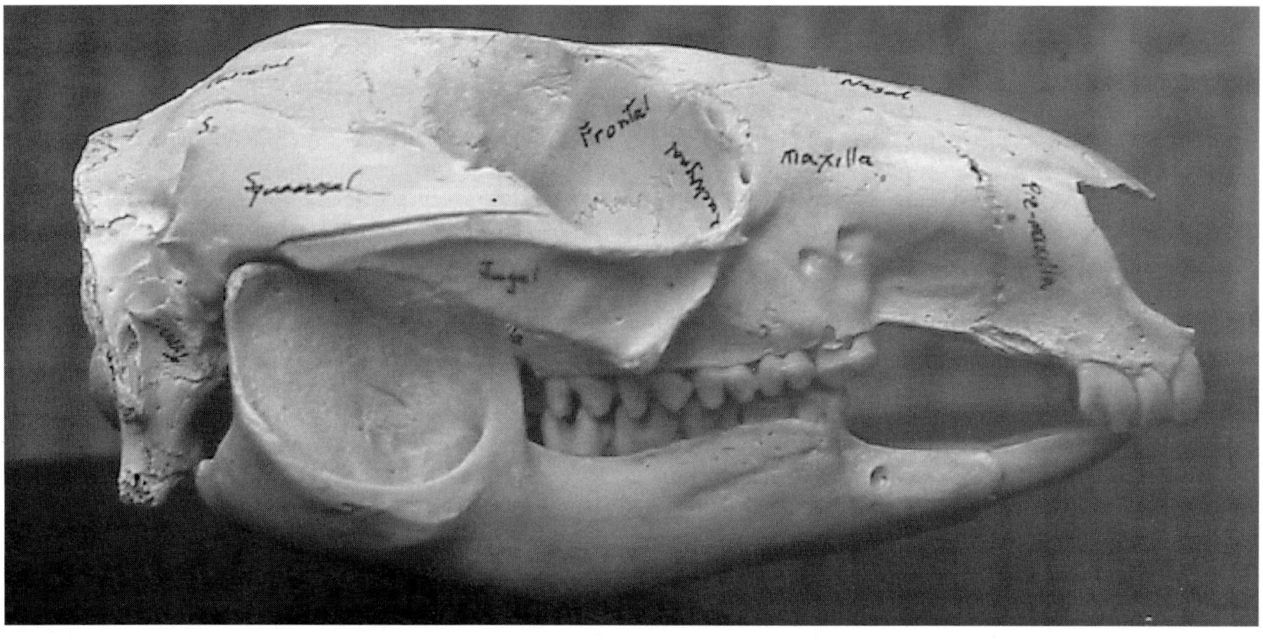

Figure 17. Skull and dentary of a modern grazing kangaroo, *Macropus rufus*. Note the elongated paroccipital processes, long masseteric process, antorbital sulcus, downward flexion of the rostrum, procumbent lower incisor.

itself derived from a hypothetical primitive member of the family Phascolarctidae, containing also koalas, wombats, diprotodon-nototheres and thylacoleons. He considered that both macropodids and potoroids were derived from primitive Phalangistini (non-ringtail possums).

Bensley (1903) presented a phylogeny of the Australian marsupials based on details of dentition and pedal morphology of extant animals (Figure 18(b)). He was certainly conscious of the value of palaeontology to such studies, commenting (p.86), 'Generally speaking the history of a fauna can only be satisfactorily determined by reference to palaeontology'. Although deposits containing Pliocene and Pleistocene marsupial fossils were well-known in Australia by that time, he lamented, '… the fossil deposits of that country have not as yet furnished ancestral or collective types.' Bensley proposed a diphyletic origin for potoroids, the Bettonginae (*Hypsiprymnodon, Bettongia, Aepyprymnus*), derived from a burramyid or phalangerid, and the Potoroinae (*Potorous-Caloprymnus*), derived from a *Gymnobelideus*-like ancestor. He considered Macropodidae to be derived from within the Potoroinae

Raven and Gregory (1946) proposed the first comprehensive phylogeny of macropodoids of the 'modern' (post World War 2) era (Figure 18(c)). Based on general morphology of extant species, their phylogeny also incorporated correlation between adaptive morphology and habitat, assuming an ancestral rainforest habitat for kangaroos. As with Bensley's phylogeny, macropodids were considered to be derived from potoroids. Macropodinae was divided into two 'sections': a basal brachyodont section containing two groups of forest-dwelling animals, a *Dorcopsulus-Dorcopsis-Dendrolagus* group and a *Thylogale-Setonix* group; a more derived hypsodont section comprising *Lagorchestes*, *Lagostrophus*, *Protemnodon*, *Macropus*, *Onychogalea* and *Petrogale*. *Thylogale* was considered to be a 'primitive forerunner' of the larger kangaroos and wallabies.

Tate (1948) considered macropodoids to be derived from phalangerids but remained uncertain as to whether potoroids or macropodids should be considered more primitive. Using general morphological data, he presented separate phylogenies for each of these groups. Within potoroids, he considered there to be three lineages: Hypsiprimnodontinae (*Hypsiprymnodon-Burramys*, *Propleopus*), Potoroinae (*Aepyprymnus*, *Bettongia*, *Potorous-Caloprymnus*) and Thylacoleontidae (Figure 19(a)). Tate considered Sthenurinae and Palorchestinae allied to Macropodinae, but could not indicate a more precise phylogenetic relationship between these three groups. Within Macropodinae he suggested three basic lineages, each with subordinate branches. The phylogenetic relationship between the *Dorcopsulus-Dorcopsis-Dendrolagus* lineage and other macropodines was also not defined (Figure 19(b)).

Pearson (1950) took into account dentition and pedal morphology as well as the female urinogenital system and cranial structure, particularly the relationship of alisphenoid-parietal and frontal-squamosal bones; he advocated the erection of two families, Potoroidae (including subfamilies Hypsiprimnodondinae and Potoroinae) and Macropodidae. A common phalangerid ancestor was postulated for both families.

Hume (1978) examined the digestive systems of macropodoids as a source of phylogenetic information but concluded, 'Rather less can be said about the phylogeny of the Macropodinae on the basis of digestive tract morphology than can be concluded from skeletal and dental characteristics.' He considered the potoroid digestive tract with its large forestomach (possibly developed primarily as a storage organ) and limited development of a midstomach, together with a well-developed caecum associated with the hindgut, to represent the primitive condition among macropodoids. *Thylogale* was considered to be a primitive macropodine on the basis of its dentition and potoroid-like stomach. Grazing macropodines were considered more derived, having developed a complex, sacculated midstomach, probably similar in function to the equine colon and enabling utilization of a high fibre diet. Langer (1979, 1980a, b) noted that the forestomach of *Hypsiprimnodon* and phalangeroids is simple in comparison to the complex, sacciform forestomach seen in potoroine species, suggesting either independent evolution of a complex stomach by macropodines or by both macropodines and potoroines.

Sanson (1978) considered grades rather than clades in an examination of mastication in Macropodidae. He proposed a basal, ancestral grade (*Wallabia*, *Dorcopsis*, *Dorcopsulus*, *Dendrolagus*, *Thylogale*, *Setonix*, *Lagostrophus*, *Protemnodon*, *Hadronomas*, *Dorcopsoides*, *Prionotemnus*, *Sthenurus*), an intermediate grade (*Petrogale*, *Troposodon*), and a derived, grazing grade (*Macropus*, *Megalaeia*, *Peradorcas*, *Onychogalea*, *Lagorchestes*, *Fissuridon*, *Procoptodon*). In a more extensive study of macropodoids, Sanson (1989) added a basal macropodid-potoroid grade, adapted for a diet of invertebrates, seeds and low plant fibre.

Archer (1984) presented a comprehensive, but speculative, phylogeny of macropodids (Figure 20) significant for its placement of Balbarinae as a basal sister-group to Macropodinae, and proposing the origin of Sthenurinae (including *Lagostrophus fasciatus* and *Troposodon*), from within Macropodinae as the sister-group to a clade containing *Petrogale*, *Peradorcas*, *Thylogale*, *Setonix*, *Lagorchestes* and *Onychogalea*. Flannery (1983b) had previously suggested the placement of *Lagostrophus* and *Troposodon* within Sthenurinae on the basis of cladistic analysis using dental and mandibular characters, although he regarded the subfamily as an early offshoot within the family Macropodidae.

Flannery (1982) examined kangaroo phylogeny using data drawn from adaptive changes of the hindlimb, followed by more comprehensive cladistic analyses of phylogeny based upon morphological data from both fossil and extant species (Flannery, 1984a, b, 1989). Flannery's work is of particular significance since he was

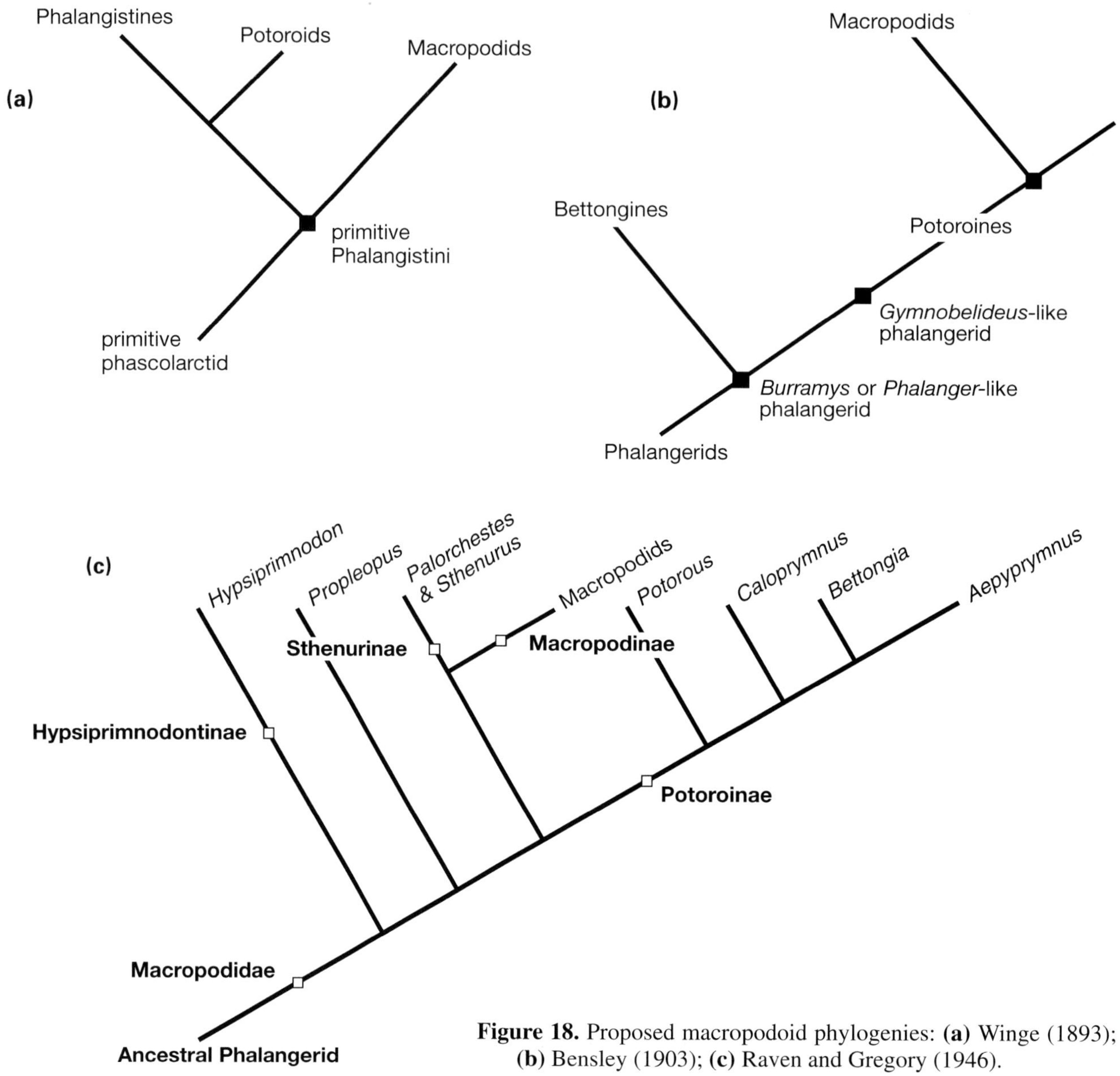

Figure 18. Proposed macropodoid phylogenies: **(a)** Winge (1893); **(b)** Bensley (1903); **(c)** Raven and Gregory (1946).

the first to include extensive data from plesiomorphic Oligo-Miocene macropodoids in a comprehensive phylogenetic analysis based on cladistic principles.

The broad phylogenetic relationships of macropodoids postulated by Flannery (1989) are indicated in Figure 21. Flannery considered macropodoids to be derived from a phalangeriform ancestral stock. Although suggesting that macropodoid differentiation occurred much earlier in the Tertiary, he proposed a differentiation of macropodoid families occurring at about 30 million years ago. A rapid radiation among potoroids soon followed, giving rise to palaeopotoroines, propleopines and hypsiprymnodontines, the latter lineage later giving rise to bulungamayines and potoroines. He considered balbarines to represent the basal macropodid lineage that subsequently gave rise to sthenurines and macropodines, the time of this differentiation estimated at about the end of the middle Miocene. In this interpretation Balbarinae thus constitutes a paraphyletic taxon.

A number of cytological studies have been undertaken in attempts to illuminate aspects of kangaroo phylogeny. Sharman (1961) recognized chromosomal affinities between *Petrogale* and *Thylogale*. Sharman *et al.* (1990) produced a phylogeny of *Petrogale* taxa based on chromosome structure, concluding that chromosomal structural rearrangements had contributed to the speciation process by increasing post-mating reproductive isolation at boundaries between parapatric taxa. Rofe (1978) revealed three different 'G'-banded karyotypes among species of *Macropus* sharing a 2n=16 chromosome number, indicating that chromosome number is not necessarily indicative of close relationship.

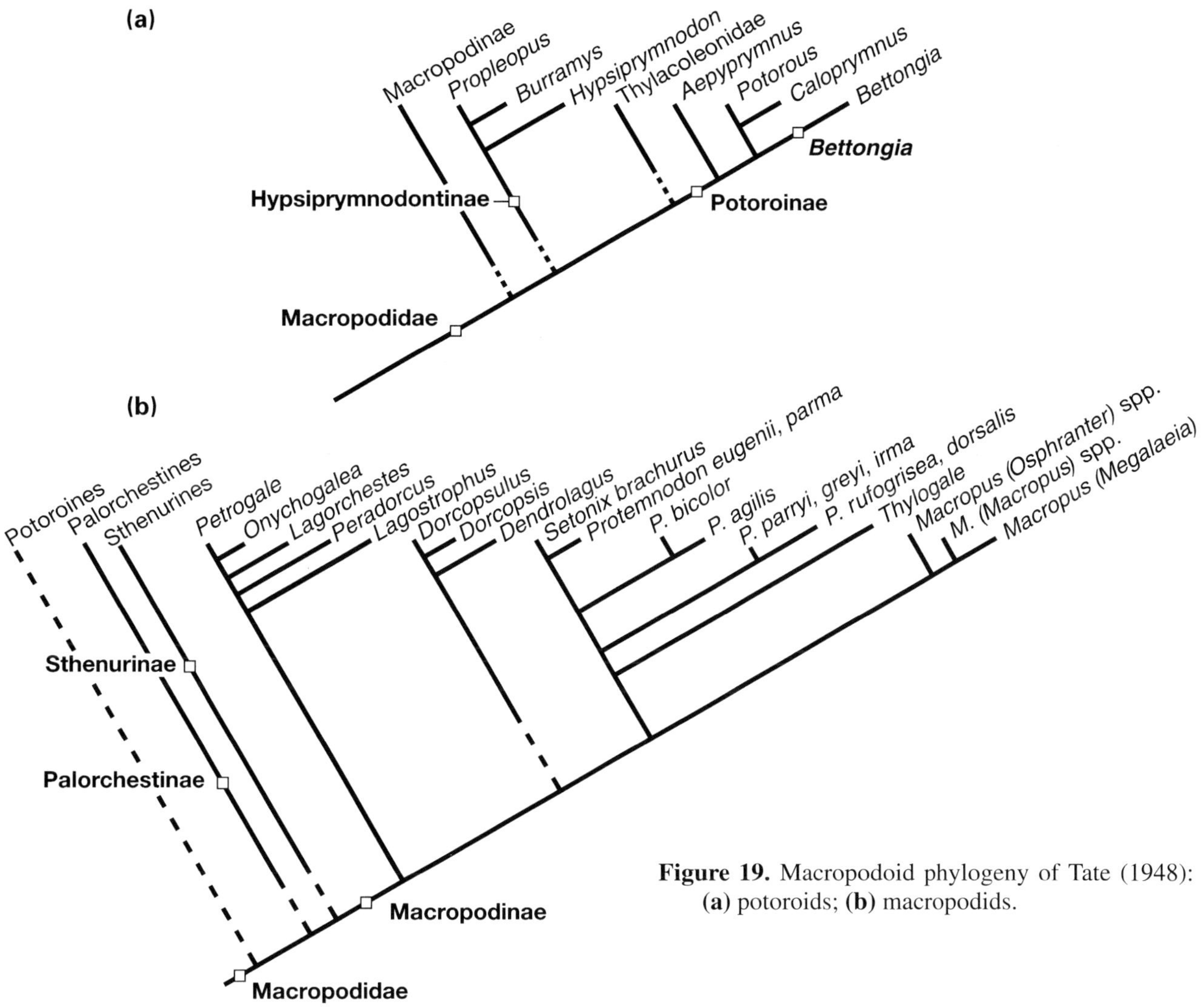

Figure 19. Macropodoid phylogeny of Tate (1948): **(a)** potoroids; **(b)** macropodids.

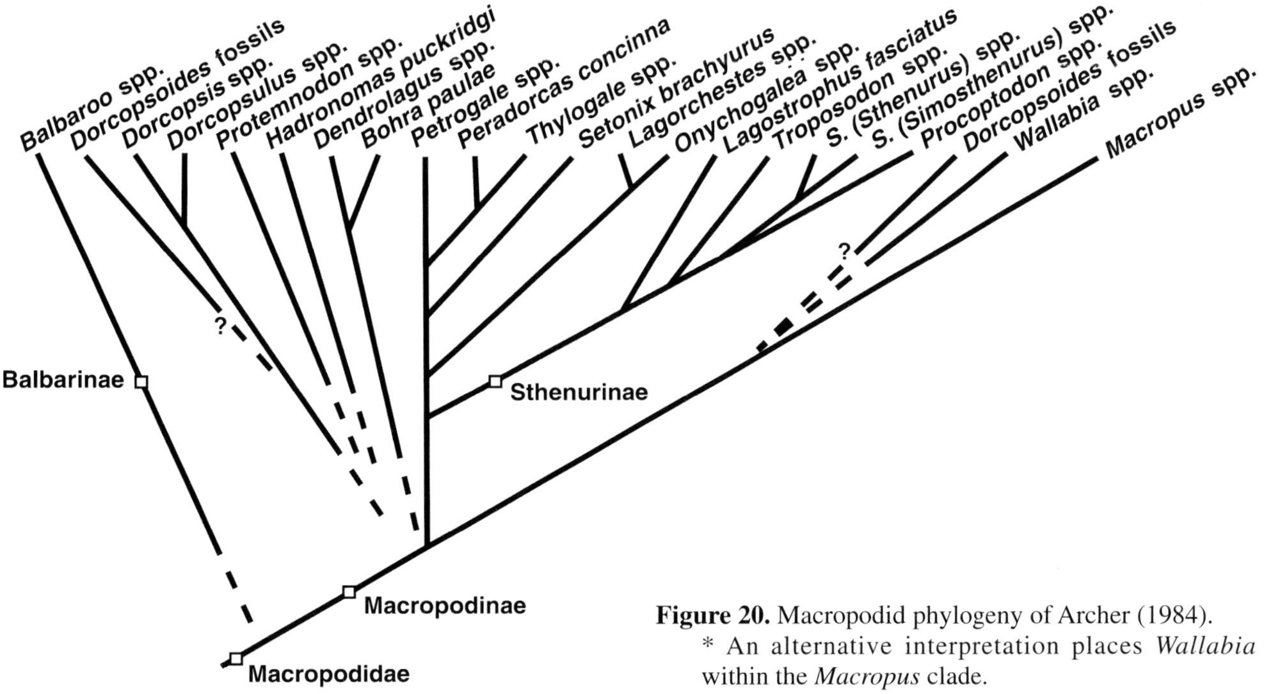

Figure 20. Macropodid phylogeny of Archer (1984). * An alternative interpretation places *Wallabia* within the *Macropus* clade.

Sharman (1989) developed a phylogeny of kangaroos based upon karyotype. Two clades were proposed, one of macropodines and the second consisting of *Lagostrophus fasciatus* as the sister taxon to potoroids with *Hypsiprymnodon* as a sister taxon to potoroines. Within Macropodinae, *Setonix brachyurus* is the sister taxon in a clade with *Thylogale*; *Petrogale*, *Lagorchestes*, *Dendrolagus*, *Onychogalea*, *Wallabia* and *Macropus (Macropus)* each form distinct clades; while *M. (Osphranter)* and *M. (Notamacropus)* together form a further clade.

Kirsch (1977) used immunological data to reconstruct phylogenetic relationships within Macropodoidea (Figure 22). In his phylogeny Potoroinae is placed as a sister group to a clade within Macropodinae containing *Wallabia* and *Macropus*. A second macropodine clade contains *Petrogale*, *Thylogale-Setonix*, *Lagorchestes-Onychogalea* with *Lagostrophus fasciatus* as the sister taxon to these.

Richardson and McDermid (1978) examined genetic relationships among Macropodidae on the basis of allozyme, cytological and immunological data (Figure 23). They placed Potoroinae as the sister group to all macropodines and indicated two series of radiations within Macropodinae. The proposed earliest radiation gave rise to three lineages: *Petrogale*, *Thylogale-Peradorcas* and *Setonix*. A later radiation gave rise to the species of *Macropus*, within which they include *M. (Wallabia) bicolor*. The species of *M. (Macropus)* were clustered together, as were the species of *M. (Prionotemnus)* and *M. (Osphranter)*. *Macropus (Megalaeia) rufus*, like *M. (Wallabia) bicolor* did not cluster with any other group. Relationships between these groups were not determined.

Baverstock *et al.* (1989, 1990) used albumin immunological data to derive a hypothesis of kangaroo phylogeny, as part of a larger study of evolutionary relationships of Australian marsupials. This study supported a macropodid-potoroid dichotomy with *Hypsiprymnodon moschatus* placed closer to potoroines than to macropodines. Surprisingly their results suggested the possibility of *Lagostrophus fasciatus* as a third major branch of macropodoids.

Westerman *et al.* (1990) applied DNA hybridization techniques to a study of 11 marsupial species representing seven major families and found that it gave rise to a phylogenetic reconstruction consistent with previously established taxonomy. They argued the value of the technique in resolving problems of marsupial taxonomy, particularly at suprageneric levels. Springer *et al.* (1990) utilized this technique in a study of the phylogeny of phalangerid marsupials and Springer and Kirsch (1991) applied it to an investigation of the radiation of diprotodontian marsupials. Results in this study suggested an early radiation of diprotodontians into three major lineages: vombatiforms, macropodoids and possums with radiation within macropodoids occurring later than within the other lineages. While the DNA hybridization data failed to resolve the basal trichotomy between the three

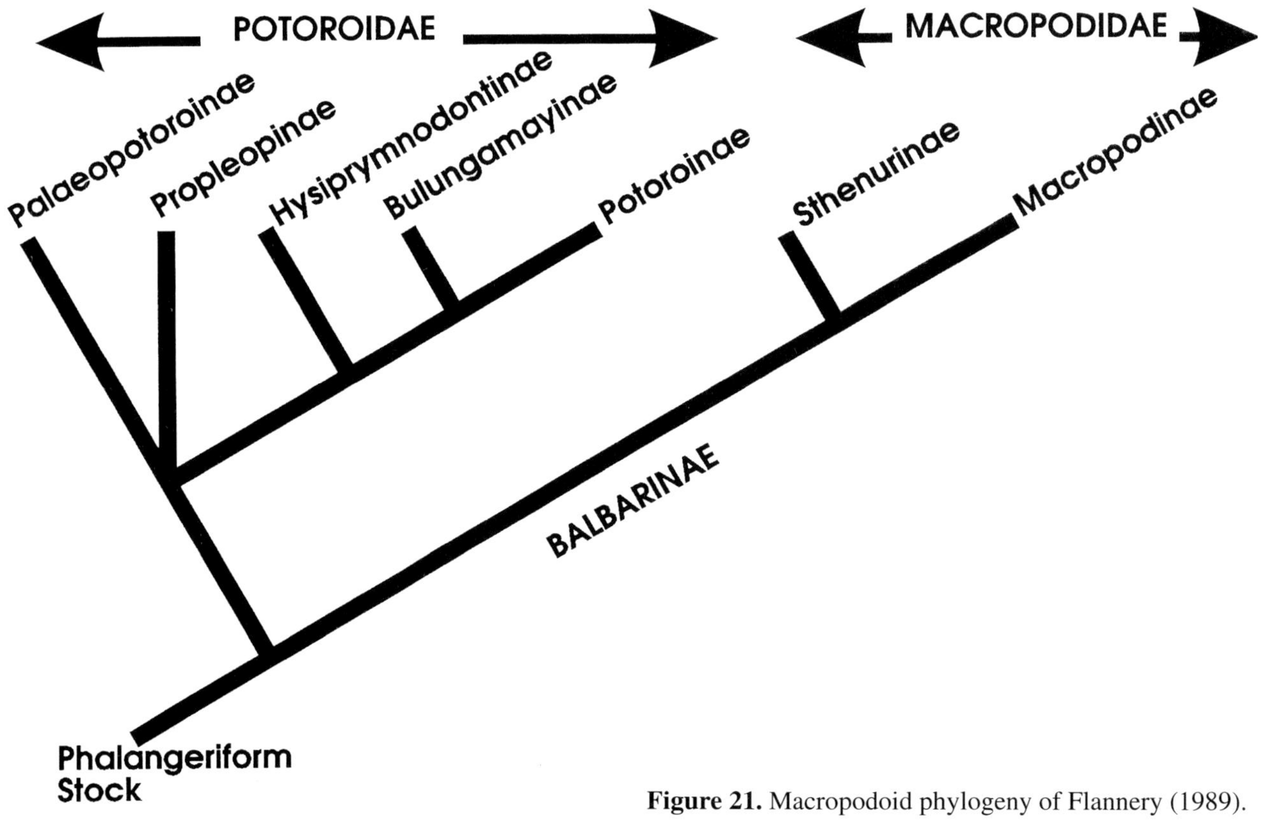

Figure 21. Macropodoid phylogeny of Flannery (1989).

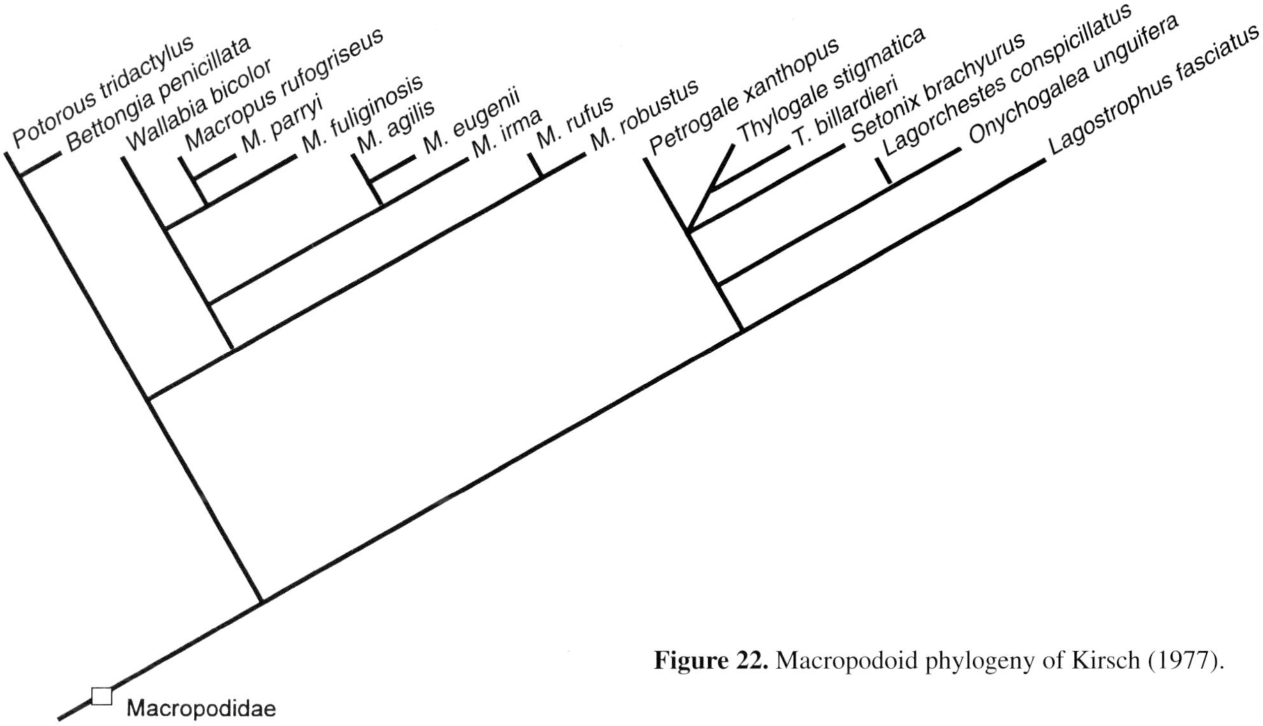

Figure 22. Macropodoid phylogeny of Kirsch (1977).

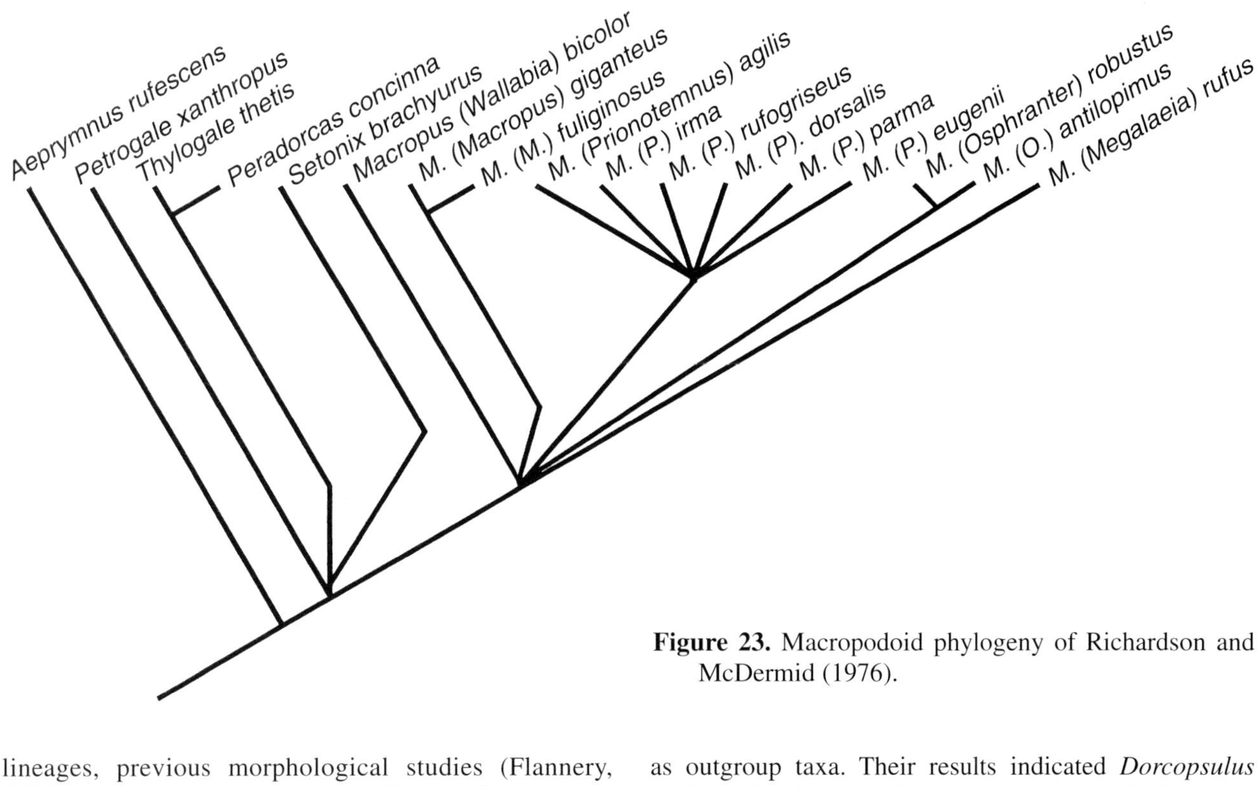

Figure 23. Macropodoid phylogeny of Richardson and McDermid (1976).

lineages, previous morphological studies (Flannery, 1987; Springer and Woodburne, 1989) also favoured the exclusion of macropodoids from a clade containing phalangerids, burramyids, acrobatids, petaurids and pseudocheirids.

Kirsch and Foeste (1995) used DNA hybridization data in a phylogenetic analysis of 11 macropodid taxa, using *Aepyprymnus rufescens* and *Trichosurus vulpecula* as outgroup taxa. Their results indicated *Dorcopsulus vanheurni* and *Setonix brachyurus* as successively closer sister groups to the remaining macropodines. Within this remainder, their results also supported a *Dendrolagus-Thylogale-Petrogale* clade, with *Macropus*, including *Wallabia bicolor* as a second clade.

Burk *et al.* (1998) sequenced 1,170 base pairs of the mitochondrial genome for 16 macropodoids. Their analy-

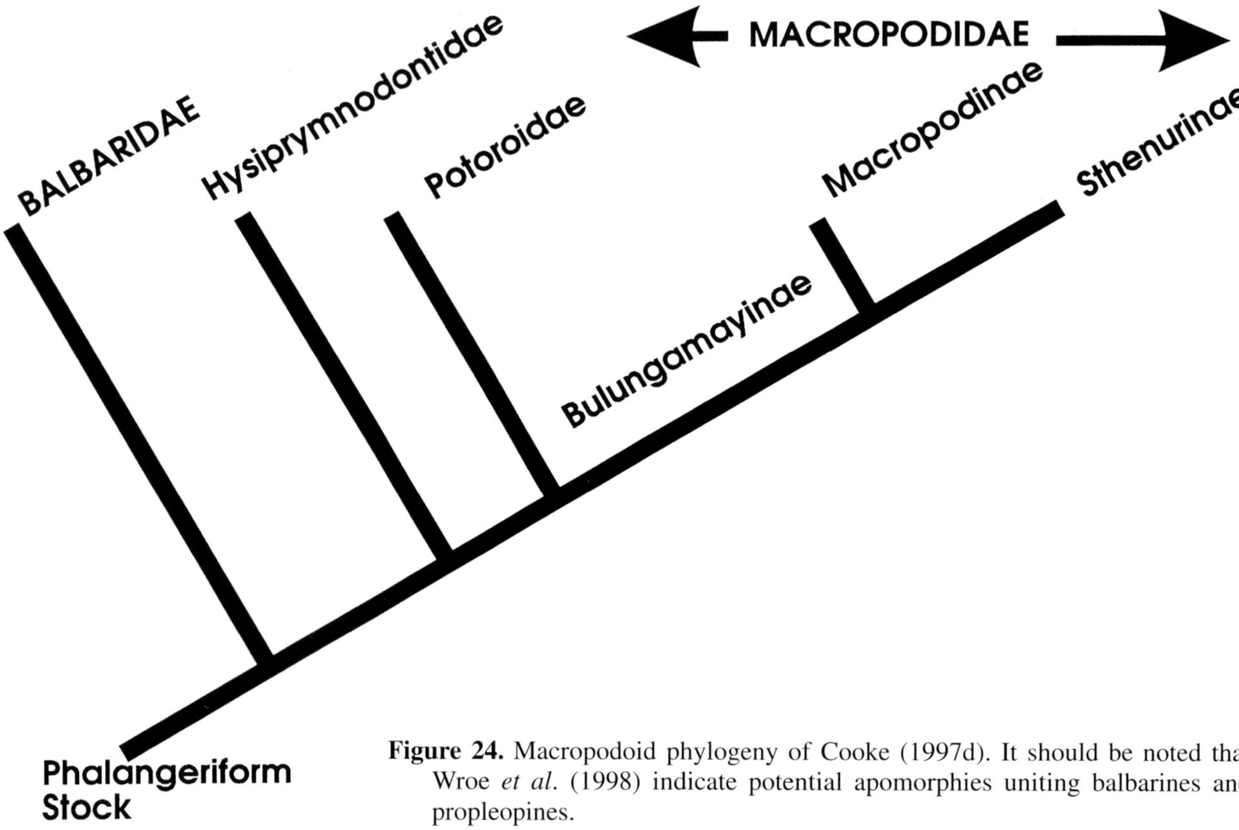

Figure 24. Macropodoid phylogeny of Cooke (1997d). It should be noted that Wroe *et al.* (1998) indicate potential apomorphies uniting balbarines and propleopines.

sis indicated monophyly of macropodids and potoroids with *Hypsiprymnodon* as a separate, basal clade. Within the larger clade potoroids and macropodids formed two distinct clades. Within the macropodid clade *Dorcopsis* and *Dorcopsulus* together formed a distinct clade with remaining macropodids, including *Lagorchestes,* forming a second clade.

A cladistic analysis by Cooke (1997d) incorporated dental and cranial information derived from plesiomorphic fossil macropodoids recovered since Flannery's (1989) analysis. Cooke revised the positions assigned to Balbarinae and Bulungamayinae by Flannery's analysis (Figure 24) with balbarines placed as a basal monophyletic group and bulungamayines as members of a large, pectinately branching clade that also included some taxa previously regarded as potoroines (*Wakiewakie lawsoni*, *Purtia mosaicus* and *Bettongia moyesi*). Bulungamayines would thus constitute a paraphyletic group. The analysis supported the monophyly of Hypsiprymnodontidae (Ride, 1993), indicated a potoroine clade restricted to extant potoroines and rendered Macropodidae (as presently understood) as paraphyletic. Kear's (1998) analysis is in general agreement with that of Cooke (1997d) but indicates bulungamayines are polyphyletic rather than paraphyletic.

Cooke's analysis further suggests that basal macropodids had squamosal-frontal contact, deeply penetrating masseteric canals, short premolars, upper canines and a tiny I_2. Elongate premolars appear in ancestors of a large potoroine/macropodid clade. Alisphenoid-parietal contact first appears at the base of the very large macropodid clade (and convergently in *Hypsiprymnodon bartholomaii* and some extant potoroines). Within the macropodid clade there is gradual modification of the elongate premolar with loss of fine ridging, later reduction in length of premolars (but secondary increase in sthenurines), flexion of the skull, decrease in the depth of masseteric penetration and partitioning of the masseteric canal from the inferior dental canal. The whole thing reeks of phyletic gradualism.

Obviously we still don't have all the answers in kangaroo phylogeny. Many areas remain unresolved, notably the ultimate origin of kangaroos which lies somewhere well before the late Oligocene. On the basis of their molecular data, Burk *et al.* (1998) estimate divergence of the *Hypsiprymnodon* lineage from other macropodids about 45 m.y.a., the divergence of macropodids and potoroids 30 m.y.a. and the divergence of the *Dorcopsis-Dorcopsulus* lineage 10 m.y.a. The 45 million year date seems closer to that suggested by the fossil record than that of Air *et al.* (1971) who proposed a differentiation of macropodoid families as much as 57 m.y.a. The only known Australian fossil mammalian assemblage of about that age is the early Eocene Tingamarra Local Fauna from Murgon, Queensland which so far has yielded a variety of mammals of uncertain affinity, but nothing even vaguely kangarooish.

So far, we do know that kangaroos had diversified into recognizable major groupings by late Oligocene and that they continued to thrive and diversify through the early Miocene. We know that the mid Miocene marks a change in the composition of the kangaroo faunal composition with the apparent rapid decline of the balbarines and the rise of macropodids, no doubt initiated by climatic change towards greater aridity beginning about that time.

To fully understand kangaroo evolutionary history we need to find fossil sites from the early Oligocene or late Eocene in order to clarify the early origins of kangaroos. Further information from Riversleigh and central Australian fossil deposits is needed to better understand the initial explosive radiation of kangaroos and to resolve questions relating to the status of balbarines, bulungamayines and hysiprymnodontids. More information from the mid to late Miocene is also required to fully understand the origins and radiation of macropodines and sthenurines. Finally, we need more molecular or biochemical studies of kangaroo phylogeny and to synthesize these with morphological studies of extant as well as fossil species.

REFERENCES

Abbie, A.A. 1939. A masticatory adaptation peculiar to some diprotodont marsupials. *Proceedings of the Zoological Society of London,* Series B, Vol. 109, pp.261–279.

Air, G.M., Thompson, E.O.P., Richardson, B.J. and Sharman, G.B. 1971. Amino-acid sequences of kangaroo myoglobulin and haemoglobin and the date of marsupial-eutherian divergence. *Nature,* Vol. 229, pp.391–394.

Aplin, K.P. 1987. Basicranial anatomy of the early Miocene diprotodontian *Wynyardia bassiana* (Marsupialia: Wynyardiidae) and its implications for wynyardiid phylogeny and classification. Pp.369–391. In Archer, M. (ed). *Possums and Opossums: Studies in Evolution.* Vol. 1. Surrey Beatty and Sons, Sydney. Pp.1–400.

Aplin, K.P. and Archer, M. 1987. Recent advances in marsupial systematics with a new syncretic classification. Pp xv–lxxii. In Archer, M. (ed). *Possums and Opossums: Studies in Evolution.* Vol. 1. Surrey Beatty and Sons, Sydney. Pp.1–400.

Archer, M. 1979. *Wabularoo naughtoni* gen. et sp. nov., an enigmatic kangaroo (Marsupialia) from the middle Tertiary Carl Creek Limestone of northwestern Queensland. Results of the Ray E. Lemley Expeditions, part 4. *Memoirs of the Queensland Museum,* Vol. 19, pp.299–307.

Archer, M. 1984. The Australian marsupial radiation. Pp.633–808. In Archer, M. and Clayton, G. (eds). *Vertebrate Zoogeography and Evolution in Australasia.* Hesperian Press, Perth. 1203pp.

Archer, M. and Bartholomai, A. 1978. Tertiary mammals of Australia: a synoptic review. *Alcheringa,* Vol. 2, pp.1–19.

Archer, M. and Flannery, T.F. 1985. Revision of the extinct gigantic rat kangaroos (Potoroidae: Marsupialia) with a description of a new Miocene genus and species and a new Pleistocene species of *Propleopus. Journal of Paleontology,* Vol. 59, pp.1331–1349.

Bartholomai, A. 1967. *Troposodon,* a new genus of fossil Macropodinae (Marsupialia). *Memoirs of the Queensland Museum,* Vol. 15, pp.21–33.

Bartholomai, A. 1973a. The genus *Protemnodon* Owen. (Marsupialia: Macropodidae) in the upper Cainozoic deposits of Queensland. *Memoirs of the Queensland Museum,* Vol. 16, pp.309–363.

Bartholomai, A. 1973b. *Fissuridon pearsoni,* a new fossil macropodid (Marsupialia) from Queensland. *Memoirs of the Queensland Museum,* Vol. 16, pp.365–368.

Bartholomai, A. 1975. The genus *Macropus* Shaw. (Marsupialia: Macropodidae) in the upper Cainozoic deposits in Queensland. *Memoirs of the Queensland Museum,* Vol. 17, pp.195–235.

Bartholomai, A. 1976. The genus *Wallabia* Trouessart (Marsupialia: Macropodidae) in the upper Cainozoic deposits of Queensland. *Memoirs of the Queensland Museum,* Vol. 17, pp.373–377.

Bartholomai, A. 1978. The fossil kangaroos. *Australian Mammalogy,* Vol. 2, pp.15–22.

Baverstock, P.R., Krieg, M. and Birrell, J. 1990. Evolutionary relationships of Australian marsupials as assessed by albumin immunology. *Australian Journal of Zoology,* Vol. 37, pp.273–287.

Baverstock, P.R., Richardson, B.J., Birrell, J. and Krieg, M. 1989. Albumin immunologic relationships of the Macropodidae (Marsupialia). *Systematic Zoology*, Vol. 38, No. 1, pp.38–50.

Bensley, B.A. 1903. On the evolution of the Australian Marsupialia: with remarks on the relationships of the marsupials in general. *Transactions of the Linnean Society of London, Zoology,* Series 2, Vol. 9, pp.83–217.

Burk, A.M., Westerman, M. and Springer, M. 1998. The phylogenetic position of the musky rat-kangaroo and the evolution of bipedal hopping in kangaroos (Macropodidae: Diprotodontia). *Systematic Biology,* Vol. 47, No. 3, pp.457–474.

Campbell, C.R. 1973. A new species of *Troposodon* Bartholomai; from the early Pleistocene Kanunka Fauna, South Australia (Macropodinae; Marsupialia). *Records of the South Australian Museum,* Vol. 16, No. 13, pp.1–18.

Case, J.A. 1984. A new genus of Potoroinae (Marsupialia: Macropodidae) from the Miocene Ngapakaldi Local Fauna, South Australia, and a definition of the Potoroinae. *Journal of Paleontology,* Vol. 58, No. 4, pp.1074–1086.

Cooke, B.N. 1992. Primitive macropodids from Riversleigh, northwestern Queensland. *Alcheringa,* Vol. 16, pp.201–217.

Cooke, B.N. 1997a. Two new balbarine kangaroos and molar evolution within the subfamily. *Memoirs of the Queensland Museum,* Vol. 41, Pt. 2, pp.269–280.

Cooke, B.N. 1997b. New Miocene bulungamayine kangaroos (Marsupialia: Potoroidae) from Riversleigh, northwestern Queensland. *Memoirs of the Queensland Museum,* Vol. 41, Pt. 2, pp.281–294.

Cooke, B.N. 1997c. Biostratigraphic implications of fossil kangaroos at Riversleigh, northwestern Queensland. *Memoirs of the Queensland Museum,* Vol. 41, Pt. 2, pp.295–302.

Cooke, B.N. 1997d. *Researches into fossil kangaroos and kangaroo evolution.* Ph.D. Thesis (unpublished), University of New South Wales. 455pp.

Cooke, B.N. 1999. *Wanburoo hilarus* gen. et sp. Nov., a

Hume, I. (eds). *Kangaroos, Wallabies and Rat-Kangaroos*. Surrey Beatty and Sons, Sydney. 835pp.

Sharman, G.B. 1961. The mitotic chromosomes of marsupials and their bearing on taxonomy and phylogeny. *Australian Journal of Zoology,* Vol. 9, pp.38–60.

Sharman, G.B. 1989. Opening address—a chromosome phylogeny of kangaroos. Pp.v–vii. In Grigg, G.G., Jarman, P. and Hume, I. (eds). *Kangaroos, Wallabies and Rat-Kangaroos*. Surrey Beatty and Sons, Sydney. 835pp.

Sharman, G.B., Close, R.L. and Maynes, G.M. 1990. Chromosome evolution, phylogeny and speciation of rock wallabies (*Petrogale*: Macropodidae). *Australian Journal of Zoology,* Vol. 37, pp.351–363.

Smith, M.J. 1983. Desert rat-kangaroo. P.192. In Strahan, R. (ed). *The Australian Museum Complete Book of Australian Mammals*. Angus and Robertson, Sydney. 530pp.

Springer, M.S. and Kirsch, J.A. 1991. DNA hybridization, the compression effect, and the radiation of diprotodontian marsupials. *Systematic Zoology,* Vol. 40, No. 2, pp.131–151

Springer, M.S., Kirsch, J.A.W., Aplin, K. and Flannery, T. 1990. Rates of single-copy DNA evolution in phalangeriform marsupials. *Journal of Molecular Evolution,* Vol. 6, pp.331–341.

Springer, M.S. and Woodburne, M.O. 1989. The distribution of some basicranial characters within the Marsupialia and a phylogeny of the Phalangeriformes. *Journal of Vertebrate Paleontology,* Vol. 9, pp.210–221.

Stirton, R.A. 1955. Late Tertiary mammals from South Australia. *Records of the South Australian Museum,* Vol. 11, pp.247–268.

Stirton, R.A. 1963. A review of the macropodid genus *Protemnodon. University of California Publications in Geological Sciences,* Vol. 44, pp.97–162.

Tate, G.H.H. 1948. Results of the Archbold Expeditions. No. 59. Studies on the anatomy and phylogeny of the Macropodidae (Marsupialia). *Bulletin of the American Museum of Natural History,* Vol. 91, pp.233–352.

Tedford, R.H. 1966. A review of the macropodid genus *Sthenurus. University of California Publications in Geological Science*, Vol. 57, pp.1–72.

Tedford, R.H. 1994. Succession of Pliocene through medial Pleistocene mammal faunas of southeastern Australia. *Records of the South Australian Museum,* Vol. 27, pp.79–93.

Vis, C.W. de 1895. A review of the fossil jaws of the Macropodidae in the Queensland Museum. *Proceedings of the Linnean Society of New South Wales*, Vol. 10, pp.75–133.

Westerman, M., Janczewski, D.N. and O'Brien, S.J. 1990. DNA-DNA hybridization studies and marsupial phylogeny. *Australian Journal of Zoology,* Vol. 37, pp.315–323.

Winge, H. 1893. The interrelationships of the mammalian genera. In Winge, H. 1941. *Monotremata, Marsupialia, Insectivora, Chiroptera, Edentata*. Vol.1. Copenhagen. (English translation by Deichmann, E. and Allen, G.M.).

Woodburne, M.O. 1967. The Alcoota Fauna, central Australia: an integrated palaeontological and geological study. *Bureau of Mineral Resources, Geology and Geophysics Bulletin,* No. 87, pp.1–187.

Woodburne, M.O. 1984. *Wakiewakie lawsoni*, a new genus and species of Potoroinae (Marsupialia: Macropodidae) of medial Miocene age, South Australia. *Journal of Paleontology,* Vol. 58, No. 4, pp.1062–1073.

Woodburne, M.O., Macfadden, B.J., Case, J.A., Springer, M.S., Pledge, N., Power, J.D., Woodburne, J.M. and Springer, K.B. 1993. Land mammal biostratigraphy and magnetostratigraphy of the Etadunna Formation (late Oligocene) of South Australia. *Journal of Vertebrate Paleontology,* Vol. 13, No. 4, pp.483–515.

Woods, J.T. 1958. The extinct marsupial genus *Palorchestes* Owen. *Memoirs of the Queensland Museum,* Vol. 13, pp.177–193.

Wroe, S. 1996. An investigation of phylogeny in the giant extinct rat kangaroo *Ekaltadeta* (Propleopinae, Potoroidae, Marsupialia). *Journal of Paleontology,* Vol. 70, No. 4, pp.681–690.

Wroe, S. and Archer, M. 1995. Extraordinary diphyodonty-related change in dental function for a tooth of the extinct marsupial *Ekaltadeta ima* (Propleopinae, Hypsiprymnodontidae). *Archives of Oral Biology,* Vol. 40, Issue 7, pp.597–603.

Wroe, S., Brammall, J. and Cooke, B.N. 1998. The skull of *Ekalatadeta ima* (Marsupialia, Hypsiprymnodontidae?): and analysis of some marsupial cranial features and a re-investigation of propleopine phylogeny, with notes on the inference of carnivory in mammals. *Journal of Paleontology,* Vol. 72, No. 4, pp.738–751.

Chapter 30
Bat Beginnings and Biogeography: The Australasian Record

Suzanne J. Hand
School of Biological, Earth and Environmental Sciences, University of New South Wales, Sydney N.S.W. 2052 Australia
E-mail: s.hand@unsw.edu.au

INTRODUCTION

Nearly 25% of all known species of mammals are bats (Figure 1). The 966 bat species recognized (Altringham, 1996) form the second largest mammalian order, outranked in number only by the rodents. Bats are most common in tropical areas where insects, fruit and flowers are most abundant, but they occur on all continents except Antarctica, with three widespread families penetrating the cool temperate zones.

DIVERSITY

Bats comprise the order Chiroptera, derived from the Greek words *chiro* meaning 'hand' and *ptera* 'wing'. They are the only mammals that have ever been able to fly. The order is usually divided into two suborders: the Megachiroptera or true fruit-bats (including flying-foxes) and the Microchiroptera, usually smaller 'insectivorous' bats.

Megachiropterans all belong to the family

Figure 1. Microchiropterans belong to 17 families of bats and, compared with megachiropterans, form a more varied group that has exploited a greater range of food and ecological niches. Generally microchiropterans are smaller than megachiropterans but the range of microchiropteran morphology is staggering. Most variation among bats of this suborder occurs in the structure of the nose-leaf and ear and, at the other end, the tail and interfemoral membrane. The microchiropteran shown here is the horseshoe bat, *Rhinolophus euryotis*, of Indonesia and Papua; from Dobson (1875).

Pteropodidae, have an Old World tropical distribution (Figure 2(a)), commonly retain a clawed phalanx on the index finger, usually rely on vision and smell rather than echolocation for navigation and food-finding, and possess teeth with a simple crown pattern for flower- and fruit-eating (Figure 3). About 175 living species are currently recognized.

Microchiropterans belong to 17 modern families, have an almost worldwide distribution (Figure 2(b)), lack an index claw, are echolocators and have diversified tooth forms that correlate with feeding habits (Figure 4). Although insectivorous, carnivorous, sanguivorous, piscivorous, nectarivorous and frugivorous microchiropterans are known, as a group they are broadly described as insectivorous. They may generally be distinguished from megachiropterans by traces at least of the insectivoran-like W-shaped labial crests on their upper molars. However, in some nectar-, blood- and fruit-eating bats of the New World (microchiropteran family Phyllostomatidae) there is a complete degeneration of the more typical microchiropteran molar cusp pattern (Koopman, 1981—see Figure 4 (d), (e)).

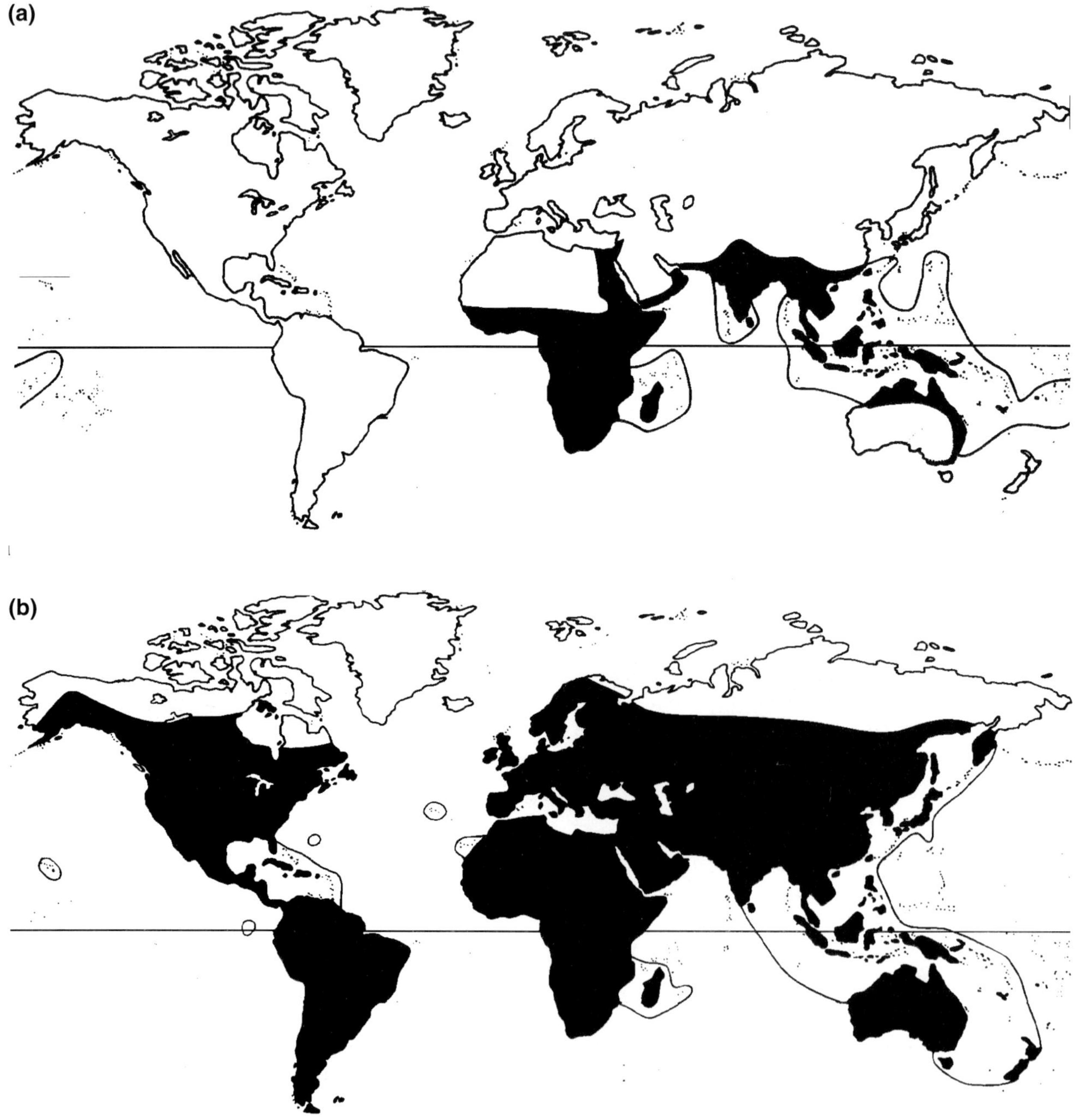

Figure 2. Overall bat distributions: **(a)** megachiropterans (family Pteropodidae) have an Old World tropical distribution; **(b)** microchiropterans (17 different families) have an almost worldwide distribution; from Yalden and Morris (1975).

DIET AND BIOLOGY

About 68% of all bats eat insects and other arthropods, with some bats eating as many as 3,000 insects (half their own body weight) in a single night (Hill and Smith, 1984). Bats help regulate insect populations in rural areas, reducing the need for widespread use of environmentally damaging insecticides. Some 30% of bats are fruit-, flower- and nectar-eaters and these pollinate or disperse the seeds of many hundreds of plant species. Fruit-eating bats carry seeds in their mouth or gut often for long distances before dropping them, hence spreading seeds over a wide radius and helping forest regrowth in cleared areas. On many oceanic islands bats are responsible for maintaining forests.

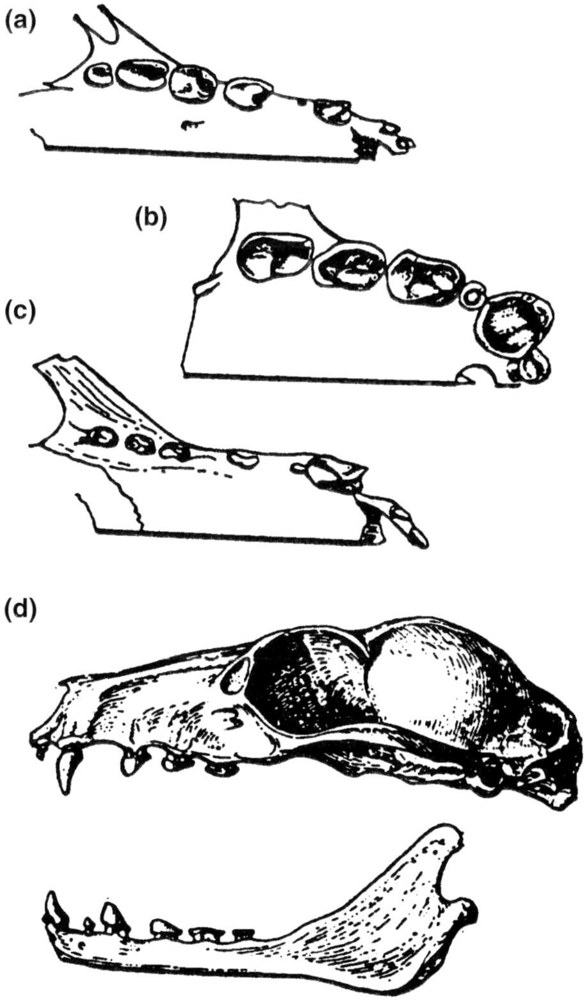

Figure 3. Megachiropterans possess teeth with a simple low-crowned pattern for flower- and fruit-eating shown in: **(a)** upper tooth row of *Pteropus* sp.; and **(b)** *Nyctimene* sp. Some long-snouted members of the suborder feed on nectar or pollen and in these bats the molars are further reduced, as in: **(c)** upper tooth row of *Melonycteris* sp.—from British Museum (Natural History) (1975); **(d)** skull of the pteropodid *Epomorphous franqueti* (from Miller, 1907).

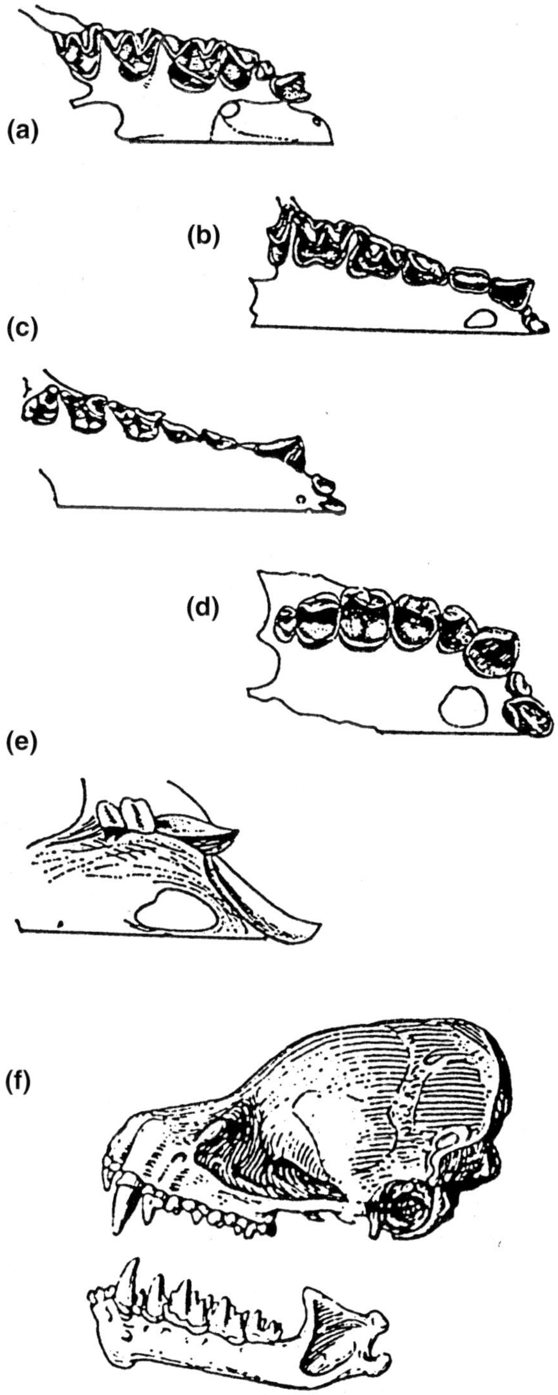

Figure 4. Microchiropterans have diversified tooth forms that correlate with feeding habits; examples are illustrated by the upper tooth row of each category. As a group, microchiropterans are broadly described as insectivorous bats, but the suborder includes bats that eat: **(a)** insects (*Rhinolophus* species and *Micronycteris* species **(b)**); **(c)** fruit (*Sturnira* species); **(d)** nectar (*Glossophaga* species); and **(e)** blood (*Desmodus*); from British Museum (Natural History, 1975); **(f)** skull of the microchiropteran *Chalinolobus tuberculatus* (from Miller, 1907).

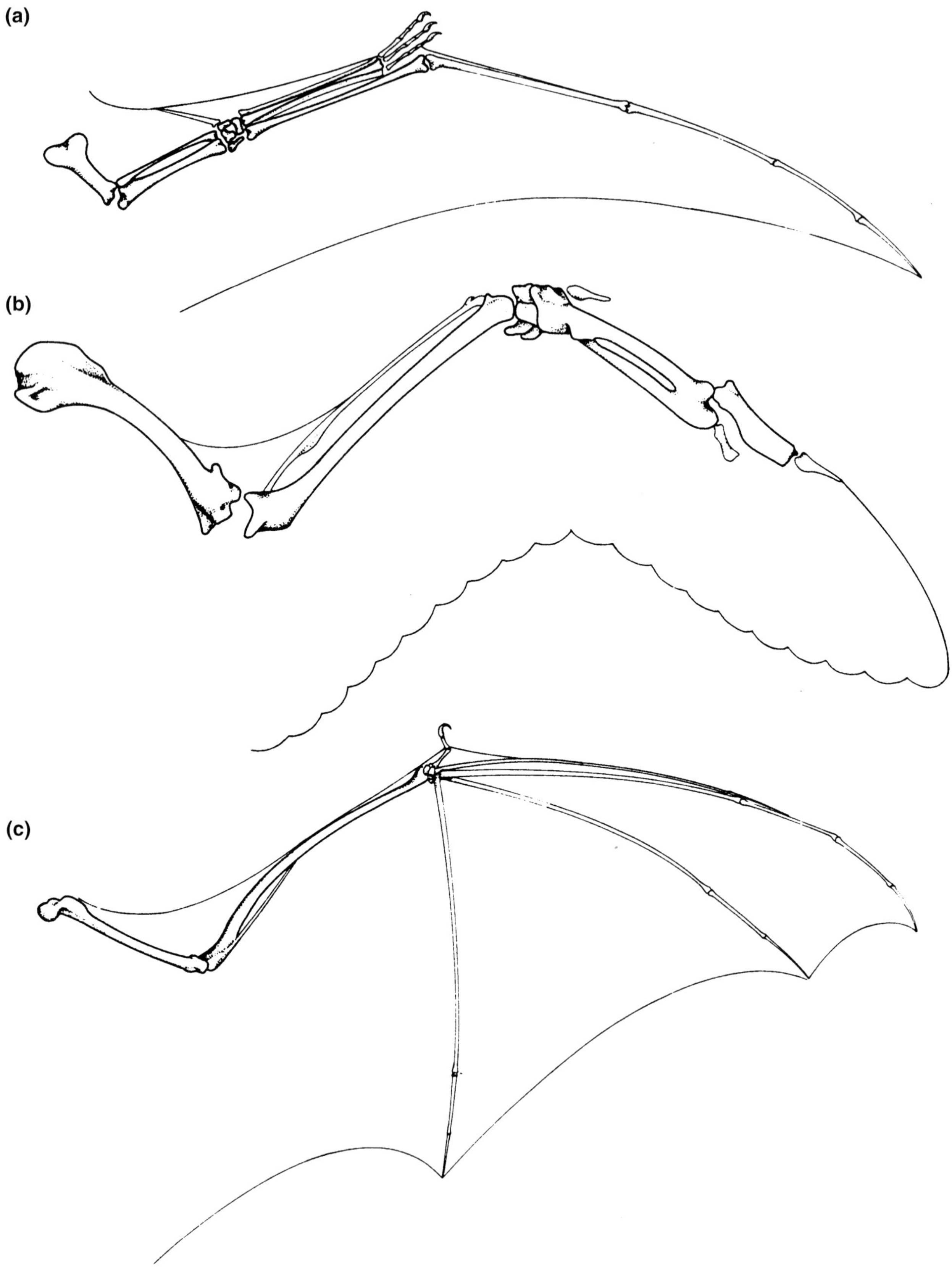

Figure 5. Comparison of wing structure showing evolutionary variations on a forelimb. In **(a)** pterosaur—the fourth digit supports the wing; in **(b)** bird—it is mostly the second, and in **(c)** bat—it is the second to fifth digits. From Hall (1984).

Among vertebrates, bats share with birds and the extinct reptilian pterosaurs the capacity for sustained flight. However, the flight apparatus in each case is quite different (Figure 5). In birds, the arm bones are elongated (but not the digits) and feathers form the primary flight apparatus. The legs and pelvis are not highly modified and birds remain capable of terrestrial locomotion. In pterosaurs and bats, the arm and finger bones are elongated and the primary flying apparatus is skin stretched over the elongated arms and fingers and attached to the sides of the body and legs. In pterosaurs it is the fourth digit that is especially elongated, whereas in bats it is digits 2 to 5. In bats the legs are an important component of wing support as well and the pelvic area and legs are modified, with the leg bones rotated 180 degrees so that the knee joints bend backwards.

Bat wings are formed by a double membrane of skin (patagium), which sandwiches elastic fibres, blood vessels and nerves. The membrane sometimes also encloses the tail (uropatagium). The shape of the wings varies in bats (Figure 6). Wing area can be large or small relative to the size of a bat and this can be expressed as a measurement of wing loading. A large bat with relatively small wings has a high wing loading and a small bat with large wing area has a low wing loading. The shape of a bat's wing can be expressed as a measurement of aspect ratio. Short, broad wings have a low aspect ratio while long, narrow wings have a high aspect ratio. Bats with long, thin wings (high aspect ratio) are generally species that forage for flying insects in relatively uncluttered areas such as over the forest canopy or water. Short, broad wings (low aspect ratio) are more common in bats that eat fruit, nectar or glean prey from vegetation or the ground (Figure 7). Bats with a high wing loading tend to fly faster than those with a low wing loading.

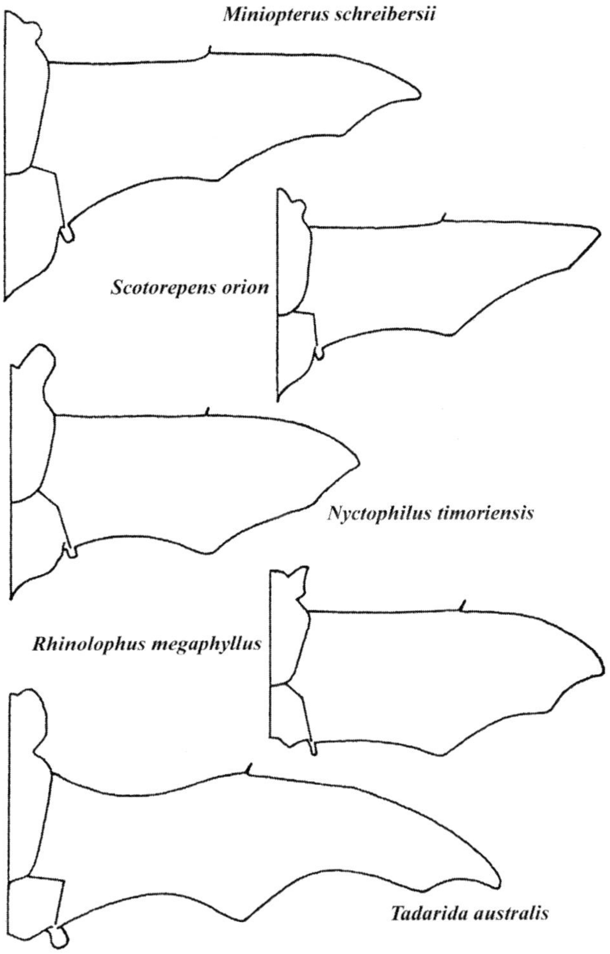

Figure 6. Outlines of the flight membranes (right hand side) for five eastern Australian bats showing the different wing shapes; from Dwyer (1965).

Bats may fly long distances nightly and/or seasonally. Some flying-foxes, such as Australia's grey-headed flying-fox *Pteropus poliocephalus*, fly over 50 km each night to reach a favourite feeding site, and seasonally may travel hundreds of kilometres to follow the flowering of certain trees (Eby, 1991). The Mexican free-tailed bat *Tadarida brasiliensis* migrates from as far as Canada to Mexico to overwinter and give birth in warmer caves in the south (Fenton, 1983). For bats, the combination of being active primarily at night and being able to fly gives them several advantages over other vertebrates. For example, bats can forage widely and rapidly and they can escape from most terrestrial predators. It also provides almost unlimited access to nearly competition-free food resources such as insects and fruit, and conserves energy and water by being active during the cooler period of the day (Altringham, 1996).

All bats have eyes that are well-adapted for seeing in dimly lit environments—none is blind. All megachiropterans and many microchiropterans use their eyes to help orient themselves, migrate long distances and find their food in the dark. Microchiropterans also use echolocation over shorter distances to help them 'see' their surrounding environment, locate prey and avoid obstacles. They do this by emitting pulses of high frequency sound (ultrasonics) and analysing the returning echos. The high frequency sound is produced in the larynx or voice box and emitted through either the mouth (in furry and smooth-faced microchiropterans) or through the nose (usually in microchiropterans with complex faces and noseleaves). Using echolocation, microchiropterans are able to detect objects less than 1 mm in diameter and can tell whether an insect is flying towards them or away, and plan their intercept accordingly. Most bat calls are above the limit of human hearing (15 kHz) but range from 9 kHz to above 200 kHz.

Bats use different kinds of echolocation depending on the habitat in which they hunt. Bats that hunt for insects in dense vegetation use ultra-high-frequency sounds to provide a very detailed and accurate sound picture of their cluttered environment. Bats that chase insects high up or in open spaces do not need such accuracy and generally emit lower-frequency calls. Bats hunting relatively large and slow-moving prey such as small vertebrates also use lower-frequency calls as well as their

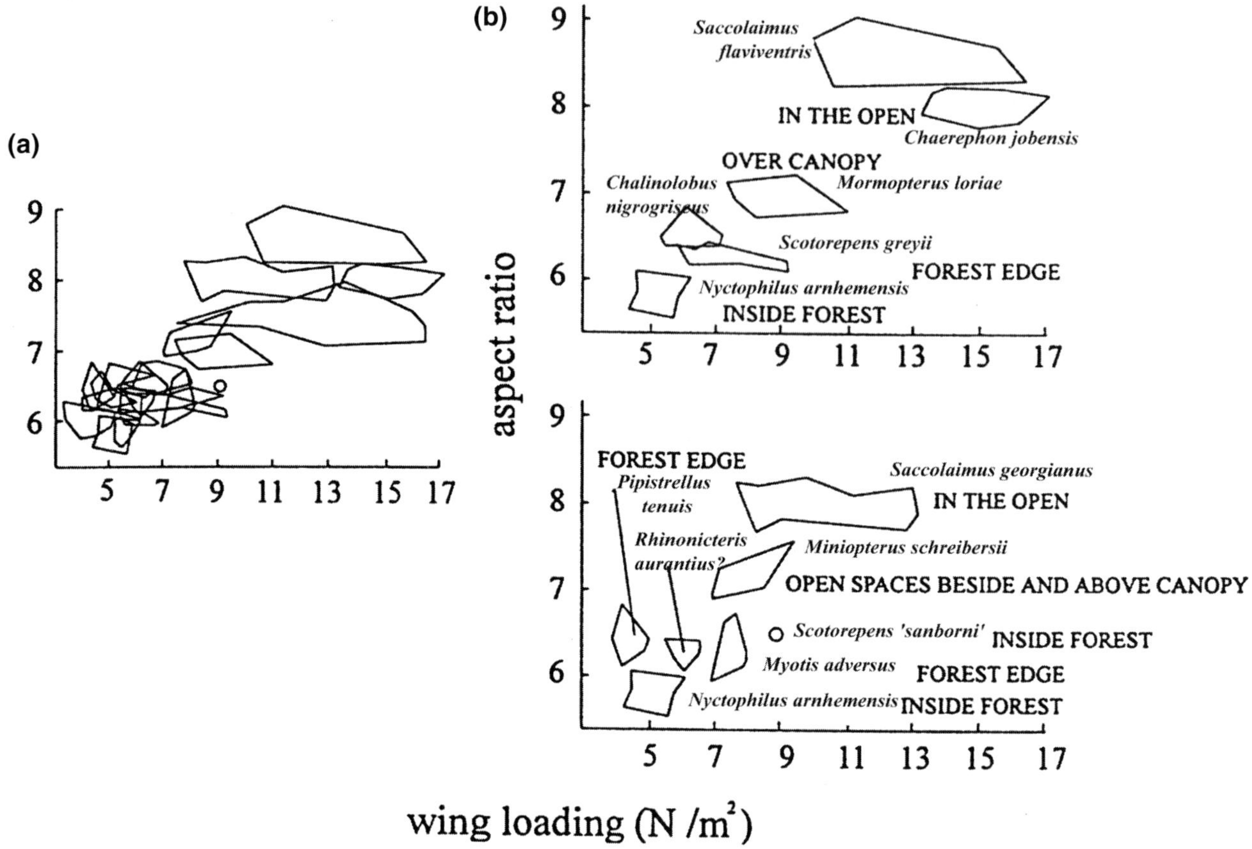

Figure 7. Aspects of flight morphology appear to affect the microhabitat in which bats forage and community structure. Here, wing loading is plotted against aspect ratio for an insectivorous bat community in an Australian mangrove forest: **(a)** shows all species in the region; **(b)** shows examples of two of the guilds studied, with the microhabitats in which each species foraged (from Altringham, 1996, after McKenzie and Rolfe, 1986).

eyes. Some echolocating bats emit their calls at a constant frequency, others use a range or sweep of frequencies (frequency modulated or FM calls), some add harmonics or overtones, and many use a combination of calls during their hunt for food (Figure 8).

Roosts are where bats spend their days. These protect them from weather and predators, and also provide energetic savings when roosting together, improved mating opportunities and maternal care, and transfer of information about foraging and roosting sites (Altringham, 1996). Some bats roost alone, others in millions. Bats choose many different kinds of day roosts (Table 1). In tropical areas, many bats use trees. Flying foxes and other megachiropterans usually make camps among the foliage (Plate 106), while some bats (both microchiropterans and small megachiropterans) make tents from leaves by biting through the midrib of large leaves to collapse them into tents. Others live under bark, in flowers, bamboo shoots and under tree roots. Many Australian bats depend on tree hollows for diurnal and maternity roosts, especially in southern Australia. New Zealand's short-tailed bat (*Mystacina tuberculata*) makes its roosts in fallen and decaying kauri and southern beech trees, using its teeth to make warm, humid burrows 100 metres long (Daniel and Williams, 1984). A cave may provide a variety of different micro-habitats (varying in temperature and humidity) that are attractive to several different bat species (Figure 9). Bats commonly also use urban and man-made roosts such as mines, tunnels, tombs, roof and other cavities in wooden buildings, cracks in stonework of buildings and bridges, expansion joints, wells, catacombs, ruins, roller doors, barns, attics, roofs and churches.

Most bats that live in cold or temperate climates hibernate during winter when temperatures fall and food becomes scarce. To conserve energy, bats enter a period of torpor during which their metabolism slows, their body temperature falls and they survive on reserves of fat they have built up during autumn. Some bats migrate to warmer areas during the winter. Most bats give birth to one or two young once a year. The gestation period varies but averages around two months. Pups are born well formed but without hair and with their eyes closed, and can weigh as much as 40% of the mother's weight. They are carried by clinging tightly to mother's breast, with special recurved milk teeth to help them hang on. Females of some species always carry their young with them, even when searching for food. Young bats stay with their mothers until they have learned to fly and are able to fend for

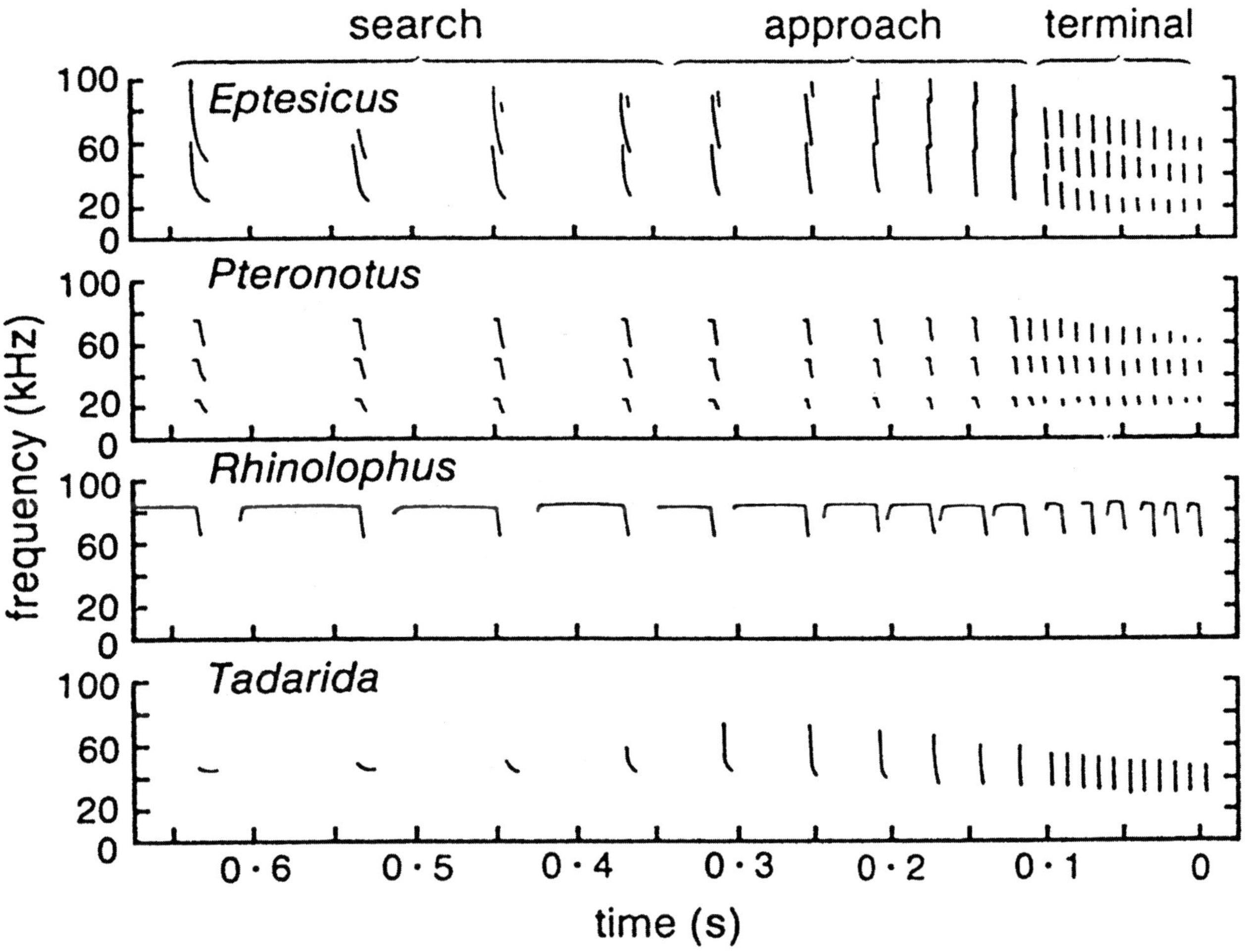

Figure 8. Diagrammatic reproductions from sonagrams of echolocation calls made by four species of bats during the hunt for insect prey. The call of each species is different, with some incorporating harmonics in addition to pure tones in their calls. As an insect is approached the bats emit more pulses of sound per second, enabling them to gather more precise information about their prey including its speed and direction of movement. From Hill and Smith (1984).

themselves. Mothers teach their young to forage, to echolocate and find alternate roost sites. In some bat species, females with young separate from males, others stay together and form huge nursery colonies.

Table 1. Types of roosts used by Australia's 90 recognized species of bats. Species numbers are minimum values. Data from Hall and Richards (1979), Hall (1984), Strahan (1995), Churchill (1998) and Duncan *et al.* (1999).

Roost type	No. of species
Shallow caves and overhangs	48
Tree hollows	47
Buildings	46
Deep caves, mines and tunnels	39
Foliage	23
Under bark	12

Food resources and roost sites (diurnal, hibernation and maternity sites) are important limiting factors that determine the distribution of living bats. The choice of roost sites by bats has also played a critical role in determining where, when and what bats have been fossilized over the last 60 million years.

EARLY BATS AND FIRST FLIGHTS

To date, more than 220 fossil bat species have been described. Many of these are from early and middle Tertiary bat-rich karst (limestone) deposits in Europe and Australia (see below), but lake deposits in Europe and North America have produced some of the oldest and best preserved fossil bats.

The world's oldest and most primitive fossil bats are from early and middle Eocene deposits in both the Northern and Southern Hemisphere. They are currently referred to nine genera: *Icaronycteris* from the United

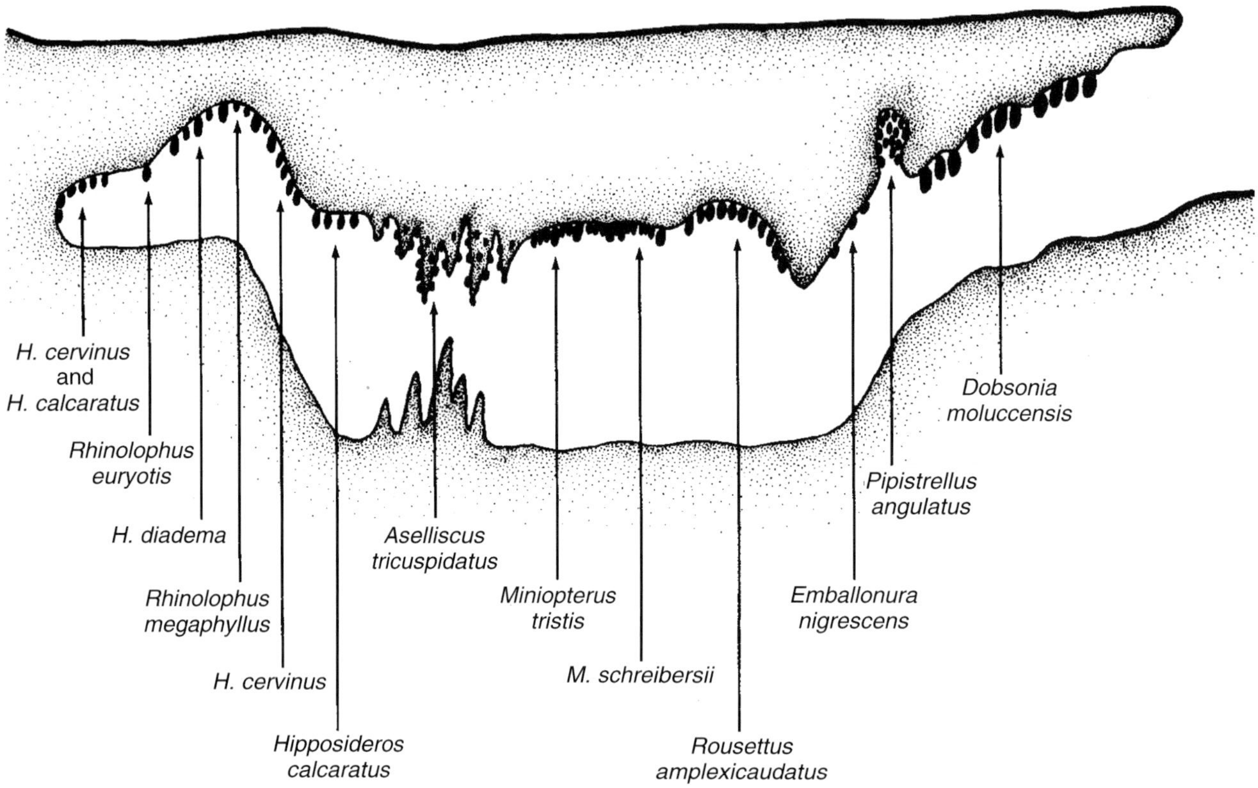

Figure 9. Diagrammatic representation of the distributions of a number of species of bats in a cave on New Ireland Island, Bismark Archipelago, New Guinea. From Hill and Smith (1984).

States and France (Jepsen, 1966; Russell *et al.*, 1973); *Palaeochiropteryx* and *Archaeonycteris* from France and Germany (Revilliod, 1917a, b, 1922; Sigé and Russell, 1980); *Ageina* from France (Russell *et al.*, 1973); *Cecilonycteris*, *Matthesia* and *Hassianycteris* from Germany (Heller, 1935; Sigé and Russell, 1980; Smith and Storch, 1981; Habersetzer and Storch, 1987; Smith and Russell, 1992); *Australonycteris* from Australia (Hand *et al.*, 1994); and *Lapichiropteryx* from central China (Tong, 1997). Other primitive early Eocene bats, too poorly represented to be named, have been found in Europe, North America and Pakistan (McKenna *et al.*, 1962; Russell and Gingerich, 1981; Russell *et al.*, 1982; Savage and Russell, 1983; Sigé and Legendre, 1983; Crochet *et al.*, 1988; Marandat, 1991).

By 50–55 m.y.a., these 'archaic' bats looked similar to modern bats, as shown by beautifully preserved, complete bat skeletons found in the 53-million-year-old shales of the Green River Formation in Wyoming, U.S.A (Jepsen, 1966, 1970; Figure 10) and the 49-million-year-old Messel deposit near Darmstadt in Germany (Habersetzer *et al.*, 1992, 1994). These archaic bats lack some of the derived features of modern bats and are referred to several families (Archaeonycterididae, Icaronycterididae, Palaeochiropterygidae and Hassianycterididae). These bats have traditionally been placed together in a separate clade or grade (e.g., Eochiroptera or Palaeochiropterygoidea in Figure 11); see a discussion of this concept by Sigé (1991a). However, Simmons and Geisler (1998) argue that the families Icaronycterididae, Archaeonycterididae, Hassianycterididae and Palaeochiropterygidae form progressively closer sister-groups to the modern microchiropteran clade (Figure 12).

Although these bats *may* lack some of the specializations found in modern microchiropterans (see Simmons and Geisler, 1998, pp.99–101), they are nonetheless 'good' bats. Adaptations for flight were already well established and evidence from their ear morphology and gut contents indicates that they used echolocation to help capture their insect prey (Habersetzer and Storch, 1989; Norberg, 1989). Habersetzer and Storch (1989) studied the functional morphology and probable ecological role of the species of archaic bats known from the Messel deposit near Darmstadt, noting that the middle Eocene bat fauna was ecologically quite diverse (Figure 13). Habersetzer *et al.* (1994) suggested that, of the species represented in the Messel deposit, those of *Archaeonycteris* were probably the most plesiomorphic, species of *Palaeochiropteryx* had a more specialized wing structure, and species of *Hassianycteris* were more modern in appearance, both dentally and osteologically.

The ancestry of the Megachiroptera is unclear. Traditionally, the order Chiroptera has been regarded as a monophyletic group (Dobson, 1875; Miller, 1907;

Figure 10. *Icaronycteris index* (Jepsen, 1966) is an archaic bat from early Eocene sediments in Wyoming, U.S.A. Beautifully preserved in shale, the few known specimens exhibit primitive traits of both modern suborders. The skeleton is shown as it was preserved but the shadowy extension of the wings and legs has been reconstructed. From Jepsen (1966).

Winge, 1942; Simpson, 1945; Van Valen, 1979), with the suborders Microchiroptera and Megachiroptera being each other's closest sister-group (Figure 14(a)). A number of recent studies of morphological, anatomical and molecular data confirm this relationship (e.g., Wible and Novacek, 1988; Mindell *et al.*, 1991; Bailey *et al.*, 1992; Stanhope *et al.*, 1992; Simmons and Geisler, 1998; Teeling *et al.*, 2000). However, some researchers (notably Smith and Madkour, 1980; Hill and Smith, 1984; Pettigrew, 1986, 1991a, b, 1994, 1995; Pettigrew and Jamieson, 1987; Pettigrew *et al.*, 1989; Rayner, 1991; Pettigrew and Kirsch, 1995) have argued that it is possible that the two suborders arose independently, from separate ancestors, and that the possession of wings has developed convergently, merely reflecting the general mammalian anatomical background from which bats developed (Figure 14(b)). Other researchers (e.g., Sigé, 1991a) suggest the possibility that the

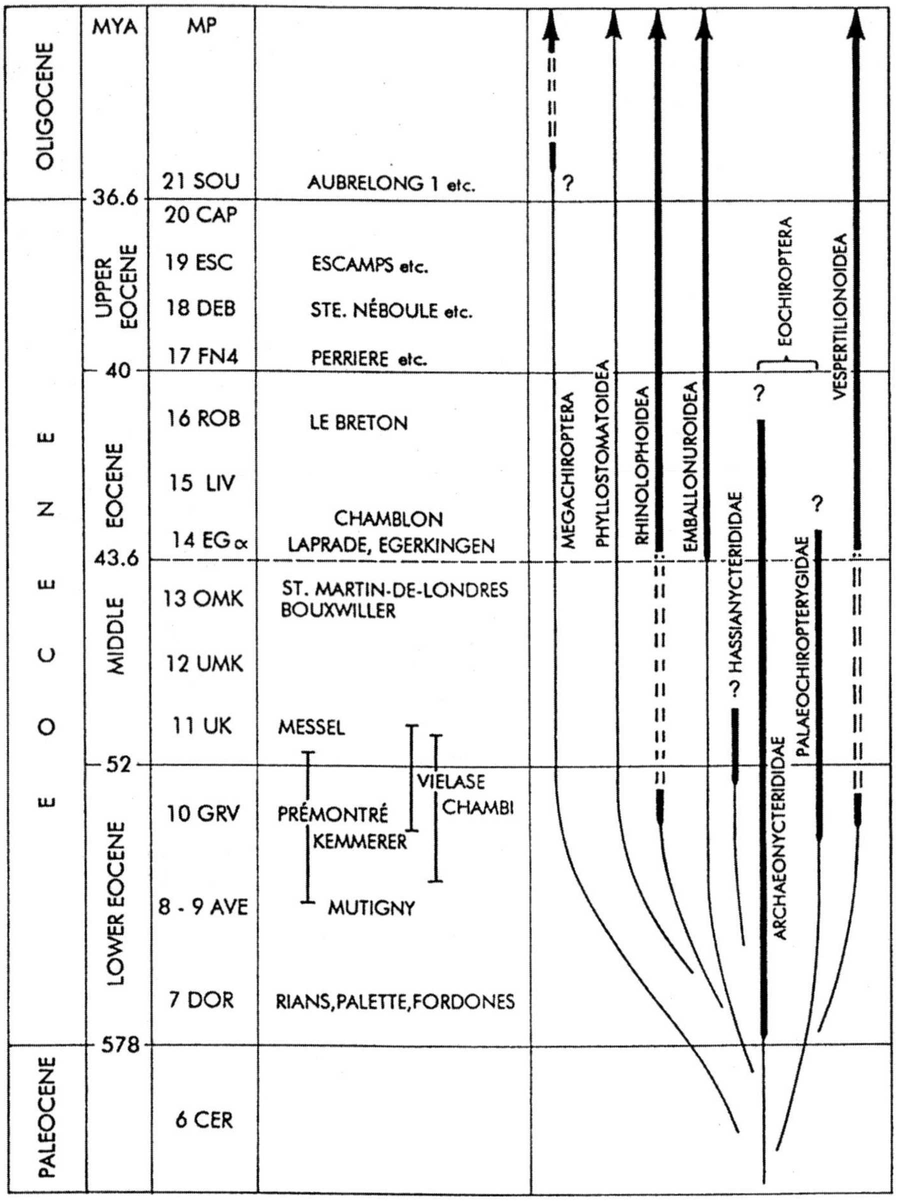

Figure 11. Summary of biogeographic data for Eocene bats and temporal distribution of major bat groups, including Eochiroptera. Modified from Sigé (1991b).

Megachiroptera may have evolved from a lineage within the Microchiroptera, for example the Rhinolophoidea (Kirsch, 1996; Kirsch and Hutcheon, 1997; Figure 15).

This hypothesis is supported by a number of subsequent studies including Teeling *et al.* (2000, 2002), Springer *et al.* (2001) and Hulva and Horacek (2002). The majority of these have been of molecular data (involving nuclear and mitochondrial genes), although Springer *et al.* (2001) include some morphological data.

The oldest megachiropteran fossil, a single premolar from the late Eocene of Thailand (Ducrocq *et al.*, 1993), appears to represent a *Rousettus*-like taxon and suggests a much older date for the divergence of megachiropterans from other bats. The more complete, but enigmatic *Archaeopteropus transiens* from an early Oligocene deposit in Italy was described by Meschinelli in 1903 from a well-preserved skeleton and poorly preserved skull. The single specimen (destroyed in World War 2) was variously regarded as an early megachiropteran (Meschinelli, 1903; Andersen, 1912; Hill and Smith, 1984; Smith and Storch, 1981; Habersetzer and Storch, 1987; McKenna and Bell, 1997) or as belonging to a more primitive grade of bats (Russell and Sigé, 1970; Smith, 1976; Sigé, 1977; Van Valen, 1979; Simmons and Geisler, 1998, pp.138–139).

Also hotly debated is the identity of the sister-group of bats. In the past, tupaiids (tree shrews), primates and dermopterans (colugos or flying lemurs) have variously been nominated as the sister-group of bats, these three groups plus bats forming the grandorder Archonta. Morphological data and some molecular data (see Simmons and Geisler, 1998) strongly support the close

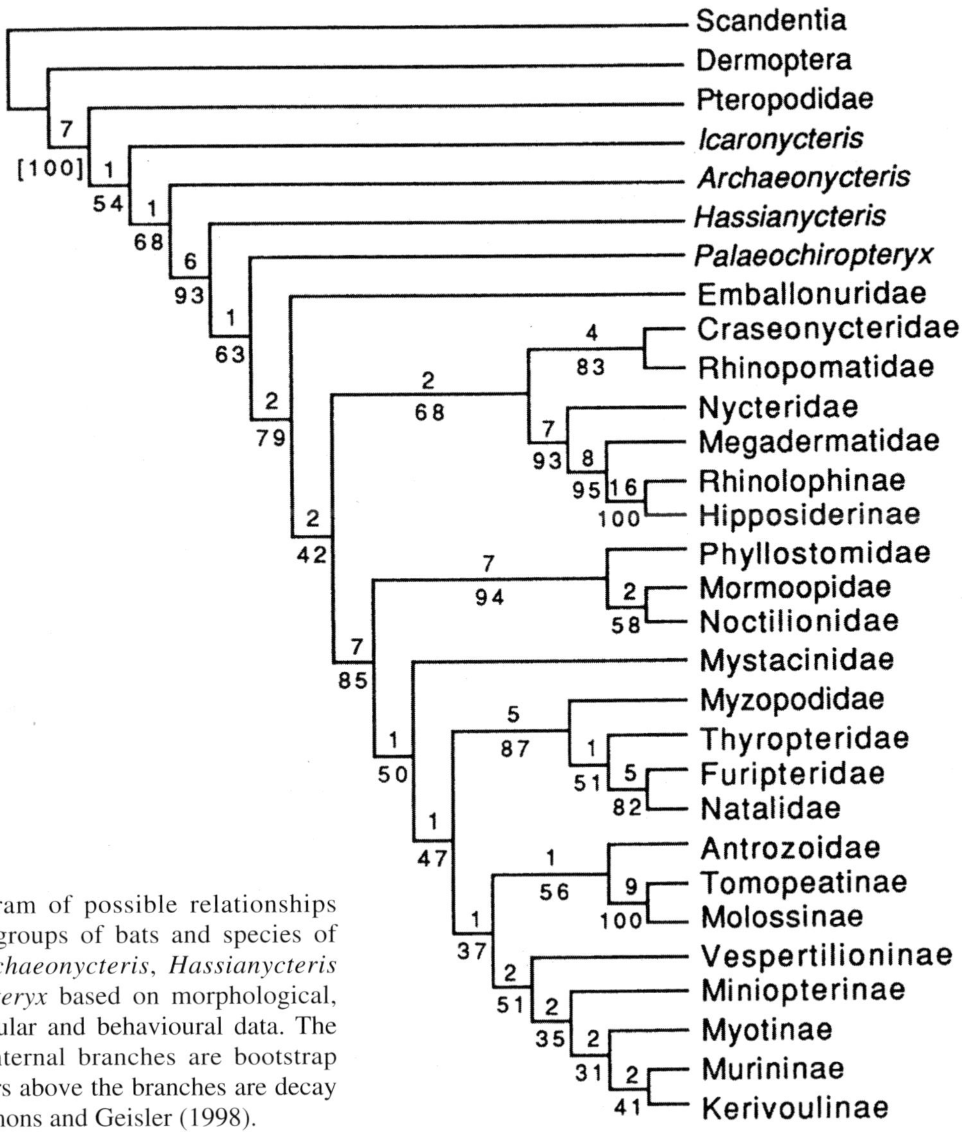

Figure 12. A cladogram of possible relationships between modern groups of bats and species of *Icaronycteris*, *Archaeonycteris*, *Hassianycteris* and *Palaeochiropteryx* based on morphological, anatomical, molecular and behavioural data. The numbers below internal branches are bootstrap values; the numbers above the branches are decay values. From Simmons and Geisler (1998).

relationship of dermopterans to bats (Wible and Novacek, 1988; Novacek, 1992, 1994; Simmons, 1993, 1994, 1995; Miyamoto, 1996) and some data sets support the concept of Archonta. However, other molecular data sets (e.g. Miyamoto *et al.*, 2000; Murphy *et al.*, 2001; Douady *et al.*, 2002; Arnason *et al.*, 2002; Van Den Bussche *et al.*, 2002) suggest alternative sister-groups for bats, including Carnivora, Insectivora, Ferungulata, Perrisodactyla, Artiodactyla + Cetacea and combinations thereof—see Figure 15.

Simmons and Geisler (1998) suggest that flight evolved in the common ancestor of Microchiroptera and Megachiroptera and that the evolution of echolocation in microchiropterans evolved in a series of evolutionary 'steps'. The earliest bats are hypothesized to have used vision for orientation and foraged by gleaning insects and perhaps fruit from foliage, bark and the ground. Later bats, including species of *Icaronycteris* and *Archaeonycteris*, developed the ability to echolocate as a tool for orientation and obstacle avoidance, and later for catching insects on the wing. The immediate ancestors of microchiropterans (including species of *Hassianycteris* and *Palaeochiropteyx*) were probably capable of aerial hawking in open areas using echolocation to detect, track and evaluate prey. This is envisaged to have been the primitive foraging strategy for bats (with gleaning, passive prey detection and perch hunting in cluttered environments secondarily derived specializations rather than retentions of primitive behaviours).

Whereas microchiropterans evolved increasingly more efficient means of capturing insect food through fine-tuning in echolocation and manoeuvrability, megachiropterans may have rapidly become strong, direct fliers with a loss (or lack) of manoeuvrability, thus further specializing the group towards being restricted to taking food in trees rather than in flight. If megachiropterans evolved from a lineage within the Microchiroptera then it suggests that the ability to echolocate may have been common to all

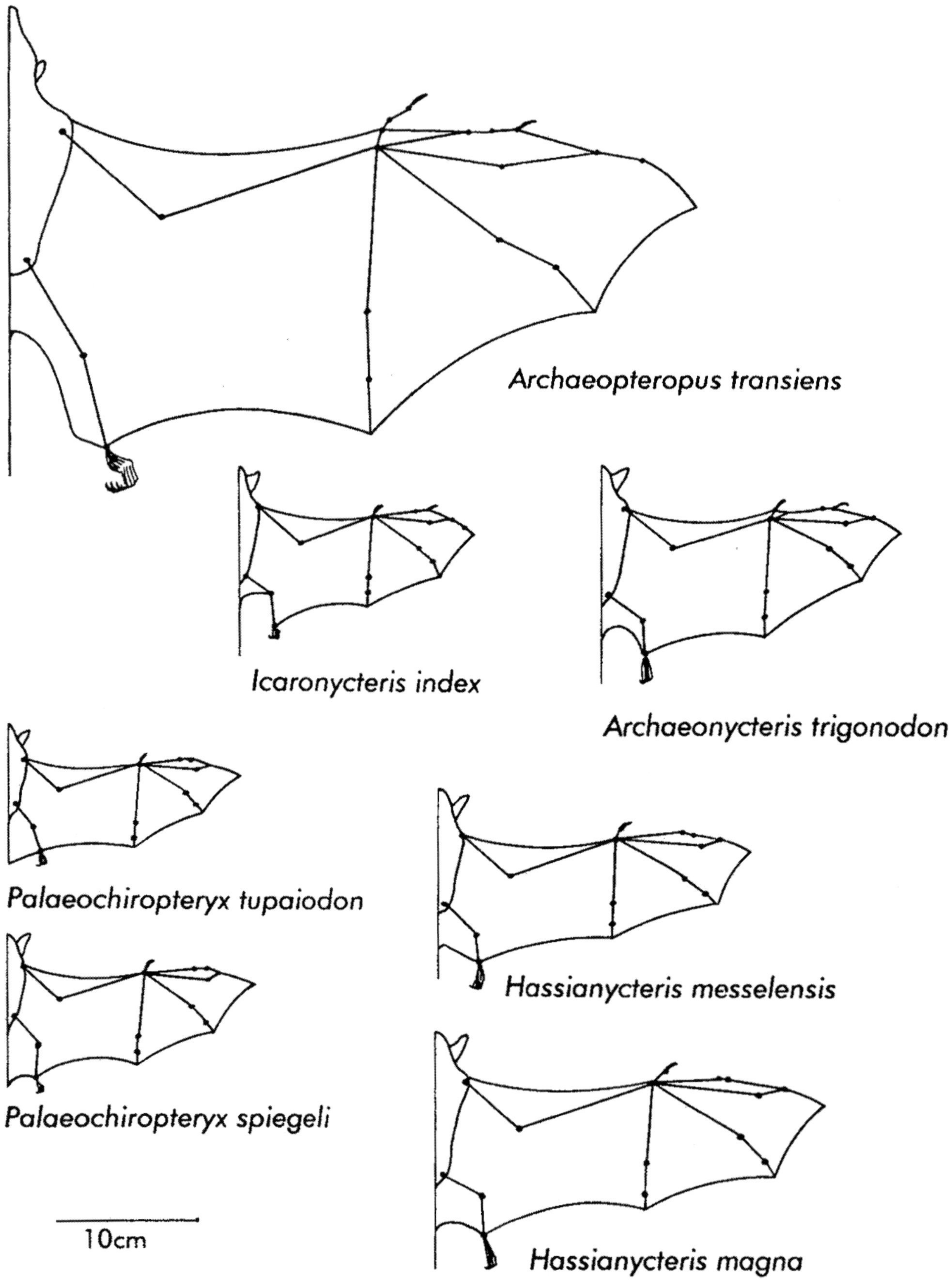

Figure 13. Reconstructions of the diverse wing shapes, presumed to reflect different foraging modes, of Eocene bats. Based on Norberg (1989) and Simmons and Geisler (1998).

Plate 109. Bat bones (skull in centre) bristle from a block of Riversleigh limestone. At least 22 species of hipposiderids have been recorded from the Tertiary cave and lake deposits at Riversleigh. Photograph R. Arnett.

Plate 111. The eastern horseshoe bat *Rhinolophus megaphyllus* inhabits forests along the eastern coast of Australia. Fossils referred to this species have been found in Pleistocene cave deposits but the family is otherwise unrepresented in the Australian fossil record. Photograph L. Hall.

Plate 110. Emballonurids or sheath-tailed bats are plain-faced, insect-eating bats that roost in rock crevices, caves, ruins, houses, trees, curled leaves and hollow logs. Some species are solitary, others, like *Taphozous georgianus* of northern Australia, roost in groups. Photograph L. Hall.

Plate 112. Wild Australian dingo from the Simpson Desert. Photograph L. Corbett.

Plate 113. Wild dingo pup (black and tan variant) from Sturt's Stony Desert. Photograph L. Corbett.

Plate 114. Free-living dingo from northern Thailand. Photograph L. Corbett.

Plate 115. Hybrid dingo (with blue heeler) bred in captivity. Photograph L. Corbett.

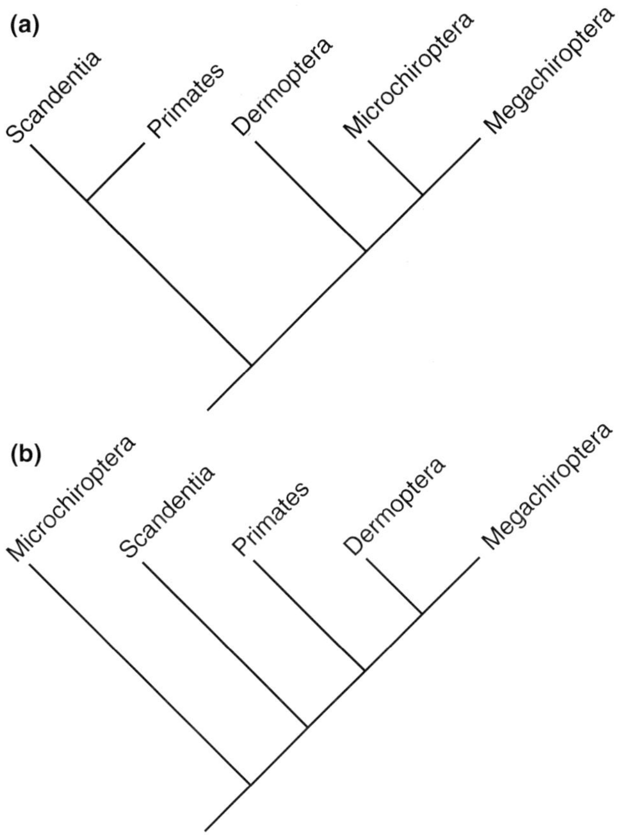

Figure 14. Two competing hypotheses concerning the phylogenetic relationships of microchiropteran bats, megachiropteran bats, primates, dermopterans (flying lemurs or colugos) and scandentians (tree shrews): **(a)** classical hypothesis; **(b)** alternative hypothesis. From Baker *et al.* (1991).

bats, but secondarily lost in megachiropterans. Most megachiropterans, with the exception of cave-dwelling pteropodids of the genus *Rousettus*, orient visually rather than acoustically. This may be related to the fact that obstacles can be detected at long distances by vision but only short distances by echolocation (Griffin, 1958). Because the mode of transmission of facultative ultrasonic calls made by bats of the genus *Rousettus* is so grossly different from that of microchiropterans (clicking of the tongue versus laryngeal mechanisms respectively), it is thought that the ability to echolocate using ultrasonics evolved separately in the *Rousettus* lineage .

The distinctive chiropteran pattern of flight musculature is basic to both suborders of bats (see Pettigrew *et al.*, 1989 on the differences between the wings of mega- and microchiropterans) and features a division of labour between large groups of muscles arising from the scapulae, vertebrae and sternum. This differs strikingly from the flight musculature of birds in which the bulk of the power of the wing-beat cycle is provided by a very few unusually large muscles originating on the greatly enlarged sternum (Vaughan, 1970a). The chiropteran flight musculature pattern enables rapid, precise control over a variety of wing movements, the basis of manoeuvrability (Vaughan, 1970a, b).

No 'missing links' between bats and their non-flying ancestor have yet been found in the fossil record. However, it is assumed that bats evolved from a small, tree-living mammal and passed through a gliding stage (perhaps like colugos) before being able to fly. As the wings and flight muscles of pre-bats increased in size and strength (Figure 16), so did the animal's overall weight and the amount of energy needed to stay aloft. To offset this increase, the pre-bat's hindquarters evidently became progressively smaller and lighter until its legs lacked the strength to support its weight. Ultimately, the ancestors of bats began to hang upside down, as all bats do today. Although their feet are relatively small, bats use their strong, sharp claws for gripping bark, stems, branches and cracks in rocks. A tendon-locking device in their legs and feet ensure that the suspended weight of a hanging bat causes the toes and claws to grip firmly even when asleep (or dead).

The oldest known bats have dilambdodont dentitions with W-shaped ectolophs on the upper molars, teeth well-suited for puncturing and slicing through the cuticle of insects. Gut contents of species of *Icaronycteris* and *Palaeochiropteryx* indicate that these early bats were eating insects; the diversity of feeding habits in the Microchiroptera, including fruit-, nectar-, fish- and blood-eating, seems to have evolved from an ancestrally insectivorous diet (Gillette, 1975).

The first representatives of modern families of microchiropteran bats also appeared in the early Eocene, overlapping widely—both geographically and temporally—with more primitive bats. A rhinolophoid and the vespertilionoid *Dizzya exsultans* are known from the late Early Eocene of Tunisia (Sigé, 1991a) and the natalid *Honrovits tsuwape* from the late Wasatchian of central Wyoming, U.S.A. (Beard *et al.*, 1992). Hooker (1996) has described a possible emballonuroid (*Eppsinycteris*) from the earliest Eocene of England. Simmons (1998) suggests that, based on phylogenetic analyses of extant and extinct bats, almost all modern bat families had probably evolved by the early Eocene.

When the main placental radiations (including that of bats) were occurring in the late Cretaceous–early Tertiary, areas of Eurasia and North America were intermittently connected as part of the Mesozoic northern supercontinent Laurasia, and Australia, Antarctica and South America comprised the last fragment of the southern supercontinent Gondwana. Africa, India and New Zealand had already split from Gondwana (Figure 17). Oceanic currents were quite different from today and climates were considerably more equable, with tropical or subtropical rainforests extending to high latitudes in both Northern and Southern Hemispheres. The presence of archaic bats in the earliest Eocene in both Northern and Southern Hemispheres indicates that bats had a worldwide distribution in the early Tertiary.

Hershkovitz (1972) and Pierson (1986) have argued, on the basis of distributions of endemic modern bat fam-

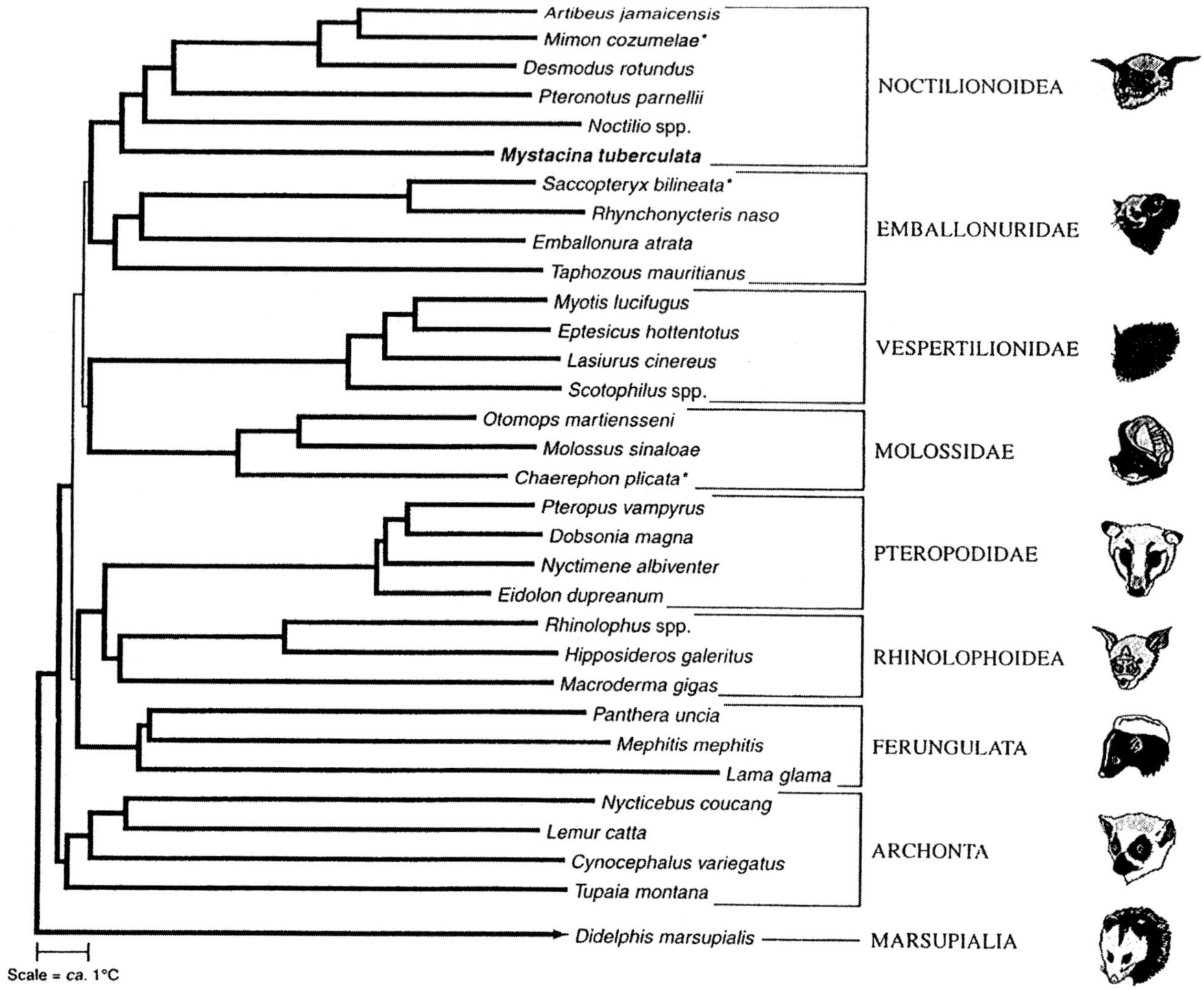

Figure 15. Cladogram of relationships of 24 bat species and eight outgroups based on DNA-hybridization comparisons and suggesting a closer relationship between Chiroptera and Ferungulata than between Chiroptera and Archonta (primates, tree shrews and colugos). From Kirsch *et al.* (1998).

ilies, for a Southern Hemisphere origin for the world's extant bat radiation. Sigé (1991a) proposed that modern bat groups evolved from isolated immigrant archaic groups in the Southern Hemisphere in the early Eocene. The rapidly expanding fossil record for bats in southern continents (Sigé, 1991a; Czaplewski, 1997; Hand, 1999) should soon help fill in the missing pieces of the four-dimensional jigsaw of early bat evolution.

THE AUSTRALIAN FOSSIL BAT RECORD

In Australia, the 90 species of living bats currently recognized (Duncan *et al.*, 1999) comprise more than a quarter of its native, non-marine mammal fauna. Both suborders and seven families are present: Pteropodidae, Emballonuridae, Rhinolophidae, Hipposideridae, Megadermatidae, Vespertilionidae and Molossidae (Table 2). Another two bat families (the Mystacinidae and Archaeonycteridae) are represented in the Australian fossil record.

No bat families and only five genera are endemic to Australia, although the Mystacinidae is endemic to the Australian region and several genera are found only in Australia and New Guinea (Table 2). Further, the taxonomy of some living Australian bats is so poorly understood that it is possible that some bats currently referred to cosmopolitan or Asian genera may represent endemic groups. The level of endemicity (at the generic level) is much lower than that recognized in Australian marsupials and rodents (see Aplin, 2006; Wroe and Archer, 2006). This perhaps reflects the ability of bats to cross water gaps, but also the morphological conservatism that characterizes them. It is not commensurate with their long history on this continent (see below).

Pre-Pleistocene records of bats in Australasia are restricted so far to continental Australia and New

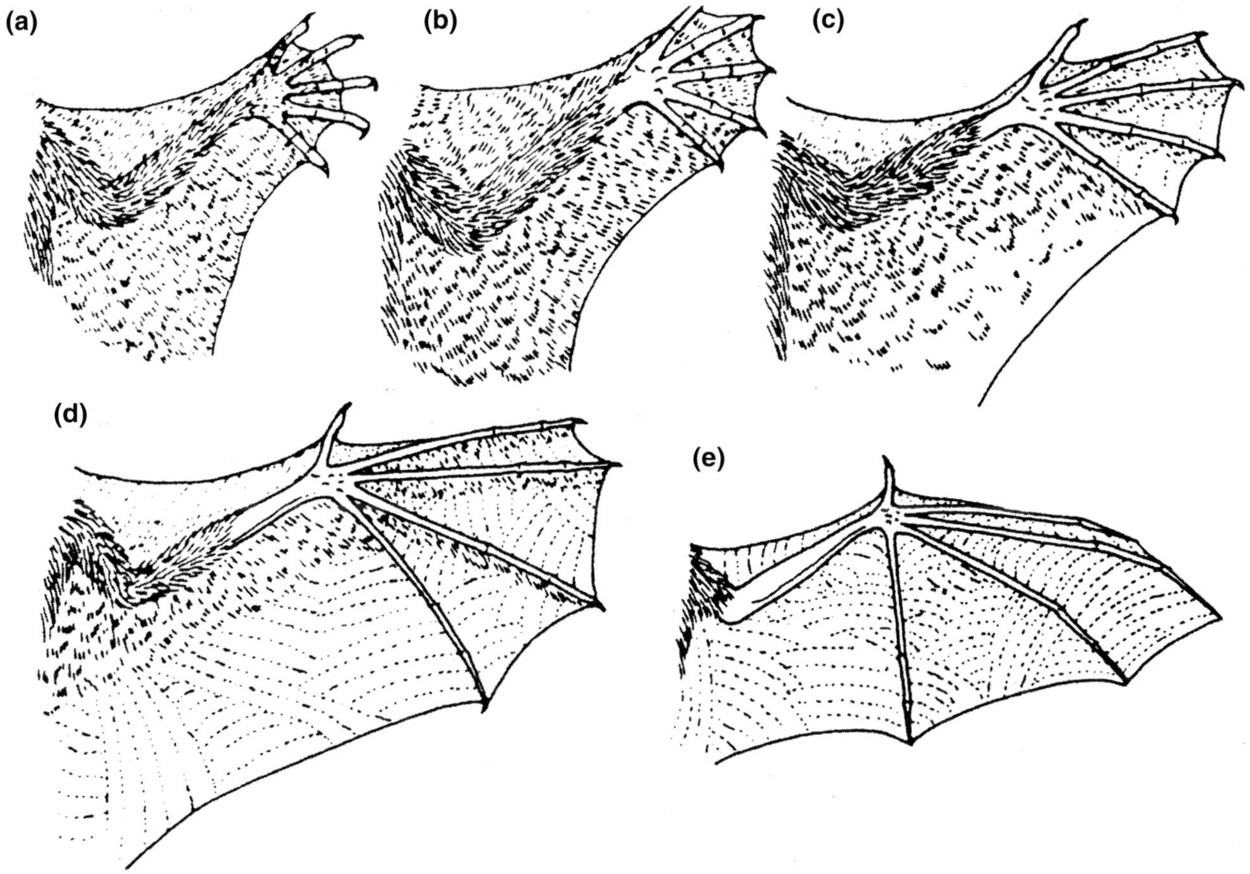

Figure 16. Hypothetical progressive stages in the development of the wing of bats: **(a)–(c)** early stages in which the emphasis may have been on gliding; **(d)** intermediate stage; **(e)** fully developed wing. After Smith (1977).

Zealand. The oldest record of bats otherwise are late Pleistocene remains from cave deposits and archaeological sites in New Guinea and islands in the south-west Pacific (Flannery, 1995a, b). The extinct species of *Nyctophilus howensis* is known from a Quaternary deposit on Lord Howe Island (McKean, 1975). From New Zealand, Worthy *et al.*, (2002) have reported, but not yet described, microchiropteran remains from an early Miocene site near St. Bathans, Otago, South Island.

The Australian fossil bat record is one of the world's oldest and best; it extends back 55 million years and records the presence of at least 50 fossil species. Early Tertiary Australian bats are archaic kinds, while, in general, middle Tertiary (Oligo-Miocene) Australian bats are ancestors or close relatives of extant Australian taxa (e.g., *Macroderma* and *Rhinonicteris* spp.), relatives of extinct and extant non-Australian taxa (e.g., *Icarops*, *Xenorhinos* and *Riversleigha* spp.), and taxa representing archaic cosmopolitan groups with no living descendants (e.g., *Petramops* and *Hydromops* spp.). Australian Pliocene bats are typically closely related to, or represent early populations of, extant Australian bats (e.g., *Taphozous* spp., *Macroderma gigas*, *Mormopterus (Micronomus)* sp.). Australian bat diversity through time is summarized in Figure 18.

ARCHAEONYCTERIDIDS

Until 1990, archaic early Eocene bats had been found mostly in the Northern Hemisphere, in areas that once formed Laurasia, but in that year a primitive Eocene bat was found in 55-million-year-old clays near Murgon, south-eastern Queensland (Hand *et al.*, 1994); see Figure 19. Murgon's *Australonycteris clarkae* is the oldest bat in the Southern Hemisphere and only slightly younger than the world's oldest bats from the Paris Basin in France (Russell *et al.*, 1982; Sigé and Legendre, 1983). *Australonycteris* (which means 'southern bat') was described on the basis of a dentary, isolated lower teeth and a periotic (ear) bone. Since then, additional fossil material from Murgon (including upper molars and post-cranial remains) clearly indicates that *Australonycteris* is a member of the family Archaeonycterididae, an archaic bat group otherwise known only from early to middle Eocene sediments in the Northern Hemisphere.

Australonycteris clarkae was a medium-sized bat (with a forearm length of 40–45 mm) and appears to have been able to echolocate. The depositional environment at the Murgon site is interpreted to have been a shallow lake or slow running stream, probably surrounded by vegetation. The tips of the molars of this early bat were curi-

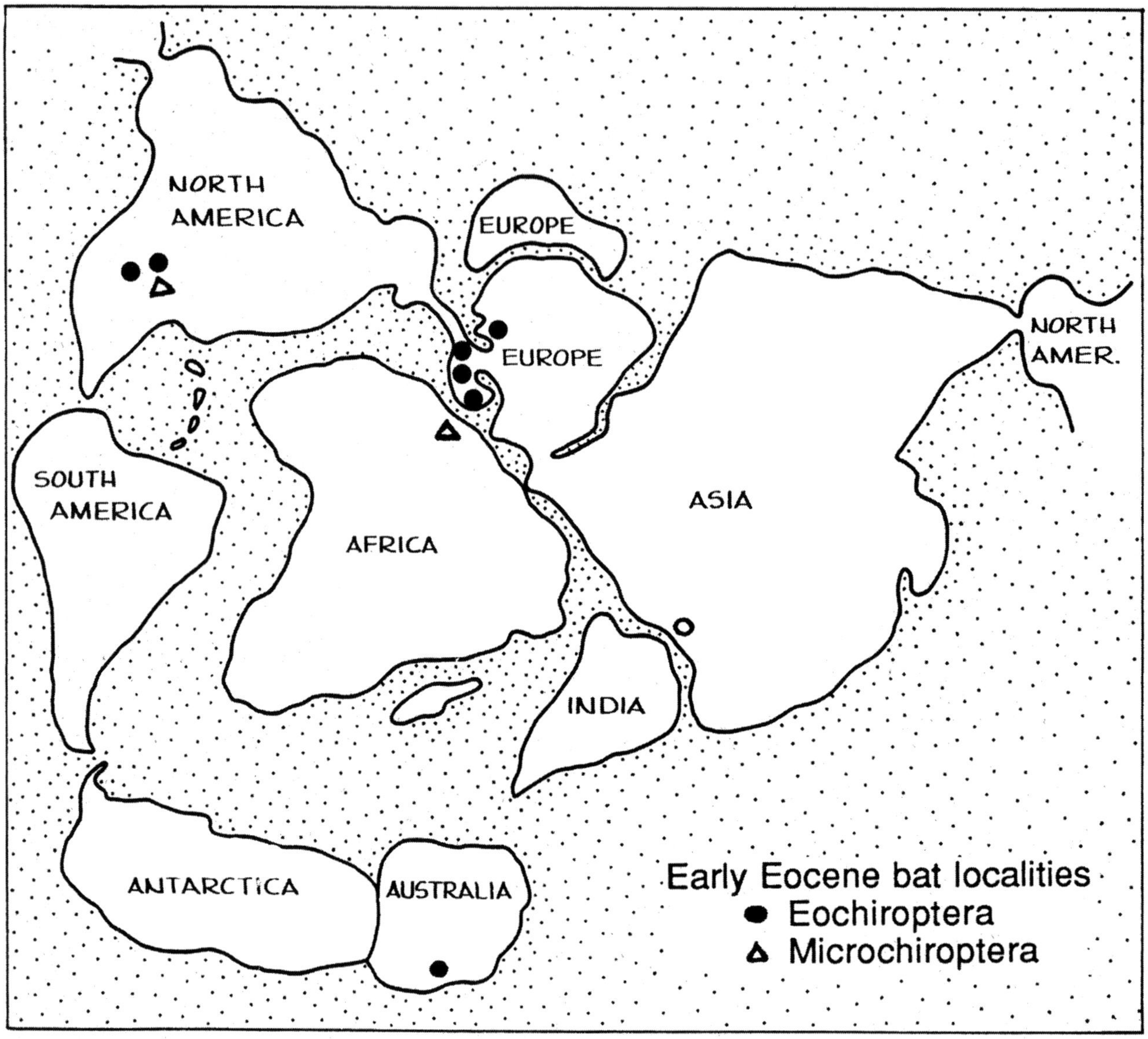

Figure 17. Palaeogeographic reconstruction of continents during the early Eocene showing fossil sites that have produced early Eocene bats. From Savage and Russell (1983).

ously worn, possibly by the abrasion of food, suggesting that *Australonycteris* was regularly carrying hard prey between its teeth. It may have foraged over water for beetles or even small fish (Hand *et al.*, 1994).

How bats first reached Australia is not yet known. They arrived before the breakup of Gondwana, but dispersal could have been through either South America or south-eastern Asia. The changes (in relationships of land masses) that have occurred in this region, over the last 80 million years, are illustrated in other chapters (Holdaway and Worthy, 2006; van Oosterzee, 2006; White, 2006) but points of special relevance to bats include: (a) in the late Cretaceous and earliest Tertiary, the Australia-New Guinea land mass was continuous with Antarctica and was separated from northern continents and Africa by the Tethys Ocean; (b) an archipelago linked South America and Antarctica, and the South-America-East Antarctica-Australian land mass appears to have been populated during that time by mammals such as monotremes, marsupials and terrestrial placentals including the condylarth-like *Tingamarra* (Woodburne and Zinsmeister, 1984; Pascual *et al.*, 1992; Godthelp *et al.*, 1992; Archer *et al.*, 1993, 1999; Woodburne and Case, 1996); (c) Australia separated completely from Antarctica in the late Eocene to early Oligocene, with the last land connection, through Tasmania, severed about 35 million years ago (Veevers, 1991).

Early Tertiary bats may have entered Australia circuitously from North America and Europe via South America and Antarctica. However, no fossil bats are yet known from Antarctica, and there is no early Tertiary record for bats in South America, despite a good pre-Oligocene micromammal record. The oldest South American bats are a late Oligocene Brazilian molossid

Table 2. Species numbers and extralimital distributions of genera of Australian bats, with data from Strahan (1995), Churchill (1998) and Duncan *et al.* (1999). Key to abbreviations: NG = New Guinea; NZ = New Zealand; Orient= Oriental region; Afric = Africa; Palae = Palaearctic region; Nearc = Nearctic region; Neotr = Neotropical region.

Genus	Australian spp no.	NG	NZ	Orient	Afric	Palae	Nearc	Neotr
Suborder Megachiroptera								
Family Pteropodidae								
Pteropus	8	+	–	+	+	–	–	–
Dobsonia	1	+	–	+	–	–	–	–
Nyctimene	2	+	–	+	–	–	–	–
Macroglossus	1	+	–	+	–	–	–	–
Syconycteris	1	+	–	–	–	–	–	–
Suborder Microchiroptera								
Family Emballonuridae								
Saccolaimus	3	+	–	+	–	–	–	–
Taphozous	5	+	–	+	+	+	–	–
Family Megadermatidae								
Macroderma	1	–	–	–	–	–	–	–
Family Hipposideridae								
Hipposideros	7	+	–	+	+	+	–	–
Rhinonicteris	2	–	–	–	–	–	–	–
Family Rhinolophidae								
Rhinolophus	4	+	–	+	+	+	–	–
Family Vespertilionidae								
Chalinolobus	5	+	+	–	–	–	–	–
Falsistrellus	2	–	–	–	–	–	–	–
Kerivoula	1	+	–	+	+	–	–	–
Miniopterus	3	+	–	+	+	+	–	–
Murina	1	+	–	+	–	–	–	–
Myotis	2	+	–	+	+	+	+	+
Nyctophilus	11	+	–	–	–	–	–	–
Pipistrellus	3	+	–	+	+	+	+	+
Scoteanax	1	–	–	–	–	–	–	–
Scotorepens	5	+	–	–	–	–	–	–
Vespadelus	9	–	–	–	–	–	–	–
Family Molossidae								
Chaerephon	1	+	–	+	+	+	–	–
Mormopterus	9	+	–	+	+	+	+	+
(*Micronomus*)		(–)	(–)	(–)	(–)	(–)	(–)	(–)
Tadarida	1	+	–	+	+	+	+	+

Mormopterus faustoi (Paula Couto, 1956; Paula Couto and Mezzalira, 1971) and some middle Miocene Colombian taxa, including the phyllostomids *Notonycteris magdalenensis* and *Tonatia* sp., a glossophagine, a phyllostomine and *Noctilio albiventris* (Czaplewski, 1997).

Some palaeogeographic reconstructions of the Australian region (e.g. Audley-Charles, 1981; Hall, 1998) postulate a complex of continental fragments between Australia and Asia during the Cretaceous and early Tertiary, perhaps acting as a chain of island 'stepping stones' for good dispersers such as bats. The palaeobotanical record shows sporadic interchange of plants between Australia and Asia since at least the Cretaceous (Truswell *et al.*, 1987).

MYSTACINIDS

There is a gap in the Australian fossil record for bats (as for all non-marine mammals) between 55 and 26 million years ago in the late Oligocene. Several sites of late Oligocene age, at Riversleigh in north-western Queensland and at one site in the Lake Eyre Basin of South Australia, have produced bat fossils. The remains are referable to modern bat families.

From late Oligocene deposits in the Lake Eyre Basin, a single bat tooth has been recovered (Figure 19). The upper molar, from the 26 million-year-old Ditjimanka Local Fauna of Lake Palankarinna, was described by Archer (1978) as a possible rhinolophid but is now understood to be a mystacinid.

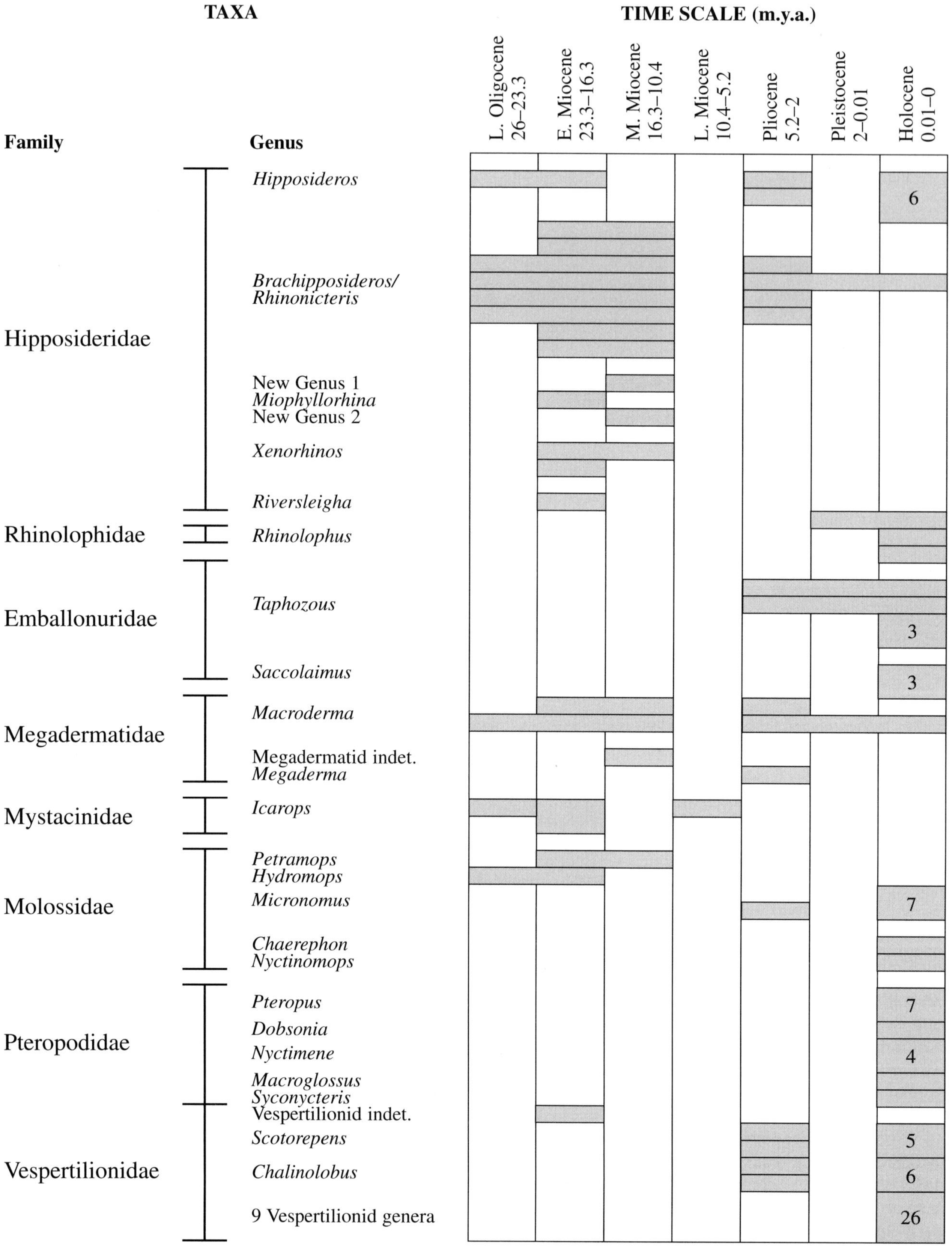

Figure 18. Australian bat diversity through time. Each block represents a distinct species unless species diversity is indicated by a number. Blocks that extend without breaks through more than one time period indicate apparently long-surviving species. Based on Hand (1999).

The only surviving member of the bat family Mystacinidae is a single, threatened species living in New Zealand's ancient Gondwanan forests. The lesser short-tailed bat *Mystacina tuberculata* (Plate 107) is world renowned for two reasons: the fact that its peculiarly broad dietary and terrestrial habits are quite unlike those of all other bats, and that its evolutionary history and relationships to other bat families are clouded in mystery. A second, larger New Zealand species (*M. robusta*) appears to have become extinct during the last 40 years (Flannery, 1987; Holdaway and Worthy, 2006).

Species of *Mystacina* have very short wings and sturdy hindlimbs and spend up to 30 percent of their time on the ground or on tree trunks and branches. They feed on ground and flying arthropods, nectar, fruit and pollen, and arguably carrion (Daniel, 1990; Lloyd, 2001). They roost in tree hollows, caves and make burrows in rotting logs. The lesser short-tailed bat is the sole pollinator of the endemic, parasitic rhizome the wood rose which produces its flowers at ground level (Eckroyd, 1993). The two *Mystacina* species were referred to their own family (New Zealand's only family of native mammals) soon after their discovery, but during the following 150 years the family has been placed in three different superfamilies (Emballonuroidea, Vespertilionoidea and Noctilionoidea).

In 1986 the semi-terrestrial *Mystacina* was interpreted by Pierson *et al.*, on the basis of immunological distance, to be a member of the South American superfamily Noctilionoidea, which dispersed from South America to New Zealand some 35 million years ago. Subsequently, Simmons (1996, 1998) found, using a 'total evidence' approach in which morphological, reproductive, behavioural, DNA and other data are included in a single data set, that mystacinids are probably basal to a large group of bats including the cosmopolitan Vespertilionidae, Molossidae and South American Nataloidea; this implies that ancestral mystacinids diverged from other bats at least 55 m.y.a. DNA-DNA hybridization studies by Kirsch *et al.* (1998) confirmed the affinity of *Mystacina* with noctilionoids, as suggested by serological studies, but suggested separation from its South American relatives occurred between 54 and 66 m.y.a. DNA sequencing studies (Kennedy *et al.*, 1999) also indicate a mystacinid-noctilionoid relationship, but differ from other molecular results in placement of *Mystacina* within the superfamily closer to phyllostomids and mormoopterids; they calculate the time of divergence of *Mystacina* from other bat taxa to be anywhere from 45 to 68 m.y.a.

In Australia, the family Mystacinidae is represented by species from several sites including Ditjimanka. More complete material belongs to several species of the new genus *Icarops* now known from early and middle Miocene Riversleigh deposits and middle Miocene limestone sediments at Bullock Creek, Northern Territory (Hand *et al.*, 1998; Figure 19).

Postcranial bones from some Riversleigh sites suggest that the terrestrial habits of the living *Mystacina* were probably already developed in its Australian Miocene ancestors. In the Miocene, *Icarops* species lived in Australian forests, probably foraging in trees and on the ground for arthropods, and perhaps over water for small fish or insects.

Figure 19. Australian fossil deposits that have produced Tertiary-aged bats. Key to Site Numbers: 1 = Murgon, Tingamarra Local Fauna, early Eocene; 2 = Lake Palankarinna, Ditjimanka Local Fauna, late Oligocene; 3 = many Riversleigh local faunas, late Oligocene to late Miocene and early Pliocene; 4 = Bullock Creek Local Fauna, middle Miocene; 5 = Hamilton Local Fauna, early Pliocene; 6 = Wellington Caves, Big Sink Local Fauna, Pliocene; 7 = Chinchilla Local Fauna, Pliocene.

The family Mystacinidae was represented in Australia at least 26 m.y.a. and survived there until at least 12 m.y.a., occurring in middle Miocene deposits at Bullock Creek in the Northern Territory. But they had disappeared, at least from the Riversleigh area, by 3–5 m.y.a.—being absent from the early Pliocene Rackham's Roost deposit (Archer *et al.*, 1994). The decline and, ultimately, extinction of this lineage in Australia may reflect the intense drying of Australia since the Late Miocene which resulted in the replacement of wet forests by open woodlands and grasslands over much of the continent (McGowran and Li, 1997; Martin, 1998). The sole surviving member of the mystacinid lineage is now restricted to New Zealand's endemic Gondwanan-type forests dominated by *Podocarpus*, *Dacrydium*, *Nothofagus* and *Agathis* (Daniel and Williams, 1984; Lloyd, 2001).

Hand *et al.* (1998) have argued that the presence of plesiomorphic mystacinids in the Australian Tertiary record most parsimoniously suggests that Australia was the immediate source of New Zealand's Quaternary mystacinids. New Zealand separated from the rest of Gondwana 80–90 million years ago (Fleming, 1979), long

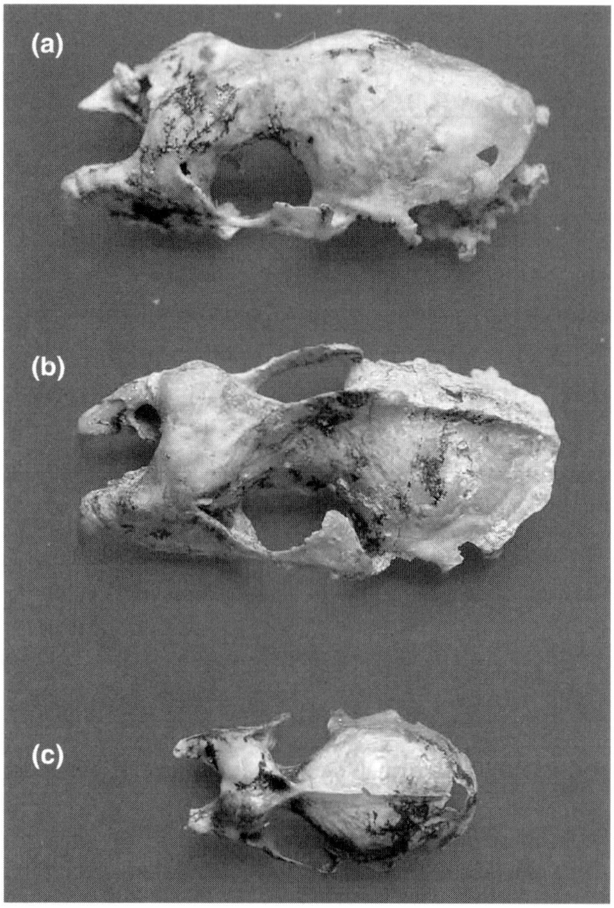

Figure 21. Skulls representing three hipposiderid genera from the early Miocene Bitesantennary Site at Riversleigh: **(a)** *Xenorhinos halli*; **(b)** *Riversleigha williamsi*; **(c)** *Rhinonicteris tedfordi*. Photograph by R. Arnett.

Brachipposideros, *Riversleigha*, *Xenorhinos*) occur in the Australian region, the early radiation of the family could have occurred there—see Figure 23. A minimum age for the early radiation of the Hipposideridae is provided by the middle Eocene age of fossil hipposiderids from Europe, such as *Pseudorhinolophus* and *Palaeophyllophora* species.

MEGADERMATIDS

Megadermatids, or false vampires, are characterized by, among other features, their large ears which are fused across the midline of the head, a large, simple noseleaf, and bifid tragus. Five living species are distributed (Figure 20(b)) in tropical areas of Africa (two species), Asia (two species) and Australia (one species). Megadermatids are not blood-drinkers like the true vampires of southern and central America, but insectivorous or facultatively carnivorous and generally have a dentition adapted for meat-eating.

The family Megadermatidae is well represented in the fossil record, being known from at least 19 fossil taxa from early Tertiary to Quaternary sediments in Eurasia, Africa and Australia. The oldest megadermatid (*Necromantis adichaster*) is known from the late Eocene of France (Weithofer, 1887).

Australia's only living megadermatid is the carnivorous ghost bat *Macroderma gigas*. *Macroderma gigas* is one of the largest microchiropterans in the world and is known to take prey up to the size of a pigeon (Douglas, 1967). Commonly this bat cohabits large caves with other species of bats, but also roosts in human-made structures such as mines and sheds. Now restricted to northern Australia, the ghost bat was once distributed also in southern and central areas of Australia, including, during the Pleistocene, the south-western corner of Western Australia (Bridge, 1975) and New South Wales (Molnar *et al.*, 1984; Dawson and Augee, 1997).

Eight Australian fossil megadermatids have been described: seven from Riversleigh Tertiary sediments and one from Wellington Caves, New South Wales (Hand, 1985, 1995, 1996; Hand *et al.*, 1988). Most appear related or referable to the endemic Australian genus *Macroderma* but several appear to represent quite different lineages (see below).

The fossil record for *Macroderma* extends from the late Oligocene to Recent, with a maximum of two species occurring syntopically during that period. *Macroderma godthelpi* is known from well-preserved maxillae, jaws and teeth from late Oligocene to middle Miocene deposits at Riversleigh (Hand, 1985). It is the smallest species of *Macroderma,* being about one half to two thirds the size of the living *M. gigas*. The slightly younger (or perhaps contemporaneous) middle Miocene species *M. malugara* (Hand, 1996) from Riversleigh appears to be approximately the same size as *M. gigas* or larger.

The oldest record for the living ghost bat (*M. gigas*) is Riversleigh's Pliocene Rackham's Roost deposit (Hand, 1996). This deposit appears to largely represent the fossilized remains of prey accumulated (3–5 m.y.a.) by an early population of *M. gigas*—it contains the finely broken up bones and teeth of vertebrates such as fish, frogs, lizards, snakes, birds, dasyurids, bandicoots, possums and other bats.

A second Pliocene species (*Macroderma koppa*) was described by Hand *et al.* (1988) based on a broken snout and many isolated teeth from the Big Sink at Wellington Caves, New South Wales (Dawson *et al.*, 1999). *Macroderma koppa* is closely related to the living *M. gigas*, but differs in having two infraorbital foramina, and lacks bony eminences on the nose-shield.

Megaderma richardsi from Riversleigh's early Pliocene Rackham's Roost deposit (Hand, 1995) is the first Australian record of this Afro-Eurasian genus, which elsewhere has a record that extends from late Oligocene to Holocene. *Megaderma richardsi* is one of the smallest megadermatids known. It exhibits a mixture of plesiomorphic and specialized features, the latter appearing to exclude it from being ancestral to any living megadermatid.

Sigé (1976) argued that the centre of radiation for the family Megadermatidae lay outside Europe, which was

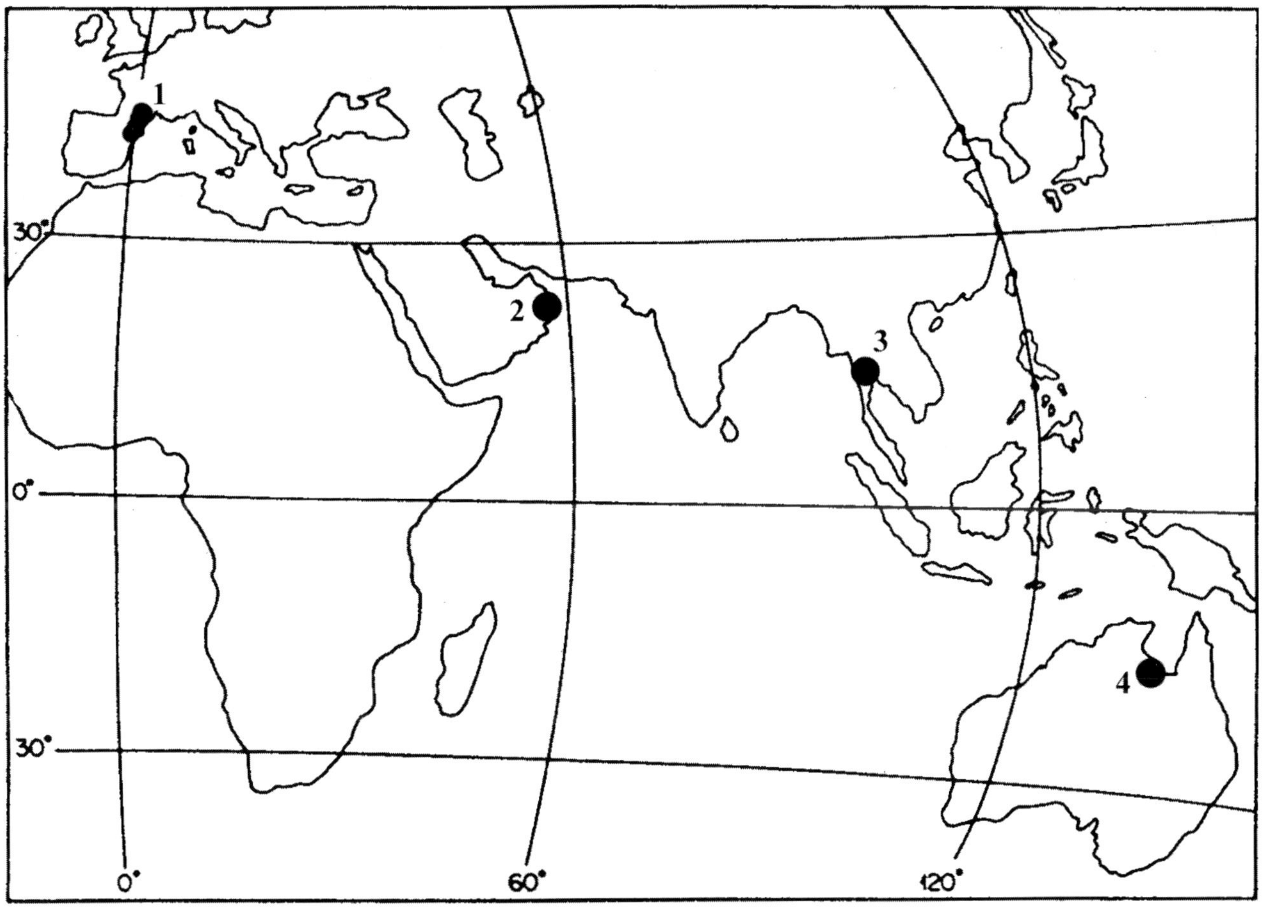

Figure 22. Known geographic distribution of the Tertiary species of *Brachipposideros*; after Sigé *et al.* (1982). Key to localities: 1 = France and Germany; 2 = Sultanate of Oman; 3 = Thailand; 4 = Riversleigh.

colonized in the middle Tertiary by relatively derived taxa (e.g., species of *Megaderma*) and Hand (1985) found that ancestors of the Australian middle Tertiary *Macroderma* radiation entered Australia before the late Oligocene. Hand (1995) speculated that *Megaderma* spp. could have entered Australia as recently as the late Miocene to earliest Pliocene, although the informally named 'Dwornamor Variant' (Hand, 1985), from the middle Miocene Gag Site, may also be a member of this lineage. In Sigé and Hand's phylogenetic and biogeographic hypotheses species of *Megaderma* and *Macroderma* are interpreted as relatively derived megadermatids (see Hand, 1985); however, Griffiths *et al.* (1992) suggest that these taxa may in fact be closer to the base of the megadermatid radiation and presumably its geographic centre.

MOLOSSIDS

Molossids are known as free-tailed or mastiff bats and are characterized by a long tail projecting beyond the edge of the tail membrane, and other features such as heavy lips and snout. They are fast, direct and relatively unmanoeuvrable fliers with long narrow wings and typically forage for flying insects in open spaces above the forest canopy or over water. Worldwide, 89 living molossid species are referred to between nine and 12 genera (Freeman, 1981; Honacki *et al.*, 1982; Legendre, 1984b; Altringham, 1996) and are distributed widely in tropical and temperate zones (Figure 20(c)).

Some 25 fossil molossids are reported from all continents except Antarctica (Legendre, 1985; Hand, 1990; Hand *et al.*, 1997; Czaplewski, 1997). The oldest are species of *Wallia* from the middle Eocene of North America (Legendre, 1985) and *Cuivierimops* from the late Eocene of Europe (Legendre and Sigé, 1984). Other early records include: a late Oligocene species from South America (Paula Couto, 1956; Paula Couto and Mezzalira, 1971); Oligo-Miocene species from Australia (Hand, 1990; Hand *et al.*, 1997); and Miocene species from Africa and Asia (Lavocat, 1961; Sigé, 1976; Legendre *et al.*, 1988).

In Australia, 11 living molossids are currently recognized: one species each of the genera *Tadarida* and *Chaerephon* and at least nine species of *Mormopterus* (*Micronomus*); the taxonomy of this last group remains confused (Duncan *et al.*, 1999). These extant molossids are distributed widely across the mainland, but not Tasmania, and all roost in tree hollows and buildings but not caves.

Australian Tertiary molossids are known from Riversleigh and also from Chinchilla, south-eastern Queensland. Several species occur in Riversleigh's Oligo-Miocene deposits, two of which have been described. *Petramops creaseri* is a plesiomorphic taxon from the middle Miocene of Riversleigh (Hand, 1990) which appears to fit near the base of the molossid family radiation (Figure 24). *Mormopterus (Hydromops) riversleighensis* from the early Miocene of Riversleigh (Hand *et al.*, 1997) belongs to the otherwise Eurasian subgenus *Hydromops*, all species of which are now extinct (Figure 24). Because other species of *Hydromops* occur in well-dated European deposits (Revilliod, 1920) as well as Miocene deposits in Thailand (Legendre *et al.*, 1988), the Australian form is helping to biocorrelate Riversleigh mammal-bearing sediments with those of Asia and Europe (Hand *et al.*, 1997).

The two Oligo-Miocene molossids from Riversleigh do not appear to be closely related to each other, nor do they appear to be ancestral to any living Australian molossid. They represent part of an older, archaic bat fauna that was widespread, and had close relatives in Eurasia and North America. Together they suggest that the modern Australian molossid bat fauna is the result of more recent colonization.

Legendre (1985) has suggested that *Mormopterus sensu lato* probably evolved from the *Cuvierimops* lineage in the late Eocene or early Oligocene and expanded wherever climates were warm (Legendre, 1984a, b, c). From this stock emerged species of *Hydromops*, *Micronomus* and *Platymops*, their three separate evolutionary paths resulting from a disjunct distribution of the ancestral group and occurred towards the end of the Oligocene (Legendre *et al.*, 1988). The ancestors of the Australian endemic *Micronomus* lineage are thought to have arrived in the Australian region from Asia by at least the late Oligocene, independently of *Petramops*, *Nyctinomus* and *Chaerephon* species (Legendre, 1984a, b, c, 1985; Hand, 1990).

The *Mormopterus* (*Micronomus*) lineage had evolved by at least the middle Pliocene, as demonstrated by isolated teeth found in Riversleigh's Pliocene Rackham's Roost Site and the Pliocene Chinchilla deposit in south-eastern Queensland (Hand *et al.*, 1999). More complete remains of extant species *Mormopterus loriae* and *Tadarida australis* are known from the Pleistocene of eastern Australia (Archer *et al.* 1984), and *Tadarida australis* is also reported as a fossil from Devil's Lair in south-western Australia (Baynes *et al.*, 1976). Today the nine species (separable on fur and penis characters) are distributed across the Australian mainland, with two *M. loriae* and *M. beccarii* found also in New Guinea and *M. beccarii* occurring on Ambon Island (McKenzie, 1995; Duncan *et al.*, 1999).

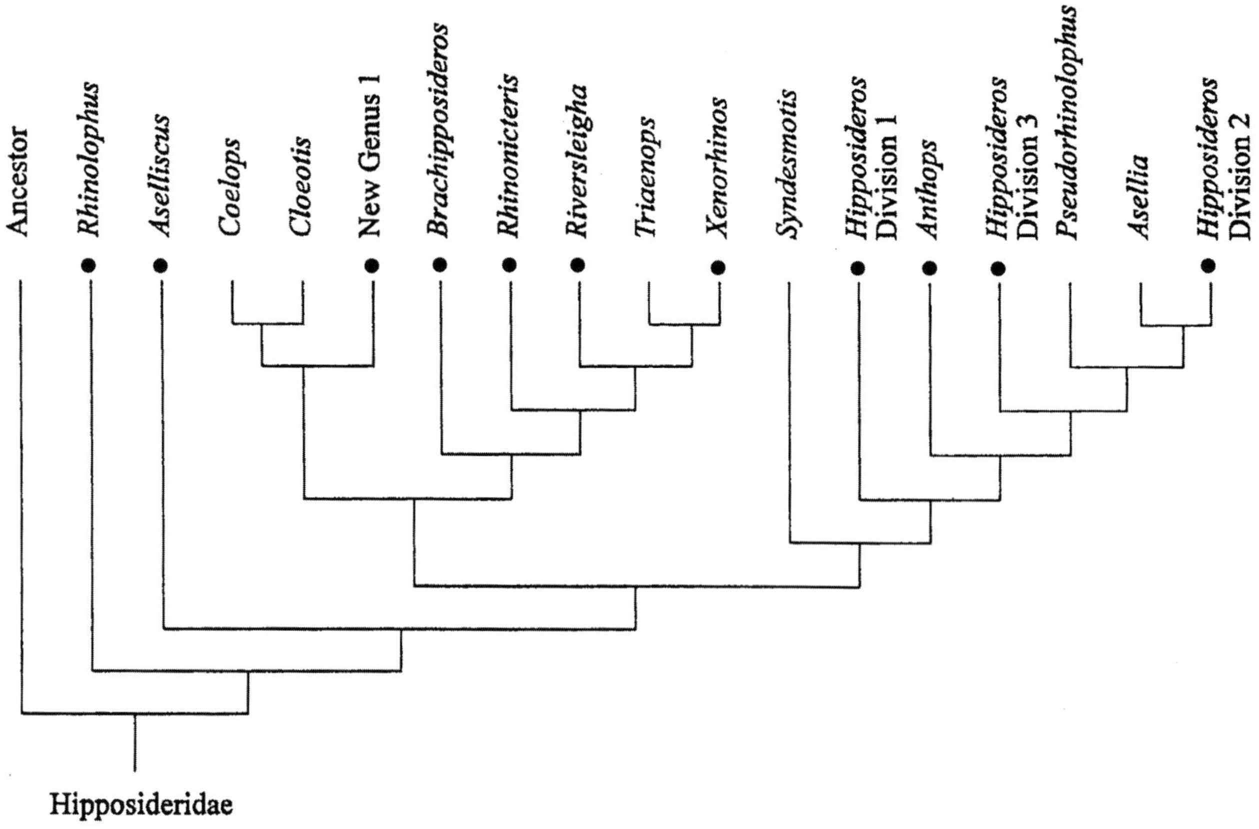

Figure 23. Cladogram of possible relationships among fossil and living hipposiderids (after Hand and Kirsch, 1998). Dots indicate taxa with representatives in the Australian region. Because many of the most plesiomorphic hipposiderids occur in the Australian region, it is possible that the family evolved and radiated there.

EMBALLONURIDS

Emballonurids or sheath-tailed bats are relatively large with flattened heads, large eyes and tails that emerge from half-way along the dorsal (upper) surface of the tail membrane (Plate 110). In flight, the tail membrane can be lengthened (and the bony tail enclosed in the membrane) by stretching the hind limbs. This enables them to manoeuvre exceptionally well in tight spaces, and gives them considerable freedom to scuttle about on cave walls and the ground. Emballonurids are today found in the tropics of both the Old and New Worlds (Figure 20(d)), where there seems to have been two distinct radiations giving rise to 51 living species.

The family dates back possibly to the earliest Eocene of England (Hooker, 1996) and certainly to the middle Eocene of Europe (Revilliod, 1920). Although there is no early or middle Tertiary record for emballonurids in the New World, they are thought to have spread to the Western Hemisphere before the middle latitude climatic deterioration in the mid to late Miocene. Legendre (1980) has suggested a minimum separation date for the Old and New World radiations as late Oligocene.

In Australia, nine living emballonurid species of the genera *Taphozous* and *Saccolaimus* roost in both caves and trees. The oldest and only fossil emballonurids known from Australia are two species of *Taphozous* from Riversleigh's Pliocene Rackham's Roost deposit, one species of which may represent the living Australian *T. hilli* from north-western Australia (Archer *et al.*, 1994). This Afro-Australasian genus has Miocene records in both Africa and Europe (Butler and Hopwood, 1957; Butler, 1969, 1978; Legendre, 1980).

VESPERTILIONIDS

Members of this family are known as common or plain-faced bats, and include the majority of living Australian microbats (41 species, Duncan *et al.*, 1999), such as the long-eared bats, broad-nosed bats and pipistrelles. They are typically small to medium-sized bats with a plain muzzle, large ears with a long tragus, long and narrow wings and a well-developed tail.

This is the largest and most widely distributed bat family, occurring from tropical to arctic zones and on many oceanic islands (Figure 2(b)). Vespertilionids have a very old record, the oldest being the middle Eocene European species of *Stehlinia* (Revilliod, 1919). Middle Tertiary records of vespertilionids are known from Europe, North America, Asia, Africa and Australia (Hand, 1984, 1999).

Of the 11 vespertilionid genera now occurring in Australia, approximately half have an Asian or worldwide distribution (Table 2). The other six genera, *Nyctophilus* (nine species), *Chalinolobus* (five species), *Vespadelus* (nine species), *Scotorepens* (five species), *Scoteanax* (one species) and *Falsistrellus* (two species), are restricted to Australia, New Zealand, New Guinea and nearby islands. Traditional understanding of the latter group has been that each lineage or genus had its closest relatives outside the Australian region (Tate, 1946; Simpson, 1961; Keast, 1972; Hand, 1984; Kitchener and Caputi, 1985); but Volleth and Tidemann (1991) suggested that, on the basis of karyological data, the Australian endemic genera (together with the New Guinean *Pharotis microtis*) form a monophyletic group, the ancestor of which possibly reached Australia in the Oligocene.

Despite their living diversity and old record elsewhere, vespertilionids have a poor fossil record in Australia. To date, the only Oligo-Miocene vespertilionid known from Australia is a single tooth, from an early Miocene Riversleigh site (Menu *et al.*, 2002). This rarity may relate to relative scarcity of vespertilionids in the Riversleigh region, or northern Australia, during the Oligo-Miocene or, possibly, preservational biases resulting from the roost and habitat preferences of such taxa.

By Pliocene time, vespertilionids had increased in number in the Riversleigh region with at least four species represented in the Rackham's Roost deposit; the fragmentary remains are possibly referable to the modern Australian genera *Chalinolobus* and *Scotorepens*, neither of which has an older record anywhere (Archer *et al.*, 1994; Hand, 1999). At least seven extant vespertilionid species are recorded from Pleistocene deposits in Australian caves (*Chalinolobus gouldii, C. morio, Miniopterus schreibersii, Nyctophilus geoffroyi, N. timoriensis* and species of *Vespadelus* and *Falsistrellus*; see Archer *et al.*, 1984 for details).

RHINOLOPHIDS

Rhinolophids or horse-shoe bats are so-named because of their elaborate U-shaped noseleaf (Plate 111). These are slow flying microbats with broad wings, well-suited for manoeuvring within the cluttered understorey of forests. All 69 species in this family are referred to the genus *Rhinolophus* and have an Old World tropical distribution. Living Australian species are confined to east coast forests. Although only two Australian species are formally recognized (*R. megaphyllus* and *R. philippensis*), molecular and morphological studies in progress indicate that at least four species are represented (Duncan *et al.*, 1999).

The fossil record of rhinolophids extends back to the Miocene of Europe and Africa (Lavocat, 1961; Sigé, 1976; 1990; Savage and Russell, 1983), but there is now doubt that Palaeogene species referred to the genus *Rhinolophus* (e.g., *R. priscus* Revilliod, 1920) are true rhinolophids (Sigé, 1990). On the basis of phylogenetic analysis of morphological and morphometric data, Bogdanowicz (1992) and Bogdanowicz and Owen (1992) argue for a south-east Asian origin for rhinolophids. These researchers propose that *R. megaphyllus* from the Malay peninsula, New Guinea, New Ireland, New Britain and Australia is the most basal member.

Rhinolophids have a poor fossil record in Australia. A

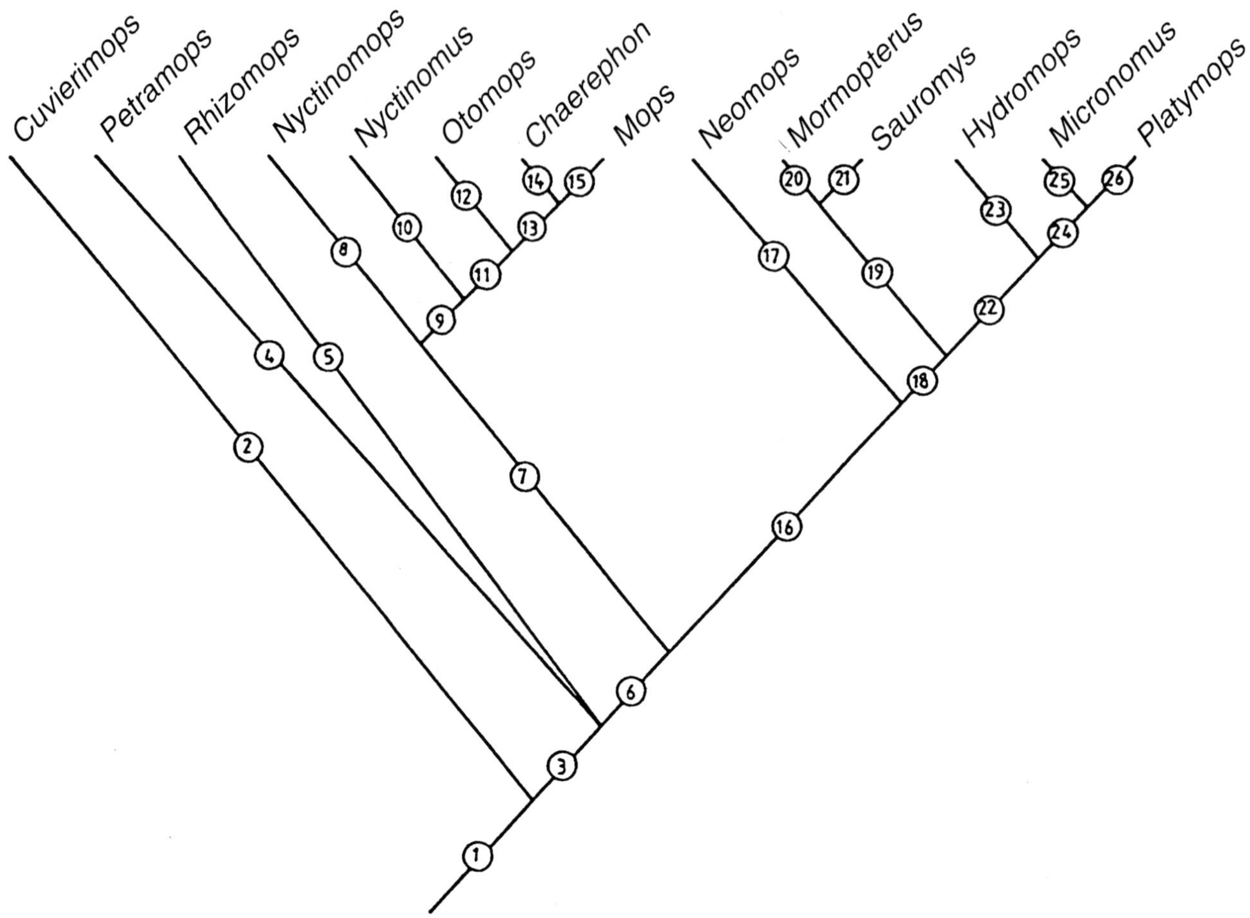

Figure 24. Cladogram of possible relationships between species of *Petramops*, *Mormopterus* and other molossids; numbers refer to possible synapomorphies. Based on Hand (1990).

possible rhinolophid from the late Oligocene Ditjimanka Local Fauna at Lake Palankarinna in central Australia (Archer, 1978) is now understood to represent a mystacinid; otherwise rhinolophids have only been recorded from Pleistocene deposits in eastern Australia: the eastern horseshoe bat, *Rhinolophus megaphyllus*, is recorded from deposits in Russenden Cave of south-eastern Queensland, and in Pyramids Cave, near Buchan, Victoria (Wakefield, 1972). It is possible that in Australian tropical palaeofaunas, such as those of Riversleigh, hipposiderids filled the rhinolophid niche (Hand, 1998b).

PTEROPODIDS

About 175 pteropodid species, distributed in tropical and subtropical areas of the Old World (Figure 2(a)), form the world's second largest bat family. In Australia, the Pteropodidae is represented by 13 living species in five genera (*Pteropus*, *Dobsonia*, *Nyctimene*, *Syconycteris* and *Macroglossus*). These are restricted mainly to coastal regions but one species, the little red flying-fox (Plate 106), extends into the interior; this species (*P. scapulatus*) has also been recorded as storm-blown waifs in New Zealand (Daniel, 1990). The range of Australia's *P. poliocephalus* extends to Victoria and South Australia giving it the most southerly distribution of any pteropodid.

Although the megachiropteran record extends back to the late Eocene in Asia (Ducrocq *et al.*, 1993) and the middle Tertiary of Africa and Europe (Sigé and Aguilar, 1987), there is no pre-Pleistocene record of this family in Australia. Extant forms are recorded from Quaternary archaeological deposits in northern Australia (White, 1967; Wakefield, 1972; Archer and Brayshaw, 1978; Aplin, 1981) and from the Pleistocene of New Guinea, the Moluccas and other islands (Menzies, 1977; Flannery *et al.*, 1974, 1983; Gillieson and Mountain, 1983; Flannery, 1995a, b; Aplin *et al.*, 1999). Kirsch *et al.* (1995) have argued that a centre of radiation for some pteropodids was in the New Guinea region and the islands to the north of Australia at least 25 million years ago. Their absence from the middle and late Tertiary mammal-rich fossil deposits of Riversleigh in northern Australia is an enigma. Competition with marsupials for food resources may have played a role in preventing their Tertiary colonization of Australia, or perhaps suitable food resources may not have been readily available in Australia until the recent past.

SUMMARY

The growing Australian fossil bat record indicates that bats entered Australia from both the north and south over a period of at least 55 million years. Initial colonization by archaic bats (e.g., *Australonycteris*) may have been from either the north or south, but at least some groups (e.g., mystacinids) probably represent vicariated Gondwanan groups. Australian hipposiderids, megadermatids, rhinolophids, emballonurids and at least some molossids (e.g., *Hydromops*) appear to be more closely related to, and are probably derived from, groups that radiated to the north of Australia. The Australian record for vespertilionids and some molossid lineages is poor, although it is possible that some of these may represent vicariant southern groups.

Phylogenetic analysis of morphological data from fossil and modern hipposiderids suggests that this family radiated in or close to northern Australia. Karyological evidence suggests that a number of modern vespertilionid genera (e.g., *Vespadelus*, *Falsistrellus*) form a monophyletic group that evidently evolved in Australia. Some bat groups appear to have had their primary centre of radiation in Australia but have living and fossil representatives more widely distributed (e.g., species of *Xenorhinus*, *Nyctophilus*, *Chalinolobus*). Others appear to be recent colonists (e.g., species of *Rhinolophus*, *Pteropus*).

Overall, components of the Australian fossil bat fauna include: taxa representing archaic cosmopolitan groups with no living descendants (e.g. species of *Australonycteris*, *Petramops*, *Hydromops*); relatives of extinct and extant non-Australian taxa (e.g., *Icarops*, *Xenorhinos* and *Riversleigha* spp.); ancestors or close relatives of extant Australian taxa (e.g., *Macroderma* and *Rhinonicteris* species); and early populations of extant Australian bats (e.g., *Taphozous* spp., *Macroderma gigas*, *Micronomus* sp.).

REFERENCES

Altringham, J.D. 1996. *Bats. Biology and Behaviour*. Oxford University Press, Oxford. 262pp.

Andersen, K. 1912. *Catalogue of the Chiroptera in the Collection of the British Museum. Volume 1. Megachiroptera*. 2nd Edit. British Museum, London.

Aplin, K.P. 1981. Faunal remains from the archaeological sites in Mangrove Creek Catchment. Pp.1–73. In Attenbrow, V. (ed). *Mangrove Creek Dam Salvage Excavation Project*. National Parks and Wildlife Service, New South Wales (Appendix 4) 2.

Aplin, K.P. 2006. Ten million years of rodent evolution in Australasia: phylogenetic evidence and a speculative historical biogeography. Pp.707–744. In Merrick, J.R., Archer, M., Hickey, G.M. and Lee, M.S.Y. (eds). *Evolution and Biogeography in Australasian Vertebrates*. Auscipub Pty. Ltd., Sydney. 942pp.

Aplin, K.P., Pasveer, J.M. and Boles, W.E. 1999. Late Quaternary vertebrates from the Bird's Head Peninsula, Irian Jaya, Indonesia, including descriptions of two previously unknown marsupial species. *Records of the Western Australian Museum, Supplement,* No. 57, pp.351–387.

Archer, M. 1978. Australia's oldest fossil bat, a possible rhinolophid. *Proceedings of the Royal Society of Queensland*, Vol. 89, p.23.

Archer, M., Arena, D.A., Bassarova, M., Black, K., Brammall, J.R., Cooke, B.N., Creaser, P.H., Crosby, K., Gillespie, A.K., Godthelp, H.J., Gott, M., Hand, S.J., Kear, B.P., Krikmann, A., Mackness, B.S., Muirhead, J., Musser, A.M., Myers, T.J., Pledge, N.S., Wang, Y. and Wroe, S. 1999. The evolutionary history and diversity of Australian mammals. *Australian Mammalogy*, Vol. 21, pp.1–45.

Archer, M. and Brayshaw, H. 1978. Recent local faunas from excavations at Herveys Range, Kennedy, Jourama, and Mount Roundback, north-eastern Queensland. *Memoirs of the Queensland Museum*, Vol. 18, p.7.

Archer, M., Clayton, G. and Hand, S.J. 1984. A checklist of Australasian fossil mammals. Pp.1027–1087. In Archer, M. and Clayton, G. (eds). *Vertebrate Zoogeography and Evolution in Australasia*. Hesperian Press, Perth. 1203pp.

Archer, M., Godthelp, H. and Hand, S.J. 1993. Early Eocene marsupial from Australia. *Kaupia*, Vol. 3, pp.193–200.

Archer, M., Hand, S.J. and Godthelp, H. 1994. *Riversleigh. The Story of Animals in Ancient Rainforests of Inland Australia.* Reed Books, Sydney. 264pp.

Arnason, U., Adegoke, J.A., Bodin, K., Born, E.W., Esa, Y.B., Gullberg, A., Nilsson, M., Short, R.V., Xu, X. and Janke, A. 2002. Mammalian mitogenetic relationships and the root of the eutherian tree. *Proceedings of the National Academy of Sciences of the United States of America*, Vol. 99, pp.8151–8156.

Audley-Charles, M.G. 1981. Geological history of the region of Wallace's line. Pp.24–35. In Whitmore, T.C. (ed). *Wallace's Line and Plate Tectonics*. Clarendon Press, Oxford. 90pp.

Bailey, W.J., Slightom, J.L. and Goodman, M. 1992. Rejection of the 'flying primate' hypothesis by phylogenetic evidence from the e-globin gene. *Science,* Vol. 256, pp.86–89.

Baker, R.J., Novacek, M.J. and Simmons, N.B. 1991. On the monophyly of bats. *Systematic Zoology*, Vol. 40, pp.216–231.

Baynes, A., Merrilees, D. and Porter, J.K. 1976. Mammal remains from the upper levels of a late Pleistocene deposit in Devil's Lair, Western Australia. *Journal of the Royal Society of Western Australia*, Vol. 58, pp.97–126.

Beard, K.C., Sigé, B. and Krishtalka, L. 1992. A primitive vespertilionid bat from the early Eocene of central Wyoming. *Comptes Rendus de l'Academie des Sciences, Paris*, Vol. 314, pp.735–741.

Bogdanowicz, W. 1992. Phenetic relationships among bats of the family Rhinolophidae. *Acta Theriologia*, Vol. 37, pp.213–240.

Bogdanowicz, W. and Owen, R.D. 1992. Phylogenetic analyses of the bat family Rhinolophidae. *Zeitschrift für Zoologische Systematik und Evolutionsforschung*, Vol. 30, pp.142–160.

Bridge, P.J. 1975. Paleo-distribution of *Macroderma gigas* in the south west of Western Australia. *Helictite*, Vol. 13, pp.34–36.

British Museum (Natural History). 1975. *Bats*. Zoology Leaflet 2.

Winnington, A.P. and Spencer, H.G. 1999. The long and short of it: branch lengths and the problem of placing the New Zealand Short-tailed Bat, *Mystacina*. *Molecular Phylogenetics and Evolution*, Vol. 13, pp.405–416.

Kirsch, J.A.W. 1996. Bats are monophyletic; megabats are monophyletic; but are microbats also? *Bat Research News*, Vol. 36, p.78.

Kirsch, J.A.W., Flannery, T.F., Springer, M.S. and Lapointe, F.-J. 1995. Phylogeny of the Pteropodidae (Mammalia: Chiroptera) based on DNA hybridisation, with evidence for bat monophyly. *Australian Journal of Zoology*, Vol. 43, No. 4, pp.395--428.

Kirsch, J.A.W. and Hutcheon, J.M. 1997. Further on the possibility that microchiropterans are paraphyletic. *Bat Research News*, Vol. 37, p.138.

Kirsch, J.A.W., Hutcheon, J.M., Byrnes, G.P. and Lloyd, B.D. 1998. Affinities and historical zoogeography of the New Zealand Short-tailed Bat, *Mystacina tuberculata* Gray 1843, inferred from DNA-hybridization comparisons. *Journal of Mammalian Evolution*, Vol. 5, pp.33–64.

Kitchener, D. and Caputi, N. 1985. Systematic revision of Australian *Scoteanax* and *Scotorepens*, with remarks on relationships to other Nycticeiini. *Records of the Western Australian Museum*, Vol. 12, pp.85–146.

Koopman, K. 1981. The distributional patterns of New World nectar-feeding bats. *Annals Missouri Botanic Gardens*, Vol. 68, pp.352–369.

Lavocat, R. 1961. Le gisement de verte'bre's mioc'enes de Beni Mellal (Maroc). Etude syste'matique de la faune des mammif'eres et conclusions generales. *Notes et Mémoirs Service Géologie Maroc*, Vol. 155, pp.1–121.

Legendre, S. 1980. Un chiroptère emballonuridé dans le Néogène d'Europe occidentale; considérations paléo-biogéographiques. *Geobios*, Vol. 13, pp.839–847.

Legendre, S. 1984a. Étude odontologique des représentants actuels du group Tadarida (Chiroptera, Molossidae). Implications phylogénetiques, systématiques et zoogéographiques. *Revue Suisse de Zoologie*, Vol. 91, pp.99–442.

Legendre, S. 1984b. Identification de deux sous-genres fossiles et compréhension phylogénique du genre *Mormopterus* (Molossidae, Chiroptera). *Comptes Rendus de l'Academie des Sciences, Paris,* Series 2, Vol. 298, pp.715–720.

Legendre, S. 1984c. Essai de biogeographie phylogénique des molossidés (Chiroptera). *Myotis*, Vol. 21-22, pp.30–36.

Legendre, S. 1985. Molossidés (Mammalia, Chiroptera) cénozoiques de l'Ancien et du Nouveau Monde; statut systématique; intégration phylogénique des données. *Neues Jahrbuch für Geologie und Palaöntologie Abhandlungen*, Vol. 170, pp.205–227.

Legendre, S., Rich, T.H.V., Rich, P.V., Knox, G.J., Punyaprasiddi, P., Trumpy, D.M., Wahlert, J. and Napawongse, P. 1988. Miocene fossil vertebrates from the Nong Hen-I (A) exploration well of Thai Shell Exploration and Production Company Limited, Phitsanulok Basin, Thailand. *Journal of Vertebrate Paleontology*, Vol. 8, pp.278–289.

Legendre, S. and Sigé, B. 1984. La place du "Vespertillion de Montmartre" dans l'histoire des chiroptères molossidés. *Actes Symposium Paléontologie G. Cuvier, Montbéliard,* pp.337–352.

Lloyd, B. 2001. Advances in New Zealand mammalogy 1990-2000: short-tailed bats. *Journal of the Royal Society of New Zealand*, Vol. 31, pp.59–81.

Marandat, B. 1991. Mammifères de l'Ilerdien moyen (Eocène Inférieur) des Corbières et du Minervois (Bas-Langedoc, France). *Palaeovertebrata*, Vol. 20, pp.55–144.

Martin, H.A. 1998. Tertiary climatic evolution and the development of aridity in Australia. *Proceedings of the Linnean Society of New South Wales*, Vol. 119, pp.115–136.

McGowran, B. and Li, Q. 1997. Australasian Cainozoic biogeography: environmental framework for a southern-temperate province on a cooling planet. *Geological Society of Australia Abstracts*, Vol. 44, pp.52–53.

McKean, J.L. 1975. The bats of Lord Howe Island with the description of a new Nyctophiline bat. *Australian Mammalogy*, Vol. 1, No. 4, pp.329–332.

McKenna, M.C. and Bell, S.K. 1997. *Classification of Mammals above the Species Level*. Columbia University Press, New York. 631pp.

McKenna, M.C., Robinson, P. and Taylor, D.W. 1962. Notes on Eocene Mammalia and Mollusca from Tabernacle Butte, Wyoming. *American Museum Novitates*, No. 2102, pp.1–33.

McKenzie, N.L. 1995. Beccari's freetail-bat *Mormopterus becarrii*. Pp.480–481. In Strahan, R. (ed). *The Mammals of Australia*. Australian Museum and Reed Books, Sydney. 756pp.

McKenzie, N.L. and Rolfe, J.K. 1986. Structure of bat guilds in the Kimberley mangroves, Australia. *Journal of Animal Ecology*, Vol. 55, pp.401–420.

Mein, P. and Ginsburg, L. 1997. Les mammifères du gisement miocène inférieur de Li Mae Long, Thaïland: systématique, biostratigraphie et paléoenvironnement. *Geodiversitas*, Vol. 19, pp.783–844.

Menu, H., Hand, S. J. and Sigé, B. 2002. Oldest Australian vespertilionid (Microchiroptera) from the Miocene of Riversleigh. *Alcheringa*, Vol. 26, pp.319–331.

Menzies, J.I. 1977. Fossil and subfossil fruit bats from the mountains of New Guinea *Australian Journal of Zoology*, Vol. 25, pp.329–336.

Meschinelli, L. 1903. Un nouvo chiropterro fossile (*Archaeopteropus transiens* Mesch.) delle ligniti de Monteviale. *Atti Reale Istituto Venteto Scienze*, Vol. 62, pp.1329–1344.

Miller, G.S.J. 1907. The families and genera of bats. *Bulletin of the United States National Museum*, Vol. 57, pp.1–282.

Mindell, D.P., Dick, C.W. and Baker, R.J. 1991. Phylogenetic relationships among megabats, microbats, and primates. *Proceedings of the National Academy of Sciences of the United States of America*, Vol. 88, pp.10322–10326.

Miyamoto, M.M. 1996. A congruence study of molecular and morphological data for eutherian mammals. *Molecular and Phylogenetic Evolution*, Vol. 6, pp.373–390.

Miyamoto, M.M., Porter, C. and Goodman, M. 2000. *c-Myc* gene sequences and the phylogeny of bats and other eutherian mammals. *Systematic Biology*, Vol. 49, pp.501–514.

Molnar, R.E., Hall, L.S. and Mahoney, J.H. 1984. New fossil localities for *Macroderma* Miller, 1906 (Chiroptera: Megadermatidae) in New South Wales and its past and present distribution in Australia. *Australian Mammalogy*, Vol. 7, pp.63–73.

Murphy, W.J., Eizirlirk, E., Johnson, W.E., Zhang, Y.P., Ryder, O.A. and O'Brien, S.J. 2001. Molecular phylogenetics and the origins of placental mammals.

Nature, Vol. 409, pp.614–618.

Norberg, U.M. 1989. Ecological determinants of bat wing shape and echolocation call structure with implications for some fossil bats. Pp.197–211. In Hanak, V., Horacek, T. and Gaisler, J. (eds). *European Bat Research 1987*. Charles University Press, Prague.

Novacek, M.J. 1992. Fossils as critical data for phylogeny. Pp.46–88. In Novacek, M.J. and Wheeler, Q.D. (eds). *Extinction and Phylogeny*. Columbia University Press, New York.

Novacek, M.J. 1994. Morphological and molecular inroads to phylogeny. Pp.85–131. In Rose, M.D. and Lauder, G.V. (eds). *Interpreting the Hierarchy of Nature: from Systematic Patterns to Evolutionary Process Theories*. Academic Press, New York.

Pascual, R., Archer, M., Ortiz Jaureguizar, E., Prado, J.L., Godthelp, H. and Hand, S.J. 1992. First discovery of monotremes in South America. *Nature*, Vol. 356, pp.704–706.

Paula Couto, C. de. 1956. Une chauvre-souris fossile des argiles feuilletées pléistocènes de Tremembé, Etat de Sao Paulo (Brésil). *Actes 4eme Congres International Quaternaire,* 1953, pp.343–347.

Paula Couto, C. de and Mezzalira, S. 1971. Nova conceituacao geochronologica de Tremembé, Estado de Sao Paulo, Brasil. *Academia Brasileira Ciencias, Anais, Supplement* 43, pp.473–488.

Pettigrew, J.D. 1986. Flying primates? Megabats have the advanced pathway from eye to mid-brain. *Science*, Vol. 231, pp.1304–1306.

Pettigrew, J.D. 1991a. Wings or brain? Convergent evolution in the origins of bats. *Systematic Zoology*, Vol. 40, pp.199–216.

Pettigrew, J.D. 1991b. A fruitful, wrong hypothesis? Response to Baker, Novacek and Simmons. *Systematic Zoology*, Vol. 40, pp.231–239.

Pettigrew, J.D. 1994. Flying DNA. *Current Biology*, Vol. 4, pp.277–283.

Pettigrew, J.D. 1995. Flying primates: crashed or crashed through? Pp.3–26. In Racey, P.A. and Swift, S.M. (eds). *Ecology, Evolution and Behaviour of Bats*. Clarendon Press, Oxford. 421pp.

Pettigrew, J.D. and Jamieson, B.G.M. 1987. Are flying foxes (Chiroptera: Pteropodidae) really primates? *Australian Mammalogy*, Vol. 10, pp.119–124.

Pettigrew, J.D., Jamieson, B.G.M., Robson, S.K., Hall, L.S., McAnally, K.I. and Cooper, H.M. 1989. Phylogenetic relations between microbats, megabats and primates (Mammalia: Chiroptera and Primates). *Philosophical Transactions of the Royal Society of London,* Series B, Vol. 325, pp.489–559.

Pettigrew, J.D. and Kirsch, J.A.W. 1995. Flying primates revisited: DNA hybridization with fractionated, GC-enriched DNA. *South African Journal of Science*, Vol. 91, pp.477–482.

Pierson, E.D. 1986. *Molecular systematics of the Microchiroptera: higher taxon relationships and biogeography*. PhD Thesis (unpublished),University of California, Berkeley, California.

Pierson, E.D., Sarich, V.M., Lowenstein, J.M., Daniel, M.J. and Rainey, W.E. 1986. A molecular link between the bats of New Zealand and South America. *Nature*, Vol. 232, pp.60–63.

Rayner, J.M.V. 1991. Complexity and a coupled system: flight, echolocation and evolution in bats. Pp.173–190. In Schmidt-Kittler, N. and Vogel, K. (eds). *Constructional Morphology and Evolution*. Springer-Verlag, Berlin. 409pp.

Remy, J.A., Crochet, J.-Y., Sigé, B., Sudre, J., de Bonis, L., Vianey-Liaud, M., Godinot, M., Hartenberger, J.-L., Lang-Badre, B. and Comte, B. 1987. Biochronologie des phosphorites de Quercy: Mise à jour des listes fauniques et nouveaux gisements de mammifères fossiles. *Münchner Geowissenschaftliche Abhandlungen A*, Vol. 10, pp.169–188.

Revilliod, P. 1917a. Fledermäuse aus der Braunkohle von Messel bei Darmstadt. *Abhandlungen der Grossherzoglich Hessischen Geologischen Landesanstalt zu Darmstadt*, Vol. 7, No. 2, pp.158–196.

Revilliod, P. 1917b. Contribution à l'étude des chiroptères des terrains tertiaires, première partie. *Mémoires de la Société Paléontologique Suisse*, Vol. 43, pp.1–58.

Revilliod, P. 1919. L'état actuel de nos connaissances sur les chiroptères fossiles. (Note preliminaire.) *Comptes Rendus de la Societé Historie Naturelle Genève*, Vol. 36, pp.93–96.

Revilliod, P. 1920. Contribution à l'étude des chiroptères des terrains tertiaires, deuxieme partie. *Mémoires de la Société Paléontologique Suisse*, Vol. 44, pp.63–129.

Revilliod, P. 1922. Contribution à l'étude des chiroptères des terrains tertiaires, troisème partie. *Mémoires de la Société Paléontologique Suisse*, Vol. 45, pp.133–195.

Russell, D.E. and Gingerich, P.D. 1981. Lipophyla, Proteutheria(?), and Chiroptera (Mammalia) from the early–middle Eocene Kuldana Formation of Kohat (Pakistan). *Contribution to the Museum of Paleontology, University of Michigan*, Vol. 25, pp.277–287.

Russell, D.E., Hartenberger, J.-L., Pomerol, Ch., Sen, S., Schmidt-Kittler, N. and Vianey-Liaud, M. 1982. Mammals and stratigraphy: the Paleogene of Europe. *Palaeovertebrata, Mémoir Extraordinaire*, pp.1–77.

Russell, D.E., Louis, P. and Savage, D.E. 1973. Chiroptera and Dermoptera of the French early Eocene. *University of California Publications in Geological Science*, Vol. 95, pp.1–55.

Russell, D.E. and Sigé, B. 1970. Révision des chiroptères lutétiens de Messel (Hesse, Allemange). *Palaeovertebrata*, Vol. 3, pp.83–182.

Savage, D.E. and Russell, D.E. 1983. *Mammalian Paleofaunas of the World*. Addison-Wesley Publishing, Reading, Massachusetts. 432pp.

Sigé, B. 1968. Les chiroptères du Miocène inférieur de Bouzigues. I. Étude systématique. *Palaeovertebrata*, Vol. 1, pp.65–133.

Sigé, B. 1976. Les Megadermatidae (Chiroptera, Mammalia) miocenes de Beni Mellal, Maroc. *Géologie Méditerranéenne*, Vol. 1, pp.71–86

Sigé, B. 1977. Les insectivores and chiroptères du Paléogène moyen d'Europe dans l'histoire des faunes de mammifères sur ce continent. *Jurij A. Orlov Memorial, Journal of the Palaeontological Society of India*, Vol. 20, pp.178–190.

Sigé, B. 1990. Nouveaux chiroptères de l'Oligocene moyen des phosphorites du Quercy, France. *Comptes Rendus de l'Academie des Sciences, Paris*, Vol. 310, pp.1131–1137.

Sigé, B. 1991a. Rhinolophoidea et Vespertilionoidea (Chiroptera) du Chambi (Eocène inférieur de Tunisie).

Aspects biostratigraphique, biogéographique and paléoécologique de l'origine des chiroptères modernes. *Neues Jahrbuch für Geologie und Palaöntologie Abhandlungen*, Vol. 182, pp.355–376.

Sigé, B. 1991b. Morphologie dentaire lactéale d'un chiroptère de l'Éocène inférieur-moyen d'Europe. *Geobios*, Vol. 13, pp.231–236.

Sigé, B. and Aguilar, J.-P. 1987. L'extension stratigraphique des megachiropteres dans le Miocene d'Europe meridionale. *Comptes Rendus de l'Academie des Sciences, Paris*, Vol. 304, pp.469–474.

Sigé, B., Hand, S.J. and Archer, M. 1982. An Australian Miocene *Brachipposideros* (Mammalia, Chiroptera) related to Miocene representatives from France. *Palaeovertebrata*, Vol. 12, pp.149–171.

Sigé, B. and Legendre, S. 1983. L'histoire des peuplements de chiropteres du bassin Méditerranéen: l'apport comparé des remplissages karstiques et des dépots fluvio-lacustres. *Mémoires Biospeléologie*, Vol. 10, pp.209–225.

Sigé, B. and Russell, D.E. 1980. Complements sure les chiroptères de l'Eocène moyen d'Europe. Les genres *Palaeochiropteryx* et *Cecilionycteris*. *Palaeovertebrata, Mémorial Jubilee of R. Lavocat*, pp.91–126.

Sigé, B., Thomas, H., Sen, S., Gheerbrandt, E., Roger, J. and Al-Sulaimani, Z. 1994. Les chiroptères de Taqah (Oligocène inférieur, Sultanat d'Oman). Premier inventaire systématique. *Münchner Geowissenschaftliche Abhandlungen A*, Vol. 26, pp.35–48.

Simmons, N.B. 1993. The importance of methods: archontan phylogeny and cladistic analysis of morphological data. Pp.1–61. In MacPhee, R.D.E. (ed). *Primates and Their Relatives in Phylogenetic Perspective*. Plenum Press, New York.

Simmons, N.B. 1994. The case for chiropteran monophyly. *American Museum Novitates*, No. 3103, pp.1–54.

Simmons, N.B. 1995. Bat relations and the origin of flight. Pp.27–43. In Racey, P.A. and Swift, S.M. (eds). *Ecology, Evolution and Behaviour of Bats*. Clarendon Press, Oxford. 421pp.

Simmons, N.B. 1996. Bat diversification and palaeobiogeography: implications of a new higher-level phylogeny. *Journal of Vertebrate Paleontology*, Vol. 16, p.66A. (abstract only)

Simmons, N.B. 1998. A reappraisal of interfamilial relationships of bats. In Kunz, T.H. and Racey, P.A. (eds). *Bat Biology and Conservation*. Smithsonian Institution Press, Washington, D.C. 365pp.

Simmons, N.B. and Geisler, J.H. 1998. Phylogenetic relationships of *Icaronycteris*, *Archaeonycteris*, *Hassianycteris* and *Palaeochiropteryx* to extant bat lineages, with comments on the evolution of echolocation and foraging strategies in Microchiroptera. *Bulletin of the American Museum of Natural History*, Vol. 235, pp.1–182.

Simpson, G.G. 1945. The principles of classification and a classification of mammals. *Bulletin of the American Museum of Natural History*, Vol. 85, pp.1–350.

Simpson, G.G. 1961. Historical zoogeography of Australian mammals. *Evolution*, Vol. 15, pp.431–446.

Smith, J.D. 1976. Chiropteran evolution. Pp.49–69. In Baker, R.J., Jones, J.K. and Carter, D.C. (eds). *Biology of Bats of the New World Family Phyllostomatidae, Part I*. Special Publication of the Museum, Texas Tech University, Lubbock. Texas.

Smith, J.D. 1977. Comments on flight and the evolution of bats. Pp.427–438. In Hecht, M.K., Goody, P.C. and Hecht, B.M. (eds). *Major Patterns in Vertebrate Evolution*. NATO Advanced Study Institutes. Series A, Life Sciences, Vol. 14. Plenum Press, New York. 908pp.

Smith, J.D. and Madkour, G. 1980. Penial morphology and the question of chiropteran phylogeny. Pp.347–365. In Wilson, D.E. and Gardner, A.L. (eds). *Proceedings of the Fifth International Bat Research Conference*. Texas Tech Press, Lubbock, Texas

Smith, J.D. and Storch, G. 1981. New middle Eocene bats from the 'Grube Messel' near Darmstadt, W-Germany (Mammalia: Chiroptera). *Senckenberg Biologie*, Vol. 61, pp.153–168.

Smith, R. and Russell, D.E. 1992. Mammifères (Marsupialia, Chiroptera) de l'Yprésian de la Belgique. *Bulletin Institut Royal des Sciences Naturelles de Belgique, Sciences de la Terre*, Vol. 62, pp.223–227.

Springer, M.S., Teeling, E.C., Madsen, O., Stanhope, M.J. and deJong, W.W. 2001. Integrated fossil and molecular data reconstruct bat echolocation. *Proceedings of the National Academy of Sciences of the United States of America*, Vol. 98, pp.6241–6246.

Stanhope, M.J., Czelusniak, Si, J.-S., Nickerson, J. and Goodman, M. 1992. A molecular perspective on mammalian evolution from the gene encoding interphotoreceptor retinoid binding protein, with convincing evidence for bat monophyly. *Molecules, Phylogeny and Evolution*, Vol 1, pp.148–160.

Strahan, R. (ed). 1995. *The Mammals of Australia*. Australian Museum and Reed Books, Sydney. 756pp.

Tate, G.H.H. 1946. Geographical distribution of bats in the Australasian archipelago. *American Museum Novitates*, No. 1323, pp.1–21.

Teeling, E.C., Madsen, O., Van Den Bussche, R.A., de Jong, W.W., Stanhope, M.J. and Springer, M.S. 2002. Microbat paraphyly and the convergent evolution of a key innovation in Old World rhinolophid microbats. *Proceedings of the National Academy of Sciences of the United States of America*, Vol. 99, pp.1431–1436.

Teeling, E.C., Scally, M., Kao, D.J., Romagnoli, M.L., Springer, M.S. and Stahn, M.J. 2000. Molecular evidence regarding the origin of echolocation and flight in bats. *Nature*, Vol. 403, pp.188–192.

Tong, Y. 1997. Middle Eocene small mammals from Liguanqiao Basin of Henan Province and Yuanqu Basin of Shanxi Province, central China. *Palaeontologia Sinica*, New Series C, 26, No. 18, pp.189–253.

Truswell, E.M., Kershaw, A.P. and Sluiter, I.R. 1987. The Australian South-east Asian connection: evidence from the palaeobotanical record. Pp.32–49. In Whitmore, T.C. (ed). *Biogeographical Evolution of the Malay Archipelago*. Clarendon Press, Oxford. 147pp.

Van Den Bussche, R.A., Hoofer, S.R. and Hansen, E.W. 2002. Characterization and phylogenetic utility of the mammalian protamine P1 gene. *Molecular Phylogenetics and Evolution*, Vol. 22, pp.333–341.

van Oosterzee, P. 2006. Drawing the Wallace Line. Pp.95–110. In Merrick, J. R., Archer, M., Hickey, G.M. and Lee, M.S.Y. (eds). *Evolution and Biogeography of Australasian Vertebrates,* Auscipub Pty. Ltd., Sydney. 942pp.

Van Valen, L. 1979. The evolution of bats. *Evolutionary*

Theory, Vol. 4, pp.104–121.

Vaughan, T.A. 1970a. The skeletal system. Pp.97–138. In Wimsatt, W.A. (ed). *Biology of Bats*. Vol. 1. Academic Press, New York. 406pp.

Vaughan, T.A. 1970b. The muscular system. Pp.139–194. In Wimsatt, W.A. (ed). *Biology of Bats*. Vol. 1. Academic Press, New York. 406pp.

Veevers, J.J. 1991. Phanerozoic Australia in the changing configuration of Proto-Pangea through Gondwanaland and Pangea to the present dispersed continents. *Australasian Systematics and Botany,* Vol. 4, pp.1–9.

Volleth, M. and Tidemann, C.R. 1991. The origin of the Australian Vespertilionine bats, as indicated by chromosomal studies. *Zeitschrift für Saeugetierkunde*, Vol. 56, pp.321–330.

Wakefield, N.A. 1972. Palaeoecology of fossil mammal assemblages from some Australian caves. *Proceedings of the Royal Society of Victoria*, Vol. 85, pp.1–26.

Weithofer, A. 1887. Zur Kenntnis der fossilen Cheiropteren der französischen Phosphorite. *Sitzungberichte der Königlich Akademie der Wissenschaften, Mathematisch-Naturwissenschafliche Klasse, Wien*, Vol. 96, No. 1, pp.341–360.

White, C. 1967. *Plateau and plain; prehistoric investigations in Arnhem Land, Northern Territory*. PhD Thesis (unpublished), Australian National University, Canberra.

White, M.E. 2006. Environments of the geological past. Pp.17–50. In Merrick, J.R., Archer, M., Hickey, G.M. and Lee, M.S.Y. (eds). *Evolution and Biogeography in Australasian Vertebrates.* Auscipub Pty. Ltd., Sydney. 942pp.

Wible, J.R. and Novacek, M.J. 1988. Cranial evidence for the monophyletic evolution of bats. *American Museum Novitates*, No. 2911, pp.1–19.

Winge, H. 1942. *The Interrelationships of Mammalian Genera. Vol. 1: Monotremata, Marsupialia, Insectivora, Chiroptera, Edentata*. C.A. Reitzels, Copenhagen. [English translation by Deichmann, E. and Allen, G.M].

Woodburne, M.O. and Case, J.A. 1996. Dispersal, vicariance, and the late Cretaceous to early Tertiary land mammal biogeography from South America to Australia. *Journal of Mammalian Evolution*, Vol. 3, pp.121–161.

Woodburne, M.O. and Zinsmeister, W.J. 1984. The first land mammal from Antarctica and its biogeographic implications. *Journal of Paleontology*, Vol. 58, pp.913–948.

Worthy, T. H., Tennyson, A.D., Jones, C. and McNamara, J.A. 2002. A diverse early-Miocene (15–20 Ma) terrestrial fauna from New Zealand reveals snakes and mammals. *IPC2002, Geological Society of Australia Abstracts,* Vol. 68, pp.174–175.

Wroe, S. and Archer, M. 2006. Origins and early radiations of marsupials. Pp.551–574. In Merrick, J.R., Archer, M., Hickey, G.M. and Lee, M.S.Y. (eds). *Evolution and Biogeography of Australasian Vertebrates.* Auscipub Pty. Ltd., Sydney. 942pp.

Yalden, D.W. and Morris, P.A. 1975. *The Lives of Bats.* David and Charles, Newton Abbot, U.K. 247pp.

Chapter 31

Ten Million Years of Rodent Evolution in Australasia: Phylogenetic Evidence and a Speculative Historical Biogeography

K.P. Aplin

National Wildlife Collection, CSIRO Sustainable Ecosystems, G.P.O. Box 284, Canberra, A.C.T. 2601 Australia
E-mail: Ken.Aplin@csiro.au

INTRODUCTION

Rodents arrived in Australasia relatively recently, in the region's long history of mammal evolution, yet today they comprise the single most speciose family of mammals both in New Guinea and in Australia. In the New Guinean region, more than 75 species occupy a wide variety of both terrestrial and arboreal niches, with the greatest diversity present in mid-montane rainforests (Flannery, 1995). In Australia, which had at least 60 native species at the time of European settlement, it was the semi-arid and arid environments that supported the highest rodent diversity (Lee *et al.,* 1981; Watts and Aslin, 1981). Many Australian rodents are now extinct or severely restricted in range, with several still regarded as critically endangered.

All Australian rodents belong to a single family, the Muridae, which has an extensive Old and New World distribution and contains more than 280 genera and almost 1,400 species (Musser and Carleton, 1993). The taxonomy of this group is complex and has been subject to major changes at the subfamilial level (see Carleton and Musser, 1984); but here I will follow the usage of Musser and Carleton (1993), who have broadened the familial concept Muridae to include many of the family-level taxa previously grouped together as Muroidea (Simpson, 1945). Under their classification, all Australian rodents belong to the single subfamily Murinae, which includes a total of more than 125 genera and 550 species distributed across the African, Eurasian and Australasian regions. Other murid subfamilies mentioned in this work are Cricetinae (Eurasia), Cricetomyinae (Africa), Dendromyinae (Africa), Otomyinae (Africa) and Sigmodontinae (Americas).

The main purpose of this chapter is to provide a synthesis of current knowledge of Australasian rodent phylogeny, classification and biogeography—drawing together information from many diverse sources including molecular biology, karyology, gross morphology, micro-anatomy and the fossil record. The phylogenetic evidence and fossil record are also examined, in the context of recent developments in knowledge of the physical history of the region, to produce a new, speculative, biogeographic synthesis for the group. Gaps in knowledge and specific issues for future research are identified.

TAXONOMIC DIVERSITY OF AUSTRALASIAN RODENTS

The wider Australasian region (including islands to the north and east of New Guinea, but not the Moluccan Islands to the west) supports 140 living species of native rodents, currently divided between 35 genera. This represents 25% of all living murine species and almost 30% of all genera. All but four of the Australasian genera are endemic; the exceptions are *Rattus* (found throughout Asia) and *Hydromys*, *Melomys* and *Pogonomelomys*, which all extend to the Moluccan Islands.

Table 1 lists all of the currently recognized genera and species of Australian native rodents. The taxonomy follows that recommended by Musser and Carleton (1993) but with the following variations or additions: (1) the use of *Abeomelomys* for *sevia* (*sensu* Menzies, 1990; returned to *Pogonomelomys* by Musser and Carleton); (2) the inclusion of all Australasian 'true-rats' in the genus *Rattus* (Musser and Carleton employ *Stenomys* for a suite of small-bodied New Guinean species: *niobe, richardsoni, vandeuseni* and *verecundus*); (3) inclusion of *Pseudomys praeconis* as a junior synonym of *P. fieldi* (following Baynes, 1990); (4) subdivision of *Melomys* into four genera: *Mammelomys, Melomys, Paramelomys* and *Protochromys* (after Menzies, 1996); (5) recognition of *Rattus omlichodes* as distinct from *R. richardsoni,* based on Flannery's (1995) observations of sympatry between the two species (he included both in *Stenomys*).

Although various fossil rodents have been described from Australia, only two of these are currently regarded as valid species and both are referred to modern genera; many more fossil species, including a number of distinctive new genera, are awaiting description. Summaries of fossil taxa currently known are provided in Tables 2 and 3.

ECOLOGICAL DIVERSITY OF AUSTRALASIAN RODENTS

The Australasian rodent fauna contains some of the smallest as well as the largest of the world's living murines. At the small end of the scale is a range of mouse-sized forms including both arid-adapted species (e.g. *Pseudomys delicatulus, P. hermansburgensis, Leggadina forresti*) and rainforest-dwelling species (e.g.

Table 1. Informal list of the genera and species of native Australasian murids, based largely on Musser and Carleton (1993), but with alterations based on other recent revisions. Taxa that have become extinct during the period of European settlement are indicated by *.

Genera	Specific name(s)
Abeomelomys	*sevia*
Anisomys	*imitator*
Chiruromys	*forbesi, lamia, vates*
Coccymys	*albidens, ruemmleri*
Conilurus	*albipes*, penicillatus*
Crossomys	*moncktoni*
Hydromys	*chrysogaster, habbema, hussoni, neobrittanicus, shawmayeri*
Hyomys	*dammermani, goliath*
Leggadina	*forresti, lakedownensis*
Leporillus	*apicalis*, conditor*
Leptomys	*elegans, ernstmayeri, signatus*
Lorentzimys	*nouhuysi*
Macruromys	*elegans, major*
Mallomys	*aroaensis, gunung, istapantap, rothschildi*
Mammelomys	*lanosus, rattoides*
Mayermys	*ellermani*
Melomys	*bougainville, burtoni, capensis, cervinipes, frigicola, leucogaster, lutillus, matambuai, rubicola, rufescens*
Mesembriomys	*gouldii, macrurus*
Microhydromys	*musseri, richardsoni*
Neohydromys	*fuscus*
Notomys	*alexis, amplus*, aquilo, cervinus, fuscus, longicaudatus*, macrotis*, mitchellii, mordax**
Parahydromys	*asper*
Paraleptomys	*rufilatus, wilhelmina*
Paramelomys	*gressitti, levipes, lorentzii, mollis, moncktoni, naso, platyops, steini, rubex*
Pogonomelomys	*bruijni, mayeri*
Pogonomys	*championi, loriae, macrourus, sylvestris*
Pseudohydromys	*asper*
Pseudomys	*albocinereus, apodemoides, australis, bolami, chapmani, delicatulus, desertor, fieldi, fumeus, fuscus, glaucus, gouldii, gracilicaudatus, hermannsburgensis, higginsi, johnsoni, laborifex, nanus, novaehollandiae, occidentalis, oralis, patrius, pilligaensis, shortridgei*
Protochromys	*fellowsi*
Rattus	*colletti, fuscipes, giluwensis, jobiensis, leucopus, lutreolus, mordax, niobe, novaeguineae, omlichodes, praetor, richardsoni, sordidus, steini, tunneyi, vandeuseni, verecundus, villosissimus*
Solomys	*ponceleti, salamonis, salebrosus, sapientis*
Uromys (Cyromys)	*imperator*, porculus, rex*
Uromys (Uromys)	*anak, boeadii, caudimaculatus, emmae, hadrourus, neobrittanicus*
Xenuromys	*barbartus*
Xeromys	*myoides*
Zyzomys	*argurus, maini, palatilis, pedunculatus, woodwardi*

Microhydromys spp., *Mayermys ellermani*). The largest species are found in montane forest habitat in New Guinea and the Solomon Islands; these attain body weights (to 2 kg) almost twice those of the largest Australian murines.

The diet of many species remains poorly known. Many are probably omnivorous. Some are distinctly carnivorous, feeding on crustaceans and fishes (e.g. *Hydromys*, *Xeromys*) or small terrestrial vertebrates (e.g. *Leptomys* spp.). Many of the Australian arid-zone species are granivorous, competing in a niche that is otherwise occupied by various birds and insects (Watts and Aslin, 1981; Watts and Kemper, 1989).

Several species are highly aquatic (*Hydromys* spp., *Crossomys moncktoni*). Some of the smallest (*Lorentzimys* spp.) and largest (*Hyomys goliath*, *Mallomys* spp.) species are arboreal and either build perched nests or inhabit tree hollows. Many of the terrestrial species occupy elaborate burrow systems, while others construct large mounds of pebbles (some *Pseudomys* spp.); arid zone species show adaptations for conserving water and energy (see review by Lee *et al.*, 1981).

Reproductive rates are generally low among Australian murines (Lee *et al.*, 1981). But exceptions are found in the *Rattus sordidus* group, where high litter sizes and rapid growth under favourable conditions in the arid zone can result in spectacular plagues or irruptions (Newsome and Corbett, 1975). Similar irruptions of the introduced house mouse (*Mus domesticus*) occur on a fairly regular basis in the major grain-growing regions of southern Australia (Singleton, 1997).

EARLY NOTIONS OF AUSTRALASIAN RODENT PHYLOGENY

Compared with the large amount of attention accorded to Australasia's unusual marsupials and monotremes during the 19th and early 20th centuries, the native rodents attracted little interest. Alston (1876) and Weber (1902) undertook some preliminary work, but Ellerman (1940–41) and Tate (1951) provided the first serious phylogenetic treatments of Australian rodents.

Ellerman (1940–41) was greatly impressed by the morphological diversity among Australasian rodents. He concluded that the group had been in Australia for a very long time, even suggesting that the Muridae might have evolved here, with subsequent dispersal *into* Asia via Sulawesi and the Philippines. From this perspective, he proposed special links between various Australasian and Asian taxa; examples include: between *Leggadina* and *Mus*; between *Melomys* and the Philippine *Apomys*; between *Mallomys* and the Floresian *Papagomys armandvillei*; between *Lorentzimys* and either *Mus* or *Haeromys* of Malaysia and Sulawesi; and between the Australasian 'water-rats' (*Hydromys, Leptomys, Xeromys*) and the Sulawesian and Philippine 'shrew-rats' (*Celaenomys, Chrotomys*). Among the New Guinean endemics, Ellerman singled out *Anisomys* and *Hyomys* as particularly isolated lineages, the former being placed in its own taxonomic 'group', Anisomyes, within the subfamily Murinae; and he placed *Macruromys* on the 'borderline' between his Murinae and Hydromyinae. Many of Ellerman's concepts persisted for some years, including the erroneous view that the *Melomys* group was close to *Rattus* (Tate, 1951; Watts, 1974).

Tate (1951) considered Australasian rodents to be the product of successive waves of migration from Asia. Among his more lasting contributions was his definition of a '*Pseudomys* complex' (including *Conilurus*, *Leggadina*, *Leporillus, Mastacomys, Mesembriomys, Notomys* and *Zyzomys*, and possibly also *Lorentzimys*) based on a suite of craniodental features. Like Ellerman, Tate placed the *Uromys* group (including *Melomys, Pogonomelomys* and *Xenuromys*) close to *Rattus*, and considered both groups to be relatively recent arrivals in the Australasian region. He also postulated links between a number of Australasian and Philippine genera, most notably between *Melomys* and *Apomys*, between *Mallomys* and the Philippine *Crateromys*, and between the Australasian 'Hydromyinae' and the Philippine 'shrew mice'. As noted above, Tate (1951) noted similarities between *Lorentzimys* of New Guinea and members of the *Pseudomys* complex of Australia (most notably *Leggadina*). However, he also saw other similarities between *Lorenztimys* and the Asian genus *Mus*.

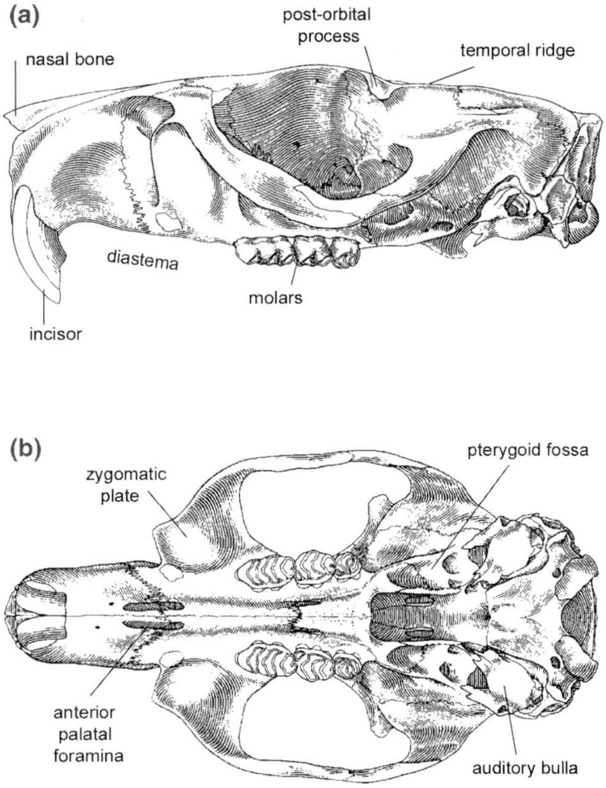

Figure 1. Cranial anatomy of a generalized murine (*Uromys neobrittanicus*), showing some of the major anatomical regions that are used to distinguish between genera and species: **(a)** lateral view; **(b)** ventral view. (Modified from Tate, 1936).

COMPARATIVE ANATOMICAL STUDIES

Several aspects of Australasian rodent anatomy have been examined in recent years, either with or without an explicit cladistic analytical framework. These are reviewed here prior to considering the phylogenetic insights derived from various biomolecular studies.

CRANIAL MORPHOLOGY

Rodents as a whole (and murines in particular) are notoriously conservative in their cranial anatomy (Figure 1). Although most genera and species can be distinguished from each other on cranial measurements, the fundamental arrangement of the bony and muscular elements, and the associated neurovascular grooves and foramina are remarkably constant in all members of the group.

Tate (1951) identified eight craniodental features that he considered diagnostic of his *Pseudomys* group. Subsequently, Martinez (cited by Lidicker and Brylski, 1987) identified one new dental and three new cranial characters common to all members of this group. Neither author considered the phylogenetic significance of these features within a cladistic framework, but seven of the nine cranial features listed by Tate and Martinez relate specifically to features of the masticatory apparatus (zygomatic plate, pterygoid fossae, bony process of the Eustachian canal, coronoid process of dentary). These features may reflect a smaller number of anatomical developments in this highly integrated functional complex.

Musser and his co-workers have identified several cranial features of interest to students of Australasian rodents. A derived carotid arterial pattern, involving reduction in the size of the stapedial artery was reported in four of the 10 genera of 'water-rats' (*Leptomys, Mayermys, Pseudohydromys* and *Xeromys*) by Musser and Heaney (1992). They also reported the same pattern in two '*Melomys*' species (*lanosus* and *rattoides*; now separated as *Mammelomys*) and in *Lorentzimys*, but noted that all other New Guinean taxa possess an ancestral arrangement of the carotid arteries, characterized by a larger stapedial artery. This character clearly represents a useful synapomorphy linking the members of each of the first two groups, but the significance of its occurrence in *Lorentzimys* is unclear.

Another important cranial feature first described by Wahlert (1974) and later surveyed by Musser (1981a) and his co-workers is a bony strut of the alisphenoid bone that separates the masticatory and buccinator rami of the mandibular nerve (Figure 2). This strut is primitively present, causing the nerve rami to exit the cranium via separate foramina, in the great majority of murids including species of all Australasian genera except *Rattus*. It is consistently absent in all *Rattus* species (Australasian and Asian) and also in a suite of closely related genera including *Bunomys* and *Bullimus* of Sulawesi and surrounding islands, *Paulomys* and *Papagomys* of Flores, *Komodomys* of Komodo and Rinca Islands, *Sundamys* of islands on the Sunda Shelf and *Berylmys* of mainland Asia.

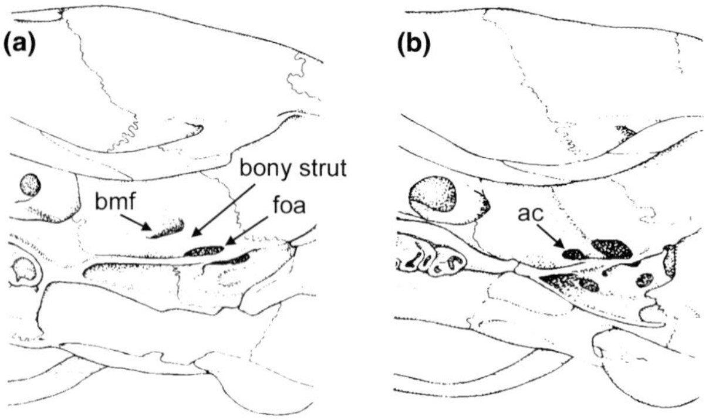

Figure 2. Infratemporal region of the skull of two murine rodents illustrating contrasting conditions of the major foramina. In **(a)**, a bony strut hides the alisphenoid canal and divides the buccinator-masticatory foramen (bmf) from the accessory foramen ovale (foa); this probably represents the primitive condition among murines. In **(b)**, the bony strut is absent, exposing the alisphenoid canal (ac) to view and leaving the foramen ovale undivided. This probably represents a derived condition, and is characteristic of the *Rattus* group of genera.

DENTAL MORPHOLOGY

Murine rodents are also conservative in dental morphology, at least in regard to dental formula and the pattern of tooth eruption and wear. In all genera there is one open-rooted, chisel-like incisor in each quadrant. These are followed by a long diastema and a short row of three (rarely one or two) molar teeth.

Dentally, murines are distinguished from all other muroids by the presence of three cusps in each of the first and second lophs of the first upper molar (see Figure 3). There are several different theories as to the origin of this pattern from one based on twinned cusps. Vanderbroek (1966) proposed that it was the central cusps that had been added, tracing the ancestry of murines back to a sigmodont theridomyid. Other authors (Miller and Gidley, 1918; Petter, 1966; Lavocat, 1967; Misonne, 1969) favoured a process involving the addition of lingual cusps, with the second row cusp (t4) added prior to the first row cusp (t1). This latter scenario is supported by the now much-improved fossil record of the earliest murines and their close relatives (Jacobs, 1977).

All murines have closed-rooted molars that often suffer extreme wear leading to near complete removal of the crown. Despite the strong selective pressure to prolong the life of the cheekteeth (and presumably, the life of the animal), no member of this subfamily seems to have ever developed continuously-growing, open-rooted molars. This is surprising in view of the prevalence of this condition in various other groups of muroid rodents.

Despite their conservatism in growth form, the molars of murines show considerable diversity in the details of cusp size, positions and connections (Figure 4 shows some of the variation found among Australasian representatives of the group). Much of this variation is fairly obviously linked to differences in dietary adaptations, such as granivory, frugivory, insectivory and piscivory.

Despite the obviously strong influence of diet on molar structure among murines, there have been several attempts to use molar characters for phylogenetic analysis. One such 'brave attempt' was made by Misonne (1969), who presented a comprehensive analysis of trends in molar evolution among extant and fossil murines. He recognized four taxonomic 'Divisions', three of which were represented in the Australasian region. His *Lenothrix-Parapodemus* Division includes many genera with 'primitive' molar characteristics such as relatively large second and third molars, non-overlapping upper molars, and well-defined, discrete cusps and a large posterior cingulum on both upper and lower molars.

The New Guinean murine fauna contains several members of Misonne's *Lenothrix-Parapodemus* Division, among them *Anisomys, Hyomys*, *Mallomys* and *Pogonomys* (including *Chiruromys*), with *Pogonomys* and *Anisomys* identified as especially primitive forms. At the comparative level, Misonne (1969) noted special similarities between *Mallomys* from New Guinea and *Papagomys armandvillei* from Flores (these taxa were regarded as congeneric by Ellerman but were separated by Tate), and he included *Hyomys* in the same 'series' as these taxa.

Misonne (1969) considered Tate's '*Pseudomys* complex' to be a regional derivative of the *Lenothrix* group and renamed it the '*Mesembriomys* series'. Of these taxa, he regarded *Mesembriomys* as having the least-modified molar pattern, with *Conilurus* only slightly more advanced. *Leporillus* was linked specifically to *Mesembriomys,* while *Zyzomys* was viewed as an 'offshoot from a *Conilurus*-type'. *Pseudomys* and *Notomys* were identified as closely allied, while '*Leggadina*' (erroneously based on *P. delicatulus* and *P. hermannsburgensis*) was placed close to *Pseudomys*. Misonne labelled the vole-like *Mastacomys* an 'aberrant genus' that had taken a very different evolutionary direction to other members of this group. *Lorentzimys* was also included in the *Mesembriomys* series, although Misonne noted certain archaic dental characters otherwise restricted to other genera (*Pogonomys*, *Carpomys* and *Batomys*).

Misonne's 'Muridae with basin-shaped molars', corresponds closely to earlier concepts of Hydromyinae (Weber, 1902; Ellerman, 1940–41) and thus includes both Australasian and Philippine taxa. *Macruromys* from New Guinea, along with *Chronomys* and *Echiothrix* from Sulawesi, were tentatively associated with this group on the basis of incipient development of basined molars.

Misonne (1969) placed all remaining Australasian rodents in his *Rattus* Division. This complex series was further divided into various 'groups', two of them of special relevance here. His *Uromys* group consisted of five Australasian genera (*Melomys, Pogonomelomys, Solomys, Uromys* and *Xenuromys*) along with the Philippine *Apomys*. Inclusion of these taxa within the *Rattus* Division was based principally on the degree of molar simplification and interdental overlap, perhaps bolstered by earlier suggestions of close ties between *Melomys* and *Rattus* (Ellerman, 1940–41; Tate, 1951). Misonne (1969) also noted special similarity in molar structure between *Uromys*, *Solomys* and *Pogonomelomys.*

Misonne's *Rattus* group included a broadly-conceived genus *Rattus*, divided into four subgenera. The subgenus *Rattus* included members of the *R. sordidus*-group, *R. tunneyi* and *R. lutreolus,* while *Stenomys* consisted of *fuscipes* (as *assimilis-greyi*), *leucopus, niobe, steini* (as *ruber*), *verecundus* and possibly *richardsoni*. Within *Stenomys*, *fuscipes* was stated to have the most conservative dental pattern, while *leucopus* was regarded as more advanced.

Crabb (1977) reviewed dental trends among Australian rodents and recognized the following five groups within *Pseudomys* based on molar morphology: (1) *P. australis, P. 'gouldii'*, *P. higginsi, P. fumeus, P. oralis* and *P. praeconis* (= *P. fieldi*); (2) *P. shortridgei* and *P. desertor*; (3) *P. novaehollandiae, P. albocinereus* and *P. occidentalis*; (4) *P. gracilicaudatus* and *P. nanus*; (5) *P. hermannsburgensis* and *P. delicatulus.* Crabb associated *Leporillus* and *Mastacomys* with his *Pseudomys* Group 1, and *Notomys* with *Pseudomys* Groups 4 and 5; but as he did not clearly distinguish ancestral from derived characters in the analysis, it is unclear which of these groups, if any, Crabb regarded as a monophyletic unit.

Musser's extensive studies of craniodental features of Indo-Malayan murines have done much to clarify generic boundaries and intergeneric affinities within this complex assemblage (Musser, 1981a, b, 1982a, b; Musser and Newcomb, 1983; Musser and Heaney, 1992). From an Australasian perspective, the most significant result has perhaps been the refutation of various supposed links between Australian and south-east Asian genera. In each case, Musser has argued convincingly that the dental similarities are the result of parallel evolution, rather than true shared ancestry.

Musser and his co-workers have also done much to clarify the generic limits of *Rattus*. In the process, they have identified numerous other generic groups, some of which bear only a superficial resemblance to the true

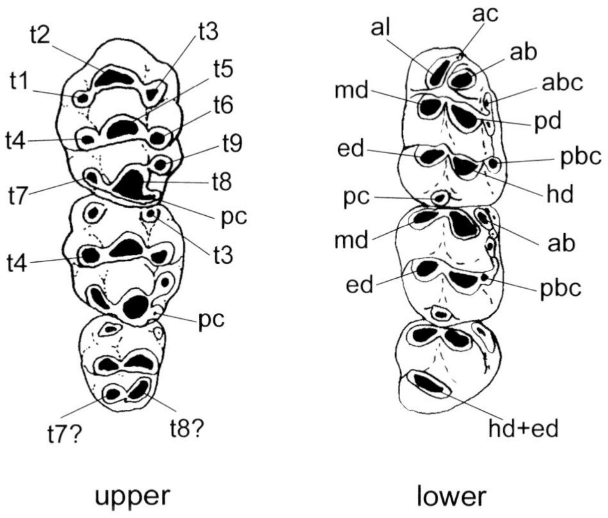

Figure 3. Diagram of the left upper and right lower molars of a generalized murine *(Pogonomys* sp.) illustrating the terminology of major cusps (slightly modified from the scheme used by Musser, 1981a). The first molar is the largest tooth in each series. Key to abbreviations:
Upper molars: major cusps are simply numbered across the laminae and are referred to with the prefix t; pc = posterior cingulum. Cusps t1, t2, and t3 comprise the 'anterocone'. The identity of the cusps making up the rear of the third molar is uncertain.
Lower molars: ab = anterobuccal cuspid; abc = anterobuccal cuspule; ac = anterocentral cuspid; al = anterolingual cuspid; ed = entoconid; hd = hypoconid; md = metaconid; pbc = posterobuccal cuspule; pc = posterior cingulid; pd = protoconid.

'rats'(e.g., *Maxomys*, *Niviventer*). However, they have also identified a suite of genera that appear to be genuinely close to *Rattus*. This group includes *Berylmys* of Indochina; *Kadarsanomys, Palawanomys* and *Sundamys* of the Sunda Shelf; *Komodomys, Papagomys* and *Paulomys* of the Nusa Tenggarran islands; *Abditomys, Bullimus, Limnomys, Tarsomys* and *Tryphomys* of the Philippines; and *Eropeplus, Lenomys, Paruromys* and *Taeromys* of Sulawesi. Members of this group share a number of dental features, including reduction of the posterior molars, increased level of upper molar overlap and reduction or loss of the posterior cingulum of the upper molars. However, unlike the good evidence for monophyly of the *Rattus* Group from cranial morphology, there is no single feature in the dentition to define the same group.

One issue that has remained unclear is the generic identity of the Australasian '*Rattus*'. In earlier works, Musser (1981b) included all of these within the genus *Rattus*. However, more recently he has resurrected the genus *Stenomys* for a group of small-bodied, slender-skulled, small-toothed species including four species from New Guinea (*niobe, richardsoni, vandeuseni* and *vere-*

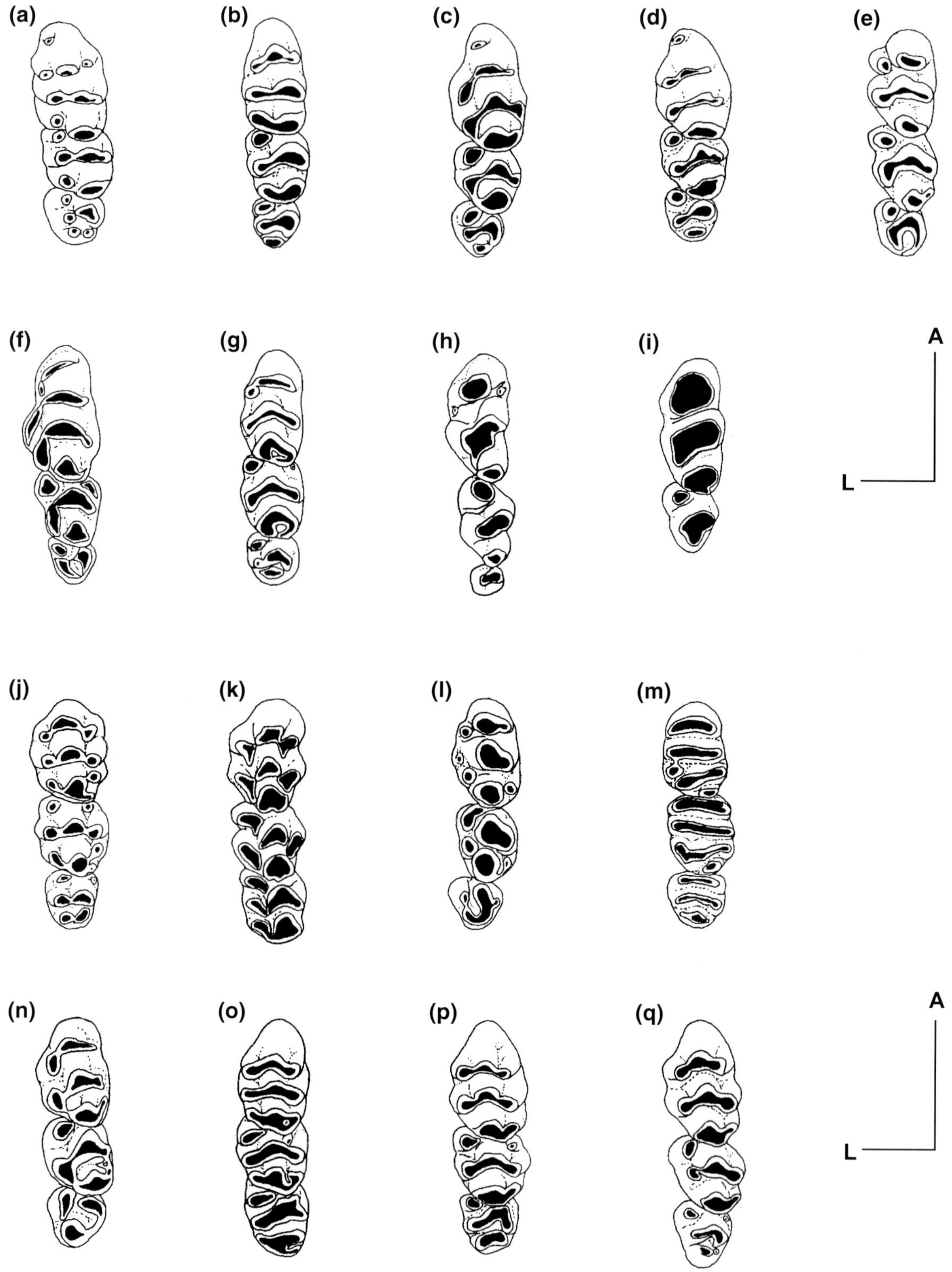

Figure 4. Variation in dental morphology among Australian and New Guinean rodents, as illustrated by a selection of 17 taxa from 16 genera. All specimens are left upper molar rows, subject to variable enlargements. **(a)** *Conilurus albipes*; **(b)** *Zyzomys argurus*; **(c)** *Pseudomys higginsi*; **(d)** *Notomys alexis*; **(e)** *Leporillus apicalis*; **(f)** *Leggadina lakedownensis*; **(g)** *Pogonomelomys mayeri*; **(h)** *Leptomys elegans*; **(i)** *Hydromys chrysogaster*; **(j)** *Pogonomys loriae*; **(k)** *Mallomys* sp.; **(l)** *Macruromys major*; **(m)** *Anisomys imitator*; **(n)** *Lorentzimys nouhuysi*; **(o)** *Hyomys goliath*; **(p)** *Rattus villosissimus*; **(q)** *Rattus leucopus*. All outlines are adapted from illustrations in Misonne (1969) except (f), which is original. Orientation: first molar at top of each series; A = anterior, L = lingual.

cundus) and one from Seram. The other Australasian species were retained provisionally within *Rattus* (Musser and Carleton, 1993). Flannery (1995) has adopted this generic scheme but recognizes one more species, *S. omlichodes* from subalpine habitats in western New Guinea.

EXTERNAL MORPHOLOGY

Several features of external morphology have been emphasized in discussions of Australasian murine evolution. One of these is the variation in number and distribution of teats, which reflects the reproductive parameter of litter size and fecundity. Among murids in general, the most widely observed mammary formula is 1 + 0 + 2 (one pectoral, no abdominal, two inguinal teats) and it is reasonable to regard this as the plesiomorphic condition. This condition is found in five New Guinean genera (*Anisomys*, *Chiruromys*, *Lorentzimys*, *Mallomys* and *Pogonomys*) but not in any Australian taxa. In all Australian murines except the native *Rattus* species, the mammary formula is reduced to 0 + 0 + 2 teats. A comparable reduced pattern is also observed in many New Guinean murines including the various genera of 'water rats', most members of the '*Uromys* group' and *Coccymys*, *Hyomys* and *Macruromys*. In *Mammelomys* a further reduction to 0 + 0 + 1 teats is seen, the lowest teat number in any murid rodent. The mammary formula in native *Rattus* species varies from the probable plesiomorphic arrangement of 1 + 0 + 2 in many of the New Guinean taxa, to a minimum of 0 + 0 + 2 in *R. vandeuseni*, and a maximum of 1 + 2 + 3 in members of the *Rattus sordidus* complex.

The morphology of the tail has been used principally to distinguish the group of 'mosaic-tailed' rats from other Australasian murines. The tail scales in members of this group are hexagonal in shape and range in form from completely flat, as in species of *Melomys*, *Paramelomys*, *Pogonomelomys, Solomys* and *Uromys*, to weakly overlapping as in *Mammelomys* and *Xenuromys*. The tail scales are also hexagonal in *Anisomys imitator*. In most other genera of murines they are ovoid or rectangular in shape, with strongly overlapping margins.

MALE REPRODUCTIVE ANATOMY

Valuable phylogenetic information has been extracted from various aspects of the gross anatomy of the male reproductive tract of myomorph rodents, most notably the glans penis with its supporting bacular apparatus (Hooper, 1958), and the accessory sex glands (Voss and Linzey, 1981). For Australasian rodents, detailed studies of the penis and baculum were undertaken by Lidicker (1968, 1973) and Lidicker and Brylski (1987), with additional information on the glans penis of Australian species supplied by Morrissey and Breed (1982) and Breed (1986). Comparative data on Asian murids are provided by Yang and Fang (1988) and Yang *et al.* (1992). Breed and Sarafis (1979) and Breed (1980, 1986) give information on the accessory sex glands of selected Australian species, but many taxa remain unstudied. The results of a comparable study of representative New Guinean taxa are currently in preparation (Breed, 2001, personal communication).

Lidicker (1968) gave detailed descriptions of the glans penis of 28 species of New Guinean rodents representing 19 genera, while Lidicker and Brylski (1987) made comparable observations on 19 Australian species representing eight genera. In both studies, a major distinction was drawn between *Rattus,* which was inferred to be a recent arrival in the region, and all other genera which were stated to share a common penile groundplan.

Lidicker's (1968) interpretation of penile evolution among New Guinean rodents shows *Anisomys* as an ancestral form. However, it is clear that this placement is based not on any independent analysis of penile characters, but rather on the presence of various primitive features in the skull and teeth of *Anisomys* (Rümmler, 1938; Ellerman, 1940–41). For this reason, Lidicker's phylogenetic tree is perhaps best treated as an unrooted network in which character polarities are unknown.

Lidicker and Brylski (1987) also peformed computer-based phenetic and cladistic analyses of a combined dataset for many Australian and a selection of New Guinean rodents; for the cladistic analysis, *Rattus* was used as an outgroup. Both methods produced broadly consistent trees in which the New Guinean taxa (including *Anisomys*) are embedded *within* the Australian assemblage which divides into two major groups, one comprising *Conilurus*, *Leporillus*, '*Mastacomys*', *Mesembriomys* and part of *Pseudomys*, and the other, *Notomys, Leggadina* and the remainder of *Pseudomys*. *Zyzomys* grouped with the first group in the cladistic analysis but with the second group in the phenetic tree. Despite this complex result, Lidicker and Brylski (1987) argued for monophyly of the Australian assemblage, explaining the inconsistent tree topology in terms of the relatively derived condition of the glans penis of *Rattus* and the presence of some of the same derivations in *Notomys* and *Leggadina.* This possibility requires further investigation with the use of appropriate outgroups; for the present, it may be more productive to examine the evolutionary polarity of some individual characters used in the analyses.

Hooper's studies of penile morphology in many different groups of myomorphs provide an appropriate foundation for assessing overall evolutionary trends (Hooper, 1959, 1960, 1962; Hooper and Musser, 1964). Hooper and Musser (1964) identified a 'complex' glans penis as the ancestral state for myomorph rodents. This includes a four-part baculum (basal element, 3 distal 'digits') and a replacement of the spongy, vascular layer by pairs of sac-like sinuses. Murines have simplified this complex pattern by the loss of the lateral bacular elements so that the lateral bacular mounds are supported by vascular tissues rather than by cartilage or bone. However, they have also made it more elaborate through the development of a second crater enclosing the urinary meatus. The distal baculum in most murid groups (including most Australasian taxa) comprises about 30–40% of the combined length. Other details of ornamentation are too variable for identification of likely ancestral states at this level.

The glans penis of *Anisomys* conforms quite well to this generalized murine pattern, as do the glandes of *Crossomys, Hydromys, Leptomys, Lorentzimys, Mallomys, Mayermys, Melomys, Neohydromys, Parahydromys, Paraleptomys, Pogonomelomys, Pogonomys, Pseudohydromys* and *Uromys* among the New Guinean genera (Lidicker, 1968), as well as *Conilurus, Leporillus, Mesembriomys, Zyzomys* and some *Pseudomys* species (*P. australis, P. fuscus, P. higginsi* and *P. nanus*) among the Australian genera (Lidicker and Brylski, 1987). Substantially different morphologies are found: in *Hyomys,* where the inner and outer craters are both very low and the distal baculum is relatively short; in *Coccymys*, where the outer crater is very high and both bacular elements are massively broadened; in *Notomys,* where the distal baculum is very short and the outer crater is moderately low; and in *Leggadina* and various *Pseudomys* species (*P. albocinereus, P. apodemoides, P. delicatulus, P. fumeus, P. hermannsburgensis* and *P. shortridgei*) where both inner and outer craters are moderately to very low and the distal baculum is usually shortened.

Other modifications that may be important indicators of phyletic groups include: the presence of internal papillae on the outer crater in the majority of the 'water rats' (*Crossomys, Hydromys, Leptomys, Parahydromys, Pseudohydromys, Mayermys* and *Neohydromys*); the low and very thin nature of the inner crater, coupled with moderately to very elongate lateral bacular processes in *Mallomys, Pogonomys* and *Lorentzimys* (the last genus differs from all others in having a straight shaft to the proximal baculum); and the placement of the dorsal papilla—positioned between the inner and outer craters in all Australian taxa and in most members of the *Uromys* group and in *Hyomys* among the New Guinean taxa, but attached to the wall of the outer crater in *Melomys platyops* and all other New Guinean taxa examined.

Morrissey and Breed (1982) and Breed (1986) reported considerable variation in penis shape and in the size and distribution of epidermal spines. In species of *Leggadina, Hydromys* and *Uromys,* and in some *Pseudomys* (*P. australis, P. desertor, P. nanus*), the penis is broader distally than at the base and the entire external surface of the shaft is spinose, often with a decrease in spine size towards the tip. *Melomys* species have a similar-shaped penis but only the distal region of the shaft is spinose. In *Notomys alexis, N. fuscus* and *N. mitchellii* the penis tapers distally and the spines are larger and even in size, while in *N. cervinus* and *N. longicaudatus* the penis is cylindrical and the spines are smaller. *Pseudomys apodemoides, P. delicatulus, P. hermannsburgensis* and *P. novaehollandiae* have a cylindrical glans penis with moderately large spines; in *P. shortridgei* the shaft is cylindrical but the spines are smaller.

The proximal baculum is bulky and broad across the base in *Melomys, Uromys* and *Hydromys*. In *Leggadina forresti, Pseudomys apodemoides, P. hermannsburgensis* and *Notomys cervinus* the baculum is less massive and narrower across the base, while in *P. australis, P. nanus* and other *Notomys* spp. (*alexis, mitchellii*) it is very slender with a narrow to very narrow base.

The accessory sex glands have received less attention. Taylor and Horner (1970, 1972) reported that coagulating glands were absent in various *Melomys* and *Pseudomys* species. This was shown to be incorrect by Breed (1986), who found coagulating glands in all genera examined (*Hydromys, Leggadina, Melomys, Notomys, Pseudomys, Uromys* and *Xeromys,*). In fact, species of *Pseudomys, Leggadina* and *Notomys* possess a full suite of accessory sex glands, although in *Notomys* spp. the seminal vesicle, coagulating, ampullary and preputial glands, as well as the dorso-lateral prostate, may be very much reduced. The ventral prostate in *Notomys* spp. is unusually large, perhaps compensating for the reduction in other glands. In *Melomys, Uromys* and *Hydromys* (and possibly also *Xeromys*), the preputial gland is absent (Breed, 1986).

SPERM MORPHOLOGY

Studies of sperm morphology have provided many important insights into the phylogeny of muroid rodents (Friend, 1936; Linzey and Layne, 1974). Work on Australian species also began early (Illisson, 1971) and progressed when taken up by Breed and his co-workers (Breed and Sarafis, 1979; Breed, 1980, 1983, 1984, 1997; Flaherty and Breed, 1983, 1987).

The majority of Australasian murines can be divided into two main groups based on the form of the sperm head (Breed, 1997); see Figure 5. Members of the first group, including *Rattus* and a suite of New Guinean genera (*Abeomelomys, Anisomys, Coccymys, Hyomys, Lorentzimys* and *Mammelomys*), have a short to elongate sperm head with a single apical hook. Members of the second group (containing most other genera) have a more complex sperm head with two accessory ventral hooks, formed by extension of the subacrosomal space. Outside the Australasian region, this complex type of sperm head has only been observed in one rodent genus, the African murine *Dasymys* (Breed 1991, 1995). As these accessory processes are absent in all other African murines examined (Breed *et al.*, 1988; Breed, 1991, 1995), it seems likely that they have evolved independently in one genus and in the Australasian rodents.

There is some variation in the form and development of accessory hooks. In species of *Hydromys, Melomys, Uromys* and *Zyzomys* the accessory hooks are relatively

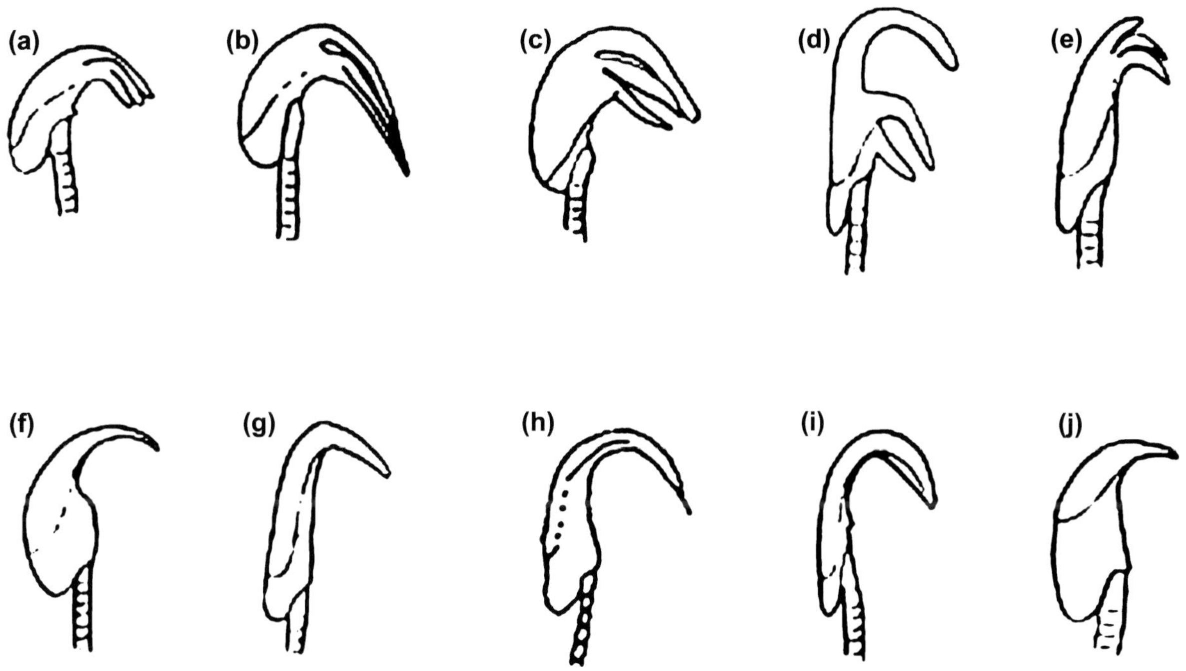

Figure 5. Variation in sperm head morphology among some Australasian and other murine rodents, illustrating the striking 'three-hooked' sperm head of members of the '*Mesembriomys* Group': **(a)** *Hydromys*; **(b)** *Uromys*; **(c)** *Notomys* and similar conditions in **(d)** the New Guinean 'old Endemic' genus *Mallomys;* and **(e)** African genus *Dasymys*. Other Australasian genera shown are: **(f)** *Anisomys*; **(g)** *Coccymys*; **(h)** *Lorentzimys*; **(i)** *Rattus*. In **(j)** the sperm head of the African 'dendromurine' genus *Acomys* is shown. Modified from Breed (1995, 1997).

short. Further reduction leading to complete loss of the hooks is observed in some *Notomys* species (*N. alexis, N. mitchellii, N. fuscus*) and in some *Pseudomys* (*P. delicatulus, P. novaehollandiae, P. pillagaensis, P. shortridgei*). Within *Pseudomys, P. shortridgei* is unique in having a simple, oval-shaped sperm head with the insertion point for the neck situated basally, rather than offset laterally as in the other species.

A broad survey of sperm morphology among New Guinean murines (Breed and Aplin, 1994) documented the presence of accessory hooks in various genera of 'water-rats' (*Leptomys, Mayermys, Neohydromys, Parahydromys, Paraleptomys* and *Pseudohydromys*) and in various species of *Melomys, Paramelomys* and *Uromys*. Accessory hooks were also observed in representatives of three of the 'Old Endemic' genera, namely *Chiruromys*, *Mallomys* and *Pogonomys,* but with considerable variation in both the length and orientation of the hooks between each of these genera. Ultrastructural examination of *Pogonomys* and *Chiruromys* sperm heads by Breed and Aplin (1994) revealed an unusual expansion of the nucleus into the accessory ventral hook in these taxa. This might indicate independent acquisition of hooks in these taxa, or may represent a further, unique development. The inner composition of the accessory ventral hook in *Mallomys* has not been determined.

Breed and Aplin (1994) reported a short, broad sperm head with a single, apical hook in *Anisomys imitator* and *Hyomys* sp., whereas species of *Abeomelomys, Coccymys, Lorentzimys, Mammelomys* and *Xenuromys* have a more elongate, falciform sperm head similar to, but not identical with, that of *Rattus*. In *Solomys salebrosus* from the Solomon Islands, the sperm head is oval-shaped, reminiscent of the condition in *Pseudomys shortridgei*. All New Guinean *Rattus* species examined had the characteristic falciform sperm head common to both Australian and Asian members of this genus.

Comparable surveys of sperm morphology among Malaysian (Breed and Yong, 1986), Indonesian (Breed, 1991), and Sulawesian-Philippine (Breed and Musser, 1991) murines failed to reveal the presence of accessory hooks in any taxon. But the discovery of hooks in the sub-Saharan *Dasymys* (Breed, 1991, 1995) is crucial as it demonstrates that even this very specific character can be subject to multiple derivations, presumably as a consequence of similar mutations and/or in response to similar selective forces.

Roldan *et al.* (1992) have postulated multiple, independent origins of the accessory ventral hooks among Australasian rodents, based largely on the suggestion that accessory hooks have evolved a number of times within the single genus *Pseudomys*. However, this argument in turn rests on the near 'basal' position in a genetic-based phylogeny (Baverstock *et al.*, 1981) of two groups with small or no hooks (i.e. *P. shortridgei* and the *P. delicatulus*-*P. novaehollandiae* group). As indicated below, subsequent immunological studies (Watts *et al.*, 1992) have failed to support this aspect of the allozyme-based phylogeny of *Pseudomys*. Moreover, the ovoid type of sperm head that Roldan *et al.* (1992) regard as ancestral for 'Muridae' is highly erratic in its taxonomic distribution. For example, it occurs in one member of the *Uromys* group (Breed and Aplin, 1994), in one species of *Taeromys* (Breed and Musser, 1991) and the Malaysian 'old endemic' *Pitecheir* sp. (Breed and Yong, 1986). Indeed, the occurrence of this sperm head type is so disjunct that Breed and Aplin (1994) speculated that it had evolved on a number of occasions within otherwise unrelated groups of murids—perhaps as a result of a simple mutation.

CYTOGENETIC, IMMUNOLOGICAL AND MOLECULAR ANALYSES

CHROMOSOMES

Chromosomes are probably more amenable to phylogenetic analysis than almost any other kind of gross or microscopic 'morphological' feature (Qumsiyeh and Baker, 1988). This is because chromosomal rearrangements represent discrete events (Robertsonian fusions/fissions, pericentric inversions, heterochromatic additions/deletions) that occur infrequently and with generally low fixation potential (Lande, 1979). However, for chromosome studies to yield well-resolved, reliable phylogenies, two essential criteria must be met. First, appropriate banding techniques (G- or R-banding; C-banding) must be performed, to identify homologous autosomal components (Baker *et al.*, 1987). Second, chromosomal rearrangements must have been sufficiently common within the group under study to provide good phylogenetic resolution, yet not so common that the history of rearrangements becomes too complex to decipher (Baverstock, 1987).

G- and C-banded karyotypes are available for a large number of Australian rodents including representatives of nearly all genera (Baverstock *et al.*, 1977a, b, 1983a, b). The karyotype of Australian *Rattus* spp. is very different from that of all other Australasian rodents, to the extent that homology cannot be identified for the majority of autosomal components. Among the remaining Australian rodents, rearrangements have not been so severe as to obscure the homology of autosomal components and this enabled Baverstock *et al.* (1983b) to reconstruct what they called the 'ancestral hydromyine' karyotype. This has a diploid number of 2n = 48 with 21 pairs of telocentric autosomes forming a graded series, 2 pairs of small submetacentric autosomes and telocentric sex chromosomes (the X relatively large). Small procentric blocks of heterochromatin occur on all autosomes, especially the smaller ones, and on the X-chromosome. The Y-chromosome is entirely C-positive. This karyotype is retained in an unmodified form in *Melomys burtoni, Hydromys chrysogaster, Mesembriomys gouldii, Conilurus penicillatus* and *Leggadina forresti,* representing each of the three major groups of Australian rodents as identified by other criteria [Tribes Uromyini (*Melomys*), Hydromyini (*Hydromys*) and Conilurini (*Conilurus, Leggadina, Mesembriomys*) *sensu* Watts and Aslin (1981)].

Various minor chromosomal rearrangements provide evidence for interspecific and, in one case, intergeneric relationships. The most important of these is a reciprocal translocation between autosome pairs 1 and 2, which occurs in all species of *Pseudomys* (including *Mastacomys fuscus*; henceforth, *P. fuscus*) and *Notomys*. As argued by Baverstock *et al.* (1983b), reciprocal translocations must arise initially as heterozygotes in which state they are known to cause severe reduction in fertility. Such mutations are thus only likely to persist where local deme size is extremely small (just a few individuals), providing an opportunity for the homozygous state first to be attained and then to proliferate without being swamped by backcrossing with the ancestral karyotype. This particular rearrangement constitutes strong evidence for monophyly of *Pseudomys* and *Notomys* and hints at some kind of dispersal or bottlenecking event in the history of this lineage.

Within the *Pseudomys-Notomys* clade, several smaller clusters of taxa are indicated by shared pericentric inversions. One inversion involving autosomal pair 8 is shared by *Pseudomys australis,* some *P. delicatulus* (Qld. but not N.T.), *P. fuscus*, *P. novaehollandiae* and *P. occidentalis*, while another involving pair 3 is restricted to *P. australis* and *P. fuscus*.

Outside the *Pseudomys-Notomys* group, the most spectacular chromosomal rearrangement occurs in the species of *Zyzomys* [data for *Z. argurus* and *Z. maini*, the latter reported as *woodwardi* by Baverstock *et al.* (1977a, 1983b)]. In these taxa, only a small fraction of the autosomal complement of 21 pairs can be matched to elements of the ancestral karyotype. This suggests either that the genome underwent a major reorganization during the origin of *Zyzomys*, or that *Zyzomys* is not derived from the same stock as the remaining Australian rodents. Unfortunately, no G-band information is available for *Leporillus*. However, based on its conventional karyotype, Baverstock *et al.* (1977a) placed *Leporillus* outside the *Pseudomys-Notomys* clade.

G-banding data are lacking for any New Guinean rodents. However, conventional karyotypes and some C-banding data are available for a significant number of taxa (Dennis and Menzies, 1979; Donnellan, 1989). Donnellan (1989) reported conventional and C-banded karyotypes for 10 species from eight genera of New Guinean murids. Karyotypes of *Mallomys aroaensis* (reported as *M. rothschildi*) and *Parahydromys asper* are indistinguishable at the gross level from the ancestral 2n = 48. *Crossomys moncktoni, Leptomys elegans* and *Uromys caudimaculatus* also have 2n = 48 but the karyotype has undergone various rearrangements to produce additional acrocentric and submetacentric autosomes, and there are more extensive C-bands including interstitial blocks. Lower diploid numbers are found in *Anisomys imitator, Lorentzimys nouhuysi* and *Pogonomys* (confirming earlier observations by Dennis and Menzies, 1979), each of which have 2n = 46. *Chiruromys* spp. have a further reduced karyotype with 2n = 44 (Dennis and Menzies, 1979; Donnellan, 1989), while *Uromys neobrittanicus* has a much lower diploid number of 32. The karyotypes of *Solomys* are reported only as a diploid number; 2n = 48 for *S. salebrosus* and 2n = 44 for *S. ponceleti* (Donnellan, 1994, personal communication). Genera for which the karyotypes are completely unknown include *Abeomelomys, Hyomys, Mammelomys, Paramelomys, Protochromys* and *Xenuromys*.

Donnellan (1989) suggested that the 2n = 46 karyotype shared by *Anisomys, Lorentzimys, Pogonomys* and *Chiruromys* might be derived from the ancestral 2n = 48 by a single, shared Robertsonian rearrangement. Alternatively, the 2n = 46 might itself be a more plesiomorphic karyotype that gave rise to the 2n = 48 pattern through fission. Given the lack of other evidence for special affinity between these taxa (see later sections), this second possibility warrants further consideration.

Musser (1990) and Rickart and Musser (1993) have put forward the notion that the 'hydromyine' karyotype might be ancestral for Murinae as a whole. Their arguments hinge on two observations. The first is that a superficially similar karyotype (predominantly telocentric with one or two small metacentrics) is found in many different genera of Asian and African murines. The second is that *Malacomys longipes*, which has a 2n = 48 karyotype very similar to that of the Australian 'hydromyines', has also been identified (Viegas-Pequignot *et al*., 1983, 1986; Viegas-Pequignot, 1985) as having an essentially plesiomorphic karyotype for Murinae as a whole. Although Musser and Rickart's arguments are unconvincing in themselves, they highlight the fact that direct chromosomal homologies have not been established between the shared ancestral 2n = 48 karyotype of the Australasian 'hydromyine' rodents and the karyotypes of other major groups of murines.

Chromosomal evolution in Australian *Rattus* has been investigated in detail by Baverstock *et al*. (1977b, 1983a), with less extensive work on some New Guinean taxa by Yosida (1973) and Dennis and Menzies (1978). Among Australian species, *R. lutreolus* alone retains the inferred ancestral karyotype (2n = 42), as determined by G-banding homology with non-Australian *Rattus* spp. Minor rearrangements have occurred in the lineages leading to *R. tunneyi* (1 pericentric inversion; 2n = 42) and *R. fuscipes* (2 Robertsonian fusions; 2n = 38), with more extensive rearrangments to *R. leucopus* (3–4 Robertsonian fusions; 2n = 34) and each of *R. sordidus* (5 Robertsonian fusions; 2n = 34), *R. colletti* (5 Robertsonian fusions, 5 fissions; 2n = 42) and *R. villosissimus* (2 Robertsonian fusions, 6 fissions; 2n = 50). The karyotypes of the latter three species have one Robertsonian fusion in common (involving pairs 2 and 4) but otherwise show entirely divergent rearrangements. Rapid chromosomal divergence involving a predominance of Robertsonian rearrangements is suggestive of major fluctuations in population size with reductions to local demes of only a few individuals (Bush *et al*., 1977; Lande, 1979). This is consistent with the known biology of these species, all of which undergo cyclic 'plaguing' followed by virtual disappearance across huge geographic ranges (Taylor and Horner, 1973). *Rattus fuscipes* and typical *R. leucopus* share one Robertsonian fusion in common (involving pairs 5 and 9) but this is lacking in the subspecies *R. l. cooktownensis*.

Conventionally stained karyotypes are available for six New Guinean *Rattus* species. New Guinean *R. sordidus* has 2n = 42 (*sordidus*) or 2n = 44 (*'gestroi'*), while New Guinean *R. leucopus* has 2n = 34 in common with southern populations of this species in Australia (Yosida, 1973; Dennis and Menzies, 1978). *Rattus niobe, R. steini* and *R. verecundus* each have 2n = 32 karyotypes, which are indistinguishable at the gross level. Dennis and Menzies (1978) suggest derivation of the 2n = 32 karyotype from the 2n = 34 karyotype of *R. leucopus* by Robertsonian fusion of pairs 6 and 9. G-banding of various New Guinean *Rattus* species is required to test this hypothesis. The karyotypes of many other New Guinean *Rattus* species are completely unknown.

Yosida (1980) and Gadi and Sharma (1983) describe G-banded karyotypes for various Asian murids. Their studies emphasize the unity of the *Rattus* Group, based on an ancestral 2n = 42 karyotype. This karyotype is retained essentially unmodified in various *Rattus* species and requires only minor modifications to produce the divergent karyotypes observed in *Bandicota, Berylmys, Leopoldamys* and *Sundamys*. Gadi and Sharma (1983) consider the *Rattus* Group karyotype to be derived from a more archaic 2n = 46 pattern that also gave rise to the divergent karyotypes of *Maxomys, Niviventer, Lenothrix, Millardia* and *Cremnomys*.

GENETIC STUDIES

Allozyme electrophoresis is an excellent tool for sibling species recognition and population-level genetics, and it can also provide useful information on overall patterns of relatedness among species and genera. This is usually done by an examination of a genetic distance matrix, but patterns of allele distribution are also amenable to cladistic analysis (Richardson *et al*., 1986).

Baverstock *et al*. (1981) analysed genetic relationships among 38 species of Australian rodents using both distance and numerical cladistic methods. Phenetic analysis of the genetic distance matrix showed a clear separation between *Rattus* and all other Australian genera and also supported a number of the traditional generic groupings, most notably *Melomys, Notomys* and *Zyzomys*. In contrast, the diverse genus *Pseudomys* failed to hold together, with several species showing closer links either to *Notomys* (*P. shortridgei*) or to *Leporillus* (*P. gracilicaudatus* and *P. nanus*). *Pseudomys* (= *Mastacomys*) *fuscus* fell within the wider *Pseudomys* group, linking specifically to *Pseudomys fumeus*, while *Conilurus* spp. and *Mesembriomys gouldii* clustered together to the exclusion of *M. macrurus*. *Uromys caudimaculatus* and *Melomys* spp. formed a fairly tight cluster, which then linked with *Leggadina forresti*. *Hydromys chrysogaster* and *Xeromys myoides* grouped distantly to each other, while *Zyzomys* spp. grouped loosely with *Notomys* spp. and *P. shortridgei*.

Baverstock *et al.*'s (1981) cladistic Wagner Tree presents a different perspective that might, in part, reflect a reduced sensitivity to variations in the rate of genetic divergence. In this analysis, *Pseudomys* is more unified although it included both *Pseudomys fuscus* (associated specifically with *P. shortridgei*) and *Leporillus apicalis* (associated with *P. gracilicaudatus*). *Zyzomys* and *Notomys* were both closely associated with *Pseudomys*, while *Conilurus* and *Mesembriomys gouldii* formed a discrete lineage, as did the species of *Melomys*. *Hydromys* and *Xeromys* clustered together but with *Mesembriomys macrurus* as well. *Uromys* and *Leggadina* occupied distinct branches of the tree, the latter sharing a short common stem with the *Hydromys*-*Xeromys* clade.

In both analyses a number of species-pairs are evident within *Pseudomys*, most notably *P. albocinereus* + *P. apodemoides*, *P. novaehollandiae* + *P. delicatulus* and *P. fieldi* (as *P. praeconis*) + *P. higginsi*. However, at a broader level, there is no clear agreement regarding fundamental divisions within the genus. Independent genetic analysis of a smaller number of taxa by Briscoe *et al.* (1981) supported the association of *P. delicatulus* with *P. novaehollandiae* and added a third species to this cluster—the newly described *P. pillagaensis*. *Pseudomys hermannsburgensis* and *P. australis* were found to be only distantly related to members of this group and to each other.

Among the Australian *Rattus*, *R. colletti* and *R. villosissimus* are genetically near identical, with no fixed allelic differences at 55 loci (Baverstock *et al.*, 1983a, 1986). *Rattus sordidus* is also close to these taxa, with only 2 fixed differences out of 55 loci examined. Of the remaining species, *R. tunneyi* appears to be closest to the *R. sordidus*-group, with *R. fuscipes* and *R. lutreolus* being more isolated from this group and from each other. *Rattus leucopus* groups with the *R. sordidus*-*R. tunneyi* group on both distance and Wagner analyses, but in each case the shared branch lengths are relatively short.

An unpublished pilot electrophoretic study involving a small number of New Guinean *Rattus* species revealed a close genetic relationship between *R. leucopus* and *R. verecundus* (Adams, 2001, personal communication), thereby casting doubt on the validity of the distinction between *Rattus* and *Stenomys* (*sensu* Musser and Carleton, 1993).

ALBUMIN IMMUNOLOGY

Immunological distance methods have provided critical insights into the phylogeny of many groups of vertebrates, including various other groups of Australian mammals (Baverstock *et al.*, 1989, 1990). Studies based on the albumin molecule have proven especially informative for elucidating intergeneric to interfamilial relationships among mammals and have revealed a tendency for a fairly even, 'clock-like' rate of mutational change.

Watts and his co-workers undertook an extensive albumin immunological study of Murinae (Watts *et al.*, 1992; Watts and Baverstock, 1994a, b, 1995a, b), based on the quantitative method of Microcomplement Fixation (MC' F). The results of this research are of great importance to our understanding of Australasian rodent evolution and are reviewed in some detail below.

Among the Australian genera, the major interest concerns the large and complex genus *Pseudomys*. Here, special affinity is demonstrated by low albumin immunological distances (AIDs) of less than 10 between: *P. laborifex* and *P. johnsoni*; *P. desertor* and *P. shortridgei*; *P. albocinereus* and *P. apodemoides*; *P. hermannsburgensis*, *P. bolami*, *P. delicatulus* and *P. novaehollandiae*; and *P. nanus* and *P. gracilicaudatus*. Considerably greater distances of around 20–40 were observed between what appear to be three major intrageneric lineages. These are: (1) the *P. laborifex* group + the *P. nanus* group; (2) the *P. apodemoides* group + *P. fumeus*; and (3) the *P. hermannsburgensis* group + the *P. shortridgei* group along with *P. fieldi* (as *P. praeconis*), *P. australis* and *P. higginsi*. *Pseudomys fuscus* forms a fourth major lineage in this cluster, while *P. occidentalis* yielded somewhat contradictory values and is not clearly associated with any other group.

The genus *Notomys* forms a monophyletic cluster independent of *Pseudomys*. Three distinct groups are evident within *Notomys*: *N. alexis* + *N. mitchellii*, *N. fuscus* and *N. cervinus*, with AIDs of 16–20 between them. *Melomys*, represented by *M. cervinipes*, *M. burtoni*, *M. capensis* and *M. rubicola*, forms a tight cluster (all AIDs less than 10), as do *Zyzomys argurus* and *Z. maini* (as *Z. pedunculatus*; AID = 4) and *Leggadina forresti* and *L. lakedownensis* (AID = 7). In contrast, the two *Mesembriomys* species do not form an exclusive cluster, *M. macrurus* falling slightly closer to *Conilurus penicillatus* (AID = 7) than to *M. gouldii* (AID = 12).

Among the New Guinean murids, intrageneric comparisons support the unity of *Pogonomys* (*P. macrourus*, *P. loriae*, *P. sylvestris*; AIDs = 13–14), but indicate wide divergences within *Uromys* (*U. caudimaculatus* to *U. anak* AID = 28; to *U. neobrittanicus* AID = 39). Within the former, broadly-conceived genus *Melomys*, a low AID was obtained between the Australian *M. cervinipes* and *M. rufescens* from New Guinea. In contrast, significantly higher distances were observed between *M. cervinipes* and species of the newly recognized genera *Paramelomys* (AIDs = 28–40) and *Mammelomys* (AID = 60). Species of the Solomon Island endemic genus

Solomys are moderately close to *Melomys cervinipes*, with AIDs of 13–16.

At the intergeneric level, the MC'F data suggest an especially close relationship between *Conilurus, Mesembriomys* and *Leporillus* (AIDs = 6–14). Somewhat further removed (AIDs = 17–30) are *Zyzomys, Uromys caudimaculatus* and *Melomys*, the last two taxa showing no special association except as members of this larger cluster. A second major group consists of the various species groups of *Pseudomys* (including *P. fuscus*), with *Notomys* slightly further removed. The two Australian genera of 'water-rats', *Hydromys* and *Xeromys*, are not clearly monophyletic, nor are they strongly associated within any other generic cluster. Distances involving *Leggadina* are similarly inconsistent but suffice to demonstrate a lack of close relationship to any other group. There is some indication of a special relationship between *Hydromys* and *Leggadina* (AIDs = 23–33).

Among the New Guinean genera, a very close relationship is revealed between *Mayermys, Neohydromys* and *Pseudohydromys* (AIDs = 0), with slightly greater distance to each of *Leptomys* and *Crossomys* (AIDs = 10). The Australian *Xeromys* also groups closely with these taxa (AIDs = 5–18), while *Hydromys* is uniformly distant from these genera (AIDs = 26–31) but close to *Parahydromys asper* of New Guinea (AID = 9). As noted earlier, *Melomys* (*sensu* Menzies, 1996) groups closely with *Solomys* (AIDs = 13–16) and is further removed from each of *Uromys* (AIDs = 25–39) and *Paramelomys* (AIDs = 28–40). *Mammelomys* is even further removed from each of *Melomys* (AID = 60) and *Uromys* (AIDs = 84–88). Among the remaining New Guinean genera, *Chiruromys* groups convincingly with *Pogonomys* (AID = 21), but *Anisomys*, *Coccymys*, *Hyomys*, *Lorentzimys*, *Macruromys* and *Mallomys* all appear to represent distinct lineages.

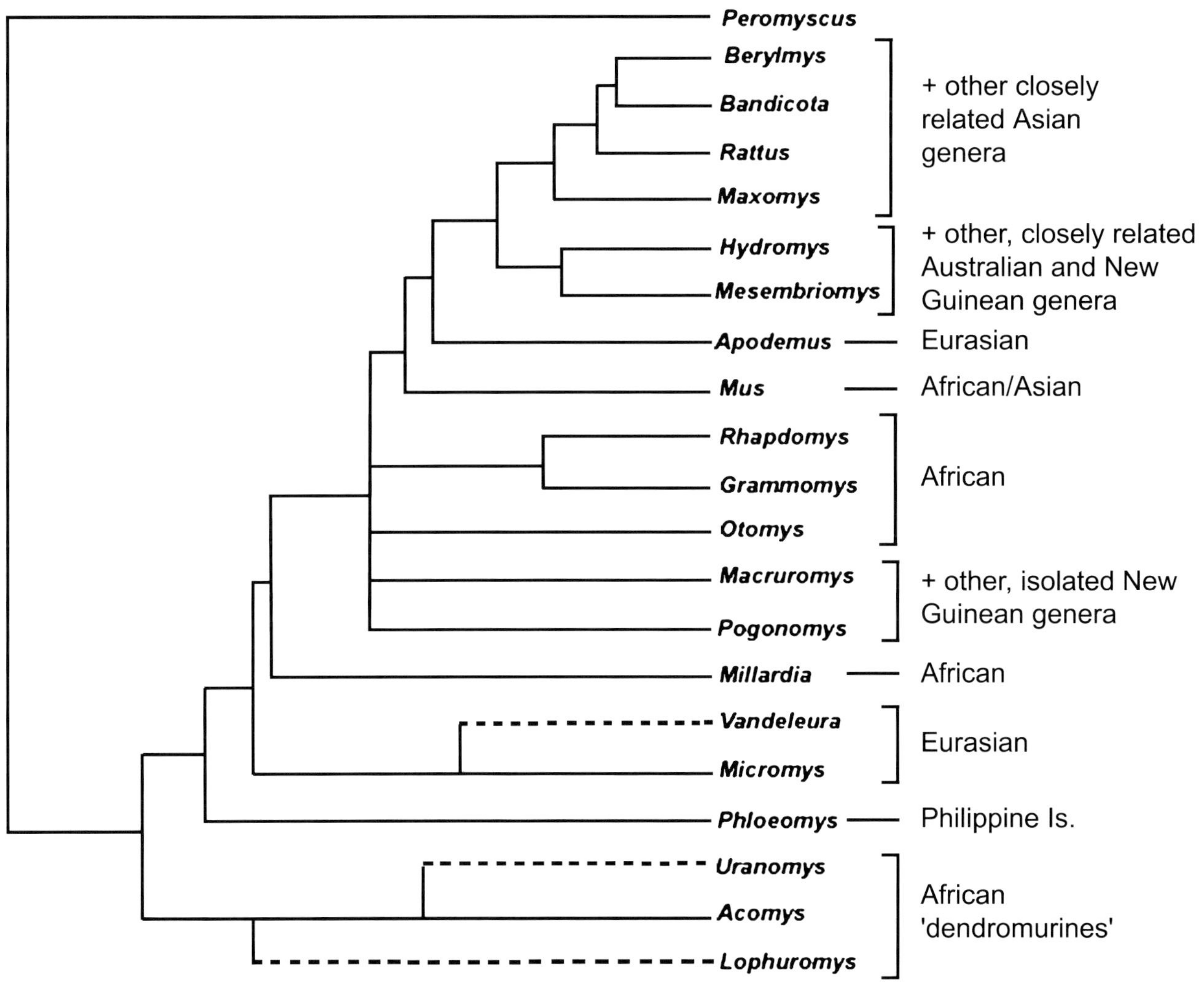

Figure 6. Broad phylogeny of murine rodents based on Microcomplement Fixation of albumin (after Watts *et al.*,1992). *Hydromys* and *Mesembriomys* are included as representatives of the '*Mesembriomys* group'. *Pogonomys* and *Macruromys* are included as representatives of the 'Old Endemic' murine assemblage in New Guinea, but they do not necessarily represent the full diversity of this group.

have dispersed outwards from New Guinea, following an earlier period of diversification within the Australasian region. Figure 7 summarizes the likely phylogenetic relationships among Australasian *Rattus* species.

THE *MESEMBRIOMYS* GROUP

The *Mesembriomys* group radiation contains at least 22 genera which are either primarily Australian in distribution (*Conilurus, Leggadina, Leporillus, Mesembriomys, Notomys, Pseudomys, Xeromys, Zyzomys*) or primarily New Guinean (*Crossomys, Hydromys, Leptomys, Mayermys, Melomys, Neohydromys, Parahydromys, Paraleptomys, Paramelomys, Pseudohydromys, Solomys, Uromys*; the poorly known *Microhydromys* presumably also belongs here). All members of the *Mesembriomys* group share the obviously derived features of accessory hooks on the sperm head and loss of all pectoral and abdominal teats; they also form a cohesive unit on immunological criteria. The common, shared karyotype may constitute further evidence of monophyly, but might also prove to be a retained ancestral character.

No murid from the immediate source area in Asia has been found to have the characteristic accessory hooks on the sperm head that might be expected of a sister taxon of the *Mesembriomys* group. This means either that the accessory hooks developed *in situ* in the Australasian region, or that the ancestral lineage became extinct in Asia after dispersal into Australasia. Several molecular studies have emphasized the long isolation of the *Mesembriomys* lineage from both Asian and African murine radiations (Watts and Baverstock, 1995a, b; Ruedas and Kirsch, 1997). These results are consistent with the idea of a relatively early dispersal of the group into, and subsequent isolation within, the Australasian region.

Among the Australian genera, two well-supported clades are identified, one containing *Mesembriomys* and *Conilurus* and the other, *Pseudomys* (including *Mastacomys*) and *Notomys*. *Leporillus* and *Zyzomys* are associated with the former group on immunological criteria, but with *Pseudomys* on genetic evidence. Chromosomal data do not help here, due to a lack of critical banding evidence in the case of *Leporillus* and because of an excessive number of confounding rearrangments in the case of *Zyzomys*. Misonne's (1969) interpretation of trends in dental morphology would group both *Leporillus* and *Zyzomys* close to *Mesembriomys* and *Conilurus*, yet Tate (1951) and Crabb (1977) both associated *Leporillus* with *Pseudomys* on essentially the same evidence. Clearly, the dental morphology needs to be reexamined within a cladistic framework. Among the remaining Australian genera, *Hydromys* and *Xeromys* are both closely related to endemic New Guinean genera and will be discussed in that context, while *Leggadina* is isolated from all other taxa.

Within *Pseudomys*, there are a number of species-groups that appear to be well-supported. These are:

(1) *P. gracilicaudatus* + *P. nanus*: special affinity is indicated by genetic and immunological data, supported by a close similarity in dental morphology.
(2) *P. delicatulus* + *P. hermannsburgensis* + *P. novaehollandiae* (+ *P. pillagaensis, P. bolami* ?): association of the three primary taxa is indicated by genetic and immunological data. Genetic data place *P. pilla-*

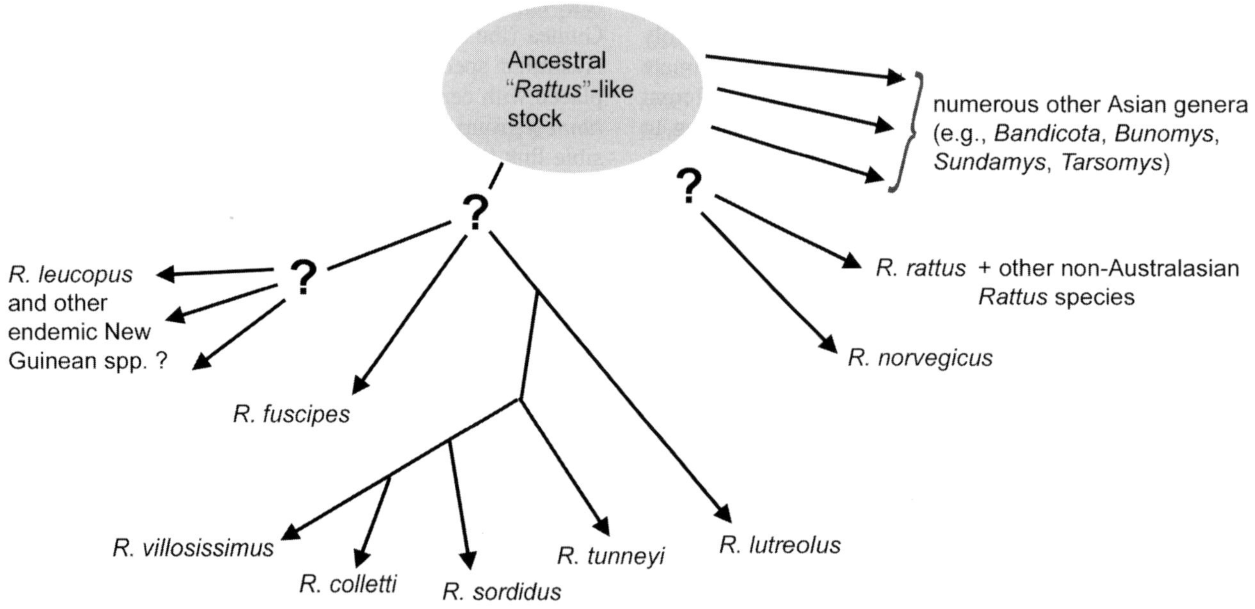

Figure 7. Schematic phylogeny of the genus *Rattus* in the Australasian region. Uncertain relationships (unresolved polychotomies) are indicated by question marks. Note that numerous New Guinean *Rattus* species remain very poorly known and could conceivably represent additional major lineages.

gaensis close to *P. delicatulus* and *P. novaehollandiae,* with *P. hermannsburgensis* somewhat further removed. Immunological data place *P. bolami* within this group. *Pseudomys delicatulus, P. novaehollandiae* and *P. pillagaensis* share a reduction in the size and number of accessory sperm head hooks. *Pseudomys delicatulus* and *P. novaehollandiae* share a likely derived glans penis morphology (shared with several other species). *Pseudomys hermannsburgensis* and *P. delicatulus* were formerly (and erroneously) associated with *Leggadina forresti* on the basis of trivial dental similarities.

(3) *P. desertor* + *P. shortridgei*: these taxa are placed close together on immunological and dental criteria. Genetic data group *P. shortridgei* with *Notomys* and *Zyzomys,* but *P. desertor* was not included in the study.

(4) *P. albocinereus* + *P. apodemoides*: these species are grouped on immunological, genetic, penile and dental criteria. Immunological evidence identifies *P. fumeus* as the next closest relative of *P. albocinereus* + *P. apodemoides*. The ?Pliocene-age fossil taxon *P. vandycki* is most similar to *P. albocinereus* in molar morphology (Godthelp, 1990).

(5) *P. johnsoni* + *P. laborifex*: these species are closely related on immunological criteria; both construct 'pebble-mounds'. *P. chapmani* probably also belongs with these taxa based on dental morphology and behaviour (Kitchener, 1980). Immunological evidence suggests that this group is related to the *P. gracilicaudatus* + *P. nanus* species-group.

The remaining species of *Pseudomys* either appear to occupy isolated positions within the genus (*P. australis, P. fieldi, P. higginsi, P. occidentalis*) or they have not been subject to critical examination (*P. glaucus, P. gouldii, P. oralis, P. patrius*). *Pseudomys fuscus* is linked to *P. fumeus* on genetic evidence but its penile morphology is more similar to that of a variety of other *Pseudomys* species including *P. nanus, P. australis* and *P. higginsi. Pseudomys fuscus* and *P. australis* share a unique pericentric inversion involving chromosome pair 3, as well as a more widely distributed inversion on pair 8. Immunological data do not support a close relationship between these taxa but place *fuscus* in a relatively isolated position within the genus.

The intrageneric phylogeny of *Notomys* is available for the few extant species only. Of these *N. alexis, N. mitchellii* and *N. fuscus* share a suite of obvious modifications in both sperm head form (reduction of accessory hooks) and male reproductive anatomy (reduction of most accessory sex glands, cryptorchid testes), while *N. cervinus* lacks these derived features and thus appears to be the sister taxon to this derived clade. Immunological evidence supports this suggestion and goes further to group *N. alexis* and *N. mitchellii* to the exclusion of *N. fuscus*. No data are available for *N. aquilo* which remains widespread in northern Australia.

Three major clades are identified among the predominantly New Guinean members of the *Mesembriomys* group. Two of these are made up of various genera of 'water-rats'. The true water-rats, *Hydromys* spp., are close to *Parahydromys* on morphological and immunological evidence, but distant from all other genera. The remaining 'water-rats' (*Crossomys, Leptomys, Mayermys, Neohydromys, Paraleptomys, Pseudohydromys* and *Xeromys*) form a second, close-knit group on immunological criteria but display considerable morphological diversity. The poorly known genus *Microhydromys* is generally included with this group due to a superficial similarity to other diminutive 'moss-mice'. However, it differs in some important respects and may not belong here. The third major clade is centred on the genera *Uromys* and *Melomys* which are tightly associated on genetic, immunological and morphological criteria. *Solomys* is close to *Melomys* on immunological and dental evidence but has a uniquely derived sperm form. *Paramelomys* appears to be somewhat further removed but shares special features of sperm and dental morphology with *Uromys* and *Melomys*. It may be the sister taxon to these genera combined.

Five other genera are traditionally associated with *Melomys* on the basis of dental morphology. *Pogonomelomys* species share a uniquely derived upper molar morphology (Figure 4) with species of both *Melomys* and *Uromys* (Misonne, 1969) and *Pogonomelomys mayeri* also shares a common penile morphology with these taxa and with *Paramelomys platyops* (Lidicker, 1968). Species of *Abeomelomys, Xenuromys* and *Mammelomys* lack the derived molar morphology of *Uromys, Melomys* and *Pogonomelomys,* but share a sperm head morphology, which is more similar to that of *Rattus* than to any member of the *Mesembriomys* group. These taxa have been grouped traditionally with *Melomys* and *Uromys* on account of their 'laminated' molar morphology and 'mosaic' tails, but immunological data place *Mammelomys* very far from *Melomys* and *Uromys. Protochromys fellowsi* also lacks the derived molar morphology of *Melomys* and *Uromys.* Immunological comparisons are not available for *Abeomelomys, Protochromys* or *Xenuromys,* nor are there any data on chromosomes or penile morphology for these taxa. Indeed, apart from the superficial resemblance to *Melomys* in dental and tail morphology, the only grounds for inclusion of these taxa within the *Mesembriomys* group is the shared derived reduction in mammary formula to 0 + 2 = 4.

The question of how each of the various lineages of the *Mesembriomys* group relates to the others cannot be answered on the basis of available evidence. Penile and dental features, as well as weak genetic evidence, suggest links between the two groups of 'water-rats'. However, immunological evidence suggests that they are, at best, distantly related and perhaps even biphyletic—the MC'F data associate the *Hydromys* group with *Leggadina* rather than with the *Leptomys-Xeromys* group (Watts and Baverstock, 1994a).

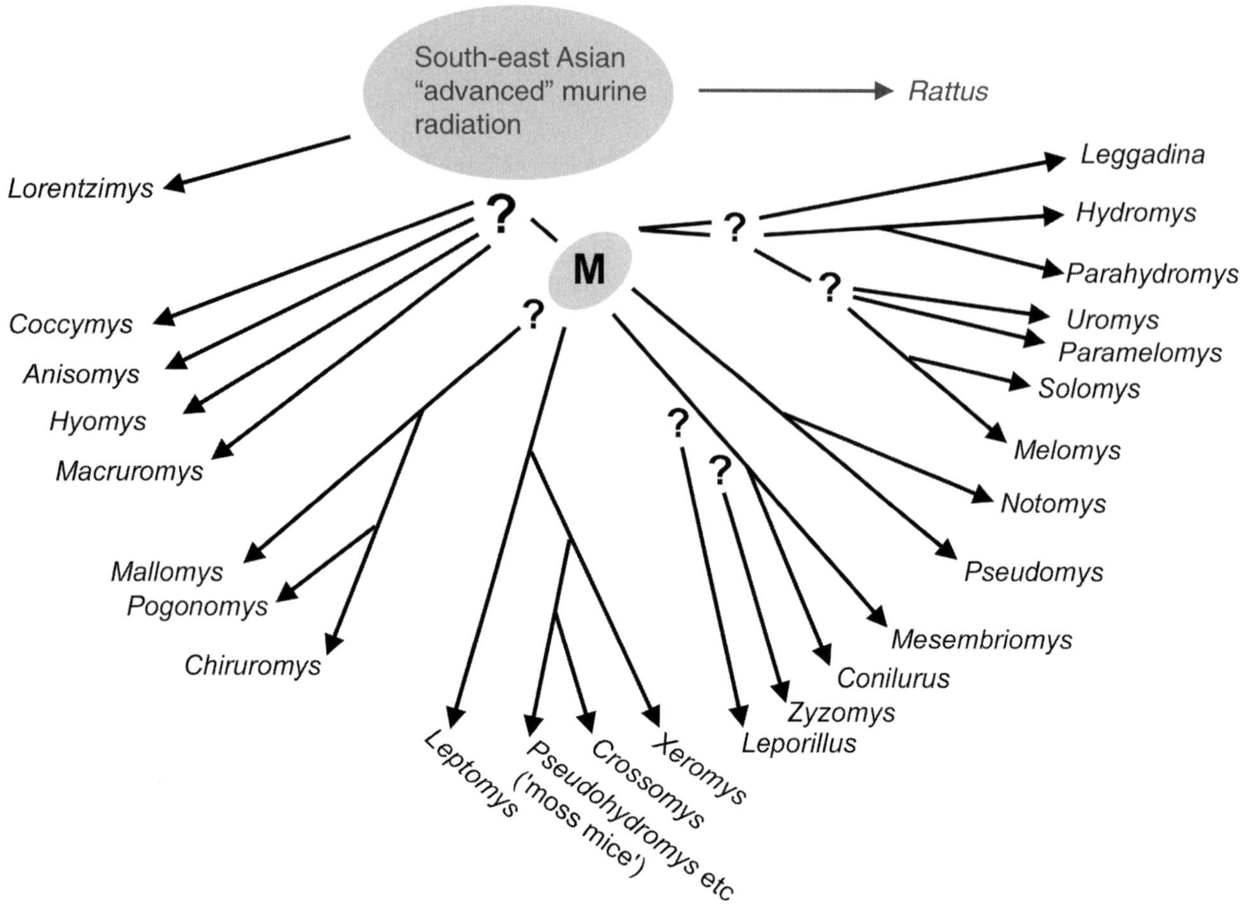

Figure 8. Schematic phylogeny of Australasian murines, based on a synthesis of all available evidence. Uncertain relationships (unresolved polychotomies) are indicated by question marks. The ancestor of the '*Mesembriomys* group' is represented by **M**. Four genera are either too poorly known to include on the phylogeny (*Microhydromys, Abeomelomys*), or are the subject of contradictory evidence (*Mammelomys, Xenuromys*). *Abeomelomys, Mammelomys* and *Xenuromys* may belong to a single lineage.

Some evidence from the male reproductive tract (broad, chunky baculum; loss of the preputial glands; presence of a second pair of ventral prostate glands) links *Hydromys* to the *Uromys-Melomys* group, but critical observations are lacking for many taxa including any of the *Xeromys-Leptomys* group. Immunological data suggest an alternative connection between the *Melomys-Uromys* group and the *Mesembriomys-Conilurus* group, also accompanied by *Zyzomys*. It is interesting to note that *Melomys, Uromys, Zyzomys* and both groups of 'water-rats' all share a sperm head form characterized by short accessory hooks; whether this is an ancestral or derived condition is unclear.

NEW GUINEAN OLD ENDEMICS

This group comprises at least eight genera, all of which are found principally, if not exclusively, in New Guinea. These are *Anisomys, Chiruromys, Coccymys, Hyomys, Lorentzimys, Macruromys, Mallomys* and *Pogonomys*. As mentioned above, one or more of *Abeomelomys*, *Mammelomys* and *Xenuromys* might also belong here rather than in the *Mesembriomys* group, but evidence on which to base this decision is lacking. With the exception of *Lorentzimys*, this assemblage corresponds to the 'New Guinean clade' identified by Watts and Baverstock (1994a) on the basis of immunological evidence. However, while they treat this group as a probable monophyletic clade, I prefer to view it as a 'grab-bag' of taxa united mainly by their lack of clear affinity to either of the preceding two groups. This hesitancy to accept the group as monophyletic relates in part to the existence of contradictory evidence, some of which provides intriguing hints of possible alternative relationships.

As indicated above, the three defining characteristic features of the *Mesembriomys* group are a reduced number of mammae, a 2n=48 karyotype and accessory sperm head hooks. Each of these characters is present in at least one taxon among the New Guinean 'Old Endemics'. Loss of pectoral mammae occurs in three of the principal genera of this group, *Coccymys, Hyomys* and *Macruromys*. In all of the others, there is one pair of pectoral mammae in addition to two inguinal pairs. The karyotypes of *Coccymys, Hyomys* and *Macruromys* unfortunately are unknown. However, *Mallomys aroaensis* has a 2n = 48 karyotype, which is at least superficially identical to the homologous, shared karyotype of the *Mesembriomys* group; and the karyotype in several other genera (*Anisomys, Chiruromys, Lorentzimys* and *Pogonomys* all with 2n = 46) is described as being readily derived from the standard karyotype (Donnellan, 1989). As noted earlier, this evidence is ambiguous, as it could be argued that the 2n = 46 karyotype is an ancestral condition from which the 2n = 48 karyotype subsequently arose by fission. Under this latter scenario, *Mallomys* might be linked phylogenetically to the *Mesembriomys* group, but no other relationship would be indicated. In either event, G-banding studies will be necessary to establish the homology of the autosomal components.

A more compelling case for links to the *Mesembriomys* group comes from sperm morphology. As noted earlier, the sperm head in *Chiruromys, Mallomys* and *Pogonomys* has two accessory ventral hooks similar to those seen in members of the *Mesembriomys* group. These hooks differ somewhat from the 'classic' *Mesembriomys* group taxa in both gross form and internal structure, and they also differ between *Mallomys* and *Pogonomys-Chiruromys*. The immunological evidence supports a relatively close relationship between *Pogonomys* and *Chiruromys,* but places these taxa closer to *Anisomys* (which has a very simple sperm head, lacking accessory hooks) and *Macruromys* (sperm unknown) than to *Mallomys.*

Watts and Baverstock (1994a) placed *Lorentzimys* in a separate group based on its degree of immunological isolation from all other taxa. As noted earlier, *Lorentzimys* is also noteworthy in combining some very primitive aspects of molar construction with a superficially *Rattus*-like sperm head morphology; at present, it is perhaps best regarded as a very distinctive member of the New Guinean 'Old Endemics'.

The many uncertainties surrounding phylogenetic patterns within this group are portrayed in Figure 8. With the possible exception of *Lorentzimys,* no member of the group appears to be sufficiently isolated in biochemical terms to qualify as a direct derivative of an 'archaic murine' radiation.

MURINE FOSSILS

The fossil record of Australasian rodents is relatively poor and incompletely analysed, but provides some useful data on the earliest appearance of several genera. By contrast, the fossil records of murines in southern Asia, Europe and Africa are relatively rich and well-studied, providing a framework of major cladogenic and dispersal events that will help to elucidate the likely history of the group in Australasia.

ASIAN, AFRICAN AND EUROPEAN RECORDS

The earliest murine fossils, referred to the genus *Antemus*, come from the Siwalik beds of Pakistan and date to ~14 m.y.a. (Jacobs, 1977, 1978; Jacobs *et al.*, 1989). A suitable ancestor for *Antemus* is provided by the 'dendromurine' genus *Potwarmus,* recorded from sediments of the Siwalik sequence back to 18.3 m.y.a. (Jacobs and Downs, 1994). More advanced murines (*Progonomys* sp.) make their appearance in the Siwalik record at 11.8 m.y.a., with the genus *Karnimata,* a possible ancestor of *Rattus* and its relatives, appearing between 10.6 and 9.5 m.y.a. (Jacobs and Downs, 1994). The genus *Mus* is first recorded in the Siwalik sequence at 5.7 m.y.a., a sudden appearance that is consistent with the idea of an African origin and subsequent migration.

Elsewhere in Asia, the fossil record is far less complete. The Li Basin fauna from Thailand has produced a possible species of *Antemus* (Jacobs *et al.,* 1989) as well as the dendromurine *Potwarmus*. The age of this fauna is variously estimated as 18–19 m.y.a. (Mein and Ginsburg, 1985) or as recent as 11 m.y.a. (Jacobs *et al.*, 1989). Chinese localities document the appearance of *Progonomys* at around 10.5 m.y.a. (Jacobs and Downs, 1994). Later faunas of late Turolean age (6.0–5.4 million) have yielded a moderately diverse murine assemblage, including an advanced species of *Karnimata* and primitive species of the extant Eurasian genera *Apodemus* and *Micromys* (Storch, 1987); Chiamanee (1998) described murine faunas of late Pliocene to middle Pleistocene age from cave deposits in Thailand. These earlier faunas contain taxa recognizably annectant to modern genera, possibly including *Rattus*, but there is no specific evidence of links with any Australasian genera.

The pre-Quaternary murine fossil record of peninsular or insular south-east Asia is extremely sparse. Musser (1982c) described *Rattus trinilensis* from sediments of probable early Pleistocene age at Trinil in Central Java. More recently, Downing *et al.* (1998) reported specimens of the extant species *Paruromys dominator* from the Pleistocene Walanae Formation of south-west Sulawesi. Sediments of late Pleistocene to Holocene age on the island of Flores contain the remains of two extinct genera (*Hooijeromys, Spelaeomys*) (Musser, 1981b). Four extinct genera, only one of which has been named (*Coryphomys buehleri*), are known from the nearby island of Timor (Hooijer, 1965; Musser, 1981b; Glover, 1986).

The African fossil record is relatively rich and well-studied (Jacobs, 1985; Winkler, 1994). A *Potwarmus*-like taxon along with other possible dendromurines are recorded in the ?late Early Miocene Gebel Zelten fauna of Libya (Winkler, 1994). Murines do not appear in the African fossil record until significantly later, around 12–11 m.y.a., at which time both *Progonomys* and

Karnimata are stated to be present (Winkler, 1994). Modern genera first appear in Africa around 5 m.y.a. Today, murines make up the dominant group of rodents in most tropical and temperate African faunas, with lesser numbers of dendromurines and cricetomyines.

The earliest fossil murines from Europe are species of *Progonomys* of early Vallesian age (11–10.5 m.y.a.) from southern France (Aguilar and Michaux, 1996) and Spain (Weerd, 1976; Fahlbusch, 1989). By lateVallesian times, murines had attained considerable diversity across western Europe (Weerd and Daams, 1978; Mein, 1984). Although some tentative evolutionary links have been made between fossil and extant genera (Michaux, 1971), the earliest, undoubted representatives of a modern genus (*Apodemus*) are of late Turolean age (6.0–5.4 m.y.a.) (Mein, 1978).

AUSTRALIAN AND NEW GUINEAN RECORDS

The pre-Quaternary fossil record of rodents in Australia is extremely poor (see Tables 2, 3). To date, no rodents have been recorded from any fossil locality of Miocene age. In the case of the abundant and diverse faunas of early to mid-Miocene age from the Lake Frome Basin and Riversleigh area, this absence is almost certainly a genuine one. However, the few known faunas of latest Miocene age (e.g., Alcoota, Sunlands) have produced relatively few small mammals, hence no great significance can be attached to the absence of any particular group (Rich *et al.*, 1991).

The arrival of rodents in Australia is commonly estimated with reference to the Hamilton Local Fauna from Grange Burn in Victoria, dated by radiometric means to 4.46 m.y.a. (Turnbull and Lundelius, 1970; Flannery *et al.*, 1992). This fauna has yielded a diverse small to medium-sized mammal fauna, but no rodents. Evidence from the faunal assemblage and from an associated microflora concur on the presence of closed wet forests in the vicinity. The microfloral evidence more specifically points to a mesotherm rainforest community, possibly with a gymnosperm overstorey including *Araucaria, Dacrycarpus, Dacrydium* and *Podocarpus* as well as one or more subcanopy layers dominated by angiosperms—but also with ferns present, including *Cyathea* (Macphail, 1996). The closest modern analogue is perhaps the Araucarian 'dry' rainforests along the coastal ranges of north-eastern Australia. Today, these dry rainforests support relatively poor mammal communities including species of *Melomys, Uromys* and *Rattus,* but no representatives of endemic genera such as *Mesembriomys* or *Pseudomys.*

Similarly, the wetter rainforest habitats of north Queensland support relatively few rodent species and only one true endemic, the recently described *Uromys hadrourus.* Why the more 'archaic' elements of the Australian rodent fauna should be absent from these habitats is unclear. Nevertheless, the possibility clearly exists that rodents were present in Australia by 4.5 million years ago, but were simply absent or poorly represented in the relictual rainforest habitats (Godthelp, 1990). The Palankarinna Local Fauna, dated by palaeomagnetic means to around 3.9 m.y.a. (Tedford *et al.,* 1992), has produced a moderately diverse vertebrate assemblage suggestive of more open, woodland habitats; the total numbers of small mammal fossils is low and the absence of rodents may be due to inadequate sampling.

Potentially the oldest Australia rodent fossils are those reported as coming from the Parwan site in Victoria, dated to 4.06–4.2 million years of age (Whitelaw, 1991). Unfortunately, these specimens have not been critically examined and may now be lost. An isolated rodent incisor from the Allingham Formation of north-east Queensland (Bluff Downs Local Fauna) may be as old as 4 million years (Archer and Wade, 1976; Hand, 1984), but this requires confirmation. The oldest, confirmed occurrence of rodents is in the Kanunka Local Fauna at around 3.4 m.y.a. Two species of 'Conilurinae' are represented but no details are yet available (Tedford *et al.,* 1992).

The most intensively studied Tertiary rodent assemblage is that derived from the late Pliocene Moorna Formation of the Murray Basin, dated at the interval 2.42–2.87 of the Gauss Chron (Whitelaw, 1991). In Crabb's (1977) analysis of this rich fauna he recorded a minimum of eight taxa including: *Leggadina* sp., four species of *Pseudomys, Leporillus* sp. cf. *L. apicalis*, a possible *Notomys*, and a species of a distinctive new genus with possible affinities to *Leggadina.* Other diverse rodent faunas from the Rackham's Roost locality at Riversleigh, north-west Queensland and from Chinchilla in south-east Queensland are currently under study. The Rackham's Roost fauna contains a minimum of 13 rodent taxa including species of *Zyzomys, Leggadina* and *Pseudomys* (Godthelp, 1987, 1994b, 1997); the Chinchilla rodents include the extinct *Pseudomys vandycki,* a possible close relative of the living *P. albocinereus* (Godthelp, 1990). Both of these faunas currently are dated by biostratigraphic correlation only and this requires independent confirmation.

Similarly diverse rodent faunas are also available from cave breccia deposits on Barrow Island, off the north-west coast of Australia and from Dingo Gap in the Napier Range, south Kimberley; these sites are currently under study by the author and A. Baynes of the Western Australian Museum. The age of the Y-Block locality on Barrow Island is dated (by magnetostratigraphy) between 2.8 and 3.6 million years. It contains four murine taxa, two of them referrable to the extant genera *Pseudomys* and *Zyzomys*, and two representing undescribed genera. One of the latter is a plesiomorphic kind of 'water-rat', with possible affinities to *Hydromys* and represented by a single lower molar. This tooth is the only Tertiary record of this highly specialized branch of the *Mesembriomys* group. The Dingo Gap fauna is undated, but thought to be

Table 2. List of Australasian rodents known only from fossil or subfossil remains. LF = Local Fauna.

Designated Taxon	Geological Interval	Site or Fauna	Published Authority
Melomys spechti	Holocene	Buka Island	Flannery and Wickler (1990)
Notomys sp.	Holocene	Flinders Range, S.A.	Watts and Aslin (1981, p.104)
Pogonomelomys sp.	Holocene	Bird's Head, Irian Jaya	Aplin *et al.* (1999)
Pseudomys vandyki	late? Pliocene	Chinchilla LF, Qld.	Godthelp (1990)
Pseudomys sp.	Holocene	Basalt Plains, Vic.	Watts and Aslin (1981, p. 175)
Pseudomys? sp.	late? Pliocene	Big Sink LF, N.S.W.	Dawson *et al.* (1999)
Rattus sanila[1]	late Pleistocene–Holocene	New Ireland	Flannery and White (1991)
Solomys spriggsarum	Holocene	Buka Island	Flannery and Wickler (1990)
Zyzomys rackhami	late? Pliocene?	Rackham's Roost LF, Qld.	Godthelp (1997)

1. Described as a subspecies of *Rattus mordax*; listed by Carleton and Musser (1984, p. 659) as a full species but without full justification. Study of new archaeological material from New Ireland supports the separation of *sanila* from *mordax*.

Table 3. Summary of Tertiary rodent-bearing fossil faunas of Australia and New Guinea. Dating method abbreviations are: M = Magnetostratigraphy; R = Radiometric; B = Biostratigraphy

Local Fauna	Age (m.y.a.)	Dating Method	Identified Murid Taxa	Ref No.
Australia				
Parwan	4.06–4.2	M	unknown, specimens possibly lost	1
Kanunka	3.4	M	Conilurinae (2 spp.)	6
Bluff Downs	3–4?	R, B	Genus indet.	3
Rackham's Roost	3?	B	*Pseudomys* (6 spp.), *Zyzomys rackhami*, *Leggadina* sp. other genera (total 13 spp.)	7
Y-Block, Barrow Island	2.8–3.6	M	*Pseudomys* (1 sp.), *Zyzomys* sp., 2 new genera (*Hydromys*-like)	9
Fisherman's Cliff	2.4–2.87	M	*Leggadina* sp., *Pseudomys* (4 spp.), *Leporillus* cf. *L. apicalis*, *Notomys* sp (?), unnamed genus	2
Dog Rocks	2.03–2.48	M	*Pseudomys* (2 spp.)	4
Bone Gulch	1.85–2.42	M	1 sp.	2
Chinchilla	2–3?	B	*Pseudomys vandyki,* other unnamed spp.	5
New Guinea				
Awe	3–3.5	R	Genus indet.	8

Reference Numbers:
1. Whitelaw, 1991; Tedford, 1994. **2.** Marshall, 1973; Crabb, 1977; Whitelaw, 1991; Tedford, 1994. **3.** Archer and Bartholomai, 1978; Mackness *et al.,* 2000. **4.** Whitelaw, 1989. **5.** Godthelp, 1990. **6.** Tedford *et al.*, 1992. **7.** Archer *et al.,* 1991; Godthelp, 1994b, 1997. **8.** Plane, 1967; Page and McDougal, 1972; Hoch and Holm, 1986. **9.** Aplin *et al.*, 2001.

late Pliocene in age. It contains several distinctive taxa but no member of the genus *Rattus*.

Only one Tertiary-aged fauna is yet known from New Guinea, the Awe Fauna from the Otibanda Formation; this has produced a single rodent incisor. The Otibanda Formation was originally dated by radiometric means and 'stage of evolution' of the diprotodontid marsupials to 5–7 m.y.a. (Plane, 1967), but the radiometric age has been revised to around 3 m.y.a. (Page and McDougal, 1972; Hoch and Holm, 1986).

The few, securely-dated Australian fossil sites of early to middle Pleistocene age are heavily biased towards larger mammals and have not as yet produced any rodent remains (Tedford, 1994). By contrast, numerous late Pleistocene and Holocene sites have produced rodent material predominantly representing extant species (Wakefield, 1960a, b; Marshall, 1973; Hope *et al.,* 1977; Baynes, 1990; Godthelp, 1994a). Many of these faunas contain species of the genus *Rattus*, which is conspicuous in its absence from all earlier faunas. The earliest known fossil remains of *Hydromys* and *Melomys* also come from the late Pleistocene. Late Pleistocene and Holocene age faunas from New Guinea likewise contain a suite of essentially modern taxa including *Rattus* spp. (White, 1972; Hope *et al*., 1993; Aplin *et al.,* 1999).

THE MOLECULAR CLOCK

Molecular clock estimates of the timing of cladogenic events are based on the concept of random mutation at the nucleotide level. While it is well known that mole-most places by volcanic or marine sediments (Hamilton, 1979; Pigram and Panggabean, 1984). Evidence of rapid uplift in the form of flights of raised coral terraces is seen on many islands (Chappell and Veeh, 1978; Pirazolli *et al.*, 1993), and it is clear that many of these islands have emerged from below the ocean only comparatively recently. However, some of the larger islands, such as Timor and Seram, as well as the smaller islands of the Sula Archipelago, have probably been at least partially emergent since early Pliocene times (Audley-Charles, 1981; Charlton, 2000). Further west, the arc of active volcanism and tectonism extends through the Lesser Sunda island chain to Java; the emergence of these islands is evidently post-Miocene (Smith *et al.*, 1994).

A key event in the biogeographic history of the region may have been the eventual closure, by tectonic activity, of the 'Indonesian Seaway', the former deep-water, high energy sea passage that connected the Indian and Pacific Oceans (Linthout *et al.*, 1997). Rapid shallowing of this passage 10–9 m.y.a. and the associated emergence of some of the modern day Moluccan Islands, may have provided the earliest stepping-stones for dispersal of many groups of plants and animals into and out of Australasia.

SEA LEVEL FLUCTUATIONS

Global sea level fluctuations during the late Cainozoic have occurred primarily as a consequence of the cyclic advance and retreat of polar and continental ice sheets. In the Australasian region, even relatively minor falls in sea level of a few tens of metres are sufficient to dramatically alter the palaeogeography, such as the emergence of an extensive land-bridge between northern Australia and southern New Guinea. Likewise in south-east Asia, periods of lower sea level are marked by a dramatic expansion of land area with the emergence of much of 'Sundaland' and the amalgamation of many of the islands of the Banda Arc into mega-islands.

Water passage depths between the Australasian continental mass and the island groups of the Banda Arc are currently far greater than 150 m—the maximum sea level depression experienced during even the most severe glaciations. However, the low gradient of much of the north-west continental shelf means that even less extreme falls in sea level are sufficient to cause a narrowing of water barriers to as little as 70 km between north-west Australia and Timor and a similar distance between the expanded western 'head' of New Guinea and the Moluccan islands of Seram and Halmahera. Periods of lowered sea level thus represent optimum times for over-water dispersal, compared with periods of high sea levels when water barriers expand to some hundreds of kilometres.

Many zoogeographers have recognized the biogeographic impact of sea level fluctuations during the Pleistocene (Schodde and Calaby, 1972; Taylor *et al.*, 1983; Flannery, 1988). However, the fact that glaciation of the polar regions had commenced far earlier, probably during the early Eocene, leading to major fluctuations in global sea level (Hayes and Frakes, 1975; Shackelton and Kennett, 1979; Shackleton, 1986), has been largely ignored by biogeographers.

The first major Neogene expansion of the Antarctic ice cap, involving the formation of an extensive sea-ice shelf, took place 14–11 m.y.a. This interval coincides with the time of closure of the Australian-Indonesian deep ocean passage (Kennett, 1977) and is marked by a progressive fall in sea level from the record high levels of 16.5–15 m.y.a. (McGowran and Li, 1994). A period of particularly intense cold 10–9 m.y.a. (Woodruff *et al.*, 1981) led to a surge in sea-ice volume and a sharp fall in sea level, perhaps to more than 50 m below current level (Vail and Hardenbol, 1979; Haq *et al.*, 1987).

The period 9–6.5 m.y.a. was one of general climatic amelioration. Sea levels probably rose over this period towards their present position, perhaps interrupted by a minor fall at ~7.5 m.y.a. (Keller and Barron, 1983). Antarctic ice began to expand once again at around 6.5 m.y.a. (Berggren and Haq, 1976), with periods of rapid fall in sea level at 5.6 million and again at 5.2 m.y.a. (Hodell *et al.*, 1986; Roberts *et al.*, 1994). This period coincides with the 'Messinian crisis' which saw the stranding and gradual drying out of the Mediterranean Basin, indicating a fall in sea level of at least 50–70 m below present level (Berggren and Haq, 1976). The volume of Antarctic ice at this time may have been 50% greater than at present, yet Northern Hemisphere polar or continental glaciation had not commenced (Hayes and Frakes, 1975; Kennett and Vella, 1979).

The period 5–3.95 m.y.a. was generally warm with seas once again returning to higher levels, although some evidence suggests a further surge in Antarctic sea-ice at 4.7–4.3 m.y.a. (Kennett and Vella, 1979) in response to 'severe cooling'. Ice retreat and minor oceanic transgression at 4.2 m.y.a. was succeeded by a brief, exceptionally warm period until 3.95 m.y.a., with Southern Ocean sea surface temperatures possibly 8–10°C warmer than at present (Ciesielski and Weaver, 1974).

A rapid return to cool conditions by 3.8 m.y.a. was followed by a gradual but progressive fall in ocean temperatures (Ciesielski and Weaver, 1974). The earliest indications of significant buildup of Arctic sea-ice date to around 3 m.y.a. (Kennett and Vella, 1979), with evidence of continental glaciation in both North America and Patagonia dating to 2.7 m.y.a. The first of the major 'Pleistocene' glaciations occurred at 2.4 m.y.a., and so of late Pliocene age (Shackelton and Hall, 1984). Other short-lived but dramatic surges in ice volume occurred at 2.1 and 2.0 m.y.a., after which the fluctuations became more regular but less extreme (Beu and Edwards, 1984). Approximately 20 full cycles of glaciation and deglaciation have occurred over the last two million years, with a marked increase in the amplitude of oscillations in the last 800,000 years.

REGIONAL MURINE BIOGEOGRAPHY

There appears to be little doubt that the earliest phase of murine evolution took place in southern Asia. Here, the excellent fossil record provides not only the earliest known murine, *Antemus chinjiensis* (at 13.75 m.y.a.), but also an excellent annectant taxon in the dendromurine *Potwarmus primitivus* (Jacobs and Downs, 1994). More advanced murines are present in the Siwalik sequence (at 11.8 m.y.a.). By 11 m.y.a., similarly advanced murine species are recorded in Africa, Europe and China, evidently reflecting a major period of dispersal (the Vallesian faunal event) involving many different mammalian groups. Barry *et al.* (1985) and Winkler (1994) have suggested a direct link between the Vallesian dispersal event and the progressive fall in global sea levels from 12.5–10 m.y.a.; but, it is likely that the dispersal was facilitated, if not induced, by the drier climatic conditions prevailing in Eurasia through much of the Vallesian and into the terminal Miocene Turolean period (Weerd and Daams, 1978).

In South-east Asia, marine regressions and depositional hiatuses are recorded in many different sedimentary basins over the period 12–7 m.y.a. (Hutchinson, 1986), and it is clear that large parts of Sundaland were emergent throughout much of this period. Murines probably dispersed widely and rapidly across the emergent platform and may also have crossed the narrowed water barriers to reach the emergent portions of the Philippines and Sulawesi.

The distinctive Philippine endemic genus *Phloeomys* (Musser and Heany, 1992) might conceivably be derived from a very early phase of dispersal of murines off the Sunda Shelf. As discussed earlier, the immunological evidence indicates that this genus is among the earliest offshoots of the Murinae, with an origin soon after the time of origin of the subfamily. However, molecular data are lacking for many of the other 'Old Endemic' taxa which form a significant component of the murine faunas in other parts of the region. These taxa are often very restricted in distribution and include many highly specialized forms of both giant and miniature size as well as frugivores, insectivores and vermivores (Rickart *et al.*, 1991). Sulawesian Old Endemics include the arboreal *Haeromys* and *Margaretomys* (Musser, 1981a, 1990), several genera of shrew-rats (*Melasmothrix, Tateomys*) and the more generalized *Lenomys meyeri* (Musser, 1969, 1981b). Mainland South-east Asia through to peninsular Malaysia and the islands of Sundaland support a variety of similarly archaic and isolated genera such as *Lenothrix, Pitecheir, Hapalomys* and *Chiropodomys.*

The Lesser Sunda island chain, representing the most likely immediate source for ancient migrants into New Guinea and/or Australia, appears to lack an Old Endemic component in its modern rodent fauna (Musser, 1981b). However, fossil material dating to Pleistocene and even Holocene times records a formerly diverse suite of endemic genera on both Flores and Timor. Two such taxa have been described, *Spelaeomys florensis* from Flores (Hooijer, 1957; Musser, 1981b) and *Coryphomys buehleri* from Timor (Hooijer, 1965), but species of other new genera await description (Musser, 1981b). Another fossil taxon from undated Quaternary sediments on Flores was described by Musser (1981b) as *Hooijeromys nusatenggara.* This species is probably related to the living *Papagomys armandvillei* and *Komodomys rintjanus,* members of the widespread group of *Rattus*-like genera. Its occurrence on Flores indicates that the radiation of the *Rattus* group of genera was formerly more diverse in Nusa Tenggaran islands, with some extinction occurring in late Pleistocene or even Holocene times.

The initial phase of murine dispersal appears to have been succeeded by regional diversification over the period 9–6 m.y.a. The main generic diversity among many of the rare and restricted south-east Asian 'Old Endemics' presumably arose during this same period, along with other genera (*Maxomys, Leopoldamys, Niviventer*), which remain abundant and widely distributed today. In South-east Asia the process of speciation may have been stimulated by a partial return to an archipelagic geography following a rise in global sea levels.

Later phases of dispersal are suggested by the relatively wide distribution of genera such as *Maxomys* and *Leopoldamys*, and by the extremely wide geographic range of the cluster of genera associated with *Rattus.* Members of this '*Rattus* group' range from the Indian subcontinent (*Bandicota, Nesokia*), through Indochina (*Berylmys*) and the islands of Sundaland (*Kadarsanomys, Leopoldamys, Palawanomys, Sundamys*), to the Philippines (*Bullimus, Tarsomys*), Sulawesi (*Bunomys, Paruromys, Taeromys*) and the Lesser Sunda Islands (*Komodomys, Papagomys, Paulomys*). *Rattus* itself occurs in all of these areas and also shows considerable regional differentiation (Musser, 1986; Musser and Holden, 1991).

Within the *Rattus* group, the majority of intergeneric distances (AIDs) fall in the range 30–40 units, although lower values are observed between some regionally-based groupings of genera (e.g. the Indian *Nesokia* + *Bandicota* – AID = 6 units). The larger distances approximate to ~5–7 m.y.a., corresponding to a period of generally lowered sea levels (6.5–5 m.y.a.). Within *Rattus,* the maximum reported molecular distance of 10 units (between *R. rattus* and *R. norvegicus*) corresponds to 1.6–1.9 m.y.a.; a similar divergence time (~2.5 m.y.a.) was suggested by analysis of a Line-1 retrotransposon subfamily (Verneau *et al.,* 1998). So dispersal and regional diversification within this highly successful genus appears to have occurred in the context of the latest Pliocene and Pleistocene sea level fluctuations and associated environmental changes.

AUSTRALASIAN MURINE BIOGEOGRAPHY

Until recently, the biogeographic history of Australasian rodents was thought to involve two waves of invasion; an earlier one leading to the 'Old Endemic' lineages of Australia and New Guinea, and a more recent one involving only *Rattus* (Simpson, 1961; Watts and Aslin, 1981; Lee *et al.*, 1981; Flannery, 1988). In each case the number of invading taxa was thought to be quite small, perhaps no more than a single species. Based on the Australian fossil record, the timing of the initial invasion was estimated as 4 m.y.a., with *Rattus* arriving considerably later—during the Pleistocene.

This review of morphological and molecular evidence suggests a far more complex history than previously suspected. Firstly, it is now clear that the diversity of murines in the contemporary New Guinean fauna is much higher than in Australia; but there is no specific evidence that the New Guinean assemblage is monophyletic, with respect to murines in potential source areas to the west. Secondly, molecular clock estimates for initial divergence of major Australasian murine clades far exceed the intervals predicted based on an arrival date of 4 m.y.a. Finally, as noted by Godthelp (1990), the presence of essentially modern murine genera in the earliest known Australian fossil rodent faunas also indicates a considerably earlier date for initial colonization.

Several biogeographic scenarios might account for these apparent conflicts between the Australian fossil record and the new molecular and morphological data. One possible solution lies in placing the initial phase of radiation outside the Australasian region entirely, perhaps in the islands of the Lesser Sunda groups, with multiple migrations into Australia and New Guinea. Flannery (1988) and Godthelp (1990) hint at a scenario of this kind, emphasizing the xeric adaptation of the majority of Australian rodents and the presence of appropriate, xeric habitats today on many of the islands of the Lesser Sunda chain. As noted by Flannery (1988), this scenario requires the subsequent, catastrophic extinction of the source populations from the Lesser Sunda islands and many of them are indeed very degraded as a result of human activity. However, the few known endemic murines from this region (e.g. *Coryphomys, Papagomys, Paulomys, Spelaeomys*) are decidely non-Australian in character, being related instead either to south-east Asian or New Guinean taxa.

An alternative solution was proposed recently by Watts and Baverstock (1996), who speculated that murids had reached the Indonesian Islands and New Guinea between 10 and 8 m.y.a. After an initial phase of rapid speciation, an ancestral species crossed from New Guinea to Australia ~5–7 m.y.a., thereafter giving rise to an endemic radiation of predominantly arid-adapted species. Other members of the Australasian clade and one or more species of *Rattus* reached Australia at a later time, probably within the last million years.

I regard the Watts and Baverstock (1996) hypothesis as the best explanation of the currently available evidence, although some revision of the chronology is needed in the light of the recalibrated molecular clock. A more detailed version of this biogeographic hypothesis (integrating all diverse lines of evidence previously discussed) is presented below.

Phase 1. Colonizing New Guinea

Murines were almost certainly widely distributed across the Sundaland platform by 10 m.y.a.; however, several lines of evidence suggest that they did not penetrate through to the Australasian region until somewhat later, perhaps around 7–9 m.y.a. Evidence for a delayed arrival of murines comes from molecular clock estimates of divergence between Australasian and other regional murines. The major clades of New Guinean and Australian murines differ both from each other and from the majority of south-east Asian murines by around 40–50 albumin units (Watts and Baverstock, 1995). This corresponds to 6.5–9.0 m.y.a., depending on the calibration values used. The New Guinean endemic genus *Lorentzimys* may present a special case. This highly specialized taxon has a mean AID of 60 from other Australasian murines yet it is no more distant to *Rattus* or to other 'outgroups', hence its greater degree of isolation cannot be attributed to an elevated rate of albumin evolution; estimates of divergence time for *Lorentzimys* fall in the range 9.5–11.0 m.y.a.

Some clues as to the possible timing and route of dispersal are provided by the palaeogeographic evidence. As discussed earlier, many of the island groups of modern day Indonesia may have remained fully submerged until the latest Miocene or Pliocene times. Possible exceptions may have included small parts of eastern Sulawesi along with the associated Sula Island group, as well as the island of Seram to the immediate west of New Guinea. This series of possible 'stepping stones', created during closure of the Indonesian Seaway, may be thought of as a northern route for dispersal through to New Guinea. By contrast, many of the land masses of the Lesser Sunda islands—an alternative southern route—appear to have emerged considerably later, during Pliocene or even Pleistocene times.

As noted above, the molecular data suggest that the *Lorentzimys* lineage diverged prior to the postulated date of entry of murines into New Guinea. The initial colonization may thus have involved at least two different species of murines. These may have arrived separately or close together in time, or *Lorentzimys* may even have arrived later in time. In any case, the isolated position of *Lorentzimys* within the Australasian murine fauna may reflect a degree of original ecological specialization that precluded any adaptive radiation of this group.

Phase 2. Radiation in the New Guinea Region

During the late Miocene times (8–6 m.y.a.) New Guinea consisted of several large land masses, surrounded by clusters of smaller islands—see Flannery (1995) for several reconstructions. All of the emergent land was relatively low compared with the mountainous terrain of today. The early murine immigrants probably dispersed rapidly throughout the various island groups and underwent an adaptive radiation to occupy a wide variety of arboreal, terrestrial and semi-aquatic niches. Later, as the more centrally placed islands coalesced, formerly distinct regional faunas blended to produce a high overall level of murine diversity including many specialized genera. Taxa that probably evolved during this initial phase of radiation include the genera *Anisomys, Hyomys*, *Macruromys* and *Mallomys* as well as stem lineages of various subgroups of the *Mesembriomys* group, namely the *Melomys-Uromys* subgroup and the *Leptomys-Xeromys* subgroup.

Phase 3. Ancestral 'Conilurins' enter Australia

The various major lineages of the *Mesembriomys* group show an unusual distributional pattern that hints at the mode of origin of the Australian murines. As discussed earlier, six major lineages can be identified within the group on a combination of immunological, chromosomal and morphological criteria, but interrelationships remain unclear; two of these lineages (the *Xeromys-Leptomys* subgroup and *Melomys-Uromys* subgroup) have endemic representatives in both Australia and New Guinea. A third (the *Hydromys-Parahydromys* subgroup) is predominantly New Guinean, with a single, shared species extending into Australia. The remaining three subgroups (*Leggadina*; the *Mesembriomys-Conilurus* subgroup; the *Pseudomys-Notomys* subgroup) are either exclusively or predominantly Australian (shared species of *Conilurus* and *Pseudomys* are found in southern New Guinea).

This pattern is best explained by derivation of the three endemic Australian lineages from a single immigrant taxon. The alternatives are to postulate two or more separate dispersal events from New Guinea, or to derive part of the *Mesembriomys* group from outside New Guinea. The first of these alternatives is certainly possible, but unlikely as it would require all three vicariant sister lineages, left behind in New Guinea, to become extinct. The second alternative, of *Mesembriomys* origins from Indonesia, conflicts with evidence from extant Nusa Tenggaran rodent faunas. For the present, the 'single New Guinean immigrant' hypothesis is preferred as the simplest one consistent with the known facts.

The postulated 6–5 m.y.a. date of entry of murines into Australia is considerably earlier than that currently favoured on the basis of the limited fossil record. However, an earlier date is strongly indicated by the relatively high levels of molecular divergence between the three Australian lineages of the *Mesembriomys* group and is further supported by palaeogeographic and biogeographic considerations.

Aplin *et al.* (1993) noted the similarity between the immunological divergences for murines and those for a number of other vicariant pairs within the Australasian marsupial fauna. Importantly, this wider list includes several groups of ringtail possums, which are extremely unlikely to cross a water barrier of any magnitude. They suggested that murines initially entered Australia as part of a more general faunal interchange during the terminal Miocene, facilitated by an exposed land connection across present day Torres Strait.

Some support for the faunal interchange idea comes from the Hamilton Local Fauna (4.46 m.y.a.). This fauna includes several marsupials with strong New Guinean affinities. Most informative of these is a species of *Dorcopsis* (Flannery *et al.*, 1992), providing the first and only record of this highly distinctive macropodid genus in an Australian fossil locality. The presence of this animal at Hamilton in Victoria clearly indicates faunal interchange sometime prior to 4.5 m.y.a., followed in this case by a shortlived expansion of *Dorcopsis* in the warm-moist rainforests that flourished throughout eastern Australia during the climatic optimum of the early Pliocene (Martin, 1991; White, 1994). The absence of murines in the Hamilton Local Fauna may say more about the pattern of habitat usage by the early 'conilurin' rodents in Australia, than it does about their time of arrival.

Phase 4. *Mesembriomys* Group Radiation within Australia

Having once arrived in Australia, murines underwent a rapid adaptive radiation. Not surprisingly, this radiation has occurred primarily within the more xeric environments, which have expanded and diversified enormously over the last four to five million years (Bowler, 1982; White, 1994). Ecological and physiological aspects of this radiation have been reviewed extensively by Lee *et al.* (1981), Watts and Aslin (1981), Watts and Kemper (1989), Dawson and Dawson (2006) and others, and are not covered here.

A noteworthy feature of the biogeography of Australian murines is the low diversity of rainforest-adapted taxa compared with either New Guinea or South-east Asia (Flannery, 1988; Godthelp, 1990). This is especially true for the *Mesembriomys-Conilurus* and *Pseudomys-Notomys* subgroups, both of which are unrepresented in these habitats within Australia (Flannery, 1988). In part this may reflect the troubled history of Australia's rainforest refugia, which have clearly been subject to repeated periods of drying and contraction, leading to local extinctions (Schodde and Calaby, 1972; Moritz *et al.*, 1997). However, the fact that the few rainforest endemics (*Melomys, Pogonomys, Uromys*), in north-east Queensland, all have their closest relatives in New Guinea suggests a recolonization of rainforest patches—by rodents from the north after local extinctions of earlier forms.

Phase 5. Faunal Exchange between Australia and New Guinea

Following the climatic amelioration at the close of the Miocene, sea levels remained relatively high and stable until ~2.4 m.y.a.—the onset of the first of the Pleistocene glaciations. Opportunities for dispersal between Australia and New Guinea were evidently limited over this period, and there are no obvious vicariant-pairs even among the wider mammalian fauna (Aplin *et al.*, 1993). By contrast, numerous sibling-pairs show molecular distances consistent with dispersal during Pleistocene times, and this includes a number of murines. These migrations presumably took place during glacial stages when lower sea level would have caused repeated exposure of Torres Strait (Doutch, 1972; Schodde and Calaby, 1972). Vegetation on the Torres land-bridge during these periods was most likely a mosaic of savanna woodlands with gallery rainforests and extensive, low-lying swamps surrounding the massive Lake Carpentaria, which covered much of the western part of the land-bridge (Torgersen *et al.,* 1988).

At least ten murine taxa appear to have crossed between Australia and New Guinea (or vice versa) at various times during the Pleistocene. Among these are several species of *Rattus*, which will be discussed separately below. Possibly the earliest of the transmigrants was the ancestor of the false water rat, *Xeromys myoides*, which has its closest relatives among a suite of highly specialized New Guinean taxa including *Crossomys moncktoni* and the various 'moss-mice' (*Pseudohydromys, Mayermys*). The molecular distance between *Xeromys* and its New Guinean relatives suggests a late Pliocene divergence, but its time of arrival in Australia is unknown. *Xeromys* has an extensive range in northern and eastern Australia, and was recently discovered in southern New Guinea; it is confined to coastal habitats including mangroves.

Uromys caudimaculatus occurs both in north-eastern Australia and in New Guinea, with possible close relatives in both areas. In New Guinea, *U. emmae* is known only from Owi Island in Geelvink Bay, Irian Jaya, while in Australia *U. hadrourus* is restricted to upland rainforest on Thornton Peak in north Queensland (Groves and Flannery, 1994). The molecular distances between *U. caudimaculatus* and other New Guinean *Uromys* (*U. anak* and *U. neobrittanicus*) are large, corresponding to a divergence date in excess of 2.5 m.y.a. (Watts and Baverstock, 1994a), hence it is possible that the *hadrourus-caudimaculatus* lineage has had a long history in Australia. Alternatively, the ancestor of *U. hadrourus* may have entered Australia separately—and much earlier than *U. caudimaculatus.* Some regional differentiation has occurred within *U. caudimaculatus*, as reflected by chromosomal differences between northern and southern populations in Australia (Baverstock *et al.,* 1977a) and between these and the New Guinean population (Donnellan, 1989), but no molecular comparisons are available.

Australian *Melomys* species are closely related to each other and to New Guinean members of this morphologically weakly differentiated cluster (Watts and Baverstock, 1994a). *Melomys* (*sensu stricto*) is well-removed from *Paramelomys* and *Uromys*, but is quite closely related to *Solomys* of the Solomon Islands. The presence of this sibling relationship to *Solomys* clearly favours New Guinea as the place of origin of *Melomys,* with a likely time of dispersal to Australia of no earlier than 0.6 m.y.a. *Melomys* species today are confined to northern, sub-tropical regions, but their remains are known from deposits of probable mid to late Pleistocene age in Victoria (Wakefield, 1960b).

The Australian *Pogonomys* has been associated with the montane New Guinean *P. loriae* group, but its precise affinities are uncertain. Within *Pogonomys,* New Guinean *P. loriae* has an AID of 13 to *P. macrourus,* corresponding to 2–2.5 m.y.a., while the AID to *Chiruromys* is 21 units, representing approximately 3–4 million years. These figures provide a rough guide to the maximum time since arrival of the genus in Australia, although this may have occurred at any time since then.

The water rat, *Hydromys chrysogaster,* is in many respects one of Australia's most characteristic murines. But the species is also widely distributed in New Guinea, along with four other species of *Hydromys* and a species of the closely related *Parahydromys.* The molecular distance between the two genera is small, corresponding to no more than 1 million years since divergence. The Australian water rat is thus clearly of New Guinean origin and may also be one of the more recent arrivals in Australia. It is interesting to note that the remaining species of *Hydromys* and *Parahydromys* are mostly found at mid- to high elevations in New Guinea, where *P. asper* has no particular association with waterways (Flannery, 1995). *Hydromys chrysogaster,* in contrast, extends to sea level and occurs on many of the smaller island groups around New Guinea; presumably it has evolved some particularly successful adaptations that have facilitated its spectacular dispersal within both the New Guinean region and beyond. Within Australia, it has a largely peripheral distribution, but extends inland along the major watercourses and also penetrates areas of more reliable rainfall such as the upland forests of north Queensland. In some areas, the species occupies coastal habitats and offshore islands.

Two distinctively Australian murine species have evidently crossed into southern New Guinea; these are *Conilurus penicillatus* and *Pseudomys delicatulus*, both of which are known from a small area of the trans-Fly plains (Flannery, 1995). Other typically northern Australian mammals including the wallabies *Thylogale stigmatica* and *Lagorchestes conspicillatus* are also recorded from this area, providing clear evidence of faunal continuity across a recently exposed, late Pleistocene Torresian plain.

Phase 6. *Rattus* Arrives

Genetic and molecular evidence clearly demonstrates the close relationship between the native *Rattus* of the Australasian region and typical *Rattus* of South-east and mainland Asia. However, it is not yet known for certain how or when *Rattus* first arrived in the region, nor has the precise relationship between the New Guinean and Australian taxa been established. A date of arrival of around 1–0.5 m.y.a. is suggested by the molecular comparisons, but these lack any precision over such short time-frames.

As discussed earlier, some evidence points to the existence of distinct Australian and New Guinean radiations, the latter with possible connections through to the Lesser Sunda Islands and the Moluccas. Within Australia, the *Rattus sordidus* group has undergone rapid speciation, largely driven or at least facilitated by chromosomal differentiation. In contrast, *R. fuscipes, R. lutreolus* and *R. tunneyi* are chromosomally more conservative but show geographic differentiation in both morphology and genetics. At least one Australian species of *Rattus, R. sordidus,* has crossed Torres Strait to colonize southern and south-eastern New Guinea, while one New Guinean species, *R. leucopus,* has successfully invaded Cape York.

Within New Guinea, many taxa occur in altitudinal replacement series (Taylor *et al.*, 1982; Flannery, 1995) and there are relatively few restricted or obviously relictual taxa. The extent of chromosomal and/or genetic differentiation remains poorly known. *Rattus leucopus,* which occurs in lowland forests both in New Guinea and in north-eastern Australia, is probably a close relative of *R. verecundus* and other montane New Guinean taxa. Like *Uromys caudimaculatus,* it shows some differentiation between southern and northern populations in north-east Queensland, indicating a not very recent arrival in Australia.

NOTES ON CLASSIFICATION

The nomenclature of Australasian rodents in most common recent usage employs a combination of subfamilial and tribal categories and is summarized in Table 5.

As noted in the introduction, Musser and Carleton's (1993) action in downgrading the 'murid' radiation as a whole to subfamilial rank within Cricetidae, clearly impacts on the continued recognition of the Hydromyinae and Murinae at that same rank. An equivalent downgrading of these taxa to tribal taxa seems possible, but new studies have already cast considerable doubt on the current tribal classification (Watts and Baverstock, 1994a).

Most critically, immunological data fail to support the distinction between Uromyini and Conilurini and also hint at a possible biphyletic origin of the tribe Hydromyini. These findings prompted Watts and Baverstock (1994a) to propose a simplified tribal classification within Hydromyinae, as follows: Anisomyini for *Anisomys, Chiruromys, Hyomys, Macruromys, Mallomys* and *Pogonomys*; Hydromyini for *Conilurus, Crossomys, Hydromys, Leggadina, Leporillus, Leptomys, Mayermys, Melomys, Mesembriomys, Microhydromys, Neohydromys, Notomys, Parahydromys, Paraleptomys, Pseudohydromys, Pseudomys, Uromys, Xeromys* and *Zyzomys;* with *Lorentzimys, Xenuromys* and *Coccymys* left unplaced, reflecting uncertainty regarding their affinities. Immunological evidence provides weak support for the monophyly of Anisomyini, but this is contradicted by evidence from sperm which appears to associate *Mallomys, Pogonomys* and *Chiruromys* with the Hydromyini (*sensu* Watts and Baverstock, 1994a; Breed and Aplin, 1994).

With these doubts hanging over the future of supra-generic nomenclature within Murinae, it would seem unwise to propose any new arrangement, and prudent to complete the process of 'taxonomic deconstruction' by abandoning Anisomyini, at least temporarily. Under this scenario, Australasian rodents could be placed in two tribes, Murini for *Rattus* and Hydromyini for what is termed the *Mesembriomys* group in the foregoing discussion. Those taxa which do not fit comfortably in either taxon can be listed as Murinae *incertae sedis*, reflecting their unresolved status. This proposed, interim arrangement is summarized in Table 6 overleaf.

FUTURE RESEARCH

This review has highlighted numerous gaps in our knowledge of the Australasian murine radiation. Many of these relate to the extraordinarily rich and diverse murine

Table 5. Conventional taxonomy of Australasian rodents (based on Watts and Kemper, 1989).

Family, Subfamily, Tribe (with authorities)	Genera Included
Family Muridae (Illiger, 1815)	
Subfamily Hydromyinae (Thomas, 1896)	
Tribe Hydromyini (Thomas, 1896; Watts and Aslin, 1981)	*Crossomys, Hydromys, Leptomys, Mayermys, Microhydromys, Neohydromys, Parahydromys, Paraleptomys, Pseudohydromys, Xeromys*
Tribe Uromyini (Watts and Aslin, 1981)	*Melomys, Pogonomelomys, Uromys*
Tribe Conilurini (Dahl, 1897; Watts and Aslin, 1981)	*Conilurus, Leggadina, Leporillus, Mesembriomys, Notomys, Pseudomys, Xenuromys, Zyzomys*
Tribe Anisomyini (Lidicker and Brylski, 1987)	*Anisomys, Chiruromys, Hyomys, Lorentzimys, Macruromys, Mallomys, Pogonomys*
Subfamily Murinae (Illiger, 1815; Thomas, 1896)	*Rattus, Stenomys*

Breed, W.G. 1991. Variation in sperm head morphology of the Murinae (Family Muridae): some taxonomic implications. Pp.937–939. In Baccetti, B. (ed). *Comparative Spermatology: 20 Years After.* Raven Press, New York. 1112pp.

Breed, W.G. 1995. Spermatozoa of murid rodents from Africa: morphological diversity and evolutionary trends. *Journal of Zoology, London,* Vol. 237, pp.625–651.

Breed, W.G. 1997. Evolution of the spermatozoon in Australasian rodents. *Australian Journal of Zoology,* Vol. 45, pp.459–478.

Breed, W.G. and Aplin, K.P. 1994. Sperm morphology of murid rodents from New Guinea and the Solomon Islands: phylogenetic implications. *Australian Journal of Zoology,* Vol. 43, pp.17–30.

Breed, W.G., Cox, G.A., Leigh, C.M. and Hawkins, P. 1988. Sperm head structure of a murid rodent from southern Africa: the red veld rat, *Aethomys chrysophilus. Gamete Research,* Vol. 19, pp.191–202.

Breed, W.G. and Musser, G.G. 1991. Sulawesi and Philippine rodents (Muridae): a survey of spermatozoal morphology and its significance for phylogenetic inference. *American Museum Novitates,* No. 3003, pp.1–15

Breed, W.G. and Sarafis, V. 1979. On the phylogenetic significance of spermatozoal morphology and male reproductive tract anatomy in Australian rodents. *Transactions of the Royal Society of South Australia,* Vol. 103, pp.127–135.

Breed, W.G. and Yong, H.S. 1986. Sperm morphology of murid rodents from Malaysia and its possible phylogenetic significance. *American Museum Novitates,* No. 2856, pp.1–12.

Briscoe, D.A., Fox, B.J. and Ingleby, S. 1981. Genetic differentiation between *Pseudomys pillagaensis* and related *Pseudomys* (Rodentia: Muridae). *Australian Mammalogy,* Vol. 4, pp.89–92.

Bush, G.L., Case, S.M., Wilson, A.C. and Patton, J.L. 1977. Rapid speciation and chromosomal evolution in mammals. *Proceedings of the National Academy of Sciences of the United States of America,* Vol. 74, pp.3942–3946.

Carleton, M.D. and Musser, G.G. 1984. Muroid rodents. Pp.289–379. In Anderson, S. and Jones, J. K. Jnr. (eds). *Orders and Families of Recent Mammals of the World.* John Wiley and Sons, New York. 686pp.

Catzeflis, F.M., Aquilar, J.-P. and Jaegar, J.-J. 1992. Muroid rodents: phylogeny and evolution. *Trends in Ecology and Evolution,* Vol. 7, pp.122–127.

Catzeflis, F.M. and Denys, C. 1992. The Afriocan *Nannomys* (Muridae): an early offshoot from the *Mus* lineage—evidence from scnDNA hybridization experiments and compared morphology. *Israel Journal of Zoology,* Vol. 38, pp.219–231.

Chappell, J. and Veeh, H.H. 1978. Late Quaternary tectonic movements and sea-level changes at Timor and Atauro Island. *Geological Society of America Bulletin,* Vol. 8, pp.356–368.

Charlton, T.R. 2000. Tertiary evolution of the Eastern Indonesian Collision Complex. *Journal of Asian Earth Sciences,* Vol. 18, pp.603–631.

Chevret, P., Denys, C., Dubois, J.-Y., Duplantier, J.-M., Fily, M.-L., Granjon, L., Jaeger, J.-J., Michaux, J. and Catzeflis, F. 1996. Evolutionary relationships of some African murid rodents (Rodentia: Mammalia): a study based on DNA/DNA hybridization. *Mammalia,* Vol. 60, p. 801.

Chevret, P., Granjon, L., Duplantier, J-M., Denys, C. and Catzeflis, F.M. 1994. Molecular phylogeny of the *Praomys* complex (Rodentia: Muridae): a study based on DNA/DNA hybridization experiments. *Zoological Journal of the Linnean Society,* Vol. 112, pp.425–442.

Chiamanee, Y. 1998. Plio-Pleistocene rodents of Thailand. *Thai Studies in Biodiversity,* Vol. 3, pp.1–303.

Ciesielski, P.F. and Weaver, F.M. 1974. Early Pliocene temperature changes in the Antarctic seas. *Geology,* 1974, pp.511–515.

Crabb, P.L. 1977. *Fossil mammals of the lower Pleistocene Moorna Sands, southwestern New South Wales with an analysis of the Australian pseudomyine murid molars.* Ph.D Thesis (unpublished), Monash University, Victoria.

Cronin, J.E. and Sarich, V.M. 1975. Molecular systematics of the New World monkeys. *Journal of Human Evolution,* Vol. 4, pp.357–375.

Dawson, L., Muirhead, J. and Wroe, S. 1999. The Big Sink Local Fauna: a lower Pliocene mammalian fauna from the Wellington Caves complex, Wellington, New South Wales. *Records of the Western Australian Museum, Supplement,* No. 57, pp.265–290.

Dawson, T.J. and Dawson, L. 2006. Evolution of arid Australia and consequences for vertebrates. Pp.51–70. In Merrick, J.R., Archer, M., Hickey, G.M. and Lee, M.S.Y. (eds). *Evolution and Biogeography of Australasian Vertebrates.* Auscipub Pty. Ltd., Sydney. 942pp.

Dennis, E. and Menzies, J.I. 1978. Systematics and chromosomes of New Guinea *Rattus. Australian Journal of Zoology,* Vol. 26, pp.197–206.

Dennis, E. and Menzies, J.I. 1979. A chromosomal and morphometric study of Papuan tree rats *Pogonomys* and *Chiruromys* (Rodentia, Muridae). *Journal of Zoology, London,* Vol. 189, pp.315–332.

Denys, C., Michaux, J., Catzeflis, F., Ducroco, S. and Chevret, P. 1995. Morphological and molecular data against the monophyly of Dendromurinae (Muridae: Rodentia). *Bonner Zoologische Beitrage,* Vol. 45, pp.173–190.

Donnellan, S.C. 1989. Phylogenetic relationships of New Guinean rodents (Rodentia: Muridae) based on chromosomes. *Australian Mammalogy,* Vol. 12, pp.61–67.

Doutch, H.F. 1972. The palaeogeography of northern Australia and New Guinea and its relevance to the Torres Strait area. In Walker, D. (ed). *Bridge and Barrier: the Natural and Cultural Heritage of the Torres Strait.* Department of Biogeography and Geomorphology, Research School of Pacific Studies, Australian National University, Canberra. 437pp.

Downing, K.F., Musser, G.G. and Park, L.E. 1998. The first fossil record of small mammals from Sulawesi, Indonesia: the large murid, *Paruromys dominator*, from the late(?) Pliocene Walane formation. Pp.105–121. In Tomida, Y., Flynn, L. J. and Jacobs, L. L. (eds). *Advances in Vertebrate Paleontology and Geochronology.* National Science Museum Monograph, No. 14, Tokyo.

Eisentraut, M. 1976. Das gaumenfaltenmuster der saugetiere und seine bedeutung fur stammesgeschichtliche und taxonomische untersuchungen. *Bonner Zoologische Monographie,* Vol. 8, p. 214.

Ellerman, J.R. 1940–41. *The Families and Genera of Living Rodents, Vols 1–2.* British Museum (Natural History), London.

Fahlbusch, V. 1989. European Neogene rodent assemblages in response to evolutionary, biogeographic, and ecologic factors. Pp.129–138. In Black, C.C. and Dawson, M.R. (eds). *Papers on Fossil Rodents, in Honor of Albert Elmer Wood.* Natural History Museum of Los Angeles County, Science Series, Los Angeles, California.

Flaherty, S.P. and Breed, W.G. 1983. The sperm head of the plains mouse, *Pseudomys australis*: ultrastructure and effects of chemical treatments. *Gamete Research,* Vol. 8, pp.231–244.

Flaherty, S.P. and Breed, W.G. 1987. Formation of the ventral hooks on the sperm head of the plains mouse, *Pseudomys australis*. *Gamete Research,* Vol. 17, pp.115–129.

Flannery, T.F. 1988. Origins of the Australo-Pacific land mammal fauna. *Australian Zoological Reviews,* Vol. 1, pp.15–24.

Flannery, T.F. 1995. *The Mammals of New Guinea.* 2nd Edition. Reed Books, Sydney. 568pp.

Flannery, T.F., Rich, T.H., Turnbull, W.D. and Lundelius, E.L. Jnr. 1992. The Macropodoidea (Marsupialia) of the Early Pliocene Hamilton Local Fauna, Victoria, Australia. *Fieldiana, (Geology),* New Series, No. 25, pp.1–37.

Flannery, T.F. and White, J.P. 1991. Animal translocations. *National Geographic Research and Exploration,* Vol. 7, pp.96–113.

Flannery, T.F. and Wickler. S. 1990. Quaternary murids (Rodentia: Muridae) from Buka Island, Papua New Guinea, with descriptions of two new species. *Australian Mammalogy,* Vol. 13, pp.127–139.

Friend, G.F. 1936. The sperm of British Muridae. *Quarterly Journal of Microscopical Science,* Vol. 78, pp.419–443.

Gadi, I.K. and Sharma, T. 1983. Cytogenetic relationships in *Rattus, Cremnomys, Millardia, Nesokia* and *Bandicota*. *Genetica*, Vol. 61, pp.21–40.

Galloway, R.W. and Kemp, E.M. 1981. Late Cenozoic environments in Australia. Pp.53–80. In Keast, A. (ed). *Ecological Biogeography of Australia.* Vol. 1. W. Junk, The Hague.

Glover, I. 1986. Archaeology in Eastern Timor, 1966–67. *Terra Australis,* 11, pp.1–241. *Department of Prehistory, Research School of Pacific Studies, Australian National University, Canberra.*

Godthelp, H.J. 1987. Rackhams Roost: the beginnings of the modern world. Pp.81–83. In Hand, S.J. and Archer, M. (eds). *The Antipodean Ark.* Angus and Robertson, Sydney. 90pp.

Godthelp, H.J. 1990. *Pseudomys vandycki*, a Tertiary murid from Australia. *Memoirs of the Queensland Museum,* Vol. 28, pp.171–173.

Godthelp, H.J. 1994a. Some rodents from the Pleistocene fluviatile deposits of the eastern Darling Downs, Queensland. *Records of the South Australian Museum,* Vol. 27, p. 220. (abstract only).

Godthelp, H.J. 1994b. A new species of the murid *Zyzomys* from the Pliocene Rackhams Roost deposit, northwest Queensland. *Records of the South Australian Museum,* Vol. 27, p. 220. (abstract only).

Godthelp, H.J. 1997. *Zyzomys rackhami* sp. nov. (Rodentia, Muridae) a rockrat from Pliocene Rackhams's Roost Site, Riversleigh, northwestern Queensland. *Memoirs of the Queensland Museum*, Vol. 41, pp.329–333.

Groves, C.P. and Flannery, T.F. 1994. A revision of the genus *Uromys* Peters, 1867 (Muridae: Mammalia) with descriptions of two new species. *Records of the Australian Museum,* Vol. 46, pp.145–169.

Haig, D.W. and Medd, D. 1996. Latest Miocene to early Pliocene bathymetric cycles related to tectonism, Puri Anticline, Papuan Basin, Papua New Guinea. *Australian Journal of Earth Sciences*, Vol. 43, pp.451–465.

Hamilton, W. 1979. *Tectonics of the Indonesian region.* United States Geological Survey Professional Paper 1079, U.S. Government Printing Office, Washington.

Hand, S.J. 1984. Australia's oldest rodents: master mariners from Malaysia. Pp.905–912. In Archer, M. and Clayton, G. (eds). *Vertebrate Zoogeography and Evolution in Australasia.* Hesperian Press, Perth. 1203pp.

Haq, B.U., Hardenbol, J. and Vail, P.R. 1987. Chronology of fluctuating sea levels since the Triassic. *Science,* Vol. 235, pp.1156–1166.

Hayes, D.E. and Frakes, L.A. 1975. General synthesis, Deep Sea Drilling Project Leg 28. *Initial Reports of the Deep Sea Drilling Project,* Vol. 28, pp.919–942.

Hill, K.C. and Gleadow, A.J.W. 1989. Uplift and thermal history of the Papuan Fold Belt, Papua New Guinea: apatite fission track analysis. *Australian Journal of Earth Sciences,* Vol. 36, pp.515–539.

Hoch, E. and Holm, P.M. 1986. New K/Ar age determinations of the Awe Fauna *Gangue*, Papua New Guinea: consequences for Papuaustralian late Cenozoic biostratigraphy. *Modern Geology,* Vol. 10, pp.181–195.

Hodell, D.A., Elmstrom, K.M. and Kennett, J.P. 1986. Latest Miocene benthic delta^{18}O changes, global ice volume, sea level and the 'Messinian salinity crisis'. *Nature,* Vol. 320, pp.411–414.

Hooijer, D.A. 1957. Three new giant prehistoric rats from Flores Lesser Sunda Islands. *Zoologische Mededelingen,* Vol. 35, pp.299–314.

Hooijer, D.A. 1965. Notes on *Coryphomys buhleri* Schaub, a gigantic murine rodent from Timor. *Israel Journal of Science,* Vol. 14, pp.128–133.

Hooper, E.T. 1958. The male phallus in mice of the genus *Peromyscus. Miscellaneous Publications of the Museum of Zoology, University of Michigan,* No. 105, pp.1–24.

Hooper, E.T. 1959. The glans penis in five genera of cricetid rodents. *Occasional Papers of the Museum of Zoology, University of Michigan,* No. 613, pp.1–10.

Hooper, E.T. 1960. The glans penis in *Neotoma* (Rodentia) and allied genera. *Occasional Papers of the Museum of Zoology, University of Michigan,* No. 618, pp.1–20.

Hooper, E.T. 1962. The glans penis in *Sigmodon, Sigmomys,* and *Reithrodon. Occasional Papers of the Museum of Zoology, University of Michigan,* No. 625, pp.1–11.

Hooper, E.T. and Musser, G.G. 1964. The glans penis in neotropical cricetines (Family Muridae) with comments on classification of muroid rodents. *Miscellaneous Publications of the Museum of Zoology, University of Michigan,* No. 123, pp.1–57.

Hope, G., Flannery, T.F. and Boeadi. 1993. A preliminary report of changing Quaternary mammal faunas in subalpine New Guinea. *Quaternary Research,* Vol. 40, pp.117–126.

Hope, J.H., Lampert, R.J., Edmondson, E., Smith, M.J. and van Tets, G.F. 1977. Late Pleistocene faunal remains from Seton rock shelter, Kangaroo Island, South Australia. *Journal of Biogeography,* Vol. 4, pp.363–385.

Hutchinson, C.S. 1986. Tertiary basins of S.E. Asia—their disparate tectonic origins and eustatic stratigraphic similarities. *Geological Society of Malaysia Bulletin,* Vol. 19, pp.109–122.

Hutchinson, C.S. 1989. *Geological Evolution of South-east Asia.* Clarendon Press, Oxford. 368pp.

Illisson, L. 1971. Comparative sperm nuclear morphology of Australian and some New Guinean Muridae. *Australian Mammal Society Bulletin,* Vol. 2, pp.232–233 (abstract only).

Jacobs, L.L. 1977. A new genus of murid rodent from the Miocene of Pakistan and comments on the origin of the Muridae. *PaleoBios,* Vol. 25, pp.1–11.

Jacobs, L.L. 1978. Fossil rodents (Rhizomyidae and Muridae) from Neogene Siwalik deposits, Pakistan. *Museum of Northern Arizona Press, Bulletin Series,* 52, pp.1–103.

Jacobs, L.L. 1985. The beginning of the age of murids in Africa. *Acta Zoologica, Fennica,* Vol. 170, pp.149–170.

Jacobs, L.L. and Downs, W.R. 1994. The evolution of murine rodents in Asia. Pp.149–156. In Tomida, Y. Li, C.K. and Setogushi, T. (eds). *Rodents and Lagomorph Families of Asian Origins and Diversification.* National Science Museum Monograph, No. 8, Tokyo.

Jacobs, L.L., Flynn, L.F. and Downs, W.R. 1989. Neogene rodents of southern Asia. Pp.157–177. In Black, C.C. and Dawson, M.R. (eds). *Papers on Fossil Rodents, in Honor of Albert Elmer Wood.* Natural History Museum of Los Angeles County, Science Series, Los Angeles, California.

Jacobs, L.L. and Pilbeam, D. 1980. Of mice and men: fossil-based divergence dates and molecular 'clocks'. *Journal of Human Evolution,* Vol. 9, pp.551–555.

Jansa, S.A., Goodman, S.M. and Tucker, P.K. 1999. Molecular phylogeny and biogeography of the native rodents of Madagascar (Muridae: Nesomyinae): a test of the single-origin hypothesis. *Cladistics,* Vol. 15, pp.253–270.

Keller, G. and Barron, J.A. 1983. Paleoceanographic implications of Miocene deep-sea hiatuses. *Geological Society of America Bulletin,* Vol. 94, pp.590–613.

Kemp, E.M. 1981. Tertiary palaeogeography and the evolution of Australian climate. Pp.31–49. In Keast, A. (ed). *Ecological Biogeography of Australia. Vol. 1.* W. Junk, The Hague.

Kennett, J.P. 1977. Cenozoic evolution of Antarctic glaciation, the circum-Antarctic ocean, and their impact on global paleoceanography. *Journal of Geophysical Research,* Vol. 82, pp.3843–3860.

Kennett, J.P. and Vella, P. 1979. Late Cenozoic planktonic foramina and paleoceanography at DSDP Site 284 in the cool subtropical South Pacific. *Initial Reports of the Deep Sea Drilling Project,* Vol. 29, pp.769–799.

Kershaw, A.P., Band, J.G., D'Costa, D.M., Edney, P.A., Peterson, J.A. and Strickland, K.M. 1991. A comparison of long Quaternary pollen records from the Atherton and Western Plains volcanic provinces, Australia. Pp.288–301. In Williams, M.A.J., De-Deckker, P. and Kershaw, A.P.(eds). *The Cainozoic in Australia.* Geological Society of Australia, Special Publication No.18. Sydney.

Kitchener, D.J. 1980. A new species of *Pseudomys* (Rodentia; Muridae) from Western Australia. *Records of the Western Australian Museum,* Vol. 8, pp.405–414.

Kitchener, D.J., Aplin, K.P. and Boeadi. 1991a. A new species of *Rattus* from Gunung Mutis, South West Timor Island, Indonesia. *Records of the Western Australian Museum,* Vol. 15, pp.445–465.

Kitchener, D.J., How, R.A. and Maharadatunkamsi. 1991b. *Paulamys* sp. cf. *P. naso* (Musser, 1981) (Rodentia; Muridae) from Flores Island, Nusa Tenggara, Indonesia—description of a modern specimen and a consideration of its phylogenetic affinities. *Records of the Western Australian Museum,* Vol. 15, pp.171–189.

Lande, R. 1979. Effective deme sizes during long-term evolution estimated from rates of chromosomal rearrangement. *Evolution,* Vol. 33, pp.234–251.

Lavocat, R. 1967. A propos de la dentition des ronguers et du probleme de l'origin des murindes. *Mammalia,* Vol. 31, pp.205–216.

Lee, A.K., Baverstock, P.R. and Watts, C.H.S. 1981. Rodents—the late invaders. Pp.1521–1553. In Keast, A. (ed). *Ecological Biogeography of Australia. Vol.3.* W. Junk, The Hague.

Li, W.-H., Tanimura, M. and Sharp, P.M. 1987. An evaluation of the molecular clock hypothesis using mammalian DNA sequences. *Journal of Molecular Evolution,* Vol. 25, pp.330–342.

Lidicker, W.Z. Jnr. 1968. A phylogeny of New Guinea rodent genera based on phallic morphology. *Journal of Mammalogy,* Vol. 49, pp.609–643.

Lidicker, W.Z. Jnr. 1973. A phenetic analysis of some New Guinea rodents. *Systematic Zoology,* Vol. 22, pp.36–45.

Lidicker, W.Z. Jnr. and Brylski, P.V. 1987. The conilurine rodent radiation of Australia, analysed on the basis of phallic morphology. *Journal of Mammalogy,* Vol. 68, pp.617–641.

Linthout, K., Helmers, H. and Sopaheluwaken, J. 1997. Late Miocene abduction and microplate migration around the southern Banda Sea and the closure of the Indonesian Seaway. *Tectonophysics,* Vol. 281, pp.17–30.

Linzey, A.V. and Layne, J.N. 1974. Comparative morphology of spermatozoa of the rodent genus *Peromyscus. American Museum Novitates,* No. 2532, pp.1–20.

Mackness, B.S., Whitehead, P.W. and McNamara, G.C. 2000. New potassium-argon basalt date in relation to the Pliocene Bluff Downs Local Fauna, northern Australia. *Australian Journal of Earth Sciences,* Vol. 47, pp.807–811.

Macphail, M.K. 1996. Neogene environments in Australia, 1. Re-evaluation of microfloras associated with important early Pliocene marsupial remains at Grange Burn, south-west Victoria. *Review of Palaeobotany and Palynology,* Vol. 92, pp.307–328.

Marshall, L.G. 1973. Fossil vertebrate faunas from the Lake Victoria region. *National Museum of Victoria, Memoirs,* No. 34, pp.151–171.

Martin, H.A. 1991. Tertiary stratigraphic palynology and palaeoclimate in the inland river systems in New South Wales. Pp.181–194. In Williams, M.A.J., De-Deckker, P. and Kershaw, A.P. (eds). *The Cainozoic in Australia.* Geological Society of Australia, Special Publication, No.18. Sydney.

Maxson, L.R., Sarich, V.M. and Wilson, A.C. 1975. Continental drift and the use of albumin as an evolutionary clock. *Nature,* Vol. 255, pp.397–400.

McGowran, B. and Li, Q. 1994. The Miocene oscillation in southern Australia. *Records of the South Australian Museum,* Vol. 27, pp.197–212.

Mein, P. 1978. A propos de *Mus gaudryi* (Mammalia, Rodentia) et des premiers *Apodemus* (*A. primaevus* et *A. gudrunae*) du Miocene superieur European. *Geobios,* Vol. 11, pp.389–392.

Mein, P. 1984. Composition quantitative des faunes de mammiferes du Miocene moyen de la region lyonnaise. *Paleobiologie Continentale,* Vol. 14, pp.339–346.

Mein, P. and Ginsburg, L. 1985. Les rongeurs miocenes de Li (Thailande). *Comptes Rendus Academie des Sciences, Paris,* Vol. 19, pp.1369–1374.

Menzies, J.I. 1990. A systematic revision of *Pogonomelomys* (Rodentia: Muridae) of New Guinea. *Science in New Guinea,* Vol. 16, pp.118–137.

Menzies, J.I. 1996. A systematic revision of *Melomys* (Rodentia: Muridae) of New Guinea. *Australian Journal of Zoology,* Vol. 44, pp.367–426.

Michaux, J. 1971. Muridae (Rodentia) Neogenes d'Europe Sud-Occidentale, evolution et rapports avec les formes actuelles. *Paleobiologie Continentale,* Vol. 2, pp.1–76.

Miller, G.S. and Gidley, J.W. 1918. Synopsis of the supergeneric groups of rodents. *Journal of the Washington Academy of Science,* Vol. 8, pp.431–448.

Misonne, X. 1969. African and Indo-Australian Muridae. Evolutionary trends. *Musee Royal de l'Afrique Centrale, Terverun, Belgique Annales, serie 8, Sciences Zoologiques,* No. 172, pp.1–219.

Moritz, C., Joseph, L., Cunningham, M. and Schneider, C. J. 1997. Molecular perspectives on historical fragmentation of Australian tropical and subtropical rainforest: implications for conservation. Pp.442–454. In Laurence, W.F. and Bieregard, R.O. (eds). *Tropical Rainforest Remnants: Ecology, Management and Conservation of Fragmented Communities.* Chicago University Press, Chicago.

Morrissey, B.L. and Breed, W.G. 1982. Variation in external morphology of the glans penis of Australian native rodents. *Australian Journal of Zoology,* Vol. 30, pp.495–502.

Musser, G.G. 1969. Results of the Archbold Expeditions. No. 91. A new genus and species of murid rodent from Celebes, with a discussion of relationships. *American Museum Novitates,* No. 2384, pp.1–41.

Musser, G.G. 1981a. Results of the Archbold Expeditions. No.105. Notes on systematics of Indo-Malayan murid rodents, and descriptions of new genera and species from Ceylon, Sulawesi and the Philippines. *Bulletin of the American Museum of Natural History,* Vol. 168, pp.225–334.

Musser, G.G. 1981b. The giant rat of Flores and its relatives east of Borneo and Bali. *Bulletin of the American Museum of Natural History,* Vol. 169, pp.67–176.

Musser, G.G. 1982a. Results of the Archbold Expeditions. No.108. The definition of *Apomys*, a native rat of the Philippine Islands. *American Museum Novitates,* No. 2746, pp.1–43.

Musser, G.G. 1982b. Results of the Archbold Expeditions. No.110. *Crunomys* and the small-bodied shrew rats native to the Philippine Islands and Sulawesi (Celebes). *Bulletin of the American Museum of Natural History,* Vol. 174, pp.1–95.

Musser, G. G. 1982c. The Trinil rats. *Modern Quaternary Research in Southeast Asia,* Vol. 7, pp.65–85.

Musser, G.G. 1986. Sundaic *Rattus*: definitions of *Rattus baluensis* and *Rattus korinchi. American Museum Novitates,* No. 2862, pp.1–24.

Musser, G.G. 1990. Sulawesi rodents: species traits and chromosomes of *Haeromys minahassae* and *Echiothrix leucura* (Muridae: Murinae). *American Museum Novitates,* No. 2989, pp.1–18.

Musser, G.G. and Carleton, M.D. 1993. Family Muridae. Pp.501–755. In Wilson, D.M. and Reeder, D.M. (eds). *Mammal Species of the World—A Taxonomic and Geographic Reference.* 2nd Edit. Smithsonian Institution Press, Washington DC. 1207pp.

Musser, G.G. and Heaney, L.R. 1992. Philippine rodents: definitions of *Tarsomys* and *Limnomys* plus a preliminary assessment of phylogenetic patterns among native Philippine murines (Murinae, Muridae). *Bulletin of the American Museum of Natural History,* Vol. 211, pp.1–138.

Musser, G.G. and Holden, M.E. 1991. Sulawesi rodents (Muridae: Murinae): morphological and geographical boundaries of species in the *Rattus hoffmanni* Group and a new species from Pulau Peleng. *Bulletin of the American Museum of Natural History,* Vol. 206. Pp.322–413.

Musser, G.G. and Newcomb, C. 1983. Malaysian murids and the giant rat of Sumatra. *Bulletin of the American Museum of Natural History,* Vol. 174, pp.327–598.

Newsome, A.E. and Corbett, L.K. 1975. Outbreaks of rodents in central Australia: origins, declines and evolutionary considerations. Pp.117–153. In Prakash, I. and Ghosh, P. (eds). *Rodents in Desert Environments.* Monographiae Biologie. W. Junk, The Hague.

Nott, J. 1994. Long-term landscape evolution in the Darwin region and its implications for the origin of landsurfaces in the north of the Northern Territory. *Australian Journal of Earth Sciences,* Vol. 41, pp.407–415.

O'hUigin, C. and Li, W.-H. 1992. The molecular clock ticks regularly in muroid rodents and hamsters. *Journal of Molecular Evolution,* Vol. 35, pp.377–384.

Page, R.W. and McDougal, I. 1972. Ages of mineralization of gold and porphyry copper deposits in the New Guinean highlands. *Economic Geology,* Vol. 67, pp.1034–1048.

Petter, F. 1966. L'origine des murides plan cricetin et plans murins. *Mammalia,* Vol. 30, pp.205–225.

Pigram, C.J. and Davies, H.L. 1987. Terranes and the accretion history of the New Guinea orogen. *Bureau of Mineral Resources, Journal of Australian Geology and Geophysics,* Vol. 10, pp.193–211.

Pigram, C.J. and Panggabean, H. 1984. Rifting of the northern margin of the Australian continent and the origin of some microcontinents in eastern Indonesia. *Tectonophysics,* Vol. 107, pp.331–353.

Pirazzoli, P.A., Radtke, U., Hantoro, W.S., Jouannic, C., Hoang, C.T., Causse, C. and Borel-Best, M. 1993. A one million-year-long sequence of marine terraces on Sumba Island, Indonesia. *Maritime Geology,* Vol. 109, pp.221–236.

Plane, M.D. 1967. Stratigraphy and vertebrate fauna of the Otibanda Formation, New Guinea. *Bureau of Mineral Resources, Geology and Geophysics Bulletin,* No. 86, pp.1–64.

Qumsiyeh, M.B. and Baker, R.J. 1988. Comparative cytogenetics and the determination of primitive karyotypes. *Cytogenetics and Cell Genetics,* Vol. 47, pp.100–103.

Rich, T.H., Archer, M., Hand, H.J., Godthelp, H., Muirhead, J., Pledge, N.S., Flannery, T.F., Woodburne, M.O., Case, J.A., Tedford, R.H., Turnbull, W.D., Lundelius,

E.L. Jnr., Rich L.S.V., Whitelaw, M.J., Kemp, A. and Rich, P.V. 1991. Australian Mesozoic and Tertiary terrestrial mammal localities. Pp.1005–1058. In Vickers-Rich, P., Monaghan, J.M., Baird, R.F. and Rich, T.H. (eds). *Vertebrate Palaeontology of Australasia.* Pioneer Design Studio and Monash University Publications Committee, Melbourne. 1437pp.

Richardson, B.J., Baverstock, P.R. and Adams, M. 1986. *Allozyme Electrophoresis: a Handbook for Animal Systematics and Population Studies.* Academic Press, Sydney. 410pp.

Rickart, E.A., Heaney, L.R. and Utzurrum, R.B. 1991. Distribution and ecology of small mammals along an elevational transect in southeastern Luzon, Philippines. *Journal of Mammalogy,* Vol. 72, pp.458–469.

Rickart, E.A. and Musser, G.G. 1993. Philippine rodents: chromosomal characteristics and their significance for phylogenetic inference among 13 species (Rodentia: Muridae: Murinae). *American Museum Novitates,* No. 3064, pp.1–34.

Roberts, A.P., Turner, G.M. and Vella, P.P. 1994. Magnetostratigraphic chronology of late Miocene to early Pliocene biostratigraphic and oceanographic events in New Zealand. *Geological Society of America Bulletin,* Vol. 10, pp.665–683.

Roldan, E.R.S., Gomendio, M. and Vitullo, A.D. 1992. The evolution of eutherian spermatozoa and underlying selective forces: female selection and sperm competition. *Biological Reviews,* Vol. 67, pp.551–593.

Ruedas, L.A. and Kirsch, J.A.W. 1997. Systematics of *Maxomys* Sody, 1936 (Rodentia: Muridae: Murinae): DNA/DNA hybridization studies of some Borneo-Javan species and allied Sundaic and Australo-Papuan genera. *Biological Journal of the Linnean Society,* Vol. 61, pp.385–408.

Rummler, H. 1938. Die Systematik und Verbreitung der Muriden Neuguineas. *Mitteilungen aus dem Zoologischen Museum in Berlin,* Vol. 23, pp.1–298.

Sarich, V.M. 1969. Pinniped origins and the rate of evolution of carnivore albumins. *Systematic Zoology,* Vol. 18, pp.286–295.

Schodde, R. and Calaby, J.H. 1972. The biogeography of the Australo-Papuan bird and mammal faunas in relation to Torres Strait. Pp.257–300. In Walker, D. (ed). *Bridge and Barrier: the Natural and Cultural Heritage of the Torres Strait.* Department of Biogeography and Geomorphology, Research School of Pacific Studies, Australian National University, Canberra. 437pp.

Shackelton, N.J. 1986. Paleogene stable isotope events. *Palaeogeography, Palaeoclimatology, Palaeoecology,* Vol. 57, pp.91–102.

Shackelton, N.J. and Hall, M.A. 1984. Oxygen and carbon isotope stratigraphy of Deep Sea Drilling Project Hole 552A: Plio-Pleistocene glacial history. *Initial Reports of the Deep Sea Drilling Project,* Vol. 81, pp.599–609.

Shackelton, N.J. and Kennett, J.P. 1979. Paleotemperature history of the Cenozoic and the initiation of Antarctic glaciation: oxygen and carbon isotope analyses in DSDP sites 277, 279 and 281. *Initial Reports of the Deep Sea Drilling Project,* Vol. 29, pp.743–755.

Shimamura, M., Yasue, H., Ohahima, K., Abe, H., Kato, H., Kishiro, T., Goto, M., Munechika, I. and Okada, N. 1997. Molecular evidence from retrotransposons that whales form a clade within even-toed ungulates. *Nature,* Vol. 388, pp.666–670.

Simpson, G.G. 1945. The principles of classification and a classification of mammals. *Bulletin of the American Museum of Natural History,* Vol. 85, pp.1–350.

Simpson, G.G. 1961. Zoogeography of Australian mammals. *Evolution,* Vol. 15, pp.431–446.

Singleton, G. 1997. Integrated management of rodents: a southeast Asian and Australian perspective. *Belgian Journal of Zoology,* Vol. 127, pp.145–156.

Smith, A.G., Smith, D.G. and Funnell, B.M. 1994. *Atlas of Mesozoic and Cenozoic Coastlines.* Cambridge University Press, Cambridge. 99pp.

Storch, G. 1987. The Neogene mammalian faunas of Ertemte and Harr Obo in Inner Mongolia (Nei Mongol), China.—7. Muridae (Rodentia). *Senckenbergiana Lethaea,* Vol. 67, pp.401–431.

Tate, G.H.H. 1936. Some Muridae of the Indo-Australian region. *Bulletin of the American Museum of Natural History,* Vol. 72, pp.501–728.

Tate, G.H.H. 1951. Results of the Archbold Expedition. No. 65. The rodents of Australia and New Guinea. *Bulletin of the American Museum of Natural History,* Vol. 97, pp.183–430.

Taylor, J.M., Calaby, J.H. and Smith, S.C. 1983. Native *Rattus*, land bridges, and the Australian region. *Journal of Mammalogy,* Vol. 64, pp.463–475.

Taylor, J.M., Calaby, J.H. and Van Deusen, H.M. 1982. A revision of the genus *Rattus* (Rodentia, Muridae) in the New Guinea region. *Bulletin of the American Museum of Natural History,* Vol. 173, pp.177–336.

Taylor, J.M. and Horner, B.E. 1970. Reproduction in the mosaic-tailed rat, *Melomys cervinipes* (Rodentia: Muridae). *Australian Journal of Zoology,* Vol. 18, pp.171–184.

Taylor, J.M. and Horner, B.E. 1972. Observations on the reproductive biology of *Pseudomys* (Rodentia: Muridae). *Journal of Mammalogy,* Vol. 53, pp.318–328.

Taylor, J.M. and Horner, B.E. 1973. Results of the Archbold Expeditions, No. 98. Systematics of Australian *Rattus* (Rodentia, Muridae). *Bulletin of the American Museum of Natural History,* Vol. 150, pp.1–130.

Tedford, R.H. 1994. Succession of Pliocene through medial Pleistocene mammal faunas of southeastern Australia. *Records of the South Australian Museum,* Vol. 27, pp.79–93.

Tedford, R.H., Wells, R.T. and Barghoorn, S.F. 1992. Tirari Formation and contained faunas, Pliocene of the Lake Eyre Basin, South Australia. *The Beagle (Records of the Northern Territory Museum of Arts and Sciences),* Vol. 9, pp.173–194.

Torgersen, T., Luly, J., DeDeckker, P., Jomes, M.R., Searle, D.E., Chivas, A.R. and Ullman, W.J. 1988. Late Quaternary environments of the Carpentaria Basin, Australia. *Palaeogeography, Palaeoclimatology, Palaeoecology,* Vol. 67, pp.445–461.

Torrance, A.W. 1997. *Biogeography and diversification of the Australian Hydromyinae.* Ph.D. Thesis (unpublished), Harvard University, Boston, Massachusetts.

Turnbull, W.D. and Lundelius, E.L.Jnr. 1970. The Hamilton fauna, a late Pliocene mammalian fauna from the Grange

Burn, Victoria, Australia. *Fieldiana (Geology),* Vol. 19, pp.1–163.

Usdin, K., Chevret, P., Catzeflis, F.M., Verona, R. and Furano, A.V. 1995. L1 (LINE-1) retrotransposable elements provide a "fossil" record of the phylogenetic history of murid rodents. *Molecular Biology and Evolution,* Vol. 12, pp.73–82.

Vail, P.R. and Hardenbol, J. 1979. Sea level changes during the Tertiary. *Oceanus,* Vol. 22, pp.71–80.

Vandebroek, G. 1966. Plans dentaires fondamentaux chez les ronguers, origine des murides. *Musee Royal de l'Afrique Centrale, Terverun, Belgique Annales, serie 8, Sciences Zoologiques,* No. 114, pp.117–152.

Verheyen, E., Colyn, M. and Verheyen, W. 1996. A mitochondrial cytochrome *b* phylogeny confirms the paraphyly of the Dendromurinae Alston, 1896 (Muridae, Rodentia). *Mammalia,* Vol. 60, pp.780–785.

Verneau, O., Catzeflis, F. and Furano, A.V. 1997. Determination of the evolutionary relationships in *Rattus* sensu lato (Rodentia: Muridae) using L1 (LINE-1) amplification events. *Journal of Molecular Evolution,* Vol. 45, pp.424–436.

Verneau, O., Catzeflis, F. and Furano, A.V. 1998. Determining and dating recent rodent speciation events by using L1 (Line-1) retrotransposons. *Proceedings of the National Academy of Sciences of the United States of America,* Vol. 95, pp.11284–11289.

Viegas-Pequignot, E. 1985. Major chromosome homeologies between Muridae and Cricetidae. *Cytogenetics and Cell Genetics,* Vol. 39, pp.258–261.

Viegas-Pequignot, E., Dutrillaux, B., Prod'Homme, M. and Petter, F. 1983. Chromosomal phylogeny of Muridae: a study of 10 genera. *Cytogenetics and Cell Genetics,* Vol. 35, pp.269–278.

Viegas-Pequignot, E., Petit, D., Benazzou, T., Prod'Homme, M., Lombard, M., Hoffschir, F., Descailleaux, J. and Dutrillaux, B. 1986. Phylogenie chromosomique chez les Sciuridae, Gerbillidae et Muridae, et etude d'especes appartenant a d'autres familles de rongeurs. *Mammalia,* Vol. 50, pp.164–202.

Voss, R. S. and Linzey, A.V. 1981. Comparative gross morphology of male accessory glands among Neotropical Muridae (Mammalia: Rodentia) with comments on systematic implications. *Miscellaneous Publications of the Museum of Zoology, University of Michigan,* No. 159, pp.1–41.

Wahlert, J.A. 1974. The cranial foramina of protrogomorphous rodents; an anatomical and phylogenetic study. *Bulletin of the Museum of Comparative Zoology*, Vol. 146, pp.363–410.

Wakefield, N.A. 1960a. Recent mammal bones in the Buchan District. Part 1. *Victorian Naturalist,* Vol. 77, pp.164–178.

Wakefield, N.A. 1960b. Recent mammal bones in the Buchan District. Part 2. *Victorian Naturalist,* Vol. 77, pp.227–240.

Watts, C.H.S. 1974. The native rodents of Australia: a personal view. *Australian Mammalogy,* Vol. 1, pp.109–116.

Watts, C.H.S. and Aslin, H.J. 1981. *The Rodents of Australia.* Angus and Robertson, Sydney. 321pp.

Watts, C.H.S. and Baverstock, P.R. 1994a. Evolution in New Guinean Muridae (Rodentia) assessed by microcomplement fixation of albumin. *Australian Journal of Zoology,* Vol. 42, pp.295–306.

Watts, C.H.S. and Baverstock, P.R. 1994b. Evolution in some South-east Asian Murinae (Rodentia), as assessed by microcomplement fixation of albumin, and their relationship to Australian murines. *Australian Journal of Zoology,* Vol. 42, pp.711–722.

Watts, C.H.S. and Baverstock, P.R. 1995a. Evolution in the Murinae (Rodentia) assessed by microcomplement fixation of albumin. *Australian Journal of Zoology,* Vol. 43, pp.105–118.

Watts, C.H.S. and Baverstock, P.R. 1995b. Evolution in some African Murinae (Rodentia) assessed by microcomplement fixation of albumin. *Journal of African Zoology,* Vol. 109, pp.423–433.

Watts, C.H.S. and Baverstock, P.R. 1996. Phylogeny and biogeography of some Indo-Australian murid rodents. Pp.47–50. In Kitchener, D.J. and Suyanta, A. (eds). *Proceedings of the First International Conference on Eastern Indonesian-Australian Vertebrate Fauna, Manado, Indonesia, November 22–26, 1994.* Western Australian Museum, Perth.

Watts, C.H.S., Baverstock, P.R., Birrell, J. and Krieg, M. 1992. Phylogeny of the Australian rodents (Muridae): a molecular approach using microcomplement fixation of albumin. *Australian Journal of Zoology,* Vol. 40, pp.81–90.

Watts, C.H.S. and Kemper, C.M. 1989. Muridae. Pp.939–950. In Walton, D.W. and Richardson, B.J. (eds). *Fauna of Australia. Vol. 1B. Mammalia.* Australian Government Printing Service, Canberra. Pp.401–1227.

Weber, M. 1902. *Der Indo-Australische Archipelago und die Geschichte seiner Tierwelt.* Jena.

Weerd, A. van de. 1976. Rodent faunas of the Mio-Pliocene continental sediments of the Teruel-Alfambra region, Spain. *Utrecht Micropaleontological Bulletin, Special Publication,* 2, pp.1–217.

Weerd, A. van de and Daams, R. 1978. Quantitative composition of rodent faunas in the Spanish Neogene and paleoecological implication. *Proceedings of the Koninklijke Nederlandse Akademie van Wetenschappen,* Series B, Vol. 81, pp.448–473.

White, J.P. 1972. Ol Tumbuna. Archaeological excavations in the eastern Central Highlands, Papua New Guinea. *Terra Australis,* 2, pp.1–176. *Department of Prehistory, Research School of Pacific Studies, Australian National University, Canberra.*

White, M.E. 1994. *After the Greening. The Browning of Australia.* Kangaroo Press, Sydney. 288pp.

Whitelaw, M.J. 1989. Magnetic polarity stratigraphy and mammalian fauna of the late Pliocene (Early Matuyama) section at Batesford (Victoria), Australia. *Journal of Geology,* Vol. 97, pp.624–631.

Whitelaw, M.J. 1991. Magnetic polarity stratigraphy of Pliocene and Pleistocene fossil vertebrate localities in southeastern Australia. *Geological Society of America, Bulletin,* Vol. 103, pp.1493–1503.

Winkler, A.J. 1994. The middle/upper Miocene dispersal of major rodent groups between southern Asia and Africa. Pp.173–184. In Tomida, Y., Li, C.K. and Setogushi, T. (eds). *Rodent and Lagomorph Families of Asian Origins and Diversification.* National Science Museum Monograph, No. 8, Tokyo.

Woodruff, F., Savin, S.M. and Douglas, R.G. 1981. Miocene stable isotope record: a detailed deep Pacific Ocean study and its paleoclimatic implications. *Science,* Vol. 212, pp.665–668.

1995a). There are also close similarities in body shape, breeding pattern, coat colour, range and social behaviours between dingoes in South-east Asia and Australia. This indicates that the dingo's general morphology has remained virtually unchanged for the past 5,500 years, although as discussed later, this situation is now rapidly changing in Australia and elsewhere through hybridization with domestic dogs. See Plates 112–114.

The eastern Asian dingo was the form that became cosmopolitan through its association with the movements of early humans as their populations expanded (Bellwood, 1978, 1984; Thorne and Raymond, 1989). It was during this expansion that dingoes were transported to Australia by Asian seafarers (Corbett, 1985); the most recent introductions were by Macassan trepangers between 90 and 350 years ago (Macknight, 1976). Dingoes probably accompanied the human migrants mainly as a source of fresh food during long sea voyages, or as guard dogs during stopovers. There were also cultural reasons for transporting dingoes (Clutton-Brock, 1977; Olsen and Olsen, 1977). For example, even today in Borneo, dogs (and dingoes) are killed, eaten and the remains buried with the dead owner as part of a funeral ceremony (Medway, 1977). Within these cultures, death is the beginning of a journey into the afterlife where man is accompanied by his dog; so perhaps this is a reason why dingoes accompanied Asian man during sea voyages.

The primitive dogs of most Pacific islands (Titcombe, 1969), the ancient Kirri dog of New Zealand (Colenso, 1877; Bay-Petersen, 1979), the Basenji in the African Congo (Coe, 1994) and the New Guinean singing dog (Troughton, 1957, Schultz, 1969) all descended from south-east Asian dingoes. According to fossil evidence, the primitive dogs of the Americas were also morphologically very similar to dingoes and they probably arrived there together with people via the Bering Strait (Olsen and Olsen, 1977). The Carolina dog is the remaining descendant of the early Amerindian canids (Brisbin and Risch, 2000).

Most of the Austronesian dingo-like dogs were closely related (Matisoo-Smith, 1998) but recent preliminary mtDNA sequencing indicates that dingoes from New Guinea and Australia have several genetic markers not found in grey wolves (*Canis lupus lupus*) or domestic dogs (Koler-Matznick *et al.,* 1998). If proven, this may be a reflection of recent dingo evolution in isolation from grey wolf populations.

AUSTRALIAN DINGO FOSSILS

Although there have been relatively few archaeological discoveries of dingo in Australia (Mulvaney, 1975), they are ubiquitous in late Holocene deposits across southern and eastern Australia in contexts dated at less than 4,000 years BP (Gollan, 1984). For example, fossil dingo remains are present in the late Holocene at McEachern Cave (Wakefield, 1967) and Gisborne Cave (McCoy, 1882) in Victoria, and Madura Cave in Western Australia (Milham and Thompson, 1976). See Table 1. The most substantial fossil is a 3,000 year old (almost) complete skeleton of a juvenile from Fromm's Landing, South Australia (Macintosh, 1964). Measurements of 80 fossil crania from caves and natural pit traps in southern and eastern Australia fall within the range of variability of modern wild dingo populations (Gollan, 1982, 1984).

Earlier (erroneous) claims that the dingo was present in Australia in the Pleistocene were based on associations between fossil dingo remains and those of extinct Australian mammals (McCoy, 1861; Krefft, 1867; Gill, 1953). Such claims are now refuted or cannot be substantiated, as they are based on stratigraphic controls or synchronicity tests on bone chemistry (Gill and Sinnott, 1973; Gollan, 1982; White and O'Connell, 1982). Other published dates of 7,030–8,600 BP (Mulvaney, 1975) are also incorrect as they are now interpreted as a mixing of upper and lower layers due to disturbance by digging and burrowing animals, and infiltration of upper deposits into lower deposits (Gollan, 1984).

There are no dingoes in the Devils Lair faunal assemblage, dated at 5,000 years BP (Balme *et al.,* 1978), which suggests that dingoes were not present prior to that date and the absence of dingoes in Tasmania is interpreted to mean that its introduction into Australia post-dated the formation of Bass Strait, some 11,000 years BP (Mulvaney, 1975).

Table 1. Dates and locations for the earliest records of fossil dingoes in Australia (adapted from Gollan, 1984).

Date (yrs BP)	Location	Context	Ref. number	Reference
3,450 ± 95	S.E. Western Australia	Madura Cave	ANU 807	Milham and Thompson (1976)
3,230 ± 100	New South Wales	Wombah Midden	Gak 568	Mulvaney (1975)
3,170 ± 94	South Australia	Fromm's Landing	NPL 29	Macintosh (1964)
2,980 ± 90	South Australia	Devon Downs	Gak 1021	Smith (1980)
2,865 ± 57	New South Wales	Capertee 3	V33	Johnson (1979)
2,200 ± 96	Nullarbor Region, Western Australia	Thylacine Hole	NSW 30	Lowry and Merrilees (1969)
1,680 ± 100	New South Wales	Burrill Lake	ANU 139	Lampert (1971)
1,020 ± 40	South Australia	Mt. Burr	Gak 426	Gollan (1982)

Similarly, given that Australia was separated from Indonesia and New Guinea by a rise in sea level 8,000–6,500 years ago (Jennings, 1971) it can be inferred that dingoes were not present on the Sunda Plain because they should have retreated to both northern and southern areas (ie. Australia) with the rise in sea level (Gollan, 1984).

In summary, the oldest dingo fossils in southern and eastern Australia are in the range 3,500–3,000 years BP. Based on the speed of radiation of foxes, feral pigs and feral cats throughout Australia in modern times, the initial radiation of dingoes throughout mainland Australia is also likely to have been rapid, especially if radiation was assisted by the Aborigines who had arrived on the continent at least 40,000 years earlier (Mulvaney, 1975; Roberts *et al.*, 1990). Gollan (1984) suggested a period of 500 years between the dingo founder population in northern Australia and their widespread appearance as fossils in southern areas of the continent. The inference is that the dingo arrived in Australia about 4,000 years ago.

INTERACTIONS WITH ABORIGINALS

The original interactions would have been with dingoes accompanying the Asian seafarers but there is evidence that many of those dingoes also returned to Asia. This conclusion is based on the distribution of a biting louse (*Heterodoxus spiniger*), found only on marsupials in Australia and modern canids in mainland Asia (Hopkins, 1949). It is suggested that the louse probably evolved after transference to the dingo from a marsupial and spread to the Asian continent on returning dingoes.

Following the discovery of eastern Australia by Captain Cook and later European settlement and inland exploration, dingoes were regularly recorded in the company of Aborigines (Breckwoldt, 1988; Walters, 1995) but there are few comments on their use by Aborigines. However, based on Aboriginal rock art and mythology and observations of early explorers (Breckwoldt, 1988), there is no doubt that dingoes played an important role in Aboriginal life. Kolig (1978) suggested that the association between dingoes and Aborigines was based on the mischievous nature of the dingo and served as a model of the need for restraint and order in Aboriginal society.

In Aboriginal mythology, Djaranging was a star-eyed dog who watched the souls of deceased Aborigines lingering at the graves of their former bodies (Harney and Elkin, 1949). Dingoes are included in Aboriginal songs such as 'The Hunters Cycle' (Harney and Elkin, 1949) and there are songs associated with the Dog Dreaming such as those sung by the Walbrai people in the Tanami Desert in central Australia (Breckwoldt, 1988). The Kundi-Djumindju clan, in the Victoria River region of northern Australia, perform a dance of how the dingo was introduced into the Dreamtime (Harney, 1951). This dance has been re-interpreted to indicate the introduction of the dingo into Australia by Asian seafarers (Rolls, 1992; Corbett, 1995a). The dingo is the chief totem of the Mildjinggi clan in Arnhem Land and there is a story and dance associated with its origin (Thompson, 1955; Wilkinson, 2000).

Although it is difficult to date rock art, the appearance of the dingo in Aboriginal art is generally consistent with the age of the oldest dingo fossils (Clarke, 1968). In Cape York, dingoes are regularly depicted in cave paintings including one of a woman suckling a dingo pup and another of dingoes chasing kangaroos that have been wounded by spears (Breckwoldt, 1988; Walters, 1995).

Some Aboriginal tribes used dingoes to hunt game, especially kangaroos and wallabies (Macropodidae), possums (Phalangeroidea) and echidnas (*Tachyglossus aculeatus*) (Finlayson, 1935; Thompson, 1949; Meggitt, 1965; Hamilton, 1972; White, 1972; Gould, 1979; Pickering, 1992) but it is inconclusive as to how effective they were. Meggitt (1965) questioned whether the dingo was an effective hunting dog based on his own observations and those of early explorers (Giles, 1889; Basedow, 1925). According to Mulvaney (1975), dingoes ate more of their (Aboriginal) master's food than they retrieved for them. Other snippets of information from anthropologists working with Aborigines indicated: that dingoes followed the scent of kangaroos but rarely killed them; that they were useful to warn of the approach of strangers; that dingoes were good for children to play with and learn about animal behaviour. Other writers recorded that Aboriginals used dingoes to locate and catch small game such as lizards, but chased the dingoes back to camp when hunting parties set out to hunt large kangaroos because dingoes disturbed the hunt in the final stages (Kolig, 1973, 1978).

Because dingoes were not always reliable hunting aids Aborigines took to domestic dogs as soon as they became available. Many pictures of Aboriginals and dogs taken around the turn of the 20th century show clearly that the dogs were domestic breeds—especially greyhound or mixed breeds involving greyhound and known as kangaroo dogs (see plates in Breckwoldt, 1988). It is likely that records this century were of Aboriginals hunting with domestic dogs or hybrids. This explains the conflicting statements about the use and value of dingoes as hunting dogs. Aboriginals may have changed some of their hunting strategies for kangaroos based on the tractable and more manageable nature of domestic dogs (White, 1972). Tasmanian Aboriginals assimilated European dogs into their hunting economy with such rapidity that by 1830, only 50 years after the introduction of domestic dogs into Tasmania, they were as ubiquitous and numerous as in mainland Aboriginal camps and were the subject of entire dances (Jones, 1970; Mulvaney, 1975).

Interestingly, there is evidence that some Aboriginal groups selected certain traits in dingoes in a manner reminiscent of the process in Asia several millennia earlier. The skeletal modification in some dingoes from archaeological burial sites in eastern Australia, as compared to wild populations, may have been due to selective breeding of camp dingoes by Aborigines (Gollan, 1984). This is supported by the fact that some Aboriginal groups had separate names for camp dingoes and wild dingoes (Hamilton, 1972). Elsewhere in Australia, however, the fact that most skeletons at burial sites are indistinguishable from skeletal deposits at non-burial sites indicates that the Aborigines regularly took dingoes from the wild. This is supported by observations of Aborigines taking pups from the wild (Finlayson, 1935) and anecdotal reports of Aborigines breaking the legs of dingoes to keep them in camp.

Some Aboriginal people suckled pups and slept with dingoes for warmth (Lumholtz, 1889) and the practice of eating dingoes was widespread. For example, it was recorded in south-east Australia (Krefft, 1866), central Australia (Basedow, 1925; Tindale, 1974) and north-west Australia (Durack, 1983). Dingo parts were used as ornaments. In central-western Queensland, knuckle bones were suspended over the ears (McCarthy, 1967) and there are examples of necklaces made from canine teeth. In central Australia Aborigines regularly traded dingo scalps to doggers (Finlayson, 1935); the author observed that scalps were still used as a form of currency in the late 1960s.

DINGO POPULATION FLUCTUATIONS SINCE EUROPEAN SETTLEMENT

Initially, dingo numbers throughout the Australian mainland would have been kept down by irregularities in food supply, by social factors and by diseases that periodically reduced populations. However, following European settlement and the development of the pastoral industry in inland Australia (about 75% of the continent) about 100 years ago (Bauer, 1983), dingo numbers rose dramatically. Food such as rabbits, stock and some macropodids became more plentiful, and dams or artesian bores increased water supply during drought. In many of the more closely settled areas of Australia, however, changes to the environment and intensive control of dingoes by shooting, trapping and poisoning largely eradicated dingoes especially in areas inside the barrier fences. Similarly, the 100-year war between pastoralism and dingoes was easily won by dingoes mainly because control techniques simply fractured packs and thereby dampened the natural population regulation in dingoes where most breeding was confined to the dominant pack female (Corbett, 1988, 1995a, b).

Ironically, however, crossbreeding with domestic dogs also increased over the same period to the extent that hybrids now occur throughout the mainland and predominantly so in southern Australia (Newsome and Corbett, 1985; Jones, 1990; Corbett, 1995a).

Earlier studies based on canonical variate analysis of skull measurements suggested that separate races of dingoes existed in Australia and Thailand and that there might be distinct types of dingoes within Australia, in particular, tropical, desert and alpine forms (Corbett, 1995a). However re-analysis of the data using a tighter definition of dingoes indicated there is only one type or form of dingo in Australia—the Australian dingo (*Canis lupus dingo*). There is no distinct 'alpine' form in the south-eastern highlands, nor desert or tropical forms (Corbett, 2001). However, dingo populations in south-eastern Australia are significantly different to dingo populations elsewhere in Australia (Figure 1). This is basically because they are hybrids as a result of their fairly long close contact with feral dogs, but it is possible that some skeletal differences may have derived from artificial selection processes by Aboriginals in earlier times.

There is no reason why hybrids will not eventually overrun dingo populations elsewhere in Australia. Even in Kakadu National Park, wild dogs that are obviously hybrids are increasingly being recorded by experienced park rangers and researchers (Corbett, 1999). By the end of the 21st century it is likely that most populations of pure dingoes will be extinct and Australia would then become a land of hybrids and feral dogs. See Plate 115.

ECOLOGICAL ROLE

INTERACTIONS WITH OTHER PREDATORS

The demise of the thylacine (*Thylacinus cynocephalus*) on the Australian mainland about 3,000 years ago (Archer, 1974; Dixon, 1989; Rounsevell and Mooney, 1995) and the Tasmanian devil (*Sarcophilus harrisii*) about 450 years ago (Jones, 1995) has been attributed to competition with dingoes according to the 'superior adaptability' hypothesis (Corbett, 1995a). This hypothesis hinges on the superior social organization of dingoes during critical periods when food supplies were scarce, widely dispersed or clumped, which usually occurs during drought or after extensive wildfire. Only dingoes form large integrated packs and cooperate to catch large prey, and to defend carcasses, water and other crucial resources. On the other hand, thylacines hunted alone or in pairs and devils were essentially solitary, so that neither could successfully compete against the weight of dingo numbers during those critical periods.

Dingoes and other wild dogs (hybrids, feral domestic dogs) may now present foxes and feral cats with a similar kind of competition especially during drought in central Australia and following extensive severe wildfire in temperate Australia. For example, during drought in central Australia in the early 1970s, foxes avoided dingoes at shared waterholes and there was an increase of cat in the dingo's diet (Corbett and Newsome, 1987; Corbett, 1995a).

PREDATION ON NATIVE FAUNA

Predation by wild dogs may have an impact on the survival of remnant populations of endangered fauna. For

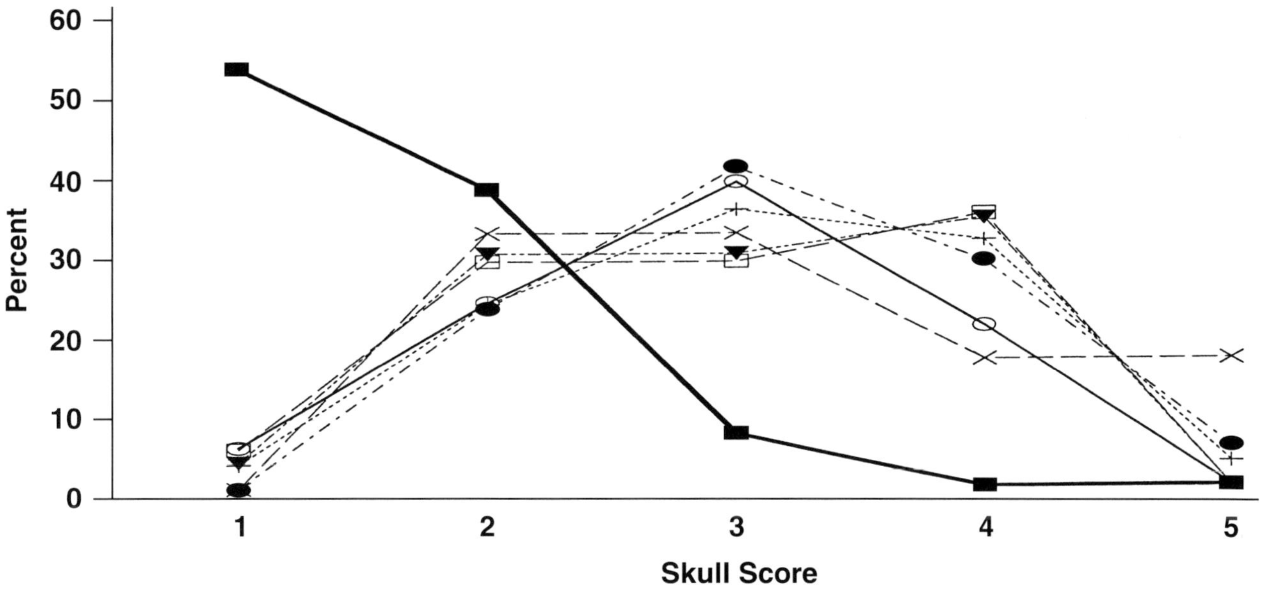

Figure 1. Distribution of skull scores of dingoes from seven major regions in Australia (from Corbett, 2001).
Key to symbols: ■ = South-eastern Highlands; ○ = Central Australia; □ = Barkly Tablelands of the Northern Territory; ▼ =Kakadu National Park; ● = Gibson Desert (south-western Northern Territory); + =Pilbara Region of north-west Western Australia; x = North-east New South Wales.
Skull scores were calculated from eight skull variables using equations provided in Newsome *et al.*, (1980), Newsome and Corbett (1982, 1985) and Corbett (1995a). Pure dingoes were defined as those animals with (1) ginger pelts, (2) skull scores >1.270, and (3) each of the eight skull variables contributing to the skull score had to be within the 95% confidence limits for dingoes (Corbett, 2001). In Corbett's (2001) study, about 74% of 180 skulls examined (from seven regions throughout Australia) were classified as dingoes and totally pure populations were not recorded anywhere. The purest dingo populations were in north, central and north-west Australia (96–74% of sampled populations were dingoes); hybrids were predominantly recorded from south-east Australia (65–48% of sampled populations were hybrids). ANOVA testing of skull scores indicated that only dingoes from the south-eastern highlands were significantly different from the other six regions—all of which had similar and even spreads of skull scores. This suggests that highland animals are either a distinct race (alpine dingoes) or hybrids. The skew to the left (towards the hybrid range of skull scores, 1.270 to –1.393) suggests there is a relatively greater proportion of hybrids in the south-eastern highlands.

example, predation by the dingo has been implicated in the extinction of the Tasmanian native-hen (*Gallinula mortierii*) from mainland Australia (Baird, 1991). In central Australia nine species of bandicoots, macropods and rat-kangaroos have become extinct since the 1930s (Newsome and Corbett, 1977; Morton, 1990). Corbett (1995a) outlined a hypothesis that described the interplay between dingoes and foxes, feral cats, rabbits, stock and fire in the first half of the 20th century. He concluded that dingoes played a dominant role in the decline and extinction of those medium-sized mammals.

Similarly, the eastern quoll (*Dasyurus viverrinus*) is believed to be extinct on mainland Australia (Godsell, 1995) and the dingo has been implicated as one factor in its demise. However, dingoes coexist with other quoll species in many regions of Australia and there is no evidence to suggest that direct predation has been a major factor. Similarly, the assertion that reduced eastern quoll populations are a result of non-target poisoning during 1080 poison baiting programmes for dingoes is based on circumstantial evidence (Fleming *et al.*, 2001).

A number of authors have suggested that dingoes may regulate prey populations. Predation by dingoes has often been cited as a cause of fluctuation or lack of fluctuation in populations of red kangaroos and emus on both sides of the dingo fence (Caughley *et al.*, 1980), red kangaroos in central Australia (Corbett and Newsome, 1987), euros (*Macropus robustus erubescens*) in north-west Western Australia (Thomson, 1992), macropods in south-eastern Australia (Newsome *et al.*, 1983), feral goats (Macknight, 1976; Parkes *et al.*, 1996), swamp wallabies in northern New South Wales (Robertshaw and Harden, 1989; Fleming, 1996) and feral pigs in Queensland (Woodall, 1983; Pavlov, 1991; Choquenot *et al.*, 1996). Although it is clear that dingo predation can affect prey populations, especially macropodids, the long-term effect of this predation and whether or not the effect is regulatory has not been experimentally assessed. The exception is one experiment, conducted in the tropical wetlands of northern Australia, to examine if dingo predation regulated populations of feral pigs (Corbett, 1995b). Data collected over seven years indicated that the dingo predation rate was negatively correlated with pig abundance—so dingoes were not regulating the feral pig population.

In conjunction with the evidence of studies of other predators in Australia (Newsome, 1990; Pech *et al.*, 1992), it appears that dingo predation alone can significantly curb prey populations only when prey numbers are initially depressed by a widespread environmental event such as drought in arid rangelands or intense wildfire in temperate forests. The theory is that prey are trapped in a 'predator pit' (Walker and Noy-Meir, 1982) where their reproductive rate is too low for the population growth rate to exceed the predation rate. At the same time, the extra food that such circumstances usually provide allows the predators to survive. For example, in arid central Australia, cattle carrion enables dingoes to survive droughts (Corbett and Newsome, 1987). The scene is thus set for predators to concentrate on a substitute prey and perhaps regulate their population after a drought.

The other environmental impacts of dingoes relate to current management practices, rather than to direct impacts by dingoes themselves. First, control measures such as poisoning may have a direct impact on non-target species, such as quolls (Fleming *et al.*, 2001). Second, reducing dingo density may result in an increase in other predators with overlapping diets. This process of substitution of predators is called 'mesopredator release' (Soulé *et al.*, 1988). For example, removing dingoes from a system where foxes also occur will result in an increase in fox numbers with consequent increased predation on mammals in the critical weight range (Denny, 1992; Smith *et al.*, 1992); however, such effects and impacts are yet to be demonstrated.

Dingoes have been present in Australia for about 4,000 years and undoubtedly have helped to shape ecosystems. However, given that essentially hybrid populations exist today in eastern Australia, there is debate as to whether or not hybrids perform the same ecological role as pure dingoes in native ecosystems. Consequently, it is unclear whether or not there have been altered predation rates on native prey and other disruptions to those ecosystems. The few available data suggest that important physiological differences do exist between dingoes and hybrids (e.g. breeding rates) where some differences lead to adverse impacts on stock (Corbett, 1995a) and, by implication, native species. Further research is required to clarify such potential impacts.

CONSERVATION

The dingo is a native Australian mammal with intrinsic and aesthetic values that the majority of the public expect to be conserved. As indicated above, the greatest threat to the survival of dingoes is hybridization with domestic dogs. There are three actions that offer the best hope of conserving dingoes.

The first relates to the identification of pure dingoes. Currently pure dingoes are identified on the basis of skull measurements. This method, however, can only be applied to adult, dead dingoes and thus is of limited use to wildlife managers and dingo breeders who need to ascertain the identity of young living animals so as to decide whether to cull or maintain them. Recent work on canine molecular genetics in Australia and overseas promises a practical method to assess the purity of wild and captive dingo populations, both living and deceased (Wayne, 1993; Vila *et al.*, 1997; Wilton, 2001). It is imperative that pre-European benchmark material is used in comparisons with modern dingo material. Such unequivocally pure dingo reference material is available in the form of fossils and cave deposits as indicated in Table 1.

The second action relates to the use of offshore islands as dingo refugia even though hybrids currently exist on all islands including Fraser Island (Woodall *et al.*, 1996). Islands such as Fraser, Melville, Bathurst and Groote offer the best hope of conserving dingoes because they: (1) contain many representative habitats of the mainland; (2) are large enough for dingoes to live and breed in partly or completely natural conditions; (3) small enough for wild populations to be screened (with DNA molecular techniques) and hybrids and feral dogs to be removed; and (4) sufficiently isolated for managers to maintain the integrity of pure dingo populations, once established, through removal of excess dingoes by culling or transportation.

The third action relates to public investment in dingoes. Dingo refugia don't necessarily need to be offshore islands. They could be 'islands' of well protected and maintained sanctuaries on the mainland, which would have the added advantage of additional habitats, such as hot deserts, not available on offshore islands. Such sanctuaries already exist on the mainland and are very successful in conserving rare and endangered marsupials and birds (Wamsley, 1998). There is no justifiable reason why enlightened and motivated people could not form companies to establish wildlife sanctuaries of adequate size for dingoes to live naturally.

In some areas where pure dingoes are locally extinct, hybrids, as top predators, may have an important, but as yet unclear, influence on the biodiversity of animal communities. They might also have an inverse density relationship with foxes and therefore be important in limiting the impact of foxes and cats on populations of small and medium-sized mammal prey. Because a dingo-sized canid has been present in Australia long enough to have affected the biodiversity of the communities in which it occurs, to completely remove wild dogs from these environments might have unforeseen impacts on biodiversity, and so some wild dogs should be retained irrespective of their genetic status.

REFERENCES

Archer, M. 1974. New information about the Quaternary distribution of the thylacine (Marsupalia, Thylacinidae) in Australia. *Journal of the Royal Society of Western Australia*, Vol. 57, pp.43–50.

Baird, R.F. 1991. The dingo as a possible factor in the disappearance of *Gallinula mortierii* from the Australian mainland. *Emu*, Vol. 91, pp.121–122.

Balme, J., Merrilees, D. and Porter, J.K. 1978. Late Quaternary mammal remains, spanning about 30,000 years from excavations in Devil's Lair, Western Australia. *Journal of the Royal Society of Western Australia*, Vol. 61, pp.33–65.

Basedow, H. 1925. *The Australian Aboriginal*. F. W. Preece and Sons, Adelaide. 422pp.

Bauer, F.H. 1983. The coming of European man. Pp.26–45. In Crook, G. (ed). *Man in the Centre: Proceedings of a Symposium held at CSIRO, Alice Springs, 3–5 April 1979.* CSIRO, Division of Groundwater Research, Perth. 232pp.

Bay-Petersen, J.A. 1979. The role of the dog in the economy of the New Zealand Maori. Pp.165–181. In Anderson, A.J. (ed). *Osteological and Archaeological Papers from the South Pacific in Honour of R. J. Scarlett. New Zealand Archaeological Association Monograph II and BAR International Series*, 62.

Bellwood, P.S. 1978. *Man's Conquest of the Pacific: the Prehistory of South-east Asia and Oceania*. Collins, Auckland. 462pp.

Bellwood, P. 1984. The great Pacific migration. Pp.80–93. In *Yearbook of Science and the Future 1984*. Encyclopaedia Britannica, Chicago.

Breckwoldt, R. 1988. *A Very Elegant Animal: The Dingo.* Angus and Robertson, Sydney. 283pp.

Brisbin, I.L. and Risch T.S. 2000. *Primitive dogs, their ecology and behaviour*. http://www.Carolinadogs.org/geninfo/primitive.html

Caughley, G., Grigg, G.C., Caughley, J. and Hill, G.J.E. 1980. Does dingo predation control the densities of kangaroos and emus? *Australian Wildlife Research,* Vol. 7, pp.1–12.

Choquenot, D., McIlroy, J. and Korn, T. 1996. *Managing Vertebrate Pests: Feral Pigs*. Bureau of Resource Sciences, Australian Government Publishing Service, Canberra. 163pp.

Clarke, G. 1968. Australian stone age. *Liber Iosepho Kostrzewski*, 1968, pp.17–18.

Clutton-Brock, J. 1977. Man-made dogs. *Science*, Vol. 197, pp.1340–1342.

Clutton-Brock, J. (ed.) 1989. *The Walking Larder: Patterns of Domestication, Pastoralism and Predation*. Unwin Hyman, London. 368pp.

Coe, C. 1994. *The Basenji*. Doral Publishing, Wilsonville, Oregon. 357pp.

Colenso, W. 1877. Notes on the ancient dog of the New Zealanders. *Transactions and Proceedings of the New Zealand Institute*, Vol. 10, pp.135–155.

Corbett, L. 1995a. *The Dingo in Australia and Asia.* University of New South Wales Press, Sydney. 200pp.

Corbett, L. 1995b. Does dingo predation or buffalo competition regulate feral pig populations in the Australian wet-dry tropics? An experimental study. *Wildlife Research*, Vol. 22, pp.65–74.

Corbett, L. 1999. Jabiluka baseline surveys: terrestrial, arboreal and volant vertebrates, and terrestrial insects. *Report to The Federal Minister for Resources and Energy*. ERA Environmental Services Pty. Ltd.

Corbett, L. 2001. The conservation status of dingoes (*Canis lupus dingo*) in Australia, with particular reference to New South Wales: threats to pure dingoes and potential solutions. Pp.10–19. In Dickman, C.R. and Lunney, D. (eds). *A Symposium on the Dingo*. Transactions of the Royal Zoological Society of New South Wales, Mosman. 110pp.

Corbett, L.K. 1985. Morphological comparisons of Australian and Thai dingoes: a reappraisal of dingo status, distribution and ancestry. *Proceedings of the Ecological Society of Australia*, Vol. 7, pp.277–291.

Corbett, L.K. 1988. Social dynamics of a captive dingo pack: population regulation by dominant female infanticide. *Ethology*, Vol. 78, pp.177–198.

Corbett, L.K. and Newsome, A.E. 1987. The feeding ecology of the dingo. III. Dietary relationships with widely fluctuating prey populations in arid Australia: an hypothesis of alternation of predation. *Oecologia*, Vol. 74, pp.215–227.

Davis, S. and Valla, F.R. 1978. Evidence of the domestication of the dog, 12,000 years ago in the Natufian of Israel. *Nature*, Vol. 276, pp.608–610.

Denny, M. 1992. *Dorrigo Management Area Fauna Impact Statement of Proposed Forestry Operations*. Forestry Commission of New South Wales, Sydney.

Dixon, J.M. 1989. Thylacinidae. Pp.549–559. In Walton, D.W. and Richardson, B.J. (eds). *Fauna of Australia. Vol. 1B. Mammalia*. Australian Government Publishing Service, Canberra. Pp.401–1227.

Durack, M. 1983. *Sons in the Saddle*. Constable, London. 442pp.

Finlayson, H.H. 1935. *The Red Centre*. Angus and Robertson, Sydney. 146pp.

Fleming, P.J.S. 1996. *Aspects of the management of wild dogs (*Canis familiaris*) in north-eastern New South Wales.* M.Sc. Thesis (unpublished), University of New England, Armidale.

Fleming, P., Corbett, L., Harden, B. and Thomson, P. 2001. *Managing the Impacts of Dingoes and Other Wild Dogs*. Bureau of Resource Sciences, Canberra. 186pp.

Fogle, B. 2003. *The Encyclopedia of the Dog*. http://www.carolinadogs.com/general.html#encyclopedia

Giles, E. 1889. *Australia Twice Traversed: the Romance of Exploration being a Narrative Compiled from the Journals of Five Exploring Expeditions into and through Central South Australia and Western Australia from 1872 to 1876.* Vol.1. Sampson Low, Marston, Searle and Rivington, London. (Reprint) (1986) Doubleday, Sydney. 320pp.

Gill, E.D. 1953. Distribution of the Tasmanian devil, the Tasmanian wolf and the dingo in southeast Australia in Quaternary time. *Victorian Naturalist*, Vol. 70, pp.86–90.

Gill, E.D. and Sinnott P.J. 1973. Age of canine teeth associated with giant extinct marsupials at Wellington Caves, New South Wales. *Mankind*, Vol. 9, pp.125–126.

Glover, I.C. 1972. *Excavations in Timor.* Ph.D. Thesis (unpublished), Australian National University, Canberra.

Godsell, J. 1995. Eastern quoll, *Dasyurus viverrinus*. Pp.70–71. In Strahan, R. (ed). *The Mammals of Australia*. Reed Books, Sydney. 756pp.

Gollan, K. 1982. *Prehistoric dingo*. Ph.D. Thesis (unpublished), Australian National University, Canberra.

Gollan, K. 1984. The Australian dingo: in the shadow of man. Pp.921–927. In Archer, M. and Clayton, G. (eds). *Vertebrate Zoogeography and Evolution in Australasia.* Hesperian Press, Perth. 1203pp.

Gould, R.A. 1979. *Living Archaeology.* Cambridge University Press, Cambridge. 270pp.

Hamilton, A. 1972. Aboriginal man's best friend? *Mankind,* Vol. 8, pp.287–295.

Harney, W.E. 1951. The dingo. *Walkabout,* 1 July, pp.36–37.

Harney, W.E. and Elkin, A.P. 1949. *Songs of the Songmen: Aboriginal Myths Retold.* Cheshire, Melbourne. 173pp.

Higham, G.H.E., Kijngam, A. and Manly, B.F.J. 1980. An analysis of prehistoric canid remains from Thailand. *Journal of Archaeological Science,* Vol. 7, pp.149–165.

Hopkins, G.H.E. 1949. The host associations of the lice of mammals. *Proceedings of the Zoological Society of London,* Vol. 119, pp.387–604.

Jennings, J.N. 1971. Sea level changes and land links. In Mulvaney, D.J. and Golsen, J. (eds). *Aboriginal Man and Environment in Australia.* Australian National University Press, Canberra. 389pp.

Johnson, I. 1979. *The getting of data. A case study from the recent industries of Australia.* Ph.D. Thesis (unpublished), Australian National University, Canberra.

Jones, E. 1990. Physical characteristics and taxonomic status of wild canids, *Canis familiaris,* from the eastern highlands of Victoria. *Australian Wildlife Research,* Vol. 17, pp.69–81.

Jones, M. 1995. Tasmanian devil *Sarcophilus harrisii.* Pp.82–84. In Strahan, R. (ed). *The Mammals of Australia.* Reed Books, Sydney. 756pp.

Jones, R. 1970. Tasmanian Aborigines and dogs. *Mankind,* Vol. 7, pp.256–271.

Kirch, P.V. and Yen, D.E. 1982. *Tikopia: The Prehistory and Ecology of a Polynesian Outlier.* Bernice P. Bishop Museum Bulletin No. 238, Honolulu.

Koler-Matznick, J., Brisbin, I.L. and McIntyre, J.K. 1998. The New Guinea singing dog: a living primitive dog. *Proceedings of the 8th International Congress of the International Council for Archaeology, August 23–24 University of Victoria, BC, Canada.*

Kolig, E. 1973. Aboriginal man's best foe. *Mankind,* Vol. 9, pp.122–123.

Kolig, E. 1978. Aboriginal dogmatics: canines in theory, myth and dogma. *Bijdragen Totdetaal Land en Volkenhunde,* Vol. 134, pp.84–115.

Krefft, G. 1866. On the vertebrated animals of the lower Murray and Darling, their habits, economy and geographical distribution. *Transactions of the Philosophical Society of New South Wales 1862–1865.*

Krefft, G. 1867. *Fossil Remains of Mammals, Birds and Reptiles from the Caves of Wellington Valley.* Vol. I. Universal Exhibition 1867, Sydney.

Lampert, R.J. 1971. *Burrill Lake and Currarong. Terra Australia I.* Australian National University Press, Canberra. 86pp.

Lowry, J.W.J. and Merrilees, D. 1969. Age of the desiccated carcase of a thylacine (Marsupialia, Dasyuridae) from Thylacine Hole, Nullarbor Region, Western Australia. *Helictite,* Vol. 7, pp.15–16.

Lumholtz, C. 1889. *Amongst Cannibals: An Account of Four Years' Travels in Australia and of Camp Life with the Aborigines of Queensland.* J. Murray, London. 395pp.

Macintosh, N.W.G. 1964. A 3,000 year old dingo from shelter 6 (Fromm's Landing, South Australia). *Proceedings of the Royal Society of Victoria,* Vol. 77, pp.498–507.

Macknight, C.C. 1976. *The Voyage to Marege: Macassan Trepangers in Northern Australia.* Melbourne University Press, Melbourne. 175pp.

Matisoo-Smith, E. 1998. The Austronesian origins of the dingo and other Pacific dogs? Evidence from mitochondrial DNA. P.190. *Proceedings of the 8th International Congress of the International Council for Archaeology, August 23–24 University of Victoria, BC, Canada.* (abstract only).

McCarthy, F.D. 1967. *Australian Aboriginal Stone Implements, including Bone, Shell and Teeth Implements.* Australian Museum, Sydney. 99pp.

McCoy, F. 1861. *On the Ancient and Recent Natural History of Victoria.* Essays for Melbourne Exhibition, Geological Survey of Victoria, Victoria.

McCoy, F. 1882. The dingo, prodromus of the palaeontology of Victoria. *Geological Survey of Victoria,* Vol. 7, pp.7–10.

Medway, L. 1977. The ancient domestic dogs of Malaysia. *Journal of the Royal Asia Society (Malaysia),* Vol. 50, pp.14–27.

Meggitt, M.J. 1965. The association between Australian Aborigines and dingoes. In Leeds, A. and Vayda, A. (eds). *Man, Culture and Animals: the Role of Animals in Human Ecological Adjustments.* American Association for the Advancement of Science, Washington. 304pp.

Milham, P. and Thompson, P. 1976. Relative antiquity of human occupation and extinct fauna at Madura Cave, southeastern Western Australia. *Mankind,* Vol. 10, pp.175–180.

Morton, S.R. 1990. The impact of European settlement on the vertebrate animals of arid Australia: a conceptual model. *Proceedings of the Ecological Society of Australia,* Vol. 16, pp.201–213.

Mulvaney, D.J. 1975. *The Prehistory of Australia.* Penguin Books, Melbourne. 327pp.

Newsome, A. 1990. The control of vertebrate pests by vertebrate predators. *Trends in Ecology and Evolution,* Vol. 5, pp.187–191.

Newsome, A.E., Catling, P.C. and Corbett, L.K. 1983. The feeding ecology of the dingo II. Dietary and numerical relationships with fluctuating prey populations in south-eastern Australia. *Australian Journal of Ecology,* Vol. 8, pp.345–366.

Newsome, A.E. and Corbett, L.K. 1977. The effects of native, feral and domestic animals on the productivity of Australian rangelands. Pp.332–356. In *The Impact of Herbivores on Arid and Semi-arid Rangelands: Proceedings of the 2nd United States/Australia Rangeland Panel, Adelaide, 1972.* Australian Rangeland Society, Perth. 376pp.

Newsome, A.E. and Corbett, L.K. 1982. The identity of the dingo. II. Hybridisation with domestic dogs in captivity and in the wild. *Australian Journal of Zoology,* Vol. 30, No. 3, pp.365–374.

Newsome, A.E. and Corbett, L.K. 1985. The identity of the dingo. III. The incidence of dingoes, dogs and hybrids and their coat colours in remote and settled regions of Australia. *Australian Journal of Zoology,* Vol. 33, No. 3, pp.363–375.

Newsome, A,E., Corbett, L.K. and Carpenter, S.M. 1980. The identity of the dingo. I. Morphological discriminants of dingo and dog skulls. *Australian Journal of Zoology*, Vol. 28, No. 4, pp.615–625.

Olsen, S.J. and Olsen, J.W. 1977. The Chinese wolf: ancestor of the New World dogs. *Science*, Vol. 197, pp.533–535.

Parkes, J., Henzell, R. and Pickles, G. 1996. *Managing Vertebrate Pests: Feral Goats*. Bureau of Resource Sciences, Australian Government Publishing Service, Canberra. 129pp.

Pavlov, P.M. 1991. *Aspects of the ecology of the feral pig (*Sus scrofa*) in semi-arid and tropical areas of eastern Australia.* Ph.D. Thesis (unpublished), Monash University, Melbourne.

Pech, R.P., Sinclair, A.R.E., Newsome, A.E. and Catling, P.C. 1992. Limits to predator regulation of rabbits in Australia: evidence from predator-removal experiments. *Oecologia*, Vol. 89, pp.102–112.

Pickering, M. 1992. Garawa methods of game hunting, preparation and cooking. *Records of the South Australian Museum*, Vol. 26, pp.9–23.

Roberts, R.G., Jones, R. and Smith, M.A. 1990. Thermoluminescence dating of a 50,000 year-old human occupation site in northern Australia. *Nature*, Vol. 345, pp.153–156.

Robertshaw, J.D. and Harden, R.H. 1989. Predation in Macropodoidea: a review. Pp.735–753. In Grigg, G., Jarman, P. and Hume, I. (eds). *Kangaroos, Wallabies and Rat-kangaroos*. Surrey Beatty and Sons, Sydney. 835pp.

Rolls, E. 1992. The disappearing howl. *The Independent Monthly*, February.

Rounsevell, D.E. and Mooney, N. 1995. Thylacine, *Thylacinus cyanocephalus*. Pp.164–165. In Strahan, R. (ed). *The Mammals of Australia*. Reed Books, Sydney. 756pp.

Schultz, W. 1969. Zur kenntis des Hallstromhundes (*Canis hallstromi*). *Sonderuck aus Zoologischer Anzeiger,* Vol. 183, pp.42–72.

Smith, A.P., Moore, D.M. and Andrews, S.P. 1992. *Proposed Forestry Operations in the Glen Innes Forest Management Area: Fauna Impact Statement*. Forestry Commission of New South Wales, Armidale. 98pp.

Smith, M.A. 1980. *Saltbush, sampling strategy and settlement pattern.* MA Thesis (unpublished), Australian National University, Canberra.

Soulé, M.E., Bolger, D.T., Alberts, A.C., Wright, J.M.S. and Hill, S. 1988. Reconstructed dynamics of rapid extinctions of chaparral-requiring birds in urban habitat islands. *Conservation Biology*, Vol. 2, pp.75–92.

Thompson, D.F. 1949. Arnhem Land: explorations among an unknown people. Part III. On foot across Arnhem Land. *Geographical Journal*, Vol. 114, pp.53–67.

Thompson, D.F. 1955. Yellow dog dingo. *Walkabout*, 1 May, pp.16–18.

Thomson, P.C. 1992. The behavioural ecology of dingoes in north-western Australia. III. Hunting and feeding behaviour, and diet. *Wildlife Research*, Vol. 19, pp.531–542.

Thorne, A. and Raymond, R. 1989. *Man on the Rim: the Peopling of the Pacific*. Angus and Robertson, Sydney. 288pp.

Tindale, N.B. 1974. *Aboriginal Tribes of Australia: their Terrain, Environmental Controls, Distribution, Limits and Proper Names*. Australian National University Press, Canberra. 404pp.

Titcombe, M. 1969. *Dog and Man in the Ancient Pacific*. Bernice P. Bishop Museum Special Publication 59, Honolulu.

Troughton, E. 1957. A new native dog from the Papuan highlands. *Proceedings of the Royal Society of New South Wales* (1955–56), pp.93–94.

Turnbull, P.F. and Reed, C.A. 1974. The fauna of the terminal Pleistocene of Palegawra Cave, a Zarzian occupation site in northeastern Iraq. *Fieldiana (Anthropology)*, Vol. 36, pp.81–146.

Vila, C.P., Savolainen, P., Maldonado, J.E., Amorim, I.R., Rice, J.E., Honeycutt, R.L., Crandall, K.A., Lundenberg, J. and Wayne, R.K. 1997. Multiple and ancient origins of the domestic dog. *Science,* Vol. 276, pp.1687–1689.

Wakefield, N.A. 1967. Preliminary report on McEacherns Cave, S.W. Victoria. *Victorian Naturalist,* Vol. 84, pp.363–383.

Walker, B.H. and Noy-Meir, I. 1982. Aspects of the stability and resilience of savanna ecosystems. Pp.556–590. In Huntly, B.J. and Walker, B.H. (eds). *Ecology of Tropical Savannas*. Springer-Verlag, New York. 669pp.

Walters, B. 1995. *The Company of Dingoes: two Decades with our Native Dog.* Australian Native Dog Conservation Society Ltd., Bargo, New South Wales. 109pp.

Wamsley, J. 1998. Rehabilitate for what? Pp.139–144. In Asher, C.J. and Bell. L.C. (eds). *Proceedings of the Workshop on Fauna Habitat Reconstruction after Mining, Adelaide, 10–11 October 1997.* Australian Centre for Mining Environmental Research, Brisbane.

Wayne, R.K. 1993. Molecular evolution of the dog family. *Trends in Genetics*, Vol. 9, pp.218–224.

White, I. 1972. Hunting dogs at Yalata: Aborigines, hunting dogs and conservation. *Mankind*, Vol. 8, pp.201–205.

White, J.P. and Allen, J. 1980. Melanesian prehistory: some recent advances. *Science*, Vol. 207, pp.728–734.

White, J.P. and O'Connell, J.F. 1982. *A Prehistory of Australia, New Guinea and Sahul*. Academic Press, Sydney. 286pp.

Wilkinson, N. 2000. http://www.dreamtimepacific.com.au/artwork/Paintings/PNW001.htm

Wilton, A. 2001. DNA methods of assessing dingo purity. Pp.49–56. In Dickman, C.R. and Lunney, D. (eds). *A Symposium on the Dingo.* Transactions of the Royal Zoological Society of New South Wales, Mosman. 110pp.

Woodall, P.F. 1983. Distribution and population dynamics of dingoes (*Canis familiaris*) and feral pigs (*Sus scrofa*) in Queensland, 1945–1976. *Journal of Applied Ecology*, Vol. 20, pp.85–95.

Woodall, P.F., Pavlov, P. and Twyford, K.L. 1996. Dingoes in Queensland, Australia: skull dimensions and the identity of wild canids. *Wildlife Research*, Vol. 23, pp.581–587.

Chapter 33

A Southern Perspective on Cetacean Evolution and Zoogeography

R. Ewan Fordyce
Department of Geology, University of Otago, Dunedin, New Zealand
E-mail: ewan.fordyce@stonebow.otago.ac.nz

INTRODUCTION

Cetaceans—whales, dolphins and porpoises (Figure 1)—are large carnivorous obligate aquatic mammals. More than 80 living species are distributed widely in global oceans and some fresh waters, with more than half this total reported from Australasia (Rice, 1998; Perrin and Brownell, 2001). The Australasian region—Australia, New Zealand and adjacent oceans—spans from tropical to sub-polar seas, encompassing several different water masses (Figure 2), and a high regional diversity is thus unsurprising. Globally, cetacean history extends more than 50 million years back to the Eocene, with significant Australasian records—more so for New Zealand than for Australia. This article reviews taxonomic, stratigraphic, evolutionary and geographic patterns amongst modern and fossil Cetacea from Australasia.

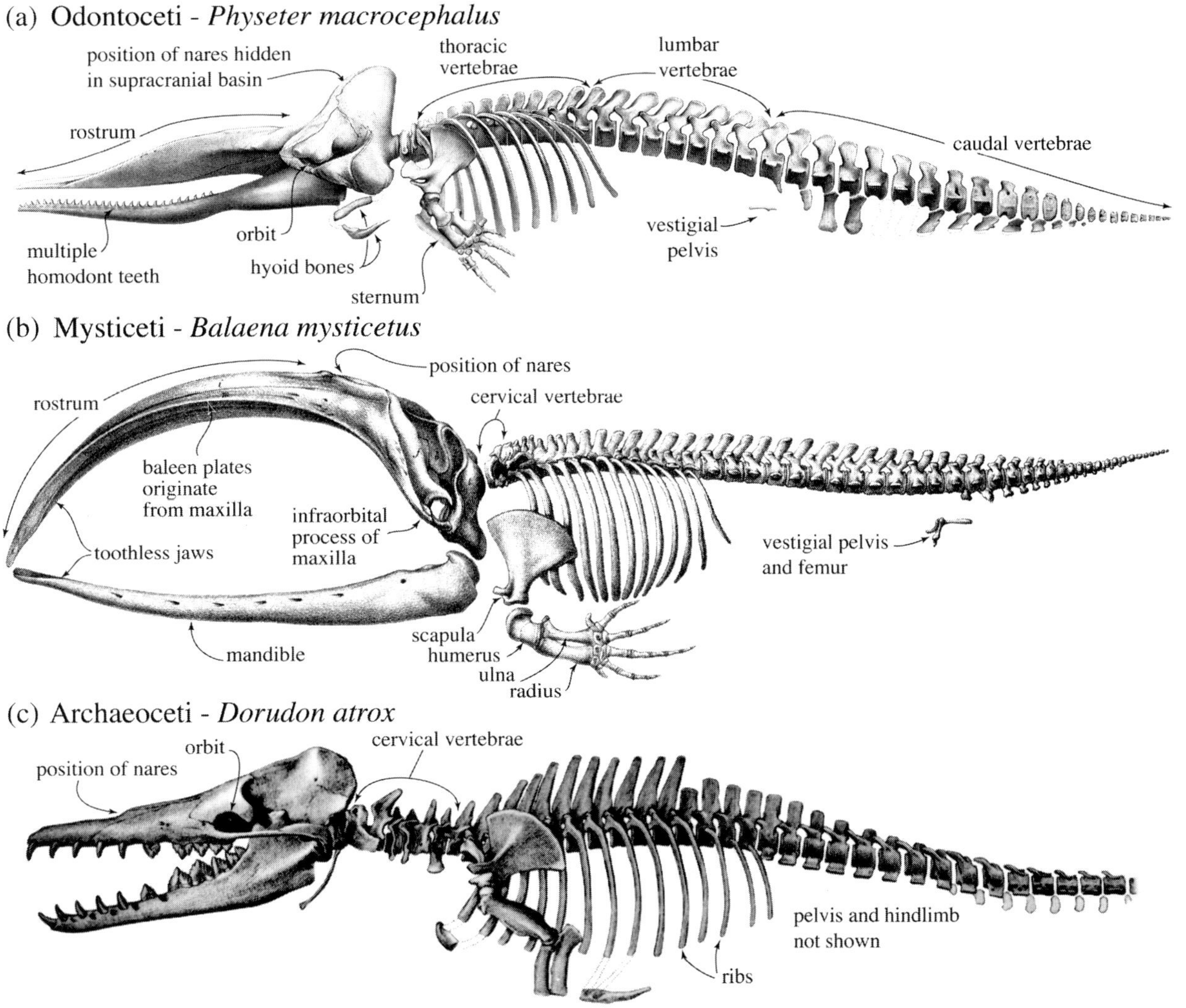

Figure 1. Examples of the cetacean skeleton: **(a)** the living sperm whale, *Physeter macrocephalus* (Odontoceti), from Van Beneden and Gervais (1868–1880); **(b)** the living bowhead whale, *Balaena mysticetus* (Mysticeti), from Van Beneden and Gervais (1868–1880); **(c)** an Eocene archaeoceti, *Dorudon atrox* (Archaeoceti), from Stromer (1908).

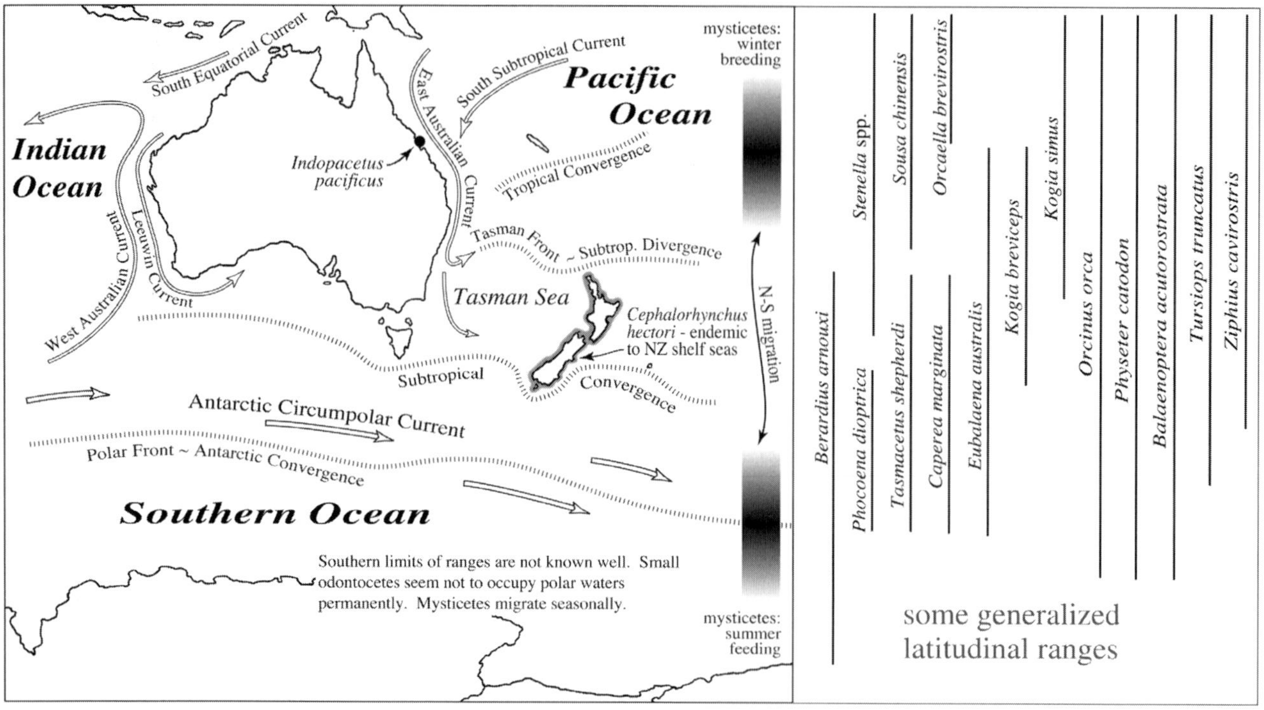

Figure 2. Oceanic features and latitudinal ranges for some living Australasian Cetacea. Cetacean ranges based on Rice (1998), Bryden *et al.* (1998) and Baker (1999).

SYSTEMATICS OF CETACEA

Studies in evolution and zoogeography need a foundation of taxonomy based on phylogenetic relationships, desirably calibrated against geological time. Concepts of relationships and stratigraphic correlation, however, are dynamic, changing with new advances. The modern taxonomy of the Cetacea—species, genera, families—originated in the pre-evolutionary classification of Linnaeus and contemporaries of the late 1700s. In the mid 1800s, family-level classification rapidly moved beyond reliance on superficial features. Cetologists such as W.H. Flower, influenced by the rise of Darwinian evolution, used anatomy at many levels (see Flower, 1868) to try to identify real groups, especially families. In modern terms, Flower aimed to identify clades; a clade is a monophyletic group containing all the descendants of the most recent common ancestor of that group. Conversely, a grade is a cluster of species which have a common pattern of organization but are not necessarily monophyletic. The older literature reveals that, despite Flower's aims, grade families were used widely in the early 20th century (e.g. Kellogg, 1928—Delphinidae; Simpson, 1945—Platanistidae), with taxa recognized by various attributes including anatomy, distribution and ecology. Recently, cladistic approaches have clarified family-level classifications for fossil and living groups (e.g. Muizon, 1991 and references therein) (Figure 3). Molecular techniques are now used widely to resolve relationships at specific and generic level (e.g. Rosel *et al.*, 1995; Dalebout *et al.*, 1998; LeDuc *et al.*, 1999; Rosenbaum *et al.*, 2000; Waddell *et al.*, 2000). Some molecular methods have not resolved higher-level branches, as the 'sperm-whales as mysticetes' contention of the mid 1990s shows (Milinkovitch, 1995; Heyning, 1997; Messenger and McGuire, 1998), but the new SINE (short interspersed element) technique (Nikaido *et al.*, 2001) shows promise to elucidate 'deep-time' events. There is still no consensus on classification of fossils (see McKenna and Bell, 1998; Fordyce and Muizon, 2001), partly because many species are named from fragmentary material of uncertain age and affinity, such as isolated teeth or fragments of postcranial skeleton. For an excellent historical account and modern summary of cetacean classification, see Rice (1998); for a current list of species, see Perrin and Brownell (2001).

The systematic section, below, reviews current understanding of relationships within the Cetacea, working mainly at family level. Groups with an Australasian record are considered in more detail than others.

LOGISTICS AND MATERIAL

The singular attributes of cetaceans greatly affect systematics. Cetology is not for the faint-hearted: dead stranded cetaceans are a fundamental source of anatomical and systematic information, yet the cetologist faces a barrage of legal, logistic, cultural and emotional hurdles in accessing material (Geraci and Lounsbury, 1993), even before physical preparation and sampling starts. Unsurprisingly, then, cetaceans are amongst the least-

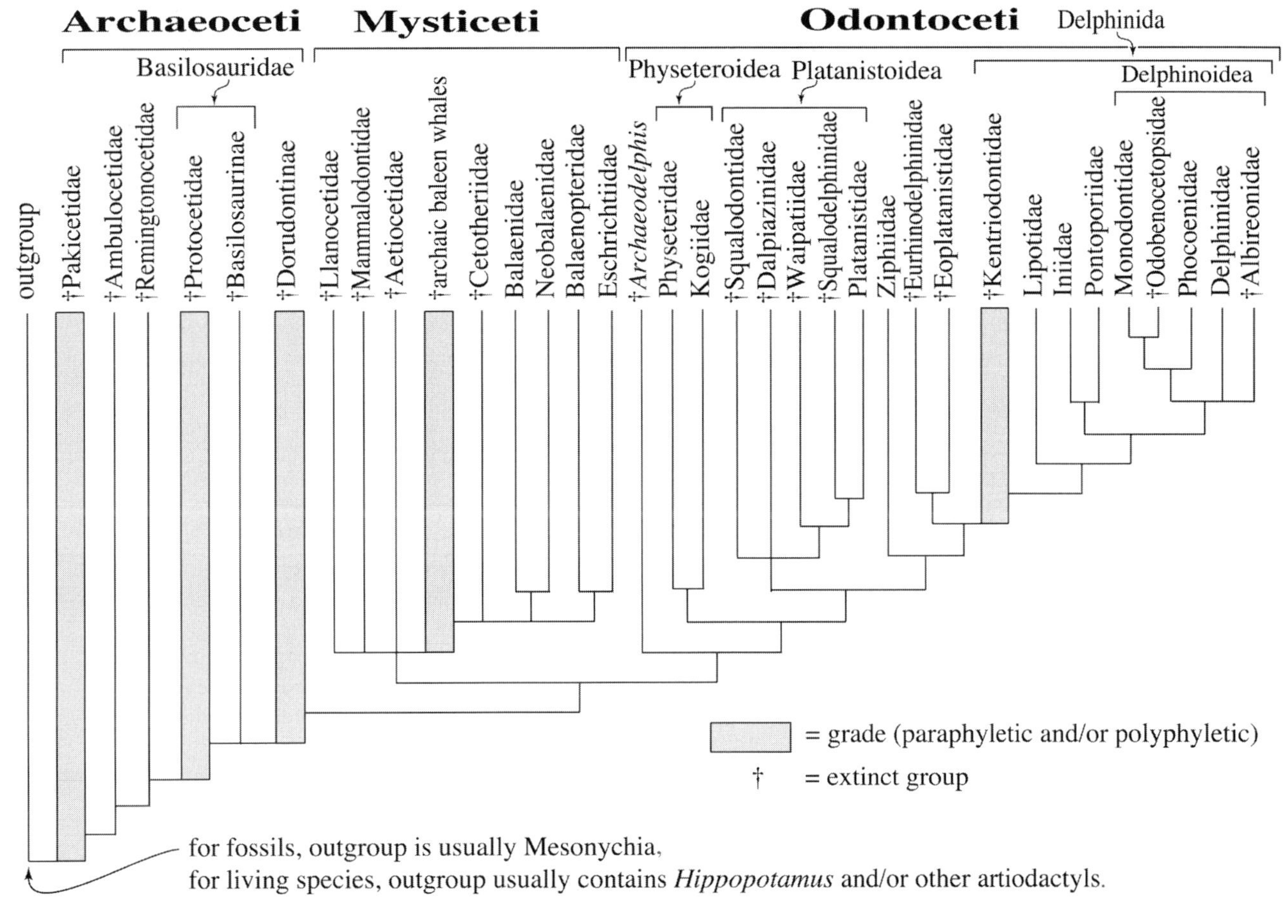

Figure 3. Current understanding of cladistic relationships within traditionally-recognized cetacean groups. Grade taxa are shaded. Archaeocete pattern based on Uhen (1998, 1999) and Luo and Gingerich (1999), odontocetes based on Muizon (1987, 1988a, 1988b, 1991), Heyning (1989) and Waddell *et al.* (2000), and mysticetes based on Geisler and Luo (1996) and McLeod *et al.* (1993).

known mammals. Fossil Cetacea are relatively rare, large, brittle and difficult to extract. They are time-consuming and costly to prepare. Extracted specimens—and modern bones—use much storage space. Only skull remains are of much use in systematics, but almost every sort of element has been used at some stage as a type-specimen. (On Australasian types, see Fordyce, 1980, 1982, 1988). The Australasian worker is remote from historically important northern centres of study, yet core literature is available in both Australia and New Zealand, and superb collections of extant cetacean skeletons are held in museums in most Australian states, and in the Museum of New Zealand. Notable fossil collections are in the Museum of Victoria and Geology Museum of the University of Otago.

STRATIGRAPHIC RECORDS AND LOCALITIES

Stratigraphic ranges outlined in Figure 4 and below revise and advance the earlier accounts of Fordyce (1982, 1984a). Figure 4 shows Cainozoic stratigraphy, lists some Australian and New Zealand cetacean-bearing strata, and shows the stratigraphic distribution of cetacean families both locally and globally based on Fordyce and Muizon (2001) and references therein. The timescale and ranges of taxa are approximate for Australian and New Zealand Tertiary strata, because of problems correlating precisely with key reference sequences in the Northern Hemisphere.

Localities for Australian fossil Cetacea are mostly in the south-east, reflecting accessible surface outcrops of marine Tertiary (Figure 5). Guides to localities include those of Mahoney and Ride (1975), Pledge and Rothausen (1977), Fordyce (1984a) and Pledge (1994). There is no significant fossil record from the continental shelf around Australia, even though Cetacea may have lived there since the Eocene. Whitmore *et al.* (1986) showed a skull, 'probably *Mesoplodon*', from the deep sea south of Tasmania.

Most localities in New Zealand occur away from the axial ranges (Figure 5); the many localities around the Waitaki Valley, on the southern margin of the Canterbury basin, reflect the presence of flat-lying and accessible

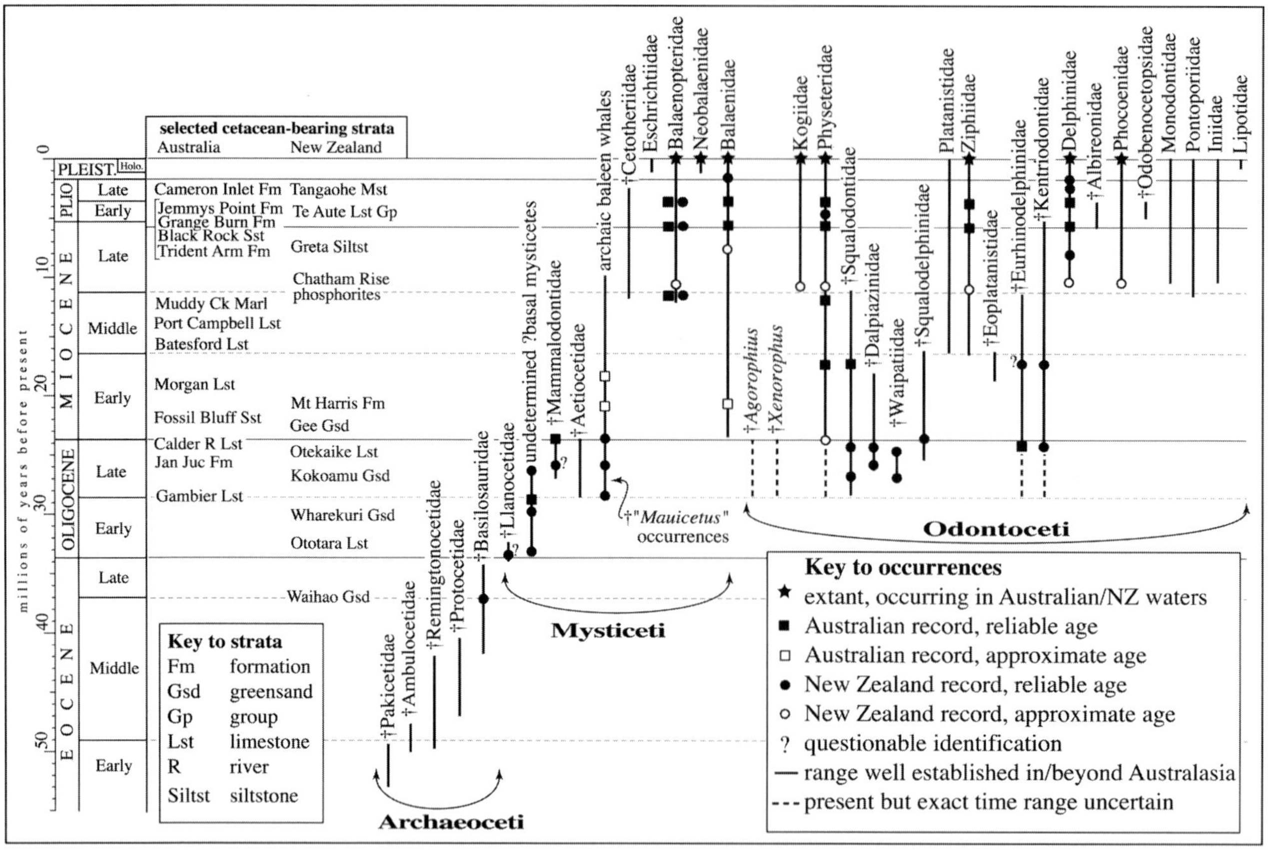

Figure 4. Stratigraphic ranges of Cetacea in Australia and New Zealand. Australasian cetacean-bearing strata follow Fordyce (1984a, 1991), with cetacean family ranges after Fordyce and Muizon (2001). Occurrences are a guide only; other records are known especially for the Oligocene. For keys to occurrences and strata abbreviations see figure boxes.

fossiliferous, bioclastic middle Tertiary sediments. Most of the important localities are in the South Island. The only specimen to have been described formally from the North Island is a delphinid jaw from Hawera (McKee and Fordyce, 1987).

PREVIOUS STUDIES

Published information abounds, with many articles produced since the 1980s. Major titles include those of Baker (1999) and Bryden *et al.* (1998), a substantial review (Bannister *et al.*, 1996), many articles in journals, and theses. Beyond regional titles, notable articles on Cetacea include those of Rice (1998) and various contributors to the edited series of Ridgway and Harrison (1994). Hershkovitz (1966) identified important early literature. The edited volume of Reynolds and Rommel (1999) contains fundamental contributions on cetacean biology. Important titles on osteology include those of Van Beneden and Gervais (1868–1880), Fraser and Purves (1960), and Kasuya (1973). Australasian fossil Cetacea have been reviewed a number of times (Mahoney and Ride, 1975; Fordyce, 1980, 1982, 1984a, 1991, 1992). Broader summaries of evolution and the fossil record include: Kellogg (1928)—outdated, yet still a classic work; Barnes *et al.* (1985); Berta and Demere (1994); Fordyce and Barnes (1994); Barnes *et al.* (1995a); Thewissen (1998); Berta and Sumich (1999) and Fordyce and Muizon (2001).

CLADE CETACEA

Cetaceans represent a clade within the Mammalia, linked by fossils to the extinct archaic ungulate group Mesonychia (O'Leary and Geisler, 1999; O'Leary and Uhen, 1999), or by molecular taxonomy to living hippopotamuses (Nikaido *et al.*, 1999, 2001). Mesonychians seem not to be related closely to hippos; which one is the sister-group for Cetacea is still unclear. Three cetacean suborders are used (Figure 1): a paraphyletic basal Archaeoceti (extinct, Eocene only), and the monophyletic living groups Odontoceti (toothed whales and dolphins) and Mysticeti (baleen whales and relatives). Both latter groups appeared about the latest Eocene (Fordyce and Muizon, 2001). Anatomical and genetic studies indicate that modern odontocetes form a clade, apparently unified by the ability to produce echolocation sounds, which are used to navigate and detect prey. Skull struc-

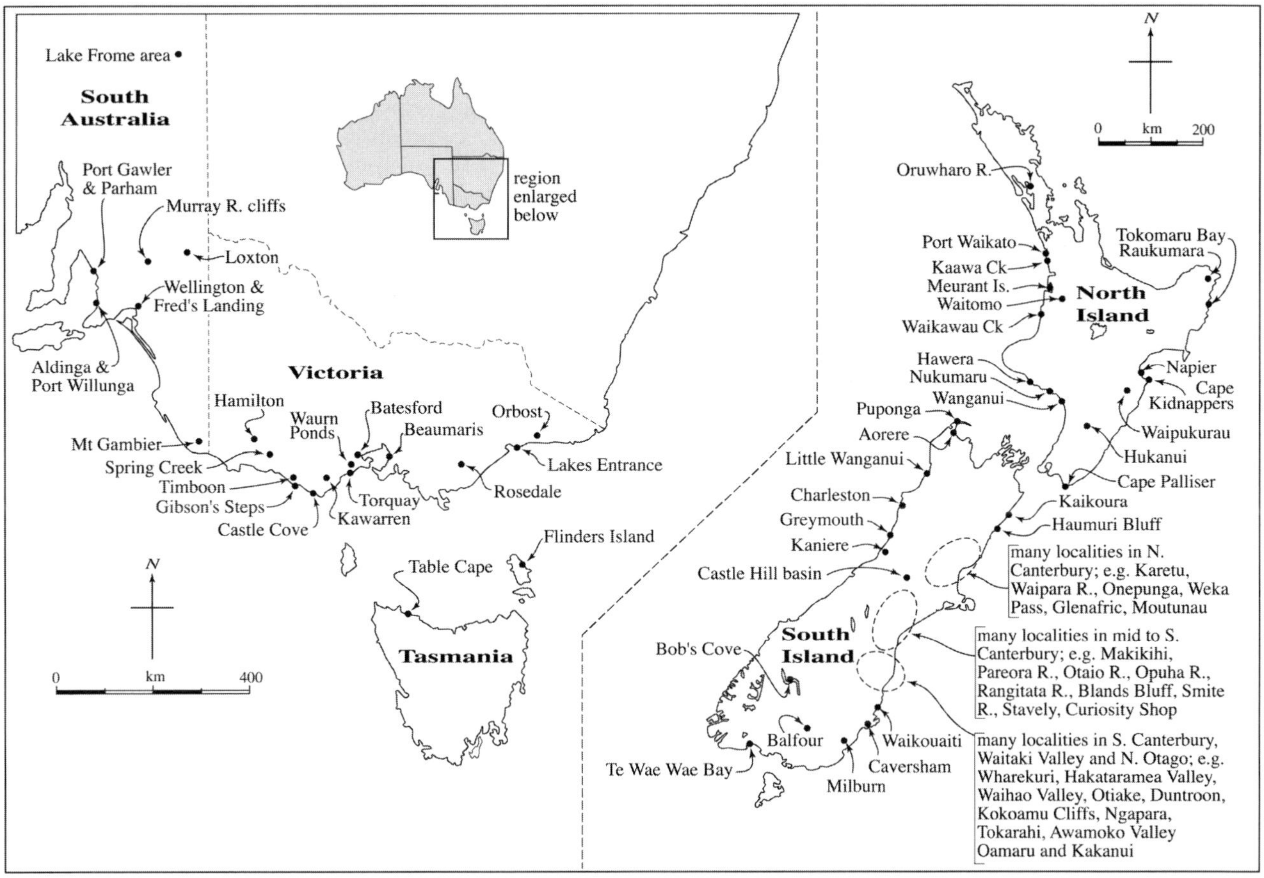

Figure 5. Significant localities for fossil Cetacea in Australia and New Zealand. Based on Fordyce (1984a, 1991).

tures in fossil odontocetes suggest that they, too, also echolocated. Conversely, the clade Mysticeti is characterized by filter-feeding, a behaviour inferred for most or all fossil mysticetes. Mysticetes are not known to echolocate, and archaeocetes show no evidence of ability to echolocate or filter-feed. Odontoceti + Mysticeti form the crown-group Cetacea (formally termed Autoceta or Neoceti; Figure 3), while archaeocetes form the stem-group Cetacea.

Figure 3 shows the present understanding of cladistic relationships within the Cetacea. This is a composite, not the result of a single cladistic analysis. Several of the fossil groups shown are paraphyletic, and will be broken into clades upon further study.

ARCHAEOCETI

Archaeocetes are extinct basal cetaceans that lack the features of odontocetes and mysticetes. The maxilla does not expand over the frontal either dorsally (as in odontocetes) or ventrally (as in mysticetes), and the parietals are well exposed at a long deep intertemporal region (Kellogg, 1936; Thewissen, 1998) (Figure 1(c)). Older literature (Miller, 1923; Kellogg, 1936) has emphasized that archaeocetes lack the 'telescoped' skull of crown-group Cetacea, but 'telescoping' is an ambiguous term best not used. Archaeocetes reportedly range from the late Early Eocene to the end of the Eocene, with no reliably-identified archaeocetes described from younger rocks. Constituent families are currently the Pakicetidae, Ambulocetidae, Protocetidae, Remingtonocetidae and Basilosauridae. Families are diagnosed on features including cusps (or denticles) on teeth, number of teeth, proportions of rostrum (or upper jaw) and cranium (or braincase), sinus fossae in the skull base, earbones, and proportions of limbs and vertebrae.

The family Pakicetidae encompasses the most-basal Cetacea. These animals, such as the small *Pakicetus inachus*, are reported only from Pakistan and India. The oldest record, at 53.5 m.y.a. (or million years ago), is *Himalayacetus subathuensis*, based on a partial jaw from India (Bajpai and Gingerich, 1998). Gingerich *et al.* (1983) suggested that these early cetaceans radiated slowly in productive shallow waters of the Tethys seaway. Slightly more-crownward archaeocetes from the eastern Tethys also include early to middle Eocene species in two other families, the Ambulocetidae and Remingtonocetidae (Thewissen, 1998). Given their archaic body form, these archaeocetes were probably amphibious to shallow-marine species unlikely to have crossed open ocean to Australasian waters.

Species of Protocetidae (early to middle Eocene; e.g. *Protocetus atavus*) occur more widely in the Tethys and middle Atlantic. Many species have been described, mostly since 1980 (see Uhen, 1999). Protocetids have a skull with an enlarged supraorbital shield, and a mandible with a large mandibular foramen, but lack complex tooth crowns or fossae for the pterygoid sinuses. In the post-cranial skeleton, hindlimbs and pelvis are present but reduced (Gingerich *et al*., 1994). Protocetids have not been reported from Australasia, but since their highest diversity is apparently in the eastern Tethys, they could occur in Eocene strata along northern Australia.

The most-crownward archaeocetes are the middle and late Eocene basilosaurids (Uhen, 1998) (Figures 1(c), 3). Like basal odontocetes and mysticetes, the basilosaurids have enlarged fossae for the pterygoid sinuses in the skull base, and cheek-teeth with multiple denticles (Figure 6(a)). The subfamily Basilosaurinae is typified by *Basilosarus cetoides* (late Eocene, western North Atlantic) which probably reached 15 m in length. A second subfamily, the Dorudontinae, includes more generalized and somewhat dolphin-like species of *Dorudon* and other genera. All are smaller than basilosaurines and lack specialized elongate vertebral bodies. Species are distinguished on size, tooth form, and limb form. Several dorudontines are equally plausible sister-taxa to the Odontoceti + Mysticeti (Uhen, 1998, 1999).

All the named basilosaurids are from the Tethys and western Atlantic, but Australasian records are also known; these comprise the oldest reported Cetacea for the Southern Hemisphere. For New Zealand, Fordyce (1985) noted teeth of a *Dorudon*-like species, and large vertebrae of an unidentified late Middle to late Eocene form, possibly a basilosaurine. Köhler and Fordyce (1997) described the partial skull and associated elements of a *Zygorhiza*-like dorudontine (Figure (6a)) from the Waihao Greensand of South Canterbury. Basilosaurid archaeocetes have not been identified from Australia, but should be expected.

CROWN-GROUP CETACEA—THE NEOCETI

Do the Odontoceti and Mysticeti share an immediate common ancestor amongst basilosaurid archaeocetes, or did each arise from a different basilosaurid archaeocete? At present, Odontoceti and Mysticeti are regarded as sister-taxa, variously termed crown-group Cetacea, or Neoceti, or Autoceta. They share features that indicate a common ancestry amongst basilosaurid archaeocetes, including details of the maxilla and facial fossa, the premaxillae and mesorostral groove, the pterygoid sinus fossa, and relationships of the bulla and periotic with adjacent skull bones. Polydonty (tooth complement exceeding the norm for mammals) and loss of tooth succession have also been cited as synapomorphies. These features need detailed analysis in terms of functional shifts during phylogeny.

Early Neoceti are known poorly because of a sparse early Oligocene global record for Cetacea in general. A global sea level fall at about the end of the early Oligocene (~29–30 m.y.a.) removed shallow-water strata including, presumably, cetacean-bearing rocks (Fordyce, 1992). Thus, the few Australasian early Oligocene Cetacea, although fragmentary, are globally quite significant (Glaessner, 1955; Fordyce, 1989b; Pledge 1994; Fordyce and Watson, 1998). Strata that have produced early Oligocene specimens include the Gambier Limestone, Ototara Limestone, Wharekuri Greensand and Amuri Limestone; some of the key fossils are mentioned below.

Mysticeti

Living mysticetes are toothless, and filter-feed small prey from the water. Hair-fringed baleen plates that hang from the upper jaw form the filtering apparatus (Figure 7(f)). The skull is greatly modified (Miller, 1923); for example, the maxilla ventrally forms an origin for baleen, the maxilla extends back below the orbit to form an infraorbital plate, and the mandibles are joined by short ligaments rather than a bony symphysis (Figure 1(b)). Many fossil species have a comparable skull structure and, although baleen preserves only rarely, these can be identified confidently as baleen whales. The most basal mysticetes have well-developed teeth in an otherwise mysticete-like skull. Mysticetes appeared shortly before the end of the Eocene, ~35 m.y.a., and radiated rapidly in the Oligocene. Limestones and greensands from New Zealand are some of the richest global sources of Oligocene mysticetes and other Cetacea, producing toothed archaeocete-like species, archaic toothed mysticetes, the *'Mauicetus'* group of early baleen whales and several other clades (Fordyce, 1992). There are four living families, of still-uncertain relationship to one another, and up to five extinct families. Some of these groups are considered in cladistic analyses (e.g. Geisler and Luo, 1996), but there is no detailed treatment of the Mysticeti as a whole.

Archaic Mysticetes

The best documented family of extinct toothed mysticetes is the Aetiocetidae, represented by small Oligocene species around the North Pacific (Barnes *et al*., 1995b). The upper jaw is flat and triangular, as in most other mysticetes, but teeth are well developed. There is no evidence of baleen. Aetiocetids have not been reported from Australasia.

Three other families usually referred to the Mysticeti each are known from a single named Austral species. First, the large latest Eocene *Llanocetus denticrenatus* Mitchell was based on a fragment of mandible from Antarctica; much of a skull is known (Fordyce, 1989a). A fragmentary toothed cetacean from the earliest Oligocene of Weston, Otago, New Zealand (the 'protosqualodont' of Keyes, 1973) may also represent *Llanocetus*. Second, *Kekenodon onamata* (Hector, 1881) typifies the Kekenodontidae (Figure 6(e)), also provisionally allocated to the Mysticeti. The only described material is a fragmentary late Oligocene specimen, long thought to be an

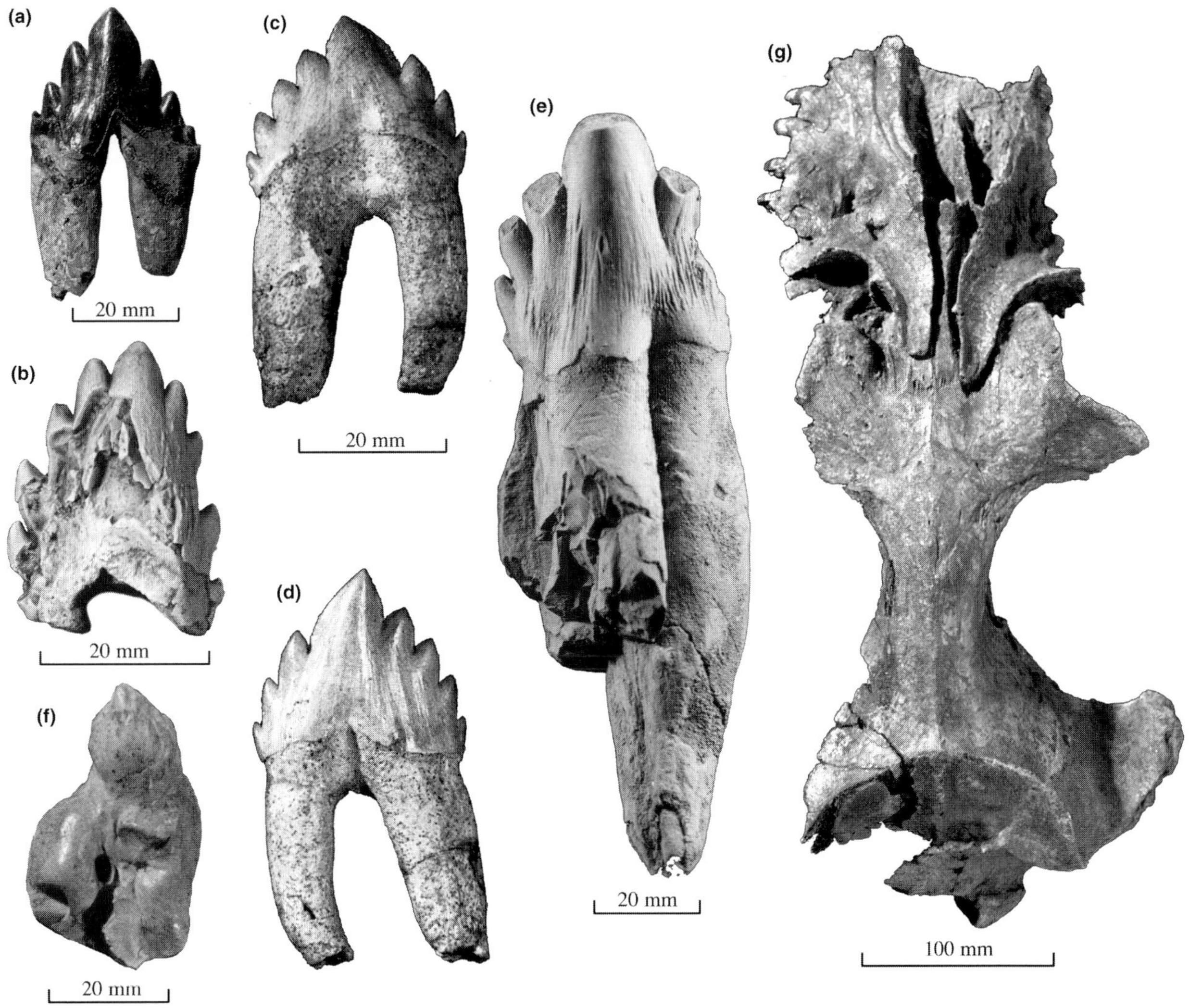

Figure 6. An archaeocete and archaic toothed Mysticeti from Australia and New Zealand. **(a)** Lingual view of tooth of cf. *Zygorhiza,* archaeocete, ?early Late Eocene, Waihao, N.Z.; Geology Museum, University of Otago. **(b)** '*Squalodon' serratus*, lingual view of single holotype tooth, late Oligocene, Karetu, N.Z.; Canterbury Museum. **(c, d)** '*Squalodon' gambierensis*, single holotype tooth, ?early Oligocene, Mt. Gambier, S.A.; University of Adelaide Department of Geology); **(c)** buccal view; **(d)** lingual view. **(e)** *Kekenodon onamata*, lingual view of tooth from holotype, late Oligocene, Wharekuri, N.Z.; Museum of N.Z. **(f)** ?*Mammalodon colliveri*, ventral view of provisionally-referred periotic, late Oligocene, Vic.; Museum of Victoria. **(g)** *Mammalodon colliveri*, dorsal view of holotype skull, late Oligocene, Vic.; University of Melbourne Geology Department.

archaeocete, from Wharekuri, Otago, New Zealand. Finally, *Mammalodon colliveri* (Pritchard, 1939) (Mammalodontidae) (Figure 6(f), (g)) is a small peculiarly short-jawed species known from a skull, jaws and many referred earbones, from Torquay, Victoria. Fragmentary *Mammalodon*-like fossils have been found in New Zealand.

Mammalodon is sufficiently important to warrant further comment and to summarize a previous account (Fordyce, 1984a): it has a small skull (condylobasal length about 600 mm; Figure 6(g)), with a distinctively short, broad rostrum carrying single-rooted cheek-teeth. These rostral features distinguish the species and, indeed, family. Bones of the rostrum articulate loosely with the cranium, as in other mysticetes. Ventrally, the rostrum is flat-based, without any evidence of baleen. Dorsally, the face is more depressed for nasofacial muscles than in dorudontines, but less so than in odontocetes, and there is no supraorbital process of the maxilla. The cranium is archaeocete-like, with a long intertemporal region. On the mandible, the symphysis is very short and was probably ligamentous, as in modern mysticetes. Skull length indicates a body size of 3–3.6 m. Like the aetiocetids of the North Pacific, *M. colliveri* in many ways straddles the gap between dorudontines and baleen whales.

Other putative archaic mysticetes are from the poorly-sampled early Oligocene. These fossils, from the Wharekuri Greensand of North Otago, include a large pathological specimen (Fordyce and Watson, 1998) and an undescribed neonate or juvenile which includes an incomplete skull. Another fragmentary jaw from New Zealand (Amuri Limestone, earliest Oligocene) may also represent an archaic mysticete (Fordyce, 1989b).

Baleen Whales

Globally, there is an extensive record of late Oligocene to Pliocene toothless fossil mysticetes which do not clearly belong in any living family. The family Cetotheriidae has long been used for these fossils (e.g. Kellogg, 1928; Barnes and McLeod, 1984; Hampe, 1996). However, 'cetothere' has been used so widely that the term obscures rather than elucidates the radiation of toothless mysticetes. The family Cetotheriidae is based ultimately on the late Miocene *Cetotherium rathkii* Brandt, which is known from an incomplete skull with a distinctive back-thrust suture between the upper jaw (rostrum) and braincase. The Cetotheriidae should be used as a clade for *C. rathkii* and relatives, rather than as a grade for archaic baleen whales; this action would exclude many fossil mysticetes from the family, including previously-reported Australasian 'cetotheres'. Notable amongst the archaic baleen whales is the South Australian early Miocene "*Aglaocetus*? sp." of Glaessner (1955)—see Figure 7(a)—for which Bearlin (1985, 1988) suggested affinities with *Parietobalaena palmeri* (middle Miocene, Maryland).

Of four species of *Mauicetus* from New Zealand, the type species, *M. parki* (Benham, 1937)—see Figure 7(c)—seems closer to balaenopterids than other mysticete groups. Marples' (1956) other species of *Mauicetus* (shown in Figure 7(d)), represent at least one other genus of uncertain family position. All these named species from New Zealand are of late Oligocene age and amongst the oldest named baleen whales, but are presently not allocated to family (Figure 4). *Mauicetus* has not been identified positively outside New Zealand.

Amongst the extant groups of mysticetes, rorquals and humpbacks—Balaenopteridae—have a significant late Miocene and younger record including Australasian fossils. The living species, such as fin whale *Balaenoptera physalus* (Figure 7(f)), are relatively large active 'gulping' predators with prominent throat grooves, in which the skull shows a distinctively depressed frontal over the orbit (Miller, 1923), in turn part of the origin for the jaw-closing temporalis muscles. Fossil balaenopterids have been reported from Australia and New Zealand (Bearlin, 1985, 1988; Fordyce, 1982, 1991), but none has yet been described formally. They include species of *Megaptera* and *Balaenoptera*, which are represented by earbones from the late Miocene and Pliocene of Victoria and Tasmania and incomplete or whole skulls from the middle Miocene or later (Figure 7(b)). The latter localities include the richest Australian fossil whale horizons, namely phosphate-rich remanié beds of latest Miocene age, at Beaumaris and Grange Burn in Victoria (Carter, 1978). A useful insight into balaenopterid taxonomy and osteology, based on Australian material, was given by Arnold *et al.* (1987).

The Family Eschrichtiidae includes a single living species, *Eschrichtius robustus*, which occurs in the North Pacific and, until recently, inhabited the North Atlantic. There are no southern records, and known fossils are less than one million years old (Barnes and McLeod, 1984). The gray whale is phenetically quite different from other living mysticetes, but is close to balaenopterids in terms of mtDNA sequence (Arnason *et al.*, 1992).

Right whales (Balaenidae) are slow-moving skim-feeders found in both hemispheres. The feeding method, using long baleen plates, is linked with a distinctive skull structure (McLeod *et al.*, 1993) which includes a narrow arched rostrum (Figure 1(b)). The oldest named fossil is the early Miocene *Morenocetus parvus* Cabrera, from Patagonia. *Morenocetus* is clearly a right whale, and balaenid origins should be sought in the Oligocene. The balaenid record is patchy in the global Miocene, but there are many records of isolated and often-worn elements in the latest Miocene and Pliocene of Australia. Published references to such fossils mention earbones (Gill, 1957)—see Figure 7(e)—which are relatively distinctive. A fragmentary presumed balaenid is known from the Miocene of South Australia (Bearlin, 1988).

The small living pygmy right whale, *Caperea marginata*, is the sole species in the family Neobalaenidae. This species occurs in cool circum-Antarctic waters, with many records from New Zealand and southern Australia. Its skeleton is divergent from that of balaenids (Beddard, 1901; Barnes and McLeod, 1984) and see Figure 7(g), although a relationship with right whales is widely accepted. The single published report of a South American fossil species has not been substantiated (Cozzuol, 1996).

Odontoceti

Living odontocetes are the toothed whales, dolphins and porpoises. Most, but not all, adults have erupted teeth, and those without erupted teeth usually have vestigial teeth. No odontocete has baleen. On the skull, each maxilla expands dorsally and posteriorly to form a supraorbital plate (Miller, 1923), making an origin for enlarged nasofacial muscles which apparently manipulate soft tissues around the nasal sinuses, between the skull and the blowhole (Mead, 1975a, Heyning, 1989). In turn, these tissues are implicated in producing the echolocation sounds which odontocetes use to navigate and detect prey (Cranford *et al.*, 1996). Most fossil odontocetes have a comparable supraorbital plate of the maxilla, implying the presence of enlarged nasofacial muscles and, thus, the ability to echolocate. Other osteological and soft-tissue features, including the basicranial sinuses and details of the ear region (Fraser and Purves, 1960; Kasuya, 1973) are also diagnostic. The oldest named and reliably dated odontocetes are from the late Oligocene (~30 m.y.a.), but older supposed odontocetes have been mentioned in abstract and, given that the Mysticeti is the sister taxon, the Odontoceti probably arose about 35 m.y.a.

Living families include the Physeteridae, Kogiidae, Platanistidae, Ziphiidae, Delphinidae, Phocoenidae,

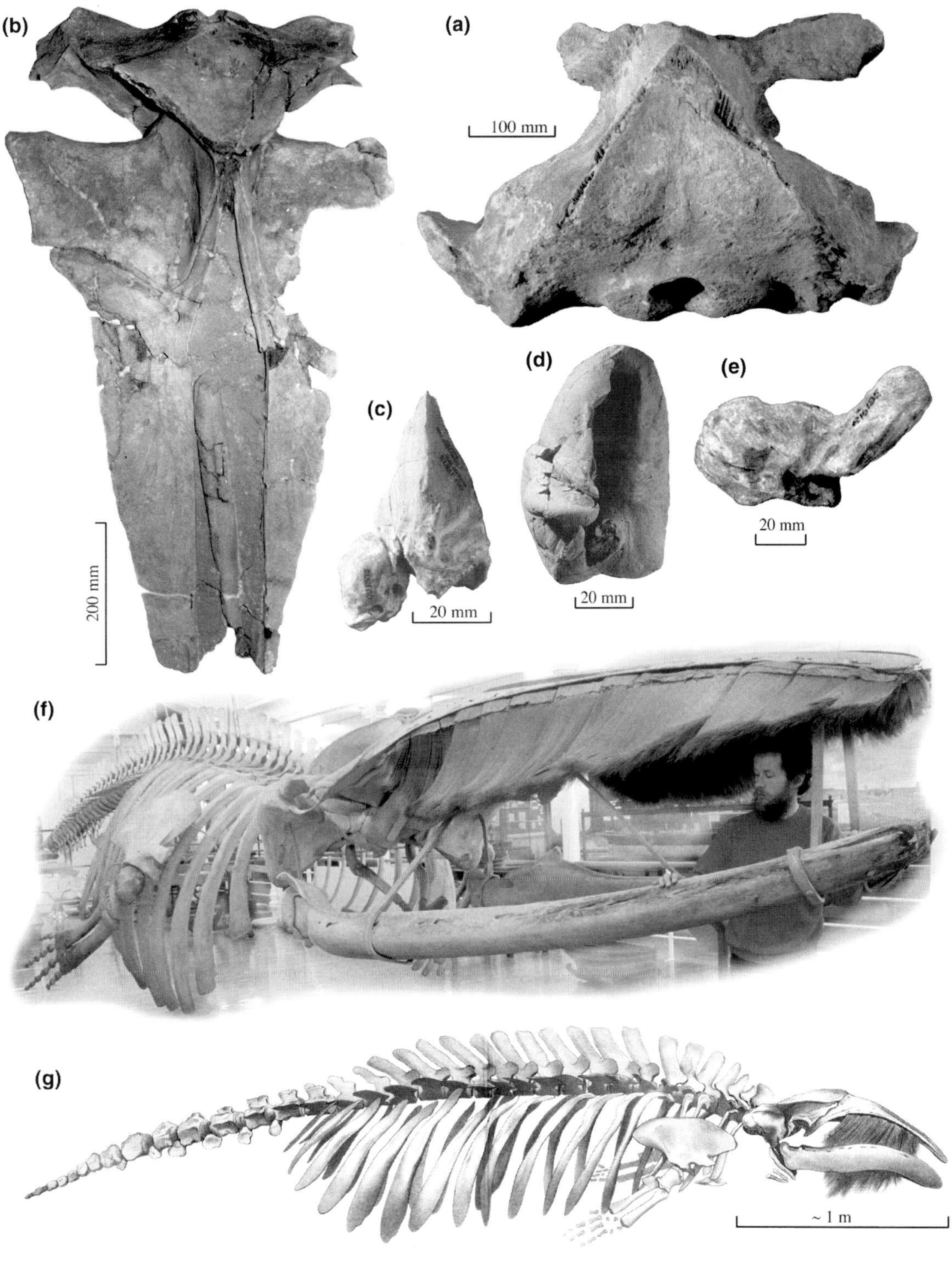

Figure 7. Baleen whales (Mysticeti) from Australia and New Zealand. **(a)** *Parietobalaena* sp. (="*Aglaocetus* ? sp. nov." of Glaessner 1955), dorsal view of incomplete skull, early Miocene, S.A.; South Australian Museum. **(b)** ?*Megaptera* of Bearlin 1985, anterodorsal view of skull, Pliocene, Vic.; Museum of Victoria. **(c)** *Mauicetus parki,* ventral view of holotype periotic, late Oligocene, Milburn, N.Z.; Geology Museum, University of Otago. **(d)** "*Mauicetus" lophocephalus*, dorsal view of holotype bulla, late Oligocene, Duntroon, N.Z.; Otago Museum. **(e)** *Eubalaena* sp., ventral view of isolated periotic, late Miocene or early Pliocene, Beaumaris, Vic., Museum of Victoria. **(f)** *Balaenoptera physalus*, anterior oblique view of mounted skeleton, extant, New Zealand; Otago Museum; photo D.V. Weston, Department of Geology, University of Otago. **(g)** *Caperea marginata*, lateral view of skeleton, extant, New Zealand; from Beddard (1901).

Monodontidae, Iniidae, Pontoporiidae, and Lipotidae, with many other extinct groups also reported (below, and Figure 3). Overall, the odontocetes are more diverse taxonomically and more disparate structurally than are the mysticetes. Some living families can be put into higher-level clusters (Figure 3 shows Physeteroidea, Platanistoidea, Delphinoidea and Delphinida), but some family-level relationships are uncertain. Cladistic approaches include those of Muizon (1987, 1988a, 1988b, 1991), Heyning (1989), and Fordyce (1994). Living species were listed by Rice (1998) (see also Perrin and Brownell, 2001).

Basal Odontocetes

Odontocetes with an archaic skull form and differentiated teeth are often placed in the widely-mentioned family Agorophiidae. Fordyce (1981) suggested that the family be used in a cladistic sense, containing only *Agorophius pygmaeus*; *A. pygmaeus* is uncertainly related to the Squalodontidae, discussed below. Amongst other archaic and enigmatic odontocetes, *Archaeodelphis patrius* Allen is structurally the most basal species. Species of the bizarre late Oligocene *Xenorophus* (see Whitmore and Sanders, 1977) are known only from the late Oligocene of the Atlantic Coastal Plain. No odontocetes of 'agorophiid' grade have been described from Australasia.

Sperm Whales

Sperm whales, Physeteridae and Kogiidae (jointly, Physeteroidea) lie near the base of the odontocete radiation. Structurally, the living sperm whales are quite disparate from other odontocetes, and they form a low-diversity clade. The huge, deep-diving, near-cosmopolitan squid-eater *Physeter macrocephalus* is the sole species of Physeteridae. The sperm whale skull has a distinctive supracranial basin (Van Beneden and Gervais, 1868–1880)—see Figures 1(a), 8(e)—which can be traced back to early Miocene fossils such as *Diaphorocetus*. Judging from the poorly-known *Ferecetotherium*, physeterids had originated by the late Oligocene. Globally, Miocene and Pliocene physeterids are represented by common fragmentary specimens, especially isolated teeth (Kellogg, 1925), with many from Australia. Despite their wide use in taxonomy, it seems that teeth are not particularly diagnostic. Chapman (1912, 1918a, 1918b) and McCoy (1879) established four new species of *Scaldicetus*, *Scaptodon* and *Physetodon* for isolated teeth from the Miocene and Pliocene of Victoria and Tasmania. Because the types are not particularly diagnostic (Figure 8(g)), the species are of uncertain relationships. Other physeterids are known from Victoria (Miocene) and Flinders Island, Tasmania (Pliocene), and Quaternary specimens occur in South Australia (Kemper *et al.*, 1996). See Figures 8(h), (i). Fragmentary physeterids from New Zealand strata include a partial rostrum from the latest Oligocene of North Otago, a ?middle–late Miocene *Scaldicetus*-like tooth dredged from the Chatham Rise (Fordyce, 1984b), and a ?late Pliocene tooth from Hawke's Bay.

Pygmy sperm whales (Kogiidae) are dramatically smaller than their close living relative, the sperm whale, and are shark-like in external form—even having a pigmented false gill slit (Caldwell and Caldwell, 1989). *Kogia* species thus provide an evolutionarily significant example of mimicry. There is a patchy fossil record back to the late or possibly middle Miocene (Fordyce and Muizon, 2001). A fossil kogiid periotic is known from Chatham Rise (Fordyce, 1984b), east of New Zealand, and kogiids might be expected from nodule beds in Victoria, Australia.

Platanistoid Dolphins

A cluster of mainly-extinct families represents the superfamily Platanistoidea, typified by the endangered long-beaked freshwater dolphin (*Platanista gangetica*) of the Indian subcontinent. Fossil relatives are marine species of *Zarhachis* and *Pomatodelphis*, from the Atlantic Coastal Plain (Kellogg, 1928; Fordyce and Muizon, 2001); these, with *Platanista*, comprise the Platanistidae. All have remarkably long rostra, and maxillary crests along the lateral margins of the face. Platanistids are not known from Australasia; earlier reports from South Australia (Tedford *et al.*, 1977) probably represent eurhinodelphinids (Fordyce, 1983).

The affinities of *Platanista* and other 'river dolphins' were long problematic until Muizon (1987, 1991, 1994) significantly modified Simpson's (1945) original concept of the Platanistoidea. Muizon excluded other living 'river dolphins' (now allied with Delphinoidea, below), but included some other hitherto problematic extinct groups: Squalodelphinidae, Squalodontidae, and possibly Dalpiazinidae. Fordyce (1994) placed the extinct Waipatiidae here also. Nikaido *et al.* (2001) placed the Platanistoidea between the Physeteridae and Ziphiidae.

Squalodelphinids appear to be the sister taxon for the Platanistidae. They are marine, mainly latest Oligocene to early Miocene, and share with the Platanistidae important synapomorphies of the ear region and scapula. Fordyce (1994) identified squalodelphinids from the New Zealand region, including the latest Oligocene *Notocetus marplesi* (Dickson, 1964) and the slightly older and fragmentary *'Microcetus' hectori* (Benham, 1935b). Previously, both were thought to be squalodontids. These named New Zealand species are only a few of the many platanistoids from the richly fossiliferous limestones and greensands of the Waitaki Valley of eastern South Island; also notable are dalpiazinids, waipatiids, squalodontids, archaic delphinoids, and possible sperm whales. The New Zealand squalodelphinids, with *Notocetus vanbenedeni* in Patagonia, indicate an early circum-Antarctic distribution for the group.

The least known of the platanistoids is the Dalpiazinidae (Muizon, 1988b), an enigmatic group including some long-beaked species previously identified as 'Acrodelphidae'. Only the European early Miocene *Dalpiazina ombonii* is named, but undescribed skulls from New Zealand probably represent late Oligocene dalpiazinids (Fordyce and Barnes, 1994).

Plate 116. A humpback whale (*Megaptera novaeangliae*) surfacing. Note the characteristic flipper—up to 5 m in length. Although generally slow-moving humpbacks are capable of bursts of speed; their feeding behaviour is versatile and includes bubblenetting, to confine a school of fish. Adults may be up to 19m long. Photograph S. Allen.

Plate 117. Humpbacks (*M. novaeangliae*) are often seen in pairs or small groups of up to 15 individuals. Females calve every 2–3 years and calves stay close to mothers for the first year. Several populations have been studied in recent decades during their long migrations between polar waters, where they feed, and tropical breeding grounds. The tail is usually lifted clear of the water before a deep dive and thousands of individuals have been photo-identified by their unique ventral tail fluke markings. Photograph S. Allen.

Plate 118. A group of humpbacks including the only known individual (in Australian populations) that is completely white. This animal was first recorded in 1991, when it was sub-adult. Late in 2003 it was involved in a collision with a yacht, but survived with apparently minor injuries. Photograph S. Allen.

Plate 119. The white humpback diving. This individual is a member of the population that migrates up and down the coast of eastern Australia. Whether it is a true albino is still unclear, but there is some indirect evidence indicating that it is. The sex has not been definitely determined—although behaviour suggests that it is a male and has now reached maturity. Photograph S. Allen.

Plate 120. Risso's dolphin (*Grampus griseus*) breaking the surface, showing the bulbous head and the prominent, tall dorsal fin. Occurs throughout Australasia. Photograph S. Allen.

Plate 121. The back of a Risso's dolphin (*Grampus griseus*) showing the body scarring characteristic of this species. This scarring is caused by the teeth of other Risso's dolphins and by squid. Older individuals become lighter in colour. Photograph S. Allen.

Plate 122. The Indo-Pacific humpbacked dolphin (*Sousa chinensis*) has a huge range in shallow, coastal waters from southern Africa eastwards to China and south to Indonesia, New Guinea and northern Australia. Adults range from 2.4–3.2 m in length and some regional variation in both morphology and colour has been noted; for example, the back is usually dark grey but may be white or pink, the undersides are light grey. The taxonomy is still controversial: there may be two species or two genetically distinct populations. Photograph S. Allen.

Plate 123. The southern right whale (*Eubalaena australis*) was another species that was intensively harvested in Australasia and numbers greatly reduced. The population is slowly recovering and individuals attract considerable media attention—especially during the migration south—when they enter estuaries such as Sydney Harbour. Here the characteristic tail is shown. Photograph S. Allen.

Plate 124. The slow-moving southern right whale (*E. australis*) has a relatively stumpy pectoral flipper—in comparison with the humpback. Photograph S. Allen.

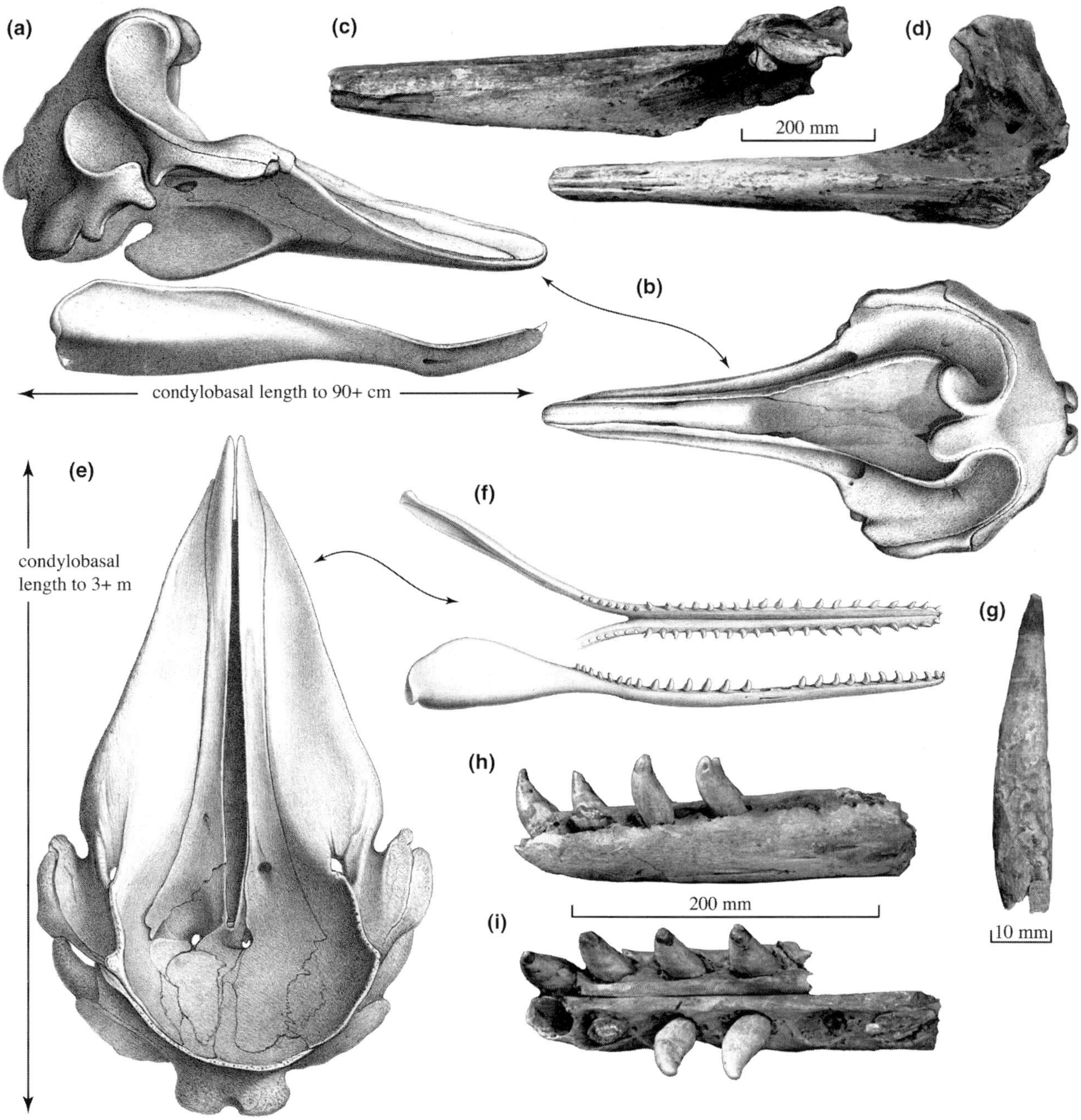

Figure 8. Beaked whales (Ziphiidae) and sperm whales (Physeteridae) including species from Australia and New Zealand. **(a, b)** *Ziphius cavirostris*, modern oceans, skull and mandibles, from Van Beneden and Gervais (1868–1880); **(a)** lateral view; **(b)** dorsal view. **(c, d)** Unidentified ziphiid (?*Mesoplodon* sp. = '*Ziphius*' of Sutherland and Kershaw, 1971), rostrum and skull base, Pliocene, Flinders Island, Tasmania; Museum of Victoria; **(c)** left lateral view; **(d)** dorsal view. **(e, f)** Sperm whale *Physeter macrocephalus*, modern oceans, skull and mandibles, from Van Beneden and Gervais (1868–1880); **(e)** dorsal view of skull; **(f)** dorsal (upper) and lateral (lower) views of mandibles. **(g)** *Scaldicetus lodgei*, holotype tooth, Hamilton, Vic.; Museum of Victoria. **(h, i)** Mandibles of indeterminate physeterid, late Miocene or early Pliocene, Beaumaris, Vic.; Museum of Victoria; **(h)** left lateral view; **(i)** dorsal view.

The literature suggests that, amongst the extinct platanistoids, the shark-toothed dolphins (Squalodontidae; late Oligocene to late Miocene) include many Australasian fossils (Keyes, 1973; Mahoney and Ride, 1975; Pledge and Rothausen, 1977). However, an apparent high diversity may be artificial, because the Squalodontidae has been used previously for many fossil cetaceans in which the teeth have multiple denticles (see Kellogg, 1923). Redefinitions of the Squalodontidae (Fordyce, 1994; Muizon, 1994) imply that Australasian squalodontids now include only '*Prosqualodon*' *hamiltoni* Benham, perhaps *Prosqualodon davidis* (Flynn,

1923) and several unnamed new species represented by well-preserved late Oligocene skulls from New Zealand (Figure 9(a)–(d), (f)).

Of the Australasian squalodontids, *Prosqualodon davidis* is by far the best-documented. The species was based on a well-preserved skull (Flynn, 1948) from Table Cape, near Wynyard, Tasmania. The type horizon was probably the Fossil Bluff Sandstone, of Longfordian age, earliest Miocene (Mahoney and Ride, 1975; Pledge and Rothausen, 1977). Regrettably, the holotype skull has been lost, although casts are held by several institutions, and Flynn's (1948) fine account of the skull allows easy comparisons with other odontocetes (Fordyce, 1984a). Most notably, the skull has a robust short rostrum with prominent coarsely ornamented and heterodont teeth (Figure 9(c)). *Prosqualodon davidis* clearly had well-developed nasofacial muscles, and was probably an echolocator. Basicranial sinus fossae are moderately developed. As in other Squalodontidae and Platanistoidea (see Muizon, 1991), the scapula lacks a coracoid process. *Prosqualodon davidis* is related closely to *P. australis* Lydekker (see also Cozzuol, 1996) from the earliest Miocene of Patagonia, and the genus is apparently represented in New Zealand by isolated teeth from the late Oligocene Waitakian Stage. These occurrences point to a circumpolar distribution. There is no consensus yet on whether *Prosqualodon* is close to *Squalodon* and relatives, or better placed in Cozzuol's (1996) family Prosqualodontidae.

Australasia has many examples of so-called squalodontids named from isolated teeth, including the Oligocene '*Squalodon' gambierensis* (Glaessner, 1955) (Figures 6(c), (d)), '*Squalodon' serratus* (Davis, 1888; see Glaessner, 1972; Figure 6(b)), *Metasqualodon harwoodi* (Sanger, 1881; see Pledge and Rothausen, 1977; Figure 9(e)), and '*Squalodon' andrewi* (Benham, 1942). Hall (1911) and Gill (1957) mentioned yet more specimens. Of these fossils, '*S.' gambierensis* and '*S.' serratus* are probably not odontocetes, while *M. harwoodi* and *S. andrewi* could be squalodontids or other odontocetes. New Zealand 'squalodontids' are equally ill-founded, including *Tangaroasaurus kakanuiensis* Benham (1935a) and *Austrosqualodon trirhizodonta* Climo and Baker (1972).

Small triangular and denticulate teeth from heterodont dolphins were long identified as from small squalodontids (Kellogg, 1923; Whitmore and Sanders, 1977). Eventually, such teeth were found in a well-preserved late Oligocene skull from New Zealand (Figure 9(g)), on which Fordyce (1994) based the new species, genus and family *Waipatia maerewhenua* (Waipatiidae), related to but separate from the squalodontids. Probably, other so-called squalodontids (e.g. *Sulakocetus*—North Pacific; ?*Microcetus*—North Sea) also belong here. Waipatiids should be expected from Australian Oligocene strata.

Beaked Whales

Beaked whales or Ziphiidae are medium- to large-sized deep-diving pelagic odontocetes that are historically well represented from New Zealand and Australia (Bannister *et al.*, 1996; Baker, 1999). Most ziphiids have greatly reduced numbers of teeth apart from prominent mandibular teeth in males which presumably are used in agonistic behaviour. The skull is deeply-basined at the face (Figure 8(a), (b)), and the rostrum may be densely ossified (Zioupos *et al.*, 1997), producing one of the densest known biological tissues. Extant ziphiids include some of the world's rarest mammals; Longman's beaked whale *Indopacetus* [or *Mesoplodon*] *pacificus* (Longman, 1926), is known from only two stranded specimens, one from Queensland and one from Somalia. Shepherd's beaked whale, *Tasmacetus shepherdi* (Oliver, 1937) is now known from strandings including New Zealand, Australia and southern South America (Baker, 1999). Members of the speciose genus *Mesoplodon* are difficult to separate morphologically, but molecular techniques should help to resolve relationships (Dalebout *et al.*, 1998).

For fossils, the oldest unambiguous ziphiids are middle Miocene. Isolated rostra are the most common fossils, with many used as the basis for new species despite their dubiously diagnostic form, and the taxonomy of fossil ziphiids needs reviewing (Mead, 1975b). Localities in South Australia, Victoria, and Tasmania (Chapman, 1918a; Glaessner, 1947; Fordyce, 1982) have yielded rostra from species of *Mesoplodon* and others related to extant ziphiids. Many of these fossils are from nodule beds. A Pliocene '*Ziphius*' from Tasmania (Sutherland and Kershaw, 1971) represents a ziphiid other than *Ziphius* (Figure 8 (c), (d)).

Deep-sea remains of ziphiids are significant. Whitmore *et al.* (1986) reported the skull of a ziphiid which was photographed in deep water in the southern Tasman Sea. New Zealand ziphiids include the worn apex of a mandible of possible middle or late Miocene age, and isolated earbones, from Chatham Rise (Fordyce and Cullen, 1979; Fordyce, 1984b) and many undescribed rostra. Associated with the Chatham Rise ziphiids is a range of delphinids, phocoenids, physeterids and balaenopterids (Fordyce and Cullen, 1979; Fordyce, 1984b).

Eurhinodelphinids

Eurhinodelphinids (Eurhinodelphidae or Rhabdosteidae of some authors) are extinct medium to long-beaked small odontocetes known from widely scattered early to middle Miocene localities (Abel, 1909, Kellogg, 1932); there are also a few late Oligocene records (Fordyce and Barnes, 1994). Genera include *Eurhinodelphis* (Figure 10(c)), *Agyrocetus,* and possibly *Squaloziphius.* Skull structure is quite specialized, so that the relationships of eurhinodelphinids are uncertain; Muizon (1988a, 1991) placed them close to the extinct *Eoplatanista* (early Miocene, Europe; Eoplatanistidae).

One of Australasia's most unusual occurrences of fossil odontocetes is that of fragmentary and indeterminate Eurhinodelphinidae from late Oligocene freshwater sediments in the Lake Frome area, South Australia. Many bones (Figure 10(a), (b)) represent at least six dolphins from different horizons, indicating significant longer-term occupation of freshwaters (Tedford *et al.*, 1977; Fordyce, 1983, 1984a; Woodburne *et al.*, 1993).

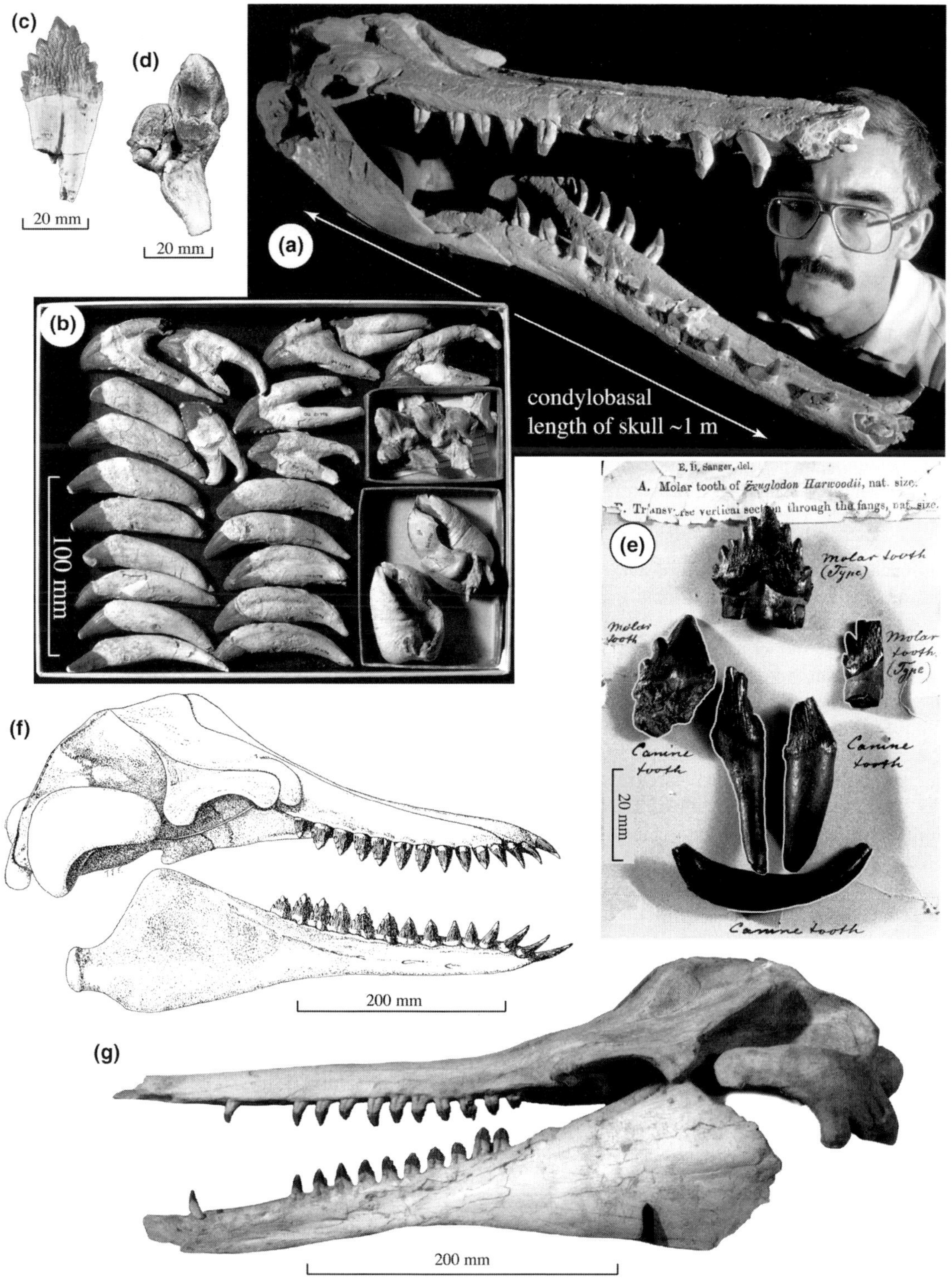

Figure 9. Platanistoid dolphins from Australia and New Zealand. **(a, b)** Un-named *Squalodon*-like squalodontid, late Oligocene, Duntroon, N.Z.; Geology Museum, University of Otago; **(a)** skull and mandibles, oblique lateral view; **(b)** teeth and earbones associated with skull of **(a)**. **(c)** Odontocete similar to *Prosqualodon davidis*, isolated tooth (figured by Hall, 1911), late Oligocene, Mt. Gambier, S.A.; Museum of Victoria. **(d)** Un-named Squalodontidae, ventral view of left periotic, middle Miocene, Batesford, Vic., Museum of Victoria. **(e)** *Metasqualodon harwoodi,* holotype teeth mounted on display card, with some outlines retouched, late Oligocene, Wellington, S.A.; South Australian Museum. **(f)** *Prosqualodon davidis*, lateral view of holotype skull and mandible, early Miocene, Tasmania; reconstruction from Flynn, 1948. **(g)** *Waipatia maerewhenua*, skull and mandibles, lateral view; late Oligocene, Duntroon, N.Z.; Geology Museum, University of Otago.

Elsewhere, eurhinodelphinids occur only in marine strata, including South America and probably New Zealand (the fragmentary *Phocaenopsis mantelli* Huxley; see Fordyce, 1982b). The South Australian specimens are too incomplete to allow useful comparison with named species of, for example, *Eurhinodelphis* (Figure 10(c)), *Argyrocetus* and *Iniopsis*. Fragments of skull indicate a size comparable with the Atlantic Miocene *Eurhinodelphis bossi* Kellogg—with a body length of about 4 m. Despite the incomplete material, this record is the best for eurhinodelphinids in Australasia.

True dolphins—Delphinoidea

Dolphins (family Delphinidae) form the most speciose group of living Cetacea. They are prominent in modern Australasian waters (Bryden *et al.*, 1998; Baker, 1999), and provide important fossil records. Delphinids are quite variable in external form, but have several consistent diagnostic skull features including asymmetrical premaxillae and extensive basicranial sinuses invading the orbit (Fraser and Purves, 1960; Barnes, 1978). Previously, the Delphinidae was used often as a grade family for diverse small odontocetes (Kellogg, 1928; Simpson, 1945), and it is only in the last three decades that the porpoises, Phocoenidae, have routinely been separated from delphinids. There is still no consensus about subdivisions within the Delphinidae. Since the 1960s and 1970s, it has been common to use subfamilies based around *Delphinus* (Delphininae), *Steno* (Steninae), *Lissodelphis* (Lissodelphinae), *Cephalorhynchus* (Cephalorhynchinae), and *Globicephala* (Globicephalinae). These groups were based largely on phenetic judgements of body profile and colour, nasofacial muscles, basicranial sinuses, and skeletal form (see Fraser and Purves, 1960; Mead, 1975a). Molecular techniques have the potential to reveal different relationships, as work on *Lagenorhynchus* (Figure 10(g)) and other delphinids shows (Cipriano, 1997; LeDuc *et al.*, 1999), although the fine study of the Irrawaddy dolphin *Orcaella brevirostris* by Arnold and Heinsohn (1996) indicates an enduring systematic role for morphology. Distribution patterns are rather well known for living Australasian delphinids, with most species ranging outside Australasia. The small Hector's dolphin, *Cephalorhynchus hectori*, is endemic to shallow New Zealand seas.

Globally, fossils with delphinid cranial features can be traced back to the late Miocene, >10 m.y.a. (Barnes, 1977). Australia's single named fossil delphinid is '*Steno*' *cudmorei* (Chapman, 1918a), which does not clearly represent the genus *Steno* (Fordyce, 1982). The holotype is a slightly worn isolated tooth which provides a poor basis for comparison, so it is hard to judge if other specimens referred to the species really are conspecific (Fordyce, 1982)—see Figure 10(d)–(f). Other Australian fossil delphinids have received brief mention (Longman, 1920; Gill, 1965), and undescribed fossil specimens are known (Fordyce, 1984a). At least three delphinids, probably all extinct, are represented by possibly middle to late Miocene periotics from phosphorites on Chatham Rise, east of New Zealand (Fordyce, 1984b), with specimens of comparable age also from Chatham Islands (Fordyce and Campbell, 1990). Undescribed fossil delphinids are known from the coast of North Canterbury, and from the Te Aute Limestone (Pliocene) at Napier. McKee and Fordyce (1987) described a Pliocene mandible, possibly representing *Delphinus* or *Stenella*, from Hawera (Figure 10(h)), and delphinid-like vertebrae are known from this region.

Two other groups are classified close to the Delphinidae: white whales, Monodontidae, and porpoises, Phocoenidae. Extant monodontids include only the beluga and narwhal of Arctic waters but, surprisingly, monodontids lived in warm equatorial waters of the eastern Pacific until the Pliocene (Barnes, 1984, Muizon, 1984). There is no evidence that monodontids ever occupied in south-west Pacific waters, but one could not fully rule out such a distribution. The Irrawaddy dolphin, *Orcaella brevirostris*, has been placed sporadically with monodontids (Kasuya, 1973; Barnes, 1984), but is conclusively a delphinid (see above).

The spectacled porpoise, *Phocoena dioptrica*, occurs sporadically in waters south of New Zealand and Australia, but otherwise phocoenids are mainly animals of the cool temperate North Pacific, south-east Pacific, Atlantic and, strangely, warm waters of the Gulf of Mexico (Barnes, 1985; Rice, 1998). Phocoenids are typically blunt-snouted, and show many diagnostic features on the skull including elevated premaxillary eminences and reduced and ventrally displaced nasals. Molecular studies (Rosel *et al.*, 1995) show relationships amongst living species, placing *P. dioptrica* close to other species of *Phocoena* rather than in a distinct genus, *Australophocoena*. As with dolphins, the globally oldest phocoenids are from the late Miocene, >10 m.y.a. (Barnes, 1985). There are no reports of Australian fossils, but a single possibly Miocene periotic is known from phosphorites on Chatham Rise, east of New Zealand (Fordyce, 1984b, 1989).

Barnes (1978) proposed that the family Kentriodontidae be used for archaic dolphins including ancestors of delphinids and possibly all other extant delphinoid families (Barnes *et al.*, 1985). Given diagnostic features such as small to moderate size, symmetrical cranium, and lack of extensive orbital extensions for the basicranial air sinuses, the Kentriodontidae is clearly a grade. Eventually, the group will be redefined as a clade typified by *Kentriodon pernix* Kellogg from the middle Miocene of the Atlantic Coastal Plain. Kentriodontids have been reported from many Miocene localities world-wide. Late Oligocene and early Miocene *Kentriodon*-like dolphins occur in New Zealand (Ichishima *et al.*, 1995), but have not yet been reported from Australia (see Figure 10(k)).

Two more extinct groups, both from the eastern subtropical to tropical Pacific, occur within the Delphinoidea. First, the porpoise-like *Albireo whistleri* Barnes, 1984 (latest Miocene) is the sole named species within the Albireonidae. Second, the bizarre *Odobenocetops peruvianus* and *O. leptodon* are uncannily like the walrus (pinniped) *Odobenus rosmarus* in skull form, with well developed and probably sexually dimorphic tusks (Muizon *et al.*, 1999). Skull form indicates regression of the nasofacial apparatus which, in other odontocetes, produces echolocation sounds, but the ear retains the distinctive characteristics of odontocetes.

Basal Delphinida

Related closely to the Delphinoidea are three families of extant 'river dolphins'. In total, these groups (Delphinoidea, Iniidae, Pontoporiidae, Lipotidae) form the Delphinida of Muizon (1988a). *Inia, Pontoporia* and fossil relatives occur only in American waters and American strata; it would be surprising to record them in the Southwest Pacific. References on osteology and palaeontology include, Kellogg (1928), Brownell (1989) and Cozzuol (1996). The oldest Pontoporiidae are probably late Middle Miocene, with Iniidae uncertainly from the late Miocene (Fordyce and Muizon, 2001). The dramatically endangered *Lipotes vexillifer* from China represents the third family, Lipotidae, which has no firm fossil record.

PATTERN AND PROCESS IN CETACEAN EVOLUTION AND RADIATIONS

DISTRIBUTIONS

Patterns of cetacean distribution have had wide treatment, from the perspectives of ecology (what governs distribution?) and historical biogeography (what are the long-term patterns, and implications for speciation and extinction?). Davies (1963) provided an early review of historical biogeography using a dispersalist framework, and the later synthesis of Gaskin (1982) is also notable; both, however, have become dated with advances in systematics (see Cipriano, 1997, Rosenbaum *et al.*, 2000). Davies' patterns can be expanded to include tropical, transequatorial/antitropical, circumpolar, endemic (including riverine/lacustrine) tropical/temperate and cosmopolitan. Stranding and sighting records for Australasian cetaceans, although biased in places by poor sampling, show some of these patterns (Bannister *et al.*, 1996; Bryden *et al.*, 1998; Baker, 1999). See Figure 2 and the following short summary.

Globally, tropical Cetacea are not as well known as temperate species. Species such as *Orcaella brevirostris* and *Sousa chinensis* occupy or range into northern Australian waters, and *Balaenoptera edeni* enters New Zealand waters. Pelagic dolphins in the genus *Stenella*, and *Kogia sima*, range into warm temperate waters but rarely further south. These species show no obvious skeletal adaptations to warm-water that might also be seen in fossils. Elsewhere, distribution patterns may indicate warm-water habits in basal archaeocetes, for these occur only in strata from the Tethys Ocean—an eastern extension of the Mediterranean, obliterated when India collided with Asia later in the Eocene.

Many Cetacea have antitropical (= transequatorial, or bipolar) distributions (Davies, 1963), with populations or sister-taxa present in the Northern and Southern Hemispheres but not in the tropics. In *Balaenoptera* spp., this includes different populations of conspecifics, while in the ziphiids *Hyperoodon* and *Berardius,* and the right whale *Eubalaena* (see Rosenbaum *et al.*, 2000), presumed sister species occur in the north and south. Many Australasian fossils belong in families, and sometimes genera, with significant northern temperate records; the fossil *Mesoplodon longirostris* is reported from southern and northern localities (Glaessner, 1947). Generally, though, taxonomic relationships and distribution patterns are too uncertain to be sure that these represent real antitropical distributions.

Some extant species are circumpolar in southern cool-temperate or sub-Antarctic waters, e.g. *Caperea marginata, Phocoena dioptrica,* and *Tasmacetus shepherdi*. Southern limits for the smaller species are not well established, but it seems that small odontocetes do not live permanently in Antarctic waters south of the Polar Front. Amongst fossils, later Oligocene species of *Prosqualodon* (Squalodontidae) occur in southern Australia, New Zealand, and Patagonia, indicating the origin of circum-Antarctic distributions by about 25 m.y.a.

A few extant small odontocetes have restricted distributions; for example, *Platanista gangetica* and other 'river dolphins', *Phocoena sinus* (Gulf of California) and *Cephalorhynchus hectori* from New Zealand. *Cephalorhynchus* is a southern circumpolar genus, in which species probably arose by allopatric speciation from a limited number of founders, as also inferred for *P. sinus*. In Australasia and elsewhere, the fossil record provides many species known from only one locality, or even one specimen. But, known biases (preservation processes, limited geographic and stratigraphic sampling) caution that such fossils could represent originally more-widespread species. The only Australasian species truly likely to have had a limited geographic range is the eurhinodelphinid from the Lake Frome area of South Australia.

Many medium-sized odontocetes have tropical/temperate distributions, ranging through most global waters other than high latitudes. In Australasia, these species include *Tursiops truncatus* (and possibly *T. aduncus*; see below), *Ziphius cavirostris* and *Pseudorca crassidens*. A few extant Cetacea seem to be cosmopolitan. For example, male sperm whales (*Physeter macrocephalus*) range globally, though females disperse less (Best, 1979). Similarly, the killer whale, *Orcinus orca*, ranges through all oceans. Wide distribution of a single species implies global gene flow, in turn limiting scope for allopatric speciation and, indeed, the widely distributed species mostly represent monospecific genera.

DIVERSITY, DISPARITY AND SPECIATION

Processes implicated in speciation include natural selection, sexual selection, coevolution, founder effects, vicariance, and hybridization. Any or all of these might

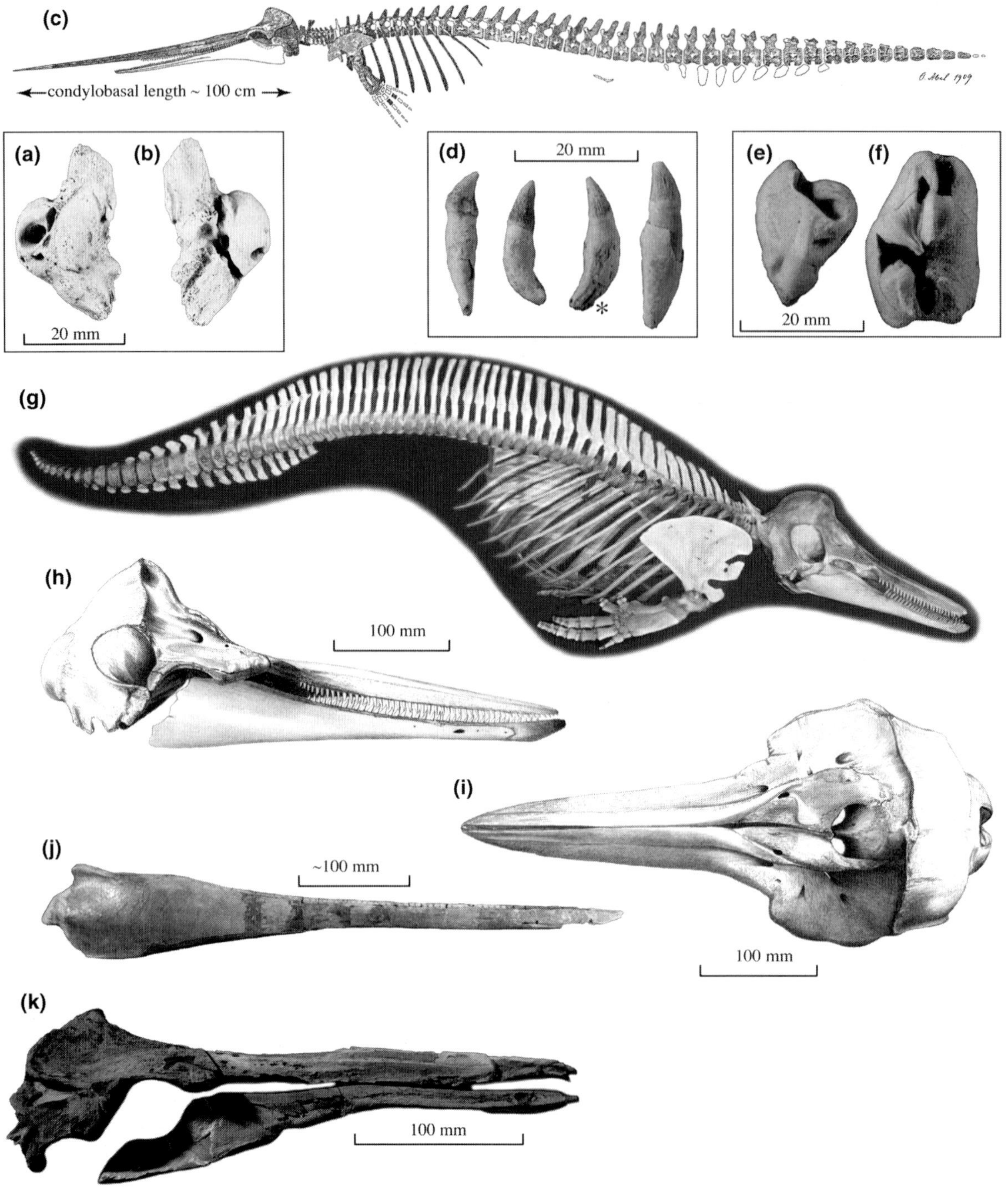

Figure 10. Eurhinodelphinidae and dolphins (Delphinoidea) including species from Australia and New Zealand. **(a, b)** periotic of indeterminate species of Eurhinodelphinidae of Fordyce 1983, late Oligocene, Lake Namba, S.A.; American Museum of Natural History; **(a)** dorsal view; **(b)** ventral view. **(c)** reconstruction of skeleton of *Eurhinodelphis*, Miocene of Belgium, from Abel (1909). **(d)** holotype tooth (marked with asterisk*) and provisionally referred teeth of '*Steno*' *cudmorei*, late Miocene or early Pliocene, Beaumaris, Vic.; Museum of Victoria. **(e, f)** earbones associated with teeth similar to those of '*Steno*' *cudmorei*, late Miocene or early Pliocene, Beaumaris, Vic.; Museum of Victoria; **(e)** periotic, dorsal view; **(f)** bulla, dorsal view. **(g)** *Lagenorhynchus obscurus*, dusky dolphin, right lateral view of skeleton, extant, New Zealand; Department of Zoology, University of Canterbury; length about 1.3 m. **(h, i)** *Stenella coeruleoalba*, modern subtropical to tropical oceans, skull and mandibles, from Gray (1846); **(h)** lateral view; **(i)** dorsal view. **(j)** *Stenella* or *Delphinus* sp., of McKee and Fordyce (1987), right mandible in lateral view; Pliocene, Hawera, N.Z. **(k)** *Kentriodon*-like dolphin, skull in right lateral view and left mandible in internal view, late Oligocene or early Miocene, Port Waikato, N.Z.; Geology Department, University of Auckland.

be invoked to explain the diversity and disparity of extant Cetacea; diversity is the number of species in a particular taxonomic group, while disparity indicates the variation in structure or basic design within a group. For organisms in general, fossils provide the only direct evidence of ancient life, but for most groups, and particularly for Cetacea, the record is usually not complete enough to reveal ancestor-to-descendant transitions (see below).

Competition through natural selection has been invoked traditionally, or at least tacitly implied, to account for speciation. For example, it appears that living balaenopterid mysticetes vary interspecifically in details of the feeding apparatus (size of rostrum, spacing of baleen), food preference and use of high-latitude feeding grounds (Laws, 1977). These patterns are consistent with the notion that balaenopterid niches overlap little, if at all. Minimal niche overlap, in turn, may imply past divergence through competition (Kawamura, 1994; Clapham and Brownell, 1996). Niche divergence (or overlap) is more difficult to assess in other pelagic Cetacea, in cryptic species and in those for which feeding behaviour is hard to monitor (i.e. most living species). Is natural selection, and correlated niche divergence, a satisfactory mechanism to explain speciation? Competition and niche separation might partly explain the high species-diversity in the Delphinidae, but vicariance, sexual selection and hybridization could also have roles. Bottlenosed dolphins indicate that explanations may be complex. These have long been thought to comprise one widespread species, *Tursiops truncatus*, but, recently, 'onshore' and 'offshore' populations have been recognized; these groups were initially characterized by behaviour and ecology, then more recently by anatomical and molecular methods (Hoelzel *et al*., 1998; Wang *et al.,* 2000). These parapatric populations probably warrant recognition as distinct species (*Tursiops truncatus, T. aduncus*).

The examples of balaenopterids and *Tursiops* remind us that Darwin regarded geography as fundamental in evolution. On this theme, Davies (1963) used allopatric speciation to explain the evolution of north-south (or antitropical) taxonomic pairs, suggesting that allopatry arose when tropical water masses expanded or contracted in response to Quaternary glaciations. Davies' speciation model was framed in terms of dispersal, although vicariance (see Zink *et al*., 2000) might equally be a mechanism. One study of the antitropical speciation concept involves Cipriano's (1997) molecular work on the delphinid genus *Lagenorhynchus*, including the Australasian *L. obscurus* shown in Figure 10(g). It seems that the six antitropical living species split at varying times, not only during the Quaternary, and that *Lagenorhynchus* is at least diphyletic, with these species representing two or more different lineages. The Austral delphinid genus *Cephalorhynchus* warrants this sort of study. *Cephalorhynchus* includes four species distributed around the Southern Ocean (Brownell and Donovan, 1988); two are allopatric (*C. hectori*, New Zealand; *C. heavisidei*, South Africa), and two partly sympatric (*C. commersoni*, Kerguelen and South America; *C. eutropia*, South America). Molecular studies could elucidate possible founder effects in speciation. Species of *Cephalorhynchus* are tacitly accepted as convergent in body form and habits with porpoises (phocoenids); how, then, is niche overlap minimized between *Cephalorhynchus* (*C. eutropia, C. commersoni*) and *Phocoena* (*P. dioptrica, P. spinnipinnis*) in South America?

Prominent display structures and other sexually dimorphic features in Cetacea indicate a role for sexual selection. Such structures include, for example, species-specific patterns of procumbent teeth in male ziphiids, enhanced body profiles (e.g. of the head, dorsal fin or genital region) in male delphinids, and greater body size (especially male sperm whales). The speciose ziphiid genus *Mesoplodon* (see Mead, 1989) is morphologically conservative, with cryptic species separated most easily on the basis of male tooth structure; presumably, sexual selection is important in the genus.

Coevolution plausibly explains the body form, and even colour pattern, of the pygmy sperm whales, *Kogia* spp. These small odontocetes are remarkably shark-like in form, even to the extent of having a pigmented 'false gill slit.' Predator-prey interaction is implied. Another possible example of coevolution related to predation involves the polar feeding/tropical breeding of mysticetes, which may be linked with reduced risk of predation from *Orcinus orca* at low latitudes (Corkeron and Connor, 1999).

FOSSIL RECORDS

No Cetacea seem to show dense evolutionary succession as seen amongst some fossil marine invertebrates. The record is too patchy to reveal rates of evolution—fast, slow, constant, varied—at the species level, and it would be hard to support claims of either gradualism or punctuated evolution. Still, the fossil record does include some speciose genera or subfamilies (e.g. *Kentriodon, Balaenoptera*, and the dorudontine archaeocetes) with species of different geological ages that may represent ancestor-to-descendant lineages. More importantly, evolution can be assessed via patterns of global fossil diversity over time, especially at generic or family level. Using this approach, the cetacean record reveals three major phases of diversification: first, the Eocene radiation of archaeocetes, with species diversity peaking in the later middle Eocene; second, the Oligocene radiation of Neoceti, with many families present by 29–30 m.y.a.; and third, the later Miocene to modern radiation of balaenopterids and, especially, delphinoids.

The magnitude of the Eocene and Oligocene radiations has only become apparent in the last two decades. Currently, archaeocetes are thought to have arisen and diversified rapidly in the Eocene, radiating into vacant 'ecospace' (O'Leary and Uhen, 1998; Thewissen, 1998).

The Oligocene was, for some time, regarded as an interval of greatly lowered diversity (see review by Fordyce, 1992). Now, it seems that the crown-group Cetacea (Odontoceti + Mysticeti) radiated rapidly between their origins at ~35 m.y.a., and the start of the Late Oligocene, 29–30 m.y.a. (Figures 3, 4), when many disparate family-level groups were well established. This Oligocene radiation involved two new groups characterized by new feeding habits—the echolocating odontocetes and the filter-feeding mysticetes—and is linked closely with the final breakup of Gondwana including the opening of the Southern Ocean, changed oceanic temperature regimes, and inferred changes in ocean ecology (see reviews by Fordyce, 1989, 1991; Fordyce and Barnes, 1994). Recently, analyses of microfossils from the Southern Ocean have supported the idea of major increase in productivity about the Eocene/Oligocene boundary (Diester-Haass *et al.*, 1996; Salamy and Zachos, 1999). The Oligocene radiation marked the disappearance of archaeocetes, which was not a true extinction as archaeocetes are a paraphyletic rather than real group. Nevertheless, the demise of archaeocetes signalled major shifts in cetacean feeding ecology.

The third radiation, in middle Miocene (12–15 m.y.a.), saw many extant families of mysticete and odontocete appear (Figure 4). Delphinoids—dolphins and relatives—in particular, radiated dramatically. Archaic cetacean groups that disappeared at about this time include the squalodontids and eurhinodelphinids. Plausibly, these events are linked with a pulse of modernization in ocean circulation (Whitmore, 1994).

The last decade has seen increased concern about climate change and extinction, including cetacean extinction (the topic of an International Whaling Commission Symposium in Hawaii in 1996). Climate change during the Quaternary involved major temperature swings between glacial and interglacial, associated with huge changes in sea level and shifts in shelf area. Changes in shelf area mean changes in habitat for neritic species, and extinction might therefore be expected. However, the record of later Pliocene and earlier Pleistocene fossils hint that at least some extant cetacean species originated before, and successfully endured, such changes. Other Pliocene assemblages, however, contain species that are now extinct. Perhaps climate change had some effect, but it was not clearly profound.

DIRECTIONS FOR THE FUTURE

Current emphasis in cetology involves especially ecology, behaviour, and conservation biology; systematics has long been a poor cousin. Modern syntheses of molecular phylogenies and the fossil record should remind us that systematic studies provide a fundamental framework for biology, whether for Cetacea or other groups. In future, combined molecular and morphometric studies of living species will help with species recognition and inferences based thereon, such as delimiting global geographic patterns and timing speciation events. The modest record of Australian fossil Cetacea probably reflects lack of study rather than reality. Eocene strata, and localities away from the south-east, deserve study, and Plio-Pleistocene strata could produce fossils useful in judging origins of extant species and in calibrating molecular clocks. Finally, living cetaceans known only or mainly from Australasia include rare forms (*Cephalorhynchus hectori*), and rather disparate species in monotypic genera (*Orcaella brevirostris, Tasmacetus shepherdi, Caperea marginata*)—all prime candidates for detailed local study with potentially exciting returns to cetacean systematics and the understanding of diversity/disparity trends.

ADDENDUM

Recent advances in cetacean systematics, including accounts of new taxa and cladistic analyses, are important to help understand Australasian material. Of note, several enigmatic toothed Oligocene Cetacea (Australian "*Squalodon*" *gambierensis*, New Zealand *Kekenodon onamata*; Figure 3 herein) probably represent a new clade of relict Archaeoceti that forms the immediate sister taxon to the living Cetacea, Neoceti (Fordyce, 2004). Amongst the Mysticeti, Australian toothed forms including *Mammalodon* are under study by Fitzgerald (2005). Perhaps some previously-enigmatic Australasian archaic species belong in newly named groups such as the Eomysticetidae and Cetotheriopsidae (see Sanders and Barnes, 2002a, b; Geisler and Sanders, 2003) or a redefined Cetotheriidae (see Bouetel, 2005).

Molecular studies show that the living pygmy right whale, *Caperea*, may be closer to *Balaenoptera* than to Balaenidae (Rychel *et al.*, 2004; Sasaki *et al.*, 2005). New Zealand fossils suggest that the right whale lineage, Balaenidae, was established by the late Oligocene (Fordyce, 2002). Within the Odontoceti, the content and relationships of the *Platanista* clade, which may include Squalodontidae, Waipatiidae and other Australasian groups, are unresolved (e.g. Geisler and Sanders, 2003; Lambert, 2005; Figure 3 herein), and southern fossils will be important in future studies. A strange toothless Pliocene dolphin from Australian Antarctic Territory, *Australodelphis mirus* (Delphinidae), represents an ecological group no longer present in high southern latitudes (Fordyce *et al.*, 2002).

ACKNOWLEDGEMENTS

Work was supported in part by a postdoctoral fellowship in the Department of Earth Sciences at Monash University (1980–1982), with the New Zealand component (1982–present) assisted extensively by the National Geographic Society and the Research Committee, University of Otago. I thank T.H. Rich and P. Vickers-Rich for facilitating Australian studies, A. Grebneff for help with New Zealand work, many colleagues for discussing cetacean systematics over the years, and two anonymous reviewers for helpful comments.

REFERENCES

Abel, O. 1909. Cetaceenstudien. I. Mitteilung: Das Skelett von *Eurhinodelphis cocheteuxi* aus dem Ober-miozän von Antwerpen. *Sitzungsberichte Mathematisch-naturwissenschaftliche Klasse, Akademie der Wissenschaften, Wien*, Vol. 118, pp.241–253.

Arnason, U., Gretarsdottir, S. and Widegren, B. 1992. Mysticete (baleen whale) relationships based upon the sequence of the common cetacean dna satellite. *Molecular Biology and Evolution*, Vol. 9, pp.1018–1028.

Arnold, P. and Heinsohn, G. 1996. Phylogenetic status of the Irrawaddy dolphin *Orcaella brevirostris* (Owen, in Gray): a cladistic analysis. *Memoirs of the Queensland Museum*, Vol. 39, pp.141–204.

Arnold, P., Marsh, H. and Heinsohn, G. 1987. The occurrence of two forms of minke whales in east Australian waters with a description of external characters and skeleton of the diminutive or dwarf form. *Scientific Reports of the Whales Research Institute, Tokyo*, Vol. 38, pp.1–46.

Bajpai, S. and Gingerich, P.D. 1998. A new Eocene archaeocete (Mammalia, Cetacea) from India and the time of origin of whales. *Proceedings of the National Academy of Sciences of the United States of America*, Vol. 95, pp.15464–15468.

Baker, A.N. 1999. *Whales and Dolphins of New Zealand and Australia. An Identification Guide*. Victoria University Press, Wellington. 133pp.

Bannister, J.L., Kemper, C.M. and Warneke, R.M. 1996. *The Action Plan for Australian Cetaceans*. Australian Nature Conservation Agency, Canberra. 242pp.

Barnes, L.G. 1977. Outline of eastern North Pacific fossil cetacean assemblages. *Systematic Zoology*, Vol. 25, pp.321–343.

Barnes, L.G. 1978. A review of *Lophocetus* and *Liolithax* and their relationships to the delphinoid family Kentriodontidae (Cetacea: Odontoceti). *Natural History Museum of Los Angeles County, Science Bulletin*, Vol. 28, pp.1–35.

Barnes, L.G. 1984. Fossil odontocetes (Mammalia: Cetacea) from the Almejas Formation, Isla Cedros, Mexico. *Paleobios*, Vol. 42, pp.1–46.

Barnes, L.G. 1985. Evolution, taxonomy and antitropical distributions of the porpoises (Phocoenidae, Mammalia). *Marine Mammal Science*, Vol. 1, pp.149–165.

Barnes, L.G., Domning, D.P. and Ray, C.E. 1985. Status of studies on fossil marine mammals. *Marine Mammal Science*, Vol. 1, pp.15–53.

Barnes, L.G., Inuzuka, N. and Hasegawa, Y. (eds). 1995a. Evolution and biogeography of fossil marine vertebrates in the Pacific realm. *The Island Arc*, Vol. 3, pp.243–537.

Barnes, L.G., Kimura, M., Furusawa, H. and Sawamura, H. 1995b. Classification and distribution of Oligocene Aetiocetidae (Cetacea; Mysticeti) from western North America and Japan. *The Island Arc*, Vol. 3, pp.392–431.

Barnes, L.G. and McLeod, S.A. 1984. The fossil record and phyletic relationships of gray whales. Pp.3–32. In Jones, M.L., Swartz, S.L. and Leatherwood, S. (eds). *The Gray Whale:* Eschrichtius robustus. Academic Press, Orlando. 600pp.

Bearlin, R.K. 1985. The morphology and systematics of Neogene Mysticetes (baleen whales) from Australia and New Zealand—a progress report. *New Zealand Geological Survey Record*, Vol. 9, pp.11–13.

Bearlin, R.K. 1988. The morphology and systematics of Neogene Mysticeti from Australia and New Zealand. *New Zealand Journal of Geology and Geophysics*, Vol. 31, p.257 (abstract only).

Beddard, F.E. 1901. Contribution towards a knowledge of the osteology of the pigmy whale (*Neobalaena marginata*). *Transactions of the Zoological Society of London*, Vol. 16, pp.87–110.

Benham, W.B. 1935a. A reptilian jaw from Kakanui, South Island, New Zealand. *Transactions of the Royal Society of New Zealand,* Vol. 65, pp.232–238.

Benham, W.B. 1935b. The teeth of an extinct whale, *Microcetus hectori* n. sp. *Transactions of the Royal Society of New Zealand,* Vol. 65, pp.239–243.

Benham, W.B. 1937. Fossil Cetacea of New Zealand II.—On *Lophocephalus,* a new genus of zeuglodont Cetacea. *Transactions of the Royal Society of New Zealand*, Vol. 67, pp.1–7.

Benham, W.B. 1942. Fossil Cetacea of New Zealand V.—*Mauicetus*, a generic name substituted for *Lophocephalus* Benham. *Transactions of the Royal Society of New Zealand*, Vol. 71, pp.260–270.

Berta, A. and Deméré, T. (eds). 1994. Contributions in marine mammal paleontology honoring Frank C. Whitmore, Jr. *Proceedings of the San Diego Museum of Natural History*, Vol. 29, pp.1–268.

Berta, A. and Sumich, J.L. 1999. *Marine Mammals: Evolutionary Biology*. Academic Press, San Diego. 460pp.

Best, P.B. 1979. Social organization in sperm whales, *Physeter macrocephalus*. Pp.227–289. In Winn, H.E. and Olla, B.O. (eds). *Behavior of Marine Animals. Volume 3: Cetaceans*. Plenum, New York. 438pp.

Bouetel, V. 2005. Phylogenetic implications of skull structure and feeding behavior in balaenopterids (Cetacea, Mysticeti). *Journal of Mammalogy*, Vol. 86, pp.139–146.

Brownell, R.L. 1989. Franciscana, *Pontoporia blainvillei* (Gervais and d'Orbigny 1844). Pp.45–67. In Ridgway, S.H. and Harrison, R.J. (eds). *Handbook of Marine Mammals. Volume 4. River Dolphins and the Larger Toothed Whales*. Academic Press, London. 442pp.

Brownell, R.L. and Donovan, G.P. (eds). 1988. Biology of the genus *Cephalorhynchus*. *Reports of the International Whaling Commission*, Special Issue 9, pp.1–344. *International Whaling Commission, Cambridge*.

Bryden, M.M., Marsh, H. and Shaughnessy, P.D. 1998. *Dugongs, Whales, Dolphins and Seals. A Guide to the Sea Mammals of Australasia*. Allen and Unwin, Sydney. 176pp.

Caldwell, D.K. and Caldwell, M.C. 1989. Pygmy sperm whale—*Kogia breviceps* (de Blainville, 1838); dwarf sperm whale—*Kogia simus* Owen, 1866. Pp.235–260. In Ridgway, S.H. and Harrison, R.J. (eds). *Handbook of Marine Mammals. Volume 4. River Dolphins and the Larger Toothed Whales*. Academic Press, London. 442pp.

Carter, A.N. 1978. The discovery of a phosphatic nodule bed at the base of the Jemmys Point Formation in Gippsland, Victoria. *Search*, Vol. 9, pp.370–372.

Chapman, F. 1912. On the occurrence of *Scaldicetus* in Victoria. *Geological Survey of Victoria, Records*, Vol. 3, pp.236–238.

Chapman, F. 1918a. On new or little known Victorian fossils in the National Museum. Part XXI.—Some Tertiary cetacean remains. *Proceedings of the Royal Society of Victoria*, Vol. 30, pp.32–43.

Chapman, F. 1918b. On an apparently new type of cetacean tooth from the Tertiary of Tasmania. *Proceedings of the Royal Society of Victoria*, Vol. 30, pp.149–152.

Cipriano, F. 1997. Antitropical distributions and speciation in dolphins of the genus *Lagenorhynchus*: a preliminary analysis. *Society for Marine Mammalogy Special Publication*, Vol. 3, pp.305–316.

Clapham, P.J. and Brownell, R.L. 1996. Potential for interspecific competition in baleen whales. *Reports of the International Whaling Commission*, Vol. 46, pp.361–367.

Climo, F.M. and Baker, A.N. 1972. A new shark-toothed dolphin (Cetacea: Squalodontidae) from the upper Oligocene of New Zealand. *Journal of the Royal Society of New Zealand*, Vol. 2, pp.61–68.

Corkeron, P.J. and Connor, R.C. 1999. Why do baleen whales migrate? *Marine Mammal Science*, Vol. 15, pp.1228–1245.

Cozzuol, M.A. 1996. The record of the aquatic mammals in southern South America. *Münchner Geowissenschaftliche Abhandlungen. A, Geologie und Paläontologie*, Vol. 30, pp.321–342.

Cranford, T.W., Amundin, M. and Norris, K.S. 1996. Functional morphology and homology in the odontocete nasal complex: implications for sound generation. *Journal of Morphology*, Vol. 228, pp.223–285.

Dalebout, M.L., Van Helden, A., Van Waerebeek, K. and Baker, C.S. 1998. Molecular genetic identification of Southern Hemisphere beaked whales (Cetacea: Ziphiidae). *Molecular Ecology*, Vol. 7, pp.687–694.

Davies, J.L. 1963. The antitropical factor in cetacean speciation. *Evolution*, Vol. 17, pp.107–116.

Davis, J.W. 1888. On fossil fish remains from the Tertiary and Cretaceo-Tertiary formations of New Zealand. *Transactions of the Royal Dublin Society*, Series 2, Vol.4, Pt. 2, pp.1–56.

Dickson, M.R. 1964. The skull and other remains of *Prosqualodon marplesi*, a new species of fossil whale. *New Zealand Journal of Geology and Geophysics*, Vol. 7, pp.626–635.

Diesterhaass, L., Robert, C. and Chamley, H. 1996. The Eocene-Oligocene preglacial-glacial transition in the Atlantic sector of the Southern Ocean (ODP Site 690). *Marine Geology*, Vol. 131, pp.123–149.

Fitzgerald, E.M.G. 2005. Toothed mysticetes (Mammalia: Cetacea) from the late Oligocene of Australia. P.25. In Uhen, M. (ed). *Evolution of Aquatic Tetrapods. Fourth Triannual Convention Abstracts*. Cranbrook Institute of Science, Miscellaneous Publications, 1. Cranbrook Institute of Science, Bloomfield Hills, Michigan.

Flower, W.H. 1868. On the osteology of the cachalot or sperm-whale (*Physeter macrocephalus*). *Transactions of the Zoological Society of London*, Vol. 6, pp.309–372.

Flynn, T.T. 1923. A whale of bygone days. *Australian Museum Magazine*, Vol. 1, pp.266–272.

Flynn, T.T. 1948. Description of *Prosqualodon davidi* Flynn, a fossil cetacean from Tasmania. *Transactions of the Zoological Society of London*, Vol. 25, pp.153–197.

Fordyce, R.E. 1980. The fossil Cetacea of New Zealand. (A catalogue of described genera and species with an annotated literature guide and reference list). *New Zealand Geological Survey Report*, Vol. 90, pp.1–60.

Fordyce, R.E. 1981. Systematics of the odontocete *Agorophius pygmaeus* and the family Agorophiidae (Mammalia: Cetacea). *Journal of Paleontology*, Vol. 55, pp.1028–1045.

Fordyce, R.E. 1982a. A review of Australian fossil Cetacea. *Memoirs of the National Museum of Victoria*, Vol. 43, pp.43–58.

Fordyce, R.E. 1982b. Redescription of early Miocene dolphin *Phocaenopsis mantelli*, Huxley 1859 (Odontoceti incertae sedis). *New Zealand Journal of Geology and Geophysics*, Vol. 24, pp.563–568.

Fordyce, R.E. 1983. Rhabdosteid dolphins (Mammalia: Cetacea) from the middle Miocene, Lake Frome area, South Australia. *Alcheringa*, Vol. 7, pp.27–40.

Fordyce, R.E. 1984a. Evolution and zoogeography of cetaceans in Australia. Pp.929–948. In Archer, M. and Clayton, G. (eds). *Vertebrate Zoogeography and Evolution in Australasia.* Hesperian Press, Perth. 1203pp.

Fordyce, R.E. 1984b. Preliminary report on cetacean bones from Chatham Rise (New Zealand). *Geologisches Jahrbuch*, Vol. D 65, pp.117–120.

Fordyce, R.E. 1985. Late Eocene archaeocete whale (Archaeoceti: Dorudontinae) from Waihao, South Canterbury, New Zealand. *New Zealand Journal of Geology and Geophysics*, Vol. 28, pp.351–357.

Fordyce, R.E. 1988. Taxonomic status of Victorian fossil whales assigned to the genus *Cetotolites* McCoy, 1879. *Memoirs of the Museum of Victoria*, Vol. 49, pp.59–65.

Fordyce, R.E. 1989a. Origins and evolution of Antarctic marine mammals. *Special Publication of the Geological Society of London*, Vol. 47, pp.269–281.

Fordyce, R.E. 1989b. Problematic early Oligocene toothed whale (Cetacea, ?Mysticeti) from Waikari, North Canterbury, New Zealand. *New Zealand Journal of Geology and Geophysics*, Vol. 32, pp.385–390.

Fordyce, R.E. 1991. A new look at the fossil vertebrate record of New Zealand. Pp.1191–1316. In Vickers-Rich, P., Monaghan, J.M., Baird, R.F. and Rich, T.H. (eds). *Vertebrate Palaeontology of Australasia.* Pioneer Design Studio and Monash University Publications Committee, Melbourne. 1437pp.

Fordyce, R.E. 1992. Cetacean evolution and Eocene/Oligocene environments. Pp.368–381. In Prothero, D. and Berggren, W.A. (eds). *Eocene-Oligocene Climatic and Biotic Evolution.* Princeton University Press, Princeton, New Jersey. 568pp.

Fordyce, R.E. 1994. *Waipatia maerewhenua*, new genus and new species (Waipatiidae, new family), an archaic late Oligocene dolphin (Cetacea: Odontoceti: Platanistoidea) from New Zealand. *Proceedings of the San Diego Museum of Natural History*, Vol. 29, pp.147–176.

Fordyce, R.E. 2002. Oligocene origins of skim-feeding right whales: a small archaic balaenid from New Zealand. *Journal of Vertebrate Paleontology*, Vol. 22, Pt. 3(Supplement), p.54a.

Fordyce, R.E. 2004. The transition from Archaeoceti to Neoceti: Oligocene archaeocetes in the Southwest Pacific. *Journal of Vertebrate Paleontology,* Vol. 24, Pt. 3(Supplement), p.59a.

Fordyce, R.E. and Barnes, L.G. 1994. The evolutionary history of whales and dolphins. *Annual Review of Earth and*

Planetary Science, Vol. 22, pp.419–455.

Fordyce, R.E. and Campbell, H.J. 1990. Fossil dolphin bones from the Chatham Islands, New Zealand. *New Zealand Natural Sciences*, Vol. 17, pp.61–65.

Fordyce, R.E. and Cullen, D.J. 1979. A Miocene ziphiid whale (Odontoceti: Cetacea) from central Chatham Rise, east of New Zealand. *New Zealand Oceanographic Institute Records*, Vol. 4, pp.45–53.

Fordyce, R.E. and Muizon, C. de. 2001. Evolutionary history of cetaceans: a review. Pp.169–234. In Buffrenil, V. de and Mazin, J. (eds). *Secondary Adaptation of Tetrapods to Life in Water.* Pfeil Verlag, München. 367pp.

Fordyce, R.E., Quilty, P.G. and Daniels, J. 2002. *Australodelphis mirus*, a bizarre new toothless ziphiid-like fossil dolphin (Cetacea : Delphinidae) from the Pliocene of Vestfold Hills, East Antarctica. *Antarctic Science*, Vol. 14, pp.37–54.

Fordyce, R.E. and Watson, A.G. 1998. Vertebral pathology in an early Oligocene whale (Cetacea: ?Mysticeti) from Wharekuri, North Otago, New Zealand. Pp.161–176. In Grimm, K.I., Grimm, M.C. and Morlo, M. (eds). *Festschift Zum 70. Geburtstag Von Karlheinz Rothausen*. Naturhistoriches Museum Mainz, Mainz. 186pp.

Fraser, F.C. and Purves, P.E. 1960. Hearing in cetaceans: evolution of the accessory air sacs and the structure of the outer and middle ear in recent cetaceans. *Bulletin of the British Museum (Natural History), Zoology*, Vol. 7, pp.1–140.

Gaskin, D.E. 1982. *The Ecology of Whales and Dolphins*. Heinemann, London. 459pp.

Geisler, J.H. and Luo, Z.X. 1996. The petrosal and inner ear of *Herpetocetus* sp (Mammalia: Cetacea) and their implications for the phylogeny and hearing of archaic mysticetes. *Journal of Paleontology*, Vol. 70, pp.1045–1066.

Geisler, J.H. and Sanders, A.E. 2003. Morphological evidence for the phylogeny of Cetacea. *Journal of Mammalian Evolution*, Vol. 10, Pt. 1, 2, pp. 23—129.

Geraci, J.R. and Lounsbury, V.J. 1993. *Marine Mammals Ashore. A Field Guide for Strandings*. Sea Grant College Program, Texas A & M University, Galveston, Texas. 305pp.

Gill, E.D. 1957. The stratigraphical occurrence and paleoecology of some Australian Tertiary marsupials. *Memoirs of the National Museum of Victoria*, Vol. 21, pp.135–203.

Gill, E.D. 1965. *Fossils of Victoria*. National Museum of Victoria, Melbourne. 24pp.

Gingerich, P.D., Raza, S.M., Arif, M., Anwar, M. and Zhou, X. 1994. New whale from the Eocene of Pakistan, and the origin of cetacean swimming. *Nature*, Vol. 368, pp.844–847.

Gingerich, P.D., Wells, N.A., Russell, D.E. and Shah, S.M. 1983. Origin of whales in epicontinental remnant seas: new evidence from the early Eocene of Pakistan. *Science*, Vol. 220, pp.403–406.

Glaessner, M.F. 1947. A fossil beaked whale from Lakes Entrance, Victoria. *Proceedings of the Royal Society of Victoria,* New Series, Vol. 58, pp.25–34.

Glaessner, M.F. 1955. Pelagic fossils (*Aturia,* penguins, whales) from the Tertiary of South Australia. *Records of the South Australian Museum*, Vol. 11, pp.353–372.

Glaessner, M.F. 1972. Redescription of the tooth of an Oligocene whale from North Canterbury, New Zealand. *Records of the Canterbury Museum*, Vol. 9, pp.183–187.

Gray, J.E. 1846. Mammalia. III. On the cetaceous animals. Pp.13–53. In Richardson, J. and Gray, J.E. (eds). *The Zoology of the Voyage of H.M.S. "Erebus and Terror", Under the Command of Captain Sir James Clark Ross, R.N., F.R.S., During the Years 1839 to 1843*. E.W. Jansen, London. 53pp.

Hall, T.S. 1911. On the systematic position of the species of *Squalodon* and *Zeuglodon* described from Australia and New Zealand. *Proceedings of the Royal Society of Victoria*, New Series, Vol. 23, pp.257–265.

Hampe, O. 1996. Ein artikuliertes Bartenwalskelett aus dem Neogen (?Undem-Schichten, Obermiozän) der Neiderrheinsichen Bucht (NW-Deutschland) und seine systematische Stellung innerhalb der Mysticeti (Mammalia: Cetacea). *Decheniana Beihefte*, Vol. 33, pp.1–75.

Hector, J. 1881. Notes on the New Zealand Cetacea, Recent and fossil (Abstract). *Transactions of the New Zealand Institute,* Vol. 13, pp.434–436.

Hershkovitz, P. 1966. Catalog of living whales. *Smithsonian Institution Bulletin*, Vol. 246, pp.1–259.

Heyning, J.E. 1989. Comparative facial anatomy of beaked whales (Ziphiidae) and a systematic revision among the families of extant Odontoceti. *Contributions in Science, Natural History Museum of Los Angeles County*, Vol. 405, pp.1–64.

Heyning, J.E. 1997. Sperm whale phylogeny revisited: analysis of the morphological evidence. *Marine Mammal Science*, Vol. 13, pp.596–613.

Hoelzel, A.R., Potter, C.W. and Best, P.B. 1998. Genetic differentiation between parapatric 'nearshore' and 'offshore' populations of the bottlenose dolphin. *Proceedings of the Royal Society of London, Series B-Biological Sciences*, Vol. 265, pp.1177–1183.

Huxley, T.H. 1859. On a fossil bird and a fossil cetacean from New Zealand. *Quarterly Journal of the Geological Society of London,* Vol. 15, pp.670–677.

Ichishima, H., Barnes, L.G., Fordyce, R.E., Kimura, M. and Bohaska, D.J. 1995. A review of kentriodontine dolphins (Cetacea: Delphinoidea: Kentriodontidae): systematics and biogeography. *The Island Arc*, Vol. 3, pp.486–492.

Kasuya, T. 1973. Systematic consideration of recent toothed whales based on the morphology of tympano-periotic bone. *Scientific Reports of the Whales Research Institute, Tokyo*, Vol. 25, pp.1–103.

Kawamura, A. 1994. A review on the feeding condition of southern baleen whales. *Reports of the International Whaling Commission*, Vol. 44, pp.261–272.

Kellogg, A.R. 1923. Description of two squalodonts recently discovered in the Calvert Cliffs, Maryland; and notes on the shark-toothed dolphins. *Proceedings of the United States National Museum*, Vol. 62, Pt. 6, pp.1–69.

Kellogg, A.R. 1925. A fossil physeteroid cetacean from Santa Barbara County, California. *Proceedings of the United States National Museum*, Vol. 66, Pt.27, pp.1–8.

Kellogg, A.R. 1928. History of whales—their adaptation to life in the water. *Quarterly Review of Biology*, Vol. 3, pp.29–76, 174–208.

Kellogg, A.R. 1932. A Miocene long-beaked porpoise from California. *Smithsonian Miscellaneous Collections*, Vol. 87, Pt. 2, pp.1–11.

Kellogg, A.R. 1936. A review of the Archaeoceti. *Carnegie*

Institute of Washington Publication, Vol. 482, pp.1–366.

Kemper, C., Pledge, N. and Ling, J.K. 1996. Subfossil evidence of strandings of the sperm whale *Physeter macrocephalus* in Gulf of St Vincent, South Australia. *Records of the South Australian Museum*, Vol. 29, pp.41–53.

Keyes, I.W. 1973. Early Oligocene squalodont cetacean from Oamaru, New Zealand. *New Zealand Journal of Marine and Freshwater Research*, Vol. 7, pp.381–390.

Köhler, R. and Fordyce, R.E. 1997. An archaeocete whale (Cetacea: Archaeoceti) from the Eocene Waihao Greensand, New Zealand. *Journal of Vertebrate Paleontology*, Vol. 17, pp.574–583.

Lambert, O. 2005. Phylogenetic affinities of the long-snouted dolphin *Eurhinodelphis* (Cetacea, Odontoceti) from the Miocene of Antwerp, Belgium. *Palaeontology*, Vol. 48, pp.653–679.

Laws, R.M. 1977. The significance of vertebrates in the Antarctic marine ecosystem. Pp.411–438. In Llano, G. (ed). *Adaptations Within Antarctic Ecosystems.* Smithsonian Institution, Washington, D.C. 1252pp.

LeDuc, R.G., Perrin, W.F. and Dizon, A.E. 1999. Phylogenetic relationships among the delphinid cetaceans based on full cytochrome B sequences. *Marine Mammal Science*, Vol. 15, pp.619–648.

Longman, H.A. 1920. Cranium of *Delphinus delphis*?. *Proceedings of the Royal Society of Queensland*, Vol. 31, p.X.

Longman, H.A. 1926. New records of Cetacea, with a list of Queensland species. *Memoirs of the Queensland Museum*, Vol. 8, pp.266–278.

Luo, Z. and Gingerich, P.D. 1999. Terrestrial Mesonychia to aquatic Cetacea: transformation of the basicranium and evolution of hearing in whales. *University of Michigan Papers on Paleontology*, Vol. 31, pp.1–98.

Mahoney, J.A. and Ride, W.D.L. 1975. Index to the genera and species of fossil Mammalia described from Australia and New Guinea between 1838 and 1969. *Western Australian Museum Special Publication*, Vol. 6, pp.1–247.

Marples, B.J. 1956. Cetotheres (Cetacea) from the Oligocene of New Zealand. *Proceedings of the Zoological Society, London*, Vol. 126, pp.565–580.

McCoy, F. 1879. *Physetodon baileyi* (McCoy). Pp.19–20. In *Geological Survey of Victoria, Prodromus of the Palaeontology of Victoria or, Figures and Descriptions of the Victorian Organic Remains, Decade 6.* Government Printer, Melbourne. 42pp.

McKee, J.W.A. and Fordyce, R.E. 1987. Dolphin mandible (Delphinidae) from the Waipipian Stage (Pliocene), Waihi Beach, Taranaki, New Zealand. *New Zealand Journal of Geology and Geophysics*, Vol. 30, pp.321–323.

McKenna, M.C. and Bell, S.K. 1998. *Classification of Mammals: Above the Species Level.* Columbia University Press, New York. 631pp.

McLeod, S.A., Whitmore, F.C. and Barnes, L.G. 1993. Evolutionary relationships and classification. Pp.45–70. In Burns, J.J., Montague J.J. and Cowles C.J. (eds). *The Bowhead Whale.* Society for Marine Mammalogy, Lawrence, Kansas. 787pp.

Mead, J.G. 1975a. Anatomy of the external nasal passages and facial complex in the Delphinidae (Mammalia: Cetacea). *Smithsonian Contributions to Zoology*, Vol. 207, pp.1–72.

Mead, J.G. 1975b. A fossil beaked whale (Cetacea: Ziphiidae) from the Miocene of Kenya. *Journal of Paleontology*, Vol. 49, pp.745–751.

Mead, J. G. 1989. Beaked whales of the genus *Mesoplodon*. Pp.349–430. In Ridgway, S.H. and Harrison, R.J. (eds). *Handbook of Marine Mammals. Volume 4. River Dolphins and the Larger Toothed Whales.* Academic Press, London. 442pp.

Messenger, S.L. and Mcguire, J.A. 1998. Morphology, molecules, and the phylogenetics of cetaceans. *Systematic Biology*, Vol. 47, pp.90–124.

Milinkovitch, M.C. 1995. Molecular phylogeny of cetaceans prompts revision of morphological transformations. *Trends in Ecology and Evolution*, Vol. 10, pp.328–334.

Miller, G.S. 1923. The telescoping of the cetacean skull. *Smithsonian Miscellaneous Collections*, Vol. 76, Pt.5, pp.1–70.

Muizon, C. de. 1984. Les vertébrés fossiles de la Formation Pisco (Pérou). Deuxieme partie: Les Odontocètes (Cetacea, Mammalia) du Pliocène inférieur de Sud-Sacaco. *Institut Français d'Étude Andines, Mémoire*, Vol. 50, pp.1–188.

Muizon, C. de. 1987. The affinities of *Notocetus vanbenedeni*, an Early Miocene platanistoid (Cetacea, Mammalia) from Patagonia, southern Argentina. *American Museum Novitates*, No. 2904, pp.1–27.

Muizon, C. de. 1988a. Les relations phylogénétiques des Delphinida (Cetacea, Mammalia). *Annales de Paléontologie*, Vol. 74, pp.159–227.

Muizon, C. de. 1988b. Le polyphylétisme des Acrodelphidae, Odontocètes longirostres du Miocène européen. *Bulletin du Muséum Nationale d'Histoire Naturelle, Paris, 4 Série*, Vol. 10 C, pp.31–88.

Muizon, C. de. 1991. A new Ziphiidae (Cetacea) from the early Miocene of Washington State (USA) and phylogenetic analysis of the major groups of odontocetes. *Bulletin du Muséum Nationale d'Histoire Naturelle, Paris, 4 Série*, Vol. 12 C, pp.279–326.

Muizon, C. de. 1994. Are the squalodonts related to the platanistoids? *Proceedings of the San Diego Museum of Natural History*, Vol. 29, pp.135–146.

Muizon, C. de, Domning, D.P. and Parrish, M. 1999. Dimorphic tusks and adaptive strategies in a new species of walrus-like dolphin (Odobenocetopsidae) from the Pliocene of Peru. *Comptes Rendus de L'Academie des Sciences Serie II Fascicule A—Sciences de la Terre et des Planetes*, Vol. 329, pp.449–455.

Nikaido, M., Matsuno, F., Hamilton, H., Brownell, R.L., Cao, Y., Ding, W., Zuoyan, Z., Shedlock, A.M., Fordyce, R.E., Hasegawa, M. and Okada, N. 2001. Retroposon analysis of major cetacean lineages: the monophyly of toothed whales and the paraphyly of river dolphins. *Proceedings of the National Academy of Sciences of the United States of America*, Vol. 98, pp.7384–7389.

Nikaido, M., Rooney, A.P. and Okada, N. 1999. Phylogenetic relationships among cetartiodactyls based on insertions of short and long interpersed elements: hippopotamuses are the closest extant relatives of whales. *Proceedings of the National Academy of Sciences of the United States of America*, Vol. 96, pp.10261–10266.

O'Leary, M.A. and Geisler, J.H. 1999. The position of Cetacea within Mammalia: phylogenetic analysis of morphological data from extinct and extant taxa. *Systematic*

Biology, Vol. 48, pp.455–490.

O'Leary, M.A. and Uhen, M.D. 1999. The time of origin of whales and the role of behavioral changes in the terrestrial-aquatic transition. *Paleobiology*, Vol. 25, pp.534–556.

Oliver, W.R.B. 1937. *Tasmacetus shepherdi*: a new genus and species of beaked whale from New Zealand. *Proceedings of the Zoological Society of London*, 1937, pp.371–381.

Perrin, W.F. and Brownell, R.L. 2001. Update of the list of recognised species of cetaceans. *Journal of Cetacean Research Management*, Vol 3 (supplement), pp.364–367.

Pledge, N.S. 1994. Cetacean fossils from the lower Oligocene of South Australia. *Records of the South Australian Museum*, Vol. 27, pp.117–123.

Pledge, N.S. and Rothausen, K. 1977. *Metasqualodon harwoodi* (Sanger, 1881)—a redescription. *Records of the South Australian Museum*, Vol. 17, pp.285–297.

Pritchard, G.B. 1939. On the discovery of a fossil whale in the older tertiaries of Torquay, Victoria. *Victorian Naturalist*, Vol. 55, pp.151–159.

Reynolds, J.E. and Rommel, S.A. (eds). 1999. *Biology of Marine Mammals*. Smithsonian Institution, Washington, D.C. 578pp.

Rice, D.W. 1998. *Marine Mammals of the World. Systematics and Distribution*. Special Publication 4, Society for Marine Mammalogy, Lawrence, Kansas. 231pp.

Ridgway, S.H. and Harrison, R. J. (eds). 1994. *Handbook of Marine Mammals. Volume 5. The First Book of Dolphins*. Academic Press, London. 416pp.

Rosel, P.E., Haygood, M.G. and Perrin, W.F. 1995. Phylogenetic relationships among the true porpoises (Cetacea: Phocoenidae). *Molecular Phylogenetics and Evolution*, Vol. 4, pp.463–474.

Rosenbaum, H.C., Brownell, R.L., Brown, M.W., Schaeff, C., Portway, V., White, B.N., Malik, S., Pastene, L.A., Patenaude, N.J., Baker, C.S., Goto, M., Best, P.B., Clapham, P.J., Hamilton, P., Moore, M., Payne, R., Rowntree, V., Tynan, C.T., Bannister, J.L. and Desalle, R. 2000. World-wide genetic differentiation of *Eubalaena:* questioning the number of right whale species. *Molecular Ecology,* Vol. 9, pp.1793–1802.

Rychel, A.L., Reeder, T.W. and Berta, A. 2004. Phylogeny of mysticete whales based on mitochondrial and nuclear data. *Molecular Phylogenetics and Evolution*, Vol. 32, pp.892–901.

Salamy, K.A. and Zachos, J.C. 1999. Latest Eocene early Oligocene climate change and Southern Ocean fertility: inferences from sediment accumulation and stable isotope data. *Palaeogeography, Palaeoclimatology, Palaeoecology*, Vol. 145, pp.61–77.

Sanders, A.E. and Barnes, L.G. 2002a. Paleontology of the late Oligocene Ashley and Chandler Bridge Formations of South Carolina, 2: *Micromysticetus rothauseni*, a primitive cetotheriid mysticete (Mammalia: Cetacea). *Smithsonian Contributions to Paleobiology*, Vol. 93, pp.271–293.

Sanders, A.E. and Barnes, L.G. 2002b. Paleontology of the late Oligocene Ashley and Chandler Bridge Formations of South Carolina, 3: Eomysticetidae, a new family of primitive mysticetes (Mammalia: Cetacea). *Smithsonian Contributions to Paleobiology*, Vol. 93, pp.313–356.

Sanger, E.B. 1881. On a molar tooth of *Zeuglodon* from the Tertiary beds on the Murray River near Wellington, S.A. *Proceedings of the Linnean Society of New South Wales*, Vol.5, pp.298–300.

Sasaki, T., Nikaido, M., Hamilton, H., Goto, M., Kato, H., Kanda, N., Pastene, L.A., Cao, Y., Fordyce, R.E., Hasegawa, M. and Okada, N. 2005. Mitochondrial phylogenetics and evolution of mysticete whales. *Systematic Biology*, Vol. 54, pp.77–90.

Simpson, G.G. 1945. The principles of classification, and a classification of mammals. *Bulletin of the American Museum of Natural History*, Vol. 85, pp.1–350.

Stromer, E. 1908. Die Archaeoceti des ägyptischen Eocäns. *Beiträge zur Geologie und Paläontologie von Österreich-Ungarn*, Vol. 21, pp.106–178.

Sutherland, L. and Kershaw, R.C. 1971. The Cainozoic geology of Flinders Island, Bass Strait. *Papers and Proceedings of the Royal Society of Tasmania*, Vol. 105, pp.151–175.

Tedford, R.H., Archer, M., Bartholomai, A., Plane, M., Pledge, N.S., Rich, T.H., Rich, P.V. and Welles, R.T. 1977. The discovery of Miocene vertebrates, Lake Frome area, South Australia. *Bureau of Mineral Resources, Journal of Australian Geology and Geophysics*, Vol. 2, pp.52–57.

Thewissen, J.G.M. (ed.) 1998. *The Emergence of Whales: Evolutionary Patterns in the Origin of Cetacea*. Plenum, New York. 477pp.

Uhen, M.D. 1998. Middle to late Eocene basilosaurines and dorudontines. Pp 29–61. In Thewissen, J.G.M. (ed). *The Emergence of Whales. Evolutionary Patterns in the Origin of Cetacea*. Plenum, New York. 477pp.

Uhen, M.D. 1999. New species of protocetid archaeocete whale, *Eocetus wardii* (Mammalia: Cetacea) from the middle Eocene of North Carolina. *Journal of Paleontology*, Vol. 73, pp.512–528.

Van Beneden, P.J. and Gervais, P. 1868–1880. *Ostéographie des Cétacés Vivants et Fossiles, Comprenant la Description et l'Iconographie du Squelette et du Système Dentaire de ces Animaux; ainsi que des documents relatifs à leur histoire naturelle*. Arthus Bertrand, Paris. 2 vols. (Text 634pp., Atlas 64 pls).

Waddell, V.G., Milinkovitch, M.C., Berube, M. and Stanhope, M.J. 2000. Molecular phylogenetic examination of the Delphinoidea trichotomy: congruent evidence from three nuclear loci indicates that porpoises (Phocoenidae) share a more recent common ancestry with white whales (Monodontidae) than they do with true dolphins (Delphinidae). *Molecular Phylogenetics and Evolution*, Vol. 15, pp.314–318.

Wang, J.Y., Chou, L.S. and White, B.N. 2000. Osteological differences between two sympatric forms of bottlenose dolphins (genus *Tursiops)* in Chinese waters. *Journal of Zoology*, Vol. 252, pp.147–162.

Whitmore, F.C. 1994. Neogene climatic change and the emergence of the modern whale fauna of the North Atlantic Ocean. *Proceedings of the San Diego Society of Natural History*, Vol. 29, pp.223–227.

Whitmore, F.C., Morejohn, G.V. and Mullins, H.T. 1986. Fossil beaked whales—*Mesoplodon longirostris* dredged from the ocean bottom. *National Geographic Research*, Vol. 2, pp.47–56.

Whitmore, F.C. and Sanders, A.E. 1977. Review of the Oligocene Cetacea. *Systematic Zoology*, Vol. 25, pp.304–320.

Woodburne, M.O., MacFadden, B.J., Case, J.A., Springer, M.A., Pledge, N.S., Power, J.D., Woodburne, J.M. and Springer, K.B. 1993. Land mammal biostratigraphy and magnetostratigraphy of the Etadunna Formation (late Oligocene) of South Australia. *Journal of Vertebrate Paleontology*, Vol. 13, No. 4, pp.483–515.

Zink, R.M., Blackwell-Rago, R.C. and Ronquist, F. 2000. The shifting roles of dispersal and vicariance in biogeography. *Proceedings of the Royal Society of London, Series B-Biological Sciences*, Vol. 267, pp.497–503.

Zioupos, P., Currey, J.D., Casinos, A. and de Buffrenil, V. 1997. Mechanical properties of the rostrum of the whale *Mesoplodon densirostris*, a remarkably dense bony tissue. *Journal of Zoology*, Vol. 241, pp.725–737.

Chapter 34
Humans Among Primates

Colin P. Groves
School of Archaeology and Anthropology, Australian National University, Canberra. A.C.T. 0200 Australia
E-mail: Colin.Groves@anu.edu.au

INTRODUCTION

The primates are arguably one of the most successful of the placental orders. Non-human primates (see Figures 1, 2) are also amongst the most adaptable and widely distributed of mammalian groups occurring naturally throughout Africa, Asia and South America and extending within the comparatively recent past into both Europe and North America.

Historically, the deep seas between the continental shelves of Asia and New Guinea-Australia have constituted one of the few long-term barriers to primate dispersal. Indeed it was not until the advent of our own species, or more pertinently of our watercraft, that

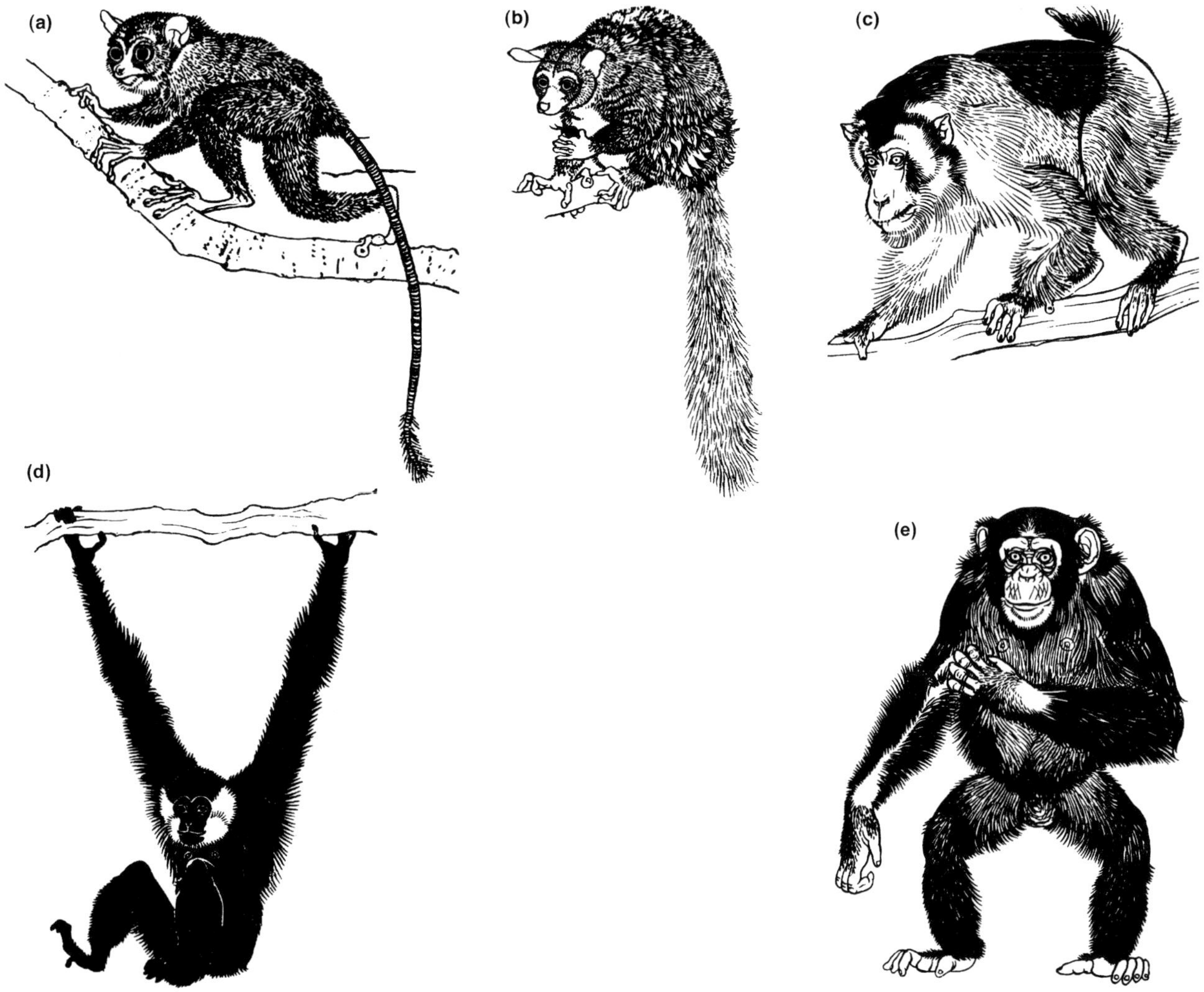

Figure 1. An assortment of living primates: **(a)** a tarsier (*Tarsius* sp.); **(b)** a galago (*Otolemur* sp.); **(c)** a macaque monkey (*Macaca nemestrina*); **(d)** a gibbon (*Nomascus leucogenys.*); **(e)** a chimpanzee (*Pan troglodytes*). Based on Le Gros Clark (1963).

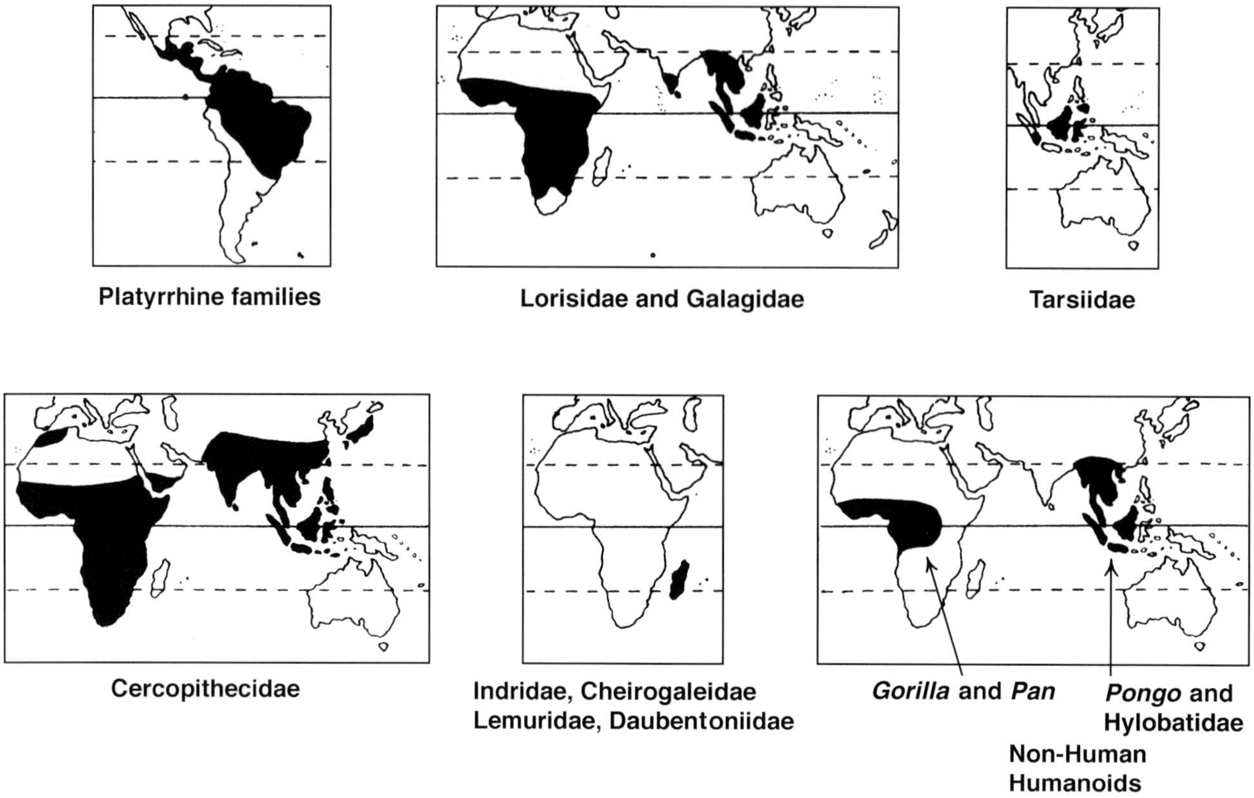

Figure 2. The family-level distribution of living primates. Humans (*Homo sapiens*) are, unfortunately for the other primates, ubiquitous. Modified from Webb *et al.* (1977).

primates of any kind crossed this natural barrier to colonize Meganesia (i.e. New Guinea, Australia and Tasmania) and wider areas of the Pacific; without watercraft primates had reached Sulawesi (Celebes), but no farther east.

Although primates have only had direct impacts on Australasia in very recent times, the whole matter of primate evolution is given broad coverage in this chapter. This is partly because the wider history of cosmopolitan primate radiations provides a valuable contrast to the almost entirely endemic mammalian radiations of Australasia.

Primate evolution is one of the most progressive and controversial areas of palaeontology. This interest is accompanied by many hypotheses on primate taxonomy. Classifications of primates, both fossil and extant, are almost as abundant as the primates themselves. Despite this diversity, few of the existing schemes are entirely satisfactory because they are based on untestable phylogenetic hypotheses or they contain too much paraphyly for cladistic palates. Accordingly, a new cladistic classification is presented in Table 1; this system is based on the phylogenetic scenario developed in the succeeding pages.

The chapter is organized around five themes: the origins and relationships of primates; the early radiation and dispersal of primates and the evolution of the extant prosimians; the origin of the higher primates and the evolution of the Old and New World monkeys; human evolution; and the course of human evolution in Australasia.

PRIMATE ORIGINS AND RELATIONSHIPS

It is important to appreciate that primates are one of the more venerable groups of living mammals. Indeed, they are presently the only extant placental group with a fossil record that may extend back into the Mesozoic Era. One implication of this antiquity is that, with more than 65 million years of independent evolution under their belts, extant primates cannot be considered closely related to any other living group of placentals.

The major hypotheses and types of evidence relating to primate origins are examined below. The evidence comes from three separate sources (fossils, comparative anatomy, molecular studies), each of which is plagued by a unique set of problems.

THE FOSSIL RECORD

The earliest known probable primate is *Purgatorius ceratops* (Figure 3(a)). This species is currently repre-

Table 1. Summary of proposed primate classification to subfamily level (based on Groves, 1989).

- Order Primates
  - Plesion* *Purgatorius*
  - Suborder Plesiadapiformes
    - Superfamily Plesiadapoidea
      - Family Plesiadapidae
      - Family Saxonellidae
      - Family Carpolestidae
    - Superfamily Paromomyoidea
      - Family Paromomyidae
        - Subfamily Paromomyinae
        - Subfamily Macromomyinae
        - Subfamily Picodontinae
      - *Incertae sedis: Navajovius*
        - *Berruvius*
  - Suborder Strepsirrhini
    - Infraorder Adapiformes
      - Family Notharctidae
      - Family Adapidae
        - Subfamily Protoadapinae
        - Subfamily Adapinae
        - Subfamily Anchomomyinae
      - Family Sivaladapidae
      - Family Petrolemuridae
      - *Incertae sedis: Azibius*
    - Infraorder Lemuriformes
      - Family Cheirogaleidae
      - Family Lemuridae
      - Family Megaladapidae
      - Family Indridae
    - Infraorder Lorisiformes
      - Family Lorisidae
      - Family Galagidae
    - Infraorder Chiromyiformes
      - Family Daubentoniidae
  - Suborder Haplorrhini
    - Plesion *Decoredon*
    - Infraorder Omomyiformes
      - Family Microchoeridae
      - Family Anaptomorphidae
        - Subfamily Teilhardininae
        - Subfamily Anaptomorphinae
        - Subfamily Tetoniinae
      - Family Omomyidae
        - Subfamily Omomyinae
        - Subfamily Washakiinae
        - Subfamily Utahiinae
        - *Incertae sedis: Donrussellia*
          - *Copelemur*
          - *Kohatius*
    - Infraorder Tarsiiformes
      - Family Tarsiidae
    - Infraorder Simiiformes
    - Plesion *Eosimias*
    - Section Platyrrhini
      - Plesion *Branisella*
      - Family Homunculidae
      - Family Cebidae
      - Family Aotidae
      - Family Pitheciidae
      - Family Atelidae
    - Section Catarrhini
      - Plesion *Pondaungia*
      - Plesion *Amphipithecus*
      - Superfamily Parapithecoidea
        - Family Parapithecidae
      - Superfamily Cercopithecoidea
        - Plesion *Prohylobates*
        - Subfamily Cercopithecinae
          - Tribe Cercopithecini
          - Tribe Papionini
        - Subfamily Colobinae
      - Superfamily Propliopithecoidea
        - Family Propliopithecidae
      - Superfamily Pliopithecoidea
        - Family Pliopithecidae
      - Superfamily Hominoidea
        - Family Proconsulidae
        - Family Hylobatidae
        - Family Oreopithecidae
        - Family Dryopithecidae
        - Family Hominidae
          - Plesion *Kenyapithecus*
          - Subfamily Ponginae
          - Subfamily Homininae

*The term 'plesion' indicates that the taxon is represented only by extinct forms and contains too little diversity to warrant the recognition of a separate suprageneric rank.

sented by a single molar from the late Cretaceous (or earliest Paleocene) Hell Creek Formation in Montana (Clemens, 1974; Bown and Rose, 1976). A second species, *Purgatorius unio* (Figure 3(c)), is found in greater abundance in the overlying Paleocene sediments of the Tullock Formation. Both species were evidently about the size of a rat. In common with many later primates, *P. unio* shows a characteristic enlargement of the central lower incisor and initial stages in the reduction and loss of the anterior premolar. In its retention of both I3 and PI, however, *P. unio* is uniquely primitive among primates. Overall, the species of *Purgatorius* appear to be sufficiently generalized to have given rise to any plesiadapiform group.

Apart from a suite of primate-flattening dinosaurs (although these may have been reworked into the deposit from older sediments), the contemporaries of *Purgatorius ceratops* included an early condylarthan, *Protungulatum donnae,* and a variety of paleoryctoid and leptictoid proteutherians. In terms of dental morphology, the species of *Purgatorius* show considerable similarities to many of these early placental groups, particularly with the diverse group of proteutherian 'insectivores'. In nearly all instances, however, these similarities are most likely due to the common retention of primitive placental dental states rather than to common descent. Clemens (1974) considered *Purgatorius* species to be too primitive in some

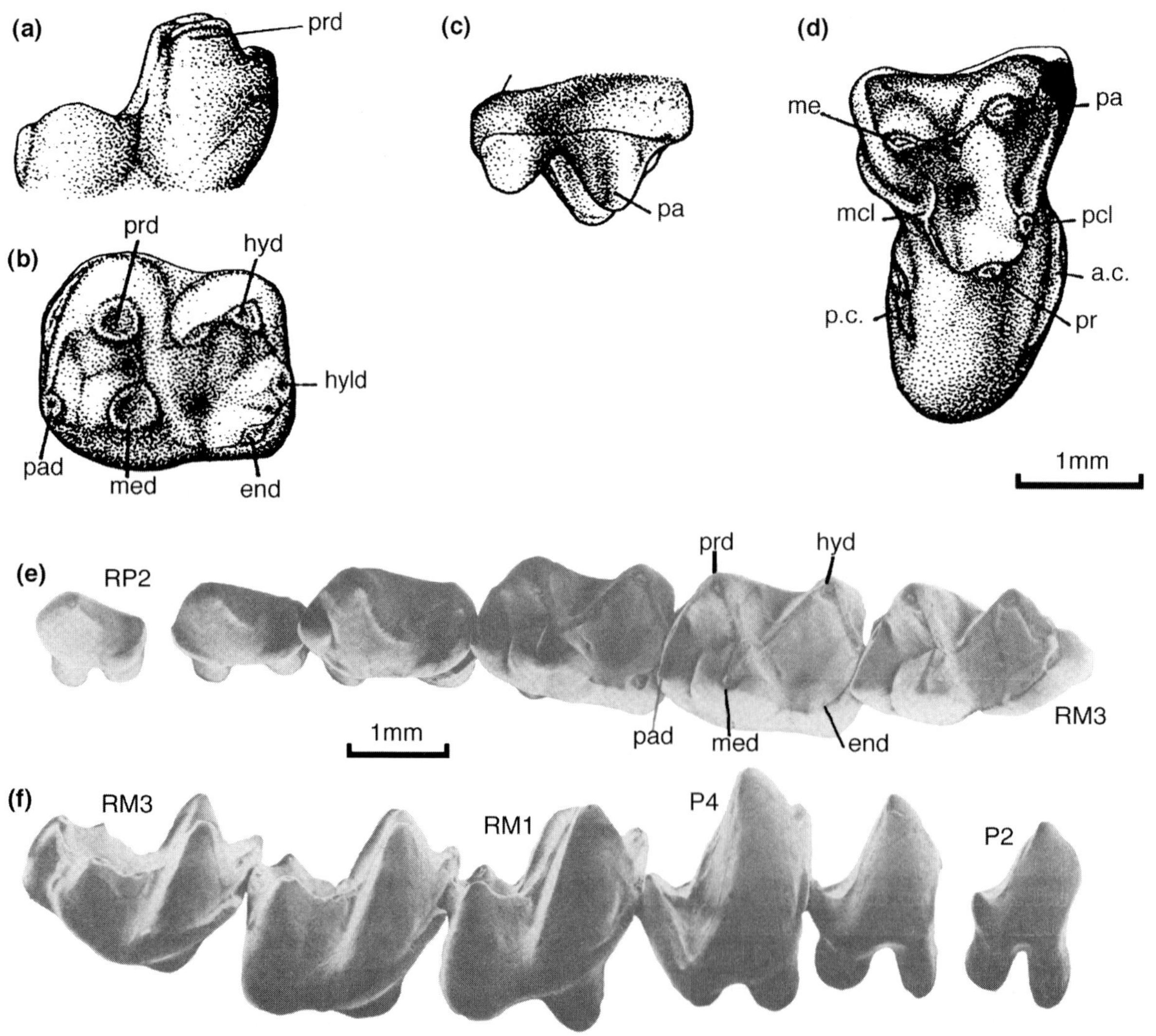

Figure 3. The teeth of the paromomyiform species of *Purgatorius*: **(a), (b)** an isolated right lower molar of the late Cretaceous North American *P. ceratops*, a contemporary of dinosaurs; **(c), (d)** an isolated RM\2 of *P. unio*, a Paleocene species from North America; **(e), (f)** occlusal and buccal views of the right P12–M/3 of *P. unio*. Modified from Szalay and Delson (1979).
Key to abbreviations: a.c. = anterior cingulum; end = entoconid; hyd = hypoconid; hyld = hypoconulid; mcl = metaconule; me = metacone; med = metaconid; pa = paracone; pad = paraconid; p.c. = posterior cingulum; pcl = protoconule; pr = protocone; prd = protoconid; P = premolar; RM = right molar; RP = right premolar.

respects to have been derived from any of the other known late Cretaceous placental groups. If this is correct then the origin of primates might be regarded as an even earlier event, one that perhaps occurred as early as 70 or 80 million years ago.

COMPARATIVE ANATOMY

In contrast to the rather uninformative fossil record, comparative anatomical studies during the last century have generated many hypotheses concerning questions of primate phylogeny. Perhaps the most influential during modern times were the ideas of William K. Gregory as outlined in his *Orders of Mammals* (Gregory, 1910). Gregory postulated a monophyletic lineage, the Archonta, in which he grouped the primates, tree-shrews (Tupaiidae), colugos (Dermoptera), elephant-shrews (Macroscelididae) and bats (Chiroptera).

Neglected for many years in favour of more conservative classifications, Gregory's Archonta has been revived recently on the basis of new anatomical and biomolecular evidence (McKenna, 1966, 1975; Szalay; 1977b; Szalay and Drawhorn, 1980). However, these authors consider the Archonta to contain only the primates, tree-shrews, colugos and bats. They place the Macroscelididae and their fossil allies elsewhere within Placentalia.

Evidence comes from a wide range of tissue systems including cranial and dental anatomy (Van Valen and Sloan, 1965; McKenna, 1966), the structure of the proximal tarsus (Szalay and Decker, 1974; Szalay, 1977b), the organization of the cephalic arterial system (Bugge, 1974), patterns of patagial muscularization (among bats and colugos) (Leche, 1885) and, most recently, from comparative serology. Despite apparent concordance of evidence, recent reanalyses of archontan affinities suggest that the question is far from resolved. There are serious claims of non-homologous elements in the auditory region (Cartmill and MacPhee, 1980) and of misinterpretation of character state polarities elsewhere within the skeletal system (Novacek, 1980; Lewis, 1983). In fact, in terms of gross morphology, there appears to be little justification for the inclusion of primates within Archonta; but, comparative morphology does not suggest any more compelling relationship between primates and any other known placental group.

The Archonta concept has fared poorly as a whole, but parts of it may be salvageable. The presumption of monophyly in the order Chiroptera (bats) has been questioned (Hill and Smith, 1984) and, if the Megachiroptera and Microchiroptera (fruit bats and insectivorous bats respectively) are independently derived from non-flying ancestors, it seems natural to ask what these ancestors could be. The affinities of Microchiroptera remain mysterious, but those of the Megachiroptera could, on a number of grounds (especially the neuroanatomy of the visual system), really be with primates (Pettigrew, 1986; Pettigrew and Jamieson, 1987). Pettigrew (1986) goes as far as to include the 'mega-bats' within the primates, and suggests that the Dermoptera might belong in the order as well. For further discussion of bats and their relationships see pp.673–705 (Hand, 2006).

MOLECULAR STUDIES

Molecular evolutionary studies, drawing on the sophisticated techniques of immunology, amino acid sequencing and, more recently, recombinant DNA and nucleotide sequencing, now constitute a major alternative to the traditional morphological approach.

As indicated above, early immunological studies (reactions between whole sera) generally supported the concept of Archonta (Gregory, 1910), at least in regard to the association of primates with tree-shrews, colugos and bats; but the same analyses also indicated fairly close ties between these groups and rodents (Rodentia), a morphologically aberrant and phyletically enigmatic placental order. In more recent years vast amounts of sequencing data (from numerous polypeptides, gene and pseudogene nucleotides), from 250 vertebrate species (including many primates), have been analysed (Goodman, 1975, 1976; Goodman *et al.*, 1982). Genealogical trees generated from these data (Figure 4) consistently associate primates with two other placental orders. One of these includes the tree-shrews and the other includes rabbits and pikas (order Lagomorpha). Beyond this immediate grouping, association is made first with rodents (confirming the earlier serological results), and then with the carnivores (order Carnivora) and ungulates (order Ungulata). Bats are placed further afield in these analyses and the results, with different techniques, are inconclusive or contradictory.

In summary, in terms of comparative morphology, no significant anatomical feature has been identified that supports an affinity between primates and any of the other groupings suggested above. The initial conclusion stands that primates as a group might well be too ancient to retain any special features that might tie them unequivocably to any other (equally ancient) placental lineage.

PALAEOECOLOGICAL CONSIDERATIONS

Speculation concerning the possible ecological circumstances of primate origins revolves around the importance of arboreal adaptations among early placentals. Several authors, including Jenkins (1974) and Kay and Cartmill (1974), consider the earliest primates to have combined both terrestrial and arboreal adaptations, possibly with a lifestyle somewhat akin to that of the pen-tailed tree-shrew, *Ptilocercus lowii*.

By contrast, Szalay and Decker (1974) consider the

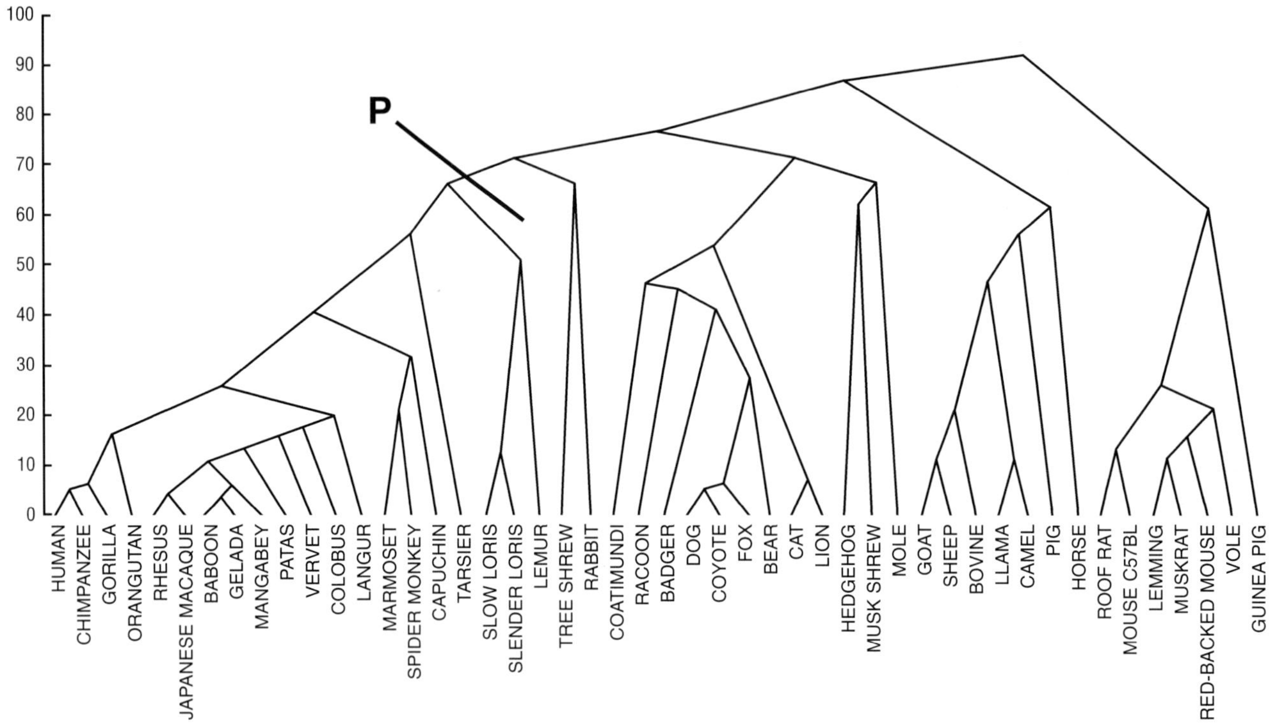

Figure 4. A molecular tree of living placental relationships suggested by the structure of their haemoglobin molecules. All groups to the left of the 'P' line are primates. Note that the closest groups to primates here are tree-shrews (tupaiids) and rabbits (lagomorphs). The ordinate scale is an estimate of millions of years based largely on palaeontological data. Based on Goodman *et al.* (1982).

earliest primates to have evolved within an arboreal milieu following the initial colonization of this niche by ancestral archontans. Furthermore, Szalay and Decker consider archontans to have been the first and, throughout much of the early Tertiary, the only placentals to have undergone significant diversification within the arboreal milieu. The future of this last hypothesis, like that of the Archonta itself, does not look particularly healthy; new biomolecular results would favour the likelihood of multiple invasions of the arboreal milieu by a variety of taxonomically diverse early placentals.

Finally, Gingerich (1976a) and Cartmill and Kay (1978) have independently questioned the arboreality of the earliest primates, citing as evidence a number of supposed terrestrial adaptations among Paleocene plesiadapiform primates. Much of their evidence, however, has itself been attacked by Szalay and Delson (1979) and Szalay (1981).

MICROSYOPIDS

Before concluding this section on primate origins there is one further group, the microsyopids (late Paleocene to Eocene) that warrants brief consideration; see Figure 5. Originally considered, on the basis of some striking dental similarities, to lie within the primate order (Szalay, 1969; Bown and Rose, 1976), microsyopids more recently have been grouped with leptictoid insectivores (Szalay, 1975, 1977a); these dental similarities are regarded as evidence of parallelism and not of phyletic affinity.

Exclusion of this group from the primates follows recognition that the microsyopids lacked one important morphological feature that all undoubted primates possess—a petrosal contribution to the auditory bulla (Szalay, 1975; MacPhee, 1979; MacPhee *et al.*, 1983). In cladistic terms, however, the microsyopid auditory region is thus merely more plesiomorphic (primitive) than that of true primates. As such there appears to be nothing about the microsyopid morphotype, either dental or basicranial, which would preclude them from occupying a structurally ancestral position with respect to primates.

Although known microsyopids occur too late in time to include the actual ancestor of the primates, the possibility that an ancestral microsyopid stock gave rise to both the later microsyopids and the primates cannot be discounted. Some researchers consider the microsyopids to be either the sister-group of the entire primate radiation, or a basal group for the plesitarsiiform radiation (Gingerich, 1976b, 1977b, 1981a; Gingerich and Schoeninger, 1977).

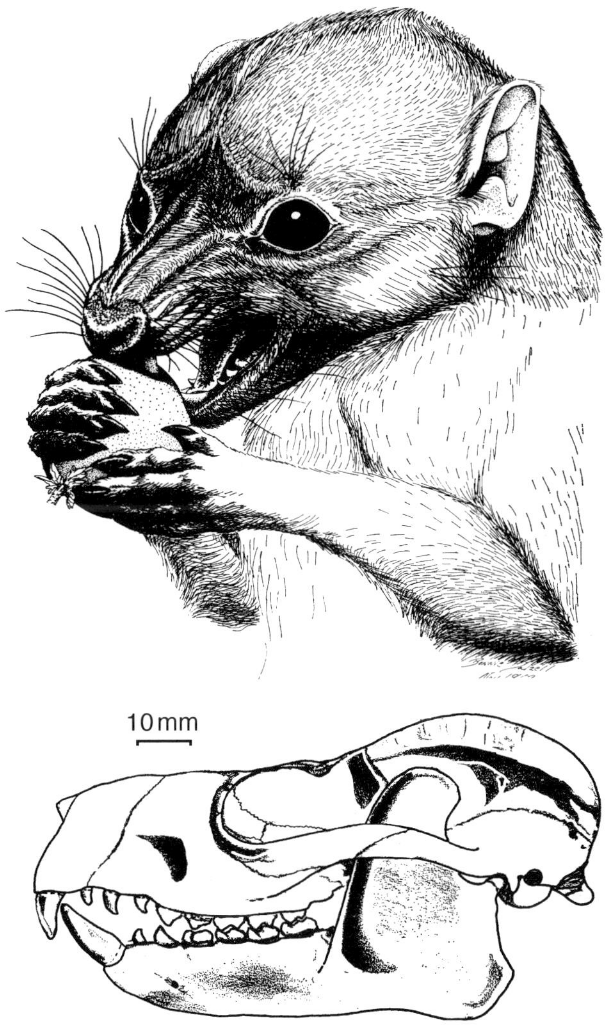

Figure 5. The microsyopids, a group of early Tertiary placentals, regarded by some workers to be the extinct plesiomorphic sister-group of primates. Others (Szalay and Delson, 1979) regard them to be near the insectivore-primate boundary. The scale relates to the skull. From Szalay (1969).

Figure 6. Species of *Plesiadapis*, early Tertiary (Paleocene and Eocene) North American and European plesiadapiform probable primates: **(a)**, **(b)** reconstructions of species of *Plesiadapis*; **(c)** a skull and dentary of *Plesiadapis tricuspidens* showing the procumbent lower first incisor and rodent-like jaw. From Wood (1976) and Gingerich (1976a, 1977a).

EARLY RADIATION AND DISPERSAL OF PRIMATES

PLESIADAPIFORMES: PLESIOMORPHIC PRIMAL PRIMATES

Although classification of the earliest Tertiary primates remains controversial, most authors favour a conservative placement of all Paleocene and some later taxa within a single higher taxon. This is variously known as the Plesiadapiformes or Paromomyiformes; and species allocated likewise vary considerably between authors. From a phylogenetic viewpoint, these arguments may be irrelevant as the group is very probably paraphyletic. Adoption of 'plesion' status for *Purgatorius* improves the situation, but the links between the remaining superfamilies (Plesiadapoidea, Paromomyoidea) are still tenuous.

Compared with all later primates, plesiadapiforms (Figure 6) are characterized by the retention of a suite of primitive features such as: lack of a postorbital bar; small body size (few were much larger than a rat); essentially unmodified molar morphology. In at least some forms there are claws on all digits and a medially placed (extrabullar) entocarotid artery (Szalay, 1973; Gingerich, 1976a). They do have, however, one derived feature in common: an enormously overdeveloped lower central incisor. Whether the procumbent form seen in the Paromomyoidea (Figure 7) is merely a modification of the more vertical type of the Plesiadapoidea (Figure 6), or is a parallel development, remains unclear.

Plesiadapiform affinities remain the subject of controversy. Gingerich and co-workers (Gingerich, 1976a, 1981a; Gingerich and Schoeninger, 1977), regarded the plesiadapiform primates to be ancestral only to the tarsiiforms amongst later groups, but later Gingerich (1989) altered his opinion. He now argued that the Plesiadapiformes was the sister-group of all primates, and proposed to remove them from the order altogether, as a new order Proprimates. Beard (1990a) criticized this move as merely a formal recognition of our ignorance of plesiadapiform affinities, but had an even more radical proposal to make. Beard (1990b) suggested that the Plesiadapiformes were not primates at all, but fossil members of the Dermoptera. A group of partial skeletons of two genera of the group, *Phenacolemur* and *Ignacius*, showed that the hands and feet had short proximal phalanges, very long intermediate phalanges, and very large terminal phalanges, obviously supporting huge claws. Such proportions are found only in the Dermoptera. Beard (1993) performed a cladistic analysis, and concluded that: the Plesiadapiformes and Micromomyiformes (a newly recognized group of small primate-like fossils) were Dermoptera; the primates should be restricted to Strepsirrhini and Haplorrhini; *Purgatorius, Berruvius* and some other Paleocene fossils should be left '*incertae sedis*'.

Although Gingerich's hypothesis was initially accepted (Schwartz *et al.,* 1978), it conflicts with much comparative anatomical evidence for the monophyly of the post-Paleocene primate radiation and with increasing immunological data (Sarich and Wilson, 1967; Hoffstetter, 1974; Goodman *et al.,* 1982). The overall conclusion places the tarsiers within the living primate radiation rather than as its sister-group. Beard's hypothesis has been more widely accepted. Indeed Martin (1990) independently proposed that the Plesiadapiformes were not primates, nor even closely related to them. Beard (1993) did not go as far as this; he placed the plesiadapids in the

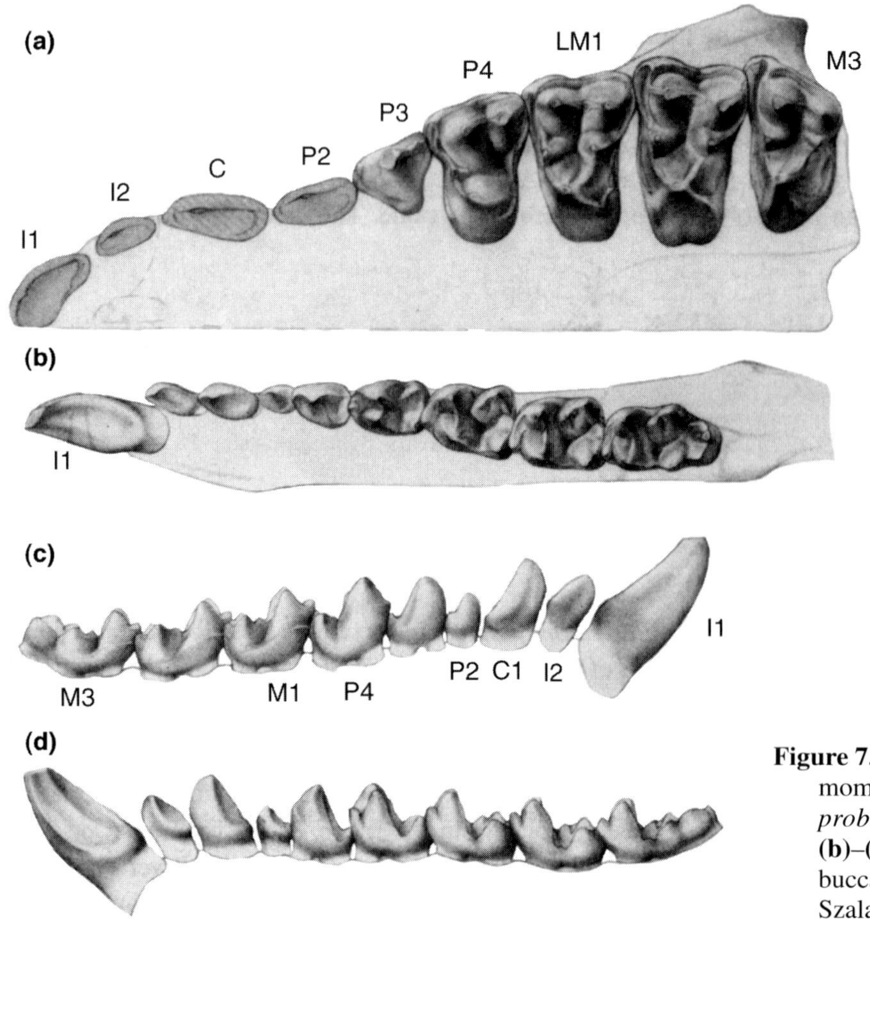

Figure 7. A Paleocene North American paromomyid (a plesiadapiform), *Plesiolestes problematicus*: **(a)** the left 1\l-M\3; **(b)**–**(d)** the right I\1–M\3 in occlusal, buccal and lingual views. Modified from Szalay (1973).

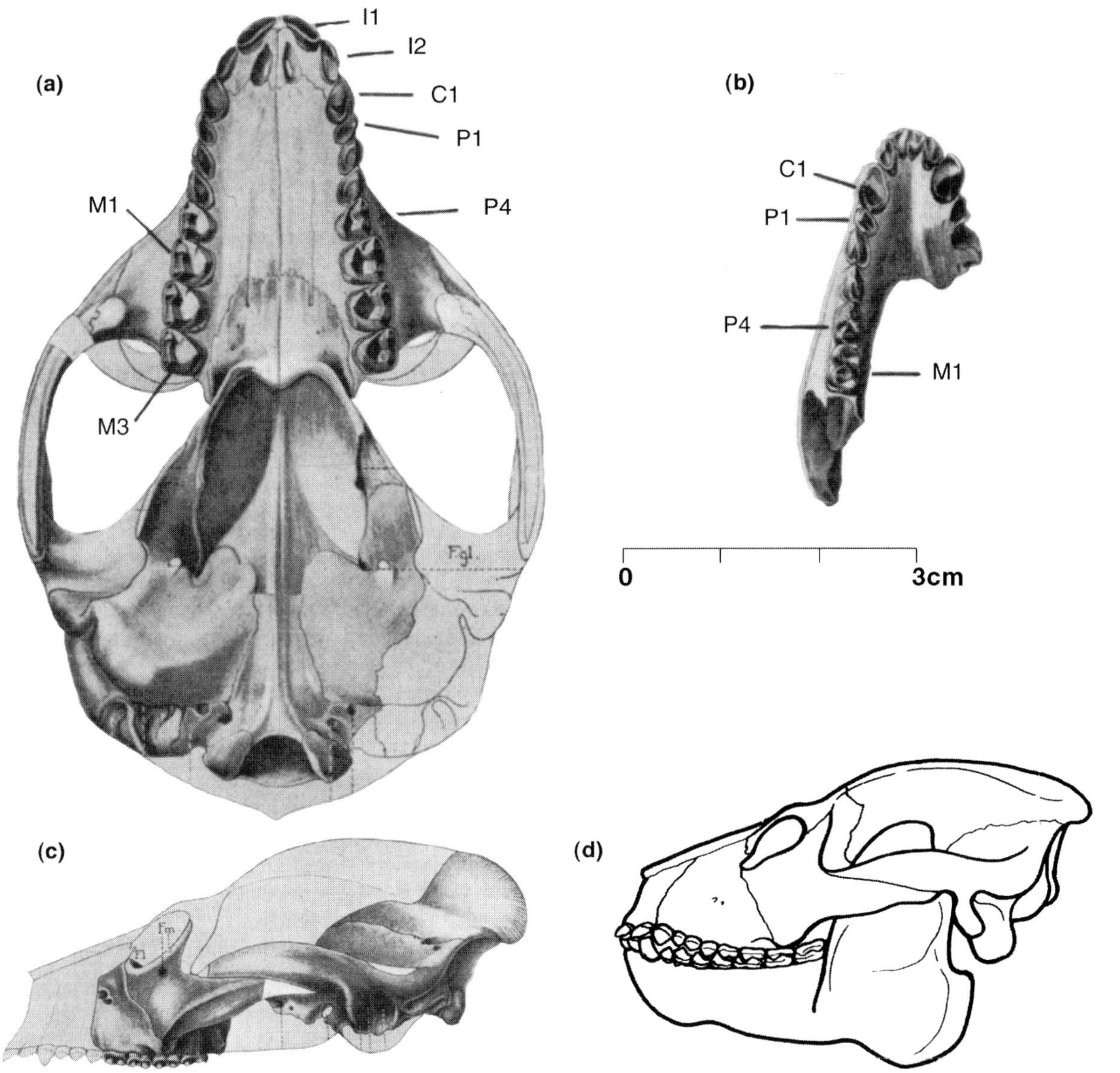

Figure 8. The skull and dentaries of the late Eocene European adapine adapid adapiform *Adapis parisiensis*. From Szalay and Delson (1979) and Gingerich (1980).

Dermoptera, but united Dermoptera and Primates together with Purgatoriidae, Palaechthonidae, Microsyopidae and Picrodontidae (none of which he assigned to any order at all) to a new mirorder, Primatomorpha.

These phylogenetic debates aside, the plesiadapiform primates present a fascinating diversity of adaptation and form. In one of several alternate classifications of the plesiadapiform primates, Szalay and Delson (1979) distinguish a total of five families split between two superfamilies, the Paromomyoidea and the Plesiadapoidea, although one family, Picrodontidae, is so clearly derived from Paromomyidae that its status should be altered. It is proposed to reduce it to subfamily level. From an Australian perspective, the most interesting of these early primates are the paromomyoids (Figure 7), a remarkably diverse group among which many species show differing degrees of enlargement of the central lower incisor, various plagiaulacoid specializations of the posterior premolars, and an occasional lowering of molar topography to a bunodont condition (Szalay, 1973). In these features the paromomyoids are similar to members of the Australian phalangeroid marsupial radiation, see pp.551–574 (Wroe and Archer, 2006) and pp.575–646 (Archer and Hand, 2006).

In terms of zoogeography, the plesiadapiform record appears to indicate an initial phase of dispersal (presumably into Europe from North America) possibly during the middle Paleocene, followed by a relatively rapid radiation (on both continents) into diverse, but largely endemic primate faunas (Walker, 1972; Hoffstetter, 1974). Two genera of plesiadapoids, *Plesiadapis* and *Chiromyoides,* have cosmopolitan distributions but are represented by differing species in each case (Gingerich, 1973). One possible paromomyoid lineage, the Navajoviini, is also reported from both continents. However, in this case there remains some doubt as to whether these minute and poorly known taxa should be retained within the primates, or, if they should, whether the two genera *Navajovius* and *Berruvius* should be associated together in the same higher-level taxon (see Szalay and Delson, 1979).

ADAPIFORMES

Soon after the close of the Paleocene, plesiadapiform primates are joined in the fossil record by a second major group of early primates, the Adapiformes. All members of this group share a number of morphologically advanced features, such as a complete postorbital bar and a fully intra-bullar internal carotid arterial system, which set them apart from their plesiadapiform ancestors (Figure 8). They are generally placed in the suborder Strepsirrhini, with the living lemurs. Although we do not know how they nourished their embryos, they do have a form of incisor occlusion that seems to foreshadow that of modern lemurs (Rosenberger *et al.,* 1985). In addition, a well-preserved pair of legs from the Eocene Messel formation was equipped with lemur-like toilet claws (von Koenigswald, 1979); this shows that proto-lemurs were around at that time, although not necessarily that they were adapiforms.

Szalay and Delson (1979) place all adapiform primates within a single family, the Adapidae, but two subfamilies are recognized (Notharctinae, Figures 9, 10; and Adapinae, Figure 8), with both groups represented in North America as well as Europe. As the Adapiformes

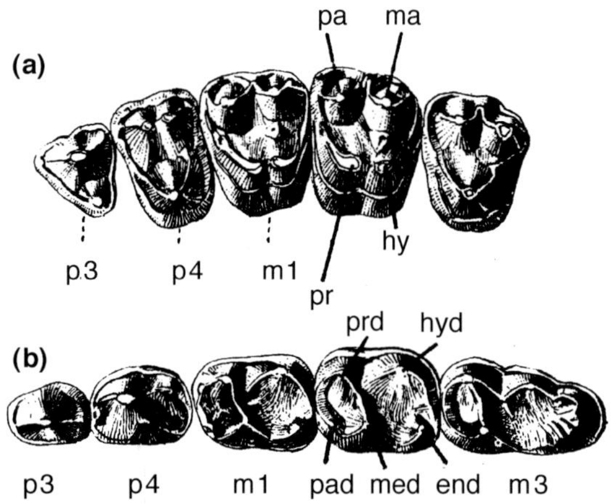

Figure 9. The teeth of the Eocene North American notharctid adapiform *Notharctus* sp. These forms had enlarging hypocones and hence were probably becoming more herbivorous. Some Eocene species of the genus, known from complete skeletons, appear to have been agile and long-distance arboreal leapers. **(a)**, **(b)** occlusal views modified from Osborn (1907).

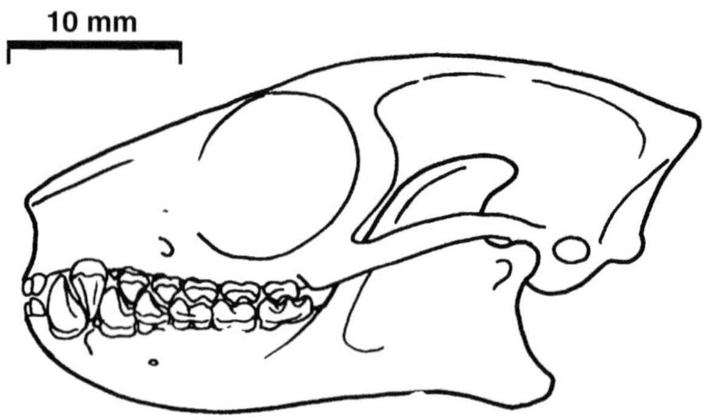

Figure 10. The skull of the middle Eocene North American notharctid *Smilodectes gracilis.* Unlike more plesiomorphic notharctids, this form had a short face. Like some of the other notharctids, it was probably a folivore. From Szalay and Delson (1979).

may well be paraphyletic, the distinct separation between the two divisions may be better recognized at family level (Adapidae, Notharctidae)—especially as there seem to be clear groupings (perhaps subfamilies) within the Adapidae (Schwartz and Tattersall, 1985). Evidence of the wrist and ankle joints supports the idea that Adapiformes are paraphyletic, with Adapidae being the sister-group of the living lemurs and lorises (Beard *et al.,* 1988). The Petrolemuridae (Szalay and Li, 1986) are difficult to associate with any other adapiform; and the origin of certain Miocene genera from India and China is at the moment impossible to pinpoint. Accordingly, it is suggested that four full families be recognized within the Adapiformes: Adapidae, Notharctidae, Petrolemuridae and Sivaladapidae. At lower taxonomic levels, only one genus, *Pelycodus,* is known from both continents (Gingerich, 1977a; Gingerich and Simons, 1977).

Although primitively long-faced with laterally facing orbits, some adapiforms, such as *Smilodectes gracilis* (Figure 10), became decidedly cat-like with foreshortened rostra and anteriorly rotated orbits, the latter feature presumably providing stereoscopic vision. Adapiform primates remained abundant throughout the duration of the Eocene and into early Oligocene, but by mid Oligocene they had all but vanished. Three genera survived into the Miocene: *Indraloris* and *Sivaladapis* from the Siwaliks of northern India, and *Sinoadapis* from China; these are closely interrelated and as long as we cannot relate them closely to any earlier group we class them as Sivaladapidae.

PROSIMIANS: SPECTRES IN THE ANTHROPOID SHADOW

Throughout virtually their entire history, adapiform primates coexisted with representatives of another major primate group, Omomyiformes. Until the idea was chal-

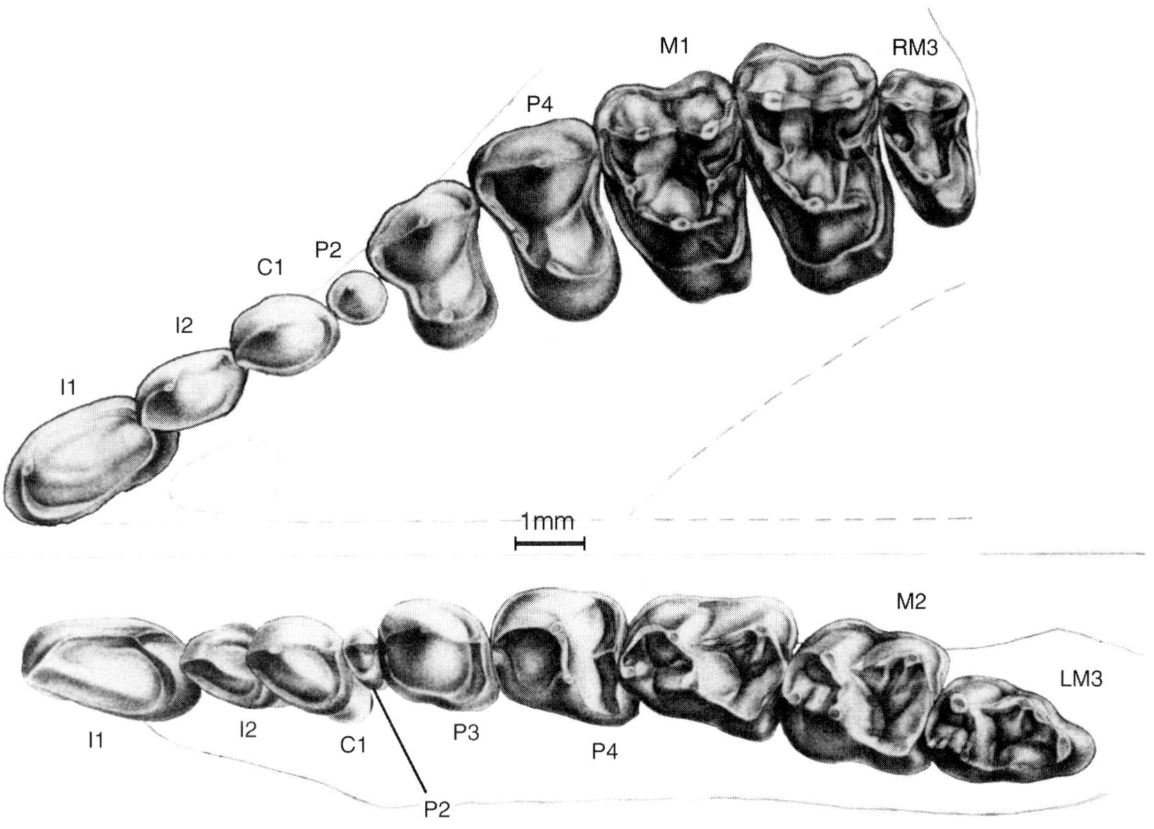

Figure 11. Teeth of a North American anaptomorphid omomyiform *Tetonius homunculus*. This is one of the few omomyids known from the skull. At about three centimetres long, the skull indicates that *T. homunculus* had a relatively large brain, large orbits and a short face. Modified from Szalay (1976).

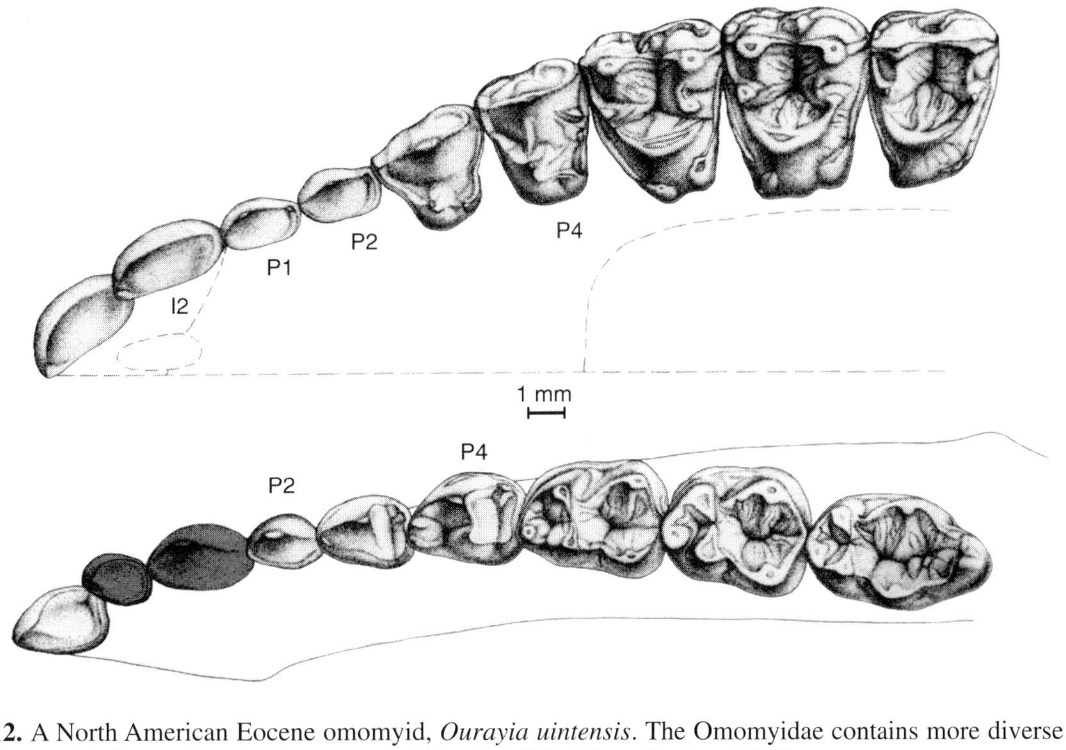

Figure 12. A North American Eocene omomyid, *Ourayia uintensis*. The Omomyidae contains more diverse and less plesiomorphic forms than the Anaptomorphidae. Modified from Szalay (1976).

Figure 13. An Eocene European microchoerid, *Necrolemur antiquus*. Well-preserved skulls and postcranial elements have enabled this reconstruction. From Wood (1976).

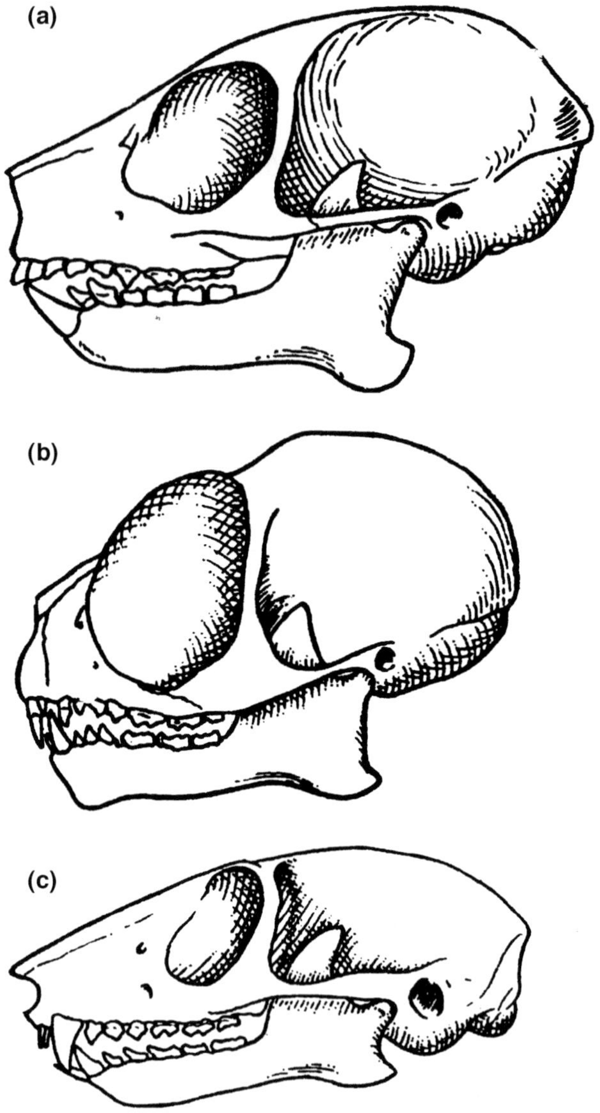

Figure 14. A comparison of the skulls and left dentaries of: **(a)** *Necrolemur antiquus*; **(b)** *Tarsius bancanus*, one of four or more living species of south-east Asian tarsiid tarsiiforms; **(c)** *Microcebus* sp., a living cheirogaleid lemuriform from Madagascar. Note the specialized and forwardly-projecting lower incisors of *Microcebus* sp., a feature characteristic of the whole infraorder. Modified from Le Gros Clark (1963).

lenged by Szalay (1976), the omomyiform primates had mostly been placed in a family, Omomyidae, in the Tarsiiformes, with a few genera actually being referred to the family Tarsiidae. Szalay showed that none of them has any convincing similarity to the Tarsiidae, beyond the general 'prosimian' grade.

This brings us to a theme of fundamental importance to an understanding of the primates in general, and the modern primates in particular: the Prosimian Concept. The 'prosimians' are, in modern primates, the lemurs, lorises and tarsiers. As early as 1918, Pocock pointed out that the tarsier has a nose like a monkey's and an upper lip that is mobile, whereas lemurs and lorises have a moist dog-like rhinarium and the upper lip is bound to the gum. On this basis he united tarsiers with monkeys, apes and people in a group Haplorrhini, separating the lemurs and lorises as Strepsirrhini. Simpson (1945) agreed that the tarsier was very likely to be close to the so-called 'higher primates' but knew the difficulties of placing Eocene (or Paleocene) fossils in one group or the other. So he opted for a 'grade' classification of primates, dividing the 'primitive' ones (lemurs, lorises and tarsiers, and most of the Palaeogene fossils) as Prosimii from the 'higher' ones, the Anthropoidea.

Although the growing evidence supporting the Strepsirrhini/Haplorrhini split has been repeatedly acknowledged (Luckett, 1976), the idea of the Prosimii/Anthropoidea split has persisted in the literature. But it is now evident that living Haplorrhini (tarsiers as well as 'Anthropoidea') share numerous derived soft-tissue characters and several involving hard tissues, including: a reduced stapedial artery; the carotid foramen posteromedial to the bulla; the olfactory process passing above the interorbital septum; and a postorbital bony wall (Cartmill, 1981; Schmid, 1981).

The Omomyiformes can all, if the material is complete enough, be seen to possess the first three of these conditions but not the fourth. Available evidence thus favours their placement within Haplorrhini, although there is some dispute whether the postorbital bony wall is homologous in tarsiers and other haplorrhines. The Omomyiformes (where material is complete) do all seem to possess an I1 that is larger than the canine,

Figure 15. Morphological diversity among living strepsirrhines: **(a)** a lemurid, *Lemur catta*; **(b)** a daubentoniid, *Daubentonia madagascariensis*; **(c)** a galagid, *Galago senegalensis*; **(d)** an indriid, *Indri indri*; **(e)** a lorisid, *Nycticebus coucang*. From Le Gros Clark (1963), Burton (1968) and Webb *et al.* (1977).

a tubular ectotympanic, and an elongated tarsal region; all characters that would support their monophyly. One family, the Microchoeridae, may retain primitive states: they have a median gap between the upper central incisors (implying, perhaps, retention of a rhinarium?), a promontory artery that is still small, and a ventrally exposed fenestra rotunda. Schmid (1981) considers this family, therefore, to be the sister-group of all other haplorrhines. On the other hand characters of more recently discovered material, especially the omomyid *Shoshomius* (Beard *et al.,* 1991), support a sister-group relationship between the Omomyidae and Tarsiidae.

There is also the incompletely described Paleocene genus *Decoredon* (Szalay *et al.,* 1986). Here it has been given the status of plesion to the haplorrhine infraorders, but Gingerich and Schoeninger (1977) maintain an adapiform ancestry for the 'Anthropoidea' as well as the Lemuriformes. Tarsiiform prosimians, according to Gingerich, arose directly from within the plesiadapoid radiation. These suggestions, regarding prosimian relationships,

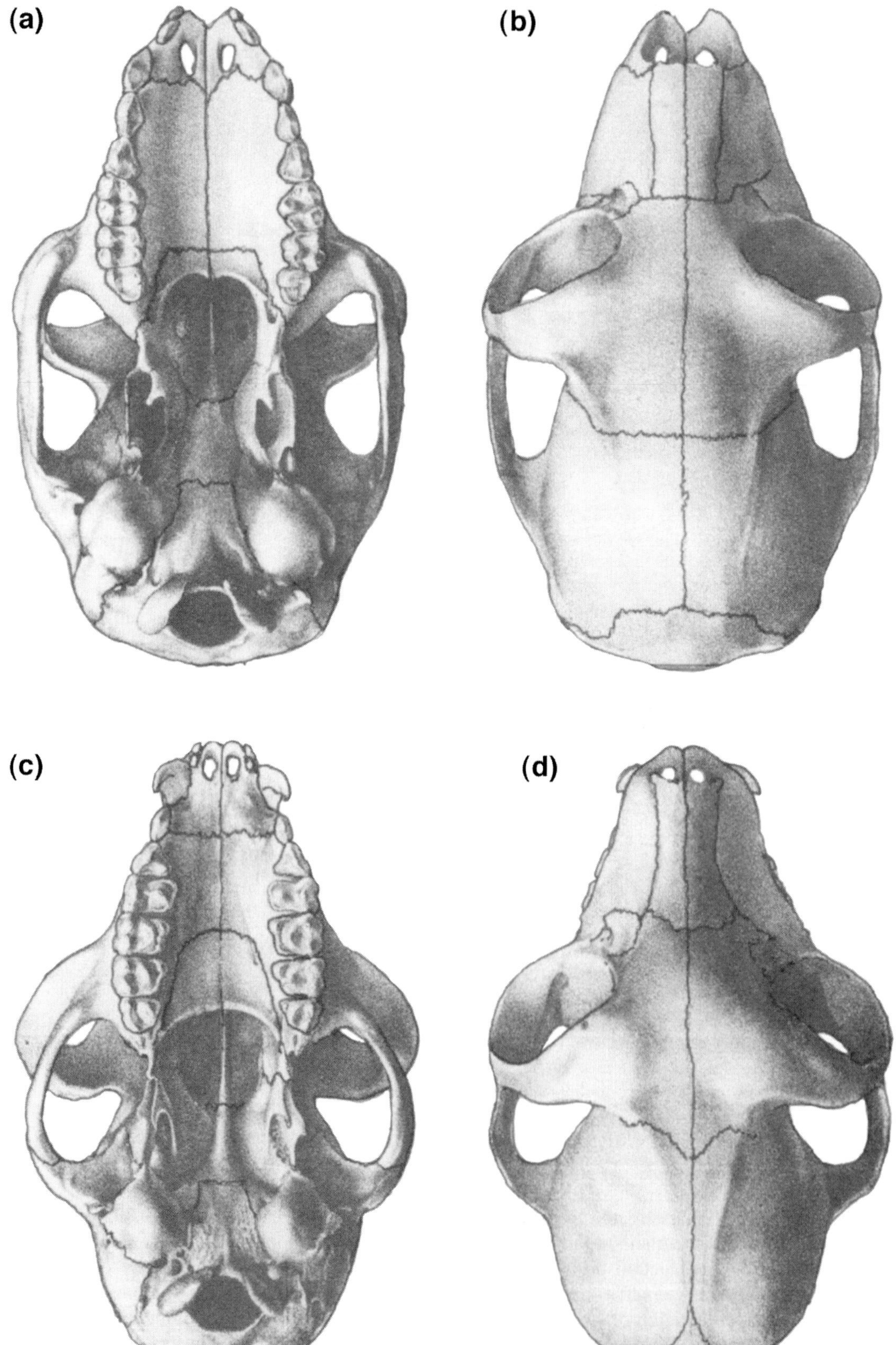

Figure 16. The skulls of living Malagasy indrids and lemurids: **(a)**, **(b)** *Indri indri*; **(c)**, **(d)** *Eulemur fulvus*—note the caniniform incisor and premolariform canine. Modified from Blainville (1839).

depend chiefly on the evidence of the anterior dentition. In plesiadapiforms and tarsiiforms the anteriormost lower incisor is often (although not invariably) enlarged and procumbent, while in Adapiformes and Lemuriformes the equivalent teeth are small, vertical and spatulate.

Independent derivation of the two extant prosimian lineages would require a remarkable degree of parallelism between these groups in many features of cranial (closure of the post-orbital bar, intra-bullar passage of the internal carotid arteries), dental (loss of I3 and P1) and post-cranial anatomy (advanced tarsal, pedal and pelvic structure). So parsimony (with all its limitations as a method) would favour the original more conservative conclusion.

Omomyiformes

This group makes its initial appearance in the fossil records of both western Europe and North America at around the start of the Eocene Epoch (Szalay, 1976; Gingerich, 1981a). Of the three families, one (Anaptomorphidae; Figure 11) has representatives in both Europe and North America, one is restricted to North America (Omomyidae), while the third (Microchoeridae) is restricted to western Europe. Within Europe, omomyiform diversity remained low and the group did not survive the Oligocene. By contrast, North America played host to a major phylogenetic radiation of these primates; but, as in Europe, they attained their greatest abundance and morphological diversity during the Eocene and Oligocene. Only one lineage, Washakiinae, represented by the highly aberrant (in both dentition and name) *Ekgmowechashala philotau,* survived into the Miocene.

Omomyiforms attained only a limited degree of variation in size and craniodental morphology. Nearly all were small, short-faced animals with large forwardly-rotated orbits (see Figure 13) and heavy, procumbent central lower incisors (Figures 11, 12). This tooth is interpreted to be an anteriorly shifted canine by Schwartz (1978).

Tarsiiformes

The living tarsiers (*Tarsius* spp.) of South-east Asia do not appear to be closely related to any of these Tertiary lineages. Accordingly, most workers now place these species in a separate family, the Tarsiidae. Presumably these truly remarkable primates represent the relicts of a once diverse, but as yet largely undiscovered, tarsiiform radiation centred within tropical Asia. Ironically, the only really convincing candidate for tarsiiform status, apart from one Eocene and one Miocene specimen (both isolated teeth) from eastern Asia, *Afrotarsius chatrathi,* is from the Oligocene Fayum of Egypt (Simons and Bown, 1985).

Lemuriformes

Lemuriform primates today occur in restricted areas of Africa and Asia (Figure 2). While this presumably represents a dramatic contraction from an historically much wider geographic range, there is as yet no evidence to suggest that lemuriform primates ever occurred in either North America or Europe.

Six distinct families are recognized within the extant lemuriform radiation. These are the Lorisidae (lorises and pottos), Galagidae (galagos or bush babies; Figure 15 (c), (d)), the Cheirogaleidae (dwarf lemurs), the Indriidae (indri and sifakas; Figure 16 (a), (b)), Megaladapidae (sportive lemur; Figure 14(c)) and the Lemuridae (lemurs; Figures 15, 16 (c), (d)). Of these only the family Lorisidae occurs on both continents. All remaining groups are today confined to the island of Madagascar and two species of Lemuridae have become naturalized on the Comoro Islands off the south-east coast of Africa. For a general introduction to lemuriform diversity and phylogeny, see Tattersall and Schwartz (1975) and Seligsohn (1977).

Lemuriforms differ most strikingly from other primates in their possession of a procumbent 'tooth-comb' (Figure 17), a result of complex modifications of the incisor and canine dentition of the lower jaw. Various aspects of dental homologies, possible function and the evolutionary derivation of the tooth-comb are discussed by Schwartz (1975) and Szalay and Seligsohn (1977).

In common with other African or Asian primate groups, the fossil record of the Lemuriformes remains

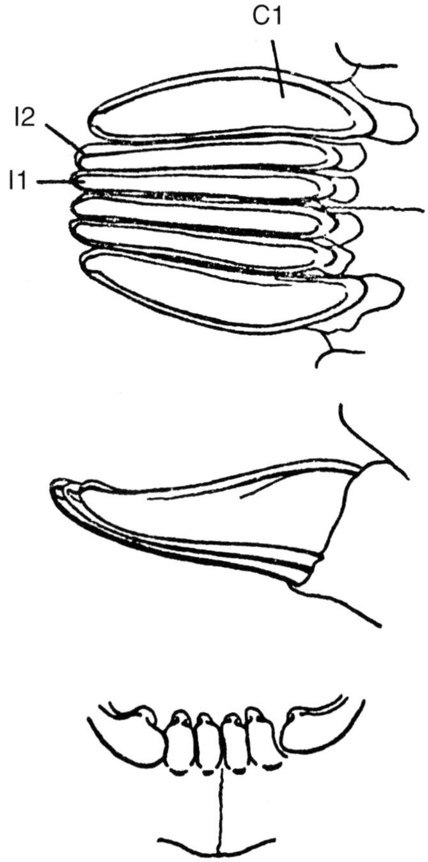

Figure 17. The specialized procumbent teeth (I/(1–2) and C/1) of the lemuriforms (here in *Eulemur fulvus*) sometimes called the strepsirrhine tooth comb, are used like a comb to groom the fur. Modified from Szalay and Seligsohn (1977).

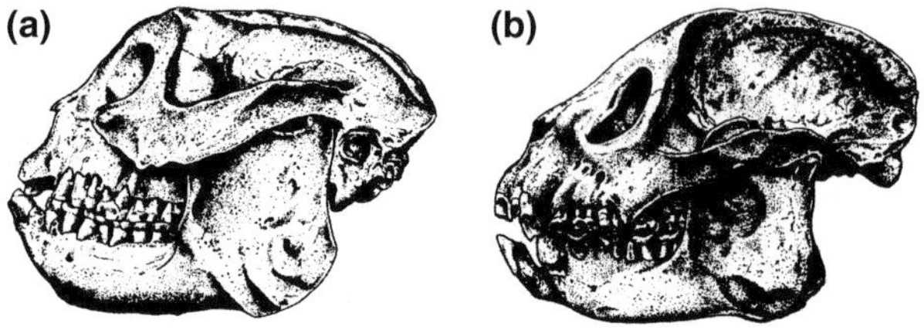

Figure 18. The skulls and dentaries of Malagasy archaeolemurine indriid lemuriforms: **(a)** *Hadropithecus stenognathus*, an extinct but Holocene form that was convergently becoming hominine in appearance; **(b)** *Archaeolemur edwardsi*, another extinct but Holocene form that was rather phalangerid-like in its dental morphology but evidently largely terrestrial in habits. From Szalay and Delson (1979).

limited (Szalay and Delson, 1979). The earliest confirmed lemuriforms, dating from the early Miocene of East Africa, belong to what is arguably the most highly derived of the lemuriform families, the Lorisidae (Szalay and Katz, 1973; Walker, 1974). There is also a single tooth provisionally referred to the Loridae, recently discovered in the Fayum Oligocene deposits of Egypt (Simons *et al.,* 1986). On this evidence major groups of the Lemuriformes appear to have differentiated from their common ancestor sometime prior to the end of the Oligocene.

The most spectacular episode of lemuriform evolution took place on Madagascar during the Neogene. Here, in the absence of many other groups of large mammals, prosimians evolved into a fantastic array of forms. In many cases these adaptations parallel developments seen elsewhere within the primates or within the wider animal kingdom. Among the more striking parallelisms include those illustrated by the decidedly anthropoid-like species of *Archaeolemur* and *Hadropithecus* (Figure 18), both referred to the subfamily Archaeolemurinae of the Indriidae; and the perhaps gibbon-like brachiating *Palaeopropithecus* and *Archaeoindris* (Palaeopropithecinae, Indriidae). Other species have no close parallels in the wider mammalian world, but instead represent wholly unique biological experiments. A good example here is the dog-sized species of *Megaladapis* (Figure 19), which combined an essentially koala-like post-cranial skeleton with a massive, decidedly pig-like skull and dentition; this is a gigantic relative of the extant, diminutive, leaf-eating *Lepilemur.*

Tragically, many of these animals appear to have become extinct when humans first colonized Madagascar before the 10th Century. Although evidence from early European explorers in the area suggests that a number of the now extinct forms, including *Megaladapis* sp., may have survived up to the 16th Century. As one of many phylogenetic tragedies related to our own evolution and dispersal, these magnificent lemuriforms apparently failed to survive their first contact with another member of their own order.

Figure 19. The extinct *Megaladapis madagascariensis*, a megaladapid lemuroid lemuriform. This browsing Malagasy primate was highly specialized, but its relationships to the extant *Lepilemur* now seem clear. Its skull is elongate, lacks permanent upper incisors and has down-turned nasals, features that led to the suggestion that it had a flexible snout. Otherwise, it was probably somewhat koala-like in appearance and habits. From Wood (1976) and Szalay and Delson (1979).

Chiromyiformes

The thoroughly peculiar aye-aye, *Daubentonia madagascariensis,* occupies a family (Daubentoniidae) and even an infraorder (Chiromyiformes) by itself (Figure 15). Although Tattersall and Schwartz (1975) suggest that it is a highly specialized derivative of the Indriidae, derived character states of the foot (Gebo, 1985) link the Indriidae with the Lemuridae alone; and retention of primitive features of the musculature, the skull and the olfactory system suggest that it is the sister-

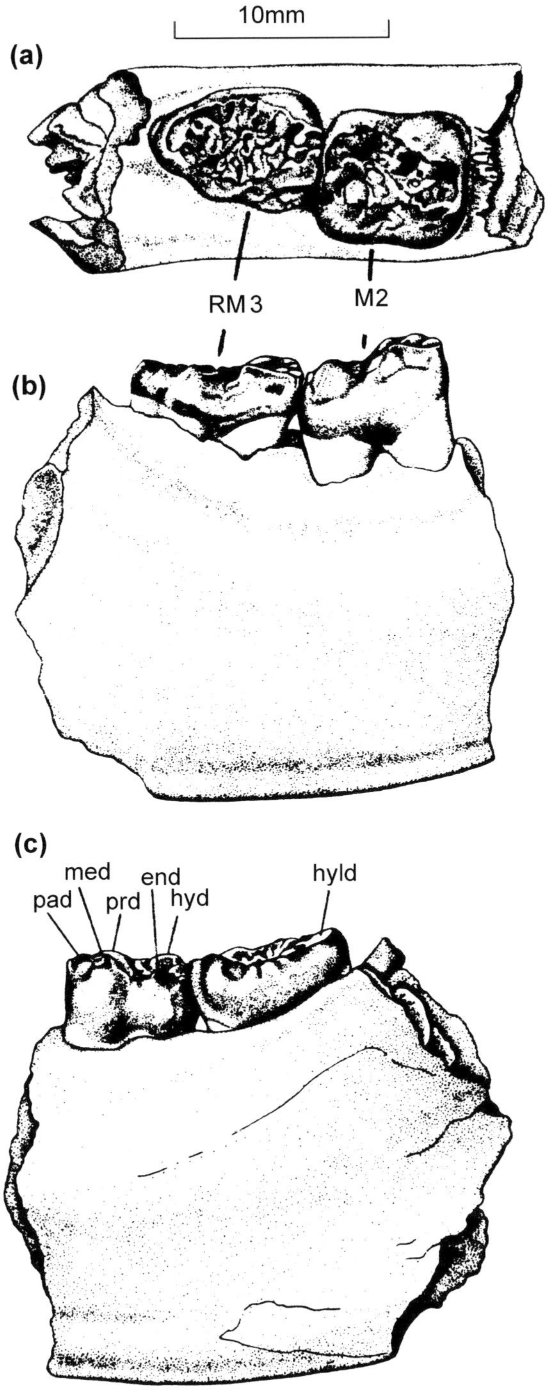

Figure 20. The late Eocene *Pondaungia cotteri* from Burma. This fragmentary right mandible represents the oldest known anthropoid primate. It was probably a fruit-eater: **(a)**–**(c)** occlusal, buccal and lingual views. Modified from Maw *et al.* (1979).

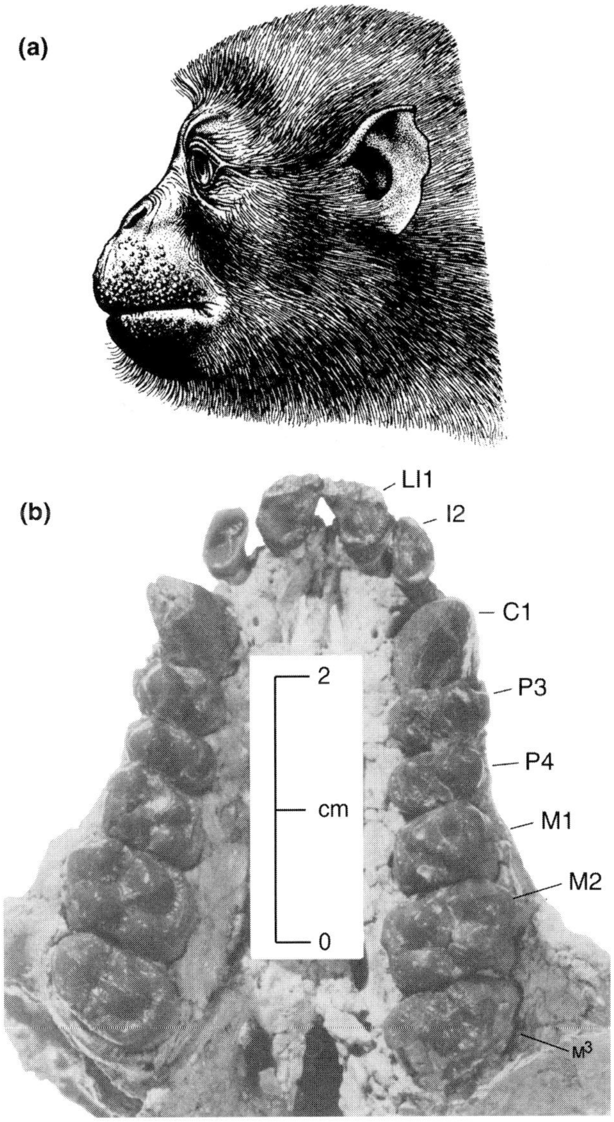

Figure 21. A middle Oligocene propliopithecid anthropoid from the Fayum of Africa, *Aegytopithecus zeuxis*: **(a)** reconstruction of lateral view of head (from Wood, 1976); **(b)** the palate with L and Ri\1-M\3—the molar morphology suggests this was a fruit-eater. Modified from Szalay and Delson (1979).

group of the other living strepsirrhines (Groves, 1974, 1989). The aye-aye is a kind of primate 'woodpecker', strikingly convergent in its craniodental and digital specializations on the phalangeroid marsupial *Dactylopsila* (Cartmill, 1974). It is one of the primate species most in danger of extinction; in its Madagascan homeland it is regarded as a creature of ill omen, and is killed on sight.

SIMIIFORM ORIGINS AND EARLY RADIATIONS

ROOTS

The term 'anthropoid', as it is currently used, is an informal name referring to either the community of higher primates in general or to great apes in particular. Hoffstetter (1974) discussed the usage of the name Anthropoidea in a formal sense, and concluded that it must be abandoned; the correct name for the non-tarsiiform, non-omomyiform haplorrhines is Simiiformes. As indicated above, the case for a tarsiiform ancestry for the 'higher primates' appears relatively secure. But where did this transformation take place? The tarsiiform fossil record is somewhat less than brilliant and, as long as omomyiforms were confused with tarsiiforms, a simiiform origin in either the Americas or the Old World seemed possible; but separation of the two infraorders deprives North America of its tarsiiforms.

An origin somewhere in the Old World appears to be more probable. But where? The earliest undoubted anthropoids come from the Far East; Burma, Thailand and China to be precise. The material in question is *Pondaungia* (Figure 20), *Amphipithecus, Siamopithecus* and *Eosimias* all considered to be middle Eocene in age (Conroy and Bown, 1976; Beard *et al.,* 1994, 1996). Unfortunately, all available specimens are too incomplete to tell us much about the relationships of these cryptic species. Ciochon *et al.* (1985) argue convincingly that they are all simiiforms, even catarrhines; while Jaeger *et al.* (1998) accept that *Amphipithecus, Pondaungia* and *Siamopithecus* are catarrhines they consider that *Eosimias* may be an omomyid.

It is not until considerably later that the first reasonable assemblages of fossil anthropoids become available. The most extensive of these comes from late Eocene and early Oligocene deposits in the Fayum Depression of present-day Egypt. Collection of fossil mammals from this highly productive region over many years has yielded a diverse suite of early anthropoids, possibly as many as 11 species in all. Although classification of these Fayum primates remains controversial, most workers agree that three major groups are represented. The first of these, the Parapithecidae, contains a suite of essentially primitive anthropoids including the relatively well known *Apidium phiomense*. Also included are the *Parapithecus* species, taxa which Simons (1974a) regarded as annectant to the Cercopithecidae but which Delson (1975a), among others, dismisses as merely convergent on later forms.

Members of the second group are placed collectively within the family Propliopithecidae. Compared with the more conservative parapithecids, propliopithecids are characterized by various craniodental specializations, at least some of which appear to foreshadow features seen among the living apes and Old World monkeys (Simons, 1965, 1974a, b, 1977; Radinsky, 1977). Probably the best known of these is *Aegyptopithecus zeuxis* (Figure 21). Szalay and Delson (1979) regard this as a distinct species among forms previously allocated to the genus *Propliopithecus*. This species was widely hailed upon its discovery as a true ancestor of the Miocene dryopithecines and, through them, of the extant great apes. A second Fayum propliopithecid, *Propliopithecus chirobates,* was originally regarded as showing a number of decidedly gibbon-like features and regarded as annectant to that group (Simons and Fleagle, 1973). It has to be noted, however, that as a result of more recent studies, most workers (Kay *et al.*, 1981) regard these propliopithecids as relatively plesiomorphic anthropoids.

Figure 22. Three distinctive types of living American platyrrhine monkeys: **(a)** a cebid, *Cebus apella*; **(b)** a callitrichine, *Callithrix jacchus*; **(c)** an atelid, *Ateles* sp. All platyrrhines have a flattened nose with a wide internarial septum. Atelids differ from cebids in having large dentaries and moderate to large hypocones on the molars; callitrichines have simple tritubercular molars, reduced to two in most species. From Le Gros Clark (1963) and Webb *et al.* (1977).

The third group is the Oligopithecidae, a possibly paraphyletic family now best known from the late Eocene levels (*Catopithecus, Proteopithecus*), but surviving into the Oligocene (*Oligopithecus*). They preserved, among other primate features, an emphasis on crest function rather than cusp function in their molar teeth.

Indeed, in virtually all respects these Oligocene anthropoids would appear to differ little from what might be expected of an all-primitive ancestral anthropoid. Where specializations are evident they appear relatively minor in scale and do not detract from the impression of these Fayum primates as basically plesiomorphic catarrhines. The position of the Oligopithecidae is unclear. Though some authors see them as the sister-group of all other Simiiformes, the latest study (Jaeger *et al.*, 1998) assigns this position to the Parapithecidae, but confirms the Oligopithecidae as the sister-group (perhaps the paraphyletic stem-group) of the other catarrhines.

SIDE-BRANCHES

There has been much discussion relating to ecological influences on anthropoid origins, including changing climates. Although relationships between ecological factors and physiological responses are rarely direct or simple, it is considered that enhanced seasonality may have been of prime importance in the development of higher primates.

Before concentrating on the Old World anthropoids, their remarkable relatives the New World Monkeys or platyrrhines are briefly considered.

Platyrrhini

The Platyrrhini or New World monkeys (Figures 22–26) are confined in their natural state to South and Central America. Szalay and Delson (1979) divide the 46 or more living species between two families, the Cebidae and Atelidae, the former containing the Callitrichinae or marmosets and the Cebinae or 'cebid' monkeys. Other authors, such as Hershkovitz (1977), favour a more extensive taxonomic breakdown of the group into additional families; fossil evidence and immunological evidence alike suggest that the platyrrhine radiation is very old, and in fact relationships between the various groups are very difficult to detect. Like Hershkovitz, Groves (1989) recognizes a number of families, these are: Homunculidae (containing the Miocene *Homunculus* only; Figure 24(c)), Callitrichidae (marmosets and the Miocene *Micodon;* Figures 22(b), 23(a)), Cebidae (capuchin, squirrel monkeys and the Oligocene or Miocene *Dolichocebus;* Figures 22(a), 23(b)), Aotidae (night monkeys and Miocene *Tremacebus;* Figure 23(d), (e)), Callicebidae (titis and perhaps the enigmatic *Xenothrix;*Figure 23(c)), Pitheciidae (sakis, uakaris and Miocene *Cebupithecia;* Figure 24(a), (b)), and Atelidae (woolly, spider and howler monkeys, and Miocene *Stirtonia*; Figure 22(c)). Schneider *et al.* (1993) and Horovitz and Meyer (1997) both used molecular data to divide the Platyrrhini into just two families, Cebidae and Atelidae, but constituted

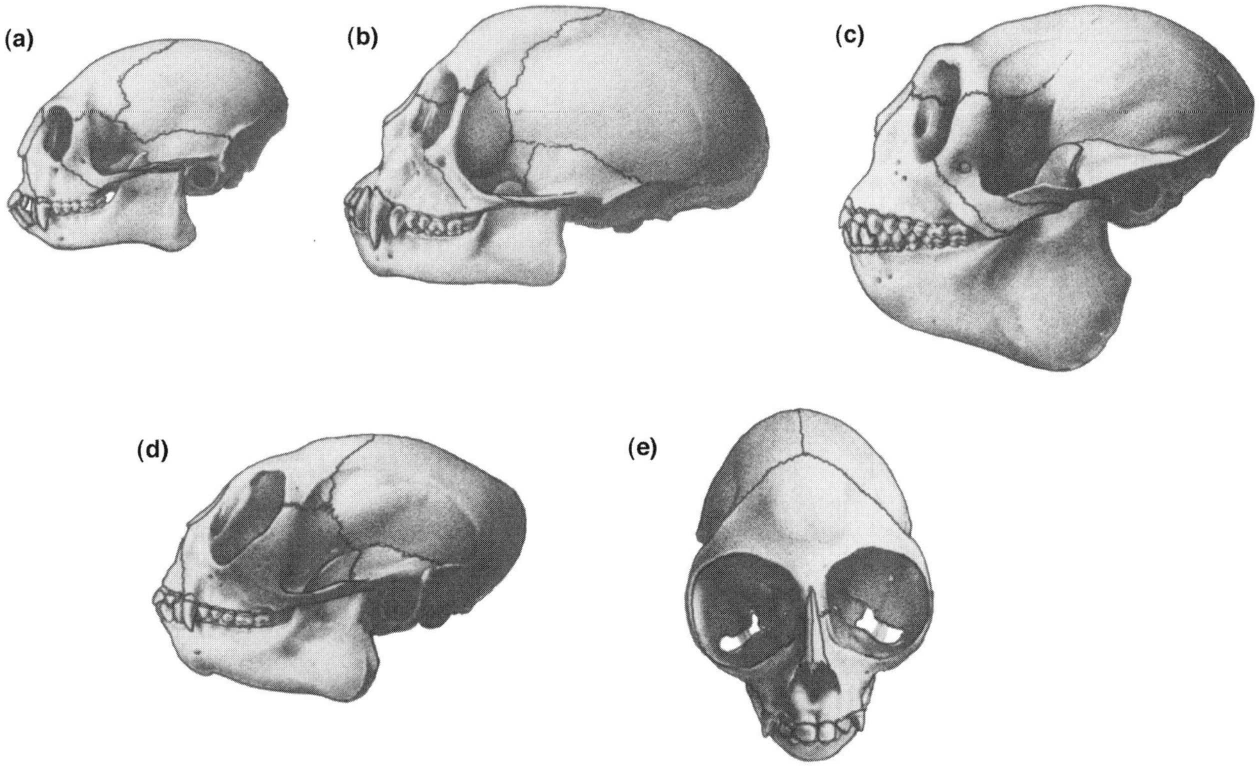

Figure 23. Skulls and dentaries of living platyrrhine primates: **(a)** a callitrichine, *Callithrix* sp.; **(b)** a cebid, *Saimiri* sp.; **(c)** a callicebin, *Callicebus* sp.; **(d), (e)** an aotid, *Aotus* sp. Modified from Blainville (1839).

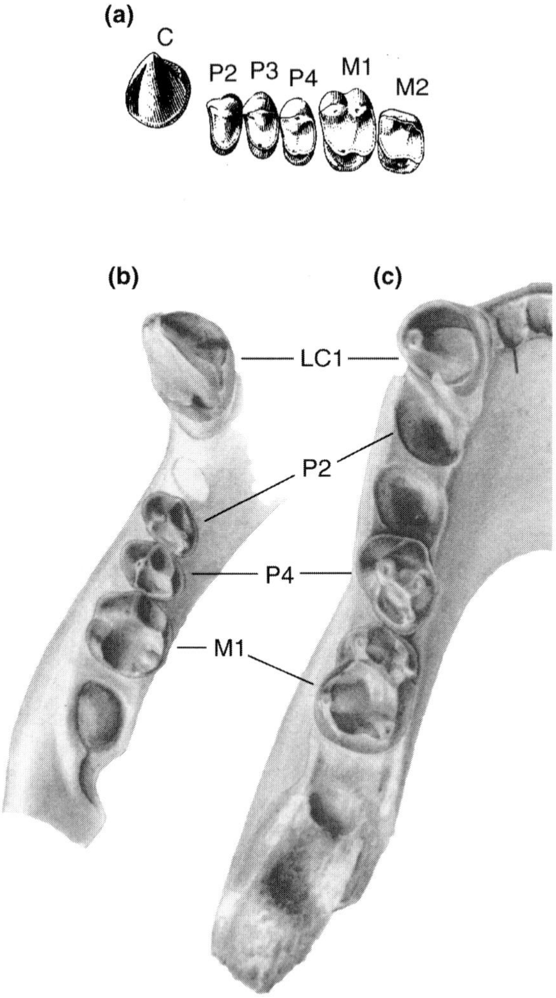

Figure 24. The teeth of platyrrhine monkeys: **(a)** the left C\1–M\2 of the Columbian Miocene pitheciid, *Cebupithecia sarmientoi;* **(b)** the left C/1 and teeth or alveoli for P/2–M/2 of *C. sarmientoi;* **(c)** the teeth or alveoli for the left I/1–M/2 of the Argentinian Miocene homunculid, *Homunculus patagonicus*. Modified from Stirton (1951).

differently from those families in Szalay and Delson (1979). The Cebidae include the Callitrichinae (marmosets), Aotinae (*Aotus*), Saimiriinae (*Saimiri*) and Celinae (*Cebus*); the Atelidae include the Atelinae (with tribes Atelini and Alouattini) and Pitheciinae (with tribes Pitheciini and Callicebini). In the opinion of Groves (2001), the affinities of *Aotus* are still not clear-cut, and the association of the Pitheciinae with the Atelinae is not absolutely definitive.

All up, platyrrhines comprise about 35% of all living primates. Although they are highly variable in feeding and locomotor habits, and occur in a wide range of different forest types, the platyrrhines are perhaps ecologically less diverse than their Old World cousins. The New World radiation does not contain any open country or savanna-dwelling representatives, although several of the forest-dwelling platyrrhine lineages, such as the Callitrichidae (and the species of *Aotus,* the night monkeys), find no close parallel among catarrhines.

The New World monkeys are universally regarded as a younger element within the South American fauna compared with, for example, the marsupials or edentates or ungulates, but exactly when and how the primates first appeared in the southlands remains unclear. The earliest known platyrrhine, a Bolivian beauty known as *Branisella boliviana* (Figure 25), is probably early Oligocene in age. But *B. boliviana* must be itself a product of some reasonable period of endemic evolution. Both palaeontological and biochemical studies suggest a late Eocene arrival of the ancestral platyrrhine stock in South America. Interestingly, this estimate corresponds closely with the known arrival time of several other external elements, the caviomorph rodents and the procyonid carnivorans. These three groups may have entered South America from the north, presumably by island-hopping across the then narrowing Panamanian Strait; but the complete absence of any anthropoid lineages in North America makes a Panamanian source for the platyrrhine primates very unlikely. An African or possibly Asian source is more probable, with dispersal taking place by chance rafting across what were, even then, formidable ocean crossings.

In keeping with the relative ecological conservatism of platyrrhines, the majority of post-Oligocene fossil platyrrhines fall within the bounds of the modern radiation (Rosenberger, 1979; see Figure 26). One possible

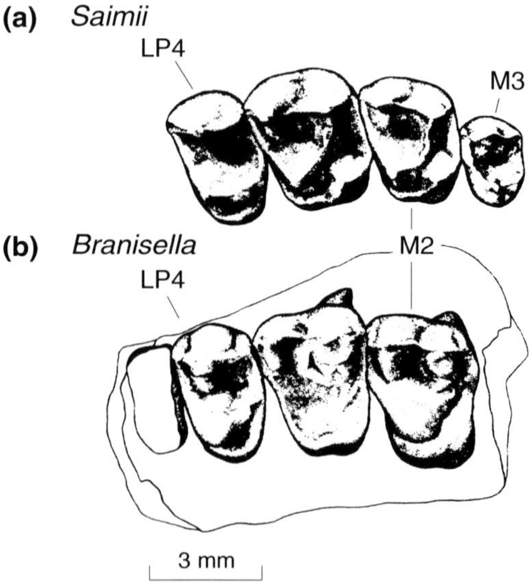

Figure 25. A comparison of the LP\1–M\3 of the living cebid *Saimiri sciurcus* with the LP\4–M\2 of the Oligocene platyrrhine *Branisella boliviana*, the oldest known platyrrhine. These forms are similar and suggest either that *B. boliviana* may be more closely related to cebids than to other platyrrhines, or else that the Cebidae are dentally plesiomorphic. Modified from Gingerich (1981b).

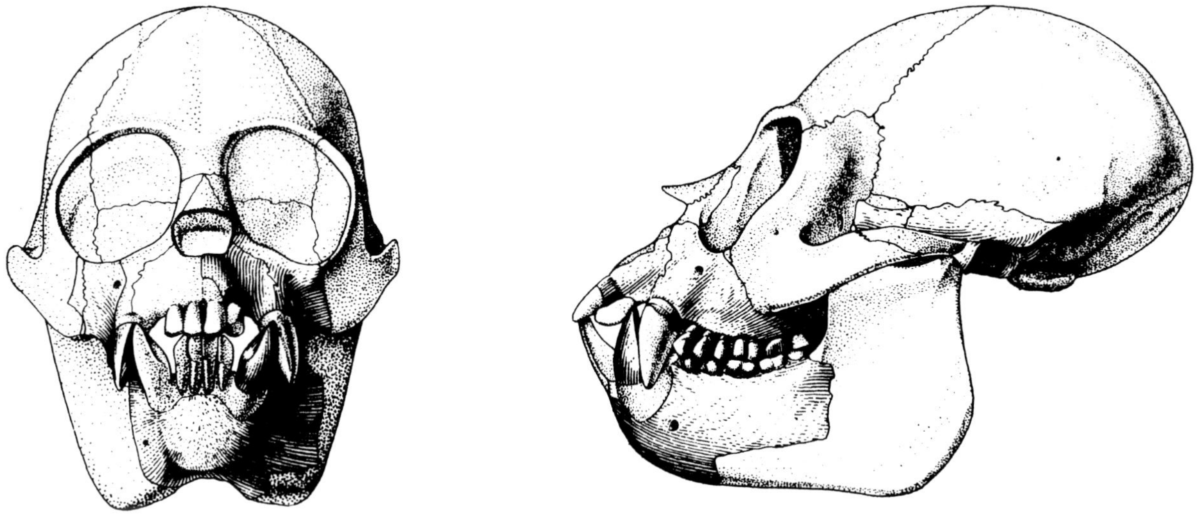

Figure 26. A middle Miocene pitheciid, *Cebupithecia sarmientoi*, from South America. Like its living relatives, it was probably a fruit- and seed-eater. From Stirton (1951).

exception to this is *Xenothrix mcgregori,* a comparatively recent but highly atypical Jamaican platyrrhine without obvious affinities to any other known lineage—except for some poorly known subfossils from Cuba and Hispaniola (Rosenberger, 1977).

LATER EVOLUTION OF THE CATARRHINES

Following the abundant diversity of the Fayum period, the fossil record of catarrhines enters a 'Dark Age' of roughly ten million years duration. When the fossil record picks up again, around the start of the Miocene, primate assemblages are dominated by recognizable members of the two extant superfamilies, the Cercopithecoidea and the Hominoidea.

CERCOPITHECOIDEA

Szalay and Delson (1979) divide known cercopithecoids into two families. The first family, the Oreopithecidae, contains only the late Miocene *Oreopithecus bambolii* (Figure 27) but Harrison (1986) argued convincingly that the oreopithecids are actually hominoids. This leaves only the Old World monkeys, which Szalay and Delson (1979) placed in a single family. The Cercopithecinae is a highly diverse group with perhaps the broadest geographic and ecological range of any primate subfamily, living or extinct (Napier, 1970; Delson, 1975a, b, 1977; Delson and Andrews, 1975). Members of this subfamily (Figures 28, 29) are typically diurnal, and differ markedly in locomotor adaptation and dietary preference from the strictly arboreal folivores of the Colobinae (Figure 30).

Despite this great diversity of life-styles, all cercopithecoids display a characteristic bilophodont dentition, where each of the molars bears two upraised transverse crests or lophs (Figure 31). Although unparalleled in primates, this dental development is duplicated within several Australian marsupial families, most notably among the macropodids or kangaroos. Lophodonty most likely provides enhanced shearing capabilities, greater wear potential and greater capability to deal with highly fibrous and/or low quality food resources (Temerin and Cant, 1983). Among primates, this dental modification is accompanied by various digestive specializations such as cheek pouches among cercopithecines (Murray, 1975) and a macropod-like stomach in colobines (Bauchop and Martucci, 1968).

PLIOPITHECOIDEA

The European mid-Miocene genera *Pliopithecus* and *Crouzelia* have at various times been thought of as either primitive gibbons or surviving propliopithecoids. Actually they are best regarded as the sister-group to the cercopithecoid/hominoid stem (Andrews, 1985). They seem to have been lightly built, arboreal, quadrupedal springers; their remains are locally abundant in western and central European deposits, but were never widespread (Ginsburg and Mein, 1980). One representative, *Laccopithecus,* has been found in China (Wu and Pan, 1984).

HOMINOIDEA

Compared with their cercopithecoid cousins, the hominoids display a relatively limited taxonomic diversity (Figure 32) and, with the exception of our own

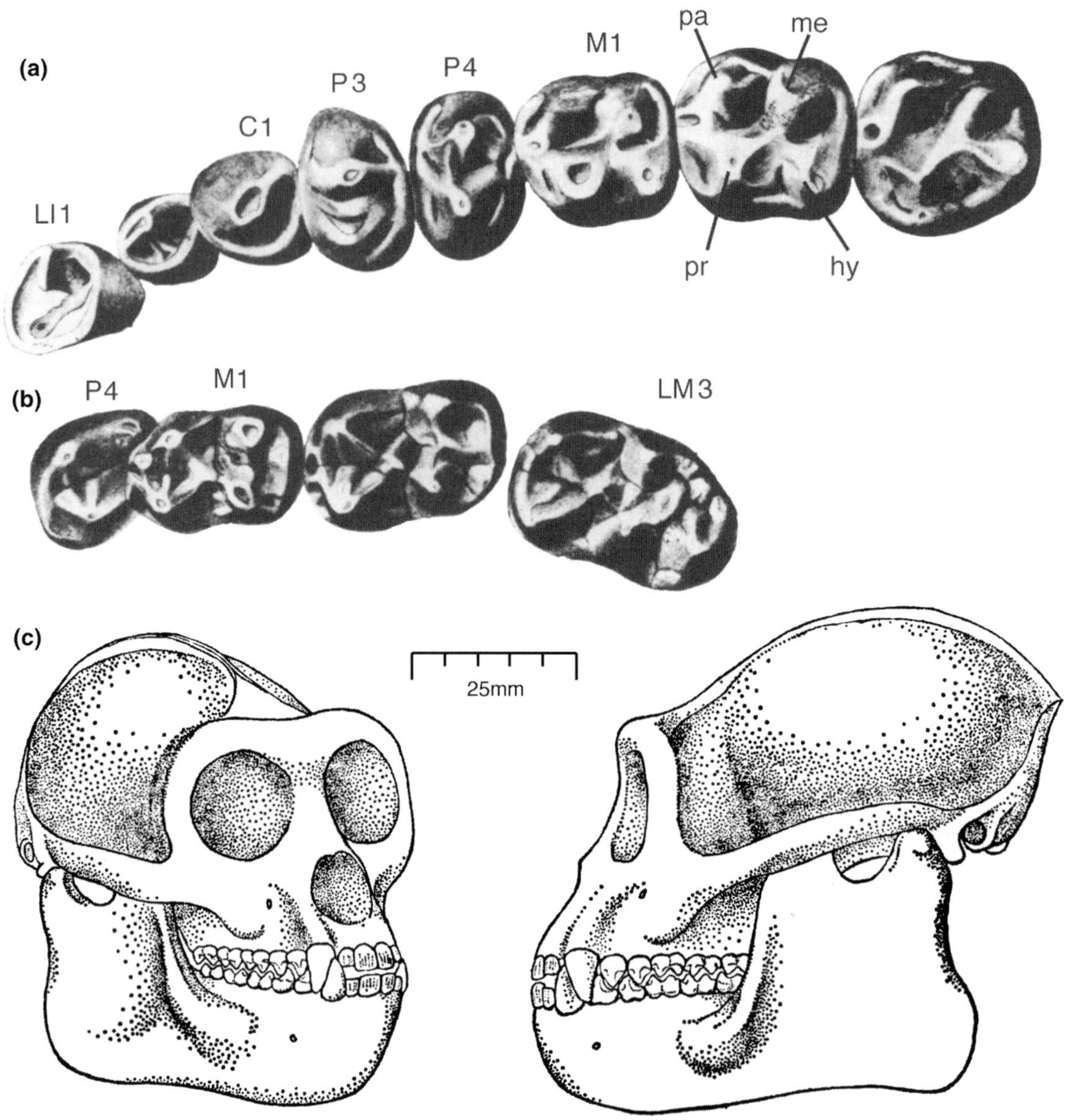

Figure 27. The late Miocene oreopithecid of Europe, *Oreopithecus bambolii*. This species was probably a forest-dwelling folivore and an agile climber: **(a)** LI\1–M\3; **(b)** LP\4MI3; **(c)** reconstruction of skull. From Hurzeler (1960) and Szalay and Delson (1979).

Figure 28. A living papionine cercopithecid, the African olive baboon (*Papio anubis*). From Le Gros Clark (1963).

species, a restricted ecological and distributional range. Morphologically, hominoids differ from Old World monkeys in having retained a relatively primitive molar series but have developed various specializations of the antemolar dentition. In the postcranium, various specializations indicate enhanced shoulder mobility and great facility for upright posture. Although previously considered as adaptations for brachiation (arm-swinging), these are better thought of as feeding adaptations (Andrews and Groves, 1975). Other features of the common hominoid ground plan include a tendency towards large body size, enlargement of the brain, variably delayed developmental rates and lowered reproductive rates.

Although several of these diagnostic features may represent allometric adjustments to increased body size,

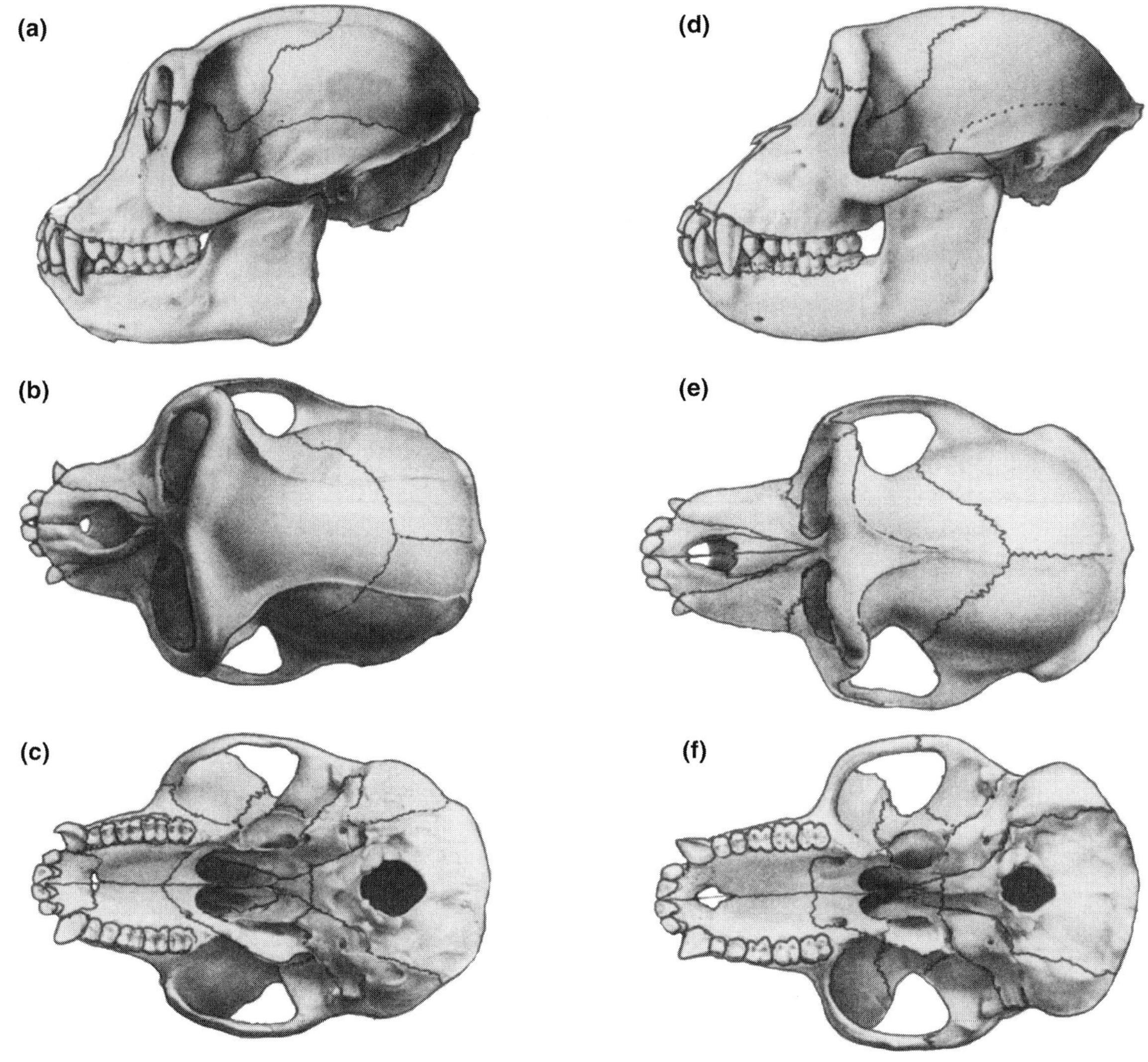

Figure 29. Skulls and dentaries of living cercopithecoids: **(a)–(c)** *Pygathrix nemacus*, an Asian colobine; **(d)–(f)** *Macaca* sp., an Asian cercopithecine. Modified from Blainville (1839).

they nevertheless reflect significant contrasts in lifestyle between the hominoids and Old World monkeys (Temerin and Cant, 1983). Humans are unquestionably the most successful hominoids and have the broadest geographic and ecological range of any known terrestrial vertebrate, living or extinct. We are also one of the most highly autapomorphic (uniquely derived) of all primates; certainly, of the Haplorrhini. This is equally true of both anatomical and behavioural features.

In traditional taxonomy, these many unique features were considered sufficient grounds upon which to distinguish the genus *Homo* and its close relatives, the australopithecines, as a separate family of hominoids. Based on purely phenetic logic, this morphological distance between humans and apes was also widely regarded as evidence for an early divergence of the two lines, possibly even predating the divergence of the great apes and the Old World monkeys. Operating within this paradigm palaeontologists have sought the roots of the human lineage among Miocene or earlier anthropoids. At various times possible special hominid affinities have been suggested for such diverse forms as *Oreopithecus bambolii,* a number of Fayum propliopithecids, the species of *Gigantopithecus* and the species of *Ramapithecus;* most of these concepts now lie behind us, although the affinities of *Oreopithecus* continue to be debated (Harrison, 1986). What stands in their place is quite a different view of human affinities.

Current Concepts of Evolution

Over 30 years ago immunological techniques (using the rates of change or mutation of blood proteins) revealed that, at a molecular level, humans were closest to chimpanzees and gorillas, then to the orangutan and gibbons. More distantly this collection of groups was related to Old World monkeys (Sarich and Wilson, 1967). Three decades and many more (confirmatory) biomolec-

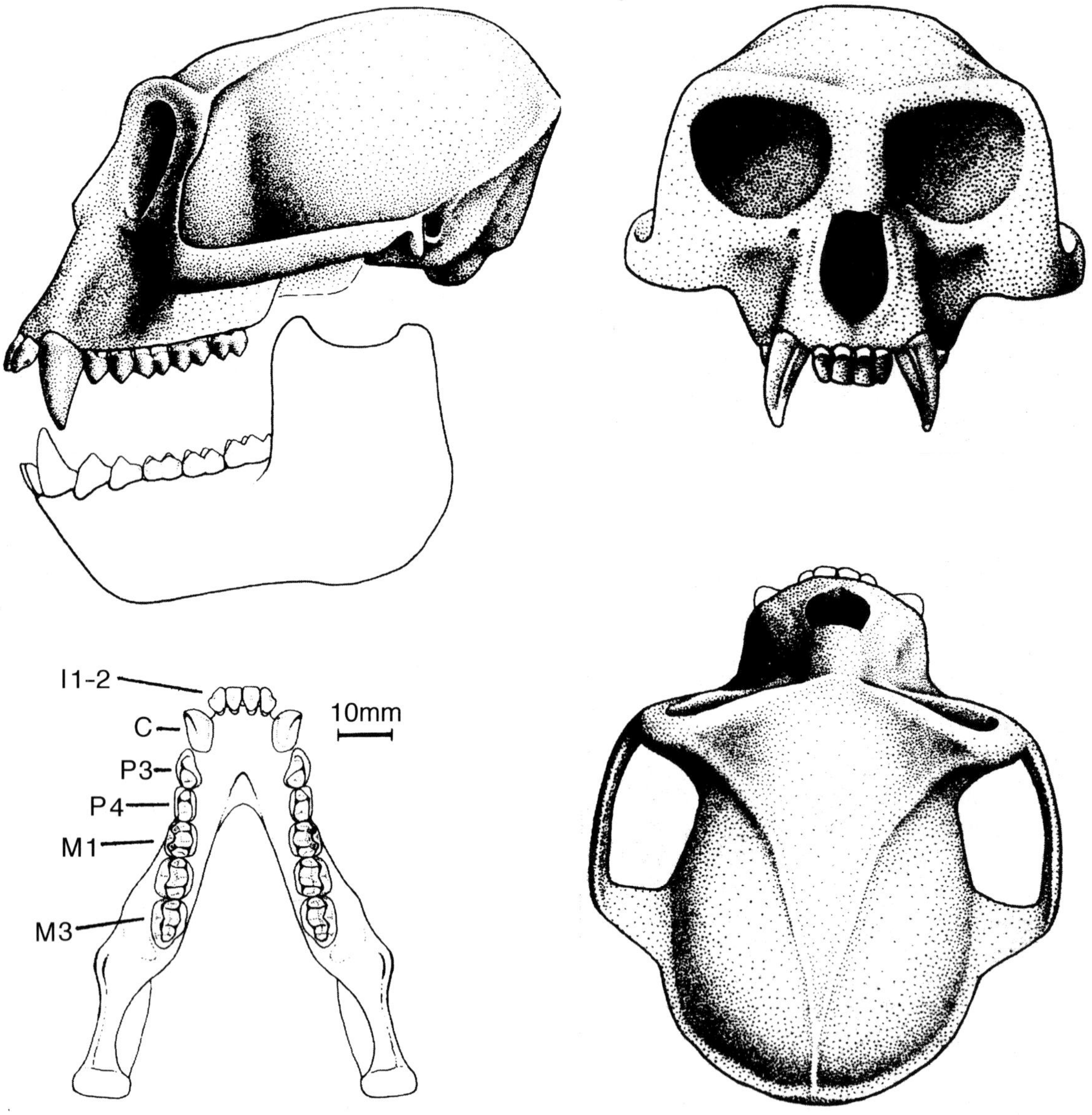

Figure 30. *Mesopithecus pentelicus*, a late Miocene Eurasian colobine. From Szalay and Delson (1979).

ular studies later, anthropologists have accepted what was, at the time of Sarich and Wilson's study (Figure 33), an almost heretical viewpoint. Humans are now widely regarded as very closely related to the African great apes, but exactly how the species of this natural group are related amongst themselves remains to be resolved. Most recent evidence places the chimps and humans together, leaving the gorilla as a slightly more distant relative; this conclusion seems to be supported by cladistic analysis of morphological characters (Groves, 1986). Looking beyond this group, affinity is found next with the orang-utan (*Pongo pygmaeus*) and then with the gibbons (Hylobatidae); so gibbons represent the earliest offshoot among living hominoids. Sarich and Wilson's original estimates of divergence times among hominoids have also been largely validated by more recent estimates (Figure 34).

A phylogenetic classification of the Hominoidea, incorporating both living and fossil forms, is included in the introduction to this chapter (see Table 1).

Hylobatidae

The living gibbons (Figures 32(a), 35) are without doubt the most highly specialized of mammalian arborealists. The 11 or so living species of gibbons are currently grouped into four subgenera within the genus *Hylobates* (Andrews and Groves, 1975; Groves, 2001). Although these are better regarded as genera, as molecular data suggest that they are more divergent from one another than is *Pan* from *Homo*! Gibbons are distributed throughout mainland and insular South-east Asia. During Pleistocene times, hylobatids of the modern type also occurred in parts of central China.

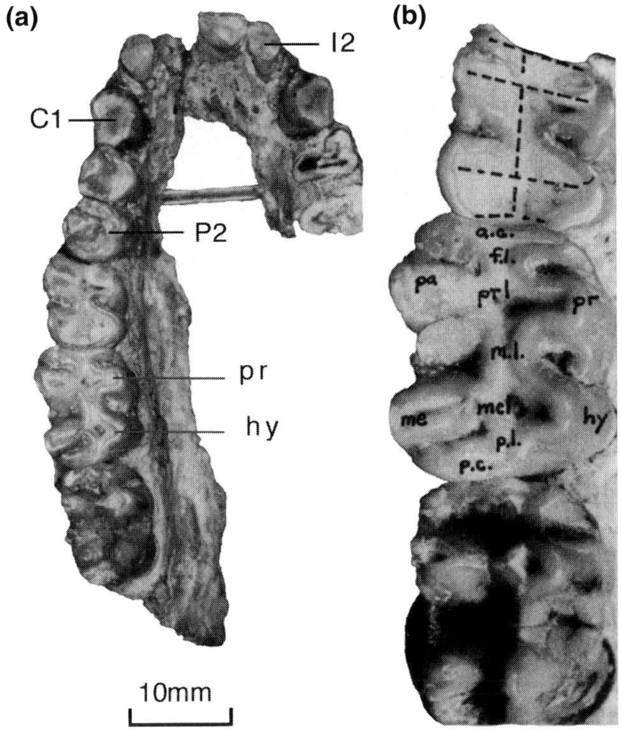

Figure 31. Plio-Pleistocene upper teeth of the African papionin *Theropithecus oswaldi*, a baboon. This dentition shows the convergence with molar teeth of kangaroos and palorchestid marsupials. Note even the anteroposterior wear on the adjoining regions of the molar teeth, a common feature of kangaroos. **(a)** a palate with teeth or alveoli representing the L and RI\(1–2), L and RC\1, L and RP\(3–4) and RM\(1–3); **(b)** an isolated RM\(1–3). The RM\2 is labelled as if it were a macropodid. Baboons are largely herbivorous and feed mainly in the open grasslands on grass, seeds and rhizomes. Modified from Szalay and Delson (1979). Key to abbreviations: a.c. = anterior cingulum; fi = forelink; hy = hypocone; me = metacone; m.l. = midlink; pa = paracone; p.c. = posterior cingulum; p.l. = postlink;. pr = protocone; prl = protoloph.

In craniodental morphology, hylobatids are little changed from the ancestral catarrhine pattern and this similarity, to primitive anthropoids, has led some authors to suggest an independent derivation of hylobatids from an early pliopithecid stock (Thenius, 1981); but their placement with hominoids is supported by immunological, chromosomal, postcranial and behavioural data.

Despite the long evolutionary history of hylobatids, no definite representatives of the family are known from any pre-Pleistocene context. One possible explanation for the scarcity of hylobatids in the fossil record is that they represent an exclusively Asian radiation. This might also account for the presence in the European and African fossil records of a number of superficially gibbon-like cercopithecoids and hominoids, which might thus be viewed as ecological vicars of (and convergent on) true Asian hylobatids.

Oreopithecidae

From a late Miocene lignite mine near Pisa, in Italy, comes a controversial set of remains, *Oreopithecus bambolii,* known by numerous teeth and a complete skeleton (Figure 27). Hurzeler (1960) saw it as directly on the human lineage; Szalay and Delson (1979) assigned it to the Cercopithecoidea. Harrison (1986) clarified the confusion about affinities by: confirming its family status (Oreopithecidae); showing that it was a side-line of the Hominoidea (neither a palaeo-monkey nor a proto-person); and supplying it with an early-to-mid Miocene East African ancestor, *Nyanzapithecus.* The most recent finding is that *Oreopithecus* had adaptations for bipedality (Rook *et al.*, 1999).

HOMINIDAE

Dryopithecines—a Paraphyletic Muddle

Along with the two extant groups of hominids (humans and apes), a third group has sometimes been recognized for all of the Miocene hominids: the Dryopithecinae. As this group almost certainly contains the ancestors of most, if not all, extant hominoids, the taxon (as defined) is clearly paraphyletic and has now been disbanded. Its components have been placed in three families: Proconsulidae (genera *Proconsul, Rangwapithecus, Limnopithecus,* and perhaps *Dendropithecus* and *Micropithecus,* all of them from the early Miocene of East Africa); Dryopithecidae (*Dryopithecus* only); and Hominidae (with *Sivapithecus* and *Gigantopithecus* placed in the Ponginae). The term 'dryopithecine' thus indicates merely a grade of hominoid evolution; although it is useful to continue to use it with this understanding, and such a classification would also omit a number of important early genera, including *Afropithecus, Heliopithecus, Morotopithecus* and *Kamoyapithecus.*

The earliest known 'dryopithecine' is *Kamoyapithecus,* from the latest Oligocene deposits in Kenya (Leakey *et al.,* 1995b); it may already have proconsulid affinities (Cameron, 1998). The next oldest come from early Miocene deposits in East Africa (Simons and Pilbeam, 1965; Pilbeam, 1969; Pickford, 1982). More than ten species may be represented in these faunas, varying in morphology from the superficially gibbon-like *Limnopithecus legetet,* to the chimpanzee-sized species of *Proconsul.* These early species occur in association with a suite of characteristically forest-dwelling vertebrates. Although some species appear to show incipient specializations toward folivory, most retain the generalized dentition of an omnivore or generalist frugivore/folivore.

Postcranial remains are known for relatively few species, with *Proconsul heseloni* being the most intensively studied. While some authors consider this species to have been a brachiator, others favour its interpretation as a quadruped similar in locomotor pattern to larger New World monkeys (Figure 36). The first genus to show definite traces of the postcranial features of the

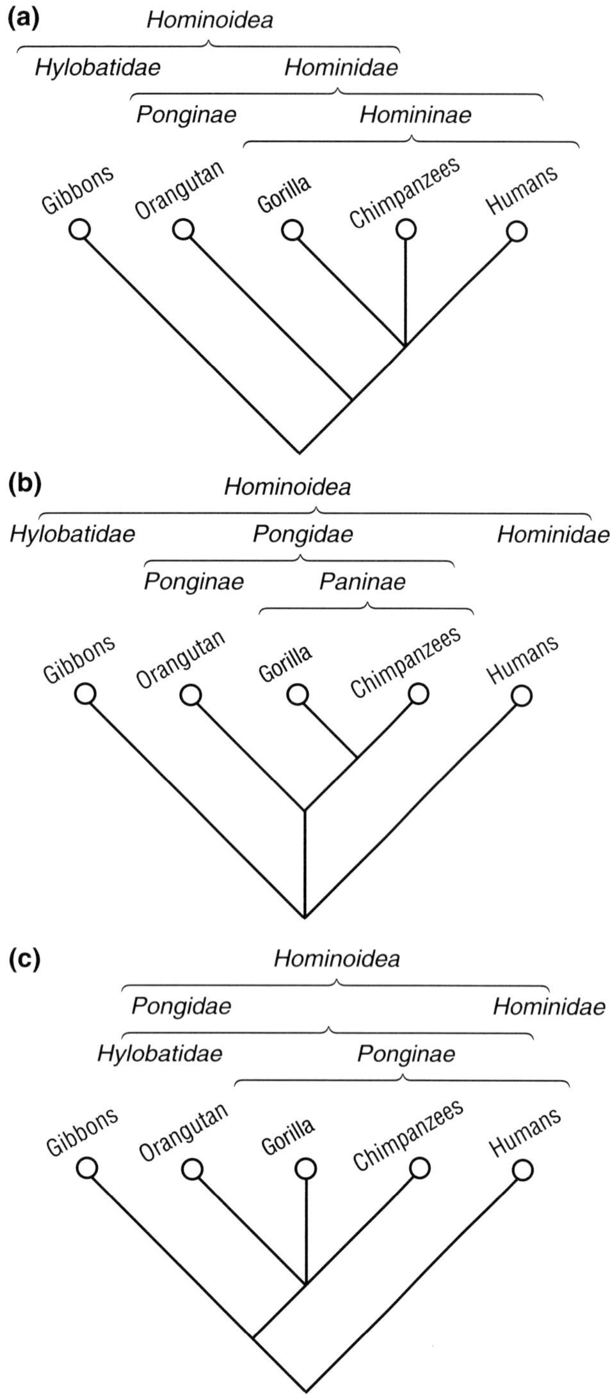

Figure 34. Alternative sister-group hypotheses involving living hominoids: **(a)** the interpretation supported by the molecular and most of the morphological data; **(b)** a less common earlier notion that gibbons, humans and pongids were equally distinct groups; **(c)** the most common view in the older literature, that humans and the apes are each other's sister-group. The author supports the first hypothesis. The data now indicate that humans and chimpanzees are each other's closest relatives and form a monophyletic group within the cluster formerly regarded as the great apes. From Patterson (1982).

million years old) were considered a range extension of *Ramapithecus* but, on the basis of more recent studies of more extensive material (Pickford, 1982), these remains have been assigned to *Kenyapithecus*. According to Andrews (1985), it corresponds very closely to the reconstructed morphotype of the Hominidae: that is, it could be the common ancestor of ourselves and the great apes. Andrews' argument is convincing and it is placed as a plesion in the Hominidae, but subsequent to *Morotopithecus*. R. Wu (1981) reported on specimens he assigned to *Ramapithecus,* as well as *Sivapithecus,* from Yunnan. These are either the two sexes, or different sized species, of *Lufengpithecus*, a genus allied to *Sivapithecus*.

Great Apes

The orangutan (genus *Pongo*) is today confined to the islands of Sumatra and Borneo. Pleistocene aged fossils from as far north as southern China, and to the south in Java, indicate a formerly much wider distribution for the genus.

In some features of dental and craniofacial morphology, *Pongo* is remarkably similar to certain species of the mid Miocene genus *Sivapithecus* (Figures 39, 40). Indeed, the similarity is such that most authors now consider the orangutan to represent the sole survivor of the once prolific Asian and eastern European *Sivapithecus-Gigantopithecus* radiation.

The African great apes *Pan troglodytes* (Figure 32(d)), *P. paniscus*, *Gorilla gorilla* and *G. beringei* (Figure 32(c)) are presently confined to areas of tropical rainforest and adjacent forest-savanna mosaic in equatorial Africa. These species are unknown from other than modern contexts. Although several authors have suggested a possible relationship between *Pan* spp. and the early Miocene *Proconsul africanus,* biomolecular evidence would suggest that this species is too early to be involved in the ancestry of any particular modern hominine lineage. The lack of a significant fossil record for either group of great apes presumably reflects the poor chances for fossilization in tropical forest environments.

Hominini

Among living primates, humans are very closely related to the African great apes. In fact, the relationship appears to be so tight that, were these matters decided on molecular evidence alone, taxonomists would probably have long since combined *Pan* and *Homo* into a single genus. That we do not adopt this arrangement reflects the fact that, despite our biochemical similarity to the apes, we as humans show some distinctive anatomical specializations that set us well apart from the other living hominids. These are particularly pronounced within the morphology of the cranium, pelvic girdle and foot as well as in external features. Evidently these aspects of our total genotype have been turning over at well above the 'average' mammalian rate, presumably because of the operation of some fairly powerful selective pressures. At any rate, we now classify ourselves in the same

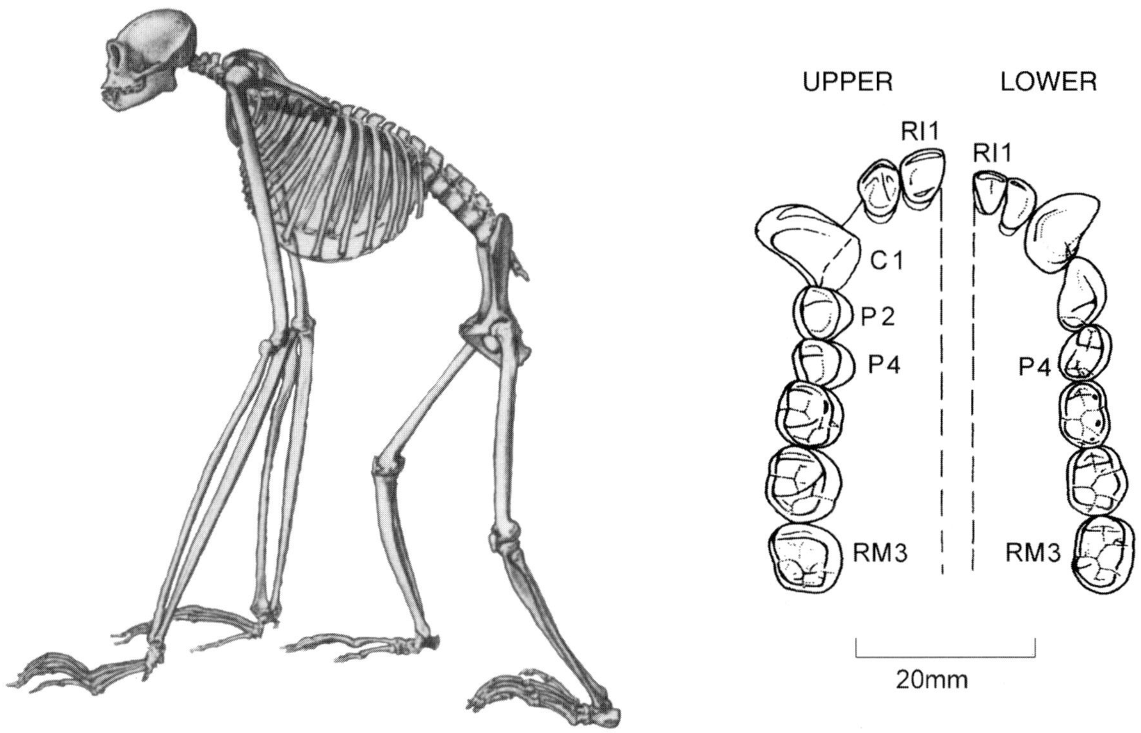

Figure 35. The skeleton and dentition of the living Asian hylobatid hominoid *Hylobates lar*. This gibbon has an essentially plesiomorphic hominoid skull and dentition but a highly specialized postcranial skeleton. Not surprisingly, they are well-known arboreal acrobats and efficient brachiators (travelling through trees swinging by the arms). Modified from Blainville (1839) and Szalay and Delson (1979).

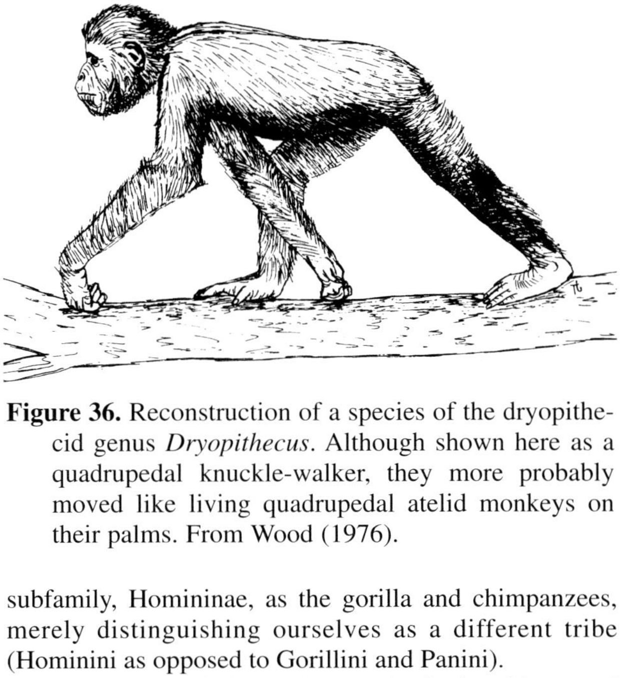

Figure 36. Reconstruction of a species of the dryopithecid genus *Dryopithecus*. Although shown here as a quadrupedal knuckle-walker, they more probably moved like living quadrupedal atelid monkeys on their palms. From Wood (1976).

subfamily, Homininae, as the gorilla and chimpanzees, merely distinguishing ourselves as a different tribe (Hominini as opposed to Gorillini and Panini).

Before examining palaeontological evidence of human evolution, let's briefly consider some of the more important processes and selective forces that may have been instrumental in the development of our own little branch of the hominid family tree.

FIRST STEPS, NEOTENY, VENTRAL VAGINAS AND BABY BRAINS

The reader could well be excused for wondering what sort of recipe this is. In fact it is some of the major processes and ingredients of which adult *Homo sapiens* are composed and which serve to mark the rapid evolutionary departure from ancestral hominines over the last five to six million years. For a detailed summary see Gould (1977).

Judging from the palaeontological evidence, the first of these to occur was the habitual adoption of an upright, bipedal gait (many apes occasionally use this gait but only Hominini do so habitually). This transition is a complex evolutionary achievement involving extensive modification of muscles and bone. Gould (1977) suggests that this was the real breakthrough for early hominines. As shall be detailed below, we know from the three to four million-year-old Laetoli trackways that probable human ancestors walked upright by that time. The same conclusion has been reached from study of the 3.2 million-year-old skeletons of the hominine dubbed *Australopithecus afarensis*. The studies of Stern and Sussman (1983) suggest that *A. afarensis* females and males were both capable of terrestrial bipedalism, although the females were also well adapted for arboreal activity; but Latimer (1991) argues that *A. afarensis* was a committed biped and that the supposed 'arboreal' features were insignificant retentions.

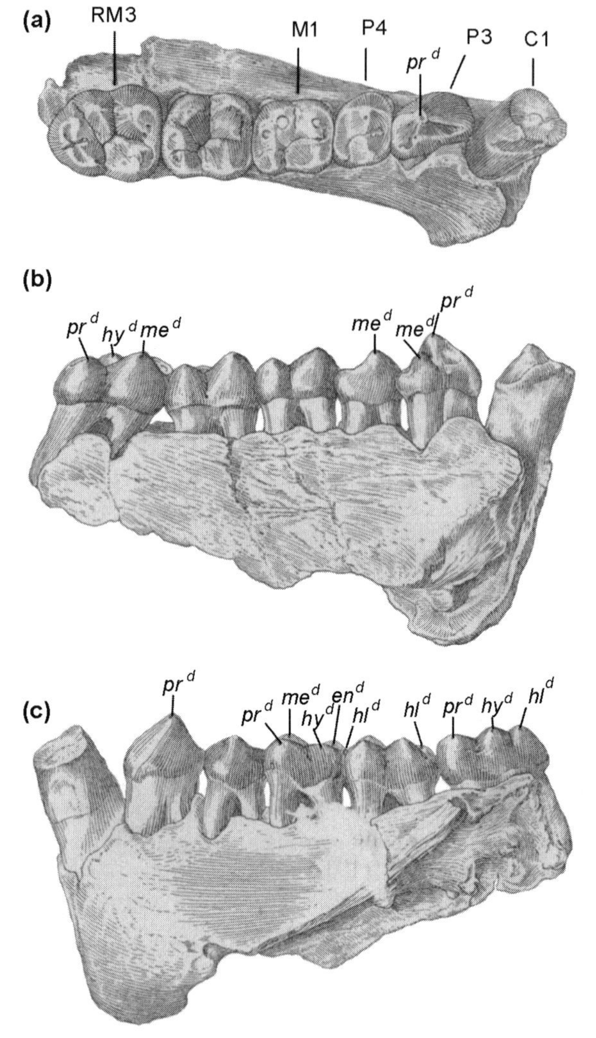

Figure 37. A left dentary fragment of the Miocene Asian pongine hominid *Sivapithecus sivalensis*. Species of this genus have thick enamel on the molar crown surfaces, a feature shared with species of *Australopithecus* and *Gigantopithecus*, indicating adaptation to feeding on hard objects such as seeds. They were probably open country primates and may have been ancestral to the pongines. Modified from Brown *et al.* (1924).

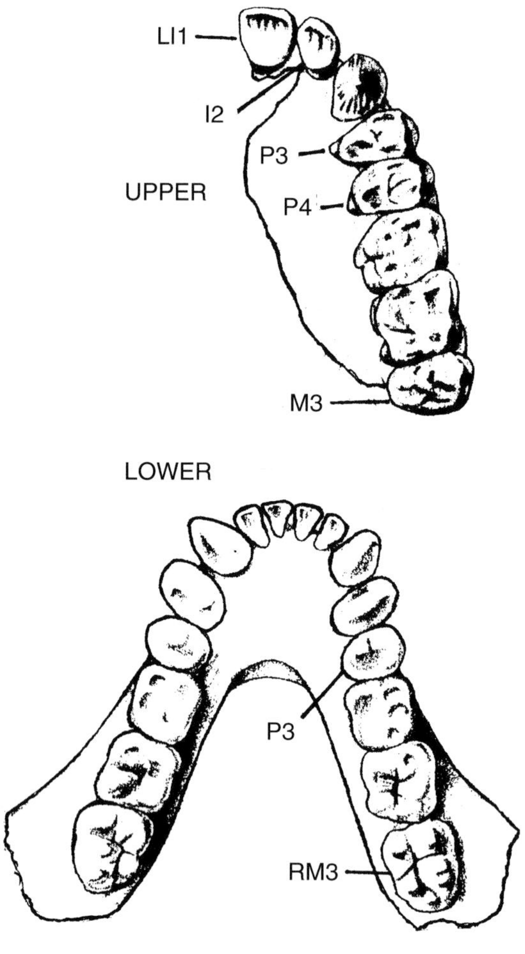

Figure 38. The teeth of a species of '*Ramapithecus*'. When first found, the fragments of dental arcades (above) were reconstructed as parabola-shaped, like those of *Homo sapiens* and the species of *Australopithecus*, rather than U-shaped like other hominids. Later, more complete material revealed the dental arcade to be V-shaped, a condition that could nevertheless have been plesiomorphic to the U-shape or the parabola shape. Modified from Zihlman and Lowenstein (1979).

Accepting the molecular clock date for the separation of the early bipeds from their non-bipedal ancestors, it appears that these modifications must have occurred in little more than two million years. Considering the morphological changes involved, this is a rather remarkable achievement.

Several attempts have been made to explain why the shift to habitual bipedalism occurred. Initially, workers such as Washburn (1978) placed the emphasis on freeing of the hands, perhaps for technological purposes. Others envisaged similar mechanisms but focused instead on seed-eating adaptations among early hominines, the hands in this case being freed for complex manipulative tasks. More recently, Lovejoy (1981) presented an elaborate hypothesis in which he identified the shift to a bipedal gait as one consequence of a complex set of behavioural innovations among proto-hominins. In essence, Lovejoy views the adoption of a bipedal gait as a means by which larger quantities of foodstuffs could be collected during foraging expeditions from a centrally located base camp.

The second process involved neoteny (a type of paedomorphosis) and retarded development, and was in all likelihood simpler to achieve than the bipedal gait but, in our case, it led to far more significant results. Neoteny is the retention of juvenile features among adults (Figures 41–44). And we are simply full of them. For example: the retention of juvenile growth rates for the brain (Figure 43); the short face (Figures 41, 42), with the protruding chin, the result of the failure of the face to

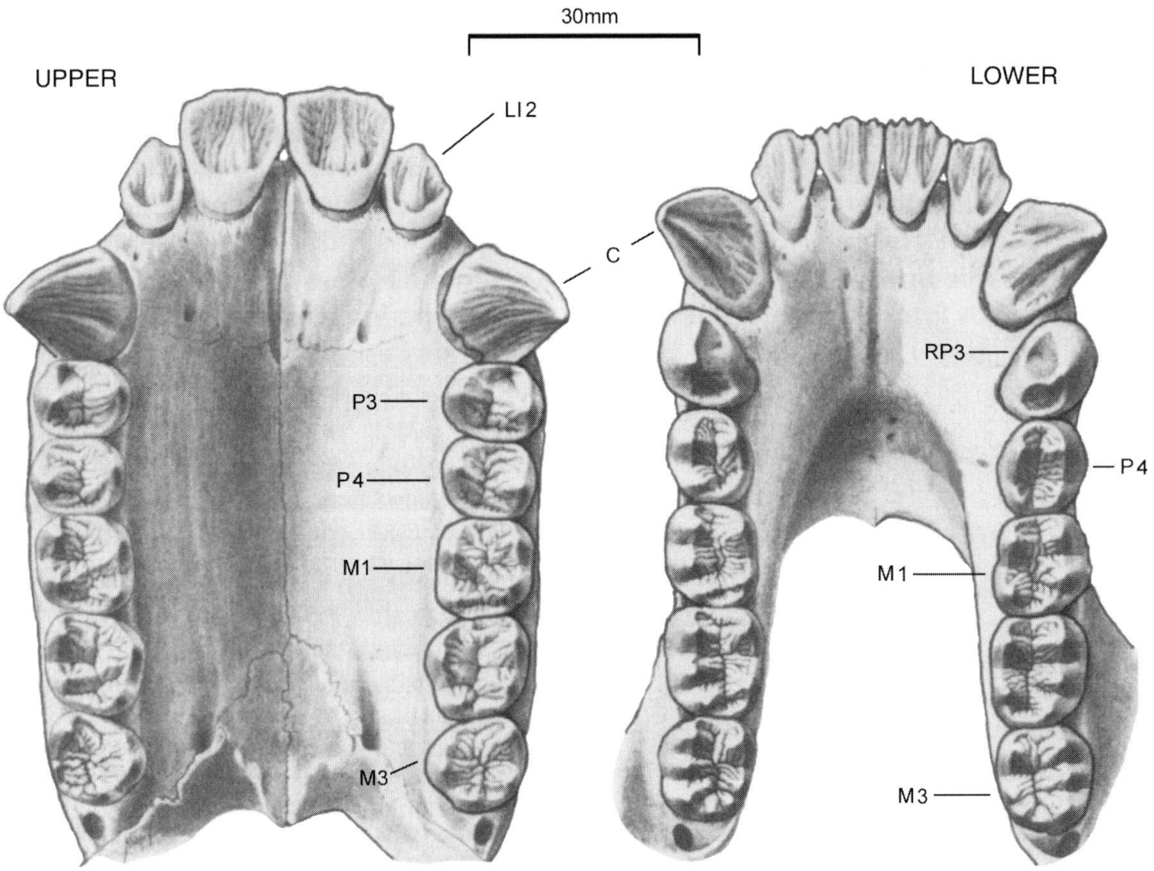

Figure 39. The dentition of a living Asian pongine, the Bornean orangutan (*Pongo pygmaeus*). Modified from Selenka (1899).

develop anteriorly and large, globular cranial vault; poorly developed brow ridges; undeveloped cranial crests; thin skull bones; small jaws (and consequent reliance on a horizontal segmentive biting weapon using incisors and small canines); the sparse body hair (Figure 44); ventral (rather than posteriorly facing) vagina; labia majora; big toe of the hind foot close to the other toes; long period of infant dependency; the long childhoods; and so on. In hominoids except humans, these features are well developed in the juveniles but lost in the adults.

The neotenic process itself is a surprisingly simple one, requiring little more than an alteration in developmental rates. Specifically, humans appear to spend a greater portion of their lives growing at juvenile rates (Figure 43). In part this would help explain our disproportionately (for our size) long lives. Primates as a whole are slow developers among mammals, and hominines especially so among primates. Humans have accentuated this to such a degree that we are in effect juveniles for almost a third of our long lives.

But why all this neoteny and retarded development? Gould (1977) believes that it evolved hand in hand with the origins of culture, the principal attraction being the provision of longer and longer periods of childhood for the assimilation of increasing loads of cultural information. As Gould notes, we are pre-eminently learning animals. If it were possible to pass on, through cultural education, the increasing loads of accumulated knowledge to each successive generation, rather than expecting each to acquire it all empirically, the species would presumably have a higher fitness. And, as all of us soon come to realize, children find it a great deal easier to assimilate information than do adults. Hence, this may be the rationalization for the selective pressure to achieve a longer period of brain growth at the juvenile rate: a prolonging of the learning process for as long as possible. As accidental byproducts of delaying the acquisition of adulthood, we have also acquired most of the neotenic features that make us distinct among primates.

FOSSIL HOMININES: FIRST FOSSIL PERSONS AND OTHER THINGS TO FIGHT ABOUT

THE 'ONLY-JUST' HOMININE

In a neat vindication of Sarich and Wilson's late date for separation of the human and chimpanzee lineages, White *et al.* (1994) discovered a very primitive hominine in 4.4 million-year-old deposits at Aramis, Ethiopia. At first they called it *Australopithecus ramidus*, but later

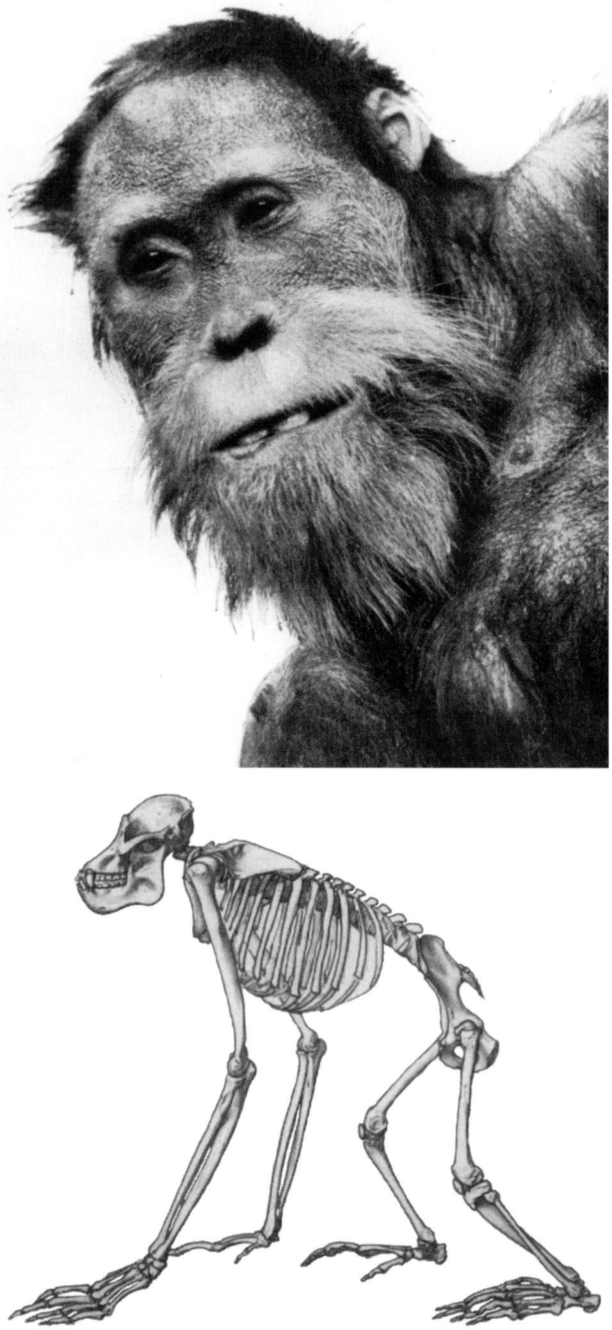

Figure 40. Although food for thought, this visage is actually of a young Sumatran orangutan (*Pongo abelii*); the skeleton is that of an adult orangutan. Modified from Blainville (1839) and Sanderson (1957).

(White *et al.*, 1995) they erected a new genus for it, *Ardipithecus*. It had larger canines than *Australopithecus*; the lower first deciduous molar was narrow, but not quite chimpanzee-like; there was no articular eminence to the jaw joint; the canine and molar enamel was thin; and the first lower premolar was partly sectorial. Despite this, its forwardly-positioned foramen magnum and some postcranial features suggest that it may already have been bipedal.

Very recently, two new candidates for 'earliest hominin' have been described: 5 million-year-old *Ardipithecus ramidus kadabba* (Haile-Selassie, 2001) and 6 million-year-old *Orrorin tugenensis* (Senut *et al.*, 2001); but both need much more evidence before the claims can be treated with confidence. The position of *Orrorin* as a basal hominin has recently been more fully argued by Pickford *et al.* (2002).

THE NEXT OLDEST HOMININE

The best known early hominine is *Australopithecus afarensis* from Laetoli (3.5 or more million years old) in Tanzania and from Hadar (3.0 to 3.3 million years old) in the Afar Triangle of Ethiopia (Figures 45–47). This species is known from a number of fairly complete fossil skeletons, including the now famous 'Lucy'; as a result it is considerably better known than even many Pleistocene taxa (Johanson and Edey, 1981).

Not surprisingly, given its great antiquity, *A. afarensis* appears to be essentially generalized and chimpanzee-like in many features of craniodental morphology. By contrast, the postcranial skeleton, although similar in size to that of a pigmy chimpanzee, is remarkably human in appearance. This is particularly striking from the waist down, leaving little doubt that *A. afarensis* was a biped (Latimer, 1991). This interpretation was dramatically confirmed through the discoveries made by Mary Leakey at Laetoli (Leakey, 1979). Here careful excavation of an ancient ashfall horizon revealed a magnificent series of criss-crossing animal trackways, among which were found several meandering trails made by a fully bipedal hominine. The age: an astonishing three to four million years. And the maker: possibly *A. afarensis* whose remains are also present at this truly remarkable site.

Of course, the *A. afarensis* concept is not without its controversial aspects. Chief among these is: do the remains all represent one species? Cladistic analysis by Groves (1989) suggested that as many as three species may be involved: one at Laetoli (the true *A. afarensis*) and two (a large and a small one) at Hadar. The small Hadar species (whose remains include 'Lucy') would be the sister-group of all other Hominini (hominins), while the large one may even be *Homo;* only the Laetoli specimen would actually belong to *Australopithecus.* More recent discoveries at Kanapoi and Allia Bay may support this idea in part: the fully bipedal species found there, *Australopithecus anamensis* (4 million years old), is said to be distinctly closer to the Laetoli sample of *A. afarensis* than to the Hadar sample (Leakey *et al.*, 1995a). Also rather more like the Laetoli fossils is the 3.0–3.5 million-year-old *Australopithecus bahrelghazali* from Chad, the first australopithecine outside East or South Africa (Brunet *et al.*, 1996).

Although the status of *A. afarensis* (if it is only one species) as the earliest really well known hominin appears secure, there are some significant differences in opinion concerning its relationships. Johanson *et al.* (1978), the

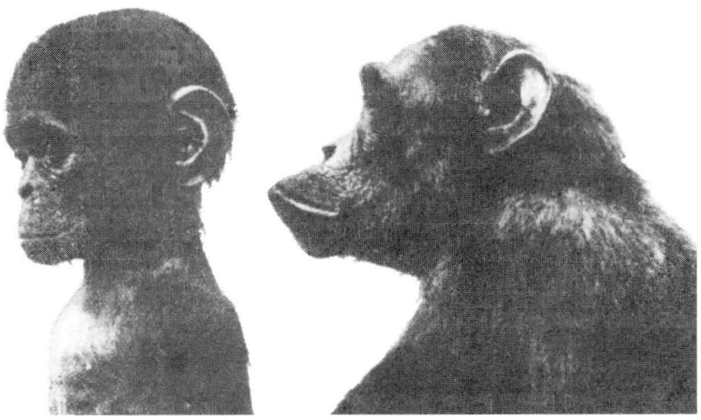

Figure 41. Most of the morphological features that distinguish us as a species from the other living primates may be the result of neoteny—a simple readjustment of the rates of development so that a species may retain, as a sexually mature adult, juvenile features of its ancestors. In contrast to the adults of non-human primates (e.g. the chimpanzee shown here), the juveniles are less hairy, have a relatively high brain weight, poorly developed brow ridges, undeveloped cranial crests, thin skull bones, flat faces, ventrally pointing vaginas, rounded bulbous crania, small jaws and teeth, and ventrally positioned foramina magna (exits from the skulls for the brain stem). All of these juvenile features are lost or profoundly changed in non-human primates as they become adults. Humans, on the other hand, retain these features as sexually mature adults. It is as if humans were primates whose general bodily developmental clock has been slowed down and overrun by relatively rapid sexual development. From Gould (1977).

describers of the species, consider it to be sufficiently generalized to have given rise to species of both *Australopithecus* and *Homo* (Figures 48, 49). Rather than erect a new generic concept for this species alone, they assign *A. afarensis* to the more plesiomorphic genus *Australopithecus* on the basis of phenetic similarity. The Leakeys (Mary and Richard), like Olson (1981) and Senut (1996), consider *afarensis* to be an amalgam of two species, one an *africanus*-like australopithecine and the other an early species of *Homo* (Leakey, 1981). Ferguson (1983) also favours the 'two species' hypothesis but identifies one of the two forms as a species of the dryopithecine genus *Sivapithecus*. Johanson and White (1979), as mentioned, attribute this variability to sexual dimorphism and associated allometric differences within a single species of hominin, and Leakey *et al.* (1995a) emphasize that the Laetoli (earlier) sample is more primitive than the later (Hadar) sample. Whatever its ultimate taxonomic status, at least some of the *afarensis* material clearly pertains to a very primitive hominin and a good structural link between the great apes and ourselves.

The successors of *A. afarensis* in Africa are both numerous and varied. For a selection of viewpoints see Robinson and Steudel (1973), Leakey (1973a, 1976), Howell (1978), Walker and Leakey (1978) or Johanson and White (1979). A major feature of hominin evolution during the late Pliocene and early Pleistocene is the presence of two contemporary lineages, one comprising the species of so-called 'australopithecines' and the other, the species of *Homo*.

AUSTRALOPITHECINES

The australopithecine species (Figures 47–51) are the more widely represented and better known of these two Pliocene hominine groups, but questions of phylogeny and classification remain highly controversial. While this is partly a function of poor age-control for many *Australopithecus*-bearing faunas (particularly South African ones), it also derives from uncertainty about the extent of size-related allometry among these early hominines. In general terms, however, most authorities agree about the existence through much of the late Pliocene of two, quite distinct, australopithecine lineages, and recently a third has been added.

The first of these, generally grouped under the name *A. africanus* (Figures 47–50), were small-bodied hominins with relatively unspecialized dentitions. Undoubted material of this species (or species group) is presently known only from southern Africa (Figure 49). Judging from its dental proportions, *A. africanus* was most likely an omnivore that included small game and carrion in its diet. Although cultural materials (stone artefacts) occur together with the remains of *A. africanus* in some South African localities, most authorities consider these to be the work of a species of *Homo,* possibly *H. habilis* (see below).

Members of the second group were substantially larger and more robust and the evidence is now good that they actually comprise a genus, *Paranthropus,* which is the sister-group of *Homo. Paranthropus* is represented by at least two species, with different forms represented in southern Africa (*P. robustus;* Figures 47–49) and East Africa (*P. boisei;* Figures 49, 50). Both species had greatly enlarged molars suitable for heavy grinding, reduced anterior teeth and massively developed masticatory musculature. These features indicated a predominantly herbivorous or possibly graminivorous diet. Interestingly, a very recently described new species, *Australopithecus garhi* from 2.5 million-year-old deposits at Bouri, Ethiopia, has equally large teeth, but is certainly not *Paranthropus* (Asfaw *et al.*, 1999).

There have been other interpretations of variation in Pliocene australopithecines. Pilbeam and Gould (1974) postulated that robust and gracile lineages could be different morphological expressions of a single species—with differences in craniodental proportions due to allometric rather than phylogenetic factors. But this hypothesis fails to account for many apparently non-allometrically related features, particularly of the basicranial region, that serve to distinguish the various 'australopithecine' species (Olson, 1981; Dean and Wood, 1982).

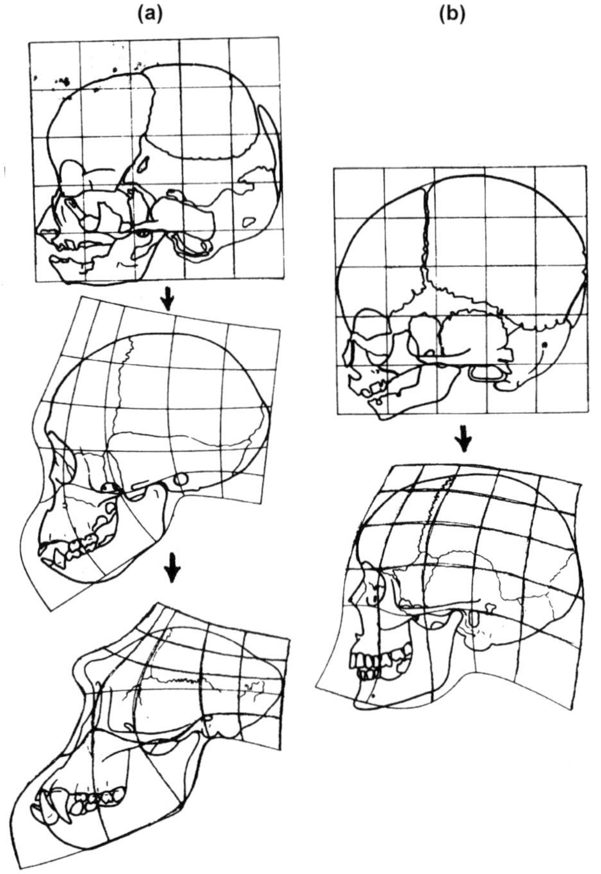

Figure 42. The ontogenetic development of a chimpanzee **(a)** and human **(b)** skull. Both start from similar foetal conditions and end with the very different adult conditions. The distortion of the grid in each series reflects the area in which the greatest ontogenetic change takes place. Note that the human undergoes much less transformation in its development and that in its adult form it more closely resembles the juvenile chimpanzee skull (middle left). Modified from Gould (1977).

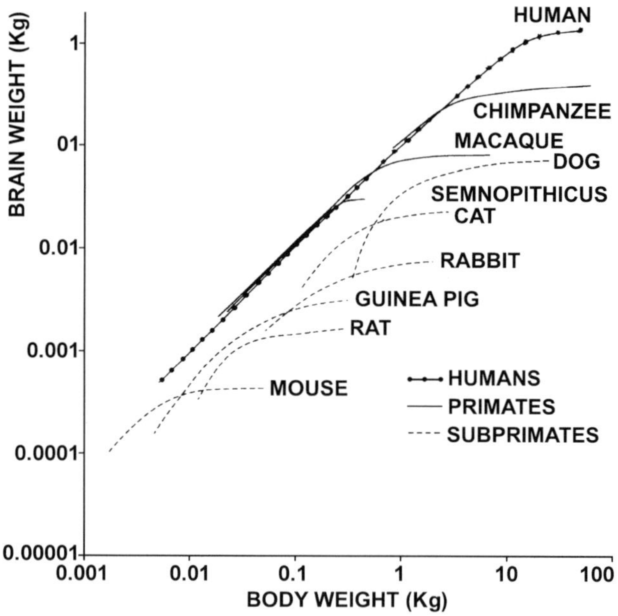

Figure 43. Ontogenetic brain–body weight curves for a number of mammals. Note that the developmental rate of increase in brain size is greatest when each animal is young and levels off as it reaches adulthood. It is probable that the enormous (relative to other primates) size of the adult human brain is in large part the result of a prolonged period of brain growth at the more rapid juvenile rate. In some primates the brain may increase by 60% in size after birth. In humans it more than quadruples in size after birth. The delay in maturation of the human brain, a probable correlate of the neotenic process, may have enabled longer periods of peak learning capacity. Modified from Gould (1977).

Furthermore, as pointed out by Wood and Stack (1980), several of the observed differences in dental proportions are precisely the opposite of allometric trends exhibited elsewhere within the Hominoidea. The newly described third genus, *Kenyanthropus* (Leakey *et al.*, 2001) is a smaller-toothed, flat-faced hominin from northern Kenya; but recently it was pointed out that the skull on which *Kenyanthropus* was described was so distorted that it may be simply 'an early variant of *A. afarensis*' (White, 2003).

The australopithecines do not appear to have undergone any significant directional evolution. Although there are few well-dated australopithecine fossils, the discovery of a fine specimen, the 'Black Skull' from West Turkana in Kenya (W.T. 17000), showed that earlier *Paranthropus* were truly more primitive than later ones and that there was some evolution within this genus. There is no evidence that australopithecines ever left their African homeland to explore the greater world.

EVOLUTION OF THE GENUS *HOMO*

Habilines

The earliest securely-dated specimens of *Homo* are a 2.3–2.5 million-year-old jaw from Urahla, in Malawi (Bromage *et al.*, 1995) and a 2.3 million-year-old maxilla from Hadar (Kimbel *et al.*, 1996). Much better known, and more complete is the now famous 1470 skull from East Turkana in Kenya (Leakey, 1973b; Figure 52). Dating around 1.8 million years, this specimen has been referred to as *Homo rudolfensis*, and it may even be descended not from *Australopithecus* but from *Kenyanthropus*. But few doubt the *Australopithecus* ancestry of the first undoubted *Homo*, *H. habilis*.

Early *Homo* differed from their australopithecine contemporaries in having a significantly expanded neurocranium, a smaller and less protrusive face and relatively smaller cheek teeth combined with an unreduced anterior dentition. Postcranial remains attributed to the so-called 'habilines' are remarkably modern in appearance and suggest a more advanced locomotor pattern than postulated for either *Australopithecus afarensis* or the later australopithecines.

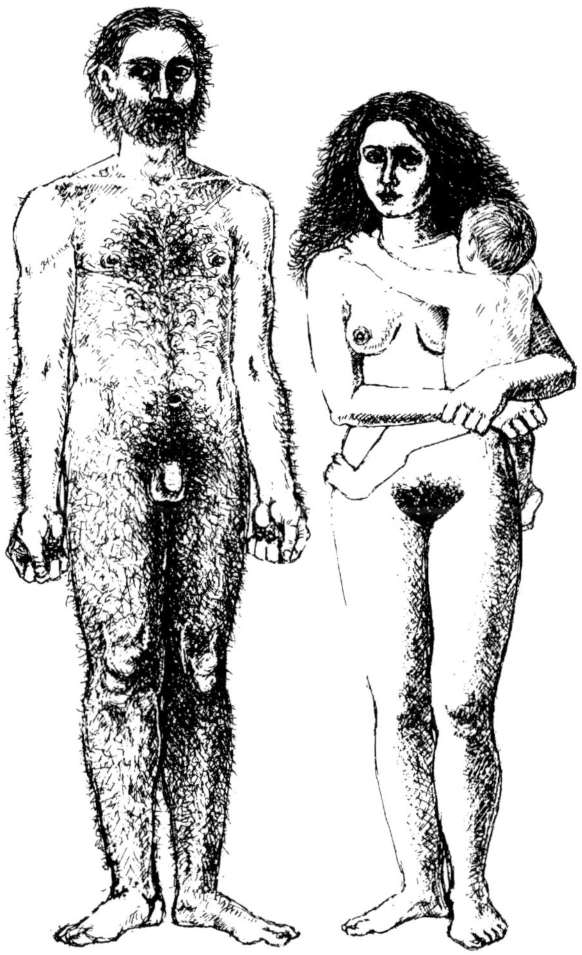

Figure 44. Humans are most conspicuously different from other primates because of their relative loss of body hair. The neotenic process may frustrate the normal mass of primate body hair from developing. However, as we age (albeit slowly relative to other primates), we do get steadily more hirsute (on all areas except the dome). From Archer (1982).

The earliest occurrences of stone artefacts in both the Omo River and Afar Triangle regions are dated to around 2.5 million years and may have been the handiwork of slightly earlier habilines. These early industries are exceedingly rudimentary, although in some instances transport of raw materials over considerable distances can be demonstrated, thus suggesting a certain degree of forward-planning in these actions. The Turkana material is coeval with a smattering of other habilines from various East African and possibly some southern African localities. Importantly, stone artefact industries accompany these remains; they show little variation between sites and throughout the duration of the late Pliocene.

Erectines

With the onset of the Pleistocene a significantly more advanced grade of *Homo* made its appearance, first in East Africa, then in various localities around the world, as the habilines began to disappear. Generally grouped as the 'erectines' (Figure 53), these early to mid Pleistocene hominines are distinguished from the earlier *habilis* by an increase in body and brain size. Probably as a consequence of their larger size, they are also on average more robust than earlier forms of *Homo,* particularly within the orofacial complex. The oldest securely dated fossils of the group come from the Lake Turkana region at about 2.0 million years old, and are referred to the species *Homo ergaster*; but there are fragments of a very small species of proto-human almost as early in China, at Longgupo in Sichuan (Huang *et al.,* 1995). On the other hand, it has been argued that the Longgupo fragment is actually from an orangutan-like form (Etler, 2000; Wu, 2000).

Hominines of this grade were apparently the first of their lineage to have wandered extensively outside Africa, presumably leaving the homeland during the earliest Pleistocene (Howells, 1980). The earliest uncontroversial finds of *Homo* outside Africa come from Dmanisi, in Georgia, dated at 1.7 million years (Gabunia *et al.*, 2000); these are ascribed to *Homo ergaster*, and may be unrelated to later European or Asian *Homo*. Unfortunately, the detailed history of the first great exodus remains unwritten; but available evidence suggests that virtually all of Europe, the Middle East and both eastern and South-east Asia had been colonized by the middle Pleistocene (Jacob, 1976). From that time on there were repeated expansions from Africa, perhaps to fill niches left vacant by local extinctions—caused by factors such as glaciation. Only the Americas and, of course, the Meganesian land mass appear to have remained beyond the ingenuity and endurance of these highly adaptable and successful explorers.

In cultural terms as well as morphologically, erectines appear to have been worlds ahead of their habiline ancestors; in fact, Wood and Collard (1999) suggest that the genus *Homo* should be restricted to this grade and not include the habilines at all. This is reflected most abundantly in the nature of their stone artefact industries, many of which include highly formalized implement types that are recognizable across broad geographic areas. Other cultural achievements assigned to this period include the mastery of fire, the cooking of food, construction of complex dwellings, the advent of big game hunting and perhaps the first artistic doodlings. This degree of cultural elaboration also implies some form of communication, perhaps even complex language.

Within their greatly expanded geographic range, these advanced populations underwent considerable regional differentiation, leading to the development of two distinctive species, each with its own subspecies: *Homo erectus* in China and Indonesia and what is, in fact, an ancestor of *Homo sapiens* in Africa and Europe, but not now classified with either of them. These forms have maintained their identities over considerable periods of time, perhaps indicating restricted levels of interregional gene flow through much of the Pleistocene. First, a species only known from Gran Dolina (Spain), nearly

Figure 45. The best known early hominin on our particular twig of the family tree, the three to four million-year-old African *Australopithecus afarensis*, which, like all later members of our family tree, walked upright. From Johanson and Edey (1981) and Archer (1982).

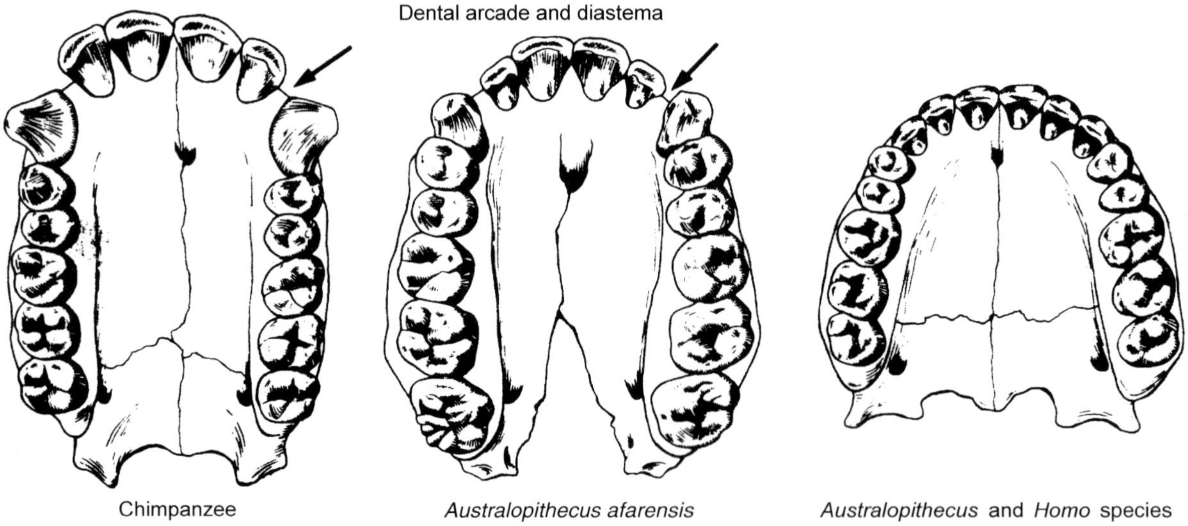

Figure 46. Palates of an ape (left), *Australopithecus afarensis* (centre) and other species of *Australopithecus* and *Homo* (right) to show the retention of a few primitive features (e.g. the diastema or space indicated by the arrow) in *A. afarensis*. From Johanson and Edey (1981).

800,000 years old, called *Homo antecessor*, is quite clearly different from *Homo erectus* (Bermúdez de Castro *et al.,* 1997). After this period, the line split in two: one line, represented by African *Homo rhodesiensis*, developed into *Homo sapiens*, the other, the European *Homo heidelbergensis*, evolved into the famous Neanderthal people, *Homo neanderthalensis*—in fact, the evolution of Neanderthals grades smoothly from primitive early *heidelbergensis* skulls like those from Petralona (Greece) and Arago (France). Meanwhile, *Homo erectus* may have survived until less than 100,000 years ago in Java.

Several workers have investigated the tempo of morphological change within *Homo* during the Pleistocene. Rightmire (1981) and Howells (1980) examined trends in cranial capacity and cheek tooth size through the course of the Pleistocene; they concluded that evolu-

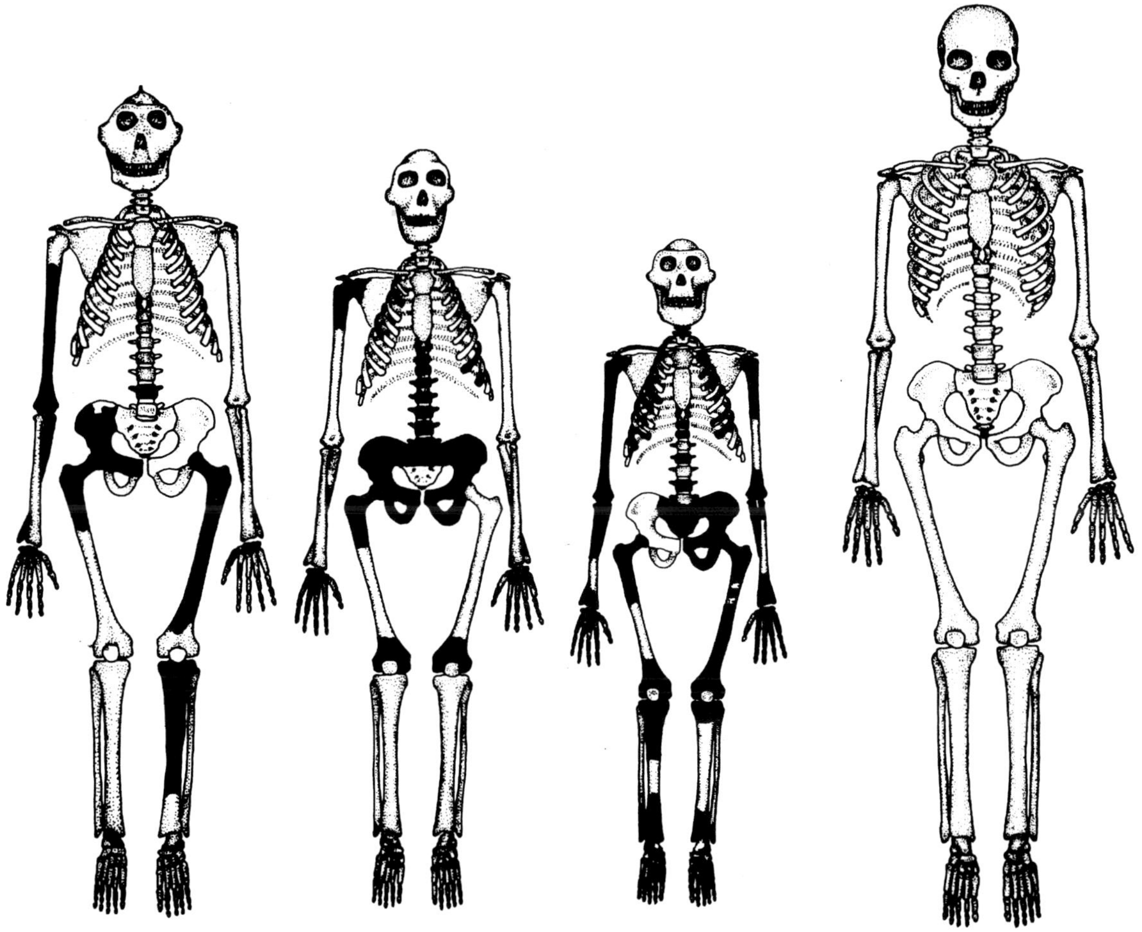

Figure 47. Skeletons of *Paranthropus*, species of *Australopithecus*, and *Homo sapiens*. The skull, dentition and skeleton of the older *A. afarensis* reveal features that indicate that this species could be structurally ancestral to the other species. Modified from Johanson and Edey (1981).

tion within the genus *Homo* followed a highly irregular pattern with little directional change evident throughout early or middle Pleistocene, followed by a distinct 'leap' between mid and late Pleistocene. Importantly, this final episode of relatively rapid morphological change appears to coincide with a period of major cultural advancement. While this apparent correlation of morphological and cultural evolution makes the non-gradualist's case an attractive one, major uncertainties in the dating of many of the key fossils leave considerable room for scepticism. Indeed Bilsborough (1976), using essentially the same materials, was able to present an equally justifiable case for gradual evolution within the *Homo* lineage. These disagreements may also be the result of failure to separate the ancestral *sapiens* from *H. erectus* fossils.

The Emergence of *Homo sapiens*

From a palaeontological viewpoint, it is difficult to separate a variety of mid to late Pleistocene *Homo* fossils into lineages and they have all sometimes been grouped as *H. heidelbergensis.* Arguing from the evidence of the time that these were most abundant in Europe, Delson (1977) postulated a European origin for *sapiens*, suggesting that this may have resulted from geographic isolation of local erectine populations in Europe during an episode of glacial advance. With later climatic amelioration, *sapiens* could have spread out into Asia and Africa, variably assimilating or replacing pre-existing erectine populations.

The view of Stringer (1984) is somewhat different: African, as well as European, middle Pleistocene fossils are to be classified as ancestral to *H. sapiens.* There is, thus, no question of *Homo erectus* evolving into *H. sapiens* as Delson (1977) postulates. In China, there appears to be a replacement, around 200,000 years ago, of *H. erectus* by *H. heidelbergensis.* According to X. Wu (1981), the Dali skull (the earliest Chinese *H. heidelbergensis*) shows signs of having evolved from Chinese *H. erectus,* whereas, according to Groves (1989), it is a descendant of European *H. heidelbergensis.* Following this theory, replacement of *erectus* by *sapiens* may have occurred somewhat later in Indonesia.

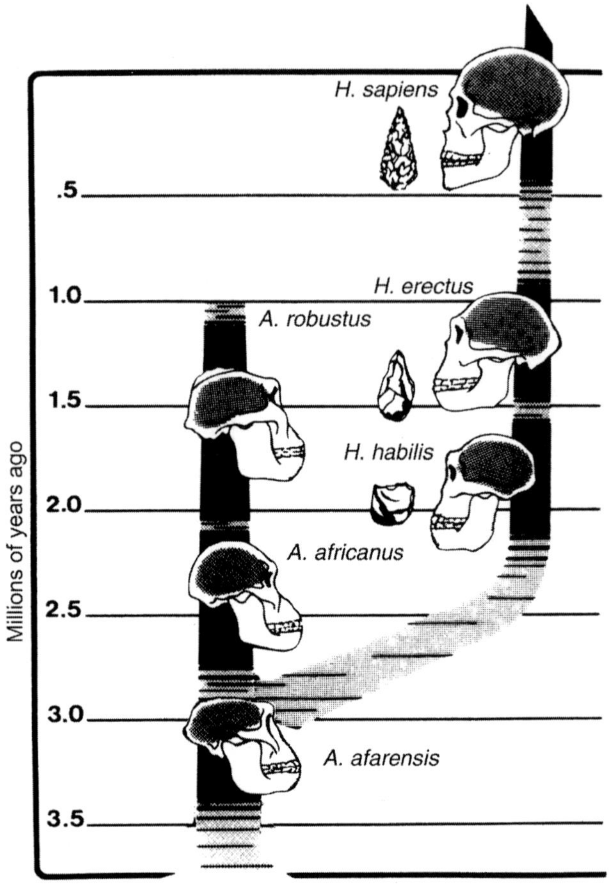

Figure 48. One view of the age and phylogenetic significance of the species of *Australopithecus* and *Homo*. The solid black parts of the column indicate the known ranges for the species. Modified from Johanson and Edey (1981).

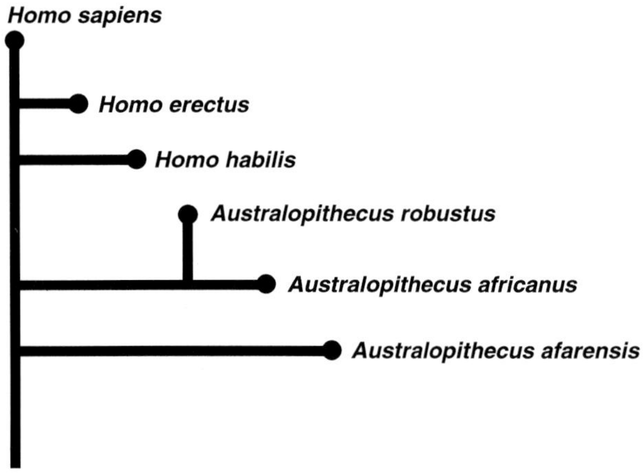

Figure 49. A cladogram involving the hominine species shown in Figure 48. From Johanson and Edey (1981).

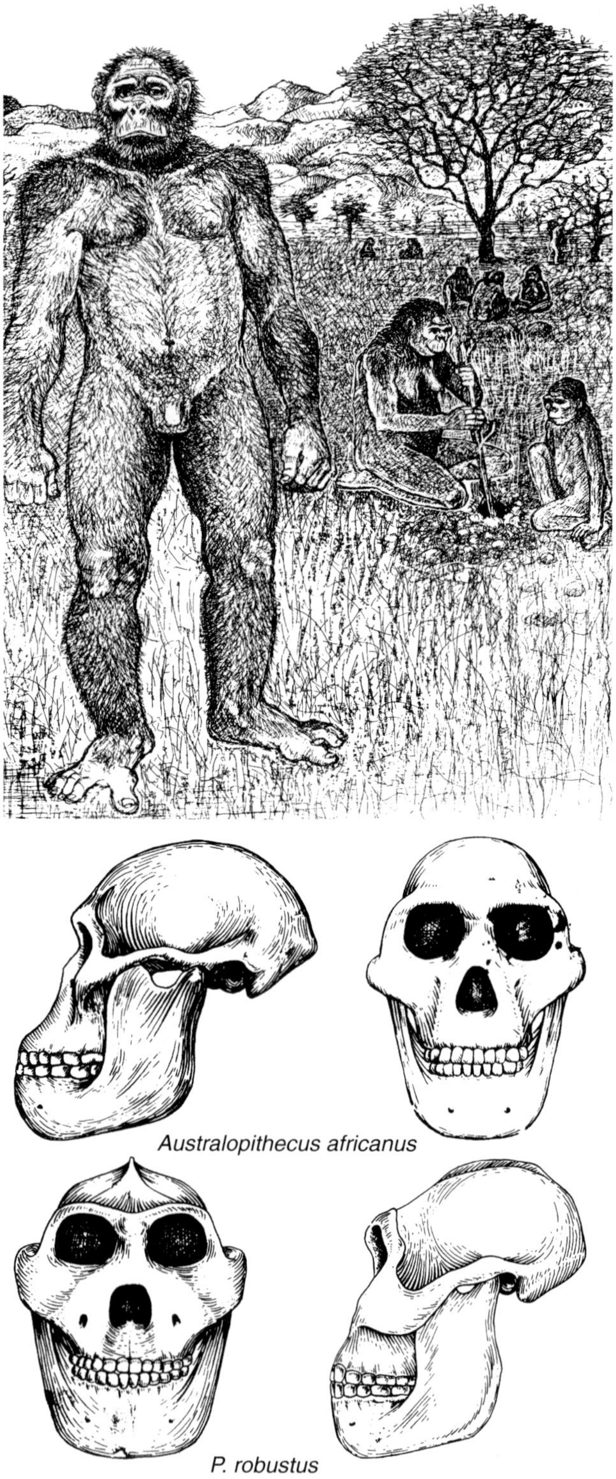

Figure 50. Reconstruction of the South African *Paranthropus robustus* from early Pleistocene, with skulls and dentaries of the African *Australopithecus africanus* and *P. robustus*. *Australopithecus africanus* is older, less specialized and more similar to *A. afarensis*. *Paranthropus robustus* is larger, had heavier teeth and a more powerful chewing musculature. From Johanson and Edey (1981) and Archer (1982).

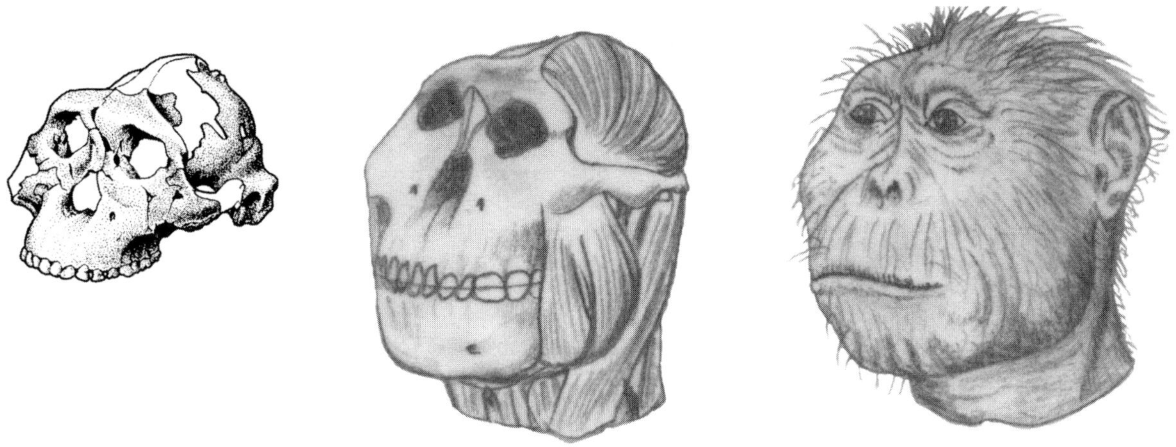

Figure 51. The acme of robusticity in an australopithecine, *Paranthropus boisei*, formerly called *Zinjanthropus boisei*. Its features are similar to those of *P. robustus* but it was even larger and the features more exaggerated. The artist-anatomist has shown two stages in the reconstruction of 'Zinj', the most robust of our close cousins. From Wood (1976).

A third view, now gaining ground, is that of Hublin (1996) that the lines leading to *Homo sapiens* and *Homo neanderthalensis* separated very early, after *H. antecessor*, and that the European fossils (so called *H. heidelbergensis*) show increasing Neanderthal features through time. This view was recently strongly vindicated by the results of a successful recovery and sequencing of mitochondrial DNA from the original Neanderthal fossil and some other examples. The Neanderthal sequence is well outside the modern range and indicates a separation of the modern and Neanderthal lineages between about 550,000 and 690,000 years ago (Krings *et al.,* 1997; Ovchinnikov *et al.*, 2000). The Neanderthals (Figure 54) were a ruggedly constructed but large-brained species that sat it out in western Europe through glacial and interglacial periods alike, living at times no more than a stone-tipped spear's throw from the mighty European ice caps (Howells, 1974; Trinkaus and Howells, 1979).

The early history of our own species (Figure 54) is now well known. Delson (1977, 1981) and others considered the earliest undoubted *Homo sapiens* to be of sub-Saharan origin but did not favour this region as the source area for the taxon—but Bräuer (1984) argues for the archaic-to-modern transition in Africa itself. The earliest specimen assignable to *Homo sapiens* is from Ethiopia and is 154,000–160,000 years old (White *et al.*, 2003); it was preceded by a series of primitive intermediates with *Homo heidelbergensis*, going back to 300,000 years ago.

Arguing chiefly from palaeontological evidence, several authors have claimed to see the presence of certain modern racial characteristics among various middle to late Pleistocene fossils. In several cases, these similarities are said to be sufficiently marked and the fossils sufficiently abundant to postulate the existence of long-standing regional 'clades', the end point in each case being one of the major living races of *Homo sapiens*. The only really well documented of these

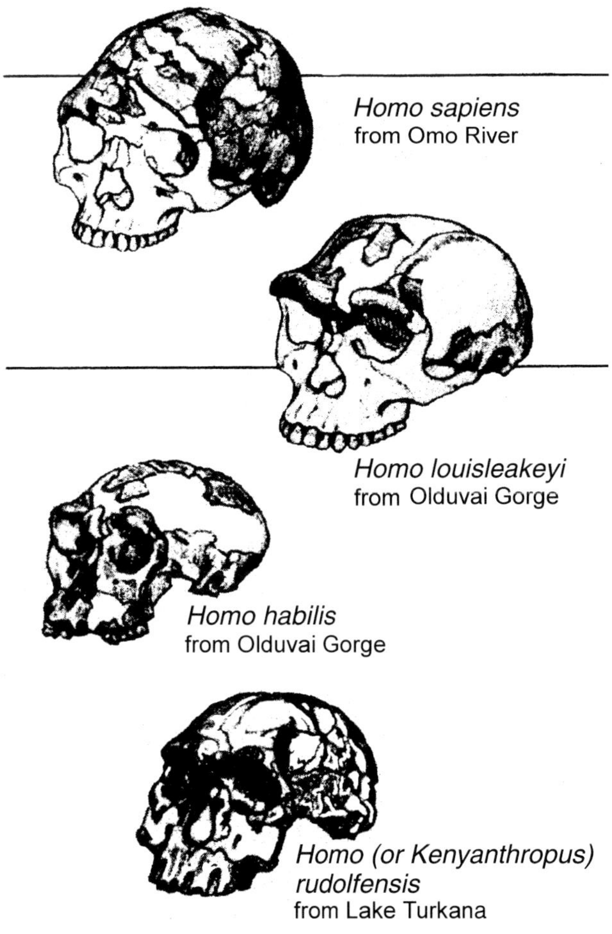

Figure 52. Representative skulls of the evolutionary stages of *Homo*. Skull 1470 is among the oldest known examples of habilines. The Omo River *H. sapiens* is among the oldest referable to our own species. Modified from Leakey (1973b).

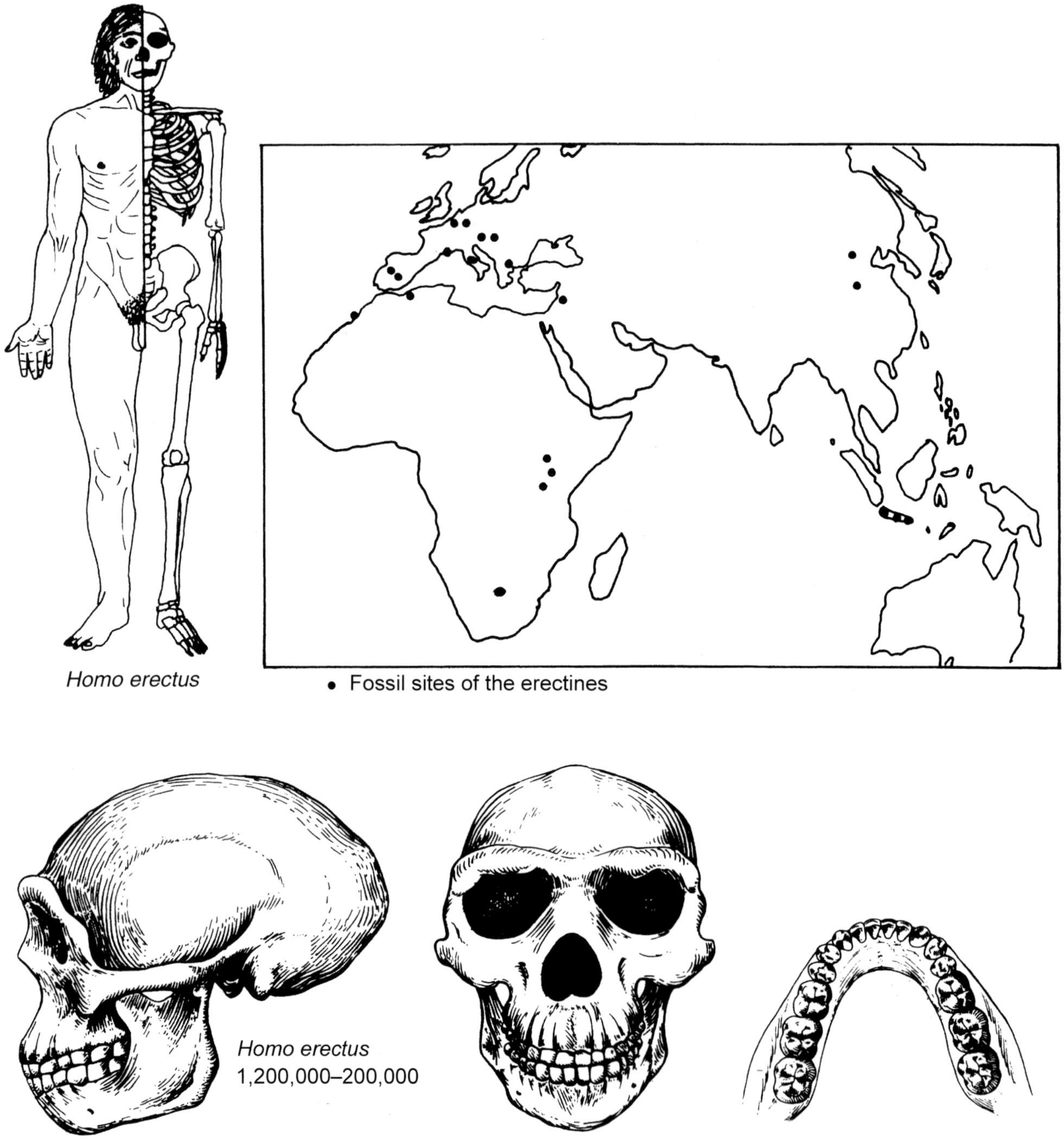

Figure 53. *Homo erectus*: skeleton and reconstruction; skull, dentary and lower dentition; and distribution of *H. erectus* and its contemporaries. From Wood (1976) and Johanson and Edey (1981).

clades is that described by Rightmire (1983) and Beaumont *et al.* (1978) for sub-Saharan Africa; but a clade postulated by Thorne and Wolpoff (1981), for Indonesian material, ending with the contemporary 'Australoids' of Australasia is still promoted by some. Thorne (1981a) and X. Wu (1981) further hinted at comparable continuity within the Chinese *Homo* record, the affinity being in this instance with modern Mongoloid populations. In both of these latter examples, the evolutionary sequences may be seen to cut across the purportive species boundary between *sapiens* and *erectus*. Both these Asian clades have been described in detail by Wolpoff *et al.* (1984).

The 'early divergence' hypothesis appears to conflict with the wealth of anatomical, genetic and reproductive data that support the essential biological unity of *Homo sapiens*. This hypothesis also appears at odds with a number of molecular studies of phyletic affinity among the living races. One such study, based on comparisons of mitochondrial DNA, dates the initial divergence of racial

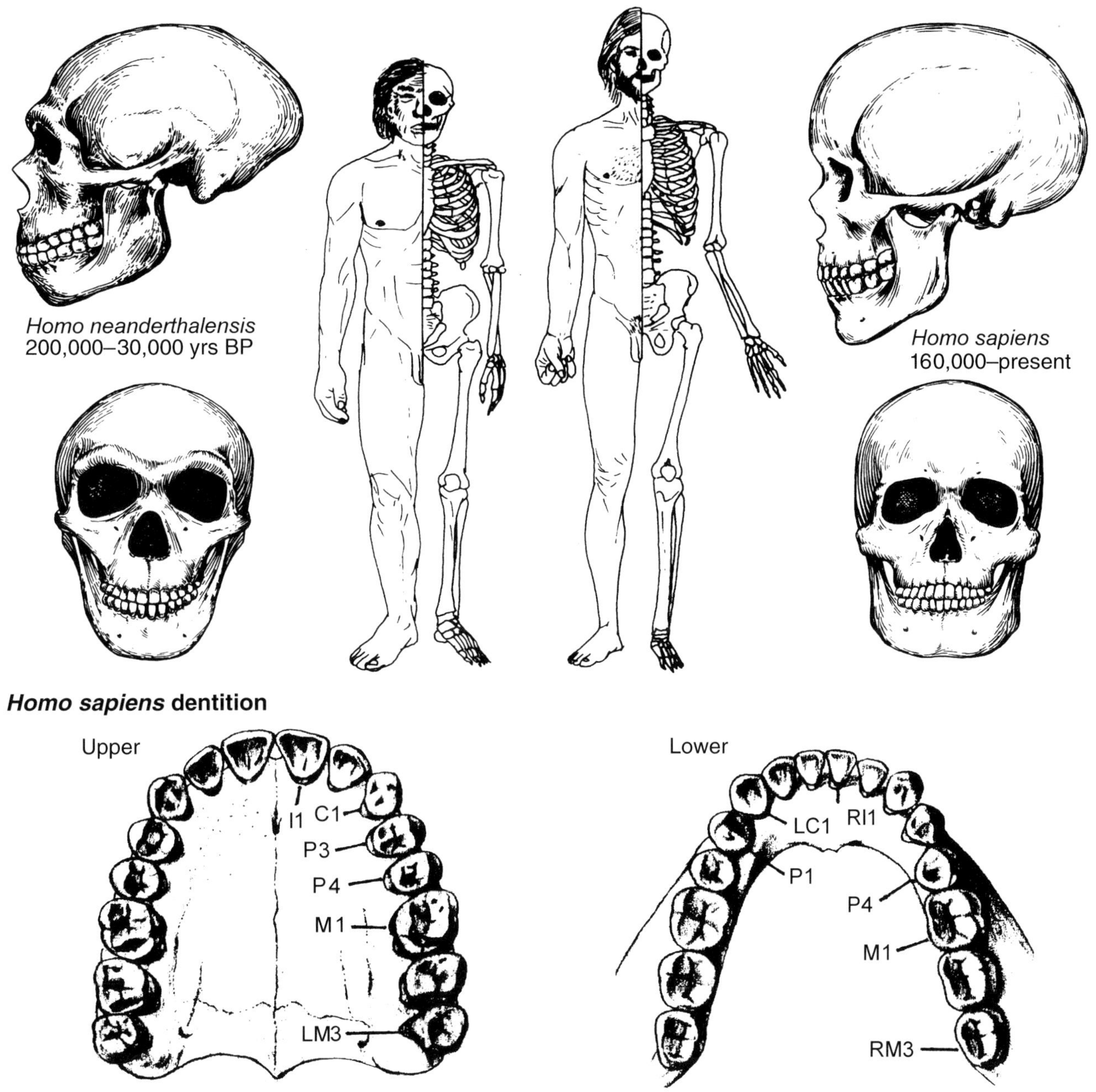

Figure 54. *Homo neanderthalensis* and *H. sapiens*: skeletons and reconstructions; skulls and dentaries; and upper and lower dentitions (of *H. sapiens*). It is unlikely that *H. sapiens* developed from *H. neanderthalensis*. It is likely that the two were regionally distinct and (in part) temporally overlapping forms. A structurally intermediate infant skeleton is known from Portugal, 25,500 years old and may be evidence that the two forms could hybridize. From Wood (1976), Zihlman and Lowenstein (1979) and Johanson and Edey (1981).

groups at around 150,000–200,000 years (Gribbin and Cherfas, 1982; Groves and Lahr, 1994), a far cry from the hundreds of thousands of years demanded by the regional clade scenario. Of course molecular evolution cannot be automatically equated with whole animal evolution. Indeed, in the strictest sense, molecular evolutionary studies provide no more than a phylogeny of molecules. Estimates of divergence times based on molecular affinity are thus of direct relevance only to the particular molecule under investigation. To return to the present example, the results of the mitochondrial DNA studies tell us only that the modern races last shared a common mitochondrial DNA structure around 250,000 years ago—the best authenticated date—but have continued to intermingle since then. Importantly, they do not tell us anything as to the reason behind this sharing of molecular structure. Certainly one possible explanation could be that the differentiation of *H. sapiens* into its modern races began around that time. However, an equally valid interpretation could be that the 250,000 year quantity represents the maximum timespan over which any one of the major racial groups have remained relatively free of genetic interchange with any other racial group.

But are these clades real? They depend on the recognition of various regional characters, that is, character states that are common to both early and modern populations within an area: to Peking Man (*Homo erectus pekinensis*) and modern mongoloids; to Java Man (*Homo erectus erectus*) and modern Australoids. The status of the features held to link ancient with recent populations has been examined by Habgood (1989) and Groves (1989), who conclude that where similarities exist, they are in primitive retentions, not derived states such as would be necessary to support an evolutionary link.

If these criticisms are valid, then *Homo erectus* would totally lose any status as the ancestor of *Homo sapiens,* Bräuer's (1984) 'Afro-European *sapiens*' model would be supported, and the mitochondrial DNA findings would revert to their definitive racial-separation status.

Finally, there is the question of whether Neanderthals contributed anything to the evolution of modern humans. Although the three examples of Neanderthal mitochondrial DNA are quite different from any modern human type (Krings *et al.,* 1997; Ovchinnikov *et al.*, 2000), there is now some evidence that there was occasional hybridization. In most of Europe, Neanderthals died out about 35,000 years ago, but in Portugal and southern Spain they survived to at least 30,000 (Hublin *et al.,* 1995), and the skeleton of what is plausibly argued to be a hybrid child was recently recovered from a site in Portugal dated to about 25,500 B.P. (Duarte *et al.,* 1999). Evidently interbreeding was a rare event, and Neanderthals have had no detectable influence on *Homo sapiens* genetics or morphology.

THE AUSTRALIAN STORY

In this final section we explore two topics relevant to the history of human populations in the Australasian Region. These are: the actual date and mode of entry of *H. sapiens* into the region; and the probable relationships of Australian Aborigines to racial groups outside Australia. Other equally fascinating topics such as the role of human activities in shaping the Australian environment lie beyond the scope of this chapter but are discussed elsewhere (see Kohen, 1995).

BIOGEOGRAPHIC BACKGROUND

Figure 55 illustrates the extent of land connections during glacial conditions involving a sea level depression of around 50 metres. Note that Australia is broadly united with both Tasmania and New Guinea into the greater land mass of Meganesia and that insular South-east Asia is similarly united into a large peninsular land mass with broad connections to mainland Asia. But even during

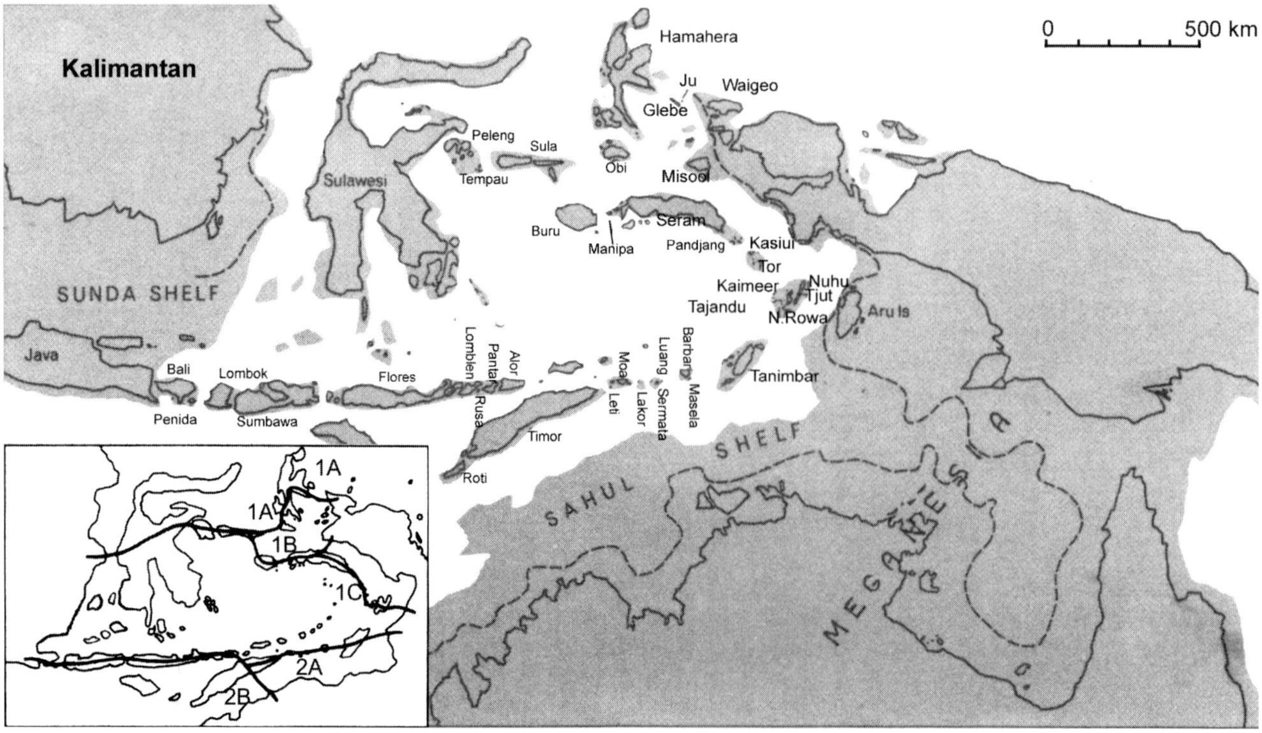

Figure 55. The area between South-east Asia and northern Australia showing (in shading) the continental shelves. The inset indicates potential dispersal routes for hominines from Sunda to Sahul. Modified from Birdsell (1977).

periods of maximal sea level depression, these enlarged land masses remained separated by a wide but island-studded expanse of deep ocean. These intervening islands and their adjoining waters are known as Wallacea—see pp.95–110 (van Oosterzee, 2006). As shown in Figure 55, these islands form a number of distinct chains, any one of which may have been used by humans for dispersal between the two regions. In all cases, however, these possible dispersal routes involve at least one open-water crossing of 65 kilometres or more (Birdsell, 1977).

The effectiveness of this deep-water passage as a barrier to dispersal between the two regions is well illustrated by the limited degree of faunistic sharing (van Oosterzee, 2006), particularly among terrestrial mammals. Indeed, for an animal the size of a human, dispersal across this zone would appear highly improbable in the absence of effective watercraft and at least rudimentary navigational skills.

Despite their many cultural achievements, pre-*sapiens* populations do not appear to have ever made the crossing into Australasia, although they appear to have got as far as Flores by mid-Pleistocene (Morwood *et al.,* 1998); this would involve at least two sea crossings of up to 10 km. Certainly there are no known pre-*sapiens* fossils from the area. But the best evidence that humans did not get to Australia until quite recently is the complete lack of any of the characteristic, middle Pleistocene stone tool industries that are found throughout South-east Asia. In general terms, there is thus little doubt that Australia remained unpopulated until after the differentiation of our own species.

THE OLDEST AUSTRALIANS AND ORIGINS

Up until about 40 years ago, most researchers were confident that Australia was unpopulated until the start of the Holocene (about 10,000 years BP). In 1964 this view was dramatically challenged by the determination of Pleistocene dates from Kennif Cave in central Queensland (Mulvaney and Joyce, 1965). Ten years later, dates of 30,000 years were available, stretching the then known limits of true *Homo sapiens* on a world-wide scale (Barbetti and Allen, 1972). Today, dates in excess of 40,000 years are a reality; convincing results come from hearths buried within an ancient depositional terrace of the Swan River near Perth. Other sites dating to greater than 30,000 years BP include the well-known middens of the Willandra and Darling Lakes systems, and the buried living surfaces at Keilor in Victoria (Bowler, 1976). Dates close to this figure are also available from Kosipe in the highlands of New Guinea (White *et al.,* 1970). Given this broad geographical spread of early sites, an initial entry date of 45,000–50,000 years BP appears reasonable, although some workers favour an earlier date (50,000–55,000 years BP) on the grounds that global sea levels dropped to an all time low over the earlier period. Water crossings would have been reduced to their narrowest extent and the probability of successful dispersals would have risen. Most recently, dates of over 60,000 years BP have been claimed for a skeleton and associated sediments from Willandra Lakes (Thorne *et al.,* 1999), although these are strongly disputed (Bowler *et al.*, 2003).

Singh *et al.* (1981) suggest an even earlier entry, in the order of 120,000 years BP, on the basis of a long pollen and carbonized-particle (CP) record from Lake George in New South Wales. This record is in fact remarkably long and complete, with a coverage extending well back into the Tertiary. It also represents a number of complete cycles of glaciation and deglaciation in south-eastern Australia.

Prior to the 120,000-year mark, these cyclic environmental changes are indicated by a regular pattern of changes within the pollen and CP curves. At the level of the last interglacial period, however, dating from 120,000 to 100,000 years ago, CP counts rise to an uncharacteristically high level. Even more surprisingly, the CP counts do not drop off during the final phase of glaciation (as they had done in all previous episodes), but instead remain uniformly high. These changes in CP abundance are mirrored by a fundamental and equally long-standing alteration of pollen spectra away from the characteristic patterns of all earlier periods. Not surprisingly, the new vegetation is not only more modern in composition but probably more fire resistant. A related finding is that in the humid tropics of north-eastern Queensland, the pollen record indicates a major change in the vegetation, from rainforest to eucalypt woodland, about 140,000 years BP.

Figure 56. Partial skull of *H. sapiens* from Lake Mungo, at about 40,000 years, is among the oldest dated human remains from Australia. It is more gracile in appearance than others of much younger age and has been interpreted by Thorne (1981a, b) as representing descendants of a Chinese delegation that came into Australia, possibly later than the ancestors of the less gracile forms. From Thorne (1981a).

Figure 57. A reconstruction of Lake Mungo about 30,000 years ago. The Aborigines who hunted and fished on the edges of this lake were similar in build to modern Aborigines. From Wood (1976).

This has also been interpreted as a consequence of Aboriginal burning (Kershaw, 1994).

Datable archaeological evidence is, at present, very much younger than Singh *et al.*'s evidence would demand, but very few deposits are known from the critical time period between 30,000 and 120,000 years, and even fewer of these have been intensively examined. Anthropological considerations likewise shed little light on this issue since both possible entry dates (60,000 and 120,000 years BP) fall well within the known temporal range of *H. sapiens* in the Old World. Recently, Kershaw (2000, unpublished data) has withdrawn his claim that the Queensland sequence is evidence of human occupation.

On one level the question of who the original Australians were has already been resolved. For, unless humans entered Australia at a very early date (more than about 250,000 years BP), there can be little doubt that the earliest colonists were fully differentiated *Homo sapiens*. On a finer level of resolution, however, we are faced with the problem of identifying the precise racial affinities and source area of Australian people within the wider regional context.

Until about 35 years ago, prior to the recovery of significant numbers of human fossils from Australia, these issues were tackled on the basis of modern racial variability and distributions (see Macintosh, 1963; Birdsell, 1967). Debate raged on how many waves of people had reached Australia and on their precise areas of origin within South-east Asia. But virtually all hypotheses considered the Australasian peoples to be derived from a once widespread 'Negritoid' population that was later overrun in most Asian areas by Mongoloid peoples. Australian Aborigines were regarded as a relict race that had changed little since the time of colonization. The fossil record at that time was very patchy and poorly dated. Although a few possibly ancient

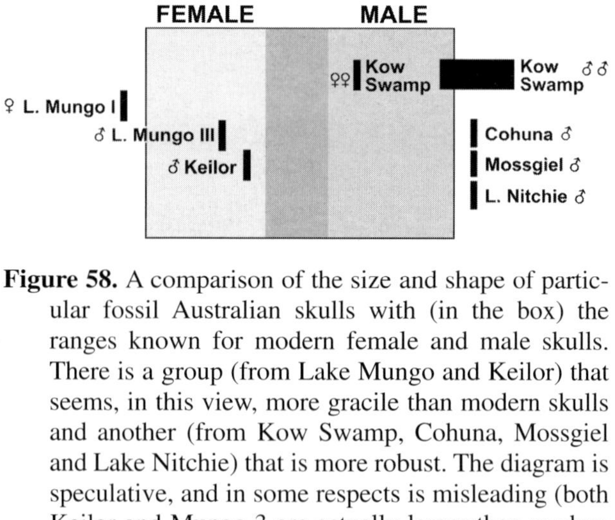

Figure 58. A comparison of the size and shape of particular fossil Australian skulls with (in the box) the ranges known for modern female and male skulls. There is a group (from Lake Mungo and Keilor) that seems, in this view, more gracile than modern skulls and another (from Kow Swamp, Cohuna, Mossgiel and Lake Nitchie) that is more robust. The diagram is speculative, and in some respects is misleading (both Keilor and Mungo 3 are actually larger than modern skulls, although gracile). From Thorne (1977).

skulls (from Talgai, Queensland), with distinct features, were known (Macintosh, 1969; Macintosh and Larnach, 1972) they were largely ignored.

Variation in Australian Populations

Since the late 1960s and early 70s the number of well-provenanced human fossils from Australia has multiplied many fold as a result of important discoveries at a number of sites. By far the most important of these were made at Lake Mungo (one of the Willandra Lakes) and at Kow Swamp, both located in the Riverina District of south-eastern Australia. From Lake Mungo, Thorne and his colleagues recovered a number of skeletons first thought to be 30,000 years BP (Bowler *et al.*, 1970, 1972; Thorne, 1976; Figures 56, 57), but now redated at 63,000 BP (Thorne *et al.*, 1999) or some 40,000 years BP (Bowler *et al.*, 2003). These individuals were essentially modern in appearance. Indeed, both male and female Mungo specimens were remarkably gracile (Figure 58).

The second series, coming from Kow Swamp, presented a much larger population of individuals (Thorne and Macumber, 1972; Thorne, 1976; Figure 59). In contrast to the Mungo specimens, this population was characterized by extreme robusticity (Figure 58), particularly within the orofacial region (Thorne, 1976). Some (although not all) specimens also displayed a peculiar flattening of the forehead. Thorne identified this and other characteristics of the Kow Swamp individuals as archaic features and an indication of the retention of more *erectus* characteristics than were present in contemporary Aborigines or in the Mungo individuals (Thorne and Wilson, 1977; Thorne and Wolpoff, 1981).

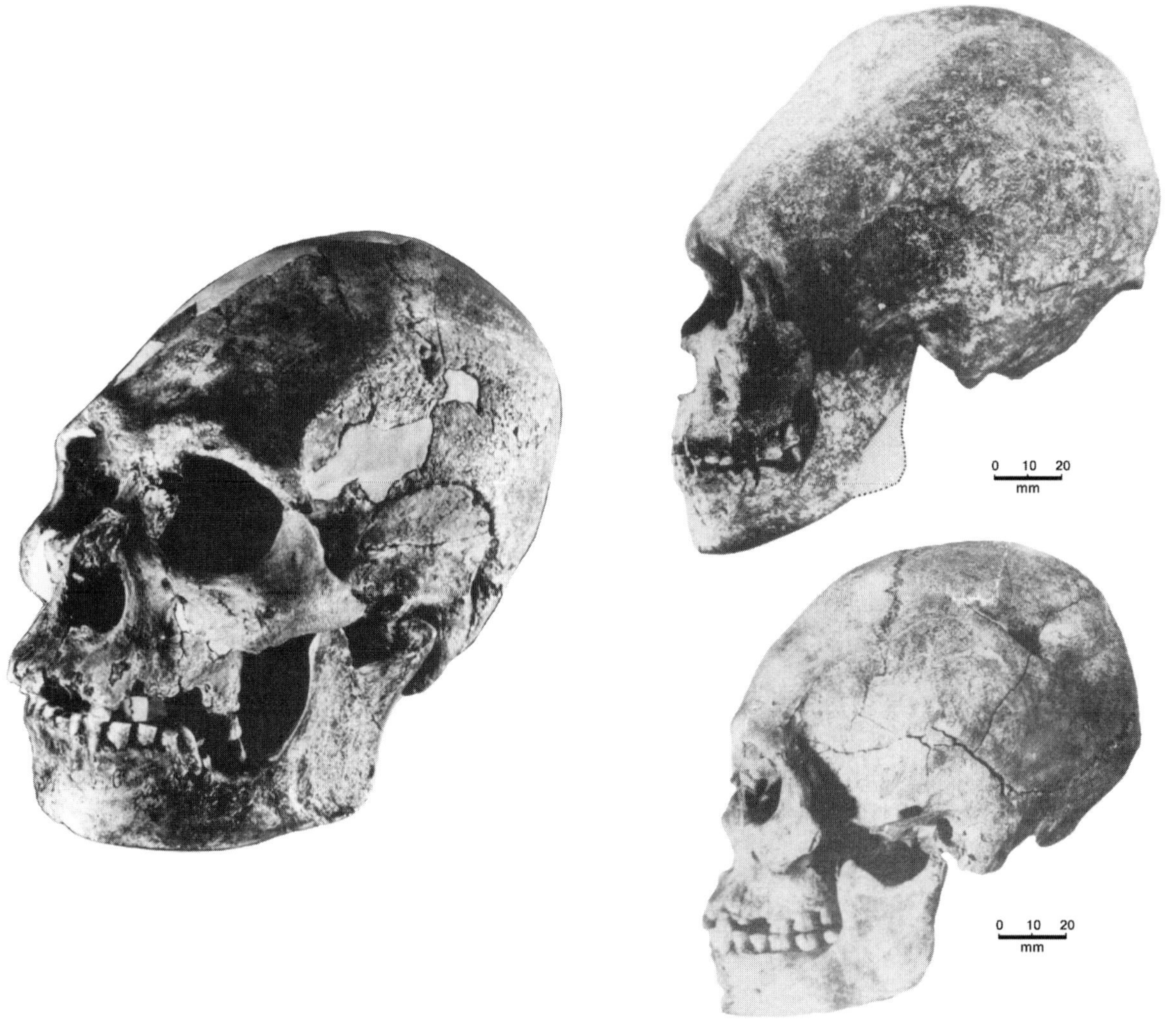

Figure 59. The more robust Australian form. Side views of male (above) and female (below) skulls from Coobool, New South Wales. Oblique view of a skull from Kow Swamp, Victoria. These people lived until 9,000–13,000 years ago and may have overlapped (or succeeded) more gracile types of the sort known from the older deposits at Lake Mungo. Robust features of the skulls shown include large and thickly-enamelled teeth and large jaws. The receding brows exhibited by many of these more robust skulls are now thought to be the result of deliberate head-deformation. From Brown (1981).

Radiocarbon dating of the Kow Swamp site dates the 'archaic' series at 13,000–9,000 years of age and therefore considerably younger than the morphologically more 'modern' Mungo specimens. To link the fossils and contemporary Aborigines as a single racial lineage required that populations had not only gone from a fully modern to a more archaic state, but that they had then reverted within a relatively short space of time. This scenario was clearly not acceptable (non-parsimonious and suggested that Australian evolution of *Homo sapiens* ran counter to accepted trends elsewhere).

Faced with this dilemma, Thorne (1977, 1981a) developed a more elaborate hypothesis in which he postulated two distinct waves of immigration into Australia. On purely evolutionary grounds he favoured an initial entry by the more archaic race, presumably sometime well prior to 30,000 years BP. The second, more modern population could then have arrived anytime after that date. Subsequently, and presumably sometime after the start of the Holocene, these populations may have merged to produce the contemporary Aboriginal morphology.

Perhaps the greatest difficulty with this hypothesis is the requirement that racial distinctiveness between Mungo and Kow Swamp populations be maintained over a very long period of time, at least 20,000 years. Moreover, a simple explanation based on possible physical isolation does not appear probable given the close geographic proximity of the two sites (although the evidence does not indicate whether the two populations overlapped temporally in the Murray Basin).

Several authors have expressed dissatisfaction with the 'two population' hypothesis. Wright (1976), for example, considered that selective forces operating on tooth size during the late Pleistocene may have been sufficient to produce the large-toothed, robustly-built morphology seen in the Kow Swamp population. Similarly, a relaxation of selective pressures, perhaps as a result of technological advancements, might have been sufficient cause for the rapid change from robust to modern morphologies at the close of the Pleistocene. Macintosh and Larnach (1976) questioned the very nature of the evidence for separate races. Instead they regarded the Mungo and Kow Swamp series to represent opposite ends of a morphological continuum observable within contemporary Aboriginal crania. At the Willandra lakes, where the now large number of human remains include both robust and gracile types, analysis by sex shows that, in the main, the 'robust' are male and 'gracile' are female (Pardoe, 1991).

Brown (1981) added a new dimension to this debate with the finding that a number of the Kow Swamp and other Pleistocene crania had been subjected in life to artificial cranial deformation. Various kinds of deformation are possible but one of the most widespread involves compression of the cranium during early life between tightly bound boards. The skull may also be shaped by hand pressure alone, or by tight binding with cloths in the absence of boards; in the Australian example, Brown suggests that a variety of techniques may have been used. But, whatever the method, characteristic results include a flattening of the forehead and a distortion of the occipital region, thus producing certain of the distinctive features of the Kow Swamp and other similar crania. More recently Antón and Weinstein (1999) have confirmed that the 'robust' Australian crania show all the characteristics of artificial deformation.

If these possibly artificially produced features are subtracted, we are left with a series of large-toothed, large-boned generally robustly constructed crania. If robusticity is to be equated with primitiveness, then the Kow Swamp

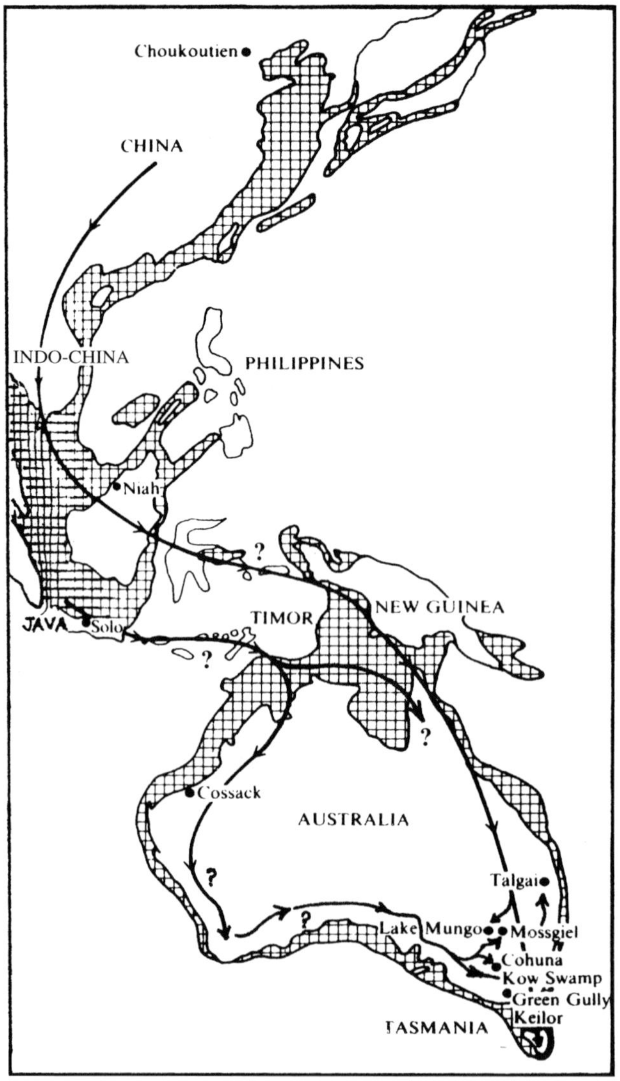

Figure 60. Recent discovery of a possibly 6,000-year-old robust-type human skull from near Cossack in Western Australia has led to the suggestion that the robust types were continentally distributed. These robust types may have originally entered Australia via Java and Timor as opposed to the more gracile types that may have entered via Kalimantan and New Guinea. Modified from Freedman and Lofgren (1979).

and other similar fossils are certainly primitive. However, as previously noted in relation to australopithecines, robusticity may also develop as a side-effect of size increase alone. The orofacial skeleton, for example, appears particularly sensitive to changes in tooth size and to the masticatory musculature required to drive it. At the most fundamenta1 level, much of the Kow Swamp morphology might be explained as the result of big teeth and big bones. And as Brown (1987) pointed out, most supposedly 'gracile' fossils, such as Mungo 3 and Keilor, have thick bones and large teeth exactly as did Kow Swamp.

Thus, large teeth and robust crania would appear to be a fairly constant feature of Pleistocene Aboriginal crania. Some, such as the undated but presumably Pleistocene Coobol Crossing series described by Brown (1981), include many individuals that are simply massive, with enormous teeth and ruggedly constructed facial regions. Another very large skull has recently turned up from a dune deposit near Cossack, north-western Western Australia. In this case the specimen may be no more than 6,000 years old (Freedman and Lofgren, 1979; Figure 60).

Clearly factors of size and robusticity have fluctuated through time and in all likelihood followed regional trends as well. If any generalization is possible, it might be that Aboriginal crania were on average larger and more robust through much of the late Pleistocene than is normal today, and this in fact seems to have been a world-wide trend. In some areas, this pattern appears to extend into the Holocene; in others a fully modern appearance appears to have been attained by the close of the Pleistocene.

Studies by Habgood (1985) and Brown (1987) have found that the distinction between 'robust' and 'gracile' types in Australia is much less convincing than it at one time appeared, and may not exist at all. A reconsideration of the whole question is overdue and there is little doubt that the Australian hominine story is shaping up as one of the world's most fascinating chapters in the evolution of modern hominids. If new fossil materials and evolutionary insights continue to appear at the rate of the last decade or two, who can predict where our understanding may stand in another ten years?

REFERENCES

Andrews, P. 1985. Family group systematics and evolution among catarrhine primates. Pp.14–22. In Delson, E. (ed). *Ancestors, the Hard Evidence: Proceedings of the Symposium at the American Museum of Natural History April 6–10, 1984, to Mark the Opening of the Exhibition'Ancestors four million years of humanity'*. A.R. Liss, New York. 366pp.

Andrews, P. and Groves, C. 1975. Gibbons and brachiation. *Gibbon and Siamang,* Vol. 4, pp.167–218.

Antón, S. and Weinstein, K.J. 1999. Artificial cranial deformation and fossil Australians revisited. *Journal of Human Evolution*, Vol. 36, pp.195–209.

Archer, M. 1982. Monkeying with the theory of evolution. *Australian Natural History,* Vol. 20, pp.364–372.

Archer, M. and Hand, S.J. 2006. The Australian marsupial radiation. Pp.575–646. In Merrick, J.R., Archer, M., Hickey, G.M. and Lee, M.S.Y. (eds). *Evolution and Biogeography of Australasian Vertebrates.* Auscipub Pty. Ltd., Sydney. 942pp.

Asfaw, B., White, T., Lovejoy, O., Latimer, B., Simpson, S. and Suwa, G. 1999. *Australopithecus garhi*: a new species of early hominid from Ethiopia. *Science*, Vol. 284, pp.629–635.

Barbetti, M. and Allen, H. 1972. Prehistoric man at Lake Mungo, by 32,000 years B.P. *Nature,* Vol. 240, pp.46–48.

Bauchop, T. and Martucci, W. 1968. Ruminant-like digestion of the Langur monkey. *Science,* Vol. 161, pp.698–699.

Beard, K.C. 1990a. Do we need the newly proposed order Proprimates? *Journal of Human Evolution*, Vol. 19, pp.817–820.

Beard, K.C. 1990b. Gliding behaviour and palaeoecology of the alleged primate family Paromomyidae (Mammalia, Dermoptera). *Nature*, Vol. 345, pp.340–341.

Beard, K.C. 1993. Phylogenetic systematics of the Primatomorpha, with special reference to Dermoptera. In Szalay, F.S., Novacek, M.J. and McKenna, M.C. (eds). *Mammal Phylogeny.* Springer Verlag, New York. 249pp.

Beard, K.C., Dagosto, M., Gebo, D.L. and Godinot, M. 1988. Interrelationships among primate higher taxa. *Nature*, Vol. 331, pp.712–714.

Beard, K.C., Krishtalka, L. and Stuck, R.K. 1991. First skulls of the early Eocene primate *Shoshonius cooperi* and the anthropoid–tarsier dichotomy. *Nature*, Vol. 349, pp.64–67.

Beard, K.C., Qi, T., Dawson, M.R., Wang, B. and Li, C. 1994. A diverse new primate fauna from middle Eocene fissure-fillings in southeastern China. *Nature*, Vol. 368, pp.604–609.

Beard, K.C., Tong, Y., Dawson, M.R., Wang, J. and Huang, X. 1996. Earliest complete dentition of an anthropoid primate from the late middle Eocene of Shanxi Province, China. *Science*, Vol. 272, pp.83–85.

Beaumont, P.B., de Villiers, H. and Vogel, J.C. 1978. Modern man in sub-Saharan Africa prior to 49,000 years B.P.: a review and evaluation, with particular reference to Border Cave. *South African Journal of Science*, Vol. 74, pp.409–419.

Bermudez de Castro, J.M., Arsuaga, J.L., Carbonell, E., Rosas, A., Martinez, I. and Mosquera, M. 1997. A hominid from the lower Pleistocene of Atapuerca, Spain: possible ancestor to Neandertals and modern humans. *Science*, Vol. 276, pp.1392–1395.

Bilsborough, A. 1976. Patterns of evolution in middle Pleistocene hominids. *Journal of Human Evolution,* Vol. 5, pp.423–439.

Birdsell, J.B. 1967. Preliminary data on the trihybrid origin of the Australian Aborigines. *Archaeology and Physical Anthropology in Oceania,* Vol. 2, pp.100–155.

Birdsell, J.B. 1977. The recalibration of a paradigm for the first peopling of greater Australia. Pp.113–167. In Allen, J., Golson, J. and Jones, R. (eds). *Sunda and Sahul: Prehistoric Studies in Southeast Asia, Melanesia and Australia*. Academic Press, London. 647pp.

Blainville, H.M.D. de. 1839. Osteographie des primates. In *Osteographie des Mammiferes. I. Primates et Secundates.* Bailliere, Paris.

Bowler, J.M. 1976. Recent developments in reconstructing late

Quaternary environments in Australia. Pp.55–77. In Kirk, R.L. and Thorne, A.G. (eds). *The Origin of the Australians*. Human Biological Series No. 6. Australian Institute of Aboriginal Studies, Canberra. 449pp.

Bowler, J.M., Johnston, H., Olley, J.M., Prescott, J.R., Roberts, R.G., Shawcross, W. and Spooner, N.A. 2003. New ages for human occupation and climatic change at Lake Mungo, Australia. *Nature*, Vol. 421, pp. 837–840.

Bowler, J.M., Jones, R.M., Allen, H. and Thorne, A.G. 1970. Pleistocene human remains from Australia: a living site and human cremation from Lake Mungo, western New South Wales. *World Archaeology.* Vol. 2, pp.39–60.

Bowler, J.M., Thorne, A.G. and Polach, H.A. 1972. Pleistocene man in Australia: age and significance of the Mungo skeleton. *Nature*, Vol. 240, pp.48–50.

Bown, T.M. and Rose, K.D. 1976. New early Tertiary primates and a reappraisal of some Plesiadapifornes. *Folia Primatologica*, Vol. 26, pp.109–138.

Braüer, G. 1984. The "Afro-European *sapiens* hypothesis" and hominid evolution in East Asia during the late middle and upper Pleistocene. *Courier Forschungsinstitut Senckenberg,* Vol. 69, pp.145–165.

Bromage, T.G., Schrenck, F. and Zonneveld, F.W. 1995. Paleoanthropology of the Malawi Rift: an early hominid mandible from the Chiwondi Beds, within Malawi. *Journal of Human Evolution*, Vol. 28, pp.71–108.

Brown, B., Gregory, W.K. and Hellman, M. 1924. On three incomplete anthropoid jaws from the Siwaliks, India. *American Museum Novitates,* No. 130, pp.1–9.

Brown, P. 1981. Artificial cranial deformation: a component in the variability in Pleistocene Australian Aboriginal crania. *Archaeology in Oceania*, Vol. 16, pp.156–167.

Brown, P. 1987. Pleistocene homogeneity and Holocene size reduction: the Australian human skeletal evidence. *Archaeology in Oceania*, Vol. 22, pp.41–67.

Brunet, M., Beauvilain, A., Coppens, Y., Heintz, E., Moutaye, A.H.E. and Pilbeam, D. 1996. *Australopithecus bahrelghugali*, une novelle espèce d'Hominidé ancion de la régón de Karo Toro (Tchad). *Comptes Rendus de l'Academie des Sciences, Paris*, Vol. 322, pp.907–913.

Bugge, J. 1974. The cephalic arterial system in insectivores, primates, rodents and lagomorphs, with special reference to the systematic classification. *Acta Anatomica,* Vol. 87, Supplement 62.

Burton, M. 1968. *University Dictionary of Mammals of the World.* Thomas Y. Crowell Co., New York.

Cameron, D.W. 1998. Anatomical variability and systematics status of the hominoids currently allocated to the African Dryopitheciinae. *Homo*, Vol. 49, pp.101–137.

Cartmill, M. 1974. *Daubentonia, Dactylopsila,* woodpeckers and klinorhynchy. Pp.655–670. In Martin, R.D., Doyle, G.A. and Walker, A.C. (eds). *Prosimian Biology: Proceedings of a Meeting of the Research Seminar in Archaeology and Related Subjects Held at the Institute of Archaeology, London University*. University of Pittsburgh Press, Pittsburgh. 983pp.

Cartmill, M. 1981. Morphology, function and evolution of the anthropoid postorbital septum. Pp.243–274. In Ciochon, R.L. and Chiarelli, A.B. (eds). *Evolutionary Biology of the New World Monkeys and Continental Drift*. Plenum Press, New York. 528pp.

Cartmill, M. and Kay, R.F. 1978. Craniodental morphology, tarsier affinities, and primate suborders. Pp.205–214. In Chivers, D.J. and Joysey, K.A. (eds). *Recent Advances in Primatology. Vol. 3. Evolution*. Academic Press, London.

Cartmill, M. and MacPhee, R.D.E. 1980. Tupaiid affinities: the evidence of the carotid arteries and cranial skeleton. Pp.205–214. In Luckett, W.P. (ed). *Comparative Biology and Evolutionary Relationships of Tree Shrews.* Plenum Press, New York.

Ciochon, R.L., Savage, D.E., Tint, T. and Maw, B. 1985. Anthropoid origins in Asia? New discovery of *Amphipithecus* from the Eocene of Burma. *Science*, Vol. 229, pp.756–759.

Clemens, W.A. 1974. *Purgatorius,* an early paromomyid primate (Mammalia). *Science*, Vol. 184, pp.903–906.

Conroy, G.C. and Bown, T.M. 1976. Anthropoid origins and differentiation: the Asian question. *Yearbook of Physical Anthropology*, Vol. 18, pp.1–6.

Dean, M.C. and Wood, B.A. 1982. Basicranial anatomy of Plio-Pleistocene hominids from East and South Africa. *American Journal of Physical Anthropology*, Vol. 59, pp.157–174.

Delson, E. 1975a. Evolutionary history of the Cercopithecidae. *Contributions to Primatology*, Vol. 5, pp.167–217.

Delson, E. 1975b. Palaeoecology and zoogeography of the Old World monkeys. Pp.37–64. In Tuttle, R. (ed). *Primate Functional Morphology and Evolution.* Mouton, The Hague. 583pp.

Delson, E. 1977. Catarrhine phylogeny and classification: principles, methods and comments. *Journal of Human Evolution*, Vol. 6, pp.433–459.

Delson, E. 1981. Paleoanthropology: Pliocene and Pleistocene human evolution. *Paleobiology*, Vol. 7,pp.298–305.

Delson, E. and Andrews, A.P. 1975. Evolution and interrelationships of the catarrhine primates. Pp.405–446. In Luckett, W.P. and Szalay, F.S. (eds). *Phylogeny of the Primates: a Multidisciplinary Approach.* Plenum Press, New York. 483pp.

Duarte, C., Maurício, J., Pettitt, P.B., Souto, P., Trinkaus, E., van der Plicht, H. and Zilhão, J. 1999. The early upper Palaeolithic human skeleton from the Abrigod Lagar Vellio (Port Royal) and modern human emergence in Iberia. *Proceedings of the National Academy of Sciences of the United States of America*, Vol. 96, pp.7604–7609.

Etler D. 2000. Implications of new fossil material attributed to Plio-Pleistocene Asian Hominidae. http://biology.windy.edu/Biol345/LECTURE19/China/longgupo.htm

Ferguson, W. 1983. An alternative interpretation of *Australopithecus afarensis* fossil material. *Primates*, Vol. 24, pp.397–409.

Freedman, L. and Lofgren, M. 1979. The Cossack skull and a dihybrid origin of the Australian Aborigines. *Nature*, Vol. 282, pp.298–300.

Gabunia, L., Vekua, A., Lordkipanidze, D., Swisher III, C.C., Ferring, R., Justus, A., Nioradze, M., Tvalchrelidze, M., Antón, S.C., Bosinski, G., Jöris, O., de Lumley, M.-A., Majsuradze, G. and Mouskhelishvili. 2000. Earliest Pleistocene hominid cranial remains from Dmanisi, Republic of Georgia: taxonomy, geological setting, and age. *Science*, Vol. 288, pp.1019–1025.

Gebo, D.L. 1985. The nature of the primate grasping foot. *American Journal of Physical Anthropology*, Vol. 67,

pp.269–277.
Gebo, D.L., MacLatchy, L., Kityo, R., Deino, A., Kingston, J. and Pilbeam, D. 1997. A hominoid genus from the early Miocene of Uganda. *Science*, Vol. 276, pp.401–404.
Gingerich, P.D. 1973. First record of the Palaeocene primate *Chiromyoides* from North America. *Nature*, Vol. 244, pp.517–518.
Gingerich, P.D. 1976a. Cranial anatomy and evolution of early Tertiary Plesiadapidae (Mammalia, Primates). *Museum of Paleontology, University of Michigan Paper, Paleontology*, 15.
Gingerich, P.D. 1976b. Paleontology and phylogeny: patterns of evolution at the species level in early Tertiary mammals. *American Journal of Science*, Vol. 276, pp.1–28.
Gingerich, P.D. 1977a. Radiation of Eocene Adapidae in Europe. *Geobios Mémoire Spécial*, Vol. 1, pp.165–182.
Gingerich, P.D. 1977b. New species of Eocene primates and the phylogeny of European Adapidae. *Folia Primatologica,* Vol. 28, pp.60–80.
Gingerich, P.D. 1980. Dental and cranial adaptations in Eocene Adapidae. *Zeitschrift für Morphologie und Anthropologie*, Vol. 71, pp.135–142.
Gingerich, P.D. 1981a. Early Cenozoic Omomyidae and the evolutionary history of tarsiiform primates. *Journal of Human Evolution*, Vol. 10, pp.345–374.
Gingerich, P.D. 1981b. Eocene Adapidae, paleobiogeography, and the origin of South American Platyrrhini. Pp.123–138. In Ciochon, R.L. and Chiarelli, B. (eds). *Evolutionary Biology of the New World Monkeys and Continental Drift*. Plenum Press, New York. 528pp.
Gingerich, P.D. 1989. New earliest Wasatchian mammalian fauna from the Eocene of northwestern Wyoming: composition and diversity in a rarely sampled high-floodplain assemblage. *University of Michigan Papers in Paleontology*, Vol. 28, pp.1–97.
Gingerich, P.D. and Schoeninger, M. 1977. The fossil record and primate phylogeny. *Journal of Human Evolution*, Vol. 6, pp.483–505.
Gingerich, P.D. and Simons, E.L. 1977. Systematics, phylogeny and evolution of early Eocene Adapidae (Mammalia, Primates) in North America. *Contributions to the Museum of Paleontology, University of Michigan,* Vol. 24, pp.245–279.
Ginsburg, L. and Mein, P. 1980. *Crouzelia rhodanica,* nouvelle espece de primate catarhinien, et essui sur la position systematique des Pliopithecidae. *Bulletin du Muséum National d'Histoire Naturelle Paris*, Vol. 40, pp.57–85.
Goodman, M. 1975. Protein sequence and immunological specificity: their role in phylogenetic studies of primates. Pp.219–248. In Luckett, W.P. and Szalay, F.S. (eds). *Phylogeny of the Primates*. Plenum Press, New York. 483pp.
Goodman, M. 1976. Towards a genealogical description of the primates. Pp.321–353. In Goodman, M. and Tashian, R.E. (eds). *Molecular Anthropology*. Plenum Press, New York.
Goodman, M., Weiss, M.L. and Czelosniak, J. 1982. Molecular evolution above the species level: branching pattern, rates, and mechanisms. *Systematic Zoology*, Vol. 31, pp.376–399.
Gould, S.J. 1977. *Ontogeny and Phylogeny*. Harvard University Press, Cambridge, Massachusetts. 501pp.
Greenfield, L.O. 1979. On the adaptive pattern of *"Ramapithecus". American Journal of Physical Anthropology*, Vol. 50, pp.527–548.
Gregory, W.K. 1910. The orders of mammals. *Bulletin of the American Museum of Natural History*, Vol.27, pp.3–524.
Gribbin, J. and Cherfas, J. 1982. *The Monkey Puzzle: a Family Tree*. Bodley Head, London. 279pp.
Groves, C.P. 1974. Taxonomy and phylogeny of prosimians. Pp.449–473. In Martin, R.D., Doyle, G.A. and Walker, A.C. (eds). *Prosimian Biology: Proceedings of a Meeting of the Research Seminar in Archaeology and Related Subjects Held at the Institute of Archaeology, London University.* University of Pittsburgh Press, Pittsburgh. 983pp.
Groves, C.P. 1986. Systematics of the great apes. Pp.187–217. In Erwin, J. and Swindler, D.R. (eds). *Comparative Primate Biology. 1. Systematics, Evolution and Anatomy*. A.R. Liss, New York.
Groves, C.P. 1989. *A Theory of Human and Primate Evolution*. Oxford University Press, Oxford. 375pp.
Groves, C.P. 2001. *Primate Taxonomy.* Smithsonian Institution Press, Washington, D.C.
Groves, C.P. and Lahr, M.M. 1994. A bush not a ladder: speciation and replacement in human evolution. *Perspectives in Human Biology,* No. 4, pp.1–11.
Habgood, P.J. 1985. The origin of the Australian Aborigines: an alternative approach and view. Pp.367–380. In Tobias, P.V. (ed). *Hominid Evolution: Past, Present and Future. Proceedings of the Taung Diamond Jubilee International Symposium, Johannesburg and Mmabatho, Southern Africa, 27th January–4th February 1985.* A.R. Liss, New York. 499pp.
Habgood, P.J. 1989. The evolution of modern humans: evidence from Australasia seen in a global context. An examination of regional continuity. Pp.245–273. In Mellars, P. and Stringer, C. (eds). *The Origins and Dispersal of Modern Humans: Behavioural and Biological Perspectives.* Cambridge University Press, Cambridge.
Haile-Selassie, Y. 2001. Late Miocene hominids from the Middle Awash, Ethiopia. *Nature*, Vol 412, pp.178–181.
Hand, S.J. 2006. Bat beginnings and biogeography: the Australasian record. Pp.673–705. In Merrick, J.R., Archer, M., Hickey, G.M. and Lee, M.S.Y. (eds). *Evolution and Biogeography of Australasian Vertebrates.* Auscipub Pty. Ltd., Sydney. 942pp.
Harrison, T. 1986. New fossil anthropoids from the middle Miocene of East Africa and their bearing on the origin of the Oreopithecidae. *American Journal of Physical Anthropology*, Vol. 71, pp.265–284.
Hershkovitz, P. 1977. *New World Monkeys (Platyrrhini).* University of Chicago Press, Chicago.
Hill, J.E. and Smith, J.D. 1984. *Bats; a Natural History.* British Museum of Natural History, London. 243pp.
Hoffstetter, R. 1974. Phylogeny and geographical deployment of the primates. *Journal of Human Biology*, Vol. 3, pp.327–356.
Horovitz, I. and Meyer, A. 1997. Evolutionary trends: the ecology of New World Monkeys inferred from a combined phylogenetic analysis of nucleo, mitochondrial and morphological data. Pp.189–224. In Givnish, T.J. and Sytsma, K.J. (eds). *Molecular Evolution and Adaptive Radiation.* Cambridge University Press, Cambridge. 621pp.
Howell, F.C. 1978. Hominidae. Pp.154–248. In Maglio, V.J. and Cooke, H.B.S. (eds). *Evolution of African Mammals.*

Harvard University Press, Cambridge, Massachusetts. 641pp.

Howells, W.W. 1974. Neandertals: names, hypotheses and scientific method. *American Anthropology*, Vol. 76, pp.24–38.

Howells, W.W. 1980. *Homo erectus:* who, when and where. *Yearbook of Physical Anthropology*, Vol. 23, pp.1–23.

Huang, W., Gochon, R., Gu, Y., Larick, R., Fang, Q., Schwarcz, H., Yonge, C., de Vos, J. and Rink, W. 1995. Early *Homo* and associated artefacts from Asia. *Nature*, Vol. 378, pp.275–278.

Hublin, J.J. 1996. The first Europeans. *Archaeology*, January/February, pp.36–44.

Hublin, J.J., Barroso Ruiz, C., Medinalara, P. Fontugene, M. and Reyss, J-L. 1995. The Moudeman site of Zafarraya (Andlucia, Spain): dating and implications on the palaeolithic peopling process of Western Europe. *Comptes Rendus de l'Academie des Sciences, Paris*, Vol. 321, pp.931–937.

Hurzeler, J. 1960. The significance of *Oreopithecus* in the genealogy of man. *Triangle*, Vol. 4, pp.164–175.

Jacob, T. 1976. Early populations in the Indonesian region. Pp.81–93. In Kirk, R.L. and Thorne, A.G. (eds). *The Origin of the Australians*. Human Biological Series No. 6. Australian Institute of Aboriginal Studies, Canberra. 449pp.

Jaeger, J-J., Aung Naing Soe, V., Aye Ko Aung, U., Benammi, M., Chaimanee, Y., Ducrocq, R-M., Than Tun, C., Tin Thein, U. and Ducrocq, S. 1998. New Myonmas middle Eocene anthropoids: an Asian origin for catarrhines? *Comptes Rendus de l'Academie des Sciences, Paris*, Vol. 321, pp.953–989.

Jenkins, F.A. Jnr. 1974. Tree shrew locomotion and the origins of primate arborealism. Pp.85–115. In Jenkins, F.A. Jnr. (ed). *Primate Locomotion.* Academic Press, New York. 390pp.

Johanson, D.C. and Edey, M. 1981. *Lucy: the Beginnings of Mankind.* Simon and Schuster, New York.

Johanson, D.C. and White, T.D. 1979. A systematic assessment of early African hominids. *Science*, Vol. 203, pp.321–330.

Johanson, D.C., White, T.D. and Coppens, Y. 1978. A new species of the genus *Australopithecus* (Primates, Hominidae) from the Pliocene of Eastern Africa. *Kirtlandia*, Vol. 28, pp.1–14.

Kay, R.F. 1981. The nut-crackers—a new theory on the adaptations of the Ramapithecinae. *American Journal of Physical Anthropology*, Vol. 55, pp.141–145.

Kay, R.F. 1982. *Sivapithecus simonsi,* a new species of Miocene hominid, with comments on the phylogenetic status of the Ramapithecinae. *International Journal of Primatology*, Vol. 3, pp.113–173.

Kay, R.F. and Cartmill, M. 1974. Skull of *Palaechthon nacimienti. Nature*, Vol. 252, pp.37–38.

Kay, R.F., Fleagle, J.G. and Simons, E.L. 1981. A review of the Oligocene apes of the Fayum Province, Egypt. *American Journal of Physical Anthropology*, Vol. 55, pp.293–322.

Kershaw, A.P. 1994. Pleistocene vegetation of the humid tropics of northeastern Queensland, Australia. *Paläogeography, Palaeoclimatology, Palaeoecology*, Vol. 109, pp.399–412.

Kimbel, W.H., Walter, R.C., Johanson, D.C., Reed, K.E., Aronson, J.L., Assefa, Z., Marean, C.W., Eck, G.G., Bobe, R., Hovers, E., Rak, Y., Vondra, C., Yemane, T., York, D., Chen, Y., Eversen, N.M. and Smith, P.E. 1996. Late Pliocene *Homo* and Oldovian tools from the Hadar Formation. *Journal of Human Evolution*, Vol. 31, pp.549–561.

Koenigswald, W. von. 1979. Ein Lemuren rest aus dem eozanen Olschiefer der Grube Messel bei Darmstadt. *Paläontologische Zeitschrift*, Vol. 53, pp.63–76.

Kohen, J.L. 1995. *Aboriginal Environmental Impacts*. University of New South Wales Press, Sydney. 160pp.

Krings, M., Stone, A., Schmitz, R.W., Krainitzki, H., Stoneking, M. and Pääbo, S. 1997. Neandertal DNA sequences and the origin of modern humans. *Cell*, Vol. 90, pp.19–30.

Latimer, B. 1991. Locomotor adaptation in *Australopithecus afarensis*: the issue of arboreality. Pp.169–179. In Coppens, Y. and Senut, B. (eds). *Origines de lu Bipedie chez Les Hominidés*. Centre National de la Recherche Scientifique, Paris.

Leakey, M.D. 1979. Footprints in the ashes of time. *National Geographic*, Vol. 155, pp.446–457.

Leakey, M.G., Feibel, C.S., McDougall, I. and Walker, A. 1995a. New four-million-year-old hominid species from Kanapoi and Allia Bay, Kenya. *Nature*, Vol. 376, pp.565–571.

Leakey, M.G., Spoor, F., Brown, F.H., Gathogo, P.N., Kiarie, C., Leakey, L.N. and McDougall, I. 2001. New hominin genus from eastern Africa shows diverse middle Pliocene lineages. *Nature*, Vol. 410, pp.433–440.

Leakey, M.G., Ungar, P.S. and Walker, A. 1995b. A new genus of large primate from the late Oligocene of Lothidok, Turkana district, Kenya. *Journal of Human Evolution*, Vol. 28, pp.519–531.

Leakey, R.E.F. 1973a. Australopithecines and hominines: a summary of the evidence from the early Pleistocene of eastern Africa. Pp.53–63. In Suckerman, L. (ed). *The Concepts of Human Evolution: Proceedings of a Symposium organised Jointly by the Anatomical Society of Great Britain and Ireland and the Zoological Society of London on 9 and 10 November, 1972.* Academic Press, London. 480pp.

Leakey, R.E.F. 1973b. Skull 1470. *National Geographic*, Vol. 143, No. 6, pp.819–829.

Leakey, R.E.F. 1976. *Australopithecus, Homo erectus* and the single species hypothesis. *Nature*, Vol. 261, pp.572–574.

Leakey, R.E. 1981. *The Making of Mankind.* Michael Joseph Limited, London. 256pp.

Leche, W. 1885. Uber die Saugetiergattung *Galeopithecus. Kungliga Svenska Vetenskapsakademiens Handlingar*, Vol. 21, No. 11, pp.1–92.

Le Gros Clark, W.E. 1963. *History of the Primates.* (8th Edition). University of Chicago Press, Chicago.

Lewis, 0.J. 1983. The evolutionary emergence and refinement of the mammalian pattern of foot architecture. *Journal of Anatomy*, Vol. 137, pp.21–45.

Lovejoy, C.O. 1981. The origin of man. *Science*, Vol. 211, pp.341–350.

Luckett, W.P. 1976. Cladistic relationships among primate higher categories: evidence of the fetal membranes and placenta. *Folia Primatologica*, Vol. 25, pp.245–276.

Macintosh, N.W.G. 1963. Origin and physical differentiation of the Australian Aborigines. *Australian Natural History*,

Vol. 14, pp.248–252.
Macintosh, N.W.G. 1969. The Talgai cranium: the value of archives. *Australian Natural History*, Vol. 16, pp.189–195.
Macintosh, N.W.G. and Larnach, S.L. 1972. The persistence of *H. erectus* traits in Australian Aboriginal crania. *Archaeology and Physical Anthropology in Oceania*, Vol. 7, pp.1–7.
Macintosh, N.W.G. and Larnach, S.L. 1976. Aboriginal affinities looked at in world context. Pp.113–126. In Kirk, R.L. and Thorne, A.G. (eds). *The Origins of the Australians*. Human Biological Series No. 6. Australian Institute of Aboriginal Studies, Canberra. 449pp.
MacPhee, R.D.E. 1979. Entotympanics, ontogeny and primates. *Folia Primatologica*, Vol. 31, pp.23–47.
MacPhee, R.D.E., Cartmill, M. and Gingerich, P.D. 1983. New Palaeogene primate basicrania and the definition of the order Primates. *Nature*, Vol. 301, pp.509–511.
Martin, R.D. 1990. *Primate Origins and Evolution*. Princeton University Press, Princeton, New Jersey.
Maw, B., Ciochon, R. and Savage, D.E. 1979. Late Eocene of Burma yields earliest anthropoid primate, *Pondaungia cotteri. Nature*, Vol. 282, pp.65–67.
McKenna, M.C. 1966. Paleontology and the origin of the primates. *Folia Primatologica*, Vol. 4, pp.1–25.
McKenna, M.C. 1975. Toward a phylogenetic classification of the Mammalia. Pp.21–46. In Luckett, W.P. and Szalay, F.S. (eds). *Phylogeny of the Primates: a Multidisciplinary Approach*. Plenum Press, New York. 483pp.
Morwood, M.J., O'Sullivan, P.B., Aziz, F. and Raza, A. 1998. Fission-crack ages of stone tools and fossils on the east Indonesian island of Flores. *Nature*, Vol. 392, pp.173–176.
Mulvaney, D.J. and Joyce, E.B. 1965. Archaeological and geomorphological investigations on Mt. Moffat Station, Queensland. *Proceedings of the Prehistorical Society*, Vol. 31, pp.147–212.
Murray, P. 1975. The role of cheek pouches in cercopithecine monkey adaptive strategy. Pp.151–194. In Tuttle, R.L. (ed). *Primate Functional Morphology and Evolution*. Mouton, The Hague. 583pp.
Napier, J.R. 1970. Paleoecology and catarrhine evolution. Pp.55–95. In Napier, J.R. and Napier, P.H. (eds). *Old World Monkeys: Evolution, Systematics and Behavior: Proceedings*. Academic Press, New York. 660pp.
Novacek, M.J. 1980. Cranioskeletal features in tupaiids and selected Eutheria as phylogenetic evidence. Pp.35–93. In Luckett, W.P. (ed). *Comparative Biology and Evolutionary Relationships of Tree Shrews*. Plenum Press, New York.
Olson, T.R. 1981. Basicranial morphology of the extant hominoids and Pliocene hominids: the new material from the Hadar Formation, Ethiopia, and its significance in early human evolution and taxonomy. Pp.99–128. In Stringer, C.B. (ed). *Aspects of Human Evolution*. Taylor and Francis, London. 233pp.
Osborne, H.F. 1907. *Evolution of Mammalian Molar Teeth to and from the Triangular Type*. Macmillan Co., London.
Ovchinnikov, I.G., Götherström, A., Romanova, G.P., Kharltonov, V.M., Lidén, K. and Goodwin, W. 2000. Molecular analysis of Neanderthal DNA from the northern Caucasus. *Nature,* Vol. 404, pp.490–493.
Pardoe, C. 1991. Competing paradigms and ancient human remains: the state of the discipline. *Archaeology in Oceania*, Vol. 26, pp.79–85.
Patterson, C. 1982. Cladistics and classification. *New Scientist*, Vol. 94, pp.303–306.
Pettigrew, J.D. 1986. Flying primates? Megabats have the advanced pathway from the eye to midbrain. *Science*, Vol. 231, pp.1304–1306.
Pettigrew, J.D. and Jamieson, B.G.M. 1987. Are flying foxes (Chiroptera: Pteropodidae) really primates? *Australian Mammalogy*, Vol. 10, pp.119–124.
Pickford, M. 1982. New higher primate fossils from the middle Miocene deposits at Majiwa and Kaloma, western Kenya. *American Journal of Physical Anthropology*, Vol. 58, pp.1–19.
Pickford, M., Senut, B., Gommery, D. and Treil, J. 2002. Bipedalism in *Orrorin tugenensis* revealed by its femora. *Comptes Rendu Palevol*, Vol. 1, pp.199–203.
Pilbeam, D.R. 1969. Tertiary Pongidae of East Africa: evolutionary relationships and taxonomy. *Yale Peabody Museum Bulletin*, Vol. 31, pp.1–185.
Pilbeam, D.R. and Gould, S.J. 1974. Size and scaling in human evolution. *Science*, Vol. 186, pp.892–901.
Pilbeam, D.R., Meyer, G.E., Badgley, C., Rose, M.D., Pickford, M.H.L., Behrensmeyer, A.K. and Ivorahim Suah, S.M. 1977. New hominoid primates from the Sivaliks of Pakistan and their bearing on hominoid evolution. *Nature*, Vol. 270, pp.689–695.
Pocock, R.I. 1918. On the external features of the lemurs and of *Tarsius. Proceedings of the Zoological Society of London*, 1918, pp.19–53.
Radinsky, L. 1977. Early primate brains: facts and fiction. *Journal of Human Evolution,* Vol. 6, pp.79–86.
Rightmire, G.P. 1981. Patterns in the evolution of *Homo erectus. Paleobiology*, Vol. 7, pp.241–246.
Rightmire, G.P. 1983. The Lake Ndutu Cranium and early *Homo sapiens* in Africa. *American Journal of Physical Anthropology*, Vol. 61, pp.245–254.
Robinson, J.T. and Streudel, K. 1973. Multivariate discriminate analysis of dental data bearing on early hominid affinities. *Journal of Human Evolution*, Vol. 2, pp.509–527.
Rook, L., Bondioli, L., Köhler, M., Moyà-Solå, S. and Macchiarelli, R. 1999. *Oreopithecus* was a bipedal ape after all: evidence from the iliac cancellous architecture. *Proceedings of the National Academy of Sciences of the United States of America*, Vol. 96, pp.8795–8799.
Rosenberger, A.L. 1977. *Xenothrix* and ceboid phylogeny. *Journal of Human Evolution*, Vol. 6, pp.561–581.
Rosenberger, A.L. 1979. Cranial anatomy and implications of *Polichocebus,* a late Oligocene ceboid primate. *Nature*, Vol. 279, pp.416–418.
Rosenberger, A.L., Strasser, E. and Delson, E. 1985. Anterior dentition of *Notharctus* and the adapid-anthropoid hypothesis. *Folia Primatologica*, Vol. 44, pp.15–39.
Sanderson, I.T. 1957. *The Monkey Kingdom: an Introduction to the Primates*. Doubleday and Co., New York. 200pp.
Sarich, V.M. and Wilson, A.C. 1967. Immunological time scale for hominid evolution. *Science*, Vol. 158, pp.1200–1205.
Schmid, P. 1981. Comparison of European nonadapids and *Tarsius*. Pp.6–13. In Chiarelli, A.B. and Corruccini, R.S. (eds). *Primate Evolutionary Biology*. Springer Verlag, Berlin.
Schneider, H., Schneider, M.P.C., Sampaio, I., Harada,

M.L., Stanhope, M., Czelusniak, J. and Goodman, M. 1993. Molecular phylogeny of the New World monkeys (Platyrrhinii, Primates). *Molecular Phylogenetics and Evolution*, Vol. 2, pp.225–242.

Schwartz, J.H. 1975. Development and eruption of the premolar region of prosimians and its bearing on their evolution. Pp.41–63. In Tattersall, E. and Sussman, R.L. (eds). *Lemur Biology.* Plenum Press, New York.

Schwartz, J.H. 1978. If *Tarsius* is not a prosimian, is it a haplorhine? Pp.195–202. In Chivers, D.J. and Joysey, K.A. (eds). *Recent Advances in Primatology. Vol. 3, Evolution.* Academic Press, London.

Schwartz, J.H. and Tattersall, I. 1985. Relationships of *Microadapis sciureus* (Stehlin, 1916), and two new primate genera from the Eocene of Switzerland. *Folia Primatologica*, Vol. 39, pp.178–186.

Schwartz, J.H., Tattersall, I. and Eldredge, N. 1978. Phylogeny and classification of the primates revisited. *Yearbook of Physical Anthropology*, Vol. 21, pp.95–133.

Selenka, E. 1899. *Menschenaffen. Studien uber Entwickelung und Schadelbau Liefg. 2. Kapitel 2. Schadel des Gorilla und Schimpanse.* Pp.93–160. Kreidel, Wiesbaden.

Seligsohn, D. 1977. Analysis of species-specific molar adaptations in strepsirhine primates. *Contributions to Primatology*, Vol. 11, pp.1–116.

Senut, B. 1996. Pliocene hominin systematics and phylogeny. *South African Journal of Science*, Vol. 92, pp.165–166.

Senut, B., Pickford, M., Gommery, D., Mein, P., Cheboi, K. and Coppens, Y. 2001. First hominid from the Miocene (Lukeino Formation, Kenya). *Comptes Rendus de l'Academie des Sciences, Paris, Sciences de la Terre et des Planetes*, Vol. 332, pp.137–144.

Simons, E.L. 1961. The phyletic position of *Ramapithecus. Postilla*, Vol. 51.

Simons, E.L. 1965. New fossil apes from Egypt and the initial differentiation of Hominoidea. *Nature*, Vol. 205, pp.135–139.

Simons, E.L. 1974a. *Parapithecus grangeri* (Parapithecidae, Old World higher primates): new species from the Oligocene of Egypt and the initial differentiation of Cercopithecoidea. *Postilla*, Vol. 166, pp.1–12.

Simons, E.L. 1974b. The relationships of *Aegyptopithecus* to other primates. *Annals Geological Survey, Egypt*, Vol. 4, pp.149–156.

Simons, E.L. 1977. *Ramapithecus. Scientific American*, Vol. 236, pp.28–35.

Simons, E.L. and Bown, T.M. 1985. *Afrotarsius chatrathi,* first tarsiiform primate (?Tarsiidae) from Africa. *Nature*, Vol. 313, pp.475–477.

Simons, E.L., Bown, T.M. and Rasmussen, D.T. 1986. Discovery of two additional prosimian primate families (Omomyidae, Lorisidae) in the African Oligocene. *Journal of Human Evolution*, Vol. 15, pp.431–437.

Simons, E.L. and Fleagle, J. 1973. The history of extinct gibbonlike primates. *Gibbon and Siamang*, Vol. 2, pp.121–148.

Simons, E.L. and Pilbeam, D.R. 1965. Preliminary revision of the Dryopithecinae (Pongidae, Anthropoidea). *Folia Primatologica*, Vol. 3, pp.81–152.

Simpson, G.G. 1945. The principles of classification and a classification of mammals. *Bulletin of the American Museum of Natural History*, Vol. 85, pp.1–350.

Singh, G., Kershaw, A.P. and Clark, R. 1981. Quaternary vegetation and fire history in Australia. Pp.23–54. In Gill, A.M., Groves, R.H. and Noble, I.R. (eds). *Fire and the Australian Biota.* Australian Academy of Science, Canberra.

Stern, J.T. and Sussman, R.L. 1983. The locomotor anatomy of *Australopithecus afarensis. American Journal of Physical Anthropology*, Vol. 60, pp.279–317.

Stirton, R.A. 1951. Ceboid monkeys from the Miocene of Colombia. *University of California, Publications in Geological Sciences*, Vol. 28, pp.315–356.

Stringer, C.B. 1984. The definition of *Homo erectus* and the existence of the species in Africa and Europe. *Courier Forschungsinstitut Senckenberg*, Vol. 69, pp.131–143.

Szalay, F.S. 1969. Mixodectidae, Microsyopidae and the insectivore-primate transition. *Bulletin of the American Museum of Natural History*, Vol. 140, pp.193–330.

Szalay, F.S. 1973. New Palaeocene primates and a diagnosis of the new suborder Paromomyiformes. *Folia Primatologica*, Vol. 19, pp.73–87.

Szalay, F.S. 1975. Where to draw the nonprimate-primate boundary. *Folia Primatologica*, Vol. 23, pp.158–163.

Szalay, F.S. 1976. Systematics of the Omomyidae (Tarsiiformes, Primates): taxonomy, phylogeny and adaptations. *Bulletin of the American Museum of Natural History*, Vol. 156, pp.157–450.

Szalay, F.S. 1977a. Phylogenetic relationships and a classification of the eutherian Mammalia. Pp.315–374. In Hecht, M.K., Goody, P.C. and Hecht, B.M. (eds). *Major Patterns in Vertebrate Evolution.* Plenum Press, New York.

Szalay, F.S. 1977b. Constructing primate phylogenies: a search for testable hypotheses with maximum empirical content. *Journal of Human Evolution*, Vol. 6, pp.3–18.

Szalay, F.S. 1981. Phylogeny and the problem of adaptive significance: the case of the earliest primates. *Folia Primatologica*, Vol. 36, pp.157–182.

Szalay, F.S. and Decker, R.L. 1974. Origins, evolution and function of the tarsus in late Cretaceous eutherians and Palaeocene primates. Pp.223–259. In Jenkins, F.A. (ed). *Primate Locomotion.* Academic Press, New York. 390pp.

Szalay, F.S. and Delson, E. 1979. *Evolutionary History of the Primates.* Academic Press, New York.

Szalay, F.S. and Drawhorn, G. 1980. Evolution and diversification of the Archonta in an arboreal milieu. Pp.133–169. In Luckett, W.P. (ed). *Comparative Biology and Evolutionary Relationships of Tree Shrews.* Plenum Press, New York.

Szalay, F.S. and Katz, C. 1973. Phylogeny of lemurs, galagos and lorises. *Folia Primatologica*, Vol. 19, pp.88–103.

Szalay, F.S. and Li, C.K. 1986. Middle Palaeocene euprimate, from southern China and the disevolution of primates in the Paleogene. *Journal of Human Evolution*, Vol. 15, pp.387–397.

Szalay, F.S., Li, C.K. and Wang, B.Y. 1986. Middle Palaeocene omomyid primate from Anhui province, China: *Decoredon anhinensis* (Xu, 1976), new combination Szalay & Li, and the significance of *Petrolemur. American Journal of Physical Anthropology*, Vol. 69, p.269.

Szalay, F.S. and Seligsohn, D. 1977. Why did the strepsirhine tooth comb evolve? *Folia Primatologica*, Vol. 27, pp.75–82.

Tattersall, I. and Schwartz, J.H. 1975. Relationships among the Malagasy lemurs, the craniodental evidence. Pp.299–312. In Luckett, W.P. and Szalay, F.S. (eds).

Phylogeny of the Primates: a Multidisciplinary Approach. Plenum Press, New York. 483pp.

Temerin, L.A. and Cant, J.G.H. 1983. The evolutionary divergence of Old World monkeys and apes. *American Naturalist*, Vol. 122, pp.335–348.

Thenius, V.E. 1981. Bemerkungen zur taxonomischen und stammesgeschichtlichen Position der Gibbons (Hylobatidae, Primates). *Sonderdruck aus Zeitschrift fur Saugetierkunde*, Vol. 46, pp.232–241.

Thorne, A., Grun, R., Mortimer, G., Spooner, N.A., Simpson, J-J., McCulloch, M., Taylor, L. and Curnoe, D. 1999. Australia's oldest human remains: age of the Lake Mungo 3 skeleton. *Journal of Human Evolution*, Vol. 36, pp.591–612.

Thorne, A.G. 1976. Morphological contrasts in Pleistocene Australians. Pp.95–112. In Kirk, R.L. and Thorne. A.G. (eds). *The Origin of the Australians.* Human Biological Series No. 6. Australian Institute of Aboriginal Studies, Canberra. 449pp.

Thorne, A.G. 1977. Separation or reconciliation? Biological clues to the development of Australian society. Pp.187–204. In Allen, J., Golson, J. and Jones, R. (eds). *Sunda and Sahul: Prehistoric Studies in Southeast Asia, Melanesia and Australia.* Academic Press, London. 647pp.

Thorne, A.G. 1981a. The arrival and adaptation of Australian Aborigines. Pp.1751–1760. In Keast, A. (ed). *Ecological Biogeography of Australia.* Junk, The Hague.

Thorne, A.G. 1981b. The centre and the edge: the significance of Australasian hominids to African palaeoanthropology. Pp.180–181. In Leakey, R.E.F. and Ogot, B.A. (eds). *Proceedings of the 8th Panafrican Congress. Prehistoric Quaternary Studies, Nairobi, September 1977.* Nairobi.

Thorne, A.G. and Macumber, P.G. 1972. Discoveries of late Pleistocene man at Kow Swamp, Australia. *Nature*, Vol. 238, pp.316–319.

Thorne, A.G. and Wilson, S.R. 1977. Pleistocene and Recent Australians: a multivariate comparison. *Journal of Human Evolution*, Vol. 6, pp.393–402.

Thorne, A.G. and Wolpoff, M.H. 1981. Regional continuity in Australasian Pleistocene hominid evolution. *American Journal of Physical Anthropology*, Vol. 55, pp.337–349.

Trinkaus, E. and Howells, I.W.W. 1979. The Neandertals. *Scientific American*, Vol. 241, pp.118–134.

van Oosterzee, P. 2006. Drawing the Wallace Line. Pp.95–110. In Merrick, J.R., Archer, M., Hickey, G.M. and Lee, M.S.Y. (eds). *Evolution and Biogeography of Australasian Vertebrates.* Auscipub Pty. Ltd., Sydney. 942pp.

Van Valen, L. and Sloan, R.E. 1965. The earliest primates. *Science*, Vol. 150, pp.743–745.

Walker, A. 1972. The dissemination and segregation of early primates in relation to continental configuration. Pp.195–218. In Bishop, W.W. and Miller, J.A. (eds). *Calibration of Homonoid Evolution.* Scottish Academic Press, Edinburgh.

Walker, A. 1974. A review of the Miocene Lorisidae of East Africa. Pp.435–437. In Martin, R.D., Doyle, G.A. and Walker, A.C. (eds). *Prosimian Biology.* Duckworth, London.

Walker, A. and Leakey, R.E.F. 1978. The hominids of East Turkana. *Scientific American*, Vol. 238, pp.54–66.

Ward, S.C. and Kimbel, W.H. 1983. Subnasal alveolar morphology and the systematic position of *Sivapithecus. American Journal of Physical Anthropology*, Vol. 61, pp.157–171.

Washburn, S.L. 1978. The evolution of man. *Scientific American*, Vol. 239, pp.146–153.

Webb, J.E., Wallwork, J.A. and Elgood, J.H. 1977. *Guide to Living Mammals.* Macmillan Press Ltd., London.

White, J.P., Crook, K.A.W. and Ruxton, B.P. 1970. Kosipe: a late Pleistocene site in the Papuan Highlands. *Proceedings of the Prehistorical Society*, Vol. 36, pp.152–170.

White, T.D. 2003. Early hominids—diversity or distortion? *Science*, Vol. 299, pp.1994–1996.

White, T.D. , Asfaw, B., DeGusta, D., Gilbert, H., Richards, G.D., Suwa, G. and Howell, F.C. 2003. Pleistocene *Homo sapiens* from middle Awash, Ethiopia. *Nature*, Vol. 423, pp.742–747.

White, T.D., Suwa, G. and Asfaw, B. 1994. *Australopithecus ramidus*, a new species of early hominid from Aramis, Ethiopia. *Nature*, Vol. 371, pp.306–312.

White, T.D., Suwa, G. and Asfaw, B. 1995. *Australopithecus ramidus*, a new species of early hominid from Aramis, Ethiopia. *Nature*, Vol. 375, p.88.

Wolpoff, M.H., Wu, X.-Z. and Thorne, A.G. 1984. Modern *Homo sapiens* origins: a general theory of hominid evolution involving the fossil evidence from East Asia. Pp.411–483. In Smith, F. and Spencer, F. (eds). *Origin of Modern Humans.* A.R. Liss, New York.

Wood, B. 1976. *The Evolution of Early Man.* Eurobook Ltd., England.

Wood, B.A. and Collard, M. 1999. The human genus. *Science*, Vol. 284, pp.65–67.

Wood, B.A. and Stack, C.G. 1980. Does allometry explain the difference between "gracile" and "robust" australopithecines? *American Journal of Physical Anthropology*, Vol. 52, pp.55–62.

Wright, R.V.S. 1976. Evolutionary process and semantics. Pp.265–274. In Kirk, R.L. and Thorne, A.G. (eds). *The Origins of the Australians.* Human Biological Series No. 6. Australian Institute of Aboriginal Studies, Canberra. 449pp.

Wroe, S. and Archer, M. 2006. Origins and early radiations of marsupials. Pp.551–574. In Merrick, J.R., Archer, M., Hickey, G.M. and Lee, M.S.Y. (eds). *Evolution and Biogeography of Australasian Vertebrates.* Auscipub Pty. Ltd., Sydney. 942pp.

Wu, R. 1981. Where did Humankind originate? On the discovery of the world's first *Ramapithecus* skull. *China Pictorial*, Vol. 7, No.387, pp.16–18.

Wu, R. and Pan, Y. 1984. A late Miocene gibbon-like primate from Lufeng, Yunnan Province. *Acta Anthropologica Sinica,* Vol. 3, pp.185–196.

Wu, X. 1981. A well-preserved cranium of an archaic type of early *Homo sapiens* from Dali, China. *Scientia Sinica*, Vol. 24, pp.530–539.

Wu, X. 2000. Longgupo hominoid mandible belongs to ape. *Acta Anthropologica Sinica*, Vol. 19, pp.1–10.

Zihlman, A.L. and Lowenstein, J.M. 1979. False start of the human parade. *Natural History*, Vol. 88, pp.86–91.

Chapter 35
Combined Keys to Extant Australasian Mammals

George M. McKay
Department of Biological Sciences, Macquarie University, N.S.W. 2109 Australia
Present Address: P.O. Box 100, Pacific Palms, N.S.W. 2420 Australia
E-mail: george.mckay@bigpond.com

INTRODUCTION

Previous chapters have discussed the origins, interrelationships (using fossil records, morphology and other evidence) and factors influencing radiations of a number of major mammal groups; in addition, mechanisms and processes contributing to recent speciation and current biogeographic patterns are considered (Aplin, 2006; Archer and Hand, 2006; Cooke, 2006; Corbett, 2006; Fordyce, 2006; Hand, 2006; Johanson, 2006; Musser, 2006; Wroe and Archer, 2006). But this chapter is intended to provide an overview of the diversity of extant faunas by way of identification keys—to familial or generic levels. These summaries are designed to be used in conjunction with detailed, fully illustrated keys for particular groups—at the generic down to specific level.

The purpose of a key is to aid the process of identification by guiding the observer through a sequence of choices that result in the elimination of unlikely alternatives. Once the final tentative identification has been made, it should be checked against a series of specimens such as in a museum, or at least by reference to a detailed description of the animal. In most instances, reference to a description is the most practical solution. A 'good' key should provide unambiguous alternatives at each decision point and should begin with more general characters, proceeding gradually to more specific details. A good key should also be easy to understand and contain a minimum of jargon. Technical terminology can, however, reduce the wordiness of descriptions. In the following keys a basic knowledge of anatomy is assumed and each time a new term is introduced it is defined briefly [in square brackets immediately following the new word].

The characteristics of orders and families used in constructing these keys are based primarily from examination of specimens in the United States National Museum, the Australian Museum and other collections and on diagnoses in Anderson and Jones (1984) and Archer and Clayton (1984). Many characters were also checked against Lawlor (1979), Jones and Baynes (1989) and Thomas (1888). Previous versions of these keys were used in teaching mammalogy at the University of Maryland and at Macquarie University. Characteristics used in the generic keys depend heavily on characters described in Churchill (1998), Flannery (1990), Jones and Baynes (1989), King (1990), Seebeck *et al.* (1990) and Smith and Winter (1984). The final section of this chapter is intended as a guide to the most useful and most recent reference works on mammals of the Australasian region and is arranged first geographically then by taxonomic grouping.

Taxonomic arrangements generally follow Wilson and Reeder (1993) except as follows: Hipposideridae and Rhinolophidae are treated as separate families; *Mastacomys* is retained for the broad-toothed rat rather than including it in *Pseudomys*; *Antechinomys* is retained for the kultarr, included by some authors in *Sminthopsis*; *Strigocuscus* is considered a synonym of *Phalanger*.

The keys in this chapter are divided into two parts. In the first part are keys based as much as possible on characters which can be seen on a living, or at least recently dead, whole animal. Where appropriate, characteristic behaviour patterns have been included. In the second part the keys are designed to aid in identifying remains based on characteristics of the skull and teeth, which tend to be the best preserved parts of mammalian carcasses. Although isolated teeth and bones other than the skull are often encountered there have been very few guides to these published (Merilees and Porter, 1979 and Triggs, 1996 are two). Where hairs are the main component of the remains, reference should be made to Brunner and Coman (1974) or Brunner *et al.* (2002).

The structure of the following keys is the standard couplet where each decision leads to another couplet, in which case the line ends in the number of the next decision or a tentative identification. In the identifications common names are given before scientific names which follow the sequence of ORDER, FAMILY (in capitals), genus (in italics). Geographic distribution of each group or subgroup is given in square brackets before the name. Abbreviations used for geographic ranges are: A = Australia, NG = New Guinea including surrounding islands, NZ = New Zealand, Tas = Tasmania, WA = Western Australia. Recently extinct taxa, identified by the symbol (†), are included as sub-fossil remains can still be found. Taxa which have been transported by humans and have established wild populations are indicated by an asterisk (*). In the whole-animal keys, the number of species currently recognized in each genus, and which are found in the region, is given in square brackets after the generic name.

The keys provide for identification of order and family for all Australasian mammals using either the whole animal or the skull. Wherever practicable, the keys to whole animals provide for further identification to the genus level. In some families, most notably the Dasyuridae, Macropodidae and Muridae, workable keys to generic level are difficult or even impossible to construct except on a local level in a restricted geographic area. For such families the genera are listed in tables with a brief description of important characters. For the bandicoots, Order Peramelemorphia, the distinction between families as they are currently recognized is based on subtle distinctions on the skull which do not adapt readily to construction of a whole-animal key. A generic key to bandicoots is provided. The skull keys do not proceed beyond the family level. For further identification of skulls, the reader should consult the relevant detailed reference works listed. Illustrations are not provided and the reader is encouraged to consult the relevant chapters on mammals in this volume or the excellent diagrams in Walton and Richardson (1989).

KEYS BASED ON EXTERNAL CHARACTERISTICS

KEY A1 TO MAMMAL ORDERS

1a. Head and body >1.5m; body streamlined, forelimbs modified as flippers without separate digits; either hindlimbs absent and tail modified as lateral flukes or hindlimbs webbed and adapted for swimming .... Whales, dolphins, dugong, seals.........2

1b. Head and body generally <1.5m; digits of manus [forefoot] and pes [hind foot] distinct even if joined by a web or membrane .... All other mammals.........4

2a. Hindlimbs absent and tail modified as lateral flukes .... Whales, dugong.........3

2b. Hindlimbs present, webbed and adapted for swimming .... Seals and sea-lions.........ORDER CARNIVORA (PINNIPEDIA)......Key A8

3a. Snout deflected ventrally with prominent rounded upper lip; no dorsal fin; swims in shallow bays and offshore areas; crops seagrasses leaving characteristic plume of silt in wake .... Dugong.........ORDER SIRENIA.........DUGONGIDAE.........*Dugong* [1]

3b. Snout may be rounded, elongated, often pointed or beaked but never deflected; dorsal fin present or absent; estuaries to open ocean; surface briefly to breathe; 'blow' [exhaled breath] of larger species often visible long before animal can be seen.................Whales.........ORDER CETACEA.........Key A9

4a. Forelimbs modified as wings with elongated metacarpals and phalanges and an extensive membrane between digits .... Bats.........ORDER CHIROPTERA.........Key A6

4b. Fore and hind limbs with many modifications but never as above 5

5a. Snout beak-like (either cylindrical or spatulate [flattened]), no teeth in adult; either feet webbed and with flattened tail and soft fur, or feet armed with long claws and fur coarse and spiny .... Monotremes.........ORDER MONOTREMATA.........KEY A2

5b. Snout not beak-like, jaws contain some teeth 6

6a. Eyes and ears vestigial; enlarged spade-like claws; keratinized nasal shield; tail reduced; desert areas; lives underground "swimming" through loose sand; no permanent burrows ....[A] Marsupial mole.........ORDER NOTORYCTEMORPHIA, NOTORYCTIDAE.........*Notoryctes*[1]

6b. Body without strong fossorial [underground] adaptations............7

7a. Hindquarters frequently appear to taper gradually into tail; urogenital and anal openings within a common cloaca; males with scrotum anterior to penis which is located in cloaca; females with mammae located in a well-defined pouch or pouch-area .... Marsupials.........8

7b. Tail generally distinct from body; cloaca generally absent, separate urogenital and anal openings; scrotum in males, if present, posterior to penis; mammae generally dispersed along ventral body wall or, if clustered, never in a pouch or pouch-area............Placentals.........10

8a. Hind feet syndactyl [pedal digits II & III united in common sheath], hallux [first digit of hind foot] if present opposable and clawless 9

8b. Hind feet never syndactyl [pedal digits II & III, if present, separate], hallux if present never both opposable and clawless; digits 5+4 or, if 5+5, hallux reduced............ ORDER DASYUROMORPHIA.........Key A3

9a. Manus with either two [II,III] or three [II-IV] functional digits bearing strong claws, remaining digits vestigial ....ORDER PERAMELEMORPHIA ....Key A4
9b. Manus with 5 functional digits usually subequal in size but frequently with digits I&II opposable to the remainder ....ORDER DIPROTODONTIA ....Key A5

10a. Large size (HB [length of head and body] >750mm) with elongated limbs and unguligrade [stand on tips of toes] stance 11
10b. Small to medium size (HB <750mm), most with generalized terrestrial/climbing adaptations although some specialization but never unguligrade 12

11a. Single functional digit on fore and hind limbs with nail enlarged as a semi-circular hoof; large herding animals with prominent head on long maned neck; long hairs on short tail............... ....[A, NZ].........Horses, donkeys.........*ORDER PERISSODACTYLA ....EQUIDAE.........*Equus*[2]
11b. Two functional weight-bearing digits on each limb; either nails enlarged as hooves or weight borne on sole-pads of toes ....*ORDER ARTIODACTYLA..................16

12a. Fur on dorsal surface with dense sharp spines; pointed snout; short rounded ears; tail inconspicuous ...[NZ].........Hedgehog.........*ORDER INSECTIVORA.........ERINACEIDAE *Erinaceus*[1]
12b. Fur without sharp spines 13

13a. Ears long, extending at least to eye if bent forward; tail shorter than ears and with tuft of fine hair; gait quadrupedal bound or hop. rhinarium [bare skin on nose] divided vertically. Either..............
... solitary except during breeding season, no permanent burrows ....[A, NZ], hare ......*Lepus*[1]
or colonial, warrens [elaborate systems of interconnecting burrows] conspicuous.....[A, NZ], rabbit...
....*Oryctolagus* [1] ....*ORDER LAGOMORPHA ....LEPORIDAE
13b. Ears not extending to eye and shorter than tail 14

14a. Incisors 3/3, canines usually prominent; equal numbers of manal and pedal digits [5+5 or 4+4]; all digits subequal in length; small to medium carnivores.........Dogs, foxes, cats, weasels......... ....*ORDER CARNIVORA [FISSIPEDIA].........Key A7
14b. Incisor teeth 1/1, canines absent; unequal numbers of manal and pedal digits, manus with vestigial pollux ....Rodents ORDER RODENTIA 15

15a. Tail uniformly furred with long hairs; [WA] Squirrels. *SCIURIDAE (Palm squirrels, *Funambulus*, escaped from Perth zoo, may be found in surrounding areas; similar escapees in Mosman, NSW apparently extinct. Eastern grey squirrel, *Sciurus carolinensis*, introduced in Melbourne and Ballarat, 19th Century, now extinct)
15b. Tail scaly and sparsely haired [A, NG,*NZ]. Rats, mice ....MURIDAE Table 1

16a. Males with head adornments, either horns or antlers 17
16b. Males without head adornments 18

17a. Medium to large animals with unbranched permanent horns on heads of males and often females ....BOVIDAE [A] Blackbuck, *Antelope*[1];.........[A] Cattle, *Bos*[2],........................[A] Buffalo, *Bubalus*[1], ...[A, NZ] Sheep *Ovis*[1], Goats *Capra*[1]; [NZ] Chamois *Rupicapra*[1], Tahr *Hemitragus* [1]
917b.Males with prominent branched antlers on head most of year or short pedicles [stumps of bone] when antlers shed; females without head adornments ....
Deer.............................CERVIDAE. [A, NG, NZ] *Axis*[2], *Cervus*[4], *Dama*[1], [NZ] *Alces*[1], *Odocoileus*[1]

18a. Very large; relatively small head on long curved neck; humped back usually visible;.................... ....[A].Camel ...CAMELIDAE ...*Camelus*[1]
18b. Medium to large; long snout with tusks [curved canine teeth] often visible; short neck; short tail with terminal tuft of hair ....[A, NG, NZ] Pigs ....SUIDAE ....*Sus*[1]

Table 1. Characteristics of genera of Muridae.

Recognition Features	Distribution	Genus	No. Species
Hindfeet fully webbed; fur very fine; ears tiny; tail with ventral crest; large size	NG	*Crossomys*	1
Hindfeet fully webbed; fur fine; medium to large size	A, NG	*Hydromys*	5
Fur sleek; hindfoot narrow; 3 molars; medium size	NG	*Leptomys*	3
Fur sleek, short, grey; eye tiny; 1 molar; small	NG	*Mayermys*	1
Fur very short, sleek, eye tiny, upper incisor grooved;	NG	*Microhydromys*	2
Fur sleek, short; eye tiny; 2 molars; tail mottled; small	NG	*Neohydromys*	1
Hindfeet partly webbed; fur coarse; 2 molars; large size	NG	*Parahydromys*	1
Fur sleek, 2 molars; medium size	NG	*Paraleptomys*	2
Fur sleek, short; eye tiny; 2 molars; small	NG	*Pseudohydromys*	2
Fur short, silky; 2 molars; small	A	*Xeromys*	1
Terminal 2/3 of tail white; upper incisors very broad	NG	*Anisomys*	1
Long, prehensile, haired tail	NG	*Coccomys*	2
Tail with crest of long black hairs; large ears	A, NG	*Conilurus*	1, †1
Prehensile tail with hexagonal scales	NG	*Chiruromys*	3
Very large; broad incisors; cheek teeth high crowned	NG	*Hyomys*	2
Small; short tail	A	*Leggadina*	2
Ears long; builds large stick nests	A	*Leporillus*	1, †1
Small; very long tail; long pointed ears	NG	*Lorentzimys*	1
Medium; tail very long; molars 3 but tiny	NG	*Macruromys*	2
Very large; fur long and coarse; molars complex	NG	*Mallomys*	4
Tail short; head broad and rounded; complex molars	A	*Mastacomys*	1
Small to medium; tail with mosaic scales, no visible hairs	A, NG	*Melomys*	17
Long hairy tail with white tip; large ears	A	*Mesembriomys*	2
Upper incisor with distinct notch; distinctive musky odour; small	*A,*NZ	*Mus*	1
Small, long hind limbs	A	*Notomys*	5, †4
Prehensile tail with hexagonal scales	NG	*Pogonomelomys*	3
Prehensile tail with rectangular scales	A, NG	*Pogonomys*	4
Small; large eyes;	A, NG	*Pseudomys*	23, †1
Fur usually with soft spines; small to medium	A, NG,*NZ	*Rattus*	15, †1
Small, fur with or without spines (included in *Rattus* by Flannery, 1990)	NG	*Stenomys*	4
Large; tail appears naked with small rectangular scales; incisors oval in cross section	A, NG	*Uromys*	4
Large; tail with large diamond-shaped scales	NG	*Xenuromys*	1
Small; tail thickened at base	A	*Zyzomys*	5

KEY A2 TO FAMILIES AND GENERA OF MONOTREMES

1a. Snout spatulate, leathery; feet webbed; tail prominent, dorso-ventrally flattened; fur fine and dense; aquatic; burrows in banks of streams, crepuscular [most active around dawn and dusk]............... ....[A].........Platypus.........ORNITHORHYNCHIDAE.........*Ornithorhynchus*[1]

1b. Snout cylindrical; feet not webbed; tail vestigial; fur coarse with spines; terrestrial; active at any time of day; gait a distinctive rolling waddle..Echidnas.........TACHYGLOSSIDAE.........2

2a. Snout short and straight or slightly up-turned ....[A, NG].........*Tachyglossus*[1]

2b. Snout long and down-curved...[NG].........*Zaglossus*[2]

KEY A3 TO FAMILIES OF DASYUROMORPH MARSUPIALS

1a. Head and body length > 750mm; body plan wolf-like, stance digitigrade; 5 clawed digits on manus and pes; tail tapering as 'extension of body'; transverse black stripes on rump.. ....[Tas].Thylacine.........†THYLACINIDAE.........*Thylacinus*[1]

1b. Head and body <750mm; hallux absent or clawless; if stripes present on rump then tail hairs longer than body hairs along entire length of tail 2

2a. Prominent transverse stripes present on rump; tail hairs longer than body hairs along entire length of tail; diurnal; feeds on termites which it harvests from their galleries................. ....[WA]..Numbat...............MYRMECOBIIDAE.........*Myrmecobius*[1]

2b. Body colour variable but never with transverse stripes; tail with short hair, occasionally with terminal tuft or dorsoventral crest of longer hairs..DASYURIDAE Table 2

Table 2. Characteristics of genera of Dasyuridae.

Recognition Features	Distribution	Genus	No. Species
Prominent ears; sparsely furred tail	A, NG	*Antechinus*	11
Long hind limbs and long brush-tipped tail	A	*Antechinomys*	1
Black crest or tuft on tail; desert	A	*Dasycercus*	2
Head and ears short; reddish brown	WA	*Dasykaluta*	1
Body with numerous white spots	A, NG	*Dasyurus*	6
Long slender tail broad dorsal stripe in 1 sp. only	NG	*Murexia*	2
Three dark dorsal stripes	NG	*Myoictis*	1
White tail tip; silvery speckled fur without dorsal stripe	NG	*Neophascogale*	1
Minute; narrower feet than *Planigale*	A	*Ningaui*	3
WA white eye-ring; heath; NT long tail, sandstone	A	*Parantechinus*	2
Black tuft on tail; arboreal in forest areas	A	*Phascogale*	2
Narrow dorsal stripe	NG	*Phascolosorex*	2
Minute; flattened head	A, NG	*Planigale*	5
Like Antechinus but often with tail fat at base	A	*Pseudantechinus*	3
Black with white patches on chest, rump	Tas	*Sarcophilus*	1
Large ears, slender feet; tail often fat at base	A, NG	*Sminthopsis*	19

KEY A4 TO THE ORDER PERAMELEMORPHIA

The characters used to separate the two families of marsupial bandicoots are small details of cranial and dental morphology. Using external characters, a generic key is most practical.

1a. Tail with distal black or white crest; ears longer than eye-ear distance 2
1b. Tail without distal crest; ears shorter than or equal to eye-ear distance 3

2a. Distal crest on tail white; 3 functional manal digits ....[A] ....*Macrotis*[2]
2b. Distal crest on tail black; 2 functional manal digits ....[A] ....†*Chaeropus*[1]

3a. Ear pointed, as long as eye-ear distance ....[A]......*Perameles* [3, †1]
3b. Ear rounded, shorter than eye-ear distance 4

4a Five upper incisors, the fifth often separated from the others by a diastema [gap] 5
4b. Four upper incisors 7

5a. Soles of feet wrinkled; auditory bullae inflated ....[A, NG] ....*Isoodon* [3]
5b. Soles of feet granulated; auditory bullae not inflated ....[NG] 6

6a. Size small [HB <300mm]; distinct dorsal stripes or white tail-tip ....*Microperoryctes* [3]
6b. Size medium [HB > 300mm] without distinctive striping ....*Peroryctes* [2]

7a. Fur soft ....[Ceram only]......*Rhynchomeles* [1]
7b. Fur spinous ....[NG, Cape York] ....*Echymipera* [5]

KEY A5 TO THE FAMILIES AND GENERA OF THE DIPROTODONTIA

1a. Tail rudimentary 2
1b. Tail prominent 3

2a. Manal digits I & II at right angles to and opposable to III–V; fur with soft fine hairs; arboreal ....
........[A] ....Koala ........PHASCOLARCTIDAE........*Phascolarctos* [1]

2b All manal digits aligned; fur coarse; terrestrial and sub-fossorial, construct prominent burrows and pathways. Either: Rhinarium naked between nostrils; fur coarse; ear relatively short..........[A]......Wombat......VOMBATIDAE......*Vombatus*[1]
or: Rhinarium haired between nostrils; fur fine; ear relatively long...[A]...Wombats...VOMBATIDAE...*Lasiorhinus*[2]

3a. Hind foot approximately same length as fore foot and with prominent opposable hallux ..........PHALANGEROIDEA 5
3b. Hind foot long and narrow, more than twice length of fore foot; hallux reduced or absent ....MACROPODOIDEA 4

4a. Hind foot with small opposable hallux; body small [HB <300mm]; tail naked and scaley......[A]......Musky rat-kangaroo..........POTOROIDAE (part) ....*Hypsiprymnodon* [1]
4b. Hind foot lacking hallux; tail furred 21

5a. Tail with lateral fringe of stiff hairs..........ACROBATIDAE 6
5b. Tail without lateral fringe of stiff hairs..........7

6a. Gliding membrane (patagium) present between fore and hind limbs ....[A] ....*Acrobates* [1]
6b. Gliding membrane absent ....[NG] ......*Distoechurus* [1]

7a. Snout elongate; tongue extensible and brush-tipped; teeth mostly vestigial; very small (<20g); body with 3 longitudinal dark stripes ....[A] ..........TARSIPEDIDAE *Tarsipes* [1]
7b. Other than above 8

8a. Digits with small claws and large apical pads; size small with membranous ears and slightly furred prehensile tail ....BURRAMYIDAE 9
8b. Digits with prominent claws 10

9a. Large sectorial premolars, much higher than molars ....[A] ....*Burramys* [1]
9b. Premolars no higher than molars.......... ....[A, NG] ....*Cercartetus* [4]

10a. Tail appears fully furred with hairs of uniform length or longer at tip but may have short inconspicuous naked tip 11
10b. Tail with hair length tapering from base to tip; often with large naked area 15

11a. Gliding membrane [patagium] present between fore and hind limbs 12
11b. Gliding membrane absent 13

12a. Ear pinna with long hair on outer surface; patagium from elbow to knee ....[A] ......PSEUDOCHEIRIDAE (part)......*Petauroides*[1]
12b. Ear pinna with short hairs on outer surface; patagium from manus to pes .... [A, NG] ....PETAURIDAE (part) ....*Petaurus*[5]

13a. Pelage of head and body alternating black and white longitudinal stripes; manal digit IV much longer than others... ....[A, NG]..........PETAURIDAE (part).........*Dactylopsila* [4]
13b. Pelage other than above; manal digit IV not elongated 14

14a. Tail fully furred and club-shaped with longer hairs near tip; prominent dark mid-dorsal stripe; venter paler than dorsum ....[A] ....PETAURIDAE (part) ....*Gymnobelideus* [1]
14b. Tail with short naked tip, otherwise with hairs of uniform length; pelage unicolour...... ....[A] .........PSEUDOCHEIRIDAE (part) ....*Hemibelideus* [1]

15a. Tail with distal 50–60% naked all round or with few short hairs..........PHALANGERIDAE (part).........16
15b. Tail with some hair on distal 40–50% 18

16a. Distal half of tail with some short dorsal hairs 17
16b. Distal half of tail entirely naked and scaly ....[A]..........*Wyulda* [1]

17a. Pelage variable in colour, often patchy but without mid-dorsal stripe; ears small and hidden in fur ....[A, NG] ......*Spilocuscus*[2]
17b. Pelage unicolour, often with dark mid-dorsal stripe; ears generally prominent ....[A, NG] ...*Phalanger* [12]

18a. Ear pinna pointed; tail with terminal hairs distinctly darker than proximal hairs.................
....[A,*NZ]...PHALANGERIDAE (part) ......*Trichosurus*[2]
18b. Ear pinna rounded; tail with terminal hairs white or same colour as proximal hairs..............
....PSEUDOCHEIRIDAE (part) 19

19a. Dorsal fur long and woolly 20
19b. Dorsal fur not long and woolly ....[A, NG] ....*Pseudocheirus* (incl. *Pseudochirulus*) [7]

20a. Dorsal fur spangled with metallic sheen and either uniformly coppery coloured or greenish
with alternating dark and pale stripes ....[A, NG] ....*Pseudochirops* [4]
20b. Dorsal fur unicolor with a dark median stripe ....[A] ....*Petropseudes* [1]

21a. Claws of manal digits II, III & IV longer
than those of I & V ....POTOROIDAE Table 3
21b. Claws of all 5 manal digits subequal in
size...MACROPODIDAE Table 4

KEY A6 TO FAMILIES OF CHIROPTERA

1a. Tail absent or vestigial or if present, not enclosed in a patagium; noseleaf and tragus absent;
manal digit II usually clawed ....PTEROPODIDAE Table 5
1b. Tail prominent and at least partially enclosed in a patagium or if absent,
noseleaf and tragus large ....MICROCHIROPTERA..........2

2a. Tail absent; noseleaf and tragus prominent ...[A]......Ghost bat......MEGADERMATIDAE......*Macroderma* [1]
2b. Tail present 3

3a. Tail protrudes >5 mm from uropatagium...4
3b. Tail entirely within uropatagium or slightly protruding...6

4a. Tail extends beyond posterior border of uropatagium ....MOLOSSIDAE Table 6
4b. Tail emerges from sheath on dorsal surface of uropatagium 5

5a. Ears and tragus rounded, manal digit II lacks phalanges............EMBALLONURIDAE.........
......NO GENERIC KEYS AVAILABLE [NG] *Emballonura* [5]; [A, NG] ....*Saccolaimus* [3], *Taphozous* [5]
5b. Ears and tragus elongated; manal digit II has 1 minute phalanx ....[NZ] MYSTACINIDAE...*Mystacina*[2]

6a. Tragus present; noseleaf absent or small ....VESPERTILIONIDAE Table 7
6b. Tragus absent; noseleaf prominent...7

7a. Toes 2–5 with 3 phalanges; noseleaf pointed
....[A, NG] ......Horseshoe bats .........RHINOLOPHIDAE ......*Rhinolophus*[4]
7b. All toes with 2 phalanges; noseleaf square ....[A, NG] ............HIPPOSIDERIDAE Table 8

KEY A7 TO THE FISSIPEDIA

1a. Manus and pes with 5 digits; body elongate; limbs short; stance plantigrade......
....[NZ]Weasel, Stoat, [A, NZ]Ferret ....*MUSTELIDAE *Mustela*[3]
1b. Manus and pes with 4 digits; stance digitigrade 2

2a. Claws retractable (do not show in footprint); snout shorter than eye-ear distance...............
....[A, NG, NZ].........Cat *FELIDAE...............*Felis* [1]
2b. Claws non-retractable (prominent in footprint); snout normally longer than eye-ear distance...
....[A] Dog, dingo, fox; [NG, NZ] dog ....*CANIDAE ......*Canis* [1], *Vulpes* [1]

Table 3. Characteristics of genera of Potoridae.

Recognition Features	Genus	No. Species
Rufous fur; hairy muzzle	*Aepyprymnus*	1
Tail either with white tip [WA islands, Tas] or black crest	*Bettongia*	3
Tail longer than head and body	*Caloprymnus*	†1
Snout pointed, slender tail	*Potorous*	3, †1

Table 4. Characteristics of genera of Macropodidae.

Recognition Features	Distribution	Genus	No. Species
Short hind feet, robust fore-limbs; arboreal	A, NG	*Dendrolagus*	9
Long slender head, tail arched when used as prop	NG	*Dorcopsis*	4
Smaller than *Dorcopsis*, tail short	NG	*Dorcopsulus*	2
Small, relatively short-tailed	A	*Lagorchestes*	2, †2
Small, transverse dark and light banding on back	A	*Lagostrophus*	1
Variable size and colour; most hop with head high	A, NG,*NZ	*Macropus*	14, †1
Slender; white shoulder and hip stripes	A	*Onychogalea*	2, †2
Long tail; hop with outstretched arms; rocky outcrops	A,*NZ	*Petrogale*	11
Short tail, short rounded ears	WA	*Setonix*	1
Small, dark; rainforest and thick vegetation	A, NG	*Thylogale*	4
Medium-large, dark; hops with head low to ground	A,*NZ	*Wallabia*	1

Table 5. Characteristics of genera of Pteropodidae.

Recognition Features	Distribution	Genus	No. Species
Small size, back furred, nostrils not tubular	NG	*Rousettus*	1
Large size, back furred, mantle same colour as body	NG	*Aproteles*	1
Large size, back furred, mantle different colour from body	A, NG	*Pteropus*	8
Medium size; back naked	A, NG	*Dobsonia*	2
Small size, long snout; no uropatagium	A, NG	*Syconycteris*	2
Small size, long snout; small uropatagium	A, NG	*Macroglossus*	1
Small size, tubular nostrils, dark dorsal stripe	A, NG	*Nyctimene*	6
Small size, tubular nostrils, no dorsal stripe	NG	*Paranyctimene*	1

Table 6. Characteristics of genera of Molossidae.

Recognition Features	Distribution	Genus	No. Species
Ears with 4 small papillae on upper margin; lips wrinkled	A, NG	*Chaerephon*	1
Face and ears less complex	A, NG	*Mormopterus*	3
Throat pouch present	A, NG	*Tadarida*	1
Ears with serrated ear margins	NG	*Otomops*	2

Table 7. Characteristics of genera of Vespertilionidae.

Recognition Features	Distribution	Genus	No. Species
Fleshy lobes at corner of mouth and base of ear	A, NG, NZ	*Chalinolobus*	7
2 pairs upper incisors, I2 minute; penis hairy	A	*Falsistrellus*	2
Long thin tragus without notch at base	NG	*Kerivoula*	1
Terminal bone of longest finger extremely long, folded back when at rest	A, NG	*Miniopterus*	6
Nostrils tubular	A, NG	*Murina*	1
Very large feet	A, NG	*Myotis*	1
Small noseleaf; ears very large	A, NG	*Nyctophilus*	8
Larger noseleaf, large ears	NG	*Pharotis*	1
Penis lacks prepuce; vulva small, longitudinal	NG	*Philetor*	1
Long thin tragus with deep notch at base	A, NG	*Phoniscus*	1
2 pairs upper incisors, inner pair not distinctly forked; 2 upper premolars	A, NG	*Pipistrellus*	7
1 pair upper incisors; no gap between I and canine; large size	A	*Scoteanax*	1
1 pair upper incisors; gap between I and canine	A, NG	*Scotorepens*	3
2 pairs upper incisors, inner pair distinctly forked; 1 upper premolar	A	*Vespadelus*	9

Table 8. Characteristics of genera of Hipposideridae.

Recognition Features	Distribution	Genus	No. Species
Noseleaf with 3 prominent dorsal projections	NG	*Aselliscus*	1
Noseleaf flat dorsally	A, NG	*Hipposideros*	11
Noseleaf rounded dorsally	A	*Rhinonycteris*	1

Table 9. Characteristics of genera of Otariidae.

Recognition Features	Distribution	Genus	No. Species
Snout relatively pointed	A, NZ	*Arctocephalus*	2
Snout blunt; males with mane pale dorsally	A	*Neophoca*	1
Snout blunt; males with dark brown mane	NZ	*Phocarctos*	1

Table 10. Characteristics of genera of Phocidae.

Recognition Features	Genus	No. Species
Long slender body, large head; wide gape; fur dense blue-black dorsally	*Hydrurga*	1
Large, fat body; small head; fur dense blue-black dorsally	*Leptonychotes*	1
Slender; small head; fur dense, brown	*Lobodon*	1
Males very large with inflatable proboscis; short hair, dark grey	*Mirounga*	1
Slender small seal; short hair, dark brown dorsally	*Ommatophoca*	1

Table 11. Characteristics of genera of Ziphiidae.

Recognition Features	Distribution	Genus	No. Species
Lower jaw longer than upper, with 2 pairs teeth	Southern oceans	*Berardius*	1
Pronounced melon; 1 pair teeth terminal	Southern oceans	*Hyperoodon*	1
Known only from skull	Tropical waters	*Indopacetus*	1
Teeth subterminal or further back on jaw	All oceans	*Mesoplodon*	7
Many small teeth behind terminal pair	Southern oceans	*Tasmacetus*	1
Short rounded beak; 1 pair terminal teeth	All oceans	*Ziphius*	1

Table 12. Characteristics of genera of Delphinidae other than *Lissodelphis*.

Recognition Features	Distribution	Genus	No. Species
Small rounded dorsal fin	All oceans	*Cephalorhynchus*	1
Distinctive black/white pattern with black stripes through eye and from lower jaw to flippers	All oceans	*Delphinus*	1
Small, blunt-headed; uniformly dark	Tropical seas	*Feresa*	1
Black with pale or white behind dorsal fin; pronounced melon	All oceans	*Globicephala*	2
Pronounced melon obscures beak; fins darker than body	All oceans	*Grampus*	1
Very small beak; medium size; white flank stripe	All oceans	*Lagenodelphis,*	1
Virtually no beak; medium size; darker above	All oceans	*Lagenorhynchus*	1
Small dolphin with bulbous head	Tropical seas	*Orcaella*	1
Distinctive black and white; males with large triangular dorsal fin	All oceans	*Orcinus*	1
Rounded forehead; white lips	All oceans	*Peponocephala*	1
Slender; rounded snout; black	All oceans	*Pseudorca*	1
Humped back; very pale pink to almost white	Tropical seas	*Sousa*	1
Slender active dolphins with pronounced patterns of striping	All oceans	*Stenella*	3
Smooth forehead without melon	All oceans	*Steno*	1
Stout beak; generally brown with indistinct patterning on sides	All oceans	*Tursiops*	2

KEY A8 TO THE PINNIPEDIA

1a. Hind feet capable of rotation for locomotion on land; ear pinnae prominent............... ....Fur seals and Sea lions ....OTARIIDAE ....Table 9
1b. Hind feet cannot be rotated; ear pinnae absent ....[A, NZ] ....Seals ......PHOCIDAE............Table 10

KEY A9 TO THE ORDER CETACEA

1a. Baleen plates in upper jaw, teeth absent; blowholes paired ....MYSTICETI...........................2
1b. Baleen absent, teeth present or absent; blowhole single ....ODONTOCETI...........................3

2a. Numerous throat grooves extending posteriorly from lower lip; dorsal fin prominent; medium to large whales: either head smooth; flippers short, tapered, seldom show flukes on diving (Rorquals *Balaenoptera*[5]) or head with many tubercles; flippers long; often show flukes (Humpback whale *Megaptera* [1]) ....BALAENOPTERIDAE

2b. Throat grooves absent; small whale with dorsal fin (*Caperea*[1]) or large whale with no dorsal fin but prominent barnacles on head......(*Eubalaena*[1]) ....Right whales.........BALAENIDAE

3a. Small to medium whales with two short throat grooves, snout prominently beaked, 1 or 2 pairs prominent teeth in lower jaw ....Beaked whales...........ZIPHIIDAE...............Table 11
3b. Throat grooves absent 4

4a. Snout blunt and squarish with lower jaw narrow and underslung [not reaching end of snout]; either large whale, snout with terminal blowhole, no true dorsal fin (*Physeter* [1]) or small whale (*Kogia* [2]) ....Sperm whales.........PHYSETERIDAE
4b. Small to medium whales and dolphins with rounded to beaked snouts but jaws not underslung........................5

5a. Dorsal fin absent 6
5b. Dorsal fin present 7

6a. Body colouring uniformly bluish to slate grey, paler below; snout rounded with prominent melon; ....[tropical waters] ....PHOCOENIDAE (part).........*Neophocoena*[1]
6b. Snout, venter and flippers white; Dorsum and flukes black; snout pointed;..................... ....[NZ, Tas].........DELPHINIDAE (part).........*Lissodelphis*[1]

7a. Snout rounded; body with distinct line between black dorsum and flukes and white venter and flippers; dorsal fin broadly triangular (female) or large and rounded (male);............ ....[subantarctic waters].........PHOCOENIDAE (part)..........*Australophocoena*[1]
7b. Snout rounded to beaked, melon prominent to absent; body pattern variable but never distinctly demarcated dorsoventral black/white ....DELPHINIDAE (part).........Table 12

KEYS TO SKULLS

PRELIMINARY KEY (B1) TO MAMMAL SKULLS

1a. Teeth absent; external nares terminal; rostrum [portion of skull anterior to the orbit] either tubular or spatulate..MONOTREMATA.........2
1b. Teeth present or absent, if absent nares dorsal to orbit and skull massive (>300mm) 4

2a. Rostrum spatulate; premaxillae separate and hooked; dentary robust... ....[A] ....Platypus .........ORNITHORHYNCHIDAE ......*Ornithorhynchus*
2b. Rostrum tubular; premaxillae joined anteriorly, dentary fragile with rudimentary coronoid and angular processes... ....TACHYGLOSSIDAE.........3

3a. Rostrum long and downcurved [NG] ....*Zaglossus*
3b. Rostrum short and straight or slightly upturned [A, NG] ....*Tachyglossus*

4a. At least two of the following characters present: large posterior palatal vacuities [gaps in the posterior half of the bony palate], nasal bones wider posteriorly than anteriorly, jugal bone [middle bone in the zygomatic arch or "cheekbone"] extends posteriorly to the point of articulation of the lower jaw, lacrimal bone extends out of orbit onto rostrum, angular process of dentary inflected [the part of the lower jaw for attachment of jaw-opening muscles; ventral to the point of articulation, is turned inwards] .... MARSUPIALIA KEY B2
4b. Alisphenoid entirely within temporal fossa, jugal if present not extending to point of articulation of the lower jaw, angular process not inflected, no palatal vacuities, nasals if present wider anteriorly than posteriorly; lacrimal generally contained within the orbit .... EUTHERIA KEY B3

KEY B2 TO THE SKULLS OF MARSUPIAL FAMILIES

1a. Polyprotodont [lower incisors 2 or 3, approximately equal in size] 2
1b. Diprotodont [first lower incisor large and directed forward] 7

2a. Upper molars with 3 or 4 major cusps in a triangular or square arrangement plus a lateral stylar shelf containing several small cusps; molars in contact with adjacent teeth 4
2b. Upper molars simplified; lacking either stylar shelf or one or more major cusps 5

3a. Molars 4, greatly simplified with either paracone or metacone absent; cheek tooth rows arched .... NOTORYCTIDAE
3b. Molars > 4, with paracone and metacone present, lacking stylar shelf; cheek tooth rows almost parallel .... MYRMECOBIIDAE

4a. Upper incisors 5, often with a diastema [large gap] between I4 and I5; auditory bullae inflated .... PERAMELIDAE
4b. Upper incisors 4 or 5 (if 5 then auditory bullae not inflated) 5

5a. Rostrum elongated with pronounced diastema between incisors, canine and premolars .... PERORYCTIDAE
5b. No large diastemata between upper teeth 6

6a. Auditory bullae inflated, or if not inflated lower molars with 4 or 5 prominent cusps; skull length < 150mm .... DASYURIDAE
6b. Auditory bullae not inflated, only 3 prominent cusps on lower molars; skull length >150mm .... THYLACINIDAE

7a. Molars tiny and peglike and pointed; skull slender; dentaries extremely slender with rudimentary processes .... TARSIPEDIDAE
7b. Molars prominent and multi-cusped 8

8a. Single upper and lower incisor only .... VOMBATIDAE
8b. Upper incisors 3; vestigial lower incisors (I2 & I3) may be present 9

9a. Dentary with masseteric fossa [hollow in outer surface of the ascending or coronoid process for attachment of the masseter muscle] extending deep into the ramus [tooth-bearing portion of lower jaw] as a canal 10
9b. Dentary without deep masseteric canal 11

10a. Lower incisors occlude with crowns of I2 and I3; I1 pointed; upper canine prominent; upper molars M3 and M4 smaller than the larger of M1 or M2 .... POTOROIDAE
10b. Lower incisors occlude on medial surface of I2 and I3; I1 broad, upper canine reduced or absent; upper molars M3 and M4 larger than the larger of M1 or M2 .... MACROPODIDAE

11a. Only one upper premolar .... PHASCOLARCTIDAE
11b. Two or three upper premolars 12

12a. Final upper premolar high-crowned, conical and aligned at an angle to the line of the molar teeth .... PHALANGERIDAE
12b. Final upper premolar not as above 13

13a. Lower molars 4 all subequal in size; upper molars with distinct crescent-shaped ridges .... PSEUDOCHEIRIDAE
13b. Lower molars 3–4, M4 if present much smaller than M1; upper molars without crescent-shaped ridges 14

14a. Molars 4/4; final upper premolar narrower anteriorly than posteriorly .... PETAURIDAE
14b. Molars 3/3 or 4/4; final upper premolar of uniform width anteriorly and posteriorly 15

15a. Molars 3/3; upper premolars P1 and P2 approximately same height as P3; upper canine strong and longer than I1; auditory bulla lacks alisphenoid component .... ACROBATIDAE
15b. Molars 4/4 or 3/3; if molars 3/3 then upper premolars P1 and P2 very much reduced; auditory bulla has typical marsupial alisphenoid component .... BURRAMYIDAE

KEY B3 TO THE SKULLS OF EUTHERIAN FAMILIES

1a. External nares terminal, directed anteriorly 2
1b. External nares on dorsal surface of cranium 24

2a. Canines absent from both upper and lower dentition; one chisel-like lower incisor; either one chisel-like upper incisor or if I2 present it is located immediately posterior to I1 3
2b. Canines variable: if upper canine absent then lower canine incisiform 5

3a. Incisors 2/1 with I2 immediately posterior to I1 .... [A, NZ]...... LEPORIDAE
3b. Incisors 1/1 .... ORDER RODENTIA...... 4

4a. Cheek teeth 5/4 (premolars 2/1, molars 3/3) .... [WA]...... SCIURIDAE
4b. Premolars absent, molars 1–3/1–3 .... [A, NG, NZ]...... MURIDAE

5a. Upper incisors absent; upper canine absent or small; lower canine incisiform 6
5b. At least 1 upper incisor present; canines variable 7

6a. Single lacrimal foramen in anterior of orbit; skull often has horns consisting of a permanent bony core and an outer keratinous sheath .... BOVIDAE
6b. Two lacrimal foramina; skulls of males typically have antlers—deciduous, often branching, bony structures growing from a permanent pedicle .... CERVIDAE

7a. Incisors 3/3, or 1/3 with upper incisor caniniform; six large selenodont or lophodont cheek teeth in each jaw; complete postorbital bar 8
7b. Incisors variable; cheek-teeth neither selenodont nor lophodont 9

8a. Incisors 3/3; canines present in some, cheek teeth lophodont .... EQUIDAE
8b. Incisors 3/3 in juveniles, 1/3 in adults; adult I3 caniniform .... CAMELIDAE

9a. Complete postorbital bar separating orbit and temporal fossa, incisors 2/2; canines not markedly longer than incisors; cheek teeth bunodont [rounded cusps]; rostrum extremely shortened; cranium expanded and rounded .... HOMINIDAE
9b. Postorbital bar never complete 10

10a. Incisors 3/3 11
10b. Incisors other than 3/3 14

11a. Canines enlarged, curved and directed laterally with prominent flattened wear surfaces forming tusks in adults; cheek teeth bunodont, increasing in size from front to rear .... SUIDAE
11b. Canines prominent but not forming tusks; final upper premolar and first lower molar carnassial shearing teeth 12

12a. Rostrum relatively long; premolars 4/4; molars 2/3 .... CANIDAE
12b. Rostrum shortened, premolars and molars fewer than above 13

13a. Premolars 3/3, molars 1/2 .... MUSTELIDAE
13b. Premolars 3/2, molars 1/1 .... FELIDAE

14a. Skulls large (total length >80mm); incisors 3/2 or 2/2; cheek teeth homodont and either conical or with anterior and posterior cusps but no medial cusps 15
14b. Skulls small; incisors variable; cheek teeth bunodont or lambdodont [cusps arranged in one or more inverted V-shaped formations] 16

15a. Upper incisors 3 .... OTARIIDAE
15b. Upper incisors 2 .... PHOCIDAE

16a. Incisors 3/2; first upper and lower incisors longer than canines .... [NZ]...ERINACEIDAE
16b. Canines large, upper incisors 2 or fewer 17

17a. Upper molars bunodont...[A, NG] .... PTEROPODIDAE
17b. Upper molars lambdodont 18

18a. Premaxilla and upper incisors absent .... MEGADERMATIDAE
18b. Premaxilla and 1 or more upper incisors present 19

19a. One lower incisor .... [NZ]...MYSTACINIDAE
19b. Two lower incisors 20

20a. Deep basisphenoid pits between auditory bullae .... EMBALLONURIDAE
20b. Not as above 21

21a. Rostrum inflated dorsally 22
21b. Rostrum not inflated 23

22a. 3 lower premolars .... RHINOLOPHIDAE
22b. 2 lower premolars .... HIPPOSIDERIDAE

23a. Upper molars with additional posterior lingual cusp .... MOLOSSIDAE
23b. Upper molars not as above .... VESPERTILIONIDAE

24a. Nasals absent; zygomatic arch robust; premaxillae angled sharply ventrally .... DUGONGIDAE
24b. Nasals present; zygomatic arch slender or incomplete; premaxillae slender and elongated .... CETACEA...25

25a. Nasals large, covering nasal passage; teeth absent 26
25b. Nasals small, not covering nasal passage; teeth present 27

26a. Posterior border of nasals and premaxillae anterior to supraorbital process of frontal; maxillae without nasal processes .... BALAENIDAE
26b. Posterior border of nasals and premaxillae posterior to supraorbital process of frontal; maxillae with nasal processes .... BALAENOPTERIDAE

27a. Teeth in upper and lower jaw >15, spatulate .... PHOCOENIDAE
27b. Teeth not spatulate 28

28a. Either 1 or 2 pairs of teeth in lower jaw only or if >2, teeth heterodont .... ZIPHIIDAE
28b. Lower teeth >2, homodont 29

29a. Lower teeth >8, upper teeth absent or vestigial .... PHYSETERIDAE
29b. Either 3 to 7 teeth in lower jaw and upper teeth absent or teeth present in both upper and lower jaw .... DELPHINIDAE

GEOGRAPHIC AND TAXONOMIC GUIDE TO PUBLICATIONS

The most important reference works for mammals of the world are Nowak (1999), which gives a short account of all the genera of living and recently extinct mammals and Anderson and Jones (1984), which contains detailed diagnostic characters for orders and families. Wilson and Reeder (1993) is an essential taxonomic and geographic reference but does not contain descriptions or diagnoses. Lawlor (1979) is a handy compact guide to the world's mammals and has good keys to the skulls.

For marine mammals, there are a number of useful worldwide guides such as King (1983) on pinnipeds, Reeves *et al.* (1992) on pinnipeds and sirenians; and Carwardine (1995) and Hoyt (1984) on cetaceans. Closer to home, Baker (1983) contains much detailed information on New Zealand and Australian distributions of whales and dolphins besides being an excellent identification guide, and Tucker (1989) is a handy guide to Australian waters. Bryden *et al.* (1998) not only has descriptions of Australasian marine mammals but also gives a summary of marine mammal biology.

For years Australia depended on the many editions of Troughton's *Furred Animals of Australia* from the first edition in 1941 to the 9th edition in 1967. The following two decades saw an increase in interest in mammals beginning with Ride (1970) and peaking with the massive collaborative efforts published as Strahan (1983), Walton and Richardson (1989) and Strahan (1995a). During this period there appeared several regional works including a reprint in 1968 of Jones (1923–25) South Australian work, Hyett and Shaw (1980) on south-eastern Australia and Green (1983) for Tasmania. A number of smaller works intended as pocket guides have been produced including Taylor (1984), Cronin (1991) and Strahan (1995b). Menkhorst (2001) has produced the first comprehensive illustrated field guide to the mammals of all of Australia with colour illustrations and some keys.

At a regional level, Tasmania is best served with useful small books by Green (1994) and Watts (1993), although Green (1983) remains a very handy guide to the skulls of Tasmanian mammals. South Australian small mammals are briefly described in Watts (1972) and Victorian mammals have received a recent comprehensive review in Menkhorst (1995).

For tracks and traces, Triggs (1996) is indispensable but Morrison (1981) still contains useful information. Both these pocket or knapsack-sized books contain useful photographs of skulls.

For Australian forms, books of more restricted taxonomic scope include Marlow (1962), on marsupials, and several works on bats and rodents. The recent work on bats by Churchill (1998) virtually supersedes all predecessors. There are, however useful descriptions and keys in Hall and Richards (1979), Parnaby (1992) and Reardon and Flavel (1987). For rodents, recent taxonomic revisions have made a number of changes, mainly additions, to the fauna but the main reference is still Watts and Aslin (1981) and Covacevich and Easton (1974) is still useful in the north-east.

New Zealand, with its predominantly introduced mammal fauna has been admirably covered by the compilation of data in King (1990), but the New Guinea region has been poorly documented. For many years researchers depended on the lists produced by Tate (1944) and Laurie and Hill (1954). Menzies and Dennis (1979) produced a small handbook of rodents, which has still not been superseded, and Ziegler (1982) attempted a checklist for the mammal fauna of the main land mass. Menzies (1991) has since produced another useful small book on the monotremes and marsupials. Flannery (1990, 1995) has produced two recent compilations, based primarily on his own explorations. The recent field guide to Papua New Guinea bats (Bonaccorso, 1998) has keys to species level; however, this region is still very poorly covered and there is enormous scope for revision particularly among the bats and rodents.

REFERENCES

Anderson, S. and Jones, J.K. Jnr. (eds). 1984. *Orders and Families of Recent Mammals of the World.* Wiley, New York. 686pp.

Aplin, K.P. 2006. Ten million years of rodent evolution in Australasia: phylogenetic evidence and a speculative historical biogeography. Pp.707–744. In Merrick, J.R., Archer, M., Hickey, G.M. and Lee, M.S.Y. (eds). *Evolution and Biogeography of Australasian Vertebrates.* Auscipub Pty. Ltd., Sydney. 942pp.

Archer, M. and Clayton, G. (eds). 1984. *Vertebrate Zoogeography and Evolution in Australasia.* Hesperian Press, Perth, W.A. 1203pp.

Archer, M. and Hand, S.J. 2006. The Australian marsupial radiation. Pp.575–646. In Merrick, J.R., Archer, M., Hickey, G.M. and Lee, M.S.Y. (eds). *Evolution and Biogeography of Australasian Vertebrates.* Auscipub Pty. Ltd., Sydney. 942pp.

Baker, A.N. 1983. *Whales and Dolphins of New Zealand and Australia.* Victoria University Press, Wellington. 133pp.

Bonaccorso, F. J. 1998. *Bats of Papua New Guinea.* University of Chicago Press. 489pp.

Brunner, H. and Coman, B.J. 1974. *The Identification of Mammalian Hair.* Inkata Press, Melbourne. 176pp.

Brunner, H., Triggs, B. and Ecobyte Pty. Ltd. 2002. *Hair ID; an Interactive Tool for Identifying Australian Mammal Hair.* CD-ROM. CSIRO, Melbourne.

Bryden, M., Marsh, H. and Shaughnessy, P. 1998. *Dugongs, Whales, Dolphins and Seals: A Guide to the Sea Mammals of Australasia.* Allen and Unwin, Sydney. 176pp.

Carwardine, M. 1995. *Eyewitness Handbooks: Whales Dolphins and Porpoises.* Stoddart, Toronto. 255pp.

Churchill, S. 1998. *Australian Bats.* Reed New Holland, Sydney. 230pp.

Cooke, B. 2006. Kangaroos. Pp.647–672. In Merrick, J.R., Archer, M., Hickey, G.M. and Lee, M.S.Y. (eds). *Evolution and Biogeography of Australasian Vertebrates.* Auscipub Pty. Ltd., Sydney. 942pp.

Corbett, L. 2006. The Australian dingo. Pp.745–753. In Merrick, J.R., Archer, M., Hickey, G.M. and Lee, M.S.Y. (eds). *Evolution and Biogeography of Australasian Vertebrates*. Auscipub Pty. Ltd. Sydney. 942pp.

Covacevich, J. and Easton, A. 1974. *Rats and Mice in Queensland*. Queensland Museum Booklet No.8, Brisbane. 72pp.

Cronin, L. 1991. *Key Guide to Australian Mammals*. Reed, Sydney. 190pp.

Flannery, T.F. 1990. *Mammals of New Guinea*. Robert Brown and Associates, Carina, Qld. 439pp.

Flannery, T.F. 1995. *Mammals of the South-West Pacific & Moluccan Islands*. Reed and Australian Museum, Sydney. 464pp.

Fordyce, R.E. 2006. A southern perspective on cetacean evolution and zoogeography. Pp.755–777. In Merrick, J.R., Archer, M., Hickey, G.M. and Lee, M.S.Y. (eds). *Evolution and Biogeography of Australasian Vertebrates*. Auscipub Pty Ltd., Sydney. 942pp.

Green, R.H. 1983. *An Illustrated Key to the Skulls of Mammals in Tasmania*. Queen Victoria Museum and Art Gallery, Launceston. 100pp.

Green, R.H. 1994. *The Fauna of Tasmania: Mammals*. Potoroo Publishing, Launceston. 56pp.

Hall, L.S. and Richards, G.C. 1979. *Bats of Eastern Australia*. Queensland Museum Booklet, No.12. Brisbane. 66pp.

Hand, S.J. 2006. Bat beginnings and biogeography: the Australasian record. Pp.673–705. In Merrick, J.R., Archer, M., Hickey, G.M. and Lee, M.S.Y. (eds). *Evolution and Biogeography of Australasian Vertebrates*. Auscipub Pty. Ltd., Sydney. 942pp.

Hoyt, E. 1984. *The Whale Watcher's Handbook*. Penguin and Madison Press, Toronto. 208pp.

Hyett, J. and Shaw, N. 1980. *Australian Mammals: A Field Guide for New South Wales, Victoria, South Australia & Tasmania*. Nelson, Melbourne. 270pp.

Johanson, Z. 2006. Origins of mammals: morphology, molecules and a petrosal or two. Pp.491–522. In Merrick, J.R., Archer, M., Hickey, G.M. and Lee, M.S.Y. (eds). *Evolution and Biogeography of Australasian Vertebrates*. Auscipub Pty. Ltd., Sydney. 942pp.

Jones, B. and Baynes, A. 1989. Illustrated keys to Australian Mammalia to generic level. Pp.1075–1171. In Walton, D.W. and Richardson, B.J. (eds). *Fauna of Australia. Vol. 1B. Mammalia*. Australian Government Publishing Service, Canberra. Pp.401–1227.

Jones, F.W. 1923–1925. *The Mammals of South Australia*. Government Printer, Adelaide. 458pp. (Reprinted 1968).

King, C.M. (ed.) 1990. *The Handbook of New Zealand Mammals*. Oxford University Press, Auckland. 600pp.

King, J.E. 1983. *Seals of the World*. British Museum (Natural History), London. 240pp.

Laurie, E.M.O. and Hill, J.E. 1954. *List of Land Mammals of New Guinea, Celebes and Adjacent Islands*. British Museum (Natural History). London. 175pp.

Lawlor, T.E. 1979. *Handbook to the Orders and Families of Living Mammals*. Mad River Press, Eureka, California. 327pp.

Marlow, B.J. 1962. *Marsupials of Australia*. Jacaranda Press, Sydney. 141pp.

Menkhorst, P.W. (ed). 1995. *Mammals of Victoria*. Oxford University Press., Melbourne. 359pp.

Menkhorst, P.W. 2001. *A Field Guide to the Mammals of Australia*. Oxford University Press, Melbourne. 269pp.

Menzies, J.I. 1991. *A Handbook of New Guinea Marsupials and Monotremes*. Kristen Press, Madang. 138pp.

Menzies, J.I. and Dennis, E. 1979. *Handbook of New Guinea Rodents*. Wau Ecology Institute Handbook No. 6. 68pp.

Merrilees, D. and Porter, J.K. 1979. *Guide to the Identification of Teeth and Some Bones of Native Land Mammals Occurring in the Extreme South West of Western Australia*. Western Australian Museum. Perth. 152pp.

Morrison, R.G.B. 1981. *A Field Guide to the Tracks & Traces of Australian Animals*. Rigby, Adelaide. 192pp.

Musser, A.M. 2006. Furry egg-layers: monotreme relationships and radiations. Pp.523–550. In Merrick, J.R., Archer, M., Hickey, G.M. and Lee, M.S.Y. (eds). *Evolution and Biogeography of Australasian Vertebrates*. Auscipub Pty. Ltd., Sydney. 942pp.

Nowak, R.M. 1999. *Walker's Mammals of the World*. 6th Edit. Johns Hopkins University Press, Baltimore. 1936pp.

Parnaby, H. 1992. An interim guide to identification of insectivorous bats of south-eastern Australia. *Technical Report of the Australian Museum*, No. 8, pp.1–33.

Reardon, T.B. and Flavel, S.C. 1987. *A Guide to the Bats of South Australia*. South Australian Museum, Adelaide. 87pp.

Reeves, R.R., Stewart, B.S. and Leatherwood, S. 1992. *The Sierra Club Handbook of Seals and Sirenians*. Sierra Club Books, San Francisco. 359pp.

Ride, W.D.L. 1970. *A Guide to the Native Mammals of Australia*. Oxford University Press, Melbourne. 249pp.

Seebeck, J.H., Brown, P.R., Wallis, R.L. and Kemper, C.M. 1990. Key to the bandicoots and bilbies of Australasia. Pp.375–376. In Seebeck, J.H., Brown, P.R., Wallis, R.L. and Kemper, C.M. (eds). *Bandicoots and Bilbies*. Surrey Beatty and Sons, Sydney. 392pp.

Smith, A. and Winter, J. 1984. A key and field guide to the Australian possums, gliders and koala. Pp.579–594. In Smith, A. and Hume, I. (eds). *Possums and Gliders*. Surrey Beatty and Sons, Sydney. 598pp.

Strahan, R. (ed). 1983. *The Australian Museum Complete Book of Australian Mammals*. Angus and Robertson, Sydney. 530pp.

Strahan, R. (ed). 1995a. *The Mammals of Australia*. Australian Museum and Reed, Sydney. 756pp.

Strahan, R. (ed). 1995b. *A Photographic Guide to Mammals of Australia*. Australian Museum and New Holland, Sydney. 134pp.

Tate, G.H.H. 1944. *A List of the Mammals of the Japanese War Area*. American Museum of Natural History, New York. Pp.5–22

Taylor, J.M. 1984. *The Oxford Guide to Mammals of Australia*. Oxford University Press, Melbourne. 148pp.

Thomas, O. 1888. *Catalogue of the Marsupialia and Monotremata in the Collection of the British Museum (Natural History)*. Trustees of the British Museum (Natural History), London. 401pp.

Triggs, B. 1996. *Tracks Scats and Other Traces: A Field Guide to Australian Mammals*. Oxford University Press, Melbourne. 340pp.

Troughton, E. 1967. *Furred Animals of Australia*. 9th Edit. Angus and Robertson, Sydney. 384pp.

Tucker, M. 1989. *Whales and Whale Watching in Australia*. Australian National Parks and Wildlife Service, Canberra. 48pp.

Walton, D.W. and Richardson, B.J. (eds). 1989. *Fauna of Australia. Vol. 1B. Mammalia.* Australian Government Publishing Service, Canberra. Pp.401–1227.

Watts, C.H.S. 1972. *Handbook of South Australian Rodents and Small Marsupials.* Field Naturalists' Society of SA (Inc.), Adelaide. 29pp.

Watts, C.H.S. and Aslin, H. 1981. *Rodents of Australia.* Angus and Robertson, Sydney. 321pp.

Watts, D. 1993. *Tasmanian Mammals—A Field Guide.* Peregrine Press, Kettering. 119pp.

Wilson, D.E. and Reeder, D.M. (eds). 1993. *Mammal Species of the World—A Taxonomic and Geographic Reference.* 2nd Edit. Smithsonian Institution Press, Washington. 1207pp.

Wroe. S. and Archer, M. 2006. Origins and early radiations of marsupials. Pp.551–574. In Merrick, J.R., Archer, M., Hickey, G.M. and Lee M.S.Y. (eds). *Evolution and Biogeography of Australasian Vertebrates.* Auscipub Pty. Ltd., Sydney. 942pp.

Ziegler, A.C. 1982. An ecological checklist of New Guinea Recent mammals. Pp.863–894. In Gressitt, J.L. (ed). *Biogeography and Ecology of New Guinea.* W. Junk, The Hague. 983pp.

Accelerated Change—the Regional Future

GEOLOGICAL TIME SCALE FOR THE PHANEROZOIC

ERA	PERIOD or SUB-PERIOD	SERIES/EPOCH	STAGE	AGE m.y.a.	EXTINCTIONS (E) ICE AGE (I)
CAINOZOIC	QUATERNARY	PLEISTOCENE	Calabrian	1.8	E, I
	TERTIARY – NEOGENE	PLIOCENE	Piacenzian	3.6	I
			Zanclean	5.3	I
		MIOCENE	Messinian	7.12	
		Late	Tortonian	11.2	
			Serravalian	14.8	
		Mid	Langhian	16.4	
			Burdigalian	20.52	
		Early	Aquitanian	23.8	
	TERTIARY – PALAEOGENE	OLIGOCENE Late	Chattian	28.5	E
		Early	Rupelian	33.7	
		EOCENE Late	Priabonian	37	
			Bartonian	41.3	
		Mid	Lutetian	49	
		Early	Ypresian	54.8	
		PALEOCENE	Thanetian	57.9	
		Late	Selandian	60.9	
		Early	Danian	65	E
MESOZOIC	CRETACEOUS	LATE	Maastrichtian	73	
			Campanian	83	
			Santonian	87	
			Coniacian	89	
			Turonian	91	
			Cenomanian	97.5	
		EARLY	Albian	108	
			Aptian	115	
			Barremian	123	
			Hauterivian	130	
			Valanginian	135	
			Berriasian	141	E
	JURASSIC	LATE	Tithonian	146	
			Kimmeridgian	151	
			Oxfordian	159	
		MIDDLE	Callovian	165	
			Bathonian	173	
			Bajocian	180	
			Aalenian	184	
		EARLY	Toarcian	190	E
			Pliensbachian	195	
			Sinemurian	202	
			Hettangian	205	E
	TRIASSIC	LATE	Rhaetian	207	
			Norian	220	
			Carnian	230	
		MIDDLE	Ladinian	235	
			Anisian	241	
		EARLY	Spathian	244	
			Nammalian	247	
			Griesbachian	251	
PALAEOZOIC	PERMIAN	LATE	Tatarian – Dorashamian	255	E
			Tatarian – Dzhulfian	259	I
			Tatarian – Midian	263	I
			Kazanian	267	I
			Ulfimian	270	I
		EARLY	Kungurian	274	I
			Artinskian	285	I
			Sakmarian	293	I
			Asselian	298	I
	CARBONIFEROUS – PENNSYLVANNIAN – SILESIAN	STEPHANIAN	Gzhelian	302	I
			Kasimovian	305	I
		WESTPHALIAN	Moscovian	311.5	
			Bashkirian	314	
		NAMURIAN	Serpukhovian	325	
	CARBONIFEROUS – MISSISSIPPIAN – DINANTIAN	VISEAN	Brigantian	328.5	
			Asbian	330.5	
			Holkerian	335	
			Arundian	341.5	
		TOURNAISIAN	Chadian	344	
			Ivorian	348	
			Hastarian	354	
	DEVONIAN	LATE	Famennian	364.5	E
			Frasnian	369	
		MIDDLE	Givetian	378	
			Eifelian	384	I
		EARLY	Emsian	399.5	I
			Pragian	404.5	I
			Lochkovian	410	
	SILURIAN	PRIDOLI		414	
		LUDLOW	Ludfordian	417	
			Gorstian	420	
		WENLOCK	Homerian	422.5	
			Sheinwoodian	425	
		LLANDOVERY	Telychian	428	I
			Aeronian	431	I
			Rhuddanian	434	E, I
	ORDOVICIAN	ASHGILL	Bolindian	443	I
		CARADOC	Eastonian	454.5	
			Gisbornian	459	
		LLANVIRN	Darriwilian	467	
			Yapeenian	471	
		ARENIG	Castlemainian	477	
			Chewtonian	481	
			Bendigonian	484.5	
		TREMADOC	Lancefieldian	486	
			Warendan	490	
	CAMBRIAN	LATE	Datsonian	491	
			Payntonian	492.5	
			Iverian	496	
			Idamean	497.5	E
		MIDDLE	Mindyallan	498.5	
			Boomerangian	500	
			Undillan	503	
			Late Templetonian-Floran	506	
			Ordian-Early Templetonian	509	
		EARLY	Atdabanian-Toyonian	528	
			Tommotian	534	
			Nemakit-Daldynian	545	

This detailed scale is based on Young and Laurie (1996) but Cambrian stages abbreviated ages are rounded to nearest 0.1 million years. More recent dating (e.g. United States Geological Survey) has resulted in some small adjustments to interval boundaries.

Chapter 36
Molecular Systematics: Implications for Phylogeny, Biogeography and Conservation

D.J. Colgan and D. O'Meally
Evolutionary Biology Unit, Australian Museum, 6 College Street, Sydney, N.S.W. 2010 Australia
E-mail: Don.Colgan@austmus.gov.au

INTRODUCTION

Advances in molecular biology over the past quarter century have been prolific and impressive. Thousands of data points can now be collected from single individuals with relative ease. Using molecular biology a relevant data type can be chosen for almost any question about evolutionary relationships and geographic distributions. However, despite their great advantages, two general points about molecular data should be recognized. First, their application must be based on an understanding of the strengths and limitations of the various types of data. Several general texts are available for more extensive background on older material (Avise, 1994; Ferraris and Palumbo, 1996; Hillis *et al.*, 1996). Second, the data are not by themselves necessarily decisive. Combining them with the results of other disciplines—including palaeontology, morphology, biogeography and natural history—can be highly illuminating.

This chapter addresses the use of molecular techniques to questions broadly divided into four overlapping categories relating to: (1) the status of species and populations; (2) evolutionary relationships among animals; (3) the geographical distribution of animals; and (4) conservation biology, incorporating applied elements of the first three.

In regard to species, we may wish to know whether morphological taxa represent multiple genetic species, the level of hybridization between species, or whether particular populations have significant genetic differentiation from other groups within a species. Intra-population level investigations examine such areas as the level of genetic variation and the type of breeding systems.

The use of molecular data to investigate the evolutionary relationships between taxa has been an active field for over 30 years but really blossomed with the development of rapid techniques for obtaining homologous sequences from different individuals. The versatility of these techniques is so great that appropriate markers can be chosen for investigating relationships within groups ranging from species to phyla.

Molecular techniques can be used at the broadest scale of biogeographic analysis, testing for instance whether a continental fauna is a single native radiation or whether it derives from single or multiple colonizations. Molecular techniques also permit very detailed biogeographic analysis if appropriately variable markers can be identified. In this emerging field, known as phylogeography, intra-specific differentiation is investigated in relation to the evolutionary history of different parts of a species' range (Avise *et al.*, 1987; Avise, 1996). Investigation of rapidly evolving sequences can be applied to identifying barriers to gene flow, establishing the provenance of colonizing populations or determining whether disjunct distributions are due to immigration or vicariance events.

Conservation genetics has the laudable aim of maximizing a taxon's chance of achieving the same evolutionary destiny as it would without human intervention. For reviews of this topic see Frankham (1995), Lambert and Millar (1995) and Frankham *et al.* (2002). Conservation geneticists may investigate whether study species have taxonomically-significant sub-structuring requiring active management or whether and/or how low heterozygosity levels in isolated populations should be augmented by active translocation. Phylogenetic information can be an important input to decision making regarding the allocation of resources within taxonomic groups or between areas that may potentially be included in nature reserves. Conservation genetics can appear marginal in an era confronted with massive extinction rates. But reductions in effective population size are correlated with population extinction (Frankham, 1998; Frankham *et al.*, 2002) and this must be addressed over the time-frame of centuries for species that survive the current crisis.

TECHNIQUES

ALLOZYME ELECTROPHORESIS

In this venerable technique, first used in population genetic studies in the mid 1960s (Shaw, 1965; Lewontin and Hubby, 1966), homogenates of tissues or blood are placed side by side on a gel and subjected to a direct electric current for a specified period. The relative mobility of proteins depends on the numbers of charged amino acids in their sequences. Changes involving a net gain or loss of charge alter the rate at which proteins move through the gel. The point to which an enzyme moves can be determined by bathing the gel in a solution containing its substrate. The enzymatic conversion of the substrate to its product is linked to another reaction where a soluble chemical is converted to an insoluble stain deposited only where the enzyme is active.

Recipes are available for visualizing numerous enzymes (Harris and Hopkinson, 1977; Richardson *et al.*, 1986; Manchenko, 1994). The most common gel medium nowadays is cellulose acetate (Richardson *et al.*, 1986). Starch and acrylamide (Brewer, 1970; Andrews, 1986) are also used as they have sieving properties permitting some changes altering the mass or three-dimensional shape of the protein to be identified even if they do not alter electric charge. Starch can be difficult to obtain, acrylamide is a proven neurotoxin and both are relatively cumbersome to use.

Allozyme electrophoresis can be too easily overlooked for more sophisticated procedures. Yet it should be seriously considered at the outset of any investigation. It provides a rapid estimate of variability at a broader sample of known loci than can presently be readily obtained with any other technique; in addition, heterozygotes are readily identified. The major disadvantages of electrophoresis are that: it cannot be directly related to DNA sequence variation; it does not always detect suitable levels of variation; it is difficult to analyze results cladistically; and it is difficult to compare accurately the results from different laboratories or experimental periods.

ISOELECTRIC FOCUSSING AND TWO DIMENSIONAL ELECTROPHORESIS

In isoelectric focussing (IEF), electrophoresis is conducted so that proteins move to the point in a pH gradient where they are neither positively nor negatively charged. Charge differences between proteins are reflected by mobility differences. The technique is more sensitive than allozyme electrophoresis, but allelic variation can be obscured by post-translational modifications. Tegelström *et al.* (1982) give experimental protocols for applying isoelectric focussing to systematics.

Future use of IEF in systematics will mostly be in 'two dimensional electrophoresis'. IEF is applied to a single sample in a tube gel. The tube is then used as the origin of a slab gel for electrophoresis that separates proteins according to mass in a second dimension. Differences between individuals are identified by comparing multiple 2-D slabs. The technique is laborious but the number of proteins that can be scored at one time is a major advantage (Goldman and O'Brien, 1993): radioactive labelling allows hundreds to be resolved on a single gel. The technique is currently receiving attention in the field of 'proteomics'.

IMMUNOLOGY (INCLUDING SEROLOGY)

Although not used widely today, immunological methods (reviewed by van Regenmortel *et al.*, 1993) have some historical significance. Generally, the strength of the cross-reactivity of antigens to a common antibody can be used to estimate genetic distances between species. The influential early work of Kirsch (1977) was based on mixtures of proteins, but more recent applications utilize purified single proteins, particularly albumin, in a technique known as micro-complement fixation (Baverstock, 1984; Baverstock *et al.*, 1989; Baverstock *et al.*, 1990).

CYTOGENETICS

Several aspects of chromosomal variation can be used as phylogenetic or biogeographic information. These include the number of chromosomes in a diploid complement, the gross morphology as determined by centromere position, the occurrence of inversions within chromosomes, translocations between non-homologous chromosomes and Robertsonian fusions where two telocentric chromosomes fuse to form a single metacentric. White's classic book (1983) is still worth reading, in conjunction with Howard and Berlocher (1998). Differential staining of chromosomal regions can be used to produce a variety of different banding patterns to allow finer-scale classification of their variation. Sessions (1996) provides a useful catalogue of protocols. Many cytogenetic studies of Australasian vertebrates have been made (Murray *et al.*, 1980; King, 1984; Rofe and Hayman, 1985; Mengden *et al.*, 1986; Sharman *et al.*, 1991; Mahony *et al.*, 1996).

DNA-DNA HYBRIDIZATION

This technique utilizes the property that the two strands of DNA molecules can be separated by heat, with the temperature required dependent on the length of the molecules, the percentage of the various bases in the molecules and the degree of non-complementarity in the two strands (Britten *et al.*, 1974). Single copy DNA (sheared to a specified length) from two species is mixed, fully denatured and the separated strands are allowed to pair up again ('annealing'). One species' DNA is radioactively labelled and included in the mix at a relatively much lower concentration so that during re-annealing its strands will most likely pair with those of the other species. The mixture is put onto a buffered hydroxy-apatite column that binds double-stranded DNA and heated. As the temperature increases, fractions are collected for determining how much radioactively-labelled unbound (i.e. single-stranded) DNA is eluted. Various measures of the melting profile (e.g. the temperature at which 50% of the radiolabelled DNA remains double-stranded) are used to compare the overall similarity of two species' genomes (Kirsch *et al.*, 1995). The technique requires a large amount of DNA and all parameters (buffers, temperature, etc.) must be controlled as strictly as possible to allow comparison between different laboratories (Lapointe and

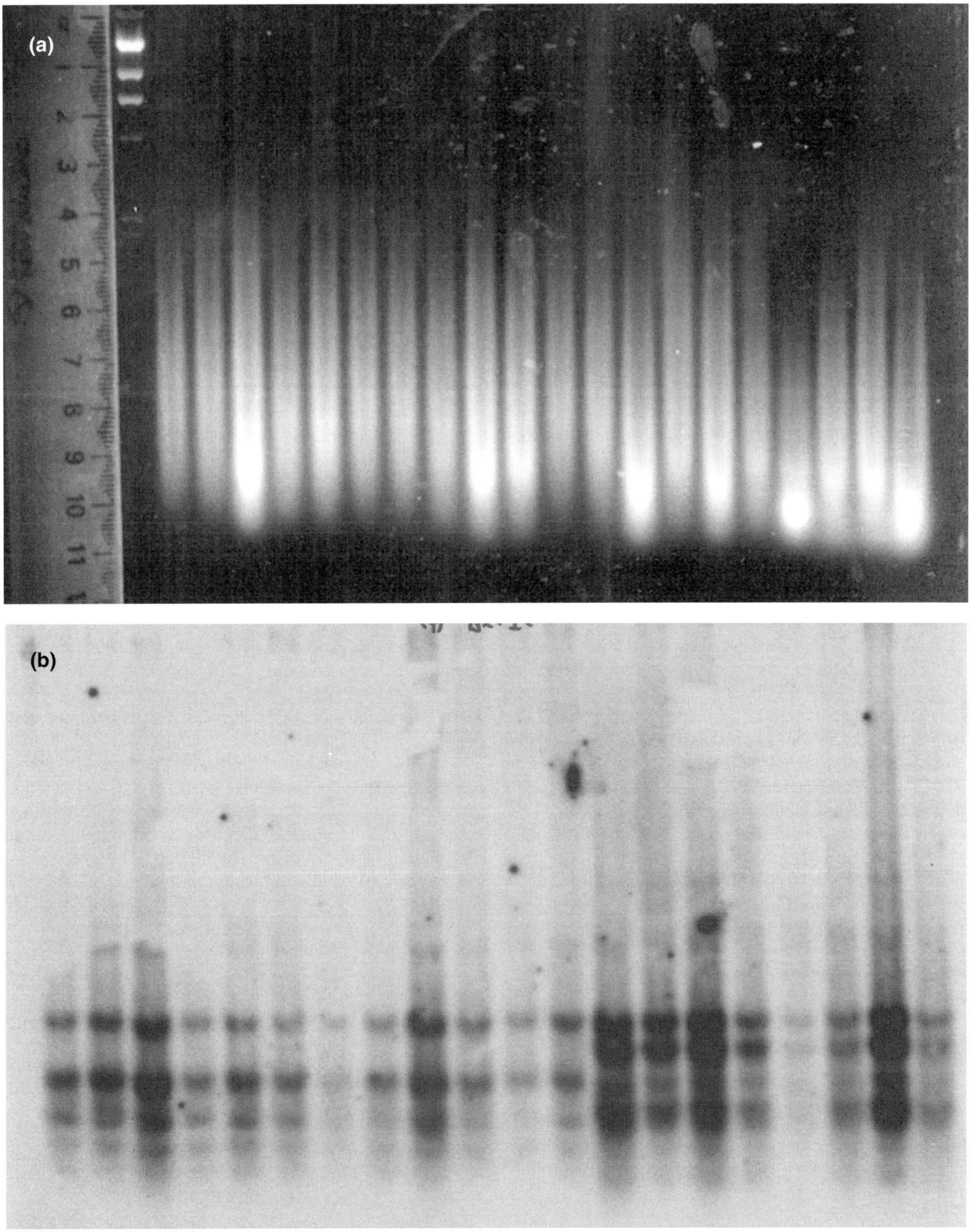

Figure 1. DNA extractions from the gemfish, *Rexea solandri*. **(a)** Samples were digested with the restriction endonuclease *Dde 1* and visualized with ethidium bromide fluorescence. The leftmost lane is a size marker, the next twelve are from locations east of Bass Strait and the remaining eight from South Australia and Western Australia. **(b)** Southern blot of (a) probed with radioactively labelled mitochondrial DNA showing distinctive patterns for individuals from east and west of Bass Strait.

Kirsch, 1995). The results are not suitable for parsimony analysis and must be treated as distances. The technique is now used only occasionally (Campeau-Péloquin *et al.*, 2001) but it has made significant contributions to avian (Sibley and Alquist, 1990) and mammalian systematics. It has been extensively used on Australasian marsupials including Springer *et al.* (1990) on Phalangeridae, Kirsch *et al.* (1990) on bandicoots, Springer *et al.* (1992) on Pseudocheiridae and Edwards and Westerman (1995) on all six families of possums and gliders. The large literature is reviewed by Kirsch *et al.* (1997). DNA-DNA hybridization remains the only available measure for summarizing overall sequence similarity.

RESTRICTION FRAGMENT LENGTH POLYMORPHISM

This technique (RFLP) depends on the fact that restriction endonucleases cut DNA at or near a specific 'recognition site', usually of four, five or six bases (occasionally eight) in length. A change in DNA sequences may cause the loss or gain of a restriction site. Southern blotting (Figure 1) is used to search for such changes in complex mixtures of DNA (often complete genomes). In this technique, the DNA is cut and run on a gel, after which it is separated into single strands and transferred by capillary action to a membrane to which it is bound—by heating (under vacuum for nitrocellulose, to prevent explosion) or UV crosslinking. The membrane is incubated with a labelled probe sequence, also separated into single-strands. Probe strands bind to complementary strands on the membrane and are detected by radioactive decay, or the chemiluminescent or colour-producing action of enzymes conjugated to biotin included in the probes (Dowling *et al.*, 1996)

This technique is laborious—one of the authors spent nearly two years in the laboratory to get 23 characters for 26 species of bats (Colgan and Flannery, 1995)—and requires a lot of DNA. Is it now generally used in combination with other techniques such as PCR or sequencing (Clegg *et al.*, 1998).

POLYMERASE CHAIN REACTION

The polymerase chain reaction (PCR) is a technique for amplifying a designated DNA region from extremely complex mixtures of sequences. It has many applications: each chapter in Innis *et al.* (1990) and Rapley (1996), for instance, provides an introduction to methods useful throughout molecular biology.

In principle, the technique has three steps. These are repeated many times on a reaction mix containing oligonucleotide 'primers' complementary to short sequences (typically 17–30 base pairs) at the ends of the region of interest, nucleotides for synthesis of new DNA strands and a DNA polymerase. This enzyme is derived from extremely thermophilic bacteria so is not destroyed by the high temperatures used during PCR.

The steps in the procedure are as follows:

(1) The DNA in the reaction is 'denatured' by heating it sufficiently to separate completely the two strands of each DNA fragment. The strands themselves are not damaged by this.

(2) The reaction is rapidly cooled to a temperature where complementary sections of DNA can again become double-stranded by annealing. Because the oligonucleotide primers are included at vastly greater concentrations than the original DNA (usually 10^7 times as many copies) annealing of the study region is almost certain to involve a single-strand of endogenous DNA and a complementary primer.

(3) The reaction temperature is raised to an intermediate level that allows the DNA polymerase to add additional nucleotides to the end of the primer ('extension') as directed by the sequence on the other strand by the principles of complementarity ($A \leftrightarrow T$, $C \leftrightarrow G$). At the end of this cycle, each copy of the study region has been duplicated. Because Taq polymerase (and indeed every known DNA polymerase) requires a double-stranded piece of DNA to initiate replication, other gene regions are not duplicated.

The procedure is repeated, with a doubling of copy numbers of the study region after each round. After numerous cycles, there are so many copies that other sequences (such as the rest of the original DNA sample) can be ignored during further procedures such as PCR-RFLP or DNA sequencing. Typical results are illustrated in Figure 2.

The critical factor for successful PCR is the design of primer pairs with sufficient specificity to prevent the amplification of sequences other than those of immediate interest and of sufficient generality to amplify these sequences from all of the individuals/species required. Programs and advice for primer design are widely avail-

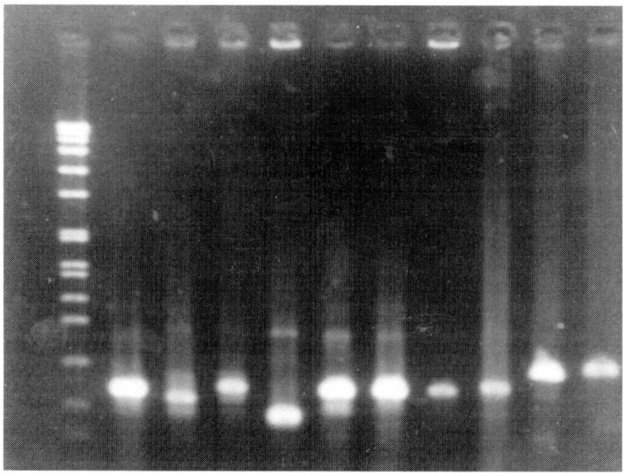

Figure 2. PCR products visualized with ethidium bromide fluorescence. The products are from the *phosphoglycerokinase* gene of various marsupials.

able, but experimentation remains the only reliable method of testing a particular oligonucleotide.

Random Amplified Polymorphic DNA

In this variant of PCR (abbreviated as RAPD), only one short primer of about ten bases, is used. The sequence of the primer is random. If by chance, there are two near sites for the primer on opposite strands then the PCR reaction will amplify the intervening sequences. Typically, multiple bands are found in the products of a RAPD reaction. Differences in the patterns from different individuals will occur if a change in one of the primer sites prevents amplification of a fragment in that individual. Because the primers are short, annealing temperatures must be greatly reduced, introducing a substantial degree of uncertainty into the reactions, with consequent problems for reproducibility. Reproducibility also depends critically on the quality of input DNA and on polymerase concentration. Applications of this technique are reviewed by Hadrys *et al.* (1992).

Amplified Restriction Fragment Length Polymorphisms

This PCR variant known as AFLP was developed by Vos *et al.* (1995). Genomic DNA is digested with two restriction endonucleases—typically one with a six base recognition sequence and one with four base recognition. The genomic fragments are ligated to oligonucleotide adaptors—i.e. short sequences complementary to the overhanging single-strands of the digestion products. A first round of PCR is performed with the 3' end of one primer corresponding to one endonuclease recognition sequence and the 3' end of the other primer to the second recognition sequence. This results in the amplification of numerous fragments. A second round of PCR is performed to reduce this number. The primers for the second reaction correspond to the oligonucleotide adaptors but have extra bases at the 3' end to increase specificity. One extra base should give a fourfold reduction in the number of amplified fragments, two extra bases a sixteen fold reduction, and so on. Generally three extra bases are used, giving about 100 fragments for a typical genome. These are detected on polyacrylamide gels through autoradiography or, more recently, by fluorescence labelling and capillary electrophoresis on automatic DNA sequencers.

AFLPs are very useful for the detection of polymorphisms—if at a particular site, a target genome has a different base to one of the extra bases in the second-round primers, then its profile will lack the corresponding band. AFLPs are more robust to variation in experimental conditions than RAPDs. They have been under-utilized in animal studies, and notably so in the Australian context.

DNA SEQUENCING

With the discovery of PCR, DNA sequence information could be obtained for homologous genes from multiple taxa, thereby permitting the widespread application of this technique to comparative biology. Currently, PCR products are sequenced using the Sanger dideoxy termination technique, either manually or more commonly on an automated sequencer (Rapley, 1996). Sequencing is carried out directly using one of the primers from the PCR reaction or a second 'nested' primer internal to the study sequence, usually adjacent to the original. Sequencing can also be carried out after cloning the PCR products. Cloning may be necessary with particular genes (for example, the internal transcribed spacers between the ribosomal genes) or because the PCR product is thought to comprise multiple sequences. With automatic sequencers, primers can be chosen to span a region of 600 or more bases. The technique will continue to play a dominant role in molecular zoology.

Compared to morphology, DNA sequence data may have a relatively high tendency to repeated independent evolution of the same character state ('homoplasy') as there are only four (or five if gaps are treated separately) possible alternatives. Conversely, morphological characters may be more likely than DNA sequence characters to be homoplasous if a particular character state is favoured by selection.

In using any DNA sequence based technique, we want to match the estimated variability of the sequence to the question being posed. Rates of evolution differ greatly between various components of the genome. Rapidly evolving genes are of restricted use for investigating ancient phylogenetic divergence, whereas conservative genes are uninformative regarding intra-specific differentiation. There can be no question that a greater range of sequences should be utilized in the future. Experience with these will give a clearer idea of what sequences to use for a particular study. Nevertheless, the following generalizations about the choice of study sequence can be made.

(1) In coding genes, there is a wide variation in the rates of evolution of sites depending on whether they represent the first, second or third position of an amino acid codon and on the function of the particular gene. In most cases, a change at the third base in a triplet does not alter the amino acid that it encodes. This is also true for two amino acids (leucine and arginine) for some changes at the first base.

(2) Most sections of mitochondrial DNA ('mtDNA') evolve quite rapidly so that its genes are generally useful for questions ranging from intra-specific to familial. Mitochondrial ribosomal DNAs can be used for ordinal level phylogenies by ignoring areas of 'uncertain alignment' (i.e. those loop regions where insertion and/or deletion of bases makes it difficult to establish evolutionary relationships of the various sequence positions). The mtDNA has a number of properties that have made it very useful to biogeography and phylogenetics (Avise *et al.*, 1987). First, it is abundant, sometimes occurring in thousands of copies per cell. Hence it was relatively easy to isolate in the early days of molecular zoology. Second, mtDNA is almost invariably inherited only from the mother, so that there

copies of the sequence. Minisatellites are scored by southern blotting using a cloned repeat sequence as a probe. In many cases it is possible to use probes from another species, not necessarily closely related. Detection is usually autoradiographic, but non-radioactive methods are available (Gissing and Crease, 1997). Because there is a great deal of variation in the number of repeats at each site, differing fragment patterns can be used to identify individuals (Burke *et al.*, 1996).

SHORT INTERSPERSED ELEMENTS

These elements ('SINES') are short (about 500 base pairs or less) and moderately to highly repeated in the genome (ten thousand to several hundred thousand copies). The copies are dispersed. SINEs can be copied to new locations by integration of DNA following transcription of an existing SINE and using the reverse trancriptase of an associated long interspersed element. Once integrated a SINE is not readily removed from the genome. Moreover, there does not appear to be a concentration of SINE insertions at 'hot spots'. Hence the positions of SINEs have high potential as phylogenetic markers (Terai *et al.*, 1998), illustrated by their success in corroborating monophyly of four tribes of cichlids in Lake Tanganyika (Takahashi *et al.*, 1998). The major problem with the technique is that current practice requires the construction and screening of genomic libraries for each tested species. It appears to us, however, that inverse PCR (Ochman *et al.*, 1990) could be adopted to circumvent most library construction after the discovery of a SINE family.

AUTOMATION

DNA microarrayers (Schema, 2000) spot tens of thousands of small individual specimens of DNA on to microscope slides or nylon membranes. In the first generation technology, these slide arrays are probed with a labelled tester DNA. Tester DNA pairs with the target DNA only where the sequences are closely matched. Unbound tester is washed from the slide and the microarray is scanned by laser confocal microscopy of the fluorescence emitted by the bound probe. Scanning requires only two to three minutes so that data collection is extremely rapid compared to previous technologies. Scoring is also much less expensive than conventional automatic sequencing.

This technology is under active development. Some second generation systems will be able to detect which of the four bases is added when an oligonucleotide primer is extended on a genomic template. This will give arrayers sufficient flexibility for two general applications to molecular systematics.

First, total DNA from a large number of different individuals would be spotted on to a microarray to be probed with a known sequence to identify whether a particular DNA sequence variant is present at a specified point. Allele-specific oligonucleotides (Dessauer *et al.*, 1996) or the oligonucleotide ligation assay could be readily adapted to array technology.

Second, as explained by Winzeler *et al.* (1998), tens of thousands of known oligonucleotide sequences from a target genome are synthesized on arrays. Separate array copies are probed with labelled test genomes (one is usually the target genome) to identify sequence differences. Using this technique, Winzeler *et al.* (1998) were able to identify 3,714 allelic variants between two strains of yeast in a single scan.

DNA arrays are expensive to make, but once available for a species (and many will be made for humans and mice in particular) it will be possible to detect differences between two other taxa by hybridization patterns. One potential difficulty is that an oligonucleotide would have to be evolutionarily conservative to hybridize to the reference species, and hence would be less likely to vary between the two test species than non-conservative oligonucleotides.

METHODS OF ANALYSIS

The methods of analysis vary with the data type and question addressed. Generally, the methods fall into three main classes—cladistic, phenetic and probabilistic; for further background on these approaches see pp.3–16 (Lee, 2006). The plethora of available software can be very useful in analyses, but cannot replace close inspection of the data. Miyamoto and Cracraft (1991) and Swofford *et al.* (1996) provide excellent surveys of various methods of phylogenetic analysis. Durbin *et al.* (1998) is more mathematical but summarizes the probabilistic treatment of biological sequence data very well and is a great introduction to the burgeoning field of bioinformatics. Most widely used packages also have comprehensive supporting material providing theoretical background. The references should be consulted to weigh the merits of the different analytical methods if they produce incongruent results on a particular dataset.

Felsenstein (2002) maintains an up-to-date annotated list of phylogeny programs at http://evolution.genetics.washington.edu/phylip/software.html; a similar list is maintained by the University of California—see University of California (2002). Many programs are freely available. Most are updated, sometimes frequently. To obtain the most recent versions, it is usually easiest to check the Felsenstein site or search the internet using the name of the program and 'download' as keywords. The packages we use most include CLUSTAL (Thompson *et al.*, 1994) for DNA sequence alignment, MacClade (Maddison and Maddison, 1992) for data editing and cladistic tree comparison, PAUP* (Swofford, 2000) for cladistic, distance and maximum likelihood methods, with particular reference to DNA sequences, PHYLIP (Felsenstein, 1989) for distance and phylogenetic analyses on a wide range of operating systems,

MEGA (Kumar *et al.*, 1993) for distance analyses of sequence data and REAP (McElroy *et al.*, 1992) for analyses of restriction endonuclease data. See the summary of the programs available in Table 1.

Any researcher using DNA sequences will need to become and remain (owing to frequent site modifications) familiar with the three major genetic databases. These have many facilities for sequence analysis, data retrieval and similarity searching. Their internet addresses are:

(1) The National Center for Biotechnology Information: http://www.ncbi.nlm.nih.gov/
(2) The DNA database of Japan: http://www.ddbj.nig.ac.jp/
(3) The EMBL—European Bioinformatics Institute: http://www.ebi.ac.uk/ebi_home.html

APPLICATIONS

In this section, representative studies that are of historical importance in Australasian molecular zoology or that illustrate the use of particular techniques on the Australian, New Guinean or New Zealand fauna are listed. One or two case studies are described in detail.

Table 1. Summary of programs for molecular systematics, together with sources and primary purpose(s)of each program.

Program	Website	Purpose/Comments
ARLEQUIN	http://acasun1.unige.ch/arlequin/	DNA/Allozyme population level work amply repays learning effort
AUTODECAY	http://www.bergianska.se/forskning.html	Bremer support (decay) indices, works as a PAUP shell.
BIOSYS	by ftp from lamar.colostate.edu in directory pub/wcb4	Population genetic analysis—dated, but still useful. FORTRAN formatting
CLUSTALX	http://www-igbmc.u-strasbg.fr/BioInfo/ClustalX/Top.html	Multiple sequence alignment
COMPONENT	http://taxonomy.zoology.gla.ac.uk/rod/cpw.html	Comparison of cladograms, tree-tree distance metrics, biogeography-area cladograms, etc.
DNASP	http://www.ub.es/dnasp/	Analysis of DNA sequence polymorphism, user friendly
CONSERVE	http://evolve.bio.ic.ac.uk/software/conserve/index.html	Evolutionary distinctiveness of species and testing of potential reserve schemes
GDA	http://alleyn.eeb.uconn.edu/gda/	Test version—has short expiration. Population genetics for frequency and sequence data. Has Monte Carlo F-statistics
GENEPOP	http://wbiomed.curtin.edu.au/genepop/	Population genetics, format conversion, Mantel tests.
HADTREE	http://imbs.massey.ac.nz/Research/MolEvol/Farside/programs.htm	Tree partition support and rejection by Hadamard transformation. DOS menu.
HENNIG 86	from Arnold Kluge at akluge@umich.edu	Maximum parsimony, command-line driven
MacCLADE	http://phylogeny.arizona.edu/macclade/macclade.html	Step matrices, dataset definition, translation, tree manipulation, sequence statistics, etc.
MEGA	http://www.megasoftware.net	Distance calculations, likelihood and parsimony methods for sequence data
MICROSAT	http://hpgl.stanford.edu/projects/microsat/	Distances based on microsatellite data
MRBAYES	http://brahms.biology.rochester.edu/software.html	Monte Casrlo Markov Chain estimates of posterior probabilities of phylogeny parameters
NONA	http://www.cladistics.com/about_nona.htm	Parsimony searches with defined character weights
PAML	http://abacus.gene.ucl.ac.uk/software/paml.html	Maximum likelihood analyses. Flexible but slow.
PAUP	http://www.sinauer.com/Titles/frswofford.htm	Many analyses, generally useful for phylogenetic analyses
PHYLIP	http://evolution.genetics.washington.edu	Phenetic, likelihood and parsimony methods for a flexible range of data
POPGENE	http://www.ualberta.ca/~fyeh/index.htm	General population genetics, easy to use
POY	by ftp from ftp.amnh.org in directory pub/people/wheeler/poy	Simultaneous alignment and phylogenetic analysis using parsimony criteria
PUZZLE	http://www.tree-puzzle.de	Maximum likelihood 'quartet puzzling' various substitution models—quick
REAP	http://bioweb.wku.edu/faculty/mcelroy/	Methods for analysis of restriction fragment data—includes Monte Carlo
TREEVIEW	http://taxonomy.zoology.gla.ac.uk/rod/treeview.html	Tree visualization, printing and format interchange
WINCLADA	http://www.cladistics.com/about_winc.htm	Interactive Shell for HENNIG, NONA, and other programs

SPECIES AND POPULATION STATUS

Many Australasian mammal species, a high proportion of birds, and substantial numbers of reptiles, amphibians and fish have been investigated to determine species status. Representative studies include Richardson and McDermid (1978) and Eldridge and Close (1992) for macropods, Colgan and Flannery (1992) for Petauridae, Colgan *et al.* (1993) for Phalangeridae, Sumner and Dickman (1998) for *Antechinus*, Southgate and Adams (1993) on the greater bilby (also studied by Moritz *et al.*, 1997), Baverstock *et al.* (1981) for Muridae, Cooper *et al.* (1998) in two species of rhinolophid bats, McDonnell *et al.* (1978), on two species of *Pseudophyryne* (Amphibia), Colgan (1997) on the green and golden bell-frog, *Litoria aurea*, Milton *et al.* (1983) and Sadlier *et al.* (1994) for skinks, Milton and Shaklee (1987) for blue grenadier fish, Lavery and Shaklee (1991) for sharks, Elliott and Ward (1994) for jackass morwong (also studied for RFLP by Grewe *et al.*, 1994), Elliott *et al.* (1995) for orange roughy. The compilation of Shaklee and Keenan (1986), though relatively old remains the best resource for fish identification using electrophoresis and one of the few synoptic studies for electrophoresis. The use of allozmes to describe a complex system of hybridization in carp gudgeon from the Murray River (Bertozzi *et al.*, 2000) shows that the technique still has much to reveal about speciation.

Direct sequencing of mtDNA PCR products for the investigation of population sub-structuring is exemplified by Edwards and Kot's study (1995) of grey-crowned babblers and Worthington Wilmer *et al.* (1994) on ghost bats, Cardinal and Christidis (1999) on bentwing bats. Eldridge *et al.* (2001b) use mtDNA sequencing to demonstrate the correctness of the original, but more recently discounted description of the purple-necked rock-wallaby as a full species. Conversely species numbers can be reduced by the application of this technique, as shown by Cooper *et al.* (2001) for bats of the genus *Myotis*. RFLP has been extensively utilized in studies of fish population structure (Ovenden, 1990; Smolenski *et al.*, 1994; Colgan and Paxton, 1997). An interesting example of its use is the demonstration on the multiple origins of parthenogenesis in geckoes (Moritz and Heideman, 1993). Microsatellites have not been used as widely as their potential would support. Spencer *et al.* (1995) isolated microsatellites from *Petrogale* and Cooper *et al.* (1997) from *Tiliqua*. More often they have been studied in relation to conservation biology (Taylor *et al.*, 1994).

PHYLOGENETIC RELATIONSHIPS

Protein electrophoretic studies with an emphasis on phylogeny include Mengden *et al.* (1986) on snakes (genus *Pseudechis*), Georges and Adams (1992) on Australian chelid turtles, Bolch *et al.* (1994) on centrolophid fishes and Christidis and Schodde (1997) on Australo-Papuan fairy wrens. The latter paper also indicates the extent of work completed on protein electrophoresis of the Australasian avifauna.

DNA sequencing studies of Australasian vertebrate phylogenetics include Krajewski *et al.* (1992) and Retief *et al.* (1995a) on dasyuroid family relationships, Armstrong *et al.* (1998) on *Antechinus* (Dasyuridae), Blacket *et al.* (1999) on Sminthopsini (Dasyuridae), Edwards *et al.* (1991) on perching birds, van Tuinen *et al.* (1998) on ratites, Brown and Toft (1999) on cockatoos, Donnellan *et al.* (1999) on gekkonoid lizards, Seddon *et al.* (1997) and Georges *et al.* (1999) on turtles, Hay *et al.* (1995) on Amphibia. The use of molecular phylogeny to map other aspects of an animal's biology can be very interesting. For instance, Fairbairn *et al.* (1998) determined, using cytochrome *b* sequences, that viviparity has arisen at least twice within the skink *Lerista*. *L. bougainvilli* contains both oviparous and viviparous populations. The geographically disjunct occurrence of the latter reproductive mode on Kangaroo Island and the Islands of eastern Bass Strait (and Tasmanian mainland) is mirrored by the different placement of individuals from these localities on phylogenetic trees. Smith *et al.* (2001) find at least two shifts in reproductive mode within the species *Saiphos equalis*.

Relationships between the Monotremata, Marsupialia and Eutherians have been addressed by many studies—see Archer and Hand (2006), Hand (2006), Johanson (2006) and Wroe and Archer (2006). Molecular aspects have recently been reviewed by Lee *et al.* (1999). Globin and protamine P1 sequences tend to the traditional Theria hypothesis that the monotremes are the sister to all other mammals. Analyses of complete mitochondrial sequences suggest that the monotremes and marsupials are sisters (Janke *et al.*, 1997). Lee *et al.* (1999) isolated a β-globin from the echidna. In all other mammals, this gene has duplicated to give one copy (ϵ-globin) whose expression is restricted to early development and one that is expressed at later stages. The echidna gene is similar in intron/exon structure to all known α-globin genes. In cladistic analyses, it is the sister to a clade comprising all other β- and ϵ-globins. No ϵ-globin has been identified in monotremes. Recent single gene sequence data (Killian *et al.*, 2001) supports the Theria hypothesis. Kathy Belov (2002, personal communication) is extending her studies of the marsupial immune system (Miller and Belov, 2001) to Monotremata in an effort to resolve the controversy.

Molecular data have been influential in discussions of the interrelationships of the seven major lineages of extant marsupials: Didelphimorphia (American opossums), Paucituberculata (family Caenolestidae), Microbiotheria (the Chilean monito del monte *Dromiciops gliroides*—synonym *australis*), Notoryctemorphia (the marsupial mole *Notoryctes*), Peramelemorphia (bandicoots and bilbies), Dasyuromorphia (the thylacine, numbat and dasyurids) and Diprotodontia (koala, wombats, macropods and phalangeroids). The most stimulating hypothesis of these relationships has, however, come from comparative morphology.

Szalay (1982) defined the Australidelphia, comprising all extant Australasian marsupials and the monito del monte *Dromiciops australis* on the basis of supposedly synapomorphic tarsal characteristics. Other living marsupials are defined as the Ameridelphia (Szalay, 1982). All studied species from this group have epididymal sperm pairing, a derived condition not found in *D. australis* (Temple-Smith, 1987; Temple-Smith, 2000, personal communication). The Ameridelphian-Australidelphian dichotomy has not been universally accepted. Reig *et al.* (1987) note that intermediate states of the tarsal characters are seen in Australian marsupials and raise other morphological objections, with respect to dental and cranial characters. Hershkovitz (1992) criticized Szalay's interpretation of the tarsal characters, claiming continuous intergradations between their alternative states, with typical 'Ameridelphian' and 'Australidelphian' states being seen in both continents.

Relevant DNA sequence data have been collected for a number of genes: 12S rDNA (Springer *et al.*, 1994; Palma and Spotorno, 1999), protamine P1 (Retief *et al.*, 1995a, b; Krajewski *et al.*, 1997), 16S rDNA (Springer *et al.*, 1998), interphotoreceptor retinoid-binding protein (eight marsupial species only—Springer *et al.*, 1998) and phosphoglycerokinase (Colgan, 1999). The best present estimate of marsupial phylogeny using molecular data should be obtained by compiling sequences for the five genes that have been scored in most of the critical taxa (12S and 16S rDNA, cytochrome *b*, protamine P1 and phosphoglycerokinase). For some 'taxa', two species in the same genus were combined if data were not available for all genes from the one species. These are indicated by an asterisk in Figure 4. For each gene about a quarter of species lack data. Areas of uncertain alignment in protamine P1 and 12S rDNA were excluded. In total, 36 taxa and 4,406 aligned bases were used in the analysis. The tree in Figure 4 was one of seven most parsimonious trees found using the default options in PAUP*. The other trees differ only near the terminals—basal clades are the same in all trees. The tree is midpoint rooted. Analyses with humans as an outgroup suggest that the root lies between *Notoryctes* and the remainder of the marsupials, with an identical ingroup topology.

The topology of the tree is remarkably concordant with morphological expectation. Among recognized morphological groups that are well supported are the Phalangeridae, Phalangeroidea, Macropodidae, Vombatoidea, *Antechinus*, *Murexia*, Dasyuridae, Dasyuroidea, Diprotodonta, Peramelemorphia, Didelphidae and Ameridelphia. The salient point regarding basal evolution is that *Dromiciops*, irrespective of the root of the tree, is sister to an Australasian clade rather than to an American clade. If the true root of the tree is between the bandicoots and the other American taxa, then the tree supports the Ameridelphia and the Australidelphia *sensu* Szalay (1982).

BIOGEOGRAPHY

Molecular studies of Australasian faunal biogeography have concentrated on phylogeography (Schneider *et al.*, 1998; Houlden *et al.*, 1999), the delineation of regional boundaries and the demonstration of dispersal. Boundaries can be quite broadly defined as, for instance, shown by the central cordillera of Papua New Guinea that separates most well-studied taxa into northern and southern populations (Colgan and Flannery, 1992; Colgan *et al.*, 1993). Bass Strait is a particularly interesting boundary. This, because of recurrent changes in sea level, has been inundated or exposed many times during the Pleistocene. It has acted therefore as both a marine and a terrestrial barrier. The eastern and western stocks of the gemfish, *Rexea solandri,* can be distinguished in the RFLP phenotype of nearly every tested restriction endonuclease (Colgan and Paxton, 1997). The fish are capable of migrating thousands of kilometres, but will not swim across the strait owing to its shallow depth so it acts as a virtually complete barrier to gene flow. The strait's function as a terrestrial barrier is illustrated by the separation of mainland and Tasmanian populations of tiger quolls using mtDNA sequencing (Firestone *et al.*, 1999). Dispersal during recent exposures makes the biogeographic history of the strait complex, however, as not all studies give a clear dichotomy (Fairbairn *et al.*, 1998). Similarly complex phylogenetic consequences of episodic inundation are shown by the freshwater Melanotaeniidae of northern Australia and New Guinea (McGuigan *et al.*, 2000).

Long-range dispersal rather than range contraction and expansion has also been revealed by sequencing (e.g. Austin (2000) on the *Candoia* boas and Burridge (2000) for the marine fish family Aplodactylidae). Fuller *et al.* (1998) investigated whether the Australian goannas (Varanidae) are a Gondwanan relict or are more recently derived from other areas of the family's distribution. Previous micro-complement fixation studies had separated the Australian species into two of the three main clades in a basal unresolved polytomy. 12S rDNA sequence data suggested that all Australian species were included in a group of Indo-Australian taxa. The data are congruent with an Asian origin of the Australian Varanidae, with the two African species representing a separate invasion from the one that led to the Australian dispersal.

The origin and history of dispersal of fruitbats (Megachiroptera) has attracted significant molecular attention; see pp.673–705 for a summary of anatomy, biology and the fossil record (Hand, 2006). Fruitbats were divided by Andersen (1912) into two subfamilies. One, the Macroglossinae, comprises the six genera of nectar feeding bats (*Eonycteris*, *Macroglossus*, *Syconycteris*, *Melonycteris* and *Notopteris* from the tropics and subtropics of the Indo-West Pacific, and the African *Megaloglossus woermanni*). The second subfamily, the Pteropodinae, contains the great majority of species. Recent molecular works generally suggest that the morphological classification requires major revision. Both DNA-DNA hybridization (Kirsch *et al.*,

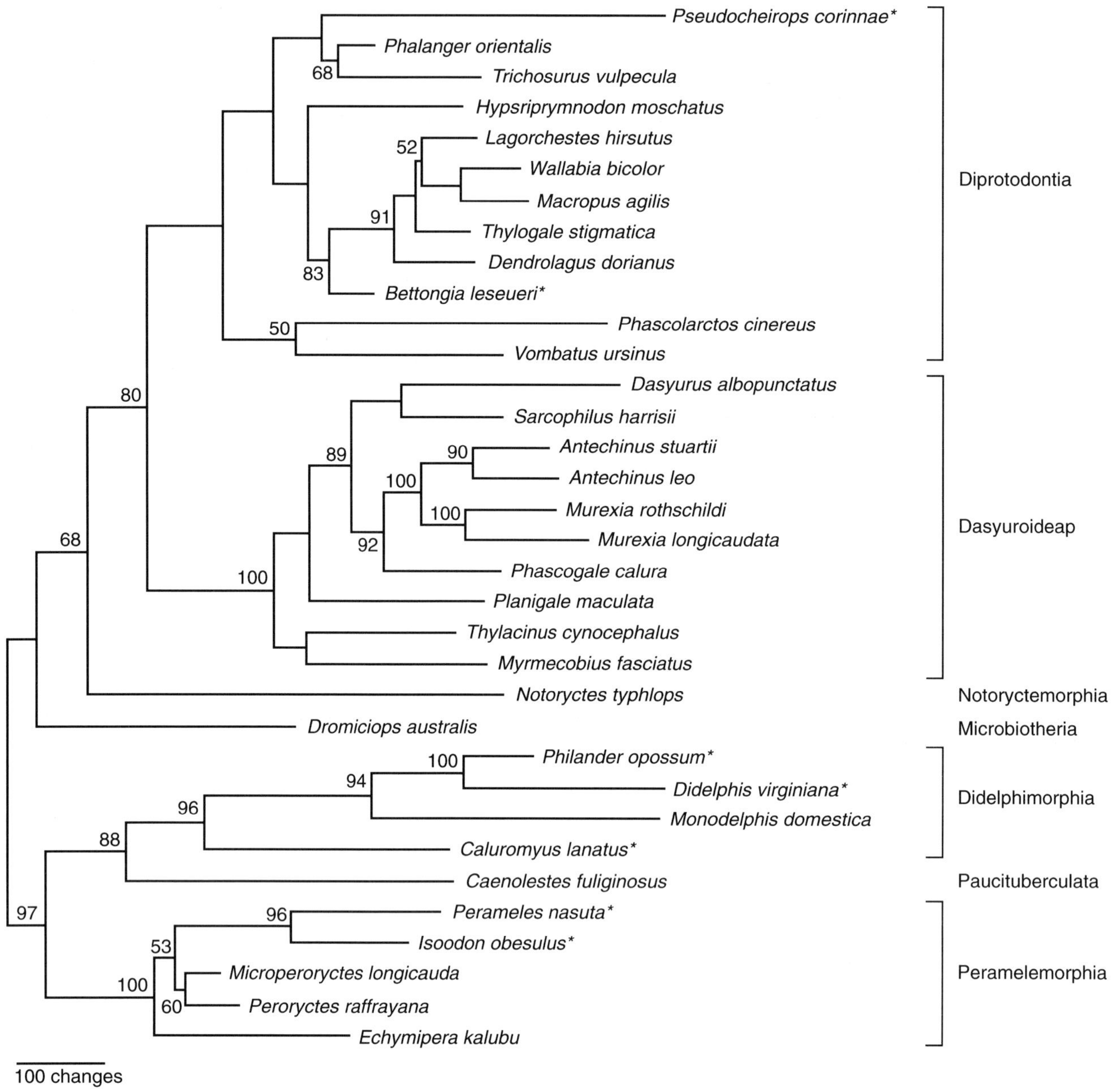

Figure 4. One of seven most parsimonious trees for the compilation of marsupial DNA sequences. The tree has a length of 7,927 steps from 1,559 parsimony informative characters and a consistency index of 0.428. Bootstrap support values are indicated near nodes. Major marsupial groups are shown at the right.

1995) and mitochondrial DNA sequencing (Hollar and Springer, 1997) suggest the existence of an African clade linked to the morphologically primitive *Rousettus*. Recently, Álvarez *et al.* (1999) have collected more mtDNA data and shown that the previously unsequenced *Eidolon* colonized the continent independently. No available analyses include all critical taxa—Álvarez *et al.* (1999) do not include the nectar-feeding *Notopteris*, enigmatic because it is the only long-tailed fruit bat. In the Australian Museum's laboratory we are completing the 12S ribosomal mtDNA dataset and compiling a comprehensive set for the nuclear gene *c*-mos. These data add weight to the hypothesis of an Indo-West Pacific origin of the group as the most basal dichotomies in the group all occur in the easternmost quarter of its distribution (Colgan and da Costa, 2002).

CONSERVATION BIOLOGY

One of the aims of conservation biology is identifying population substructuring that is significant for the long-term preservation of all genetic variability in the species. Recently, assessments of significance have often been made in terms of evolutionarily significant units (ESUs) sometimes defined as groups that are 'reciprocally monophyletic for mtDNA alleles and show significant divergence of allele frequencies at nuclear loci' (Moritz, 1994). Whilst this is an operational definition that has many advantages, it is too narrow. For instance, Firestone *et al.* (1999) did not find reciprocal monophyly between the two mainland ranges of the tiger quolls (New South Wales and Victoria versus far north Queensland). The latter populations do not form an ESU under the strict definition, but are clearly an evolutionary separate trajectory and would be so irrespective of European disruption of the species' range.

Although protein electrophoresis may not be sufficiently sensitive (Leung *et al.*, 1993), DNA-based techniques have been fruitful for estimation of the amount of genetic variation in endangered or vulnerable populations. Notable demonstrations of reduced variation include minisatellite analysis of koalas (Timms *et al.*, 1993), microsatellite analysis of the small single remnant population of the critically endangered northern hairy-nosed wombat (Taylor *et al.*, 1994), haplotype diversity in Southern right whales apparently lost through extensive hunting in the 19th century (Baker *et al.*, 1999) and allelic variation at the multiple histocompatibility locus in the bush rat *Rattus fuscipes* in Australian mainland and island populations (Seddon and Baverstock, 1999). In this species, there were 36 alleles at the locus over all samples. Most of the islands were, however, depauperate, 13 of 14 sampled having only one or two alleles. The study of Fowler *et al.* (1998), using RAPDs, found a lower level of variability in southern, as compared to northern, populations of the koala raising concerns as to population bottlenecks in the former areas.

Translocations of individuals to increase heterozygosity or reoccupy a habitat must pay close attention to the provenance of individuals. If selection is common against migrants from a differently adapted local population, as suggested by Cooper (2000) for the bandicoot *Isoodon obesulus*, then options for translocation will be limited. Preferably, before translocation, the populations that might be donors of recent natural migrants should be identified by microsatellite surveys as developed by Eldridge *et al.* (2001a) for rock-wallabies.

Genetic Criteria for Reserve Selection

The allocation of certain areas as nature reserves by legislative protection has an important role in conservation biology. Any serious attempt to preserve 'ecosystem' diversity is critically dependent on such reservation. Nevertheless, the alienation of these lands from other uses imposes an opportunity cost. Consequently, great care is necessary in the selection of particular areas for inclusion in reserves. Many criteria are brought to these decisions since each reserve must be thought of as an element in an overall system designed with respect to comprehensive preservation of all the ecosystems. There is never sufficient information about the species in an ecosystem for conservation measures to be tailored individually for the requirements of each. Consequently, 'habitat protection' is usually assumed to be the most efficient method of preserving as many species as possible. Operationally, 'protection' is defined with respect to the requirements of one or several species taken as exemplars. Genetic criteria have been generally ignored in the past because of the paucity of applicable data. Moritz and Faith (1998) make a promising start to using genetic information from multiple taxa in comparing distinct geographic areas. Their 'phylogenetic diversity' approach is particularly useful, if the members of the group have restricted geographic distributions. Frankham *et al.* (2002) also consider the genetic implications of restricted or fragmented ranges.

The authors are conducting an investigation aimed at ranking sites for reservation priority based on the combination of genetic data for different taxonomic groups of herpetofauna (Plate 126). We are using protein electrophoresis and SSCP/DNA sequencing of cytochrome b for the following species: *Amphibolurus nobbi*, *Bassiana platynota*, *Ctenotus robustus*, *Ct. taeniolatus*, *Egernia whitii*, *E. striolata*, *Eulamprus quoyii*, *Hemiergis decresiensis*, *Oedura lesueurii*, *O. tryoni* and *Pseudemoia pagenstecheri*. The study area is concentrated on the north-western slopes of New South Wales. In the interim bioregionalization scheme proposed by Thackway and Cresswell (1995), the range is principally the Nandewar and New England Tablelands, with some extension into the New South Wales North Coast region. Less than 5% of the land in the first two regions is protected.

Various metrics have been suggested for ranking of particular species or populations. (reviewed in Krajewski, 1994; Crozier, 1997; Petit *et al.*, 1998). In general, we suggest calculating overall metrics by combining (usually summing or averaging) values or numerical rankings for the individual groups. Two metrics are illustrated here.

(1) The average of the genetic distances between a given population and all other populations within a species or species group. This was expressed as a fraction of the maximum value found for any population within the particular species. These values were added for all species present at a site and divided by the number of such species to give an overall value to each site.
(2) The heterozygosity of a population at a site—expressed as a fraction of the maximum observed average heterozygosity in a population of the given species, with the overall value being calculated as in (1).

The metrics are illustrated in Figures 5 and 6. Sample points with high values on these genetic criteria are indicated by name. Some areas have high values on both metrics (e.g. the Mount Kaputar area), whereas other sites with high values on one metric rank lower with the other. This suggests that combining or weighting metrics might be difficult—although this is already recognized as a general problem in conservation biology (Freitag and Van Jaarsveld, 1997). The metrics illustrated here, and indeed all others presently available, are subject to potentially severe distortions due to low sample sizes of some species within sites, or of low numbers of sites for a species.

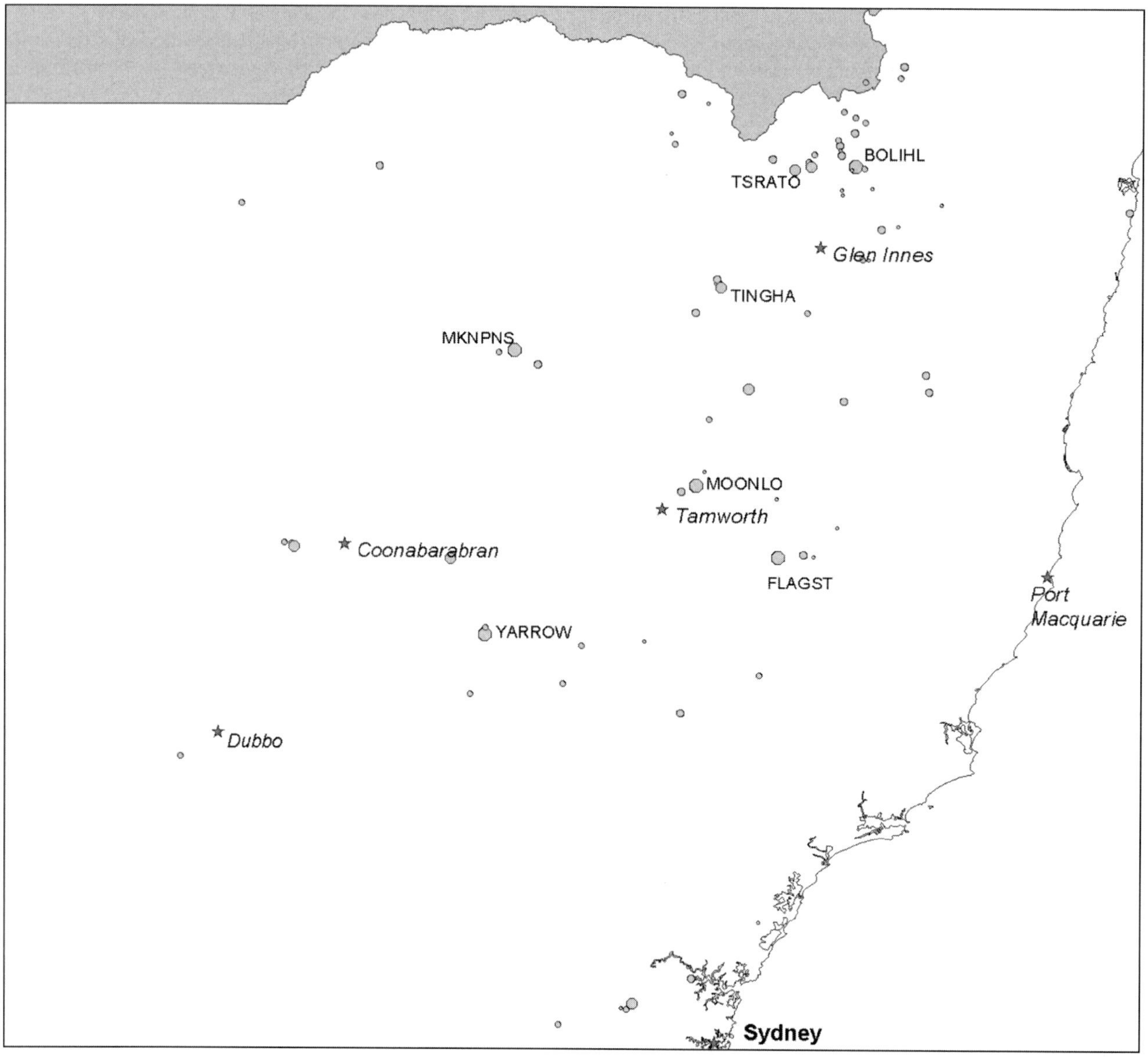

Figure 5. The values of metric (1)—genetic distance, at study sites in the Northern Tablelands. The sizes of the circles are proportional to the site's value. The maximum value is found at Mount Kaputar National Park (MKNPNS). Other notable sites include The Flags/Walcha (FLAGST, value 0.98), Bolivia Hill (BOLIHL-0.97), Moonbi (MOONLO-0.81), Yarrowick (YARROW-0.70), Tingha (TINGHA-0.64) and Torrington State Recreation Area (TSRATO-0.60)

Many high-ranking sites are included in nature reserves. Others, such as The Flags, Bolivia Hill and Moonbi (for genetic distance) or Apsley Falls and Gregadoo (for heterozygosity) are not. When additional reserves in the area are considered, the value of these sites on genetic criteria is a significant argument for their protection.

CONCLUSIONS

This chapter has summarized the techniques used in molecular systematics and reviewed their application to evolutionary studies and the conservation of Australian vertebrate fauna. In the intermediate future, such studies will predominantly use variants of the polymerase chain reaction and DNA sequencing. Yet until the use of DNA array technology becomes routine in systematics, protein electrophoresis will continue to be useful. The possibilities of other techniques, such as cytogenetics, should also be remembered. In the longer term, DNA array technology will dominate systematic investigations at all levels, from the provenance of individuals, to the population that has to be assessed for conservation importance, to higher order phylogenetic relationships. Then, conceivably, programmatic biodiversity survey with standardized arrays

Figure 6. The values of metric (2)—genetic heterozygosity, at study sites in the Northern Tablelands. The sizes of the circles are proportional to the site's value. The maximum values are found at Kingsgate (KINGS) and Mount Kaputar National Park (MKNPMY). Other notable sites are Marramarra National Park (MURRA, value 0.83), Apsley Falls (APSLEY-0.83), Gregadoo (GREGAD-0.83), Llangothlin (LLANGO-0.81) and Bonshaw (BONSHA-0.62).

could allow scientists finally to have a comprehensive description of animal life.

ACKNOWLEDGEMENTS

We wish to thank the late Ken and Yasuko Myer for their generous support of the Evolutionary Biology Unit, The Ian Potter Foundation for support of our marsupial studies and the N.S.W. State Government for Policy Enhancement funding of the 'Genetic Criteria Project'. Appreciation is expressed to K. Belov, P. Temple-Smith and P. Rankin for providing unpublished observations and the photograph respectively. Don Colgan wishes to dedicate the chapter to Drs M. da Cruz, M. Dexter and F. Vostaganis and the staff of the neurosurgery wards of Westmead Hospital.

REFERENCES

Álvarez, Y., Juste, J.B., Tabares, E., Garrido-Pertierra, A., Ibáñez and Bautista, J.M. 1999. Molecular phylogeny and morphological homoplasy in fruitbats. *Molecular Biology and Evolution*, Vol.16, pp.1061–1067.

Andersen, K. 1912. *Catalogue of the Chiroptera in the*

Collections of the British Museum. I. Megachiroptera. British Museum (Natural History), London. 854pp.

Andrews, A.T. 1986. *Electrophoresis: Theory, Techniques, and Biochemical and Clinical Applications.* Second Edition. Clarendon Press, Oxford. 452pp.

Archer, M. and Hand, S.J. 2006. The Australian marsupial radiation. Pp.575–646. In Merrick, J.R., Archer, M., Hickey, G.M. and Lee, M.S.Y. (eds). *Evolution and Biogeography of Australasian Vertebrates.* Auscipub Pty. Ltd., Sydney. 942pp.

Armstrong, L.A., Krajewski, C. and Westerman, M. 1998. Phylogeny of the dasyurid marsupial genus *Antechinus* based on cytochrome-*b*, 12S -rRNA and protamine -P1 genes. *Journal of Mammalogy*, Vol. 79, pp.1379–1389.

Austin, C.C. 2000. Molecular phylogeny and historical biogeography of Pacific island boas (*Candoia*). *Copeia,* 2000, pp.341–352.

Avise, J.C. 1994. *Molecular Markers, Natural History and Evolution.* Chapman and Hall, NewYork. 511pp.

Avise, J.C. 1996. Toward a regional conservation genetics perspective: Phylogeography of faunas in the southeastern United States. Pp.431–470. In Avise, J.C. and Hamrick, J.L. (eds). *Conservation Genetics, Case Histories from Nature.* Chapman and Hall, New York. 512pp.

Avise, J.C, Arnold, J., Ball, R.M., Bermingham, E., Lamb, T., Neigel, J.E., Reeb, C.A. and Saunders, N.C. 1987. Intraspecific phylogeography: the mitochondrial DNA bridge between population genetics and systematics. *Annual Review of Ecology and Systematics*, Vol. 18, pp.489–522.

Bagley, M.J., Medrano, J.F. and Gall, G.A.E. 1997. Polymorphic molecular markers from anonymous nuclear DNA for genetic analysis of populations. *Molecular Ecology*, Vol. 6, pp.309–320.

Baker, C.S., Patenaude, N.J., Bannister, J.L. and Kato, H. 1999. Distribution and diversity of mtDNA lineages among southern right whales (*Eubalaena australis*) from Australia and New Zealand. *Marine Biology*, Vol. 134, pp.1–7.

Baverstock, P.R. 1984. The molecular relationships of Australasian possums and gliders. Pp.1–8. In Smith, A.P. and Hume, I.D. (eds). *Possums and Gliders.* Surrey Beatty and Sons and Australian Mammal Society, Sydney. 598pp.

Baverstock, P.R., Krieg, M. and Birrell, J. 1990. Evolutionary relationships of Australian marsupials as assessed by albumin immunology. *Australian Journal of Zoology*, Vol. 37, pp.273–287.

Baverstock, P.R., Richardson, B.J., Birrell, J. and Krieg, M. 1989. Albumin immunological relationships of the Macropodidae (Marsupialia). *Systematic Zoology,* Vol. 38, pp.38–50.

Baverstock, P.R., Watts, C.H.S., Adams, M. and Cole, S.R. 1981. Genetical relationships among Australian rodents (Muridae). *Australian Journal of Zoology*, Vol. 29, pp.289–303.

Bertozzi, T., Adams, M. and Walker, K.F. 2000. Species boundaries in carp gudgeons (Eleotrididae: *Hypseleotris*) from the River Murray, South Australia: evidence for multiple species and extensive hybridization. *Marine and Freshwater Research*, Vol. 51, No. 8, pp.805–815.

Blacket, M.J., Krajewski, C., Labrinidis, A., Cambron, B., Cooper, S. and Westerman, M. 1999. Systematic relationships within the dasyurid marsupial tribe Sminthopsini—a multigene approach. *Molecular Phylogenetics and Evolution*, Vol. 12, pp.140–155.

Bolch, C.J.S., Ward, R.D. and Last, P.R. 1994. Biochemical systematics of the marine fish family Centrolophidae (Teleostei: Stomateoidei) from Australian waters. *Australian Journal of Marine and Freshwater Research,* Vol. 45, No. 7, pp.1157–1172.

Brewer, G.J. 1970. *An Introduction to Isozyme Techniques.* Academic Press, New York. 186pp.

Britten, R.J., Graham, D.E. and Neufeld, B.R. 1974. Analysis of repeating DNA sequences by re-association. Pp.363–418, Vol. 29, Part E. In Grossman, L. and Moldave, K. (eds). *Methods in Enzymology.* Academic Press, New York.

Brown, D.M. and Toft, C.A. 1999. Molecular systematics and biogeography of the cockatoos (Psittaciformes: Cacatuidae). *Auk*, Vol. 116, pp.141–157.

Bruford, M.W., Cheesman, D.J., Coote, T., Green, H.A., Haines, S.A., O'Ryan, C. and Williams, T.R. 1996. Microsatellites and their application to conservation genetics. Pp.278–297. In Smith, T.B. and Wayne, R.K. (eds). *Molecular Genetic Approaches in Conservation.* Oxford University Press, New York. 483pp.

Burke, T., Hanotte, O. and Van Pijlen, I. 1996. Minisatellite analysis in conservation genetics. Pp.251–277. In Smith, T.B. and Wayne, R.K. (eds). *Molecular Genetic Approaches in Conservation.* Oxford University Press, New York. 483pp.

Burridge, C.P. 2000. Molecular phylogeny of the Aplodactylidae (Perciformes: Cirrhitoidea), a group of Southern Hemisphere marine fishes. *Journal of Natural History*, Vol. 34, pp.2173–2185.

Campbell, N.J.H., Harris, F.C., Elphinstone, M.S. and Baverstock, P.R. 1995. Outgroup heteroduplex analysis using temperature gradient gel electrophoresis: high resolution, large scale screening of DNA variation in the mitochondrial DNA control region. *Molecular Ecology,* Vol. 4, pp.407–418.

Campeau-Péloquin, A., Kirsch, J.A.W., Eldridge, M.D.B. and Lapointe, F.-J. 2001. Phylogeny of the rock-wallabies, *Petrogale* (Marsupialia: Macropodidae) based on DNA/DNA hybridisation. *Australian Journal of Zoology*, Vol. 49, pp.463–486.

Cardinal, B.R. and Christidis, L. 1999. Mitochondrial DNA and morphology reveal three geographically distinct lineages of the large bentwing bat (*Miniopterus schreibersii*) in Australia. *Australian Journal of Zoology*, Vol. 48, pp.1–19.

Christidis, L. and Schodde, R. 1997. Relationships within the Australo-Papuan fairy-wrens (Aves: Malurinae): an evaluation of the utility of allozyme data. *Australian Journal of Zoology*, Vol. 45, pp.113–129.

Clegg, S.M., Hale, P. and Moritz, C. 1998. Molecular population genetics of the red kangaroo (*Macropus rufus*). *Molecular Ecology,* Vol. 7, pp.679–686.

Colgan, D.J. 1990. Systematic evidence from gene duplication and regulation. Pp.145–158. In Ladiges, P.Y. and Martinelli, L.W. (eds). *Plant Systematics in the Age of Molecular Biology.* CSIRO Publishing, Melbourne. 163pp.

Colgan, D.J. 1997. Electrophoretic variation in the Green and Golden Bell Frog, *Litoria aurea. Australian Zoologist*, Vol. 30, pp.170–177.

Colgan, D.J. 1999. Phylogenetic studies of marsupials based on phosphoglycerate kinase DNA sequences. *Molecular Phylogenetics and Evolution,* Vol. 11, pp.13–26.

Colgan, D.J. and da Costa, P. 2002. Megachiropteran evolution studied with 12s rDNA and *c*-Mos DNA sequences. *Journal of Mammalian Evolution*, Vol. 9, pp.3–22.

Colgan, D.J. and Flannery, T.F. 1992. Biochemical systematic studies in the genus *Petaurus* (Marsupialia: Petauridae). *Australian Journal of Zoology*, Vol. 40, pp.245–256.

Colgan, D.J. and Flannery, T.F. 1995. A phylogeny of Indo-West Pacific Megachiroptera based on ribosomal DNA. *Systematic Biology*, Vol. 44, pp.209–220.

Colgan, D.J., Flannery, T.F., Trimble, J. and Aplin, K. 1993. Electrophoretic analysis of the systematics of the *Phalanger orientalis* (Marsupialia) species complex in Papua New Guinea and the Solomon Islands. *Australian Journal of Zoology*, Vol. 41, pp.355–378.

Colgan, D.J. and Paxton, J.R. 1997. Biochemical systematics and recognition of a western stock of the common gemfish, *Rexea solandri* (Scombroidea: Gempylidae), in Australia. *Marine and Freshwater Research*, Vol. 48, No. 2, pp.103–118.

Cooper, M.L. 2000. Random amplified polymorphic DNA analysis of southern brown bandicoot (*Isoodon obesulus*) populations in Western Australia reveals genetic differentiation related to environmental variables. *Molecular Ecology*, Vol. 9, pp.469–479.

Cooper, S.J.B., Bull, C.M. and Gardner, M.E. 1997. Characterization of microsatellite loci from the socially monogamous lizard *Tiliqua rugosa* using a PCR-based isolation technique. *Molecular Ecology,* Vol. 6, pp.793–795.

Cooper, S.J.B., Day, P.R., Reardon, T.B. and Schulz, M. 2001. Assessment of species boundaries in Australian *Myotis* (Chiroptera: Vespertilionidae) using mitochondrial DNA. *Journal of Mammalogy*, Vol. 82, pp.328–338.

Cooper, S.J.B., Reardon, T.B. and Skilins, J. 1998. Molecular systematics of Australian rhinolophid bats (Chiroptera: Rhinolophidae). *Australian Journal of Zoology*, Vol. 46, pp.203–220.

Crozier, R.H. 1997. Preserving the information content of species: genetic diversity, phylogeny and conservation worth. *Annual Review of Systematics and Ecology*, Vol. 28, pp.243–268.

Dessauer, H.C., Reeder, T.W., Cole, C.J. and Knight, A. 1996. Rapid screening of DNA diversity using dot-blot technology and allele-specific oligonucleotides: maternity of hybrids and unisexual clones of hybrid origin (lizards, *Cnemidophorus*). *Molecular Phylogenetics and Evolution*, Vol. 6, pp.366–372.

Donnellan, S.C., Hutchinson, M.N. and Saint, K.M. 1999. Molecular evidence for the phylogeny of Australian gekkonoid lizards. *Biological Journal of the Linnean Society*, Vol. 67, pp.97–118.

Dowling, T.E., Moritz, C., Palmer, J.D. and Rieseberg, L.H. 1996. Nucleic acids II: restriction site analysis. Pp.249–320. In Hillis, D.M., Moritz, C. and Mable, B. (eds). *Molecular Systematics.* Second Edition. Sinauer Associates, Sunderland, Massachusetts. 655pp.

Drouin, G., Prat, F., Ell, M. and Clarke, G.D.P. 1999. Detecting and characterizing gene conversions between multigene family members. *Molecular Biology and Evolution*, Vol. 16, pp.1369–1390.

Durbin R., Eddy, S., Krogh, A. and Mitchison, G. 1998. *Biological sequence analysis: probabilistic models of proteins and nucleic acids*. Cambridge University Press, New York. 356pp.

Edwards, D. and Westerman, M. 1995. The molecular relationships of possums and gliders families as revealed by DNA-DNA hybridisation. *Australian Journal of Zoology*, Vol. 43, pp.231–240.

Edwards, S.V., Arctander, P. and Wilson, A.C. 1991. Mitochondrial resolution of a deep branch in the genealogical tree for perching birds. *Proceedings of the Royal Society of London,* Series B, Vol. 243, pp.99–107.

Edwards, S.V. and Kot, M. 1995. Comparative methods at the species level: geographic variation in morphology and group size in grey-crowned bablers (*Pomatostomus temporalis*). *Evolution*, Vol. 49, pp.1134–1146.

Eldridge, M.D.B. and Close, R.L. 1992. Taxonomy of rock-wallabies, *Petrogale* (Marsupialia: Macropodidae). I. A revision of the eastern *Petrogale* with the description of three new species. *Australian Journal of Zoology*, Vol. 40, pp.605–625.

Eldridge, M.D.B., Kinnear, J.E. and Onus, M.L. 2001a. Source populations of dispersing rock-wallabies (*Petrogale lateralis*) identified by assignment tests on multilocus genotypic data. *Molecular Ecology*, Vol. 10, pp.2867–2876.

Eldridge, M.D.B., Wilson, A.C.C., Metcalfe, C.J., Dollin, A.E., Bell, J.N., Johnson, P.M., Johnston, P.G. and Close, R.L. 2001b. Taxonomy of rock-wallabies, *Petrogale* (Marsupialia: Macropodidae). III. Molecular data confirms the specific status of the purple-necked rock-wallaby (*Petrogale purpureicollis* Le Souef). *Australian Journal of Zoology*, Vol. 49, pp.323–343.

Elliott, N.G., Haskard, K. and Koslow, J.A. 1995. Morphometric analysis of orange roughy (*Hoplostethus atlanticus*) off the continental slope of southern Australia. *Journal of Fish Biology*, Vol. 46, pp.202–220.

Elliott, N.G. and Ward, R.D. 1994. Enzyme variation in Jackass Morwong, *Nemadactylus macropterus* (Schneider, 1801) (Teleostei: Cheilodactylidae), from Australia and New Zealand waters. *Australian Journal of Marine and Freshwater Research*, Vol. 45, No. 1, pp.51–67.

Estoup, A., Cornuet, J.-M., Rousset, F. and Guyomard, R. 1999. Juxtaposed microsatellite systems as diagnostic markers for admixture: theoretical aspects. *Molecular Biology and Evolution*, Vol. 16, pp.898–908.

Fairbairn, J., Shine, R., Moritz, C. and Frommer, M. 1998. Phylogenetic relationships between oviparous and viviparous populations of an Australian lizard (*Lerista bougainvilli*, Scincidae). *Molecular Phylogenetics and Evolution*, Vol. 10, pp.95–103.

Felsenstein, J. 1989. PHYLIP Phylogeny Inference Package (Version 3.2). *Cladistics*, Vol. 5, pp.164–166.

Felsenstein, J. 2002. Phylogeny programs. http://evolution.genetics.washington.edu/phylip/software.html

Ferraris, J.D. and Palumbo, S.R. 1996. *Molecular Zoology: Advances, Strategies, and Protocols***.** Wiley-Liss, New York. 580pp.

Firestone, K.B., Elphinstone, M.S., Sherwin, W.B. and Houlden, B.A. 1999. Phylogeographic structure of tiger quolls *Dasyurus maculatus* (Dasyuridae: Marsupialia), an endangered carnivorous marsupial. *Molecular Ecology,* Vol. 8, pp.1613–1625.

Fowler, E.V., Hoeben, P. and Timms, P. 1998. Randomly amplified polymorphic DNA variation in populations of eastern Australian koalas, *Phascolarctos cinereus*. *Biochemical Genetics*, Vol. 36, pp.381–393.

Frankham, R. 1995. Conservation genetics. *Annual Review of Genetics*, Vol. 29, pp.305–327.

Frankham, R. 1998. Inbreeding and extinction: island populations. *Conservation Biology*, Vol. 12, pp.665–675.

Frankham, R., Ballou, J.D. and Briscoe, D.A. 2002. *Introduction to Conservation Genetics*. Cambridge University Press, Cambridge. 617pp.

Freitag S. and Van Jaarsveld, A.S. 1997. Relative occupancy, endemism, taxonomic distinctiveness and vulnerability: prioritising regional conservation actions. *Biodiversity and Conservation*, Vol. 6, pp.211–232.

Fuller, S., Baverstock, P. and King, D. 1998. Biogeographic origins of goannas (Varanidae): a molecular perspective. *Molecular Phylogenetics and Evolution*, Vol. 9, pp.294–307.

Georges, A. and Adams, M. 1992. A phylogeny for Australian chelid turtles based on allozyme electrophoresis. *Australian Journal of Zoology*, Vol. 40, pp.453–476.

Georges, A., Birrell, J., Saint, K.M., McCord, W. and Donellan, S.C. 1999. A phylogeny for side-necked turtles (Chelonia: Pleurodira) based on mitochondrial and nuclear gene sequence variation. *Biological Journal of the Linnean Society*, Vol. 67, pp.213–246.

Gissing, G.J. and Crease, T.J. 1997. A simple and effective chemiluminescent DNA fingerprinting technique using digoxigenin-labelled minisatellite probes. *Canadian Journal of Zoology*, Vol. 75, pp.817–820.

Goldman, D. and O'Brien, S.J. 1993. Two-dimensional protein electrophoresis in phylogenetic studies. Pp.113–121. In Zimmer, E.A., White, T.J., Cann, R.L. and Wilson, A.C. (eds). *Molecular Evolution: Producing the Biochemical Data*. Academic Press, San Diego. 725pp.

Grewe, P.M., Smolenski, A.J. and Ward, R.D. 1994. Mitochondrial DNA diversity in jackass morwong (*Nemadactylus macropterus*: Teleostei) from Australian and New Zealand waters. *Canadian Journal of Fisheries and Aquatic Sciences*, Vol. 51, pp.1101–1109.

Groth, J.G. and Barrowclough, G.F. 1999. Basal divergences in birds and the phylogenetic utility of the nuclear RAG-1 gene. *Molecular Phylogenetics and Evolution*, Vol. 12, pp.115–123.

Hadrys, H., Balick, M. and Shierwater, B. 1992. Applications of random amplified DNA (RAPD) in molecular ecology. *Molecular Ecology*, Vol. 1, pp.55–63.

Hand, S.J. 2006. Bat beginnings and biogeography: the Australasian record. Pp.673–705. In Merrick, J.R., Archer, M., Hickey, G.M. and Lee, M.S.Y. (eds). *Evolution and Biogeography of Australasian Vertebrates*. Auscipub Pty. Ltd., Sydney. 942pp.

Harris, H. and Hopkinson, D.A. 1977. *Handbook of Enzyme Electrophoresis in Human Genetics*. North-Holland, Amsterdam.

Hay, J.M., Ruvinsky, I., Hedges, S.B. and Maxson, L.R. 1995. Phylogenetic relationships of amphibian families inferred from DNA sequences of mitochondrial 12S and 16S ribosomal RNA genes. *Molecular Biology and Evolution*, Vol. 12, pp.928–937.

Hayashi, K. 1996. PCR SSCP: Single-strand conformation polymorphism analysis of PCR products. Pp.14–22. In Landegren, U. (ed). *Laboratory Protocols for Mutation Detection*. Oxford University Press, Oxford. 192pp.

Hershkovitz, P. 1992. Ankle bones: the Chilean opossum *Dromiciops gliroides* Thomas, and marsupial phylogeny. *Bonner Zoologische Beiträge*, Vol. 43, pp.181–213.

Hillis, D.M., Moritz, C. and Mable, B.K. (eds). 1996. *Molecular Systematics*. Second Edition. Sinauer Associates, Sunderland, Massachusetts. 655pp.

Hollar, L.J. and Springer, M.S. 1997. Old World fruitbat phylogeny: evidence for convergent evolution and an endemic African clade. *Proceedings of the National Academy of Sciences of the United States of America*, Vol. 94, pp.5716–5721.

Houlden, B.A., Costello, B.H., Sharkey, D., Fowler, E.V., Metzer, A., Ellis, W., Carrick, F., Baverstock, P.R. and Elphinstone, M.S. 1999. Phylogeographic differentiation in the mitochondrial control region in the koala *Phascolarctos cinereus* (Goldfuss 1817). *Molecular Ecology*, Vol. 8, pp.999–1011.

Howard, D.J. and Berlocher, S.H. 1998. *Endless Forms: Species and Speciation*. Oxford University Press, New York. 470pp.

Innis, M.A., Gelfand, D.H., Sninsky, J.J. and White, T.J. 1990. *PCR Protocols, a Guide to Methods and Applications*. Academic Press, San Diego. 482pp.

Janke, A., Xu, X. and Arnason, U. 1997. The complete mitochondrial genome of the wallaroo (*Macropus robustus*) and the phylogenetic relationships among Monotremata, Marsupialia, and Eutheria. *Proceedings of the National Academy of Sciences of the United States of America*, Vol. 94, pp.1276–1281.

Jin, L., Macaubas, C., Hallmayer, J., Kimura, A. and Mignot, E. 1996. Mutation rate varies among alleles at a microsatellite locus. *Proceedings of the National Academy of Sciences of the United States of America*, Vol. 93, pp.15285–15288.

Johanson, Z. 2006. Origins of mammals: morphology, molecules and a petrosal or two. Pp.491–522. In Merrick, J.R., Archer, M., Hickey, G.M. and Lee, M.S.Y. (eds). *Evolution and Biogeography of Australasian Vertebrates*. Auscipub Pty. Ltd., Sydney. 942pp.

Killian, J.K., Buckley, T.R., Stewart, N., Munday, B.L. and Jirtle, R.L. 2001. Marsupials and Eutherians reunited: genetic evidence for the Theria hypothesis of mammalian evolution. *Mammalian Genome*, Vol. 12, pp.513–517.

King, M. 1984. Karyotypic evolution in *Gehyra* (Gekkonidae: Reptilia) IV. Chromosome change and speciation. *Genetica*, Vol. 64, pp.101–114.

Kirsch, J.A.W. 1977. The comparative serology of Marsupialia, and a classification of marsupials. *Australian Journal of Zoology*, Supplementary Series 52, pp.1–152.

Kirsch, J.A.W., Flannery, T.F., Springer, M.S. and Lapointe, F.-J. 1995. Phylogeny of the Pteropodidae (Mammalia: Chiroptera) based on DNA hybridisation, with evidence for bat monophyly. *Australian Journal of Zoology*, Vol. 43, pp.395–428.

Kirsch, J.A.W., Lapointe, F.-J., and Springer, M.S. 1997. DNA studies of marsupials and their implications for metatherian classification. *Australian Journal of Zoology*, Vol. 45, pp.211–280.

Kirsch, J.A.W., Springer, M.S., Krajewski, C., Archer, M., Aplin, K. and Dickerman, A.W. 1990. DNA/DNA hybridisation studies of the carnivorous marsupials I. The intergeneric relationships of bandicoots (Marsupialia, Perameloidea). *Journal of Molecular Evolution*, Vol. 30, pp.434–448.

Krajewski, C. 1994. Phylogenetic measures of biodiversity: a comparison and critique. *Conservation Biology,* Vol. 69, pp.33–39.

Krajewski, C., Buckley, M. and Westerman, M. 1997. DNA phylogeny of the marsupial wolf resolved. *Proceedings of the Royal Society of London,* Series B, Vol. 264, pp.911–917.

Krajewski C., Driskell, A.C., Baverstock, P.R. and Braun, M.J. 1992. Phylogenetic relationships of the thylacine (Mammalia: Thylacinidae) among dasyuroid marsupials: evidence from cytochrome *b* DNA sequences. *Proceedings of the Royal Society of London,* Series B, Vol. 250, pp.19–27.

Kumar, S.K., Tamura, K. and Nei, M. 1993. *MEGA, Molecular Evolutionary Genetic Analysis.* 1. 01. Pennsylvania State University, University Park.

Lambert, D.M. and Millar, C.D. 1995. DNA science and conservation. *Pacific Conservation Biology,* Vol. 2, pp.21–38.

Lapointe, F.-J., and Kirsch, J.A.W. 1995. Estimating phylogenies from lacunose distance matrices, with special reference to DNA hybridisation data. *Molecular Biology and Evolution*, Vol.12, pp.266–284.

Lavery, S. and Shaklee, J.B. 1991. Genetic evidence for separation of two sharks, *Carcharhinus limbatusi* and *C. tilstoni* from northern Australia. *Marine Biology*, Vol. 108, pp.1–4.

Lee, M.-H., Shroff, R., Cooper, S.J.B. and Hope, R. 1999. Evolution and molecular characterization of a β-globin gene from the Australian echidna *Tachyglossus aculeatus* (Monotremata). *Molecular Phylogenetics and Evolution*, Vol.12, pp.205–214.

Lee, M.S.Y. 2006. Systematics and the Australasian vertebrate fauna. Pp.3–16. In Merrick, J.R., Archer, M., Hickey, G.M. and Lee, M.S.Y. (eds). *Evolution and Biogeography of Australasian Vertebrates.* Auscipub Pty. Ltd., Sydney. 942pp.

Leung, L.K.-P., Dickman, C.R. and Moore, L.A. 1993. Genetic variation in fragmented populations of an Australian rainforest rodent, *Melomys cervinipes. Pacific Conservation Biology,* Vol.1, 58–65.

Lewontin, R.C. and Hubby, J. 1966. A molecular approach to the study of genetic heterogeneity in natural populations. II. Amounts of variation and degree of heterozygosity in natural populations of *Drosophila pseudoobscura. Genetics*, Vol. 54, pp.595–609.

Lisitsyn, N., Lisitsyn, N. and Wigler, M. 1993. Cloning the differences between two complex genomes. *Science*, Vol.259, pp.946–951.

Lunt, D.H., Hutchinson, W.F. and Carvalho, G.R. 1999. An efficient method for PCR-based isolation of microsatellite arrays (PIMA). *Molecular Ecology,* Vol. 8, pp.891–894.

Maddison, W.P. and Maddison, D.R. 1992. *MacCLADE, Version 3.0.* Sinauer Associates, Sunderland, Massachusetts. 398pp.

Mahony, M.J., Norris, R.M. and Donnellan, S.C. 1996. Karyotypes of South-west Pacific ranid frogs. *Australian Journal of Zoology*, Vol. 44, pp.119–128.

Manchenko, G.P. 1994. *Handbook of Detection of Enzymes on Electrophoretic Gels.* CRC Press, Boca Raton, Florida. 341pp.

McDonnell, L.J., Gartside, D.F. and Littlejohn, M.J. 1978. Analysis of a narrow hybrid zone between two species of *Pseudophryne* (Anura: Lepidodactylidae) in south-eastern Australia. *Evolution*, Vol. 32, pp.602–612.

McElroy, D., Moran, P., Bermingham, E. and Kornfeld, I. 1992. REAP: an integrated environment for the manipulation and phylogenetic analysis of restriction data. *Journal of Heredity*, Vol. 83, pp.157–158.

McGuigan, K., Zhu, D., Allen, G.R. and Moritz, C. 2000. Phylogenetic relationships and historical biogeography of melanotaeniid fishes in Australia and New Guinea. *Marine and Freshwater Research*, Vol. 51, No. 7, pp.713–723.

Mengden, G.A., Shine, R. and Moritz, C. 1986. Phylogenetic relationships within the Australasian venomous snakes of the genus *Pseudechis*. *Herpetologica*, Vol. 42, pp.215–229.

Miller, R.D. and Belov, K. 2001. Immunoglobulin genetics of marsupials. *Developmental and Comparative Immunology*, Vol. 24, pp.485–490.

Milton, D.A., Hughes, J.M. and Mather, P.B. 1983. Electrophoretic evidence for the specific distinctness of *Egernia modesta* and *E. whitii* (Lacertilia: Scincidae). *Herpetologica*, Vol.39, pp.100–105.

Milton, D.A. and Shaklee, J.B. 1987. Biochemical genetics and population structure of Blue Grenadier, *Macruronus novaezelandiae* (Hector) (Pisces: Merluccidae), from Australian waters. *Australian Journal of Marine and Freshwater Research*, Vol. 38, No. 6, pp.727–742.

Miyamoto, M.M. and Cracraft, J. 1991. *Phylogenetic Analysis of DNA Sequences*. Oxford University Press, New York. 358pp.

Moritz, C. 1994. Defining 'Evolutionary Significant Units' for conservation. *Trends in Ecology and Evolution,* Vol. 9, pp.373–375.

Moritz, C. and Faith, D. 1998. Comparative phylogeography and the identification of genetically divergent areas for conservation. *Molecular Ecology,* Vol. 7, pp.419–429.

Moritz, C. and Heideman, A. 1993. The origin and evolution of parthenogenesis in *Heteronotia binoei* (Gekkonidae): reciprocal origins and diverse mitochondrial DNA in western populations. *Systematic Biology*, Vol. 42, pp.293–306.

Moritz, C., Heideman, A., Geffen, E. and McRae, P. 1997. Genetic population structure of the bilby, *Macrotis lagotis*, a marsupial in decline. *Molecular Ecology*, Vol. 6, pp.925–936.

Murray, J.D., Sharman, G.B., McKay, G.M. and Calaby, J.H. 1980. Karyotypes, constitutive heterochromatin and taxonomy of ringtail opossums of the Genus *Pseudocheirus* (Marsupialia: Petauridae). *Cytogenetics and Cell Genetics*, Vol. 27, pp.73–81.

Nekrutenko, A., Makova, K.D., Chesser, R.K. and Baker, R.J. 1999. Representational difference analysis to distinguish cryptic species. *Molecular Ecology*, Vol. 8, pp.1235–1238.

Ochman, W.H., Medhora, M.M., Garza, D. and Hartl, D.L. 1990. Amplification of flanking sequences by inverse PCR. Pp.219–227. In Innis, M.A., Gelfand, D.H., Sninsky, J.J. and White, T.J. (eds). *PCR Protocols: A Guide to Methods and Applications.* Academic Press, San Diego. 482pp.

Ovenden, J.R. 1990. Mitochondrial DNA and marine stock

assessment: a review. *Australian Journal of Marine and Freshwater Research*, Vol. 41, No. 6, pp.835–853.

Palma, R.E. and Spotorno, A.E. 1999. Molecular systematics of marsupials based on the rRNA 12S mitochondrial gene: the phylogeny of Didelphimorphia and of the living fossil microbiotheriid *Dromiciops gliroides* Thomas. *Molecular Phylogenetics and Evolution*, Vol. 13, pp.525–535.

Petit, R.J., El Mousadik, A. and Pons, O. 1998. Identifying populations for conservation on the basis of genetic markers. *Conservation Biology,* Vol. 12, pp.844–855.

Powell, L.M. 1990. RNA processing: Apo-B. Pp.237–244. In Innis, M.A., Gelfand, D.H., Sninsky, J.J. and White T.J. (eds). *PCR Protocols, a Guide to Methods and Applications*. Academic Press, San Diego. 482pp.

Prychitko, T.M. and Moore, W.S. 1997. The utility of DNA sequences of an intron from the â-fibrinogen gene in phylogenetic analyis of woodpeckers (Aves: Picidae). *Molecular Phylogenetics and Evolution*, Vol. 8, pp.193–204.

Rapley, R. 1996. *PCR Sequencing Protocols*. Humana Press, Totowa, New Jersey. 221pp.

Rassmann, K., Schlötterer, C. and Tautz, D. 1991. Isolation of simple sequence loci for use in polymerase chain reaction-based DNA fingerprinting. *Electrophoresis*, Vol. 12, pp.113–118.

Reig, O.A., Kirsch, J.A.W. and Marshall, L.G. 1987. Systematic relationships of the living and neocenozoic American "opossum-like" marsupials (suborder Didelphimorphia), with comments on the classification of these and of the Cretaceous and Paleogene New World and European metatherians. Pp.1–89. In Archer, M. (ed). *Possums and Opossums*: *Studies in Evolution*. Vol. 1. Surrey Beatty and Sons, Sydney. Pp.1–400.

Retief, J.D., Krajewski, C., Westerman, M. and Dixon, G.H. 1995a. The evolution of protamine P1 genes in dasyurid marsupials. *Journal of Molecular Evolution*, Vol. 41, pp.549–555.

Retief, J.D., Krajewski, C., Westerman, M., Winkfein, R.J. and Dixon, G.H. 1995b. Molecular phylogeny and evolution of protamine P1 genes. *Proceedings of the Royal Society of London,* Series B, Vol. 259, pp.7–14.

Richardson, B.J., Baverstock, P.R. and Adams, M. 1986. *Allozyme Electrophoresis*. Academic Press, Sydney. 410pp.

Richardson, B.J. and McDermid, E.M. 1978. A comparison of genetic relationships within the Macropodidae as determined by allozyme, cytological and immunological data. *Australian Mammalogy*, Vol. 2, pp.43–51.

Rico, C., Rico, I. and Hewitt, G. 1996. 470 million years of conservation of microsatellite loci among fish species. *Nature Genetics*, Vol. 13, pp.549–557.

Rofe, R.H. and Hayman, D.L. 1985. G-banding evidence for a conserved complement in the Marsupialia. *Cytogenetics and Cell Genetics,* Vol. 39, pp.40–50.

Sadlier, R.A., Colgan, D.J. and Shea, G.M. 1994. Taxonomy and distribution of the Australian scincid lizard *Saproscincus challengeri* and related species in south-eastern Australia. *Memoirs of the Queensland Museum*, Vol. 34, pp.139–158.

Schema, M. (ed). 2000. *Microarray Biochip Technology.* BioTechniques Books, Westborough, Massachusetts. 298pp.

Schneider, C.J., Cunningham, M. and Moritz, C. 1998. Comparative phylogeography and the history of endemic vertebrates in the wet tropics rainforests of Australia. *Molecular Ecology,* Vol. 7, pp.487–498.

Seddon, J.M. and Baverstock, P.R. 1999. Variation on islands: multiple histocompatibility (*Mhc*) polymorphisms in populations of the Australian bush rat. *Molecular Ecology,* Vol. 8, pp.2071–2079.

Seddon, J.M., Georges, A., Baverstock, P.R. and McCord, W. 1997. Phylogenetic relationships of chelid turtles (Pleurodira: Chelidae) based on mitochondrial 12S rRNA gene sequence variation. *Molecular Phylogenetics and Evolution*, Vol. 7, pp.55–61.

Sessions, S.K. 1996. Chromosomes: molecular cytogenetics. Pp 121–169. In Hillis, D.M., Moritz, C. and Mable, B. (eds). *Molecular Systematics*. Second Edition. Sinauer Associates, Sunderland, Massachusetts. 655pp.

Shaklee, J.B. and Keenan, C.P. 1986. A practical laboratory guide to the techniques and methodology of electrophoresis and its application to fish fillet identification. *Australia CSIRO Marine Laboratories Report,* No. 177, pp.1–59.

Sharman, G.B., Close, R.L and Maynes, G.M. 1991. Chromosome evolution, phylogeny, and speciation of rock-wallabies. *Australian Journal of Zoology*, Vol. 37, pp.351–363.

Shaw, C.R. 1965. Electrophoretic variation in enzymes. *Science*, Vol. 149, pp.936–943.

Sibley, C.G. and Ahlquist, J.E. 1990. *Phylogeny and Classification of Birds: A Study in Molecular Evolution*. Yale University Press, New Haven, Connecticut. 976pp.

Smith, S.A., Austin, C.C. and Shine, R. 2001. A phylogenetic analysis of variation in reproductive mode within an Australian lizard (*Saiphos equalis*, Scincidae). *Biological Journal of the Linnean Society*, Vol. 74, pp.131–139.

Smolenski, A.J., Ovenden, J.R. and White, R.W.G. 1994. Preliminary investigation of mitochondrial DNA variation in jack mackerel (*Trachurus declivis*, Carangidae) from south-eastern Australian waters. *Australian Journal of Marine and Freshwater Research*, Vol. 45, No. 4, pp.495–505.

Southgate, R. and Adams, M. 1993. Genetic variation in the greater bilby (*Macrotis lagotis*). *Pacific Conservation Biology,* Vol. 1, pp.46–52.

Spencer, P.B.S., Odorico, D.M., Jones, S.J., Marsh, H.D. and Miller, D.J. 1995. Highly variable microsatellites in isolated colonies of the rock-wallaby (*Petrogale assimilis*). *Molecular Ecology,* Vol. 4, pp.523–525.

Springer, M., Kirsch, J.A.W., Aplin, K. and Flannery, T.F. 1990. DNA hybridisation, cladistics and the phylogeny of phalangerid marsupials. *Journal of Molecular Evolution*, Vol. 30, pp.298–311.

Springer, M., McKay, G., Aplin, K. and Kirsch, J.A.W. 1992. Relations among ringtail possums (Marsupialia: Pseudocheiridae) based on DNA-DNA hybridisation. *Australian Journal of Zoology*, Vol. 40, pp.423–435.

Springer, M.S., Westerman, M., Kavanagh, J.R., Burk, A., Woodburne, M.O., Kao, D.J. and Krajewski, C. 1998. The origin of the Australasian marsupial fauna and the phylogenetic affinities of the enigmatic monito del monte and marsupial mole. *Proceedings of the Royal Society of London,* Series B, Vol. 265, pp.2381–2386.

Springer, M., Westerman, M. and Kirsch, J.A.W. 1994. Relationships among orders and families of marsupials based on 12S ribosomal DNA sequences and the timing of the marsupial radiation. *Journal of Mammalian Evolution*, Vol. 2, pp.85–115.

Steelman, J.T., Zarfoya, R., Meyer, A. and Karl, S.A. 1998. Multilocus phylogeny of cichlid fishes (Pisces: Perciformes): evolutionary comparison of microsatellite and single-copy nuclear loci. *Molecular Biology and Evolution*, Vol. 15, pp.798–808.

Sumner, J. and Dickman, C.R. 1998. Distribution and identity of species in the *Antechinus stuartii-A. flavipes* group (Marsupialia: Dasyuridae) in south-eastern Australia. *Australian Journal of Zoology*, Vol. 46, pp.27–41.

Sunnucks, P., Wilson, A.C.C., Beheregaray, L.B., Zenger, K., French, J. and Taylor, A.C. 2000. SSCP is not so difficult: the application and utility of single-stranded conformational polymorphism in evolutionary biology and molecular ecology. *Molecular Ecology*, Vol. 9, pp.1699–1710.

Swofford, D.L. 2000. *PAUP*: Phylogenetic Analysis Using Parsimony*. Version 4.0. Laboratory of Molecular Systematics. Smithsonian Institution, Washington, D.C. and Sinauer Associates, Sunderland, Massachusetts.

Swofford, D.L., Olsen, G.J., Waddell, P.J., and Hillis, D.M. 1996. Phylogenetic inference. Pp.407–514. In Hillis, D. M., Moritz, C. and Mable, B. (eds). *Molecular Systematics*. Second Edition. Sinauer Associates, Sunderland, Massachusetts. 655pp.

Szalay, F. S. 1982. A new appraisal of marsupial phylogeny and classification. Pp.621–640. In Archer, M. (ed). *Carnivorous Marsupials*. Vol. 2. Royal Zoological Society of New South Wales, Sydney. pp.397–804.

Takahashi, K., Terai, Y., Nishida, M. and Okada, N. 1998. A novel family of short interspersed repetitive elements (SINEs) from cichlids: the patterns of insertion of SINEs at orthologous loci support the proposed monophyly of four major groups of cichlid fishes in Lake Tanganyika. *Molecular Biology and Evolution*, Vol. 15, pp.391–407.

Taylor, A.C., Sherwin, W.B. and Wayne, R.K. 1994. Genetic variation of microsatellite loci in a bottlenecked species: the northern hairy-nosed wombat *Lasiorhinus krefftii*. *Molecular Ecology*, Vol. 3, pp.277–290.

Tegelström, H., Wyöni, P.-E. and Ryttman, H. 1982. Thin layer isoelectric focusing as a tool for higher category systematics. *Annales Zoologicae Fennici*, Vol 19, pp.203–207.

Temple-Smith, P. 1987. Sperm structure and marsupial phylogeny. Pp.171–193. In Archer, M. (ed). *Possums and Opossums: Studies in Evolution*. Vol. 1. Surrey Beatty and Sons, Sydney. 400pp.

Terai, Y., Takahashi, K. and Okada, N. 1998. SINE cousins: the 3'-end tails of the two oldest and distantly related families of SINEs are descended from the 3' ends of LINEs with the same genealogical origin. *Molecular Biology and Evolution*, Vol. 15, pp.1460–1471.

Thackway, R. and Cresswell, I.D. (eds). 1995. *An Interim Biogeographic Regionalisation for Australia: a Framework for Setting Priorities in the National Reserve System Cooperative Program*. Version 4.0. Australian Nature Conservation Agency, Canberra. 88pp.

Thompson, J.D., Higgins, D.G. and Gibson, T.J. 1994. CLUSTAL W: improving the sensitivity of progressive multiple sequence alignment through sequence weighting, position specific gap penalties and weight matrix choice. *Nucleic Acids Research*, Vol. 22, pp.4673–4680.

Timms, P., Kato, J., Maugeri, M. and White, N. 1993. DNA fingerprint analysis of a free-range koala population. *Biochemical Genetics*, Vol. 31, pp.363–373.

University of California. 2002. Phylogenetics Software Resources. http://www.ucmp.berkeley.edu/subway/phylo/phylosoft.html

Van Regenmortel, M.H.V., Joisson, C. and Wetter, C. 1993. Comparative immunological methods. Pp.130–140. In Zimmer, E.A., White, T.J., Cann, R. L. and Wilson, A.C. (eds). *Molecular Evolution: Producing the Biochemical Data*. Academic Press, San Diego. 725pp.

Van Tuinen, M., Sibley, C.G. and Hedges, S.B. 1998. Phylogeny and biogeography of ratite birds inferred from DNA sequences of the mitochondrial ribosomal genes. *Molecular Biology and Evolution*, Vol. 15, pp.370–376.

Vos, P., Hogers, R., Bleeker, M., Reijans, M., van der Lee, T., Hornes, M., Frijters, A., Pot, J. and Peleman, J. 1995. AFLP: a new technique for DNA fingerprinting. *Nucleic Acids Research*, Vol. 21, pp.4407–4414.

Wallace, B., Ugozzoli, L., Reyes, A.A. and Lowery, J. 1996. LCR: ligase chain reaction. Pp.101–104. In Landegren U. (ed). *Laboratory Protocols for Mutation Detection*. Oxford University Press, Oxford. 192pp.

Weber, J.L. and Wong, C. 1993. Mutation of human short tandem repeats. *Human Molecular Genetics*, Vol. 2, pp.1123–1128.

White, M.J.D. 1983. *Animal Cytology and Evolution*. Cambridge University Press, Cambridge. 455pp.

Winzeler, E.A, Richards, D.R., Conway, A.R., Goldstein, A.L., Kalman, S., McCullogh, M.J., McCusker, J.H., Stevens, D.A., Wodicka, L., Lockhart, D.J. and Davis, R. W. 1998. Direct allelic variation scanning of the yeast genome. *Science*, Vol. 281, pp.1194–1197.

Worthington Wilmer, J., Moritz, C., Hall, L. and Toop, J. 1994. Extreme population structuring in the threatened ghost bat, *Macroderma gigas*: evidence from mitochondrial DNA. *Proceedings of the Royal Society of London*, Series B, Vol. 257, pp.193–198.

Wroe. S. and Archer, M. 2006. Origins and early radiations of marsupials. Pp.551–574. In Merrick, J.R., Archer, M., Hickey, G.M. and Lee, M.S.Y. (eds). *Evolution and Biogeography of Australasian Vertebrates*. Auscipub Pty. Ltd., Sydney. 942pp.

Chapter 37
The Role of GIS in Systematic Conservation Planning

Paul Flemons and Gerasimos Cassis
Centre for Biodiversity and Conservation Research, Australian Museum, 6 College Street, Sydney, N.S.W. 2010 Australia
E-mail: Paul.Flemons@austmus.gov.au

INTRODUCTION

Geographic Information Systems (GIS) have played a key role in the development of systematic conservation planning. Theories of conservation biology that relate to opportunities and constraints of the real world can now be tested and applied. GIS facilitates the process by which conservation biologists and resource planners can systematically select and design reserve systems.

In a strict sense GIS is a computer information system that describes places on the Earth's surface. In reality GIS has a much broader range of applications. In terms of conservation planning, GIS has the following key attributes: (i) computer hardware; (ii) computer software; (iii) geographically referenced data; and (iv) technical personnel. See Figure 1. Maps and spatial analysis have been used for over a century to plan land use. In the 1960s GIS technology became available for the same purpose. By the 1990s GIS had evolved from a research tool for mapping and modelling species distributions into a fundamental component of biodiversity assessment and conservation planning.

The capacity to collect and collate data, derive models based on observed patterns in time and space, and predict outcomes are crucial to conservation planning. Data and models inform the process of targeting areas for conservation. GIS contributes to this process through a broad range of applications, including: modelling habitat for rare and endangered species (Catling *et al.,* 1998; Carroll *et al.,* 1999); creating priority lists of ecoregions on a global basis (WWF, 2000); identifying hot spots of species richness (Gaston and David, 1994) and endemism (Myers *et al.,* 2000); establishing species range sizes (Gaston, 1996a); modelling of species distributions (Ferrier and Watson, 1997); demonstrating biodiversity surrogates (Humphries *et al.,* 1995; Balmford *et al.,* 1996b; Andersen, 1997); and for identifying gaps between land areas that are rich in biodiversity and areas that are managed for conservation (Scott and Csuti, 1997).

This chapter examines the role of GIS in the conservation planning process, focusing on selection of survey sites, the use of surrogates to measure and model biodiversity, and the selection and design of reserve systems.

SITE SELECTION

Biodiversity data underpin conservation planning. Generally, there are three types of biodiversity data: genetic, species and ecosystem. Genetic data require the extraction and analysis of DNA material from individual organisms. Collecting species data involves the taxonomic identification of each individual organism. Ecosystem data (usually flora-related) require the grouping or classification of species data. All three of these data types rely on site based collection or observation of organisms through biological survey.

Biological surveys are ideally designed to maximize the representativeness of data in a cost-effective manner. Representativeness in this context refers to the extent to which variation in environmental conditions is sampled for the area of interest. Environmental conditions include temperature, rainfall, radiation, aspect, slope, soil, geology and altitude. Evidence suggests that environmental variation is a major determinant of species diversity (Faith and Walker, 1996a), therefore by maximizing the representation of environmental variation, sampling of species diversity is also maximized.

The importance of obtaining survey data that are representative cannot be understated. Conservation planning outcomes for any area are limited by the representativeness of the available data. Much existing data are biased by collector preference for particular locations or environments, or because access constraints or ease of collection have meant that roadside environments are over represented. GIS plays an important role in overcoming these biases by providing the tools for designing representative biological surveys.

Simple mapping of existing survey/collection sites can quickly give an idea of geographic gaps in collections (Figure 2). When existing sites are mapped along with GIS data layers such as roads, rivers, and cleared land, a survey can be designed that will improve the geographic coverage of data whilst minimizing survey costs. This does not necessarily maximize the representativeness of a survey however, as geographic space is only one variable that describes environmental variation. The survey sites selected may be heavily skewed towards environments that determine the location of roads and agricultural lands, or to environments already adequately sampled.

A more meaningful approach is to use spatial representations of environmental variables. The complexity of the data can be reduced by classifying each variable into discrete classes (e.g. rainfall ranging from 20 mm to 140 mm may be grouped into 3 classes: 20–60, 60–100, 100–140).

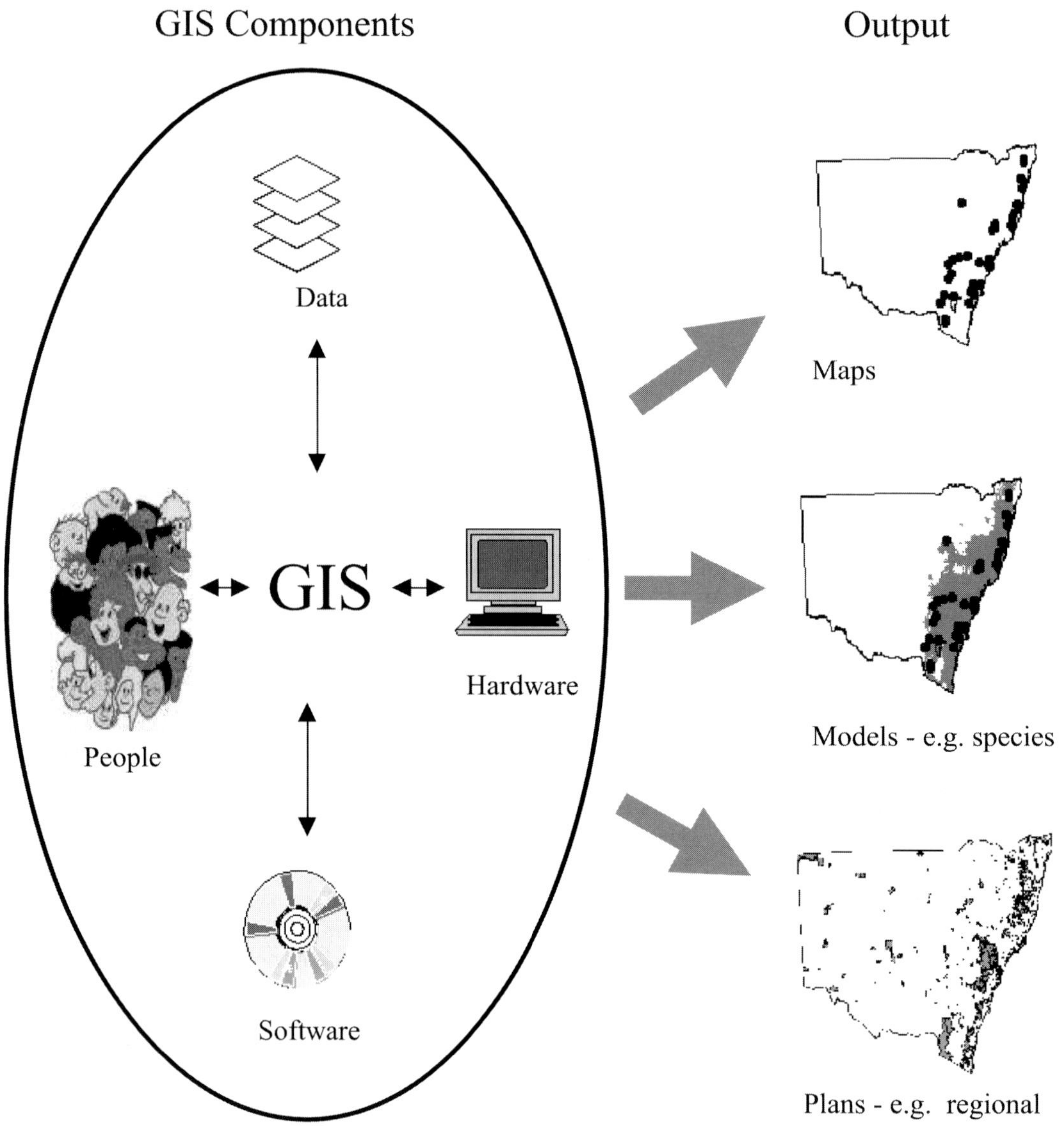

Figure 1. The components of GIS and some outputs, using New South Wales as an example.

All the classified variables are then combined to form environmental units (unique combinations of environmental variables). By ensuring that each of these units is sampled, it is assumed that the representation of environmental variation is broadly sampled. In such a survey design the representativeness of the data is dependent on how finely the environmental variables are classified and how many sites can be surveyed. If there are more environmental units than sites then some must remain unsampled. Conversely, fewer units may enable all units to be sampled but may not adequately represent environmental variation in the study area. Any classification is likely to be arbitrary, driven by the desire to sample all units equally or because it appears adequate, making selection of the 'best' classification impossible.

An approach that overcomes these problems involves the retention of each of the environmental variables as continuous (data as originally recorded). An ordination of all the available environmental data for the study area is used to create a multi dimensional environmental space (Figure 3). This 'maps' the sites in the study area in geometrical space, where the distance between points in the space is directly related to the environmental differences between sites. The distance in the environmental space is correlated to species-composition differences (Faith and Walker, 1996a). This means that sites that are close together in the environmental space are likely to have more similar species compositions than those further apart. Therefore the representativeness of a group of sites is reflected in the degree to which the sites span the environmental space.

A measure of the minimum average distance in environmental space from a point or points to all other points can be measured through the p-median index (Faith and Walker, 1996a). This is based on a method for optimally locating objects in a space. The p-median index can pro-

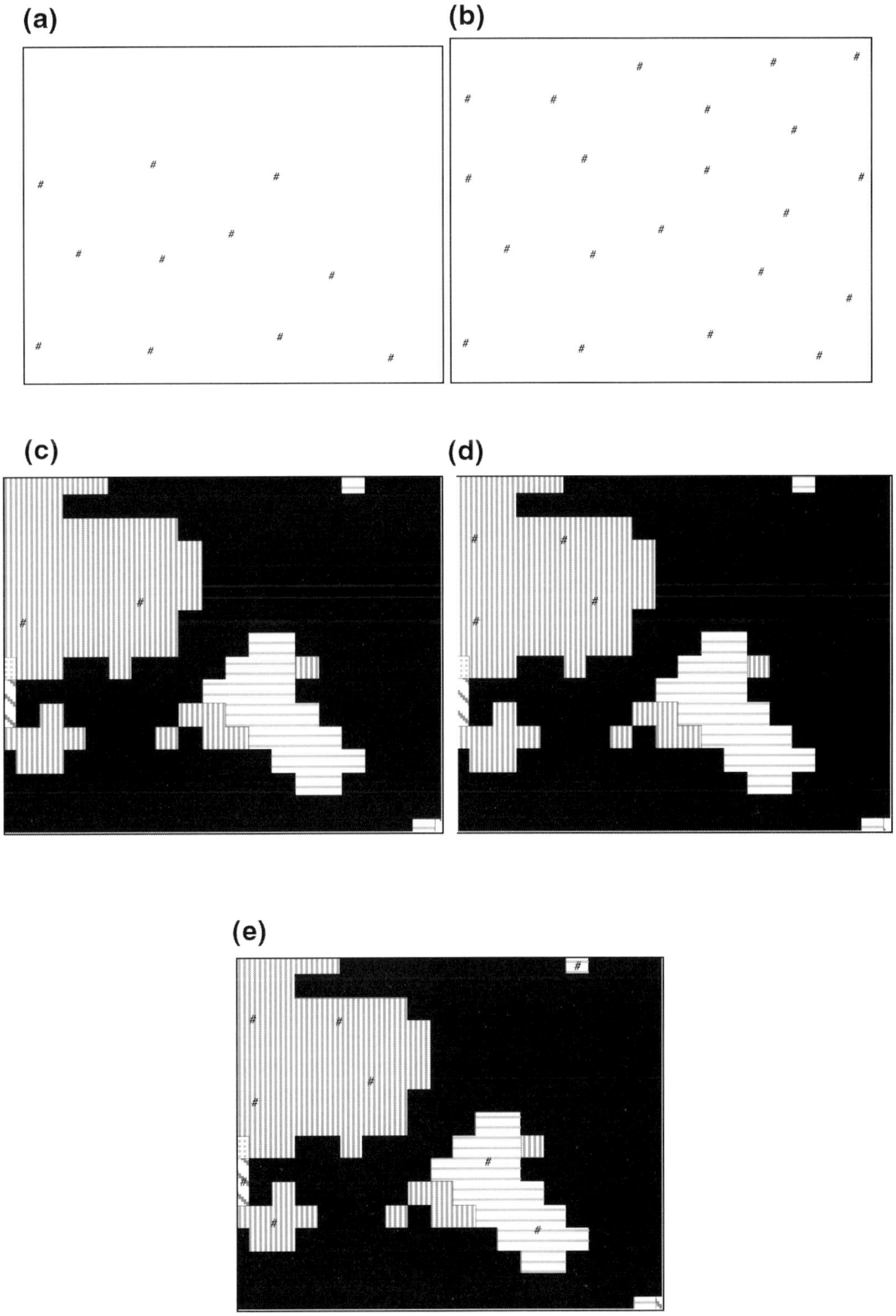

Figure 2. The problem associated with selecting sites using only geographical space can be seen here. Each box represents the same view of a study area showing: **(a)** 11 existing survey sites; **(b)** 11 existing survey sites plus 10 new survey sites chosen only on the basis of attempting to represent the geographic variation within the study area; **(c)** 11 existing sites displayed with environmental variation (represented by discrete classes) within the study area; **(d)** 11 existing survey sites plus the additional 10 sites chosen in (b) and showing environmental variation; and, **(e)** 11 existing sites plus 10 additional sites chosen on the basis of representing the environmental and geographic variation in the study area.

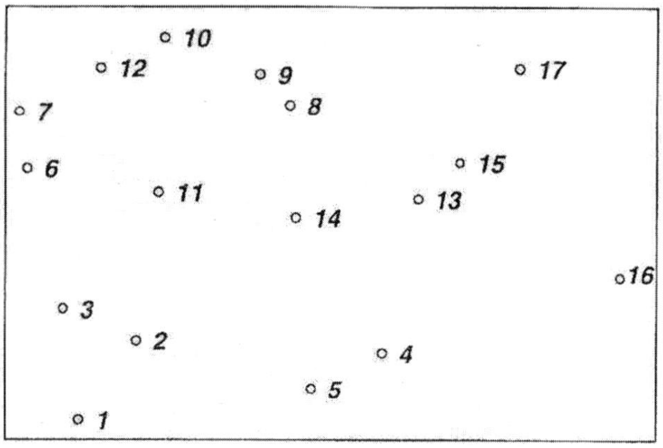

Figure 3. A hypothetical 2-dimensional environmental space; from Faith and Walker (1996a).

vide a single quantitative measure of survey adequacy and has recently been integrated into a GIS environment (N.S.W. N.P.W.S., 1998). This enables the existing survey adequacy to be visualized as a GIS layer, which in turn makes it possible to identify areas that are poorly surveyed. In selecting a new site, the p-median index for each candidate site is calculated (smaller p-median values indicate better coverage) and displayed according to the magnitude of the p-median values. This allows the user to quickly identify those sites that would most improve the representativeness of the survey data. GIS layers of roads and tenure can then be used in assisting site selection. Each time a new site is added p-median values are recalculated and candidate sites are redisplayed.

MEASURING AND MODELLING BIODIVERSITY

Biodiversity is a systems concept that refers to the diversity of all living organisms; the genetic diversity they contain, the species they belong to and the ecosystems of which they are part. This hierarchical definition does not provide an operational measure for all biodiversity, so it is necessary to use one component of biodiversity to represent (or act as a surrogate for) other components of biodiversity. In using surrogates it is important to establish a relationship between the surrogate and the components of biodiversity you want to represent.

There are two types of biodiversity surrogates, taxonomic and spatial. The taxonomic surrogate uses a few components of biodiversity (e.g. a few species of vertebrates) to represent many components of biodiversity (e.g. all vertebrates) or all biodiversity (e.g. genetic, species and ecosystem diversity). Taxonomic surrogates are generally applied on a site basis or through area-based summaries. They do not involve extrapolation over areas where there are no biological survey data. The spatial surrogate fills the geographic gaps not covered by biological survey. This is done by extrapolating the biological component from point based information to create a contiguous estimate (or surface) of biodiversity information across an area of interest (e.g. planning area). Spatial surrogates can be modelled distributions of biodiversity features (e.g. several species of large carnivore), remotely mapped land characteristics (e.g. vegetation mapped from air photos or satellite imagery) or abiotic (non-biological) classifications of environmental variables such as climate, topography and soils. Faith and Walker (1996a), Ferrier and Watson (1997) and Williams (2000) give detailed treatments on biodiversity surrogates.

GIS techniques can be used to develop biodiversity surrogates for conservation planning applications, by using site-based data obtained through biological surveys. These tools and techniques can be classified into two groups; the first involves measuring biodiversity, the second involves extrapolation or modelling of existing data to make predictions about where else a feature may be found.

MEASURING WITH TAXONOMIC SURROGATES

Species are the most commonly used and convenient unit for measuring biodiversity. They provide a simple, discrete, and easily recognizable unit for which considerable volumes of data already exist. Characters are also potential units for biodiversity measurement. Characters are defined as 'any feature or attribute of an organism which forms the basis for comparison' (Lincoln *et al.*, 1982) and may be morphological, genetic or behavioural. Examples of characters are presence or absence of hairs on the thorax of an insect, the relative lengths of antennal segments, courtship display of bower birds or nucleotide sequences (Crandall, 1998). Some biodiversity measures, such as richness, could equally use characters or species as the unit of biodiversity. However, in comparison to species, the data for characters are more difficult to assemble in a comparative sense, and deriving new information requires more effort than is the case for species.

Measures of biodiversity are generally used as a means of reporting on the biodiversity of an area. They include species richness, endemism and rarity. Species richness is a count of the number of species recorded in each grid cell or planning unit. It is the most frequently and widely applied biodiversity measure (Gaston, 1996b) and has voluminous coverage in the literature, by both proponents and critics (Williams *et al.*, 1996). In its favour is its simplicity and the fact that it appears to be correlated with character diversity (Williams *et al.*, 1996). Negative aspects of species richness are that it is highly scale-dependent and has no scope for differentiating between species. This may result in rare or endemic species being poorly represented if conservation priority setting is based on species richness alone.

Rarity is a function of the number of organisms and their geographic distribution, whereas endemism refers to

Plate 125. Percichthyids (in the genera *Maccullochella* and *Macquaria*) have been subject to more intense molecular investigations than most other Australian freshwater fishes. Aside from clarifying taxonomy and identifying regional genetic stocks, recent DNA analyses have been used to estimate levels of genetic variability in surviving wild stocks of endangered species. Photograph J. Merrick.

Plate 126. *Ctenotus taeniolatus*, one of the species investigated in the 'Genetic Criteria for Reserve Selection' project. Photograph P. Rankin, Sydney.

Plate 127. There have been more molecular studies on dolphins than on other groups of marine mammals. Reasons for these investigations include: clarifying taxonomy where external variation is apparent in disparate regional populations; estimating levels of genic exchange between adjacent populations; demonstrating levels of intra-populational genetic variation; elucidating kinship between and within sub-groups of a local population, in relation to behaviour. Photograph S. Allen.

Plate 128. A crabeater seal (*Lobodon carcinophaga*) from the pack ice of Antarctica. This species consumes krill. Photograph R. Harcourt.

Plate 129. Australian fur seals (*Arctocephalus pusillus doriferus*) underwater. This subspecies was heavily culled in the 19th century, but recent evidence indicates that southern Australian populations are recovering. Photograph R. Harcourt.

Plate 130. This Australian sea lion (*Neophoca cinerea*) mother and pup are from Kangaroo Island, South Australia. Photograph S. Allen.

Plate 131. A young female Australian sea lion hauled out on a Kangaroo Island beach. Photograph S. Allen.

species that have restricted geographic ranges regardless of their abundance. Rarity is significantly affected by scale. Species that are considered rare at the global scale (because of their restricted distribution) may be abundant at the local level and thus difficult to be thought of as rare. Therefore at the global scale such species are both rare and endemic, whereas at the local scale they are only endemic. Alternatively some species may occur over a large geographic range but be uncommon and these may be considered rare but not endemic. Whilst endemic species have distributions that are restricted to a particular area (such as an island or ecoregion), narrow range endemics are restricted to very small areas, regardless of their abundance.

One of the first uses of biodiversity surrogates for planning purposes has been the identification of biodiversity hot spots. Hot spot analysis involves the identification and prioritization of high value areas in terms of biodiversity measures such as species richness, endemism, numbers of rare or threatened species and intensity of threat (Reid, 1998). Endemism and rarity hot spots are areas that contain a relatively large number of endemic or rare species. The first application of the hot spots concept used a measure of endemism combined with level of threat (Myers, 1990). Since then a plethora of papers has been published on hot spots (Gaston and David, 1994; Prendergast *et al.*, 1993). A recent case study identified 25 global hot spots that comprise only 1.4% of Earth's land surface but contain 35% of all species from four vertebrate groups (Myers *et al.*, 2000).

Indicator, focal or umbrella taxa are often used as surrogates for a larger part or all of biodiversity. These indicator taxa are often based, on what data are available and not on evidence of suitability. They are often charismatic and usually vertebrates or higher plants. They are generally chosen because they have broad habitat or large area requirements or because they have the best data coverage. The assumption is made that, because of these requirements, a large proportion of biodiversity will be represented by the umbrella taxon. Unfortunately the literature does not support the use of indicator taxa as surrogates for overall biodiversity (Lindenmayer *et al.*, 1991; Flather *et al.*, 1997; Margules and Pressey, 2000). The use of vertebrate taxa as surrogates for invertebrates in particular has been found to be unreliable (Kerr, 1997), although there is some evidence for partial surrogacy (Flather *et al.*, 1997). Emphasis has moved towards the use of a suite or umbrella of indicator species (Lambeck, 1997).

Though richness is generally measured in terms of species there is considerable interest in being able to use higher taxon richness as a surrogate for species richness. The basis for higher taxon surrogacy is that higher taxon richness is predictive of species richness and that obtaining the data to map higher taxon distributions is cheaper than for species (Gaston and Williams, 1993). There is some support for the concept of higher taxon surrogacy (Williams and Gaston, 1994) at local (Balmford *et al.*, 1996b) and regional scales (Balmford *et al.*, 1996a), with the genus-group being more suitable than family-group (Balmford, *et al.*, 1996b). It is unclear how useful the higher taxon approach is for conservation planning and this is reflected in the paucity of literature on its application. Its attraction for conservation planning lies in its potential for including a broad range of organisms represented by surveyed data. This is preferable to the indicator type surrogate, which relies on the existence of questionable (or unknown) correlations between the distribution of different taxonomic groups.

Using WORLDMAP

The use of GIS tools and techniques for measuring biodiversity is typified by the WORLDMAP (Williams and Humphries,1996) software package. It is based on the application of a regular grid over an area of interest (Figure 4(a)). Each cell of the grid is referenced to a database that is populated with point locations of species data

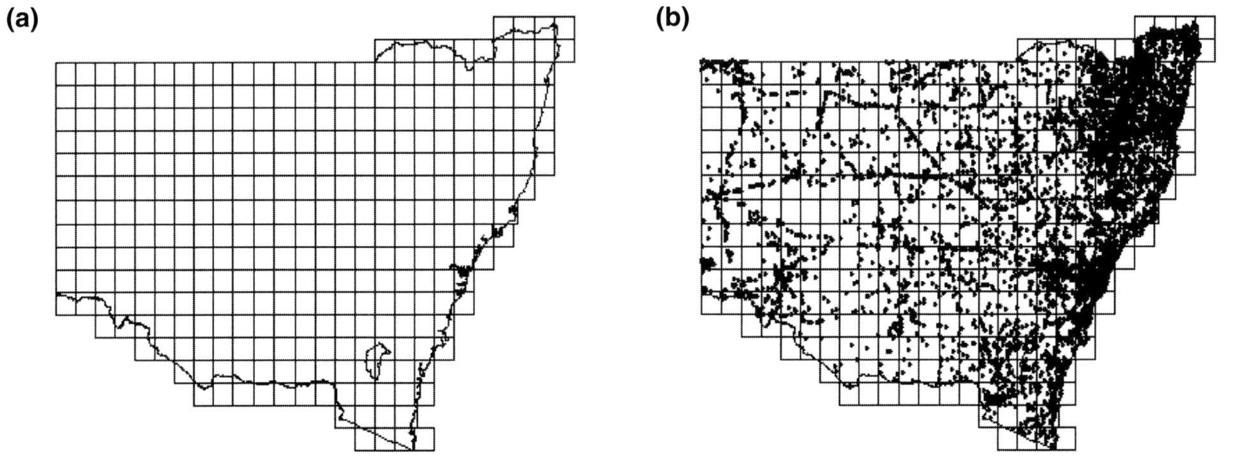

Figure 4. How WORLDMAP works—using the State of New South Wales as an example: **(a)** a regular grid is applied over an area; **(b)** species location records are then linked to each grid cell to enable biodiversity measures to be calculated for each grid cell.

(Figure 4(b)). This allows a broad range of algorithms to be applied cell by cell to create a spatial distribution of various biodiversity measures such as those described above. Varying the grid cell size allows the user to strike a balance between the scale of the analysis and the computing time required to carry it out. It also provides a means of assessing the influence of scale on observed patterns. This approach makes no attempt to extrapolate spatially from existing data.

A strength of WORLDMAP lies in its capacity to assess taxonomic surrogates. It has been used for assessing a wide range of surrogates including species richness, endemism, indicator species and higher taxon surrogacy (Gaston and Williams, 1993; Williams and Gaston, 1994; Balmford *et al.*, 1996a; Balmford *et al.*, 1996b; Csuti *et al.*, 1997). It provides a simple, fast and robust tool for analyzing patterns and relationships in the distribution of individual species, groups of taxa and overall biodiversity. It is a useful research tool for priority setting at a very broad scale (continental and global level). However it is rarely an effective planning tool at local or regional scales, because it requires relatively comprehensive and consistent data coverage for the area of interest. Each grid cell must contain a large number of species records and different species. This is generally achievable only at the continental and global scale. It has been used in smaller areas (e.g. United Kingdom, and Oregon, U.S.A.; see Figure 5) but only when they have been heavily sampled. Once the grid cell size starts decreasing, less and less species records fall within each cell. So for sparsely sampled areas, such as many developing countries, the density of species records is insufficient to support a comprehensive analysis at the local or regional scales and often even at the national scale.

MODELLING WITH SPATIAL SURROGATES

Knowing the location of biodiversity hot spots is in itself not sufficient for the purposes of conservation planning. Local and regional conservation planning requires detailed and comprehensive data on relevant species distributions and the spatial distribution of biodiversity as a whole. Most areas of the world, particularly at the local and regional level, have sparse and incomplete biodiversity survey data. Informed decision making in these areas requires the ability to extrapolate from existing data.

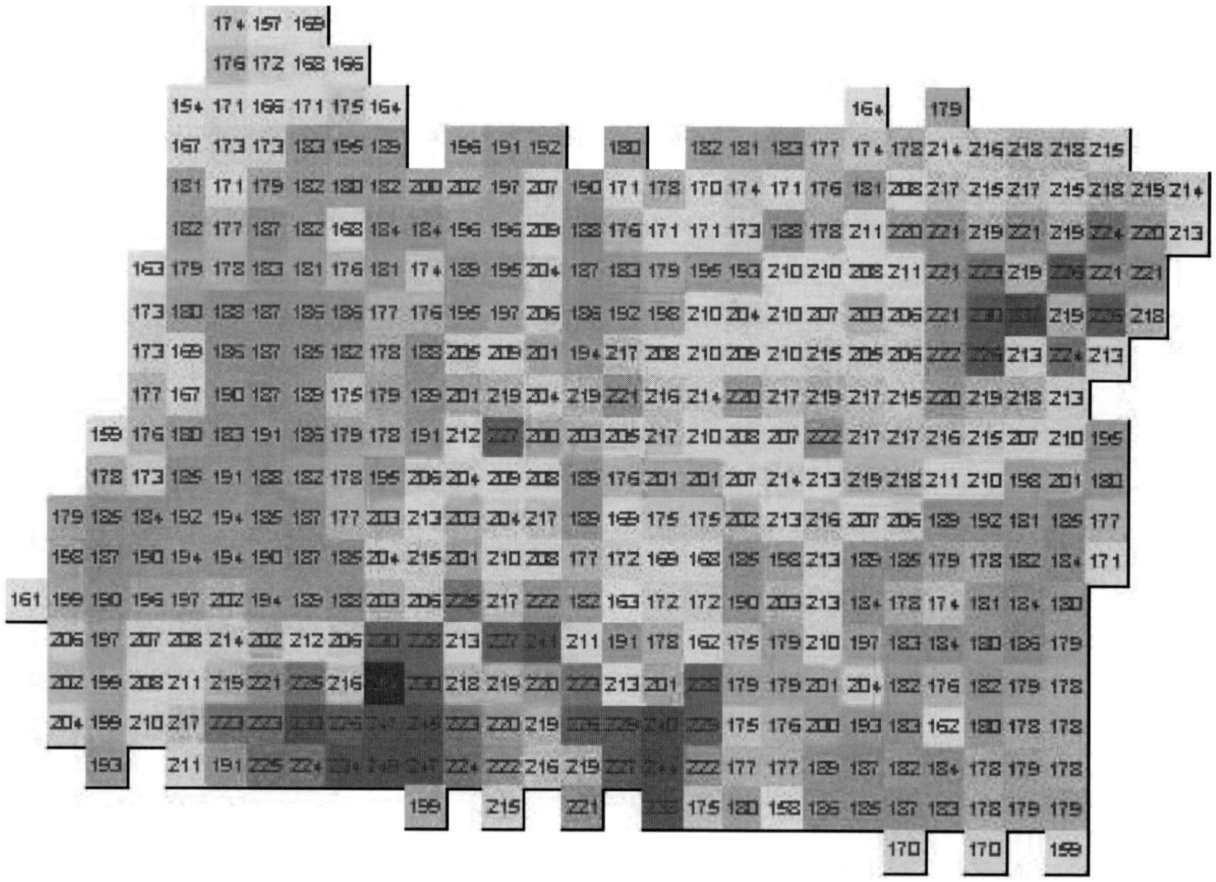

Figure 5. WORLDMAP database (from Williams and Humphries, 1996) for the species of vertebrates in the state of Oregon (U.S.A), showing species richness. Values in each cell correspond to the number of species found within the area covered by that grid cell. Total number of species in this WORLDMAP database for this area is 426. The dark shaded areas, those with high species richness values, are referred to as hot spots of species richness.

Selecting reserves that adequately represent significant species and overall biodiversity is only possible if data layers are available that accurately describe the distribution of these features across the entire study area. Such data layers can only be obtained through the use of spatial surrogates. Spatial surrogates can be grouped into three categories: species modelling, mapped land classes, and environmental classifications and ordinations.

Species Modelling

Species modelling is used to create spatial surrogates that define the distribution of suitable habitat for individual species. These species models, in the form of GIS layers, are crucial to conservation planning as they provide an explicit means of assessing whether existing reserve systems adequately represent the habitat required for long-term viability of a species. This is particularly important for rare or threatened species. Species modelling of one group of taxa has been shown to have potential as a surrogate for other taxa (Ferrier and Watson, 1997). It has also been shown to be a superior surrogate for distribution of biodiversity than other surrogates such as vegetation mapping and environmental ordinations (Ferrier and Watson, 1997).

Modelling techniques that are routinely employed include profile matching, generalized linear modelling (GLM), generalized additive modelling (GAM), and genetic algorithms (Figure 6).

Profile matching

Profile matching was one of the first species modelling techniques (Busby, 1985). It uses the values of each climatic variable from each site to construct a profile of the range of values for each variable. This technique provides a means of displaying the predicted distribution based on core subsets of the biological data (e.g. points falling within say 5–95% or 25–75% of the range of the data). Output is in the form of a binary map and there is no measure of probability of species occurrence on a cell by cell basis. Profile matching, available through a number of different software programs such as BIOCLIM or DOMAIN(Carpenter *et al.*, 1993) is simple and easy to use, and has been applied to the modelling of elapid snake distributions (Nix, 1986), rare bats (Clague *et al.*, 1999) and invertebrates (Ponder *et al.*, 2001). The BIOCLIM approach can use variables other than climatic ones, such as soil depth, soil moisture and slope. For a comparative summary of this modelling process see Figure 6.

GLM and GAM

These two modelling techniques are variations on the same theme. They are forms of multiple regression suitable for use with presence/absence data (systematic data that indicates if a species is present or absent at a survey site) (Crosbie and Hinch, 1985). GLM uses linear functions of the predictor variables whilst GAM uses a smoothing function. GLM was the first of these techniques used for species modelling (Austin *et al.*, 1984; Nicholls, 1989; Lindenmayer *et al.*, 1991; Pearce *et al.*, 1994; Lechmere-Oertel and Cowling, 1999). The non-linear nature of variable responses led to GAM gaining favour due to its ability to fit smoothing functions to the data. GAM was first used in modelling plant species (Yee and Mitchell, 1991) and has since been used widely for creating species models for use in regional conservation planning (Carroll *et al.*, 1999; N.S.W. N.P.W.S., 1999a). For a comparative summary of these two techniques see Figure 6.

GARP (Genetic Algorithm for Rule-set Production)

Genetic algorithms are inspired by the concept of evolution by natural selection. These algorithms evolve solutions to problems in an analogous way to organismal evolution (Stockwell, 2000b). GARP uses site based biological data and environmental data layers to produce species distribution maps and models. A major advantage of GARP is its ability to produce probability based models with all types of data, including presence only data. GARP has been used in modelling invertebrate distributions for identifying endemism hot spots (Australian Museum, 2000) and is available as an interactive modelling tool on the internet (Stockwell, 2000a). For a comparative summary of this modelling process see Figure 6.

Care has to be taken in the use of species modelling. The use of poor data, biological or environmental, that is biased or unrepresentative invariably leads to models that are biased and unrepresentative (Maddock and Benn, 2000). The scale of environmental data used can also influence the quality of species models. Ferrier and Watson (1997) found that the use of finer-scale environmental data produced better species models than when using presence/absence data. When presence only data was used there was no significant improvement in the predictive capacity of the models.

The most accurate means of accuracy assessment is to use ***a posteriori*** ground truthing. However, it is not often used for testing models because of its prohibitive costs in time and money. Accuracy assessment is often done by randomly splitting the data into two; one half is used for creating the models and the other half for testing the results. Expert knowledge has also been used for assessing accuracy and improving species and asssemblage models (N.S.W. N.P.W.S., 1999a).

Mapped Land Classes

This approach generally involves the mapping of single attribute, such as vegetation mapping (Ferrier and Watson, 1997; Oliver *et al.*, 1998) or forest types (Ferrier and Watson, 1997) or land systems (Pressey and Logan, 1998), datasets using remote sensing. Vegetation mapping is the most commonly used type of mapped land class. It is assumed that land classes are a surrogate for biodiversity, so by maximizing the diversity of classes reserved, conservation of biodiversity is maximized. The use of mapped vegetation classes has considerable support (Noss, 1990; Scott and Jennings, 1994; Scott and Csuti, 1997; Maddock and Du Plessis, 1999). This approach assumes that biodiversity at lower organizational levels (genetic and species) is embraced at higher levels (ecosystem) and that the presence of intact indigenous vegetation implies that natural ecological processes are functioning efficiently (Noss, 1990).

Environmental Classifications and Ordinations

These techniques are also used to derive surrogates of biodiversity on the assumption that environmental variation is correlated with biodiversity. Environmental classifications are produced by numerical clustering of abiotic environmental data (resulting in environmental domains) or by simple combination of broad classes (environmental units). Environmental ordination creates a multi-dimensional environmental space for the area of interest. The number of dimensions is equal to the number of environmental variables employed. Any point in the environmental space can be related to a point or area on the ground. The environmental space is then used as a surrogate for biodiversity (Faith and Walker, 1996b). Maximizing the representation of environmental space theoretically maximizes representation of biodiversity. These biodiversity surrogates perform reasonably for vertebrates and higher plants but perform poorly for invertebrates (Ferrier and Watson, 1997; Ferrier *et al.*, 1999).

The reality of reserve selection (see below) is that no one surrogate is sufficient to represent overall biodiversity. It is likely that even a combination of the best surrogates are inadequate for truly representing overall biodiversity. It is a matter of using the best available data and technology to develop suitable datasets for use in reserve selection and design.

RESERVE SELECTION AND DESIGN

Reserves are the central plank of strategies designed to conserve biodiversity. Until recently reserves were mostly selected in an ad-hoc fashion (Pressey and Tully, 1994). Often they have been established for recreational or scenic value or because they were on economically unproductive land. In addition, the reserve systems they contributed to were poorly designed, not considering factors such as proximity, connectivity or configuration. The consequence of this situation is that reserves do not represent an area's biodiversity, nor maintain the persistence or viability of that biodiversity. In response to the need for more systematic reserve selection and design, two new fields of conservation biology research emerged during the 1980s, namely reserve selection and reserve design. GIS has been an important component of the development of these fields, providing a mechanism by which theories could be tested and applied to real world data.

SELECTION TENETS

In the early 1990s the term 'reserve selection' (Pressey, 1990) became synonymous with research that is focused on using computer algorithms to select parcels of land for protection. The algorithms are designed to maximize the representation of biodiversity features within a reserve system or network, whilst minimizing the land required and therefore the lost opportunities for human activities. Through the 1990s more novel algorithms were introduced (Church *et al.*, 1994), others were refined (Pressey *et al.*, 1997) and assessed (Virolainen *et al.*, 1999; Rodrigues *et al.*, 2000). The basic premise of these algorithms is to select a subset of land units, from a larger available set, that represents all features (species or other biodiversity surrogate) within the least number of units or area of land.

Central to reserve selection are the concepts of representativeness, complementarity and irreplaceability; these are explained as follows. Representativeness simply means that biodiversity in a region's reserve system should be representative of that region's overall biodiversity. Complementarity proposes that, in adding an area to the reserve system, the selection of that area should be based on it having biodiversity features most complementary to the existing reserve system. The next site chosen would be the next most complementary site and so on. This is an efficient mechanism for ensuring that the minimum necessary area is used to represent all biodiversity features. These biodiversity features may be species or any other component of biodiversity. In the planning process the concept of complementarity is applied to selection of areas for reservation so as to maximize the representativeness of the reserve system and ensure the inclusion of areas of irreplaceable features. This is constrained within the current socio-economic environment by available land, competing land uses and political climate. Irreplaceability refers to the importance of a site in the context of its potential contribution to a representative reserve system and the extent to which the options for a representative reserve system are lost if that site is lost (Ferrier *et al.*, 2000).

An additional major consideration in selecting reserves is that of species turnover (β diversity). Recent research has shown that species turnover, both spatially (Ferrier *et al.*, 1999) and temporally (Rodrigues *et al.*, 2000) has implications for reserve selection. Spatial turnover refers to the rate of change in species composition at a site with changes in geographic distance between sites. The theory predicts that as distance between sites increases the number of species common to the sites will decrease. This distance has been shown to be much less for invertebrates than it is for vertebrates (Ferrier *et al.*, 1999). Also, invertebrates have been shown to exhibit high spatial turnover within the same vegetation community/habitat type (Ferrier *et al.*, 1999). The implication of this is that reserve systems need to sample vegetation communities over their geographic ranges if biodiversity is to be effectively represented in reserves.

Temporal turnover refers to the site based turnover in species composition over time. The minimum set

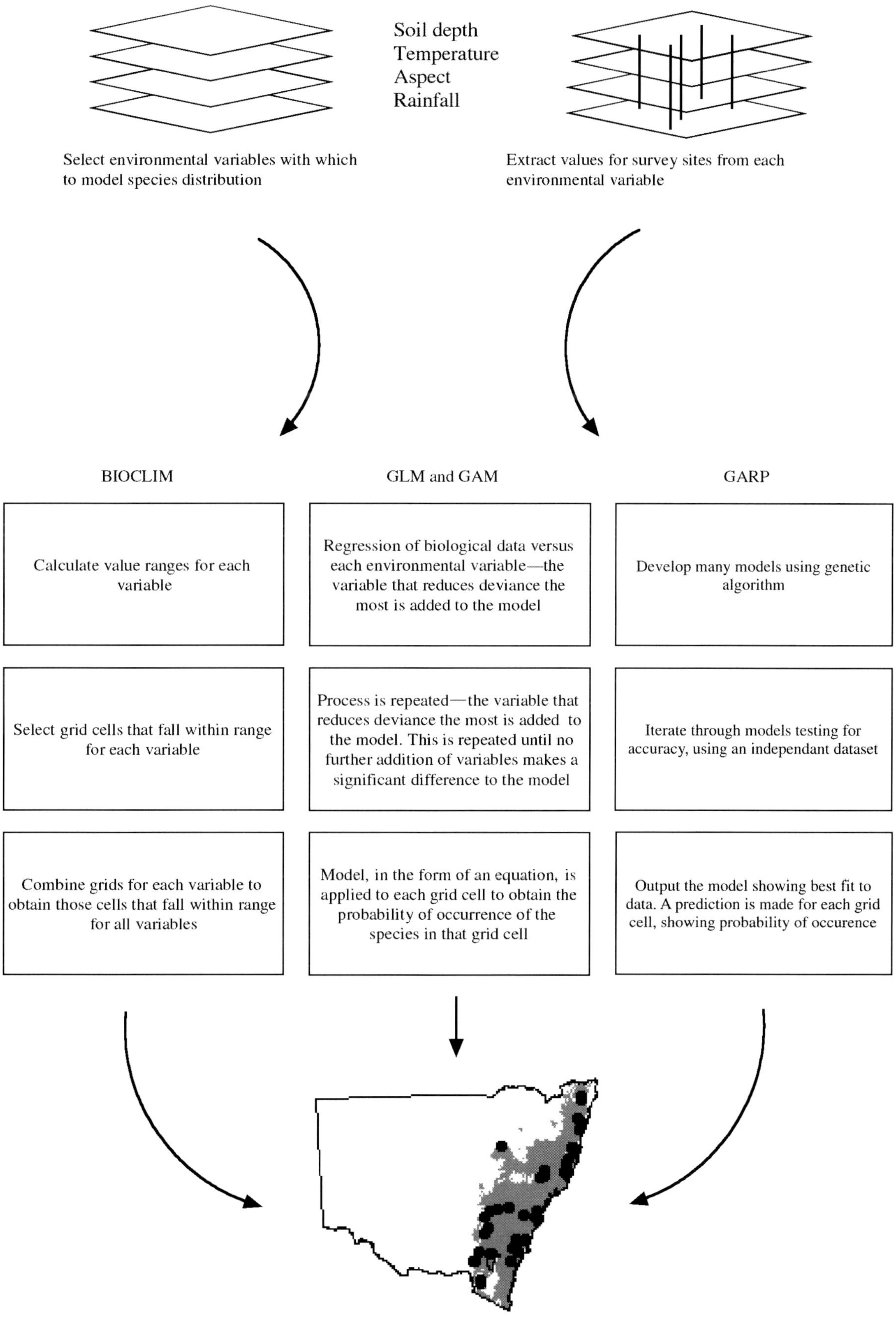

Figure 6. Generalized comparison of species modelling processes.

strategy (the selection of the least number of sites/areas possible to represent all known biodiversity features) used in reserve selection algorithms to efficiently (in the least area) represent biodiversity, is not always effective over time (Rodrigues *et al.*, 2000). This is because a minimum set solution does not consider design issues such as size and shape of reserves and the maintenance of connectivity between reserves.

DESIGN ISSUES

Reserve design needs to take into account many issues relating to ecological and evolutionary processes. This is essential to convert the selection of areas for reservation based on biological patterns into reserve systems that are representative and capable of ensuring the persistence of biodiversity. The spatial nature of reserve design means that protection of these processes needs to be based on spatial characteristics such as size, configuration, connectivity and heterogeneity.

In many cases where reserves are isolated from other natural areas by human development (agriculture, urban), reserves in effect become islands. Species area relationship (island biogeography theory) predicts that the smaller an island is the less number of species it will support—see pp.459–476 (Ford, 2006). This would suggest that a single large reserve would be better than several small reserves for conserving biodiversity. In reality the situation is not that simple and has been the subject of considerable debate in the literature (Diamond, 1975; Gilpin and Diamond, 1980; Simberloff, 1982; Ford, 2006). The simplistic notion of large being better than small is confounded by issues such as heterogeneity and quality of habitat. Smaller reserves with a range of good quality habitats can be just as important to a reserve system as large areas, more so if the large areas are homogeneous or contain poor quality habitat.

The potential effects of isolation on species is significant in terms of likelihood of extinction. A species is often made up of a number of separate populations linked by dispersal (metapopulations). Whilst dispersal links (such as corridors) are intact each population is able to be recolonized if local extinction occurs. Where a reserve becomes isolated recolonization cannot occur. If the reserves within a reserve system become isolated, the probability of species extinction through sequential local extinctions increases (Frankham *et al.*, 2002).

Reserve design should be based on patterns of biodiversity, derived from biodiversity data and surrogates. To achieve persistence the configuration and connection of reserves must include sufficient area of a broad range of habitat. Optimally, reserve systems should be designed in a regional (political or ecological or a combination of these) context, with each region having its own particular issues that need to be addressed.

ROLE OF GIS

Reserve selection algorithms have been transformed by integration into GIS-based decision support systems (Margules and Pressey, 2000; Groves *et al.*, 2000). Originally research tools, this integration has enabled them to become powerful conservation planning tools. Reserve selection and reserve design can now be carried out within an integrated software environment. The GIS acts as an interface by which reserve selection scenarios can be interactively and iteratively developed enabling different land allocation decisions to be visually and statistically compared. The keys to these interactive systems are algorithms that calculate near optimal (full representation in least area) solutions in very short time periods. This makes it possible to recalculate the values of irreplaceability and complementarity each time a new land parcel is selected for addition to, or removal from, the reserve system. Interest groups can now negotiate reserve selection and design through interaction with GIS decision support.

Recent improvements in some decision support systems (N.S.W. N.P.W.S., 1999b) now provide for improved consideration of size and connectivity in the selection and design of reserves. Measures of geographical and environmental spread are included. This enables better consideration of the sampling of ecosystems across their geographic range, addressing issues such as species turnover. New work (Scotts and Drielsma, 2003) provides an improved means of identifying the most suitable location for conserving habitat corridors—considered to be valuable conservation tools (Beier and Noss, 1998). The importance of each grid cell in connecting the overall landscape is calculated by using GIS layers to estimate the cost to an animal of moving through each grid cell. For example, a cleared paddock would be expected to have a much higher cost than a forested area.

A field that has been neglected in terms of its application in GIS based decision support systems (DSS) is the influence of climate change. Software is available that enables the effects of various climate change scenarios on the distribution of biodiversity (Sutherst and Maywald, 2000) to be examined. This is essential if they are to be used in selecting and designing suitable reserve systems for long-term conservation of biodiversity. Without considering climate change, reserve systems are likely to be unable to allow species room to adjust their ranges as habitat changes in response to climate change.

DATA TYPES

The amount and detail of data used in reserve selection and design may vary and will determine how comprehensive a planning project will be. The types of data used fall into four categories:

(i) Planning units—these are the building blocks for new reserves. They can be administrative, natural or arbitrary divisions of the landscape. Examples are forest compartments, squares, hexagons, catchments and local government areas. They need to be small enough to act as a viable planning unit and large enough to be significant from a biodiversity perspective. Landscape processes and land ownership/management implications also need to be considered. These layers are captured or created, from other layers, in the appropriate scale and format.

(ii) Biodiversity data—these data describe the distribution of biodiversity in the area of interest. They include a range of biodiversity surrogates such as species richness and species models, vegetation mapping and environmental domains. Species models are generally developed for rare or threatened species and other species for which adequate data are available for modelling, usually vertebrates and higher plants. The use of any particular type of biodiversity surrogate is determined by availability of suitable data, which are in turn determined by available resources for data collection (Table 1). Point based biodiversity data (e.g. species occurrence records) are used to develop surrogates for use in planning. They are also used in raw form for reference purposes.

(iii) Socio-economic data—these data are extremely important to the systematic conservation planning process. If comprehensive they should provide detailed data for each planning unit in terms of potential production and development. These data can then be used alongside biodiversity data to help design a reserve system that enables biodiversity conservation targets to be met whilst minimizing lost production and development opportunities.

(iv) Infrastructure data—these include roads, rivers, administrative boundaries etc. that provide context for the visualization and negotiation of reserve selection and design.

Integrated GIS decision support systems (N.S.W. N.P.W.S., 1999b) were used recently for selecting and designing reserves in the south-eastern forests of Australia. The biodiversity surrogates used included species modelling for vertebrates and higher plants, endemism layers for vertebrates, plants and invertebrates and forest ecosystems. Socio-economic information on wood supply, tourism and employment impacts was also considered. The planning unit layer was made up of a combination of forest compartments and sub-catchments. Considering the contribution made by existing reserves, irreplaceable planning units were identified. The complementarity principle was then used to select additional units for addition to the reserve system. Each time a unit was added to or subtracted from the reserve system the complementarity value for each unit was recalculated.

Table 1. Using any particular surrogate involves a trade-off between the precision of the surrogate as a means of measuring biodiversity, and the cost of deriving that surrogate (Gaston,1996b).

Advantage: precision as a measure of character diversity	**A scale of surrogacy for character diversity**	**Advantage: ease (/cost) of measurement**
Low	(Ecosystems)	High
↓	Landscapes	↑
↓	Land classes	↑
↓	Species assemblages	↑
↓	Higher taxa	↑
↓	Species	↑
High	(Characters)	Low

CONSERVATION PLANNING FRAMEWORKS

GIS has enabled the development of a number of approaches to conservation planning. These approaches or frameworks use a broad range of GIS applications. They aim to improve biodiversity conservation planning activities at the regional, national and global level. Each of these frameworks has differing strengths and weaknesses and needs to be seen as complementary rather than opposing. They have arisen for different reasons and seek to achieve different goals. The most notable of these frameworks are discussed below.

HOT SPOTS

There are various measures of biodiversity that have been used for defining hot spots (Reid,1998), and the literature is brimming with debate on the merits and shortcomings of each of these (Prendergast *et al.*, 1993; Kerr, 1997; Reid, 1998). Regardless of which measure is used, the identification of global hot spots of biodiversity serves as a means of flagging those areas most deserving of conservation effort. Several systematic and comprehensive global assessments of hot spots have been carried out—two are listed below.

(a) A consortium of non-government organizations identified 25 global hot spots (Myers *et al.*, 2000), based on the criteria of species endemism and degree of threat. These hot spots identify areas where conservation efforts would achieve the greatest return in terms of biodiversity conservation.

(b) The Global 200 is an initiative of the World Wildlife Fund (WWF, 2000). In this case the focus is on representing the world's diverse ecosystems. This was done by choosing an outstanding ecoregion (or hot spot) from within each major habitat type on the basis of their 'species richness, species endemism, unique higher taxa, unusual ecological or evolutionary phenomena, and global rarity of major habitat types'.

SYSTEMATIC CONSERVATION PLANNING (SCP)

This approach has been developed and refined in Australia and its influence was evident in the Australian Government's recent Regional Forestry Agreements (RFA) process (Margules and Pressey, 2000). The development of this particular framework was influenced strongly by the ad-hoc nature of conservation planning in New South Wales (Pressey, 1990; Pressey and Tully, 1994), where until the RFAs, reserves were heavily skewed towards land that is rugged, inaccessible and undesirable for agriculture (Pressey and Tully, 1994).

There are six stages in this framework:

(1) Measure and map biodiversity
(2) Identify conservation goals for the planning region
(3) Review existing reserves
(4) Select additional reserves
(5) Implement conservation actions on the ground
(6) Manage and monitor reserves

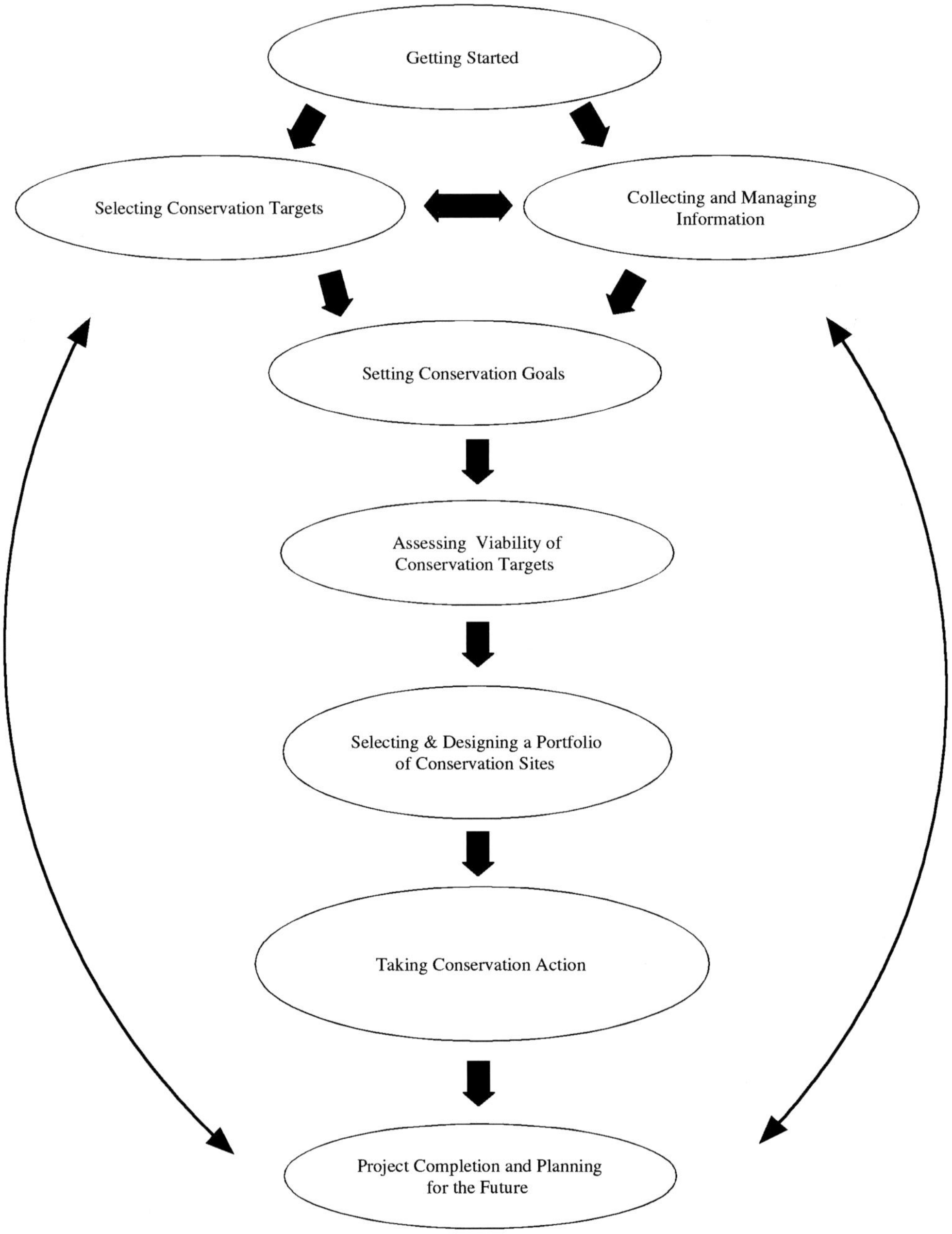

Figure 7. The Nature Conservancy's Ecoregional Planning Process (Groves *et al.*, 2000)

This framework provides planners at all levels of action from local to global with a flexible but rigorous approach to biodiversity conservation. The flexible nature of this framework enables the scope and detail of the data to be the determining factor in the scale and scope of the planning outcomes rather than the approach itself. This contrasts with the other three frameworks mentioned here which all have predefined scales at which they are applied, fitting the data to the scale. Also this approach does not have an active conservation programme associated with it, instead it provides a blueprint of concepts and processes that should be addressed in any conservation planning exercise.

GAP ANALYSIS

This framework was developed by the U.S. Fish and Wildlife Service as a research project and then transferred to the National Biological Service for coordination as a national program of conservation assessment. As a collaborative approach it consists of more than 200 organizations including local, state and federal governments, universities, special interest groups and businesses. The objectives of the GAP analysis program as stated in the GAP handbook (Scott and Jennings, 1994) are:

(1) 'To determine the representation of each mapped element (land cover types and animal species as identified in the land cover and animal modelling chapters) in different categories of land stewardship (management authority) and management status.'
(2) 'To interpret the results from Objective 1 in a way that is useful for land stewards in their development of land use and management plans that conserve or restore those biotic elements or avoid causing further impairment.'
(3) 'To provide the analysis and data sets in suitable form for additional modelling, analyses, biodiversity assessment, and planning activities by a broad group of users.'

No explicit mention is made of attempting to systematically design a reserve system, rather the GAP framework is more about capturing data in a systematic manner so that it can be used in land use planning and management by a broad range of users. There is evidence (Scott and Csuti, 1997) indicating that, as the data become more comprehensive and reserve selection and design techniques move from research to operational, GAP will become more outcome focused.

ECOREGIONAL CONSERVATION PLANNING (ECP)

The Nature Conservancy is a privately funded conservation agency that has developed a framework (see Figure 7) for conservation planning based on ecoregions. Detailed in a practitioners handbook (Groves *et al.*, 2000), this approach is similar to the SCP approach in its systematic nature and embracing of the latest science for guiding principles (e.g. the complementarity approach to reserve selection). ECP is the only framework of these four that has dedicated funding for the purchase of lands for inclusion in ecoregion based reserve systems. ECP sets goals and prioritizes potential reserve sites based on biodiversity values and greatest threat.

CONCLUSIONS

A major component of our efforts to conserve biodiversity, centres around the identification and reservation of lands of high conservation value. Competing land uses such as agriculture, housing and industry put pressure on the selection and design of these reserves. Conservation of biodiversity must be maximized whilst being mindful of minimizing the economic impact on local, regional and national communities. This is best achieved by using scientifically based and systematically implemented processes. Systematic conservation planning provides the conceptual framework for such processes and GIS provides the mechanism for implementing them.

The ability of GIS-based decision support systems to facilitate a consultative (involving all relevant parties—business, indigenous, environmental, local communities) approach that makes best use of available data (both biodiversity and socio-economic) has revolutionized our approach to conservation planning. Future work should focus on developing and testing biodiversity surrogates and improving the ability of the decision support systems to incorporate configuration and climate factors into selecting and designing reserve systems.

REFERENCES

Andersen, A.N. 1997. Measuring invertebrate biodiversity: surrogates of ant species richness in the Australian seasonal tropics. *Memoirs of the Museum of Victoria*, Vol. 56, No. 2, pp.355–359.

Austin, M.P., Cunningham, R.B. and Fleming, P.M. 1984. New approaches to direct gradient analysis using environmental scalars and statistical curve-fitting procedures. *Vegetatio*, Vol. 55, pp.11–27.

Australian Museum. 2000. *Evaluation of effectiveness of derived forest ecosystems as surrogates for invertebrate biodiversity, and identification of hotspots of invertebrate endemism.* Australian Museum Report for the Resource and Conservation Assessment Council.

Balmford, A., Green, M.J.B. and Murray, M.G. 1996a. Using higher-taxon richness as a surrogate for species richness: I. Regional tests. *Proceedings of the Royal Society of London,* Vol. 263, pp.1267–1274.

Balmford, A., Jayasuriya, A.H.M. and Green, M.J.B. 1996b. Using higher-taxon richness as a surrogate for species richness: II. Local applications. *Proceedings of the Royal Society of London,* Vol. 263, pp.1571–1575.

Beier, P. and Noss, R.F. 1998. Do habitat corridors provide connectivity? *Conservation Biology*, Vol. 12, No. 6, pp.1241–1252.

Busby, J.R. 1985. *Bioclimate Prediction System Users Manual, Version 1.2.* Bureau of Flora and Fauna, Canberra.

Carpenter, G., Gillison, A.N. and Winter, J. 1993. DOMAIN: a flexible modelling procedure for mapping potential distributions of plants and animals. *Biodiversity and Conservation*, Vol. 2, pp.667–680.

Carroll, C., Zielinski, W.J. and Noss, R.F. 1999. Using presence-absence data to build and test spatial habitat models for the Fisher in the Klamath Region, U.S.A. *Conservation Biology*, Vol. 13, No. 6, pp.1344–1359.

Catling, P.C., Bart, R.J. and Forrester, R.I. 1998. Models of the distribution and abundance of ground-dwelling mammals in the eucalypt forests of south-eastern New South Wales. *Wildlife Research*, Vol. 25, pp.449–466.

Church, R.L., Stoms, D.M. and Davis, F.W. 1994. Reserve selection as a maximal covering location problem. *Biological Conservation*, Vol. 76, pp.105–112.

Clague, C.I., Coles, R.B., Whybird, O.J., Spencer, H.J. and Flemons, P. 1999. The occurrence and distribution of the tube-nosed insectivorous bat (*Murina florium*) in Australia. *Proceedings of the Linnean Society of New South Wales,* Vol. 121, pp.175–191.

Crandall, K.A. 1998. Conservation phylogenetics of Ozark crayfishes: assigning priorities for aquatic habitat protection. *Biological Conservation*, Vol. 84, pp.107–117.

Crosbie, S.F. and Hinch, G.N. 1985. An intuitive explanation of generalised linear models. *New Zealand Journal of Agricultural Research,* Vol. 28, pp.19–29.

Csuti, B., Polasky, S., Williams, P.H., Pressey, R.L., Camm, J.D., Kershaw, M., Kiester, A.R., Downs, B., Hamilton, R., Huso, M. and Sahr, K. 1997. A comparison of reserve selection algorithms using data on terrestrial vertebrates in Oregon. *Biological Conservation,* Vol. 80, pp.83–97.

Diamond, J.M. 1975. The Island dilemma: lessons of modern biogeographic studies for the design of natural reserves. *Biological Conservation,* Vol. 7, pp.129–146.

Faith, D.P. and Walker, P.A. 1996a. Environmental diversity: on the best-possible use of surrogate data for assessing the relative biodiversity of sets of areas. *Biodiversity and Conservation,* Vol. 5, pp.399–415.

Faith, D.P. and Walker, P.A. 1996b. How do indicator groups provide information about the relative biodiversity of different sets of areas?: on hotspots, complementarity and pattern-based approaches. *Biodiversity Letters,* Vol. 3, pp.18–25.

Ferrier, S., Cassis, G.A., Gray, M.R. and Wilkie, L. 1999. Spatial patterns of species turnover in ground dwelling invertebrates in north-east New South Wales: implications for selection of forest conservation reserves. In Ponder, W. and Lunney, D. (eds). *The Other 99%: the Conservation and Biodiversity of Invertebrates.* Royal Zoological Society of New South Wales, Sydney. 454pp.

Ferrier, S., Pressey, R.L. and Barrett, T.W. 2000. A new predictor of the irreplaceability of areas for achieving a conservation goal, its application to real-world planning, and a research agenda for further refinement. *Biological Conservation,* Vol, 93, pp.303–325.

Ferrier, S. and Watson, G. 1997. *An Evaluation of the Effectiveness of Environmental Surrogates and Modelling Techniques in Predicting the Distribution of Biological Diversity.* Environment Australia, Canberra. 193pp.

Flather, C. H., Wilson, K. R., Dean, D. J. and McComb, W. C. 1997. Identifying gaps in conservation networks: of indicators and uncertainty in geographic-based analyses. *Ecological Applications,* Vol. 7, No. 2, pp.531–542.

Ford, H.A. 2006. Island biogeography: as illustrated by birds in the Australasian region. Pp.459—476. In Merrick, J.R., Archer, M., Hickey, G.M. and Lee, M.S.Y. (eds). *Evolution and Biogeography of Australasian Vertebrates.* Auscipub Pty. Ltd., Sydney. 942pp.

Frankham, R., Ballou, J.D. and Briscoe, D.A. 2002. *Introduction to Conservation Genetics.* Cambridge University Press, Cambridge. 617pp.

Gaston, K.J. 1996a. Species-range-size distributions: patterns, mechanisms and implications. *Tree,* Vol. 11, No. 5, pp.197–201.

Gaston, K.J. 1996b. Species richness: measure and measurement. Pp.77–113. In Gaston, K.J. (ed). *Biodiversity: a Biology of Numbers and Difference.* Blackwell Science, Cambridge. Massachusetts. 396pp.

Gaston, K.J. and David, R. 1994. Hotspots across Europe. *Biodiversity Letters,* Vol. 2, pp.108–116.

Gaston, K.J. and Williams, P.H. 1993. Mapping the world's species-the higher taxon approach. *Biodiversity Letters,* Vol. 1, pp.2–8.

Gilpin, M.E. and Diamond, J.M. 1980. Subdivision of nature reserves and the maintenance of species diversity. *Nature,* Vol. 285, p.567.

Groves, C., Valutis, L., Vosick, D., Neely, B., Wheaton, K., Touval, J. and Runnels, B. 2000. *Designing a Geography of Hope: a Practitioner's Handbook to Ecoregional Conservation Planning.* The Nature Conservancy, United States of America.

Humphries, C.J., Williams, P.H. and Vane-Wright, R.I. 1995. Measuring biodiversity value for conservation. *Annual Review of Ecological Systematics,* Vol. 26, pp.93–111.

Kerr, J.T. 1997. Species richness, endemism, and the choice of areas for conservation. *Conservation Biology,* Vol. 11, No. 5, pp.1094–1100.

Lambeck, R.J. 1997. Focal species: a multi-species umbrella for nature conservation. *Conservation Biology,* Vol. 11, No. 4, pp.849–856.

Lechmere-Oertel, R.G. and Cowling, R.M. 1999. Predicting the distribution of fynbos and succulent karoo biome boundaries and plant communities using generalised linear models and geographic information systems. *South African Journal of Botany,* Vol. 65, No. 1, pp.89–96.

Lincoln, R.J., Boxshall, G.A. and Clark, P.F. 1982. *A Dictionary of Ecology, Evolution and Systematics.* Cambridge University Press, Cambridge. 298pp.

Lindenmayer, D.B., Cunningham, R.B., Tanton, M.T., Nix, H.A. and Smith, A.P. 1991. The conservation of arboreal marsupials in the montane ash forests of the Central highlands of Victoria, South East Australia:III. The habitat requirements of leadbeater's possum, *Gymnobelideus leadbeateri* and models of the diversity and abundance of arboreal marsupials. *Biological Conservation*, Vol. 56, pp.295–315.

Maddock, A. and Benn, G.A. 2000. Identification of conservation-worthy areas in Northern Zululand, South Africa. *Conservation Biology,* Vol. 14, No. 1, pp.155–166.

Maddock, A. and Du Plessis, M.A. 1999. Can species data only be appropriately used to conserve biodiversity? *Biodiversity and Conservation,* Vol. 8, pp.603–615.

Margules, C.R. and Pressey, R.L. 2000. Review article: systematic conservation planning. *Nature*, Vol. 405, pp.243–253.

Myers, N. 1990. The biodiversity challenge: expanded hotspots analysis. *The Environmentalist*, Vol. 10, No. 4, pp.243–256.

Myers, N., Mittermeier, R.A., Mittermeier, C.G., da Fonseca, G.A.B. and Kent, J. 2000. Biodiversity hotspots for conservation priorities. *Nature*, Vol. 403, pp.853–858.

Nicholls, A.O. 1989. How to make biological surveys go further with generalised linear models. *Biological Conservation*, Vol. 50, pp.51–75.

Nix, H.A. 1986. A biogeographic analysis of Australian elapid snakes. In Longmore, R. (ed). *Atlas of Elapid Snakes of Australia*. Australian Government Publishing Service, Canberra. 115pp.

Noss, R.F. 1990. Indicators for monitoring biodiversity: a hierarchical approach. *Conservation Biology*, Vol. 4, No. 4, pp.355–364.

N.S.W. N.P.W.S. 1998. *Vertebrate fauna survey*. New South Wales National Parks and Wildlife Service Report for the Resource and Conservation Assessment Council.

N.S.W. N.P.W.S. 1999a. *Modelling areas of habitat significance for vertebrate fauna and vascular flora in north east NSW*. New South Wales National Parks and Wildlife Service Report for the Resource and Conservation Assessment Council.

N.S.W. N.P.W.S. 1999b. *Development of C-Plan Functionality to guide achievement of Spatial Configuration Objectives*. New South Wales National Parks and Wildlife Service Report for the Resource and Conservation Assessment Council.

Oliver, I., Beattie, A.J. and York, A. 1998. Spatial fidelity of plant, vertebrate, and invertebrate assemblages in multiple-use forest in eastern Australia. *Conservation Biology*, Vol. 12, No. 4, pp.822–835.

Pearce, J.L., Burgman, M.A. and Franklin, D.C. 1994. Habitat selection by helmeted honeyeaters. *Wildlife Research*, Vol. 21, pp.53–63.

Ponder, W.F., Carter, G.A., Flemons, P. and Chapman, R.R. 2001. Evaluation of museum collection data for use in biodiversity assessment. *Conservation Biology*, Vol. 15, No. 3, pp.648–657.

Prendergast, J.R., Quinn, R.M., Lawton, J.H., Eversham, B.C. and Gibbons, D.W. 1993. Rare species, the coincidence of diversity hotspots and conservation strategies. *Nature*, Vol. 365, pp.335–337.

Pressey, R.L. 1990. Reserve selection in New South Wales: where to from here? *Australian Zoologist*, Vol. 26, Pt. 2, pp.70–75.

Pressey, R.L. and Logan, V.S. 1998. Size of selection units for future reserves and its influence on actual vs targeted representation of features: a case study in western New South Wales. *Biological Conservation*, Vol. 85, pp.305–319.

Pressey, R.L., Possingham, H.P. and Day, J.R. 1997. Effectiveness of alternative heuristic algorithms for identifying indicative minimum requirements for conservation reserves. *Biological Conservation*, Vol. 80, pp.207–219.

Pressey, R.L. and Tully, S.L. 1994. The cost of ad hoc reservation: a case study in western New South Wales. *Australian Journal of Ecology*, Vol. 19, pp.375–384.

Reid, W.V. 1998. Biodiversity hotspots. *Tree*, Vol. 13, No. 7, pp.275–280.

Rodrigues, A.S.L., Gregory, R.D. and Gaston, K.J. 2000. Robustness of reserve selection procedures under temporal species turnover. *Proceedings of the Royal Society of London*, Vol. 267, pp.49–55.

Scott, J.M. and Csuti, B. 1997. Gap Analysis for biodiversity survey and maintenance. In Reaka-Kudla, M.L., Wilson, D.E., and Wilson, E.O. (eds). *BioDiversity II: Understanding and Protecting our Biological Resources*. Joseph Henry Press, Washington, D.C.

Scott, J.M. and Jennings, M.D. 1994. *A Handbook for Gap Analysis*. Idaho Cooperative Fish and Wildlife Research Unit, University of Idaho, Moscow, Idaho.

Scotts, D. and Drielsma, M. 2003. Developing landscape frameworks for regional conservation planning; an application integrating fauna spatial data and ecological principles. *Pacific Conservation Biology*, Vol. 8, pp.235–254.

Simberloff, D.S. 1982. Big advantages of small refuges. *Natural History*, Vol. 91.

Stockwell, D. 2000a. Biodiversity Species Workshop. http://biodi.sdsc.edu/bsw_home.html.

Stockwell, D. 2000b. GARP Modelling System User's Guide and Technical Reference. http://biodi.sdsc.edu/Doc/GARP/Manual/manual.html.

Sutherst, R.W. and Maywald, G.F. 2000. *CLIMEX: Predicting the Effects of Climate on Plants and Animals.* CSIRO Publishing, Melbourne.

Virolainen, K.M., Virola, T., Suhonen, J., Kuitunen, M., Lammi, A. and Siikamaki, P. 1999. Selecting networks of nature reserves: methods do affect the long-term outcome. *Proceedings of the Royal Society of London*, Vol. 266, pp.1141–1146.

Williams, P. 2000. Biodiversity and WORLDMAP. http://www.nhm.ac.uk/science/projects/worldmap/

Williams, P.H. and Gaston, K.J. 1994. Measuring more of biodiversity: can higher-taxon richness predict wholesale species richness? *Biological Conservation*, Vol. 67, pp.211–217.

Williams, P.H. and Humphries, C.J. 1996. WORLDMAP and prioritisation for conservation: integration of systematic data for conservation evaluation. Pp.98–99. In Jermy, A. C., Long, D., Sands, M.J.S., Stork, N.E. and Winser, S. (eds). *Biodiversity Assessment: a Guide to Good Practice*. Department of the Environment, London.

Williams, P.H., Prance, G.T., Humphries, C.J. and Edwards, K.S. 1996. Promise and problems in applying quantitative complementary areas for representing the diversity of some Neotropical plants (families Dichapetalaceae, Lecythidaceae, Caryocaraceae, Chrysobalanaceae and Proteaceae). *Biological Journal of the Linnean Society*, Vol. 58, pp.125–157.

WWF. 2000. World Wildlife Fund. http://www.wwf.org.

Yee, T.W. and Mitchell, N.D. 1991. Generalised additive models in plant ecology. *Journal of Vegetation Science*, Vol. 2, pp.587–602.

Table 1. List of pinniped species found in the Australasian region (modified from Shaughnessy, 1999).

Family	Scientific Name	Common English Name	Breeding Area
Phocidae	*Mirounga leonina*	southern elephant seal	subantarctic islands
	Hydrurga leptonyx	leopard seal	Antarctica (sea ice)
	Lobodon carcinophagus	crabeater seal	Antarctica (sea ice)
	Leptonychotes weddelli	Weddell seal	Antarctica (fast ice)
	Ommatophoca rossii	Ross seal	Antarctica (sea ice)
Otariidae	*Phocarctos hookeri*	New Zealand sea lion	New Zealand (subantarctic islands and mainland)
	Neophoca cinerea	Australian sea lion	mainland Australia
	Arctocephalus forsteri	New Zealand fur seal	mainland Australia/NZ/subantarctic islands
	Arctocephalus pusillus doriferus	Australian fur seal*	mainland Australia
	Arctocephalus gazella	Antarctic fur seal	subantarctic islands
	Arctocephalus tropicalis	subantarctic fur seal	subantarctic islands

* A subspecific designation is given to distinguish these populations from the Cape fur seals (*A. pusillus pusillus*) of southern Africa.

time and space. In addition, the availability of broad-scale satellite imagery has improved to the point where biologists can use bio-physical measures of ocean productivity to understand the relationship between pinnipeds and their marine environment (McConnell *et al.*, 1992; McCafferty *et al.*, 1998, 1999; Bornemann *et al.*, 2000; Guinet *et al.*, 2001).

The main technological advance enabling biologists to collect data on seals' at-sea behaviour was the development of a data-archiving tag known as the Time-Depth Recorder, commonly abbreviated to 'TDR' (Gentry and Kooyman, 1986b). The original devices were large, cumbersome and collected data on depth and time using a mechanical clock and inscribing trace. Nevertheless, they permitted a first glimpse into the clandestine life of diving animals (Kooyman, 1965; Kooyman *et al.*, 1976; Gentry and Kooyman, 1986b). By the late 1980s, microprocessor-controlled TDRs were collecting data on the at-sea behaviour of seals and other marine animals (Gentry and Kooyman, 1986a; Le Boeuf *et al.*, 1989; Goebel *et al.*, 1991; Hindell *et al.*, 1991b; Bengston and Stewart, 1992; Boness *et al.*, 1994; Boyd and Croxall, 1996). Technology has continued to improve to the point where today's TDRs have not only increased archiving capacity, but can also record a variety of additional variables including temperature, light levels, conductivity, swim velocity, heart rate and stomach temperature. In addition, other devices such as animal-mounted, miniature video cameras, electronic compasses, accelerometers and inclinometers can provide more precise behavioural information (Davis *et al.*, 1999).

It is not our intention to review the vast literature describing the applications and data recorded by TDRs and other behavioural data archival tags; we have devoted this chapter to the description of techniques available for the study of the spatial use of the sea by pinnipeds (i.e., marine movements). Many of these techniques can enhance information collected from TDRs, or they can be used independently to provide other types of data. Here we provide detailed descriptions of the applicability and efficiency of each technique using recent examples of research on three Australasian pinnipeds. These techniques include satellite telemetry, geo-location, acoustic tracking and radio tracking. We also discuss the future of potential technological advances in these techniques to continually improve the information available on this group of marine mammals.

SATELLITE TRACKING

PLATFORM TRANSMITTER TERMINALS (PTTs)

One of the more common methods employed to locate seals at sea is satellite telemetry, which requires a platform transmitter terminal (PTT) to be attached to the animal (Plate 132). PTTs transmit a UHF (ultra-high frequency) pulse to satellites orbiting the Earth. The system employed by ecologists for tracking animals is the ARGOS satellite-based system, which was implemented in 1979 by the French space agency (CNES), U.S. National Oceanic and Atmospheric Administration (NOAA) and the U.S. National Aeronautics Space Administration (NASA) (Argos, 1989). The system works in four simple stages: 1) the ARGOS transmitter is deployed on the seal and automatically sends a message at a pre-determined rate to multiple low-Earth orbit satellites; 2) these messages are then relayed from satellite to a ground station; 3) which then automatically forwards the data to the ARGOS processing centre in France. The data are then processed to calculate the location of the transmitter and; 4) the results are delivered to the researcher. Locations can vary in accuracy and are divided into quality classes by ARGOS. These are: **Z** = invalid location; **A/B** = no estimate of accuracy; **0** = >1000 m; **1** = >350 to <1000 m; **2** = >150 to <350 m; and **3** = <150 m (Argos, 1989).

PTTs can vary in complexity with the simplest device containing only the transmitter, battery and antenna all housed in a waterproof casing. Seals spend much of their time submerged during their foraging trips and signal transmission at these times is a waste of battery power. Another way to save on battery power and costly satellite time (Table 2) is to inhibit the number of signals emitted while the animal is on land or ice. These type of transmissions are controlled by a conductivity switch which, when not activated by salt water for a predetermined period, inhibits the signal (i.e., the 'haul-out threshold').

Other possible modifications to PTTs for pinnipeds may include modified TDRs with a pressure transducer that suspends signal transmission while the animal is diving and restarts transmission as the seal approaches the surface (McConnell *et al.*, 1994). Attaching archival tags, such as TDRs to collect dive behaviour, swim speeds, light intensity and water temperature and then transmit those data to the satellites can further increase the data collected for each deployment. However, there are limitations in the amount of supplementary data that can be transferred via satellites due to restricted transmission rates.

However, the size (i.e., frontal surface area) and weight of the instruments, caused by the requirement for large batteries (e.g., up to 400 g), can cause excessive drag on the seal while swimming. Even with the continued miniaturization of electronic components, large devices can limit the number of species and age-classes on which instruments can be deployed (see also Boyd *et al.*, 1991; Culik *et al.*, 1994; Walker and Boveng, 1995; Boyd *et al.*, 1997; Harcourt and Davis, 1997 for results on the effects of deploying instruments on marine animals).

In addition, an important consideration when using PTTs is that large samples (i.e., >10 PTT units) are uncommon due to the cost of the devices themselves and the use of the satellites (Table 2). Often the most expensive component of studies employing PTTs is the charge incurred by using the ARGOS satellite system rather than the initial cost of the devices. One advantage of this high cost is that data retrieval occurs whether or not the instrument itself is recovered. When the population to be studied is in an extremely remote, or difficult to access area (e.g., pack ice or islands at high latitudes) or the species being studied is elusive and difficult to re-catch, time and money do not need to be invested to retrieve archived data (this is the case when using TDRs). Researchers therefore need to assess not only the minimum sample size required to answer their proposed questions, but also must determine the magnitude of difficulty in accessing the data.

FUR SEALS AND FISHERIES: USING SATELLITES TO ASSESS OVERLAP

After the initiation of commercial sealing in the late 18th century, Australian fur seal populations were severely reduced, although the exact degree of the reduction is unknown (Warneke and Shaughnessy, 1985). The Australian fur seal has been protected for most of the last century and its numbers are now increasing (Warneke and Shaughnessy, 1985; Shaughnessy *et al.*, 1995;

Table 2. Approximate costs per unit (1999–2000 prices in Australian dollars*) and additional costs (e.g., data retrieval, necessary equipment, etc.), for a range of remote data-collection devices. Examples of unit sources are given (for a more comprehensive list, see the website http://www.biotelem.org/). Figures quoted should only be used as relative guides, since actual market costs fluctuate rapidly.

Unit Type	Unit/Cost ($)	Other Costs	Source
PTT	1,700–3,000	data retrieval (30/day/unit)	Sirtrack, N.Z.; Microwave Telemetry, U.S.A.; GFT, Germany (unit); Argos, France (data retrieval)
GPS§	400–600§	unit retrieval (manual costs)	B. Philips, Australian Antarctic Division; private development
geo-locating TDR	2,000	unit retrieval	Wildlife Computers, U.S.A.; Vemco Ltd., Canada; Lotek Marine Technologies, Canada
acoustic transmitter	1,075 (depth modulated)	47,000 (4-buoy array); unit retrieval	Vemco Ltd., Canada; Lotek Marine Technologies, Canada; HTI, U.S.A.
VHF radio	220	1,500 (scanner & aerial) 9,200 (data logger)	numerous sources
mini-CTD	400–600	unit retrieval	Star-Oddi, Iceland
route recorder	500–600†	unit retrieval	IEI‡, Italy; F. Bonadonna, CNRS, France

* Some AUD prices are converted from USD at a rate of 0.60 AUD/USD.
§ Projected cost.
† Non-commercial construction cost.
‡ Instituto per l'Elaborazione dell'Informazione, Pisa.

Arnould and Littnan, 2000). They are now found throughout south-eastern Australian coastal waters and as far north as Jervis Bay, N.S.W. (Shaughnessy, 1999). As is the case with many seal species, the Australian fur seal has been suggested to be a major competitor with fisheries, in particular with the South-east Trawl Fishery (SETF) (Anon., 1998, 1999). Several other smaller fisheries occur within the range of Australian fur seals including the South-east Non-Trawl Fishery (SENTF), the Southern Squid Jig Fishery (SSJF), and the Jack Mackerel Fishery (JMF) (Anon., 1998, 1999).

To determine if overlap (i.e., using the same resources, as opposed to 'competition', which is 'an interaction in which one organism consumes a resource that would have been available to, and might have been consumed by, another'—Begon *et al.*, 1986, p. 199) exists between seals and the fisheries two basic criteria must be met: (1) overlap in prey species eaten or harvested (Harwood, 1992; Fea and Harcourt, 1997), and (2) overlap in areas fished (Harwood, 1992; Harcourt and Davis, 1997). With regard to criterion number one, Australian fur seals do eat commercially hunted fish. Indeed, 60% of the prey species of Australian fur seals (Gales *et al.*, 1993; Gales and Pemberton, 1994; Littnan, unpublished data) are fished commercially (Anon., 1998, 1999). However, the effect this overlap may have on the subsistence of both fur seals and the fisheries is unclear. Further research is required to determine annual, seasonal and spatial variation in Australian fur seal diet, fur seal prey consumption rates, and population dynamics of both fur seals and fish stocks.

The second part of the equation is obtaining information on the overlap between areas used by seals and commercial fishing grounds (Harwood, 1992; Harcourt and Davis, 1997). This is important for managing fish stocks for sustainable use while ensuring the persistence of seal populations. TDRs and PTTs have been deployed on animals in Tasmanian and Victorian waters to determine foraging locations and dive behaviour (Arnould, Harcourt, Hindell, Kirkwood, Littnan, Pemberton, unpublished data). Much of the work has been done with adult female fur seals for a number of reasons. First, they return to the colony to suckle their pups and so expensive instruments may be recovered. Second, females are significantly smaller and therefore, more tractable and more easily handled than males. Third, and most importantly, females provide information on the reproductive performance (and thus trends) of the population.

A majority of dives recorded from female fur seals on Kanowna Island (see Figure 1) were to 65–85 metres, suggesting that animals feed primarily on the continental shelf within Bass Strait (Arnould, Hindell, Littnan, unpublished data). This is supported by data from ten females fitted with PTTs: Kanowna Island (n = 3), The Skerries (n = 1) and Tenth Island (n = 6). Figure 1 shows a composite of these data expressed as the amount of time spent in hours per 20 km x 20 km grid cell. Data from Kanowna Island and The Skerries (Arnould, Harcourt, Littnan, unpublished data) show females foraging within Bass Strait between Tasmania and Victoria, but also along the east coast of Victoria near Lakes Entrance. One female even foraged along the N.S.W. coast as far north as Merimbula. The Tenth Island females foraged entirely within southern Bass Strait and within 200 km of the colony (Hindell, Pemberton, McCarrey, unpublished data). In addition, four adult males have been tracked from Seal Rocks in northern Bass Strait (Kirkwood, Gales, Lynch, Dann, 2000, personal communication). These seals foraged mostly in western Bass Strait, where water depths are <100 m, and >20 km from any land (Kirkwood, 2000, personal communication).

With the limited data available, it appears that seals foraging in Bass Strait and Tasmanian waters may have some spatial overlap with the relatively small Jack Mackerel Fishery. The greatest region of overlap is along the east coast of Victoria, near Lakes Entrance, where the continental slope occurs very close to shore (see Irvine *et al.*, 1997). It may be that some fur seals in colonies to the north of eastern Bass Strait forage in areas where there is substantial commercial fishing, but much more information is still required to comment on the degree of overlap between Australian fur seals and Australian commercial fisheries.

GEO-LOCATION

LOCATIONS AT SEA (FROM TEMPERATURE AND LIGHT RECORDINGS)

An alternative to satellite location of animals at sea is to record environmental parameters that can be processed later to provide an indication of where the seal was at the time that the recordings were made. In the late 1980s and early 1990s, this technique began by using water temperatures that were recorded concurrently with depth data in early TDRs (Hindell, 1991; Slip *et al.*, 1994). Comparisons of temperatures when the animal is at the surface with concurrent sea surface temperature data (generally derived from satellite imagery) can be used to estimate in which general oceanographic regions (e.g., south of the Antarctic Polar Front—Hull, 1999) the animals were foraging. In deeper diving animals, these locations can be refined by including temperature data at a range of depths, for example 100, 200 and 500 m, and relating them to oceanographic temperature-depth profiles (Hindell *et al.*, 1991a). These additional data enable animals to be located to more precise areas of the ocean (Slip, 1997). The precision of these estimates depends on the magnitude of variation in the vertical thermal structure of the water column and on the detail and accuracy of the oceanographic data. Unfortunately, the latter are sparse for some regions of the ocean and for some times of the year. Nonetheless, this technique remains a powerful analytical tool in marine ecology.

Plate 132. A Time-Depth Recorder (TDR) and Platform Transmitter Terminal (PTT) attached to a male fur seal Photograph R. Kirkwood.

Plate 133. Attaching an acoustic transmitter to a female Weddell seal Photograph T. Dorr and R. Harcourt.

Plate 134. A southern elephant seal female at Macquarie Island with a geo-locating Time-Depth Recorder (TDR) and VHF transmitter (with antenna) attached. Photograph C. Bradshaw.

Plate 135. The single species of sirenid represented in Australasia is the Dugong or sea cow (*Dugong dugon*). Aside from inclusion in keys, this group of marine mammals is not discussed elsewhere in this volume; but conservation of dugongs, as well as the related manatees elsewhere, is of major concern due to human impacts on shallow coastal waters and reduced populations from previous over-hunting. Photograph L. Bejder.

Plate 136. The dugong is herbivorous and Australian populations are largely restricted to coastal shallows or estuaries where there are extensive seagrass beds; it is a slow swimmer and recruitment rates are low. This adult is a member of the large population in Shark Bay, Western Australia. Photograph L. Bejder

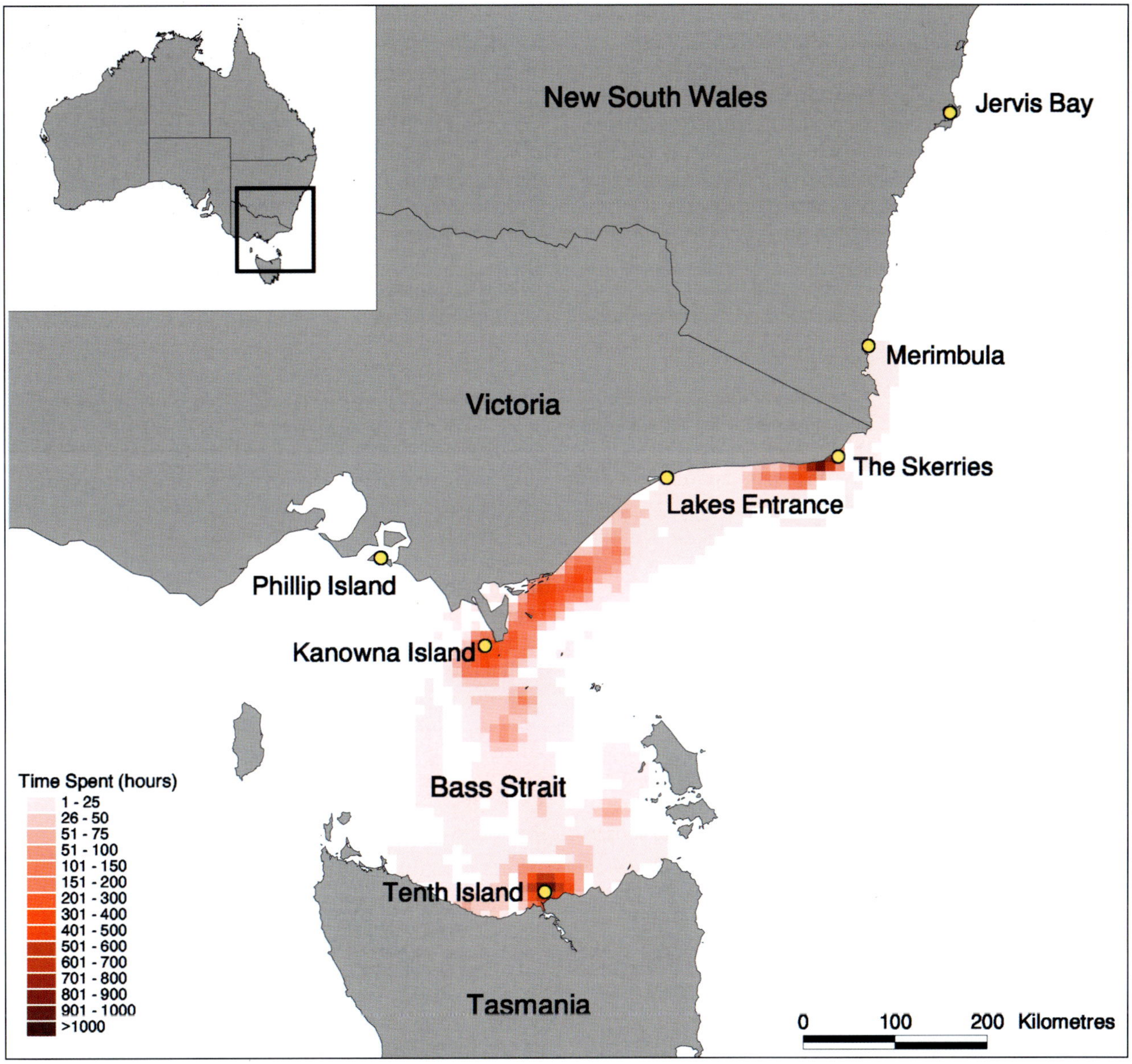

Figure 1. Plot of the time spent at sea by Australian fur seals equipped with satellite transmitters. Animals were captured and released from The Skerries ($n = 1$; 21 Jun.–27 Aug. 1999), Kanowna Island ($n = 3$; 27 May–18 Nov. 1999) (Littnan, Harcourt, Arnould, unpublished data) and Tenth Island ($n = 6$; 18 Dec. 1996–26 Feb. 1997) (Hindell, Pemberton, McCarrey, unpublished data).

An alternative parameter used to locate animals is light intensity (DeLong *et al.*, 1992). Again, this is done conventionally in association with a TDR, which is programmed to record light intensity at set time intervals, or when the animal is within a set distance from the surface. The basic data of light intensity, time and depth can be processed to calculate the time of sunrise and sunset for each day that the animal has spent at sea. Standard equations for solar navigation (Nautical Almanac Office, 1991) can then convert this information into estimates of latitude and longitude for that day.

The use of light levels to estimate position on the Earth's surface is not as accurate as locations derived from satellite telemetry. Even at their very best, locations will be accurate only to within approximately one degree of latitude (i.e., 111 km), and often they are less accurate than this (Hill, 1994). One problem is estimating the time of sunrise and sunset from light data recorded on actively swimming and diving animals. The resultant light curves are rarely smooth, and the algorithms employed to smooth and interpolate between the points often have trouble identifying the time of sunrise or sunset with sufficient accuracy. Another source of error occurs when an animal moves several hundred kilometres in a day. Movements to the east or west will artificially increase or decrease the apparent day length, and subsequently provide inaccurate locations. Perhaps the main limitation to the method is the occurrence of the equinoxes. During an equinox, day length is the same over the entire globe, making accurate assessment of latitude impossible, although longitude is not affected. This means that there are generally several weeks each year for which latitudinal fixes cannot be made. Polar regions offer special problems because they have extended periods of complete daylight, or complete darkness, which again prevents the derivation of a location. Likewise, equatorial regions with their limited variation in day length provide few navigational cues. Geo-location is therefore most effective when used in mid-latitudes between 10 and 60 degrees.

Some studies have attempted to validate geo-location from light levels against known at-sea locations either from ships (DeLong *et al.*, 1992), or from animals carrying both geo-location TDRs and satellite transmitters (van den Hoff, unpublished data). The validation studies have confirmed that, under ideal conditions, locations at sea derived from light levels are accurate to ±100 km, and that biases increase with increasing latitude. These validations are a crucial component of geo-location studies as they determine the appropriate scale for the subsequent analysis of location data, and for the calculation of adjustment algorithms.

Despite the problems associated with light levels in estimating the location of animals at sea, this technique has been used successfully on a number of species (Goldsworthy *et al.*, 1997; Stewart, 1997; Hull, 1999). There are two prerequisites for the successful use of light data:

(1) A suitable question: studies need to address questions that are concerned with meso-scale events (i.e., in the order of $\geqslant 100$ km^2). Ecological variables that operate on these scales are events such as ocean currents, frontal systems, bathymetry and some types of fisheries activity.
(2) A suitable species: studies need to focus on species that cover a large area of ocean (Stewart and DeLong, 1993). As the best accuracy possible is to within one degree of latitude or longitude, species that do not move >100 km during a foraging trip will reveal little of value. Alternatively, animals that travel too far too quickly, such as far-ranging sea birds, will have large errors associated with their locations. Ideal species are those that make foraging trips of more than 10 days, range over at least several hundred kilometres, and do not move more than 200 km per day.

ADVANTAGES OF GEO-LOCATION

Geo-location systems have several advantages over existing telemetric systems. Telemetric devices have antennae that need to leave the water before they can transmit a signal, which restricts where on the body the devices can be placed, and may increase the drag while the animal is submerged and swimming (see Boyd *et al.*, 1991; Walker and Boveng, 1995; Harcourt and Davis, 1997). The relatively compact size of light-archiving units reduces the amount of drag, thus permitting deployment of the devices on small marine animals such as penguins and small fur seals (Hull, 1997). TDRs with light-archiving capability record depth and water temperature (and sometimes other parameters such as swimming speed), enabling all of these data to be referenced spatially and greatly enhancing the usefulness of conventional diving behaviour data. To collect similar information using radio or satellite telemetry requires much larger units incorporating additional sensors and archival capacity, or even two separate units (e.g., a transmitter and a TDR shown in Plate 134). Both of these options increase size and drag and restrict both the choice of species and duration of deployments. However, the single biggest advantage to geo-location TDRs is cost. The initial purchase of a satellite transmitter can be up to twice or three times that of a TDR, and then there is an additional daily charge while the units transmit (Table 2). There is therefore a trade-off between using satellite transmitters and getting more high-quality locations per animal (but for fewer animals) without the burden of unit retrieval, or using light-archivers and getting fewer, less accurate locations per animal but from a far greater number of animals after unit retrieval. The resolution to this dilemma lies in the nature and scale of the questions being asked, and the probability of retrieving instruments after initial deployment.

ELEPHANT SEALS FORAGING IN THE SOUTHERN OCEAN

One species that is particularly well suited to geo-location is the southern elephant seal. This species has a circumpolar distribution, breeding on subantarctic islands, and dispersing into the Southern Ocean for periods of up to eight months at a time (Hindell and Burton, 1988; Hindell *et al.*, 1991a; McConnell *et al.*, 1992; Fedak and McConnell, 1993; Figure 2). Geo-location has been used to track these animals from several of their major breeding sites, and has revealed much about their migration patterns (Bester and Pansegrouw, 1991; Jonker and Bester, 1998). Adult seals migrate twice each year: the first migration is after the breeding season when the animals need to re-build their fat reserves for the coming moult. During this time, they can disperse up to 2,500 km from the breeding site. In females, this post-breeding migration lasts for a mean of 70 days, and in males it lasts for about 90 days (Le Boeuf and Laws, 1994). For elephant seals that breed on subantarctic Macquarie Island, males tend to concentrate their post-breeding foraging effort over the Antarctic continental shelf, and to a lesser extent over subantarctic ridges such as the Campbell Plateau (Slip *et al.*, 1994). Dispersal of adult females is

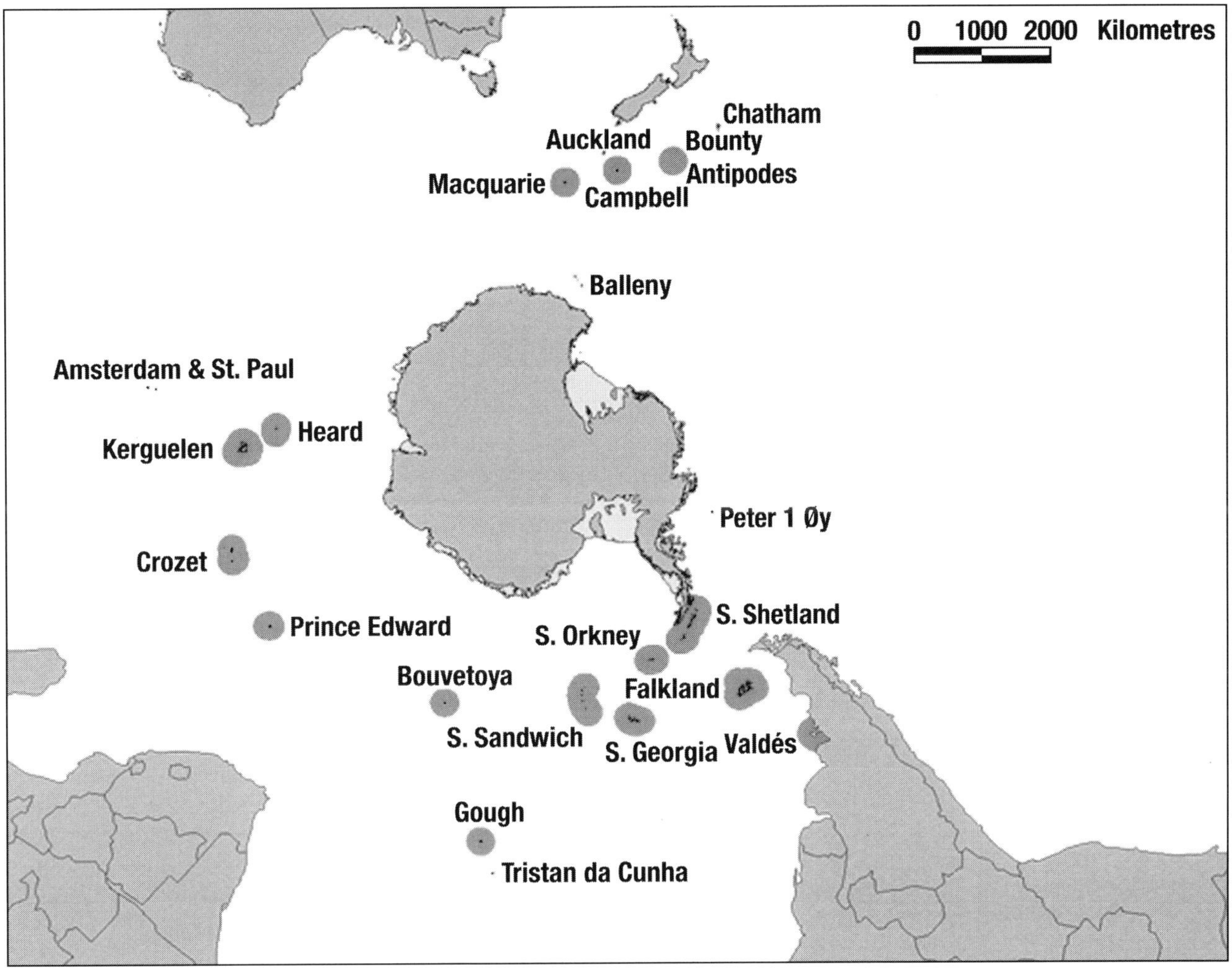

Figure 2. The breeding distribution of southern elephant seals (*Mirounga leonina*) in the Southern Ocean (after King, 1983; Ling and Bryden, 1992; Slade, 1997).

more variable and often more pelagic; females have been located in a range of marine habitats from the Antarctic continental shelf, to oceanic waters north of the sub-antarctic convergence (McConnell *et al.*, 1992; Slip *et al.*, 1994). The second migration is longer (up to 8 months in females and 6 months in males) and takes place after the annual moult. General patterns of dispersal seem to be the same as during the post-breeding migration, although less is known about this phase (Slip *et al.*, 1994).

Due to the wide-ranging and protracted nature of their migrations, elephant seals are ideal subjects for questions that are directed towards meso-scale events. One example of this type of study is recent work using geo-locating TDRs, which aims to quantify the physical and biological oceanographic characteristics that determine the foraging grounds, and ultimately the foraging success, of adult females (Bradshaw, Hindell, Michael, unpublished data). A study on female foraging ecology began in the summer of 1999–2000. Here, 17 female elephant seals from Macquarie Island were instrumented with TDRs (Plate P2) and the data collected confirmed that there was considerable individual variation in the areas of the Southern Ocean used by these seals (Figure 3). Animals from Macquarie Island moved to the west, the north-east (adjacent to the Campbell Plateau), to the south-west into the Antarctic pack ice, and >2,000 km to the east.

Although helpful in indicating general dispersal patterns, the plotted tracks of seal locations themselves yield little quantitative information. A more informative way to represent these data is to calculate the time that the animals actually spent in each part of the ocean. For elephant seals tracked with geo-location, an appropriate spatial resolution is reflected by a raster grid consisting of 100–200 km^2 cells. Presenting the data for all 17 seals as 'time spent' per grid cell produces a much clearer picture of their spatial use (Figure 4). There is a clear trend for the seal activity to be concentrated along a band extending in an approximately north-west to south-east direction, and within the band there are regions of high use. Similar regions of high use are associated with the Antarctic continent, and near

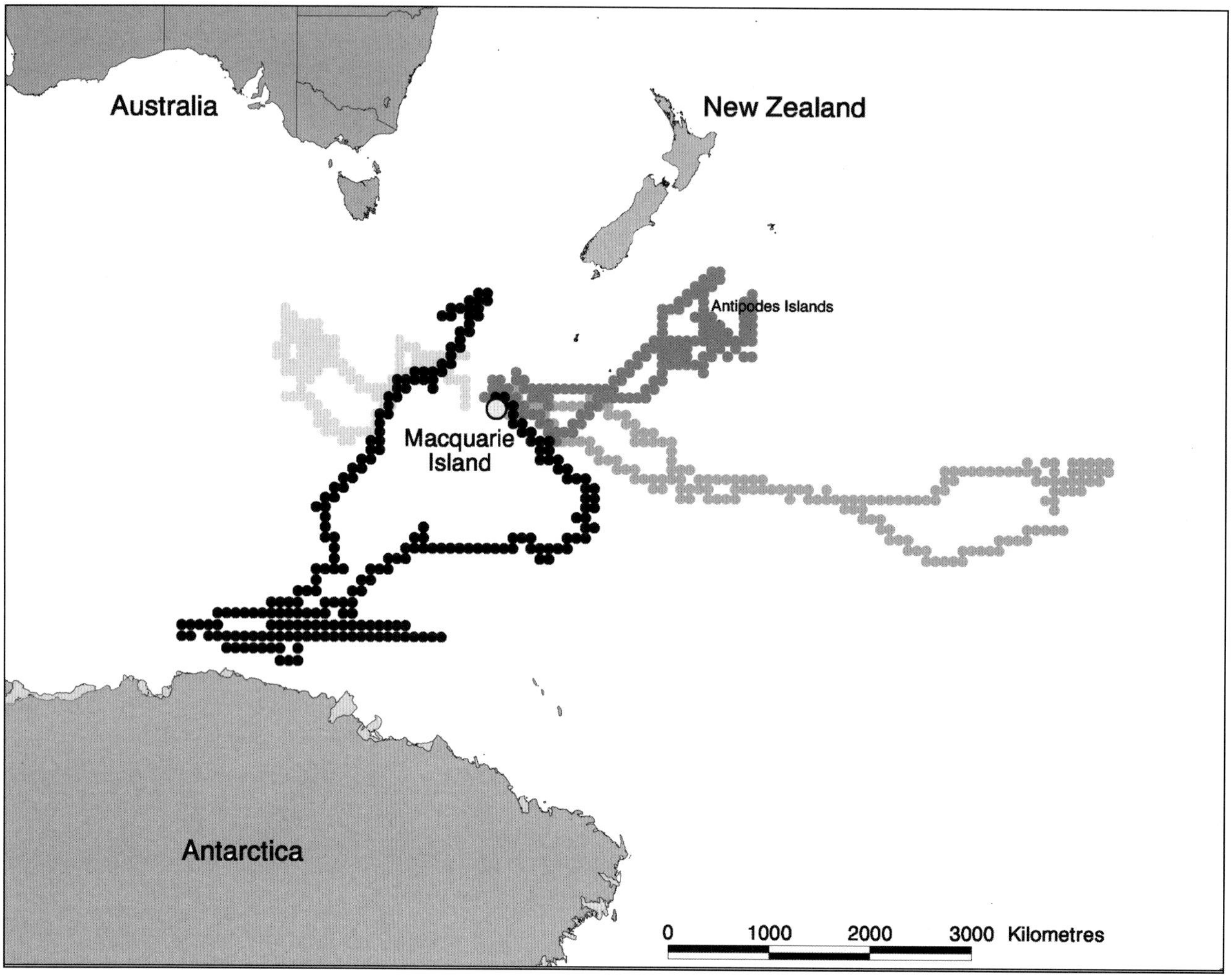

Figure 3. Plot of the four extreme female elephant seal tracks made in the 1999–2000 post-lactation foraging trip (Oct. 1999–Jan. 2000).

the Antipodes Islands (Figure 3). This figure represents all the data combined over a 100-day period, but it is also helpful to represent the data on more restricted time scales (Figure 5). This representation indicates that the areas of the Southern Ocean visited and time spent there differ throughout the spring and summer, reflecting the outward part of their migrations, the time spent at remote feeding grounds, and the return phase.

How top predators respond to variation in the intensity and location of food patches in their environment will, in the long term, influence population trends (Testa *et al.*, 1990; Boyd and Arnbom, 1991; Burton, 1998). Spatial and temporal distribution data from seals can be incorporated into models that help us interpret how they vary their movements in relation to prey abundance. An extension of these models would allow the prediction of elephant seal behaviour as a function of a range of physical and biological ocean variables (Croxall *et al.*, 1988; Whitehead *et al.*, 1990; Testa *et al.*, 1991; Trillmich and Ono, 1991; Burton, 1998; McCafferty *et al.*, 1998, 1999; Bornemann *et al.*, 2000; Guinet *et al.*, 2001).

At this time it is still difficult to estimate the configuration of oceanographic features, but we do have access to some data that are at least partially related to the biological productivity of the Southern Ocean (Vincent *et al.*, 1991; Ansorge *et al.*, 1999). Measurements of sea-surface temperature, bathymetry, ocean currents and ocean colour (a proxy for chlorophyll A, produced by phytoplankton) are made through satellite imagery of the ocean surface at a variety of temporal and spatial scales. Assessing these properties at scales similar to those produced by the remote data for elephant seals at sea can produce models that best explain the observed patterns of seal distribution and foraging behaviour. Spatial data for all input variables are best organized and presented using geographic information systems—see pp.873–887 (Flemons and Cassis, 2006)—and associated using standard spatial correlations (Odland, 1988), or more recently, artificial neural networks (Bradshaw *et al.*, 2002). Such analyses are starting to provide powerful predictive tools that describe elephant seal distribution under various oceanographic scenarios (Bradshaw, Hindell, unpublished data). The analyses will

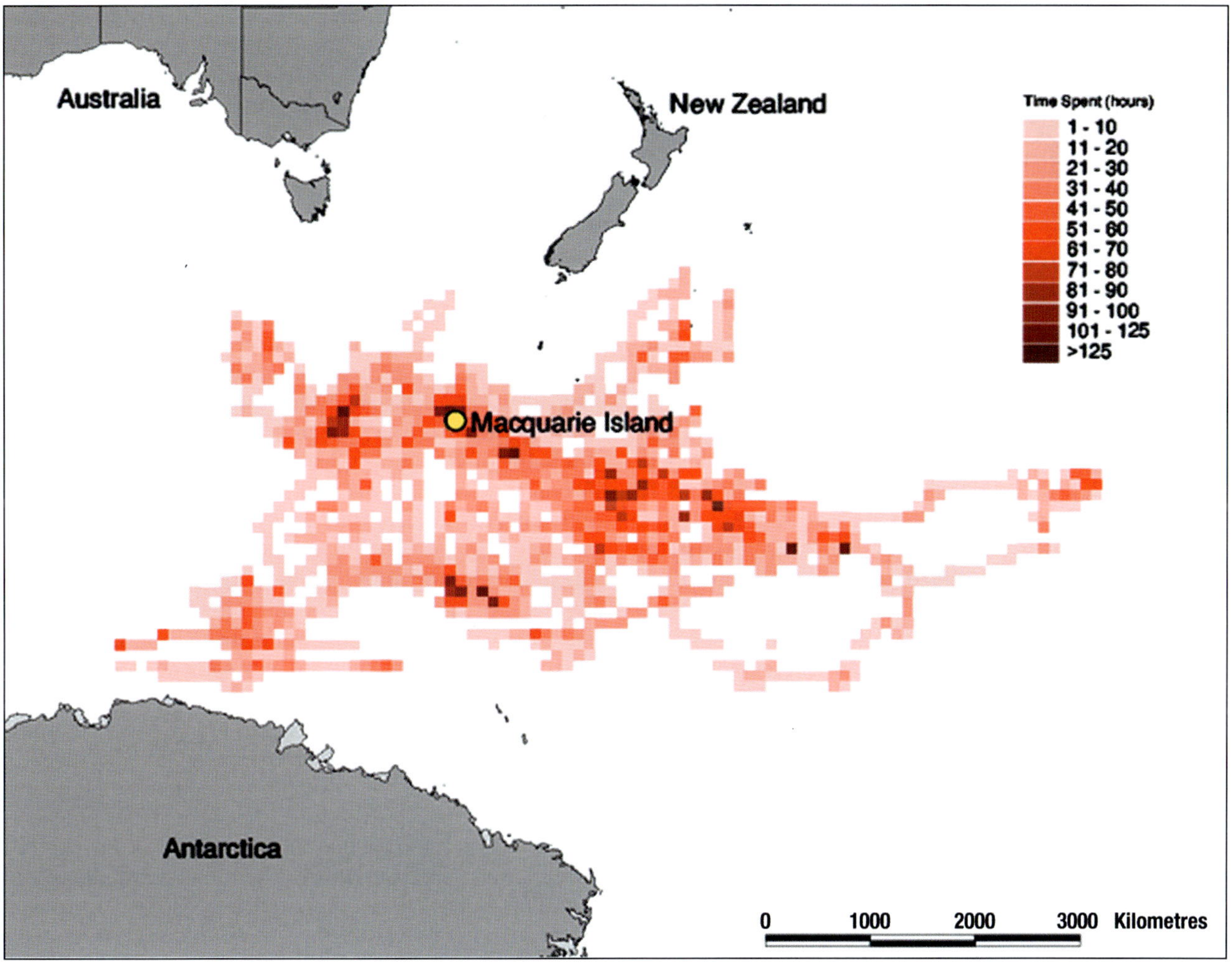

Figure 4. Plot of the time spent at sea for 17 female southern elephant seals during the 1999–2000 post-lactation foraging trip (Oct. 1999–Jan. 2000). Darker shades indicate more time spent in those grid cells.

be made even more powerful by incorporating behavioural data recorded simultaneously with location data that can provide indices of foraging effort (such as number of dives, total time spent at depth, and total vertical distance) at the same scales. Although in its infancy, the integrative modelling approach to marine predator studies is an exciting, new development that will shed light on linkages between biology and the environment that could not be explored in the past.

ACOUSTIC TRACKING

MEASURING BEHAVIOUR IN THREE DIMENSIONS

TDRs have revolutionized our understanding of the ecology of diving animals (Gentry and Kooyman, 1986a). Yet despite their technical sophistication, to date they have one major limitation: dive profiles for records produced by archival tags are limited normally to two dimensions: time and depth. However, diving animals inhabit a three-dimensional world (four dimensions if time is included). Although satellite telemetry allows us to visualize movements over broad spatial scales, it cannot provide information on fine-scale directional changes, that is, within individual dives. Although archival tags record other parameters, such as swim velocity, that provide a mechanism for calculating dive angle and activity changes in different phases of each dive (Ponganis *et al.*, 1990; Thompson and Fedak, 1993; Boyd *et al.*, 1995; Horning and Trillmich, 1997), most still do not provide information on finer-scale activity; however, for further discussion of route recorders, see the Other Techniques section (p.901).

To counter this problem, alternative methods of studying dive behaviour of seals are available. One method employs an acoustic tracking system that enables the reconstruction of the movements of individual seals in four dimensions (the horizontal X and Y coordinates, the vertical Z coordinate (or depth and time) (Wartzok *et al.*, 1992a, b; Harcourt *et al.*, 2000). The degree of precision with this technique is high (within 1–2 m^3), and it can be used with a number of animals simultaneously. Therefore, it allows for the measurement of changes in the underwater behaviour of individual seals relative to each other, and is thus a powerful tool in the study of social behaviour.

FORAGING OF WEDDELL SEALS UNDER ANTARCTIC ICE

Acoustic tracking has been used on Weddell seals (Hindell *et al.*, 1999; Harcourt *et al.*, 2000) which give birth and suckle their pups on fast ice at the coastal margins of Antarctica during the Austral spring (Castellini *et al.*, 1992a). Females leave the water along tide-cracks in the ice to give birth and feed their pups during the five- to six-week lactation period (Tedman and Green, 1987). During this time, territorial males spend the majority of their time under the ice competing for, or maintaining, territories (Kaufman *et al.*, 1975; Siniff *et al.*, 1977; Bartsh *et al.*, 1992; Harcourt *et al.*, 1998). Because females are concentrated along the cracks, the movements of males are also localized nearby. Females dive frequently towards the end of lactation but are constrained by the limited number of gaps through the ice surface to the air, and also by the need to return to their pup at the breeding site. Males that successfully hold territories rarely stray more than a kilometre from the tide crack (Harcourt *et al.*, 1998); therefore, it is possible to use acoustic tracking on many free-ranging animals *in situ*.

Movements under the ice have been monitored using a Cabled Acoustic Positioning System (CAPS) manufactured by Vemco Limited (Canada). An acoustic transmitter is attached to the seal using quick-dry epoxy glue (Plate 133). The attached device transmits a series of ultrasonic signals at a single, predetermined frequency in the range 60–80 kHz every 1.5 seconds. The signal contains four pulses, the first three occur at fixed time intervals and the last varies with depth. Only transmitters using frequencies >60 kHz are used currently because it has been observed that these frequencies produce no noticeable effects on the behaviour of seals and should be beyond the upper range of their hearing sensitivity (Wartzok *et al.*, 1992b). Very High Frequency (VHF) radio transmitters (Sirtrack 150–152 MHz) are attached simultaneously to the dorsal surface of the seal, both for in-water time budgets (see below) and to find animals that might leave the study area. An array of three or four omnidirectional hydrophones (Vemco VH65, 50–80 kHz) each connected to an Ultrasonic Acoustic Receiver (Vemco VR20) is placed around the area where the density of seals is highest (Figure 6). Hydrophones are placed through 80 mm diameter holes drilled in the sea-ice (thickness 1.4–2.2 m), or alternatively, for studies in ice-free areas, they can be connected to moored buoys to float on open water (see below). The cables to the hydrophones are suspended inside three-metre-long PVC pipes to hold the hydrophone in place and also to protect the cable and prevent it from being frozen permanently into the ice. Each hydrophone must be held securely at approximately the same depth and well clear of the undersurface of the ice to avoid echoes and reverberations. A four-part, diamond-shaped array gives a larger area of maximum precision, although a three-part triangular array still provides an area of coverage of >200,000 m^2. The distance between receivers is critical to the area to be tracked; typically, a distance of 400–500 m. Each receiver is connected via communication cables (length 308 m) to a central base station. Radio links between buoys are also possible.

Locations of each animal can then be determined using custom software, such as POSITION V3.07© (Vemco Ltd., Canada) which calculates the X and Y coordinates of the transmitter based on differences in the arrival time of pulses at each of the three hydrophones, and determines depth (± 1 m) from the modulating changes in pulse interval with depth (Figure 6). Within the area under the array, the X, Y and Z values are accurate to approximately ± 1–2 m^3. Outside the array area precision is lower, but measurement for animals up to 1 km from the area can be made with satisfactory precision. Transmitters have a custom depth range, and for deep diving Weddell seals (Kooyman, 1981), the maximum possible of 0–680 m was chosen.

The procedure for calculating locations requires the central base-station to interrogate each of the buoys in turn, and each interrogation takes a number of seconds, as several transmitted pulses are required to calculate a position. This equates to a sampling rate of one location every 15 or 30 seconds, depending on the number of animals being recorded. For fine-scale movements, minimum sampling intervals (e.g., a rule-of-thumb is a minimum of 20 sampling points per dive) are required to prevent artefacts arising in the dive profile as a result of an infrequent sampling regime (Boyd, 1993; Wilson *et al.*, 1995; Hindell and Lea, 1998). In this case, detailed analysis is restricted to dives exceeding five minutes in duration.

The data output collected using POSITION consists of a number of files containing all locations for all animals at each period sampled. Erroneous positions are occasionally recorded, presumably due to echoes and reverberations in the acoustic signals. Therefore, data are always filtered to remove consecutive X, Y and Z positions that require a swim velocity >4 ms^{-1}, twice the modal velocity of adult Weddell seals (Castellini *et al.*, 1992b, Davis *et al.*, 1999, Harcourt *et al.*, 2000).

It is only possible to display TDR dive profiles in two dimensions. However, for three-dimensional profiles several perspectives are needed to view the nuances of each dive. Figure 7 illustrates two dives by a female Weddell seal with the three-dimensional dive rotated three times through 90° in order to see the dive from several angles. In Figure 7(a), a female 'moving' dive, the female dives rapidly to approximately 150 m and then swims more slowly in a large loop before returning rapidly to the surface. Her entry and exit holes in the ice are in different positions. In Figure 7(b), a female 'flat' dive, the same female again swims rapidly to approximately 150 m, but there her velocity falls to zero ms^{-1}. After spending considerable time (approximately 9 minutes of a 19-minute dive) waiting in this place, the female returns rapidly to the surface. In this dive, the entry and exit holes are identical. From our knowledge of the bathymetry under the array, it is known that the females were close to, if not on, the bottom for most of these dives. As females may be diving to feed (Testa *et al.*, 1989; Hindell *et al.*, 1999) these two dives may illustrate alternate foraging strategies. Moving dives (Figure 7(a)) may indicate a seal actively searching for sedentary prey, while flat dives (Figure 7(b)) may illustrate a 'sit and wait' strategy as would be expected when the animal is preying on sparse, active prey (Thompson *et al.*, 1993). As has been shown recently with diving male Weddell seals (Harcourt *et al.* 2000), two-dimensional TDR profiles of these two dives would indicate that they were comparable; it is only when the third dimension is added that they become so distinct.

This illustrates that, for a single animal, three-dimensional dive profiles can enhance our understanding of foraging tactics of marine predators. There are limitations in the system given the large-scale movements by most marine predators, as described elsewhere in this chapter. It may also be possible to apply this technology to other seal species not restricted to the fast ice. Identifying areas of preferred foraging by other means (e.g., satellite, geolocation), and then establishing a floating acoustic array on the ocean surface above these areas could provide finer-scale details on the foraging habits of many fur seal species. It may even be possible to track individuals of seal prey species simultaneously to provide concurrent distributional and behavioural data on both the predators and their prey. Given the fine-scale (10–100 m) at which patchiness in the distribution of many prey species occurs (Zamon *et al.*, 1996), the insights offered by this type of system in appropriate environments will enhance our understanding of behaviour at broader scales. Interpreting dive behaviour becomes simultaneously more complex and more realistic when the entire three-dimensional path is apparent.

RADIO TRACKING

VERY HIGH FREQUENCY (VHF) TELEMETRY

Very High Frequency (VHF) radio tags have been employed for many years to monitor movements of wild animals ranging in size from large arthropods (e.g., giant weta, Sirtrack, New Zealand) to the great whales (Goodyear, 1993). VHF transmitters work by emitting a series of pulsed signals at very high frequencies (30–240 MHz) via a transmitting antenna. These signals are received by another, directionally sensitive antenna, which is connected to a receiver within line-of-sight of the transmitter (i.e., there must be a clear path between the transmitter and receiver) (McDonald and Amlaner, 1980). The directional properties of the receiving antenna allow the location of the animal to be calculated by taking bearings on the signal from at least two places and noting where the bearings intersect (i.e., 'triangulation'). Yagi, loop or Adcock antennae are required for taking bearings (see Kenward, 1987).

VHF signals do not transmit through the air-water interface, so locating marine animals can only occur when the animal is ashore or has breached the surface. Errors arise in four main ways: (1) system error (antennae flex and twist in the wind, especially when mounted on aircraft or other vehicles); (2) animal movement while bearings are being taken (a particularly important problem for triangulation of marine animals that breach the surface only briefly); (3) topographical error arising from reflection and refraction of the signal creating false bearings (McDonald and Amlaner, 1980); and (4) determining direction depends on the human ear or some other measure of signal strength, and these are often not very accurate over distances of several hundred metres.

In Australia, wildlife transmitters operate on frequencies between 150–152 MHz. Transmitters for diving animals such as seals usually have the batteries and electronic parts embedded in an epoxy resin to prevent leakage and implosion due to pressure at depth, and have an attached whip antenna (plastic-coated, stainless-steel cable approximately 25–30 cm long). For these sealed units, an external, magnet-reed switch activates and de-activates the transmitters. For most purposes, VHF units transmit at a fixed rate, and the current drain and hence the life of the transmitter is a function of battery size and the pulse rate. For example, with a pulse rate of 40 pulses per minute (ppm) the average current drain may be 30 microamps, whereas with a pulse rate of 60 ppm, the current drain would be approximately 45 microamps. Since seals are large animals, the current drain is not usually a problem as large, long-life batteries can be packaged readily with the transmitter and still not approach the 'rule-of-thumb' for wildlife tracking that the package not exceed 2–3% of the animal's weight (Kenward, 1987). For example, using VHF transmitters with a pulse rate of 40 ppm and a battery life of 9.6 months requires a minimum weight of 40 g. These units have been attached to fur seal females weighing 30–90 kg (i.e., 0.04–0.13% of body weight).

The maximum range for a signal is a function of the power of the transmitter, whether it is a one- or two-stage transmitter, and a number of other factors including the length and position of the transmitter antenna, the receiver sensitivity, the receiver antenna gain, the height above the ground or sea of both the transmitter and the receiving system, the humidity and the topography (Kenward, 1987). A quarter-wavelength antenna with a good Earth plane is the optimal solution for maximizing the tracking range of a transmitter, although it is not always realized. VHF operates on 'line-of-sight', although VHF waves can be detected slightly over the horizon, by a factor of 1.4 in

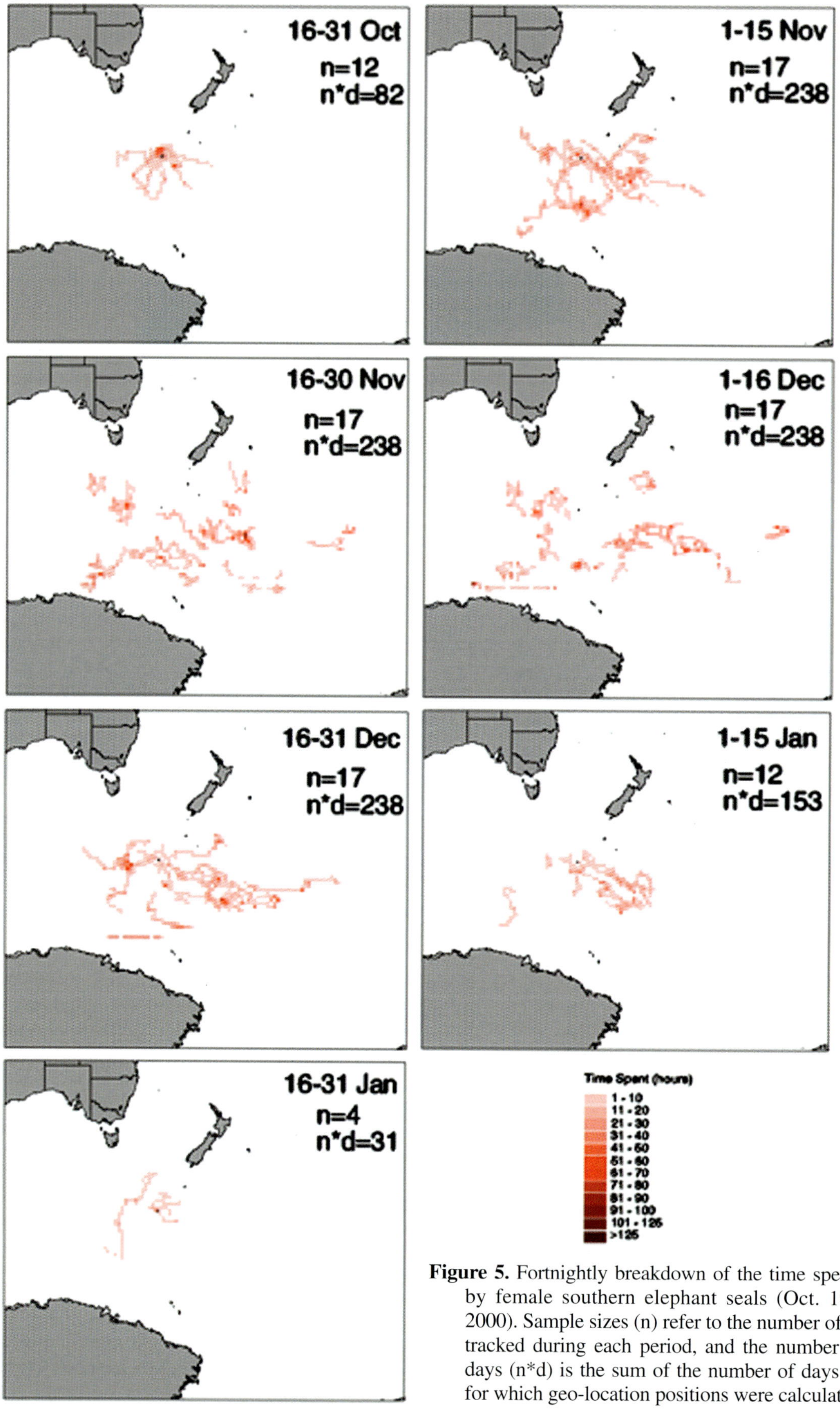

Figure 5. Fortnightly breakdown of the time spent at sea by female southern elephant seals (Oct. 1999–Jan 2000). Sample sizes (n) refer to the number of females tracked during each period, and the number of seal-days (n*d) is the sum of the number of days per seal for which geo-location positions were calculated.

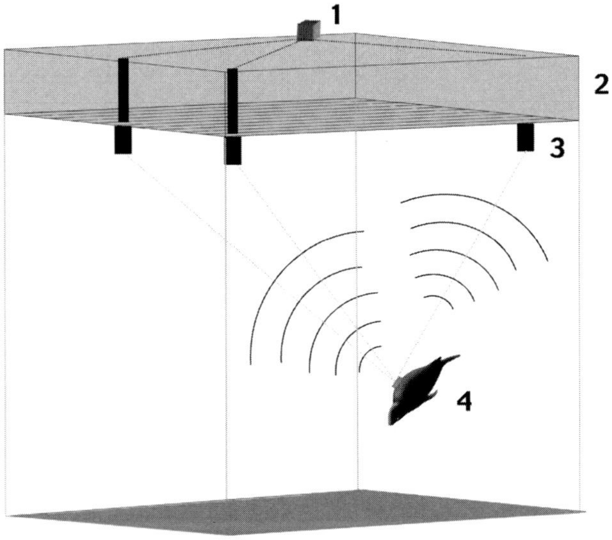

Figure 6. Schematic representation of an acoustic hydrophone array embedded within the sea ice. The Ultrasonic Acoustic Receiver (1), sea ice (2), omnidirectional hydrophones (3) and seal with acoustic transmitter attached (4) are shown.

ideal propagation conditions (Kenward, 1987). Thus, there are severe limitations in tracking marine mammals from land-based stations. At sea level the effective range is only 5–20 km, and even with an elevation of 100 m this increases to only about 33 km (Thompson, 1993).

USING VHF TELEMETRY TO TRACK SEALS

Early attempts to determine where fur seals and other otariids forage at sea included aircraft flying transects over areas where seals equipped with VHF transmitters were foraging (see Antonelis *et al.*, 1990; Loughlin *et al.*, 1987; Goebel *et al.*, 1991). This method is effective, but extremely expensive, allows only for the tracking of a few animals at any one time, and is constrained by the range of the aircraft. With the advent of effective satellite telemetry, except in particular circumstances (e.g., fjords and other semi-enclosed areas), VHF transmitters are rarely used today to determine foraging areas of seals; but they are still used widely for measuring the presence of instrumented individuals at breeding colonies.

An example of the value of VHF telemetry in current research is the measurement of foraging trip duration. Otariid seals alternate time at sea feeding with time ashore suckling their pups, so the proportion of the time spent foraging has provided a surrogate measure of foraging effort (Gentry and Kooyman, 1986a; Boyd *et al.*, 1991; Gales and Mattlin, 1997; Goldsworthy *et al.*, 1997; Gentry, 1998). An automated receiver with an omnidirectional whip antenna that scans a range of frequencies and then archives the data measures the time ashore relative to the time at sea for large numbers of females. Due to the line-of-sight limitation for VHF transmitters, deploying medium power transmitters (with a range of 2–3 km) and a strategically-placed receiver at a colony allows accurate records to be obtained with relative ease. It is important when setting up a system such as the one described that the signal strength be checked throughout the colony area because shadow effects from large objects (e.g., boulders, outcroppings, etc.) can mask signals and thereby produce misleading attendance patterns.

Similar methods have been used to determine haul-out patterns and activity budgets of phocid seals. Siniff *et al.* (1969) first employed VHF transmitters in the 1960s to record the proportion of time individual Weddell seals spent hauled out on the ice during the breeding season. A similar system has been used with success on breeding Weddell seals (Harcourt, Hindell, Waas, Bell, unpublished data). VHF transmitters do have a tendency to drift in frequency when used in severe cold such as that found in Antarctica, so caution is required when establishing an automated receiving station.

Finally, VHF transmitters are of immense worth when animals need to be found again after a period at sea. Therefore, VHF transmitters are most often used in combination with other archival tags such as TDRs so that the devices can be recovered when animals come ashore in large colonies or to inaccessible areas (Gales and Mattlin, 1997). Under these circumstances, they act as an effective and inexpensive form of insurance.

OTHER TECHNIQUES

GLOBAL POSITIONING SYSTEM (GPS)

The Global Positioning System is a worldwide, satellite-based, radio-navigation system developed by the U.S. Department of Defence (DOD) (Anon., 1996; Dana, 1999). The GPS includes a constellation of 21 operating and three spare satellites that orbit the earth. A ground-based GPS receiver calculates the time it takes for individual signals to arrive from at least three satellites to the receiver to compute a two-dimensional, horizontal fix (latitude and longitude) given an assumed height (with marine mammals assumed to be sea level). The acquisition (i.e., the detection of satellite signals) of four satellites can determine three-dimensional positions and time, while five or more can provide position, time, redundancy and the certainty of the greater position fix (Dana, 1999). There are two services available in the GPS, the Precise-Positioning Service (PPS), and the Standard-Positioning Service (SPS). The PPS is a highly accurate military positioning, velocity, and timing service that is under the control of the DOD and unavailable to the public. The SPS is a positioning service available to all GPS users on a continuous, worldwide basis with no direct charge. The SPS was, however, degraded intentionally by the DOD by use of Selective Availability (time varying bias). Most receivers were capable of using the SPS signal and its predictable

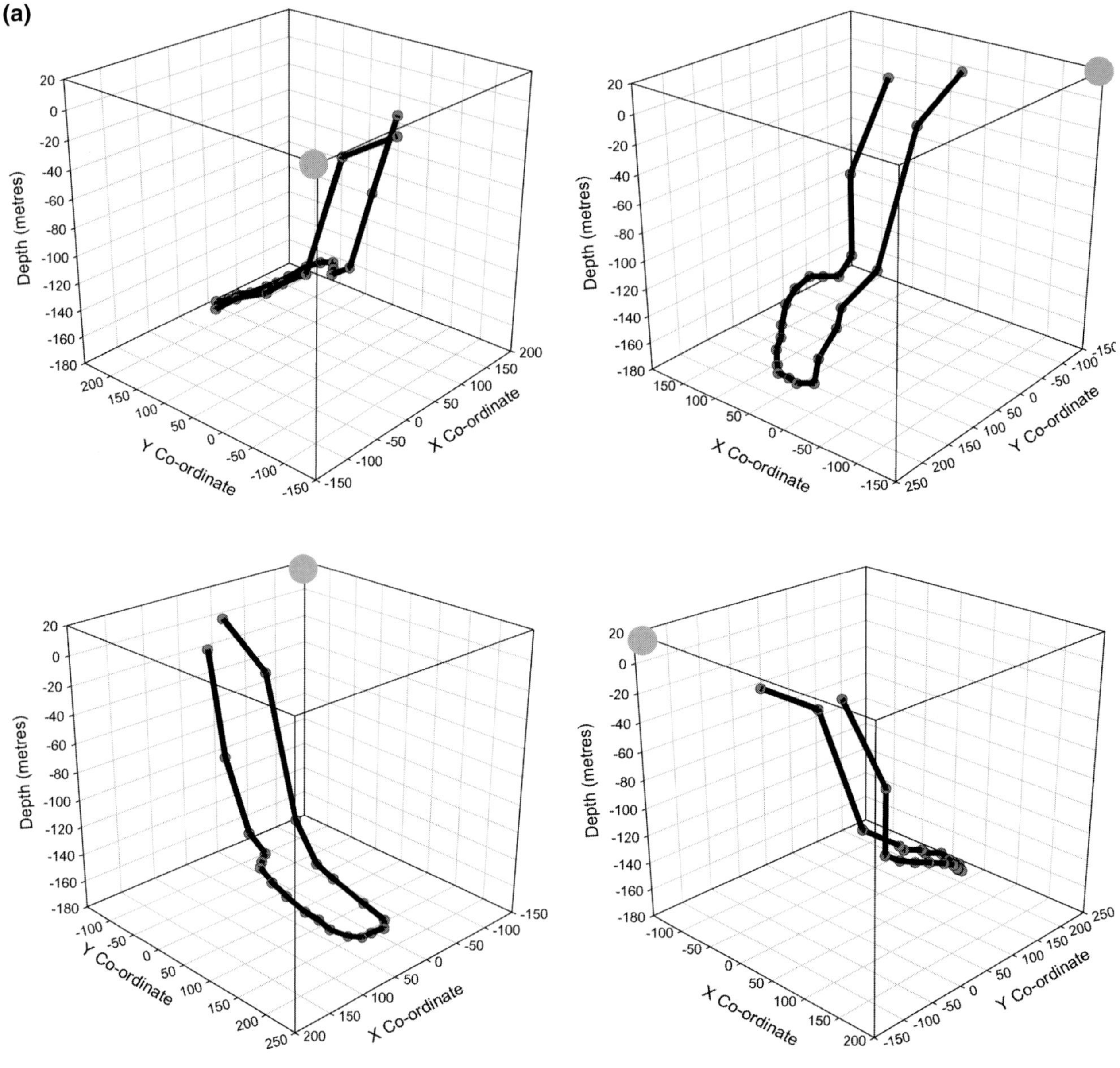

Figure 7. Multiple, three-dimensional views of two types of female Weddell seal dives (during the lactation): **(a)** a female 'moving' dive; and **(b)** a female 'flat' dive.

accuracy was 100 m (horizontal), 156 m (vertical), and 340 nanoseconds (time). In May 2000, the DOD lifted the restrictions on the PPS service and its predictable accuracy is now 22 m (horizontal), 27.7 m (vertical), and 22 nanoseconds (time) (Anon., 2000).

GPS as a method for studying seals is limited by the cost, size and speed of satellite acquisition by the commercially available devices (Lotek, 1996). Most GPS devices in the domain of biotelemetry were developed for large terrestrial mammals since these taxa could carry the excessive bulk of the units. Consequently, GPS units were impractical for all but the largest marine mammals due to drag and weight. However, a far more important limitation, has been the time that it took GPS devices to lock onto (i.e., acquire) the necessary number of satellites to predict accurately a location (e.g., up to ten minutes). Smaller (~6 g), faster units have now been produced and these, with further improvements in battery efficiency and reduced bulk, could be attached to a seal without compromising swimming efficiency (e.g., GXB2000 GPS receiver, Sony Corporation). Acquisition times can now be as short as two seconds, which reduces the constraints imposed by short between-dive intervals at the surface. However, acquisition times of this interval represent only the best possible scenarios; more realistic acquisition times range from 20–40 seconds. The satellites also send regular updates on the satellite orbits and positions (ephemeris data) and a receiver must update these stored

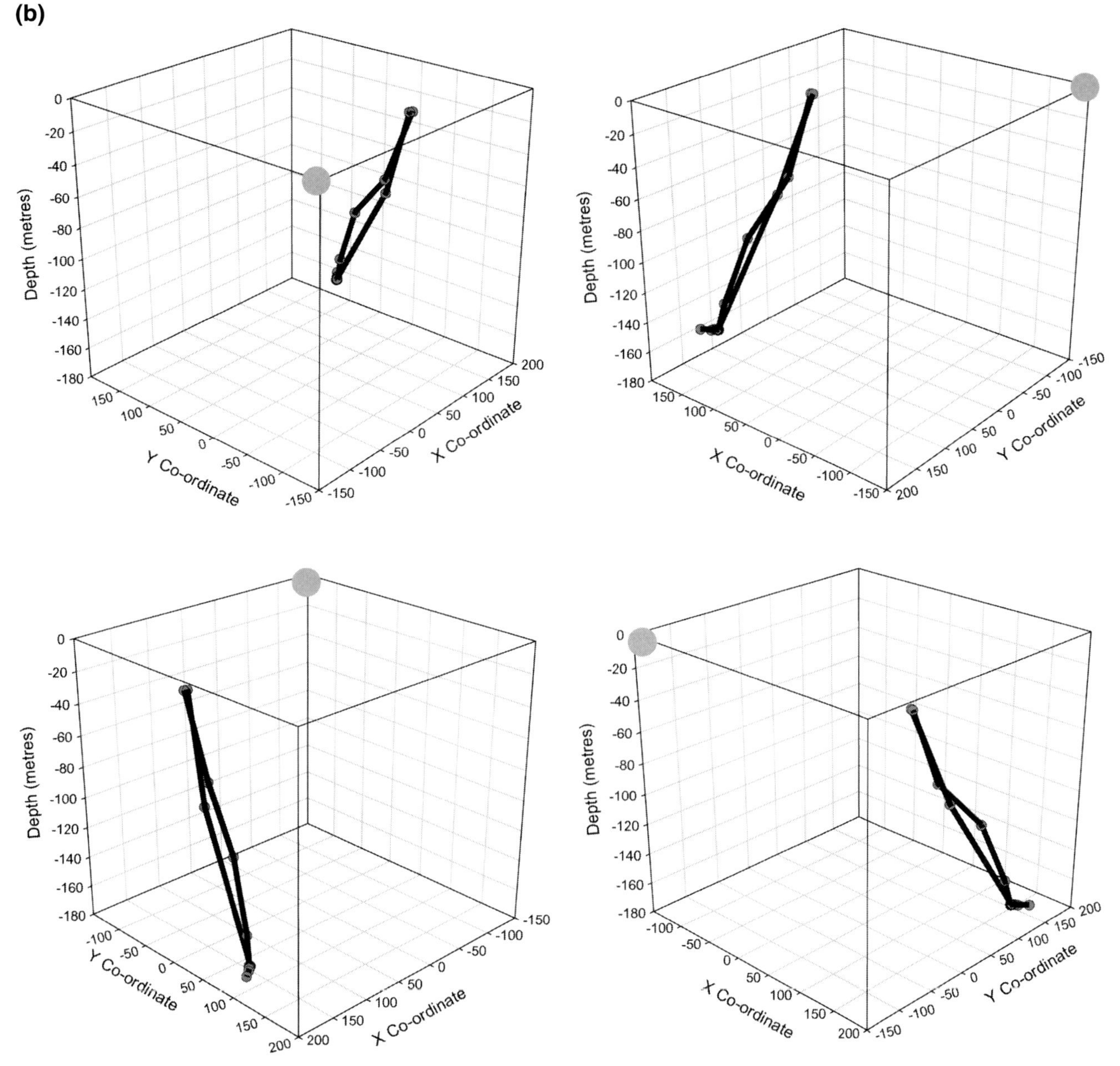

signals approximately every four hours at the most (Dana, 1999). This adds to the time required to acquire all the necessary data to provide an accurate location.

However, the improvements in acquisition time and the associated reduction in power demand combined with lower costs (Table 2) and greater accuracy (<100 m) compared to those of PTTs (~500 m) make GPS devices one of the best candidates for future studies of seal movements. Apart from power drain, the greatest drawback is the need to recover the device in order to download the archived data. This could be avoided by using a GPS archival tag combined with a PTT or, more economically, a VHF transmitter, to download remotely the data upon the seal's return to land or ice. GPS will be able to answer the same questions as PTTs, including location and rate of travel, but with far greater accuracy, far more locations per day and at much less expense (Table 2). Of course, the main limitation of GPS units on some pinniped species would be the recovery of the units themselves. Pack-ice seals such as crabeater, leopard and Ross seals are not easily re-captured after the deployment of units. Therefore, relying on archival tags for location data in these species is not advised.

CONDUCTIVITY-TEMPERATURE-DEPTH (CTD) ARCHIVAL TAGS

To advance both the precision of models combining oceanographic variables and biological activity as well as our basic understanding of ocean structure, there is an increasing emphasis on precise measurement of oceanographic variables while at sea (Ancel *et al.*, 1992; McCafferty *et al.*, 1999). Therefore, more sensitive and complex communications equipment is needed. Another type of archival data tag is the miniature Conductivity-Temperature-Depth (CTD) sensor (Figure 8) analogous to the larger oceanographic CTDs available for oceanographic sampling. These are miniature archival tags designed mostly for deployment on fish species, but have been deployed on both larger and smaller taxa (ICES Anadromous and Catadromous Fish Committee, 1997; Freire and González-Gurriarán, 1998; Sturlaugsson and Johannsson, 1998). The available parameters for measurement include temperature, light, pressure (depth), salinity (conductivity) and tilt angle.

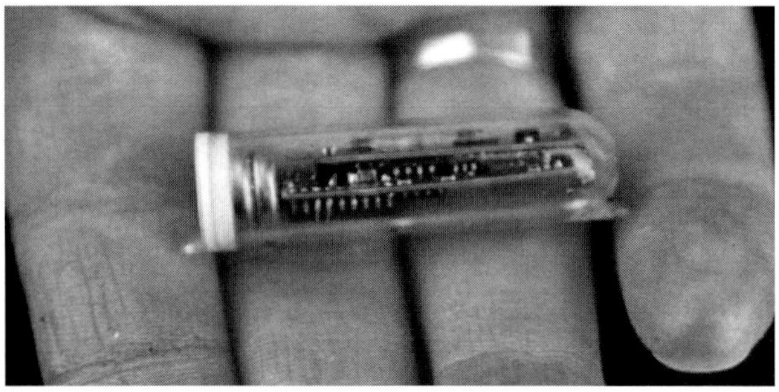

Figure 8. A mini Conductivity-Temperature-Depth (CTD) archival tag (Data Storage Tag, Star-Oddi Marine Device Manufacturing, Reykjavik, Iceland). Reproduced with permission from Star-Oddi.

ROUTE RECORDERS

Another type of device, known as a 'route' or 'direction' recorder, was developed to track homing pigeons (Bramanti *et al.*, 1988; Papi *et al.*, 1991). However, it has now been applied to sea birds (Dall'Antonia *et al.*, 1995; Benvenuti *et al.*, 1998), green turtles (Luschi *et al.*, 1996), and fur seals (Georges *et al.*, 2000). These devices detect and archive the headings of an animal during its movement path by measuring the angle between the horizontal component of the Earth's magnetic field and the instrument's main axis at regular intervals. Prototypes were equipped with a compass and a transducer that converted the angular values to electrical resistance values (Bramanti *et al.*, 1988; Dall'Antonia *et al.*, 1992); but more recent route recorders have been improved with the addition of a microprocessor to control other components (Benvenuti *et al.*, 2001). With further technological developments, these devices may provide more fine-scale information on pinniped movements unavailable from the other methods described in this chapter.

FUTURE DEVELOPMENT OF ARCHIVAL TAGS

The real advantage in the improvements in archival tag technology is the precision with which one can measure a series of oceanographic variables (McCafferty *et al.*, 1999). There is growing interest in the use of marine mammals as sampling units for physical oceanography data collection (see Ancel *et al.*, 1992; Wilson *et al.*, 1994; Weimerskirch *et al.*, 1995). Of primary importance to both biologists and to oceanographers contemplating using marine animals for this role is the precise measurement of temperature (McCafferty *et al.*, 1999). Sensors now available can measure temperature accurately to 0.001°C within fractions of a second. The data are useful to describe the temperature gradients in which animals select to forage and the physical data are sufficiently precise to model oceanographic trends in time and space.

The real test for scientists will be the development of multi-platform measurement devices that satisfy the requirements of both biologists and physico-chemical oceanographers. Individual ecological laboratories are now in the process of developing such systems to attract the interest of oceanographers to invest in the continued development of these systems and their applications.

CONCLUSIONS AND MANAGEMENT CONSIDERATIONS

In this chapter, we have described a variety of techniques available for the remote study of pinniped movement during the essentially clandestine phase of their life history—foraging at sea. Within the last 20 years, there has been impressive development in technology allowing for the collection of data that now begin to describe the nuances of these fascinating predators. Of course, many of the methods described herein are not restricted to the study of marine mammals such as pinnipeds. In fact, most of the methods were developed for deployment on terrestrial organisms. However, the most exciting aspect to the understanding of at-sea behaviour of marine mammals is that we are now beginning to understand the complexities of their behaviour. In addition, we now have the means to begin the immense task of collecting information that can unlock the biophysical secrets of the oceans. Applying archival measuring devices to organisms that must live and feed within these vast regions is a promising means to achieve these goals.

The application of a particular method of remotely tracking seals depends on the research questions to be addressed for each specific study. The temporal or spatial scale of concern, the target species, and the domain of research (behaviour, population dynamics, management, etc.) will dictate to what extent the specific advantages and disadvantages of each technique affect a researcher's choice. We have summarized in Table 3 the type of information one can expect to collect for each of the main methods described in this chapter. The main consideration appears to be the spatial scale at which one intends to investigate behaviour. Fine-scale movements can be achieved by acoustic arrays and perhaps GPS satellite telemetry, whereas more broad-scale questions should be addressed using PTT, geo-location or even VHF telemetry. Of course, with continued advances in technology the options available to researchers will undoubtedly increase.

One of the most practical management applications for telemetric data of seal distribution, foraging effort and at-sea behaviour is the delineation of regions of potential conflict between these predators and human industry. Domestic and international policies are evolving toward the recognition of sustainable practices in fisheries, with the need for increased information on ecosystem relationships among constituent species (Staples, 1996; Caton and McLoughlin, 2000). Recognizing that single-species management fails to embrace a realistic perspective of ecological relationships, management is promoting the concept of multi-species models to account for predator-prey relationships, by-catch issues and complex interactions within marine food webs (National Research Council, 1999). Fisheries managers must pursue the objectives of long-term, ecological sustainability and the reduction of impacts on non-target species (Fisheries Management Act, 1991). To achieve these objectives it is essential to have reliable data on the distribution, abundance, consumption rates, growth, biomass and behaviour of all species involved (National Research Council, 1999). Establishing areas of spatial and dietary overlap between pinnipeds and Australasian commercial fishing industries is the first step in prescribing management restrictions or recommendations to achieve the sustainability of marine living resource use. Currently, multi-species management is in its infancy within the fisheries sector, but more information is required to develop the necessary models.

The example of Australian fur seals tracked within and around Bass Strait has shown that there may be some overlap with breeding females and many commercial fishing zones (Figure 1), although more data are required to elucidate the amount of overlap and possible competition. With fewer commercial fisheries operating in regions where southern elephant seals forage, the potential for overlap is less, although the spatial scale at which one assesses the degree of competition must concur with how animals exploit this area of ocean. In addition to assessing diet composition, distribution and spatial use, spatially-allocated indices of foraging (i.e., dive behaviour data from TDRs) are required to delineate the more important regions of the Southern Ocean required to maintain healthy populations of these marine predators. Of course, seasonal and annual changes in ocean resources must be assessed continually to update biophysical models.

Another practical application of telemetric data can be the establishment of marine reserves (Clark and Dingwall, 1985; Cooper and Ryan, 1994; Scott, 1994; Dingwall, 1995; Shaughnessy, 1999). Without a detailed understanding of the biological communities that are supported in highly-productive areas of the ocean, it is impossible to isolate and protect sensitive regions. Pinnipeds are important higher predators within a number of marine ecosystems. Determining where their important feeding areas are located contributes to our knowledge of the importance of these regions as areas of high biological productivity (Croxall and Wace, 1995). It makes sense that pinnipeds must locate and exploit areas of high productivity to survive; therefore, protecting such areas from over-exploitation by human industry will contribute to the protection of entire biological communities. The literature available on this type of approach is vast and does not fall within the context of this chapter. However, improving methods required to track and monitor higher predators such as pinnipeds is an important step in the conservation, management and theoretical understanding of ocean life.

Table 3. Movement and dive parameters available from satellite telemetry, geo-locating TDRs, acoustic tracking and VHF transmitters.

Parameter	PTT	Geo-locating TDRs	Acoustic	VHF
Horizontal Distance				
–horizontal distance	yes	broad-scale	fine-scale	yes
–bearing	broad-scale	broad-scale	fine-scale	no
–large-area coverage	yes	yes	no	no
Diving				
–start-end distance of dive	only with additional sensors	no	yes	no
–depth	only with additional sensors	yes	yes	no
–dive duration	only with additional sensors	yes	yes	yes
–distance during dive	only with velocity meter	only with velocity meter	yes	no
–shape of dive	limited	limited	yes	no
–ascent/descent rate	partial	yes	yes	no
–directional changes	broad-scale	broad-scale	within dives	between dives
–velocity of dive phases	partial	only with velocity meter	yes	no
–conspecific interactions	broad-scale	broad-scale	yes	no

ACKNOWLEDGEMENTS

Funding for the research projects mentioned in this chapter was provided by the following: Antarctica New Zealand, Antarctic Science Advisory Committee (ASAC), Australian Geographic Society, Australian Research Council (ARC), Department of Natural Resources (Victoria), Linnean Society of New South Wales, Macquarie University, Natural Sciences and Engineering Research Council of Canada (NSERC), PADI Foundation, Parks Victoria, P. Rowsthorn, Sea World Research and Rescue Foundation Inc. (SWRRI), University of Tasmania and Waikato University. We thank members of the 52nd and 53rd Australian National Antarctic Research Expeditions (ANARE), S. Allen, P. Barker, D. Bell, M. Biuw, H. Burton, M. Chambers, T. Dorr, G. Enbom, I. Field, J. Harrington, A. Irvine, S. Jadhav, D. Kamien, C. McKinley, C. McMahon, E. Meiling, K. Michael, P. Mitchell, T. Mitchell, A. Morrissey, C. Nave, B. Priest, Q. Smith, D. Thompson, M. Tierney, E. Turner, J. Tyler, J. Tripovich, J. Waas, D. Watts, A. Welling, K. Wheatley, K. Willis, S. Wilson and S. Winter for analytical, logistical or field support. Thanks to J. Arnould, P. Dann, N. Gales, R. Kirkwood, M. Lynch, S. McCarrey, D. Pemberton for providing unpublished data, and to F. Bonadonna, M.-A. Lea, and T. Potts who provided technical information for specific sections of the chapter. Star-Oddi Marine Device Manufacturing (Reykjavik, Iceland) provided the image of the mini CTD archival tag. Many thanks to R. Kirkwood and an anonymous reviewer for detailed comments on the manuscript.

REFERENCES

Ancel, A., Kooyman, G.L., Ponganis, P.J., Gendner, J-P., Lignon, J., Mester, X., Huin, N., Thorson, P.H., Robisson, P. and Le Maho, Y. 1992. Foraging behaviour of emperor penguins as a resource detector in winter and summer. *Nature,* Vol. 360, pp.336–339.

Anon. 1996. *Fact Sheet: U.S. Global Positioning System Policy.* Office of Science and Technology Policy, National Security Council, Washington, D.C.

Anon. 1998. *Scalefish Fishery Policy Document.* Department of Primary Industry and Fisheries, Tasmania.

Anon. 1999. *Australian Fisheries Management Authority Annual Report 1998–99.* Canberra.

Anon. 2000. *President Clinton: Improving the Civilian Global Positioning System (GPS).* Press Release, Office of the Press Secretary, The White House, Washington, D.C.

Ansorge, I.J., Froneman, P.W., Pakhomov, E.A., Lutjeharms, J.R.E., Perissinotto, R. and van Ballegooyen, R.C. 1999. Physical-biological coupling in the waters surrounding the Prince Edward Islands (Southern Ocean). *Polar Biology,* Vol. 21, pp.135–145.

Antarctic Treaty (Environment Protection) Act. 1980. *Schedule 1. Convention for the conservation of Antarctic seals.* Act 103 of 1980 as amended (18 October 1999). Prepared by the Office of Legislative Drafting, Attorney-General's Department, Canberra. pp.41–51.

Antonelis, G.A., Stewart, B.S. and Perryman, W.F. 1990. Foraging characteristics of female northern fur seals (*Callorhinus ursinus*) and California sea lions (*Zalophus californianus*). *Canadian Journal of Zoology,* Vol. 68, pp.150–158.

Argos, 1989. *User's manual for the Argos System.* Argos CLS, 176, Toulouse.

Arnould, J.P.Y and Littnan, C.L. 2000. Pup production and breeding areas of Australian fur seals (*Arctocephalus pusillus doriferus*) at Kanowna Island and The Skerries in northeastern Bass Strait. *Australian Mammalogy,* Vol. 22, pp.51–54.

Bartsch, S.S., Johnston, S.D. and Siniff, D.B. 1992. Territorial behaviour and breeding frequency of male Weddell seals (*Leptonychotes weddelli*) in relation to age, size, and concentrations of serum testosterone and cortisol. *Canadian Journal of Zoology,* Vol. 70, pp.680–692.

Begon, M., Harper, J.L. and Townsend, C.R. 1986. *Ecology. Individuals, Populations, and Communities.* Sinauer Associates, Sunderland, Massachusetts. 876pp.

Bengston, J.L. and Stewart, B.S. 1992. Diving and haulout behavior of crabeater seals in the Weddell Sea, Antarctica, during March 1986. *Polar Biology,* Vol. 12, pp.635–644.

Benvenuti, S., Bonadonna, F., Dall'Antonia, L. and Gudmundsson, G.A. 1998. Foraging patterns of breeding thick-billed murres (*Uria Lomvia*) in Iceland, as revealed by a direction recorder. *Auk,* Vol. 115, pp.57–66.

Benvenuti, S., Bonadonna, F., Dall'Antonia, L. and Ribolini, A. 2001. Homing and foraging flights tracked by bird-born data loggers. *Proceedings of Fifth European Conference on Wildlife Telemetry. 25–30 August 1996.* Strasbourg, France.

Best, P.B. and Shaughnessy, P.D. 1979. An independent account of Captain Benjamin Morrell's sealing voyage to the south-west coast of Africa and the Antarctic, 1828/29. *Fisheries Bulletin of South Africa,* Vol. 12, pp.1–19.

Bester, M.N. 1980. Population increase in the Amsterdam Island fur seal *Arctocephalus tropicalis* at Gough Island. *South African Journal of Zoology,* Vol. 15, pp.229–234.

Bester, M.N. 1990. Population trends of subantarctic fur seals and southern elephant seals at Gough Island. *South African Journal of Antarctic Research,* Vol. 20, pp.9–12.

Bester, M.N. and Pansegrouw, H.M. 1991. Ranging behaviour of southern elephant seal cows from Marion Island. *South African Journal of Science,* Vol. 88, pp.574–575.

Boness, D.J., Bowen, W.D. and Oftedal, O.T. 1994. Evidence of a maternal foraging cycle resembling that of otariid seals in a small phocid, the harbor seal. *Behavioral Ecology and Sociobiology,* Vol. 34, pp.95–104.

Bonner, W.N. 1982. *Seals and Man: a Study of Interactions.* University of Washington Press, Seattle. 170pp.

Bonner, W.N. and Laws, R.M. 1964. Seals and sealing. Pp.163–190. In Priestly, R.E., Adie, R.J. and Robin, G.D.Q. (eds). *Antarctic Research: a Review of British Scientific Achievement in Antarctica.* Butterworths, London. 360pp.

Bornemann, H., Kreyscher, M., Ramdohr, S., Martin, T., Carlini, A., Sellmann, L. and Plötz, J. 2000. Southern elephant seal movements and Antarctic sea ice. *Antarctic Science,* Vol. 12, pp.3–15.

Boyd, I.L. 1993. Selecting sampling frequency for measuring diving behaviour. *Marine Mammal Science,* Vol. 9, pp.424–430.

Boyd, I.L. and Arnbom, T. 1991. Diving behaviour in relation to water temperature in the southern elephant seal: foraging implications. *Polar Biology*, Vol. 11, pp.259–266.

Boyd, I.L. and Croxall, J.P. 1996. Dive durations in pinnipeds and seabirds. *Canadian Journal of Zoology*, Vol. 74, pp.1696–1705.

Boyd, I.L., Lunn, N.J. and Barton, T. 1991. Time budgets and foraging characteristics of lactating Antarctic fur seals. *Journal of Animal Ecology*, Vol. 60, pp.577–592.

Boyd, I.L., McCafferty, D.J. and Walker, T.R. 1997. Variation in forage effort by lactating Antarctic fur seals: response to simulated increased foraging costs. *Behavioral Ecology and Sociobiology*, Vol. 40, pp.135–144.

Boyd, I.L., Reid, K. and Bevan, R.M. 1995. Swimming speed and allocation of time during the dive cycle in Antarctic fur seals. *Animal Behaviour*, Vol. 50, pp.769–784.

Bradshaw, C.J.A., Davis, L.S., Purvis, M., Zhou, Q. and Benwell, G.L. 2002. Using artificial neural networks to model the suitability of coastline for breeding by New Zealand fur seals. *Ecological Modelling*, Vol. 148, pp.111–131.

Bradshaw, C.J.A., Lalas, C. and Thompson, C.M. 2000. Clustering of colonies in an expanding population of New Zealand fur seals (*Arctocephalus forsteri*). *Journal of Zoology, London*, Vol. 250, pp.105–112.

Bramanti, M., Dall'Antonia, P. and Papi, F. 1988. A new technique to monitor the flight paths of birds. *Journal of Experimental Biology*, Vol. 134, pp.467–472.

Burton, H.R. 1998. Is survivorship of southern elephant and Weddell seal pups related to the Antarctic Circumpolar Wave? *SCAR VII International Biological Symposium. Christchurch, New Zealand, 1–5 Sep. 1998.* (Poster).

Castellini, M.A., Davis, R.W. and Kooyman, G.L. 1992a. Annual cycles of diving behavior and ecology of the Weddell seal. *Bulletin of the Scripps Institute of Oceanography*, Vol. 28, pp.1–54.

Castellini, M.A., Kooyman, G.L. and Ponganis, P.J. 1992b. Metabolic rates of freely diving Weddell seals: correlations with oxygen stores, swim velocity and diving duration. *Journal of Experimental Biology*, Vol. 165, pp.181–194.

Caton, A. and McLoughlin, K. (eds). 2000. *Fishery Status Reports 1999: Resource Assessments of Australian Commonwealth Fisheries*. Bureau of Rural Sciences, Canberra.

Chapman, D.C. 1981. Evaluation of marine mammal population models. Pp.277–296. In Fowler, C.W. and Smith, T.D. (eds). *Dynamics of Large Mammal Populations*. John Wiley and Sons, New York. 477pp.

Clark, M.R. and Dingwall, P.R. 1985. *Conservation of Islands in the Southern Ocean*. International Union for the Conservation of Nature and Natural Resources, Gland, Switzerland.

Cooper, J. and Ryan, P.G. 1994. *Management plan for the Gough Island Wildlife Reserve*. Report for the World Wide Fund for Nature, Government of Tristan da Cunha, Edinburgh (Tristan da Cunha).

Croxall, J.P. and Wace, N. 1995. Interactions between marine and terrestrial ecosystems. Pp.115–120. In Dingwall, P.R. (ed). *Progress in Conservation of the Subantarctic Islands*. Proceedings of the SCAR/IUCN Workshop on Protection, Research and Management of Subantarctic Islands, Paimpont, France, 27–29 April 1992. Conservation of the Southern Polar Region No. 2. IUCN, Gland, Switzerland.

Croxall, J.P., McCann, T.S., Prince, P.A. and Rothery, P. 1988. Reproductive performance of seabirds and seals at South Georgia and Signy Island, South Orkney Islands, 1976–1987: implications for Southern Ocean monitoring studies. Pp.261–285. In Sahrhage, D. (ed). *Antarctic Ocean and Resource Availability*. Springer-Verlag, Berlin.

Culik, B.M., Bannach, R. and Wilson, R.P. 1994. External devices on penguins: how important is shape? *Marine Biology*, Vol. 118, pp.353–357.

Dall'Antonia, L., Dall'Antonia, P., Benvenuti, S., Ioalé, P., Massa, B. and Bonadonna, F. 1995. The homing behaviour of Cory's shearwaters (*Calonectris diomedea*) studied by means of a direction recorder. *Journal of Experimental Biology*, Vol. 198, pp.359–362.

Dall'Antonia, L., Dall'Antonia, P. and Ribolini, A. 1992. Flight path reconstruction of birds by a route recorder. Pp.544–549. In Mancini, P., Fioretti, S., Cristalli, C. and Bedini, R. (eds). *Biotelemetry XII, Proceedings of the Twelfth International Symposium on Biotelemetry*. Litografia Felici, Pisa, Italy.

Dana, P.H. 1999. *Global Positioning System overview*. The Geographer's Craft Project, Department of Geography, University of Colorado, Boulder.

Davidson, J. 1984. *The Prehistory of New Zealand*. Longman Paul, Auckland. 270pp.

Davis, R.W., Fuiman, L.A., Williams, T.M., Collier, S.O., Hagey, W.P., Kanatous, S.B., Kohin, S. and Horning, M. 1999. Hunting behaviour of a marine mammal beneath the Antarctic fast-ice. *Science*, Vol. 283, pp.993–996.

DeLong, R.L., Stewart, B.S. and Hill, R.D. 1992. Documenting migrations of northern elephant seals using day length. *Marine Mammal Science*, Vol. 8, pp.155–159.

Dingwall, P.R., (ed). 1995. *Progress in Conservation of the Subantarctic Islands*. Proceedings of the SCAR/IUCN Workshop on Protection, Research and Management of Subantarctic Islands, Paimpont, France, 27–29 April 1992. Conservation of the Southern Polar Region No. 2. IUCN, Gland, Switzerland.

Fea, N. and Harcourt, R. 1997. Assessing the use of faecal and regurgitate analysis as a means of determining the diet of New Zealand fur seals. Pp.143–150. In Hindell, M. and Kemper, C. (eds). *Marine Mammal Research in the Southern Hemisphere. Volume 1: Status, Ecology and Medicine*. Surrey Beatty and Sons, Sydney. 186pp.

Fedak, M.A. and McConnell, B.J. 1993. Observing seals by satellite. Open ocean behaviour of southern elephant seals. *NERC News*, April, pp.26–28.

Fisheries Management Act. 1991. Act no. 162 of 1991 as amended (14 July 1999). Prepared by the Office of Legislative Drafting, Attorney-General's Department, Canberra.

Flemons, P. and Cassis, G. 2006. The role of GIS in Systematic Conservation Planning. Pp.873–887. In Merrick, J.R., Archer, M., Hickey, G.M. and Lee, M.S.Y. (eds). *Evolution and Biogeography of Australasian Vertebrates*. Auscipub Pty. Ltd., Sydney. 942pp.

Freire, J. and González-Gurriarán, E. 1998. New approaches to the behavioural ecology of decapod crustaceans. Pp.123–372. In Lagardère, J-P., Bégout Anras, M.L. and Claireaux, G. (eds). *Advances in Invertebrates and Fish Telemetry*. Kluwer Academic Publishers, Belgium.

Gales, N.J. and Costa, D.P. 1997. The Australian sea lion: a

Philosophical Transactions of the Royal Society of London, Series B, Vol. 279, pp.67–79.

Payne, M.R. 1979. Growth in the Antarctic fur seal *Arctocephalus gazella*. *Journal of Zoology, London*, Vol. 187, pp.1–20.

Ponganis, P.J., Ponganis, E.P., Ponganis, K.V., Kooyman, G.L., Gentry, R.L. and Trillmich, F. 1990. Swimming velocities in otariids. *Canadian Journal of Zoology*, Vol. 68, pp.2105–2112.

Richards, R. 1982. *Whaling and Sealing at the Chatham Islands*. Roebuck, California. 89pp.

Richards, R. 1994. The upland seal of the Antipodes and Macquarie Islands: a historian's perspective. *Journal of the Royal Society of New Zealand*, Vol. 24, pp.289–295.

Robinson, A.C. and Dennis, T.E. 1988. The status and management of seal populations in South Australia. Pp.87–110. In Augee, M.L. (ed). *Marine Mammals of Australasia. Field Biology and Captive Management*. Royal Zoological Society of New South Wales, Mosman. 140pp.

Roux, J-P. 1987. Recolonisation processes in the subantarctic fur seal, *Arctocephalus tropicalis*, on Amsterdam Island. Pp.189–194. In Croxall, J.P. and Gentry R.L. (eds). Status, Biology, and Ecology of Fur Seals. *National Oceanic and Atmospheric Administration Technical Report NMFS 51. National Marine Fisheries Service, Cambridge, Massachusetts.*

Scott, J. 1994. *Marine Conservation at Macquarie Island. A Marine Conservation Strategy and an Account of the Marine Environment*. Parks and Wildlife Service, Tasmania, Hobart.

Shaughnessy, P.D. 1998. Continued increase in the population of Antarctic fur seals, *Arctocephalus gazella*, at Heard Island, Southern Ocean. *Marine Mammal Science*, Vol. 14, pp.384–389.

Shaughnessy, P.D. 1999. *The Action Plan for Australian Seals*. Environment Australia, Canberra. 116pp.

Shaughnessy, P.D. and Goldsworthy, S.D. 1990. Population size and breeding season of the Antarctic fur seal *Arctocephalus gazella* at Heard Island—1987/88. *Marine Mammal Science*, Vol. 6, pp.292–304.

Shaughnessy, P.D., Testa, J. and Warneke, R.M. 1995. Abundance of Australian furseal pups, *Arctocephalus pusillus doriferus*, at Seal Rocks, Victoria, 1991–92 from Petersen and Bayesian estimators. *Wildlife Research*, Vol. 22, pp.625–632.

Siniff, D.B., DeMaster, D.P. and Hofman, R.J. 1977. An analysis of the dynamics of a Weddell seal population. *Ecological Monographs*, Vol. 47, pp.319–335.

Siniff, D.B., Testa, J.R. and Keuchl, V.B. 1969. Population studies of Weddell seals at McMurdo Station. *Antarctic Journal of the United States*, Vol. 4, pp.120–121.

Slade, R.W. 1997. Genetic studies of the southern elephant seal *Mirounga leonina*. Pp.11–29. In Hindell, M.A. and Kemper, C. (eds). *Marine Mammal Research in the Southern Hemisphere. Volume 1: Status Ecology and Medicine*. Surrey Beatty and Sons, Sydney. 186pp.

Slip, D.J. 1997. Diving and foraging behaviour of juvenile southern elephant seals from Heard Island. Pp.114–124. In Hindell, M.A. and Kemper, C. (eds). *Marine Mammal Research in the Southern Hemisphere. Volume 1: Status Ecology and Medicine*. Surrey Beatty and Sons, Sydney. 186pp.

Slip, D.J., Hindell, M.A. and Burton, H.R. 1994. An overview of diving behaviour of southern elephant seals from Macquarie Island. Pp.253–270. In Le Boeuf, B.J. and Laws, R.M. (eds). *Elephant Seals. Population Ecology, Behavior and Physiology*. University of California Press, Berkeley. 414pp.

Smith, I.W.G. 1989. Maori impact on the marine megafauna: pre-European distributions of New Zealand sea mammals. Monograph 17, pp.76–108. In Sutton, D.G. (ed). *Saying So Doesn't Make it So. Papers in Honour of B. Foss Leach*. New Zealand Archaeological Association, Dunedin, New Zealand.

Smith, R.I.L. 1988. Destruction of Antarctic terrestrial ecosystems by a rapidly increasing fur seal population. *Biological Conservation*, Vol. 45, pp.55–72.

Staples, D. 1996. Indicators of Sustainable Fisheries Development. Pp.719–725. In Hancock, D.A., Smith, D.C., Grant, A. and Beumer, J.P. (eds). *Second World Fisheries Congress: Developing and Sustaining World Fisheries Resources*. CSIRO Australia, Canberra.

Stewart, B.S. 1997. Ontogeny of differential migration and sexual segregation in northern elephant seals. *Journal of Mammalogy*, Vol. 78, pp.1101–1116.

Stewart, B.S. and DeLong, R.L. 1993. Seasonal dispersion and habitat use of foraging northern elephant seals. Pp.171–194. In Boyd, I.L. (ed). *Marine Mammals: Advances in Behavioural and Population Biology*. Symposium of the Zoological Society of London, Oxford University Press, Oxford. 404pp.

Strange, I.J. 1973. The silent ordeal of a south Atlantic archipelago. *Natural History*, Vol. 82, pp.30–39.

Sturlaugsson, J. and Johansson, M. 1998. Migration study of wild sea trout (*Salmo trutta* L.) in SE-Iceland: depth movements and water temperature recorded by data storage tags in freshwater and marine environment. *Proceedings of the Fifth European Conference on Wildlife Telemetry, 25–30 August 1996*. Strasbourg, France.

Tedman, R. and Green, B. 1987. Water and sodium fluxes and lactational energetics in suckling pups of Weddell seals (*Leptonychotes weddellii*). *Journal of Zoology, London*, Vol. 212, pp.29–42.

Testa, J.W., Hill, S.E.B. and Siniff, D.B. 1989. Diving behavior and maternal investment in Weddell seals (*Leptonychotes weddellii*). *Marine Mammal Science*, Vol. 5, pp.399–405.

Testa, J.W., Oehlert, G., Ainley, D.G., Bengston, J.L., Siniff, D.B., Laws, R.M. and Rounsevell, D. 1991. Temporal variability in Antarctic marine ecosystems: periodic fluctuations in the phocid seals. *Canadian Journal of Fisheries and Aquatic Sciences*, Vol. 48, pp.631–639.

Testa, J.W., Siniff, D.B., Croxall, J.P. and Burton, H.R. 1990. A comparison of reproductive parameters among three populations of Weddell seals (*Leptonychotes weddelli*). *Journal of Animal Ecology*, Vol. 59, pp.1165–1175.

Thompson, D. and Fedak, M. 1993. Cardiac responses of grey seals during diving at sea. *Journal of Experimental Biology*, Vol. 174, pp.139–164.

Thompson, D., Hiby, A.R. and Fedak, M.A. 1993. How fast should I swim: behavioural implications of diving physiology. Pp.349–368. In Boyd, I.L. (ed). *Marine Mammals: Advances in Behavioural and Population Biology*. Symposium of the Zoological Society of London, Oxford

University Press, Oxford. 404pp.

Thompson, P.M. 1993. Harbour seal movement patterns. Pp.225–239. In Boyd, I.L. (ed). *Marine Mammals: Advances in Behavioural and Population Biology*. Symposium of the Zoological Society of London, Oxford University Press, Oxford. 404pp.

Trillmich, F. and Ono, K.N. (eds). 1991. *Pinnipeds and El Niño: Responses to Environmental Stress*. Springer-Verlag, Berlin.

Vincent, W.F., Howard-Williams, C., Tildesley, P. and Butler, E. 1991. Distribution and biological properties of oceanic water masses around the South Island, New Zealand. *New Zealand Journal of Marine and Freshwater Research*, Vol. 25, pp.21–42.

Walker, B.G. and Boveng, P.L. 1995. Effects of time-depth recorders on maternal foraging and attendance behavior of Antarctic fur seals (*Arctocephalus gazella*). *Canadian Journal of Zoology*, Vol. 73, pp.1538–1544.

Warneke, R.M. 1982. The distribution and abundance of seals in the Australasian region, with summaries of biology and current research. Pp.431–475. In *Mammals in the Sea. Volume 4: Small Cetaceans, Seals, Sirenians, and Otters*. FAO Fisheries Series 5, Food and Agriculture Organisation of the United Nations, Rome.

Warneke, R.M. and Shaughnessy, P.D. 1985. *Arctocephalus pusillus*, the South African and Australian fur seal: taxonomy, evolution, biogeography and life history. Pp.53–57. In Ling, J.K. and Bryden, M.M. (eds). *Studies of Sea Mammals in South Latitudes*. South Australian Museum, Adelaide. 132pp.

Wartzok, D., Elsner, R., Stone, H., Kelly, B. and Davis, R. 1992b. Under-ice movements and the sensory basis of hole finding by ringed and Weddell seals. *Canadian Journal of Zoology*, Vol. 70, pp.1712–1722.

Wartzok, D., Sayegh, S., Stone, H., Barchak, J., Barnes, W. 1992a. Acoustic tracking system for monitoring under-ice movements of polar seals. *Journal of the Acoustic Society of America*, Vol. 92, pp.682–687.

Weimerskirch, H., Wilson, R.P., Guinet, C. and Koudil, M. 1995. Use of seabirds to monitor sea-surface temperatures in the Southern Ocean. *Marine Ecology Progress Series*, Vol. 126, pp.299–303.

Whitehead, M.D., Johnstone, G.W. and Burton, H.R. 1990. Annual fluctuations in productivity and breeding success of Adélie penguins and fulmarine petrels in Prydz Bay, East Antarctica. Pp.214–223. In Kerry, K.R. and Hempel, G. (eds). *Antarctic Ecosystems. Ecological Changes and Conservation*. Springer-Verlag, Berlin. 427pp.

Wildlife Protection (Regulation of Exports and Imports) Act. 1982. Act no. 149 of 1982 as amended (24 January 2000). Prepared by the Office of Legislative Drafting, Attorney-General's Department, Canberra.

Wilkinson, I.S. and Bester, M.M. 1990. Continued population increase in fur seals, *Arctocephalus tropicalis* and *A. gazella*, at the Prince Edward Islands. *South African Journal of Antarctic Research*, Vol. 20, pp.58–63.

Wilson, R.P., Culik, B.M., Bannasch, R. and Lage, J. 1994. Monitoring Antarctic environmental variables using penguins. *Marine Ecology Progress Series*, Vol. 106, pp.199–202.

Wilson, R.P., Putz, K., Charrassin, J.B., Lage, J. 1995. Artifacts arising from sampling interval in dive depth studies of marine endotherms. *Polar Biology*, Vol. 15, pp.575–581.

Zamon, J.E., Greene, C.H., Meir E., Demer, D.A., Hewitt, R.P. and Sexton, S. 1996. Acoustic characterization of the three-dimensional prey field of foraging chinstrap penguins. *Marine Ecology Progress Series*, Vol. 131, pp.1–10.

Index

This comprehensive listing has been designed for quick, easy use. Where subjects are referred to in several ways, or under different headings, cross-referencing is provided. Some terms and names occur many times in different chapters, but page numbers listed are restricted to initial usage, major references or detailed explanations.

In general, organisms are only listed down to generic level except where a particular species is used to demonstrate a significant point; in very speciose groups, such as birds, designation may only be to family level. The emphasis is on terms used in the text and usually only the current accepted scientific names are included; older alternative names (now invalidated or placed in synonomy) are not listed unless subjected to on-going controversy—although they may be incorporated in tables. Where a number of topics are included under a generic name the sections are listed in alphabetical order, so that species names may be interspersed with subject headings.

Only the major geological time intervals are listed—for details of sub-interval nomenclature, sequences and ages, or regional names, see the summary table on each section title page.

B

L

P

Q

U

NOTES

NOTES

These local stages (often referred to in older literature) are summarized for direct comparison with the current standard scale. Boundaries of some intervals are subject to controversy and, in the Quaternary of New Zealand, only the two initial stages of the Wanganui and Hawera Series are listed; however, a number of other stages have been recognized.

Adapted from the following sources:

Archer, M. and Clayton, G. (eds). 1984. *Vertebrate Zoogeography and Evolution in Australasia.* Hesperian Press, Perth. 1203pp.

Suppate, R.P., Stevens, G.R. and Te Punga, M.T. (eds). 1978. *The Geology of New Zealand.* Volume 1. Government Printer, Wellington. 343pp.

REGIONAL GEOLOGICAL STAGES FOR THE CAINOZOIC

ERA	PERIOD or SUB-PERIOD		SERIES/EPOCH		STAGE	AGE m.y.a.	AUSTRALIAN STAGE	NEW ZEALAND STAGE
CAINOZOIC	QUATERNARY		PLEISTOCENE		Calabrian	1.8	Weikooian	Porikan, Hautawan
	TERTIARY	NEOGENE	PLIOCENE		Piacenzian	3.6	Yatalan	Waitotaran; Waipipian
					Zanclean	5.3	Kalimnan	Opoitian
			MIOCENE	Late	Messinian	7.12	Cheltenhamian	Kapitean
					Tortonian	11.2	Mitchellian	Tongaporutuan
				Mid	Serravalian	14.8	Balcombian	Waiauan Lillburnian
					Langhian	16.4	Batesfordian	Clifdenian
				Early	Burdigalian	20.52	Longfordian	Altonian, Awamoan, Hutchinsonian
					Aquitanian	23.8	Janjukian	Otaian
		PALAEOGENE	OLIGOCENE	Late	Chattian	28.5		Waitakian
				Early	Rupelian	33.7	Aldingan	Duntroonian; Whaingaroan
			EOCENE	Late	Priabonian	37		Runangan
				Mid	Bartonian	41.3		Kaiatan; Bortonian
					Lutetian	49		Porarian
				Early	Ypresian	54.8	Wangerripian	Heretaurigian; Mangaorapan
			PALEOCENE	Late	Thanetian	57.9		Waipawan; Teurian
					Selandian	60.9		
				Early	Danian	65		

GEOLOGICAL TIME SCALE FOR THE PHANEROZOIC

ERA	PERIOD or SUB-PERIOD	SERIES/EPOCH	STAGE	AGE m.y.a.	EXTINCTIONS (E) ICE AGE (I)
CAINOZOIC	QUATERNARY	PLEISTOCENE	Calabrian	1.8	E, I
	TERTIARY – NEOGENE	PLIOCENE	Piacenzian	3.6	I
			Zanclean	5.3	I
		MIOCENE – Late	Messinian	7.12	
			Tortonian	11.2	
		MIOCENE – Mid	Serravalian	14.8	
			Langhian	16.4	
		MIOCENE – Early	Burdigalian	20.52	
			Aquitanian	23.8	
	TERTIARY – PALAEOGENE	OLIGOCENE – Late	Chattian	28.5	
		OLIGOCENE – Early	Rupelian	33.7	E
		EOCENE – Late	Priabonian	37	
		EOCENE – Mid	Bartonian	41.3	
			Lutetian	49	
		EOCENE – Early	Ypresian	54.8	
		PALEOCENE – Late	Thanetian	57.9	
			Selandian	60.9	
		PALEOCENE – Early	Danian	65	E
MESOZOIC	CRETACEOUS	LATE	Maastrichtian	73	
			Campanian	83	
			Santonian	87	
			Coniacian	89	
			Turonian	91	
			Cenomanian	97.5	
		EARLY	Albian	108	
			Aptian	115	
			Barremian	123	
			Hauterivian	130	
			Valanginian	135	
			Berriasian	141	E
	JURASSIC	LATE	Tithonian	146	
			Kimmeridgian	151	
			Oxfordian	159	
		MIDDLE	Callovian	165	
			Bathonian	173	
			Bajocian	180	
			Aalenian	184	
		EARLY	Toarcian	190	E
			Pliensbachian	195	
			Sinemurian	202	
			Hettangian	205	E
	TRIASSIC	LATE	Rhaetian	207	
			Norian	220	
			Carnian	230	
		MIDDLE	Ladinian	235	
			Anisian	241	
		EARLY	Spathian	244	
			Nammalian	247	
			Griesbachian	251	

This detailed scale is based on Young and Laurie (1996) but Cambrian stages abbreviated ages are rounded to nearest 0.1 million years. More recent dating (e.g. United States Geological Survey) has resulted in some small adjustments to interval boundaries.

ERA	PERIOD or SUB-PERIOD	SERIES/EPOCH	STAGE	AGE m.y.a.	EXTINCTIONS (E) ICE AGE (I)
PALAEOZOIC	PERMIAN	LATE	Tatarian – Dorashamian	255	E
			Tatarian – Dzhulfian	259	I
			Tatarian – Midian	263	I
			Kazanian	267	I
			Ulfimian	270	I
		EARLY	Kungurian	274	I
			Artinskian	285	I
			Sakmarian	293	I
			Asselian	298	I
	CARBONIFEROUS	PENNSYLVANIAN – SILESIAN – STEPHANIAN	Gzhelian	302	I
			Kasimovian	305	I
		PENNSYLVANIAN – SILESIAN – WESTPHALIAN	Moscovian	311.5	
			Bashkirian	314	
		PENNSYLVANIAN – SILESIAN – NAMURIAN	Serpukhovian	325	
		MISSISSIPPIAN – DINANTIAN – VISEAN	Brigantian	328.5	
			Asbian	330.5	
			Holkerian	335	
			Arundian	341.5	
		MISSISSIPPIAN – DINANTIAN – TOURNAISIAN	Chadian	344	
			Ivorian	348	
			Hastarian	354	
	DEVONIAN	LATE	Famennian	364.5	E
			Frasnian	369	
		MIDDLE	Givetian	378	I
			Eifelian	384	I
		EARLY	Emsian	399.5	I
			Pragian	404.5	I
			Lochkovian	410	
	SILURIAN	PRIDOLI		414	
		LUDLOW	Ludfordian	417	
			Gorstian	420	
		WENLOCK	Homerian	422.5	
			Sheinwoodian	425	
		LLANDOVERY	Telychian	428	I
			Aeronian	431	I
			Rhuddanian	434	E, I
	ORDOVICIAN	ASHGILL	Bolindian	443	I
		CARADOC	Eastonian	454.5	
			Gisbornian	459	
		LLANVIRN	Darriwilian	467	
			Yapeenian	471	
		ARENIG	Castlemainian	477	
			Chewtonian	481	
			Bendigonian	484.5	
		TREMADOC	Lancefieldian	486	
			Warendan	490	
	CAMBRIAN	LATE	Datsonian	491	
			Payntonian	492.5	
			Iverian	496	
			Idamean	497.5	E
		MIDDLE	Mindyallan	498.5	
			Boomerangian	500	
			Undillan	503	
			Late Templetonian-Floran	506	
			Ordian-Early Templetonian	509	
		EARLY	Atdabanian-Toyonian	528	
			Tommotian	534	
			Nemakit-Daldynian	545	

Biogeo

Changes in Thinking

When attempting to explain the biogeography of extant faunas, it is necessary to be aware of recent advances in knowledge of processes, key factors or mechanisms that influence the distributions of living organisms. These include:

(1) The variety and combinations of dispersal and colonization mechanisms now known to have occurred, together with recognition that island biogeography differs markedly from the continental situation.

(2) The increasing resolution of molecular techniques to detect intraspecific variation, indicating the relationships (derivation) of regional populations or local stocks.

(3) The growing evidence for local rapid evolutionary responses to changed conditions (e.g. bird bills and preferred seed size in the Galapagos Islands), which can change from year to year.

(4) The realization that, superimposed on the main glaciation cycles of the Pleistocene, significant and short-term (within a few years) climatic fluctuations (both regional and local), have occurred over the last 100,000 years; these changes appear to be associated with disruptions of major oceanic circulation systems and altered heat exchange patterns.

(5) The acknowledgement of the extent of the impacts of human activities—especially translocations with the resulting bio-invasive processes. A single non-indigenous organism can directly or indirectly cause a total transformation of an ecosystem—devastating endemic populations.

RECOMMENDED READING

Augee, M.L. and Fox, M.D. 2000. *Biology of Australia and New Zealand.* Pearson Education, Sydney. 100pp.

Briggs, J.C. 1995. *Global Biogeography.* Elsevier, Amsterdam. 452pp.

Chatterjee, S. and Scotese, C.R. 1999. The breakup of Gondwana and the evolution and biogeography of the Indian Plate. Pp.189–217. In Sahni, A. and Loyal, R. (eds). *Gondwanaland Assembly: Current Issues and Problems.* Indian National Science Academy, New Delhi.

Cox, C.B. and Moore, P.D. 2000. *Biogeography: an Ecological and Evolutionary Approach.* 6th Edition. Blackwell Science, Malden, Massachusetts. 298pp.

Craw, R.C., Grehan, J.R. and Heads, M.J. 1999. *Panbiogeography: Tracking the History of Life.* Oxford University Press, New York. 229pp.

Humphries, C.J. and Parenti, L.R. 1999. *Cladistic Biogeography: Interpreting Patterns of Plant and Animal Distributions.* 2nd Edition. Oxford University Press, Oxford. 187pp.

MacDonald, G. 2001. *Biogeography: Space, Time and Life.* 1st Edition. J. Wiley and Sons, New York. 480pp.

McDowall, R.M. 2004. What biogeography is: a place for process. *Journal of Biogeography,* Vol. 31, pp.345–351.

Spellerberg, I.F., Sawyer, J.W.D. and Whitten, T. 1999. *An Introduction to Applied Biogeography.* Cambridge University Press, Cambridge. 257pp.

Waters, J.M. and Burridge, C.P. 1999. Extreme intraspecific mitochondrial DNA sequence divergence in *Galaxias maculatus* (Osteichthys: Galaxiidae), one of the world's most widespread freshwater fish. *Molecular Phylogenetics and Evolution,* Vol. 11, pp.1–12.